JOHNNIE MAE BERRY LIBRARY
CINCINNATI STATE
3520 CENTRAL PARKWAY
CINCINNATI, OHIO 45223-2690

For Reference

Not to be taken from this room

WILEY ENCYCLOPEDIA OF
BIOMEDICAL ENGINEERING

VOLUME 1

WILEY ENCYCLOPEDIA OF BIOMEDICAL ENGINEERING

Editor
Metin Akay

Editorial Board
Henrietta Galiana
McGill University

Yongmin Kim
University of Washington

Larry McIntire
Georgia Institute of Technology

Banu Onaral
Drexel University

Linda S. Powers
Utah State University

Jose Principe
University of Florida

Toshiyo Tamura
China University

Kamil Ugurbil
University of Minnesota

Herbert Voigt
Boston University

Wolf W. von Maltzahn
Rensselaer Polytechnic Institute

Area Editors
Paolo Bonato
Harvard Medical School

Eugene N. Bruce
University of Kentucky

Sergio Cerutti
Polytechnic University, Milan, Italy

Li-Shan Chou
University of Oregon

Patricia Connolly
University of Strathclyde in Glasgow

Atam Dhawan
New Jersey Institute of Technology

Olaf Dössel
Universität Karlsruhe, Germany

Andrew Karduna
University of Oregon

Michael C. K. Khoo
University of Southern California

Michael Korenberg
Queen's University

Andrew Laine
Columbia University

Z.-P. Liang
University of Illinois at Urbana-Champaign

Andy Marsh
National Technical University of Athens

Roberto Merletti
Politecnico di Torino

Karen Moxon
Drexel University

Robert F. Munzner
Regulatory Affairs Consultant

Barbara Oakley
Oakland University

George Pins
Worcester Polytechnic Institute

Peter Rolfe
University of Genova

Chiarra Sabatti
University of California at Los Angeles

Editorial Staff
Vice President, STM Books: **Janet Bailey**
Sponsoring Editor: **George J. Telecki**
Assistant Editor: **Cassie Craig**

Production Staff
Director, Book Production and Manufacturing:
Camille P. Carter
Managing Editor: **Shirley Thomas**
Production Editor: **Kristen Parrish**
Illustration Manager: **Dean Gonzalez**

WILEY ENCYCLOPEDIA OF
BIOMEDICAL ENGINEERING

VOLUME 1

Metin Akay, Editor

The *Wiley Encyclopedia of Biomedical Engineering* is available online at
http://www.mrw.interscience.wiley.com/ebe

A John Wiley & Sons, Inc., Publication

Copyright © 2006 by John Wiley & Sons, Inc. All rights reserved.

Published by John Wiley & Sons, Inc., Hoboken, New Jersey.

Published simultaneously in Canada.

No part of this publication may be reproduced, stored in a retrieval system, or transmitted in any form or by any means, electronic, mechanical, photocopying, recording, scanning, or otherwise, except as permitted under Section 107 or 108 of the 1976 United States Copyright Act, without either the prior written permission of the Publisher, or authorization through payment of the appropriate per-copy fee to the Copyright Clearance Center, Inc., 222 Rosewood Drive, Danvers, MA 01923, (978) 750-8400, fax (978) 750-4470, or on the web at www.copyright.com. Requests to the Publisher for permission should be addressed to the Permissions Department, John Wiley & Sons, Inc., 111 River Street, Hoboken, NJ 07030, (201) 748-6011, fax (201) 748-6008, or online at http://www.wiley.com/go/permission.

Limit of Liability/Disclaimer of Warranty: While the publisher and author have used their best efforts in preparing this book, they make no representations or warranties with respect to the accuracy or completeness of the contents of this book and specifically disclaim any implied warranties of merchantability or fitness for a particular purpose. No warranty may be created or extended by sales representatives or written sales materials. The advice and strategies contained herein may not be suitable for your situation. You should consult with a professional where appropriate. Neither the publisher nor author shall be liable for any loss of profit or any other commercial damages, including but not limited to special, incidental, consequential, or other damages.

For general information on our other products and services or for technical support, please contact our Customer Care Department within the United States at (800) 762-2974, outside the United States at (317) 572-3993 or fax (317) 572-4002.

Wiley also publishes its books in a variety of electronic formats. Some content that appears in print may not be available in electronic formats. For more information about Wiley products, visit our web site at www.wiley.com.

Library of Congress Cataloging-in-Publication Data:

Wiley encyclopedia of biomedical engineering/Metin Akay, editor-in-chief
    p. cm.
  Includes index.
  ISBN-13: 978-0-471-24967-2
  ISBN-10: 0-471-24967-X (cloth : set)
    –ISBN-13: 978-0-471-74037-7
      ISBN-10: 0-471-74037-3 (cloth : v. 1)
    –ISBN-13: 978-0-471-74039-1
      ISBN-10: 0-471-74039-X (cloth : v. 2)
    –ISBN-13: 978-0-471-74038-4
      ISBN-10: 0-471-74038-1 (cloth : v. 3)
    –ISBN-13: 978-0-471-74040-7
      ISBN-10: 0-471-74040-3 (cloth : v. 4)
    –ISBN-13: 978-0-471-74041-4
      ISBN-10: 0-471-74041-1 (cloth : v. 5)
    –ISBN-13: 978-0-471-74042-1
      ISBN-10: 0-471-74042-X (cloth : v. 6)
1. Biomedical engineering–Encyclopedias.   I. Title: Encyclopedia of biomedical engineering.
  [DNLM: 1. Biomedical Engineering–Encyclopedias.   QT 13 W676
2006]
R856.A3.W55 2006
610'.2803–dc22

                                                                                   2006001110

Printed in the United States of America

10 9 8 7 6 5 4 3 2 1

PREFACE

The integration of engineering, mathematics, and the computational and information sciences and technologies with biology, the life and cognitive sciences, medicine, and health care is defining the new biomedical frontier. This development is marked by innovations that are affecting, at an unprecedented pace and in profound ways, the social, industrial, and economic bases of our global society. This sense of urgency permeates the health care landscape. Concurrently, biomedical engineering is reorganizing its knowledge base while expanding its scope. As the biological sciences make phenomenal strides, the necessity for systems-integrated and quantitative approaches, the hallmark of engineering, becomes more critical in tackling complex biological problems and health care issues. This trend fuels the desire to elucidate the relationships between molecular, functional, and systemic levels, across a multiplicity of scales and many degrees of complexity.

Recent advances in computing power and nanotechnology are transforming the processes of discovery, learning, and communication and hence are reshaping the interaction between users and knowledge repositories such as encyclopedias. The compilation, delivery, and dissemination of the intellectual content are no longer confined by the constraints of the hard copy. The most important feature of the hard copy and its electronic embodiment will be to present the content of the encyclopedia from different perspectives such as the physiological system (e.g., cardiovascular, respiratory, etc.), the level of organization ranging from molecular interactions to the analysis of the health care system, or the tools and techniques.

The *Wiley Encyclopedia of Biomedical Engineering* is a living and evolving repository of the biomedical engineering knowledge base. To represent the vast diversity of the field and its multi- and cross-disciplinary nature and serve the BME community, the scope and content are comprehensive including the material directed to the needs of readers ranging from undergraduate and graduate students, post-docs, research scientists, and experts.

As a primer, educational material, technical reference, and research and development resource, the *Wiley Encyclopedia of Biomedical Engineering*, in terms of its intellectual substance and rigor, is peer reviewed. For the online version, mechanisms will be in place to upgrade and update the content and to capture and assimilate emerging trends. The contributors have been invited from diverse disciplinary groups representing academia, industry, and private and government organizations. The make-up of participants represents the geographical distribution of BME activity around the world.

I am confident that the Encyclopedia of Biomedical Engineering will become the unique biomedical enginering resource that contributes to the organization of the BME knowledge domain and facilitates its growth and development.

I am grateful for Prof. Banu Onaral for her strong support of this initiative. I would like to thank Mr. George Telecki for his support and Ms. Cassie Craig for her strong support, help, and hard work during the entire time of editing this encyclopedia. Working in concert with the Editorial Board Members and Area Editors and the contributors really helped me in developing the content and in managing the peer-review process. I am grateful to them.

Finally, many thanks to my wife, Dr. Yasemin M. Akay, for her unmatched care, love and devotion to her family, and to our son, Altug R. Akay, for their support, encouragement and patience. They have been my driving source. It was indeed a pleasure working with her during the weekends and holidays to finish this encyclopedia.

The encyclopedia is dedicated to the memory of my late brother, Cetin Akay, who dedicated his short, but meaningful life to the well-being and happiness of others. May God bless his soul.

Metin Akay
February 2006
Scottsdale, Arizona

CONTRIBUTORS

ALI E. ABBAS, *University of Illinois at Urbana-Champaign, Urbana, Illinois,* Bayesian Analysis; Bioinformatics; Entropy; Hidden Markov Models; Markov Chains

SHIMON ABBOUD, *Tel Aviv University, Tel Aviv, Israel,* Electrical Impedance Technique for Cryosurgery Monitoring

BRUCE ABERNETHY, *University of Hong Kong, Hong Kong,* Gait Patterns

PETER ACHERMANN, *University of Zurich, Zurich, Switzerland,* Sleep

EHSAN AFKHAMI, *Boston University, Boston, Massachusetts,* Proteomics

ATA AKIN, *Boğaziçi University, Bebek, Istanbul, Turkey,* Bio-Optical Signals

SOLANGE AKSELROD, *Tel Aviv University, Tel Aviv, Israel,* Heart Rate Variability (HRV): Nonlinear HRV; Heart Rate Variability (HRV); Hypertension: Time-Frequency Analysis for Early Diagnosis

DAVID ALGOOD, *University of Pittsburgh, Pittsburgh, Pennsylvania,* Rehabilitation Engineering: An Overview

ABDALHAMEED ALKHATEEB, *King Abdulazziz University, Jeddah, Saudi Arabia,* Medical Device Acquisition

KYLE D. ALLEN, *Rice University, Houston, Texas,* Temporomandibular Joint Disc

ALEJANDRO J. ALMARZA, *Rice University, Houston, Texas,* Temporomandibular Joint Disc

GIL ALTEROVITZ, *Massachusetts Institute of Technology, Cambridge, Massachusetts,* Proteomics

JOSÉ ÁLVAREZ-GÓMEZ, *Polytechnical University of Cartagena, Cartagena, Spain,* Anesthesia Machines

Y. H. AN, *Medical University of South Carolina, Charleston, South Carolina and Clemson University, Clemson, South Carolina,* Bone, Mechanical Testing of

A. SERMET ANAGUN, *Osmangazi University, Bademlik, Eskisehir, Turkey,* Sleep Apnea: Detection and Classification of Infant Patterns

I. A. ANDERSON, *University of Auckland, Auckland, New Zealand,* Force Measurement

JAMES M. ANDERSON, *Case Western Reserve University, Cleveland, Ohio,* Foreign Body Reaction; Sensor Biocompatibility and Biofouling in Real-Time Monitoring

LARS ARENDT-NIELSEN, *Aalborg University, Aalborg, Denmark,* Muscle Pain

SERDAR ARITAN, *Hacettepe University, Hacettepe, Turkey,* Bulk Modulus

DOUGLAS ARM, *Cytori Therapeutics, Inc., San Diego, California,* Autologous Platelet-Based Therapies

MARIA TERESA ARREDONDO, *Universidad Politecnica de Madrid, Life Supporting Technologies, Ciudad Universitaria, Madrid, Spain,* Home Telehealth: Telematics-Supported Services; Telemedicine in Emergency Medical Services

M. ARSIERO, *University of Bern, Bern, Switzerland,* Biological Neuronal Networks, Modeling of

ANTHONY ATALA, M.D. *Wake Forest University School of Medicine, Winston-Salem, North Carolina,* Tissue Engineering of Kidney, Bladder, and Urethra

KYRIACOS A. ATHANASIOU, *Rice University, Houston, Texas and University of Texas - Health Science Center, Houston, Texas,* Knee Meniscus, Biomechanic of; Musculoskeletal Cell Mechanics; Temporomandibular Joint Disc

ANDRÉ E. AUBERT, *University Hospital Gasthuisberg, K.U. Leuven, Leuven, Belgium,* Bioenergetics and Systemic Responses to Exercise; Electrodes; Space Physiology

NIZAMETTIN AYDIN, *Bahcesehir University, Istanbul, Turkey,* Venous Air Embolism: Detection via Wavelet Transform

FABIO BABILONI, *University of Rome, Rome, Italy,* Electroencephalography (EEG): Inverse Problems

PAUL BACH-Y-RITA, *University of Wisconsin, Madison, Wisconsin,* Human Nervous System, Noninvasive Coupling of Electronically Generated Data into

DAN L. BADER, *University of London, London, United Kingdom,* Pressure Sensors

M. SAFWAN BADR, *Wayne State University, Detroit, Michigan,* Upper Airway Mechanics

ER-WEI BAI, *Shanghai University, Shanghai, China,* Computed Tomography

ROBERT BAIER, *State University of New York, Buffalo, New York,* Biocompatibility of Engineering Materials

JAMES BAISH, *Bucknell University, Lewisburg, Pennsylvania,* Thermal Conductivity Measurement of Biomaterials

TADEJ BAJD, *University of Ljubljana, Ljubljana, Slovenia,* Surface Electrostimulation Electrodes

JAKE D. BALLARD, *Lockheed Martin/KAPL, Niskayuna, New York,* Nanophase Materials

JOSEPH BARILLARI, *Harvard University, Cambridge, Massachusetts,* Proteomics

GRACE T. BARTOO, *Decus Biomedical, LLC, San Carlos, California,* Clinical Trials; Risk Management

GIUSEPPE BASELLI, *Politecnico di Milano, Milano, Italy,* Closed-Loop System Identification

JAMES B. BASSINGTHWAIGHTE, *University of Washington, Seattle, Washington,* Capillary Permeability

JASON H. T. BATES, *University of Vermont, Burlington, Vermont,* Pulmonary Mechanics

REBECCA BECK, *University College Dublin and The National Rehabilitation Hospital, Dublin, Ireland,* Muscle Fiber Conduction Velocity

FRANK BECKERS, *University Hospital Gasthuisberg, K.U. Leuven, Leuven, Belgium,* Bioenergetics and Systemic Responses to Exercise; Electrodes; Space Physiology

JULIAN BEDOYA, *Texas A&M University, College Station, Texas,* Vascular Mechanics

KHOSROW BEHBEHANI, *University of Texas, Arlington, Texas,* Heart Rate Variability (HRV): Sleep Disordered Breathing

ANDRES BELALCAZAR, *University of Minnesota, Minneapolis,* Plethysmography

BESIM BEN-NISSAN, *University of Technology, Sydney, Australia,* Bioactive Glasses and Glass Ceramics

LISA BENSON, *Clemson University, Clemson, South Carolina,* Accelerometers

FRANÇOIS BERTHIAUME, *Massachusetts General Hospital, Boston, Massachusetts,* Tissue-Engineered Liver

JASMINE BHATHENA, *McGill University, Montreal, Québec, Canada,* Microencapsulation

VASUDA BHATIA, *Jawaharlal Nehry Centre for Advanced Scientific Research, Bangalore, India,* Photoconducting Devices

PAOLO BIANCHINI, *University of Genoa, Genoa, Italy,* Multiphoton Microscopy

CONTRIBUTORS

LYNNE E. BILSTON, *University of NSW, Sydney, Australia*, Viscoelasticity

ANAND SINGH BISEN, *Indiana University, Purdue University, Indianapolis, Indiana*, Distributed Processing

KATARZYNA BLINOWSKA, *Warsaw University, Warszawa, Poland*, Electroencephalography (EEG)

PAOLO BOLZERN, *Politecnico di Milano, Milano, Italy*, Closed-Loop System Identification

MICHAEL L. BONINGER, *University of Pittsburgh, Pittsburgh, Pennsylvania*, Rehabilitation Engineering: An Overview; Wheelchair Engineering

MARK BORDEN, *Director of Biomaterials Research, Interpore Cross International, Irvine, California*, Resorbable Materials in Orthopedic Surgery

CONNIE BORROR, *Arizona State University, Phoenix, Arizona*, Probability Distributions

RICCARDO BOSCOLO, *University of California, Los Angeles*, Independent Component Analysis

BARBARA D. BOYAN, *Georgia Institute of Technology, Atlanta, Georgia and University of Hamburg-Eppendorf, Hamburg, Germany*, Platelet-Rich Plasma in Bone Repair

J. BRABYN, *The Smith-Kettlewell, Eye Research Institute*, Sensory Aids

STEVEN L. BRESSLER, *Florida Atlantic University, Boca Raton, Florida*, Event-Related Potentials

ALEXANDER M. BRONSTEIN, *Technion–Israel Institute of Technology, Haifa, Israel*, Blind Source Separation

MICHAEL M. BRONSTEIN, *Technion–Israel Institute of Technology, Haifa, Israel*, Blind Source Separation

KARL BROWN, *University of Pittsburgh, Pittsburgh, Pennsylvania*, Spinal Cord Stimulation Systems

MELISSA BROWN, *Duke University, Durham, North Carolina*, Vascular and Capillary Endothelium

MARIJN E. BRUMMER, *Emory University School of Medicine, Atlanta, Georgia*, NMR Imaging

JOHN E. BRUNELLE, *Pegasus Biologics, Irvine, California*, Arthroscopic Fixation Devices

CHRISTOPHER D. BUCKLEY, *University of Birmingham, Birmingham, United Kingdom*, Cell Adhesion Molecules: Conversational Signalers

DAVID M. BUDGETT, *University of Auckland, Auckland, New Zealand*, Extracorporeal Electrodes

TODD BUERSMEYER, *Georgetown University Medical Center, Washington, DC*, Computer Aided Surgery

OMRAN BUKHRES, *Indiana University, Purdue University, Indianapolis, Indiana*, Biological Database Integration; Distributed Processing; Information Retrieval in Biomedical Research

KAREN BURG, *Clemson University, Clemson, South Carolina*, Tissue Engineering

RALPH BUSCHBACHER, *Indiana University, Indianapolis, Indiana*, Electrophysiology

GREG BUSH, *University of New Brunswick, New Brunswick, Canada*, Myoelectric Control of Powered Upper-Limb Prostheses

ROBERT BUTERA, *Georgia Institute of Technology, Atlanta, Georgia*, Nernst Potential

VALENTINA CAORSI, *University of Genoa, Genoa, Italy*, Multiphoton Microscopy

ALICIA L. CARLSON, *The University of Texas of Austin, Austin, Texas*, Epithelial Pre-cancers and Cancers, Optical Technologies for Detection and Diagnosis of

ANDRES E. CASTELLANOS, *Drexel University College of Medicine, Philadelphia, Pennsylvania*, Robotic Surgery

PAOLO CASTIGLIONI, *Fondazione Don Carlo Gnocchi ONLUS, Milano, Italy*, Arterial Blood Pressure Processing

JUAN JOSÉ RAMOS CASTRO, *Universitat Politecnica de Catalunya, Barcelona, Spain*, Noise in Instrumentation

HENRY A. CATHERINO, *U.S. Army Research, Development and Engineering Command, Warren, Michigan*, Electrochemical Biosensors

M. CENK ÇAVUŞOĞLU, *Case Western Reserve University, Cleveland, Ohio*, Medical Robotics in Surgery

A. ENIS ÇETIN, *Bilkent University, Ankara, Turkey*, Image Coding; Vector Quantization

RAKIÉ CHAM, *University of Pittsburgh, Pittsburgh, Pennsylvania*, Rehabilitation Biomechanics

F. H. Y. CHAN, *The University of Hong Kong, Hong Kong*, Evoked Potentials, Adaptive Filtering of

GREGORY CHAN, *The Chinese University of Hong Kong, Hong Kong*, Heart Sounds and Stethoscopes

WARREN C. W. CHAN, *University of Toronto, Toronto, Ontario, Canada*, Nanoparticles in Biomedical Photonics

ELIANA CHAVES, *University of Pittsburgh, Pittsburgh, Pennsylvania*, Rehabilitation Engineering: An Overview

J. D. Z. CHEN, *University of Texas Medical Branch, Galveston, Texas*, Electrogastrography (EGG)

S. C. CHEN, *University of New South Wales, Sydney, Australia*, Functional Optical Imaging of Intrinsic Signals in Cerebral Cortex

WEI CHEN, *University of Minnesota Medical School, Minneapolis, Minnesota*, Brain Function, Magnetic Resonance Imaging of

NEIL S. CHERNIACK, *New Jersey Medical School, Newark, New Jersey*, Pulmonary Medicine, Applications of Biomedical Engineering to

BERNARDINO CHIAIA, *Politecnico, Torino, Italy*, Respiration Measurements

MIREYA FERNÁNDEZ CHIMENO, *Universitat Politecnica de Catalunya, Barcelona, Spain*, Electrical Safety; Noise in Instrumentation

GIUSEPPE CHIRICO, *University of Milano-Bicocca, Milano-Bicocca, Italy*, Multiphoton Microscopy

B. DEVIKA CHITHRANI, *University of Toronto, Toronto, Ontario, Canada*, Nanoparticles in Biomedical Photonics

FRANCISCO J. CHORRO, *Hospital Clínico Universitario de Valencia, Valencia, Spain*, Ectopic Activity

DAVID CHRISTINI, *Weill Medical College of Cornell University, New York, New York*, Cardiac Arrhythmia

CHRISTODOULOS CHRISTODOULOU, *University of Cyprus, Nicosia, Cyprus*, Ultrasonic Imaging of Carotid Atherosclerosis

CONSTANCE R. CHU, *University of Pittsburgh, Pittsburgh, Pennsylvania*, Tissue-Engineered Cartilage

ARTHUR CIARKOWSKI, *Center for Devices and Radiological Health / FDA, Rockville, Maryland*, Risk Management for Medical Devices

MEHMET CILINGIROGLU, *Baylor College of Medicine, Houston, Texas*, Cardiac Hypertrophy

KEVIN CLEARY, *Georgetown University Medical Center, Washington, DC*, Computer Aided Surgery

S.L. CLOHERTY, *University of Newcastle, Callaghan, Australia and University of New South Wales, Sydney, Australia*, Electrical Activity in Cardiac Tissue, Modeling of; Functional Optical Imaging of Intrinsic Signals in Cerebral Cortex

MAURICE E. COHEN, *California State University, Fresno, California and University of California, San Francisco, California*, Chaos; Neural Networks

GERY COLOMBO, *Balgrist University Hospital, Zürich, Switzerland and Hocoma AG, Volketswil, Switzerland*, Gait Retraining After Neurological Disorders

STEVEN A. CONRAD, *Louisiana State University Health Sciences Center, Shreveport, Louisiana and Louisiana Tech University,*

Ruston, Louisiana, Extracorporeal Membrane Oxygenation; Hemodialysis and Hemofiltration; Sepsis

RORY A. COOPER, *University of Pittsburgh, Pittsburgh, Pennsylvania,* Rehabilitation Engineering: An Overview; Wheelchair Engineering

ROSEMARIE COOPER, *University of Pittsburgh, Pittsburgh, Pennsylvania,* Rehabilitation Engineering: An Overview; Wheelchair Engineering

THOMAS CORFMAN, *University of Pittsburgh, Pittsburgh, Pennsylvania,* Rehabilitation Engineering: An Overview

RUSSELL CRAWFORD, *Swinburne University of Technology, Melbourne, Australia,* Adhesion of Bacteria

FU-ZHAI CUI, *Tsinghua University, Beijing, China,* Tissue-Engineered Bone

DAVID CUNNINGHAM, *Abbott Laboratories, Abbott Park, Illinois,* Microfluidics; Technology Assessment of Medical Devices

AMY DE JONGH CURRY, *The University of Memphis, Memphis, TN,* Defibrillation

ALBERT DAHAN, *Leiden University Medical Center (LUMC), Leiden, The Netherlands,* Hypercapnia

P. DARAS, *Informatics & Telematics Institute, C.E.R.T.H., Thessaloniki, Greece,* Deformable Objects, Interactive Simulation of; Haptic Interaction in Medical Virtual Environments

DAVID Z. D'ARGENIO, *University of Southern California, Los Angeles, California,* Pharmacokinetic and Pharmacodynamic Control

ALIREZA DARVISH, *University of North Carolina at Charlotte, Charlotte, North Carolina,* Maximum Likelihood Estimation; Neural Networks: Applications in Biomedical Engineering

J. ANDREW DAUBENSPECK, *Dartmouth College, Lebanon, NH,* Respiratory Related Evoked Potentials

SHASHANK DAVE, *Indiana University, Indianapolis, Indiana,* Electrophysiology

LINDA R. DAVRATH, *Tel Aviv University, Tel Aviv, Israel,* Heart Rate Variability (HRV); Hypertension: Time-Frequency Analysis for Early Diagnosis

J. C. DE MUNCK, *VU Medical Center, Amsterdam, The Netherlands,* Forward Problem

SANJUKTA DEB, *King's College London, London, United Kingdom,* Orthopedic Bone Cement

JACK C. DEBES, *John Fiske Brown Associates Inc., Forensic Engineers, Solana Beach, California,* Stress

M. ELENA DEL-CAMPO-ADRIÁN, *Polytechnical University of Cartagena, Cartagena, Spain,* Assistive Technology

FRANCISCO DEL POZO, *Universidad Politécnica de Madrid, Madrid, Spain,* Diabetes Care, Biomedical and Information Technologies for; Telemedicine: Teleconsultation between Medical Professionals; Telemedicine: Ubiquitous Patient Care

SEMAHAT S. DEMIR, *The University of Memphis, and The University of Tennessee, Health Science Center, Memphis, Tennessee,* Cardiac Action Potentials; Careers; Education

GREGORY M. DEMYASHEV, *Swinburne University of Technology, Melbourne, Australia,* Carbyne-Containing Surface Coatings

JAYDEV P. DESAI, *Drexel University, Philadelphia, Pennsylvania,* Robotic Surgery

MICHAEL S. DETAMORE, *Rice University, Houston, Texas,* Temporomandibular Joint Disc

STEVEN DEUTSCH, *Pennsylvania State University, University Park, Pennsylvania,* Flow Measurement

ATAM DHAWAN, *New Jersey Institute of Technology, Newark, New Jersey,* Radon Transform

MARCO DI RIENZO, *Fondazione Don Carlo Gnocchi ONLUS, Milano, Italy,* Arterial Blood Pressure Processing

ALBERTO DIASPRO, *University of Genoa, Genoa, Italy,* Confocal Microscopy; Multiphoton Microscopy; Optical Microscopy

G. V. DIMITROV, *Bulgarian Academy of Sciences, Sofia, Bulgaria,* EMG Modeling

N. A. DIMITROVA, *Bulgarian Academy of Sciences, Sofia, Bulgaria,* EMG Modeling

DAN DING, *University of Pittsburgh, Pittsburgh, Pennsylvania,* Rehabilitation Engineering: An Overview; Wheelchair Engineering

MINGZHOU DING, *University of Florida, Gainesville, Florida,* Event-Related Potentials

JONATHAN B. DINGWELL, *University of Texas, Austin, Texas,* Lyapunov Exponent

DRAGO DJORDJEVIĆ, *University of Belgrade, Serbia, Montenegro,* Ionic Channels

S. DOKOS, *University of New South Wales, Sydney, Australia,* Electrical Activity in Cardiac Tissue, Modeling of

TAMMY HAUT DONAHUE, *Michigan Technological University, Houghton, Michigan,* Meniscal Replacements

SILVANO DONATI, *University of Pavia, Pavia, Italy,* Photomultipliers

KARIN DORMAN, *Iowa State University, Ames, Iowa,* Trees, Evolutionary

EDWARD R. DOUGHERTY, *University of Texas, Houston, Texas,* Genomic Networks: Statistical Inference from Microarray Data

MIKE R DOUGLAS, *University of Birmingham, Birmingham, United Kingdom,* Cell Adhesion Molecules: Conversational Signalers

MANUEL VARGAS DRECHSLER, *Universitat Politecnica de Catalunya, Barcelona, Spain,* Noise in Instrumentation

AARON J. DULGAR-TULLOCH, *General Electric Global Research Center, Niskayuna, New York,* Nanophase Materials

LOUIS-GILLES DURAND, *Institut de Recherches Cliniques de Montréal, Montreal, Canada,* Aortic Stenosis and Systemic Hypertension, Modeling of

PIOTR DURKA, *Warsaw University, Warszawa, Poland,* Electroencephalography (EEG)

MITRA DUTTA, *University of Illinois at Chicago, Chicago, Illinois,* Optoelectronics

TOURADJ EBRAHIMI, *EPFl-STI-ITS-LTS, Lausanne, Switzerland,* Human Brain Interface: Signal Processing and Machine Learning

CHRISTOPHER N. EICHELBERGER, *University of North Carolina at Charlotte, Charlotte, North Carolina,* Back-Propagation; Genetic Algorithms; Software Engineering

KOST ELISEVICH, *Henry Ford Hospital, Detroit, Michigan,* Hippocampus

SUNNY ELOOT, *Ghent University, Ghent, Belgium,* Artificial Kidney, Modeling of Transport Phenomena in

PAUL ENCK, *University Hospitals, Tübingen, Germany,* Electromyography (EMG) of Pelvic Floor Muscles

JOHN ENDERLE, *University of Connecticut, Storrs, Connecticut,* Eye Movements

KEVIN ENGLEHART, *University of New Brunswick, New Brunswick, Canada,* Myoelectric Control of Powered Upper-Limb Prostheses

ROGER M. ENOKA, *University of Colorado, Boulder, Colorado,* Motor Unit

JEREMIE EVEN, *University of Illinois at Urbana-Champaign, Urbana, Illinois,* Hidden Markov Models; Markov Chains

SHULAMIT EYAL, *Tel-Aviv University, Tel-Aviv, Israel,* Heart Rate Variability (HRV): Nonlinear HRV

B. MURAT EYÜBOĞLU, *Middle East Technical University, Ankara, Turkey,* Electrical Impedance Imaging, Injected Current; Electrical Impedance Plethysmography; Magnetic Resonance Current Density Imaging; Magnetic Resonance—Electrical Impedance Tomography

CONTRIBUTORS

VITO FANELLI, *Ospendale S. Giovanni Battista, Italy,* Respiration Measurements

MARIO FARETTA, *IFOM-IEO Campus for Oncogenomics, Milan, Italy,* Confocal Microscopy

DARIO FARINA, *Aalborg University, Denmark, Italy,* Myoelectric Manifestations of Muscle Fatigue; Surface Electromyography (EMG) Signal Processing

STEFANIA FATONE, *Northwestern University, Chicago, Illinois,* Orthotics

FEDERICO FEDERICI, *University of Genoa, Genoa, Italy,* Confocal Microscopy

FRANCESCO FELICI, *Istituto Universitario di Scienze Motorie, Rome, Italy,* Muscle, Skeletal

ENRIQUE FERNÁNDEZ, *Universitat Politècnica de Catalunya, Barcelona, Spain,* Bioactive Bone Cements

MIREYA FERNANDEZ, *Universitat Politecnica de Catalunya, Barcelona, Spain,* Electromagnetic Interference and Compatibility

ANDREA FERRARI, *Ospendale S. Giovanni Battista, Italy,* Respiration Measurements

ÁNGEL FERRERO, *Hospital Clinico Universitario, Valencia, Spain,* Ischemia

JOSÉ MARÍA FERRERO, JR., *Universidad Politécnica de Valencia, Valencia, Spain,* Ectopic Activity; Ischemia

LUISA FILIPPONI, *Swinburne University of Technology, Hawthorn, Australia,* Cell Patterning

DAVID M. FINDLAY, *University of Adelaide, Adelaide, Australia,* Bone Resorption

GERALD FISCHER, *University for Health Sciences Medical Informatic and Tecnnology (UMIT), Austria,* Electrocardiogram (ECG) Mapping

DAN FLETCHER, *UC Berkeley, Berkeley, California,* Nanometer-Scale Probes

STEPHEN FLORCZYK, *SUNY Downstate Medical Center, Brooklyn, New York,* Ethical Issues in Biomedical Research

ARNOLD A. FONTAINE, *Pennsylvania State University, University Park, Pennsylvania,* Flow Measurement

BELMA FORD, *University of North Carolina, Charlotte, North Carolina,* Protein Structure, Folding, and Conformation

KENNETH R. FOSTER, *University of Pennsylvania, Philadelphia, Pennsylvania,* Radiofrequency Energy, Biological Effects of

DIMITRIOS I. FOTIADIS, *University of Ioannina, Ioannina, Greece,* Clinical Decision Support Systems; Elasticity; Electrocardiogram (ECG): (Automated Diagnosis); Intelligent Patient Monitoring; Wearable Medical Devices

BRIAN FRICKE, *University of Missouri-Kansas City, Kansas City, Missouri,* Dentin–Enamel Junction of Human Teeth

CARLO FRIGO, *Polytechnic university of Milan, Milano, Italy,* Neuromuscular Coordination in Gait, EMG Analysis of

BINGMEI M. FU, *The City College of the City University of New York, New York,* Microvessel Permeability; Transport Across Endothelial Barriers, Modeling of

FLORIN FULGA, *Swinburne University of Technology, Hawthorn, Australia,* Biomolecular Layers: Quantification of Mass and Thickness

STEFAN M. GABRIEL, *Innovative Spinal Technologies, Mansfield, Massachusetts,* Arthroscopic Fixation Devices

DAMIEN GARCIA, *Institut de Recherches Cliniques de Montréal, Montreal, Canada,* Aortic Stenosis and Systemic Hypertension, Modeling of

J. GARCIA-CASADO, *CI2B, Universidad Politécnica de Valencia, Spain,* Intestinal Motility

STEVEN GARD, *Northwestern University, Chicago, Illinois,* Prosthetic Devices and Methods

ROGER GAUMOND, *Pennsylvania State University, University Park, Pennsylvania,* Medical Device Acquisition

NEVZAT G. GENÇER, *Middle East Technical University, Ankara, Turkey,* Electrical Impedance Tomography, Induced Current

ÖMER NEZIH GEREK, *Anadolu University, Eskişehir, Turkey,* Image Coding; Vector Quantization

MAYSAM GHOVANLOO, *North Carolina State University, Raleigh, North Carolina,* Transcutaneous Magnetic Coupling of Power and Data

STAVROULA GIANNOULI, *Pi-Medical Ltd., Athens, Greece,* Computer Assisted Radiation Therapy (CART)

IZABELLA A. GIERAS, *American College of Clinical Engineering, Plymouth Meeting, Pennsylvania,* American College of Clinical Engineering

RONALD MARK GILLIES, *WorleyParsons Advanced Analysis Group, Sydney, Australia,* Mechanical Testing

FRANCISCO J. GIL-SÁNCHEZ, *Polytechnical University of Cartagena, Cartagena, Spain,* Anesthesia Machines

M. GIUGLIANO, *Brain Mind Institute – EPFL, Lausanne, Switzerland and University of Bern, Bern, Switzerland,* Biological Neuronal Networks, Modeling of; Substrate Arrays of Microelectrodes for *in vitro* Electrophysiology

CONSTANTINE GLAROS, *University of Ioannina, Ioannina, Greece,* Wearable Medical Devices

NEIL GLOSSOP, *Traxtal Technologies, Bellaire, Texas,* Computer Aided Surgery

MASSIMILIANO GOBBO, *Università degli Studi di Brescia, Brescia, Italy,* Mechanomyography

YORGOS GOLETSIS, *University of Ioannina, Ioannina, Greece,* Clinical Decision Support Systems

ENRIQUE J. GÓMEZ, *Universidad Politécnica de Madrid, Madrid, Spain,* Diabetes Care, Biomedical and Information Technologies for; Telemedicine: Teleconsultation between Medical Professionals; Telemedicine: Ubiquitous Patient Care

TOMÁS GÓMEZ-CÍA, *University of Seville, Seville, Spain and Virgen del Rocío University Hospital, Seville, Spain,* Skin Lesions: Burns

JULIO GOMIS-TENA, *Universidad Politécnica de Valencia, Valencia, Spain,* Ectopic Activity

MIGUEL ÁNGEL GARCÍA GONZÁLEZ, *Universitat Politecnica de Catalunya, Barcelona, Spain,* Noise in Instrumentation

SHAI GOZANI, *NeuroMetrix, Inc., Waltham, Massachusetts,* F-Wave

BERNHARD GRAIMANN, *University of Technology Graz, BCI-Lab, Graz, Austria,* EEG-Based Brain-Computer Interface System

THOMAS GRAVEN-NIELSEN, *Aalborg University, Aalborg, Denmark,* Muscle Pain

BONNIE GRAY, *Simon Fraser University, Burnaby, Canada,* Photodiodes

MARTIN A. GREEN, *University of New South Wales, Sydney, Australia,* Photovoltaic Cells

WARREN M. GRILL, *Duke University, Durham, North Carolina,* Nerve Stimulation; Neuromuscular Stimulation

SVERRE GRIMNES, *University of Oslo, Oslo, Norway,* Bioimpedance

GARRETT GRINDLE, *University of Pittsburgh, Pittsburgh, Pennsylvania,* Spinal Cord Stimulation Systems

NORBERTO M. GRZYWACZ, *University of Southern California, Los Angeles, California,* Biological Neural Control

JIANJUN GUAN, *University of Pittsburgh, Pittsburgh, Pennsylvania,* Soft Tissue Scaffolds

ANTONIO GUILLAMÓN-FRUTOS, *Polytechnical University of Cartagena, Murcia, Spain,* Fatigue

SONGFENG GUO, *University of Pittsburgh, Pittsburgh, Pennsylvania,* Spinal Cord Stimulation Systems; Wheelchair Engineering

LEONTIOS J. HADJILEONTIADIS, *Aristotle University of Thessaloniki, Thessaloniki, Greece,* Bioacoustic Signals; Higher-Order Statistics

L. E. HALLUM, *University of New South Wales, Sydney, Australia,* Functional Optical Imaging of Intrinsic Signals in Cerebral Cortex

NOAM HAREL, *University of Minnesota Medical School, Minneapolis, Minnesota,* Brain Function, Magnetic Resonance Imaging of

FRANCIS X. HART, *University of the South, Sewanee, TN,* Tissues, Electric Properties of

DAVID R. HAYNES, *University of Adelaide, Adelaide, Australia,* Bone Resorption

BIN HE, *University of Minnesota, Minneapolis, Minnesota,* Electrocardiogram (ECG): Inverse Problem; Electroencephalography (EEG): Inverse Problems

R. M. HEETHAAR, *VU Medical Center, Amsterdam, The Netherlands,* Forward Problem

JENNI HEINO, *Helsinki University of Technology, Espoo, Finland,* Near-Infrared Spectroscopic Imaging

RAPHAEL HEINZER, *Brigham and Women's Hospital, Boston, Massachusetts,* Sleep Apnea Syndrome

TIMOTHY HENNING, *Abbott Laboratories, Abbott Park, Illinois,* Technology Assessment of Medical Devices

M. ELENA HERNANDO, *Universidad Politécnica de Madrid, Madrid, Spain,* Diabetes Care, Biomedical and Information Technologies for; Telemedicine: Teleconsultation between Medical Professionals; Telemedicine: Ubiquitous Patient Care

WALTER HERZOG, *University of Calgary, Calgary, Alberta, Canada,* Articular Cartilage

MICHAEL R. HILLMAN, *Wolfson Centre, Royal United Hospital Bath, Somerset, United Kingdom,* Assistive Robotics

HIIE HINRIKUS, *Tallinn Technical University, Tallinn, Estonia,* Electromagnetic Waves; Microwave Imaging

ULRICH HOFFMANN, *EPFl-STI-ITS-LTS, Lausanne, Switzerland,* Human Brain Interface: Signal Processing and Machine Learning

NEVILLE HOGAN, *Massachusetts Institute of Technology, Cambridge, Massachusetts,* Robotic Rehabilitation Therapy

CURTIS F. HOLMES, *Greatbatch Incorporated, Clarence, New York,* Batteries for Implantable Biomedical Applications

M. P. HORAN, *Medical University of South Carolina, Charleston, South Carolina,* Bone, Mechanical Testing of

JUNICHI HORI, *Niigata University, Niigata, Japan,* Electroencephalography (EEG): Inverse Problems

FERNANDO HORNERO, *Hospital General Universitario, Valencia, Spain,* Atrial Fibrillation and Atrial Flutter

JINGTAO HUANG, *Louisiana Tech University, Ruston, Louisiana,* Obstructive Sleep Apnea: Electrical Stimulation Treatment

BERNARD HUDGINS, *University of New Brunswick, New Brunswick, Canada,* Myoelectric Control of Powered Upper-Limb Prostheses

DONNA L. HUDSON, *University of California-San Francisco, Fresno, California,* Medical Expert Systems; Pattern Classification

MARK S. HUMAYUN, *University of Southern California, Los Angeles, California,* Semiconductor-Based Implantable Prosthetic Devices

J. D. HUMPHREY, *Texas A&M University, College Station, Texas,* Vascular Growth and Remodeling

GEORGE K. HUNG, *Rutgers University, Piscataway, New Jersey,* Oculomotor Control

KEVIN HUNG, *The Chinese University of Hong Kong, Shatin, Hong Kong,* Telemedicine

HELMUT HUTTEN, *University of Technology, Graz, Austria,* Cardiac Pacemakers

NED H. C. HWANG, *National Health Research Institutes, Zhunan, Taiwan,* Cardiac Valves

WILLIAM A. HYMAN, *Texas A&M University, College Station, Texas,* Medical Devices, Design and Modification of

RISTO J. ILMONIEMI, *Helsinki University of Technology, Espoo, Finland and Nexstim Ltd., Helsinki, Finland,* Transcranial Magnetic Stimulation

OLGA A. IMAS, *Medical College of Wisconsin, Milwaukee, Wisconsin,* Coherence

EILEEN INGHAM, *University of Leeds, Leeds, United Kingdom,* Allogeneic Cells and Tissues

MICHAEL F. INSANA, *University of Illinois, Urbana, Illinois,* Elasticity Imaging; Ultrasonic Imaging

CINTHIA ITIKI, *University of Sao Paolo, Sao Paolo, Brazil,* Discrete Fourier Transform

ELENA P. IVANOVA, *Swinburne University of Technology, Melbourne, Australia,* Adhesion of Bacteria

LAWRENCE D. JACOBS, *University of Chicago Hospitals, Chicago, Illinois,* Echocardiography

MICHEL Y. JAFFRIN, *Technological University of Compiegne, Compiegne Cedex, France,* Plasma Separation and Purification by Membrane

NORIAH JAMAL, *University of Malaya, Kuala Lumpur, Malaysia,* Mammography

JENS INGEMANN JENSEN, *University of Rochester School of Medicine and Dentistry, Rochester, New York,* Hypoxia

TENG-FEI JIANG, *Tsinghua University, Beijing, China,* Tissue-Engineered Bone

JACINTO JIMÉNEZ-MARTÍNEZ, *Polytechnical University of Cartagena, Cartagena, Spain,* Anesthesia Machines

JAMES A. JOHNSON, *The University of Western Ontario, London, Ontario, Canada,* Computer Aided Design

C.W. JONES, *Prince of Wales Hospital, Sydney, Australia; University of Western Australia, Perth, Australia and University of Sydney, Sydney, Australia,* Sutures

JULIAN R. JONES, *Imperial College London, London, United Kingdom,* Scaffolds for Cell and Tissue Engineering

AMY JORDAN, *Brigham and Women's Hospital, Boston, Massachusetts,* Sleep Apnea Syndrome

PETER JORDAN, *Weill Medical College of Cornell University, New York, New York,* Cardiac Arrhythmia

EMIL JOVANOV, *University of Alabama in Huntsville, Huntsville, Alabama,* Multimodal Presentation of Biomedical Data; Virtual Instrumentation

ALEKSANDAR JOVIČIĆ, *University of Illinois at Urbana-Champaign, Urbana, Illinois,* Parametric Adaptive Identification and Kalman Filter; State-Space Methods

TIMO KAJAVA, *Helsinki University of Technology, Espoo, Finland,* Near-Infrared Spectroscopic Imaging

RUBEN Y. KANNAN, *University College London, London, United Kingdom,* Vascular Networks

DIMITRIOS G. KATEHAKIS, *Foundation for Research and Technology – Hellas, Heraklion, Crete, Greece,* Electronic Health Record

TOIVO KATILA, *Helsinki University of Technology, Espoo, Finland,* Near-Infrared Spectroscopic Imaging

RAIN KATTAI, *Tallinn Technical University, Tallinn, Estonia,* Bio-Optics: Optical Measurement of Pulse Wave Transit Time

J. LAWRENCE KATZ, *University of Missouri-Kansas City, Kansas City, Missouri,* Dentin

ANNMARIE KELLEHER, *University of Pittsburgh, Pittsburgh, Pennsylvania,* Wheelchair Engineering

D. J. KELLY, *Trinity College, Dublin, Ireland,* Stents

ERNST KENNDLER, *University of Vienna, Vienna, Austria,* Capillary Electrophoresis

STEPHEN KERCEL, *University of New England, Brunswick, Maine,* Human Nervous System, Noninvasive Coupling of Electronically Generated Data into

OMAR S. KHALIL, *Abbott Laboratories, Abbott Park, Illinois,* Metabolites, NonInvasive Optical Measurements of

WILLIAM KING, *Abbott Laboratories, Abbott Park, Illinois,* Technology Assessment of Medical Devices

J. H. KINNEY, *Lawrence Livermore National Laboratory, Livermore, California,* Cortical Bone Fracture

RICHARD KNIGHT, *University of Leeds, Leeds, United Kingdom,* Allogeneic Cells and Tissues

VIVIAN KNIGHT, *David Geffen School of Medicine at UCLA, Los Angeles, California,* DNA Sequencing

GEORGE K. KNOPF, *The University of Western Ontario, London, Ontario, Canada,* Computer Aided Design; Data Visualization

ANDRÉ FABIO KOHN, *University of São Paulo, São Paulo, Brazil,* Autocorrelation and Cross Correlation Methods

PETER KOKOL, *University of Maribor, Maribor, Slovenia,* Knowledge Engineering; Software Agents

XUAN KONG, *NeuroMetrix, Inc., Waltham, Massachusetts,* F-Wave

GEORGE KONTAXAKIS, *Universidad Politécnica de Madrid, Madrid, Spain,* Positron Emission Tomography (PET)

ALICIA M. KOONTZ, *VA Pittsburgh HealthCare System, Pittsburgh, Pennsylvania and University of Pittsburgh, Pittsburgh, Pennsylvania,* Rehabilitation Biomechanics

MICHAEL J. KORENBERG, *Queen's University, Kingston, Ontario, Canada,* Gene Expression Profiles, Nonlinear System Identification In

RAVI KOTHARI, *IIT Hauz Khas, New Delhi, India,* Data Mining

KALLE KOTILAHTI, *Helsinki University of Technology, Espoo, Finland,* Near-Infrared Spectroscopic Imaging

D. KOUTSONANOS, *Informatics & Telematics Institute, C.E.R.T.H., Thessaloniki, Greece,* Haptic Interaction in Medical Virtual Environments

WADE F. KRAUSE, *University of Texas Health Science Center, San Antonio, Texas,* Platelet-Rich Plasma in Bone Repair

HERMANO I. KREBS, *Massachusetts Institute of Technology, Cambridge, Massachusetts and Weill Medical College of Cornell University, White Plains, New York,* Robotic Rehabilitation Therapy

J. YASHA KRESH, *Drexel University College of Medicine, Philadelphia, Pennsylvania,* Robotic Surgery

J. J. KRUZIC, *Oregon State University, Corvallis, Oregon,* Cortical Bone Fracture

OMAR YUSEF KUDSI, *Brigham and Women's Hospital, Harvard Medical School, Boston, Massachusetts,* Skin

ANDREW M. KWARCIAK, *University of Pittsburgh, Pittsburgh, Pennsylvania,* Wheelchair Engineering

EFTHYVOULOS KYRIACOU, *Cyprus Institute of Neurology and Genetics, Nicosia, Cyprus,* Ultrasonic Imaging of Carotid Atherosclerosis

PABLO LAGUNA, *Zaragoza University, Spain,* Electrocardiogram (ECG) Signal Processing

C. LALLY, *Dublin City University, Glasnevin, Dublin, Ireland,* Stents

TANIA LAM, *Balgrist University Hospital, Zürich, Switzerland and University of British Columbia, Vancouver, BC, Canada,* Gait Retraining After Neurological Disorders

LUIGI LANDINI, *University of Pisa, Pisa, Italy,* Cellular and Molecular Imaging

JAANUS LASS, *Tallinn Technical University, Tallinn, Estonia,* Bio-Optics: Optical Measurement of Pulse Wave Transit Time

GIANLUCA LAZZI, *North Carolina State University, Raleigh, North Carolina,* Transcutaneous Magnetic Coupling of Power and Data

JENNIE B. LEACH, *University of Maryland, Baltimore, Maryland,* Tissue-Engineered Peripheral Nerve

MARIA J. LEDESMA-CARBAYO, *Universidad Politecnica Madrid, Madrid, Spain,* Cardiac Imaging

J. MICHAEL LEE, *Dalhousie University, Halifax, Nova Scotia, Canada,* Tissue Mechanics

JEN-SHIH LEE, *University of California-San Diego, La Jolla, California and University of Virginia, Charlottesville, Virginia,* American Institute for Medical and Biological Engineering

CECILE LEGALLAIS, *Technological University of Compiegne, Compiegne Cedex, France,* Plasma Separation and Purification by Membrane

PETER A. LEWIN, *Drexel University, Philadelphia, Pennsylvania,* Piezoelectric Devices in Biomedical Applications

NIANHUA LI, *Indiana University, Purdue University, Indianapolis, Indiana,* Biological Database Integration; Distributed Processing; Information Retrieval in Biomedical Research

ARISTIDIS LIKAS, *University of Ioannina, Ioannina, Greece,* Clinical Decision Support Systems; Electrocardiogram (ECG): (Automated Diagnosis); Intelligent Patient Monitoring; Wearable Medical Devices

K. LIM, *University of Auckland, Auckland, New Zealand,* Force Measurement

ZHIYUE LIN, *University of Kansas Medical Center, Kansas City, Kansas,* Electrogastrography (EGG)

JOHN M. LIPCHITZ, *Innovative Spinal Technologies, Mansfield, Massachusetts,* Arthroscopic Fixation Devices

TARMO LIPPING, *Tampere University of Technology, Pori, Finland,* Digital Filters; Higher-Order Spectral Analysis

JING LIU, *Chinese Academy of Sciences & Tsinghua University, Beijing, China,* Bioheat Transfer Model

LEI LIU, *University of Illinois at Urbana-Champaign, Urbana, Illinois,* Bioinformatics

YANG LIU, *Indiana University, Purdue University, Indianapolis, Indiana,* Biological Database Integration

WENTAI LIU, *University of California, Santa Cruz, California,* Semiconductor-Based Implantable Prosthetic Devices

GEOFFERY R. LOCKWOOD, *Queen's University, Kingston, Ontario, Canada,* Ultrasonic Transducers for Medical Imaging

CHRISTOPH H. LOHMANN, *University of Hamburg-Eppendorf, Hamburg, Germany,* Platelet-Rich Plasma in Bone Repair

CHRISTOS LOIZOU, *Intercollege Limassol Campus, Limassol, Cyprus,* Ultrasonic Imaging of Carotid Atherosclerosis

QUAN LONG, *Imperial College, London, United Kingdom,* Blood Flow Simulation, Patient-Specific *in-vivo*

GUY S. LONGOBARDO, *New Jersey Medical School, Newark, New Jersey,* Pulmonary Medicine, Applications of Biomedical Engineering to

KOK-KEONG LOO, *Brunel University, Uxbridge, United Kingdom,* Wide Area Networks

EDMUND F. LOPRESTI, *AT Sciences, Pittsburgh, Pennsylvania,* Cognitive Assistive Technology

NOAH LOTAN, *Technion – Israel Institute of Technology, Haifa, Israel,* Artificial Blood

N.H. LOVELL, *University of New South Wales, Sydney, Australia and Australian Technology Park, Eveleigh, Australia,* Electrical Activity in Cardiac Tissue, Modeling of; Functional Optical Imaging of Intrinsic Signals in Cerebral Cortex

DONGHUI LU, *Shanghai University, Shanghai, China,* Computed Tomography

WILLIAM W. LU, *University of Hong Kong, Hong Kong,* Gait Patterns

CLIFFORD K. C. LUI, *The Chinese University of Hong Kong, Hong Kong,* Wireless Biomedical Sensing

B. L. LUK, *City University of Hong Kong, Hong Kong,* Wireless Biomedical Sensing

LARS LÜNENBURGER, *Balgrist University Hospital, Zürich, Switzerland and University of British Columbia, Vancouver, BC, Canada,* Gait Retraining After Neurological Disorders

KENNETH R. LUTCHEN, *Boston University, Boston, Massachusetts,* Lung Tissue Viscoelasticity

MAI LY, *University of New South Wales, Sydney, Australia,* Wound Healing

TING MA, *The Chinese University of Hong Kong, Shatin, NT, Hong Kong,* Biometrics

JASON MAES, *Michigan Technological University, Houghton, Michigan,* Meniscal Replacements

MALIKA MAHOUI, *Indiana University, Purdue University, Indianapolis,* Information Retrieval in Biomedical Research

ATUL MALHOTRA, *Brigham and Women's Hospital, Boston, Massachusetts,* Sleep Apnea Syndrome

ROBERT MALKIN, *Duke University, Durham, North Carolina, and Engineering World Health, Memphis, Tennessee,* Defibrillation; Health Care Technology for the Developing World

CLAUDIA MANFREDI, *Universita degli Studi di Firenze, Firenze, Italy,* Voice Analysis

KEEFE B. MANNING, *Pennsylvania State University, University Park, Pennsylvania,* Flow Measurement

ALI J. MARIAN, *Baylor College of Medicine, Houston, Texas,* Cardiac Hypertrophy

DONALD E. MARLOWE, *U.S. Food and Drug Administration, Rockville, Maryland,* Biomedical Products, International Standards for

LUIS MARTÍ-BONMATÍ, *Hospital Universitari Dr. Peset, Valencia, Spain,* NMR Imaging

FERRAN SILVA MARTÍNEZ, *Universitat Politecnica de Catalunya, Barcelona, Spain,* Electrical Safety; Electromagnetic Interference and Compatibility

FRANCISCO MARTÍNEZ-GONZALEZ, *Polytechnical University of Cartagena, Murcia, Spain,* Fatigue

SYLVAIN MARTEL, *Ecole Polytechnique Montreal, Montreal, Quebec, Canada,* Micromechanical Devices

COLIN MARTIN, *North Glasgow University Hospitals, Glasgow, Scotland,* Ionizing Radiation, Biological Effects of; Radiation Safety

J. L. MARTINEZ-DE-JUAN, *CI2B, Universidad Politécnica de Valencia, Spain,* Intestinal Motility

S. MARTINOIA, *University of Genova, Genova, Italy,* Substrate Arrays of Microelectrodes for *in vitro* Electrophysiology

ØRJAN G. MARTINSEN, *University of Oslo, Oslo, Norway,* Bioimpedance

GIUSEPPE MARUCCIO, *National Nanotechnology Laboratory of CNR-INFM, Lecce, Italy,* Molecular Electronics

SHARMEEN MASOOD, *Imperial College, London, United Kingdom,* Blood Flow Measurement

CHRISTOS V. MASSALAS, *University of Ioannina, Ioannina, Greece,* Elasticity

JEAN L. MCCRORY, *University of Pittsburgh, Pittsburgh, Pennsylvania,* Rehabilitation Biomechanics

HUGH J. MCDERMOTT, *The University of Melbourne, East Melbourne, Australia,* Cochlear Implants

W. BARRY MCKAY, *Baylor College of Medicine, Houston, Texas,* Spasticity and Upper Motor Neuron Dysfunction

CRAIG MCLACHLAN, *BioQ Devices Pty Ltd, Brisbane, Australia and Reperfusion Pty Ltd, Sydney, Australia,* Stenosis and Thrombosis

JACQUELYN MCMILLAN, *Georgia Institute of Technology, Atlanta, Georgia,* Platelet-Rich Plasma in Bone Repair

KALJU MEIGAS, *Tallinn Technical University, Tallinn, Estonia,* Bio-Optics: Optical Measurement of Pulse Wave Transit Time

WILLIAM W. MELEK, *Alpha Global IT Inc., North York, Ontario, Canada,* Fuzzy Neural Networks

JAMES MELROSE, *Royal North Shore Hospital & University of Sydney, St. Leonards, Australia,* Adhesion of Cells to Biomaterials; Extracellular Matrix

YITZHAK MENDELSON, *Worcester Polytechnic Institute, Worcester, MA,* Pulse Oximetry

ROBERTO MERLETTI, *Politechnic of Torino, Torino, Italy,* Electromyography (EMG) of Pelvic Floor Muscles; Myoelectric Manifestations of Muscle Fatigue

ROBERT MERRIFIELD, *Imperial College, London, United Kingdom,* Blood Flow Simulation, Patient-Specific *in-vivo*

MATEJ MERTIK, *University of Maribor, Maribor, Slovenia,* Knowledge Engineering

CLARK MEYER, *Texas A&M University, College Station, Texas,* Vascular Mechanics

WILLIAM C. MEYERS, *Drexel University College of Medicine, Philadelphia, Pennsylvania,* Robotic Surgery

LAMPROS MICHALIS, *University of Ioannina, Ioannina, Greece,* Electrocardiogram (ECG): (Automated Diagnosis)

ALEX MIHAILIDIS, *University of Toronto, Toronto, Ontario, Canada,* Cognitive Assistive Technology

M. MIHELIN, *University Institute of Clinical Neurophysiology, Ljubljana, Slovenia,* Electromyography (EMG), Needle

DAMIJAN MIKLAVČIČ, *University of Ljubljana, Ljubljana, Slovenia,* Electroporation; Tissues, Electric Properties of

ZINA BEN MILED, *Indiana University, Purdue University, Indianapolis, Indiana,* Biological Database Integration; Distributed Processing; Information Retrieval in Biomedical Research

MART MIN, *Tallinn University of Technology, Tallinn, Estonia,* Biomedical Electronics

MARTIN MINTCHEV, *University of Calgary, Alberta, Canada,* Sampling Theorem and Aliasing in Biomedical Signal Processing

NATASA MITIK-DINEVA, *Swinburne University of Technology, Melbourne, Australia,* Adhesion of Bacteria

RICHARD MOFFITT, *Georgia Tech University, Atlanta, Georgia,* Microarray Data Analysis

MARTA MONSERRAT, *Universidad Politécnica de Valencia, Valencia, Spain,* Ectopic Activity

FULGENCIO MONTILLA, *Universidad Politécnica de Valencia, Valencia, Spain,* Ischemia

JAMES E. MOORE, JR., *Texas A&M University, College Station, Texas,* Vascular Mechanics

RAJSHREE MOOTANAH, *Anglia Ruskin University, Chelmsford, Essex, United Kingdom,* Pressure Sensors

PIETRO G. MORASSO, *University of Genova, Genova, Italy,* Human Motion Analysis

DAVID MORATAL, *Universitat Politècnica de València, Valencia, Spain,* NMR Imaging

MICHAEL R. MORENO, *Texas A&M University, College Station, Texas,* Vascular Mechanics

DAISUKE MORI, *Texas A&M University, College Station, Texas,* Vascular Mechanics

EVELYN MORIN, *Queen's University, Kingston, Canada,* Myoelectric Signal Processing

J. W. MORLEY, *University of New South Wales, Sydney, Australia,* Functional Optical Imaging of Intrinsic Signals in Cerebral Cortex

JENNIFER MORSE, *University of San Diego, San Diego, California,* Electric Shock

MICHAEL MORSE, *University of San Diego, San Diego, California,* Electric Shock

KONSTANTINOS MOUSTAKAS, *Informatics & Telematics Institute, Thessaloniki, Greece*, Deformable Objects, Interactive Simulation of; Haptic Interaction in Medical Virtual Environments

ROBERT MUNZNER, *Medical Device Consultant, Schuyler, Virginia*, Medical Devices: Regulations, Codes, and Standards

DAVID J. MURRAY-SMITH, *University of Glasgow, Scotland, United Kingdom*, Neuromuscular Systems

SVERRE MYHRA, *University of Oxford, Yarnton, United Kingdom*, Atomic Force Microscopy

DAVID J. NAGEL, *The George Washington University, Washington, DC*, Microsensors and Nanosensors

KAYVAN NAJARIAN, *University of North Carolina, Charlotte, North Carolina*, Back-Propagation; Genetic Algorithms; Maximum Likelihood Estimation; Neural Networks: Applications in Biomedical Engineering; Protein Structure, Folding, and Conformation; Software Engineering

R. K. NALLA, *Intel Corporation, Chandler, Arizona*, Cortical Bone Fracture

K. S. NARAYAN, *Jawaharlal Nehry Centre for Advanced Scientific Research, Bangalore, India*, Photoconducting Devices

HOMER NAZERAN, *University of Texas, El Paso, Texas*, Heart Rate Variability (HRV): Sleep Disordered Breathing

ROBERT M. NEREM, *Institute of Bioengineering and Bioscience, Georgia Institute of Technology, Atlanta, Georgia*, Tissue Engineering Blood Vessels

CHRISTA NEUPER, *University of Technology Graz, BCI-Lab, Graz, Austria*, EEG-Based Brain-Computer Interface System

KWAN-HOONG NG, *University of Malaya, Kuala Lumpur, Malaysia*, Mammography

DAVID NICKERSON, *The University of Auckland, Auckland, New Zealand*, Cardiac Electromechanical Coupling

ANDREW NICOLAIDES, *Cyprus Institute of Neurology and Genetics, Nicosia, Cyprus*, Ultrasonic Imaging of Carotid Atherosclerosis

DAN NICOLAU, *Swinburne University of Technology, Hawthorn, Australia*, Biocomputation; Cell Patterning; Biomolecular Layers: Quantification of Mass and Thickness

DAN NICOLAU, JR., *University of Melbourne, Melbourne, Australia and Swinburne University of Technology, Melbourne, Australia*, Biocomputation

POUL M. F. NIELSEN, *University of Auckland, Auckland, New Zealand*, Extracorporeal Electrodes

G. NIKOLAKIS, *Informatics & Telematics Institute, C.E.R.T.H., Thessaloniki, Greece*, Haptic Interaction in Medical Virtual Environments

I. NISSILÄ, *Helsinki University of Technology, Espoo, Finland*, Near-Infrared Spectroscopic Imaging

TOMMI NOPONEN, *Helsinki University of Technology, Espoo, Finland*, Near-Infrared Spectroscopic Imaging

ŽELJKO OBRENOVIĆ, *Centrum voor Wiskunde en Informatica, Amsterdam, Netherlands*, Virtual Instrumentation; Multimodal Presentation of Biomedical Data

MARI OGIUE-IKEDA, *University of Tokyo, Tokyo, Japan*, Bioelectricity and Biomagnetism

BRENDA M. OGLE, *Mayo Clinic College of Medicine, Rochester, Minnesota*, Xenotransplantation

HIROYUKI OHSHIMA, *Tokyo University of Science, Tokyo, Japan*, Electrophoresis

YOSHIWO OKAMOTO, *Chiba Institute of Technology, Chiba, Japan*, Electrocardiogram (ECG): Inverse Problem

ERIK OLOFSEN, *Leiden University Medical Center (LUMC), Leiden, The Netherlands*, Hypercapnia

CLAUDIO ORIZIO, *Università degli Studi di Brescia, Brescia, Italy*, Mechanomyography

A. ORTMANN, *University of Pittsburgh, Pittsburgh, Pennsylvania*, Sensory Aids

BOONE B. OWENS, *University of Minnesota, Minneapolis, Minnesota*, Batteries for Implantable Biomedical Applications

LARRY D. PAARMANN, *Wichita State University, Wichita, Kansas*, Analog to Digital Conversion

MÓNICA PADILLA, *University of Southern California, Los Angeles, California*, Biological Neural Control

MICHAEL R. PAGNOTTO, *University of Pittsburgh, Pittsburgh, Pennsylvania*, Tissue-Engineered Cartilage

C. PALMER, *University of Pittsburgh, Pittsburgh, Pennsylvania*, Sensory Aids

YVONNE PAMULA, *Women's and Children's Hospital, Adelaide, Australia*, Heart Rate Variability (HRV): Sleep Disordered Breathing

STAVROS M. PANAS, *Aristotle University of Thessaloniki, Thessaloniki, Greece*, Bioacoustic Signals; Higher-Order Statistics

DORIN PANESCU, *Refractec, Irvine, California*, Medical Device Industry

CHRISTINE A. PANGBORN, *Rice University, Houston, Texas*, Knee Meniscus, Biomechanics of

MARIOS PANTZIARIS, *Cyprus Institute of Neurology and Genetics, Nicosia, Cyprus*, Ultrasonic Imaging of Carotid Atherosclerosis

ATHANASSIOS PAPADOPOULOS, *University of Ioannina, Ioannina, Greece*, Clinical Decision Support Systems

COSTAS PAPALOUKAS, *University of Ioannina, Ioannina, Greece*, Electrocardiogram (ECG): (Automated Diagnosis)

JEANETTE PAPP, *David Geffen School of Medicine at UCLA, Los Angeles, California*, DNA Sequencing

GIANFRANCO PARATI, *University of Milano-Bicocca and S.Luca Hospital, Milano, Italy*, Arterial Blood Pressure Processing

GRACE E. PARK, *Purdue University, West Lafayette, Indiana*, Cartilage Scaffolds

KYUNGSOO PARK, *Yonsei University, Seoul, Korea*, Pharmacokinetic and Pharmacodynamic Control

TOOMAS PARVE, *Tallinn University of Technology, Tallinn, Estonia*, Biomedical Electronics

MAGDA PASSATORE, *Universita' Di Torino, Torino, Italy*, Muscle Sensory Receptors

CONSTANTINOS S. PATTICHIS, *University of Cyprus, Nicosia, Cyprus*, Ultrasonic Imaging of Carotid Atherosclerosis

MARIOS S. PATTICHIS, *University of New Mexico, Albuquerque, New Mexico*, Ultrasonic Imaging of Carotid Atherosclerosis

ABHIJIT PATWARDHAN, *University of Kentucky, Lexington, Kentucky*, Respiratory Sinus Arrhythmia

NATAŠA PAVŠELJ, *University of Ljubljana, Ljubljana, Slovenia*, Tissues, Electric Properties of

ANTONIO PEDOTTI, *Polytechnic university of Milan, Milano, Italy*, Neuromuscular Coordination in Gait, EMG Analysis of

THOMAS PENZEL, *Hospital of Philipps-University, Marburg, Germany*, Sleep Laboratory

GERT PFURTSCHELLER, *University of Technology Graz, BCI-Lab, Graz, Austria*, EEG-Based Brain-Computer Interface System

CHANDLER A. PHILLIPS, *Wright State University, Dayton, Ohio*, Haptic Devices and Interfaces; Human Factors Engineering

ITZIK PINHAS, *Tel Aviv University, Tel Aviv, Israel*, Heart Rate Variability (HRV); Hypertension: Time-Frequency Analysis for Early Diagnosis

JEFFREY L. PLATT, *Mayo Clinic College of Medicine, Rochester, Minnesota*, Xenotransplantation

ROBI POLIKAR, *Rowan University, Glassboro, New Jersey*, Pattern Recognition

BOHDAN POMAHAC, *Brigham and Women's Hospital, Harvard Medical School, Boston, Massachusetts*, Skin

J. L. PONCE, *Hospital Universitario "La Fe" de Valencia, Spain*, Intestinal Motility

LAURA A. POOLE-WARREN, *University of New South Wales, Sydney, Australia*, Wound Healing

P. W. F. POON, *National Cheng Kung University, Taiwan*, Evoked Potentials, Adaptive Filtering of

VINCENZO POSITANO, *CNR Institute of Clinical Physiology of Pisa, Pisa, Italy*, Cellular and Molecular Imaging

PETRA POVALEJ, *University of Maribor, Maribor, Slovenia*, Knowledge Engineering

MARCO POZZO, *Karolinska Institutet, Stockholm, Sweden*, Electromyography (EMG), Electrodes and Equipment For

MANUEL PRADO, *University of Seville, Sevilla, Spain*, Simulation Languages

NATHAN PRAHLOW, *Indiana University, Indianapolis, Indiana*, Electrophysiology

SATYA PRAKASH, *McGill University, Montreal, Quebec, Canada*, Microencapsulation

P. J. PRENDERGAST, *Trinity College, Dublin, Ireland*, Stents

VASILIOS C. PROTOPAPPAS, *University of Ioannina, Ioannina, Greece*, Elasticity; Intelligent Patient Monitoring

MARKO PUC, *University of Ljubljana, Ljubljana, Slovenia*, Electroporation

ANDREW J. PULLAN, *University of Auckland, Auckland, New Zealand*, Extracorporeal Electrodes

W. QIU, *State University of New York, Plattsburgh, New York*, Evoked Potentials, Adaptive Filtering of

MICHAL M. RADAI, *Tel Aviv University, Tel Aviv, Israel*, Electrical Impedance Technique for Cryosurgery Monitoring

DEJAN RAKOVIĆ, *University of Belgrade, Serbia, Montenegro*, Ionic Channels

MARCO RAMONI, *Harvard Medical School and Harvard Partners Center for Genetics and Genomics, Boston, Massachusetts*, Proteomics

VITO MARCO RANIERI, *Ospendale S. Giovanni Battista, Italy*, Respiration Measurements

DON M. RANLY, *Georgia Institute of Technology, Atlanta, Georgia*, Platelet-Rich Plasma in Bone Repair

TONG KAI YU RAYMOND, *The Hong Kong Polytechnic University, Hung Hom, Hong Kong*, Functional Electrical Stimulation (FES) for Stroke Rehabilitation

JOHN M. REID, *Drexel University, Philadelphia, Pennsylvania*, Piezoelectric Devices in Biomedical Applications

JAVIER REINA-TOSINA, *University of Seville, Seville, Spain*, Skin Lesions: Burns

IOANNIS T. REKANOS, *Technological and Educational Institute of Serres, Serres, Greece*, Bioacoustic Signals; Higher-Order Statistics

ANDREW J. RENTSCHLER, *University of Pittsburgh, Pittsburgh, Pennsylvania*, Rehabilitation Engineering: An Overview; Intelligent Mobility Aids

DANIEL W. REPPERGER, *Air Force Research Laboratory, Dayton, Ohio*, Evolutionary Algorithms; Haptic Devices and Interfaces; Human Factors Engineering

DAVID B. REYNOLDS, *Wright State University, Dayton, Ohio*, Human Factors Engineering

IEAD REZEK, *University of Oxford, Oxford, United Kingdom*, Multivariate Biomedical Signal Processing

ELAHEH RHABAR, *Texas A&M University, College Station, Texas*, Vascular Mechanics

BRYCE S. RICHARDS, *Australian National University, Canberra, Australia*, Photovoltaic Cells

REBECCA RICHARDS-KORTUM, *Rice University, Houston, Texas*, Epithelial Pre-cancers and Cancers, Optical Technologies for Detection and Diagnosis of

ROBERT RIENER, *Balgrist University Hospital, Zürich, Switzerland and Swiss Federal Institute of Technology, (ETH), Zürich, Switzerland*, Gait Retraining After Neurological Disorders

JEVGENI RIIPULK, *Tallinn University of Technology, Tallinn, Estonia*, Microwave Imaging

ROSS RINALDI, *National Nanotechnology Laboratory of CNR-INFM, Lecce, Italy*, Molecular Electronics

R. O. RITCHIE, *University of California, Berkeley, California*, Cortical Bone Fracture

PERE J. RIU, *Universitat Politecnica de Catalunya, Barcelona, Spain*, Electrical Safety; Electromagnetic Interference and Compatibility

GIUSEPPE RIVA, *Istituto Auxologico Italiano, Milan, Italy and Università Cattolica del Sacro Cuore, Milan, Italy*, Virtual Reality

LAURA M. ROA, *University of Seville, Seville, Spain*, Simulation Languages; Skin Lesions: Burns

SILVESTRO ROATTA, *Universita' Di Torino, Torino, Italy*, Muscle Sensory Receptors

JOAQUIN ROCA-DORDA, *Polytechnical University of Cartagena, Murcia, Spain*, Anesthesia Machines; Assistive Technology; Fatigue

JOAQUÍN ROCA-GONZÁLEZ, *Polytechnical University of Cartagena, Murcia, Spain*, Anesthesia Machines; Assistive Technology; Fatigue

BLANCA RODRÍGUEZ, *Universidad Politécnica de Valencia, Valencia, Spain and Tulane University, New Orleans, Louisiana*, Ischemia

KRISTINA M. ROPELLA, *Marquette University, Milwaukee, Wisconsin*, Coherence

JAMES A. ROWLEY, *Wayne State University, Detroit, Michigan*, Upper Airway Mechanics

PAUL RUFFIN, *Redstone Arsenal, Alabama*, Fiber Optic Sensors

ZIAD SAAD, *National Institute of Mental Health, Bethesda, Maryland*, Hilbert Transform

SUBRATA SAHA, *SUNY Downstate Medical Center, Brooklyn, New York*, Ethical Issues in Biomedical Research

MESUT SAHIN, *Louisiana Tech University, Ruston, Louisiana*, Obstructive Sleep Apnea: Electrical Stimulation Treatment

JAVIER SAIZ, *Universidad Politécnica de Valencia, Valencia, Spain*, Ectopic Activity; Ischemia

HIROMI SAKAI, *Waseda University, Tokyo, Japan*, Blood Substitutes

RODNEY SALO, *Guidant Corporation, St. Paul, Minnesota*, Biomedical Electronics

MAR SANEIRO-SILVA, *Polytechnical University of Cartagena, Cartagena, Spain*, Assistive Technology

ARCHANA SANGOLE, *McGill University, Montreal, Quebec, Canada*, Data Visualization

VITTORIO SANGUINETI, *University of Genova, Genova, Italy*, Human Motion Analysis

MARIA FILOMENA SANTARELLI, *CNR Institute of Clinical Physiology of Pisa, Pisa, Italy*, Cellular and Molecular Imaging

ANDRES SANTOS, *Universidad Politecnica Madrid, Madrid, Spain*, Cardiac Imaging

PAOLA SBRICCOLI, *Istituto Universitario di Scienze Motorie, Rome, Italy*, Muscle, Skeletal

KATJA SCHENKE-LAYLAND, *David Geffen School of Medicine, Los Angeles, California*, Tissue Engineering of Cardiac Tissues

F. B. SCHREUDER, *Prince of Wales Hospital, Sydney, Australia and St Lukes Hospital, Hand Unit, Sydney, Australia*, Sutures

ZVI SCHWARTZ, *Georgia Institute of Technology, Atlanta, Georgia; University of Texas Health Science Center, San Antonio, Texas and Hebrew University Hadassah, Jerusalem, Israel*, Platelet-Rich Plasma in Bone Repair

L. KEITH SCOTT, *Louisiana State University Health Sciences Center, Shreveport, Louisiana*, Sepsis, Extracorporeal Membrane Oxygenation

KATHERINE SEELMAN, *University of Pittsburgh, Pittsburgh, Pennsylvania*, Sensory Aids

ALEXANDER M. SEIFALIAN, *University College London, London, United Kingdom*, Vascular Networks

MASAKI SEKINO, *University of Tokyo, Tokyo, Japan*, Bioelectricity and Biomagnetism

JAMES L. SHERLEY, *Massachusetts Institute of Technology, Cambridge, Massachusetts*, Biochemical Pathways Research

ARTHUR SHERWOOD, *National Institute on Disability and Rehabilitation Research, Washington, DC*, Spasticity and Upper Motor Neuron Dysfunction

ADRIAN C. SHIEH, *Rice University, Houston, Texas*, Musculoskeletal Cell Mechanics

ZVIKA SHINAR, *Tel Aviv University, Tel Aviv, Israel*, Heart Rate Variability (HRV)

RICHARD W. SIEGEL, *Rensselaer Polytechnic Institute, Troy, New York*, Nanophase Materials

RICARDO SILVA, *Pennsylvania State University, University Park, Pennsylvania and Simon Bolivar University, Caracas, Venezuela*, Medical Device Acquisition

DAN T. SIMIONESCU, *Clemson University, Clemson, South Carolina*, Artificial Heart Valves

RICHARD C. SIMPSON, *University of Pittsburgh, Pittsburgh, Pennsylvania*, Intelligent Mobility Aids

SARIT SIVAN, *Technion – Israel Institute of Technology, Haifa, Israel*, Artificial Blood

MOHANASANKAR SIVAPRAKASAM, *University of California, Santa Cruz, California*, Semiconductor-Based Implantable Prosthetic Devices

HAMID SOLTANIAN-ZADEH, *Henry Ford Hospital, Detroit, Michigan*, Hippocampus

SANJAY P. SOOD, *C-DAC School of Advanced Computing, Quatre Bornes, Mauritius*, Telesurgery

LEIF SÖRNMO, *Lund University, Sweden*, Electrocardiogram (ECG) Signal Processing

DONALD M. SPAETH, *University of Pittsburgh, Pittsburgh, Pennsylvania*, Wheelchair Engineering

PAULETTE SPENCER, *University of Missouri-Kansas City, Kansas City, Missouri*, Dentin

OWEN C. STANDARD, *The University of New South Wales, Sydney, Australia*, Alumina

JOHN J. STANKUS, *University of Pittsburgh, Pittsburgh, Pennsylvania*, Soft Tissue Scaffolds

DUŠAN STARČEVIĆ, *University of Belgrade, Belgrade, Serbia and Montenegro*, Multimodal Presentation of Biomedical Data; Virtual Instrumentation

GREGOR STIGLIC, *University of Maribor, Maribor, Slovenia*, Software Agents

DUŠAN M. STIPANOVIĆ, *University of Illinois at Urbana-Champaign, Urbana, Illinois*, Parametric Adaptive Identification and Kalman Filter; State-Space Methods

ULRICH A. STOCK, *Charité University and Heart Center Brandenburg Bernau, Berlin, Germany*, Tissue Engineering of Cardiac Tissues

PAUL R. STODDART, *Swinburne University of Technology, Melbourne, Australia*, Adhesion of Bacteria

MICHAEL G. STRINTZIS, *Informatics & Telematics Institute, Thessaloniki, Greece*, Deformable Objects, Interactive Simulation of; Haptic Interaction in Medical Virtual Environments

MICHAEL A. STROSCIO, *University of Illinois at Chicago, Chicago, Illinois*, Optoelectronics

G. J. SUANING, *University of Newcastle, Callaghan, Australia and University of New South Wales, Sydney, Australia*, Functional Optical Imaging of Intrinsic Signals in Cerebral Cortex

BÉLA SUKI, *Boston University, Boston, Massachusetts*, Lung Tissue Viscoelasticity

SETSUO TAKATANI, *Tokyo Medical and Dental University, Tokyo, Japan*, Artificial Heart

TOSHIYO TAMURA, *Chiba University, Chiba, Japan*, Temperature Sensors

DALIN TANG, *Worcester Polytechnic Institute, Worcester, Massachusetts*, Flow in Healthy and Stenosed Arteries

TOGAWA TATSUO, *Waseda University, Tokorozawa, Japan*, Temperature Sensors

ALEXANDER L. TAUBE, *Swinburne University of Technology, Melbourne, Australia*, Carbyne-Containing Surface Coatings

DAWN TAYLOR, *Case Western Reserve University, Cleveland, Ohio*, Neural Control of Assistive Technology

XIAO-FEI TENG, *The Chinese University of Hong Kong, Shatin, Hong Kong*, Blood Oxygen Saturation Measurements

CHUCK THOMAS, *Clemson University, Clemson, South Carolina*, Tissue Engineering

ARTHUR R. TILFORD, *Member of the Institute of Electrical and Electronics Engineers (IEEE), Member of the Association for the Advancement of Medical Instrumentation (AAMI), Fellow of the Royal Society of Medicine (RSM)*, Homecare

K. H. TING, *The University of Hong Kong, Hong Kong*, Evoked Potentials, Adaptive Filtering of

TATSUO TOGAWA, *Waseda University, Tokorozawa, Saitama, Japan*, Biomedical Transducers

ERAN TOLEDO, *Tel Aviv University, Tel Aviv, Israel*, Heart Rate Variability (HRV); Hypertension: Time-Frequency Analysis for Early Diagnosis

GEORGIA D. TOURASSI, *Duke University Medical Center, Durham, North Carolina*, Computer Assisted Radiology (CAR)

BEATRIZ TRÉNOR, *Universidad Politécnica de Valencia, Valencia, Spain*, Ischemia

JOŽE TRONTELJ, *University Institute of Clinical Neurophysiology, Ljubljana, Slovenia*, Electromyography (EMG), Needle

GEORGE A. TRUSKEY, *Duke University, Durham, North Carolina*, Vascular and Capillary Endothelium

MANOLIS TSIKNAKIS, *Foundation for Research and Technology – Hellas, Heraklion, Crete, Greece*, Electronic Health Record

EISHUN TSUCHIDA, *Waseda University, Tokyo, Japan*, Blood Substitutes

DIMITRIOS TZOVARAS, *Informatics & Telematics Institute, C.E.R.T.H., Thessaloniki, Greece*, Deformable Objects, Interactive Simulation of; Haptic Interaction in Medical Virtual Environments

KENJI UCHINO, *The Pennsylvania State University, University Park, Pennsylvania*, Piezoelectric Actuators

SHOOGO UENO, *University of Tokyo, Tokyo, Japan*, Bioelectricity and Biomagnetism

KÂMIL UĞURBIL, *University of Minnesota Medical School, Minneapolis, Minnesota*, Brain Function, Magnetic Resonance Imaging of

KAMIL ULUDÁ, *Max Plank Institute for Biological Cybernetics, Tübingen, Germany*, Brain Function, Magnetic Resonance Imaging of

MARIA FERNANDA CABRERA UMPIERREZ, *Universidad Politecnica de Madrid, Life Supporting Technologies, Ciudad Universitaria, Madrid, Spain*, Telemedicine in Emergency Medical Services

CESARE USAI, *CNR, National Research Council, Genoa, Italy*, Confocal Microscopy; Multiphoton Microscopy; Optical Microscopy

MAX VALENTINUZZI, *Universidad Nacional de Tucumán, Tucumán, Argentina,* Plethysmography

MIGUEL A. VALERO, *Universidad Politécnica de Madrid, Madrid, Spain,* Home Telehealth: Telematics-Supported Services

ANA VALLÉS-LLUCH, *Universitat Politècnica de València, Valencia, Spain,* NMR Imaging

NENS VAN ALFEN, *Radboud University Nijmegen Medical Center, Nijmegen, The Netherlands,* Electrodiagnosis in Neuromuscular Disorders

PIERRE-FRANCOIS VAN DE MOORTELE, *University of Minnesota Medical School, Minneapolis, Minnesota,* Brain Function, Magnetic Resonance Imaging of

PAUL VAN DE VOORDE, *University Hospital Gasthuisberg, K.U. Leuven, Medtronic, Belgium,* Electrodes

PASCAL VERDONCK, *Ghent University, Ghent, Belgium,* Artificial Kidney, Modeling of Transport Phenomena in

BART VERHEYDEN, *University Hospital Gasthuisberg, K.U. Leuven, Leuven, Belgium,* Bioenergetics and Systemic Responses to Exercise; Electrodes; Space Physiology

MATEJA VERLIC, *University of Maribor, Maribor, Slovenia,* Software Agents

IVAN VESELY, *University of Southern California, Los Angeles, California,* Heart Valve Tissue Engineering

JEAN-MARC VESIN, *EPFl-STI-ITS-LTS, Lausanne, Switzerland,* Human Brain Interface: Signal Processing and Machine Learning

ERIK VIIRRE, *University of California, San Diego, California,* Cognitive Systems

EBERHARD O. VOIT, *Georgia Institute of Technology and Emory University, Atlanta, Georgia,* Biochemical Processes/Kinetics

GABRIELA VOSKERICIAN, *Case Western Reserve University, Cleveland, Ohio, and Proxy Biomedical Ltd., Galway, Ireland,* Foreign Body Reaction; Sensor Biocompatibility and Biofouling in Real-Time Monitoring

WILLIAM R. WAGNER, *University of Pittsburgh, Pittsburgh, Pennsylvania,* Soft Tissue Scaffolds

MARY P. WALKER, *University of Missouri-Kansas City, Kansas City, Missouri,* Dentin–Enamel Junction of Human Teeth

CHARLES S. WALLACE, *Duke University, Durham, North Carolina,* Vascular and Capillary Endothelium

PETER WALSH, *BioQ Devices Pty Ltd, Brisbane, Australia and WorleyParsons Services Ltd, Sydney, Australia,* Stenosis and Thrombosis

W.R. WALSH, *Prince of Wales Hospital, Sydney, Australia,* Sutures

GE WANG, *Shanghai University, Shanghai, China,* Computed Tomography

GUOXING WANG, *University of California, Santa Cruz, California,* Semiconductor-Based Implantable Prosthetic Devices

MAY WANG, *Georgia Tech University, Atlanta, Georgia,* Microarray Data Analysis

XIAODONG WANG, *Columbia University, New York, New York,* Genomic Networks: Statistical Inference from Microarray Data

YONG WANG, *University of Missouri-Kansas City, Kansas City, Missouri,* Dentin

DENHAM S. WARD, *University of Rochester School of Medicine and Dentistry, Rochester, New York,* Hypoxia

SUSAN A. WARD, *University of Leeds, Leeds, United Kingdom,* Exercise Hyperpnea

CHARLES L. WEBBER, JR., *Loyola University Medical Center, Maywood, Illinois,* Recurrence Quantification Analysis; Ventilatory Pattern Variability in Mammals

THOMAS J. WEBSTER, *Brown University, Providence, Rhode Island,* Cartilage Scaffolds

ANDREW S. WECHSLER, *Drexel University College of Medicine, Philadelphia, Pennsylvania,* Robotic Surgery

JOACHIM WEGENER, *Technische Universität Hamburg-Harburg, Hamburg, Germany,* Cell Surface Interactions

BRUCE WEST, *US Army Research Office, Research Triangle Park, North Carolina,* Complexity, Scaling, and Fractals in Biological Signals

BRIAN J. WHIPP, *University of Leeds, Leeds, United Kingdom,* Exercise Physiology

JOHN M. WHITELOCK, *University of New South Wales, New South Wales, Sydney, Australia,* Adhesion of Cells to Biomaterials; Extracellular Matrix

LORI L. WICKHAM, *John Fiske Brown Associates Inc., Forensic Engineers, Solana Beach, California,* Stress

MATTHEW WILKINSON, *VA Pittsburgh HealthCare System, Pittsburgh, Pennsylvania,* Rehabilitation Biomechanics

ANDY P. WILLIAMS, *University of Illinois at Urbana-Champaign, Urbana, Illinois,* Bayesian Analysis

ERIK J. WOLF, *University of Pittsburgh, Pittsburgh, Pennsylvania,* Rehabilitation Engineering: An Overview

MARK WONG, *University of Texas - Health Science Center, Houston, Texas,* Temporomandibular Joint Disc

ANDREW WOOD, *Swinburne University of Technology, Melbourne, Australia,* Nonionizing Radiation

BRADFORD J. WOOD, *National Institutes of Health Clinical Center, Bethesda, Maryland,* Computer Aided Surgery

TERRY O. WOODS, *FDA Center for Devices & Radiological Health, Rockville, Maryland,* MRI Safety

YUN XU, *Imperial College, London, UK,* Blood Flow Simulation, Patient-Specific *in-vivo*

ESSA YACOUB, *University of Minnesota Medical School, Minneapolis, Minnesota,* Brain Function, Magnetic Resonance Imaging of

YUSHENG YANG, *VA Pittsburgh HealthCare System, Pittsburgh, Pennsylvania and University of Pittsburgh, Pittsburgh, Pennsylvania,* Rehabilitation Biomechanics

GUANG-ZHONG YANG, *Imperial College, London, United Kingdom,* Blood Flow Measurement, Blood Flow Simulation, Patient Specific *in-vivo*

JIA-LIN YANG, *University of New South Wales, New South Wales, Australia,* Cancer

WENBING YAO, *Brunel University, Uxbridge, United Kingdom,* Wide Area Networks

MARTIN L. YARMUSH, *Shriners Hospital for Children, Boston, Massachusetts,* Tissue-Engineered Liver

JOHN T. W. YEOW, *University of Waterloo, Waterloo, Ontario, Canada,* Biomedical Sensors; Fuzzy Neural Networks

STUART (SHIZHUO) YIN, *The Pennsylvania State University, University Park, Pennsylvania,* Fiber Optic Sensors

LESLIE YING, *University of Wisconsin-Milwaukee, Milwaukee, WI,* Phase Unwrapping

LUNG YIP, *The Chinese University of Hong Kong, Hong Kong,* Heart Sounds and Stethoscopes

H. O. YLÄNEN, *Åbo Akademi University, Turku, Finland,* Bioactive Glasses and Glass Ceramics

YAN YU, *University of New South Wales, New South Wales, Australia,* Cancer

LYLE D. ZARDIACKAS, *University of Mississippi Medical Center, Jackson, Mississippi,* Stainless Steels for Implants; Titanium and Titanium Alloys

JOSEPH P. ZBILUT, *Rush University Medical Center, Chicago, Illinois,* Recurrence Quantification Analysis; Ventilatory Pattern Variability in Mammals

XIN-YU ZHANG, *The Chinese University of Hong Kong, Hong Kong,* Heart Sounds and Stethoscopes

YAN ZHANG, *The Chinese University of Hong Kong, Shatin, NT, Hong Kong,* Biometrics

YUAN-TING ZHANG, *The Chinese University of Hong Kong, Shatin, NT, Hong Kong,* Biometrics; Blood Oxygen Saturation Measurements; Heart Sounds and Stethoscopes; Telemedicine; Wireless Biomedical Sensing

MING HAO ZHENG, *University of Western Australia, Nedlands, Australia,* Cellular Engineering

XIAOBO ZHOU, *Texas A&M University, College Station, Texas,* Genomic Networks: Statistical Inference from Microarray Data

XIAOHONG JOE ZHOU, *University of Illinois Medical Center, Chicago, Illinois,* Diffusion Tensor Imaging

MINGCUI ZHOU, *University of California, Santa Cruz, California,* Semiconductor-Based Implantable Prosthetic Devices

XIAOHAN ZHU, *University of Minnesota Medical School, Minneapolis, Minnesota,* Brain Function, Magnetic Resonance Imaging of

MICHAEL ZIBULEVSKY, *Technion–Israel Institute of Technology, Haifa, Israel,* Blind Source Separation

EMILY ZIPFEL, *University of Pittsburgh, Pittsburgh, Pennsylvania,* Spinal Cord Stimulation Systems; Wheelchair Engineering

SHARON ZLOCHIVER, *Tel Aviv University, Tel Aviv, Israel,* Electrical Impedance Technique for Cryosurgery Monitoring

MACHIEL J. ZWARTS, *Radboud University Nijmegen Medical Center, Nijmegen, The Netherlands,* Electrodiagnosis in Neuromuscular Disorders

A

ACCELEROMETERS

LISA BENSON
Clemson University
Clemson, South Carolina

1. INTRODUCTION

Acceleration is the rate of change of either the magnitude or the direction of the velocity of an object, and it is measured in units of length per time squared (i.e., m/s^2) or units of gravity (g) (1 g = 9.81 m/s^2). Devices that measure acceleration, or accelerometers, are used in high-performance devices such as missile guidance systems, airbag deployment systems in automobiles, in handheld electronics, as well as in biomedical applications such as motion analysis or assessment of physical activity.

In the most general terms, an accelerometer consists of a mass, spring, and damper. The acceleration of the mass is measured by the deformation of the spring, and oscillation of the spring is controlled by the damper. The acceleration of the mass should not affect the actual acceleration of the object being monitored, so size is an important design criterion. The earliest devices that were designed in the 1930s were based on strain gages in a Wheatstone bridge configuration on a small frame supporting a mass weighing over 3 pounds (1). Current accelerometers are typically electronic devices based on piezoelectric or semiconductor technology. Depending on the device and its design, an accelerometer can measure very small accelerations caused by vibration (micro g), or very large accelerations during motion or impact (thousands of g's).

2. THEORY

Most accelerometers contain a known mass (sometimes referred to as a seismic mass or a proof mass) and some means for measuring either the resultant force or the displacement when the mass is accelerated. These types of accelerometers work by the principle of Newton's second law of mechanics, which states that the force ($\boldsymbol{F}$) acting on an object that is accelerating will be proportional to its acceleration ($\boldsymbol{a}$), and the proportionality constant is the mass (m) of the object: $\boldsymbol{F} = m\boldsymbol{a}$. An example of a simple accelerometer is a mass attached to a spring and ruler, where the deflection of the spring is proportional to the force acting on it. As the force is directly proportional to the acceleration of the mass, acceleration in the direction of the ruler can be calculated by measuring the spring deflection. Most electronic accelerometers use a similar principle, measuring the displacement of the mass in the desired direction and producing an output signal proportional to the acceleration.

Piezoelectric materials, which develop an electric charge when deformed, were used to make some of the first modern accelerometers in the 1940s and 1950s and are the most common and cost-effective types of accelerometers being used today. Such a device is typically comprised of a piezoelectric crystal supporting a small mass. Acceleration produces a force proportional to the mass, which causes a change in the electrical output of the piezoelectric crystal.

Piezoresistive accelerometers measure the force produced by the acceleration of an internal mass by the change in resistance of the material (often silicon based) when a force is applied to it. Recently developed thermal accelerometers do not use a mass/spring configuration, but instead sense acceleration by changes in heat transfer between micromachined elements within a small, insulated space.

3. PERFORMANCE

The optimum performance of an accelerometer is based on the linearity of its response, its frequency bandwidth, its sensitivity to acceleration, and minimal cross-sensitivity. All electronic devices have some useable range in which their response is linear (i.e., the electrical output is directly proportional to the parameter being measured). Mass-spring-damper accelerometers have a natural frequency at which they oscillate that is particular to their design and construction, also called the resonant frequency. The usable frequency range is the flat area of the frequency response curve, which is a plot of the deviation in output versus frequency. Frequencies below about one-third to one-half of the resonant frequency are generally within the range in which the device response is linear, and this range is referred to as bandwidth (see Fig. 1). Sensors that can sense constant accelerations are said to

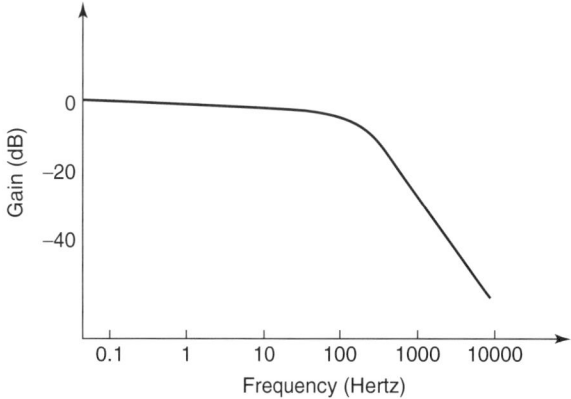

Figure 1. Gain (dB) vs frequency (Hz), also known as a Bode plot. The frequency response of any electronic device, plotted as gain (ratio of output to input) versus the frequency at which it is operating, shows its bandwidth, or range in which its output is a direct response to the input, as well as how it behaves at the cutoff frequency. Gain is reported in units of decibels (dB) for mathematical purposes (db = 20 log$_{10}$ X). If the gain equals unity (output equals the input), gain = 0 dB. Every order of magnitude change in gain equals 20 dB. The device in this figure has a linear response from the DC range up to 100 Hz.

have a "DC response" (0 Hz). Those that are functional below several hundred Hz are referred to as "low frequency" sensors. The upper limit of the bandwidth is determined by the physical properties of the device such as the dimensions of the mass and beam, the location of the mass on the beam, and the elastic modulus of the material that it is made from. The lower limit is governed by the amplifier and electronic response of the device. These parameters are all optimized to provide the widest possible bandwidth or range of operation for a specific device.

The sensitivity of the accelerometer is defined as the ratio of the change in output to the relative change in acceleration. As sensitivity is determined by the same physical properties as bandwidth, there is a tradeoff between sensitivity and bandwidth for most devices (except thermal accelerometers). A small mass usually means a lower sensitivity, which is typical for most high-frequency accelerometers. Accelerometers generally have an output that is some voltage that is proportional to acceleration, reported as a ratio of the output to input. Output units are typically mV/V/G. As the output is dependent on the input, or excitation voltage, care must be taken to provide a properly regulated input voltage.

Accelerometers are designed to be sensitive to accelerations only in a specified direction. However, the accelerometer will also sense accelerations in other directions to a certain degree, which is called cross-sensitivity or transverse sensitivity, which can contribute to errors in measurement and is usually quantified and reported by device manufacturers. All devices have some sensitivity to fluctuations in temperature that may affect their performance, which is also reported by the manufacturer, and the thermal environment should be taken into consideration when choosing a device for a particular application.

Electronic devices can be damaged by overloading or by accelerating it beyond its normal range of operation. Some devices have overload protection built in to their design, which will protect against accelerations during typical handling. However, a trade-off exists between protective mechanisms and damping of the device response, so these mechanisms have limitations. Dropping a device can expose it to accelerations of up to 400 g, which is difficult to accommodate for in a low-range device. The overrange limit is usually specified by the manufacturer.

As gravity is acceleration toward the center of the Earth, accelerometers that have a DC response are sensitive to gravity. An accelerometer at rest or in steady-state motion (i.e., not subject to other accelerations caused by motion) with its sensitive axis pointing toward the center of the Earth will have an output equal to 1 g. This property is commonly used to calibrate the gain and zero offset of an accelerometer by positioning the sensitive axis at a known angle relative to vertical and measuring the output (2).

High-frequency accelerometers, such as piezoelectric accelerometers, have a small physical size to ensure a high natural frequency (above 10 kHz). They are used in applications such as vibration and impact monitoring. Devices that provide the best DC response include piezoresistive, capacitive, force balance, and thermal accelerometers. These devices would be used in navigation and robotic applications as tilt sensors, constant acceleration measurements, or sensors taking measurements over prolonged periods. Table 1 summarizes some typical performance characteristics of devices used for biomedical applications.

4. PHYSICAL FEATURES

Semiconductor accelerometers are often MEMS, which are a combination of mechanical and electrical components made from silicon using micromachining processes.

Table 1. Device Specifications and Features for Accelerometers used in Typical Biomedical Applications, as Listed by the Manufacturers

Application	Motion Analysis	Tilt Sensor, Motion Tracking	Vibration	Impact Testing
Manufacturer and device model number	Entran Inc., EGAXT3	Analog Devices Inc., ADXL311	Kistler Instrument Co., 8704B50	Endevco Inc., 2255B-01
Range	±10 g	±2 g	±50 g	±50,000
Overrange limit	±10,000 g	3500 g	±100 g	±200,000
Bandwidth	0 to 250 Hz	6 kHz	3 to 7000 Hz	10 to 1200 Hz
Natural frequency	600 Hz	10 kHz	54 kHz	300 kHz
Sensitivity	6 mV/g	167 mV/g	100 mV/g	0.1 mV/g
Current		0.4 mA	2–20 mA	2–10 mA
Excitation voltage	15 VDC	5 VDC	18–30 VDC	18–24 VDC
Non-linearity	±1% full scale	±0.2% full scale	±1% full scale	<2.5%
Transverse sensitivity	3% max	±2%	1.5% (3% max)	<5%
Thermal sensitivity	±2.5% per 50°C	−0.025% per °C	−0.06% per °C	10 g per °C
Temperature range		−40 to +105°C	−55 to +100°C	−55 to +125°C
Zero offset	±15 mV	1.5 V	11 V	8.5–11.5 VDC
Size	Triaxial: 0.5″ cube	Smallest available: 4 × 4 × 1.45 mm	0.5″ hex × 1″ ht	0.3″ hex × 0.5″ ht
Weight	Triaxial: 6 g	<1 g	8.6 g	2 g

Note: See Bibliography for list of manufacturers' web sites. Specifications are nominal or typical values, and are listed here for reference purposes only for illustrating the differences between devices designed for a variety of applications. Many more designs and specifications are available from these and other manufacturers.

Piezoelectric devices are usually machined and packaged with necessary electronic components for filtering and amplifying their output, and are therefore referred to as Integrated Electronics Piezo Electric (IEPE) devices.

Semiconductor accelerometers measure acceleration in a variety of ways. Piezoresistive devices have resistive elements embedded in the spring components, which sense deflection as a change in resistance. Piezoelectric devices have a layer of material on the springs that responds to deflection with a change in electrical output. Others work like plate capacitors, with the silicon being machined into tiny comb-like structures that move with the attached mass when they are accelerated. The change in the distance between the combs causes a change in the electrical output. Capacitive devices have advantages of low power consumption, a wide range of operation, and low thermal sensitivity. Some designs combine these technologies to form a force-balanced capacitive accelerometer, typically used in automotive technology. These devices measure the force required to maintain a proof mass in a central, preset position. One drawback of this design is that because they are closed-loop accelerometers, the feedback loop slows down the response time, resulting in a smaller bandwidth than other devices. However, they have other features that are desirable for high-precision applications, such as accuracy and high signal-to-noise ratios.

Thermal accelerometers have recently been developed that sense acceleration by changes in heat transfer between micromachined elements within a small, insulated space on a silicon wafer. Some rely on thermal sensing for a seismic mass; as the mass is accelerated toward a heat source within the device, heat flows from the source to the mass and is sensed as acceleration. Other thermal accelerometers have no mass or spring elements, and therefore they are not subject to the typical trade-off between size/weight and sensitivity. They are made using MEMS technology, and are more cost effective than other mass and spring MEMS systems because of their ease of manufacturing. They are highly shock-resistant (reportedly up to 50,000 g) and are less prone to noise and drift than other MEMS devices. The main disadvantages of thermal devices are their limited frequency response (usually less than 100 Hz) and sensitivity of some designs to the temperature of their environment (3).

All electronic devices require power and a means to measure their output. The power requirements for accelerometers vary depending on the type of device. In general, piezoelectric devices require a current in the milliamp range, and have a preamplifier and signal conditioners packaged within the device. Especially for small devices, this input can have a significant thermal effect on the device, and therefore they require a warm-up period before use.

Accelerometers are usually packaged in a case that allows them to be firmly adhered to the object being measured in order to minimize the effects of damping or vibration. The sensitive axis is indicated on the outside of the package. Multiple axis accelerometers are available, either as multiple single-axis devices within one package, or as an inherently multiple-axis device (4). Usually, the axes are aligned orthogonally (at right angles to each other), and the multiple axes are indicated on the package. The outer case is usually metal, such as stainless steel, titanium, or a composite, which serves to protect and seal internal components. Accelerometers come in a variety of forms and sizes, such as through-hole and surface-mount packages for use on a circuit board (Small Outline Integrated Circuits, or SOICs). Others are in a self-contained form that can be attached directly to a data acquisition system or in a package that can be attached to the object whose acceleration is being measured. The forms typically include a tapped hole or stud (threaded or unthreaded), so they can be bolted securely to the object being monitored. Adhesives such as hot glue or double-sided tape can also be used to hold the device in position. Depending on the inherent stiffness of the device, the amount of torque applied when bolting it in place can change the output. Care must be taken to ensure consistent mounting torque when taking measurements over time or when comparing measurements between locations or devices.

Actual accelerometer sizes can be as small as a few hundred microns for MEMS devices, but the final size of the product is larger, typically several millimeters, because it is packaged together with the necessary electronics to amplify and process the output. The mass of the device should be small relative to the mass of the object being monitored. Otherwise, the behavior of the object may be affected, a condition referred to as mass loading. For some applications where size constraint is critical, or for high-frequency measurements, a miniature device may be required, generally considered to weigh less than 100 grams. Although these devices use advanced manufacturing methods and can be highly sensitive and accurate, there are trade-offs between size and stability, ruggedness and cost (5).

5. APPLICATIONS

The measurement of the acceleration of limb segments has been used for human motion analysis, motion detection, vibration, and impact studies. To measure the linear acceleration as well as the angular acceleration and angular velocity of the center of mass of a body segment in two or three dimensions, groupings of six to twelve linear accelerometers can be placed on a body segment such as the upper body (6) or the lower limb (7). However, it has been found that because of the magnitude of experimental errors, accelerometers alone may not be a reliable method of calculating limb kinematics, such as position and orientation (8). As semiconductor and MEMS technology develop, it is likely that accelerometers will become smaller, more sensitive, and more accurate. In fact, the performance capabilities of these devices has greatly improved over the past decade, and it is possible that this will lead to more widespread use for human motion analysis (9).

Vibration studies have been conducted to examine the shock-absorbing properties of joints and soft tissue structures such as the spine (10,11) and lower leg (12,13). Accelerometers can quantify the accelerations being sensed by these parts of the body and the shock attenuation that takes place within the soft tissue structures. This information is useful for establishing the long-term effects of exposure to low-level vibrations such as lower back pain and other musculoskeletal disorders. Although industrial vibration sensors require a high bandwidth, those used for human body vibration monitoring are typically below 60 Hz. Studies of accelerations caused by impact during normal walking and running have found accelerations of up to 10 g during running, with a frequency range of up to 100 Hz, which has significance in the area of athletic shoe design, as well as the study of tissue damage caused by high impact.

Detection of motion and the initiation of movement have been investigated by researchers for position feedback in systems being investigated to restore gait for paraplegics (14) and stroke victims (15). With functional neuromuscular stimulation (FNS), the lower limbs are stimulated to move, and accelerometers provide feedback on body segment and joint positions. Another application of accelerometers in motion detection is for monitoring balance during walking (16,17). Accelerometers and gyroscopes, which measure angular velocity, can be combined to form an inertial sensing system that works similarly to the human vestibular system, providing complete information on linear and angular accelerations (18). Several researchers and manufacturers are investigating ways to combine three-axis accelerometers and gyroscopes into a six-degree of freedom inertial sensing device.

One of the most common biomedical applications of accelerometers is in the area of physical activity monitoring. The use of accelerometer-based monitors is gaining widespread acceptance as a means to monitor physical activity of patients undergoing rehabilitation or treatment for mobility-related disorders, as well as activity of the general public (19). Accelerometer-based pedometers log the number of acceleration spikes identified as "steps," and are readily available to consumers through several manufacturers. Other accelerometer systems are used by physical activity researchers to discern not only the number of movements, but the type of activity, such as sitting, standing, lying down, and dynamic activity (20). These are sometimes referred to as activities of daily living (ADL) monitors. The accelerometer unit is strapped somewhere on the trunk and is wired to a separate data logger, which stores and displays the cumulative data. These typically use a time-frequency analysis to establish the duration of the different activities. The latest research efforts in physical activity monitoring are focused on discriminating between different dynamic activities, such as walking and stair climbing, which cannot be achieved with currently available devices.

Human motion, biofeedback, and wireless computer interfaces all make use of accelerometers as tilt sensors, which measure the direction of an object relative to the ground. Future applications include tilt sensors for easier manipulation of handheld devices such as cell phones and personal digital assistants (PDAs). Researchers are investigating data entry to a device by a combination of tilting it in different directions and allowing multiple devices to communicate by tapping them together like champagne glasses (21), which holds promise for use in for adaptive devices for the disabled and feedback devices for rehabilitation.

BIBLIOGRAPHY

1. P. L. Walter, *The History of the Accelerometer*. Bay Village, OH: Sound and Vibration, 1997, pp. 16–22.
2. R. Moe-Nilssen, A new method for evaluating motor control in gait under real-life environmental conditions. Part 1: The instrument. *Clin. Biomech.* 1998; **13**:320–327.
3. MEMSIC, Moving with heat. *Eur. Semiconduct.* 2001; March:63–67.
4. J. C. Lotters, W. Olthuis, P. H. Veltink, and P. Bergveld, On the design of a triaxial accelerometer. *J. Micromechan. Microengineer.* 1995; **5**:128–131.
5. J. G. Pierson (2001). *Miniature Sensors: When, Where and Why Should We Use Them?* (online). Available: http://www.entran.com/TechTipPart1.htm.
6. A. J. V. D. Bogert, L. Read, and B. M. Nigg, A method for inverse dynamic analysis using accelerometry. *J. Biomech.* 1996; **29**(7):949–954.
7. J. Bussmann, L. Damen, and H. Stam, Analysis and decomposition of signals obtained by thigh-fixed uni-axial accelerometry during normal walking. *Med. Biol. Eng. Comput.* 2000; **38**(6):632–638.
8. D. Giansanti, V. Macellari, G. Maccioni, and A. Cappozzo, Is it feasible to reconstruct body segment 3-D position and orientation using accelerometric data? *IEEE Trans. Biomed. Engineer.* 2003; **50**(4):476–483.
9. N. Yazdi, F. Ayazi, and K. Najafi, Micromachined inertial sensors. *Proc. IEEE* 1998; **86**(8):1640–1659.
10. M. H. Pope, D. G. Wilder, and M. L. Magnusson, A review of studies on seated whole body vibration and low back pain. *Proc. Inst. Mech. Eng. [H]* 1999; **213**(6):435–446.
11. C. Rubin, M. Pope, J. C. Fritton, M. Magnusson, T. Hansson, and K. McLeod, Transmissibility of 15-hertz to 35-hertz vibrations to the human hip and lumbar spine: determining the physiologic feasibility of delivering low-level anabolic mechanical stimuli to skeletal regions at greatest risk of fracture because of osteoporosis. *Spine* 2003; **28**(23):2621–2627.
12. M. A. Lafortune, M. J. Lake, E. Hennig, Transfer function between tibial acceleration and ground reaction force. *J. Biomech.* 1995; **28**(1):113–117.
13. W. Kim, A. S. Voloshin, S. H. Johnson, and A. Simkin, Measurement of the impulsive bone motion by skin-mounted accelerometers. *J. Biomech. Eng.* 1993; **115**:47–52.
14. P. H. Veltink and H. B. K. Boom, 3D movement analysis using accelerometry - theoretical concepts. In: A. Pedotti et al., eds., *Neruoprosthetics - from Basic Research to Clinical Applications, Biomedical and Health Research Program (BIOMED) of the European Union, Concerted Action: Restoration of Muscle Activity through FES and Associated Technology (RAFT)*. New York: Springer Verlag, 1996, pp. 317–326.
15. P. H. Veltink, P. Slycke, J. Hemssems, R. Buschman, G. Bultstra, and H. Hermens, Three dimensional inertial sensing of foot movements for automatic tuning of a two-channel implantable drop-foot stimulator. *Med. Eng. Phys.* 2003; **25**(1):21–28.
16. R. Moe-Nilssen, A new method for evaluating motor control in gait under real-life environmental conditions. Part 2: Gait analysis. *Clin. Biomech.* 1998; **13**:328–335.
17. S. J. Richerson, L. W. Faulkner, C. J. Robinson, M. S. Redfern, and M. C. Purucker, Acceleration threshold detection during

short anterior and posterior perturbations on a translating platform. *Gait Posture* 2003; **18**(2):11–19.
18. L. C. Benson, and M. Laberge, Measurement of knee kinematics and kinetics using inertial sensors. *ASME BED Advances Bioengineer.*, in press.
19. G. J. Welk, S. N. Blair, K. Wood, S. Jones, and R. W. Thompson, A comparative evaluation of three accelerometry-based physical activity monitors. *Med. Sci. Sports Exer.* 2000; **32**(9):S489–S497.
20. K. Aminian, P. H. Robert, E. E. Buchser, B. Rutschmann, D. Hayoz, and M. Depairon, Physical activity monitoring based on accelerometry: validation and comparison with video observation. *Med. Biol. Eng. Comput.* 1999; **37**:304–308.
21. L. Hardesty, Just tilt to enter text. *Technol. Rev.* 2003; **106**(9):24.

READING LIST

Some very useful resources for selecting an accelerometer can be found on the manufacturers' websites.

Analog Devices, Inc., Norwood, MA. Available: *http://www.analog.com/*.

Endevco Corp, San Juan Capistrano, CA. Available: *http://www.endevco.com/*.

Entran Devices, Inc., Fairfield, NJ. Available: *http://www.entran.com*.

Kistler Instrument Corp, Amherst, MA. Available: *http://www.kistler.com*.

Measurement Specialties, Inc., Fairfield, NJ. Available: *http://www.msiusa.com*.

Memsic, Inc., North Andover, MA. Available: *http://www.memsic.com*.

Motorola, Inc., Schaumburg, IL. Available: *http://e-www.motorola.com*.

PCB Piezotronics, Inc., Depew, NY. Available: *http://www.pcb.com*.

Crossbow Technology, Inc., San Jose, CA. Available: *http://www.xbow.com*.

ADHESION OF BACTERIA

NATASA MITIK-DINEVA
PAUL R. STODDART
RUSSELL CRAWFORD
ELENA P. IVANOVA
Swinburne University of Technology
Melbourne, Australia

1. INTRODUCTION

Bacterial adhesion plays a pivotal role in natural, industrial, and clinical environments. The consequences of bacterial adhesion can be beneficial or deleterious depending on the situation. For instance, bioreactor and biofilter systems rely on the adhesion and growth of bacterial cells on support materials for effective operation. Biofilms are often used in wastewater treatment for degradation of soluble waste. In marine environments, on the other hand, irreversible attachment and the subsequent growth of bacteria on surfaces can affect the attachment of other organisms, leading to corrosion or to the fouling of heat exchangers and ship's hulls. Significant consequences can develop from adhesion to other nonbiological or synthetic surfaces (e.g., colonization of surfaces in food and water industries) or to biological surfaces (e.g., infections of tissues or prosthetic implants) (1–3), which are just a few examples of the diverse results of bacterial attachment and growth on different surfaces. Most of the surfaces studied for bacterial adhesion have been chosen on the basis of application needs (e.g., glass and metal surfaces) or because of symbiotic and pathogenic interactions (4).

As a result of the enormous diversity of both bacteria and substrata, bacterial adhesion is far from being fully understood. The attachment process is complex and becomes further complicated if changes in cell metabolism subsequently occur. However, progress has been made in understanding the factors that influence adhesion on biological and nonbiological surfaces and the properties of bacteria that facilitate their attachment. Intensive studies over the last three decades have focused on two major aspects of bacterial attachment, namely physicochemical and biological. As microbial cell surfaces are chemically and structurally complex and heterogeneous, the physicochemical approach to explaining the microbial adhesive interactions is complicated. On the other hand, the biological drivers of bacterial adhesion are often quite specific to the structure of the individual bacterium and are, therefore, difficult to generalize.

The vast majority of our present knowledge regarding bacteria reflects the average properties and behavior of very large assemblies of the cells attached to surfaces. Indeed, if bacterial adhesion is followed by colonization, then the eventual result is the establishment of structured communities referred to as biofilm. The colonization of biological and nonbiological surfaces and the specific role of bacterial biofilms have received considerable attention over the last decade. After a brief overview of novel techniques that have recently been applied to studies of bacterial adhesion, this chapter will describe the general principles that apply to mechanisms of bacterial adhesion. Recent progress in understanding the driving forces of bacterial adhesion and the impact of the surface on metabolic activity and basic biofilm characteristics will also be discussed (5).

1.1. Definition

The word adhesion is derived from the Latin *adhaesio*, or *adhaerere* (i.e. to stick to), and has a wide range of definitions depending on it's usage; for example: (i) the property of remaining in close proximity such as that resulting from the physical attraction of molecules to a substance, or the molecular attraction existing between the surfaces of contacting bodies; (ii) the stable joining of parts to each other, which may occur abnormally; and (iii) a fibrous band or structure by which parts abnormally adhere.

In the context of bacterial adhesion, this term can be defined as the discrete and sustained association between a bacterium and a surface, or substratum (6), which is synonymous with the English word "attachment" as something that attaches one thing to another.

2. TECHNIQUES FOR STUDYING BACTERIAL ATTACHMENT

Recent advances in microscopy have provided a way to bypass the difficulties of single-cell observations. Adhesion studies have been facilitated by the emergence of several experimental techniques that can probe the interactions of single cells with their neighboring surfaces or other cells. Among these single-cell probing and manipulation techniques, the most prominent are atomic force microscopy (AFM) mapping and manipulation of single cells on surfaces and scanning confocal laser microscopy (SCLM) (7).

The AFM technique enables the high-resolution imaging of the native bacterial cell surface in three dimensions, without staining or shadowing, by mechanically scanning a sharp tip mounted on a flexible cantilever over the sample surface. Razatos (8,9) developed an AFM-based methodology for directly measuring the interaction forces between AFM cantilevers with standard silicon nitride (Si_3N_4) tips and bacterial lawns immobilized onto flat glass substrates. When microspheres are glued onto standard tips, the probes can be modified to study the interaction of bacteria with various planar surfaces of interest (e.g., mica, hydrophilic glass, hydrophobic glass, polystyrene, and Teflon).

SCLM has become a standard technique for obtaining high-resolution images at various depths in a sample, which allows 3-D images to be reconstructed from successive focal planes. The SCLM method requires minimal sample preparation and, thus, provides an opportunity to study the organism itself, the surrounding environment, metabolic activity, and genetic control within the biofilm. The SCLM transforms the optical microscope into an analytical spectrofluorimeter (10).

X-ray photoelectron spectroscopy (XPS) has also found application in recent studies (11). This technique provides semiquantitative information on the elemental composition of the outer 1–5 nm of a surface, together with some basic chemical information (e.g., the relative concentration of an element in different functional groups). Some modern XPS instruments provide a chemical mapping capability with 3–5 micron resolution. The data is often useful for indicating the relative concentration of different elements or functional groups on a surface after exposure to different experimental conditions (12). Sample preparation requires some care, as samples must be vacuum compatible.

3. PHYSICOCHEMICAL ASPECTS OF BACTERIAL ADHESION

Although the forces involved in bacterial adhesion are not fully understood, it is evident from numerous studies of different systems that bacterial adhesion basically depends on physicochemical and biological parameters. These two issues will be discussed in this and the following chapter, respectively. In addition, environmental factors can also play a significant role in bacterial adhesion. For example, the number of attached cells tends to decrease as the temperature is reduced from room temperature down to approximately 3°C. On the other hand, pH changes between 4 and 9 do not significantly affect the adhesion process. The reader is referred to McEldowny and Fletcher (13) and Pasmore et al. (14) for more information on this subject.

One set of parameters depends on the surface characteristics: surface/substratum wettability, roughness, ligand density, dipole moment, and charge density. For example:

- Roughness—rough irregular surfaces can provide better and more stable initial attachment than a flat surface. However, it is not yet known how space limitations affect the bacterial attachment.
- Surface wettability—materials with different wettability have been used, such as glasses, metals, and polymers. Some material, such as polystyrene, have been chemically treated in order to create a series of substrata with different wettabilities, which has demonstrated that hydrophobic substrata are generally better suited to bacterial attachment.

It is clear that the surface characteristics of both the substrate and the bacteria must play a role in the attachment process. Although the underlying physicochemical interactions must also be common to both surfaces, studies have tended to focus on the effects of variations in the properties of either one or the other component of this essentially binary system. The discussion in this section reflects this somewhat artificial dichotomy, but concludes with an attempt to synthesize the results of a number of studies into an overall picture of bacterial attachment. At the same time, it must be noted that the biological surface is capable of subtle and complex responses to stimuli provided by the substrate, which leads to a rich phenomenology of adhesive interactions that is discussed in the subsequent sections below.

3.1. Wettability of the Substratum Surface

Numerous studies have attempted to correlate bacterial adhesion with the physicochemical properties of substrata. As previously mentioned, bacterial attachment has been found to be more effective on hydrophobic substrata with respect to numbers of attached cells, rate of attachment, and strength of binding (15,16). The "hydrophobic effect" refers to the proclivity of one nonpolar molecule for another nonpolar molecule over water. When two hydrophobic surfaces approach one another in an aqueous environment, inverse-ordered layers of water are displaced. The entropy increase during the displacement process creates energetically favorable conditions for adhesion (17). However, some other findings have demonstrated similar adhesion on hydrophilic and hydrophobic surfaces or even better adhesion on hydrophilic substrata [reviewed by Mittelman (6)]. Hydrophilic surfaces that are highly hydrated can be more resistant to bacterial adhesion because of adsorbed water that must be displaced before adhesion can occur. As a result, the effectiveness of the adhesion will depend on the difference between the bacterium ↔ substratum attraction forces and the adsorbed water ↔ substratum attraction forces.

3.1.1. Examples of Bacterial Adhesion on Hydrophilic/Hydrophobic Surfaces.
Chavant (18) have investigated the surface physicochemical properties of *Listeria monocytogenes* LO28 under different conditions (temperature and growth phase) during bacterial adhesion on hydrophilic (stainless steel) and hydrophobic (polytetrafluoroethylene—PTFE) surfaces. Stainless steel and PTFE were selected because of their common use in food-processing plants and because they have different physicochemical characteristics. The growth temperature and the phase of growth may influence the cell wall composition and thereby modify the surface electrical properties, hydrophobicity, and electron donor or electron acceptor character of the bacteria.

The affinity of bacterial cells, as determined by the comparison of two different solvents, was higher for the electron acceptor solvent and weaker for the electron donor solvent under all conditions of the study, indicating a strong electron donor nature and weak electron acceptor nature of the bacteria. Lewis acid-base interactions indicated that the cells had basic surface properties that correlated with their hydrophilic characteristics. The results showed that the hydrophilicity varied with the temperature and growth phase. The bacterial cells were always negatively charged and possessed hydrophilic surface properties, which were negatively correlated with growth temperature. As previously described by Smoot and Pierson (19), hydrophilicity increased as the growth temperature decreased, except at 37°C and 20°C for mid-log-phase cells, where the cells shared the same properties. Hydrophilicity was greater for mid-log-phase cells than for stationary-phase cells at 37°C, but this trend was the opposite at 20°C and 8°C. At low temperatures, *L. monocytogenes* was strongly hydrophilic in either growth phase, which suggests some modifications in cell wall composition. It is well known that some bacteria maintain an optimal degree of membrane fluidity by modifying the cell wall lipid composition with temperature and cell growth. Smoot and Pierson (19) assumed that membrane modifications probably occurred in *L. monocytogenes* LO28 because the hydrophilicity varies with temperature and growth phase, but these modifications do not influence the global charge of the cells, which were always electronegatively charged under all of the conditions studied. The results of the electrophoretic mobility tests were surprising, because bacterial cells generally have an isoelectric point of between pH 2.0 and 3.5 (20). Nevertheless, this cell surface property seems to be characteristic of *L. monocytogenes*, with several authors reporting the same result for strains of *L. monocytogenes* from different origins (17–22). The electronegative global charge over the pH range studied might indicate the presence of compounds with a very low pKa on the cell surface. A statistical analysis of the variances showed that temperature and the nature of the surface were the main factors affecting adhesion and colonization.

Consequently, better adhesion and colonization were observed on the stainless steel, confirming the hydrophilic character of *L. monocytogenes* LO28 and the importance of this property in these processes. Moreover, no bacterial mat could be formed at 8°C on PTFE, and the initial adherent population decreased during the first days of the experiment, which suggests that colonization on a hydrophobic surface is very difficult or even impossible. The combination of these two antagonist parameters—a hydrophilic cell envelope and a hydrophobic surface—leads to a very significant decrease of the colonization power of the strain. It is noteworthy that the presence of similar initial adherent populations does not imply the same kinetics of colonization for the two surfaces, which suggests that adhesion and colonization phases could require different molecular adaptations. Significant detachment occurred at 37°C on PTFE, but only after the surface was completely colonized. It seems that the rapid colonization of the surface resulted in the formation of cellular aggregates based on a relatively small number of adherent cells. Eventually, the aggregates became too voluminous and sloughed off because of the weak interactions of the few adherent cells with the hydrophobic PTFE. It is interesting to note that a phase of detachment took place after 1 day at 20°C on stainless steel, but this result was transitory and the surface was almost completely colonized again 24 h later. On the other hand, the same bacterium at 20°C colonized PTFE more slowly, but without detachment. At the end of one week, a 3-D structure with voluminous aggregates was visible. The pace of colonization may explain the significant difference in biofilm stability observed on PTFE between 37°C and 20°C.

These dynamics may be critical for the dissemination of micro-organisms and contamination of surfaces, such as occurs in the food industry, for example. From a practical point of view, the results suggest that the use of PTFE (hydrophobic) surfaces in cold rooms might minimize the development of *L. monocytogenes* biofilms in food plants. Many studies have indicated that *L. monocytogenes* can grow or survive on fresh or processed produce, including asparagus, broccoli, cauliflower, and leafy vegetables. Outbreaks caused by *L. monocytogenes*-contaminated foods and the serious illnesses and fatalities that occur in susceptible individuals illustrate the importance of understanding the fundamental mechanisms that initially allow the bacterium to associate with produce and how it remains attached. Gorski (23) have developed a model system with cut radish tissue to study the molecular mechanisms involved in *L. monocytogenes* adhesion to plant surfaces. The results indicated that *L. monocytogenes* may use different attachment factors at different temperatures and that temperature should be considered an important variable in studies of the molecular mechanisms of *Listeria* fitness in complex environments.

3.1.2. Examples of Bacterial Adhesion on Self-assembled Monolayers.
Weincek and Fletcher (24) investigated in detail the attachment of the gram-negative bacterium, *Pseudomonas* sp., to hydrophobic and hydrophilic self-assembled monolayer (SAM) surfaces. Long-chain alkanethiols, $HS(CH_2)_nX$, adsorb from solution onto gold

substrata and form oriented, well-ordered, stable monolayers. The sulfur moieties of the alkanethiol bind to the gold, causing the molecules to pack tightly in an orientation nearly perpendicular to the surface of the gold film. For an alkanethiol of sufficient chain length ($n \approx 10$), the resulting single-component monolayer consists of an ordered organic film with wetting properties that are determined by the terminal functional group, X. Several studies of the surface properties of self-assembled monolayers have focused on single- and mixed-component SAMs with methyl and hydroxyl functional groups.

The controlled adhesion of uropathogenic *Escherichia coli* for rapid screening of potential inhibitors of bacterial adhesion and or quantitative evaluation of the efficacy of the inhibitors has been investigated using arrays of alkanethiolate SAMs on gold (25). The SAMs presented mixtures of alpha-D-mannopyranoside (a ligand that promotes the adhesion of uropathogenic *Escherichia coli* by binding to the FimH proteins on the tip of type I pili) and tri(ethylene glycol) moieties (organic groups that resist nonspecific adsorption of proteins and cells). With this approach, the SAMs provided surfaces for studying the adhesion of uropathogenic *E. coli* to specific ligands, while ensuring excellent resistance to nonspecific adhesion. Using arrays of mannoside-presenting SAMs, inhibitors of bacterial adhesion were easily screened by observing the number of bacteria that adhered to the surface of the SAMs in the presence of inhibitors. The properties of SAMs, when combined with the convenience and standardization of a microtiter plate, make arrays of SAMs a versatile tool that can be applied to high-throughput screening of inhibitors of bacterial, viral, and mammalian cell adhesion and of strongly binding ligands for proteins.

Generally, studies have shown that hydroxyl-terminated SAMs are wetted by water and hexadecane, whereas methyl-terminated SAMs are hydrophobic and oleophobic. It appears that two basic attachment mechanisms are relevant. Adsorption on hydrophobic surfaces is rapid with strong binding forces, whereas adhesion to hydrophilic surfaces followed the model of reversible and irreversible adhesion proposed by Marshall (26) and can be described by DLVO theory (see below). Initially, a weak and reversible stage of the adhesion was observed at separation distances of several nanometers, at which point the bacterium can be removed by shear forces or desorb spontaneously. At a later stage, this attachment can be converted into irreversible adhesion by synthesis of extracellular biopolymers or by stabilization of conformational changes in existing polymers. These polymers bridge separation distances of less than 1 nm, displacing the adsorbed water or neutralizing the electrostatic repulsion.

3.2. Properties of the Bacterium Surface

3.2.1. Hydrophobic/Hydrophilic Characteristics of the Bacterial Cell Surface. The cell surface reactivity of the cyanobacterium *Calothrix* sp. strain KC97, an isolate from the Krisuvik hot spring, Iceland, has been investigated in terms of its proton-binding behavior and charge characteristics by using acid-base titrations, electrophoretic mobility analysis, and transmission electron microscopy (27).

As the cyanobacterial surface is composed of an intricate array of complex biological molecules, resulting in a high degree of chemical heterogeneity in the cell envelope, electronic and steric interactions within this complex arrangement can be quite large, and thus the pKa of a functional group will be influenced by the group's location within this highly reactive meshwork. Any given functional group type is likely to exhibit a range of pKa that is dependent on its distribution throughout the cell envelope.

It was shown in this study that the three main functional groups (carboxyl, phosphoryl, and amine) present on whole, intact filaments produced six potentially discrete binding sites. Carboxylic functional groups deprotonate over a low pH range and predominantly exhibit pKa between 2 and 6, which is likely to be representative of carboxyl groups contained within the galacturonic acid of the sheath. The phosphoryl groups commonly exhibit intermediate pKa between 5.4 and 8. Amine groups commonly exhibit pKa between 8 and 11. Hydroxyl groups, such as those abundant on the neutral sugars, do not usually deprotonate until the pH is 12. Carboxyl groups are found in abundance in cell wall constituents such as diaminopimelic and glutamic acids in the peptide cross-linkages of peptidoglycan molecules. Additional carboxyl groups may be provided by fatty acids such as hydroxypalmitic acid, a component of the lipid A fraction rarely detected in the lipopolysaccharide (LPS) of most gram-negative bacteria but common in most cyanobacteria. On the other hand, carboxylated teichoic acids, covalently bound to peptidoglycan, do not occur within the cell walls of cyanobacteria. Other phosphoryl-containing polymers in the cell envelope include LPS, although phosphate levels in the LPS of cyanobacteria are generally low. Lipids present in the cell wall may also exhibit phosphoryl groups, such as phosphatidylglycerol and phospha-tidylcholine, which are found in the cell wall of the cyanobacterium *Synechocystis* sp. strain PCC 6714. Electrophoretic mobility (EPM) analysis of isolated sheath fragments indicated that the sheath exhibited a net surface charge that is close to neutral at pH 5.5. This net neutrality is consistent with a low concentration of functional groups on the sheath and implies that the dominant, negatively charged carboxyl and positively charged amine groups, present at natural pH, must occur in approximately equal proportions. The slight increase in the negative charge of the sheath with increasing pH was suggested to be predominantly a result of deprotonation of carboxylic and phosphoric groups to form negatively charged species, but may also be contributed to by deprotonation of amine groups to form electroneutral species. EPM analysis of fragmented filaments showed that they separate into a highly electronegative fraction suggested to be unsheathed cells and a weakly electronegative fraction representative of sheathed filaments. Additional evidence for the charge distribution between the sheath and cell wall was clearly illustrated by TEM analysis, in which the highly reactive cell wall exhibited extensive heavy metal staining in comparison with the weak staining of the less reactive sheath. Whatever their exact distribution, the cyanobacterial envelope clearly contains an abundance of proton-binding sites, which exhibit a net negative charge when deprotonated and thus display a high reactivity with cationic species in the surrounding liquid (27).

A number of attempts have been made to find a general pattern for modifying hydrophobicity and surface charge by external treatment of the cells. Castellanos (28) investigated two *Azospirillum* species, *A. brasilense* and *A. lipoferum*, but were not able to identify a general pattern. However, they did find that various chemicals, temperatures, and enzymatic treatments affect the cell-surface hydrophobicity and charge in the different strains.

3.3. Theoretical Approaches to Bacterial Adhesion

Initially, two different approaches existed for describing the physicochemical aspects of microbial adhesion: the DLVO (Derjaguin, Landau, Verwey, and Overbeek) approach and the thermodynamic theory (29).

3.3.1. DLVO Theory. In describing the interaction energies between the two surfaces, the classic DLVO theory developed by Derjaguin, Landau, Verwey, and Overbeek takes into account the attractive van der Waals and repulsive electrostatic interactions. According to this theory, two distances of separation exist. At larger separation distances, the bacterium is only weakly held near the surface by van der Waals forces. Shear forces can easily remove it and closer approach is inhibited by electrostatic repulsion. Once this repulsion is overcome by a bacterium, it may be bound at the closer separation distance, where the attractive forces are strong and adhesion becomes irreversible.

The DLVO theory was used to explain the impact of electrolyte concentration and net surface charge on bacterial adhesion in laboratory experiments. It was demonstrated that attachment numbers increase with increasing ionic strength of the medium [Weerkamp (30)]. The theory predicts that adhesion will be inhibited when both bacteria and substrata bear significant negative surface charges.

The classic DLVO theory has been extended to include short-range acid-base interactions (29). The acid-base interactions are based on electron-donating and electron-accepting interactions between polar moieties in aqueous solutions. These interactions have an enormous influence compared with the electrostatic and van der Waals interactions. However, they are also short-range interactions (less than 5 nm), which means that they only become operative when the interacting surfaces are in close proximity.

3.3.2. Thermodynamic Theory. The second theoretical approach that has been used to explain or predict the mechanism of bacterial adhesion is based on thermodynamic principles. This theory interprets the attachment as a spontaneous decrease in the free energy of the system. To confirm this concept, contact angle measurements of liquids on test substrata and bacterial cell surfaces have been carried out in laboratory experiments (31). According to this theory, microbial adhesion generally has the nature of weak, reversible, secondary minimum adhesion, which can develop into irreversible primary minimum adhesion. As expected, this progression is much easier in high ionic strength suspensions and when the micro-organisms involved have surface appendages (32).

3.3.3. Tentative Scenario for Initial Bacterial Attachment at Nanometer Proximity. On approaching a surface, a bacterium experiences a sequential chain of interactions. The first attractive and repulsive forces that act between two surfaces occur at distances of more than 50 nm, where van der Waals forces are considered to be attractive, although rather weak.

Electrostatic forces play a major role at distances of 10–20 nm. Electrostatic interactions are attractive when the surfaces have opposite net surface charges and repulsive when surfaces have like charges. However, in many natural environments, repulsive forces are reduced as the ionic strength of the medium increases. For example, the electrolyte concentration of seawater is sufficient to eliminate the electrostatic repulsion barrier.

Interfacial water may be a barrier to adhesion at distances of 0.5–2 nm, although it can be removed by hydrophobic interactions. Furthermore, at separation distances of less than 1 nm, hydrogen bonding, cation bridging, and receptor-ligand interactions become important (6). For additional information on this specific topic, the reader is referred to Ref. 32.

3.3.4. Application of the Adhesion Theories. Difficulties exist in the application of the thermodynamic and DLVO theories in bacterial adhesion because of the chemical complexity of bacterial surfaces and substrata. The structural and chemical diversity of bacterial cells, together with the lack of finite boundaries because of fimbriae, stalks, flagella, or long-chain biopolymers, frustrates the use of a reductionist, theoretical approach. During adhesion, bacteria employ several stages of interaction, each involving different molecular components of the bacterial or substratum surfaces. Nonpolar (hydrophobic) groups on fimbriae, lipopolysaccharides, or outer membrane proteins may assist the bacterium to approach the surface more closely, which may cause conformational changes in other bacterial surface polymers, thereby exposing further functional groups for stronger attractive interactions. The production of additional adhesive polymers can also be triggered at this time, which is an important issue for the thermodynamic theory, which can only be applied to processes in an equilibrium situation. The bacterial production of extracellular biopolymers is irreversible at the initial stages of adhesion and therefore disturbs the necessary equilibrium. At this point, sufficient evidence exists to confirm that some aspects of bacterial adhesion can be described by physicochemical approaches. However, as the complexity of the cell surface appendages increases, the applicability of this approach decreases (29,32).

4. BIOLOGICAL ASPECTS OF BACTERIAL ADHESION

Depending on whether the whole cell surface or just one specific molecular group is interacting with the surface, bacterial cell adhesion can be considered as specific or nonspecific. In both cases, the previously mentioned cell appendages play an important role in sensing proximity and attaching to the surface (33).

Nonspecific bacterial adhesion is typically involved in the colonization of inert surfaces, as opposed to specific bacterial adhesion, which is typical for attachment to biological surfaces. In the case of specific bacterial adhesion, only certain highly localized stereochemical molecular groups on the microbial surface interact and make contact with the surroundings. The specific binding mechanism can be used to differentiate between stereo-specific "lock and key" or "receptor-ligand" interactions in the adhesion process. When discussing bacterial adhesion on biological surfaces, it is essential to keep in mind that the host cells are not inert surfaces, but play an active role in the process through the activation of specific genes (34,35). These factors, which depend on the particular characteristics of individual biological systems, are discussed in this section.

4.1. Adhesins

One set of biological parameters derives from inherent bacterial characteristics, such as surface charge, bacterial shape, or the presence of adhesins, assembled into hairlike appendages known as fimbria or pili, which extend out from the bacterial surface. In other cases, the adhesins are directly associated with the microbial cell surface (so-called nonpilus adhesins) (36). Recent studies have shown that bacteria express a wide variety of adhesins in order to establish adhesion to host cells. The type of adhesin used is dependent on the strain, environmental factors, and the receptors expressed by the host cell. These adhesive interactions have developed in symbiotic and pathogenic relationships to ensure the recognition of certain surfaces and binding sites via the signaling system of surface components or receptors. Some examples of bacterial appendages that are involved in cellular adhesion include:

- P pili and type 1 pili, such as those observed on the surface of *Escherichia coli* and *Enterobacteriaceae*. Eleven genes for P pili and seven genes for type 1 pili are required for the expression and assembly of these organelles. They mediate binding to mannose-oligosaccharides and represent important virulence determinants.
- Type IV pili have been implicated in a variety of functions such as adhesion to host-cell surfaces, twitching motility, modulation of target cell specificity, and bacteriophage adsorption. They are found on bacteria such as *Pseudomonas aeruginosa*, pathogenic *Neisseria*, *Moraxella bovis*, *Vibrio cholerae*, and enteropathogenic *Escherichia coli*.
- Curli are thin, irregular, and highly aggregated surface structures mainly produced in *E. coli* and *Salmonella enteritidis*. They mediate binding to a variety of host proteins including fibrinogen, plasminogen, and human contact proteins.
- Flagella are large complex protein assemblages spanning out from the bacterial wall.

Some specific adhesins, such as lectins and type I fimbriae, can also serve as mediators of adhesion via specific functional groups on the surface of the biomaterials. For example, dental implants are rapidly coated by *Streptococcus sanguis*, which produce a sialic acid-binding protein. Piliated strains of *Pseudomonas aeruginosa* easily adhere to unworn contact lenses in contrast to nonpiliated mutants (2). In the case of implanted biomaterials, conditioning films composed of polysaccharides or proteinaceous components in human fluids (blood, urine, bile, saliva, etc.) can greatly facilitate specific bacterial adhesion. For example, in *P. aeruginosa*, flagellar expression is directly associated with adhesion to mucin, as shown in patients with cystic fibrosis. Evidence exists that fibrinogen, albumen, and mucin all serve to enhance adhesion of *Staphylococcus* to polyethylene, nylon, and polyvinylchloride surfaces, whereas plasma and serum albumin appear to inhibit the adhesion (6).

4.1.1. Hydrophobins: Eukaryotic Homologs of Adhesins.

Similar mechanisms of specific adhesion have been reported for eukaryotic micro-organisms, such as yeasts and fungi. In filamentous fungi, adhesion is facilitated by small amphipathic proteins known as hydrophobins. Hydrophobins show specific physicochemical characteristics, such as surface activity and self-assembly, and have the ability to adhere efficiently and stably to natural and artificial surfaces. For example, an SC3 hydrophobin coating on Teflon can withstand treatment with the boiling detergent sodium dodecyl sulfate. On assembly on the surface, an amphipathic protein layer is formed that changes the wettability of the solid from hydrophobic to hydrophilic and vice versa. To explain the specificity of hydrophobins, several functions have been suggested, ranging from protein immobilization to surface modification. In nature, hydrophobins are commonly found in emergent structures of filamentous fungi, aerial hyphae, spores, and infection structures, where they form a hydrophobic, water-repellent layer. The hydrophobin coating protects the fungal structures both from water ingress and from dessication (37).

The role of hydrophobins in the attachment of pathogenic fungi to the surfaces of plants has been investigated by Nakari-Setala (38). The properties of the yeast *Saccharomyces cerevisiae* were modified by the expression on the cell surface of a hydrophobin from the filamentous fungus *Trichoderma reesei*, which resulted in an increase in surface hydrophobicity (more precisely, a decrease in cell surface hydrophilicity). The hydrophobin-expressing strain also displayed slightly lower negative surface charge, which is expected to contribute to a reduction in the electrostatic repulsion between the cell and the negatively charged supports. Indeed, both steady-state and kinetic experiments suggested that hydrophobin display assists in the first steps of adsorption. The authors noted that hydrophobin-producing yeast cells did not form more or larger cell aggregates than the parent strain and, therefore, flocculation of the cells could not be correlated with increased adsorption.

Cell surface expression of proteins has been exploited in various ways, including the display of peptide or antibody libraries for rapid screening and engineering of ligands and binders or of enzyme activities. It has also been used to improve the immobilization capacity of cells. Native surface proteins may be used to target the protein of interest at the cell surface.

4.1.2. Fimbriae. To study the role of type 1 fimbriae in *E. coli* interactions with the mussel *Mytilus hemocytes*, a wild-type strain expressing type 1 fimbriae (MG155) and an unfimbriated mutant (AAEC072) were used to infect hemocyte monolayers in both artificial seawater (ASW) and hemolymph serum at 18°C (39). Of the two strains, it was shown that the fimbriated variety associates more readily with hemocytes—both in ASW and hemolymph serum, and after incubation for 60 and 120 min. Moreover, the presence of hemolymph serum greatly increased the association of the fimbriated strain, but did not affect the association of the unfimbriated mutant. The adherence of the fimbriated strain in hemolymph serum was four to six times greater than that of AAEC072, whereas in ASW the number of adherent fimbriated bacteria was only about twice the number of adherent unfimbriated cells. These data demonstrate that type 1 fimbriae play a role in the surface interactions between *E. coli* and mussel hemocytes and suggest that the mussel hemocytes express receptors for type 1 fimbriae.

4.2. Extracellular Biopolymers

Extracellular biopolymers comprise extracellular polysaccharides that can interact with hydrophilic surfaces, lipopolysaccharides that can interact with hydrophobic surfaces, and proteins that often react in specific attachment mechanisms (40). The composition of extracellular polysaccharides varies considerably depending on the bacterium habitat. For example, bacterial alginate (β-1, 4-D-mannuronic, and L-guluronic acids) produced by the opportunistic pathogen *P. aeruginosa* during its colonization of lungs has a binding capacity to facilitate attachment to the epithelial cells of the respiratory tract. In contrast, the alginate produced by the river dwelling *P. fluorescens* inhibits adhesion. These physicochemical properties of the alginates are caused by different degrees of acetylation and different ratios of mannuronic to guluronic acid (41).

A growing body of evidence suggests that surfaces can induce the expression of genes responsible for the production of lateral flagella, polysaccharides, and extracellular proteins including enzymes. For example, in *P. aeruginosa*, the specific activity of the *algC* reporter gene product was 19-fold higher in biofilm bacteria than in those bacteria remaining suspended in the surrounded environment (42). It was found that initial adsorption was independent of *algC* promoter activity, but 15 minutes after attachment to a glass surface, the activity of *algC* promoter was detected. Goldberg (43) reported that the *algC* gene encodes a bifunctional enzyme, which is involved in alginate and O-antigen synthesis in lipopolysaccharides. Results of another laboratory study have indicated that the production of extracellular polysaccharides increased after sand was added to free-living bacteria (44).

A mechanism for detachment of cells from surfaces also exists through, for example, the synthesis of an alginate lyase. Production of this enzyme can result in degradation of the alginate followed by increased cell detachment (45).

Two surface proteins of *E. faecalis*, namely the aggregation substance (Agg) and enterococcal surface protein (Esp), are reported to be associated with infections, suggesting that these proteins may increase the ability of this micro-organism to adhere. It was shown that the presence of the Esp gene is associated with the capacity of *E. faecalis* to form a biofilm on polystyrene. Later on, Waar (12) demonstrated that growth in the presence of ox bile stimulates the adhesion of some *E. faecalis* strains, as up to twice as many bacteria adhered at the stationary phase. These results indicate that growth in bile changes the adhesion properties of *E. faecalis* and thereby possibly increases its capacity to adhere and infect. Analysis of the cell surface properties revealed that the strains were generally more hydrophobic and zeta potentials were more negative after growth in the presence of ox bile, concurrent with decreases in N/C and P/C elemental surface concentration ratios and an increase in O/C surface concentration ratio [except for OG1X (pAM373)]. These changes may be due to the expression of different surface proteins in response to bile salts present in the growth medium. A negative surface charge is often associated with the presence of oxygen-rich functional groups, whereas a more positive charge is associated with the presence of nitrogen-rich groups (43). In keeping with this trend, all enterococcal strains possessing a higher number of oxygen-rich groups (and a similar or slightly lower number of nitrogen-rich groups) after growth in the presence of ox bile have a more negative zeta potential. On the other hand, strain OG1X (pAM373) has a lower number of oxygen-rich groups after growth in the presence of ox bile, but this strain gains negative charge through a relatively large loss of positively charged, nitrogen-rich groups. In oral streptococci, lower N/C elemental surface concentration ratios and higher O/C surface concentration ratios have been associated with decreased expression of proteinaceous, fibrillar structures on the surface and a loss of cell surface hydrophobicity (11).

5. IMPACT OF SURFACES ON METABOLIC ACTIVITY OF ATTACHED BACTERIA

5.1. Morphological Change

Elucidation of the mechanisms responsible for surface-induced morphological changes in bacterial cells has a potentially wide impact on the fundamental understanding of bacterial adhesion and biofilm formation, as well as on industrial and medical applications, which offer the potential to manipulate morphology for favorable outcomes.

The adhesion capabilities of *Sulfitobacter* spp. and *Staleya guttiformis* and the morphological changes of their respective vegetative cells during adhesion on surfaces of the polymer, poly-*tert*-butylmethacrylate (P*t*BMA), of varying hydrophobicity, have been studied using atomic force microscopy (46). The results revealed that two type strains, namely *S. pontiacus* and *S. brevis*, fail to attach to P*t*BMA, while the vegetative cells of *S. mediterraneus*, which do attach to the P*t*BMA surface, began the conversion to coccoid form after 24 hours. Figure 1 shows the various dynamic stages of formation of the coccoid body [i.e., vegetative cells with subpolar flagella (Fig. 1a), the beginning of formation of the coccoid body (Fig. 1b), and coccoid bodies with sizes ranging from 0.9–1.2 μm (Fig. 1c)]. In contrast to *Sulfitobacter* spp., type strain *Staleya guttiformis* DSM 11458^T formed a multilayered biofilm (Fig. 2a) producing extracellular polysaccharides (Fig. 2b). The cellular surface morphology presented a mucoid shell encapsulating the entire cell.

The presence of coccoid forms in nonaging cultures of *Sulfitobacter mediterraneus* during the 24 hours of adhesion on P*t*BMA polymeric surfaces deserves special attention. Previously, coccoid body formation was reported in aging cultures (46). This process was followed by a number of significant changes, including retraction of the protoplast away from the cell wall to one side of the cell; an increase in cell surface hydrophobicity; a decrease in lipid and RNA content; changes in fatty acid composition; and use of reserves such as poly β-hydroxybutyrate. In the case of bacterial adhesion onto P*t*BMA surfaces, the latter probably triggered the conversion of vegetative cells into coccoid forms (47). The precise mechanism of this change remains unclear.

5.2. Antibacterial Activity

As we have seen, bacterial surface properties are rather divergent, depending on the taxonomic affiliation of the bacterium. The metabolic status of the bacteria is controlled by the substratum surface, which can also affect the bacterial surface properties. The adhesive behavior of two strains of *Pseudoalteromonas* sp. has been studied on surfaces with different hydrophobicities (47). These bacteria secrete antimicrobial substances that allow the level of metabolic activity to be inferred after attachment. The hydrophobicity (measured by means of water contact angles) and chemical functionality of the surfaces were manipulated with photo- and thermo-chemistry of diazo-naphto-quinone/novolak—a polymeric system commonly used as a photoresist material in semiconductor manufacturing. In general, the number of attached bacterial cells,

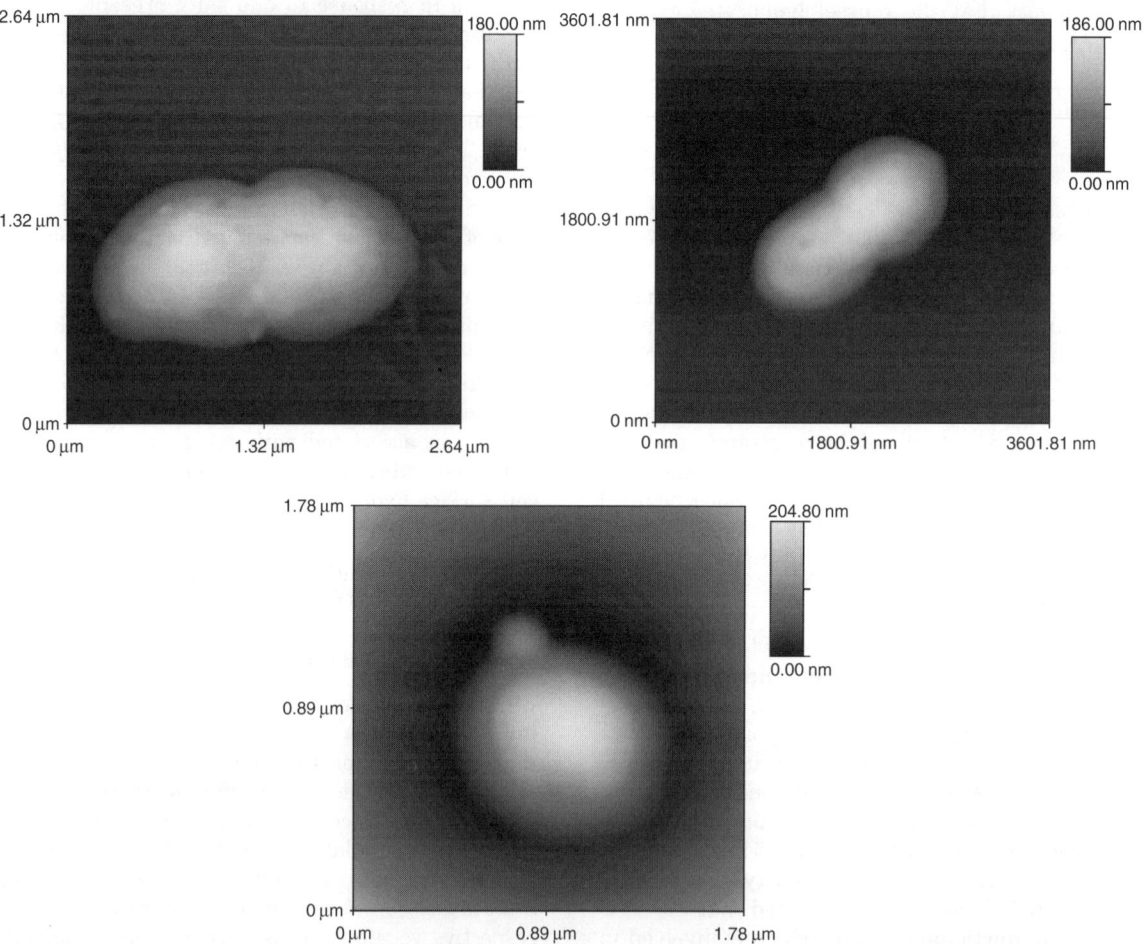

Figure 1. Conversion of vegetative cells of *S. mediterraneus* ATCC 700856^T into coccoid forms after attachment to P*t*BMA, 24 h: (a) vegetative cells with subpolar flagella; (b) initial step toward coccoid body formation; (c) coccoid form of *S. mediterraneus* ATCC 700856^T (47). (This figure is available in full color at http://www.mrw.interscience.wiley.com/ebe.)

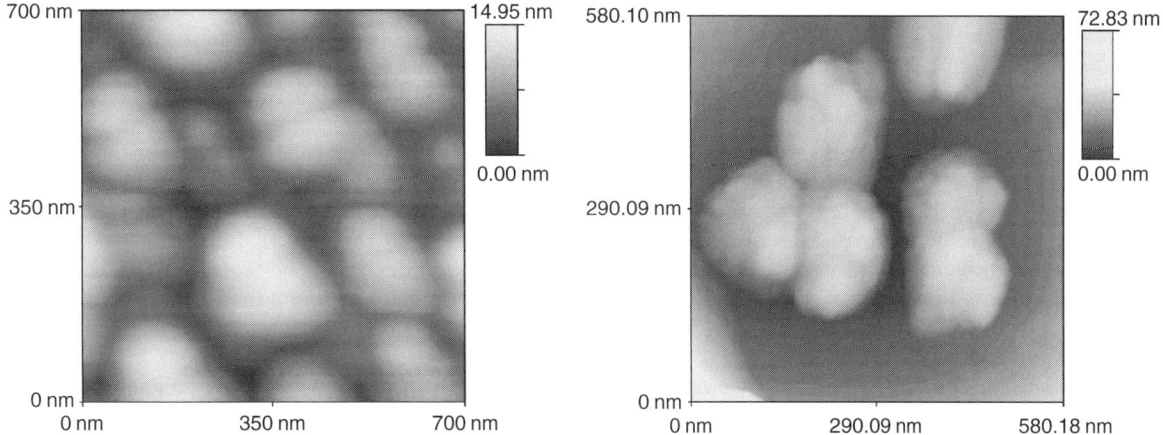

Figure 2. Adhesion behavior of *Staleya guttiformis* on P*t*BMA, 72 h: (a) multilayer biofilm formation; (b) expression of extra-cellular polysaccharides by cells (47). (This figure is available in full color at http://www.mrw.interscience.wiley.com/ebe.)

along with the level of antimicrobial activity, was greater on the unexposed polymeric films and less on exposed surfaces with the same level of hydrophobicity. An exception occurred on exposed, low-energy polymeric surfaces, where antimicrobial activity reached its highest level, which suggests that the chemical characteristics of the polymeric films are a major factor controlling the level of antimicrobial activity. The exposed films of photoresist are enriched by carboxylic residues, which induce the attached bacterial cells to synthesize antimicrobial substances more actively. Treatment with trypsin showed that these antimicrobial substances may inhibit the growth of *Staphylococcus epidermidis* via a proteinaceous factor. It is known that the most common protein-secretion pathway in prokaryotes involves the *sec* secretion system. This system recognizes a characteristic N-terminal secretion signal, consisting of a stretch of hydrophobic residues flanked by basic residues, which is removed during secretion. However, the substrate requirements for this secretion pathway are still unclear. This study suggested that the surface hydrophobicity controlled the number of cells on the surface, but it was also found that the relative concentration of nitrogen (decreasing with exposure energy delivered to the photopolymer surface) independently influences the antimicrobial activity of these marine bacteria (47).

5.3. Antibiotic Resistance of Biofilm Bacteria

On the other hand, bacterial adhesion and subsequent biofilm formation on biological surfaces may result in enhanced antibiotic resistance. This resistance is caused by bacterial phenotypic changes within the biofilm and by suppression of antibiotic activity by the extracellular enzymes produced by the biofilm itself. Moreover, biofilm bacteria produce antigens that stimulate the host immune system to produce antibodies. Although bacteria residing in the biofilm are often resistant to these mechanisms, the immune response may cause damage to surrounding tissue. Numerous examples of diseases caused by the formation of biofilm exist, like otitis media, periodontitis, pneumoniae, and, of course, implant infections (48).

6. BASIC BIOFILM CHARACTERISTICS

Although biofilm structures have been recognized for some time, we are just beginning to understand the formation process at the molecular level. Recent studies indicate that biofilms are constantly changing structures that pass through several stages, namely initiation, maturation, maintenance, and dissolution. In order to form a 3-D structure, bacteria must be able to attach to, move on, and sense the surface. It is now clear that the natural assemblages of bacteria in the biofilm itself function as a cooperative system with extremely complex control mechanisms upregulated by 15 proteins and downregulated by 30 proteins (48,49). The biofilm does not necessarily involve one bacterial type. In general, a few bacterial types are involved, and each of them plays a separate role in order to maintain a stable community.

6.1. Biofilm Initiation

The initiation process begins at a point when bacteria sense and respond to favorable environmental conditions, such as nutrients, iron, oxygen, specific osmolarity, temperature, or even the surface texture. These conditions can trigger a transition from a planktonic (free-swimming) bacterial form to a complex biofilm. This development will continue as long as fresh nutrients are provided. Environmental stimuli vary among bacteria. Although some bacteria, like *Vibrio cholerae* or *Escherichia coli*, demand very specific nutrient media, others, like *Pseudomonas aeruginosa* or *Pseudomonas fluorescens*, form biofilm under any conditions. Planktonic bacteria that can easily be triggered into forming biofilm under appropriate conditions are called wild type, whereas those that are unable to form biofilm are described as surface attachment defective (SAD) types. Some of the SAD mutants undergo initial attachment normally, but cannot perform the further stages in the biofilm development process (50).

Multiple genetic pathways regulate biofilm formation. The information gathered thus far suggests that the expression of surface structures may change, depending on the environmental conditions to which bacteria are exposed. For example, in *P. fluorescens*, the Clp protease participates in biofilm formation, but after the initial process additional protein synthesis is no longer required. In the case of this bacterium, therefore, the earliest stages of biofilm development can be divided into two phases: in the first phase, the initial interaction with the abiotic surface requires new protein synthesis, whereas in the second phase, the short-term maintenance of the attached cells does not require synthesis of new proteins. Additional cytoplasmic proteins located on the surface of the bacterial cell are also thought to be important for these bacteria to attach to abiotic substrates (51).

In *P. aeruginosa*, flagella and type IV pili seem to play an important role in early biofilm development The organism has been observed to swim along the surface as if it is scanning for an appropriate location for initial contact. After the initial contact has been made, the bacterium continues to move, but now it uses twitching motility rather than swimming. Twitching motility is strictly dependent on type IV pili and is much slower than flagella-dependent swimming. Nevertheless, by extending or retracting their pili, the bacteria can push or pull themselves across the surface. This type of motility is only active when cells are in contact with other cells, suggesting community behavior. Similarly, in *V. cholerae*, flagella and type IV pili are important in the initial stage. However, they are not essential, because even the SAD mutants can form biofilms. Their progress is delayed by 1–2 days; but once the biofilm forms, its architecture is identical to that of the wild type (52,53).

In gram-positive bacteria, such as *Streptococcus epidermidis*, *Staphylococcus aureus*, or *Enterococcus* spp. the initial attachment seems to be mediated by a number of factors, including uncharacterized surface proteins, extracellular proteins, capsular polysaccharide/adhesin, and a cell-surface-localized autolysin. After cell-surface interactions, these bacteria initiate cell-cell interactions that involve the production of the intercellular adhesin ICA-polysaccharide.

6.2. Biofilm Maturation

In order to maintain a newly established biofilm, bacteria undergo further maturation and the formation of characteristic biofilm architectures that are often associated with the production of signaling molecules, including short peptides, cyclic dipeptides (CDPs), fatty acid derivates, together with the commonly reported acyl-homoserine lactones (HSLs) and exopolysaccharides (EPSs). Acyl homoserine lactones (the so-called quorum-sensing molecules) are synthetized after the initial adhesion and are essential for cell-cell communication, which is, in turn, important for establishing a well-organized surface community. This results in characteristic biofilm architectures referred to as "mushroom" or "tower" structures. These different forms contain fluid-filled channels to supply nutrients and oxygen and remove metabolic waste. Studies of *P. aeruginosa* biofilm formation showed that the removal of the genes necessary for synthesis of HSLs resulted in biofilms forming without the usual well-spaced microcolonies, which means that the mutants are capable of early cell-surface interactions, but are unable to form biofilms that have the specific architecture essential for biofilm sustainability. Like the acyl-HSLs, EPSs seem to play a specific role in establishing stable interactions between cells and surfaces, and between cells themselves. For example, a strain of *V. cholerae* defective in EPS synthesis can hardly establish initial attachment. Even those cells that successfully attached were unable to develop architecture specific to the wild type. The EPS complex provides strong adhesive forces between the cell and the attachment surface, and probably slows down the penetration of antibiotic or enzymes from surface predators such as protozoans and microalgae (53).

If the environment becomes depleted in nutrients, or if temperature or pH change, bacteria may detach from the biofilm and return to their planktonic form. Some authors believe that the loss of EPS, or the production of enzymes such as alginate lyase, may play a role in detachment. However, in general, very little is known about the pathways involved in this process. A better understanding of bacterial detachment will contribute significantly to the prevention of undesirable adhesion. For additional information on the specific topic of biofilms and biofilm characteristics, the reader is referred to some excellent reviews (50–53).

7. CONCLUSION

Despite important advances, many fundamental aspects of bacterial adhesion, especially regarding the initial stages of adhesion and further bacterial biofilm development, remain far from fully resolved. The adhesion of bacterial cells, starting from attachment of single cells while they are exploring the "suitability" of the surface, is a key point in the chain of events leading to the formation of biofilms. Once aggregated, the biofilm cells trigger efficient "self-protection" mechanisms that shield the cells from the outer environment. A thorough understanding of bacterial adhesion and the mediating surface properties will have a significant impact on the prevention of biofilm formation (antibacterial surfaces) and on the use of biofilms (in bioremediation).

BIBLIOGRAPHY

1. C. J. Whittaker, C. M. Klier, and P. E. Kolenbrander, Mechanisms of adhesion by oral bacteria. *Annu. Rev. Microbiol.* 1996; **50**:513–552.

2. M. R. Parsek and P. K. Singh, Bacterial biofilms: an emerging link to pathogenesis. *Annu. Rev. Microbiol.* 2003; **57**: 677–701.

3. Y. Benito, C. Pin, M. L. Marin, M. L. Garcia, M. D. Selgas, and C. Clasas, Cell surface hydrophobicity and attachment of pathogenic and spoilage bacteria to meat surfaces. *Meat Sci.* 1997; **45**:419–425.

4. W. Chen, F. Brühlmann, R. D. Richins, and A. Mulchandani, Engineering of improved microbes and enzymes for bioremediation. *Curr. Opin. Biotechnol.* 1999; **10**:137–141.

5. A. Filloux and I. Vallet, Biofilm: set-up and organization of a bacterial community. *Med. Sci (Paris)* 2003; **19**:77–83.
6. M. W. Mittelman, Adhesion to biomaterials. In: *Bacterial Adhesion: Molecular and Ecological Diversity*. New York: Wiley, 1996.
7. J. Ubbink and P. Schär-Zammaretti, Probing bacterial interactions: integrated approaches combining atomic force microscopy, electron microscopy and biophysical techniques. *Micron* 2005; **36**:293–320.
8. A. Razatos, Y. L. Ong, M. M. Sharma, and G. Georgiou, Molecular determinants of bacterial adhesion monitored by atomic force microscopy. *Proc. Natl. Acad. Sci. USA* 1998; **95**:11059–11064.
9. A. Razatos, Application of atomic force microscopy to study initial events of bacterial adhesion. *Methods Enzymol.* 2001; **337**:276–285.
10. D. E. Caldwell, D. R. Korber, and J. R. Lawrence, Confocal laser microscopy and digital image analysis in microbial ecology. *Advan. Microbial Ecol.* 1992; **12**:1–67.
11. H. C. van der Mei, J. de Vries, and H. J. Busscher, X-ray photoelectron spectroscopy for the study of microbiol cell surfaces. *Surface Sci. Rep.* 2000; **39**:1–24.
12. K. Waar, H. C. van der Mei, H. J. Harmsen, J. E. Degener, and H. J. Busscher, Adhesion to bile drain materials and physicochemical surface properties of *Enterococcus faecalis* strains grown in the presence of bile. *Appl. Environ. Microbiol.* 2002; **68**:3855–3858.
13. S. McEldowney and M. Fletcher, Effect of pH, temperature, and growth conditions on the adhesion of a gliding bacterium and three nongliding bacteria to polystyrene. *Microbial Ecol.* 1988; **16**:183–195.
14. M. Pasmore, P. Todd, B. Pfiefer, M. Rhodes, and C. N. Bowman, Effect of polymer surface properties on the reversibility of attachment of *Pseudomonas aeruginosa* in the early stages of biofilm development. *Biofouling* 2002; **18**:65–71.
15. R. J. Doyle and M. Rosenberg, *Microbial Cell Surface Hydrobicity*. Washington, DC: American Society for Microbiology, 1990.
16. H. J. Busscher, J. Sjollema, and H. C. van der Mei, Relative importance of surface free energy as a measure of hydrophobicity in bacterial adhesion to solid surfaces. In: *Microbial Cell Surface Hydrophobicity*. Washington, DC: American Society for Microbiology, 1990.
17. M. C. M. van Loosdrecht and A. J. B. Zehnder, Energetics of bacterial adhesion. *Experiencia* 1990; **46**:817–822.
18. P. Chavant, B. Martinie, T. Meylheuc, M. N. Bellon-Fontaine, and M. Hebraud, *Listeria monocytogenes* LO28: surface physicochemical properties and ability to form biofilms at different temperatures and growth phases. *Appl. Environ. Microbiol.* 2002; **68**:728–737.
19. L. M. Smoot and M. D. Pierson, Influence of environmental stress on the kinetics and strength of attachment of *Listeria monocytogenes* Scott A to Buna-N rubber and stainless steel. *J. Food Prot.* 1988; **61**:1286–1292.
20. R. T. Briandet, C. Meylheuc, C. Maher, and M. N. Bellon-Fontaine, *Listeria monocytogenes* Scott A: cell surface charge, hydrophobicity, and electron donor and acceptor characteristics under different environmental growth conditions. *Appl. Environ. Microbiol.* 1999; **65**:5328–5333.
21. N. Mozes and P. G. Rouxhet, Methods for measuring hydrophobicity of micro-organisms. *J. Microbiol. Methods* 1987; **6**:99–112.
22. A. A. Mafu, D. Roy, J. Goulet, and L. Savoie, Characterization of physicochemical forces involved in adhesion of *Listeria monocytogenes* to surfaces. *Appl. Environ. Microbiol.* 1991; **57**:1969–1973.
23. L. Gorski, J. D. Palumbo, and R. E. Mandrell, Attachment of *Listeria monocytogenes* to radish tissue is dependent upon temperature and flagellar motility. *Appl. Environ. Microbiol.* 2003; **69**:258–266.
24. K. M. Weincek and M. Fletcher, Bacterial adhesion to hydroxyl- and methyl-terminated alkanethiol self-assembled monolayers. *J. Bacteriol.* 1995; **177**:1959–1966.
25. X. Qian, S. J. Metallo, I. S. Choi, H. Wu, M. N. Liang, and G. M. Whitesides, Arrays of self-assembled monolayers for studying inhibition of bacterial adhesion. *Anal. Chem.* 2002; **74**:1805–1810.
26. K. C. Marshall, Biofilms: an overview of bacterial adhesion, activity, and control at surfaces. *ASM News* 1992; **58**:202–207.
27. V. R. Phoenix, R. E. Martinez, K. O. Konhauser, and F. G. Ferris, Characterization and implications of the cell surface reactivity of *Calothrix* sp. strain KC97. *Appl. Environ. Microbiol.* 2002; **68**:4827–4834.
28. T. Castellanos, F. Ascencio, and Y. Bashan, Cell-surface hydrophobicity and cell-surface charge of *Azospirillum* spp. *FEMS Microbiol. Ecol.* 1997; **24**:159–172.
29. H. J. Busscher and H. C. van der Mei, Physico-chemical interactions in initial microbial adhesion and relevance for biofilm formation. *Adv. Dent. Res.* 1997; **11**:24–32.
30. A. H. Weerkamp, H. M. Uyen, and H. J. Busscher, Effect of zeta potential and surface energy on bacterial adhesion to uncoated and saliva-coated human enamel and dentin. *J. Dent. Res.* 1988; **67**:1483–1487.
31. M. Fletcher and K. C. Marshall, Bubble contact angle method for evaluation substratum interfacial characteristics and its relevance to bacterial attachment. *Appl. Env. Microbiol.* 1982; **44**:184–192.
32. R. Bos, H. C. Van der Mei, and H. J. Busscher, Physico-chemistry of initial microbial adhesive interactions—its mechanisms and methods for study. *FEMS Microbiol. Rev.* 1999; **23**:179–230.
33. H. Ton-That and O. Schneewind, Assembly of pili on the surface of *Corynebacterium diphtheriae*. *Mol. Microbiol.* 2003; **50**:1429–1438.
34. K. Triandafillu, D. J. Balazs, B. O. Aronsson, P. Descouts, P. Tu Quoc, C. van Delden, H. J. Mathieu, and H. Harms, Adhesion of *Pseudomonas aeruginosa* strains to untreated and oxygen-plasma treated poly(vinyl chloride) (PVC) from endotracheal intubation devices. *Biomaterials* 2003; **24**:1507–1518.
35. S. Laarmann and M. A. Schmidt, The *Escherichia coli* AIDA autotransporter adhesin recognizes an integral membrane glycoprotein as receptor. *Microbiology* 2003; **149**:1871–1882.
36. G. E. Soto and S. J. Hultgren, Bacterial adhesins: common themes and variations in architecture and assembly. *J. Bacteriol.* 1999; **181**:1059–1071.
37. H. J. Hektor and K. Scholtmeijer, Hydrophobins: proteins with potential. *Curr. Opin. Biotechnol.*, 2005; **16**:434–439.
38. T. Nakari-Setala, J. Azeredo, M. Henriques, R. Oliveira, J. Teixeira, M. Linder, and M. Penttila, Expression of a fungal hydrophobin in the *Saccharomyces cerevisiae* cell wall: effect on cell surface properties and immobilization. *Appl. Environ. Microbiol.* 2002; **68**:3385–3391.
39. L. Canesi, G. Gallo, M. Gavioli, and C. Pruzzo, Bacteria-hemocyte interactions and phagocytosis in marine bivalves. *Microsc. Res. Tech.* 2002; **57**:469–476.

40. J. H. Ryu, H. Kim, and L. R. Beuchat, Attachment and biofilm formation by *Escherichia coli* O157:H7 on stainless steel as influenced by exopolysaccharide production, nutrient availability, and temperature. *J. Food Prot.* 2004; **67**(10): 2123–2131.
41. E. Cotni, A. Flaibani, M. O'Regan, and I. W. Sutherland, Alginate from *Pseudomonas fluorescencs* and *P. putida*: production and properties. *Microbiology* 1994; **140**:1125–1132.
42. D. G. Davies and G. G. Geesey, Regulation of the alginate biosynthesis gene *algC* in *Pseudomonas aeruginosa* during biofilm development in continuous culture. *Appl. Environ. Microbiol.* 1995; **61**:860–867.
43. J. B. Goldberg, W. L. Gorman, J. L. Flynn, and D. E. Ohman, A mutation in *algN* permits trans activation of alginate production by *algT* in *Pseudomonas* species. *J. Bacteriol.* 1993; **175**:1303–1308.
44. P. Vandevivere and D. L. Kirchman, Attachment stimulates exopolysaccharide synthesis by a bacterium. *Appl. Environ. Microbiol.* 1993; **59**:3280–3286.
45. A. Boyd and A. M. Chakrabarty, Role of alginate lyase in cell detachment of *Pseudomonas aeruginosa*. *Appl. Environ. Microbiol.* 1994; **60**:2355–2359.
46. E. P. Ivanova, D. K. Pham, J. P. Wright, and D. V. Nicolau, Detection of coccoid forms of *Sulfitobacter mediterraneus* using atomic force microscopy. *FEMS Microbial Lett.* 2002; **214**:177–181.
47. E. P. Ivanova, D. V. Nicolau, N. Yumoto, T. Taguchi, K. Okamoto, and S. Yoshikawa, Impact of the conditions of cultivation and adsorption on antimicrobial activity of marine bacteria. *Marine Biol.* 1998; **130**:545–551.
48. M. E. Davey and G. A. O'Toole, Microbial biofilms: from ecology to molecular genetics. *Microbiol. Mol. Biol. Rev.* 2000; **64**:847–867.
49. K. Sauer and A. K. Camper, Characterization of phenotypic changes in *Pseudomonas putida* in response to surface associated growth. *J. Bacteriol.* 2001; **183**:6579–6589.
50. G. A. O'Toole and R. Kolter, Flagellar and twitching motility are necessary for *Pseudomonas aeruginosa* biofilm development. *Mol. Microbiol.* 1998; **30**:295–304.
51. G. A. O'Toole and R. Kolter, Initiation of biofilm formation in *Pseudomonas fluorescens*. *Mol. Microbiol.* 1998; **28**:449–461.
52. L. A. Pratt and R. Kolter, Genetic analyses of bacterial biofilm formation. *Curr. Opin. Microbiol.* 1999; **2**:598–603.
53. G. O'Toole, H. B. Kaplan, and R. Kolter, Biofilm formation as microbial development. *Annu. Rev. Microbiol.* 2000; **54**:49–79.

ADHESION OF CELLS TO BIOMATERIALS

JOHN M. WHITELOCK
University of New South Wales
Sydney, Australia

JAMES MELROSE
Royal North Shore Hospital &
University of Sydney
St. Leonards, Australia

1. INTRODUCTION

The adhesion of cells to biomaterials is a complex multistep process involving the binding and often denaturation of proteins and glycoproteins on the surface of a synthetic polymer, ceramic or metal, followed by the interaction of those adsorbed molecules with specific receptors on the surface of the cell. The proteins that adsorb to the surface of most biomaterials are found in the extracellular milieu of many connective tissue types in the body and are known collectively as extracellular matrix (ECM) molecules. They consist of proteins, such as the collagens; proteoglycans like the chondroitin sulfate proteoglycan, aggrecan or heparan sulfate proteoglycans, such as perlecan; and other structural ECM glycoproteins. These molecules have diverse functions yet often contain homologous modules or domains and will only be discussed here in relation to their roles in cell adhesion to biomaterials.

2. EXTRACELLULAR MATRIX ADSORPTION TO BIOMATERIALS

The relative degree of denaturation of a specific ECM molecule when it binds to a biomaterial surface is an important determinant on the subsequent cell-binding properties of a treated surface. For example, the chemistry of the bulk polymer is known to have effects on the adhesion of cells because it affects the way cell-adhesive ECM proteins are adsorbed to biomaterials as a consequence of bulk-transfer phenomena because of concentration-dependent effects, which can be critical in systems where growth factor signals are being delivered in a complex biological fluid such as plasma to encourage the growth of endothelial cells on biomaterial surfaces (1). Another factor that controls the adsorption of ECM molecules is the contact angle of the polymer at its surface, which is directly related to the type of polymer (i.e., polyurethane versus polyvinyl alcohol) and its surface chemistry. Generally, the more hydrophobic the biomaterial surface and higher the contact angle, the more protein that adheres to the surface but the more denatured it tends to be. The relationship between the nativity and the quantity of the adsorbed proteins both affect the amount of cell adhesion to biomaterials (2). This principle has been examined with methacrylate polymers and endothelial cell adhesion (3) and has been used to treat surfaces with compounds that effect the surface hydrophobicity, which alters the adsorption of collagen type I and fibronectin creating a predetermined pattern of adhered mammalian cells (4). Nonadhesive surface treatments, such as the addition of polyethylene oxide (PEO), have been shown to exert their effects because they are relatively resistant to the adsorption of proteins (5). A future important area of research in tissue engineering will be the use of appropriate adhesive and nonadhesive signals to achieve correct alignment and patterning of cells leading to correct tissue and organ structure (6,7). An example of appropriate signals is the combination of gas plasma treatment of polymers that have been engineered to contain PEO (8), which results in a mostly nonadhesive surface except where vitronectin or other adhesive proteins have adhered to areas where the surface has been modified by the gas plasma treatment and subsequently results in the adhesion of cells to specific areas of biomaterial surfaces and the generation of patterned multicellular constructs in a predefined arrangement. Polymerization initiated by light has also been used to modify a surface to control the adsorption of proteins and subsequent adhesion of cells (9).

3. POLYMER MODIFICATION TO ENCOURAGE ADSORPTION OF ECM PROTEINS

The most common polymer that is used to grow cells in the tissue culture laboratory is synthesized from polystyrene and is modified by glow discharge methodology (10), or treatment with an acidic solution (11), which make the surface more positively charged and, hence, hydrophilic, which improves the attachment of cells. Untreated polystyrene does not adhere cells well when compared with these more wettable surfaces (12). A major reason that gas plasma treatment has been successful in promoting the adhesion and growth of cells in culture is that they are grown in the presence of serum usually purified from either newborn or fetal cows. The major cell-adhesive components in serum are fibronectin and vitronectin; studies have shown that the adsorption of these molecules from serum (and hence culture medium) is more effective on the more hydrophilic surfaces (13). Furthermore, studies have also demonstrated that vitronectin is more important than fibronectin for the adhesion of cells in the presence of serum (14). Vitronectin and fibrinogen are the most important cell-adhesive molecules in the initial response of the body to biomaterials. After surgery, where a device or synthetic biomaterial is inserted, local injury occurs leading to the leakage of plasma from the circulation, the generation of serum as a thrombus, and its subsequent adsorption onto the surface of the biomaterial. Vitronectin and fibrinogen are both present in plasma at appreciable levels, the latter plays a major role in thrombus stabilization and platelet adhesion (15). Vitronectin is a 7.4 kDa protein, which is present as a monomer in the circulation at concentrations of approximately 400 µg/ml (16). When it binds to the surface of platelets (17) or biomaterials (18), vitronectin denatures, which exposes cryptic sites in the protein and unmasks new biological activities such as its ability to bind to cells. The denatured form of vitronectin also has the ability to self-polymerize to form large aggregates, which causes layering of the protein on the implant surface, further increasing its biological effects. Denatured vitronectin displays a higher affinity for heparin, which promotes the neutralization of antithrombin III activity, suggesting that vitronectin has some regulatory role to play in fibrinolysis (16).

3.1. Cell Attachment to ECM Molecules

The cell attachment site in vitronectin consists of an Arginine-Glycine-Aspartic acid (RGD) sequence in the connecting domain toward its N-terminus (19). The same RGD tripeptide sequence is responsible for the cell-adhesive properties of fibronectin and other ECM molecules such as thrombospondin and osteopontin (20). Fibronectin has also been used to modify the surface of many polymers including tissue culture polystyrene, where it is used to promote the *in vitro* growth of endothelial cells and has also been shown to increase the adhesion of bone cells to polyvinylidene fluoride (PVDF) (21).

Many cells also display adhesive properties to other ECM molecules and experiments have been performed to identify the specific peptide sequences responsible for adhesion. The glycine-phenylalanine-hydroxyproline-glycine-glutamate-arginine (GFOGER, single amino acid code) sequence found in triple helical native type I collagen is one such example, which interacts with the alpha-2 beta-1 integrins (22). Collagens (23), collagen peptide mimetics (24), and collagen cross-linked with chitosan (25) have all been used to modify the surfaces of biomaterials to improve cell adhesion or tissue biocompatibility. Additional adhesive epitopes with the sequences YIGSR and SIKVAV have been identified in the beta 1 chain of laminin and the long arm of the alpha chain of laminin (26). Both sequences are implicated in endothelial cell adhesion to basement membranes as well as other biological processes involving cellular attachment, such as the spread of cancer cells, which facilitates their invasion and metastasis (27). Other cell-adhesive peptides originating from protein sequences identified in laminin have been conjugated to chitosan membranes to promote cell attachment for tissue-engineering applications (28). Additional peptide sequences have been implicated in cell adhesion, but these remain illusory and largely uncharacterized because their sequences either do not align with known cell-attachment sequences or binding activity is apparently because of protein-folding affects on the tertiary structure of a putative cell-binding peptide motif. For example, the protein core of the basement membrane heparan sulfate proteoglycan, perlecan has adhesive properties for endothelial and smooth muscle cells on tissue culture polystyrene and polyurethane (29). This cell-binding site on perlecan has been shown to involve the beta 1 integrin family of molecules and has been localized to various domains of the perlecan molecule (30,31). However, such binding sites on the perlecan core protein do not contain linear peptide sequences with homology to RGD, YIGSR, or other known cell-attachment sequences suggesting that these cell interactive properties of perlecan may be caused by multiple binding epitopes possibly arranged in a specific but yet to be determined three-dimensional conformation. Other confounding factors in many experiments performed to assess the adhesion of cells to biomaterial surfaces are attributable to the presence of vitronectin from serum in the system or the time allowed for the assay to be performed, which may allow some cells to synthesize their own endogenous adhesive ECM proteins. The presence of serum has been overcome in certain systems with serum specifically immuno-depleted of the major adhesive protein, vitronectin (32), or by the use of serum-free conditions using defined media; however, many cell types bind relatively poorly under such conditions.

4. MODIFICATION OF POLYMERS WITH CELL ADHESIVE FRAGMENTS

The RGD tripeptide has been used to investigate and control the adhesion of cells to many different types of polymers used for both clinical and nonclinical applications (33).

These studies have included extensive investigations on the surface density (concentration), orientation, and biological activity of peptides containing RGD sequences together with investigations to examine other peptide sequences or functional groups. This method allows for covalent coupling to the biomaterial surface, and accommodates certain spatial arrangements or conformations required by key binding moieties. In many of the studies, the aim has been to encourage endothelial cell attachment to the polymers used in the synthesis of vascular grafts, as it has been recognized that the presence of a confluent and functional endothelial cell monolayer reduces platelet adhesion, controlled smooth muscle hyperplasia, and leads to long-term patency of vessels. The RGD peptide has been used in modified copolymer blends (34), as well as in covalent attachment to the polymer surface (35). Silicon surfaces have been coated with linear RGD-containing peptides to encourage bone-cell adhesion (36), whereas cyclic RGD peptides have been grafted onto poly-methyl methacrylate (PMMA) to encourage bone-cell adhesion and growth (37). Although the initial data obtained by many of these investigators demonstrated that these short fragments can be used to adhere cells to biomaterials in the short term, the process of cell adhesion, maintenance of cell phenotype, and sustained growth on biomaterials in the longer term involves many signals, some of which are cell-type and substrate specific, and cannot be totally mimicked by the simple modification of the biomaterial surface with such short peptides.

5. CELL ADHESION RECEPTORS

The integrity of a tissue relies on the adhesion of cells with other cells and the surrounding ECM and involves four major families of molecules, (i) the cadherins, (ii) selectins, (iii) cell-adhesion molecules (CAM) that are members of the immunoglobulin superfamily, and (iv) the integrin superfamily of cell surface receptors. All four types of molecules except the CAMs require a divalent cation for activity and only the integrin family are involved in interactions between cells and the ECM. Cadherins tend to be involved in interactions involving the same counter-receptor on another similar cell type. Selectins are also involved in cell–cell adhesion and involve the recognition of specific glycosylation structures present on the cell surface, such as the blood group antigens and other related structures, such as the sialyl Lewis X and A antigens (38). As a result of the large number of these types of ligands on the surface of circulating blood cells, they tend to be involved in the extravasation of leukocytes and other inflammatory cells through the endothelial lining of blood vessels. In some instances, it may be desirable in tissue-engineering applications to actually prevent or inhibit cell adhesion such as platelets and inflammatory cells in vascular grafts or orthopaedic implants, which can result in localized tissue damage or loosening of the implant.

6. CELL ADHESION MOLECULES INVOLVED IN BINDING TO THE ECM—THE INTEGRINS

Integrins are cell surface glycoprotein receptors that are involved in the attachment of cells to ECM molecules (39), and as we have discussed previously, ECM molecules are molecules that adsorb to the surface of biomaterials from the plasma or are synthesized and secreted by cells in the surrounding connective tissues. Therefore, it follows that the integrins are the major cell-adhesive components involved in the adhesion of cells to biomaterial surfaces. There are 18 different types of alpha subunits and eight different types of beta subunits that bind to each other in a noncovalent heterodimeric fashion that relies on the presence of divalent cations to give 24 characterized and distinct integrins (40,41). Both subunits have large extracellular N-terminal domains and short intracellular C-terminal domains that are linked to cytoskeletal components of the cell, such as actin, by various accessory proteins such as talin or paxillin (see Fig. 1). When they bind to the appropriate protein sequences in the ECM, they tend to cluster on the cell membrane, which has a positive effect on the formation of other linkages with the ECM and leads to the formation of a concentrated region of integrins close to the ECM with links to cytoskeletal proteins. These links have been termed focal adhesions, which all together provide for swift and specific two-way communication between the extracellular and intracellular

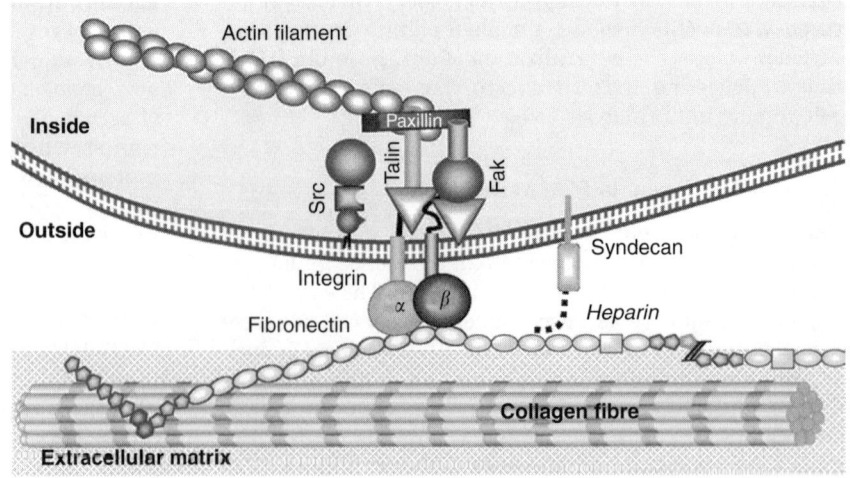

Figure 1. Schematic illustration of interactions between the collagen fibrillar network, structural glycoproteins of the ECM, cell-surface integrins, and HS proteoglycans (syndecan) and resultant signaling events through the cell cytoskeleton where actin filaments play a major role in signal transduction. The linking of the ECM and cell cytoskeleton is achieved in various ways, including the one illustrated where the ECM protein fibronectin binds both to collagen and to ECM receptors such as integrins and syndecan. The integrins are linked to actin inside the cell through a focal adhesion complex comprised of proteins like focal adhesion kinase (FAK), Src family kinases, talin, and paxillin. Reproduced with permission (42).

environments. This phenomenon has been termed inside-out signaling if an interaction within the cell results in an effect on the extracellular state of the cell and outside-in signaling if the interaction commences in the extracellular environment and results in intracellular signaling (43). Cell adhesion is a classic example of outside-in signaling in that the formation of focal adhesions, as described above, has the primary response of altering cell shape. These shape changes have been shown to have major effects on the cell, enabling it to sense the surrounding environment. This process has been termed mechanosensation or tensegrity and is facilitated by sensory cues on the cell from the surrounding ECM by the tensional forces exerted on the cells actin cytoskeleton (44,45). The result of these interactions can be reflected in the increased migration of cells through the ECM, as well as the possibility of proliferative signals and other intracellular signaling events that may lead to a change in the synthesis of ECM by the cell and the maintenance of tissue homeostasis (46).

The assembly of focal adhesions also leads to the activation of an enzyme called focal adhesion kinase (FAK), which phosphorylates tyrosine residues on proteins and can bind to other intracellular proteins, which in turn can initiate a cascade of intracellular events leading to signal transduction and DNA and protein synthesis (see Fig. 1) (47).

7. INTEGRIN—ECM BINDING SPECIFICITY

The interactions between integrins and the various ECM ligands generally are of low affinity and usually in the micromolar to nanomolar scale range and display considerable overlap in specificity (see Table 1). For example, $\alpha 1\beta 1$ $\alpha 2\beta 1$ both bind to collagens type I and IV, as well as to laminin type-I. It has been suggested that the binding epitope of laminin-I involves the sequence YGYYGDALR, which also inhibits the binding of endothelial cells to collagen type IV. A related sequence in type IV collagen, FYFDLR also inhibits the binding of the endothelial cells to collagen type I, type IV, and laminin-I (49). The overlap in binding specificity between the integrins and their interactive ECM molecules has changed over time to cope with the fact that some tissues have specific ECM components such as the basement membrane composed of type IV collagen and not type I collagen, as found elsewhere, and cartilage that is composed of type II collagen and not type I. With a certain amount of overlap, the repertoire of integrins required by different cell types for ECM interactions is kept to a more manageable number. If a certain cell (e.g., endothelial cell) expresses a subset of alpha and beta integrin subunits, it will still have the ability to adhere to type IV collagen or laminin in the basement membrane, whereas it is of little biological significance that it can also bind to collagen type I because its presence in the basement membrane is limited. Besides fibronectin, the laminins are the most promiscuous group of ECM molecules with respect to binding with various integrins. This promiscuity has an advantage in a cell in that, because of the low-affinity reactions of integrin-ECM binding, cells have the ability to partially bind and explore their surrounding ECM, which has been termed the velcro effect, without which cells would become irreversibly stuck to their surrounding ECM. If the number of low-affinity binding reactions increases to an amount whereby focal adhesions cluster and cause a strong association with the ECM, then cells become firmly adhesive. When the cells need to migrate or proliferate, they need to become less adhesive by reducing the number of these binding events involved in focal adhesion. *In vitro*, this can be promoted by chelation of available cations, which has downline effects on integrin structure and their interactive properties with ECM components.

8. QUANTIFICATION OF CELL ADHESION

Various methods exist available for the quantitation of the cell adhesion, which occurs to a particular surface. Some of these methods include the use of histochemical dyes (e.g., crystal violet or methylene blue) that stain proteins or other structural components of the cell (50). The challenge with these methods is the adsorption of the stain onto the biomaterial surface itself to such a degree that significant interference occurs that may override any specific signal obtained by the adherance of the cells to the biomaterial. Also, the ECM produced by the cells as they are cultured may also take up stain and thus be a source of interference because of nonspecific binding effects by the histochemical dyes used (51). Alternative methods have used radio labeling of cells with radioactive isotopes such as ^{35}S methionine (52). These molecules reflect the production of proteins by cells and are therefore related to the metabolic activity occurring within the cell during the assay, which varies considerably from cell type to cell type, which affects the amount of time that the cell-adhesion step of the assay should be carried out for. For example, primary cells have a relatively lower metabolic rate than transformed or continuous cell lines, so cell adhesion assays using transformed cells should only have a 4-hour cell-adhesive step, whereas primary cells can be left for 16

Table 1. Extracellular Matrix Molecules and Integrins

ECM molecule	Integrins involved in binding
Collagen I	$\alpha 1\beta 1, \alpha 2\beta 1, \alpha 3\beta 1, \alpha 10\beta 1, \alpha 11\beta 1$
Collagen IV	$\alpha 1\beta 1, \alpha 2\beta 1$
Fibronectin	$\alpha 3\beta 1, \alpha 4\beta 1, \alpha 5\beta 1, \alpha 8\beta 1, \alpha 4\beta 7, \alpha V\beta 1,$ $\alpha V\beta 3, \alpha V\beta 6, \alpha 1\beta 6, \alpha 2\beta 8$
Vitronectin	$\alpha V\beta 1, \alpha V\beta 3, \alpha V\beta 5, \alpha 1\beta 8$
Laminin – 1	$\alpha 1\beta 1, \alpha 2\beta 1, \alpha 3\beta 1, \alpha 6\beta 1, \alpha 7\beta 1, \alpha 6\beta 4$
Laminin – 2	$\alpha 6\beta 1$
Laminin – 4	$\alpha 6\beta 1$
Laminin – 5	$\alpha 6\beta 1$
Thrombospondin	$\alpha V\beta 3, \alpha 2\beta 1, \alpha 8\beta 1, \alpha 9\beta 1$
Osteopontin	$\alpha V\beta 3, \alpha 9\beta 1$
Bone sialoprotein 1	$\alpha V\beta 3$
Gelatin	$\alpha V\beta 3$
Tenascin – C	$\alpha 9\beta 1$
Epiligrin	$\alpha 3\beta 1$
Nidogen / Entactin	$\alpha 3\beta 1$
Fibrinogen	$\alpha V\beta 3$

Source: Refs. 40,41,48.

hours to ensure maximal adhesion. Careful interpretation of data from different cell lines is mandatory, and if the adhesion to different biomaterial surfaces is to be measured, only the results from one cell type should be compared within each experiment. An alternative assay that uses centrifugal force to measure the interaction between the cell and a biomaterial surface has been developed, and is especially useful for high-throughput applications (24). These assays are based on the use of polystyrene 96 well plates, which work well when one is examining the adhesive capacity of various ECM molecules on that synthetic polymer. The assay becomes more challenging when examining different synthetic polymer substrata.

9. FUTURE DIRECTIONS

In the context of this chapter (i.e., cell attachment to biomaterials), it is important to understand how cells normally use various binding modules on specific biomolecules to interact with ECM components in normal and pathological processes in connective tissues. This knowledge may provide clues and be subsequently "de-engineered" for the development of strategies to encourage the adherence of cells to biomaterials with the maintenance of cellular viability and expression of an appropriate cell phenotype. A number of so-called matricellular proteins have important roles to play in the regulation of cell adhesion in the ECM, but their full potential has yet to be realized in the area of cell attachment to biomaterials (53–56). Matricellular proteins are secreted multidomain macromolecules that are responsible for molecular interactions both at the cell surface and within the ECM, thus, they may be useful in novel engineering applications. Such interactions with the matricellular proteins modulate cell function but do not appear to contribute directly to the organization or physical properties of fibrils or laminae within tissues, thus, the matricellular proteins are distinct from other structural or adhesive glycoproteins of the ECM, such as fibronectin and laminin (56). Some members of the thrompospondins and matrilin families of ECM proteins are also members of the matricellular protein family, which include thrombospondin-5 (COMP, cartilage oligomeric protein) and matrilin-1 (CMP, cartilage matrix protein) (56), which have a limited tissue distribution whereas other members of the matricellular proteins (e.g., osteopontin) have a more widespread distribution in connective tissues. In some cases, the peptide sequences on the ECM components, which are responsible for cell binding, have been determined. Osteopontin (secreted phosphoprotein-1, OPN) is an ECM protein containing the well known RGD cell attachment sequence, which interacts with the $\alpha V \beta 3$ integrin to promote cell attachment and migration (57,58). OPN promotes osteoclast binding to resorptive sites through the $\alpha V \beta 3$ integrin and stimulates osteoclastic resorption of calcified cartilage during growth-related functional remodeling (57,58). Chondroadherin is a small leucine-rich protein, that promotes the attachment of chondrocytes to plastic via the $\alpha 2 \beta 1$ integrin (59). Anchorin CII (annexin V) functions as a collagen receptor and has a key role in the anchorage of the chondrocyte to type II collagen in the ECM and in the mineralization of calcified cartilage (60,61). Furthermore, other nonproteinaceous ECM components have a role to play in the regulation of cellular attachment. Decorin, a small chondroitin sulphate (CS) or dermatan sulphate (DS)-substituted small leucine-rich proteoglycan (SLRP), has been shown to interfere with cell attachment to thrombospondins via a KKTR attachment site on the matricellular protein (62,63), and with the attachment of fibroblasts to fibronectin by blocking the RGD cell attachment site (64,65). The glycosaminoglycan (GAG) side chain of the SLRPs have been shown to be responsible for this effect (66). It should also be noted that L-iduronate containing GAGs such as DS have been shown to potently inhibit fibroblast proliferation (67,68) emphasizing the importance of an integrated approach and a full understanding of the biology of cell attachment before developing any prospective engineering applications in this area.

Although technically outside the scope of this chapter, understanding how cells are prevented from attaching to normal ECM components and biomaterials is nevertheless an important consideration in many biomedical and tissue-engineering applications and undoubtedly will be an important area of future investigation. Thus, tissue engineering will employ the appropriate use of adhesive and nonadhesive signals to achieve correct alignment and patterning of cells leading to correct tissue and organ structure (6,7). Hyaluronan (HA), a $(1\text{-}20 \times 10^6 \text{ Da})$ polyanion copolymer GAG composed of linear arrays of the repeating disaccharide $[-\beta 1,3\text{-GlcNAc-}\beta 1,4\text{-GlcA-}]_n$, is an important biomolecule in this regard during embryonic development and displays traits that may be exploitable in future tissue-engineering applications. HA has diverse functional roles to play in cellular attachment and cellular migration (leucocyte trafficking, metastasis), but also displays antiadhesive properties. HA has extensive water regain, space-filling, and viscoelastic properties that result in the expansion of tissue volume and the maintenance of a loose ECM through which cells can readily migrate, and also displays biophysical properties that protect vital internal organs from physical damage. These are important considerations in prospective tissue-engineering applications. Furthermore, HA has matrix and cell regulatory properties that are attributable to its interactive properties with ECM proteoglycans of the aggrecan superfamily and its participation in signal-transduction events through cell surface HA receptors such as CD-44, receptor for HA-mediated motility (RHAMM), and lymphatic vessel endothelial cell HA receptor (LYVE)-1. HA can also serve as a cell attachment scaffold and influence cellular migration in developmental processes and wound healing and thus may have properties that are exploitable in tissue-engineering applications. One such application involves an HA-binding peptide (Pep-1), based on the HA-binding motif of BRAL-1 a brain specific link-protein, which is a small glycoprotein that displays affinity for HA and binds specifically to soluble, immobilized and cell-associated forms of HA but also displays interactive properties with HA-binding proteoglycans (69), which almost completely inhibits the binding of leucocytes to HA via its

CD-44 receptor, which normally acts as a homing receptor during inflammatory conditions. Pep-1 is currently being evaluated for its ability to prevent leucocyte trafficking in inflammatory disorders (69).

Recent advances in the characterization of biomolecules using techniques such as tandem mass spectrometry (MS-MS) and matrix-assisted laser desorption ionization time of flight mass spectroscopy (MALDI-TOF MS) will undoubtedly assist in the identification of prospective biomolecules that will be produced by advanced recombinant techniques. Furthermore, increasing sophistication in molecular modeling techniques will aid in the definition of the conformational requirements of cell-binding modules and aid in the design of such functional proteins. We are, therefore entering an exciting era in the area of biomolecule and polymer design and its future application in tissue engineering.

BIBLIOGRAPHY

1. P. A. Underwood, J. M. Whitelock, P. A. Bean, and J. G. Steele, Effects of base material, plasma proteins and FGF2 on endothelial cell adhesion and growth. *J. Biomater. Sci. Polym. Ed.* 2002; **13**(8):845–862.
2. Y. Ikada, Surface modification of polymers for medical applications. *Biomaterials* 1994; **15**:725–736.
3. P. B. van Wachem, C. M. Vreriks, T. Beugeling, J. Feijin, A. Bantjes, J. P. Detmers, and W. G. van Aken, The influence of protein adsorption on interactions of cultured human endothelial cells with polymers. *J. Biomed. Mat. Res.* 1987; **21**:701–718.
4. J. L. Dewez, J. B. Lhoest, E. Detrait, V. Berger, C. C. Dupont-Gillain, L. M. Vincent, Y. J. Schneider, P. Bertrand, and P. G. Rouxhet, Adhesion of mammalian cells to polymer surfaces; from physical chemistry of surfaces to selective adhesion on defined patterns. *Biomaterials* 1998; **19**:1441–1445.
5. L. G. Cima, Polymer substrates for controlled biological interactions. *J. Cell. Biochem.* 1994; **56**:155–161.
6. K. E. Healy, A. Rezania, and R. A. Stile, Designing biomaterials to direct biological responses. *Ann. N Y Acad. Sci.* 1999; **875**:24–35.
7. L. G. Griffith, Emerging design principles in biomaterials and scaffolds for tissue engineering. *Ann. N Y Acad. Sci.* 2002; **961**:83–95.
8. M. B. Olde Riekerink, M. B. Claase, G. H. Engbers, D. W. Grijpma, and J. Feijen, Gas plasma etching of PEO/PBT segmented block copolymer films. *J. Biomed. Mater. Res.* 2003; **65A**(4):417–428.
9. K. M. DeFife, M. S. Shive, K. M. Hagen, D. L. Clapper, and J. M. Anderson, Effects of photochemically immobilized polymer coatings on protein adsorption, cell adhesion, and the foreign body reaction to silicone rubber. *J. Biomed. Mater. Res.* 1999; **44**(3):298–307.
10. C. Amstein and P. Hartman, Adaptation of plastic surfaces for tissue culture by glow discharge. *J. Clin. Micro.* 1975; **2**:46–54.
11. G. Hannan and B. McAuslan, Immobilised serotonin: a novel substrate for cell culture. *Exp. Cell Res.* 1987; **171**:153–163.
12. T. G. van Kooten, H. T. Spijker, and H. J. Busscher, Plasma-treated polystyrene surfaces: model surfaces for studying cell-biomaterial interactions. *Biomaterials* 2004; **25**(10):1735–1747.
13. J. G. Steele, G. Johnson, W. D. Norris, and P. A. Underwood, Adhesion and growth of cultured human endothelial cells on perfluorosulphonate: role of vitronectin and fibronectin in cell attachment. *Biomaterials* 1991; **12**:531–539.
14. J. G. Steele, G. Johnson, and P. A. Underwood, Role of serum vitronectin and fibronectin in adhesion of fibroblasts following seeding onto tissue culture polystyrene. *J. Biomed. Mat. Res.* 1992; **26**:861–864.
15. W. B. Tsai, J. M. Grunkemeier, and T. Horbett, Human plasma fibrinogen adsorption and platelet adhesion to polystyrene. *J. Biomed. Mater. Res.* 1999; **44**(2):130–139.
16. K. T. Preissner, R. Wassmuth, and G. Muller - Berghaus, Physicochemical characterisation of human S-protein and its function in the blood coagulation system. *Biochem. J.* 1985; **231**:349–355.
17. D. Seiffert and N. V. Wagner, Evidence for a specific interaction of vitronectin with arginine: effects of reducing agents on the expression of functional domains and immunoepitopes. *Biochimie* 1997; **79**:205–210.
18. P. A. Underwood, J. G. Steele, and B. A. Dalton, Effects of polystyrene surface chemistry on the biological activity of solid phase fibronectin and vitronectin, analysed with mouse antibodies. *J. Cell. Sci.* 1993; **104**:793–803.
19. S. Suzuki, M. D. Pierschbacher, E. G. Hayman, K. Nguyen, Y. Ohgren, and E. Ruoslahti, Domain structure of vitronectin. *J. Biol. Chem.* 1984; **259**:15307–15314.
20. X. Sun, K. Skorstengaard, and D. F. Mosher, Disulfides modulate RGD-inhibitable cell adhesive activity of thrombospondin. *J. Cell Biol.* 1992; **118**(3):693–701.
21. D. Klee, Z. Ademovic, A. Bosserhoff, H. Hoecker, G. Maziolis, and H. J. Erli, Surface modification of poly(vinylidenefluoride) to improve the osteoblast adhesion. *Biomaterials* 2003; **24**(21):3663–3670.
22. J. Emsley, C. G. Knight, R. W. Farndale, and M. J. Barnes, Structure of the integrin alpha2beta1-binding collagen peptide. *J. Mol. Biol.* 2004; **335**(4):1019–1028.
23. D. F. Sweeney, R. Z. Xie, M. D. Evans, A. Vannas, S. D. Tout, H. J. Griesser, G. Johnson, and J. G. Steele, A comparison of biological coatings for the promotion of corneal epithelialization of synthetic surface *in vivo*. *Invest. Ophthalmol. Vis. Sci.* 2003; **44**(8):3301–3309.
24. C. D. Reyes and A. J. Garcia, Engineering integrin-specific surfaces with a triple-helical collagen-mimetic peptide. *J. Biomed. Mater. Res.* 2003; **65A**(4):511–523.
25. X. H. Wang, D. Li, W. J. Wang, Q. L. Feng, F. Z. Cui, Y. X. Xu, X. H. Song, and M. van der Werf, Crosslinked collagen/chitosan matrix for artificial livers. *Biomaterials* 2003; **24**(19):3213–3220.
26. N. Alminana, M. R. Grau-Oliete, F. Reig, and M. P. Rivera-Fillat, *In vitro* effects of SIKVAV retro and retro-enantio analogues on tumor metastatic events. *Peptides* 2004; **25**(2):251–259.
27. K. Yamamura, M. C. Kibbey, S. H. Jun, and H. K. Kleinman, Effect of Matrigel and laminin peptide YIGSR on tumor growth and metastasis. *Semin. Cancer Biol.* 1993; **4**(4):259–265.
28. M. Mochizuki, Y. Kadoya, Y. Wakabayashi, K. Kato, I. Okazaki, M. Yamada, T. Sato, N. Sakairi, N. Nishi, and M. Nomizu, Laminin-1 peptide-conjugated chitosan membranes as a novel approach for cell engineering. *FASEB J.* 2003; **17**(8):875–877.
29. J. M. Whitelock, L. D. Graham, J. Melrose, A. D. Murdoch, R. V. Iozzo, and P. A. Underwood, Human perlecan immunopurified from different endothelial cell sources has different

adhesive properties for vascular cells. *Matrix Biol.* 1999; **18**(2):163–178.

30. J. C. Brown, T. Sasaki, W. Gohring, Y. Yamada, and R. Timpl, The C-terminal domain V of perlecan promotes beta1 integrin-mediated cell adhesion, binds heparin, nidogen and fibulin-2 and can be modified by glycosaminoglycans. *Eur. J. Biochem.* 1997; **250**(1):39–46.

31. B. Schulze, T. Sasaki, M. Costell, K. Mann, and R. Timpl, Structural and cell-adhesive properties of three recombinant fragments derived from perlecan domain III. *Matrix Biol.* 1996; **15**(5):349–357.

32. P. A. Underwood, P. A. Bean, S. M. Mitchell, and J. M. Whitelock, Specific affinity depletion of cell adhesion molecules and growth factors from serum. *J. Immunol. Methods* 2001; **247**:217–224.

33. U. Hersel, C. Dahmen, and H. Kessler, RGD modified polymers: biomaterials for stimulated cell adhesion and beyond. *Biomaterials* 2003; **24**(24):4385–4415.

34. D. A. Wang, L. X. Feng, J. Ji, Y. H. Sun, X. X. Zheng, and J. H. Elisseeff, Novel human endothelial cell-engineered polyurethane biomaterials for cardiovascular biomedical applications. *J. Biomed. Mater. Res.* 2003; **65A**(4):498–510.

35. H. B. Lin, C. Garcia-Echeverria, S. Asakura, W. Sun, D. F. Mosher, and S. L. Cooper, Endothelial cell adhesion on polyurethanes containing covalently attached RGD-peptides. *Biomaterials* 1992; **13**(13):905–914.

36. A. R. El-Ghannam, P. Ducheyne, M. Risbud, C. S. Adams, I. M. Shapiro, D. Castner, S. Golledge, and R. J. Composto, Model surfaces engineered with nanoscale roughness and RGD tripeptides promote osteoblast activity. *J. Biomed. Mater. Res.* 2004; **68A**(4):615–627.

37. M. Kantlehner, P. Schaffner, D. Finsinger, J. Meyer, A. Jonczyk, B. Diefenbach, B. Nies, G. Holzemann, S. L. Goodman, and H. Kessler, Surface coating with cyclic RGD peptides stimulates osteoblast adhesion and proliferation as well as bone formation. *Chembiochem.* 2000; **1**(2):107–114.

38. A. Varki, Selectin ligands. *Proc. Natl. Acad. Sci. USA* 1994; **91**(16):7390–7397.

39. E. Ruoslahti, Integrins. *J. Clin. Invest.* 1991; **87**(1):1–5.

40. K. M. Hodivala-Dilke, A. R. Reynolds, and L. E. Reynolds, Integrins in angiogenesis: multitalented molecules in a balancing act. *Cell Tissue Res.* 2003; **314**(1):131–144.

41. J. Ivaska and J. Heino, Adhesion receptors and cell invasion: mechanisms of integrin-guided degradation of extracellular matrix. *Cell Mol. Life Sci.* 2000; **57**(1):16–24.

42. I. D. Campbell, Modular proteins at the cell surface. *Biochem. Soc. Trans.* 2003; **31**(6):1107–1114.

43. F. G. Giancotti and E. Ruoslahti, Integrin signaling. *Science* 1999; **285**(5430):1028–1032.

44. D. E. Ingber, Tensegrity I. Cell structure and hierarchical systems biology. *J. Cell. Sci.* 2003; **116**(7):1157–1173.

45. N. Wang, J. P. Butler, and D. E. Ingber, Mechanotransduction across the cell surface and through the cytoskeleton. *Science* 1993; **260**(5111):1124–1127.

46. M. Chiquet, A. S. Renedo, F. Huber, and M. Fluck, How do fibroblasts translate mechanical signals into changes in extracellular matrix production? *Matrix Biol.* 2003; **22**(1):73–80.

47. R. L. Juliano, Signal transduction by cell adhesion receptors and the cytoskeleton: functions of integrins, cadherins, selectins, and immunoglobulin-superfamily members. *Annu. Rev. Pharmacol. Toxicol.* 2002; **42**:283–323.

48. K. Anselme, Osteoblast adhesion on biomaterials. *Biomaterials* 2000; **21**(7):667–681.

49. P. A. Underwood, F. A. Bennett, A. Kirkpatrick, P. A. Bean, and B. A. Moss, Evidence for the location of a binding sequence for the alpha 2 beta 1 integrin of endothelial cells, in the beta 1 subunit of laminin. *Biochem. J.* 1995; **309**(3):765–771.

50. P. A. Underwood, J. G. Steele, B. A. Dalton, and F. A. Bennett, Solid-phase monoclonal antibodies. A novel method of directing the function of biologically active molecules by presenting a specific orientation. *J. Immunol. Methods* 1990; **127**(1):91–101.

51. P. A. Underwood and P. A. Bean, Extracellular matrix interferes with colorimetric estimation of cell number. *In vitro Cell Dev. Biol. Anim.* 1998; **34**(3):200–202.

52. B. A. Dalton, M. Dziegielewski, G. Johnson, P. A. Underwood, and J. G. Steele, Measurement of cell adhesion and migration using phosphor-screen autoradiography. *Biotechniques* 1996; **21**(2):298–303.

53. P. Bornstein and E. H. Sage, Matricellular proteins: extracellular modulators of cell function. *Curr. Opin. Cell. Biol.* 2002; **14**:608–616.

54. P. Bornstein, Thrombospondins as matricellular modulators of cell function. *J. Clin. Invest.* 2001; **107**:929–934.

55. P. Bornstein, Cell-matrix interactions: the view from the outside. *Methods Cell Biol.* 2002; **69**:7–11.

56. J. C. Adams, Thrombospondins: multifunctional regulators of cell interactions. *Annu. Rev. Cell Dev. Biol.* 2001; **17**:25–51.

57. S. Shibata, K. Fukuda, S. Suzuki, T. Ogawa, and Y. Yamashita, In-situ hybridisation and immunohistochemistry of bone sialoprotein and secreted phosphoprotein-1 (osteopontin) in the developing mouse mandibular and condylar cartilage compared with limb buds. *J. Anat.* 2002; **200**:309–320.

58. H. Sugiyama, M. Imada, A. Sasaki, Y. Ishino, T. Kawata, and K. Tanne, The expression of osteopontin with condylar remodeling in growing rats. *Clin. Orthod. Res.* 2001; **4**:194–199.

59. L. Camper, D. Heinegard, and E. Lundgren-Akerlund, Integrin alpha 2 beta 1 is a receptor for the cartilage matrix protein chondroadherin. *J. Cell Biol.* 1997; **138**:1159–1167.

60. J. Mollenhauer, J. A. Bee, M. A. Lizarbe, and K. von der Mark, Role of anchorin CII, a 31,000-mol-wt membrane protein, in the interaction of chondrocytes with type II collagen. *J. Cell Biol.* 1984; **98**:1572–1579.

61. K. von der Mark and J. Mollenhauer, Annexin V interactions with collagen. *Cell Mol. Life Sci.* 1997; **53**:539–545.

62. M. Winnemoller, P. Schon, P. Vischer, and H. Kresse, Interactions between thrombospondin and the small proteoglycan decorin: interference with cell attachment. *Eur. J. Cell Biol.* 1992; **59**:47–55.

63. B. Merle, L. Malaval, J. Lawler, P. Delmas, and P. Clezardin, Decorin inhibits cell attachment to thrombospondin-1 by binding to a KKTR-dependent cell adhesive site present within the N-terminal domain of thrombospondin-1. *J. Cell Biochem.* 1997; **67**:75–83.

64. M. Winnemoller, G. Schmidt, and H. Kresse, Influence of decorin on fibroblast adhesion to fibronectin. *Eur. J. Cell Biol.* 1991; **54**:10–17.

65. K. Lewandowska, H. U. Choi, L. C. Rosenberg, L. Zardi, and L. A. Culp, Fibronectin-mediated adhesion of fibroblasts: inhibition by dermatan sulfate proteoglycan and evidence for a cryptic glycosaminoglycan-binding domain. *J. Cell Biol.* 1987; **105**:1443–1454.

66. D. Bidanset, R. Lebaron, L. Rosenberg, and J. Murphy-Ullrich, Regulation of cell substrate adhesion: effects of small galactosaminoglycan-containing proteoglycans. *J. Cell Biol.* 1992; **118**:1523–1531.

67. D. Bidanset, R. Lebaron, L. Rosenberg, and J. Murphy-Ullrich, Regulation of cell substrate adhesion: effects of small galactosaminoglycan-containing proteoglycans. *J. Cell Biol.* 1992; **118**:1523–1531.
68. G. Westergren-Thorsson, S. Persson, A. Isaksson, P. O. Onnervik, A. Malmstrom, and L. A. Fransson, L-iduronate-rich glycosaminoglycans inhibit growth of normal fibroblasts independently of serum or added growth factors. *Exp. Cell Res.* 1993; **206**:93–99.
69. M. E. Mummert, M. Mohamadzadeh, D. I. Mummert, N. Mizumoto, and A. Takashima, Development of a peptide inhibitor of hyaluronan-mediated leukocyte trafficking. *J. Exp. Med.* 2000; **192**:769–779.

FURTHER READING

The following review has some excellent information on the specific structures employed for cell-ECM interactions. J. C. Adams (2002). *Molecular organisation of cell-matrix contacts: essential multiprotein assemblies in cell and tissue function.* (online). Available: *http://www-ermm.cbcu.cam.ac.uk/02004039h.htm.*

The WWW Virtual Library of Cell Biology contains links to websites on cell adhesion, ECM, the cytoskeleton and cell motility. (online). Available: *http://www.vlib.org/Science/Cell_Biology/.*

The Medscape Website has information about preclinical and clinical studies related to various fields including cell-extracellular matrix research. (online). Available: *http://www.medscape.com/.*

The Wound Healing Society web-site has links to relevant journals and other links to cell-matrix research related to wound healing. (online). Available: *http://www.woundheal.org.*

ALLOGENEIC CELLS AND TISSUES

RICHARD KNIGHT
EILEEN INGHAM
University of Leeds
Leeds, United Kingdom

1. CELLS AND TISSUES

All of the organs and tissues in the human body develop from a single fertilized egg, the zygote. The zygote is described as being totipotent because it is ultimately able to differentiate into all the cell types that constitute the adult body. During a 9-month gestation period, this single cell divides repeatedly through a highly organized and controlled development process to produce many billions of daughter cells. The genetic material (DNA) of the zygote is retained in all of the daughter cells during the development process [with the exception of mature red cells (no DNA) and mature lymphocytes (loss of some DNA during immunoglobulin and receptor gene rearrangements)]. During development, cells differentiate as different proteins are expressed in cascades that regulate the DNA in each cell, restricting the transcription of genes in different cells, which enables cells in different areas of the developing embryo to differentiate into the highly specialized cells that will eventually make up the adult body. As these cells continue to differentiate and grow according to their newly developed function, they produce the extracellular matrix (ECM) that comprises the bulk of most tissues, such as bone, cartilage, muscle fibers, and many other structural and nonstructural components. The cell is therefore the basic unit of life and a fundamental component of all tissues. 216 different cell types are identified in the human body, including mesenchymal cells such as fibroblasts, chondrocytes, osteoblasts, adipose cells, muscle cells, and cardiomyocytes; blood cells, including erythrocytes, white cells or leucocytes including neutrophils, eosinophils, basophils, mononuclear phagocytes, and lymphocytes; different types of nerve cells such as astrocytes and glial cells; different types of epithelial cells such as keratinocytes of the skin, urothelial cells of the bladder, intestinal epithelial cells; different types of glandular secretory cells such as beta cells in the pancreas that make and secrete insulin; and highly specialized cells found in particular organs.

The ECM is composed of fibrous proteins such as collagen (mainly collagen type I) and elastin, glycosaminoglycans (chains of carbohydrate molecules), and proteoglycans (proteins and carbohydrates). The abundance, organization, and proportions of the macromolecular components vary according to the tissue type and determine the mechanical properties of the tissue in which they are situated. Skin and blood vessels are resilient because of numerous elastic fibers; tendons have great tensile strength owing to collagen; and bone is incompressible and rigid because it is a calcified collagen matrix. The ECM is not a static structure, but is fully dynamic in that it is constantly being remodeled by the cells that create it. ECM is constantly being degraded and replaced with new ECM, enabling the structure and function of the ECM to adapt with changes in the physiology of the tissue (e.g., build-up of muscle in athletes). Furthermore, the ECM supports specialized neuronal and muscle cells in various organs. On the level of gross anatomy, cells and fibers form fascia, tendons, cartilage, and bone that support the organs of the body. The ECM also provides an arena for cellular defense systems. Specialized "professional" phagocytic white cells of the immune system roam connective tissue to find foreign and injured cells and other matter that may develop in the tissues. Connective tissue also provides avenues for communication and supply within the body as both the circulatory system and the peripheral nervous system run through the connective tissue compartments of each organ.

The third and final component of the tissues is a fluid. This fluid is composed of the fluid phase of blood/lymph and fluid that constitutes part of the tissue itself. Most tissues are supplied with blood vessels and lymph vessels. The blood brings nutrients to the cells in the tissue and oxygen bound to hemoglobin in the red blood cells (erythrocytes). Water and nutrients diffuse from the blood into the tissues, and the excess fluid is collected into lymph vessels and eventually returned to the blood. Along with proportions of ECM components varying between tissue types, so does the proportion of cells, ECM, and fluid. For example, if one considers the blood as a tissue, no ECM exists at all and instead it comprises solely of cells and fluid. In contrast, epithelia, although supported by a basement membrane composed of ECM, are themselves composed solely of cells.

2. HISTOCOMPATIBILITY

The development and maintenance of all the cells and tissues in an individual is regulated by their DNA. DNA exists as a double-stranded helix in the nucleus of all cells throughout the body and is gathered in sequences called genes, which encode the proteins that make up both cellular structures and functions and the ECM. Every cell contains a full set of DNA and, consequently, it is the switching on and off of specific genes in different cells in different parts of the body that accounts for the differences in cell function. The DNA is split into 23 pairs of chromosomes, and specific genes are always found on the same chromosome in different individuals. The vast majority of gene sequences are more or less conserved between individuals of the same species, and to some extent between individuals of closely related species. Hence, in humans, the proteins that make up the extracellular matrix, such as the collagen and elastin molecules, in one individual will be the same as the collagen molecules in another, unrelated individual.

However, some genes exist that are highly polymorphic within a species. The most highly polymorphic genes are the genes that encode for proteins that are expressed on the surface of living cells, which are responsible for preventing the free transplantation of tissues and cells from one individual to another. These genes and proteins are known as histocompatibility genes and proteins, and a basic understanding of these is essential to appreciate why, when foreign tissues and organs containing living cells are transplanted from one individual to another (excluding monozygotic twins, which have identical genes), the donated tissue or organ is recognized as foreign to the host and is thus attacked by the immune system or "rejected."

In humans, the genes that code for the histocompatibility proteins are found within a set of genes called the major histocompatibility complex (MHC) of genes. The ones that are relevant here are those that are classified as Class I and Class II (MHC Class I and MHC Class II). Three Class I genes and three Class II genes exist on each of the pair of chromosome 6, making six genes for Class I and six genes for Class II. As one of each chromosome is from each parent, every individual inherits three Class I and three Class II genes from their father and three Class I and three Class II genes from their mother. The genes are "expressed" in a codominant manner, which means that the proteins coded by both the paternal and maternal genes are expressed on the cell surface (Fig. 1).

In the human, the proteins that are expressed on the cell surface are called human leukocyte-associated or HLA proteins because they were first discovered on the leukocytes in the blood. The three Class I genes on each chromosome code for HLA-A, HLA-B, and HLA-C proteins. The three Class II genes on each chromosome code for HLA-DR, HLA-DP, and HLA-DQ proteins (Fig. 1).

An extremely important feature of the genes that code these proteins is their polymorphism within the human population, which means that different individuals have slightly different DNA codes for each protein, known as alleles. In humans, different alleles are given different numbers, for example, HLA-B17 or HLA B27. Numerous different alleles exist for each Class I and Class II protein in the human population, indeed it is probable that the full extent of polymorphisms that exist for each gene is not yet known. This feature, together with the fact that the alleles are expressed codominantly, means that every individual has a "set" of alleles and proteins that are essentially unique to them (unless they have an identical twin or a sibling who has inherited the same chromosome from both mother and father). For example, an individual may be HLA-A2 and A7, HLA-B17 and B27, HLA-C6 and C8, and so on. Another individual may have a totally different set of alleles and proteins, or individuals may have some of the same alleles and proteins and only one or two that are different. However, the extreme polymorphism means that it is almost impossible that two random individuals will have identical sets of the alleles and proteins. It is analogous to a credit card or bank account number.

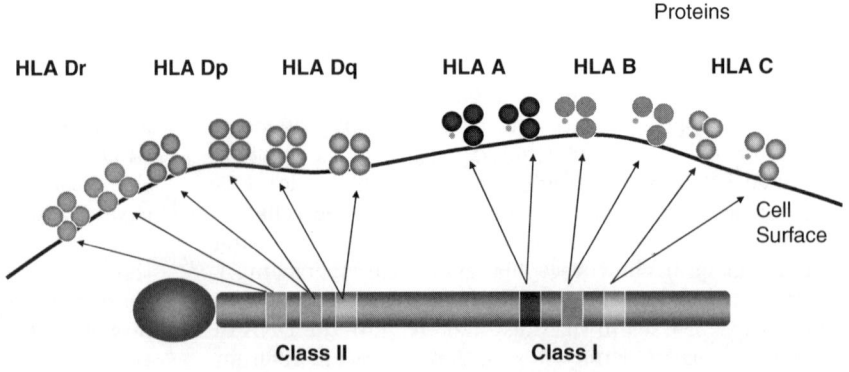

Figure 1. MHC Genes and Proteins. In humans, the genes that code for the histocompatibility proteins are found within a set of genes called the major histocompatibility complex (MHC) of genes (MHC Class I and MHC Class II). Three Class I genes and three Class II genes exist on each of the pair of chromosome 6, making six genes for Class I and six genes for Class II. As one of each chromosome is from each parent, every individual inherits three Class I and three Class II genes from their father and three Class I and three Class II genes from their mother. The genes are "expressed" in a codominant manner, which means that the proteins coded by both the paternal and maternal genes are expressed on the cell surface. (This figure is available in full color at http://www.mrw.interscience.wiley.com/ebe.)

An individual's MHC is unique to them. It is what makes them "immunologically self." These proteins have very important functions in the immune response, which is explained in more detail below.

3. BASIC IMMUNOLOGY

The primary function of the immune system is to preserve "self" by recognizing any substance that is "non-self" that manages to gain entry into the body and bringing about its elimination from the body. Hence, in a healthy individual, the immune system protects an individual from environmental agents by eliminating harmful foreign substances. Such foreign substances include pathogenic organisms, such as certain bacteria, viruses, and fungi. However, the immune system will also attack potentially beneficial nonself material such as tissues, organs, and cells obtained from another individual. It is this response that causes the greatest problem when attempting to transplant cells and tissues between individuals.

3.1. Innate and Adaptive Immunity

Any immune response essentially has two phases: first, recognition of the pathogen or foreign material; and second, a reaction against it to eliminate it. The types of reaction that can be generated are often divided into two categories: innate and adaptive immune responses (1). Both are dependent on the activities of white blood cells, or leukocytes. Innate immunity involves leukocytes that act in a nonspecific manner and that respond to "danger" signals such as mediators of inflammation released by damaged cells, which include phagocytic cells called neutrophils and mononuclear phagocytes (monocytes in blood, macrophages in tissues) that recognize micro-organisms and their products via nonspecific "pattern" receptors, ingest, kill, and catabolize them. These cells use a primitive recognition system that allows them to bind to a variety of foreign molecules and act as the first line of defense while the adaptive response is initiated. Initiation of an adaptive or specific immune response requires that foreign molecules are taken up by specialized cells of the innate immune systems called immunostimulatory dendritic cells, which are related to macrophages and can present foreign material to lymphocytes.

The adaptive immune response depends on lymphocytes (B-cells and T-cells) that can provide life-long immunity (1). Foreign material that stimulates an adaptive immune response is called an immunogen (generates immune response) or antigen. B-cells combat foreign material through the interaction of highly specific cell surface molecules called immunoglobulins (Ig), which specifically recognize foreign molecular shapes on the surface of antigens, called epitopes. Once this interaction takes place, more Ig of the same specificity is secreted in a form called antibody. The antibodies then bind to the specific epitope that stimulated their production, which can inhibit the activities of the antigen or facilitate phagocytosis by neutrophils and macrophages. This process is known as humoral immunity. Similarly, T-cells recognize epitopes derived from antigens via membrane-bound proteins called T-cells receptors (TCRs), which are structurally related to Ig. However, T-cells do not recognize and bind epitopes on antigens directly, but instead recognize short peptide epitopes that are bound to MHC molecules on the surface of other cells (1). T-cells have a range of activities including control of B-cell responses, activation of phagocytic cells to aid destruction of ingested material (T-helper cells, Th), and recognition and destruction of self cells infected by viruses (cytotoxic T-cells, Tc). Th cells recognize peptide epitopes presented by MHC Class II molecules and cytotoxic Tc cells recognize peptide epitopes presented by MHC Class I molecules.

3.2. Humoral Immunity

The antibody molecule has two separate functions: one is to bind specifically to the epitope that elicited the immune response; the other is to recruit other cells and molecules to destroy the antigen once the antibody is bound to it, for example, binding of antibody marks antigens for destruction by phagocytes. Antibodies are basically Y-shaped molecules. The two arms of the Y end in regions that are highly variable between different antibody molecules and are called the V or variable regions. The end of the V region interacts with the epitope of the antigen and is unique to each specific antibody molecule. The stem of the Y is much more conserved and is referred to as the C or constant region, and this end interacts with leukocytes and other proteins. Five different classes of antibody or immunoglobulin exist and these classes are referred to as IgM, IgG, IgD, IgA, and IgE. The classes differ in the structure of their C region and have different roles, but IgG is the most abundant class in the blood. The basic structure of IgG is shown in Fig. 2. Each antibody has two identical antigen-binding sites, meaning that one antibody can bind two identical epitopes. Antibodies are highly specific and will only bind to a given shape of epitope. Hence, in order to cope with the millions of foreign substances that may gain entry into the body, the immune system has a massive repertoire of antibodies (circa 10^7–10^8) that are able to bind with every conceivable epitope of antigen that it might encounter (1).

The humoral immune response begins with recognition of a specific antigen by binding to a B-cell via its surface antibody receptor (clonal selection), which activates the B-cell and ultimately results in cell division and differentiation (clonal expansion). As the dividing B-cells differentiate, some develop into "memory cells" and some into cells called plasma cells. The plasma cells are terminally differentiated cells that have a short life span and secrete large amounts of the specific antibody to combat the antigen. Should the immune system encounter the same antigen again, many, many more B-cells (the memory B-cells) now exist of the correct specificity to recognize and respond to it. The memory cells become activated and rapidly clonally expand to generate more memory cells and plasma cells. This memory response is much faster and more potent (in terms of levels of antibodies produced) than when the antigen was originally encountered, enabling the immune system to eradicate the foreign material much more rapidly, leading to immunity from the

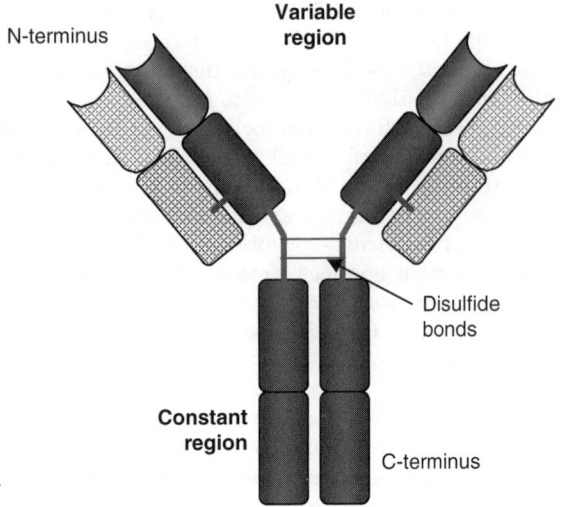

Figure 2. Immunoglobulins released as antibody by B-cells specifically bind to foreign antigens to aid identification and phagocytosis of foreign material. The two arms of the Y end in regions that are highly variable between different antibody molecules and are called the V or variable regions. The end of the V region interacts with the epitope of the antigen and is unique to each specific antibody molecule. The stem of the Y is much more conserved and is referred to as the C or constant region, and this end interacts with leukocytes and other proteins. Each antibody has two identical antigen-binding sites, meaning that one antibody can bind two identical epitopes. Dark shading shows heavy chains; lighter shading shows light chains. (This figure is available in full color at http://www.mrw.interscience.wiley.com/ebe.)

effects of the antigen in the body. The first response to the antigen is called the primary immune response, and the memory cell response in known as the secondary immune response. The response "adapts" to combat the antigen.

Another aspect of the humoral immune response is the complement system. The complement system consists of a system of proteins in serum that is normally biologically inactive. When "activated," the proteins react in a cascade of biochemical reactions that result in the generation of mediators of inflammation, which attract and activate phagocytic cells and can lyse target cells by inserting a hole in the target cell membrane. The binding of IgG or IgM antibody molecules to antigens can activate the complement cascade. The activities of complement are said to "complement" the activities of antibodies (2).

3.3. Cell-Mediated Immunity

The reactions of T-cells in the adaptive immune response are generally referred to as cell-mediated immunity. As mentioned above, T-cells are specific but do not recognize native antigens. The T-cell receptor recognizes a combination of a peptide epitope derived from a foreign antigen presented to it on the surface of another body cell in association with an MHC molecule (3). Two subpopulations of T-cells exist, the helper T-cells and the cytotoxic T-cells. Helper T-cells express a marker that is used to identify them called CD4 (cluster of differentiation molecule 4) and are often called CD4+ T-cells. The marker that is used to recognize cytotoxic T-cells is called CD8 (CD8+ T-cells). CD4+ T-cells recognize a combination of an MHC Class II molecule and peptide epitope, whereas CD8+ T-cells recognize a combination of MHC Class I and peptide epitope.

T-cells are educated in how to recognize foreign epitopes during their development in a special organ called the thymus gland during early life. This process is very complex, involving a process of positive and negative selection that is beyond the scope of this chapter. The process ensures that only T-cells that express a receptor that recognizes self-MHC plus foreign epitope are present within the immune system. Any T-cells that develop a receptor that can bind very tightly (and therefore potentially be activated by) a self-MHC molecule in combination with a self-epitope are killed, which ensures that the T-cells of the immune system only recognize and respond to non-self epitopes. As with B-cells and antibodies, the immune system generates millions of different clones of CD4+ and CD8+ T-cells, with each clone expressing a specific T-cell receptor.

These mature T-cells recognize "altered self." A given T-cell receptor will bind loosely to the self-MHC molecule that it has been educated to recognize. This process is not, however, sufficient to activate the T-cell to respond. If the self-MHC is altered by the binding of a foreign peptide epitope, the molecular shape is altered, which will enable the T-cell receptor to bind tightly, stimulating a response.

All cells in the body except red blood cells express MHC Class I molecules. Hence, any cell can present an epitope in association with an MHC Class I molecule to the respective CD8+ T-cell. Such epitopes are usually derived from microbial proteins that are present within the cytoplasm of cells, such as virus proteins. MHC Class II molecules are, however, only normally expressed on the surface of specialized cells that are called antigen-presenting cells (dendritic cells, macrophages, and B-cells). These cells have the capacity to take up foreign proteins or organisms from outside the cell, internalize them, and then present peptide epitopes derived from them on the cell surface in combination with MHC Class II molecules for the respective CD4+ T-cells to recognize.

In order to activate mature but naive T-cells (T-cells that have not previously responded to an antigen), a complex series of molecular signals are required. For CD4+, these signals can only be delivered via specialized immunostimulatory dendritic cells that have presented the complementary epitope in combination with an MHC class II molecule and also express "co-stimulatory" molecules (Fig. 3). In order to activate a naive CD8+ T-cell, presentation of epitope via MHC Class I is required plus a chemical signal in the form of a cytokine called interleukin-2 (IL-2) produced by an activated CD4+ T-cell. Once activated, T-cells, like B-cells, undergo clonal expansion and differentiation into memory cells and "effector" cells. The effector cells must recognize a cell expressing the same MHC molecule and epitope in order to exert their effects (efferent recognition). In this way, the reactions of the T-cells are targeted to cells that contain their specific foreign antigen. However, for effector cells, the requirements for recognition are much less stringent and they can exert their effects on any cells expressing the correct

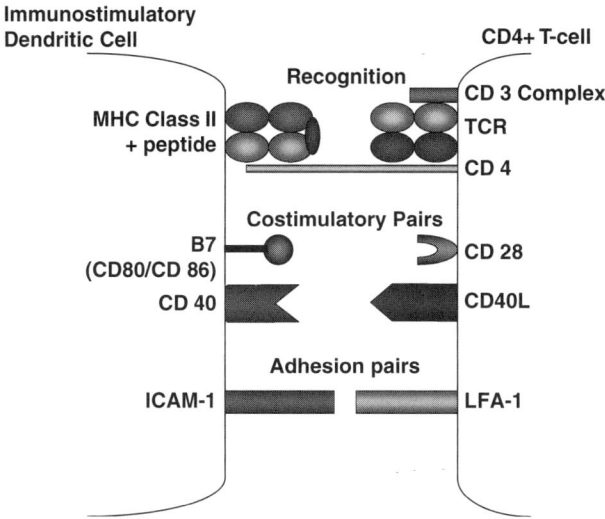

Figure 3. Key cell surface interactions leading to CD 4+ T-cell activation and cytokine secretion. In order to activate mature but naive T-cells (T-cells that have not previously responded to an antigen), a complex series of molecular signals are required. For CD4+, these signals can only be delivered via specialized immunostimulatory dendritic cells that have presented the complementary epitope in combination with an MHC class II molecule. (This figure is available in full color at http://www.mrw.interscience.wiley.com/ebe.)

MHC molecule and epitope. In this respect, CD4+ T-cells are still limited to reacting with MHC Class II expressing cells (mononuclear phagocytes, dendritic cells and B-cells). In the efferent response, CD4+ T-cells release soluble factors (cytokines) that activate the target cell. If the target cell is a macrophage, the macrophages will be activated to destroy the foreign material they have ingested. One particular cytokine that is very important in this respect is interferon gamma (IFN-γ), which activates macrophages and stimulates them to increase their killing capability via the production of free oxygen radicals and enzymes (4), which helps macrophages to kill particularly persistent pathogens that they have phagocytosed. When an effector CD8+ T-cell recognizes its target cell, it reacts by killing the cell. By killing the target cell, the life cycle of any virus that has infected the cell is cut short, and the virus is "killed."

Hence, T-cells protect the body against microbial pathogens that live inside body cells; however, as indicated below, the consequences of the way that T-cells recognize foreign antigens are that they also react aggressively against foreign cells and tissues.

4. TRANSPLANTATION

4.1. Relationship Between Donor and Recipient

An *autograft* describes the transplant of tissue from one body site to another body site on the same individual. This transfer is autogeneic, meaning that the donor and recipient is the same individual and therefore the donor tissue is seen as "self" and, consequently, the immune system does not react to it. An example is the skin grafting of patients with burns. If the skin on the arms, upper trunk, or face is burnt and lost, but the lower body is unaffected, the burns surgeon will take skin from the unaffected area (e.g., buttocks, thighs; donor site) and transplant it to the burned area.

An *isograft* describes the transplant of tissue between individuals of identical genotype. The donor tissue is described as isogeneic or syngeneic, which describes cells of the same genotype but from different individuals. An example would be tissue transplanted between monozygotic (identical) twins. In addition, when animals are used for scientific experimentation, they are inbred in order to produce genetically identical animals in order to reduce variability. Again, the tissues are accepted because the cells are genetically identical and are therefore seen as "self."

An *allograft* describes tissue transplanted between unrelated individuals of the same species. The donor tissue is described as allogeneic and is genetically different from the recipient and as such is "non-self." Such tissue will be recognized as foreign by the recipient and rejected by either "acute" or "chronic rejection" depending on how closely matched genetically the individuals are. In clinical transplantation, in which allogeneic tissues and organs from living or cadaveric donors are transplanted, they only survive because of the treatment of the recipient with immunosuppressive drugs.

A *xenograft* describes tissue transplanted between different species. The tissue is described as xenogeneic and is genetically very different and consequently "non-self." Such tissue will be recognized as foreign by the recipient and rejected by a mechanism referred to as hyperacute rejection. An illustration of tissue transfer can be seen in Fig. 4.

4.2. Immune Mechanisms and Allograft Rejection

When allogeneic cells, tissues, or organs are transplanted, the graft is usually "rejected" either rapidly, as in acute rejection, or over many years, as in chronic rejection. Both are mediated by T-cells. Which mechanism occurs depends on how closely genetically matched, in respect of the MHC genes and proteins, the recipient and host cells are. Acute rejection occurs when implanted tissues are poorly matched causing a rapid immune response that rejects the tissue in a matter of months.

The MHC proteins present on the surface of allogeneic cells are responsible for stimulating T-cells and bringing about rejection. As indicated above, in the reaction against microbial antigens, T-cells recognize microbial peptide epitopes (8–12 amino acids in length) associated with self-MHC molecules (5). It is believed that, for a given individual, the same T-cell that recognizes a self-MHC plus foreign epitope will also have the capability of recognizing an allogeneic MHC protein, which is not difficult to understand if one considers that a microbial peptide of only 8–12 amino acids in length is sufficient to alter the shape of a self-MHC molecule to enable recognition. What the T-cell recognizes is "altered self." In essence, an allogeneic MHC molecule that differs by a few amino acids appears like "altered self." It is estimated that around 10% of

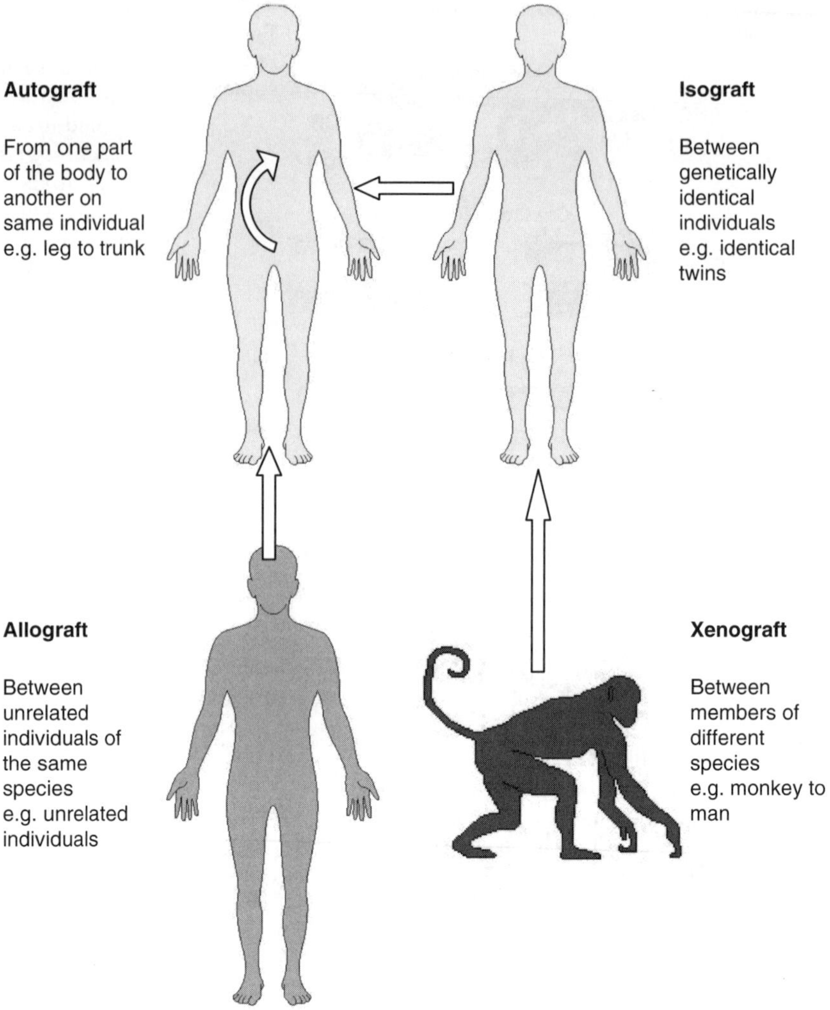

Figure 4. Tissue Transfer between individuals and species. An autograft describes the transplant of tissue from one body site to another body site on the same individual. An isograft describes the transplant of tissue between individuals of identical genotype. An allograft describes tissue transplanted between unrelated individuals of the same species. A xenograft describes tissue transplanted between different species. (This figure is available in full color at http://www.mrw.interscience.wiley.com/ebe.)

Autograft
From one part of the body to another on same individual e.g. leg to trunk

Isograft
Between genetically identical individuals e.g. identical twins

Allograft
Between unrelated individuals of the same species e.g. unrelated individuals

Xenograft
Between members of different species e.g. monkey to man

T-cells will recognize allogeneic MHC molecules in this way (6).

Therefore, when allogeneic tissues or organs are transplanted, the allogeneic cells activate the recipient T-cells and initiate an immune response. The basic rules described above, however, still hold true. The recipient CD4+ T-cells must be stimulated by specialized immunostimulatory dendritic cells expressing MHC Class II molecules and recipient CD8+ T-cells must recognize the allogeneic MHC Class I proteins plus receive the appropriate IL-2 signal from the CD4+ T-cell in order to clonally expand and give rise to memory and effector cells. As most tissue and organ transplants contain a population of immunostimulatory dendritic cells, the requirements for recognition and activation of the recipient T-cells are present within the transplanted allogeneic tissues. The effector cells then recognize the allogeneic MHC proteins and exert their effects on the cells expressing them. CD8+ Tc cells then kill the foreign cells expressing allogeneic MHC Class I and CD4+ Th cells recognize allogeneic APCs expressing allogeneic MHC Class II, and produce cytokines that activate macrophages to destroy the foreign tissue cells via the production of toxic oxygen radicals and enzymes. Ultimately, the grafted tissue becomes necrotic and dies (Fig. 5). In this scenario, the mechanisms of afferent recognition that initiate the response that leads to rejection is referred to as the ***direct pathway of alloantigen recognition*** (7).

In addition, the recipient T-cells can recognize the transplanted allogeneic tissue cells via a second mechanism called the ***indirect pathway of alloantigen recognition*** (8). In this mechanism, allogeneic MHC proteins are shed by the donor tissue cells and are taken up by recipient immunostimulatory dendritic cells. The recipient dendritic cells can then present peptide epitopes derived from the foreign MHC molecules in combination with self-MHC Class II molecules to recipient CD4+ Th cells. The activated CD4+ T-cells then release IL-2, which activates CD8+ Tc cells that have directly recognized the allogeneic MHC Class I proteins on the donor cells (Fig. 6).

4.3. Immune Mechanisms and Xenograft Rejection

When xenogeneic tissues or organs are transplanted, the graft is usually rejected extremely quickly by a mechanism referred to as hyperacute rejection, which is mediated by preformed antibodies active against the implanted tissue. This reaction is so severe that it can begin within

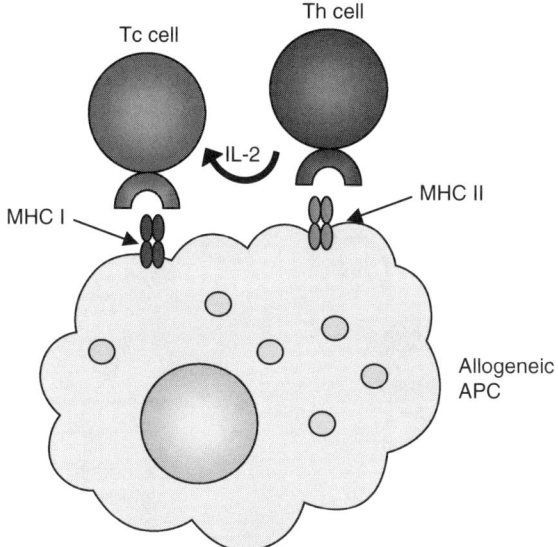

Figure 5. Direct antigen recognition. As most tissue and organ transplants contain a population of immunostimulatory dendritic cells, the requirements for recognition and activation of the recipient T-cells are present within the transplanted allogeneic tissues. Consequently, recipient Th cells activate recipient Tc cells to destroy allogeneic tissue. The effector cells then recognize the allogeneic MHC proteins and exert their effects on the cells expressing them. CD8+ Tc cells then kill the foreign cells expressing allogeneic MHC Class I and CD4+ Th cells recognize allogeneic APCs expressing allogeneic MHC Class II, and produce cytokines that activate macrophages to destroy the foreign tissue cells via the production of toxic oxygen radicals and enzymes. (This figure is available in full color at http://www.mrw.interscience.wiley.com/ebe.)

minutes of the implant being transplanted. Humans have high-titres of antibodies to a sugar epitope present on bacteria that live in the human gut, which is similar in shape to a sugar residue called galα1-3gal (α-gal) (9). Unfortunately, all mammalian species, with the exception of humans and old world monkeys, express this sugar on the surface of their cells. Humans do not have the sugar because, during evolution, they have lost the enzyme that makes it. Therefore, should a tissue or organ containing α-gal positive cells be implanted into a human, the antibodies quickly bind to the epitope expressed on the xenogeneic endothelial cells in the blood vessels of the tissue, resulting in the rapid activation of the complement cascade, leading to inflammation, blood clotting, and blockage of the vessels resulting in hypoxia followed by death of the tissue caused by the lack of oxygen.

4.4. Blood Transfusion

One tissue exists that can be transplanted without evoking a rejection response among the recipient T-cells: the transplantation of blood, as in blood transfusion. Mature red blood cells (or erythrocytes) are not living cells in the strictest sense because they lack a nucleus. As a consequence, erythrocytes do not express any MHC molecules on their surface and cannot, therefore, be recognized by T-lymphocytes. They do, however, possess other surface molecules that can vary between individuals in the human population and, thus, act as antigens. Of these, the ABO antigen system is of primary importance. The epitopes of the antigens concerned occur on many cell types in addition to erythrocytes and are located on the carbohydrate units of glycoproteins. Individuals can express either A, B, or AB antigens on their erythrocytes (blood group antigens), or they may have neither A nor B, in which case they are designated blood group O. Humans develop antibodies to the blood group antigens that they do not posses, again because of stimulation by similar epitopes expressed by bacteria in their gut. Hence, individuals who are Group B develop anti-A antibodies. Individuals who are blood group A develop anti-B antibodies (Table 1). However, all people are tolerant to the O blood cells, and so individuals with blood type O are universal donors with respect to the ABO system. Failure to match these major blood types results in hyperacute rejection of the transfused blood, which in this situation can produce a very rapid and potentially fatal systemic lysis of the red cells as the new blood is transfused.

4.5. Tissue Typing

Tissue typing is a means of identifying the genetic similarities between donor and recipient prior to transplantation to ensure as close a match as possible in order to minimize the immune reaction. The perfectly matched donor and recipient would be isogeneic (e.g., monozygotic twins). However, this situation is rare, and in all other cases, major or minor histocompatibility differences exist between the donor and recipient. If the MHC Class I and Class II genes and proteins do not match well, a much greater chance exists that the recipient will rapidly reject the donated tissue, as in acute rejection. The chances of completely matching two individuals at random are extremely remote (10). Good organ graft survival is obtained when the donor and recipient share only the same MHC class II antigens, especially HLA-DR, because these are the antigens that directly activate the recipient's CD4+ T-cells.

A number of different techniques are used to identify the antigens on the cells. Serological tests are quickest and easiest, taking only a few hours, and can be performed while the donor tissue is preserved on ice. Typically, blood samples are exposed to commercially available monoclonal antibodies that bind to specific histocompatibility antigens. Depending on the tissue type of the patient, the antibodies may or may not bind to the cells. Binding can then be visualized and quantified by colorimetric techniques. More recently, sensitive and accurate typing has been achieved using the polymerase chain reaction (PCR) to identify MHC genes in the nucleic acid of donors and recipients. PCR uses nucleic acid-binding markers to specifically bind to gene sequences if they are present within an individual's genetic material. The mixed lymphocyte reaction (MLR) is an older technique that can be used as an *in vitro* correlate of the direct pathway of alloantigen recognition in the transplantation reaction to test the responsiveness of recipient lymphocytes to MHC proteins expressed on donor dendritic cells. The MLR involves

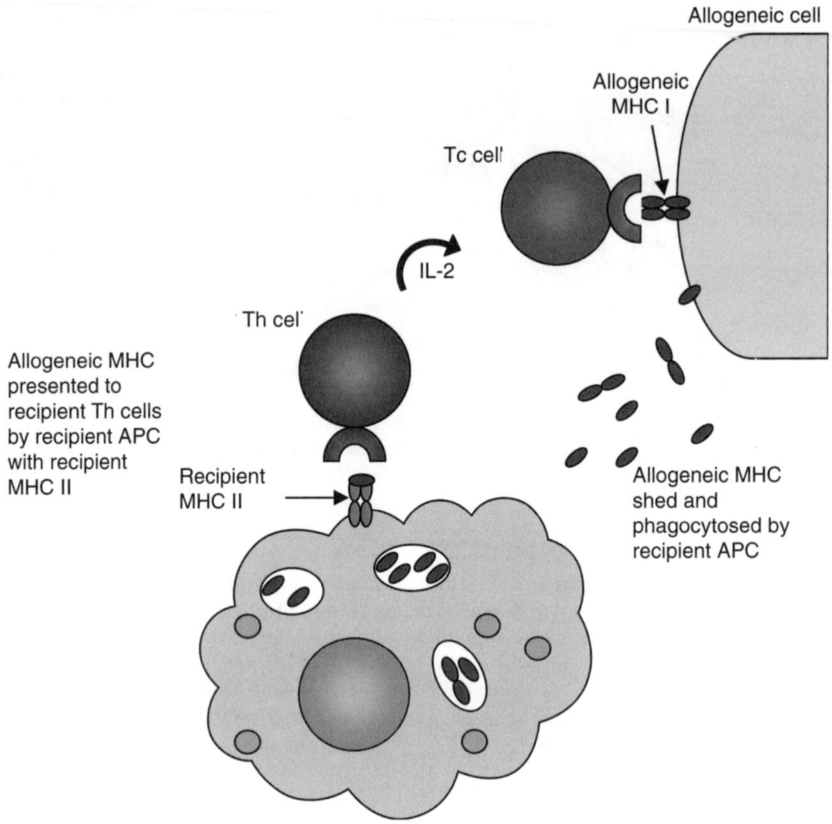

Figure 6. Indirect antigen recognition. In this mechanism, allogeneic MHC proteins are shed by the donor tissue cells and are taken up by recipient immunostimulatory dendritic cells. The recipient dendritic cells can then present peptide epitopes derived from the foreign MHC molecules in combination with self-MHC Class II molecules to recipient CD4+ Th cells. The activated CD4+ T-cells then release IL-2, which activates CD8+ Tc cells that have directly recognized the allogeneic MHC Class I proteins on the donor cells. (This figure is available in full color at http://www.mrw.interscience.wiley.com/ebe.)

Table 1. Cross-Matching the ABO Blood Group Types

Blood Group	Genotypes	Antigens	Antibodies to ABO in Serum
A	AA, AO	A	Anti-B
B	BB, BO	B	Anti-A
AB	AB	A and B	None
O	OO	O	Anti-A and Anti-B

mixing populations of mononuclear cells (lymphocytes, dendritic cells, and mononuclear phagocytes) from two allogeneic individuals. The "donor's" cells are irradiated to prevent cell proliferation while maintaining metabolic activity and are termed the stimulators, and the "recipients" cells are termed responders. The T-cells of the responder population react with the allogeneic MHC proteins on the surface of stimulator immunostimulatory dendritic cells, causing the T-cells to rapidly proliferate, which can be assayed by measuring tritiated thymidine incorporation into the DNA of the dividing cells. The level of T-cell proliferation can be correlated to the degree of MHC mismatch between the two cell sources: Closely matched cells show a minimal T-cell proliferation whereas mismatched cells show a high T-cell proliferation. Low recipient antidonor MLR responses are associated with excellent transplant survival. However, the MLR test takes 5–7 days to complete, making it impractical in most clinical organ transplantation because organs from dead donors cannot be preserved for more than 24–48 h.

In clinical practice, very little difference in survival rates between matched and mismatched tissues can be observed (10,11). Consequently, only minimal tissue typing is routinely carried out because of the added complications brought about by immunosuppression (which is given to all transplant recipients regardless of how closely matched their tissues are), the fact that some individuals show differing responses to transplants and immunosuppression, and further complications from infections and other factors.

4.6. Immunosuppression

The first two successful allogeneic kidney transplants in the world in 1959 used sublethal total-body irradiation of the recipient to prevent rejection. Irradiation targets rapidly dividing cells, such as responding lymphocytes, and is therefore an effective form of immunosuppression. The kidney grafts themselves were donated by fraternal twins and survived for 20 and 26 years without any maintenance immunosuppression (10). Although these grafts were exceptionally successful, total-body irradiation was soon abandoned because of the exorbitant risks involved and pharmacological alternatives were developed. Nowadays, recipients are routinely given immunosuppressive drugs to suppress the activities of the T-cells and improve the chances of graft survival. Standard clinical practices use drugs to nonspecifically suppress the activity of the immune system as a whole. However, this method has a major disadvantage in that the graft recipient is left extremely vulnerable to infection because the defenses that protect us and help us deal with everyday infections are knocked-out in the process, which means that otherwise minor infections can rapidly become life-threatening to transplant recipients. Three nonspecific agents are most widely used clinically: corticosteroids, cyclosporin, and azathioprine (12). Steroids have anti-inflammatory properties and suppress activated macrophages, interfere with APC function, and reduce the expression of MHC antigens. Cyclosporin is a fungal macrolide produced by soil organisms that interferes with the expression of IL-2 receptors on T-cells undergoing activation. Azathioprine is a 6-mercaptopurine analogue that incorporates into the DNA of dividing cells to prevent further proliferation. These agents can be effective when used alone, although they typically require high doses that increase the likelihood of adverse toxic effects. They are known to have a synergistic effect when used in combination, as they interfere with different stages of the rejection pathway, which means individual doses can be lowered, thereby reducing the toxic side effects. Clinical results since the introduction of cyclosporin are very good; however, the expected life of a kidney transplant is 7–8 years because of the problem of chronic rejection and the long-term use of drugs is still associated with adverse effects. New, more selective agents are under development, for example, using monoclonal antibodies against lymphocyte surface molecules to eliminate specific cell types or block their function; however, these remain in development.

4.7. Implications for the Future

Despite the undoubted success of clinical transplantation during the last century, a need to research alternative strategies for the treatment of individuals with degenerative or traumatized organs and tissues remains. Even with the most favorable tissue-matching, the survival rate after 10 years for most transplanted tissues and organs is only approximately 50% (13). In addition, simply not enough donors exist to match the number of people requiring replacement tissues or organs. With an increasingly aging population in the western world, the gap in number between potential donors and those waiting is increasing annually, and tens of thousands of people die while waiting for a transplant organ to become available. Hence, the past 10 years has seen a growth of activity in the area of regenerative medicine encompassing tissue engineering, which seeks to develop a new generation of therapeutic options.

5. TISSUE ENGINEERING

5.1. Basic Principles

Tissue engineering is a rapidly developing field that combines biological and engineering principles in an attempt to create replacement organs or structures from cells and biological or synthetic scaffolds to restore, maintain, or improve tissue function. Although the technology is still in its infancy, the potential of this field is vast, and the research is driven by the intrinsic problems of current techniques for repair and replacement of tissues, such as limited donor availability and poorly compatible materials.

A tissue-engineered construct (TEC) can include cells only, a scaffold only, or a combination of the two. In considering the use of a scaffold, depending on the nature of the tissue being replaced, the addition of cells may or may not be necessary. It is thought that once implanted, the scaffold would gradually be degraded as host cells infiltrate the construct. As they degrade it, however, it is thought that the infiltrating cells would then gradually replace the scaffold by producing an appropriate extracellular matrix (ECM) so that, ultimately, the implanted TEC will be completely replaced with new living tissue capable of growth, repair, and regeneration that is ideally identical to the native tissue. The addition of cells before the TEC is implanted could be beneficial to the functioning of certain tissues as host cells gradually infiltrate it. Such cells could increase the rate at which new replacement ECM is laid down and also add mechanical strength and function to the construct. Four key processes that occur during the *in vitro* or *in vivo* phases of tissue formation and maturation have been identified (14):

1. Cell proliferation and differentiation;
2. Extracellular matrix production;
3. Degradation of the scaffold; and
4. Remodeling and potentially growth of the tissue.

Once combined in the correct manner, a dynamic and interactive cell phenotype and ECM could be produced that would ultimately develop a microarchitecture comparable with a native tissue (15). Multiple challenges for the tissue engineering approach to tissue replacement exist. The physical and biological properties of the natural tissue must be recapitulated, and any synthetic materials used must be tolerated by the body and adequately degraded as the new tissue develops and remodels. The basic principle involves seeding cells onto a matrix scaffold. However, a number of ways of creating a suitable matrix scaffold and many potential cell sources exist.

5.2. Tissue Engineering Scaffolds

A number of different biodegradable materials exist with minimal immunogenicity currently being investigated. These scaffolds are usually designed to gradually degrade over time and be replaced with new host tissue and cells but must possess sufficient structural integrity to temporarily withstand functional loading *in vitro* and *in vivo*. Consequently, it must be ensured that materials used are nonimmunogenic, nontoxic, and are adequately reabsorbed by the body. A variety of scaffold materials have been examined for use in tissue engineering applications including fibrin gel (16), and collagen (17,18). Synthetic materials that have been investigated include polyglycolic acid (PGA), polylactic acid (PLA), and polyhydroxyalkanoate (19–21).

An alternative to synthetic scaffolds is the use of acellular xenogeneic or allogeneic tissue. This strategy would provide a ready-made scaffold, which already possesses optimal mechanical and biochemical qualities and characteristics. A distinct advantage in that there should be minimal host response to a naturally occurring ECM also exists (22). Methods for producing acellular matrices typically use a combination of enzymes and detergents to remove the cells from tissues (23,24). The difficulty with this technique, however, lies in the decellularization process: all of the cells must be completely removed, while not damaging the tissue matrix so that mechanical and biochemical properties are maintained as much as possible. However, questions remain as to the suitability of unfixed animal sources for human implantation because of the risks of transmitting animal viruses and prions, which are difficult to screen for and adequately remove if detected (25,26).

It is possible that a matrix may not be required at all for repairing damaged tissues. Simply injecting desirable cells into the body at a distant site from the damaged tissue may be sufficient. This method of cell delivery has been used to inject mesenchymal stem cells into infarcted myocardium where they have been shown to engraft into the damaged tissue and differentiate into myocardial cells that begin to recover the damage and initiate repair (27–30).

5.3. Cell Type: Terminally Differentiated Cells vs. Stem Cells

When considering the cell type, two choices of cell that can be investigated primarily exist: terminally differentiated cells and stem cells. Terminally differentiated cells are the cells that would be found in the native tissue that is being repaired or replaced. These cells are generally highly specialized cells whose function is specific to the tissue in which they are located. In theory, these cells would be the ideal cells of choice; however, shortcomings exist, including the fact that these cells generally have a low proliferation rate as their main function is synthetic rather than proliferative. When they are encouraged to proliferate, for example, when transferred to *in vitro* conditions, they can only undergo a limited number of cell doublings before they become senescent (31), which is a consequence of the mechanism by which DNA is copied during cell proliferation whereby cells lacking a particular enzyme (telomerase) suffer from a shortening of their DNA with each round of replication until the cells are no longer able to proliferate, and, as this point approaches, cells gradually lose functionality and senesce. This process is particularly problematic here because in a clinical setting, ideally a relatively small biopsy would have to be taken and the cells from it expanded *in vitro* so sufficient cell numbers would be available for seeding the TEC. The reduced replication capacity of these cells would make this difficult. Furthermore, they might not naturally be migratory cells and, therefore, seeding the TEC would also be difficult as the cells might only sit on the surface rather than fully penetrate the tissue.

Stem cells have several advantages over terminally differentiated cells. These cells are a type of precursor cell that are able to differentiate into multiple cell types and furthermore can, in theory, proliferate indefinitely in culture, without any loss of function. Consequently, cells could be harvested from a common source (e.g., bone marrow) and expanded and differentiated *in vitro* into the most desirable cell type needed for creating the tissue construct. Alternatively, they could be seeded in their undifferentiated state, and a bioreactor used to culture the TEC, which would hopefully provide the adequate mechanical signals to induce appropriate differentiation *in situ*. A further option would be to implant the stem cells in their native state and allow the biochemical and mechanical influences *in vivo* to promote the appropriate differentiation and functioning of the cells. Stem cells essentially come in two forms: embryonic stem (ES) cells and adult stem (AS) cells (Table 2). ES cells are derived from the inner cell mass of developing embryos (32). These cells are the ultimate stem cell as these cells differentiate and proliferate to produce all the cells in the human body. Consequently, they are referred to as being "totipotent," as one cell can produce progeny of multiple phenotypes (in contrast, terminally differentiated cells can only produce direct copies of themselves). AS cells can be derived from multiple sources in the fully grown adult and are thought to exist as part of a repair mechanism for repairing damaged cells and tissues (33). The most widespread source being investigated is the bone marrow, where the AS cells are called mesenchymal stem cells (MSC) (34). Other sources include fat (adipose) tissue, muscle, liver (hepatic), and brain (35–38). Compared with ES cells, it is believed that AS cells possess a more limited differentiation capacity (multipotent) that is limited to the cell types from the tissue in which they are found, for example, MSC are believed to produce only mesenchymal cell types such as bone, cartilage, and fat. However, more recent research suggests that there may be functional overlap that enables MSC to produce neural cells and hepatic cells (39,40).

Several advantages and disadvantages exist to the use of these different types of stem cells, but the main drawback to the use of ES cells is an ethical one. Even if ethical issues did not exist, a major disadvantage of these cells is

Table 2. Comparison of Embryonic and Adult Stem Cells

	Embryonic Stem Cells	Adult Stem Cells
Advantages	Can be propagated indefinitely.	Autologous derivation.
	Pluripotent differentiation capacity.	Respond to developmental signals in the injury environment and differentiate.
	Absence of any potential effects of age.	Easy to obtain.
Disadvantages	Need to be grown on mouse feeder cells.	Not pluripotent.
	Can give rise to teratocarcinomas.	May be donor-age-associated changes in proliferative and differentiative capacity.
	Epigenetic instability.	Current culture conditions do not maintain multipotency.
	Allogeneic derivation.	

that they would always be of allogeneic derivation and would therefore suffer from the associated complications once implanted. Other disadvantages include epigenetic instability and the ability of the cells to spontaneously form teratocarcinomas *in vivo*. Although their differentiation capacity is reduced when compared with ES cells, AS cells can be derived autologously, and furthermore, because they are derived from adults, no ethical issues exist. Unfortunately, their proliferation capacity *in vitro* does seem to be limited, which may be because of the patients from whom these cells are being harvested, as it is believed that as the donor age increases, the number and functional capacity of AS cells is reduced (41). However, it could simply be a reflection of inadequate culture conditions that it may be possible to overcome with further more basic research.

5.4. Cell Source: Autologous vs. Allogeneic

In any tissue engineering strategy, an appropriate source of suitable cells is required that have similar properties to the cells of the native tissue. It may be that multiple cell types are required (i.e., endothelial and smooth muscle cells for blood vessels) in order that an endothelial lining is established to prevent thrombus formation, where smooth muscle cells can penetrate and remodel the tissue.

The primary consideration when using cells for a TEC is to decide whether the cells will be derived from the person who will ultimately receive the graft (i.e., autologous cells) or whether the cells must be derived from another individual (i.e., allogeneic cells). In an ideal situation, the cells used would be autologous and taken from the tissue that is being replaced, as upon implantation of the TEC, these cells would be recognized as self and, consequently, an immune reaction against the cells within the TEC would not occur. However, because the aim of tissue engineering is to repair or replace a damaged tissue, then it is likely that the cells themselves will be in some way diseased or damaged. Consequently, in any tissue engineering or cell therapeutic strategy, the decision must be made as to whether it would be better to isolate autologous cells from another site in the individual (adult stem cells from the patient's bone marrow) or to use allogeneic cells. A very important aspect to consider will be the time required to culture autologous cells prior to the implantation of the TEC: It may take several weeks to obtain sufficient numbers of autologous cells from a small biopsy; however, allogeneic cells could be obtained and grown in advance and stored ready for use in any individual.

The potential for an adverse immune response against allogeneic cells used in tissue engineering and regenerative medicine strategies has become an issue for debate and controversy. On one hand, from an immunological perspective, it is difficult to perceive of a situation in which allogeneic cells would not eventually be recognized and rejected by the recipient in the absence of some form of immunosuppression. On the other hand, the practical and commercial advantages of being able to use allogeneic cells in tissue engineering applications are immense, which has driven developments in their use in particular applications. The important issue is really whether the benefits of using allogeneic cells in tissue engineering and regenerative medicine approaches will be greater than any potential adverse effects of the immunological response to the cells, which may well vary considerably with the particular application.

A school of thought exists that allogeneic cells of nonhematopoietic bone marrow origin, such as the stromal or mesenchymal stem cells, will not be recognized and rejected by the immune system because they do not normally express MHC Class II proteins or other costimulatory molecules necessary for the activation of naive CD4+ T-cells. However, it is clear from the literature that this is not the case.

Historically, as the mechanisms of immunological rejection of tissue and organ grafts were being elucidated during the 1970s and 1980s, it was initially thought that allogeneic graft rejection was a result solely of the direct pathway of alloantigen recognition: So-called "passenger" allogeneic dendritic cells (expressing MHC Class II and costimulatory molecules) within the graft were implanted with the tissue and elicited an immune response by directly interacting with recipient T-cells (42). This view was strongly supported by experiments at the time that showed that *in vitro* one-way MLR reactions using allogeneic cells derived from donor tissue that had been treated to remove passenger dendritic cells did not cause proliferation of responder T-cells (43). However, around the same time, *in vivo* experiments clearly showed that allogeneic tissues devoid of passenger dendritic cells could still result in rejection (44,45). Thus, it was hypothesized that alloantigens were sloughing off the implanted tissues, and being presented by recipient dendritic cells to CD4+ T-cells in the draining lymphoid tissues, resulting in the generation of "effector" CD4+ T-cells, as shown in Fig. 7. Host Langerhans cells, which colonize the epithelial grafts, could subsequently present epitopes of allogeneic MHC or nonMHC antigens to the effector T-cells at the implant site (45), resulting in rejection of the tissue via what came to be referred to as the indirect pathway of alloantigen recognition (8). Overwhelming evidence now exists that this pathway operates in clinical transplantation (46,47).

It should be noted, however, that in the "rejection" of allogeneic cells, such as allogeneic keratinocyte sheets, and other nonvascularised tissue-engineered constructs containing allogeneic cells, the rejection process may not necessarily result in the overt pathology that is observed with vascularized tissue and organ grafts. The rejection of vascularized organs and tissues, such as split skin allografts, is characterized by the involvement of the donor vascular endothelium. Following transplantation, the blood supply to the transplanted tissue is cut off resulting in some degree of tissue hypoxia. Following the first few days, recipient endothelial cells are attracted into the grafted tissue and reestablish the blood supply by linking with vessels in the grafted tissue. Effector CD4+ T-cells generated in the afferent phase of the response and, indeed, antibodies gain access back to the transplanted tissue via the vascular system. It is not until the blood supply is reestablished, therefore, that the rejection response ensues. The immune response is largely directed toward

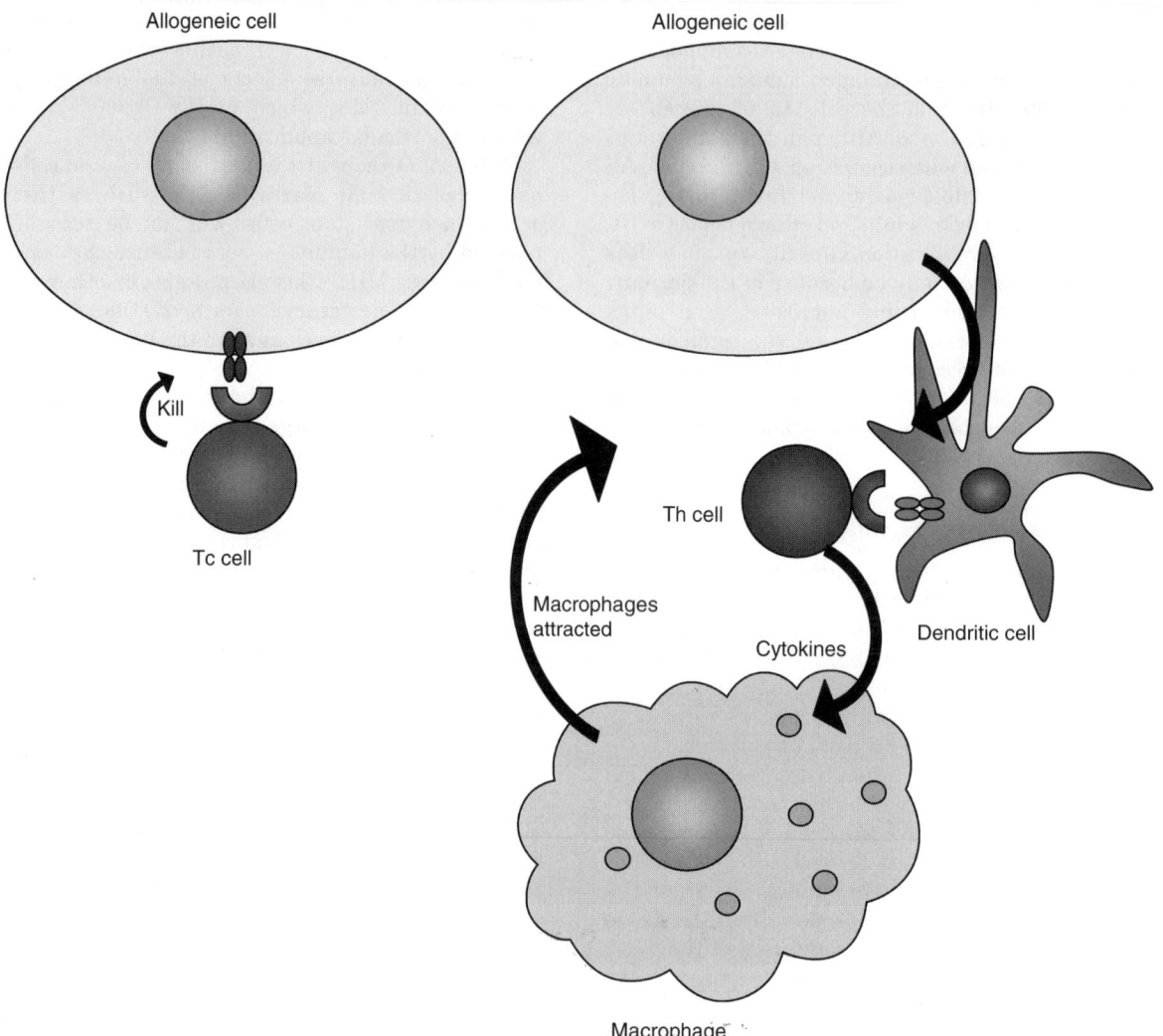

Figure 7. The efferent immune response to allogeneic cells. Allogeneic MHC molecules slough off donor cells and are presented to recipient Th cells by dendritic cells. The Th cells attract macrophages by releasing cytokines and activate Tc cells, which then cause apoptosis of allogeneic cells. (This figure is available in full color at http://www.mrw.interscience.wiley.com/ebe.)

the allogeneic endothelium, because it is the first target that the effector cells encounter, leading to vessel occlusion and tissue necrosis. In contrast, allogeneic keratinocyte sheets are not vascularized. Consequently, donor keratinocyte cells will not be rejected in this typical pathological manner. The encounter between allogeneic keratinocytes/donor Langerhans cells and effector T-cells will be dependent on the normal migratory pathways of these cells through the epidermis, leading to the gradual "removal" of the cells over time (48). The allogeneic cells may survive for sufficient duration to produce growth factors and other molecules that promote the reestablishment of a normal healthy epithelium, which would otherwise take much longer to form.

It is interesting to speculate on the likely consequence of the implantation of a 3D tissue-engineered construct comprised of a scaffold and allogeneic stromal cells that has been grown *in vitro*. Removal from the optimal culture environment would result in some degree of construct hypoxia, and surgical implantation would lead to local acute inflammation. Increased permeability of the surrounding recipient tissue would provide a route for the sloughing of donor cells/antigens and transport to the lymphoid tissues for afferent induction of CD4+ T-cells via the indirect pathway. In order for the effector CD4+ T-cells/APCs to gain access to the allogeneic cells in the construct, the construct would need to have a local blood supply and the endothelial cells would need to be activated to express adhesion molecules to allow adhesion and migration of the cells into the construct. The initial inflammatory process and construct hypoxia would attract recipient inflammatory and endothelial cells into the site where the construct was located, and neovascularization would commence in a manner analogous to the normal wound healing process. However, the neovasculature, being of recipient origin, would not be a target for immune attack, but would merely provide a conduit for the effector cells to gain access to the construct. The extent and duration of the inflammatory response would be dependent on the "quality" of the construct with respect to materials used for the scaffold, degree of hypoxia, and so on. In the worst-case scenario, if the neovasculature was activated over several days, allowing effector T-cells and APCs access to the allogeneic cells in the construct, one would expect to see the development of a chronic inflammatory infiltrate, characterized by CD4+ T-cells and macrophages and necrosis of the construct after about 12–14 days. Conversely, if the construct stimulated minimal acute inflammation, any neovasculature may not express the adhesion molecules necessary for the recruitment of effector cells into the construct once generated. The likely consequence would be that T-cells and APCs would only gain access to the allogeneic cells "by chance" via normal migratory pathways (in skin, for example, skin-homing T-cells, Langerhans cells). In this theoretical best-case scenario, allogeneic cells might then be eliminated gradually over prolonged periods of time without any overt tissue pathology.

Consideration of the above may help to explain the clinical success of the only currently licensed tissue-engineered product, which uses allogeneic cells in a skin substitute, namely Apligraf. Apligraf consists of living allogeneic fibroblasts and structural proteins. The lower dermal layer combines bovine type 1 collagen and human allogeneic fibroblasts, which produce additional matrix proteins. The upper epidermal layer is formed by promoting human allogeneic keratinocytes first to multiply and then to differentiate to replicate the architecture of the human epidermis. Apligraf is not designed as nor licensed for use as a tissue graft for permanent acceptance, but is used solely as temporary dressing to restore homeostasis and healing of chronic nonhealing ulcers and skin wounds. The initial presence of the cells protects the wound and produces growth factors and other molecules that aid and promote the host repair mechanisms to begin to replace the damaged area. Such tissue-engineered constructs do not contain donor endothelial cells and any vascularization that occurs will be with recipient endothelial cells, which will not be targets for the rejection response. When used in the treatment of acute wounds, the allogeneic cells in Apligraf have been shown to persist for around 4 weeks, after which they could no longer be detected. Loss of allogeneic cells was associated with some degree of vascularization below the Apligraf, and an intense cellular infiltrate. Furthermore, the Apligraf exhibited no features of engraftment, but resulted in wound healing in 4–9 weeks (49).

In practice, therefore, the issues of the potential immune response to allogeneic cells depends on the exact nature and purpose of the TEC and have to be balanced against the required function of the cells *in situ* and the location in which the TEC is to be implanted.

5.5. Allogeneic Mesenchymal Stem Cells

It has been suggested that allogeneic mesenchymal stem cells (MSC) have some unusual immunological properties. The introduction of allogeneic MSC into fetal sheep before and after the expected development of immunological competence has shown that the cells were capable of migration across the peritoneal cavity and engrafted and persisted in multiple tissues for as long as 13 months after transplantation (50,51). Similarly, intravenous infusion of autologous and allogeneic MSCs combined with hematopoietic stem cells in an irradiated nonhuman primate model (baboon) has shown that MSC were capable of homing to the bone marrow and persisted for over a year (52). A control animal that was not irradiated and infused with allogeneic MSC had cells present in its bone marrow up to 442 days post infusion, leading to speculation that allogeneic MSC are not recognized and rejected by the immune system.

In addition, some groups have carried out experiments to investigate the effect of MSC on mixed lymphocyte reactions (MLR). Experiments by Klyushnenkova et al. (53) and Le Blanc et al. (54) suggested that human MSC may suppress activated T-cells, because the addition of MSC to an MLR resulted in near complete suppression of T-cell proliferation whether the MSC were added to the MLR at the initiation of culture or midway through a seven-day culture period. It has also been shown that although proliferation activity in the MLR was suppressed with the addition of between 10,000 and 40,000 MSC, the addition of just 10–1000 MSC led to a less consistent suppression or a marked lymphocyte proliferation (54). However, no other cell source was used as a control for these experiments and, therefore, no means exist to verify these results were specific to MSC or a general phenomenon of the cell culture system. More recent experiments by Bartholomew et al. (55) showed that T-cell suppression was reduced when MSC were added midway through an MLR and could be partially reversed using interleukin-2 (IL-2).

Although data of this nature raises the possibility that human MSC may possess immunosuppressive effects, which may render them either 'immunoprivileged' or perhaps immunosuppressive *in vivo*, overall the data remain inconclusive. If allogeneic MSC are indeed immunosuppressive, this would raise additional concerns regarding their clinical use. If allogeneic MSC became infected with a virus or became neoplastic once transplanted, the MSC would be exempt from the body's immune surveillance systems, with potentially disastrous consequences. Obviously, this area requires extensive rigorous investigation.

MSC do not normally express MHC II and costimulatory molecules, and therefore cannot induce proliferation of T-cells directly themselves *in vitro*. The *in vivo* data indicating that allogeneic MSC survive following infusion can be explained by the mode of delivery, injection as a cell suspension intravenously or per-orally both avoid stimulation of inflammation and the generation of "danger" signals. The true test of immunogenicity of allogeneic MSC will be delivery *in vivo* in a 3D TEC.

If allogeneic MSC are immunogenic, it is still possible that allogeneic MSC could be useful clinically. If implanted with minimal "danger signals" and inflammation, the MSC might persist long enough to provide some healing responses and initiate regeneration of the surrounding tissue in advance of the recipient cell infiltration, which offers great potential benefits for tissue engineering purposes as MSC from any donor could be used to create the ideal off-the-shelf tissue engineering product suitable for any recipient.

6. FUTURE DEVELOPMENTS IN ALLOGENEIC TRANSPLANTATION

6.1. Tolerance Induction

It has been recognized for some years that if allogeneic stem cells are implanted into a fetal animal sufficiently early during development, then when the animal reaches adulthood, it will be tolerant to any tissues transplanted from a donor isogeneic with the transplanted cells (56), because if implanted before full immunocompetence is developed, the alloantigens from the tissues will be seen as self and consequently any T-cells that react against the alloantigens will be killed in the thymus, as happens with any self-reacting T-cells during development. The concept of inducing "tolerance" to allogeneic cells and tissues in the adult human has been a major area of investigation over the past 50 years. It is based simply on the idea that deletion or regulation of the T-cells that respond to the allogeneic MHC proteins will result in immune acceptance of an allogeneic tissue.

In some experimental models in which tolerance occurs spontaneously, or under the appropriate conditions of immunosuppression, the immune response is too weak to eliminate the donor immunostimulatory dendritic cells and regresses. Collapse of the response occurs when the activated host T-cell clones reach their limit of proliferation and are exhausted and die by incompletely understood mechanisms (57). This process is referred to as clonal exhaustion-deletion and is the seminal mechanism of organ engraftment and of acquired tolerance (58,59). However, the exhaustion is never complete and, consequently, maintenance of the variable deletional tolerance achieved at the outset depends on the persistence of residual stimulatory donor dendritic cells (60). In experimental models in which organ allograft rejection by the unmodified recipient is the normal outcome, the passenger dendritic cell-induced response can be reduced enough to be deleted by treating the recipient prior to transplantation with total lymphoid irradiation, conventional immunosuppressive therapy, or anti-T-cell monoclonal antibodies (61). The immune response can also be brought into the deletable range by immunosuppression given after transplantation. However, if the immunosuppression is too much, the capacity to exhaust T-cell clones is also affected. When immunosuppression is reduced later, in an initially overtreated recipient, the undeleted donor-specific clones may recover along with the return of immune reactivity. Graft survival is then dependent on permanent immunosuppression.

A range of other experimental approaches to induce transplantation tolerance exist, including the establishment of mixed chimerism, reprogramming via allogeneic dendritic cells, and induction of regulatory T-cells. A full discussion of these approaches is beyond the scope of this chapter but has been recently reviewed (61).

6.2. New Therapeutic Agents for Immunosuppression

Various new agents are becoming available or under development as adjunctive immunosuppression or to assist in the induction of tolerance. These agents are outlined in Table 3.

Table 3. The Most Recent Therapies Under Development (62)

T-cell depletion or TCR signal transduction	Anti-CD3 immunotoxin
	CD52-specific monoclonal antibody (Campath-1H)
	CD45RB-specific monoclonal antibody
Costimulatory blockade	CD40-specific monoclonal antibody (CD154:CD40 pathway)
	B7-specific monoclonal antibodies. CTLA4Ig, LEA29Y (CD28:B7 pathway)
	Cytokine receptor signalling (JAK 3 kinase inhibitors)
B-cell depletion	CD20-specific monoclonal antibody
Lymphocyte trafficking	LFA-1-specific monoclonal antibody
	Sphingosine-1-phosphate receptor modulator, FTY720 (causes lymphocyte sequestration in lymphoid tissue)
	CXCR3/CCR1/5 antagonists (inhibits lymphocyte trafficking to rejection site)

Monoclonal antibodies can be used to specifically target T-cells, and this has been investigated for some time (63). Targeted depletion of the cells or blocking of the antigen receptors (TCRs) using monoclonal antibodies to T-cell-specific cell surface markers, such as CD45 and CD52, prior to the implantation of allogeneic cells or tissues could prevent their interaction with the foreign tissue, preventing acute rejection (64,65). The costimulatory system is a critical aspect of the immune response, as without costimulatory molecules being present at the implantation site, activated Th cells are unable to directly react with allogeneic tissue. Hence, blocking the biochemical pathways that lead to the production of these molecules, such as the CD40 or B7 pathways, would be another means to prevent T-cells from interacting with the allogeneic tissue (66–68). B-cells represent the first step in the adaptive immune response and generate antibodies against foreign tissues. Consequently, depletion of these cells using a specific monoclonal antibody, such as one directed against the cell surface marker CD20, would reduce the adaptive immune response (69,70). Another potential target is lymphocyte trafficking: Throughout the immune response, APCs and T-cells need to travel to the site of the implanted tissue in order bring about acute rejection of implanted allogeneic tissues. Therefore, by targeting the receptors that are responsible for this trafficking, such as the sphingosine-1-phosphate receptor, LFA-1, and CXCR3, the cells can be sequestered and their movements inhibited (71–73).

Overall, in the short term, it is necessary to develop better immunosuppressive agents in order to reduce the side effects of some of the currently available drug therapies and also as an alternative to those patients who can become unresponsive to current agents. In the longer term, the development of novel agents that work on specific areas of the immune response may help in the induction of tolerance so that maintenance immunosuppression can ultimately be stopped altogether.

BIBLIOGRAPHY

1. C. A. Janeway and P. Travers, *Immunobiology: The Immune System in Health and Disease*, 2nd ed. Edinburgh, UK: Churchill Livingstone, 1996.
2. S. K. A. Law and K. B. M. Reid, *Complement*, 1st ed. Oxford, UK: IRL Press, 1988.
3. A. G. Schrum, L. A. Turka, and E. Palmer, Surface T-cell antigen receptor expression and availability for long-term antigenic signaling. *Immunolog. Rev.* 2003; **196**:7–24.
4. A. Billiau, H. Heremans, K. Vermeire, and P. Matthys, Immunomodulatory properties of interferon-γ. *Ann. NY Acad. Sci.* 1998; **856**:22–32.
5. H. O. McDevitt, Discovering the role of the major histocompatibility complex in the immune response. *Annu. Rev. Immunol.* 2000; **18**:1–17.
6. E. J. Suchin, P. B. Langmuir, E. Palmer, M. H. Sayegh, A. D. Wells, and L. A. Turka, Quantifying the frequency of alloreactive T cells in vivo: new answers to an old question. *J. Immunol.* 2001; **166**:973–981.
7. A. Warrens, G. Lombardi, and R. Lechler, MHC and alloreactivity: presentation of major and minor histocompatibility antigens. *Transplant. Immunol.* 1994; **2**:103–107.
8. P. Hornick and R. Lechler, Direct and indirect pathways of alloantigen recognition: relevance to acute and chronic allograft rejection. *Nephrol. Dialysis Transplant.* 1997; **12**:1806–1810.
9. U. Galili, M. R. Clark, S. B. Shohet, J. Buehler, and B. A. Macher, Evolutionary relationship between the natural anti-Gal antibody and the Galα1-3Gal epitope in primates. *Proc. Natl. Acad. Sci. USA* 1987; **84**:1369–1373.
10. F. H. J. Claas, M. K. Dankers, M. Oudshoorn, J. J. van Rood, A. Mulder, D. L. Roelen, R. J. Duquesnoy, and I. I. N. Doxiadis, Differential immunogenicity of HLA mismatches in clinical transplantation. *Transplant. Immunol.* 2005; **14**:187–191.
11. T. E. Starzl, The saga of liver replacement, with particular reference to the reciprocal influence of liver and kidney transplantation (1955–1967). *J. Am. Coll. Surg.* 2002; **195**:587–610.
12. V. S. Gorantla, J. H. Barker, J. W. Jones, K. Prabhune, C. Maldonado, and D. K. Granger, Immunosuppressive agents in transplantation: mechanisms of action and current anti-rejection strategies. *Microsurgery* 2000; **20**:420–429.
13. UK Transplant (2005). Transplant activity in the UK 2004–2005. (online). Available: http://www.uktransplant.org.uk/.
14. E. Rabkin and F. J. Schoen, Cardiovascular tissue engineering. *Cardiovasc. Pathol.* 2002; **11**:305–317.
15. E. Rabkin, S. P. Hoerstrup, M. Aikawa, J. E. Mayer, and F. J. Schoen, Evolution of cell phenotype and extracellular matrix in tissue-engineered heart valves during in vitro maturation and in vivo remodeling. *J. Heart Valve Dis.* 2002; **11**:308–314.
16. S. Jockenhoevel, K. Chalabi, J. S. Sachweh, H. V. Groesdonk, L. Demircan, M. Grossmann, G. Zund, and B. J. Messmer, Tissue engineering: complete autologous valve conduit—a new moulding technique. *Thorac. Cardiovasc. Surg.* 2001; **49**:287–290.
17. M. Rothenburger, P. Vischer, W. Völker, B. Glasmacher, E. Berendes, H. H. Scheld, and M. Deiwick, In vitro modelling of tissue using isolated vascular cells on a synthetic collagen matrix as a substitute for heart valves. *Thorac. Cardiovasc. Surg.* 2001; **49**:204–209.
18. P. M. Taylor, S. P. Allen, S. A. Dreger, and M. H. Yacoub, Human cardiac valve interstitial cells in collagen sponge: a biological three-dimensional matrix for tissue engineering. *J. Heart Valve Dis.* 2002; **11**:298–307.
19. G. Zund, T. Breuer, T. Shinoka, P. X. Ma, R. Langer, J. E. Mayer, and J. P. Vacanti, The in vitro construction of a tissue engineered bioprosthetic heart valve. *Eur. J. Cardiovasc. Thorac. Surg.* 1997; **11**:493–497.
20. T. Shinoka, D. Shum-Tim, P. X. Ma, R. E. Tanel, N. Isogai, R. Langer, J. P. Vacanti, and J. E. Mayer, Creation of viable pulmonary artery autografts through tissue engineering. *J. Thorac. Cardiovasc. Surg.* 1998; **115**:536–546.
21. R. Sodian, S. P. Hoerstrup, J. S. Sperling, S. H. Daebritz, D. P. Martin, A. M. Moran, B. S. Kim, F. J. Schoen, J. P. Vacanti, and J. E. Mayer, Early in vivo experience with tissue-engineered trileaflet heart valves. *Circulation* 2000; **102**(Suppl III):22–29.
22. S. F. Badylak, The extracellular matrix as a scaffold for tissue reconstruction. *Sem. Cell Develop. Biol.* 2002; **13**:377–383.
23. A. Bader, T. Schilling, O. E. Teebken, G. Brandes, T. Herden, G. Steinhoff, and A. Haverich, Tissue engineering of heart valves—human endothelial cell seeding of detergent acellularized porcine valves. *Eur. J. Cardiovasc. Thorac. Surg.* 1998; **14**:279–284.
24. C. Booth, S. A. Korossis, H. E. Wilcox, K. G. Watterson, J. N. Kearney, J. Fisher, and E. Ingham, Tissue engineering of cardiac valve prostheses I: development and histological characterization of an acellular porcine scaffold. *J. Heart Valve Dis.* 2002; **11**:457–462.
25. C. Patience, Y. Takeuchi, and R. A. Weiss, Infection of human cells by an endogenous retrovirus of pigs. *Nat. Med.* 1997; **3**:282–286.
26. R. Knight and S. Collins, Human prion dieases: cause, clinical and diagnostic aspects. *Contribut. Microbiol.* 2001; **7**:68–92.
27. J. G. Shake, P. J. Gruber, W. A. Baumgartner, G. Senechal, J. Meyers, J. M. Redmond, M. F. Pittenger, and B. J. Martin, Mesenchymal stem cell implantation in a swine myocardial infarct model: engraftment and functional effects. *Ann. Thorac. Surg.* 2002; **73**:1919–1926.
28. S. Tomita, D. A. G. Mickle, R. D. Weisel, Z-Q. Jia, L. C. Tumiati, Y. Allidina, P. Liu, and R-K. Li, Improved heart function with myogenesis and angiogenesis after autologous porcine bone marrow stromal cell transplantation. *J. Thorac. Cardiovasc. Surg.* 2002; **123**:1132–1140.
29. K. H. Grinnemo, A. M†nsson, G. Dellgren, D. Klingberg, E. Wardell, V. Drvota, C. Tammik, J. Holgersson, O. Ringd,n, C. Sylv,n, and K. Le Blanc, Xenoreactivity and engraftment of human mesenchymal stem cells transplanted into infarcted rat myocardium. *J. Thorac. Cardiovasc. Surg.* 2004; **127**:1293–1300.
30. L. C. Amado, A. P. Saliaris, K. H. Schuleri, M. St John, J-S. Xie, S. Cattaneo, D. J. Durand, T. Fitton, J. Q. Kuang, G. Stewart, S. Lehrke, W. W. Baumgartner, B. J. Martin, A. W. Heldman, and J. M. Hare, Cardiac repair with intramyocardial injection of allogeneic mesenchymal stem cells after myocardial infarction. *Proc. Natl. Acad. Sci.* 2005; **102**:11474–11479.
31. L. Hayflick, The limited in vitro lifetime of human diploid cell strains. *Experiment. Cell Res.* 1965; **37**:614–636.
32. J. A. Thomson, J. Itskovitz-Eldor, S. S. Shapiro, M. A. Waknitz, J. J. Sweirgiel, V. S. Marshall, and J. M. Jones,

Embryonic stem cell lines derived from human blastocysts. *Science* 1998; **282**:1145–1147.
33. T. C. Mackenzie and A. W. Flake, Human mesenchymal stem cells persist, demonstrate site-specific multipotential differentiation, and are present in sites of wound healing and tissue regeneration after transplantation into fetal sheep. *Blood Cells Molecules Dis.* 2001; **27**:601–604.
34. M. F. Pittenger, A. M. Mackay, S. C. Beck, R. K. Jaiswal, R. Douglas, J. D. Mosca, M. A. Moorman, D. W. Simonetti, S. Craig, and D. R. Marshak, Multilineage potential of adult human mesenchymal stem cells. *Sci.* 1999; **284**:143–147.
35. J. T. Williams, S. S. Southerland, J. Souza, A. F. Calcutt, and R. G. Cartledge, Cells isolated from adult human skeletal muscle capable of differentiating into multiple mesodermal phenotypes. *Am. Surg.* 1999; **65**:22–26.
36. B. E. Petersen, Hepatic "stem" cells: coming full circle. *Blood Cells Molecules Dis.* 2001; **27**:590–600.
37. P. A. Zuk, M. Zhu, H. Mizuno, J. Huang, J. W. Futrell, A. J. Katz, P. Benhaim, H. P. Lorenz, and M. H. Hedrick, Multi-lineage cells from human adipose tissue: implications for cell-based therapies. *Tissue Eng.* 2001; **7**:211–228.
38. D. A. Peterson, Stem cells in brain plasticity and repair. *Curr. Opin. Pharmacol.* 2002; **2**:34–42.
39. B. E. Petersen, W. C. Bowen, K. D. Patrene, W. M. Mars, A. K. Sullivan, N. Murase, S. S. Boggs, J. S. Greenberger, and J. P. Goff, Bone marrow as a potential source of hepatic oval cells. *Science* 1999; **284**:1168–1170.
40. C. Svendsen, A. Bhattacharyya, and Y-T. Tai, Neurons from stem cells: preventing an identity crisis. *Nat. Rev.* 2001; **2**:831–834.
41. S. C. Mendes, J. M. Tibbe, M. Veenhof, K. Bakker, S. Both, P. P. Platenburg, F. C. Oner, J. D. de Bruijn, and C. A. van Blitterswijk, Bone tissue-engineered implants using human bone marrow stromal cells: effect of culture conditions and donor age. *Tissue Eng.* 2002; **8**:911–920.
42. K. J. Lafferty, S. J. Prowse, C. J. Simeonovic, and H. S. Warren, Immunobiology of tissue transplantation: a return to the passenger leukocyte concept. *Annu. Rev. Immunol.* 1983; **1**:143–173.
43. E. Ingham, J. B. Matthews, J. N. Kearney, and G. Gowland, The effects of variation of cryopreservation protocols on the immunogenicity of allogeneic skin grafts. *Cryobiology* 1993; **30**:443–458.
44. J. W. Fabre and P. R. Cullen, Rejection of cultured keratinocyte allografts in the rat. *Transplantation* 1989; **48**:306–315.
45. J. Auböck, E. Irschick, N. Romani, P. Kompatscher, R. Hopfl, M. Herold, G. Schuler, M. Bauer, C. Huber, and P. Fritsch, Rejection, after a slightly prolonged survival time, of langerhans cell-free allogeneic cultured epidermis used for wound coverage in humans. *Transplantation* 1988; **45**:730–737.
46. N. Suciu-Foca, Z. Liu, A. I. Colovai, P. Fisher, E. Ho, E. F. Reed, E. A. Rose, R. E. Michler, M. A. Hardy, P. Cocciolo, F. Gargano, and R. Cortesini, Indirect T-cell recognition in human allograft rejection. *Transplant. Proc.* 1997; **29**:1012–1013.
47. H. Auchincloss Jr, In search of the elusive Holy Grail: the mechanisms and prospects for achieving clinical transplantation tolerance. *Am. J. Transplant.* 2001; **1**:6–12.
48. A. Brain, P. Purkis, P. Coates, M. Hacket, H. Navsaria, and I. Leigh, Survival of cultured allogeneic keratinocytes transplanted to deep dermal bed assessed with probe specific for Y chromosome. *Brit. Med. J.* 1989; **298**:917–919.

49. M. Griffiths, N. Ojeh, R. Livingstone, R. Price, and H. Navsaria, Survival of apligraf in acute human wounds. *Tissue Eng.* 2004; **10**:1180–1195.
50. J. P. Rubin, S. R. Cober, P. E. M. Butler, M. A. Randolph, G. S. Gazelle, F. L. Ierino, D. H. Sachs, and W. P. A. Lee, Injection of allogeneic bone marrow cells into the portal vein of swine *in utero*. *J. Surg. Res.* 2001; **95**:188–194.
51. K. W. Liechty, T. C. Mackenzie, A. F. Shaaban, A. Radu, A. B. Moseley, R. Deans, D. R. Marshak, and A. W. Flake, Human mesenchymal stem cells engraft and demonstrate site-specific differentiation after in utero transplantation in sheep. *Nat. Med.* 2000; **6**:1282–1286.
52. S. M. Devine, A. M. Bartholomew, N. Mahmud, M. Nelson, S. Patil, W. Hardy, C. Sturgeon, T. Hewett, T. Chung, W. Stock, D. Sher, S. Weissman, K. Ferrer, J. D. Mosca, R. J. Deans, A. B. Moseley, and R. Hoffman, Mesenchymal stem cells are capable of homing to the bone marrow of non-human primates following systemic infusion. *Experiment. Hematol.* 2001; **29**:244–255.
53. E. N. Klyushnenkova, V. I. Shustova, J. D. Mosca, A. B. Moseley, and K. R. McIntosh, Human mesenchymal stem cells induce unresponsiveness in pre-activated, but not naive alloantigen-specific T cells. *Experiment. Hematol.* 1999; **27**(Suppl I):122.
54. K. Le Blanc, L. Tammik, B. Sundberg, S. Haynesworth, and O. Ringden, Mesenchymal stem cells inhibit and stimulate mixed lymphocyte cultures and mitogenic responses independently of the major histocompatibility complex. *Scand. J. Immunol.* 2003; **57**:11–20.
55. A. Bartholomew, C. Sturgeon, M. Siatskas, K. Ferrer, K. McIntosh, S. Patil, W. Hardy, S. Devine, D. Ucker, R. Deans, A. Moseley, and R. Hoffman, Mesenchymal stem cells suppress lymphocyte proliferation in vitro and prolong skin graft survival in vivo. *Experiment. Hematol.* 2002; **30**:42–48.
56. R. E. Billingham, L. Brent, and P. B. Medawar, 'Actively acquired tolerance' of foreign cells. *Nature* 1953; **172**:603–606.
57. S. Qian, L. Lu, F. Fu, Y. Li, W. Li, T. E. Starzl, J. J. Fung, and A. W. Thomson, Apoptosis within spontaneously accepted mouse liver allografts. *J. Immunol.* 1997; **158**:4654–4661.
58. T. E. Starzl, A. J. Demetris, N. Murase, S. Ildstad, C. Ricordi, and M. Trucco, Cell migration, chimerism, and graft acceptance. *Lancet* 1992; **339**:1579–1582.
59. T. E. Starzl and R. M. Zinkernagel, Transplantation tolerance from a historical perspective. *Nat. Rev. Immunol.* 2001; **1**:233–239.
60. G. Mathe, J. L. Amiel, L. Schwarzenberg, J. Choay, P. Trolard, M. Schneider, M. Hayat, J. R. Schlumberger, and C. Jasmin, Bone marrow graft in man after conditioning by antilymphocytic serum. *Brit. Med. J.* 1970; **2**:131–136.
61. H. Waldmann and S. Cobbold, Exploiting tolerance processes in transplantation. *Science* 2004; **305**:209–212.
62. R. I. Lechler, M. Sykes, A. W. Thomson, and L. A. Turka, Organ transplantation—how much of the promise has been realized? *Nat. Med.* 2005; **11**:605–613.
63. R. J. Benjamin and H. Waldmann, Induction of tolerance by monoclonal antibody therapy. *Nature* 1986; **320**:449–451.
64. C. K. Asiedu, S. S. Dong, A. Lobashevsky, S. M. Jenkins, and J. M. Thomas, Tolerance induced by anti-CD3 immunotoxin plus 15-deoxyspergualin associates with donor-specific indirect pathway unresponsiveness. *Cell. Immunol.* 2003; **223**:103–112.
65. A. Vathsala, E. T. Ona, S. Y. Tan, S. Suresh, H. X. Lou, C. B. Casasola, H. C. Wong, D. Machin, G. S. Chiang, R. A. Danguilan, and R. Calne, Randomized trial of Alemtuzumab

for prevention of graft rejection and preservation of renal function after kidney transplantation. *Transplantation* 2005; **80**:765–774.
66. C. P. Larsen, J. M. Austyn, and P. J Morris, The role of graft-derived dendritic leukocytes in the rejection of vascularized organ allografts. *Ann. Surg.* 1990; **212**:308–315.
67. T. Shiraishi, Y. Yasunami, M. Takehara, T. Uede, K. Kawahara, and T. Shirakusa, Prevention of acute lung allograft rejection in rat by CTLA4Ig. *Am. J. Transplant.* 2002; **2**:223–228.
68. D. C. Borie, P. S. Chengelian, M. J. Larson, M. S. Si, R. Paniagua, J. P. Higgins, B. Holm, A. Campbell, M. Lau, S. Zhang, M. G. Flores, G. Rousvoal, J. Hawkins, D. A. Ball, E. M. Kudlacz, W. H. Brissette, E. A. Elliott, B. A. Reitz, and R. E. Morris, Immunosuppression by the JAK3 inhibitor CP-690,550 delays rejection and significantly prolongs kidney allograft survival in nonhuman primates. *Transplantation* 2005; **79**:791–801.
69. Y-G. Yang, E. deGoma, H. Ohdan, J. L. Bracy, Y. Xu, J. Iacomini, A. D. Thall, and M. Sykes, Tolerization of anti-galα1-3gal natural antibody-forming B cells by induction of mixed chimerism. *J. Experiment. Med.* 1998; **187**:1335–1342.
70. H. Ohdan, K. G. Swenson, H. Kitamura, Y-G. Yang, and M. Sykes, Tolerization of Galα1,3Gal-reactive B cells in pre-sensitized α1,3-galactosyltransferase-deficient mice by nonmyeloablative induction of mixed chimerism. *Xenotransplantation* 2001; **8**:227–238.
71. J. E. K. Hildreth, V. Holt, J. T. August, and M. D. Pescovitz, Monoclonal antibodies against porcine LFA-1: species cross-reactivity and functional effects of α-subunit-specific antibodies. *Molec. Immunol.* 1989; **26**:883–895.
72. J. A. Belperio, M. P. Keane, M. D. Burdick, J. P. Lynch, D. A. Zisman, Y. Y. Xue, K. Li, A. Ardehali, D. J. Ross, and R. M. Strieter, Role of CXCL9/CXCR3 chemokine biology during pathogenesis of acute allograft rejection. *J. Immunol.* 2003; **171**:4844–4852.
73. Y. Y. Lan, A. D. Creus, B. L. Colvin, M. Abe, V. Brinkmann, P. T. H. Coates, and A. W. Thomson, The sphingosine-1-phosphate receptor agonist FTY720 modulates dendritic cell trafficking *in vivo*. *Am. J. Transplant.* 2005; **5**:2649–2659.

READING LIST

N. Mahmud, W. Pang, C. Cobbs, P. Alur, J. Borneman, R. Dodds, M. Archambault, S. Devine, J. Turian, A. Bartholomew, P. Vanguri, A. MacKay, R. Young, and R. Hoffman, Studies of the route of administration and role of conditioning with radiation on unrelated allogeneic mismatched mesenchymal stem cell engraftment in a nonhuman primate model. *Experiment. Hematol.* 2004; **32**:494–501.

ALUMINA

Owen C. Standard
The University of New South Wales
Sydney, Australia

1. DEFINITION

Alumina (Al_2O_3) is a ceramic oxide material that has been used as a biomedical implant material since the early 1970s. In materials science and related fields, alumina refers to chemical compounds containing mainly, but not necessarily restricted to, aluminum and oxygen. Depending on the context, alumina can refer to one of many compounds including amorphous hydrous and anhydrous oxides, crystalline hydroxides and oxides, and aluminas containing small amounts of alkali or alkali earth oxides (1). Thermodynamically, the most stable form of alumina is Al_2O_3 of hexagonal crystal structure. This crystallographic phase is called corundum (designated as α-Al_2O_3) and is the usual type of alumina encountered in biomedical engineering applications, being used in either polycrystalline form (hereafter referred to as *alumina*) or single crystal form (hereafter referred to as *sapphire*).

2. PREPARATION

The principal mineralogical source of alumina for ceramic manufacturing purposes is bauxite (1). Bauxite is a naturally occurring, heterogeneous mixture composed primarily of one or more aluminum hydroxide minerals (usually gibbsite, boehmite, and/or diaspore) plus impurities of silica, iron oxide, titania, and aluminosilicates in minor amounts. Alumina is refined from bauxite by the Bayer process. This process involves crushing mined bauxite ore to a fine powder and then digesting it in an aqueous solution of caustic soda at $\sim 250°C$ and ~ 3 atm to form water-soluble aluminum hydroxide. The impurities remain insoluble and are removed from the solution usually by settling or filtration, producing the by-product red mud. The solution is then cooled to room temperature, which results in the precipitation of aluminum hydroxide. The aluminum hydroxide is collected, washed in water, dried, and lastly calcined at $\sim 1100°C$ to produce alumina powder. Alumina powder produced by this method contains typically (in weight percent) $>99.0\%$ Al_2O_3, 0.02–0.12% SiO_2, 0.03–0.06% Fe_2O_3, 0.01–0.05% CaO, and 0.01–0.5% Na_2O. The grain size of the powder ranges from ~ 0.02 μm to tens of millimeters depending on the Bayer processing conditions. Alumina powders for ceramic manufacturing applications have average particle sizes of typically $\sim 0.1–0.5$ μm and are refined by powder comminution and size classification methods to control the particle size, particle size distribution, and extent and nature of particle agglomeration.

3. FABRICATION OF ALUMINA COMPONENTS

Polycrystalline alumina components are fabricated from high-purity, submicron alumina powders using conventional ceramic processing routes. A schematic summarizing the main routes is given in Fig. 1. Alumina starting powder having desired purity and particle characteristics is first formed into its initial shape by one of several forming techniques. Dry pressing involves the compaction of powder at high pressure ($\sim 50–100$ MPa) in a high-strength steel die by means of hydraulically driven plungers (2). In cold isostatic pressing, powder is sealed in an elastomeric mould and immersed in hydraulic fluid contained in a pressure chamber; then the pressure of the

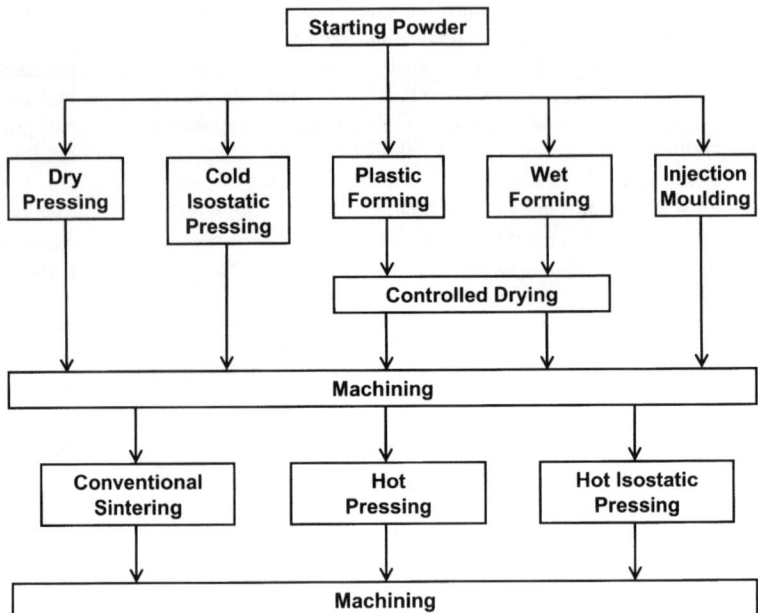

Figure 1. Summary of conventional routes used for the fabrication of engineering ceramics.

fluid is increased, typically up to 200 MPa or more, to compact isostatically the powder. In both dry pressing and cold isostatic pressing, organic lubricants usually are used to minimise bulk density variations that occur in the powder compact due to interparticle friction (this occurs in dry pressing and cold isostatic pressing) and particle-die wall friction (this occurs in dry pressing only).

By the addition of a suitable amount of liquid (usually water) and an appropriate organic plasticizer, the starting powder can be made into a plastic mass that can be readily shaped without breaking (2). The main plastic forming technique for small-sized components of uniform cross-section is ram extrusion. In this technique, plastic material is deaired in a cylindrical chamber and then forced though a die located at one end of the chamber by means of a hydraulically driven ram piston. The plastic mass is compacted as it passes through the die, with the cross-sectional shape of the extruded material being determined by the shape of the die. A wide variety of solid or hollow cross sections can be made by this method. The formed component is cut to the required shape and then dried in a controlled manner to remove the liquid.

Injection moulding is an advanced plastic forming technique used to make complex shapes that are difficult to make by the other more conventional methods or that are difficult to machine (2). Dry ceramic starting powder is mixed with organic binders and special plasticizers/lubricants. The mixture is softened by heating and then formed into small pellets by extrusion through a multiple-orifice die. After cooling, the pellets are loaded into the heated chamber of an injection moulder such that the organic materials soften and melt. The hot mixture is forced by a hydraulic piston through a channel (called the *sprue*) and into a cooled metal mould, where it is then allowed to solidify. After removal from the mould, the formed component is broken from the sprue piece and trimmed of any flashing. The organic material is then removed from the ware by controlled heating, often under reduced pressure, prior to sintering.

Wet-forming techniques involve the dispersion of the starting powder in a liquid (usually water). Through pH control and the addition of small amounts of soluble polyelectrolytes, the surface charge of the oxide can be modified such that the resultant electrostatic repulsion between particles is sufficient to separate and disperse the individual particles in the liquid (2). This mechanism is known as *deflocculation*. Optimal deflocculation greatly retards and, in some cases, prevents sedimentation of the particles and greatly reduces the amount of liquid needed to make a suspension of suitable (low) viscosity. The main wet-forming technique for small components is slip casting. This involves pouring the deflocculated suspension (also known as a *slip*) into a microporous mould of plaster of Paris or, more recently, polymer. The porosity of the mould provides a capillary suction pressure of ~ 0.1–0.2 MPa, which draws liquid from the suspension into the mould, causing particles to be deposited at the mould surface. This dewatering process results in the progressive buildup of a layer of particles on the walls of the mould and continues as long as slip remains in the mould. The cast article assumes the outer configuration of the mould. After casting, the formed component is allowed to dry slowly in the mould to a point at which it is sufficiently rigid for it to be removed from the mould without damage. The component is then dried in a controlled manner to remove the remaining liquid. Other, less common wet-forming methods that utilize deflocculated suspensions include pressure casting and tape casting.

Components formed by the above methods consist of particles packed together in a close packed arrangement (2). For approximately equiaxed particles having a narrow particle size distribution, this close packing gives typically ~ 40–50 volume% of apparent porosity in the formed component. The particles at this stage are held together by weak van der Waals forces; mechanical interlocking and friction between particles; and residual surface water, organic molecules, and salt bridges concentrated at the contact points. Organic binders are also incorporated during forming to facilitate machining of the powder compact when dry.

After forming, the resultant powder compact is heat-treated at an elevated temperature to promote bonding between the particles. This process is called *sintering* (3). While forming provides the basic shape, sintering determines the characteristics of the final component, including the dimensions, microstructure (sizes and distribution of phases, including porosity), and properties, including bulk density, stiffness, strength, hardness, wear resistance, and corrosion resistance. In alumina, sintering of particles involves diffusion of Al and O ions to the contact points between particles and is driven thermodynamically by the decrease in surface free energy achieved by the reduction in free surface area as the particles bond. Like any diffusion process, this requires an activation energy, and sintering, in general, occurs at temperatures above ~ 0.5–0.7 of the melting point (in K) of the material; alumina has a melting point of 2045°C (2318 K) and is sintered at typically ~ 1500–1700°C (1773–1973 K). In general, the finer the initial particle size, the greater the driving force for sintering and hence the lower the temperature required. Concurrent with interparticle bonding are changes in the particle shape and thus the pore morphology. In alumina and most ceramics in general, sintering is accompanied by a reduction and, in some cases, a complete elimination of initial pore volume. For the latter, this means that a component containing 50 volume % of porosity after forming will shrink by 50 volume % during sintering. This equates to a linear shrinkage of approximately 17%. Linear shrinkages of 15–25% are typical for high-purity, single-phase, engineering ceramics, and control of this shrinkage is critical in ensuring dimensional accuracy of the final component.

The latter stage of sintering occurs at porosity levels less than ~ 5–10 volume% and is accompanied usually by grain growth in which larger particles (now referred to as *grains*) grow at the expense of smaller ones (3). This process can also result in the pores that were initially located between the grains becoming trapped within the growing grains. The resultant intragranular porosity is virtually impossible to remove by further heat treatment and, as examined later, degrades the mechanical properties of the ceramic. Commercial alumina components usually contain approximately 0.05 wt% MgO, which promotes sintering and limits grain growth such that the porosity is able to be completely eliminated (4). Also, components for critical load-bearing applications can be either hot pressed (pressed in a die during sintering) or, as in total hip replacements, hot isostatically pressed (i.e., sintered in a sealed chamber under a high gas pressure) to completely remove porosity and to reduce grain growth. The resultant ceramic has a microstructure that consists of alumina grains of the order of ~ 2–5 μm in size with (ideally) zero residual porosity. In applications requiring accurate dimensional tolerances and/or low roughness surface finishes, diamond machining and polishing of sintered components are done. However, the procedures are expensive and time-consuming.

Although the microstructure and properties of the final component are developed during sintering, they can be critically affected by deficiencies in any of the stages of fabrication. In particular, heterogeneities in the starting powder (e.g., wide-ranging particle sizes or hard particle agglomerates) and the formed powder compact (e.g., bulk density variations) can result in nonuniform sintering and grain growth behavior, which leads to final sintered microstructures that have nonuniform grain size and contain residual porosity. As examined below, such microstructural defects are undesirable because they compromise the mechanical and other properties of the ceramic.

4. MATERIAL PROPERTIES

Typical material properties of high-purity and high-density alumina are summarized in Table 1 (1,5). Compared with other materials, alumina has a high melting point, very high compressive strength, very high hardness, high stiffness, exceptional wear resistance, and very high corrosion resistance. These properties derive directly from the ceramic's having very strong ionic bonding between

Table 1. Room-Temperature Material Properties of High-Purity, High-Density Polycrystalline Alumina

Property	General (1,5)	ISO6474 (6)
Chemical Properties		
Purity (weight %)	>99.0%	>99.5
Atomic Bonding	ionic	
Crystal Type	hexagonal	
Space Group	$D6_3d$	
a Lattice Parameter	0.4758 nm	
c Lattice Parameter	1.2991 nm	
Physical Properties		
Melting Point (°C)	2045	
Theoretical Density (g.cm^{-3})	3.98	
Bulk Density (g.cm^{-3})	>3.85	>3.90
Porosity (%)	<5	0
Average Grain Size (μm)	~ 1–7	<4.5
Coefficient of Thermal Expansion (0–200°C)	6.5×10^{-6} °C^{-1}	
Mechanical Properties		
Young's Modulus (GPa)	366	380
Poisson's Ratio	0.26	
Tensile Strength (MPa)	310	
Compressive Strength (MPa)	3790	
Flexural Strength (MPa)	~ 450–550 MPa	>400
Biaxial Flexural Strength (MPa)		250
Fracture Toughness (MPa.m$^{0.5}$)	4–5	
Microhardness (GPa)	22–23	23
Impact Strength (cm.MPa)		>40
Wear Resistance (mm^3/h)		0.01
Corrosion Resistance (mg/m^2.d)		<0.1

the aluminum and oxygen ions that make up the crystalline lattice structure. As examined later, the high degree of structural and chemical stability of alumina gives the ceramic excellent biocompatibility, which is classified as being bioinert. This combination of properties has resulted in the extensive use of alumina in load-bearing tribological orthopedic applications. The specific requirements of alumina for this purpose are standardized by the International Standards Organization in ISO6474(6) and are listed in Table 1. However, the relatively low tensile strength, flexural strength, and fracture toughness of alumina limit its structural applications to compressive loading situations, such as, most notably, femoral and acetabular components of total hip replacements as well as dental implants.

The mechanical properties of alumina, and oxide ceramics in general, are diminished significantly by decreased chemical purity, increased porosity, and increased grain size. Control of the processing conditions during manufacture, therefore, is critical to the development of optimal microstructures and, ultimately, to the resultant properties of the final component. The effects of chemical purity, porosity, and grain size on relevant material properties will now be examined.

Common impurities in alumina starting powders include Fe_2O_3, Na_2O, CaO, MgO, and SiO_2 (7). The limit of solid solubility of SiO_2 in Al_2O_3 is ~ 0.03 wt% and SiO_2 dissolved in the Al_2O_3 grains segregates strongly toward the grain boundaries. During sintering, SiO_2 present in excess of this limit, but still in relatively small amounts of the order of less than 1 wt%, reacts with other impurity oxides, such as Na_2O, CaO, and MgO, which also have low solubilities in Al_2O_3 and segregate preferentially to the grain boundaries. This segregation results in the formation of a thin (~ 10 nm) layer of aluminosilicate glass that wets the grain boundaries. The effect of this glassy phase is to promote grain growth at the expense of densification during sintering and to promote fatigue failure through subcritical crack growth in service (8). Accordingly, these impurities must be kept to a minimum (<0.1 wt%) in order to maintain the high density and small grain size necessary for biomedical-grade alumina.

The presence of porosity in a sintered ceramic component significantly diminishes the strength of the component because the pores reduce the effective cross-sectional area over which the applied load acts; the pores also serve as stress concentrators that effectively amplify the load at each pore (9). As the volume fraction (p) of porosity increases, the strength of the component (σ_p) diminishes according to the following empirical relationship:

$$\sigma_p = \sigma_o e^{-Bp},$$

where σ_o is the strength of the component in the absence of porosity and B is a constant that depends on the morphology and distribution of the pores. This relationship indicates that, as the amount of porosity in a component decreases, the strength exponentially increases. This highlights one of the fundamental objectives in the manufacture of structural ceramics, which is to fabricate components

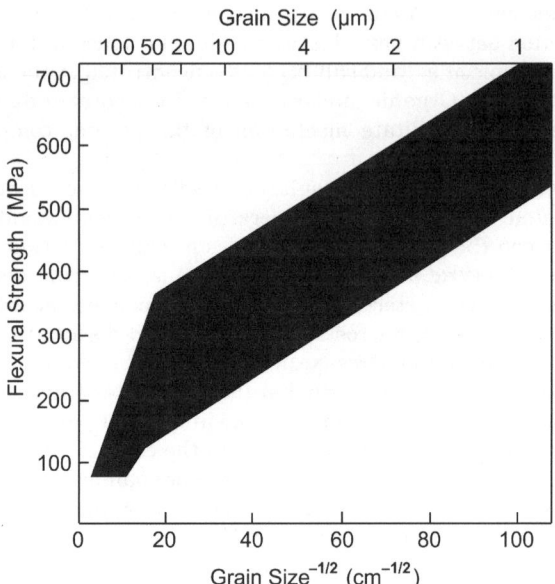

Figure 2. Diagram showing the room-temperature dependency of flexural strength versus grain size for high-density, high-purity polycrystalline alumina. The shaded region encompasses the experimental data presented by Rice (10).

that have the highest bulk density possible and, preferably, that are fully dense (i.e., contain no porosity).

The flexural strength of high-density alumina at room temperature increases with the inverse square root of grain size as indicated by the data shown in Fig. 2 (10). For ceramics loaded in tension or bending, fracture usually initiates at preexisting flaws in the structure. Such flaws are usually surface cracks that are introduced by grain boundary grooving during sintering, by grinding and polishing operations during machining, or by accidental impact damage in service. Also, differential cooling of a component from the sintering temperature inevitably causes differential contraction, which can result in microcracking, especially close to the surface. Such cracks usually do not extend deeper than approximately one grain diameter before they impinge upon another grain and stop (11). Consequently, these flaws scale approximately with the grain size of the ceramic. During loading of a ceramic, the applied stress is effectively amplified at the crack and, as the load increases, the stress at the crack tip increases. The stress generated at the crack tip is a complex quantity and is a function of the type of loading, sample configuration, and crack geometry. If the stress at the crack tip exceeds the maximal strength of the atomic bonding, then the crack will become unstable and will propagate through the material. According to the Griffith criterion for brittle fracture (12), the tensile strength (σ) at which a brittle material fractures decreases as the crack size (σ) increases:

$$\sigma = \frac{K_{Ic}}{Y\sqrt{\pi a}},$$

where K_{Ic} is the fracture toughness of the material (a material property) and Y is a dimensionless constant (~ 1)

that depends on the sample shape, the crack geometry, and the size of the crack relative to the sample size. Since the crack size is approximately equal to the grain size, the strength also scales with the inverse square root of grain size, thus accounting for the trend shown in Fig. 2.

When the strength of a set of nominally identical ceramic components is measured, the scatter in the size of the cracks produces considerable variation in the strength values (13). In particular, the strength that can be safely used in service (i.e., that below which will give an acceptable probability that failure will not occur) is considerably less than the average strength. Furthermore, as the component size increases, the probability of its containing a larger crack increases and so too does its probability of failure. Weibull statistics are used to characterize the distribution of strength data measured for a material and enable the probability of failure of a component to be calculated as a function of applied load and component volume. The Weibull analysis also enables the calculation of a safety factor, which is defined as the mean strength divided by the maximal allowable design stress that will ensure a specified probability of survival of a given component. Furthermore, load-bearing alumina components are now usually proof tested after fabrication to ensure further that they will not fail in service. Proof testing involves overloading the ceramic, in a similar configuration to how it would be loaded in service, at a load equal usually to the maximum service load multiplied by the safety factor. The component is considered safe to use if it does not fail what is known as the *proof test*. The proof load for alumina femoral heads used in total hip replacements (see below) is of the order of 50 times the average patient body weight, and the current fracture rate of current components is 0.004% (14).

Related to the fracture behavior of alumina is high-cycle fatigue, which is the failure of components due to cyclic loading at stresses well below those usually necessary to cause brittle fracture. The effect of cycling is to open and close repeatedly the tip of a particular preexisting crack such that the atomic bonds at the crack tip break, causing the crack to grow a small amount, typically the order of an atomic spacing, during each cycle. With continued cycling, the crack progressively grows until it reaches the critical size necessary for the stress intensity at the crack tip (given by the Griffith equation) to exceed (at the maximal load) the fracture toughness of the material and thus cause brittle fracture. The number of cycles to failure decreases as the maximal load, load amplitude, and initial crack size each increase. In aqueous environments such as the human body, crack growth in alumina is exacerbated by the presence of glassy silicate phases at the grain boundaries—water reacts with Si–O bonds of any glassy phases at the crack tip effectively reducing the energy needed for crack propagation. This mechanism is known as *stress corrosion* and highlights the important need for alumina starting powders used in the manufacture of biomedical load-bearing alumina components, such as those used in total hip prostheses, to be of high purity. Alumina components meeting ISO6474 specifications contain sufficiently small amounts (<0.5 wt%) of impurity oxides such that stress corrosion is not a problem and, as a consequence, the design lifetimes for cyclic fatigue of current components are very long.

The grain size of alumina also strongly influences the wear behavior of the ceramic. As illustrated in Fig. 3a, the application of a contact load on a ceramic surface generates stresses immediately below the contact point, which result in elastic and plastic deformation of the contact region (15). If the tensile strength of the material is exceeded, microcracking occurs in the deformed region. As examined above, microcracking can initiate from stress concentrators, such as preexisting microcracks and grain boundaries. In the case of wear, this situation is complicated by frictional and tangential forces associated with the contact load sliding on the surface. As illustrated in Fig. 3b, the friction between the two surfaces introduces a shear component to the loading such that the tensile stress is decreased at the leading edge of the contact and is increased at the trailing edge. Therefore, crack initiation and propagation are enhanced at the trailing edge of the sliding load. Although not examined here, the pattern of cracking below the contact load is related to the grain size, hardness, and fracture toughness of the ceramic and manifests as one (or a combination) of the following characteristic crack types: cone, radial, median, half-penny, and lateral.

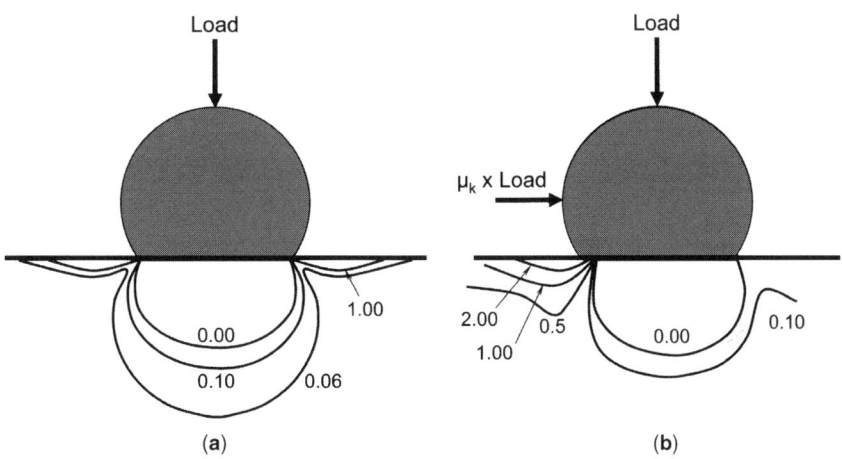

Figure 3. Diagram illustrating the contours of constant tensile stress generated immediately below the surface of a material subjected to (a) a stationary normal contact load and (b) a sliding normal contact load (15). (The force required for sliding is equal to normal load times the coefficient of dynamic friction (μ_k)).

At the beginning of wear, the microfracture is transgranular (i.e., within grains) and typically is confined to the small region immediately adjacent to the each crack and to fine scratches caused by wear debris and asperities on the counterface (16). This regime is known as *mild wear* and is characterized by low wear rates that are either approximately constant or increasing slowly. Wear surfaces in this regime usually have an almost polished appearance. As wear proceeds, the extent of subsurface fracture increases such that the cracks progressively link up, causing material to detach from the surface, with the volume of detached material being significantly greater than the apparent contact area. The surface thus becomes rough and irregular and grain pullout, intergranular fracture (i.e., between grains), and surface delamination are usually apparent. This wear regime is known as *severe wear* and is characterized by increased values of surface roughness and coefficients of friction and significantly greater wear rates.

The wear debris can be ejected from between the sliding surfaces as they move past each other, or crushed into finer particles, or can remain between the two surfaces. The latter results in an undesirable situation known as *third body wear* in which the wear debris generate high-stress contact points at the sliding surfaces that, in turn, generate further, more rapid, wear. However, for alumina, (fine) wear debris can react with water vapor to produce aluminum hydroxide, which then forms a tribofilm on the surface. Tribofilms fill surface porosity and roughness and lower the coefficient of friction (17). However, it is likely that this mechanism is not of great significance to today's alumina-on-alumina hip replacements because the amount of wear debris produced is exceedingly small. Also, articulation of the joint and lubrication of the joint by synovial and other physiological fluids effectively removes wear debris from between the articulating surfaces before it can build up in sufficient amounts to form appreciable tribofilms. These fluids also play an important role in lubricating alumina–alumina and alumina–ultrahigh molecular-weight polyethylene (UHMWPE) total hip replacements: alumina is hydrophilic and consequently water and biological macromolecules, such as proteins, adsorb onto the surface to create a lubricating film. Compared with dry conditions, this lubricating film significantly diminishes the coefficient of friction and wear rate of both alumina-alumina and alumina-UHMWPE wear couples, often by as much as a factor of five to ten.

Like the requirements for brittle fracture, the ability of a ceramic to resist fracture during wear is favored by small grain size, minimization of the extent and size of structural flaws, such as microcracks and pores, and elimination of internal tensile stresses (18). As shown in Fig. 4, the time (or number of cycles) that it takes for the wear mechanism to change from mild to severe decreases with increasing grain size. In the case of alumina used for biomedical tribological applications, the data further highlight the need to have a small grain size in the ceramic in order to avoid severe wear. A small grain size is also important for when severe wear does begin because it reduces the size of the individual particles that are produced

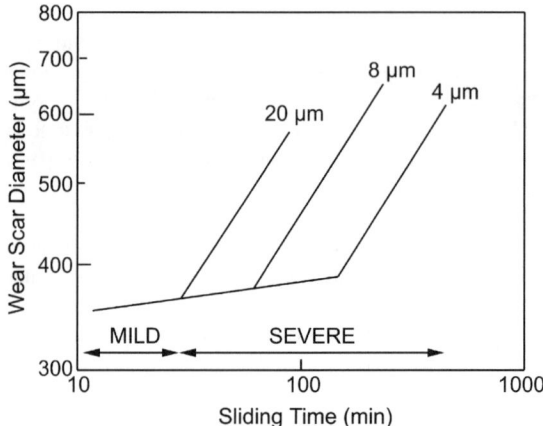

Figure 4. Diagram of wear scar diameter versus sliding time for high-density, high-purity polycrystalline alumina. The transition time at which wear mechanism changes from mild to severe increases with decreasing grain size (data are from Cho et al. (18)).

by intergranular cracking, thus reducing the severity of any third body wear that subsequently occurs.

5. BIOCOMPATIBILITY

The biocompatibility of alumina was well established in the late 1960s and early 1970's by *in vitro* cellular studies and *in vivo* animal studies. Alumina has been demonstrated to be nonirritating, noncytotoxic, and noncarcinogenic (19). *In vitro*, alumina surfaces support the normal biochemical and biological functions of attachment, migration, and proliferation of numerous cell types such as fibroblasts, osteoblasts, osteoclasts, and epithelial cells. Also, particles and fibers of alumina are generally noncytotoxic and are well tolerated in soft tissues *in vivo*. Recent studies have shown that, despite alumina's being bioinert, cellular activity on alumina of nanometer grain sizes (<100 nm) is enhanced compared with conventional alumina (20). The reasons for this are unclear but may relate to the finer scale of surface roughness arising from decreased surface grain size and decreased diameter of surface pores, which are closer to the size range of the focal contacts involved in cell attachment (focal contacts are the points of the cell cytoskeleton that attach to a substrate via specific proteins adsorbed on the surface). Alternatively, increased wettability associated with nanoscale surface roughness may increase protein adsorption and hence enhance cell interaction with the surface. Related to this could be an inherently greater reactivity of the surface derived from a higher proportion of unsaturated ionic bonds.

Alumina bioceramics are classified as being bioinert because they elicit minimal response from living tissues, both soft and osseous, in which they are implanted (19). This bioinertness is a direct consequence of the very high chemical and physical stability of the ceramic: alumina undergoes very little chemical or physical change during long-term exposure to the physiological environment. Unlike bioactive ceramics (such as hydroxyapatite and specific compositions of bioactive glass) and resorbable ceramics (such as tricalcium phosphate and specific compositions of bioactive glass), leaching of soluble species from conventional biomedical alumina is not detectable by spectroscopic microanalysis techniques and the surface does not exhibit any degree of surface activity with respect to contacting cells. Following hemostasis and acute inflammation immediately after surgery, the wound-healing response involves the deposition of a disorganized fibrous matrix of mesenchymal tissue, chiefly collagen, at the wound site to form a soft tissue callus. This soft tissue callus becomes progressively denser and, depending on the type of tissue in which the alumina is implanted, is converted to either organized soft tissue or trabecular and/or lamellar bone. In either case, the long-term tissue reaction to alumina is minimal and is characterized by the material's being encapsulated by a thin layer of fibrous tissue composed chiefly of collagen. Chronic inflammation, as typified by the proliferation of macrophages, mononuclear cells, lymphocytes, and giant cells, and the resultant formation of granulation tissue at the implant site, is usually minimal or nonexistent.

In the case of alumina implanted in bone, the space between the implant and bone is progressively filled during the healing process, usually by bone and/or fibrous tissue (19). The formation of fibrous tissue rather than bone around an implant may be a specific response determined by the properties of the implant or a result of this and other factors including excessive tissue damage during implantation (e.g., bone necrosis caused by excessive heating of the bone during drilling or cutting); movement of the implant during healing; and deprivation of nutrients, including oxygen. In the case of good apposition of osseous tissue adjacent to alumina, the thickness of the interfacial fibrous tissue layer decreases from ∼100 μm a few weeks after surgery to an equilibrium value of ∼25–50 μm after ∼6 months, the latter values tending to increase with micromotion of the implant. Areas of direct contact between bone and alumina may exist but, unlike bioactive ceramics, a specific chemical bond does not form, due to the inherent bioinertness of the ceramic. A typical interface between alumina and osseous tissue is shown in Fig. 5.

6. BONE INGROWTH INTO POROUS ALUMINA

Bone ingrowth into macroporous, roughened, or textured surfaces is used as a means of fixing orthopedic and dental

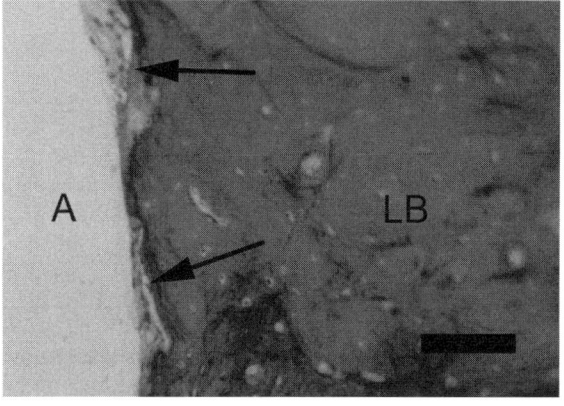

Figure 5. Section of high-density, high-purity polycrystalline alumina (A) implanted for 12 weeks in a rabbit femur (stained with Von Kossa and toluidine blue) showing lamellar bone (LB) separated from the alumina (A) by fibrovascular tissue (indicated by arrows). Scale bar is equal to 50 μm. (This figure is available in full color at http://www.mrw.interscience.wiley.com/ebe.)

prostheses into the skeletal system. Stabilization results from ingrowth of osseous tissue into the surface of the implant, thereby mechanically interlocking the implant in the host bone (21). The ability of bone to grow into a porous implant is dependent on the pore size and the amount and size of interconnecting spaces between the pores. Numerous histomorphometry and mechanical pushout studies have demonstrated that bone ingrowth is facilitated by pore diameters in the range of ∼100–600 μm (21). Pores smaller than approximately 50 μm tend to become filled with fibrovascular tissue rather than bone, and bone ingrowth occurs only in pores larger than approximately 50 to 100 μm. As the pore size increases above ∼600 μm diameter, osseous tissue is unable to fill the pores completely and instead increasing amounts of fibrovascular are usually present. Also, in order for bone to infiltrate pores deeper in the implant, the interconnecting pores must be larger than ∼50–100 μm in diameter. As shown in Fig. 6, bone ingrowth into deep alumina pores occurs by the progressive migration of a tissue front consisting of undifferentiated mesenchymal cells, fibrovascular tissue, osteoid, woven bone, and marrow (22). Behind this front, woven bone is remodeled to produce a thin layer of lamellar bone along the pore wall and a central lumen of marrow.

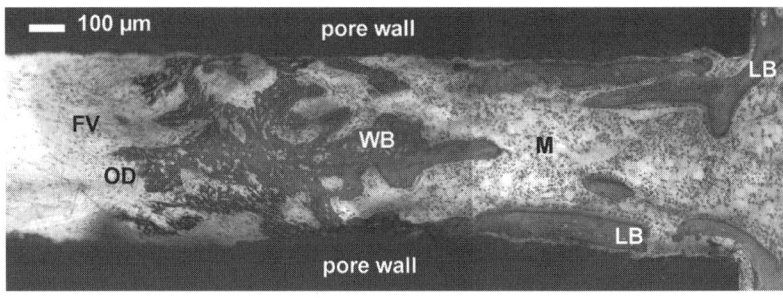

Figure 6. Longitudinal section of a cylindrical pore (600 μm diameter) in high-purity polycrystalline alumina implanted in the medullary canal of a rat femur for 4 weeks (stained with methylene blue and basic fuchsin) (22). Bone ingrowth into the tube (from the pore entrance at the right) consists of an advancing tissue front composed of loose fibrovascular tissue (FV), osteoid (OD), and woven bone (WB). Behind the advancing tissue front to the pore entrance, tissues are progressively remodeled into lamellar bone (LB) that lines the tube and marrow (M) which fills the tube lumen. (This figure is available in full color at http://www.mrw.interscience.wiley.com/ebe.)

An advantage of using porous surfaces to fix prostheses in the skeletal system is that mechanical interlocking of bone and the implant results in high interfacial strengths. The larger interfacial area distributes the applied load more evenly and minimizes localized stresses, which may otherwise cause tissue necrosis. A disadvantage of using porous surfaces is that it takes a period of time after implantation for the bone to grow into the porous structure. During this time, the implant should not be loaded and therefore the patient must be immobilized. Loading and/or movement of the implant during this time may seriously disrupt bone ingrowth, resulting instead in fibrous tissue ingrowth and consequently poor fixation. This can be overcome by the use of mechanical fixation techniques, such as the use of screws and plates, to complement the long-term fixation by bone ingrowth to give immediate postoperative stability.

7. BIOMEDICAL APPLICATIONS

Alumina has been used as a biomedical implant material since the early 1970s. Due to its unique combination of excellent biocompatibility, high compressive strength and stiffness, very high corrosion resistance, exceptional wear resistance, relatively low tensile strength, and low fracture toughness (the latter two being characteristic of ceramics in general), alumina has been used extensively in articulating, compressive-type, load-bearing, orthopedic applications (mainly total hip prostheses) and in dental applications.

7.1. Orthopedic Applications

High-density polycrystalline alumina is used extensively as one or both bearing surfaces in total hip prostheses. Current designs typically consist of a high-density polycrystalline alumina femoral head fitted to a titanium or cobalt-chromium alloy femoral stem. The stem is fixed in the medullary canal of the femur by cementation using polymethylmethacrylate or by a press-fit combined with subsequent bone ingrowth into a porous coating or bone attachment onto a bioactive coating. The latter two techniques are known as cementless fixation and, in some implant designs, are combined in the form of a porous layer (usually sintered beads or woven mesh of the same metal as the stem) for mechanical osseointegration with the exposed pore surfaces, which are coated by a bioactive material (usually hydroxyapatite) to promote osteoconduction into the pores and to give direct bone apposition onto the pore surfaces. The alumina head articulates against an acetabular cup made of either UHMWPE or alumina. The acetabular cup of either material is fitted into a titanium or cobalt-chromium alloy cup liner that is fixed in the acetabulum by the same techniques as used for the stem component as well as by bone screws anchored into the acetabular bone. Some designs employ a sandwich design for the acetabular component consisting of a layer of polyethylene interposed between the metallic liner and the ceramic cup; the polyethylene acts to absorb impact loads, thereby minimizing their transmission to the acetabular bone.

Provided that the alumina is of high density and has a small grain size (i.e., it satisfies ISO6474 requirements specified in Table 1), surfaces having very low surface roughness can be fabricated. Compared with conventional metal head/UHMWPE cup configurations, the low surface roughness as well as the inherent hydrophilicity and related lubricity of the alumina head result in significantly less wear of the UHMWPE (23). UHMWPE wear debris has been implicated in the long-term clinical failure of both cementless and cemented prostheses. Failure occurs by aseptic loosening, which occurs as a result of bone resorption and osteolysis at the bone–prothesis interface (24,25). While the pathogenesis of aseptic loosening is not clearly understood, there is evidence to suggest that it is triggered by phagocytosis of UHMWPE wear debris by macrophages, which, in turn, release cytokines and other cell mediators to recruit osteoclasts, thereby promoting bone resorption.

Replacing the UHMWPE with high-density polycrystalline alumina further improves the tribological performance of total hip prostheses. The coefficient of friction of alumina against alumina decreases with time, and the wear rate correspondingly diminishes toward that of a natural joint (23). The linear wear rate of today's alumina/alumina total hip prostheses is of the order 0.001 mm/year, which is considerably less than that of UHMWPE/alumina (0.1 mm/year) and UHMWPE/metal (0.2–0.5 mm/year) (26). The very low rates of clinical wear of alumina are due, in part, to the wear mechanisms being in the mild regime such that wear involves effectively only relief polishing. Examination by scanning electron microscopy (SEM) of tissues retrieved from around noncemented alumina-on-alumina total hip prostheses typically shows the presence of alumina wear particles of 0.1–0.5 μm diameter located either within phagocytic cells or in the extracellular space (27). Reaction to this debris is usually minimal, consisting of either no obvious pathology or small areas of necrosis; unlike UHMWPE debris, there is little in the way of giant cell reactions. This is attributed to the relatively small amount of wear debris rather than to a greater tolerance for the wear debris, although some studies have suggested that alumina wear debris is inherently less biologically active than UHMWPE wear debris (28). In cases of massive alumina wear, for example, with poorly aligned prostheses, significant tissue necrosis and associated osteolytic reaction can occur.

The first alumina/alumina total hip prostheses in the early 1970s suffered high rates of clinical failure by excessive wear and catastrophic fracture arising largely from the inadequate properties of the ceramic, namely, excessive residual porosity, large grain size, and wide grain size distribution. However, better knowledge of the underlying materials science, improvements in manufacturing technology, and more stringent quality control have significantly improved the compositional, physical, and mechanical properties of the ceramic. These developments are reflected in the introduction of ISO6474 in 1979 and its subsequent revisions in 1989 and 1994 (6). As indicated by reviews of clinical data of alumina/alumina total hip prostheses, the incidence of implant failure arising from excessive wear or catastrophic fracture has significantly diminished with the technological improvements in the manufacturing, and hence performance, of alumina (29,30). In particular, the current standard introduced in 1994 requires alumina for load-bearing implants to be completely densified (i.e., must contain 0% porosity) and to have a fine grain size of <4.5 µm. The intention of this demand is obvious: to improve the reliability of alumina components by increasing the maximal strength, decreasing the surface roughness, and decreasing the wear rate. However, the only feasible way to achieve this demand is to hot isostatically press the components, which requires highly specialized and expensive facilities. Today, all alumina components manufactured for total hip prostheses are hot isostatically pressed.

Alumina components satisfying the requirements of the current ISO6474 standard (6) and implanted in correct alignment usually show very low wear rates (0.001 mm/year), low clinical failure rates ($\sim$ 1%/year), and exceedingly low incidences of fracture (0.004%/year) (26). Incidences of excessive wear and, in severe cases, chipping and catastrophic fracture are attributable usually to biomechanical problems, such as misalignment of the prosthesis in the initial implantation, ball contact with the unpolished rim of the acetabular component, or repeated linear impact contact. The latter two situations are caused usually by microseparation of the head and cup arising from problems such as excessively short neck of the femoral stem (for example, as a result of revision surgery), incorrect vertical placement of the acetabular cup, or progressive vertical movement of the acetabular cup due to pelvic bone osteolysis.

Other clinical orthopedic applications of polycrystalline alumina that have been developed include components in knee joint, shoulder joint, elbow joint, and ankle joint prostheses. Like total hip prostheses, these all utilize alumina as bearing surfaces for articulation. However, due to size constraints, complexity of shape, and the difficulty in avoiding tensile loading components, these applications remain largely experimental and have not been commercially developed to any major extent.

7.2. Dental Applications

Polycrystalline alumina and sapphire have been used as endosteal dental implants to replace extracted teeth, both mandibular and maxilliary, immediately or soon after extraction. They consist of a form of plug that fills the endosteal defect and provides an abutment for the later fitting of a dental crown or a dental bridge. The desirable combination of high compressive strength, excellent (bioinert) biocompatibility, and outstanding wear resistance makes alumina an attractive material for this application. Although sapphire endosteal implants have been used clinically since the early 1980s, they are susceptible to fracture if excessive shear forces are imposed (31). As a consequence, titanium alloy components tend to be favored at present.

A wide variety of endosteal alumina plug designs exists including threaded screws, tapered cylinders, stepped cylinders, dimpled cylinders, and circumferentially grooved cylinders. All are designed to provide optimal load transfer to the implant. Long-term osseointegration with the bone in which they are implanted is promoted by specific surface macrotopography, such as roughened or porous surfaces, as well as the application of bioactive coatings of hydroxyapatite, which facilitate attachment of osseous tissue directly with the surface. Implanted endosteal implants are not mechanically loaded for a period of 3 to 6 months in order to allow osseointegration to occur such that the implant is stable. During this time, the implant remains sealed by the gingival tissues that were closed at the time of implantation. After the period of osseointegration, the gingival tissues are resected to expose the implant such that the abutment connection and dental crown or bridge can be fitted.

Polycrystalline alumina and sapphire dental implants elicit favorable hard-tissue and soft-tissue biocompatibility for implantation in the mouth. In addition to good mechanical osseous integration in bone, which is necessary for stable fixation, the gingival epithelium forms a biological seal where the implant exits the gum, a case that is similar to that of natural teeth. Some implant designs utilize a circumferential groove around the implant to promote this epithelial seal. In a properly sealed implant, the microflora that exist below the gingiva are no different from these observed for natural healthy teeth.

BIBLIOGRAPHY

1. W. H. Gitzen, *Alumina as a Ceramic Material*. Columbus, OH: The American Ceramic Society, 1970.
2. M. N. Rahaman, *Ceramic Processing and Sintering*. New York: Marcel Dekker, 1995.
3. R. M. German, *Sintering Theory and Practice*. New York: Wiley, 1996.
4. S. J. Bennison and M. P. Harmer, History of the role of MgO in the sintering of α-Al_2O_3. *Ceram. Trans.* 1990; **7**: 13–49.
5. R. Morrell, *Handbook of Properties of Technical and Engineering Ceramics. Part 2. Data Sheets. Group A – High-Alumina Ceramics*. Her Majesty's Stationary Office: London, 1985.
6. *ISO 6474 Implants for Surgery – Ceramic Materials Based on High Purity Alumina*, International Standards Organisation: Geneva, 1994.
7. W. D. Kingery (Editor), *Advances in Ceramics, Volume 10. Structure and Properties of MgO and Al_2O_3 Ceramics*. American Ceramic Society Inc: Columbus, OH, 1984.
8. B. J. Dalgleish and R. D. Rawlings, A comparison of the mechanical behaviour of alumina in air and simulated body environments. *J. Biomed. Mater. Res.* 1981; **15**:527–542.
9. M. Barsoum, *Fundamentals of Ceramics*. McGraw Hill: New York, 1997.
10. R. W. Rice, Effects of environment and temperature on ceramic tensile strength-grain size relations. *J. Mater. Sci.* 1997; **32**:3071–3087.
11. R. W. Davidge, *Mechanical Behaviour of Ceramics*. Cambridge University Press: Cambridge, 1979.
12. A. A. Griffith, The phenomena of rupture and flow in solids. *Philos. Trans. Royal Soc. Lond.* 1920; **A221**:163–198.
13. J. B. Wachtman, *Mechanical Propertries of Ceramics*. New York: Wiley, 1996.

14. H. G. Pfaff, Ceramic component failure and the role of proof testing. *Clin. Orthop. Rel. Res.* 2000; **379**:29–33.
15. B. R. Lawn, *Fracture of Brittle Solids – Second Edition*. Cambridge: Cambridge University Press, 1993.
16. A. Blomberg, M. Olsson, and S. Hogmark, Wear mechanisms and tribo mapping Al_2O_3 and SiC in dry sliding. *Wear* 1994; **171**:77–89.
17. Y. Wang and S. M. Hsu, The effects of operating parameters and environments on the wear and wear transition of alumina. *Wear* 1996; **195**:90–96.
18. S. J. Cho, B. J. Hockey, B. R. Lawn, and S. J. Bennison, Grain size and R-curve effects in the abrasive wear of alumina. *J. Am. Ceram. Soc.* 1989; **72**:1249–1252.
19. P. Griss and G. Heimke, Biocompatibility of high density alumina and its applications in orthopedic surgery. In D. F. Williams (ed.), *Biocompatibility of Clinical Implant Materials*. Boca Raton, FL: CRC Press, 1981, pp. 155–198.
20. T. J. Webster, C. Ergun, R. H. Doremus, R. W. Siegel, and R. Bizios, Enhanced osteoclast-like cell functions on nanophase ceramics. *Biomaterials* 2001; **22**:1327–1333.
21. H. Schliephake, F. W. Neukam, and D. Klosa, Influence of pore dimensions on bone ingrowth into porous hydroxylapatite blocks used as bone graft substitutes: A histomorphometric study. *Int. J. Oral Maxillofac. Surg.* 1991; **20**:53–58.
22. M. B. Pabbruwe, O. C. Standard, C. C. Sorrell, and C. R. Howlett, Use of an intramedullary in vivo model to study bone formation and maintenance in ceramic porous domains. *J. Biomed. Mater. Res.* 2004; **68A**:305–313.
23. B. D. Ratner, A. S. Hoffman, F. J. Schoen, and J. E. Lemons (eds.), *Biomaterials Science. An Introduction to Materials in Medicine, Second Edition*. Amsterdam: Elsevier, 2004.
24. S. Corbett, M. Hukkanen, P. Hughes, and J. Polak, Aseptic loosening of total hip prostheses. pp. 178–179 in M. Hukkanen, P. Hughes, and J. Polak J (eds.), *Nitric Oxide in Bone and Joint Disease*. Cambridge University Press: Cambridge, 1998.
25. S. Goodman, P. Aspenberg, Y. Song, D. Regula, and L. Lidgren, Polyethylene and titanium alloy particles reduce bone formation. *Acta Orthop. Scand.* 1996; **67**:599–605.
26. G. M. S. Willmann, Ceramic femoral head retrieval data. *Clin. Orthop. Rel. Res.* 2000; **379**:22–28.
27. A. Hatton, J. E. Nevelos, A. A. Nevelos, R. E. Banks, J. Fisher, and E. Ingham, Alumina-alumina artificial hip joints. Part I: A histological analysis and characterisation of wear debris by laser capture microdissection of tissues retrieved at revision. *Biomaterials*. 2002; **23**:3429–3440.
28. D. Granchi, G. Ciapetti, I. Amato, S. Pagani, E. Cenni, and L. Savarino, Influence of alumina and ultra-high molecular weight polyethylene particles on osteoblast-osteoclast cooperation. *Biomaterials*. 2004; **25**:4037–4045.
29. C. Heisel, M. Silva, and T. Schmalzried, Bearing surface options for total hip replacement in young patients. *J. Bone Jt. Surg.* 2003; **85-A**:1366–1379.
30. A. Toni, A. Sudanese, S. Terzi, M. Tabarroni, F. Calista, and A. Giunti, Ceramics in Total Hip Arthroplasty. In D. L. Wise, D. J. Trantolo, D. E. Altobelli, M. J. Yaszemski, J. D. Gresses, and E. R. Schwartz. (eds.), *Encyclopedic Handbook of Biomaterials and Bioengineering. Part A: Materials*. New York: Marcel Dekker, 1995, pp. 1510–1544.
31. T. I. Berge and A. G. Gronningsaeter, Survival of single crystal sapphire implants supporting mandibular overdentures. *Clin. Oral Implants Res.* 2000; **11**:154–162.

AMERICAN COLLEGE OF CLINICAL ENGINEERING

IZABELLA A. GIERAS
American College of Clinical Engineering
Plymouth Meeting, Pennsylvania

The American College of Clinical Engineering (ACCE) was established with a special focus on promoting the profession of clinical engineering in the United States and in the international arena, http://www.ACCEnet.org. With members across the globe, the ACCE is the only professional society for clinical engineers with international recognition. Clinical engineers are healthcare professionals who work in hospitals, medical device industry, research and development, academic institutions, consulting, regulatory agencies, and other medical technology-based institutions. Their main focus is to apply engineering and management background to enhance patient care. With this focus in mind, clinical engineers work very closely with biomedical engineering technologists, physicians, clinicians, information technologists, risk managers, systems engineers, construction engineers, procurement, and other healthcare professionals. The ACCE's multidisciplinary activities engage the organization and its members on local, state, national, and international levels. These activities include conferences, educational workshops, teleconferences, partnerships, and cosponsorships with professional organizations and societies as well as diverse medical device industry initiatives that are all further fostering the profession of clinical engineering.

1. ESTABLISHMENT

In 1990, the ACCE, the first organization focused exclusively on clinical engineering was established in the State of Washington. The first Board of Directors was formed and charged with developing the mission and definition of a clinical engineer. The ACCE's mission is specific to its dedication to the clinical engineering profession and is well reflected throughout all of the organization's activities. The Board of Directors and its founding members approved a four-point mission that includes: *To establish a standard of competence and to promote excellence in clinical engineering practice*; *To promote safe and effective application of science and technology in patient care*; *To define the body of knowledge on which the profession is based*; and *To represent the professional interests of clinical engineers* (1). The definition of a clinical engineer was formulated in 1991. Dyro (2) describes it as follows: "A clinical engineer is a professional who supports and advances patient care by applying engineering and managerial skills to healthcare technology."

During the ACCE's formative years, the organization established the *Code of Ethics* and the *Bylaws*. The Code of Ethics is a separate document from its mission and is adhered to by each member of the college. The Bylaws were created with the inception of the organization and have been since revised. They provide guidance to the ACCE Board of Directors and its members on how the college is managed.

2. MEMBERSHIP

Each ACCE member plays a vital role in the betterment of the healthcare system, leading to a safe and effective healthcare environment. The ACCE has had a steady growth to over 250 members in the United States and abroad, including members in Canada, Argentina, Egypt, Germany, China, South Africa, and other parts of the world.

2.1. Categories

The ACCE has five membership categories based on the existing Bylaws: Individual, Fellow, Emeritus, Associate, and Candidate, each with different approval criteria. The new member applications are reviewed by the ACCE Membership Committee, and a formal recommendation is submitted for review and approval by the ACCE Board of Directors. International applicants are also reviewed by the ACCE International Committee.

2.2. Benefits

The ACCE offers a wide variety of membership benefits and opportunities for its members to enhance their clinical engineering careers and shape the medical device industry. Examples include access to a national and international network of practicing clinical engineers, ACCE newsletter, monthly teleconferences, diverse educational opportunities, sponsorship of professional awards, partnerships with professional organizations, as well as other opportunities within the healthcare environment.

2.3. Board

The ACCE is managed by the Board of Directors, which consists of ten members: president, president-elect, vice president, secretary, treasurer, immediate past president, and four members at large. The ACCE Executive Board consists of the president, president-elect, vice president, secretary, treasurer, and immediate past president. The Executive Board meetings provide an avenue to review the current status of all ACCE activities and to set the agenda for the full board meetings. Each board member has his/her own unique tasks that are reported on during the board meetings, held via teleconference calls. Most of the communication between board members takes place by electronic mail.

2.4. Committees

The ACCE Board of Directors works closely with several ACCE committees on different clinical engineering activities. Based on the current ACCE Bylaws, the organization has two types of committees: standing and general. The ACCE has two standing committees: a membership and a nomination committee. Members of these two committees need to be ACCE members in good standing at the membership level of Individual, Fellow, or Emeritus (3). The general committees comprise ACCE members and non-members. The chairs of these committees are appointed by the president, with the approval of the Board. The ACCE has several general committees, including Advocacy, Body of Knowledge, Education, International, Professional Practices, Strategic Development, and Symposium Planning. Each committee has its unique mission and reports directly to the ACCE Board of Directors.

Assistance needed for special projects such as development or review of white papers and responses to guidance documents has led to the formation of task force groups. The ACCE has three active task force groups: Health Insurance Portability and Accountability Act (HIPAA), Integrating the Healthcare Enterprise (IHE), and Medical Errors. Some of the different committees and task force groups will be highlighted throughout this article.

2.5. Secretariat

The continued growth of the organization and its expansion in membership, activities, and partnerships is supported by the ACCE Secretariat, which was established in 2000. The Secretariat provides services on behalf of the ACCE, the ACCE Healthcare Technology Foundation, and the Clinical Engineering Certification Program. The work of the Secretariat is instrumental in providing administrative and secretarial support in processing organizational membership applications, managing membership databases, coordinating teleconference meeting schedules, and processing organizational mailing for each organization (4).

2.6. Annual Membership Meetings

The different professional conferences are a perfect avenue for the clinical engineering community to get together and discuss the latest events in healthcare. They serve as an ideal location for ACCE activities. One such activity is the ACCE annual membership meeting. This meeting is filled with rich agendas focusing on the review of the ACCE's annual activities, Advocacy Awards, ACCE Board elections, and outlook on the organization's future, concluding with an open and interactive discussion on any issues and concerns related to the ACCE membership and the clinical engineering profession. Other activities include special board meetings and committee meetings. These meetings usually take place before the annual membership meeting and serve as a preparatory time to finalize the membership meeting agenda.

3. ADVOCACY ACTIVITIES

3.1. Advocacy Committee

The ACCE Advocacy Committee plays an important role in promoting the clinical engineering profession to clinical engineers and other healthcare professionals. One way such promotion is occurring is with the tools and materials developed by the committee members for the clinical engineers to use in promoting the profession to high-school and college students as well as to other healthcare and non-healthcare professionals.

In addition, the ACCE has established an Advocacy Awards Program. The advocacy awards recognize clinical engineers for their excellence and advocacy efforts in the profession. The Advocacy Awards Program includes the *Thomas O'Dea Advocacy Award*, *Professional Achievement in Technology/Professional Development Award*, *Professional Achievement in Management/Managerial Excellence Award*, *DEVTEQ Patient Safety Award*, and the *Challenge Award*.

In 2003, the ACCE introduced the *Best Student Paper Award*. This award recognizes a work of clinical engineering that was either published in a professional journal or presented at a professional conference by a student pursuing the field of clinical or biomedical engineering. The award provides the student with an introduction to the college and its benefits by awarding the student with a 1-year ACCE membership.

The 2004 marked an important year for the ACCE Advocacy Awards Program when the ACCE Board of Directors established the *ACCE Lifetime Achievement Award*. The recipient is a person who has greatly contributed to the profession of clinical engineering and whose professional accomplishments are well recognized in the healthcare community.

4. PROFESSIONAL RELATIONSHIPS

4.1. Biomedical Equipment Technologists

Clinical engineers working in hospitals have a close relationship with biomedical equipment technologists (BMETs). These persons are responsible for the direct support, service, and repair of medical equipment in the hospital (5). The close affiliation between these two professionals has led to a relationship between the ACCE and the Medical Equipment and Technology Association (META).

4.2. Information Technologists

The design of modern medical technologies has become more integrated, resulting in new professional relationships. The ACCE has been working closely with the Healthcare Information and Management Systems Society (HIMSS) on collaborative efforts between clinical engineers and information technologists (IT), focusing on the management of healthcare technologies. The ACCE is a regular cosponsor, presenter, and participant at the Annual HIMSS Conference and Exhibition.

4.3. Regulatory Agencies

The ACCE has several professional relationships with regularity agencies, for example, the Food and Drug Administration (FDA) and Joint Commission on Accreditation of Healthcare Organizations (JCAHO). The ACCE has designated an ACCE representative to the FDA Medical Device Industry Coalition (FMDIC) group. The coalition was formed to promote communication, education, and cooperation between the regulators and the regulated industry (6). The ACCE has also reviewed and commented on FDA, JCAHO, and other guidance documents. In 2004, the ACCE Board of Directors approved the formation of a subcommittee under the Advocacy Committee that will focus on reviewing industry-based white papers, guidance documents, and endorsements, representing the ACCE in the medical device arena.

4.4. International Professional Organizations

The ACCE has established a new international collaboration with the Italian Association of Clinical Engineers (Associazione Italiana Ingegneri Clinici, AIIC). The ACCE was invited by AIIC to present at the Heath Technology Assessment International (HTAi) meeting in Rome, Italy, in the summer of 2005 on *Clinical Engineering's Role in HTA: US Perspective*. The presentation evoked enthusiasm in clinical engineering and the interest in pursuing an international clinical engineering certification program (7). The ACCE continues its successful relationships with the World Health Organization (WHO), Pan American Health Organization (PAHO), and ORBIS on international activities with special focus on the Advanced Clinical Engineering Workshops (ACEW) and other initiatives in medical technology management and patient safety. The ACCE is also involved with the International Federation for Medical and Biological Engineering (IFMBE) and its Clinical Engineering Division.

4.5. Other Professional Organizations

The ACCE has successful affiliations with other professional organizations and has been involved in cosponsorships at many of the professional organizations' annual conferences and meetings. The ACCE has established a professional relationship with the Association for the Advancement of Medical Instrumentation (AAMI) and works closely with the AAMI in organizing an annual educational symposium held at the AAMI Annual Conference and Expo.

In 2001, a collaboration of ACCE members, the ACCE Board of Directors, and AAMI developed an award, through the AAMI Foundation, to recognize the longtime ACCE leader, Robert L. Morris. The first recipient was Robert L. Morris who was granted the award posthumously in 2001. Five recipients including Dr. Morris have now been recognized with this special award, which is presented each year at the AAMI Annual Conference and Expo. Some recipients have worked closely with Dr. Morris, and all have worked to improve health conditions through the application of health technology (8).

Other professional relationships include the American Institute for Medical and Biological Engineering (AIMBE), the American Society for Healthcare Engineering (ASHE), ECRI (formerly the Emergency Care Research Institute), and the Institute of Electrical and Electronics Engineers–Engineering in Medicine and Biology Society (IEEE-EMBS).

5. PROFESSIONAL INITIATIVES

5.1. ACCE Visibility in Healthcare

The ACCE's professional partnerships, medical device industry, and healthcare-based initiatives have increased the ACCE's visibility and public awareness of clinical engineering. The organization has been closely involved in medical device security and IHE endeavors. Both initiatives have had a positive impact on the organization and its members and will lead to streamlining medical technologies in the healthcare environment, increasing patient and caregiver safety.

5.2. Medical Device Security Initiatives

The new era of interconnected and computer-based medical devices makes these devices vulnerable to security breaches (9). Healthcare providers and recently clinical engineers are racing against time to ensure the integrity, availability, and confidentiality of information maintained and transmitted by these technologies are not compromised as stated by Grimes (10). ACCE members are involved in several medical device security initiatives and hold positions in workgroups involved in addressing issues related to healthcare securities. In 2005, the ACCE presented a symposium on *Information Security for Medical Technology* at the AAMI Annual Conference and Expo. The ACCE was also involved in the release of the *Information Security for Biomedical Technology: A HIPAA Compliance Guide*, which will be discussed later in this article.

5.3. IHE Initiatives

With the ever increasing complexity and interoperability of medical technologies, the need for seamless flow of data is essential for all healthcare providers. Keeping interoperability in mind, a joint effort of the Radiological Society of North America (RSNA) and the HIMSS started the IHE initiative in 1999. The IHE focuses on a strong relationship between healthcare professionals and medical device industry in improving the exchange of healthcare information (11).

In 2005, the ACCE was appointed as the sponsor of the IHE Domain for Patient Care Devices (PCD). The PCD initiative focuses on implementing standards for effective and efficient communication of patient information throughout the healthcare systems. The PCD undertaking provides the necessary building blocks for the development of Electronic Health Records (EHRs), integrating medical device data (12). In the fall of 2005, the ACCE hosted the formation meeting for the PCD. This brought medical device experts, clinicians, information technology leaders, and clinical engineers to participate in the shaping of the medical device industry. Participants discussed the current state of the medical device industry as related to its integration and infrastructure to help define priorities for subsequent implementation (13).

The ACCE is looking forward to pursuing these efforts and targeting standardization of information flow from medical devices to EHRs. This initiative aims at improving patient safety; assisting clinicians, physicians, and administrators in providing better care; eliminating medical errors; and leading to medical technology interoperability.

6. EDUCATION ACTIVITIES

The ACCE places a special focus on promoting and enhancing the profession of clinical engineering by providing its members with many educational activities. Education is one of the core objectives within the college, meeting the interests and needs of the members and the profession. Many tasks are in collaboration with other professional organizations.

6.1. Teleconference Program

The ACCE provides a comprehensive educational teleconference program on topics that are most pertinent to clinical engineers and healthcare professionals. The program consists of ten teleconferences that are administered once a month for the duration of 1 hour, including a 10–15-minute question-and-answer session.

6.2. Symposia

For the past 8 years, the ACCE has organized a symposium on topics relevant to the clinical engineering community. The symposium is held in conjunction with the AAMI Annual Conference and Expo. The first symposium was entitled *The Future of Clinical Engineering*. The symposium focused on the future of the clinical engineering profession and the resources needed to support its ongoing evolution.

Over the years, the ACCE was instrumental in designing symposiums on topics such as *Clinical Engineering and Information Systems (IS)*, *Medical Telemetry*, *HIPAA*, and most recently *Information Security for Medical Technology*.

6.3. Advanced Clinical Engineering Workshops (ACEWs)

The ACCE engages in national and international education focused on healthcare technologies. The organization's international and education committees, in collaboration with sponsoring international organizations such as PAHO, WHO, International Aid (IA), and ORBIS, have been involved in teaching educational workshops for over 10 years. The purpose of such workshops is to provide clinical engineers and healthcare professionals in other countries with an introduction to the U.S. clinical engineering management systems and their working methods. The workshops are ideal for exchanging information and ideas essential to clinical engineering in the country where the workshops are held. The previous workshops have taken place in countries across the globe, including Brazil, Columbia, Mexico, South Africa, Russia, and more (14).

6.4. Infratech

The infratech listserv was created in 1999 by the WHO and PAHO in the form of an Internet-based discussion group, to exchange information on healthcare infrastructure and technology. The ACCE is the coordinator of this very successful venture. Infratech holds membership from countries across the world. Daily discussion focuses on technical and management-related issues ranging from parts, manuals, repairs to other more complex technology management-related questions, and comments from its members.

7. ACCE HEALTHCARE TECHNOLOGY FOUNDATION

At the end of 2002, the ACCE Healthcare Technology Foundation (AHTF) was inaugurated and registered in the State of Pennsylvania. It is the wisdom of ACCE senior members and their years of professional experience that contributed to the successful establishment of the Foundation. The purpose of the Foundation is as follows: *The Improving Healthcare Delivery by Promoting the Development and Application of Safe and Effective Healthcare Technologies Through the Global Advancement of Clinical Engineering Research, Education, Practice and Other Related Activities* (15).

The Foundation now plays an important role in enhancing ACCE's mission and contributes to building the legacy on which the clinical engineering field of practice is evolving. The AHTF fundraising activities support many successful programs. The programs focus on furthering the profession of clinical engineering and include certification for clinical engineers, public awareness, clinical alarms management and integration, clinical engineering excellence institute, benchmarking on management of medical devices, and patient safety (15).

8. CLINICAL ENGINEERING CERTIFICATION PROGRAM

The ACCE has been actively involved in the development of the Clinical Engineering Certification Program that is now offered under the administration of the Healthcare Technology Certification Commission and the U.S. Board of Examiners for Clinical Engineering Certification. The program is currently sponsored by the ACCE Healthcare Technology Foundation. The first certification examination under this administration was given in November 2003. The ACCE fully endorses the clinical engineering certification program, which is an indication of a practicing clinical engineer's current professional competence. The certification program provides a tremendous sense of personal attainment and is well respected within the healthcare environment.

The body of knowledge for the clinical engineering profession was adopted and defined by the ACCE on the basis of the performance domain on which the certification examinations are structured. Dyro (2) adds that the ACCE has been charged with periodically updating the body of knowledge for the Board of Examiners. The ACCE is updating the existing body of knowledge, which will be reflected in future certification examinations.

9. PUBLICITY AND PUBLICATIONS

The ACCE and its members have been involved in many professional publications on clinical engineering and healthcare technology with some of this work highlighted in this article. The ACCE has issued five informative brochures to help the public learn more about the organization and its activities as follows: *The American College of Clinical Engineering*, *What's a Clinical Engineer*, *Providing Clinical Engineering Support to Countries Around the World*, *A Guide to Clinical Engineering Certification*, and the *Clinical Engineering and Information Technology*. These documents serve as educational and promotional tools to sustain the ACCE and the clinical engineering profession.

In 2001, the ACCE published an important white paper focused on patient safety entitled *Enhancing Patient Safety—The Role of Clinical Engineering*. The paper describes the unique role clinical engineering plays in enhancing patient safety, with a special focus on medical technology as it is applied in the healthcare delivery system (16). The ACCE continues to play an active role in different patient safety initiatives and plans to update the existing white paper with new developments within the clinical engineering profession.

The ACCE's involvement in international activities led to the publication on *Guidelines for Medical Device Donations*, which helps to facilitate donations of medical equipment to third-world countries. The document is also referenced in lectures provided by the ACEW faculty.

The ACCE in collaboration with one of its members and the ECRI, an independent, nonprofit health services research agency, created a CD-ROM, *Information Security for Biomedical Technology: A HIPAA Compliance Guide*. It serves as a tool to healthcare organizations to identify and address security issues. The guide comprises applications and resources on the overview of the HIPAA Security Rule and other safety measures healthcare institutions need to undertake to sustain a secure patient care environment (17).

9.1. ACCE News

In addition to the professional publications, the ACCE has an official newsletter, *ACCE News*, which is published bimonthly. The newsletter contains articles on important developments within the ACCE, success stories in clinical engineering, regular columns, and other practical information. The newsletter's content is valuable to all healthcare technology professionals, consultants, researchers, and profession's advocates.

9.2. ACCE Website

In the summer of 2005, the ACCE launched a new and improved ACCE website, *http://www.ACCEnet.org*. The website presents a highly professional image providing ACCE present and future members with a comprehensive and easy-to-navigate webpage. The website content ranges from ACCE Bylaws, ACCE Mall, Publications, and References, to a restricted area for ACCE Members Only.

9.3. ACCE Membership Survey

In its constant strive for improvement, the ACCE has introduced an online membership survey that allows for a quick and user-friendly submittal of comments and suggestions from ACCE members. A summary of the survey is published every year in the *ACCE News*. ACCE Board and Committee chairs are assigned the task of reviewing the feedback and taking appropriate action as suggested by the ACCE members to sustain continued enhancements to the organization, its members, and its diverse activities.

9.4. Conferences

The ACCE thrives on its resourceful membership and partnerships with other professional organizations. The organization's members are also members of other professional societies and organizations, and such relationships facilitate closer collaborations on healthcare-related issues. The organization has designated members to act as liaisons between the ACCE and other professional organizations. The ACCE has participated in numerous society meetings as a presenter, moderator, organizer, sponsor, and exhibitor.

Acknowledgment

The author would like to acknowledge the professional sources which served as references for the content of this work as well as the different professionals who have been involved in its review.

BIBLIOGRAPHY

1. ACCE, *The American College of Clinical Engineering*. Plymouth Meeting, PA: ACCE, 2005.
2. J. F. Dyro, *Clinical Engineering Handbook*. San Diego: Elsevier Academic Press, 2004.
3. ACCE. (2005). Home page. (Online). ACCE Bylaws. Available: *http://www.accenet.org/default.asp?page=governance&section=bylaws*
4. ACCE Secretariat, *Notification on new Secretariat*. ACCE membership e-mail, 2004.
5. ACCE, *What's a Clinical Engineer*. Plymouth Meeting, PA: ACCE, 2005.
6. FMDIC. (2004). Home page. (Online). FMDIC—FDA Medical Device Industry Coalition. Available: *http://www.fmdic.org/*.
7. ACCE News, US Government Shows Increased Interest in Healthcare IT. 2005; **15**:4.
8. AAMI. (2005). Home page. (Online). AAMI Awards. Available: *http://www.aami.org/awards/index.html*.
9. ACCE, Information Security for Medical Technology, AAMI Annual Conference and Expo, Tampa, FL, 2005.
10. S. Grimes, Security: A new clinical engineering paradigm. *IEEE Eng. Med. Biol. Mag.* 2004; July/August:82.
11. IHE. (2005). Home page. (Online). Available: *http://ihe.net*.
12. ACCE, ACCE-IHE Sponsor Memorandum of Understanding, 2005.
13. ACCE. (2005). Home page. (Online). ACCE Headlines: ACCE and IHE Launch Patient Care Devices Domain. Available: *http://www.accenet.org*.
14. ACCE. (2005). Home page. (Online). ACCE International Committee. Available: *http://www.accenet.org/default.asp?page=about&section=international*.
15. AHTF. (2005). Home page. (Online). ACCE Healthcare Technology Foundation. Available: *http://www.acce-htf.org/*.
16. ACCE. (2001). Home page. (Online). Enhancing Patient Safety—The Role of Clinical Engineering. Available: *http://www.accenet.org/downloads/ACCEPatientSafetyWhitePaper.pdf*.
17. ECRI. (No date). Home page. (Online). Information Security for Biomedical Technology: A HIPAA Compliance Guide. Available: *http://www.ecri.org/Products_and_Services/Products/HIPAA_Compliance_Guide/Default.aspx*.

AMERICAN INSTITUTE FOR MEDICAL AND BIOLOGICAL ENGINEERING

JEN-SHIH LEE
University of California–
San Diego
La Jolla, California
and
University of Virginia
Charlottesville, Virginia

1. INTRODUCTION

The Age of Biotechnology has been upon us for two to three decades (1). The health-care cost, at an increasing trend, has risen to 13% of our total national gross product. Invasive diagnostic methods have been replaced with unprecedented advancements in noninvasive medical imaging. Many medical devices are coming out of the market to significantly improve the quality of life of patients. Detailed mapping of the human genome was completed in 2002, 5 years ahead of the projected date.

Our federal government has made tremendous investments in biomedical research. The budget of the National Institutes of Health (NIH) was doubled over a 5-year period from 1998 to 2003 and reached a level of $28 billion in 2004. Many laws enacted by the Congress have impacted the nation's economy and health-care delivery. To compete more effectively in a health-care environment that is moving toward managed care and cost containment, many medical device companies grew in the 1990s through acquisition and expansion. As an example, growing from a company with sales of steerable catheters, Boston Scientific set out to become the largest corporation in the world devoted to minimally invasive medicine. The ambitious Medtronic sets the goal of manufacturing pacemakers at the rate of one implant every 7 seconds. Promoting the increase in knowledge in medical and biological engineering and its utilization lead to the formation of new professions in engineering, medicine, and biology and the establishment of new professional societies. As these inevitable developments proceed, we must ask ourselves how best we can carry out the advancements as a nation for the benefit of mankind. The public needs to be better informed, because these developments have already impacted the life of people and the economy of the nation.

2. FORMATION OF AIMBE

In a 1988 meeting at the National Science Foundation (NSF), Dr. John White, the Assistant Director for Engineering, stated that the bioengineering community would never have any influence in Washington, D.C. unless it could unify. This comment led to a proposal to the NSF, jointly submitted by the Alliance for Engineering in Medicine and Biology (AEMB) and the U.S. National Committee on Biomechanics (USNCB), in which funding was sought to support a process that would lead to a unification of the community.

This proposal to the NSF was funded, and a joint AEMB/USNCB task force was established in 1989. Co-chaired by Arthur Johnson from AEMB and Robert Nerem from USNCB, this task force conducted a series of meetings and workshops in Washington, D.C. over the next 2 years, with the first workshop being held in August 1989 at Crystal City, Virginia. Out of this process came a recommendation to establish the American Institute for Medical and Biological Engineering (AIMBE), and in October 1991, the AIMBE became an official organization with 501c3 status (2).

3. MISSIONS AND ORGANIZATION OF AIMBE

To unify the medical and biological engineering community to address public policy issues, the AIMBE has the following missions:

- To establish a clear and comprehensive identity for the field of medical and biological engineering
- To promote public awareness of medical and biological engineering
- To establish liaisons with government agencies and other professional groups
- To improve intersociety relations and cooperation within the field of medical and biological engineering
- To serve and promote the national interest in science, engineering, and education
- To recognize individual and group achievements and contributions to the field of medical and biological engineering

The AIMBE was structured to be inclusive for the fulfilling of its missions, without competing with the professional societies involved in bioengineering activities and holding scientific and technical conferences. The structure established included the following four constituents. First, individual membership was to be available for those elected as a Fellow of AIMBE. These AIMBE Fellows were to be represented through the College of Fellows. Second, was the membership of societies being represented through the Council of Societies (COS). Third, was the membership of universities active in bioengineering, and these were organized into the Academic Council. Finally, it was desired to involve industry, and the Industry Council was formed.

The Executive Officers overseeing AIMBE consist of the President, President-Elect, Treasurer-Secretary, and Past-President. These Officers, Executive Director, Second Past-President, four Vice Presidents At-Large, Chair of College of Fellows, Council of Societies, Academic Council and Industry Council, and Chair of three standing committees make up the Board of Directors to govern all activities of AIMBE and to oversee the work of the College of Fellows and the three Councils.

4. ANNUAL MEETING OF AIMBE

To provide a forum on AIMBE's activities and to showcase the development of medical and biological engineering is the Annual Meeting, which was first held in February 1992 in Washington, D.C. This meeting is continuing on an annual basis with a 1-day scientific symposium, sponsored by the College of Fellows, a ceremony to induct the newly elected Fellows, and a 1-day series of business meetings focused on public policy and other issues of interest to AIMBE's constituents. The first two are held at the National Academy of Sciences.

To exemplify the topics covered by the scientific symposium, this author chooses the one organized by Janie Fouke (3) for the 2003 Annual Meeting and entitled "Bioengineering Education in 21st Century." Here is the way she described the three sessions of the symposium:

> The first session was devoted to contemporary topics of emerging importance: nanotechnology and molecular imaging. What are the enabling platform technologies and what advances are on the horizon? As educators and practitioners, we need to learn these so that we can maintain the most modern curriculum. The second session asked industry to keep us on track. Certainly the research and education that is the foundation to an engineering education has to be responsive to the market place. Finally, a session was devoted to pedagogy and the setting for learning. From the integration of research and education and the voice that professional educators raise to advise the engineers, to public scientific literacy in a less formal and more assessable medium such as a museum, the opportunities as we continue to define our discipline are substantial, as are the challenges.

The scientific symposium for 2004 addressed, marvelously, this topic "Imaging and Bioengineering, Partners for the Future." The Program Chair Frank Yin indicated that, "One focus of the symposium will be on new frontiers of imaging and bioengineering. The role of both disciplines and the interactive and symbiotic relations will be highlighted. Another focus will be on best practices for translating discoveries from the laboratory to industry and the clinic" (4). NIH Director Elias Zerhouni was the keynote speaker. He talked about the NIH Roadmap for Medical Research, the importance of multidisciplinary science and technology teams to solve problems in biology and medicine, and the need to effectively translate research to patient applications. He challenged medical and biological engineers to take a leadership role in the development of emerging technologies and the establishment of multidisciplinary collaborations for the advancement of health care (5).

For the business meetings, there were workshops on "Global Issues in Medical Engineering," "Translational Research: Speeding the Journey From Research to Commercialization," "Effective Partnering for Integrative Biomedical Engineering Projects," and "Developing a Consensus in Public Policy Issues Facing Medical and Biological Engineering Societies." The Council and College business meetings provided the opportunity for the participants to deliberate public policy strategies.

5. PRESIDENCIES OF AIMBE

The first president of AIMBE was Robert M. Nerem who served from October 1991 until the Annual Meeting in

1994. During his tenure, a proposal was submitted to The Whitaker Foundation for funds to allow for the establishment of a Washington, D.C. office and the hiring of an Executive Director. With the award from The Whitaker Foundation, a search was initiated, and in 1994 Pierre Galletti, AIMBE's second president, appointed Kevin O'Connor as AIMBE's Executive Director. The current Executive Director is Patricia Ford-Roegner.

Since Robert Nerem and Pierre Galletti, the following persons have served as president:

- Jerome S. Schultz, University of Pittsburgh, 1995–1996
- Winfred M. Phillips, University of Florida, 1996–1997
- Larry V. McIntire, Rice University (now at Georgia Tech), 1997–1998
- William R. Hendee, Medical College of Wisconsin, 1998–1999
- John H. Linehan, Marquette University/the Whitaker Foundation, 1999–2000
- Shu Chien, University of California, San Diego, 2000–2001
- Peer M. Portner, Stanford University, 2001–2002
- Buddy D. Ratner, University of Washington, 2002–2003
- Arthur J. Coury, Genzyme Corporation, 2003–2004
- Don P. Giddens, Georgia Tech, 2004–2005
- Thomas R. Harris, Vanderbilt University, 2005–2006
- Herbert F. Voigt, Boston University, 2006–2007

These people, as well as many others, have provided leadership to AIMBE, all of whom represent literally a "who's who" of bioengineering.

The activities being pursued by the President and Board of Directors of AIMBE are carried out with consensus relevant to the College, Councils, and their memberships. To highlight the actions taking by AIMBE in 2003–2004, President Art Coury sent letters to all U.S. Senators endorsing a bill to limit liability of medical practitioners to reasonable levels. This bill had already been passed in the House of Representatives. With a similar mandate, he wrote on behalf of AIMBE to the House and Senate Appropriations Committees petitioning for a National Institute of Biomedical Imaging and Bioengineering (NIBIB) funding of $350 million for the fiscal year 2004. Members of the AIMBE Board also met with the Directors of NIH and its National Institute of Biomedical Imaging and Bioengineering (NIBIB) to discuss policy and funding strategies (6).

Operationally for the same year, Coury appointed an ad hoc Communications Committee, which addressed shortcomings such as the content of the Annual Meeting and website and information disseminated to new Fellows. A Vice-President level committee was appointed in conjunction with COS Chair to help organize and coordinate several consensus conferences. During the tenure of Coury as the President, the Industry Council had new corporations signing on as members, COS and the Industry Council proposed several initiatives in public policy and education that would benefit from the new support, and the Academic Council provided ongoing service by ranking institutions in medical and biological engineering, addressing standardization of terminology in the field, and considering timely topics such as the effects of the War on Terrorism on international students.

6. COLLEGE OF FELLOWS

The College is established to represent the collective consideration by AIMBE Fellows on all aspects of the field of medical and biological engineering. Fellows are elected for their distinguished contributions in research, industrial practices, and/or education in medical and biological engineering. The Chair of the College leads the committee that plans the scientific symposium at the Annual Meeting.

A total of 202 medical and biological engineers were inducted in 1992 and 1993 as the Founding Fellows of AIMBE. By 2005, the College grew to over 1000 Fellows, which is about $2 \sim 3\%$ of the total number of individuals active in medical and biological engineering. The talent and reputation of these Fellows make them an extraordinary resource for AIMBE. Foreign Fellows make up a small fraction of the total membership of the College. In 2004, Honorary Fellows were established by AIMBE. The first group includes Dr. Earl E. Bakken, founder of Medtronic and humanitarian; Dr. Arnold Beckman, founder of Beckman Instruments and philanthropist; and Senator David Durenberger, congressional leader in health-care legislation and leading strategist dealing with issues important to health care. In 2005, Dr. Norman E. Borlaug, architect of the "Green Revolution," which enhances food production to eliminate hunger in the developing world, and Nobel Prize winner, was inducted as an Honorary Fellow.

7. COUNCIL OF SOCIETIES

AIMBE established its COS to serve the interests of Society Members in AIMBE. These interests include:

- Providing a collaborative forum for the establishment of Society Member positions on issues affecting the field of medical and biological engineering
- Fostering intersociety dialogue, harmony, and cooperation to provide a cohesive public representation for medical and biological engineering
- Effecting a means for coordinating activities of Member Societies with academia, government agencies and research laboratories, the health-care sector, industry, the public and private biomedical communities, and the AIMBE

Societies are eligible for membership in AIMBE if they have substantial and continuing professional interest in the field of medical and biological engineering and in the missions of AIMBE.

COS now represents 19 professional societies with a combined membership of more than 50,000 scientists and engineers in the field of medical and biological engineering. Representatives from the Member Societies run the affairs of COS.

COS surveyed the members of Member Societies in 2002 to identify the national agenda that AIMBE and COS should consider. From the resulting responses of more than 400, COS identified three focus areas for AIMBE actions:

- The development of a national consensus from engineering, scientific, and medical societies to address fundamental policy issues of government in healthcare management and technology
- The creation of a more-enlightened public on the importance of medical and biological engineering
- The enhancement of biomedical research to better treat patients

COS has formed a National Affairs Working Group to develop an implementation plan for consideration by the AIMBE Board. In the meantime, COS is working to open up a National Consensus Conference with participation from all Member Societies, their individual members, and other constituents of AIMBE. Recently, Jason Rivkin joined AIMBE as the Communication Director who will work closely with the Member Societies to develop public consensus on a variety of issues facing the disciplines of medical and biological engineering.

As a Member Society, the Biomedical Engineering Society (BMES) has collaborated with the AIMBE in several areas. Many BMES members serve on the Board of AIMBE and participate in AIMBE affairs. Recognizing the importance of the AIMBE Annual Meeting, the BMES leadership decided to hold the spring meeting of the BMES Board of Directors in conjunction with the AIMBE Annual Meeting. AIMBE has enabled BMES to bring public affairs matters of importance to the BMES membership in a timely fashion. As biomedical engineering and bioengineering continue to expand as career options for our brightest university students, BMES is working closely with AIMBE to educate the public and industry about the singularly important role bioengineers can play in health care.

8. ACADEMIC COUNCIL

The Academic Council is established to address issues of particular concern to academic programs in medical and biological engineering. The Academic Council shall provide a forum for positions taken by academic programs and deliberate on issues such as definitions, standards, and accreditation of academic programs. Academic institutions are eligible for membership in AIMBE if they have substantial and continuing professional interest in the field of medical and biological engineering.

The Academic Council is composed of the representatives of some 85 education programs, biomedical engineering tends to be the numerically dominant discipline, but biological engineering and several others participate. The Academic Council meets twice yearly. Agenda items have concentrated on the following:

- Accreditation issues including help for new programs seeking accreditation
- Curricula
- Programming for educational topics at national meetings
- The relationships of educational programs to NIBIB
- Job placement for bioengineers

The Academic Council offers a place to review these topics and plan for discussions and representations dealing with them. The Academic Council is an excellent example of how the AIMBE acts as a "congress" of units and groups and seeks to focus opinion and effort on topics of overriding mutual interest. The Academic Council has served as the catalyst for building an agenda for enhancing the collaboration among the council members. One example illustrating this catalytic role is the Forum on "Effective Partnering for Integrative Biomedical Engineering Projects" organized by Eugene Eckstein for the 2004 AIMBE Annual Meeting.

9. INDUSTRY COUNCIL

The founding fathers envisioned the AIMBE as an organization, that would play a pivotal role in fostering the understanding and adoption of new biomedical technology for the advancement of health care around the world. It was realized that this could not be accomplished without the full support of the industry community. The path to innovation begins often in the research laboratories but leads ultimately to the commercialization of technology. Without a shared dialogue involving industry, academia, and government, this vision could not be realized. Industrial organizations are eligible for membership in AIMBE if they have substantial and continuing professional interest in the field of medical and biological engineering.

The Industry Council was established within AIMBE to bring the voice of industry into the deliberations of AIMBE's Board of Directors as well as the greater membership of the organization and to engage in dialogue, which ultimately supports and accelerates innovation. AIMBE assists industry in many ways. By electing members of the College of Fellows who are from industry and incorporating them into the membership, leadership, working groups, and events, the AIMBE assures that the diversity of opinions is captured in the dialogue and positions of AIMBE. The AIMBE often plays a role in providing the leadership and impartiality in communicating issues of public policy such that the voice of industry emanates from a sound scientific foundation.

The AIMBE continues to reach out to industry and is strengthening efforts to involve industry in the sponsorship of AIMBE events and in joining with other AIMBE Councils to create a unified voice. As an example, four of the Directors of the 2004 AIMBE Board are from industry. In the 2004 Annual Meeting, a forum was organized by Vincent De Caprio for the AIMBE to engage dialogue and action along the topic of accelerating the translation of research into products and services that will improve health and wellness. It is hoped that the work sponsored by the Industry Council and the AIMBE will lead to streamlined approval processes by the government and greater and more efficient utilization of capital and products of the highest performance and safety.

10. PAST, PRESENT, AND FUTURE OF AIMBE

The AIMBE's achievements in fostering medical and biological engineering include:

- Starting under the leadership of Pierre Galleti, the AIMBE joined with other organizations in advocating what became the Biomaterials Availability Act, signed into law in 1998.
- Working with the Academy of Radiology Research, the AIMBE promoted the idea of forming a new institute at NIH, and in 2000, the NIBIB was established as the 19th Institute of NIH.
- Providing recommendations and support during the formative stages of NIBIB such as the search for the NIBIB Director and the formation of Advisory Committees.
- Participating together with the Academy of Radiology Research and the NIH at a Congressionally mandated, working-group review of grant transfers from established Institutes of NIH to NIBIB.
- Co-sponsoring the successful symposia organized by the Bioengineering Consortium (BECON) of NIH since 1997. The BECON symposium is to foster collaborations and transdisciplinary initiatives among the biological, medical, physical, engineering, and computational sciences.

Also commented by Buddy Ratner at the 12th Annual Meeting (7), the AIMBE is embarking on aggressive strategies to impact policy in the bioengineering aspects of medical devices, clinical therapies, and research. The AIMBE established Strategic Working AIMBE Teams (SWATs) to address issues that confront our field. To educate and inform the public and government policy makers about the value of biomedical technology, a platform in the form of the **AIMBE Hall of Fame** is established to exemplify achievements in medical and biological engineering and highlight their benefits to saving lives and the health and well-being of people.

The AIMBE is an organization of volunteers. Many consider their efforts with the AIMBE to be personally rewarding because many fine people have also chosen to work toward this end. Their attitude of contributing to society in an upbeat, committed manner permeates the AIMBE's activities. The membership of the AIMBE often provides grassroots forums, which bring important societal issues to the local level, where legislators listen most.

The generous funding in its early years provided by The Whitaker Foundation, the NSF, and other donors allows the AIMBE to have an office at Pennsylvania Avenue in Washington, D.C. with two full-time staff members. With much more to be accomplished, the AIMBE definitely needs more resources to support the volunteers in carrying out the AIMBE's public advocacy activities, to enhance the communication between AIMBE and the 50,000 medical and biological engineers, and to involve them in public advocacy for and public education in medical and biological engineering. In 2002, the Board of Directors voted for a capital campaign to raise $1 million as the endowment base for expanding the operation and support of the AIMBE. With generous donations from The Wallace H. Coulter Foundation, Biomet Corporation, and several generous individuals including Earl E. Bakken, Shu Chien, Jen-shih Lee, Dane Miller, and the Board of Directors, a dedicated Development Committee achieved about half of the campaign's goal by 2003 and the campaign was officially kicked off that year.

As the President of AIMBE for 2004–2005, Don Giddens worked with the Board and staff to reengineer AIMBE's headquarters, to develop a strategic plan for presentation to the membership, and to pursue the capital campaign **"Fund for Excellence."** Subsequently Giddens announced in the 2005 Annual Meeting that The Wallace H. Coulter Foundation has just made a $1 million challenge grant to the Fund for Excellence. The Coulter Foundation has committed to promoting biomedical engineering with an emphasis on the impact of the field on health care of patients. Their vision is that unless biomedical research is translated into commercial products, this impact is limited. Thus, The Coulter Foundation made this challenge grant, which is a one-for-one match to other gifts raised by AIMBE, to assist the AIMBE and its constituents in reducing barriers to innovation, in advancing translational research, in assuring a smooth transition between university research and industrial adaptation, and in continuing to attract young people into biomedical engineering.

Under the presidencies of Tom Harris and Herb Voigt in coming years, the AIMBE will continue its "reengineering" effort with key activities to include the following:

- **Annual Meeting:** The program of the Annual Meeting will be expanded to achieve active participation by all constituencies of the AIMBE and to renew its emphasis on issues important to public policy.
- **Finance:** The Development Committee is working to generate revenue for the AIMBE to pursue its ambitious program of advocacy for medical and biological engineering.
- **Council of Societies:** The Council of Societies works to bring the multiplicative effect of the Member Societies to bear on public policy in medical and biological engineering. A major initiative of this Council will be the development of a consensus research agenda that is important to the growth of medical and biological engineering, the health of people, and the economy of the nation.
- **Academic Council:** It has been invited to seek means by which the great achievements listed in the AIMBE Hall of Fame might be found useful in university's curricula and be employed by students to develop case studies in medical and biological engineering. Under consideration by the Council is a program session at the Annual Meeting to display or communicate university-based innovations in

medical and biological engineering that are moving toward commercialization.
- **Industry Council:** This Council is seeking new industrial memberships to significantly enhance industry–academic–society interchange within the AIMBE and beyond. This Council has taken the lead to develop forums and venues examining barriers to innovation and seeking ways to enhance innovation.

In conclusion, the author echoes with Art Coury the firm belief that "AIMBE is an organization with staggering potential, which is exploiting only a fraction of its capability. Even so, its achievements and reputation are substantial; ever growing and well justify the vision of its Founding Fellows. The ongoing challenge to AIMBE is to activate and energize the unused potential of the membership of its constituents to more fully achieve AIMBE's missions of benefiting our society and nation by influencing public policy in medical and biological engineering."

Acknowledgments

The author would like to thank (in alphabetical order) Drs. Michael Ackerman (Chair of COS, 2003–2005), Kryiacos Athanasiou (President of BMES, 2004–2005), Art Coury (AIMBE President, 2003–2004), Vincent De Caprio (Chair of Industry Council, 2004–2006), Eugene Eckstein (Chair of Academic Council, 2003–2004), Don Giddens (President, 2004–2005), Thomas Harris (President, 2005–2006 and Chair of Academic Council, 1998–1999), Robert Nerem (Founding President), and Herbert Voigt (President, 2006–2007) for their input to this article.

BIBLIOGRAPHY

1. J. Rifkin, *The Biotech Century*. New York: Tarcher/Putnam, 1998.
2. AIMBE website. Available: *http://www.aimbe.org*.
3. J. Fouke, Welcome from the Program Chair. *Proc. 2003 AIMBE Annu. Meeting.* p. 2.
4. F. Yin, Welcome from the Program Chair. *Proc. 2004 AIMBE Annu. Meeting.* p. 2.
5. E. Zerhouni, The NIH Roadmap. *Science.* 2003; **302**:63–64.
6. R. I. Pettigrew, Senate 2005 NIBIB Budget Testimony. Available: *http://www.nibib.nih.gov*.
7. B. Ratner, President's Message. *Proc. 2003 AIMBE Annu. Meeting.* p. 1.

ANALOG TO DIGITAL CONVERSION

LARRY D. PAARMANN
Wichita State University
Wichita, Kansas

The process of converting an analog, or continuous time, signal into a digital signal is known as analog-to-digital conversion (ADC). Many signals are inherently analog. Examples of analog signals obtained from a transducer abound, such as those obtained from a microphone, a pressure transducer, and a temperature transducer. Other analog signals are inherently electrical, such as those obtained from biomedical electromyography (EMG) or electroencephalography (EEG) electrodes.

In many cases, it is deemed desirable to process a signal by a microprocessor or computer to extract information, or to store the signal in some digital storage medium such as a hard drive, a floppy disk, or a CD ROM. Also, some display devices may require the input to be in digital signal format. If the original signal is in analog format, in such cases, it is necessary to convert the analog signal into an equivalent digital signal. This article describes the process of ADC.

Not all biomedical research, of course, involves ADC. But it frequently does. When ADC is involved, sometimes researchers report the analog-to-digital methodology in detail, whereas sometimes detail is omitted but some method of performing the conversion is implied. For example, a recent issue of the *IEEE Transactions on Biomedical Engineering*, selected at random, gives examples of where researchers have either explicitly given details on the analog-to-digital system that they used in obtaining data for the research reported in the paper or have at least mentioned that ADC was used (1–7). Also, a recent issue of the *Annals of Biomedical Engineering*, selected at random, gives similar examples (8–10). These examples include the analysis of EEG and EMG signals, voice impairment detection, and several cardiac studies.

1. INTRODUCTION

A digital computer can store or process only discrete numbers. In addition, because those numbers are stored or manipulated in digital storage registers internal to the computer, those numbers can only have a certain degree of precision, which is described in terms of the maximum number of binary bits. Therefore, no matter how accurate and precise an ADC is, the analog signal must be represented as a sequence of numbers (samples) and those numbers inherently must have finite precision to be compatible with a digital microprocessor or computer. Hence, the analog signal, independent variable must be sampled only at discrete points to have a sequence of numbers, and the analog signal dependent variable must be quantized to be compatible with the inherently finite precision registers of the microprocessor or computer. Both of these issues, sampling and quantization, introduce error in the digital signal compared with the original analog signal. Under certain conditions that will be discussed in this article, sampling of the independent variable can be accomplished such that no information is lost. Quantization, on the other hand, introduces unrecoverable loss of information. This loss of information, known as quantization error, is reduced as the number of bits in an ADC is increased. This result leads to two fundamental areas of advancement in ADC technology: increasing the maximum sampling rate and increasing the number of resolution bits. A third related area of advancement is hardware implementation to realize higher sampling rates and larger numbers of bits.

In the next section, basic types of ADCs are briefly reviewed. Two types, the successive approximation converter and the flash converter, are presented with a little more detail. A third type, because of its uniqueness, is presented in a later section.

Also, the sampling theorem is presented in this article. Although the sampling theorem indicates that, under certain conditions, an analog signal is fully recoverable from discrete-time, unquantized, samples of it, it does not indicate how to accomplish the feat. And in practice, the samples are also quantized, and therefore, the quantization noise prevents full recovery of the original analog signal. Nonetheless, in practice, low levels of error are achievable.

Furthermore, aliasing is described. And quantization noise is briefly described. The advantages of oversampling, which is sampling at a rate higher than the Nyquist rate specified by the sampling theorem, are presented. The delta-sigma ADCs are also described. These converters are more readily integrated as they do not require precision components, and they can obtain the higher number of bits required for increased precision. Finally, some comments are offered to assist the biomedical engineer in selecting an ADC method and some basic parameters.

2. ADC TYPES

ADCs are implemented in a variety of ways. Two well-known techniques are the successive approximation ADC and the flash ADC. They will be considered here. Delta-sigma ADCs will be considered later. Other converter types include ramp converters, pipelined converters, charge redistribution converters, and so on (11–14).

2.1. Successive Approximation ADC

The successive approximation ADC is illustrated in Fig. 1. In the figure, $x_a(t)$ is the input analog signal. In many practical applications, the input analog signal may be filtered by a low-pass filter, which reduces the signal bandwidth and thereby lowers the required Nyquist rate and prevents aliasing. The design of the analog low-pass filter is dependent on the desired passband specifications, basic filter type, and transition band and stopband requirements (15). The output of the analog low-pass anti-aliasing filter in Fig. 1 is denoted as $x(t)$, and its bandwidth is now, because of the low-pass filter, known. For example, analog speech signals may well contain significant spectral energy to 10 kHz and beyond, yet it is known that most of the intelligence in conversational speech is contained below 3 or 4 kHz, and therefore, an analog low-pass anti-aliasing filter can significantly reduce the required Nyquist rate.

As can be seen in Fig. 1, the low-pass filter output signal to be digitized, $x(t)$, is applied as the input to a sample-and-hold circuit. The sample-and-hold circuit, in addition to that shown in the figure, has a clock input at the sample rate. That is, if $x(t)$ is to be sampled every T seconds, the clock input to the sample-and-hold circuit will have a frequency of $1/T$ Hz. At the time instance $t = nT$, the output of the sample-and-hold circuit will be $x[n] = x(nT)$, and it will hold that value constant until $t = (n+1)T$; at which time, it changes to $x[n+1] = x((n+1)T)$. The output of the sample-and-hold circuit is discrete time, but x is still continuous; that is, it has not yet been quantized to discrete amplitude levels.

The output of the sample-and-hold circuit is one of the two inputs to the comparator circuit. The comparator circuit, in addition to the two inputs shown in Fig. 1, has a clock input at a higher rate than that used by the sample-and-hold circuit. The clock used by the comparator is also used by the rest of the circuitry shown in Fig. 1: It may be at the frequency of the system clock, but it must be, at a minimum, at a frequency of $(N+1)/T$ Hz, where N is the number of conversion bits of the ADC.

Any ADC is designed for a certain range of amplitude values of $x[n]$, such as 0 to 5 V, or -1 V to $+1$ V. In general, this range may be expressed as V_{R_1} to V_{R_2}. In Fig. 1, the digital-to-analog converter (DAC) is set up so that when the N-bit sample output is midrange between its minimum and maximum digital values, the analog output of the DAC will be midrange between V_{R_1} and V_{R_2}. When the ADC begins a new conversion cycle, it begins with the N-bit sample output midrange between its minimum and maximum digital values, and the current sampled/held value of $x[n]$ is compared with the midrange value at the output of the DAC. If $x[n]$ is the larger of the two, the output of the comparator will be a logical one; if it is the smaller, the output of the comparator will be a logical zero. If the output of the comparator is a logical one, the digital

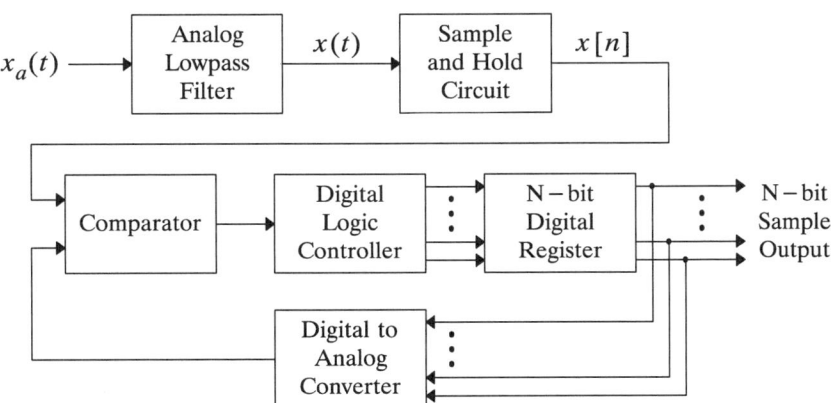

Figure 1. Block diagram of the successive approximation ADC.

logic controller will adjust the N-bit digital register so that the N-bit sample output is moved up halfway between its starting value and the maximum N-bit digital number; if the output of the comparator is a logical zero, the N-bit register will move down halfway between its starting value and the minimum N-bit digital number.

The change in the N-bit sample output will result in a change in the output of the DAC. Then a change in the output of the comparator may or may not result: The N-bit sample output moves up, or moves down, but this time about an amount only half as much as the first move. This process continues until the output of the DAC is as close to the amplitude value of $x[n]$ as the number of bits of the ADC, N, will allow. This process may take N clock periods. After N clock periods and the output of the DAC has converged as close as possible to the value of $x[n]$, the ADC flags the processor that an N-bit sample is ready to be read. After the sample has been read, the ADC resets to its initial value and waits for a new held value input from the sample and hold circuit.

2.2. Flash ADC

The flash ADC is illustrated in Fig. 2. In the figure, $x_a(t)$ is the analog input signal. As with the successive approximation ADC, a low-pass analog filter is included as an antialiasing filter. The low-pass filter output signal to be digitized, $x(t)$, is applied as the input to a sample-and-hold circuit. The sample-and-hold circuit, in addition to that shown in the figure, has a clock input at the sample rate. That is, if $x(t)$ is to be sampled every T seconds, the clock input to the sample-and-hold circuit will have a frequency of $1/T$ Hz. At a sample point, all $2^N - 1$ comparators simultaneously produce outputs, comparing the input analog signal amplitude value with a string of reference values derived from V_{R_1} and V_{R_2}, the minimum and maximum values that the ADC is designed to accept on the input. Depending on which comparator outputs are logical ones and which outputs are logical zeros, the digital logic controller determines the N-bit sample output. Note that this process requires only one system clock cycle. For this reason, a flash ADC is generally used where high sample rates are required, such as in image processing.

The comparator threshold voltage values, V_{T_1} through $V_{T_{2^N-1}}$, are usually obtained by a string of resistors across the reference voltages V_{R_1} and V_{R_2}. For example, if the input signal voltage sample is below V_{T_1}, the N-bit sample output would be all zeros. If the input signal voltage sample is above V_{T_1} but below V_{T_2}, only the least significant bit of the N-bit sample output would be a one. If the input signal voltage sample is above $V_{T_{2^N-1}}$, then the N-bit sample output would be all ones.

Because the number of comparators required is $2^N - 1$, a flash converter can only be reasonably used when a small number of bits is adequate. For example, a 5-bit ADC would require 31 comparators.

3. THE SAMPLING THEOREM

The sampling theorem, according to Shannon (16), establishes that, under certain conditions, there is no loss of information in samples of an analog signal compared with the analog signal itself. The theorem may be stated as follows. Given an analog signal, $x(t)$, that is bandlimited such that

$$X(j\Omega) \equiv 0 \qquad \forall |\Omega| \geq \Omega_{max}, \qquad (1)$$

and samples of $x(t)$, denoted as $x[n]$, are obtained as shown by the sample and hold circuits in Figs. 1 and 2, then $x(t)$ may be recovered from $x[n]$ as follows:

$$x(t) = \sum_{n=-\infty}^{\infty} x[n] \frac{\sin(\pi(t-nT)/T)}{\pi(t-nT)/T}, \qquad (2)$$

if and only if $T \leq \pi/\Omega_{max}$. $\square$

The minimum sampling rate for reconstruction from the samples is denoted the Nyquist rate (17)

$$f_{NR} = \Omega_{max}/\pi,$$

and the maximum sampling period is as follows:

$$T_{NR} = \pi/\Omega_{max}. \qquad (3)$$

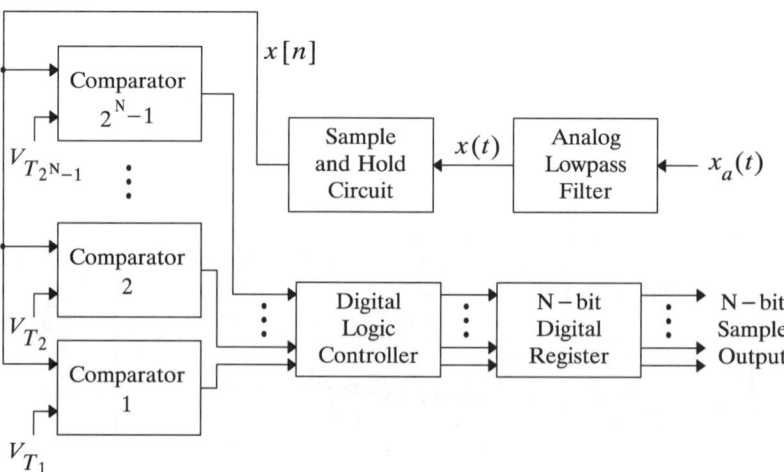

Figure 2. Block diagram of the flash ADC.

Note that $x(t)$ is expressed in Equation 2 as an expansion of the form

$$x(t) = \sum_{n=-\infty}^{\infty} a_n \varphi_n(t),$$

where $\{\varphi_n(t)\}$ are the basis functions and $\{a_n\}$ are the expansion coefficients. Various basis functions could be used, but those used here

$$\frac{\sin(\pi(t-nT)/T)}{\pi(t-nT)/T}$$

are known as Cardinal functions. The beauty of the representation in Equation 2 is that the expansion coefficients are simply the time-domain samples of the signal $x(t)$.

The sampling theorem may be proved as follows. Assuming that conditions are such that the Fourier transform of $x(t)$ exists and that the signal is bandlimited as expressed in Equation 1, $x(t)$ may be expressed as follows:

$$x(t) = \frac{1}{2\pi} \int_{-\infty}^{\infty} X(j\Omega) e^{j\Omega t} \, d\Omega$$

$$= \frac{1}{2\pi} \int_{-\Omega_{max}}^{\Omega_{max}} X(j\Omega) e^{j\Omega t} \, d\Omega.$$

But because $T \leq \pi/\Omega_{max}$ it follows that $\Omega_{max} \leq \pi/T$. Therefore, $x(t)$ may be expressed as follows:

$$x(t) = \frac{1}{2\pi} \int_{-\pi/T}^{\pi/T} X(j\Omega) e^{j\Omega t} \, d\Omega. \quad (4)$$

Let $X_p(j\Omega)$ be defined as the periodic extension of $X(j\Omega)$ with period $2\pi/T$

$$X_p(j(\Omega + 2\pi k/T)) = X(j\Omega) \quad \forall k,$$

where k is an integer. Because $X_p(j\Omega)$ is periodic, it may be expressed by a Fourier series as follows:

$$X_p(j\Omega) = \sum_{k=-\infty}^{\infty} \alpha_k e^{jkT\Omega}, \quad (5)$$

where the Fourier series coefficients may be computed as follows:

$$\alpha_k = \frac{T}{2\pi} \int_{-\pi/T}^{\pi/T} X_p(j\Omega) e^{-jkT\Omega} \, d\Omega.$$

However, because $X_p(j\Omega) = X(j\Omega)$ for $-\pi/T \leq \Omega \leq \pi/T$, it follows that

$$\alpha_k = \frac{1}{2\pi} \int_{-\pi/T}^{\pi/T} X(j\Omega) e^{-jkT\Omega} \, d\Omega. \quad (6)$$

Letting $t = -kT$ in Equation 4

$$x(-kT) = \frac{1}{2\pi} \int_{-\pi/T}^{\pi/T} X(j\Omega) e^{-jkT\Omega} \, d\Omega. \quad (7)$$

Comparing Equations 6 and 7, note that $\alpha_k = Tx(-kT)$. Substituting $\alpha_k = Tx(-kT)$ into Equation 5:

$$X_p(j\Omega) = \sum_{k=-\infty}^{\infty} Tx(-kT) e^{jkT\Omega}. \quad (8)$$

Finally, recalling that $X_p(j\Omega) = X(j\Omega)$ for $-\pi/T \leq \Omega \leq \pi/T$, substituting Equations 8 into 4 and simplifying results in Equation 2.

4. ALIASING

Aliasing occurs if the sampling rate is less than the Nyquist rate. Theoretical analysis of aliasing is facilitated if the concept of ideal sampling is introduced. Let $x(t)$ be the bandlimited input analog signal that is to be sampled, and let this signal be multiplied by a periodic train of impulses spaced T seconds apart

$$x_s(t) = x(t) s(t),$$

where

$$s(t) = \sum_{n=-\infty}^{\infty} \delta(t - nT).$$

Therefore,

$$\begin{aligned} x_s(t) &= x(t) \sum_{n=-\infty}^{\infty} \delta(t - nT) \\ &= \sum_{n=-\infty}^{\infty} x(nT) \delta(t - nT) \end{aligned} \quad (9)$$

Taking the continuous-time Fourier transform of Equation 9 results in the following (17):

$$X_s(j\Omega) = \frac{1}{T} \sum_{k=-\infty}^{\infty} X(j(\Omega - k\Omega_s)), \quad (10)$$

where $\Omega_s = 2\pi/T$ is the sampling frequency in radians per second. If the input signal is bandlimited as specified in Equation 1 and T is less than T_{NR}, as specified in Equation 3, then it follows that

$$X_s(j\Omega) = \frac{1}{T} X(j\Omega) \quad \forall |\Omega| < \Omega_{max},$$

and therefore, $x(t)$ is recoverable from $x_s(t)$ by low-pass filtering. Clearly, if the sampling rate does not satisfy the Nyquist rate, then aliasing will occur and $x(t)$ will not be fully recoverable from $x_s(t)$. In fact, it is not difficult to show that a frequency component at $\Omega_1 > \Omega_{max}$ (see Equation 10) will appear as a frequency component at $\Omega_1 - \Omega_s/2$.

5. QUANTIZATION NOISE

One potential source of error, namely aliasing, was previously discussed. Another source of error is quantization error, or quantization noise as it is commonly called. No matter what type of ADC is used, and no matter how precisely implemented the ADC may be, quantization error is inherent in the ADC process. The input analog signal has a continuously variable amplitude, whereas the digital output of the ADC has only discrete amplitude values. Properly implemented, the ADC digital output sample value, or rather its analog equivalent, will be closer to the analog input value at that sample time than any of the other possible digital sample values available, but it will not be without error.

Therefore, if $x_q[n]$ is denoted as the quantized output sample value, then it may be represented as follows:

$$x_q[n] = x[n] + q[n],$$

where $x[n]$ is the input sample value (discrete time but x is still a continuous variable, and $x[n]$ is the output of the sample-and-hold circuit), and $q[n]$ is the quantization error introduced by the quantizer. Assuming that the input sample value $x[n]$ is somewhere between the values that the ADC is designed for, i.e., between V_{R_1} and V_{R_2}, the error $q[n]$ must be somewhere between $\pm \Delta/2$, where Δ is the step size of the quantizer

$$\Delta = \frac{V_{R_2} - V_{R_1}}{2^N}, \tag{11}$$

where N is the number of bits in the quantizer. Assuming that the analog input signal is sufficiently busy, e.g., not a constant, it is reasonable to assume, as it usually is, that the quantization error will be uncorrelated with the input signal and uniformly distributed between $\pm\Delta/2$. Therefore, the variance of the quantization error may be computed as follows:

$$\sigma_q^2 = \frac{1}{\Delta} \int_{-\Delta/2}^{\Delta/2} q^2 \, dq = \frac{\Delta^2}{12} = \frac{(V_{R_2} - V_{R_1})^2}{12 \times 2^{2N}}. \tag{12}$$

Because it is also usually assumed that the quantization error is uncorrelated with itself for a sufficiently busy input signal, which also is a reasonable assumption, it follows that the power spectral density (PSD) of the quantization error may be expressed as follows:

$$S_q(e^{j\omega}) = \sigma_q^2 = \frac{\Delta^2}{12}; \tag{13}$$

that is, the PSD is flat.

5.1. Example 1

If it is assumed that the input signal $x[n]$ is uniformly distributed between V_{R_1} and V_{R_2} and $V_{R_2} = -V_{R_1}$, which is not usually the case but serves as an example here, then similar to the above,

$$\sigma_x^2 = \frac{(V_{R_2} - V_{R_1})^2}{12} = \frac{(2^N \Delta)^2}{12} = \frac{2^{2N}}{12} \Delta^2.$$

The signal-to-noise ratio (SNR) for the above example would be as follows:

$$\mathrm{SNR} = 2^{2N},$$

or in terms of decibels:

$$\mathrm{SNR}_{\mathrm{dB}} = 10 \log_{10}(\mathrm{SNR}) = 20N \log_{10}(2)$$
$$\cong 6.02 N.$$

It is noted here that the $\mathrm{SNR}_{\mathrm{dB}}$ increases 6.02 dB for every 1-bit increase in the quantizer. But also, with the above assumptions, the actual $\mathrm{SNR}_{\mathrm{dB}}$ may be readily calculated. For example, an 8-bit quantizer will have $\mathrm{SNR}_{\mathrm{dB}} = 48$ dB, and a 16-bit quantizer will have $\mathrm{SNR}_{\mathrm{dB}} = 96$ dB. □

6. OVERSAMPLING

Oversampling is the term that indicates that the sample rate exceeds the Nyquist rate. The oversampling ratio, denoted O_R, is defined as follows:

$$O_R = f_s / f_{NR};$$

that is, the oversampling ratio is the actual sample rate divided by the Nyquist rate. If the output of the ADC, no matter whether the ADC is a successive approximation ADC, a flash ADC, or any other type of ADC, denoted as $x_q[n]$, is the input to a digital low-pass filter, which is assumed to be ideal here for convenience, with a normalized cutoff frequency of $\omega_c = \pi/O_R$, the effective cutoff frequency of the low-pass filter in terms of continuous-time frequencies would be

$$\Omega_c = \frac{\pi}{O_R} \times \frac{1}{\pi} \times \frac{2\pi f_s}{2} = \pi f_{NR} = \Omega_{max}.$$

Because $x_q[n] = x[n] + q[n]$, it follows that the $x[n]$ component of $x_q[n]$ will pass through the digital low-pass filter unaffected because $\Omega_c = \Omega_{max}$. However, the quantization noise $q[n]$ will be affected. As noted, the power spectral density of the quantization error may be expressed as given in Equation 13. Therefore, the output of the digital low-pass filter because of the $q[n]$ component of $x_q[n]$ on the input, denoted as $q_{lp}[n]$, will have the following PSD:

$$S_{q_{lp}}(e^{j\omega}) = \begin{cases} \sigma_q^2, & |\omega| < \pi/O_R \\ 0, & |\omega| > \pi/O_R \end{cases}$$
$$= \begin{cases} \frac{\Delta^2}{12}, & |\omega| < \pi/O_R \\ 0, & |\omega| > \pi/O_R \end{cases}.$$

Therefore, the power in the quantization error at the output of the digital low-pass filter will be

$$P_{q_{lp}} = \frac{\sigma_q^2}{O_R} = \frac{\Delta^2}{12 O_R}. \tag{14}$$

As noted, the $x[n]$ component of $x_q[n]$ will pass through the digital low-pass filter unaffected. Therefore, the signal power at the output of the digital low-pass filter will not be a function of the oversampling ratio. The output of the digital low-pass filter may be denoted as follows:

$$x_{lp}[n] = x[n] + q_{lp}[n].$$

Because the quantization power is a function of the oversampling ratio, the signal to quantization noise ratio, SNR, will be a function of the oversampling ratio. Every time O_R is doubled, the SNR will improve by 3 dB. Or, every time O_R is increased by a factor of 10, the SNR will improve by 10 dB. This improvement is modest, but it does improve.

6.1. Example 2

Suppose a 16-bit ADC is designed for a zero-mean input bounded by ± 5 V. Suppose that the input is uniformly distributed such that the average signal power, on a 1-Ω

basis, is 6.25 W. The ADC bin size is $\Delta = 152.6\,\mu\text{V}$. Therefore, when sampling at the Nyquist rate, the quantization noise power would be 1.94 nW. The SNR would be 3.22×10^9, or 95 dB. If the sampling rate was increased such that $O_R = 2$, then SNR = 98 dB. If the sampling rate was increased such that $O_R = 4$, then SNR = 101 dB. If the sampling rate was increased such that $O_R = 10$, then SNR = 105 dB. If the sampling rate was increased such that $O_R = 64$, then SNR = 113 dB. □

The bin size for a conventional ADC is given by Equation 11 and is inversely related to the number of bits. The variance of the quantization error is given by Equation 12. As the number of bits is increased, the quantization noise power goes down, and therefore, the SNR goes up. The quantization noise power for an ADC that includes oversampling followed by digital low-pass filtering and often downsampling is given by Equation 14. Substituting Equation 11 into Equation 14,

$$P_{q_{lp}} = \frac{\sigma_q^2}{O_R} = \frac{(V_{R_2} - V_{R_1})^2}{12 \times 2^{2N} O_R}. \quad (15)$$

It is easy to equate Equation 15 to an equivalent ADC without oversampling to obtain the equivalent number of bits. Every time O_R doubles, the number of equivalent bits increases by 0.5. Or, every time O_R increases by a factor of 16, the number of bits increases by 2 bits.

6.2. Example 3

Consider the scenario given in Example 2. The quantization noise power for the conventional Nyquist sampled ADC is 1.94 nW, and the corresponding SNR is 3.22×10^9, or 95 dB. If the sampling rate was increased such that $O_R = 2$, then SNR = 98 dB. It implies that the equivalent noise power would be 990.6 pW. From Equation 15, it is implied that the equivalent number of bits would be 16.5. If the sampling rate was increased such that $O_R = 4$, then SNR = 10 dB, which implies an equivalent number bits equal to 17. If the sampling rate was increased such that $O_R = 10$, then SNR = 105 dB and the equivalent numbers of bits would be 17.7. If the sampling rate was increased such that $O_R = 64$, then SNR = 113 dB and the equivalent number of bits would be 19. □

7. DELTA-SIGMA ADC CONVERTERS

Delta-sigma ADC converters accomplish two important and perhaps initially seemingly contradictory things. First, they are better suited for integrated circuit (IC) technology. Whether it is desired to implement a stand-alone, general-purpose ADC, or whether it is desired to include an ADC in a mixed-signals, application-specific, embedded system IC, delta-sigma ADCs have desirable implementation characteristics. Most importantly, they do not require precision components. The threshold voltages that are required for successive approximation ADCs. flash ADCs (see Figs. 1 and 2 and accompanying text), and other common ADC types are usually obtained by having a string of precision resistors connected across the reference voltages V_{R_1} and V_{R_2}. Precision resistors are difficult components to realize in current IC technology, and therefore, they are usually external components to the IC. This result is undesirable because of the expense and because of the physical space requirements. Therefore, delta-sigma ADCs are attractive for many commercially available ADC ICs. They are also attractive for embedded system ICs, such as those found in many cell phones, and other highly integrated systems. Secondly, where high-precision (many bits) ADC is required, without special requirements as to a high level of integration, such as in ADCs used in professional audio recording studios, delta-sigma ADCs have advantages.

Delta-sigma conversion has the above advantages because of the manner in which it achieves a high SNR. It does so by the use of oversampling and signal processing to shape the quantization noise spectrum such that most quantization noise power is pushed to the high frequencies located beyond the signal band. The resultant signal plus noise may then be low-pass filtered to remove most quantization noise. This processing is so effective that often the basic quantizer can be simply a binary decision (a threshold detector) and yet achieve the equivalent of, for example, 24 bits of precision.

The cost for the performance obtainable from a delta-sigma ADC is a very high O_R, a complex circuit design to achieve the desired performance, and stability issues caused by the necessary internal feedback loops. Delta-sigma ADC development is ongoing, but significant accomplishments have been achieved with many commercial applications in place. High-speed IC development, as basic device dimensions in IC technology continue to decrease, has fueled development of delta-sigma ADCs.

In the presentation below, conventional delta-sigma ADCs of first and second order will be covered first. Then, some brief information will be given of more advanced delta-sigma architectures (18,19).

7.1. First-Order Delta-Sigma ADC

The basic block diagram of a first-order delta-sigma ADC is shown in Fig. 3. Note that the input signal $x[n]$, as is the case for the ADCs shown in Figs. 1 and 2, is a discrete-time signal, but the amplitude has not been quantized; the amplitude is continuous. The signal $x[n]$ is obtained from the analog input $x_a(t)$ by first filtering with an analog low-pass filter and then processing with a sample-and-hold circuit, again as shown in Figs. 1 and 2. As shown in Fig. 3, the input to the integrator is the difference between the input sample $x[n]$ and the output of the DAC. The DAC need not include any smoothing, but need only simply to convert the digital output of the quantizer to its equivalent analog level so that subtraction from $x[n]$ is feasible. Therefore, the input to the integrator, just like $x[n]$, is discrete time but with a continuous amplitude. Note that this signal is discrete time in the sense that amplitude changes only occur at discrete times, but it is continuous time in the sense that it is defined for all time. The output of the integrator (an analog circuit), however, because of the nature of integration, will not be discrete time. Therefore, the quantizer, or ADC, shown at the output of the integrator converts the output signal from the integrator

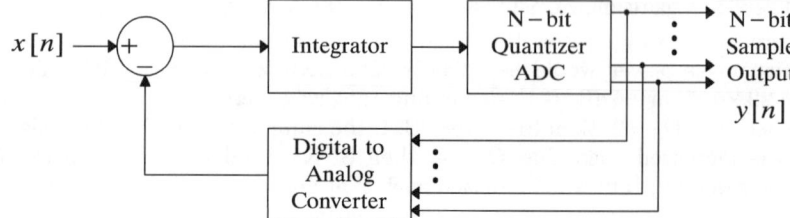

Figure 3. Block diagram of the first-order delta-sigma ADC.

into a digital signal. The ADC must discretize both time and amplitude. Therefore, the ADC must include a sample-and-hold circuit as well as the quantizer. It must also include a clock running at the same frequency as the input samples $x[n]$. For simplicity, the block is simply shown as an N-bit quantizer ADC. Note that this circuit is inherently a mixed-signals circuit: It includes both analog and digital circuitry. At first glance, because the circuit includes an ADC within it, it may seem to be unnecessarily complex. But the resultant SNR far exceeds what can be accomplished by the internal ADC. So much so, as mentioned before, that the internal ADC may be a simple binary device, at each clock cycle simply indicating whether the output of the integrator is positive (a logical one) or negative (a logical zero). Note that, as mentioned earlier for oversampling, the circuit may be followed by low-pass filtering and downsampling.

Note that the output of the integrator need only be known at the discrete-time values. Therefore, it significantly simplifies analysis if a discrete-time equivalent to Fig. 3 is used. Also, for simplicity, it is very helpful to linearize the internal ADC; without this step, the entire circuit is nonlinear. Therefore, for analysis, the discrete-time and linearized circuit in Fig. 4 is preferred. Note in the figure that a one-time-step delay is assumed for the DAC, and that the integrator is represented in discrete-time form with a transfer function of $1/(1-z^{-1})$.

From Fig. 4, using superposition for the linearized circuit, the signal transfer function (STF) can easily be shown to be

$$\text{STF} = \frac{Y_x(z)}{X(z)} = 1, \quad (16)$$

where $y_x[n]$ is the output from the input $x[n]$ only, and the noise transfer (NTF) is

$$\text{NTF} = \frac{Y_q(z)}{Q(z)} = 1 - z^{-1}, \quad (17)$$

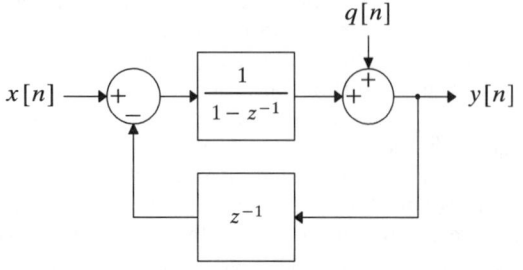

Figure 4. Discrete-time equivalent first-order delta-sigma ADC.

where $y_q[n]$ is the output from the quantization noise $q[n]$ only. From Equations 16 and 17, it follows that

$$y[n] = y_x[n] + y_q[n] = x[n] + q_1[n]$$
$$= x[n] + q[n] - q[n-1],$$

where $q_1[n] = y_q[n] = q[n] - q[n-1]$ is the quantization noise of the overall delta-sigma quantizer. From Equation 17,

$$Q_1(z) = (1-z^{-1})Q(z). \quad (18)$$

From Equation 18, the PSD of the quantization noise may be expressed as follows:

$$S_{q_1}(e^{j\omega}) = |1-e^{-j\omega}|^2 S_q(e^{j\omega})$$
$$= |1-e^{-j\omega}|^2 \sigma_q^2 = 4\sigma_q^2 \sin^2(\omega/2). \quad (19)$$

From Equation 19, the noise shaping that the first-order delta-sigma quantizer has accomplished is apparent. At low frequencies ($\omega < \pi/3$), $S_{q_1} < \sigma_q^2$. However, at $\omega = \pi$, $S_{q_1} = 4\sigma_q^2$. Therefore, the noise power has been shaped such that the most power is at higher frequencies.

On the normalized radian frequency scale, the maximum signal frequency is $\omega_{max} = \pi/O_R$, and it follows that the noise power in the signal band may be computed as

$$\sigma_{q_1}^2 = \frac{1}{\pi}\int_0^{\omega_{max}} S_{q_1}(e^{j\omega})d\omega = \frac{4\sigma_q^2}{\pi}\int_0^{\omega_{max}} \sin^2(\omega/2)d\omega$$
$$= \frac{2\sigma_q^2}{\pi}[\omega_{max} - \sin(\omega_{max})]. \quad (20)$$

7.2. Example 4

From Equation 20, it is noted that, if $O_R = 1$ (no oversampling) and therefore $\omega_{max} = \pi$, that $\sigma_{q_1}^2 = 2\sigma_q^2$. Thus, there is nothing to be gained from a delta-sigma data converter unless it is combined with oversampling. If $O_R = 10$, then $\sigma_{q_1}^2 = 3.27\times 10^{-3}\sigma_q^2$. If $O_R = 20$, then $\sigma_{q_1}^2 = 4.11\times 10^{-4}\sigma_q^2$. If $O_R = 40$, then $\sigma_{q_1}^2 = 5.14\times 10^{-5}\sigma_q^2$. If $O_R = 80$, then $\sigma_{q_1}^2 = 6.43\times 10^{-6}\sigma_q^2$. Therefore, it is observed that the in-band noise power is approximately divided by eight every time the oversampling rate is doubled. □

Substituting $\omega_{max} = \pi/O_R$ into Equation 20 results in

$$\sigma_{q_1}^2 = \frac{2\sigma_q^2}{\pi}[\pi/O_R - \sin(\pi/O_R)]. \quad (21)$$

As noted, $y[n] = x[n] + q_1[n]$, and because $x[n]$ is bandlimited to a normalized π/O_R, the $x[n]$ component of $y[n]$ will pass through a digital low-pass filter with $\omega_c = \pi/O_R$ unaffected (ideally). Therefore, the signal power at the output of the digital low-pass filter will not be a function of the oversampling ratio. However, as noted in Equations 20 and 21, and as illustrated in Example 4, the quantization noise power at the output of the low-pass filter will be affected. The output of the digital low-pass filter may be denoted as follows:

$$y_{lp}[n] = x[n] + q_{lp}[n].$$

Because the quantization power is a function of the oversampling ratio, the SNR will be a function of the oversampling ratio. Every time O_R is doubled, the SNR will improve by about 9 dB. Or, every time O_R is increased by a factor of 10, the SNR will improve by about 30 dB. Compare these figures with that obtained for simple oversampling of 3 dB and 10 dB.

7.3. Example 5

Suppose a 16-bit ADC serves as the internal quantizer in a first-order delta-sigma data converter. Suppose that the input is uniformly distributed such that the average signal power, on a 1-Ω basis, is 6.25 W. Also suppose that the quantization noise power of the internal quantizer is 1.94 nW. If the sampling rate was such that $O_R = 10$, then SNR = 120 dB. Comparing this result with that of Example 2, it is noted that the SNR here is superior to that achieved by oversampling alone with an O_R of 64. □

Substituting Equation 12 into Equation 21,

$$\sigma_{q_1}^2 = \frac{2\sigma_q^2}{\pi}[\pi/O_R - \sin(\pi/O_R)]$$
$$= 2\frac{(V_{R_2} - V_{R_1})^2}{12 \times 2^{2N}\pi}[\pi/O_R - \sin(\pi/O_R)]. \quad (22)$$

By equating Equation 22 to an equivalent non-delta-sigma ADC without oversampling, the equivalent number of bits of the first-order delta-sigma data converter may be computed. As noted in Example 4, the in-band noise power is approximately divided by eight every time the oversampling ratio is doubled. Therefore, every time the oversampling ratio for a first-order delta-sigma data converter is doubled, the equivalent number of bits increases by 1.5.

7.4. Example 6

Consider the scenario given in Examples 2, 3, and 5. The SNR for a Nyquist sampled conventional ADC is 95 dB. If simple oversampling is applied, as in Examples 2 and 3, with an oversampling ratio of 10, the SNR increases to 105 dB and the equivalent number of bits is 17.7. If, as in Example 5, a first-order delta-sigma data converter is used with an oversampling ratio of 10, the SNR increases to 120 dB. From Equation 22, the equivalent number of bits is 20. □

7.5. Second-Order Delta-Sigma ADC

The discrete-time, and linearized, equivalent block diagram of a conventional second-order delta-sigma ADC is shown in Fig. 5. From Fig. 5, using superposition for the

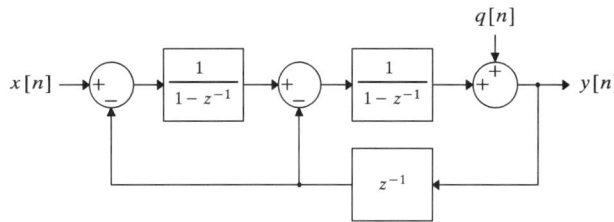

Figure 5. Discrete-time equivalent second-order delta-sigma ADC.

linearized circuit, the STF can be shown to be

$$\text{STF} = \frac{Y_x(z)}{X(z)} = 1, \quad (23)$$

where $y_x[n]$ is the output from the input $x[n]$ only, and the NTF is

$$\text{NTF} = \frac{Y_q(z)}{Q(z)} = (1 - z^{-1})^2, \quad (24)$$

where $y_q[n]$ is the output from the quantization noise $q[n]$ only. From Equations 23 and 24, it follows that

$$y[n] = y_x[n] + y_q[n] = x[n] + q_1[n]$$
$$= x[n] + q[n] - 2q[n-1] + q[n-2],$$

where $q_1[n] = y_q[n] = q[n] - 2q[n-1] + q[n-2]$ is the quantization noise of the overall delta-sigma quantizer. From Equation 24,

$$Q_1(z) = (1 - z^{-1})^2 Q(z). \quad (25)$$

From Equation 25, the PSD of the quantization noise may be expressed as follows:

$$S_{q_1}(e^{j\omega}) = |1 - e^{-j\omega}|^4 S_q(e^{j\omega})$$
$$= |1 - e^{-j\omega}|^4 \sigma_q^2 = 16\sigma_q^2 \sin^4(\omega/2). \quad (26)$$

From Equation 26, the noise shaping that the second-order delta-sigma quantizer has accomplished is apparent. At low frequencies ($\omega < \pi/3$), $S_{q_1} < \sigma_q^2$. However, at $\omega = \pi$, $S_{q_1} = 16\sigma_q^2$. Therefore, the noise power has been shaped such that the most power is at higher frequencies.

On the normalized radian frequency scale, the maximum signal frequency is $\omega_{max} = \pi/O_R$, and it follows that the noise power in the signal band may be computed as

$$\sigma_{q_1}^2 = \frac{1}{\pi}\int_0^{\omega_{max}} S_{q_1}(e^{j\omega})d\omega =$$
$$= \frac{16\sigma_q^2}{\pi}\int_0^{\omega_{max}} \sin^4(\omega/2)d\omega \quad (27)$$
$$= \frac{\sigma_q^2}{\pi}[6\omega_{max} - 8\sin(\omega_{max}) + \sin(2\omega_{max})].$$

7.6. Example 7

From Equation 27, it is noted that if $O_R = 1$ (no oversampling) and therefore $\omega_{max} = \pi$, that $\sigma_{q_1}^2 = 6\sigma_q^2$. Therefore, as was noted for the first-order delta-sigma ADC, there is nothing to be gained from a delta-sigma data converter unless it is combined with oversampling. If $O_R = 10$, then $\sigma_{q_1}^2 = 1.925 \times 10^{-4} \sigma_q^2$. If $O_R = 20$, then $\sigma_{q_1}^2 = 6.07 \times 10^{-6} \sigma_q^2$. If $O_R = 40$, then $\sigma_{q_1}^2 = 1.901 \times 10^{-7} \sigma_q^2$. If $O_R = 80$, then $\sigma_{q_1}^2 = 5.92 \times 10^{-9} \sigma_q^2$. Therefore, it is observed that the in-band noise power is approximately divided by 32 every time the oversampling rate is doubled. □

Substituting $\omega_{max} = \pi/O_R$ into Equation 27 results in

$$\sigma_{q_1}^2 = \frac{\sigma_q^2}{\pi} [6\pi/O_R - 8 \sin(\pi/O_R) + \sin(2\pi/O_R)]. \quad (28)$$

As noted, $y[n] = x[n] + q_1[n]$, and because $x[n]$ is bandlimited to a normalized π/O_R, the $x[n]$ component of $y[n]$ will pass through a digital low-pass filter with $\omega_c = \pi/O_R$ unaffected (ideally). Therefore, the signal power at the output of the digital low-pass filter will not be a function of the oversampling ratio. However, as noted in Equations 27 and 28, and as illustrated in Example 7, the quantization noise power at the output of the low-pass filter will be affected. The output of the digital low-pass filter may be denoted as follows:

$$y_{lp}[n] = x[n] + q_{lp}[n].$$

Because the quantization power is a function of the oversampling ratio, the SNR will be a function of the oversampling ratio. Every time O_R is doubled, the SNR will improve by about 15 dB. Or, every time O_R is increased by a factor of 10, the SNR will improve by about 50 dB. Compare these figures with that obtained for simple oversampling of 3 dB and 10 dB, and with that obtained by the first-order delta-sigma ADC of 9 dB and 30 dB.

7.7. Example 8

Suppose a 16-bit ADC serves as the internal quantizer in a second-order delta-sigma data converter. Suppose that the input is uniformly distributed such that the average signal power, on a 1-Ω basis, is 6.25 W. Also suppose that the quantization noise power of the internal quantizer is 1.94 nW. If the sampling rate was such that $O_R = 10$, then SNR = 132 dB. Comparing this result with that of a first-order delta-sigma converter, it is noted that the SNR here is about the same as that achieved by a first-order delta-sigma converter with an O_R of 26. □

Substituting Equation 12 into Equation 28,

$$\sigma_{q_1}^2 = \frac{(V_{R_2} - V_{R_1})^2}{12 \times 2^{2N} \pi} [6\pi/O_R - 8 \sin(\pi/O_R) + \sin(2\pi/O_R)]. \quad (29)$$

By equating Equation 29 to an equivalent non-delta-sigma ADC without oversampling, the equivalent number of bits of the second-order delta-sigma data converter may be computed. As noted in Example 7, the in-band noise power is approximately divided by 32 every time the oversampling ratio is doubled. Therefore, every time the oversampling ratio for a second-order delta-sigma data converter is doubled, the equivalent number of bits increases by 2.5.

7.8. Example 9

Consider the scenario given in Examples 2, 3, 5, and 8. The SNR for a Nyquist sampled conventional ADC is 95 dB. If simple oversampling is applied, as in Examples 2 and 3, with an oversampling ratio of 10, the SNR increases to 105 dB and the equivalent number of bits is 17.7. If, as in Example 5, a first-order delta-sigma data converter is used with an oversampling ratio of 10, the SNR increases to 120 dB and the equivalent number of bits is 20. If, as in Example 8, a second-order delta-sigma data converter is used with an oversampling ratio of 10, the SNR increases to 132 dB. From Equation 29, the equivalent number of bits is 22. □

7.9. Comparisons for First- and Second-Order Delta-Sigma ADCs

In this section, performance comparisons are made, for SNR and equivalent number of bits, for first-order and second-order delta-sigma ADCs compared with a conventional ADC that includes oversampling. The reference comparison is with a conventional ADC with N bits that is Nyquist sampled. For various values of the oversampling ratio O_R, the improvement in the SNR, in decibels, is reported, as is the equivalent increase in the number of bits.

For simple oversampling, as has been briefly presented, Equation 30 computes the improvement in SNR, in decibels, that is achieved by oversampling, based on Equation 14

$$\text{SNR}_{dB} = 10 \log_{10}(O_R). \quad (30)$$

The resultant equivalent number of added bits, caused by simple oversampling, based on Equation 15, is as follows:

$$\text{bits} = \frac{\log(O_R)}{2 \log(2)}.$$

For the first-order delta-sigma ADC, which has been briefly presented, Equation 31 computes the improvement in SNR, in decibels, that is achieved as a function of O_R, based on Equation 21:

$$\text{SNR}_{dB} = -10 \log_{10}[(2/\pi)(\pi/O_R) - \sin[\pi/O_R]]. \quad (31)$$

The resultant equivalent number of added bits for the first-order delta-sigma ADC with oversampling ratio O_R, based on Equation 22, is as follows:

$$\text{bits} = \frac{\log(\pi) - \log[2(\pi/O_R - \sin[\pi/O_R])]}{2 \log(2)}.$$

For the second-order delta-sigma ADC, which has been briefly presented, Equation 32 computes the improvement in SNR, in decibels, that is achieved as a function of O_R, based on Equation 28:

$$\text{SNR}_{dB} = -10 \log_{10}[(1/\pi)(6\pi/O_R) - 8 \sin[\pi/O_R] + \sin[2\pi/O_R]]. \quad (32)$$

The resultant equivalent number of added bits for the second-order delta-sigma ADC with oversampling ratio O_R, based on Equation 29, is as follows:

$$\text{bits} = \frac{\log(\pi) - \log[6\pi/O_R - 8\sin(\pi/O_R) + \sin(2\pi/O_R)]}{2\log(2)}.$$

The comparison results are summarized in Table 1. In each column labeled SNR_{dB}, the improvement in SNR is reported as a function of O_R. In each column labeled bits, the equivalent number of added bits is reported as a function of O_R. It should, however, be noted that the basic quantization noise that is used for comparison is that of a conventional ADC with N bits that is Nyquist sampled. That quantization noise is denoted as $q[n]$, its PSD is assumed to be flat (i.e., it is uncorrelated with itself except for a lag of zero), and its variance is denoted as σ_q^2. It has been throughout this discussion assumed that the internal quantizer has the same quantization noise characteristics, which of course may not be true. However, the assumption does allow for comparisons with practical significance.

In Table 1, OS denotes a conventional ADC with oversampling, DS1 denotes a first-order delta-sigma ADC, and DS2 denotes a second-order delta-sigma ADC. Note that the data in the table imply that for a desired number of equivalent bits for an ADC, there are several ways of achieving the desired results. For example, if it is desired to obtain 12 bits of resolution, this may be achieved by (1) a conventional ADC designed for 12 bits sampled at the Nyquist rate; (2) a conventional ADC designed for 10 bits sampled at 20 times the Nyquist rate; (3) a first-order delta-sigma ADC that includes an internal conventional ADC designed for 4 bits, sampled at 60 times the Nyquist rate; and (4) a second-order delta-sigma ADC that includes an internal binary quantizer sampled at 40 times the Nyquist rate.

The fact that it is possible to achieve a significant number of bits of resolution with a delta-sigma ADC using only a binary internal quantizer is a compelling reason to use such an engineering solution in many applications. It lends itself readily to IC implementation without the need for external precision components, except for if the internal ADC is multibit. The internal clock speed may be high, but developments in IC technology have greatly facilitated internal speed capabilities.

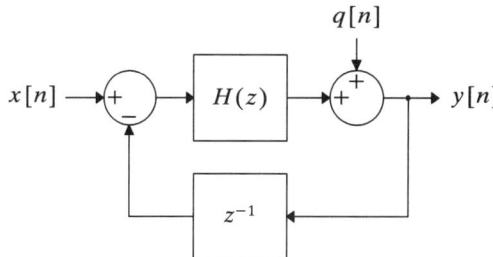

Figure 6. A generalization of Fig. 4 that allows for higher orders.

7.10. Higher Order Delta-Sigma Architectures

For stability reasons, conventional delta-sigma ADCs, such as the first-order and second-order ones discussed here, cannot be implemented beyond order two. However, other architectures have been proposed for higher orders. One well-received approach is the MASH architecture (18,19). Another example is the redundant signed digit (RSD) approach (20,21) This is area of research and development is active. One example will be given here (22).

A simple modification of Fig. 4 is shown in Fig. 6, where the integrator transfer function $1/(1-z^{-1})$ has been replaced with the general transfer function $H(z)$. Because an integrator is a low-pass function, we might expect that $H(z)$ would need to be low pass as well, and such is the case. For an arbitrary $H(z)$, it can be shown that

$$\text{STF} = \frac{Y_x(z)}{X(z)} = \frac{H(z)}{1 + z^{-1}H(z)} \qquad (33)$$

and

$$\text{NTF} = \frac{Y_q(z)}{Q(z)} = \frac{1}{1 + z^{-1}H(z)}. \qquad (34)$$

Table 1. Performance Comparison Among Simple Oversampling, a First-Order Delta-Sigma ADC, and a Second-Order Delta-Sigma ADC

	OS		DS1		DS2	
O_R	SNR_{dB}	(bits)	SNR_{dB}	(bits)	SNR_{dB}	(bits)
1	0	N	-3.0	-0.5	-7.8	-1.3
2	$+3.0$	$+0.5$	$+4.4$	$+0.7$	$+3.4$	$+1.1$
4	$+6.0$	$+1.0$	$+13.0$	$+2.2$	$+17.5$	$+2.9$
6	$+7.8$	$+1.3$	$+18.2$	$+3.0$	$+26.2$	$+4.3$
8	$+9.0$	$+1.5$	$+22.0$	$+3.6$	$+32.3$	$+5.4$
10	$+10.0$	$+1.7$	$+24.8$	$+4.1$	$+37.2$	$+6.2$
20	$+13.0$	$+2.2$	$+33.9$	$+5.6$	$+52.2$	$+8.7$
30	$+14.8$	$+2.5$	$+39.1$	$+6.5$	$+61.0$	$+10.1$
40	$+16.0$	$+2.7$	$+42.9$	$+7.1$	$+67.2$	$+11.2$
50	$+17.0$	$+2.8$	$+45.8$	$+7.6$	$+72.1$	$+12.0$
60	$+17.8$	$+3.0$	$+48.2$	$+8.0$	$+76.0$	$+12.6$
70	$+18.5$	$+3.1$	$+50.2$	$+8.3$	$+79.3$	$+13.2$
80	$+19.0$	$+3.2$	$+51.9$	$+8.6$	$+82.2$	$+13.6$
90	$+19.5$	$+3.2$	$+53.5$	$+8.9$	$+84.8$	$+14.1$
100	$+20.0$	$+3.3$	$+54.8$	$+9.1$	$+87.1$	$+14.4$

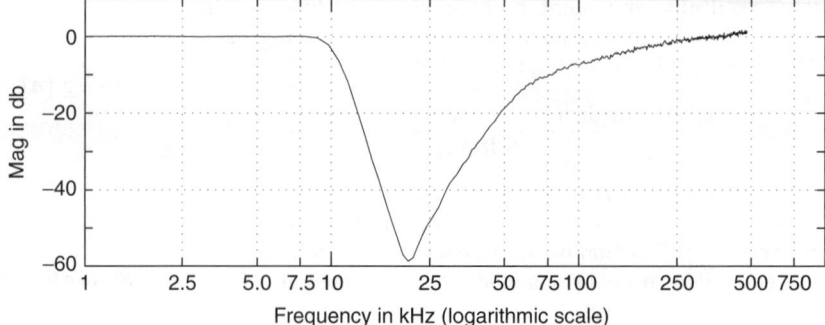

Figure 7. Spectrum of the output of a fourth-order Chebyshev type II design for a random input.

As was noted for the first-order and second-order delta-sigma ADCs, the NTF is a high-pass response shaping the quantization noise spectrum to higher frequencies beyond the signal band. Suppose it is desired for the NTF given in Equation 34 to be a fourth-order Chebyshev type II high-pass response. This condition allows for a systematic design procedure. Given the desired NTF, the required $H(z)$ may be obtained from Equation 34. Once $H(z)$ has been determined, the design of the delta-sigma ADC shown in Fig. 6 is complete. The low-pass characteristic of the STF may be verified from Equation 33.

The results of such a design are shown in Figs. 7 and 8. The design parameters are as follows: The internal quantizer is binary, the desired signal band is dc to 20 kHz, the sampling rate is 1 MHz (i.e., $O_R = 25$), the 3-dB cutoff frequency of the Chebyshev type II high-pass filter is about 68 kHz, the designed stopband attenuation of the high-pass response is -60 dB, and the stopband edge frequency is 20 kHz. Note that the output of the quantizer shown in Fig. 6 is a binary bit stream.

When the input $x[n]$ is a bandimited random signal, produced by a white noise process driving a tenth-order Butterworth low-pass filter with a 3-dB cutoff frequency of 10 kHz, the spectrum of the output bit stream is as shown in Fig. 7. Note that the random input spectrum is -60 dB at 20 kHz, as determined by the characteristics of the Butterworth filter, and beyond which the noise spectrum dominates as observed in Fig. 7. Therefore, in Fig. 7, below 20 kHz there is signal-plus-noise, and above 20 kHz is shown the quantization noise spectrum.

To better isolate the signal and noise in-band, a very narrow signal spectrum is desired, and this is often accomplished with a sinewave input, which is illustrated in Fig. 8. In Fig. 8, the input $x[n]$ is a 1-kHz sinewave. The spectrum of the output bit stream is shown in Fig. 8. The quantization noise is about -72 dB relative to the signal power. From Fig. 8, $S_x(e^{j\omega}) = \delta(\omega)$, and from

$$\sigma_x^2 = \frac{1}{\pi} \int_0^\pi S_x(e^{j\omega}) \, d\omega$$

(see Equation 20), $\sigma_x^2 = 1/\pi$. Because σ_x^2 is the power in $x[n]$, a sinewave, $x[n]$ must have a peak value of $\sqrt{2/\pi}$. Therefore, $V_{R_2} = -V_{R_1} = \sqrt{2/\pi}$. From Equation 12 and from Fig. 8,

$$\sigma_q^2 = 6.3 \times 10^{-8} = \frac{(V_{R_2} - V_{R_1})^2}{12 \times 2^{2N}} = \frac{8}{12\pi \times 2^{2N}}. \quad (35)$$

From Equation 35, it follows that the equivalent number of bits, N, is about 12 bits. Similar results may be obtained by a Butterworth–MASH architecture (23).

8. SELECTION CRITERIA FOR AN ADC CONVERTER

Several considerations are necessary for the proper selection and use of an ADC. First are basic considerations as to the environment in which it is to be used. These considerations are concerned with whether a particular converter will work properly with the computer and system already in place. The converter must mechanically fit, have appropriate voltages available for its use, have cable or bus connections of the appropriate type, and have an analog input voltage range that is appropriate.

Second, does the converter have the needed number of bits and will it operate at the desired sample rate? Often in a particular application it may not be necessary to know the technical details of how the converter works, that is,

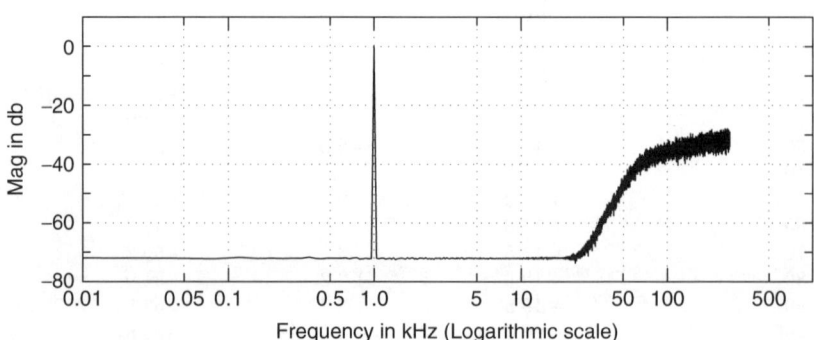

Figure 8. Spectrum of the output of a fourth-order Chebyshev type II design for a 1-kHz sinewave input.

for example, whether it is a successive approximation converter or some type of delta-sigma converter, but it will almost always be of significant importance as to the number of bits and the sample rate (sometimes called the word rate). The number of bits is related to the SNR. From Example 1,

$$\mathrm{SNR_{dB}} = 20 N \log_{10}(2),$$

where N is the number of bits, or the inverse may be stated as follows:

$$N = \frac{\mathrm{SNR_{dB}}}{20 \log_{10}(2)}. \qquad (36)$$

These equations are true no matter what type of converter is used. The particular application will determine the number of bits, or equivalently, the $\mathrm{SNR_{dB}}$. For example, if the sampled data are going to be analyzed for some artifact that may be 60 dB below the maximum average power level of the input, the $\mathrm{SNR_{dB}}$ of the converter needs to be significantly greater that 60 dB. That is, from Equation 36, more than ten bits is required in the ADC. In addition, the maximum sample rate (word rate) of the converter needs to be at least as high as the Nyquist rate (twice as high as the highest frequency component in the input signal).

Third, is the software control for the ADC sufficiently flexible so that the converter will accomplish what is needed? Will the software allow a flexible selection of sample rates, or are only a few sample rates available? Is the input analog voltage range under software control?

Fourth, what sampling frequency accuracy is required in the specific application? Sampling frequencies are typically derived from a system clock and obtained by dividing the clock frequency by some integer power of two. If the clock frequency is in error from its nominal value by $x\%$, then the sampling period T will also be in error by about $x\%$. Critical time values in the data analysis, and critical frequency values, will be in error by $x\%$. If the precise value of the clock frequency is known, then so will be the precise value of the sampling period, and the clock frequency error can be readily adjusted for in the data analysis.

This discussion provides a brief indication of the criteria needed for the proper selection and use of an ADC. Technical details have been provided in earlier sections of this article.

BIBLIOGRAPHY

1. Y. Zhong, H. Wang, K. H. Ju, K.-M. Jan, and K. H. Chon, Nonlinear analysis of the separate contributions of autonomic nervous systems to heart rate variability using principal dynamic modes. *IEEE Trans. Biomed. Eng.* 2004; **51**(2):255–262.

2. J. Bourien, J. J. Bellanger, F. Bartolomei, P. Chauvel, and F. Wendling, Mining reproducible activation patterns in epileptic intracerebral EEG signals: application to interictal activity. *IEEE Trans. Biomed. Eng.* 2004; **51**(2):304–315.

3. J. Tang and S. T. Acton, Vessel boundary tracking for intravital microscopy via multiscale gradient vector flow snakes. *IEEE Trans. Biomed. Eng.* 2004; **51**(2):316–324.

4. E. A. Goldstein, J. T. Heaton, J. B. Kobler, G. B. Stanley, and R. E. Hillman, Design and implementation of a hands-free electrolarynx device controlled by neck strap muscle electromyographic activity. *IEEE Trans. Biomed. Eng.* 2004; **51**(2):325–332.

5. E. Entcheva, Y. Kostov, E. Tchernev, and L. Tung, Fluorescence imaging of electrical activity in cardiac cells using an all-solid-state system. *IEEE Trans. Biomed. Eng.* 2004; **51**(2):333–341.

6. B. A. Schnitz, D. Xu Guan, and R. A. Malkin, Design of an integrated sensor for *in vivo* simultaneous electrocontractile cardiac mapping. *IEEE Trans. Biomed. Eng.* 2004; **51**(2):355–361.

7. J. I. Godino-Llorente and P. Gómez-Vilda, Automatic detection of voice impairments by means of short-term cepstral parameters and neural network based detectors. *IEEE Trans. Biomed. Eng.* 2004; **51**(2):380–384.

8. H. A. Al-Nashash, J. S. Paul, W. C. Ziai, D. F. Hanley, and N. V. Thakor, Wavelet entropy for subband segmentation of EEG during injury and recovery. *Ann. Biomed. Eng.* 2003; **31**(6):653–658.

9. J. Zhao, J. Yang, L. Vinter-Jensen, F. Zhuang, and H. Gregersen, Biomechanical properties of esophagus during systemic treatment with epidermal growth factor in rats. *Ann. Biomed. Eng.* 2003; **31**(6):700–709.

10. D. R. McGaughey, M. J. Korenberg, K. M. Adeney, A. D. Collins, and G. J. M. Aitken, Using the fast orthogonal search with first term reselection to find subharmonic terms in spectral analysis. *Ann. Biomed. Eng.* 2003; **31**(6):741–751.

11. D. F. Hoeschele, *Analog-to-Digital and Digital-to-Analog Conversion Techniques*, 2nd ed. New York: John Wiley & Sons, 1994.

12. M. Gustavsson, J. J. Wikner, and N. Tan, *CMOS Data Converters for Communications*. Boston, MA: Kluwer Academic Publishers, 2000.

13. P. G. A. Jespers, *Integrated Converters: D to A and A to D Architectures, Analysis and Simulation*. New York: Oxford University Press, 2001.

14. R. J. van de Plassche, *CMOS Integrated Analog-to-Digital and Digital-to-Analog Converters*, 2nd ed. Boston, MA: Kluwer Academic Publishers, 2003.

15. L. D. Paarmann, *Design and Analysis of Analog Filters: A Signal Processing Perspective*. Boston, MA: Kluwer Academic Publishers, 2001.

16. R. A. Haddad and T. W. Parsons, *Digital Signal Processing: Theory, Applications, and Hardware*. New York: Computer Science Press, 1991.

17. A. V. Oppenheim and R. W. Schafer, with J. R. Buck, *Discrete-Time Signal Processing*, 2nd ed. Upper Saddle River, NJ: Prentice-Hall, 1999.

18. S. R. Norsworthy, R. Schreier, and G. C. Temes, eds., *Delta-Sigma Data Converters: Theory, Design, and Simulation*. New York: Wiley-IEEE Press, 2001.

19. J. C. Candy and G. C. Temes, eds., *Oversampling Delta-Sigma Data Converters: Theory, Design, and Simulation*. New York: Wiley-IEEE Press, 2001.

20. B. Ginetti, P. G. A. Jespers, and A. Vandemeulebroecke, A CMOS 13-b cyclic RSD A/D converter. *IEEE J. Solid-State Cir.* 1992; **27**(7):957–965.

21. Y. H. Atris and L. D. Paarmann, Hybrid RSD-cyclic-sigma-delta analog-to-digital converter architecture. Proc. Int. Sig. Proc. Conf., Dallas, TX, 2003.

22. L. D. Paarmann and S. Darsi, A cascode architecture which implements a fourth-order Chebyshev type II highpass NTF in delta-sigma data converters. Proc. Int. Sig. Proc. Conf., Dallas, TX, 2003.

23. L. D. Paarmann and V. Subbarao, A MASH architecture which implements a two-stage Butterworth highpass NTF in delta-sigma data converters. Proc. Int. Sig. Proc. Conf., Dallas, TX, 2003.

ANESTHESIA MACHINES

JOAQUIN ROCA-DORDA
JOSÉ ÁLVAREZ-GÔMEZ
JOAQUÍN ROCA-GONZÁLEZ
JACINTO JIMÉNEZ-MARTÍNEZ
FRANCISCO J. GIL-SÁNCHEZ
Polytechnical University of Cartagena
Cartagena, Spain

1. INTRODUCTION

Since the introduction of general anesthesia in 1846 (October 16th, the "ether day"), surgical procedures may be carried out while the patient remains insensitive to pain and other external stimuli (1). As a result of the simplicity of the systems used for the delivery of the first anesthetic agents, the use of anesthesia spread quickly among the medical community. As new gaseous agents were introduced, the equipment required for delivering them started to be considered as part of the standard medical equipment, so that when the first vaporized agents came onto the scene, the medical community was ready for the adoption of this kind of equipment (2). A similar timeline was followed after the introduction of the first intravenous agent in 1934, leading to the modern infusion systems in current use worldwide. As with many other manufactured products, the machines used for anesthesia have changed as technology developed within the industrial and military fields. In this sense, it is easy to understand that modern anesthesia machines include accurate electronic sensors, complex embedded computers, and high-precision mechanic elements in order to ease the operative procedures to the anesthesiologists and, at the same time, grant the security of the patient.

This work will focus on different technological aspects involved in the design and selection of new equipment for anesthesia delivery. With this orientation in mind, the authors have centered this review on anesthesia machines for delivering vaporized and intravenous agents, because of the recent growth of this later method.

1.1. Anesthesia Machines in the Operating Room

Although anesthesia has been generally described as the part of the medical profession that ensures that the patient's body remains insensitive to pain and other stimuli during surgical operations, the currently accepted role for the anesthesiologist includes all of the procedures that have to be performed to offer a completely balanced anesthesia (3). In this sense, anesthesia is understood as a patient care within four different domains: sedation (reversible patient unconsciousness), relaxation (temporal reduction of the motoric functions of the patient in order to ease the surgical procedures), analgesia (insensitivity to pain), and respiration (granting the respiratory function in order to avoid permanent damage to the different tissues), which has led medical device manufacturers to develop complex machines for anesthesia delivery and patient monitoring.

Figure 1, which shows a typical setting found in many operating rooms, includes the equipment required for the delivery of vaporized and intravenous agents in order to allow the proper sedation method for each patient and surgical procedure. The main elements of the anesthesia delivery system illustrated include the gas supply system, the gas mixing subsystem, the vaporizer for inhaled agents, the mechanical ventilator, the breathing circuit of the patient, the absorber for CO_2 removal, the infusion pumps required for intravenous anesthesia, and the monitoring subsystem for patient and equipment supervision (4). In order to cover each system in depth (as allowed by the length of the article and the knowledge of the authors), the next sections will review the elements listed above.

2. ANESTHESIA MACHINES FOR INHALED AGENTS DELIVERY

The design of anesthesia machines, as they are known today, can be considered as a direct result of almost one century of technological advances in the medical field. On the one hand, the fabric masks used for the application of ether or chloroform were abandoned on behalf of vaporizers able to supply accurate concentrations of the vaporized agents. On the other hand, the first mechanical ventilators introduced at the beginning of the twentieth century took advantage of the advances generated within the military and chemical industries, leading to the modern ventilation systems.

This section will try to provide a review of the different elements constituting these devices from an operationally centered approach. In this sense, the reader is encouraged to consult Fig. 2 in order to follow the outline of this review.

2.1. Principles of Operation

According to the definition included in the European Standard EN740 (Clause 3.1), an anesthesia machine is "a system for the administration of inhalation anesthesia which includes one or more actuator modules, their particular monitoring and alarm modules, and essential hazard protection modules." A simpler, yet complete, definition of an anesthesia machine may be stated as a device that delivers a precisely-known but variable gas mixture, including anesthetizing and life-sustaining gases. In this sense, anesthesia units dispense a mixture of gases and vapors of known concentrations in order to control the level of consciousness or analgesia of the patient undergoing surgery.

As this work falls beyond the scope of clinical anesthesia but within the technological aspects of its delivery, the reader should understand the basics of the administration of inhaled anesthetics, so a basic description is enclosed below (5).

ANESTHESIA MACHINES 71

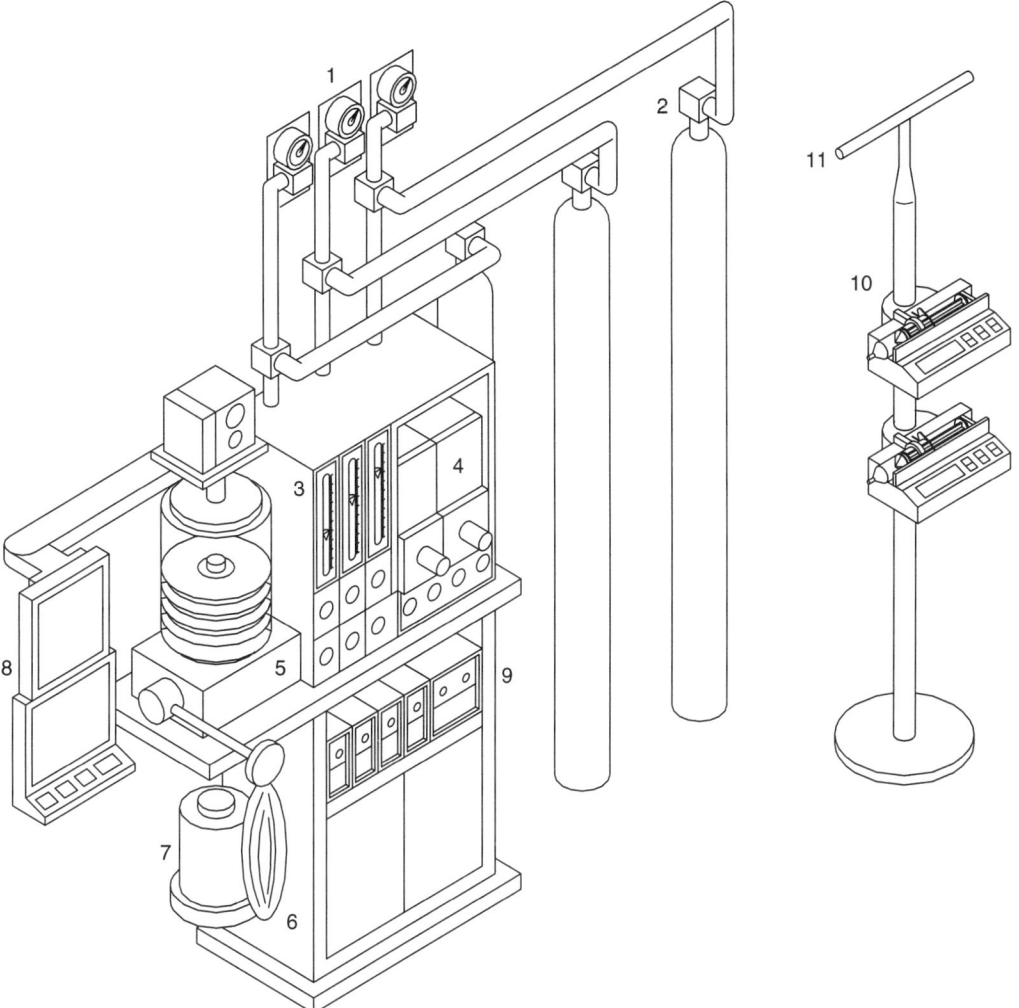

Figure 1. Simplified view of the anesthesia machines in the operating room. (1) Central gas supply (oxygen, nitrous oxide, and air), (2) high-pressure gas cylinders, (3) gas flowmeters and mixing controls, (4) anesthetic agent vaporizers, (5) mechanical ventilator, (6) breathing reservoir bag, (7) absorber for carbon dioxide removal, (8) patient and machine monitors, (9) monitoring amplifier modules, (10) infusion pumps, and (11) standing pole.

Anesthesia is achieved by administering a mixture of O_2, the vapor of a volatile liquid halogenated hydrocarbon anesthetic, and, if necessary, N_2O and other gases. As spontaneous breathing is often depressed by anesthetic agents and by muscle relaxants administered in conjunction with them, respiratory support is usually necessary to deliver the breathing gas to the patient.

For these purposes, the anesthesia machine must perform the following functions:

- Assuring the proper oxygen (O_2) flow delivery to the patient.
- Vaporizing the volatile anesthetic agent and blending it into a gas mixture with O_2, nitrous oxide (N_2O), other medical gases, and air.
- Granting the ventilation of the patient by controlling spontaneous ventilation and using mechanical assistance if needed.

- Minimize the anesthesia-related risk to the patient and the clinical personnel.

At this point, providing the reader with an overview of the basic machine operation becomes mandatory, as it will ease further discussions of the different components of these machines.

The gas flow supplied by either the pipelines (2) or the security high-pressure cylinders (1) is regulated at the flowmeters (5) and mixed in the common gas manifold entering the vaporizer (8), where this mixture is vaporized with the anesthetic agent used. This fresh gas flow is then sent to the patient through the breathing circuit (12), that also collects the expired gas in order to process it through the circuit selected (15). In either case, the gas will pass through the absorber (14) in order to remove carbon dioxide before returning to the inspiratory branch. If mechanical ventilation is used, the ventilator (18) sets the inspiratory and expiratory cycles according to the control

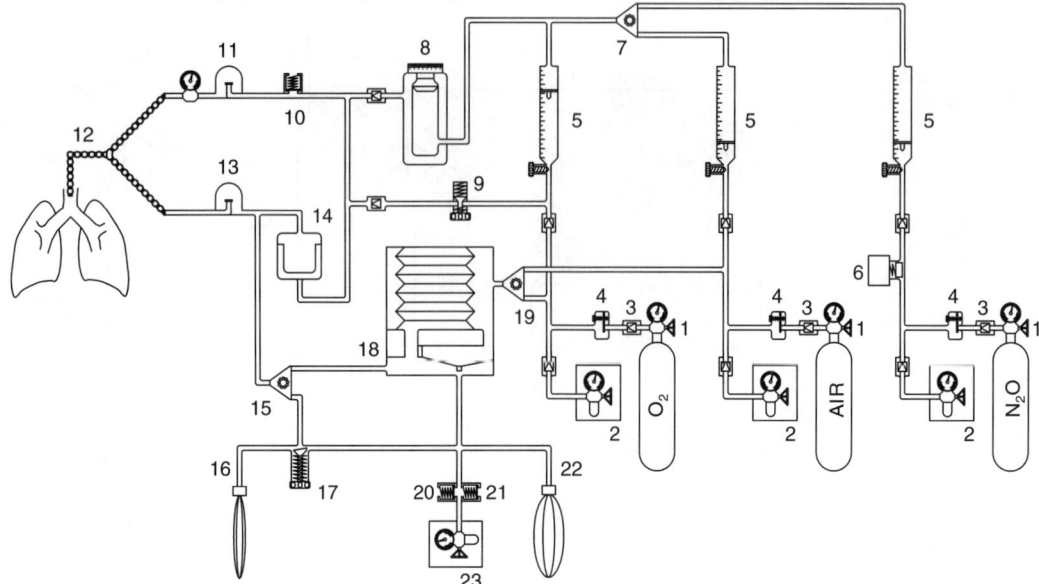

Figure 2. Schematic diagram of an anesthesia machine for the delivery of inhaled agents. (1) High-pressure gas cylinder, (2) central gas supply outlet, (3) unidirectional-flow valve, (4) pressure regulator, (5) gas flowmeter, (6) fail-safe device, (7) carrier gas selector, (8) vaporizer, (9) oxygen flush valve, (10) fresh gas flow positive pressure relief valve, (11) unidirectional inspiratory valve, (12) patient breathing circuit, (13) unidirectional expiratory valve, (14) carbon dioxide absorber, (15) mechanical ventilation or spontaneous breathing circuit selector, (16) breathing reservoir, (17) adjustable pressure limiting valve or pop-off valve, (18) mechanical ventilator, (19) ventilator driving gas selector, (20) scavenging gas positive pressure relief valve, (21) scavenging gas negative pressure relief valve, (22) scavenging reservoir bag, and (23) central vacuum inlet.

adopted. In the case of spontaneous ventilation, the exhaled gas is scavenged through an adjustable pressure limiting valve (APL) to the available waste gas removal system (23).

From the description related above, the anesthesia machine may be understood as the ensemble of the following subsystems:

- Gas supply
- Flow Regulators
- Vaporizer
- Breathing system
- Scavenging system

2.2. Gas Supply System

As mentioned above, anesthesia machines do not just administer the anesthetic agents but also life-sustaining gases, such as oxygen and nitrous oxide (that may be substituted by medical air or helium, among others), so that dedicated systems have been developed for precisely supplying the proper concentration of these gases to the patient. In this sense, the gas supply system relies on three different components: the gas source, the flow regulators, and the associated safety devices.

2.2.1. Gas Sources. The gases commonly used in anesthesia (oxygen, nitrous oxide, and compressed air) are under high pressure and may be piped in from a central storage area or used directly from nearby compressed gas cylinders in case of central supply failure (4–9).

2.2.1.1. Central Gas Supply. As medical gases are used in most of the hospital's rooms besides the operating rooms, it is typical to provide a central storage area for the whole installation. Both oxygen (produced by fractional distillation of liquid air) and nitrous oxide (produced by heating of ammonium nitrate in the presence of iron) are stored in liquid form in high-pressure tanks placed outside the facility in order to minimize the risk of combustion, whereas the medical air is produced and distributed from a compressor plant on-site. These gases are usually supplied at 345 KPa (50 psi) after a two-stage regulation from the nominal pressure of the tanks. The wall outlets (2) are usually suited with primary and secondary check valves (that prevent reverse flow of gases from machine to pipeline or atmosphere), a pressure regulator, and a filter for the removal of the impurities.

Besides the fact that wall connectors should be color-coded for each gas according to national standards, the piping connections for O_2 and N_2O can be accidentally interchanged during installation or repair of medical gas systems, which may lead to harmful conditions. In order to minimize these risks, several regulations have proposed the use of specially designed connection systems, such as in the United States, where the Diameter Index Safety System (DISS) introduced by the Compressed Gas Association (CGA) has been adopted as a clinical standard.

This system, designed to operate below 1380 KPa (200 psi), establishes a standard for non-interchangeable, removable connections for use with medical gases, vacuum (suction), and evacuation service. Each DISS connection consists of a body adaptor, nipple, and nut. As the diameter of the body adaptor increase/decrease proportionally, only properly mated and intended parts fit together to permit thread engagement. In place of diameter indexing, oxygen (DISS 1240) has been assigned the long-established 9/16″-18 thread connection as its safety standard. The DISS system is used on gas pressure regulator outlets, wall outlets, anesthesia equipment, and respiratory equipment.

2.2.1.2. Gas Cylinders. In case of central supply failure, the anesthesia machine should be fitted with a backup source of medical gases in order to grant continuous ventilation to the patient, which has become mandatory in most countries where the use of backup gas cylinders is specifically included within the regulations related to anesthesia machines. For this purpose, it is advisable to include cylinders for oxygen and nitrous oxide delivery.

These compressed gas cylinders (which are mounted on yokes attached to the anesthesia machine) use a filter, an unidirectional flow check valve, a pressure regulator, and gauge. The pressure regulator (4) is needed to set the gas pressure below the pipeline supply pressure (310 KPa for the cylinders) in order to prevent recirculation of the gas from the cylinders to the central supply system. Aside from this pressure regulator, cylinders usually include additional security features such as a safety relief device consisting of a frangible disc that bursts under extreme pressure, a fusible plug made of Wood's metal that has a low melting point, a safety relief valve that opens at extreme pressure, or a combination there of.

Safety is assured by means of color codes identifying the gas contained by the cylinders and a hardware identifying system such as the American Standard Safety System (ASSS) and Pin Index Safety System (PISS). Although the ASSS applied to cylinders having threaded gas ports is mostly used for distinguishing between flammable and nonflammable gases, PISS, a subsystem of the first, has become mandatory within the medical field. This system relies on the use of holes drilled in the face of the cylinder valve face and a yoke with pins matching the holes in the valve face for each specific gas (six hole-pin combinations exist).

2.2.2. Safety Devices. In order to prevent damage associated with hypoxic ventilation, several safety devices are included within the anesthesia machine. The hypoxic guard system links the controls of O_2 and N_2O in order to avoid the administration of hypoxic gas mixtures (mixtures containing less than 25% oxygen). This system is complemented by means of the so-called fail-safe device (6), which shuts off the nitrous oxide supply when the oxygen pressure at the flowmeter falls below a certain threshold value, which typically ranges from 69 KPa (10 psi) to 138 KPa (20 psi). Additionally, an oxygen flush valve (9) must be included in order to allow the rapid (35–75 L/min) washout of the breathing circuit in case of emergency, as this valve directly injects oxygen into the patient without passing through any kind of vaporizer (4–9).

2.3. Flow Regulators

Since the introduction of the "Rotameters" of Küpers by Neu in 1910, anesthesia machines have included independent flow controls for each of the medical gases used in order to cover the requirements of the anesthesiologist for precisely controlling the amount of each gas flowing into the breathing circuit attached to the patient. These flow meters (5) typically consist of a glass tube in which a floating conical element rotates at different heights as a function of the flow streaming out from the meter. Although modern machines have included electronic flowmeters based on different sensing principles (ultrasound Doppler, electromagnetic sensing, etc.) and digital displays, it is advisable to include at least one conventional glass flowmeter in order to allow operation even when electrical power fails (8,9).

Regulations, such as American ASTM F1850 or the European EN740, specify the minimum requirements regarding the use of flowmeters in anesthesia machines. These regulations establish that a single control located next to a flow indicator should be included for each gas. The control knob associated with the oxygen flow control should be uniquely shaped in order to ease fast operation in case of emergency. The flowmeter should be protected against damage caused by excessive rotation by means of a valve stop or some other mechanism. Although the location of the oxygen flowmeter within the flowmeter is different in the United States (right) and United Kingdom (left), it is assumed worldwide that the oxygen flow going out of the flowmeter has to join the common gas stream downstream of the other gases.

Additional components such as a carrier gas circuit selector (7) or a ventilator driving gas selector (19) may be included in order to offer more features.

2.4. Vaporizer System

In order to deliver most of the inhaled anesthetic agents through the breathing circuit, these liquid substances must be vaporized into the carrier gas stream. To achieve this goal, special devices have been developed and, today, are considered one of the most important elements found in anesthesia machines.

A vaporizer enriches the carrier gas mixture with a vapor fraction of the volatile agent by means of different principles, leading to different families of these devices such as those known as variable bypass, heated blender, measured flow, draw-over, and the recently introduced injectors (5). The most common are shown in Fig. 3.

2.4.1. Variable Bypass. The variable bypass vaporizer is the most commonly used in today's machines for the vaporization of many agents such as enflurane, isoflurane, halothane, and sevoflurane. This type of vaporizer receives this name because a variable shunt valve (c in Fig. 3) regulates the proportion of gas flowing into the vaporization chamber and into the mixing chamber (b). As the gas flowing out of the vaporizer chamber is mixed with

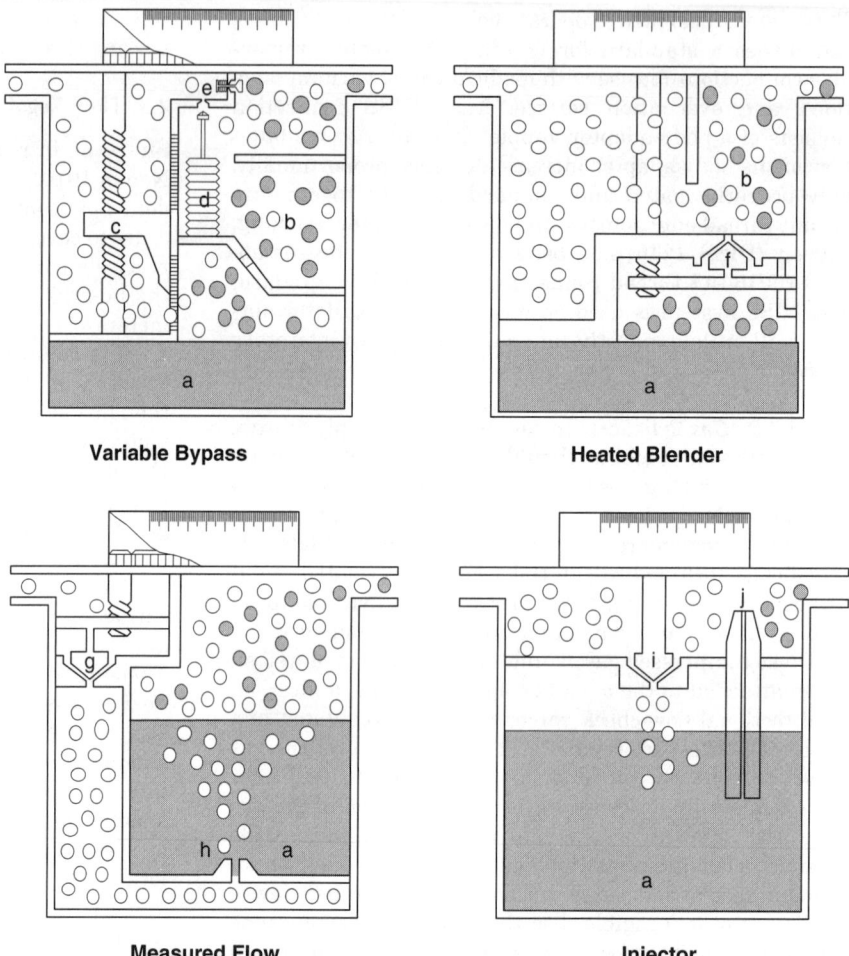

Figure 3. Idealized views of different types of vaporizers. (a) Liquid agent, (b) mixing chamber, (c) bypass valve, (d) temperature compensation bellows, (e) pressure relief valve, (f) feedback-controlled metering valve, (g) constant flow valve, (h) gas mixture bubbler, (i) bypass valve, and (j) injector.

the bypassed gas stream, the concentration of the agent in the gas mixture is directly related to the splitting ratio of the valve. Temperature compensation bellows (d) are included in order to compensate for the effect of temperature changes that affect the equilibrium vapor pressure above the agent. As this kind of vaporizer is able to deliver accurate concentrations of the anesthetic agent, specific designs and calibration methods are used for each type of liquid agent.

2.4.2. Heated Blender. Initially introduced for use with desflurane, the liquid agent is heated within a sump chamber before entering the mixing chamber through an adjustable feedback-controlled metering valve (f) that regulates the vapor stream flow.

2.4.3. Measured Flow. Also known as the copperketlle of flowmeter-controlled vaporizers, these devices are not able to deliver accurate concentrations of the liquid agent because they are not calibrated, which is because of the vaporization control implemented, which is based on a constant flow of carrier gas (g) bubbling up (h) through the liquid agent. Other types of similar vaporizers exist such as the flow-over (where the carrier gas flows over the surface of the liquid agent instead of bubbling through it) and the draw-over vaporizers (employed in the military filed and in situations where pressurized gas sources are not available).

2.4.4. Injector. This recently introduced system uses a valve (i) to regulate the amount of fresh gas flowing into a pressurized chamber where the liquid agent is stored. As the pressure of this chamber increases, the agent is forced up through the injector nozzle (j) where it is atomized within the fresh gas flow because of the Venturi effect. As no vaporization occurs (just atomization of the liquid agent), temperature compensation is not required.

In addition to the technology adopted, regulations established by the competent authorities have made mandatory the use of a vaporizer interlock that grants that only one vaporizer of the set installed at the machine may work at a given time in order to prevent multiple agent vaporization into the patient circuit. The vaporizers should include a liquid agent level in order to prevent overfilling and a keyed-filler device in order to avoid filling a vaporizer with the wrong agent.

2.5. Breathing System

Once the anesthetic agent is vaporized at the desired concentration, the gas mixture has to be administered to the patient in order to get the desired therapeutic effect and the proper ventilation. For this purpose, anesthesia machines are connected to the patient by the so-called breathing circuits, made out of black antistatic rubber or polypropylene (despite the fact of not withstanding autoclaving, this type is currently the most frequently used because of its low cost and hygiene, as it is designed for one-time use only). Transnational regulations have set a worldwide standard coaxial connector of 22/15 mm diameter for the connection of the patient's circuit to the common gas outlet.

2.5.1. Breathing Circuits. Breathing circuits are often classified either as open systems or closed systems. In open systems, the fresh gas flow is administered to the patient before being scavenged, whereas in closed systems, the exhaled gases are processed in order to recycle them, reducing the total amount of agent required (in order to reduce anesthesia costs and staff exposure to the agents, as proposed by the NIOSH). As these types may present different variations, a further classification groups the breathing circuits as open (where no rebreathing occurs at all and the gases are scavenged directly), semiopen (although no rebreathing occurs, gases are scavenged through the reservoir bag), semiclosed (which present a partial rebreathing of the exhaled agents and use a reservoir bag), and closed (where complete rebreathing occurs through the reservoir bag). Although the circle system is the one most commonly used, non-rebreathing breathing circuits (NRB or "T-piece" systems) are widely used, so a deeper discussion is provided below (6).

2.5.1.1. Circle System. The breathing circuit receives the vapor-enriched gas mixture from the vaporizer outlet and sends it to the patient circuit (12) through the unidirectional inspiratory valve (11), installed to prevent rebreathing from the patient to this branch of the circuit. The inspired branch is completed with a security overpressure valve (10) installed in order to release the gas out of the circuit in case the pressure of the gas mixture going into the patient circuit exceeds the threshold of 12.5 kPa (125 cm H_2O) established by the regulating authorities.

The exhaled gases return through the unidirectional expiratory valve (13) flowing into the absorber (14), which is installed to remove carbon dioxide in order to recycle the breathing gas. In the function of the circuit selected (15), part of the exhaled gases will cycle through the ventilator (18) or the spontaneous breathing circuit formed by the breathing reservoir bag (16) and the adjustable pressure limiting valve (APL) (17) before returning to the absorber inlet. It should be said that the actual design of the machine differs from the theoretical view of Fig. 2 as both branches of the circuit (inspiratory and expiratory) share a common gas outlet (CGO) where the patient circuit is connected through a coaxial assembly.

The reader should be aware that the "Bain" circuit is a modification of the Mapleson Type "D" circuit, where the fresh gas flow enters the circuit through the inner conduit of the coaxial tubing used.

2.5.1.2. Non-Rebreathing Circuits. These circuits allow oxygen, nitrous oxide, and the volatile agent to flow in and out of the patient before being released to the scavenging system required to avoid the contamination of the atmosphere at the operating room.

Besides being simple to operate, the agent is wasted, so the costs associated with the anesthetic procedure are increased. In NRB circuits, unidirectional valves (11 and 13) are not used because both branches of the respiratory circuit use different tubing. As rebreathing does not occur, no need exists for chemical removal of the exhaled C_{O_2}, so absorbers are not used. The five types of NRB most common in the operating rooms were classified by Mapleson, who identified them with a letter (A to E), a system that has been followed from that time by other authors, such as Willis et al. in 1975 when they introduced type F. These circuits, which are shown in Fig. 4, differ on the location of the inlets for the fresh gas flow and the exhaled gases, and in the use of the breathing bag.

2.5.2. Carbon Dioxide Absorber. Circle systems require the use of chemical elements for carbon dioxide removal from the exhaled gases before recycling them back to the inspiratory limb of the breathing circuit. These absorbers present in the form of a canister containing pellets of the absorbing compound, which is placed within a shell connected inline with the circle system. Although several compounds have been developed since the adoption of this technique (initially introduced in the military field for the regeneration of the air in submarines and in the mining industry), most of them are based on soda lime absorption of carbon dioxide in a three-stage process depicted in Fig. 5.

Carbon dioxide is hydrated with the water vapor leading to the formation of carbonic acid (i). Then, the acid reacts with the sodium hydroxide (which is a base) in a neutralization process that generates a salt (sodium carbonate) and a water-releasing energy in the form of heat (ii). Finally, this salt reacts with the calcium hydroxide (soda lime) leading to the precipitation of another salt (calcium hydroxide) and sodium hydroxide (iii). The final balance indicates that only one part of carbon hydroxide is consumed per part of CO_2 absorbed, leading to a part of the precipitated salt and a part of water so that no harmful byproducts are generated (theoretically). Note that soda lime pellets do not only contain sodium hydroxide but also potassium hydroxide, which leads to similar reactions for that compound.

Aside from the activators (sodium hydroxide and potassium hydroxide), soda lime pellets also contain other substances as hardeners (such as silica and kieselguhr) and indicators. This last type of additive, which is included in order to give a visual cue of the exhaustion of the absorptive properties of the pellets, react to an increase of the pH produced by an accumulation of carbonic acid that cannot be absorbed by the hydroxides, changing the color of the pellets (6).

Despite the obvious advantages of this absorbent, it presents some drawbacks that have favored the introduction of new substances. On the one hand, as the sevofluorane is unstable in soda lime, a substance known as "Compound A" may be generated, endangering the life of the patient, because this substance is lethal at

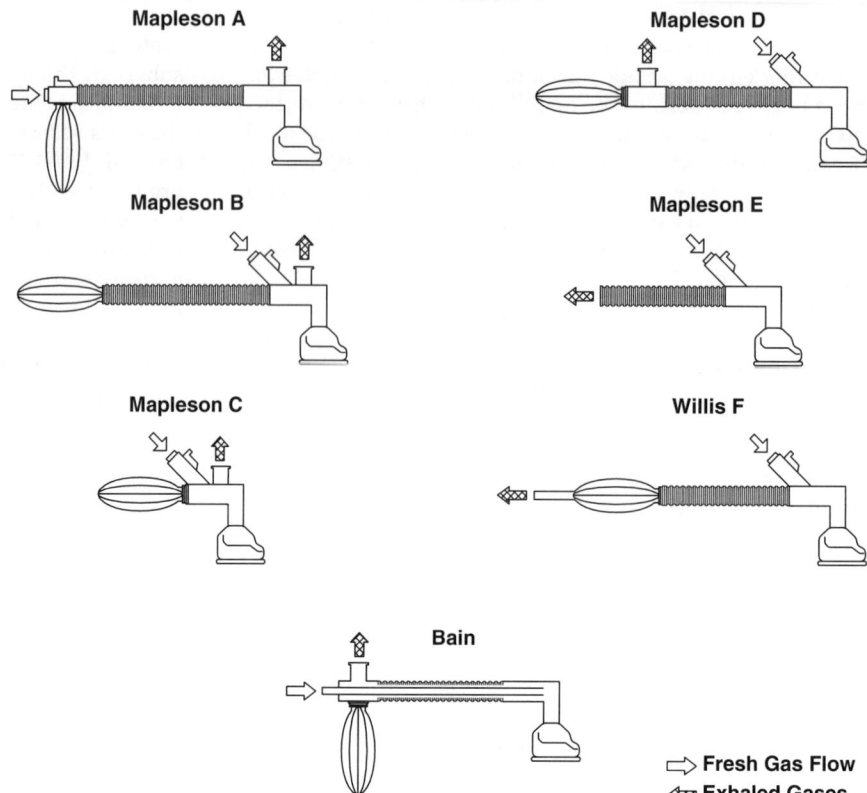

Figure 4. Non-rebreathing breathing (NRB) circuits.

130–340 ppm, and values around 25–50 ppm (which have been associated with renal injury in rats) have been extensively reported within normal clinical procedures, which does not just happen with this agent as the soda lime is totally incompatible with the trichloroethylene, generating a cranial neurotoxin (dichloroacetylene) and a potent pulmonary irritant (phosgene). On the other hand, Carbon monoxide (CO) is produced with many agents (desflurane, enflurane, isoflurane, halothane, and sevoflurane) because of the strong bases used as activators (KOH, NaOH). In order to overcome these problems and reduce the risk to the patient, alternative substances such as the barium hydroxide ($Ba(OH)_2$) have been recently introduced. The mechanism of the carbon dioxide absorption for this base is shown in Fig. 6.

In this case, a two-way removal of the carbon dioxide from the exhaled gas flow exists as one-tenth of this gas is removed by the hydrated barium hydroxide, releasing the amount of water required to produce the carbonic acid required for the absorption of the remaining nine-tenths of the carbon dioxide. As a result of these reactions, only water and two precipitated salts are generated (calcium and barium carbonates).

i) $CO_2 + H_2O \longrightarrow H_2CO_3$
ii) $H_2CO_3 + 2NaOH \longrightarrow Na_2CO_3 + 2H_2O + Energy$
iii) $Na_2CO_3 + Ca(OH)_2 \longrightarrow CaCO_3\downarrow + 2NaOH$

Balance $CO_2 + Ca(OH)_2 \Longrightarrow CaCO_3\downarrow + H_2O + Energy$

Figure 5. Carbon dioxide absorption by soda lime.

2.5.3. Ventilation. As mentioned before, ventilation of the patient should be granted for the supply of the life-sustaining gases and the vaporized anestethic agent, which is usually done in two ways, depending on the ventilatory assistance strategy adopted. If the patient presents a healthy condition and the surgical procedure does not interfere with the function of the respiratory system, the anesthesiologist in charge of the sedation may opt for keeping spontaneous ventilation throughout the procedure instead of appealing to mechanical ventilation.

If the first strategy is adopted, the patient will breath spontaneously inhaling the gas mixture through the inspiratory limb of the circuit before exhaling the mixture to the expiratory one. As the gas is expired, it enters into the reservoir bag (16) connected to the pop-off or adjustable pressure limiting valve (17), which is sensitive to the pressure of the exhaled stream.

In case this pressure exceeded the preset level, the APL valve will open, sending a fraction of the exhaled gases directly to the scavenging system. If mechanical ventilation is used, a similar valve located at the ventilator will release the excess pressure to the scavenging system, assuring the proper operation of the machine. As mechanical ventilation is reviewed in a dedicated chapter of this encyclopedia, the reader is encouraged to consult that work in order to get further information on these devices, which cannot be included on this chapter for obvious reasons.

2.6. Scavenging System

Once the gas mixture has been released by the ventilation circuit, it must be properly disposed in order to avoid the

i) $Ba(OH)_2 \cdot 8H_2O + CO_2 \longrightarrow BaCO_3 \downarrow + 9H_2O + Energy$
ii) $9H_2O + 9CO_2 \longrightarrow 9H_2CO_3$
iii) $9H_2CO_3 + 9Ca(OH)_2 \longrightarrow 9CaCO_3 \downarrow + 18H_2O + Energy$

Figure 6. Carbon dioxide absorption by barium hydroxide and soda lime.

risks associated with the contamination of the operating room's atmosphere. For this purpose, most of the hospitals count on central scavenging systems based on vacuum pipelines.

In order to guard the patient airway from possible suction by the scavenging system, modern systems include pressure relief valves. If the suction pressure was greater than the pressure of the exhaled gas stream, the negative pressure relief valve will open, letting the air from the operating room flow into the scavenging system until the pressure of the line reaches the preset value of the valve. In the opposite case, when the suction pressure was not enough to assure the disposal of the exhaled gases, a positive pressure relief valve will open, ejecting the exhaled gas mixture into the atmosphere of the operating room. This system, based around valves between the anesthesia machine and the scavenging system, is known as a closed system since the introduction of the first open systems in recent years. In these other systems, which have become popular, the exhaled gases are scavenged from a dedicated reservoir bag without requiring pressure relief valves.

3. INFUSION PUMPS IN ANESTHESIA

From the first experiences of Sir Christopher Wren (who has been credited as the first physician to administer intravenous drugs for anesthetic purposes in 1655) and the development of the hypodermic syringe by Alexander Wood in 1855, anesthesiologists had to wait until 1934 for the presentation of the first known intravenous agent (the thiopentone, first administered by Lundy & Waters) in order to replace the inhaled anesthetic agents used at that time (ether and chloroform) for general anesthesia. What was initially introduced as the final solution for anesthesia turned into tragedy because of the lack of knowledge of the drugs pharmacokinetics, to the point that, when a large number of patients injured at Pearl Harbor were lost because of problems with the induction and maintenance dosing, this agent started to be described as "an ideal method of euthanasia."

With the development of the pharmacokinetic modeling techniques, traditional dosing in the form of discrete boluses was substituted by total intravenous anesthesia (or TIVA) where the drugs are administered at a constant rate in function of the required dosage. Again, although this method proved useful for some drugs of simple kinetics (one compartment models), the performance at everyday clinical procedures was limited with other drugs that required continuous intervention by the anesthesiologist in order to correct the administration according to the clinical signs related to depth of anesthesia (10). In order to overcome this limitation, model predictive controllers were proposed to drive the infusion of the hypnotic drug in function of the predicted behavior of the drug within the patient through the use of accurate pharmacokinetic models. The first device approved for clinical use, which was introduced in 1996 for the administration of propofol, used an embedded computer to simulate the plasmatic concentration of the drug in order to adjust the required perfusion rate by means of an analytical solution of the associated three-compartment model (11,12).

On the one hand, this type of administration, also known as target-controlled infusion (TCI), has renewed the hope in intravenous anesthesia, and has recently encouraged the development of new depth of anesthesia measurement methods suitable for closed-loop administration of the intravenous agents (13). An example of the different infusion schemes (bolus administration, TIVA, and TCI) for propofol may be seen in the simulations depicted in Fig. 7 (14).

On the other hand, the stability of the sedation level offered by TCI systems has led to the development of intravenous anesthesia machines that include all of the classic systems except the vaporizer, resulting in very compact units specially suitable for portable or in-vehicle operation, particularly useful in military and humanitarian missions.

3.1. Architecture of Anesthesia Infusion Pumps

As infusion pumps are widely use for purposes other than anesthesia delivery, this encyclopedia will cover most of the different technological topics involved in a chapter dedicated to these devices; nevertheless, as the equipment specifically designed for the delivery of intravenous anesthetics and muscle relaxants presents several aspects not found in other pumps, a slight review of the internal architecture of these devices is included below. In order to guide this review, the reader is encouraged to consult Fig. 8, which depicts a general overview of these pumps.

The pump mechanics connect the stepper motor shaft with the syringe plunger through the transmission gear, which includes a clutch in order to ease the syringe loading and disposal procedures. The stepper motor, so called because it is able to turn in precise angular steps, is driven by the embedded controller of the pump, which triggers the motor driver in order to supply the required stimuli for turning one angular step.

Both rotary encoders are placed at the motor and the transmission shafts, providing the feedback signals required to assess the actual advance of the syringe plunger, which are also useful for the detection of the clutch state (as transmission declutch should generate an alarm when infusing).

The pumping pressure may be monitored by a strain gauge sensor located at one end of the slider antagonist spring, once conditioned by the proper signal conditioning circuitry. The other sensors located at the pump mechanics are intended for the identification of the syringe type through the measurement of the diameter (which is done by means of a linear potentiometer designed to wrap

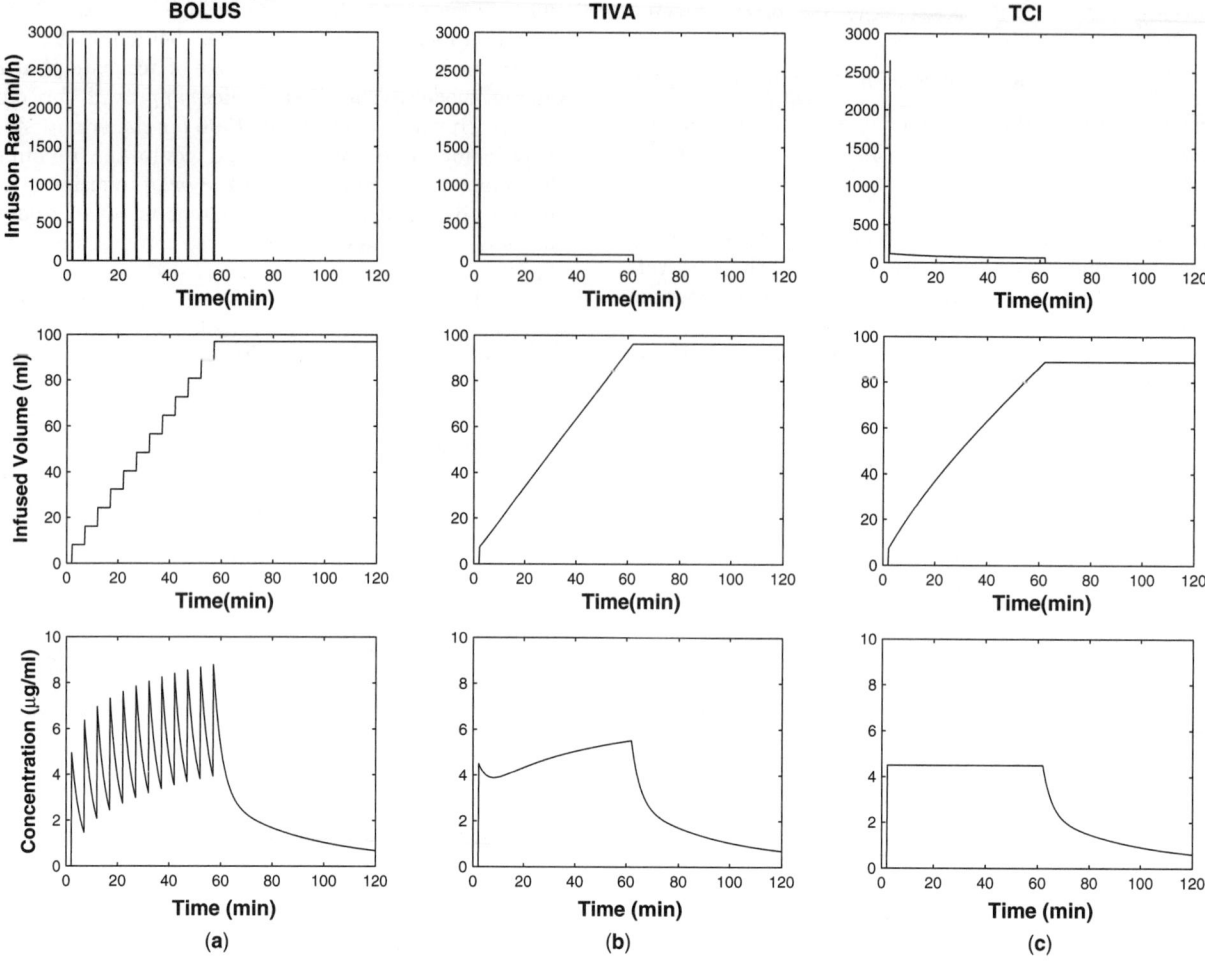

Figure 7. Different schemes used in infusion pumps for intravenous anesthesia delivery. (a) Bolus administration, (b) total intravenous anesthesia, and (c) target-controlled infusion. Although the total infused volume is almost the same (around 90 ml) for the three plots, the advantages of TCI are obvious as no overshoot is produced and the drug concentration in plasma is kept maximally stable during surgery. Simulations made by the authors for a 45-year-old male patient of 70 Kg and Marsh kinetic set for propofol under STANPUMP (14).

around the pumps outer cylinder) and the length (estimated in function of the initial slider position) of the selected syringe.

As pump operation should be granted in adverse situations, a high number of security elements have to be included in this basic design. First of all, an isolated power source (which is always included in order to comply with the required isolation barrier stated by the medical devices regulations) charges the emergency battery for overcoming power losses and allowing portable operation. The watchdog circuit should continuously monitor the output of this power supply in order to trigger the power-loss and low-battery alarms. This circuit is also used for the supervision of the integrity of the embedded processor, as a system breakdown should be notified immediately, as it is one of the most serious failures. Memory integrity, usually checked through the memory map checksum, and parity calculation should be included in order to assess the integrity of the machine code stored for the pump operation.

In order to manage the different alarm conditions, a priority system should be included (as a watchdog alarm should be more important than a notification of the near end of infusion condition). In fact, part of the alarming system should be implemented in hardware (through a sequential logic circuit designed around custom ASICs or specially programmed PLDs), rather than in software, in order to react to embedded processor malfunction. This subsystem should include a nurse call function in order to allow the remote notification of the alarming condition.

As happens with many other medical devices, serial ports are included in order to provide remote operation of the pump and data logging functions. On the one hand, these interfaces (which usually conform with the EIA RS232-C, RS-422, or RS-485 standards) should be properly isolated in order to reduce the risk of electric shock to the patient and the anesthesiologists. On the other hand, the integrity of the transmitted data should be granted by means of standard cyclic redundancy checksum methods such as CRC-16. Finally, additional modules, such as TCI controllers that connect to the main processor, should include their own security features in order to check the

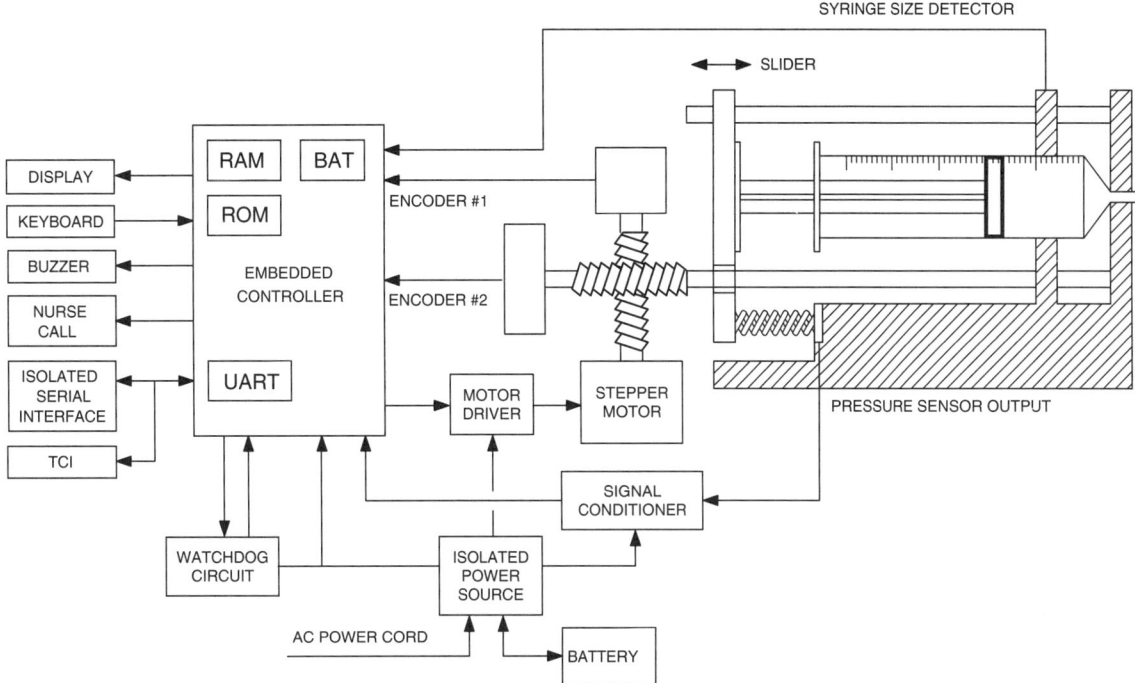

Figure 8. General infusion pump architecture.

internal condition and notify the main processor in case of failure.

3.2. Characteristics and Specifications

Although it might be difficult to state what the main characteristics are that should be considered when evaluating different infusion pumps for anesthesia, a complete list should include at least all of the following:

Infusion Range: Maximum and minimum values of the infusion rate. These values are especially important when designing new control strategies and should be included in the pumps model in order to limit the controller output to physically achievable values (e.g., it would be easy to get a flash induction by setting the infusion rate to infinite, although the pump mechanics would never be able to reach that speed). On the other hand, controllers should ignore negative perfusion speeds because these devices are intended just to pump the drug into the plasma but not to extract it (as this is physically impossible). Typical ranges cover from 0–0.1 to 1200 ml/h, although the same infusion pump may present different ranges for each one of the supported syringe types.

Bolus Rate: Maximum and minimum flow rates allowed for bolus injection. As this parameter depends on the supported syringe types, a value should be included for each one of them. The maximum values usually coincide with the maximum infusion rates.

Bolus Limit: Maximum volume to be infused by a single bolus. As the bolus injection is limited by the bolus rate, the actually infused volume per each bolus will be the result of the product of the bolus rate per the bolus injection time. Usual values are limited to 25 ml in order to minimize the risk of overdosing.

Volumetric Accuracy: Although this parameter depends on the different type of syringes supported, it has to be taken into account in order to reduce the mismatch between the actually infused volume and the one demanded by the controller. It should be noted that this value is often expressed graphically through the so-called "trumpet diagrams" that represent the evolution of the volumetric error along time in order to consider the transient time required by the mechanical elements of the pump.

Critical Volume: In order to grant the safety of the patient, pump manufacturers have to include the maximum volume that may be automatically injected by the pump in case of a single fault condition. The electronic surveillance systems included in current machines limit this value to 0.5 ml, which is small enough when compared with dangerous doses.

Maximum Pumping Pressure Limit: As an excessive pumping pressure could cause severe effects on the condition of the patient, pressure sensors are built-in within the pumps mechanics in order to trigger overpressure and occlusion alarms (typically set around 280 mmHg and 650 mmHg, respectively).

Resolution: Minimum change at the output rate. As most of the pumps are intended to cover a wide range of infusion rates, the actual resolution offered by the pump is usually presented as a function of the actual infusion rate, so that lower rates present smaller steps among possible rate values than at higher rates. Typical values start at 0.1 ml/h for the lowest infusion rates (below 99 ml/h) and extend up to 10 ml/h (above 1000 ml/h).

Update Rate: As mentioned before, the pumps mechanical elements present a significant response time in order to grant the accuracy of the infusion rate. In order to completely define the controller's specifications, the actual pump update rate should be included for manual and remote operation modes, as additional time delays are frequently introduced by the serial communications interface. As in many cases linear operation is not supported, the pump has to be stopped in order to change the infusion rate and then started again, the whole cycle time should be considered for the purposes listed above.

Purge Limit: In order to grant the proper administration of the intravenous agents, the air in the infusion lines should be removed in order to avoid administration of bubbles that would endanger the patient. The purge volume is usually limited in order to increase the manual control of the process and to avoid the waste of the agents used. Typical values are around 2 ml.

3.3. Infusion Control for Accurate Dosing

As pointed out previously, intravenous anesthesia has gained wider acceptance because of the advances introduced on dosing as pharmacokinetic models have been developed for the agents used. This section will describe the basic principles of the modeling technique involved in the operation of model-driven infusion pumps, as described in Fig. 9.

The infusion of an intravenous drug may be modeled as a mass flow (I) entering into the blood plasma, diffusing toward the different tissues perfused by the blood flow, which behave as reservoirs storing part of the drug before turning it back to the blood plasma. Finally, the drug is washed out from the plasma through liver and renal clearances, which may be modeled mathematically through the use of compartmental modeling techniques by associating a virtual compartment of known volume to the plasma (C_1 in Fig. 9) and several adjacent compartments to the rest of the tissues (C_2 and C_3). The diffusive processes may be modeled by first-order systems of gain and time constants derived after the so-called rate microconstants that model the intercompartmental diffusion.

Although early designs appealed to numerical integration techniques in order to obtain discrete approximations to the analytical solution of the differential equation set associated with the model description (illustrated in Fig. 9), current designs use the algorithm proposed by Bailey and Shafer (11) in order to maximize the accuracy of the predicted concentrations.

Target-controlled infusion is achieved, as seen in Fig. 10, by means of linear interpolation caused by the linearity of the system. Suppose the pump is about to calculate the rate needed to hold the target concentration (C_T) for the next ΔT seconds. At this point, the model is simulated assuming an input infusion rate equal to the maximum rate of the pump (I_{MAX}), resulting in a maximum value for the concentration for that time (C_{MAX}). This process is simulated again but supposing an input rate equal to the minimum rate offered by the pump (I_{min}). It is possible to find the linear relationship among the perfusion rate and the final concentration at that certain point, which enables the calculation of the exact rate required to hold the desired concentration. Finally, the model is simulated considering this rate in order to obtain the state variables (C_1, C_2, and C_3) required for future predictions.

The reliability of this system is limited by the proximity of the patient's model described by the kinetic set considered and the actual model of the patient. In order to overcome these limitations, models are derived after considering a high number of clinical samples, which limits the number of models available for each drug. It should be said that, in this work, the target concentrations have been described for the plasma, but actual devices offer the possibility of driving the infusion to reach certain concentration levels at the effect site (a virtual compartment receiving the drug from the plasma, which is supposed to be the actual origin of the therapeutic effect of the anesthetic agent) (15,16).

4. MONITORING

In addition to the main components described above (gas delivery, vaporizer, ventilator, and breathing circuit),

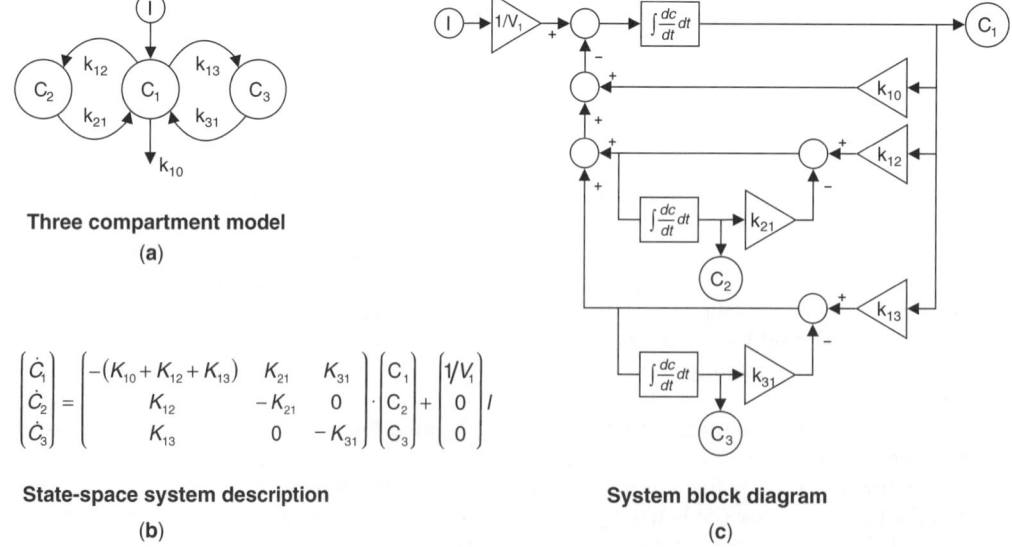

Figure 9. Pharmacokinetic models used for target-controlled infusion.

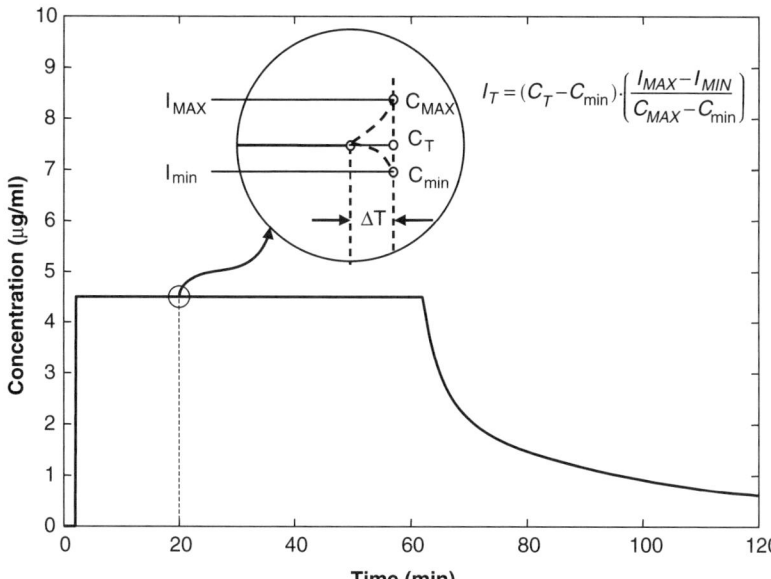

Figure 10. Target-controlled infusion (TCI) of propofol.

anesthesia machines usually include a monitoring subsystem specifically designed for the supervision of the state of the patient and the proper function of the whole system in order to:

1. Improve the security of the patient as a whole by preventing potentially harmful physiological conditions and detecting malfunctions in the normal operation of the anesthesia delivery unit.
2. Collect several physiological measures, which are considered to be interesting by the specialist in charge of the anesthetic procedure.
3. Check the integrity of the different components involved in anesthesia administration and taking corrective actions in response to system malfunctions.
4. Study the degree of change on a certain indicator in order to analyze trends and provide data required for the forecast of the patient evaluation (prognosis) during the surgical procedure.
5. Validate the impact of an specific therapy on the physiological state of the patient in order to give personalized anesthetic care.

Despite the potential offered by current systems, the information gathered by the monitoring equipment should be analyzed by the specialist in charge of the anesthesia as a whole, along with the clinical observation of the patient, in order to bear in mind those possible harmful conditions that may not be automatically detected.

4.1. Anesthesia Delivery System Supervision

System failures or malfunctions (such as hypoxic gas mixture administration, poor ventilation of the lungs because of low-volume gas mixture supply, or misconnections in the administration piping, overdosing, etc.). seriously endanger patients undergoing the anesthetic procedure. In order to avoid the undesired effects of these failures, granting the proper operation of the anesthesia delivery unit, several variables should be monitored, such as (7):

- Inspired oxygen concentration.
- Anesthetic vapor concentration.
- Carbon dioxide concentration.
- Air pressure.
- Exhaled gas volume.
- Manually operated valves and regulators set-points.

4.1.1. Gaseous Concentrations. The measurement of the concentration of the inhaled and exhaled gas mixtures is important to grant the administration of the proper proportion of medical gases and anesthetic agents, which provides timely information required to supervise the integrity and proper operation of the respiratory circuit and the anesthesia machine.

As the anesthesiologist has to grant the proper oxygenation of the body tissues throughout the procedure, oxygen concentration has to be continuously monitored in order to reduce the risk of administration of hypoxic or hyperoxygenated gas mixtures. Delivery of inappropriate gas mixtures should trigger the proper alarming mechanism. Although most of the devices limit the minimum rate of oxygen to a minimum of a 25%, inspired oxygen concentration has to be sensed at the inhalatory branch of the breathing circuit by means of transducers based on different principles of operation such as electrochemical, galvanic, polarographic, or paramagnetic elements of proved immunity to other gases.

Concentration of the other gaseous components present in the gas mixture (such as C_{O_2} and vaporized agents) is sensed using other techniques based on the optical properties of the different substances such as mass spectrometry, Raman spectrometry, and infrared spectrometry, the latter being the most widely used because of size, reliability, and life span. These sensors sample the gas flow at different points depending on the technology used. The most widely used systems are those sampling centrally (known as mainstream sensors) or laterally (known as side-stream sensors). Central sampling systems, which analyze the gas concentration through a miniature optic sensing element wrapped around the endotracheal tube, are mainly used for the assessment of the CO_2 concentration, as these devices are intended for the analysis of just one type of gas. The lateral or side-stream system (which is suited with a water trap to avoid the possible damage caused by condensed water vapor), takes samples of the gas flow near the endotracheal tube at a rate ranging from $50\,mL \cdot min^{-1}$ (from children) up to $250\,mL \cdot min^{-1}$ (from adults), analyzing the whole composition of the gas mixture before discarding or returning the gas samples.

4.1.2. Ventilation. Patient ventilation should be continuously evaluated in order to grant the proper respiratory function along the anesthetic procedure. Mechanical ventilation should be assessed by means of clinical observation along with capnography and the measurement of the volume, respiratory rate, and the maximum and minimum values of the airway pressure. In addition to these critical parameters, the anesthesia machine should monitor eventual disconnections at any point of the breathing circuit as well as limit the maximum pressure of the flow delivered to the patient in order to avoid barotrauma.

Capnography has spread within the anesthesiologist community because of the information that can be extracted from this measure of the evolution of the CO_2 concentration. If central sampling systems are used, it is possible to plot volumetric capnography as the concentration of the exhaled CO_2 and the exhaled volume, which provides an indirect measure of the alveolar and arterial pressure of the CO_2. The resulting plot may be studied to detect the changes associated with the respiratory function of the patient.

The measurement of the flowing and the exhaled volumes (which is performed at the respiratory branch of the breathing circuit returning from the patient) is essential for the evaluation of spontaneous or mechanically assisted ventilation. Respiratory gas flow may be measured at one end of a one-way valve by means of flowmeters based on rotors, neumotachographic, ultrasound, or differential temperature sensors.

As the measurement of airway pressure should be available at every ventilator, anesthesia machines usually include built-in sensors located near the ventilating device. This design ensures that the measured pressure reflects not only that attributed to the patient's airway but also the effect of the ventilator, its circuit, and the endotracheal tube, which must be taken into account in order to discard the alterations devoid of the patients state. The course of the pressure wave may help in monitoring the effect of the ventilation volume and rate on the pressure of the airways, at the same time enabling the detection of an obstruction in the patient's airways.

Knowing the amount of work required for reaching the pressure needed to trigger a change in lung volume is useful for setting the level of effort that will avoid the onset of fatigue and, thus, respiratory failure, as the ventilation work can be considered to be the result of overcoming the elastic forces and flow resistances associated with the respiratory effort. This value is easily measured from the volume-pressure loop when mechanical ventilation is applied as the ventilator develops all of the work. In case of spontaneous respiration, the measurement is more complicated and the use of esophageal pressure is required.

4.2. Patient Monitoring

As happens in most of the medical procedures, anesthesia should minimize the patient's perception of pain while preserving the normal functionality of all of the body systems. In order to grant the integrity of the patient, certain physiological parameters should be monitored to correct possible alterations before they lead to permanent cellular damage. In this sense, monitoring the oxygenation, ventilation, circulation, temperature, and neurological function of the patient is mandatory when performing anesthetic procedures. Among others, typical anesthetic protocols include the acquisition of hemodynamic, respiratory, and neurological variables, which will be discussed in the next sections.

4.2.1. Temperature. As body temperature is altered by the anesthetic procedure because of the induced changes on the adjustment thresholds of the patient system responsible for temperature regulation, it is usually monitored by the specialist in charge of the anesthesia, because of the simplicity of the measurement setting. On the one hand, it is known that decreased body temperature may lead to alterations of the state of the patient such as extending of the anesthetic effect, decreasing blood coagulation capability, and increasing the risk of post-surgical infections. On the other hand, in some surgical procedures, it is useful to keep body temperature slightly inferior to normal in order to protect brain functions when performing operations involving reduced blood perfusion to the brain.

4.2.2. Electrocardiogram (ECG, EKG). The electrocardiogram monitors the electrical activity of the heart recorded at the surface of the skin. Anesthesia monitors include a variable number of electrodes, ranging from a minimum of three up to 12 electrodes, which are used to obtain the three bipolar derivations from the Einthoven triangle (I, II, III), the three unipolar derivations (aVL, aVR, and aVF), and the six precordial derivations proposed by Wilson (V1 to V6; V5 being one of the most frequently used for the evaluation of the left ventricle function). ECG signal bandwidth is usually limited to 0.5–30 Hz for this purpose, which is quite low when compared with the standard bandwidth used for diagnostic purposes (0.05–150 Hz). Nevertheless, the advances in the technology involved have led to extending the typical bandwidth to the diagnostic range in order to monitor the ST segment for the detection of intraoperatory ischemia (17).

4.2.3. Blood Pressure.
The pressure originated by the mechanical contraction of the ventricles is usually monitored for the evaluation of the functional integrity of the cardiovascular system. For this purpose, several invasive and noninvasive methods have been developed in order to obtain the mean, systolic, and diastolic arterial pressures, as well as the heart rate.

4.2.3.1. Noninvasive Blood Pressure (NIBP).
Among the noninvasive methods used, oscillatory devices are the ones most commonly employed. These methods use an air-pressurized cuff with a built-in pressure sensor, which is wrapped around one of the limbs (generally an arm or leg, although it is possible to use wrist and finger cuffs) in order to detect the blood pressure of the peripheral arteries. The monitor inflates and deflates the cuff at programmable time periods (typically 1, 2, 3, 4, 5, 10, and 15 minutes) in order to acquire the blood pressure measure as needed by the specialist, at the same time that protracted inflated periods are avoided to reduce the risk of ischemia or brain damage. The measurement range is variable as a function of the patients age, with typical the ranges of 25–260 mmHg for adults, 20–195 mmHg for children, and 15–145 mmHg for babies, whereas the initial pressure ranges from 185, 150, and 120 mmHg for the same cases, although sometimes it is calculated as 25–40 mmHg above the last measured mean systolic value. In order to avoid cuff overpressure, the inflation mechanism usually includes security valves that release excess pressure in order to avoid harmful conditions for the patient.

4.2.3.2. Invasive Blood Pressure.
In order to obtain the pulse wave, arterial blood pressure must be acquired invasively through the use of a catheter-mounted pressure sensor, which is introduced within the blood flow through a puncture in a peripheral artery, although better signals may be obtained with larger arteries (at the cost of increased risk) (18).

The pressure sensor, which is usually built around a strain gauge bridge or a piezoelectric sensor, must present a useful range from -40 to $350\,mmHg$, a maximum allowed variation of 5% (equivalent to $\pm 2\,mmHg$) and a recommended gain of $5\,uV/V/mmHg$ (although typical on-screen resolution is limited to $1\,mmHg$). The bandwidth required for sampling is limited to the tenth harmonic of the pulse wave, taking the typical value of $20\,Hz$ associated with a $2\,Hz$ (or 120 beats per minute) pulse wave.

4.2.3.3. Pulmonary Artery Pressure (PAP).
Special measurement techniques have been developed for the pulmonary artery pressure, as its systolic, diastolic, and mean values have been related to the precapillary resistance and the pulmonary artery occluding pressure (PAOP), an indicator of the hemodynamic state of the left heart cavities. Monitored with a flow-directed pulmonary artery catheter, the typical values range from 15–25 mmHg (systolic) to 0–8 mmHg (diastolic) for the right ventricle and 15–25 mmHg (systolic), 10–20 mmHg (mean) to 8–15 mmHg (diastolic) for the pulmonary artery, whereas the PAOP ranges from 6 mmHg to 12 mmHg.

4.2.4. Hemodynamic Parameters.
Cardiac output, which represents a measure of the volume of blood impelled by the heart over a 1-minute period, may be calculated as the systolic volume multiplied by the heart rate. This measure has been related to the oxygen and nutrient requirements of the different body tissues as the heart pumps more or less blood as a function of those requirements. The usual measurement method involves the use of a catheter inserted into the pulmonary artery, suited with a thermistor at the tip. This temperature-sensing element is intended to record the temperature gradient appearing after the rapid injection of a known volume bolus of a cold saline solution (between 0–25°C). This thermodilution method (19) uses the formula proposed by Steward–Hamilton in order to calculate the cardiac output as a function of the area under the curve associated with the temporal temperature evolution. As the snapshot value of this parameter varies significantly with time, it is normal to give an averaged value obtained after several trials, taking typical values ranging from $4\,L/min$ to $12\,L/min$. Additionally, this parameter may be normalized to body surface area in order to provide a new parameter known as Cardiac Index, which ranges from $2\,L/min/m^2$ to $5\,L/min/m^2$.

As happens with the left ventricle, which may be fairly estimated by the joint study of the PAOP and the cardiac output, the right ventricular function can be evaluated through the observation of the cardiac output and the central venous pressure (CVP), as CVP is a measure of the pressure in the right atrium, which reflects changes in right ventricular end-diastolic pressure. Aside from these measures, the ejection fraction is also used for the evaluation of ventricular function, as it is related to the amount of blood impelled by the heart on every contraction as estimated through the thermodilution method.

Many other hemodynamic parameters derived after these measures exist that ease a complete evaluation of the cardiac function such as the ejection volume and the systemic and pulmonary vascular resistances (Table 1).

4.2.4.1. Esophageal Doppler.
This method is used to estimate the cardiac output derived after the measurement of the blood flow at the descending aorta by means of esophageal probes equipped with doppler transducers, which enable the recording of the blood flow over time. The blood volume ejected by the heart may be calculated as the product of the valve area multiplied by the integral of the flow over time; once this value is known, aortic output can be evaluated as the product of this volume multiplied by the heart rate, presenting a typical value close to 70% of the cardiac output. This method presents some technical drawbacks related to the estimation of the aortic section area and the alignment of the doppler beam in the direction of the blood flow. A variation of this technique may be found in transesophageal echocardiography, which has increased in popularity among anesthesiologists (20). This procedure offers real-time images that may be processed to assess the cardiac anatomy, possible alterations, and the result of a surgical procedure as well as the size of the ventricular cavities, among other measures, such as the systolic and diastolic areas that are frequently used

Table 1. Different Hemodynamic Parameters Usually Calculated by Anesthesia Monitors

Parameter Calculation	Acronym	Normal Range	Units
Cardiac Index $CI = CO/BS$	CI	2,4–4,2	$L \cdot min^{-1} \cdot m^{-2}$
Stroke Volume $SV = CO * 1000/PR$	SV	50–110	$mL \cdot beat^{-1}$
Stroke Volume Index $SVI = SV/BS$	SVI	30–65	$mL \cdot beat^{-1} \cdot m^{-2}$
Systemic Vascular Resistance $SVR = (ARTm - CVP) * 80/CO$	SVR	800–1500	$din \cdot seg^{-1} \cdot cm^{-5}$
Pulmonary Vascular Resistance $PVR = (PAPm - PAOP) * 80/CO$	PVR	150–250	$din \cdot seg^{-1} \cdot cm^{-5}$
Right Ventricular Stroke Work Index $RVSWI = 1.36(PAPm - CVP)SVI/100$	RVSWI	5–10	$g \cdot m^{-1} \cdot m^{-2}$
Left Ventricular Stroke Work Index $LVSWI = 1,36(ARTm - PAOP)SVI/100$	LVSWI	45–60	$g \cdot m^{-1} \cdot m^{-2}$

CO: Cardiac Output ($L \cdot min^{-1}$); BS: Body surface (m^2); PR: Pulse Rate (b.p.m); ARTm: Arterial Pressure (mean); PAPm: Pulmonary Artery Pressure (mean)

for calculating the preload and the ventricular ejection fraction.

4.2.5. Pulse Oximetry and Mixed Venous Saturation.

4.2.5.1. Pulse Oximetry. Pulse oximeters provide noninvasive information for monitoring the arterial oxygen saturation, measured as a function of the hemoglobin light absorption ratios at different wavelengths. In practice, the algorithms commercially implemented are based on empirical models of the absorption ratios of light emitted by an LED or laser source at 660 nm and 940 nm placed within a clamp-shaped enclosure specially designed to be fitted at the finger or earlobes of the patient. The measured value (which typically ranges from 40% to 100%) is severely affected by sensor displacement and external light sources (e.g., fluorescent lighting), so that powerful artifact algorithms are used in order to increase the reliability of the procedure.

4.2.5.2. Mixed Venous Oxygen Saturation. As mentioned above, the cardiac output measures the volume of blood flow ejected by the heart per time unit, although it is impossible to asses if this volume is enough to meet the oxygen demands of the whole body by the sole use of this parameter. In order to cover this gap, venous saturation, a measure of the balance between oxygen supply and the demands imposed by the different body tissues, has been widely adopted. This technique uses an optical fiber bundled into a light-emitting catheter that is inserted into the pulmonary artery in order to measure the ratio of reflected light. As light is reflected by the reduced hemoglobin in the blood flow, this ratio constitutes a valuable indicator of the mixed venous saturation, which is usually expressed as a percentage typically ranging from 65% to 85%.

4.2.6. Neurological Monitoring.
Central nervous system monitoring became widespread in the early 1990s, as it proved essential for the prevention and identification of brain ischemia, not just in neurosurgery but also in heart and carotid artery surgery. The equipment developed since that time has helped research the effect of different drugs on the brain, as well as the quantification of possible brain lesions that enable personalized health care for the patient. The observed changes of the parameters acquired by the monitor reflect changes in the function or blood flow of the specific area of the central nervous system under study during surgery. Electrical activity of the brain is usually monitored through the electroencephalogram (EEG), although anesthesiologists have recently increased their interest in the electrocardiograms and the evoked potentials because of the changes associated to different depth of anesthesia levels.

As the EEG, which reflects the electrical activity of the neurons at the brain cortex, is sensitive to different physiopathological conditions as well as to different drugs, it has been proposed for the evaluation of the central effects of different anesthetic drugs. For this purpose, specific algorithms and methods have been developed in order to analyze this signal discarding all of the non-relevant information. The typical procedure follows a simple scheme: Once the signal is amplified and filtered (generally to a 0.1–30 Hz bandwith), time-domain (average amplitude, zero crossing rate, aperiodic analysis, burst suppression ratio, etc.) and frequency-domain (spectral analysis, spectral margin, bispectral analysis, etc.) measures are performed.

When the surgical procedure enables the placement of the recording electrodes directly over the brain cortex, electrocardiograms are recorded in order to provide further information about the brain function (21). As the evoked potentials, triggered by external stimuli (visual, audible, motoric, etc.), are affected by hypnotic drugs as a function of their site of effect, they have been proposed for the evaluation of the hypnotic state (e.g., as some hypnotics may inhibit the peripheral nerve conduction, the amplitude and latency of the evoked potential may be dimmed when compared with the baseline values) (22).

4.2.7. Brain Hemodynamics.
One of the most frequent causes of nervous system damage is the insufficient blood flow to the brain. In addition to transcranial doppler and thermodilution methods, today, with the spread of the

Xenon gas as an anesthetic agent, functional magnetic resonance imaging techniques enable the precise measurement of central blood flow.

4.2.8. Neuromuscular Blockade Monitoring. Patient relaxation (temporal lack mobility), achieved by certain anesthetic agents but mostly by muscle relaxants, should be monitored in order to obtain the proper dosing regime, as different surgical procedures require different levels of muscular blockade. This function is evaluated by the measurement of the exerted force after a muscle contraction is triggered when electrical stimuli (usually in the form of simple spikes, double bursts, train of four (TOF), or post titanic bursts) are applied to a nerve.

5. FINAL WORDS

As happens in many other medical fields, anesthesia has adopted different technological advances as they became available. Evolution of anesthetic equipment in response to trends imposed by the clinical community, the manufacturers, and the regulating agencies will continue. In this regard, the current trends for the design of these machines may develop to cover the following topics:

Multimodal anesthesia delivery units: The integration of infusion pumps with anesthesia machines as a built-in module may not seem significant at first, but it is very important because it enables one to monitor the infusion parameters and predicted concentration values along with the rest of the physiological indicators, which are essential for influencing proper dosing. Some manufacturers have gone further, developing the latest generation of these machines to include all of the elements required for both kinds of anesthetic agents (inhalation and injectable). These designs seem to point out the future direction of these machines because of their versatility and adaptability to each patient and surgical procedure.

Personalized anesthesia delivery: Parameters acquired by the monitoring equipment offer a measure of the different aspects of the depth of anesthesia of the patient. The current trend integrates this data along with the specific effects of the different drugs (as well as the possible interactions) in order to provide the proper dose of the different agents to the patient, leading to a personalized anesthesia administration. This goal may ultimately lead to the introduction of efficient closed-loop systems for the administration of the agents, adjusting the dose to the actual requirements of each patient (22,23).

Anesthesia delivery in Magnetic Resonance Imaging: The requirements imposed by the MRI equipment and its environment have led the medical device manufacturers to develop machines specifically designed to comply with the electromagnetic restrictions in this field. In this sense, magnetic shielding and interference reduction techniques are not the only modifications to ordinary designs, but also the location of the machine with respect to the imaging equipment and the adaptation of the peripheral devices such as infusion pumps.

Depth of anesthesia: Cortical components of anesthesia (those relate to the anesthetic effects on the brain cortex, which is related to the cognitive process) are measured through the bispectral index (BIS), which is currently available in the form of stand-alone monitors or integrated within the monitoring equipment of the anesthesia machines. Additionally, subcortical components are even more important because they are involved in the regulation of the motoric and the autonomic stability, which grants antinociception (related to the nociceptive process of the nervous system that leads to a proper degree of analgesia). The role of antinoception is essential for surgery as it may reduce negative consequences immediately and over the long-term.

Integration with information systems: As with other medical devices, the integration of anesthesia machines within the hospital information system has become mandatory. Anesthetic equipment is connected to other devices through an information system that enables remote monitoring and recording, thus most manufacturers are offering this option. The recent introduction of ubiquitous computing and spot technologies (Internet-enabled devices) that have been favored by the spread of wireless networks, based on Bluetooth and IEEE 802.11, may open the doors to the next generation of anesthesia machines.

BIBLIOGRAPHY

1. E. Rahardjo, Ether, the anesthetic from 19th through 21st century. *International Congress Series* 2002; **1242**:51–55.

2. J. Lagunilla, J. Diz, A. Franco, and J. Alvarez, Changes in the social position of anesthesiologists in the 19th and 20th centuries. *International Congress Series* 2002; **1242**:337–342.

3. N. Webel, B. Harrison, and P. Southorn, Anesthesia origins of the intensive care physician. *International Congress Series* 2002; **1242**:613–617.

4. J. A. Dorsch and S. E. Dorsch, *Understanding Anesthesia Equipment*, 4th ed. Baltimore, MD: Williams and Wilkins, 1998.

5. ECRI. (2002). *Healthcare Product Comparison System*: *Anesthesia Units* (online). Available: *http://www.ecri.org*.

6. M. P. Dosch. (2002). *The Anesthesia Gas Machine*. University of Detroit Mercy Graduate Program in Nursing (online). Available: *http://www.udmercy.edu/crna/agm/*.

7. A. W. Paulsen, Chap. 84:Essentials of Anesthesia Delivery. In: J. D. Bronzino, ed., *The Biomedical Engineering Handbook*, 2nd ed. Boca Raton, FL: CRC Press, 2000.

8. J. C. Otteni, J. B. Cazalaà, and P. Feiss, Appareil d'anesthésie: systèmes d'alimentation en gaz frais. 1. Systèmes mécaniques avec rotamètres et vaporisateurs. *Ann. Fr. Anesth. Réanim.* 1999; **18**:956–975.

9. J. C. Otteni, J. Ancellin, J. Cazalaà, and P. Feiss, Appareil d'anesthésie: systèmes d'alimentation en gaz frais. II. Systèmes électroniques. *Ann. Fr. Anesth. Réanim.* 1999; **18**:976–986.

10. I. Smith, Inhalation versus intravenous anesthesia for day surgery. *J. Amb. Surg.* 2003; **10**:89–94.

11. J. M. Bailey and S. L. Shafer, A simple analytical solution to the three-compartment pharmacokinetic model suitable for CCIPs. *IEEE Trans. BME* 1991; **38**:522–525.

12. J. B. Glen, The development of 'Diprifusor': a TCI System for Propofol. In: *Anesthesia*. Oxford, UK: Blackwell Science, 1998 (53S1); pp. 13–21.

13. W. M. Haddad, T. Hayakawa, and J. M. Bailey, Adaptive control for non-negative and compartmental dynamical systems with applications to general anesthesia. *Int. J. Adapt. Control Signal Proc.* 2003; **17**:209–235.
14. S. L. Shafer, *STANPUMP freeware*. Palo Alto, CA: Anesthesiology Service PAVAMC, 1998.
15. J. R. Jacobs, Algorithm for optimal linear-model based control with application to pharmacokinetic model-driven drug delivery. *IEEE Trans. BME* 1990; **37**:107–109.
16. R. N. Upton, Relationships between steady state blood concentrations and cardiac output during intravenous infusions. *Biopharm. Drug Disposition* 2000; **21**:69–76.
17. M. J. London, M. Hollenberg, M. G. Wong et al., Intraoperative myocardial ischemia. Localization by continuous 12-lead electrocardiography. *Anesthesiology* 1988; **69**:232–241.
18. F. H. Van Bergen, D. S. Weatherhead, A. E. Treloar et al., Comparison of direct and indirect methods of measuring arterial blood pressure. *Circulation* 1954; **10**:481–450.
19. G. Fegler, Measurement of cardiac output in anaethetised animals by a athermodilution method. *Q. J. Exp. Physiol.* 1954; **39**:153–164.
20. K. H. Sheikh, N. P. De Brujin, J. S. Rankin et al., The utility of transesophageal echocardiography and Doppler color-flow imaging in patients undergoing cardiac valve surgery. *J. Am. Coll. Cardiol.* 1990; **15**:33–70.
21. R. A. W. Lehman, Intraoperative cortical stimulation and recording. In: G. B. Russell and L. D. Rodichock, eds., *Primer of Intraoperative Neurophysiologic Monitoring*. Boston, MA: Butterworth-Heinemann, 1995, pp. 195–204.
22. M. Mahfouf, A. Asbury, and D. Linkens, Physiological modelling and fuzzy control of anesthesia via vaporisation of isoflurane by liquid infusion. *Int. J. Simulation* (ISSN 1473-804x online, 1473-8031 print) 2001; **2**(1):55–66.
23. L. Vefghi and D. Linkens, Internal representation in neural networks used for classification of patient anesthetic states and dosage. *Comput. Meth. Progr. Biomed.* 1999; **59**: 75–89.

FURTHER READING

C. Thornton and R. M. Sharpe, Evoked responses in anesthesia. *Br. J. Anesth.* 1998; **81**:771–781.

AORTIC STENOSIS AND SYSTEMIC HYPERTENSION, MODELING OF

DAMIEN GARCIA
LOUIS-GILLES DURAND
Institut de Recherches Cliniques
de Montréal
Montreal, Quebec, Canada

1. INTRODUCTION

Aortic stenosis (AS) is the most common cardiovascular disease after hypertension and coronary artery disease and is the most frequent cause of valvular replacement in developed countries (1,2). This disease refers to the narrowing of the aortic valve opening during left ventricular (LV) ejection (Fig. 1a). This can be caused by a congenital abnormality of the valve (for example, the valve could have only two cusps instead of three). The most common cause of AS today is, however, the calcification of the valve cusps induced by a progressive degeneration of leaflet tissue (senile degenerative stenosis) (3). The narrowing of the valve aperture induces an obstruction to blood flow from the left ventricle to the aorta resulting in an increase in LV afterload (Fig. 1b). When AS becomes severe, symptoms such as shortness of breath, chest pain, and dizziness may occur and survival is markedly reduced (3). Once patients develop symptoms, prompt aortic valve replacement is generally needed. In patients with AS, the prevalence of systemic hypertension ranges from 30 to 40 % (4,5).

Systemic hypertension (HPT) is due to abnormally high blood pressure in the systemic arteries, i.e., in the vessels that carry blood from the heart to the organs. It is generally brought on by two interrelated physiological factors (6): (1) a reduction in the caliber of small arteries or arterioles with an ensuing increase in systemic vascular resistance and mean blood pressure, and (2) a reduction in the arterial compliance (ability of an artery to distend with increasing transmural pressure) with a resulting increase in pulse pressure (systolic minus diastolic blood pressure). When AS coexists with HPT, the left ventricle faces a double pressure overload (valvular and vascular). Consequently, symptoms of AS develop at a lesser degree of valvular obstruction in hypertensive compared with normotensive patients (4). In the presence of AS and/or HPT, a compensatory concentric hypertrophy of the left ventricle appears which contributes to preserve an adequate cardiac performance. Concentric hypertrophy is primarily characterized by an LV wall thickening, as new contractile-protein units are generated in parallel to the existing ones, whereas the LV cavity volume generally remains unchanged (Fig. 1b). It thus compensates for the increased LV wall stress and tends to maintain a normal cardiac output (7,8). In the clinical situation of coexistent valvular and vascular overloads, it is difficult to differentiate the events caused by AS from those caused by HPT. A detailed understanding of the respective impacts of AS and HPT on the LV function would help to predict whether aortic valve replacement and/or antihypertensive medical treatment would be beneficial in this particular situation. It is, however, difficult to perform a comprehensive analysis of the interaction between different pathologies in the context of a clinical study because this approach usually requires a large number of physiological measurements in a large cohort of patients. In addition, these measurements are often difficult to achieve, if at all possible, in patients. The use of mathematical or numerical models may overcome this dilemma.

Several theoretical cardiovascular models have been proposed to analyze the effect of vascular properties and/or AS on the LV function (9–17). Most of these models contain, however, numerous independent input parameters so that their potential application in the clinical setting may be very limited. Moreover, only a few numerical models include the effect of AS (11,12,14). But these latter assume a linear relationship between the transvalvular flow rate and the pressure difference across the AS. Such a

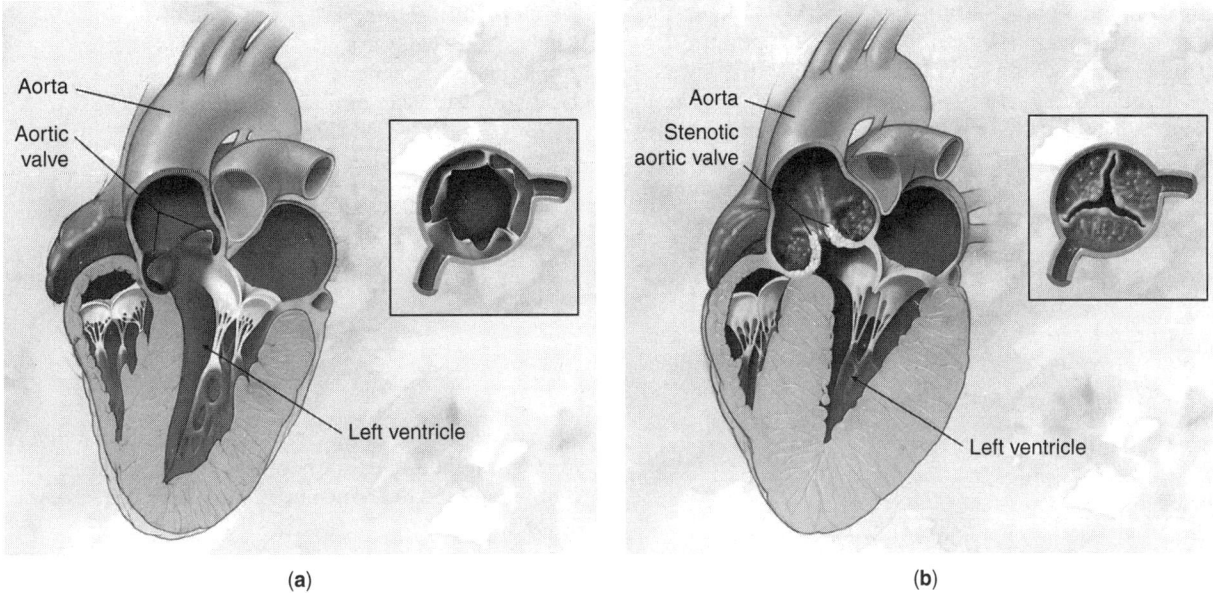

Figure 1. Schema of the heart during ejection. (a) without aortic stenosis, the aortic valve is fully opened; (b) in the presence of aortic stenosis, the calcified aortic valve cannot open fully, which causes an obstruction to blood flow from the left ventricle to the aorta and produces a transvalvular flow jet. Note the increase in left ventricular wall thickness. From Nishimura (3) with permission. (This figure is available in full color at http://www.mrw.interscience.wiley.com/ebe.)

representation is inaccurate since the flow-pressure relationship in AS is by far non-linear. To overcome these limitations, we thus developed an explicit analytical representation of the AS hemodynamics (18), which was incorporated in a simple cardiovascular model based on the so-called three-element windkessel model (19). The resulting ventricular-valvular-vascular mathematical model (V^3 model) and its main features will be depicted in this chapter. In the next section, we first describe the AS hemodynamics and the inherent hypotheses used for the derivation of the corresponding analytical model. The ventricular and the vascular models are then briefly explained before introducing the V^3 model. We then describe some findings obtained with the numerical V^3 model and their clinical implications. We finally conclude by presenting potential future improvements of the mathematical V^3 model.

2. THEORETICAL MODELS

2.1. Description of the Flow across the Aortic Stenosis

For convenience, we define left ventricular (LV) systole as the LV ejection period, i.e., the period where the transvalvular flow rate (Q) is strictly positive. The flow pattern across an aortic stenosis (AS) is very similar to the one occurring in orifice plates used as differential-pressure flow metering devices (20). It is mainly characterized by a flow contraction as far as the vena contracta, followed by an abrupt expansion (Fig. 2). The vena contracta corresponds to the location where the cross-sectional area of the jet is minimal (location 2, Fig. 2). This area is called the effective orifice area (*EOA*) of the valve. Within the contraction region, upstream from the vena contracta, some static pressure is converted to dynamic pressure. Flow contraction is a stable process with virtually no energy loss (21). Beyond the vena contracta, the fluid decelerates as the area occupied by the throughflow increases to fill the cross-section of the ascending aorta. The jet is rapidly lost in a region of turbulent mixing which involves

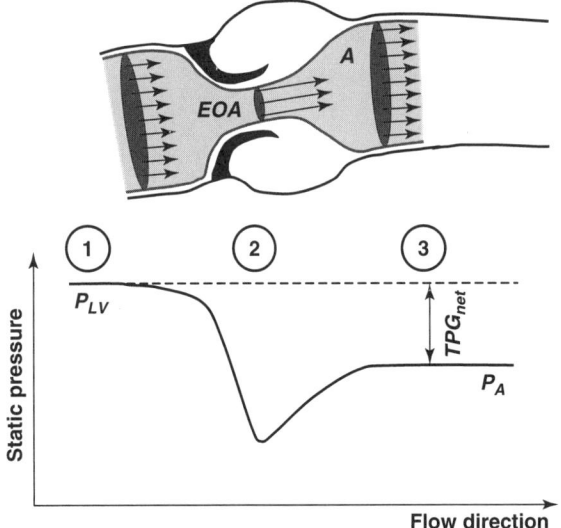

Figure 2. Schema of the flow across an aortic valve during systole and corresponding pressure field along the flow axis. Locations 1, 2, and 3 correspond to the detachment of the flow from the left ventricular outflow tract, the vena contracta, and the location where pressure is recovered. TPG_{net} is the net transvalvular pressure gradient and *EOA* is the valvular effective orifice area. P_{LV} = left ventricular pressure, P_A = aortic pressure, A = cross-sectional area of the ascending aorta. (This figure is available in full color at http://www.mrw.interscience.wiley.com/ebe.)

significant fluid energy dissipation. In this region, the static pressure increases until it reaches a maximum beyond the location where the reattachment of the flow occurs. The difference between LV pressure and recovered aortic pressure is called the net transvalvular pressure gradient (location 3, Fig. 2) or TPG_{net} (22). Note that cardiologists rather use the expression "pressure gradient" where a physicist would properly utilize "pressure difference," reserving the term "gradient" to express the rate of change in pressure per unit distance (23). To remain consistent with the clinical terminology, the term "pressure gradient" will be used in this chapter. The following section thoroughly describes the theoretical development of the instantaneous TPG_{net} expression as a function of transvalvular flow rate (Q) and effective orifice area (EOA). The mathematical derivation is adapted with permission from our previous work (18).

2.2. The Pressure-Flow Relationship in Aortic Stenosis

According to the aforementioned properties of the transvalvular flow, we first suppose that the fluid is ideal (i.e., incompressible and nonviscous) upstream from the vena contracta. Second, we postulate that the valve opens and closes instantaneously and that its EOA remains constant throughout systole. Third, we assume that the velocity profiles are flat (plug flow) within the throughflow. Finally, for simplicity's sake, we suppose that the respective cross-sectional areas of the LV outflow tract (upstream section, location 1, Fig. 2) and of the ascending aorta (downstream section, location 3, Fig. 2) are equal and noted A.

Because gravitation has no significant effect on the transvalvular flow and blood is an incompressible fluid, the generalized Bernoulli equation used along the axial streamline linking LV outflow tract (location 1) with ascending aorta (location 3) yields TPG_{net}:

$$TPG_{net} = P_{LV} - P_A = P_1 - P_3 = \frac{1}{2}\rho(U_3^2 - U_1^2) + \rho \int_1^3 \frac{\partial U}{\partial t} dl + E_L, \quad (1)$$

where P, U and ρ are the pressure, the velocity, and the density of the fluid, respectively. Coordinate l is the curvilinear coordinate along the considered streamline and the subscripts refer to the location number. E_L is the total energy loss induced by the flow expansion. Because the cross-sectional areas are similar in locations 1 and 3 and velocity profiles are flat, $U_1 = U_3$. Therefore,

$$TPG_{net} = \rho \int_1^3 \frac{\partial U}{\partial t} dl + E_L. \quad (2)$$

According to the conservation of mass, the transvalvular flow rate Q can be written as $Q = A(l)U(l)$, where $A(l)$ is the cross-sectional area of the through-flow at location l. If we further assume that $A(l)$ is not time-dependent, Equation 2 therefore yields

$$TPG_{net} = \rho \frac{\partial Q}{\partial t} \int_1^3 \frac{1}{A(l)} dl + E_L. \quad (3)$$

To obtain the complete expression of TPG_{net}, one has to know E_L. Recalling that no energy loss occurs upstream from the vena contracta, E_L appears in the generalized Bernoulli equation applied between locations 2 and 3:

$$P_2 - P_3 = \frac{1}{2}\rho(U_3^2 - U_2^2) + \rho \int_2^3 \frac{\partial U}{\partial t} dl + E_L. \quad (4)$$

Using the mass conservation, as above, this can be rewritten as

$$P_2 - P_3 = \frac{1}{2}\rho Q^2 \left(\frac{1}{A^2} - \frac{1}{EOA^2}\right) + \rho \frac{\partial Q}{\partial t} \int_2^3 \frac{1}{A(l)} dl + E_L. \quad (5)$$

For a control volume Ω that is fixed and non-deforming and whose boundary is noted Γ, Newton's second law of motion applied to an incompressible flow can be written as (24)

$$\rho \int_\Omega \frac{\partial U}{\partial t} d\Omega + \rho \int_\Gamma UU \cdot n d\Gamma = \sum F, \quad (6)$$

where U is the fluid velocity vector, n is the outward-pointing normal, and F are the forces that act on what is contained in Ω. Let's consider the fluid contained in the control volume Ω delimited by the aortic wall and the respective cross-sections at location 2 and location 3 (Fig. 2). Since the action of gravity and the wall shear forces are negligible here, the only axial forces acting on the fluid contained in Ω are the inlet and outlet pressure forces. Because velocities are uniform in locations 2 and 3, it follows from Equation 6 applied to Ω:

$$\rho \int_\Omega \frac{\partial U}{\partial t} d\Omega - \rho U_2^2 EOA + \rho U_3^2 A = (P_2 - P_3)A. \quad (7)$$

Again, the conservation of mass leads to

$$\rho \int_2^3 \frac{\partial Q}{\partial t} dl + \rho Q^2 \left(\frac{1}{A} - \frac{1}{EOA}\right) = (P_2 - P_3)A. \quad (8)$$

From the combination of Equations 5 and 8, one can get the energy loss:

$$E_L = \rho \frac{\partial Q}{\partial t} \left(\frac{1}{A}\int_2^3 dl - \int_2^3 \frac{1}{A(l)} dl\right) + \frac{1}{2}\rho Q^2 \left(\frac{1}{EOA} - \frac{1}{A}\right)^2. \quad (9)$$

Finally, Equations 2, 9 give the expression of TPG_{net}:

$$TPG_{net} = \rho \frac{\partial Q}{\partial t} \left(\frac{L_{23}}{A} + \int_1^2 \frac{1}{A(l)} dl\right) + \frac{1}{2}\rho Q^2 \left(\frac{1}{EOA} - \frac{1}{A}\right)^2, \quad (10)$$

where $L_{23} = \int_2^3 dl$ is the recovery length, i.e., the distance separating the vena contracta (location 2, Fig. 2) from the location where static pressure is totally recovered (location 3, Fig. 2). As shown by Equation 10, TPG_{net} is governed by the local inertia (involving $\partial Q/\partial t$) and by the convective inertia (involving Q^2). The first term in brackets, related to the local inertia, is purely associated to the geometry of the flow jet and is homogeneous to the inverse of a length. We therefore define the parameter λ, homogeneous to a length, as follows:

$$\frac{1}{\lambda} = \frac{L_{23}}{A} + \int_1^2 \frac{1}{A(l)} dl, \quad (11)$$

so that Equation 10 is reduced to

$$TPG_{net} = \rho \frac{1}{\lambda} \frac{\partial Q}{\partial t} + \frac{1}{2} \rho Q^2 \left(\frac{1}{EOA} - \frac{1}{A}\right)^2. \quad (12)$$

It can be noticed from Equation 11 that when EOA [i.e., $A(l=2)$] converges toward zero, $1/\lambda$ tends toward $+\infty$. On the contrary, when $EOA = A$ (no stenosis), locations 1, 2, and 3 are superimposed and $1/\lambda$ thus equals zero. The expression $1/\lambda$ is exclusively dependent upon the flow jet geometry and more precisely upon EOA and A, so that a dimensional analysis (25) gives the following relationship:

$$\frac{\sqrt{A}}{\lambda} = f\left(\frac{A}{EOA}\right). \quad (13)$$

A simple type of functions f defined on $[1, +\infty]$ that meets the two aforementioned boundary conditions [i.e., $f(x) \to +\infty$ when $x \to +\infty$ and $f(1) = 0$] is the following:

$$\frac{\sqrt{A}}{\lambda} = \alpha \left(\frac{A}{EOA} - 1\right)^\beta, \quad (14)$$

where α and β are two strictly positive constants to be determined. Integrating Equation 14 in Equation 12, TPG_{net} becomes

$$TPG_{net} = \rho \frac{1}{\sqrt{A}} \frac{\partial Q}{\partial t} \alpha \left(\frac{A}{EOA} - 1\right)^\beta + \frac{1}{2} \rho Q^2 \left(\frac{1}{EOA} - \frac{1}{A}\right)^2. \quad (15)$$

It is difficult to solve α and β from a purely analytical development. We therefore determined these two constants by means of *in vitro* experiments performed in a mock flow circulation model. Nine orifice plates, simulating several grades of aortic stenosis severity, were tested under numerous physiological pulsatile flow conditions. A minimization method provided $\alpha = 2\pi$ and $\beta = 1/2$ (18). From those results, the instantaneous TPG_{net} is finally expressed as

$$TPG_{net} = 2\pi \rho \frac{\partial Q}{\partial t} \sqrt{\frac{1}{EOA} - \frac{1}{A}} + \frac{1}{2} \rho Q^2 \left(\frac{1}{EOA} - \frac{1}{A}\right)^2. \quad (16)$$

When using the expression of the energy loss coefficient ($E_L Co$) defined as $EOA \times A / (A - EOA)$ (26), TPG_{net} can be written as

$$TPG_{net} = P_{LV} - P_A = \frac{2\pi\rho}{\sqrt{E_L Co}} \frac{\partial Q}{\partial t} + \frac{\rho}{2 E_L Co^2} Q^2. \quad (17)$$

Equation 17 shows that, for a given transvalvular flow rate $Q(t)$, the pressure difference between the left ventricle and the aorta, i.e., TPG_{net}, is dependent upon a unique valvular parameter, namely $E_L Co$. This equation was validated with bioprosthetic heart valves in the abovementioned *in vitro* model and was shown to predict accurately the instantaneous TPG_{net} measured by micromanometer-tipped Millar catheters. We invite the reader to refer to (18) for a detailed description of the complete protocol.

2.3. The Left Ventricular Pressure-Volume Relationship

Pressure-volume graphs are commonly used to assess the inotropic state of the left ventricle (27). When tracing LV pressure (in mmHg) as a function of LV cavity volume (in mL), a complete loop, called the pressure-volume loop (PV loop), is described. LV stroke work is the work of the left ventricle during each heart beat and is represented by the area contained within the PV loop (Figs. 3 and 11). LV stroke work has been shown to effectively characterize the outcome of patients with AS (28). If the loading conditions on the heart are pharmaceutically (e.g., by administration of phenylephrine) or mechanically (e.g., by occlusion of the inferior vena cava) modified, while myocardial contractility remains unchanged, a series of PV loops is obtained (Fig. 3). In a physiological range, the top left corners of the PV loops may be connected by a regression line (Fig. 3) whose slope is called the maximal elastance (E_{max}). The intercept of this line with the abscissa volume axis is the unloaded volume (V_0). It should be noted that the linear approximation often leads to negative V_0 (29). Hence, V_0 has no physical or physiological meaning but should rather be considered as a virtual parameter

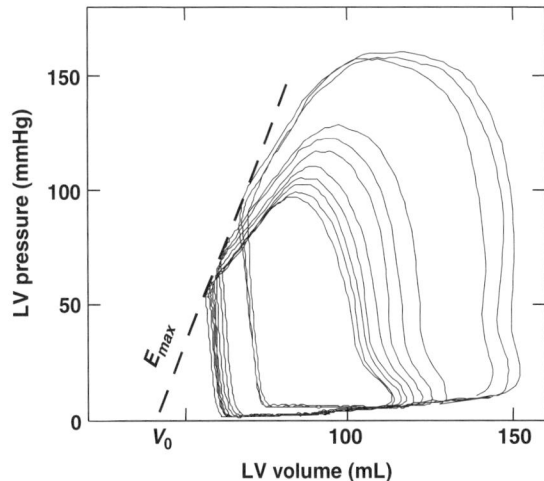

Figure 3. Series of left ventricular pressure-volume loops obtained in a patient with aortic stenosis. E_{max} is the maximal elastance. V_0 is the left ventricular unloaded volume. From Dekker et al. (36) with permission.

computed by extrapolation. This linear characteristic remains, however, largely appropriate over a large physiological or pathophysiological range. In the 1970s, Suga et al. (30) defined the time-varying LV elastance as LV pressure (P_{LV}) divided by LV cavity volume (V) decremented by V_0:

$$E(t) = \frac{P_{LV}(t)}{V(t) - V_0}. \qquad (18)$$

Interestingly, it has been shown by Senzaki et al. that the elastance waveform, when normalized with respect to its amplitude (E_{max}) and time to peak value (T_{Emax}), is somewhat similar in the normal or diseased human hearts despite the presence of differences with regard to etiology of heart disease, LV myocardial contractility and loading conditions (31). Fig. 4 depicts the normalized LV elastance (E_N) as a function of normalized time, as measured by Senzaki et al. in patients and normal subjects. Due to the universal property of the LV normalized elastance, LV pressure can be related to LV volume by means of only three independent LV parameters, namely E_{max}, T_{Emax}, and V_0, as follows:

$$E_{max} E_N(t/T_{Emax}) = \frac{P_{LV}(t)}{V(t) - V_0}. \qquad (19)$$

2.4. The Three-Element Windkessel Model of the Peripheral System

The three-element windkessel (WK3) model is a lumped model that has been proven to simulate adequately the hemodynamic characteristics of the peripheral system (32,33). It includes three independent vascular parameters: the characteristic impedance of the proximal aorta (Z_0); the arterial compliance (C); and the systemic vascular resistance (R). The resistance R reproduces the hydraulic resistance in the small arteries and arterioles, the compliance C reflects the ability of the large arteries to distend with increasing inner pressure, and the characteristic impedance Z_0 mainly relates the transvalvular

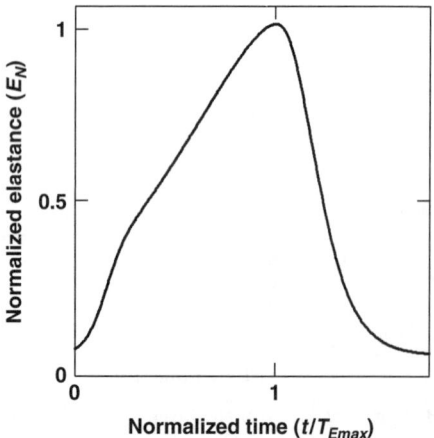

Figure 4. Normalized left ventricular elastance as a function of normalized time (dimensionless).

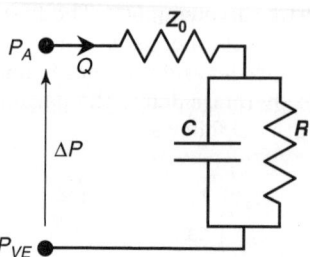

Figure 5. Electrical representation of the three-element windkessel model. P_A = aortic pressure, P_{VE} = central venous pressure, Q = transvalvular flow rate. Z_0 = aortic characteristic impedance, R = systemic vascular resistance, C = arterial compliance.

flow rate (Q) to the aortic pressure in the high-frequency range (highly pulsatile range). An electrical analog model of the WK3 is illustrated in Fig. 5. In this figure, P_A and P_{VE} represent the aortic pressure and the central venous pressure, respectively. Using complex notation, the peripheral impedance is written as

$$Z^* = Z_0 + \frac{R}{1 + j\omega RC}, \qquad (20)$$

where ω is the pulsation frequency and j is the unit complex number. Then, using Ohm's law ($\Delta P = P_A - P_{VE} = Z * Q$), one obtains

$$j\omega \Delta P + \frac{\Delta P}{RC} = \frac{Z_0 + R}{RC} Q + Z_0 j\omega Q, \qquad (21)$$

which may be rewritten, in the time domain ($j\omega$ is changed to $\partial/\partial t$), as

$$\frac{\partial P_A(t)}{\partial t} + \frac{P_A(t)}{RC} = \frac{Z_0 + R}{RC} Q(t) + Z_0 \frac{\partial Q(t)}{\partial t} + \frac{P_{VE}}{RC}. \qquad (22)$$

We note t_0 the time at which LV ejection begins and T the cardiac period. The analytical resolution of this linear differential equation yields the aortic pressure (P_A) within the time interval [$t_0; t_0 + T$]:

$$P_A(t) = e^{-(t-t_0)/RC} \left(\int_{t_0}^{t} \frac{Q^*(\tau)}{Ce^{-(\tau - t_0)/RC}} d\tau + DP_A \right), \qquad (23)$$

where Q^* is related to Q as follows:

$$Q^*(\tau) = \frac{Z_0 + R}{R} Q(\tau) + Z_0 C \frac{\partial Q(\tau)}{\partial \tau} + \frac{P_{VE}}{R}, \qquad (24)$$

and where the diastolic blood pressure (DP_A) can be determined using the periodic property of P_A and is given by

$$DP_A = \int_{t_0}^{t_0 + T} \frac{Q^*(t)e^{(t-t_0)/RC}}{Ce^{T/RC} - C} dt. \qquad (25)$$

For a given central venous pressure, the WK3 representation described by Equations 23–25 thus relates aortic pressure to transvalvular flow rate using only three vascular parameters (Z_0, R, and C).

2.5. The V^3 (Ventricular-Valvular-Vascular) Model

2.5.1. Derivation of the V^3 Model. The ventricular-valvular-vascular V^3 model results from the coupling of the abovementioned LV time-varying elastance, TPG_{net} equation and WK3 model (Fig. 6). The sum $\partial(\text{Equation 17})/\partial t$ + (Equation 17)/(RC) yields

$$\left(\frac{\partial P_{LV}}{\partial t} + \frac{P_{LV}}{RC}\right) - \left(\frac{\partial P_A}{\partial t} + \frac{P_A}{RC}\right)$$
$$= \frac{2\pi\rho}{\sqrt{E_L C_0}}\left(\frac{\partial^2 Q}{\partial t^2} + \frac{1}{RC}\frac{\partial Q}{\partial t}\right) + \frac{\rho}{2}\frac{Q}{E_L C_0^2}\left(2\frac{\partial Q}{\partial t} + \frac{Q}{RC}\right). \quad (26)$$

The expression $(\partial P_A/\partial t + P_A/RC)$ in Equation 26 is expressed as a function of Q using Equation 22, and P_{LV} is replaced using Equation 19. Throughout ejection, transvalvular flow rate can be written as: $Q(t) = -\partial V(t)/\partial t$. Therefore, Equation 26 becomes

$$\frac{2\pi\rho}{\sqrt{E_L C_0}}\frac{\partial^3 V(t)}{\partial t^3} = a_3(t)\frac{\partial^2 V(t)}{\partial t^2} + a_2(t)\frac{\partial V(t)}{\partial t}$$
$$+ a_1(t)V(t) + a_0(t),$$

where

$$a_0(t) = V_0 \frac{E_{\max}}{T_{E\max}}\frac{\partial E_N(\hat{t})}{\partial \hat{t}} + V_0 E_{\max}\frac{E_N(\hat{t})}{RC} + \frac{P_{VE}}{RC},$$
$$a_1(t) = -\frac{E_{\max}}{T_{E\max}}\frac{\partial E_N(\hat{t})}{\partial \hat{t}} - E_{\max}\frac{E_N(\hat{t})}{RC}, \quad (27)$$
$$a_2(t) = \frac{\rho}{2RCE_L C_0^2}\frac{\partial V(t)}{\partial t} - \frac{Z_0 + R}{RC} - E_{\max}E_N(\hat{t}),$$
$$a_3(t) = \frac{\rho}{E_L C_0^2}\frac{\partial V(t)}{\partial t} - \frac{2\pi\rho}{RC\sqrt{E_L C_0}} - Z_0,$$

where $\hat{t}$ is the normalized time ($t/T_{E\max}$). At the onset of the ejection (at $t = t_0$), LV volume V is equal to the LV end-diastolic volume ($V(t_0) = LVEDV$) and $Q(t_0) = 0$. LV ejection begins when LV pressure reaches aortic pressure and TPG_{net} thus equals zero. According to Equation 17, $\partial Q/\partial t$ is therefore also equal to 0 at the ejection onset. Because $Q(t) = -\partial V(t)/\partial t$ during LV ejection, the initial conditions are therefore

$$V(t_0) = LVEDV; \frac{\partial V}{\partial t}(t_0) = 0; \frac{\partial^2 V}{\partial t^2}(t_0) = 0. \quad (28)$$

The 3rd-order nonlinear differential equation 27, with the corresponding initial conditions (Equation 28), describes thoroughly the LV volume during ejection under the conditions that the ventricular, valvular, and vascular properties are given. Table 1 summarizes the independent input parameters necessary for solving the V^3 model. The mathematical V^3 model has been validated in patients during surgery, before and after aortic valve replacement, as described in details in (19).

2.5.2. Numerical Computation. Time reference ($t = 0$) is fixed at the onset of the isovolumic LV contraction. An arbitrary diastolic pressure (DP_A) is chosen and t_0 (onset of ejection) is then determined from Equation 19 such that it satisfies the following condition: $E_{\max}E_N(t_0/T_{E\max}) = DP_A/(LVEDV - V_0)$. LV cavity volume during ejection, $V(t)$, is then calculated from Equation 27 using an explicit Runge–Kutta method starting from t_0 until $\partial V(t)/\partial t$ reaches zero. Transvalvular flow rate during LV ejection is $Q(t) = -\partial V(t)/\partial t$ and is further assumed to be zero throughout the rest of the cardiac cycle (no aortic regurgitation). Aortic pressure is then deduced from Equations 23–25 and a second DP_A is therefore obtained. If the difference between the two DP_A exceeds a given relative error, a new iteration is performed using the latest DP_A value until the desired precision is reached. Knowing $V(t)$, LV pressure during isovolumic contraction, ejection, and

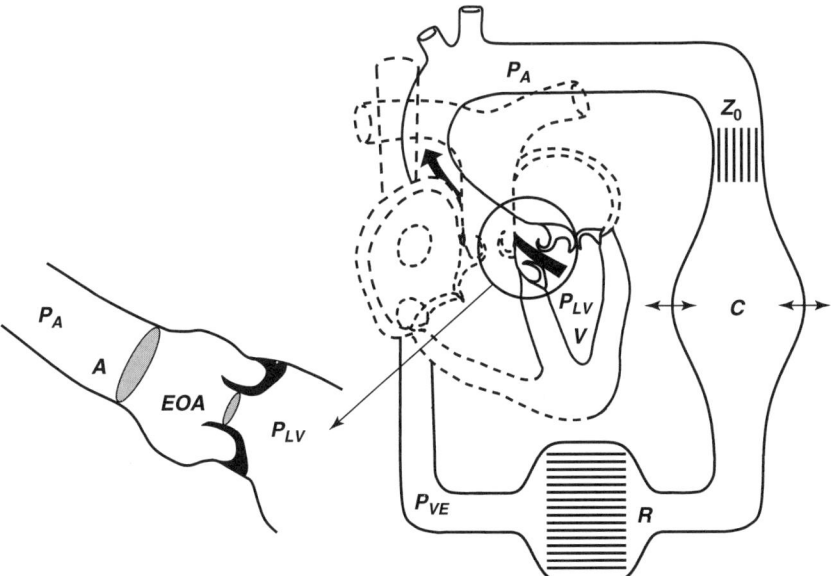

Figure 6. Schematic representation of the V^3 model. V = left ventricular cavity volume, P_{LV} = left ventricular pressure, P_A = aortic pressure. See also legends of Figs. 2 and 5. From Garcia et al. (19) with permission.

Table 1. List of the Cardiovascular Input Parameters Required for the Resolution of the V3 Model [Values in the right column are typical physiological values used for the simulations with a heart rate of 70 beats per minute. Emax (mmHg/mL), R (mmHg.s/mL) and C (mL/mmHg) were adjusted according to the desired hemodynamic conditions as explained in the text (see also Table 3). EOA was varied from 4 cm² (no aortic stenosis) down to 0.5 cm² (severe stenosis).]

Ventricular parameters		
Left-ventricular end-diastolic volume	LVEDV	150 mL
Unloaded volume	V_0	15 mL
Maximal elastance	E_{max}	Adjusted for stroke volume (70 mL)
Time to maximal elastance	T_{Emax}	0.33 s
Vascular parameters		
Aortic characteristic impedance	Z_0	0.07 mmHg s/mL
Systemic vascular resistance	R	Adjusted for blood pressure level
Total arterial compliance	C	Adjusted for blood pressure level
Central venous pressure	P_{VE}	5 mmHg
Valvular parameters		
Effective orifice area	EOA	From 4 down to 0.5 cm2
Aortic cross-sectional area	A_A	5 cm²

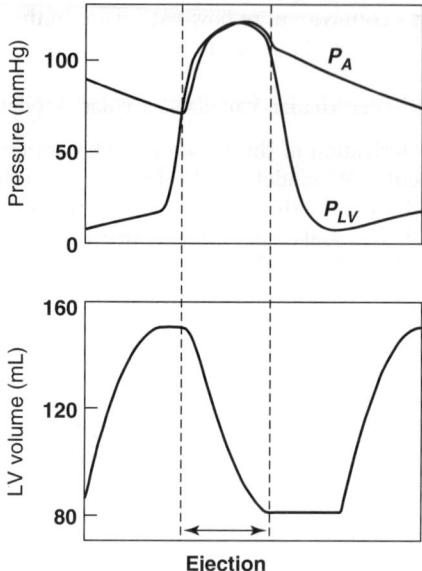

Figure 7. Top—left ventricular (P_{LV}) and aortic pressure (P_A) waveforms simulated with the V³ model under normal conditions. Bottom—corresponding simulated left ventricular cavity volume. Time scale covers more than one cardiac cycle.

isovolumic relaxation is finally calculated by means of Equation 19 and is linearly extrapolated during the LV filling. LV volume during LV filling is extrapolated using a 2nd-order polynomial so that its temporal derivative is zero at the end of diastole. A detailed Matlab program for the resolution of the V³ model is available online in (34). Figure 7 shows normal physiological waveforms simulated with the V³ model (normotensive condition, normal aortic valve, and normal cardiac conditions). It should be noted that the V³ model in the present form may exclusively simulate LV volume and pressure for the periods of ejection and isovolumic contraction and relaxation. Their waveforms during LV filling are thereby extrapolated. Thus, the V³ model as presented here cannot be used to analyze LV diastolic dysfunction.

3. APPLICATIONS

The V³ model provides a potentially useful tool for simulating the effects of AS on pressure and transvalvular flow waveforms with different grades of AS severity. To date, no explicit theoretical model has been shown to reflect accurately the cardiovascular hemodynamics in presence of AS (19). As for the V³ model, it may explicitly and correctly describe some cardiovascular features if only a few cardiovascular parameters are known, and as importantly, its validity has been tested in patients with AS. In clinical practice, AS is graded as mild if $EOA > 1.5$ cm², moderate if $EOA > 1.0$ and ≤ 1.5 cm², or severe if $EOA \leq 1.0$ cm² (Table 2). When stenosis is severe and cardiac output is normal, the mean TPG_{net} is generally > 50 mmHg (1 mmHg ≈ 133 Pa) (35). Because it is cost-effective, completely non-invasive, and rapid, echocardiography is presently the most commonly applied modality in clinical cardiology for establishing the diagnosis of aortic stenosis. Echocardiography allows one to display high-quality images of the heart as well as the transvalvular blood velocities (Fig. 8). Doppler echocardiography is mainly used to determine EOA by means of the continuity equation (equation of mass conservation). The peak velocity across the valve (at location 2, Fig. 2) measured by Doppler echocardiography has also been shown to be predictive of symptom onset in AS patients (Fig. 8) (1). When the clinical and echocardiographic data yield conflicting diagnoses or when the echocardiographic data are unconvincing, invasive measurements by cardiac catheterization are needed and max or mean TPG_{net} are measured

Table 2. Classification of Hypertension and Aortic Stenosis Severity According to the European Society of Hypertension (44) and to the American College of Cardiology/ American Heart Association (35) (Systolic and diastolic pressures are in mmHg. EOA = effective orifice area.)

	Hypertension		Aortic Stenosis
	Systolic	Diastolic	EOA
Normal	120–129	80–84	≥ 3 cm²
High normal	130–139	85–89	
Mild	140–159	90–99	> 1.5 cm²
Moderate	160–179	100–109	1.0–1.5 cm²
Severe	≥ 180	≥ 110	≤ 1.0 cm²

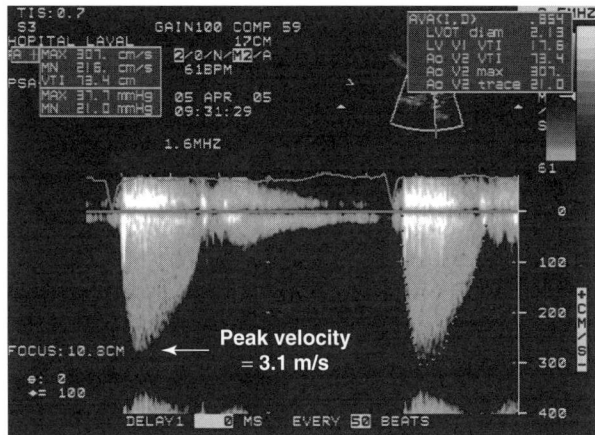

Figure 8. Continuous-wave Doppler recording of an aortic stenosis jet in a patient with severe aortic stenosis ($EOA = 0.85\,cm^2$). (This figure is available in full color at http://www.mrw.interscience.wiley.com/ebe.)

(Fig. 9), among other parameters. When severity of AS becomes significant, pressure loss due to the aperture narrowing of the valve induces an increase in LV cavity pressure and LV stroke work. If concomitant HPT is present, effects of AS on LV function are worsened due to the further vascular overload. The following paragraphs will show how AS may affect LV pressure and more specifically LV stroke work, by using the numerical V^3 model. The additional effect of concomitant HPT will then be described.

3.1. Simulations of Flow and Pressure Waveforms with Aortic Stenosis

The ejection fraction is defined as the amount of blood ejected divided by the amount of blood contained in the left ventricle at end of diastole. In absence of serious LV dysfunction, ejection fraction is usually normal (50–60 %) in patients with AS (7). For a normal average stroke volume of 70 mL, $LVEDV$ is therefore maintained at ~ 150 mL. In addition, typical values are assumed for V_0 (15 mL), T_{Emax} (0.33 s), A (5 cm^2), Z_0 (0.07 mmHg.s/mL)

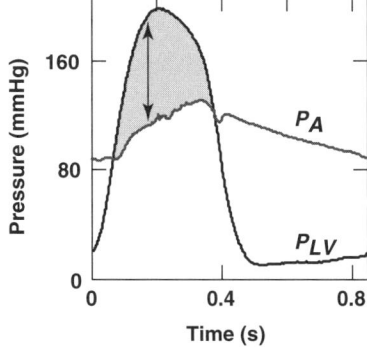

Figure 9. Typical left ventricular (P_{LV}) and aortic pressure (P_A) waveforms in a patient with severe aortic stenosis. The arrow represents the peak TPG_{net} (net transvalvular pressure gradient). During a clinical examination, mean TPG_{net} is calculated as the gray area divided by ejection time. (This figure is available in full color at http://www.mrw.interscience.wiley.com/ebe.)

and P_{VE} (5 mmHg) to achieve the simulations (19,36). The heart rate was fixed at 70 beats per minute and maximal elastance (E_{max}) was adjusted so that stroke volume equals 70 mL (normal outflow condition). For example, the respective calculated values for E_{max} without AS and with very severe AS ($EOA = 0.5\,cm^2$) were 1.63 and 2.35 mmHg/mL. Total vascular resistance (R) and arterial compliance (C) were also adjusted to obtain normotensive conditions (systolic/diastolic pressures = 120/80 mmHg): their respective values were 1.07 mmHg.s/mL and 2.05 mL/mmHg. Table 1 summarizes the chosen values for the input cardiovascular parameters. Figure 10 illustrates three simulations achieved with the V^3 model: (1) no AS ($EOA = 4\,cm^2$); (2) moderate AS ($EOA = 1\,cm^2$); and (3) severe AS ($EOA = 0.5\,cm^2$). This figure shows that simulated LV and aortic pressure waveforms are very similar to those observed in patients (see Fig. 9 for comparison). Whereas LV peak pressure has a normal value (120 mmHg) without AS, it may be as high as 200 mmHg with very severe AS, even under normotensive conditions. It has been reported that ejection time lengthens (37,38) and that peak transvalvular flow rate occurs later in ejection with increasing AS severity (39,40). The simulated instantaneous transvalvular flow rates are very consistent with these observations as shown in Fig. 10. Figure 11 illustrates the PV loops corresponding to the conditions of Fig. 10. Stroke work ($SW = P_{LV}dV$, represented by the inner area of PV loops) increases significantly when EOA decreases, i.e., when severity of AS increases. Figure 12 shows more accurately how LV stroke work varies with AS severity. It can be observed that LV stroke work remains relatively stable around the value of 1 J when $EOA > 1\,cm^2$, which means that AS does not greatly affect the LV pump when graded as mild or even moderate. But when AS is severe ($EOA < 1\,cm^2$), a small decrease in EOA induces a drastic increase in LV stroke work, and it is precisely in this range that patients with AS generally develop symptoms.

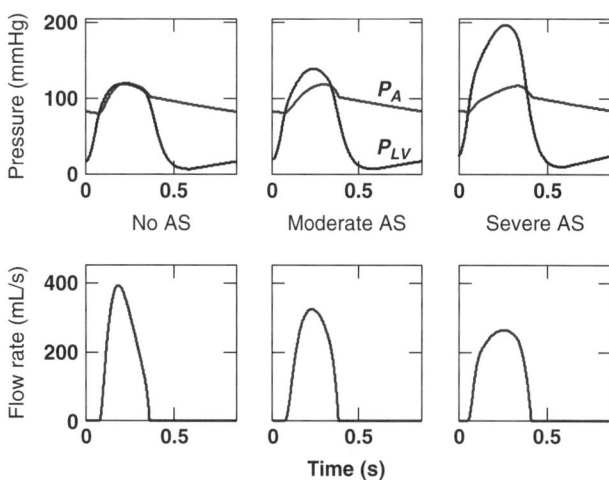

Figure 10. Pressure and transvalvular flow waveforms simulated with the V^3 model. Note that simulated pressure waveforms with severe aortic stenosis are similar to those measured in patient (Fig. 9). (This figure is available in full color at http://www.mrw.interscience.wiley.com/ebe.)

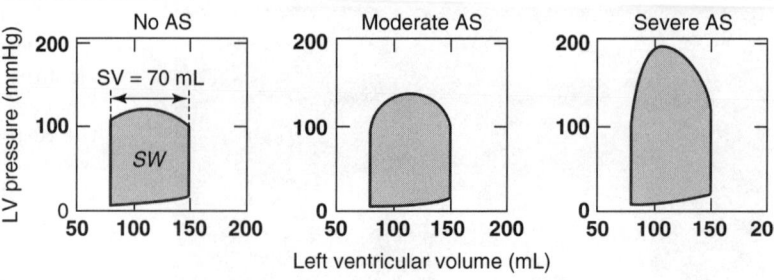

Figure 11. Simulated left ventricular pressure-volume loops for the hemodynamic conditions illustrated in Fig. 10. Left ventricular stroke work (SW) is represented by the inner area. SV is stroke volume (70 mL). (This figure is available in full color at http://www.mrw.interscience.wiley.com/ebe.)

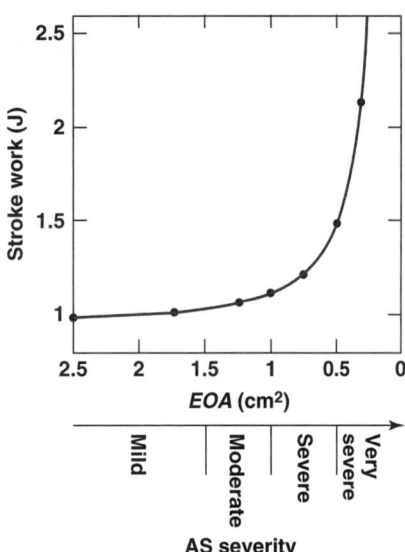

Figure 12. Left ventricular stroke work as a function of aortic stenosis severity. One dot characterizes one simulation performed with the V^3 model. (This figure is available in full color at http://www.mrw.interscience.wiley.com/ebe.)

3.2. Simulations in the Presence of Aortic Stenosis and Coexistent Systemic Hypertension

Left ventricular pressure overload caused by AS or systemic arterial HPT generally results in LV concentric hypertrophy, which has been shown to be a strong independent risk factor for morbidity and mortality (41,42). Systemic hypertension (HPT) has a prevalence of 30–40 % in patients with AS (4,43). The resulting vascular overload in such patients adds further to the valvular overload, which increases the LV afterload and affects LV function and patient outcome. The V^3 model may help to quantify the respective impacts of AS and HPT on LV function. For this purpose, we simulated the combined effect of AS and HPT on the LV stroke work. Aortic stenosis severity was varied from mild to severe ($EOA = 1.75$ to $0.5\,\text{cm}^2$), and for each degree of AS severity, blood pressure level was progressively increased from normotensive conditions to severe HPT (systolic/diastolic pressures = 120/80 to 190/115 mmHg). For each degree of HPT, systemic vascular resistance (R) and arterial compliance (C) were adjusted to obtain the desired systolic and diastolic aortic pressures (Table 3). Systolic and diastolic aortic pressures were chosen according to the classification of blood pressure levels published by the European Hypertension Society (44) (Table 2). The values found for R and C were comparable to those measured in hypertensive patients (45,46). As mentioned previously, $E_{\max}$ was adjusted for each simulation so that stroke volume was equal to 70 mL. The chosen values of the other input cardiovascular parameters are listed in Table 1. Figure 13 illustrates four simulations obtained with a normal valve and a moderate aortic stenosis ($EOA = 1\,\text{cm}^2$) with and without moderate hypertension (170/105 mmHg). The peak LV pressure is largely increased when hypertension coexists with AS (no AS = 120, moderate AS = 139, moderate HPT = 169, moderate AS + HPT = 185 mmHg). It should also be noted that moderate AS has a small impact on LV stroke work in comparison with moderate HPT (no AS = 1.00, moderate AS = 1.11, moderate HPT = 1.37, moderate AS + HPT = 1.47 J). The following valvular parameters: peak TPG_{net}, mean TPG_{net}, and peak jet velocity (Table 4) were very consistent with those of patients reported in the literature (1,22). As an example, mean TPG_{net} and peak jet velocity with an EOA of $0.5\,\text{cm}^2$ were found to be 55 mmHg and 5.3 m/s, respectively, which represent typical values for patients with severe AS (1). As expected, because cardiac flow conditions were fixed, these parameters were only dependent upon AS severity and were not influenced by the degree of HPT (Table 4). The theoretical influence of

Table 3. Values of Total Peripheral Resistance (R) and Arterial Compliance (C) Utilized for Simulating the Different Blood Pressure Levels in This Study

Blood Pressure Level	Arterial Pressure (mmHg)		R (mmHg.s/mL)	C (mL/mmHg)
	Systolic	Diastolic		
Normal	120	80	1.07	2.05
High normal	135	87	1.21	1.47
High normal	135	87	1.21	1.47
Mild	150	95	1.35	1.23
Moderate	170	105	1.53	0.98
Severe	190	115	1.71	0.83

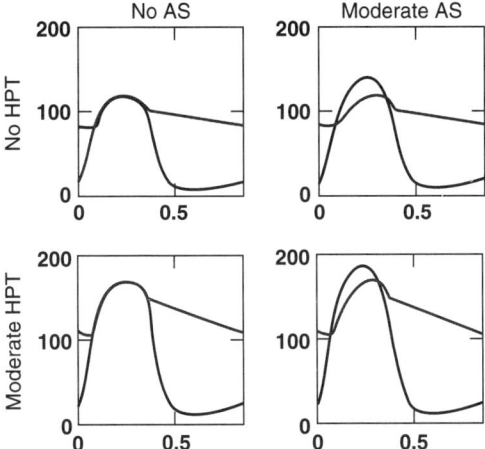

Figure 13. Simulated left ventricular (P_{LV}) and aortic (P_A) pressures in normal aortic valve (no AS) and moderate aortic stenosis (AS) with and without moderate hypertension (HPT). Pressures are in mmHg, time is in s. (This figure is available in full color at http://www.mrw.interscience.wiley.com/ebe.)

AS/HPT on LV stroke work is represented on Fig. 14. This graph shows that LV stroke work is shifted upward from one blood pressure level to the following, independently of the AS grade AS. Thus, HPT has a quasi-linear effect on LV work. By contrast, an increase in AS severity from mild to moderate ($EOA > 1\,cm^2$) has a very small impact on LV stroke work. The latter, however, increases noticeably when AS becomes severe ($EOA < 1\,cm^2$).

3.3. Potential Clinical Implications

According to current guidelines, the decision to replace a stenotic valve is mainly based on EOA and presence of symptoms (35). Unfortunately, there are often discrepancies between the AS severity and the symptomatic status. Some patients are indeed symptomatic although they have only a moderate AS, whereas others remain asymptomatic despite the presence of severe AS. Our simulations may in part explain these discrepancies. According to our numerical results, a patient having moderate AS with coexistent moderate HPT may have higher LV afterload than a normotensive patient with severe AS. Accordingly, HPT is a well-established cardiovascular risk factor. Its impact on the clinical outcome of patients with AS, however, is still unknown (47), but it has been recently

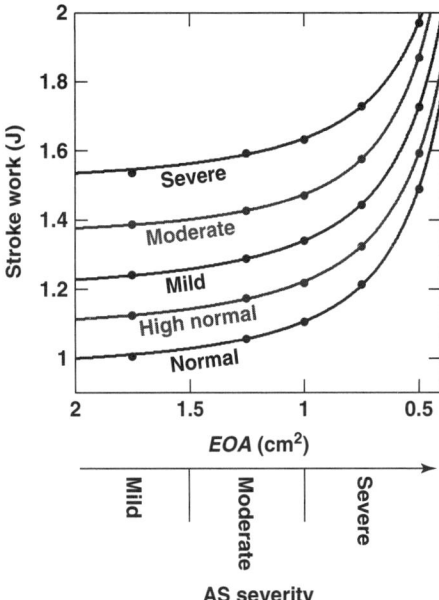

Figure 14. Left ventricular stroke work as a function of aortic stenosis severity for normal arterial pressure and different grades of hypertension. One dot characterizes one simulation performed with the V^3 model. (This figure is available in full color at http://www.mrw.interscience.wiley.com/ebe.)

reported that in hypertensive symptomatic AS patients, symptoms of AS develop at a relative earlier stage of the disease, with larger EOA (4). This is very likely due to the additional LV afterload induced by HPT itself, because our simulations tended to show that even mild HPT may greatly influence LV stroke work and therefore LV function in AS patients. Thus, antihypertensive medication should be initiated soon after aortic valve replacement in patients with coexistent HPT. As an example, Fig. 15 illustrates a hypothetical case of a patient with moderate HPT (systolic/diastolic pressures = 170/105 mmHg), severe AS ($EOA = 0.6\,cm^2$), and normal output flow conditions (stroke volume = 70 mL, heart rate = 70 bpm), whose left ventricle develops a stroke work of 1.75 J. In such a patient, aortic valve replacement alone would reduce LV stroke work to 1.4 J. This is identical to the work done by the left ventricle in presence of severe AS (with an EOA of $0.6\,cm^2$) under normotensive conditions. Thus, this patient would not fully benefit from aortic valve replacement if no antihypertensive medication was prescribed, since LV

Table 4. Hemodynamic Valvular Parameters Obtained with the V^3 Model for Given Degrees of Aortic Stenosis Severity Under Normal Flow Conditions and Different Levels of Arterial Blood Pressure (stroke volume = 70 mL, heart rate = 70 bpm) (Values are represented as mean ± standard deviation.)

Aortic Stenosis	EOA	Mean TPG_{net} (mmHg)	Peak TPG_{net} (mmHg)	Peak Velocity (m/s)
No AS	$4\,cm^2$	0.8 ± 0.7	11.7 ± 2.0	1.1 ± 0.03
Mild	$1.75\,cm^2$	4.5 ± 0.5	18.1 ± 2.3	2.1 ± 0.02
Moderate	$1.25\,cm^2$	9.7 ± 0.7	22.0 ± 2.3	2.7 ± 0.01
Moderate to severe	$1.0\,cm^2$	15.2 ± 0.8	27.2 ± 1.0	3.2 ± 0.01
Severe	$0.75\,cm^2$	26.4 ± 0.6	42.8 ± 0.7	3.9 ± 0.01
Very severe	$0.5\,cm^2$	55.4 ± 1.3	84.7 ± 1.1	5.3 ± 0.03

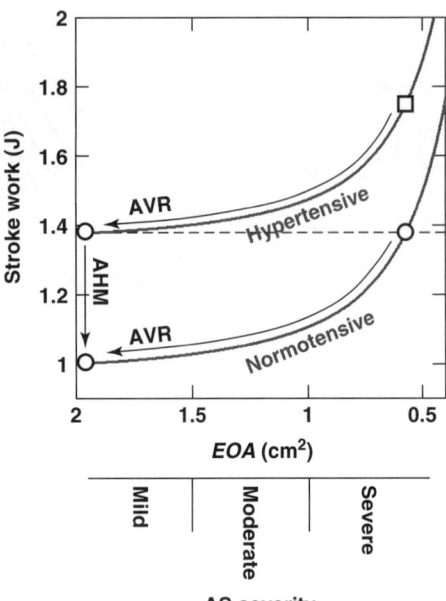

Figure 15. Respective outcome of aortic valve replacement (AVR) and antihypertensive medication (AHM) in a patient (represented by a square) with severe aortic stenosis and moderate systemic hypertension. (This figure is available in full color at http://www.mrw.interscience.wiley.com/ebe.)

stroke work would remain abnormally high. This could explain why short-term clinical results vary considerably from one patient to another after aortic valve replacement. This surgery removes the valvular component of the LV overload but not its vascular component related to HPT. Hence, patients with AS and concomitant HPT may have only minimal LV functional outcome if aortic valve replacement alone is carried out. The influence of HPT on the development of LV overload and symptoms in patients with AS is often underestimated in the clinical practice, while the V^3 model shows that HPT, even at mild degree, is an important determinant of LV stroke work in patients with AS.

4. FUTURE INVESTIGATIONS

The V^3 model has been proved to predict accurately LV pressure, aortic pressure, and LV volume waveforms in patients with AS (19). Contrary to previously published mathematical models of the cardiovascular circulation, the V^3 model contains a realistic representation of the transvalvular pressure-flow relationship. Because the cardiovascular input parameters are all measurable *in vivo*, it may also enlighten clinical inquiries with regard to the effects of some cardiovascular diseases on the LV function. In addition, future updates could be made to extend its clinical applications. As an example, combining the V^3 model with the lumped representation of left coronary circulation developed by Judd and Mates (48) could allow the examination of the impact of AS and HPT on coronary blood flow, and would help to clarify the occurrence of LV failure in patients with AS and/or HPT. Indeed, the left coronary circulation is essentially governed by LV pressure and by the left coronary inlet pressure (49). Also, in the current version of the V^3 model, *EOA* is supposed constant throughout ejection. Whereas this hypothesis is adequate in normal and mildly or moderately calcified aortic valves, it has been reported that *EOA* may vary notably during ejection in patients with severe AS (39,50). Time-dependence of *EOA* may significantly influence TPG_{net} and LV pressure waveforms and consequently LV stroke work. The influence of time-varying *EOA* on LV stroke work could be studied by writing *EOA* as a function of transvalvular flow. This could be performed for example by means of a simple spring-damper representation of the valve aperture or by using a Lagrangian dynamic leaflet model as that proposed by Fenlon and David (51). Despite its relative simplicity, the V^3 model may therefore allow the investigation of complex cardiovascular interactions. In conclusion, mathematical cardiovascular models, such as the V^3 model, may answer important clinical questions. It is, however, essential to experimentally validate the model per se and its findings before any clinical application.

Acknowledgments

The authors gratefully thank Dr. Lysanne Goyer and Dr. Lyes Kadem, from the Institut de Recherches Cliniques de Montréal, for having carefully reviewed this chapter.

BIBLIOGRAPHY

1. D. M. Shavelle and C. M. Otto, Aortic stenosis. In: Crawford MH, Dimarco JP, ed., Cardiology. London: Mosby, 2000: 9.1–9.9.
2. P. Tornos, [New aspects in aortic valve disease]. *Rev. Esp. Cardiol.* 2001; **54**(Suppl 1):17–21.
3. R. A. Nishimura, Cardiology patient pages. *Aortic valve disease. Circulation* 2002; **106**(7):770–772.
4. F. Antonini-Canterin, G. Huang, E. Cervesato, P. Faggiano, D. Pavan, R. Piazza, and G. L. Nicolosi, Symptomatic aortic stenosis: Does systemic hypertension play an additional role? *Hypertension* 2003; **41**(6):1268–1272.
5. M. Briand, J. G. Dumesnil, L. Kadem, A. G. Tongue, R. Rieu, D. Garcia, and P. Pibarot, Reduced systemic arterial compliance impacts significantly on LV afterload and function in aortic stenosis: Implications for diagnosis and treatment. *J. Am. Coll. Cardiol.* 2005; **46**(2):291–298.
6. M. E. Safar, B. I. Levy, and H. Struijker-Boudier, Current perspectives on arterial stiffness and pulse pressure in hypertension and cardiovascular diseases. *Circulation* 2003; **107**(22):2864–2869.
7. K. E. Berkina and S. G. Ball, Essential hypertension: The heart and hypertension. *Heart* 2001; **86**(4):467–475.
8. W. Grossman, D. Jones, and L. P. McLaurin, Wall stress and patterns of hypertrophy in the human left ventricle. *J. Clin. Invest.* 1975; **56**(1):56–64.
9. D. H. Fitchett, LV-arterial coupling: interactive model to predict effect of wave reflections on LV energetics. *Am. J. Physiol.* 1991; **261**(4 Pt 2):H1026–H1033.
10. R. R. Ha, J. Qian, D. L. Ware, J. B. Zwischenberger, A. Bidani, and J. W. Clark, An integrative cardiovascular model of the standing and reclining sheep. *Cardiovasc. Eng.* 2005; **5**(2):53–75.

11. T. Korakianitis and Y. Shi, Numerical simulation of cardiovascular dynamics with healthy and diseased heart valves. *J. Biomech.* 2005; doi:10.1016/j.jbiomech.2005.06.016.

12. J. K. Li, J. Y. Zhu, and M. Nanna, Computer modeling of the effects of aortic valve stenosis and arterial system afterload on left ventricular hypertrophy. *Comput. Biol. Med.* 1997; **27**(6):477–485.

13. P. Segers, N. Stergiopulos, and N. Westerhof, Quantification of the contribution of cardiac and arterial remodeling to hypertension. *Hypertension* 2000; **36**(5):760–765.

14. B. W. Smith, J. G. Chase, R. I. Nokes, G. M. Shaw, and G. Wake, Minimal haemodynamic system model including ventricular interaction and valve dynamics. *Med. Eng. Phys.* 2004; **26**(2):131–139.

15. N. Stergiopulos, J. J. Meister, and N. Westerhof, Determinants of stroke volume and systolic and diastolic aortic pressure. *Am. J. Physiol.* 1996; **270**(6 Pt 2):H2050–H2059.

16. M. Ursino, Interaction between carotid baroregulation and the pulsating heart: A mathematical model. *Am. J. Physiol.* 1998; **275**(5 Pt 2):H1733–H1747.

17. M. Zacek and E. Krause, Numerical simulation of the blood flow in the human cardiovascular system. *J. Biomech.* 1996; **29**(1):13–20.

18. D. Garcia, P. Pibarot, and L. G. Durand, Analytical modeling of the instantaneous pressure gradient across the aortic valve. *J. Biomech.* 2005; **38**(6):1303–1311.

19. D. Garcia, P. J. Barenbrug, P. Pibarot, A. L. Dekker, F. H. van der Veen, J. G. Maessen, J. G. Dumesnil, and L. G. Durand, A ventricular-vascular coupling model in presence of aortic stenosis. *Am. J. Physiol. Heart Circ. Physiol.* 2005; **288**(4): H1874–H1884.

20. A. J. Ward-Smith, Internal fluid flow. The fluid dynamics of flow in pipes and ducts. Oxford: Clarendon Press, 1980.

21. D. S. Miller, Internal flow systems. second ed. Bedford: BHR, 1996.

22. H. Baumgartner, T. Stefenelli, J. Niederberger, H. Schima, and G. Maurer, "Overestimation" of catheter gradients by Doppler ultrasound in patients with aortic stenosis: a predictable manifestation of pressure recovery. *J. Am. Coll Cardiol.* 1999; **33**(6):1655–1661.

23. J. D. Thomas, and Z. B. Popovic, Intraventricular pressure differences: A new window into cardiac function. *Circulation* 2005; **112**(12):1684–1686.

24. B. R. Munson, D. F. Young, and T. H. Okiishi, Viscous flow in pipes. Fundamentals of fluid mechanics. Second edition. New York: John Wiley & Sons, Inc., 1994: 455–547.

25. A. A. Sonin, A generalization of the Pi-theorem and dimensional analysis. *Proc. Natl. Acad. Sci. U S A.* 2004; **101**(23): 8525–8526.

26. D. Garcia, P. Pibarot, J. G. Dumesnil, F. Sakr, and L. G. Durand, Assessment of aortic valve stenosis severity: A new index based on the energy loss concept. *Circulation* 2000; **101**(7):765–771.

27. D. Burkhoff, I. Mirsky, and H. Suga, Assessment of systolic and diastolic ventricular properties via pressure-volume analysis: A guide for clinical, translational, and basic researchers. *Am. J. Physiol. Heart Circ. Physiol.* 2005; **289**: H501–H512.

28. J. Bermejo, R. Odreman, J. Feijoo, M. M. Moreno, P. Gomez-Moreno, and M. A. Garcia-Fernandez, Clinical efficacy of Doppler-echocardiographic indices of aortic valve stenosis: A comparative test-based analysis of outcome. *J. Am. Coll. Cardiol.* 2003; **41**(1):142–151.

29. D. A. Kass, and W. L. Maughan, From 'Emax' to pressure-volume relations: A broader view. *Circulation* 1988; **77**(6):1203–1212.

30. H. Suga, K. Sagawa, and A. A. Shoukas, Load independence of the instantaneous pressure-volume ratio of the canine left ventricle and effects of epinephrine and heart rate on the ratio. *Circ. Res.* 1973; **32**(3):314–322.

31. H. Senzaki, C. H. Chen, and D. A. Kass, Single-beat estimation of end-systolic pressure-volume relation in humans. A new method with the potential for noninvasive application. *Circulation* 1996; **94**(10):2497–2506.

32. R. Fogliardi, M. Di Donfrancesco, and R. Burattini, Comparison of linear and nonlinear formulations of the three-element windkessel model. *Am. J. Physiol.* 1996; **271**(6 Pt 2): H2661–H2668.

33. N. Westerhof, G. Elzinga, and P. Sipkema, An artificial arterial system for pumping hearts. *J. Appl. Physiol.* 1971; **31**(5):776–781.

34. D. Garcia, Personal home page. http://garciadam.free.fr. 2006.

35. R. O. Bonow, B. Carabello, A. C. De Leon, L. H. Edmunds, Jr., B. J. Fedderly, M. D. Freed, W. H. Gaasch, C. R. McKay, R. A. Nishimura, P. T. O'Gara, R. A. O'Rourke, S. H. Rahimtoola, J. L. Ritchie, M. D. Cheitlin, K. A. Eagle, T. J. Gardner, A. Garson, Jr., R. J. Gibbons, R. O. Russell, T. J. Ryan, and S. C. Smith, Jr. Guidelines for the management of patients with valvular heart disease: executive summary. A report of the American College of Cardiology/American Heart Association Task Force on Practice Guidelines (Committee on Management of Patients with Valvular Heart Disease). *Circulation* 1998; **98**(18):1949–1984.

36. A. L. Dekker, P. J. Barenbrug, F. H. van der Veen, P. Roekaerts, B. Mochtar, and J. G. Maessen, Pressure-volume loops in patients with aortic stenosis. *J. Heart Valve Dis.* 2003; **12**(3):325–332.

37. R. J. Bache, Y. Wang, and J. C. Greenfield, Jr. Left ventricular ejection time in valvular aortic stenosis. *Circulation* 1973; **47**(3):527–533.

38. P. Kligfield, P. Okin, R. B. Devereux, H. Goldberg, and J. S. Borer, Duration of ejection in aortic stenosis: effect of stroke volume and pressure gradient. *J. Am. Coll. Cardiol.* 1984; **3**(1):157–161.

39. L. M. Beauchesne, R. deKemp, K. L. Chan, and I. G. Burwash, Temporal variations in effective orifice area during ejection in patients with valvular aortic stenosis. *J. Am. Soc. Echocardiogr.* 2003; **16**(9):958–964.

40. J. Chambers, R. Rajani, M. Hankins, and R. Cook, The peak to mean pressure decrease ratio: a new method of assessing aortic stenosis. *J. Am. Soc. Echocardiogr.* 2005; **18**(6): 674–678.

41. K. Yamamoto, Q. N. Dang, Y. Maeda, H. Huang, R. A. Kelly, and R. T. Lee, Regulation of cardiomyocyte mechanotransduction by the cardiac cycle. *Circulation* 2001; **103**(10): 1459–1464.

42. B. H. Lorell and B. A. Carabello, Left ventricular hypertrophy: Pathogenesis, detection, and prognosis. *Circulation* 2000; **102**(4):470–479.

43. G. E. Pate, Association between aortic stenosis and hypertension. *J. Heart Valve Dis.* 2002; **11**(5):612–614.

44. Guidelines committee. 2003 European Society of Hypertension-European Society of Cardiology guidelines for the management of arterial hypertension. *J. Hypertens.* 2003; **21**(6):1011–1053.

45. D. Chemla, I. Antony, Y. Lecarpentier, and A. Nitenberg, Contribution of systemic vascular resistance and total arterial compliance to effective arterial elastance in humans. *Am. J. Physiol. Heart Circ. Physiol.* 2003; **285**(2):H614–H620.
46. D. Chemla, J. L. Hebert, C. Coirault, K. Zamani, I. Suard, P. Colin, and Y. Lecarpentier, Total arterial compliance estimated by stroke volume-to-aortic pulse pressure ratio in humans. *Am. J. Physiol.* 1998; **274**(2 Pt 2):H500–H505.
47. J. Bermejo, The effects of hypertension on aortic valve stenosis. *Heart* 2005; **91**(3):280–282.
48. R. E. Mates and R. M. Judd, Models for coronary pressure-flow relationships. In: Sideman S., Beyar R., ed., Interactive phenomena in the cardiac system. New York: Plenum Press, 1993: 153–161.
49. J. I. Hoffman and J. A. Spaan, Pressure-flow relations in coronary circulation. *Physiol. Rev.* 1990; **70**(2):331–390.
50. M. Arsenault, N. Masani, G. Magni, J. Yao, L. Deras, and N. Pandian, Variation of anatomic valve area during ejection in patients with valvular aortic stenosis evaluated by two-dimensional echocardiographic planimetry: Comparison with traditional Doppler data. *J. Am. Coll. Cardiol.* 1998; **32**(7):1931–1937.
51. A. J. Fenlon and T. David, Numerical models for the simulation of flexible artificial heart valves: part I–computational methods. *Comput. Methods Biomech. Biomed. Engin.* 2001; **4**(4):323–339.

ARTERIAL BLOOD PRESSURE PROCESSING

MARCO DI RIENZO
PAOLO CASTIGLIONI
Fondazione Don Carlo Gnocchi
ONLUS
Milano, Italy

GIANFRANCO PARATI
University of Milano-Bicocca
and S.Luca Hospital
Milano, Italy

1. INTRODUCTION

The arterial blood pressure wave is the result of the interaction between the heart ventricles ejecting a given volume of blood during the contraction and the arterial vascular system, which receives the ejected blood. At each beat, an incident blood pressure wave propagates from the heart to the periphery, where it is reflected by bifurcations and arteriolar beds. The resulting blood pressure waveform therefore consists of the summation of incident and reflected waves, whose magnitude and timing determine the final pulse contour. For these reasons and for the energy dissipation, amplitude, and shape of each arterial blood pressure wave change while traveling from the aorta to the periphery. These changes depend on the vascular district (pulmonary, systemic, coronary beds), and on the distance from the heart where the measure is performed (1). Additionally, arterial blood pressure is also characterized by an important beat-by-beat modulation imposed by the need to guarantee at any moment a proper blood flow supply in response to the multiple endogenous and exogenous challenges occurring in daily life. Part of blood pressure variability also depends on other factors, e.g., the interaction between cardiovascular and respiratory systems. Moreover, because excessive blood pressure deviations from the optimal value may be harmful for the organism, a number of biological control mechanisms also impinge on the cardiovascular system to limit blood pressure variability within a safety range. Age and specific pathologies may determine differences in the shape of the blood pressure wave and in its beat-to-beat fluctuations (2).

Arterial blood pressure can be measured by a variety of devices based on different technologies. Available instrumentation differs according to the site of measurement (from distal or proximal arteries), invasiveness, and continuity of the monitoring.

This chapter contains descriptions of (1) signal processing techniques for the analysis of data derived from devices providing a discontinuous measure of blood pressure, (2) techniques for the analysis of continuous arterial blood pressure recordings, (3) the main methods for the analysis of blood pressure variability, and (4) techniques for the estimation of the baroreflex function from the analysis of spontaneous fluctuations of blood pressure and heart rate.

2. ARTERIAL BLOOD PRESSURE PROCESSING FROM DISCONTINUOUS MEASUREMENTS

Noninvasive devices based on the arm cuff allow the sporadic assessment of systolic and diastolic blood pressures (i.e., the maximum and minimum values in the blood pressure wave), mean blood pressure, and heart rate. This simple and low-cost methodology is often used to obtain also an intermittent assessment of blood pressure values over the 24 hours in ambulant subjects. In this instance the systems are programmed to repeat the measurements over time at a rate that may vary from hours to minutes. While providing important clinical information, the low sampling rate does not allow the assessment of fast components of blood pressure variability, which are known to convey important information on the mechanisms controlling the cardiovascular system (see next sections). From the data obtained by these devices, the overall 24h mean and standard deviation are typically computed, being the latter taken as a global index of blood pressure variability. Obviously the time interval between blood pressure measurements influences the accuracy of the estimates. It has been shown that while one pressure assessment every 30 minutes may suffice for a correct assessment of the 24h blood pressure mean, the measuring interval should not exceed 10 minutes for estimating the 24h blood pressure standard deviation (3). An additional quantification of 24h variability is based on computation of the difference between the average blood pressure values observed during the day and night. More sophisticated methods are based on harmonic analysis of the blood pressure series. These methods include the single cosinor method (which estimates the single cosine function best fitting the 24h blood pressure profile) (4), the estimation of the first four harmonics of the Fourier spectrum (5), or the assessment of the cusum statistics (6).

The ambulatory blood pressure monitoring through intermittent arm-cuff devices is often performed to determine the optimal dosage of antihypertensive drugs. For this reason, further indexes have been proposed to quantify the effects of drugs on the 24 h blood pressure profile (7). Estimation of these indexes is based on the evaluation of two 24 h blood pressure recordings performed before and after the antihypertensive treatment. The *smoothness index* provides a quantification of the homogeneity of the blood pressure reduction following the drug administration. This index is calculated by first computing the hourly profile of the blood pressure reduction induced by the drug. The smoothness index is obtained as the ratio between mean and standard deviation of this profile (8) Another index for assessing the effects of drugs is the *trough-to-peak ratio* (9). It is estimated as the ratio of the reduction in blood pressure at the end of the interval between doses (trough) and the reduction in blood pressure at the time of the maximal effect of the drug (peak).

3. ARTERIAL BLOOD PRESSURE PROCESSING FROM CONTINUOUS MEASUREMENTS

The entire informative content of the arterial blood pressure signal can only be retrieved through continuous measurements of the whole waveform over time. In the past, the continuous assessment of arterial blood pressure was possible by intra-arterial techniques requiring the insertion of a catheter into an artery (10). The recent development of noninvasive devices based on finger plethysmography or applanation tonometry has offered a noninvasive alternative to the intra-arterial approach. In particular, devices for measuring blood pressure at the finger level allow the continuous recording of arterial blood pressure to be performed even for 24 hours in ambulant subjects (11). When the recording time is limited to a few minutes, devices based on applanation tonometry can be also used to detect the arterial blood pressure waveform at different superficial arteries (12).

In this section, the following aspects of the analysis of continuous arterial blood pressure waveforms will be described: (1) the application of transfer functions to estimate the shape of the pressure wave at arteries different from the measurement site; (2) the beat-by-beat extraction of traditional parameters from the blood pressure wave; and (3) the estimation of characteristics of the vascular bed and cardiac output through the pulse contour technique.

The analysis of the changes over time of the above parameters and their use to assess the baroreflex function will be illustrated in later sections.

3.1. Transfer Function between Measurement Sites

A limit of most devices for the continuous blood pressure recording is that they detect the signal from peripheral arteries (finger, radial or carotid arteries). As mentioned in the introduction, however, the shape of the blood pressure waveform is distorted while traveling from the heart to the periphery of the vascular system. Particularly for

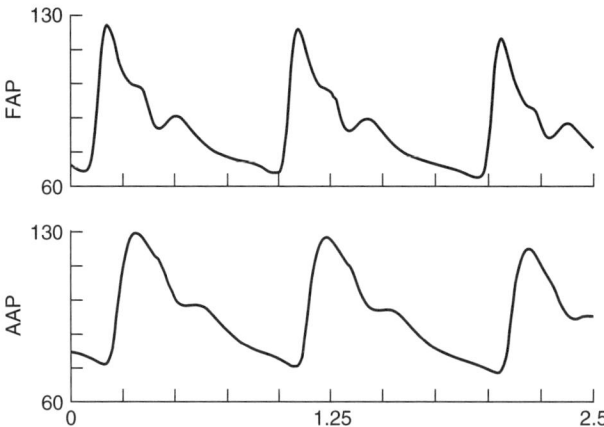

Figure 1. Noninvasive finger arterial pressure (FAP) and the corresponding aortic arterial pressure (AAP) obtained from FAP by application of filters based on the transfer function technique.

the systemic vascular bed, this phenomenon results in a progressive increase in the systolic value and a progressive reduction in the diastolic value while the pressure wave moves from the aorta to the small arteries. Thus, most of the recorded pressure waveforms are actually modified by these aberrances (Fig. 1).

To alleviate this problem, procedures have been proposed in order to estimate a proximal waveform by the application of specific filters to distal pressure signals. Details on these filters are available in literature to transform femoral into aortic (13), carotid into aortic (14), radial into aortic (15,16), finger into brachial (17,18) and finger into aortic (19) blood pressure waveforms.

The above filters have been designed by considering the inverse of the transfer functions describing the relation between central and peripheral waveforms. These transfer functions have been experimentally determined by simultaneously measuring arterial pressure at a proximal and a distal site in selected groups of subjects, and by estimating the relevant parameters through the application of Fourier analysis (18), the identification of a modified windkessel model of the arterial vessel (19), or by the application of autoregressive modelling techniques (16).

This approach has been shown to provide good estimates of parameters of the aortic waveform, like mean, maximum and minimum values of the pulse (20), while poor results have been obtained on the estimation of specific features of the aortic wave related to the reflection of the pressure wave (21,22).

3.2. Main Parameters of the Blood Pressure Wave

From continuous blood pressure recordings, possibly after application of the filtering procedure illustrated in the previous subsection, a number of parameters can be derived from each pulse wave. As a first step of the parameter extraction, a segmentation of the continuous signal into individual heart beats should be performed. When a simultaneous electrocardiographic recording is available, start and end of the nth heart beat are defined by the occurrence of the nth and $(n+1)^{\text{th}}$ R peaks in the ECG signal. Coherently, systolic and diastolic blood pressures are defined as the maximum and minimum blood pressure values occurring within the nth heart beat. According to this definition, the nth diastolic blood pressure precedes the nth systolic blood pressure. If a simultaneous ECG recording is not available, a fiducial point within the pressure wave must be taken to identify the start of each beat. Possible fiducial points can be the systolic peak, the foot of the wave, the point of maximal derivative in the upslope phase of the pulse waveform (Fig. 2).

From each pulse wave, besides SBP and DBP, other parameters can also be derived: pulse pressure (defined as the difference between systolic and diastolic pressures of the same beat); mean blood pressure (defined as the average of all blood pressure values observed in a beat); pulse interval (defined as the duration of the pulse wave). The latter is estimated as the time interval between consecutive systolic peaks or diastolic minima and is commonly taken as a quantification of the heart interval when the RR interval cannot be measured.

In case of use of systolic peak as a marker for segmentation, a "derivative and threshold" algorithm can be used for the identification of the systolic maximum. In this case systolic blood pressure can be identified as the greatest blood pressure value occurring within a data window of fixed length (i.e., few hundredths of milliseconds), after the first derivative of the signal exceeded an adaptive threshold (Fig. 3).

In some instances pressure waves may display uncommon contours and slightly different definitions of the derived parameters are preferred. This is the case for bisferiens pulses, i.e., blood pressure waves that show two distinct peaks during the phase of systolic ejection and the maximal pressure value alternates between these peaks (see Fig. 4, left panel). Although the definition of

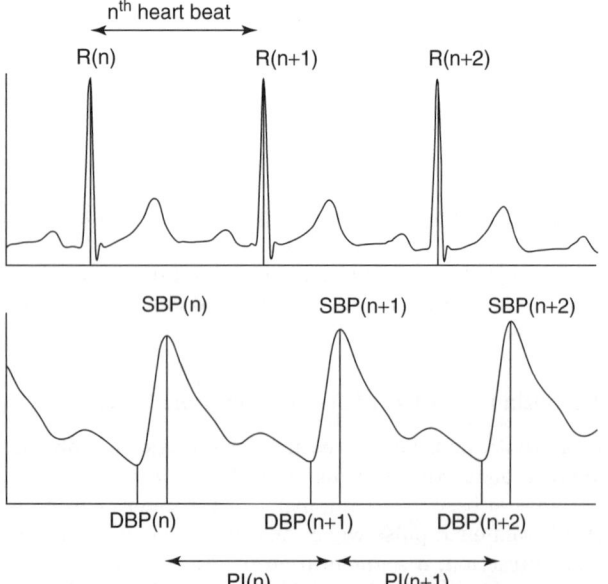

Figure 2. Definition of systolic blood pressure (SBP), diastolic blood pressure (DBP), and pulse interval estimated as interval between systolic peaks.

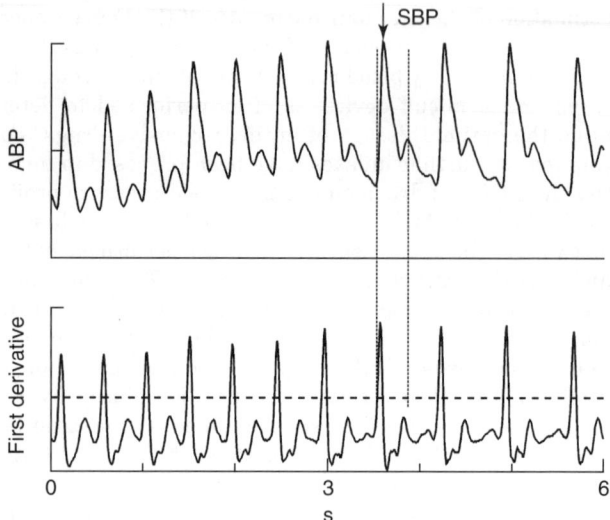

Figure 3. Example of "derivative and threshold" algorithm for identifying the systolic blood pressure peak in each heart beat. The original signal (upper trace) is differentiated (lower trace) and an adaptive threshold (dashed line) is computed as a percentage of the standard deviation of the derived signal in the last two minutes. Systolic blood pressure is the highest blood pressure value occurring within a data window of fixed length (in this example, equal to 350 ms) after the derived signal exceeded the threshold.

systolic blood pressure does not change (it remains the maximal pressure value in the wave), pulse should be measured by considering the first peak of the pair. This gives a measure of the wave duration closer to the R-R interval. Another exception is represented by blood pressure waves where the lowest blood pressure value does not occur just before the upslope phase of the next beat (this pattern may be observed in case of marked wave reflection). In this instance diastolic blood pressure is defined as the last relative minimum before the next systolic phase (Fig. 4, right).

When the central blood pressure wave is available (through a direct recording or as a result of filtering of a peripheral wave), another useful parameter that can be derived is the augmentation pressure AP, defined as the boost of pressure from the systolic shoulder (i.e., the inflection point during the ejection phase) to the systolic pressure peak (Fig. 5). Augmentation pressure can also be expressed as percentage of pulse pressure by the augmentation index, AI. Since the augmentation pressure depends on amplitude and timing of the reflected waves, AI is used to characterize the arterial stiffness (23).

3.3. Other Parameters from the Pulse Contour Analysis

Because the shape of the blood pressure wave is the result of the coupling between the ventricular activity and the vascular bed, several methods have been proposed to derive information on the heart and vascular functions from the analysis of the blood pressure pulse contour. In one application, the possibility of investigating alterations in arterial mechanical properties through analysis of the

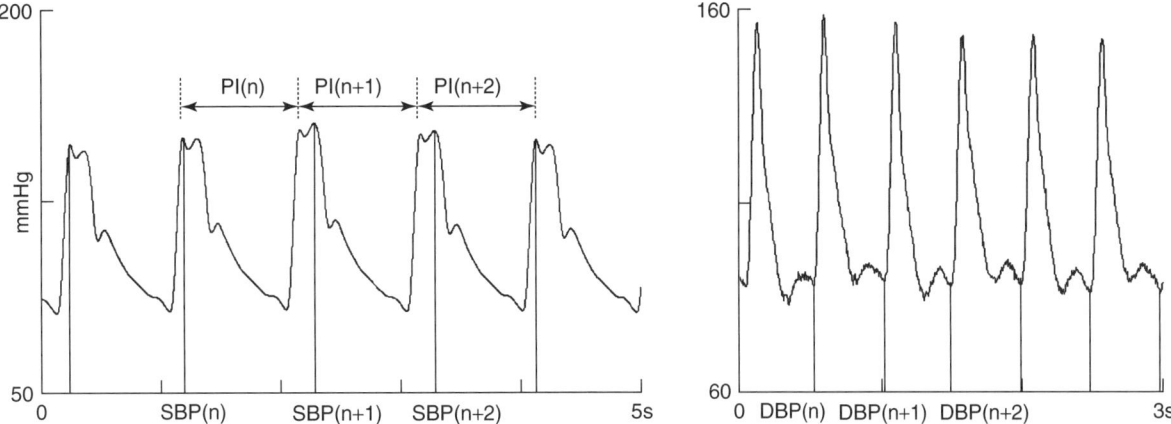

Figure 4. Definitions of pulse interval PI (left panel) or diastolic blood pressure DBP (right panel) in case of bisferiens pulses or when the minimum of the blood pressure wave does not coincide with the end of the diastole.

diastolic portion of the brachial or radial wave has been suggested (24). The method is based on fitting the parameters of a 4-element windkessel model (25) to the diastolic decay (Fig. 6). This is done by finding the six parameters A_i ($i = 1,\ldots,6$) that describe the exponential pressure decay during diastole, $P(t)$, as

$$P(t) = A_1 e^{-A_2 t} + A_3 e^{-A_4 t} \cos(A_5 t + A_6)$$

The four windkessel elements are related to the A_i coefficients. This method provides estimates of the capacitive compliance CC (ratio between volume fall and pressure fall during the diastolic decay) and oscillatory compliance OC (ratio between volume and pressure changes around the exponential pressure decay during diastole).

Other methods have also been proposed to estimate stroke volume and cardiac output from the shape of the blood pressure wave (26). One of the most frequently used is the model-flow method, based on an extended 3-element windkessel model (27). The model consists of two elements which nonlinearly depend on the input arterial pressure wave, $P(t)$ (the characteristic impedance of proximal aorta, Z_0, and the arterial compliance C) and a third element represented by the time-varying peripheral resistance $R(t)$ (Fig. 7). On the base of the measured central blood pressure wave $P(t)$, the "model flow" method provides an estimate of Z_0 and C, from which the blood flow $F(t)$ and resistances R are derived. The estimation of impedance and compliance is based on the arctangent model of aortic

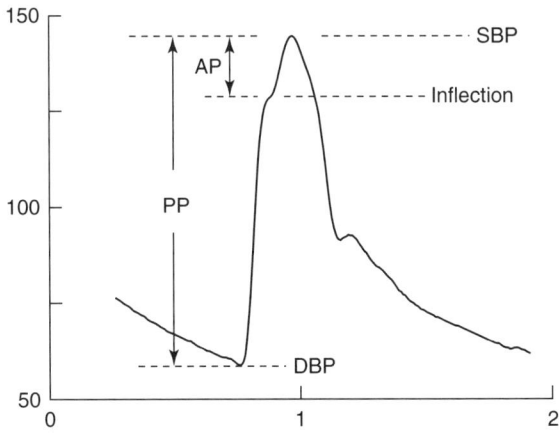

Figure 5. Augmentation pressure (AP) in a central blood pressure wave is calculated as the increment of blood pressure from the inflection point to the systolic peak (SBP).

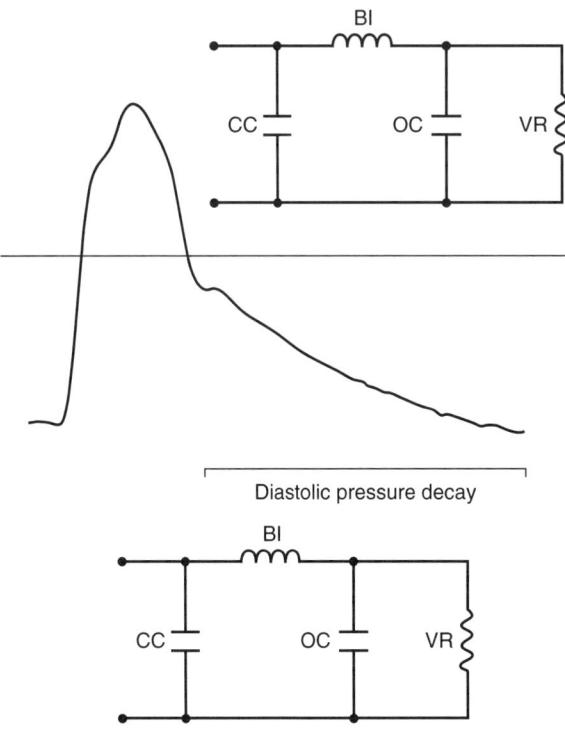

Figure 6. Windkessel model for the estimation of capacitive compliance (CC), oscillatory compliance (OC), systemic vascular resistance (VR), and blood inertia (BI) from analysis of the pressure decay during diastole. (This figure is available in full color at http://www.mrw.interscience.wiley.com/ebe.)

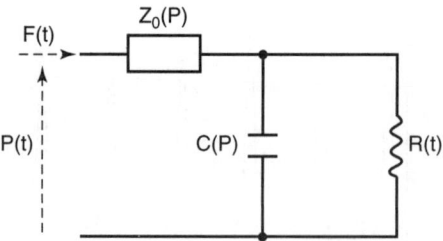

Figure 7. Extended windkessel model for estimating blood flow F from the measured blood pressure P; Z_0 and C are aortic characteristic impedance and arterial compliance, both nonlinearly dependent on blood pressure P; R is the time-varying peripheral resistance.

mechanics (28), which nonlinearly relates the thoracic aortic cross-sectional area to blood pressure, $A(P)$. From this relationship, the aortic compliance per unit area, C', is derived as: $C' = dA(P)/dP$. The arterial compliance C is obtained as $C = L \times C'$, where L is the aortic length, which depends on patient height and weight. Characteristic impedance Z_0 is calculated as $Z_0 = \sqrt{\rho/(AC')}$, where ρ is the blood density. Given an input pressure waveform, aortic systolic inflow is determined by the time constant $Z_0 \times C$. Stroke volume is computed by integrating the model flow during systole, and cardiac output by multiplying stroke volume with instantaneous heart rate. This method cannot provide accurate absolute values of cardiac output, but relative changes in cardiac output can be tracked with good precision (29).

4. BEAT-BY-BEAT BLOOD PRESSURE VARIABILITY

The blood pressure waveform changes continuously over time because of the varying needs of the different vascular districts, the effects of external perturbations, and the autonomic and humoral regulation of circulation. Therefore, several signal processing procedures have been developed to quantify the beat-by-beat dynamics of blood pressure parameters such as systolic blood pressure, diastolic blood pressure, and heart rate (derived as the reciprocal of the pulse interval). In specific instances, blood pressure may also be analyzed in concomitance with other signals, such as the ECG and signals related to the respiratory function (e.g., thoracic movements, oro-nasal airflow, oxygen saturation). Most of the available procedures for the assessment of blood pressure dynamics are based on the decomposition of the overall pressure variability in frequency components as obtained by the spectral analysis.

A common requirement characterizing standard spectral methods is that the input signal must be evenly sampled over time. Unfortunately, the parameters derived from the blood pressure waves are events occurring once per beat, and thus they are irregularly sampled at a frequency equal to the instantaneous heart rate. Although specific methods for the analysis of unevenly sampled time series have been proposed (30,31), two simpler approaches are generally used. The first one does not requires any manipulation of the original series; it simply assumes that the series is evenly sampled at a frequency equal to the mean heart rate. The spectral distortion resulting from this assumption is negligible if the heart rate does not change markedly in the analyzed segment of data. The second approach consists in the interpolation of the original beat-by-beat series followed by an even resampling of the interpolated signal.

Concerning the choice of the components of blood pressure variability to be considered, attention is usually focused on spectral components with period shorter than 30 s, which appear to reflect the action of the mechanisms controlling the cardiovascular system and the influence of respiration (32,33). In particular, it has been shown that the spectral powers in the so-called high frequency (HF: 0.15–0.40 Hz) band is caused by the interaction of respiration with the thorax hemodynamics while it has been hypothesized that the power in the low frequency (LF: 0.04–0.15 Hz) band may be due to central oscillators modulating the sympathetic efferences on the heart and vasculature (34) or to a resonance in the baroreflex loop (35).

The most popular spectral methods for the analysis of spontaneous cardiovascular variability are based on the fast Fourier transform, FFT (36,37) or on the autoregressive, AR, modelling (38). These methods yield similar spectra when FFT is used with some degree of smoothing and the AR modeling is applied with a sufficiently high model order (Fig. 8). However, specific properties make

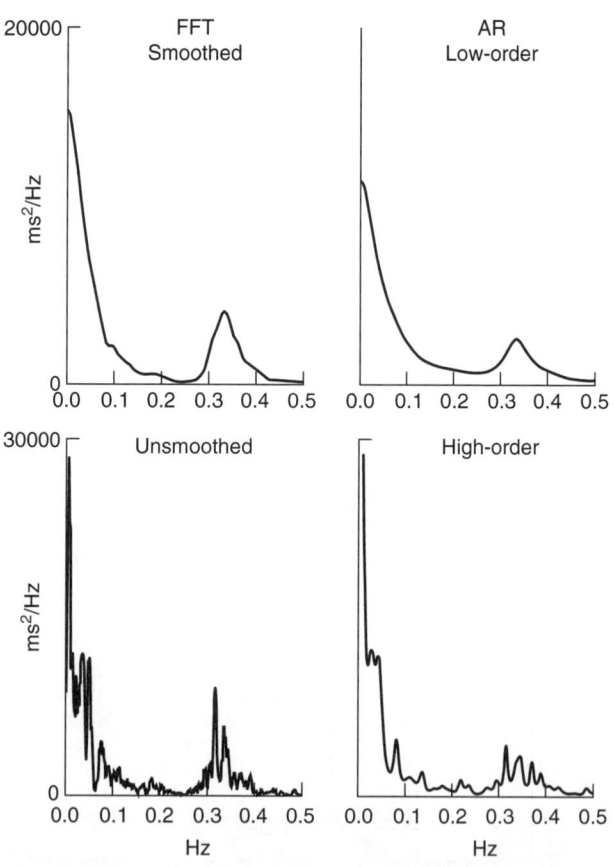

Figure 8. Spectra of the same 10-min recording of beat-to-beat pulse interval data analysed by FFT (left) with (upper panel) or without (lower panel) smoothing, or by AR modeling (right) with low (upper panel) or high (lower panel) model order.

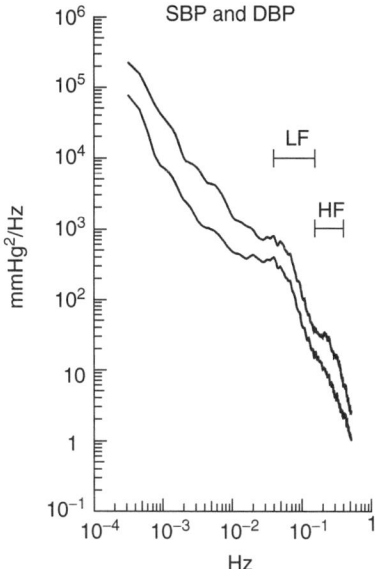
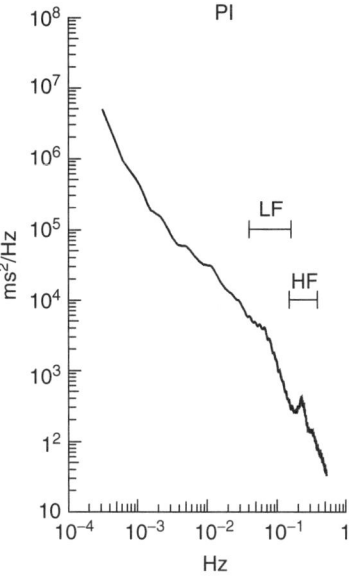

Figure 9. Broadband spectra of systolic blood pressure (left panel, upper trace), diastolic blood pressure (left panel, lower trace), and pulse interval (right panel) obtained from a 24-hour recording; position of low- and high-frequency bands, LF and HF, are shown.

each approach appropriate in special applications. For instances, AR spectra generally have higher-frequency resolution, making the AR modelling preferable when only short segments of data are available. Moreover, the AR approach makes it easy the analytical identification of power and central frequency of the spectral peaks (39). The FFT method is preferable to analyze spectra with the power spread over a wide range of frequencies (Fig. 9).

When the interest is to focus on the global evaluation of the spectral characteristics of a long-term blood pressure recording, the so called broadband analysis can be performed. This approach is usually based on the evaluation of a single spectrum from the whole recording followed by the application of smoothing procedures to increase the consistency of the spectral estimate (37,40; Di Rienzo et al., 1995). The broadband approach was used to show that, like the R-R interval time series (41), also beat-by-beat blood pressure profiles are characterized by the so-called $1/f$ trend (42), i.e., by fluctuations having a power that decreases with the frequency according to the $1/f^\beta$ power law, where the exponent β can be identified by the regression line fitting the slower components of the spectrum plotted in a log-log scale.

Since the biological mechanisms controlling the cardiovascular system simultaneously modulate blood pressure and also other variables, such as heart rate, quantification of the coupling between variables is usually performed to get information on the efficiency of the cardiovascular control. The degree of linear correlation between variables is usually quantified by the coherence function:

$$\gamma(f) = \frac{G_{XY}(f)}{\sqrt{G_{XX}(f)G_{YY}(f)}}$$

where X and Y are the two input time series, $\gamma(f)$ is the coherence function, $G_{XY}(f)$ is the cross-spectrum, and $G_{XX}(f)$ and $G_{YY}(f)$ are the spectra of X and Y.

From the above formula, the coherence squared modulus, $|\gamma(f)|^2$, and phase are usually derived (36). Pairs of variables are considered to be coupled if the value of $|\gamma(f)|^2$ is above a certain threshold, which may depend on the degrees of freedom of the estimation procedure and on the features of the input signals. In literature, 0.5 is the most common value used as a threshold in the analysis of cardiovascular data, although criticisms may be raised of such an arbitrary choice (43). Figure 10 shows typical coherence squared modulus and phase between systolic blood pressure and pulse interval and between systolic and diastolic blood pressure in a healthy subject at rest.

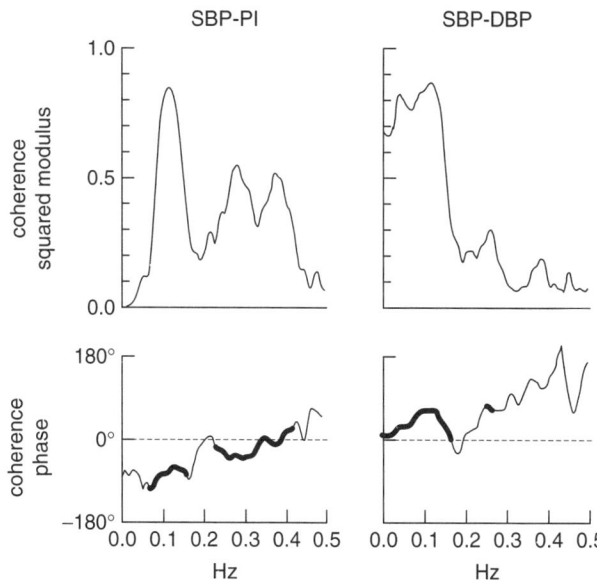

Figure 10. Squared modulus and phase of the coherence function between systolic blood pressure and pulse interval (left) and between systolic and diastolic blood pressures, from a 10-min recording in a young healthy subject at rest. Solid dots (●) show phase estimates obtained when the corresponding squared coherence modulus was greater than 0.3. A negative phase implies the SBP oscillation to lead the PI or DBP oscillation.

Typically, the modulus of coherence between systolic blood pressure and pulse interval (the reciprocal of heart rate) shows two peaks: around the respiratory frequency and at 0.1 Hz. These peaks indicate that respiratory oscillations and LF powers are linearly correlated. By contrast, coherence is remarkably low at lower frequencies, suggesting that oscillations of longer periods are uncoupled, or coupled through nonlinear mechanisms (36,44). Conversely, the modulus of coherence between systolic and diastolic blood pressure is high only at the slowest components of the spectrum (from 0.1 Hz toward the lower frequencies) while at the respiratory frequency the coherence modulus is relatively low. This can be explained by considering the relationships between blood pressure and heart interval as described by the windkessel model, and by considering the fast vagally mediated baroreflex control of heart rate (44). Indeed, SBP changes at the respiratory frequency are usually followed by reflex parallel changes in pulse interval because of the buffering action exerted by the baroreflex. This reflex change in the beat length modifies the duration of the pressure decay during diastole and weakens the SBP-DBP coupling at the respiratory frequencies. The example of Fig. 10 shows that the SBP oscillations lead the PI oscillations and the DBP oscillations lead the SBP oscillations in the LF band.

5. ASSESSMENT OF BAROREFLEX FUNCTION

The arterial blood pressure homeostasis is importantly controlled by the arterial baroreflex. This biological mechanism counteracts deviations of blood pressure from the reference set point by modulating heart rate, peripheral vascular tone, and other cardiovascular variables.

An impairment of the baroreflex function may result in a significant disregulation of blood pressure. This may become manifest as sudden pressure drops on shifting from a supine to a standing position or as aberrant pressure rises under stressful behavioral conditions. Baroreflex dysfunction may occur in different pathological conditions, including hypertension, primary autonomic failure, acute myocardial infarction, congestive heart failure, renal failure, diabetes mellitus, and a number of neurodegenerative diseases. In these instances the identification of a baroreflex impairment has been suggested to have diagnostic and prognostic relevance (45–47).

The efficiency of the baroreflex function is commonly estimated by evaluating the sensitivity of the baroreflex control of the heart (BRS), i.e., by quantifying the capability of the baroreflex to modulate the heart rate in the attempt to buffer a unitary change in blood pressure. Traditionally, the BRS assessment is performed in a laboratory environment through techniques that provide isolated evaluations of the reflex heart rate changes following (1) blood pressure changes artificially induced by external interventions on the subject such as injections of vasoactive drugs or (2) external manipulations of carotid baroreceptors, e.g., by a neck-chamber device, or (3) voluntary changes in the respiratory pattern, e.g., through the Valsalva maneuver. These techniques allow an assessment of BRS under standardized and controlled conditions. However, they suffer from two major limitations (48): first, the required "artificial" stimulations may interfere with the baroreflex function, and second, the isolated assessment these approaches yield do not allow an adequate characterization of the dynamic feature of the baroreflex control in daily life. The last limitation is particularly important. Indeed, it is known that in daily life BRS changes over time as the result of continuous influences exerted by central and peripheral inputs with the purpose to adapt the cardiovascular system to specific behavioral tasks. For example, during physical exercise or under emotional stress, blood pressure should rise from its reference level. Given the high efficiency of the baroreflex (49), this pressure rise can be obtained only through a reduction of the baroreflex gain (50). Indeed, should the baroreflex gain remain stable at its maximal level, no pressure increase would be allowed.

A significant advancement in the assessment of the baroreflex function is represented by the development of techniques that quantify the "spontaneous" sensitivity of the baroreflex control of the heart. These techniques, sometimes referred to as "modern techniques," do not require any external intervention on the subject under evaluation. Moreover, most of them can be used also to investigate the dynamic features of baroreflex modulation in daily life. These "modern techniques" derive information on the baroreflex function from the joint analysis of spontaneous systolic blood pressure and heart rate beat-by-beat variability as obtained by time or frequency domain algorithms, by statistical procedures, or through the identification of mathematical models (Fig. 11) (40,44,51–61). All these approaches are based on the following rationale. Since the baroreflex continuously modulates the heart rate to buffer the spontaneous SBP changes occurring in daily life, it actually produces a certain level of coupling between specific components of blood pressure and heart rate variabilities. A proper assessment of this coupling is thus expected to provide information on the baroreflex function.

5.1. Techniques for the Assessment of Spontaneous Baroreflex

Hereafter, the features of the most employed techniques for spontaneous baroreflex analysis will be described.

5.1.1. Time Domain Techniques.
5.1.1.1. The Sequence Technique. This technique is based on the automatic scanning of beat-by-beat SBP and RRI (or PI) series searching for spontaneous sequences of three or more consecutive beats in which a progressive rise in systolic blood pressure is followed, with a delay of zero, one, or two beats, by a progressive increase in R-R interval (+RRI/+SBP sequences) or by a progressive decrease in systolic blood pressure and shortening in R-R interval (−RRI/−SBP sequences). The slope of the regression line between SBP and RRI values forming the sequence is estimated and taken as an index of the sensitivity of arterial baroreflex modulation of heart rate (Fig. 12) (40,48,62–64). About 80 spontaneous sequences of 4-beat length can typically be observed per each hour

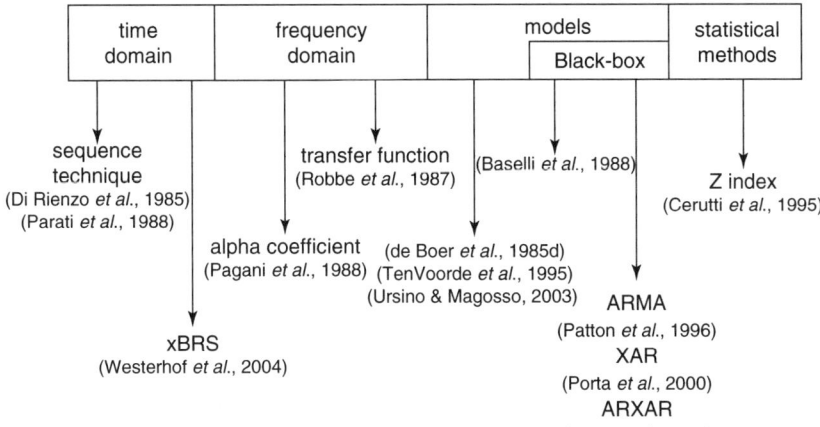

Figure 11. Scheme of available techniques for estimating BRS (40,44,51–61).

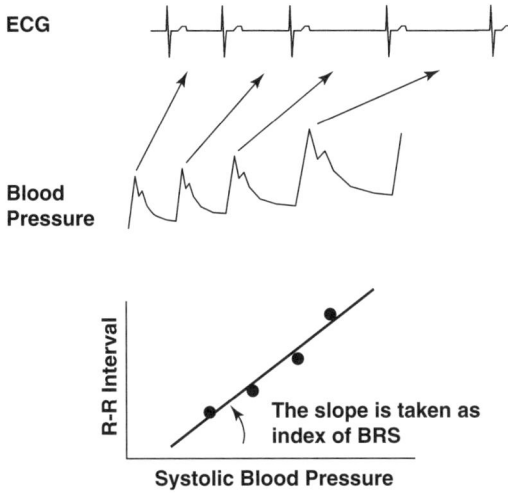

Figure 12. Scheme of the sequence technique: from (62), with permission, © 2001 IEEE and (63) with permission from IOS press.

during daily activities, thus making this method suitable to provide a detailed profile of BRS modulation over time. The specificity of this technique was demonstrated by showing a drastic reduction in the number of spontaneous sequences (−86%) after surgical opening of the baroreflex loop (64).

5.1.1.2. x-BRS. This approach is based on the estimation of the cross-correlation between very short (10-sec) segments of SBP and RRI data. For any SBP segment, the cross-correlation is repeatedly estimated by considering sliding 10-sec segments of RRI with a time delay with respect to SBP up to 5 sec. The SBP and RRI segments showing the maximal correlation coefficient are selected and the slope of the regression line obtained from the corresponding SBP and RRI values is taken as an index of BRS, provided the probability of being a random regression is lower than 1% (52).

5.1.2. Frequency Domain Techniques.
5.1.2.1. Alpha Coefficient. This approach is based on the splitting of the beat-by-beat SBP and RRI series into short data segments, each lasting a few minutes (usually 2–5), followed by estimation of the power spectra for each segment and by the computation of the following indexes:

$$\alpha_{LF} = (\text{RRI Power}_{LF}/\text{SBP Power}_{LF})^{1/2}$$

$$\alpha_{HF} = (\text{RRI Power}_{HF}/\text{SBP Power}_{HF})^{1/2}$$

where RRI Power$_{LF}$, SBP Power$_{LF}$, RRI Power$_{HF}$, and SBP Power$_{HF}$ are the RRI and SBP powers estimated in the LF and HF frequecy bands, respectively. With the hypothesis that SBP-RRI coherence may reflect the degree of baroreflex–induced coupling between SBP and RRI, α_{LF} and α_{HF} are taken as indexes of BRS whenever the squared modulus of the RRI-SBP coherence is greater than 0.5 (53,65).

5.1.2.2. Transfer Function. In this case, after having split the SBP and RRI series into short data segments, the SBP spectrum and the cross-spectrum between SBP and RRI are estimated. The transfer function between the output RRI and the input SBP is then evaluated as the ratio between the cross-spectrum and the SBP spectrum. The values of the transfer function modulus in the LF and HF frequency regions are taken as BRS estimates (54).

Because blood pressure and heart rate recordings performed in spontaneous conditions include frequent SBP and RRI rhythmic oscillations in the LF and HF frequency band, also the above techniques can be used for the assessment of the dynamic features of BRS.

5.1.3. Statistical Techniques.
5.1.3.1. Z-Method. This approach is based on the computation of the statistical dependence between SBP and RRI values, as obtained by considering the Z-coefficient (66). The slope of the regression line estimated from the SBP and RRI values showing over-threshold z-coefficient values is taken as an index of BRS (61).

5.1.4. Model-Based Techniques.
Another approach for estimating the baroreflex sensitivity is based on the mathematical modeling of the cardiovascular regulatory mechanisms. In this case the recorded SBP and RRI beat-by-beat series are used to identify the coefficients of

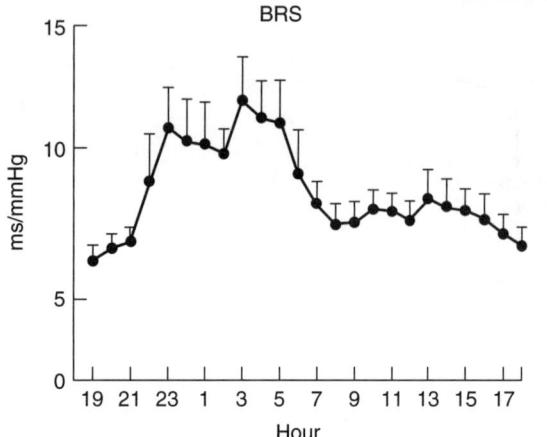

Figure 13. BRS profile over the 24 hours in a group of 8 young healthy subjects: (63) with permission from IOS press.

a preselected mathematical model of the baroreflex. Once the model has been identified, the transfer function, and hence the gain, delay, and time constant of the baroreflex control can be derived from the equations of the model. The accuracy of the estimates so obtained depends on how well the mathematical model fits the physiological complexity of the cardiovascular control mechanisms. So far a number of different models have been proposed (13,44,56). In many instances a "black-box" modeling of baroreflex control has been used, sometimes including also respiratory signals. This class of models includes dynamic adjustment models (57); autoregressive-moving average, ARMA, models (58); exogenous models with autoregressive input, XAR (59); bivariate autoregressive models with two exogenous inputs, ARXAR (60).

This approach to the assessment of the baroreflex function is particularly intriguing and challenging because, provided a suitable model is identified, it may offer the possibility of addressing aspects of the baroreflex control of the circulation, in addition to BRS, that can usually only be investigated through complex experimental procedures in a laboratory setting.

5.2. Dynamic Assessment of Spontaneous Baroreflex

Figure 13 shows an example of BRS assessment over the 24 hours (sequence technique) in a group of eight young subjects. Application of these procedures demonstrated that in young subjects BRS is significantly lower during the day than in the night time during sleep, in line with the hypothesis of an inverse relation between BRS and mental arousal or physical activity (51).

Figure 14 illustrates another example of the capability of the sequence technique and alpha coefficient to track fast changes of BRS. The figure shows BRS estimates obtained in a healthy volunteer undergoing a sequence of maneuvers (changes of posture, mental stress, incremental exercise) that are known to activate the autonomic cardiac modulation at different levels. BRS displays

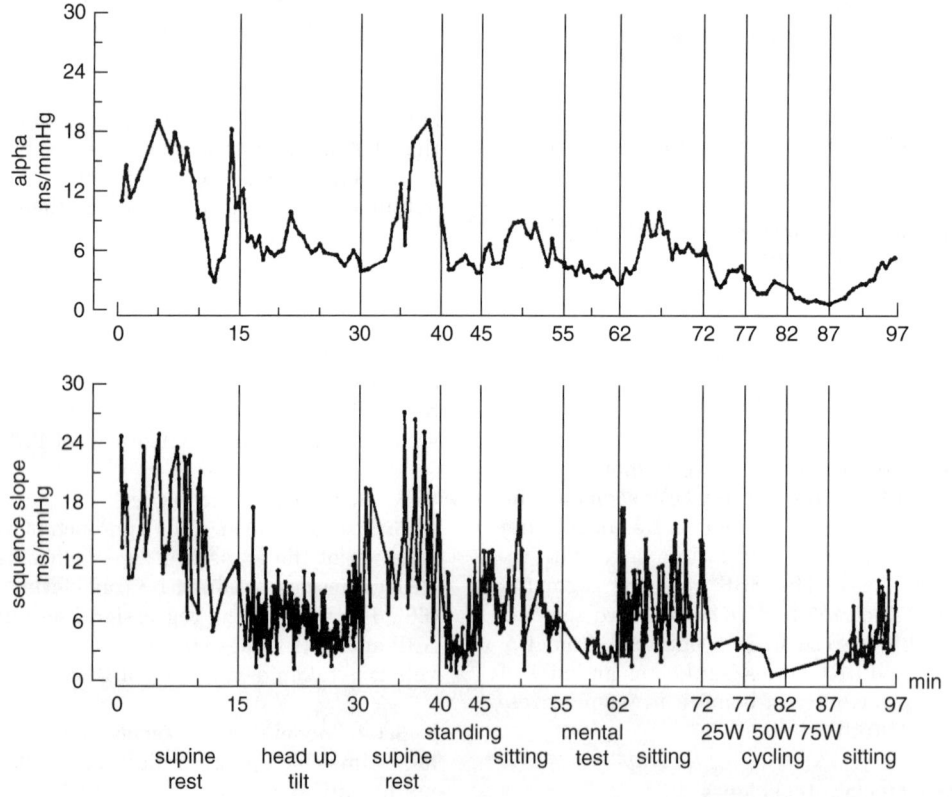

Figure 14. Time course of BRS estimated by alpha (upper panel) and sequence (lower panel) techniques during a series of manoeuvres: (63) with permission from IOS press.

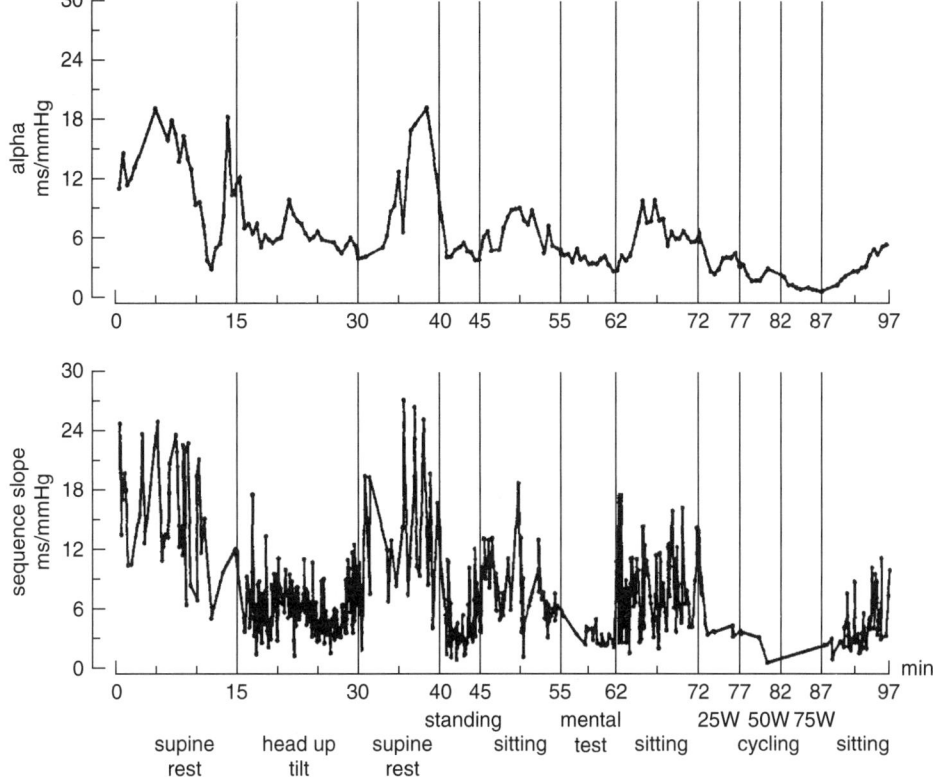

Figure 15. Profile of BRS during 24 hours (left) and its power spectrum (right). From (62) with permission © 2001 IEEE.

continuous changes from one condition to the next one. The highest values were observed while the subject was supine at rest (lowest sympathetic tone and highest vagal tone). The BRS estimates decreased during tilt, standing, and mental stress and dropped further during physical exercise, in parallel with the sympathetic activation (62,63). As a methodological remark, the overall similarity in the BRS estimates obtained by the sequence and the alpha techniques should be noted from the figure. This indicates that in spite of the marked differences in the algorithms used by these methods, they actually provide an uniquivocal quantification of the baroreflex function.

Concerning the long-term description of BRS dynamics, it was shown (67) that over the 24 h the spectral characteristics of BRS follow a $1/f$ trend (Fig. 15) and this finding opens the question on whether such a behavior of the BRS dynamics may be one of the cofactors responsible for the $1/f$ pattern also observed in blood pressure and heart rate spectra (50).

BIBLIOGRAPHY

1. W. W. Nichols and M. F. O'Rourke, (2005). *McDonald's Blood Flow in Arteries*. 5th edition. New York: Oxford University Press.
2. J. T. Shepherd and F. M. About, (1983). *Handbook of Physiology*, IV ed., pp. 755–794. Bethesda, MD: American Physiological Society.
3. M. Di Rienzo, G. Grassi, A. Pedotti, and G. Mancia, Continuous vs intermittent blood pressure measurements in estimating 24-hour average blood pressure. *Hypertension* 1983; **5**:264–269.
4. F. Halberg, Chronobiology. *Annu. Rev. Physiol.* 1969; **31**:675–725.
5. N. P. Chau, J. M. Mallion, G. R. de, E. Ruche, J. P Siche, O. Pelen, and G. Mathern, Twenty-four-hour ambulatory blood pressure in shift workers. *Circulation* 1989; **80**:341–347.
6. A. Stanton, J. Cox, N. Atkins, K. O'Malley, and E. O'Brien, Cumulative sums in quantifying circadian blood pressure patterns. *Hypertension* 1992; **19**:93–101.
7. S. Omboni, G. Parati, and G. Mancia, The trough:peak ratio and the smoothness index in the evaluation of control of 24 h blood pressure by treatment in hypertension. *Blood. Press. Monit.* 1998; **3**:201–204.
8. G. Parati, M. Di Rienzo, G. Bertinieri, G. Pomidossi, R. Casadei, A. Groppelli, A. Pedotti, A. Zanchetti, and G. Mancia, Evaluation of the baroreceptor-heart rate reflex by 24-hour intra-arterial blood pressure monitoring in humans. *Hypertension*. 1988; **12**:214–222.
9. J. A. Staessen, L. Thijs, G. Bijttebier, D. Clement, E. T. O'Brien, P. Palatini, J. L. Rodicio, J. Rosenfeld, and R. Fagard, Determining the trough-to-peak ratio in parallel-group trials. Systolic Hypertension in Europe (SYST-EUR) Trial Investigators. *Hypertension*. 1997; **29**:659–667.
10. A. T. Bevan, A. J. Honour, and F. H. Stott, Direct arterial pressure recording in unrestricted man. *Clin. Sci.* 1969; **36**:329–344.
11. B. P. Imholz, G. J. Langewouters, G. A. van Montfrans, G. Parati, G. J.van, K. H. Wesseling, W. Wieling, and G. Mancia,

Feasibility of ambulatory, continuous 24-hour finger arterial pressure recording. *Hypertension* 1993; **21**:65–73.

12. R. Kelly, C. Hayward, J. Ganis, J. Daley, A. Avolio, and M. O'Rourke, Noninvasive registration of the arterial pressure pulse waveform using high-fidelity applanation tonometry. *J. Vasc. Med. Biol.* 1989; **1**:142–149.

13. D. H. Fitchett, Aortofemoral transfer function: A method to determine the instantaneous aortic valve gradient in aortic valve stenosis. *J. Am. Coll. Cardiol.* 1993; **22**: 1909–1914.

14. M. Karamanoglu, and M. P. Feneley, Derivation of the ascending aortic-carotid pressure transfer function with an arterial model. *Am. J. Physiol.* 1996; **271**:H2399–H2404.

15. C. J. Chen, E. Nevo, B. Fetics, P. H. Pak, F. C. Yin, W. L. Maughan, and D. A. Kass, Estimation of central aortic pressure waveform by mathematical transformation of radial tonometry pressure. Validation of generalized transfer function. *Circulation* 1997; **95**:1827–1836.

16. B. Fetics, E. Nevo, C. H. Chen, and D. A. Kass, Parametric model derivation of transfer function for noninvasive estimation of aortic pressure by radial tonometry. *IEEE Trans. Biomed. Eng.* 1999;**46**: 698–706.

17. W. J. Bos, G. J van, G. A. van Montfrans, A. H. van den Meiracker, and K. H. Wesseling. Reconstruction of brachial artery pressure from noninvasive finger pressure measurements. *Circulation* 1986; **94**:1870–1875.

18. P. Gizdulich, A. Prentza, and K. H. Wesseling, Models of brachial to finger pulse wave distortion and pressure decrement. *Cardiovasc. Res.* 1997; **33**:698–705.

19. M. Karamanoglu, and M. P. Feneley, On-line synthesis of the human ascending aortic pressure pulse from the finger pulse. *Hypertension* 1997; **30**:1416–1424.

20. A. L. Pauca, M. F. O'Rourke, and N. D. Kon, Prospective evaluation of a method for estimating ascending aortic pressure from the radial artery pressure waveform. *Hypertension* 2001; **38**:932–937.

21. S. A. Hope, D. B. Tay, I. T. Meredith, and J. D. Cameron, Use of arterial transfer functions for the derivation of aortic waveform characteristics. *J. Hypertens.* 2003; **21**:1299–1305.

22. S. C. Millasseau, S. J. Patel, S. R. Redwood, J. M. Ritter, and P. J. Chowienczyk, Pressure wave reflection assessed from the peripheral pulse: is a transfer function necessary? *Hypertension* 2003; **41**:1016–1020.

23. M. F. O'Rourke and G Mancia, Arterial stiffness. *J. Hypertens.* 1999; **17**:1–4.

24. J. N. Cohn, S. Finkelstein, G. McVeigh, D. Morgan, L. LeMay, J. Robinson, and J. Mock, Noninvasive pulse wave analysis for the early detection of vascular disease. *Hypertension* 1995; **26**:503–508.

25. R. Goldwyn, and T. Watt, Arterial pressure pulse contour analysis via a mathematical model for the clinical quantification of human vascular properties. *IEEE Trans. Biomed. Eng.* 1967; **14**:11–17.

26. C. Cerutti, M. P. Gustin, P. Molino, and C. Z. Paultre, Beat-to-beat stroke volume estimation from aortic pressure waveform in conscious rats: comparison of models. *Am. J. Physiol. Heart Circ. Physiol.* 2001; **281**:H1148–H1155.

27. K. H. Wesseling, J. R. Jansen, J. J. Settels, and J. J. Schreuder, Computation of aortic flow from pressure in humans using a nonlinear, three-element model. *J. Appl. Physiol.* 1993; **74**:2566–2573.

28. G. J. Langewouters, K. H. Wesseling, and W. J. Goedhard, The pressure dependent dynamic elasticity of 35 thoracic and 16 abdominal human aortas in vitro described by a five component model. *J. Biomech.* 1985; **18**:613–620.

29. W. J. Stok, F. Baisch, A. Hillebrecht, H. Schulz, M. Meyer, and J. M. Karemaker, Noninvasive cardiac output measurement by arterial pulse analysis compared with inert gas rebreathing. *J. Appl. Physiol.* 1993; **74**:2687–2693.

30. P. Laguna, G. B. Moody, and R. G. Mark, Power spectral density of unevenly sampled data by least-square analysis: performance and application to heart rate signals. *IEEE Trans. Biomed. Eng.* 1998; **45**:698–715.

31. B. J. TenVoorde, J. C. Faes, and O. Rompelman, Spectra of data sampled at frequency-modulated rates in application to cardiovascular signals: Part 1. Analytical derivation of the spectra. *Med. Biol. Eng. Comput.* 1994; **32**:63–70.

32. G. Mancia, G. Parati, P. Castiglioni, and M. Di Rienzo, Effect of sinoaortic denervation on frequency-domain estimates of baroreflex sensitivity in conscious cats. *Am. J. Physiol.* 1999; **276**:H1987–H1993.

33. Task Force of the European Society of Cardiology and the North American Society of Pacing and Electrophysiology. Heart rate variability: Standards of measurement, physiological interpretation and clinical use. *Circulation* 1996; **93**:1043–1065.

34. N. Montano, T. Gnecchi-Ruscone, A. Porta, F. Lombardi, A. Malliani. and S. M. Barman. Presence of vasomotor and respiratory rhythms in the discharge of single medullary neurons involved in the regulation of cardiovascular system. *J. Auton. Nerv. Syst.* 1996; **57**:116–122.

35. R. W. deBoer, J. M. Karemaker, and J. Strackee, Hemodynamic fluctuations and baroreflex sensitivity in humans: a beat-to-beat model. *Am. J. Physiol.* 1987; **253**:H680–H689.

36. R. W. de Boer, J. H. Karemaker, and J. Strackee, Relationships between short-term blood-pressure fluctuations and heart-rate variability in resting subjects. I: A spectral analysis approach. *Med. Biol. Eng. Comput.* 1985a; **23**:352–358.

37. R. E. Challis, and R. I., Kitney, Biomedical signal processing (in four parts). Part 3. The power spectrum and coherence function. *Med. Biol. Eng. Comput.* 1991; **29**:225–241.

38. G. Baselli, S. Cerutti, S. Civardi, D. Liberati, F. Lombardi, A. Malliani, and M. Pagani, Spectral and cross-spectral analysis of heart rate and arterial blood pressure variability signals. *Comput. Biomed. Res.* 1986; **19**:520–534.

39. S. J. Johnsen, and N. Andersen, On power estimation in maximum entropy spectral analysis. *Geophysics* 1978; **43**:681–690.

40. M. Di Rienzo, G. Bertinieri, G. Mancia, and A. Pedotti, A new method for evaluating the baroreflex role by a joint pattern analysis of pulse interval and systolic blood pressure series. *Med. Biol. Eng. Comput.* 1985; 23 Suppl.**1**:313–314.

41. M. Kobayashi and T. Musha, 1/f fluctuation of heartbeat period. *IEEE Trans. Biomed. Eng.* 1982; **29**:456–457.

42. M. Di Rienzo, G. Parati, A. Pedotti and P. Castiglioni, (1997). 1/f modeling of blood pressure and heart rate spectra. In *Frontiers of Blood Pressure and Heart Rate Analysis* Amsterdam: IOS Press, pp. 45–53.

43. L. Faes, G. D. Pinna, A. Porta, R. Maestri, and G. Nollo, Surrogate data analysis for assessing the significance of the coherence function. *IEEE Trans. Biomed. Eng.* 2004; **51**:1156–1166.

44. R. W. de Boer, J. M. Karemaker, J. and Strackee J. Relationships between short-term blood-pressure fluctuations and heart-rate variability in resting subjects. II: A simple model. *Med. Biol. Eng. Comput.* 1985b; **23**:359–364.

45. Y. Katsube, H. Saro, M. Naka, B. H. Kim, N. Kinoshita, Y. Koretsune, and M. Hori, Decreased baroreflex sensitivity in patients with stable coronary artery disease is correlated with the severity of coronary narrowing. *Am. J. Cardiol.* 1996; **78**:1007–1010.
46. M. T. La Rovere, G. D. Pinna, R. Maestri, A. Mortara, S. Capomolla, O. Febo, R. Ferrari, M. Franchini, M. Gnemmi, C. Opasich, P. G. Riccardi, E. Traversi, and F. Cobelli, Short-term heart rate variability strongly predicts sudden cardiac death in chronic heart failure patients. *Circulation* 2003; **107**:565–570.
47. T. G. Farrell, O. Odemuyiwa, Y. Bashir, T. R. Cripps, M. Malik, D. E. Ward, and A. J. Camm, Prognostic value of baroreflex sensitivity testing after acute myocardial infarction. *Br. Heart J.* 1992; **67**:129–137.
48. G. Parati, R. M. Di, and G. Mancia, How to measure baroreflex sensitivity: From the cardiovascular laboratory to daily life. *J. Hypertens.* 2000; **18**:7–19.
49. A. C. Guyton, T. G. Coleman, A. W. Cowley, Jr., J.F. Liard, R. A. Norman, Jr. and R. D. Manning, Jr. Systems analysis of arterial pressure regulation and hypertension. *Ann. Biomed. Eng.* 1972; **1**:254–281.
50. K. H. S. J. Wesseling, (1995). Baromodulation explains short-term blood pressure variability. In: *Psychophysiology of Cardiovascular Control. Models, Methods and Data.* New York: Plenum, pp. 69–97.
51. G. Parati, S. Omboni, D. Rizzoni, E. Gabiti-Rosei, and G. Mancia, The smoothness index: A new, reproducible and clinically relevant measure of the homogeneity of the blood pressure reduction with treatment for hypertension. *J. Hypertens.* 1988; **16**:1685–1691.
52. B. E. Westerhof, J. Gisolf, W. J. Stok, K. H. Wesseling, and J. M. Karemaker. Time-domain cross-correlation baroreflex sensitivity: Performance on the EUROBAVAR data set. *J Hypertens.* 2004; **22**:1371–1380.
53. M. Pagani, V. Somers, R. Furlan, S. Dell'Orto, J. Conway, G. Baselli, S. Cerutti, P. Sleight, and A. Malliani, Changes in autonomic regulation induced by physical training in mild hypertension. *Hypertension* 1988; **12**:600–610.
54. H. W. Robbe, L. J. Mulder, H. Ruddel, W. A. Langewitz, J. B. Veldman, and G. Mulder, Assessment of baroreceptor reflex sensitivity by means of spectral analysis. *Hypertension.* 1987; **10**:538–543.
55. B. J. TenVoorde, J. C. Faes, T. W. J. Janssen, G. J. Scheffer, and Rompelman O. Respiratory modulation of blood pressure and heart rate studied with a computer model of baroreflex control. In: M. Di Rienzo, G. Mancia, G. Parati, A. Pedotti, and A. Zanchetti eds., *Computer analysis of cardiovascular signals.* Amsterdam: IOS Press, 1995, pp. 119–134.
56. M. Ursino and E. Magosso, Short-term autonomic control of cardiovascular function: A mini-review with the help of mathematical models. *J. Integr. Neurosci.* 2003; **2**:219–247.
57. G. Baselli, S. Cerutti, S. Civardi, A. Malliani, and M. Pagani, Cardiovascular variability signals: towards the identification of a closed-loop model of the neural control mechanisms. *IEEE Trans. Biomed. Eng.* 1988; **35**:1033–1046.
58. D. J. Patton, J. K. Triedman, M. H. Perrott, A. A. Vidian, and J. P. Saul, Baroreflex gain: Characterization using autoregressive moving average analysis. *Am. J. Physiol.* 1996; **270**:H1240–H1249.
59. A. Porta, G. Baselli, O. Rimoldi, A. Malliani, and M. Pagani, Assessing baroreflex gain from spontaneous variability in conscious dogs: role of causality and respiration. *Am. J. Physiol. Heart Circ. Physiol.* 2000; **279**:H2558–H2567.
60. G. Nollo, A. Porta, L. Faes, G. M. Del, M. Disertori, and F. Ravelli, Causal linear parametric model for baroreflex gain assessment in patients with recent myocardial infarction. *Am. J. Physiol. Heart Circ. Physiol.* 2001; **280**:H1830–H1839.
61. C. Cerutti, M. Ducher, P. Lantelme, M. P. Gustin, and C. Paultre, Assessment of spontaneous baroreflex sensitivity in rats a new method using the concept of statistical dependence. *Am. J. Physiol.* 1995; **268**:R382–R388.
62. M. Di Rienzo, P. Castiglioni, G. Mancia, A. Pedotti, and G. Parati, Advancements in estimating baroreflex function. *IEEE Eng. Med. Biol. Mag.* 2001; **20**:25–32.
63. M. Di Rienzo, P. Castiglioni, G. Parati, G. Mancia, and A. Pedotti, The wide-band spectral analysis: A new insight into long term modulation of blood pressure, heart rate and baroreflex sensitivity. In: M. Di Rienzo, G. Mancia, G. Parati, A. Pedotti A, and A. Zanchetti eds., *Computer Analysis of Cardiovascular Signals.* Amsterdam: IOS Press, 2001, pp. 67–74.
64. G. Bertinieri, M. Di Rienzo, A. Cavallazzi, A. U. Ferrari, A. Pedotti, and G. Mancia. Evaluation of baroreceptor reflex by blood pressure monitoring in unanesthetized cats. *Am. J. Physiol.* 1988; **254**:H377–H383.
65. G. Parati, A. Frattola, R. M. Di, P. Castiglioni, A. Pedotti, and G. Mancia, Effects of aging on 24-h dynamic baroreceptor control of heart rate in ambulant subjects. *Am. J. Physiol.* 1995; **268**:H1606–H1612.
66. M. Ducher, C. Cerutti, M. P. Gustin, and C. Z. Paultre, Statistical relationships between systolic blood pressure and heart rate and their functional significance in conscious rats. *Med. Biol. Eng. Comput.* 1994; **32**:649–655.
67. M. Di Rienzo, P. Castiglioni, G. Parati, A. Frattola, G. Mancia, and A. Pedotti, Effects of 24-h modulation of baroreflex sensitivity on blood pressure variability. Computers in Cardiology 1993, 551-554. 1993. Loa Alamitos (CA), IEEE Computer Society Press. Ref Type: Conference Proceeding

ARTHROSCOPIC FIXATION DEVICES

STEFAN M. GABRIEL
JOHN M. LIPCHITZ
Innovative Spinal Technologies
Mansfield, Massachusetts

JOHN E. BRUNELLE
Pegasus Biologics
Irvine, California

1. INTRODUCTION

The surgical art of tissue repair and reconstruction has evolved over the course of the twentieth century to include procedures that are done with increasingly less trauma to structures that are not at the repair site. The goal of this evolution in technique is to improve surgical outcomes by decreasing pain and healing time by reducing the cutting and retraction of tissue merely to create access to a repair site. Endoscopic surgery—the performance of surgical techniques within a cavity or organ of the body through a relatively small access channel—is the area of surgical specialty that encompasses these procedures. One area of surgical specialty within endoscopic surgery is that of arthroscopic surgery. Arthroscopic surgery is endoscopic surgery that is performed on or within the joints of the body. In the United States, arthroscopy is now one of the most common surgical procedures, and an estimated 2 million repairs were done arthroscopically in 2001; (1).

The developments of early medical pioneers make today's endoscopic and arthroscopic techniques possible. Endoscopes and arthroscopes are essentially tubes that transmit light along their lengths and into the body location where they are placed as well as allowing light to travel back out, providing external images of the internal body structures. The first recorded creation and use of an endoscope was by Danish surgeon H. C. Jacobaeus in 1910 (2). His work was followed by the work of another Danish surgeon Severin Nordentoft, who described his own examination of the knee joint with arthroscopic equipment he had built himself (2). Japanese surgeon K. Takagi performed arthroscopy in 1918 (2); Swiss surgeon Eugen Bircher reported on 18 diagnostic knee arthroscopies in 1921 (3); and the first reference to arthroscopy in the United States was by P. Kreuscher in 1925 (2). Another Japanese surgeon, M. Watanabe, continued and built on Takagi's device development work in the 1950s leading to the first production arthroscope in 1959 (4).

Arthroscopes and techniques of arthroscopic visualization allowed the effective use of additional devices for surgical resection and repair. In most cases, these devices had to be specially designed or redesigned to pass through small access channels (typically no larger than around 10 mm to 12 mm in diameter). The shafts of the insertion instruments used to advance and deploy the devices also have to be long enough and strong enough to allow them to be used within the body while the surgeon's hands remain outside the body. In this way, surgical techniques including repair and fixation of tissue are done within the body through relatively small incisions with visual feedback provided via arthroscope.

Arthroscopic fixation devices are available for use in all of the major joints of the body. Fixation of tendinous or ligamentous tissue to other tendinous or ligamentous tissue or bone allows torn and damaged structures at the joint to be repaired, reducing pain and restoring function to injured or unstable joints. The following sections of this article contain more detailed descriptions of devices used for fixation in the knee, shoulder, and other joints, as well as discussions of design aspects and of the materials used in the devices.

2. APPLICATION AREAS

2.1. Knee

Motion and stability at the knee joint are controlled by the articulation of the femur (thigh bone) and tibia (shin bone); the articulation of the patella (knee cap) and femur; the restraints of the anterior and posterior cruciate ligaments (ACL and PCL) in the central region and the medial and lateral collateral ligaments on either side; and the quadriceps tendons, hamstring tendons, and other tendons from surrounding muscles (5). The medial meniscus and lateral meniscus aid in load-bearing on either side of the articulation between the femur and the tibia. If one or more of these structures are damaged, the pain or loss of function or stability at the knee can be debilitating.

2.1.1. Meniscal Repair. Surgery to repair damage to a meniscus is the most prevalent arthroscopic repair procedure performed in the United States today (1). Meniscal tissue does not heal well on its own, and so, even today, arthroscopic removal of all or part of the meniscus is most often performed to treat pain or instability because of meniscal damage. However, an increasing trend toward repairing meniscal tears to restore function exists. The development of small polymer blocks connected by fine suture, small absorbable polymer screws, small staples, and barbed dart-like devices have spurred this trend by providing the surgeon with meniscal repair options. The goal of these devices is to repair and restore the function of the meniscus by fixing one part of the meniscus to another. They are designed to attach to the meniscal tissue on either side of a tear, span and close the tear in the meniscus, and hold the two sides in apposition to aid healing.

An early arthroscopic meniscal fixation device is the T-fix (Smith & Nephew Endoscopy, Andover, MA). The T-fix allowed the introduction of the fixation bars into and through the meniscus on either side of a tear, but then relied on the surgeon to tie, advance, and secure knots at the repair site from outside the knee. Tying, advancing, and securing knots arthroscopically can be a difficult and time-consuming part of a surgical procedure. Subsequent arthroscopic meniscal fixation devices address this concern via designs that do not rely on sutures to span and hold the tear closed or that provide a fixed cable strand or pretied sliding/locking knots between fixation members. A representative listing of meniscal arthroscopic fixation devices is shown in Table 1 (6–11).

2.1.2. Anterior Cruciate Ligament Repair. In the knee, repair of the ACL is the next-most prevalent arthroscopic procedure (1). Unlike the majority of meniscal procedures, however, in which the meniscus is removed, the primary aim of arthroscopic ACL procedures is replacement of the ligament structure and restoration of its function, which is predominantly done by using tissue grafts taken either from the patient (autograft) or another donor (allograft) in the form of hamstrings or patellar tendons.

These grafts are secured at the origin and insertion sites on the femur and tibia using fixation devices specially designed to hold tissue within holes prepared in each bone. In this way, the fixation devices maintain the graft in position as it spans the femoro-tibial articulation at the joint, ideally in the same location and in the same way as the ACL did prior to its damage or rupture. The successful healing process results in the incorporation of the graft into the femur and tibia at either end, and the formation of a ligamentous structure along the graft that is essentially a new ACL. A representative listing of arthroscopic fixation devices used in ACL repair is shown in Table 2 (6–11).

2.2. Shoulder

Motion and stability at the shoulder joint are controlled by the articulation of the humerus (upper arm bone) and

Table 1. Arthroscopic Meniscal Repair Fixation Devices (6–11)

Device Name	Description	Material	Company
Meniscal Dart	Double-barbed shaft that pierces through and spans tear	Absorbable polymer (PLA)	Arthrex, Naples, FL
Meniscal Staple	Staple that spans tear	Absorbable copolymer (PGA/PLA)	Arthrotek/Biomet, Warsaw, IN
BioDart	Dart that pierces and spans tear	Absorbable polymer (PLA with TCP)	Biocomposites, Wilmington, NC
Contour Meniscus Arrow	Shaft that pierces and spans tear	Absorbable polymer (PLA)	Linvatec, Largo, FL
BioStinger	Shaft that pierces and spans tear	Absorbable polymer (PLA)	Linvatec, Largo, FL
RapidLoc	Spike with button piece attached via suture and sliding/locking knot (spike pierces and fixes on one side of tear, suture spans tear and button fixes on other side of tear)	Absorbable polymers (PDS & PLA)	Mitek, J&J DePuy, Norwood, MA
Clearfix Screws	Headless screws that are rotated in and span tear	Absorbable polymer (PLA)	Mitek, J&J DePuy, Norwood, MA
Meniscal Repair System	Tie-bars—angled 'T' ends fix device on either side of tear, center link spans tear	Nonabsorbable (prolene) and Absorbable (PDS) polymers	Mitek, J&J DePuy, Norwood, MA
T-fix	'T' bars with suture passing through—one 'T' fixed on either side of tear, suture tied by surgeon spans tear	Nonabsorbable polymer (polyacetal)	Smith & Nephew Endoscopy, Andover, MA
FasT-Fix	Pair of 'T' bars with pretied sliding/locking knot suture in between—one 'T' fixed on either side of tear, suture pulled tight spans tear	Nonabsorbable (polyacetal) and absorbable (PLA and PLA with HA) polymers	Smith & Nephew Endoscopy, Andover, MA
Polysorb Meniscal Staple	Barbed posts connected by fixed, short length of braided cable	Absorbable polymer	U.S. Surgical, North Haven, CT

Note: See Materials section for discussion of materials used.

glenoid of the scapula (shoulder blade); the restraints of the capsule, which includes ligaments from the four surrounding rotator cuff muscles (supra-spinatus, infra-spinatus, subscapularis, and teres minor); and the labrum, which stabilizes and aids in distributing the contact load (5). If one or more of these structures are damaged, the pain or loss of function or stability at the shoulder can be a severe limitation to activities of daily living.

Arthroscopic procedures to repair damaged structures in the shoulder are second in prevalence only to arthroscopic procedures to address meniscal problems in the knee (1). Arthroscopic shoulder procedures can be roughly divided into two main categories: procedures to restore function by repairing damage to rotator cuff structures and procedures to restore stability by repairing damage to other capsular structures such as the labrum.

The devices specific to rotator cuff repair and specific to instability repair are similar in many aspects of design and function but differ in overall size. The arthroscopic suture anchors used to repair rotator cuff damage fix tendons back to the predominantly softer cancellous bone of the proximal humerus. As a result, they are generally relatively large with wide thread forms to enhance fixation.

The devices used for repairing capsular structures to address instability fix ligamentous or labral tissue back to the predominantly harder cortical bone of the glenoid of the scapula. They are generally smaller and shorter with shallower thread forms or ribbed shafts for fixation into the cortical bone. Tissue anchors are most prevalent for instability applications, but they have also been successfully used for rotator cuff repair.

As noted above for arthroscopic meniscal repair, the arthroscopic use of many current suture anchors requires the surgeon to tie, advance, and secure knots at the repair site from outside the joint, which can be a difficult and time-consuming part of the procedure. Use of suture anchors for arthroscopic fixation also requires the passage of suture limbs through ligamentous or tendinous tissue from outside the joint. This process can also be a challenging aspect of the procedure. Tissue anchors address this challenge via designs that do not rely on sutures, and some suture anchors address this challenge via designs that provide features for locking the suture without the need for knots (knotless designs). Yet other suture anchors address this challenge via designs that include pretied sliding/locking knots between the anchor and fixation members that stay on the opposite side of the tissue. A representative listing of arthroscopic fixation devices used for shoulder repair is shown in Table 3 (6–11).

It should be noted that the devices presented above for meniscal and ACL repair in the knee are designed to address the relatively specific needs of securing meniscal

Table 2. Arthroscopic ACL Repair Fixation Devices (6–11)

Device Name	Description	Material	Company
Bio-Interference Screw	Interference screw—screw is advanced into bone hole with graft in place pressing graft against the hole and holding it firmly	Absorbable polymer (PLA)	Arthrex, Naples, FL
Delta Tapered Bio-Interference Screw	Interference screw	Absorbable copolymer (PLA/PGA)	Arthrotek/Biomet, Warsaw, IN
BiLok	Interference screw	Absorbable polymer (with TCP)	Biocomposites, Wilmington, NC
EndoPearl	Spherical bead provides anchoring for suture above an interference screw to help hold a graft within the femoral bone tunnel	Absorbable polymer (PLA)	Linvatec, Largo, FL
BioScrew	Interference screw	Absorbable polymer (PLA)	Linvatec, Largo, FL
Guardsman	Interference screw	Titanium alloy (Ti6Al4V)	Linvatec, Largo, FL
SoftSilk	Interference screw	Titanium alloy (Ti6Al4V)	Smith & Nephew Endoscopy, Andover, MA
EndoFix and EndoFix L	Interference screws	Titanium alloy (Ti6Al4V) and absorbable polymer (L) (PLA)	Smith & Nephew Endoscopy, Andover, MA
Endobutton	Button device to provide anchoring for suture against the outer femoral cortex to help hold a graft within the bone tunnel	Titanium alloy (Ti6Al4V)	Smith & Nephew Endoscopy, Andover, MA
RCI	Interference screw	Titanium alloy (Ti6Al4V)	Smith & Nephew Endoscopy, Andover, MA
BioRCI and BioRCI HA	Interference screws	Absorbable polymers (PLA and PLA with HA)	Smith & Nephew Endoscopy, Andover, MA
Biosteon	Interference screw	Absorbable polymer	Stryker, Kalamazoo, MI

Note: See Materials section for discussion of materials used.

tissue to itself and of securing graft tissue within femoral and tibial bone tunnels. The devices for arthroscopic shoulder repair, however, are more generally applicable to any situation in which ligamentous or tendinous tissue is to be fixed to bone. For this reason, many of the devices listed in Table 3 have been used successfully to arthroscopically repair damage at other joints in the body. The criterion for use of these devices in the hip, elbow, wrist, or ankle is the relative size and density of the bone at the joint, which then governs the size and type of anchor used. The application of arthroscopic devices in these other joints is presented and discussed in more detail below.

2.3. Other Joints

As in the knee and shoulder, the motion and stability at the elbow, wrist, ankle, and hip joints are controlled by the articulation of the bones at the joint and the surrounding musculo-tendinous and ligamentous restraints (5). Again, as in the knee and shoulder, if one or more of these structures are damaged, the pain or loss of function or stability at the joint can be severely detrimental to quality of life.

2.3.1. Elbow. The elbow is a hinge joint in which the humerus and ulna meet that bends and straightens the arm. The articular surfaces of the elbow are connected together by a capsule that is comprised of the following ligaments: anterior, posterior, ulnar collateral, and radial collateral. Common injuries requiring arthroscopic fixation devices include tennis elbow and torn ulnar collateral ligament reconstruction.

2.3.2. Hand and Wrist. The bones of the hand are divided into three segments. The first segment, the Carpus (wrist bones), is composed of eight bones arranged in an upper and lower row. The second segment, the Metacarpus (bones of the palm), is comprised of five bones. The third segment, the phalanges (bones of the fingers), is composed of 14 bones, two for the thumb and three for each finger. The wrist is comprised of 27 articular surfaces between the radius and ulna, carpal bones, and metacarpals. Examples of typical procedures in the hand include ulnar and radial collateral ligament reconstruction of the thumb, collateral ligament repair of the wrist, scapholunate ligament reconstruction, and repair of metacarpal fractures.

2.3.3. Foot and Ankle. The ankle is a hinge joint at which the foot (talus) is attached to the ends of the tibia and fibula and allows for up-and-down motion of the foot. The anterior tibiofibular ligament connects the tibia to the fibula. The lateral collateral ligaments attach the fibula to the calcaneus, which allows outside stability of the ankle. Medial stability is provided by the deltoid ligaments on the inside of the ankle, which connect the tibia to the talus and calcaneus. Although ankle joint arthroscopy was

Table 3. Arthroscopic Shoulder, Elbow, Wrist, Ankle, and Hip Repair Fixation Devices (6–11)

Device Name	Description	Material	Company
Corkscrew and BioCorkscrew	Threaded (screw-type) suture anchors primarily for rotator cuff repair	Titanium alloy (Ti6Al4V) and absorbable polymer (PLA)	Arthrex, Naples, FL
TissueTak and TissueTak II	Tack-type tissue anchors with ribbed shaft and fixed head	Absorbable polymer (PLA)	Arthrex, Naples, FL
FASTak and FASTak II	Threaded suture anchors	Titanium alloy (Ti6Al4V)	Arthrex, Naples, FL
Bio-Phase	Push-in (nonthreaded) suture anchor	Absorbable copolymer (85%PLA/15%PGA)	Arthrotek/Biomet, Warsaw, IN
Harpoon and Mini-Harpoon	Push-in suture anchors (4.4 mm and 2.9 mm dia.)—wing features splay out within bone after insertion (primarily for rotator cuff, 'mini' primarily for instability)	Stainless Steel	Arthrotek/Biomet, Warsaw, IN
Contour Labral Nail	Tack-type tissue anchor with angled-over head primarily for instability repair	Absorbable polymer (PLA)	Linvatec, Largo, FL
UltraFix RC, MiniMite and MicroMite	Push-in suture anchors (2.9 mm, 2.3 mm, and 1.5 mm dia.)—wing features splay out within bone after insertion	Stainless Steel	Linvatec, Largo, FL
Super Revo, Revo, and Mini-Revo	Threaded suture anchors (5.0 mm, 4.0 mm, and 2.7 mm dia.)	Titanium alloy (Ti6Al4V)	Linvatec, Largo, FL
BioTwist RC	Threaded tissue anchor primarily for rotator cuff repair	Absorbable polymer (PLA)	Linvatec, Largo, FL
UltraSorb RC	Push-in suture anchor—reorientation of anchor after insertion into bone resists pull-out	Absorbable polymer (PLA)	Linvatec, Largo, FL
GII, Rotator Cuff, MiniAnchor, SuperAnchor	Push-in suture anchors—nitinol barbs deflect during insertion and then spread out within bone	Titanium alloy (Ti6Al4V) with nitinol barbs	Mitek, J&J DePuy, Norwood, MA
Knotless, Bioknotless, Bioknotless RC	Push-in, fixed-loop suture anchors—capture of loop in bone hole within anchor feature after passage through tissue eliminates knot-tying	Titanium alloy (Ti6Al4V) with nitinol barbs, Absorbable polymer (PLA)	Mitek, J&J DePuy, Norwood, MA
Fastin, Fastin RC	Threaded suture anchors (3.0 mm and 5.0 mm)	Titanium alloy (Ti6Al4V)	Mitek, J&J DePuy, Norwood, MA
Panalok, Panalok RC	Push-in suture anchors—reorientation of anchor after insertion into bone resists pull-out	Absorbable polymer (PLA)	Mitek, J&J DePuy, Norwood, MA
RotorloC	Push-in suture anchor—reorientation of anchor with rotation after insertion into bone resists pull-out	Absorbable polymer (PLA)	Smith & Nephew Endoscopy, Andover, MA
TAG Rod and TAG Wedge	Push-in suture anchors (3.0 mm and 3.7 mm dia.)	Nonabsorbable (polyacetal) and absorbable (PLA) polymers	Smith & Nephew Endoscopy, Andover, MA
Fastenator	Push-in suture anchor—wing features splay upon deployment after insertion into bone	Titanium alloy (Ti6Al4V)	Smith & Nephew Endoscopy, Andover, MA
Suretac, Suretac II and Suretac III	Tack-type tissue anchors with ribbed shaft and fixed head	Absorbable polymer (polyglyconate)	Smith & Nephew Endoscopy, Andover, MA
TwinFix Ti	Threaded suture anchors (2.8 mm, 3.5 mm, 5.0 mm, and 6.5 mm dia.)	Titanium alloy (Ti6Al4V)	Smith & Nephew Endoscopy, Andover, MA
TwinFix AB and PMR	Threaded suture anchors (5.0 mm, and 6.5 mm dia.)	Absorbable (PLA) and nonabsorbable (polyacetal) polymers	Smith & Nephew Endoscopy, Andover, MA
TwinFix Quick-T	Threaded suture anchors with 'T'-bar connected via suture loop with pretied sliding/locking knot—anchor is advanced through tissue and into bone leaving 'T' above tissue; suture tightened to hold tissue down	Titanium alloy (Ti6Al4V) with nonabsorbable polymer 'T' (PEEK)	Smith & Nephew Endoscopy, Andover, MA
Statak and Bio-Statak	Threaded suture anchors (2.5 mm, 3.5 mm, 5.0 mm (regular & Bio-), and 5.2 mm dia.)	Titanium alloy (Ti6Al4V), Absorbable polymer (PLA)	Zimmer, Warsaw, IN

Note: See Materials section for discussion of materials used.

adopted after knee and shoulder arthroscopy, it is considered standard procedure in many situations. Typical surgical repairs of the ankle and foot requiring arthroscopic fixation devices include Hallux Valgus reconstruction (deviation of the great toe toward the fibular border of the foot), lateral instability, medial instability, Achilles tendon reconstruction or repair, metatarsal ligament repair, and mid-foot reconstruction.

Arthroscopic repair procedures performed in the elbow, foot, ankle, hand, and wrist use much the same types of devices as in the knee and shoulder but are much less prevalent than those in the knee and shoulder. Repairing damage in these smaller joints poses a special challenge to arthroscopic surgeons because of their small size. Not only must instruments such as cannulae and scopes be smaller than in the knee and shoulder to aid in visualization, the anchors must be smaller also, but not compromise the strength of repair throughout the healing period.

3. DEVICE DESIGN

Numerous designs of arthroscopic devices exist to allow repair of ligaments, tendons, joint capsule, labrum, or other tissue that has become torn away from a bone origin or insertion site or that has a tear within it. In general, these designs can be characterized as belonging in one of a number of groups of designs. Within a given design group, the devices share the same design approach. The material from which a device is made, or the length, diameter, or width of a particular feature of a particular type of design, can vary within a group. These intragroup variations can be determined by application of a particular device to one joint or another or use of a device in one type of bone or another.

3.1. Tissue-to-Tissue Repair

Devices for tissue-to-tissue repair must hold one part of the tissue in apposition to another part across a tear. The most prevalent use for this type of repair device is for meniscal repair in the knee (Table 1). Devices for this application can basically be divided into three design groups.

3.1.1. Tissue-to-Tissue Repair—Trans-Tear Devices.
One group consists of the post or shaft devices that are characterized by a central body that is inserted into the tissue on one side of a tear and that goes across the tear to be inserted and fixed within the tissue on the opposite side of the tear. The Contour Menicus Arrow (Linvatec, Largo, FL), a barbed polymer shaft, is an example of this type of design.

3.1.2. Tissue-to-Tissue Repair—Trans-Tear Devices with Suture.
A second group of designs relies on suture to pass through the tissue and across the tear, but the two parts of the device held together by the suture are positioned against the outer surfaces of the tissue on either side of the tear. Variations of this design include adjustable-length suture loops with pretied locking knots for suture tensioning and securement. An example of this type of design is the FasT-Fix (Smith & Nephew Endoscopy, Andover, MA) (Fig. 1). By providing the pretied knot and locking suture capability, this particular design addresses

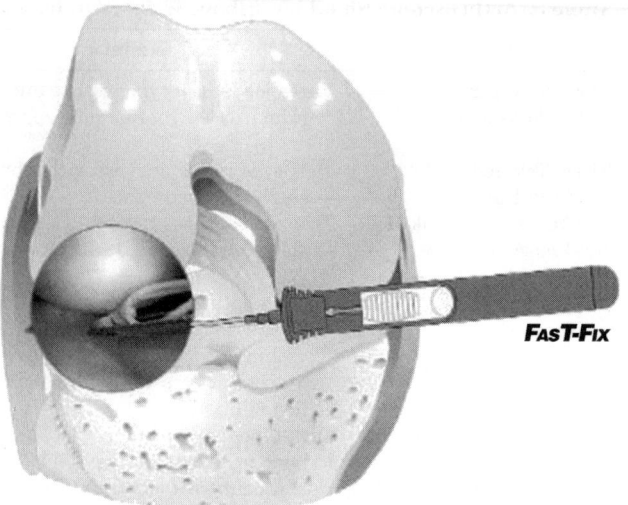

Figure 1. Arthroscopic fixation device with insertion instrument used for meniscal repair in the knee. (Shown is the FasT-Fix device on its insertion instrument schematically shown during a repair procedure.) The circular image on the left shows the view of the repair in progress as seen through an arthroscope. (This figure is available in full color at http://www.mrw.interscience.wiley.com/ebe.)

knot-tying and suture handling, which is one of the major difficulties inherent in using suture-based repair devices.

3.1.3. Tissue-to-Tissue Repair—Tear-Spanning Devices.
The third group of tissue-to-tissue repair devices consists of devices that are fixed in the tissue on either side of a tear, but instead of passing through the tissue they span the tear by passing around it. The Meniscal Staple (Arthrotek/Biomet, Warsaw, IN) is an example of this type of design.

Design considerations for tissue-to-tissue repair devices include (1) a small cross-section (or fine suture) so as to keep the size of the hole in the tissue created by the repair device (or suture) as small as possible, (2) features such as barbs, or other means of maintaining secure fixation of the parts of a device within the tissue on either side of a tear, and (3) use of relatively soft materials (polymers), and reduction or elimination of parts of the device that might protrude from the tissue repair site, which is done to decrease the chances for damage to surrounding structures such as cartilage at a joint.

3.2. Repair of Tissue within Bone

Devices for tissue within bone repair must hold tissue (usually ligamentous) within a bone tunnel for eventual attachment of the tissue onto the bone. The most prevalent use for this type of repair device is for ligament repair in the knee (Table 2), although they can and have also been used for tendon repair (specifically biceps tendon repair) in the shoulder. Devices for this type of repair fall into two categories, interference screws and fixation components, which rest on the outer surface of the cortex near a joint and secure the ligament within a bone tunnel via transcortical sutures.

3.2.1. Repair of Tissue within Bone—Interference Screws.
Interference screws are designed with and without rounded heads and manufactured from polymer or metal materials. In ligament repair using interference screws, a bone tunnel is created at the site of desired reattachment of the ligament to the bone. The diameter of this hole is such that it is smaller by at least a few millimeters than the combined dimension of the ligament graft plus the diameter of the screw. One end of the graft is placed in the bone tunnel and the screw is advanced until it is also within the tunnel. In this position, the interference between the screw, ligament graft, and bone tunnel provides the radial force necessary to resist axial loads on the graft during healing and thereby hold the graft in place in the tunnel.

The threads and heads of interference screws press against the ligament graft with substantial force both during insertion and after implantation. As a result, the thread forms are relatively shallow, and the thread tip is blunt and rounded (Fig. 2), which allows the screw to press against and hold the ligament graft in place without cutting or damaging it. Different sized grafts are used for different patients, so length and diameter vary within product lines and among designs. The insertion torque can also be relatively large, so size restrictions and material properties also determine acceptable design and material combinations. The Delta (Arthrotek/Biomet, Warsaw, IN) and RCI and BioRCI (Smith & Nephew Endoscopy, Andover, MA) devices are examples of interference screw designs.

3.2.2. Repair of Tissue within Bone—Trans-Cortical Fixation.
The second group of devices for tissue within bone repair are those that rely on trans-cortical suture or cable. In these devices, one end of a suture or cable loop is fixed to the ligament graft and passes through the far end of the bone tunnel out of the bone. There, the other end is fixed to a button or bar that rests on the outer surface of the bone cortex.

Small size and low profile are requirements for the button or bar portion of these devices, which allows passage of the device through a relatively small hole at the end of the bone tunnel for techniques that require it. It also reduces the chances for tissue damage or irritation on the outer bone surface. These requirements, however, compete with the need for strength to hold the ligament graft in place via the suture or cable loop. The Endobutton (Smith & Nephew Endoscopy, Andover, MA) is an example of this type of design for which adequate strength is obtained in a relatively small perforated bar by making it out of metal.

3.3. Repair of Tissue onto Bone
Suture anchors and tissue anchors are examples of arthroscopic fixation devices used to repair tissue onto bone. They can be further divided into screw-in and push-in design groups. Design variations within these groups are determined by the type and size of bone into which they are to be placed. A number of these devices are listed in Table 3.

3.3.1. Tissue Anchors.
Tissue anchors are characterized by shafts that pass through the tissue to be repaired and are inserted into bone. Radial compression or catching features on the shafts then hold them within bone. Head features on tissue anchors allow them to act as tacks or nails, bringing and holding the tissue down onto the bone. The benefit of tissue anchor designs in arthroscopic repair is that no suture exists to take care of during insertion and tissue attachment. The drawback is that the head feature protrudes above the repaired tissue where it can possibly irritate other tissues around the repair site.

3.3.1.1. Threaded Tissue Anchors.
Threaded-shaft, or screw-in tissue anchors have been designed for use in relatively soft bone at the rotator cuff reattachment site in the shoulder. The relatively wide, thin thread profile of these devices allows them to withstand a greater force for pull-out than shallower, thicker thread profiles would. The BioTwist RC (RC for rotator cuff) (Linvatec, Largo, FL) is an example of a threaded tissue anchor. The lower strength required for the relatively soft bone at the rotator cuff site in the shoulder allows this design to be manufactured from an absorbable polymer.

3.3.1.2. Push in Tissue Anchors.
Push-in tissue anchors rely on the radial compression of the bone around the shaft interacting with features on the shaft to resist pull-out forces. Suretac (Smith & Nephew Endoscopy, Andover, MA) (Fig. 3) with its ribbed shaft and head with spikes on the underside is a good example of a push-in tissue anchor. It is intended for use in relatively hard bone and the ribbed shaft design allows it to interact with the inner surface of a slightly undersized hole prepared in the bone. As with other push-in tissue anchors, for a given material

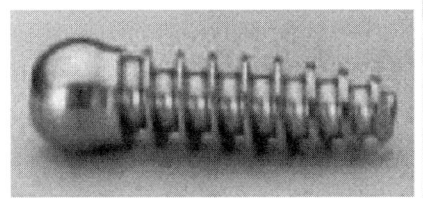

Figure 2. Interference screws used for arthroscopic ACL fixation and repair in the knee. (Shown are the metal RCI (left) and absorbable polymer BioRCI HA (right) designs.) (This figure is available in full color at http://www.mrw.interscience.wiley.com/ebe.)

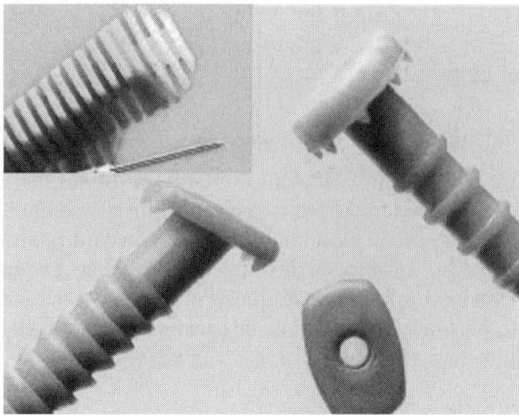

Figure 3. Tissue anchors [Suretac II (right) and Suretac III (lower left and center)] and the handle of the insertion instrument (upper left) used for instability repair in the shoulder as well as to fix tissue to bone at many other repair sites. (This figure is available in full color at http://www.mrw.interscience.wiley.com/ebe.)

such as absorbable polymer, it can be made smaller than screw-in anchors for the same intended uses, because it needs only to withstand the compressive forces of implantation and retention within a bone hole, and the tensile and bending forces from the tissue it is holding in place. In contrast, a screw-in anchor must also be able to withstand the torque during insertion.

3.3.2. Suture Anchors. The designs for suture anchors—devices to hold suture in place in the bone—are perhaps the most varied. They can be divided into five different design groups, two groups of threaded designs, and three groups of push-in designs. Additional groups may exist in the future as well, as material and manufacturing innovations are developed that allow additional means of fixing suture into bone.

3.3.2.1. Threaded Suture Anchors. The first two groups of suture anchors are threaded designs. The first of these groups is represented by threaded or screw-in anchors with a hole or holes, or other suture-retaining means. Many variations within this group exist and it contains perhaps the largest number of devices, including shorter, smaller-diameter, metal screws that can carry only one small-diameter suture for use in relatively hard bone and in relatively small joint spaces. Longer, larger-diameter, polymer and absorbable polymer screws that can carry two larger-diameter sutures for use in softer bone and in relatively larger joint spaces such as the shoulder also exist. Additional variations include multi-lead and multi-height thread designs. The TwinFix Ti, AB, and PMR (Smith & Nephew Endoscopy, Andover, MA) anchors (Fig. 4) are examples of these threaded anchors.

The TwinFix Quick-T (Smith & Nephew Endoscopy, Andover, MA) anchor (Fig. 5) is an example of a suture anchor in the second group of threaded anchors. This design features a threaded anchor that is held in the bone in the same way as all of the other threaded suture anchors, but is placed into the bone in the same way as a tissue anchor, through the tissue to be repaired. A second com-

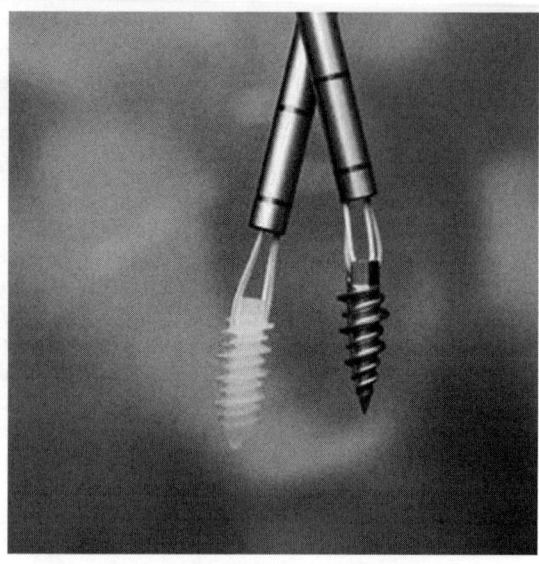

Figure 4. Screw-in suture anchors used for rotator cuff repair in the shoulder as well as to fix tissue to bone at many other repair sites. Shown are the devices just off the ends of their insertion instruments. An absorbable anchor (TwinFix AB) is shown on the left and a titanium alloy anchor (TwinFix Ti) is shown on the right. (This figure is available in full color at http://www.mrw.interscience.wiley.com/ebe.)

ponent remains above the tissue and is used to bring the tissue and hold it down on the bone via suture connecting it to the threaded component in the bone.

3.3.2.2. Push-in Suture Anchors. The third, fourth, and fifth groups consist of push-in suture anchors. The third group of designs contains push-in anchors that rely on fixation within the bone through passive features such as ribs, wings, or barbs along or protruding from the body of the anchor. As the anchor is pushed into a hole in the bone, the features are deflected at least partially out of the way. When a force tries to pull the anchor back out of the bone, the passive features interact with the bone to resist this

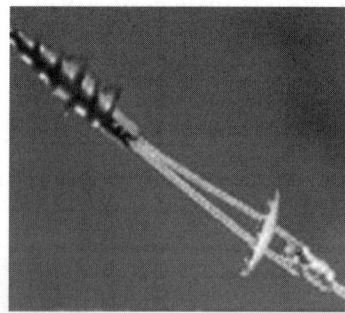

Figure 5. An arthroscopic fixation device (TwinFix Quick T) used for rotator cuff repair in the shoulder as well as to fix tissue to bone at many other repair sites. The titanium alloy suture anchor shown in the upper left is screwed through the tissue to be reattached and into the bone. The nonabsorbable polymer (PEEK) T-bar shown in the lower right is advanced and held tight against the tissue with the aid of a pretied sliding-locking knot shown in the far lower right corner. (This figure is available in full color at http://www.mrw.interscience.wiley.com/ebe.)

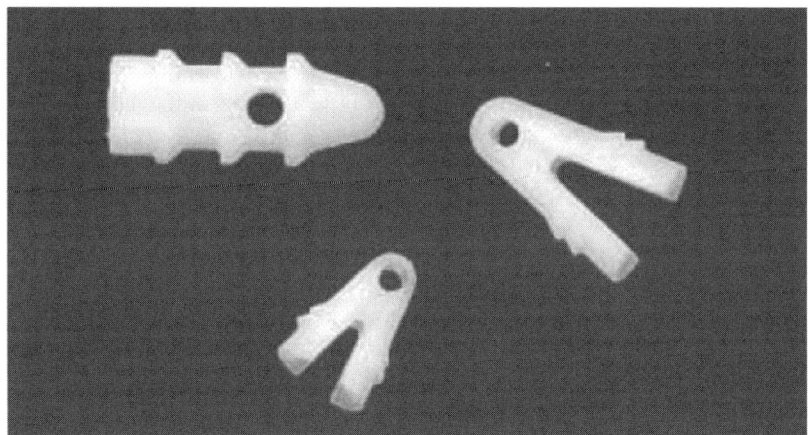

Figure 6. Push-in suture anchors [TAG rod (upper left) and TAG wedges] used for instability repair in the shoulder as well as to fix tissue to bone at many other repair sites.

pull-out. The GII (Mitek, J&J DePuy, Norwood, MA) anchor and the TAG (Smith & Nephew Endoscopy, Andover, MA) anchors (Fig. 6) are examples of push-in anchors with passive fixation design features.

The fourth group consists of push-in anchors with active fixation design features. The Fastenator (Smith & Nephew Endoscopy, Andover, MA) anchor is a good example of this type of design. The body of the anchor is designed as a tube with axial slits. After the body of the anchor is inserted into the bone, the insertion tool is used to compress the anchor body axially, which causes the walls of the tube to deform radially outward to form winglike features that help to hold the anchor in the bone.

The fifth and final group consists of push-in anchors that rely on reorientation of the anchor within the bone for resistance to pull-out. The Panalok (Mitek, J&J DePuy, Norwood, MA) anchor is a good example of this type of design. The elongated and tapered body of the anchor is mounted onto the end of a flexible inserter shaft and is inserted into a hole in the bone with one end going in first. After releasing the anchor from the end of the inserter shaft, tensioning the suture jams the body of the anchor somewhat sideways in the hole in the bone. As with other anchors relying on reorientation for fixation in the bone, the bone must be relatively soft to allow some reorientation and proper fixation of the anchor within the hole.

3.4. Overall Design Considerations

Suture anchor designs are generally the same for different applications across different joints, as long as the type of bone, the type of tissue, and the space available is similar. Devices used in the wrist and ankle, for example, are similar to those used in the shoulder, but are smaller (usually 3 mm or less) and typically come preloaded with only one suture, which is also smaller (usually) United States Pharmacopoeia (USP) size 2-0. For example, the V-Tak (Arthrex, FL) is used for ligament repair of the thumb and wrist and is 2.2 mm in diameter and 7 mm long (compared with a 7.5 mm dia. and 15 mm length typical of a threaded suture anchor used for rotator cuff repair in the shoulder). Other design features such as angled wedges, ribs for push-in anchors, wings or barbs that deploy, and radial compression push-in anchors are also used in many different applications and are sized according to the joint space.

Anchors may come unloaded or preloaded with braided nonabsorbable suture (polyester or high-strength ultra-high-molecular-weight polyethylene) ranging from USP designation size 3 (0.6–0.7 mm dia.) to size 2-0 (0.3–0.4 mm dia.). Mitek, J&J DePuy also offer the option of braided Panacryl suture, which is an absorbable PLA, PGA copolymer suture.

The smallest screw-in anchors are typically titanium because the small size of the anchors require increased material properties to prevent failure during insertion. Several companies such as Smith and Nephew, Inc, Mitek, J&J DePuy, and Arthrex offer complete lines of titanium screw-in suture anchors. Several types of absorbable poly lactic acid (poly lactide) screw-in anchors are available for those that prefer not leaving an implant embedded in the bone for an extended period of time. The benefits of absorbable anchors are discussed later in the material(s) section. Some anchors are manufactured from nonabsorbable polymers such as Smith and Nephew's polyetheretherketone (PEEK) Quick-T anchor or Mitek, J&J DePuy's polyethylene ROC-EZ anchor.

A summary of some of the arthroscopic fixation devices that have been used in arthroscopic hip, elbow, wrist, and ankle repairs is given in Table 3. Of the devices listed, the smaller-diameter anchors are most frequently used in the elbow, wrist, hand, and ankle, and the larger ones are most frequently used in the shoulder.

Within the push-in suture anchor design groups, a number of design variations exist that have been implemented to aid in suture management, especially regarding the elimination of knot-tying. Some designs employ a fixed-length suture loop that is passed through the tissue and then around the anchor before inserting the anchor into the bone. Others feature crimping or other constricting means to hold the suture once the anchor is fixed and the desired suture/tissue tension is obtained. Finally, others have pretied sliding/locking knots that are engaged after anchor placement and tissue tensioning.

Regardless of how a given design attempts to address the difficulties of suture management in general, and knot-tying in particular, the choice is still up to the surgeon. He or she must decide whether an anchor allows them to adequately repair the tissue at a desired location on the bone. In some cases, the added benefits of simplified suture management may be outweighed by added complexity or an increase in the inconsistency of the repair technique.

4. BIOMATERIALS

An extensive assortment of devices is used for arthroscopic tissue fixation, and material selection can be a challenge as it greatly depends on intended function and available technology. The process of selection of materials covers a variety of basic considerations, including: biocompatibility, including cytotoxicity, surface morphology, and degradation characteristics; functional requirements, such as physical, chemical, mechanical, and biological properties; and processing issues, such as feasibility, quality control, sterilization, and packaging. Currently, devices typically are fabricated from metals, inorganic fillers, polymers, and tailored composites thereof. Recent advances in materials research are focused in developing materials that mimic the behavior of the tissue being replaced, and materials that are quickly and completely replaced by natural, viable tissue as part of the healing and repair response. This focus has spurred investigation into new absorbable materials especially and critical review of these materials. This is true as the industry becomes more aware of the long-term biological implications of these materials, which is especially important for those devices implanted into bone. Most procedures require the removal or alteration of bone at the surgical site, which can ultimately adversely affect the structural integrity of the joint.

4.1. Metals

Metals such as stainless steel and titanium alloys are most suited for use in tissue fixation devices, specifically those implanted into bone, for their initial mechanical properties and biocompatibility. These materials are generally well-tolerated in the physiological environment, and ideal for mechanical load realized during implantation, and can be feasibly processed into a variety of devices. Although metallic implants may provide strong initial fixation, they still may require a secondary operation for removal once the bone is healed, and present the risk of adverse tissue response. In addition, metallic implants can distort magnetic resonance image (MRI) data, interfering with a surgeons' ability to properly assess the progress of the repair or to diagnose any future injury at the implant site.

4.1.1. Stainless Steel.
The austenitic stainless steels, particularly the 316 series, are widely used for implants because they are nonmagnetic and possess superior corrosion resistance (12). The inclusion of molybdenum in these materials provides enhanced resistance to pitting corrosion in saline environments. More specifically, the American Society for Testing and Materials (ASTM) recommend grades 316L and 316LS rather than 316, as these materials contain less carbon (0.03% as opposed to 0.08%) and are better chemically balanced for resistance to pitting corrosion. A wide range of properties exists for these materials depending on heat treatment (annealing to obtain softer materials) or cold working (for greater strength and hardness) (13). Caution must be taken in designing with these materials as even the most corrosion-resistant stainless steels can potentially corrode inside the body in highly stressed and oxygen-depleted regions, and therefore are typically only used in temporary implants.

4.1.2. Titanium and its Alloys.
Titanium and its alloys have been used extensively as implantable biomaterials as they exhibit excellent properties for fixation applications. They are biocompatible, exhibit excellent corrosion resistance, and are biomechanically advantageous to stainless steel because of their strength-to-weight ratio and lower modulus, which reduces the effects of stress shielding in surrounding bone. Several grades of unalloyed commercially pure titanium exist for surgical implant applications, distinguished by their impurity contents (oxygen, iron, nitrogen), which should be carefully controlled. Oxygen in particular has a great influence on ductility and strength (13).

The addition of alloying elements to titanium allows for a wide range of properties. In particular, a titanium (Ti), aluminum (Al), Vanadium (V) alloy Ti-6Al-4V is widely used in implants. The primary alloying elements are aluminum (5.5–6.5%) and vanadium (3.5–4.5%), resulting in a material with excellent strength characteristics and corrosion resistance (13). Titanium derives its resistance to corrosion by the formation of a solid oxide layer that forms a thin adherent film and passivates the material (13).

Manufacturing titanium implants presents an array of challenges as the material is very reactive at high temperature and burns readily in the presence of oxygen. High-temperature processing requires an inert atmosphere and any hot working or forging operation should be carried out below 925°C. Machining at room temperature also has its problems, as the material tends to gall or seize cutting tools. The use of very sharp tools at slow speeds is an effective manufacturing technique, as well as electrochemical machining (12).

4.2. Polymers

Polymeric biomaterials have been used extensively in fixation devices as an effective alternative to metal implants, offering a wide array of unique properties. These materials exhibit good biocompatibility, adequate mechanical properties, and can be manufactured and sterilized using conventional methods. The increased use of polymers in arthroscopic fixation devices in recent years has produced implants with considerable advantages over their metallic counterparts. Although these materials do not retain the mechanical properties of metals, they are desirable because the enormous array of fabrication options can produce materials with characteristics that are unique and beneficial for arthroscopic applications. The spectrum of biomedical polymers has grown in recent years because variables such as chemical composition, crystallization polymerization methods and parameters, inherent viscosity, residual monomer, and molecular weight distribution have been optimized. In addition, tailoring methods such as compounding and advances in copolymers/terpolymers have resulted in more application-specific materials. Improvements in implant fabrication processes (e.g., injection molding) have played a crucial role in the development of these materials and greatly influence final material properties, functional behavior of the implant, and *in vitro/in vivo* characteristics.

In addition to processing and manufacturing advantages, the employment of polymeric materials maintain a biological advantage in comparison with metals because their elastic moduli more closely match that of bone, limiting the effect of stress shielding, a phenomenon caused by a modulus mismatch between the implant material and the surrounding bone that can result in bone necrosis and bone resorption by osteoclasts. The use of polymeric materials also reduces the risk of long-term adverse reactions because of ion release, eliminating the need for a secondary procedure to remove the implant. They are radiolucent, will not distort MRI data, and reduce the complexity of secondary and revision procedures at the surgical site. Table 4 displays the flexural moduli of several polymers in comparison with metals and bone (12–15).

In general, polymeric biomaterials can be categorized as nonabsorbable; those polymers that permanently remain within in the body as an inert object, and bioabsorbable; polymers that degrade over time via hydrolysis, and are eventually absorbed by the body and disposed of through natural metabolic pathways.

4.2.1. Nonabsorbable Polymers.

4.2.1.1. Polyacetal (POM). Polyacetals, properly referred to as polyoxymethylene (POM), are biocompatible, semi-crystalline linear thermoplastic materials that are suitable for implant applications. They exhibit excellent mechanical, thermal, and chemical properties, attributed to a stiffened main backbone chain and a regular, highly crystalline structure (12,13). This material is essentially inert *in vivo*, as it possesses tremendous chemical and moisture resistance. Notable physical properties include yield strength, stiffness, hardness, toughness, dimensional stability, creep and fatigue resistance, and low coefficient of friction. Polyacetals are produced by reacting formaldehyde and can be easily processed by conventional melt processing techniques and machined from stock shapes. These materials are typically sterilized by steam, or chemical sterilization techniques such as ethylene oxide, as they do not retain their properties in the presence of radiation.

4.2.1.2. Polyethylene (PE). Polyethylene is a linear thermoplastic polyolefin generally available in low-density, high-density, and ultra-high-molecular-weight grades, all with various medical applications. High-density polyethylene (HDPE) has had limited but effective use in fixation devices because it exhibits good wear properties and is suitable for living hinge applications. HDPE is synthesized from ethylene, which is polymerized at high temperature and high pressure to produce a highly crystalline, high-density linear polymer that exhibits good mechanical properties (yield strength, elastic modulus, hardness) as well as superior creep and wear properties, and good chemical and hydrolysis resistance (14). Polyethylene can be easily processed using conventional extrusion and injection molding processes. Polyethylene is sterilzed using both chemical techniques such as ethylene oxide and via radiation, although sterilization at high radiation dosages ($>10^6$ Gy) can cause this polymer to become hard and brittle because of random chain scission and crosslinking (12).

4.2.1.3. Polyetheretherketone (PEEK). Polyetheretherketone (PEEK) has increasingly gained attention as an implant material that has successfully been used in bone fixation applications. This semi-crystalline material exhibits excellent biocompatibility and is widely regarded as a high-performance material with properties exceeding most engineering thermoplastics. The chemical composition of PEEK consists of two repeating ether groups and a ketone group combined by phenyl groups. Available in both virgin and carbon-filled resins, PEEK can be processed by conventional melt processing techniques. PEEK has excellent chemical and hydrolysis resistance; displays superb mechanical performance in its tensile, flexural, creep, fatigue, and impact properties; and is functional at high temperatures (260°C). The use of PEEK also has benefits in that this polymer can be sterilized by a variety of techniques as it has excellent radiation resistance.

4.2.2. Absorbable Polymers.
Polymers derived from aliphatic polyesters, specifically the family of polylactic acid, or polylactide, (PLA), polyglycolic acid, or polyglycolide, (PGA), and Copolymers of lactic and glycolic acids (PLGA) are bioabsorbable, and have been increasingly used in fixation devices. The main mode of degradation for these polymers is hydrolysis, a mechanism that breaks down the implant into natural metabolic intermediates, which are eventually absorbed by the body. An implant that degrades over time is physiologically advantageous, as load transfer from the implant to the healing tissue can increase the quality of the repair. An extended period of release of degradation products may also be advantageous to reduce any inflammatory response that might occur in reaction to an acute release of a relatively

Table 4. General Properties of Biocompatible Materials Compared with Cortical and Cancellous Bone (12–15)

Material	Young's Modulus (GPa)
Cortical Bone	17.0
Cancellous Bone	0.1
Stainless Steel	200.0
Titanium, Ti6Al4V	110.0
Polyacetal (POM)	3.0
Polyethylene (HDPE)	1.7
Polyetheretherketone (PEEK)	4.0
PEEK, 30% carbon filled	13.0
Poly glycolide (PGA)	7.0
Poly L-lactide (PLLA)	2.7
Poly D,L-lactide (PDLA)	1.9
Poly D,L-lactide-co-glycolide (PDLGA)	2.0

large volume of materials. The degradation rate can be controlled depending on chemical composition, molecular weight, molecular weight distribution, residual monomer, and processing conditions. These polymers are readily available and able to be fabricated into a variety of shapes and forms using conventional melt processing techniques.

The addition of inorganic fillers such as tri-calcium phosphate (TCP) and hydroxyapatite (HA) that contain minerals found in natural bone matrix may also increase the potential for a device inserted into bone to be integrated into and replaced by the surrounding material during and after healing. Anchoring devices made from absorbable polymers in combination with absorbable sutures such as those made from polydioxanone (PDS) allow completely absorbable constructs to be used in arthroscopic repairs.

4.2.2.1. Polylactide (PLA). Lactide is the cyclic dimer of lactic acid that exists as two optical isomers, D and L. L-lactide is the naturally occurring isomer. This material exhibits high tensile strength and low elongation, making it suitable for load-bearing applications. Poly L-lactides are typically about 30–50% crystalline and generally require 3–5 years to be completely absorbed by the body (15).

D,L-lactide is a synthetic blend of D-lactide and L-lactide. Poly D,L-lactide is an amorphous polymer arranged in a random order of both isomeric forms of lactic acid. This material has a lower tensile strength, higher elongation, and a much faster absorption time (<1 year) than Poly L-lactide. The glass transition temperature of this material is about 55–60°C, and proper handling is essential, as implants of this material will deform above this temperature. Copolymers of L-lactide and D,L-lactide have been produced to decrease crystallinity and increase the degradation time of Poly L-lactide (15).

4.2.2.2. Polyglycolide (PGA). Polyglycolide is the simplest linear aliphatic polyester and is synthesized from the dimerization of glycolic acid. PGA is a highly crystalline material (45–55%), with a high melting point with a total degradation time of approximately 6–12 months. PGA has been incorporated into medical devices for several load-bearing applications as it exhibits excellent initial mechanical properties. However, this material loses strength very quickly and is typically copolymerized with other absorbable materials to tailor mechanical properties and increase its strength retention (15). Devices are generally not made exclusively out of PGA, which is most likely due to its relatively rapid degradation and the accompanying release of acidic degradation products into the surrounding tissue within a short time after placement of the device and healing of the tissue.

4.2.2.3. Poly lactide-co-glycolide (PLGA). Polylactide and polyglycolide are often copolymerized to obtain properties of both polymers, or to alter the properties of each polymer. Copolymers of glycolide and both L-lactide and D,L-lactide have been developed for the medical device and drug delivery industries. The copolymerization of glycolide and D,L-lactide will increase the degradation rate from either polymers depending on the copolymer ratio. These materials are amorphous and typically exhibit a glass transition temperature of about 50–55°C, and therefore implants of this material are temperature-sensitive and require a controlled environment to prevent deformation. These materials are typically used in applications where fast implant degradation is desired and strength retention is not required past 6 weeks.

5. SUMMARY AND FUTURE DIRECTIONS

The proper combination of design and material has resulted in devices that are effective in surgical repair performed through 10-mm diameter (and smaller) openings. The continuing desire of the surgeon to perform surgery through a smaller and smaller access opening along with the desire of the patient to be left with less and less evidence of surgery (from pain, rehabilitation, and cosmetic perspectives) will continue to drive the development of smaller and smaller fixation devices and the tools to use them, which will ensure the continued evolution of arthroscopic surgery.

Mechanical arthroscopic devices similar to the screws for both relatively soft cancellous bone and relatively hard cortical bone, and the push-in anchors primarily for relatively hard bone in small joints will continue to be widely and successfully used. The current trend away from metallic and nonabsorbable materials and toward absorbable materials will increase. Absorbable materials that include biologically active agents will be developed to increase the speed of healing.

In challenging repair scenarios, and especially for cases of avascular tissue (such as cartilage) that does not heal readily by itself, biological augmentation to mechanical repair devices will be critical to ultimate success. Materials and devices will be developed to ensure desirable properties in the tissue that is created during healing. In the field of arthroscopic repair, as in other areas of surgical intervention, fixation devices incorporating the best designs and materials will be developed to try to achieve the ideal result, namely a repair site that is indistinguishable from one that was never damaged.

BIBLIOGRAPHY

1. Medtech Insight, LLC, U.S. Markets for Sports Medicine Devices and Products. Report #A330, 2003.
2. C. W. Kieser and R. W. Jackson, Severin Nordentoft: The first arthroscopist. *Arthroscopy* 2001; **17**:532–535.
3. C. W. Kieser and R. W. Jackson, Eugen Bircher (1882–1956) the first knee surgeon to use diagnostic arthroscopy. *Arthroscopy* 2003; **19**:771–776.
4. R. W. Jackson, Quo venis quo vadis: The evolution of arthroscopy. *Arthroscopy* 1999; **15**:680–685.
5. H. Gray, *Gray's Anatomy*. Philadelphia, PA: Running Press, 1974.
6. Stryker. (2004). (online). Available: *http://www.stryker.com*.
7. Arthrex. (2004). (online). Available: *http://www.arthrex.com*.
8. Biomet. (2004). (online). Available: *http://www.biomet.com*.

9. Conmed. (2004). (online). Available: *http://www.conmed.com*.
10. J&J DePuy Mitek. (2004). (online). Available: *http://www.jnj.com*.
11. FDA CDRH 510(k). (2004). (online). Available: *http://www.accessdata.fda.gov/scripts/cdrh/cfdocs/cfPMN/pmn.cfm*.
12. J. B. Park and R. S. Lakes, Biomaterials, *An Introduction*. New York: Plenum Press, 1992.
13. J. D. Bronzino, *The Biomedical Engineering Handbook*. Boca Raton, FL: CRC Press, 1995.
14. McGraw-Hill. *Modern Plastics Encyclopedia Handbook*. New York: McGraw-Hill, 1994.
15. J. C. Middleton and A. J. Tipton, Synthetic biodegradable polymers as medical devices. *Med. Plast. Biomater.* 1998; Mar/Apr.

ARTICULAR CARTILAGE

WALTER HERZOG
University of Calgary
Calgary, Alberta, Canada

1. INTRODUCTION

Articular cartilage is a thin (about 1–6 mm in human joints) layer of fibrous connective tissue covering the articular surfaces of bones in synovial joints (Fig. 1). It consists of cells (2–15% in terms of volumetric fraction) and an intercellular matrix (85–98%) with a 65–80% water content. Articular cartilage is a viscoelastic material and, in conjunction with synovial (joint) fluid, allows for virtually frictionless movement (coefficients of friction from 0.002–0.05) of the joint surfaces. Osteoarthritis is a joint disease that is associated with a degradation and loss of articular cartilage from the joint surfaces and a concomitant increase in joint friction causing pain and disability, particularly in the elderly population. The primary functions of articular cartilage include force transmission across joints, distribution of articular forces so as to minimize stress concentrations, and provision of smooth surface for relative gliding of joint surfaces. In most people, articular cartilage fulfills its functional role for decades, although the incidence of osteoarthritis in North America is about 50% among people of age 60 and greater.

2. STRUCTURE

Articular cartilage is structurally heterogeneous, and material properties change as a function of depth. Although these changes are continuous, articular cartilage is typically divided into four structural zones (Fig. 2):

– Superficial zone,
– Middle (or transitional) zone,
– Deep (or radial) zone, and
– Calcified zone.

The superficial zone is the thinnest, most superficial region that forms the gliding surface of joints. It contains a surface layer (lamina splendens) of about 2 μm thickness, which is made up of randomly aligned collagen fibrils, and a deep layer consisting of collagen fibrils aligned parallel to the cartilage surface following the so-called "split line" pattern (1), which follows the direction of normal joint movement.

The collagen fibrils in the superficial zone show a wave-like pattern referred to as crimp. This waving or crimping shows dips and ridges in the micrometer range (2), thus the surface, although apparently smooth to touch and with a low friction coefficient, is not quite smooth microscopically speaking.

The deep layer of the superficial zone contains articular cartilage cells (chondrocytes) that are flat (3), and

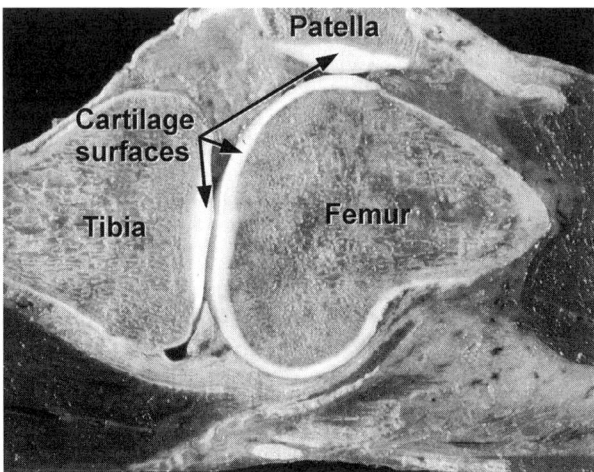

Figure 1. Sagittal plane section through a human knee showing the femur, tibia, and patella and the associated articular cartilage.

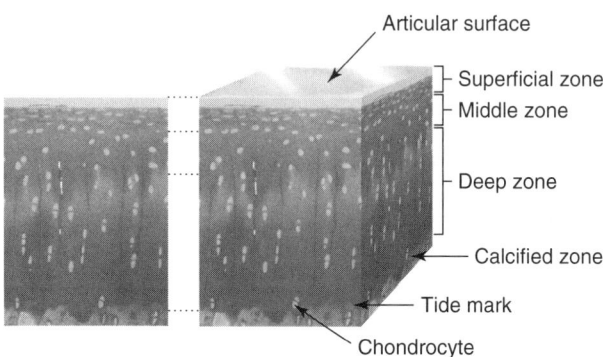

Figure 2. The four zones of articular cartilage. The superficial zone provides the sliding surface of joints with collagen fibrils aligned parallel to the surface, and the flat, metabolically relatively inactive cells. The middle (or transitional) zone contains collagen fibrils that are oriented randomly, and the cells are nearly spherical. The deep (or radial) zone contains collagen fibrils that are oriented perpendicular to the subchondral bone (and articular surface), and the cells are typically aligned in radial columns. The calcified zone provides a mechanical transition that separates the relatively soft cartilage tissue from the stiff subchondral bone.

metabolically relatively inactive (4), as evidenced by a low content of mitochondria, golgi organs, and endoplasmic reticula. The deep layer of the superficial zone also contains little proteoglycan but has the highest water concentration (about 80%) of all zones, as water content decreases with depth to a value of about 65% in the deep zone (5,6).

The middle (or transitional) zone is typically thicker than the superficial zone. Collagen fibrils have a greater diameter in this zone than the superficial zone and are oriented in a nearly random fashion. Proteoglycan content is greater and aggregate complexes are larger than in the superficial zone. Chondrocytes are nearly spherical in this zone and contain great numbers of mitochondria, golgi bodies, and a vast endoplasmic reticulum network, suggesting that cells in this zone are metabolically more active than those of the superficial zone.

The deep (or radial) zone contains the largest diameter collagen fibrils that are oriented perpendicular to the subchondral bone and the cartilage surface. Water content is lowest and proteoglycan content is typically highest in this zone. The chondrocytes tend to be aligned in radial columns and they contain intracytoplasmic filaments, glycogen granules, endoplasmic reticula, and golgi bodies suggesting great protein synthesis activity. Benninghof (7) proposed that collagen fibers form arcades that extend from the deep to the superficial zone, and recent studies confirmed that collagen fibers, indeed, might be continuous through the various cartilage zones (8).

The calcified zone provides a mechanical transition that separates the relatively soft cartilage tissue and the stiff subchondral bone. It is characterized by hydroxyapatite, an inorganic constituent of bone matrix. The calcified zone is separated from the deep (radial) zone by the "tidemark," an undulating line of a few micrometers thickness. Collagen fibers from the deep zone cross the tidemark and anchor into the calcified zone, thereby providing strong adhesion of cartilage to bone. The calcified zone contains metabolically active chondrocytes, serves for structural integration and has been considered important for nutrition, and cartilage repair developing from the underlying bone (3).

The structural differences across the various zones of articular cartilage have been thought to be the primary cause for the anisotropy of articular cartilage (e.g., (9)) and have motivated structural models of transverse isotropic cartilage models (10).

3. COMPOSITION

Articular cartilage consists mostly (85–98%) of matrix and a sparse population of cells (chondrocytes about 2–15% in terms of volumetric fraction), and it is avascular, aneural, and alymphatic.

3.1. Cells

Chondrocytes are metabolically active cells that are responsible for the synthesis and degradation of the matrix. They are isolated, lie in lacunae, and receive nourishment through diffusion of substrates, which is thought to be facilitated by cycling loading/unloading that is common for many articular joint surfaces. As described above, the volumetric fraction, shape, and metabolic activity of cells varies as a function of location, and also varies across cartilages in different joints and even within the same joint but different locations (e.g., (11–13)). It is accepted that normal loading of articular cartilage produces deformations in chondrocytes and the corresponding cell nuclei (14,15), and these deformations have been thought responsible for the biosynthetic activity of cells; in general, static, long-lasting loads have been associated with a tissue degrading response whereas dynamic loading of physiological magnitudes has been related to positive adaptive responses.

Chondrocytes are softer than the surrounding extracellular matrix by a factor of about 1000. Therefore, one would expect them to be deformed much more during loading than the matrix, and might even expect them to collapse. However, chondrocytes are surrounded by a protective cover, consisting of a pericellular matrix and a pericellular capsule (16). The chondrocyte, its pericellular matrix and capsule, constitute the chondron, the primary functional and metabolic unit of cartilage. It has been suggested that the chondron acts hydrodynamically to protect the chondrocytes from excessive stresses and strains during cartilage compression. Chondrocytes in different types of cartilages have different functional and metabolic roles. In articular cartilage, they specialize in producing type II collagen and proteoglycan.

4. MATRIX

The intercellular **matrix** of articular cartilage is largely responsible for the functional and mechanical properties associated with this tissue. It consists of structural macromolecules and tissue fluid. Fluid comprises between 65–80% of the wet weight of articular cartilage, and its volumetric fraction decreases from the superficial to the deep zone. Macromolecules comprise 20–40%, and they are produced by the chondrocytes. Of the macromolecules, collagens contribute about 50% of the tissue dry weight, with proteoglycans and noncollagenous proteins/glycoproteins contributing approximately 30–35% and 15–20%, respectively (17). The interactions of tissue fluid with the structural macromolecules give articular cartilage its specific mechanical and electrostatic properties. As the distribution, orientation, and density of these macromolecules changes with cartilage depth, so do the functional and material properties of the tissue.

At least 18 different types of **collagen** exist, but in articular cartilage, type II collagen is by far the most abundant (about 80–85% of all collagens), but other types (V, VI, IX, X, and XI) are also found and have been associated with specific functional roles (18,19). Collagen molecules are comprised of three α-chains that are interwoven in a helical configuration. Each chain has high hydroxylysine content and covalently bound carbohydrates that make it adhere readily with proteoglycans.

Collagens have a high tensile stiffness [about 2–46 MPa for uncross linked type I collagen from rat tail tendon and

380–770 MPa when cross linked, (20)], but because of their fibrillar structure are typically thought to have negligible compressive strength, as they are assumed to buckle when subjected to compressive loading (21,22). However, some scientists think of collagen fibrils like reinforced inclusions, similar to steel rods embedded in concrete, with appreciable compressive capabilities, and thereby contributing substantially to withstand compressive forces (23). Collagens form a structural network that gives cartilage its tensile strength, and because of the characteristic orientation of the fibrillar network, they are associated with providing resistance to compressive loading as fluid pressurization tends to load the collagen network in tension. Collagen fibrils are cross linked for further strength (20), and are connected to proteoglycans via molecular chains developing from glycosaminoglycans and polysaccharides. Thus, collagens are intimately associated with other macromolecules and so provide for a tough tissue that, despite its thinness, can withstand high repetitive loading for a lifetime.

Proteoglycans are large molecules composed of a central core protein with glycosaminoglycan side chains covalently attached. The protein core makes up about 10% and the side chains about the remaining 90% of the molecular weight of proteoglycan. The glycosaminoglycan side chains contain sugars on a protein core that have a negative electrostatic charge. These negatively charged molecules repel each other and attempt to occupy as much volume as possible. Proteoglycans are kept from dissolving into the fluid and being swept away by their attachment to the stretched collagen network. The negative charges are thus forced to stay in close proximity, and when subjected to external compressive loads, proteoglycans are further compressed, and the repulsive forces increase from their natural pre-tensed state.

Articular cartilage contains large aggregating proteoglycans (aggrecan and versican) and small interstitial proteoglycans (biglycan, decorin, fibromodulin, and lumican). The large proteoglycans contribute 50–85% of the total proteoglycan content. Aggrecan is the major proteoglycan; it consists of a core protein and up to 150 chondroitin and keratin sulphate chains (Fig. 3). The core protein's N-terminal G1 domain interacts with link proteins and hyaluronan, and these three components form stable macromolecular complexes (Fig. 3).

Changes in proteoglycan structure and decreased density often accompany articular cartilage degeneration and aging (24,25). These changes are typically accompanied by a corresponding increase in water content, a decreased stiffness, and a reduced resistance to withstand mechanical loading.

Noncollagenous proteins play a role in the assembly and integrity of the extracellular matrix, and form links between the chondrocytes and the matrix. Noncollagenous proteins include adhesive glycoproteins such as fibronectin, thrombospondin, chondroadherin, and other matrix proteins such as the link protein, cartilage matrix oligomeric protein, cartilage matrix protein, and proline arginine-rich and leucine-rich repeat proteins. The detailed function of many of these noncollagenous proteins is currently not well understood.

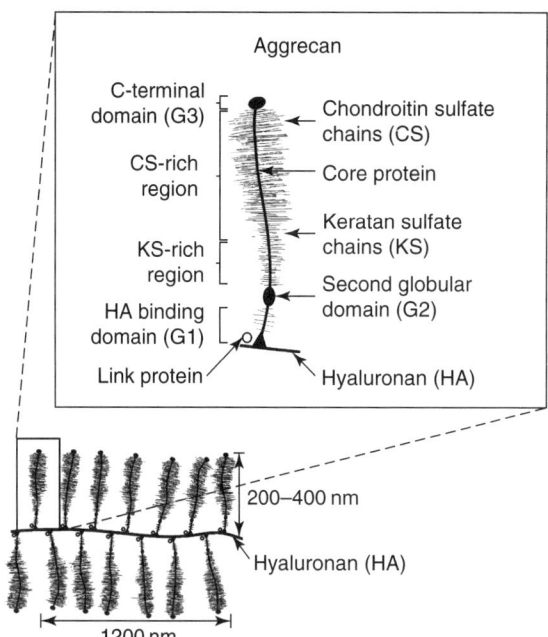

Figure 3. Macromolecular aggregate formed by aggrecan molecules (inset blown up) binding to a chain of hyaluronan through a link protein. The aggrecan molecule consists of a core protein with several domains: hyaluronan-binding G1 domain, G2 domain, keratin sulfate-rich region, chondroitin sulfate-rich region, and C-terminal domain, G3.

Articular cartilage tissue **fluid** consists of water and dissolved gas, small proteins, and metabolites. Water is the most abundant component of cartilage and accounts for 65–80% of its wet weight. Fluid is moved around in articular cartilage and is closely associated with the synovial fluid of the joint.

Fluid in articular cartilage is intimately associated with the proteoglycan network, which, to a certain degree, hinders its movement. This arrangement makes cartilage sponge-like in that water is restrained, but with pressure caused by cartilage loading, will flow with the pressure gradient. When modeling articular cartilage, fluid flow is described by permeability, which is a measure of how easily a fluid flows through a porous material. Permeability is inversely proportional to the force required for a fluid to flow at a given speed through the tissue. Permeability in cartilage is low, therefore fluid flow is typically small, and any substantial exchange of tissue fluid and synovial fluid, or any significant loss of fluid caused by cartilage compression takes a long time, in the order of minutes. Compared with the typical loading cycle of cartilage in a joint, for example, about 0.5 s of loading during a step cycle, fluid flow is very slow. Therefore, cartilage maintains its stiffness well when loaded, and fluid takes up a big part of the forces in the tissue during physiological (quick) loading cycles, thereby presumably protecting the matrix, and, in turn, the cells, from stresses and strains during normal everyday loading cycles.

Synovial fluid provides lubrication to joints. It is transparent, alkaline, and viscous and is secreted by the synovial membranes that are contained in joints. It is a dialysate of blood containing hyaluronic acid. Synovial fluid is viscous and contains a glycoprotein called lubricin, which is found on the articular surfaces of joints but not within the cartilage tissue. Cartilage surfaces are physically separated by a 0.5–1.0 µm of synovial fluid film. Synovial fluid is essential for the mechanical sliding characteristics of joints, and permits the diffusion of gases, nutrients, and waste products, thereby making it essential for the health of the tissue.

5. MECHANICAL PROPERTIES

Articular cartilage is a passive structural tissue whose primary functions are mechanical: smooth sliding surface, and transmission and distribution of forces across joints. Therefore, in order to understand the functional properties of this tissue, its mechanical (material) properties need to be known, which is, insofar, difficult as articular cartilage structure and composition varies from one joint to the next, its properties also change with location within a joint and even across its depth. Nevertheless, many attempts have been made to elucidate the compressive, tensile, and shear properties of articular cartilage using a variety of engineering testing approaches.

5.1. Compressive Properties

Compressive properties of articular cartilage have not been obtained for the whole tissue attached to its native bone, although indentation testing (Fig. 4a) has been used to evaluate compressive properties at specific locations using exposed joint surfaces (e.g., (26)), or in patients using indentation devices that can be introduced arthroscopically into joints (e.g., (27,28)). Typically, material properties are determined for articular cartilage explants that are subjected to unconfined or confined compression testing (Fig. 4b and 4c).

Compressive properties, elastic modulus, and stiffness of a given sample of articular cartilage vary with depth in the tissue. For example, Schinagl et al. (9) report a more than 20-fold increase in the compressive modulus across the full thickness of bovine articular cartilage tested in unconfined compression. Compressive strength is directly related to proteoglycan concentration, and so is the tissue's compressive stiffness (Fig. 5) (29). As proteoglycan concentration is known to increase with depth, one would expect this result. However, even a five-fold increase in proteoglycan concentration is only associated with a doubling or tripling of the compressive modulus (Fig. 5), thus it appears that the great changes in stiffness observed experimentally in different cartilage layers must have an additional explanation. One such explanation might be the orientation of the collagen fibrillar network and assuming that the collagen network might help resist compressive loading.

5.2. Tensile Properties

Articular cartilage functions as a compressive load absorber. Nevertheless, when loaded, tensile forces are thought to act on the collagen network, and thus an understanding of the tensile properties might be important

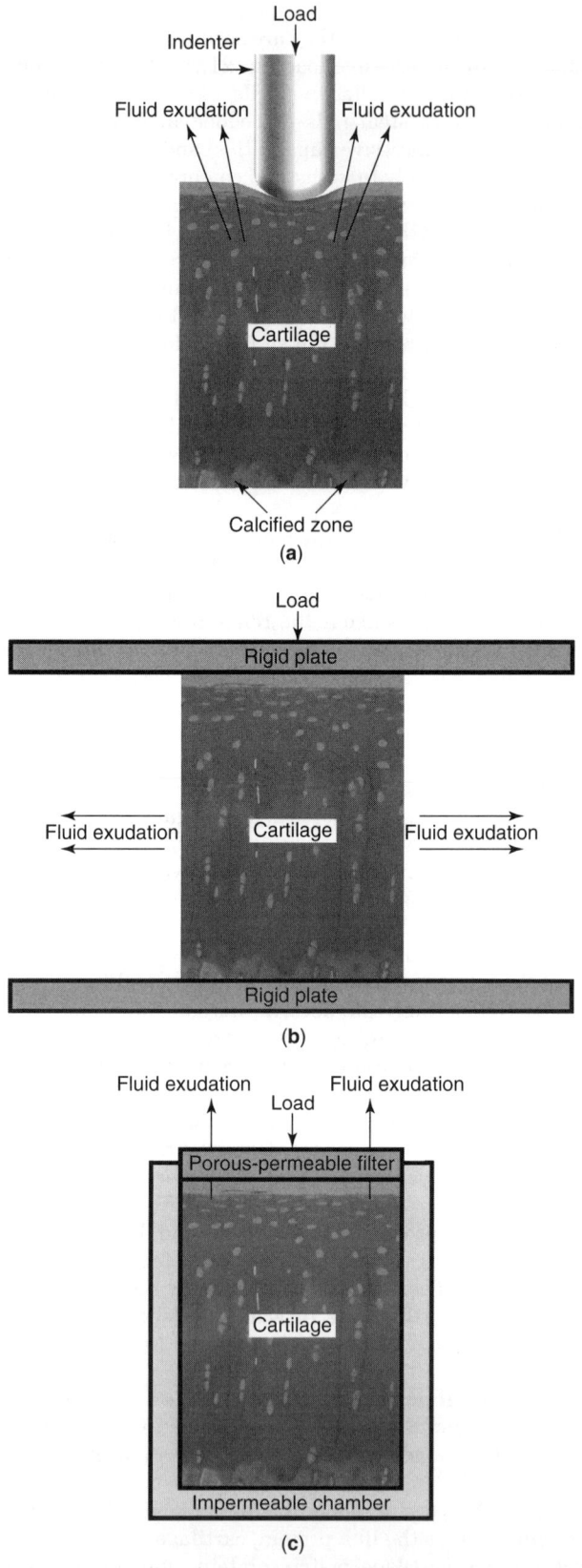

Figure 4. Schematic illustration of the three most commonly used techniques for determining articular cartilage mechanical properties: (a) indentation testing, (b) unconfined compression, and (c) confined compression.

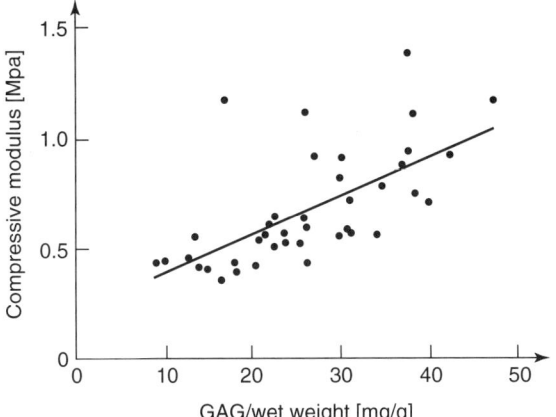

Figure 5. Compressive modulus of elasticity as a function glycosaminoglycan (GAG) concentration. As GAG concentration increases with cartilage depth, so does the tissue's compressive stiffness [adapted from (29), with permission].

in understanding the functioning of cartilage *in vivo*. Tensile strength in articular cartilage is highest in the superficial zone and decreases continuously with increasing depth in the tissue (Fig. 6) (30), likely because of variations in the orientation of collagen fibrils, the cross linking of collagen fibrils, and the ratio of collagen to proteoglycan. Also, tensile properties are direction-dependent. When testing occurs along the split line, that is along the predominant orientation of the long axis of collagen fibrils, tensile strength and the tensile modulus are greater than when testing is performed perpendicular to the split line (Fig. 6).

5.3. Shear Properties

Shear properties in articular cartilage are not well understood, and shear testing is associated with numerous problems. For example, the smoothness of the surface layer makes it difficult to apply shear loads without slippage. Shear testing has become important for identifying flow-independent viscoelasticity of the tissue, as ideally small deformation shear testing is accomplished with a constant volume and no fluid flow. Shear resistance is thought to come primarily from the collagen network, and consistent with that idea, the shear modulus is decreased with decreasing collagen content as may occur in cartilage diseases, such as osteoarthritis. Also, as the proteoglycan network is thought to prestress the collagen network, loss of prestress caused by loss of proteoglycans has also been found to decrease the dynamic and equilibrium shear properties (31).

5.4. Viscoelasticity

If a material's response to a constant force or deformation varies in time, its mechanical behavior is said to be viscoelastic, in contrast to an elastic material's response that does not vary in time. Articular cartilage is known to be viscoelastic (Fig. 7); however, the origin for this property remains controversial. Most people agree that one aspect of the viscoelastic behavior is associated with fluid flow, and some believe that a flow-independent viscoelasticity also exists that resides in the matrix proper. As fluid flow is proportional to the pore pressure gradient in water, it can be described by the coefficient for hydraulic permeability. The inverse of this coefficient gives a measure of the material's resistance to fluid flow. As proteoglycans attempt to restrain fluid flow, permeability is smallest in areas with high proteoglycan concentration (deep layers), and cartilage is most permeable in zones of low proteoglycan concentration (surface layer).

6. BIOMECHANICS

In its role of force transmission and force distribution across joints, and providing a smooth surface for virtually frictionless gliding, articular cartilage material properties are of great importance. However, articular cartilage is

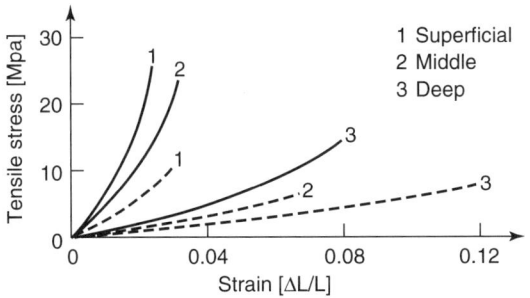

Figure 6. Stress-strain curves for tensile testing of articular cartilage specimens from the superficial (1), middle (2), and deep zone (3), and along the long axis of the collagen fibrils (solid lines) and perpendicular to the long axis of the collage fibrils (dashed lines). Tensile strength decreases continuously from the surface to the deep zone and is greater along the collagen fibril direction than perpendicular to it (adapted from (30), with permission).

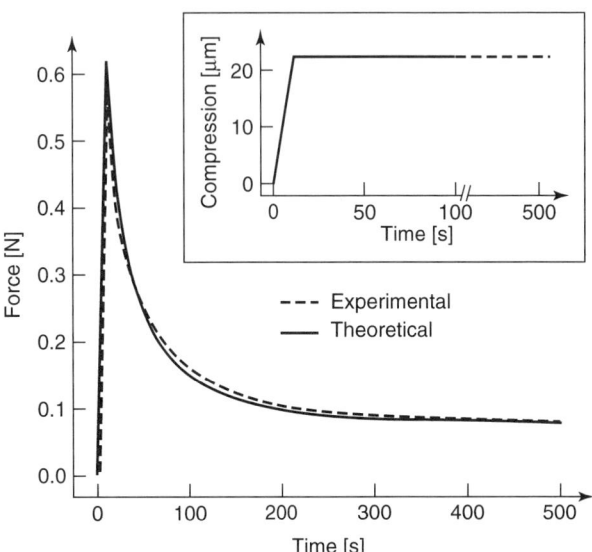

Figure 7. Experimental and theoretical force-relaxation curve for articular cartilage exposed to unconfined ramp loading (as shown in the inset). Articular cartilage exhibits a typical viscoelastic response that is associated with fluid flow, and possibly also with an inherent viscoelasticity of the matrix component.

only one component of a joint, and in order to understand articular cartilage biomechanics, its functional requirements within the "organ" joint need to be known. Malfunctioning of one component of a joint, for example, a breakdown of the articular cartilage, loss of a guiding ligament, or rupture of a meniscal inclusion, will affect all structures of a joint, and thus focusing on a single component, might not do justice to the intricate functionality of that component in its native environment.

When a joint moves, the articular surfaces slide relative to one another, the size and location of the contact area changes, and the contact pressure distribution is affected by the changing surface geometry and the variable forces applied to the joint. Furthermore, for a given joint configuration, that is a given orientation of the articular surfaces, the contact area increases with increasing force transmission across the joint. For example, the cat patellofemoral joint contact area increases with increasing force potential of the quadriceps muscles, thereby providing a natural way for keeping average pressure between the contacting cartilage surfaces low when forces become big. Similarly, when forces transmitted across the patellofemoral joint are increased by a factor of five (from 100–500N), contact area increases by a factor of about four (from about 7 to about 30 mm^2) whereas the mean contact pressure only increases by a factor of about 0.6 (from 7.7 to 12.9 MPa) (Fig. 8). Therefore, contact pressures on the articular surfaces of joints are kept small with increasing forces by increasing contact areas for joint angles where large muscular forces are possible, and by increasing contact area as a function of the applied forces through the viscoelastic properties of articular cartilage.

The loads transmitted across joint surfaces are not well known, as most of the loading in joints is caused by muscles and muscle forces are hard to measure *in vivo* and cannot be estimated accurately and reliably. However, Rushfeldt et al. (32,33) developed techniques to measure local hip joint contact pressures by mounting pressure transducers on an endoprosthesis that replaced the natural femoral head. They measured average articular surface pressures in the joint of about 2–3 MPa, with peak pressures reaching 7 MPa, for loads equivalent to the single stance phase of walking (2.6 times body weight). Peak hip joint pressures reached were in the range of 18 MPa and were obtained for walking downstairs and rising from a chair (34). Using telemetry-based force measurements from patients with total hip joint replacements, Bergman et al. (35) measured resultant hip forces of 2.8 to 4.8 times body weight for walking at speeds ranging from 1 to 5 km/h. Very fast walking or slow running increased these forces to about 5.5 times body weight, whereas stumbling (unintentionally) gave temporary peak resultant hip joint forces of 7.2 and 8.7 times body weight in two patients. No corresponding joint pressure or force data are available for intact human joints during normal everyday movements.

Similarly, direct pressure measurements in an intact diarthrodial joint during unrestrained movements are not available from animal experimentation. However, quadriceps forces and the corresponding knee kinematics have been measured in freely walking cats. Joint pressure distributions could then be obtained using Fuji pressure-sen-

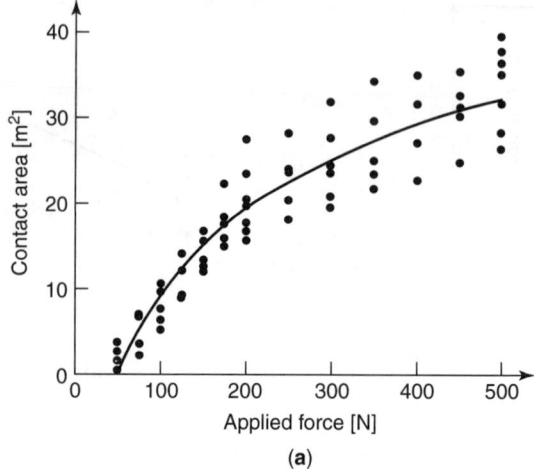

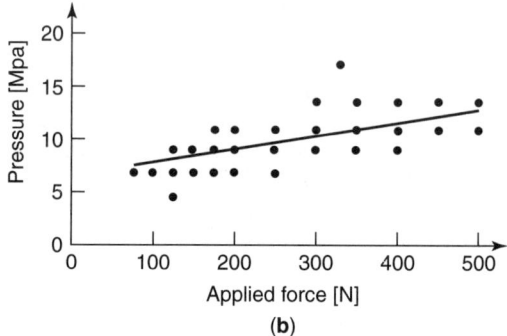

Figure 8. Patellofemoral contact area (a) and peak pressure (b) transmitted across the joint. Contact area increases quickly with increasing force, especially in the range of physiological loading (0–200 N), whereas peak pressures are less sensitive to changes in joint loading.

sitive film in the anesthetized animals while reproducing the joint angles and muscle forces observed during locomotion. Average peak pressures obtained by using this approach were 5.7 MPa in the patellofemoral joint. When the quadriceps muscles were fully activated (through electrical stimulation), median and peak pressures in the cat patellofemoral joint reached 11 and 47 MPa, respectively, thereby indicating that joint pressures in normal every day movements are much smaller than can be produced by maximal muscle contraction (36).

The forces transmitted by articular cartilage across leg joints are in the order of several times body weight for normal every day tasks such as walking, getting up from a chair, and walking up or down stairs. The corresponding articular surface pressures range from about 2 to 10 MPa, but maximal muscle contraction can produce peak pressures in knee articular cartilage of almost 50 MPa. Such high pressures, when applied to articular cartilage explants through confined or unconfined compression, cause cell death and matrix damage, and when applied to the whole joint through impact (peak pressures attained within 0.5–3 ms), damage occurs as well. Muscular loading of the intact joint does not cause articular cartilage damage, thus indicating that the time of load application (typically >100 ms for human muscles), joint integrity, and articular surface movements during force production provide a safe environment for high cartilage loading, which is lost when testing articular cartilage explants *in vitro*.

7. OSTEOARTHRITIS

Osteoarthritis is a joint degenerative disease that affects about 50% of all people above the age of 60 in North America. It is associated with a thinning and local loss of articular cartilage from the joint surfaces, osteophyte formations at the joint margins, swelling of the joint, and pain. The causes for osteoarthritis are not known, but some of the risk factors include age, injury, and obesity. Osteoarthritis is often combined with a loss of muscle mass and movement control, as well as a decrease in strength.

It is well accepted that physiological loading of articular cartilage is essential for its health and integrity. Animal models of joint disruption through strategic cutting of ligaments, or removal of menisci from knees, have caused joint degeneration, presumably because of the altered loading conditions of the articular surfaces, although very hard to quantify *in vivo*. For example, when cutting the anterior cruciate ligament of the knee in a dog, cat, or rabbit unilaterally, an unloading of the experimental and an overloading of the contralateral side occurs that lasts for a few weeks (Fig. 9). This intervention is associated with an instability of the knee, ostephyte formation at the joint margins, thickening of the medial aspect of the joint capsule and the medial collateral ligament, and an increased thickness, water content, and softness of the articular cartilage (37,38). In the long-term, anterior cruciate ligament transection in the dog and cat have been shown to lead to bona fide osteoarthritis, as evidenced by full thickness loss of the articular cartilage in some load-bearing regions of the joint, leading to an increase in joint friction, decreased range of motion, pain, and a loss of muscle mass and strength in muscles crossing the knee (13,39–41).

Contact pressure measurements in intact cat patellofemoral joint that were loaded by muscular contraction showed that for a given amount of force transmitted through the joint, contact areas are increased and peak contact pressures are decreased by 50% in joints with early osteoarthritis compared with normal controls (38). These results can be explained readily with the increased cartilage thickness and decreased stiffness observed in this phase of cartilage degeneration, and theoretical models with the altered geometries and functional properties have confirmed the experimental results. In late osteoarthritis, articular cartilage becomes thinner and harder than normal, and forces are transmitted across a decreased contact area and with peak pressures that are increased compared with the normal joint (42).

It has been demonstrated that loading of articular cartilage affects the biological response of chondrocytes, which in turn affects the health and integrity of the cartilage matrix. Although the results are controversial, and no unified consensus exists on what loading is "good" and "bad" for cartilage health, it has been demonstrated that long, static loading is associated with an increase in cartilage degrading enzymes, whereas intermittent physiological loading appears to strengthen the cartilage matrix through the formation of essential matrix proteins (e.g., (19,43)).

One of the proposed pathways for transmitting load signals across the cell to the genome has been the cytoskeleton, specifically the integrin network. Integrins are transmembrane extracellular matrix receptors that have been shown to transmit forces from the matrix to the cytoskeleton of chondrocytes and vice versa. The mechanical signals transmitted by integrins may result in biochemical responses of the cell through force-dependent release of chemical second messengers, or by force-induced changes in cytoskeletal organization that activate gene transcription (44,45). However, a direct link between force transmission in integrins and a corresponding biosynthetic response has yet to be demonstrated.

Another potential cytoskeletal pathway for force transmission is actin. Guilak (14) demonstrated that chondrocytes and their nuclei deformed when articular cartilage was loaded (15% compression) in an unconfined configuration. When the actin cytoskeleton was removed from the preparation and the same loading was applied to the cartilage, chondrocyte deformation remained the same, whereas nuclear deformation was changed, demonstrating an intimate connection between the actin cytoskeleton and cell nuclei. Although these results do not prove that actin works as a mechanical signaling pathway, it demonstrates that mechanical linkages exist that affect nuclear shape and might affect gene transcription.

A clinical observation in patients with osteoarthritis is that they often have diminished strength in the musculature surrounding the arthritic joint (46–48), that maximal muscle contractions are not possible because of reflex inhibitions (e.g., (49)), and that muscle contraction patterns change from normal for everyday movements (47).

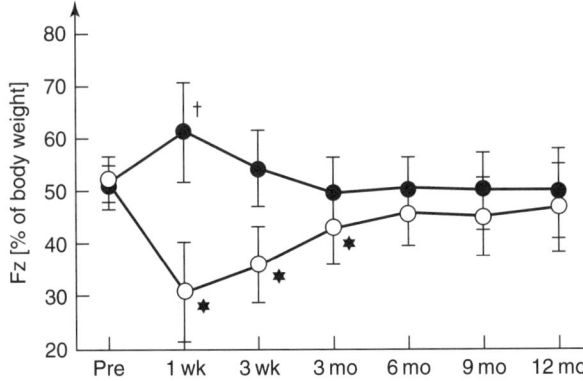

Figure 9. Vertical ground reaction forces, F_z, normalized to body weight, as a function of time following unilateral anterior cruciate ligament (ACL) transection in cats (n = 7). "Pre" indicates testing before ACL transection, 1 wk, 3 wk, ..., 12 mo indicate 1 week, 3 weeks, ...12 months post ACL transection. The vertical ground reaction forces are significantly (*) decreased in the ACL-transected (open symbols) compared with the contralateral (closed symbols) hind limb for 3 months post ACL transection; thereafter, the differences are not statistically significant. One week following ACL transection, the vertical ground reaction force in the contralateral leg is greater (†) than in the pretransection condition.

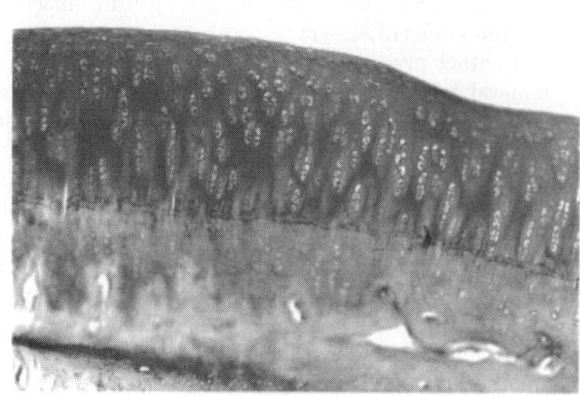
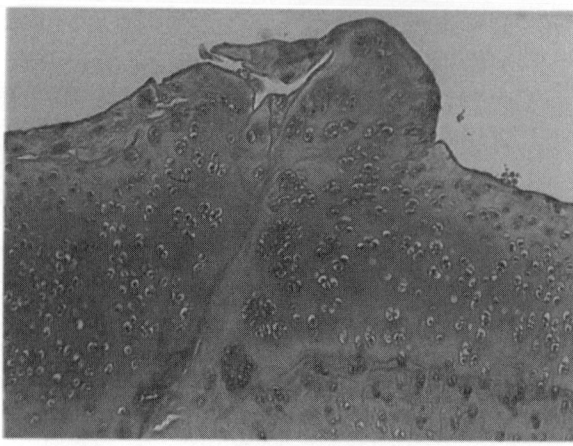

Figure 10. Histological sections of articular cartilage from the femoral condyle of a normal and an experimental rabbit. The experimental rabbit received a single injection of botulinum toxin type-A into the knee extensor muscles causing weakness, and was sacrificed four weeks following the injection. The Mankin scores for the normal and experimental cartilage were 0 (normal) and 12 (signs of degeneration), indicating that muscle weakness may be a risk factor for joint degeneration. (This figure is available in full color at http://www.mrw.interscience.wiley.com/ebe.)

Furthermore, it is generally accepted that muscles provide the primary loading of joints, with weight forces and external forces playing a relatively smaller role. Nevertheless, the role of muscles in the development and progression of osteoarthritis has largely been neglected, and some basic question should be addressed in the future, including the role of muscle activation patterns and muscle weakness as risk factors for osteoarthritis.

In studying osteoarthritic patients, only weak associations have been found between radiographic signs of joint disease and joint pain and disability (50,51), whereas quadriceps weakness has been found to be one of the earliest and most common symptoms reported by patients with osteoarthritis (49,52–54). Furthermore, quadriceps weakness has also been shown to be a better predictor of disability than radiographic changes or pain (46). Therefore, suggestions have been made that muscle weakness might represent an independent risk factor for cartilage degeneration, joint disease, and osteoarthritis. Recent evidence from an animal model of muscle weakness supports this idea showing that articular cartilage is degenerated following just four weeks of muscle weakness in knees that were otherwise not compromised (Fig. 10). Based on the evidence that cartilage degeneration and osteoarthritis are associated with changed loading of joints, that mechanical loading influences the biosynthetic activity of chondrocytes, and that muscle coordination patterns and muscle strength are intimately associated with cartilage disease, it is suggested that conservative treatment modalities, such as appropriate movement biomechanics, exercise, and strengthening, might represent promising approaches for preventing cartilage disease and osteoarthritis.

BIBLIOGRAPHY

1. W. Hultkrantz, Über die Spaltrichtungen der Gelenkknorpel. *Verh. Anat. Ges.* 1898; **12**:248–256.
2. D. Dowson, Biotribology of natural and replacement of synovial joints. In: V. C. Mow, A. Ratcliffe, and S. L.-Y. Woo, eds., *Biomechanics of Diarthrodial Joints*. New York: Springer Verlag, 1990, pp. 305–345.
3. E. Hunziker, Articular cartilage structure in humans and experimental animals. In: K. E. Peyron, J. G. Schleyerback, and V. C. Hascall, eds., *Articular Cartilage and Osteoarthritis*. New York: Raven Press, 1992, pp. 183–199.
4. M. Wong, P. Wuethrich, P. Eggli, and E. Hunziker, Zone-specific cell biosynthetic activity in mature bovine articular cartilage: a new method using confocal microscopic stereology and quantitative autoradiography. *J. Orthop. Res.* 1996; **14**(3):424–432.
5. A. Maroudas, Biophysical chemistry of cartilaginous tissues with special reference to solute and fluid transport. *Biorheology*. 1975; **12**:233–248.
6. P. A. Torzilli, Influence of cartilage conformation on its equilibrium water partition. *J. Orthop. Res.* 1985; **3**:473.
7. A. Benninghof, Form und Bau der Gelenkknorpel in ihren Beziehungen zur Funktion. In: Z. Zellforsch, ed., *Der Aufbau des Gelenkknorpel in seinen Beziehungen zur Funktion*. 1925; pp. 783–862.
8. H. Notzli and J. Clark, Deformation of loaded articular cartilage prepared for scanning electron microscopy with rapid freezing and freeze-substitution fixation. *J. Orthop. Res.* 1997; **15**:76–86.
9. R. M. Schinagl, D. Gurskis, A. C. Chen, and R. L. Sah, Depth-dependent confined compression modulus of full-thickness bovine articular cartilage. *J. Orthop. Res.* 1997; **15**:499–506.
10. S. Federico, A. Grillo, G. La Rosa, G. Giaquinta, and W. Herzog, A transversely isotropic, transversely homogeneous microstructural-statistical model of articular cartilage. *J. Biomech.* 2005; **38**:2008–2018.
11. R. S. Stockwell, *Biology of Cartilage Cells*. Cambridge: Cambridge University Press; 1979.
12. I. H. M. Muir, Proteoglycans as organisers of the intercellular matrix. *Biochem. Soc. Trans.* 1983; **11**(6):613–622.
13. A. L. Clark, T. R. Leonard, L. Barclay, J. R. Matyas, and W. Herzog, Opposing cartilages in the patellofemoral joint adapt

differently to long-term cruciate deficiency: chondrocyte deformation and reorientation with compression. *Osteoarthritis Cartilage* doi:10.1016/j.joca.2005.07.010, 2005.
14. F. Guilak, Compression-induced changes in the shape and volume of the chondrocyte nucleus. *J. Biomech.* 1995; **28**:1529–1541.
15. F. Guilak, A. Ratcliffe, and V. C. Mow, Chondrocyte deformation and local tissue strain in articular cartilage: a confocal microscopy study. *J. Orthop. Res.* 1995; **13**:410–421.
16. C. A. Poole, R. T. Gilbert, D. Herbage, and D. J. Hartmann, Immunolocalization of type IX collagen in normal and spontaneously osteoarthritic canine tibial cartilage and isolated chondrons. *Osteoarthritis Cartilage.* 1997; **5**:191–204.
17. J. A. Buckwalter, E. B. Hunziker, L. C. Rosenberg, R. Coutts, M. Adams and D. Eyre, Articular cartilage: Composition and structure. In: S. L.-Y. Woo and J. A. Buckwalter, Eds., *Injury and Repair of the Musculoskeletal Soft Tissues.* American Academy of Orthopaedic Surgeons: Park Ridge, 1991; pp. 405–425.
18. J. T. Thomas, S. Ayad, and M. E. Grant, Cartilage collagens: strategies for the study of their organisation and expression in the extracellular matrix. *Ann. Rheum. Dis.* 1994; **53**:488–496.
19. E. M. Hasler, W. Herzog, J. Z. Wu, W. Muller, and U. Wyss, Articular cartilage biomechanics: theoretical models, material properties, and biosynthetic response. *Crit. Rev. Biomed. Eng.* 1999; **27**(6):415–488.
20. G. D. Pins, E. K. Huang, D. L. Christiansen, and F. H. Silver, Effects of static axial strain on the tensile properties and failure mechanisms of self-assembled collagen fibers. *J. Appl. Polymer Sci.* 1997; **63**:1429–1440.
21. L. P. Li, M. D. Buschmann, and A. Shirazi-Adl, A fibril reinforced nonhomogeneous poroelastic model for articular cartilage: inhomogeneous respones in unconfined compression. *J. Biomech.* 2000; **33**:1533–1541.
22. J. Soulhat, M. D. Buschmann, and A. Shirazi-Adl, A fibril-network-reinforced biphasic model of cartilage in unconfined compression. *J. Biomech. Eng.* 1999; **121**:340–347.
23. J. Z. Wu and W. Herzog, Elastic anisotropy of articular cartilage is associated with the microstructures of collagen fibers and chondrocytes. *J. Biomech.* 2002; **35**:931–942.
24. M. W. Lark, E. K. Bayne, J. Flanagan, C. F. Harper, L. A. Hoerrner, N. I. Hutchinson, I. I. Singer, S. A. Donatelli, J. R. Weidner, H. R. Williams, R. A. Mumford, and L. S. Lohmander, Aggrecan degradation in human cartilage. Evidence for both matrix metalloproteinase and aggrecanase activity in normal, osteoarthritic, and rheumatoid joints. *J. Clin. Invest.* 1997; **100**:93–106.
25. J. A. Buckwalter, K. E. Kuettner, and E. J. Thonar, Age-related changes in articular cartilage proteoglycans: electron microscopic studies. *J. Orthop. Res.* 1985; **3**:251–257.
26. A. L. Clark, L. D. Barclay, J. R. Matyas, and W. Herzog, In-situ chondrocyte deformation with physiological compression of the feline patellofemoral joint. *J. Biomech.* 2003; **36**:553–568.
27. J. H. Dashefsky, Arthroscopic measurement of chondromalacia of patella cartilage using a microminiature pressure transducer. *Arthroscopy* 1987; **3**:80–85.
28. T. Lyyra, J. Jurvelin, P. Pitkänen, U. Väätäinen, and I. Kiviranta, Indentation instrument for the measurement of cartilage stiffness under arthroscopic control. *Med. Eng. Phys.* 1995; **17**:395–399.
29. R. L. Sah, A. S. Yang, A. C. Chen, J. J. Hant, R. B. Halili, M. Yoshioka, D. Amiel, and R. D. Coutts, Physical properties of rabbit articular cartilage after transection of the ACL. *J. Orthop. Res.* 1997; **15**:197–203.
30. G. E. Kempson, The tensile properties of articular cartilage and their relevance to the development of osteoarthrosis. Orthopaedic Surgery and Traumatologie. *Proceedings of the 12th International Society of Orthopaedic Surgery and Traumatologie, Tel Aviv Exerpta Medica, Amsterdam.* 1972; 44–58.
31. W. Zhu, V. C. Mow, T. J. Koob, and D. R. Eyre, Viscoelastic shear properties of articular cartilage and the effects of glycosidase treatments. *J. Orthop. Res.* 1993; **11**:771–781.
32. P. D. Rushfeldt, R. W. Mann, and W. H. Harris, Improved techniques for measuring *in vitro* the geometry and pressure distribution in the human acetabulum-I. Ultrasonic measurement of acetabular surfaces, sphericity and cartilage thickness. *J. Biomech.* 1981; **14**:253–260.
33. P. D. Rushfeldt, R. W. Mann, and W. H. Harris, Improved techniques for measuring *in vitro* the geometry and pressure distribution in the human acetabulum. II Instrumented endoprosthesis measurement of articular surface pressure distribution. *J. Biomech.* 1981; **14**:315–323.
34. R. W. Mann, Comment on "an articular cartilage contact model based on real surface geometry", Han Sang-Kuy, Salvatore Federico, Marcelo Epstein and Walter Herzog. *J. Biomech.* 2005; **38**:1741–1742.
35. G. Bergmann, F. Graichen, and A. Rohlmann, Hip joint loading during walking and running, measured in two patients. *J. Biomech.* 1993; **26**(8):969–990.
36. E. M. Hasler and W. Herzog, Quantification of *in vivo* patellofemoral contact forces before and after ACL transection. *J. Biomech.* 1998; **31**:37–44.
37. W. Herzog, M. E. Adams, J. R. Matyas, and J. G. Brooks, A preliminary study of hindlimb loading, morphology and biochemistry of articular cartilage in the ACL-deficient cat knee. *Osteoarthritis Cartilage* 1993; **1**:243–251.
38. W. Herzog, J. Z. Wu, T. R. Leonard, E. Suter, S. Diet, C. Muller, and P. Mayzus, Mechanical and functional properties of cat knee articular cartilage 16 weeks post ACL transection. *J. Biomech.* 1998; **31**:1137–1145.
39. K. D. Brandt, S. L. Myers, D. Burr, and M. Albrecht, Osteoarthritic changes in canine articular cartilage, subchondral bone, and synovium fifty-four months after transection of the anterior cruciate ligament. *Arthritis Rheum.* 1991; **34**:1560–1570.
40. K. D. Brandt, E. M. Braunstein, D. M. Visco, B. O'Connor, D. Heck, and M. Albrecht, Anterior (cranial) cruciate ligament transection in the dog: A bona fide model of osteoarthritis, not merely of cartilage injury and repair. *J. Rheumatol.* 1991; **18**:436–446.
41. D. Longino, T. Butterfield, and W. Herzog, Frequency and length dependent effects of Botulinum toxin-induced muscle weakness. *J. Biomech.* 2005; **38**:609–613.
42. W. Herzog and S. Federico, Considerations on joint and articular cartilage mechanics. *Biomechan. Modeling Mechanobiol.* 2005; in press.
43. V. C. Mow, N. M. Bachrach, L. A. Setton, and F. Guilak, Stress, strain, pressure and flow fields in articular cartilage and chondrocytes. In: V. C. Mow, F. Guilak, R. Tran-Son-Tray, and R. M. Hochmuth eds., *Cell Mechanics and Cellular Engineering.* New York: Springer Verlag, 1994, pp. 345–379.
44. D. Ingber, Integrins as mechanochemical transducers. *Cell Biol.* 1991; **3**:841–848.
45. A. Ben-Ze'ev, Animal cell shape changes and gene expression. *BioEssays.* 1991; **13**:207–212.

46. C. Slemenda, K. D. Brandt, D. K. Heilman, S. Mazzuca, E. M. Braunstein, B. P. Katz, and F. D. Wolinsky, Quadriceps weakness and osteoarthritis of the knee. *Ann. Intern. Med.* 1997; **127**:97–104.
47. M. V. Hurley, The role of muscle weakness in the pathogenesis of osteoarthritis. *Rheumat. Disease Clin. N. Am.* 1999; **25**(2):283–298.
48. C. Slemenda, D. K. Heilman, K. D. Brandt, B. P. Katz, S. Mazzuca, E. M. Braunstein, and D. Byrd, Reduced quadriceps strength relative to body weight. A risk factor for knee osteoarthritis in women? *Arthritis Rheum.* 1998; **41**:1951–1959.
49. M. V. Hurley and D. J. Newham, The influence of arthrogenous muscle inhibition on quadriceps rehabilitation of patients with early unilateral osteoarthritic knees. *Br. J. Rheumatol.* 1993; **32**:127–131.
50. A. A. Claessens, J. S. Schouten, F. A. van den Ouweland, and H. A. Valkenburg, Do clinical findings associate with radiographic osteoarthritis of the knee? *Ann. Rheumatic. Diseases* 1990; **49**:771–774.
51. T. E. McAlindon, C. Cooper, J. R. Kirwan, and P. A. Dieppe, Determinants of disability in osteoarthritis of the knee. *Ann. of Rheumat. Diseases* 1993; **52**:258–262.
52. N. M. Fisher, D. R. Pendergast, G. E. Gresham, and E. Calkins, Muscle rehabilitation: its effect on muscular and functional performance of pateients with knee osteoarthritis. *Arch. Phys. Med. Rehabil.* 1991; **72**:367–374.
53. N. M. Fisher, G. E. Gresham, M. Abrams, J. Hicks, D. Horrigan, and D. R. Pendergast, Quantitative effects of physical therapy on muscular and functional performance in subjects with osteoarthritis of the knees. *Arch. Phys. Med. Rehabil.* 1993; **74**:840–847.
54. N. M. Fisher, S. C. White, H. J. Yack, R. J. Smolinski, and D. R. Pendergast, Muscle function and gait in patients with knee osteoarthritis before and after muscle rehabilitation. *Disabil. Rehabil.* 1997; **19**(2):47–55.

ARTIFICIAL BLOOD

Sarit Sivan
Noah Lotan
Technion – Israel Institute of Technology
Haifa, Israel

1. INTRODUCTION

Blood substitute, artificial blood, or artificial red blood cells are commonly employed names defining "synthetic or hybrid materials prepared in the laboratory to replace red blood cells for use in transfusion and other applications where red blood cells are normally required" (1,2). Mostly, these materials are required as a resuscitation fluid when extensive blood loss occurred following accidental injuries or during elective surgery. Additional applications are envisaged in cell cultures for production of biopharmaceuticals, as well as in tissue engineering.

When these materials are used as resuscitation fluids, they are administered directly into the blood stream and are expected to remain in circulation for periods of no more than up to a few days. In these cases, their most critical functional characteristic is the ability to bind oxygen and to deliver it to the tissues. It is most desirable that this function be performed efficiently, even when the patient inhales plain air, rather than pure oxygen or oxygen-enriched air.

The quest for a resuscitative fluid continues to be in the focus of activity of many researchers. The limitations involved in using human blood and the wide range of applications of blood substitutes have fueled a worldwide search for safe, effective, and universal blood substitute materials. The global market for artificial blood is estimated at $30 billion dollars annually in the United States alone (1,2). It is estimated that approximately 14 million units of blood were transfused in the United States in 2001.

In view of the above, and especially because of the concern related to HIV, the last 30 years witnessed an intensive effort to develop a resuscitation fluid that is capable to mimic the O_2-carrying characteristics of natural blood, yet is free of all the limitations associated with the use of this natural material (2–56). Such a fluid should have been called "artificial O_2-carrier." Yet, in the accepted terminology, such a fluid is called "artificial blood" or "blood substitute," although it does not substitute for all the functions of blood.

Basic requirements from any resuscitative fluid include the ability to carry oxygen from the lungs to vital tissues and organs, the ability to transport carbon dioxide from these tissues and organs back to the lungs, and the capability to travel through the circulatory system with minimal side effects. It must also be stable, sterile, nonimmunogenic, biocompatible, and have appropriate rheologic properties.

Two major routes in the development of blood substitutes have been undertaken during the years. The first proposed the use of perfluorinated solvents (3–9). However, in view of their typical oxygen-binding characteristics, perfluorocarbons can be used only when the patient is provided with essentially pure oxygen and are totally inefficient when the patient consumes regular air. The second route involved the use of isolated human Hb, either cellular (encapsulated Hb, within vesicles) or acellular, as oxygen-carrying materials (10–56). As Hb has a high colloidal activity, a high oxygen capacity, and also a high blood compatibility, it turned out to be the basic component for a potential blood substitute material.

However, cell-free Hb tends to dissociate into its individual alpha-beta dimers, and the latter are rapidly eliminated from the circulatory system causing kidney damage. In addition, cell-free Hb lacks an essential effector, 2,3-diphosphoglycerate (DPG). In the absence of the latter, Hb binds oxygen, yet prevents its appropriate release to the tissues.

In order to use Hb, it has to be either molecularly modified or encapsulated. Chemical modifications of Hb have been devised to prevent its dissociation into dimers, to decrease its high oxygen affinity and to increase the retention time in circulation. This route of research requires the use of specifically designed cross-linker reagents necessary to modify the Hb protein.

Taken in part from the Ph.D. thesis of S. Sivan, Technion – Israel Institute of Technology, Haifa, Israel (2000).

To avoid the problems associated with blood transfusion and to meet the need for a prehospital resuscitation fluid, acellular Hb-based solutions are being extensively developed as clinical blood substitutes. Currently, several Hb-based formulations have entered FDA Phase II or III clinical trials (34).

In this series of accounts (32,33), we adduce results of studies addressed to the design and synthesis of NAD-modified Hb for the purpose of developing an oxygen-carrying material. The modifying reagent is aimed at the 2,3-DPG pocket of Hb, in order to mimic the effect of 2,3-DPG.

The choice of NAD-based reagents in general and oxidized-NAD (o-NAD) in particular is based on the following considerations: (a) NAD-based molecules are biocompatible, therefore decreasing possible side effects; (b) under physiological conditions, these molecules are negatively charged and, therefore, will be easily attracted to the positively charged 2,3-DPG pocket of Hb, leading to homogeneous products; (c) the flexible backbone and the four aldehydic groups of o-NAD will help to ensure the attachment of this molecule to the nucleophilic functional groups in the 2,3-DPG pocket of Hb; (d) these molecules are abundant and relatively cheap, which are important features for their use in a large-scale process.

With molecular engineering techniques, we have developed a new family of Hb-modifying reagents. It is based on nicotinamide-adenine dinucleotide (NAD), in which the two sugar rings are opened by periodate oxidation, leading to formation of four aldehydic groups (32,33). These cross-linkers, specifically, o-NAD, contain two phosphate groups and a flexible backbone and, as such, may serve as 2,3-DPG analogues.

Encouraged by the modeling results (33), o-NAD was reacted with deoxyHb. The cross-linked product mixture, HbNAD, was then purified, characterized, and further polymerized to obtain a final product, poly-HbNAD. Both HbNAD and poly-HbNAD solutions were characterized in terms of their oxygen affinity and oxygen-binding and releasing properties, as well as their viscosity and colloidal osmotic (oncotic) activity.

HbNAD and poly-HbNAD were found to have relatively low to normal oxygen-delivering capacity at 37°C (P_{50} values of 37.2 and 24.1 mmHg, respectively). Poly-HbNAD was found to be a mixture characterized by molecular masses ranging from 64 kD to 500 kD and a normal COP value (13.4 mmHg obtained for a 6% preparation).

2. MATERIALS

Human Hb, ferrous A_0 (Hb A_0), and oxidized beta-NAD were purchased from Sigma Chemical Co. (USA). Preswollen, microgranular anion exchanger diethylaminoethyl cellulose (DE-52) was obtained from Whatman Inc. (USA) and low-molecular-weight protein standards from New England BioLabs Inc. (USA). The HP-GPC column (Fractogel TSK-G3000) was obtained from TosoHaas (USA), and the C_4 reversed-phase column (Vydac 214TP^{TM}) was a product of Vydac (USA). All other reagents used were of analytical grade.

3. METHODS

3.1. Spectrophotometry

UV/Visible absorption spectra were measured with a Spectronic-2000 spectrophotometer (Bausch & Lomb, Rochester, NY) or a HP 8452A Diode-Array Spectrophotometer (Hewlett Packard, USA) supported by the HP 89532A UV-Visible Operating Software. Both were equipped with 1-cm light path-length quartz cuvettes.

3.2. High-Performance Liquid Chromatography (HPLC)

High-performance liquid chromatography (HPLC) experiments were performed with a system equipped with a model 2152 solvent-delivery controller (LKB, Sweden), a model 2150 dual-piston pump (LKB, Sweden), a model 2156 solvent conditioner, and a variable wavelength detector (Knauer, Germany) equipped with a 1-cm light path flow cell (10-µl internal volume) or by using a HPLC Model 1100 system (Hewlett-Packard, USA) equipped with a diode array detector and an automatic autosampler. With the latter, data was collected and analyzed with the HP_chemstation Software, version A.05.01.

3.3. Size Exclusion Chromatography (SEC)

Size exclusion chromatography (SEC) was used to estimate the molecular weight distribution. Protein solutions were chromatographed on a size exclusion column (7.5 mm i.d. × 600 mm TSK3000SW, TosoHaas, USA). Elution was carried out at a flow rate of 1 ml/min, at room temperature, with 1 M $MgCl_2$ or with 0.02 M Tris pH 8.0 buffer, also containing 50 mM NaCl. Column outflow was monitored at 280 nm. A calibration curve (log MW vs. elution volume), produced from nonheme proteins ranging from 13.7 kD to 440 kD, was used throughout.

3.4. Reversed-Phase Chromatography (RPC)

Separation of the globin chains was carried out by reversed-phase chromatography (RPC). Samples were loaded onto a Vydac 214TP^{TM} C_4 reversed-phase column (4.6 mm i.d. × 250 mm, silica, particle size 5 µ, pore diameter 300 angstrom), and eluted by employing a gradient of 35–60% acetonitrile in H_2O, also containing 0.1% TFA. Flow rate of 1 ml/min was used over a period of 60 minutes. Readings were taken at 220 nm and also at 576, 540, and 404 nm.

3.5. Ion Exchange Chromatography (IEC)

Protein solutions were loaded onto an anion exchange column (10 mm i.d. × 250 mm) packed with diethylaminoethyl-cellulose (DE-52, Whatman) and eluted by an increasing piecewise linear gradient of 30–500 mM NaCl. The column outflow was monitored spectrophotometrically at 280 nm, 404 nm, 540 nm, and 576 nm.

3.6. Sodium Dodecyl Sulfate Polyacrylamide Gel Electrophoresis (SDS-PAGE)

SDS-PAGE was performed essentially as reported (57), with 12% polyacrylamide slab gels. Gels were stained with coomassie blue R-250 and destained by a mixture solution of 45% methanol and 7.5% acetic acid. Molecular weight and the amount of the proteins present in each spot were determined by scanning the gel with a Bio-Rad GS-690 densitometer (Bio-Rad Laboratories, USA).

3.7. Viscometry

Measurements were carried out in a Cannon–Ubbelhode semimicro dilution type, size 50 capillary viscometer (Cannon Instrument Co., USA), at 37°C, with a 10 mM phosphate-buffered saline (PBS) pH 7.4 as solvent.

The following definitions were used:

Relative viscosity (η_r): $\eta_r = t/t_0$, where t and t_0 are the time required for the solution and solvent, respectively, to flow through the capillary.

Specific viscosity (η_{sp}): $\eta_{sp} = \eta_r - 1 = (t-t_0)/t_0$.

Intrinsic viscosity ($[\eta]$): $[\eta] = (\eta_{sp}/C)_{c=0}$, where C is the solute concentration.

Intrinsic viscosity, $[\eta]$, was determined by measuring relative viscosities (η_r) at a number of solute concentrations, C, by plotting the data as η_{sp} vs. C, and, finally, by extrapolating the best-fit line to infinite dilution (i.e., to the intercept with the y-axis).

The reduced viscosity, η_{red}, is related to C by a linear relationship as follows:

$$\text{Reduced viscosity } (\eta_{red}): \quad \eta_{red} = \eta_{sp}/C = aC + b.$$

The dynamic viscosity of a fluid, μ, is related to the concentration of the solute, C, and to the relative viscosity, η_r, by a second-order polynomial equation, as follows:

$$\text{Relative viscosity } (\eta_r): \quad \eta_r = \mu/\mu_{solvent} = aC^2 + bC + 1.$$

$$\text{Dynamic viscosity } (\mu): \quad \mu = \alpha_2 C^2 + \alpha_1 C + \alpha_0.$$

Where:

$\alpha_2 = a\mu_{solvent}$;
$\alpha_1 = b\mu_{solvent}$;
$\alpha_0 = \mu_{solvent}$.

3.8. Osmometry

Measurements were carried out essentially as reported earlier (58). 2 ml samples, in a 10 mM PBS buffer pH 7.4, were injected to a Wescor Model 4100 colloid osmometer (Wescor Inc., USA), equipped with a 32 kD cutoff membrane. Calibration of the osmometer was carried out with 2 ml of a 5% human albumin solution in saline (standard solution), at room temperature. For this solution, the colloid osmotic pressure (COP) obtained was 19.8 mmHg (expected COP value is 19.5 ± 1 mmHg).

3.9. Hemoglobinometry

This section addresses itself to the determination of the concentration of solutions of Hb and of its derivatives by means of chemical and spectrophotometric methods, as well as to the characterization of the oxygen-binding properties of these materials (59).

3.9.1. Cyanomethemoglobin Method (59).
Hb solution (0.02 ml) is added to 5 ml of working reagent (prepared from 200 mg $K_3Fe(CN)_6$, 50 mg KCN, and 140 mg KH_2PO_4 pH 7–7.4, all dissolved in 1 liter) and allowed to stand for 3–5 minutes. The absorbance value is then read at 540 nm, using water or the working reagent as the blank. Hb content is then calculated using the following equation:

$$c(\text{g/liter}) = A(540)_{Hb^+ - CN^-} \times F \times M / \varepsilon_{Hb^+ - CN^-} \times l. \quad (1)$$

In this equation, $A(540)_{Hb^+ - CN^-}$ is the absorbance measured at 540 nm; F = dilution factor (251); M = molecular weight of Hb monomer (16114.5); $\varepsilon_{Hb^+ - CN^-}$ = millimolar extinction coefficient of Hb^+-CN^- monomer at 540 nm (11.00 $mM^{-1} cm^{-1}$); and l = light path length in cm.

3.9.2. Multiwavelength Spectrophotometry.
The concentrations of oxyHb and metHb were assessed using the multiwavelength spectrophotometry approach as described below:

A. OxyHb concentration was determined using the millimolar extinction coefficients of $\varepsilon = 13.8$ $mM^{-1} cm^{-1}$ at 541 nm, or $\varepsilon = 14.6$ $mM^{-1} cm^{-1}$ at 577 nm. These coefficients are on a heme basis (60)

B. MetHb (Hb^+) content as a percent of total Hb was assessed by the method described by Asakura et al. (61). According to this method, metHb fraction was calculated using the following equation:

$$\text{MetHb}(\%) = 100 * (15.5 - 0.33R)/(12.1 + 3.97R), \quad (2)$$

where R is the absorption ratio A_{578}/A_{630}.

3.10. Hb-Oxygen Dissociation Analysis

Measurements were performed with a model B Hemox-analyzer (TCS-Medical products Co., USA) equipped with an optical cell containing an oxygen electrode and a thermistor. Oxygen affinity and cooperativity were outlined in terms of oxygen pressure at half saturation (P_{50}) and Hill coefficient, using the Hemox Analytical Software (HAS).

Measurements were performed at two stages. First, Hb solution (0.5 mM, on a heme basis) was fully deoxygenated with nitrogen at 37°C (± 0.1°C) in a 0.1 M KH_2PO_4 buffer pH 7.0, also containing 0.5 mM EDTA. Then, the sample was slowly oxygenated with air, while measuring the oxygen saturation as a function of the oxygen partial pressure.

PO_2 was also determined, using the StatProfile-5 Blood Analyzer (*Nova* Biomedical, USA). Hb samples, dissolved in 0.1 M Bis Tris pH 7.4, containing 0.15 M NaCl and 5 mM EDTA, were first deoxygenated by sodium-dithionite, and the measurements were repeated for each sample with increasing amounts of oxygen. All measurements were taken at 37°C.

Oxygen saturation (%) and metHb content (%) were measured by the IL-482 CO-Oximeter (Instrument Laboratories, USA).

4. RESULTS

4.1. Preparation of HbNAD

4.1.1. Procedure. Hb A_0 solution (1.55×10^{-6} moles in 50 mM Bis Tris buffer, pH 7.0) was first deoxygenated for 6 hours under nitrogen in the presence of 10 μl of caprilic alcohol. Separately, o-NAD solution (1.55×10^{-6} moles in 100 μl 50 mM Bis Tris buffer, pH 7.0) was also deoxygenated for 1 hour under nitrogen and then mixed with the deoxygenated Hb A_0 solution. The reaction was allowed to proceed for 24 hours at 4°C under nitrogen. Subsequently, reduction was carried out at 4°C overnight, under nitrogen, with $NaBH_4$ (3×10^{-5} moles in 50 mM Bis Tris buffer, pH 7.0, deoxygenated for 1 hour under nitrogen). Finally re-dilution and concentration of the reaction mixture was carried out in a Centricon-10 filter (Millipore, USA), by centrifugation at 5000 g for 2 hours at 4°C either in 0.2 M glycine buffer pH 8.0, or 10 mM PBS buffer pH 7.4, or 0.1 M Bis Tris buffer pH 7.4 also containing 0.15 NaCl and 5 mM EDTA, or in 1 M $MgCl_2$ according to subsequent needs. As a result, a HbNAD-containing mixture is obtained. In this study, this mixture is referred to as HbNAD.

4.1.2. Size Exclusion Chromatography (SEC). High-salt solutions, like magnesium chloride, are known to completely dissociate native Hb into its monomers, whereas in cross-linked Hb, this is fully prevented (12). This fact was used in this study to separate covalently cross-linked Hb from the unmodified Hb by passage through a SEC column in 1 M $MgCl_2$.

Samples of HbNAD, (20 μl, 0.1 mg/ml in 1 M $MgCl_2$) and of unmodified Hb (20 μl, 0.1 mg/ml in 1 M $MgCl_2$) were diluted (100 times) with 1 M $MgCl_2$ and analyzed on a SEC column, also equilibrated with 1 M $MgCl_2$. Separations were carried out at room temperature at a flow rate of 1 ml/min. The column outflow was monitored at 280 nm.

The results obtained (Fig. 1) indicate that unmodified Hb yields a single peak eluting at 25 ml (peak 3), which corresponds to a MW of 16 kD. HbNAD, on the other hand, shows a different elution pattern, containing three main peaks; these are eluted at 25 ml (peak 3), 23.5 ml (peak 2), and 21.5 ml (peak 1), indicating MW components of 16, 32, and 48 kD, respectively. The 16 kD peak represents the free, unmodified monomers; the 32 kD peak represents the cross-linked dimer (alpha-alpha or beta-beta), and the 48 kD peak represents a triply cross-linked entity. On comparison of these two chromatograms and integration of the area under the peaks, the yield of the cross-linking process is estimated at about 75–80%.

4.1.3. Optimization of the Cross-Linking Procedure. Attempts were made to reach for the optimal reaction conditions in terms of molar ratio of Hb to cross-linker. An optimal reaction is considered as the one for which maximal modification occurs specifically on the beta chains,

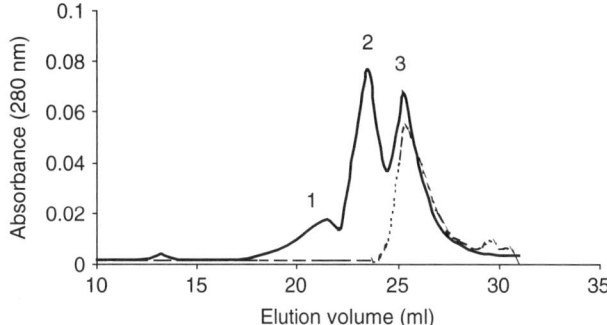

Figure 1. Size exclusion chromatography profile of: (—) modified Hb solution (HbNAD, 20 μl, 0.1 mg/ml in 1 M $MgCl_2$), prepared as indicated in Procedure section of Results; (- - -) unmodified Hb solution (20 μl, 0.1 mg/ml in 1 M $MgCl_2$). Samples were loaded to a column (7.5 × 600 mm, TSK3000SW, TosoHaas, USA) also equilibrated with 1 M $MgCl_2$. Separations were performed at room temperature at a flow rate of 1 ml/min. The column outflow was monitored at 280 nm. Peaks (1), (2), and (3) represent fractions of molecular masses of 48, 32, and 16 kD, respectively.

which will consequently lead to the disappearance of the latter accompanied with minimal (if any) disappearance of the alpha chains. Also, the appearance of new products is expected to complement only for the disappearance of the beta chains.

Attempts to optimize the cross-linking reaction were made by carrying out the reaction at various Hb (tetramer basis):o-NAD molar ratios. All other reaction parameters (i.e., temperature, environment, and duration of reaction) were kept constant.

Experiments were carried out at 1:1, 1:2, 1:3, 1:4, and 1:10 molar ratios of Hb (tetramer basis) to o-NAD, respectively. The reaction products were analyzed on a Vydac 214TP™ C_4 reversed-phase column, as described in the RPC section.

The results obtained are presented in Fig. 2. The elution volume of the globin chains (alpha and beta), as well as of the heme group, were in accordance with previously published work, using the same column under similar conditions (62). Significantly, the peaks indicating the presence of higher molecular weight fractions (elution volumes between 22 and 27 ml) confirm that modification of Hb took place at all molar ratios. Disappearance of the beta peak and appearance of new peaks (modified Hb; i.e., HbNAD) occurs, which suggests that the modification took place at several sites of the Hb molecule. Also, a significant new peak (eluted at about 27 ml) is observed at 1:4 and 1:10 ratio. This peak is significantly smaller at lower ratios.

The optimal ratio of Hb to o-NAD was delineated at about 1:3.5 (Fig. 3). We note that maximal difference of about 40% between the beta and alpha peaks is achieved at this molar ratio. Also, at this molar ratio, the formation of new products complements for the disappearance of beta chains only. Above this molar ratio, the alpha chains are also modified. At molar ratios of 1:3.5 and above, the conversion of native Hb to cross-linked Hb is up to 90%. Following this analysis, all further cross-linking processes were performed at 1:3.5 molar ratio of

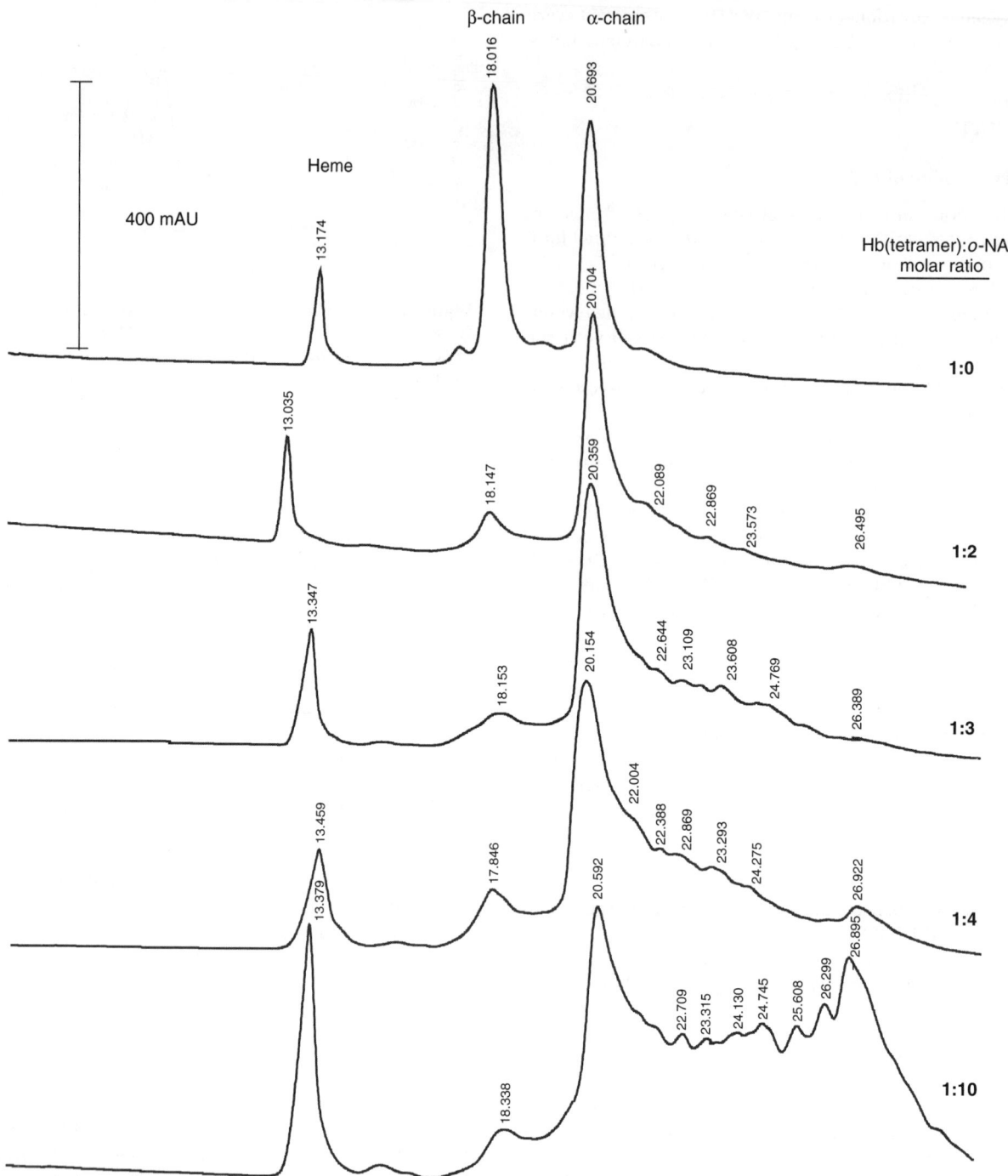

Figure 2. Optimization of cross-linking reaction in terms of molar ratio of Hb (tetramer-basis) to cross-linker (o-NAD), as indicated. The separations were performed on a Vydac 214TP^{TM} C$_4$ reversed-phase column as described in the RPC section. Monitoring was performed at 220 nm. The numbers on the peaks indicate the pertinent elution volume. (This figure is available in full color at http://www.mrw.interscience.wiley.com/ebe.)

Hb (tetramer basis) to o-NAD, using the reaction conditions described in Table 1.

4.2. Preparation of Polymerized HbNAD (Poly-HbNAD)

Poly-HbNAD was prepared by reacting 1 ml of HbNAD solution (0.1 g/ml, in standard kidney dialysis buffer, pH 7.4, deoxygenated for 2 hours under nitrogen in the presence of 10 μl of caprilic alcohol) with 0.75 mg glutaraldehyde in the same buffer (also deoxygenated for 1 hour under nitrogen). The polymerization was allowed to proceed at 25°C for 3 hours under nitrogen. Quenching of the remaining aldehydic groups was carried out for 1 hour using ethanolamine (0.06 mmoles) also dissolved in a standard kidney

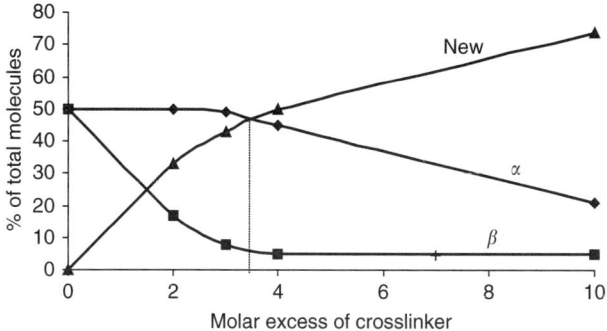

Figure 3. Optimization of the cross-linking reaction in terms of percentage of total molecules (alpha + beta chains) as a function of molar excess of cross-linker to Hb tetramer. The molar ratio of 1:3.5 (Hb:cross-linker) is marked as dashed line.

dialysis buffer, pH 7.4. Subsequently, reduction of the products obtained was carried out for 2 hours using $NaBH_4$ (3.1×10^{-5} moles dissolved in 1 mM NaOH and deoxygenated for 1 hour under nitrogen). Finally, concentration of the poly-HbNAD-containing mixture was carried out with a Centricon-10 filter (Millipore, USA) at 5000 g for 2 hours at 4°C either in 10 mM PBS buffer pH 7.4 or in 0.1 M Bis Tris containing 0.15 M NaCl and 5 mM EDTA, pH 7.4, or in 0.02 M Tris buffer pH 8.0 also containing 0.05 NaCl, according to subsequent needs. As a result of this procedure, poly-HbNAD solution was obtained.

4.3. Purification and Characterization of HbNAD

4.3.1. Ion Exchange Chromatography (IEC). Separation of the individual components comprising the HbNAD solution was carried out by IEC (IEC section) and by heat treatment (Heat treatment section). The results obtained are presented in this section and in the next section, respectively.

The HbNAD sample (3 mg in 0.5 ml of 0.2 M glycine buffer pH 8.0, containing 30 mM NaCl) was analyzed by IEC. A piecewise linear gradient of 30–500 mM NaCl was employed as the eluting solvent, with a flow rate of 0.3 ml/min at room temperature. The fractions collected were analyzed for their absorption at 576, 540, 404, and 280 nm.

The results (readings at 404 nm) presented in Fig. 4 suggest that the cross-linking process does not yield a single product, but a mixture of at least five of them. Fractions eluted at 12–30 ml (peak 1), 100–130 ml (peak 2), 180–210 ml (peak 3), 230–290 ml (peak 4), and 310–330 ml (peak 5) were individually pooled and diafiltrated against

Table 1. Optimal Conditions for the Reaction of Hb with o-NAD

Parameter	Value
Total Hb concentration (g/dL)	5–10
MetHb concentration (% of total Hb)	Below 10
Deoxygenation under nitrogen (hours)	5–6
Molar ratio (Hb tetramer:o-NAD)	1:3.5
pH	6.8–7.0
Temperature (°C)	4°C
Reaction duration (hours)	3–24

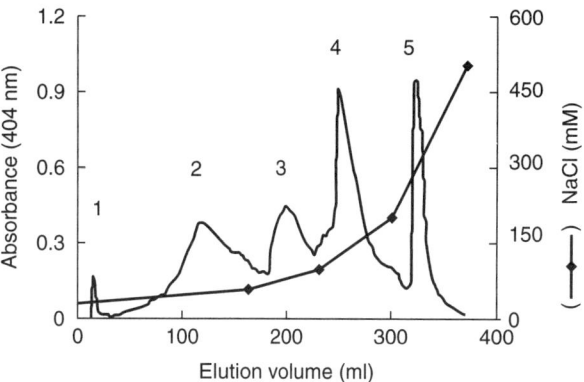

Figure 4. Ion exchange chromatography profile of the cross-linked products, using a DEAE-cellulose (DE-52) column (10 × 250 mm). Separation was performed at room temperature using a piecewise linear gradient of 30–500 mM NaCl in 0.2 M glycine buffer, pH 8.0, at a flow rate of 0.3 ml/min. The column outflow was monitored at 404 nm.

20 volumes of 0.1 M Bis Tris buffer, pH 7.4, also containing 0.15 M NaCl and 5 mM EDTA, using a Centricon-10 filter (Millipore, USA). Subsequently, each fraction was characterized by SDS-PAGE and reversed-phase chromatography (SDS-PAGE and RPC sections to follow), as well as for its oxygen affinity (P_{50}), viscosity, and metHb content (Physico-Chemical Properties of HbNAD and Poly-HbNAD section).

4.3.2. Heat Treatment. One of the most convenient methods for separating cross-linked from non cross-linked Hb is by heat treatment (63,64).

Heat treatment of Hb-containing solutions followed essentially the procedure described earlier (63) with slight modifications. Sodium-dithionite (0.121 g) was slowly added to HbNAD solution (0.1 g dissolved in 10 ml 10 mM PBS buffer pH 7.4). The reaction mixture was sealed into airtight vials and heated to 80°C for 1 hour.

Recovery of the heat-treated sample was carried out by centrifuging at 5000 g for 10 minutes. The supernatant was re-diluted and then concentrated in 0.1 M Bis Tris buffer, pH 7.4 also containing 0.15 M NaCl and 5 mM EDTA.

HbNAD and Hb solutions were tested spectrophotometrically and by SDS-PAGE for their molecular integrity. Absorption spectra of unmodified and modified Hb were recorded and compared (Fig. 5). One can see that the

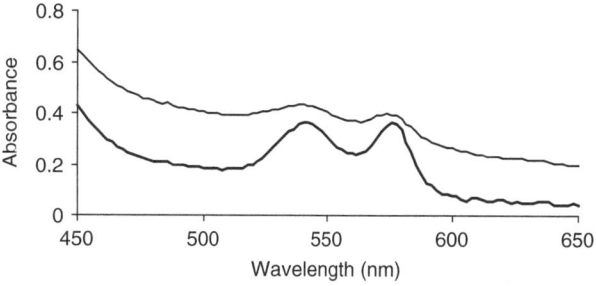

Figure 5. Absorption spectra of unmodified Hb (−) and heat-treated HbNAD (−).

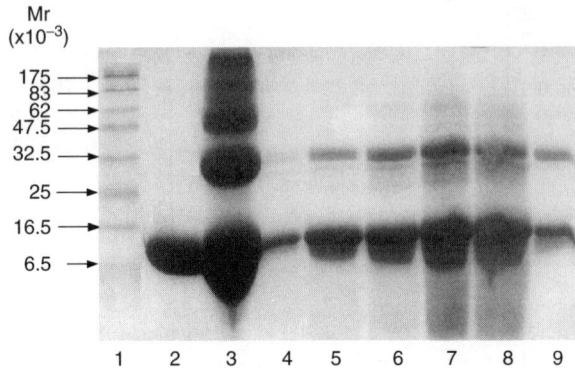

Figure 6. SDS-PAGE analysis of cross-linked products obtained by separation on DEAE-cellulose (IEC section) and after the heat treatment process (Heat treatment section). Lane (1) - molecular markers (15 μl, New England BioLabs, USA); lane (2) - unmodified Hb (50 μg); lane (3) – HbNAD (170 μg); lanes (4–8) – fractions collected as peaks 1–5, respectively, (see Fig. 4), for amounts loaded of 19, 25, 39, 66, and 62 μg, respectively; lane (9) – HbNAD after heat-treatment (22 μg). Electrophoresis was performed on a 12% SDS-polyacrylamide gel, at 30 mA for 45 minutes at room temperature, as described in SDS-PAGE section. (This figure is available in full color at http://www.mrw.interscience.wiley.com/ebe.)

absorption maxima at 540 and 576 nm, exhibited by HbNAD after heat treatment, are similar to those obtained with the unmodified Hb. Also, the electrophoretic analysis shows that bands obtained for samples that were not heat-treated also appear in the heat-treated sample (Fig. 6, lane 9). These results show that no deterioration of HbNAD took place during the heat treatment process.

4.3.3. SDS-PAGE. Samples obtained after separation by IEC and by the heat treatment processes (IEC and Heat treatment sections, respectively) were analyzed by SDS-PAGE. The results obtained are presented in Fig. 6. Unmodified Hb (lane 2) yields a single band, characteristic to the free chains of Hb, each of 16 kD. Samples of modified Hb (lanes 3–9) clearly point to the presence of cross-linked products with higher molecular weights (multiples of 16 kD), indicating that multiple cross-bridges have been formed during the cross-linking reaction.

The samples eluted on IEC (lanes 4–8, corresponding to peaks 1–5, respectively, in Fig. 4) show two dominant bands, one related to the free Hb chains of 16 kD, and the other to the cross-linked dimer of 32 kD. Additional bands exist between these two bands, suggesting possible modifications, with no actual bridging of the dimers. Higher molecular weight bands suggest that cross-linking also occurred at multiple sites of Hb simultaneously. The heat-treated sample (lane 9) shows only bands that are multiples of 16 kD, two of which are dominant: of 16 and 32 kD. The relative amounts of proteins in the individual bands were assessed by densitometry, from which the yield of the reaction was determined to be about 75%, which agrees with previous results obtained by SEC (SEC section).

These results and the ones obtained in the Heat treatment section suggest that the heat treatment process is indeed efficient in precipitating the noncross-linked Hb (i.e., only products containing intramolecular linkages but not intermolecular ones). This process is considered as a purification step in the recovery of cross-linked Hb. The yield of the product obtained is about 80%.

4.3.4. Reversed-Phase Chromatography (RPC). RPC was used to locate the modification site on the Hb molecule. Samples of unmodified Hb, HbNAD, as well as fractions eluted from the DEAE-cellulose column (IEC section) and heat-treated ones (Heat treatment section) were analyzed with a Vydac C_4 reversed-phase column. Separations were performed essentially as described in the RPC section, and readings were taken at 220 nm. The results obtained are depicted in Fig. 7. We note that unmodified Hb (Fig. 7a) shows three distinct bands, which correspond to the heme group and to the beta and alpha chains of Hb, respectively. These results agree with previously reported results (65). Moreover, for all the reaction products (Fig. 7b–g), a drastic decrease in the intensity of the beta chains peak is observed, whereas the alpha chains peak remained virtually unchanged, which is clear evidence that the modification specifically occurred on the beta chains of the Hb molecule. This decrease in the content of beta chains is complemented by the appearance of new peaks emerging at higher elution volumes. The chromatogram representing the first peak eluted from the DEAE-cellulose (Fig 7b) indicates that this material is also modified at the beta chain, which is in contradiction to previously published works, in which the first eluted peak is believed to correspond to unmodified Hb (11,20,37). One possible explanation is that, under these conditions, beta chains of unmodified Hb underwent a slight modification i.e., attachment of o-NAD moieties (but not a cross-linking), which changed their hydrophobicity, but not their molecular weight, which is demonstrated by the 16 kD peak (Fig. 1).

The decrease in the intensity of the beta peak in the modified sample relative to that in the unmodified Hb is expressed in a change of the ratios of the areas under the peaks from 1:1 (beta:alpha) in the unmodified Hb to 1:9 in the modified sample (HbNAD), suggesting that up to 90% of the Hb is modified.

4.4. Characterization of Poly-HbNAD

A sample from the polymerization reaction mixture (0.2 ml from a reaction solution of 0.01 g/ml, dissolved in 0.02 M Tris buffer, pH 8.0, also containing 0.05 M NaCl) was analyzed by SEC, using a column, also equilibrated with the same buffer (SEC section). Separation was carried out at room temperature at a flow rate of 1 ml/min. The column outflow was monitored at 280 nm.

The results obtained are presented in Fig. 8. Based on the molecular weight calibration curve, we note that polymerization of HbNAD yielded a mixture of products ranging from 64 kD to about 500 kD. Four major fractions were collected: (I) MW of 64 kD; (II) MW of 128 kD; (III) of 192–256 kD and (IV) above 320 kD. A rough estimation, based on the area under the peaks, reveals that some 40% of the sample is in the form of tetramers (fraction I), 40% is in the form of oligomers (fractions II and III), and 20% are in the form of polymers (fraction IV). Based on these results, we conclude that the yield of the polymerization reaction is 60%, with some 40% of the molecules remaining in the form of tetramers.

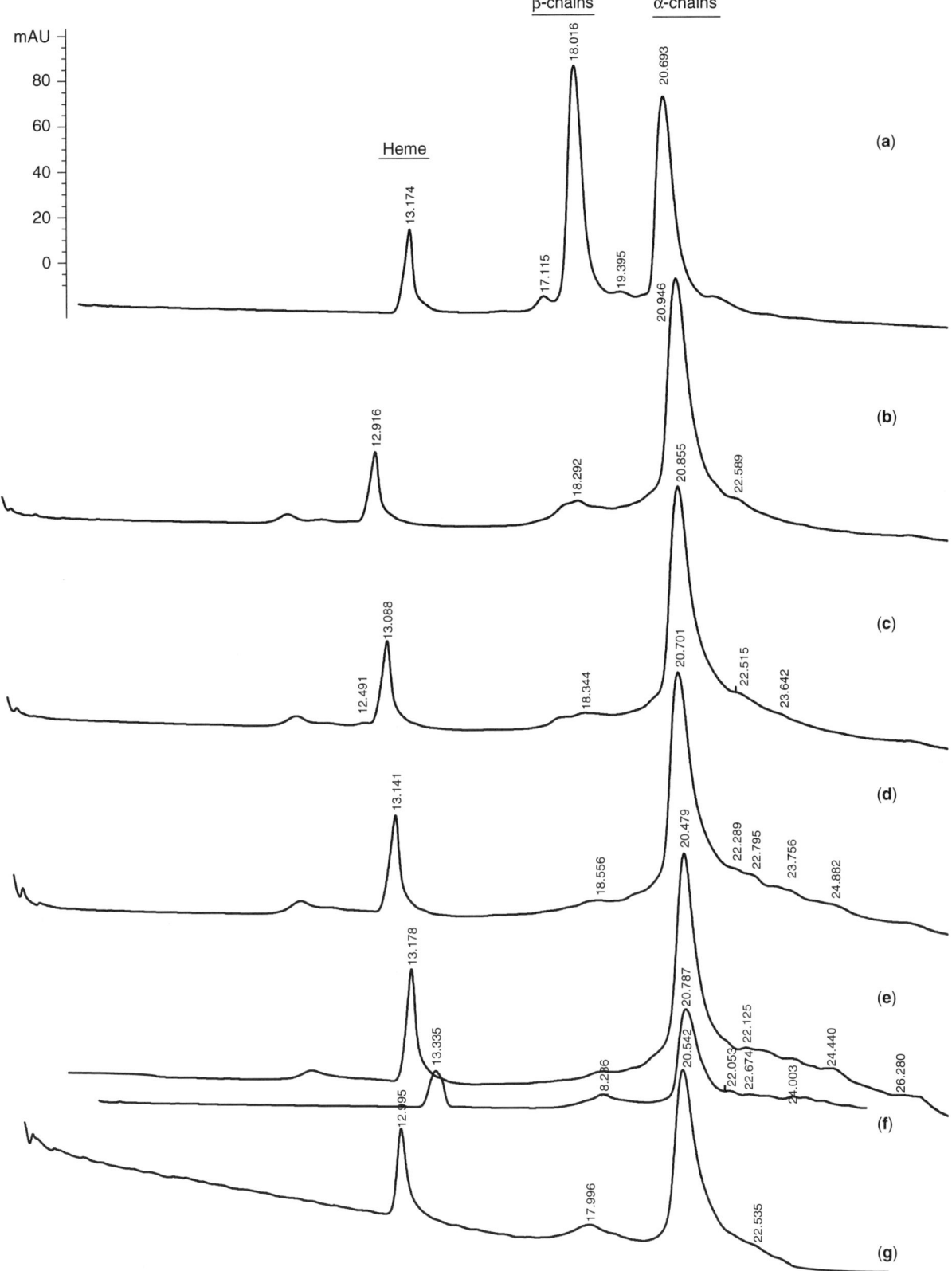

Figure 7. Reversed-phase chromatographic elution profile of: (a) - unmodified Hb; (b–f) - peaks eluted from DEAE-cellulose column (fractions 1–5, respectively, Fig. 4); and (g) - heat-treated HbNAD. Separations were performed on a Vydac 214TP™ C₄ reversed-phase column by employing a linear gradient of 35–60% acetonitrile in H_2O, also containing 0.1% TFA as the eluting solvent, at a flow rate of 1 ml/min, over a period of 60 minutes, as indicated in the RPC section. The column outflow was monitored at 220 nm.

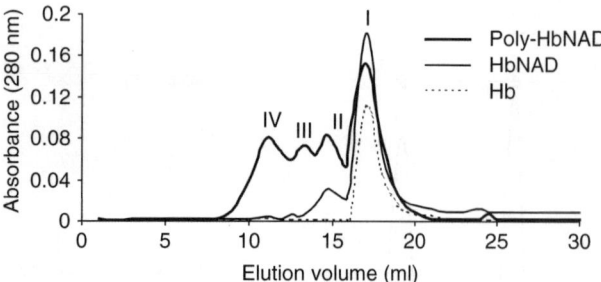

Figure 8. Size-exclusion chromatography profile of (- - - - -) unmodified Hb; HbNAD (—), and poly-HbNAD (—). Samples, dissolved in 0.02 M Tris buffer, 0.05 M NaCl, pH 8.0, were loaded onto a TSK3000SW column (7.5 × 600 mm, TosoHaas, USA) also equilibrated with the same buffer. Separations were performed at room temperature at a flow rate of 1 ml/min. The column outflow was monitored at 280 nm. Peaks I–IV are mentioned in the text.

4.5. Physico-Chemical Properties of HbNAD and Poly-HbNAD

4.5.1. Oxygen-Binding Properties. Samples of unmodified Hb, HbNAD, and poly-HbNAD were analyzed for their oxygen-binding properties, as described in the Hb-Oxygen Dissociation Analysis section. The pertinent oxygen-binding curves and P_{50} values are presented in Fig. 9 as well as in Tables 2 and 3. Oxygen equilibrium data (Fig. 9) were also plotted according to the Hill equation (Equation 3), and the results obtained are presented in Fig. 10.

$$\log Y/(1-Y) = n \log PO_2, \quad (3)$$

where Y is the fractional saturation of Hb with oxygen, and n, the slope of this line, is a measure for subunit cooperativity of the Hill coefficient. The x-intercept of this line at Y = 0.5 gives the P_{50} value.

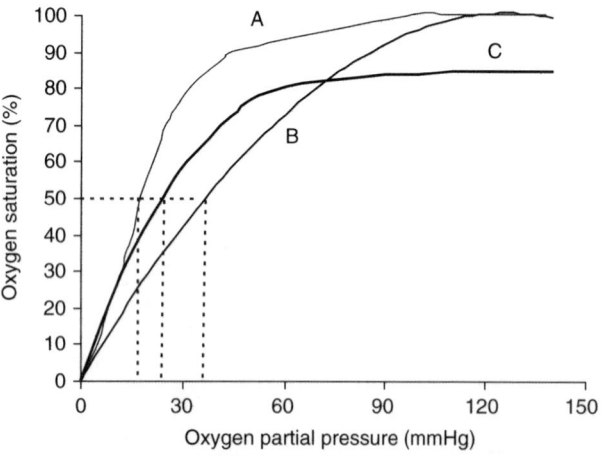

Figure 9. Oxygen-binding curves of: (A) unmodified Hb (sample #2 in Table 2); (B) HbNAD (sample #3 in Table 2), and (C) poly-HbNAD (sample #9 in Table 2). Curves were obtained with an IL-482 CO-Oximeter (International Labs, USA) at 37°C (±0.1°C) in 0.1 M Bis-Tris, pH 7.4, also containing 0.15 M NaCl and 5 mM EDTA. The broken lines at 50% O_2-saturation give the corresponding P_{50} values. (This figure is available in full color at http://www.mrw.interscience.wiley.com/ebe.)

The P_{50} value obtained for the HbNAD preparation is high (37.2 mmHg), indicating low affinity to oxygen. P_{50} obtained for poly-HbNAD (24.1 mmHg) is close to that of whole blood (26.6 mmHg). Also, hyperbolic rather than sigmoidal curves were obtained for HbNAD and poly-HbNAD, with Hill coefficients of 1.95 and 1.07, respectively (Fig. 10 and Oxygen-Binding Properties section).

4.5.2. Viscosity.

4.5.2.1. Intrinsic Viscosity. The viscosity of Hb, HbNAD, and poly-HbNAD in 10 mM PBS buffer, pH 7.4, at 37°C was measured as described in Viscometry section, and the results obtained are depicted in Fig. 11. As shown, the intrinsic viscosity of Hb and HbNAD are not much different from one another, with values of 3.6 and 4.3 ml/g, respectively. The intrinsic viscosity of poly-HbNAD is higher, 9.5 ml/g, as expected for a higher molecular weight material.

4.5.2.2. Dynamic Viscosity. In order to compare the viscosities of Hb, HbNAD, and poly-HbNAD to that of whole blood, the apparent dynamic viscosity of these compounds was estimated, which was achieved as indicated in the Viscometry section, using the value $\mu_0 = 0.69$ cP at 37°C (66). The data obtained are shown in Fig. 12a, and these were fitted by second-order polynomial, as indicated on this figure.

In order to compare these data with the data for blood and blood plasma, the curves in Fig. 12a were extrapolated to the physiological range of Hb concentration (100–150 g/L), using the equation derived in Fig. 12a. The extrapolated curves are shown in Fig. 12b, indicating that the pertinent dynamic viscosities are 1–1.2 cP for Hb, 3–5.8 cP for HbNAD, and 13.9–30 cP for poly-HbNAD. For comparison, the corresponding value for plasma is 1 cP and for blood with hematocrit of 45% is 4 cP, both at 37°C (39,67).

4.5.3. Colloidal Osmotic Pressure (COP). HbNAD (7% (w/v) in 0.1 M Bis Tris buffer, pH 7.4, also containing 0.15 M NaCl and 5 mM EDTA) and poly-HbNAD (6% (w/v) in the same buffer) were characterized for their colloidal osmotic pressure (COP), as indicated in the Osmometry section. The values thus obtained were 24.8 mmHg and 13.4 mmHg, respectively.

5. DISCUSSION

5.1. Microscopic Characteristics

DeoxyHb was reacted with o-NAD to produce a new cross-linked product of Hb i.e., HbNAD. This reaction was validated chromatographically (SEC section). The reaction yield was estimated at above 75% using both SEC (SEC section, Fig. 1) and RPC (RPC section, Fig. 2). SDS-PAGE revealed bands corresponding to molecular masses of 16 kD for the unreacted subunit and also multiples thereof (Fig. 6), thus indicating the formation of cross-linked Hb. Heat treatment experiments of HbNAD revealed a stable

Table 2. Summary of the Oxygen-Binding Properties for the Materials Studied

Sample #	Oxygen Carriers	P_{50} (mmHg)	Hill Coefficient (n)	MetHb % of Total Hb
1	Human blood	26.6	2.7	Up to 5%
2	Unmodified Hb	19.6	2.4	15–20
3	HbNAD	37.2	1.95	25–30
4–8	Peaks b-f (DEAE-Cellulose**)	2.9, 4.2, 4.3, 3.8, and 7.1, respectively	1.0–1.1	ND*
9	Poly-HbNAD	24.1	1.07	25–30

*ND – not determined.
**Peaks b–f correspond to peaks 1–5, respectively, eluted from DEAE-cellulose column.

material (Fig. 5, Fig. 6), suggesting that sterilization is amenable in a large-scale process.

A molar ratio of Hb:o-NAD of 1:3.5 was found to be optimal for the cross-linking reaction, because at this ratio the reaction occurs specifically on the beta chains at maximal rate (Fig. 2, Fig. 3), which is demonstrated by maximal disappearance of the free beta chains, accompanied by minimal disappearance of the alpha chains, and a relatively high-yield formation of new products. At molar ratios above 1:3.5, the alpha chains were also cross-linked. Accordingly, a molar ratio of 1:3.5 was used in the subsequent experiments. Moreover, based on a previous work by Bakker et al. (17), other parameters associated with the cross-linking reaction were kept unchanged.

HbNAD was polymerized to meet physiological requirements associated with colloidal osmotic pressure. The polymerized material, poly-HbNAD, consists of a mixture of modified intermolecularly cross-linked Hb and is characterized by molecular masses ranging from 64 to 500 kD (Fig. 8). At this range of molecular masses, retention of the poly-HbNAD within the blood circulation is achieved. This reaction was proved to be reproducible with a yield of about 60%.

5.2. Oxygen-Binding Properties

Oxygen-binding properties were measured for Hb, HbNAD, and poly-HbNAD. As shown in Fig. 9, the P_{50} of unmodified Hb (19.6 mmHg) is low, suggesting a very high affinity to oxygen, which is expected, because of the absence of the 2,3-DPG effector. On the other hand, P_{50} obtained for the HbNAD preparation is high (37.2 mmHg), indicating that the modification influenced the oxygen-binding properties of Hb in a way that the oxygen equilibrium is shifted toward the low-affinity, T-form conformation of Hb. The fact that cross-linking reagents, which are also DPG analogues, decrease the overall affinity was previously studied (14,37). Oxygen affinity characteristic of poly-HbNAD is close to that of whole blood, with P_{50} of 24.1 mmHg. Also, hyperbolic rather than sigmoidal curves were obtained for HbNAD and poly-HbNAD with Hill coefficients of 1.95 and 1.07, respectively (Fig. 10). The fact that further polymerization increases the oxygen affinity and decreases oxygen-binding cooperativity is well documented (13,16,17,19,39) and is most likely caused by the increased steric hindrance imposed by multiple intermolecular covalent cross-links within the cross-linked Hb and mainly within the polymerized preparations.

5.3. Rheological Properties

Hb solutions generally possess advantageous rheological flow properties because these acellular solutions are more Newtonian in nature than blood (68). The total Hb concentration of these solutions and the degree to which the contained materials are polymerized directly affect the viscosity of the final preparations.

Viscosity of a Newtonian fluid is fully characterized by its dynamic (absolute) and kinematic viscosity. When a complete rheological characterization of a nonNewtonian

Table 3. Macroscopic Properties of HbNAD and Poly-HbNAD

Analytical Parameter	Whole Blood	Hb	HbNAD	Poly-HbNAD
Total Hb content (%)	10–15%	10	7	6
pH at 37°C	7.2	7.4	7.45	7.45
MetHb content (% of total Hb)	3–5	15–20	25–30	25–30
Intrinsic viscosity (ml/g)	- - - -	3.8	4.2	9.3
Dynamic viscosity (cP) (within Hb concentration range of 10–15 g/dL)	Whole blood—4.07 cP Plasma—1 cP	1–1.22	3–5.8	13.9–30
P_{50}	26.6	19.6	37.2	24.1
Hill coefficient (n)	2.7	2.4	1.95	1.07
COP (mmHg)	Normal plasma—22–25	- - - -	24.8 (7%)*	13.4 (6%)*

*Numbers in parantheses indicate the concentration (w/v) of the sample analyzed.

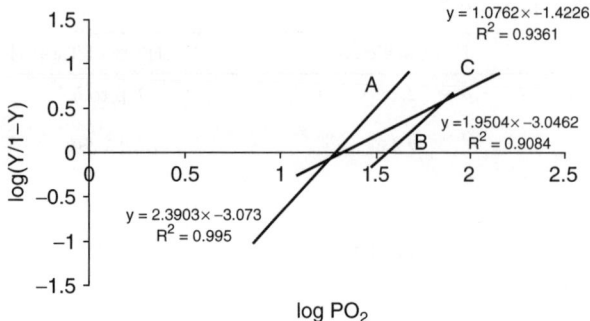
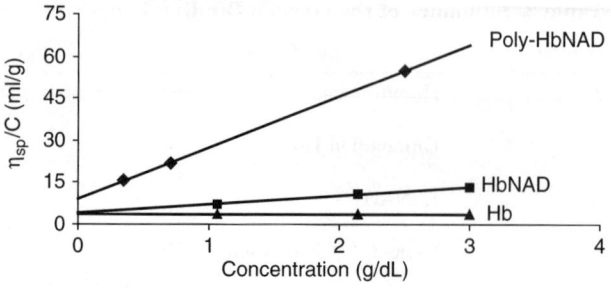

Figure 10. Hill plot of: (A) Hb (sample #2 in Table 2); (B) HbNAD (sample #3 in Table 2); and (C) poly-HbNAD (sample #9 in Table 2). The slopes values are the Hill coefficient, n.

Figure 11. Reduced viscosity of Hb [taken from (66)], HbNAD, and poly-HbNAD, at 37°C, in 10 mM phosphate-buffered saline, pH 7.4. Intrinsic viscosity is the intercept with the Y-axis, where $C=0$.

fluid is required, a stress-strain plot must be determined. Such a complete description is beyond the scope of this study. In the current analysis, relative, reduced, specific, and intrinsic viscosities were used to study both Newtonian and nonNewtonian fluids.

The concentration-dependent reduced viscosity of Hb, HbNAD, and poly-HbNAD at 37°C in 10 mM phosphate buffered saline pH 7.4, is shown in Fig. 11. The intrinsic viscosity value was obtained by extrapolating the linear regression of the reduced viscosity to infinite dilution (i.e., the intercept with the y-axis, where $C=0$). As shown, the rheological behavior of Hb and HbNAD are quiet similar, with intrinsic viscosity values of 3.6 and 4.3 ml/g, respectively. The intrinsic viscosity of poly-HbNAD (9.45 ml/g) is higher, as expected.

The intrinsic viscosity is described by the coefficient of the linear term (b) (Viscometry section), whereas coefficient (a) describes the nonlinear behavior and is related to internal protein-protein interactions within the solution (69). As expected, the nonlinear behavior of poly-HbNAD is dominant (Fig. 11), and therefore, it is impossible to reduce the high-concentration viscosity by only reducing the intrinsic viscosity, represented by the coefficient (b) (Viscometry section).

The dynamic viscosity is well-defined when the rheology of the solution is Newtonian. Poly-HbNAD solution, unlike the Hb one, is less Newtonian, as its concentration

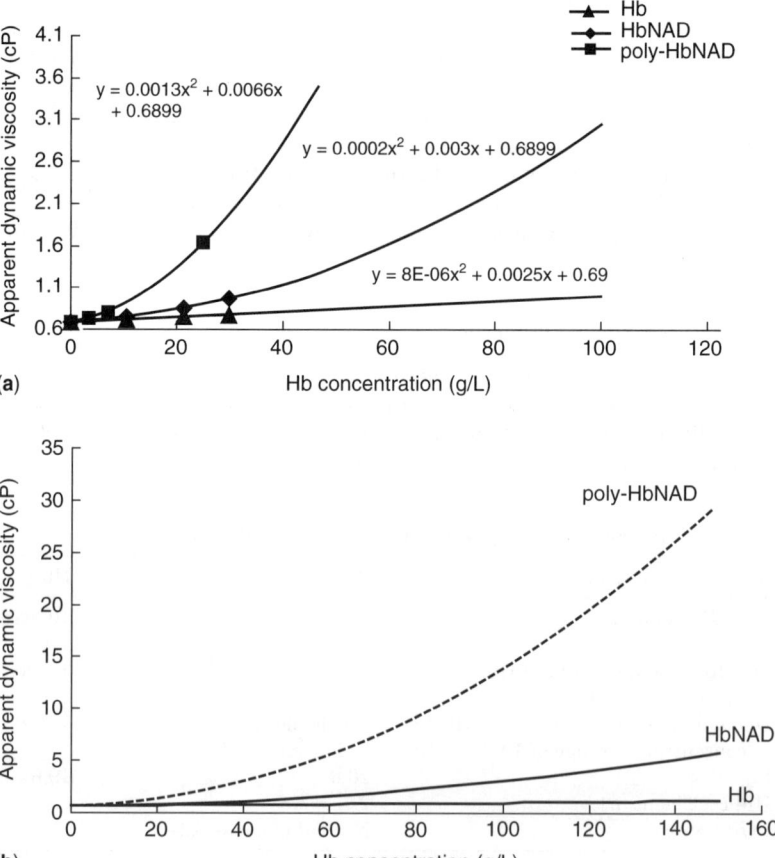

Figure 12. Apparent dynamic viscosity curves of Hb, HbNAD, and poly-HbNAD. Calculated dynamic viscosity values are shown on graph (a). Results of extrapolation to higher Hb concentrations are presented on an extended Hb concentration scale on graph (b).

increases. The term dynamic viscosity will be used here to describe an apparent viscosity for both the Newtonian and the nonNewtonian fluids.

In order to compare the viscosity of Hb, HbNAD, and poly-HbNAD with that of whole blood, the dynamic viscosity was estimated, as described in the Viscometry section. For comparison, a physiological range of 1–4 cP (plasma-whole blood, respectively) was also delineated (67).

In order to estimate the dynamic viscosity behavior of Hb, HbNAD, and poly-HbNAD at higher Hb concentrations (physiological range of 100–150 g/L), the graphs obtained at lower concentrations were extrapolated. The results obtained (Fig. 12) suggest that the dynamic viscosity values obtained for both Hb and HbNAD are within the physiological range (1–1.22 cP and 3–5.8 cP, respectively), whereas that of poly-HbNAD is much higher (13.9–30 cP).

The results, thus obtained, suggest that the next stage design of poly-HbNAD should also include the optimization of this process in terms of choosing those fractions of poly-HbNAD (I to IV, Fig. 8), which also meet with the physiological viscosity limitations.

A complete study of the rheological behavior of the poly-HbNAD preparation was not carried out. For the aim of future design of poly-HbNAD, a stress-strain plot of this preparation must also be determined.

5.4. Osmotic Pressure Properties

The relationship between Hb concentration and the COP depends on the number of colloidal particles and is not linear, tending to deviate to higher COP values at higher protein concentrations. The COP value obtained for HbNAD is similar to that of normal plasma (24.8 mmHg compared with 22–25 mmHg, respectively, Table 3). The only difference is that this value (24.8 mmHg) was obtained for a 7% (w/v) preparation of HbNAD, whereas the physiological Hb concentration is 15% (w/v); in other words, the COP value obtained for HbNAD corresponds to only half the normal Hb concentration in blood.

Consequently, one of the benefits of Hb polymerization is to lower the hyperoncotic nature (excessive COP) of highly concentrated Hb solutions (39). Poly-HbNAD had a significantly lower COP of only 13.4 mmHg at a 6% (w/v) preparation.

Our results of COP of poly-HbNAD versus concentration of Hb are similar to previously reported data for Hb concentrations of upto 7% (w/v) (17). Based on this similarity, it is expected that the COP values of poly-HbNAD, at a physiological Hb concentration (11%, w/v), will also be similar to the pertinent value obtained by Bakker et al. (17), which is 27 mmHg.

6. CONCLUSIONS

One of the basic approaches used in this thorough study was the exploitation of molecular engineering methodologies for the design (32,33) and later synthesis of novel cross-linking reagents for hemoglobin (Hb). This approach proved to be very fruitful, because, from the results obtained, it is evident that poly-HbNAD possesses a physiological oxygen-carrying capacity and oncotic characteristics similar to whole blood. These properties make poly-HbNAD a potential viable candidate as a resuscitation fluid.

Acknowledgment

Taken in Part from the Ph.D. Thesis of S. Sivan, Technion-Israel Institute of Technology, Haifa, Israel (2000). This research was supported in part by grants from the Leonard and Diane Sherman Research Fund, and from the Fund for the Promotion of Research at the Technion (to N.L.), as well as by the fellowship from the Forchheimer Foundation (to S.S.). All are gratefully acknowledged.

We would like to thank Prof. Toshio Asakura from the department of pediatrics at the Children's Hospital of Philadelphia for his generous help with the oxygen equilibrium experiments, Mr. Willy Piva for his help with the HPLC runs, Dr. Helen Rosenberg from the Hemodynamic Laboratory of the Rambam Hospital for her kind assistance with the osmometry experiments, and Prof. Rosa Azhari from the Ort-Braude College at Karmiel for her help throughout this study. One of the authors (S.S) would also like to thak Dr. Uri Shavit from the Dept. of Agriculture Engineering at the Technion, for his assistance in the viscosity studies.

BIBLIOGRAPHY

1. J. S. Jahr, S. B. Nesargi, K. Lewis, and C. Johnson, Blood substitutes and oxygen therapeutics: an overview and current status. *Am. J. Ther.* 2002; **9**:437–443.
2. D. Hunkeler, A. Cherrington, A. Prokop, and R. Rajotte, Bioartificial organs: glossary of terms. *Ann. New York Acad. Sci.* 2001; **944**:7–17.
3. S. Matsumoto and Y. Kuroda, Perfluorocarbon for organ preservation before transplantation. *Transplantation* 2002; **27**:1804–1809.
4. S. A. Gould, A. L. Rosen, L. R. Schgal, H. L. Sehgal, L. A. Langdale, L. M. Krause, C. L. Rice, and W. H. Chamberlin, Fluosol-DA as a red cell substitute in acute anemia. *N. Engl. J. Med.* 1986; **314**:1653–1656.
5. R. M. Kacmarek, Liquid ventilation. *Respir. Care Clin. N. Am.* 2002; **8**:187–209.
6. R. P. Geyer, Whole animal perfusion with fluorocarbon dispersion. *Fed. Proc. Fed. Am. Soc. Exp. Biol.* 1970; **29**:1758–1763.
7. J. G. Riess and M. P. Krafft, Fluorinated materials for *in vivo* oxygen transport (Blood Substitutes), diagnosis and drug delivery. *Biomaterials* 1998; **19**:1529–1539.
8. D. Wong and N. Lois, Perfluorocarbons and semifluorinated alkanes. *Semin. Ophthalmol.* 2000; **15**:25–35.
9. S. Rudiger, U. Gross, and E. Kemnitz, Perfluorocarbons - useful tools for medicine. *Eur. J. Med. Res.* 2000; **5**:209–216.
10. R. Benesch and R. E. Benesch, Intracellular organic phosphates as regulators of oxygen release by hemoglobin. *Nature* 1969; **221**:618–622.
11. R. Benesch, R. E. Benesch, S. Yung, and R. Edalji, Hemoglobin covalently bridged across the polyphosphate binding site. *Biochem. Biophys. Res. Commun.* 1975; **63**:1123–1129.

12. R. Benesch and R. E. Benesch, Preparation and properties of hemoglobin modified with derivatives of pyridoxal. *Methods Enzymol.* 1981; **76**:147–159.
13. J. C. Hsia, D. L. Song, L. T. L. Wong, and S. S. Er, Molecular determination of o-Raffinose-polymerized human hemoglobin. *Biomat. Art. Cells and Immob. Biotech.* 1992; **20**:293–296.
14. R. E. Benesch, S. Yung, T. Suzuki, and C. Bauer, Pyridoxal compounds as specific reagents for the α and β N-termini of hemoglobin. *Proc. Nat. Acad. Sci. USA.* 1973; **70**:2595–2599.
15. R. E. Benesch and S. Kwong, Bis-pyridoxal phosphates: a new class of specific intramolecular crosslinking agents for hemoglobin. *Biochem. Biophys. Res. Commun.* 1988; **156**:9–14.
16. R. E. Benesch and S. Kwong, Hemoglobin tetramers stabilized by a single intramolecular cross-link. *J. Protein Chem.* 1991; **10**:503–510.
17. J. C. Bakker, G. A. M. Berbers, W. K. Bleeker, P. J. den Boer, and P. T. M. Bisselles, Preparation and characterization of crosslinked and polymerized hemoglobin solutions. *Biomat. Art. Cell Immob. Biotechnol.* 1992; **20**:233–241.
18. P. Bender, Hemoglobin-based oxygen carriers. *Biotech. Lab. Int.* 1999; **4**:16–19.
19. E. Bucci, A. Razynsak, B. Urbaitis, and C. Fronticelli, Pseudo cross-link of human hemoglobin with mono-(3,5-dibromosalicyl)fumarate. *J. Biol. Chem.* 1989; **264**:6191–6195.
20. E. Bucci, C. Fronticelli, A. Razynska, V. Militello, R. Koehler, and B. Urbaitis, Hemoglobin tetramers stabilized with polyaspirins. *Biomat. Art. Cells Immob. Biotech.* 1992; **20**:243–252.
21. T. M. S. Chang, Hemoglobin corpuscles. (Report of a research project of B.Sc. Honors Physiology, McGill University, 1957, pp. 1–25. Medical Library, McIntyre Building McGill University, 1957), reprinted in *Artific. Red Blood Cells Artific. Org.* 1988; **16**:1–9.
22. T. M. S. Chang, Blood substitutes: principles, methods, products and clinical trials. In: T. M. S. Chang, ed., *Modified Hemoglobin-Based Blood Substitutes: Crosslinked, Recombinant and Encapsulated Hemoglobin*. New York: Karger-Lands Systems, 1998.
23. T. M. S. Chang, Recent and future developments in modified hemoglobin and microencapsulated hemoglobin as red blood cell substitutes. *Art. Cells, Blood Subs. Immob. Biotechnol.* 1997; **25**:1–24.
24. R. Chatterjee, E. V. Welty, R. Y. Walder, S. L. Pruitt, P. H. Rogers, A. Arnone, and J. A. Walder, Isolation and characterization of a new hemoglobin derivative cross-linked between the α chains (lysine $99\alpha_1 \rightarrow$ lysine $99\alpha_2$). *J. Biol. Chem.* 1986; **261**:9929–9937.
25. C. P. Stowell, Hemoglobin-based oxygen carriers. *Curr. Opin. Hematol.* 2002; **9**:537–543.
26. E. Dellacherie, M. Grandgeorge, F. Prouchayret, and G. Fasan, Hemoglobin linked to polyanionic polymers as potential red blood cell substitute. *Biomat. Art. Cells Immob. Biotechnol.* 1992; **20**:309–317.
27. L. Djordjevich and I. F. Miller, Synthetic erythrocytes from lipid encapsulated hemoglobin. *Exp. Hamatol.* 1980; **8**:584–592.
28. A. G. Fallon and R. L. Schnaare, Cross-linked hemoglobin as a potential membrane for an artificial red blood cell. *Art. Cells Blood Subs. Immob. Biotechnol.* 2001; **29**:285–296.
29. W. J. Fantl, L. R. Manning, H. Ueno, A. Di Donato, and J. M. Manning, Properties of carboxymethylated cross linked hemoglobin A. *Biochemistry* 1987; **26**:5755–5761.
30. M. C. Farmer and B. P. Gaber, Liposome-encapsulated hemoglobin as an artificial oxygen-carrying system. *Methods Enzymol.* 1987; **149**:184–200.
31. A. S. Rudolph, Encapsulated hemoglobin: current issues and future goals. *Art. Cells Blood Sub. Immob. Biotechnol.* 1994; **22**:347–360.
32. S. Sivan, Molecular engineering of bioactive materials with defined specificity: design and synthesis of hemoglobin derivatives, Ph.D. thesis, Technion – Israel Institute of Technology, Haifa, Israel, 2000.
33. S. Sivan and N. Lotan, Molecular engineering of proteins with predetermined function: part I: design of a hemoglobin-based oxygen carrier. *Biomolecular Engineer.* 2003; **20**:83–90.
34. D. R. Spahn, Current status of artificial oxygen carriers. *Adva. Drug Deliv. Rev.* 2000; **40**:143–151.
35. F. J. Carmichael, A. C. Ali, J. A. Campbell, S. F. Langlois, G. P. Biro, A. R. Willan, C. H. Pierce, and A. G. Greenburg, A phase I study of oxidized raffinose cross-linked human hemoglobin. *Crit. Care Med.* 2000; **28**:2283–2292.
36. K. Hanazawa, H.Ohzeki, H. Moro, S. Eguchi, T. Nakajima, T. Makifuchi, K. Miyashita, M. Nishiura, and H. Naritomi, Effects of partial blood replacement with pyridoxylated hemoglobin polyoxyethylene conjugate solution on transient cerebral ischemia in gerbil. *Art. Cells, Blood Subs. Immob. Biotechnol.* 1997; **25**:104–155.
37. E. Ilan, P. G. Morton, and T. M. S. Chang, Bovine Hemoglobin anaerobically reacted with divinylsulfone: a potential source for hypothermic oxygen carriers. *Biomat. Art. Cells Immob. Biotechnol.* 1992; **20**:263–275.
38. Y. Iwashita, A. Yabuki, K.Yamaji, K. Iwasaki, T. Okami, C. Hirati and K. Kosaka, A new resuscitation fluid "Stabilized Hemoglobin": preparation and characteristics. *Artific. Cells Artific. Org.* 1988; **16**:271–280.
39. P. E. Keipert and T. M. S. Chang, Pyridoxylated-polyhemoglobin solution: a low viscosity oxygen-delivering blood replacement fluid with normal oncotic pressure and long-term storage feasibility. *Biomat. Artif. Cells Artif. Org.* 1988; **16**:185–196.
40. P. E. Keipert, Properties of chemically cross linked hemoglobin solutions designed as temporary oxygen carriers. *Adv. Exp. Med. Biol.* 1992; **37**:453–464.
41. D. D. Powanda and T. M. S. Chang, Cross-linked polyhemoglobin-superoxide dismutase-catalase supplies oxygen without causing blood-brain barrier disruption or brain edema in a rat model of transient global brain ischemia-reperfusion. *Art. Cells Blood Subs. Immob. Biotechnol.* 2002; **30**:23–37.
42. J. E. Squires, Artificial blood. *Science* 2002; **295**:1002–1005.
43. D. Looker, D. Abbott-Brown, P. Cozart, S. Durfee, S. Hoffman, A. J. Mathews, J. Miller-Roehrich, S. Shoemaker, S. Trimble, G. Fermi, N. H. Komiyama, K. Nagai, and G. Stetler, A human recombinant hemoglobin designed for use as a blood substitute. *Nature (Lond).* 1992; **356**:258–260.
44. L. R. Manning, S. Morgan, R. C. Beavis, B. T. Chait, J. M. Manning, J. R. Hess, M. Cross, D. L. Currell, M. A. Marini, and R. M. Winslow, Preparation, properties and plasma retention of hemoglobin derivatives: comparison of uncrosslinked carboxymethylated hemoglobin with crosslinked tetrameric hemoglobin. *Proc. Natl. Acad. Sci. USA.* 1991; **88**:3329–3333.
45. M. A. Marini, G. L. Moore, S. M. Christensen, R. M. Fishman, R. G. Jesse, F. Medina, S. M. Snell, and A. Zegna, Re-examination of the polymerization of pyridoxylated hemoglobin with glutaraldehyde. *Biopolymers* 1990; **29**:871–882.
46. M. Marta, M. Patamia, A. Lupi, M. Antenucci, M. Di Iorio, S. Romeo, M. Petruzzelli, M. Pomponi, and B. Giardina, Bovine hemoglobin cross-linked through the beta chains. *J. Biol. Chem.* 1996; **271**:7473–7478.

47. A. O. Schubert, J. F. Hara, R. J. Przbelski, J. E.Tetzlaff, K. E. Marks, E. Mascha, and A. C. Novick, Effect of diaspirin cross-linked hemoglobin (DCLHb HemAssist) during high blood loss surgery on selected indices of organ function. *Art. Cells Blood Subs. Immob. Biotechnol.* 2002; **30**:259–283.

48. K. W. Olsen, Q. Y. Zhang, H. Huang, G. K. Sabaliauskas, and T. Yang, Stabilities and properties of multilinked hemoglobin. *Biomat. Art. Cells Immob. Biotechnol.* 1992; **20**:283–285.

49. S. A. Shoemaker, M. J. Gerber, G. L. Evans, L. E. Archer-Paik, and C. H. Scoggin, Initial clinical experience with a rationally designed, genetically engineered recombinant human hemoglobin. *Art. Cells, Blood Subs. Immob. Biotechnol.* 1994; **22**:457–465.

50. H. Sakai, M. Yuasa, H. Onuma, S. Takeoka, and E. Tsuchida, Synthesis and physicochemical characterization of a series of hemoglobin-based oxygen carriers: objective comparison between cellular and acellular types. *Bioconjug. Chem.* 2000; **11**:56–64.

51. K. D. Vandegriff and R. M. Winslow, Hemoglobin-based blood substitutes. *Chem. Industry* 1991; **14**:497–501.

52. K. D. Vandegriff, D. F. H. Wallach, and R. M. Winslow, Encapsulation of hemoglobin in non-phospholipid vesicles. *Art. Cells, Blood Subs. Immob. Biotechnol.* 1994; **22**:849–854.

53. A. Pocker, Synthesis of 2-nor-2-formylpyridoxal 5′-phosphate, a bifunctional reagent specific for the cofactor site in proteins. *J. Org. Chem.* 1973; **38**:4295–4299.

54. D. Pliura, Selective crosslinking of hemoglobins by oxidized, ring-opened saccharides, U.S. Patent 5,532,352, 1994.

55. W. P. Yu and T. M. S. Chang, Preparation and characterization of hemoglobin nanocapsules: 1. Evaluation of polymer composite in the nanocapsule formulation. *Art. Cells, Blood Subs. Immob. Biotechnol.* 1998; **26**:69.

56. S. H. Zuckerman, M. P. Doyle, R. Gorozynski, and G. J. Rosenthal, Preclinical biology of recombinant human hemoglobin, rHb1.1. *Art. Cells, Blood Subs. Immob. Biotechnol.* 1998; **26**:231–257.

57. U. K. Laemmli, Cleavage of structural proteins during the assembly of the head of bacteriophage T4. *Nature* 1970; **227**:680–685.

58. J. W. Prather, K. A. Gaar, and A. C.Guyton, Direct continuous recording of plasma colloid osmotic pressure of whole blood. *J. Appl. Physiol.* 1968; **24**:602.

59. L.Tentorii and M. Salvati, Hemoglobinometry in human blood. *Methods Enzymol.* 1981; **76**:707–715.

60. E. Antonini and M. Brunori. Hemoglobin. *Annu. Rev. Biochem.* 1970; **39**:977–1042.

61. T. Asakura, O. Tsuyoshi, S. Friedman, and E. Schwartz, Abnormal precipitation of oxyhemoglobin S by mechanical shaking. *Proc. Natl. Acad. Sci. USA*. 1974; **71**:1594–1598.

62. B. Masala and L. Manca, High-performance liquid chromatography of globin chains in the identification of human globin gene abnormalities. *Biophys. Chem.* 1990; **37**:225–230.

63. T. N. Estep, M. K. Bechtel, T. J. Miller, and A. Bagdasarian, Virus inactivation in hemoglobin solutions by heat. *Biomat. Art. Cells Art. Org.* 1988; **16**:129–134.

64. M. Farmer, A. Ebeling, T. Marshall, W. Hauck, C. S. Sun, E. White, and Z. Long, Validation of virus inactivation by heat treatment in the manufacture of diaspirin crosslinked hemoglobin. *Biomat. Art. Cells Immob. Biotechnol.* 1992; **20**:429–433.

65. J. B. Shelton, J. R. Shelton, and W. A. Schroeder, High performance liquid chromatography of globin chains on a large-pore C_4 column. *J. Liquid Chromatogr.* 1984; **7**:1969–1977.

66. N. Lotan, personal communication.

67. W. W. Nichols, *McDonald's Blood Flow in Arteries: Theoretic, Experimental and Clinical Principles*. London: Arnold, 1998.

68. S. Usami and S. Chien, Hemoglobin solutions as plasma expanders: effect on blood viscosity. *Proc. Soc. Exp. Biol. Med.* 1971; **136**:1232–1236.

69. M. Blank, Molecular association and viscosity of hemoglobin solutions. *J. Theor Biol.* 1984; **108**:55–64.

ARTIFICIAL HEART

SETSUO TAKATANI
Tokyo Medical and Dental
University
Tokyo, Japan

1. INTRODUCTION

The artificial heart (AH) is a mechanical pump that partially or completely replaces the pumping function of the native heart. Two types of AH exist depending on use. The first is called ventricular assist device (VAD) and consists of a single pump to help the pumping function of the left or right ventricle, or both ventricles by combining two VADs without removing the native heart. The second AH is called total artificial heart (TAH) and is comprised of two pumps to maintain both lung circulation and systemic circulation after removal of the native heart. Around the world, over 100,000 patients per year are in end-stage heart failure that may be saved by heart transplantation, but only about 4000 donor hearts are available per year for transplantation. The patients who cannot benefit from heart transplantation may be saved by either VAD or TAH temporarily or indefinitely. Of those who may be saved by circulatory assistance, approximately 70–80% suffer from left or right ventricular failure requiring VAD, whereas the rest have biventricular failure requiring biventricular assist device (BVAD) or TAH.

2. AH SYSTEM

Figure 1 shows the basic configuration of the AH system. The basic components of the AH system are (1) pumping chamber, (2) actuator and energy converter, (3) controller and CPU (central processing unit), and (4) energy source and transmission system. In the case of TAH, two pumps are used to replace left and right heart function. As an energy source, electrical energy in the form of chemical batteries is supplied using a direct percutaneous wire connection or transcutaneous high-frequency transduction method and is converted into mechanical energy using an energy converter such as an electrical motor. The actuator translates the rotating motion of the motor into a linear motion using mechanical systems such as cam, crank-shaft, roller screw, or impeller. The actuator then delivers mechanical power to the blood pump. The device or biological information obtained using various sensors

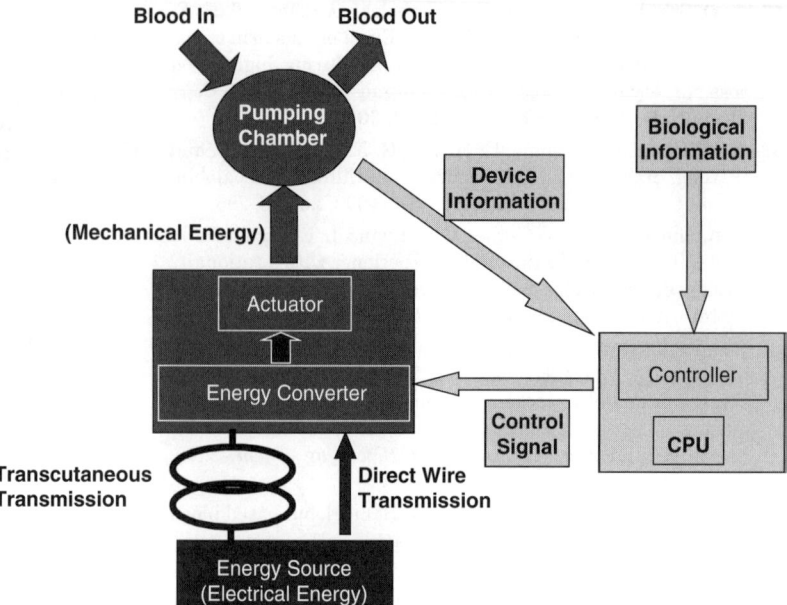

Figure 1. Configuration of the aritificial heart including (1) pumping chamber, (2) actuator and energy converter, (3) controller and CPU, and (4) energy source and energy transmission system. (This figure is available in full color at http://www.mrw.interscience.com/ebe.)

are input into the controller/CPU to appropriately control the motor speed and to meet the varying needs of the biological system.

2.1. Pulsatile and Nonpulsatile (or Continuous Flow) Pumps

2.1.1. Pulsatile pump. Two types of blood pumps exist, pulsatile and continuous flow, based on the flow pattern generated by the device. Employing a prosthetic one-way heart valve, one each in the inflow and outflow port of the pump chamber, generates the pulsatile flow like the native heart. Fig. 2 shows models of pulsatile devices that are in use including (1) hemispherical diaphragm type, (2) sac type, (3) concentric tube type, and (4) pusher-plate type. Pump models 1–3 have been proposed to use the pneumatic or hydraulic power, where pressurized air, fluid, or vacuum alternately enters the chamber to actuate the flexible membrane and hence to displace the blood pooled on the other side of the membrane. These pumps have been used for applications from temporary extracorporeal to semi-completely implantable permanent uses with the patient being tethered to an external control-drive console. The pneumatically operated Thoratec VAD (1,2) is an example of the sac-type pump used for extracorporeal support of left, right, or both ventricles. In the inflow and outflow port of the Thoratec VAD, Bjork-Shiley tilting disk valves are installed to create a unidirectional pulsatile flow. The intracorporeal CardioWest TAH (3–5) uses two of the hemispherical diaphragm-type blood pumps driven pneumatically by the extracorporeal console and supports the lung and systemic circulation.

The pusher-plate-type pump was designed to use an electromechanical actuator and to achieve a completely implantable system. In addition, in the pusher-plate design, the plate movement can be easily monitored using sensors such as a Hall effect device so as to precisely control the pusher-plate movement. With the hemispherical diaphragm and sac-type pump, it is difficult to predict the movement of the diaphragm or sac. The diaphragm and sac do not move in a predictable manner like the pusher-plate. Currently available clinical VADs such as Novacor (6,7) and HeartMate-I (8–20) both employ the pusher-plate-type pump coupled with an electromechanical actuator. The Novacor VAD combines a solenoid and a double counter-lever mechanism to squeeze the sac-type pump, whereas the HeartMate-I employs a high-torque low-speed dc motor in combination with a face cam mechanism to displace the pusher-plate. The completely implantable AbioCor TAH (11,12) that has moved into clinical trial in 2001 is a compact, one-piece totally implantable system comprising left and right pusher-plate-type pumps and an electro-hydraulic actuator accommodated between the two pumps. The AbioCor TAH has four tri-leaflet polyurethane valves built into the inflow and outflow ports of the left and right pumps.

2.1.2. Continuous flow pump. Continuous flow pumps offer features such as (1) compact design, (2) simple control, (3) valve-less design, and (4) cost-effectiveness in comparison with the pulsatile devices. Three types of continuous flow pumps exist, including (1) axial, (2) centrifugal, and (3) mixed flow or diagonal, as shown in Fig. 3.

2.1.2.1. Axial flow pump. In the axial flow design, the inflow and outflow direction are the same. Within a straight conduit, a flow straightener, an impeller, and a diffuser are lined up to push the blood in the axial direction. The axial flow device can be made extremely small with its diameter as small as 3–4 mm and length of 10–100 mm. As a result of their small size, they are suited for intracorporeal implantation as well as for paracorporeal insertion through a peripheral vessel into the cardiac chamber. The axial pumps require a high impeller speed usually at around 15,000 to 20,000 revolutions per minute to generate the flow required to assist the left ventricle.

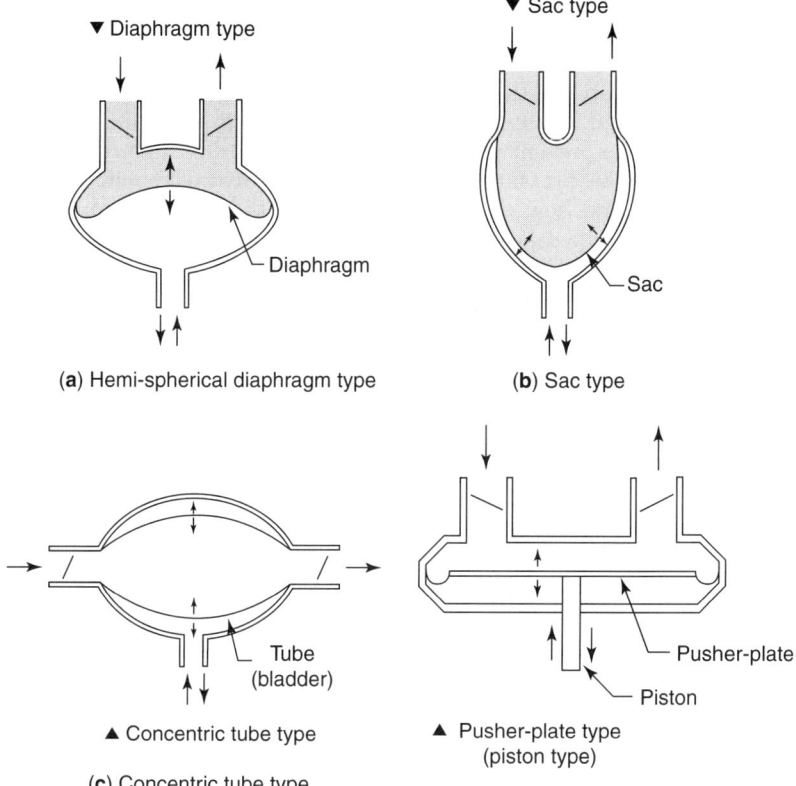

Figure 2. Pulsatile pump models; (a) hemispherical diaphragm type, (b) sac type, (c) concentric tube type, and (d) pusher-plate type.

Axial flow devices, however, are not suitable when high afterload pressure is exerted on the pump, causing back flow through the pump. The MicroMed DeBakey VAD (13–15) and Jarvik 2000 Flowmaker (16–19), which have been in clinical applications since early 2000, are of axial type and are called the second-generation AH in reference to the first-generation AH of the pulsatile blood pumps. In these pumps, the rotating impeller is supported with pivot bearings at both ends for stable rotation.

The MicroMed DeBakey VAD has been implanted in approximately 300 patients as a bridge to transplantation in the United States and Europe. The Jarvik 2000 Flowmaker is designed to be inserted into the left ventricular apex and connected to the descending aorta through a 12 mm Dacron graft. The percutaneous cable connected to the belt-worn controller and battery exits through the abdominal wall or alternatively via a skull-mounted pedestal especially for permanent use (Fig. 4). A patient implanted with the Jarvik 2000 Flowmaker celebrated 5 years of survival in June 2005. Although possible wear in the bearing support material may not warrant the durability of these devices beyond 5–10 years, it is interesting to know how much longer the Jarvik 2000 AH will continue to work.

The Impella Recover (20–23) incorporates a tiny axial pump at the catheter tip that can be inserted through the peripheral femoral artery without opening the chest for cardiology intervention or postoperative recovery of the heart. It has been implanted in approximately 200 patients in Europe during the post-operative recovery period for an average duration of 7 days showing remarkable prognosis.

2.1.2.2. Centrifugal pump. In the centrifugal pump, the outflow port direction is perpendicular to the inflow direction, this feature making the device suitable for high afterload operation such as in cardiopulmonary bypass. The incoming blood is directed in the radial direction by the centrifugal action of the impeller. The impeller rotates usually at around 2000–3000 revolutions per minute, much lower than that of the axial pumps to provide comparable flow to the axial device. The centrifugal blood pumps came into clinical environment as cardiopulmonary bypass pumps to replace roller pumps for their easier handling, safer operation, and lower cost. So, in the earlier models of centrifugal pumps, the impellers were directly mounted on the motor shaft with a shaft-seal preventing leakage of blood, or supported by the mechanical bearings immersed in the blood. Thus, the earlier centrifugal pumps were limited in their usage to 24–48 h because of blood leak from the shaft seal or blood cell damage and blood clot formation around the area of the blood-immersed mechanical bearings.

The longevity and reliability of the earlier centrifugal pump models was improved by making them into seal-less structures by suspending the impeller in the blood through the use of blood-immersed pivot bearings or contact-free magnetic bearings. The Gyro (24,25) pump achieved suspension of the impeller using pivot bearings made of ceramic and the centrifugal blood pumps such as DuraHeart (26,27) incorporated the magnetic suspension mechanism to achieve mechanical contact-free operation. The magnetic levitation was accomplished by feedback control of the electromagnets with the impeller position

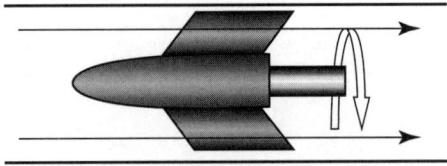

(a) Axial flow pump

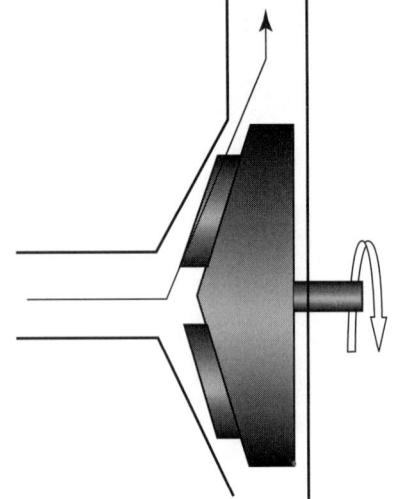

(b) Centrifugal pump

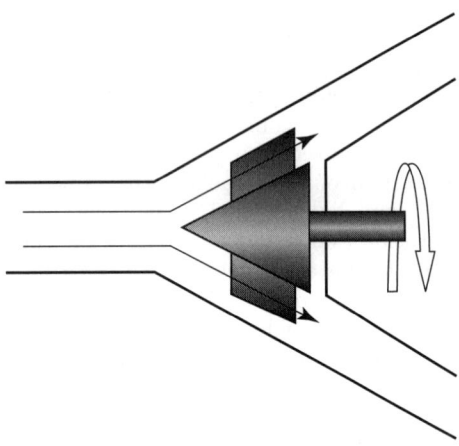

(c) Mixed flow or diagonal pump

Figure 3. Three types of continuous flow pumps; (a) axial pump, (b) centrifugal pump, and (c) mixed flow or diagonal pump.

continuously monitored by the position sensor. Usually, a sophisticated mechanism including multi-axial control is required to obtain stable suspension of the impeller without any mechanical contact. To solve this problem, Asama et al. proposed a simple magnetic levitation by combining two-degree active levitation in the x- and y-axes with a passive attractive force generated by a set of permanent magnets (Fig. 5) (28–30). The concept of magnetic levitation has also been incorporated in the axial device called INCOR, which moved into clinical trials in Europe (31).

Another mechanical contact-free operation realized in centrifugal pumps is hydrodynamic bearing, which is based on the fluid pressure generated in the narrow gap between the stationary pump housing and rotating impeller, repelling the impeller back into the opposite direction to prevent direct contact of the impeller. Figure 6 shows the pressure distribution in the hydrodynamic bearing and how the fluid pressure in the narrow and wedge-shaped gap created by shifting of the impeller with respect to the center of rotation acts on the rotor to correct its position (32,33). The VentrAssist (24–36) and CorAide (37–39) centrifugal VAD both incorporated the concept of hydrodynamic bearing to achieve contact-free operation of the impeller. The hydrodynamic bearing does not require feedback control of the electromagnets as required in the magnetic bearing mechanism, but its unique pump housing and impeller design attains a low cost and simple device structure for long-term reliable and safe usage.

2.1.2.3. Mixed flow or diagonal pump. The third continuous flow pump is the mixed flow pump obtained by combining the features of the centrifugal and axial flow blood pumps. It can be thus made as small as the axial flow device, but it can be operated at lower impeller speeds close to that of the centrifugal pump to achieve similar performance. It is also called the diagonal pump, because the blood flows in the diagonal direction in comparison with the axial direction of the axial pump and radial direction of the centrifugal pump. The diagonal pumps are suited for applications such as extracorporeal membrane oxygenation (ECMO) where high afterload pressure, sometimes around 300 mmHg, is generated. The DELTASTREAM (40,41) diagonal pump is an example that has been used in Europe for extracorporeal perfusion.

The continuous flow devices with mechanical contact-free operation are named the third-generation AH in comparison with the second-generation AH that has blood-immersed mechanical contact bearings. The third-generation AH is expected to last longer than ten years and is intended for permanent application or destination therapy (DT) in patients who are ineligible for heart transplantation.

2.2. Design Consideration

For successful and efficient development of AH, interdisciplinary collaboration across various areas such as clinical and basic medicine, mechanical, electronic, material, and control engineering is important. AH design must achieve (1) compact design, (2) efficient and reliable control, (3) long-term durability, (4) tissue and blood compatibility, and (5) efficient energy-saving operation.

2.2.1. Compact design. For implantable devices, the limited space inside the body varying from person to person must be well understood in designing a system that does not affect the function of the other organs in the vicinity of the implanted AH. The external pump dimension and its shape are important for it to fit in the available space after removal of the native heart or to implant it heterotopically without removing the heart. The direction and length of the inflow and outflow ports of the TAH are

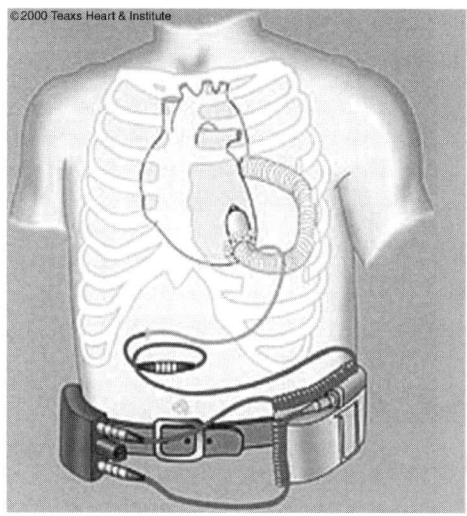
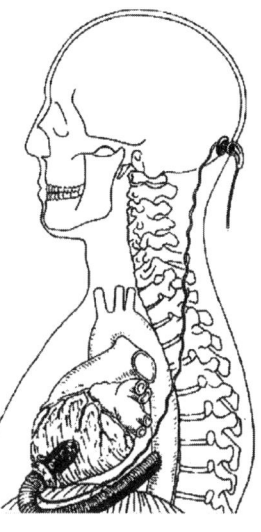

Figure 4. Configuration of the Jarvik 2000 Flowmaker showing that (left) the percutaneous cable connected to the belt-worn controller and battery exits through the abdominal wall or (right) alternatively via a skull-mounted pedestal especially for permanent use. (This figure is available in full color at http://www.mrw.interscience.com/ebe.)

important in order not to create an obstruction to blood flow because of kinking of the tubes connecting the blood pump to the remnant great vessels or atria. A compact pump design is the first priority for anatomical compatibility as well as surgical implantation.

Historically, a human cadaver was used to do a fitting study as part of the design evaluation (42,43). However, the cadaver data misled the anatomical evaluation because of shrinkage of the tissue and organs by dehydration after death. To overcome the shortcomings of the cadaver model, CT (computer tomography) and MRI (magnetic resonance imaging) data have been used to construct the 3-D computer graphics data and to evaluate the anatomical compatibility of a device on the computer. CAD/ACM is now becoming more and more accepted for precision design and manufacturing of the device before spending time and money for tedious trial and error processes in device construction. A computer-assisted surgical navigation system is also becoming a valuable tool for proper implantation of the device based on the preassembled implantation scheme.

The design of the totally implantable TAH AbioCor (Fig. 7) was based on the 26 heart transplant recipients performed at Baylor College of Medicine (44,45). After removal of the native heart prior to implantation of the donor heart, available space in the pericardial sac, direction and length between the pump model and remnant atria and great vessels were measured to design a one-piece compact TAH system. Since July 2001, AbioCor TAH has been implanted in 14 end-stage cardiac patients who were ineligible for heart transplantation and who might not survive more than 30 days. Prior to implantation, a so-called AbioCor fit analysis was usually performed in every patient using the patient's CT data together with the 3-D CAD image of the AbioCor TAH (46,47). So, advanced computer technology was assisting clinical medicine that had been based on human experience.

2.2.2. Control of AH.

2.2.2.1. Control modes for pulsatile pump. Currently used control modes include the fixed pulse rate mode with fill-limited operation, variable pulse rate mode, and demand mode synchronized to the external signal such as the patient ECG. All these methods can mimic the Frank–Starling like intrinsic regulation of the cardiac output. When the return to the pump increases, the pump output can increase. The fixed pulse rate with fill-limited operation can attain Frank–Starling-like control by setting the pulse rate at a high level so as to vary the beat-to-beat fill volume depending on the filling rate. When the fill volume reaches the full level, the pump rate will be set at a higher level to assure fill-limited operation. The variable pulse rate mode also known as full-fill/full empty mode can be implemented when the pump fill signal is available (48–50). This mode of operation is included in Novacor and HeartMate-I VAD with the pusher-plate position signal obtained using a Hall effect sensor. The pump rate, thus, varies automatically depending on the fill rate. When the pump fills faster, the pump rate automatically increases to deliver more flow to the peripheral organs. The patients implanted with Novacor and HeartMate-I VAD when put on an exercise machine for rehabilitation experienced a better response with the full-fill/full empty control through the automatic increase of the pump flow. The third mode is the synchronous mode where the patient's ECG signal can be used to trigger the pump ejection (51). This mode, however, has not been used because of difficulty in obtaining a stable ECG signal to synchronize the pump operation on a chronic base.

2.2.2.2. Control modes for continuous flow pumps. Control of continuous flow pumps is based on the head pressure vs. pump flow characteristics of the pump (Fig. 8). The impeller speed is adjusted manually knowing the relation between the head pressure vs. pump flow to maintain near normal cardiac output and blood pressure. The pump flow is usually measured using a flow sensor attached to the outflow tubing or based on the intrinsic motor current characteristics. Although the demand mode based on the patient heart rate has been suggested based on the relation obtained prior to the study, again difficulty in obtaining a stable ECG and defining an appropriate control algorithm are the problems that must be

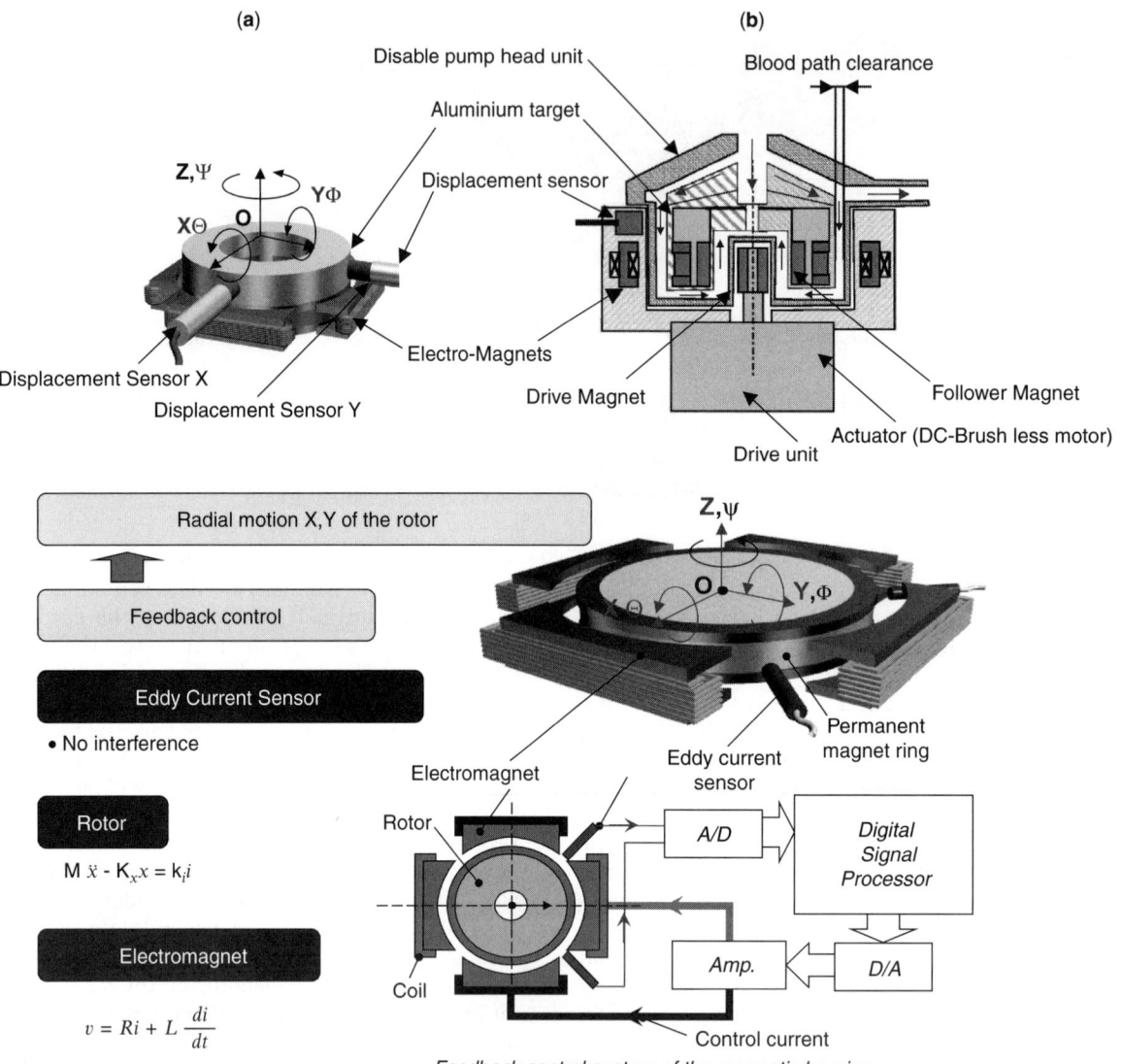

Figure 5. Schematic diagram of a simple two-degree of freedom magnetic bearing system (a) and magnetically levitated centrifugal blood pump (b). Also shown is the control system (bottom). (This figure is available in full color at http://www.mrw.interscience.com/ebe.)

overcome. In the actual operation, emphasis has been placed on how to prevent the suction effect by the continuous flow device in the pulsating ventricle. The mismatch in the venous return to the ventricle and suction by the pump will collapse the ventricle and the inflow tip may suck in the endo-cardial tissue resulting in possible tissue damage. If the bypass flow information is available, the suction effect may be easily detected, but long-term usage of the flow sensor is difficult, because of stability problems and a lack of *in vivo* calibration of the sensor. Pump flow and suction phenomena can be detected from the motor current level and waveform without relying on external flow or pressure sensors. When suction is detected, the motor speed can be programmed to vary and to release the suction in the ventricle. The MicroMed DeBakey VAD incorporated the suction detection scheme based on the pattern recognition of various events that are correlated with pump status and showed favorable outcomes (52).

2.2.2.3. Control of TAH. In controlling TAH, in addition to individual pump control, left-right flow balance is an important issue. Too little right pump output will cause venous congestion and low left pump output, and too much right pump output with respect to the left pump capacity will cause lung edema. Thus, balance of the right and left pump output control is an important issue for patient survival. In the case of the clinical TAH, pneumatic Cardio-West TAH operates the left and right pumps at the same rate, ejecting simultaneously, with both pumps hopefully in the fill-limited mode. The drive pressure and fill vacuum of each pump are independently adjusted to maintain the left atrial pressure within an acceptable level less than 10–15 mmHg or so, while delivering the required left

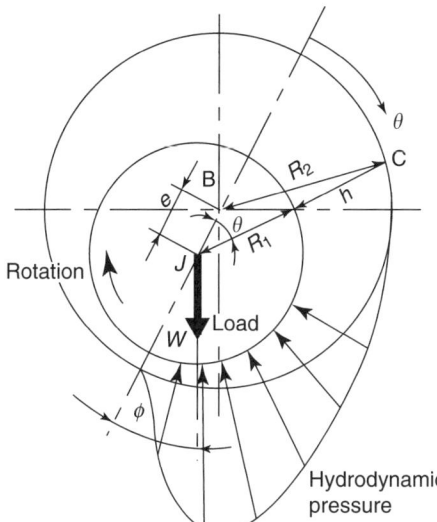

Figure 6. A hydrodynamic bearing system showing how the pressure generated in the gap will act on the impeller for contact-free operation. The figure shows that as the impeller position shifted, the hydrodynamic pressure is generated in the wedged area as if the impeller rotation forces and squeezes the fluid into a narrow wedged area. The generated hydrodynamic pressure is highest in the lower corner where the gap is the smallest and looks as if a compressed fluid exists because of impeller rotation.

pump flow. In the one-piece TAH such as the AbioCor, left-right operation is somewhat complex and a high-level control logic is built into the device. In the AbioCor TAH, the left and right pumps are operated in the alternate ejection mode with a single actuator shared by the two pumps. A unique left-right flow balance mechanism reduces the right pump maximum stroke volume as the left atrial pressure increases above a certain level. The left atrial pressure is sensed in a chamber attached to the inlet port of the left pump. When the left atrial pressure exceeds a certain level and pushes back the atrial chamber diaphragm, the hydraulic fluid behind the left atrial chamber is shifted toward the right pump to limit the right full stroke to a lower level, thus reducing the right pump output. In addition to the unique left-right flow balance, another advantage of this mechanism is that it can eliminate the need for a compliance chamber, which is required for a displacement-type pump when it is completely contained within the body. The compliance chamber or volume compensation device is needed to compensate for the volume change occurring on the blood side when a displacement-type pump is implanted within the body.

2.2.2.4. Control problems common to VAD and TAH.
A common to control problem in VAD and TAH using either the pulsatile or continuous flow blood pumps is the fact that both systems are operated in the open-loop control mode without any physiological feedback regulation. The intrinsic Starling-like control is the only control accepted in current clinical AHs. No reliable flow and pressure sensor exist that can be used in the blood for an indefinite duration with pronounced accuracy. Drift and calibration problems associated with pressure and flow sensors make their use in clinical settings difficult for prolonged duration. In addition, it is difficult to quantitatively evaluate the adequacy of the cardiac output requirement in a living system. The AH result may provide the cardiac output requirement data, which could not be obtained with the normal circulatory system. In order to evaluate the adequacy of pump output, the mixed venous hemoglobin oxygen saturation has often been measured using a Swan-Gans catheter inserted in the pulmonary artery or by attaching an optical sensor to the right pump of the TAH (53,54). However, long-term and continuous pump output control based on the mixed venous hemoglobin oxygen saturation level has not yet been demonstrated. Nevertheless, the long-term survival of the AbioCor TAH patients has proved that the simple, intrinsic Starling-like control without any active biological feedback regulation works in humans for a prolonged time providing sufficient improvement in the quality of life (QOL) as measured by patient mobility and activity level.

As no neural feedback and hormonal control mechanism exist attached to the AH system, pump ejection triggered by neural information or by specific hormonal levels might improve the performance of the device and QOL of the patients. To accomplish the neural feedback regulation, we must design and develop a sensor that can measure the nerve activity over a prolonged time as well as a control algorithm based on the nerve activity. Improvements in sensor technology together with nano- and micro-fabrication technology may lead to a bionic heart that can respond to autonomic nervous activity reflecting even the emotional state.

2.2.3. Durability. The durability of the AH system depends on the whole range from individual component and material durability to the durability of an entire system. Our native heart pumps approximately 100,000 times per day without any rest, which amounts to about 200 million cycles for a projected 5 year continuous operation. In order to meet 5 year continuous operation in biological environment, not only the pump components, such as flexing diaphragm and actuator, but also the peripheral components such as inflow and outflow prosthetic valves must meet repeated durability testing in a biological environment. The biological environment with an average temperature of 37°C at a 100% humidity level increases the deterioration rate of the implanted materials including polymers, metals, plastics, and all sorts of electronic components. In addition to mechanical durability of the pump components, charge-discharge cycle of implanted rechargeable batteries, durability of electrical cables used to carry electrical power and control signals, durability of the controller electronics, and fluid sealing property of the feed-through cable line are all problem areas that must be addressed when AH system durability is considered. Knowing these difficult problems in achieving long-term durability in biological environments, the technology is making day-to-day progress, and it has been reported that patients implanted with Novacor and HeartMate⁻I VADs have survived over 2 years (55,56). More recently, the patient implanted with a Jarvik 2000 Flowmaker has cleared the milestone of 5 year survival in June 2005

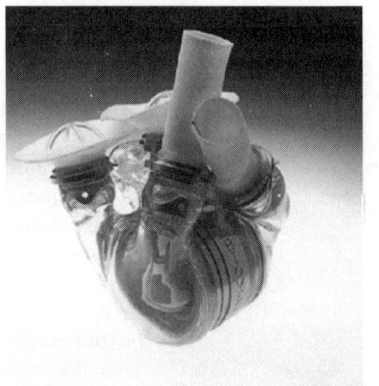
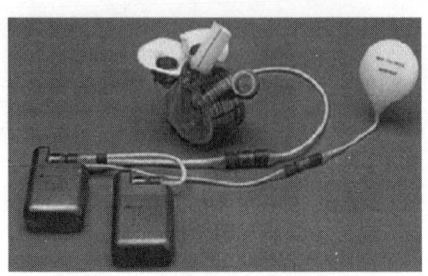
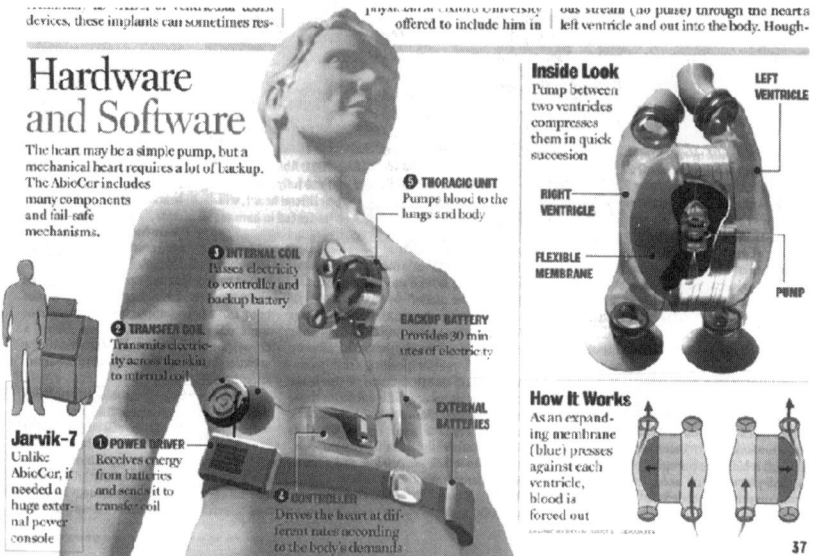

Figure 7. Completely implantable AbioCor TAH (top left) and implanted components (top right). Inside look of the AbioCor TAH and how it is implanted inside the body (bottom). (This figure is available in full color at http://www.mrw.interscience.com/ebe.)

(57). It may be concluded that the continuous flow devices with simple design and fewer components without requiring inflow and outflow valves as required for pulsatile devices may be better suited for permanent support of patients who are ineligible for heart transplantation.

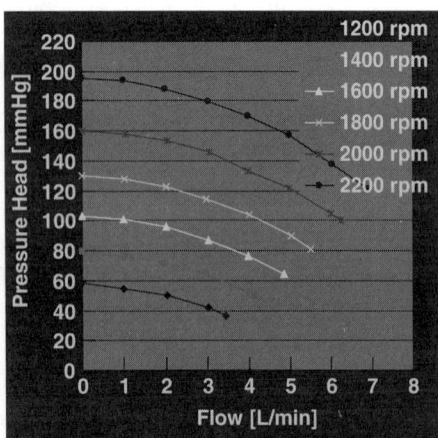

Figure 8. Pressure and flow characteristics of a centrifugal blood pump. (This figure is available in full color at http://www.mrw.interscience.com/ebe.)

The applicability of the reduced pulsatile flow in patients must await further clinical results.

2.2.4. Anti-thrombogenic performance. Of all the patients who have died after prolonged survival with the VAD or TAH, probably 80% are lost because of thromboembolic complications. A blood clot formed in the pumping chamber, possibly with associated components, is released into the blood stream creating emboli in the arteries causing infarction in peripheral organs. Particularly when thrombi are released into cerebral circulation, neurologic complications can be fatal or cause permanent damage in various motor and sensory systems. Chemical, physical, electrical, and surface properties of the material; fluid dynamic characteristics of blood, including blood viscosity, laminar or turbulent flow patterns, or streak lining; and hematological factors including coagulation and the activation state of the blood can lead to such an event. Blood clot formation in the blood chamber has been one of the major problems in the development of reliable AH systems since the early stages of AH development and are still without concrete solution.

Although problems associated with thrombus formation at the blood contacting surface of the AH has not yet been solved, two approaches exist that have made long-term application of the devices possible in human patients: smooth surface and rough or textured surface. In the smooth surface approach, segmented polyurethane or metallic surface such as mirror-finished titanium alloy with the surface roughness controlled to within less than a micron have been employed in implantable VADs such as Novacor, and in extracorporeal devices such as Thoratec, Toyobo VADs together with the administration of anticoagulants such as heparin (58–62). Also, most of the continuous flow devices used a smooth mirror finished surface of the titanium providing a nonreactive surface to obtain long-term thrombus-free operation. As mentioned before, the Jarvik 2000 Flowmaker employed a smooth titanium surface as the blood contacting surface to accomplish 5 year successful operation in human patients.

On the other hand, the textured surface is employed in the HeartMate-I (Fig. 9) (63,64). In the textured surface, with exposure to the blood, the blood proteins are trapped in the rough surface and fill the textured roughness of about 5–60 microns to form a pseudo neointimal layer making it appear like its own surface, which is called the biolization process to attain clot-free long-term operation. Hence, in the textured surface, anticoagulation therapy is not usually enforced thus reducing complications such as bleeding and multi-organ failure secondary to infusion of anticoagulants as required in the smooth surface device. Although the textured surface has been applied also in the continuous flow devices, favorable results have not been reported maybe because of differences in the surface shear rate between the pulsatile and continuous flow devices. Stabilization of the polymer surface through employment of nano-level molecular modification as well as formation of compound material consisting of metal and polymer might attain a better anti-thrombogenic performance of the AH.

2.2.5. Energy source and energy transmission. Electrical energy in the form of chemical batteries is the most reliable source of energy for the implanted AH system. A direct wire connection can be made between the external battery and the implanted blood pump in the form of a percutanous cable. As the skin entrance point of the percutaneous wire could cause irritation eliciting an inflammatory response, which could invite infection when a cable movement exists because of body motion. Thus, the entrance or exit site in the abdominal area may be prone to more body movement inviting infection. From the point of view of long-term stable operation, the approach taken by the Jarvik 2000 Flowmaker tunneling the percutaneous cable to be connected to the skull-mounted pedestal behind the ear may have been a good approach, because the tissue-device interface is not subject to movement as in the abdominal area (65). A hydro-oxy-apatite skin button has been specially made for skull mount to make stable tissue adhesion and easier connection to the exterior power source.

As an alternative to the percutaneous cable, a transcutaneous energy transmission system (TETS) can completely isolate the implanted components from outside of the body (66,67). The electrical energy available in the chemical batteries is first converted into high-frequency ac power at around 100–200 KHz and then transmitted across the skin using a coil couple with the primary coil being outside the body on the skin and the secondary one beneath the skin. The shape of the secondary coil is usually made into a cone type so as to make a mound when implanted in the subcutaneous space and to prevent the dislocation of the primary coil over the skin. The efficiency of the coil transformer through the skin could be important to minimize heat dissipation and hence tissue necrosis. The energy transmission efficiency of the TETS has been increased to greater than 80% to ensure safe and efficient operation of the implanted AH.

When the TETS is used for energy transmission across the skin, the secondary rechargeable battery is usually implanted to operate the implanted AH in case the primary coil of the TETS fixed on the skin is accidentally dislocated or when the patients desire to take a bath or shower. The high-energy density battery such as Lithium-ion battery or Nickel-metal hydride can be implanted to supply energy to the implanted AH. The charge-discharge cycle, energy density, sealing of the package in the biological environment is also an important aspect in designing a safe and reliable system for prolonged operation in human patients.

In the LionHeart VAD CUBS trial performed in Europe, the TETS remarkably reduced the infection rate to improve the performance of the VAD system (68,69). In the AbioCor TAH, the TETS helped improve patient mobility and quality of life in those patients who are ineligible for heart transplantation (46,47). The surgical complication in TETS implantation must be cleared through training of the surgeons as well as awareness of the patients implanted with the device.

3. EVOLUTION OF AH USAGE

The progress in AH technology and in clinically effective use of the devices has brought out various options for

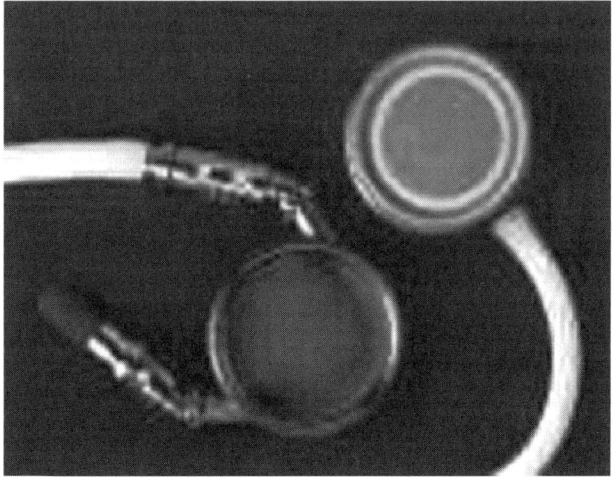

Figure 9. Textured blood contacting surface of the HeartMate I VAD. (This figure is available in full color at http://www.mrw.interscience.com/ebe.)

treating end-stage heart failure patients. The applications of the VADs include temporary support for acute cardiogenic shock, as a bridge to recovery (BTR)/therapy in idiopathic cardiomyopathy, as a bridge to transplantation (BTT) as a most common usage, and, more recently, destination therapy (DT) for those who are not eligible for heart transplantation. The support duration of the both VADs and TAH has also been extended from a few hours to several years to permanent application until expiration. Following the enrichment of devices, it is important to use the device appropriately for each patient with proper timing.

3.1. Temporary Extracorporeal

In the early stage of clinical application of AH, VAD usage was intended for temporary support of the heart until its recovery. As early as 1963, the first clinical VAD application had been performed by Dr. DeBakey whose patient was successfully weaned from the heart-lung machine and recovered thereafter, followed by additional six cases until 1968 (70). Particularly, in postcardiotomy patients who could not be weaned from the cardiopulmonary bypass following open heart surgery, extracorporeal pneumatic VADs had been used to maintain circulation as well as to unload the LV function of the patients while waiting for recovery. The postcardiotomy support of the failing heart instigated more open-heart surgery as well as development of better VAD systems. Although pulsatile devices had been used primarily, easy-to-handle and low-cost centrifugal pumps for temporary applications started to come into the market in the 1980s. With the progress in support techniques together with device improvement, efficient and reliable support became possible yielding more and more successful cases, which lead to the next-generation support, which is called bridge to transplantation.

3.2. Bridge to Transplantation (BTT)

This concept dates back to 1969 when the first TAH was used to support the circulation of the patient before finding an appropriate donor heart (71). The second BTT with the TAH was performed in 1981 by Akutsu who originated the artificial heart research with Kolff in 1957 (72). These two events probably established the concept of BTT, particularly the usage of the TAH system. The VAD bridging to heart transplantation was for the first time performed in 1978 (73), and followed with more applications in the late 1980s when the NIH implantable VAD projects had been completed. The implantable Novacor VAD and pneumatic HeartMate-I were the devices that helped patients improve the circulatory status as they prepared for transplantation of the hearts. More frequent usage of these devices bridging the patients to heart transplantation together with the emergence of an electrical version of the HeartMate-I in the early 1990s allowed the patient for the first time to be discharged while waiting for heart transplantation (74,75). This event really revolutionized the application of VAD, because the previous VADs had required patients to stay within the hospital prior to heart transplantation. The patient implanted with the electrical VAD became more mobile with a rechargeable battery carried in the shoulder bag eliminating a bulky console, which limited their mobility and activity. The evolution in medical technology came to improve the patient QOL, which was questioned in the early 1980s with implantation of the pneumatic TAHs in half a dozen patients who were restricted inside the hospital until their expiration.

3.3. Destination Therapy (DT)

The destination therapy with AH was first demonstrated in 1982 when the patient who was ineligible for heart transplantation volunteered for implantation of the pneumatic TAH, Jarvik 70, and survived for a duration of 112 days (76). This event was followed with five additional implantations allowing a longest survival of 622 days. Although the survival benefit was demonstrated with the TAH, the patients had to be kept inside the hospital because of tethering to the external pneumatic drive console. The quality of life for the patients was questioned, and this process pressured the development of the completely implantable system. The causes leading to the death of these patients were mainly micro-thrombi that caused neurologic complications and infection through the skin entrance site. As a result of these outcomes, the usage of pneumatic TAH was stopped until the completely implantable TAH will become available at a later time.

Meanwhile, the VAD outcomes had been improving with more and more patients being bridged to heart transplantation. However, the donor availability had become an important issue in the United States and Europe. Approximately 2500 donors in the United States and 1500 in Europe had been available, but the number of patients requiring donor hearts was increasing every year. Also, an age limitation existed according to which people older than 60 years were not eligible for heart transplantation. These restrictions pushed forward to start off what is called destination therapy by the VAD for those who are ineligible for heart transplantation. The efficacy of semi-completely implantable electrical HeartMate-I VAD for permanent circulatory support of the patients who were ineligible for heart transplantation had been evaluated randomly against the conventional maximal drug therapy from 1999 to 2001. The trial was called REMATCH (Randomized Evaluation of the Mechanical Assistance for the Treatment of Congested Heart Failure) and compared the survival rate and the QOL of the heart failure patients between the mechanical circulatory support and maximal drug therapy (63). End-stage cardiac patients were randomly assigned to either the mechanical circulatory support therapy or the conventional maximal drug therapy. This study was supported by the National Heart, Lung, and Blood Institute (NHBLI) and half a dozen of the participating medical centers and Thoratec Inc. supplied all the devices used for the trial, whereas the routine and specific medical expenses were shared by Medicare insurance and the participating medical centers. The results indicated the two-year survival benefit of approximately 28% by the VAD over 8% of the conventional medical therapy. Moreover, the patient QOL with the VAD was substantially improved over the medical therapy as evaluated through patient mobility as demonstrated, for example, by climbing up and down the stairway. The device failure, infection, and thromboembolic complications were the major events terminating the VAD patients. This study promoted the FDA approval of the permanent use of HeartMate-I–VE in end-stage heart failure patients ineligible for heart transplantation. The FDA approved the device for DT use in 2002 and the medical community for reimbursement in 2003. The recent design improvements to newer HeartMate-I–XVE proved to meet the design requirements for DT application. The mechanical failures of the HeartMate-I–VE included inflow valve dysfunction, percutaneous lead breakage, diaphragm fractures or punctures, bearing failures, outflow graft erosion, and pump disconnects.

Following this study, the completely implantable Lion-Heart made by Arrow International Inc. took place in Europe called CUBS (Clinical Utility Baseline Study) trial (68). The LionHeart with the energy delivered by the TETS lowered the infection rate to 0.17 per patient year in comparison with 0.60 per patient year in the REMATCH trial with a percutaneous cable.

In 2001, the completely implantable electrical TAH whose development was initiated in 1988 became available for clinical study and the FDA approved the pilot trial for those who were ineligible for heart transplantation and who might not survive for another 30 days (46,47). This study looked to be the advanced version of the trial performed in 1982 using an externally tethered pneumatic TAH. The electrical energy was transcutaneously induced inside the body to power the implanted electro-hydraulic TAH. The survival benefit, QOL of the patients in comparison with the Jarvik 70 pneumatic TAH had been obvious, whereas the thromboembolic complications anatomical fit were the problems that still required improvement. The AbioCor II with 30% reduction in its overall volume and a passive fill electromechanical system instead of an active fill electrohydraulic system is now ready for another trial with better performance.

3.4. Bridge to Recovery (BTR)

From the long-term VAD trials in Europe and Japan, previously thought unrecoverable DCM (dilated cardiomyopathy) patients started to show signs of recovery (77). Spreading use of VADs for applications such as combined therapy with beta-agonist clenbuterol or stem cell transplantation has revealed some DCM patients recover heart function after an extended period of support (78,79). Particularly in Japan where donor heart availability is very scarce (27) and also in the United States and Europe donor availability is declining every year, the combined use of VADs with drug therapy and cell implantation can promote recovery of the myocardium allowing removal of the VADs. In order to promote combined therapy, reliability and safety of the VADs must be improved, particularly mechanical durability, anti-thrombogenic performance, and infection control. A smaller-sized, compact device without inflow and outflow valves such as axial flow and centrifugal pumps with a mechanical contact-free feature in combination with the maximal drug therapy or cell and tissue implantation may bridge to more recovery of the hearts without requiring the donor hearts.

4. FUTURE DIRECTION

The advance in the field of AH technology and in management of end-stage heart failure patients in the last 5 years have made great progress in the recognition of AH, both VAD and TAH, as a possible alternative therapy to heart transplantation. This achievement is a blessing for patients who are facing death because of failure of cardiac function but who are ineligible for transplantation for various reasons.

In the next 5 years, the third-generation devices that are currently being (or will soon be) tested for safety and efficacy in clinical trials will move into routine clinical application and will have more advanced performance in comparison with the second-generation devices available today. The second-generation devices, such as MicroMed DeBakey VAD and Jarvik 2000 Flowmaker, succeeded in reducing the bulky sizes of the first-generation devices. The use of the Jarvik 2000 Flowmaker in DT application, as of June 2005, cleared 5 years of continuous pumping in the patient. It was shown that the reduced pulsatility with continuous flow device in humans works for 5 years and the patient was discharged home for routine life. The possible 10-year survival of the patient in 2010 with the Jarvik 2000 Flowmaker should mark a milestone and bring firm belief in AH technology. All should look forward to hearing such news. As the second model of the completely implantable AbioCor TAH, 30% downsized version of the first AbioCor TAH is just now on the corner to be tested in the human subject. Also, surgeons and researchers have just launched a study testing the feasibility of continuous flow TAH by combining two of the MicroMed DeBakey and Jarvik 2000 VADs in animals (80,81). A completely depulsed mode of circulation that can be derived from continuous flow TAH can be studied soon in long-term support of congested heart failure patients in comparison with the conventional pulsatile TAH such as AbioCor TAH.

Hopefully in 2010, the third-generation devices such as DuraHeart or INCOR that are on the horizon will eventually prove to provide extremely prolonged longevity passing the performance of the second-generation devices. The smaller VADs based on the continuous flow technology should be applicable for treatment of congenital as well as acquired heart failure of children and infants (82–84). The constant problem of biocompatibility must be addressed in the next 5 years by using new technologies such as nano-surface modulation. As for energy, based on the recent evolution of batteries and energy transmission techniques in industrial technology, the improvement in QOL for device recipients will be expected. The control of the AH to meet peripheral organ needs must be addressed to achieve an optimal control strategy. To this end, reliable biosensors and control algorithm should be put into practical use. In 2010, we should have greater advancements in information technology that transmit patient information from the home to the medical center for rapid diagnosis and for providing proper care.

Recently, the working group of the USA NHLBI (National Heart, Lung, and Blood Institute) summarized strategies and recommendations for next-generation AHs targeted for permanent use (85). According to this summary, the limitations of currently available AHs are (1) risk of thromboembolic events, (2) risk of infection, especially because of percutaneous control or drive lines, (3) limited durability, (4) large device size, (5) limited physiological control strategies during prolonged use, (6) uncertainty about the long-term consequences of nonpulsatile flow, (7) adverse effects on the gastrointestinal system with abdominal wall device placement, (8) substantial invasive surgery, and (9) high costs. This summary clearly suggests the needs for development of next-generation totally implantable, easier to implant, more reliable, and more efficient AHs. To meet this need requires the multidisciplinary collaborative efforts of bioengineers, cardiologists, cardio-thoracic surgeons, biomaterial scientists, biologists, other scientists, and industrial support. Most of all, the cardiologists who provide the initial treatment and referral to cardiovascular surgeons for implantation of the device must be fully informed about the capabilities, limitations, and future trends of the VADs. In addition, NHLBI's effort in establishing a database and clinical coordinating center to manage a registry of patients receiving a mechanical circulatory support device will also be helpful in identifying problems and establishing a guideline for the proper use of AHs. With all the effort targeted to identify and to solve the problems, survival and improved QOL of patients with AHs that approach those of the transplant recipients will be achieved.

AHs should also be made a routine medical procedure and technology that everyone can afford, not just the rich man's survival tool. Furthermore, much of the data summarized from heart failure patients who receive VADs and TAHs will indicate the route for development of guidelines that establish the standard for patient selection and for the next VADs and TAHs.

BIBLIOGRAPHY

1. Thoratec Corp. (2005). (online). Available: www.thoratec.com.
2. M. A. Sobieski, T. A. George, and M. S. Slaughter, The Thoratec mobile computer: initial in-hospital and outpatient experience. *ASAIO J.* 2004; **50**:373–375.
3. J. G. Copeland, R. G. Smith, F. A. Arabia, P. E. Nolan, D. McClellan, P. H. Tsau, G. K. Sethi, R. K. Bose, M. E. Banchy, D. L. Covington, and M. J. Slepian, Total artificial heart bridge to transplantation: a 9-year experience with 62 patients. *J. Heart Lung Transplant.* 2004; **23**(7):823–831.
4. A. El-Banayosy, L. Arusoglu, M. Morshuis, L. Kizner, G. Tenderich, P. Sarnowski, H. Milting, and R. Koerfer, CardioWest total artificial heart: bad Oeynhausen experience. *Ann. Thorac. Surg.* 2005; **80**(2):548–552.
5. Syncardia. (2005). (online). Available: www. Syncardia.com.
6. J. Lee, P. J. Miller, H. Chen, M. G. Conley, J. L. Carpenter, J. C. Wihera, J. S. Jassawalla, and P. M. Portner, Reliability model from the in vitro durability tests of a left ventricular assist system. *ASAIO J.* 1999; **45**(6):595–601.
7. Y. C. Yu, J. R. Boston, M. A. Simaan, P. J. Miller, and J. F. Antaki, Pressure-volume relationship of a pulsatile blood pump for ventricular assist device development. *ASAIO J.* 2001; **47**:293.
8. D. B. Gernes, W. F. Bernhard, W. C. Clay, C. W. Sherman, and D. Burke, Development of an implantable, integrated electrically powered ventricular assist system. *Trans. ASAIO* 1983; **29**:546.
9. O. H. Frazier, T. J. Myers, and B. Radovancevic, The HeartMate left ventricular assist system: overviews and 12-year experience. *Tex. Heart Inst. J.* 1998; **25**:265–271.
10. O. H. Frazier, E. A. Rose, M. C. Oz, W. Dembitsky, P. McCarthy, B. Radovancevic, V. Poirier, and K. A. Dasse, Multicenter clinical evaluation of the HeartMate vented electric left ventricular assist system in patients awaiting heart transplantation. *J. Thorac. Cardiovasc. Surg.* 2001; **122**:1186–1195.
11. No authors listed. AbioCor totally implantable artificial heart. How will it impact hospitals? *Health Devices* 2002; 31(9):332–341.
12. O. H. Frazier, R. D. Dowling, L. A. Gray Jr., N. A. Shah, T. Pool, and I. Gregoric, The total artificial heart: where we stand. *Cardiology* 2004; **101**(1-3):117–121.
13. G. P. Noon, D. Morley, S. Irwin, and R. Benkowski, Development and clinical application of the MicroMed DeBakey VAD. *Curr. Opin. Cardiol.* 2000; **15**:166–171.
14. M. E. DeBakey, A miniature implantable axial flow ventricular assist device. *Ann. Thorac. Surg.* 1999; **68**:637–640.
15. D. J. Goldstein, M. Zucker, L. Arroyo, D. Baran, P. M. McCarthy, M. Loebe, and G. P. Noon, Research correspondence: safety and feasibility trial of the MicroMed DeBakey ventricular assist device as a bridge to transplantation. *J. Am. Coll. Cardiol.* 2005; **45**(6):962–963.
16. R. Jarvik, V. Scott, M. Morrow, and E. Takecuhi, Belt worn control system and battery for the percutaneous model of the Jarvik 2000 heart. *Artif. Organs* 1999; **23**(6):487–489.
17. S. Westaby, R. Jarvik, A. Freeland, D. Pigott, D. Robson, S. Saito, P. Catarino, and O. H. Frazier, Postauricular percutaneous power delivery for permanent mechanical circulatory support. *J. Thorac. Cardiovasc. Surg.* 2002; **123**:977–983.
18. O. H. Frazier, T. J. Myers, I. D. Gregoric, T. Khan, R. Delgado, M. Croitoru, K. Miller, R. Jarvik, and S. Westaby, Initial clinical experience with the Jarvik 2000 implantable axial-flow left ventricular assist system. *Circulation* 2002; **105**:2855–2860.
19. M. P. Siegenthaler, J. Martin, A. van de Loo, T. Doenst, W. Bothe, and F. Beyersdorf, Implantation of the permanent Jarvik-2000 left ventricular assist device: a single-center experience. *J. Am. Coll. Cardiol.* 2002; **39**:1764–1772.
20. Impella CardioSystems AG. (2005). (online). Available: www.impella.com.
21. T. Schmidt, J. Siefker, S. Spiliopoulos, and O. Dapunt, New experience with the paracardial right ventricular axial flow micropump impella elect 600. *Eur. J. Cardio-thoracic. Surg.* 2003; **24**:307–308.
22. M. J. Jurmann, H. Siniawski, M. Erb, T. Drews, and R. Hetzer, Initial experience with miniature axial flow ventricular

22. assist devices for postcardiotomy heart failure. *Ann. Thorac. Surg.* 2004; **77**:1642–1647.
23. M. P. Siegenthaler, K. Brehm, T. Strecker, T. Hanke, A. Notzold, M. Olschewski, M. Weyand, H. Sievers, and F. Beyersdorf, The Impella Recover microaxial left ventricular assist device reduces mortality for postcardiotomy failure: a three-center experience. *J. Thorac. Cardiovasc. Surg.* 2004; **127**:812–822.
24. Y. Ohara, I. Sakuma, K. Makinouchi, G. Damm, J. Glueck, K. Mizuguchi, K. Naito, K. Tasai, Y. Orime, S. Takatani et al., Baylor Gyro Pump: a completely seal-less centrifugal pump aiming for long-term circulatory support. *Artif. Organs* 1993; **17**(7):599–604.
25. Y. Nose and K. Furukawa, Current status of the gyro centrifugal blood pump: development of the permanently implantable centrifugal blood pump as a biventricular assist device (NEDO project). *Artif. Organs* 2004; **28**(10):953–958.
26. Terumo Corp. (2005). Press release on Terumo Corp. (online). Available: www.terumo.co.jp.
27. S. Takatani, H. Matsuda, A. Hanatani, C. Nojiri, K. Yamazaki, T. Motomura, K. Ohuchi, T. Sakamoto, and T. Yamane, Mechanical circulatory support devices (MCSD) in Japan: current status and future directions. *J. Artif. Organs* 2005; **8**:13–27.
28. J. Asama, T. Shinshi, H. Hoshi, S. Takatani, and A. A. Shimokohbe, New design for a compact centrifugal blood pump with a maganetically levitated rotor. *ASAIO J.* 2004; **50**(6):550–556.
29. H. Hoshi, H. Kataoka, K. Ohuchi, J. Asama, T. Shinshi, A. Shimokohbe, and S. Takatani, Magnetically suspended centrifugal blood pump with a radial magnetic driver. *ASAIO J.* 2005; **51**(1):60–64.
30. H. Hoshi, J. Asama, T. Shinshi, K. Ohuchi, T. Mizuno, H. Arai, A. Shimokohbe, and S. Takatani, Disposable, Maglev centrifugal blood pump: design and in vitro performance. *Artif. Organs* 2005; **29**(7):520–526
31. C. Schmid, T. D. T. Tjan, C. Etz, C. Schmidt, F. Wenzelburger, M. Willhelm, M. Rothenburger, G. Drees, and H. H. Scheld, First clinical experience with the INCOR left ventricular assist device. *J. Heart Lung Transplant* 2005; **24**:1188–1194.
32. H. Kataoka, Y. Kimura, H. Fujita, and S. Takatani, Measurement of the rotor motion and corresponding hemolysis of a centrifugal blood pump with a magnetic and hydrodynamic bearing. *Artif. Organs* 2005; **29**(7):547–556.
33. R. Holmes, The vibration of a rigid shaft on short sleeve bearings. *J. Mech. Eng. Sci.* 1960; **2**(4):337–341.
34. Ventracor. (2005). (online). Available: www.ventracor.com.
35. P. A. Watterson, J. C. Woodard, V. S. Ramsden, and J. A. Reizes, VentrAssist hydrodynamically suspended, open, centrifugal blood pump. *Artif. Organs* 2000; **24**(6):475–477.
36. D. S. Esmore, D. Kaye, R. Salamonsen, M. Buckland, M. Rowland, J. Negri, Y. Rowley, J. Woodard, J. R. Begg, P. Ayre, and F. L. Rosenfeldt, First clinical implant of the VentrAssist left ventricular assist system as destination therapy for end-stage heart failure. *J. Heart Lung Transplant.* 2005; **24**(8):1150–1154.
37. Arrow International Inc. (2005). (online). Available: www.arrowintl.com.
38. Y. Ochiai, L. A. Golding, A. L. Massiello, A. L. Medvedev, R. L. Gerhart, J. F. Chen, M. Takagaki, and K. Fukamachi, In vivo hemodynamic performance of the Cleveland Clinic CorAide blood pump in calves. *Ann. Thorac. Surg.* 2001; **72**(3):747–752.
39. K. Fukamachi, New technologies for mechanical circulatory support: current status and future prospects of CorAide and MagScrew technologies. *J. Artif. Organs* 2004; **7**(2):45–57.
40. C. Gobel, A. Arvand, R. Eilers, O. Marseille, C. Bals, B. Meyns, W. Flameng, G. Rau, and H. Reul, Development of the MEDOS/HIA DeltaStream extracorporeal rotary blood pump. *Artif. Organs* 2001; **25**(5):358–365.
41. S. Beholz, M. Kessler, R. Tholke, and W. F. Konertz, Priming Reduced Extracorporeal Circulation Setup (PRECiSe) with the DeltaStream diagonal pump. *Artif. Organs* 2003; **27**(12):1110–1115.
42. J. Urzua, O. Sudilovsky, T. Panke, R. J. Kiraly, and Y. Nose, Preliminary report. Anatomic constraints for the implantation of an artificial heart. *J. Surg. Res.* 1974; **17**: 262–268.
43. L. K. Fujimoto, G. Jacobs, J. Przybysz, S. Collins, T. Meaney, W. Smith, R. Kiraly, and Y. Nose, Human thoracic anatomy based on computed tomography for development of a totally implantable left ventricular assist system. *Artif. Organs* 1984; **8**:436–444.
44. A. S. Shah, M. Shiono, T. Jikuya, S. Takatani, M. E. Sekela, G. P. Noon, J. B. Young, Y. Nose, and M. E. DeBakey, Intra-operative determination of mediastinal constraints for a total artificial heart. *ASAIO Trans.* 1991; **37**(2):76.
45. M. Shiono, G. P. Noon, K. R. Hess, S. Takatani T. Sasaki, Y. Orime, J. B. Young, Y. Nose, and M. E. DeBakey, Anatomic constraints for a total artificial heart in orthotopic heart transplant recipients. *J. Heart Lung Transplant.* 1994; **13**:250–262.
46. R. D. Dowling, L. A. Gray Jr., S. W. Etoch, H. Laks, D. Marelli, L. Samuels, J. Entwistle, G. Couper, G. J. Vlahakes, and O. H. Frazier, Initial experience with the AbioCor implantable replacement heart system. *J. Thorac. Cardiovasc. Surg.* 2004; **127**(1):131–141.
47. R. D. Dowling, L. A. Gray Jr., S. W. Etoch, H. Laks, D. Marelli, L. Samuels, J. Entwistle, G. Couper, G. J. Vlahakes, and O. H. Frazier, The AbioCor implantable replacement heart. *Ann. Thorac. Surg.* 2003; **75**(6 Suppl):S93–S99.
48. S. Takatani, H. Harasaki, S. Suwa, S. Murabayashi, R. Sukalac, G. Jacobs, R. Kiraly, and Y. Nose, Pusher-plate type TAH system operated in the left and right free-running variable rate mode. *Artif. Organs* 1981; **5**(2):132–142.
49. R. Kiraly, G. Jacobs, J. Urzua, and Y. Nose, Performance analysis of pneumatically driven blood pumps. *Ann. Biomed. Eng.* 1976; **4**:6.
50. C. S. Kwan-Gett, M. J. Crosby, A. Shoenberg, S. C. Jacobsen, and W. J. Kolff, Control systems for artificial hearts. *Trans. Am. Soc. Artif. Intern. Organs* 1968; **14**:284.
51. B. Vajapeyam, B. C. McInnis, R. L. Everett, A. K. Vakamudi, and T. Akutsu, A microprocessor based control system for the artificial heart. *Trans. Am. Soc. Artif. Intern. Organs* 1981; **25**:379.
52. M. Vollkron, H. Schima, L. Huber, R. Benkowski, G. Morello, and G. Wieselthaler, Development of a suction detection system for axial blood pumps. *Artif. Organs* 2004; **28**(8):709–716.
53. S. Takatani, H. Noda, H. Takano, and T. Akutsu, Optical sensor for hemoglobin content and oxygen saturation: application to artificial heart research. *Trans. Am. Soc. Artif. Intern. Organs* 1988; **34**(3):808–812.
54. T. H. Stanley, J. Volder, and W. J. Kolff, Extrinsic artificial heart control via mixed venous blood gas tension

analysis. *Trans. Am. Soc. Artif. Intern. Organs* 1973; **19**: 258–261.

55. O. H. Frazier and R. M. Delgado, Mechanical circulatory support for advanced heart failure. *Circulation* 2003; **108**: 3064–3068

56. U.S. Department of Health and Human Services. 2002 Annual Report: The US Organ Procurement and Transplantation Network and The Scientific Registry of Transplant Recipients (Transplant Data 1992–2001). Washington, DC: U.S. Department of Health and Human Services, 2002, p. 602.

57. S. Westaby, A. P. Banning, R. Jarvik, O. H. Frazier, D. W. Pigott, X. Y. Jin, P. A. Catarino, S. Saito, D. Robson, A. Freeland, T. J. Myers, and P. A. Poole-Wilson, First permanent implant of the Jarvik 2000 Heart. *Lancet* 2000; **356**:900–903.

58. I. D. Bella, F. Pagani, C. Banfi, E. Ardemagni, A. Capo, C. Klersy, and M. Vigano, Results with the Novacor assist system and evaluation of long-term assistance. *Eur. J. Cardio-Thorac. Surg.* 2000; **18**:112–116.

59. P. M. Portner, P. G. M. Jansen, P. E. Oyer, D. R. Wheeldon, and N. Ramasamy, Improved outcomes with an implantable left ventricular assist system: a multicenter study. *Ann. Thorac. Surg.* 2001; **71**:205–209.

60. J. T. Strauch, D. Spielvogel, P. L. Haldenwang, R. K. Correa, R. A. deAsla, P. E. Siessler, D. A. Baran, A. L. Gass, and S. L. Lansman, Recent improvements in outcome with the Novacor left ventricular assist device. *J. Heart Lung Transplant.* 2003; **22**:674–680.

61. T. Mussivand, R. Hetzer, E. Vitali, B. Meyns, P. Noihomme, R. Koerfer, A. El-Banayosy, E. Wolner, G. Wieselthaler, B. Reichart, P. Uberfuhr, R. Halfmann, and P. Portner, Clinical results with an ePTFE inflow conduit for mechanical circulatory support. *J. Heart Lung Transplant* 2004; **23**:1366–1370.

62. H. Takano and T. Nakatani, Ventricular assist systems: experience in Japan with Toyobo pump and Zeon pump. *Ann. Thorac. Surg.* 1996; **61**:317–322.

63. E. A. Rose, A. C. Gelijns, A. J. Moskowitz, D. F. Heitjan, L. W. Stevenson et al., Long-term use of a left ventricular assist device for end-stage heart failure. *N. Engl. J. Med.* 2001; **345**:1435–1443.

64. R. D. Dowling, S. J. Park, F. D. Pagani, A. J. Tector, Y. Naka, T. B. Icenogle, V. L. Poirier, and O. H. Frazier, HeartMate VE LVAS design enhancements and its impact on device reliability. *Eur. J. Cardio-Thorac. Surg.* 2004; **25**: 958–963.

65. S. Westaby, R. Jarvik, A. Freeland, D. Pigott, D. Robson, S. Saito, P. Catarino, and O. H. Frazier, Postauricular percutaneous power delivery for permanent mechanical circulatory support. *J. Thorac. Cardiovasc. Surg.* 2002; **123**:977–983.

66. J. C. Schuder, J. E. Stephenson, and B. A. Rowley, Excitation of a large coil set for the transport of energy into chest. *Eng. Med. Biol.* 1962; **15**:57.

67. C. Sherman, W. Clay, K. Dasse, and B. Daly, Energy transmission across intact skin for powering artificial internal organs. *Trans. Am. Soc. Artif. Intern. Organs* 1981; **27**:137–141.

68. LionHeart CUBS trial website. (2005). (online). Available: www.Scienceblog.com/community.

69. A. El-Banayosy, L. Arusoglu, L. Kizner, M. Morshuis, G. Tenderich, W. E. Pae Jr., and R. Korfer, Preliminary experience with the LionHeart left ventricular assist device in patients with end-stage heart failure. *Ann. Thorac. Surg.* 2003; **75**(5):1469–1475.

70. DeBakey M.E. Left ventricular pump for cardiac assistance. Clinical experience. *Am. J. Cardiol* 1971; **27**:3–11.

71. D. A. Cooley, D. Liotta et al., Orthotopic cardiac prosthesis for two-stage cardiac replacement. *Am. J. Cardiol.* 1969; **24**: 723–730.

72. T. Akutsu and W. J. Kolff, Permanent substitutes for valves and hearts. *Trans. Am. Soc. Artif. Intern. Organs* 1957; **4**:230.

73. P. E. Oyer, E. B. Stinson, P. M. Portner, A. K. Ream, and N. E. Shumway, Development of a totally implantable, electrically actuated left ventricular assist system. *Am. J. Surg.* 1980; **140**:17.

74. O. H. Frazier, T. J. Myers, and B. Radovancevic, The HeartMate left ventricular assist system: overviews and 12-year experience. *Tex. Heart Inst. J.* 1998; **25**:265–271.

75. O. H. Frazier, E. A. Rose, M. C. Oz, W. Dembitsky, P. McCarthy, B. Radovancevic, V. Poirier, and K. A. Dasse, Multicenter clinical evaluation of the HeartMate vented electric left ventricular assist system in patients awaiting heart transplantation. *J. Thorac. Cardiovasc. Surg.* 2001; **122**: 1186–1195.

76. W. C. DeVries, The permanent artificial heart. Four case reports. *J. Am. Med. Assoc.* 1988; **259**:849–859.

77. J. Muller, G. Wallukat, Y. Weng, M. Dandel, S. Speigelsberger, S. Semarau, K. Brandes, M. Loebe, R. Meyer, and R. Hetzer, Treatment of idiopathic dilated cardiomyopathy by insertion of a left ventricular support system. *Heart Replacement Artif. Heart* 1996; **6**:281–291.

78. H. Tsuneyoshi, W. Oriyanhan, H. Kanemitsu, R. Shiina, T. Nishina, S. Matsuoka, T. Ikeda, and M. Komeda, Does the beta2-agonist clenbuterol help to maintain myocardial potential to recover during mechanical unloading? *Circulation* 2005; **30**(112):I51–I56.

79. J. K. Hon and M. H. Yacoub, Bridge to recovery with the use of left ventricular assist device and clenbuterol. *Ann. Thorac. Surg.* 2003; **75**:S36–S41.

80. O. H. Frazier, E. Tuzun, W. Cohn, D. Tamez, and K. A. Kadipasaoglu, Total heart replacement with dual centrifugal ventricular assist devices. *ASAIO J.* 2005; **51**(3):224–229.

81. R. Benkowski, G. Morello, S. Alexander, E. Tuzun, J. Conger, K. Kadipasaoglu, C. Gemmato, W. Cohn, O. H. Frazier, G. P. Noon, and M. E. DeBakey, Development of a bi-ventricular replacement system using two axial flow pumps. Proceedings of the 13th Congress of the International Society for Rotary Blood Pumps, 2005; **80**.

82. J. T. Baldwin, NHLBI's program for the development of pediatric circulatory support devices. Proceedings of the 13th Congress of the International Society for Rotary Blood Pumps, 2005: **43**.

83. B. W. Duncan, M. Lorenz, M. W. Kopcak, K. Fukamachi, Y. Ootaki, H. M. Chen, P. A. Chapman, S. J. Davis, and W. A. Smith, The PediPump: a new ventricular assist device for children. *Artif. Organs* 2005; **29**(7):527–30.

84. H. Borovetz, S. Badylak, R. Boston, R. Kormos, M. Kameneva, M. Simaan, T. Snyder et al., The Pedi Flow pediatric ventricular assist device. Proceedings of the 13th Congress of the International Society for Rotary Blood Pumps, 2005; **44**.

85. National Heart, Lung, and Blood Institute, Information for Researchers. (2005). (online). Available: www.nhlbi.gov/meetings/workshops/nextgen-vads.htm.

ARTIFICIAL HEART VALVES

DAN T. SIMIONESCU
Clemson University
Clemson, South Carolina

1. INTRODUCTION

The cardiovascular system is a closed circulatory system that ensures blood flow through the human body. Blood circulates through two systems; the first is responsible for pushing blood from the heart to the lungs to capture oxygen, and the second is responsible for distributing oxygenated blood to body organs. This system is composed of a double pump with branched tubes that leads blood from the heart to the organs (arteries) and returns it to the heart (veins). The heart is essentially a muscular pump with four rooms, two for receiving incoming blood (atria) and two for pushing blood toward the organs (ventricles). In order to maintain unidirectional flow of blood, 3-D open-close "gateways" (valves) are optimally located between atria and ventricles, and between the ventricles and the emerging arteries. The workload of heart valves is nothing short of extraordinary. The valves open and close once per second totaling more than 3 billion times in a lifetime. These movements take place under constant pressure and in flowing blood, a very viscous fluid rich in minerals, proteins, lipids, and cells.

Valves may progressively become defective and critically influence the performance of the heart. These defects are collectively named valve diseases. As the properties of normal and diseased valves are not sufficiently understood, no drugs are available to treat and cure valve diseases. The only curative solution is to perform open-heart surgery to replace diseased valves with artificially engineered devices. These valve substitutes are nonliving materials fashioned in the form of valves that reestablish initial mechanical functions and restore the functionality of the heart. However, these devices may eventually fail because of imperfections in design, composition, or biocompatibility. Once a device shows signs of failure, the patient requires a second open-heart surgery for replacement of the defective device. Reoperations are generally risky and pose additional problems for children who naturally outgrow their implants.

Thus, the long-term goal of biomedical engineering is to find better methods for treating valvular diseases and thus impact lives of millions of patients worldwide. This daunting effort brings together a multidisciplinary team of specialists in medicine, biology, engineering, and mechanics. This chapter describes the structure, function, biology, and pathology of heart valves; information on current replacement devices; and the necessary prerequisites for constructing an "ideal" replacement valve. Ongoing research is aimed at improving existing devices by enhancing biocompatibility, as well as pioneering work on novel tissue-engineering approaches, which would facilitate complete regeneration of valve tissues.

2. NATURAL HEART VALVES ARE SPECIALIZED CARDIOVASCULAR TISSUES

The mechanical action of the human heart is similar to that of the two-cylinder engine. Four natural heart valves maintain the direction of blood flow within the circulatory system (Fig. 1). The tricuspid and mitral valves (collectively called the atrioventricular valves) open to allow blood to fill the ventricular cavities. These two "admission" (inflow) valves are connected to the subjacent heart muscle via fibrous extensions (chordae tendinae). When the inflow valves are closed, blood is forced to flow through the aortic and pulmonary valves (collectively called the semilunar valves). These two "ejection" (outflow) valves open in response to blood flow and immediately close after ejection. Although not initially apparent, a structural and functional correlation between the heart muscle and the valves exists (Fig. 1). A 3-D network of extra cellular matrix (mainly collagen fibers), formally known as the "cardiac skeleton," maintains the spatial structure and function of the heart muscle. This cardiac skeleton includes large structures (macro-skeleton), such as the heart valves, the chordae tendinae, and the septum, and a fibrous "micro-skeleton" of connective tissue, which

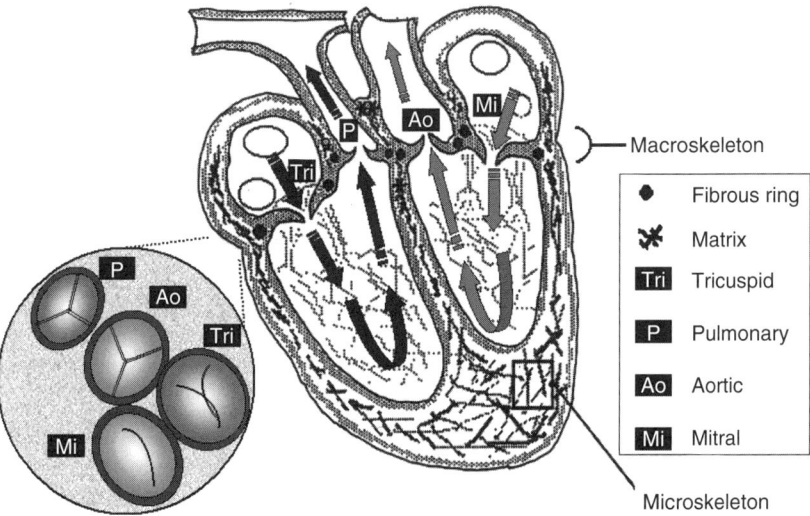

Figure 1. Functional anatomy of the heart valves. Longitudinal section diagram depicting venous blood (dark blue arrows) collecting in the right ventricle via the tricuspid valve (Tri) and sent through the pulmonary valve (P) to the lungs. Oxygenated blood (red arrows) returns to the left ventricle via the mitral valve (Mi) and is directed to the organs via the aortic valve (Ao). Insert at left shows a cross section through the heart at the level of the four heart valves. (This figure is available in full color at http://www.mrw.interscience.wiley.com/ebe.)

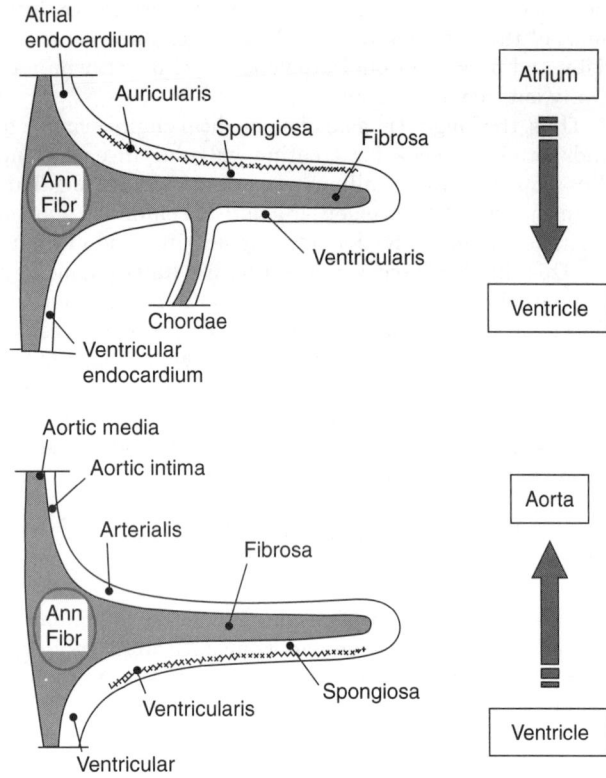

Figure 2. General structure of the heart valves. Representative schematic depicting a simplified cross section through an atrioventricular valve (top) and a semilunar valve (bottom); direction of blood flow is shown by thick arrow. Note the central supporting structure (fibrosa) inserted in the fibrous ring (Ann. Fibr.) and the presence of shock-absorbing layers (spongiosa), which are structural extensions of the cardiac endocardium.

comprises the entire collagenous network that encloses heart muscle fibers. Besides providing structural support for the cells, the cardiac skeleton allows a mechanical "coupling" function [i.e., the concomitant contraction of all muscle fibers during ventricular contraction (systole)] (1).

The heart valves are structurally composed of thin sheets of tissue (cusps or leaflets), which represent anatomical extensions of the cardiac endocardium (Fig. 2). These cusps are strengthened with a hydrophobic, collagenous, load-bearing tissue layer (fibrosa) that maintains continuity with the fibrous ring into which it is inserted (2). Elastic fibers that provide resilience and recoil are predominantly present in the inflow layers of the heart (e.g., the ventricularis of the semilunar valves) (3). Between these two layers is a loose connective tissue comprised predominantly of hydrophilic, shock-absorbing glycosaminoglycans (spongiosa) (4). This gel-like matrix acts as a space-filler and a lubricant to permit continuous bending, rearranging, straightening, and rotating of structural fibers with minimal friction (5).

In addition to these explicit biochemical and 3D constructs, specific cells exhibiting metabolic and contractile properties are also found in each heart valve, which include matrix-producing cells known as fibroblasts, active valvular interstitial cells known as myofibroblasts, and endothelium and resident macrophages (Fig. 3). Valvular cells exhibit high metabolic activities related to matrix homeostasis (synthesis and degradation of collagen and proteoglycans) as well as reactivity toward vasoactive compounds such as epinephrine and angiotensins. Therefore, this multivariate cell design clearly indicates that specific cellular activities characterize the form and function of each valve, which permits continuous adaptation to subtle hemodynamic modifications (6). In conclusion, heart valves are unique, specialized components of the cardiac connective tissue network that rely on delicately balanced homeostatic activities for their mechanical and biological functions.

3. HEART VALVE DISEASES ARE "CURED" BY SURGICAL REPLACEMENT

Valvular pathology is an important aspect of cardiovascular diseases. Mechanically speaking, valve dysfunction can be expressed as either imperfect closure (insufficiency) or incomplete opening (stenosis). In the majority

Figure 3. Cell types and distribution in heart valves. Representative schematic depicting a simplified cross section through an atrioventricular valve (top) and a semilunar valve (bottom); direction of blood flow is shown by arrow. Myocytes and smooth muscle cells populate the base of the valve indicating a "transition area" from a muscular structure to a specialized fibrous connective tissue. Fibroblasts and interstitial cells mainly populate the load-bearing valve structures; valves are fully covered by a continuous layer of endothelial cells.

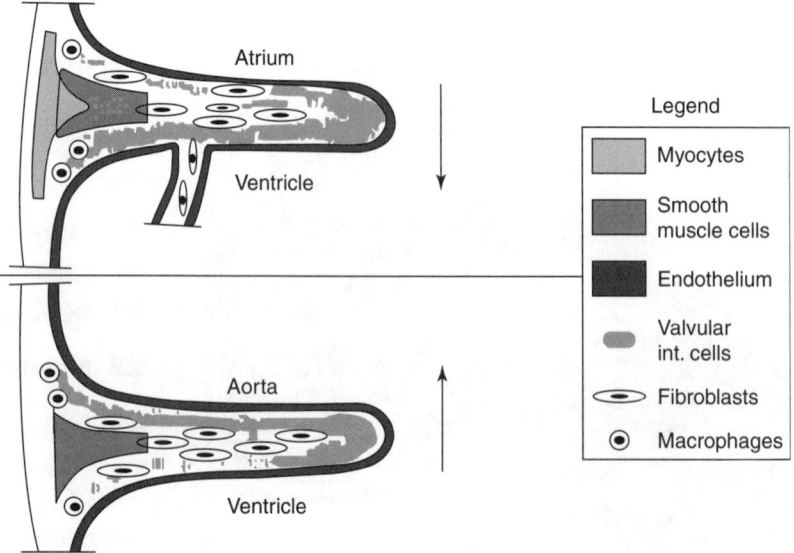

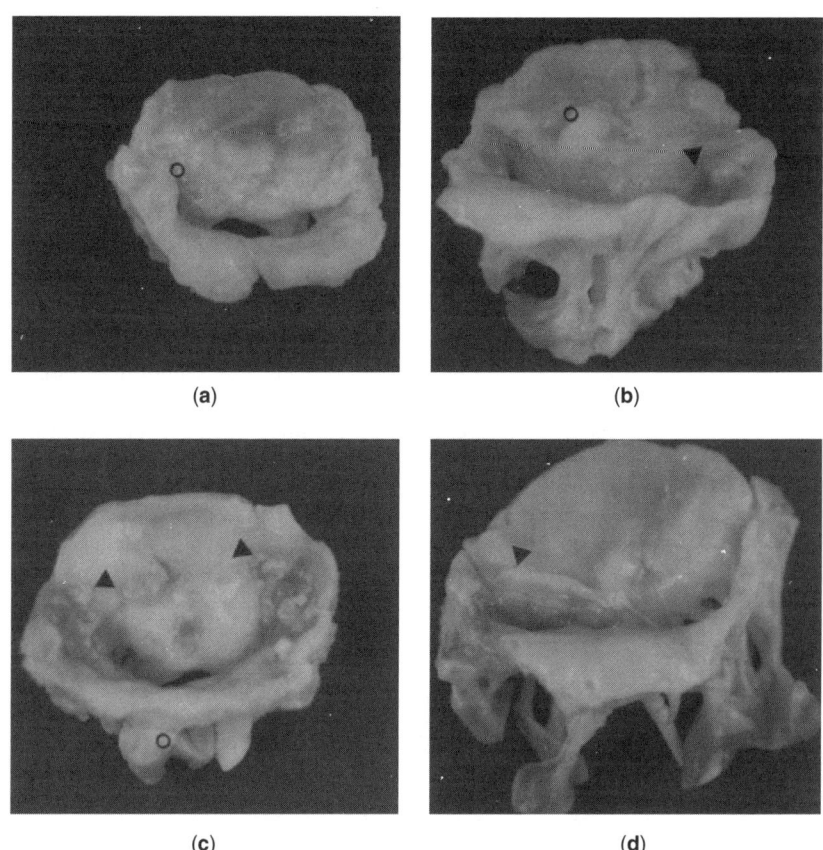

Figure 4. Pathology of human mitral valves (viewed from the atrial side). Photo depicting (a) an apparently normal valve, with some signs of leaflet thickening (o); (b) stenotic valve with signs of thickening (o); and extensive calcium deposits (▲) in both (c) and (d) with signs of chordae thickening. (This figure is available in full color at http://www.mrw.interscience.wiley.com/ebe.)

of pathological cases, valvular tissue appears either excessively deformed (floppy) or very thick and rigid, characterized by leaflet calcification (Fig. 4). The importance of the structural/functional connection of the heart muscle and the valves is clearly demonstrated by known pathologic situations. For example, in aortic stenosis, the heart muscle attempts to compensate for deficiencies in blood flow by growing in mass (hypertrophy). Consequently, regression of myocardial mass after valve replacement is one of the major clinical indicators of effective treatment of valvular pathology (7).

Valvular diseases associate with cell activation and alterations in metabolism of collagen and proteoglycans. These structural changes induce functional modifications that eventually lead to malfunction. The causes of valvular diseases are typically associated with congenital defects, atherosclerosis, infections, or postrheumatic episodes (6). However, the mechanisms of onset and progression of valvular diseases are largely unknown, specifically because of the lack of adequate experimental models.

Heart valve diseases progress rapidly and may become fatal. As the majority of heart valve disease is irreversible, its progression cannot be prevented by pharmacological treatments. Moreover, damaged heart valves lack the ability to spontaneously regenerate (8). The only treatment option presently available is open-heart surgery, in which the diseased valve is removed and replaced with an artificial device. Although effective, this procedure is quite traumatic for the patient. For a fortunate few, however, small areas of the valve can be reconstructed or repaired using biomaterials, obviating the need for major surgery to install new artificial heart valves. Pathological heart valves exhibit significantly altered mechanical and biological properties that eventually require surgical replacement with artificial devices. For an excellent update on heart valve pathology, please refer to the work of F. J. Schoen (6).

4. ARTIFICIAL HEART VALVES ARE EXCELLENT MECHANO-MIMETIC DEVICES

Heart valve replacement, pioneered in the early 1970s, is now a routine surgical procedure that employs devices made of nonliving, nonresorbable biomaterials for substitution of the valvular mechanical functions (9). Replacement of heart valves offers an excellent improvement in the quality of life for thousands of patients and can be considered one of the major accomplishments of biomedical engineering. It is estimated that more than 275,000 replacement heart valves are implanted annually worldwide; thus the social and economic impact of heart valves research and development is considerable (10).

Engineered devices, or mechanical valves, are used to replace diseased human heart valves in approximately 50% of the cases. Valves made of processed biological tissues are used in an additional 45% of cases. Pulmonary autograft valves (whereby the patient's own pulmonary valve is transplanted into the aortic position) and human cryopreserved allograft valves represent the remainder of implanted valves. Autografts and allografts exhibit excellent durability after implantation, but are not readily available for all patients (11).

Mechanical heart valves are constructed from rigid supporting materials and mobile components whose designs range from that of a "caged ball" device to designs encompassing free-moving tilting discs or flaps with restricted movement (Fig. 5). The components ensuring proper opening and closing of valvular orifices are made of inert, biocompatible materials such as pyrolitic carbon, polyester (Dacron)-covered polymers, or metals (12). Blood pressure differences within the chambers of the heart enable these mechanical valves to properly function. For example, for a caged ball valve inserted in the aortic position, contraction of the left ventricle will create sufficient blood pressure to thrust the ball toward the opposite end of the cage, thus allowing blood to flow around the ball. Immediately after systole (ventricular contraction), the ventricular pressure decreases and the ball is pushed toward the circular valve base, closing the orifice completely. However, imperfect blood flow through some of these devices can create nonphysiological patterns that may induce the formation of blood clots. Furthermore, the surfaces exposed directly to blood flow are not entirely antithrombogenic. For these reasons, in order to avoid complications caused by thrombus formation, patients with mechanical heart valves are required to take anticoagulation medications for the rest of their lives (13). As such, because patients may have possible episodic internal bleeding, mechanical valves are of limited use for pregnant women or women considering pregnancy and patients with coagulation diseases.

Biological heart valves (bioprostheses) are made of porcine aortic valves or bovine pericardial sheets that are mounted on adequate supports (stents) to mimic the valvular architecture, or they are left unmounted (stentless) (Fig. 5) (14). To understand how these devices are created, a typical procedure for creating a pericardium-derived heart valve is shown in Fig. 6. Bovine pericardium (the external fibrous sac that encloses the heart) is collected at the slaughterhouse, chemically stabilized, cut into three leaflets, and mounted on a three-legged (stented) ring using surgical sutures. The final product is adapted with a circular ring at the base that allows the surgeon to secure the device to the human fibrous ring.

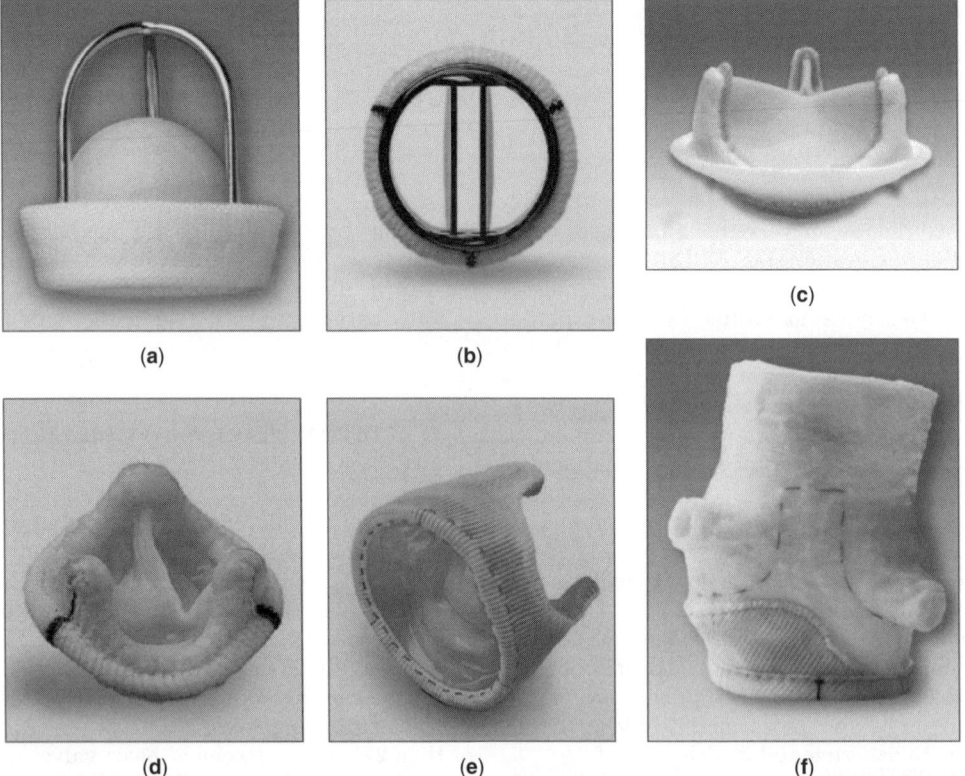

Figure 5. Artificial heart valves. Photo showing mechanical valves including devices that employ a caged ball (a) or two hinged leaflets (b) for functioning. Tissue valves are either made from bovine pericardium (c), porcine heart valves mounted on stents (d), or left unmounted (e and f). Valves in (a), (c)–(e) are shown in closed position, and valve (b) in a partially open position. Images (a), (c), and (f) were provided courtesy of Edwards Lifesciences, Inc. Models depicted are: Starr-Edwards® Ball Valve® (a), Carpentier-Edwards® PERIMOUNT Magna® Pericardial Bioprosthesis (c), and Edwards Prima Plus® Stentless Bioprosthesis (f). Copyright Edwards Lifesciences, Inc. 2005. All rights reserved. Images (b), (d), and (e) were provided courtesy of St. Jude Medical, Inc. Models depicted are: St Jude Medical® Regent Valve (b), Biocor Aortic (d), and Toronto SPV (e). Copyright St. Jude Medical, Inc. 2005. All rights reserved. (This figure is available in full color at http://www.mrw.interscience.wiley.com/ebe.)

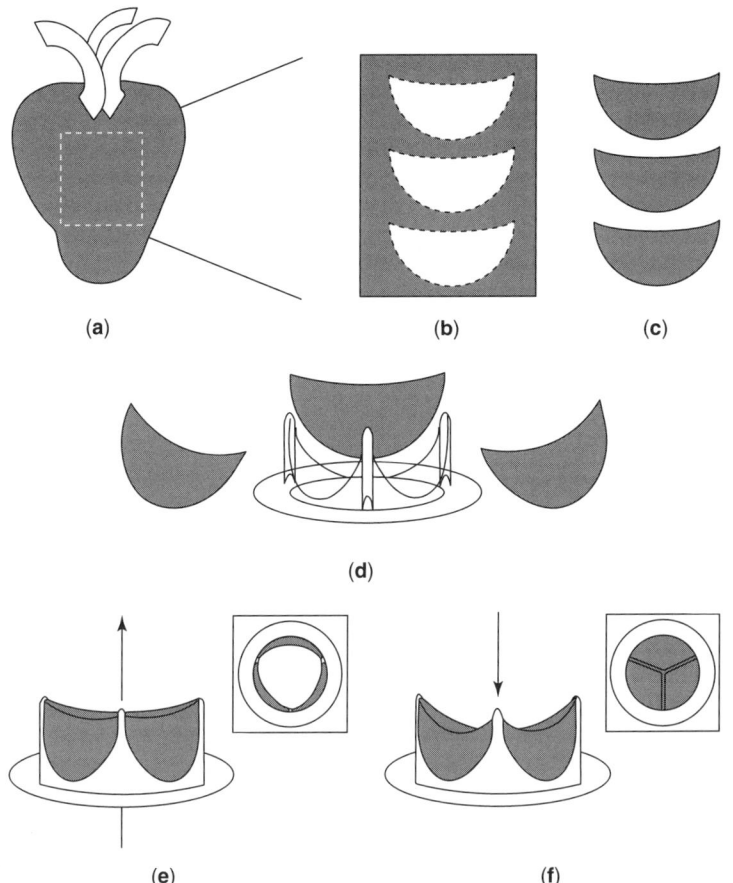

Figure 6. Artificial heart valve from bovine pericardium. Diagram showing the steps involved in creating a heart valve device. Bovine pericardium is collected (a) then cross-linked in glutaraldehyde (b), fashioned into three leaflets (c), mounted on a stent (d), and fitted with a sewing ring. While opening, blood flow (arrow) pushes leaflets toward the outside (e) and creates an almost circular orifice (insert). Upon closure, leaflets are pushed toward the center of the valve (f) and complete closure is ensured by central coaptation of leaflets (insert).

Similar to natural valves, the function of bioprostheses is pressure-driven. When the blood pressure below the valve is greater than the pressure above, the leaflets are forced toward the outside, thus reversing curvature and causing the valve to open. When the blood pressure above the valve is greater than the pressure below, the valve closes by pushing the leaflets toward the center. As the total surface of the leaflets exceeds that of the orifice, the leaflets overlap in the center (coaptation) and allow for complete closure of the orifice, without backflow (regurgitation).

Cross-species implantation of animal tissues is clearly prone to severe immune rejection and rapid tissue degeneration. For this reason, bovine or porcine tissues are treated with glutaraldehyde, a water-soluble cross-linker, which almost completely reduces tissue antigenicity (15). In addition, glutaraldehyde devitalizes tissues and kills all resident cells, prevents degradation by host enzymes, and sterilizes the tissue for implantation. Interestingly, the use of glutaraldehyde was inspired from its use in microscopy (as a fixative) and as a hospital sterilant (in the form of Cidex, a commonly used bactericidal liquid sterilization medium for heat-sensitive materials) (16). Tissue-derived heart valves are less thrombogenic than their mechanical counterparts and do not require long-term anticoagulation. For an excellent overview of different models and properties of heart valves, please refer to the work of J. Butany et al. (14).

Overall, mechanical and tissue-derived heart valves perform exceptionally well as devices intended to open and close valvular orifices, and thus could be considered as admirable mechano-mimetic replacements.

5. ARTIFICIAL HEART VALVES HAVE A LIMITED BIOLOGICAL DURABILITY

Artificial heart valves function quite effectively for many years after implantation, but their long-term durability is quite limited. Clinical follow-ups indicate that more than 50% of patients with artificial valve implants develop complications within 10 years (6). This disturbing trend suggests that the majority of implanted valves would have to be explanted after 20 years. From the perspective of the valvular patients, a second open-heart surgery to retrieve and replace the defective device is prone to high clinical risks, and therefore undesirable. The lack of long-term valve durability is especially important in the case of pediatric patients, where supplemental surgical procedures are required to accommodate the natural growth of the patients. The choice between bioprosthetic and mechanical valves depends on specific patient characteristics. Mechanical valves are more durable but require life-long anticoagulant therapy, whereas bioprostheses tend to deteriorate more rapidly because of degenerative processes but do not require anticoagulation therapy. In general,

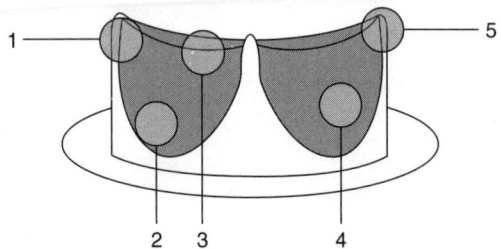

Figure 7. Pathology of tissue heart valves. Diagram showing typical aspects observed in explanted devices, including (1) tissue ruptures, (2) abrasions, (3) thickening of the tissue, (4) calcification, and (5) host tissue overgrowth.

tissue valves are implanted in elderly patients (60–65 years or older), which have a lower tendency to calcify tissue implants and a mean life expectancy of 10 to 15 years (17). Conversely, more mechanical valves are implanted in younger patients and children.

Analysis of explanted valves revealed that the principal causes of device failure are thrombosis, degeneration, and lack of full integration into host tissue, perivalvular leaks, host tissue overgrowth, and susceptibility to infections. Complications related to thrombogenicity are the major problems of mechanical valves, whereas tissue-derived valves are mainly affected by the degeneration and calcification of the tissue components (18). For example, explanted pericardial valves reveal tissue abrasion and erosion, tearing and perforations, deformations and loss of bending stiffness, along with deposition of calcium minerals (Fig. 7).

Tissue degeneration in artificial heart valves occurs as a result of the interplay between host-related factors and implant-related factors. As a response to the implant, the host selects one or more defense mechanisms that correspond to the degree of activation induced by the implant. Implantation of tissue heart valves induces a chronic foreign-body response, a low degree of immune reaction, activation of the coagulation cascade, and inflammatory response. These reactions are translated into deposition of fibrin immunoglobulins and complement infiltration of activated macrophages and lymphocytes and growth of host fibrous tissue over nonmoving parts of the implant (18).

The main pathways of bioprosthetic heart valve failure are structural damage and calcification of the tissue component. Analysis of explanted (failed) bioprosthetic heart valves typically reveals the coexistence of structural damage and calcification. Structural damage may be pure (noncalcific), stress-induced disruption of fiber architecture (19), or mediated by enzymatic degradation (20). Mechanical deformation was shown to accelerate calcification and, conversely, calcific deposits significantly alter mechanical properties (21). These processes may be synergistic, but convincing evidence also exists that each may occur independently. Many improvements in design and geometry of the leaflets have reduced, but not fully eliminated, the incidence of mechanical damage. For an excellent overview of heart valve design issues, please refer to the work of I. Vesely (22).

Calcium deposition in tissue-derived artificial heart valves is one of the major causes of their clinical failure. For this reason, noteworthy efforts have been made to elucidate the mechanisms of valve calcification and to implement treatments that would prevent this unwarranted side effect (23). Calcification typically appears as granular, fibrillar, or lamellar deposits, which greatly reduce tissue mechanical properties and can effectively contribute to erosion, abrasions, tearing, and perforations. The major factors involved in calcification are young age, hypercholesterolemia, and flexural stress, implant composition, and chemical pretreatments of the biological tissues. Calcification may occur through passive mechanisms (direct deposition of calcium and phosphate) but host cells, via remodeling mechanisms involving protease-mediated degeneration of matrix components and deposition of bone proteins, can mediate it (24). Mechanisms underlying calcification of tissue-derived artificial heart valves in patients are still not fully understood, although tissue composition and the use of glutaraldehyde appear to be two essential factors. In most cases analyzed thus far, the mineral phase is composed of bone-like hydroxyapatite associated with collagen, elastin, and chemically devitalized cells. Ongoing research focuses on prevention of calcification in tissue-derived artificial heart valves by attempts to remove or extract cells, by structural modification of collagen and elastin, and by stabilization or addition of natural calcification inhibitors (25,26).

Glutaraldehyde induces adequate preservation of collagenous structures, but also triggers severe cell alterations, which result in the formation of cell debris that resemble matrix vesicles formed by bone cells. For this reason, treatments are under study to extract all cellular components before implantation (23). Residual loosely bound or unreacted aldehyde groups may also be involved in tissue calcification. Neutralization of free aldehyde groups with compounds that possess reactive amines such as glutamine, glycine, homocysteic acid, and lysine has been shown to reduce calcification in animal models. In addition to simple aldehyde neutralization, amino-oleic acid, a treatment recently implemented in clinical use, may also hinder calcification by mechanisms related to reduction of calcium diffusion through tissues (27). Additionally, amino-biphosphonates (28) may also act as crystal growth inhibitors to hinder calcification. A different approach involving the use of stabilization chemistries, which do not employ glutaraldehyde, is also under investigation. These chemistries include carbodiimides, dye-mediated photo fixation, and epoxy-based crosslinkers (29). Tissues cross-linked with nonglutaraldehyde agents do not calcify in experimental models to the same extent as glutaraldehyde-fixed tissues, thus directly linking the presence of glutaraldehyde as one direct cause of tissue calcification. Extensive research on alternative cross-linking methods is currently being pursued, but none have yet reached clinical use.

These results indicate that mechanical and biological factors contribute significantly to failure of artificial heart valves. Although they perform well mechanically, artificial heart valves do not possess sufficient biological assets to fulfill the requirements of true biomimetic devices. Their lack of biologic properties may account for their limited durability after implantation. As artificial heart valves lack live cells, they are deficient in their ability to maintain and adapt the valvular matrix composition or to maintain an adequate calcium homeostasis. In the absence of live cells, degenerative processes induced by mechanical fatigue, proteolytic enzymes, and calcium deposition slowly erode the structural components and lead to progressive valve deterioration. In addition, the lack of an intact layer of endothelial cells may allow free influx of blood components and could also contribute to device-related thrombogenicity. Finally, the majority of current artificial heart valves, except newer stentless valves, do not maintain their cardiac skeleton-valve continuity and thus may progressively impair heart function.

In conclusion, pathological artificial valves exhibit altered mechanical properties, similar to human pathological valves. However, exciting research is currently ongoing to improve these devices and increase their biological durability. It is expected that efforts targeted toward understanding the pathology of artificial heart valves will facilitate greater insights into human valvular pathology.

6. THE "IDEAL" ARTIFICIAL HEART VALVE

One of the greatest biomedical engineering challenges today is to develop an implantable device that resists the natural conditions to which heart valves are subjected, without eliciting host reactions that would impair their function. This research field is very exciting and poses abundant questions whose answers will positively impact the lives millions of patients. Specifically, these natural conditions affecting the lifespan of an artificial heart valve include their durability to 40 million beats per year for about 80 years (summing up to about 3.2 billion cycles in a lifetime) under a cyclical pressure of about $200 \, g/cm^2$ in a very aggressive, corrosive fluid of high viscosity. Other research endeavors include testing valves within elevated concentrations of calcium and phosphate, proteins, lipids, enzymes, and cells, in the presence of potentially destructive defense mechanisms such as the immune system, coagulation cascade, inflammation, encapsulation, and calcification (30). This ongoing research has initially concluded that the ideal artificial heart valve should fulfill the following specific clinical, mechanical, and biological prerequisites. These prerequisites include, but are not limited to:

1. Full valve closure and release, and an internal orifice area close to the natural valve;
2. Limited resistance to blood flow while moving, and laminar flow through valve;
3. Mechanically durable, resistant to wear, maintain properties throughout lifetime;
4. Perfect integration into host tissue with minimal healing responses;
5. Chemically stable, nonleachable, noncytotoxic components;
6. Nonhemolytic, nonthrombogenic, substrate not supportive of infections;
7. Nonimmunogenic, noninflammatory, noncalcifying;
8. Long-term shelf life without changes in properties and sterility;
9. Continuity with the cardiac skeleton;
10. Ease of implantation, preferably using minimally invasive surgery;
11. Available in all sizes and reasonably priced.

To assess these properties, artificial heart valves are rigorously tested *in vitro* and *in vivo*, before being approved for human use. *In vitro* testing includes analysis of flow patterns by computer modeling and evaluation of hydrodynamic functions using ventricle simulators (31). In addition, valve components are routinely tested for mechanical properties such as elasticity, tensile strength, and wear resistance. For long-term durability evaluation, valves are subjected to fatigue testing at accelerated rates (up to 20 times faster that normal heart rate) (32). *In vivo* tests include subdermal implantation of tissue valve components for calcification studies (33) and valve replacement in sheep (34).

Currently, no artificial heart valve device, either mechanical or tissue-derived, fulfills all those previously described prerequisites. No material produced by the human imagination or through cutting-edge technology available can fully reproduce the complexity of the natural heart valve. However, the possibilities for creating such a naturally complex construct are endless if biomaterials or approaches can be developed that, instead of waging war with biology, will promote perfect integration within local and systemic physiologic systems. One such promising avenue is the regenerative medicine/tissue engineering approach.

7. REGENERATIVE MEDICINE APPROACHES TO HEART VALVE REPLACEMENT

The field of regenerative medicine is based on the innovative and visionary principle of using the patient's own cells and extracellular matrix components to restore or replace tissues and organs that have failed. This regenerative approach is a derivative of reconstructive surgery, where surgeons use the patient's tissues to rebuild injured or aging body parts. Modern approaches to heart valve regenerative medicine include several research methodologies, collectively known as tissue engineering. The most intensely researched approaches are (1) the use of decellularized tissues as scaffolds for *in situ* regeneration, (2) construction of tissue equivalents in the laboratory before implantation, and (3) use of scaffolds preseeded with stem cells.

A widespread approach is to use native (uncrosslinked) decellularized valves obtained from processed human or animal tissues. Once implanted, decellularized tissues are expected to provide proper environment and sufficient stimuli for host cells to infiltrate, remodel, and eventually regenerate the valvular tissue. As antigenic determinants are mainly concentrated on the surface of cells, removal of the original cells (decellularization) is necessary for avoiding complications related to immune rejection, which can be satisfactorily attained using combinations of detergents and enzymes, leaving behind 3-D scaffolds comprising apparently intact matrix molecules. Experimental data obtained with decellularized porcine valves yielded spectacular results (35). *In vitro* hydrodynamic performance was excellent, and valves performed well after implantation in sheep for 5 months as replacement of pulmonary valves. Explanted valves showed good repopulation of porcine scaffolds with fibroblast-like cells as well as an almost complete re-endothelialization of valve surfaces. These encouraging studies prompted several small-scale studies in humans using a commercially available decellularized porcine heart valve. However, despite the initial enthusiasm, the results of a small study in children were catastrophic.

In a breakthrough and intrepid study, a group of cardiac surgeons from Austria reported that three out of four children implanted with decellularized porcine heart valves had died within 1 year after implantation because of tissue rupture, degeneration, and calcification of the implant (36). Analysis of explanted decellularized porcine heart valves revealed lack of cell repopulation or endothelialization. The collagen matrix induced a severe inflammatory response and encapsulation of the graft and also served as a substrate for development of numerous calcification sites. The same group of researchers also recently identified the presence of the porcine cell-specific disaccharide galactose-alpha-1-3-galactose (alpha-Gal epitope) in decellularized porcine heart valves as the possible source of immunogenicity (37). Similarly, decellularized blood vessels also elicited inflammatory responses (38) and failed to maintain patency (39). Evidently, more basic studies are required for further development of these products as well as to reevaluate the relevance of animal models. Several academic groups, as well as a group of medical device companies, continue to pursue research and development of decellularized cardiovascular tissues (40–42).

A second approach involves construction of tissue equivalents in the laboratory prior to implantation, with the expectation that assembling the tissue-engineered valvular constructs from appropriate cells, and synthetic or natural matrix components, would create mechanically competent, nonthrombogenic, living tissues capable of adaptation and growth. Compared with decellularized tissues, this approach provides better control of the device properties before implantation. However, because of the enormity of this task, and because of possible early clinical failures of decellularized tissues, researchers in the field of heart valve tissue regeneration have taken a cautious, stepwise approach.

The *in vitro* assembly of heart valve tissue from its individual components was pioneered by I. Vesely et al. (43,44). In this ingenious approach, a chemically cross-linked nonbiodegradable glycosaminoglycan gel was seeded with vascular cells and cultured *in vitro*, which resulted in the formation of a thin sheet of elastin at the interface between cells and the glycosaminoglycan gel. The collagen structural component was created by fibroblast-mediated compaction of soluble collagen, with the expectation that *in vitro* assembly of these building blocks would, at some point, create a valve-like structure. For more details, please refer to the chapter on HEART VALVE TISSUE ENGINEERING in this Encyclopedia.

In an exemplary collaborative effort, clinicians, basic scientists, polymer chemists, and biomedical engineers focused on creating functional tissue-engineered heart valves *in vitro* using biodegradable scaffolds (45). These efforts included scaffold design and characterization, optimization of cell sources, finding adequate mechanisms for cell delivery, and optimizing *in vitro* culture of the seeded scaffolds, culminating with the surgical replacement of heart valves in sheep. A wide variety of scaffolding models created from biodegradable polymers have been tested for heart valve tissue engineering, including polyglycolic acid, polylactic acid, polycaprolactone, and biodegradable elastomers, which are manufactured into nonwoven textiles in shapes and conformations that mimic the natural architecture of the heart valve. These scaffolds are then seeded with cells that can be obtained from (1) fully differentiated cells such as myofibroblasts and endothelial cells derived from systemic arteries, or (2) pluripotent stem cells derived from adipose tissue, bone marrow, or peripheral blood (46). As mature cells have a limited lifespan, an attractive cell source was stem cells for heart valve tissue engineering. Recently, in a landmark experiment, mesenchymal stem cells obtained from ovine bone marrow were seeded onto biodegradable scaffolds and the constructs were cultured *in vitro* before implantation as a valve in the pulmonary position of sheep (47) for up to 8 months. Tissue-engineered valves performed well hemodynamically, with signs of slow degradation of the scaffolds, and concomitant deposition of new extracellular matrix. Moreover, cell types resembled those present in natural heart valves. Overall, these exciting results hold great promise for truly effective regenerative approaches to treatment of heart valve disease.

8. FUTURE AND PERSPECTIVES

By virtue of its clinical applications, biomedical engineering has matured into an interdisciplinary medical specialty, which necessitates intertwining expertise from numerous fields, including surgery, pathology, cell and molecular biology, matrix biochemistry, engineering, and mechanics. In anticipation of major scientific breakthroughs in this field, surgeons will continue to treat these diseases by first reestablishing the mechanical functions of the heart using artificial valves. The yellow brick road leading to effective treatments of valvular disease continues to present multiple challenges. Three extremely exciting lines of investigation are:

A. Finding causes and developing nonsurgical therapy approaches for valvular disease. To achieve this goal, more detailed molecular, biochemical, and cellular studies are needed on normal and diseased human valves as well as development of more clinically relevant animal models for valve diseases.
B. Improvement of current artificial devices. Regarding mechanical valves, it would be a milestone of success to drastically reduce the incidence of

thrombogenicity caused by valve replacement with mechanical devices. For tissue valves, more effective stabilization and anticalcification treatments are required to reduce the incidence of tissue degeneration and extend device durability. More studies on implant-host tissue interactions are needed to develop innovative materials that elicit minimal healing responses that are caused by surgically invasive and inherently traumatic valve replacement techniques. Finally, a need to develop minimally invasive implantation approaches that would reduce the need for open-heart surgery for valve replacement exists.

C. Regenerative medicine approaches. Heart valve tissue engineering is only in the nascent stage of research. Therefore, to succeed, it is imperative that we fully understand the structural-functional properties of normal heart valves and to determine how the valvular cells remodel the tissue. As stem cell research offers a marvelous potential for effective regeneration of heart valve tissue by differentiation into the "proper" cells, insights into embryological development are also essential (8).

9. CONCLUSIONS

Heart valves are connective tissues that maintain their remarkable mechanical and biological functions under austere biomechanical and biochemical conditions within the human body. As heart valve injuries lead to diseases that cause progressive valve degeneration, open-heart surgery to replace diseased tissue with an engineered device is, at present, the only curative method for correcting these altered mechanical and biological properties. Mechanical and tissue-derived artificial heart valves successfully reestablish mechanical functions but do not replenish the biological requisites for long-term durability and eventually fail. Remarkable research efforts are underway to develop better devices as well as to open new and exciting avenues for tissue engineering approaches, which could potentially facilitate a complete regeneration of valve tissues.

BIBLIOGRAPHY

1. J. M. Icardo and E. Colvee, Atrioventricular valves of the mouse: III. Collagenous skeleton and myotendinous junction. *Anat. Rec.* 1995; **243**(3):367–375.
2. I. Vesely and R. Noseworthy, Micromechanics of the fibrosa and the ventricularis in aortic valve leaflets. *J. Biomech.* 1992; **25**(1):101–113.
3. I. Vesely, The role of elastin in aortic valve mechanics. *J. Biomech.* 1998; **31**(2):115–123.
4. J. Lovekamp and N. Vyavahare, Periodate-mediated glycosaminoglycan stabilization in bioprosthetic heart valves. *J. Biomed. Mater. Res.* 2001; **56**(4):478–486.
5. J. J. Lovekamp, D. T. Simionescu, J. J. Mercuri, B. Zubiate, M. S. Sacks, and N. R. Vyavahare, Stability and function of glycosaminoglycans in porcine bioprosthetic heart valves. *Biomaterials*, in press.
6. F. J. Schoen, Cardiac valves and valvular pathology: update on function, disease, repair, and replacement. *Cardiovasc. Pathol.* 2005; **14**(4):189–194.
7. X. Y. Jin and J. R. Pepper, Do stentless valves make a difference? *Eur. J. Cardiothorac. Surg.* 2002; **22**(1):95–100.
8. E. Rabkin-Aikawa, J. E. Mayer, Jr., and F. J. Schoen, Heart valve regeneration. *Adv. Biochem. Eng. Biotechnol.* 2005; **94**:141–179.
9. M. I. Ionescu, B. C. Pakrashi, D. A. Mary, I. T. Bartek, and G. H. Wooler, Replacement of heart valves with frame-mounted tissue grafts. *Thorax* 1974; **29**(1):56–67.
10. Y. S. Morsi, I. E. Birchall, and F. L. Rosenfeldt, Artificial aortic valves: an overview. *Int. J. Artif. Organs* 2004; **27**(6):445–451.
11. F. J. Schoen and R. J. Levy, Founder's Award, 25th Annual Meeting of the Society for Biomaterials, perspectives. Providence, RI, April 28–May 2, 1999. Tissue heart valves: current challenges and future research perspectives. *J. Biomed. Mater. Res.* 1999; **47**(4):439–465.
12. J. Butany, M. S. Ahluwalia, C. Munroe, C. Fayet, C. Ahn, P. Blit, C. Kepron, R. J. Cusimano, and R. L. Leask, Mechanical heart valve prostheses: identification and evaluation. *Cardiovasc. Pathol.* 2003; **12**(6):322–344.
13. D. Horstkotte, Prosthetic valves or tissue valves–a vote for mechanical prostheses. *Z. Kardiol.* 1985; **74** (Suppl 6):19–37.
14. J. Butany, C. Fayet, M. S. Ahluwalia, P. Blit, C. Ahn, C. Munroe, N. Israel, R. J. Cusimano, and R. L. Leask, Biological replacement heart valves. Identification and evaluation. *Cardiovasc. Pathol.* 2003; **12**(3):119–139.
15. M. Dahm, M. Husmann, M. Eckhard, D. Prufer, E. Groh, and H. Oelert, Relevance of immunologic reactions for tissue failure of bioprosthetic heart valves. *Ann. Thorac. Surg.* 1995; **60**(2 Suppl):S348–S352.
16. D. T. Cheung, N. Perelman, E. C. Ko, and M. E. Nimni, Mechanism of crosslinking of proteins by glutaraldehyde III. Reaction with collagen in tissues. *Connect. Tissue Res.* 1985; **13**(2):109–115.
17. A. M. Borkon, L. M. Soule, K. L. Baughman, W. A. Baumgartner, T. J. Gardner, L. Watkins, V. L. Gott, K. A. Hall, and B. A. Reitz, Aortic valve selection in the elderly patient. *Ann. Thorac. Surg.* 1988; **46**(3):270–277.
18. F. J. Schoen and C. E. Hobson, Anatomic analysis of removed prosthetic heart valves: causes of failure of 33 mechanical valves and 58 bioprostheses, 1980 to 1983. *Hum. Pathol.* 1985; **16**(6):549–559.
19. M. S. Sacks and F. J. Schoen, Collagen fiber disruption occurs independent of calcification in clinically explanted bioprosthetic heart valves. *J. Biomed. Mater. Res.* 2002; **62**(3):359–371.
20. D. T. Simionescu, J. J. Lovekamp, and N. R. Vyavahare, Extracellular matrix degrading enzymes are active in porcine stentless aortic bioprosthetic heart valves. *J. Biomed. Mater. Res. A* 2003; **66**(4):755–763.
21. I. Vesely, J. E. Barber, and N. B. Ratliff, Tissue damage and calcification may be independent mechanisms of bioprosthetic heart valve failure. *J. Heart Valve Dis.* 2001; **10**(4):471–477.
22. I. Vesely, The evolution of bioprosthetic heart valve design and its impact on durability. *Cardiovasc. Pathol.* 2003; **12**(5):277–286.
23. D. T. Simionescu, Prevention of calcification in bioprosthetic heart valves: challenges and perspectives. *Expert Opin. Biol. Ther.* 2004; **4**(12):1971–1985.
24. A. Simionescu, D. Simionescu, and R. Deac, Biochemical pathways of tissue degeneration in bioprosthetic cardiac

valves. The role of matrix metalloproteinases. *Asaio. J.* 1996; **42**(5):M561–M567.
25. N. Vyavahare, M. Ogle, F. J. Schoen, and R. J. Levy, Elastin calcification and its prevention with aluminum chloride pretreatment. *Am. J. Pathol.* 1999; **155**(3):973–982.
26. C. M. Giachelli, Ectopic calcification: gathering hard facts about soft tissue mineralization. *Am. J. Pathol.* 1999; **154**(3):671–675.
27. J. P. Gott, C. Pan, L. M. Dorsey, J. L. Jay, G. K. Jett, F. J. Schoen, J. M. Girardot, and R. A. Guyton, Calcification of porcine valves: a successful new method of antimineralization. *Ann. Thorac. Surg.* 1992; **53**(2):207–215; discussion 216.
28. R. J. Levy, G. Golomb, J. Wolfrum, S. A. Lund, F. J. Schoen, and R. Langer, Local controlled-release of diphosphonates from ethylenevinylacetate matrices prevents bioprosthetic heart valve calcification. *Trans. Am. Soc. Artif. Intern. Organs* 1985; **31**:459–463.
29. M. A. Moore, PhotoFix: unraveling the mystery. *J. Long Term Eff. Med. Implants* 2001; **11**(3–4):185–197.
30. T. Gudbjartsson, S. Aranki, and L. H. Cohn, Mechanical/bioprosthetic mitral valve replacement. In: L. H. Cohn and L. H. Edmunds, Jr., eds., *Cardiac Surgery in the Adult*. New York: McGraw-Hill, 2003, pp. 951–986.
31. A. P. Yoganathan, Z. He, and S. Casey Jones, Fluid mechanics of heart valves. *Annu. Rev. Biomed. Eng.* 2004; **6**:331–362.
32. K. Iwasaki, M. Umezu, K. Iijima, and K. Imachi, Implications for the establishment of accelerated fatigue test protocols for prosthetic heart valves. *Artif. Organs* 2002; **26**(5):420–429.
33. F. J. Schoen, G. Golomb, and R. J. Levy, Calcification of bioprosthetic heart valves: a perspective on models. *J. Heart Valve Dis.* 1992; **1**(1):110–114.
34. M. F. Ogle, S. J. Kelly, R. W. Bianco, and R. J. Levy, Calcification resistance with aluminum-ethanol treated porcine aortic valve bioprostheses in juvenile sheep. *Ann. Thorac. Surg.* 2003; **75**(4):1267–1273.
35. S. Goldstein, D. R. Clarke, S. P. Walsh, K. S. Black, and M. F. O'Brien, Transpecies heart valve transplant: advanced studies of a bioengineered xeno-autograft. *Ann. Thorac. Surg.* 2000; **70**(6):1962–1969.
36. P. Simon, M. T. Kasimir, G. Seebacher, G. Weigel, R. Ullrich, U. Salzer-Muhar, E. Rieder, and E. Wolner, Early failure of the tissue engineered porcine heart valve SYNERGRAFT in pediatric patients. *Eur. J. Cardiothorac. Surg.* 2003; **23**(6):1002–1006; discussion 1006.
37. M. T. Kasimir, E. Rieder, G. Seebacher, E. Wolner, G. Weigel, and P. Simon, Presence and elimination of the xenoantigen gal (alpha1, 3) gal in tissue-engineered heart valves. *Tissue Eng.* 2005; **11**(7–8):1274–1280.
38. F. Sayk, I. Bos, U. Schubert, T. Wedel, and H. H. Sievers, Histopathologic findings in a novel decellularized pulmonary homograft: an autopsy study. *Ann. Thorac. Surg.* 2005; **79**(5):1755–1758.
39. M. A. Sharp, D. Phillips, I. Roberts, and L. Hands, A cautionary case: the SynerGraft vascular prosthesis. *Eur. J. Vasc. Endovasc. Surg.* 2004; **27**(1):42–44.
40. H. E. Wilcox, S. A. Korossis, C. Booth, K. G. Watterson, J. N. Kearney, J. Fisher, and E. Ingham, Biocompatibility and recellularization potential of an acellular porcine heart valve matrix. *J. Heart Valve Dis.* 2005; **14**(2):228–236; discussion 236–237.
41. E. Rieder, G. Seebacher, M. T. Kasimir, E. Eichmair, B. Winter, B. Dekan, E. Wolner, P. Simon, and G. Weigel, Tissue engineering of heart valves: decellularized porcine and human valve scaffolds differ importantly in residual potential to attract monocytic cells. *Circulation* 2005; **111**(21):2792–2797.
42. D. T. Simionescu, Q. Lu, Y. Song, J. Lee, T. N. Rosenbalm, C. Kelley, and N. R. Vyavahare, Biocompatibility and remodeling potential of pure arterial elastin and collagen scaffolds. *Biomaterials* 2006; **27**(5):702–713.
43. A. Ramamurthi and I. Vesely, Evaluation of the matrix-synthesis potential of crosslinked hyaluronan gels for tissue engineering of aortic heart valves. *Biomaterials* 2005; **26**(9):999–1010.
44. Y. Shi, A. Ramamurthi, and I. Vesely, Towards tissue engineering of a composite aortic valve. *Biomed. Sci. Instrum.* 2002; **38**:35–40.
45. S. P. Hoerstrup, R. Sodian, S. Daebritz, J. Wang, E. A. Bacha, D. P. Martin, A. M. Moran, K. J. Guleserian, J. S. Sperling, S. Kaushal, J. P. Vacanti, F. J. Schoen, and J. E. Mayer, Jr., Functional living trileaflet heart valves grown in vitro. *Circulation* 2000; **102**(19 Suppl 3):III44–III49.
46. F. W. Sutherland and J. E. Mayer, Jr., Tissue engineering for cardiac surgery. In: L. H. Cohn and L. H. Edmunds, Jr., eds., *Cardiac Surgery in the Adult*. New York: McGraw-Hill, 2003, pp. 1527–1536.
47. F. W. Sutherland, T. E. Perry, Y. Yu, M. C. Sherwood, E. Rabkin, Y. Masuda, G. A. Garcia, D. L. McLellan, G. C. Engelmayr, Jr., M. S. Sacks, F. J. Schoen, and J. E. Mayer, Jr., From stem cells to viable autologous semilunar heart valve. *Circulation* 2005; **111**(21):2783–2791.

ARTIFICIAL KIDNEY, MODELING OF TRANSPORT PHENOMENA IN

SUNNY ELOOT
PASCAL VERDONCK
Ghent University
Ghent, Belgium

1. BACKGROUND

1.1. Function of Healthy Kidneys

The urinary system consists of two kidneys, which filter blood and deliver the produced urine into the two ureters. From the ureters, the urine is passed to the urinary bladder, which is drained, via the urethra during urination. The kidneys are bean-shaped organs of about 11 cm long, 4 to 5 cm wide, and 2 to 3 cm thick and lie bilaterally in the retroperitoneum in the abdominal cavity. The smallest functional unit of the kidney is the uriniferous tubule, each containing a nephron and a collecting tubule. Approximately 1 to 1.3 million nephrons exist in each kidney. One nephron is composed subsequently of a vascular part (glomerulus), a drainage part (Bowman's capsule), a proximal tubule, Henle's loop, and a distal tubule (Fig. 1). Several nephrons are drained by one collecting tubule, which enlarges downstream until it becomes a duct of Bellini and perforates the renal papilla (1).

The major function of the kidneys is removing toxic byproducts of the metabolism and other molecules smaller than 69000 Da (i.e., smaller than albumin) by filtration of the blood flowing through the glomerulus. They also

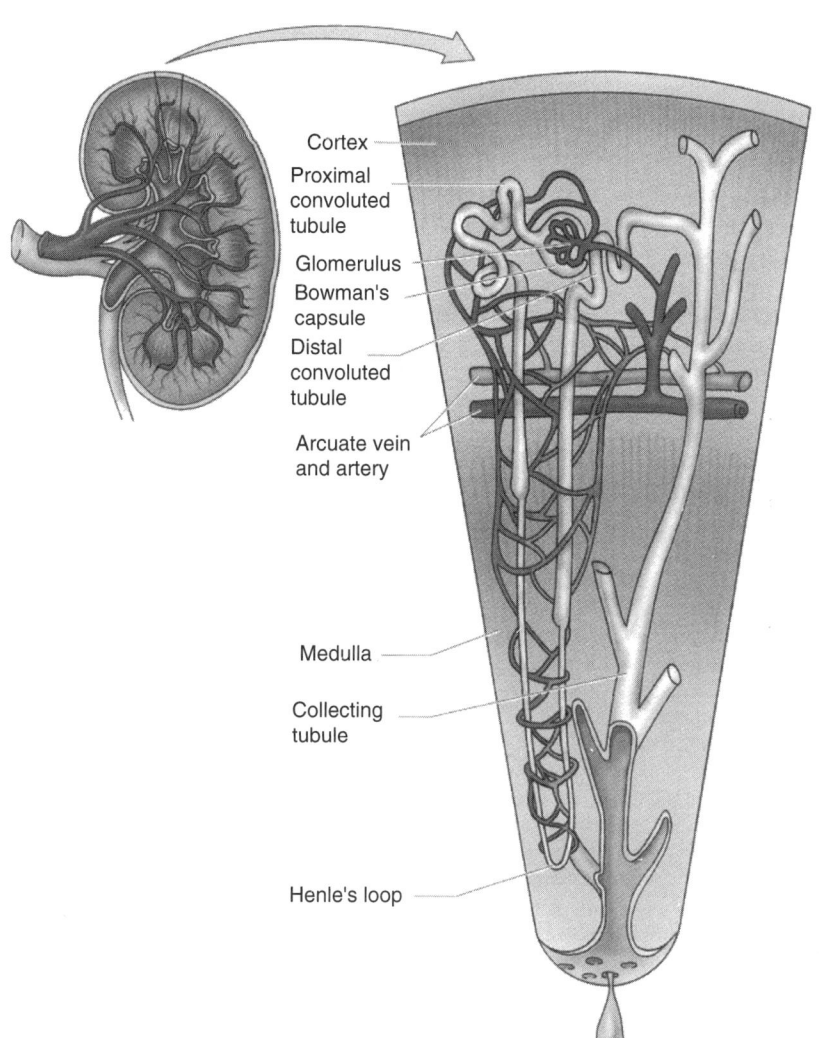

Figure 1. The uriniferous tubule, the smallest functional unit of the kidney.

regulate body fluid composition and volume. Specifically, resorption of salts (Na^+, K^+, Cl^-), glucose, creatine, proteins, and water takes place in the tubular parts. As a result of these eliminating and conserving functions, the kidneys also contribute to the regulation of the blood pressure, hemodynamics, and the acid-base balance of the body. Additionally, kidneys have an endocrine function: They produce the hormones renin, erythropoietin, and prostaglandines[1] and help in converting Vitamin D to dihydroxycholecalciferol, a substance that controls calcium transport (1).

1.2. The Uremic Syndrome

Renal insufficiency can be subdivided into three categories according to the duration that the kidneys lose their ability to purify the blood: acute (hours to days) (2), subacute (months), and chronic (years) renal failure. In contrast with the subacute and chronic form, acute renal failure is often reversible. The uremic syndrome is the result of the retention of compounds, normally cleared by healthy kidneys, and of a disorder in the hormonal and enzymatic homeostasis (3). As renal failure progresses, glomerular filtration rate as well as the amount of functional nephrons decreases. The main causes of end-stage renal disease are diabetes and hypertension, whereas the most important symptoms are found in the cardiovascular (4–6), neurological (7,8), hematological (9,10), and immunological (11–13) status.

1.3. Renal Replacement Therapies

The diagnosis of chronic renal failure is based on the indication of a decreased renal function or a disorder in urine sedimentation. In daily practice, creatinine[2] clearance (95 ± 20 ml/min for women and 120 ± 25 ml/min for men) (14) is used as a measure of the glomerular ultrafiltration rate and quantifies the remaining renal function. Ureum[3] clearance is, in contrast with the creatinine clearance, strongly dependent on the protein intake, the catabolic state of the patient, and the urine flow rate.

---

[1]Derivatives of essential fatty acids to maintain homeostasis
[2]An endproduct of the muscular metabolism
[3]A product of the amino acid catabolism

Nevertheless, the increase of serum ureum is a useful additional marker of chronic renal failure.

As renal replacement therapy, three modalities are available: kidney transplantation from cadaver or living donors, peritoneal dialysis, and hemodialysis.

1.3.1. Transplantation. The introduction of the end-to-end anastomosis technique (15), the revelation of the secrets of the HLA (Human Leukocyte Antigen) (16,17), and the availability of immunosuppressiva (18) opened the way to successful transplantations. The implantation of the donor kidney usually occurs in one of the fossae iliacae[4]. The venous anastomosis consists of a side-to-end connection of the vena renalis with the vena iliaca communis/externa. The arteria renalis is anastomosed with the arteria iliaca interna (end-to-end) or with the arteria iliaca communis/externa (side-to-end). Last but not least, the donor ureter must be fixed at the supralateral side of the bladder roof after performing an antireflux channel at the bladder wall.

Transplantation has to deal with two main problems: organ rejection and donor shortage. Possible rejections of transplanted organs can be subdivided into two groups: hyperacute rejections, which are serological processes based on preformed antibodies, and the rejections caused by cellular reactions between T-lymphocytes and HLA. The latter can be acute or chronic.

Potential living kidney donors must have a bilateral renal function, tissue histocompatibility, and may not have any systemic disease. Kidneys from brain death donors can only be used if the donor did not have vascular diseases, diabetes, or malignancies. Furthermore, because of different selection criteria, age and cardiac situation, and health economic reasons, only a small percentage of patients get on the waiting list. And because of the shortage of kidney donors, only a small percentage of patients on the waiting list are actually transplanted (25% for West Europe).

In spite of those drawbacks, kidney transplantation can be called *the* solution for chronic renal failure with immense advantages for the patient: no limitation concerning water intake, less restricted diet, no suffering from anemia, normalization of the bone metabolism, and return to a dynamic life with a social and professional reintegration.

1.3.2. Peritoneal Dialysis. With peritoneal dialysis (19), a mostly hypertonic glucose dialysis fluid (20) is injected in the peritoneal cavity by means of a permanent peritoneal catheter (21–25). The peritoneal cavity is an intra-abdominal space that is surrounded by a serous membrane called the peritoneum (1–1.5 m^2). It is a semipermeable membrane that contains mesothelial cells on an interstitium that consists of connective tissue with capillaries and lymphatic vessels. In between the mesothelial cells, intercellular gaps (range 5 nm) are responsible for the major solute transport between the dialysis fluid and the blood in the capillaries (14). Ultrafiltration is mainly achieved by the osmotic effect of the glucose, poly-glucose, or other osmotic agents in the dialysate, and takes place through the very small pores (range 0.5 nm), also called aquaporins (26). Solute clearance and ultrafiltration are mainly determined by the permeability of the peritoneal membrane. This characteristic is patient-dependent and can be calculated by the peritoneum equilibration test (PET) (27).

After a dwell time of 2–14 hours, the peritoneal cavity is drained and filled again by means of a cycling machine or manually by gravimetrical flow. In order to assist the patient with the fluid exchanges, different user-friendly and automated machines are available as well as devices to assure right, safe, and sterile connections with the bags or the machine (e.g., UV Flash™—Baxter, Chicago, IL; stay.safe®—Fresenius Medical Care, Germany).

Peritoneal dialysis can be performed continuously [CAPD = Continuous Ambulatory Peritoneal Dialysis (28), CCPD = Continuous Cyclic Peritoneal Dialysis (29,30)], or intermittently (DAPD = Daytime Ambulatory Peritoneal Dialysis, IPD = Intermittent Peritoneal Dialysis (31), NIPD = Nightly Intermittent Peritoneal Dialysis).

1.3.3. Hemodialysis. Hemodialysis is a blood purifying therapy in which the blood of a patient is circulated through an artificial kidney, also called hemodialyzer. This process is realized in an extracorporeal circuit (Fig. 2) where one or two needles (or catheters) can be used as the patient's vascular access. A general hemodialysis therapy lasts about 9–15 hours a week, mostly spread over three sessions. It can take place in the hospital, in a low-care unit, or at home.

Two types of hemodialyzers are in use: plate and hollow fiber dialyzers (32). In a plate dialyzer, membrane sheets are packed together and blood and dialysate flow in subsequent layers. The priming volume is around 30% larger than in a hollow fiber dialyzer. The latter (Fig. 3) consists of thousands of small capillaries (inner diameter in the range of 200 µm and wall thickness of 8–40 µm). Blood flows inside the capillaries whereas dialysate flows counter-currently around them. Typical blood flow rates are in the range of 200 to 350 ml/min (33), whereas dialysate flows are preferably twice the blood flow (34). Besides the

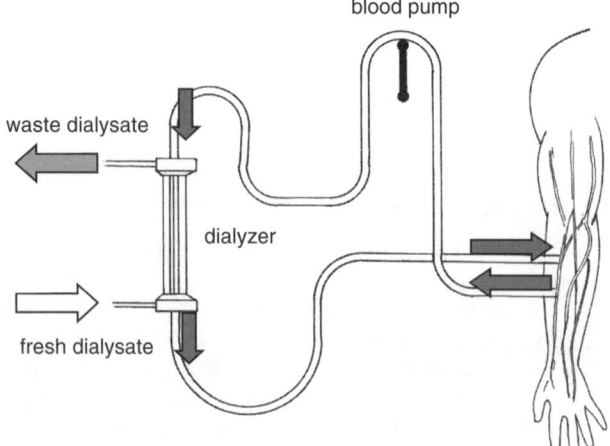

Figure 2. The extracorporeal circuit in hemodialysis.

---
[4]Cavity at the intestinal bone

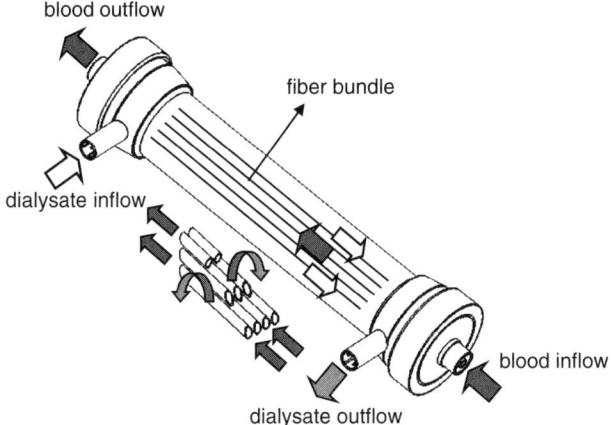

Figure 3. The hollow fiber dialyzer.

advantage of a small blood volume, these dialyzers suffer from problems like clotting in and clogging of the capillaries. With respect to the membrane characteristics, distinction can be made between low-, medium-, and high-flux dialyzers on one hand (ultrafiltration coefficient lower than 15, between 15 and 40, and higher than 40 mL/h/mmHg, respectively), and low- and high-area dialyzers on the other (membrane surface lower and higher than 1.5 m^2, respectively).

Already, from the start of hemodialysis, the challenge for nephrologists was to obtain an adequate vascular access. The Quinton–Scribner shunt (35) with the use of an external access is, today, if used, only used in patients with acute renal failure and important vascular problems. More often, catheters are used for acute short phase of renal failure. The original subcutaneous internal arterio-venous fistula, described by Brescia and Cimino (36), between the arteria radialis and the vena cephalica is still the most successful angioaccess method (37). In the latter, arterial flow and pressure dilates the vein, facilitating repetitive puncture. In case vessel conditions are inadequate or fail to dilate (10–30% of the patients), bridge grafts between an artery and a suitable vein are used. Several types of graft material are used, including autologous veins (38–41), allografts (42), and synthetic grafts (43,44). As more elderly people with peripheral vascular disease are recruited on dialysis, the central venous catheter, which was initially introduced for acute dialysis, is gaining popularity in long-term dialysis treatment (45–47).

As hemodialysis implies a repeated and compulsory contact of blood with foreign materials, biocompatibility problems are unavoidable. Traditionally, biocompatibility is defined as the absence of functional or biochemical reaction during or after the contact of the body, a body fluid or an organ with an artificial device or a foreign material (48,49). Dialysis-related biocompatibility problems are mainly because of the intermittent nature, the application of high blood flows, and the use of dialysis fluid and of semipermeable membranes. They can be summarized as problems related to clotting phenomena (50–52), complement and leukocyte activation (53,54), susceptibility to bacterial (55) and tuberculosis infection (56), leaching (57), surface alterations (58), allergic reactions (59,60),

shear (61), and inverse transfer of electrolytes (62) or endotoxins from the dialysate toward the blood (63).

Besides the problems related to the vascular access and the biocompatibility, patients on chronic hemodialysis may also suffer from cardiac, hematological, coagulation, acid-base, pulmonary, infectious, dermatological, neurological, nutritional and gastrointestinal, ocular, endocrine, and sexual problems (64).

2. BIOPHYSICS OF A HEMODIALYZER

In hemodialysis therapy, the dialyzer succeeds in purifying the blood and extracting the excess water caused by basic transport phenomena, such as diffusion, ultrafiltration, and osmosis. As transport takes place between the blood and dialysate compartment over a semipermeable membrane, fluid characteristics and membrane properties should also be considered.

2.1. Blood Characteristics

2.1.1. Blood Constitution and Major Functions. An average adult has a total blood volume of about 5l, which is approximately 7% of total body weight. Blood is a dark red, viscous, slightly alkaline suspension (pH 7.4) of cells—erythrocytes (red blood cells), leukocytes (white blood cells), and thrombocytes (platelets)—suspended in a fluid (plasma). The volume fraction of cells (45% for male, 43% for female) is better known as the hematocrit (14).

The main *functions* of blood include transportation of nutrients from the gastrointestinal system to all cells of the body and, subsequently, delivering waste products of these cells to organs for elimination. Oxygen (O_2) is carried from the lungs to all cells of the organism by the hemoglobin in the erythrocytes, whereas carbon dioxide (CO_2) is transported back to the lungs for elimination, both by the hemoglobin and by the plasma. Besides nutrients, the bloodstream also transports numerous other metabolites, cellular products, and electrolytes. Additionally, blood has a function of regulating the body temperature and maintaining the acid-base and osmotic balance of the body fluids.

Plasma consists of water (90%), proteins (9%) and inorganic salts, ions, nitrogens, nutrients, and gases (1%) (14). Several plasma proteins exist with different origin and function (e.g., albumin (69000 Da), α- and β-globulins (80000-1E+6 Da), γ-globulins, clotting proteins, complement proteins (C1 to C9), and plasma lipoproteins).

Erythrocytes (14) are nonnucleated, biconcave-shaped disks, 7.5 μm in diameter and 1–2 μm thick. Their large surface-volume proportion benefits the exchange of gases. Erythrocytes are packed with hemoglobin, a large protein (68000 Da) composed of four polypeptide chains, which are covalently bound to an iron-containing heme. In regions of high oxygen concentration, the hemoglobin part releases CO_2 while the iron binds to O_2. *Leukocytes* (14) use the bloodstream as a means for traveling and only fulfill their function after diapedesis[5]. Within the bloodstream,

---
[5]Leaving the blood vessels and entering the surrounding connective tissue

leukocytes are round, whereas they are pleomorphic in connective tissue. Their main function is to defend the human body against foreign substances. They can be classified into two main groups: granulocytes (60–70% neutrophiles, 4% eosinophiles, and 1% basophiles) and agranulocytes (20–25% lymphocytes and 3–8% monocytes). *Thrombocytes* are small (2–4 μm in diameter), disk-shaped, nonnucleated cell fragments containing several tubules and granules. They function in limiting hemorrhage of blood vessel endothelium in case of injury (14).

2.1.2. Blood Rheology. Blood is a nonNewtonian fluid characterized by a nonlinear relationship between shear stress τ (Pa) and strain rate $\gamma = \partial u/\partial y$ (1/s) (65):

$$\tau = \mu \cdot \left(\frac{\partial u}{\partial y}\right)^m = \mu \cdot \gamma^m, \quad (1)$$

with μ the dynamic viscosity (Pa.s), u the velocity in axial direction (m/s), y the direction perpendicular to the flow direction (m), and m a coefficient equal to unity for Newtonian fluids and smaller than 1 for shear thinning fluids like blood (–).

The shear thinning behavior as well as the dependence of the blood viscosity μ on the hematocrit H and the plasma viscosity μ_p, is described among others by Quemada (66) (Fig. 4):

$$\mu = \frac{\mu_p}{(1 - 1/2k.H)^2}. \quad (2)$$

Parameter k is function of the intrinsic viscosities $k_0(H)$, characterizing the red blood cell aggregation at zero shear

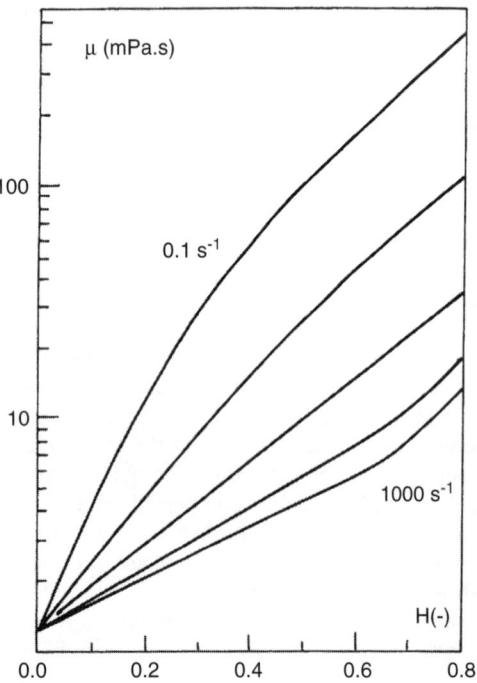

Figure 4. Dynamic viscosity as a function of hematocrit, described by Quemada (66).

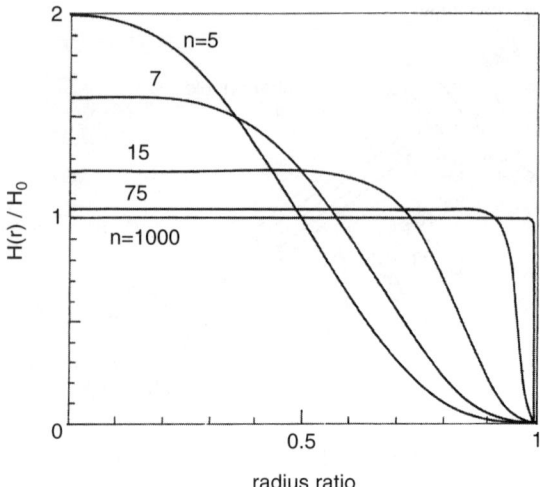

Figure 5. The radial variation of the hematocrit over the fiber radius, described by Lerche and Oelke (68): nonuniformity of cell distribution increases with decreasing n values.

stress, $k_\infty(H)$, describing the orientation and deformation of red blood cells at important shear stress, and the shear rate γ (66). For a fixed hematocrit, viscosity decreases with increasing shear rate, whereas for a fixed shear rate, viscosity increases with hematocrit.

Blood flowing through small capillaries exhibits a redistribution of the red blood cells creating a plasma-skimming layer that can be observed near the wall while red blood cells are concentrated in the center. Gaehtgens (67) described the effect of this nonuniform cell distribution on the flow by defining an apparent blood viscosity μ_{app} (Pa.s) for use in the Haegen–Poiseuille equation, describing laminar flow in a circular tube (65):

$$Q = \frac{1}{\mu_{app}.L} \cdot \frac{\pi.d^4}{128} \cdot \Delta P, \quad (3)$$

with Q the flow rate (m^3/s) through a tube with diameter d (m) and ΔP the pressure drop over the tube length L (m).

The radial variation of the hematocrit was deduced by Lerche and Oelke (68) using a parameter n that describes the degree of plasma skimming: nonuniformity of cell distribution increases with decreasing n (Fig. 5):

$$H(r) = H_0 \cdot \left(\frac{-n.(n+1).(n-1)}{2}\right) \cdot \left[\frac{r^n}{n} - \frac{2.r^{n-1}}{n-1} + \frac{r^{n-2}}{n-2} - \frac{2}{n.(n-1).(n-2)}\right], \quad (4)$$

with H_0 the mean hematocrit (–), r the relative position in the capillary (–), and n the dimensionless Lerche parameter.

2.2. Dialysis Fluid Characteristics

The hemodialysis fluid should be considered as a temporary extension of the patient's extracellular fluid because of the bidirectional transport process when blood and dialysate are flowing through the dialyzer. Therefore, the

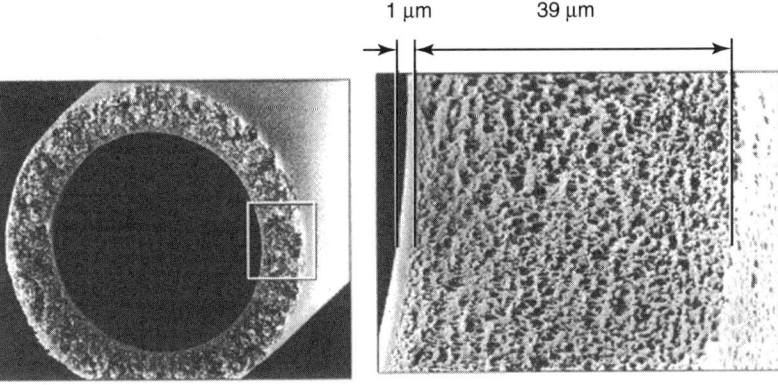

Figure 6. Synthetic polysulphone membrane.

composition of dialysis fluid is critical in achieving the desired blood purification and body fluid and electrolyte homeostasis. It contains reverse osmosis water, dextrose, and different electrolytes like calcium-, magnesium-, potassium-, and sodium-chloride and sodium acetate or sodium bicarbonate. The latter two fulfill the function of dialysate buffer, responsible for the correction of metabolic acidosis in the uremic patient. Hydrogen ions (H^+) are, soon after their production, buffered by plasma bicarbonate and can only be removed by the diffusive flux of alkaline from the dialysate into the blood replacing the blood buffers (69).

Besides the chemical composition, the physical and microbiological characteristics are important. As the use of highly permeable membranes in hemodialysis is responsible for backfiltration or backdiffusion[6], toxic and pyrogenic substances can move from the dialysate toward the blood, resulting in febrile reactions (63).

Today, the composition of dialysis fluid is prescribed for each single patient to individualize the dialysis therapy according to the personal needs (70). The actual dialysis machines guarantee accurate proportioning of treated water and concentrated salts, continuous monitoring of the final composition, and a constant maintenance of the required conductivity values (71).

The hemodialysis system is the endpoint of a hydraulic circuit where tap water is changed into reverse osmosis water through water supply, water pretreatment, water purification (72), and dialysis fluid preparation. The pretreatment consists of flowing tap water through filters, softeners, carbon filters, and microfilters. The subsequent treatment concerns flow through one or two reverse osmosis membranes (73) and a deionizer (74), closing the purification chain with ultrafiltration and submicrofiltration.

2.3. Membrane Properties

Hemodialysis membranes vary in chemical compositional structure, transport properties, and biocompatibility. Polymers can be categorized in three major groups (75,76): regenerated and modified cellulose membranes, and synthetic membranes. Regenerated cellulose membranes replaced collodion, the first polymer to be used as an artificial membrane, and showed a better performance and mechanical stability. Cuprophan, for example, is a polysaccharide with the same chemical but with other physical characteristics than the original cellulose because of a chemical modification. These membranes are very hydrophilic and form a hydrogel when absorbing water. Solute diffusion occurs through highly water-swollen amorphous regions.

Examples of synthetically modified cellulose are cellulose (di) (tri) acetate and hemophan. In the first, three hydroxyl groups are changed by an acetate group making it more hydrophobic than cellulose. With hemophan, 1% of the hydrogen (H) in the hydroxyl (OH) groups is changed by an amino ligand. The majority of cellulose and modified cellulose membranes have a thickness of 5–11 µm and a surface of 0.8–2.5 m².

Polysulphone (PS), polyamide (PA), and polyacrylonitrile polyvinylchloride copolymer (PAN-PVC) are membranes prepared from synthetic-engineered thermoplastics and are hydrophobic, asymmetric, and anisotropic with solid structures and open void spaces (76). These membranes are also characterized by a thin skin layer, determining the hydraulic permeability and solute retention properties, and a bulk spongy region providing mechanical strength (Fig. 6). Synthetic materials are usually less activating complement cascade and are less restrictive to the transport of middle and large molecules. The AN69 (acrylonitrile) is different from the other synthetic membranes because of its symmetric structure. The well-chosen proportion of the hydrophilic sulphonate groups and the hydrophobic nitrile groups makes it a membrane with good permeability and biocompatible characteristics.

As a result of varying polymer compositions, membranes with the same polymer names may differ in their hemocompatibility, flux properties, and adsorption characteristics (77,78). Proteins like beta2-microglobulin, fibrinogen and coagulation factors, complement proteins, or hormones like parathormone and erythropoietin are differently adsorbed by dialysis membranes, which contributes to the removal characteristics (79).

2.4. Basic Transport Phenomena

Diffusion of uncharged molecules in an isothermal system refers to the transport driven by a concentration

---

[6]Filtration and diffusion from the dialysate compartment toward the blood compartment

difference, and described by the law of Adolf Fick (80):

$$J = -D_S . A . \frac{\Delta C}{\Delta x}, \quad (5)$$

with J the net flux (mol/s), D_S the solute diffusivity (m²/s) being a unique property of the solute-solvent at a specific temperature, A the area of diffusion (m²), and $\Delta C/\Delta x$ the concentration difference (mol/m³) over membrane thickness (m).

Ultrafiltration is a mode of convective transport with a pressure difference as driving force. As solutes are conveyed by the fluid, it can be seen as passive transport of solutes. A general equation for ultrafiltration is given by Darcy's law:

$$J_u = h_m . A . \Delta P, \quad (6)$$

with J_u the volumetric flux (m³/s), h_m the hydraulic permeability (m/Pa.s), A the area of ultrafiltration (m²), and ΔP the pressure difference (Pa).

Osmosis can be described as diffusive transport. The difference with diffusion, however, is that the dissolved particles cannot pass the membrane (e.g., proteins). Thus, water passes the membrane in the opposite direction to tend to equalize the concentrations. The osmotic pressure $\Delta \pi$ is given by the expression of Van't Hoff (81):

$$\Delta \pi = \sigma . R . T . \Delta C, \quad (7)$$

with σ the reflection coefficient of the membrane (−), R the universal gas constant (8.314 J/mol.K), T the absolute temperature (K), and ΔC the concentration gradient (mol/l).

In hemodialysis, diffusion is the major transport phenomenon, while the term hemofiltration is used for the therapy in which solutes are mainly cleared by convection (82,83). In the latter, the excess water and vital solute removal is counterbalanced by adding a dilution fluid at the dialyzer inlet [predilution (84)], halfway to the blood trajectory in the dialyzer (e.g., Nephros, NY), or at the dialyzer outlet [post dilution technique (85)]. In hemodiafiltration therapy, toxic agents are removed by a combination of diffusion and convection resulting in a better clearance of high-molecular-weight (HMW) solutes (MW > 12000 Da) while maintaining the performance for low-molecular-weight (LMW) solutes (MW < 500 Da) (86).

2.5. Mass Transfer in Hemodialyzers

The practical application of the diffusion law (Equation 5), requires the definition of different coefficients that can help in either dialyzer design or clinical practice. From this point of view, the overall mass transfer coefficient K_0 (m/s) can be defined transforming Equation 5 into:

$$J = -K_0 . A . \Delta C. \quad (8)$$

The reciprocal of K_0 can be seen as the resistance to diffusive transport, R_0, which is the sum of blood-side, membrane, and dialysate-side resistances R_B, R_M, and R_D, respectively (87). Furthermore, as the mass transfer coefficient K_0 for radial diffusive mass transfer is equal to $D_S/\Delta x$ (solute diffusivity/traveled distance), the total resistance R_0 can be written as:

$$R_0 = R_B + R_M + R_D = \frac{\Delta x_B}{D_B} + \frac{\Delta x_M}{D_M} + \frac{\Delta x_D}{D_D}. \quad (9)$$

D_B, D_M, and D_D represent the diffusivities in blood, membrane, and dialysate. Δx_B and Δx_D symbolize a characteristic distance for diffusion in the blood and dialysate domain, whereas Δx_M is the membrane thickness.

From Equation 9 it can be observed that dialyzer efficiency can be best increased by reducing the largest resistance. The blood and dialysate-side resistances are mainly covered by the diffusion distance from the main fluid stream to and from the membrane, Δx_B and Δx_D. The membrane resistance, however, depends on membrane thickness, Δx_M, as well as on the diffusivity in the membrane, D_M, varying with the chemical composition of it.

The *diffusive dialysance* D (ml/min) is defined as the change in solute content in the blood inflow per unit of concentration driving force (88):

$$D = \frac{Q_{Bi} \cdot (C_{Bi} - C_{Bo})}{C_{Bi} - C_{Di}} = \frac{Q_{Di} \cdot (C_{Do} - C_{Di})}{C_{Bi} - C_{Di}}, \quad (10)$$

with Q_{Bi} the inlet blood flow rate (ml/min) and C_{Bi}, C_{Bo}, C_{Di}, and C_{Do} the blood inlet and outlet concentrations, and the dialysate inlet and outlet concentrations, respectively. As the dialysate inlet concentration is zero in the case of hemodialysis, Equation 10 can be simplified to the definition of the *diffusive clearance* K (ml/min), a definition that is analogous to the physiological kidney clearance (89):

$$K = \frac{Q_{Bi} \cdot (C_{Bi} - C_{Bo})}{C_{Bi}} = \frac{Q_{Di} \cdot (C_{Do} - C_{Di})}{C_{Bi}}. \quad (11)$$

In case ultrafiltration takes place, the clearance K is increased by the net contribution of ultrafiltration Q_{UF} (ml/min) to the flux:

$$K' = \frac{Q_{Bi} \cdot (C_{Bi} - C_{Bo})}{C_{Bi}} + Q_{UF} \cdot \frac{C_{Bo}}{C_{Bi}} = K + Q_{UF} \cdot \frac{C_{Bo}}{C_{Bi}}. \quad (12)$$

As these relations hold for aqueous solutions, a correction factor should be added, counting for the heterogeneous nature of blood. The influence of the hematocrit H (−), plasma water, and solute protein binding is considered by replacing Q_{Bi} by Q_E in the conventional formulas (90):

$$Q_E = Q_{Bi} \cdot \left[F_P - \frac{H}{100} \cdot (F_P - F_R \cdot k' \cdot \varphi) \right], \quad (13)$$

where F_P is the plasma water fraction, F_R the red blood cell water fraction, k' the equilibrium distribution coefficient, and φ the red blood cell water fraction that participates in solute transfer during blood flow through the dialyzer.

In clinical practice, clearance index, Kt/V_{urea} equal to 1.2–1.4 is used as the gold standard for adequate dialysis

(91,92). This indicator is larger for better clearance, K, longer dialysis time, t, or for a smaller patient distribution volume, V_{urea}. In general, an increase of Kt/V_{urea} by 0.1 is associated with a substantially decreased risk of death from cardiac, cerebrovascular, and infectious diseases (93). Kt/V_{urea}, however, measures only removal of low-molecular-weight substances, which occurs predominantly by diffusion, and does not consider clearance of larger molecules. Babb et al. (94) introduced middle-molecular-weight solutes (500–12000 Da) (86), playing an important role in uremic toxicity, especially in processes related to inflammation, atherogenesis,[7] and malnutrition. Moreover, they defined their clearance as the product of overall mass transfer coefficient K_0 and membrane area A, the proportion factor in Equation 8. Both described parameters (i.e., Kt/V_{urea} and K_0A) are linked by the Michaels equation (87), stating that diffusive clearance K is a function of blood and dialysate flow rates and of the dialyzer specific parameter K_0A.

Besides the mass transfer to and from the patient, described by dialysance D or clearance K, a transfer of water toward or from the dialysate compartment to control the patient's distribution volume also exists. In analogy with Darcy's law (Equation 6), the *ultrafiltration coefficient* K_{UF} (ml/min.mmHg) can be defined as (95):

$$K_{UF} = \frac{Q_{UF}}{\Delta P - \pi} = \frac{Q_{UF}}{TMP}, \quad (14)$$

with Q_{UF} the ultrafiltration flow rate (ml/min) and ΔP the hydraulic pressure difference (mmHg) between blood and dialysate compartment. The latter can be defined as the sum of transmembrane pressure (TMP) and oncotic pressure π exerted by the proteins present at dialyzer blood side. Although low-flux dialyzers were originally designed as diffusive exchangers (96), high-flux dialyzers have the therapeutic advantage of an increased solute removal by ultrafiltration. Their open pore structure results in high rates of small molecule diffusion (97) and middle molecule diffusion and convection (96,97).

Backfiltration may occur whenever the TMP becomes negative (98). The existence and importance of backfiltration during high-flux hemodialysis have been extensively demonstrated performing hydrostatic and oncotic pressure measurements (63,99–102). The main problem related to backfiltration is the bacterial contamination by liquid bicarbonate concentrate and the passage of endotoxins toward the blood compartment (63). Ronco (103), however, demonstrated the positive influence of high forward filtration in the proximal and backfiltration in the distal segment of the dialyzer for the removal of large molecules.

After the membrane is exposed to proteins, diffusive transport as well as hydraulic permeability decreases significantly because of protein adsorption (58). Moreover, these plasma proteins exert an oncotic pressure of 20–30 mmHg opposing the applied hydrostatic pressure (32,104). Furthermore, the ultrafiltration flow deviates from linearity for high-TMP values because of concentration polarization of high-molecular-weight substances in the blood that are not freely filtrated through the membrane pores (104,105).

3. HEMODIALYZER MODELING: A REVIEW

The flow distribution in the blood and dialysate compartment of a hollow fiber dialyzer is an important determinant of the overall mass transfer efficiency. A uniform flow distribution benefits local mass transfer, and any mismatch caused by nonuniform flow in the blood or dialysate compartment results in a worse uremic solute removal from the blood (106,107).

3.1. Flow Visualization and Modeling

Medical imaging techniques are a useful tool to visualize the overall flow in the blood and dialysate compartment of hemodialyzers. The use of different methods like CT (Computerized Tomography) and MRI (Magnetic Resonance Imaging) is described in literature. Besides experimental and theoretical approaches, computational fluid dynamics (CFD) can be seen as a third dimension in engineering. The cornerstones of CFD are the fundamental governing equations of fluid dynamics: continuity, momentum, and energy equations. These mathematical statements are based on the fundamental physical principles of mass conservation, Newton's second law, and energy conservation (65), respectively, which are usually partial differential equations in their most general form. By replacing these differential equations with numbers that are advanced in space or time, a final numerical description of the complete flow field of interest is obtained. However, the numerical results should at least be validated with analytical solutions or experimental measurements (108).

3.1.1. Blood Flow Distribution.
Nordon and Schindhelm (109) investigated, using a finite volume method, the parameters influencing blood flow distribution in a 2-D axisymmetrical model of a hollow fiber cell separator system. In the inlet and outlet manifolds, the continuity and Navier–Stokes equations (conservation of mass and momentum) were solved for steady incompressible laminar flow, whereas the hollow fibers were modeled as a porous medium. In the upstream manifold, boundary layer separation occurred at the point of channel divergence causing the formation of a separation bubble. Moreover, it was found that the uniformity of the porous medium flow is mainly influenced by the radial-to-axial hydraulic permeability ratio. Using magnetic resonance Fourier velocity imaging to determine blood and dialysate flow distributions simultaneously, Zhang et al. (110) found a uniform blood and nonuniform dialysate flow distribution.

3.1.2. Dialysate Flow Distribution.
Osuga et al. (111) determined dialysate pressure isobars in a low-flux hollow fiber dialyzer by combining the results of MRI and CFD of a contrast solution injection in the dialysate flow. The latter was regarded as a porous medium flow in the CFD model, depending on the radial-to-axial hydraulic permeability ratio, which was determined comparing the results

---
[7]Start of degeneration of the inner vessel wall

of both techniques. The *a priori* made assumption that the ultrafiltration flux into the dialysate can be ignored was supported by the finding that the isobars had no steep radial gradient such that a homogeneous flow pattern could be assumed. Using x-ray CT, however, Takesawa et al. (112) found a radially distributed dialysate flow caused by the breaking and twisting of fiber bundles made of cellulose membrane, resulting in preferential flow channels.

While striving for a homogeneous dialysate distribution, different dialyzer designs were developed and evaluated with medical flow imaging. Using a helical CT scan, Ronco et al. evaluated the use of spacing filaments (113) and hollow fibers with a Moiré-structured wave design (114). They found those techniques effective in preventing fiber twisting, thereby resulting in an idealized dialysate flow pattern without dialysate channeling. Using MRI, Poh et al. (115) found the design of flow baffles in combination with spacing filaments promoting flow uniformity even more. Besides those experimental techniques, CFD also plays an important role in the development of new dialyzer designs. Karode and Kumar (116) showed with a 3-D steady-state CFD model that spacing filaments in flat dialyzers are very effective in the promotion of fluid mixing while reducing concentration polarization by increasing the shear rate at the membrane surface (117). As a consequence, whether the dialysate flow distribution is homogeneous is mainly influenced by the dialyzer design with respect to fiber twisting.

Besides the macroscopic computation of blood and dialysate flow in a dialyzer using porous media, it should be noted that some important aspects like ultrafiltration and related fluid properties, convection-diffusion, concentration polarization, and protein adsorption at the membrane should be investigated with microscopic models.

3.1.3. Ultrafiltration. Ultrafiltration, generated by osmotic and hydrostatic pressure drops over a porous membrane, was originally numerically described by Kedem and Katchalsky (118) using thermodynamics of irreversible processes. The membrane properties were described in terms of filtration, reflection, and permeability. Instead of using the complex formulas describing irreversible processes, the Maxwell–Stefan approach, assuming all diffusion vectors as a linear function of the flux, can be used, as this theory is especially suited to describe multicomponent mass transport (119). Rather than using thermodynamics, Kargol (120) redefined those coefficients assuming a membrane with randomly distributed pore sizes.

Using a theoretical model validated with experiments, Wüpper et al. (121) described the profile of ultrafiltration and concentration along the axis of high-flux hollow fiber dialyzers. They found that the hydrostatic pressure profile could be approximated as linear even in the presence of a nonlinear concentration profile for impermeable solutes. As a result, changes in fiber radius and membrane permeability can be studied using the Darcy (Equation 6) and Haegen–Poiseuille laws (Equation 3). Karode (122) derived analytical expressions for the pressure drop in a permeable tube as a function of wall permeability, channel dimensions, axial position, and fluid properties by differentiating the Haegen–Poiseuille formula and applying it locally to infinitesimal sections.

3.1.4. Concentration-Polarization. In the presence of ultrafiltration, particles within the main stream are subjected to a drag force and accumulate near the membrane surface, whereas the accumulated particles tend to migrate to the feeding stream driven by the concentration gradient (Equation 5). The boundary layer concentration modifies the solute or solvent properties like viscosity, density, and solute molecular diffusivity (123). As a result of particle accumulation, ultrafiltration flow decreases with time. A steady-state value, described as a function of the concentration ratio near the membrane and in the main stream, and of the mass transfer coefficient, was derived by Michaels (124) using the one-dimensional convection-diffusion equation. Backfiltration, caused by the oncotic effect, induces a shear stress, which was incorporated by Zydney and Colton (125) by using the shear-induced hydrodynamic diffusion coefficient analyzed experimentally by Eckstein et al. (126). Moreover, the shear stress, maximal at the membrane on the fluid-like concentrated layer, causes this layer to get fluidized. This shear-induced hydrodynamic diffusion process was implemented by Romero and Davis (127) taking into account the two-dimensional characteristics of ultrafiltration by integrating the axial momentum equation into the one-dimensional convection-diffusion equation. Lee and Clark (128) used the two-dimensional convection-diffusion equation to describe the ultrafiltration flow decline caused by concentration polarization. They found that concentration as well as boundary layer thickness increases with axial distance but decreases for higher diffusion coefficients and axial velocities.

Membrane fouling by protein deposition decreases ultrafiltration until a quasisteady flux is obtained, which is determined by the balance between the drag force on the proteins and the intermolecular repulsive interactions between the proteins in the bulk and those in the boundary layer (129). In order to investigate membrane fouling, Ho and Zydney (130) developed a combined pore-blockage and cake-filtration model for bovine serum albumin. The model was recently found effective for a variety of proteins (131).

For computational modeling of such a thin boundary layer in which the solute concentration changes intensively, a very dense grid should be used. To enable the use of a large grid, Miranda and Campos (123) applied a simple natural logarithmic variable transformation in the solute transport equation attenuating the concentration derivatives inside the boundary layer. Other studies (132,133) refer to the use of CFD to model concentration polarization in the fluid phase adjacent to the membrane without taking into account the selective permeation through the membrane fluid phase.

3.1.5. Multiphase Flow. The unidirectional shear flow of highly concentrated fluid-particle suspensions, showing particle migration from regions of high shear to regions of low shear, has been investigated using a two-dimensional (134) and axisymmetrical (135) numerical model. The suspension, treated as a Newtonian fluid, is modeled using the momentum and continuity equation, whereas the particle motion is governed by a modified transport equation accounting for the effects of shear-induced particle migrations. Although the parameters of rigid neutrally buoyant particles (135) are far from matching those of the deformable red blood cells, the numerical results, in good agreement with analytical predictions of Phillips et al. (136), may contribute to a better understanding of the possible local variations of the hematocrit.

3.2. Physical Models Describing Mass Transport

In case diffusion is the dominant transport phenomenon, Jaffrin et al. (137) showed that the assumption of a linear concentration profile in the boundary layers is only valid for LMW solutes. In the case of larger molecules, it should rather be assumed exponential. Grimsrud and Babb (138) and Colton et al. (139) investigated the concentration profile for diffusion in blood flowing through two infinite flat plates and at the entrance region, respectively, assuming zero ultrafiltration. Berman (140) solved the Navier–Stokes equations asymptotically for small constant ultrafiltration values. As the assumption of a zero or constant ultrafiltration velocity was not realistic, Ross (141) analyzed the mass transport in a laminar Newtonian fluid flow through a permeable tubular membrane. He used an ultrafiltration velocity depending on hydrostatic and osmotic pressure differences while the solute transport was determined by the reflection coefficient of the membrane. He found that the exit concentration decreases as membrane permeability, fiber diameter, and fiber length increase. In addition, the change of concentration profile by varying radial transport is most expressed at 0% and 60% radial distance.

In several hemodiafiltration studies (121,137,142–147), transport phenomena have been described based on the assumption of a constant ultrafiltration flow (142–145), and neglecting the accumulation of partially rejected solutes at the membrane wall (142,144). Jaffrin (146) and Zydney (148) found that the convective contribution of HMW solutes partially rejected by the membrane is larger than expected and, in the case of dominant ultrafiltration, is independent of the sieving coefficient. Others (137,143,144,146) investigated the solute transport accounting for the module geometry, the membrane properties, and the operating conditions, but without considering the effective ultrafiltration profile along the dialyzer length or the change of the mass transfer coefficient. On the other hand, Legallais et al. (149) incorporated those aspects and the concentration polarization aspect in a one-dimensional transport model for hemodiafilters using Newtonian fluids and assuming the desired ultrafiltration rate. Pressure in both compartments was assumed to drop according the Haegen–Poiseuille equation applied to an equivalent cylindrical tube (150) and using local fluid viscosities.

For the investigation of mass transport in dialyzers, most authors neglect or simplify the ultrafiltration aspect, whereas others (141,149) take into account a linear pressure distribution over the length of the dialyzer, determining the ultrafiltration profile.

4. A NUMERICAL MODEL OF THE OVERALL BLOOD AND DIALYSATE FLOW IN A HEMODIALYZER

4.1. Introduction

The main objective of the present study was to assess the flow distribution in hemodialyzers by combining experimental and numerical techniques (151). A CFD model was developed for the simulation of the flow distributions in a low-flux dialyzer, providing detailed quantitative information about local velocities and fiber bundle permeabilities. Numerical results were further validated using single-photon emission-computed tomography (SPECT) imaging, a different imaging technique then those previously described in literature. The combination of the SPECT images and the CFD simulations are used to assess details of flow and mass transport in hemodialyzers (see next section).

4.2. Geometry in the CFD Model

As a result of the fact that the fiber compartment is modeled as a porous medium, blood and dialysate flow are investigated using separate models. The blood flow in the dialyzer housing is modeled with a three-dimensional finite volume model (Fluent 6, Sheffield, UK). As a result of symmetry reasons, only a quarter part of the dialyzer needed to be simulated. As dialysate inlet and outlet nozzles are unsymmetrically placed on the dialyzer housing, a three-dimensional CFD model of the entire dialyzer was developed (Fluent 6, Sheffield, UK).

4.3. Governing Equations and Fluid and Porous Medium Properties

In the inlet and outlet nozzles, and in the manifolds and distribution rings for the blood and dialysate compartment, respectively, the continuity and Navier–Stokes equations are used describing conservation of mass and momentum. The fiber compartment is modeled as a porous medium using Darcy's law with corresponding permeability proportion ratio in the different flow directions. The permeability characteristics of the fiber bundle are obtained from *in vitro* tests in which pressure drop and flow measurements were performed for the cases of flow inside the hollow fibers (blood compartment model), flow around the fibers in axial direction, and flow around the fibers in radial direction (dialysate compartment model). In the latter, an axial-to-radial permeability ratio of 22 was found.

In both CFD models, a parabolic velocity profile at the inlet nozzle and a zero pressure at the corresponding outlet nozzle are prescribed. Besides the symmetry boundaries, all other boundaries are defined as walls where no slip is assumed.

Blood was modeled as a Newtonian fluid with a density of $1054 \, \text{kg/m}^3$ and a dynamic viscosity of $2.96 \, \text{mPa.s}$ derived from (152):

$$\mu = \mu_p \cdot \exp(2.31 \cdot H), \quad (15)$$

with plasma viscosity μ_p equal to $1.3 \, \text{mPa.s}$ and the hematocrit of anemic blood H 35%. As blood thickening because of ultrafiltration occurs along the dialyzer length, blood viscosity was assumed to increase linearly up to 4.10 mPa.s, which corresponds to an overall ultrafiltration flow of 2 L/h (152).

From bicarbonate dialysate samples, taken *in vivo* from the dialyzer supply and drain, dynamic viscosity and density were determined using a capillary Ubbelohde viscosimeter (Schott, Germany) and a density-hydrometer-aerometer (Assistant, Germany), respectively. As it was found that both properties are not influenced by the dialysis session, dialysate flow is assumed as an incompressible, isothermal laminar Newtonian flow with a constant viscosity (0.687 mPa.s) and density (1008 kg/m³) at body temperature.

4.4. Numerical Results

A homogeneous *blood* flow distribution is found in the fibers, while vortices are remarked at the inlet manifold. Although vortices are also observed in the dialysate compartment at the transition of the inlet nozzle with the *dialysate* distribution ring, an axisymmetrical flow pattern is observed around the fibers (Fig. 7).

4.5. Validation Using Medical Imaging Technique

A new *in vitro* setup was built to visualize blood and dialysate flow through hemodialyzers using SPECT imaging (151). Using an upstream overflow reservoir, a steady-state flow was accomplished in the compartment under study, either the blood or the dialysate compartment. A computer-controlled injection system using an electronic valve was used for radioactive bolus injection in the dialyzer supply tubing. The hemodialyzer was centrally positioned in between the two- and three-headed gamma camera for blood and dialysate flow visualizations, respectively. At the dialyzer outlet, flow rates were measured gravimetrically.

For the *blood flow measurements*, radioactive transport through the semipermeable membrane was avoided by injecting 10 ml boluses of 200000 Da 99m-Technetium-labeled MAA (MacroAggregated Albumin). With the dialyzer fixed inside a two-headed gamma camera, as close as possible to one head to maximize spatial resolution, planar dynamic 2-D images were taken of the bolus passage. Assuming axisymmetrical flow in the blood compartment, the planar images were considered as taken at each view angle. After multiplication of the images over 360° and after performing a filtered backprojection (153), a three-dimensional dynamic image was obtained.

Using water instead of blood in the dialyzer blood compartment, flow rates were adjusted to account for the difference in dynamic viscosity and density. To obey dimensional similarity between *in vivo* blood and *in vitro* water flow, Reynolds numbers Re ($-$) were kept equal in both models (65):

$$\mathrm{Re} = \frac{\rho.Q.d}{\mu.\varepsilon.A_f} = \text{constant}, \qquad (16)$$

with ρ fluid density (kg/m³), Q flow rate (m³/s), d fiber diameter (m), μ dynamic viscosity (Pa.s), ε porosity ($-$), and A_f gross frontal area (m²). Flow rates were adjusted with a factor of 2.75 to account for the difference in density and viscosity. As a consequence, to simulate a blood flow of 300 ml/min, a water flow rate of 109 ml/min was used in the *in vitro* setup. During investigation of the flow pattern in the blood compartment, the dialysate side was filled with fresh tap water at room temperature and hermetically closed.

As was also found in the numerical study, a homogeneous velocity profile was found in the fiber compartment. Activity accumulation was noticed in both manifolds and might be because of the adhesion of macroaggregates to the plastic dialyzer capsule. Moreover, these accumulation sites correspond with regions of low blood velocity and possible blood clotting.

For the *dialysate flow visualizations*, 10 ml boluses of 99m-Tc-DMSA (Dimercaptosuccinic Acid, MW 281 Da) tracer were injected, while pressurized air was forced simultaneously and counter-currently through the blood compartment. The latter technique was useful to avoid water filtration and Technetium diffusion through the dialyzer membrane, and yet no air bubbles were forced through the membrane.

3-D acquisitions of the dialysate compartment were made by rotating the three-headed gamma camera over 120° in 12° increments. For each angular position, 2-D planar intensity pictures of the bolus passage through the dialysate compartment were taken. By evaluating the images of the 10 measurements at different time steps, and by performing a filtered backprojection (153), a dynamic 3-D image was constructed of the bolus propagation in the dialysate compartment.

In contrast with the numerical results, SPECT imaging of dialysate flow resulted in a skewed velocity profile with a maximum near the nozzles side housing, while a minimum value was located down the dialyzer axis (at 23–25 mm vertical distance from the inlet nozzle) (Fig. 7). As a consequence, at the opposite site of inlet and outlet nozzles, nonefficient sites with respect to mass transport are observed (106). Furthermore, a nonhomogeneous velocity distribution was also observed in the plane perpendicular to the plane along the nozzles, which implies that the

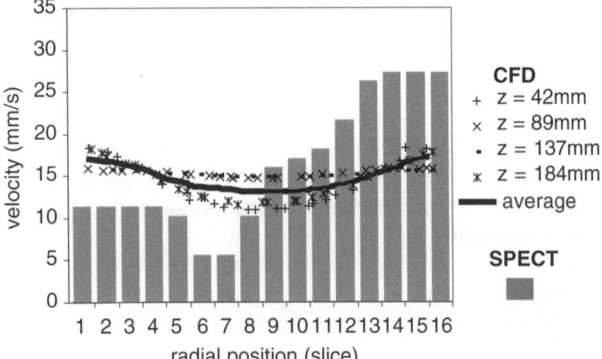

Figure 7. Radial distribution of axial dialysate velocity as computed with CFD for four different axial z positions compared with average. The CFD results are also compared with mean bolus peak velocity (bars) as measured with SPECT.

nonhomogeneous velocity distribution is not only caused by the flow distribution near the dialyzer inlet but also to the preferential flow channeling resulting from fiber twisting. As a result, the discrepancy between the CFD and SPECT results might be attributed to the optimistic assumption in the CFD model of a homogeneous radial permeability over an entire cross section.

4.6. Conclusion

In order to define regions characterized by diminished mass transfer efficiency, the existence of nonuniform flow in the blood and dialysate compartment was investigated by computational fluid dynamics validated with SPECT imaging technique. With both techniques, a fully homogeneously distributed blood flow was found, whereas a discrepancy was observed for the dialysate flow in the case a constant fiber bundle permeability was modeled numerically. The SPECT results can be applied for validation of the CFD model with respect to the fiber bundle permeabilities, such that the validated CFD model can be further used for new dialyzer design and optimization.

5. A NUMERICAL MODEL FOR THE BLOOD/DIALYSATE INTERFACE

5.1. Introduction

Although dialyzer manufacturers only provide information about their products as a black box, this study aimed at optimizing dialyzer geometry by looking more in detail at transport processes and fluid properties inside the dialyzer, using computational fluid dynamics (CFD). Therefore, a microscopic numerical model was developed to describe flow and mass transport in a single dialyzer fiber. After model calibration and validation, the impact on solute removal of a variable fiber length and diameter was assessed for small and middle molecules. Also, the model is used to examine the effect of dialysate flow maldistribution by implementing the experimental SPECT results of the prior section.

5.2. Model Geometry

A three-dimensional finite volume microscopic model of the blood-dialysate interface over the complete length of a dialyzer was developed (Fluent 6, Sheffield, UK) (154). Assuming the fibers spaced in a hexagonal lattice and based on symmetry, a twelfth part of one single fiber is

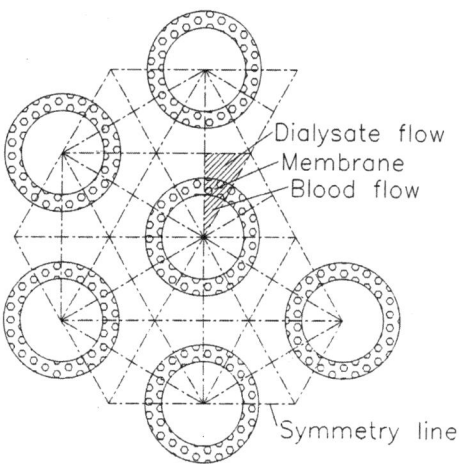

Figure 8. Hexagonal lattice in a cross section of the hollow fiber dialyzer.

isolated (Fig. 8). The parameter settings of the three-dimensional unit module (Fig. 9) were assessed for a high-flux polysulphone dialyzer (Fresenius F60, Bad Homburg, Germany), characterized by a fiber inner diameter of 200 μm, membrane thickness of 40 μm (1 μm inner layer and 39 μm bulk layer—Fig. 6), and dialysate compartment dimensions (maximum radius 230 μm) calculated from fiber density. The membrane module has an active length of 230 mm whereas inlet and outlet tubes (each 12.5-mm long) are integrated in both fluid compartments simulating the manifold and potting region. The mesh generation is performed using Gambit (Fluent, Sheffield, UK). The different domains are filled with hexahedron volume elements and local grid refinements were performed near the blood-membrane and membrane-dialysate interfaces. For the implementation in the numerical model, properties of the three compartments, blood, dialysate, and the semipermeable membrane in between, are derived from literature and experimental investigations.

5.3. Fluid Characteristics

As ultrafiltration takes place over the length of the dialyzer, water is removed from the *blood*, resulting in a decrease of the plasma volume fraction, such that the hematocrit as well as the blood viscosity value is augmented. The influence on viscosity of the local shear rate and hematocrit is taken into account using Quemada's

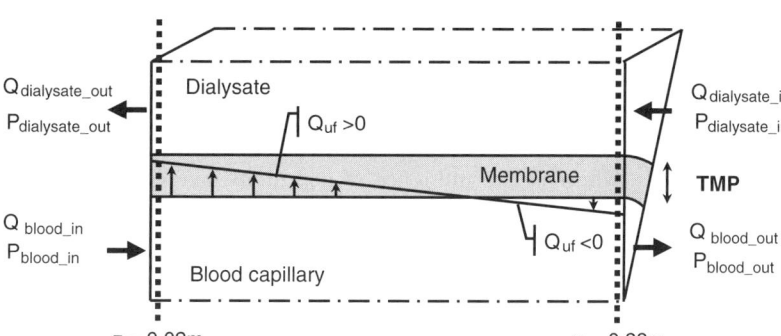

Figure 9. 3-D visualization of an isolated unit showing blood, membrane, and dialysate compartment.

equation (Equation 2). To describe the radial viscosity variation caused by cell redistribution, the formula proposed by Lerche (Equation 4) is used. The parameter n in this equation was derived as follows. First, Poiseuille's law (Equation 3) was used in a horizontal fiber to define the relation between flow Q and pressure drop ΔP as a function of blood viscosity μ. Using the semiempirical viscosity model of Haynes, the apparent blood viscosity μ_{app} (Pa.s) was calculated for a given plasma viscosity μ_p (Pa.s), hematocrit value H $(-)$, and for the proposed fiber geometry (155):

$$\mu_{app} = \mu_p \cdot \frac{1}{\left[1 - \left(1 - \frac{\mu_p}{\mu_c}\right) \cdot \left(1 - 2 \cdot \frac{\delta}{d}\right)^4\right]}, \quad (17)$$

with $\mu_c = \exp(0.48 + 2.35H)$ (mPa.s), parameter $\delta = 2.03 - 2H$ (μm), and d the fiber diameter (μm).

Secondly, a three-dimensional numerical CFD model was developed of the same fiber (given length L). Using the Navier–Stokes equations and the viscosity model of Quemada (Equation 2) combined with the Lerche formula (Equation 4), the parameter n could be determined in an iterative way. The iteration was stopped if, for a given flow, the obtained pressure drop was similar to the one found in the theoretical model using Haynes' formula (Equation 17).

This process was repeated with a constant fiber diameter of 200 μm and for each flow-pressure set, because each velocity profile corresponds with another hematocrit distribution and, as a consequence, with another n value. n was then fitted and was found to vary hyperbolically from 400 to 185 in the cross-sectional hematocrit range 25–50% (Fig. 10). An overview of the algorithm, used for the calculation of local hematocrit and viscosity, is given in Fig. 11.

As plasma density (1030 kg/m^3) differs from the density of platelets and blood cells (1090 kg/m^3), the density of blood ρ_{blood} (kg/m^3) varies with the local hematocrit H $(-)$:

$$\rho_{blood} = 1030 \cdot (1 - H) + 1090 \cdot H. \quad (18)$$

For the *bicarbonate dialysis fluid*, viscosity and density, as found with the earlier described *in vivo* tests, are used.

5.4. Membrane Properties

The permeability characteristics of the membrane were obtained from laboratory tests in which a dialysate flow was forced through the membrane (inflow in blood inlet

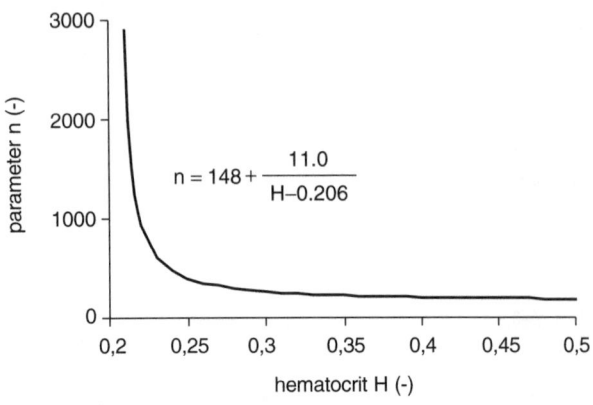

Figure 10. The Lerche parameter n as a function of hematocrit.

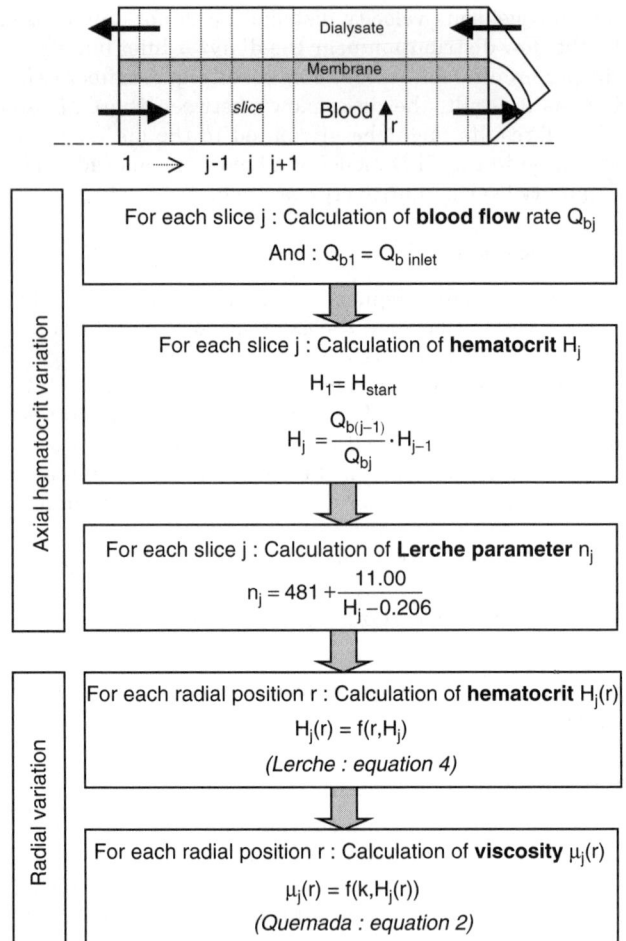

Figure 11. Algorithm for the calculation of local hematocrit and viscosity values.

and outlet and outflow from dialysate inlet and outlet). The ultrafiltration coefficient, K_{UF}, was calculated from flow, Q_{UF}, and transmembrane pressure, TMP, measurements using Equation 14. Furthermore, the overall hydraulic membrane permeability k_m (m^2/s.Pa) was derived from the ultrafiltration coefficient, K_{UF} (m^3/s.Pa), membrane surface, A (m^2), and membrane thickness, d_m (m):

$$k_m = K_{UF} \cdot \frac{d_m}{A}. \quad (19)$$

The tests were done for forward and backfiltration using sterile dialyzers (overall permeability 7950 nm^2/s.Pa) as well as samples in which a protein layer was induced on the membrane (overall permeability 3650 nm^2/s.Pa) simulating a clinical session (156). The permeability of a sterile inner layer k_i (thickness d_i) and of a bulk layer k_b (thickness d_b) were implemented as a series of two resistances, whereas the influence of a protein layer on the overall membrane permeability was incorporated as a higher resistive inner layer:

$$\frac{d_m}{k_m} = \frac{d_b}{k_b} + \frac{d_i}{k_i}. \quad (20)$$

5.5. Governing Equations

In the blood and dialysate compartment, conservation of mass and momentum are described by the three-dimensional steady incompressible continuity and Navier–Stokes equations, using the local and constant viscosity and density for blood and dialysis fluid, respectively (65):

$$\begin{cases} \nabla(\overline{u}) = 0 \\ \rho \cdot \frac{\partial \overline{u}}{\partial t} + \frac{1}{2} \cdot \rho \overline{\nabla} \overline{u}^2 + \rho \cdot (\overline{\nabla} \times \overline{u}) \times \overline{u} + \nabla p = \nabla \overline{\tau}, \end{cases} \quad (21)$$

where ∇ represents the divergence $\nabla = (\frac{\partial}{\partial x}, \frac{\partial}{\partial y}, \frac{\partial}{\partial z})$, $\overline{u}$ represents the velocity vector $\overline{u} = (u, v, w)$, ρ the density of the fluid, and the stress tensor is defined as with p the pressure, a unit tensor, and $\vec{\tau}$ a symmetrical tensor without diagonal sum.

The transmembrane water transport, a function of the membrane permeability and the local oncotic pressure, is described by the Darcy equation for porous media (65):

$$\begin{cases} \nabla \overline{u} = k \cdot \Delta p = 0 \\ \overline{u} = k \cdot \nabla p, \end{cases} \quad (22)$$

where Δ represents the Laplace operator $\Delta = \frac{\partial^2}{\partial x^2} + \frac{\partial^2}{\partial y^2} + \frac{\partial^2}{\partial z^2}$ and k is defined as the hydraulic membrane permeability tensor. The use of Darcy's law is allowed because the Reynolds number is only in the range E-10–E-9, using the equation for porous media (157).

Knowing the velocities in all nodes of the finite volumes, the mass transfer can be calculated with the stationary convection-diffusion equation in the absence of a source or sink reaction:

$$S \cdot (\overline{u} \bullet \overline{\nabla} C) - \overline{\nabla}(D_S \cdot \overline{\nabla} C) = 0. \quad (23)$$

D_S represents the solute diffusion coefficient (m²/s) and C the solute concentration (mol/m³). Although this equation is valid for all domains, note that the diffusion coefficient D is different for the three domains. Furthermore, although the sieving coefficient S (–) can be eliminated in the blood and dialysate domain, S can deviate from unity in the membrane domain (104,158,159). A sieving coefficient equal to unity corresponds to unhindered solute transport through the membrane, whereas S equal to zero implies that the membrane is impermeable to the considered solute (e.g., large proteins like albumin). In the present study, solutes of distinct molecular weight were investigated. Urea (MW60) was used as marker for small water-soluble solutes, whereas Vitamin B12 (MW1355) and inulin (MW5200) were used as middle molecule markers (160). The choice of the uremic solutes was mainly driven by the available clearance data from the manufacturer and by the fact that all three solutes have a sieving coefficient equal to unity for the considered high-flux polysulphone membrane.

5.6. Model Boundary Conditions

In the blood and dialysate compartment, a constant inlet velocity is given, whereas outlet conditions are specified either as outlet pressures or as a flow percentage distribution in both compartments to apply the desired ultrafiltration flow. These boundary conditions comply with the mass balance and the zero diffusion flux for all variables. The information at both outlets is derived from the calculated values at internal mesh points.

Oncotic pressure, which is exerted by the plasma proteins and opposes the hydrostatic transmembrane pressure, is implemented as a discontinuous local pressure drop at the skin-bulk interface. As a result of the difficulty to induce *in vitro* a stable protein layer on the membrane (156), initial oncotic pressure from literature (25 mmHg) was used (32,104) rather than their own *in vitro* derived values. Moreover, as hemoconcentration takes place in axial direction, the oncotic pressure is varying with hematocrit.

As the smallest blood-membrane-dialysate entity is isolated, all other boundaries are symmetry planes. Through such a symmetry plane, convective and diffusive fluxes are nonexistent. As a consequence, the normal velocity component and the normal diffusive gradient is zero. Moreover, the shear stress in a symmetry plane is zero.

5.7. Calibration and Validation of the Diffusivities

Although the solute diffusion coefficients in blood and dialysate were known from literature, membrane diffusivity was derived from the inlet and outlet blood (C_{Bi} and C_{Bo}) and dialysate (C_{Di} and C_{Do}) concentrations. Blood concentrations (mol/l) were assessed from Equation 11, where the manufacturer typically reports K values for given blood/dialysate flow combinations (e.g., 300/500 ml/min) (161,162). Dialysate concentrations (mol/l) were calculated from the mass balance in the dialyzer, which is a function of blood and dialysate flows Q_B and Q_D (ml/min) (87):

$$(C_{Bi} - C_{Bo}) \cdot Q_B = (C_{Do} - C_{Di}) \cdot Q_D. \quad (24)$$

Although the concentration difference between blood and dialysate, ΔC, will decrease exponentially along the dialyzer length, a linear approximation is allowed for low and middle molecules (137):

$$\frac{d(\Delta C)}{dz} = \frac{\Delta C_i - \Delta C_o}{L}, \quad (25)$$

with ΔC_i and ΔC_o the blood-dialysate concentration difference at the blood inlet and outlet, respectively.

By multiplying both terms with the mass flux J (mol/s), as defined by Fick's law (Equation 5), and after integration of Equation 25 and solving it for the mass flux J, clearance K can then be written as a function of the mass transfer coefficient K_0 (m/s), which is the reciprocal of total resistance R_0, and of the logarithmic mean concentration difference ΔC_{lm} (87):

$$K = \frac{K_0 \cdot A}{C_{Bi}} \cdot \Delta C_{lm} = \frac{1}{R_0} \cdot \frac{A}{C_{Bi}} \\ \cdot \frac{(C_{Bi} - C_{Do}) - (C_{Bo} - C_{Di})}{\ln\left(\frac{C_{Bi} - C_{Do}}{C_{Bo} - C_{Di}}\right)}. \quad (26)$$

Furthermore, using Equation 9, membrane diffusivity D_M (m²/s) can be derived from total resistance R_0 and the convective mass transfer coefficients $1/R_B$ and $1/R_D$ (m/s). As Δx_B and Δx_D were not *a priori* known, the diffusion coefficient in the membrane for a particular solute was derived iteratively until the clearance as found with the simulations matches the manufacturer's data for a Q_B/Q_D ratio equal to 250/500 ml/min. The power of the numerical model (160) was checked performing simulations for a Q_B/Q_D equal to 300/500 ml/min, and comparing the numerically derived clearance value with the manufacturer's data (161,162).

5.8. Numerical Results

Assuming a constant blood and dialysate inlet flow of 250 and 500 ml/min, respectively, blood and dialysate outlet pressures of 10 kPa and 5 Pa, respectively, and an initial oncotic pressure of 3.33 kPa (32,104), the pressure distribution renders an overall ultrafiltration flow of 45 ml/min while no backfiltration occurs (Figs. 12 and 13).

As blood, with an initial viscosity of 3 mPa.s, flows through the dialyzer, the water removal causes hemoconcentration. As a consequence, the hematocrit shows an axial variation from its initial value of 30% at blood entrance up to 42% at the outlet, resulting in a mean viscosity increase from 3 mPa.s to 4.5 mPa.s. The plug flow of blood cells at the axis (maximum viscosity 7.5–11.8 mPa.s) and the plasma layer near the membrane wall (viscosity 1.3 mPa.s) demonstrates the radial variation of the blood viscosity (Fig. 14).

The oncotic pressure, varying with the local hematocrit, increases from its initial value 3.33 kPa (25 mmHg) up to 4.20 kPa (31 mmHg). As a result of this oncotic pressure opposing the hydraulic driving pressure, ultrafiltration flow is decreased by 28%.

The shear stress, zero at blood and dialysate axes, is maximal at the blood-membrane interface decreasing from 0.97 Pa at blood inlet to 0.78 Pa at the outlet, whereas it is slightly increasing at the dialysate-membrane interface from 0.23 Pa up to 0.26 Pa at dialysate outlet.

As a result of ultrafiltration, one may expect a deviation from the linear flow-pressure drop profile described by Poiseuille (Equation 3) as well as from the parabolic velocity profile. Nevertheless, for an ultrafiltration flow of 45 ml/min in a dialyzer module of 230 mm in length, the pressure distribution in the blood compartment deviates only slightly from linearity (maximum 0.28–0.33% at blood inlet and outlet, respectively) (Fig. 13), while the same is true for the parabolic velocity profile ($R^2 = 0.997$–0.993 at blood inlet and outlet, respectively) (Fig. 15).

With respect to mass transfer, Fig. 16 shows the concentration profiles for urea, Vitamin B12, and inulin in a standard F60 dialyzer with overall blood and dialysate flows of 250 and 500 ml/min, respectively. The radial concentration variation, most pronounced for inulin, illustrates the lower diffusivities in blood and dialysate compared with urea. Comparing outlet with inlet solute concentrations, clearances of 213, 126, and 61 ml/min were found for urea, Vitamin B12, and inulin, respectively, corresponding to the data as provided by the manufacturer.

5.9. Experimental Validation of the Numerical Model

In order to validate the numerical model, an *ex vivo* study was performed using the clinical dialysis set up. As blood substitution fluid, bovine blood was collected during slaughtering and was prevented from coagulation by adding heparin. Blood was flowing through the dialyzer at 37°C using a Bellco Formula dialysis machine, which allows adjusting blood, dialysate, and ultrafiltration flow.

The tests consisted of flow and pressure measurements as well as blood property measurements (viscosity,

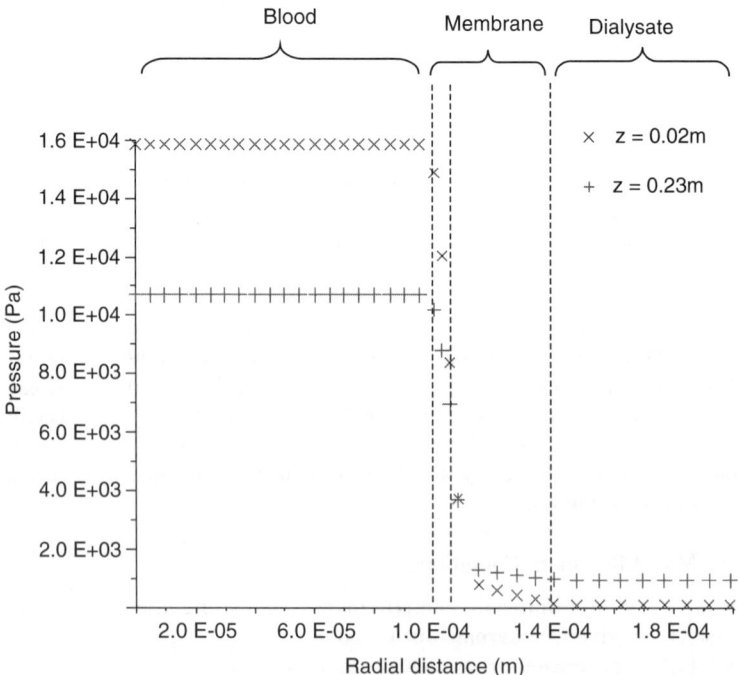

Figure 12. Radial pressure distribution at blood inlet ($\times$) and outlet ($+$) section. (This figure is available in full color at http://www.mrw.interscience.wiley.com/ebe.)

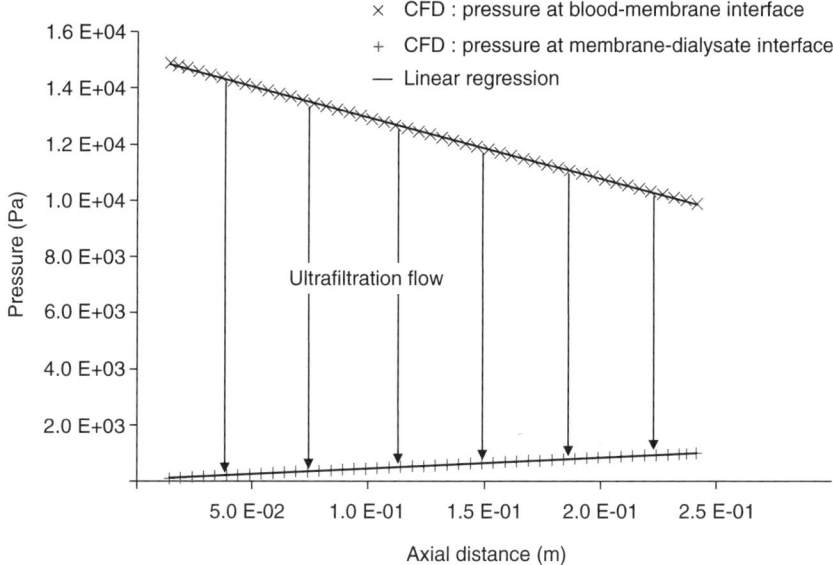

Figure 13. Axial pressure distribution in blood (×) and dialysate (+) compartment. (This figure is available in full color at http://www.mrw.interscience.wiley.com/ebe.)

hematocrit, pH, and oxygen saturation). The pressure was measured at all dialyzer inlets and outlets with fluid-filled pressure transducers (Ohmeda, Belgium), while downstream, blood and dialysate flows were measured gravimetrically. The viscosity measurements of the bovine blood samples were done with a plate and cone viscosimeter (Rheolyst AR 1000-N, TA Instruments, UK) that allows deriving dynamic viscosity from the torque corresponding with the cone's angular velocity. Using this technique, the shear-thinning behavior of blood was adequately visualized and a mean viscosity value was derived for a fixed shear rate. For each flow setting, samples were taken from the up and downstream blood line in order to set hematocrit, oxygen saturation, and pH in the medical laboratory.

As blood pH remained stable (7.34 ± 0.05) during the entire test session, blood acidification did not take place and addition of a buffer (like bicarbonate powder) was unnecessary. Oxygen saturation was also constant during the tests, such that extra oxygenation was superfluous. Limited discrepancies were found between viscosity and hematocrit variations with the *ex vivo* study and the numerical model (10% and 20%, respectively). The important discrepancy found for the pressure drop in blood compartment was caused by the formation *ex vivo* of an important protein layer on the membrane, and the blockage of fibers using highly concentrated bovine blood. Using Poiseuille's law, the theoretical pressure drops are in good agreement with those found in the CFD model.

5.10. Parameter Study

With the extended finite volume model, a parameter study was performed varying fluid flows (blood and dialysate flows of 150–350 ml/min and 300–800 ml/min, respectively), dialyzer geometry (axial as well as radial dimensions), or initial blood properties (hematocrit and plasma viscosity).

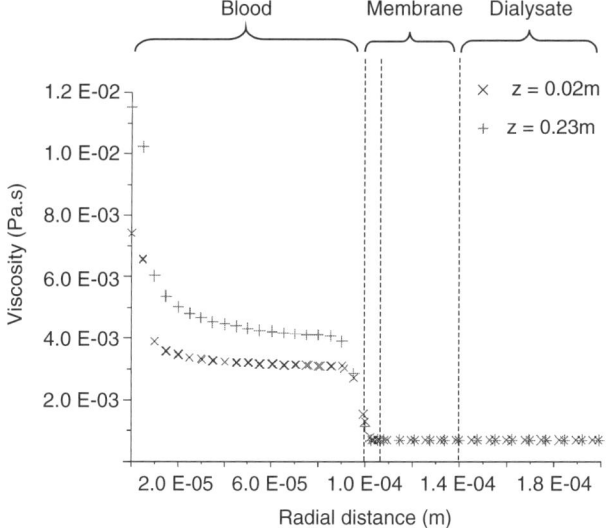

Figure 14. Radial viscosity distribution at blood inlet (×) and outlet (+) section. (This figure is available in full color at http://www.mrw.interscience.wiley.com/ebe.)

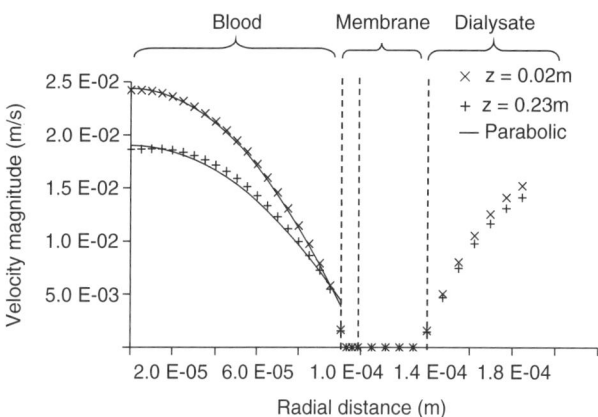

Figure 15. Spatial velocity profile in blood and dialysate compartment at blood inlet (×) and outlet (+) section.

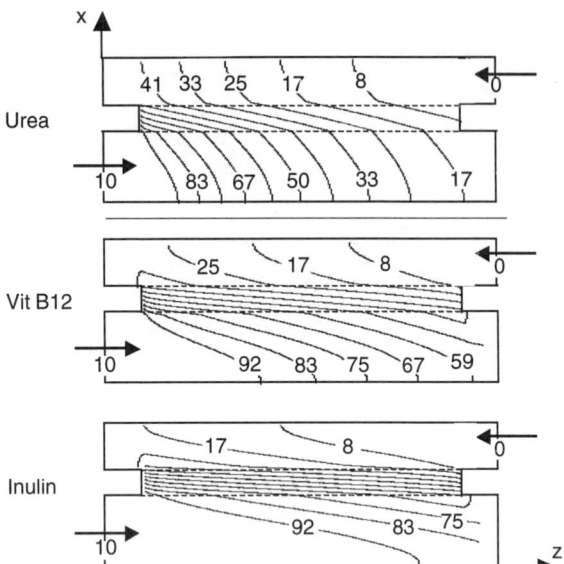

Figure 16. Concentration profiles in a dialyzer fiber (diameter 200 μm and length 230 mm) for urea (upper panel), Vitamin B12 (middle panel), and inulin (lower panel). The relative blood start concentration was 100, whereas blood and dialysate flows were 250 and 500 ml/min, respectively.

Increasing blood and dialysate flow rate while keeping the geometrical and blood characteristics constant, the blood viscosity and hematocrit percentage variations are reduced exponentially. Moreover, hematocrit increase is even more expressed than viscosity increase if blood flow rates over 220 ml/min are considered. With respect to fiber length, blood viscosity as well as hematocrit increases are smaller for shorter fibers (180 mm) compared with the standard ones (230 mm). Further lengthening of fibers (280 mm), however, does not change blood property variations significantly. Varying the radial dimensions (fiber inner diameter 150–200–250 μm), blood property variations are from the same magnitude of those found in the case of axial variation. The value of the initial plasma viscosity has only influence on the blood viscosity variation, whereas hematocrit varies independently. However, if blood samples with a more physiological hematocrit (43% instead of 30% for anemic blood) are considered, similar blood thickening can be remarked while flowing through a dialyzer.

Finally, the impact of dialyzer dimensions on dialyzer clearance was studied by changing fiber diameter and length in a wide range, while keeping a constant fluid velocity. Clearances were found enhanced by 13% (urea), 50% (Vitamin B12), and 89% (inulin) for a fiber twice as long as a standard one, and by 5.5% (Vitamin B12) and 21% (inulin) for a fiber diameter of 150 μm instead of 200 μm. In general, the impact of fiber dimensions was more pronounced for the middle molecules compared with urea.

5.11. Impact of Flow Distribution on Mass Transport

The impact of flow maldistributions on solute transfer efficiency was evaluated numerically by implementing the local velocities as obtained from the SPECT measurements (Previous section). Mass transfer calculation was performed for urea and Vitamin B12 in the case of maximum (20.5 mm/s) and minimum dialysate flow velocity (4.3 mm/s), instead of the mean dialysate velocity of 12.1 mm/s. Meanwhile, blood flow was considered homogeneously distributed, and a Poiseuille velocity profile with mean velocity of 17.3 mm/s was used at the blood inlet, corresponding to a uniform overall blood flow of 300 ml/min.

Although the observed maximum dialysate velocity resulted in a clearance increase of 1.3% and 12% for urea and Vitamin B12, the minimum dialysate velocity deteriorated solute removal by 12% and 28%, respectively. By calculating the clearances for intermediate blood-dialysate flow combinations, and by integration over the cross section, an overall decrease in solute removal efficiency by 3.8% and 4.4% was obtained for urea and Vitamin B12, respectively.

5.12. Conclusion

The presented numerical model incorporates blood, dialysate, and membrane flow in hollow fiber dialyzers allowing an accurate investigation of the fluid properties and the presence and localization of backfiltration. The hydraulic permeability of the dialyzer is based on a more accurate method than in previous *ex vivo* studies, and blood is modeled as a nonNewtonian fluid with properties varying in radial as well as axial direction. Furthermore, the model allows calculating concentration profiles inside the dialyzer and related solute clearances. After validation of the CFD model, different parameters such as geometrical and fluid properties were investigated.

6. SUMMARY

Hemodialysis is the most frequently used renal replacement therapy for patients suffering from chronic renal failure. During hemodialysis, the patient's blood is flowing through an extracorporeal circuit using a hemodialyzer. The main function of a dialyzer is purifying the blood by means of basic transport phenomena: diffusion and ultrafiltration. Different dialyzer modalities are available with different geometries and membrane characteristics.

To have better insight in the interaction of pressure, concentration, and flow profiles, a review is given about the overall flow distribution on one hand and more local phenomena, like concentration polarization and mass transport, on the other. As most manuscripts report one specific topic, a description of dialyzer flow taking into account all present transport phenomena with consecutive fluid property variations is not yet available.

To optimize mass transport, it is important that blood as well as dialysate flow show a uniform flow profile. In order to find ineffective regions with respect to dialyzer flow, a CFD model was developed and validated using SPECT imaging. The local fluid velocities were further implemented in a microscopic model of the blood/dialysate interface, providing velocity, pressure, and concentration profiles. Furthermore, the influence of pressure distributions and ultrafiltration profiles on fluid properties and solute clearances was investigated performing a parameter study changing geometry, flow, and initial fluid properties.

BIBLIOGRAPHY

1. L. P. Gartner and J. L. Hiatt, *Color Textbook of Histology*. Philadelphia, PA: W.B. Saunders Company, 1997.
2. C.M. Kjellstrand and P. T. Brendan, Acute renal failure. In: C. Jacobs, C. M. Kjellstrand, K. M. Koch, and J. F. Winchester, eds., *Replacement of Renal Function by Dialysis*. 4th ed. Dordrecht, the Netherlands: Kluwer Academic Publisher, 1996, pp. 821–862.
3. R. Vanholder, R. De Smet, P. Vogeleere, C. H. Hsu, and S. M. Ringoir, The uraemic syndrome. In: C. Jacobs, C. M. Kjellstrand, K. M. Koch, and J. F. Winchester, eds., *Replacement of Renal Function by Dialysis*. 4th ed. Dordrecht, the Netherlands: Kluwer Academic Publisher, 1996, pp. 103–113.
4. J. M. Lazarus, E. G. Lowrie, C. L. Hampers, and J. P. Merrill, Cardiovascular disease in uremic patients on hemodialysis. *Kidney Int. Suppl.* 1975; **2**:167–175.
5. G. Mall, M. Rambausek, A. Neumeister, S. Kollmar, F. Vetterlein, and E. Ritz, Myocardial interstitial fibrosis in experimental uremia-implications for cardiac compliance. *Kidney Int.* 1988; **33**:804–811.
6. P. S. Parfrey, J. D. Harnett, S. M. Griffiths, M. H. Gault, and P. E. Barre, Congestive heart failure in dialysis patients. *Arch. Intern. Med.* 1988; **148**:1519–1525.
7. C. L. Fraser and A. I. Arieff, Nervous system complications in uremia. *Ann. Intern. Med.* 1988; **109**:143–153.
8. P. L. Kimmel, G. Miller, and W. B. Mendelson, Sleep apnea syndrome in chronic renal disease. *Am. J. Med.* 1989; **86**:308–314.
9. V. Pavlovic-Kentera, G. K. Clemons, L. Djukanovic, and L. Biljanovic-Paunovic, Erythropoietin and anemia in chronic renal failure. *Exp. Hematol.* 1987; **15**:785–789.
10. E. Bogin, S. G. Massry, J. Levi, M. Djaldeti, G. Bristol, and J. Smith, Effect of parathyroid hormone on osmotic fragility of human erythrocytes. *J. Clin. Invest.* 1982; **69**:1017–1025.
11. R. Vanholder and S. Ringoir, Infectious morbidity and defects of phagocytic function in end-stage renal disease: a review. *J. Am. Soc. Nephrol.* 1993; **3**:1541–1554.
12. S. L. Lewis and D. E. Van Epps, Neutrophil and monocyte alterations in chronic dialysis patients. *Am. J. Kidney Dis.* 1987; **9**:381–395.
13. S. E. Goldblum and W. P. Reed, Host defenses and immunologic alterations associated with chronic hemodialysis. *Ann. Intern. Med.* 1980; **93**:597–613.
14. A. C. Guyton, *Textbook of Medical Physiology*, 7th ed. Philadelphia, PA: WB Saunders Company, 1986.
15. A. Carrel, La technique opératoire des anastomoses vasculaires et la transplantation des viscères. *Lyon Méd.* 1902; **98**:859–864.
16. B. A. Cunningham, The structure and function of histocompatibility antigens. *Sci. Am.* 1977; **237**:96–107.
17. C. M. Zmijewski, Human leukocyte antigen matching in renal transplantation: review and current status. *J. Surg. Res.* 1985; **38**:66–87.
18. B. D. Kahan, C. T. van Buren, S. M. Flechner, M. Jarowenko, T. Yasumura, A. J. Rogers, N. Yoshimura, S. LeGrue, D. Drath, and R. H. Kerman, Clinical and experimental studies with cyclosporine in renal transplantation. *Surgery* 1985; **97**:125–140.
19. C. M. Mion, Continuous peritoneal dialysis. In: C. Jacobs, C. M. Kjellstrand, K. M. Koch, and J. F. Winchester, eds., *Replacement of Renal Function by Dialysis*. 4th ed. Dordrecht, the Netherlands: Kluwer Academic Publisher, 1996, pp. 562–602.
20. M. Feriani, C. Ronco, and G. La Greca, Solutions for peritoneal dialysis. In: C. Jacobs, C. M. Kjellstrand, K. M. Koch, and J. F. Winchester, eds., *Replacement of Renal Function by Dialysis*. 4th ed. Dordrecht, the Netherlands: Kluwer Academic Publisher, 1996, pp. 103–113.
21. H. Tenckhoff and H. Schechter. A bacteriologically safe peritoneal access device. *Trans. Am. Soc. Artif. Intern. Organs* 1968; **14**:181–187.
22. R. Khanna, S. Izatt, D. Burke, R. Mathews, S. Vas, and D. G. Oreopoulos, Experience with the Toronto Western Hospital permanent peritoneal catheter. *Perit Dial Bull.* 1984; **4**:95.
23. D. Kim, D. Burke, S. Izatt, R. Mathews, G. Wu, R. Khanna, S. Vas, and D. G. Oreopoulos, Single- or double-cuff peritoneal catheters? A prospective comparison. *Trans. Am. Soc. Artif. Intern. Organs* 1984; **30**:232–235.
24. S. R. Ash, Chronic peritoneal dialysis catheters: overview of design, placement, and removal procedures. *Semin. Dial.* 2003; **16**:323–334.
25. J. W. Moncrief, R. P. Popovich, L. J. Broadrick, Z. Z. He, E. E. Simmons, and R. A. Tate, The Moncrief-Popovich catheter. A new peritoneal access technique for patients on peritoneal dialysis. *Asaio J.* 1993; **39**:62–65.
26. B. Rippe, G. Stelin, and B. Haraldsson, Computer simulations of peritoneal fluid transport in CAPD. *Kidney Int.* 1991; **40**:315–421.
27. Z. J. Twardowski, K. D. Nolph, R. Khanna, B. F. Prowant, C. P. Ryan, H. L. Moore, and M. P. Nielsen, Peritoneal equilibration test. *Perit Dial Bull.* 1987; **7**:138–147.
28. R. P. Popovich, J. W. Moncrief, K. D. Nolph, A. J. Ghods, Z. J. Twardowski, and W. K. Pyle, Continuous ambulatory peritoneal dialysis. *Ann. Intern. Med.* 1978; **88**:449–456.
29. D. Nakagawa, C. Price, B. Stinebaugh, and W. Suki, Continuous cycling peritoneal dialysis: a viable option in the treatment of chronic renal failure. *Trans. Am. Soc. Artif. Intern. Organs* 1981; **27**:55–57.
30. J. A. Diaz-Buxo, C. D. Farmer, P. J. Walker, J. T. Chandler, and K. L. Holt, Continuous cyclic peritoneal dialysis: a preliminary report. *Artif. Organs* 1981; **5**:157–161.
31. P. D. Doolan, W. P. Murphy Jr., R. A. Wiggins, N. W. Carter, W. C. Cooper, R. H. Watten, and E. L. Alpen, An evaluation of intermittent peritoneal lavage. *Am. J. Med.* 1959; **26**:831–844.
32. N. A. Hoenich, C. Woffindin, and C. Ronco, Haemodialysers and associated devices. In: C. Jacobs, C. M. Kjellstrand, K. M. Koch, and J. F. Winchester, eds., *Replacement of Renal Function by Dialysis*. 4th ed. Dordrecht, the Netherlands: Kluwer Academic Publisher, 1996, pp. 188–230.
33. J. P. Van Waeleghem, M. M. Elseviers, and E. J. Lindley, Management of the vascular access in Europe. Part I - A study of centre based policies. *EDTNA/ERCA J* 2000; **26**:28–33.
34. J. E. Sigdell and B. Tersteegen, Clearance of a dialyzer under varying operating conditions. *Artif. Organs* 1986; **10**:219–225.
35. W. Quinton, D. Dillard, and B. H. Scribner, Cannulation of blood vessels for prolonged hemodialysis. *Trans. Am. Soc. Artif. Intern. Organs* 1960; **6**:104–113.
36. M. J. Brescia, J. E. Cimino, K. Appel, and B. J. Hurwich, Chronic hemodialysis using venipuncture and a surgically created arteriovenous fistula. *N. Engl. J. Med.* 1966; **275**:1089–1092.

37. W. H. Bay, S. Van Cleef, and M. Owens, The hemodialysis access: preferences and concerns of patients, dialysis nurses and technicians, and physicians. *Am. J. Nephrol.* 1998; **18**:379–383.
38. J. Valenta, J. Bilek, and K. Opantmry, Autogenous saphenous vein grafts as secondary vascular access for hemodialysis. *Dial Transpl.* 1985; **14**:567–571.
39. M. C. Coburn and W. I. Carney Jr., Comparison of basilic vein and polytetrafluoroethylene for brachial arteriovenous fistula. *J. Vasc. Surg.* 1994; **20**:896–902.
40. A. Hatjibaloglou, D. Grekas, N. Saratzis, A. Megalopoulos, I. Moros, D. Kiskinis, and V. Dalainas, Transposed basilic vein-brachial arteriovenous fistula: an alternative vascular access for hemodialysis. *Artif. Organs* 1992; **16**:623–625.
41. H. D. Dardik, I. M. Ibrahim, S. Sprayregen, and I. I. Dardik, Clinical experience with modified human umbilical cord vein for arterial bypass. *Surgery* 1976; **79**:618–624.
42. M. Haimov and J. H. Jacobson, 2nd, Experience with the modified bovine arterial heterograft in peripheral vascular reconstruction and vascular access for hemodialysis. *Ann. Surg.* 1974; **180**:291–295.
43. L. Flores, I. Dunn, E. Frumkin, R. Forte, R. Requena, J. Ryan, M. Knopf, J. Kirschner, and B. S. Levowitz, Dacron arteriovenous shunts for vascular access in hemodialysis. *Trans. Am. Soc. Artif. Intern. Organs.* 1973; **19**:33–37.
44. J. H. Tordoir, J. M. Herman, T. S. Kwan, and P. M. Diderich, Long-term follow-up of the polytetrafluoroethylene (PTFE) prosthesis as an arteriovenous fistula for haemodialysis. *Eur. J. Vasc. Surg.* 1988; **2**:3–7.
45. N. H. Shusterman, K. Kloss, and J. L. Mullen, Successful use of double-lumen, silicone rubber catheters for permanent hemodialysis access. *Kidney Int.* 1989; **35**:887–890.
46. A. H. Moss, C. Vasilakis, J. L. Holley, C. J. Foulks, K. Pillai, and D. E. McDowell, Use of a silicone dual-lumen catheter with a Dacron cuff as a long-term vascular access for hemodialysis patients. *Am. J. Kidney Dis.* 1990; **16**:211–215.
47. R. Uldall, M. DeBruyne, M. Besley, J. McMillan, M. Simons, and R. Francoeur, A new vascular access catheter for hemodialysis. *Am. J. Kidney Dis.* 1993; **21**:270–277.
48. S. Ringoir and R. Vanholder, An introduction to biocompatibility. *Artif. Organs.* 1986; **10**:20–27.
49. R. Vanholder and S. Ringoir, Bioincompatibility: an overview. *Int. J. Artif. Organs* 1989; **12**:356–365.
50. J. H. Joist and D. G. Pennington, Platelet reactions with artificial surfaces. *ASAIO Trans.* 1987; **33**:341–344.
51. J. W. Marshall, D. J. Ahearn, R. J. Nothum, J. Esterly, K. D. Nolph, and J. F. Maher, Adherence of blood components to dialyzer membranes: morphological studies. *Nephron* 1974; **12**:157–170.
52. D. J. Lyman, K. Knutson, B. McNeil, and K. Shibatani, The effects of chemical structure and surface properties of synthetic polymers on the coagulation of blood. IV. The relation between polymer morphology and protein adsorption. *Trans. Am. Soc. Artif. Intern. Organs* 1975; **21**:49–54.
53. L. S. Kaplow and J. A. Goffinet, Profound neutropenia during the early phase of hemodialysis. *JAMA* 1968; **203**:1135–1137.
54. P. R. Craddock, J. Fehr, A. P. Dalmasso, K. L. Brighan, and H. S. Jacob, Hemodialysis leukopenia. Pulmonary vascular leukostasis resulting from complement activation by dialyzer cellophane membranes. *J. Clin. Invest.* 1977; **59**:879–888.
55. W. F. Keane, F. L. Shapiro, and L. Raij, Incidence and type of infections occurring in 445 chronic hemodialysis patients. *Trans. Am. Soc. Artif. Intern. Organs* 1977; **23**:41–47.
56. M. C. Belcon, E. K. Smith, L. M. Kahana, and A. G. Shimizu, Tuberculosis in dialysis patients. *Clin. Nephrol.* 1982; **17**:14–18.
57. J. Bommer and E. Ritz, Spallation of dialysis materials-problems and perspectives. *Nephron* 1985; **39**:285–289.
58. T. Bosch, B. Schmidt, W. Samtleben, and H. J. Gurland, Effect of protein adsorption on diffusive and convective transport through polysulfone membranes. *Contrib. Nephrol.* 1985; **46**:14–22.
59. F. Villarroel and A. A. Ciarkowski, A survey on hypersensitivity reactions in hemodialysis. *Artif. Organs* 1985; **9**:231–238.
60. J. T. Daugirdas and T. S. Ing, First-use reactions during hemodialysis: a definition of subtypes. *Kidney Int.* 1988; **24**:S37–S43.
61. E. F. Leonard, C. Van Vooren, D. Hauglustaine, and S. Haumont, Shear-induced formation of aggregates during hemodialysis. *Contrib. Nephrol.* 1983; **36**:34–45.
62. C. Quereda, L. Orofino, R. Marcen, J. Sabater, R. Matesanz, and J. Ortuno, Influence of dialysate and membrane biocompatibility on hemodynamic stability in hemodialysis. *Int. J. Artif. Organs* 1988; **11**:259–264.
63. U. Baurmeister, J. Vienken, and V. Daum, High-flux dialysis membranes:endotoxin transfer by backfiltration can be a problem. *Nephrol. Dial Transplant* 1989; **4**:89–93.
64. N. Lameire and R. L. Mehta, *Complications in dialysis*. New York: Basel, Marcel Dekker, 2000.
65. J. R. Welty, C. E. Wicks, R. E. Wilson, and G. L. Rorrer, *Fundamentals of Momentum, Heat, and Mass Transfer*. New York: John Wiley & Sons, 2001.
66. D. Quemada, General features of blood circulation in narrow vessels. In: C. M. Rodkiewicz, ed., *Arteries and Arterial Blood Flow: Biological and Physiological Aspects*. Vienna: Springer-Verlag, 1983.
67. P. Gaehtgens, Flow of blood through narrow capillaries: rheological mechanisms determining capillary hematocrit and apparent viscosity. *Biorheology* 1980; **17**:183–189.
68. D. Lerche and R. Oelke, Theoretical model of blood flow through hollow fibres considering hematocrit-dependent, non-Newtonian blood properties. *Int. J. Artif. Organs* 1990; **13**:742–746.
69. C. Ronco, A. Fabris, and M. Feriani, Hemodialysis fluid composition. In: C. Jacobs, C. M. Kjellstrand, K. M. Koch, and J. F. Winchester, eds., *Replacement of Renal Function by Dialysis*. 4th ed, Dordrecht, the Netherlands: Kluwer Academic Publisher, 1996, pp. 256–276.
70. J. P. Merrill, E. Schupak, E. Cameron, and C. L. Hampers, Hemodialysis in the Home. *JAMA* 1964; **190**:468–470.
71. E. J. Serfass and V. H. Troutner, Portable dialysate supply system. in US, Milton Roy Comp. 1970.
72. B. J. Canaud and C. M. Mion, Water treatment for contemporary haemodialysis. In: C. Jacobs, C. M. Kjellstrand, K. M. Koch, and J. F. Winchester, eds., *Replacement of Renal Function by Dialysis*. 4th ed. Dordrecht, the Netherlands: Kluwer Academic Publisher, 1996, pp. 231–255.
73. R. F. Madsen, B. Nielsen, O. J. Olsen, and F. Raaschou, Reverse osmosis as a method of preparing dialysis water. *Nephron* 1970; **7**:545–558.
74. J. J. Petrie, R. Fleming, P. McKinnon, R. J. Winney, and J. Cowie, The use of ion exchange to remove aluminum from water used in hemodialysis. *Am. J. Kidney Dis.* 1984; **4**:69–74.
75. H. Klinkmann and J. Vienken, Membranes for dialysis. *Nephrol. Dial Transplant* 1995; **10**:39–45.

76. C. K. Colton and M. J. Lysaght, Membranes for hemodialysis. In: C. Jacobs, C. M. Kjellstrand, K. M. Koch, and J. F. Winchester, eds., *Replacement of Renal Function by Dialysis*. 4th ed. Dordrecht, the Netherlands: Kluwer Academic Publisher, 1996, pp. 103–113.

77. J. Floege, C. Granolleras, G. Deschodt, M. Heck, G. Baudin, B. Branger, O. Tournier, B. Reinhard, G. M. Eisenbach, L. C. Smeby, K. M. Koch, and S. Shaldon, High-flux synthetic versus cellulosic membranes for beta 2-microglobulin removal during hemodialysis, hemodiafiltration and hemofiltration. *Nephrol. Dial Transplant* 1989; **4**:653–657.

78. G. Lonnemann, K. M. Koch, S. Shaldon, and C. A. Dinarello, Studies on the ability of hemodialysis membranes to induce, bind, and clear human interleukin-1. *J. Lab. Clin. Med.* 1988; **112**:76–86.

79. A. Fujimori, H. Naito, and T. Miyazaki, Adsorption of complement, cytokines, and proteins by different dialysis membrane materials: evaluation by confocal laser scanning fluorescence microscopy. *Artif. Organs* 1998; **22**: 1014–1017.

80. W. J. Beek, K. M. K. Muttzall, and J. W. van Heuvel, *Transport Phenomena*, 2nd ed. Chichester, UK: John Wiley & Sons, 1999.

81. J. H. Byrne and S. G. Schultz, *An Introduction to Membrane Transport and Bioelectricity: Foundation of General Physiology and Electrochemical Signalling*, 2nd ed. New York: Raven Press, 1994.

82. A. Geiger, A method of ultrafiltration in vivo. *J. Physiol.* 1931; **71**:111–120.

83. P. Kramer, W. Wigger, J. Rieger, D. Matthaei, and F. Scheler, Arteriovenous haemofiltration: a new and simple method for treatment of over-hydrated patients resistant to diuretics. *Klin. Wochenschr.* 1977; **55**:1121–1122.

84. L. W. Henderson, A. Besarab, A. Michaels, and L. W. Bluemle, Blood purification by ultrafiltration and fluid replacement (diafiltration). *Trans. Am. Soc. Artif. Intern. Organs* 1967; **16**:216.

85. E. Quellhorst and E. Plashues, Ultrafiltration: eliminatation harnpflichtiger substanzen mit hilfe neuartiger membranen, In: P. Ditrich and F. Skrabel, eds., *Aktuelle Probleme der Dialyseverfahren und der Niereninsuffizienz*. Friedberg: Bindernagel, 1971.

86. R. Vanholder, R. De Smet, G. Glorieux, A. Argiles, U. Baurmeister, P. Brunet, W. Clark, G. Cohen, P. P. De Deyn, R. Deppisch, B. Descamps-Latscha, T. Henle, A. Jorres, H. D. Lemke, Z. A. Massy, J. Passlick-Deetjen, M. Rodriguez, B. Stegmayr, P. Stenvinkel, C. Tetta, C. Wanner, and W. Zidek, Review on uremic toxins: classification, concentration, and interindividual variability. *Kidney Int.* 2003; **63**:1934–1943.

87. J. A. Sargent and F. A. Gotch, Principles and biophysics of dialysis. In: C. Jacobs, C. M. Kjellstrand, K. M. Koch, and J. F. Winchester, eds., *Replacement of Renal Function by Dialysis*. 4th ed. Dordrecht, the Netherlands: Kluwer Academic Publishers, 1996, pp. 34–102.

88. A. V. Wolf, D. G. Remp, J. E. Kiley, and G. D. Currie, Artificial kidney function; kinetics of hemodialysis. *J. Clin. Invest.* 1951; **30**:1062–1070.

89. H. W. Smith, *The Kidney: Structure and Function in Health and Disease*. New York: Oxford University Press, 1951.

90. A. W. B. Morcos and A. R. Nissensen, Erythropoietin and high-efficiency dialysis. In: J. P. Bosch, ed., *Contemporary Issues in Nephrology vol. 27 - Hemodialysis: High-Efficiency Treatments*. New York: W.B. Saunders Company, 1993.

91. F. A. Gotch and J. A. Sargent, A mechanistic analysis of the National Cooperative Dialysis Study (NCDS). *Kidney Int.* 1985; **28**:526–534.

92. J. A. Sargent, Shortfalls in the delivery of dialysis. *Am. J. Kidney. Dis.* 1990; **15**:500–510.

93. W. E. Bloembergen, D. C. Stannard, F. K. Port, R. A. Wolfe, J. A. Pugh, C. A. Jones, J. W. Greer, T. A. Golper, and P. J. Held, Relationship of dose of hemodialysis and cause-specific mortality. *Kidney Int.* 1996; **50**:557–565.

94. A. L. Babb, R. P. Popovich, T. G. Christopher, and B. H. Scribner, The genesis of the square meter-hour hypothesis. *Trans. Am. Soc. Artif. Intern. Organs* 1971; **17**:81–91.

95. N. J. Ofsthun and J. K. Leypoldt, Ultrafiltration and backfiltration during hemodialysis. *Artif. Organs* 1995; **19**:1143–1161.

96. N. J. Ofsthun and A. L. Zydney, Importance of convection in artificial kidney treatment. In: K. Maeda and T. Shinzato, eds., *Effective Hemodiafiltration: New Methods*. Basel: Karger Publisher, 1994, pp. 54–70.

97. T. Kanamori, K. Sakai, T. Awaka, and M. Fukuda, An improvement on the method of determining the solute permeability of hollow-fiber dialysis membranes photometrically using optical fibers and comparison of the method with ordinary techniques. *J. Membr. Sci.* 1994; **88**:159–165.

98. C. Ronco, Backfiltration: a controversial issue in modern dialysis. *Int. J. Artif. Organs* 1988; **11**:69–74.

99. M. Schmidt, C. A. Baldamus, and W. Schoeppe, Backfiltration in hemodialysers with high permeable membranes. *Blood Purif.* 1984; **2**:108–114.

100. S. W. Hyver, J. Petersen, and J. Cajias, An in vivo analysis of reverse ultrafiltration during high-flux and high-efficiency dialysis. *Am. J. Kidney Dis.* 1992; **19**:439–443.

101. J. K. Leypoldt, B. Schmidt, and H. J. Gurland, Measurement of backfiltration rates during hemodialysis with highly permeable membranes. *Blood Purif.* 1991; **9**:74–84.

102. J. K. Leypoldt, B. Schmidt, and H. J. Gurland, Net ultrafiltration may not eliminate backfiltration during hemodialysis with highly permeable membranes. *Artif. Organs* 1991; **15**:164–170.

103. C. Ronco, Backfiltration in clinical dialysis: nature of the phenomenon, mechanisms and possible solutions. *Int. J. Artif. Organs* 1990; **13**:11–21.

104. L. W. Henderson, Biophysics of ultrafiltration and hemofiltration. In: C. Jacobs, C. M. Kjellstrand, K. M. Koch, and J. F. Winchester, eds., *Replacement of Renal Function by Dialysis*. 4th ed. Dordrecht, the Netherlands: Kluwer Academic Publisher, 1996, pp. 114–145.

105. N. J. Ofsthun, C. K. Colton, and M. J. Lysaght, Determination of fluid and solute removal rates during hemofiltration. In: L. W. Henderson, E. A. Quellhorst, C. A. Baldamus, and M. J. Lysaght, eds., *Hemofiltration*. Berlin: Springer-Verlag, 1986, pp. 17–39.

106. C. Vander Velde and E. F. Leonard, Theoretical assessment of the effect of flow maldistributions on the mass transfer efficiency of artificial organs. *Med. Biol. Eng. Comput.* 1985; **23**:224–229.

107. A. Frank, G. G. Lipscomb, and M. Dennis, Visualization of concentration fields in hemodialyzers by computed tomography. *J. Membr. Sci.* 2000; **175**:239–251.

108. P. Verdonck, The role of computational fluid dynamics for artificial organ design. *Artif. Organs* 2002; **26**:569–570.

109. R. E. Nordon and K. Schindhelm, Design of hollow fiber modules for uniform shear elution affinity cell separation. *Artif. Organs* 1997; **21**:107–115.

110. J. Zhang, D. L. Parker, and J. K. Leypoldt, Flow distributions in hollow fiber hemodialyzers using magnetic resonance Fourier velocity imaging. *Asaio J.* 1995; **41**:M678–M682.

111. T. Osuga, T. Obata, H. Ikehira, S. Tanada, Y. Sasaki, and H. Naito, Dialysate pressure isobars in a hollow-fiber dialyzer determined from magnetic resonance imaging and numerical simulation of dialysate flow. *Artif. Organs* 1998; **22**:907–909.

112. S. Takesawa, M. Terasawa, M. Sakagami, T. Kobayashi, H. Hidai, and K. Sakai, Nondestructive evaluation by x-ray computed tomography of dialysate flow patterns in capillary dialyzers. *ASAIO Trans.* 1988; **34**:794–799.

113. C. Ronco, M. Scabardi, M. Goldoni, A. Brendolan, C. Crepaldi, and G. La Greca, Impact of spacing filaments external to hollow fibers on dialysate flow distribution and dialyzer performance. *Int. J. Artif. Organs* 1997; **20**:261–266.

114. C. Ronco, A. Brendolan, C. Crepaldi, M. Rodighiero, P. Everard, M. Ballestri, G. Cappelli, M. Spittle, and G. La Greca, Dialysate flow distribution in hollow fiber hemodialyzers with different dialysate pathway configurations. *Int. J. Artif. Organs* 2000; **23**:601–609.

115. C. K. Poh, P. A. Hardy, Z. Liao, Z. Huang, W. R. Clark, and D. Gao, Effect of spacer yarns on the dialysate flow distribution of hemodialyzers: a magnetic resonance imaging study. *Asaio J.* 2003; **49**:440–448.

116. S. K. Karode and A. Kumar, Flow visualization through spacer filled channels by computational fluid dynamics I. Pressure drop and shear rate calculations for flat sheet geometry. *J. Membr. Sci.* 2001; **193**:69–84.

117. A. R. Dacosta, A. G. Fane, C. J. D. Fell, and A. C. M. Franken, Optimal channel spacer design for ultrafiltration. *J. Membr. Sci.* 1991; **62**:275–291.

118. O. Kedem and A. Katchalsky, Thermodynamic analysis of the permeability of biological membranes to nonelectrolytes. *Biochim. Biophys. Acta* 1958; **27**:229–246.

119. S. W. B. De Lint, Transport of electrolytes through ceramic nanofiltration membranes. In: *Chemical Engineering*. Twente: University of Twente, 2003.

120. A. Kargol, A mechanistic model of transport processes in porous membranes generated by osmotic and hydrostatic pressure. *J. Membr. Sci.* 2001; **191**:61–69.

121. A. Wupper, F. Dellanna, C. A. Baldamus, and D. Woermann, Local transport processes in high-flux hollow fiber dialyzers. *J. Membr. Sci.* 1997; **131**:181–193.

122. S. K. Karode, Laminar flow in channels with porous walls, revisited. *J. Membr. Sci.* 2001; **191**:237–241.

123. J. M. Miranda and J. B. Campos, An improved numerical scheme to study mass transfer over a separation membrane. *J. Membr. Sci.* 2001; **188**:49–59.

124. A. S. Michaels, New separation tehcnique for the CPI. *Chem. Eng. Prog.* 1968; **64**:31–40.

125. A. L. Zydney and C. K. Colton, A concentration polarization model for the filtrate flux in cross-flow microfiltration of particulate suspensions. *Chem. Eng. Prog.* 1986; **47**:1–21.

126. E. C. Eckstein, D. G. Bailey, and A. H. Shapiro, Self-diffusion of particles in shear flow of a suspension. *J. Fluid Mech.* 1974; **79**:191–208.

127. C. A. Romero and R. H. Davis, Global-model of cross-flow microfiltration based on hydrodynamic particle diffusion. *J. Membr. Sci.* 1988; **39**:157–185.

128. Y. Lee and M. M. Clark, A numerical model of steady-state permeate flux during cross-flow ultrafiltration. *Desalination* 1997; **109**:241–251.

129. S. P. Palecek and A. L. Zydney, Intermolecular electrostatic interactions and their effect on flux and protein deposition during protein filtration. *Biotechnol. Prog.* 1994; **10**:207–213.

130. C. C. Ho and A. L. Zydney, A combined pore blockage and cake filtration model for protein fouling during microfiltration. *J. Coll. Interf. Sci.* 2000; **232**:389–399.

131. L. Palacio, C. C. Ho, and A. L. Zydney, Application of a pore-blockage cake-filtration model to protein fouling during microfiltration. *Biotechnol. Bioeng.* 2002; **79**:260–270.

132. D. Bhattacharyya, S. L. Back, and R. I. Kermode, Prediction of concentration polarisation and flux behaviour in reverse osmosis by numerical analysis. *J. Membr. Sci.* 1990; **48**:231–262.

133. C. Rosen and C. Tragardh, Computer simulation of mass transfer in the concentration boundary layer over ultrafiltration membranes. *J. Membr. Sci.* 1993; **85**:139–156.

134. K. Zhang and A. Acrivos, Viscous resuspension in fully developed laminar pipe flows. *Int. J. Multiphase Flow* 1994; **20**:579–591.

135. M. Hofer and K. Perktold, Computer simulation of concentrated fluid-particle suspension flows in axisymmetric geometries. *Biorheology* 1997; **34**:261–279.

136. R. J. Phillips, R. C. Armstrong, and R. A. Brown, A constitutive equation for concentrated suspensions that accounts for shear-induced particle migration. *Phys. Fluids* 1992; **4**:30–40.

137. M. Y. Jaffrin, L. H. Ding, and J. M. Laurent, Simultaneous convective and diffusive mass transfers in a hemodialyser. *J. Biomech. Eng.* 1990; **112**:212–219.

138. L. Grimsrud and A. L. Babb, Velocity and concentration profiles for laminar flow of a newtonian fluid in a dialyser. *Chem. Eng. Prog.* 1966; **62**:20–31.

139. C. Colton, K. Smith, P. Stroeve, and E. Merrill, Laminar flow mass transfer in a flat duct with permeable walls. *Am. Inst. Chem. Eng.* 1971; **17**:773–780.

140. A. Berman, Laminar flow in channels with porous walls. *J .Appl. Phys.* 1953; **24**:1232–1235.

141. S. M. Ross, Mathematical-model of mass-transport in a long permeable tube with radial convection. *J. Fluid Mech.* 1974; **63**:157–175.

142. M. Abbas and V. P. Tyagi, On the mass transfer in a circular conduit dialyzer when ultrafiltration is coupled with dialysis. *Int. J. Heat Mass Transfer* 1988; **31**:591–602.

143. M. Y. Jaffrin, B. B. Gupta, and J. M. Malbrancq, A one-dimensional model of simultaneous hemodialysis and ultrafiltration with highly permeable membranes. *J. Biomech. Eng.* 1981; **103**:261–266.

144. J. E. Sigdell, Calculation of combined diffusive and convective mass transfer. *Int. J. Artif. Organs* 1982; **5**:361–372.

145. A. Werynski and J. Waniewski, Theoretical description of mass transport in medical membrane devices. *Artif. Organs* 1995; **19**:420–427.

146. M. Y. Jaffrin, Convective mass transfer in hemodialysis. *Artif. Organs* 1995; **19**:1162–1171.

147. A. Wupper, D. Woermann, F. Dellanna, and C. A. Baldamus, Retrofiltration rates in high-flux hollow fiber hemodialyzers: analysis of clinical data. *J. Membr. Sci.* 1996; **121**:109–116.

148. A. L. Zydney, Bulk mass transport limitations during high flux hemodialysis. *Artif. Organs* 1993; **17**:919–924.

149. C. Legallais, G. Catapano, B. von Harten, and U. Baurmeister, A theoretical model to predict the in vitro performance of hemodiafilters. *J. Membr. Sci.* 2000; **168**:3–15.

150. N. Hosoya and K. Sakai, Backdiffusion rahter than backfiltration enhances endotoxin transport through highly permeable dialysis membranes. *Trans. Am. Soc. Artif. Intern. Organs* 1990; **36**:311–313.

151. S. Eloot, Y. D'Asseler, P. De Bondt, and P. Verdonck, Combining SPECT medical imaging and computational fluid dynamics for analyzing blood and dialysate flow in hemodialyzers. *Int. J. Artif. Organs*, in press.

152. L. F. Mockros and R. Leonard, Compact cross-flow tubular oxygenators. *Trans. Am. Soc. Artif. Intern. Organs* 1985; **31**:628–633.

153. A. C. Kak and M. Slaney, *Principles of Computerized Tomographic Imaging*. New York: IEEE Press, 1988.

154. S. Eloot, D. De Wachter, I. Van Tricht, and P. Verdonck, Computational flow modeling in hollow-fiber dialyzers. *Artif. Organs* 2002; **26**:590–599.

155. M. F. Kiani and A. G. Hudetz, A semi-empirical model of apparent blood viscosity as a function of vessel diameter and discharge hematocrit. *Biorheology* 1991; **28**:65–73.

156. S. Eloot, D. De Wachter, J. Vienken, R. Pohlmeier, and P. Verdonck, In vitro evaluation of the hydraulic permeability of polysulfone dialysers. *Int. J. Artif. Organs* 2002; **25**:210–216.

157. J. Bear, *Dynamics of Fluids in Porous Media*. New York: American Elsevier, 1972.

158. C. K. Colton, L. W. Henderson, C. A. Ford, and M. J. Lysaght, Kinetics of hemodiafiltration. I. In vitro transport characteristics of a hollow-fiber blood ultrafilter. *J. Lab. Clin. Med.* 1975; **85**:355–371.

159. L. W. Henderson, C. K. Colton, and C. A. Ford, Kinetics of hemodiafiltration. II. Clinical characterization of a new blood cleansing modality. *J. Lab. Clin. Med.* 1975; **85**:372–391.

160. S. Eloot, J. Vierendeels, and P. Verdonck, Optimisation of dialyser performance using a three-dimensional finite volume model, *submitted for publication*, 2005.

161. Fresenius, Product brochure: Hemoflow F-series Fresenius Polysulfone Capillary Dialysers.

162. C. Ronco, Hemofiltration and hemodiafiltration. In: *Contempory Issues in Nephrology*, vol. 27. New York: Churchill-Livingstone, 1993, pp. 119–133.

ASSISTIVE ROBOTICS

MICHAEL R. HILLMAN
Wolfson Centre, Royal United
Hospital Bath
Somerset, United Kingdom

1. INTRODUCTION

For many years, and particularly since the first electric wheelchair in the 1950s, disabled people have benefited from the use of mechanical and electrical technology to provide powered mobility. The dream of assistive robotics is to extend that enablement to the area of powered manipulation. Assistive robotics is a combination of a technology and an application area. As such, overlap exists both with the wider field of robotics and with assistive technology. In attempting to discuss assistive robotics, it is wise to start from an official definition. The Robot Institute of America defined a robot as "a re-programmable, multifunctional manipulator designed to move material, parts, tools or specialized devices through variable programmed motions for the performance of a variety of tasks."

Although this definition was obviously intended for industrial robots, it identifies the key features of programmability, flexibility, and movement. Robotics has obviously moved on from that early definition. Although robots were initially employed as handling machines in factories, their application is now much wider. In 1987, the Department of Trade and Industry in the United Kingdom launched an advanced robotics initiative to encourage the wider use of robotics in areas other than factories. They used this definition of advanced robotics: "The integration of enabling technologies and attributes embracing manipulators, mobility, sensors, computing (Intelligent Knowledge Based Systems, Artificial Intelligence) and heirarchical control to result ultimately in a robot capable of autonomously complementing man's endeavours in unstructured and hostile environments."

This definition is very wide-ranging, but one of the key features is the "integration of technologies." The aims of this integration of advanced technologies are to produce devices that can both operate autonomously and in environments that may be unstructured or hostile.

Many examples of advanced robotics now exist. What is important is not just the level of technology, but that the robots have moved out of the factory and into the wider, sometimes hostile, and certainly unstructured world. The examples cover a wide area and include devices for the exploration both of Mars and the earth's oceans as well as more domestic applications such as filling a car with fuel, mowing the lawn, and cleaning the floor. The use of robotics in medical and associated applications is another major area in which robots are coming "out of the factory."

Robotics has been used in a number of medical applications. Literally at the "cutting edge" is the application of robotic technology in surgery, for example, in orthopedics, neurosurgery, and endoscopy. Robot-mediated therapy (or robotic rehabilitation therapy) is the application of robotics in the therapeutic rehabilitation of those with various neurological conditions, particularly those who have had strokes. Our interest is the application of robotics to assistive technology. Assistive technology (AT) is defined in the King's Fund consultation (March 14, 2001) as "a product or service designed to enable independence for older or disabled people," which may range from something as "simple" as a modified cup to a complex electromechanical device.

Many of the uses of robotics in AT are either in a vocational environment or as aids to daily living. However, other areas exist either where robotic technology has been applied or might be applied:

- Mobility - increasing the usefulness of more traditional powered wheelchairs.
- Prosthetics and orthotics - robotics has already been applied in this area.

- Education - modern robotic toys might have an impact as well as equipment more specifically aimed at the disabled.
- Communication - advanced computer technology is already widely in use, but there is the scope of incorporating more mechanical technologies as well.

2. POTENTIAL USES

Many people including the report of Prior (1) have surveyed the potential uses of robotics to assist people with physical disabilities, and the following areas have been identified.

- Eating and drinking
- Personal hygiene – washing, shaving, applying make up
- Work and leisure – particularly computer use, equipment such as hi-fi and video systems, and games
- Mobility – opening doors, windows
- General reaching – up to shelves, down to the floor

These areas are all valuable. Many nonrobotic devices exist that are dedicated to specific tasks, some designed for people with disabilities, others readily available on the general market. In this case, the choice and installation of such devices (in consultation with the user) predetermines what tasks he or she can carry out. Most assistive robots, however, are designed as a general-purpose tool and intended to be used as the user desires, rather than for any predetermined task. It is possible to construct lists of specific tasks for which an assistive robot might be used. Some tasks will no longer be relevant as technology progresses. For example, many early workstation robots demonstrated the manipulation of 5.25" floppy disks. Similarly, tasks will come into the list, for example, bringing a personal digital assistant (PDA) close to a user. The tasks will also vary from country to country depending on cultural norms. Independence only comes when the user can decide at any time what activity they would like to be involved in. A good example is the user who gained great satisfaction from using his independence to open his Christmas presents.

3. HISTORICAL HIGHLIGHTS

The term robot was first used in Capek's *Rossums Universal Robots* in the 1920s, from the Czech word for a slave. The film Metropolis in 1927 represented the familiar humanoid robot loved by science fiction. In practical terms, robotics can be dated back to the DeVilbiss paint sprayer in the 1930s. By 1950, the concept of robots was well established in science fiction, particularly in Asimov's exemplary book *I, Robot* (2). In the 1950s, the Unimation robot company was formed. 750,000 robots are now in industrial use.

Most reviews of rehabilitation robotics cite the work at the CASE Institute of Technology in the early 1960s (3) as the first application of robotics technology to a

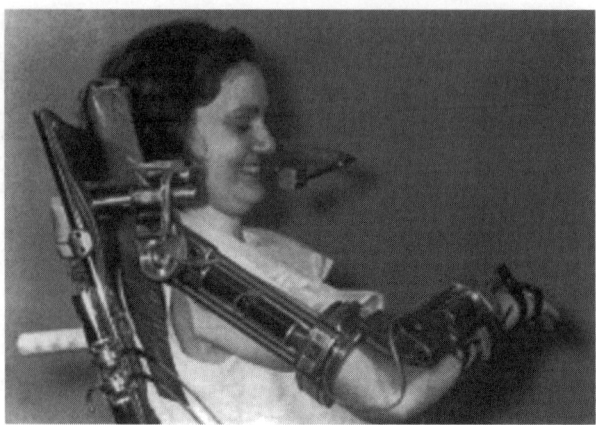

Figure 1. Rancho Golden Arm.

rehabilitative manipulator, which was a powered orthosis with four degrees of freedom. The exoskeletal structure supported the user's paralyzed arm while performing prerecorded manipulative tasks, these sequences being taught by an able-bodied assistant during training. Interestingly, when so much current work uses electrical actuators, this used pneumatic actuators with closed-loop position control achieved using incremental encoders.

Another early project was the Rancho Los Amigos *Golden Arm* (Fig. 1) (3), which was a seven-degree-of-freedom battery-powered electric orthosis. Several versions were built, and at least one was wheelchair-mounted. It was controlled using a form of joint-by-joint control; during evaluation, this control was found to be unintuitive.

In considering these two early devices, it is instructive to put them in the context of the technology of the day. The early 1960s was a time when the integrated circuit had just been invented, and ten years before the microprocessor. Computers were beginning to come down in size from room size to a more compact "cabinet."

4. WORKSTATION-BASED ROBOTS

Apart from the two prosthetic devices mentioned above, work in the more specific area of assistive robotics started in the mid-1970s. One of the earliest projects was the workstation-based system designed by Paeslack and Roesler (4) in Heidelberg, Germany. The purpose designed, five degree of freedom manipulator was placed in a specially adapted desktop environment, using rotating shelf units.

Another early workstation system was that of Seamone and Schmeisser (5) at the Johns Hopkins University, supported by the Veterans Administration in the United States from 1974. The arm of this system was based around an electrically powered prosthetic arm, mounted on a horizontal track. Various items of equipment (e.g., telephone, book rest, computer disks) were laid out on the simple, but cleverly designed, workstation table and could be manipulated by the arm using preprogrammed commands. The system thus required that items be in precisely known positions as there were no sensors on the arm. User input was by simple scanning switch selection of routines on a simple LED display.

Figure 2. DeVAR IV workstation Veteran's Administration.

In France, an early project was the Spartacus robot (6), based around a large, relatively high-power manipulator from the nuclear industry. The table-mounted arm was able to reach down to the floor or up to a shelf. This project is of particular significance in that it led to the Manus project in Holland and the Master project in France.

In any review of work in rehabilitation robotics there must be recognition of the continuing work at Stanford University, initiated by Leifer in the Department of Mechanical Engineering, with Van der Loos at the Palo Alto VA Centre. They built four generations of DeVAR (Desktop Vocational Assistive Robot) systems (7,8). DeVAR III was a tabletop system laid out for daily living tasks, whereas DeVAR IV was used in a vocational environment.

The DeVAR IV system (Fig. 2) used the Puma 260 robot, a standard industrial manipulator, mounted upside down on an overhead track, thus making much better use of the available space. This highlights a problem of using a commercially available robot in that the work environment has to be tailored around the arm. How this result is achieved can be crucial to how successful the system is. The usefulness of the system depends on how many "tasks" can be laid out within the work environment. Whenever a certain task is not available to the robot or the user, the usefulness of the system comes to a halt and a colleague has to be called in to intervene. Obviously, in an office environment, this solution can be allowed for, but the ability to work independently has been compromised.

Besides the engineering work, this project is notable for its use in a real working environment on an 8-hour-a-day basis at the Pacific Gas and Electric Company by Bob Yee. The Stanford group has also done a lot of work to justify the high cost of such a system ($50–$00,000) in terms of financial savings relative to the cost of employing a human assistant.

Whereas the Stanford system used the Puma robot, initially designed for an industrial application, many other projects have been based around the RT series robot. The RT series was sold by UMI (Universal Machine Intelligence, UK). One significant use of the RT series robot was by the Master project (9) in France. Their approach to maximizing the workspace was to mount the arm at the back of the workstation, which gave good visibility of the whole work area. In setting up any workstation, it is vital that the arm does not obscure visibility of the environment, in this case either the large vertical column or the bulk of the arm or gripper in different orientations.

The RT robots were used as the basis for the RAID project (10), funded by the European Commission. The RAID project, as with all projects under the TIDE initiative (Telematics for the Integration of Disabled and Elderly people), was collaborative and multinational. Among the partners were Oxford Intelligent Machines (OxIM, UK), who were then the manufacturers of the RT robots, and the Master project team. The RT series robot was set up in an extended workstation. The work space of the basic robot is extended by, in this case, mounting the robot on a horizontal track enabling it to retrieve paper work etc from shelving units. The outcome of the RAID project was commercialized by OxIM (who have now ceased trading) and Afma Robots in France.

4.1. Feeding

In terms of numbers sold, probably the most successful rehabilitation robot is the Handy 1 (Fig. 3) (11), sold by Rehab Robotics (UK). The project originated from the Masters Degree project of Topping at the University of Keele (UK). Topping had a neighbor, a young boy named Peter, who had severe problems eating. He used a cheap educational robot to produce a device that allowed Peter to eat independently, at the speed he chose, for the first time. The company has sold at least 250 systems with very positive feedback of its effectiveness. One of the strengths of this project, and reasons for its success, is that it stemmed from the real problems of an individual client. Another feature is that it originally made no attempt to be multi-functional. However, more recently, extensions of the system have been developed for it to be used for applying makeup, painting, washing, and shaving.

The Handy 1 is essentially a feeding aid based around a robot. By comparison, the Winsford feeder (RTD-ARC, New Jersey) has been available for a long time as a feeding aid, but only recently has been promoted as being a robotic device. With around 2000 units having been sold, its commercial impact is obviously greater than the Handy 1. The MySpoon (Secom Co. Ltd, Tokyo, Japan) is available in Japan and is similar in concept. In the United Kingdom, the Neater Eater (Buxton, UK) was initially

Figure 3. Handy 1 feeding robot M Topping/Rehab Robotics.

designed as a purely manual device to assist those with a tremor to eat, using a damped arm. More recently, a powered, programmable device has emerged that may be considered to be a robotic device.

5. WHEELCHAIR-MOUNTED ROBOTS

If a mobile robot can be seen as a mechanical servant or slave, the concept of mounting a manipulator onto a wheelchair provides what may be termed a third arm.

One very early wheelchair-mounted robot was designed by Carl Mason (12) at the VA Prosthetics Center in New York. Mechanically, it was a well-engineered system, able to reach from the floor to ceiling. Apparently, however, it was not rigid enough, making it difficult to control and the simple hook-end effector (as used by many prosthetic hand wearers) was not very successful. The control was quite basic, being operated on a joint-by-joint basis that has been found to be unintuitive and time-consuming to control.

By comparison, placing a simple educational robot on a wheelchair tray is the most basic of arrangements. Zeelenberg's son was diagnosed as having Muscular Dystrophy. His parents wanted him to make the fullest development of his abilities and skills. His father obtained an educational robot and simply mounted it on the wheelchair tray (13). As Muscular Dystrophy (Duchenne) is characterized by a general weakness and muscle wasting, it is often possible to use a small push-button controller, which was chosen for the input device. This choice would not pretend to be a technically sophisticated device but derived out of a real need and close collaboration between the developer and user. Among the uses of the robot are opening the door, moving chess pieces, and using the telephone. One reason for highlighting this project is that out of this work came the Manus project.

Manus (Fig. 4) (14) is a sophisticated robotic manipulator able to be mounted to a number of different wheelchairs. It has seven degrees of freedom, as well as a simple gripper. The extra degree of freedom extends its vertical range, allowing it to fold compactly at the side of the wheelchair. The mounting of the arm, protruding from the side of the chair, raises the crucial issue for all wheelchair-mounted robots of integrating the arm with the wheelchair, not least to ensure no unacceptable increase in overall width or compromise of the stability of the wheelchair occurs. The work started as far back as 1984 involving collaboration between IRV (Institute for Rehabilitation Research), in Hoensbroek, led by Hok Kwee, the Institute for Applied Physics and the TNO Product Centre in Delft, and the Netherlands Institute of Preventive Health Care. It is now manufactured by Exact Dynamics and is on the prescription list of the Dutch government.

Many have been sold to rehabilitation centers with much development going on around it, but more importantly with significant sales to end-users. Manus is seen as the standard against which other assistive robotic systems are measured. Further development of Manus is being carried out by both the manufacturers and also under the European Commanus project (15).

The other wheelchair-mounted manipulator that is available commercially is the Raptor (Fig. 5) (16), which is currently being produced by Phybotics. It makes an interesting comparison with Manus. Whereas Manus is a relatively high cost, sophisticated device, the Raptor has introduced compromises to bring the cost down to about a third of that of Manus. In particular, it has only four degrees of freedom. It will be very interesting to see how these two devices perform commercially and in terms of their effectiveness, with their differences in cost and functionality. It will also be interesting to see how the larger American market affects the viability.

A unique approach to wheelchair-mounted manipulators came from Hennequin in the United Kingdom. He

Figure 4. Manus ARM. With permission Exact Dynamics. (This figure is available in full color at http://www.mrw.interscience.wiley.com/ebe.)

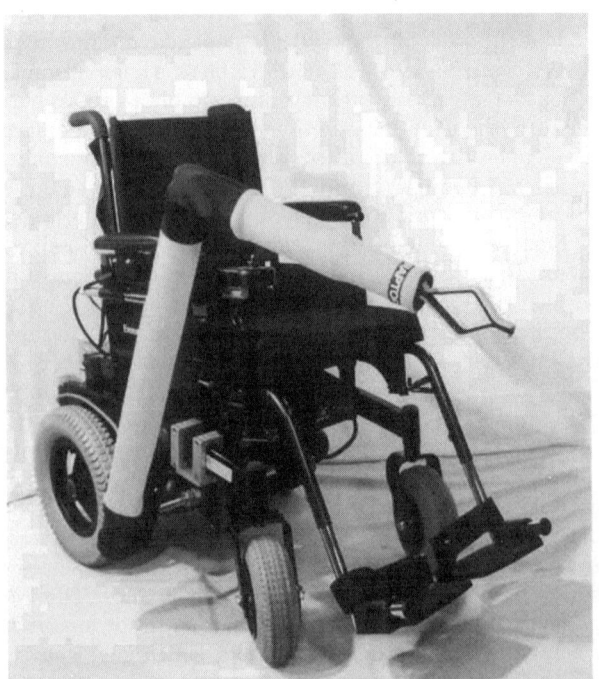

Figure 5. Raptor. With permission Phybotics.

was best known for his Spitting Image satirical puppets, which appeared on television in the United Kingdom. The puppets used a pneumatic air muscle, which were used for the drive motors of the Inventaid (17) wheelchair-mounted robot. He claimed a high power-to-weight ratio for the air muscles. He also claimed that it was simple enough to be maintained by a backstreet fitter.

6. MOBILE ROBOTS

Compared with the workstation and wheelchair-based devices, the number of mobile assistive robots is very small and the commercial impact has been negligible. One of the best known is probably the MoVAR system (18) from Stanford University, which is essentially a DeVAR on wheels. The mobile base had sophisticated omnidirectional "mechanum" wheels. The MoVAR was controlled from a console with several monitor screens giving feedback to the user from an on-board camera as well as a map of the environment and a control environment. It was a very capable system, but at the time it was not packaged in a way that would appeal to a general user. With advances in computer technology, the idea could be revisited and a more marketable product achieved.

As the founder of the Unimation Company, Engelberger has been called the father of robotics. He has had an interest in service robotics and particularly in medical applications. He proposed (19) the use of his Helpmate robot as a fetch and carry robot for a disabled person. However, although the Helpmate has been successfully used for moving supplies around a hospital, it has not been used successfully in a rehabilitation application. The cluttered environment of a home is not appropriate for such a mobile robot.

More recently, the KARES II robot system (20) has been developed at KAIST in Korea. This project is a wide-ranging project investigating various modes of user control including the use of visual servoing, an eye mouse, and a haptic suit as well as the design of the robot arm itself. The arm has been mounted in a number of different configurations, but primarily on a remote-controlled mobile base.

A different approach was investigated at the Bath Institute of Medical Engineering with their Wessex robot (Fig. 6) (21). Having identified the shortcomings of a fixed-site robotic workstation in a domestic environment, a nonpowered mobile base was designed. Although a workstation system works well in a vocational environment, it may be very restricting in a domestic environment where different tasks are normally carried out in different rooms of the home, for example, to wash in the bathroom, to listen to music in the living room, and to eat in the kitchen or dining room. The mobile base was intended to be moved from room to room by a carer or might be clipped to the front of a wheelchair.

7. BODY-WORN ROBOTS

The ultimate body-worn robot is what is termed an exoskeleton, a "suit" that can greatly enhance a person's

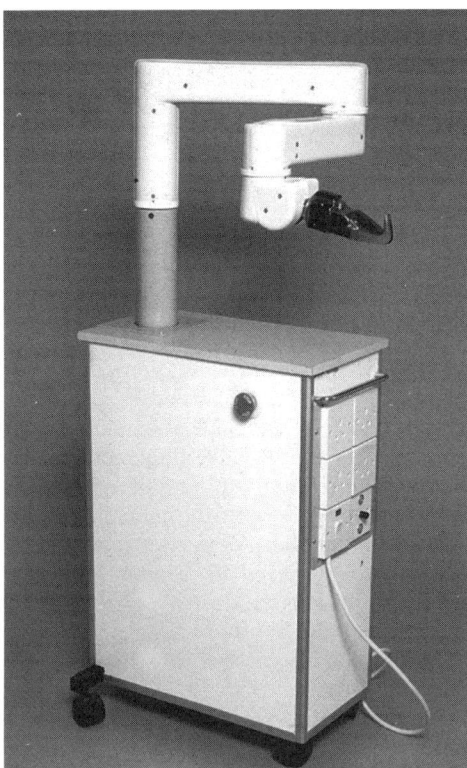

Figure 6. Wessex robot. With permission Bath Institute of Medical Engineering.

physical abilities. Interest exists, mainly in the United States and Japan, in such systems both for the elderly and disabled as well as for the military. A system that is claimed to be nearing commercial availability for people with disabilities is the Hybrid Assistive Limb-5 (HAL-5), designed by Sankai of the University of Tsukuba (22). The earlier HAL-3 suit consisted of frames to support both the user's legs and has motors at knee and hip joints. The latest version adds an upper arm exoskeleton. EMG signals are detected, aiming to detect the user's movement intentions, and are input to a computer system, which controls the motors. The system is also taught the user's normal stride pattern. For all body-worn systems, a major constraining factor is the storage of energy; HAL-5 quotes a usable 2 h 40 min using NiMH cells.

It is clear from the early work, mentioned above, that the more traditional fields of prosthetics and orthotics have been closely associated with rehabilitation robotics. It is useful at this stage to define a prosthesis as an artificial limb (although the term can also be used for an internal organ or joint) and an orthosis as a device to support or control part of the body.

7.1. Upper Limb

The early work at CASE and Rancho Los Amigos (mentioned above) were orthotic systems. More recently, the Mulos project, funded under the EU TIDE funding initiative (23) was a powered upper-limb orthotic system. Rahman et al. at the AI duPont Institute (Wilmington, DE) have been intimately involved in the rehabilitation robotics field. They have been involved in the design of powered and nonpowered arm orthoses. In particular, their antigravity arm orthosis (24) is noteworthy, although, being a balanced system with no external power supply, it may not come within our definition of a robotic system.

Although a lot of commercial work in prosthetic arms and hands exists, very little of this work has used robotic technology, but rather has been a development from existing technologies. However, one computer-controlled upper-arm prosthesis is the Utah/MIT artificial arm and dextrous hand developed by Jacobsen (25). Another long-standing project is the work that originated with the Southampton Hand (26) at Southampton University, UK and progressed at the Nuffield Orthopaedic Centre in Oxford, UK. Initially a complex five-fingered hand, the mechanism has been simplified, while retaining the capabilities of forming the hand into several functional configurations both for precision and for power. As a continuation of this work, the ToMPAW project (27) combines the earlier Leverhulme hand prosthesis with a prosthetic arm developed at Edinburgh, UK.

Two main issues in powered prosthetics and orthotics exist. One is the issue of miniaturization and the other is the issue of power. The problem of miniaturization is particularly critical for a hand prosthesis, where the complete system has to fit within the outline of a human hand. The problems of integrating a robotic system onto a wheelchair have already been noted. The problems of prosthetics are obviously at a far greater level of magnitude. Although a hand prosthesis and an upper-arm orthosis require different levels of power, both present problems in how to store the energy to give a day's use of the device before recharging is required. With a hand prosthesis, the requirement is to fit batteries with sufficient capacity within the hand. For an upper-arm prosthesis, the energy requirement is far greater. Although the volume/mass constraints are not so difficult, for a truly portable system, it is still a major problem area. Besides electrical batteries, compressed CO_2 has also been used as a power supply.

7.2. Lower Limb

Blatchford's Intelligent Knee prosthesis (Basingstoke, UK), however, is a nonpowered device. It uses sensors to regulate the swing of the knee, dependent on the rate of walking and other programmable values. Although it is a passive rather than an active device, it is truly a robot, although nowhere in Blatchford's publicity is the word "robot" used. Similarly, Otto-Bock's C-Leg (Duderstadt, Germany) senses the position in the walking cycle, knee angle position, and ground force reactions to adjust hydraulic valves to provide the required damping.

8. SMART WHEELCHAIRS

Within mainstream robotics, a major area of both research and commercial application is that of automatic guided vehicles (AGV). This technology obviously has potential in addressing the mobility needs of people with a wide range of disabilities. All powered wheelchairs operate in two degrees of freedom, conventionally under the direct control of the user, and so are not that much different from a two-degree-of-freedom telemanipulator. Modern powered wheelchairs will often add other powered functions such as leg raisers and seat tilt. Most modern powered wheelchairs also have programmable controllers. Such chairs may be technically sophisticated but would not make the claim of being robotic. As soon as sensors and a control system that can react to the output of those sensors are added, the wheelchair becomes an AGV. Such devices are normally referred to as smart wheelchairs. Smart wheelchairs use sensors to detect objects in the environment. On-board processing of the constantly changing and moving relative environment can allow the user such functions as to track a wall, go through a door, or dock at a table or desk.

One approach is to adapt a standard commercial base. For example, the CALL Centre in Edinburgh (UK), have many years experience in this area. In their initial work (28) they used a standard electric wheelchair to produce a smart wheelchair for children and teenagers. In their latest smart wheelchair, the Smart Controller acts as if it was a second joystick plugged into the DX (Dynamic, New Zealand) wheelchair bus system. Various Smart Wheelchair tools can be easily selected in different combinations to suit the pilot and environment. Alternatively, it is possible to build a wheelchair that is "smart" from the outset. Such an approach was used by the CEC TIDE-funded OMNI project (29), which was an omnidirectional wheelchair integrated with autonomous control features.

The big difference between an AGV and a smart wheelchair is that a powered wheelchair is not normally required to be completely autonomous. The issue is how to handle the conflicts between when the user should have an appropriate level of control of the chair and when the smart processor will take over, and vice versa.

A different approach to mobility comes from Dean Kamen who invented the iBOT wheelchair. Although the chair may be driven in a conventional way, gyroscopic sensors allow the chair to balance on two wheels or to climb stairs. The safety issues of relying on gyroscopes and processors to provide the basic stability of the device are paramount. In common with other safety critical "fly by wire" systems, multiple redundancy is used. A commercial company, Independence Technology (US) has recently received FDA approval for the iBOT in the United States. Since the first Everest and Jennings powered wheelchair in the 1950s, the traditional design has changed little; perhaps the first truly novel product for many years is the iBOT wheelchair.

Not all assistance to mobility implies that the person needs to be transported by the device. In 1977, Meldog (30) was developed at the Mechanical Engineering Labs at Tsukuba Science City in Japan. It provided mobility for a blind person by guiding them around city streets, downloading a basic map, and using landmark sensors. It would function in much the same way as a guide dog would be used in other cultures. With the increasing miniaturization of electronics and GPS positioning, it is possible that the same functionality could be obtained today on a body-worn device without the problems of kerbs and steps.

Figure 7. Cambridge University Educational Robot. Photograph with permission Bath Institute of Medical Engineering.

9. ROBOTICS IN SPECIAL NEEDS EDUCATION

The use of robots in education for those with physical and learning disabilities has received attention over the years. For example, a young child learns much from simple play activities. However, if he has impaired mobility, these opportunities are greatly restricted. The Cambridge/CUED robot (Fig. 7) (31), based on a UMI RTX robot with a vision recognition system, allowed the child to interact with his environment in various ways ranging from simply dropping a toy brick onto a drum to painting or playing board games.

Many electronic toys are now cheaply available on the mainstream market, which may provide special benefits for the disabled. However, AnthroTronix (Maryland) are developing what they describe as telerehabilitation tools to motivate and integrate therapy, learning, and play for children with disabilities.

10. ROBOTICS IN COMMUNICATIONS

It has already been identified that most communication aids are not robots, which certainly applies when communication is through an acoustic medium. Communication can, however, use other senses—especially visual, but also tactile.

The Dexter hand (32) is intended to act as a finger spelling hand for those who are deaf/blind. Such people use finger spelling for communication, but this language is not widely known by the general population. Dexter enables someone to input text, for example, at a keyboard that is then converted to finger spelling by Dexter, to communicate with a person who uses finger spelling.

One difficulty that many physically disabled people encounter is the desire to read a book, magazine, or newspaper. Page turning is one of the tasks that comes highest on the list of priorities for an assistive robot. It is also one of the most difficult, although ways to achieve it exist. Several page-turners exist on the market, most of them bulky and expensive. They are often not very effective and limited in what they can achieve, which would seem to be a prime area for the application of robotic technologies. For what is essentially a single function device, cost may be a constraint, but if it was reliable and effective, a large market would exist for such a device.

11. A REVIEW OF THE TECHNOLOGY

11.1. Commercial Device or Custom Build?

Many assistive robots have used commercially available robots. Some of these have been industrial robots, some sold for educational purposes, and others designed at least partly with rehabilitation applications in mind.

The use of industrial robots presents several problems. Industrial robots tend to be large, high load, and expensive; often they are non-backdrivable or rigid. For a robot, which needs to operate in the close environment of a user, these traits limit the choice to the range of industrial robots that are small and relatively low-powered. The advantages of using an industrial robot are that such devices are well-developed, highly specified, and reliable, allowing the research and design to swiftly progress to more user-related issues. One compact industrial robot that has been widely used is the Puma 260 arm, used, for example, of the Stanford DeVAR systems (7,8).

Educational robots have been used by groups starting in this field of work, but the lack of robustness and reliability of such devices means that they have not been successful in the long term. However, in many instances, they have proved a valuable introduction before progressing to a purpose designed device.

The RTX robot (9,10) was designed by Tim Jones of UMI (Universal Machine Intelligence, UK) in 1985. One of the early application areas promoted by the manufacturers was its use in rehabilitation (other application areas were laboratory use and for educational purposes). The original RTX robot was followed by higher-powered versions, the RT100 and the RT200. The RT series has been widely used in assistive robotics as described in earlier sections of this article. The Manus robot (14,15), described above and designed specifically for a wheelchair-mounted application, has recently been used by several research groups in different mounting configurations.

For an assistive robot, the design and configuration of the manipulator needs to be appropriate to the environment and likely tasks. Many robot configurations are available (Fig. 8), some more appropriate for an assistive robot than others. The Puma is a revolute configuration, as is the Manus; the RT series are jointed cylindrical configuration robots, also known as modified SCARA. These two configurations have proved particularly appropriate for rehabilitation applications.

In assistive robotics, the integration of the robot in the environment is critical. The environment is normally centered around human users with their sensate and manipulative skills. For a workstation-based system, it is possible to arrange the workstation around the robot. However, in order to better integrate a robot into the environment, whether a workstation, wheelchair, or mobile

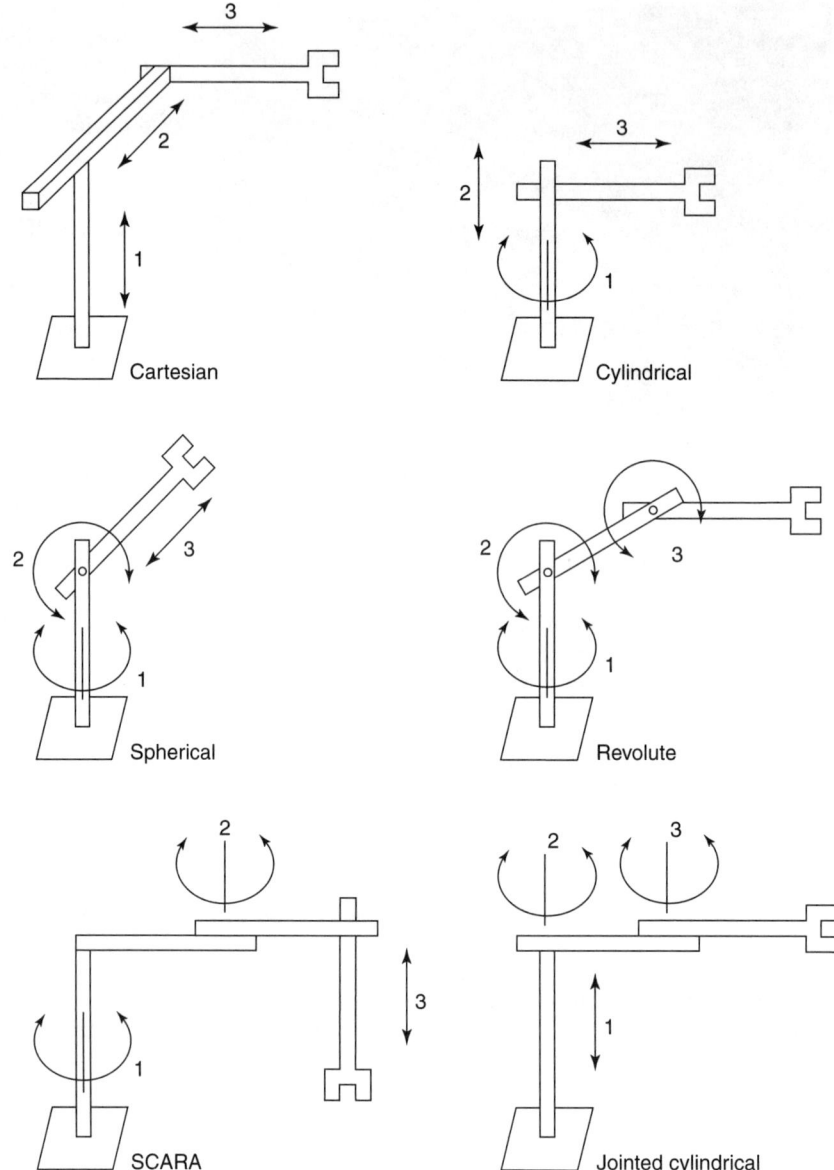

Figure 8. Robot configurations. With permission M. Hillman, PhD thesis.

base, an optimum solution may be to custom-design and build a robot arm.

As an example, the Neil Squire Foundation designed their Regenesis manipulator (Fig. 9) (33) to best suit the environment and tasks. Their manipulator was based on a horizontal beam, around which an extending arm could translate and rotate, which gave access to a large working volume and could be mounted at the back of a desk or across a bed, for example.

Besides integration, other reasons for custom building a robotic device exist. In the case of a device aimed for a specific task, such as feeding, it may be more appropriate to design a system particularly for that task. Although a successful feeding device may use a commercially available robot arm (the Handy 1 device started with a simple educational robot), a more appropriate approach may be to start with the specific movements required to bring food to a person's mouth. Also, the advantage of not being reliant on the continuing availability of a commercial device exists.

Another area that is more important for an assistive robot to be used in a domestic environment, compared with a purely functional industrial robot, is the aesthetic design. An assistive robot becomes both a consumer item and an extension of the user's style and personality. In his Master of Philosophy thesis, Pullin raised the question of what the appearance of a robotic device "says" (Fig. 10) - mechanical or macabre?, traditional or technological?, laboratory or lounge? Similar issues have been raised by Kwee (34).

A custom-built device, produced in small numbers, is likely to be more expensive than one made for a wider market in higher volumes. This disadvantage may be countered by optimizing the design appropriate to the specific system requirements. Any finished device needs to be of a quality and reliability comparable with the best commercial standards.

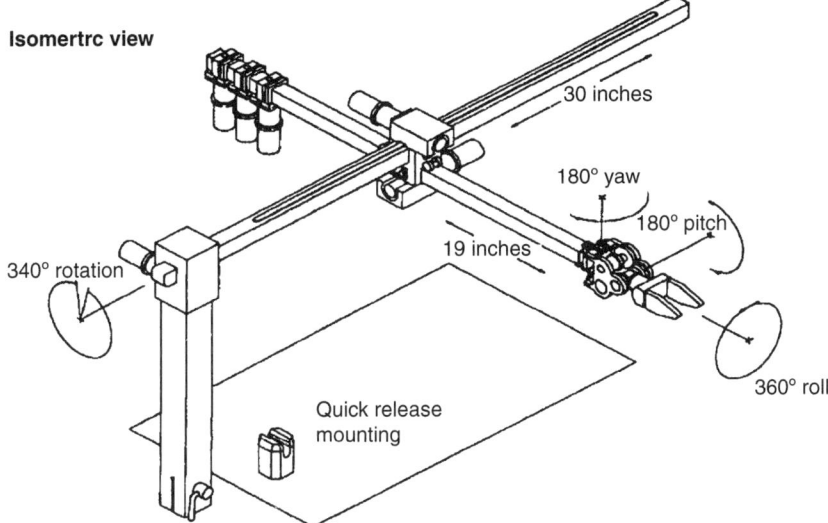

Figure 9. Regenesis workstation.

Mechanical or macabre?

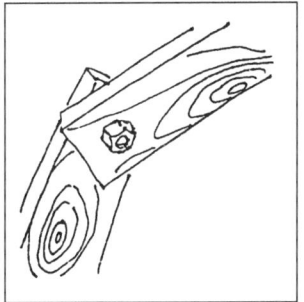
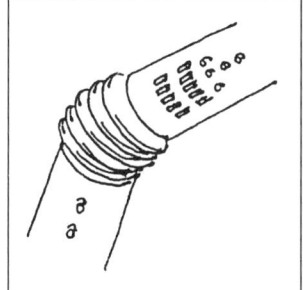

Traditional or technological?

Laboratory or lounge?

Figure 10. Questions of perception, identity, and taste. With permission G M Pullin.

11.2. Arm Design

To design and build a robotic device is not a trivial exercise, but offers benefits for a closely integrated system. It is beyond the scope of this article to describe in detail the design of a robot arm, but just to touch on the various areas that must be considered, and highlighting those of specific relevance to assistive robotics. Many good robot reference books exist; a good grounding in engineering design, from both a mechanical and electronics point of view, is essential.

The starting point is to understand what is required of the robot; the robot will be required to work closely with a human user, whose abilities and needs are central. The requirements of the robot will include:

- the movement patterns required,
- the environment in which it will be working (workstation/mobile base/wheelchair/unstructured human centred environment),
- the quantitative specification (load, range, stiffness, velocity).

With this information, and bearing in mind the other considerations detailed above, a decision can be made as to whether to purchase a commercial robot or proceed with a custom build.

Some basic design decisions will be made early on. Probably the most fundamental is what robot configuration to use and the number of degrees of freedom. Although the revolute and jointed cylindrical configurations have been widely used in assistive robots, others might be considered. The revolute configuration is appropriate because it is closest to a human arm. The jointed cylindrical configuration is also similar to a human arm; as most of the movements are in a horizontal plane, lower powered motors can be used. To allow the end effector to be oriented freely in any position in space requires a six-degree-of-freedom system (not including any end effector movements). Depending on the environment, it may be possible to simplify the system to less than 6 degrees. If the arm is mounted on a wheelchair or mobile base, the base may provide 2 degrees of freedom, but, in most instances, it is better for all six degrees of freedom to be within the arm itself.

Most assistive robots use electrical motors, which has proved to be the most appropriate, although other power devices have been used. Hydraulic systems may be used where high force/torque is required, but this case is rare with assistive robots and the necessary safely considerations. Pneumatic systems have been particularly used for prosthetic systems. The compliance can be appropriate in this situation, although is less helpful in a more structured workstation environment.

A robot may be described as a series of fixed elements joined by rotational or linear actuators. The motors to drive the actuators may be mounted in the base or adjacent to the actuators. Placing the motors in the base gives a lighter-weight upper arm, but with a more complex drive train. The drive train may use pulleys, cables, or rotating shafts. Mounting the motors adjacent to the actuators is simpler but with more weight in the upper arm, the structure is not so efficient. Even with the motor adjacent to the actuator, some gearing will be required, whether pulleys, an integrated gearhead or spur, bevel, or worm and wheel gears. Many different materials may be considered for the arm structure, including aluminium, carbon or glass fiber, plastics, or some composite structure. The aim is to achieve maximum stiffness and strength with the most economical use of materials, for low cost, weight, and cross section. The cross-sectional "bulk" is particularly relevant for an assistive robot where the user requires to see the end effector clearly without moving his or her body or trunk.

11.3. Electronics

Although a robotic device can be controlled by a simple series of electrical switches controlling power to the motors, the assumption is that most systems will be more sophisticated. The discussion below assumes an electrically powered robot, although many of the basic systems will also apply to a pneumatic- or hydraulic-powered robot.

The basic electronic components for a robot are as follows.

- Power supply
- Central processor
- Communications bus
- Motor control and drive
- Sensors

The robot will be either battery-powered or mains-powered. If mains-powered, it must be transformed or rectified to appropriate voltage levels for both the motor drive and processor functions. Efficient use of power is an important issue for wheelchair-mounted robots where the power will be taken from the wheelchair power supply; wheelchair users are very concerned about the range of their wheelchairs. In the case of a robot on a mobile base, power consumption is also an important issue.

The central processor is essentially a computer without a keyboard input device or monitor display system. It may be as small as a credit card or as large as a PC motherboard. For a simple device, a microcontroller-based circuit may be appropriate. With increasing computing power, what may have seemed impractical a few years ago may now require just an average processor system. Of particular importance are the input and output capabilities of the processor to communicate with the user interface and the robot hardware.

The processor will communicate with other subsystems (particularly motor control boards and sensor devices) along a bus system. Depending on the physical location of these subsystems, the bus will need to communicate over a distance of anything up to several meters. The choice of bus will depend on the distance and amount and speed of data (e.g., position information, control commands, video picture) to be transferred. Within a robot, a wired bus is likely, although a radio-frequency bus may have application. For a short-distance bus, either a serial or parallel system may be appropriate. For longer distances, the bus is likely to be a serial system because of the considerations of feeding the cables through the arm structure. Of particular note is the M3S bus, a development of the CAN bus standard for assistive devices with extra safety lines (35).

The motor control electronics may either be implemented on board(s) close to the central processor or distributed through the arm, which requires the provision of appropriate voltage levels to the motor and feeding back the motor speeds and displacements. Motor control chips and boards may be used that take a position/speed command and drive the motors based on feedback from a position or speed encoder. The use of such integrated control systems is easier to implement than programming the low-level motor control on the central processor.

11.4. Sensors

If the motor position and speed encoders give proprioceptive information internal to the robot, sensing the external environment gives exteroceptive information. A wide range of sensing devices exist that could give useful information. Some of the most likely to be used in an assistive robot are:

- Vision systems, probably based around a CCD array. A web cam is probably the cheapest device that might be considered.
- Ultrasonic range and movement sensors.
- Infra-red sensors.
- Touch sensors. These sensors will often be integrated into the end effector or gripper. This category includes simple switches, whisker switches, force-sensitive resistors, and slip sensors.

11.5. Software

The choice of programming language and environment will depend on the processor system being used as well as the speed of processing required. In general, a low-level language such as C/C++ will be used for low-level control where speed of processing is important. Forth is one language that was designed specifically for control applications. For higher level processing (such as processing user commands), a high-level language such as Basic, Java, or Paschal may be preferable.

Although many robot programming languages exist, several rehabilitation-friendly programming languages have been developed for the RT robots, although different drivers could allow the software to be applied to other robots. One example is the Cambridge University Robot Language (CURL) (36).

Well-written structured robot control software will have an architecture relating the different levels of control. Several architectures are used for robot control (37). A typical simplified architecture might be to divide the software into:

- Execution – Low-level drivers for motors and so on, collecting data from sensors.
- Coordination - Planning of movements.
- Organization - Input of controls from user. Degrees of intelligent action depending on environmental events.

11.6. Human-Machine Interface

A mobile assistive robot may be envisaged as a "slave"—it can be instructed to fetch an item from the kitchen or to place a book on the shelf. If the technology is adequate for it to operate autonomously, the instructions can be in what is virtually a natural language, for example, "Grip the red mug in front of you," or simply "reach down to the floor." A similar approach can be taken to a workstation environment. As the environment is more structured, it is easier for the robot to operate autonomously.

By comparison, a wheelchair-mounted manipulator may be seen as a "third arm" and would normally be controlled in some direct way, for example, to move the arm forward or to grip an object. It is one of the great challenges to come up with a control system that can even begin to compete with the way in which those without disabilities freely move their arms and hands. Even without the impairments that develop from a disabling condition, it is a challenge to come up with a control system that allows a human to control a robot. For example, a simple joystick input has two degrees of freedom whereas a robot is likely to have six degrees of freedom. Although joysticks with more than two degrees of freedom (even up to six) are available, they are unlikely to be appropriate for disabled people with movement impairment of their hands.

This distinction need not be hard and fast. It is obviously possible to drive a mobile robot around the home in the same way one would drive a remote-controlled car, perhaps with the benefit of an on-board camera. Similarly, task-type commands can be given to a wheelchair-mounted manipulator. Two approaches, however, can be identified from the user's point of view, whether to "drive" the arm or to command it at a "higher" level.

The role of the human-machine interface is to act as a link between the human and the machine. It must therefore take into account the abilities of the user and the requirements of the machine. For a flexible system, different user abilities must be able to be used as input. In fact, for a truly flexible system, it should be possible to control a range of different robotic devices.

Assistive robots are normally used by people with physical impairments as a result of diseases and conditions such as spinal cord injuries, motor neuron disease, multiple sclerosis, and muscular dystrophy. In general, these conditions lead to a reduced ability (movement, force, accuracy, speed) to control an input device. Taking an example of a two-degree-of-freedom input, a joystick may be considered. Many disabled people will be able to use a hand-operated joystick, although often with a reduced level of control. Sometimes software can provide a filter to detect the user's intention from an input that may be compromised by tremor or spasticity. If hand movement cannot be used, alternatives such as head movement (measured using an ultrasonic position detection system or a physical joystick), a chin-operated joystick, or an eye position tracking system might be used. As stated above, it is a challenge to control a six-degree-of-freedom robot with a two-degree-of-freedom input. This challenge may be overcome by changing between different modes, which each control two degrees of freedom. Alternatively, the joystick may be used as a switch menu. If a two-degree-of-freedom system is not possible, two or even just one switch scanning can be used to select items from a menu. Besides physical switches, "Sip Puff" and EMG switches may be used. Some users may have limited reach and force capability but be able to access a larger number of switches. The first Manus systems were designed for people with muscular dystrophy and controlled by a 16-key input device.

For those with very limited physical ability, voice control may be considered. Although voice control may be used to directly control the arm (for example, forward/faster/slower/stop), problems exist because of the necessary time delays to process the voice command. Significant safety issues exist, for example, a system trained to stop with the word "stop" may not recognize the word "stop" said in a stressed voice associated with a hazardous situation. More appropriate for voice control is to select items from a menu or to issue natural language command. Voice control may also be used to augment joystick control, for example, to switch between different control modes.

If a user is controlling a two-degree-of-freedom "robot" using a joystick, the only feedback to the user may be the physical position of the joystick and the physical movement of the "robot." In most cases, in assistive robots, some feedback of status information to the user is necessary. This feedback might be in an audible fashion (spoken text, "beeps") but is more likely to be visual. The simplest system, used on the Manus robot at one time, was a simple 6-segment LED display on the arm itself to indicate the control mode. More normally, systems use a 2-D display of varying sizes, which may be lines of text or images. Large displays present a lot of information but may visually block the user's view and are likely to be more expensive.

Given the limited user control ability, maximum use should be made of all sources of control information. A combination of switch control, head position, and voice control may be all used for optimum control of the robot.

Other external sources of information should be used. A system controlled directly by a user relies purely on the visual feedback of the human user "in the loop." Use of sensors allows the robot to move autonomously or semi-autonomously (38). Sensor information can come from a device that gives a 2-D or 3-D view of the environment (for example, vision systems, ultrasonic range finders). Alternatively, more localized information, for example, from gripper-mounted force or slip sensors, can be valuable in semi-autonomously guiding a robot to grip an item (39). The other source of information can be a database storing known information on the environment. This information may be input manually to the system, acquired by a vision (or similar) system, or acquired as the user interacts with the environment and objects within it using the robot (40).

11.7. Gripper

A robot arm is of limited ability unless it has a "tool" at its end. A manipulating tool is normally referred to as a gripper, but more generally it may be referred to as an end-effector. A robot may be designed to use just one end-effector, whether a very specific tool or a general-purpose gripper. Alternatively, a robot may use interchangeable end-effectors for different tasks.

Most assistive robots take the approach of using a simple but effective gripper, which will often be a two-jaw system, with just one degree of freedom, to open or close. The Manus robot (14) uses freely swiveling jaws to conform better to a range of objects. More complex designs may be envisaged to replicate the functionality of the human hand. However, besides the mechanical complexity, the user control aspects are a critical aspect of the usefulness of such devices. Any useful, multi-degree-of-freedom end-effector will almost always need to be a semi-autonomous device, making much use of sensor systems both on the end-effector itself as well as possibly remote cameras. An advanced prosthetic hand may be used as a robot gripper, but, in practice, this approach has not proved successful. Although much beneficial crossover exists between robot grippers and prosthetic hands, the requirements of each application will require a tailored design. For any gripper, the minimum level of sensing will be the control of gripping force, which may be simply limited to a predetermined level, or more adjustable control allowed.

The use of interchangeable or alternative end-effectors has been a part of the Master (9) and RAID systems (10). Besides a gripper, a suction tool has proved useful for turning pages and handling paper.

11.8. Safety

The normal approach to robot safety in an industrial environment is to completely separate the robot from any personnel, which is done using steel cages, safety interlocks, and sensors to detect the presence of any humans. In an assistive robotics setting, one has to take a completely different approach; the robot is intended to work in the close environment of the user, and sometimes to physically interact with him or her. Moreover, the human users have limited ability to move away from a robot that might potentially harm them. In some cases, it is appropriate to separate the user from the robot, for example, under a testing or training situation. For example, the Spartacus robot (6) was a relatively high-power device and early training of users was done with the arm behind a clear screen.

More normally, the approach to safety is a combination of sensors and a robot that is physically unable to apply a dangerous level of force. Touch sensors on the arm can detect when the arm touches a person (or other obstruction). If voice recognition software is in use, the word "stop" is a powerful safety instruction as long as it is robustly recognized. The safest system is one that is less able to cause any harm, which will be through the use of low-power motors. An arm that is light will have low inertia. A related approach is to have slip elements between the motor and actuator that will "fail" when too high a load is applied. Sometimes the danger may be through what the robot is holding, for example, a sharp knife or vessel with boiling liquid in it.

For all software-controlled systems, it is important for the software to be as robust and reliable as possible. The safest software is that which has been carefully designed, is well structured, and extensively tested. Many protocols for writing and testing safe software exist. One approach used for safety-critical software on aircraft, for example, is to have redundant processors where the results from several processors acting in parallel are compared. Given the cost constraints of assistive robots, a fully redundant system may not be appropriate, but there are elements of this approach that might be considered. Overall safety can also be given by a watch dog system that will close down the system if a software "crash" is detected. However, it is not guaranteed that removing power from the system will necessarily put it into a safe state.

Looking at the safety of the overall system, the "human in the loop" must also be considered. A disabled user might go into spasm causing an unwanted movement, or his hand might slip off a joystick. It is important that it is easy both operationally and conceptually for the user to bring the robot to a safe stop. It is not inconceivable that a user might want to deliberately "self-harm" with a robot. Ultimately, with such a system, a residual risk will always exist.

11.9. Evaluation

In designing an evaluation study of an assistive robot, the most important questions to ask are what the purpose of the evaluation is and what type of output information is required? Hammel (41) identifies the use of evaluations in the following roles:

- Conceptual brainstorming
- Clinical feasibility testing
- Viability testing - Evaluation in everyday performance contexts

In each case, the study can be separated into goals, methods, and lessons learned.

All items of assistive technology must be developed in close collaboration with potential users. Such users are not just disabled people, but also domestic carers and care professionals. Valuable feedback may be in an anecdotal form, "I like this feature," "It's too slow," and so on. It may be in a more quantitative form such as scoring a particular feature of the robot or potential task. A valid comparison is to look at the response of people to the use of a robot system when they are first introduced to the concept and after practical experience with a system.

Currently, most studies comprise just one or a few users, and can be formally presented as case studies (42). One of the most comprehensive case studies of an individual user is the work of Eftring (43), who studied in depth the use of Manus by one disabled user. The Manus is believed to have over 200 end users using their assistive robot on a daily basis. When a robot has been widely used, as with the Manus robot, a study of the cumulative experience of a large number of users would be valuable.

A greater degree of valid quantitative data can be obtained by studying, for example, the time to carry out a particular task. In evaluating the Handy 1 feeding robot (44), Pinnington and Hegarty measured the change in weight of children eating with the Handy 1 robot. Various numerical scales are also used by occupational therapists to quantify functional tasks, which may be relevant to the study of assistive robots.

In occupational therapy, dozens of standardized assessment tests exist. Their use allows the comparison of scores. Schuyler (45) explored the use of the Jebsen Hand Test, the Box and Blocks Test (BBT), and the Minnesota Rate of Manipulation Test (MMRT). Jebsen times the manipulation of a range of household items; BBT was developed for the assessment of the gross manual dexterity of patients with cerebral palsy, and the patient is asked to move 1 inch cubes from one side of a box to another; MMRT involves the manipulation of two-sided draughts/checkers pieces. Schuyler found that the occupational therapy assessment tests identified could be administered to measure the manipulation skill of a disabled person operating an assistive robot.

If the aim of the evaluation is to demonstrate to an outside group the benefits of a robot system, a study of the costs and potential financial benefits of a robot can be carried out. A lot of work was carried out by the Stanford group using a single-subject case study to demonstrate the potential financial savings in attendant time by using DeVAR (46). Benefits can also be illustrated by quality of life scores.

11.10. Commercial/Marketing Issues

The real benefit of rehabilitation robotics is in devices being readily available on the open market, so the number of systems sold commercially is of paramount importance. A survey in 2003 (47) showed the AfMaster system (at $50,000) being the only workstation system commercially available. Wheelchair-mounted robots are represented by the Manus and Raptor robots. Manus is believed to have sold over 200 units at a cost (dependent on training and support package) of $35,000, whereas at that time, less than 20 Raptors had been sold at $12,000. An interesting comparison exists between Manus and Raptor, whether the technically and functionally superior Manus or the lower-cost Raptor with the benefit of the huge U.S. market will make the greatest impact. Cost is obviously a major constraining factor in the growth of the assistive robotics market, but reduced costs should not be at the expense of adequate functionality.

As an expensive item of assistive technology, the breakthrough for Manus is that it is now on the list of equipment prescribable by the Dutch government. This situation has been achieved by a clear demonstration of the cost effectiveness of Manus over many years experience and close work between Exact Dynamics, manufacturers of Manus, and RTD, the Dutch prescribing agency (48).

Feeders are an interesting lower-cost area. The Winsford feeder has been available for many years and sold 2000 units at $2499; it is not actively promoted as being a robotic device. The Handy 1 is the device that is most obviously marketed as being robotic, but its cost is highest at $6300. Other feeding "robots" are the Neater ($3600) and the MySpoon ($3200).

12. ALTERNATIVES TO ROBOTICS IN REHABILITATION

Before concluding this survey, the alternatives to using robots in rehabilitation should be considered. These alternatives should be seen not as competition but as complementary. Most assistive robots aim to be multifunctional, but stand-alone devices, whether sold specifically for the disabled market or mainstream market, can have elements of the same functionality. One area of growth at the moment is the integration of different technological devices and approaches, particularly within a smart house environment.

Today, many activities can be carried out on a computer without the need to interact with the real world. Examples are computer art and music, computer chess, and the whole area of 3-D computer games and virtual reality.

Animals are often used in rehabilitation. Dogs are particularly used to guide the blind, but may also be used to assist those with mobility and hearing impairments. Monkeys have also been used, although they are more difficult to train. Human carers will always be important. Against the issues of independence must be balanced the need for human interaction and companionship. However sophisticated a technical system might be, it is unlikely to match the abilities of a human.

In parallel with the development of robotic devices for rehabilitation, much research is continuing into the origins and treatment of debilitating diseases and conditions.

13. CONCLUSIONS

However good the research may be, success is ultimately measured in assistance being given to patients and disabled people in real life. Research must be seen as a stepping stone to commercial products, which will be through devices being sold and bought commercially (whether through private, institutional, or state funding). For devices to succeed commercially, the correct balance between function and cost must be achieved.

Although the use of robots in industry is well-established and a growing market exists for service robots (including robots to cut the lawn and to clean the floor, as well as in physical rehabilitation), the uptake of robots as assistive devices has been disappointing. A major factor is that assistive robotics is being asked to perform a much more general purpose task than any of the examples given above. Objections and reservations on the use of robotics in assistive technology, expressed by both users and carers, include:

- High cost: High cost is often cited as being a reason for the slow uptake of assistive robotics, although other assistive solutions can be equally expensive. The high cost problem is more often the cost that is being asked, for what is currently a limited functionality.
- Low functionality: This obstacle is a complicated balance against cost. Often the functionality has been compromised by both the demands for a low unit cost and restrictions on the development resources.
- Demands on the user: Most assistive robotic devices operate on the basis of an intelligent user in the loop of control, which allows the robot to operate in an unstructured environment, but demands that the user, often a disabled person with limited functional mobility, has to provide many commands through a low communication "bandwidth."
- Obtrusiveness of the technology: For adequate reach and lifting capability, a robotic device will be large, which is particularly a problem for use on a wheelchair, which is a very restricted environment. The weight of a robot makes it difficult for a carer to move the device about.
- Fear (perceived) of a robot: Although some users are attracted by the technologically advanced label of a robot, others see this technology as a threat.

However, experience with volunteer users and those who use the technology in their day-to-day lives, suggest that a real potential benefit exists to be realized. The success of Handy 1 and Manus show the way to a time when assistive robotics may be as accepted as powered wheelchairs.

One way in which the field of assistive robotics will expand is with a move away from the traditional idea of a robot "arm" to the concept of using robotic technologies—sometimes as a multifunctional device, or sometimes as a single function tool—in the most appropriate fashion. Increasingly, robotic technology will be part of a wider smart environment. A convergence will also occur between devices developed for people with disabilities and mainstream consumer products, particularly in the area of electronics.

This article has consistently referred to such devices as "robots," but does it matter whether it is a robot—should the "R" word be used? Commercially, the word robot may be used positively to give an impression of technical sophistication. On the other hand, many consumers are frightened by the word robot. However, ultimately and finally the most important aspect is the benefit of disabled people and their carers.

BIBLIOGRAPHY

1. S. D. Prior, An electric wheelchair mounted arm—a survey of potential users. *J. Med. Eng. Technol.* 1990; **14**(4)143–154.
2. I. Asimov, *I, Robot*. New York: Gnome Press, 1950.
3. L. Leifer, Rehabilitative robotics. *Robot. Age* 1981;May/June:4–15.
4. V. Paeslack and H. Roesler, Design and control of a manipulator for tetraplegics. *Mechanism Machine Theory* 1977; **12**:413–423.
5. W. Seamone and G. Schmeisser, Evaluation of the JHU/APL robot arm workstation. In: R. Foulds, ed., *Interactive Robotic Aids*. New York: World Rehabilitation Fund Monograph #37, 1986, pp. 51–53.
6. H. H. Kwee, M. Tramblay, R. Barbier, M. Dupeyroux, M. F. Vinceneux, P. Semoulin, and S. Pannier, First experimentation of the Spartacus telethesis in a clinical environment. *Paraplegia* 1983; **21**:275–286.
7. J. Hammel, K. Hall, D. Lees, L. Leifer, M. Van der Loos, I. Perkash, and R. Crigler, Clinical evaluation of a desktop robotic assistant. *J. Rehabil. Res. Develop.* 1989; **26**(3):1–16.
8. M. Van der Loos, VA/Stanford rehabilitation robotics research and development program: lessons learned in the application of robotics technology to the field of rehabilitation. *IEEE Trans. Rehabil. Eng.* 1995; **3**(1):46–55, 1995.
9. J. M. Detriche, B. Lesigne, T. Bernard et al., Development of a workstation for handicapped people including the robotized system Master. Proc. ICORR '91, Atlanta, GA, 1991:1–16.
10. T. Jones, RAID—Towards greater independence in the office and home environment. Proc. ICORR '99, Stanford, CA, 1999:201–206.
11. M. Topping, Handy 1 - A robotic aid to independence for severely disabled people. In: M. Mokhtari, ed., *Integration of Assistive Technology in the Information Age*. Amsterdam, the Netherlands: IOS Press, 2001, pp. 142–147.
12. C. P. Mason and E. Peizer, Medical manipulator for quadriplegic. Proc. Int'l Conf. on Telemanipulators for the Physically Handicapped, IRIA, 1978.
13. A. P. Zeelenberg, Domestic use of a training robot-manipulator by children with muscular dystrophy. In: R. Foulds, ed., *Interactive Robotic Aids*. New York: World Rehabilitation Fund Monograph #37, 1986, pp. 29–33.
14. H. H. Kwee, J. J. Duimel, J. J. Smits, A. A. Tuinhof de Moed, J. A. van Woerden, L. W. v.d. Kolk, and J. C. Rosier, The Manus wheelchair-borne manipulator: system review and first results. Proc. IARP Workshop on Domestic and Medical & Healthcare Robotics, Newcastle, UK, 1989: 385–396.
15. H. G. Evers, E. Beugels, and G. Peters, MANUS towards a new decade. In: M. Mokhtari, ed., *Integration of Assistive Technology in the Information Age*. Amsterdam, the Netherlands: IOS Press, 2001, pp. 155–161.
16. R. M. Mahoney, The Raptor wheelchair robot system. In: M. Mokhtari, ed., *Integration of Assistive Technology in the Information Age*. Amsterdam, the Netherlands: IOS Press, 2001, pp. 135–141.
17. R. D. Jackson, Robotics and its role in helping disabled people. *Eng. Sci. Ed. J.* 1993; **2**(6):267–272.
18. M. Van der Loos, S. Michalowski, and L. Leifer, Design of an omnidirectional mobile robot as a manipulation aid for the severely disabled. In: R. Foulds, ed., *Interactive Robotic Aids*. New York: World Rehabilitation Fund Monograph #37, 1986, pp. 61–63.

19. J. Engelberger, *Robotics in Service*. Cambridge, MA: The MIT Press, 1989.
20. Z. Bien, D-J. Kim, D. H. Stefanov, J-S. Han, H-S. Park, and P-H. Chang, Development of a novel type rehabilitation robotic system KARES II. In: *Universal Access and Assistive Technology*. London: Springer, 2002, pp. 201–212.
21. M. Hillman and A. Gammie, The Bath Institute of Medical Engineering assistive robot. Proc. ICORR '94 Wilmington, DE, 1994:211–212.
22. E. Guizzo and H. Goldstein, The rise of the body bots. *IEEE Spectrum* 2005; **42**(10)(INT):42–48.
23. G. R. Johnson, D. A. Carus, G. Parrini, S. Scattareggia Marchese, and R. Valeggi, The design of a five-degree-of-freedom powered orthosis for the upper limb. *Proc. Instn. Mech. Engrs. Part H* 2001; **215**(3):275–284, 2001.
24. T. Rahman, W. Sample, R. Seliktar, M. Alexander, and M. Scavina, An anti-gravity arm orthosis for people with muscular weakness. In: M. Mokhtari, ed. *Integration of Assistive Technology in the Information Age*. Amsterdam, the Netherlands: IOS Press, 2001 pp. 31–36.
25. S. C. Jacobsen, D. F. Knutti, R. T. Johnson, and H. H. Sears. Development of the Utah artificial arm. *IEEE Trans. Biomed. Eng.* 1982; **BME-29**(4):249–269.
26. P. J. Kyberd, C. Light, P. H. Chappell, J. M. Nightingale, D. Whatley, and M. Evans, The design of anthropomorphic prosthetic hands: a study of the Southampton Hand. *Robotica* 2001; **19**(6):593–600.
27. A. S. Poulton, P. J. Kyberd, and D. Gow, Progress of a modular prosthetic arm. In: S. Keates, P. Langdon, P. J. Clarkson, and P. Robinson, eds., *Universal Access and Assistive Technology*. London: Springer, 2002, pp. 193–200.
28. P. D. Nisbet, J. P. Odor, and I. R. Loudon, The CALL Centre Smart Wheelchair. Proc. First Int'l Workshop on Robotic Applications in Medical and Healthcare, Ottawa, Canada, 1988:9.1–9.10.
29. H. Hoyer, U. Borgolte, and R. Hoelper, An omnidirectional wheelchair with enhanced comfort features. Proc. ICORR 97, Bath, UK, 1997:31–34.
30. S. Tachi, K. Tanie, K. Komoriya, and M. Abe, Electrocutaneous communication in a guide dog robot (MELDOG). *IEEE Trans. Biomed. Eng.* 1985; **32**:461–469.
31. W. S. Harwin, A. Ginige, and R. D. Jackson, A potential application in early education and a possible role for a vision system in a workstation based robotic aid for physically disabled persons. In: R. Foulds, ed., *Interactive Robotic Aids*. New York: World Rehabilitation Fund Monograph #37, 1986, pp. 18–23.
32. D. Gilden and D. Jaffe, Dexter, a robotic hand communication aid for deaf-blind. *Int. J. Rehabil. Res.* 1988; **11**(2):188–189.
33. W. M. Cameron, Manipulative appliance development in Canada. In: R. Foulds, ed., *Interactive Robotic Aids*. New York: World Rehabilitation Fund Monograph #37, 1986, pp. 24–28.
34. H. H. Kwee, Rehabilitation robotics - softening the hardware. *IEEE Eng. Med. Biol.* 1995; May/Jun:330–335.
35. S. Linnman, M3S: the local network for electric wheelchairs and rehabilitation equipment. *IEEE Trans. Rehabil. Eng.* 1996; **4**(3):188–192.
36. J. L. Dallaway et al. An interactive robot control environment for rehabilitation applications. *Robotica* 1993; **11**:541–551.
37. S. G. Tzafestas and N. I. Katevas, Control architectures of intelligent mobile robots. In: N. I. Katevas, ed., *Mobile Robots in Healthcare*. Amsterdam, the Netherlands: IOS Press, 2001.
38. W. A. McEachern, J. L. Dallaway, and R. D. Jackson, Sensor-based modification of manipulator trajectories for application in rehabilitation robotics. *Trans. Inst. MC* 1995; **17**(5):272–280.
39. R. G. Gosine, W. S. Harwin, L. J. Furby, and R. D. Jackson, An intelligent end-effector for a rehabilitation robot. *J. Med. Eng. Technol.* 1989; **13**(1/2):37–43.
40. J. L. Dallaway and A. J Tollyfield, Task-specific control of a robotic aid for disabled people. *J. Microcomp. Applicat.* 1990; **13**:321–335.
41. J. M. Hammel, The role of assessment and evaluation in rehabilitation robotics research and development: moving from concept to clinic to context. *IEEE Trans. Rehabil. Eng.* 1995; **3**(1):56–61.
42. N. Evans, M. Hillman, and R. Orpwood, The role of user evaluation in designing robotics. *Brit. J. Ther. Rehabil.* 2002; **9**(12):485–489.
43. H. Eftring, *The Useworthiness of Robots for People with Physical Disabilities*. Lund, Sweden: CERTEC, 1999.
44. L. L. Pinnington and J. R. Hegarty, Preliminary findings on the everyday use of a robotic aid to eating. *Clin. Rehabil.* 1994; **8**:258–265.
45. J. L. Schuyler and R. M. Mahoney, Assessing human–robotic performance for vocational placement. *IEEE Trans. Rehabil. Eng.* 2000; **8**(3):394–404.
46. J. M. Hammel, M. Van der Loos, and I. Perkash, Evaluation of a vocational robot with a quadriplegic employee. *Arch. Phys. Med. Rehabil.* 1992; **73**:683–693.
47. M. R. Hillman, Rehabilitation robotics from past to present - a historical perspective. In: Z. Z. Bien and D. Stefanov, eds., *Advances in Rehabilitation Robotics*. New York: Springer, 2004.
48. G. W. Romer, H. Stuyt, G. Peters, and K. van Woerden, Processes for obtaining a "Manus" (ARM) robot within the Netherlands. In: Z. Z. Bien and D. Stefanov, eds., *Advances in Rehabilitation Robotics*. New York: Springer, 2004.

ASSISTIVE TECHNOLOGY

JOAQUÍN ROCA-DORDA
M. ELENA DEL-CAMPO-ADRIÁN
JOAQUÍN ROCA-GONZALEZ
MAR SANEIRO-SILVA
Polytechnical University of Cartagena
Cartagena, Spain

1. INTRODUCTION

In order to guarantee the inclusion of people with disabilities, experts working in different disciplines such as engineering, medicine, and psychology conceived this new field of technological development, which is today considered to be part of biomedical engineering. Assistive technology pursues inclusion of the users by means of devices, techniques, and services that enable them to overcome the limitations induced by their disabilities. This article has been conceived by the authors in order to introduce the field of disability to novice readers.

1.1. Disabled and Elderly People in the Present Society

The capability of people for living a normal life, as members of the human specie, depends on their genetic inheritance and their relationships within the physical and social environments they are immersed in.

On the one hand, if a subject presents a genetic predisposition, several disabilities may appear, limiting his ability to perform certain tasks (e.g., Huntington disease). On the other hand, if the evolution of a subject through his whole life differs from what is naturally expected, because of accidental circumstances (e.g., road traffic accident), than additional disabilities might appear (e.g., spinal cord injury, palsy). Finally, the natural degenerative process that takes place as the subject gets older may produce similar effects (e.g., hearing impairment). In both groups, disabled and elderly people, inclusion is threatened because of the loss of autonomy and the lack of ability to perform certain tasks (Fig. 1).

At this point, it should be considered that the most optimistic statistics show that more than 500 million people are affected by some kind of disability all over the world (80% in developing countries). Besides, an average prevalence rate of disabilities of 25% of the population may be found in many countries, but just 10% suffer from limitations along their daily life routine. However, the impact of these limitations on the social inclusion of people with disabilities differ from country to country, depending on external factors such as the degree of social support offered by the community as well as the socioeconomic and technological levels of the country. In this sense, it is admitted that around 350 million of people with disabilities in the world cannot have access to proper social, health, and technological support, as it is happening in third world countries and in some developing countries, where prevalence rates as high as a 50% of the population may be found. Moreover, around 10 million refugees, of those spread all over the world, are affected by some kind of disability.

Obviously, the situation in the United States significantly differs from that of developing countries, as it may be found after analyzing the data from different population databases (1,2), as well as those from the national census (3). According to these sources, the disabled population in 2000 was estimated to be close to 50 million people, that is, slightly greater than 18% of the population of the United States in the same year (over 53 million people and 18.2% in 2005). If this analysis is centered on the disabled population of working age, only 56.6% had a job, whereas 77.2% of the nondisabled people had a job (4).

If a similar analysis is performed on the data from the European Union for that same year, the disabled population may be found to be close to 38 million (58% are women and 70% elderly), a value slightly greater than a 10% of the EU population (5). At this point, population aging should be considered a major concern for the quality of life of future generations. As an example, the average life expectancy in the United States has increased from 46 years in 1900 to 77 years in 2002. In the world scenario, the elderly population (close to 600 million today) is expected to reach 2 billion by 2050 (6).

As both disabled and elderly people may be affected by deficiencies and handicaps, public and private organizations should consider them as a whole when developing policies and programs toward their social inclusion.

1.2. Deficiency, Disability, and Handicap

As these three concepts are often misused as synonyms, the authors have included their definitions in order to explain the differences between these terms.

- Deficiency: lack or alteration of a physiological structure or psychological function.
- Disability: lack of capacity (because of a deficiency) to perform a task that is considered normal according to age, sex, and social situation.
- Handicap: adverse situation for an individual that makes achievement unusually difficult or impossible, as a result of a deficiency or disability.

1.3. Right to Accessibility, Social Integration, and Self-Care

Inclusion of people with disabilities in all areas of society is usually remarked as one of the main goals for most of the international organizations working for the advance of people with disabilities. Current trends in support policies have left the traditional paternal-like care schemes in favor of new ones based on a modern concept for disability.

In this sense, it is now widely accepted that disabled persons have the inherent right to respect for their human dignity and have the same fundamental rights as their fellow citizens of the same age. They are entitled to the measures designed to enable them to become as self-reliant as possible and to develop their capabilities and skills.

For this purpose, social integration should be reached by proper training of the remaining functionalities of the user, as well as the adaptation of the environment and the tools required by the different tasks to be performed. In other words, social inclusion may be assured after the user's accessibility is guaranteed. At this point, accessibility may be defined as the ability of products, services, and environments to be used by these people. A more detailed

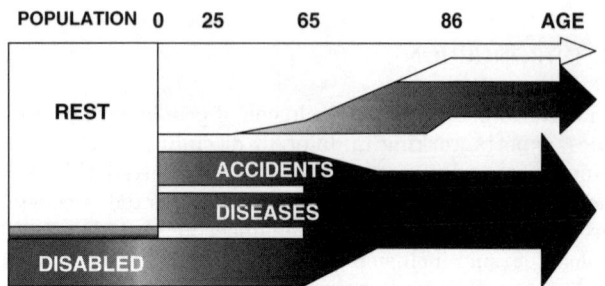

Figure 1. Evolution of disabled population and aging.

definition of this term may be extracted after that provided for software accessibility, which may be stated as:

"Accessibility refers to the ability of products and environments to be used by people. In this particular context, accessibility is used to refer to the ability of standard application software to be accessed and used by people with disabilities. Although the way people access the software may vary, a program is accessible to an individual if the individual is able to use it to carry out all of the same functions and to achieve the same results as individuals with similar skill and training who do not have a disability (7)."

Besides, it should be said that it is an integral part of usability, as access results in the ability to use a product, program, or service. The concept of usability was extensively used by software developers since the early days of computer programming. In this sense, usability was considered to be the ease with which a user can learn to operate, prepare inputs for, and interpret outputs of a system or component.

The term usability was initially defined in 1998 for the computers systems in ISO 9241-11(1998) "Ergonomic requirements for office work with visual display terminals, VDTs," and later revised in order to include all kind of products (ISO/DIS 20282, Usability of everyday products).

In these standards, usability is defined as "the extent to which a product can be used by specified users to achieve specified goals witch effectiveness, efficiency, and satisfaction in a specific context of use."

Three additional terms exist related to usability, also defined by ISO 9241:

- Effectiveness: "The accuracy and completeness with which specified users can achieve specified goals in particular environments."
- Efficiency: "The resources expended in relation to the accuracy and completeness of goals achieved."
- Satisfaction: "The comfort and acceptability of the work system to its users and other people affected by its use."

In a similar manner, ISO ISO TS 16071 "Ergonomics of Human-System Interaction-Guidance of Software Accessibility," defined accessibility as "the usability of a product, service, environment, or facility by people with the widest range or capabilities."

Obviously, different disabilities require different accessibility solutions. Regarding this last fact, a review of the different disabilities is included in the next subsection.

1.4. Classification of Disabilities and Emerging Disabilities

The simplest classification of disabilities relates these to the physiological structures affected. From this point of view, the disabilities may be:

- Physical: This kind of disability is originated by deficiencies in the bone, muscle, or neuromotor systems, which affect mobility and manipulation.
- Cognitive: This group includes those disabilities whose origin lies on deficiencies at brain level, affecting the intellectual capacity. Among others, reasoning, language, and communication may be affected.
- Sensorial: These deficiencies are those affecting organs and tissues involved in sensorial perception (vision, hearing, touch).
- Multiple: Occasionally, these alterations may appear simultaneously, leading to multiple disabilities.

In order to settle a unifying framework for classifying the consequences of disease, the World Health Organization (WHO) introduced, in 1980, the International Classification of Impairments, Disabilities, and Handicaps (ICIDH) for trial purposes. After 9 years of international revision efforts, the World Health Assembly on May 22, 2001 approved the International Classification of Functioning, Disability, and Health (ICF) (8). This classification of health and health-related domains (domains that help describe changes in body function and structure) focuses on both capacity and performance levels of the subject.

In this regard, the level of capacity of a subject evaluates what a person with a health condition can do in a standard environment, whereas the level of performance is related to what he or she actually does in his or her usual environment. For this purpose, ICF proposes up to a hundred descriptors for the classification of the different disabilities in terms of the affected functions and anatomical structures. At this point, it should be said that the benefits of ICF extend further than just being a mere classification methodology.

On the one hand, during the classification process, it is possible to extract valuable information related to the possibilities of rehabilitation and interaction within the environment, leading to the identification of those aspects that should be modified in order to guarantee the inclusion of people with disabilities. On the other hand, it is also possible to identify the so-called emerging disabilities, those appearing after the introduction of new demographic (i.e., population aging and migration), social (i.e., the advancements in health care), and even technological frameworks.

It should be said that because the development of new technologies (e.g., Internet) also leads to new activities, jobs, and situations, it may also give rise to new challenges for the inclusion of the disabled and the elderly. Besides, it may also lead to the introduction of new terms such as driver disabilities (those affecting the ability for driving) or e-disabilities (those related to the use of computers and information technologies) that did not exist before.

2. ASSISTIVE TECHNOLOGY

As mentioned before, disabilities appear after the development of conflicts between the capabilities of the user and his or her environment; so both factors should be taken into account when looking forward to a solution based on the design of dedicated systems and new social and technological resources.

For this purpose, two different approaches are usually followed. In some occasions, custom technological solutions have to be developed, whereas in other cases, existing technological tools have to be adapted to the capabilities of the subject and the requirements imposed by the task to be performed (hence the introduction of the term adaptive technology). Thus, it is easy to come out to the idea that technology either "assists" (i.e., helps the user) or "adapts" the available resources (i.e., so that the user can use them), leading to an initial definition for the main goal of assistive technology (AT).

2.1. Definition of Assistive Technology

Since the early attempts for helping people with disabilities through technology, many terms have been used for describing this engineering discipline. Among others, prosthetics and orthopedics, rehabilitation technology (RT) or rehabilitation engineering (RE) (9), assistive technology (AT), and biomedical engineering of the disability have been proposed throughout history.

In our opinion, this last term provides a holistic definition of this scientific field at present, but it has not been widely adopted because of its inherent complexity. This situation is similar to that faced by Enderle (10) when setting the non-interchangeability of the terms assistive technology and rehabilitation engineering after defining the first one as "a result of the rehabilitation engineering activities; as the health care is for the medicine practice."

It is usually admitted that the first definition for rehabilitation engineering was given by James Reswick in 1982 as "the application of the science and technology on the solution of the problems of people with disabilities." In 1988, the term assistive technology is introduced in the third paragraph of the U.S. Technology-Related Assistance for Individuals with Disabilities Act (PL100-407), which has become known as the Assistive Technology Act. This act was later modified by the PL101-476 Federal Regulations for the Individuals with Disabilities Education Act (PL 101-476) to make it more applicable to children with disabilities, leading to the finally adopted definition for this term: "Assistive technology means any item, piece of equipment or product system, whether acquired commercially off-the-shelf, modified, or customized, that is used to increase, maintain, or improve the functional capabilities of children with disabilities" (11).

This definition was later revised in order to include the different fields of engineering involved (12): "Assistive technologies include mechanical, electronic, and microprocessor-based equipment, non-mechanical and non-electronic aids, specialized instructional materials, services, and strategies that people with disabilities can use either to (a) assist them in learning, (b) make the environment more accessible, (c) enable them to compete in the workplace, (d) enhance their independence, or (e) otherwise improve their quality of life. These may include commercially available or home-made devices that are specially designed to meet the idiosyncratic needs of a particular individual." As this last definition includes materials, services, and strategies, it is clear that the scope of this discipline extends further than just the mere use, research, and development of technical aids.

2.2. From Technical Aids to Assistive Technology

Traditionally, the term technical aids referred to that set of tools designed for compensating the user's disability or handicap by means of function substitution or augmentation. Thus, it seems clear that this term made reference to physical and technological objects, leaving apart the social, formative, and organizational aspects of the disability.

However, in the last decades, the AT has been coined as a new philosophy able to promote the support and assistance of people with disabilities, from a holistic scope of the problem. As far as this goes, this technology branch is configured as a complex infrastructure made up by technical aids, but also by any service, programs, or other item that helps to avoid or reduce the disadvantages of people with disabilities toward their social inclusion. For this purpose, a great interdisciplinary labor is required; implying varied kinds of professionals in the fields of medicine, engineering, biology, and other fields as well (politics, socio-economic, etc.) (13).

According to the International Organization for Standardization, technical aids (TA) are defined as those "products, instruments, technical devices or systems to be used specifically by disabled people or/and elderly people, available on the market to prevent, compensate, or shorten a handicap." These products and devices are usually classified according to the guidelines proposed by international standards such as the ISO 9999 or "Classification of technical aids for persons with disabilities."

The first publication of ISO 9999 was in April 1992 with an integral approval by CEN of ISO 9999 as EN 29999:1994. At the moment, the latest (third) edition published is the ISO 9999:2002, and the ISO/TC173 SC2 is working on the revision of this document (fourth edition). This international standard establishes a classification of technical aids for persons with disabilities, especially produced or generally available. It is restricted to technical aids intended mainly for the use by an individual. Technical aids used by a person with a disability and that require the assistance of a helper for their operation are included in the classification, but not those used by professionals only.

Basically, TAs were conceived to act as interfaces between the disabled user and the standard resources (tools, services, etc.) designed for nondisabled users, however, without thinking of the difficulties that its use would produce on users with disabilities.

The progressive social sensibilization with regard to disability ended up influencing the design parameters of not just technical aids but also those considered for the introduction of new tools or devices intended for everyday use.

This process began with the introduction of the classic concept of ergonomic design (i.e., centered design on the average user) and evolved through custom and adapted design (i.e., adapted to the user differences) toward the so-called universal design or design for all, used at present.

In this sense, this last design philosophy recommends that all the objects, tools, and services had to be designed more usable for more people (as will be discussed in depth later in this article). Apart from improving accessibility of the disabled (i.e., as no adaptation should be required), the application of this principle increases usability by general users.

At present, universal design and usability criteria are applied in all the fields of technology in order to eliminate barriers and handicaps.

In this regard, as a need for quantifying usability exists, an extended concept for it has been defined by Roca-Dorda et al. (14). So, it was suggested that usability of all devices (as well as products and assistive services) should include at least four independent components identified as:

- Intrinsic Usability (IU): IU is offered by the product or systems to the nontrained user (because of its hard-soft design, functional organization, etc.).
- Training Usability (TU): TU is the component of usability that can be improved by proper training of the user for device operation.
- Adaptive Usability to Diversity (AUD): AUD is related to the device ability to maintain a minimum usability range, when it is used by different disabled users with slightly different characteristics. It may be found as the quotient between the usability range (of the device) and the diversity range (of the users).
- Adaptive Usability to Evolution (AUE): AUE is related to the capability of the device to adapt to a degenerative evolution of the remaining functionalities of the disabled user. It may be calculated as the quotient between the usability range (of the device) and the degenerative evolution range (of the users).

All of these components may be combined in a linear expression in order to get a unified measure for usability:

$$\text{Usability} = K1 \cdot IU + K2 \cdot TU + K3 \cdot AUD + K4 \cdot AUE$$

$$K1 + K2 + K3 + K4 = 1 \text{ and } 0 \leq Ki \leq 1, \text{ for } i = 1, 2, 3, 4,$$

where K1 to K4 are weight constants whose value must be defined according to the intended use for the device.

In this sense, as K4 affects the AUE, it should be increased for elderly people and other users affected by degenerative processes. In a similar way, K3 should be increased if the device is installed in a public space, in order to be used by a diverse disabled population, as it modifies the AUD. Additionally, K2 should also have to be increased if people with cognitive disabilities are considered as potential users. On the other hand, if the device under evaluation should offer a high grade of usability at first sight (e.g., an automatic teller machine located at an international airport), K1 should be increased, which would allow untrained users to operate the device efficiently.

In short, the balance among these constants should be chosen as a function of both the final audience of the product as well as of the environmental conditions of its operation.

For instance, the study of the usability of different devices used as computer mice, which is being carried out by the authors, centers on intrinsic usability evaluation. For this purpose, a custom tracking task (15) has been chosen to measure the spatiotemporal accuracy of the pointer-to-target response.

On the one hand, as this task requires many of the resources available to the user at different levels (physical, sensorial, and cognitive) (16), the intrinsic usability may be evaluated. On the other hand, as performance improves after the first time the tracking task is used, training usability may be also evaluated. Similar testing methodologies may be defined for other kinds of products and services, as happens in the evaluation of the usability of graphical user interfaces (GUIs) (17).

Besides the above application, measuring usability may be valuable for:

- Adequate selection (i.e., the best device).
- Proper configuration (i.e., the best settings).
- Measuring user's adaptation (i.e., the results of training or rehabilitation).

2.3. AT Resources

In order to get to the final inclusion of the disabled, assistive technology makes use of different resources, as follows:

- AT Products and Devices (AP & AD): AP & AD are those technological resources, of diverse technical complexity, that help disabled people to minimize their handicaps, thus increasing their quality of life. These may be either custom-designed (i.e., specially conceived for disabled people) or commercially available (i.e., designed for a general audience).
- AT Services (AS): AT services are those services covering different issues related to the care and support of the disabled, such as social health and education, among many others. These services are usually oriented toward disabled users, their families, and the professionals involved in their social inclusion.

In order to provide a deeper overview of these products and services, the next sections will introduce general classification for both kinds of resources.

2.4. Classification of Assistive Products and Devices

If a top to bottom approach is used, APs and ADs may be classified according to different factors such as their technological level (i.e., how complex), design characteristics (i.e., custom vs. general design), the disability affecting the user, and the type of aid provided by them (10,18).

2.4.1. Technological Level.
From this point of view, devices and products can be classified as:

- No-tech products: Simple objects used in everyday life by common people (e.g., book-stand, toilet seat raiser).
- Low-tech Products: Those products derived after the adaptation of existing simple tools (early concept boards, dressing aids, etc.).
- Medium-tech Products: Products and devices of certain technological complexity that are specially conceived for AT use (e.g., wheel chairs).
- High-tech Products: These complex systems include innovative technological resources from different

fields of technological innovation such as information and communication technologies, robotics, and biomedical engineering.

2.4.2. Design Characteristics. According to the design criterion, assistive devices and products can be:

- A consequence of a specific development in AT field (i.e., custom products).
- A result of the adaptation of standard devices (i.e., non-AT devices).
- Derived after adapting existing assistive products.
- Off-the-self assistive products (i.e., ready-made assistive devices).

2.4.3. Kind of Disability. In this sense, assistive products and devices may be classified as a function of the kind of disability affecting the user. The most classical classification groups them as devices and products for:

- Physical disabilities.
- Psycho-cognitive disabilities.
- Sensory disabilities.

2.4.4. Type of Aid Provided. From this point of view, assistive devices and products can be classified according to the way their aid is provided, aside from their technological foundation.

- Alternative aids: Alternative aids are those aids offering alternative ways for performing tasks (e.g., pushing a button instead of pulling a handle).
- Augmentative aids: The operation principle of these devices relies on the augmentation or amplification of the remaining functionalities of the disabled (e.g., magnifying glasses, hearing aids, crutches, orthopedic harnesses).
- Substitutive aids: In these devices, the damaged or limited body functions are substituted by another of the remaining functionalities (19) (e.g., using Braille lines as computer monitors and strobes lights as phone ring indicators).

2.4.5. Device Role. ISO 99999 proposes the classification of APs and ADs into the following groups according to their role:

- Therapy and training aids.
- Orthosis and prosthesis.
- Aids for hygiene and personal care.
- Aids for personal mobility and transportation.
- Aids for home care.
- Furniture, households, and buildings.
- Aids for communication, information, and signaling.
- Aids for object handling.
- Tools, machines, and environmental adaptations.
- Aids for leisure and spare-time activities.

At this point, the reader should have noticed the variety and diversity of assistive products and devices available for disabled people. In order to complete this review, the next section will introduce the different families of assistive products and devices.

2.5. A Brief Overview of Assistive Products

This section systematically reviews the different families of assistive products, according to the kind of disability they are intended for.

This discussion only covers the most representative products for each family, so the reader is encouraged to review some other references (10,18,20–22).

2.5.1. Devices and Products for Physical Disabilities. Under this category, most products and devices related to mobility and manipulation are reviewed, except for those related to prostheses and orthoses (as these are extensively treated in other chapters of this encyclopedia). This family of assistive resources includes devices and products for:

- Mobility and transport: Wheelchairs (motorized or not), special vehicles, and adaptations for cars; lifting platforms and cranes; standing aids, walkers, rollators, and canes.
- Hygiene and personal care: Toilet seats and stands, bath chairs, lifts and lying supports, etc.
- Household tasks: Feeding (from spoons and nonspill glasses to robotic arms for the severely disabled), house cleaning and labor, etc.
- Augmentative and alternative communication (AAC): This family includes those devices intended for helping people with limited speech or writing capabilities.
- Computer access: As many disabled users cannot operate computers through standard input/output peripherals, special devices have been designed for this purpose. Among many others, disabled people may use dedicated keyboards (extended, reduced, or programmable), mouse emulators, head-mounted gaze trackers, voice recognition systems, scanning access systems activated by switches (from the simplest electromechanical ones to those activated after biosignal processing (e.g., EMG and EOG switches), brain computer interfaces, and also auxiliary elements (keyboard guards, head pointers, disk insertion tools, etc.).
- Autonomy enhancers: Robotic assistants, page-turners, graspers, adaptable pointers (mechanical and lasers), and home control systems.
- Leisure time accessories.

2.5.2. Devices and Products for Psycho-Cognitive Disabilities. In this field, AP may cover the different functions in order to overcome the limitations of some individuals in learning processes (interpretation and settling of relationships among abstract concepts), execution of complex tasks, language comprehension, and so on.

For this purpose, several APs are available, such as:

- Task sequencers: These APs (configured around PCs or PDAs), break down complex tasks into simple operations that the disabled person is supposed to execute one at a time (in order to do this, a graphical interface offers the disabled a visual cue of the task required to be completed before receiving further instructions for the execution of the next operation).
- Relational association devices: Built around graphical pads, these products are used within early stimulation training programs, or with users of very low cognitive resources.
- Concept communicators: Often used within the field of AAC, these devices can help people, enabling them to understand what is said to them and say or write down what they want. These devices are structured around portable communication boards or concept pads that enable communication through ideographic or symbolic methods (23).

2.5.3. Devices and Products for Visual Disabilities. These products are intended for helping people with sensorial disabilities affected by blindness or reduced vision. Among many others, vision impaired people may use:

- Implants: These devices enable the generation of luminous impressions directly over the brain as a result of electrical stimulation at different levels of the vision system (retinal or cortical electrode arrays). At present, edge detection and contrast enhancement systems are setting the basis for the future direct vision prosthesis.
- Complementary mobility aids: This family of devices ranges from the simplest object detectors that work under the principle of sensory substitution (tactile or auditory feedback) to the more recently introduced environment visualizers that provide a 3-D "sonic image" of the environment surrounding the disabled user.
- Reading aids: Reading aids include such devices as screen readers and screen explorers (that transcribe the contents and structure of the computer screen into speech), Braille lines, screen magnifiers, optic aids (lens and magnifying glasses), augmented reality devices, and so on.
- Writing aids: Writing aids include devices such as typewriters, Braille keyboards, Braille printers, and so on.

2.5.4. Devices and Products for the Hearing Impaired. Apart from cochlear implants that turn sounds into electrical impulses applied at the auditory nerve by means of internal electrodes (which are covered in depth by other articles in this Encyclopedia), other devices and products (hearing aids, etc.) may be considered for these sensorial disabled people.

- General communication tools: These devices consist of hearing aids (mostly digital and programmable built around Digital Signal Processors, DSPs), transmitters (used in classrooms and conference halls), light signalers (used for translating sound warnings into visual cues appearing on the screen), and tactile indicators for the hearing impaired (e.g., vibration units used at mobile phones).
- Telephony: Telephony devices include telephony aids (amplifiers, coils) and text telephones (TDD/TTY) (composed by an alphanumeric keyboard and a small screen that enables the composition of texts to be sent through a modem). Text telephony relay services enable the hearing impaired to perform calls to standard telephony users. In these services, a specially trained communications assistant relays the message by reading the text message to the hearing person at the other end. Then relays the hearing person's spoken words by typing them back to the TDD/TTY user. Fully automatic systems based on speech recognition and syntheses are currently under research.

Finally, it should be said that, aside from the type of disability being considered, the resources described above may be integrated within more complex systems that overcome the limitations of these products when operated in a stand-alone environment (24).

2.5.5. Devices and Products for the Disabled and the Elderly. As previously discussed, disabled and elderly people may be affected by the same kind of deficiencies (e.g., elderly people are usually affected by mobility reduction and hearing impairment). In order to overcome these handicaps, both user groups may use similar solutions such as:

- Home control systems: By means of these environmental control systems, it is possible to activate, deactivate, and adjust the operation of the electric appliances and even the communications, comfort, and security systems of the user's home. These systems include different facilities such as sensing (light, temperature, flooding, fire, gas, surveillance, etc.), control (door openers, blind drivers, gas and water valves, etc.), and communications (telephony, Internet services, etc.). In order to connect the different components of these systems, dedicated buses (using coaxial or twisted-pair cables), power lines (carrier currents over 100 kHz), or wireless links (infrared, radio frequency, Bluetooth, WiFi, etc.) are used. This technology has evolved significantly since its introduction back in the 1970s to the present-day systems such as the EIB/KNX bus, although classical systems such as the popular X-10 are still being used. With the introduction of additional services (telemedicine, telecare, and telework) and technologies (ubiquitous computing, intelligent sensors, and fiber-optic networks) (25), these systems have turned to be the most powerful tools for guaranteeing the autonomy and the integration of the disabled and elderly (26). An application of home control technologies for the disabled and the elderly may be found in Fig. 2, which shows a prototype for a voice-controlled motorized

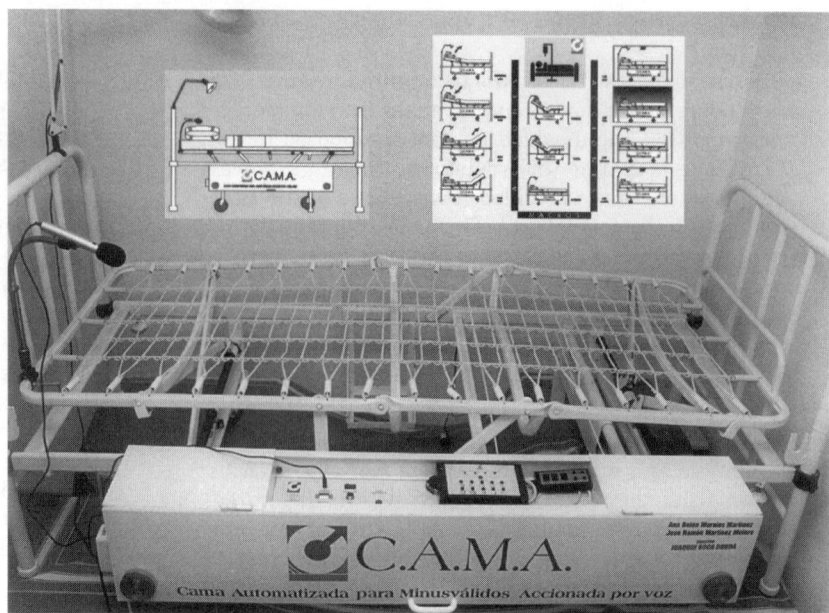

Figure 2. Voice controlled hospital bed for the disabled and the elderly. This prototype was presented by the authors in the 3rd International Congress on Project Engineering held in Barcelona in 1996, and awarded with the Spanish Institute of Social Services (IMSERSO) Prize to the "Best Project on Disability." (This figure is available in full color at http://www.mrw.interscience.wiley.com/ebe.)

hospital bed that included environmental control of the patient's room.

- Security control systems: Usually integrated within home control systems, security control systems make use of the existing sensors to generate the proper alarm signals after receiving information about the activity of the surrounding environment of the disabled, such as smoke or fire presence, activation of electric appliances, flooding detection, intrusions, and so on.
- Remote assistance or telecare: These systems (that may be integrated within home control systems) provide remote assistance, which is vital for preserving the autonomy of the disabled and the elderly without affecting the security and self-care of these people (27).
- Telemedicine: Remote monitoring of the health state of the users (through phone, GSM, Internet, etc.) is usually offered by health providers from private enterprises and some government organizations.
- Telework: The possibility of working from the home (also known as telecommuting) by means of computer access systems and Internet-related tools (e-mail, ftp, www, etc.) has opened a new front for the social integration of the disabled.
- Teaching, distant, and self-paced education: The introduction of e-learning systems and the advances in telecommunications have enhanced the access to education, the most powerful means of integration.

2.5.6. Other Devices and Products. Aside from the products and devices previously reviewed, APs and ADs exist that are specially oriented for the workplace, adapted furniture (for both work and living), adapted transport and public facilities, and urban and architectonic accessibility, which should be included as additional categories. An example of many of the above-mentioned devices may be seen in Fig. 3.

2.6. Assistive Product Selection

Adequate selection of technological aids and APs (from the myriad of them commercially available) starts with a multidisciplinary evaluation of the user that has to be carried out by different actors such as supporting professionals (physicians, rehabilitators, psychologists, educators, and engineers), the user itself, and his/her family. This process must begin by making an evaluation of the disfunctionalities, limitations, and difficulties of the user, as well as for the specific requirements imposed by his/her habitual environment (home, work, leisure).

Once these requirements, as well as the capacities and limitations of the subject, have been identified, the typical methodology follows the next steps:

A. First, the different solutions offered by different commercially available APs are analyzed in terms of efficiency, adaptability, quality/price, and so on.

B. A trained advisor is recalled to make an initial selection of the most appropriate technological aid.

C. The degree of adaptation of the selected product to the user's particularities is studied for a period of time long enough to observe if further adaptations are required.

D. These results are then cross-compared with those offered by the different adaptations initially selected. At this point, the satisfaction of the end users should be a primary concern. Once this stage is finished, the proper technology suitable for that disabled user should be available.

On the one hand, this methodology should be adopted from a clear understanding of the subject-environment interrelationships (Fig. 4 illustrates the actions of the rest of the actors involved in the process of selection and evaluation of the proper AT resource). On the other hand, it should not be forgotten that the basic problem to be solved by AT will be that of getting the successful adaptation of

Figure 3. Diverse assistive products. (This figure is available in full color at http://www.mrw.interscience.wiley.com/ebe.)

the user within the different tasks, (despite the limitations imposed by the disfunctionalities and the resources required by these tasks).

With this last statement in mind, it is easy to understand why APs should be organized in a manner that minimizes the "load" induced into the user during the operation, in order to reduce the probability of rejection or "under-exploitation" of these by the users. Regarding this last idea, if this dilemma is centered within the design and technological aspects of the APs (despite the sociological and legal issues), the problem may be approached from the points of view offered by engineering psychology, ergonomics, or human factors in design and engineering (28,29). In this sense, the identification and quantification of the human factors involved in the operation of the APs should not be forgotten when considering different AT solutions for overcoming the limitations imposed by different disabilities.

From this point of view, it is usual to consider four different areas related to the requirements or "load factors" imposed by the different tasks:

- Physical requirements (physical exertion, movement, manipulation, etc.).
- Cognitive requirements (reasoning, analysis, and memory capabilities, etc.).
- Linguistic requirements (message and signaletics comprehension capabilities).
- Temporal requirements (time required for completing the task).

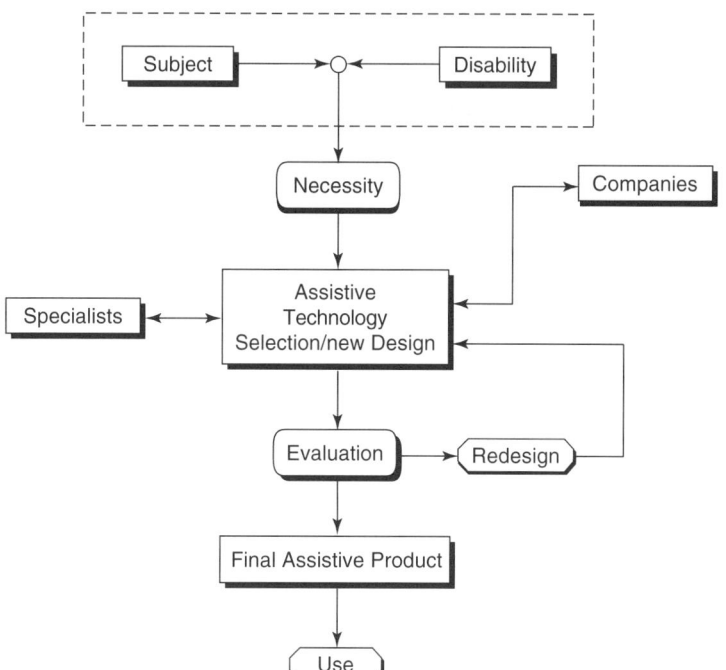

Figure 4. Actors involved in assistive product design and selection.

This set of efforts will then turn into the "negative resources" required for the competition of the task that shall be compensated by the user. Besides, the disabled may rely on a resource element acting as an "aid factor" (when this is positive) that may be found in his/her desire for completing the task. However, this motivating factor may also act as a "negative" factor as a result of a prior bad experience that may have caused the user to be demotivated because of the lack of information or training or another psychological factor. From this point of view, it is clear that the evaluation of the expected success of the user in the performance of a certain task should begin with the evaluation of the requirements described above.

For this purpose, Baker (30) introduced his own equation within a classic application prepared by him for the AAC. On his own, King adapted this first equation in order to evaluate the success in the use of AT products and services (31). In this sense, this major reference offers remarkable indications for increasing the satisfaction on the assistive technology use (SATU) by means of the proper maximization of the motivation factor (M), while the load factors (P, C, L, and T) are minimized accordingly:

$$SATU = \frac{M}{P+C+L+T}$$

Load Factors: $P = $ Physical, $C = $ Cognitive,

$L = $ Linguistic, $T = $ Temporal

$M = $ Motivation $(0 - 100), P + C + L + T = 100$.

Aside from this approximation, other factors such as the "transparency" of the AP should be considered for the estimation of the success with use. The concept of transparency was initially introduced within the field of the AAC for describing the efficiency of a certain symbol or icon for describing a certain topic (32). When adapted for AP evaluation, it summarizes the principle that as the AP operation turns simpler, more evident, and intuitive, the better this AP performs. In this sense, when principle is applied to AT, two additional terms are derived:

- Transparency: A single AP (or a complex assistive system) is said to be transparent (or offers transparency) when its operating logic (and the signaletics used), as well as other ergonomic aspects related to design, are evident and unequivocally perceptible. So the user should not need further information, or additional training, in order to use it properly. Under these conditions, its operation is said to be transparent.
- Translucency: This term is reserved for those AP or systems that do no offer a clear understanding of the operating logic or signaletics (icon meaning, control function, etc.) at first sight, but that will be easy to operate once the user has received a short training period about the criteria adopted for the design (why some controls are on the left and others on the right, which color stands for the maximum on a color scale, etc.).
- Opacity: This term is reserved for those AP or systems of bad design, where the user has no option other than memorizing the meaning of each control and indicator (as well as the sequence of operation of the device) in order to properly operate it, because no way of establishing a reasonable operative principle exists. These designs force the user to take long training sessions, and successful use cannot be guaranteed in the case of cognitive disfunctionalities such as happens with the elderly because of memory limitations. It is evident that opacity should be avoided in order to increase the accessibility and thus the fulfillment of an assistive product, as the maximum transparency should be pursued at the design stage (for both hardware and software products).

2.7. Classification of Assistive Technology Services

With assistive technology services, a whole range of social, health, and education services supporting the disabled and their families (as well as the professionals of this sector) should be considered. Generally, these services are offered by the governments, but also by private organizations such as associations and foundations for the support of the disabled, that provide resources for:

- Evaluation of the disfunctionalities.
- Rehabilitation.
- Technical support.
- Information and documentation.
- Search and selection of APs.
- Education and training on the use of APs.
- Management and funding for acquisition or renting (of equipment and services).
- Research.
- Professional training.
- Spread of knowledge and support for education.

Today, it is customary to incorporate these services within centers for the information and documentation, databases, web pages and Internet portals, virtual knowledge spaces, and research centers (18,21,33–36).

Education (at any level, from kindergarten to postgraduate levels at universities) should not be forgotten, as this is the field with more potential regarding the integration of the disabled, and it is the field where the AT resources may be used with the greatest efficiency for the improvement of the quality of life of the users.

2.8. Effects of Assistive Technology to the Users

Today, social development is growing in such a fast way that is very difficult to chase, even for people considered "nondisabled" within the current social standards. To be considered a "citizen" (whose rights and needs are taken into account), it is necessary to participate in each and every social task (work, education, transport, communication, leisure time). Such a social conceptualization is a real handicap for disabled people. They are far from reaching those perfection criteria.

As a consequence of their disability, these people have functional, physical, and psychological limitations preventing or shortening their participation in the different social tasks. The present AT development is provoking the progressive "fall down" of integration barriers, thanks to the advance of the assistive product and services. If it is analyzed from social relations and a personal growth point of view, AT can really constitute a fundamental tool that allows disabled people to participate in the society in which they live. AT has allowed disabled people to overcome the barriers imposed by their functional limitations and to have access to activities that would otherwise be impossible for them. The immediate consequence of such an achievement is a greater participation in society, a decrease of discrimination and social exclusion, which improves personal development, a fundamental requisite for a good interaction and adaptation to the environment.

However, apart from these "material" results, the advancement in the field of technological support leads to a parallel development of the self-concept and self-esteem in disabled people. Personal autonomy allowed them to be loosed from dependency and to reach freedom. Individuality and personal sufficiency are fundamental concepts for effective and emotional development. Social acceptance is important, but it is no less important than the autonomy of the subject when facing others.

Psychological equilibrium is based on a good emotional state and living in an environment that is adapted to the individual needs. From a human point of view, AT support can enable the subject to develop adapted alternative answers in order to eliminate architectonical or technological barriers and negative and paternalist social attitudes in their environment. The psychological equilibrium also helps to overcome personal barriers converting his/her disability in one more of his/her personal characteristics, which can convert his/her disability into his/her ability to participate, transforming disability into a rising positive value for society.

In short, AT support allows disabled people to feel (and to be in fact) his/her own life protagonist, and one citizen more, through his/her social participation.

3. ASSISTIVE PRODUCTS AS A GUARANTEE FOR ACCESSIBILITY

Prior to the definition of assistive technology, the so-called technical aids (as defined as hardware devices or software applications that allow a disabled person to access and to use standard equipment) were born to compensate the user's disfunctionalities, by means of technological resources.

3.1. Assistive Products as User Interfaces

As discussed before, the ability of subjects to perform common tasks will depend on both their functional capacities and the resources (tools, services, legal and economical frame, etc.) offered to them. From this point of view, the resources offered by the society to their users will act as operative interfaces (OI), which insert between users and the activities or tasks that characterize this society, as

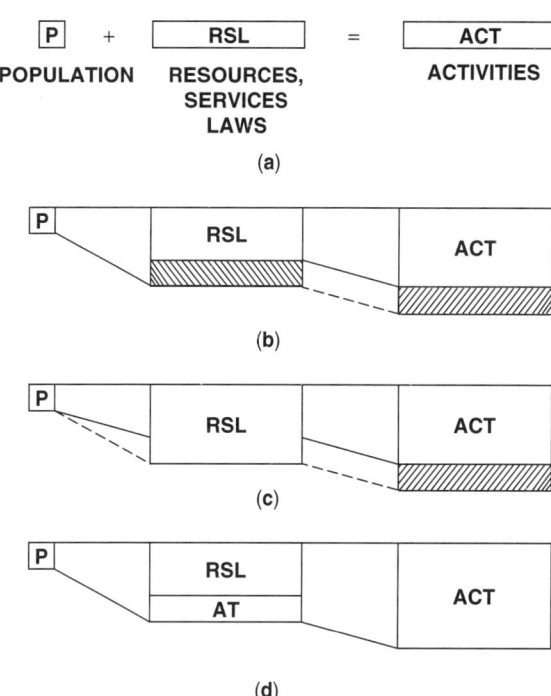

Figure 5. Assistive technology as an interface between people and activities. (a) Man-Resource-Task system. (b) The lack of resources inhibit the access to certain activities. (c) Effect of the lack of capabilities. (d) Assistive technology enables the access to the resources, and therefore to the activities.

seen in Fig. 5a. These elements (users, interfaces, and tasks) may be bundled into a set that can be configured as a complex Man-Resources-Task (MRT) system.

Within this logic, different issues should be analyzed:

- The particular characteristics of the task to be performed: These characteristics define the demand of operational capacities and resources required for performing the task, within a certain MRT system, such as happens in Elemental Resource Models (ERM) (29).
- The capabilities and skills of the user: When analyzing the capability of a subject to perform a certain task, both physiological capabilities and psychological capabilities should be considered. It is also mandatory to include the expected effects of therapy and rehabilitation.
- The resources available to users: The totality of the tools, techniques, and services that people may use in order to perform a certain task. At this point, legal issues and regulations as well as social aids should be included.

If this MRT approach is followed, a disability for performing a certain task may appear because of:

- The lack of resources: In this case, a trained user with no disfunctionalities is not able to perform a task because of a "failure in the interface" to the resources as seen in Fig. 5b.

- The lack of capacities: If the user does not have the required physiological or psychological capabilities, a "failure in the user" appears, as seen in Fig. 5c. If this happens, the existing OI should be redesigned in order to compensate the disabilities of the user, making him able to perform the task.

In this sense, assistive devices and products add to operative interfaces in order to compensate the user disabilities and assure access to the resources (see Fig. 5d). Thus, AT can be considered to act by breaking those barriers preventing disabled users from performing certain tasks.

3.2. The Functional and Cognitive Barriers of Access

From the point of view of access, a barrier can be defined as any handicap caused by the disabilities of the user that prevents him/her for the manipulation of an object or the execution of a certain task.

In the case of the physical or the sensorial disabled, a functional barrier is said to exist (e.g., this barrier prevents a tetraplegic from operating the controls of a video monitor and a blind user from reading the text on a screen).

When psycho-cognitive disabilities are considered, a cognitive barrier is said to exist (e.g., this barrier prevents a cognitive disabled user to remember a sequence of operations required for formatting a text).

Consequently, assistive devices and products should be designed in order to break (i.e., by augmentation of the user's abilities) or jump over (i.e., providing an alternative method for executing the task) these barriers.

3.3. Evolution of Technical Aids

Although the concept of technological aids seems to be relatively new, their main objective, expressed in terms such as to help the user to do, technologically, and in an efficient way, a function for which the user is not able, is as old as the hills. In this sense, it is widely known that men throughout history have applied the best of their technological knowledge to improve their own life conditions. Throughout the history of society, different solutions were given to different needs according to the available technologies.

From those tools initially developed in the Stone Age such (e.g., stone axes and arrowheads) to those used at present (e.g., wrenches and screwdrivers), a similar distance exists to that among the first technical aids (e.g., wooden prostheses and canes) and the most recently developed ones.

In the old times, the first prostheses (new Latin, from Greek, addition, from *prostithenai* to add to, from *pros* − in addition to + *tithenai* to put − more at) or technological devices were introduced in order to substitute a lost member by an artificial part. This fact is also true for orthoses (from Greek: *orthosis* straightening, from *orthoun* to straighten, from *orthos*), which are technological devices designed to fulfill the function of a handicap member or limited sense.

On the one hand, this way of acting at first produced very primitive devices such as walking sticks, crutches, hooks, and wooden or pottery dentures that would evolve to reach more complex forms. On the other hand, the technological development may be responsible for the compensation or reduction in a disability, what contributes to coin the need of new tasks, actions or "emerging activities" whose fulfillment will not take too much time to being considered as a standard characteristic of the inclusion of the user into a fully-developed society.

Accordingly, a revealing example is that of the car: Since its early booming in the United States during the first quarter of the last century, cars have become indispensable devices for both work and leisure. People who cannot drive cars may be considered disabled too, at present. Other examples could be the phone, computer, and the like.

In a few words, in all society and in all time, the right to carry on new "emerging" activities may drive us to the birth of a "new disability" produced by the social and technological advance. Obviously, the solution for these "new disabilities" will come from the application of newer technological methods and resources.

3.4. The Integrating Effect of New Technologies

Derived from the above-mentioned effects, it can be deduced that the booming of new technological resources will contribute to guarantee that disabled and elderly people may have new and better man-task interfaces, which may favor their integration. At this point of the review, it is obvious that new technologies may be adopted in several fields of AT. In this sense, the authors have found it useful to include a short review about the application of these resources in different assistive products and services.

3.4.1. Communication. With the advancement of technology, it was possible to evolve from the early communicators (ideograms printed on cardboard cards) to electrical communication boards (backlighted slide holders). The introduction of personal computers made possible the development of programmable on-screen communicators (fully customizable to the user's needs) and voice output devices (speech synthesizers) that are currently used in AAC.

3.4.2. Mobility, Transportation, and Manipulation. As in other fields, mobility technology has taken advantage of the development of electrical, electronic, and robotic technologies to improve the quality of life of disabled users. This fact made it possible to build electrical-powered wheelchairs, which were first controlled by electrical switches, then were voice-operated, and recently, even self-guided. A similar pathway has been followed in the field of manipulation, where biosignal-controlled prostheses and robotic manipulators are being currently developed.

3.4.3. Hearing Aids. These aids evolved from the early developed ear trumpets and auricles, which were able to amplify sounds by means of sound resonant cavities. The popularization of telephony and radio contributed to the advancement of electronics, leading to the introduction of early electronic sound amplifiers, which can be considered the predecessors of those ones used at present. Today, these devices offer fully programmable frequency response (i.e., each frequency band is amplified as required by the residual hearing capabilities of the user), embedding dedicated digital signal processors (DSPs) for this purpose. Besides, the early bone conduction devices that were used to transmit sound vibrations through the teeth to the skull have been replaced by cochlear implants, which directly stimulate the acoustic nerve by means of electrical impulses.

3.4.4. Blindness and Visual Impairment. In order to improve the vision of the visually impaired, different solutions have been proposed through the ages. In this sense, as soon as optical magnifiers were introduced, so were glasses and special lenses for the correction and augmentation of the visual capabilities of the users. Today, video-based magnifiers, screen readers (text-to-speech screen analyzers), and optical character recognition (OCR) techniques are widely used. The forthcoming solutions for the blind will use electronic cameras and sensors for feeding electronic circuits for the generation of electric stimuli applied directly at either the retina or the brain cortex by means of microelectrode arrays.

3.4.5. Independent Living, Telecare, and Telehealth. The successive apparition of new technologies such as environmental control, mobile telephony, and Internet has lead to the improvement of the autonomy and self-care of the disabled and the elderly.

3.4.6. Access to Computer. It is an admitted fact that computer systems can be used as technical aids providing the augmentative, adaptive, or substitutive resources required for compensating the deficiencies of some functionalities. In practice, this example is one of the clearest of how the AT can introduce a new way for the social integration of disabled and elderly people (37). The techniques currently used are derived after those experimentally tested with complex computer systems during the early 1970s. People with disabilities had to wait until the boom of compatible personal computers (1980s) to take advantage of these systems. The use of these techniques was favored by the popularization of computer use, turning them into new home appliances. In short, all of these applications cited above were made possible after the introduction of innovative low-cost technological resources derived after the advancements of the information and communication technologies. This fact has lead the authors to propose a case study related to this topic, as an example of the innovation process in AT.

3.5. Functional and Cognitive Barriers Analysis

The first step toward the final design of an assistive product should identify the actual access barriers imposed by the user's disabilities. Obviously, for this purpose, many approaches are followed in function of the type of disability and the different environmental conditions. As a result of space constraints, the authors have chosen to

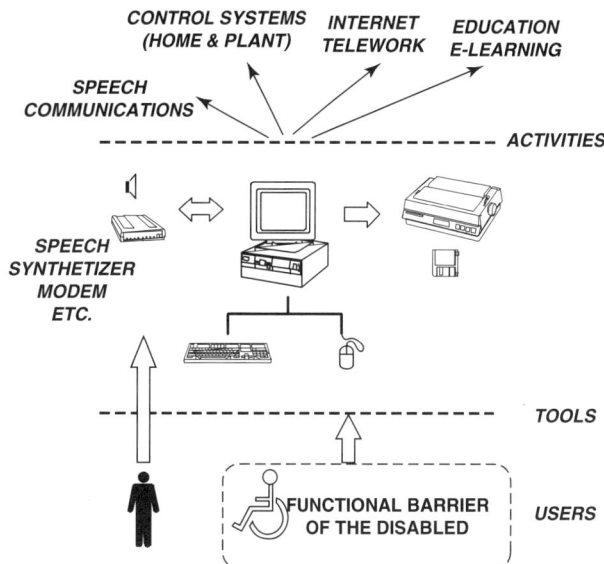

Figure 6. The functional barrier of the disabled.

center this discussion on a particular case study related to computer access systems, included in the section entitled Computer Access for the Motor Disabled as a Case Study.

3.5.1. Analyzing Barriers on Computer Access Systems. From the very beginning, researchers, rehabilitators, and disabled supporting organizations came to be aware of the existence of an important barrier in the communications between man and computer, which has defined the requirements to be fulfilled by Human-Computer-Interfaces (HCI) (38). This obstacle, also known as functional barrier (shown in Fig. 6), lies on the idea that computer systems have been designed for and by users who normally can use their standard input and output units (screen, keyboards, mice, etc.), and quoting M. Battro: "If simple actions such as reading a screen, or push a keyboard are inhibited by a motor or neurological deficiency, can be an obstacle for the professional ability and social integration of the person" (39).

Besides, in the early years, computer systems were not considered to be used by people suffering from cognitive-type disabilities, because of the complexity of the primitive user interfaces used at that time (punched card inputs, printer outputs and complex command syntax, etc.). This situation changed at the popular level after the emerging of graphic user interfaces, which favored the migration from uncomfortable command line interfaces (e.g., DOS) into friendly user interfaces (e.g., Windows). Simultaneously, different research teams all over the world were betting for using computers for the early stimulation of people with disabilities and the development of AAC systems. At this point, some cognitive-disabled people started to be considered as potential computer users because they could take advantage of the usage of these systems. This situation led to the settling of a wider definition of the so-called access barrier, as functional and cognitive barriers, which could be formulated starting from these two considerations:

- For some users, simple actions such as reading on a screen or typing on a keyboard can be inhibited by the motor or sensorial deficiencies.
- For others, the capacity to follow an access sequence or understand the instructions provided by a GUI can be blocked by cognitive or memory deficiencies.

The diversity of users imposed by the different type of disabilities (cognitive, sensorial, physical, etc.) (40,41), the degree of affection, and the concurrency of multiple disabilities leads to different situations (41,42) to be considered.

3.5.2. Sensory Barriers. In relation to computer access, three different cases should be analyzed.

- Vision impairment: Visually impaired people can suffer from visual loss to total blindness. In these users, the functional barrier limits the access to output devices (e.g., screen). In some cases, access to input devices is also affected, as happens with keyboards, because slight visual deficiencies may make the identification of the characters in connection to each key difficult.
- Hearing impairment: These deficiencies, from the medium levels of hypoacusia to complete deafness, may cause problems with specific output units (sound cards), when sound signaling or voice output are provided by the software. In some occasions, problems of indirect type should be taken into account with some input units (e.g., acoustic feedback when hitting a key or clicking an object).
- Handicaps in the touch sense: These disabled people have full access to computer systems but not those using substitutive aids that rely on this sense, such as happens with Braille input/output devices.

3.5.3. Psychological Barriers. Three cases should be considered:

- Cognitive deficiency: Cognitive deficiencies range from simple problems with the system operation and the access logic to the complexity of software and the comprehension of system messages. In these cases, GUI may be improved up to a limit imposed by the degree of the cognitive deficiency of the user and the complexity of the task.
- Other deficiencies that affect the behavior (e.g., an anxious and nervous user can hardly access to the computer).
- Language deficiencies (of interpretative and constructive type): These deficiencies may introduce some difficulties for the interpretation of the messages generated by the software being executed.

3.5.4. Physical Barriers (Type I). Within this group, those deficiencies related to the physical elements involved in speech communication should be considered. These elements affect the phonation, the verbalization, and the language capabilities (neither interpretative nor constructive). These people with disabilities are only affected when using some "special" input units (speech recognition devices).

3.5.5. Physical Barriers (Type II). This group is related with those deficiencies affecting the motor skills of the users. In this sense, different levels of motor disfunctionalities should be considered from skill deficiencies (poor accuracy of movements), limited strength, to complete limb impairment (of the upper, lower or both limbs) or any other kind of alternative movement. In any case, the access barrier basically centers on the impossibility for the manipulation, which provokes serious problems with the input units (mouse, keyboard, and joysticks) or even with the mechanical environment (power plugs, on/off switches, disk handling, printer paper, etc.) (42).

3.5.6. Barriers in Concurrent or Multiple Disabilities. It is worth mentioning that some pathologies may produce concurrent disabilities, so that many of the above limitations can appear simultaneously.

3.5.7. Barriers for the Elderly. Concurrencies of a lesser multiplicity and with more reduced degrees of affection are usually produced by the age, affecting elderly people gradually suffering from hypoacusia, visual loss, reduction of motor abilities, response speed, interpretation capability, sequencing of operations, and so on. All of these items will create difficulties for a group of people that, in the next years, will need the computer to reach their autonomy.

Finally, aging will make all of us become affected by some sort of disability. In this sense, the authors have taken the license to include a reflection about a famous quotation of a French poet of the nineteenth century (43). According to Charles Baudelaire, *"we all spend our life making a corpse out of ourselves..."*, but these days it seems obvious that if we live long enough, we will all spend our life making a disabled person... that will need to use computers.

4. COMPUTER ACCESS FOR THE MOTOR DISABLED AS A CASE STUDY

As a result of shortage of space, an overview (as a case study) will be presented to illustrate the evolution of assistive technology used for guaranteeing computer access to the motor disabled people. The advances introduced since the early adoption of hand-made aids has led to the complex systems available today, as has happened with some other solutions for other disabilities such as blindness and hearing impairment.

4.1. Early Attempts

In the beginning, taking over the use of standard input devices (keyboard and mouse) was attempted through intuitive and very low technological solutions:

4.1.1. Start Points Concerning the Keyboard. Several characteristics exist in a standard keyboard that are able to build up difficulties for motor disabled people, although they handle some functional abilities with their upper limbs, and even their fingers and hands:

- Difficult handle of some keys (e.g., Ctrl + C, Caps).
- Automatic repetition (e.g., typing "aaaa" instead of "a").
- Proximity and key-size (e.g., typing "st" instead of "at" on qwerty keyboards).

In the beginning, several low-tech devices such as retention clips (for keystroke of capital letters) and keyboard guards were used in the early 1980s to allow some users with imprecise control in their hands and upper limbs the access to the computer keyboard. They basically consist of a drilled shield (metal or plastic) placed over the keyboard, in order to serve as guides for the fingers when typing on the keys. When the users could not move their fingers, or the strength was not enough to ease the activation, a hand palm stick was used for pushing the right key and not the wrong one.

Luckily for this kind of disabled users, adapted keyboards with more room between keys, of, bigger size, and with reduced force requirements were available. On the other hand, at present, many operative systems include their own software accessibility resources that can cut down most of the above-mentioned problems and others in relation to doubled push keys, automatic repetition, and so on.

In the case of the most severe motor disabled (also tetraplegic affected), the first solution used for computer access relied on pointing devices, such as head-pointer sticks, or "Licornium" (10,18), which substituted the hand and palm stick into another that was on the user's head with a helmet mount. This device was used to push on the keys, one after the other. The long experience on this one (which can produce an important amount of eye-related problems) was followed by the emerging of head-mounted optical pointers and other more sophisticated techniques. Furthermore, both devices, key guards and head sticks, were also used at the same time.

The head stick (still keeps on enabling computer access in countries and communities with low income and technological resources) presented serious deficiencies, including:

- Slow operation.
- Need of a very good head control movement.
- Unjustified, but very real, social prejudices drawing the other peoples' attention (it should be remembered that this goes against inclusion).

4.1.2. Start-Points Concerning the Mouse. As in the case of the keyboard, several characteristics of this device may also make access for the motor disabled difficult, although they may have some residual functionality in their upper limbs and even in fingers and hands. In this sense, the following difficulties may be found:

- Need of high-precision hand and wrist movement (especially for graphics and design software packages).
- Need of fast operations (double-click selections).
- Need of "simultaneous" operation (click-and-drag operations).

Although, in the long run, nearly all drivers of these input devices, and even some operative systems, include software solutions for these problems, the initial solutions for disabled users (with any residual control on the upper limbs) were focused on the use of joysticks and other adapted mice. At presents, it is possible to select among these kinds of devices (e.g., trackballs, joysticks, touch pads) those that best suit the special characteristics of the disabled users.

4.2. Access Techniques for the Motor Disabled

Obviously, the fast evolution of technology soon permitted both the development of the most sophisticated and effective solutions as well as the definition and development of the basic techniques in this field. In a related manner, several techniques such as detection, general access, acceleration, and reduction of involuntary activations were developed.

4.3. Detection Techniques

Logically, in order to assure the right efficiency of a computer access assistive product in the worst cases, it is necessary to have, at least, a Boolean or logical signal (on/off, 0/1, true/false) able to control the assistive product working as a user interface. This signal must be generated consciously by the disabled user. Depending on the residual functionalities and their anatomic location, multiple options will be presented (head movement, eye-lid closure, etc.), which may justify the existence of many kinds of switches or sensors and different detention techniques.

On the one hand, in switch-controlled computer access processes, a closed relationship is established between the movement accuracy of the user and the final success of the task. In fact, not only in the AT field, but also in the operation of all kinds of industrial home switch-activated controls, we could be dealing with a "switching ability." This ability can be developed through adequate training.

On the other hand, it is obvious that, generally, the capacity of performing a fast movement toward a target switch brings about the inaccuracy of these movements. Fitts and Peterson (44) worked out a generic equation for this and other cases of human-machine interfaces. Such an equation can be adopted as a tool for the evaluation of the level of computer access at the very first stages of the AT selection for guaranteeing the operation by the disabled user with switches (31).

$$\mathrm{MT} = a + (b \bullet \mathrm{Log}\, 2(2D/W)),$$

where MT is the time required from the very start of the movement of the limb used for the access until the target used for switching is reached; D is the distance as far as the target switch; W stands for the switches width (square section of the target active area); and a and b are empirically derived constants after population studies.

After working with this formula, other useful relationships were deduced:

- When the anatomic movement speed increases, the targeting accuracy decreases.
- The time required for hitting the target increases when the sensitivity or the target switch resolution decreases.
- The necessary time of targeting activation goes up when the switch selectivity decreases.

At the same time, special attention should be taken when selecting from different switches (45). In relation to the different kinds of switches for this purpose, a great variety exists, such as:

4.3.1. Contact Switches. Here, one can include all kinds of simple switches whose action implies a real and effective anatomical contact (as long and strong as necessary):

A. Electromechanical switches: The simple use of these switches, on-off type with a normally open (NO) or normally closed (NC) contact set, was the earliest, most common, and cheapest solution. This kind of switch would be suitable whenever the disabled user may carry on an adequate residual functionality (an anatomical movement of at least 0.3 to 2 cm and a strength between 10 and 80 gr). When the switch changes from an NO position to an NC position, a logical signal is generated and applied to hardware or software in use. In this case, it would be necessary to take into account the switching noise problems to avoid the fact that some multiple rising edges (switch bounces) can be interpreted as other switch activations, producing undesired multiple selections (the use of Hall-effect switches is free of switching noises). The mistake of always choosing the most sensitive switch should not be made. Generally, an excessive sensibility (need of lesser activation effort) and the selection of a more reduced displacement length (need of a lesser anatomic movement) can make easier the undesired switch activation because of tremors and spastic involuntary movements.

B. Pneumatic switches: In this case, the residual functions to be used should be those related to breathing, so the activation required of the user, the possibility of taking or exhaling air through a small tube or a mouth-tip (Puff/Sip switch). The required positive or negative pressure can be very varied within some H2O cms. The expertise of the professional in charge of the training and rehabilitation reveals that, sometimes, the disabled user and more often his or her family (especially for very young users) can reject this kind of switch for both hygienic and esthetic reasons. A derivation is the blow switch (Squeeze ball switch), in which it is not necessary to place into the mouth.

C. Piezo-film sensors: These sensors are the most modern contact switches and are based on piezoelectric effects. In these sensors, the digital output signal is noise-free. On the other hand, the piezo-film can be acquired in rolls or big size sheets and easily cut and conformed with a simple pair of scissors and a heat gun, which, together with its flexibility, allows adapting and placing the switch on any anatomic area of the user or any nearby equipment.

D. Zero-force switches: Several kinds of switches exist that are able to detect the simple touch on a sensitive surface. When using these switches, no effort or minimal movements are needed, being used then with such small residual functions.

4.3.2. Contactless Switches. When we work with users with scare residual functionalities, that are unable to carry on, the necessary strength for activating the switch, several sensors can be used, such as:

A. Proximity sensors: Proximity sensors are based on the effect of the dielectric properties of the skin and muscular tissues over the electric field around the sensing element.

B. Optic detectors: Optic detectors are based on cut-off or reflection of an infrared beam over some user's anatomic area.

C. Ultrasound detectors: They work in a similar way as the ones above, but for an ultrasound beam instead of an infrared one.

D. Special-type detectors: A great variety of these detectors exist, organized as high-tech assistive products, able to determine the eye position or the eye gaze through different techniques (artificial vision included).

4.3.3. Biosignal Detection. The idea of using this modality is as old as the very same electromedical technology; hence, the use of different biosignals should be taken into account:

A. Electromyographic (EMG) signals: The EMG signals represent the currents generated by the ionic flow, which get through the muscle cells. The measure and register of these action potentials constitute the basis of the electromyographic techniques. By nature, the EMG signal is representative for the muscle activity and provides useful information about the muscle tissues and nerves that activate them. For its use in technical aids, they are used more than generalized noninvasive methods, surface EMG (sEMG) signals in the range of 0.1 mV to 5 mV with a bandwidth between 10–20 and 500 Hz (46). In many cases for on-off applications in AT it will be enough with the use of Ag or Ag-AgCl electrodes, similar to those used for ECG. The location and placement of the electrodes will be determined depending on every user's characteristics. Generally, a training period, through biofeedback techniques, can develop better user control. In some occasions, it is possible to use a technique, which has been named, "false targeting" (47). This technique consists, basically, of training the user, in order to try to have a go with a movement with an area of his or her anatomy in direction to the false target location (even in the case of the user's impossibility to fulfill this movement). The electrodes should be set in another area associated to secondary muscles structures where there may be a response (even involuntary or reflex modes) as a consequence of a previous unsuccessful movement proved. On the other hand, several authors (46,48) have done a survey dealing with the relationship between the sEMG signal and the related strength, coming to the conclusion that the RMS value of this signal has a closed relationship with the exerted muscle strength. Otherwise, in the frequency domain, it is possible to check the existence of exerted muscle strength—associated shifts, for the median frequency in the Power Spectral Density (PSD), which allows the implementation of low-cost detection systems (47,49), as can be seen in Fig. 7c. On the other hand, not only the RMS value (Fig. 7a) but also their derivate and median frequencies (in Fig. 7b) provide useful information. Besides, it is worthy to observe (in Figs. 7a and 7b) the logical anticipative condition of the sEMG. It is easy to implement a cheap EMG switch, starting from the RMS value (or better its derivate); comparing it with a reference threshold voltage (Vref), a digital signal can be obtained. The adjusting of the reference threshold permits an additional sensibility control (Fig. 7c). Other more powerful solutions are also possible, either by using EMG signals (49) or other biosignals such as the electro-oculographic activity (EOG) and even after using artificial vision techniques.

B. Electroencephalographic (EEG) signals: In this case, the signal generated by the brain activity is used. In the beginning, some trials were based on the simple quantification of the rhythms (delta, theta, alpha, and beta), but the characteristics of the EEG signal (ranging from 5 to 300 µV across a 0.01–150 Hz bandwidth) and the necessary equipment, postponed the advances in this field. At present, and after a long time, the possibility of a direct computer interface is greatly latent in successful outcomes such as the Brain-Computer Interface (BCI) of the Technological University of Graz in Austria and others (50). Among other aspects of the brain activity, alpha rhythm desynchronization related to movement planning or event-related desynchronization (ERD) changes on the mu-rhythm; EEG patterns from evoked potentials and pattern recognition over the brain activity distribution related to orders have been considered. All the development of the BCI lies on the fact that different mind activities are associated to different neuronal activities patterns, included in EEG potential distributions, which can be quantified and

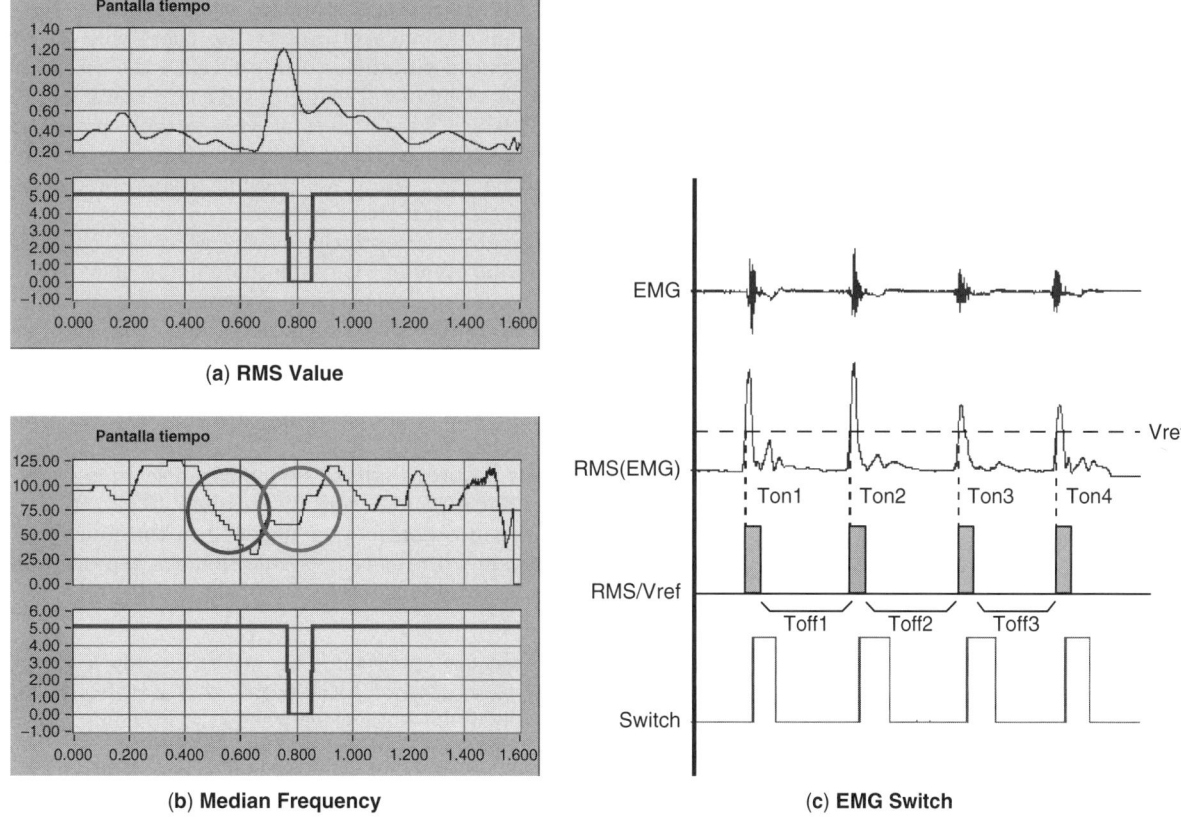

Figure 7. Electromyographic signal (EMG) as switch input. (This figure is available in full color at http://www.mrw.interscience.wiley.com/ebe.)

recognized. This system permits the control of a screen cursor, as well as the option selections in software, only by thinking in making a movement. From here onward, several devices have been developed, such as a virtual keyboard and prosthesis controls (51,52). Here, this information is not analyzed in full detail, as, in several articles of this Encyclopedia, a more detailed analysis is provided.

4.4. General Computer Access Techniques

Once a digital signal is voluntarily generated by the disabled user, certain techniques should be defined with the aim to control the computer. The highly different characteristics of each user make necessary the availability of more diverse computer access strategies. So, whereas the nondisabled user can access the computer systems using standard input interfaces, the motor disabled people must do it through adapted or alternative input interfaces (AIIs).

Furthermore, for some disabilities (e.g., blindness), people will require the use of substitutive output interfaces (SOIs) such as screen talkers. Other people with disabilities (e.g., motor disabled people) can need other kinds of input interfaces (e.g., voice recognition, switches). Here, the concern is with the case of the motor disabled people, where misleading of voice recognition access can occur (18,53).

According to the different characteristics of the users, several options can be considered:

4.4.1. Direct Access with Standard Devices. Those people with disabilities and some control of the upper limbs, but with slight skill, deficiencies can have access through simple means such as: (1) with standard devices, (but with a low efficiency and many mistakes; (2) with standard devices and low-tech aids (guards, sticks, etc.).

4.4.2. Direct Access by Adapted Devices. The users with more disfunctionalities, but maintaining a certain upper limbs control, can access through adapted devices such as special and alternative mouse and keyboards.

4.4.3. Direct Selection Access. This level includes disabled people with the capacity to operate with a conventional or adapted mouse, but not with a conventional keyboard. In this case, the answer lies in a mixed solution—direct access for the mouse and, through direct selection, for the keyboard—using the mouse as an element to select, directly, characters and commands on a virtual keyboard (on-screen keyboard).

4.4.4. Encoding Access. In this case, the necessary number of selections (in both modes, direct and scanning, as will be seen later) shortens thanks to encoding actions with generally two icons or digits (such as number–number, character–character, and color–color) and their combinations. On the other hand, the number of codes cannot be very high because the user must memorize them, which may be difficult for the cognitive disabled or elderly people.

(If possible, a code board can be displayed on screen.) It is also possible to work using a "serial coding" scheme such as the Morse code.

4.4.5. Switches and Scanning Access. On the one hand, some motor disabled people have a unique and very limited functionality (e.g., head movement, open or close eyelid, thumb movement), so this unique functionality must be used for activating only one or two switches (only one is common). On the other hand, from this unique functionality, it is not possible to generate, directly, so many different signals as those of a standard keyboard or mouse. For this reason, the only way to guarantee the computer access will lie on high-tech assistive devices and products such as scanning systems, which are able to offer the user the possibility of selecting, in a sequential way (and because of the activation of an unique switch) a determined option out of the many possible.

The first switch-scanning systems were set from author or dedicated software, which read a serial port (or a mouse button) to detect the external switch activation. The control software, generally written in machine code, was totally specific, both in the early personal computers, (1977–1980: Pet 2001, Commodore; TRS-80 Tandy Radio Shack, Apple I & II, Sinclair), and also later for the first PC-compatibles under DOS that were unable to work in multitasking mode (without using terminate-and-stay-resident programming techniques). Under these conditions, it was problematic to reach a simultaneous operation of standard software and the control scanning software, which did not permit the use of a standard word processor, databases, and educational software to the users with disabilities. This initial situation quickly focused the researchers on the development of external keyboard and mouse emulators, which were able to provide to these ports the same signals as the standard devices. In this way, and in a truly transparent mode, it was possible to simultaneous run any standard commercial software.

Finally, the evolution of the personal computers, the generalization of PC compatibility, and above all, the development of the operative systems such as Windows, able to bear a multitask functioning, made possible a good operation free of collision problems. At present, it is possible, without any problem, to display a virtual keyboard on the screen where the scanning takes place or on which the direct selection is done.

By nature, a typical switch-based scanning access process (for AAC and work integration (49,53), etc.) presents an organization as that depicted in the functional block diagram with the components that are exposed in the following lines (Fig. 8):

- Scanning screen: This screen is only a graphic display (monitor, LEDs panel, etc.) in which the different available options (characters, commands, etc.) are being shown and enabled sequentially.
- Scanning generator: It consists of an internal clock that marks the scanning speed and highlights sequentially, and one-to-one, the different scanning options on the screen.

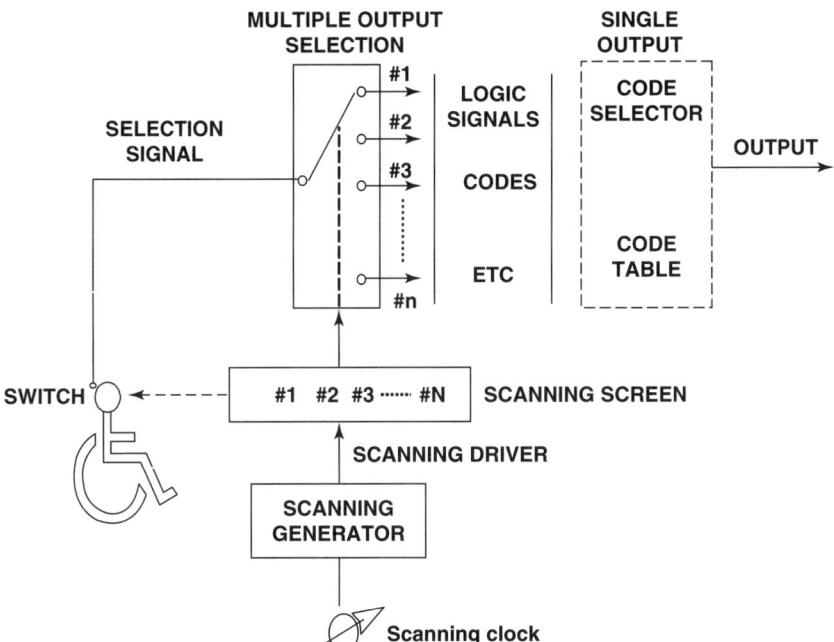

Figure 8. Elements of a scanning access system.

- Scanning selector: This device, with a logic structure similar to a multiplexer, sequentially "switches" (clock-driven scanning) the only input coming from the switch for different characters and code generation.

On synchronizing both (the scanning screen and the selector), any signal introduced on the unique input (by the switch) will lead to the exit corresponding to the selected option on the screen. During the operation, the user will observe the option selection menu on the scanning screen. When he or she checks that the desired option is selected (e.g., highlighted) the switch is activated (with his or her residual functionalities), which produces a logical signal that may be decoded by the system bringing along the computer execution of the highlighted option.

Supposing that an orthogonal column and row scanning, also known as 2-D scanning, were adopted (Fig. 9), the scanning starts by highlighting each of the column screen, and once one of them is selected, a new scanning (row scanning) will let us select the chosen option. Obviously a switch scan system has an inconvenience: its slowness (although this may be the only access via for people with these disabilities). Otherwise, its scanning speed must be adapted to the user's response. For this reason, it is important to know the different techniques that can be used:

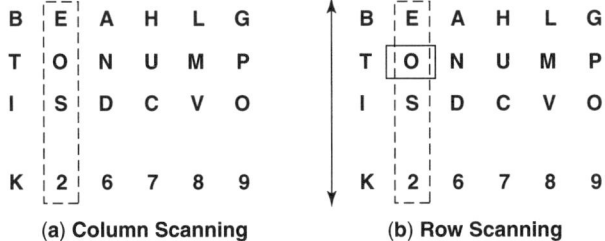

Figure 9. 2-D or column/row scanning.

A. Scanning techniques: The efficiency of the scanning access that must be applied to a person with a particular disability lies in selecting the control and organization of the most suitable scanning to this user:

- Automatic scanning: Automatic scanning is also called clock-driven scanning. In this case, an internal clock enables each next option to select, following a software established order. The timing of an internal clock lets us use the most appropriate scanning rate for each user.
- Direct scanning: In this scanning, a first click of the switch decides the direction in which the next scanning will move, and then the next scanning will make the selection.
- Step scanning or pseudo-scanning: In this case, an internal clock does not exist; an isolated switch click "moves" the scanning pointer to the next programmed location. A time without selection or one double click makes the selection with this option.
- Inverse scanning: In this case, the scanning starts with pushing the switch and holding it, the selection takes place when the switch is let go.
- Two-speed scanning: Here, the scanning rate is quick or slow, whenever the scanning pointer is far or near in relation to the target option.

B. Scanning formats: An account of the "geometrical scanning organization" and the option screen distribution can be remarked (18):

- Linear scanning: In this case, the options to select are placed on a linear distribution (1-D scanning).
- Rotary scanning: In this scanning (1-D), the options are placed on a circular structure, so when the latter option has been scanned, the first option is newly scanned.

- Orthogonal scanning: File-row scanning or vice versa: Here, the option set can be organized from a matrix vector 2-D. To select a specific option (i.e., a file scanning is done), and when one of the files has been selected, then a row scanning takes place, until the desired individual element can be selected.
- Group-item scanning: It is implemented over a 2-D structure, where the options (items) have been distributed, forming, at the same time, 2-D (groups or zones). During the use, the first step is to scan the groups, and when one of them is selected, the file-row scanning is started in it; until the chosen item is reached.
- Angular-radial scanning: This scanning has a similar behavior to that of file-row scanning but for a circular structure. In it, it is also possible to organize here the sector-item philosophy.
- Nested-scanning or double scanning: In case of very complex control, any of the elements (groups, files, or rows) may contain inputs that lead to another scanning screen. In this way, it is possible to work out very complex structures but with the inconvenience of the transparency loss.

C. Scanning improvement: The slow process that characterizes the scanning selection technology (compared with direct selection), strongly recommends looking for improvement techniques for this. To ensure a quicker data input, it is possible to make use of these basic techniques:

- Bistable keys: Here, once one option has been selected, the remaining options alter their function as far as the initial option is reselected (similar to capital letters operation).
- Hidden options: When using this technique, some options are not scanned, reaching an invisible stage on the screen because they have been disabled through a previous selection.
- Interruption of the sequencing: It is possible to spot along the scanning screen (i.e., to one-third or one-quarter of this) exit or home return areas, which permit the users a fast recovery from possible mistakes, preventing unnecessary scanning cycles.
- Nonselection outputs: In the double scanning cases or with a circular structure, automatic outputs ending of a subscaning (when two scanning cycles without selections are done) must be provided.
- Optimized item distribution: Normally, when the user has taken a selection on the screen, the scanning pointer must be restarted automatically home. So the scanning item distribution (keys, mouse controls, etc.) plays an important role in the access optimization; here, several aspects related to scanning operation, ergonomics, and labeling should be considered, for which the structures of virtual keyboards and scanning screens must be easily defined and reprogrammed. To optimize the scanning, shortening the time until a new selection, it is possible to place the most frequent characters or commands in the left upper corner of the scanning screen (for the case of 2-D file-column scanning). The greatest difficulty lies in identifying the items that should be used in every case, logically, depending on the language, the software in use (as well as how and what for this is used).
- Scanning distribution prediction: A more advanced technique consists of analyzing, in real time (statistics or best AI techniques), the frequency when a character can appear behind the previous one (or a pair of them). As a consequence, the first ten positions of the left upper angle of the on-screen scanning keyboard will come to be updated, in real time, during the operation process. In these positions, the most likely ten characters will be placed considering their probability degree (predictive scanning or forecasting scanning). This technique allows a clear improvement in relation to the task time.
- Encoding: The scanning can be simplified using encoding techniques and two independents selection zones, one for each code digit (or even using only one switch and a serial coding scheme such Morse code). Encoding may not be affordable by some cognitive disabled or elderly people.

At last, it may be necessary to handle scanning systems able to generate the same signals (at electric, logic, and code levels) that are made by these standard devices. Only in this way it will be possible to avoid the use of standard devices. In this respect, several possibilities exist, including:

- External emulators: At first, using external independent emulators (see Figs. 10 and 11) organized a module with a reduced scanning screen (LEDs, LCD, etc.) placed next to the screen computer.
- Internal emulators: After the multitasking became available, the computational and screening resources of the PC were in use. Here, the virtual keyboards will be pointed out on the screen. These devices can work by scanning or direct selection access with a click of the conventional mouse. At present, many operation systems include this as a standard option.

4.5. Acceleration and Optimization Techniques

The most important handicap for any of these users working with a computer access assistive product (in scanning or selection) is its very reduced working speed, especially in editing for word-processing and AAC tasks (23). This problem, which was detected early on (54), can be softened if one:

- Avoids, as much as possible, the use of scanning. A whole-body motor impaired user can access a PC through a switch scanning selection system, either head-controlled or a direct selection assistive device, with infrared or laser pointer on a glasses-frame, or

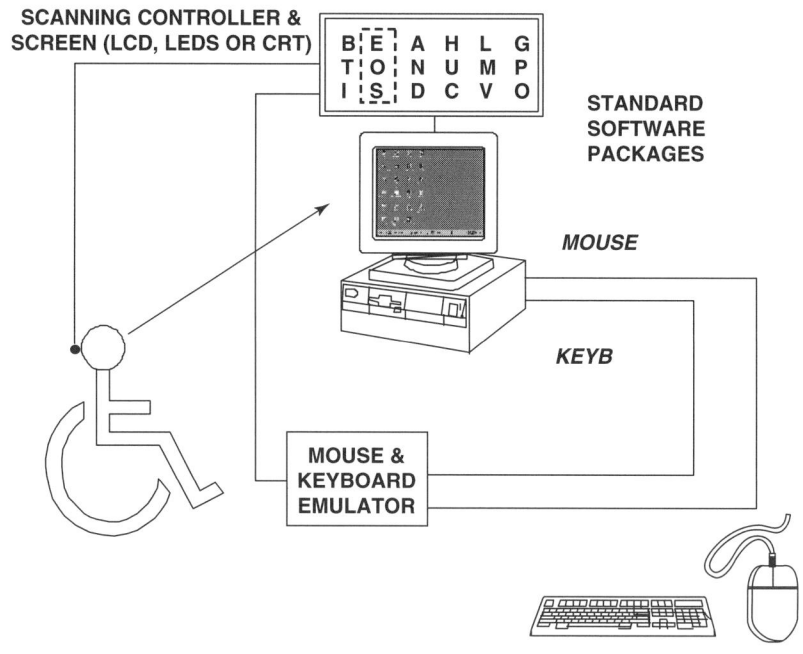

Figure 10. Scanning keyboard and mouse emulator. (This figure is available in full color at http://www.mrw.interscience.wiley.com/ebe.)

by means of a head movement mouse control or other high-tech system such as "gaze detection", and so on.

- Optimizes the scanning by selecting a distribution and set of character and commands, suitable to the disabled user's characteristics and the proper task demands. Hence, reprogrammable and reconfigurable scanning are recommended.

Besides, it is possible to use strategies to reduce the necessary selection numbers to finish the task, increasing the working speed, which is especially useful in applications like word processing, AAC, and so on. For this purpose, several acceleration techniques are available, including:

A. Automatic punctuation: If the software detects the text ending (. , ; : etc.), it must automatically insert a blank (if capital letters follow a full stop (.), an indent in the beginning of a paragraph, etc).

B. Abbreviating: Adding an abbreviation file to the software, as soon as an abbreviation is identified, it will be automatically replaced by its full meaning (words or even sentences). This method (also known as compression-expansion) is limited by the user-handicaps to memorize a huge number of codes.

C. Word prediction: In this case, when a word is totally completed, the most used following word is automatically written (with letters in different colors). If the user waits for a short time without making any selection, the suggested word disappears, restarting the scanning; otherwise the word is inserted and the process continues.

D. Word complementation or anticipation: This technique is suitable both for words and commands. In the

Figure 11. Girl with a disability using a combined keyboard and mouse emulator. (This figure is available in full color at http://www.mrw.interscience.wiley.com/ebe.)

former, as the user writes the different characters of a word, he or she is offered to write the full most frequent word (or command label), with a direct and unique selection.

E. Built-in sentences: It is possible to use sentences (previously recorded in a file) selected from a scanning. This technique is very useful in AAC.

F. Situational adaptation: Sentences or word file sets can be arranged or organized in relation to different situations in use (i.e., at home, work, shopping, a business letter). So the user must first select the situational area where he or she must operate, discharging the corresponding file load.

In order to evaluate the efficiency of these and other methods, special techniques have been developed (55).

4.6. Involuntary Activations Minimization Techniques

Working with scan-switch techniques, both the switch hardware interface and the specific software must provide precise and well-defined functioning characteristics. The first condition here is the perfect synchronization of the very complex system user-switch-interface-software-computer in order to ensure that the switch action will provoke, in the PC, the right action and not other. Under these work conditions, the user will be included in a servo system with an optical and acoustic feedback system between the switch action and the task required.

Whenever the system is synchronized and the switch is only activated in the precise moment the computer access is taken for granted. Unlucky and very commonly, in some physical disabilities, such as cerebral palsy, the voluntary movements (VM) of the muscle mass are blended together with others of spasmodic or involuntary type movements (IM) that can originate in the computer, mistaking selections during the scanning access (56,57).

On the other hand, one should take into account that the different users with different switches placed on different anatomical areas can show very different types and rates of IMs related to very diverse pathologies and diagnostics. In an initial classification of these perturbations, those that are spastic, atetoid, ataxic, coreic, and even PK tremors can be considered.

In general, the AMs act as artifacts during the access, originating involuntary switch activations, which impedes the use of these assistive products, as happens with different kinds of tremors (58,59), as in Parkinson's and other pathologies. The situation produced by the IMs corresponds to that shown in Fig. 12. As one can see when IMs do not exist (E4), the signal produced by VM action (E1) is synchronized with the help of the software and optical feedback watching the scanning screen (E3). This action brings along a corresponding signal related to the desired option (S1). This signal, applied to the interface will generate the (S2) signal (which will originate the realization of the desired option S4), displaying the information (S3) that will act as an optical feedback (E2) on the PC screen and reporting to the user the right execution. On the other hand, when IMs do exist, they provoke undesired action (E4), applied to the control system out

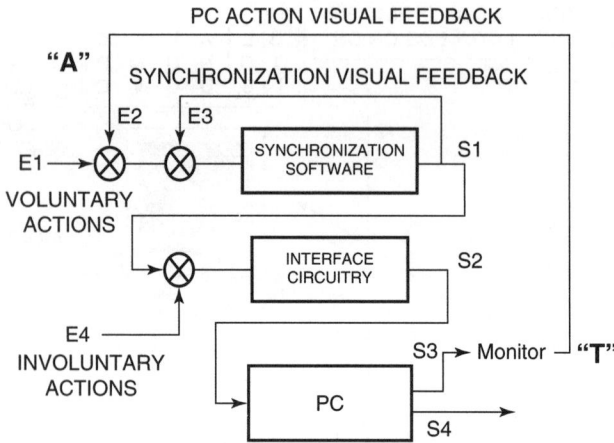

Figure 12. Effect of the involuntary activations of the selection switch.

of synchronism, originating in the PC the fortuitous execution of undesired actions. If, additionally, it is taken into account that during the immediate correction process, it is not possible to guarantee the non appearance of other undesired actions originated by IMs, it is easy to come to the conclusion that the presence of IMs (especially in editing applications), can make the task time increase. The importance of being able to identify the IMs actions is clearly evident, rejecting them before they can be processed by the software.

So, to solve these problems, at first empiric methods were used based on experience rather than justified hypothesis. Further studies developed different solutions based on e-analysis of the fundaments of this process (55,59–61). In this respect, the applied solutions can be catalogued in:

- Passive: Reject: In this method, the involuntary switch activations will be rejected, after being identified as IMs originated through switch signal analysis or also over the muscle activity (EMG).
- Active or statically adjusted: Here, the mechanic features of the switch are modified (in factories) so as to make it less sensitive to the effect of IMs (having a limited efficiency).
- Mixed or dynamically adjusted: The features of the switch are dynamically modified in real time depending on the analysis results (haptic devices).

From these possibilities, several kinds of solutions have been developed, such as sensibility adjustment and adaptive dynamic control of the switch; timing debugging and filtering of the generated signals; as well as EMG pattern recognition.

4.6.1. Switch Sensibility Adjustment. Some designers thought that by increasing the switch activation strength (hardness), or also the switch activation displacement, it would be possible to minimize the IMs effects so that they could not activate the switch. All of this is a little fuzzy and has reached a poor efficiency.

4.6.2. Adaptive Dynamic Control. In this case, the device (e.g., mouse) was structured as a feedback system with adequate sensors and actuators (magnetic brakes, torque engines, etc.) that work in the same way as a braking servo system, within a real-time adjustment, of the device friction effect, in relation to the intensity and the frequency characteristics of the user's tremor.

4.6.3. Timing Debugging. This technique starts from the idea that, generally, many of the involuntary actions (IMs hold) produce "faster" logic signals than those voluntary actions VMs generated; for this, all the state changes (in ON) that do not last a minimum time (TON) or keep (in OFF) between them a minimal time difference (TOFF) will come to be ignored.

This system is appealing (because of the ease of implementation using a simple algorithm on the scanning software), but the results were never adequate for extreme cases. In 1998, a research team defined a variation in this sense (the structure for low-cost time depuration), based on the use of three time parameters (ON time, OFF time, and elimination time) and two different techniques (adding and eliminating) for the switch digital signal treatment, which allowed improvements of nearly 13% in the false action (VMs originated) rejection (62).

4.6.4. Filtering Techniques. The application over the input device signals (mouse) of frequency-domain analysis and filtering techniques shows effectiveness in the case of Parkinson's disease (tremors from 4 to 12 Hz). These devices use different filtering algorithms directly over a standard or adapted mouse. Generally, it is compulsory to use multimodal techniques. So, band rejected filtering, BR (for tremor in the 3–20 Hz band); Fourier Linear Combinators, FLC (for high-frequency tremors); Weight Frequency Linear Combinators, WFLC (for periodical artifacts of unknown frequency and amplitude); as well as other variations (for essential type tremors) (63,64).

4.6.5. EMG Pattern Recognition. The signals generated by switches, joystick, or mouse correspond to the complex function of transference of a biological system (brain, nervous system, motor units, muscle fibers, etc.). This system attacks another simpler electro-mechanical system (the switch), which is why it seems logical to admit that it will be more effective in the identification of involuntary actions (IMs produced), directly through the identification of patterns in the EMG activity (that is to say in the biological system).

For the development of study in this field, the simultaneous and synchronized acquisition of both signals (sEMG and switch) must be guaranteed. So, for example, for the study of an involuntary activation to be done with one's foot, it is possible to use two electrodes (1 cm diameter with inter-electrode distance of 2.5 cm, both placed over the *M. Peroneus Tertius*. Moreover, a structure can be used organized as it is shown in Fig. 13 where the computer acts as a controller of the data acquisition and a "synchronizer" element; originating acoustic and lightening signals (loudspeaker tones and color changes on the PC screen) with a previously programmed cadence and lasting. These signals will be understood as a stimulus by the person to be analyzed that must "respond" by generating synchronized switch actions, repeating the programmed sequence. All the signals (excitation, stimulus, switch, and sEMG signals) are monitored by the PC (65).

As can be seen in Fig. 14, not making either "attention-related mistakes" or involuntary actions, the ideal responding signal (Fig. 14b, switch action) caused by VMs, activation (Fig. 14c, EMG ideal) must follow, in spite of the unavoidable delays, the cadence of the stimuli signal (Fig. 14a).

When the system is used on people highly affected by spasmodic movements, it is possible to obtain registers of the type of those included in Fig. 14d (switch signal) affecting involuntary activations (Fig. 14e, EMG signal).

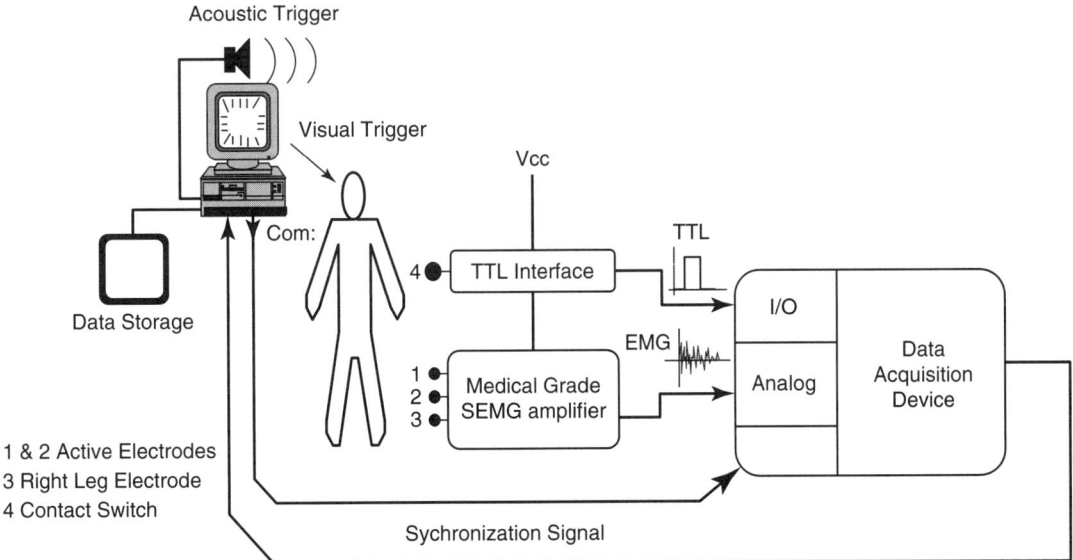

Figure 13. Synchronization of the three signals (sEMG, switch and, stimulus). (This figure is available in full color at http://www.mrw.interscience.wiley.com/ebe.)

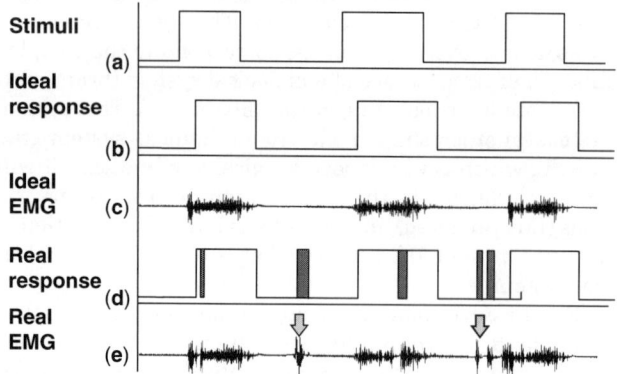

Figure 14. Effect of the involuntary movements on the sEMG and switch response after stimuli.

In this analysis, several anomalies can be identified, such as:

- Failures in ON. They have their origin when the spasm of the user produces an undesired switch release.
- Pulses in OFF. These pulses appear when the user's spasm produces an undesired switch activation.

The sEMG signal characteristic that are useful to develop this identification correspond with:

- Transitions of the d/dt of the RMS value;
- Shifts of the median frequency (in the PSD); and
- Temporal patterns associated with the comparison of the RMS value with a certain threshold.

Once done, the IMs identification (on the EMG) of the activation of the switch can be rejected before they are processed by the scanning software. Unfortunately, this technique will have a relatively high cost (computational and in equipment) (56).

4.7. Accessibility to Electronics Contents

Once the different techniques, used for guaranteeing computer access, have been reviewed, the contents provided by these systems should be analyzed. In fact, this review of AT resources would not be completed without making reference to the web access, which, at present, is one of the greatest resources for assistive services.

Today, the diffusion of the Internet, and of its contents and varied use (which make it widely used to buy, sell, learn, teach, keep relations, communicate, and inform) clearly define a powerful application for disabled and elderly people. So, on a web page, both text and graphical contents must be clearly accessible. Otherwise, web accessibility must tend to get an accessible design on pages and portals not depending on the hardware and software in use, at the same time as other requirements are fulfilled (66,67).

On the one hand, the use of tools and methods (Bobby (68), Taw (69), html Validator, CSS Validator), the comparison of usability with "text-only" web browsers, and of course the "human evaluation" permit one to classify the contents of the pages organized in three levels (A, AA, AAA) depending on the fulfillment of three accessibility priorities; if they all are fulfilled, then it is possible to obtain total information access.

On the other hand, the consideration of the conceptual accessibility, especially for the people with cognitive disabilities, must be highlighted. With the spread of Internet usage, and the rise in the number of portals and online databases, multiple projects have focused on the enhancement of web accessibility (70,71).

Apart from the use of a system-independent design, several recommendations have been made for web page developers (72–74), including:

- Provide texts equivalent for each of the nontext elements.
- Make explicit the meaning of the acronyms and abbreviations to enable the synthesizer operations.
- Provide the user's control over the changes done on the screen.
- Make use of "work modes" repetitive that do not change its logic through the www browser (especially for people with cognitive disabilities and the elderly).

5. EVOLUTION OF ASSISTIVE TECHNOLOGY

Throughout history, the relationship of the society and people with disabilities has greatly improved. Initially, society adopted good-hearted, charitable, philanthropic, and paternal attitudes toward disability (via the participation of the church, the crown, the wealthy social classes, etc.) followed by the state actions justified by the public or company-like deeds (war-crippled and labor handicap people, etc.). Only later did it reach the present position, on a whole aiming for the complete guarantee of the natural rights of all people as members of a fair, democratic, and equal society.

Gradually, along this evolution, important landmarks such as the following ones have taken place:

- Basic care in shelters;
- "Primitive" prostheses and orthoses;
- Early social rules;
- Rehabilitation of acquired disabilities (wars, labor activities);
- Early attempts of social and labor integration;
- Start of disability prevention;
- Early education of people with disabilities;
- Beginning of the modern technical aids and the assistive services;
- Early stimulation and rehabilitation;
- Emerging of the accessibility concept;
- Advanced technological aids;
- Autonomy concept emerges;
- New concept of accessibility (total accessibility);

- Development of the general rules for a more adequate design of the environment;
- Development of the concept of usability; and
- Universal design or "design for all."

Further information may be found in the works of Pedrotty and Brian who have carried out an excellent review of the different stages of the evolution of this matter along three different periods (before 1900, 1900–1972, and 1973 to present) (18).

All this helped to make up the AT as it is known today and also what it will be in future. Unfortunately, only two basic concepts of the present AT are mentioned here for space purposes.

5.1. Total Accessibility

According to this new concept (that emerges complementing that of accessibility), it must be guaranteed that the physical environment (architectonic, urban, services related, etc.), the society and legal rules, as well as the operative environment (tools, systems, resources) must be established permitting everybody to live equally and in the most independent way as possible, integrating the varied needs and rights of people (in laws) and offering them tools and resources that can be used by everybody. The objective of total accessibility comes to lie in the fulfillment of a physical, legal, and operative environment that is comfortable and safe, and that everybody can enjoy, including people with disabilities.

At the same time, within the total accessibility concept, and in the attempt to create a technological order able to aim the research and development, the initial conditions of universal design (or design for all) were established.

5.2. Universal Design

It is because of the concept of total accessibility and with the aim of guaranteeing the right interaction with the user and his or her environment that makes equipment, objects, services, and tools necessary that can be developed as valid designs for most of the users (whether disabled or not).

It must not be forgotten that the most important belonging of a human being is his or her diversity. Additionally, the life of the human being suffers from changes, including the simple day-by-day passing of its biologic and time cycle. Finally, it should also be considered that the functional capacities of the user may also suffer extreme changes because of indirect circumstances far from the "natural" evolution of its biologic life, such as accidents, illnesses, and so on, which is (the users' diversity) what provoked the emerging of a new scope of the design.

The universal design, as defined by Ron Mace in the 1970s, was initially referred only to the requirements for architectural design guaranteeing a proper accessibility to all kinds of buildings and facilities. Finally, this philosophy would come to spread to all the fields of design.

5.2.1. Characteristics of Universal Design.
As far as the above-mentioned goes, it is clearly explained that the new sight of the design (aiming at the establishment of technology today), will not only affect the present resources of the AT but also the most varied kinds of products, objects, tools, systems, and services related to the production sectors, which until now was the so-called "general" or "big audience."

To reach the above-mentioned goals, the universal design fulfills two basic principles:

- Rejecting the simplistic classification of the population in people with and without disability.
- Resolving the technological problems related to the total accessibility, in a general way, making use only of special arrangements in adequate and limited cases.

Hence, new bases (mentioned below) were born for the design, affecting a whole spread of technological fields (75–77):

- Products cannot be created without thinking of the diversity of users.
- The environment must be adapted to the user's needs.
- It must be known that every subject, as a member of the specie, can be very diverse according to their physical, psychic, cultural, religious, and social characteristics.
- The abilities and characteristics of every human being are different and are modified throughout their lives.

Within the logic of the universal design (or design for all), the products, equipment, and assistive developed techniques will have to fulfill some conditions (78–80), including:

- Indistinct and slightly differentiated (for people with or without disabilities).
- Flexibility in the use.
- Logics of simplified and intuitive use.
- Systems tolerant to mistakes.
- Easy perception of the information (signals and GUIs, adapted and friendly).
- Need of a low physical effort for its fulfillment.
- Setting and size that allow the nearby and use.
- Possibility of including connections for other external interfaces.

On the other hand, the assistive products, structured following this philosophy, must be simple, being able to be adapted to the capacities of interaction of the users. At the same time, they must have a "transparent" conception showing at once what they are like and what they are used for within their design, including methods for making evidence, through a constant feedback, the relationship between the user's action and the reaction of the product or service. Finally, the agents involved in the design must assure the functioning of the equipment is adjusted to the previous experience and the expectations of the user.

5.2.2. Agents Involved in Universal Design. The emerging of the concept of universal design and for reasons of its own operability as an integrating technological resource are justified not only on the capacity to encourage research in the field of AT but also on the capacity of stimulating the introduction of these techniques in the general public.

To summarize, this technology will enable resolution of the problems motivated by the early customized development of the AT products, (76,79–82) including:

- Design of individualized and much differentiated solutions.
- Impossibility of industrial manufacturing in batches.
- Lack of interest of the enterprises (for economical reasons).
- Expensive solutions, of difficult maintenance and availability.
- Lack of stimuli to the development in I+D in the field of the assistive products.
- Lack of stimuli on the manufacturing and trade sector of products for the AT.

Under these conditions, it is clear enough that for the success of universal design, as it is shown in Fig. 4, will have to involve different agents such as:

- Users (in relation to a single user or in associations);
- Researchers and designers.
- Manufacturers (both of assistive products and also of great consumption products).
- The public administration in its different services (not only the assistance services but also those related to disability).

On the other hand, it must be also taken into account that the above-mentioned agents have a direct influence on the goals of universal design in many different ways. So, the consumers must think that the product that he or she acquires also must fulfill their needs in the future and outline this to the manufacturer. The designer will have to know and consider the needs of the consumer, users, and organizations to carry out the product design. The user, in his or her diversity, will press so that the product design (and the environment) bears the user's diversity. The enterprise will make the product possible and will have to help the users and organizations to take part of the stages of the design. The administration will have to be responsible for the fulfillment of the equality in opportunities (and the universal accessibility) for every member of the society. Last but not least, they all (the organizations of people with disabilities, as well as consumers, designers, and administration) will be responsible for the promotion and fulfillment of the use and spread of the universal design (83).

5.3. The Future of Assistive Technology

It is possible to foresee the future evolution of assistive technology in relation to its three main components:

- The population of disabled people, as well as those with integration and autonomy problems.
- The technology available to solve these limitations.
- The social structures at legislative, economic, and social assistance levels.

If an analysis of the evolution of the interactions within these agents is carried out, it will be possible to outline the future evolution of the AT scenario (31,84). If the current technological and social scenarios are analyzed in order to detect the different factors affecting the prospective evolution of the assistive technology, several topics should be studied in depth, with the most important ones described below.

5.3.1. New Needs Imposed by Emerging Disabilities. Under the recently introduced term of emerging disabilities, a whole new set of deficits, alterations, and disorders not previously considered as disabilities are included. These "new disabilities" are a direct result of the great socio-economic changes introduced by the scientific and technological advances that have taken place in recent years.

As it is known, any emerging process presents some basic difficulties, such as defining its extent and the associated problems or adopting a concept and a denomination, that have complicated the adoption of a rigorous classification specifying these new alterations. Consistently, as happens in many other fields of science, it shall be necessary to start from their consequences in order to identify the causes prior to naming them. In order to set an example, the next processes are included for being susceptible to be included within those emerging disabilities, according with what is understood this topic stands for.

- Alterations associated and originated with the aging process: It is well-known that the population of the world has largely surpassed what was considered as the life expectancy during the decade before, as it is evident that the number of people going over 65 years is increasing year after year and it is not uncommon for someone to reach 80 years of age, which has lead to the apparition of new disabling situations related to the process of aging, thus not considered at that moment because most people did not reach that age before.
- Survival of people affected of prenatal and perinatal disabilities: The number of people born with severe disabilities (not supported by the medical care available years before) that survive to the adult age has increased because of the advances in science and technology at both preventive and assistive levels.
- New diseases associated with new lifestyles: Stress and anxiety are becoming more frequent among healthy people in the developed countries. The frantic and competitive rhythm in which the activities of these countries are performed drains most of the resources of the human body (which has not apparently changed when observed from the evolution time scale) in order to get the required efficiency. This

fact has led to limit situations threatening the survival of the individuals.
- New tasks and roles in modern times: The necessity for performing new tasks and assuming new roles, because of the social and technical changes introduced in everyday life, may induce a new generation of people with disabilities. Those people cannot take advantage of the new technological developments offered to the society.

Assistive technology should face this challenges without forgetting that, for these people suffering from emerging disabilities, their resources (i.e., assistive products and services) will be the unique bridge enabling these people to get out from exclusion, live a successful independent life, and get the desired social insertion, turning them into one of the most valuable elements of the society in their time.

5.3.2. Welfare State Support. Social and economic policies developed by the different countries on a worldwide scale justify their primary objectives for reaching great economical development for sharing those benefits with their citizens, giving them a better quality of life, including the improvements of the social and health coverage offered by the governments.

This social economy has evolved from the political economy that has been ruling the economic policies since the French Revolution, when the principles of liberty, equality, and fraternity spread among all modern democrat constitutions, favoring associations for the support of the socially excluded.

Later on, as the activities carried out by the citizens put pressure on the political powers, these political powers became aware of the necessity for the formal regulation of these activities and gave way to the different social structures with the required resources for guaranteeing the social and work inclusion of the socially excluded.

Unfortunately, most of the time, the trade-off among the cost of these services and the tax pressure has limited the extent of these actions. Besides, the population of the world has experienced great changes in both size and nature, which has led to the increase of the number of groups that composed it as certain population categories have increased the number of individuals, as happens with people with disabilities and the elderly, whose integration, autonomy, and quality of life have increased significantly. These population groups still constitute a minority today, in fact, it is possible that in only 25 years these groups may extend to nearly 50% of the population.

At this point, it seems obvious that any country "advanced and in progress" should be interested in the necessities of this population, increasing the funding associated to their social programs. For this purpose, policies shall include the access to AT for these people in order to keep them as active elements of the society for longer periods of time.

5.3.3. Overview of the Future of Assistive Technology. The future of assistive technology [that will have to look forward to multidisciplinary solutions (18,31)] will run in parallel with the future development of science and technology (31).

It is then possible to foresee that the new techniques on microintegration (for both electronic and mechanical microsystems or MEMS); the advent of faster processors; the gain in knowledge on biosignals, new materials and sensors (including those biocompatible), nanotechnologies and robotics, and so-called artificial intelligence; and the advances in biomedical engineering in general (and those in genetic engineering in particular) will significantly influence the future of assistive technology; especially on the development of:

- Gene therapy.
- Methods for the early detection of disabilities.
- Direct or "mind" control for ortheses and prostheses.
- New implants and biosystems.
- Adaptive and intelligent human control interfaces.
- Virtual environments, augmented reality.
- Intelligent task sequencers for home and work use.
- Artificial language, speech generation, and recognition.
- Control systems for autonomy, job and home, based on intelligent environments.
- Advanced robotic systems for the mobility and environmental manipulation.
- Assistant robots for the home, work, and leisure.
- Advanced transduction technology.
- New models for user characterization.
- Intelligent assistive products, able to detect changes (in the user and the environment) in order to automatically adapt to new conditions.

From the particular point of view of computer access and the information and communication technologies, it seems that it will be significantly developed with the nonstop diffusion of ubiquitous computing and communication, wearable computers, augmented or virtual reality, and direct access systems such as brain computer interfaces. Most of these topics are being covered by the scientific community, and promising results are being presented year after year in dedicated journals and conferences (85,86).

On the other hand, it is possible that the adoption of the principles of universal design on a global scale will attain the two main objectives of this movement: The usability of most of the products will be increased so they will become accessible for all the citizens and this adoption will impel the research and manufacturing sectors of AT to make developments available for the general public as standard resources for mass-produced goods.

6. LEGISLATION

As it was previously remarked, the most important tool in the social integration of people with disabilities is the existence of legal rules that may assure their rights. Hence, in this section is displayed a (unfortunately brief and incomplete) revision of the evolution of the legal rules related to disability.

Historically, it is illogical to make reference to international judicial aspects about disability without mentioning its beginnings, in the period from 1983 to 1992 of the United Nations Organization, UNO, after the declaration of 1981 as the International Year of the Disabled People. This initiative was the origin to take into account the difficulties, problems, and discriminations that people were suffering.

In the United States, the first actions to get legislation had their roots in the Veterans Administration, after the end of Word War II, within the frame of urgent measures developed for the attention of veterans. In 1973, the Federal Rehabilitation Act was promulgated, which enabled the foundation of the National Institute for Handicapped Research and some other specialized centers in engineering of rehabilitation that can be considered as the root of the development of the assistive technology.

Later, the Rehab Act of 1986 was edited followed by the Technology-Related Assistance for Individuals with Disabilities Act of 1988. It is in 1990 when the U.S. Congress enacted the Law of the North American Disabled People (ADA-Access and Opportunities: A Guide to Disability Awareness), on which the partial legislations have been based that have taken place in other countries.

The United States shows the most complete and advanced legislation in this respect, going further than the 1990 legislation (ADA) and the Individuals with Disabilities Education Act Amendments of 1997, and in 1998, enacting legislation about AT (Assistive Technology Act, ATA), taking the Law of 1988 as a base (87–89). The application of the ATA extents from the District of Columbia to Puerto Rico and other places such as Samoa and the Mariana Islands (Commonwealth of the Northern Mariana).

The ATA foresees that the states should be able to respond, with the AT resources, to the needs of the citizens with disabilities and indicates that the states, within the following ten years, could develop AT programs devoted to cover these needs.

As far as the European Union (EU) goes, through the Amsterdam Treaty, article 13 about discrimination, which certain groups of people do not have to suffer, was included in the Treaty of the European Union (90). In spite of all this, some countries in the EU have a specific legislation about nondiscrimination with disability. Only some countries have a partial legislation to this respect, such as: The United Kingdom (Disability Discrimination Act (1995), Part III, Access to Goods and Services) (91), Ireland, Sweden (together with that of the British act, they both mention in their legislation the nondiscrimination of people with disabilities considering the lack of equipment or adaptation of overcoming barriers of all kinds). France and Italy also changed their laws that discriminated to people with disabilities. In the constitutions of Germany, Finland, Greece, and Spain, specific references about nondiscrimination of disabled people do not exist.

In particular, in Spain, the Spanish Constitution of 1978, in article 49, remarks the specific fulfillment for forces in office to carry along policies of integration for people with disabilities, and in its article 9.2 focuses on the "promotion of the conditions so that the freedom and equality of the people and groups where they are integrated, come to be real and effective." It is also foreseen, in article 53, the judicial care that these people must have, when infringing their rights so as not to be discriminated for their disability; hence, they can call for them both at the Ordinary Court and the Constitutional Court. On the other hand, Italy and Holland, in their respective constitutions, at the beginning only mentioned the integration as a duty and a right but without specifying or including these people.

Other countries have made laws about nondiscrimination to improve the lifestyle of the people with disabilities, such as Canada, Australia, New Zealand, and South Africa.

At present, many international institutions and countries, following the previous examples, are proceeding to update their legal rules.

BIBLIOGRAPHY

1. Cornell University. (2005). National Institute on Disability and Rehabilitation Research. Disability: An Online Resource for U.S. Disability Statistics. (online). Available: http://www.DisabilityStatistics.org.
2. Centers for Disease Control and Prevention. (2005). U.S. Department of Health and Human Services. National Center for Health Statistics. (online). Available: http://www.cdc.gov/nchs/.
3. U.S. Comerse. (2005). U.S. Census Bureau. (online). Available: http://www.census.gov/.
4. A. Houtenville (2002). Disability and Employment in the USA: National Overview based on 2000 Census, *Disability World-U.S. Department of Education National Institute on Disability and Rehabilitation Research (NIDRR)*. (online). Available: http://www.disabilityworld.org/09-10_02/employment/overview.shtml.
5. European Union. (2005). The European Union Online. (online). Available: http://europa.eu.int.
6. U.N. Division. (2005). Disability Statistics. (online). Available: http://unstats.un.org/unsd/disability/introduction.asp.
7. G. C. Vanderheiden, Making software more accessible for people with disabilities: a white paper on the design of software application programs to increase their accessibility for people with disabilities. *ACM SIGCAPH Comput. Physically Handicapped* 1993; **47**:2–32.
8. International Classification of Functioning, Disability and Health. (2005). ICF Hypertext, Searchable On-line Version. (online). Available: http://www3.who.int/icf/onlinebrowser/icf.cfm.
9. C. Robinson, Chapter 135: Rehabilitation engineering, science and technology. In J. Bronzino, ed., *Handbook of Biomedical Engineering*. Boca Raton, FL: CRC Press, 1995.
10. J. D. Enderle, S. M. Blanchard, and J. D. Bronzino, *Introduction to Biomedical Engineering*. San Diego, CA: Academic Press, 2000.
11. *Federal Register*. 1991; **101–476**:41272.
12. A. E. Blackhurst and E. A. Lahm, Technology and exceptionality foundations. In J. Lindsey, ed., *Technology and Exceptional Individuals*. Austin, TX: Pro Ed, 2000.
13. AT Network. (2005). AT Network-Assistive Technology Resources. (online). Available: http://atnet.org/resources/.

14. J. Roca-Dorda, J. Roca-González, and M. E. Del-Campo-Adrián, Metodología de evaluación de la usabilidad en periféricos orientados a discapacitados físicos: resultados preliminaries. Proceedings of the XXI Congreso Anual de la Sociedad Española de Ingeniería Biomédica CASEIB'2003, Mérida, Spain, 2003: 129–132.
15. J. Roca, L. Roa, E. Del-Campo, and J. Roca, Jr., Tracking task for the evaluation of technological aids for the disabled used in computer access. Proceedings of the 2nd European Medical and Biological Engineering Conference EMBEC'02, Vienna, Austria, 2002: 1726–1727.
16. R. D. Jones, Chapter 149: Measurement of-sensory-motor control performance capacities: tracking tasks. In: J. Bronzino, ed., *Handbook of Biomedical Engineering*. Boca Raton, FL: CRC Press, 2000.
17. C. Stephanidis and D. Akoumianakis, Managing accessibility guidelines during user interface design. In: J. Abascal and C. Nicolle, eds., *Inclusive Design Guidelines for HCI*. London: Taylor & Francis, 2001.
18. D. P. Bryant and B. R. Bryant, *Assistive Technology for People With Disabilities*. Boston, MA: Pearson Allyn & Bacon, 2002.
19. K. A. Kaczmarek, Chapter 143: Sensory augmentation and substitution. In: J. Bronzino, ed., *Handbook of Biomedical Engineering*. Boca Raton, FL: CRC Press, 2000.
20. J. Vidal-García-Alonso, J. Prat-Pastor, C. Rodríguez-Porrero-Miret, J. Sánchez-Lacuesta, and P. Vera-Luna, *Libro Blanco de I+D+I al Servicio de las Personas con Discapacidad y las Personas Mayores*. Valencia, Spain: Instituto de Biomecánica de Valencia (IBV), 2003.
21. J. Lubin. (2005). Disability Information and Resources. (online). Available: *http://www.makoa.org*.
22. D. Hobson and E. Trefler, Rehabilitation engineering technologies: principles of application. In: J. Bronzino, ed., *Handbook of Biomedical Engineering*. Boca Raton, FL: CRC Press, 1995.
23. B. Romich, G. Vanderheiden, and K. Hill, Augmentative and alternative communication. In: J. D. Bronzino, ed., *Biomedical Engineering Handbook*. Boca Raton, FL: CRC Press, 2000.
24. R. Lubinski and D. J. Higginbotham, *Communication Technologies for the Elderly: Vision, Hearing and Speech*. San Diego, CA: Singular Press, 1997.
25. C. F. Valdivieso, Contribución al Desarrollo de Sensores y Redes de Sensores de Fibra Óptica para Aplicaciones Domóticas. Ph.D. thesis Public University of Navarra (Spain)-Electronic Technology, 2003.
26. J. Roca-Dorda, J. A. Vera, and M. Jiménez, EIBUS as a key technology for integrating people with disabilities: a case study. Proceedings of the Scientific Conference of EU & Technology Workshop, EIB EVENT 2000, Munich, Germany, 2000.
27. S. Brownsell, D. Bradley, and J. Porteus, *Assistive Technology and Telecare: Forging Solutions for Independent Living*. Bristol, UK: The Policy Press, 2003.
28. M. S. Sanders and E. J. McCormick, *Human Factors in Engineering and Design*. New York: McGraw-Hill Science/Engineering/Math, 1993.
29. G. V. Kondraske, A working model for human system-task interfaces. In: J. Bronzino, ed., *Handbook of Biomedical Engineering*. Boca Raton, FL: CRC Press, 1995.
30. B. R. Baker, Using images to generate speech. In: E. P. Glinert, ed., *Visual Programming Environments: Applications and Issues*. Los Alamitos, CA: IEEE Computer Society Press, 1990.
31. T. W. King, *Assistive Technology: Essential Human Factors*. Needham Heights, MA: Pearson Allyn & Bacon, 1999.
32. D. R. Fuller and L. L. Lloyd, A study of physical and semantic characteristics of a graphic symbol system as predictors of perceived complexity. *Augment. Altern. Común.* 1987; **3**: 26–35.
33. J. Lubin (2001). All Disability Links. (online). Available: *http://www.eskimo.com/~jlubin/disabled/all.htm*.
34. College of Engineering University of Wisconsin-Madison. (2005). Trace Research & Development Center. (online). Available: *http://www.tracecenter.org/*.
35. Disabled Living Foundation. (2005). DLF. (online). Available: *http://www.dlf.org.uk*.
36. CEAPAT. (2005). Centro Estatal de Autonomía Personal y Ayudas Técnicas (Spain). (online). Available: *http://www.ceapat.org*.
37. J. J. Lazzaro, Opening doors for the disabled. *BYTE* 1990; **15**(8):258–268.
38. K. J. Maxwell, Human-computer interface design issues. In: J. Bronzino, ed., *Handbook of Biomedical Engineering*. Boca Raton, FL: CRC Press, 2000.
39. M. Battro, *Computación y Aprendizaje Especial*. Buenos Aires: El Ateneo, 1986.
40. J. Scoin, Desarrollo de software y hardware, para personas con discapacidad. *Novática* 1990; **17**(90):33–38.
41. G. C. Vanderheiden, White paper on the design of software application programs to increase their accessibility for people with disabilities. Trace R & D Center. University of Wisconsin-Madison, Madison, Wisconsin, 1994.
42. J. J. Roca-Dorda, J. M. Fernández-Meroño, and L. M. Tomás-Balibrea, "SISTCOM": integrated system of multifunctional communication for motor disabled people, development Project. Proceedings of the 1993 IEEE International Conference of System, Man and Cybernetics: System Engineering in the Service of Humans, Le Touquet, France, 1993: 539–542.
43. C. Baudelaire, *Les Fleurs du Mal*. 1857.
44. P. Fitts and J. Peterson, Information capacity of discrete motor responses. *J. Experiment. Psychol.* 1964; **67**(2): 103–112.
45. E. Lancioni, M. F. O'Reilly, N. N. Singh, D. Oliva, G. Piazzolla, P. Pirani, and J. Groeneweg, Evaluating the use of multiple microswitches and responses for children with multiple disabilities. *J. Intell. Disabil. Res.* 2002; **46**(4):346.
46. J. V. Basmajian and C. J. Luca, *Muscles Alive: Their Functions Revealed by Electromyography*. Baltimore, MD: Lippincott, Williams & Wilkins, 1985.
47. J. Roca-Dorda, J. M. Fernández-Meroño, and J. A. Villarejo-Mañas, Captador para discapacitados utilizando señales de EMG. Proceedings of the I Congreso Latinoamericano de Ingeniería Biomédica MAZATLAN'98, Mazatlán, Mexico, 1998: 524–527.
48. J. Wilen, S. A. Sisto, and S. Kirshblum, Algorithm for the detection of muscle activation in surface electromyograms during periodic activity. *Ann. Biomed. Eng.* 2002; **30**:97–106.
49. J. Roca-Dorda and J. A. Villarejo, A new option to conventional switches for the motor disabled. Proceedings of the Fifth conference of the European Society for Engineering and Medicine/Sixth International Symposium of Biomedical Engineering and Telemedicine ESEM'99-CASEIB'99, Barcelona, Spain, 1999: 103–106.
50. J. Wolpaw, D. McFarland, D. Neat, and C. Forneris, An EEG-based brain-computer interface for cursor control. *Electroencephalogr. Clin. Neurophysiol.* 1991; **78**:252–259.

51. B. Obermaier, G. R. Muller, and G. Pfurtscheller, "Virtual keyboard" controlled by spontaneous EEG activity. *IEEE Trans. Neural Syst. Rehábil. Eng.* 2003; **11**(4):422–426.
52. T. M. Vaughan, W. J. Heetderks, L. Trejo, W. Z. Rymer, M. Weinrich, M. M. Moore, A. Kubler, B. H. Dobkin, N. Birbaumer, E. Donchin, E. W. Wolpaw, and J. R. Wolpaw, Brain computer interface technology: a review of the second international meeting. *IEEE Trans. Neural Syst. Rehabil. Eng.* 2003; **11**(2):104–109.
53. J. R. Berliss, P. A. Borden, and G. C. Vanderheiden, *Trace Resourcebook: Assistive Technologies for Communication Control and Computer Access*, 1989–90 ed. New York: Demos Medical Publishing, 1990.
54. G. C. Vanderheiden, A high efficiency flexible keyboard input acceleration technique: speedkey. Proceedings of the Second International Conference on Rehabilitation Engineering, Ontario, Canada, 1984: 353–354.
55. C. Riviere and N. Thakor, Modeling and canceling tremor in human-machine interfaces. *IEEE Eng. Med. Biol. Mag.* 1996; **15**(3):29–36.
56. J. Roca-Dorda, J. A. Villarejo-Mañas, and J. Roca-González, Los movimientos involuntarios como fuente de accionamientos indeseados en las ayudas tenológicas: problemática de su supresion, Proceedings of the International Congress of New Technologies and Special Education Necessities, Murcia, Spain, 2000: 41–49.
57. C. Riviere and P. Khosla, Augmenting the human-machine interface: improving manual accuracy. *Proc. IEEE Intl. Conf. Robot. Automát.* 1997; **4**:3546–3550.
58. R. J. Elble and W. C. Koller, *Tremor* (Johns Hopkins Series in Contemporary Medicine and Public Health). Baltimore, MD: Johns Hopkins University Press, 1990.
59. C. Riviere and N. Thakor, Modeling and canceling tremor in human-machine interfaces. *IEEE Eng. Med. Biol. Mag.* 1996; **15**(3):29–36.
60. R. Edwards and A. Beuter, Indexes for identification of abnormal tremor using computer tremor evaluation systems. *IEEE Trans. Biomed. Eng.* 1999; **46**(7):895–898.
61. A. Barrientos, R. González, A. Mora, and J. Martínez, Ayudas al diagnóstico, evaluación y seguimiento de la discapacidad por temblor. Proceedings of the Congreso Iberoamericano IBERDISCAP 2000, 2000: 241–244.
62. J. Roca-Dorda, J. A. Villarejo-Mañas, and J. Roca-González, Captador para ayudas técnicas, a partir de señal de emg, con depuración de accionamientos involuntarios por la hipótesis temporal extendida. *Revista Mex. Ingenier. Bioméd.* 2000; **21**(3):82–88.
63. P. Feys, A. Romberg, J. Ruutiainen, A. Davies-Smith, R. Jones, C. A. Avizzano, M. Bergamasco, and P. Ketelaer, Assistive technology to improve PC interaction for people with intention tremor. *J. Rehábil. Res. Dev.* 2001; **38**(2):235–243.
64. A. Barrientos, R. González, I. Lafoz, P. Portero, J. Martínez, A. Mora, and M. Ferre, "ACORTE" un sistema de ayuda al acceso al ordenador para personas con temblor. Proceedings of the Congreso Iberoamericano IBERDISCAP 2000, 2000: 311–314.
65. J. Roca-Dorda, J. Villarejo-Mañas, and J. Roca-González, La señal de EMG superficial: una alternativa en los captadores para ayudas tecnológicas con depuración de accionamientos involuntarios. Proceedings of the Congreso Iberoamericano IBERDISCAP 2000, 2000: 115–118.
66. G. C. Vanderheiden, Cross-modal access to current and next-generation internet-fundamental and advanced topics in Internet accessibility. *Technol. Disabil.* 1998; **8**(3):115–126.
67. World Wide Web Consortium (W3C). (2005). Web Accessibility Initiative (WAI). (online). Available: *http://www.w3.org/WAI/*.
68. Watchfire Corporation. (2005). Bobby Online Free Portal. (online). Available: *http://bobby.watchfire.com/bobby/html/en/index.jsp*.
69. Fundación CTIC. (2005). TAW. Test de Accesibilidad Web. (online). Available: *http://www.tawdis.net/*.
70. J. G. Abascal, Accessibility to databases. Proceedings of the COST219 Seminar: Databases and Information Systems For People with Disability, Madrid, Spain, 1996: 37–41. Available: *http://www.stakes.fi/cost219/DATABA95.DOC*.
71. Adaptive Technology Resource Centre. (2005). Best Practices in Web-based Instruction. (online). Available: *http://snow.utoronto.ca/best/*.
72. Access-Board. (2005). Web-based Intranet and Internet Information and Applications (1194.22). (online). Available: *http://www.access-board.gov/sec508/guide/1194.22.htm*.
73. C. Egea-García, Diseño accesible de páginas web. Proceedings of the TECNONEET 2000 1st International Congress of New Technologies and Special Educational Needs, Murcia, Spain, 2000: 37–41.
74. European Comisión. (2005). Eeurope 2005: An Information Society for All. (online). Available: *http://europa.eu.int/information_society/eeurope/2005/all_about/action_plan/index_en.htm*.
75. G. C. Vanderheiden, Guidelines for the design of telecommunication products to make them more accessible and compatible for people with disabilities. Proceedings of the Rehabilitation Engineering and Assistive Technology Society of North America 1997 Conference, Pittsburgh, PA, 1997: 506–509.
76. North Carolina State University. (2005). Center for Universal Design. (online). Available: *http://www.design.ncsu.edu/cud*.
77. G. C. Vanderheiden, Fundamentals and priorities for design of information and telecommunication Technologies. In: W. F. E. Preiser and E. Ostroff, eds., *Universal Design Handbook*. New York: McGraw-Hill Professional, 2001.
78. G. C. Vanderheiden, Development of generic accessibility/ability usability design guidelines for electronic and information technology products. Proceedings of the 1st International Conference on Universal Access in Human-Computer Interaction (HCI), New Orleans, LA, 2001.
79. C. Rodríguez-Porrero-Miret, P.Vera-Luna, J. Vidal-García-Alonso, and J. Sánchez-Lacuesta, Introducción a las tecnologías al servicio de la discapacidad y las personas mayores. In: J. Vidal-García-Alonso, J. Prat-Pastor, C. Rodríguez-Porrero-Miret, J. Sánchez-Lacuesta, and P. Vera-Luna, eds., *Libro Blanco de I+D+I al Servicio de las Personas con Discapacidad y las Personas Mayores*. Valencia, Spain: Instituto de Biomecánica de Valencia (IBV), 2003. Available: *http://www.ibv.org/liberia/AdaptingShop/usuario/productos/fichaproducto2.asp?idProducto=16*.
80. G. C. Vanderheiden and J. Tobias, Universal design of consumer products: current industry practice and perceptions. Proceedings of the XIVth Triennial Congress of the International Ergonomics Association and 44th Annual Meeting of the Human Factors and Ergonomics Society, vol. 6. San Diego, CA, 2000: 19–22.
81. G. C. Vanderheiden and J. Tobias (2005). Barriers, Incentives and Facilitators for Adoption of Universal Design Practices By Consumer Product Manufacturers. (online). Available: *http://www.tracecenter.org/docs/hfes98_barriers/barriers_incentives_facilitators.htm*.

82. G. C. Vanderheiden and J. Tobias (2005). Universal Design of Consumer Products: Current Industry Practice and Perceptions. (online). Available: *http://www.tracecenter.org/docs/ud_consumer_products_hfes2000/*.
83. Telematic Applications Programme, Design for All and ICT Business Practice: Addressing the Barriers. European Comisión DG XII-CE, Report No. 98.70.022, 1998.
84. Halftheplanet Foundation. (2005). Disability Information for People With Disabilities, Disabled Veterans and the Public. (online). Available: *http://www.halftheplanet.org*.
85. W. Goodman (2005). Predicting the Future of Assistive Technology. (online). Available: *http://www/halftheplanet.com/departments/technology/the_future.html*.
86. G. M. Craddock, L. P. McCormack, R. B. Reilly, and H. T. Knops, *Assistive Technology-Shaping the Future*. Amsterdam, The Netherlands: IOS Press, 2003.
87. Trace Research & Development Center. (2001). Laws, Regulations and Other Governmental Efforts. (online). Available: *http://www.tracecenter.org/legal/*.
88. U.S. Department of Justice. (2005). Information for Individuals and Communities-Disabilities. (online). Available: *http://www.usdoj.gov/disabilities.htm*.
89. NCD. (2005). National Council on Disability. (online). Available: *http://www.ncd.gov*.
90. European Union. (2005). European Disability Forum (online). Available: *http://www.edf-feph.org/*.
91. Disability Unit. (2005). Department for Work and Pensions (dwp), Disability, Home Page. (online). Available: *http://www.disability.gov.uk/index.html*.

ATOMIC FORCE MICROSCOPY

SVERRE MYHRA
University of Oxford
Yarnton, United Kingdom

1. INTRODUCTION – THE BIOENGINEERING CONTEXT

Atomic force microscopy (AFM) (1) is now the most widely used member of a large family of related techniques known as Scanning probe microscopy (SPM)—the acronyms will be taken to refer to the technique, microscopy, as well as the instrument, microscope. During the last decade, SPM has arguably been the main contributor to the dramatic increase in knowledge about surfaces and interfaces on the nano- and meso-scale. Thus, SPM is not only becoming a tool of the trade in surface science and technology in the research laboratory, but also in the industrial setting. Two other trends have played coincident and synergistic roles with that of SPM. Advances in the life sciences, in depth and breadth, have increasingly been based on interrogation and manipulation of biosystems at the molecular level. Also, nanotechnology has joined biotechnology at the emergent scientific and technological frontier, and as the likely driver of future postindustrial economies. In the near term, it is likely that the first breakthroughs will take place at the nano-at-bio intersection, with SPM and numerically intensive computation as principal ingredients in the enabling technologies.

The Scanning tunneling microscope (STM) was the first member of the family. The STM was invented at the IBM Rüschlikon Laboratory by Binnig et al. in 1982 (2), and earned the discoverers the Nobel Prize for physics. The unequivocal demonstration of single-atom resolution in images obtained from the 7 × 7 reconstructed Si (111) surface was a defining moment in the history of STM, and indeed in surface science. The excitement surrounding those developments is reflected in papers in two issues of the IBM J. Res. Develop. (3). After the arrival of the AFM in 1986, many other members of the increasingly large SPM family were demonstrated, based on sensing of various tip-to-surface interactions, and many novel operational modes were developed and described for each member of the family. First-generation commercial instruments came on the market in the late 1980s. Second-generation general purpose multimode, multitechnique modular instruments became available in the early 1990s. More recently, manufacturers have begun to cater to niche markets with instruments that are optimized for particular classes of applications. For instance, the major manufacturers now offer instruments that are intended principally for analysis of biosystems in biocompatible fluids.

The essential elements of an SPM system are shown schematically in Fig. 1. They consist of a sharp tip and a surface or interface, where the tip and the surface are connected by an interaction having strength and range. Both tip and surface have geometrical and physicochemical attributes. Extent of localization of a measurement

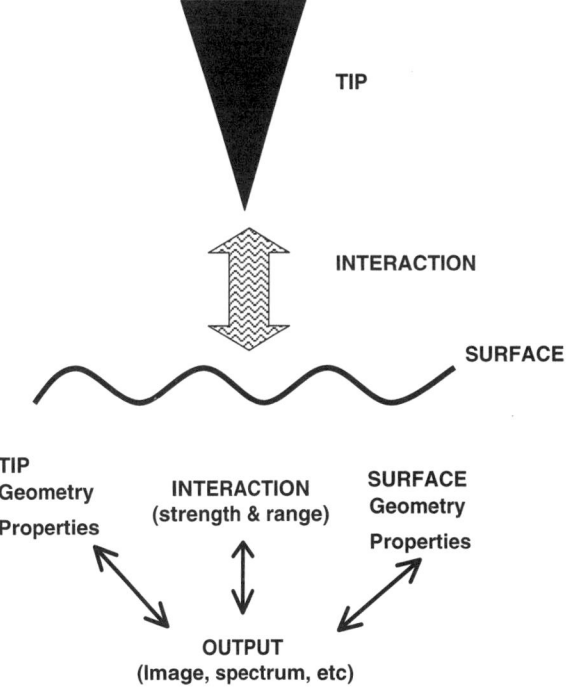

Figure 1. Schematic of the essential elements of the SPM system. The schematic suggests that if adequate knowledge exists about any two of the three elements, then the third element can be subjected to investigation. Accordingly, the system is far richer than what is implied by its name. Conversely, it will be clear that the output is a 'convolution' of the geometries and properties of the three elements.

will depend on the sharpness of the tip and on the range of the interaction(s). In the case of STM, the sharpness is defined by a single atom at the tip apex whereas the characteristic range of the interaction is determined by the spatial decay of an atomic wavefunction (<0.1 nm) (2). In the case of the AFM, on the other hand, the effective radius of curvature of the tip is rarely much less than a few nm, and the net interaction is usually dominated by the van der Waals force (1–2 nm range). Thus, single molecule resolution can be obtained with the AFM only in favorable cases. When the interaction is dominated by the longer-range electrostatic or magnetic interactions, then the localization is correspondingly degraded.

Another measure of the merit of an analytical technology is that of the characteristic 'interaction volume' (i.e., the volume of space from which the information is obtained). The interaction volumes of STM and AFM can be thought of as being those of a single atom ($<10^{-30}$ m^3) and of a single molecule ($<10^{-28}$ m^3), respectively (4). However, if the specimen surface is 'soft,' as in the case of biomaterials, then the tip-to-surface interaction will cause deformation and the interaction volume may be rather greater. Nevertheless, SPM allows interrogation with truly exceptional spatial resolution.

The AFM began life as a meso-scale imaging tool, with lateral and vertical spatial resolution in the low nm range. A particular merit is that image quality is essentially independent of surface conductivity (as distinct from the STM). As well, unlike the electron-optical and photon-optical imaging methods, image quality does not depend on pretreatment of the specimen in order to enhance the contrast. On the other hand, the AFM does not offer analytical compositional information. As described more fully below, the AFM in its various operational modes, aside from topographical imaging, has turned out to be a rich and varied source of information about those attributes of surfaces and interfaces that cannot readily be obtained with 'conventional' microscopy techniques. The functionality of the AFM is substantially unaffected by operation in air or a fluid environment. Thus 'living' biosystems can be probed nondestructively *in vitro* in biocompatible fluids. Therein is the principal attraction of AFM in a biomolecular and biomedical context.

2. INSTRUMENTAL ESSENTIALS

Detailed accounts of the technicalities of AFM instrumentation can be found in the literature (4,5). The basic elements of a generic current-generation AFM are shown in Fig. 2. The probe is at the heart of the system. It consists of a force-sensing/imposing lever and an integral tip at the free end of the lever. The interaction between tip and surface causes deformation of the lever. That deformation is 'sensed' by an optical lever system where a collimated beam from a laser diode is incident on the top surface of the lever at the location of the tip. The optical beam is reflected and then detected by a quad-segment position-sensitive photodiode (PSPD). If the deformation is that of simple deflection, analogous to that of a diving board, arising from out-of-plane force components, then the deflected light beam will be detected by the top-bottom segments. If, on the other hand, the lever undergoes torsional deformation in response to in-plane force components acting at the apex of the tip, then the deflected light beam will be detected by left-right segments of the PSPD. The angular sensitivity of the detection system is in the range 10^{-6}–10^{-7} radians. As the deflection of the cantilever, in terms of z-displacement at its free end, is proportional to the angular change, then the top-bottom PSPD signal is a

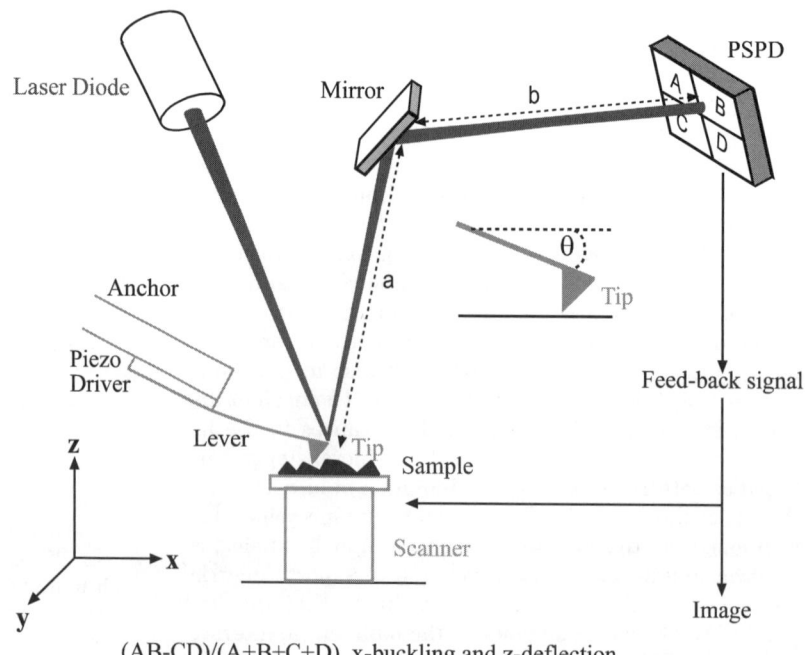

Figure 2. Schematic of AFM system. The segments of the photodetector are labeled A through D. The respective signals can be manipulated so as to detect either simple bending or torsional deformation of the lever. The piezo-electric driver can be activated in order to engage the tapping mode. The specimen is translated in x-y-z space by the piezo-electric scanner. (This figure is available in full color at http://www.mrw.interscience.wiley.com/ebe.)

(AB-CD)/(A+B+C+D) x-buckling and z-deflection
(AC-BD)/(A+B+C+D) y-torsional deflection

measure of the z-excursion of the tip from an arbitrary set-point. As the tip is rigidly attached to the lever, then the z-excursion of the tip will translate to an equal bending of the lever. The system is quasistatic, and the net force acting between tip and surface will be balanced by the force imposed by the lever.

Most AFM instruments have adopted the scheme whereby the specimen is translated while the probe is held stationary. Localization and translation in x-y-z space is effected by a stage consisting of piezo-electric elements. (A piezo-electric material has the property that direct correspondence exists between an applied electric field and dimensional change. The field derives from application of a voltage. Typical figures of merit for a piezo-electric stage range from a few to several hundred nm of dimensional change per volt applied to the stage element). A high-resolution stage may have a maximum x-y field of view of less than $1 \times 1 \, \mu m^2$, whereas other stages may have fields of view of more than $100 \times 100 \, \mu m^2$; the latter capability is useful when a large area needs to be surveyed, but extreme resolution is not required. A raster is generated by application of ramp or sawtooth signals to the x-y elements of the scanning stage. The resolution for a particular stage is generally on the order of 10^{-5} times the field of view. Thus, the smallest raster increment ranges from less than 0.1 to more than 10 nm, depending on the stage and on the number of pixels in the image. The z-range of a stage is generally about one-tenth of the maximum field of view, and defines the 'depth of focus,' whereas the resolution along the z-direction is correspondingly better (typically 0.01 nm) than in the x-y plane.

The signals from the quad segments are conditioned (i.e., buffered, amplified, and filtered). The top-bottom difference signal is compared against a set-point, when the instrument is operated in the *constant force mode* (in actuality, the mode is maintaining constant lever deflection). Any deviation from the set-point, deriving from a change in z-height at a particular location on the surface will then generate a feedback signal. That signal is managed by an electronic feedback loop, the output from which is used to control the height of the z-stage so as to take the deviation back to zero. The change in z-stage height required to maintain constant force constitutes the information that is used to generate a contour map of the surface (in this case, a map where the contours correspond to the z-stage extension required to maintain constant lever force). The implicit assumption for the interpretation of the resultant topographic image is that the surface is much stiffer than the lever (i.e., the lever is the only compliant element). If the surface is relatively flat, such as in the case of a cleaved crystalline face, then the feedback loop can be turned off, and the z-height information is obtained directly from the difference signal from the top-bottom segments (the image is then obtained in the *constant height mode*).

System control and data acquisition, processing, and handling is almost entirely software driven in current-generation instruments. First-generation instruments were highly susceptible to acoustic and mechanical vibrational noise, thermal drift, and electronic noise. A great deal of clever engineering has substantially overcome those problems in current-generation instruments. Likewise, rates of data acquisition are now much higher, and software packages are more user-friendly and incorporate powerful image and data processing tools.

3. THE PROBE

The tip is the primary sensor of interactions with the surface and defines the location of the measurement, whereas the lever is the force transducer. The quality of the probe is defined by a number of parameters including:

- Radius of curvature at the tip apex, R_{tip}, (rarely less than 2 nm, and up to 50 nm for a 'standard' probe).
- The aspect ratio of the tip, A_r, usually taken as the ratio of tip height to half width, defines the steepest gradient of surface features that can be traced (specially 'sharpened' tips may have an aspect ratio of 10, whereas routine tips may have a figure of near unity).
- Tip height, h, will affect the range of height variations that can be probed reliably; it will also affect the sensitivity to torsional forces (tip height may range from <3 to $>15 \, \mu m$).
- Lever stiffness arises from its geometry and the elastic modulus of the material, will determine its response to forces being sensed and will affect the range of forces being imposed by the tip on the surface (the normal force constant, k_N, ranges from 0.001 to 100 N/m, whereas the torsional force constant, k_T, tends to be higher by factors of 10–100).

The relationship between force sensed/imposed by the lever and displacement, d, of the tip, with respect to the z-axis, is given by

$$\text{Force [in nN]} = k \text{ [in N/m]} \times d \text{ [in nm]}.$$

Thus, a lever with a force constant of 1 N/m will impose/sense a force of 1 nN, if a z-deflection of 1 nm exists. One might think that a soft lever in combination with a z-resolution of 0.01 nm would allow force sensitivity in the sub-pN range. Unfortunately, thermal fluctuations impose a limit of 1–10 pN on resolution. A generic lever has dimensions of ca. $200 \, \mu m$ long, by $30 \, \mu m$ wide, by $2 \, \mu m$ thick. SEM images of a beam-shaped probe with a sharpened conical tip are shown in Fig. 3.

The probe is microfabricated from doped Si or from nominally stoichiometric Si_3N_4. A V-shaped two-beam geometry is generally adopted when torsional rigidity is required, whereas a single-beam 'diving board' configuration is preferable when quantitative lateral force measurements are to be obtained. In many cases, especially when biosystems are probed, the surface chemistry of the tip will be an important factor. Clean Si or Si_3N_4 will be coated by a thin native oxide and will thus present a hydrophilic surface (thus, the tip can be functionalized by biochemical agents that will couple to an oxide). However, probes stored in air are generally hydrophobic because of adsorbed hydrocarbon contamination. Special purpose

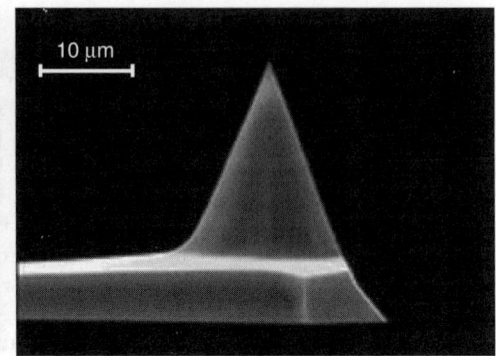

Figure 3. SEM images of a beam-shaped AFM probe and of the tip.

probes are now available for particular applications (e.g., Au coatings are preferred for thiol coupling, a nanotube or graphite spike can be grown on the tip apex in cases where extreme tip sharpness and high aspect ratio are required).

The choice of probe depends on the type of measurement (e.g., contact mode imaging requires a soft lever, whereas intermittent contact mode imaging is commonly carried out with a stiffer lever), the roughness and hardness of the sample, and the desired resolution (hard and flat specimens are suitable for short and sharp tips and stiff levers). Biomaterials are 'soft' objects in the context of AFM analysis (the plasma membrane of a live cell has an effective Young's Modulus of less than 1 MPa). Accordingly, it is necessary to work with a lever with k_N in the range 0.001–0.1 N/m in order to avoid excessive tip indentation. As-received probes will come with the manufacturer's nominal figures of merit. In many cases, it is necessary to determine the characteristics of particular probes. The actual parameters for a particular probe can be calibrated/measured by one of several methods described in the literature (6). The probe is a consumable item, because of wear, contamination, or accidental damage. The cost ranges from a few dollars/probe for the 'standard' varieties to more than $100 for a special purpose probe.

4. OPERATIONAL MODES

4.1. Contact Mode Imaging

Short-range interatomic interactions at the point of tip-to-surface contact are now balanced by the quasistatic bending of the lever. As the force constants of lattice potentials are in the range of 10^2–10^3 N/m, whereas that of the lever is typically 0.01–1 N/m, the compliance of the system is principally confined to that of the lever. However, biomaterials are 'soft' with effective force constants comparable with that of the lever. Thus, the surface will become deformed by force imposed by the tip, leading to an extended area of contact and a corresponding degradation of lateral resolution.

4.2. Intermittent Contact Mode Imaging

The lever is now stimulated by excitation of a piezoelectric actuator at the anchor point to oscillate at, or near, its free-running resonance frequency (from <10 to ca. 500 khz depending on the stiffness of the lever). The tip is then located in the z-direction with respect to the surface so that the tip enters the short-range force field of the surface at the point of closest approach. The effect is to turn the probe into a damped and driven oscillator. The damping causes a decrement in amplitude of oscillation and a change in phase with respect to the phase of the driving signal and the free-running amplitude of the probe. Those changes depend on the strength of the short-range interactions at 'contact'; the respective decrement and shift constitute variables that can be compared against a set-point. In many cases, the intermittent-contact AC mode (also known as 'tapping mode^{TM}') will result in better image quality for 'soft' specimens. The mode has two distinct advantages. The effect of lateral forces is substantially eliminated (important because biomaterials tend to have low resistance to shear stress). Also, because the interaction now derives from an impulse action at one extreme of the oscillatory motion of the tip, then the inertia of the sample will resist deformation. However, the resolution of AC mode imaging depends on the stiffness of the lever (related to the free-running frequency) and the width of the resonance envelope. The latter is severely degraded in water, and soft levers are preferred for analysis of live cells. Although AC mode imaging has many advantages, *in vitro* imaging of cells is generally carried out in the contact mode (7–10).

4.3. Lateral Force Imaging/Analysis

In-plane as well as out-of-plane force components will act on the tip at the point of contact with the surface. The former will exert a friction force on the tip in the direction of travel. If the fast-scan raster direction is perpendicular to the long axis of the lever, then the lateral friction force will cause a torsional deformation of the lever that can be sensed by the signal on the left-right PSPD segments. To a first approximation, the effect is similar to that of macroscopic friction between two objects in sliding contact (the lateral force is independent of relative speed, and independent of the 'contact' area). The relationship between lateral force, F_L, being sensed, and normal force, F_N, being imposed by the lever, is given by:

$$F_L = \mu(F_N + F_A),$$

where μ is the coefficient of friction and F_A is the force of adhesion between tip and surface. In the single asperity regime, when a sharp tip is sliding across a hard surface, the relationship is more complex (11,12). A surface may be laterally differentiated by virtue of variation in surface chemistry (the differentiation may not manifest itself in the topographical contrast). A chemical contrast will result from chemical differentiation that manifests itself as change in adhesive force. Hence, the mode is often called chemical force microscopy [or friction force microscopy (FFM)] (13). The LFM mode is particularly useful for delineating phase-separated polymer surfaces, as shown in Fig. 4 (14).

4.4. Force Versus Distance (F-d) Analysis

Quasistatic F-d analysis can be undertaken by holding the tip at a particular x-y location far away from the surface. The sample is then driven toward the tip at a rate that is slow in comparison with the mechanical response of the system. The net force is sensed during the approach, contact, and retraction parts of the cycle. Two idealized response curves are shown in Fig. 5, with stage travel and lever deflection plotted on the horizontal and vertical axes, respectively. The vertical units can be converted into force sensed/applied by the lever by the simple expediency of multiplying the deflection by k_N. The curve in Fig. 5a represents the case when both tip and surface are incompressible, and when the surface is covered with a thin adsorbed aqueous film. The various segments represent:

Approach half-cycle:

- AB - tip and surface are well-separated, no interaction;
- BC - the tip senses the attractive interaction from the meniscus layer, the force constant of interaction exceeds k_N, and the tip snaps into contact with the 'hard' surface. A regime of instability exists because of the force constant of interaction being greater than that of the lever;
- CD - tip and surface are incompressible, and the stage travel distance must therefore be equal to the lever deflection.

Retract half-cycle:

- DE - the system retraces itself because all deformations/deflections are elastic;
- EF - the meniscus interaction has increased because of capillary action, and a greater lift-off instability/discontinuity will occur (than for the snap-on);
- FA - return to large separation and no interaction.

This kind of curve is representative of events for an air-ambient instrument because of adsorbed moisture. The meniscus interaction is generally a nuisance feature in that it will mask other surface mechanical effects. The meniscus can be eliminated by carrying out F-d analysis under water (or some other fluid ambient). Alternatively, in the case of an instrument operated within a vacuum envelope the aqueous phase will be pumped away. The 'hard' surface F-d curve is used to calibrate the detection system so that a measurable detector response can be related accurately to the lever deflection. Applications of F-d analysis in cellular biology, biomolecular interactions, and protein folding are now well-established. Several reports have described current state of the art (7,13,15).

The schematic curve in Fig. 5b shows lever deflection as a function of tip indentation of the specimen surface, or separation from the surface (i.e., the difference between stage travel and lever deflection), and thus illustrates other surface mechanical aspects of the system that are accessible to F-d analysis. The shape of the curve assumes that at no stage is the force constant of interaction greater than k_N (hence, instabilities at snap-on and lift-off are suppressed). The information content of the generic F-d curve in Fig. 5b may be summarized as follows:

- The force constant of interaction, k_i, is simply the slope of the curve, $k_N z_L/|z_d - z_L|$, where z_d and z_L refer to stage travel and lever deflection, respectively.

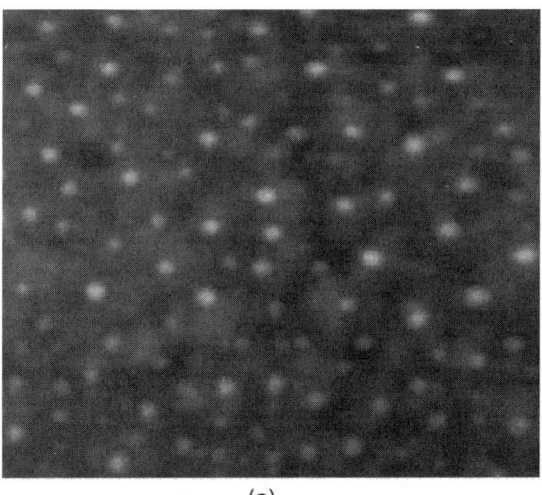

(a)

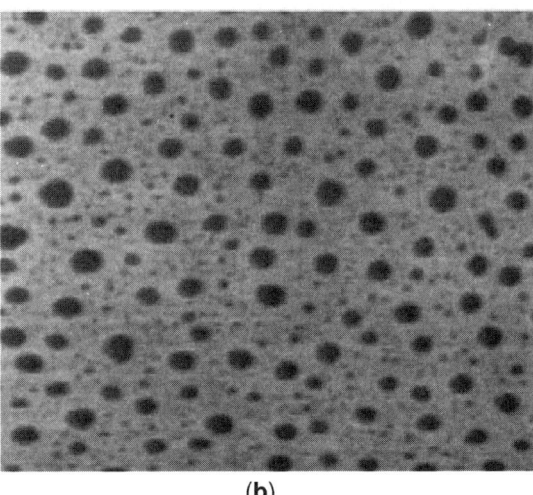

(b)

Figure 4. Topographic (A) and lateral force (B) image of a phase-separated polymer blend film (PMA/PMMA at 95:5 nominal weight fraction). The images show that the phase structure is more clearly delineated in the LFM than in the topographic imaging mode (14).

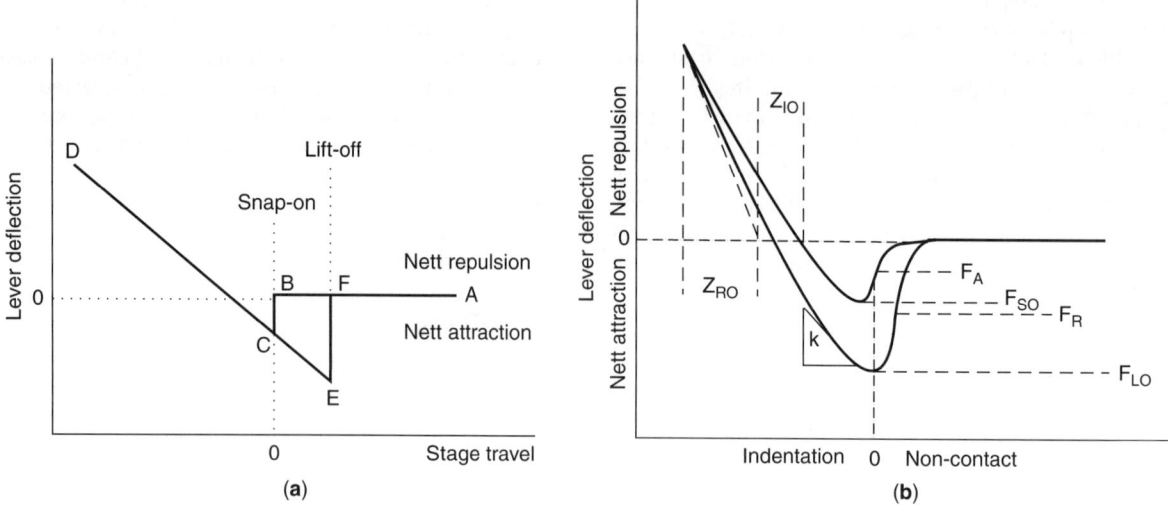

Figure 5. Generic outcomes of F-d analysis. (a) The system consists of an incompressible tip and a surface, with the surface being covered by an adsorbed aqueous film. (b) The surface is compliant so that the tip indents the surface. The surface undergoes deformation in response to the force being applied by the lever, in combination with partial elastic recovery.

- The forces at 'contact' on approach and retract, F_A and F_R, may be defined at the inflection points where the force constants of interaction are the greatest.
- The snap-on and lift-off forces, F_{SO} and F_{LO}, are measured at the points of greatest net attraction. The latter is generally taken to be the force of adhesion.
- The distances z_{RO} and z_{LO} are measures of the elastic recovery and the plastic indentation, both at zero lever loading.
- The extent of hysteresis in the system is given by the area enclosed by the approach and retract curves.

The parameters defined above cannot readily be related to the familiar macroscopic definitions of mechanical properties (e.g., hardness, adhesion, Young's modulus, tensile strength, flexural strength, etc.) unless the system can be specified further (e.g., tip shape, contact area, surface free energy of tip, etc.) and unless additional assumptions are made (homogeneity, isotropy, surface topography, etc.).

A consequence of the discussion above is the need to match the force constant of the lever to that of the interaction being investigated in order to extract maximum information. If a mismatch exists, then either the lever will be the only compliant element, and no information is obtained about the surface, or the surface will be the only compliant element, and the deflection of the lever is not measurable.

4.5. Colloidal Probe Analysis

Some AFM-based work on biomolecular interactions has been carried out by sensing forces in the sub-nN range versus distance (i.e., standard F-d analysis), between a functionalized Si or Si_3N_4 tip and a functionalized surface. More often, however, a microbead is attached at the location of the tip, so as to produce a 'colloidal' probe; that arrangement has a number of merits for investigations of intermediate- and short-range interactions (16). The geometry of the probe is then known with greater relative certainty than in the case of the tip. Likewise, the surface chemistry of a bead can be prepared with greater flexibility and reliability than in the case of a tip; a wide range of functionalized microspheres/beads can be obtained from several suppliers. Also, the much greater surface area will effectively amplify the strength of interaction and thus improve the signal-to-noise ratio. On the other hand, greater uncertainty as to the number of interacting species will occur, and lateral spatial resolution is substantially degraded in comparison with a sharp probe tip. In the case of a functionalized tip, with a radius of curvature less than 50 nm, the number of interacting species may be less than ten, thus allowing single-event-binding forces to be estimated from a histogram of the data (17–19).

5. ELEMENTS OF METHODOLOGY

5.1. Probe Functionalization

Two main generic coupling schemes exist. The more popular scheme is based on formation of a thiol self-assembled monolayer on an Au-coated substrate [e.g., Ulman (20)]. The modified substrate is stable chemically and relatively robust mechanically. Thiols with a great variety of terminal functionalities are available commercially. Silane coupling is the main alternative route [e.g., Sigueira Petri et al. (21)]. The surface is now prepared with silanol groups, which act as sites of attachment for any one of a great variety of silane-coupling agents with different terminal functionalities. A variation on the thiol theme involves attachment of an amine-terminated polyethylene glycol spacer to the probe, followed by attachment of an antibody to the amine group (22).

5.2. Ambient Conditions

Cleanliness and control of ambient conditions are important issues for any surface-specific analytical technique. When the surface is extremely reactive or susceptible to contamination, then UHV is the only acceptable environment for specimen preparation and analysis. However, in less demanding circumstances, or when the specimen is incompatible with UHV (such as live cells *in vitro*), then analysis in air, in an inert gaseous envelope, or in a fluid environment is common practice. Indeed, the overwhelming majority of AFM instruments are intended for, and used in, air or fluid ambient environments. In the case of analysis of biomaterials, then the ability to have full functionality in a biocompatible fluid is one of its unique strengths.

5.3. Static Conditions

The principal variables requiring control are those of temperature and fluid volume. If the fluid cell consists of a trapped droplet, then frequent replenishment of the reservoir is required. Thus, imaging conditions will need to be re-established at regular intervals (typically 30 mins), but then opportunity also exists to re-establish optimum temperature. A larger fluid cell, such as a culture dish, with a volume of some 5 ml or more, will have a longer life-time with respect to evaporative losses, and the greater thermal inertia will promote temperature stability. Long-term stability over some hours will require replenishment, however. The optimum temperature can then be re-established by total replacement of the media. Although flow-through replacement from an external reservoir is another option, the imaging conditions are likely then to be affected by fluid convection when a 'soft' lever is being used. Another alternative is that of continuous heating of the cell by a hot stage; the disadvantages then being associated with thermal contraction/expansion and with thermal convection currents in the fluid.

5.4. Injection of Reactants for Dynamic Studies

In the case of slow dynamics being investigated, reactants may be introduced when replenishment or replacement of media occurs. More rapid injection and mixing is required if fast dynamics is being investigated. Access to the fluid cell is generally restricted by the compact design of most instruments. Accordingly a flow-through cell arrangement may be the better choice when interruption to the imaging conditions cannot be tolerated, or when a particular field of view needs to be tracked continuously.

6. CELLS

Interrogation of systems that consist of live cells *in vitro* has traditionally been based on the mature photon optical techniques. Spatial resolution has, until recently, been limited by diffraction and aperture effects, but the availability of spectroscopic information has continued to add value to the methodologies. Moreover, the advent of laser illumination has given us optical tweezers and scissors, thus providing the means for microscale manipulation of cells. Likewise, high-quantum-yield solid-state detection of photons and digitization of images have added new dimensions to data acquisition and processing. More recently, it has been demonstrated that the diffraction limit can be beaten, and near-field imaging [near-field scanning optical microscopy (NSOM), also known as SNOM] can offer optical resolution of 30–50 nm.

The electron-optical techniques offer considerably greater lateral spatial resolution, <10 nm for the current generation of field-emission low-voltage scanning electron microscopes (FESEM), and sub-nm point-to-point resolution for high-resolution transmission electron microscopes [(HR)TEM]. Both families provide analytical information in various operational modes, whereas the TEM offers structural information from electron diffraction. However, the electron-optical techniques are incompatible with a cell being in a biocompatible environment and remaining viable.

Although AaFM is a relatively recent entry into the field of cellular biology, it has gained increasing importance as a complementary tool for visualization and characterization of cells and their dynamics. The pioneering work was carried out in the early 1990s by Henderson

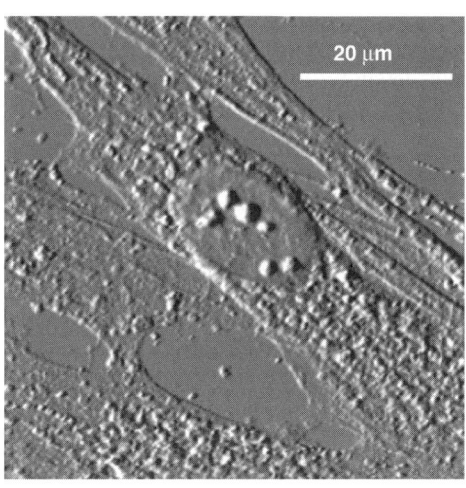
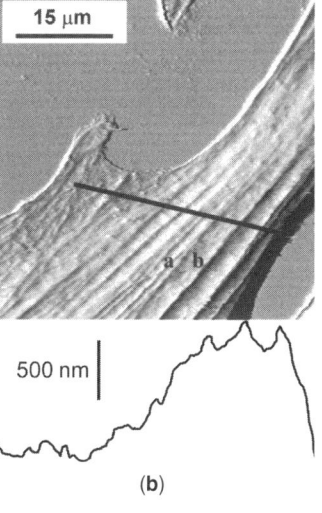

Figure 6. Contact mode images of fixed (a) and live (b) human fibroblasts. The contour line illustrates the resolution available routinely with a large field of view scanner. The locations 'a' and 'b' refer to F-d analysis of the mechanical properties [adapted from Bushell et al. (25)].

et al. (23,24). The current state of the art can be illustrated by some examples from the literature.

6.1. Fixed or Dehydrated Cells

When a cell is fixed, through cross-linking of the plasma membrane, or dehydrated, it becomes a 'hard' object. Consequently, it can be imaged in air by routine AFM procedures; a typical contact mode image of a fixed cell is shown in Fig. 6a (25). However, SEM/TEM methodologies are then richer sources of information, and are generally to be preferred.

6.2. Live Cells *InVitro*

Live cells on untreated cover slips in a biofluid rarely produce acceptable contact mode images, because of poor adherence to the substrate, gross fouling of the tip, and destructive tip-to-membrane interactions. On the other hand, cells grown on surface-treated culture dishes offer greatly improved imaging conditions. Continuous scanning at linear scan speeds of 150 µm/s and at force loadings in the low nN range can now be carried out over several hours without any apparent damage to viable cells. A typical example of a contact mode image of a lamellipodial region of a fibroblast is shown in Fig. 6b; the contour line reveals intracellular structure with a z-resolution in the low nm range.

6.3. Cell Dynamics (Slow)

The AFM is uniquely capable of meso-scale visualization of the temporal evolution of systems consisting of living cells in a biocompatible fluid. Distinctly different methodologies exists for 'slow' (>10 min) and 'fast' (<10 min) dynamics. The acquisition of an image takes typically 1–3 min. Thus, sequential imaging over a particular field of view can track cell dynamics *in vitro* on the time scale of some minutes. A soft lever ($k_N < 0.01$ N/m) in combination with a low applied force (<1 nN) will enhance information obtained from the 'softer' elements of the cell, whereas a stiffer lever and greater applied force will deform the plasma membrane and enhance visualization of the less compressible cytoskeletal and intracellular structures. The more informative studies have exploited the latter strategy in order to gain insight into cytoskeletal dynamics (24–27). Tapping mode imaging has proven to be particularly useful for probing the internal structure of living cells (8). An example of investigations of slow intracellular dynamics is shown in a sequence of images in Fig. 7 (25), where the intracellular nucleation and growth over a period of 3 h of formazan crystals is apparent. The crystals derive from enzymatic conversion of a tetrazolium salt during the MTT assay of viable cells.

6.4. Cell Dynamics (Fast)

Biological activity on the subsecond time scale can be observed and analyzed by AFM by monitoring the deflection of a lever held stationary in the x-y plane and sensed in the constant height mode (i.e., where the time constant of the feedback loop is longer than that of the biological response mechanism). The tip is simply landed at an appropriate location predetermined from an image. The x-y scan function is deactivated, and the dynamic response is monitored through the z-deflection of the lever. A soft lever is most appropriate, because the probe is ideally a passive participant in the temporal evolution. The method has been deployed with considerable success in the case of cardiomyocytes (28,29). Figure 8 (29) illustrates fast dynamics of spontaneously beating cardiomyocyte cells. In principle, dynamic data can be acquired at rates that approach the lowest mechanical or electronic frequency modes of the system, say 10 khz.

6.5. F-d Measurements

The interaction between a tip and a surface is a function of the force imposed by the lever, and manifests itself as a force acting on the tip and being sensed by the lever. The system consists of two compliant elements—the lever and the surface—whereas all other components of the system are assumed to be rigid. The forces will cause bending of the lever and indentation of the surface by the tip. If the shape of the tip, the spring constant of the lever, the applied force, and the depth of indentation are known, then an effective Young's Modulus for the surface can be calculated. The details of the procedure have been described elsewhere (25,30,31). Adhesion between tip and surface manifests itself as an attractive force causing deflection of the lever at the point of 'lift-off' in the F-d curve. As the strength of adhesion is a reflection of the local surface chemistry, then mapping of the surface adhesion of living cells and other biomaterials may provide valuable additional information (32).

The effective Young's Modulus of a supported section of the plasma membrane is in the range of 1–10 kPa, whereas the corresponding value for a membrane more strongly supported by the cytoskeletal structure is in the range of 15–50 kPa. The effective modulus of fixed cells is higher by an order of magnitude or more. The results in Fig. 9 (25) illustrate typical outcomes of F-d analysis carried out *in vitro* on living fibroblasts. A standard probe with $k_N = 0.03$ N/m and a pyramidal tip shape, nominal radius of curvature of 40 nm, and an aspect ratio of 0.7, was used for those measurements. The data illustrate tip indentation and reveal differences in stiffness measured at different locations ('a' and 'b') on the plasma membrane. The horizontal axis refers to z-stage travel, whereas the vertical axis shows lever deflection in the z-direction.

7. DNA IMAGING

Controlling and manipulating DNA molecules are potentially important enabling nano-scale technologies [e.g., DNA-based molecular computation (33) and conductive molecular wires (34)]. DNA-based crystalline structures or networks derive from self-assembly in response to functionalization or by tailoring of the aqueous chemical environment. The outcomes are essentially two-dimensional structures supported on flat substrates (e.g., mica, polymer films, or glass).

Single DNA molecules can be stretched by several methods, and thus made more accessible to detailed analysis of

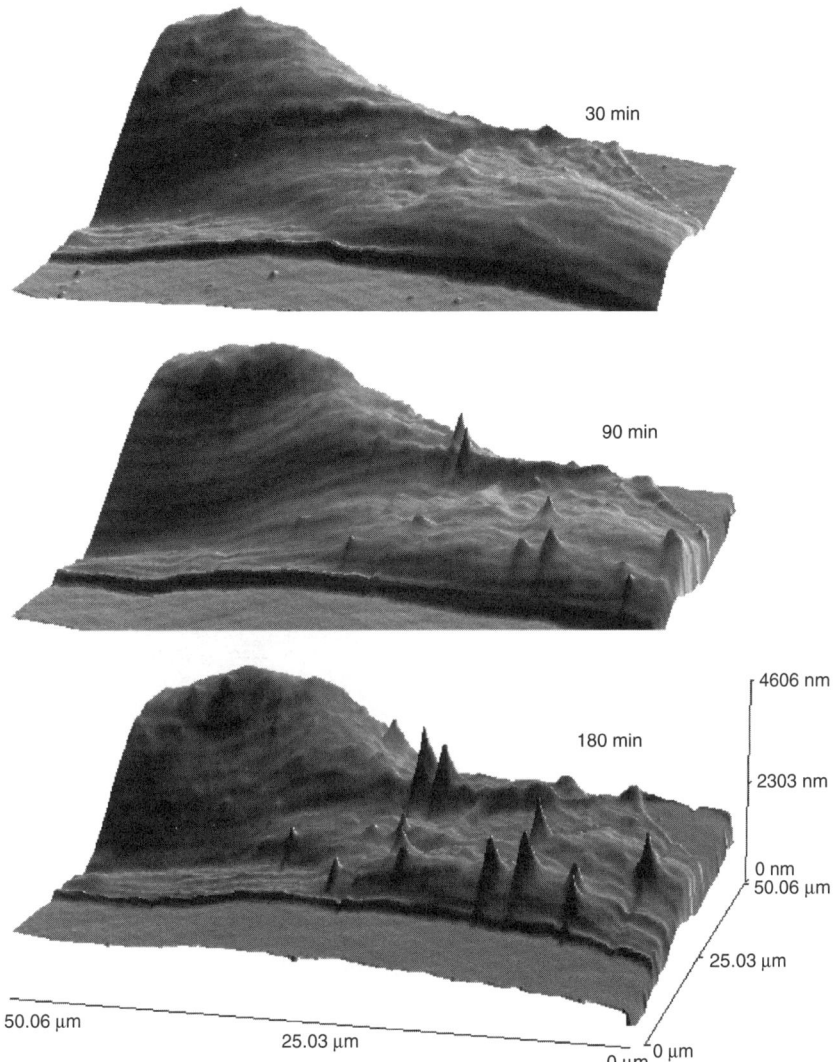

Figure 7. A sequence of contact mode images obtained with a soft lever over 3 h showing intracellular nucleation and growth of formazan crystals within a living cell (deriving from application of the MTT assay) [Adapted from Bushell et al. (25)].

structure and functionality. Figure 10 shows AFM tapping mode images of stained lambda-phage DNA stretched by a spin-coating method (35). In combination with fluorescence analysis by an attached SNOM stage (Scanning near-field microscopy), the method has the merit of discriminating between features that are specific to an intact DNA molecule and those that develop from debris. In a variation on that theme, DNA in suspension is deposited on a polymer film, whereupon stretching is effected by directional micro-pipetting (36). The air/water interface will thus constitute the mobile anchor point. The results of AFM tapping mode imaging and height contour analysis are shown in Fig. 11 (36).

Double-crossover (DX) DNA molecules can form relatively rigid structures; such structures could, in principle, provide the skeleton for assembly of bio-organic periodic structures. Moreover, the assembly of two-dimensional DX-based crystals resemble that of a tiling sequence, which has attributes common with those of a Turing Machine. Accordingly, the self-assembly process and outcome have relevance for the design of molecular processing devices. Contact mode images obtained in isopropanol of two-dimensional DNA crystals deposited onto cleaved mica are shown in Fig. 12 (33).

The electrical properties of aperiodic self-assembled DNA networks have been investigated recently. The images in Fig. 13 show the topographical features of a network, as well as an electrical conductivity map (34). The latter is obtained in the contact mode with a conducting probe acting as a traveling electrode [also known as ECM (electrical conductivity microscopy)]. The current flow between the tip and a gold electrode at the edge of the network is sensed as a function of location; the bright contrast in the image represents high conductivity.

8. BIOMOLECULAR BINDING AND RECOGNITION

Measurement of biomolecular interactions by AFM is straightforward in principle. Thus, the holy grail of gaining detailed and specific insight into molecular recognition at the level of the single event should be within our grasp. The apex of the tip is functionalized with a particular molecular species (e.g., biotin), whereas a countermolecule

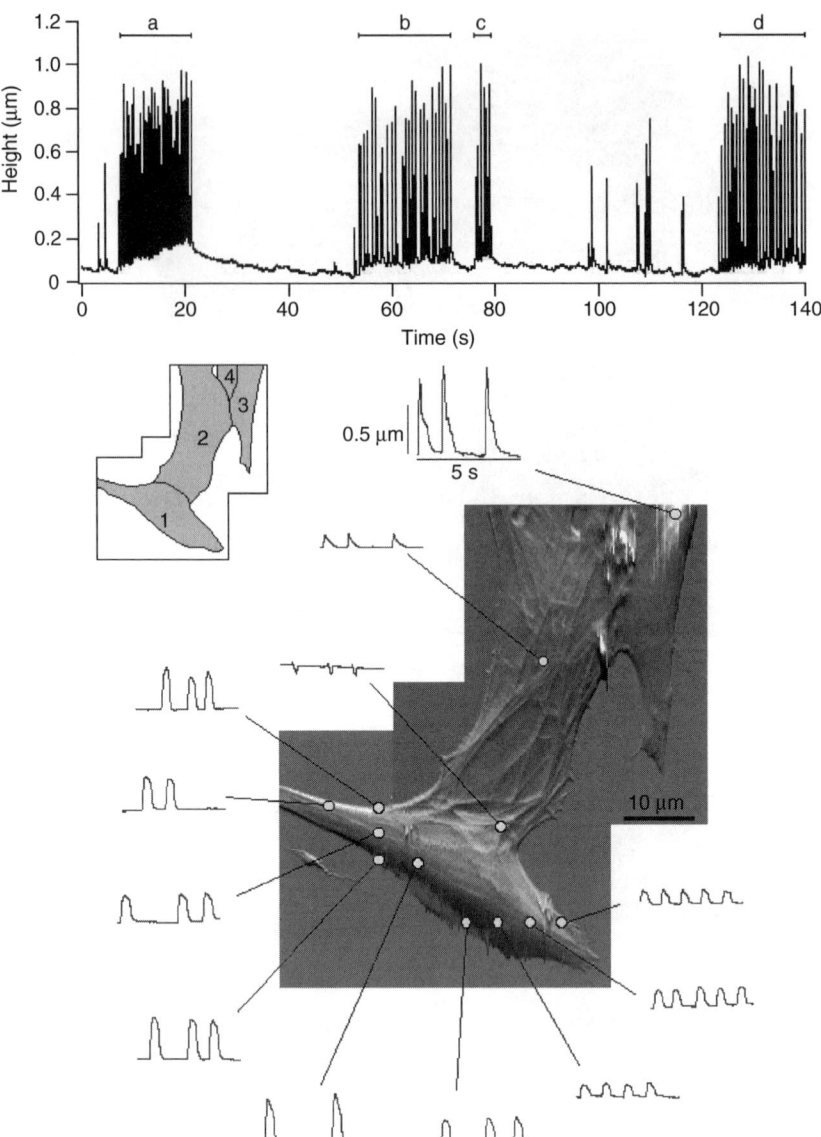

Figure 8. The trace in the upper panel illustrates irregular beat of a live cardiomyocyte. The image and traces in the lower panel demonstrate that both amplitude and frequency depend on the location within a single cell [Adapted from Domke et al. (29)].

(e.g., avidin) is attached to a substrate. The AFM is then operated in the F-d mode whereby the two species are brought into contact, causing recognition to take place. The probe is then retracted and the force of 'adhesionv,' corresponding to rupture of the bond, is inferred from the deflection of the lever. Typical rupture forces are in the range of 10–200 pN (17–19,22,37), which is well within the resolution of the instrument, when a soft lever is being

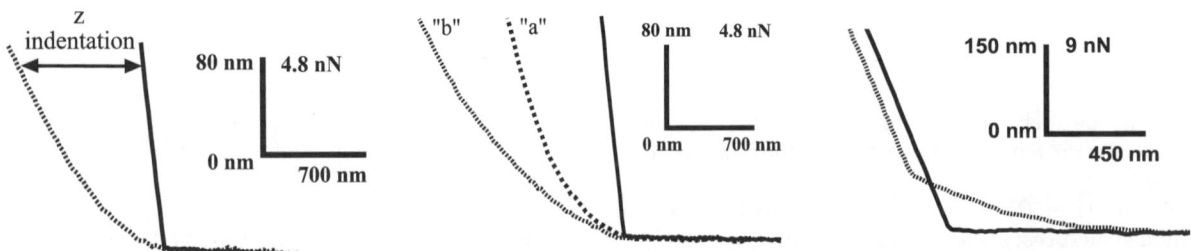

Figure 9. F-d analysis of fibroblasts [adapted from Bushell et al. (25)]. The solid line constitutes a calibration of the system where the tip is interacting with the incompressible substrate. Thus, lever deflection is then equal to z-stage displacement. The curves in the middle diagram show lever response at two different locations, 'a' and 'b' in Fig. 6b, corresponding to locations where, respectively, the plasma membrane is supported by submembrane cytoskeletal structure, and a location without cytoskeletal support. The broken curve in the diagram on the right illustrates compression of an unsupported part of the cell between the tip and the hard substrate.

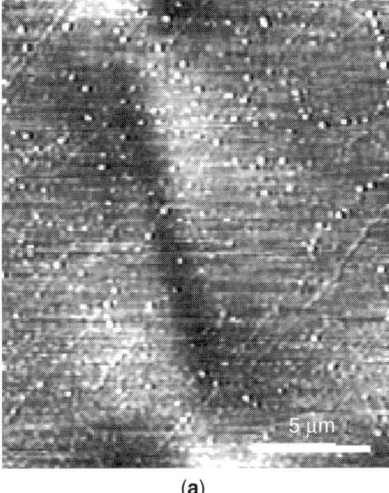

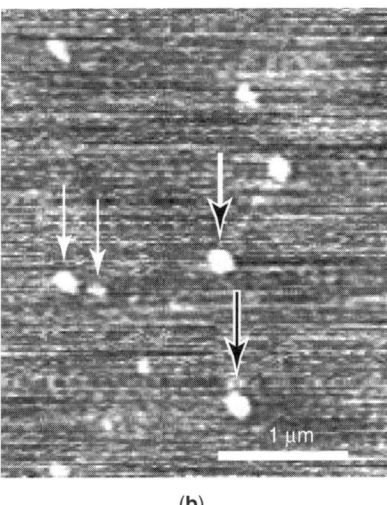

Figure 10. Topographic images of lambda-phage DNA. Top image shows stretched DNA on an Mg^{2+} treated mica substrate. The stretching was the outcome of the sample, at a dilution of 4.5 µg/ml, being spun at 4700 rpm for 20 s. The lower high-resolution image shows two kinds of features. Correlation with a SNOM image allows adventitious debris (bright arrows) to be discriminated from stretched DNA (dark arrows) [Adapted from Yoshino et al. (35)].

used. The in-principle arrangement is illustrated schematically in Fig. 14.

In practice, a great deal more is behind biomolecular interactions, not only at the conceptual level, but also with regard to methodological issues. The highlights will be summarized below, and additional details are in the literature (17–19,22,37). The idea of a biomolecular 'interaction' must be given additional consideration (38), as described below in summary form.

8.1. Biomolecular 'Interactions'

8.1.1. Van der Waals Interaction.
This interaction will always be present for all systems. It is short-range, 1–2 nm; the effective force constant of interaction will be greater than that of a 'soft' lever, causing snap-on to occur. The functional dependence, strength, and range can be inferred from the F-d approach curve.

8.1.2. Capillary Meniscus Force.
Moisture will condense on surfaces exposed to humid air; the layer of moisture will interact with the tip and give rise to a capillary force. The strength of that force will depend on the thickness of adsorbed layer, the surface chemistry of the tip, and the geometry of the tip. In most cases, the capillary force is much greater than any other contribution to adhesive interaction. However, the capillary force is eliminated if F-d analysis is carried out under water.

8.1.3. Double-Layer Interaction.
Biospecies in solution will present a positive or negative surface charge depending on solution pH with respect to the iso-electric point (IEP). Counterions or coions will cause partial screening and result in the formation of a diffuse double-layer. Accordingly, electrostatic interaction will occur with a characteristic range of 10–30 nm. The functional dependence, range, and strength of the interaction can be determined from features in the approach curve and then used to infer characteristics of the system in the context of one or more descriptions of colloid interaction, such as the DLVO model (39,40).

8.1.4. Hydrophobic/Hydrophilic Interactions.
Features in the F-d curves are often encountered that cannot readily be explained by the conventional descriptions of biocolloids in solution or through known short-range interactions. Forces of intermediate range caused by hydrophobicity/philicity have been proposed, although their theoretical justification is less than satisfactory.

8.1.5. Solvation Force.
Ordering will take place in the fluid phase when it is constrained between two surfaces in

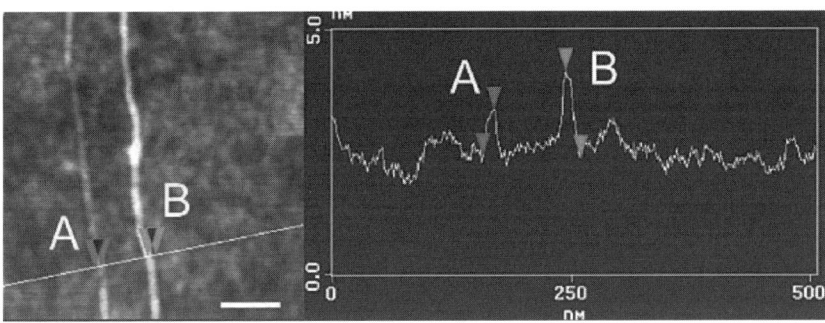

Figure 11. Tapping mode topographic image of straightened lambda-phage DNA fixed on a polymer substrate (left). The scale bar represents 100 nm. The contour line (right) shows an apparent diameter of ca. 0.9 nm at point A, consistent with the diameter of a single double-stranded DNA molecule [Adapted from Nakao et al. (36)].

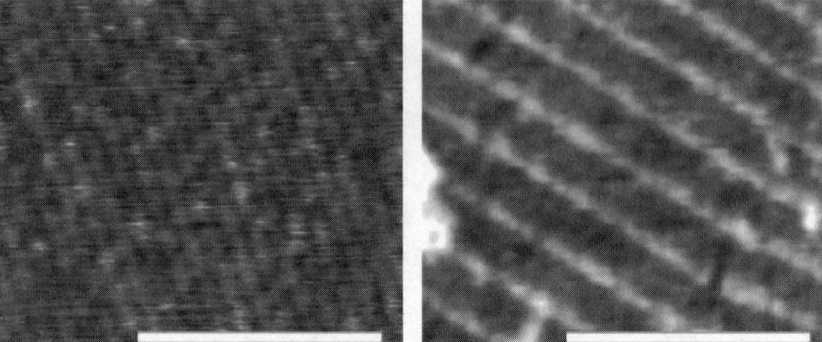

Figure 12. Two-dimensional DNA DX crystal structures. DOA-E and DAE-O structures are shown in left and right images, respectively; see Ref. 33 for detailed description. The scale bars are 300 nm.

close proximity. During approach, such ordering may give rise to oscillations in the force being sensed by the tip.

8.1.6. Adhesion. Adhesion is conventionally taken as the force required to separate the tip from the surface during probe retraction. Strength of adhesion will depend on the net strength of short-range interactions. It will also depend on the contact area, and thus on the extent of indentation by the tip. When 'soft' objects (i.e., functionalized surfaces) come into contact, then compression, deformation, and structural rearrangement will occur, which are synergistically dependent on short-range interactions (16). Accordingly, no satisfactory method exists for determining adhesion per unit area from analysis of F-d data.

8.1.7. Effects of Loading Rate. It has been demonstrated that the force corresponding to bond disruption depends on the loading rate (41). Such a dependence is thought to derive from the probability of thermal rupture being described by one or more Boltzmann-type factors, which are describing a dissociation process that is proceeding through multiple potential minima, and where the dissociative pathway, therefore, will be dependent on the rate of force loading. Accordingly, a unique measurement of bond rupture may require an accurate description of the loading rate. A related effect would develop from the kinetics of molecular conformational change that is generally taking place during bond disruption. At the present time, such effects cannot be satisfactorily accounted for.

One of the defining experiments is shown in Fig. 15 (22). An amine-terminated polyethylene glycol spacer is first attached to the functionalized probe, followed by attachment of an antibody to the amine group. The antibody can then be 'towed' across a substrate that has been functionalized sparsely with the antigen. Simultaneous measurement of the lateral force component acting on the

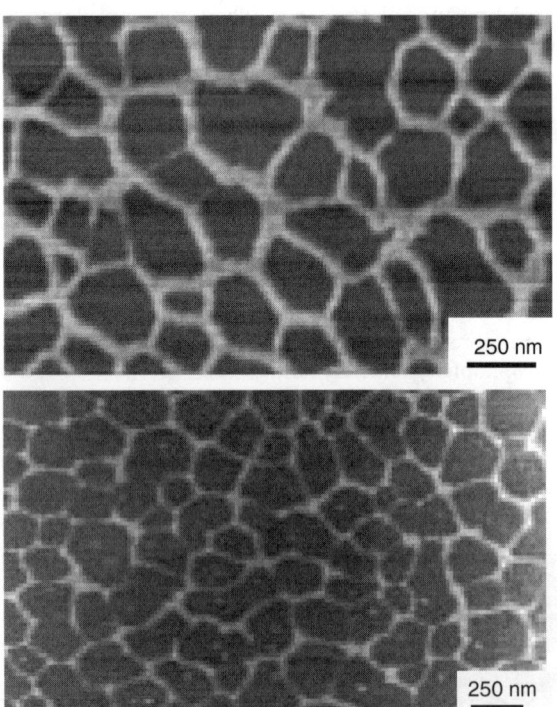

Figure 13. Self-assembled DNA networks on a mica substrate (poly(dG)·poly(dC)/5U where 1U = 50 ng/μl). Topography is revealed by tapping mode imaging (upper image), whereas the electrical conductivity is revealed by contact mode ECM (lower image) [Adapted from Cai et al. (34)].

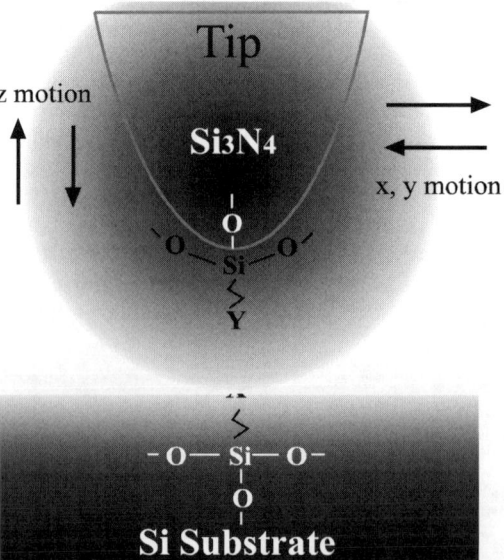

Figure 14. Schematic illustration of AFM-based measurement of biomolecular interactions. Tip and substrate are functionalized via silane coupling. A particular biomolecule is then attached to the tip, whereas a countermolecule is attached to the substrate.

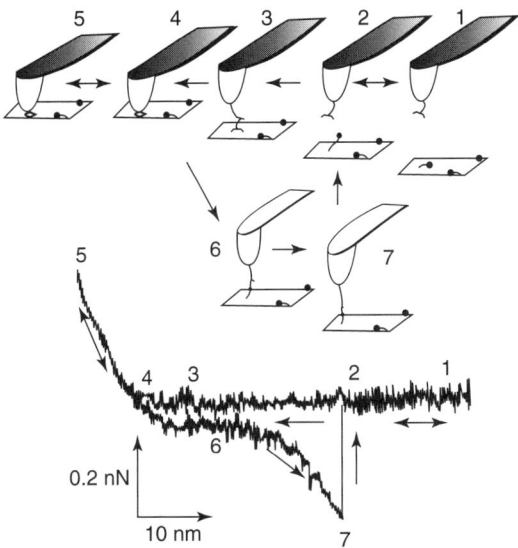

Figure 15. The schematic sequence shows (1–5) approach of the antibody and attachment to the antigen, followed by (6) retraction of the probe and stretching of the tether and bond. Finally, the antibody-antigen bond is ruptured at (7) because of the lever-imposed force of ca. 200 pN. Actual data in the form of an F-d curve are shown below [Adapted from Hinterdorfer et al. (22)].

lever will then allow localization in the x-y plane of single recognition events.

9. PROTEIN (UN)FOLDING

Each long-chain molecule has its own unique folded native structure; the structure will depend on the environment and on the location of the biomolecule in its functional life cycle. As a result of the enormous number of degrees of freedom, it is a daunting task to calculate its minimum energy configuration. Nevertheless, a polypeptide chain will find its folded global minimum energy configuration in a remarkably short time (Levinthal's paradox) (42,43). Currently favored explanations revolve around there being an identifiable directed pathway through the multidimensional potential landscape. One possible approach to gaining insight into the problem is to reverse-engineer the folding process, namely to induce unfolding.

It has recently been demonstrated that an AFM operated in the F-d mode can shed considerable light on the problem (44). A folded protein is attached to a substrate without being denatured. The probe tip, possibly functionalized, is then attached to a reactive site through trial and error. The probe is then withdrawn and a force is applied, causing extension initially, and then subsequently to cause a sequence of unfolding stages. A typical outcome is shown in Fig. 16 (44).

10. BIOSENSING BASED ON THE AFM PLATFORM

Biosensing based on molecular recognition has attracted attention and resources because of the potential high specificity and sensitivity. Few of the traditional technologies have proven to be compatible with the requirements of molecularly specific sensing. Thus, novel principles and technologies, such as those that are AFM-based, are being explored. However, cost, lack of stability, and degraded performance in the nonlaboratory environment have so far hampered commercialization (45).

The force-sensing lever is at the heart of the SPM system. It responds to external quasistatic forces through deflection, which can be detected routinely with a resolution of 10^{-6}–10^{-7} rad. Alternatively, it may respond to mass-loading through a change in its characteristic mechanical vibrational mode(s). Therein are the principal ingredients for a class of novel biosensors based on SPM technologies.

10.1. Lever as a Femtogram Microbalance Sensor

A stiff microfabricated lever will have an as-received lowest order resonance frequency of 10^5–10^6 Hz. Following functionalization on both sides, it will present specific receptor surface sites to species in solution. Selective biomolecular adsorption from solution will then add mass to the lever, causing a shift in the resonance frequency, detectable by phase-sensitive methods, and is given by

$$\Delta m = \frac{k}{4\pi^2}\left[\frac{1}{f_1^2} - \frac{1}{f_2^2}\right],$$

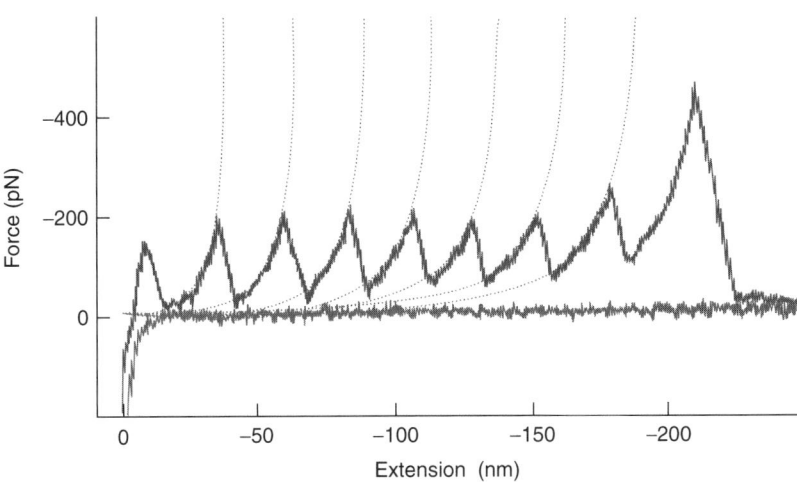

Figure 16. A force versus extension graph of an octameric TI 127 polyprotein being pulled by an AFM probe. The curve represents a sequence of continuous extension curves (rising portions of the trace) and sudden unfolding events [Adapted from Best and Clarke (44)]. (This figure is available in full color at http://www.mrw.interscience.wiley.com/ebe.)

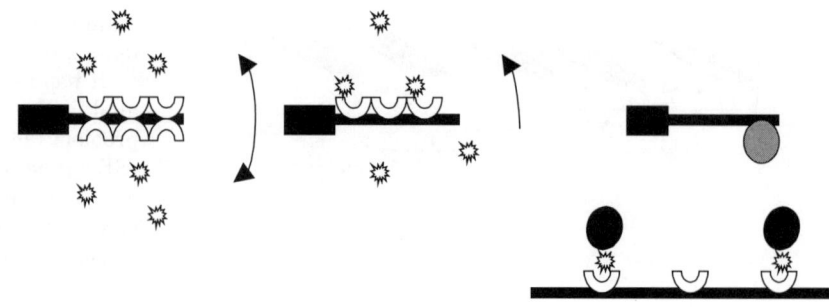

Figure 17. Schematic illustrations of SPM-based biosensor devices. The concepts are based on (a) adsorption of species in solution onto functionalized lever; (b) adsorption of species in solution onto one side of functionalized lever; (c) the use of magnetic interaction to image and selectively remove adsorbed species from a substrate, where the strength of magnetic force can be 'tuned' to that of a particular binding force.

(a) Change in resonance frequency with change in adsorbed mass

(b) change in deflection with change in mass

(c) Imaging and selective removal of adsorbed species with magnetic interaction

where k is the spring constant and f_1 and f_2 refer to the resonance frequency before and after addition of mass. The arrangement is illustrated in Fig. 17a. The response is quantifiable, and information on the kinetics of adsorption can be obtained. The system has been demonstrated as a humidity sensor (46).

10.2. Lever as a Detector of Shear Stress Deriving from Biomolecular Adsorption

The microfabricated lever is now functionalized on only one side. Thus, differential biomolecular adsorption from solution will occur on the side that presents the preferred receptor sites. The adsorbed species will generate a differential shear stress, causing bending of the lever. The change in shear stress is given by

$$\Delta\sigma = \frac{Et^2 \Delta d}{3(1-v)L^2},$$

where E is Young's modulus, t is lever thickness, Δd is lever deflection, v is Poisson's ratio, and L is lever length. The angular deflection, as a function of concentration of adsorbed species, can readily be measured with an optical lever method (as in the case of the AFM). The inferred sensitivity is on the order of a few hundred adsorption events. The arrangement is shown schematically in Fig. 17b. Several devices have been demonstrated, including an artificial nose (47) and a hydrogen/mercury array sensor (48).

10.3. Force-Sensing through Magnetic Dipole-Dipole Interaction

Sensor specificity and sensitivity can be obtained by reconfiguring the AFaM for detection of magnetic forces. The relevant biospecies are now attached to a functionalized super-paramagnetic microbead. The biomolecular recognition takes place at receptor sites on a functionalized substrate, where the species in solution have been adsorbed and immobilized by specific binding. The substrate is then imaged at a distance by a force-sensing lever with a ferromagnetic particle attached to its free end. The strength of magnetic dipole-dipole interaction depends on the separation of the imaging and stationary beads and manifests itself as a deflection of the lever. When the strength of magnetic interaction exceeds that of biomolecular binding, the bead will be removed from the substrate, and will disappear from the magnetic image. Thus, the arrangement makes available single-event sensitivity as well as quantitative measurement of the binding force. The system has been demonstrated in two recent studies (49,50). The arrangement is illustrated schematically in Fig. 17c.

11. ANALYSIS AND MANIPULATION OF NATURAL NANO-STRUCTURES

Naturally occurring nanostructures have so far not attracted much attention, but are rich sources of products that meet specifications imposed by natural selection. Although the pharmaceutical industry has long recognized the value of natural compounds, the emerging industries based on nanotechnology have so far made little use of 'free' technology that has been 'invented' over evolutionary time-scales in response to the imperatives of species adaption and survival. Characteristics of naturally occurring nanostructured arrays, produced by natural selection, ought to be of great scientific and technological relevance as these may provide ideal models or templates for manmade technologies and devices. Studies of this kind go back to the work of Bernhard et al. some 40 years ago (51). They observed that certain insects exhibited ordered hexagonal close-packed nm-size protuberances on their corneal surfaces. The authors suggested that these 'corneal nipple arrays' had an antireflective function for light in the visible spectrum. The optical properties were thought to act as a survival mechanism whereby insects could evade predators through improved camouflage. With the ready availability of high-resolution imaging techniques, such as the AFM, the field is now ready for exploitation.

An AFM contact mode image of a regular array of nanostructures on the dorsal and ventral wing sections of a cicada (*Psaltoda claripennis*) is shown in Fig. 18. The structures exhibit close-packed hexagonal structure with a spacing of ca. 225 nm, and a comparable peak-to-valley depth. The hypothetical function as an antireflection coating was tested by progressive removal of the structures.

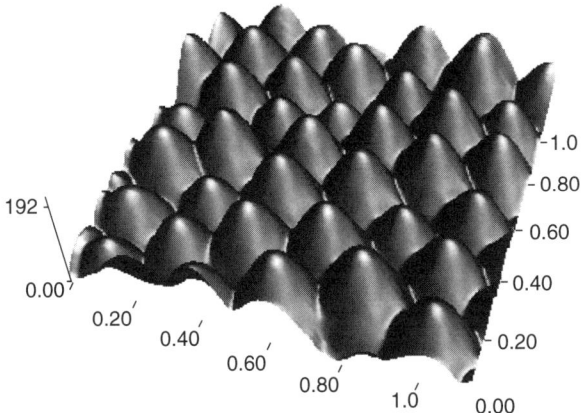

Figure 18. An AFM contact mode image shows a regular array of close-packed structures on the wings of a cicada. The spacing is ca. 250 nm. (This figure is available in full color at http://www.mrw.interscience.wiley.com/ebe.)

The AFM was then operated as a nanomachining tool whereby a stiff probe was rastered over a particular field of view at a force loading of ca. 1 μN. The effect was to remove the structures in a controlled manner. The change in reflectivity was monitored by measuring *in situ* the intensity of transmitted white light. The results are shown in Figs. 19 and 20 (52).

Evolution is an efficient process in the sense that particular survival attributes can sometimes meet more than one requirement. In the case of the arrays on the wings on cicadas, good evidence exists, from F-d analysis, for the surfaces being hydrophobic. Hydrophobicity, in combination with the geometry of the array, has the effect of producing a dirt-repellent surface, as well as being an example of natural stealth technology (52).

12. FUTURE PROSPECTS

Current SPM instrumentation is a great deal more versatile and user-friendly than in the past. Nevertheless, effective and efficient exploitation of the technique and of the methodologies requires considerable hands-on commitment and investment by the user. Although SPM has unique and impressive strengths, especially as a tool for nonintrusive interrogation of biosystems, it also has limitations and weaknesses. Future prospects are very much conditioned by the desire to build on those strengths while circumventing the limitations.

The dynamic response of current-generation instruments is essentially limited by the mechanical response of the lever. The principal resonance mode ranges from less than 10 khz (soft levers for contact mode imaging and F-d analysis) to ca. 500 khz (for noncontact and intermittent contact modes). Therein lies the limitation on rates of data acquisition. Moreover, the electronic and mechanical platform has been designed with that limitation in mind. It is likely that next-generation instruments will be based on levers with smaller physical dimensions, and thus higher resonance modes, and with matching characteristics for the mechanical and electronic platform. Thus, faster data acquisition rates will be available, allowing for better dynamical response and better noise rejection.

Most SPM instruments are currently stand-alone facilities (although UHV instruments are often integrated with related surface analytical techniques within a common envelope). Greater capability for integration of SPM with related and complementary techniques is undoubtedly a future trend. Indeed low-grade AFM stages are now available for attachment as a screw-on 'objective lens' for standard optical microscopy, and most biological SPM instruments are now purposefully designed for being mated to an inverted optical microscope. Increasing use of spectroscopic fingerprinting, as well as use of quantitative analysis, will most likely see a trend toward its integration with SPM methods, in particular with near-field scanning optical microscopy.

Nano/meso-scale interrogation of biosystems *in vitro* will place greater demand on control of the ambient environment. Although a great deal of work has been carried out on the design of special-purpose fluid cells and their

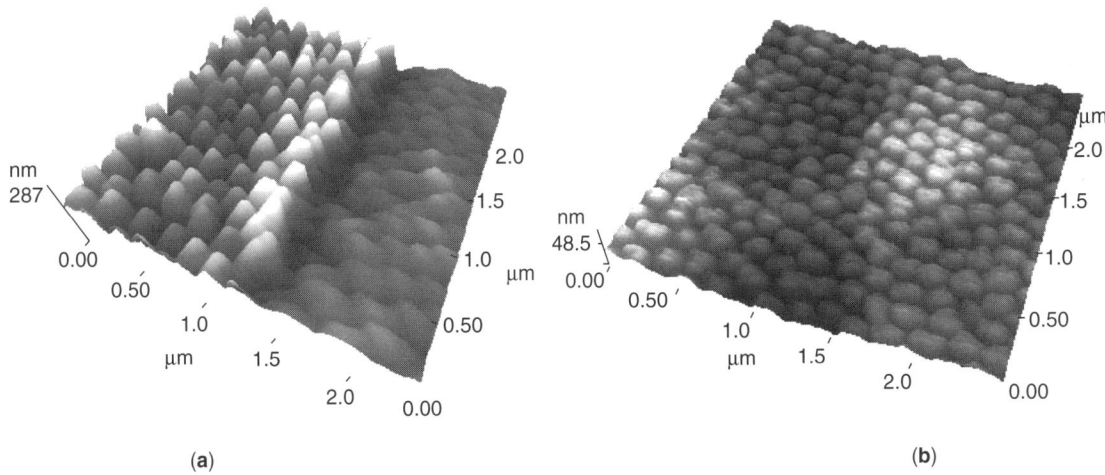

Figure 19. The images show the outcome of nanomachining by AFM the nipple array on a cicada wing. The image on the right shows the remaining parts of the structures after the top 200 nm has been removed.

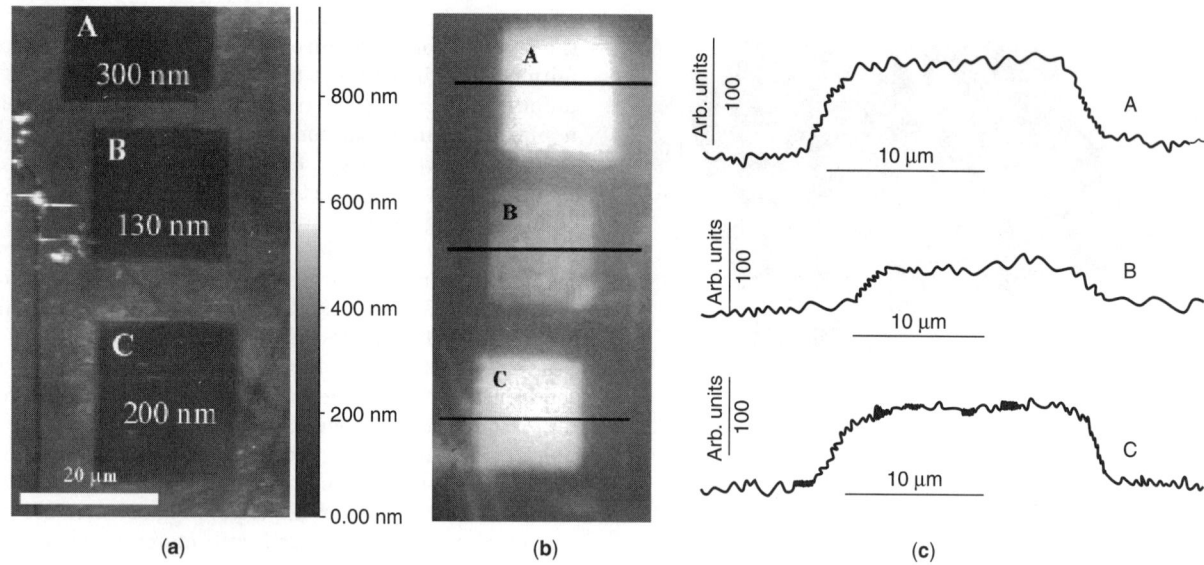

Figure 20. Demonstration of natural stealth technology. The data show that increasing extents of removal of the nipple array (A) are correlated with increasing transmission of white light (B and C).

integration with SPM, much remains to be done in order to improve the flexibility of such platforms.

Most of the mainstream instruments come with well-developed software for instrument control, data acquisition, and image processing. However, additional development of software packages for data processing, especially for lateral force and F-d analysis, will enhance and facilitate interpretation of primary data.

In conclusion, it is clear that SPM has contributed greatly to our understanding of nano- and meso-scale structure, chemistry, and processes relevant to surface and interface science and technology. In comparison with instrumentation that was available at the beginning of the 1990s, we have come a long way. SPM is no longer an esoteric technique for a few dedicated specialists, but rather a mainstream tool-of-the-trade for thousands of practitioners around the world. The exponential growth phase for SPM is most likely nearing the end, but one can confidently predict that a steep linear growth curve will take us into the future.

Acknowledgements

Some of the work was funded in part by the Australian Research Council and by the US DOD DARPA. The work and insight described above is the result of the diligent and creative efforts of all past and present members and colleagues of the SPM Group at Griffith University and at other institutions.

BIBLIOGRAPHY

1. G. Binnig, C. F. Quate, and C. Gerber, Atomic force microscope. *Phys. Rev. Lett.* 1986; **56**:930–933.
2. G. Binnig, H. Röhrer, C. Gerber, and E. Weibel, Surface studies by scanning tunnelling microscopy. *Phys. Rev. Lett.* 1982; **49**:57–61.
3. See two special issues of *IBM J. Res. Develop.* 1986; **30**:July/September.
4. S. Myhra, Introduction to scanning probe microscopy. In: J. C. Rivière and S. Myhra, eds., *Handbook of Surface and Interface Analysis*. New York: Dekker, 1998.
5. D. Sarid, *Scanning Force Microscopy*. New York: Oxford University Press, 1991.
6. C. T. Gibson, G. S. Watson, and S. Myhra, Scanning force microscopy - calibrative procedures for 'best practice'. *Scanning* 1997; **19**:564–581.
7. C. A. J. Putman, K. O. van der Werf, B. G. de Grooth, N. F. van Hulst, and J. Greve, Viscoelasticity of living cells allows high resolution imaging by tapping mode atomic force microscopy. *Biophys. J.* 1994; **67**:1749–1753.
8. C. Le Grimelec, E. Lesniewska, M.-C. Giocondi, E. Finot, and J.-P. Goudonnet, Simultaneous imaging of the surface and submembraneous cytoskeleton in living cells by tapping mode atomic force microscopy, *Academie des Sciences Biophysique* 1997; **320**:637–643.
9. V. Vié, M.-C. Giocondi, E. Lesniewska, E. Finot, J.-P. Goudonnet, and C. Le Grimelec, Tapping-mode atomic force microscopy on intact cells: optimal adjustment of tapping conditions by using the deflection signal. *Ultramicroscopy* 2000; **82**: 279–288.
10. G. R. Bushell, C. Cahill, S. Myhra, and G. S. Watson, Analysis of human fibroblasts by atomic force microscopy. In: P. C. Braga and D. Ricci, eds., *Methods in Molecular Biology*, vol. 244. Totowa, NJ: Humana Press, 2003.
11. C. T. Gibson, G. S. Watson, and S. Myhra, Lateral force microscopy—quantitative procedures. *Wear* 1997; **213**: 72–77.
12. R. W. Carpick and M. Salmeron, Scratching the surface: fundamental investigations of tribology and atomic force microscopy. *Chem. Rev.* 1997; **97**:1163–1194.
13. A. Noy, C. D. Frisbie, L. F. Rozsnyai, M. S. Wrighton, and C. M. Lieber, Chemical force microscopy: exploiting chemically-modified tips to quantify adhesion, friction, and functional group distributions in molecular assemblies. *J. A. Chem. Soc.* 1995; **117**:7943–7951.

14. W.-K. Li, Surface morphology of immiscible polymer blend using scanning force microscopy. *Polymer Testing* 2004; **23**:101–105.
15. N. A. Burnham and R. J. Colton, Measuring the nanomechanical properties and surface force of materials using an atomic force microscope. *J. Vac. Sci. Technol. B* 1989; **7**:2906–2913.
16. G. S. Watson, J. A. Blach, C. Cahill, S. Myhra, D. V. Nicolau, D. K. Pham, and J. Wright, Poly(amino acids) at Si-oxide interfaces – bio-colloidal interactions, adhesion and 'conformation'. *Coll. Polym. Sci.* 2003; **282**:56–64.
17. J. Zlatanova, S. M. Lindsay, and S. H. Leuba, Single molecule force spectroscopy in biology using atomic force microscopy. *Prog. Biophys. Mol. Biol.* 2000; **74**:37–61.
18. O. H. Willemsen, M. M. E. Snel, A. Cambi, B. G. de Groot, and C. G. Figdor, Biomolecular interactions measured by atomic force microscopy. *Biophys. J.* 2000; **79**:3267–3281.
19. S. W. Wong, E. Joselevich, A. T. Woolly, C. L. Cheung, and C. M. Lieber, Covalently functionalized nanotubes as nanometre-sized probes in chemistry and biology. *Nature* 1998; **394**:52–55.
20. A. Ulman, ed., *Self-Assembled Monolayers of Thiols*. San Diego, CA: Academic Press, 1998, and references cited therein.
21. D. F. Siqueira Petri, G. Wenz, P. Schunk, and T. Schimmel, An improved method for the assembly of amino-terminated monolayers on SiO_2 and the vapour deposition of gold layers. *Langmuir* 1999; **15**:4520–4523, and references cited therein.
22. P. Hinterdorfer, W. Baumgartner, H. J. Gruber, K. Schilcher, and H. Schindler, Detection and localization of individual antibody-antigen recognition events by atomic force microscopy. *Proc. Natl. Acad. Sci. USA* 1996; **93**:3477–3481.
23. E. Henderson, P. G. Haydon, and D. S. Sakaguchi, Actin filament dynamics in living glial cells imaged by atomic force microscopy. *Science* 1992; **257**:1944–1946.
24. V. Parpura, P. G. Haydon, and E. Henderson, Three-dimensional imaging of living neurons and glia with the atomic force microscope. *J. Cell Sci.* 1993; **104**:427–432.
25. G. R. Bushell, C. Cahill, F. M. Clarke, C. T. Gibson, S. Myhra, and G. S. Watson, Imaging and force-distance analysis of human fibroblasts *in vitro* by atomic force microscopy. *Cytometry* 1999; **36**:254–264.
26. F. Braet, C. Saynaeve, R. de Zanger, and E. Wisse, Imaging surface and submembraneous structures with the atomic force microscope: a study on living cancer cells, fibroblasts and macrophages. *J. Microsc.* 1998; **190**:328–338.
27. C. Rotsch and M. Radmacher, Drug-induced changes of cytoskeletal structure and mechanics in fibroblasts: an atomic force microscopy study, *Biophys. J.* 2000; **78**:520–535.
28. S. G. Shroff, D. R. Saner, and R. Lal, Dynamic micromechanical properties of cultured rat atrial myocytes measured by atomic force microscopy. *Am. J. Physiol.* 1995; **269**:C286–C292.
29. J. Domke, W. J. Parak, M. George, H. E. Gaub, and M. Radmacher, Mapping the mechanical pulse of single cardiomyocytes with the atomic force microscope. *Eur. Biophys. J.* 1999; **28**:179–186.
30. J. A. A. Crossley, C. T. Gibson, L. D. Mapledoram, M. G. Huson, S. Myhra, D. K. Pham, C. J. Sofield, P. S. Turner, and G. S. Watson, Atomic force microscopy analysis of wool fibre surfaces in air and under water. *Micron* 2000; **31**:659–667.
31. J. Blach, W. Loughlin, G. Watson, and S. Myhra, Surface characterization of human hair by atomic force microscopy in the imaging and F-d modes, *Intl. J. Cosm. Sci.* 2000; **23**:165–174.
32. C. D. Frisbie, L. F. Rozsnyai, A. Noy, M. S. Wrighton, and C. M. Lieber, Functional group imaging by chemical force microscopy. *Science* 1994; **265**:2071–2074.
33. E. Winfree, F. Liu, L. A. Wenzler, and N. C. Seeman, Design and self-assembly of two-dimensional DNA crystals. *Nature* 1998; **394**:539–544.
34. L. Cai, H. Tabata, and T. Kawai, Self-assembled DNA networks and their electrical conductivity. *Appl. Phys. Lett.* 2000; **77**:3105–3106.
35. T. Yoshino, S. Sugiyama, S. Hagiwara, D. Fukushi, M. Shichiri, H. Nakao, J.-M. Kim, T. Hirose, H. Muramatsu, and T. Ohtani, Nano-scale imaging of chromosomes and DNA by scanning near-field optical/atomic force microscopy. *Ultramicroscopy* 2003; **97**:81–87.
36. H. Nakao, H. Hayashi, T. Yoshino, S. Sugiyama, K. Otobe, and T. Ohtani, Development of novel polymer-coated substrates for straightening and fixing DNA. *Nanoletters* 2002; **2**:475–479.
37. H. Takano, J. R. Kenseth, S.-S. Wong, J. C. O'Brien, and M. D. Porter, Chemical and biochemical analysis using scanning force microscopy. *Chem. Rev.* 1999; **99**:2845–2890.
38. J. N. Israelachvili, *Intermolecular and Surface Forces*. New York: Academic Press, 1992.
39. B. V. Derjaguin and L. Landau, Theory of the stability of strongly charged lyophobic sols and of the adhesion of strongly charged particles in solutions of electrolytes. *Acta Physicochem. URSS* 1941; **41**:633–662.
40. E. J. W. Verway and J. Th. G. Overbeek, In: *Theory of the Stability of Lycophobic Colloids*. New York: Elsevier, 1948.
41. J. A. Fritz, G. Katopodis, F. Kolbinger, and D. Anselmetti, Force mediated kinetics of single P. selectin/ligand complexes observed by atomic force microscopy. *Proc. Natl. Acad. Sci. USA* 1998; **95**:12283–12288.
42. C. Levinthal, Are there pathways for protein folding? *J. Chim. Phys.* 1968; **65**:44–45.
43. U. Mayor, C. M. Johnson, V. Daggett, and A. R. Fersht, Protein folding and unfolding in microseconds to nanoseconds by experiment and simulation. *Proc. Natl. Acad. Sci. USA* 2000; **97**:13518–13522.
44. R. B. Best and J. Clarke, What can atomic force microscopy tell us about protein folding? *Chem. Commun.* 2002: 183–192.
45. C. R. Lowe, Chemoselective biosensors. *Curr. Opin. Chem. Biol.* 1999; **3**:106–111.
46. R. Berger, Ch. Gerber, H. P. Lang, and J. K. Gimzewski, Micromechanics: a toolbox for femtoscale science: 'towards a laboratory on a tip'. *Microelectronic Eng.* 1997; **35**:373–379.
47. H. P. Lang, M. K. Baller, R. Berger, Ch. Gerber, J. K. Gimzewski, F. M. Battiston, P. Fornaro, J. P. Ramseyer, E. Meyer, and H. J. Güntherodt, An artificial nose based on a micromechanical cantilever array. *Anal. Chim. Acta* 1999; **393**:59–65.
48. C. L. Britton, R. L. Jones, P. I. Oden, Z. Hu, R. J. Warmack, S. F. Smith, W. L. Bryan, and J. M. Rochelle, Multiple-input microcantilever sensors. *Ultramicroscopy* 2000; **82**:17–21.
49. D. R. Baselt, G. U. Lee, M. Natesan, S. W. Metzger, P. E. Sheehan, and R. J. Colton, A biosensor based on magnetoresistance technology. *Biosensors and Bioelectronics*, 1998; **13**:731–739.
50. R. G. Rudnitsky, E. M. Chow, and T. W. Kenny, Rapid biochemical detection and differentiation with magnetic force microscope cantilever arrays. *Sensors and Actuators A* 2000; **83**:256–262.

51. C. G. Bernhard, W. H. Miller, and A. R. Møller, The insect corneal nipple array. A biological broad-band impedance transformer that acts as an anitireflection coating. *Acta Physiol. Scand.* 1965; **63**:1–79.
52. G. S. Watson, Unpublished results. Nathan Australia: Griffith University, 2005.

ATRIAL FIBRILLATION AND ATRIAL FLUTTER

FERNANDO HORNERO
Hospital General Universitari
Valencia, Spain

1. ATRIAL FIBRILLATION AND FLUTTER

Normal heart activation is characterized by regular alternation between depolarization/repolarization and rest. Activation originates in the right atrium, around the sinus node area, and spreads until the atria are completely activated. Once activation is complete, atrial tissue is refractory and a period of rest necessarily precedes the next activation cycle. This sequence is caused by pacemaker cells with discharge rates that respond to neural and humoral stimuli, thus allowing adaptation of heart function to physiologic demands. However, this normal sequence is not so simple to sustain, as demonstrated by the frequent appearance of reentrant tachycardias. Rapid, complete, uniform activation of all atrial tissue is important for rhythm stability. Preferential conduction pathways have been long recognized in the atria, despite the absence of bundles of specialized conduction like the His–Purkinje network of the ventricles.

An electrocardiogram (EKG) is the electrical representation of the depolarization/repolarization of the myocardial cells, that is to say of the mechanical process of contraction of the chambers of the heart. Electrocardiographically, a cardiac cycle begins in the sinoatrial node, spreads to the auricles (wave P in the EKG), and finally crosses the atrioventricular node and goes on to the ventricles (complex QRS in the EKG). Sinus rhythm is the normal physiological rhythm of the heart (Fig. 1). Arrhythmia is any heart rhythm that is not the sinus rhythm.

Arrhythmias can originate in the atria (supraventricular arrhythmias) or in the ventricles of the heart (ventricular arrhythmias). Atrial fibrillation and atrial flutter are two types of supraventricular arrhythmia very frequent in clinical practice. All arrhythmias are diagnosed by means of EKG, attending to morphological and regularity criteria. An EKG can be obtained from the skin of the patient or from inside the heart by means of catheters (electrograms).

2. ATRIAL FIBRILLATION

Atrial fibrillation (AF) is the most frequent arrhythmia, affecting more than 5 million people in the United States, Europe, and Japan, including more than 2 million in the United States alone (1). Its medical management and treatment are still unsatisfactory, which explains the increased interest and numerous clinical investigations at present. AF is one of the potentially large cardiovascular markets for which medical devices and drugs that provide effective cures or treatment still do not exist.

2.1. Concept

AF is a type of atrial arrhythmia that is defined electrocardiographically as a fast atrial rhythm (between 400 and 700 beats/min) that is chaotic and disorganized, without the capacity to generate effective atrial contractions. AF is easily recognizable in a conventional EKG as an irregular ventricular rhythm, with an electric baseline in which it is difficult to distinguish electrical atrial activity (Fig. 2). AF is characterized electrocardiographically by two criteria:

1. The presence of f waves that replace the P sinoatrial wave, and
2. Irregular ventricular rhythm, because of the different degrees of penetration of the f waves in the atrioventricular node. All f waves do not cross the atrioventricular node; most of them are blocked in different ways.

2.2. Prevalence

In the general population, AF frequency increases slightly with age. Its prevalence is less than 0.5% up to 50 years

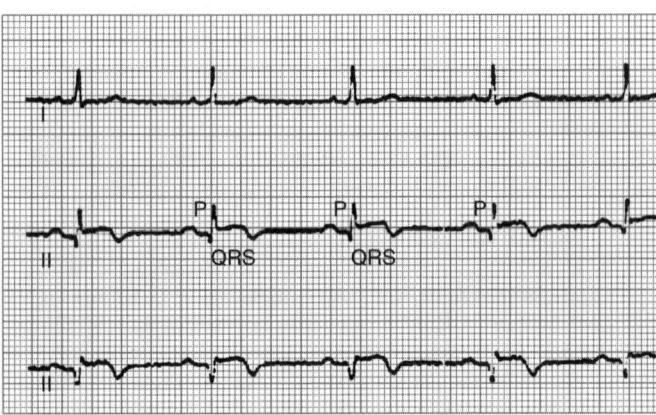

Figure 1. Normal sinus rhythm record in EKG leads I-II-II, with P waves and QRS ventricular complexes.

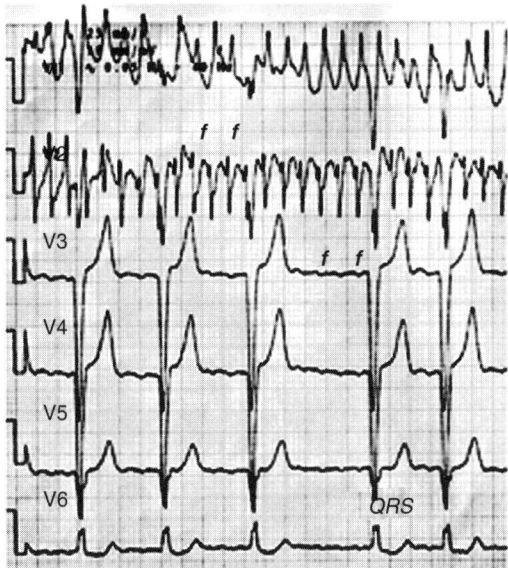

Figure 2. Representative example of EKG recordings during atrial fibrillation epicardial (heart surface) and surface electrograms. At the top, two recordings of epicardial electrograms obtained directly from heart surface; auricular rate is between 330 and 315 beats/min. The rest of the leads correspond with electrograms recorded from surface of the patient, with ventricular rate between 58 and 100 beats/min. Electrical activity of the atrium is detected electrocardiographically as small irregular baseline undulations of variable amplitude and morphology, called f waves, at a rate of 350 to 600 beats/min. The ventricular response is grossly irregular.

of age, reaches 3% at 75, and affects up to 9% of the population over 80 years old (2–4). It has been demonstrated that 4% of the cardiac disorders and 40% of the patients with heart failure have AF. On the other hand, it is very common in patients with mitral valve stenosis, with a prevalence of up to 80% in people over 60 years old (5). It is also frequent in patients with Wolf–Parkinson–White syndrome (preexcitation syndrome).

2.3. Pathogenic Mechanism—Electrophysiological Substrate

The mechanisms of AF have been the subject of speculation for many years. Two theories or mechanisms have been classically proposed: (1) increase of automatism, involving one or more fast discharging automatic focuses, and (2) a reentry mechanism, involving one or more circuits in the atria. Recently, other possible mechanisms have been described as the source of AF, such as rotors, which are a spiral wave turning without a central obstacle, but on the basis of excitability gradients that make activation advance slower in the center than at the periphery of the rotor (6). Anatomic or functional reentry mechanisms, ectopic atrial triggers, and possibly new vortex concepts also integrate ideas of 3D activation. At present, it is understood that AF could respond to different mechanisms simultaneously, and these mechanisms are not necessarily mutually exclusive. Atrial mapping of AF in experimental models and in studies on human using cartography systems are fundamental for understanding the pathophysiological mechanism of this arrhythmia.

2.4. Reentry Mechanism—The Multiple Wavelet Theory

Reentry is the mechanism of tachycardia in supraventricular tachycardias, such as AF and atrial flutter. A fixed anatomic obstacle can support the reentrant circuit but is not an essential requirement for all forms of reentries. In functionally determined reentry, propagation occurs through relatively refractory tissue without predetermined circuit and an absence of a fully excitable gap exists. Several requirements exist for the initiation and maintenance of this type of reentry. Initiation of a circular movement tachycardia requires a unidirectional conduction block in one limb of a circuit. Unidirectional block may occur as a result of acceleration of the heart rate or block of a premature impulse that impinges on the refractory period of the pathway. Slow conduction is usually required for both initiation and maintenance of a circus movement tachycardia.

In 1959, Moe and Abildskov proposed the multiple wavelet theory that the perpetuation of AF depends on the continuous and random propagation of various individual wavelets through the atria (7). The grossly irregular wavefront becomes fractionated as it divides about islets or strands of refractory tissue, and each of the daughter wavelets may now be considered as independent offspring. Such a wavelet may accelerate or decelerate as it encounters tissue in a more or less advanced state of recovery. It may become extinguished as it encounters refractory tissue; it may divide again or combine with a neighbor; it may be expected to fluctuate in size and change in direction. Its course, although determined by excitability or refractoriness of surrounding tissue, would appear to be as random as Brownian motion. Fully developed fibrillation would then be a state in which many such randomly wandering wavelets coexist (7). A few years later, Moe et al. (8) confirmed this hypothesis using a computer model, and noted the importance of a critical atrial mass and a short refractory period in the sustainability of this arrhythmia.

Later, an experimental study in dogs by Allessie et al. (9) provided evidence of the reentrant nature of AF, with the estimation that a critical number of three–six simultaneous wavelets was required to maintain this arrhythmia. In 1991, Cox et al. (10), using high-density epicardial mapping of the free wall of the right atrium in patients with Wolff–Parkinson–White syndrome, confirmed the presence during AF of multiple wavelets fleeting in appearance and location. Microreentry or focal automaticity was not observed. The importance of multiple wandering wavelets in perpetuation of AF was confirmed by the successful development of a surgical treatment of AF, named Maze procedure (11). In fact, multiple atrial incisions, dividing the atria into smaller segments, interrupt all possible reentrant wavelets.

The wavelength concept is defined as the product of refractory period and conduction velocity. Therefore, it expresses the distance traveled by the depolarization wave during the refractory period. Allessie et al. (9) confirmed this concept by using epicardial electrodes in dogs, demonstrating that the slower the conduction velocity and the shorter the refractory period, the more likely it is that reentry will occur, and very short wavelengths facilitate more complex forms of reentry, such as fibrillation versus flutter. As for AF a critical number of wandering wavelets are required, the wavelength is important for perpetuation of fibrillation. If the wavelength is relatively long, a fewer number of waves can circulate through the atria, and AF tends to terminate spontaneously. Conversely, short tissue wavelength tends to favor the onset and perpetuation of AF and may be caused by fibrosis and inflammation (decreasing conduction velocity), increased parasympathetic activity and thyrotoxicosis (shortening the refractory period), ischemia (decreasing both), and stretch, which in some experimental studies has been shown to shorten refractoriness.

Garrey hypothesized almost a century ago that a critical mass existed, above which AF was sustained and below which it could be prevented (12). In 2005, Byrd et al. supports this hypothesis (13), increasing tissue surface area correlating significantly with the probability of sustaining AF. The concept of critical mass expresses that for maintenance of AF, a critical mass of myocardial tissue is required, as larger tissue masses allow a greater space available for the wavelets to circulate. Direct proof that perpetuation of AF is dependent on a critical mass is provided by the fact that in different animal species larger hearts fibrillate longer than small hearts (14) and that, in a given animal, AF is less stable than ventricular fibrillation. In humans, atrial size has long been known to be critical in the ability to generate AF, and it has been shown to correlate with increased vulnerability. The importance of atrial enlargement may explain the propensity for AF to occur in valvular disease and cardiac failure.

In patients with AF, an increased dispersion of refractoriness has been observed (15). This heterogeneity of refractoriness may provide the setting for unidirectional block when an extrasystole developing in a zone with short refractory periods fails to excite an area with long refractory periods. This increased dispersion of refractoriness may be the substrate for the enhanced inducibility and spontaneous occurrence of AF.

All these concepts must be integrated together, and the probability of sustained AF depends of the increasing tissue area, width, and weight and decreasing effective refractory periods and wavelengths.

2.5. AF Because of a Rapidly Firing Focus

Recently, clinical and experimental studies (16–19) have shed light on the electrophysiological mechanisms responsible for the onset of AF. To date, on the basis of the results of these studies, one can schematically distinguish three different types of AF initiation. First, AF can occur in patients who have other forms of supraventricular tachycardia. The disappearance of AF in most patients after radio-frequency ablation of these tachycardias confirms their role in AF onset (tachycardia-induced tachycardia). The mechanism involved may be related to the increase in atrial pressure caused by the occurrence of atrial systole during tachycardia when the atrioventricular valves are closed or at least not fully open. The augmentation of atrial pressure causes atrial stretch, which prolongs atrial refractoriness (15) and, more importantly, increases atrial dispersion of refractoriness (20). A second interesting mode of AF initiation (focal AF) has recently been identified by the Bordeaux group (18,21). In this group of patients, the initiation and maintenance of AF is because, as confirmed by radio-frequency ablation, of a rapidly discharging focal tachycardia that is underlying and apparently driving the AF. Patients with focal AF are typically younger, without structural heart disease, with frequent runs of atrial tachycardia that may degenerate in episodes of AF. The pulmonary veins, particularly the left and right superior veins, are the source of these focal drivers in the majority of cases (22). The predominant distribution of foci in the superior veins matches the prevalent extension of myocardial sleeves over the orifices of these structures. The mechanism underlying the focal arrhythmia may be abnormal automaticity or triggered activity because this type of arrhythmia could not be induced by programmed electrical stimulation. These clinical observations seem to confirm the original hypothesis by Sherf (23) that AF may result from a single focus firing at such a rapid rate that the remainder of the atria cannot follow synchronously. The third and, probably, the most frequent mode of onset of AF is caused by single or multiple extrasystoles, more often particularly earlier (P wave on T wave) or developing in critical areas. Haissaguerre et al. (21) showed that in most patients the pulmonary veins are the source of the premature beats triggering AF. Thus, in the pulmonary veins may be present either a focal driver, continually firing during a paroxysm of AF so that when firing stops, AF ceases, or a focal trigger, with single or multiple extrasystoles favoring reentrant beats and thus the onset of AF. In a minority of patients, the ectopic beats initiating AF were found in the right atrium along the crista terminalis and close to the coronary sinus ostium (18,21).

2.6. Spiral Wave Theory

The development of the spiral wave theory has provided a new model for the fundamental properties of cardiac reentry (6). According to this theory, reentry is maintained through the ability of circulating spiral waves to perpetuate in media with sufficient excitability to support the angle of spiral curvature. Evidence has been provided for a nonactivated, excitable core at the center of reentry circuits during AF in contrast to the constantly excited and refractory tissue predicted by leading circle theory.

2.7. Computer Modeling of AF

Theoretical models are a powerful tool for studying and predicting electrophysiological aspects of the AF, created by different designs under different tissue characteristics or environment conditions (24–27).

Several computer models of electrical propagation in the atria have been developed recently (28–30). These models differ in the accuracy of the representation of electrophysiologic and anatomic details. To simulate signals with sufficient duration for appropriate analysis, a computer model of the human atria was developed by Virag et al. in which the geometry, derived from magnetic resonance imaging data, was represented by a 3D monolayer (28). Other work using a 2D computer model of AF is based on realistic cellular ionic properties and reproduces atrial electrical dysfunction that provides a favorable substrate for the arrhythmias (31).

Models of cardiac ablation have been proposed and some experimentally validated (24,26). The finite element method specifically allows the calculation of the temperature distribution by solving the bioheat equation. This method has been previously used to study, for example, the heating pattern of long electrodes (32), the effect of the dispersive electrode location during a radio-frequency cardiac ablation (33), to investigate the thermal-electrical behavior of atrial tissue during radio-frequency heating (25), and to understand the evolution of temperature distributions in the esophagus during percutaneous catheter ablation (34).

2.8. Etiology

AF can appear in almost all cardiac disorders, with rheumatic valvulophaty, especially in mitral stenosis (it represents 20–30% of the total AF); atherosclerotic cardiopathy (30–40% of total AF); systemic arterial hypertension (10% of the total); and hyperthyroidism (3% of the total) as the most frequent. On occasions, the arrhythmia is a short crisis of paroxysmal AF, generally in the absence of organic cardiac disease in relation with dysfunctions of the autonomous vague system or sympathicotonie, ingestion of alcoholic drinks, extreme physical effort, or as a consequence of a paroxysmal supraventricular tachycardia.

2.9. Clinical

The symptoms are those of fast and irregular heart rhythm. Besides the sensation of palpitations, it can show precordial pain (angina), paroxysmal dyspnea, or signs of low cerebral output. The undesirable effects of AF are in relation with: (1) fast and irregular heart rhythm that can originate precordial pain, signs of low output, and restoration or aggravation of heart failure; (2) absence of atrial contraction, which may be serious in patients with diastolic ventricular dysfunction, and (3) presentation of embolisms, especially systemic, more often in valvular patients than in other etiologies.

2.10. Classification

AF appears in many clinical and electrocardiographic forms, and happens in the presence or absence of cardiac disease. It is therefore difficult to establish an agreed terminology to classify it.

AF appears in three clinical forms:

- Sporadic. This form occurs in cases of isolated and autolimited AF, unleashed by an intermittent cause, generally extracardiac (alcohol, vagotonia, hyperthyroidism, etc.), generally in patients without cardiac diseases, and corresponds to approximately 10% of AF cases.
- Paroxysmal. This form occurs in approximately 25% of cases. It includes patients with recurrent paroxysms of AF (sometimes several daily episodes) with or without cardiac diseases, and can be present for many years. Cases of AF related to dysfunctions of the autonomous system are included in this form. In those of vagal origin, the crisis begins coinciding with a slowing heart rate and normally appears during the night, after a period of rest or during digestion. In 75% of cases, AF alternates with atrial flutter. In cases of sympathicotonie origin, the episodes coincide with an acceleration of the heart rate and normally appear by day, during physical exercise, or when experiencing emotions, but not during rest. In these cases, AF alternates more with atrial tachycardias than with atrial flutter. According to Gotfredsen's report (35), a third of the paroxysmal AF cases turn into chronic AF in 27 years of follow-up.
- Chronic. This form consists of 65% of AF cases. Normally, this type occurs in valvular and atherosclerotic patients, although on occasion in the absence of apparent cardiac diseases, especially in the elderly.

Sopher and Camm (17) have proposed a classification of AF, by the name of "the three P," based on the time of its evolution and stability: *P*aroxysmal AF when the duration is short (generally less than two days) and the arrhythmia disappears spontaneously or after administration of a drug; *P*ersistent AF when it lasts more than two days, and in general a solution is obtained more by electrical cardioversion than with medication; and *P*ermanent or chronic AF when the attempts to reestablish sinoatrial rhythm have failed and the patient is considered to be a chronic case.

The most recent classification of the American College of Cardiology/American Heart Association/European Society of Cardiology defined AF as permanent or chronic when the arrhythmia is considered to be irreversible and the habitual measures have failed to achieve cardioversion (36). A period of 7 days is proposed to differentiate the paroxysmal form (<7 days) from the permanent (>7 days) (37). Therefore, paroxysmal AF is defined as arrhythmia in episodes of less than 7 days duration that includes periods of sinus rhythm. Persistent AF is arrhythmia that needs specific intervention to reestablish sinus rhythm. Paroxysmal and persistent AF can be recurrent in time, until the form of continuous or permanent AF is reached.

Electrophysiological atrial mapping has been used by different authors to differentiate types of AF. Wells (38,39) identified four types with the help of bipolar atrial electrograms:

- In type I AF, the atrial wall was activated by a single wavefront propagating uniformity without significant conduction delay.
- During type II AF, the activation patterns were characterized either by a single wavefront associated with areas of slow conduction or multiple lines of block of conduction, or by the presence of two wavelets.

- In type III AF, the atrial wall was activated by three or more wavelets associated with areas of slow conduction and multiple lines of conduction block.
- In type IV, electrogram alters fragments type II and III.

2.11. Electrocardiographic Diagnosis

AF is characterized by two principal factors: the presence of f wave that replace the P sinus wave, and irregular ventricular response, because of the different degrees of penetration of the f waves in the atrioventricular node (Fig. 2). The f waves, whose frequency ranges between 400 and 700 per min, have changeable amplitude and morphology, being more visible in V1 and V2. Cases with small f waves (<1 mm of height) are fundamentally because of atrial fibrosis; in some cases with severe fibrosis, the f waves can completely disappear. When the atrial activation is organized but irregular, it corresponds to an intermediate situation between flutter and fibrillation (fibrillo-flutter or *coarse* fibrillation).

2.11.1. QRS Complex and Types of Ventricular Response.
The morphology of the QRS complexes of the driven f waves is usually similar to that which exists in sinoatrial rhythm. The frequency of spontaneous ventricular rate is usually high (from 120 to 160 per min) under normal conditions. Low frequencies are shown in patients that take drugs.

2.12. Treatment of AF

In treating patients with AF, decision-making is complicated by the wide spectrum of clinical presentation of the arrhythmia and the tremendous heterogeneity of the afflicted population. Currently, pharmacologic treatment is the first line of therapy. Drugs with a negative cronotropic response on the atrioventricular node are used for ventricular rate control; cardioversion or antiarrhythmic agents are used for restoration or maintenance of sinus rhythm. Whether it is preferable to treat patients with AF with rate control alone vs repeated attempts to restore and maintain sinus rhythm remains controversial (40). Drug treatment involves medications that restore normal sinus rhythm (including procainamide, quinidine, disopyramide, and amiodarone); drugs that control the ventricular pace, which is associated with the rapid beating felt by some patients (including beta blockers, calcium channel blockers, and digoxin); and anticoagulants with warfarin and antiplatelet to reduce the risk of blood clot formation and stroke associated with AF. Patient responses to these drug regimens vary widely and several of the medications carry significant side effects, whereas none cure the underlying problem. Given these shortcomings of pharmacologic therapy, major interest exists in developing nonpharmacologic treatment options for the management of AF.

Ablative therapy for the management of arrhythmias is based on the observation that most arrhythmias develop from a focal origin or are critically dependent on conduction through a defined anatomic structure. If those critical regions are irreversibly damaged or destroyed, then the arrhythmia should no longer occur spontaneously or with provocation. Focal endocardial injury could be achieved by a controlled delivery of destructive energy through a catheter. Ablation generally involves an electrophysiologist using a mapping procedure to identify certain problematic electrical points or foci that are causing the arrhythmia. Then the physician directs energy (generally radio-frequency generated) through catheters to destroy the tissue that is the source of the arrhythmia.

The only current effective method of curing AF is a surgical procedure called the Maze procedure. It was developed approximately 10 years ago and involves resecting the atria. The concept is to prevent a critical mass of contiguous atrial tissue to sustain the AF. Strategic placement of incisions in both atria, as a "maze," stops the formation and conduction of reentrant circuits or ectopic electrical impulses, and channels the normal electrical impulse in one direction from the sinus node to the atrioventricular node. Scar tissue generated by the incisions permanently blocks the abnormal paths that cause the AF, thus eradicating the arrhythmia. However, Maze procedure is so lengthy, complex, and traumatic for patients that few of the procedures are performed.

Different alternative therapies can be used in the clinical practice.

2.13. Therapies for Ventricular Rate Control

The goal of nonpharmacologic therapies designed to slow the ventricular rate during AF is to modify normal atrioventricular conduction by targeting the His bundle, the atrioventricular node, or the inputs to the atrioventricular node. They may be considered in individuals with chronic AF who fail to respond to or cannot tolerate standard pharmacologic atrioventricular node blockade. Until data from a randomized trial directly comparing medical therapy vs atrioventricular node ablation with pacemaker implantation vs atrioventricular node modification become available, all nonpharmacologic treatment approaches should be considered as a second-line therapy for rate control in AF, reserved for cases refractory to drug therapy. Ablating the atrioventricular node essentially disconnects the electrical pathway and therefore requires placement of a permanent pacemaker at the time of the procedure to provide a steady heartbeat. Although these procedures often diminish the symptoms of AF, it does not cure the condition (in fact, the atria continue to fibrillate) nor does it diminish the increased risk of stroke.

2.13.1. Atrioventricular Node Ablation.
Catheter ablation of the atrioventricular node using radio-frequency energy, with the goal of producing complete atrioventricular block, was introduced in 1987 (41). With the use of large-tip (4-mm length) ablation catheters, a success rate of 90% has been achieved using a right-sided heart transvenous approach. Following successful atrioventricular node ablation, the patient is rendered pacemaker-dependent. Anticoagulation should continue if the atrium remains in permanent AF.

2.13.2. Atrioventricular Node Modification.
The most important disadvantage of atrioventricular node ablation is the need for permanent pacing after a successful procedure. A technique that could slow atrioventricular conduction without creating high-degree atrioventricular block would be attractive (42). However, because induction of high-grade atrioventricular block can occur in up to 20% of treated patients, it seems appropriate currently to limit this procedure to symptomatic patients who would be otherwise considered for atrioventricular node ablation and pacemaker implantation (43).

2.14. Therapies for Conversion of AF to Sinus Rhythm—Cardioversion

In patients with apparent chronic AF in whom pharmacologic rate control and anticoagulation are usually applied, it may be reasonable to attempt restoring sinus rhythm with cardioversion at least once. Three different types exist.

2.14.1. High-Energy, External.
The technique for this routine procedure has been thoroughly reviewed elsewhere (44). The overall success rate with synchronized direct current cardioversion exceeds 80%. Factors that decrease transthoracic impedance, thereby increasing current flow and probability of successful cardioversion, include the following: large paddle diameter (8 to 12 cm); use of a saline solution containing a cutaneous coupling agent; firm contact pressure; and possibly repeated attempts. Success is decreased by long duration of AF and the use of antiarrhythmic agents that increase defibrillation threshold (i.e., amiodarone). Major complications from external cardioversion are rare, although ventricular arrhythmias, systemic embolism, and depression of left ventricular contractility have been reported.

2.14.2. High-Energy, Internal.
This technique may be considered when conversion of AF to sinus rhythm is desirable, but external cardioversion has failed. Energy was delivered between a quadripolar catheter in to the mid-right atrium and a backplate or cutaneous patch.

2.14.3. Low-Energy, Internal: the Implantable Atrial Defibrillator.
Intracardiac energy delivery can result in effective cardioversion using only a small fraction of the energy required for external cardioversion (45), which is possible because only 4% of the current from an external shock traverses the myocardium, the remainder following extracardiac pathways through the chest. After early animal studies demonstrated the feasibility of intracardiac low-energy ($<5 J$) cardioversion of AF, three major advances have increased the applicability of this technique to humans: use of large surface area electrodes; use of a defibrillation vector to encompass both atria (vector from the right atrium to the distal coronary sinus); and biphasic waveform shocks. Implantable atrial defibrillators are currently undergoing testing in preliminary clinical trials examining efficacy, safety, and tolerability.

2.15. Therapies to Prevent AF Recurrences/Maintain Sinus Rhythm

2.15.1. Atrial Pacing in the Prevention of AF.
Current understanding of the role of pacing in the prevention of AF derives primarily from patients with sick sinus syndrome, a population with a high prevalence of this arrhythmia. In patients with symptomatic sick sinus syndrome, a lower rate of progression to chronic AF has been demonstrated when atrial-based pacing (AAI or DDD mode) rather than single-chamber ventricular pacing (VVI mode) is used. Over a follow-up period of 3 to 5 years, four of the largest retrospective studies of pacing in sick sinus syndrome determined the incidence of AF to be between 4% and 13% with AAI/DDD pacing as compared with 18–47% with VVI pacing (46,47). The results of these studies, however, need to be interpreted with caution as several of them were fraught with potential bias that tended to favor the outcome of the AAI or DDD group.

2.15.2. Catheter ablation for AF.
The impressive success of catheter ablation of supraventricular tachycardia has precipitated considerable interest in the development of a similar curative technique for AF. Symptomatic patients with chronic AF as well as patients with intractable paroxysmal AF who have failed to respond to medical therapy would be candidates for this procedure. Most attempts thus far have been focused in creating linear endocardial lesions in the atria, similar to the surgical Maze procedure, via catheter radio-frequency ablation. Encouraging preliminary results have been reported (22,48–50). A successful result will probably require ablation of both atria. However, there have been reports of successful AF ablation in which radio-frequency energy lesions were created only in the right atrium. Some of the current concerns with this investigational procedure include the associated high level of radiation exposure and the thromboembolic potential after extensive ablation in the left atrium. Successful results with catheter ablation have also been reported in patients with paroxysmal AF apparently triggered by or coexisting with other "primary" atrial arrhythmias (such as atrial tachycardia and flutter) (50). In these cases, ablation of the primary arrhythmia substrate seemed to eliminate recurrences of AF, at least in the short term. However, uncertainty remains concerning the requisite number of lesions, their optimal location, and the need for continuous lines. Indeed, focal ablation has been proposed as an alternative approach on the basis of the demonstration that ectopic beats originating within or at the ostium of the pulmonary veins may be the source of paroxysmal and even persistent AF (19,21). Despite high acute success rates, the feasibility of this technique is limited by the difficulty in mapping the focus if the patient is in AF or has no consistent firing, the frequent existence of multiple foci causing high recurrence rates, and an incidence of pulmonary veins narrowing as high as 42% (19). A recent article by Doll et al. (51) documents a 1% incidence of esophageal perforation with intraoperative radio-frequency ablation of AF. The esophageal injuries resulted from the application of a heat-based energy source to the left atrial endocardium. In each case, the esophagus,

which courses posterior to the left atrium, suffered a burn with resulting esophageal perforation. They note that this complication has occurred with unipolar radio frequency and microwave energies. It is likely that collateral damage in general and esophageal injury in particular will occur occasionally with any heatbased.

2.15.3. Surgical therapy. Diverse surgical procedures have been described for the treatment of the AF. The left atrial isolation (52) and the Corridor procedure (53) isolate electrically the ventricles of the atria to reach a regular rhythm. Their main limitation is that the atria follow in AF, maintaining the stroke risk. The only effective method of curing AF is the Maze procedure, described in 1991 by Cox et al. (11). Development of this surgical technique was based on the evidence supporting wavelet reentry as the pathogenetic mechanism of AF and the requirement of a critical mass of contiguous atrial tissue for perpetuation and maintenance of the arrhythmia. Precisely placed atrial incisions can create linear barriers to electrical conduction. These barriers can prevent propagation of the reentrant wavelets in the atrial tissue and theoretically should inhibit arrhythmia recurrence in patients with paroxysmal AF or interrupt chronic AF. The Maze procedure interrupts all potential reentrant circuits in the atria, maintaining atrioventricular synchrony and preserving atrial transport function. A series of incisions were performed in the atria. Both atrial appendages were removed and the pulmonary veins were isolated. As a critical relationship exists between the size of the macroreentrant circuits, the distance between the Maze suture lines, and the effectiveness of the procedure in curing AF, this same concept explains why the Maze procedure fails when performed in extremely large atria. In large atria, it is recommended that we use atrial muscle resection. Other authors have report isolated atrial size reduction procedure with variables results (54,55).

Long experience indicated that the Maze operation is highly successful in restoring sinus rhythm, 90%. The late recurrence of AF has been under 7%. However, postoperatively, 9% of the patients required permanent pacing for sinus node dysfunction (56,57). Although effective, this long operation is rather complex and causes severe damage to the atria. Recently, an intraoperative catheter ablation technique mimicking the Maze procedure has been proposed (58,59). Creating the Maze pattern with energy ablation rather than incisions has simplified the procedure and increased its application (60–63). The goals of these new alternative energy sources are to provide a less invasive approach, shorten the operating time, and simplify the operation. Significant economic resources are being spent on research and development for alternative energy sources to perform Maze procedures. These energy sources include cryogens, radio frequency (unipolar and bipolar), microwave, ultrasonic, and laser (60,64–69). All result in damage to the myocardial tissue, hopefully causing a transmural scar, which does not conduct electricity. Traditional cryotherapy systems used nitrogen for cooling ($-70°C$) (70,71), but newer instruments use argon or helium, which allows lower temperatures to be reached ($-186°C$) (72–74). Radio frequency may be unipolar or bipolar, with bipolar probes having a reduced risk of damaging adjacent structures such as the esophagus (68,75,76). Irrigated radio-frequency probes have greater efficiency because the cooling effect on the surface of the tissue drives the focus of energy deeper into the tissue and prevents char accumulation on the surface (77,78). Microwave energy creates deeper lesions than does radio-frequency energy in a similar length of time and may have more potential for epicardial application (79). Laser and ultrasound energy are still relatively new energy sources, but both may produce transmural lesions even through epicardial fat (65,80–82). A plethora of devices exist that have been developed and even more lesion patterns exist. Provided the ablation creates transmural lesions and complete conduction block, little difference probably exists between the types of energy used. Several shortcomings exist compared with the standard cut-and-sew Maze. The cost is high because most of the probes are disposable. For example, the cost of a bipolar radio-frequency ablation probe is about $2,000 US dollars compared with $50 US dollar cost of the suture required for the cut-and-sew Maze and the cryoprobes are reusable. Local and regional tissue damage exist, which can be wider and deeper than the standard cut-and-sew method, particularly with unipolar radio-frequency devices that rely on time of application to control the depth of the lesion.

3. ATRIAL FLUTTER

Atrial flutter has long been considered a reentrant arrhythmia, but it is only recently that the full structure of the right atrial circuit was understood (83). Today, flutter is considered a type of macroreentrant atrial tachycardia.

3.1. Concept

Concept is an organized and regular fast atrial rhythm (between 200 and 300 beats/min) that originates atrial waves with no isoelectric baseline among them (F waves of flutter). These waves are fundamentally of two morphologies, Figs. 3 and 4: negative in DII, DIII, and aVF (common flutter) and positive in DII, DIII, and aVF (atypical flutter). Differentiation between atrial flutter and atrial tachycardia depends on a rate cut-off around 240–250 beats/min and the presence of an isoelectric baseline between atrial deflections in atrial tachycardia but not in atrial flutter.

3.2. Classification

Electrophysiological studies have shown that the simple EKG definition includes tachycardias using a variety of reentry circuits. The reentry circuits often occupy large areas of the atrium and are referred to as macroreentrant. The precise type of flutter and, in particular, dependence on a defined isthmus is an important consideration for catheter ablation but does not alter the initial approach to management. The atrial flutter is classified as:

1. Isthmus-dependent atrial flutter refers to circuits in which the arrhythmia involves the cava-tricuspid isthmus (CTI). The most common patterns include a tachycardia showing a counterclockwise rotation

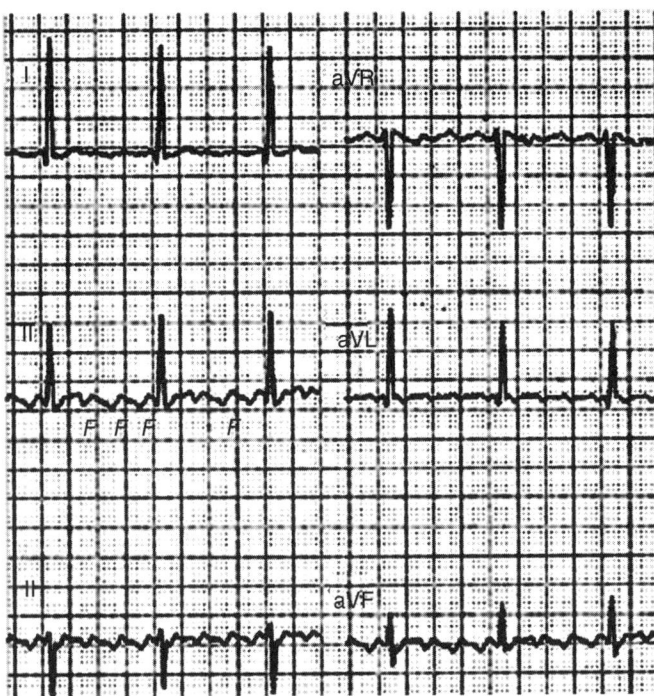

Figure 3. Typical atrial flutter with a rate of 300 beats/min, and the ratio of flutter waves to conducted ventricular complexes is 4:1. Observe the typical morphology in "sawtooth," negative in DII, DIII, and aVF.

around the tricuspid valve denominated typical or type I flutter. A less common pattern involves clockwise rotation around the tricuspid annulus (i.e., reverse typical flutter). Typical atrial flutter is the most common type of macroreentrant atrial tachycardia. It is defined as flutter with an atrial frequency between 240–340 beats/min. In this atrial flutter, activation of the right atrium is reentrant, bounded anteriorly by the tricuspid orifice, and posteriorly by a combination of anatomical obstacles (orifices of the superior and inferior vena cava and the Eustachian bridge) and functional barriers (the region of the crista terminalis). Transverse block may be fixed in some patients and of a functional nature in others, and it occurs in the region between the vena cava, possibly because of anisotropy. The direction of activation in the circuit is in descent in the case of anterior and lateral walls and in ascent in the septal and posterior walls of the right atrium (counterclockwise reentry). The superior pivot point is not well defined (84). Current data suggest that in most cases it includes the right atrial roof anterior to the superior vena cava orifice, including the initial portions of Bachmann's bundle. The inferior pivot point is the inferior CTI area, bounded anteriorly by the inferior part of the tricuspid orifice and posteriorly by the inferior vena cava orifice and its continuation in the Eustachian bridge. This area has been also called the subeustachian isthmus, inferior isthmus, or simply flutter isthmus. Complete transection or ablation of this isthmus interrupts and prevents typical atrial flutter (85,86).

Other flutter types do not have any direction. The opposite direction of activation, descending the septum and ascending the anterior (clockwise reentry), occurs in 10% of clinical cases and characterizes reverse typical atrial flutter (87). Other CTI-dependent flutter circuits are flutters that may also occur as double-wave or lower-loop reentry. Double-wave reentry is defined as a circuit in which two flutter waves simultaneously occupy the usual flutter pathway. Lower-loop reentry is defined as a flutter circuit in which the reentry wavefront circulates around the inferior vena cava because of conduction across the crista terminalis.

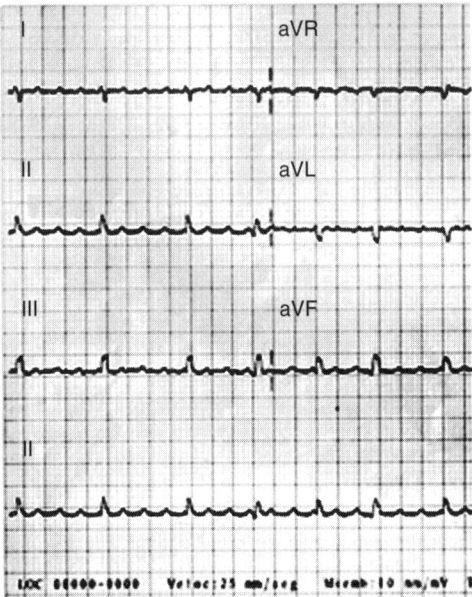

Figure 4. Atypical atrial flutter with an atrial rate of 240 beats/min and a variable ratio of flutter waves to conducted ventricular complexes, alternating between 2:1 and 3:1 atrioventricular conduction. Morphology of the F waves is positive in DIII, DIII, and aVF.

2. Noncavotricuspid isthmus-dependent atrial flutters caused by macroreentry circuits that do not use the CTI are less common than CTI-dependent atrial flutter (88). They are another type of atrial flutter that does not have the atrial frequency of the typical flutter or the localization in the right atrium. As they do not complete a reentry mechanism, in them the treatment is not very established. Most are related to an atrial scar that creates a conduction block and a central obstacle for reentry. Prior cardiac surgery involving the atrium, such as repair of congenital heart disease, mitral valve surgery, or the atrial Maze procedure, is a common cause. The resulting arrhythmias are referred to as lesion-related macroreentrant atrial tachycardias.

3.3. Pathogenic Mechanism—Electrophysiological Substrate

It is possible that all types of flutter begin with a premature atrial extrasystole, which on finding a zone of the auricles with unidirectional blockade initiates a circular movement (reentry). When the conduction velocity is slow or the circuit is long, it is easier for the arrhythmia to be perpetuated. Lewis and Rosenblueth believed that it was necessary for an anatomic obstacle (cava or pulmonary veins) to exist, around which the circular movement was perpetuated (89,90). The recent works of Allesie et al. have demonstrated that an anatomical obstacle is not necessary and that the circuit can either be exclusively in the left auricle or in the right. In some cases, it is possible that the maintenance of the arrhythmia is caused by the same mechanism involved in its initiation or by an ectopic focus that discharges around 300 impulses per minute.

Typical flutter has been better defined as a right atrial macroreentrant tachycardia using a circuit bound anteriorly by the tricuspid ring and posteriorly by a mixed obstacle made by the superior and inferior vena cava orifices and a line of functional block related to the terminal crest (91). The flutter circuit has an anatomical isthmus between the inferior vena cava and the low tricuspid ring that is probably also the most sensitive target for treatment action.

Understanding of the myocardial abnormalities underlying atrial flutter and fibrillation will be necessary to improve this long-term outlook.

3.4. Prevalence

Prevalence is much less frequent than AF (relationship 1 to 10), and normally it is a temporal autolimited arrhythmia.

3.5. Etiology

Etiology is because of practically the same reasons that can cause AF.

3.6. Clinical Manifestations

Patients with atrial flutter commonly present acute symptoms of palpitations, dyspnea, fatigue, or chest pain. However, this arrhythmia may also be presented with more insidious symptoms or conditions, such as exercise-induced fatigue, worsening heart failure, or pulmonary disease. The atrial flutter shows up clinically in three different forms: (1) temporal or sporadic flutter, at a certain frequency the atrial flutter is a temporal arrhythmia before the development of AF; (2) paroxysmal flutter, which is a more infrequent form than paroxysmal AF, although still relatively frequent—flutter and fibrillation often alternate in the same patient when the paroxysmal arrhythmia is of vagal origin, but atrial flutter is infrequent in a state of excess stimulus of the simpathicotonie tone; and (3) chronic flutter; which is not very frequent. With time, it can change to AF or to sinoatrial rhythm. The F-wave flutter causes mechanical contractile activity; the possibility of embolism is rare. The poor clinical tolerance depends fundamentally on the high ventricular rate.

3.7. Electrocardiographic Diagnosis

3.7.1. F waves of flutter. Typical atrial flutter (counterclockwise) shows a characteristic EKG "sawtooth" pattern present in leads II, III, or aVF, which consists of a downsloping segment, followed by a sharper negative deflection, then a sharp positive deflection with a positive "overshoot" leading to the next downsloping plateau. The relative size of each component can vary markedly. Lead V1 often shows a positive deflection, but biphasic or negative deflections can be seen in some cases. Leads I and aVL characteristically show low-voltage deflections. Reverse typical atrial flutter (clockwise reentry) can be recognized with a high degree of reliability in the presence of broad, positive deflections in the inferior leads, although morphologies similar to that of typical atrial flutter have been reported. Wide negative deflections in V1 may be the most specific diagnostic sign. Reverse typical atrial flutter can produce other atypical patterns that need atrial mapping for precise diagnosis of the mechanism. In some cases, discrete P waves are difficult to identify, possibly because of extensive atrial scar. Definitive diagnosis requires intracardiac mapping.

3.7.2. Ventricular response and morphology of the QRS. In the typical flutter, often up to 300/min, one of every two F waves is usually stopped in the atrioventricular node, so that the ventricular response is regular and around 150/min. Nevertheless, they can find different degrees of blockade, 3×1 (rarely), 4×1, 6×1, or more, and often different types of variable blockade with Wenckebach's phenomenon, especially with the effect of certain drug depressors of the atrioventricular conduction. The morphology of QRS is usually narrow, similar to the normal sinus rhythm. When it is wide, it is because of conduction aberrancy or conduction by an anomalous path (image of intraventricular blocking).

3.8. Treatment of Flutter

In approximately 60% of patients, atrial flutter occurs as part of an acute disease process, such as exacerbation of pulmonary disease, postoperative cardiac or pulmonary surgery, or during acute myocardial infarction. If the patient survives the underlying disease process, then chronic therapy for the arrhythmia is usually not required after sinus rhythm is restored. The acute treatment of atrial flutter might include the initial use of electrical pacing, direct current or chemical cardioversion, or atrioventricular nodal-blocking agents (sotalol, ibutilide, amiodarone, etc.). Acute therapy for patients with atrial flutter depends on clinical presentation.

3.9. External Direct Current Cardioversion

If the patient presents with acute hemodynamic collapse or congestive heart failure, then emergent direct current-synchronized shock is indicated. The success rate for external direct current cardioversion for patients with flutter is between 95% and 100%. Conversion can often be achieved with relatively small amounts of energy (i.e., 5 to 50 J), especially when biphasic wave forms are used.

3.10. Atrial Overdrive Pacing

The use and efficacy of rapid atrial pacing to terminate atrial flutter has been long established, and a comprehensive review showed a cumulative success rate of 82% (range from 55% to 100%) (92). It is important to recognize that atrial overdrive pacing may result in the induction of sustained AF.

3.11. Catheter Ablation of the Flutter

Successful ablation is dependent on identifying a critical portion of the reentry circuit where it can be interrupted with either one or a line of radio-frequency applications.

3.11.1. Catheter Ablation of the Cavotricuspid Isthmus for Isthmus-Dependent Flutter.
This isthmus, a relatively narrow part of the circuit between the tricuspid annulus and the inferior vena cava, has become the established target for typical flutter ablation. A technique for placing lesions at the isthmus to block the atrial flutter circuit and cure patients with atrial flutter is available (83,86,93). Complete bidirectional isthmus block is the final goal of flutter ablation, which has to be assessed, after flutter interruption, by pacing both sides of the ablation line while recording electrogram sequences from the opposite right atrial wall and the isthmus itself. Radio-frequency ablation after a mean follow-up of 21 months showed only 36% of patients treated with drugs compared with 80% of those treated with catheter ablation remained in sinus rhythm (94). Success is great in terms of prevention of flutter recurrence; however, a 30% incidence of AF during follow-up casts a large shadow on long-term prognosis (95). The incidence of AF after successful ablation of the CTI flutter circuit varies, depending on the presence of AF before ablation. For patients with a history of only atrial flutter, the occurrence of AF over a follow-up of 18 months was only 8%, whereas AF recurred in 86% of those in whom AF predominated prior to ablation.

3.11.2. Catheter Ablation and Mapping of Noncavotricuspid Isthmus-Dependent Flutter.
Ablation of nonCTI-dependent flutter can be substantially more difficult than for CTI-dependent flutter.

3.11.3. Ablation of the Atrioventricular Node.
In the cases with severe deterioration, bad hemodynamic tolerance, or unsuccessful ablation procedure, one should consider the ablation of the atrioventricular node with radio frequency and to implant a permanent transvenous pacemaker.

BIBLIOGRAPHY

1. W. B. Kannel, P. A. Wolf, E. J. Benjamin, and D. Levy, Prevalence, incidence, prognosis, and predisposing conditions for atrial fibrillation: population-based estimates. *Am. J. Cardiol.* 1998; **82**(8A):2N–9N.
2. C. D. Furberg, B. M. Psaty, T. A. Manolio, J. M. Gardin, V. E. Smith, and P. M. Rautaharju, Prevalence of atrial fibrillation in elderly subjects (the Cardiovascular Health Study). *Am. J. Cardiol.* 1994; **74**(3):236–241.
3. E. J. Benjamin, D. Levy, S. M. Vaziri, R. B. D'Agostino, A. J. Belanger, and P. A. Wolf, Independent risk factors for atrial fibrillation in a population-based cohort. The Framingham Heart Study. *JAMA* 1994; **271**(11):840–844.
4. W. A. Wattigney, G. A. Mensah, and J. B. Croft, Increased atrial fibrillation mortality: United States, 1980–1998. *Am. J. Epidemiol.* 2002; **155**:819–826.
5. W. M. Feinberg, J. L. Blackshear, A. Laupacis, R. Kronmal, and R. G. Hart, Prevalence, age distribution, and gender of patients with atrial fibrillation. Analysis and implications. *Arch. Intern. Med.* 1995; **155**(5):469–473.
6. J. Jalife, J. M. Davidenko, and D. C. Michaels, A new perspective on the mechanisms of arrhythmias and sudden cardiac death: spiral waves of excitation in heart muscle. *J. Cardiovasc. Electrophysiol.* 1991; **2**(Suppl S):133–152.
7. G. K. Moe and J. A. Abildskov, Atrial fibrillation as a self-sustaining arrhythmia independent of focal discharge. *Am. Heart J.* 1959; **58**:59–70.
8. G. K. Moe, W. C. Rheinboldt, and J. A. Abildskov, A computer model of atrial fibrillation. *Am. Heart J.* 1964; **64**(67):200–220.
9. M. A. Allessie, F. I. Bonke, and F. J. Schopman, Circus movement in rabbit atrial muscle as a mechanism of tachycardia. II. The role of nonuniform recovery of excitability in the occurrence of unidirectional block, as studied with multiple microelectrodes. *Circ. Res.* 1976; **39**(2):168–177.
10. J. L. Cox, T. E. Canavan, R. B. Schuessler, M. E. Cain, B. D. Lindsay, C. Stone, P. K. Smith, P. B. Corr, and J. P. Boineau, The surgical treatment of atrial fibrillation. II. Intraoperative electrophysiologic mapping and description of the electrophysiologic basis of atrial flutter and atrial fibrillation. *J. Thorac. Cardiovasc. Surg.* 1991; **101**(3):406–426.
11. J. L. Cox, The surgical treatment of atrial fibrillation. IV. Surgical technique. *J. Thorac. Cardiovasc. Surg.* 1991; **101**(4):584–592.
12. W. F. Garrey, The nature of fibrillary contraction of the heart: its relation to tissue mass and form. *Am. J. Physiol.* 1914; **33**:397–414.
13. G. D. Byrd, S. M. Prasad, C. Ripplinger, T. R. Cassilly, R. B. Schuessler, J. P. Boineau, and R. Damiano, Importance of geometry and refractory period in sustaining atrial fibrillation. Testing the critical mass hypothesis. *Circulation* 2005; **112**:(suppl I):I-7–I-13.
14. E. N. Moore and J. F. Spear, Natural occurrence and experimental initiation of atrial fibrillation in different animal species. In: H. E. Kulbertus, S. B. Olsson, and M. Schlepper, eds., *Atrial Fibrillation*. Molndal, Suecia: AB Hassell, 1982.

15. S. Kaseda and D. P. Zipes, Contraction-excitation feedback in the atria: a cause of changes in refractoriness. *J. Am. Coll. Cardiol.* 1988; **11**(6):1327–1336.
16. S. A. Chen, C. T. Tai, W. C. Yu, Y. J. Chen, C. F. Tsai, M. H. Hsieh, C. C. Chen, V. S. Prakash, Y. A. Ding, and M. S. Chang, Right atrial focal atrial fibrillation: electrophysiologic characteristics and radiofrequency catheter ablation. *J. Cardiovasc. Electrophysiol.* 1999; **10**(3):328–335.
17. M. Haissaguerre, F. I. Marcus, B. Fischer, and J. Clementy, Radiofrequency catheter ablation in unusual mechanisms of atrial fibrillation: report of three cases. *J. Cardiovasc. Electrophysiol.* 1994; **5**(9):743–751.
18. P. Jais, M. Haissaguerre, D. C. Shah, S. Chouairi, L. Gencel, M. Hocini, and J. Clementy, A focal source of atrial fibrillation treated by discrete radiofrequency ablation. *Circulation* 1997; **95**(3):572–576.
19. S. A. Chen, M. H. Hsieh, C. T. Tai, C. F. Tsai, V. S. Prakash, W. C. Yu, T. L. Hsu, Y. A. Ding, and M. S. Chang, Initiation of atrial fibrillation by ectopic beats originating from the pulmonary veins. Electrophysiological characteristics, pharmacological responses, and effects of radiofrequency ablation. *Circulation* 1999; **100**:1879–1886.
20. T. Satoh and D. P. Zipes, Unequal atrial stretch in dogs increases dispersion of refractoriness conducive to developing atrial fibrillation. *J. Cardiovasc. Electrophysiol.* 1996; **7**: 833–842.
21. M. Haissaguerre, P. Jais, D. C. Shah, A. Takahashi, M. Hocini, G. Quiniou, S. Garrigue, A. Le Mouroux, P. Le Metayer, and J. Clementy, Spontaneous initiation of atrial fibrillation by ectopic beats originating in the pulmonary veins. *N. Engl. J. Med.* 1998; **339**(10):659–666.
22. M. Haissaguerre, L. Gencel, B. Fischer, P. Le Metayer, F. Poquet, F. I. Marcus, and J. Clementy, Successful catheter ablation of atrial fibrillation. *J. Cardiovasc. Electrophysiol.* 1994; **5**(12):1045–1052.
23. D. Sherf, Studies on auricular tachycardia caused by acotine administration. *Proc. Exp. Biol. Med.* 1947; **64**: 233–240.
24. D. Panescu, J. G. Whayne, S. D. Fleischman, M. S. Mirotznik, D. K. Swanson, and J. Webster, Three-dimensional finite element analysis of current density and temperature distributions during radiofrequency ablation. *IEEE Trans. Biomed. Eng.* 1995; **42**:879–890.
25. E. Berjano and F. Hornero, Thermal-electrical modeling for epicardial atrial radiofrequency ablation. *IEEE Trans. Biomed. Eng.* 2004; **51**:1348–1357.
26. A. V. Sahidi and P. Savard, A finite element model for radiofrequency ablation of the myocardium. *IEEE Trans. Biomed. Eng.* 1994; **42**:963–968.
27. V. Jacquemet, N. Virag, Z. Ihara, L. Dang, O. Blanc, S. Zozor, J. Vesin, L. Kappenberger, and C. Henriquez, Study of unipolar electrogram morphology in a computer model of atrial fibrillation. *J. Cardiovasc. Electrophysiol.* 2003; **14**:S172–S179.
28. N. Virag, V. Jacquemet, C. Henriquez, S. Zozor, O. Blanc, J. Vesin, E. Pruvot, and L. Kappenberger, Study of atrial arrhythmias in a computer model based on magnetic resonance images of human atria. *Chaos* 2002; **12**:754–763.
29. D. Harrild and C. Henriquez, A computer model of normal conduction in the human atria. *Circ. Res.* 2000; **87**:e25–e36.
30. E. Vigmond, R. Ruckdeschel, and N. Trayanova, Reentry in a morphologically realistic atrial model. *J. Cardiovasc. Electrophysiol.* 2001; **12**:1046–1054.
31. J. Beamont, N. Davidenko, J. M. Davidenko, and J. Jalife, Spiral waves in two-dimensional models of ventricular muscle: formation of a stationary core. *Biophys. J.* 1998; **75**:1–14.
32. I. D. McRury, D. Panescu, M. A. Mitchell, and D. E. Haines, Nonuniform heating during radiofrequency catheter ablation with long electrodes: monitoring the edge effect. *Circulation* 1997; **96**(11):4057–4064.
33. M. K. Jain and P. D. Wolf, A three-dimensional finite element model of radiofrequency ablation with blood flow and its experimental validation. *Ann. Biomed. Eng.* 2000; **28**:1075–1084.
34. E. Berjano and F. Hornero, What affects esophageal injury during radiofrequency ablation of the left atrium? An engineering study based on finite-element analysis. *Physiol. Meas.* 2005; **26**:837–848.
35. J. Godtfredsen, Atrial fibrillation: course and prognosis. A follow-up study of 1212 cases. In: H. E. Kulbertus, S. B. Olson, and M. Schlepper, eds., *Atrial Fibrillation*. Stockholm, Sweden, Astra Cardiovascular, 1982.
36. V. Fuster, I. Ryden, R. W. Asigner, D. S. Cannom, H. Crijns, R. L. Frye, J. L. Halperin, G. N. Kay, W. Klein, S. Lévy, R. L. McNamara, E. N. Prytowsky, L. S. Wann, and D. G. Wyse, ACC/AHA/ESC Guidelines for the management of patients with atrial fibrillation. *JACC* 2001; **38**(4):1231–1265.
37. S. Levy, P. Novella, P. Ricard, and F. Paganelli, Paroxysmal atrial fibrillation: a need for classification. *J. Cardiovasc. Electrophysiol.* 1995; **6**:69–74.
38. J. L. J. Wells, R. B. Karp, N. T. Kouchoukos, W. A. MacLean, T. N. James, and A. L. Waldo, Characterization of atrial fibrillation in man: studies following open heart surgery. *Pacing Clin. Electrophysiol.* 1978; **1**(4):426–438.
39. K. T. Konings, C. J. Kirchhof, J. R. Smeets, H. J. Wellens, O. C. Penn, and M. A. Allessie, High-density mapping of electrically induced atrial fibrillation in humans. *Circulation* 1994; **89**(4):1665–1680.
40. The Atrial Fibrillation Follow-up Investigation of Rhythm Management (AFFIRM) Investigators, A comparison of rate control and rhythm control in patients with atrial fibrillation. *N. Engl. J. Med.* 2002; **347**(23):1825–1833.
41. S. K. Huang, S. Bharati, A. R. Graham, M. Lev, F. I. Marcus, and R. C. Odell, Closed chest catheter desiccation of the atrioventricular junction using radiofrequency energy—a new method of catheter ablation. *J. Am. Coll. Cardiol.* 1987; **9**(2):349–358.
42. B. D. Williamson, K. C. Man, E. Daoud, M. Niebauer, S. A. Strickberger, and F. Morady, Radiofrequency catheter modification of atrioventricular conduction to control the ventricular rate during atrial fibrillation. *N. Engl. J. Med.* 1994; **331**(14):910–917.
43. G. K. Feld, R. P. Fleck, O. Fujimura, D. L. Prothro, T. D. Bahnson, and M. Ibarra, Control of rapid ventricular response by radiofrequency catheter modification of the atrioventricular node in patients with medically refractory atrial fibrillation. *Circulation* 1994; **90**(5):2299–2307.
44. R. E. Kerber, External direct current cardioversion-defibrillation. In: D. P. Zipes and J. Jalife, eds., *Cardiac Electrophysiology from Cell to Bedside*. Philadelphia, PA: WB Saunders, 1990.
45. S. Saksena, P. P. Tarjan, S. Bharati, D. Boveja, D. Cohen, T. Joubert, and M. Lev, Low-energy transvenous ablation of the canine atrioventricular conduction system with a suction electrode catheter. *Circulation* 1987; **76**(2): 394–403.

46. S. Saksena, V. S. Prakash, and M. R. Hill, Prevention of recurrent atrial fibrillation with chronic dual-site right atrial pacing. *J. Am. Coll. Cardiol.* 1996; **28**:687–694.
47. P. Mabo, V. Paul, W. Jung, J. Cementy, A. Bouhour, and J. C. Daubert, Biatrial synchronous pacing for atrial arrhythmia prevention: the SYMBIAPACE study. *Eur. Heart J.* 1999; **20**(suppl):4.
48. A. Elvan, H. P. Pride, J. N. Eble, and D. P. Zipes, Radiofrequency catheter ablation of the atria reduces inducibility and duration of atrial fibrillation in dogs. *Circulation* 1995; **91**(8):2235–2244.
49. J. F. Swartz, G. Pellersels, J. Silvers, et al., A catheter-based curative approach to atrial fibrillation in humans. (Abstract). *Circulation* 1994; **18**:I:335.
50. C. Pappone, G. Oreto, F. Lamberti, G. Vicedomini, M. L. Loricchio, S. Shpun, M. Rillo, M. P. Calabro, A. Conversano, S. A. Ben-Haim, R. Cappato, and S. Chierchia, Catheter ablation of paroxysmal atrial fibrillation using a 3D mapping system. *Circulation* 1999; **100**(11):1203–1208.
51. N. Doll, M. Borger, A. Fabricius, S. Sthepan, J. Gummert, F. W. Mohr, J. Hauus, H. Kottkamp, and G. Hindricks, Esophageal perforation during left atrial radiofrequency ablation: is the risk too high? *J. Thorac. Cardiovasc. Surg.* 2003; **125**:836–842.
52. J. M. Williams, R. M. Ungerleider, G. K. Lofland, and J. L. Cox, Left atrial isolation: new technique for the treatment of supraventricular arrhythmia. *J. Thorac. Cardiovasc. Surg.* 1980; **80**:373–380.
53. G. M. Guiraudon, C. S. Campbell, D. L. Jones, J. L. McLellan, and J. L. McDonald, Combined sino-atrial node atrio-ventricular node isolation: a surgical alternative to His ablation in patients with atrial fibrillation. *Circulation* 1985; **72**(suppl III) (III):220.
54. N. M. Sankar and A. E. Farnsworth, Left atrial reduction for chronic atrial fibrillation associated with mitral valve disease. *Ann. Thorac. Surg.* 1998; **66**(1):254–256.
55. F. Hornero, I. Rodriguez, J. Buendía, M. Bueno, M. J. Dalmau, S. Canovas, O. Gil, R. Garcia-Fuster, and J. A. Montero, Atrial remodeling after mitral valve surgery in patients with permanent atrial fibrillation. *J. Card. Surg.* 2004; **19**(5):376–382.
56. J. L. Cox, R. B. Schuessler, D. G. Lappas, and J. P. Boineau, An 8 1/2-year clinical experience with surgery for atrial fibrillation. *Ann. Surg.* 1996; **224**(3):267–273.
57. S. M. Prasad, H. S. Maniar, C. J. Camillo, R. B. Schuessler, J. P. Boineau, T. M. Sunt III, J. L. Cox, and R. Damiano, The Cox maze III procedure for atrial fibrillation: long-term efficacy in patients undergoing lone versus concomitant procedures. *J. Thorac. Cardiovasc. Surg.* 2003; **126**(6):1822–1828.
58. H. T. Sie, W. P. Beukema, M. A. Ramdat, A. Elvan, J. Ennema, M. Haalebos, and H. J. Wellens, Radiofrequency modified maze in patients with atrial fibrillation undergoing concomitant cardiac surgery. *J. Thorac. Cardiovasc. Surg.* 2001; **112**:249–256.
59. F. Hornero, J. A. Montero, S. Canovas, and M. Bueno, Biatrial radiofrequency ablation for atrial fibrillation: epicardial and endocardial surgical approach. *Interact. Cardiovasc. Thorac. Surg.* 2002; **2**:101–105.
60. F. Hornero, I. Rodriguez, M. Bueno, J. Buendia, M. J. Dalmau, S. Canovas, O. Gil, R. Garcia-Fuster, and J. A. Montero, Surgical ablation of permanent atrial fibrillation by means of maze-radiofrequency. Mid-term results. *J. Card. Surg.* 2004; **9**(5):383–388.
61. S. Grumbrecht, J. Neuzner, and H. F. Pitschner, Interrelation of tissue temperature versus flow velocity in two different kinds of temperature controlled catheter radiofrequency energy applications. *J. Interv. Card. Electrophysiol.* 1998; **2**(2):211–219.
62. S. Benussi, S. Nascimbene, E. Agricola, G. Calori, S. Calvi, A. Caldarola, M. Oppizzi, V. Casati, C. Pappone, and O. Alfieri, Surgical ablation of atrial fibrillation using the epicardial radiofrequency approach: mid-term results and risk analysis. *Ann. Thorac. Surg.* 2002; **74**:1050–1057.
63. J. Melo, P. Adragao, J. Neves, M. M. Ferreira, A. Timoteo, T. Santiago, R. Ribeiras, and M. Canada, Endocardial and epicardial radiofrequency ablation in the treatment of atrial fibrillation with a new intra-operative device. *Eur. J. Cardiothorac. Surg.* 2000; **18**:182–186.
64. J. L. Cox and N. Ad, New surgical and catheter-based modifications of the maze procedure. *Semin. Thorac. Cardiovasc. Surg.* 2000; **12**(1):68–73.
65. J. Ninet, X. Roques, R. Seitelberger, C. Deville, J. L. Pomar, J. Robin, O. Jegaden, F. Wellens, E. Wolner, G. Vedrinne, R. Gottardi, J. Orrit, M. A. Billes, D. A. Hoffmann, J. L. Cox, and G. L. Champsaur, Surgical ablation of atrial fibrillation with off-pump, epicardial, high-intensity focused ultrasound: results of a multicenter trial. *J. Thorac. Cardiovasc. Surg.* 2005; **30**:803–809.
66. R. Damiano, Alternative energy sources for atrial ablation: judging the new technology. *Ann. Thorac. Surg.* 2003; **75**(2):329–330.
67. S. Nath and D. E. Haines, Biophysics and pathology of catheter energy delivery systems. *Prog. Cardiovasc. Dis.* 1995; **37**(4):185–204.
68. S. L. Gaynor, M. D. Diodato, S. M. Prasad, Y. Ishii, R. B. Schuessler, M. S. Bailey, N. R. Damiano, J. B. Bloch, M. R. Moon, and R. Damiano, A prospective, single-center trial of a modified Cox maze procedure with bipolar radiofrequency ablation. *J. Thorac. Cardiovasc. Surg.* 2004; **128**(4):535–542.
69. M. Knaut, S. G. Spitzer, L. Karolyi, H. H. Ebert, P. Richter, S. M. Tugtekin, and S. Schuler, Intraoperative microwave ablation for curative treatment of atrial fibrillation in open heart surgery—the MICRO-STAF and MICRO-PASS pilot trial. MICROwave application in surgical treatment of atrial fibrillation. MICROwave application for the treatment of atrial fibrillation in bypass surgery. *J. Thorac. Cardiovasc. Surg.* 1999; **47**(Suppl):379–384.
70. W. L. Holman, M. Ikeshita, J. M. Douglas, P. K. Smith, and J. L. Cox, Cardiac cryosurgery: effects of myocardial temperature on cryolesion size. *Surgery* 1983; **93**(2):268–272.
71. W. L. Holman, M. Ikeshita, J. M. Douglas, P. K. Smith, G. K. Lofland, and J. L. Cox, Ventricular cryosurgery: short-term effects on intramural electrophysiology. *Ann. Thorac. Surg.* 1983; **35**(4):386–393.
72. J. J. Bredikis and A. J. Bredikis, Surgery of tachyarrhythmia: intracardiac closed heart cryoablation. *Pacing Clin. Electrophysiol.* 1990; **13**(12 Pt 2):1980–1984.
73. A. Skanes, G. Klein, A. D. Krahn, and R. Yee, Cryoablation: potentials and pitfalls. *J. Cardiovasc. Electrophysiol.* 2004; **15**(Suppl. 1):S28–S34.
74. E. Manasse, F. Gaita, S. Ghiselli, A. Barbone, L. Garberoglio, E. Citterio, D. Ornaghi, and R. Gallotti, Cryoablation of the left posterior atrial wall: 95 patients and 3 years of mean follow-up. *Eur. J. Cardiothorac. Surg.* 2003; **24**(5):731–740.
75. M. Gillinov, G. Pettersson, and T. Rice, Esophageal injury during radiofrequency ablation of atrial fibrillation. *J. Thorac. Cardiovasc. Surg.* 2001; **122**:1239–1240.
76. O. G. Anfinsen, E. Kongsgaard, A. Foerster, J. P. Amlie, and H. Aass, Bipolar radiofrequency catheter ablation creates

confluent lesions at larger interelectrode spacing than does unipolar ablation from two electrodes in the porcine heart. *Eur. Heart J.* 1998; **19**(7):1075–1084.

77. H. Nakagawa, W. S. Yamanashi, J. V. Pitha, M. Arruda, X. Wang, K. Ohtomo, K. J. Beckman, J. H. McClelland, R. Lazzara, and W. M. Jackman, Comparison of in vivo tissue temperature profile and lesion geometry for radiofrequency ablation with a saline-irrigated electrode versus temperature control in a canine thigh muscle preparation. *Circulation* 1995; **91**(8):2264–2273.

78. M. Güden, B. Akpinar, I. Sanisoglu, E. Sagbas, and O. Bayindir, Intraoperative saline-irrigated radiofrequency modified maze procedure for atrial fibrillation. *Ann. Thorac. Surg.* 2002; **74**:S1301–S1306.

79. M. Knaut, S. M. Tugtekin, S. G. Spitzer, and V. Gulielmos, Combined atrial fibrillation and mitral valve surgery using microwave technology. *Semin. Thorac. Cardiovasc. Surg.* 2003; **14**(3):226–231.

80. J. H. Levine, J. C. Merillat, M. Stern, H. F. Weisman, A. H. Kadish, E. N. Moore, J. F. Spear, J. Fonger, and T. Guarnieri, The cellular electrophysiologic changes induced by ablation: comparison between argon laser photoablation and high-energy electrical ablation. *Circulation* 1987; **76**(1):217–225.

81. D. Keane and J. N. Ruskin, Linear atrial ablation with a diode laser and fiberoptic catheter. *Circulation* 1999; **100**(14):e59–e60.

82. T. Ohkubo, K. Okishige, Y. Goseki, T. Matsubara, K. Hiejima, and C. Ibukiyama, Experimental study of catheter ablation using ultrasound energy in canine and porcine hearts. *Jap. Heart J.* 1998; **39**(3):399–409.

83. F. G. Cosio, F. Arribas, M. Lopez-Gil, and H. D. Gonzalez, Atrial flutter mapping and ablation II. Radiofrequency ablation of atrial flutter circuits. *Pacing Clin. Electrophysiol.* 1996; **19**(6):965–975.

84. F. Arribas, M. Lopez-Gil, A. Nuñez, and F. Cosio, The upper link of the common atrial flutter circuit. *Pacing Clin. Electrophysiol.* 1997; **20**:2924–2929.

85. M. D. Lesh, G. F. Van Hare, L. M. Epstein, A. P. Fitzpatrick, M. M. Scheinman, R. J. Lee, M. A. Kwasman, H. R. Grogin, and J. C. Griffin, Radiofrequency catheter ablation of atrial arrhythmias. Results and mechanisms. *Circulation* 1994; **89**(3):1074–1089.

86. F. G. Cosio, M. Lopez-Gil, A. Goicolea, F. Arribas, and J. L. Barroso, Radiofrequency ablation of the inferior vena cava-tricuspid valve isthmus in common atrial flutter. *Am. J. Cardiol.* 1993; **71**(8):705–709.

87. J. E. Olgin, J. M. Kalman, M. Chin, C. Stillson, M. Maguire, P. Ursel, and M. D. Lesh, Electrophysiological effects of long, linear atrial lesions placed under intracardiac ultrasound guidance. *Circulation* 1997; **96**(8):2715–2721.

88. F. Cosio, A. Goicolea, M. Lopez-Gil, F. Arribas, J. L. Barroso, and R. Chicote, Atrial endocardial mapping in the rare form of atrial flutter. *Am. J. Cardiol.* 1991; **66**:715–720.

89. T. Lewis, Observations upon flutter and fibrillation: part IV. Impure flutter: theory of circus movement. *Heart* 1920; **7**:293–331.

90. A. Rosenblueth and J. García Ramos, Estudios sobre el flútter y la fibrilación. II. La influencia de los obstáculos artificiales en el flútter auricular experimental. *Arch. Inst. Cardiol. Mex.* 1947; **17**:1–19.

91. F. G. Cosio, F. Arribas, M. Lopez-Gil, and J. Palacios, Radiofrequency catheter ablation of atrial flutter circuits. *Arch. Mal. Coeur Vaiss.* 1996; **89**(Spec No 1):75–81.

92. D. E. Pittman, T. C. Gay, I. I. Patel, and C. R. Joyner, Termination of atrial flutter and atrial tachycardia with rapid atrial stimulation. *Angiology* 1975; **26**(11):784–802.

93. B. Fischer, P. Jais, D. Shah, S. Chouairi, M. Haissaguerre, S. Garrigues, F. Poquet, L. Gencel, J. Clementy, and F. I. Marcus, Radiofrequency catheter ablation of common atrial flutter in 200 patients. *J. Cardiovasc. Electrophysiol.* 1996; **7**(12):1225–1233.

94. A. Natale, K. H. Newby, and E. Pisano, Prospective randomized comparison of antiarrhythmic therapy versus first-line radiofrequency ablation in patients with atrial flutter. *J. Am. Coll. Cardiol.* 2000; **35**:1898–1904.

95. F. Philippon, V. J. Plumb, A. E. Epstein, and G. N. Kay, The risk of atrial fibrillation following radiofrequency catheter ablation of atrial flutter. *Circulation* 1995; **92**(3):430–435.

AUTOCORRELATION AND CROSSCORRELATION METHODS

ANDRÉ FABIO KOHN
University of São Paulo
Sao Paulo, Brazil

1. INTRODUCTION

Any physical quantity that varies with *time* is a **signal**. Examples from physiology are being an electrocardiogram (ECG), an electroencephalogram (EEG), an arterial pressure waveform, and a variation of someone's blood glucose concentration along time. In Fig. 1, one one can see examples of two signals: an electromyogram (EMG) in (a) and the corresponding force in (b). When a muscle contracts, it generates a force or torque while it generates an electrical signal that is the EMG (1). The experiment to obtain these two signals is simple. The subject is seated with one foot strapped to a pedal coupled to a force or torque meter. Electrodes are attached to a calf muscle of the right foot (*m. soleus*), and the signal is amplified. The subject is instructed to produce an alternating pressure on the pedal, starting after an initial rest period of about 2.5 s. During this initial period, the foot stays relaxed on the pedal, which corresponds to practically no EMG signal and a small force because of the foot resting on the pedal. When the subject controls voluntarily the alternating contractions, the random-looking EMG has waxing and waning modulations in its amplitude while the force also exhibits an oscillating pattern (Fig. 1).

Biological signals vary as time goes on, but when they are measured, for example, by a computerized system, the measures are usually only taken at pre-specified times, usually at equal time intervals. In a more formal jargon, it is said that although the original biological signal is defined in continuous time, the measured biological signal is defined in discrete time. For **continuous-time signals**, the time variable t takes values either from $-\infty$ to $+\infty$ (in theory) or in an interval between t_1 and t_2 (a subset of the real numbers, t_1 indicating the time when the signal

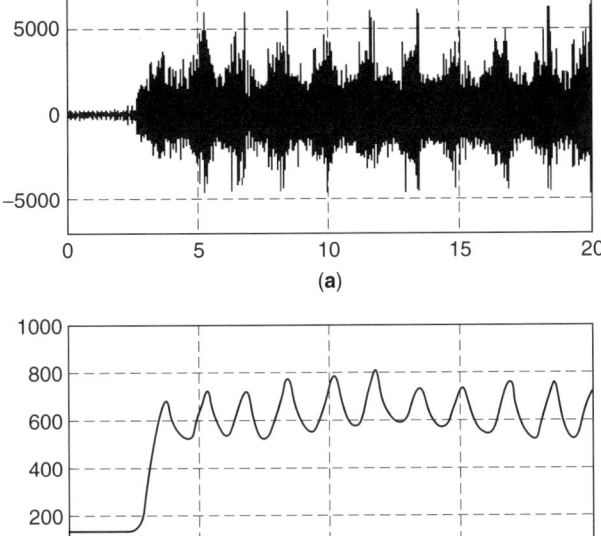

Figure 1. Two signals obtained from an experiment involving a human pressing a pedal with his right foot. (a) The EMG of the soleus muscle and (b) the force or torque applied to the pedal are represented. The abscissae are in seconds and the ordinates are in arbitrary units. The ordinate calibration is not important here because in the computation of the correlation a division by the standard deviation of each signal exists. Only the first 20 s of the data are shown here. A 30 s data record was used to compute the graphs of Figs. 2 and 3.

started being observed in the experiment and t_2 the final time of observation). Such signals are indicated as $y(t)$, $x(t)$, $w(t)$, and so on. On the other hand, a **discrete-time signal** is a set of measurements taken sequentially in time (e.g., at every millisecond). Each measurement point is usually called a sample, and a discrete-time signal is indicated by by $y(n)$, $x(n)$, and $w(n)$, where the index n is an integer that points to the order of the measurements in the sequence. Note that the time interval T between two adjacent samples is not shown explicitly in the $y(n)$ representation, but this information is used whenever an interpretation is required based on continuous-time units (e.g., seconds). As a result of the low price of computers and microprocessors, almost any equipment used today in medicine or biomedical research uses digital signal processing, which means that the signals are functions of discrete time.

From basic probability and statistics theory, it is known that in the analysis of a *random variable* (e.g., the height of a population of human subjects), the mean and the variance are very useful quantifiers (2). When studying the linear relationship between two random variables (e.g., the height and the weight of individuals in a population), the correlation coefficient is an extremely useful quantifier (2). The correlation coefficient between N measurements of pairs of random variables, such as the weight w and height h of human subjects, may be estimated by

$$\rho = \frac{\frac{1}{N}\sum_{i=0}^{N-1}(w(i)-\bar{w})\cdot(h(i)-\bar{h})}{\sqrt{\left(\frac{1}{N}\sum_{i=0}^{N-1}(w(i)-\bar{w})^2\right)\cdot\left(\frac{1}{N}\sum_{i=0}^{N-1}(h(i)-\bar{h})^2\right)}}, \quad (1)$$

where $[w(i), h(i)]$, $i = 0, 1, 2, \ldots, N-1$ are the N pairs of measurements (e.g., from subject number 0 up to subject number $N-1$); $\bar{w}$ and $\bar{h}$ are the mean values computed from the N values of $w(i)$ and $h(i)$, respectively. Sometimes ρ is called the *linear correlation coefficient* to emphasize that it quantifies the degree of linear relation between two variables. If the correlation coefficient between the two variables w and h is near the maximum value 1, it is said that the variables have a strong positive linear correlation, and the measurements will gather around a line with positive slope when one variable is plotted against the other. On the contrary, if the correlation coefficient is near the minimum attainable value of -1, it is said that the two variables have a strong negative linear correlation. In this case, the measured points will gather around a negatively sloped line. If the correlation coefficient is near the value 0, the two variables are not linearly correlated and the plot of the measured points will show a spread that does not follow any specific straight line. Here, it may be important to note that two variables may have a strong nonlinear correlation and yet have almost zero value for the linear correlation coefficient ρ. For example, 100 normally distributed random samples were generated by computer for a variable h, whereas variable w was computed according to the quadratic relation $w = 300*(h-\bar{h})^2 + 50$. A plot of the pairs of points (w, h) will show that the samples follow a parabola, which means that they are strongly correlated along such a parabola. On the other hand, the value of ρ was 0.0373. Statistical analysis suggests that such a low value of linear correlation is not significantly different to zero. Therefore, a near zero value of ρ does not necessarily mean the two variables are not associated with one another, it could mean that they are nonlinearly associated (see Extensions and Further Applications).

On the other hand, a *random signal* is a broadening of the concept of a random variable by the introduction of variations along time and is part of the theory of random processes. Many biological signals vary in a random way in time (e.g., the EMG in Fig. 1) and hence their mathematical characterization has to rely on probabilistic concepts (3–5). For a random signal, the mean and the autocorrelation are useful quantifiers, the first indicating the constant level about which the signal varies and the second indicating the statistical dependencies between the values of two samples taken at a certain time interval. The time relationship between two random signals may be analyzed by the cross-correlation, which is very often used in biomedical research.

Let us analyze briefly the problem of studying quantitatively the time relationship between the two signals shown in Fig. 1. Although the EMG in Fig. 1a looks erratic, its amplitude modulations seem to have some periodicity. Such slow amplitude modulations are sometimes called the "envelope" of the signal, which may be estimated by smoothing the absolute value of the signal. The force in Fig. 1b is much less erratic and exhibits a clearer oscillation. Questions that may develop regarding such signals (the EMG envelope and the force) include: what periodicities are involved in the two signals? Are they the same in the two signals? If so, is there a delay between the two oscillations? What are the physiological interpretations? To answer the questions on the periodicities of each signal, one may analyze their respective autocorrelation functions, as shown in Fig. 2. The autocorrelation of the absolute value of the EMG (a simple estimate of the envelope) shown in Fig. 2a has low-amplitude oscillations, those of the force (Fig. 2b) are large, but both have the same periodicity. The much lower amplitude oscillations in the autocorrelation function of the absolute value of the EMG when compared with that of the force autocorrelation function reflects the fact that the periodicity in the EMG amplitude modulations is masked to a good degree by a random activity, which is not the case for the force signal. To analyze the time relationship between the EMG envelope and the force, their cross-correlation is shown in Fig. 3a. The cross-correlation function in this figure has the same period of oscillation as that of the random signals. In the more refined view of Fig. 3b, it can be seen that the peak occurring closer to zero time shift does so at a negative delay, meaning the EMG precedes the soleus muscle force. Many factors, experimental and physiologic, contribute to such a delay between the electrical activity of the muscle and the torque exerted by the foot.

Signal processing tools such as the autocorrelation and the cross-correlation have been used with much success in a number of biomedical research projects. A few examples will be cited for illustrative purposes. In a study of absence epileptic seizures in animals, the cross-correlation between waves obtained from the cortex and a brain region called the subthalamic nucleus was a key tool to show that the two regions have their activities synchronized by a specific corticothalamic network (6). The cross-correlation function was used in Ref. 7 to show that insulin secretion by the pancreas is an important determinant of insulin clearance by the liver. In a study of preterm neonates, it was shown in Ref. 8 that the correlation between the heart rate variability (HRV) and the respiratory rhythm was similar to that found in the fetus. The same authors also employed the correlation analysis to compare the effects of two types of artificial ventilation equipment on the HRV-respiration interrelation.

After an interpretation is drawn from a cross-correlation study, this signal processing tool may be potentially useful for diagnostic purposes. For example, in healthy subjects, the cross-correlation between arterial blood pressure and intracranial blood flow showed a negative peak at positive delays, differently from patients with a malfunctioning cerebrovascular system (9).

Next, the step-by-step computations of an autocorrelation function will be shown based on the known concept of correlation coefficient of statistics. Actually, different, but related, definitions of autocorrelation and cross-correlation exists in the literature. Some are normalized versions of others, for example. The definition to be given in this section is not the one usually studied in undergraduate engineering courses, but is being presented here first because it is probably easier to understand by readers from other backgrounds. Other definitions will be presented in later sections and the links between them will be readily apparent. In this section, the single term *autocorrelation* shall be used for simplicity and, later (see Basic Definitions), will present some of the more precise names that have been associated with the definition presented here (10,11).

The approach of defining an autocorrelation function based on the cross-correlation coefficient should help in the understanding of what the autocorrelation function tells us about a random signal. Assume that we are given a random signal $x(n)$, with n being the counting variable: $n = 0, 1, 2, \ldots, N - 1$. For example, the samples of $x(n)$ may have been measured at every 1 ms, there being a total of N samples.

The mean or average of signal $x(n)$ is the value $\bar{x}$ given by

$$\bar{x} = \frac{1}{N} \sum_{n=0}^{N-1} x(n), \qquad (2)$$

and gives an estimate of the value about which the signal varies. As an example, in Fig. 4a, the signal $x(n)$ has a mean that is approximately equal to 0. In addition to the

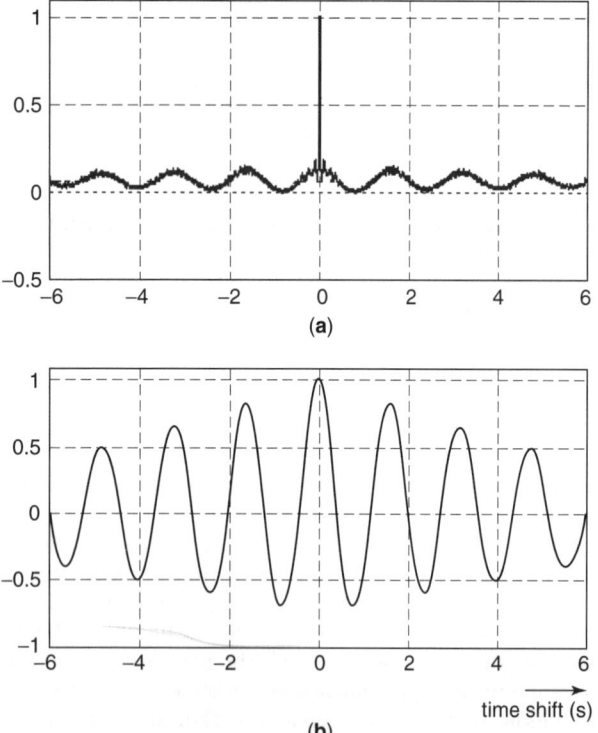

Figure 2. Autocorrelation functions of the signals shown in Fig. 1. (a) shows the autocorrelation function of the absolute value of the EMG and (b) shows the autocorrelation function of the force. These autocorrelation functions were computed based on the correlation coefficient, as explained in the text. The abscissae are in seconds and the ordinates are dimensionless, ranging from −1 to 1. For these computations, the initial transients from 0 to 5 s in both signals were discarded.

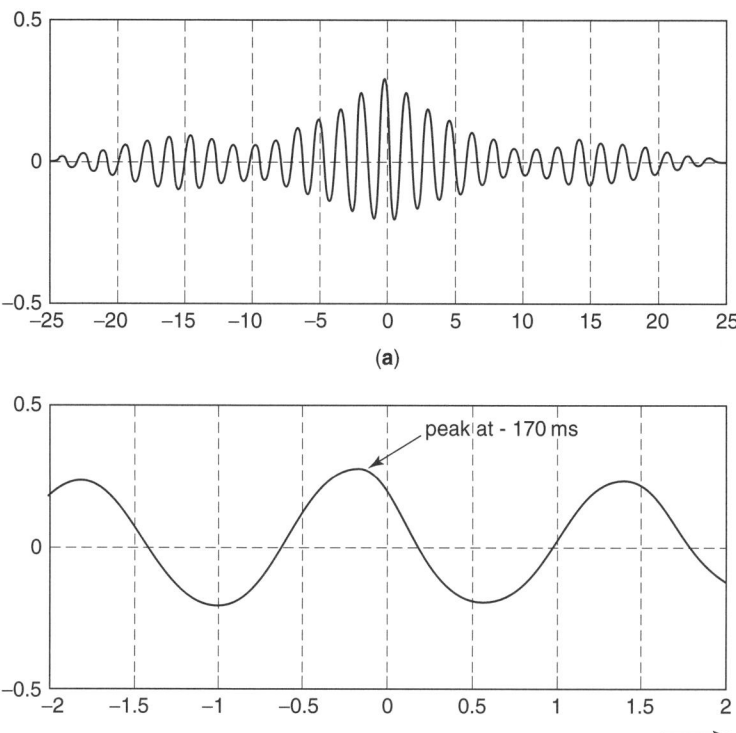

Figure 3. Cross-correlation between the absolute value of the EMG and the force signals shown in Fig. 1. (a) shows the full cross-correlation and (b) shows an enlarged view around abscissa 0. The abscissae are in seconds and the ordinates are dimensionless, ranging from −1 to 1. For this computation, the initial transients from 0 to 5 s in both signals were discarded.

mean, another function is needed to characterize how $x(n)$ varies in time. In this example, one can see that $x(n)$ has some periodicity, oscillating with positive and negative peaks repeating approximately at every 10 samples (Fig. 4a). The new function to be defined is the autocorrelation $\rho_{xx}(k)$ of $x(n)$, which will quantify how much a given signal is similar to time-shifted versions of itself (5). One way to compute it is by using the following formula based on the definition (Equation 1)

$$\rho_{xx}(k) = \frac{\frac{1}{N}\sum_{n=0}^{N-1}(x(n-k)-\bar{x})\cdot(x(n)-\bar{x})}{\frac{1}{N}\sum_{n=0}^{N-1}(x(n)-\bar{x})^2}, \qquad (3)$$

where $x(n)$ is supposed to have N samples. Any sample outside the range $[0, N-1]$ is taken to be zero in the computation of $\rho_{xx}(k)$.

The computation steps are as follows:

- Compute the correlation coefficient between the N samples of $x(n)$ paired with the N samples of $x(n)$ and call it $\rho_{xx}(0)$. The value of $\rho_{xx}(0)$ is equal to 1 because any pair is formed by two equal values [e.g., $[x(0),x(0)], [x(1),x(1)],\ldots,[x(N-1),x(N-1)]$] as seen in the scatter plot of the samples of $x(n)$ with those of $x(n)$ in Fig. 5a. The points are all along the diagonal, which means the correlation coefficient is unity.
- Next, shift $x(n)$ by one sample to the right, obtaining $x(n-1)$, and then determine the correlation coefficient between the samples of $x(n)$ and $x(n-1)$ (i.e., for $n=1$, take the pair of samples $[x(1),x(0)]$, for $n=2$ take the pair $[x(2),x(1)]$ and so on, until the pair $[x(N-1),x(N-2)]$). The correlation coefficient of these pairs of points is denoted $\rho_{xx}(1)$.
- Repeat for a two-sample shift and compute $\rho_{xx}(2)$, for a three-sample shift and compute $\rho_{xx}(3)$, and so on. When $x(n)$ is shifted by 3 samples to the right (Fig. 4b), the resulting signal $x(n-3)$ has its peaks and valleys still repeating at approximately 10 samples, but these are no longer aligned with those of $x(n)$. When their scatter plot is drawn (Fig. 5b), it seems that their correlation coefficient is near zero, so we should have $\rho_{xx}(3) \approx 0$. Note that as $x(n-3)$ is equal to $x(n)$ delayed by 3 samples, the need exists to define what the values of $x(n-3)$ are for $n=0,1,2$. As $x(n)$ is known only from $n=0$ onwards, we make the three initial samples of $x(n-3)$ equal to 0, which has sometimes been called in the engineering literature as zero padding.
- Shifting $x(n)$ by 5 samples to the right (Fig. 4c), generates a signal $x(n-5)$ still with the same periodicity as the original $x(n)$, but with peaks aligned with the valleys in $x(n)$. The corresponding scatter plot (Fig. 5c) indicates a negative correlation coefficient.
- Finally, shifting $x(n)$ by a number of samples equal to the approximate period gives $x(n-10)$, which has its peaks (valleys) approximately aligned with the peaks (valleys) of $x(n)$, as can be seen in Fig. 4d. The corresponding scatter plot (Fig. 5d) indicates a positive correlation coefficient. If $x(n)$ is shifted by multiples of 10, there will again be coincidences between its peaks and those of $x(n)$, and again the correlation coefficient of their samples will be positive.

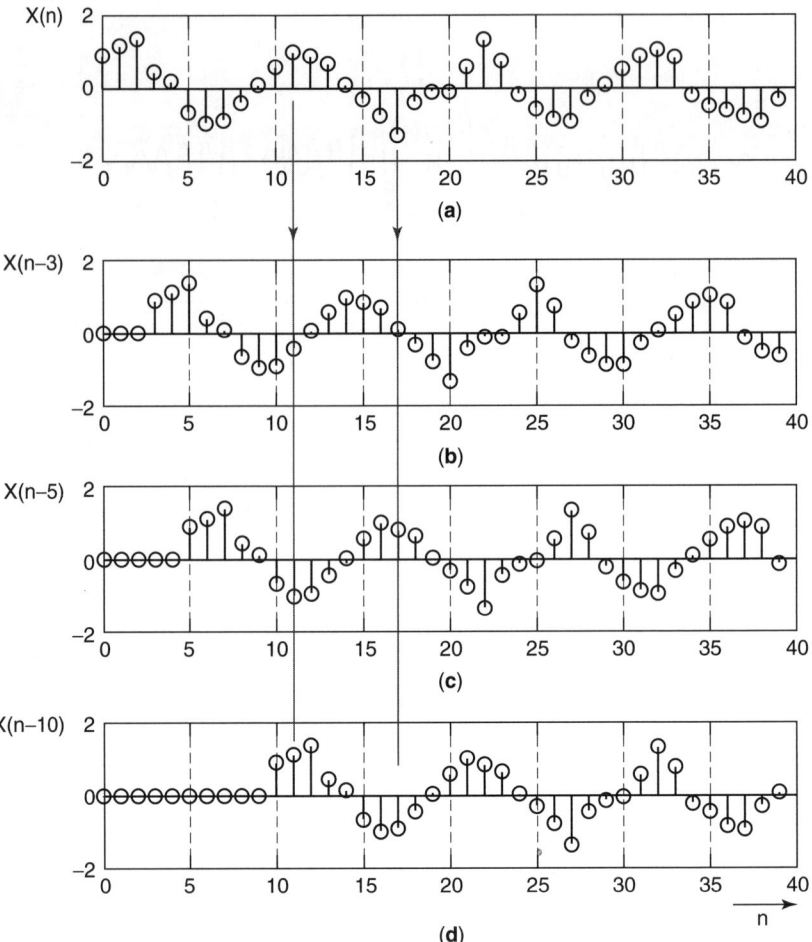

Figure 4. Random discrete-time signal $x(n)$ in (a) is used as a basis to explain the concept of autocorrelation. In (b)–(d), the samples of $x(n)$ were delayed by 3, 5, and 10 samples, respectively. The two vertical lines were drawn to help visualize the temporal relations between the samples of the reference signal at the top and the three time-shifted versions below. (This figure is available in full color at http://www.mrw.interscience.wiley.com/ebe.)

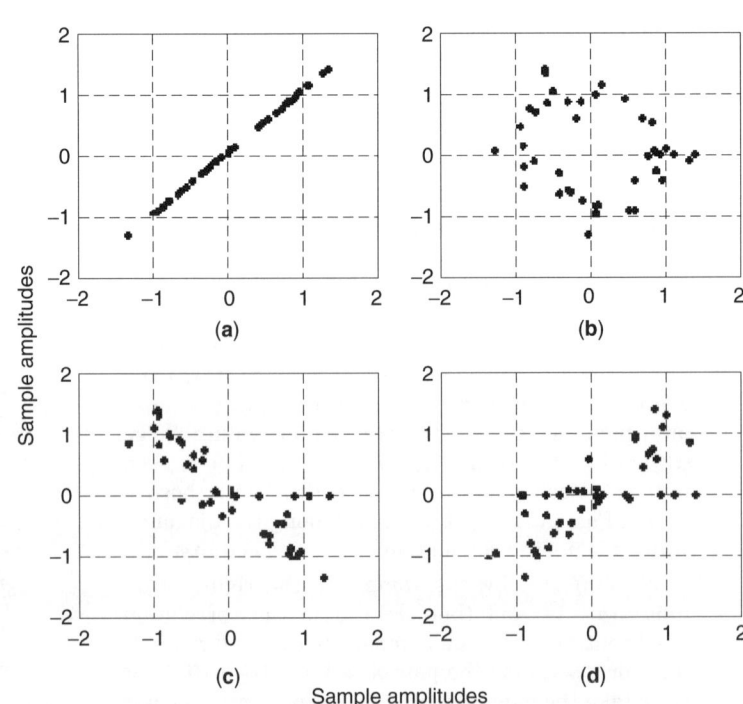

Figure 5. Scatter plots of samples of the signals shown in Fig. 4: (a) $x(n)$ and $x(n)$, (b) $x(n)$ in the abscissa and $x(n-3)$ in the ordinate, (c) $x(n)$ in the abscissa and $x(n-5)$ in the ordinate, (d) $x(n)$ in the abscissa and $x(n-10)$ in the ordinate. The computed values of the correlation coefficient from (a)–(d) were 1, −0.19, −0.79, and 0.66, respectively.

Collecting the values of the correlation coefficients for the different pairs $x(n)$ and $x(n-k)$ and assigning them to $\rho_{xx}(k)$, for positive and negative shift values k, the autocorrelation shown in Fig. 6 is obtained. In this and other figures, the hat "∧"over a symbol is used to indicate estimations from data, to differentiate from the theoretical quantities, for example, as defined in Equations 14 and 15. Indeed, the values for $k=0$, $k=3$, $k=5$ and $k=10$ confirm the analyses based on the scatter plots of Fig. 5. The autocorrelation function is symmetric with respect to $k=0$ because the correlation between samples of $x(n)$ and $x(n-k)$ is the same as the correlation between $x(n)$ and $x(n+k)$, where n and k are integers. This example suggests that the autocorrelation of a periodic signal will exhibit peaks (and valleys) repeating with the same period as the signal's. The decrease in subsequent peak values away from time shift $k=0$ is because of the finite duration of the signal, which requires the use of zero-padding in the computation of the autocorrelation (this process will be dealt with in section 6.1 later in this chapter). In conclusion, the autocorrelation gives an idea of the similarity between a given signal and time-shifted replicas of itself. A different viewpoint is associated with the following question: Does the knowledge of the value of a sample of the signal $x(n)$ at an arbitrary time $n=L$, say $x(L)=5.2$, give some "information-" as to the value of a future sample, say at $n=L+3$? Intuitively, if the random signal varies "slowly," then the answer is yes, but if it varies "fast," then the answer is no. The autocorrelation is the right tool to quantify the signal variability or the degree of "information" between nearby samples. If the autocorrelation decays slowly from value 1 at $k=0$ (e.g., its value is still near 1 for $k=3$), then a sample value 5.2 of the given signal at $n=L$ tells us that at $n=L+3$ the value of the signal will be "near" the sample value 5.2, with high probability. On

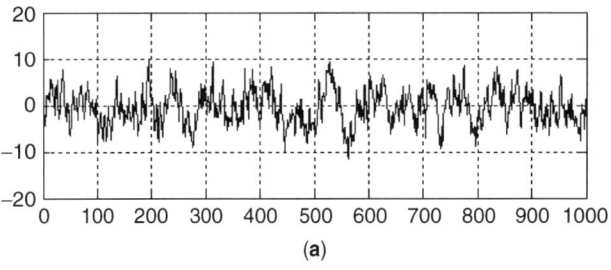

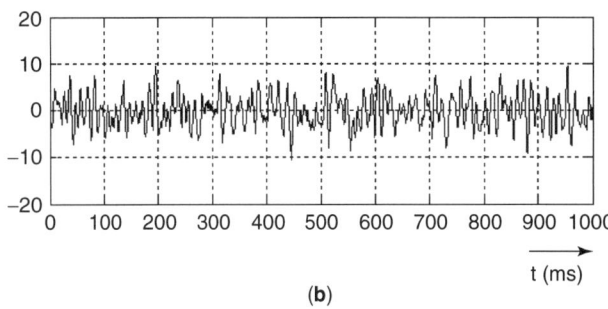

Figure 7. Two random signals measured from two different systems. They seem to behave differently, but it is difficult to characterize the differences based only on a visual analysis. The abscissae are in miliseconds.

the contrary, if the autocorrelation for $k=3$ is already near 0, then the value of the signal at $n=L+3$ has little or no relation to the value 5.2 attained three units of time earlier. In a loose sense, one could say that the signal has more memory in the former situation than in the latter. The autocorrelation depends on the independent variable k, which is called "lag," "delay," or "time shift" in the literature. The autocorrelation definition given above was based on a discrete-time case, but a similar procedure is followed for a continuous-time signal, where the autocorrelation $\rho_{xx}(\tau)$ will depend on a continuous-time variable τ.

Real-life signals are often less well-behaved than the signal shown in Fig 4a, or even those in Fig. 1. Two signals $x(t)$ and $y(t)$ shown in Fig. 7a and 7b, respectively, are more representative of the difficulties one usually encounters in extracting useful information from random signals. A visual analysis suggests that the two signals are indeed different in their "randomness," but it is certainly not easy to pinpoint in what aspects they are different. The respective autocorrelations, shown in Fig. 8a and 8b, are monotonic for the first signal and oscillatory for the second. Such an oscillatory autocorrelation function would mean that two amplitude values in $y(t)$ taken 6 ms apart (or a time interval between about 5 ms and 10 ms) (see Fig. 8b) would have a negative correlation, meaning that if the first amplitude value is positive, the other will probably be negative and vice-versa (note that in Equation 3 the mean value of the signal is subtracted). The explanation of the monotonic autocorrelation will require some mathematical considerations, which will be presented later in this chapter in section 3. An understanding of the ways different autocorrelation functions may occur could be important in discriminating between the behaviors of a biological system subjected to two different experimental conditions, or between normal and pathological cases. In

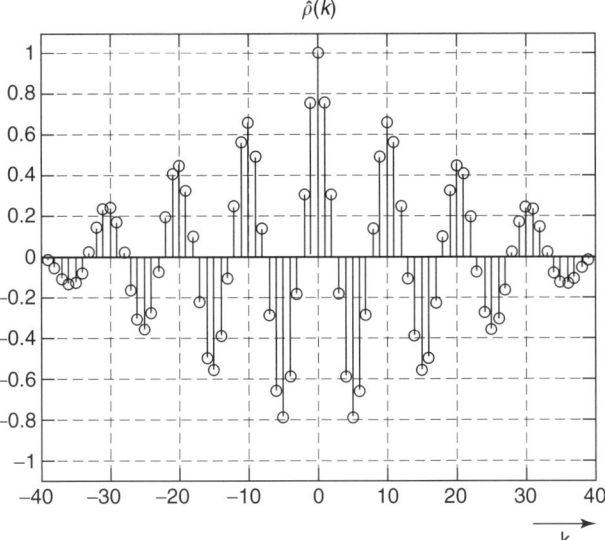

Figure 6. Autocorrelation (correlation coefficients as a function of time shift k) of signal shown in Fig. 4a. The apparently rhythmic behavior of the signal is more clearly exhibited by the autocorrelation, which indicates a repetition at every 10 samples.

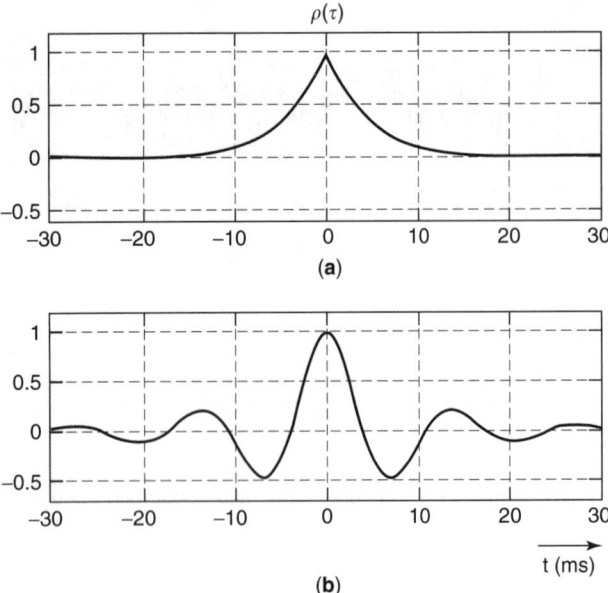

Figure 8. The respective autocorrelation functions of the two signals in Fig. 7. They are quite different from each other: in (a) the decay is monotonic to both sides from the peak value of 1 at $\tau = 0$, and in (b) the decay is oscillatory. These major differences between the two random signals shown in Fig. 7 are not visible directly from their time courses. The abscissae are in miliseconds.

addition, the autocorrelation is able to uncover a periodic signal masked by noise (e.g., Fig. 1a and Fig. 2a), which is relevant in the biomedical setting because many times the biologically interesting signal is masked by other ongoing biological signals or by measurement noise. Finally, when validating stochastic models of biological systems such as neurons, autocorrelation functions obtained from the mathematical model of the physiological system may be compared with autocorrelations computed from experimental data obtained from the physiological system, see, for example, Kohn (12).

When two signals x and y are measured simultaneously in a given experiment, as in the example of Fig. 1, one may be interested in knowing if the two signals are entirely independent from each other, or if some correlation exists between them. In the simplest case, there could be a delay between one signal and the other. The cross-correlation is a frequently used tool when studying the dependency between two random signals. A (normalized) cross-correlation $\rho_{xy}(k)$ may be defined and explained in the same way as we did for the autocorrelation in Figs. 4–6 (i.e., by computing the correlation coefficients between the samples of one of the signals, $x(n)$, and those of the other signal time-shifted by k, $y(n-k)$) (5). The formula is the following:

$$\rho_{xy}(k) = \frac{\frac{1}{N}\sum_{n=0}^{N-1}(x(n)-\bar{x})\cdot(y(n-k)-\bar{y})}{\sqrt{\left(\frac{1}{N}\sum_{n=0}^{N-1}(x(n)-\bar{x})^2\right)\cdot\left(\frac{1}{N}\sum_{n=0}^{N-1}(y(n)-\bar{y})^2\right)}}, \quad (4)$$

where both signals are supposed to have N samples each. Any sample of either signal outside the range $[0, N-1]$ is taken to be zero in the computation of $\rho_{xx}(k)$ (zero padding).

It should be clear that if one signal is a delayed version of the other, the cross-correlation at a time shift value equal to the delay between the two signals will be equal to 1. In the example of Fig. 1, the force signal at the bottom is a delayed version of the EMG envelope at the top, as indicated by the cross-correlation in Fig. 3.

The emphasis in this chapter is to present the main concepts on the auto and cross-correlation functions, which are necessary to pursue research projects in biomedical engineering.

2. AUTOCORRELATION OF A STATIONARY RANDOM PROCESS

2.1. Introduction

Initially, some concepts from random process theory shall be reviewed briefly, as covered in undergraduate courses in electrical or biomedical engineering, see, for example, Peebles (3) or Papoulis and Pillai (10). A **random process** is an infinite collection or ensemble of functions of time (continuous or discrete time), called sample functions or realizations (e.g., segments of EEG recordings). In continuous time, one could indicate the random process as $X(t)$ and in discrete time as $X(n)$. Each sample function is associated with the outcome of an experiment that has a given probabilistic description and may be indicated by $x(t)$ or $x(n)$, for continuous or discrete time, respectively. When the ensemble of sample functions is viewed at any single time, say t_1 for a continuous time process, a random variable is obtained whose probability distribution function is $F_{X,t_1}(\alpha_1) = P[X(t_1) \leq \alpha_1]$, for $\alpha_1 \in \mathbb{R}$, and where $P[.]$ stands for probability. If the process is viewed at two times t_1 and t_2, a bivariate distribution function is needed to describe the pair of resulting random variables $X(t_1)$ and $X(t_2)$: $F_{X,t_1,t_2}(\alpha_1,\alpha_2) = P[X(t_1) \leq \alpha_1, X(t_2) \leq \alpha_2]$, for α_1, $\alpha_2 \in \mathbb{R}$. The random process is fully described by the joint distribution functions of any N random variables defined at arbitrary times $t_1, t_2, \ldots, N$, for arbitrary integer number N

$$P[X(t_1) \leq \alpha_1, X(t_2) \leq \alpha_2, \ldots, X(t_N) \leq \alpha_N], \quad (5)$$
$$\text{for } \alpha_1, \alpha_2, \ldots, \alpha_N \in \mathbb{R}.$$

In many applications, the properties of the random process may be assumed to be independent of the specific values $t_1, t_2, \ldots, t_N$, in the sense that if a fixed time shift T is given to all instants t_i $i=1,\ldots,N$, the probability distribution function does not change:

$$P[X(t_1+T) \leq \alpha_1, X(t_2+T) \leq \alpha_2, \ldots, X(t_N+T) \leq \alpha_N]$$
$$= P[X(t_1) \leq \alpha_1, X(t_2) \leq \alpha_2, \ldots, X(t_N) \leq \alpha_N]. \quad (6)$$

If Equation 6 holds for all possible values of t_i, T and N, the process is called **strictsense stationary** (3). This class of random processes has interesting properties, such as:

$$E[X(t)] = \mu = \text{constant } \forall t \quad (7)$$
$$E[X(t+\tau) \cdot X(t)] = function(\tau) \quad (8)$$

The first result in Equation 7 means that the mean value of the random process is a constant value for any time t. The second result in Equation 8 means that the second-order moment defined on the process at times $t_2 = t + \tau$ and $t_1 = t$, depends only on the time difference $\tau = t_2 - t_1$ and is independent of the time parameter t. These two (Equations 7 and 8) are so important in practical applications that whenever they are true, the random process is said to be **wide-sense stationary.** This definition of stationarity is feasible to be tested in practice, and many times a random process that satisfies Equations 7, 8 is simply called "stationary" and otherwise it is simply called "nonstationary." The autocorrelation and cross-correlation analyses developed in this chapter are especially useful for such stationary random processes. All random processes considered in this chapter will be wide-sense stationary.

In real-life applications, the wide-sense stationarity assumption is usually valid only approximately and only for a limited time interval. This interval must usually be estimated experimentally (4) or be adopted from previous research reported in the literature. This assumption certainly simplifies both the theory as well as the signal processing methods. In the present text, it shall be assume that all random processes are wide-sense stationary. Another fundamental property that shall be assumed is that the random process is **ergodic**, meaning that any appropriate time average computed from a given sample function converges to a corresponding expected value defined over the random process (10). Thus, for example, for an ergodic process (Equation 2) would give useful estimates of the expected value of the random process (Equation 7), for sufficiently large values of N. Ergodicity assumes that a finite set of physiological recordings obtained under a certain experimental condition should yield useful estimates of the general random behavior of the physiological system, under the same experimental conditions and in the same physiological state, which is of utmost importance because in practice all we have is a sample function and from it have to estimate and infer things related to the random process that generated that sample function.

A **random signal** may be defined as a sample function or realization of a random process. Many signals measured from humans or animals exhibit some degree of unpredicatibility, and may be considered as random. The sources of unpredictability in biological signals may be associated with (1) a large number of uncontrolled and unmeasured internal mechanisms, (2) intrinsically random internal physicochemical mechanisms, and (3) a fluctuating environment. When measuring randomly varying phenomena from humans or animals, only one or a few sample functions of a given random process are obtained. Under the ergodic property, appropriate processing of a random signal may permit the estimation of characteristics of the random process, which is why random signal processing techniques (such as auto and cross-correlation) are so important in practice.

The complete probabilistic description of a random process Equation 5 is impossible to obtain in practical terms. Instead, first and second moments are very often employed in real-life applications. Examples of these applications are the mean, the auto correlation, and cross-correlation functions (13), which are the main topics of this chapter, and the auto and cross-spectra. The auto-spectrum and the cross-spectrum are functions of frequency, being related to the auto and cross-correlation functions via the Fourier transform.

Knowledge of the basic theory of random processes is a pre-requisite for a correct interpretation of results obtained from the processing of random signals. Also, the algorithms used to compute estimates of parameters or functions associated with random processes are all based on the underlying random process theory.

All signals in this chapter will be assumed to be real and originating from a wide-sense stationary random process.

2.2. Basic Definitions

The mean or expected value of a *continuous-time* random processes $X(t)$ is defined as

$$\mu_x = E[X(t)], \quad (9)$$

where the time variable t is defined on a subset of the real numbers and $E[]$ is the expected value operation. The mean is a constant value because all random processes are assumed to be wide-sense stationary. The definition above is a mean calculated over all sample functions of the random process.

The autocorrelation of a *continuous-time* random processes $X(t)$ is defined as

$$R_{xx}(\tau) = E[X(t+\tau) \cdot X(t)], \quad (10)$$

where the time variables t and τ are defined on a subset of the real numbers. As was mentioned before, the nomenclature varies somewhat in the literature. The definition of autocorrelation given in Equation 10 is the one typically found in engineering books and papers. The value $R_{xx}(0) = E[X^2(t)]$ is sometimes called the average total power of the signal and its square root is the "root mean square" (RMS) value, employed frequently to characterize the "amplitude" of a biological random signal such as the EMG (1).

An equally important and related second moment is the autocovariance, defined for continuous time as

$$C_{xx}(\tau) = E[(X(t+\tau) - \mu_x) \cdot (X(t) - \mu_x)] = R_{xx}(\tau) - \mu_x^2 \quad (11)$$

The autocovariance at $\tau = 0$ is equal to the variance of the process and is sometimes called the average ac power (the average total power minus the square of the dc value):

$$C_{xx}(0) = \sigma_x^2 = E[(X(t) - \mu_x)^2] = E[X^2(t)] - \mu_x^2 \quad (12)$$

For a stationary random process, the mean, average total power and variance are constant values, independent of time.

The autocorrelation for a *discrete-time* random process is

$$R_{xx}(k) = E[X(n+k) \cdot X(n)] = E[X(n-k) \cdot X(n)], \quad (13)$$

where n and k are integer numbers. Any of the two expressions may be used, either with $X(n+k) \cdot X(n)$ or with $X(n-k) \cdot X(n)$. In what follows, preference shall be given to the first expression to keep consistency with the definition of cross-correlation to be given later.

For discrete-time processes, an analogous definition of autocovariance follows:

$$C_{xx}(k) = E[(X(n+k) - \mu_x) \cdot (X(n) - \mu_x)] = R_{xx}(k) - \mu_x^2, \quad (14)$$

where again $C_{xx}(0) = \sigma_x^2 = E[(X(k) - \mu_x)^2]$ is the constant variance of the stationary random process $X(k)$. Also $\mu_x = E[X(n)]$, a constant, is its expected value. In many applications, the interest is in studying the variations of a random processes about its mean, which is what the autocovariance represents. For example, to characterize the variability of the muscle force exerted by a human subject in a certain test, the interest is in quantifying how the force varies randomly around the mean value, and hence the autocovariance is more interesting than the autocorrelation.

The independent variable τ or k in Equations 10, 11, 13 or 14 will be called, interchangeably, *time shift*, *lag*, or *delay*.

It should be mentioned that some books and papers, mainly those on time series analysis (5,11), define the *autocorrelation function* of a random process as (for discrete time):

$$\rho_{xx}(k) = \frac{C_{xx}(k)}{\sigma_x^2} = \frac{C_{xx}(k)}{C_{xx}(0)} \quad (15)$$

(i.e., the autocovariance divided by the variance of the process). It should be noticed that $\rho_{xx}(0) = 1$. Actually, this definition was used in the Introduction of this chapter. The definition in Equation 15 differs from that in Equation 13 in two respects: The mean of the signal is subtracted and a normalization exists so that at $k = 0$ the value is 1. To avoid confusion with the standard engineering nomenclature, the definition in Equation 15 may be called the *normalized autocovariance*, *the correlation coefficient function*, or still the *autocorrelation coefficient*. In the text that follows, preference is given to the term **normalized autocovariance**.

2.3. Basic Properties

From their definitions, the autocorrelation and autocovariance (normalized or not) are **even** functions of the time shift parameter, because, $X(t+\tau) \cdot X(t) = X(t-\tau) \cdot X(t)$ and $X(n+k) \cdot X(n) = X(n-k) \cdot X(n)$ for continuous- and discrete-time process, respectively. The property is indicated below only for the discrete-time case (for continuous-time, replace k by τ):

$$R_{xx}(k) = R_{xx}(-k), \quad (16)$$

and

$$C_{xx}(k) = C_{xx}(-k), \quad (17)$$

as well as for the normalized autocovariance:

$$\rho_{xx}(k) = \rho_{xx}(-k). \quad (18)$$

Three important inequalities may be derived (10) for both continuous-time and discrete-time random processes. Only the result for the discrete-time case is shown below (for continuous-time, replace k by τ):

$$|R_{xx}(k)| \leq R_{xx}(0) = \sigma_x^2 + \mu_x^2 \quad \forall k \in \mathbb{Z} \quad (19)$$

$$|C_{xx}(k)| \leq C_{xx}(0) = \sigma_x^2 \quad \forall k \in \mathbb{Z}, \quad (20)$$

and

$$|\rho_{xx}(k)| \leq 1 \quad \forall k \in \mathbb{Z}, \quad (21)$$

with $\rho_{xx}(0) = 1$.

These relations say that the maximum of either the autocorrelation or autocovariance occurs at lag 0.

Any discrete-time ergodic random process without a periodic component will satisfy the following limits:

$$\lim_{|k| \to \infty} R_{xx}(k) = \mu_x^2 \quad (22)$$

and

$$\lim_{|k| \to \infty} C_{xx}(k) = 0 \quad (23)$$

and similarly for continuous-time processes by changing k for τ. These relations mean that two random variables defined in $X(n)$ at two different times n_1 and $n_1 + k$ will tend to be uncorrelated as they are farther apart (i.e., the "memory" decays when the time interval k increases).

A final, more subtle, property of the autocorrelation is that it is positive semi-definite (10,14), expressed here only for the discrete-time case:

$$\sum_{i=1}^{K} \sum_{j=1}^{K} \alpha_i \alpha_j R_{xx}(k_i - k_j) \geq 0 \quad for \; \forall K \in \mathbb{Z}^+, \quad (24)$$

where $\alpha_1, \alpha_2, \ldots, \alpha_K$ are arbitrary real numbers and $k_1, k_2, \ldots, k_K \in \mathbb{Z}$ are any set of discrete-time points. This same result is valid for the autocovariance (normalized or not). This property means that not all functions that satisfy Equations 16 and 19 can be autocorrelation functions of some random process, they also have to be positive semi-definite.

2.4. Fourier Transform of the Autocorrelation

A very useful frequency-domain function related to the correlation/covariance functions is the power spectrum S_{xx} of the random process X (continuous or discrete-time), defined as the Fourier transform of the autocorrelation function (10,15). For continuous time, we have

$$S_{xx}(j\omega) = Fourier\ transform\ [R_{xx}(\tau)], \quad (25)$$

where the angular frequency ω is in rad/s. The average power P_{xx} of the random process $X(t)$ is

$$P_{xx} = \frac{1}{2\pi} \int_{-\infty}^{\infty} S_{xx}(j\omega)\,d\omega = R_{xx}(0). \quad (26)$$

If the average power in a given frequency band $[\omega_1, \omega_2]$ is needed, it can be computed by

$$P_{xx} = \frac{1}{\pi} \int_{\omega_1}^{\omega_2} S_{xx}(j\omega)\,d\omega. \quad (27)$$

For discrete time

$$S_{xx}(e^{j\Omega}) = Discrete\ Time\ Fourier\ transform\ [R_{xx}(k)], \quad (28)$$

where Ω is the normalized angular frequency given in rad ($\Omega = \omega \cdot T$, where T is the sampling interval). In Equation 28 the power spectrum is periodic in Ω, with a period equal to 2π. The average power P_{xx} of the random process $X(n)$ is

$$P_{xx} = \frac{1}{2\pi} \int_{-\pi}^{\pi} S_{xx}(e^{j\Omega})\,d\Omega = R_{xx}(0). \quad (29)$$

Other common names for the power spectrum are power spectral density and autospectrum. The power spectrum is a real **non-negative** and even function of frequency (10,15), which requires a positive semi-definite autocorrelation function (10), which means that not all functions that satisfy Equations 16 and 19 are valid autocorrelation functions because the corresponding power spectrum could have negative values for some frequency ranges, which is absurd.

The power spectrum should be used instead of the autocorrelation function in situations such as: (1) when the objective is to study the bandwidth occupied by a random signal, and (2) when one wants to discover if there are several periodic signals masked by noise (for a single periodic signal masked by noise the autocorrelation may be useful too).

2.5. White Noise

Continuous-time white noise is characterized by an autocovariance that is proportional to the Dirac impulse function:

$$C_{xx}(\tau) = \Psi \cdot \delta(\tau), \quad (30)$$

where Ψ is a positive constant and the Dirac impulse is defined as

$$\delta(\tau) = 0\ for\ \tau \neq 0 \quad (31)$$

$$\delta(\tau) = \infty\ for\ \tau = 0, \quad (32)$$

and

$$\int_{-\infty}^{\infty} \delta(\tau)\,d\tau = 1. \quad (33)$$

The autocorrelation of continuous-time white noise is:

$$R_{xx}(\tau) = \Psi \cdot \delta(\tau) + \mu_x^2, \quad (34)$$

where μ_x is the mean of the process.

From Equations 12 and 30 it follows that the variance of the continuous-time white process is infinite (14), which indicates that it is not physically realizable. From Equation 30 we conclude that, for any time shift value τ, no matter how small, the correlation coefficient between any value in $X(t)$ and the value at $X(t+\tau)$ would be equal to zero, which is certainly impossible to satisfy in practice because of the finite risetimes of the outputs of any physical system. From Equation 25 it follows that the power spectrum of continuous-time white noise (with $\mu_x = 0$) has a constant value equal to Ψ at all frequencies. The name white noise comes from an extension of the concept of "white light," which similarly has constant power over the range of frequencies in the visible spectrum. White noise is non-realizable, because it would have to be generated by a system with infinite bandwidth.

Engineering texts circumvent the difficulties with the continuous-time white noise by defining a **band-limited white noise** (3,13). The corresponding power spectral density is constant up to very high frequencies (ω_c), and is zero elsewhere, which makes the variance finite. The maximum spectral frequency ω_c is taken to be much higher than the bandwidth of the system to which the noise is applied. Therefore, *in approximation*, the power spectrum is taken to be constant at all frequencies, the autocovariance is a Dirac delta function, and yet a finite variance is defined for the random process.

The utility of the concept of white noise develops when it is applied at the input of a finite bandwidth system, because the corresponding output is a well-defined random process with physical significance (see next section).

In discrete time, the white-noise process has an autocovariance proportional to the unit sample sequence:

$$C_{xx}(k) = \Psi \cdot \delta(k), \quad (35)$$

where Ψ is a finite positive real value, and

$$R_{xx}(k) = \Psi \cdot \delta(k) + \mu_x^2, \quad (36)$$

where $\delta(k) = 1$ for $k = 0$ and $\delta(k) = 0$ for $k \neq 0$. The discrete-time white noise has a finite variance, $\sigma^2 = \Psi$ in Equation 35, is realizable and it may be synthesized by taking a sequence of independent random numbers from an arbitrary probability distribution. Sequences that have independent samples with identical probability distributions are usually called **i.i.d.** (independent identically distributed). Computer-generated "random" (pseudo-random) sequences are usually very good approximations to a white-noise discrete-time random signal, being usually of zero mean and unit variance for a normal or Gaussian distribution. To achieve desired values of Ψ in Equation 35 and μ_x^2 in Equation 36 one should multiply the (zero mean, unit variance) values of the computer-generated white sequence by $\sqrt{\Psi}$ and sum to each resulting value the constant value μ_x.

3. AUTOCORRELATION OF THE OUTPUT OF A LINEAR SYSTEM WITH RANDOM INPUT

In relation to the examples presented in the previous section, one may ask how may two random processes develop such differences in the autocorrelation as seen in Fig. 8. How may one autocorrelation be monotonically decreasing (for increasing positive τ) while the other exhibits oscillations? For this purpose, how the autocorrelation of a signal changes when it is passed through a time-invariant linear system will be examined.

If a continuous-time linear system has an impulse response $h(t)$ and a random process $X(t)$ with an autocorrelation $R_{xx}(\tau)$ is applied at its input the resultant output $y(t)$ will have an autocorrelation given by the following convolutions (10):

$$R_{yy}(\tau) = h(\tau) * h(-\tau) * R_{xx}(\tau). \quad (37)$$

Note that $h(\tau) * h(-\tau)$ may be viewed as an autocorrelation of $h(\tau)$ with itself and, hence, is an even function.

Taking the Fourier transform we conclude Equation 37, it is that the output power spectrum $S_{yy}(j\omega)$ is the absolute value squared of the frequency response function $H(j\omega)$ times the input power spectrum $S_{xx}(j\omega)$:

$$S_{yy}(j\omega) = |H(j\omega)|^2 \cdot S_{xx}(j\omega), \quad (38)$$

where $H(j\omega)$ is the Fourier transform of $h(t)$, $S_{xx}(j\omega)$ is the Fourier transform of $R_{xx}(\tau)$, and $S_{yy}(j\omega)$ is the Fourier transform of $R_{yy}(\tau)$.

The corresponding expressions for the autocorrelation and power spectrum for the output signal from a discrete-time system are (15)

$$R_{yy}(k) = h(k) * h(-k) * R_{xx}(k) \quad (39)$$

and

$$S_{yy}(e^{j\Omega}) = |H(e^{j\Omega})|^2 \cdot S_{xx}(e^{j\Omega}), \quad (40)$$

where $h(k)$ is the impulse (or unit sample) response of the system, $H(e^{j\Omega})$ is the frequency response function of the system, and $S_{xx}(e^{j\Omega})$ and $S_{yy}(e^{j\Omega})$ are the discrete-time Fourier transforms of $R_{xx}(k)$ and $R_{yy}(k)$, respectively. $S_{xx}(e^{j\Omega})$ and $S_{yy}(e^{j\Omega})$ are the input and output power spectra, respectively. As an example, suppose that $y(n) = [x(n) + x(n-1)]/2$ is the difference equation that defines a given system. This is an example of a finite impulse response (FIR) system (16) with impulse response equal to 0.5 for $n = 0, 1$ and 0 for other values of n. If the input is discrete-time white noise with unit variance, then from Equation 39 the output autocorrelation is a triangular sequence centered at $k = 0$, with amplitude 0.5 at $k = 0$, amplitude 0.25 at $k = \pm 1$ and 0 elsewhere.

If two new random processes are defined as $U = X - \mu_x$ and $Q = Y - \mu_y$, it follows from the definitions of autocorrelation and autocovariance that $R_{uu} = C_{xx}$ and $R_{qq} = C_{yy}$. Therefore, applying Equation 37 or Equation 39 to a system with input U and output Y, similar expressions to Equations 37 and 39 are obtained relating the input and output autocovariances, shown below only for the discrete-time case:

$$C_{yy}(k) = h(k) * h(-k) * C_{xx}(k). \quad (41)$$

Furthermore, if $U = (X - \mu_x)/\sigma_x$ and $Q = (Y - \mu_y)/\sigma_y$ are new random processes, we have $R_{uu} = \rho_{xx}$ and $R_{qq} = \rho_{yy}$. From Equations 37 or 39 similar relations between the normalized autocovariance functions of the output and the input of the linear system are obtained shown below only for the discrete-time case (for the continuous-time, use τ instead of k):

$$\rho_{yy}(k) = h(k) * h(-k) * \rho_{xx}(k). \quad (42)$$

A monotonically decreasing autocorrelation or autocovariance may be obtained (e.g., Fig. 8a), when, for example, a white noise is applied at the input of a system that has a monotonically decreasing impulse response (e.g., of a first-order system or a second-order overdamped system). As an example, apply a zero-mean white noise to a system that has an impulse response equal to $e^{-\alpha t} \amalg(t)$, where $\amalg(t)$ is the Heaviside step function ($\amalg(t) = 1, t \geq 0$ and $\amalg(t) = 0$, $t < 0$). From Equation 37 this system's output random signal will have an autocorrelation that is

$$R_{yy}(\tau) = e^{-\alpha\tau} \amalg(\tau) * e^{\alpha\tau} \amalg(-\tau) * \delta(\tau) = \frac{e^{-\alpha|\tau|}}{8}. \quad (43)$$

This autocorrelation has its peak at $\tau = 0$ and decays exponentially on both sides of the time shift axis, qualitatively following the shape seen in Fig. 8a. On the other hand, an oscillatory autocorrelation or autocovariance, as seen in Fig. 8b, may be obtained when the impulse response $h(t)$ is oscillatory, which may occur, for example, in a second-order underdamped system that would have an impulse response $h(t) = e^{-\alpha t} \cos(\omega_0 t) \amalg(t)$.

4. CROSS-CORRELATION BETWEEN TWO STATIONARY RANDOM PROCESSES

The presentation up to now was developed for both the continuous-time and discrete-time cases. From now on the expressions for the discrete-time case will only be presented.

4.1. Basic Definitions

Two signals are often recorded in an experiment from different parts of a system because an interest exists in analyzing if they are associated with each other. This association could develop from an anatomical coupling (e.g., two interconnected sites in the brain (17)) or a physiological coupling between the two recorded signals (e.g., heart rate variability and respiration (18)). On the other hand, they could be independent because no anatomical and physiological link exists between the two signals. Start with a formal definition of independence of two random processes $X(n)$ and $Y(n)$.

Two random processes $X(n)$ and $Y(n)$ are **independent** when any set of random variables $\{X(n_1), X(n_2), \ldots, X(n_N)\}$ taken from $X(n)$ is independent of another set of random variables $\{Y(n'_1), Y(n'_2), \ldots, Y(n'_M)\}$ taken from $Y(n)$. Note that the time instants at which the random variables are defined from each random process are taken arbitrarily, as indicated by the set of integers n_i, $i = 1, 2, \ldots, N$ and n'_i, $i = 1, 2, \ldots, M$, with N and M being arbitrary positive integers. This definition may be useful when conceptually it is known beforehand that the two systems that generate $X(n)$ and $Y(n)$ are totally uncoupled. However, in practice, usually no *a priori* knowledge about the systems exists and the objective is to discover if they are coupled, which means that we want to study the possible association or coupling of the two systems based on their respective output signals $x(n)$ and $y(n)$. For this purpose, the definition of independence is unfeasible to test in practice and one has to rely on concepts of association based on second-order moments.

In the same way as the correlation coefficient quantifies the degree of linear association between two random variables, the cross-correlation and the cross-covariance quantify the degree of linear association between two random processes $X(n)$ and $Y(n)$. Their cross-correlation is defined as:

$$R_{xy}(k) = E[X(n+k) \cdot Y(n)], \quad (44)$$

and their cross-covariance as

$$C_{xy}(k) = E[(X(n+k) - \mu_x) \cdot (Y(n) - \mu_y)] \\ = R_{xy}(k) - \mu_x/\mu_y. \quad (45)$$

It should be noted that some books or papers define the cross-correlation and cross-covariance as $R_{xy}(k) = E[X(n) \cdot Y(n+k)]$ and $C_{xy}(k) = E[(X(n) - \mu_x) \cdot (Y(n+k) - \mu_y)]$, which are time-reversed versions of the definitions above Equations 44 and 45. The distinction is clearly important when viewing a cross-correlation graph between two experimentally recorded signals $x(n)$ and $y(n)$, coming from random processes $X(n)$ and $Y(n)$, respectively. A peak at a positive time shift k according to one definition would appear at a negative time shift $-k$ in the alternative definition. Therefore, when using a signal processing software package or when reading a scientific text, the reader should always verify how the cross-correlation was defined. In Matlab (MathWorks, Inc.), a very popular software tool for signal processing, the commands xcorr(x,y) and xcov(x,y) use the same conventions as in Equations 44 and 45.

Similarly to what was said before for the autocorrelation definitions, texts on time series analysis define cross-correlation as the normalized cross-covariance:

$$\rho_{xy}(k) = \frac{C_{xy}(k)}{\sigma_x \sigma_y} = E\left[\frac{(X(n+k) - \mu_x)}{\sigma_x} \cdot \frac{(Y(n) - \mu_y)}{\sigma_y}\right] \quad (46)$$

where σ_x and σ_y are the standard deviations of the two random processes $X(n)$ and $Y(n)$, respectively.

4.2. Basic Properties

As $X(n+k) \cdot Y(n)$ is in general different from $\pm[X(n-k) \cdot Y(n)]$, the cross-correlation and the cross-covariance have a more subtle symmetry property than the autocorrelation and autocovariance:

$$R_{xy}(k) = R_{yx}(-k) = E[Y(n-k) \cdot X(n)]. \quad (47)$$

It should be noted that the time argument in $R_{yx}(-k)$ is negative *and* the order xy is changed to yx. For the cross-covariance, a similar symmetry relation applies:

$$C_{xy}(k) = C_{yx}(-k) \quad (48)$$

and

$$\rho_{xy}(k) = \rho_{yx}(-k). \quad (49)$$

For the autocorrelation and autocovariance, the peak occurs at the origin, as given by the properties in Equations 19 and 20. However, for the crossed moments, the peak may occur at any time shift value, with the following inequalities being valid:

$$|R_{xy}(k)| \leq \sqrt{R_{xx}(0)R_{yy}(0)} \quad (50)$$

$$|C_{xy}(k)| \leq \sigma_x \sigma_y \quad (51)$$

$$|\rho_{xy}(k)| \leq 1. \quad (52)$$

Finally, two discrete-time ergodic random processes $X(n)$ and $Y(n)$ without a periodic component will satisfy the following:

$$\lim_{|k| \to \infty} R_{xy}(k) = \mu_x \mu_y \quad (53)$$

and

$$\lim_{|k| \to \infty} C_{xy}(k) = 0, \quad (54)$$

with similar results being valid for continuous time. These limit results mean that the "effect" of a random variable taken from process X on a random variable taken from process Y decreases as the two random variables are taken farther apart. In the limit they become uncorrelated.

4.3. Independent and Uncorrelated Random Processes

When two random processes $X(n)$ and $Y(n)$ are independent, their cross-covariance is always equal to 0 for any time shift, and hence the autocorrelation is equal to the product of the two means:

$$C_{xy}(k) = 0 \quad \forall k \tag{55}$$

and

$$R_{xy}(k) = \mu_x \mu_y \quad \forall k. \tag{56}$$

Two random processes that satisfy the two expressions above are called **uncorrelated** (10,15) whether they are independent or not. In practical applications, one usually has no way to test for the independence of two random processes, but it is feasible to test if they are correlated or not. Note that the term uncorrelated may be misleading because the cross-correlation itself is not zero (unless one of the mean values is zero, when the processes are called orthogonal), but the cross-covariance is.

Two independent random processes are always uncorrelated, but the reverse is not necessarily true. Two random processes may be uncorrelated (i.e., have zero cross-covariance) but still be statistically dependent, because other probabilistic quantifiers (e.g., higher-order central moments (third, fourth, etc.)), will not be zero. However, for Gaussian (or normal) random processes, a one-to-one correspondence exists between independence and uncorrelatedness (10). This special property is valid for Gaussian random processes because they are specified entirely by the mean and second moments (autocovariance function for a single random process and cross-covariance for the joint distribution of two random processes).

4.4. Simple Model of Delayed and Amplitude-Scaled Random Signal

In some biomedical applications, the objective is to estimate the delay between two random signals. For example, the signals may be the arterial pressure and cerebral blood flow velocity for the study of cerebral autoregulation, or EMGs from different muscles for the study of tremor (19). The simplest model relating the two signals is $y(n) = \alpha \cdot x(n - L)$ where $\alpha \in \mathbb{R}$, and $L \in \mathbb{Z}$, which means that y is an amplitude-scaled and delayed (for $L > 0$) version of x. From Equation 46 $\rho_{xy}(k)$ will have either a maximum peak equal to 1 (if $\alpha > 0$) or a trough equal to -1 (if $\alpha < 0$) located at time shift $k = -L$. Hence, peak location in the time axis of the cross-covariance (or cross-correlation) indicates the delay value.

A slightly more realistic model in practical applications assumes that one signal is a delayed and scaled version of another but with an extraneous additive noise:

$$y(n) = \alpha x(n - L) + w(n). \tag{57}$$

In this model, $w(n)$ is a random signal caused by external interference noise or intrinsic biological noise that cannot be controlled or measured. Usually, one can assume that $x(n)$ and $w(n)$ are signals from uncorrelated or independent random processes $X(n)$ and $W(n)$. Let us determine $\rho_{xy}(k)$ assuming access to $X(n)$ and $Y(n)$. From Equation 45,

$$C_{xy}(k) = E[(X(n+k) - \mu_x) \cdot (\alpha(X(n-L) - \mu_x)] \\ + E[(X(n+k) - \mu_x) \cdot (W(n) - \mu_w)], \tag{58}$$

but the last term is zero because $X(n)$ and $W(n)$ are uncorrelated. Hence,

$$C_{xy}(k) = \alpha C_{xx}(k+L). \tag{59}$$

As $C_{xx}(k+L)$ attains its peak value when $k = -L$ (i.e., when the argument $(k+L)$ is zero) this provides a very practical method to estimate the delay between two random signals: Find the time-shift value where the cross-covariance has a clear peak. If an additional objective is to estimate the amplitude-scaling parameter α, it may be achieved by dividing the peak value of the cross-covariance by σ_x^2 (see Equation 59).

Additionally, from Equations 46 and 59:

$$\rho_{xy}(k) = \frac{\alpha C_{xx}(k+L)}{\sigma_x \sigma_y}. \tag{60}$$

To find σ_y, we remember that $C_{yy}(0) = \sigma_y^2$ from Equation 12. From Equation 57 we have

$$C_{yy}(0) = E[\alpha(X(n-L) - \mu_x) \cdot \alpha(X(n-L) - \mu_x)] \\ + E[(W(n) - \mu_w) \cdot (W(n) - \mu_w)], \tag{61}$$

where again we used the fact that $X(n)$ and $W(n)$ are uncorrelated. Therefore,

$$C_{yy}(0) = \alpha^2 C_{xx}(0) + C_{ww}(0) = \alpha^2 \sigma_x^2 + \sigma_w^2, \tag{62}$$

and therefore

$$\rho_{xy}(k) = \frac{\alpha C_{xx}(k+L)}{\sigma_x \sqrt{\alpha^2 \sigma_x^2 + \sigma_w^2}}. \tag{63}$$

When $k = -L$, $C_{xx}(k+L)$ will reach its peak value equal to σ_x^2, which means that $\rho_{xy}(k)$ will have a peak at $k = -L$ equal to

$$\rho_{xy}(-L) = \frac{\alpha \sigma_x}{\sqrt{\alpha^2 \sigma_x^2 + \sigma_w^2}} = \frac{\alpha}{|\alpha|} \frac{1}{\sqrt{1 + \frac{\sigma_w^2}{\alpha^2 \sigma_x^2}}}. \tag{64}$$

Equation 64 is consistent with the case in which no noise $W(n)$ ($\sigma_w^2 = 0$) exists because the peak in $\rho_{xy}(k)$ will equal $+1$ or -1, if α is positive or negative, respectively. From Equation 64, when the noise variance σ_w^2 increases, the peak in $\rho_{xy}(k)$ will decrease in absolute value, but will still occur at time shift $k = -L$. Within the context of this example, another way of interpreting the peak value in the normalized cross-covariance $\rho_{xy}(k)$ is by asking what fraction $\Gamma_{x \to y}$ of Y is because of X in Equation 57, meaning the ratio of the standard deviation of the term $\alpha X(n-L)$ to the total standard deviation in $Y(n)$:

$$\Gamma_{x \to y} = \frac{|\alpha|\sigma_x}{\sigma_y}. \quad (65)$$

From Equation 60

$$\rho_{xy}(-L) = \frac{\alpha \sigma_x}{\sigma_y}, \quad (66)$$

and from Equations 65 and 66 it is concluded that the peak size (absolute peak value) in the normalized cross-covariance indicates the fraction of Y because of the random process X:

$$|\rho_{xy}(-L)| = \Gamma_{x \to y}, \quad (67)$$

which means that when the deleterious effect of the noise W increases (i.e., when its variance increases), the contribution of X to Y decreases, which by Equations 64 and 67 is reflected in a decrease in the size of the peak in $\rho_{xy}(k)$.

The derivation given above showed that the peak in the cross-covariance gives an estimate of the delay between two random signals linked by model Equation 57 and that the amplitude-scaling parameter may also be estimated.

5. CROSS-CORRELATION BETWEEN THE RANDOM INPUT AND OUTPUT OF A LINEAR SYSTEM

If a discrete-time linear system has an impulse response $h(n)$ and a random process $X(n)$ with an autocorrelation $R_{xx}(k)$ is applied at its input, the cross-correlation between the input and the output processes will be (15):

$$R_{xy}(k) = h(-k) * R_{xx}(k), \quad (68)$$

and if the input is white the result becomes $R_{xy}(k) = h(-k)$.

The Fourier Transform of $R_{xy}(k)$ is the cross power spectrum $S_{xy}(e^{j\Omega})$ and from Equation 68 and the properties of the Fourier transform an important relation is found:

$$S_{xy}(e^{j\Omega}) = H^*(e^{j\Omega}) \cdot S_{xx}(e^{j\Omega}). \quad (69)$$

The results in Equations 68 and 69 are frequently used in biological system identification, when the system linearity may hold. Therefore, if the input signal is white, one may obtain an estimate of $R_{xy}(k)$ from the measurement of the input and output signals. Following Equation 68 the only thing to do is invert the time axis (what is negative becomes positive, and vice-versa) to get an estimate of the impulse response $h(k)$. In the case of nonwhite input, it is better to use Equation 69 to obtain an estimate of $H(e^{j\Omega})$ by dividing $S_{xy}^*(e^{j\Omega})$ by $S_{xx}(e^{j\Omega})$ (4,21) and then obtain the estimated impulse response by inverse Fourier transform.

5.1. Common Input

Let us assume that signals x and y, recorded from two points in a given biological system, are used to compute an estimate of $R_{xy}(k)$. Let us also assume that a

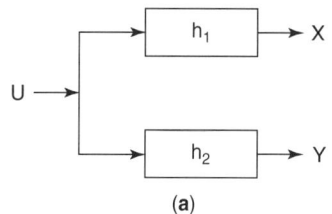

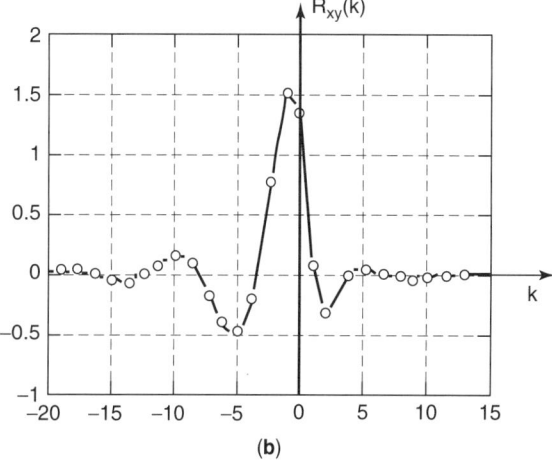

Figure 9. (a) Schematic of a common input U to two linear systems with impulse responses h_1 and h_2, the first generating the output X and the second the output Y. (b) Cross-correlation between the two outputs X and Y of a computer simulation of the schematic in (a). The cross-correlation samples were joined by straight lines to improve the visualization. Without additional knowledge, this cross-correlation could have come from a system with input x and output y or from two systems with a common input, as was the case here.

clear peak appears around $k = -10$. One interpretation would be that signal x "caused" signal y, with an average delay of 10, because signal x passed through some equivalent (yet unknown) sub-system to generate signal y, and $R_{xy}(k)$ would follow from Equation 68. However, in biological systems, one notable example being the nervous system, one should never discard the possibility of a common source exerting effects on two subsystems whose outputs are the measured signals x and y. This situation is depicted in Fig. 9a, where the common input is a random process U that is applied at the inputs of two subsystems, with impulse responses h_1 and h_2. In many cases, only the outputs of each of the subsystems X and Y (Fig. 9a) can be recorded, and all the analyses are based on their relationships. Working in discrete time, the cross-correlation between the two output random processes may be written as a function of the two impulse responses as

$$R_{xy}(k) = h_1(k) * h_2(-k) * R_{uu}(k), \quad (70)$$

which is simplified if the common input is white:

$$R_{xy}(k) = h_1(k) * h_2(-k). \quad (71)$$

Figure 9b shows an example of a cross-correlation of the outputs of two linear systems that had the same random signal applied to their inputs. Without prior information (e.g., on the possible existence of a common input) or additional knowledge (e.g., of the impulse response of one of the systems and $R_{uu}(k)$) it would certainly be difficult to interpret such a cross-correlation.

An example from the biomedical field shall illustrate a case where the existence of a common random input was hypothesized based on empirically obtained cross-covariances. The experiments consisted of evoking spinal cord reflexes bilaterally and simultaneously in the legs of each subject, as depicted in Fig. 10. A single electrical stimulus, applied to the right or left leg, would fire an action potential in some of the sensory nerve fibers situated under the stimulating electrodes (st in Fig. 10). The action potentials would travel to the spinal cord (indicated by the upward arrows) and activate a set of motoneurons (represented by circles in a box). The axons of these motoneurons would conduct action potentials to a leg muscle, indicated by downward arrows in Fig. 10. The recorded waveform from the muscle (shown either to the left or right of Fig. 10) is the so-called H reflex. Its peak-to-peak amplitude is of interest, being indicated as x for the left leg reflex waveform and y for the right leg waveform in Fig. 10. The experimental setup included two stimulators (indicated as **st** in Fig. 10) that applied simultaneous trains of 500 rectangular electrical stimuli at 1 per second to the two legs. If the stimulus pulses in the trains are numbered as $n=0$, $n=1,\ldots, n=N$ (the authors used $N=500$), the respective sequences of H reflex amplitudes recorded on each side will be $x(0), x(1), \ldots, x(N)$ and $y(0), y(1), \ldots, y(N)$, as depicted in the inset at the lower right side of Fig. 10. The two sequences of reflex amplitudes were analyzed by the cross-covariance function (22).

Each reflex peak-to-peak amplitude depends on the upcoming sensory activity discharged by each stimulus and also on random inputs from the spinal cord that act on the motoneurons. In Fig. 10, a "**U?**" indicates a hypothesized common input random signal that would modulate synchronously the reflexes from the two sides of the spinal cord. The peak-to-peak amplitudes of the right- and left-side H reflexes to a train of 500 bilateral stimuli were measured in real time and stored as discrete-time signals x and y. Initial experiments had shown that a unilateral stimulus only affected the same side of the spinal cord. This finding meant that any statistically significant peak in the cross-covariance of x and y could be attributed to a common input. The first 10 reflex amplitudes in each signal were discarded to avoid the transient (nonstationarity) that occurs at the start of the stimulation.

Data from a subject are shown in Fig. 11, the first 51 H-reflex amplitude values shown in Fig. 11a for the right leg, and the corresponding simultaneous H reflex amplitudes in Fig. 11b for the left leg (R. A. Mezzarane and A. F. Kohn, unpublished data). In both, a horizontal line shows the respective mean value computed from all the 490 samples of each signal. A simple visual analysis of the data probably tells us close to nothing about how the reflex amplitudes vary and if the two sides fluctuate together to some degree. Such quantitative questions may be answered by the autocovariance and cross-covariance functions. The right- and left-leg H-reflex amplitude normalized autocovariances, computed according to Equation 15, are shown in Fig. 12a and 12b, respectively. The value at time shift 0 is 1, as expected from the normalization. Both autocovariances show that for small time shifts, some degree of correlation exists between the samples because

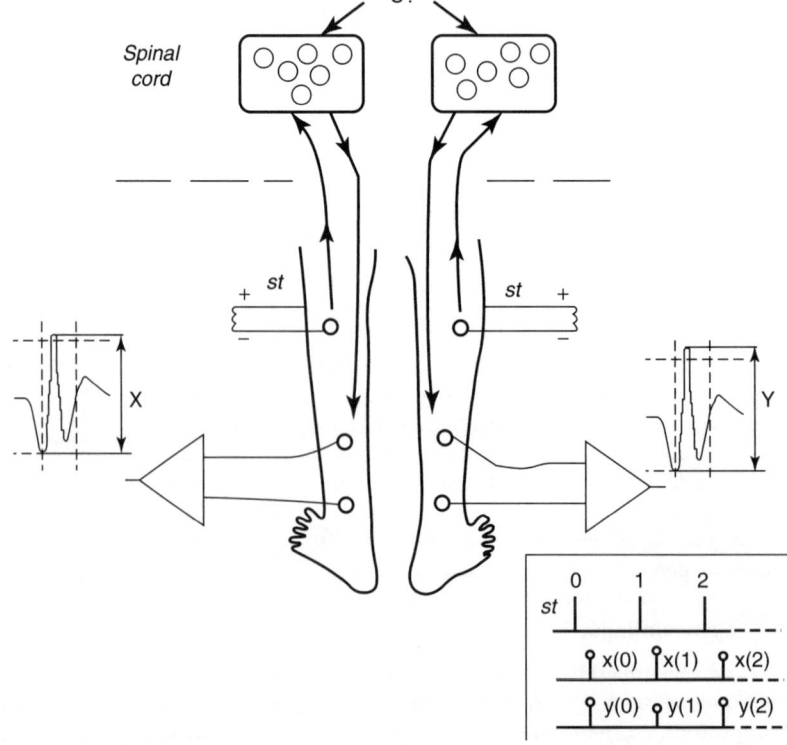

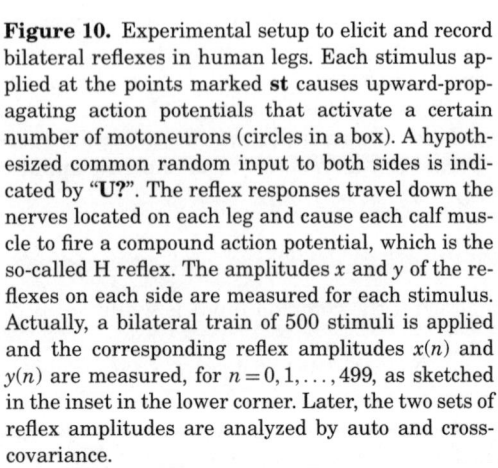

Figure 10. Experimental setup to elicit and record bilateral reflexes in human legs. Each stimulus applied at the points marked **st** causes upward-propagating action potentials that activate a certain number of motoneurons (circles in a box). A hypothesized common random input to both sides is indicated by "**U?**". The reflex responses travel down the nerves located on each leg and cause each calf muscle to fire a compound action potential, which is the so-called H reflex. The amplitudes x and y of the reflexes on each side are measured for each stimulus. Actually, a bilateral train of 500 stimuli is applied and the corresponding reflex amplitudes $x(n)$ and $y(n)$ are measured, for $n=0,1,\ldots,499$, as sketched in the inset in the lower corner. Later, the two sets of reflex amplitudes are analyzed by auto and cross-covariance.

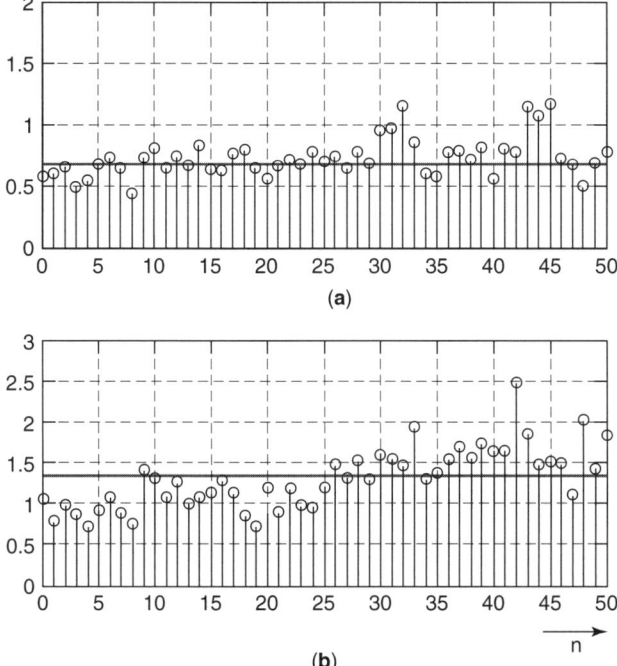

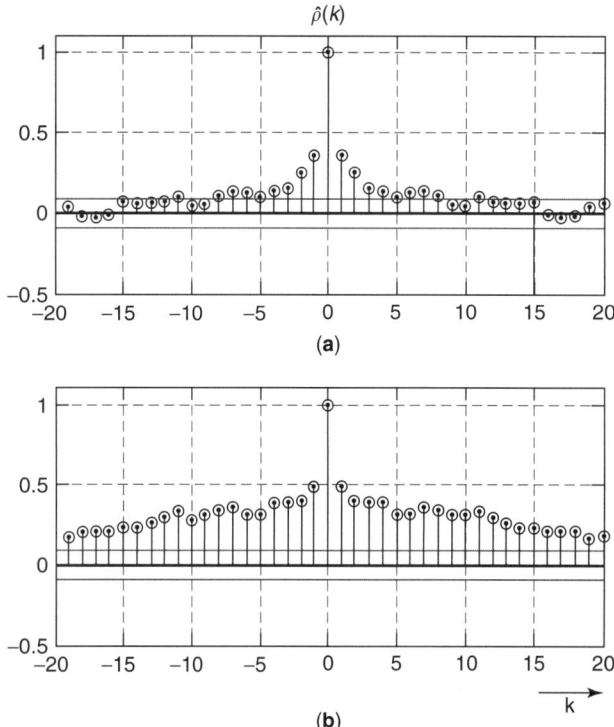

Figure 11. The first 51 samples of two time series $y(n)$ and $x(n)$ representing the simultaneously-measured H-reflex amplitudes from the right (a) and left (b) legs in a subject. The horizontal lines indicate the mean values of each complete series. The stimulation rate was 1 Hz. Ordinates are in mV. (This figure is available in full color at http://www.mrw.interscience.wiley.com/ebe.)

the autocovariance values are above the upper level of a 95% confidence interval shown in Fig. 12 (see subsection Hypothesis testing). This fact supports the hypothesis that randomly varying neural inputs exists in the spinal cord that modulate the excitability of the reflex circuits. As the autocovariance in Fig. 12b decays much slower than that in Fig. 12a, it suggests that the two sides receive different randomly varying inputs. The normalized cross-covariance, computed according to Equation 46, is seen in Fig. 13. Many cross-covariance samples around $k = 0$ are above the upper level of a 95% confidence interval (see subsection Hypothesis testing), which suggests a considerable degree of correlation between the reflex amplitudes recorded from both legs of the subject. The decay on both sides of the cross-covariance in Fig. 13 could be because of the autocovariances of the two signals (see Fig. 12), but this issue will only be treated later. Results such as those in Fig. 13, also found in other subjects (22), suggested the existence of common inputs acting on sets of motoneurons at both sides of the spinal cord. However, as the peak of the cross-covariance was lower than 1 and as the autocovariances were different bilaterally, it can be stated that each side receives common sources to both sides plus random inputs that are uncorrelated with the other side's inputs. Experiments in cats are being pursued, with the help of the cross-covariance analysis, to unravel the neural sources of the random inputs found to act bilaterally in the spinal cord (23).

Figure 12. The respective autocovariances of the signals shown in Fig. 11. The autocovariance in (a) decays faster to the confidence interval than that in (b). The two horizontal lines represent a 95% confidence interval. (This figure is available in full color at http://www.mrw.interscience.wiley.com/ebe.)

6. ESTIMATION AND HYPOTHESIS TESTING FOR AUTOCORRELATION AND CROSS-CORRELATION

This section will focus solely on discrete-time random signals because, today, the signal processing techniques are almost all realized in discrete-time in very fast computers.

If *a priori* knowledge about the stationarity of the random process exists a test should be applied for stationarity (4). If a trend of no physiological interest is discovered in the data, it may be removed by linear or nonlinear regression. After a stationary signal is obtained, then the problem is to estimate first and second moments as presented in the previous sections.

Let us assume a stationary signal $x(n)$ is known for $n = 0, 1, \ldots, N - 1$. Any estimate $\Theta(x(n))$ based on this signal would have to present some desirable properties, derived from its corresponding estimator $\Theta(X(n))$, such as unbiasedness and consistency (15). Note that an estimate is a particular value (or a set of values) of the estimator when a given sample function $x(n)$ is used instead of the whole process $X(n)$ (which is what happens in practice). For an **unbiased estimator** $\Theta(X(n))$, its expected value is equal to the actual value being estimated θ:

$$E[\Theta(X(n))] = \theta. \qquad (72)$$

For example, if we want to estimate the mean μ_x of the process $X(n)$ using $\bar{x}$ Equation 2, the unbiasedness means that the expected value of the estimator has to equal μ_x. It

276 AUTOCORRELATION AND CROSSCORRELATION METHODS

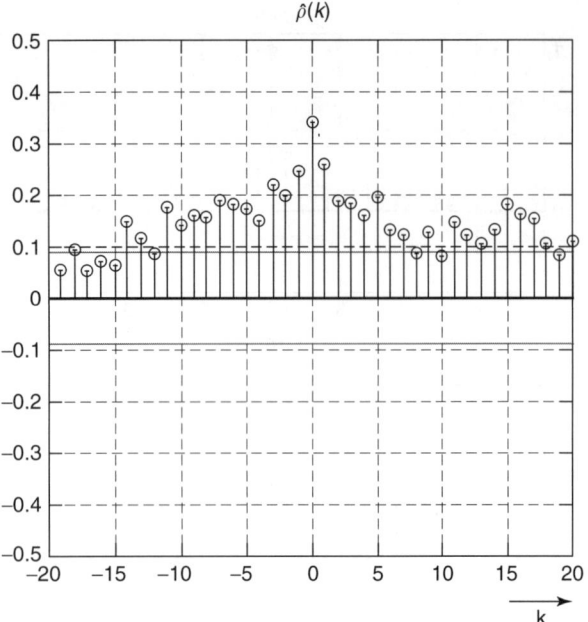

Figure 13. Cross-covariance between the two signals with 490 bilateral reflexes shown partially in Fig 11. Only the time shifts near the origin are shown here because they are the most reliable. The two horizontal lines represent a 95% confidence interval. (This figure is available in full color at http://www.mrw.interscience.wiley.com/ebe.)

is easy to show that the finite average $[X(0) + X(1) + \cdots + X(N-1)]/N$ of a stationary process is an unbiased estimator of the expected value of the process:

$$E[\Theta(X(n))] = E\left[\frac{1}{N}\sum_{n=0}^{N-1} X(n)\right] = \frac{1}{N}\sum_{n=0}^{N-1} E[X(n)]$$

$$= \frac{1}{N}\sum_{n=0}^{N-1} \mu_x = \mu_x. \quad (73)$$

If a given estimator is unbiased, it still does not assure its usefulness, because if its variance is high it means that for a given signal $x(n)$ - a single sample function of the process $X(n)$ - the value of $\Theta(x(n))$ may be quite far from the actual value θ. Therefore, a desirable property of an unbiased estimator is that its variance tends to 0 as the number of samples N tends to ∞, a property called **consistency**:

$$\sigma_\Theta^2 = E[(\Theta(X(n)) - \theta)^2]\underset{N\to\infty}{\longrightarrow} 0. \quad (74)$$

Returning to the unbiased mean estimator Equation 2, let us check its consistency:

$$\sigma_\Theta^2 = E[(\Theta(X(n)) - \mu_x)^2]$$

$$= E\left[\left(\frac{1}{N}\sum_{n=0}^{N-1}(X(n) - \mu_x)\right)^2\right], \quad (75)$$

which, after some manipulations (15), gives:

$$\sigma_\Theta^2 = \frac{1}{N}\sum_{k=-N+1}^{N-1}\left(1 - \frac{|k|}{N}\right) \cdot C_{xx}(k). \quad (76)$$

From Equation 23 it is concluded that the mean estimator is consistent. It is interesting to observe that the variance of the mean estimator depends on the autocovariance of the signal and not only on the number of samples, which is because of the statistical dependence between the samples of the signal, quite contrary to what happens in conventional statistics, where the samples are i.i.d..

6.1. Estimation of Autocorrelation and Autocovariance

Which estimator should be used for the autocorrelation of a process $X(n)$? Two slightly different estimators will be presented and their properties verified. Both are much used in practice and are part of the options of the Matlab command xcorr.

The first estimator will be called the *unbiased autocorrelation estimator* and defined as

$$\hat{R}_{xx,u}(k) = \frac{1}{N - |k|}\sum_{n=0}^{N-1-|k|} X(n+|k|)X(n); \quad (77)$$

$$|k| \leq N - 1.$$

The basic operations are sample-to-sample multiplications of two time-shifted versions of the process and arithmetic average of the resulting samples, which seems to be a reasonable approximation to Equation 13. To confirm that this estimator is indeed unbiased, its expected value shall be determined:

$$E[\hat{R}_{xx,u}(k)] = \frac{1}{N-|k|}\sum_{n=0}^{N-1-|k|} E[X(n+|k|)X(n)]; \quad (78)$$

$$|k| \leq N-1,$$

and from Equations 13 and 16 it can be concluded that the estimator is indeed unbiased (i.e., $E[\hat{R}_{xx,u}(k)] = R_{xx}(k)$). It is more difficult to verify consistency, and the reader is referred to Therrien (15). For a Gaussian process $X(n)$, it can be shown that the variance indeed tends to 0 for $N \to \infty$, which is a good approximation for more general random processes. Therefore, Equation 77 gives us a consistent estimator of the autocorrelation. When applying Equation 77 in practice, the available signal $x(n), n = 0, 1, \ldots, N-1$, is used instead of $X(n)$ giving a sequence of $2N - 1$ values $\hat{R}_{xx,u}(k)$.

The *unbiased autocovariance estimator* is similarly defined as

$$\hat{C}_{xx,u}(k) = \frac{1}{N-|k|}\sum_{n=0}^{N-1-|k|}(X(n+|k|)-\bar{x})\cdot(X(n)-\bar{x}); \quad (79)$$

$$|k| \leq N-1,$$

where $\bar{x}$ is given by Equation 2. Similarly, $\hat{C}_{xx,u}(k)$ is a consistent estimator of the autocovariance.

Nevertheless, one of the problems with these two estimators defined in Equations 77 and 79 is that they may not obey Equations 19, 20 or the positive semi-definite property (14,15).

Therefore, other estimators are defined, such as the biased autocorrelation and autocovariance estimators $\hat{R}_{xx,b}(k)$ and $\hat{C}_{xx,b}(k)$

$$\hat{R}_{xx,b}(k) = \frac{1}{N} \sum_{n=0}^{N-1-|k|} (X(n+|k|)X(n); \quad |k| \leq N-1 \quad (80)$$

and

$$\hat{C}_{xx,b}(k) = \frac{1}{N} \sum_{n=0}^{N-1-|k|} (X(n+|k|) - \bar{x}) \cdot (X(n) - \bar{x});$$

$$|k| \leq N-1. \quad (81)$$

The only difference between Equations 77 and 80, or Equations 79 and 81 is the denominator, which means that $\hat{R}_{xx,b}(k)$ and $\hat{C}_{xx,b}(k)$ are biased estimators. The zero padding used in the time-shifted versions of signal $x(n)$ in Fig. 4b–d may be interpreted as the cause of the bias of the estimators in Equations 80 and 81 as the zero samples do not contribute to the computations in the sum but are counted in the denominator N. The bias is defined as the expected value of an estimator minus the actual value, and for these two biased estimators the expressions are:

$$Bias[\hat{R}_{xx,b}(k)] = -\frac{|k|}{N} R_{xx}(k); \quad |k| \leq N-1 \quad (82)$$

and

$$Bias[\hat{C}_{xx,b}(k)] = -\frac{|k|}{N} C_{xx}(k); \quad |k| \leq N-1. \quad (83)$$

Nevertheless, when $N \to \infty$, the expected values of $\hat{R}_{xx,b}(k)$ and $\hat{C}_{xx,b}(k)$ equal the theoretical autocorrelation and autocovariance (i.e., the estimators are asymptotically unbiased). Their consistency follows from the consistency of the unbiased estimators. Besides these desirable properties, it can be shown that these two estimators obey Equations 19 and 20 and are always positive semi-definite (14,15).

The question is then, which of the two estimators, the unbiased or the biased, should be choosen? Usually, that choice is less critical than the choice of the maximum value of k used in the computation of the autocorrelation or autocovariance. Indeed, *for large k* the errors when using Equations 80 and 81 are large because of the bias, as seen from Equations 82 and 83, whereas the errors associated with Equations 77 and 79 are large because of variance of the estimates, due to the small value of the denominator $N - |k|$ in the formulas. Therefore, the user should try to limit the values of $|k|$ to as low a value as appropriate for the application, for example, $|k| \leq N/10$ or $|k| \leq N/4$, because near $k=0$ the estimates have the highest reliability. On the other hand, if the estimated autocorrelation or autocovariance will be Fourier-transformed to provide an estimate of a power spectrum, then the biased autocorrelation or autocovariance estimators should be chosen to assure a non-negative power spectrum.

For completeness purposes, the normalized autocovariance estimators are given below. The unbiased estimator is:

$$\hat{\rho}_{xx,u}(k) = \frac{\hat{C}_{xx,u}(k)}{\hat{C}_{xx,u}(0)}; \quad |k| \leq N-1, \quad (84)$$

where $\hat{\rho}_{xx,u}(0) = 1$. An alternative expression should be employed if the user computes first $\hat{C}_{xx,u}(k)$ and then divides it by the variance estimate $\hat{\sigma}_x^2$, which is usually the unbiased estimate in most computer packages:

$$\hat{\rho}_{xx,u}(k) = \left(\frac{N}{N-1}\right) \frac{\hat{C}_{xx,u}(k)}{\hat{\sigma}_x^2}; \quad |k| \leq N-1. \quad (85)$$

Finally, the biased normalized autocovariance estimator is:

$$\hat{\rho}_{xx,b}(k) = \frac{\hat{C}_{xx,b}(k)}{\hat{C}_{xx,b}(0)}; \quad |k| \leq N-1. \quad (86)$$

Assuming the available signal $x(n)$ has N samples, all the autocorrelation or autocovariance estimators presented above will have $2N-1$ samples.

An example of the use of Equation 84 was already presented in the subsection "Common input" above. The two signals, right- and left-leg H-reflex amplitudes, had $N = 490$ samples, but the unbiased normalized autocovariances shown in Fig. 12 were only shown for $|k| \leq 20$. However, the values at the two extremes $k = \pm 489$ (not shown) surpassed 1, which is meaningless. As mentioned before, the values for time shifts far away from the origin should not be used for interpretation purposes. A more involved way of computing the normalized autocovariance that assures values within the range $[-1,1]$ for all k can be found on page 331 of Priestley's text (14).

From a computational point of view, it is usually not recommended to compute directly any of the estimators given in this subsection, except for small N. As the autocorrelation or autocovariance estimators are basically the discrete-time convolutions of $x(n)$ with $x(-n)$, the computations can be done very efficiently in the frequency domain using an FFT algorithm (16), with the usual care of zero padding to convert a circular convolution to a linear convolution. Of course, for the user of scientific packages such as Matlab, this problem does not exist because each command, such as xcorr and xcov, is already implemented in a computationally efficient way.

6.2. Estimation of Cross-Correlation and Cross-Covariance

The concepts involved in the estimation of the cross-correlation or the cross-covariance are analogous to those of the autocorrelation and autocovariance presented in the previous section. An important difference, however, is that the cross-correlation and cross-covariance do not have even symmetry. The expressions for the unbiased and biased estimators of cross-correlation given below are in a form appropriate for signals $x(n)$ and $y(n)$, both defined for $n = 0, 1, \ldots, N - 1$. The unbiased estimator is

$$\hat{R}_{xy,u}(k) = \begin{cases} \frac{1}{N-|k|} \sum_{n=0}^{N-1-|k|} X(n+k)Y(n); & 0 \leq k \leq N-1 \\ \frac{1}{N-|k|} \sum_{n=0}^{N-1-|k|} X(n)Y(n+|k|); & -(N-1) \leq k < 0, \end{cases} \quad (87)$$

and the biased estimator is

$$\hat{R}_{xy,b}(k) = \begin{cases} \frac{1}{N} \sum_{n=0}^{N-1-|k|} X(n+k)Y(n); & 0 \leq k \leq N-1 \\ \frac{1}{N} \sum_{n=0}^{N-1-|k|} X(n)Y(n+|k|); & -(N-1) \leq k < 0. \end{cases} \quad (88)$$

In the definitions above, $\hat{R}_{xy,u}(k)$ and $\hat{R}_{xy,b}(k)$ will have $2N - 1$ samples. Both expressions could be made more general by having $x(n)$ with N samples and $y(n)$ with M samples ($M \neq N$), resulting in a cross-correlation with $N + M - 1$ samples, with k in the range $[-N, M]$. However, the reader should be cautioned that the outputs of algorithms that were initially designed for equal-sized signals may require some care. A suggestion is to check the outputs with simple artificially generated signals for which the cross-correlation is easily determined or known.

The corresponding definitions of the unbiased and biased estimators of the cross-covariance, for two signals with N samples each, are

$$\hat{C}_{xy,u}(k) = \begin{cases} \frac{1}{N-|k|} \sum_{n=0}^{N-1-|k|} (X(n+k) - \bar{x})(Y(n) - \bar{y}); \\ \quad 0 \leq k \leq N-1 \\ \frac{1}{N-|k|} \sum_{n=0}^{N-1-|k|} (X(n) - \bar{x})(Y(n+|k|) - \bar{y}); \\ \quad -(N-1) \leq k < 0 \end{cases} \quad (89)$$

and

$$\hat{C}_{xy,b}(k) = \begin{cases} \frac{1}{N} \sum_{n=0}^{N-1-|k|} (X(n+k) - \bar{x})(Y(n) - \bar{y}); \\ \quad 0 \leq k \leq N-1 \\ \frac{1}{N} \sum_{n=0}^{N-1-|k|} (X(n) - \bar{x})(Y(n+|k|) - \bar{y}); \\ \quad -(N-1) \leq k < 0 \end{cases} \quad (90)$$

where $\bar{x}$ and $\bar{y}$ are the means of the two processes $X(n)$ and $Y(n)$. The normalized cross-covariance estimators are given below with the $\odot$ meaning either b or u (i.e., biased or unbiased versions):

$$\hat{\rho}_{xy,\odot}(k) = \frac{\hat{C}_{xy,\odot}(k)}{\sqrt{\hat{C}_{xy,\odot}(0) \cdot \hat{C}_{yy,\odot}(0)}}. \quad (91)$$

If the user first computes $\hat{C}_{xy,u}(k)$ and then divides this result by the product of the standard deviation estimates, the factor $N/(N-1)$ should be used in case the computer package calculates the unbiased variance (standard deviation) estimates

$$\hat{\rho}_{xy,u}(k) = \left(\frac{N}{N-1}\right) \frac{\hat{C}_{xy,u}(k)}{\hat{\sigma}_x \hat{\sigma}_y}, \quad (92)$$

which may be useful, for example, when using the Matlab command xcov, because none of its options include the direct computation of an unbiased and normalized cross-covariance between two signals x and y. The Equation 92 in Matlab would be written as:

$$\hat{\rho}_{xy,u}(k) = xcov(x, y, \text{'unbiased'})$$
$$* N/((N-1) * std(x) * std(y)). \quad (93)$$

Note that Matlab will give $\hat{\rho}_{xy,u}(k)$ (or other alternatives of cross-covariance or cross-correlation) as a vector with $2N - 1$ elements (e.g., with 979 points if $N = 490$) when the two signals x and y have N samples. For plotting the cross-covariance or cross-correlation computed by a software package, the user may create the time shift axis k by $k = -(N-1) : (N-1)$ to assure the correct position of the zero time shift value in the graph. In Matlab, the values of k are computed automatically if the user makes [C,k] = xcov(x,y,'unbiased'). If the sampling frequency f_s is known and the time shift axis should reflect continuous time calibration, then vector k should be divided by f_s.

The Equation 93 was employed to generate Fig. 13, which is the normalized unbiased estimate of the cross-covariance between the reflex amplitudes of the right and left legs of a subject, as described before.

Similarly to what was said before for the autocorrelation and autocovariance, in practical applications only the cross-covariance (or cross-correlation) samples nearer to the origin $k = 0$ should be used for interpretation purposes. This is because the increase in variance or bias in the unbiased or biased cross-covariance or cross-correlation estimates at large values of $|k|$.

6.3. Hypothesis Testing

One basic question when analyzing a given signal $x(n)$ is if it is white (i.e., if its autocovariance satisfies Equation 35. The problem *in practical applications* is that the autocovariance computed from a sample function of a white-noise process will be nonzero for $k \neq 0$, contrary to what would be expected for a white signal. This problem is typical in statistics, and it may be shown that a 95% confidence interval for testing the whiteness of a signal with N samples is given by the bounds $\pm 1.96/\sqrt{N}$ (11) when either of the normalized autocovariance estimators is used Equation 86, or Equation 84 for $k \ll N$), which comes from the fact that each autocovariance sample is asymptotically normal or Gaussian with zero mean and unit variance, and hence the range ± 1.96 corresponds to a probability of 0.95.

When two signals are measured from an experiment, the first question usually is if the two have some correlation or not. If they are independent or uncorrelated, theoretically their cross-covariance is zero for all values of k.

Again, *in practice*, this result is untenable as is exemplified in Fig. 14, which shows the normalized cross-covariance of two *independently generated* random signals $x(n)$ and $y(n)$. Each computer-simulated signal, $x(n)$ and $y(n)$, had $N=1000$ samples and was generated independently from the other by a filtering procedure applied on normally distributed random samples. It can be shown (11) that under the hypothesis that two random signals are *white*, a 95% confidence interval for testing their uncorrelatedness is given by $\pm 1.96/\sqrt{N}$ when the normalized cross-covariance formula Equation 91 is used, either for the biased case or for the unbiased case if $k \ll N$. This confidence interval is drawn in Fig. 14, even without knowing beforehand if the signals are indeed white. Several peaks in the cross-covariance oscillations are clearly outside the confidence interval, which could initially be interpreted as a statistical dependence between the two signals. However, this cannot be true because they were generated independently. The problem with this apparent paradox is that the cross-covariance is also under the influence of each signal's autocovariance.

One approach to the problem of testing the hypothesis of independence of two arbitrary (nonwhite) signals is the **whitening** of both signals before computing the cross-covariance (11). After the whitening, the $\pm 1.96/\sqrt{N}$ confidence interval can be used. The whitening filters most used in practice are inverse filters based either on autoregressive (AR) or autoregressive moving average (ARMA) models. In the present example, two AR models shall be obtained for each of the signals, first determining the order by the Akaike information criterion (24) and then

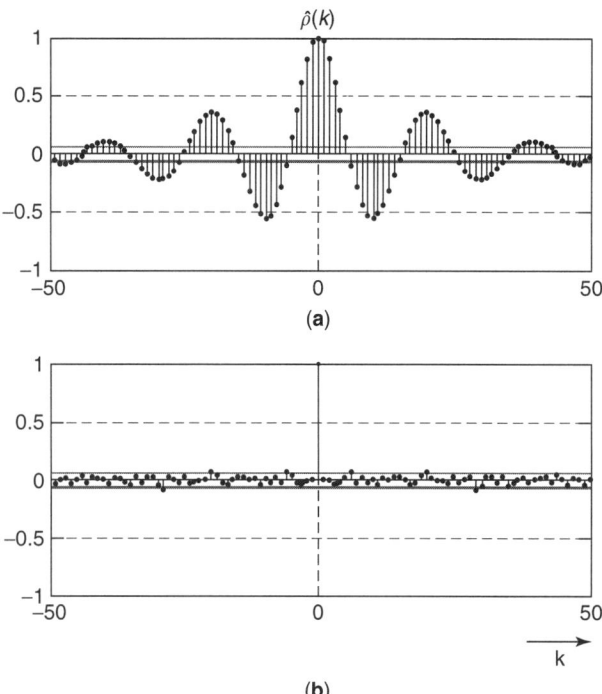

Figure 15. The autocovariance of signal $x(n)$ that generated the cross-covariance of Fig. 14 is shown in (a). It shows clearly that it is not a white signal because many of its samples fall outside a 95% confidence interval (two horizontal lines around the horizontal axis). (b) This signal was passed through an inverse filter computed from an AR model fitted to $x(n)$, producing a signal $x_o(n)$ that can be accepted as white as all the autocovariance samples at $k \neq 0$ are practically inside the 95% confidence interval. (This figure is available in full color at http://www.mrw.interscience.wiley.com/ebe.)

estimating the AR models by the Burg method (25,26). Each inverse filter is simply a moving average (MA) system with coefficients equal to those of the denominator of the transfer function of the respective all-pole estimated AR model. Figure 15a shows the autocovariance of signal $x(n)$, which clearly indicates that $x(n)$ is nonwhite, because most samples at $k \neq 0$ are outside the confidence interval. As $N=1000$, the confidence interval is $\pm 6.2 \times 10^{-2}$, which was drawn in Fig. 15a and 15b as two horizontal lines. Figure 15b shows the autocovariance of the signal $x_o(n)$ at the output of the respective inverse filter that has $x(n)$ at its input. Similarly, Fig. 16a shows the autocovariance of signal $y(n)$ and Fig. 16b the autocovariance of the output $y_o(n)$ of the corresponding inverse filter that has $y(n)$ at its input. The two autocovariances in Fig. 15b and Fig. 16b show that, except for the sample at $k=0$, which is equal to 1, the other samples stay within the confidence band, so that the hypothesis of white $x_o(n)$ and $y_o(n)$ can be accepted.

Next, the cross-covariance between $x_o(n)$ and $y_o(n)$ was computed and is shown in Fig. 17. Now, practically all the samples are contained in the confidence interval and the hypothesis that the two random signals $x(n)$ and $y(n)$ are uncorrelated (and independent, if they are Gaussian) can be accepted. For the few samples outside the confidence band, the user should use the fact that at each time shift k,

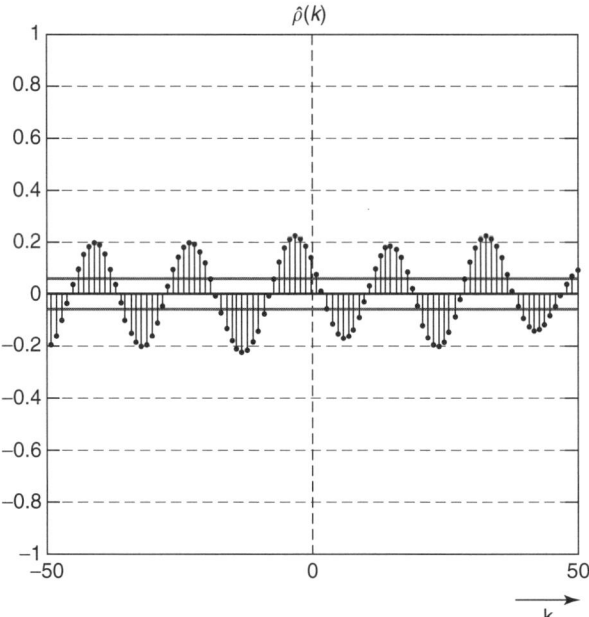

Figure 14. Cross-covariance of two signals $x(n)$ and $y(n)$, which were generated independently, with 1000 samples each. The cross-covariance shows oscillations above a 95% confidence interval indicated by the two horizontal lines, which would suggest a correlation between the two signals, but this conclusion is false because they were generated independently. (This figure is available in full color at http://www.mrw.interscience.wiley.com/ebe.)

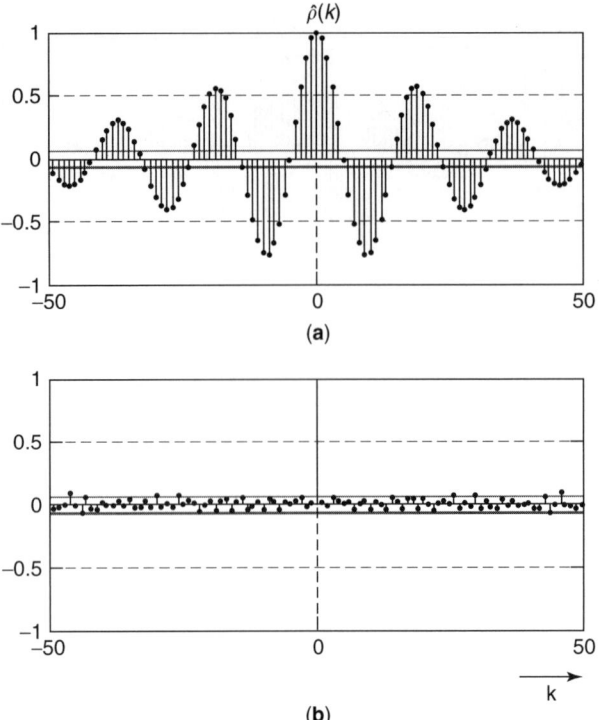

Figure 16. The autocovariance of signal $y(n)$ that generated the cross-covariance of Fig. 14 is shown in (a). It shows clearly that it is not a white signal because many of its samples fall outside a 95% confidence interval (two horizontal lines around the horizontal axis). (b) This signal was passed through an inverse filter computed from an AR model fitted to $y(n)$, producing a signal $y_0(n)$ that can be accepted as white as all the autocovariance samples at $k \neq 0$ are practically inside the 95% confidence interval. (This figure is available in full color at http://www.mrw.interscience.wiley.com/ebe.)

5% of the times a cross-covariance value could exist outside the 95% confidence interval, for independent signals. Besides this statistical argument, the biomedical user should also consider if the time shifts associated with samples outside the confidence band have some physiological meaning. Additional issues on testing the hypothesis of independency of nonwhite signals may be found in the text by Brockwell and Davis (11).

Let us apply this approach to the bilateral reflex data described in subsection "Common input." Note that the autocovariance functions computed from experimental data, shown in Fig. 12, indicate that both signals are not white because samples at $k \neq 0$ are outside the confidence band. Therefore, to be able to properly test whether the two sequences are correlated, the two sequences of reflex amplitudes shall be whitened and the cross-covariance of the corresponding whitened sequences computed. The result is shown in Fig. 18, where one can see that a sample exists at zero time shift clearly out of the confidence interval. The other samples are within the confidence limits, except for two that are slightly above the upper limit line and are of no statistical or physiological meaning.

From the result of this test the possibility that the wide peak found in Fig. 13 is an artifact can be discarded and one can feel rather confident that a common input acting synchronously (peak at $k = 0$) on both sides of the spinal cord indeed exists.

An alternative to the whitening approach is to compute statistics based on surrogate data generated from each of the two given signals $x(n)$ and $y(n)$ (27). Each surrogate signal $x_i(n)$ derived from $x(n)$, for example, would have an amplitude spectrum $|X_i(e^{j\Omega})|$ equal to $|X(e^{j\Omega})|$, but the phase would be randomized. Here $X(e^{j\Omega}) = DiscreteTimeFourierTransform[x(n)]$. The phase would be randomized with samples taken from a uniform distribution between $-\pi$ and π. The corresponding surrogate signal would be found by inverse Fourier Transform. With this Monte Carlo approach (27), the user will have a wider applicability of statistical methods than with the whitening method. Also, in case high frequencies generated in the whitening approach cause some masking of a genuine correlation between the signals, the surrogate method is a good alternative.

Finally, returning to the problem of delay estimation presented earlier, in section 4.4 it should be pointed out that several related issues were not covered in this chapter. For example, the standard deviation or confidence interval of the estimated delay may be computed to characterize the quality of a given estimate (4). A modified cross-covariance, based on a minimum least square method, may optimize the estimation of the time delay (20). Alternatively, the model may be different from Equation 57, (e.g., one of the signals is not exactly a delayed

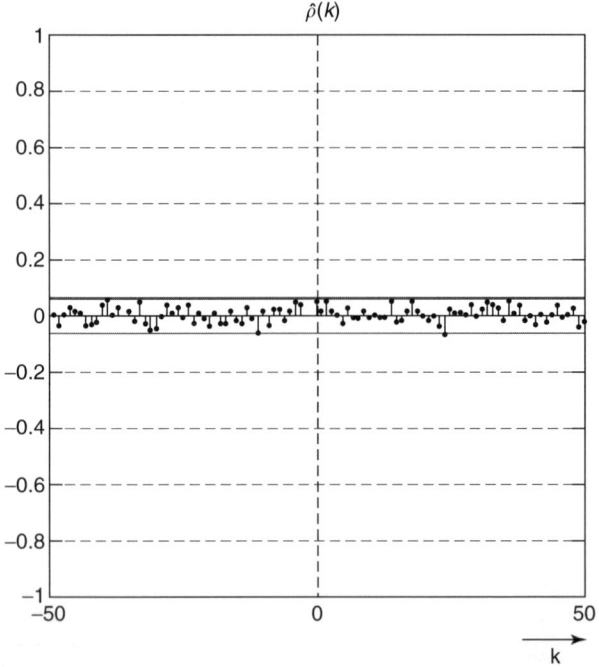

Figure 17. The cross-covariance estimate of the white signals $x_0(n)$ and $y_0(n)$ referred to in Figs. 15 and 16. The latter were obtained by inverse filtering the nonwhite signals $x(n)$ and $y(n)$ whose estimated cross-covariance is shown in Fig. 14. Here, it is noted that all the samples fall within the 95% confidence interval (two horizontal lines around the horizontal axis), which suggests that the two signals $x(n)$ and $y(n)$ are indeed uncorrelated. (This figure is available in full color at http://www.mrw.interscience.wiley.com/ebe.)

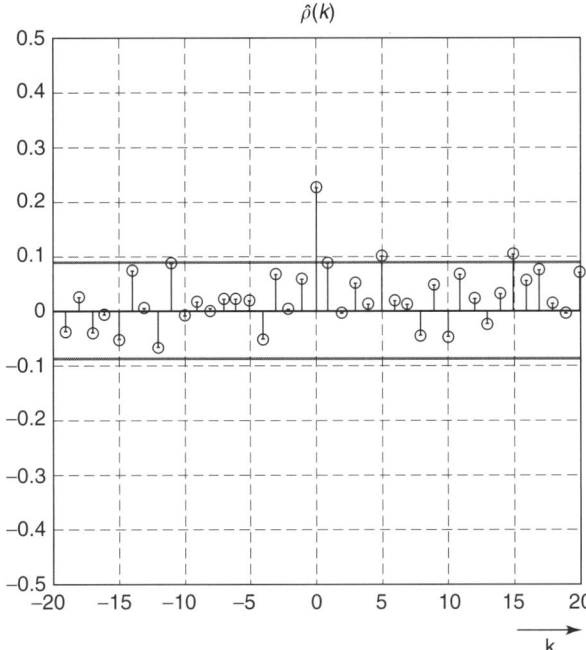

Figure 18. Cross-covariance computed from the whitened bilateral reflex signals shown in Fig. 11. Their original cross-covariance can be seen in Fig. 13. After whitening the two signals, the new cross-covariance shown here indicates that a statistically significant peak at time shift zero indeed exists. The two horizontal lines represent a 95% confidence interval. (This figure is available in full color at http://www.mrw.interscience.wiley.com/ebe.)

version of the other with added noise). One possible model would be $y(n) = \alpha q(n-L) + w(n)$, where $q(n) = x(n) * h(n)$. Here, $h(n)$ is the impulse response of a linear system that changes the available signal $x(n)$ into $q(n)$. The second available signal, $y(n)$, is a delayed and amplitude-scaled version of $q(n)$ with additive measurement noise. The concepts presented in section 5 earlier can be applied to this new model.

7. EXTENSIONS AND FURTHER APPLICATIONS

In spite of the widespread use of the autocorrelation and cross-correlation functions in the biomedical sciences, they have some intrinsic limitations. These limitations should be understood so that more appropriate signal processing tools may be employed or developed for the problem at hand. A few of the limitations of the auto and cross-correlation functions will be mentioned below together with some alternative signal processing tools that have been presented in the literature.

The autocorrelation and cross-correlation should not be applied to nonstationary random processes because the results will be difficult to interpret and may induce the user to errors. When the nonstationarity is because of a varying mean value, a subtraction of the time-varying mean will result in a wide-sense stationary random process (and hence the autocorrelation and cross-correlation analyses may be used). If the mean variation is linear, a simple detrending operation will suffice. If the mean varies periodically, a spectral analysis may indicate the function to be subtracted. If the nonstationarity is because of time-varying moments of order higher than 1, it is more difficult to transform the random process into a stationary one. One of the approaches of dealing with these more complicated cases is the extension of the ideas of deterministic time-frequency analysis to random signals. An example of the application of such an approach to EEG signals may be found in Ref. 28.

When two stationary signals are, respectively, the input and the output of a linear system, the cross-correlation and the cross-spectrum presented above are very useful tools. For example, if the input is (approximately) white, the cross-correlation between the input and output provide us the impulse response of the system. A refinement is the use of the coherence function, which is the cross-spectrum divided by the square root of the product of the two signal autospectra (11,21). The squared magnitude of the coherence function is often used in biomedical signal processing applications (29,30). However, when significant *mutual interactions* exists between the two signals x and y, as happens when two EEG signals are recorded from the brain, it is very relevant to analyze the direction of the influences ($x \to y$; $y \to x$). One approach described in the literature is the partial directed coherence (31).

Several approaches have been proposed in the literature for studying nonlinear relations between signals. However, each proposed method works better in a specific experimental situation, contrary to the unique importance of the cross-correlation for quantifying the linear association between two random signals. The authors shall refer briefly to some of the approaches described in the literature in what follows.

In the work of Meeren et al. (32), the objective was to study nonlinear associations between different brain cortical sites during seizures in rats. Initially, a *nonlinear correlation coefficient* h^2 was proposed as a way to quantify nonlinear associations between two random variables, say x and y. The coefficient h^2 ($0 \leq h^2 \leq 1$) estimates the proportion of the total variance in y that can be predicted on the basis of x by means of a nonlinear regression of y on x (32). To extend this relationship to signals $y(n)$ and $x(n)$, the authors computed h^2 for every desired time shift k between $y(n)$ and $x(n)$, so that a function $h^2(k)$ was found. A peak in $h^2(k)$ indicates a time lag between the two signals under analysis.

The concept of *mutual information* (33) from information theory (34) has been used as a basis for alternative measures of the statistical association between two signals. For each chosen delay value the mutual information (MI) between the pairs of partitioned (binned) samples of the two signals is computed (35). The MI is non-negative, with MI = 0 meaning the two signals are independent, and attains a maximum value when a deterministic relation exists between the two signals. This maximum value is given by $-\log_2 \varepsilon$ bits, where ε is the precision of the binning procedure (e.g., for 20 bins, $\varepsilon = 0.05$, and therefore MI ≤ 4.3219 bits). For example, Hoyer et al. (18) have employed such measures to better understand the statistical structure of the heart rate variability of cardiac patients and to evaluate the level of respiratory heart rate modulation.

Higher-order cross-correlation functions have been used in the Wiener–Volterra kernel approach to characterize a nonlinear association between two biological signals (36,37).

Finally, the class of oscillatory biological systems has received special attention in the literature with respect to the quantification of the time relationships between different rhythmic signals. Many biological systems exist that exhibit rhythmic activity, which may be synchronized or entrained either to external signals or with another subsystem of the same organism (38). As these oscillatory (or perhaps chaotic) systems are usually highly nonlinear and stochastic, specific methods have been proposed to study their synchronization (39,40).

BIBLIOGRAPHY

1. R. Merletti and P. A. Parker, *Electromyography*. Hoboken, NJ: Wiley Interscience, 2004.
2. J. H. Zar, *Biostatistical Analysis*. Upper Saddle River, NJ: Prentice-Hall, 1999.
3. P. Z. Peebles, Jr., *Probability, Random Variables, and Random Signal Principles*, 4th ed., New York: McGraw-Hill, 2001.
4. J. S. Bendat and A. G. Piersol, *Random Data: Analysis and Measurement Procedures*. New York: Wiley, 2000.
5. C. Chatfield, *The Analysis of Time Series: An Introduction*, 6th ed., Boca Raton, FL: CRC Press, 2004.
6. P. J. Magill, A. Sharott, D. Harnack, A. Kupsch, W. Meissner, and P. Brown, Coherent spike-wave oscillations in the cortex and subthalamic nucleus of the freely moving rat. *Neuroscience* 2005; **132**:659–664.
7. J. J. Meier, J. D. Veldhuis, and P. C. Butler, Pulsatile insulin secretion dictates systemic insulin delivery by regulating hepatic insulin extraction in humans. *Diabetes* 2005; **54**:1649–1656.
8. D. Rassi, A. Mishin, Y. E. Zhuravlev, and J. Matthes, Time domain correlation analysis of heart rate variability in preterm neonates. *Early Human Develop.* 2005; **81**:341–350.
9. R. Steinmeier, C. Bauhuf, U. Hubner, R. P. Hofmann, and R. Fahlbusch, Continuous cerebral autoregulation monitoring by cross-correlation analysis: evaluation in health volunteers. *Crit. Care Med.* 2002; **30**:1969–1975.
10. A. Papoulis and S. U. Pillai, *Probability, Random Variables and Stochastic Processes*. New York: McGraw-Hill, 2001.
11. P. J. Brockwell and R. A. Davis, *Time Series: Theory and Models*. New York: Springer Verlag, 1991.
12. A. F. Kohn, Dendritic transformations on random synaptic inputs as measured from a neuron's spike train — modeling and simulation. *IEEE Trans. Biomed. Eng.* 1989; **36**:44–54.
13. J. S. Bendat and A. G. Piersol, *Engineering Applications of Correlation and Spectral Analysis*, 2nd ed. New York: John Wiley, 1993.
14. M. B. Priestley, *Spectral Analysis and Time Series*. London: Academic Press, 1981.
15. C. W. Therrien, *Discrete Random Signals and Statistical Signal Processing*. Englewood Cliffs, NJ: Prentice-Hall, 1992.
16. A. V. Oppenheim, R. W. Schafer, and J. R. Buck, *Discrete-Time Signal Processing*. Englewood Cliffs: Prentice-Hall, 1999.
17. R. M. Bruno and D. J. Simons, Feedforward mechanisms of excitatory and inhibitory cortical receptive fields. *J. Neurosci.* 2002; **22**:10996–10975.
18. D. Hoyer, U. Leder, H. Hoyer, B. Pompe, M. Sommer, and U. Zwiener, Mutual information and phase dependencies: measures of reduced nonlinear cardiorespiratory interactions after myocardial infarction. *Med. Eng. Phys.* 2002; **24**:33–43.
19. T. Muller, M. Lauk, M. Reinhard, A. Hetzel, C. H. Lucking, and J. Timmer, Estimation of delay times in biological systems. *Ann. Biomed. Eng.* 2003; **31**:1423–1439.
20. J. C. Hassab and R. E. Boucher, Optimum estimation of time delay by a generalized correlator. *IEEE Trans. Acoust. Speech Signal Proc.* 1979; **ASSP-27**:373–380.
21. D. G. Manolakis, V. K. Ingle, and S. M. Kogon, *Statistical and Adaptive Signal Processing*. New York: McGraw-Hill, 2000.
22. R. A. Mezzarane and A. F. Kohn, Bilateral soleus H-reflexes in humans elicited by simultaneous trains of stimuli: symmetry, variability, and covariance. *J. Neurophysiol.* 2002; **87**:2074–2083.
23. E. Manjarrez, Z. J. Hernández-Paxtián, and A. F. Kohn, A spinal source for the synchronous fluctuations of bilateral monosynaptic reflexes in cats. *J. Neurophysiol.* 2005; **94**:3199–3210.
24. H. Akaike, A new look at the statistical model identification. *IEEE Trans. Automat. Control* 1974; **AC-19**:716–722.
25. M. Akay, *Biomedical Signal Processing*. San Diego, CA: Academic Press, 1994.
26. R. M. Rangayyan, *Biomedical Signal Analysis*. New York: IEEE Press - Wiley, 2002.
27. D. M. Simpson, A. F. Infantosi, and D. A. Botero Rosas, Estimation and significance testing of cross-correlation between cerebral blood flow velocity and background electro-encephalograph activity in signals with missing samples. *Med. Biolog. Eng. Comput.* 2001; **39**:428–433.
28. Y. Xu, S. Haykin, and R. J. Racine, Multiple window time-frequency distribution and coherence of EEG using Slepian sequences and Hermite functions. *IEEE Trans. Biomed. Eng.* 1999; **46**:861–866.
29. D. M. Simpson, C. J. Tierra-Criollo, R. T. Leite, E. J. Zayen, and A. F. Infantosi, Objective response detection in an electroencephalogram during somatosensory stimulation. *Ann. Biomed. Eng.* 2000; **28**:691–698.
30. C. A. Champlin, Methods for detecting auditory steady-state potentials recorded from humans. *Hear. Res.* 1992; **58**:63–69.
31. L. A. Baccala and K. Sameshima, Partial directed coherence: a new concept in neural structure determination. *Biolog. Cybernet.* 2001; **84**:463–474.
32. H. K. Meeren, J. P. M. Pijn, E. L. J. M. van Luijtelaar, A. M. L. Coenen, and F. H. Lopes da Silva, Cortical focus drives widespread corticothalamic networks during spontaneous absence seizures in rats. *J. Neurosci.* 2002; **22**:1480–1495.
33. N. Abramson, *Information Theory and Coding*. New York: McGraw-Hill, 1963.
34. C. E. Shannon, A mathematical theory of communication. *Bell Syst. Tech. J.* 1948; **27**:379–423.
35. N. J. I. Mars and G. W. van Arragon, Time delay estimation in non-linear systems using average amount of mutual information analysis. *Signal Proc.* 1982; **4**:139–153.
36. P. Z. Marmarelis and V. Z. Marmarelis, *Analysis of Physiological Systems. The White-Noise Approach*. New York: Plenum Press, 1978.

37. P. van Dijk, H. P. Wit, and J. M. Segenhout, Dissecting the frog inner ear with Gaussian noise. I. Application of high-order Wiener-kernel analysis. *Hear. Res.* 1997; **114**:229–242.
38. L. Glass and M. C. Mackey, *From Clocks to Chaos. The Rhythms of Life*. Princeton, NJ: Princeton University Press, 1988.
39. M. Rosenblum, A. Pikovsky, J. Kurths, C. Schafer, and P. A. Tass, Phase synchronization: from theory to data analysis. In: F. Moss and S. Gielen, eds., *Handbook of Biological Physics*. Amsterdam: Elsevier, 2001, pp. 279–321.
40. R. Quian Quiroga, A. Kraskov, T. Kreuz, and P. Grassberger, Performance of different synchronization measures in real data: a case study on electroencephalographic signals. *Phys. Rev. E* 2002; **65**:041903/1–14.

AUTOLOGOUS PLATELET-BASED THERAPIES

DOUGLAS ARM
Cytori Therapeutics, Inc.
San Diego, California

1. OVERVIEW

Healing and fusion in challenging bone grafting procedures is not guaranteed using the "gold standard" of autogenous iliac crest bone, and clinicians are increasingly interested in tissue engineering solutions using cells and signaling factors to enhance both overall success rate and time to fusion. The potential role in wound repair from application of cytokines contained in platelets was initially described more than 20 years ago. A rapid, simple preparation method to perioperatively separate and collect the platelet-rich fraction was developed a decade later, expanding the potential range of clinical applications. Numerous additional point-of-care techniques have since been introduced, potentially providing a cost-effective alternative to single recombinant proteins. As with many emerging technologies, clinicians are confronted with a frequently conflicting array of largely anecdotal claims for clinical utility. With a focus on structural bone graft repair, this chapter will present the therapeutic basis for clinical use of topically applied autogenous growth factor preparations and summarize existing experimental and application-specific clinical outcomes reports.

2. INTRODUCTION

Regardless of the specific application, tissue repair requires all three essential elements of the Triad of Tissue Regeneration (Fig. 1): an appropriate matrix, responding cells capable of forming the desired phenotype, and the required signaling molecules. Physicians must either ensure these elements are available via the surrounding host tissue or provide those that are lacking as part of the treatment modality. In orthopedics, for example, autologous bone harvested from the patient's own iliac crest has maintained its status as the gold standard in spinal bone grafting applications in large part because of its ability to provide all three essential elements to the graft site. The use of autologous bone, however, even with rigid internal fixation, does not guarantee a solid posterolateral fusion (1–15). A number of risk factors, including smoking, diabetes, prior pseudarthrosis, multilevel fusion, and osteoporosis, can result in a failed fusion. In addition, harvesting iliac crest autograft results in significant residual pain in up to 30% of patients (16); requires an additional surgical incision, thus increasing the risk of infection; and, depending on the patient's age and build, can be of limited quantity and questionable quality.

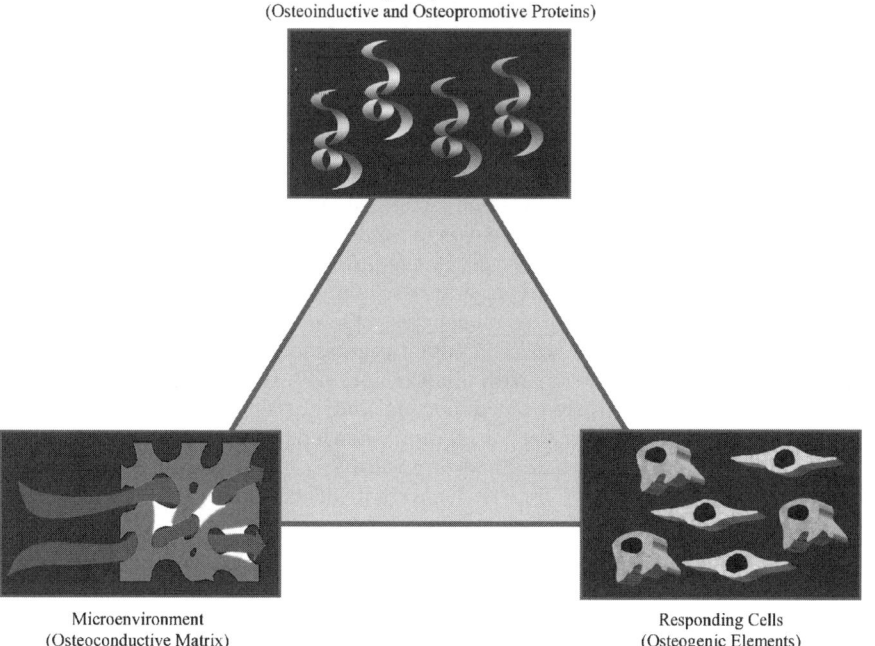

Figure 1. The Triad of Tissue Regeneration is composed of the three essential elements required for successful bone formation: a stable microenvironment consisting of an osteoconductive matrix, the right combination of osteoinductive and osteopromotive signaling molecules, and the appropriate numbers of responding osteogenic cells. These elements may be provided by the host tissue or implanted in the graft as part of a tissue engineering solution. (This figure is available in full color at http://www.mrw.interscience.wiley.com/ebe.)

In an effort to minimize or eliminate the need to harvest the patient's own bone, numerous new osteoconductive synthetic and allogeneic biomaterials have become available in the last decade. Although bone graft substitutes can provide a matrix for cellular ingrowth/ongrowth or act as scaffold for new bone formation, they lack the biological molecules and cells present in the normal healing environment. Under certain conditions, such as in a mechanically stable environment in close proximity to viable host bone, osteoconductive matrices have proven to be as effective as autograft because the host tissue can provide sufficient responding cells and signaling factors (17–23). The ability of the host tissue to consistently contribute these bone promoting cells and proteins, and produce a solid fusion mass in a challenging environment like a posterolateral graft site after implantation solely with an osteoconductive scaffold, however, has not been reliably demonstrated.

To address this issue, an industry-wide effort has been made to provide a bioactive "tissue engineering" solution by providing cells or proteins in addition to an appropriate matrix. These approaches have ranged from single recombinant factors to purified or concentrated cocktails, the use of intraoperatively harvested marrow cells to culture expanded progenitor cell populations, and combinations of the above with a wide array of carrier materials and structures. Great strides have been made in the last 40 years since Marshall Urist first proposed the existence of a bone morphogenetic protein (24) to identify, isolate, characterize, and elucidate the function of a myriad of growth factors and cytokines that influence the cascade of bone formation (25–27). The most well-studied proteins individually documented to affect tissue healing include Platelet-Derived Growth Factor (PDGF), Fibroblast Growth Factor (FGF), Insulin-like Growth Factor (IGF), Epidermal Growth Factor (EGF), Vascular Endothelial Growth Factor (VEGF), and the most widely studied, the Transforming Growth Factor-Beta (TGF-β) family, which includes the Bone Morphogenetic Proteins (BMPs).

3. RATIONALE FOR AN AUTOLOGOUS GROWTH FACTOR ALTERNATIVE

The realization that specifics of both the bone and tissue healing processes are not yet completely understood has led to an effort to examine alternative methods other than the application of single exogenous factors for providing signaling molecules to a graft or wound site. Although it is well known that receptors from many different cytokines are expressed in the lineage progression from an uncommitted progenitor cell to a mature cellular phenotype (28), the exact time each protein is turned on, for how long, at what dosage, and how they interact with each other remains undetermined. As a result, a growing perception exists that rather than rely on a single factor to consistently affect the entire cascade of events that encompass the desired tissue regeneration process, a combination of factors that act in concert to emulate the natural process of healing may be a more desirable approach.

In looking to the natural healing response for a potential therapy approach, it is well understood that one of the most immediate responses of the body to a traumatic event is the migration of platelets and white blood cells to the injury site. Platelets and white blood cells contain a number of growth factors that have been demonstrated to have active roles during various stages of the healing cascade, including PDGF, TGF-β, VEGF, EGF, IGF, and bFGF (29). A partial list of factors contained within these formed elements is shown in the text box below. The complex interaction of these and other factors chemotactically recruits mesenchymal cells to the local site, and subsequently causes their proliferation and maturation into matrix-forming cells (30–32).

Growth Factors and Cytokines Contained In Platelets and White Blood Cells

- PDGF-AA, BB, AB
- TGF-β1, β2, α
- IGF-I, IGF-II
- VEGF
- EGF
- bFGF
- PD-ECGF
- HGF
- Osteocalcin
- Fibronectin
- Fibrinogen
- BMP-2, BMP-4
- Platelet Factor 4
- P-Selectin
- Connective-tissue activating peptide III
- Lipoxygenases
- Antiplasmin
- Histamine
- EPEA
- LMW-EMA, HMA-EMA
- 12-HETE, 15-HETE
- PGE1, PGE2, PGA2, PGJ2, PGF1a, 2a

Again using bone formation as an example, different factors act at different stages of the bone healing process. The actions of some of the most important of these factors are summarized in Table 1 (33). It is important to be aware that many of these studies reporting the effects of growth factors present in platelets and white cells focus on the properties of individual growth factors and the effects that they elicit *in vitro* using cell populations of a single type. However, studies examining the effects of individual growth factors, either *in vitro* or *in vivo*, do not include all of the mechanisms that cause a specific effect to be observed *in vivo*. Frequently, it is not the individual growth factor that is responsible for the effects observed in a study, but the interaction between factors and with their target tissues (34). As will be discussed in more detail below, growth factors act synergistically in the processes of healing (35).

PDGF is found in abundance in platelets; is also produced by monocytes, macrophages, and endothelial cells (36), and is a potent regulator of bone cells by itself and in concert with other factors (27). It is released from platelets during blood clotting and, therefore, its presence shortly after the initial injury has led to the conclusion that PDGF plays an important role early in the initiation of bony healing (35). PDGF exerts its effects through receptor proteins that ultimately activate genes controlling DNA synthesis, cell replication, and the production of various proteins. PDGF has been shown to chemotactically attract and induce mitogenesis in osteoblasts and bone progenitor cells, increase cellular differentiation, enhance angiogenesis,

Table 1. Documented Functions of Some of the Growth Factors Contained within Platelets and White Blood Cells

Platelet Factor	Also Found in	Function	Effects
PDGF	Monocytes	• Stimulates chemotaxis and proliferation of osteoblasts • Promotes adult osteoclastic bone resorption • Induces proliferation in pluripotential and progenitor stem cells • Promotes cartilage formation in RC cells in culture • Increases collagen and noncollagen protein synthesis	• Increases callus formation at fracture site • Role in regulation of bone regeneration
	Macrophages Endothelial cells		
TGF-β1	Osteoblasts	• Stimulates osteoblast proliferation • Stimulates chondrocyte proliferation • Stimulates matrix synthesis	• Increases bone formation *in vitro* and *in vivo* • Increased callus formation *in vivo*
TGFβ-2	Chondrocytes Bone matrix		
IGF-I	Osteoblasts	• Stimulates chondrocyte proliferation • Stimulates collagen synthesis • Stimulates noncollagen synthesis • Decreases collagen degradation • Stimulates osteoblast proliferation	• Increases woven bone formation
	Cartilage		
IGF-II	Osteoblasts	• Stimulates chondrocyte proliferation • Increases bone collagen synthesis • Decreases collagen degradation • Stimulates osteoblast proliferation	• Increases woven bone formation
VEGF	Multiple tissues	• New vessel formation • Increased vascular permeability	• Stimulates new vessel growth
bFGF	Megakaryocytes	• Supports smooth muscle cell proliferation • Stromal cell mitogen	• Stimulates bone formation
EGF	Platelet Cytoskeleton	• Stimulates periosteal bone formation • Stimulates endosteal bone resorption	• Stimulates bone growth

Reprinted with permission from Krause et al. (33).

activate macrophages, and stimulate synthesis of connective tissue matrix components such as collagen (37–39).

Both TGF-β1 and TGF-β2 have been shown to be present in platelets, as well as macrophages, chondrocytes, and osteoblasts (25). TGF-β proteins act by binding and activating receptors, which subsequently activate Smad proteins. The Smad proteins mediate the gene responses to TGF-β (40). TGF-β has been shown to be present in normal human fracture healing and plays a number of diverse functions in bone formation (41,42), which include enhancing angiogenesis, stimulating mesenchymal stem cell proliferation, promoting mesenchymal cell commitment to the osteoblast lineage, amplifying synthesis of extracellular matrix by osteoblasts and their precursors (43), and inducing osteoclast apoptosis (44). Other reports suggest that TGF-β may affect bone healing by augmenting the osteoinductive activity of BMPs (45).

Basic Fibroblast Growth Factor (bFGF) is present in platelets, monocytes, macrophages, osteoblasts, and chondrocytes, as well as many other tissues. Through tyrosine kinase activity, it stimulates fibroblast, osteoblast, chondrocyte, and endothelial cell proliferation (46,47) and mediates angiogenesis (48). Basic FGF has been shown to be present in callus and periosteum in the early stages of fracture repair (49).

Vascular Endothelial Growth Factor (VEGF) is synthesized by activated platelets and megakaryocytes (50,51) and is also produced by osteoblasts (52,53). Studies investigating the angiogenic properties of VEGF illustrate that it supports capillary growth during wound repair by increasing endothelial cell migration to the wound site, and by subsequently increasing the general angiogenic response. In addition to its crucial role in the stimulation of neovascularization (54), VEGF induces osteoblastic migration, differentiation, and alkaline phosphatase expression *in vitro* (55) and has been shown to be an important factor in both the angiogenesis and osteogenesis that occurs during spinal fusion (56).

Insulin-like Growth Factors I and II (IGF-I and IGF-II) are signaling molecules that have also been shown to play a role in skeletal growth and development. IGF-I appears to play a greater role in growth-promoting activity and bone regeneration compared with IGF-II. Platelets also maintain IGF receptors (57), and their α-granules contain IGF-Binding Protein-3 (58). IGF-I stimulates bone growth directly by increasing osteoblastic bone formation and indirectly by increasing osteoprogenitor cell replication, which ultimately increases the number of functional osteoblasts. Through an enzymatic cascade mediated by growth hormone, complexes of IGF-I and its binding proteins in harvested platelets stimulate cellular mitogenesis, matrix synthesis, and noncollagenous protein production, all of which ultimately lead to bone growth (59).

Of recent note, Sipe et al. (60) used immunohistochemistry to identify BMPs 1 through 6 within human fetal megakaryocytes, the cells that give rise to platelets. The possibility of platelets containing BMPs, contributing to their osteoinductive potential, was the focus of this study. Further study using the Western blotting technique on lysed human platelets isolated from serum identified both BMP-2 and BMP-4 within the lysate. The authors suggested that the presence of significant amounts of BMP-2 and BMP-4 in platelets may contribute to their role in bone formation and repair (60).

Interpreting the mechanism of any single exogenous factor as it relates to a complex system of many factors may not be feasible. TGF-β, for example, has been shown to enhance and suppress differentiation and bone formation at different levels when studied individually. How TGF-β at various dosages influences healing in a bony environment when many other factors are also present, as is the case naturally, is not uniquely known. Within the stages of the bone formation process defined as angiogenesis, chemotaxis, mitogenesis, differentiation, and mineralization, all of the factors discussed above have been shown to interact in synergistic and modulatory manners. For example, Asahara et al. observed synergy between VEGF and bFGF in vascularization (30). Progenitor cell migration has been enhanced with the combination of PDGF and TGF-β in comparison with either factor by itself (26,32). PDGF, TGF-β, and EGF have a documented synergistic effect on cell proliferation by Slater et al. (61), whereas Kasperk described synergy between FGF, TGF-β, and IGF in enhancing osteoblastic differentiation (35). To illustrate the complex interactions between these various factors, it has been suggested that an increase in VEGF following trauma amplifies angiogenesis and endothelial cell proliferation, which in turn increase IGF-I synthesis (62). IGF-I then increases osteoprogenitor cell proliferation, differentiation, and matrix synthesis (63).

4. OVERVIEW OF CLINICAL PREPARATION TECHNIQUES

As applying all of these factors exogenously at the appropriate dosages at the right time for the correct duration is not yet practical, the use of formed blood elements as carriers of growth factors makes theoretical sense. Several methods have been developed over the last decade or so to perioperatively isolate and concentrate either autologous platelets or autologous buffy coat (platelets and white cells). These methods fall generally into one of three categories: (1) traditional cell separators or autotransfusion devices; (2) dedicated low-volume centrifuge systems; and (3) a combined approach using both centrifugation and ultrafiltration. A description and comparison of these approaches are presented below.

4.1. Cell Separators

Standard cell separators, frequently used in spinal procedures or other large orthopedic cases for autotransfusion, have been employed to fractionate and collect the platelets from the whole blood for over a decade. The latest machines from leading manufacturers all have sequestration programs, including Haemonetics Cell Saver 5, Medtronic Sequestra, Sorin-Dideco Compact Advanced, Cobe BRAT II, and Fresenius C.A.T.S. Depending on brand, software revision, and desired product composition, some of the sequestering programs are mostly automated and others are more manual. Each manufacturer has recommended protocols for platelet sequestration that concentrate the platelets from two to three times baseline (64). The sequestration protocol is, in general, a two-stage procedure based on standard continuous-flow Latham or Baylor centrifuge bowl technology to separate the blood fractions, with the exception of Fresenius, which uses a continuous-flow disk separation technology. A hard, or fast, spin is used to separate out the platelet-poor plasma (PPP). A soft, or slow, spin is then used to better separate the platelets from the denser RBC fraction and collect the platelet-rich plasma (PRP).

Traditional cell separators allow a fairly large volume of product to be obtained, but must start with a full unit of blood because of the size of the bowls. The need to draw off 450 ml of blood volume can intimidate inexperienced anesthesiologists despite the fact that the majority of the unused fractions are returned to the patient. Although the more manual cell separators allow the knowledgeable, vigilant user flexibility in tailoring a product for a specific patient's or physician's needs in terms of platelet count and product volume, it also creates concern that this subjectivity may create unwanted product variability. The dedicated time required to harvest the PRP by the technician and the potentially increased cost associated with the operator's time are both further drawbacks of the traditional cell separator approach.

4.2. Dedicated Centrifuges

Recognizing the need for an alternative technique to collect PRP from lower volumes in a more portable and straightforward manner, a number of devices based primarily on clinical laboratory centrifuges were developed. Leading manufacturers and models are DePuy Symphony™ II (originally licensed from Harvest Technologies) (Fig. 2a), Biomet's GPS2™ (Fig. 2b), and Medtronic's Magellan™ (Fig. 2c). Although they differ to some degree in disposable configuration and automation, these devices all use a fixed starting volume of blood in the disposable

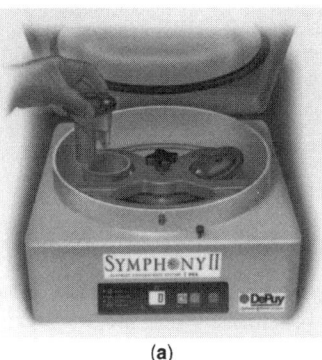 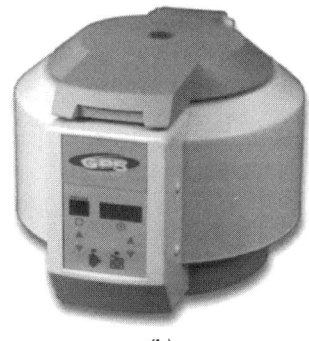 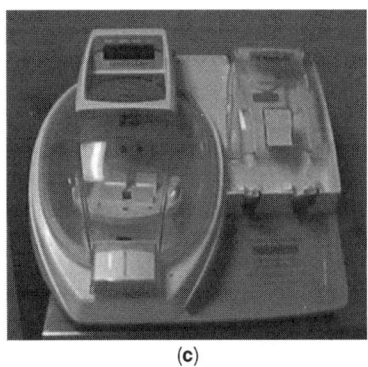

Figure 2. Leading available centrifuge systems dedicated to producing platelet-rich plasma: (a) Symphony™ II from DePuy Acromed, Raynham, MA; (b) GPS™ from Biomet, Warsaw, IN; and (c) Magellan™ from Medtronic, Minneapolis, MN. (This figure is available in full color at http://www.mrw.interscience.wiley.com/ebe.)

chamber and collect about 65–80% of the available platelets. The Symphony II harvests the platelets in the form of a platelet pellet. This pellet, with essentially no volume, is then resuspended or diluted with PPP to the user's specification, which calculates to approximately 5–6 ml of PRP with a 4- to 6-fold increase in platelet and, hence, growth factor concentratsion when starting with 60 ml of anticoagulated blood (manufacturer's literature). With the GPS2 system, no plasma removal or operator-dependent dilution is necessary as it is designed to automatically collect 6 ml of PRP with a 6- to 8-fold increase in platelets from each disposable unit (65). GPS2 also captures white cells, increasing them by approximately 2- to 4- times baseline. In both of these machines, it is possible to run two disposables at once, so approximately 10–12 ml of PRP with the concentration levels described above can be produced from 120 ml of anticoagulated blood in approximately 15–20 minutes. Magellan processes a maximum of 60 ml of blood at a time, but is capable of running three sequential processes with the same disposable set. Its yield appears to be comparable with Symphony, with 6 ml of PRP produced at a 6-fold platelet and growth factor concentration increase (manufacturer's literature). It is important to note for all of these devices that, although higher PRP volumes are possible from a single disposable by applying additional plasma, it is only at the expense of the growth factor concentrations.

4.3. Combined Centrifugation and Ultraconcentration

In an effort to provide a more consistent and highly concentrated autologous buffy coat, Interpore Cross licensed and further developed a proprietary technology termed Autologous Growth Factors® (AGF®). AGF uses the buffy coat produced from centrifuged blood separation as a raw material and passes it through an UltraConcentrator® that functions by filtering out water and low-molecular-weight solutes through the hollow fiber membranes.

Fig. 3 summarizes the classic intraoperative technique used to create the AGF gel. Using the optimized cell separator protocols, up to 180 ml of buffy coat can be collected from a single unit (450–500 ml) of blood. As not all platelets and white cells (only about 65%) are collected on any single pass through the centrifuge bowl, the reconstituted platelet-poor blood can be processed through the bowl several times. With this methodology, over 90% of the available platelets in the starting unit can be collected, with a resulting platelet concentration 2.5–3 times above patient baseline. Depending on the volume of buffy coat required, this sequestration step can take 15 minutes for 60 ml of buffy coat and up to 45 minutes for 180 ml. The remaining RBC and PPP are immediately available for reinfusion to the patient, minimizing the total blood volume loss.

The buffy coat is then processed through the fibers of the UltraConcentrator to reduce the volume by two-thirds and obtain as much as 60 cc of AGF. These fibers have an effective pore size of about 30,000 Daltons, small enough to retain virtually all of the formed elements, growth factors, and plasma proteins including fibrinogen. The volume reduction translates into a corresponding three-fold increase in platelet and white cell concentrations in AGF over the starting buffy coat, therefore approximately 6 to 8 times over native baseline blood levels. As the molecular weight of fibrinogen is 340 kDa and, thus, is retained within the perfusate, this ultraconcentration correlates directly to a three-fold increase in the fibrinogen concentration.

To address the clinical issues associated with the user-intensive classic technique, including blood draw volume requirements and operator variability, Interpore Cross recently released Access® (Fig. 4), a fully automated device that combines the centrifugation and ultraconcentration functions in a single housing. Although still providing the same composition of endproduct as the classic AGF method, Access® also enables the user to start with as little as 100 ml of blood when only small volumes of AGF are required, but maintains the flexibility to still process a full unit in a single disposable set.

4.4. Gel Preparation

Regardless of the processing method, once the PRP or concentrated buffy coat is prepared, it is transferred onto the sterile field. There it can be mixed with a small volume of thrombin, creating a gel containing the patient's own growth factors. The thrombin can be obtained from several potential sources. Most typically, in the United States,

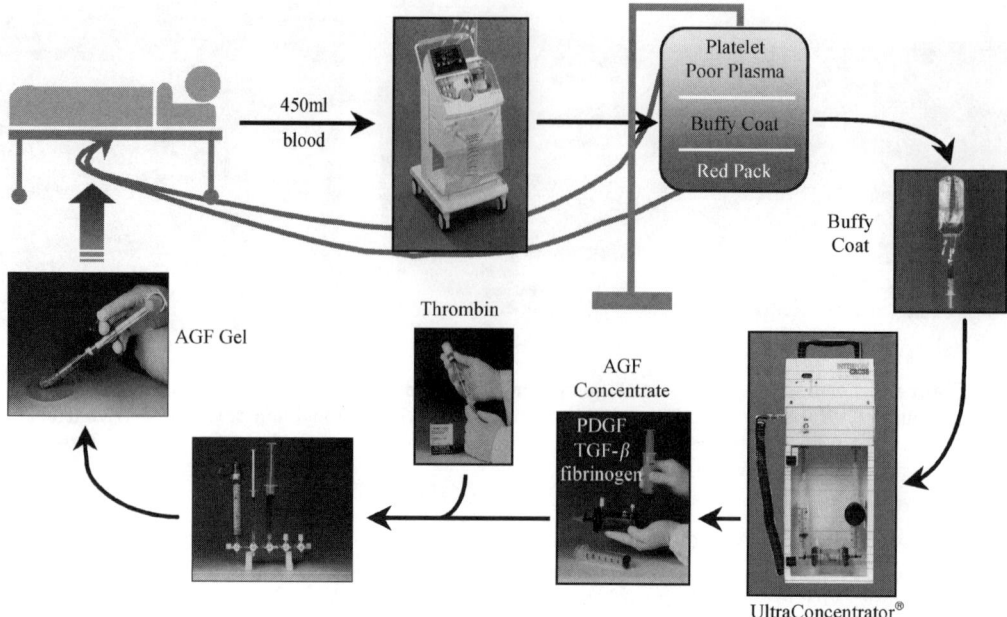

Figure 3. The classic method of preparing Autologous Growth Factors® (AGF)®. Using a standard cell separator, a unit of blood is fractionated into platelet-poor plasma, buffy coat rich in platelets and white cells, and the red cell pack. Up to 180 ml of buffy coat can be generated, and the remaining platelet-poor plasma and red cells can be returned to the patient. The buffy coat is then further processed through a proprietary UltraConcentrator® (Interpore Cross) to one-third of the original buffy coat volume. This concentrated buffy coat, termed AGF, is then mixed with a small volume of thrombin and the desired bone graft, creating a gelled composite that has elevated growth factor and fibrinogen levels. (This figure is available in full color at http://www.mrw.interscience.wiley.com/ebe.)

thrombin purified from a bovine source is used. Bovine thrombin is convenient and available off-the-shelf, and it has the advantage of being easily diluted to various concentrations. In previously available less-pure formulations, however, bovine thrombin had been suspected of creating antibodies to Factor V, resulting in inhibition of coagulation with repeat topical thrombin usage. Only an ultra-pure form of bovine thrombin, Thrombin JMI (King Pharmaceuticals, Bristol, TN) is now available.

Other available alternatives include pooled human allogeneic thrombin, available in some clinical markets outside of the United States, and various products and techniques to produce autologous thrombin from the patient's own blood directly. Although the pooled allogeneic human thrombin offers the same off-the-shelf convenience of bovine thrombin, its availability is limited, the cost is high, and it has the same potential drawbacks inherent to all allogeneic tissues. The techniques used to create an autologous thrombin are typically performed in the operating theater from the patient's own blood or plasma. One method involves inducing and then compressing a clot from non-anticoagulated blood, in the process collecting the exudate containing activated thrombin. This method has been found to be quite user- and patient-subjective, and frequently results in a thrombin preparation that is inadequate to create a gel. Simple disposable cartridges such as the Thrombin Activation Device™ (Interpore Cross) that remove fibrinogen from plasma and then activate the thrombin have made the process more consistent, but these products still depend on the patient's native blood composition.

The mechanism of action when exogenous thrombin is added directly follows the natural coagulation cascade (Fig. 5). Briefly, the thrombin cleaves the inactive fibrinogen into fibrin monomer strands and also activates Factor XIII, another plasma protein. Active Factor XIII will combine with calcium ions present from within the platelets and from the calcium chloride diluent and cause the fibrin strands to crosslink into a polymer matrix (66). The platelets will begin to degranulate and release some of their contents. Some growth factors remain in latent form attached to the fragmented platelet membranes and fibrin clot, and are only released and activated on later consumption of the fragments by macrophages.

For bone grafting applications, standard ultra-pure bovine thrombin from the hospital pharmacy is reconstituted to a concentration of 200 units per ml in a saline or calcium chloride solution. An advantage of suspending the thrombin in 10% calcium chloride is that it will counter the effects of the citrate-based anticoagulant. Shortly before graft implantation, the PRP or AGF concentrate and thrombin are combined in a 10:1 volume ratio using several potential methods. Figure 6a shows the use of an applicator tip to mix AGF, thrombin, and graft material in a sheet configuration, whereas a manifold/graft delivery syringe (Fig. 6b) combination can yield a cylindrical graft, convenient for delivery into the posterolateral gutters. The gel can also be formed directly into a cage filled

Figure 4. The recently released Access® System from Interpore Cross, Irvine, CA, is a dedicated machine that automates the classic AGF method detailed in Figure 9-3 and allows processing of as little as 100 ml of blood. (This figure is available in full color at http://www.mrw.interscience.wiley.com/ebe.)

with graft (Fig. 6c) or sprayed directly onto the site as a hemostatic agent (Fig. 6d). The volume ratio of autologous concentrate to graft material is approximately equal, with 60 ml of gel able to cover 60–90 ml of graft, depending on the specific graft composition. As the graft material will soak up site blood, and thus dilute the growth factor concentration, it is recommended that preparing the gel/graft composite be done on the back table prior to implantation. This activity will also facilitate handling of the graft during implantation.

The concentrations and volumes used have been optimized to ensure maximum gel strength and appropriate gelling speed. The speed at which gelling of the concentrate occurs is dependent on the concentration of active thrombin, not volume. The 200 units/ml recommended will cause gelling in 30–60 seconds, allowing time for the cells and growth factors to disperse throughout the bone graft. If faster gelling is desired, for example, if the gel is being applied topically, full-strength (1000 units/ml) thrombin is typically used.

Fibrinogen is a critical factor in the coagulation cascade and is the key element in determining gel strength of AGF or PRP upon the addition of exogenous thrombin (66). The 10:1 thrombin volume ratio is used to minimize the

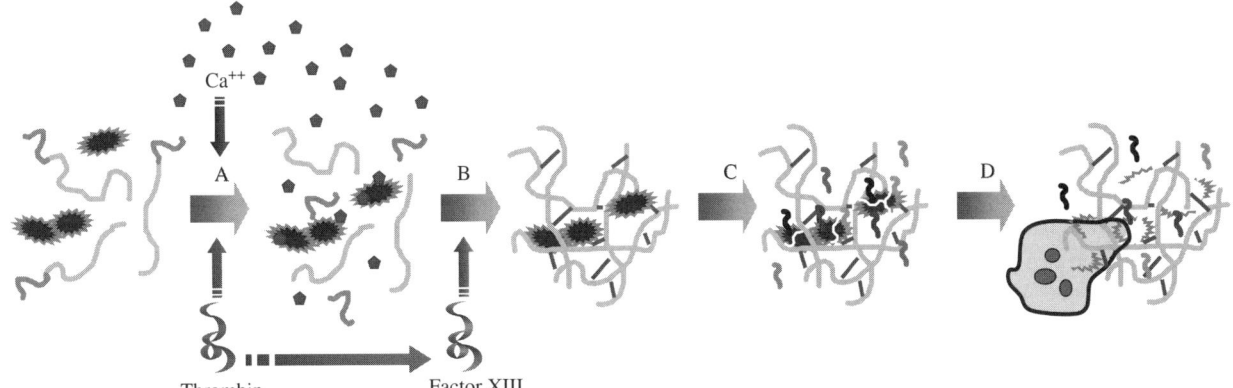

Figure 5. The mechanism of growth factor release from a platelet or buffy coat gel on the addition of thrombin-includes (a) thrombin-cleaving fibrinogen strands into fibrin; (b) crosslinking of the fibrin strands into an interwoven mesh and (if necessary) activation of the platelets; (c) degranulation of the platelets and initial release and activation of some growth factors and cytokines from the platelets and white cells; and (d) subsequent release of latent growth factors attached to cell membrane fragments on consumption by macrophages. (This figure is available in full color at http://www.mrw.interscience.wiley.com/ebe.)

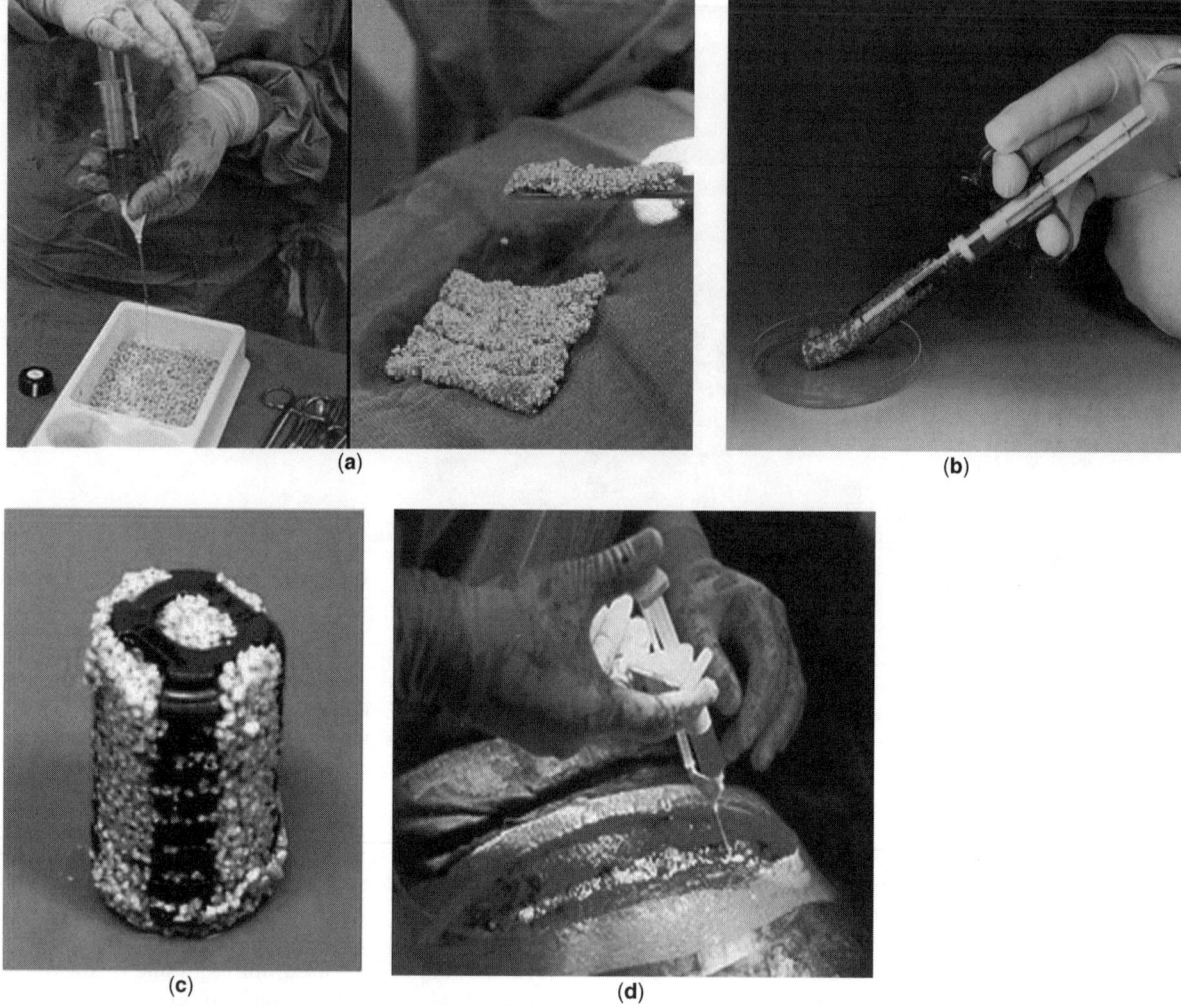

Figure 6. Methods of mixing autologous gels with bone graft material: (a) dual-cannula applicator; (b) Graft Delivery Syringe; (c) directly within an interbody cage; and (d) topically sprayed *in situ*. (This figure is available in full color at http://www.mrw.interscience.wiley.com/ebe.)

dilution of the fibrinogen concentration. Adding more thrombin will not quicken the gel formation and will only serve to dilute the fibrinogen and weaken the gel. It should be noted that none of the centrifuge-only techniques currently used clinically can increase the fibrinogen concentration. The fibrinogen in PRP or unfiltered buffy coat produced solely by a centrifugation technique is at baseline levels and, depending on the application, has made for an inconsistent gel (67). Gelling inconsistencies are a major reason why platelet gels directly from cell separators have not been widely accepted in the orthopedic bone grafting arena, where a strong gel is beneficial for handling, graft stability, and a more sustained growth factor delivery profile, as is discussed below in greater detail.

5. MECHANISM OF GROWTH FACTOR RELEASE

The initiation of wound repair is a highly complex and interdependent mechanism involving many factors. Each factor plays multiple roles in the process and also depends on the activities of the others to fulfill its own roles. Fibrin is the naturally optimized carrier for delivery of hematogenic tissue growth factors. Platelets and the coagulation proteins, including fibrinogen, combine to form a fibrin gel in a process that initially is rapidly potentiated by their properties, but that then just as quickly exhibits powerful inhibitory activities that limit the eventual extent of gel formation. Platelets adhere to fibrin strands while dispersing tissue growth factors and other active cytokines throughout the forming fibrin matrix. Shortly thereafter, contractile proteins inside the platelets bring about clot retraction, drawing the fibrin strands closer together and causing an expression of serum that bathes the surrounding tissues with growth factors.

Signaling molecules released during the coagulation process immediately attract white cells in great numbers, including macrophages that begin breakdown of the coagulum via plasmin clipping of the fibrin strands. As these cells digest the clot, growth factors are released into the surrounding tissue along with additional growth factors contained within and synthesized by neutrophils and macrophages themselves (68,69). A substantial proportion of the growth factors are in latent form, and these are only released and converted to an active state upon exposure to plasmin (70).

6. MODEL OF BONE REGENERATION

A schematic of the mechanism of action of the concentrated combination of growth factors contained within AGF or PRP is shown in Fig. 7. Building upon current understanding of single growth factors and their role in osteogenic regulation, Marx et al. eloquently described this general model for the role that growth factors released from platelets play in bone regeneration (71). Bone regeneration begins with the degranulation of platelets at the graft site that, in turn, releases PDGF, TGF-β, IGF, and all the other growth factors. Through chemotaxis and mitogenesis, they cause an increase in the number of pre-osteoblastic stem cells at the site, as well as influence the differentiation of these stem cells into mature osteoblasts. Early in the healing process, osteoblasts begin to produce bone matrix, and fibroblasts lay down collagen matrix at the graft site. Capillary angiogenesis is also initiated to support the growing graft, and within the first 14–17 days, the graft is completely permeated by capillaries (72).

With PRP, platelets survive in the wound for only approximately five days, so the direct effects of platelets and their growth factors are short-lived. Although the platelets have a short lifespan in the wound, the activity of platelet-derived growth factors causes the healing sequence to continue beyond the first couple of days after the initial injury. Growth factor-stimulated healing continues in part because of the platelet-induced increase in osteoblasts. Recalling the triad of tissue regeneration, this increase in the number of appropriate responding cells is critical for successful bone formation. These osteoblasts continue to secrete TGF-β, IGF, and BMPs into the osteoid matrix long after the platelets have ceased functioning at the wound site. Platelets also indirectly stimulate healing by inducing the chemotaxis and activation of macrophages, which become an important source of growth factors after the third day. During the first four weeks, woven bone forms and vascularization continues to increase within the graft. As a result, macrophages leave the graft site because the oxygen gradient needed to maintain their activity is no longer present. Over time, BMPs and other factors cause the woven bone to be replaced with mature bone.

Even at similar platelet concentrations, the mechanism of action of AGF differs somewhat from centrifuged PRP because of several factors including the initial inclusion of concentrated leukocytes (WBCs) within the gel, the elevated fibrinogen levels, and a corresponding decrease in the clot retraction phenomena. White cells are part of the natural healing response, and, in addition to the antibacterial and anti-inflammatory role they play in normal wound healing, they also contribute significant quantities of growth factors and cytokines that may enhance the

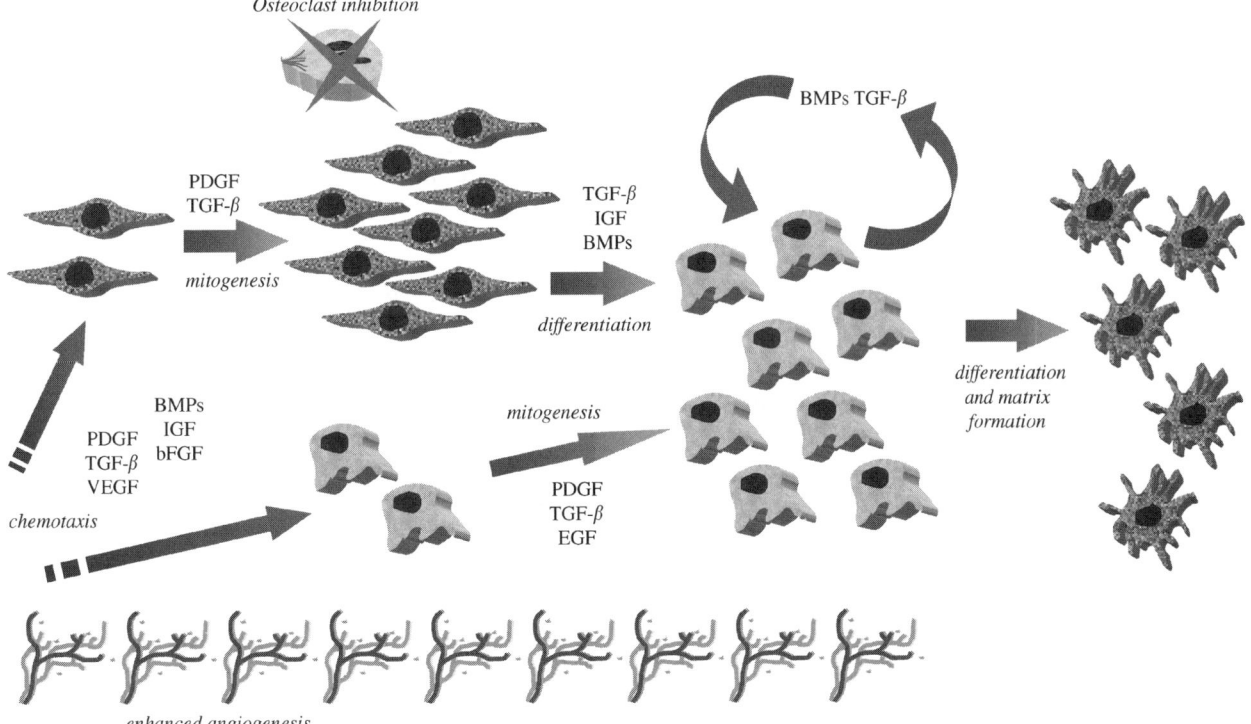

Figure 7. Schematic of a conceptual model of enhanced bone regeneration from multiple topically applied growth factors delivered by buffy coat concentrates includes increased recruitment and proliferation of progenitor cells and osteoblasts from the surrounding host tissue, enhanced revascularization, continued differentiation within the orthotopic environment into active osteoblasts, and their subsequent transition to mature osteocytes. (This figure is available in full color at http://www.mrw.interscience.wiley.com/ebe.)

healing effect. Several studies have documented stores of VEGF within white cells (68,69), and Zimmerman et al. observed significantly greater delivery of PDGF and TGF-β in concentrates containing WBCs (73). The increased delivery of TGF-β, due in part to the elevated leukocyte concentration, is thought to be further beneficial for graft preservation because of its suppressive effect on osteoclast activity (74,75).

The elevated fibrinogen within AGF imparts a number of benefits. The ability to create a firm, reliable gel that holds the graft together, facilitating handling and minimizing graft migration, has already been described. Further benefits are equally valuable, if not as readily observed visually. After implantation, it provides effective fixation of tissue, graft, and growth factors at the local site. AGF acts as a fibrin gel to fixate the bone graft material at the arthrodesis site in contact with the patient's bony structures providing stability and proximity, both known to enhance bony healing. Also, because AGF fills the spaces between the bone graft material, it prevents inflow of systemic blood into the bone graft material with consolidation of mostly red cell clot. Ordinarily, the red cell clot would need to be lysed by the normal healing mechanism before the ingrowth of blood vessels and the initiation of the bony healing cascade occurs; so, by eliminating this step, a potential jump start to the healing process can be envisioned.

For topical applications of growth factors to improve outcomes, the signaling process must be sustained for many days, which is especially true in patients with a healing impairment or where new tissue growth must be obtained over substantial distances, such as in bone fusion beds. Logic dictates that the *in situ* duration of elevated growth factor preparations directly influences the number of osteogenic repair cells recruited, the magnitude of their replication, and their rate of maturation. Gels with only a brief presence will not be as effective at these critical functions or at achieving fusion (76).

At normal platelet and fibrinogen concentrations, a PRP coagulum will quickly begin to extrude serum as it shrinks. The gel will only be half its original size in two hours, and eventually reaches one quarter or less of its original volume (77). Increasing platelet concentration, necessary to provide therapeutic levels of growth factors, both accelerates and magnifies this clot retraction effect (78) so that, as Marx described, gels formed from plasma with normal range fibrinogen are removed in three to five days (71,79).

AGF limits this clot retraction phenomenon by three mechanisms. First, the three-fold increase in fibrinogen, fibronectin, and Factor XIII levels as a result of UltraConcentration produces a relatively dense fibrin gel with greater mechanical strength than a gel formed from the original buffy coat, which, in turn, makes the gel more resistant to the platelet contractile forces. Second, a three-fold increase in total protein content means a much higher percentage of the residual plasma water is bound to proteins, which reduces the proportion of water in the coagulum available for serous extrusion. Finally, the literature has shown that platelets may become activated during centrifugation, and that the activated platelets tend to form both platelet-platelet and platelet-white cell

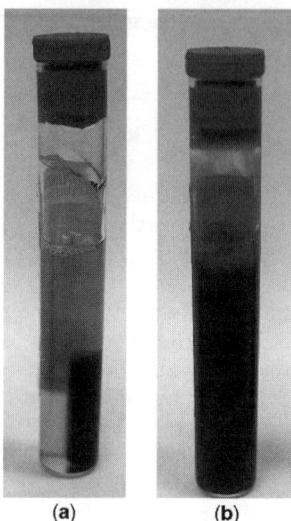

Figure 8. Example of the clot retraction phenomenon in two gels, both with platelet concentrations of 1,000,000/µl but different fibrinogen levels. The gel with baseline fibrinogen (a) exhibited significantly greater retraction and serous expression at 6 hours than the one with a three-fold increase in fibrinogen (b) (82). (This figure is available in full color at http://www.mrw.interscience.wiley.com/ebe.)

aggregates (78,80,81). The shear forces generated in the UltraConcentrator contribute to further this aggregation. These aggregates are much less efficient at pulling the fibrin mesh together because most of their energy is expended pulling on each other. The cumulative effect is that AGF releases less than 10% percent of its volume in clot retraction, compared with the 75% from PRP gels (Fig. 8) (82). The AGF gel maintains its presence and therefore this sustained release for about 14 days, comparable with the length of time required for revascularization of a sizable bone graft implant to be complete.

Kevy and Jacobson have argued that this activation and aggregation causes all growth factors to be released from the platelets immediately and thus not deliverable to the field (83). Other preclinical and clinical evidence, including a careful examination of Kevy's own data, suggests this is not the case (84). Provided the platelets are not intentionally or unintentionally fully degranulated during preparation of PRP or AGF, the activation state of the platelets prior to mixing with thrombin is immaterial in terms of growth factor delivery. The addition of thrombin to all topically applied PRPs as well as AGF begins the degranulation of all the platelets in the concentrate, after which the growth factor are delivered over time as directed by the *in situ* environment. It is important to reiterate that implantation of a material containing prereleased or quickly released factors does not appear to correspond to the timeframe for revascularization and optimal bone formation.

7. CLINICAL REQUIREMENTS

The advantages and drawbacks of each system must be weighed by the physician in choosing the most appropriate method within the bounds of the requirements for each specific case. Regardless of preparation technique, however, the objectives for an autologous gel therapy for orthopedic indications are similar. Effective levels of growth factors need to be delivered to and retained at the graft site, and they should be released in a sustained manner until revascularization has been completed. These sustained release requirements can vary from indication to indication, however. The 3–5 day growth factor delivery from a PRP gel with native levels of fibrinogen may be appropriate for a soft tissue or small-volume application where the gel is applied in a thin layer and vascularization can be quite rapid, but insufficient for a larger bone grafting procedure. The gel should provide fixation of the tissue and the graft *in situ*, and provide effective hemostasis and act as a fluid seal. The gel should also facilitate the handling characteristics of the graft composite, minimizing migration of graft materials during implantation. In addition, a sufficient volume of gel must be provided for all procedures. In today's clinical environment, the process must also be simple, reliable, and safe, and be of reasonable cost for the benefits delivered. Finally, presented and published scientific evidence specific for each clinical approach should be convincing. All of these objectives should be considered by the surgeon when evaluating which of the techniques are most appropriate for their patients.

To be effective, autologous growth factor preparations must reliably yield wound repair cytokine concentrations at a therapeutic dose. The principle that a minimum platelet concentration of 1,000,000 per μl is required to elicit a universally beneficial effect was established by Knighton for soft tissue (31,85). Although early work by Marx suggested that lower concentrations might be useful in small fusion beds (71), he has recently affirmed the need for 1,000,000 platelets per μl to routinely assure enhancement of bone fusion (86).

8. PRECLINICAL CHARACTERIZATION AND EFFICACY

Although the evidence from the growth factor literature combined with the methodology of autologous blood-derived therapies makes a compelling story for enhanced promotion of bone formation, it is the direct evaluation of each technique and its application in appropriate animal models and clinical indications that will determine its efficacy. Published or presented results from preclinical studies as well as clinical experiences and trials are detailed in the following sections. It is important to reiterate that the results from one preparation cannot be unilaterally applied to other platelet or buffy coat gels.

8.1. *In Vitro* Studies

Although most cell separator manufacturers have had their own protocols for the preparation of PRP, these were not designed for the collection of buffy coat for processing into AGF, as defined by the orthopedic clinical requirements above. Further, some user subjectivity exists on these manufacturer protocols, as evidenced by the data collected from a leading platelet gel institution. This analysis revealed both a wide-scattered distribution of platelet concentrations in the PRP and a median value only slightly above patient baseline, with numerous instances of platelet counts below baseline (Fig. 9). This data was the impetus leading to the development of recommended protocols for the optimal sequestration of buffy coat for AGF preparation. The yields from these protocols have been presented (87) and, in all cases, compare favorably to the manufacturer's protocols (64) as shown in Table 2.

Initial efforts confirmed that AGF contains significantly elevated levels of platelets, active growth factors, and fibrinogen. Cell counts were made on blood, buffy coat, and AGF using a Coulter AcT-10 hematology analyzer. Initial measurements of PDGF, TGF-β, VEGF, bFGF, and EGF were made using ELISA kits (R&D Systems), and revealed increases corresponding to the increase in platelet concentration (Fig. 10) (88,89). Quantitative fibrinogen was determined using an optical density change calibrated to a known standard, and, as expected, an increase in concentration in AGF compared with blood and buffy coat reflecting the volume reduction was observed (Fig. 11) (89). Additional evaluation of PDGF content in the PPP and the filtrate following ultraconcentration was made to ensure minimal amounts of growth factors were lost during AGF preparation (Fig. 12).

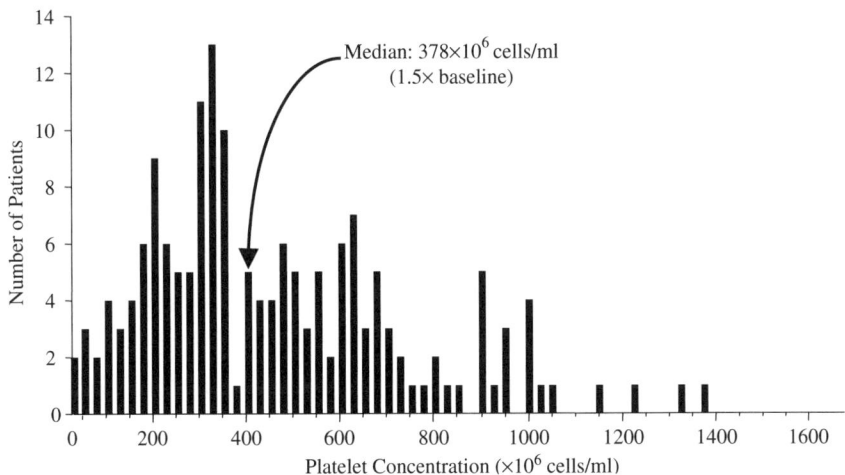

Figure 9. Distribution of platelet counts in the platelet-rich plasma fraction using manufacturer's cell separator protocols over a 9-month timeframe at a leading early platelet gel institution. With a median level just above baseline, the observed variability led to less subjective, more quantifiable, optimized protocols for the preparation of AGF from buffy coat.

Table 2. Comparison of Manufacturer Protocols with Optimized Buffy Coat Sequestration Protocols

	Platelet concentration ($\times 10^6/\mu l$)	
Device	Manufacturer	ICI Optimized
Medtronic Sequestra	607	670
Haemonetics Cell Saver 5	358	760
Sorin Compact Advanced	613	815
Cobe BRAT 2	504	900
Fresenius C.A.T.S.	422 (96 ml)	435 (180 ml)

In all cases, the optimized protocols result in a higher platelet concentration (87).

A growing number of laboratory studies have demonstrated direct and dose-dependent effects of extended exposure to platelets and their releasates on the recruitment and proliferation of osteoblasts and uncommitted progenitor cells (61,90–94). In a landmark study on human fetal osteoblast cultures, Slater et al. concluded that the marked increases observed in proliferative activity and the continued differentiated activities, including matrix formation and mineral deposition of osteoblast-like cells in the platelet-supplemented cultures, provide evidence to support the proposal that the substances released by platelets may play a role in and have clinical applications in fracture healing (61). This group also evaluated buffy coat and AGF in their culture model, and preliminary findings demonstrated a significant proliferative response compared with the controls, but also that while the AGF cultures maintained their osteoblastic activity as measured by alkaline phosphatase, the buffy coat cultures had lost about 30% of their activity over time (Fig. 13) (90).

Recently, Haynesworth et al. presented an *in vitro* study looking at the effects of PRP on mesenchymal stem cell proliferation, chemotaxis, and differentiation (91). The platelet releasates were documented to mitogenically increase stem cell numbers and stimulate their chemotactic migration in a dose-dependent manner, without affecting the cells' osteogenic potential. However, the releasates did not support osteogenic differentiation in this culture model. Similar dramatic effects have been documented for stromal stem cells by Lucarelli et al. (92), trabecular bone-derived osteoblasts by Gruber et al. (93), and osteoblast precursor cells by Weibrich et al. (94).

8.2. Animal Models

Most of the *in vivo* evaluations of AGF to date have been in spinal applications. Walsh et al. studied the effects of AGF in combination with different graft materials in a posterolateral sheep spinal fusion model (95). An instrumented fusion at L3–L4 was created in 24 aged sheep, and grafted bilaterally with either autograft alone, autograft with AGF, Pro Osteon 500R with AGF, or Pro Osteon 500R with AGF and aspirated marrow (Fig. 14) (95). Fusion rates as determined by x-ray and CT scans, mechanical analysis, histology, and immunohistochemistry were used to evaluate the fusion at 6 months. Fusion rates and mechanical stiffness are shown in Table 3. The addition of AGF to autograft increased the fusion rate compared with autograft alone, whereas the combination of Pro Osteon, AGF, and marrow demonstrated equivalent fusion rates and a trend toward increased stiffness compared with autograft. In this challenging model, the AGF + Pro Osteon

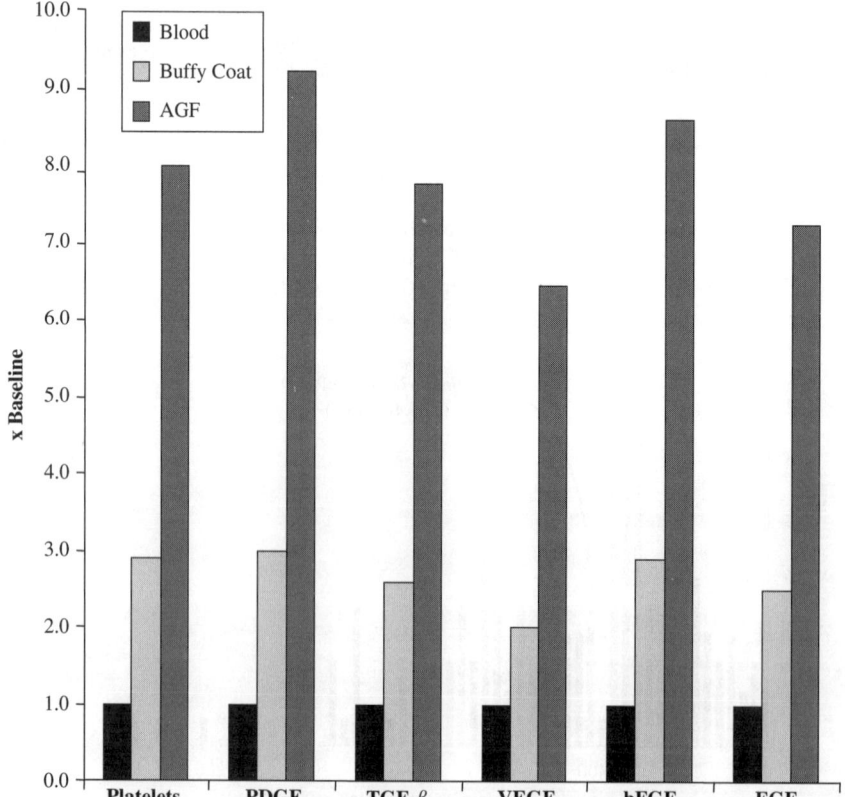

Figure 10. Relative levels of platelets and several growth factors from blood, buffy coat, and AGF samples. As expected, the increase in growth factor content as determined by ELISA corresponds to the observed measurement of the platelet concentration. (This figure is available in full color at http://www.mrw.interscience.wiley.com/ebe.)

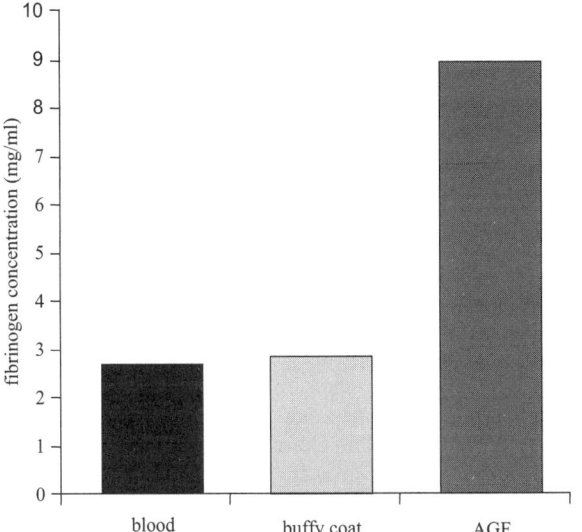

Figure 11. Fibrinogen levels in blood, buffy coat, and AGF samples. Note that no increase in fibrinogen is observed following centrifugation, only upon the removal of excess water through ultraconcentration. (This figure is available in full color at http://www.mrw.interscience.wiley.com/ebe.)

group alone was not equivalent to the other groups, perhaps because of the reduced numbers of host progenitor cells available in these aged animals. In an earlier posterolateral fusion study examining Pro Osteon 500R with and without AGF, this group evaluated the fusion site at three months using immunohistochemistry and found increased biological activity of many growth factors at the AGF-treated fusion site, including not only PDGF, TGF-β, bFGF, and IGF, but also BMP2 and BMP7 (96), which lends credence to the concept of increased recruitment and proliferation of osteoblasts by AGF, with their subsequent expression of various BMPs at the appropriate times in the healing cascade.

Several studies have examined the use of AGF in an interbody fusion model. Allen et al. concluded that the combination of AGF and Pro Osteon 200R was as effective as autograft at producing interbody fusion in disc levels treated with a threaded interbody fusion cage (97). Five skeletally mature sheep underwent lateral discectomy and interbody fusion with custom BAK/Proximity® threaded fusion cages. Two nonadjacent thoracic disc levels (T10–11; T12–13) were used in each animal; one level was grafted with autologous corticocancellous bone, whereas the other was treated with resorbable Pro Osteon® 200R supplemented with AGF. Animals were euthanized 4 months postoperatively, and the thoracic spine graded as either fused or not fused based on radiography and computed tomography. Nondestructive compression and torsional biomechanical tests were conducted, and the stiffness and range of motion were calculated for each specimen. Sections embedded in PMMA were microradiographed and then stained with MacNeals tetrachrome. The fractional areas of bone, cartilage, and fibrous tissue were measured within the central region of the fusion cage. In this study, as in the Walsh study, the increase in platelets and growth factors was measured to be over six times baseline levels. Radiography and axial CT revealed solid bridging in only one of four autograft specimens compared with three of four Pro Osteon-AGF specimens (Figs. 15a and 15b). Histomorphometry confirmed that the nature of the tissue within the fusion site was qualitatively and quantitatively similar in the two groups, consisting of mature mineralized bone interspersed with islands of endochondral ossification, fibrocartilage, and fibrous tissue. Mechanical test data showed a consistent trend toward higher stiffness values in the Pro Osteon-AGF specimens, but these differences did not reach statistical significance with the limited sample size.

An interbody fusion study of similar design was conducted by Sefter et al. in which 14 mature Alpine goats received various treatments at the C3–C4 and C5–C6 levels, including Rabea Cages packed with either local autogenous bone or AGF + Pro Osteon (98). At four months, radiographic analysis from three independent observers and biomechanical analysis indicated no statistical difference in fusion rates between autograft and AGF + Pro Osteon.

Two recent studies have been presented to date on the use of DePuy's Symphony system in the spine. In Sethi et al., platelet gels concentrated to 1 million platelets per microliter were added to either 1.5 cc (reduced volume) or

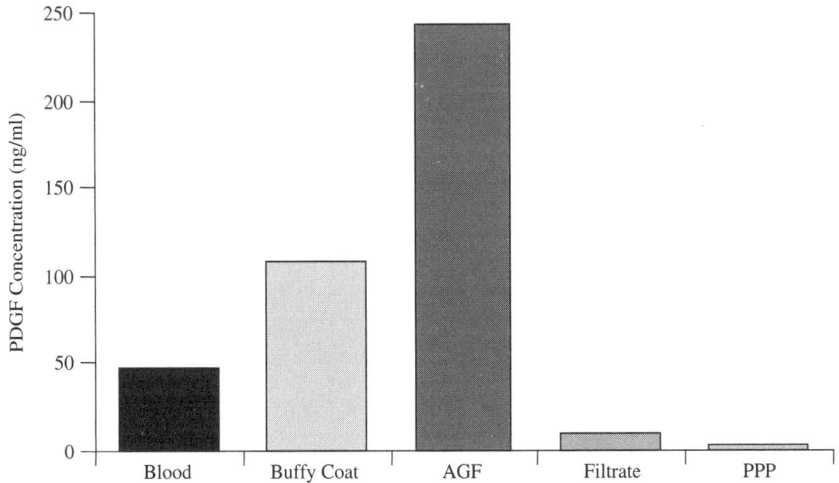

Figure 12. PDGF was assayed in the all fractions during classic AGF preparation to illustrate that most of the growth factors from the starting blood are collected in the buffy coat and AGF, and to illustrate that minimal quantities of growth factors are lost in the filtrate. (This figure is available in full color at http://www.mrw.interscience.wiley.com/ebe.)

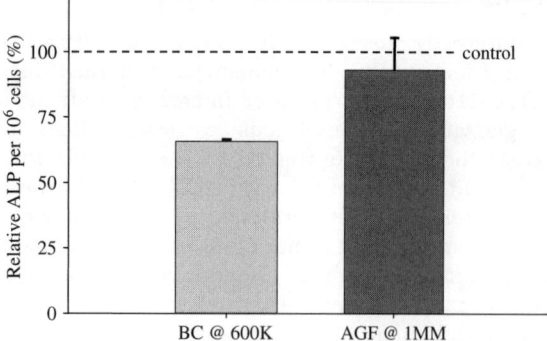

Figure 13. Relative alkaline phosphatase activity in fetal osteoblasts after culture with unfiltered buffy coat and AGF. Although increased proliferation was observed in both groups, maintenance of osteoblastic function was only observed in the AGF group. (This figure is available in full color at http://www.mrw.interscience.wiley.com/ebe.)

3.0 cc (normal volume) of autograft in an L5–L6 intertransverse rabbit fusion model (99). Analysis at five weeks revealed no difference in flexion ($3.5 \pm 1.8°$ with platelets vs. $3.4 \pm 1.0°$ without) or fusion rate (both 86%) upon the addition of the platelets to the normal graft volume group. In the reduced-volume group, however, flexion decreased from $6.7 \pm 3.7°$ to $4.6 \pm 2.6°$ and the fusion rate doubled from 29% to 58%, indicating that concentrated platelets may be of significant benefit when graft quantity is limited.

More recently, Li et al. compared an osteoconductive scaffold (β-tricalcium phosphate) with and without PRP produced from Symphony to autograft in a 3-level porcine ALIF model (100). Brantigan cages were used, and the fusion mass explanted at 3 months. Both the β-TCP alone and the β-TCP with PRP yielded 10% new bone, compared with 42% for the autograft group, indicating that, in this ALIF model, TCP loaded with PRP did not show better results than using TCP alone, and both had a poorer outcome than that of autograft. Speculation about possible

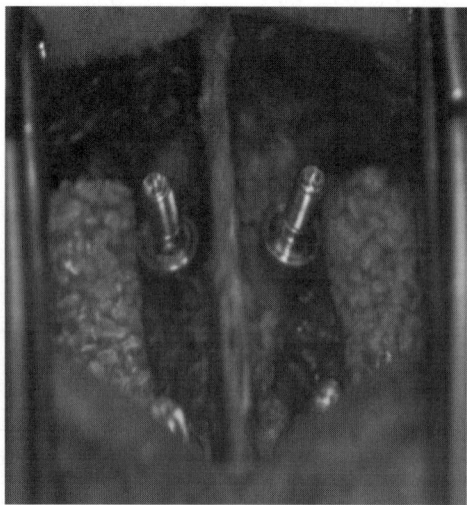

Figure 14. Posterolateral composite graft consisting of AGF, Pro Osteon 500R, and marrow cells, implanted in an aged sheep model. From Walsh et al. (95). (This figure is available in full color at http://www.mrw.interscience.wiley.com/ebe.)

Table 3. Fusion Rates and Mechanical Stiffness of the Graft Site in an Aged Sheep Model (95)

Group	Fusion Rate	Stiffness (N/mm)
500R + AGF	16% (1/6)	750.5 ± 145.6
Autograft	67% (4/6)	830.2 ± 129.7
Autograft + AGF	100% (6/6)	950.5 ± 140.8
500R + AGF + Marrow	67% (4/6)	1094.2 ± 330.9

reasons for the lack of efficacy include suboptimal volume of PRP added to the cage or fibrous tissue enhancement because of micromotion of the segments studied.

Outside the spine, the ability of AGF to enhance bone ingrowth into a chamber implanted bilaterally into the proximal tibiae of athymic rats was evaluated by Siebrecht et al. (101). The chambers in this well-established model allow tissue ingrowth from one end of a long cylindrical space, so that ingrowth distances along the cylindrical axis can be measured (102). The use of athymic animals allowed human blood to be used to make the AGF. Seventeen rats received pairs of titanium bone chambers filled with Pro Osteon 200 implanted bilaterally in the proximal tibiae. In one chamber in each rat, the hydroxyapatite was impregnated with the AGF and thrombin to form a gel. The rats were sacrificed after 4 weeks, and tissue (bone + soft tissue) and bone ingrowth were measured as ingrowth distances on histological slides. As shown in Fig. 16, AGF more than doubled the bone ingrowth distance, from 0.8 mm in controls to 1.8 mm, and also significantly increased the total tissue ingrowth distance by two-thirds, from 2.9 mm in controls to 4.1 mm.

A second nonspine study worthy of mention showed that the use of AGF with morselized bone allograft increases the strength of early fixation and the osseointegration of noncemented implants (103). This study used eight dogs to examine the effect of allograft chips with or without AGF on bone ingrowth across 2 mm into plasma-sprayed cylindrical titanium implants. By adding AGF to allograft, ultimate strength significantly increased by 50% (1.24 MPa to 1.88 MPa, $p < 0.05$). Furthermore, AGF significantly increased bone in contact with the implant from 7 to 11% ($p < 0.05$). The AGF-treated group also demonstrated increases in interface stiffness and energy to failure; but this trend was not statistically significant. No evidence existed of increased bone graft resorption with AGF, in contrast to a previous study with bone graft and exogenous OP-1 (104). In a further study by this group that highlights the point that different preparations cannot be judged together, centrifuged PRP with allograft did not show any increase in ultimate shear strength, energy absorption, and apparent stiffness (105).

9. ORTHOPEDIC CLINICAL EXPERIENCE AND OUTCOMES

The concept of applying platelet releasates to enhance wound healing has been discussed since the 1980s (31,85), and platelet gels prepared intraoperatively from the cell separators have been used clinically in soft tissue for the better part of a decade (106–111). Oral/maxillofacial surgery has been the area of greatest use, with additional

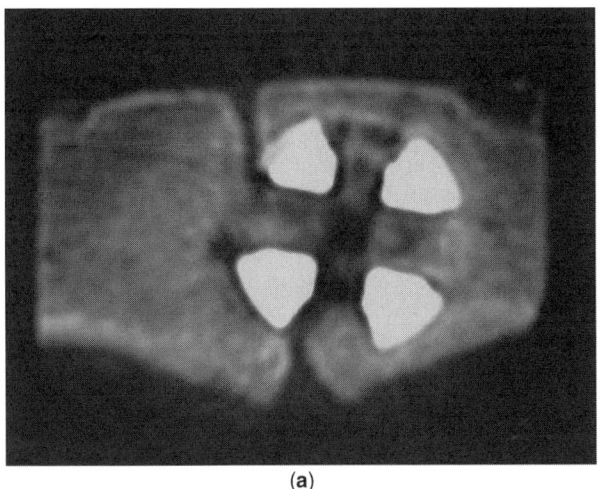 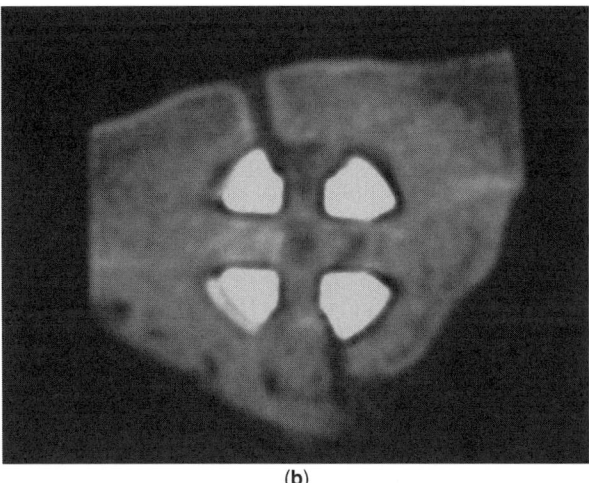

Figure 15. CT images from Allen et al. (97) indicating bridging bone through an intervertebral cage containing autograft (a) or Pro Osteon 200R and AGF (b).

usage in plastics, wound care, obstetrics, ENT, cardiothoracic surgery, general surgery, and orthopedics.

Anecdotal results were positive, although a definitive clinical study was not published until 1998, when Marx et al. examined the effects of small volumes of PRP on graft enhancement in mandibular defect reconstruction (71). At the 6-month interval, radiographic and histomorphometric studies were done to examine the maturity and density of the grafts, respectively. The results showed that PRP caused the graft to assume a greater rate of maturation and a greater bone density compared with controls, indicating that PRP is an effective clinical method for mandibular bone graft incorporation enhancement.

The effectiveness of PRP gels in orthopedic soft tissue applications to promote wound healing and decrease blood loss have recently been demonstrated by Mooar et al. (112) and Pritchard (113). In 106 total knee arthroplasty cases, Mooar applied PRP gel from the Medtronic Sequestra to cut bone surfaces, the synovium, and the lining of the wound at closure after tourniquet release. They documented an earlier function range of motion (increase of 6.1°), a smaller drop in patient hemoglobin post-op (2.68 g/dl vs. 3.12 g/dl), and a reduced hospital stay (4.04 days vs. 5.07 days) in the patients that received the PRP gel compared with their historical controls. Further, IV narcotic use was halved (17 mg/day vs. 34.5 mg/day), and oral narcotic requirements also decreased (1.84 pills/day vs. 2.72 pills/day) (112). Pritchard applied PRP to 28 patients undergoing a mini-open rotator cuff repair, with an additional 29 patients serving as untreated controls (113). At 8 weeks post-op, he reported that 57% of PRP-treated patients had a pain score ≤ 2 and met the functional range of motion landmarks required for release from physical therapy, which was significantly greater than the 21% of untreated patients who met the same criteria at 8 weeks (113).

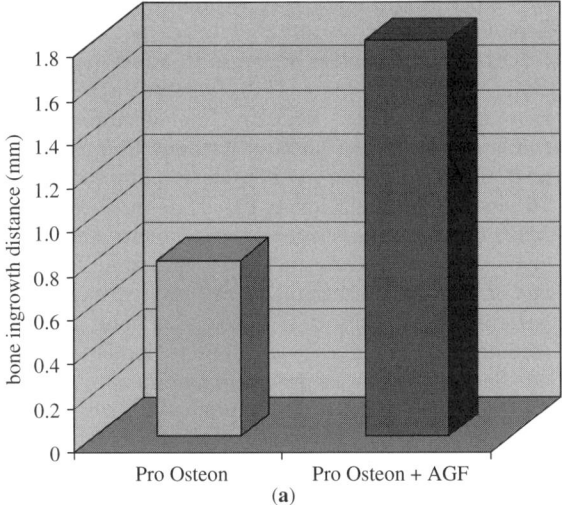 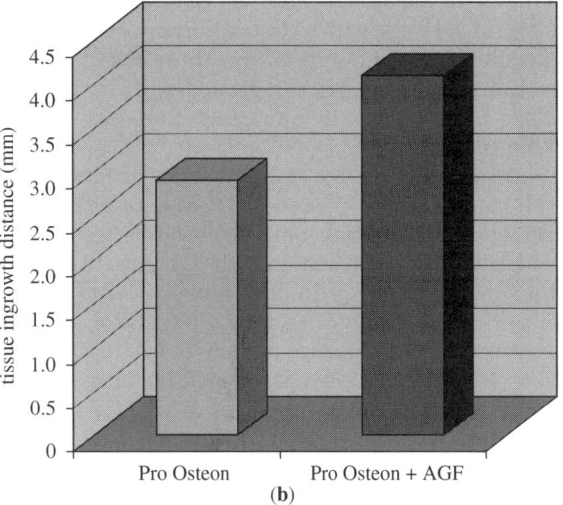

Figure 16. Results from Siebrecht et al. (101) indicating more than double new bone ingrowth (a) as well as increased total tissue ingrowth (b) in grafts containing AGF and Pro Osteon 200, compared with Pro Osteon 200 by itself. (This figure is available in full color at http://www.mrw.interscience.wiley.com/ebe.)

PRP gels have seen limited use and have not been widely studied clinically in orthopedic bone grafting applications, however. In contrast, since its 1999 launch in the United States, AGF has been used in over 50,000 bone grafting cases. Both retrospective analyses and prospective studies have examined the clinical effect of AGF in a variety of orthopedic applications, either in conjunction with an appropriate trabecular graft material as a substitute for harvesting iliac crest bone or as an enhancer to potentially increase the fusion rate or decrease the time to fusion when used with autograft (114–128). Case reports and initial analyses have been favorable in a number of various indications, including posterolateral and interbody adult spinal fusion, scoliosis, revision total joint arthroplasty, delayed and nonunions, and complex foot and ankle fusion.

The first preliminary bone grafting results with AGF were published by Lowery et al. in a nonrandomized, retrospective clinical series of 19 patients (114). AGF was used as an adjunct to autologous bone and coralline hydroxyapatite (Pro Osteon® 500) in anterior and posterolateral fusions. No pseudarthroses were identified at 6 to 18 months postoperatively with a large majority of levels graded as fused radiographically. Visual confirmation of fusion was obtained in five patients who underwent adjacent level fusion or had their hardware removed.

Several studies in which surgeons prospectively tracked AGF patients have been presented at major peer-reviewed clinical and scientific conferences. Lettice et al. evaluated AGF in instrumented posterolateral lumbar arthrodeses (115). Preliminary results based on serial radiographs of 34 patients with at least 5 months follow-up suggested that the combination of AGF and autograft produced a more rapid maturation of the posterolateral graft compared with historical controls of autograft with allogeneic demineralized bone matrix gel. Kucharzyk et al. tracked 42 patients who had undergone an average 3-level instrumented lumbar fusion and were grafted with AGF, Pro Osteon, and local laminar and spinal process bone (116). Solid fusion was observed radiographically in 95% (40/42) of patients, with a clinical success rate of 93%. When compared with the literature, these results were similar to instrumented fusions grafted with iliac crest autogenous bone without the increased morbidity at the harvesting site.

In a retrospective analysis of AGF used as an enhancer, Hee et al. independently examined 23 patients with AGF, autograft, and TLIF instrumentation who underwent 1- or 2-level fusions and compared it with 111 patients from their historical controls with just autograft and TLIF (117). Upon x-ray at only 4 months, 70% (16/23) of the AGF patients showed evidence of bony fusion, compared with 36% (40/111) of controls. At 6 months, all patients who would ultimately heal were fused with AGF (22/23, 96%), compared with 64% (71/111) of controls. Differences at both of these time points were significant ($p < 0.05$). The one pseudarthrosis in the AGF group was a chronic smoker with a failed previous lumbar fusion at L4–L5 and also presented with DDD and spinal stenosis at L5–S1. She had a TLIF at L5–S1 and posterolateral fusion from L4–S1, but still developed a pseudarthrosis. At 24 months, 22/23 (96%) of the AGF group remained fused. Although this result was not ultimately significant over the control autograft+TLIF group (104/111, 94%), the clinical benefit of a decreased time to healing to both patient and the health-care system is significant. Another important aspect of this study is that the platelet, PDGF, and TGF-beta information were quantified, verifying proper preparation of the AGF. They reported a platelet count of 1.3 million ($\sim 5 \times$ increase), a $6 \times$ increase in PDGF, and a $4 \times$ increase in TGF-β (117).

Bose et al. published a retrospective analysis of AGF used with local bone and Pro Osteon, reporting on his first 60 patients undergoing interbody fusion or posterolateral intertransverse fusion (118). No iliac crest bone grafting was used. Early clinical outcomes indicated solid or maturing fusions in 58 of 60 patients. No adverse effects were observed that could be attributed to AGF, and the authors concluded that AGF represents an economical and readily available autologous source of growth factors to enhance bone healing.

Similarly, a recent presentation by Broom retrospectively studied 177 patients who had 1–2-level instrumented lumbar spine fusions, no previous lumbar surgery, and no iliac crest bone graft at the time of their fusion (119). Most patients presented with degenerative spondylolisthesis (74) patients, isthmic spondylolisthesis (36) patients, and stenosis and instability (61) patients. Local bone from spinous process and lamina was used along with cortical demineralized allograft and AGF. Patients were analyzed radiographically using A/P, lateral, and F/E x-rays at an average of 15 months follow-up. Nine patients were judged to have pseudarthroses, a nonunion rate of 5% (9/177). On further analysis of the pseudarthroses, seven occurred in patients with isthmic spondylolisthesis, where it was also noted that fractures of the sacral screws were prevalent. Excluding this group, pseudarthrosis rate in the remaining patients was 1.4% (2/141), leading the author to conclude that the combination of local bone, demineralized cortical allograft, and AGF has been shown to yield a high fusion rate without the use of iliac crest bone graft.

Other retrospective analyses in adult spinal fusion include those by Waldrip (120), Logan et al. (121), and Kurica et al. (122). Waldrip has documented a 97% radiographic fusion success without iliac crest bone for posterolateral fusion in 103 elderly patients (average age 76 years, average 3 levels fused) using a combination of AGF, bone marrow aspirate, local bone, and either allograft chips or ProOsteon (120). Logan et al. found that the use of AGF and resorbable coralline hydroxyapatite in 15 patients as an adjunct to autologous bone yielded an extremely high fusion rate and solid fusion before 24 months postoperatively, an improvement on fusion rates with autologous bone alone in the published literature (121). In a preliminary study, Kurica et al. demonstrated the ability of a bone graft replacement consisting of AGF with Pro Osteon 500R in instrumented posterolateral lumbar fusion to be effective in producing new bone and a subsequent solid fusion as an extender of autograft (122). On one side of the posterolateral fusion in 22 patients, AGF was mixed with resorbable Pro Osteon, whereas autologous bone taken from the iliac crest was mixed with DBM gel and placed on the contralateral side. At a mean follow-up of 17.5 months, radiographic evaluation showed solid fusion in 91.0% (20/22) of the AGF and ceramic sides and in 81.8% (18/22) of the autograft and DBM sides.

Rohmiller et al. demonstrated that grafts consisting of local bone, Pro Osteon 500R, and AGF were safe and effective alternatives to autogenous crest or allograft bone in deformity correction (123). Twenty-one patients who underwent instrumented posterior spinal fusion were followed for a minimum follow-up of 24 months, and all patients were judged to have a stable spinal fusion by clinical and radiographic analyses.

Positive retrospective results have also been reported in long bone nonunions by Jimenez and Anderson (124). AGF, morselized allograft, and electrical stimulation were used on 109 long bone nonunions in 107 patients with an average nonunion duration of 12 months. 105 of 109 (96%) healed clinically and radiographically, and the four initial failures were successfully revised with AGF, indicating that this approach appears to be a safe and efficacious alternative to autologous bone grafting in severe nonunions. In the only direct clinical comparison of two different methodologies, Grant and Jacobus studied both AGF (Interpore Cross) and PRP (Harvest SmartPReP™) mixed with morselized local dead bone graft material and fixation in high-risk midfoot fusion procedures (125). They found a total fusion rate of 91% with AGF compared with 71% with PRP (125). Like PRP, AGF has been shown to reduce postoperative blood loss as well. Birch et al. randomized 70 patients undergoing posterior instrumented lumbar fusion to receive AGF gel on all the raw surfaces of their wounds prior to closure, or no spray. AGF reduced postoperative blood loss by 34% compared with controls, whereas in cases when intraoperative blood loss was greater than 500 ml, the difference was even greater (44%) (126).

Although these retrospective or prospectively tracked studies provide useful information and offer insight into the efficacy of AGF in multiple potential applications, more rigorous masked, prospective studies in these fields are needed, and are in fact underway. In spinal applications, these studies seek to evaluate the use of AGF both in combination with autogenous iliac crest bone and as a substitute for autograft. These studies are critical for the long-term acceptance of AGF as a biologic enhancer of bone regeneration. Jenis et al. (127) randomized 37 patients undergoing 1 or 2 level L3–S1 anterior/posterior interbody fusion for DDD/spondylolisthesis to one of two graft treatment groups: iliac crest autograft (22) patients, or AGF combined with cancellous allograft (15) patients. The operative technique consisted of an ALIF approach in which the endplates were preserved, a titanium mesh cage with the graft material implanted in the interbody space, and standard posterior pedicle screw fixation. A/P, lateral, and dynamic flexion/extension radiographs were taken at 6 weeks and 3, 6, 12, and 24 months, and as well as a 6-month reconstructed CT image. Images were evaluated by an independent reviewer. Clinical outcome data was independently collected based on VAS pain scores, Oswestry, and SF-36 questionnaires. With an average follow-up of 17 months, 85% of the autograft group and 89% of the allograft/AGF group were considered radiographically fused. Using strict criteria, 56% of the patients in both groups were considered fused based on the CT reconstruction. No significant differences in clinical outcomes between the treatment groups were observed; decreases in VAS pain scores and Oswestry disability scores, and the increase in SF-36 scores were all identical. Additional benefits were lack of postoperative donor site morbidity and improved graft handling characteristics (127).

In the treatment of 57 delayed and nonunions, Volgas found that the complication rate, and possibly time to healing, were less using AGF and allograft compared with autograft, with similar success rates (128). 25 patients received iliac crest bone graft and 32 patients received AGF and cancellous allograft chips. Postoperative pain averaged 6/10 on day two with autograft versus 4.7/10 on day two with AGF/allograft. Radiographic time to union was 194 days in the autograft group and 119 days in allograft/AGF group, although outliers existed that may explain the differences. 3/20 (15%) of autograft patients and 6/31 (19%) of allograft/AGF patients required re-operation for further nonunion, a statistically insignificant difference. SF-36 scores (33 for autograft and 32 for allograft/AGF) and days in the hospital (2.9 for autograft and 2.7 for autograft) were also not significantly different. Dr. Volgas concluded that the complication rate, short- and long-term pain, and perhaps even time to healing are less using AGF and the failure rates of the two procedures are similar (128).

Some authors (129,130) have reported no measurable additive effect from the application of AGF to autograft with instrumented posterolateral fusions where the control group of autograft alone demonstrated success rates well above 90%, not a surprising outcome. As shown by Hee et al., however, the clinical benefit of AGF in such settings is a significant acceleration of time to healing (117), a considerable benefit to patients, surgeons, and the health-care system at large. One report, from Weiner and Walker, has suggested that AGF may inhibit bone graft repair in the setting of noninstrumented posterolateral fusion (131). However, substantial flaws exist in their analysis (132). They claimed a 91% radiographic fusion success in the control group with autograft alone, a surprisingly anomalous outcome compared with the rest of the literature where only 60–70% success has been seen with noninstrumented fusion (133–141) and one for which they offer neither explanation nor discussion. In comparison, the AGF group outcome of 62% fusion is not out of line with consensus results for this type of procedure. In addition, their proposed theoretical basis for "inhibition" by AGF—that it interferes with BMP function—is solely drawn from a pair of in vitro studies where single rhPDGF and rhTGF-β were employed with rhBMPs (142,143). Other studies contradict their hypothetical basis, reporting that TGF-β demonstrated enhanced bone formation with both BMP-2 and OP-1 compared with either rhBMP alone (144,145). Interaction between two individual recombinant growth factors at pharmacological doses is unlikely to reflect the myriad subtle and intricate interactions that occur between autogenous BMPs and the numerous factors contained in AGF. At most, this report suggests that a proper prospective, randomized, controlled study of AGF in noninstrumented posterolateral spinal fusion is desirable.

10. CONCLUSIONS

The autologous nature of AGF or PRP is attractive to clinicians and patients alike because it provides a source of multiple osteopromotive growth factors and is nonimmunogenic. Only with the advent of dedicated systems offering consistently elevated platelets, white cells, and, in the case of AGF, fibrinogen has the orthopedic market started to explore and embrace this autologous, multiple growth factor solution to enhancement of bone regeneration. The hypothesized mechanism of action of blood-derived growth factors, based on studies of individual and multiple exogenous factors, has begun to be confirmed through *in vitro* studies evaluating the direct application of platelets and buffy coat on progenitor and bone cells, and, with *in vivo* animal models, demonstrating increased bone formation, fusion rates, and biological activity. Initial clinical studies that have followed proper AGF preparation techniques have reported excellent results in spinal fusion and other applications. AGF mixed with synthetic or allogeneic grafts have demonstrated fusion or nonunion healing results equivalent to the gold standard of autologous grafts and reduced or eliminated the need to harvest bone from the patient's hip, and the addition of AGF to autogenous bone grafts have shown evidence of decreased time to fusion. As always, longer-term follow-up and the publication of indication-specific prospective, masked clinical studies is critical for the widespread acceptance of autologous gel therapies. In addition to more rigorous trials evaluating their efficacy in bone repair and regeneration, future research directions may focus on the use of these gels as carriers of other biologic cells and molecules. Their ability to gel *in situ*, encourage vascularization and tissue ingrowth, and degrade in a matter of days to weeks may make them ideal candidates for regeneration of intervertebral disc nuclei, repair of meniscal or cartilage lesions, or orthopedic soft tissue applications.

Acknowledgement

I would like offer my sincere appreciation to a number of colleagues who have recognized the potential of the novel concept of autologous therapy in orthopedics, and through their efforts have contributed to its foray into the clinic. In particular, I would like to thank Andrew Hood, Michael Ponticiello, and Dr. Edwin Shors (Interpore Cross), Drs. Barbara Boyan and Jacquelyn McMillan (Georgia Institute of Technology), Dr. William Walsh (University of New South Wales), Dr. Matthew Allen (Syracuse University), Dr. Joan Bechtold (Minnesota Orthopedics Research Foundation), Dr. Per Aspenberg (Linkopings Universitet), Joel Higgins (Biomet), and James Giuffre (International Spinal Development & Research Foundation).

Further, I would specifically like to acknowledge and thank Andrew Hood, Barbara Boyan, and James Giuffre for their contributions to this chapter.

BIBLIOGRAPHY

1. H. S. An, K. Lynch, and J. Toth, Prospective comparison of autograft vs. allograft for adult posterolateral lumbar spine fusion: differences among freeze-dried, frozen, and mixed grafts. *J. Spinal Disord*. 1995; **8**(2):131–135.

2. P. Axelsson, R. Johnsson, B. Stromqvist, M. Arvidsson, and K. Herrlin, Posterolateral lumbar fusion. Outcome of 71 consecutive operations after 4 (2–7) years. *Acta Orthop. Scan*. 1994; **65**(3):309–314.

3. M. Bernhardt, D. E. Swartz, P. L. Clothiaux, R. R. Crowell, and A. A. White III, Posterolateral lumbar and lumbosacral fusion with and without pedicle screw internal fixation. *Clin. Orthop*. 1982; (284):109–115.

4. F. B. Christensen, K. Thomsen, S. P. Eiskjaer, J. Gelenick, and C. E. Bunger, Functional outcome after posterolateral spinal fusion using pedicle screws: comparison between primary and salvage procedure. *Eur. Spine. J*. 1998; **7**(4):321–327.

5. M. Deguici, A. J. Rapoff, and T. A. Zdeblick, Posterolateral fusion for isthmic spondylolisthesis in adults: analysis of fusion rate and clinical results. *J. Spinal Disord*. 1998; **11**(6):459–464.

6. J. C. France, M. J. Yaszemski, W. C. Lauerman, J. E. Cain, J. M. Glover, K. J. Lawson, J. D. Coe, and S. M. Topper, A randomized prospective study of posterolateral lumbar fusion. Outcomes with and without pedicle screw instrumentation. *Spine* 1999; **24**(6):553–560.

7. C. G. Greenough, M. D. Peterson, S. Hadlow S, and R. D. Fraser, Instrumented posterolateral lumbar fusion. Results and comparison with anterior interbody fusion. *Spine* 1998; **23**(4):479–486.

8. S. S. Jorgenson, T. G. Lowe, J. France, and J. Sabin, A prospective analysis of autograft versus allograft in posterolateral lumbar fusion in the same patient. A minimum of 1-year follow-up in 144 patients. *Spine* 1994; **19**(18):2048–2053.

9. J. A. McCulloch, Uninstrumented posterolateral lumbar fusion for single level isolated disc resorption and/or degenerative disc disease. *J. Spinal Disord*. 1999; **12**(1):34–39.

10. K. M. Parker, S. E. Murrell, S. D. Boden, and W. C. Horton, The outcome of posterolateral fusion in highly selected patients with discogenic low back pain. *Spine* 1996; **21**(16):1909–1917.

11. H. Pihlajamaki, O. Bostman, M. Ruuskanen, P. Myllynen, J. Kinnunen, and E. Karaharju, Posterolateral lumbosacral fusion with transpedicular fixation: 63 consecutive cases followed for 4 (2–7) years. *Acta Orthop. Scan*. 1996; **67**(1):63–68.

12. J. D. Rompe, P. Eysel, and C. Hopf, Clinical efficacy of pedicle instrumentation and posterolateral fusion in the symptomatic degenerative lumbar spine. *Eur. Spine J*. 1995; **4**(4):231–237.

13. C. L. Schnee, A. Freese, and L. V. Ansell, Outcome analysis for adults with spondylolisthesis treated with posterolateral fusion and transpedicular screw fixation. *J. Neurosurg*. 1997; **86**(1):56–63.

14. J. S. Thalgott, R. C. Sasso, H. B. Cotler, M. Aebi, and S. H. LaRocca, Adult spondylolisthesis treated with posterolateral lumbar fusion and pedicular instrumentation with AO DC plates. *J. Spinal Disord*. 1997; **10**(3):204–208.

15. K. Thomsen, F. B. Christensen, S. P. Eiskjaoe, E. S. Hansen, S. Fruensgaard, and C. E. Bunger, 1997 Volvo Award winner in clinical studies. The effect of pedicle screw instrumentation on functional outcome and fusion rates in posterolateral lumbar spinal fusion: a prospective, randomized clinical study. *Spine* 1997; **22**(24):2813–2822.

16. M. W. Chapman and E. M. Younger, Morbidity at bone graft donor sites. *J. Orthop. Trauma* 1989; **3**:192–195.
17. E. C. Shors, The development of coralline porous ceramic graft substitutes. In: C. T. Laurencin, ed. *Bone Graft Substitutes*. West Conshohocken, PA: ASTM International, 2003, pp. 271–288.
18. R. W. Bucholz, A. Carlton, and R. Holmes, Interporous hydroxyapatite as a bone graft substitute in tibial plateau fractures. *Clin. Orthop. Rel. Res.* 1989; **240**:53–62.
19. E. C. Shors, Bone graft substitutes: clinical studies using coralline hydroxyapatite. In: G. H. I. M. Walenkamp, ed. *Biomaterials in Surgery*. Stuttgart, Germany: Georg Thieme Verlag, 1998, pp. 83–89.
20. S. W. Wolfe, L. Pike, J. F. Slade III, and L. D. Katz, Augmentation of distal radius fracture fixation with coralline hydroxyapatite bone graft substitute. *J. Am. Hand Surg.* 1999; **24**:816–827.
21. F. Rahimi, B. T. Maurer, and M. G. Enzwiler, Coralline hydroxyapatite: a bone graft alternative in foot and ankle surgery. *J. Foot Ankle Surg.* 1997; **36**:192–203.
22. R. B. Irwin, M. Bernhard, and A. Biddinger, Coralline hydroxyapatite as a bone substitute in orthopedic oncology. *Am. J. Orthop.* 2001; **30**:544–550.
23. J. S. Thalgott, K. Fritts, J. M. Giuffre, and M. Timlin, Anterior fusion of the cervical spine with coralline hydroxyapatite. *Spine* 1999; **24**:1295–1299.
24. M. R. Urist, Bone formation by autoinduction. *Science* 1965; **150**:893–899.
25. M. E. Joyce, S. Jingushi, S. P. Scully, and M. E. Bolander, Role of growth factors in fracture healing. *Prog. Clin. Biol. Res.* 1991; **365**:391–416.
26. M. Lind, Growth factor stimulation of bone healing. Effects on osteoblasts, osteomies, and implants fixation. *Acta Orthop. Scand. Suppl.* 1998; **283**:2–37.
27. S. B. Trippel, R. D. Coutts, T. M. Einhorn, G. R. Mundy, and R. G. Rosenfeld, Growth factors as therapeutic agents. *J. Bone Joint Surg.* 1996; **78A**(8):1272–1286.
28. A. I. Caplan, The mesengenic process. *Clin. Plast. Surg.* 1994; **21**(3):429–435.
29. K. L. Kaplan, M. J. Broekman, A. Chernoff, G. R. Lesznik, and M. Drillings, Platelet alpha-granule proteins: studies on release and subcellular localization. *Blood* 1979; **53**:604–618.
30. T. Asahara, C. Bauters, L. P. Zheng, S. Takeshita, S. Bunting, N. Ferrara, J. F. Symes, and J. M. Isner, Synergistic effect of vascular endothelial growth factor and basic fibroblast growth factor on angiogenesis in vivo. *Circulation* 1995; **92**(9 Suppl):II365–II371.
31. D. R. Knighton, T. K. Hunt, K. K. Thakral, and W. H. Goodson III, Role of platelets and fibrin in the healing sequence: an in vivo study of angiogenesis and collagen synthesis. *Ann. Surg.* 1982; **196**(4):379–388.
32. M. Lind, B. Deleuran, K. Thestrup-Pedersen, K. Soballe, E. F. Eriksen, and C. Bunger, Chemotaxis of human osteoblasts. Effects of osteotropic growth factors. *APMIS* 1995; **103**(2):140–146.
33. W. F. Krause, J. McMillan, C. Lohmann, Z. Schwartz, D. M. Arm, and B. D. Boyan, Platelet-rich plasma—a review of its components and use in bone grafting. *Clin. Orthop. Rel. Res.*, in submission.
34. D. L. Hwang, L. J. Latus, and A. Lev-Ran, Effects of platelet-contained growth factors (PDGF, EGF, IGF-I, and TGF-beta) on DNA synthesis in porcine aortic smooth muscle cells in culture. *Exp. Cell. Res.* 1992; **200**:358–360.
35. C. H. Kasperk, J. E. Wergedal, S. Mohan, D. L. Long, K. H. W. Lau, and D. J. Baylink, Interactions of growth factors present in bone matrix with bone cells: effects on DNA synthesis and alkaline phosphatase. *Growth Factors* 1990; **3**:147–158.
36. R. Ross, E. W. Raines, and D. F. Bowen-Pope, The biology of platelet-derived growth factor. *Cell* 1986; **46**(2):155–169.
37. C. P. Kiritsy, A. B. Lynch, and S. E. Lynch, Role of growth factors in cutaneous wound healing: a review. *Crit. Rev. Oral Biol. Med.* 1993; **4**(5):729–760.
38. T. J. Nash, C. R. Howlett, C. Martin, J. Steele, K. A. Johnson, and D. J. Hicklin, Effect of platelet-derived growth factor on tibial osteotomies in rabbits. *Bone* 1994; **15**(2):203–208.
39. M. Centrella, T. L. McCarthy, and E. Canalis, Platelet-derived growth factor enhances DNA and collagen synthesis in osteoblasts-enriched cultures from fetal rat parietal bone. *Endocrinology* 1989; **125**(1):13–19.
40. J. Massague, TGF-beta signal transduction. *Ann. Rev. Biochem.* 1998; **67**:753–791.
41. J. G. Andrew, J. Hoyland, S. M. Andrew, A. J. Freemont, and D. Marsh, Demonstration of TGF-beta 1 mRNA by in situ hybridization in normal human fracture healing. *Calcif. Tissue Int.* 1993; **52**(2):74–78.
42. W. R. Gombotz, S. C. Pankey, L. S. Bouchard, D. H. Phan, and P. A. Puolakkainen, Stimulation of bone healing by transforming growth factor-β released from polymeric or ceramic implants. *J. Appl. Biomat.* 1994; **5**:141–150.
43. M. Centrella, T. L. McCarthy, and E. Canalis, Transforming growth factor beta and remodeling of bone. *J. Bone Joint Surg.* 1991; **73A**(9):1418–1427.
44. T. Murakami, M. Yamamoto, K. Ono, M. Nishikawa, N. Nagata, K. Motoyoshi, and T. Akatsu, Transforming growth factor-beta1 increases mRNA levels of osteoclastogenesis inhibitory factor in osteoblastic/stromal cells and inhibits the survival of murine osteoclast-like cells. *Biochem. Biophys. Res. Commun.* 1998; **252**(3):747–752.
45. M. Centrella, M. C. Horowitz, J. M. Wozney, and T. L. McCarthy, Transforming growth factor-beta gene family members and bone. *Endocr. Rev.* 1994; **15**(1):27–39.
46. E. Canalis and L. G. Raisz, Effect of fibroblast growth factor on cultured fetal rat calvaria. *Metabolism* 1980; **29**:108–114.
47. T. Nakamura, Y. Hara, M. Tagawa, T. Yuge, H. Fukuda, and H. Nigi, Recombinant human basic fibroblast growth factor accelerates fracture healing by enhancing callus remodeling in experimental dog tibial fracture. *J. Bone Miner. Res.* 1998; **13**(6):942–949.
48. M. Arras, W. D. Ito, D. Scholz, B. Winkler, J. Schaper, and W. Schaper, Monocyte activation in angiogenesis and collateral growth in the rabbit hindlimb. *J. Clin. Invest.* 1998; **101**(1):40–50.
49. S. Jingushi, S. P. Scully, M. E. Joyce, Y. Sugioka, and M. E. Bolander, Transforming growth factor-beta 1 and fibroblast growth factors in rat growth plate. *J. Orthop. Res.* 1995; **13**(5):761–768.
50. U. Wartiovaara, P. Salven, H. Mikkola, R. Lassila, J. Kaukonen, V. Joukov, A. Orpana, A. Ristimaki, M. Heikinheimo, H. Joensuu, K. Alitalo, and A. Palotie, Peripheral blood platelets express VEGF-C and VEGF which are released during platelet activation. *Thromb. Haemost.* 1998; **80**:171–175.
51. R. Mohle, D. Green, M. A. S. Moore, R. L. Nachman, and S. Rafii, Constitutive production and thrombin-induced release of vascular endothelial growth factor by human megakaryocytes and platelets. *Proc. Nat. Acad. Sci. USA* 1997; **94**:663–668.
52. J. M. Schlaeppi, S. Gutzwiller, G. Finkenzeller, and B. Fournier, 1,25-Dihydroxyvitamin D3 induces the expression of

vascular endothelial growth factor in osteoblastic cells. *Endocr. Res.* 1997; **23**(3):213–229.

53. S. Harada, S. B. Rodan, and G. A. Rodan, Expression and regulation of vascular endothelial growth factor in osteoblasts. *Clin. Orthop. Rel. Res.* 1995; **313**:76–80.

54. N. Ferrara, K. Houck, L. Jakeman, and D. W. Leung, Molecular and biological properties of the vascular endothelial growth factor family of proteins. *Endocr. Rev.* 1992; **13**(1):18–32.

55. V. Midy and J. Plouet, Vasculotropin/vascular endothelial growth factor induces differentiation in cultured osteoblasts. *Biochem. Biophys. Res. Commun.* 1994; **199**(1):380–386.

56. X. Guo, J. C. Y. Cheng, L. P. Law, and P. H. Chow, Vascular endothelial growth factor (VEGF) involvement in the endochondral ossification process during spinal fusion. *Trans. Scoli Res. Soc.* 2000.

57. K. Hartmann, T. G. Baier, R. Loibl, A. Schmitt, and D. Schonberg, Demonstration of type I insulin-like growth factor receptors on human platelets. *J. Recept. Res.* 1989; **9**(2):181–198.

58. E. M. Spencer, A. Tokunaga, and T. K. Hunt, Insulin-like growth factor binding protein-3 is present in the alpha-granules of platelets. *Endocrinology* 1993; **132**(3):996–1001.

59. J. M. Hock, M. Centrella, and E. Canalis, Insulin-like growth factor I has independent effects on bone matrix formation and cell replication. *Endocrinology* 1988; **122**:254–260.

60. J. B. Sipe, C. A. Waits, B. Skikne, M. Imkie, Dhanyamraju, and H. C. Anderson, The presence of bone morphogenetic proteins (BMPs) in megakaryocytes and platelets. 24th Amer. Soc. Bone Miner. Res., 2002.

61. M. Slater, J. Patava, K. Kingham, and R. S. Mason, Involvement of platelets in stimulating osteogenic activity. *J. Orthop. Res.* 1995; **13**(5):655–663.

62. J. Street, D. Winter, J. H. Wang, A. Wakai, A. McGuinness, and H. P. Redmond, Is human fracture hematoma inherently angiogenic? *Clin. Orthop. Rel. Res.* 2000; **378**:224–237.

63. W. V. Giannobile, R. A. Hernandez, R. D. Finkelman, S. Ryan, C. P. Kiritsy, M. D'Andrea, and S. E. Lynch, Comparative effects of platelet-derived growth factor-BB and insulin-like growth factor-I, individually and in combination, on periodontal regeneration in Macaca fascicularis. *J. Periodontal Res.* 1996; **31**(5):301–312.

64. T. J. Hannon, G. Polston, W. J. Pekarske, W. Carnivali, N. Wall, and D. Leivers, *Determination of Platelet Yields from Platelet Rich Plasma for Five Autotransfusion Machines*. Cardiothoracic Research and Education Foundation, 1999.

65. B. L. Eppley, J. E. Woodell, and J. Higgins, Plastelet quantification and growth factor analysis from platelet-rich plasma (PRP): implications for wound healing. *Plastic Reconstruc. Surg.*, in press.

66. C. L. Lake, Normal hemostasis. In: C. L. Lake and R. A. Moore, eds. *Blood*. New York: Raven Press, 1995.

67. D. Green, *Proc. Am. Acad. Clin. Perf.* 1997; **18**:74–76.

68. N. S. Taichman, S. Young, A. T. Cruchley, P. Taylor, and E. Paleolog, Human neutrophils secrete vascular endothelial growth factor. *J. Leukoc. Biol.* 1997; **62**(3):397–400.

69. M. Gaudry, O. Bregerie, V. Andrieu, J. El Benna, M. A. Pocidalo, and J. Hakim, Intracellular pool of vascular endothelial growth factor in human neutrophils. *Blood* 1997; **90**(10):4153–4161.

70. Y. Sato, F. Okada, M. Abe, T. Seguchi, M. Kuwano, S. Sato, A. Furuya, N. Hanai, and T. Tamaoki, The mechanism for the activation of latent TGF-beta during co-culture of endothelial cells and smooth muscle cells: cell-type specific targeting of latent TGF-beta to smooth muscle cells. *J. Cell Biol.* 1993; **123**(5):1249–1254.

71. R. E. Marx, E. R. Carlson, R. M. Eichstaedt, S. R. Schimmele, J. E. Strauss, and K. R. Georgeff, Platelet-rich plasma: growth factor enhancement for bone grafts. *Oral Surg. Oral Med. Oral Pathol. Oral Radiol. Endod.* 1998; **85**(6):638–646.

72. R. E. Marx, W. J. Ehler, and M. Peleg, Mandibular and facial reconstruction: rehabilitation of the head and neck cancer patient. *Bone* 1996; **19**(Suppl 1):595–625.

73. R. Zimmermann, R. Jakubietz, M. Jakubietz, E. Strasser, A. Schlegel, J. Wiltfang, and R. Eckstein, Different preparation methods to obtain platelet components as a source of growth factors for local application. *Transfusion* 2001; **41**:1217–1224.

74. J. M. Lane, BMPs: why are they not in everyday use? *J. Bone Joint Surg.* 2001; **83A**(Suppl 1 Pt 2):S161–S163.

75. S. Mohan and D. J. Baylink, Bone growth factors. *Clin. Orthop. Rel. Res.* 1991; **263**:30–43.

76. O. Grundnes and O. Reikeras, The importance of the hematoma for fracture healing in rats. *Acta Orthop. Scand.* 1993; **64**:340–342.

77. S. M. Tilkian, M. B. Conover, and A. G. Tilkian, eds. *Clinical Implications of Laboratory Tests*, 3rd ed. St. Louis, MO: CV Mosby Co., 1983, p. 371.

78. G. V. R. Born, Functional physiology of platelets. In: R. Biggs, ed. *Human Blood Coagulation, Haemostasis and Thrombosis*, 2nd ed. Oxford, UK: Blackwell Scientific Publications, 1976, p. 187.

79. G. P. McNicol and A. S. Douglas, The fibrinolytic enzyme system. In: R. Biggs, ed. *Human Blood Coagulation, Haemostasis and Thrombosis*, 2nd ed. Oxford, UK: Blackwell Scientific Publications, 1976, p. 413.

80. K. Gutensohn, A. Alisch, W. Krueger, N. Kroeger, and P. Kuehnl, Extracorporeal plateletpheresis induces the interaction of activated platelets with white blood cells. *Vox. Sang.* 2000; **78**:101–105.

81. J. Zeller, An electronmicroscopic study of microaggregates in ACD stored blood. Trans III Congress on Thrombosis and Haemostasis, 1972: 264.

82. A. G. Hood and G. D. Reeder, Unpublished data, 2003.

83. S. V. Kevy and M. S. Jacobson, Comparison of methods for point of care preparation of autologous platelet gel. *J. Extra Corpor. Technol.* 2004; **36**(1):28–35.

84. S. V. Kevy, M. S. Jacobson, and R. Lazar, Quantitative and qualitative analysis of autologous platelet products: a comparative study. Proc. Bone Summit, Cleveland, OH, May 2004, poster #135.

85. D. R. Knighton, Classification and treatment of chronic nonhealing wounds: successful treatment with autologous platelet-derived wound healing factors. *Ann. Surg.* 1986; **204**(3):322–330.

86. R. E. Marx, Platelet-rich plasma (PRP): what is PRP and what is not PRP? *Implant Dentistry* 2001; **10**(4):225–228.

87. G. D. Reeder, A. G. Hood, and P. S. Potter, Optimizing platelet yields from perioperative pheresis devices. *Proc. Am. Acad. Cardiovasc. Perf.* 1999.

88. D. M. Arm, G. L. Lowery, A. G. Hood, and E. C. Shors, Characterization of an autologous buffy coat gel containing multiple growth factors. 45th Ann Orthop Res Soc, 1999: 604.

89. D. M. Arm, M. Ponticiello, and E. C. Shors, *Autologous Growth Factors: Characterization and Clinical Use*. Australian Spine Society, 2001.

90. D. M. Arm, A. G. Hood, and R. S. Mason, Unpublished data, 1998.

91. S. E. Haynesworth, S. Kadiyala, L.-N. Liang, T. Thomas, and S. P. Bruder, Mitogenic stimulation of human mesenchymal stem cells by platelet releasate suggests a mechanism for enhancement of bone repair by platelet concentrate. 48th Ann Orthop Res Soc, 2002.

92. E. Lucarelli, A. Beccheroni, D. Donati, L. Sangiorgi, A. Cenacchi, A. M. Del Vento, C. Meotti, A. Z. Bertoja, R. Giardino, P. M. Fornasari, M. Mercuri, and P. Picci, Platelet-derived growth factors enhance proliferation of human stromal stem cells. *Biomaterials* 2003; **24**(18):3095–3100.

93. R. Gruber, F. Varga, M. B. Fischer, and G. Watzek, Platelets stimulate proliferation of bone cells: involvement of platelet-derived growth factor, microparticles and membranes. *Clin. Oral Impl. Res.* 2002; **13**(5):529–535.

94. G. Weibrich, S. H. Gnoth, M. Otto, T. E. Reichert, and W. Wagner, Growth stimulation of human osteoblast-like cells by thrombocyte concentrates in vitro. *Mund. Kiefer Ges.* 2002; **6**(3):168.

95. W. Walsh, S. Nicklin, A. Loefler, Y. Yu, and D. Arm, Autologous growth factors (AGF) and spinal fusion. 47th Ann Orthop Res Soc, 2001: 951.

96. W. R. Walsh, A. Loefler, D. M. Arm, R. E. Stanford, J. Harrison, and D. H. Sonnabend, Growth factor gel and a resorbable porous ceramic for use in spinal fusion. 45th Ann Orthop Res Soc, 1999: 270.

97. M. J. Allen, J. E. Schoonmaker, F. Li, N. R. Ordway, D. M. Arm, and H. A. Yuan, Interbody fusion with a resorbable bone substitute supplemented with Autologous growth factors (AGF): evaluation in a sheep model. 48th Ann Orthop Res Soc, 2002.

98. J. C. Sefter, B. W. Cunningham, and P. C. McAfee, Autologous growth factor versus autogenous iliac graft for anterior cervical interbody arthrodesis. Int. Mtg. Adv Spine Tech E-poster, 2001: 71.

99. P. Sethi, J. Miranda, J. Grauer, S. Friedlaender, S. Kadiyala, and T. Patel, The use of platelet concentrate in posterolateral fusion: biomechanical and histologic analysis. 28th Int Soc Study Lumbar Spine, 2001: 21.

100. H. Li, X. Zou, Q. Xue, N. Egund, M. Lind, and C. Bunger, Anterior interbody lumbar fusion with carbon fiber cage loaded with bioceramics and platelet rich plasma. An experimental study on pigs. *Eur. Spine J.* 2004; Jan: 17.

101. M. A. N. Siebrecht, P. P. DeRooij, D. M. Arm, M. L. Olsson, and P. Aspenberg, Platelet concentrate increases bone ingrowth into porous hydroxyapatite. *Orthopedics* 2002; **25**(2):169–172.

102. P. Aspenberg and J.-S. Wang, A new bone chamber used for measuring osteoconduction in rats. *Eur. J. Exp. Musculoskel. Res.* 1993; **2**:69–74.

103. T. B. Jensen, J. E. Bechtold, X. Chen, L. Kidder, and K. Soballe, Autologous Growth Factors™ (AGF™) in combination with morselized bone allograft improves implant fixation. 48th Ann Orthop Res Soc, 2002.

104. T. B. Jensen, S. Overgaard, M. Lind, O. Rahbek, C. Bunger, and K. Soballe, Osteogenic protein 1 device increases bone formation and bone graft resorption around cementless implants. *Acta Orthop. Scand.* 2002; **73**(1):31–39.

105. T. B. Jensen, O. Rahbek, S. Overgaard, and K. Soballe, Platelet rich plasma and fresh frozen bone allograft as enhancement of implant fixation. An experimental study in dogs. *J. Orthop. Res.* 2004; **22**(3):653–658.

106. A. G. Hood, Perioperative autologous sequestration III: a new physiologic glue with wound healing properties. *Trans. Am. Acad. Cardiovasc. Perf.* 1993; **14**:126–128.

107. R. E. Marx, D. M. Green, and B. Klink, Platelet gel as an intraoperatively procured platelet-based alternative to fibrin glue. *Plast. Reconstr. Surg.* 1998; **101**(4):1161–1162.

108. D. H. Whitman, R. L. Berry, and D. M. Green, Platelet gel: an autologous alternative to fibrin glue with applications in oral and maxillofacial surgery. *J. Oral Maxillofac. Surg.* 1997; **55**(11):1294–1299.

109. E. Anitua, Plasma rich in growth factors: preliminary results of use in the preparation of future sites for implants. *Int. J. Oral Maxillofac. Implants* 1999; **14**:529–535.

110. S. Bhanot and J. C. Alex, Current applications of platelet gels in facial plastic surgery. *Fac. Plastic Surg.* 2002; **18**(1):27–33.

111. R. Marx, Healing enhancement of skin graft donor sites with PRP. Trans. 82nd Ann Ame Acad Oral Max Surg, September 2000.

112. P. A. Mooar, M. J. Gardner, P. R. Klepchick, and H. H. Sherk, The efficacy of autologous platelet gel in total knee arthroplasty. Trans. 67th Amer Acad Orthop Surg. PE 148, 2000.

113. J. Pritchard, Platelet-rich plasma in mini-open rotator cuff repair. 16th Ann Biomet Hip and Knee Update, Lake Arthur, Lousiana, November 2003.

114. G. L. Lowery, S. Kulkarni, and A. E. Pennisi, Use of Autologous Growth Factors in lumbar spinal fusion. *Bone* 1999; **25**(S2):47S–50S.

115. J. J. Lettice, T. A. Kula, J. Kelley, and M. McCort, Supplemental Autogenous Growth Factors in spinal reconstructive surgery: technique and results. Trans 34th Scoli Res Soc, 1999.

116. D. W. Kucharzyk and G. Alavanja, The role of coralline hydroxyapatite with Autologous Growth Factors in instrumented lumbar spinal fusions. Int Mtg Adv Spine Tech E-poster 41, 2001.

117. H. T. Hee, M. E. Majd, R. T. Holt, and L. Myers, Do Autologous Growth Factors enhance transforaminal lumbar interbody fusion? *Eur. Spine J.* 2003; **12**:400–407.

118. B. Bose and M. A. Malzarini, Bone graft gel: Autologous Growth Factors used with autograft bone for lumbar spine fusions grafted with AGF, Pro Osteon, and local bone. *Adv. Therapeut.* 2002; **19**(4):170–175.

119. M. J. Broom and M. Scherb, Use of autologous growth factors in posterior lumbar spine fusion. Spinal Skeletal Solutions, Maui, Hawaii, January 2004.

120. R. C. Waldrip, Bone grafting composites with Autologous Growth Factors in posterior lumbar fusion. Spinal Skeletal Solutions, Maui, Hawaii, January 2004.

121. J. B. Logan, D. D. Dietz, and J. M. Guiffre, Autologous bone augmented with Autologous Growth Factors and resorbable porous ceramic for posterolateral lumbar fusion: a preliminary report. Int Mtg Adv Spine Tech, 2002.

122. K. L. Kurica, K. Booton, and J. M. Giuffre, Autologous Growth Factors and resorbable porous ceramic without bone graft for instrumented posterolateral lumbar fusion. Int Mtg Adv Spine Tech, 2002.

123. M. T. Rohmiller, G. A. Mencio, and N. E. Green, Use of coralline hydroxyapatite and autogenous growth factors to achieve spinal fusion in adolescents undergoing posterior spinal fusion for correction of scoliosis. Scoli Res Soc, 2002.

124. M. L. Jimenez and T. L. Anderson, The use of allograft, platelet derived growth factors, and internal bone stimulation for treating recalcitrant nonunions. Am Assoc Orthop Surg, 2003: 16.

125. W. P. Grant and D. Jacobus, Autologous growth factors in Charcot reconstruction. Amer Podiatric Medical Assn, 2003.

126. N. Birch, D. Noyes, and M. Shaw, Autologous Growth Factor (AGF) and thrombin spray used to reduce post-operative surgery. Trans ISSLS. poster 351, 2004.

127. L. G. Jenis, R. J. Banco, and B. Kwon, A prospective study of Autologous Growth Factors® (AGF®) in lumbar interbody fusion. Spinal Skeletal Solutions, Maui, Hawaii, Jan 2004; *Amer. Acad. Orthop. Surg. Spine J.*, in press.

128. D. Volgas, A randomized, controlled prospective trial of allograft with Autologous Growth Factors versus iliac crest bone graft for nonunions and delayed unions. *Orthop. Trauma Assoc.*, submitted.

129. F. Castro, Activated growth factors in spinal fusions. 18th Ann North Amer Spine Soc, 2003.

130. Y. Anekstein, S. Glassman, R. Puno, and L. Carreon, Platelet gel (AGF) fails to increase fusion rate. 18th Ann North Amer Spine Soc, 2003.

131. B. K. Weiner and M. Walker, Efficacy of autologous growth factors in lumbar intertransverse fusions. *Spine* 2003; **28**(17):1968–1971.

132. D. M. Arm, *Spine* 2004; **29**(8):946–948.

133. G. W. Wood II, R. J. Boyd, T. A. Carothers, F. L. Mansfield, G. R. Rechtine, M. J. Rozen, and C. E. Sutterlin III, The effect of pedicle screw/plate fixation on lumbar/lumbrosacral autogenous bone graft fusions in patients with degenerative disc disease. *Spine* 1995; **20**(7):819–830.

134. M. Lorenz, M. Zindrick, P. Schwaegler, L. Vrbos, M. A. Collatz, R. Behal, and R. Cram, A comparison of single-level fusions with and without hardware. *Spine* 1991; **16S**(8):S455–S458.

135. J. S. Fischgrund, M. Mackay, H. N. Herkowitz, R. Brower, D. M. Montgomery, and L. T. Kurz, Degenerative spondylolisthesis with spinal stenosis: a prospective, randomized study comparing decompressive laminectomy and arthrodeses with and without spinal instrumentation. *Spine* 1997; **22**(24):2807–2812.

136. J. Mochida, K. Suzuki, and M. Chiba, How to stabilize a single level lesion of degenerative lumbar spondylolisthesis. *Clin. Orthop.* 1999; **368**:126–134.

137. R. A. McGuire and G. M. Amundson, The use of primary internal fixation in spondylolisthesis. *Spine* 1993; **18**(12):1662–1672.

138. S. A. Grubb and H. J. Lipscomb, Results of lumbrosacral fusion for degenerative disc disease with and without instrumentation. Two- to five-year follow-up. *Spine* 1992; **17**(3):349–355.

139. M. Bernhardt, D. E. Swartz, P. L. Clothiaux, R. R. Crowell, and A. A. White III, Posterolateral lumbar and lumbrosacral fusion with and without pedicle screw internal fixation. *Clin. Orthop.* 1992; **284**:109–115.

140. S. M. Mardjetko, P. J. Connolly, and S. Shott, Degenerative lumbar spondylosis: a meta-analysis of literature 1970–1993. *Spine* 1994; **20**(Suppl):S2256–S2265.

141. T. C. Patel, A. R. Vaccaro, E. Truumees, J. S. Fischgrund, H. N. Herkowitz, and A. Hilibrand, Two-year follow-up on patients in a pilot safety and efficacy study of OP-1 (rhBMP-7) in posterolateral lumbar fusion as a replacement for iliac crest autograft. 18th Ann North Amer Spine Soc, 2003.

142. L. J. Marden, R. S. Fan, G. F. Pierce, A. H. Reddi, and J. O. Hollinger, Platelet-derived growth factor inhibits bone regeneration induced by osteogenin, a bone morphogenetic protein, in rat craniotomy defects. *J. Clin. Invest.* 1993; **92**:2897–2905.

143. S. E. Harris, L. F. Bonewald, M. A. Harris, M. Sabatini, S. Dallas, J. Q. Feng, N. Ghosh-Choudhury, J. Wozney, and G. R. Mundy, Effects of transforming growth factor β on bone nodule formation and expression of bone morphogenetic protein 2, osteocalcin, osteopontin, alkaline phosphatase, and type I collagen mRNA in long-term cultures of fetal rat calvarial ostcoblasts. *J. Bone Miner. Res.* 1994; **9**(6):855–863.

144. N. Duneas, J. Crooks, and U. Ripamonti, Transforming growth factor-beta 1: induction of bone morphogenetic protein genes expression during endochondral bone formation in the baboon, and synergistic interaction with osteogenic protein-1. *Growth Factors* 1998; **15**:259–277.

145. X. Si, Y. Jin, and L. Yang, Induction of new bone by ceramic bovine bone with recombinant human bone morphogenetic protein 2 and transforming growth factor beta. *Int. J. Oral Max. Surg.* 1998; **27**(4):310–314.

FURTHER READING

C. A. Kirker-Head, T. N. Gerhart, S. H. Schelling, G. E. Hennig, E. Wang, and M. E. Holtrop, Long-term healing of bone using recombinant human bone morphogenetic protein-2. *Clin. Orthop. Rel. Res.* 1995; **318**:222–230.

S. D. Cook, G. C. Baffes, M. W. Wolfe, T. K. Sampath, and D. C. Rueger, Recombinant human bone morphogenetic protein-7 induces healing in a canine long bone segmental defect. *Clin. Orthop. Rel. Res.* 1994; **301**:302–312.

S. D. Cook, T. E. Dalton, E. H. Tan, T. S. Whitecloud III, and D. C. Rueger, In vivo evaluation of recombinant human osteogenic protein (rhOP-1) implants as a bone graft substitute for spinal fusion. *Spine* 1994; **19**:1655–1663.

S. D. Boden, J. H. Schimandle, and W. C. Hutton, The use of an osteoinductive growth factor for lumbar spinal fusion. Part II: Study of dose, carrier, and species. *Spine* 1995; **20**:2626–2632.

S. D. Boden, T. A. Zdeblick, H. S. Sandhu, and S. E. Heim, The ese of rhBMP-2 in interbody fusion cages. Definitive evidence of osteoinduction in humans: a preliminary report. *Spine* 2000; **25**:376–381.

J. K. Burkus, M. F. Gornet, C. A. Dickman, and T. A. Zdeblick, Anterior lumbar interbody fusion using rhBMP2 with tapered interbody cages. *J. Spinal Disord. Tech.* 2002; **15**(5):337–349.

S. D. Boden, Bone substitutes in orthopedics: 2002 and beyond. 115th Meeting of the American Orthopedic Association, Symposium #3, 2002.

G. J. Martin, S. D. Boden, M. A. Morone, and P. A. Moskovitz, Posterolateral intertransverse process spinal fusion arthrodeses with rhBMP-2 in a non-human primate. Important lessons learned regarding dose, carrier, and safety. *J. Spinal Disord.* 1999; **12**:179–186.

S. D. Boden, Bioactive factors for bone tissue engineering. *Clin. Orthop. Rel. Res.* 1999; **367S**:S84–S94.

H. Kaneko, T. Arakawa, H. Mano, T. Kaneda, A. Ogasawara, M. Nakagawa, Y. Toyama, Y. Yabe, M. Kumegawa, and Y. Hakeda, Direct stimulation of osteoclastic bone resorption by bone morphogenetic protein (BMP)-2 and expression of BMP receptors in mature osteoclasts. *Bone* 2000; **27**(4):479–486.

M. Kanatani, T. Sugimoto, H. Kaji, T. Kobayashi, K. Nishiyama, M. Fukase, M. Kumegawa, and K. Chihara, Stimulatory effect of bone morphogenetic protein-2 on osteoclast-like cell formation and bone-resorbing activity. *J. Bone Miner. Res.* 1995; **10**(11):1681–1690.

J. Alexander and C. Branch, Recombinant human bone morphogenic protein-2 in a posterior lumbar interbody fusion construct: 2-year clinical and radiologic outcomes. *Proc. 17th Ann. NASS/Spine J.* 2002; **2**:47S–128S.

D. R. Knighton, Stimulation of repair in chronic nonhealing cutaneous ulcers using platelet-derived wound healing formula. *Surg. Gynecol. Obstet.* 1990; **170**(1):56–60.

BACK-PROPAGATION

KAYVAN NAJARIAN
CHRISTOPHER N.
EICHELBERGER
University of North Carolina at
Charlotte
Charlotte, North Carolina

You will say, ever since such and such a time you have been going downhill, you have been feeble, you have done nothing. Is that entirely true?
Vincent van Gogh, Letter to Theo van Gogh (July 1880) (1)

1. INTRODUCTION

Many problems in computational sciences, and by implication biomedical engineering, are too difficult for conventional, direct analysis: They are non linear; or they require extensive *a priori* knowledge about the distribution of values; or they cannot guarantee reasonable results. Artificial neural networks (ANNs), first proposed by McCulloch and Pitts in 1943 (2) and described elsewhere within this volume, can be a useful family of tools to help address such problems. Many different ANNs exist, one of the most useful types are the multilayer, feed-forward sigmoid neural networks trained under back-propagation. These networks and this training method are the focus of this chapter.

Two key components make up any successful ANN: the nature of the network (chiefly, the repeating computational unit, the neuron); and the method by which the network is trained, by which the connections between neurons are strengthened or weakened to help produce the desired behavior. An abundance of literature exist on both of these subjects; this treatment deals directly with a very useful, specific intersection: Multilayer feed-forward neural networks trained under back-propagation.

The chapter begins with a brief introduction to the theory and nature of these ANNs, which leads to a review of how these networks are used in practice, and then to a review of key applications of these networks and training algorithm in biomedical engineering. Each of these topics is considered in a separate section that follows.

2. THEORY

For a more detailed discussion of ANNs in general, please see the corresponding chapter elsewhere within this encyclopedia. What follows is the briefest possible review of the mechanics of ANNs, so as to provide adequate context for the ensuing presentation of multilayer, feed-forward ANNs trained using back-propagation.

2.1. Background

All ANNs are directed graphs of simple computing units, or *neurons*. A single neuron is depicted in Fig. 1, labeled Y_0. The following elements are significant:

- Inputs: The left-hand column of boxes represent inputs. These inputs may come directly from cases presented to the network, or they may be the outputs from other neurons. Note that the top input has a constant value of 1.0; this node is a *bias* node.
- Weights: Each input's signal is attenuated by a weight, some $w_{i \to 0}$.
- Activation: The neuron, Y0, converts the sum of the weighted input signals to a single output signal via an *activation function*. That is, $y_0 = f(\sum_i x_i w_{i \to o})$, for some suitable f to be considered shortly.
- Output: The single output signal, y_0, may either be read as a terminal output (a prediction for some property for which you have trained the network), or it may be an intermediate output value that is sent, attenuated by its own set of weights, to other neurons for continued processing.

When these neurons are connected together to form networks, they are trained to adjust the connection weights so that they can map a series of input patterns to a series of output patterns[1]. When done well, this process is more than rote memorization, because the combination of variation in the cases used to train the network with the training algorithm often produces an ANN that generalizes significant properties of the larger system.

Clearly, the knowledge contained in any ANN is embodied in both the shape of the network (its *topology*) and in the connection weights themselves. The network topology is generally less fluid than are the connection weights, which undergo many refinements during training. Each network type has one or more weight-training algorithms.

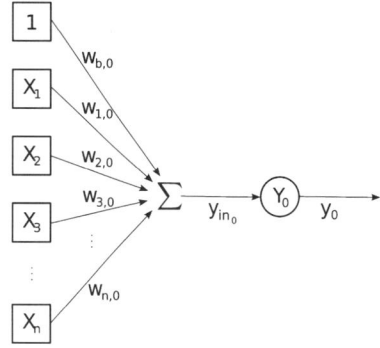

Figure 1. A generic neuron.

---
[1]ANNs can perform more tasks than this, but they are outside the scope of this presentation. Please refer to the longer, more general discussion of ANNs in another chapter in this encyclopedia.

One of the earliest and simplest neurons, the perceptron, could be used to address some interesting problems (3), but as Minsky and Papert pointed out (4), the perceptron is rather severely limited.

An ANN exists, however, that addresses these concerns, and allows the practitioner to address very difficult, non linear problems: the multilayer, feed-forward ANN.

2.2. Multilayer, Feed-Forward Networks

The architecture is simple (see Fig. 2). One input layer and one output layer exist. Note that the output layer can consist of multiple neurons. One or more hidden layers also exist; that is, layers that are internal to the network, and produce no output that is visible from outside of the ANN. Figure 2 only shows one hidden layer, but many may exist. There is not necessarily any limitation on the number of neurons that may be in a hidden layer, nor is it necessary for the graph to be fully connected, although it is customary. Many attempts have been made to try to define rules to suggest how many hidden layers, and how many neurons per hidden layer, to use in a network for a given purpose; see the discussion of Probably Approximately Correct (PAC) Learning for more details on this subject.

These networks are sometimes referred to as "sigmoid" networks, a title that highlights the activation function. That is, once the inputs have been weighted and summed, the function that computes the neuron's output is the sigmoid function; see Equation 1.

$$f(x) = \frac{1}{1+e^{-x}}. \qquad (1)$$

Note: It is not strictly necessary for the activation function to be the sigmoid function for this class of ANN to provide all of its desirable properties. The constraints on the activation function that are relevant to back-propagation are as follows:

- The function must be differentiable, and hence, continuous.
- The function ought to be defined for the entire real number line.
- The function ought to return values on $[-1, +1]$ in the case of bi-polar data representation, or $[0, 1]$ in the case of binary data representation.

Other common choices for the activation function for these networks include *arctangent* and *hyperbolic tangent*.

With increased capability comes increased complexity: A multilayer, feed-forward network requires a considerably more elaborate weight-training scheme than does a perceptron. Fortunately, an elegant solution exists: back-propagation.

2.3. Gradient Descent

Back-propagation is a gradient-descent method. Like other gradient-descent methods, it is easy to visualize. Consider an intuitive, visual treatment of gradient descent first, and then a more detailed mathematical presentation of the same topic.

When you first create an artificial neural network, and initialize the weights to some values, the network will probably not perform well. That is, the mean-squared error (MSE) between the actual outputs and the desired outputs will be high. Many different combinations of possible weights exists, most of which will produce slightly different MSE terms. If you were trying to pick the optimal value for a single weight in the network, you might generate a graph like that in Fig. 3, in which the x-axis is the value of the weight you are adjusting, and the y-axis is the MSE of the network for each value of that weight. Clearly, the lowest MSE value is located at point A on the plot; if you wished the network to perform best with all other weights frozen, you would set your target weight to the value corresponding to point A.

To identify the lowest MSE value requires having the entire plot, but this is time-consuming to create, even for a single weight value. Instead, gradient-descent methods make a simplifying assumption: Wherever your initial, typically random, value starts, always head *downhill*. If you begin at point a on the plot, and can recognize that the slope of the plot at that point is negative, then you should adjust your weight value to follow that slope: Follow the curve downhill, using relatively small steps. Eventually, you may well end up at point A, the lowest MSE on the plot, without having first generated the entire plot.

An inherent risk of gradient-descent methods such as back-propagation is that, because they rely on only local data to adjust weights, they can become stuck on *local minima*. Assume that your network started at point b on the same plot, and moved downhill, which might lead you to identify point B as the best network weight configuration, because all paths leading away from point B entail

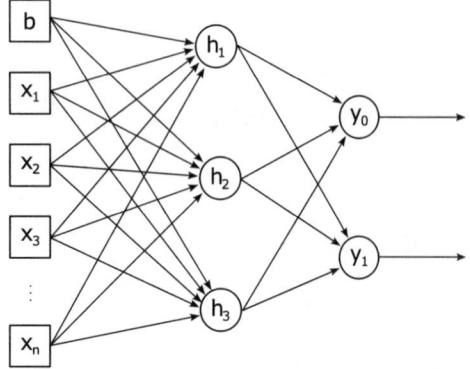

Figure 2. The multilayer, feed-forward ANN.

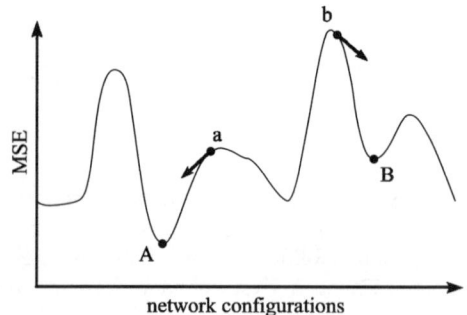

Figure 3. Gradient descent.

moving uphill. Knowing the shape of the entire plot, one can identify point A as the global minimum.

Remember that a local minimum may be good enough: The example in Fig. 3 suggests that points A and B are significantly different, which may not be the case. If the MSE at point A is 1.1×10^{-8}, and the MSE at point B is 1.6×10^{-8}, then both points may represent acceptable solutions. The effort spent to refine the solution must be commensurate with the requirements of your application.

In more formal terms, consider what it means to train the network to reduce the total mean-squared error of an ANN by following the gradient. As the name implies, the error is computed for the output neurons, and then allowed to propagate backwards through the network. The error, E, for the entire network can be expressed in Equation 2 as:

$$E = \frac{1}{2} \sum_j (t_j - y_j)^2. \qquad (2)$$

The direction *downhill* is defined by the gradient of the curve; the weights should be adjusted so that the total error is reduced, which means moving against the partial derivative of MSE with respect to the single weight one wishes to adjust. Remember that to fight cycles in weight training, a learning rate, α, is often useful. Hence, the full weight update is computed by:

$$\Delta w_{i \to j} = -\alpha \frac{\partial E}{\partial w_{i \to j}}. \qquad (3)$$

Equation 3 is computed differently for an output neuron than for an internal node. Consider Fig. 4 throughout the following discussion.

To update $w_{i \to j}$, a weight attenuates the connection from intermediate node H_i to output node Y_j. This weight is updated according to Equation 4, where t_j is defined as the known correct value for output neuron j, and y_{in_j} is the sum of weighted inputs on which Y_j's activation function acts.

$$\begin{aligned}\Delta w_{i \to j} &= -\alpha \frac{\partial E}{\partial w_{i \to j}} \\ &= -\alpha \left(\frac{\partial E}{\partial y_j} \cdot \frac{\partial y_j}{\partial w_{i \to j}} \right) \\ &= \alpha (t_j - y_j) f'(y_{in_j}) h_i. \end{aligned} \qquad (4)$$

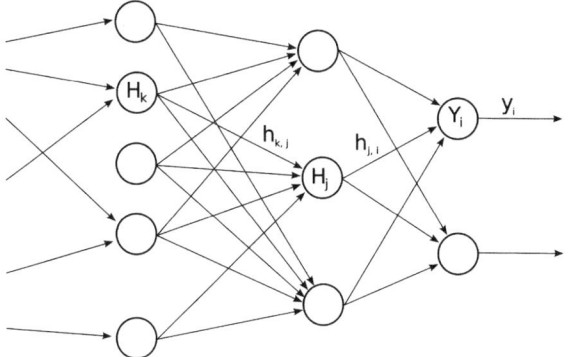

Figure 4. Example network to be trained under back-propagation.

Given two internal nodes, $H_k \to H_i$, and the corresponding connection weight $w_{k \to i}$, the adjustment of this weight is defined by:

$$\begin{aligned}\Delta w_{k \to i} &= -\alpha \frac{\partial E}{\partial w_{k \to i}} \\ &= -\alpha \left(\frac{\partial E}{\partial h_i} \cdot \frac{\partial h_i}{\partial w_{k \to i}} \right) \\ &= -\alpha \left(-\sum_n \delta_n w_{i \to n} \right) f'(h_{in_i}) h_k. \end{aligned} \qquad (5)$$

It is customary to define an intermediate function, δ, when describing the back-propagation algorithm, defined in Equation 6. Using this definition, it is possible to summarize the entire back-propagation scheme using the following simple equation: $\Delta w_{a \to b} = \alpha \delta_b h_a$.

$$\delta_b = \begin{cases} (t_b - y_b) f'(y_{in_b}) & \text{if node } b \text{ is an output neuron} \\ \left(-\sum_n \delta_n w_{b \to n} \right) f'(h_{in_b}). & \text{if node } b \text{ is an internal neuron} \end{cases} \qquad (6)$$

It becomes clear why the activation function needs to be differentiable for a back-propagation ANN: The first derivative of the activation function is part of the equation to compute weight adjustment.

The main event loop for a multilayer, feed-forward ANN trained under back-propagation looks something like this:

1. Initialization: Create the network topology and initialize the weights to small, random values.
2. Training: Present all training instances to be learned which are the known cases. In response to each known case, compute the error between the ANN outputs and the known correct values; use this error to update the network weights. One complete iteration through the training instances is termed an *epoch*.
3. Testing: Present test instances to the network. The set of test instances is disjoint with the set of training instances; that is, none of the test instances were used to train the network. The error rate is computed on the body of test instances. No network weights are adjusted during this phase.
4. Terminating condition: If the error rate is sufficiently low, freeze the network weights. Otherwise, return to the training phase described earlier. Continue looping until the terminating condition is satisfied.

3. PRACTICE

The preceding section has described some of the background and motivation behind back-propagation and multilayer, feed-forward (sigmoid) ANNs. This section briefly presents examples of how these ideas are used, and risks that need to be managed for these ANNs to be effective.

3.1. Debugging Back-Propagation Networks

Training ANNs under back-propagation is not without risks, the most significant of which are failure of the network weights to converge, the tendency of the method to settle on sub optimal local minima, and overfitting. In this section, each of these risks is presented in conjunction with ways each might be managed.

3.1.1. Convergence.
Weights adjustments computed using back-propagation will not necessarily converge; when this happens, your never achieve a useful ANN. If the error (cost) function has areas that are significantly flat, for example, the gradient may be too small to encourage the network to reach any minimum in an acceptable amount of time. In contrast, an error function that has many tightly packed peaks and valleys may consistently encourage weight adjustments that are too large, in effect jumping from hillside to hillside rather than rolling downhill.

A few strategies exists to help encourage convergence, but they all approach the problem from one of three perspectives: Change the error landscape to make it friendlier to back-propagation; modify the network topology; modify the back-propagation algorithm itself to make it behave better on the existing error landscape. Each approach is briefly considered here.

3.1.1.1. Changing the Error Landscape.
Ideally, the cost function would be smooth and bowl-shaped, akin to $y = x^2$, providing back-propagation with a single, easily-attainable global minimum. In practice, the error landscape rarely is so accommodating, but it may be possible to influence its shape so as to make weight training easier and faster.

The following issues are important to consider both as you design the initial network and training scheme and when you are trying to fight with a network that will not converge:

- Represent data in a way that is meaningful. This topic is considered in greater detail in the chapter on artificial neural networks, but is also of significance here. Data representation affects the shape of the error landscape significantly: In the simplest case, consider the choice between a binary and bipolar representation; each will produce a different series of peaks and troughs in terms of MSE. But data representation can have a more subtle effect on the fitness landscape. Consider an application in which a sequence of nucleotides is the input pattern that maps to a binary indicator of whether that location is a probable splice site. One could naively choose to represent each nucleotide as a single input neuron in which $A = -1.0$, $C = -0.33$, $G = 0.33$, $T = 1.0$. This representation is dangerous, because the ordering is arbitrary, and denies the system the ability to differentiate, for example, between the pyrimidines and purines. A better representation might use multiple input neurons per nucleotide, perhaps dedicating one to the type of nucleotide (pyrimidine, purine) and one to the presence or position of an amino group. Perhaps a scheme that used four input neurons per nucleotide would be appropriate, allowing one to specify the degree of similarity or separation among A, C, G, and T more elegantly. Careful design and experimentation are key to finding an effective data representation.

- Clean data. This effect is very similar to that of data representation. Care must be taken to ensure that the distribution of values within a given input neuron is meaningful: If 96% of the values are on $[-1.0, -0.9]$, and only 4% are on $[-0.9, 1.0]$, the network is unlikely to benefit from this input. Generally, exists to ensure that a meaningful map exists between the raw data value and the range chosen for the input neuron. Normalization can be important when preparing data for analysis. Also, one should be aware that large volumes of inconsistent data (wherein one input pattern maps to two significantly different output patterns) may also inhibit convergence. Know the data and the application domain, and tailor the network and its training accordingly.

- Use a different cost function. MSE is not the only cost function, although it is very common. Instead of MSE, a network could use a linear error cost function, or one based on some estimate of entropy. It is possible to hand-craft a novel cost function that is specific to each application. Each of these choices, however, will influence the performance of back-propagation differently, and ought to be approached with some trepidation.

- Modify the activation function. As back-propagation relies on the gradient of the error function with respect to each weight, and that gradient in turn depends on the partial derivative of the activation function, it is easy to see why adjusting the activation function can influence the rate of convergence (5).

- Introduce noise into the training cases. Introducing noise is a way to introduce minor variation into test cases. If, for example, one is training a network to learn the ubiquitous XOR function, they have only 4 test cases by default, which means that they may require many iterations of back-propagation before their network converges. Inflating the number of training examples by adding minor variations on the real test cases can reduce the number of epochs required to reach convergence, because each minor variation essentially jogs the weight Training, helping to prevent too much cyclic weight adjustment.

- Tailor the test cases. Although 10-fold cross-validation may be a useful way to estimate the average performance of a network topology and weight-training scheme, completely random selection of the training and testing instances may prevent the network from converging quickly. The data may oversample some conditions; this oversampling may bias the weight adjustment so that it is more difficult to reach a stable set of weights. Again, the safe rule is: craft the data and network to be complementary, not conflicting.

3.1.1.2. Modify the Network Topology.
Selecting a suitable network topology the number of input, hidden, and

output neurons and their connections is a topic well beyond the scope of this chapter, but one should understand that it is a crucial factor in the rate of convergence of their network. In specific, one should refer to the chapter on Probably Approximately Correct (PAC) Learning elsewhere in this volume, as it contains key concepts on how much information a network of a given size ought to be able to contain.

Many interesting algorithms exist that change the network over time to help convergence. Some grow the network, whereas others start with a large network, and prune nodes that contribute little.

3.1.1.3. Modify Back-Propagation. Many variations on the back-propagation training algorithm exist such as Quickprop (6); SuperSAB (7); and RPROP (8). Generally, these methods focus on helping back-propagation networks converge faster. Quickprop, for example, remembers the weight adjustments and the partial derivatives at each node per weight, and uses these data to compute a coarse estimate of the second derivative of error with respect to each weight change (5). Being able to predict higher-order derivatives allows quickprop to make more intelligent decisions about step size and direction based on models of where the function is heading in a larger sense. Most of these back-propagation variants also use some form of momentum, covered in the next section.

3.1.2. Local Minima. Even when a network's weights converge to a single, stable configuration, they may not settle on a *good* configuration. As discussed previously, local minima are associated with gradient-descent weight-training algorithms: They occur when the initial network weights are positioned so that strictly following the gradient prevents one from reaching the best possible network configuration. A few common ways exists to help reduce the risk of getting stuck on a local minimum:

- Biased weight initialization. Clearly, the choice of the initial connection weights defines where one begins gradient descent with back-propagation. If one can choose their starting point intelligently, they improve their chances of reaching a global (or at least better) optimum. Nguyen–Widrow initialization often is a useful method of selecting good initial weights that is based on a quick study of the distribution of values among the inputs (9).
- Momentum. Originally, it was that network training algorithms were subject to cyclic updates. The learning rate, α, mitigates this effect. A side-effect of the learning rate, however, is that if it declines too rapidly, the ANN may more easily get stuck on a local optimum in the fittness landscape. Intuitively, one might observe that, having been making updates in one useful direction, the system ought to recognize the value of having the improvement proceed along that direction, which is accomplished by introducing *momentum*. In back-propagation, the momentum term is the product of a momentum rate, $\mu \in [0, 1]$, and the magnitude of the preceding weight update. This momentum term is added to the change in weights for the current time-step: $\Delta w_{i \to j}(t) = \alpha \delta_j y_i + \mu \Delta w_{i \to j}(t - 1)$.
- Simulated annealing. This term refers to a process in metallurgy in which hot metals are gradually cooled, allowing the material to reach a more stable state. In gradient-descent methods, this process can be implemented in multiple ways: It can be used to reduce the gradient-descent step size over time from a large initial value; periods of simulated annealing can be interleaved with periods of back-propagation (10); or simulated annealing can be used instead of back-propagation.
- Genetic algorithms. Back-propagation may be slow or may fail to converge when the error landscape is highly multimodal, that is, if many local minima exist that are unacceptable, with significant barriers separating them from each other. GAs can perform well in exploring such a fitness landscape and can be used either with back-propagation or in lieu of gradient-descent methods altogether.

3.1.3. Overfitting. ANNs are susceptible to overfitting, that is, shaping the network so that it learns the test data to the exclusion of the more general principal the test data are to represent. Generally, overfitting occurs when too few training instances are used to communicate a concept that is much larger.

For example, assume you were trying to teach an ANN how to discriminate between cars and airplanes based on a few well-understood properties. If you present the network with too few training data, then the network may well over-exaggerate the importance of properties that coincidentally help to explain the difference in the training data, such as paint color. If most of the cars in your test data sample are red, the the network training method may reward weights that over-emphasize the importance of the color red in determining whether a vehicle (instance) is a car.

Overfitting is insidious, because the fitness of an ANN on training data will generally increase with the number of epochs, yet at some point on the curve, training will influence the connection weights in excess of the real, useful information content of the training data. When this happens, the network begins to be overfitted. Unfortunately, the accuracy of the training data will likely continue to increase with additional epochs of training, which means that as a researcher and a practitioner, you have a responsibility to monitor (and publish) not only those networks that produce impressive training accuracies, but also to watch the gap between the training and testing accuracy values. When either the gap begins to widen significantly or the testing error begins to increase (see Fig. 5), it may suggest that you are overfitting the training data.

4. APPLICATIONS IN BIOMEDICAL ENGINEERING

Artificial neural networks trained under back-propagation are remarkably effective general problem solvers. To highlight this property, consider the following list of real

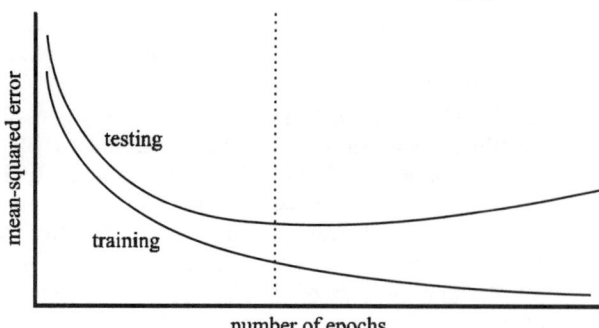

Figure 5. Curve showing a network becoming overfitted.

applications, and note that generally these applications are open-ended, difficult applications that are typified by noisy or incomplete data, which makes these ANN systems particularly valuable tools.

4.1. Predicting Medical Outcomes

When a physician treats a shallow, open scalp wound, decisions must be made, but they are reasonably well understood and non controversial; no artificial intelligence is required to assist either the physician or the patient.

When an oncologist is considering treatment options for a patient with an internal tumor, many very difficult decisions must be make, which is an example where ANNs may contribute. As part of the treatment plan, there will have been numerous tests performed on the patient and the tumor; the medical record already contains information such as base demographics and family and medical histories. Billing records at the hospital provide historical data about diagnosis codes, procedure codes, and outcomes. Radiology has large volumes of images on this and other cases. All of these data are candidates for analysis within a neural network. The process of putting a neural network in place might resemble the following:

1. What is the purpose of predicting outcomes? Commonly overlooked, this issue is the most important to address in the entire effort. You might decide to build a model of treatment success to help suggest treatment plans for new patients; you might decide to build a model to help train residents to understand what constitutes real-life best practice in oncology; you might build a model to suggest preventive care programs that could benefit the local community; and so on. If you can build a reasonably accurate model of medical outcomes, it can be put to many uses.

2. What are the data to be analyzed? Experts healthcare providers and researchers surely have a good intuitive understanding of the data elements that are important; these elements are likely to be the core set of inputs that you provide to the network. Your application will help shape the outputs: Are you interested in whether the patient is alive through the end of treatment? through the end of one year? through the end of five years?

3. Prepare your data set for analysis, which means collecting all of the data from all sources: electronic records, papers, interviews, and so forth. This preperation also includes reviewing the data set manually; perhaps all of your patients are females, in which case that field adds no value to the analysis.

4. Construct an initial network and begin training it. The details will depend on what software package you are using (or have written). Software such as Matlab, for example, will automate many of the topology and training decisions. For other packages, you will have to make your own choices as to the number of layers, the number of nodes per layer, how to partition the test cases between training and testing, how many epochs to allow back-propagation to run, and so on.

5. Review. If the network predictions are accurate only 60% of the time, you will probably have to revert to an earlier step, such as reconsidering your data, representation, topology, or training method. Even if you have acceptable predictive accuracy, you probably want to experiment to see how good a network you can build.

6. Deploy the network. Select the best performing network among all of the models you built and put it in production use. As new data or procedures or ideas become available, you may want to revise the analysis and compare the two.

For more detailed examples of how neural networks have been used in predicting medical outcomes, see Refs. 11 and 12.

4.2. Image Processing

Computer tomography and photographic archiving and control systems (PACS) make an overwhelming volume of data readily available to biomedical practitioners and researchers. Unfortunately, much of these data must still be analyzed by humans. The opportunity to shunt some of these analysis tasks to hardware and software systems is significant. Not only can automatic analysis be faster, but cases exist where it may be more reliable, and at a minimum more objective. Consider these image processing tasks where a neural network might be applicable:

- Tumor identification. Where *precisely* on the radiological film does the tumor end and healthy tissue end? This task is an inordinately complicated to perform manually.
- Estimating the condition of a prosthesis. As an artificial joint ages, its condition deteriorates. Medical images, when properly analyzed, may be able to pinpoint not merely the adequacy of the device, but may be able to identify key points of failure that are not otherwise obvious.
- Pathology. Microscopic images can also be good candidates for automatic analysis, whether your application is geared toward identifying deviant cells or estimating cell counts.

The process for conducting an image processing project using a neural network is very similar to that for predicting medical outcomes. (See the preceding example.) Clearly, very significant issues concerning the data representation exist.

For more detailed examples of how neural networks have been used in image processing, see Refs. 13–15.

4.3. Control Systems

One criticism of ANNs is that they operate like black boxes: They accept inputs and produce outputs, but the intermediate processing occurs in a manner that is not easily understandable. This description, of course, is remarkably similar to a controller: It is essentially a black box that produces an output from some series of inputs.

Neural networks have proven to be useful at modeling nonlinear control systems. The applications are considerable, including such possibilities as:

- Controlling the infusion rate of a drug to ensure that the level of that drug in the blood matches the prescription.
- Allowing a patient to control artificial limbs, or a surgeon to teleoperate.
- Planning the delivery of radiation to a patient in a way that maximizes the effectiveness of the dose, while reducing the risk to the healthy tissues.
- Route shaping for network packets so that the most critical information is sure to arrive in a timely manner, less important operational data are shunted to slower paths.

The process for conducting an image processing project using a neural network is also very similar to that for predicting medical outcomes. (See the example above.)

For more detailed examples of how neural networks have been used in control systems, see Refs. 16–18.

5. CONCLUSIONS

Good reason exists for multilayer, feed-forward ANNs trained under back-propagation (or its variants) being so popular: Few other methods can hope to solve difficult, non linear problems with noisy data nearly as effectively. Yet with increased ability comes increased complexity; fine-tuning these ANNs can be time-consuming and tedious. As with any tool, a better understanding of the background theory, current practice, and real biomedical applications should help contribute to putting this technology to good use.

BIBLIOGRAPHY

1. V. van Gogh, Letter to Theo van Gogh. written July 1880 in Cuesmes. Translated by Mrs. Johanna van Gogh-Bonger. *The Complete Letters of Vincent van Gogh* 1991; **22**:133.
2. W. S. McCulloch and W. Pitts, A logical calculus of the ideas immanent in nervous activity. *Bull. Math. Biophys.* 1943; **5**:115–133.
3. F. Rosenblatt, The perceptron, a probabilistic model for information storage and organisation in the brain. *Psycholog. Rev.* 1958; **62**:386.
4. M. Minsky and S. Papert, *Perceptrons: An Introduction to Computational Geometry*, 3rd ed. Cambridge, MA: MIT Press, 1988.
5. S. E. Fahlman, An empirical study of learning speed in back-propagation networks. Technical Report Computer Science Technical Report, 1988.
6. S. Fahlman, Faster-learning variations on back-propagation: an empirical study. In: D. S. Touretzky, G. E. Hinton, and T. J. Sejnowski, eds., *Proceedings of 1988 Connectionist Models Summer School*. Los Altos, CA: Morgan Kaufmann Publishers, 1988, pp. 38–51.
7. T. Tollenaere, Supersab: fast adaptive back propagation with good scaling properties. *Neural Networks* 1990; **13**:561–573.
8. M. Riedmiller and H. Braun, A direct adaptive method for faster backpropagation learning: the RPROP algorithm. Proceedings of the IEEE International Conference on Neural Networks, San Francisco, CA, 1993:586–591.
9. D. Nguyen and B. Widrow, Improving the learning speed of 2-layer neural networks by choosing initial values of the adaptive weights. *Proc. Intl. Joint Conf. Neural Networks* 1990; **3**:21–26.
10. B. E. Rosen and J. M. Goodwin, Training hard to learn networks using advanced simulated annealing methods. SAC '94: Proceedings of the 1994 ACM Symposium on Applied Computing, New York, 1994:256–260.
11. M. H. Schwartz, R. E. Ward, C. Macwilliam, and J. J. Verner, Using neural networks to identify patients unlikely to achieve a reduction in bodily pain after total hip replacement surgery. *Med. Care* 1997; **35**(10):1020–1030.
12. C. M. Ennett and M. Frize, Weight-elimination neural networks for mortality prediction in coronary artery surgery. *IEEE Trans. Inform. Technol. Biomed.* 2003; **2**:86–92.
13. T. Kondo and A. S. Pandya, Medical image recognition by using logistic gmdh-type neural networks. SICE 2001. Proceedings of the 40th SICE Annual Conference. International Session Papers, July 25–27, 2001:259–264.
14. R. I. Chaplin, R. M. Hodgson, and S. Gunetileke, Towards the use of problem knowledge in training neural networks for image processing tasks. *Image Proc. Applications* Seventh International Conference (Conf. Publ. No. 465). 1999; **1**:62–66.
15. M. S. Unluturk and J. Saniie, Deconvolution neural networks for ultrasonic testing. *IEEE Proc. Ultrason. Sympos.* 1995; **1**:715–719.
16. P. J. Antsaklis, Neural networks for control systems. *IEEE Trans. Neural Networks* 1990; **1**(2):242–244.
17. J. Rosen, M. B. Fuchs, and M. Arcan, Performances of hill-type and neural network muscle models toward a myosignal-based exoskeleton. *Comput. Biomed. Res.* 1999; **32**:415–439.
18. Q. Li, C. L. Teo, A. N. Poo, and G. S. Hong, Response of a feedback system with a neural network controller in the presence of disturbances. 1991 IEEE International Joint Conference on Neural Networks, November 18–21, 1991, vol. 2, pp. 1560–1565.

READING LIST

L. Fausett, *Fundamentals of Neural Networks: Architectures, Algorithms, and Applications*. Upper Saddle River, NJ: Prentice-Hall, 1994.

R. Caruana, S. Lawrence, and C. L. Giles, Overfitting in neural networks trained with backpropagation, conjugate gradient, and early stopping. In: T. K. Leen, T. G. Dietterich, and V. Tresp, eds., *Advances in Neural Information Processing Systems 13*. Cambridge, MA: MIT Press, 2001.

BATTERIES FOR IMPLANTABLE BIOMEDICAL APPLICATIONS

CURTIS F. HOLMES
Greatbatch Incorporated
Clarence, New York

BOONE B. OWENS
University of Minnesota
Minneapolis, Minnesota

1. INTRODUCTION

Two categories of "medical batteries" exist, those that are external to the human body when in use (1) and those that are internal to the human body when in use. Figure 1 illustrates the variety of implantable devices, not all of which contain batteries. Batteries for external medical devices were recently reviewed by Passerini and Owens (2) and by Holmes (3). The present chapter focuses on batteries used to power implantable medical devices such as pacemakers and neurostimulators. Therefore, the term "medical battery" in this chapter will exclude those batteries used in the many external medical devices that include, for example, hearing aids, portable defibrillators, glucose monitors, iontophoretic drug delivery systems, external stimulators for relief of pain, and even wheelchairs.

Implantable medical devices may be either passive or active with respect to their need for electrical power. Passive devices are those that do not require electrical power for operation, and include artificial hip replacements and heart valves. Perhaps the most familiar active device is the heart pacemaker. An important characteristic of active devices is that battery failure terminates the operation of the device. Therefore, it is imperative that a highly reliable, safe battery be incorporated into active medical devices, especially if they are life-supporting devices. However, failure of the active medical device does not necessarily mean the battery failed. For example, an internal short in the device circuitry can result in rapid battery depletion, but does not indicate a battery design or manufacturing defect.

Nearly 50 years have lapsed since a battery-powered pacemaker was first implanted in a human, in 1958 in Sweden (4). Since that time, significant developments in battery technology have played a key role in enabling the successful development of the small, highly reliable medical devices of the twenty-first century. Implantable devices have been developed to treat cardiac arrhythmias (bradycardia, tachycardia, and fibrillation), chronic pain, epilepsy, hearing loss, obesity, scoliosis, and bone fracture. Over 8 million people with medical disabilities have been treated with a variety of small battery-powered devices that were surgically implanted into their bodies.

Historically, the most widely used implantable battery is the Li/I_2 pacemaker battery. As of the year 2005, it is estimated that more than 500,000 lithium battery-powered pacemakers are implanted annually. The first human implant of a Li/I_2 battery-powered pacemaker took place in 1972, in Florence, Italy. Within just a few years, this type of battery displaced the mercury-zinc cell as the preferred power source for cardiac pacemakers. It is estimated that for the 5-year period from 2000 through 2004, 3 million lithium batteries were implanted in pacemakers alone.

The size of an implantable medical device and its operating service life are important properties. The battery is a key to determining both of these properties. A smaller battery will reduce overall device size, but it will also cause the device to have a shorter service life. Likewise, a larger battery will allow a longer service life, but at the cost of a bigger device, a fundamental trade-off. One outcome is a large incentive for the device designer to use the highest energy density battery available. In practice, the final design is a compromise between device size and service longevity.

Appendix A presents both illustrations of cell designs and descriptions of chemistries, as presently incorporated into commercial batteries by Greatbatch, Inc. This information was abstracted, and paraphrased from their website and is reprinted by permission of Greatbatch, Inc., 2005. The reader should note in particular that although the external appearance of the implantable lithium batteries are very similar, they have highly significant differences in the internal design of features such as cathode and electrolyte materials, current collectors, separators, and individual electrode areas and thicknesses.

The *Handbook of Batteries* (5) provides an in-depth description of all types of batteries as well as the principles of operation of batteries. The reader is encouraged to refer to chapter 1 of that publication for a thorough description of the fundamental concepts of battery design and operation.

A limited number of reference materials exist that deal specifically with medical batteries. The reader is referred to *Batteries for Implantable Biomedical Devices* (1) as well as other articles by Owens (6) and Holmes (3,7). The book edited by Ellenbogen et al. (8) includes a chapter on power sources. In addition, the reader may refer to websites of the manufacturers of medical batteries (Litronic, Greatbatch, Medtronic, and Quallion; see Table 1) for further information.

Salkind et al. (9) provide an overview of electrically-driven implantable prostheses, including pacemakers, defibrillators, bone growth, hearing devices, neurostimulators, and drug delivery systems. Recent clinical trials report the use of total artificial hearts (TAH) for recipients awaiting availability of organs for transplant (Abiomed website, 2005).

1.1. Battery Specifications

Medical batteries are unique because of the following:

- Operation is in an isothermal environment at a constant temperature of 37°C.
- In a life-sustaining device, battery failure may have serious health consequences. The battery (and

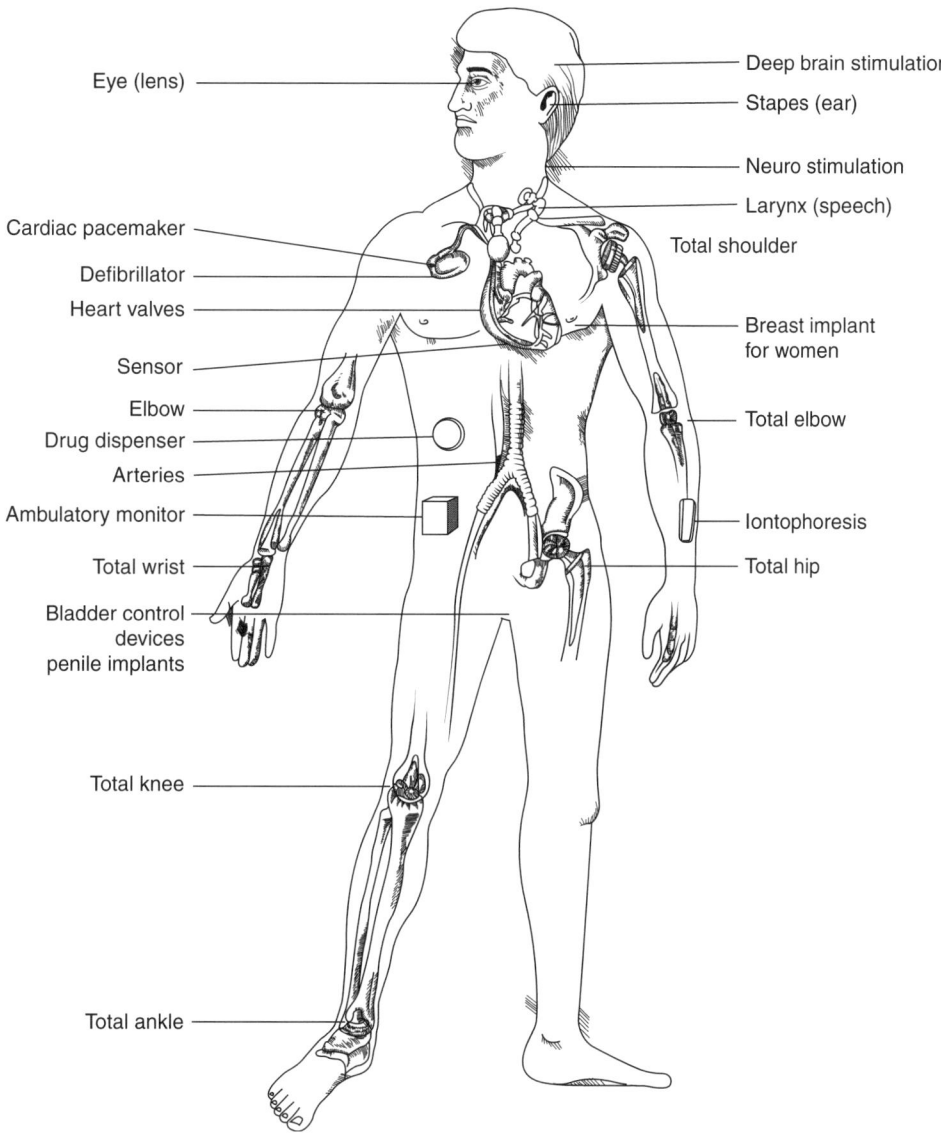

Figure 1. Implantable medical devices (1).

therefore the device) should be replaced prior to total battery depletion. An intrinsic end-of-service indicator is useful.

- Battery replacement requires surgical intervention.
- Battery size (especially volume) may dominate the device size.
- Battery shape may determine the device thickness.
- Battery is hermetically sealed.

As a result of the constraints imposed by the implantable medical devices, the batteries that are used to power them are often custom-designed for the specific models of

Table 1. Major Suppliers of Implantable Device Batteries

Battery Manufacturer	Cell Chemistry					
	Li/I_2	$Li/SOCl_2$	$LiMnO_2$	Li/CF_x	Li/SVO	Li-Ion
Greatbatch Inc.,[a]	X		X	X	X	X
Litronik GmbH, (Biotronik Inc.)[b]	X		X			
Medtronic Inc.[c]	X	X	X		X	X
Quallion LLC,[d]				X		X

[a] http://www.greatbatch.com/.
[b] http://www.biotronik.de/.
[c] http://www.medtronic.com/.
[d] http://www.quallion.com/.

devices. In any case, extensive testing is necessary before a particular battery is considered as qualified for use in a specific device.

The power requirements for implantable devices vary considerably, both between the devices used for different applications and between various periods within the use of a specific device. For example, a pacemaker (PG) typically has a background current of 10 µA. Even frequent pacing will only increase this current by 20–30% in most patients. Hence, the low-rate capability of the Li/I_2 battery is well-suited to this application. Similarly, modern defibrillators (ICD) also only need a background current of about 10 µA. However, when they are called on to defibrillate the heart, they need to deliver a 40 J shock within about 10 s, which, in turn, requires the battery to quickly deliver as much as 10 A for a few seconds. The power requirements of these two cardiac devices can be met with lithium batteries of nominally the same capacity (1–2 Ah), but because of the difference in the power requirements, the internal designs and the battery chemistries are very different. The figures in Appendix A illustrate some of the design differences between the Li/I_2 pacemaker battery and the LiSVO defibrillator battery.

Some terms used in discussing batteries are summarized in Table 2, which uses the example of the lithium/iodine battery to illustrate the definitions. The Li/I_2 battery designation indicates the anode is lithium and the cathode is iodine, which may be further simplified to Li/I, or it can be expanded to include the electrolyte LiI as in $Li/LiI/I_2$. All of these are simplified designations because the electrolyte and the cathode contain a number of chemical species that also change during discharge of the cell.

The battery is the component in a medical device that provides the necessary power for operation throughout the desired service life. The energy content of the battery limits this service life. One of the most important characteristics of the battery is its energy density, which is the ratio of its energy content to its volume, usually expressed in units of Wh/cc for small cells. Reddy and Linden (5) review the principles and methods for calculating energy density as well as other battery performance characteristics.

Table 2. Cell Components (Example is the Li Iodine Pacemaker Cell)

ANODE: Negative electrode, where the oxidation reaction takes place.

$$Li = Li^+ + e^-$$

CATHODE: Positive electrode, where the reduction reaction takes place.

$$I_2 + 2e^- = 2I^-$$

ELECTROCHEMICAL CELL: The fundamental element that contains chemical energy in a form to be converted to electrical energy by electrochemical or Redox reaction, such as a container filled with Li and I, arranged so as to prevent direct chemical combination reaction, but to readily permit the cell reaction providing electrical power to an external load.

ELECTROLYTE: A material that is both an ionic conductor and an electronic resistor, such as LiI solid salt. The electrolyte provides for the transport of ions between the anode and the cathode during cell operation, to maintain electroneutrality through the cell. The electrolyte interfaces both the anode and the cathode.

CELL DESIGNATION: The usual method is Anode/Cathode or Anode / Electrolyte / Cathode, as in

$$Li/I_2 \quad \text{or} \quad Li/LiI/I_2$$

SEPARATOR: A solid material that prevents direct contact of the anode and the cathode, to avoid internal shorting of the cell. If the electrolyte is solid as is LiI, the solid electrolyte serves the dual function of electrolyte and separator. More generally, the electrolyte is a liquid, and so a microporous polymer membrane serves as separator, while also permitting the liquid electrolyte to permeate the separator.

CURRENT COLLECTOR: The current collector is an electronic conductor (usually an inert metal) that is in intimate contact with the electroactive materials so that when the cell reaction occurs, the electrons transport through the anode current collector, to the external circuit, and then return to the cell through the cathode current collector. The current collector must maintain a high electronic conductivity and be chemically stable in contact with the active materials of the electrodes, not corroding or undergoing chemical or electrochemical reactions during the lifetime of the battery. With stainless steel (SS) as the current collector for a pacemaker cell, the designation is

Negative Terminal/SS/Li/LiI/I_2/SS/Positive Terminal

BATTERY: The battery is the sealed package that contains stored chemical energy, in a form that is readily converted to electrical energy, delivered through the positive and the negative terminals of the battery. The battery may contain just a single cell, or it may contain a multitude of cells arranged in a combination of series and parallel connections. The capacity and the voltage of the battery are determined by the number of cells and their arrangement.

PRIMARY BATTERY: A battery that is designed to be discharged one time, and is not to be recharged. It provides power for a single discharge cycle, and then must be replaced. In medical devices, usually the device is surgically removed and replaced with a new device. Ideally, this will be after 5 or 10 years.

SECONDARY BATTERY: A battery that may be recharged for continued use. The number of times the battery can be discharged and then recharged is its cycle-life, which is a highly variable term, depending on many parameters including the duty cycle, environmental conditions, and depths of discharge.

2. MEDICAL BATTERIES

Implantable device batteries fall into one of four categories, with respect to the power level and the service lifetimes required by the medical device, which will be referred to as Type 1 through Type 4 Batteries. The highest power level needed by the device determines whether the battery is designed for low, medium, or high rate (Table 3).

> **Type 1. Low-rate (µW) primary batteries** for devices such as the PG.
>
> **Type 2. Intermediate-rate (mW) primary batteries** for devices such as neurostimulators and some advanced PGs.
>
> **Type 3. High-rate (W) primary batteries** for devices that require a low-rate monitoring current (10 µA) in combination with the ability to deliver intermittent high-power pulses (1–10 A), such as in the ICD.
>
> **Type 4. Rechargeable batteries** for devices that require high power for short times and that can then be recharged *in situ*, for an extended number of cycles, such as in the total artificial heart (TAH) or left ventricular assist device (LVAD), to extend the service lifetime.

2.1. Type 1. Low-Rate (µW) Primary Batteries

Implantable biomedical devices that require microwatts of power, either continuously or intermittently include:

BGS: Bone Growth Stimulators
IGS: Implantable Gastric Stimulator
IHD: Implantable Hearing Device
PG: Cardiac pacemaker or Implantable Pulse Generator

Historically, the Li/I_2 system has been the battery of choice because it is very predictable and can reliably deliver µW of power continuously for many years. Furthermore, this battery can include an intrinsic end-of-service indicator (EOSI) to give early warning of need for replacement. This indicator is usually based on the load voltage or the impedance. In either case, the device continuously monitors the chosen parameter and then signals when the condition has been met for some period of time. The parameter is chosen so that ample time still exists for the patient to have the device changed before the power runs out, assuming normally frequent physician follow-up visits.

Cell	OCV	Typical EOSI	Impedance
$Li/LiI/I_2$	2.801 V	2.0 V	10–10,000 Ω

Table 3. Battery Chemistries Used in Implantable Device Batteries

Cell Design	Cell Chemistry					
	Li/I_2	$Li/SOCl_2$	$LiMnO_2$	Li/CF_x	Li/SVO	Li-Ion
Primary Batteries						
● Low Rate	BGS			IHD		
	IGS			PG		
	IHD					
	PG					
● Medium Rate		NRS		BGS	ICT	
				DDS		
				ICT		
				ISP		
				NRS		
● High Rate			ICD		ICD	
Secondary Batteries						
● Low Rate						
● Medium Rate						CI
						ICT
						IHD
						NRS
● High Rate						TAH
						LVAD

BGS: Bone Growth Stimulators
CI: Cochlear implants
DDS: Drug delivery system
ICD: Cardioverter-defibrillator
ICT: Incontinence Devices
IGS: Implantable Gastric Stimulator
IHD: Implantable hearing device
ISP: Insulin Pumps
LVAD: Left ventricular assist device
NRS: Neurostimulator (neuromuscular stimulators, spinal cord stimulators, deep brain stimulation systems)
PG: Cardiac pacemaker
TAH: Total artificial heart

Figure 2. Size reduction in PG, from 1965 to 1980. (This figure is available in full folor at http://www.mrw.interscience.wiley.com/ebe.)

The Li/I$_2$ battery replaced the earlier Zn/HgO cell in the mid-1970s for cardiac pacemakers. The major application of the Li/LiI/I$_2$ battery continues to be the PG.

Figure 2 illustrates the size reduction in PGs from 1965 up to 1980, as a result of advances in battery technology concurrent with advances in PG and lead designs. The first PG contains 10 mercury-zinc cells that dominate the space. The other two devices are hermetically sealed and contain Li/I$_2$ batteries.

An example of an implantable hearing device (IHD) is shown in Fig. 3. The placement of this IHD in a skull cavity is shown in Fig. 4, and graphically illustrates the need for a safe and reliable battery, free of problems such as electrolyte leakage, corrosion, and volume change (swelling) during discharge.

It can be a long and slow process to develop a new medical device, even if it uses a well-established battery. The IHD device is presently in clinical tests in the United States. About 20 devices have been implanted to date. The

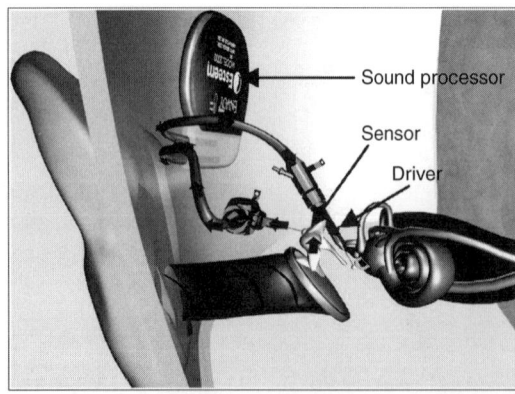

Figure 3. Implantable hearing device. The Envoy® system fits in the skull, and leads connect to transducer in middle ear. (This figure is available in full color at http://www.mrw.interscience.wiley.com/ebe.)

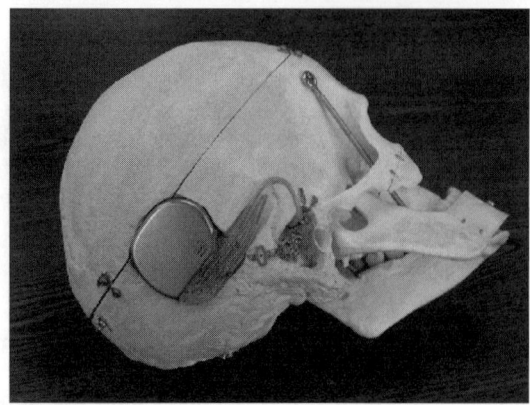

Figure 4. Placement of Envoy® hearing device in skull cavity. (This figure is available in full color at http://www.mrw.interscience.wiley.com/ebe.)

device needs very little power, about 50 to 100 μW. A standard Li/I2 pacemaker cell provides this power, with a capacity of 800 mAh. The device requires a nominal current draw of 20 to 30 μA and has a longevity of about 4 years. Surgical replacement is required after that time, because of the depletion of the battery capacity. Further information is available from the website www.EnvoyMedical.com.

A second battery chemistry that is also used in Type 1 applications is the Li/CF$_x$ system, which has the advantage of lower internal impedance that allows higher power pulses, for example, during device interrogation or programming.

2.2. Type 2. Intermediate-Rate (mW) Primary Batteries

The following biomedical devices require milliwatts of power either continuously or intermittently:

BGS: Bone Growth Stimulators
DDS: Drug Delivery System
ICT: Incontinence Devices
IHD: Implantable Hearing Device
ISP: Insulin Pumps
NRS: Neurostimulator (neuromuscular stimulators, spinal cord stimulators, deep-brain stimulation systems)

Note that the same type of device (BGS or IHD) may be listed under low-power and medium-power devices. For example, although PG are shown only as a low-power device, the more advanced models with longer-range telemetry will require higher power, intermittently.

The mW power levels are not readily provided by the Li/I$_2$ system because of its high internal resistance. Li batteries that use liquid electrolytes have better rate capabilities and are used for these applications. The cell systems include:

Cell	OCV	Typical EOSI
Li/CF$_X$	3.2 V	2.5 V
Li/SOCl$_2$	3.65 V	3.0 V
Li/MnO$_2$	3.5 V	2.0 V

The impedance of these cells is very design-dependent, and can range from a few ohms to several thousand ohms. These cells resemble the standard Li/I_2 cells pictured in the figures in this chapter in outward appearance, but they differ significantly in internal construction.

2.3. Type 3. High-Rate (W) Primary Batteries

The implantable cardiac defibrillator (ICD) needs to deliver a power pulse of 40 J energy to the heart in a matter of a few milliseconds, which requires a power output of 40 kW over about a 1 ms period. No implantable battery can directly deliver such power, so it is delivered by low-impedance capacitors charged to a very high voltage (typically about 700 V) via a high-voltage transformer circuit connected to the battery.

The battery that powers the charging circuitry is designed to provide the energy in a matter of 5 to 10 s, which requires an output current of about 1 to 10 A out of a battery with a rated capacity of a few Ah, a very high rate of discharge for the battery. However, this output is only part of the power that the battery is expected to provide.

The ICD requires a battery with two opposing power capabilities. The battery must function as a PG battery and provide about 10–100 μW of power continuously for a period of 5 years. The battery must also provide a power pulse of 1–10 W at any time during this 5-year period. The pulse lasts about 5–10 s and may be repeated up to four times during a single fibrillation episode. After defibrillation is successful, the device returns to the monitoring mode until the next event. Battery and device design determine the nominal performance range of the device. Patient condition and frequency of tachyarrythmia events will determine the actual operating life of the device, under ideal conditions.

The ICD battery must include the following characteristics:

1. Use a cell system that is inherently capable of rapid electrode kinetics.
2. Internal cell design maximizes the electrode/electrolyte interfacial area.
3. Maintain low impedance during use time, up to at least 5 years.
4. No voltage delay during high-rate operation.
5. Capable of discharging at 1–5 Amps, safely.
6. Withstand abuse tests including external or internal shorting.

Careful design and testing are especially necessary with this type of battery. If a pacemaker cell is shorted, the current may increase to 10 mA and run the cell down over a period of days or months. If an ICD battery with an internal resistance of a few tenths of an ohm is shorted, the current may peak at 30 A, with significant heating of the battery. Thus, safety features in high-rate cells are of utmost importance, and are built into the battery through careful design. One important safety feature often used is a separator material that will melt and become nonporous, and hence shut down the cell if either an internal or external short occurs.

Two Li battery systems that incorporate liquid electrolytes are used for this application. The cell systems are as follows:

Cell	OCV
Li/MnO_2	3.5 V
$Li/AgV_2O_{5.5}$	3.2–2.2 V

The battery chemistry most widely used is the LiSVO system ($Li/AgV_2O_{5.5}$). It is manufactured by both Medtronic, Inc. and Greatbatch, Ltd. The Li Manganese Oxide defibrillator battery has been produced by Litronik, a division of Biotronik (10).

2.4. Type 4. Rechargeable Batteries

Among the devices for which rechargeable batteries are being used or considered are the following low- to medium-rate applications:

- CI: Cochlear Implants
- ICT: Incontinence Devices
- IHD: Implantable Hearing Device
- NRS: Neurostimulator (neuromuscular stimulators, spinal cord stimulators, deep-brain stimulation systems)

Higher-rate designs have been produced for the following devices:

- LVAD: Left Ventricular Assist Device
- TAH: Total Artificial Heart

The total artificial heart (TAH) and the left ventricular assist device (LVAD) require continuous power at a level of 10 to 40 W. Thus, for a battery pack to run for 24 h, it would have to contain 250 to 500 Wh of energy. The most energetic rechargeable Li-ion battery may contain 200 Wh/kg, so this battery would have to weigh 2 or 3 kg. This size would still have to be recharged daily and is too large for implanting. Therefore, the designs of TAH and LVAD have an external battery pack. Inductive coupling through the skin is used to get power transmission to the implanted device. A small battery in the device is used as an emergency backup source capable of operating the device for 30 min to 1 h.

3. RECENT DEVELOPMENTS

Advances are being made to address limitations of present systems and to increase energy density and longevity (11–13). Two such developments are now in clinical use in implantable devices. Gan et al. at Greatbatch, Inc. have developed a unique cathode system combining the CF_x and SVO chemistries (11). Merritt et al. at Medtronic (12) have reported a hybrid cathode system of another design in which the CF_x and SVO active materials are mixed before forming the cathode pellet. These hybrid batteries provide higher current delivery capability than that of the lithium/iodine battery while showing comparable energy density.

Work is ongoing to improve both primary and secondary battery systems used for implantable medical devices. Although the improvements in the primary systems are most likely to come from the few companies who supply or develop these systems for their own use (Table 1), a lot of the improvements in the secondary systems are likely to come from outside the medical device industry by people such as those who develop consumer product batteries.

4. SUMMARY

Battery-powered implantable devices have been used to treat human illnesses since the implantation of the first successful pacemaker in 1960. Primary lithium batteries have been used in these devices since 1972. Today, a remarkable variety of implantable devices include neurostimulators, drug delivery systems, implantable cardioverter/defibrillators, gastric stimulators, implantable hearing assist devices, left ventricular assist devices, and the totally artificial heart. These devices are all powered by lithium primary or lithium-ion rechargeable batteries.

Acknowledgment

The authors wish to acknowledge the contributions provided by the reviewers of this paper, and especially the suggestions and clarifications from Dr. Darrel F. Untereker. Figures 3 and 4 are included with the permission of Envoy Medical Corp, (website http://www.envoymedical.com). Appendix A is included with the permission of Greatbatch, Inc. (website http://www.greatbatch.com).

BIBLIOGRAPHY

1. B. B. Owens, In: B. B. Owens, ed., *Batteries for Implantable Biomedical Devices*. New York: Plenum Publishing, 1986, p. 37.
2. S. Passerini, and B. B. Owens, Medical batteries for external medical devices. Proceedings of the 10th International Meeting on Lithium Batteries, Como, Italy, (2000). *J. Power Sources* 2001; **97-98**:750–754.
3. C. F. Holmes, *Electrochemical Power Sources and the Treatment of Human Illness*. Interface, The Electrochemical Society, 2003, pp. 26–29.
4. Pacing, *Pacing Clin. Electrophysiol.* 2003; (1 Pt 1):114–124.
5. D. Linden and T. B. Reddy, In: D. Linden and T. B. Reddy, eds., *Handbook of Batteries*, 3rd ed. New York: McGraw-Hill Publishers, 2002.
6. B. B. Owens, Medical batteries. In: H. Vankatasetty, ed., *Lithium Battery Technology and Applications*. New York: Wiley-Interscience, 1984.
7. C. F. Holmes, *The Role of Electrochemical Power Sources in Modern Health Care*. Interface, The Electrochemical Society, 1999, pp. 32–34.
8. K. Ellenbogen, G. Kay, and B. Wilkoff, Power systems for implantable pacemakers and ICDs. In: *Clinical Cardiac Pacing and Defibrillation*, 2nd ed. Philadelphia, PA: WB Saunders Co, 2000, pp. 167–193.
9. A. J. Salkind, et al. Key events in the evolution of implantable pacemaker batteries. In: B. B. Owens, ed., *Batteries For Implantable Biomedical Devices*. New York: Plenum Publishing, 1986, p. 37.
10. J. Drews, R. Wolf, G. Fehrmann, and R. Staub, The 6 volt battery for implantable cardioverter/defibrillators. *Biomed. Tech. (Berl)*. 1998; **43**(1-2):2–5.
11. H. Gan, A. Shah, Y. Zhang, R. Rubino, S. Davis, and E. Takeuchi, Advanced battery technology for medium rate implantable medical devices using unique laminated cathode construction. Abstract No. 247. 208th Meeting of The Electrochemical Society, Los Angeles, California, October 2005.
12. D. R. Merritt, W. G. Howard, C. L. Schmidt, and P. M. Skarstad, *Hybrid cathode battery for implantable medical devices*. Abstract No. 831. 208th Meeting of The Electrochemical Society, Los Angeles, California, October 2005.
13. H. Yumoto, N. Bourgeon, T. Tan, and H. Tsukamoto, $Li_{1+x}V_3O_8$ and $(CF)_n$ hybrid cathode materials for ICD batteries. Abstract No. 209. 208th Meeting of The Electrochemical Society, Los Angeles, California, October 2005.

APPENDIX A. GREATBATCH IMPLANTABLE BATTERIES

Greatbatch, Inc. has been producing lithium batteries for implantable biomedical applications for over 35 years. Different shapes, sizes, and chemistries are available in a range of power capabilities. Greatbatch (GB) offers standard cell designs and also develops proprietary products specific to customer applications. Custom pin configurations for electrical connections are also available in standard and custom designs.

The following pages (taken verbatim, or with minor paraphrasing, from the Greatbatch website, www.greatbatch.com), present details on various primary battery systems in current use in implantable devices. (Reprinted by permission of Greatbatch, Inc., 2005):

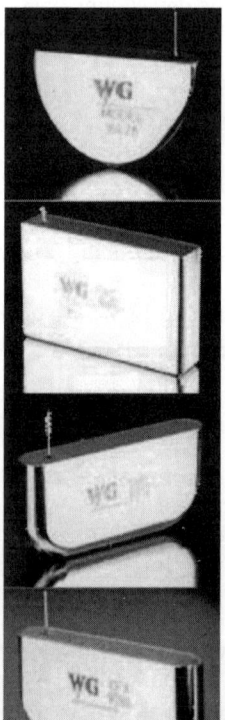

Lithium iodine (Li/I_2) cells

High rate Lithium silver vanadium oxide (Li/SVO)

Medium rate Lithium silver vanadium oxide (Li/SVO)

Figure A.1. Implantable batteries.

A.1. LITHIUM IODINE (Li/I₂) CELLS

Low Rate: Li/I₂ cells are designed to deliver current drains in the microampere range, reliably over long periods of time. They are available in a variety of sizes, shapes, and capacities. The smallest standard GB- Li/I₂ cell is 4-mm thick and has a nominal capacity of 0.42 ampere hours.

A.2. LITHIUM SILVER VANADIUM OXIDE (Li/SVO)

These solid cathode cells are designed for either high-rate or medium-rate applications. The high-rate designs are capable of ampere-level current pulses, whereas the medium-rate cells are designed for milliampere rate discharge currents. Although the battery chemistries are

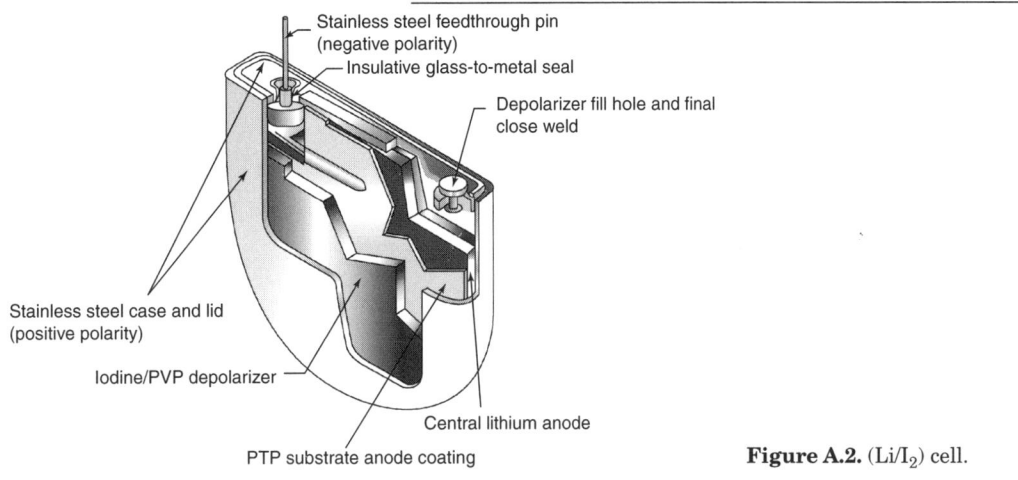

Figure A.2. (Li/I₂) cell.

- Low rate cell designed for implantable pacemaker applications
- Construction in a hermetic glass-to-metal sealed 304L stainless-steel case
- Current drains in the microamp range
- Optimum operating temperature 37°C
- High energy density
- Storage temperature range −40°C to +52°C

the same, the internal design of the hardware optimizes the performance to the application requirements.

A.2.1. High-Rate Li/SVO Cells

- High-rate cell designed for implantable defibrillator applications

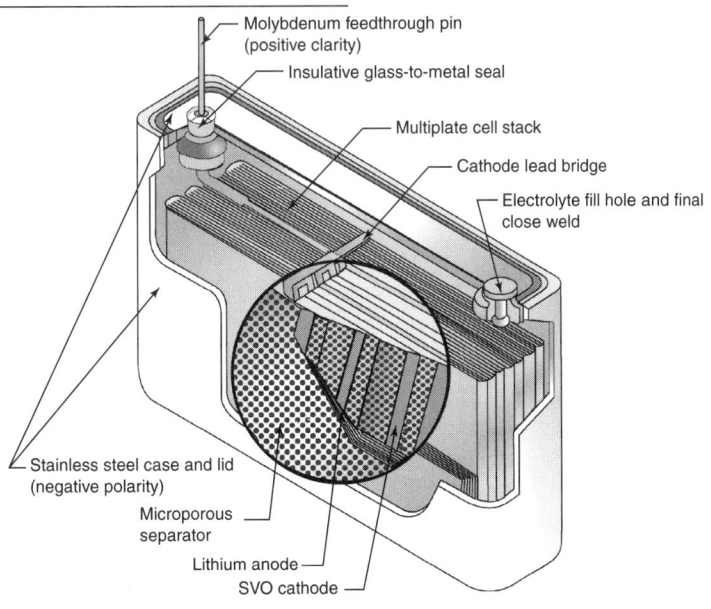

Figure A.3. High-rate Li/SVO

- Multiplate construction in a hermetic glass-to-metal sealed 304L stainless-steel or titanium case
- Current drains in the 0.5. to 2 ampere range
- Cells are capable of continuous operation at a nominal cell temperature of 37°C
- High energy density and long shelf life
- Storage temperature range −40°C to +58°C

A.2.2. Medium-Rate Li/SVO Cells

These cells are generally capable of current drains in the milliampere range and are available in standard shapes. These cells have found use in a number of implantable applications such as neurostimulators, atrial defibrillators, and drug-infusion devices.

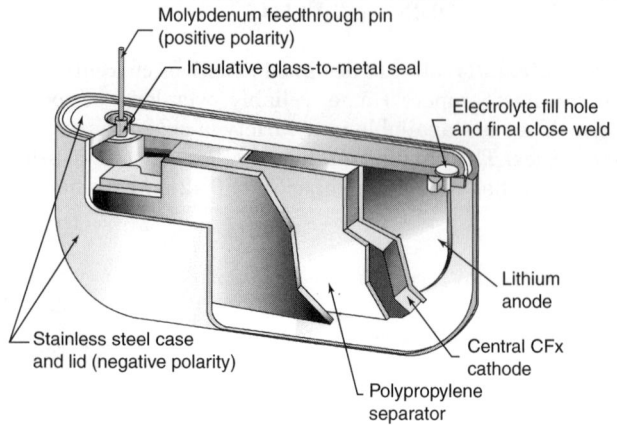

Figure A.5. Li/CF$_x$ cell.

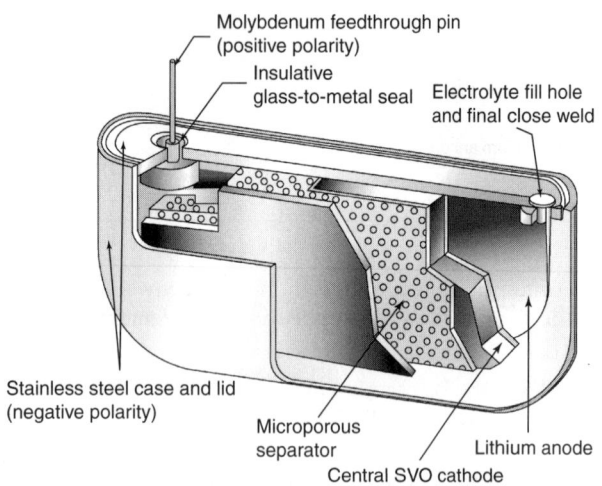

Figure A.4. Medium-rate Li/SVO cell.

- Medium-rate cell designed for implantable neurostimulators and drug-infusion devices to handle moderate-rate pulse currents (50 mA)
- Construction in a hermetic glass-to-metal sealed stainless-steel or titanium case
- High energy density
- Cells are capable of continuous operation at a nominal cell temperature of 37°C
- Long shelf life
- Storage temperature range −40°C to +58°C

A.3. LITHIUM CARBON MONOFLUORIDE (Li/CF$_x$)

This system is characterized by a relatively flat voltage discharge profile, extremely low internal impedance, and low overall cell weight. This system finds use in drug-infusion pumps, neurostimulators, and most recently in pacemakers.

- Medium-rate, low-impedance cell designed for implantable neurostimulator and pacemaker applications
- Construction in a hermetic glass-to-metal sealed stainless-steel or titanium case
- Designed to handle currents in the microamp to milliamp range
- Cells are capable of continuous operation at a nominal cell temperature of 37°C
- High energy density and long shelf life
- Storage temperature range −40°C to +52°C

A.4. LITHIUM-ION CELLS (Li-ION)

Lithium-ion secondary batteries have recharging capability that makes them a good solution for implantable devices such as neurostimulators that have power requirements that may exceed the total power available from primary cells.. Secondary cells also excel in applications where power requirements change based on the user's activity or device programming.

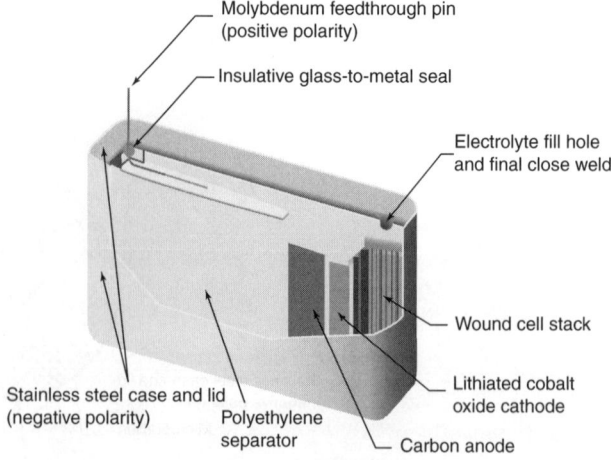

Figure A.6. Li-Ion cell.

BAYESIAN ANALYSIS

ANDY P. WILLIAMS
ALI E. ABBAS
University of Illinois at Urbana-Champaign
Urbana, Illinois

1. INTRODUCTION

Consider the following question. If you knew that 4 out of every 60 men with heart disease also had high cholesterol, what would you believe to be the probability that any man known to have high cholesterol would also have heart disease? Consider further; that you could conduct a test to determine whether a patient had a particular condition, but the test could not predict perfectly. It might result in an occasional false positive, or it might miss a positive altogether. What would you believe to be the probability that an individual who tested positive had the condition?

Quantifying uncertainty has been a prime interest for centuries, essentially as far back as games of chance. Probability theory rose to prominence in the seventeenth century with the works of Blaise Pascal and others, addressing specifically the odds associated with a roll of a die. One question that grew out of such concerns was the following: Given the number of times an event has happened and failed, determine the chance that the probability of it happening in a single trial lies between two named probabilities.

The answer to that particular question was proposed in the eighteenth century by the Reverend Thomas Bayes, and the approach (and therefore the theory and rules that it is founded on) has taken his name in honor of the discovery [a reprint of Bayes's work appeared in Barnard and Bayes (1)]. In general, Bayesian analysis refers to the process of assessing the current state of knowledge, gathering new data, and updating the analysis to incorporate both the prior (called so because it is prior to collecting anything other than the background information) assessment and the new data. The process has a wide range of applications, including scientific inquiry, clinical testing, public policy, and developing new e-mail spam filters.

2. PROBABILITY THEORY

To present Bayesian analysis, we must begin by laying out the basic building blocks that lead up to it. Probability is the fundamental building block from which Bayesian analysis is founded. We begin by defining an event. An event is a particular situation, described completely enough to allow for clarity and agreement on its occurrence once it has occurred. An event may be a future occurrence (such as the winner of an election) or a current circumstance regardless of whether it is currently known or unknown (such as the number of men in the room with you). In probability theory, events are discussed at length, because they are a convenient manner in which to describe some occurrence. Events, though, may also be considered to be a special case of a proposition, which is any logical statement.

Probability is a measure pertaining to your belief about the plausibility of an event or a proposition. Many have attempted to define probability based on the long-run frequency of a particular outcome given repeated independent revelations. Although we will not go into the details of this definition's shortcoming, it shall suffice to say that this definition fails when discussing events which will occur only once or the plausibility of a proposition such as "It is sunny at the moment." The more general definition of probability, the one adhered to by Bayesian analysts, is essentially that probability is a degree of belief about a proposition, given my current state of information. If my current information is represented by &, and the proposition is E, then my probability $p(E \mid \&)$ is my degree of belief about the plausibility of proposition E, given that I know &. Choosing to select a long-run fraction as the degree of belief ignores all other information that may be available besides the long-run fraction.

In 1946, Cox (2) developed a set of axioms that required that probability be divisible, comparable, consistent, and make common sense. The axioms required that the following functional relationships must exist between logical probabilities:

$$Axiom\,1 \quad P(A|\&) = f(p(\bar{A}|\&)), \qquad (1)$$

$$Axiom\,2 \quad p(A,B|\&) = g(p(A|\&), p(B|A,\&)), \qquad (2)$$

where f and g are some deterministic functions. The first axiom supports the notion that the plausibility of a proposition should tell us something about the plausibility of its negation. That is, the plausibility of a proposition can be presented as a function of the probability of the counter-proposition. The second supports the notion that the plausibility of the conjunction of two propositions is a function of the probability of one proposition and the probability of the second given that the first is true. These axioms resulted in functional equations with one unique nontrivial set of laws to which probability must adhere [see Jaynes (3) for more details]. These laws are

$$Law\,1 \quad p(A|\&) + p(\bar{A}|\&) = 1, \qquad (3)$$

$$Law\,2 \quad p(A,B|\&) = p(A|\&)p(B|A,\&)$$
$$= p(B|\&)p(A|B,\&). \qquad (4)$$

The first law supports additivity, meaning that for some proposition A, the sum of the probability of A and the probability of not A should total to the certain probability 1. The second law is more commonly presented in the following form:

$$p(A|B,\&) = \frac{p(B|A,\&)p(A|\&)}{p(B|\&)}, \qquad (5)$$

which relates the conditional probabilities of two events, using the unconditional probability of each. This form is the one referred to as Bayes's Rule.

Bayes's Rule allows us to relate the probabilities in a way that is useful for a wide variety of applications. Before advancing further it is important to clearly define the notions of conditional probability, unconditional probability, and independence. In the strictest sense, there is no such thing as unconditional probability, because there will always be a background state of information, which is represented by &.

An example of a state of information at a particular instant in time could be the belief that each side of a die will be equally likely to land on any given roll, leading to a probability of one sixth for getting a "3" on a particular roll of one die. We may all agree on this fact, and it becomes part of our &. However, knowledge that the die has a bent corner may lead us to assign a different probability for "3" on a particular role. In this latter case, our state of information has changed.

The rules also imply that the probability of an event is the sum of the joint probabilities of the event with a set of mutually exclusive and collectively exhaustive events, or

$$p(B|\&) = \sum_i p(B, A_i|\&) = \sum_i p(B|A_i, \&)p(A_i|\&). \quad (6)$$

This implies that Bayes's Rule can also be expressed as

$$p(A|B, \&) = \frac{p(B|A, \&)p(A|\&)}{p(B|A, \&)p(A|\&) + p(B|\bar{A}, \&)p(\bar{A}|\&)}. \quad (7)$$

3. INDEPENDENCE

Now we introduce the concepts of independence and conditioning. Two events A and B are independent given the background state & if

$$p(A, B|\&) = p(A|\&)p(B|\&). \quad (8)$$

The expression $p(A,B \mid \&)$ is called the joint probability of A and B given &, meaning the probability that both will occur. From the laws presented earlier, it follows that independence also implies that

$$p(A|\&) = p(A|B; \&). \quad (9)$$

Independence, therefore, implies that knowing the outcome of the event B provides no information that would cause us to change our belief about the probability of A. Generally, a set of events is said to be pairwise independent if

$$p(A_m, A_n|\&) = p(A_m|\&)p(A_n|\&) \; \forall m \neq n \quad (10)$$

and is said to be mutually independent if, for every subset of events from A,

$$p(A_{n_1}, A_{n_2}, \ldots, A_{n_k}|\&) = p(A_{n_1}|\&)p(A_{n_2}|\&) \ldots p(A_{n_k}|\&). \quad (11)$$

When two events are not independent, they are referred to as dependent. As the opposite of independence, dependence implies that knowing the outcome of a particular event has bearing on the probability assigned to the other. It is important to note that dependence does not imply causality. Dependence only speaks to the informational relationship between two events.

For those probabilities that are dependent on another piece of information, we will call the probability $p(A \mid B, \&)$ the conditional probability. Alternatively, the conditional probability will also be called a posterior probability, because it reflects a probability determined posterior to incorporating some relevant information. In general, the probability of a sequence of dependent events is denoted as

$$\begin{aligned}p(A_1, A_2, \ldots, A_k|\&) &= p(A_1|\&)p(A_2|A_1, \&) \\ p(A_3|A_2, A_1, \&) &\ldots p(A_k|A_{k-1}, \ldots, A_1, \&).\end{aligned} \quad (12)$$

For those familiar with Markov chains, when each state is conditioned only on the previous state, the system may be represented as a Markov chain, where

$$\begin{aligned}p(A_1, A_2, \ldots, A_k|\&) &= p(A_1|\&)p(A_2|A_1, \&) \\ p(A_3|A_2, \&) &\ldots p(A_k|A_{k-1}, \&).\end{aligned} \quad (13)$$

Markov chains establish a relationship where each state is dependent only on the previous state.

3.1. Example: Bayes' Rule in Action and the Associative Logic Error

Suppose that a man stands trial for murder. It is known that the murderer lived in a particular city with 1,000,000 men. DNA evidence has been collected from a suspect that matches that of the individual who stands trial. It is shown that the probability of two DNA profiles matching is 1/10,000. Finally, it is believed that the testing procedure for DNA is highly accurate, and that the probability that the murderer's DNA will match that which is found is 99%. If a man is pulled at random and found to be a match, what is the probability that the man is the murderer?

$$p(Match|\&)p(Murderer|Match, \&)$$
$$= p(Murderer|\&)p(Match|Murderer, \&)1/10,000 * X$$
$$= 1/1,000,000 * 99/100.$$

The answer is that X = 99/10,000 – less than 1%. Although intuition may lead us to believe it should be higher (after all, the test is highly accurate), we must be careful to understand that $p(Match \mid Murderer, \&)$ does not equal $p(Murderer \mid Match, \&)$. This error is often called the associative logic error or the prosecutor's fallacy.

3.2. Example: Sensitivity and Specificity

A primary issue in medical testing is the concern over the sensitivity and specificity of a test. Sensitivity is the probability that the test will come up positive, "+", given that the patient is sick, $p(+ \mid Sick, \&)$. Bayes's Rule shows us that sensitivity is equal to true positives over true positives plus false negatives:

$$\frac{p(+, Sick|\&)}{p(+, Sick|\&) + p(-, Sick|\&)} = \frac{p(Sick|+, \&)p(+|\&)}{p(Sick|\&)}$$
$$= p(+|Sick, \&). \quad (14)$$

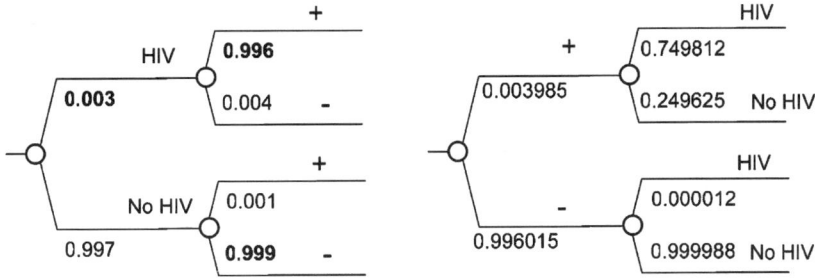

Figure 1. Probability trees for test 1.

Specificity is the probability that the test will come back negative given that the patient is not sick, $p(-|\overline{Sick},\&)$. Specificity is therefore true negatives over true negatives plus false positives:

$$\frac{p(-,\overline{Sick}|\&)}{p(+,\overline{Sick}|\&)+p(-,\overline{Sick}|\&)}=\frac{p(\overline{Sick}|-,\&)p(-|\&)}{p(\overline{Sick}|\&)}$$
$$=p(-|\overline{Sick},\&). \quad (15)$$

Consider that there are two possible tests for the HIV virus. A blood bank uses the tests on incoming blood samples before accepting the blood. Test 1 has a sensitivity of 99.6% and a specificity of 99.9%. Test 2 has a sensitivity of 99.7% and a specificity of 98.5%. Given that the base rate for HIV is 0.3%, the prior probabilities for the first test can be represented as in Fig. 1.

In Fig. 1, the values in bold represent the known values for the prior $p(HIV|\&)$, the sensitivity $p(+|HIV,\&)$, and the specificity $p(-|NoHIV,\&)$. The rest of the probabilities have been filled in based on the information provided by using Bayes's rule. For example, the probability of having HIV given that the test says positive is the product of $p(HIV|\&)$ and $p(+|HIV,\&)$ divided by $p(+|\&)$, or $0.749812 = 0.003 * 0.996/0.003985$.

For the second test:

The blood bank is looking to determine whether the new test, test 2, is better than test 1. One can easily see in Fig. 2 that the second test is more sensitive and is therefore more likely to catch that the sample has the virus, leading to a lower chance of accepting tainted blood. The tradeoff is that the test is less likely to correctly state that the virus is not there, meaning that more samples will be turned down when there was no reason to turn them down.

Both tests have extremely low probabilities of error in the case of a negative test result $p(HIV|-,\&)$, but the probability of making an error on a positive test varies considerably. Notice (in Fig. 1) that for test 1, a positive test result resulted in a 75% chance that the individual has the HIV virus. Meanwhile, a positive result from test 2 (in Fig. 2) yielded only a 16.7% chance of having the HIV virus.

Figure 3 highlights how the predictive capabilities of the test are closely related to the specificity. Notice how quickly the probability that a positive test truly indicates the HIV virus drops as the specificity gets lower. This highlights how one may believe that a test is highly accurate even though the positive test only has a 16.7% chance of being associated with a person afflicted with the virus. Note that the perceptions about the test's accuracy fall prey to the base-rate fallacy. People tend to lose track of population statistics, in this case, that the overall incidence rate of HIV is very low, which leads to the low value for $p(HIV|+,\&)$.

In the example, test 2 might be more appropriate for a blood bank testing the blood after it is received; if we assume that the cost of making an error when the virus is present (infecting another individual) is much higher than making an error when it is not (discarding perfectly good blood). Test 1 may make more sense for individual testing, because a positive test result might have grave consequences for the individual.

4. BAYESIAN INFERENCE WITH CONTINUOUS DISTRIBUTIONS

As we saw in the previous section, when two events are relevant (not independent), the prior belief $p(A|\&)$ is related to the posterior belief $p(A|B,\&)$ through Bayes's Rule. In this section, we will introduce Bayesian inference with continuous distributions. Thus, the process involves taking prior beliefs in the form of a probability distribution and modifying those beliefs on the revelation of other events or the introduction of new data, to get a posterior

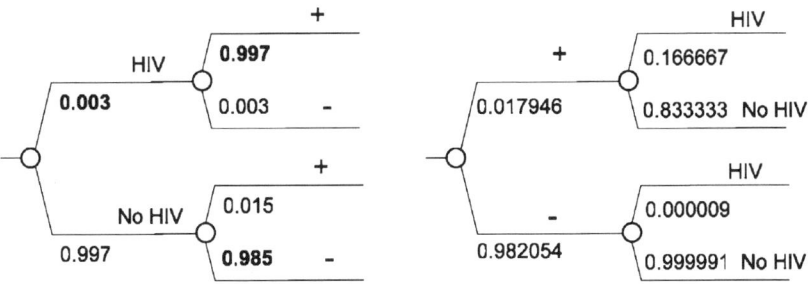

Figure 2. Probability trees for test 2.

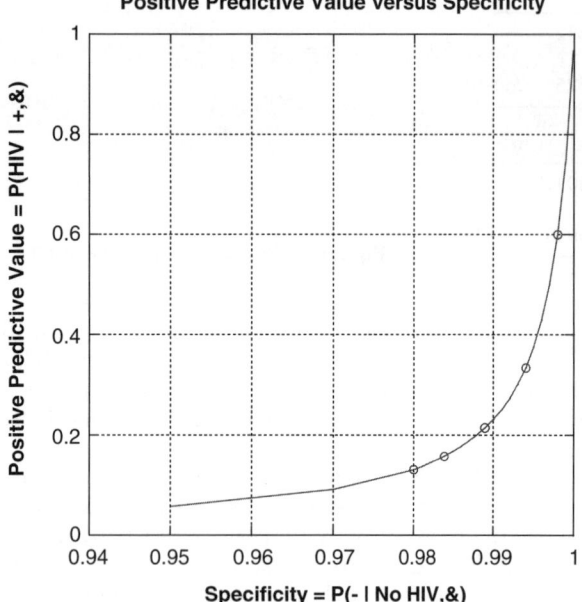

Figure 3. Predictive value versus specificity. (This figure is available in full color at http://www.mrw.interscience.wiley.com/ebe.)

probability distribution. One very common example of Bayesian inference is taking a population statistic (say mean and variance), collecting new data via a survey or experiment, and arriving at a new population statistic based on the experiment AND the prior statistic.

The basic premise of Bayesian inference is as follows. Suppose there is an unknown value θ. Before any observation outside of the background level of information, our belief about θ can be represented as $p(\theta\,|\,\&)$. In cases where θ is a continuous random variable, $p(\theta\,|\,\&)$ is actually a probability distribution, equal to

$$p(\theta|\&). \quad (16)$$

Suppose now that an observation X is made that is relevant to θ. Additionally, we can state that the probability distribution of X depends on θ in some known way, and this can be represented (in the continuous case) by the function

$$p(X|\theta,\&). \quad (17)$$

We wish to arrive at a posterior probability density function (pdf) for θ, given the prior belief and the new data. As we have seen, Bayes's Rule is the tool of choice, and from its application, we arrive at

$$p(\theta|X,\&) = \frac{p(X|\theta,\&)p(\theta|\&)}{p(X|\&)}, \quad (18)$$

where $p(X\,|\,\theta,\&)$ is the density function with respect to X and $p(\theta\,|X,\&)$ is a density function with respect to θ. Note also that, because we are discussing continuous functions, the unconditional probability can be determined via integrating the conditional probabilities such that

$$p(X|\&) = \int p(X|\theta,\&)d\theta. \quad (19)$$

Note that this is merely the continuous form of the previously mentioned formula

$$p(X|\&) = \sum_i P(X|\theta_i,\&)P(\theta_i|\&). \quad (20)$$

This allows us to represent Bayes's Rule as

$$p(\theta|X,\&) = \frac{p(X|\theta,\&)p(\theta|\&)}{\int p(X|\theta,\&)d\theta}. \quad (21)$$

For most purposes, though, it is most convenient to think of $p(X\,|\,\theta,\&)$ as a function of θ rather than X. This is because X is an observed value (fixed) and θ is our unknown. When the prior is expressed as a function of θ, it is called the likelihood function, expressed as

$$p(X|\theta,\&) = l(\theta|X,\&). \quad (22)$$

The likelihood function is the cornerstone of the analysis that we will undertake from here on out. The likelihood function expresses the likelihood of θ, given that the data X were observed. Consequently, we know that the prior likelihood of X is the integral of the prior likelihood for each possible state θ, implying

$$p(X|\&) = \int p(X|\theta,\&)d\theta = \int l(\theta|X,\&)d\theta, \quad (23)$$

which allows us to express Bayes's Rule in a new form, which is

$$p(\theta|X,\&) = \frac{p(\theta|\&)l(\theta|X,\&)}{\int l(\theta|X,\&)d\theta}. \quad (24)$$

The term $l(\theta|X,\&)/\int l(\theta|X,\&)d\theta$ can be thought of as the standardized likelihood, because it is scaled such that the volume under the surface is unity. This new form of Bayes's Rule allows us to compute the posterior probability density function as a product of the standardized likelihood (which is a function of θ) and the prior probability density function.

The challenge in using the theory described here, as with any statistical modeling, is in the ability to relate the likelihood in terms of a function, as well as the ability to model observations as a function. A word of caution should be offered that adopting particular constraints, such as assuming normality, may lead to poor models even if the method is used properly.

5. CONJUGATE PRIORS

Because of this computational complexity associated with calculating the posterior distribution function given any arbitrarily chosen prior and likelihood, it is often convenient to choose priors of a specific form. The most common strategy is to choose a prior such that the posterior will have the same functional form given the distribution of the likelihood. Recall that the following relationship holds:

$$p(\theta|X,\&) \propto l(\theta|X,\&)p(\theta|\&). \quad (25)$$

For a given likelihood function $l(\theta\,|\,X,\&)$, a class of prior distributions is said to form a conjugate family if the posterior density is in the same class as the prior. The convenience of using conjugate priors is that the likelihood function will update the distribution's parameters without changing its functional form. The use of conjugate priors greatly simplifies the computational complexity of the analysis, but one must be careful not to betray the details of the problem, which is being analyzed to reduce complexity.

Suppose that you have an unknown parameter θ for which your prior belief can be represented by a normal distribution with mean θ_0 and variance σ_0^2, such that

$$\theta \sim N(\theta_0, \sigma_0^2), \quad (26)$$

and an observation x can be made which is also normally distributed such that

$$x \sim N(\theta, \sigma^2), \quad (27)$$

where the mean equals the parameter of interest θ. For this example, we will assume that θ_0, σ_0^2, and σ^2 are known. If this is the case, then the following are true:

$$p(\theta|\&) = (2\pi\sigma_0^2)^{-\frac{1}{2}} E^{-\frac{1}{2}\frac{(\theta-\theta_0)^2}{\sigma_0^2}}, \quad (28)$$

$$l(\theta|X,\&) = (2\pi\sigma^2)^{-\frac{1}{2}} E^{-\frac{1}{2}\frac{(x-\theta)^2}{\sigma^2}}, \quad (29)$$

$$p(\theta|X,\&) \propto p(X|\theta,\&)p(\theta|\&)$$
$$\propto E^{-\frac{1}{2}\theta^2\left(\frac{1}{\sigma_0^2}+\frac{1}{\sigma^2}\right)+\theta\left(\frac{\theta_0}{\sigma_0^2}+\frac{x}{\sigma^2}\right)}. \quad (30)$$

If we represent the posterior variance as

$$\sigma_1^2 = \left(\frac{1}{\sigma_0^2} + \frac{1}{\sigma^2}\right)^{-1} \quad (31)$$

and the mean as

$$\theta_1 = \sigma_1^2\left(\frac{\theta_0}{\sigma_0^2} + \frac{x}{\sigma^2}\right), \quad (32)$$

then

$$p(\theta|X,\&) \propto E^{-\frac{1}{2}\left(\frac{\theta^2}{\sigma_1^2}\right)+\theta\left(\frac{\theta_1}{\sigma_1^2}\right)}, \quad (33)$$

which, upon adding the constant $-1/2\,\theta_1^2/\sigma_1^2$ into the exponent, and normalizing so that the density integrates to unity is equal to

$$p(\theta|X,\&) \propto (2\pi\sigma_1^2)^{-\frac{1}{2}} E^{-\frac{1}{2}\frac{(\theta-\theta_1)^2}{\sigma_1^2}}. \quad (34)$$

Therefore, the posterior is also a normal distribution

$$\theta|x \sim N(\theta_1, \sigma_1^2). \quad (35)$$

Thus, we have a concise and clear manner in which to update a normal prior via a normal likelihood function. The result is a posterior that is also normal, with $\sigma_1^2 = 1/(1/\sigma^2 + 1/\sigma_0^2)$ and $\theta_1 = \sigma_1^2(\theta_0/\sigma_0^2 + x/\sigma^2)$. Similarly, if x represents a set of observations rather than one specific

Table 1. Some Common Conjugate Priors

Likelihood	Conjugate Prior	Posterior
$l(\theta\|X) = p(X\|\theta)$	$p(\theta)$	$p(\theta\|X)$
Normal	Normal	Normal
Binomial	Beta	Beta
Poisson	Gamma	Gamma
Multinomial	Dirichlet	Dirichlet

observation, the posterior remains a normal with

$$\sigma_1^2 = \left(\frac{1}{\sigma_0^2} + \frac{n}{\sigma^2}\right)^{-1}, \quad (36)$$

$$\theta_1 = \sigma_1^2\left(\frac{\theta_0}{\sigma_0^2} + \sum_i \frac{x_i}{\sigma^2}\right). \quad (37)$$

Conjugate priors greatly simplify the assessment and calculation process. Table 1 provides common conjugate priors.

5.1. Example: A Normally Distributed Prior and Likelihood

Suppose that two separate tests are performed to determine the level of blood urea nitrogen (BUN). Upon initial diagnosis, the belief about the BUN could be represented by a normal distribution with a mean of 20 mg and a variance of 5 mg. Thus, our prior belief can be represented by N(20,5). A set of laboratory tests were conducted that suggested that the BUN level was a normal distribution with a mean of 16 mg and a variance of 4 mg.

If we use the method presented above, then we see that the posterior belief can be represented by $\theta|x \sim N(\theta_1, \sigma_1^2)$, where

$$\sigma_1^2 = 1/(1/\sigma^2 + 1/\sigma_0^2) = 1/(1/5 + 1/4) = 2.2, \quad (38)$$

$$\theta_1 = \sigma_1^2(\theta_0/\sigma_0^2 + x/\sigma^2) = 2.2 * (20/5 + 16/4) = 17.8. \quad (39)$$

Thus, the posterior belief about the BUN is $\theta|x \sim N(17.8, 2.2)$.

5.2. Example: A Gamma Prior with Poisson Likelihood

Suppose that a particular drug is believed to have some probability of an adverse effect. Consider also that the likelihood of an adverse effects increases as the number of doses taken increases. In such a scenario, the probability distribution for an adverse reaction is typically modeled with a Poisson distribution:

$$Poisson(n|\lambda, t, \&) = \frac{e^{-\lambda t}(\lambda t)^n}{n!}, \quad (40)$$

where n is the number of incidents, λ is the rate of incidents per dose, and t is the number of doses. This Poisson distribution represents the likelihood that "n" adverse effects will occur within "t" doses. One would like to arrive at a posterior probability, after a specific number of tests, which can then be used for any hypothesis testing or decision making that is relevant to the drug's interaction likelihood. The natural choice for the prior and posterior probability is the Gamma distribution, because it is the conjugate prior to the Poisson. The Gamma distribution represents a belief about the rate of incidence given a particular number of incidents within a specific number of doses. The Gamma prior takes the following form:

$$Gamma(\lambda|N, T, \&) = \frac{e^{-\lambda T}(\lambda t)^{N-1} T}{(N-1)!}, \quad (41)$$

whereas the posterior takes the form:

$$Gamma(\lambda|N, T, n, t, \&)$$
$$= \frac{e^{-\lambda(T+t)}(\lambda(T+t))^{N+n-1}(T+t)}{(N+n-1)!}. \quad (42)$$

Thus, the posterior is of the same family as the prior, whereas the parameters of the Gamma distribution have shifted from N, T to $N+n, T+t$.

5.3. Example: A Beta Prior with Binomial Likelihood

Suppose that a particular drug is believed to succeed in curing an individual a specific portion of the time, but that portion is currently unknown. We would like to perform testing on this drug to establish a posterior probability that can then be used for any hypothesis testing or decision making that is relevant to the drug's use. These tests will each result in a pass/fail determination, so therefore the likelihood function is a binomial distribution. The natural choice for the prior and posterior probability is the Beta distribution, because it is the conjugate prior to the binomial. The binomial distribution and the Beta prior take the following forms:

$$b(n|N, p, \&) = \binom{N}{n} p^n (1-p)^{N-n}, \quad (43)$$

$$Beta(p|\alpha, \beta, \&) = \frac{1}{\beta(\alpha, \beta)} p^{\alpha-1}(1-p)^{\beta-1}, \quad (44)$$

where $\beta(r, s)$ is the Beta function with parameters r and s. The posterior probability is also a Beta distribution, with the new form being

$$Beta(p|\alpha, \beta, n, N, \&)$$
$$= \frac{1}{\beta(\alpha+n, \beta+N-n)} p^{n+\alpha-1}(1-p)^{N-n+\beta-1}. \quad (45)$$

Much like with the Gamma example, the Beta prior and binomial trial lead to a Beta posterior. The new distribution reflects the prior belief, as well as the evidence from the data inherent in the binomial likelihood function. The parameters of the Beta distribution have shifted from α, β to $\alpha+n, \beta+N-n$.

Consider the following example, based on Howard (4). Consider the long-run fraction of "Heads" in tosses of a coin. Suppose we have no further information about the coin and assign a Beta(50,50) as our prior distribution for the long-run fraction. This particular Beta gives us a mean of 0.5 and a variance of 0.0025. Now suppose the coin is flipped 100 times and 54 of the tosses yield a heads. The likelihood function is represented by the binomial distribution of Bernoulli trials with $r = 54$ and $N = 100$. As the binomial likelihood is conjugate to the Beta prior, the posterior probability for the long-run fraction after observing the trials is therefore Beta(104, 96). The new mean is now 0.52. The belief about the long-run fraction of the coin is no longer symmetric as before, but it may or may not lead to a different course of action. In the remaining section, we will address this question of changes in belief, by examining the issue of hypotheses testing and decision making from a Bayesian perspective.

6. HYPOTHESIS TESTING

Classic statistics places a large emphasis on hypothesis testing. Particularly in the area of scientific inquiry, the tendency is to test hypothesis in the traditional sense, e.g., to determine whether two treatments will yield different results or whether an observed entity is a member of a classification based on specific parameters. In classic statistics, testing is usually performed on a null hypothesis H_0 (that which must be disproved) and the alternative hypothesis H_1 (which is typically "not H_0"). The goal is to investigate the probability of getting a set of data given the hypothesis. In the Bayesian view, the goal is to investigate the likelihood of various hypotheses given a set of data (and a prior belief).

The challenge with classic hypothesis testing is that it is highly sensitive to the choice of the null hypothesis. It is important to note that in classic hypothesis testing, there is an asymmetry; you do not necessarily come to the same conclusion if you switch the null and alternative hypotheses. In classic statistics, a decision regarding what test to use on a hypothesis is made to balance the probability of a type I error (rejecting a true null hypothesis) and a type II error (failing to reject a false null hypothesis). Due to the nature of the null/alternative hypotheses approach, it is common that the smaller the probability of a type I error, the larger the probability of the type II error, and vice versa.

In Bayesian hypothesis testing, the question of testing is one of determining the likelihood of various hypotheses, and is only relevant when different conclusions will lead to different actions on the part of the tester. Otherwise, there is no reason to perform a test and the prior belief about each hypothesis will be maintained. We begin by assuming that there are two hypotheses H_0 and H_1. It is also assumed that an action B_0 is best if H_0 is true and action B_1 is best if H_1 is true. Therefore, our problem has two parts. The first step is to determine the probability of each case being true: $p(H_0|\&)$ and $p(H_1|\&)$. The second step is to determine the consequences (gains or losses) associated with successes and errors.

Typically, we refer to the null and alternative hypothesis as

$$H_0 : \theta \in \Theta_0, \quad (46)$$

$$H_1 : \theta \in \Theta_1. \quad (47)$$

In the Bayesian approach, our goal is to calculate the posterior probabilities given the test data (X), which are

$$p_0 : p(\theta \in \Theta_0 | X, \&), \qquad (48)$$

$$p_1 : p(\theta \in \Theta_1 | X, \&), \qquad (49)$$

where $p_0 + p_1$ sum to one only when the hypotheses are mutually exclusive and collectively exhaustive. The posterior probabilities provide us with a means through which to judge the likelihood of each hypothesis. Posterior odds, p_0/p_1, are the ratio of the likelihood of the null hypothesis to the likelihood of the alternative hypothesis. If the ratio is low, odds are in favor of the alternative; likewise, if the odds are high, then the null hypothesis is viewed as more likely.

As we have shown, to arrive at a posterior probability, we will need the prior probabilities

$$\pi_0 = p(\theta \in \Theta_0 | \&) \qquad (50)$$

$$\pi_1 = p(\theta \in \Theta_1 | \&) \qquad (51)$$

and the likelihood function. In this situation, it is preferred to use, instead of the likelihood function, the likelihood ratio

$$\frac{p(x|\theta \in \Theta_0, \&)}{p(x|\theta \in \Theta_1, \&)} = \frac{l(\theta \in \Theta_0 | X, \&)}{l(\theta \in \Theta_1 | X, \&)}. \qquad (52)$$

The likelihood ratio (in classic and Bayesian statistics) is viewed as the odds in favor of H_0 and H_1 given the data X. Recall also that the posterior odds is the product of the prior odds and the likelihood ratio.

We may also choose to work with the Bayes's factor, which is the ratio of the posterior odds to the prior odds. When the two hypotheses are mutually exclusive and collectively exhaustive, $\pi_0 = 1 - \pi_1$ and $p_0 = 1 - p_1$, and the Bayes's factor equals the likelihood ratio. In this situation, the Bayes's factor is, as mentioned, the odds in favor of H_0 over H_1 given the data.

When the hypotheses are not so simple, which would indicate either overlapping hypotheses or a hypothesis consisting of a set of possibilities, the Bayes's factor must be computed as follows:

$$B = \frac{p_0/p_1}{\pi_0/\pi_1} = \frac{\int_{\theta \in \Theta_0} l(\theta \in \Theta_0 | X, \&) p(\theta | \&)/\pi_0 d\theta}{\int_{\theta \in \Theta_1} l(\theta \in \Theta_1 | X, \&) p(\theta | \&)/\pi_1 d\theta}. \qquad (53)$$

This factor does not provide the interpretation given by the likelihood ratio, although it may still yield some insight into the weight of the evidence as it relates to the hypothesis. One highly useful interpretation of the Bayes's factor is that it is the amount by which the beliefs were altered by the experiment or data. Together, the Bayes's factor and the posterior odds provide a powerful method by which hypotheses can be tested.

6.1. Example: Testing a Simple Hypotheses — A Fair Die

Consider the following example, based on an example found in von Winterfeldt and Edwards (5). Suppose the fairness of a die was called into question. It is believed that either it produces 6 s with a frequency of 1/6 (H_0) or 1/5 (H_1). The ratio of H_0/H_1, on the basis of observing a 6, is (1/6)/(1/5) = 5/6. Likewise, the ratio of H_0/H_1, on the basis of observing anything other than a 6, is (5/6)/(4/5) = 25/24. Although the prior odds are your prerogative, let us assume that you feel that the die is twice as likely to be fair as unfair, and thus the prior odds equal 2. If the die results in the following sequence, what is the new likelihood ratio?

$$4, 6, 2, 4, 1, 3, 6, 2, 5, 6, 1, 3, 5, 4.$$

For this example, the posterior odds are

$$\frac{p_0}{p_1} = 2 * \left(\frac{5}{6}\right)^3 * \left(\frac{25}{24}\right)^{11} = 1.8, \qquad (54)$$

and the Bayes's factor is

$$\frac{p_0/p_1}{\pi_0/\pi_1} = \left(\frac{5}{6}\right)^3 * \left(\frac{25}{24}\right)^{11} = 0.9. \qquad (55)$$

The die is still more likely to be fair than unfair, but the die is deemed less likely to be fair than was originally believed. Additional tests may lead to the alternative hypothesis being accepted.

6.2. Example: Testing a Simple Null Hypothesis—To Treat or Not?

Recall the clinical test for BUN. The posterior belief about the BUN was $\theta|x \sim N(17.8, 2.2)$. Suppose that the patient in question is currently undergoing dialysis, and that a BUN of 19 or higher would indicate a need to have dialysis performed twice as often, or face possible kidney failure. If we assume that the cost of kidney failure is ten times the cost of increased dialysis, should the patient have dialysis performed more often?

To guide our decision making, we must first determine the likelihood that the BUN is greater than 19. Our null hypothesis is that the patient has a BUN of 19 or higher. We begin by calculating the area under the normal curve that is greater than 19:

$$p_0 = p_{\theta|x}(\theta \geq 19) = p\left(Z \geq \frac{19 - 17.8}{\sqrt{2.2}}\right) \qquad (56)$$
$$= 1 - \Phi(0.809) = 0.21,$$

where $Z = \frac{\theta - 17.8}{\sqrt{2.2}}$ is a standard normal random variable and Φ is the cumulative distribution function of the standard normal distribution.

The posterior probability is interpreted to mean that there is a 21% likelihood that the BUN is greater than 19. If we establish a threshold, based on the costs given above as

$$\frac{1}{1+10} = 0.091, \qquad (57)$$

then we see that, because the $p_0 = 0.21$ is greater than the threshold (0.091), one would choose to increase the rate of dialysis. Notice that if the costs were equal rather than differing by a power of ten, the threshold would have been

0.5, and the hypothesis would be rejected and the dialysis treatment rate would not be increased.

As a matter of completeness of analysis, we can see that the posterior odds would be $0.21/(1 - 0.21) = 0.266$. The prior odds would have been 2.05, leaving a Bayes's factor of 0.13, indicating that the collected data were in favor of rejecting the null hypothesis. This is intuitive given that the prior mean was 20.

6.3. Bayesian View on the Point Null Hypothesis

The examples of hypothesis testing provided above were of the two more common types. The first is the "simple" hypothesis that tests two competing independent hypotheses whose likelihoods sum to one. The other is called a "one-tailed" test due to the fact that we are looking to determine whether a value is greater than or equal to some threshold hypothesized value. The other common scenario is that of the "two-tailed" hypothesis test, or the point hypothesis. In this scenario, the null hypothesis is that a parameter will be exactly equal to a provided value.

This approach is problematic for the Bayesian, because priors are typically continuous distributions and will therefore assign a likelihood of zero to any exact value. Classic statisticians admit that it will essentially never be the case that some $\theta = \theta_0$. The Bayesian can take several approaches to this problem. The first is to translate the problem to a one-tailed problem whenever possible. The second would be to change the hypothesis from an exact value to an appropriately small range of values $(\theta_0 - \varepsilon, \theta_0 + \varepsilon)$. The challenge with this approach is determining just how small the range of values should be.

A third approach to this problem requires that the Bayesian not use a continuous distribution as the prior. Other options include discrete and hybrid distributions that would assign a specific probability π_0 to the likelihood of achieving the exact value, and distribute the rest of the probability $\pi_1 = 1 - \pi_0$ across a range of values that might appear in the test.

After the third approach, if we define the spread distribution as $\zeta(\theta)$, then the prior is

$$\pi(\theta) = \delta_{\theta_0}\pi_0 + \pi_1\zeta(\theta), \tag{58}$$

where δ_{θ_0} is the Dirac delta with mass at θ_0. The posterior probability is then

$$\pi(\theta = \theta_0|x, \&) = \frac{p(x|\theta_0, \&)\pi_0}{\pi_0 p(x|\theta_0, \&) + \pi_1 \int p(x|\theta, \&)\zeta(\theta)d\theta}. \tag{59}$$

The Bayes's factor for this type of hypothesis is

$$B = \frac{p_0/p_1}{\pi_0/\pi_1} = \frac{p(x|\theta_0, \&)}{\int p(x|\theta, \&)\zeta(\theta)d\theta}. \tag{60}$$

6.4. Example: Testing a Point Hypothesis and Lindley's Paradox

Consider the following example, in the vein of Lindley's Paradox (6), as explained in Ref. 7. Suppose that a sampling $x = (x_1, x_2, \ldots, x_n)$ is taken from a distribution such that $\bar{x} = N(\theta, \varphi)$, with a known variance φ. We assume also that the spread distribution $\zeta(\theta)$ is also normal, with $N(\mu, \psi)$. It will be assumed that $\mu = \theta_0$ because we are assuming that the mean is the most likely value.

By writing $\bar{x} = (\bar{x} - \theta) + \theta$, we can see that the marginal density for $\bar{x}$ is $\int p(\bar{x}|\theta, \&)\zeta(\theta)d\theta$. Thus, because $x - \theta \sim N(0, \varphi/n)$ and $\theta \sim N(\theta_0, \psi)$, $\bar{x}$ has a density of $N(\theta_0, \psi + \varphi/n)$.

The Bayes's factor B is then

$$B = \frac{p(\bar{x}|\theta_0, \&)}{\int p(\bar{x}|\theta, \&)\zeta(\theta)d\theta} = \sqrt{(1 + n\psi/\varphi)}$$
$$* E^{-\frac{1}{2}z^2(1+\varphi/n\psi)^{-1}}, \tag{61}$$

where Z is again the standard normal. The posterior probability is

$$p_0 = \frac{1}{1 + ((1-\pi_0)/\pi_0)B^{-1}} = 0.40. \tag{62}$$

Lee (7) shows that if $\pi_0 = 1/2$ and $\varphi = \psi$, then the values $Z = 1.96$ and $n = 15$ (n being the number of samples) give rise to a Bayes's factor of 0.66 and a posterior probability of 0.4.

Notice that, from a classic hypothesis testing standpoint, a Z value of 1.96 would arise with a probability of 0.05, which would lead to a rejected hypothesis at the 5% level. Meanwhile, the posterior probability indicates that there is a 0.40 likelihood that the null hypothesis is true. Even more interestingly, as n reaches 100, the posterior probability increases to 72.5%. The Bayes's factor is an increasing function in n. As $n \to \infty$, $B \to \infty$ and because p_0 is of the order $1/\sqrt{n}$, we also note that $p_0 \to 1$. Notice that as n increases, the evidence gets stronger for the Bayesian; yet the classic statistician would continue to reject the null hypothesis when $Z = 1.96$, regardless of the sample size.

Lindley's Paradox shows that a two-tailed significance test may reveal significance at the 0.05 level, whereas the posterior probability (even with quite small priors) is as high as 0.95. Lindley (6) points out that the "common-sense interpretations" are in direct conflict and should cast serious doubt on use of significance levels when testing point null hypotheses.

7. SUMMARY

Classic statistics, the "frequentist" version of statistics based on long-run fractions, has enjoyed a mislabeling for a considerable time. It is important to note that "classic" statistics is considerably younger than the Bayesian view of statistics. Statistical tools such as p values, significance levels, and confidence intervals grew out of the frequentist statistical camp, but all have their parallels in Bayesian statistics. Many agree that the Bayesian approaches, which focus on beliefs and observed data, rather than expected results of repeated trials, are more natural. It is only through our training (most likely in our university courses) that we are taught to consider the classic methods as natural.

BIBLIOGRAPHY

1. G. A. Barnard and T. Bayes, Studies in the history of probability and statistics: IX. Thomas Bayes's essay toward solving a problem in the doctrine of chances. *Biometrika*. 1958; **45**(3/4):293–315.
2. R. T. Cox, Probability, frequency, and reasonable expectation. *Am. J. Phys.* 1946; **14**:1–13.
3. E. T. Jaynes, *Probability Theory: The Logic of Science*. Cambridge, U.K.: Cambridge University Press, 2003.
4. R. A. Howard, Decision analysis: Perspectives on inference, decision, and experimentation. *Proc. IEE*. 1970; **58**(5):823–834.
5. D. von Winterfeldt and W. Edwards, *Decision Analysis and Behavioral Research*. Cambridge, U.K.: Cambridge University Press, 1986.
6. D.V. Lindley, A statistical paradox. *Biometrika*. 1957; **44**: 187–192.
7. P. M. Lee, *Bayesian Statistics: An Introduction*. London: Edward Arnold, 1989.

BIOACOUSTIC SIGNALS

LEONTIOS J. HADJILEONTIADIS
STAVROS M. PANAS
Aristotle University of Thessaloniki
Thessaloniki, Greece

IOANNIS T. REKANOS
Technological and Educational Institute of Serres
Serres, Greece

1. INTRODUCTION

Bioacoustic signals (BAS) are defined as a family of signals that includes the sounds produced by the human body, whereas, in general, the study of sounds produced by living organisms is referred to as bioacoustics (1). As BAS are the audible outcome emitted from the human body, they possess valuable diagnostic information regarding the functionality of the human organs involved in the sound production mechanisms.

An everyday-life example of BAS is the pressure signals produced from the beating of the heart, namely heart sounds (HS). To obtain HS, a stethoscope is typically placed on the chest above the heart. The end of the stethoscope has a diaphragm that bends in response to sound; the sound waves travel through the tubes of the stethoscope to the clinician's ears. The clinician listens to these sounds to determine certain aspects of the mechanical functioning of the heart (mostly whether the heart valves are functioning properly or not), of which the clinician has been trained to detect. This measurement is qualitative. A microphone can also be used to pick up sound. Microphones convert the mechanical energy of sound into electric energy. The resulting analog electric signal can be digitized and analyzed quantitatively with computer-based algorithms.

Consequently, proper extraction and use of the diagnostic information of the digitized BAS can transform the art of BAS auscultation into a scientific discipline. With proper signal processing methods, the BAS could be converted from acoustic vibrations inside the human organism into graphs and parameters with diagnostic value, resulting in novel diagnostic tools that objectively track the characteristics of the relevant pathology and assist the clinicians in everyday practice.

Apart from HS, the BAS family also includes lung sounds (LS), bowel sounds (BS), and joint sounds (JS), with LS and HS being the most often used BAS in everyday clinical practice, although, recently, sufficient research effort has been placed on BS and JS analysis, facilitating their adoption by the physicians as diagnostic means.

2. DEFINITION, CATEGORIZATION, AND MAIN CHARACTERISTICS OF BAS

2.1. Lung Sounds

2.1.1. Definition and Historical Background. From the time of ancient Greeks and their doctrine of medical experimentation until at least the 1950s, the LS were considered as *the sounds originating from within the thorax* and they were justified mainly on the basis of their acoustic impression. For example, the writings of the Hippocratic School, in about 400 B.C., describe the chest (lung) sounds as splashing, crackling, wheezing, and bubbling sounds emanating from the chest (2). An important contribution to the qualitative appreciation of LS and HS was the invention of the stethoscope by René Theophil Laënnec in 1816. Laënnec's gadget, which was originally made of wood, replaced the "ear-upon-chest" detection procedure enhancing the emitted LS and HS (3). Unfortunately, it took over a century for quantitative analysis to appear. Attempts for a quantitative approach date to 1930, but the first systematic, quantitative measurement of their characteristics (i.e., amplitude, pitch, duration, and timing in controls and in patients) is attributed to McKusick et al., in 1953 (4); the door to the acoustic studies in medicine was finally opened.

2.1.2. Categorization and Main Characteristics. The LS are divided into two main categories [i.e., *normal* LS (NLS) and *abnormal* LS (ALS)]. The NLS are certain sounds heard over specific locations of the chest during breathing in healthy subjects. The character of the NLS and the location at which they are heard defines them. Hence, the category of the NLS includes the following types (5):

- *tracheal* LS (TLS), heard over the trachea having a high loudness,
- *vesicular* LS (VLS), heard over dependent portions of the chest, not in immediate proximity to the central airways,
- *bronchial* LS (BLS), heard in the immediate vicinity of central airways, but principally over the trachea and larynx,

- *bronchovesicular* LS (BVLS), which refers to NLS with a character in between VLS and BLS heard at intermediate locations between the lung and the large airways, and
- *normal crackles* (NC), inspiratory LS heard over the anterior or the posterior lung bases (6,7).

The ALS consist of LS of a BLS or BVLS nature that appear at typical locations (where VLS are the norm). The ALS are categorized between *continuous adventitious sounds* (CAS) and *discontinuous adventitious sounds* (DAS) (8), and include the following types (9–11):

- *wheezes* (WZ), musical CAS that occur mainly in expiration and invariably associated with airway obstruction, either focal or general,
- *rhonchi* (RH), low-pitched sometimes musical CAS that occur predominantly in expiration, associated more with chronic bronchitis and bronchiectasis than with asthma,
- *stridors* (STR), musical CAS that are caused by a partial obstruction in a central airway, usually at or near the larynx,
- *crackles*, discrete, explosive, nonmusical DAS, further categorized between:

  • *fine crackles* (FC), high-pitched exclusively inspiratory events that tend to occur in mid-to-late inspiration, repeat in similar patterns over subsequent breaths and have a quality similar to the sound made by strips of Velcro being slowly pulled apart; they result from the explosive reopening of small airways that had closed during the previous expiration, and
  • *coarse crackles* (CC), low-pitched sound events found in early inspiration and occasionally in expiration as well, develop from fluid in small airways, are of a "popping" quality, and tend to be less reproducible than the FC from breath to breath,

- *squawks* (SQ), short, inspiratory wheezes that usually appear in allergic alveolitis and interstitial fibrosis (12), predominantly initiated with a crackle, caused by the explosive opening and fluttering of the unstable airway that causes the short wheeze, and
- *friction rub* (FR), DAS localized to the area overlying the involved pleura and occur in inspiration and expiration when roughened pleural surfaces rub together, instead of gliding smoothly and silently.

As the preceding paragraph has demonstrated, it is not difficult to provide evidence that the LS are directly related to the condition of the pulmonary function. The variety in the categorization of LS implies changes in the acoustic characteristics either of the source or the transmission path of the LS inside the lungs because of the effect of a certain pulmonary pathology. It is likely that the time- and frequency-domain characteristics of the LS signals reflect these anatomical changes (13).

In particular, the time-domain pattern of the NLS resembles a noise pattern bound by an envelope, which is a function of the flow rate (13). Tracheal sounds have higher intensity and a wider frequency band (0–2 kHz) than the chest wall sounds (0–600 Hz) and contain more acoustic energy at higher frequencies (14). The CAS time-domain pattern is a periodic wave that may be either sinusoidal or a set of more complex, repetitive sound structures (13). In the case of WZ, the power spectrum contains several peaks ("polyphonic" WZ), or a single peak ("monophonic" WZ), usually in the frequency band of 200–800 Hz, indicating bronchial obstruction (13). Crackles have an explosive time-domain pattern, with a rapid onset and short duration (13). It should be noted that this waveform may be an artifact of high-pass filtering (15). Their time-domain structural characteristics (i.e., a sharp, sudden deflection usually followed by a wave) provide a means for their categorization between FC and CC (16), as it is shown in Fig. 1.

For an extensive description and a variety of examples regarding the LS structure and characteristics, the reader should refer to (13).

2.2. Heart Sounds

2.2.1. Definition and Historical Background.
HS are defined as *the repetitive "lub-dub" sounds of the beating of the heart* (17). Heart auscultation followed similar pathways with lung auscultation because of the topological coexistence of the heart with lungs. Hippocrates (460–377 B.C.) was familiarized with heart auscultation, and he may have used HS for diagnostic purposes (18). Nevertheless, it took almost two thousand years for re-evaluation of HS by William Harvey (1578–1657), and three hundred years more, with the contribution of Laënnec's stethoscope (1816), for Dr. Joseph Skoda (1805–1881) first to describe the cardiac sounds and murmurs, by pinpointing their locations and defining the clinical auscultatory signs that have allowed the noninvasive diagnosis of cardiac pathology via auscultation (18).

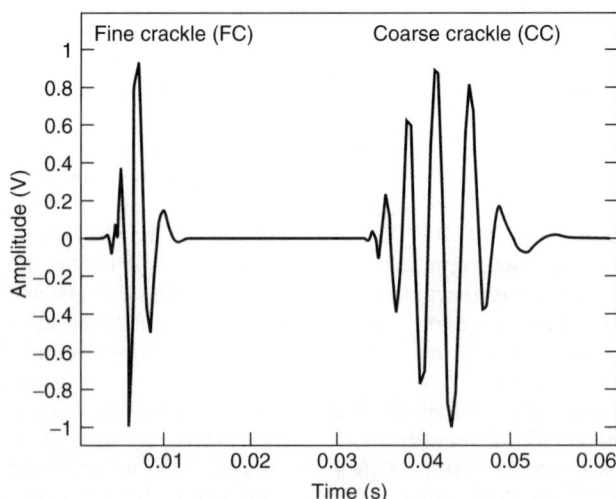

Figure 1. An example of the crackles morphology in the time-domain used for their categorization into fine and coarse crackles.

2.2.2. Categorization and Main Characteristics. The HS are named according to their sequence of occurrence and are originated at specific points in the cardiac cycle (18). In particular, the HS are categorized as follows:

- *first heart sound*, or S_1, occurs at the beginning of ventricular systole when ventricular volume is maximal and is considered as normal HS,
- *second heart sound*, or S_2, occurs at the end of ventricular systole and is considered as normal HS too,
- *third heart sound*, or S_3 or "S_3 gallop," occurs just after the S_2, as a result of decreased ventricular compliance or increased ventricular diastolic volume, and is a sign of congestive heart failure,
- *fourth heart sound*, or S_4 or "S_4 gallop," occurs just before the S_1, as a result of decreased ventricular compliance or increased volume of filling, and is a sign of ventricular stress, and
- *murmurs*, sustained noises that are audible during the time periods of systole, diastole, or both, associated with backward regurgitation, forward flow through narrowed or deformed valve, a high rate of blood flow through normal or abnormal valves, vibration of loose structures within the heart (chordaetendineae), and continuous flow through A-V shunts. They are further categorized between:

  - *systolic murmurs* (SM), sustained noises that are audible between S_1 and S_2, categorized as *early* SM, which begin with S_1 and peak in the first third of systole; *mid* SM (or "ejection" murmur), which begin shortly after S_1, peak in mid-systole, and do not quite extend to S_2; *late* SM, which begin in the later one-half of systole, peak in the later third of systole, and extend to S_2; and *pansystolic murmurs*, which begin with S_1 and end with S_2, thus heard continuously at about constant amplitude throughout systole, and
  - *diastolic murmurs* (DM), sustained noises that are audible between S_2 and the next S_1, categorized as *early* DM, which begin with S_2 and peak in the first third of diastole; *mid* DM, which begin after S_2 and peak in mid-diastole; *late* DM, which begin in the latter one-half of diastole, peak in the later third of diastole, and extend to S_1; and *pandiastolic murmurs*, which begin with S_2 and extend throughout the diastolic period.

The acoustic content of HS is concentrated in low frequencies (19). In fact, over 95% of the acoustic energy of S_1 and 99% of the one of S_2 is concentrated under the 75 Hz (20). The average spectrum decays exponentially from its peak at approximately 7 Hz and often contains one or more shallow and wide peaks (21). The S_2 contains much more energy in the lower frequencies than S_1 does. In addition, the frequency content of most cardiac murmurs is also in the low range (20). The two most common HS (i.e., S_1 and S_2) have much longer duration and thus a lower pitch than crackles, and both of them are short-lived compared with NLS and CAS (17), which contributes to the facilitation of their separation from the LS. Nevertheless, HS from patients with irregular cardiac rhythms (transient character) or loud murmurs tend to overlap the LS more and thus are less separable.

2.3. Bowel Sounds

2.3.1. Definition and Historical Background. Bowel sounds are defined as *the sounds heard when contractions of the lower intestines propel contents forward* (22). Knowledge of BS has advanced little since Cannon's pioneering work in 1902 (23), which used the sounds as a way of studying the mechanical activity of the gastrointestinal tract. Clinical tradition assesses the BS in a "passive" way (i.e., by tracing not their presence but their absence) because the latter is an indicator of intestinal obstruction or ileus (paralysis of the bowel) (22). This lack of interest in "active" abdominal auscultation is due, in part, to its lack of support in scientific fact and definitely not because of its lack of diagnostic information. Bowel sound patterns in normal people have not been clearly defined, as only a small number of them have been studied (24–28). In addition, the trivial signal processing methods that have been involved (29–31) have also been a problem. The vague notion about the usefulness of the BS in clinical practice that has been cultivated all these years, along with the lack of a worldwide-accepted BS categorization, has resulted in a reduced interest in their processing and evaluation, even today.

2.3.2. Categorization and Main Characteristics. Up to now, no reference of what can be considered as normal bowel sound activity, and no physiological understanding of the significance of different types of BS exist. Attempts to describe the acoustic impression of normal BS, such as "rushes," "gurgles," etc., fail because of their subjectivity bias. Sporadic works in the literature try to analyze the time- and frequency-domain characteristics of BS, defining as normal BS those with a frequency content in the range of 100–1000 Hz, with durations within a range of 5–200 msec, and with widely varying amplitudes (32). An analysis of the "staccato pop," which is one of the most common BS characteristic of the colon, proves that it has a frequency content between 500 and 700 Hz and a duration of 5 to 20 msec (32).

Recent work based on the differences seen between the sound-to-sound intervals of BS from different pathologies (i.e., irritable bowel syndrome, Crohn's disease, and controls) introduces a time-domain-based tool for relating BS to the associated bowel pathology (33). In a similar vein, the approaches of BS presented in this chapter contribute to the provision of an extended, accurate, and objective alternative to the current BS processing and evaluation status.

2.4. Joint Sounds

2.4.1. Definition and Historical Background. JS are defined as the sounds heard during the functioning of a joint. In particular, the sound heard when our knuckles crack results when a finger joint is extended almost to the end of its range. The joint, surrounded by a lubricating fluid, is encased in a capsule. At times, we intentionally or unintentionally extend the joint so far that the gas dissolved in the fluid spontaneously separates from the solution, forming a small bubble, and making a cracking sound. Not until the fluid reabsorbs the gas can the sound be reproduced. On the other hand, the sound heard when we do deep-knee bends is a snapping sound, produced when our tendons, which are merely the fibers that connect muscles and bones, elastically snap into new positions as our joints move under stress. Because the tendons shift back and forth with the movement of the joint, no waiting time exists before this snapping sound can be reproduced.

Another interesting joint that produces JS is the temporomandibular (jaw) joint (TMJ), which is the "ball and socket" joint that allows the lower jaw to open, close, and move sideways when chewing and speaking. Located about one centimeter in front of the ears, they are the only joints in the head. The ball is technically known as the "condyle" of the joint, and it rotates in a cuplike depression (the socket) technically known as the "fossa." Although the joint looks like it is attached directly to the sinuses, it is actually separated from them by soft tissue ligaments that entirely enclose the joint. The "meniscus" is a disk of cartilage that lives in the space between the condyle and the fossa and is capable of moving forward and backward as the jaw opens and closes. The condyle and the fossa are each covered with a thin layer of nonmovable cartilage of their own. All three layers of cartilage help to provide smooth, frictionless surfaces for comfortable joint operation. TMJ sounds (TMJS) are produced in the form of clicking or crepitation. In many cases, the cartilage that separates the ball from the socket tends to tear and displace so that it bunches up in front of the ball. When opening the jaw wide, the ball moves forward pushing the bunched up cartilage in front of it. At some point in this forward movement, the ball jumps over the mass of cartilage snapping back hard onto the bone on the other side causing a loud pop or clicking sound. Sometimes, the bones are forced into such close approximation that you might hear a grinding noise (crepitus) when opening or closing (34).

Robert Hooke is credited with the first auscultation of a joint in the Seventeenth century and for the first suggestion that joint noises could be used as diagnostic sign in patients suffering from painful joints. Heuter described, in 1885, a "myo-dermato-osteophone" to localize loose bodies in joints (35). In 1902, Blodgett reported on auscultation of the knee, with attention to sounds of apparently normal joints, the change in sounds with repetitive motions, and reproducibility over several days (36). He notes an age-related increase of sound and the relative absence of sound when an effusion was present. Bicher, in 1913, reported that each type of meniscal injury emitted a distinctive sound signal (37). A paper on the *Value of Joint Auscultation* was published in 1929 by Walters, in which joint sounds recorded from 1600 patients were graded and correlated with presumed pathology (38). The results were said to help in the diagnosis of painful joints, with limited physical findings and normal "skiagrams." These studies were expanded by Erb, who used a contact microphone to reduce extraneous sounds (39). His was the first attempt at electronic and graphic recording of the knee joint sound (KJS) signal. A "cardiophone" was used by Steindel to listen to sounds at several locations (40), and he appears to be the first to have recorded joint angles and the first to include filtering to remove noise and improve the signal. He classified the JS based on the pitch, amplitude, and sequence of the sounds, which in many cases were correlated with pathology demonstrated by operative findings. In 1953, Peylan reported on a study of 214 patients with several types of arthritis using a regular and an electronic stethoscope (41). He believed that he could distinguish between periarticular sounds, osteoarthritis, and rheumatoid arthritis, but did not present evidence of any "blind" evaluation of this ability. Fischer and Johnson showed, in 1960, that sound signals could be detected in rheumatoid arthritis before x-ray changes were observable (42). Chu et al. at Ohio began the true scientific analysis of JS in a series of papers dealing with methods to reduce skin friction and ambient noise; the use of statistical parameters, such as the autocorrelation function of the KJS signal for classification of signals into categories, such as rheumatoid arthritis, degenerative arthritis, and chondromalacia patella; and the relationship between signal acoustic power and articular cartilage damage (surface roughness) (43,44). They also pointed to a number of artifacts (local and ambient) in the acoustic recording of KJS. A good understanding of the nature of the KJS signals, their diagnostic potentials, as well as the problems encountered in practice may be obtained from the work of Mollan et al. in Northern Ireland (45–49).

2.4.2. Categorization and Main Characteristics. From the aforementioned JS, the TMJS and KJS are the most important aspects. According to Widmalm et al. (50), the TMJS are categorized in five types of clicking, i.e.,:

- *Types* 1, 2, *and* 3: short-duration signals with well localization in the time-frequency domain with a single energy peak in the frequency range below 600 Hz, from 600 Hz to 1200 Hz, and above 1200 Hz, respectively.
- *Types* 4 *and* 5: long-duration crepitation signals with nonlocalization in the time-frequency domain. Type 4 sounds contain all of the energy in the frequency region below 600 Hz, whereas type 5 sounds have energy both below 600 Hz and above 600 Hz.

Similar to the case of BS, the categorization of KJS is implied through the related knee pathology. As Mollan et al. note (49), patients with meniscal injuries produce characteristic KJS signals, and that alterations in normal joint crepitus may be a useful indicator of early cartilage degeneration. They report the largest KJS signal on the affected side, with a large displacement, and repetitive appearance at the same knee angle in cycles. In addition, bizarre, irregular KJS signals are noticable in the range of 300 Hz to 600 Hz, produced by degenerate articular cartilage. Furthermore, Frank et al. (51) report sharp bursts (click) in the KJS signal for the case of meniscal lesions, with short-duration energy in the range of 0–200 Hz. Mild chondromalacia (softening of the cartilaginous surface of the inner side of the patella, namely the chondrol) is seen in the KJS signal as long-duration activity in the range of 0–300 Hz. Finally, KJS signals associated with severe chondromalacia are of relatively low frequencies (i.e., 0–100 Hz) because of loose cartilage tissue existing between the rubbing surfaces.

3. RECORDING STANDARDS OF BAS

3.1. Recording Sensor Types

The BAS auscultation is traditionally performed by means of the stethoscope. Listening to BAS by means of a stethoscope involves several physical phenomena, such as vibrations of the chest wall that are converted into pressure variations of the air in the stethoscope, and these pressure variations are then transmitted to the eardrum. However, the stethoscope cannot be used in the quantitative analysis of BAS, because it does not provide any means of signal recording. In addition, it presents a selective behavior in its frequency response instead of a flat one (at least in the area of interest, 40–4000 Hz).

One basic category of recording devices of BAS signals refers to microphones. Whatever the type of microphone, it always has a diaphragm, like in the human ear, and the movement of the diaphragm is converted into an electric signal. Two major microphone approaches exist: the "kinematic" approach, which involves the direct recording of chest-wall movement ("contact sensor") and the "acoustic" approach, which involves the recording of the movement of a diaphragm exposed to the pressure wave induced by the chest-wall movement ("air-coupled sensor"). The chest-wall movements are so weak that a free-field recording is not possible; it is essential to couple the diaphragm acoustically with the chest wall through a closed-air cavity. Whatever the approach, kinematic or acoustic, vibrations must be converted into electric signals using transduction principles of (52): electromagnetic induction (movement of a coil in a magnetic field induces an electric current through the coil); condenser principle (changing the distance between the two plates of a charged capacitor induces a voltage fluctuation); and piezoelectric effect [bending of a crystal (rod, foil) for the induction of an electric charge on the surface].

The other category of recording devices includes piezoelectric accelerometers. A piezoelectric accelerometer applies the piezoelectric effect in such a way that the output voltage is proportional to the acceleration of the whole sensor. Early applications used heavy-weight sensors with high sensitivity and good signal-to-noise ratio. The disadvantages because of its heavy mass are mechanical loading of the surface wall, difficulties with attachment, and a low resonance frequency (well within the band of interest). A piezoelectric accelerometer of very low mass (1 g) has been applied successfully in the past, yet it may be so fragile that routine clinical applications may be difficult (52).

Although both condenser microphones and piezoelectric contact sensors are displacement receivers, the waveforms that they deliver are different because of the coupling differences. Selection criteria of a device should also include size, average lifetime, and maintenance cost. Both sensors possess some disadvantages [i.e., piezoelectric sensors are very sensitive to movement artifacts, for example, by the connecting wire, their characteristics depend on the static pressure against the body surface (53) and they are brittle, whereas condenser microphones need mounting elements that change the overall characteristics of the sound transduction (52)].

3.2. Recording Procedures and Considerations

The most common bandwidth for LS is from 60–100 Hz to 2 kHz when recorded on the chest and from 60–100 Hz to 4 kHz when recorded over the trachea. In addition, recording of adventitious sounds on the chest requires a bandwidth from 60–100 Hz to 6 kHz (52). For HS, the useful bandwidth is from 20 Hz to 150 Hz, whereas for BS, the useful bandwidth is similar to the LS recorded from the chest (i.e. 60–2000 Hz). As a result, sampling frequencies of 5 kHz to 10 kHz and of 1 kHz to 2 kHz are sufficient enough for the acquisition of all LS and HS signals, respectively.

The KJS bandwidth is controversial, and different ranges have been reported in the literature. In particular, Chu et al. mention that the frequency content of the that KJS extends to 20 kHz (43,44) whereas Mollan's work indicates that the KJS signal is predominantly low-frequency in nature (45–49). They also point out the importance of the lower frequencies, which are missed by acoustic microphones but are recorded by contact sensors. The use of the latter obviates the need to record the KJS signals in an anechoic or soundproof chamber, or the need to employ a differential microphone pair for noise cancellation, as in Chu's experimental setup (43,44). Contact sensors are insensitive to background noise and small in size in order to be securely fixed to minimize skin friction (51). Furthermore, multiple sensors can be employed to localize defects by multichannel KJS signal analysis (49).

For the case of TMJS, recent studies (50,54) indicate that the analysis of TMJS needs to cover the whole audible range (from 20 Hz to 20 kHz). Consequently, sampling frequency of 48 kHz is the most preferable. Electret microphones, like the SONY ECM 77, have a flat frequency response from 40 Hz up to 20 kHz in free-space and are usually preferred in the recordings of almost all types of BAS.

For the processing of BAS, an analogue system is required, which consists of a sensor, an amplifier, and filters that condition the signal prior to analogue-to-digital (A/D) conversion (usually with a 12- or 16-bit resolution). A combination of low-pass filters (LPF) and high-pass filters (HPF) in cascade is usually applied. The purpose of using a HPF is to reduce the heart, muscle, and contact noises. The LPF is needed to eliminate aliasing. The amplifier is used to increase the amplitude of the captured signal so that the full A/D converter range can be optimally used, and sometimes to adjust the impedance of the sensor.

Table 1 tabulates the recommendations regarding the acquisition of the LS signals. These recommendations can easily be adapted to the rest of the BAS, taking into account the aforementioned characteristics per signal category.

Table 1. Summary of Recommendations for the Case of LS Signal Acquisition with Piezoelectric or Condenser Sensors.

Sensor specifications	
Frequency response	Flat in the frequency range of the sound. Maximum deviation allowed 6 dB
Dynamic range	>60 dB
Sensitivity	Must be independent of frequency, static pressure, and sound direction
Signal-to-noise ratio	>60 dB ($S = 1\,\text{mV} \cdot \text{Pa}^{-1}$)
Directional characteristics	Omnidirectional
Coupling	
Piezoelectric contact	
Condenser air-coupled	Shape: conical; depth: 2.5–5 mm; diameter at skin: 10–25 mm; vented
Fixing methods	
Piezoelectric	Adhesive ring
Condenser	Either elastic belt or adhesive ring
Noise and interferences	
Acoustic	Shielded microphones; protection from mechanical vibrations
Electromagnetic	Shielded twisted pair or coaxial cable
Amplifier	
Frequency response	Constant gain and linear phase in the band of interest
Dynamic range	>60 dB
Noise	Less than that introduced by the sensor
High-pass filtering	Cut-off frequency 60 Hz; roll-off >18 dB $\cdot$ octave^{-1}; phase as linear as possible; minimized ripple
Low-pass filtering	Cut-off frequency above the upper frequency of the signal; roll-off >24 dB $\cdot$ octave^{-1}; minimized ripple

Source: Ref. 52.

4. BAS ANALYSIS

The original method of BAS interpretation is auscultation of the sound signals by the physician. Modern computer-based analysis extend BAS capabilities by supplying information not directly from the sound perception. Nevertheless, auscultation is still a widespread technique, because it involves the use of the stethoscope. To this end, in most cases, the agreement of an automatic method with physician's interpretation is a basic criterion for its acceptance.

The BAS analysis includes three main categories: noise reduction-detection, modeling, and feature extraction-classification.

4.1. Noise Reduction Detection

Objective analysis of BAS requires a preprocessing procedure regarding the elimination of possible noise effect in the recorded signal, which mainly refers to the elimination of the impact of the HS on the LS recordings, elimination of the background noise from BAS recordings or detection of BAS with transient character, such as DAS or BS bursts, and reduction of interference in different representations of BAS, like the TMJS or HS time-frequency representation. The above cases are met in the routine recordings of BAS; hence, advanced signal processing techniques are needed in order to efficiently circumvent these noise effects.

4.1.1. HS Interference Elimination.
Heartbeating produces an intrusive quasi-periodic interference that masks the clinical auscultative interpretation of LS. This is a serious noise effect that creates major difficulties in LS analysis because of the fact that the intensity of HS is three- to ten-fold that of LS over the anterior chest and one- to three-fold that of LS from the posterior bases (55). This high relative intensity of the HS can easily saturate the analog amplifiers or the A/D converter, leading to a truncated signal and artifacts, especially in the case of children and infants, where the heart rate is high and its HS are loud (56,57). As it is already mentioned, the main frequency components of the HS are in the range of 20 Hz to 150 Hz (58,59) and overlap with the low-frequency components of breath sound spectrum in the range of 20 Hz to 2000 Hz (14).

Two basic approaches for heart noise reduction were followed in the literature: high-pass (HP) linear filtering, with a cutoff frequency varying from 50 Hz to 150 Hz, and adaptive filtering. The first approach (14,52) effectively reduces the HS noise, but, at the same time, it degrades the respectively overlapped frequency region of breath sounds. The second approach uses the theory of adaptive filtering (60) with a reference signal highly correlated to the noise component of the input signal. To this vein, a signal preprocessing system with varying amplifier gain using an adaptive filter (61) was initially proposed. In addition, a portable breath sound analysis system that uses an adaptive filter based on the Least Mean Square (LMS) algorithm for removing HS interferences (62) was suggested. In both cases, HS signals recorded on the patient's heart location were used as the reference input for adaptive filtering, which inevitably included LS as well. Electrocardiogram (ECG) signal information was also used as the reference signal for LMS-based adaptive filtering to reduce HS (57,63), using hundreds of taps (i.e., 1000 and 300, respectively) resulted, however, in a high adaptation time. The use of reduced order Kalman filtering (ROKF) for HS reduction was proposed by (64). In this work, an autoregressive model was fitted to the HS segments free of respiratory sounds (i.e., breath hold segments including HS) under the adoption of three assumptions (i.e., heart and respiratory sounds are mutually uncorrelated; these sounds have additive interaction; and the prior and subsequent HS are linearly related to the HS corrupted by the respiratory signal). However, these prior assumptions make the method inefficient in practical implementations, and it was tested only with synthesized data and not actual LS; in addition, this ROKF-based approach exhibits increased computational cost.

Kompis and Russi (65) proposed a modification to the adaptive LMS algorithm (58), which combines the advantages of adaptive filtering with the convenience of using only a single microphone input, eliminating the need of recording a reference signal. Their modification consisted of adding a LPF with a cutoff frequency of 250 Hz in the error signal path to the filter. However, their results showed that HS were still clearly audible because of the improper identification of the HS segments within a long sound recording, achieving a moderate heart noise reduction by 24% up to 49%.

In the same perspective, yet with much better HS noise reduction results, another single recording adaptive noise cancellation (ANC) technique based on fourth-order statistics (FOS) was proposed by Hadjileontiadis and Panas (66). The ANC-FOS algorithm is an adaptive de-noising tool that employs a reference signal $z(k)$ highly correlated with the HS noise. The ANC-FOS algorithm analyzes the incoming LS signal $x(k)$ to generate the reference signal $z(k)$, by detecting the real location of the HS in the incoming LS signal. Therefore, the ANC-FOS scheme initially applies on $x(k)$ a peak detection algorithm, namely *localized reference estimation* (LOREE) algorithm (66), which searches for the true locations of HS noise, based on amplitude, distance, and noise-reduction percentage criteria. Its output, $z(k)$, is a localized signal with precise tracking of the first (S_1) and second (S_2) heartbeats, highly correlated with heart noise and with no extra recording requirement as Iyer's method requires (58). The basic characteristic of the ANC-FOS is that, during the adaptation procedure, its update equation consists only of the fourth-order cumulants of the incoming and reference signals, and it is not affected by Gaussian uncorrelated noise. This observation reveals the enhancement achieved in the robustness of an ANC scheme when FOS are employed in its structure. Comparing the ANC-FOS algorithm with the adaptive schemes proposed by Iyer et al. (58) and Kompis with Russi (65), it can be deduced that, unlike the latter, the ANC-FOS algorithm finds the true locations of S_1 and S_2 without assuming any similarity in the two HS. In addition, neither Iyer's nor Kompis' methods take into account any nonlinear transformation that might be necessary to correctly account for nonlinearities in the system, as the ANC-FOS algorithm does because of the employment of FOS. Furthermore, although Kompis' method, unlike Iyer's, avoids the need of extra recording, it results in moderate noise reduction percentages compared with that of the ANC-FOS scheme. Finally, both Iyer's and Kompis' methods are sensitive to additive Gaussian noise, whereas the ANC-FOS algorithm is insensitive to Gaussian uncorrelated noise and independent of it, because it is based on FOS, which are zero for Gaussian processes. An example of the output of the ANC-FOS, when applied on breath sound recordings contaminated with HS, is depicted in Fig. 2. From this figure, the efficiency of the ANC-FOS in deteriorating the HS noise leaving the LS signal unaffected is evident.

Recently, a bandpass-filtered version of the original signal was used as the reference input for an ANC-Recursive Least Squares (ANC-RLS) filtering technique (67). Although the performance of this method is promising

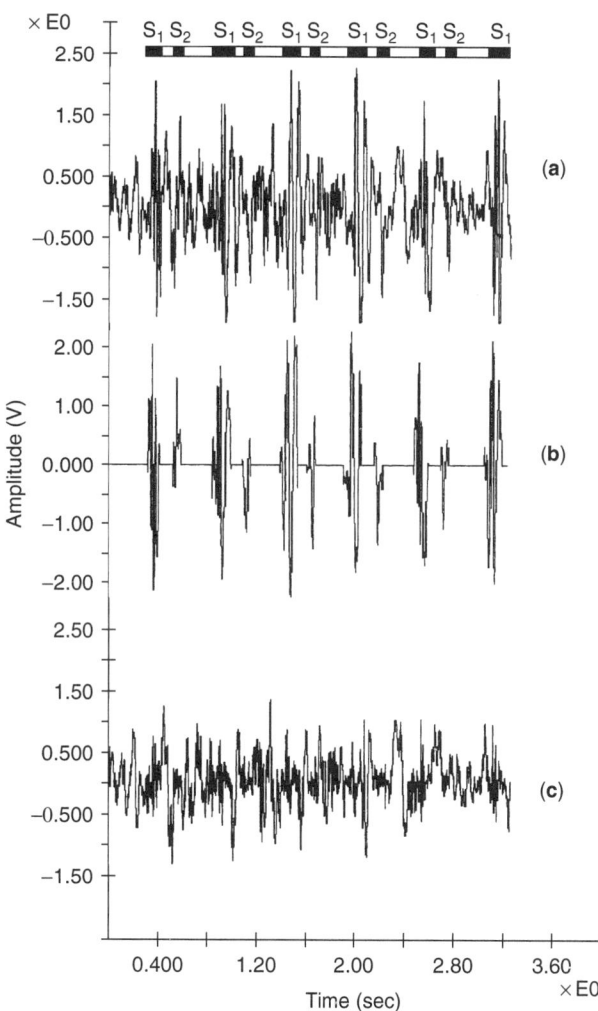

Figure 2. (a) Time waveform of 3.27-s raw contaminated LS recorded from a subject during normal breathing. (b) Time waveform of the corresponding reference signal [LOREE output (66)]. (c) Time waveform of adaptively filtered LS [ANC-FOS output (66)]. The small, black lines in the input tracing indicate the presence of the HS noise (S_1 and S_2 correspond to first and second HS, respectively).

both qualitatively and quantitatively, its computational load is quite high.

Multiresolution analysis, based on the discrete wavelet transform (WT), has also been used to form filtering techniques for the HS reduction from LS (68,69). In particular, in (69), an ANC-WT filter is proposed based on the fact that explosive peaks in the time-domain (HS peaks) have large signal amplitudes over many WT scales. On the contrary, "noisy" background (LS) dies out swiftly with increasing scale. The definition of "noise" is not always clear. For example, in this case, the HS peaks are temporary considered as the "desired" signal in order to be isolated from the LS (background noise); after their isolation, the residual (i.e., the de-noised LS) is now the desired signal and the isolated HS peaks are the noise. Hence, it is better to view an N-sample signal as being noisy or incoherent relative to a basis of waveforms if it does not correlate well with the waveforms of the basis (70). From this notion, the separation of HS from LS becomes a matter of breath sounds coherent structure extraction. The basic characteristic of the ANC-WT scheme is that, unlike most of the adaptive-filtering techniques, it does not need any reference signal for its adaptive performance. Comparing the ANC-WT algorithm (69) with the adaptive schemes proposed by Iyer et al. (59), Kompis and Russi (65), and the previously described ANC-FOS (66), it can be noticed that the ANC-WT scheme possesses all the advantages of the ANC-FOS algorithm over Iyer's and Kompis' methods, such as finding the true locations of S_1 and S_2 without assuming any similarity in the two HS, avoiding the need of extra recording, and resulting in high noise reduction. In addition, the ANC-WT scheme performs better than the ANC-FOS scheme when the additive noise has an impulsive character (e.g., friction sound caused by movement, impulsive ambient noise, etc). On the other hand, the ANC-FOS scheme performs better than the ANC-WT scheme when the input LS signal is contaminated by additive Gaussian uncorrelated noise (66,69). Although the ANC-WT algorithm results in smaller reduction percentages (84.1%) than the ANC-FOS algorithm ($> 90\%$), its "reference-free" structure equalizes this loss characterizing both of them as attractive noise reduction schemes. An example of the performance of the ANC-WT when applied on breath sound recordings contaminated with HS is depicted in Fig. 3. From this figure, the adaptation efficiency of the ANC-WT to HS noise reduction only is evident.

4.1.2. Background Elimination-BAS Detection. The elimination of background noise in the BAS recordings is a major issue to be considered prior to their diagnostic analysis, because the presence of background noise severely influences the clinical auscultative interpretation of BAS. The elimination of noise results in de-noised BAS that provide a more reliable and accurate characterization of the associated pathology.

The noise sources in the case of the BS recordings include instrumentation noise introduced during the recording process, sounds from the stomach, as well as cardiac and respiratory sounds, which occur mostly in the case of infants. In order to yield a successful BS classification, an effective reduction of noise from the contaminated BS

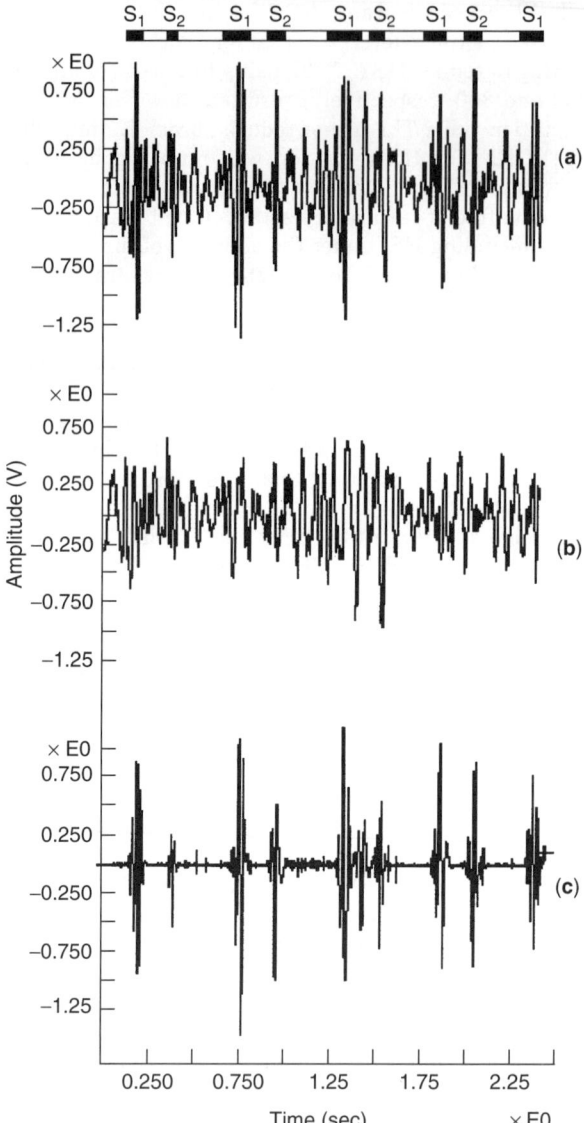

Figure 3. (a) Time waveform of 2.48-s raw contaminated LS recorded from a subject during normal breathing. (b) Time waveform of adaptively filtered LS [stationary ANC-WT output (69)]. (c) Time waveform of the estimated HS [nonstationary ANC-WT output (69)]. The small, black lines in the input tracing indicate the presence of the HS noise (S_1 and S_2 correspond to first and second HS, respectively).

signal is required, because the presence of the noise introduces pseudoperiodicity, masks the relevant signal, and modifies the energy distribution in the spectrum of BS. From the available studies related to BS analysis, only a few are focused on the BS de-noising process in order to extract their original structure before any further diagnostic evaluation (30,71). Unfortunately, the method used for noise reduction in (30) was based only on assumptions and general descriptions of the noise characteristics, forming a static, rather than a dynamic, noise reduction scheme. In addition, the noise is manually extracted, after subjective characterization and localization of the noise presence in the histogram of the recorded BS. The

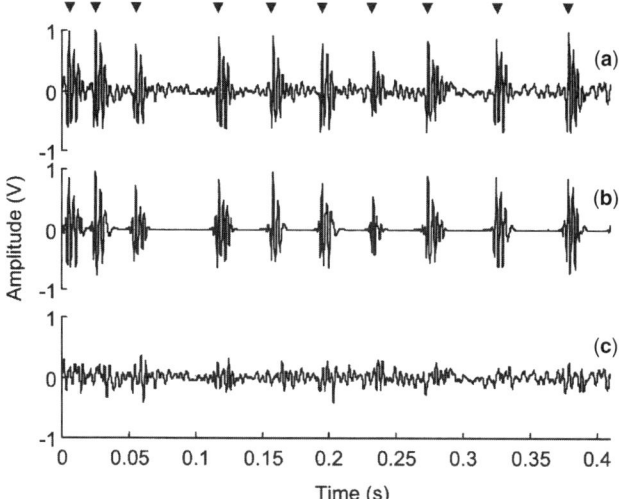

Figure 4. The performance of the ANC-WT filter when applied on recorded BS from a patient with irritable bowel syndrome (72). (a) The originally acquired BS signal. The arrowheads in the input tracing indicate the events of interest that correspond to BS. (b) The de-noised BS [nonstationary output of the ANC-WT filter (72)]. (c) The estimated background noise [stationary output of the ANC-WT filter (72)].

second de-noising method of BS, used by Mansy and Sandler (71), is based on adaptive filtering (60) and eliminates the HS noise from BS recorded from rats. Although it results in satisfactory enhancement of BS, it still requires careful construction of a noise reference signal, and it needs an empirical set-up of its adaptation parameters.

The previously described ANC-WT has been successfully applied to circumvent this problem (72). By a simple modification in its input (BS instead of LS), it results in efficient de-noising of BS signals, as it is apparent from Fig. 4.

Comparing the ANC-WT algorithm (72) with the adaptive scheme proposed by Mansy and Sandler (71), it can be noticed that, unlike the ANC-WT scheme, Mansy and Sandler's method is a "nonreference-free" scheme, because, apart from the BS recordings, it requires a reference signal for its performance. In addition, their scheme constructs a heart sound template, with averaged HS from previous successive heartbeats, to estimate the characteristics of the heart noise. On the contrary, it is not required by the ANC-WT scheme to describe the characteristics of the aggregated interference. Moreover, Mansy and Sandler's scheme is only tested on rats for the construction of the heart noise template, using their structural characteristics. On the contrary, the ANC-WT scheme is implemented in BS analysis derived from humans, with or without any gastrointestinal pathology. Finally, its implementation, unlike Mansy and Sandler's scheme, does not require empirical definition of a set of parameters, which is always prone to subjective human judgment.

As it was mentioned in the description of the LS characteristics in the LS section of this chapter, the DAS behave as a nonstationary explosive noise superimposed on breath sounds. The DAS are rarely normal, so their presence indicates an underlying pathological malfunction, which means that their separation from the background sound (i.e., the vesicular lung sounds or VLS) could reveal significant diagnostic information. In order to achieve automated separation, the nonstationarity of DAS must be taken into account. Thus, neither HP filtering, which destroys the waveforms, nor level slicing, which cannot overcome the small amplitude of FC, are adequate for this task. Application of time-expanded waveform analysis in crackle time-domain analysis (16,73) results in separation; it is, however, time-consuming, with large interobserver variability. Nonlinear processing, proposed by Ono et al. (74), and modified by Hadjileontiadis and Panas (75), obtains more accurate results, but requires empirical definition of the set of parameters of its *Stationary-nonstationary* (ST-NST) filter. To more efficiently overcome this problem (i.e., to achieve reliable automated separation of three types of DAS, namely fine crackles (or FC), coarse crackles (or CC), and squawks (or SQ), from VLS), a series of works have been proposed in the literature based on WT, fuzzy logic (FL), and fractal dimension (FD).

In particular, the WT-based method for automated separation of DAS from VLS refers to a form of the ANC-WT scheme adapted for this task. As the problem addressed here deals with the separation of the nonstationary part of the LS from the stationary part, the *ANC-WT scheme is renamed to wavelet transform-based ST-NST* (WTST-NST) filter, originally introduced by Hadjileontiadis and Panas (76). The only changes needed to the ANC-WT scheme refer to the input signal, which is now the breath sounds (DAS and VLS), and to the two outputs (nonstationary and stationary) corresponding to the separated DAS and the background VLS, respectively. Comparing the results of the WTST-NST algorithm (76) with those derived from the application of the nonlinear scheme (ST-NST) proposed by Ono et al. (74) at the same breath sounds [analytically shown in Table II in (76)], it is clear that the WTST-NST filter performs better than the ST-NST filter in all cases of each type of DAS, which is a result of the empirical definition of the parameters employed in the performance of the ST-NST filter, depending on the characteristics of the input signal. The WTST-NST filter may be applied to all types of DAS requiring neither empirical definition nor adaptive updating of its parameters according to the characteristics of the input signal. In addition, the WTST-NST filter, unlike the ST-NST filter, separates the whole DAS, without leaving the later parts (i.e., the part of DAS that follows the initial peak). Regarding the performance of the WTST-NST filter with that of the rest of separation techniques mentioned before, the WTST-NST filter clearly overcomes the disadvantages of HP filtering, because it results in nondestroyed DAS and VLS; unlike a level slicer it detects all DAS, even those with small amplitude, and unlike time-expanded waveform analysis (16,73), it results in accurate, fast, and objective results, regardless of DAS type and without employing any human intervention.

Although the WTST-NST filter obtains the most accurate separation results among all other methods reported in the recent literature, it cannot serve as the optimum solution when the real-time analysis of the LS is the primary aim. To this end, enhanced *real-time* DAS detectors based on FL have been proposed. The first one, namely *generalized FST-NST* (GFS-NST) filter, is based on a generalized version of the *fuzzy logic-based ST-NST* (FST-NST) filter proposed by Tolias et al. (77), and uses one *fuzzy inference system* (FIS) dedicated to the estimation of the stationary part of the signal to provide a reference to the second FIS that estimates the nonstationary part (78). In an alternative approach in developing fuzzy models for real-time separation of DAS from VLS, the model generation process is based on the *Orthogonal Least Squares* (OLS) concept (79), providing structure and parameter identification, as well as performing input selection. In particular, the parameters of the resulting models are calculated in a forward manner without requiring an iterative training algorithm. Therefore, the method does not suffer from drawbacks inherent to gradient-based techniques, such as tapping to local minima and extensive training time. This approach uses two FISs that operate in parallel and form the *OLS-based fuzzy filter* (OLS-FF) (80). In general, the OLS-FF improves the structure employed in (78), introducing a more flexible architecture. As a result, it activates the optimum employed fuzzy rules, uses a lower number of rules, and comprises the most appropriate inputs for each of them, which is a clear advantage over the previous work (78), because the latter uses a complete rule base and consequent parts with fixed number of terms. In addition, unlike the GFST-NST filter, the OLS-FF does not require a training phase for the estimation of the optimum model parameters, because they are calculated on the spot. Although the OLS-FF exhibits an inferior performance compared with the WTST-NST filter (76), it satisfies the real-time implementation issue. However, the GFST-NST filter (78) performs better than the OLS-FE by 1% up to 4% for the cases of CC and SQ, and by 14% for the case of FC as a trade-off between the structural simplicity of the OLS-FF. On the contrary, the OLS-FF requires a significantly smaller (by 62%) computational load than the GFST-NST filter, improving the procedure of clinical screening of DAS under a real-time implementation context (80,81). Figure 5 illustrates

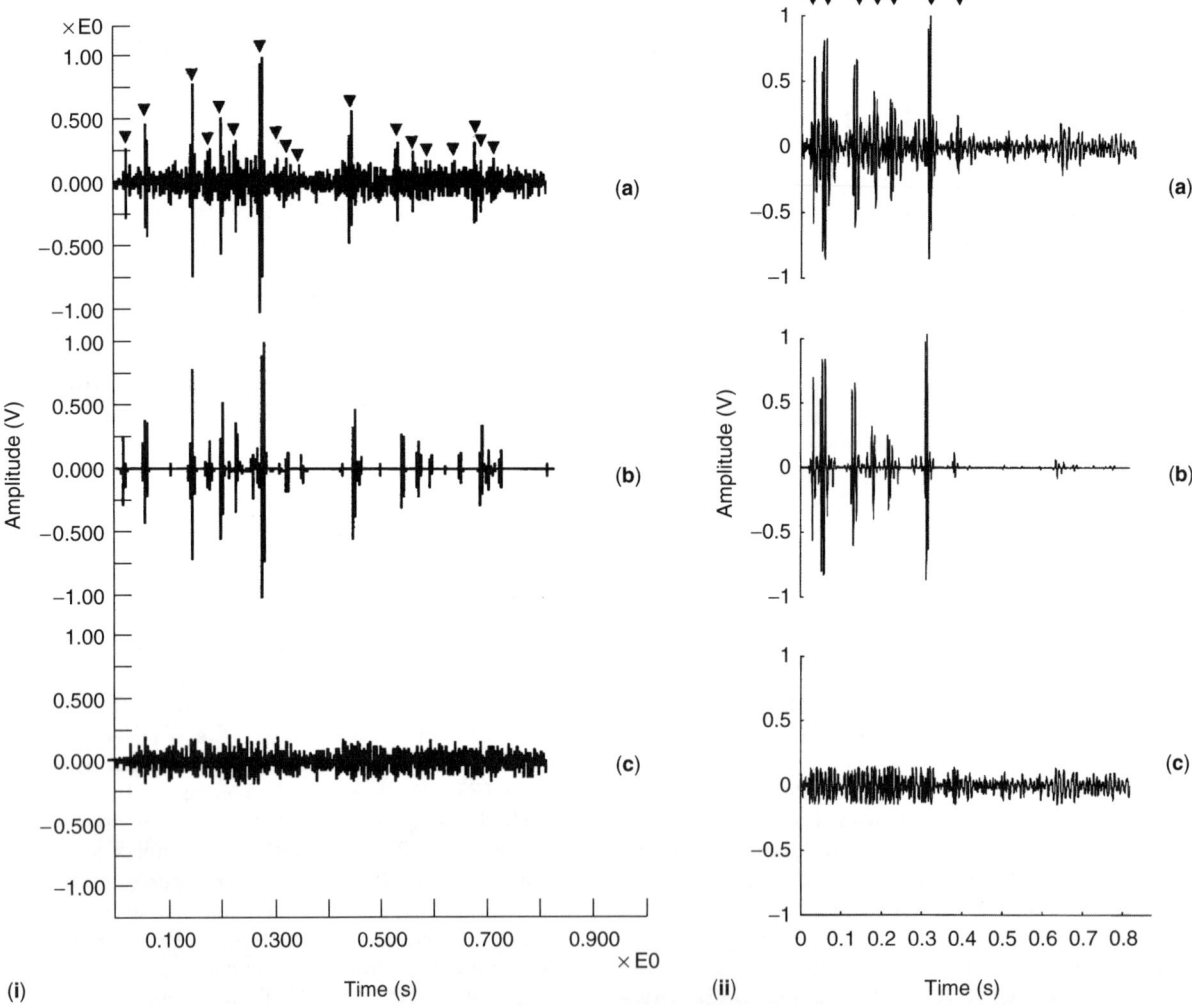

Figure 5. Examples from the performance of (i) the WTST-NST filter (76) on FC and (ii) the GFST-NST filter (78) on CC. In both subfigures, (a)–(c) correspond to the input signal, the estimated denoised BAS, and the estimated background noise, respectively. Furthermore, the arrowheads in both input tracings indicate the events of interest that correspond to BAS.

some examples of the performance of the WTST-NST and the GFST-NST filters.

Recently, an FD-based approach has been proposed to address the detection of DAS in the LS signal and of BS in the background noise (82). In particular, this technique forms an FD-based detector (FDD), which assesses the complexity of the sound recordings in the time-domain in order to efficiently detect the time location and duration of the nonstationary LS and BS. As nonstationarity is the key feature for the detection of these sound recordings, its direct relationship with the signal complexity makes the FD measure a possible tool for their detection. Compared with the previous approaches (75–78,80,83), the FDD is more attractive because of the simplicity in the evaluation of the FD measure, because the FDD is a time-domain FD algorithm that operates directly on the signal and not on state space, and hence it has fast computational implementation.

4.1.3. Reduction of Interference in TMJS and HS Time-Frequency Representation.
Time-frequency signal analysis has recently experienced an accelerating development of interest. Although spectrogram and Wigner–Ville distribution [both members of Cohen's Class of Distributions (84)] have been used in tracking the frequency content of nonstationary signals, they both exhibit liabilities, such as failure to resolve signal components close in frequency and production of interference (cross terms) because of interactions between signal components, respectively. To this end, Williams and Jeong have proposed a *reduced interference distribution* (RID) (85,86), aiming at achieving high-resolution time-frequency distributions within Cohen's class with much reduced cross-term or interference activity. The suppression of cross terms was achieved by attenuating distant terms in the ambiguity-domain, at the same time retaining a number of other desirable mathematical properties, which are not exhibited by other members of Cohen's class (85,86). The RID-based analysis of TMJS (50,86–88) provided the successful categorization of TMJS to five types described in the Joint Sounds section of this chapter.

In a similar vein, Wood and Barry and their colleagues (89–92) used the binomical form of the RID in the HS analysis in order to clearly track the alterations of the dynamic changes in physiological properties caused by pathology. Guo et al. (93,94) devised the Bessel kernel that generally falls into the RID class, exhibiting further advantages in the HS analysis, particularly of valve sounds. They have also found that, by increasing the computational burden and relaxing some of the constraints on the distribution, like the time support constraint such that the kernel is integrating (or summing) over a larger range of signal values, the noise is further suppressed (93,94).

4.2. Modeling

Modeling of the system that produces the BAS contributes to the understanding of their production mechanisms. The way the pathology affects these mechanisms is reflected in the modeling parameters adopted, which is a very important issue in the analysis of BAS, because efficient modeling could lead to an objective description of the changes a disease imposes to the production or transmission path of the BAS.

LS originated inside the airways of the lung are modeled as the input to an all-pole filter, which describes the transmission of LS through the parenchyma and chest wall structure (95). The output of this filter is considered to be the LS recorded at the chest wall. As it is already mentioned in previous subsections, the recorded LS also contain heart sound interference, the reduction of which has been thoroughly addressed in the Noise Reduction-Detection section of this chapter. Muscle and skin noise, along with instrumentation noise, are modeled as an additive Gaussian noise. With this model, given a signal sequence of LS at the chest wall, an *autoregressive* (AR) analysis based on *third-order statistics* (TOS), namely AR-TOS, can be applied to compute the model parameters. Therefore, the source and transmission filter characteristics can be separately estimated, as it is thoroughly described in (81,96). The profound motivations behind the use of the AR-TOS model is the suppression of Gaussian noise, because TOS of Gaussian signals are identically zero. Hence, when the analyzed waveform consists of a nonGaussian signal in additive symmetric noise (e.g., Gaussian), the parameter estimation of the original signal with TOS takes place in a high-SNR domain, and the whole parametric presentation of the process is more accurate and reliable (97,98). The model used for the LS originated inside the airways considers the LS source as the output from an additive combination of three kinds of noise sequences (14). The first sequence (periodic impulse) describes the *CAS sources*, because they have characteristic distinct pitches, and they are produced by periodic oscillations of the air and airway walls (see also the Categorization and Main Characteristics subsection of Lung Sounds) (9,17). The second sequence (random intermittent impulses) describes the *crackle sources*, because they are produced by sudden opening/closing of airways or bubbling of air through extraneous liquids in the airways, both phenomena associated with sudden intermittent bursts of sounds energy (see also the Categorization and Main Characteristics subsection of Lung Sounds) (9,17). Finally, the third sequence (white nonGaussian noise) describes the *breath sound sources*, because they are produced by turbulent flow in a large range of airway dimensions (see also the Categorization and Main Characteristics subsection of Lung Sounds) (9,17). The estimation of the AR-TOS model input (LS source) can be derived from the prediction error by means of inverse filtering (96,99).

As it was described in the Lung Sounds section, among the different kinds of BAS recorded from the human body, some exhibit sharp peaks or occasional bursts of outlying observations that one would expect from normally distributed signals, such as the DAS and some types of impulsive BS. In addition, these BAS occur with a different *degree of impulsiveness*. As a result of their inherent impulsiveness, these BAS have a density function that decays in the tails less rapidly than the Gaussian density function (100). Modeling based on *alpha-stable distribution* is appropriate for enhanced description of impulsive processes (100).

Furthermore, under the nonGaussian assumption in the context of analyzing impulsive processes, among the various distribution models that were suggested in the past, the *alpha*-stable distribution is the only one that is motivated by the *generalized central limit theorem*. This theorem states that *the limit distribution of the sum of random variables with possibly infinite variances is stable distribution* (101). Stable distributions are defined by the stability property, which says that *a random variable, X, is stable if and only if the sum of any two independent random variables with the same distribution as X also has the same distribution* (101). Following this approach, impulsive LS (i.e., DAS) and BS with explosive character can be modeled by means *of alpha-stable distribution, symmetrical alpha-stable* ($S\alpha S$) *distribution*, and *lower-order statistics* (LOS) (102). From the estimated parameters of the ($S\alpha S$) distribution of the analyzed impulsive BAS with the $\log|S\alpha S|$ method (100) (because the fractile method (103) works only for $\alpha \geq 0.6$, and the regression method (104), needs empirical parameter setting), it is derived that when the contaminated (without de-noising) impulsive BS are analyzed, the values of α are large enough (near 1.5), whereas when the AR-estimated inputs of the de-noised data are analyzed, these values decrease dramatically. This fact is because of the increase of impulsiveness of the analyzed signal, since the vesicular sound and the background noise are suppressed after de-noising. In fact, the impulsive source sound, initiated inside the lung or the bowel, reaches the surface with different signal characteristics, indicating a shifting toward the Gaussian distribution because of the superimposed Gaussian noise. Generally, the values of dispersion γ are low enough, indicating small deviation around the mean, whereas the location parameter a has mean values around zero, indicating low mean and median for $1 < \alpha \leq 2$ and $0 < \alpha < 1$, respectively (102). The *covariation coefficient* λ_{SQ-FC} calculated for the cases of the sound sources of SQ and FC (102) shows an almost 50% correlation between the SQ and FC, confirming the accepted theory that SQ are produced by the explosive opening, because of a FC, and decaying fluttering of an unstable airway (9,17), which proves that $S\alpha S$ distribution- and LOS-based modeling of impulsive BAS provide a measure of their impulsiveness, and, at the same time, reveals the underlying relationships between the associated production mechanisms and pathology.

4.3. Feature Extraction Classification

The characteristics of BAS provide a variety of structural features that are directly connected to the underlined pathology. The correlation of these features with parameters derived from advanced processing of BAS provides a means for their efficient classification and, consequently, for the classification of the associated pathology.

The LS that are characterized as musical are the CAS [i.e., wheezes (WZ), rhonchi (RH), and stridors (STR)] described the Categorization and Main Characteristics subsection of Lung Sounds. The musical character of these LS is reflected to the power spectrum through distinct frequency peaks (14). As the second-order statistics suppress any phase information, they fail to detect and characterize the nonlinear interactions of the distinct harmonics. Spectral-based methodologies have been widely applied to frequency-domain analysis of musical LS (mainly wheezes) (14,105,106). They, however, do not take into account the nonlinearity and nonGaussianity of the analyzed processes. On the contrary, *higher-order statistics* (HOS) preserve the phase character of the signals (97) and have been used as a useful tool for the detection of nonlinearity and deviation from Gaussianity of the signals. In this way, the process of musical LS is expanded to more accurate estimations of their true character. In particular, based on the results presented in (107), it can be concluded that a strong quadratic phase coupling exists among the distinct harmonics of the musical HS along with a strong nonGaussianity of the analyzed processes for the case of increased pathology.

Furthermore, bispectral analysis of HS (108) indicate that more efficient discrimination of cardiac pathologies could be established in the bispectrum-domain [higher (>2)-order spectrum] rather in the power spectrum one, which can be used for further elicitation of the diagnostic character of HS.

Moreover, bispectral analysis of BS (109) shows the ability of bispectrum to differentiate the BS recorded from a patient with bowel polyp before and after the polypectomy, hence to reflect the anatomical changes of the bowel. In addition, in (109), HOS-based analysis of BS recorded from controls shows a strong deviation from normality and linearity; for BS recorded from patients with diverticular disease, it shows increased nonGaussianity and decreased nonlinearity; for BS recorded from patients with ulcerative colitis, it reveals a strong deviation from normality and linearity; and for BS recorded from patients with irritable bowel syndrome, it results in increased nonGaussianity and decreased nonlinearity (109). These results show that HOS provide a new perspective in the computerized analysis of BS, defining an enhanced field of new features, which reinforce the diagnostic character of BS.

Considering the nonGaussianity and the impulsiveness of crackles and artifacts, we could model them as $S\alpha S$-distributed processes, in the way it is described in the Modeling section of this chapter. By estimating the characteristic exponent α for each category (crackles, background noise, and artifacts) using the $\log|S\alpha S|$ method (100), a classification criterion could be established based on the estimated *alpha* values. This approach has been adopted in (110), which deduces that the FC follow approximately the Cauchy distribution (mean $\alpha = 0.92$, which is close to $\alpha = 1.0$ and holds in the case of the Cauchy distribution), whereas the CC deviate both from Cauchy and Gaussian distributions (mean $\alpha = 1.33$). The artifacts have high impulsiveness (mean $\alpha = 0.0218$). The events that resemble the vesicular sound are seen as background noise and are modeled as Gaussian processes (values of $\alpha > 1.8$ or very close to 2.0; a value that holds in the case of the Gaussian distribution).

Higher-order crossings (HOC) [i.e., the zero-crossing counts resulting from the application of a bank of filters to a time series (111)] have also been applied in the classification analysis of BAS, because HOC constitute a domain by itself, which "sits" between the time- and spectral-domains (111). Consequently, the use of the *white noise test* (WNT) (i.e., the measurement of the distance of the signals under consideration from white Gaussian noise) and the *HOC-scatter plot* (HOC-SP), which is a scattergram that depicts a pair of simple HOC (D_j, D_k) (111), can be used for the BAS classification. In particular, the WNT is employed so the two categories of the signals, for example, FC and CC, are indirectly compared with each other. The ith order of HOC that results in the maximum distance between the two categories, defined as the maximum distance from the lower curve of one category to the upper curve of the other at the ith order, provides with a linear discrimination in the HOC domain. In particular, by plotting D_i versus $D_{i \pm m}$, where m usually equals 1, 2, or 3 (for $i \geq 5$), the resulted HOC-SP reveals two *linearly* separated classes. As a result of the linear character of the separation, a simple single-layer perceptron neural network (112) can be used for decision on class membership.

Results from the HOC-based analysis of crackles (113), initially de-noised with the WTST-NST filter (76), are illustrated in Fig. 6. From Fig. 6a, it is clear that the order, which provides with the maximum distance between FC and CC, the so-called "opened-eye," equals 12 (denoted by an arrow). This distance is reflected to the corresponding HOC-SP (D_{12}, D_{13}) of Fig. 6b, where FC and CC are concentrated in two distant, linearly separated classes. These results indicate that the HOC-based indices result in fast, easy, and efficient discrimination of crackles, providing with a simple decision rule for categorizing FC and CC.

Further application of HOC-based classification of BAS includes discrimination between normal LS and abnormal bronchial LS heard over the trachea (tracheobronchial LS) (114); de-noised BS recorded from patients with different pathologies of the large bowel, such as ulcerative colitis and diverticular disease; and de-noised BS recorded from patients with small volume ascites and controls (115). In all referenced cases, a linear separation is feasible, exhibiting the potentiality of the HOC-based analysis to provide an efficient classification domain in a diversity of BAS types.

Classification of TMJS from joint with and without effusion using spectrogram-based analysis is presented in (116) where it is concluded that joints with effusion can be identified through the unstable sound patterns as they are revealed in the time-frequency domain. In this way, screening of patients with suspected effusion is much more simple and less expensive than the current use of magnetic resonance imaging.

The way the head tissues affect spectral characteristics of TMJS and how these differences (because of different positioning of sensors) can be used in the localization of source are explored in (117). They conclude that spectral analysis of bilateral electronic TMJS recordings is of diagnostic value when bilateral clicking is heard at auscultation and can help to avoid diagnosing a silent joint as clicking.

Wavelet transform-based representations of TMJS are used in (118) to visually differentiate between controls and patients with reducing displaced disks. A 3×7 biorthogonal spline WT is used to create three-dimensional time-frequency graphs of the TMJS from each category, resulting in distinct visual differences between patients and controls (118).

In parallel to the aforementioned approaches, a number of research efforts have been placed on the computerized classification of TMJS, especially in the RID-analysis domain. In particular, replacement of the visual analysis

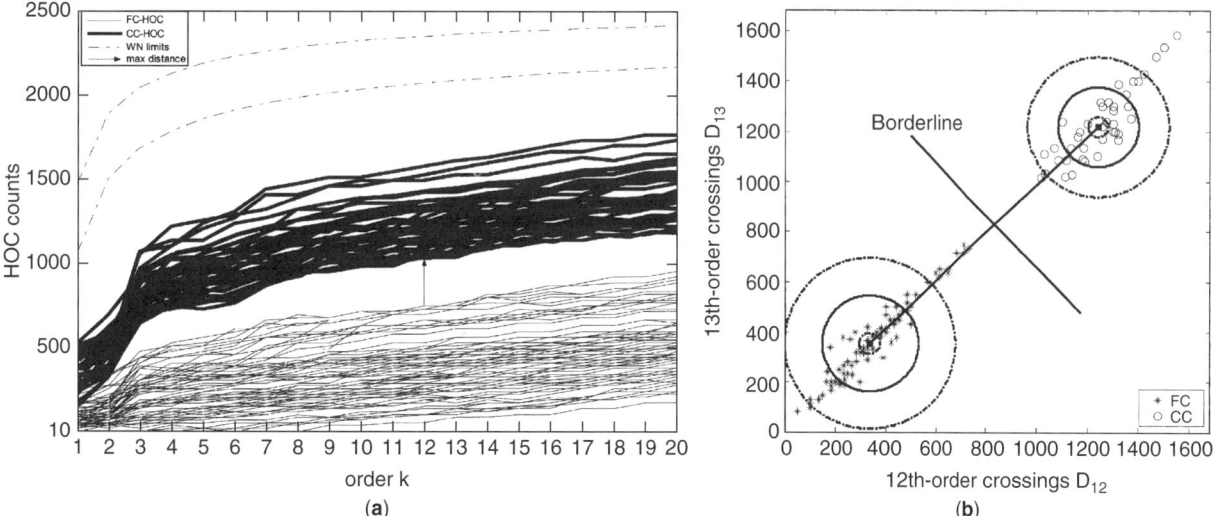

Figure 6. (a) Results from the WNT for FC and CC with the HOC order ranging from 1 to 20. The estimated HOC sequences of FC (FC-HOC) and CC (CC-HOC) are compared with the white noise (WN) limits. The maximum distance between FC-HOC (most upper) and CC-HOC (most lower) lines is denoted with an arrow located at the 12th HOC order. (b) Scatter plot of 12th-order (D_{12}) vs. 13th-order (D_{13}) crossings of FC (stars) and CC (circles). The ■ symbol denotes the center of each group (FC and CC), whereas borderline is the estimated linear border that clearly discriminates the two groups. The solid and dash-dotted circles correspond to [mean value ± (standard deviation + standard error of the mean)] of the distance from cluster center, both for FC and for CC.

of RIDs of the TMJS (an expensive and error-prone procedure) with neural networks has been proposed in (119). Adaptive Gabor transforms (120) and the third-order Rényi number of the RIDs were also used for discrimination among the five TMJS types (see the Joint Sounds section of this chapter), especially for the differentiation between the clickings and the crepitations (120). Yang et al. (121) used neural networks-, nearest linear combination-, and nearest constrained linear combination-based classifiers to automatically distinguish between types 1–3 of the TMJS classes (see the Joint Sounds section of this chapter). Finally, an extension of Yang's work, employing concepts of time-shift invariance with and without scale in variance, is presented in (122), which shows that the classifier performance is significantly improved when scale invariance is omitted.

Spectrogram-based analysis of KJS from patients with end-stage osteoarthritis, patellofemoral joint disorders, and lateral meniscal bucket handle tears is presented in (123). They have found a distinct shift of the spectrum to lower frequencies after intra-articular injection of Hyaluronic acid to the knee, establishing an efficient way of measuring the improvement of the lubricative functions through the spectral characteristics of KJS.

The use of matching pursuit method (124) in creating high-resolution and interference-free time-frequency representations of KJS is presented in (125). They involve time-frequency features, such as instantaneous energy, instantaneous energy spread, instantaneous frequency, and instantaneous frequency spread, in order to classify KJS from normal and abnormal knees. They report classification accuracy up to 86%, exhibiting high sensitivity in screening patellofemoral articular cartilage disorders, such as chondromalacia patella. Furthermore, they employ Hough (126) and Radon (127) transforms in order to extract patterns/signatures from the time-frequency domain, which could aid in better screening and understanding of how the movement of knee affects KJS patterns.

The issues presented here adopt the same endeavor: exploitation and surfacing of the valuable diagnostic information that exists in bioacoustic signals. Under this common aim, the different approaches of de-noising, detection, modeling, feature extraction, and classification described here introduce new pathways in the analysis of BAS, toward a noninvasive, efficient, and objective evaluation and understanding of the human body functioning.

BIBLIOGRAPHY

1. Houghton Mifflin Company (2000). *The American Heritage Dictionary of the English Language*, 4th ed. (online). Available: http://www.dictionary.com/cgi-bin/dict.pl?term=bioacoustics.
2. V. S. McKusick, *Cardiovascular Sound in Health and Disease*. Baltimore, MD: Williams & Wilkins, 1958, p. 3.
3. J. Rapoport, Laënnec and the discovery of auscultation. *Israel J. Med.* 1986; **22**:597–601.
4. V. S. McKusick, *Cardiovascular Sound in Health and Disease*. Baltimore, MD: Williams & Wilkins, 1958, p. 13.
5. S. S. Kraman, Vesicular (normal) lung sounds: how are they made, where do they come from and what do they mean? *Semin. Respir. Med.* 1985; **6**:183–191.
6. R. E. Thacker and S. S. Kraman, The prevalence of auscultatory crackles in subjects without lung disease. *Chest* 1982; **81**(6):672–674.
7. P. Workum, S. K. Holford, E. A. Delbono, and R. L. H. Murphy, The prevalence and character of crackles (rales) in young women without significant lung disease. *Amer. Rev. Respir. Dis.* 1982; **126**(5):921–923.
8. A. J. Robertson, Rales, ronchi, and Laënnec. *Lancet* 1957; **1**:417–423.
9. S. S. Kraman, *Lung Sounds: An Introduction to the Interpretation of Auscultatory Findings*. Northbrook, IL: American College of Chest Physicians, 1983, pp. 14–21.
10. R. L. H. Murphy, Discontinuous adventitious lung sounds. *Semin. Respir. Med.* 1985; **6**:210–219.
11. D. W. Cugell, Lung sound nomenclature. *Am. Rev. Respir. Dis.* 1987; **136**:1016.
12. J. E. Earis, K. Marsh, M. G. Rearson, and C. M. Ogilvie, The inspiratory squawk in extrinsic allergic alveolitis and other pulmonary fibroses. *Thorax* 1982; **37**(12):923–936.
13. N. Gavriely and D. W. Cugell, *Breath Sounds Methodology*. Boca Raton, FL: CRC Press, 1994, p. 2.
14. N. Gavriely, Y. Palti, and G. Alroy, Spectral characteristics of normal breath sounds. *J. Appl. Physiol.* 1981; **50**:307–314.
15. T. Katila, P. Piirila, K. Kallio, E. Paajanen, T. Rosqvist, and A. R. A. Sovijarvi, Original waveform of lung sound crackles: a case study of the effect of high-pass filtration. *J. Appl. Physiol.* 1991; **71**(6):2173–2177.
16. R. L. H. Murphy, S. K. Holford, and W. C. Knowler, Visual lung-sound characterization by time-expanded wave-form analysis. *New Eng. J. Med.* 1978; **296**:968–971.
17. N. Gavriely and D. W. Cugell, *Breath Sounds Methodology*. Boca Raton, FL: CRC Press, 1994, p. 8.
18. B. Erickson, *Heart Sounds and Murmurs: A Practical Guide*, 3rd ed. St. Louis, MD: Mosby-Year Book, 1997, p. 2.
19. R. C. Cabot and H. F. Dodge, Frequency characteristics of heart and lung sounds. *JAMA* 1925; **84**:1793–1795.
20. N. Gavriely and D. W. Cugell, *Breath Sounds Methodology*. Boca Raton, FL: CRC Press, 1994, p. 116.
21. A. P. Yoganathan, R. Gupta, J. W. Miller, F. E. Udwardia, W. H. Corcoran, R. Sarma, J. L. Johnson, and R. J. Bing, Use of the fast Fourier transform for frequency analysis of the first heart sound in normal man. *Med. Biol. Eng.* 1976; **1**:69–81.
22. Academic Medical Publishing & Cancer WEB. (1997). *Online Medical Dictionary* (online). Available: http://www.graylab.ac.uk/cgi-bin/omd?bowel+sounds.
23. W. B. Cannon, Auscultation of the rhythmic sounds produced by the stomach and intestine. *Am. J. Physiol.* 1905; **13**:339–353.
24. G. E. Horn and J. M. Mynors, Recording the bowel sounds. *Med. Biol. Eng.* 1966; **4**:205–208.
25. W. C. Watson and E. C. Knox, Phonoenterography: the recording and analysis of BS. *Gut.* 1967; **8**:88–94.
26. D. Dalle, G. Devroede, R. Thibault, and J. Perrault, Computer analysis of BS. *Comp. Biol. Med.* 1975; **4**:247–256.
27. J. P. Politzer, G. Devroede, C. Vasseur, J. Gerand, and R. Thibault, The genesis of bowels sounds: influence of viscus and gastrointestinal content. *Gastroenterology* 1976; **71**:282–285.
28. J. Weidringer, S. Sonmoggy, B. Landauer, W. Zander, F. Lehner, and G. Blumel, BS recording for gastrointestinal

29. S. Michael and M. Redfern, Computerized Phonoenterography: the clinical investigation of a new system. *J. Clin. Gastroenterol.* 1994; **18**(2):139–144.

30. H. Yoshino, Y. Abe, T. Yoshino, and K. Ohsato, Clinical application of spectral analysis of BS in intestinal obstruction. *Dis. Col. Rect.* 1990; **33**(9):753–757.

31. R. H. Sandler, H. A. Mansy, S. Kumar, P. Pandya, and N. Reddy, Computerized analysis of bowel sounds in human subjects with mechanical bowel obstruction vs. ileus. *Gastroenterology* 1993; **110**(4):A752.

32. D. Bray, R. B. Reilly, L. Haskin, and B. McCormack, Assessing motility through abdominal sound monitoring. In: B. Myklebust and J. Myklebust, eds., Proc. 19th Annu. Int. Conf. IEEE/EMBS, Chicago, IL: IEEE Press, 1997, pp. 2398–2400.

33. B. L. Craine, M. L. Silpa, and C. J. O'Toole, Enterotachogram analysis to distinguish irritable bowel syndrome from Crohn's disease. *Dig. Dis. Sci.* 2001; **46**(9):1974–1979.

34. M. S. Spiller, (2000) TMJ (online). Available: *http://www.doctorspiller.com/TMJ.htm*.

35. C. Heuter, *Grundiss der Chirurfie,* 3rd ed. Leipzig, Germany: FCW Vogel, 1885.

36. W. E. Blodgett, Auscultation of the knee joint. *Boston Med. Surg. J.* 1902; **146**(3):63–66.

37. E. Bircher, Zur diagnose der meniscusluxation und des meniscusabrisses. *Zentralbl. Chir.* 1913; **40**:1852–1857.

38. C. F. Walters, The value of joint auscultation. *Lancet* 1929; **1**:920–921.

39. K. H. Erb, Uber die moglichkeit der registrierung von gelenkgerauschen. *Deutsche Ztschr. Chir.* 1933; **241**:237–245.

40. A. Steindler, Auscultation of joints. *J. Bone Int. Surg.* 1937; **19**:121–124.

41. A. Peylan, Direct auscultation of the joints (Preliminary clinical observations). *Rheumatism* 1953; **9**:77–81.

42. H. Fischer and E. W. Johnson, Analysis of sounds from normal and pathologic knee joints. Proc. 3rd Int. Congr. Phys. Med., 1960: 50–57.

43. M. L. Chu, I. A. Gradisar, M. R. Railey, and G. F. Bowling, Detection of knee joint diseases using acoustical pattern recognition technique. *J. Biomechan.* 1976; **9**:111–114.

44. M. L. Chu, I. A. Gradisar, and R. Mostardi, A non-invasive electroacoustical evaluation technique of cartilage damage in pathological knee joints. *Med. Biologic. Engineer. Comput.* 1978; **16**:437–442.

45. R. A. B. Mollan, G. C. McCullagh, and R. I. Wilson, A critical appraisal of auscultation of human joints. *Clin. Orthopaed. Related Res.* 1982; **170**:231–237.

46. W. G. Kernohan and R. A. B. Mollan, Microcomputer analysis of joint vibration. *J. Microcomp. Applicat.* 1982; **5**:287–296.

47. R. A. B. Mollan, W. G. Kernohan, and P. H. Watters, Artefact encountered by the vibration detection system. *J. Beomechan.* 1983; **16**(3):193–199.

48. W. G. Kernohan, D. E. Beverland, G. F. McCoy, S. N. Shaw, R. G. H. Wallace, G. C. McCullagh, and R. A. B. Mollan, The diagnostic potential of vibration arthrography. *Clinic. Orthopaed. Related Res.* 1986; **210**:106–112.

49. G. F. McCoy, J. D. McCrea, D. E. Beverland, W. G. Kernohan, and R. A. B. Mollan, Vibration arthrography as a diagnostic aid in diseases of the knee. *J. Bone Joint Surg.* 1987; **B**(2):288–293.

50. S. E. Widmalm, W. J. Williams, R. L. Christiansen, S. M. Gunn, and D. K. Park, Classification of temporomandibular joint sounds based upon their reduced interference distribution. *J. Oral Rehab.* 1996; **23**:35–43.

51. C. B. Frank, R. M. Rangayyan, and G. D. Bell, Analysis of knee joint sound signals for non-invasive diagnosis of cartilage pathology. *IEEE Eng. Med. Biol. Mag.* 1990; **9**(1):65–68.

52. L. Vannuccini, J. E. Earis, P. Helistö, B. M. G. Cheetham, M. Rossi, A. R. A. Sovijärvi, and J. Vanderschoot, Capturing and preprocessing of respiratory sounds. *Eur. Respir. Rev.* 2000;**10**(77):616–620.

53. M. J. Mussell, The need for standards in recording and analysing respiratory sounds. *Med. Biol. Eng. Comput.* 1992; **30**:129–139.

54. S. E. Widmalm, W. J. Williams, and B. S. Adams, The wave forms of temporomandibular joint sounds clicking and crepitation. *J. Oral Rehab.* 1996; **23**:44–49.

55. V. K. Iyer, P. A. Ramamoorthy, and Y. Ploysongsang, Quantification of heart sounds interference with lung sounds. *J. Biomed. Eng.* 1989; **11**:164–165.

56. N. J. McLellan and T. G. Barnett, Cardiorespiratory monitoring in infancy with acoustic detector. *Lancet* 1983; **2**(8364):1397–1398.

57. A. Tal, I. Sanchez, and H. Pasterkamp, Respisonography in infants with acute bronchiolitis. *Am. J. Dis. Child.* 1991; **145**:1405–1410.

58. V. K. Iyer, P. A. Ramamoorthy, H. Fan, and Y. Ploysongsang, Reduction of heart sounds from lung sounds by adaptive filtering. *IEEE Trans. Biomed. Eng.* 1986; **33**(12):1141–1148.

59. Y. Ploysongsang, V. K. Iyer, and P. A. Ramamoorthy, Characteristics of normal lung sounds after adaptive filtering. *Am. Rev. Respir. Dis.* 1989; **139**:951–956.

60. B. Widrow, J. R. Glover, J. McCool, M. J. Kaunitz, C. S. Williams, R. H. Hearn, J. R. Zeidler, E. Dong, and R. C. Goodlin, Adaptive noise canceling: principles and applications. *Proc. IEEE* 1975; **63**(12):1692–1716.

61. L. Yang-Sheng, L. Wen-Hui, and Q. Guang-Xia, Removal of the heart sound noise from the breath sound. Proc. 10th Annu. Int. Conf. IEEE/EMBS 1988: 175–176.

62. L. Guangbin, C. Shaoqin, Z. Jingming, C. Jinzhi, and W. Shengju, The development of a portable breath sounds analysis system. Proc. 14th Annu. Int. Conf. IEEE/EMBS 1992: 2582–2583.

63. L. Yip and Y. T. Zhang, Reduction of heart sounds from lung sound recordings by automated gain control and adaptive filtering techniques. Proc. 23rd Annu. Int. Conf. IEEE/EMBS 2001: 2154–2156.

64. S. Charleston and M. R. Azimi-Sadjadi, Reduced order Kalman filtering for the enhancement of respiratory sounds. *IEEE Trans. Biomed. Eng.* 1996; **43**(4):421–424.

65. M. Kompis and E. Russi, Adaptive heart-noise reduction of lung sounds recorded by a single microphone. Proc. 14th Annu. Int. Conf. IEEE/EMBS 1992: 691–692.

66. L. J. Hadjileontiadis and S. M. Panas, Adaptive reduction of heart sounds from lung sounds using fourth-order statistics. *IEEE Trans. Biomed. Eng.* 1997; **44**(7):642–648.

67. J. Gnitecki, Z. Moussavi, and H. Pasterkamp, Recursive least squares adaptive noise cancellation filtering for heart sound reduction in lung sounds recordings. Proc. 25th Annu. Int. Conf. IEEE/EMBS 2003: 2416–2419.

68. I. Hossain and Z. Moussavi, An overview of heart-noise reduction of lung sound using wavelet transform based filter. Proc. 25th Annu. Int. Conf. IEEE/EMBS 2003: 458–461.

69. L. J. Hadjileontiadis and S. M. Panas, A wavelet-based reduction of heart sound noise from lung sounds. *Int. J. Med. Infor.* 1998; **52**:183–190.
70. R. Coifman and M. V. Wickerhauser, Adapted waveform 'denoising' for medical signals and images. *IEEE Eng. Med. Biol. Soc.* 1995; **14**(5):578–586.
71. H. A. Mansy and R. H. Sandler, Bowel-sound signal enhancement using adaptive filtering. *IEEE Eng. Med. Biol.* 1997; **16**(6):105–117.
72. L. J. Hadjileontiadis, C. N. Liatsos, C. C. Mavrogiannis, T. A. Rokkas, and S. M. Panas, Enhancement of bowel sounds by wavelet-based filtering. *IEEE Trans. Biomed. Eng.* 2000; **47**(7):876–886.
73. R. Loudon and R. L. Murphy, Jr., Lung sounds. *Amer. Rev. Respir. Dis.* 1984; **130**:663–673.
74. M. Ono, K. Arakawa, M. Mori, T. Sugimoto, and H. Harashima, Separation of fine crackles from vesicular sounds by a nonlinear digital filter. *IEEE Trans. Biomed. Eng.* 1989; **36**(2):286–291.
75. L. J. Hadjileontiadis and S. M. Panas, Nonlinear separation of crackles and squawks from vesicular sounds using third-order statistics. Proc. 18th Annu. Int. Conf. IEEE/EMBS 1996: 2217–2219.
76. L. J. Hadjileontiadis and S. M. Panas, Separation of discontinuous adventitious sounds from vesicular sounds using a wavelet-based filter. *IEEE Trans. Biomed. Eng.* 1997; **44**(12):1269–1281.
77. Y. A. Tolias, L. J. Hadjileontiadis, and S. M. Panas, A fuzzy rule-based system for real-time separation of crackles from vesicular sounds. Proc. 19th Annu. Int. Conf. IEEE/EMBS. 1997: 1115–1118.
78. Y. A. Tolias, L. J. Hadjileontiadis, and S. M. Panas, Real-time separation of discontinuous adventitious sounds from vesicular sounds using a fuzzy rule-based filter. *IEEE Trans. Biomed. Eng.* 1998; **2**(3):204–215.
79. S. Chen, C. F. N. Cowan, and P. M. Grant, Orthogonal least squares learning algorithm for radial basis functions network. *IEEE Trans. Neural Networks* 1991; **2**:302–309.
80. P. A. Mastorocostas, Y. A. Tolias, J. B. Theocharis, L. J. Hadjileontiadis, and S. M. Panas, An orthogonal least squares-based fuzzy filter for real-time analysis of lung sounds. *IEEE Trans. Biomed. Eng.* 2000; **47**(9):1165–1176.
81. L. J. Hadjileontiadis, Y. A. Tolias, and S. M. Panas, Intelligent system modeling of bioacoustic signals using advanced signal processing techniques. In: C. T. Leondes, ed., *Intelligent Systems: Technology and Applications*, vol. 3. Boca Raton, FL: CRC Press, 2002, pp. 103–156.
82. L. J. Hadjileontiadis and I. T. Rekanos, Detection of explosive lung and bowel sounds by means of fractal dimension. *IEEE Signal Proc. Lett.* 2003; **10**(10):311–314.
83. J. Hoevers and R. G. Loudon, Measuring crackles. *Chest* 1990; **98**:1240–1243.
84. L. Cohen, Generalized phase-space distribution function. *J. Math. Phys.* 1966; **7**:781–786.
85. W. J. Williams and J. Jeong, Reduced interference time-frequency distributions. In: B. Boashash, ed., *Time-Frequency Signal Analysis Methods and Applications*. New York: Longman and Cheshire/Wiley Halsted Press, 1992, pp. 75–97.
86. W. J. Williams, Reduced interference distributions: biological applications and interpretations. *Proc. IEEE* 1996; **84**(9):1264–1280.
87. S. E. Widmalm, W. J. Williams, and C. Zheng, Time frequency distributions of TMJ sounds. *J. Oral Rehab.* 1991; **18**:403–412.
88. S. E. Widmalm, P. L. Westesson, S. L. Brooks, M. P. Hatala, and D. Paesani, Temporomandibular joint sounds: correlation to joint morphology in fresh autopsy specimens. *Amer. J. Orthodont. Dentofac. Orthoped.* 1992; **101**:60–69.
89. J. C. Wood and D. T. Barry, Radon transformation of time-frequency distributions for analysis of multicomponent signals. *IEEE Trans. Signal Proc.* 1994; **42**(11):3166–3177.
90. J. C. Wood and D. T. Barry, Time-frequency analysis of the first heart sound. *IEEE Eng. Med. Biol. Mag.* 1995; **14**(2):144–151.
91. J. C. Wood et al. Differential effects of myocardial ischemia on regional first heart sound frequency. *J. Appl. Physiol.* 1994; **36**(1):291–302.
92. J. C. Wood and D. T. Barry, Time-frequency analysis of skeletal muscle and cardiac vibrations. *Proc. IEEE* 1996; **84**(9):1281–1294.
93. Z. Guo, L.-G. Durand, and H. C. Lee, Comparison of time-frequency distribution techniques for analysis of simulated Doppler ultrasound signals of the femoral artery. *IEEE Trans. Biomed. Eng.* 1994; **41**:332–342.
94. Z. Guo, L.-G. Durand, and H. C. Lee, The time-frequency distributions of nonstationary signals based on a Bessel kernel. *IEEE Trans. Signal Proc.* 1994; **42**:1700–1707.
95. V. K. Iyer, P. A. Ramamoorthy, and Y. Ploysongsang, Autoregressive modeling of lung sounds: characterization of source and transmission. *IEEE Trans. Biomed. Eng.* 1989; **36**(11):1133–1137.
96. L. J. Hadjileontiadis and S. M. Panas, Autoregressive modeling of lung sounds using higher-order statistics: Estimation of source and transmission. In: A. Petropulu, ed., Proc. IEEE Signal Processing Workshop on Higher-Order Statistics '97 (SPW-HOS '97), Banff, Alberta, Canada: IEEE Signal Processing Society, 1997, pp. 4–8.
97. C. L. Nikias and A. P. Petropulu, *Higher-Order Spectra Analysis: A Nonlinear Signal Processing Framework*, 1st ed. Englewood Cliffs, NJ: Prentice-Hall, 1993, chaps. 1–2.
98. A. Swami, J. M. Mendel, and C. L. Nikias, *Higher-Order Spectral Analysis Toolbox*, 3rd ed. Natick, MA: The Mathworks, 1998, chap. 1.
99. L. J. Hadjileontiadis, Analysis and processing of lung sounds using higher-order statistics-spectra and wavelet transform, Ph.D. dissertation, Aristotle University of Thessaloniki, Thessaloniki, Greece, 1997, pp. 139–175.
100. C. L. Nikias and M. Shao, *Signal Processing with Alpha-Stable Distributions and Applications*, 1st ed. New York: Wiley & Sons, 1995, chaps. 1–7.
101. W. Feller, *An Introduction to Probability Theory and Its Applications*, vol. 2. New York: Wiley & Sons, 1996.
102. L. J. Hadjileontiadis and S. M. Panas, On modeling impulsive bioacoustic signals with symmetric α-Stable distributions: application in discontinuous adventitious lung sounds and explosive bowel sounds. Proc. 20th Annu. Int. Conf. IEEE/EMBS, 1998: 13–16.
103. J. H. McCulloch, Simple consistent estimators of stable distribution parameters. *Commun. Statist. Simula.* 1986; **15**(4):1109–1136.
104. I. A. Koutrouvelis, An iterative procedure for the estimation of the parameters of stable laws. *Commun. Statist. Simula.* 1981; **10**(1):17–28.
105. S. K. Chowdhury and A. K. Majumder, Digital spectrum analysis of respiratory sound. *IEEE. Trans. Biomed. Eng.* 1981; **28**(11):784–788.

106. T. R. Fenton, H. Pasterkamp, A. Tal, and V. Chernick, Automated spectral characterization of wheezing in asthmatic children. *IEEE Trans. Biomed. Eng.* 1985; **32**(1):50–55.
107. L. J. Hadjileontiadis and S. M. Panas, Nonlinear analysis of musical lung sounds using the bicoherence index. Proc. 19th Annu. Int. Conf. IEEE/EMBS. 1997: 1126–1129.
108. L. J. Hadjileontiadis and S. M. Panas, Discrimination of heart sounds using higher-order statistics. Proc. 19th Annu. Int. Conf. IEEE/EMBS, 1997: 1138–1141.
109. C. N. Liatsos, L. J. Hadjileontiadis, C. C. Mavrogiannis, T. A. Rokkas, and S. M. Panas, On revealing new diagnostic features of bowel sounds using higher-order statistics. *Digestion* 1998; **59**(3):694.
110. L. J. Hadjileontiadis, A. J. Giannakidis, and S. M. Panas, α-Stable modeling: a novel tool for classifying crackles and artifacts. Proc. 25th Annu. Int. Lung Sounds Association Conference ILSA 2000, H. Pasterkamp, ed., Chicago, IL, 2000.
111. B. Kedem, *Time Series Analysis by Higher Order Crossings*. New York: IEEE Press, 1994, chaps. 1,4, and 8.
112. L. J. Hadjileontiadis, T. P. Kontakos, C. N. Liatsos, C. C. Mavrogiannis, T. A. Rokkas, and S. M. Panas, Enhancement of the diagnostic character of bowel sounds using higher-order crossings. Proc. 1st Joined Annu. Int. Conf. IEEE BMES/EMBS, 1999: 1027.
113. L. J. Hadjileontiadis, Discrimination analysis of discontinuous breath sounds using higher-order crossings. *Med. Biol. Eng. Comput.* 2003;**41**:445–455.
114. L. J. Hadjileontiadis, C. D. Saragiotis, and S. M. Panas, Discrimination of lung sounds using higher-order crossing. Proc. 24th Annu. Int. Lung Sounds Association Conference ILSA 1999. P. Wichert, ed., Marburg, Germany, 1999.
115. C. Liatsos, L. J. Hadjileontiadis, C. Mavrogiannis, D. Patch, S. M. Panas, and A. K. Burroughs, Bowel sounds analysis: a novel noninvasive method for diagnosis of small-volume ascites. *Dig. Dis. Sci.* 2003; **48**(8):1630–1636.
116. T. Sano, S. E. Widmalm, P. L. Westesson, T. Yamaga, M. Yamamoto, K. Takahashi, K. I. Michi, and T. Okano, Acoustic characteristics of sounds from temporomandibular joints with and without effusion: an MRI study. *J. Oral Rehabil.* 2002; **29**(2):161–166.
117. S. E. Widmalm, W. J. Williams, B. K. Ang, and D. C. McKay, Localization of TMJ sounds to side. *J. Oral Rehabil.* 2002; **29**:1–7.
118. J. Radke, R. Garcia, Jr., and R. Ketcham, Wavelet transform of TM joint vibrations: a feature extraction tool for detecting reducing displaced disks. *Cranio* 2001; **19**(2):84–90.
119. M. L. Brown, W. J. Williams, and S. E. Widmalm, Automatic classification of temporomandibular joint sounds. Proc. Artificial Neural Network in Engineering (ANNIE) Conf. 1994: 725–730.
120. M. L. Brown, W. J. Williams, and A. O. Hero, III, Non-orthogonal Gabor representations of biological signals. *Proc. IEEE-SP Int. Conf. Acoust. Speech Sig. Proc.* 1994; **4**:305–308.
121. K. P. Yang, D. Djurdjanovic, K. H. Koh, W. J. Williams, and S. E. Widmalm, Automatic classification of the temporomandibular joint sounds using scale and time-shift-invariant representation of their time-frequency distributions. Proc. IEEE Int. Symp. Time-Frequency and Time-Scale Analysis, Pittsburgh, PA, 1998: 265–268.
122. D. Djurdjanovic, S. E. Widmalm, W. J. Williams, C. K. Koh, and K. P. Yang, Computerized classification of temporomandibular joint sounds. *IEEE Trans. Biomed. Eng.* 2000; **47**(8):977–984.
123. I. Matsuura and M. Naito, An analysis of knee joint sound in patients with OA, PF disorders and meniscal lesions (efficacy of HA administration). *Med. Bull. Fukuoka Univ.* 2000; **27**(2):87–92.
124. S. G. Mallat and Z. Zang, Matching pursuit with time-frequency dictionaries. *IEEE Trans. Signal Process.* 1993; **41**:3397–3415.
125. S. Krishnan, (2002). *VAG Project.* (online) Available: *http://www.ee.ryerson.ca/~krishnan/person/VAG.html.*
126. P. V. C. Hough, Methods and means for recognition complex pattern, U.S. Patent 3,069,654, 1962.
127. T. S. Durrani and D. Bisset, The Radon transform and its properties. *Geophys.* 1984; **49**:1180–1187.

BIOACTIVE BONE CEMENTS

Enrique Fernández
Universitat Politècnica de Catalunya
Barcelona, Spain

1. INTRODUCTION

Bioactive bone cements is just a topic of the fascinating field of *biomaterials*. Before commenting on the concept of "bioactive cements," this review begins with some words from Professor Larry L. Hench who, in 1998, wrote: *"It is time to accept that the revolution of the last 30 years, the revolution of replacement of tissues by transplants and implants has run its course. It has led to a remarkable increase in the quality of life for millions of patients ... However, continuing the same approach of the last 30 years is not likely to reach a goal of 20–30 years implants survivability, a requirement of our ageing population. We need a change in emphasis in biomaterials research; in fact, we need a new revolution"* (1). The words of Professor Jonathan Black can also help to understand where exactly we are now and what kind of new approaches we should be looking for; in 1999 he wrote: *"Since my retirement ... this ... marks the end of a 30-year career as student, researcher and teacher in this field. It has been an interesting time, with gains and losses, both personally and for society. As I move on to other pursuits, I wish the very best to those who continue. May your efforts and insights continue to produce improvements in the human condition"* (2).

In order to improve our understanding of today's biomaterials field, it is interesting to note that Professor Hench (1) defined biomaterials as *"man-made materials to interface with living, host tissues,"* whereas Professor Black (2) expanded this definition to include *"materials of natural or man-made origin used to direct, supplement, or replace the functions of living tissue."* A consensus definition is that of Professor David F. Williams in 1987 (3) i.e., *"a material intended to interface with biological systems to evaluate, treat, augment, or replace any tissue, organ, or function of the body."*

The above definitions lead to the concept of biocompatibility that Professor Black solves in the following manner

(2): *"The real issue of biocompatibility is not whether there are adverse biological reactions to a material, but whether that material performs satisfactorily (that is, in the intended fashion) in the application under consideration, and can be considered a successful biomaterial."* According to this concept, biomaterials are today classified as: (a) *inert* [i.e., implantable materials that elicit little or no host response (local and systemic response, other than the intended therapeutic response, of living systems to the material)]; (b) *interactive* (i.e., implantable materials designed to elicit specific, beneficial responses, such as ingrowth, adhesion, etc); (c) *viable* (i.e., implantable materials, incorporating or attracting living cells at the implantation site, that are treated by the host as normal tissue matrices and are actively resorbed or remodeled); and (d) *replant* (i.e., implantable materials consisting of native tissue, cultured *in vitro* from cells obtained previously from the specific implant patient).

According to Prof. Black (2), a replant material (implantable, live tissue with identical genetic code and immunological determinants of the recipient patient) *"represents the fulfilment of the original search for biocompatibility: implantable materials demonstrating harmonious interaction,"* which is exactly the same idea reported by Prof. Hench (1): *"The message for the next millennium ... We need to shift the emphasis of biomaterials research towards assisting or enhancing the body's own reparative capacity ... The proposal is that we seek biomaterials that behave in a manner equivalent to an autograft, i.e. a regenerative allograft."*

The above conclusion helps to clarify the history of biomaterials where in the past the key question was the removal of tissue, in the present is the replacement of tissues, and in the future will be the regeneration of tissues. If we consider the present [i.e., the replacement of tissues by implants (*manmade materials to interface with living, host tissues*)], such biomaterials have the advantages of availability, reproducibility, and reliability (GMPs, standards, and regulations). On the other hand, the main reported disadvantage relates to the interfacial stability with host tissue (i.e., fixation of the implant) (1). Some partial solutions have promoted morphological fixation (press-fitting), cement fixation, biological fixation (porous ingrowth), or the trendy approach of Prof. Hench (1), the *bioactive fixation* (i.e., interfacial bonding of an implant to tissue through a biologically active *hydroxyapatite* layer on the implant surface). In this sense, materials showing bioactive fixation are called *bioactive implants*. A more general definition is that a bioactive material is a biomaterial that is designed to elicit or modulate biological activity (2). However, the term bioactive is often related to the presence of the bone-like mineral phase hydroxyapatite. In this sense, *bioactive bone cements* are those cementitious materials (calcium phosphate based) that through setting reactions form a scaffold of interconnecting crystals of bone apatite-like mineral phase.

In this chapter, the authors will approach this topic to give the reader some insights on the evolution of *bioactive bone cements* since their discovery in 1985 by Brown and Chow (4) until the present time, where *minimally invasive surgery* (5,6) and *tissue engineering* (7) are the key issues controlling further developments.

2. THE CONCEPT OF CEMENT MATERIAL

Cements are made up of a powder and a liquid phase that, on mixing, form slurry-like materials that set with time to a solid body. Bioactive bone cements are made up of a mixture of calcium phosphates (8) and an aqueous-base liquid solution (see Fig. 1). The physical, the chemical, or the mechanical properties $P(x_i,t)$ that characterize the setting of these materials can be approximated by mathematical functions $f(x_i,t)$ depending on the experimental factors x_i and the time t (see Equation 1). This procedure has been often used in experimental design (9) to plan for new product development processes in bioactive bone cement's research (10).

$$P(x_i,t) \equiv f(x_i,t). \quad (1)$$

3. THE CONCEPT OF SETTING

The setting of bioactive bone cements is a continuous process that involves dissolution and precipitation chemical reactions (11). During these processes, new crystals are formed; with time, the crystals grow and entangle, so the initial cement-like material loses its viscofluid properties into a solid body (12). The entire process is called *the setting of cement*. However, in bioactive bone cements, it is often distinguished into the *setting* and the *hardening* of the cement. The *setting* is identified to the period of time from the initial powder and liquid mixture until the cement starts to lose the viscofluid properties. On the other hand, the *hardening* makes reference to the period between the setting (as defined before) and the full transformation of the cement into a solid body, when no further chemical reactions are observed (see Fig. 1).

3.1. Characterization of the Setting Period

Some physical properties that are often measured to characterize the setting period are the *swelling time* $S(x_i)$, the *initial setting time* $I(x_i)$, and the *final setting time* $F(x_i)$ (see Fig. 1). The swelling time S, also called the cohesion time (13), is defined as the time needed, after mixing the cement's powder and liquid phase into a homogeneous paste, not to observe any cement disintegration upon early contact with liquid phases (water, Ringer's solution, simulated body fluid or blood) (14–16) (see Fig. 2). The initial setting time I, which marks the start of the setting period, has been identified to the maximum time the surgeon has to implant the homogeneous cement paste into the body without affecting the microstructure of the cement and, as a result, its final properties after *hardening* (17,18). Similarly, the final setting time F, which marks the end of the setting and the start of the hardening period, has been defined as the maximum time the surgeon must wait before closing and stressing the wound without affecting the mechanical stability and the end properties of the cement implant (17,18).

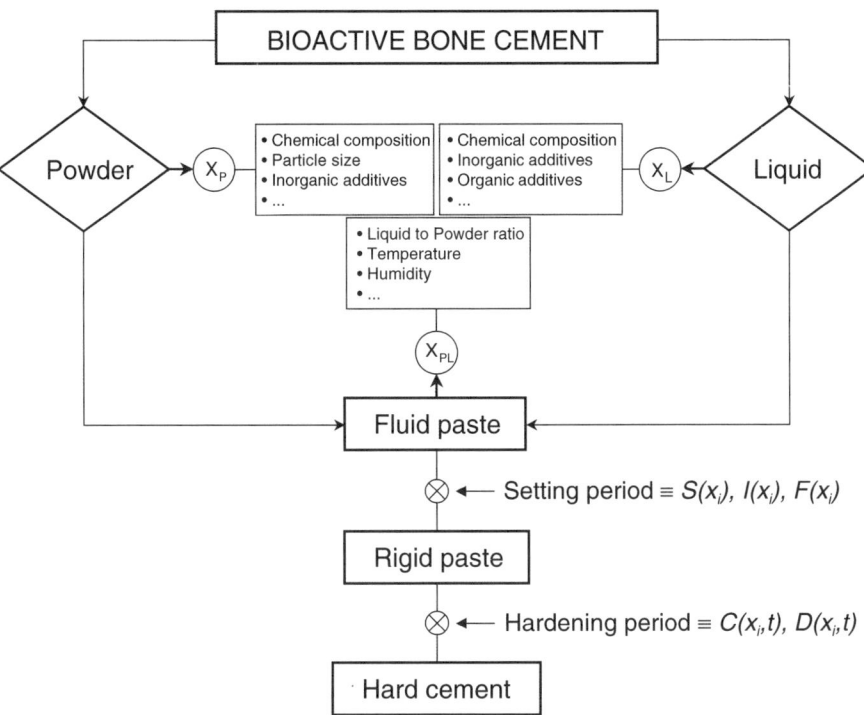

Figure 1. Block diagram of bone cement's technology. Some factors affecting the powder (X_P), the liquid (X_L), and the mixing (X_{PL}) are listed. Distinction between the setting and the hardening periods are clearly stated. (Note: S, I, and F are the swelling, the initial, and the final setting times, respectively; these characteristic times can be affected by X_P, X_L, or X_{PL} factors. C and D are the compressive and the diametral-tensile strength, respectively; these properties are also affected by the above factors but also change with time t. See the text for further details.)

The above definitions are used in experimental design to optimize research or commercial bioactive bone cements. In this sense, researchers are looking for bioactive bone cements with swelling times approaching zero ($S \to 0$) (15,16) or at least with swelling times being much lower than the initial setting time ($S \ll I$) (19,20), otherwise cement disintegrates upon contact with body fluids. Optimum values for the initial and final setting times in clinical procedures have been reported to be $4 < I(\min) < 8$ and $10 < F(\min) < 15$ (18,21).

The setting times are measured using the standards ASTM C191-92 (*Vicat* needles) (22) or ASTM C266-89 (*Gillmore* needles) (23) (see Fig. 3). Basically, both standards measure a static pressure. The procedure is as follows: (a) a needle of certain diameter and weight is rested, at control times, onto a flat surface of certain volume of

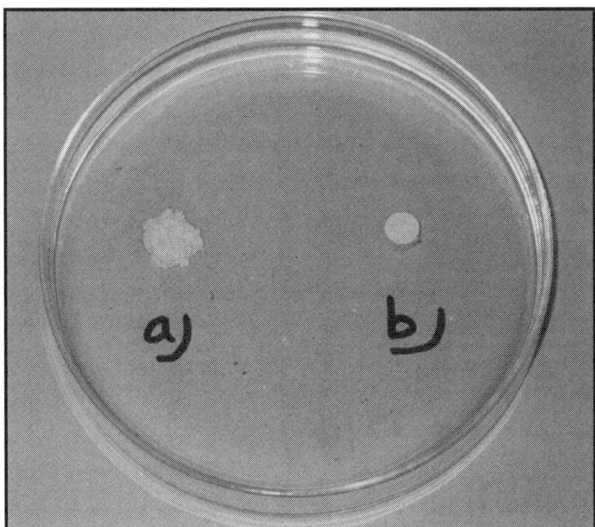

Figure 2. The cement sample on the left shows disintegration upon early contact with the liquid phase; after the swelling time, the cement on the right shows no disintegration. (This figure is available in full color at http://www.mrw.interscience.wiley.com/ebe.)

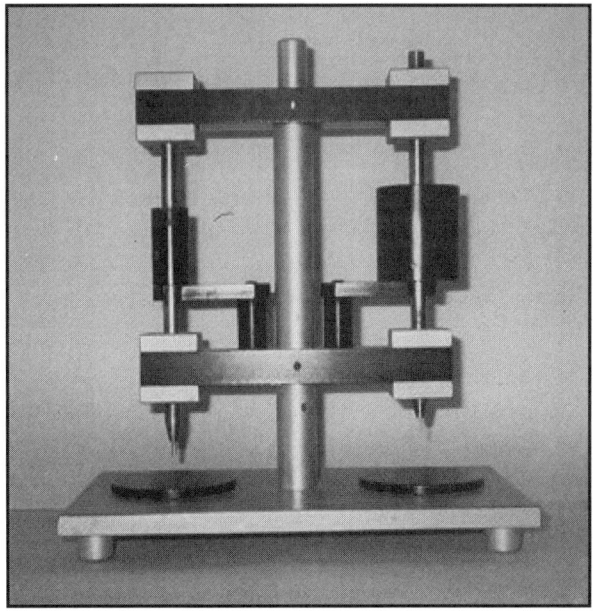

Figure 3. Gillmore needles according to the standard ASTM C266-89. The light (heavy) needle on the left (right) is used to measure the initial setting time (final setting time).

cement; (b) when no visual mark is let down onto the surface by the resting needle, the cement is considered to be set. Whereas the *Vicat* method only uses a needle to determine *the final setting time F*, the *Gillmore* method uses two different needles: one for the *initial setting time I* and the other for the *final setting time F*. Most researchers prefer the *Gillmore* method because it also gives a measure for *I*. The equivalent static pressures applied by the *Gillmore* needles are 0.3 MPa for *I* and 5 MPa for *F* (23). It is found experimentally that *F* doubles the *I* time ($F \approx 2I$) (19).

The *Vicat* and *Gillmore* methods have been used since the discovery of calcium phosphate cements (11,24). However, these methods have not avoided criticism because of their inherent subjectivity (visual inspection) and to the relative value that setting times have in minimally invasive surgery procedures, such as spinal surgery (*vertebro-* or *kyphoplasties*), where the cement should be injected far before the completion of the *initial setting time I*; otherwise it is impossible to inject the cement into the implant site. For this reason, some authors have started to record the *setting period* in a continuous manner (25,26) as well as to point out that the lack of injectability of the present bioactive bone cements is the key issue to be solved in today's minimally invasive surgery (10,27).

3.2. Characterization of the Hardening Period: Mechanical Point of View

The hardening period (transition from viscous to solid state) is characterized by measuring the evolution of some mechanical property. For convenience, the *compressive strength* $C(x_i, t)$ or the *diametral tensile strength* $D(x_i, t)$ are often used (28). Most authors use the compressive strength and follow the standard *ISO 9917-1* (29) for dental cements. Basically, compressive strength samples of 6 mm of diameter and 12 mm of height are made using teflon or stainless-steel molds. Then, samples are allowed

Figure 4. Cement samples are tested under compression in an electromechanical testing machine. The maximum compressive strength at failure is recorded against the reaction time to obtain the hardening curve [i.e., $C(x_i, t)$]. (This figure is available in full color at http://www.mrw.interscience.wiley.com/ebe.)

to set for different fixed times at near *in vivo* conditions (i.e., 37°C and 100% humidity, by immersion in a liquid solution (water, Ringer's solution, or simulated body fluid). Then, the cement samples are immediately tested under compression in an electromechanical testing machine at a normal cross-head speed of 1 mm/min, and the maximum compressive strength before fracture is recorded (30) (see Fig. 4).

The optimum procedure to characterize the hardening of bioactive bone cements is by recording the whole *hardening curve* (i.e., the compressive strength $C(x_i, t)$ for a fixed and controllable number of experimental factors x_i as a function of time t), from after the final setting time F (if possible) until saturation with time (i.e., no further evolution of the compressive strength) (30–35) (see Fig. 5).

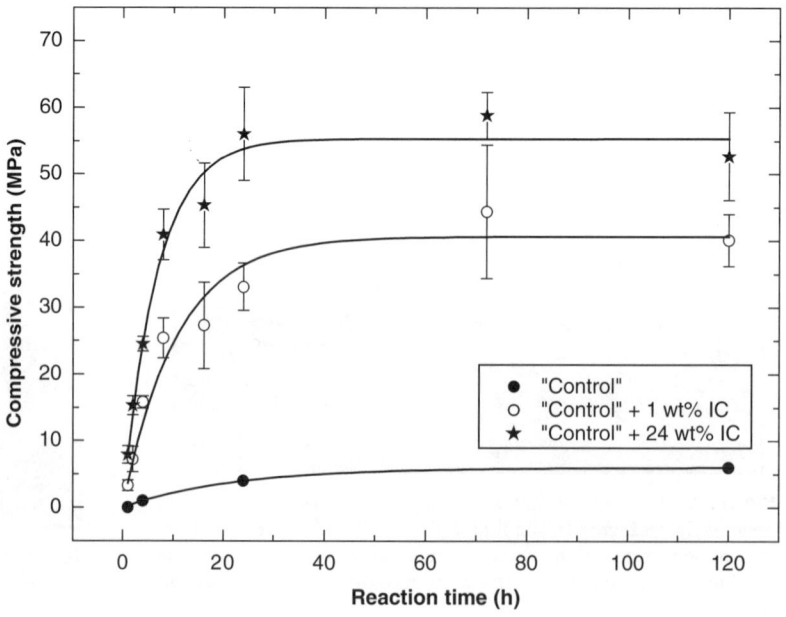

Figure 5. Hardening curves obtained for a bioactive bone cement after modification with iron citrate (adapted from Ref. 35, with kind permission of Springer Science and Business Media).

This way, relevant information is obtained about the chemical and physical mechanisms that are involved during the hardening period. These mechanisms are confirmed by other characterization techniques (*Scanning Electron Microscopy* or *X-ray Diffraction*) (31–35), which is commented later on.

Moreover, it is the only way two different cements (i.e., cements made with different experimental factors x_i) can be objectively compared, and it is the only way to study seriously the influence of several experimental factors x_i on the mechanical properties. Following this procedure, some authors have recorded the hardening curve and modeled the hardening process of bioactive bone cements following Equation 2 (31–35) (see Fig. 5).

$$C(x_i, t) = C_0(x_i)^* [1 - \exp(-t/\tau_c(x_i))] \quad (2)$$

The information obtained through the mechanical characteristic time constant $\tau_c(x_i)$, which is a function of the experimental factors x_i, has linked the microstructural development of the crystal scaffold network, responsible for the mechanical strength, to the chemical reactivity of the powder cement's phases (31–35). Moreover, this approximation has put forward the relevance of kinetic *versus* thermodynamic studies.

Unfortunately, since the discovery of bioactive bone cements (4,11) most authors have reported the compressive strength at only certain times (28,36) (i.e., after 24 h of setting) in accordance to the standard of dental cements (29), which has been the consequence of traditionally looking for bioactive bone cements passing certain "required" threshold to be considered for further optimization. In this sense, it is not rare to find in the past literature certain discrepancies among authors claiming for high [i.e., 90 Mpa (28,36,37), 55 Mpa (38,39), 40 Mpa (4,12)] or low [i.e., 5–10 Mpa (40)] compressive strength values, all suitable values for the claimed bone applications.

Whatever the strength threshold, experience has shown that bioactive bone cements (i.e., a ceramic material) should be intended for nonload bone bearing applications (6). For this reason today, strength values around the compressive strength of cancellous bone (i.e., 10–30 MPa) are considered enough for these materials, and research has been focused on more relevant problems such as their injectability (5,6,10,41–44) or their stability and transformation into bone tissue *in vivo* (45,46).

3.3. Characterization of the Hardening Period: Microstructural Point of View

Bioactive bone cements are intended for bone filling applications. It is important to remember that the trendy approach of biomaterials for this century is the replant-type materials or tissue engineering (see the Introduction). In this sense, a key issue in bioactive cements' research has been how to improve the osteointegration and further transformation of the cement implant into real bone tissue. Some authors have used the concept of *osteotransductive bone cements* to explain that the microstructure of these materials (network of chemically entangled apatite-like crystals) are slowly resorbed and replaced by new bone tissue without loss of mechanical stability during transformation (47).

Since the discovery of bioactive bone cements (4,11), researchers have followed by scanning electron microscopy (SEM) the evolution of the crystal-network structure developed during the cement's hardening, something like a function $\Omega(x_i, t)$, which obviously has not been measured but has been visually observed and reported (see Fig. 6). The information gained on the hardening behavior of bioactive bone cements has been tremendous. It has been possible to observe how the microstructure changed completely by only adding, for example, minor additions of citric acid (32) or polymeric additives (48) or how the porosity developed within the cement (34,48,49) (see Fig. 7). Most important, these observations have helped to explain the short-range and long-range mechanical stability of these materials during setting. The processes of dissolution, nucleation, precipitation, growth, and entanglement of apatite-like crystals have been perfectly reported (32,33,48,50) (see Fig. 8). This information has also served to point out the relevance of kinetic *versus* thermodynamic studies in calcium phosphate cements.

3.4. Characterization of the Hardening Period: Chemical Point of View

The chemistry (thermodynamics) of calcium phosphate cements is well known and for a long time has been the basis for further material's development. In the literature, many good reviews exist that explain the main facts of chemical cement's technology. Interested readers should address them (4,6,8,11,51,52). In particular, concepts such as *solubility isotherms*, *singular points*, *supersaturated solution*, *dissolution*, and *precipitation* should be well understood. Although this knowledge has served to identify those thermodynamically optimum calcium phosphate powder mixtures (6,51,52), as well as the expected crystallization steps, in practice, the actual evolution of the cement's chemistry has been followed by *X-ray diffraction* (XRD) (12,31–34,50,53).

Some authors have followed at the same time the evolution of the compressive strength (as reported before) and the evolution of the chemical phases disappearing (dissolution of cement's powder phases) or appearing (precipitation of new crystal phases) during the setting/hardening process. Similarly, the *extent of chemical reaction* [i.e., $R(x_i, t)$] has been calculated for both the dissolved or the new precipitated phases according to Equation 3, and comparison with Equation 2 has led to very interesting results and conclusions (31,32,34).

$$R(x_i, t) = R_0(x_i)^* [1 - \exp(-t/\tau_r(x_i))] \quad (3)$$

For example, it has been made clear throughout the values of $\tau_c(x_i)$ and $\tau_r(x_i)$ that the cement's strengthening and the extent of chemical reaction (chemical reactivity) are directly related (i.e., both are different measures (observations) of the same interrelated processes (dissolution, nucleation, precipitation, growth, and entanglement of apatite-like crystals) (31,32,34). It has also been observed that the dissolution of the main cement's powder reactant, accounted by XRD through the time constant

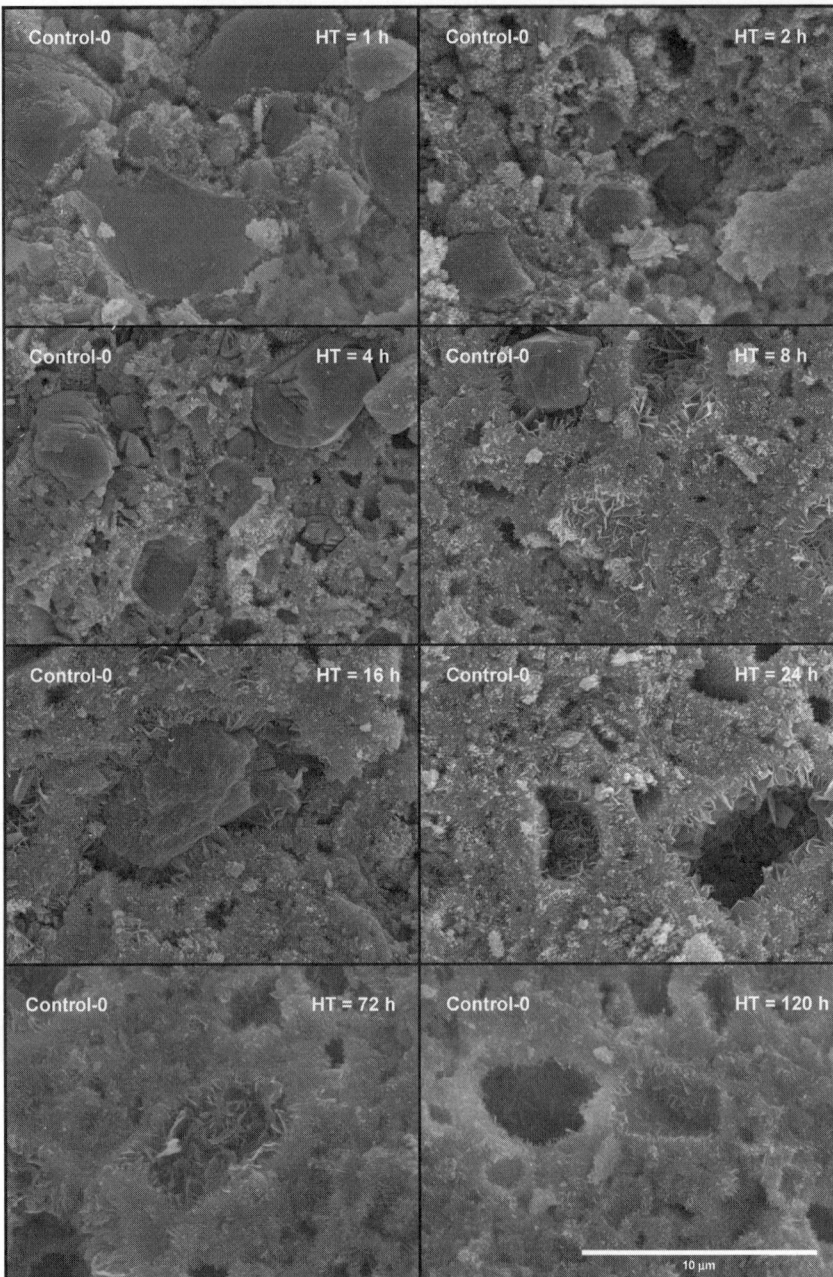

Figure 6. Scanning electron pictures, taken at different hardening times, showing the evolution of the microstructure of a bioactive bone cement (reprinted from Ref. 34, with kind permission of Elsevier).

$\tau_r^d(x_i)$, is directly followed by the precipitation of the apatite phase [i.e., $\tau_r^p(x_i) \geq \tau_r^d(x_i)$], which means that some grade of dissolution is needed before precipitation can occur, which also agrees with the fact that some precipitation is also needed before some mechanical strength can be measured [i.e., $\tau_c(x_i) \geq \tau_r^p(x_i)$] (31,32,34). Most complex analyses have also clarified, with the supporting help of SEM microstructure observations, when the chemical reactions were being controlled by the available surface area for dissolution of the main cement powder reactant or by diffusion through a shell of precipitated apatite crystals surrounding the main reactant particles (32,33,50) (see Fig. 9).

In general, these studies have put forward the relevance of the kinetic *versus* the thermodynamic approximation to the study of the setting of bioactive (calcium phosphate-based) cements. Moreover, the understanding that cement's properties can be modulated, according to Equation 1, by selecting a proper set of experimental factors x_i has been the start of a new revolution in the experimental design of bioactive bone cements. In the next section, the attention is put on these experimental factors.

4. FACTORS AFFECTING CEMENT'S PROPERTIES

Equation 1 makes clear that any measurable cement property (at some time t) is a complicated function depending on several experimental factors x_i (controlled or not).

Some of the main factors x_i affecting the powder phase (and so the end-setting properties of the cement)

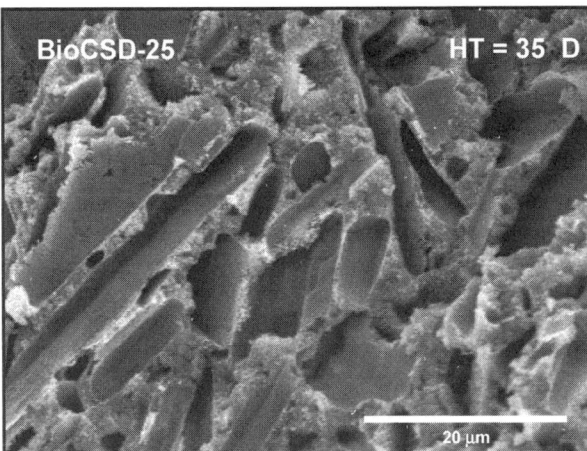

Figure 7. This bioactive bone cement, named BioCSD-25, developed macroporosity after 35 days of setting when modified with calcium sulphate dihydrate crystals (adapted from Ref. 34, with kind permission of Elsevier).

are: (a) the chemical nature of the reactants (6,52), (b) the number and the relative weight proportion of the reactants (19,30,52); (c) the particle size of the reactants (33,44); and (d) the presence of some minor additives (accelerators or retarders) (44,54). Similarly, the main factor affecting the aqueous liquid phase is its chemical purity because of the presence of some additives (accelerators, retarders, fluidifiers, etc.) (20,44,48). The main factor affecting the mixing is the liquid-to-powder (L/P) ratio, acting on both the setting times and the cement's rheology (44,55); other factors are the temperature (i.e., at higher temperatures cements set faster) and the humidity (43,56,57) (see Fig. 1).

It is evident from the above description that cement optimization is an iterative and complicate process. For example, optimum bioactive bone cement could be defined as that cement that shows no swelling, short setting times, maximum strength, good injectability, or the ability to develop porosity. However, experience shows (as an example) that in order to shorten setting times and increase the ultimate strength, it is good to have reactants with small particle size (33), which is understandable because from thermodynamics small particle size (high surface area) show high chemical reactivity (faster dissolution, supersaturation is attained earlier and, as a result, precipitation).

However, if the L/P ratio is maintained constant the swelling often increases (19), which is also understandable because if surface area is increased, more liquid is adsorbed onto the particles and so the cement paste is dryer (19), which makes the cement impossible to inject despite the fact that it could show higher strength at saturation (19,33). This example is clear on how factors x_i, interact between them in an experimental design to favor certain properties (setting times or strength) in front of others (injectability or porosity). For this reason, bioactive bone cements should be optimized having in mind its

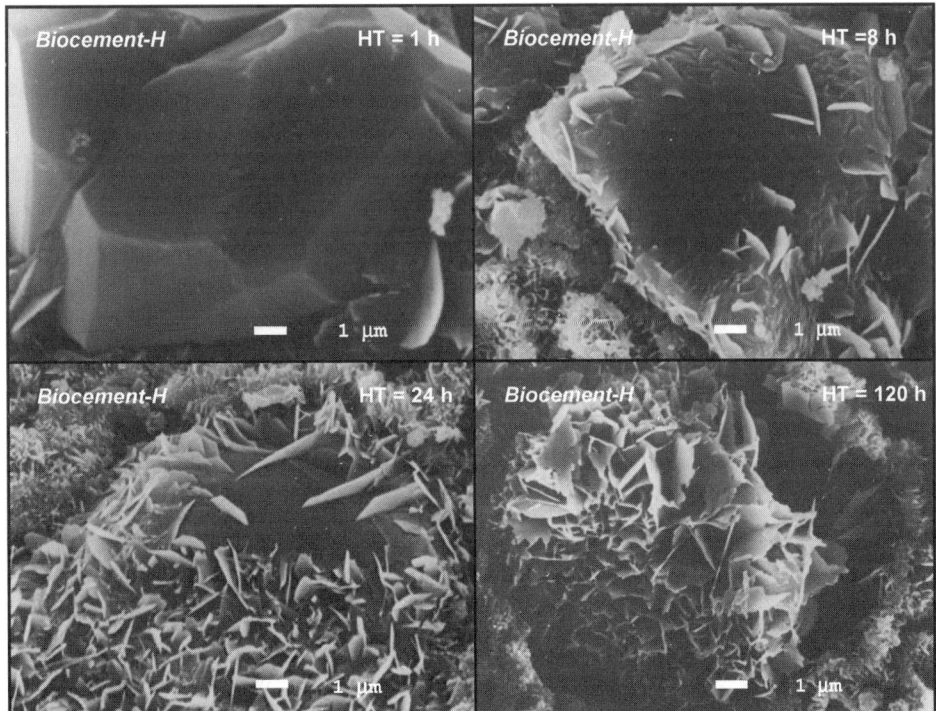

Figure 8. Scanning Electron Microscopy pictures showing different stages of *Biocement-H*$^{©}$'s setting (left-top: dissolution of α-tricalcium phosphate particles after 1 h of setting; right-top: nucleation and growth of apatite crystals after 8 h; left-bottom: further surface-control growth after 24 h; right-bottom: further diffusion-control growth after 120 h) (reprinted from Ref. 48, with kind permission of Elsevier).

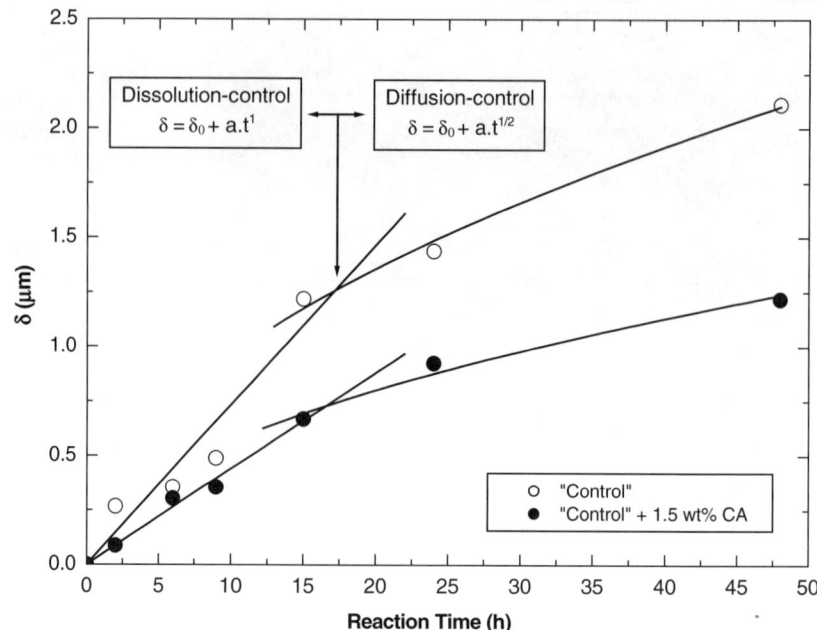

Figure 9. Bioactive bone cement modified with citric acid. Certain models determine when the setting reaction was controlled by dissolution or by diffusion (adapted from Ref. 32).

final clinical application, which is why the opinion of the final user (i.e., the surgeon (clinical procedure)) is so important and should be considered from the start of any cement product development.

A critical assessment on how to handle these technological issues to develop more efficient calcium phosphate bone cements has been recently published (58); interested readers should carefully look at it to see that critical properties, such as the injectability, are today open questions where several researchers are focusing their efforts. It is expected that more publications will appear soon on this direction.

5. FINAL REMARK

It should be mentioned that the term *bioactive bone cements* makes reference in this review to those cementitious materials made exclusively by bioactive materials (i.e., calcium phosphate cements). However, it has been suggested by Prof. Oonishi (59) that three types of bioactive bone cements exist: (a) *bioactive bone cements* where the whole material is bioactive [i.e., calcium phosphate cements (the focus of the present review)]; (b) *surface-bioactive bone cements*, where, for example, a bioactive filler particle is added to a nonbioactive matrix (i.e., hydroxyapatite added to polymeric cements such as poly-methylmethacrylate); and (c) *interface-bioactive bone cements* where a bioactive material is placed between the bone and the nonbioactive cement (i.e., layers of hydroxyapatite granules between bone cement). Those readers interested in surface-bioactive bone cements can also look at the review made by Dr. Harper (60).

Acknowledgment

I appreciate the improvements made on this review by Prof. Dr. Marc Bohner (Robert Mathys Foundation, Bischmattstrasse 12, CH-2544 Bettlach, Switzerland). I thank the *Ministerio de Educación y Ciencia* and the *Generalitat de Catalunya* (Spain) for funding my research through projects MAT2005-02778 and 2005SGR-00732, respectively.

BIBLIOGRAPHY

1. L. L. Hench, Biomaterials: a forecast for the future. *Biomaterials* 1998; **19**:1419–1423.
2. J. Black, In: *Biological Performance of Materials: Fundamentals of Biocompatibility*, 3rd ed. New York: Marcel Dekker, 1999.
3. D. F. Williams, ed., Definitions in biomaterials. Proceedings of a Consensus Conference of the European Society for Biomaterials, Chester, UK, March 3–5, 1986, Elsevier Science Publishers B.V., Amsterdam, 1987.
4. W. E. Brown and L. C. Chow, Dental restorative cement pastes, U.S. Patent 4,518,430, May 21, 1985.
5. M. Mushipe, Injectable micro-particles and pastes for minimally invasive orthopaedic surgery. MD Technology Watch Series, Article 2, August 2003.
6. M. Bohner, Physical and chemical aspects of calcium phosphates used in spinal surgery. *Eur. Spine J.* 2001; **10**:S114–S121.
7. J. E. Barralet, L. Grover, T. Gaunt, A. J. Wright, and I. R. Gibson, Preparation of macroporous calcium phosphate cement tissue engineering scaffold. *Biomaterials* 2002; **23**:3063–3072.
8. K. De Groot, ed., In: *Bioceramics of Calcium Phosphates*. Boca Raton, FL: CRC Press, 1983.
9. G. E. P. Box, W. G. Hunter, and J. S. Hunter, In: *Statistics for Experimenters: An Introduction to Design, Data Analysis, and Model Building*. New York: John Wiley & Sons, 1978.
10. M. Bohner and G. Baroud, Injectability of calcium phosphate pastes. *Biomaterials* 2005; **26**:1553–1563.
11. W. E. Brown and L. C. Chow, A new calcium phosphate water-setting cement. In: *Cements Research Progress*. P. W. Brown,

ed., Westerville, OH: American Ceramic Society, 1986, pp. 351–379.

12. M. P. Ginebra, E. Fernández, E. A. P. De Maeyer, R. M. H. Verbeeck, M. G. Boltong, J. Ginebra, F. C. M. Driessens, and J. A. Planell, Setting reaction and hardening of an apatitic calcium phosphate cement. *J. Dent. Res.* 1997; **76**(4):905–912.

13. I. Khairoun, M. G. Boltong, F. C. M. Driessens, and J. A. Planell, Effect of calcium carbonate on the compliance of an apatitic calcium phosphate bone cement. *Biomaterials* 1997; **18**:1535–1539.

14. E. Fernández, M. G. Boltong, M. P. Ginebra, F. C. M. Driessens, O. Bermúdez, and J. A. Planell, Development of a method to measure the period of swelling of calcium phosphate cements. *J. Mater. Sci. Lett.* 1996; **15**:1004–1005.

15. K. Ishikawa, Y. Miyamoto, M. Kon, M. Nagayama, and K. Asaoka, Non-decay type fast-setting calcium phosphate cement: composite with sodium alginate. *Biomaterials* 1995; **16**:527–532.

16. M. Takechi, Y. Miyamoto, K. Ishikawa, M. Yuasa, M. Nagayama, M. Kon, and K. Asaoka, Non-decay type fast-setting calcium phosphate cement using chitosan. *J. Mater. Sci. Mater. Med.* 1996; **7**:317–322.

17. F. C. M. Driessens, M. G. Boltong, O. Bermúdez, and J. A. Planell, Formulation and setting times of some calcium orthophosphate cements: a pilot study. *J. Mater. Sci. Mater. Med.* 1993; **4**:503–508.

18. M. P. Ginebra, E. Fernández, M. G. Boltong, O. Bermúdez, J. A. Planell, and F. C. M. Driessens, Compliance of an apatitic calcium phosphate cement with the short-term clinical requirements in bone surgery, orthopaedics and dentistry. *Clin. Mater.* 1994; **17**:99–104.

19. E. Fernández, F. J. Gil, M. P. Ginebra, F. C. M. Driessens, J. A. Planell, and S. M. Best, Production and characterization of new calcium phosphate bone cements in the $CaHPO_4$-α-$Ca_3(PO_4)_2$ system: pH, workability and setting times. *J. Mater. Sci. Mater. Med.* 1999; **10**:223–230.

20. I. Khairoun, F. C. M. Driessens, M. G. Boltong, J. A. Planell, and R. Wenz, Addition of cohesion promotors to calcium phosphate cements. *Biomaterials* 1999; **20**:393–398.

21. I. Khairoun, M. G. Boltong, F. C. M. Driessens, and J. A. Planell, Limited compliance of some apatitic calcium phosphate bone cements with clinical requirements. *J. Mater. Sci. Mater. Med.* 1998; **9**:667–671.

22. Standard Test Method for Time of Setting of Hydraulic Cement Paste by Vicat Needle, ASTM C191-92, *Annual Book of ASTM Standards*, vol. 04.01: *Cement, Lime, Gypsum*. Philadelphia, PA: ASTM, 1993, pp. 158–160.

23. Standard Test Method for Time of Setting of Hydraulic Cement Paste by the Gillmore Needles, ASTM C266-89, *Annual Book of ASTM Standards*, vol. 04.01: *Cement, Lime, Gypsum*. Philadelphia, PA: ASTM, 1993, pp. 189–191.

24. Y. Takezawa, Y. Doi, S. Shibata, N. Wakamatsu, H. Kamemizu, T. Goto, M. Iijima, Y. Moriwaki, K. Uno, F. Kubo, and Y. Haeuchi, Self-setting apatite cement. II. Hydroxyapatite as setting accelerator. *J. Japan Soc. Dent. Mater. Devices* 1987; **6**(4):426–431.

25. M. Nilsson, J. Carlson, E. Fernández, and J. A. Planell, Monitoring the setting of calcium-based bone cements using pulse-echo ultrasound. *J. Mater. Sci. Mater. Med.* 2002; **13**:1135–1141.

26. J. Carlson, M. Nilsson, E. Fernández, and J. A. Planell, An ultrasonic pulse-echotechnique for monitoring the setting of $CaSO_4$-based bone cement. *Biomaterials* 2003; **24**:71–77.

27. G. Baroud, M. Bohner, P. Heini, and T. Steffen, Injection biomechanics of bone cements used in vertebroplasty. *Biomed. Mater. Eng.* 2004; **14**(4):487–504.

28. O. Bermúdez, M. G. Boltong, F. C. M. Driessens, and J. A. Planell, Compressive strength and diametral tensile strength of some calcium-orthophosphate cements: a pilot study. *J. Mater. Sci. Mater. Med.* 1993; **4**:389–393.

29. ISO 9917-1. Dentistry. Water-based cements. P. I: Powder/liquid acid-base cements. Geneva: ISO, 2003.

30. E. Fernández, F. J. Gil, S. M. Best, M. P. Ginebra, F. C. M. Driessens, and J. A. Planell, Improvement of the mechanical properties of new calcium phosphate bone cements in the $CaHPO_4$-α-$Ca_3(PO_4)_2$ system: compressive strength and microstructural development. *J. Biomed. Mater. Res.* 1998; **41**:560–567.

31. E. Fernández, M. P. Ginebra, M. G. Boltong, F. C. M. Driessens, J. Ginebra, E. A. P. De Maeyer, R. M. H. Verbeeck, and J. A. Planell, Kinetic study of the setting reaction of calcium phosphate bone cement. *J. Biomed. Mater. Res.* 1996; **32**:367–374.

32. S. Sarda, E. Fernández, M. Nilsson, M. Balcells, and J. A. Planell, Kinetic study of citric acid influence on calcium phosphate bone cements as water-reducing agent. *J. Biomed. Mater. Res.* 2002; **61**:653–659.

33. M. P. Ginebra, F. C. M. Driessens, and J. A. Planell, Effect of the particle size on the micro and nanostructural features of a calcium phosphate cement: a kinetic analysis. *Biomaterials* 2004; **25**:3453–3462.

34. E. Fernández, M. D. Vlad, M. M. Gel, J. López, R. Torres, J. V. Cauich, and M. Bohner, Modulation of porosity in apatitic cements by the use of α-tricalcium phosphate-calcium sulphate dihydrate mixtures. *Biomaterials* 2005; **26**:3395–3404.

35. E. Fernández, M. D. Vlad, M. Hamcerencu, A. Darie, R. Torres, and J. López, Effect of iron on the setting properties of α-TCP bone cements. *J. Mater. Sci.* 2005; **40**:3677–3682.

36. F. C. M. Driessens, M. G. Boltong, O. Bermúdez, J. A. Planell, M. P. Ginebra, and E. Fernández, Effective formulations for the preparation of calcium phosphate bone cements. *J. Mater. Sci. Mater. Med.* 1994; **5**:164–170.

37. B. R. Constantz, Formulation for in situ prepared calcium phosphate minerals. European Patent 416,761, March 13, 1993.

38. B. R. Constantz, I. C. Ison, M. T. Fulmer, R. D. Poser, S. T. Smith, M. VanWagoner, J. Ross, S. A. Goldstein, J. B. Jupiter, and D. I. Rosenthal, Skeletal repair by in situ formation of the mineral phase of bone. *Science* 1995; **267**:1796–1799.

39. E. Fernández, J. A. Planell, S. M. Best, and W. Bonfield, Synthesis of dahllite through a cement setting reaction. *J. Mater. Sci. Mater. Med.* 1998; **9**:789–792.

40. Y. Yin, F. Ye, S. Cai, K. Yao, J. Cui, and X. Song, Gelatin manipulation of latent macropores formation in brushite cement. *J. Mater. Sci. Mater. Med.* 2003; **14**:255–261.

41. I. Khairoun, M. G. Boltong, F. C. M. Driessens, and J. A. Planell, Some factors controlling the injectability of calcium phosphate bone cements. *J. Mater. Sci. Mater. Med.* 1998; **9**:425–428.

42. M. P. Ginebra, A. Rilliard, E. Fernández, C. Elvira, J. San Roman, and J. A. Planell, Mechanical and rheological improvement of a calcium phosphate cement by the addition of a polymeric drug. *J. Biomed. Mater. Res.* 2001; **57**:113–118.

43. S. Sarda, E. Fernández, J. Llorens, S. Martinez, M. Nilsson, and J. A. Planell, Rheological properties of an apatitic bone cement during initial setting. *J. Mater. Sci. Mater. Med.* 2001; **12**:905–909.

44. G. Baroud, E. Cayer, and M. Bohner, Rheological characterization of concentrated aqueous β-tricalcium phosphate suspensions: the effect of liquid-to-powder ratio, milling time, and additives. *Acta Biomaterialia* 2005; **1**(3):357–363.
45. T. Saito, Y. Kin, and T. Kosbino, Osteogenic response of hydroxyapatite cement implanted into the femur of rats with experimentally induced osteoporosis. *Biomaterials* 2002; **23**:2711–2716.
46. M. Bohner, T. Theiss, D. Apelt, W. Hirsiger, R. Houriet, G. Rizzoli, E. Gnos, C. Frei, J. A. Auer, and B. von Rechenberg, Compositional changes of a dicalcium phosphate dihydrate cement after implantation in sheep. *Biomaterials* 2003; **24**:3463–3474.
47. F. C. M. Driessens, J. A. Planell, M. G. Boltong, I. Khairoun, and M. P. Ginebra, Osteotransductive bone cements. *Proc. Inst. Mech. Eng.* 1998; **212**(H6):427–435.
48. E. Fernández, S. Sarda, M. Hamcerencu, M. D. Vlad, M. Gel, S. Valls, R. Torres, and J. Lopez, High-strength apatitic cement by modification with superplasticizers. *Biomaterials* 2005; **26**:2289–2296.
49. S. Sarda, M. Nilsson, M. Balcells, and E. Fernández, Influence of surfactant molecules as air-entraining agent for bone cement macroporosity. *J. Biomed. Mater. Res.* 2003; **65A**:215–221.
50. M. P. Ginebra, E. Fernández, F. C. M. Driessens, and J. A. Planell, Modelling of the hydrolysis of α-tricalcium phosphate. *J. Am. Ceram. Soc.* 1999; **82**(10):2808–2812.
51. E. Fernández, F. J. Gil, M. P. Ginebra, F. C. M. Driessens, J. A. Planell, and S. M. Best, Calcium phosphate bone cements for clinical applications. I: Solution chemistry. *J. Mater. Sci. Mater. Med.* 1999; **10**:169–176.
52. E. Fernández, F. J. Gil, M. P. Ginebra, F. C. M. Driessens, J. A. Planell, and S. M. Best, Calcium phosphate bone cements for clinical applications. II: Precipitate formation during setting reactions. *J. Mater. Sci. Mater. Med.* 1999; **10**:177–183.
53. E. Fernández, F. J. Gil, S. Best, M. P. Ginebra, F. C. M. Driessens, and J. A. Planell, The cement setting reaction in the $CaHPO_4$-α-$Ca_3(PO_4)_2$ system: an X-ray diffraction study. *J. Biomed. Mater. Res.* 1998; **42**:403–406.
54. Q. Yang, T. Troczynski, and D. Liu, Influence of apatite seeds on the synthesis of calcium phosphate cement. *Biomaterials* 2002; **23**:2751–2760.
55. J. Friberg, E. Fernández, S. Sarda, M. Nilsson, M. P. Ginebra, S. Martínez, and J. A. Planell, An experimental approach to the study of the rheology behaviour of synthetic bone calcium phosphate cements. *Key Engineering Materials* 2001; **192–195**:777–780.
56. M. P. Ginebra, M. G. Boltong, E. Fernández, J. A. Planell, and F. C. M. Driessens, Effect of various additives and temperature on some properties of an apatitic calcium phosphate cement. *J. Mater. Sci. Mater. Med.* 1995; **6**:612–616.
57. M. P. Ginebra, E. Fernández, F. C. M. Driessens, M. G. Boltong, J. Muntasell, J. Font, and J. A. Planell, The effects of temperature on the behaviour of an apatitic calcium phosphate cement. *J. Mater. Sci. Mater. Med.* 1995; **6**:857–860.
58. M. Bohner, U. Gbureck, and J. E. Barralet, Technological issues for the development of more efficient calcium phosphate bone cements: a critical assessment. *Biomaterials* 2005; **26**:6423–6429.
59. H. Oonishi, Bioactive bone cements. In: *The Bone-Biomaterial Interface*. J. E. Davies, ed., Toronto, Canada: University of Toronto Press, 1991, pp. 321–333.
60. E. J. Harper, Bioactive bone cements. *Proc. Instn. Mech. Engrs.* 1998; **212**(H):113–120.

BIOACTIVE GLASSES AND GLASS CERAMICS

BESIM BEN-NISSAN
University of Technology
Sydney, Australia

H. O. YLÄNEN
Åbo Akademi University
Turku, Finland

1. INTRODUCTION

When a person has a joint pain, he or she is mainly concerned with relieving the pain and returning to a healthy and functional lifestyle. Degeneration and diseases often make surgical repair or replacement necessary, which usually requires replacement of skeletal parts, including knees, hips, finger joints, elbows, vertebrae, teeth, and repair of the mandible. The worldwide biomaterials market is valued at over US$24,000M. Revenues generated by sales of total orthopedic products worldwide exceeded US$13,000M in 2000 (1). Expansion in these areas is expected to continue because of a number of factors, including the aging population, an increasing preference by younger to middle-aged candidates for undergoing surgery, improvements in technology and lifestyle, a better understanding of body functionality, improved aesthetics, and the need for better function.

Biomaterial by definition is "a nondrug substance suitable for inclusion in systems that augment or replace the function of bodily tissues or organs." From as early as a century ago, artificial materials and devices have been developed to a point where they can replace various components of the human body. These materials are capable of being in contact with bodily fluids and tissues for prolonged periods while eliciting little, if any, adverse reactions.

Even in the preliminary stages of this field, surgeons and engineers identified materials and design problems that resulted in premature loss of implant function through mechanical failure, corrosion, or inadequate biocompatibility of the component. For the success of any implant, the key factors identified in a biomaterial are its biocompatibility and biofunctionality. Ceramics, and, in certain applications, bioactive glasses and glass ceramics, are ideal candidates with respect to the above functions, except for their brittle behavior.

In this chapter, we review general definitions of glass; types of glass presently available and used bioactive glasses; and glass ceramics, their preparation methods, properties, and applications. Classification of biomaterials and bioactivity will be introduced to cover the development and the progress of the commercially available and currently investigated bioglasses and glass ceramics, their chemistry, bioactivity, theories behind their bonding within a physiological environment, their preparation methods, and their applications in the biomedical field.

2. HOW CAN GLASS AND GLASS CERAMICS BE DEFINED?

Glass has been known to humankind for thousands of years. Obsidian, a natural glass formed from silicate magma, was known to prehistoric people long before how to make glass was discovered. The Phoenicians are thought by many to have been the first people to make glass.

Glass is made from the molten product of oxides; the molten material is cooled rapidly to prevent crystallization or devitrification. A hard, brittle, and amorphous material is produced.

Glass can now be manufactured in such a way that the properties of the glass can be predicted and therefore controlled. Much of this control comes from the use of appropriate raw materials. The choice of materials is usually based on their glass-making properties, as mentioned below.

2.1. Glass-formers

Glass-formers are oxides that can be made into a glass without the addition of any other oxide, although very high temperatures are required to initially melt the oxide. The most common glass-former is silica (SiO_2), usually obtained from sand; other examples of glass-formers are B_2O_3 and P_2O_5.

2.2. Modifiers

The major groups of compounds usually added to silica are called modifiers. As the name suggests, they change or modify the properties of the glass-forming oxides. They may also be used to prevent defects in the final glass product. Two types of modifiers are generally added to glass-forming oxides: fluxes and stabilizers.

Fluxes are chemical components that, when added to glass-forming oxides, change the underlying properties of the oxides. For example, fluxes can lower the melting point of glass-forming oxides. Examples of common fluxes are sodium oxide (Na_2O) and potassium oxide (K_2O). Particularly when used in conjunction with boric oxide (B_2O_3), these oxides can lower the viscosity of the glass, allowing the compounds to move with a greater degree of freedom (increased fluidity). Modifiers such as calcium oxide (CaO), magnesium oxide (MgO), and aluminum oxide (Al_2O_3) are known as *stabilizers*. They can be used to prevent the crystallization of oxides as well as to improve the chemical durability of glass. Crystallization may be undesirable in certain applications because of its effect on light scattering, hence reducing transparency. Like fluxes, stabilizers may also affect the working temperature of glass-formers.

2.3. Melting and Refining Agents

Small bubbles are undesirable in the manufacture of glass as they substantially affect the properties of the glass. To reduce the number of bubbles, the glass is said to be refined via the addition of compounds such as sodium sulphate, sodium nitrate, sodium chloride, calcium fluoride, and carbon. However, the purity and close control of additives during the synthesis of glasses are important areas for the biomedical applications of these materials, because of biocompatibility and toxicity issues.

3. TYPES OF GLASSES

The chemical composition of the glass can be used to give the glass specific properties. Table 1 and Table 2 provide an indication of some of the different types of glasses made in this fashion, along with the desired properties for a range of engineering applications.

3.1. Soda-Lime Glasses

The presence of soda (Na_2O) in glass lowers the melting point of the glass, and the lime (CaO) keeps the glass from crystallizing.

3.2. Borosilicate Glasses

Boron oxide (B_2O_3) acts as both a modifier and a glass former; it produces a glass with a low coefficient of thermal expansion, which results in a glass that is better equipped to deal with thermal shock. Pyrex® is the common trade name for borosilicate glasses and is often used in areas where temperature differences are a problem. The higher presence of alumina (Al_2O_3) is intended to prevent crystallization and to improve chemical durability and the hardness of the glass.

3.3. Aluminosilicate Glasses

One particular type of aluminosilicate glass is used in the production of E-glass fibers (also contains CaO). Aluminosilicate glasses are hard, usually have a good chemical resistance, and do not devitrify readily. They also have high-heat shock resistance and can withstand heat even better than borosilicate glasses.

Table 1. Types of Glasses, Showing their Chemical Composition in Weight Percent

Component	Soda-Lime Glass	Lead Glass	Borosilicate Glass	Alumino Silicate Glass	High-Silica Glass Vycor®	Bioglass® 45S5
SiO_2	70–75	53–68	73–82	57	96	45
Na_2O	12–18	5–10	3–10	1.0	—	24.5
K_2O	0–1	1–10	0.4–1	—	—	—
CaO	5–14	0–6	0–1	5.5	—	24.5
PbO	—	15–40	0–10	—	—	—
B_2O_3	—	—	5–20	4.0	3	—
Al_2O_3	0.5–2.5	0–2	2–3	20.5	—	—
MgO	0–4	—	—	12.0	—	—
P_2O_5	—	—	—	—	—	6

Table 2. Major Components of Various Glass Products

Type	Major Components, Weight Percent						Requirements
	SiO_2	Na_2O	CaO	Al_2O_3	B_2O_3	MgO	
Window	72	14	10	1	—	2	High durability
Plate (arch)	73	13	13	1	—		High durability
Container	74	15	5	1	—	4	Easy workability, chemical resistance
Light bulbs	74	16	5	1	—	2	Easy workability
Fibre (elect.)	54	—	16	14	10	4	Low alkali
Pyrex®	81	4	—	2	12	—	Low thermal expansion, low ion exchange
Fused silica	99	—	—	—	—	—	Very low thermal expansion

3.4. Lead Glasses

Lead glass is often called crystal glass because the improved machinability allows the glass to be engraved more easily. It also gives the glass a heaviness and blue appearance.

A high refractive index is one of the most important properties of this glass, which gives brilliance when properly cut or graved. As the lead oxide (PbO) acts as a flux and a modifier, the melting point and shaping (hot working) temperature of the glass are lowered to acceptable levels. Another useful application for lead glasses is in radiation shielding.

Some of the older terms associated with glass are flint and crown glasses. The term flint glass was originally used to describe lead glass, because flint was used as a source of good-quality silica free from color. It is now more loosely used to describe glasses with good color. Crown glasses are alkali-lime-silica based, for example, soda-lime glass.

3.5. Glass Ceramics

A glass ceramic is initially a glass in which, at some stage, the formation of nuclei is enhanced by using specific compositions, which are self-nucleating, or by the addition of a nucleating agent (2). The resulting material contains very small crystals. Various factors influence the glass ceramics' final properties of crystalline phase, crystal orientation, grain size, intergranular bonding, percentage of crystallinity, and distribution of any remaining glassy phase. In the past, we have been successful in controlling these factors by controlling the base composition, the choice of nucleant (nucleating agent), and an appropriate heat-treatment schedule (2). Early work on glass ceramics was concentrated in the lithia-silica (Li_2O-SiO_2) system, and at a later stage, alumina was introduced to destabilize the basic composition (Li_2O-SiO_2-Al_2O_3). A polymorph of $LiAlSiO_4$, β-spodumene, is precipitated to occupy most of the volume of the glass ceramic, for example, in a Pyroceram® system. In this system, Al^{3+} substitutes for Si^{4+} in the network structure and Li^+ is held nearby to maintain the charge balance. The preceding system was a MgO-Al_2O_3-SiO_2 system, in which MgO replaces the lithia completely. Nucleation is achieved by TiO_2, ZrO_2, and SnO_2. In another composition, Na_2O is used to replace the lithia.

Glasses and glass ceramics, as we will demonstrate in the following sections, are widely investigated and produced as bioactive or surface-active biomaterials with CaO and P_2O_5 additions to their base compositions. One of the main advantages of phosphate-based materials is their chemical relationship with carbonated apatite, one of the main constituents of bone and teeth.

The structures of phosphate glass ceramics and glasses are based on networks of corner-sharing phosphate tetrahedra. Apatite-mullite glass ceramics based on SiO_2-Al_2O_3-P_2O_5-CaO-CaF_2 compositions have been developed and have been observed to form fluorapatite and mullite with a specific heat-treatment procedure.

3.6. Machinable Glass Ceramics

The base composition of the glass ceramics (MgO-Al_2O_3-SiO_2) can be changed by replacing Li_2O with a mixture of K_2O and MgF_2 to improve machinability. Glass ceramics are produced by reheating these specific glasses in the temperature range of 650–1150°C to induce a randomly oriented dispersion of tetra-silicic-mica crystals. These crystals have the formula $KMg_{2.5}Si_4O_{10}F_2$, a structure similar to the tri-silicic mica fluorophlogopite, $KMg_3AlSi_3O_{10}F_2$. The structure is, therefore, analogous to the natural mica mineral phlogopite. Cleavage or rotation in the K^+ planes is relatively easy, and because the crystals in the glass ceramic are in random orientations, the propagating cracks in the material are continuously deflected in different directions, which results in the propagation energy being quickly absorbed. The fracture paths follow the mica-cleavage planes or the mica-glass interfaces, removing very small fragments in the process, so that a good machined finish is easily obtained (2).

3.7. Bioglasses and Glass Ceramics

Since the discovery of the bioglasses, which bond to living tissue (Bioglass®), by Hench and Wilson (3), various kinds of bioactive glasses and glass ceramics with different functions, such as high mechanical strength, high machinability, and fast setting ability, have been developed. Bioactive is the name given to a range of biocompatible materials that bond relatively fast to hard (bone) and soft (skin) tissues without generating any adverse reaction. The glasses that have been investigated for implantation are primarily based on silica (SiO_2), which may contain small amounts of other crystalline phases. The most prominent and successful application is Bioglass®, which is described in detail in various comprehensive reviews (4–6). The first generation bioactive glass compositions lie in the system Na_2O-CaO-P_2O_5-SiO_2. The first development of such a bioglass began in 1971 when 45S5 Bioglass® was proposed with a composition of 45% SiO_2, 24.5% CaO, 24.5% Na_2O, and 6% P_2O_5 by weight (7). Hench (4) and Vrouwenvelder et al. (8) suggested that Bioglass® 45S5 has greater osteoblastic activity than hydroxyapatite, which is attributed to a rapid exchange of alkali ions with hydronium ions at the surface, which in turn leads to the formation of a silica-rich layer over a period of time. This layer allows for the migration of Ca^{2+} and PO_{43}^- ions to the silica-rich surface where they combine with soluble calcium and phosphate ions from the solution, where the formation of an amorphous CaO-P_2O_5 layer takes place. This layer undergoes crystallization upon the interaction of OH^-, CO_3^{2-}, and F^- from solution. A similar phenomenon has been observed in bioglass with similar compositions by other researchers (9). Li et al. (10) prepared glass ceramics from a similar composition with differing degrees of crystallinity and found that the amount of glassy phase remaining directly influences the formation of an apatite layer, with total inhibition when the glassy phase constitutes less than about 5 weight percent (wt%).

As a result of their surface activity, these specific glasses (for example, Bioglass®) have been accepted as bioactive biomaterials and have found applications in non-load-bearing conditions. Bioglasses® have been used successfully in clinical applications as artificial middle-ear bone implants and alveolar ridge maintenance implants (3).

By using a specific heat-treatment method, a bioactive glass with reduced alkaline oxide content can be produced, containing precipitated crystalline apatite within the glass. The resultant glass ceramic, which was named Ceravital®, showed a high mechanical strength but lower bioactivity than Bioglass®.

In 1982, Kokubo et al. (11) produced a glass ceramic containing oxyfluorapatite $Ca_{10}(PO_4)_6(OH,F_2)$ and wollastonite ($CaO \cdot SiO_2$) in an MgO-CaO-SiO_2 glassy matrix, which was named A-W glass ceramic (Cerabone® A-W). In the earlier stages, it was reported that the A-W glass ceramic spontaneously bonded to living bone without forming fibrous tissue around them.

A bioactive and machinable glass ceramic named Bioverit®, containing apatite and phlogopite $(Na,K)Mg_3(AlSi_3O_{10})(F)_2$, has also been developed. It has been used in clinical applications, such as the artificial vertebra.

3.8. General Concepts in Bioceramics

A biomaterial that is designed to have or modulate biological activity is defined as a bioactive material (12). When exposed to body fluids, bioactive materials develop an adherent interface with the host tissue. In many cases, this interfacial strength is equivalent to or even greater than the strength of the material itself or the tissue bonded to the bioactive implant (13). The interfacial bonding is mostly formed between the bioactive implant and bone, but some specialized bioactive materials elicit the property of bonding to soft tissue as well (14). Naturally, depending on the material's characteristics, different bioactive materials show different bioactivity if tested *in vitro* or *in vivo*. When a manmade material is placed within the human body, tissue reacts to the implant in a variety of ways, depending on the material type. It has been accepted that no foreign material placed within a living body is completely compatible. The only substances that conform completely are those manufactured by the body itself (autogenous), and any other substance that is recognized as foreign initiates some type of reaction (host-tissue response). The mechanism of tissue interaction (if any) depends on the tissue response to the implant surface. In general, a biomaterial may be described or classified into one of three groups, represented by the tissues' responses: bioinert, bioresorbable, and bioactive, which are well covered in a range of excellent review papers (3–5,13).

The term bioinert refers to any material that, once placed within the human body, has minimal interaction with its surrounding tissue. Examples of such materials are stainless steel, titanium, alumina, partially stabilized zirconia, and ultra-high molecular weight polyethylene. Generally, a fibrous capsule might form around bioinert implants, hence its biofunctionality relies on tissue integration through the implant.

Bioactive refers to a material that, upon being placed within the human body, interacts with the surrounding bone and, in some cases, even soft tissue. This interaction occurs through a time-dependent kinetic modification of the surface, triggered by its implantation within the living bone. An ion-exchange reaction between the bioactive implant and surrounding body fluids results in the formation of a biologically active carbonate apatite (CHAp) layer on the implant that is chemically and crystallographically equivalent to the mineral phase of bone. Prime examples of these materials are synthetic hydroxyapatite $[Ca_{10}(PO_4)_6(OH)_2]$ (7), glass ceramic A-W (6,8,9), and bioglass® (10).

Bioresorbable refers to a material that, upon placement within the human body, starts to dissolve (resorbed) and is slowly replaced by advancing tissue (such as bone). Common examples of bioresorbablematerials are β-tricalcium phosphate $[Ca_3(PO_4)_2]$, calcium carbonate, calcium oxide, gypsum, and a range of new-generation polymers (for example, polylactic-polyglycolic acid copolymers).

4. BIOACTIVE BIOMATERIALS

The pioneer in the field of bioactive biomaterials, Professor Hench, describes two distinct classes of bioactive materials according to their biological behavior: class A (osteopromotive or osteoinductive) and class B (osteoconductive) (15,16). On the surface of a bioactive material representing class A *a process occurs whereby a biological surface is colonized by osteogenic stem cells free in the defect environment as a result of surgical intervention.* Thus, class A bioactive materials show both an extracellular and intracellular response at the interface and are, accordingly, osteoproductive materials. Class B bioactive materials, however, allow only bone ingrowth, therefore showing osteoconductivity (15). According to this A/B class system, the different bioactive materials can be divided into subgroups: synthetic hydroxyapatite (HA) (class B), some glass ceramics (class A), and bioactive glasses (class A). As a result of the chemical reactions

Table 3. Some Comparative Properties of Various Bioactive Materials

Biomaterial	Classification of Bioactivity	Shear Strength (MPa)	Type of the Failure
Bioactive glass	A	29.8	Cohesive
Hydroxyapatite	B	19.6	Cohesive
Titanium	Inert	1.9	Interfacial
Zirconia	Inert	1.3	Interfacial

occurring on the surface of bioactive materials, a bond is formed between the material surface and host tissue. The bioactivity of a material somehow can be carried out by measuring the strength of the interfacial bond. In order to minimize the gap between the implant and host bone, in 1992, Andersson et al. (17) introduced a method of using conical implants for press-fit insertion. At predetermined time intervals, the animals were sacrificed, the bone blocks containing the implants were harvested, and the cylindrical bones were opened using a diamond drill. The cones were then mechanically pushed out from the holes and the maximum force at failure was measured. The strength of the bonding was expressed as the maximum strength at failure per unit contact area. According to a comparative study (18), the interfacial strengths of various bioactive materials correlate well with the above-mentioned classification of the bioactive materials (Table 3). In the case of bioactive materials, the failure is usually caused by fracture of the material or the bone. However, the separation of an inert implant from the host bone occurs at the interface (17,19). It is important to note that many investigators employed their own protocols and settings such as the shape, size, location, insertion techniques of the implants, and testing methods that the load for failure at push-out or pull-out methods cannot be easily compared with each other.

In general, the nature and rate of the bioactivity depend on the nature of chemical reactions occurring on the material's surface. As a result of the high interfacial strength formed between the implant and host tissue, bioactive materials are widely used to enhance the biological fixation of prostheses to bone as an implant or a coating.

4.1. Bioactive Glass Ceramics

Bone is a natural composite consisting of 60–70 wt% calcium phosphate, a mineral analogous to hydroxyapatite and 6–7 wt% collagen fibers. The rest of the bone consists of water. The composite structure of bone has an important advantage from a mechanical point of view: The numerous interfaces between the different components of the material make it a perfect crack-arrester, resulting in the unique strength of the skeleton material (20).

In the 1980s and 1990s, various groups including Gross et al., Bromer et al., Kitsugi et al., Nakamura and Yamamuro, and Kokubo et al. introduced a range of phosphate-based glass and glass ceramics (16,21–30). Kokubo and Nakamura and their colleagues gradually heated glass powder in the system $MgO\text{-}CaO\text{-}P_2O_5\text{-}CaF_2\text{-}SiO_2$ up to 1050°C. As a result, a material consisting of crystalline wollastonite ($CaO \cdot SiO_2$) and oxyfluoroapatite ($Ca_{10}(PO_4)_6(O,F_2)$) within a homogenous glassy phase was obtained (21,22). As stated earlier, this bioactive glass ceramic was called A-W, derived from the names of the crystalline phases. The two-phase structure of the bioactive A-W glass ceramic (A-W-GC) was reported to resemble the composite structure of bone and can be machined into various shapes with diamond tools. The bending strength of the A-W-GC is almost double that of dense hydroxyapatite and even higher than that of human cortical bone (23). During the 1970s and 1980s, other groups of researchers introduced different glass ceramics (Ceravital®, Bioverit I–III®, Ilmaplant®, BAS-O) (24–27). All these glass ceramics consist of apatite crystals or apatite/wollastonite/phlogopite crystals within a homogenous glassy matrix. Of all glass ceramics, it was reported that A-W-GC elicits the highest mechanical strength and bioactivity (class A) (11,16).

The bioactivity of glass ceramics is believed to be caused by the dissolution of calcium from wollastonite or the glassy phase. The bonding between glass ceramic and bone is formed by the precipitation of the dissolved calcium from the material and phosphate originating from the body fluid (28,29). The silicate ions probably provide nucleation sites for the calcium phosphate formation (22,28). Table 4 (23,31) presents the compositions and some mechanical properties of glass ceramics and sintered synthetic HA. The varying bioactivities of different glass ceramics may be explained by different calcium contents in the materials. Rigid nondegradable bioactive glass ceramics can be manufactured with various methods for different applications (29). They are widely used in orthopedics, odontology, and in head and neck surgery as prostheses, spacers, or granulated defect fillers.

With the development of next-generation bioactive glass ceramics, questions were raised regarding their appropriateness as a coating material. In early experiments, the bioactive glass ceramic was fixed on the load-bearing implant with various techniques (30,32). In 1993, Takatsuka et al. (33) in turn modified A-W-GC, which was successfully sintered on Ti-alloy because of the similar coefficients of thermal expansion. The bioactive coating showed bonding strengths comparable with A-W-GC implants, which were used as the control. A-W-GC coatings have also demonstrated encouraging results in various other studies (34–36). As mentioned earlier, it is believed that the glassy part of the material is mainly responsible for the ion dissolution from the material and, consequently, the chemical reactions occurring on the material surface. Thus, one part of the coating substance is resorbed and the mechanical properties of the material change. Kitsugi et al.'s studies (36) suggest that the use of A-W-GC-coated metal implants should be limited to short-term implantation only, because of the risk of fracture of the coating layer.

Table 4. The Compositions and Some Mechanical Properties of Glass Ceramics and Synthetic Sintered Hydroxyapatite in Clinical Use

Composition (mass %)	Ceravital®	A/W-CG®	Implant®L1	Bioverit®
Na$_2$O	5–10	0	4.6	3–8
K$_2$O	0.5–3.0	0	0.2	3–8
MgO	2.5–5.0	4.6	2.8	2–21
CaO	30–35	44.7	31.9	10–34
Al$_2$O$_3$	0	0	0	8–15
SiO$_2$	40–50	34	44.3	19–54
P$_2$O$_5$	10–15	16.2	11.2	2–10
CaF$_2$	0	0.5	5.0	3–23 (F)
Phase(s)	Apatite	Apatite (Ca10(PO4)6(O,F2))	Apatite	Apatite
		β-Wollastonite (CaO·SiO2)	β-Wollastonite	Phlogopite glass (Na,K)Mg3(AlSiO10)F2)
	Glass	Glass	Glass	
Property				
Compressive strength (MPa)	500	1080	—	500
Bending strength (MPa)	100–150	215	160	100–160

Modified from Kokubo et al. (23) and Hench and West (31).

4.2. Bioactive Glasses

The concept of a strong bonding between bone and synthetic materials by chemical reactions occurring on a glass surface was first proposed in 1969 (31,37). The innovation concerned the chemical reactivity of the surface of a silica-based material, which had the amorphous structure of silicate glass. The biomaterial, a bioactive glass, which was appropriately named Bioglass®, was introduced by Hench in the early 1970s (38). In fact, bioactive glasses can be considered precursors to all bioactive ceramics. A major characteristic of glasses is the amorphous structure of the material. In general, the structure of silicate glass is based on the SiO$_4$ tetrahedron (Fig. 1). The tetrahedra are only linked to the oxygen ions at the corners. In crystalline silica, the tetrahedra are regulary arranged, which is characteristic of all crystalline material (Fig. 2a). In the structure of silica glass, the tetrahedra are present but they are no longer regularly arranged (amorphous structure); however, as in crystalline quartz, each of the oxygen ions still connects two tetrahedra (Fig. 2b). A silica glass, however, has a more open structure, which enables the accommodation of cations (as stated earlier, referred to as network modifiers), thus providing an option to manufacture a wide range of silicate glasses. The presence of some cations in the glass (for example, Na$^+$, K$^+$, and Ca^{2+}) results in disruption of the continuity of the glassy network caused by the breaking of some of the Si–O–Si bonding, leading to a formation of nonbridging oxygen ions (Fig. 2c). Bioactive glasses resemble ordinary soda-lime silica glass. However, the composition of bioactive glasses has one main difference compared with ordinary glass. In ordinary glass, the amount of the network former SiO$_2$ is >65% and the network modifier part consists mainly of Na$_2$O (<14%) and CaO (10%). Bioactive glasses, in turn, contain significantly less network former, which is replaced mainly by metal oxides (network modifiers).

The chemical characteristics of bioactive glasses are as follows: The amount of SiO$_2$ is between 45 and 60 wt%, the Na$_2$O or K$_2$O content is high, and the ratio of Ca/P is relatively high (13).

As a result, the surface of a bioactive glass is highly reactive when exposed to body fluids. In general, the complex reactions occurring at the surface of the glass under *in vitro* or *in vivo* conditions based on various studies (13,39–49) can be summarized as:

- Exchange of Na$^+$ and K$^+$ with H$^+$ or H$_3$O$^+$ from solution (*in vivo* or *in vitro*).
- During this exchange, the other constituents of the glass are unaffected.
- Dissolution of the silica network caused by the attack of H$^+$ ions, resulting in breaking of the Si–O–Si bonds and formation of SiOH and Si(OH)$_4$ groups at the surface of the glass. This process is dependent on the silica content of the glass, and the possible explanations for this will be discussed in the next section.
- Condensation and repolymerization, resulting in a SiO$_2$-rich layer on the glass surface, which is depleted in alkalis and alkaline-earth cations:

$$Si-OH + OH-Si \rightarrow Si-O-Si + H_2O.$$

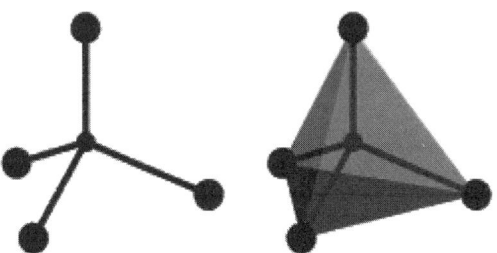

Figure 1. Diagrammatic representation of a basic building block on which the crystalline and glassy structures of silica are based. Silicon ion is bonded to four oxygen ions forming a silicon (SiO$_4$) tetrahedron.

Figure 2. (a) Two-dimensional diagrammatic representation of a crystalline SiO_2 (quartz). The SiO_4 tetrahedra are regularly ordered, which is characteristic of a crystal. (b) Two-dimensional diagrammatic representation of glassy SiO_2. The SiO_4 tetrahedra are no longer regularly arranged; the structure is more open. (c) Two-dimensional diagrammatic representation of silicate glass. The open structure of glass is able to accommodate cations. Some cations, such as Na^+, K^+, and Ca^{2+} (black circles), produce nonbridging oxygen ions (semicircles).

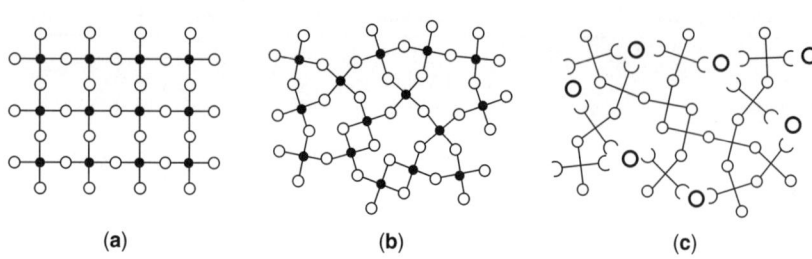

- Migration of Ca^{2+} and PO_4^{3-} groups to the surface through the Si-rich layer, forming a calcium phosphate-rich layer on the top of the Si-rich layer. (However, the fine details of the calcium phosphate formation are still only partially understood, which will be discussed in the next section.)
- Growth of the calcium phosphate layer by incorporation of soluble calcium and phosphate from the solution.
- Formation of a polycrystalline apatite layer, by incorporation of OH^-, CO_3^{2-}, or F^- anions from solution to form a mixed hydroxyl, carbonate, and fluorapatite layer on the uppermost portions of the bioactive glass.
- Incorporation of organic components.

The formation of an Si-rich layer and calcium phosphate layer on bioactive glass is well established and reported (42–47). Figure 3 shows EDAX line-profiles of Si, Ca, and P scanned over a cross section of a bioactive glass sphere at 21 days of immersion in simulated body fluid (SBF) (48).

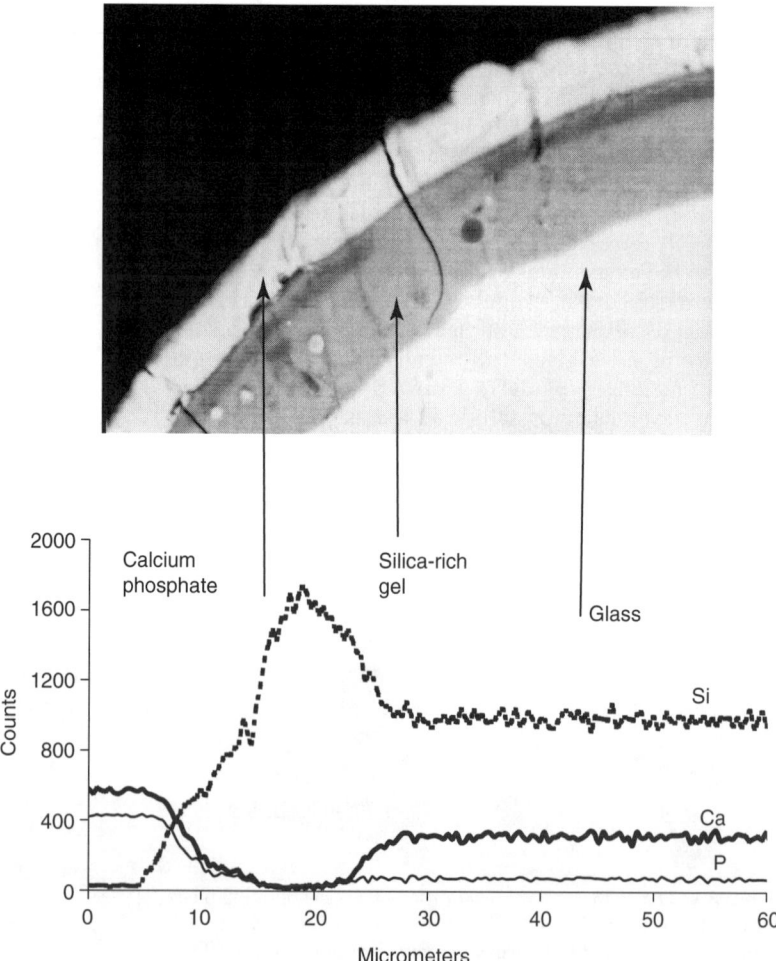

Figure 3. EDAX line-profiles of Si, Ca, and P scanned over a cross section of a bioactive glass sphere after 21 days of immersion in simulated body fluid (SBF). The line-profiles demonstrate bulk glass, a silica-rich layer, and a calcium phosphate layer (magnification × 2000).

The Si-rich layer is clearly seen on the white bulk glass and the calcium phosphate is seen as a light uppermost layer on the Si-rich gel. However, several factors contributing to the formation of the reaction layers exist. Today, the *in vitro* bioactivity of bioactive glasses can be controlled by the oxide composition of the glass (48).

5. BIOACTIVITY AND THE INTERFACIAL STRENGTH

Bonding of glasses to bone has been demonstrated for a certain compositional range of bioactive glasses containing SiO_2, Na_2O, CaO, and P_2O_5 in specific proportions. As stated above, three compositional changes differentiate them from soda-lime-silica glasses: high Na_2O and CaO content, less than 60% SiO_2, and a high CaO/P_2O_5 ratio. These compositional features create highly reactive surfaces when exposed to an aqueous medium. However, in bioactive glasses, the amount of SiO_2, which is only about 45–60%, and hence repeated hot working, can easily cause problems in the form of phase separation and crystallization of the glassy material (38,50). Crystallization of the material will reduce the rate of bioactivity of the glass (51), and partial crystallization leads to a glassy phase of incontrollable composition. Crystallization of a bioactive glass can be controlled by its chemical composition (49,52).

A new generation of bioactive glasses in the system, $Na_2O-K_2O-MgO-CaO-B_2O_3-P_2O_5-SiO_2$, can be repeatedly heated without the risk of devitrification (53). Thus, for instance, microspheres can be manufactured and sintered into porous implants of different shapes and sizes (54). The porosity of a bioactive glass body does not only noticeably increase the total reacting surface of the glass, but also allows a three-dimensional formation of the healing bone tissue. The porosity and the mechanical strength of the bioactive glass implant can be controlled with different sintering temperatures and times (55). For the best mechanical strength of the sintered implant, the glass must retain its amorphous structure during the heat treatment. The development of third-generation bioactive glasses at Åbo Akademi University, Turku, Finland has enabled spinning of high-quality thin (20–30 μm) bioactive glass fibers. From these fibers, a variety of bioactive glass fabrics can be made. Moreover, an impulse laser beam can be used to melt a fine powder made from the new bioactive glass. As a result, a glass coating on a titanium implant can be relatively easily produced. This coating consists of very small glass droplets firmly attached to the titanium. The glass shows excellent bioactivity in spite of the repeated high temperature procedures involved in the process (56,57).

5.1. Bioactivity of the Glass

As with other bioactive materials, bioactive glasses can be divided into the two categories mentioned earlier: osteopromotive or osteoinductive (class A) and osteoconductive (class B). Class A bioactive glasses rapidly release soluble silicic acid, and consequently Ca^{2+}, resulting in fast precipitation of calcium phosphate on the depleted Si-rich glass surface. Class B bioactive glasses, in turn, possess a low rate of network dissolution (16,58). Glasses with a high rate of dissolution contain less than 55 wt% SiO_2 (58). The higher the concentration of SiO_2 in the glass, the lower the rate of network dissolution and formation of an Si-rich layer. When the concentration of SiO_2 exceeds about 60 wt%, the rate of Si-rich gel formation on the glass surface is so low that it has no practical significance on the bioactivity. A reasonable explanation was given by Strnad (59) and at a later stage by Karlsson and Ylänen (60), who showed that the bioactivity of glass is based on the mean number of nonbridging oxygen ions in the silica tetrahedron. Instead of sharing a corner with another tetrahedron, the charge of the oxygen ion in the corner is balanced by a network modifier anion, that is Na^+, K^+, or Ca^{2+}. When exposed to body fluid, because of a rapid exchange of these anions for H^+ or H_3O^+ from the solution, a hydration of the gel structure ($\equiv$Si–(OH) instead of $\equiv$Si–O$^-$ Na^+, K_+) occurs. In silicate glasses, each silicon is bonded to four oxygen atoms, thus the number of nonbridging oxygen ions in the tetrahedron can take any value between 0 and 4. Number 0 represents a fully polymerized, three-dimensional network of silica tetrahedra; number 4 a dissolved SiO_4^{4-} ion. In order to get bioactivity (SiO_2 less than 60 wt%), the number must be greater than 2.6. The surfaces of glasses with the concentration of SiO_2, 50–55 wt% contain 2–3 nonbridging oxygen ions, resulting in hydration of two or three oxygen ions of each silica tetrahedron, that is, formation of $=Si(OH)_2$ or $-Si(OH)_3$. The influence of nonbridging oxygens on the bioactivity of the glass has also been studied recently by Serra et al. (61,62). The number of nonbridging oxygen ions as a function of SiO_2 and Na_2O content in a glass is presented in Table 5.

During the dissolution of the glass, the silica network is partially broken and SiOH and $Si(OH)_4$ groups are found on the uppermost layer of the glass. The higher the number of nonbridging oxygen ions in the gel, the higher the bioactivity of the glass. Finally, if none of the four oxygen ions is bridged, a totally dissolved monomeric SiO_4^{4-} ion is formed. In this case, however, the concentration of SiO_2 needs to be very low (SiO_2 less than 40 wt%) and it is questionable whether obtaining a glass phase of this composition is possible (39). In summation, the crucial factor controlling the bioactivity of a glass is said to be the formation of the hydrated Si-rich gel on the glass surface for which the SiO_2 content should be 50–60 wt%.

5.2. The Interfacial Strength of the Bonding Between Bioactive Glass and Bone

Several studies have concluded that the interfacial chemical reactions occurring on the bioactive glass surface result in a unique and firm bonding of bone to the glass

Table 5. Number of Nonbridging Oxygen Ions as a Function of SiO_2 and Na_2O Content in a Glass (wt%)

Q^n	SiO_2	Na_2O	Non-bridging O^-
4	100	0	0
3	66.6	33.3	1
2	50	50	2
1	40	60	3
0	33.3	66.6	4

implant. The strength of the bonding has been emphasized by mechanical testing in which the interfacial strengths of different biomaterials are compared with each other by determination of the push-out to failure forces. Andersson et al. (63) showed that true bioactive glasses bond to host bone through a firm calcium phosphate layer. However, glasses with appropriate amounts of SiO_2 (<60 wt%) but containing Al^{3+} >2.5 wt% showed only a low interfacial strength. Titanium cones were used as the reference in the same study. The bonding strength of titanium implants was, however, only about one-tenth of that of the bioactive glass implants (63). Glass containing both greater than 60 wt% SiO_2 and greater than 2.5 wt% Al^{3+} showed no bone contact at all. By using the same method, Niki et al. (18) showed the same difference in bonding strength between bioactive glass implants and titanium implants. The bonding of bioactive glass to bone is based on an even dissolution of material and formation of a calcium-phosphate layer connecting the glass implant to the host bone. The difference in the type of the interfacial bonding between hydroxyapatite and bone can also be seen by comparing the bonding strengths of hydroxyapatite implants with those of bioactive glass (18,64,65). In most studies, the interfacial bonding strength of bioactive glass to bone has been superior compared with that of hydroxyapatite. This finding supports the principle of classifying the bioactive materials in two distinct classes according to their bioactivity. Theoretically, the weakest point of the interface between bioactive glass and bone is the silica-rich gel. However, according to mechanical testing, the push-out failure by the maximum force occurs in most of the cases by fracture of the bone close to the glass implant (19,63).

The valuable property of bioactive glasses being able to bond firmly to bone through chemical reactions and to ultimately be replaced by bone allows them to be used for medical applications. Most importantly, the constituents in bioactive glass are physiological chemicals found in the body: silicon, sodium, potassium, magnesium, oxygen, calcium, and phosphorus. During the bonding and formation of bone, the concentration of the specific elements never increases to a level that could disturb the adjacent tissues (65,66). The use of bioactive glass as an implant material or in manufacturing medical devices is, however, limited by the mechanical properties of glass. Glass is brittle and, accordingly, cannot be applied in places where load-bearing properties are needed. Glass can be cast to plates, rods, etc., or it can be formed by sawing or grinding cast rods to rigid medical devices. Alternatively, glass can be used as a filler material in the form of particulate. The use of bioactive glass as a bioactive coating on mechanically stronger implant materials is also an option that has recently been widely studied.

6. APPLICATION OF BIOACTIVE GLASSES AND GLASS CERAMICS

6.1. Maxillofacial and Ear, Nose, and Throat Applications

The composition of this first bioactive glass, Bioglass®, and the composition of S53P4, which has been widely used

Table 6. Compositions of Bioglass® and S53P4 (wt%)

Component	Bioglass®	S53P4
SiO_2	45	53
Na_2O	24.5	23
CaO	24.5	20
P_2O_5	6	4

clinically, is presented in Table 6. Although the SiO_2 content in this Bioglass® is relatively low, it has been reported to have good bioactive properties.

The glass can be cast into shaped implants for use as small medical devices in places subject to only minor mechanical loading. The glass is relatively soft and thus appropriate for microsurgical drilling techniques (67). Bioglass® has been used to replace small bones in the middle ear and as cone-shaped implants to fill defects in the mandible and maxilla. It also elicits a property of bonding to soft tissue, which has opened new possibilities for wider application, for example, in ear, nose, and throat surgery. In the treatment of profound deafness, various electronic components are used within appropriate devices. Electrodes attached to the electronic devices outside and inside the middle ear are coated with Bioglass® (67). The device is inserted so that a part is in contact with bone, resulting in a firm anchoring. The other part of the bioactive glass-coated anchor passes through the eardrum and bonds to the soft tissue, thus providing a seal between the inner and outer ear. Stanley et al. (68) introduced the concept of filling a hole in the jawbone after extracting a tooth in order to prevent resorption of the bone. For this procedure, conical implants are made using injection molding. For the best possible fit of the conical implant, a mating drill bit is used to prepare the bone for the implant. Solid-shaped bioactive glass of a different composition (S53P4) has also been used in the treatment of facial injuries to replace the bone that supports the eye (69,70). The clinical results of Aitasalo et al. (70) suggest that thin, slowly resorbable bioactive glass provides a promising option for the reconstruction of orbital floor defects.

In odontology, bioactive glass particles are widely used to fill defects associated with periodontal disease, for example, the loss of bone surrounding teeth (58).

Some recent studies show promising results from using crushed bioactive glass or paste as obliteration material in frontal sinusitis and in special types of rhinitis (70–74). The findings of Stoor et al. (73,74) suggest that the encouraging results of using bioactive glass crush in ear, nose, and throat surgery are partially because of the antibacterial effects of bioactive glass paste on oral micro-organisms. Sintering of bioactive glass microspheres enables the manufacture of rigid porous bioactive glass bodies. The rate of bone ingrowth into the three-dimensional porosity of the bioactive glass material is significantly higher compared with the similar porosity of porous titanium (75).

6.2. Bone Graft Applications

Using animal models, solid-shaped bioactive glass implants have been investigated for the reconstruction of deep osteochondral defects (joint surface defects, which penetrate the bone under the joint cartilage) (76,77). According to these studies, the bioactive glass implants bonded to bone, but the rate of hyaline-like cartilage formation on the joint surface was only minimal, or at best moderate. The use of porous bioactive glass implants resulted in more promising results (54).

Critical size defect can be defined as the smallest size defect that cannot heal spontaneously during the lifetime of a clinical animal. A number of clinical trials showed that an 8-mm diameter defect created in the calvaria of Sprague–Dawley rats did not heal after a 12-week period. In contrast, many long bones contain a primary nutrient artery. No primary nutrient artery exists in human calvaria. The results of Lee et al. (78) have indicated that calcium phosphate glass (bioglass) can affect the differentiation and calcification of the pre-osteolastic (MC3T3-E1) cells *in vitro* and promotes new bone formation in the calvarial defects.

Using bioactive glass as granules, it is reported to enhance bonding because of the chemistry associated with the bioactivity, reactivity because of increased surface area, and the method of application (58,79–81). Bioactive glass granules have been reported as providing a promising option to be used alone as a filler material in bone defects or to fill the gap around the implants (82–84). However, according to Virolainen et al. (85), the bone-forming capacity of bioactive glass particulate seems to be lower compared with an autogenous bone graft.

6.3. Bioglasses in *In Situ* Radiotherapy and Hyperthermia

One of the most common approaches in cancer treatment is the removal of the diseased parts; however, recovery or return of full function is unfortunately seldom achieved. Noninvasive treatment techniques in which only the cancer cells are destroyed were introduced in the mid-1980s. In 1987, microspheres of $17Y_2O_3$-$19Al_2O_3$-$64SiO_2$ (mol%) glass, 20–30 µm in diameter, were shown to be effective for *in situ* radiotherapy of liver cancer (86,87). ^{89}Yttrium in this glass is nonradioactive, but can be activated by neutron bombardment, to ^{90}Y, which is a β-emitter with a half-life of 64.1 hours. The microspheres are usually injected into diseased liver through the hepatic artery and entrapped in small blood vessels, which block the blood supply to the cancer and directly irradiate the cancer with β-rays. As the β-ray transmits living tissue only 2.5 mm in diameter and the glass microspheres have high chemical durability, the surrounding normal tissue is hardly damaged by the β-rays.

These glass microspheres are already used clinically in Australia, Canada, and the United States and are commercially available. The content of Y_2O_3 in the microsphere is, however, limited to only 17 mole%, as they are prepared by conventional glass melting techniques. Recently, Kawashita et al. successfully prepared pure Y_2O_3 polycrystalline microspheres 20–30 µm in diameter by a high-frequency induction thermal plasma melting technique (88) (Fig. 4). It was reported that they observed higher chemical durability than the Y_2O_3-containing glass microspheres. It was further reported that these ceramic

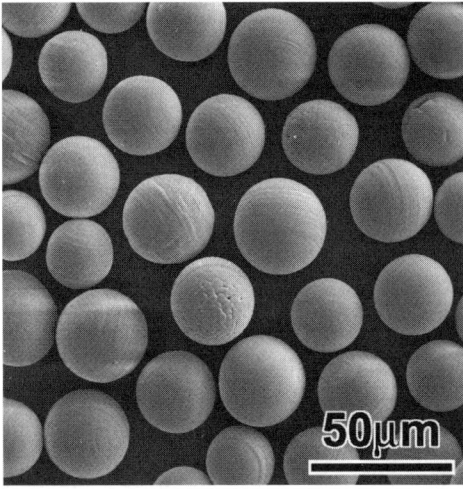

Figure 4. SEM image of Y_2O_3 microspheres for radiotherapy applications [after Kawashita et al. (88)].

microspheres are more effective for *in situ* radiotherapy of cancer.

Oxygen is known to be poorly supplied to cancerous cells to produce lactic acid, and hence can be destroyed around 43°C, whereas the normal living cells can be kept alive even around 48°C. If ferri- or ferromagnetic materials are implanted around cancers and placed under an alternating magnetic field, it is expected that locally heated cancer cells can be destroyed by magnetic hysteresis loss of the ferri- or ferromagnetic materials.

Kawashita et al. (88) prepared ferrimagnetic glass-ceramic compositions containing 36 wt% of magnetite (Fe_3O_4) particles 200 nm in size within a CaO-SiO_2 matrix. It was reported that cancerous cells in the medullary canal of a rabbit tibia were completely destroyed when this glass ceramic was inserted into the tibia and placed under an alternating magnetic field of 300 Oe with 100 kHz (89). This kind of invasive treatment, however, cannot be applied to humans, because cancer cells metastasize. In the case of humans, ferri- or ferromagnetic material must be injected into the cancer in a form of microsphere of 20–30 µm in diameter through blood vessels similar to the radioactive microspheres. For this purpose, the heat-generating efficiency of the ferrimagnetic material must be further increased.

Recently, microspheres of 20–30 µm in diameter in which number of layers of magnetite particles of 50 nm were deposited on silica microspheres were produced. The technique involved deposition of FeO(OH) from a solution and its subsequent transformation into Fe_3O_4 by a specific heat treatment at 600°C under CO_2-H_2 gas atmosphere (90). The heat-generating efficiency of this material was reported to be about four times that of the glass ceramic described above.

7. CONCLUDING REMARKS

As discussed in this chapter, the properties of biomaterials in general, bioactive glasses, and glass ceramics are strongly influenced by the synthesis methods employed, the chemistry and the compounds used, and any other thermal processes used. All of these factors contribute to their final structure and hence to their long-term performance as bioglasses. In the early 1970s, bioceramics were employed to perform biologically inert roles, such as providing parts for bone replacement. The realization that cells and tissues in the body perform many other vital regulatory and metabolic roles has highlighted the limitations of synthetic materials as tissue substitutes. The demands placed on bioceramics have changed from maintaining an essentially physical function without eliciting a host response to providing a more integrated interaction with the host, which has been accompanied by increasing demands on medical devices to improve the quality of life, as well as to extend its duration.

Bioceramics can be used as body interactive materials, helping the body to heal, or promoting regeneration of tissues, thus restoring physiological functions. This approach could be further explored in the development of next-generation bioceramics incorporating biogenic materials with a widened range of applications. Recently, the tissue-engineering field has been directed to take advantage of the combined use of living cells and tridimensional ceramic scaffolds to deliver vital cells to a patient's damaged site. Feasible and productive strategies have been aimed at combining a relatively traditional approach such as bioceramics implants with the acquired knowledge applied to the field of cell growth and differentiation of osteogenic cells.

A common characteristic of bioactive glasses and bioactive ceramics is a time-dependent kinetic modification of the surface that occurs after implantation. The surface forms a carbonate hydroxyapatite (hydroxylcarbonate apatite) layer. Results of various studies show that bioactive glass affects the osteogenesis by increasing collagen synthesis and calcification of the extracellular matrix *in vitro* and promotes new bone formation in calvarial defects in the Sprague–Dawley rats. Bonding of glass ceramic to bone has been observed to occur between two weeks and two months after implantation. Clinical studies are promising and their use in the orthopedic field is increasing. Glass ceramics are known to have high bending strength and compressive strength, and it has been reported that they could be used in the reconstruction of acetabular defects. Bioglass® and bioactive glasses are currently filling a gap in surgical orthopedic and maxillofacial applications. However, the challenge of providing safe and efficacious glass and glass ceramics with the required properties and an acceptable biocompatibility level remains. As the field of biomaterials finds increasing applications in cellular and tissue engineering, it will continue to be used in new ways as part of the most innovative therapeutic strategies.

BIBLIOGRAPHY

1. *The Worldwide Orthopaedic Market – 2000–2001, Dorland's Biomedical*. Philadelphia, PA: Knowledge Enterprises, Inc., 2001.
2. P. W. McMillan, *Glass Ceramics*. London, UK: Academic Press, 1964.
3. L. L. Hench and J. Wilson, Surface active materials. *Science* 1984; **226**:630–636.
4. L. L. Hench, Bioactive ceramics. In: P. Ducheyne and J. E. Lemons, eds., *Bioceramics: Materials Characteristics vs. in vivo Behavior*, vol. 523, New York: New York Academy of Science, 1988, pp. 54–71.
5. T. Kokubo, Recent progress in glass-based materials for biomedical applications, *J. Ceramic. Soc. Japan* 1991; **99**:965–973.
6. T. Kokubo, Novel biomaterials derived from glasses. In: W. Soga and A. Kato, eds., *Ceramics: Towards the 21st Century*, *J. Ceramic. Soc. Japan* 1991, pp. 500–518.
7. L. L. Hench, R. J. Splinter, W. C. Allen, and T. K. Greenlee, Bonding mechanisms at the interface of ceramic prosthetic materials. *J. Biomed. Mater. Res. Symp.* 1972; **2**:117–141.
8. W. C. A. Vrouwenvelder, C. G. Groot, and K. de Groot, Histological and biochemical evaluation of osteoblasts cultured on bioactive glass, hydroxylapatite, titanium alloy, and stainless steel. *J. Biomed. Mater. Res.* 1993; **27**:465–475.
9. O. H. Andersson and I. Kangasniemi, Calcium phosphate formation at the surface of bioactive glass *in vitro*. *J. Biomed. Mat. Res.* 1991; **25**:1019–1030.
10. P. Li, Q. Yang, F. Zhang, and T. Kokubo, The effect of residual glassy phase in a bioactive glass-ceramic on the formation of its surface apatite layer *in vitro*. *J. Mater. Sci.: Mater. Med.* 1992; **3**:452–456.
11. T. Kokubo, M. Shigematsu, Y. Nagashima, M. Tashiro, T. Nakamura, T. Yamamuro, and S. Higashi, Apatite-wollastonite containing glass-ceramic for prosthetic application. *Bull. Inst. Chem. Res.* 1982; **60**:260–268.
12. D. F. Williams, ed., Definitions in biomaterials. Proceedings of a Consensus Conference of European Society for Biomaterials, Chester, United Kingdom, 1986.
13. L. L. Hench, Bioceramics: from concept to clinic. *J. Am. Ceram. Soc.* 1991; **74**(7):1487–1510.
14. J. Wilson, Bonding of bioactive ceramics in soft tissues, paper presented at *7th CIMTEC World Ceramic Congress*, Montecatini Terme, Italy, 1990.
15. L. L. Hench, Bioactive ceramics: theory and clinical applications. In: Ö. H. Andersson and A. Yli-Urpo, eds., *Bioceramics*, vol. 7, Oxford: Butterford-Heinemann, 1994.
16. L. L. Hench, Glass and genes: a forecast to future. *Glastech Ber. Glass Sci. Technol.* 1997; **70**:439–452.
17. Ö. H. Andersson, G. Liu, K. Kangasniemi, and J. Juhanoja, Evaluation of the acceptance of glass in bone. *J. Mat. Sci. Mat. Med.* 1992; **3**:145–150.
18. M. Niki, G. Ito, T. Matsuda, and M. Ogino, Comparative push-out data of bioactive and non-bioactive materials of similar rugosity. In: J. E. Davies, ed., *Bone-Material Interface*. Toronto, Canada: University of Toronto Press, 1991.
19. R. Z. LeGeros and J. P. LeGeros, Dense hydroxyapatite. In: L. L. Hench and J. Wilson, eds., *An Introduction to Bioceramics*, vol. 1, Singapore: World Scientific, 1993.
20. J. Cook and J. E. Gordon, A mechanism for the control of crack propagation in all-brittle systems. *Proc. Roy. Soc. London* 1964; A282.
21. T. Kokubo, S. Ito, S. Sakka, and T. Yamamuro, Formation of high-strength bioactive glass-ceramic in the system $MgO-CaO-SiO_2-P_2O_5$. *J. Mater. Sci.* 1986; **21**:536–540.
22. T. Nakamura and T. Yamamuro, Development of a bioactive ceramic, A/W glass-ceramic. In: P. Ducheyne and

23. D. Christiansen, eds., *Bioceramics*, vol. 6, Oxford: Butterworth-Heinemann, 1993.
23. T. Kokubo, Bioactivity of glasses and glass-ceramics. In: P. Ducheyne, T. Kokubo, and C. A. Van Blitterswijk, eds., *Bone-Bonding*. Leiderdorp, The Netherlands: Reed Healthcare Communications, 1992.
24. B. A. Blencké, H. Brömer, and K. K. Deutscher, Compatibility and long-term stability of glass-ceramic implants. *J. Biomed. Mater. Res.* 1978; **12**(3):307–316.
25. W. Vogel and W. Höland, Development, structure, properties and application of glass-ceramics for medicine. *J. Non-Cryst. Solids* 1990; **123**:349–353.
26. G. Berger, R. Sauer, G. Steinborn, F. G. Wihsmann, V. Thieme, S. T. Kohler, and H. Dressel, Clinical application of surface reactive apatite/wollastonite containing glass-ceramics. In: O. V. Mazurin, ed., *Proc. XV International Congress on Glass*, Nauka, Leningrad, 1989.
27. V. Pavek, Z. Novak, Z. Strnad, D. Kudrnova, and B. Navratilova, Clinical application of bioactive glass-ceramic BAS-O for filling cyst cavities in stomatology. *Biomaterials* 1994; **15**(5): 353–358.
28. T. Kokubo, Surface chemistry of bioactive glass-ceramics. *J. Non-Cryst. Solids* 1990; **120**:138–151.
29. T. Kokubo, Bioactive glass-ceramics: properties and applications. *Biomaterials* 1991; **12**(2):155–163.
30. V. Strunz, M. Bunte, U. M. Gross, K. Manner, H. Brömer, and K. Deutscher, Coating of metal implants with the bioactive glass-ceramics Ceravital. *Dtsch. Zahnarztl. Z* 1978; **33**(12): 862–865.
31. L. L. Hench and J. K. West, Biological applications of bioactive glasses. *Life Chem. Rep.* 1996; **13**:187–241.
32. G. A. Fuchs and K. Deutscher, Glass-ceramic coated implants. A simple model for a loaded hip prosthesis with a bioactive interface. *Arch. Orthop. Trauma Surg.* 1981; **98**(2):121–126.
33. K. Takatsuka, T. Yamamuro, T. Kitsugi, T. Nakamura, T. Shibuya, and T. Goto, A new bioactive glass-ceramic as a coating material on titanium alloy. *J. Appl. Biomater.* 1993; **4**(4): 317–329.
34. K. Ido, Y. Matsuda, T. Yamamuro, H. Okumura, M. Oka, and H. Takagi, Cementless total hip replacement. Bioactive glass ceramic coating studied in dogs. *Acta Orthop. Scand.* 1993; **64**(6):607–612.
35. Z. L. Li, T. Kitsugi, T. Yamamuro, Y. S. Chang, Y. Senaha, H. Takagi, T. Nakamura, and M. Oka, Bone-bonding behavior under load-bearing conditions of an alumina ceramic implant incorporating beads coated with glass-ceramic containing apatite and wollastonite. *J. Biomed. Mater. Res.* 1995; **29**(9): 1081–1088.
36. T. Kitsugi, T. Nakamura, M. Oka, Y. Senaha, T. Goto, and T. Shibuya, Bone-bonding behavior of plasma-sprayed coatings of Bioglass®, AW-glass ceramic, and tricalcium phosphate on titanium alloy. *J. Biomed. Mater. Res.* 1996; **30**(2):261–269.
37. L. L. Hench and H. A. Paschall, Direct chemical bond of bioactive glass-ceramic materials to bone and muscle. *J. Biomed. Mater. Res.* 1973; **7**(3):25–42.
38. L. L. Hench, R. J. Splinter, W. C. Allen, and T. K. Greenlee, Bonding mechanisms at the interface of ceramic prosthetic materials. *J. Biomed. Mater. Res. Symp.* 1971; **2**:117–141.
39. P. Ducheyne, S. Brown, N. Blumenthal, L. Hench, A. Krajewski, G. Palavit, A. Ravaglioli, S. Steineman, and S. Windeler, Bioactive glasses, aluminum oxide, and titanium. Ion transport phenomena and surface analysis. *Ann. NY Acad. Sci.* 1988; **523**:257–261.
40. L. L. Hench and Ö. H. Andersson, Bioactive glasses. In: L. L. Hench and J. Wilson, eds., *An Introduction to Bioceramics*, vol. 1, Singapore: World Scientific, 1993.
41. R. D. Rawlings, Bioactive glasses and glass-ceramics. Review paper. *Clin. Mater.* 1993; **14**:155–179.
42. L. L. Hench, C. J. Pantano, Jr, P. J. Buscemi, and D. C. Greenspan, Analysis of bioglass fixation of hip prostheses. *J. Biomed. Mater. Res.* 1977; **11**(2):267–282.
43. K. H. Karlsson, Ö. H. Andersson, A. Yli-Urpo, and K. Kangasniemi, *Physical Properties and Biomedical Behavior of Phosphate Opal Glasses*. Berlin: Kurtzreferate, DKG-DGG, 1987.
44. Ö. H. Andersson, K. H. Karlsson, K. Kangasniemi, and A. Yli-Urpo, Models for physical properties and bioactivity of phosphate opal glasses. *Glastech. Ber.* 1988; **61**:300–305.
45. E. Schepers, P. Ducheyne, and M. De Clercq, Interfacial analysis of fiber-reinforced bioactive glass dental root implants. *J. Biomed. Mater. Res.* 1989; **23**(7):735–752.
46. Ö. H. Andersson and I. Kangasniemi, Calcium phosphate formation at the surface of bioactive glass in vitro. *J. Biomed. Mater. Res.* 1991; **25**(8):1019–1030.
47. K. Ohura, T. Nakamura, T. Yamamuro, Y. Ebisawa, T. Kokubo, Y. Kotoura, and M. Oka, Bioactivity of $CaO-SiO_2$ glasses added with various ions. *J. Mater. Sci. Mater. Med.* 1992; **3**:95–100.
48. H. O. Ylänen, K. H. Karlsson, A. Itälä, and H. T. Aro, Effect of immersion in SBF on porous bioactive bodies made by sintering bioactive glass microspheres. *J. Noncryst. Solids* 2000; **275**:107–115.
49. S. Hayakawa, K. Tsuru, and C. Ohtsuki, Mechanism of apatite formation on a sodium silicate class in a simulated body fluid. *J. Am. Ceram. Soc.* 1999; **82**(8):2155–2160.
50. Ö. H. Andersson, Glass transition temperature of glasses in the $SiO_2-Na_2O-CaO-P_2O_5-Al_2O_3-B_2O_3$ system. *J. Mat. Sci. Mat. Med.* 1992; **3**:326–328.
51. O. Peitl Filho, G. P. LaTorre, and L. L. Hench, Effect of crystallization on apatite-layer formation of bioactive glass 45S5. *J. Biomed. Mater. Res.* 1996; **30**(4):509–514.
52. M. Brink, Bioactive glasses with a large working range, Ph.D. thesis, Åbo Akademi University, Åbo, Finland, 1997.
53. H. E. Arstila, L. Vedel, L. Hupa, H. Ylänen, and M. Hupa, Measuring the devitrification of bioactive glasses. *Key Engineer. Mater.* 2004; **254–256**:67–70.
54. H. O. Ylänen, T. Helminen, A. Helminen, J. Rantakokko, K. Mattila, K. H. Karlsson, and H. T. Aro, Porous bioactive glass matrix in reconstruction of articular osteochondral defects. *Ann. Chir. Gyn.* 1999; **83**(3):237–245.
55. L. Fröberg, L. Hupa, and M. Hupa, Porous bioactive glasses with controlled mechanical strength. *Key Engineer. Mater.* 2004; **254–256**:973–976.
56. N. Moritz, E. Vedel, H. Ylänen, M. Jokinen, M. Hupa, and A. Yli-Urpo, Creation of bioactive glass coating on titanium by local laser irradiation, Part I: Optimisation of the processing parameters. *Key Engineer. Mater.* 2003; **240–242**:221–224.
57. E. Vedel, N. Moritz, H. Ylänen, M. Jokinen, A. Yli-Urpo, and M. Hupa, Creation of bioactive glass coating on titanium by local laser irradiation, Part II: Effect of the irradiation on the bioactivity of the glass. *Key Engineer. Mater.* 2003; **240–242**:225–228.
58. L. L. Hench, Bioceramics. *J. Am. Ceram. Soc.* 1998; **81**(7): 1705–1727.
59. Z. Strnad, Role of the glass phase in bioactive glass-ceramic. *Biomaterials* 1992; **13**(5):317–321.

60. K. H. Karlsson and H. O. Ylänen, Porous bone implants. In: P. Vincenzini, ed., *Materials in Clinical Applications, Advances in Science and Technology 28. 9th Cimtec-World Forum on New Materials*, Florence, Italy, 1998.

61. J. Serra, P. González, S. Liste, S. Chiussi, B. León, M. Pérez-Amor, H. O. Ylänen, and M. Hupa, Influence of the non-bridging oxygen groups on the bioactivity of silicate glasses. *J. Mater. Sci. Mater. Med.* 2002; **13**:1221–1225.

62. J. Serra, P. Gonzáles, S. Liste, C. Serra, S. Chiussi, B. León, M. Pérez-Amor, H. O. Ylänen, and M. Hupa, FTIR and XPS studies of bioactive silica based glasses. *J. Noncryst. Solids* 2003; **332**:20–27.

63. Ö. H. Andersson, K. H. Karlsson, and K. Kangasniemi, Calcium-phosphate formation at the surface of bioactive glass in vivo. *J. Non-Cryst. Solids* 1990; **119**:290–296.

64. T. Fujiu and M. Ogino, Difference of bone bonding behavior among surface active glasses and sintered apatite. *J. Biomed. Mater. Res.* 1984; **18**(7):845–859.

65. J. Wilson, G. H. Pigott, F. J. Schoen, and L. L. Hench, Toxicology and biocompatibility of bioglasses. *J. Biomed. Mater. Res.* 1981; **15**(6):805–817.

66. W. Lai, P. Ducheyne, and J. Garino, Removal pathway of silicon released from bioactive glass granules *in vivo*. In: R. Z. LeGeros and J. P. LeGeros, eds., *Bioceramics*, vol. 11, New York: World Scientific, 1998, p. 383.

67. J. Wilson, A. Yli-Urpo, and R.-P. Happonen, Bioactive glasses: clinical applications. In: L. L. Hench and J. Wilson, eds., *An Introduction to Bioceramics*, vol. 1, Singapore: World Scientific, 1993, p. 63.

68. H. R. Stanley, M. B. Hall, A. E. Clark, C. J. King, III, L. L. Hench, and J. J. Berte, Using 45S5 bioglass cones as endosseous ridge maintenance implants to prevent alveolar ridge resorption: a 5-year evaluation. *Int. J. Oral Maxollofac. Implants* 1997; **12**(1):95–105.

69. E. Suominen and J. Kinnunen, Bioactive glass granules and plates in the reconstruction of defects of the facial bones. *Scand. J. Plast. Reconstr. Surg. Hand Surg.* 1996; **30**(4):281–289.

70. K. Aitasalo, J. Suonpää, I. Kinnunen, and A. Yli-Urpo, Reconstruction of orbital floor fractures with bioactive glass (S53P4). In: H. Ohgushi, G. W. Hastings, and T. Yoshikawa, eds., *Bioceramics*, vol. 12, Singapore: World Scientific, 1999, p. 49.

71. J. Suonpää, J. Sipilä, K. Aitasalo, J. Antila, and K. Wide, Operative treatment of frontal sinusitis. *Acta Otolaryngol. Suppl. (Stockh)* 1997; **529**:181–183.

72. M. Peltola, J. Suonpää, K. Aitasalo, M. Varpula, A. Yli-Urpo, and R. P. Happonen, Obliteration of the frontal sinus cavity with bioactive glass. *Head Neck* 1998; **20**(4):315–318.

73. P. Stoor, E. Söderling, and J. I. Salonen, Antibacterial effects of a bioactive glass paste on oral microorganisms. *Acta Odontol. Scand.* 1998; **56**(3):161–165.

74. P. Stoor, E. Söderling, and R. Grenman, Interactions between the bioactive glass S53P4 and the atropic rhinitis-associated microorganism Klebsiella ozaenae. *J. Biomed. Mater. Res.* 1999; **48**(6):869–874.

75. H. O. Ylänen, Bone ingrowth into porous bodies made by sintering bioactive glass microspheres, Ph.D. thesis, Åbo Akademi University, Åbo, Finland, 2000.

76. J. T. Heikkilä, A. J. Aho, A. Yli-Urpo, Ö. H. Andersson, H. J. Aho, and R. P. Happonen, Bioactive glass versus hydroxyapatite in reconstruction of osteochondral defects in the rabbit. *Acta Orthop. Scand.* 1993; **64**(6):678–682.

77. E. Suominen, A. J. Aho, E. Vedel, I. Kangasniemi, E. Uusipaikka, and A. Yli-Urpo, Subchondral bone and cartilage repair with bioactive glasses, hydroxyapatite, and hydroxyapatite-glass composite. *J. Biomed. Mater. Res.* 1996; **32**(4):543–551.

78. Y.-K. Lee, J. Song, H. J. Moon, S. B. Lee, K. M. Kim, S. H. Choi, and R. Z. LeGeros, In Vitro and in vivo evaluation of non-crystalline calcium phosphate glass as a bone substitute. *Key Engineer. Mater.* 2004; **254–256**:185–188.

79. E. Schepers, M. De Clercq, P. Ducheyne, and R. Kempeneers, Bioactive glass granulate material as a filler for bone lesions. *J. Oral Rehabil.* 1991; **18**(5):439–452.

80. A. M. Gatti and D. Zaffe, Long-term behavior of active glasses in sheep mandibular bone. *Biomaterials* 1991; **12**(3):345–350.

81. A. J. Garcia and P. Ducheyne, Numerical analysis of extracellular fluid flow and chemical species transport around and within porous bioactive glass. *J. Biomed. Mater. Res.* 1994; **28**(8):947–960.

82. A. M. Gatti, G. Valdre, and A. Tombesi, Importance of microanalysis in understanding mechanism of transformation in active glassy biomaterials. *J. Biomed. Mater. Res.* 1996; **31**(4):475–480.

83. J. T. Heikkilä, H. J. Aho, A. Yli-Urpo, R. P. Happonen, and A. J. Aho, Bone formation in rabbit cancellous bone defects filled with bioactive glass granules. *Acta Orthop. Scand.* 1995; **66**(5):463–467.

84. H. Oonishi, L. L. Hench, J. Wilson, F. Sugihara, E. Tsuji, S. Kushitani, and H. Iwaki, Comparative bone growth behavior in granules of bioceramic materials of various sizes. *J. Biomed. Mater. Res.* 1999; **44**(1):31–43.

85. P. Virolainen, J. Heikkilä, A. Yli-Urpo, E. Vuorio, and H. T. Aro, Histomorphometric and molecular biologic comparison of bioactive glass granules and autogenous bone grafts in augmentation of bone defect healing. *J. Biomed. Mater. Res.* 1997; **35**(1):9–17.

86. H.-M. Kim, F. Miyaji, T. Kokubo, S. Nishiguchi, and T. Nakamura, Graded surfaces structure of bioactive titanium prepared by chemical treatment. *J. Biomed. Mater. Res.* 1999; **45**(2):100–107.

87. G. J. Ehrhardt and D. E. Day, Therapeutic use of ^{90}Y microspheres. *Int. J. Radiation Appl. Inst. -Part B. Nucl. Med. Biol.* 1987; **14**(3):233–242.

88. M. Kawashita, T. Kokubo, and Y. Inoue, Preparation of Y_2O_3 microspheres for *In Situ* radiotherapy of cancer. In: H. Ohgushi, G. W. Hastings, and T. Yoshikawa, eds., *Bioceramics*, vol. 12, Singapore: World Scientific, 1999, pp. 555–558.

89. M. Ikenaga, K. Ohura, T. Yamamuro, Y. Kotoura, M. Oka, and T. Kokubo, Localized hyperthermic treatment of experimental bone tumours with ferromagnetic ceramics. *J. Orthop. Res.* 1993; **11**(6):849–855.

90. M. Kawashita, M. Tanaka, T. Kokubo, T. Yao, S. Hamada, and T. Shinjo, Preparation of magnetite microspheres for hyperthermia of cancer. S. Brown, I. Clarke, and P. Williams, eds., *Bioceramics*, vol. 14, Switzerland: Trans Tech Pub., 2002, pp. 645–648.

READING LIST

L. L. Hench, Materials characteristics versus *in vivo* behavior. In: P. Ducheyne and J. E. Lemons, eds., *Bioceramics. J. Ann. NY Acad. Sci.* 1988; **523**:54.

BIOCHEMICAL PATHWAYS RESEARCH

JAMES L. SHERLEY
Massachusetts Institute of
Technology
Cambridge, Massachusetts

1. A BRIEF PERSPECTIVE ON BIOCHEMICAL PATHWAYS RESEARCH

All living organisms, whether existing as single-cellular or more complex multicellular forms, perform the same basic chemical reactions for life. These include reactions for energy transformation, chemical transformations (metabolism), synthesis of cellular materials (anabolism), and breakdown of cellular materials (catabolism) (1). These chemical processes occur in living cells through serial chemical reactions that are conceptually organized as biochemical pathways. Such pathways for processes like electron transport, photosynthesis, glycolysis, lipid biosynthesis, amino acid biosynthesis, and purine catabolism are the focus of introductory textbooks (2,3) familiar to every beginning student of biochemistry and medicine. In addition to essential biochemical pathways found in all cells, a vast array of pathways exist that are unique to different species (e.g., plant pigment biosynthesis) and to different cell types within multicellular organisms (e.g., neurotransmitter biosynthesis in neurons).

Although the substrate, metabolite, and product biomolecules of chemical reactions in cells are typically used to illustrate biochemical pathways, the field of biochemistry is more concerned with the chemistry-specific biopolymers that catalyze their reactions. Protein enzymes are the main biopolymers responsible for catalyzing cellular biochemical reactions. In general, unlike chemical catalysts, which can increase the rate of reactions between diverse types of molecules (e.g., acid-base catalysis; metal surface catalysis), enzymes are efficient catalysts for usually only one type of chemical reaction between specific biomolecules.

Catalyst-chemistry specificity is a unique feature of biochemical pathways. This development in Earth chemistry may have been the key evolutionary event that led to the emergence of living organisms. Catalyst-chemistry specificity makes possible the assembly of chemical systems in which chemical reactions are partitioned in time and space by the dynamic production or localization of the catalysts. Encapsulation of such systems may have been the beginning of living cells capable of regulation of chemical reactions in a manner that led to self-duplication and other properties ascribed to living organisms.

Much of the early biochemical pathway research was concerned not with the origins of life, but instead with the elucidation of the nature of the chemical reactions of life and properties of the responsible enzyme catalysts. Chemical reactions found to be essential for life were intensely studied, because they were likely to be important effectors on human health, disease, and death. Many enzyme catalysts have been investigated for their potential to be therapeutics, diagnostics, and targets for drug development. This research yielded a wealth of knowledge and understanding about cellular biochemical reactions, their substrates, their cofactors, their products, their organization into pathways, and the physicochemical properties of the enzymes that catalyzed them. Much of this knowledge, and the principles of biochemical pathway research developed in its pursuit, can be found in well-known textbooks of biochemistry (2,3).

Often, in modern discussions of the contributions of the field of biochemistry to our understanding of cellular function and the future of this field of investigation, classic intermediary metabolism is the primary focus, which is a gross oversight of the breadth of biochemical pathway research. The field of biochemical research encompasses more than just the anabolism and catabolism of small metabolites. All enzyme-based reactions that lead to chemical changes in biomolecules fall under the rubric of biochemistry (Fig. 1). Therefore, many aspects of molecular biological research also fall well within the realm of biochemical investigation, and in many cases have their beginning there as well. Typically, macromolecular synthesis processes that preserve, translate, and modify genetic information, like DNA replication, RNA transcription, RNA editing, protein translation, protein modification, and signal transduction, are considered to be in the realm of molecular biology. However, these essential cellular processes are all accomplished by the action of enzymes often functioning in the context of highly regulated molecular ensembles and machines, which catalyze chemical reactions that either extend biopolymers or modify them with small covalent biomolecules (e.g., phosphates). As will be discussed later in greater detail, the biochemical underpinning of molecular biology is often overlooked, especially in terms of how macromolecular synthesis and modification reactions are integrated with other small biomolecule pathways (e.g., nucleotide metabolism) (4).

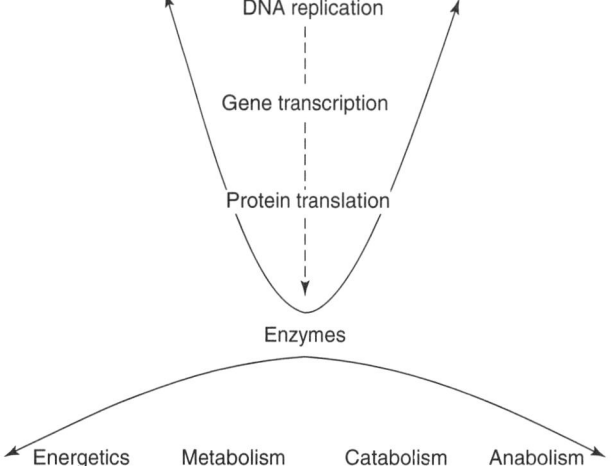

Figure 1. Enzymes at the integration point of molecular and biochemical processes in cells. Enzymes are both the components and products of macromolecular synthesis pathways. Thus, biochemical pathway research encompasses molecular processes, as well as the biochemistry of small biomolecules. (This figure is available in full color at http://www.mrw.interscience.wiley.com/ebe.)

2. BIOMEDICAL APPLICATIONS OF BIOCHEMICAL PATHWAY RESEARCH

Biochemical pathway research has had its greatest impact in biomedicine in the areas of disease diagnosis and drug development. Biomedicine is distinct from medicine *per se*. Medicine is largely an applied discipline in which a vast store of empirical information regarding the nature of human disease and health is integrated with an exact practiced method to effect improvements in the health of injured or ailing individuals. In contrast, biomedicine is a discipline in which knowledge from life sciences research is sought that can be used to better inform the empiricism of medical practice and thereby yield significant advances in medical care.

During the twentieth century, few, if indeed any, areas of life sciences discovery contributed more to advances in biomedicine than biochemical pathway research. One of the best illustrations of its great impact is research to elucidate biochemical defects caused by inborn errors of metabolism (5). Long before the genetic basis for many of these inherited childhood illnesses were known, the missing or defective enzymes responsible for specific defects in biosynthesis or catabolism were identified. This research showed that debilitating and life-threatening diseases could result from defects in single enzymes that catalyzed a specific reaction in biochemical pathways. These studies discovered enzymes and biochemical pathways that were essential for human health and life.

Research on inherited biochemical pathway defects also contributed greatly to the evolution of two major concepts in medicine and biomedicine, respectively. First, it became clear that changes in specific metabolite concentrations could signal alterations in their respective biochemical pathways and, therefore, could be used to diagnose a disease and provide clues to disease etiology. Second, enzymes in defective biochemical pathways identified as responsible for a disease process could be targeted with drugs to restore normal pathway function and health. These two concepts are core principles for modern medical diagnoses and drug discovery, respectively. The first-year medical student is already well-versed in the battery of metabolite tests that are performed for patients on admission to evaluate the status of several important biochemical processes; and specific signs and symptoms lead to additional evaluations of metabolites that provide information of specific suspected defects in biochemical pathways that govern health. The majority of drugs in our vast arsenal of pharmaceuticals that act against cell and tissue processes have a basis as effectors of well-defined biochemical pathways; and the guiding principle of modern-day pharmaceutical discovery is to first identify an important enzyme involved in a disease mechanism and then target it for development of therapeutic effectors of its action.

3. CHALLENGES FOR APPLICATION OF BIOCHEMICAL PATHWAY RESEARCH TO BIOMEDICAL ENGINEERING

Whereas the great impact of biochemical pathway research in advances in biomedicine is self-evident, its role in biomedical engineering is less clear. To begin this discussion, it is important to consider the areas of biomedical engineering for which biochemical research is likely to be most relevant, based on whether biochemical pathway research can enable biomedical engineering. Two subfields of biomedical engineering seem most appropriate for this: metabolic engineering and cell and tissue engineering.

Metabolic engineering, or biochemical engineering as it is also called, is a recently established discipline that has the goal of elucidation and manipulation of metabolic pathways (6). Its primary focus is manipulation of metabolic pathways in microbial organisms to improve their efficiency for production of industrial compounds, including pharmaceuticals. It is instructive that the motivation for establishment of this field is recent advances in genomics, not biochemistry. Much of the engineering in this field is in the form of genetic manipulation of enzymes responsible for biochemical processes that can yield industrial and pharmaceutical products. Although metabolic engineering does have the potential to impact medicine indirectly through designing better methods for production of pharmaceuticals, it does not have a goal of elucidation of the biological processes on which the produced drugs act. As many of the methods of metabolic engineering are designed for application to microbial systems, they may not be readily transferable to biomedical engineering efforts that have the goal of direct manipulation of human cell and tissue physiology.

Cell and tissue engineering is more in sync with the areas of biomedical engineering that seek to elucidate and manipulate human cells and tissues for the purpose of improving health. Much of previous cell and tissue engineering has been in the area of development of prostheses and tissue-compatible materials, including acellular biological substrates for tissue grafts like skin. Knowledge from cellular biochemistry has not been a major factor in the development of these applications. Even theoretical approaches for direct cell engineering have largely taken the form of genetic engineering to alter cell function, instead of regulation of biochemical pathways.

As the field of biomedical engineering continues its growth into the areas of cell and tissue engineering, knowledge of biochemical pathways will play a more important role. Biochemical reactions are the ultimate readout of all information programmed in the genome. All changes in gene expression programs are reflected in cellular changes that reconfigure a complex four-dimensional system of small biomolecules with time-varying states that are unbound, transiently noncovalently or covalently bound to reactive enzyme surfaces, transiently noncovalently interacting with specific target molecules, covalently bound to macro- or micromolecules, or nonspecifically interacting with molecules. Theoretically, detailed exact knowledge of kinetics and regulation of each cellular enzyme, exact monitoring of the concentrations of small biomolecules in each of the described states, and rates of chemical change for all involved components could be used to specify the gene expression status, physiological state, or pathological state of cells and tissues. The goal of cell engineering for biomedicine is to use such information to design approaches to (1) predict cell behavior; (2) manipulate cell behavior; (3) induce repair of diseased cells and tissues; and (4) provide better technologies for interrogating cell and tissue function for the purpose of engineering.

A major barrier to infusion of concepts from biochemical pathway research into cell and tissue engineering is that the current body of knowledge from biochemical pathway research is not suitably formulated for productive cell and tissue engineering for medicine. The generation of ideas and information from biochemical pathway research was not performed with engineering principles in mind. The information in biochemistry textbooks is a summarization of data derived from many different investigators working over the years with species as diverse as bacteria, rats, and humans. Although the main outline of essential biochemical pathways are very well conserved, important differences exist even for the same chemical reaction coordinate that make it impractical to consider all sources of data as universally applicable to human cell physiology, health, and disease. Moreover, it is woefully inadequate for engineering design.

Many examples exist in which differences in detailed aspects of bacterial versus human biochemical pathways manifest as significant differences in biological effects. In many cases, these detailed differences have been significant enough to support the development of effective antimicrobials for medicine. For example, several folate antagonists are potent inhibitors of the bacterial enzyme dihydrofolate reductase (DHFR), which is essential for nucleotide biosynthesis, but, at similar concentrations, have little effect on the analogous human enzyme. However, at higher doses, these drugs are used as a cancer chemotherapy agent, because actively proliferating tumor cells are more sensitive to DHFR inhibition than normal tissues that have a smaller fraction of cycling cells. Similarly, compared with single-celled organisms, multicellular organisms have more enzyme isoforms that exhibit similar enzyme-chemistry specificity, but differences in regulation. For example, the rate-limiting enzyme for guanine ribonucleotide metabolism, inosine-5′-monophosphate dehydrogenase (IMPDH), has a single form in bacteria, four different genes in some yeasts (7), and two different genes in mammals (8), and the different isoforms of enzyme produced differ in cellular regulation and enzyme kinetics properties (8).

Thus, the current body of biochemical pathway data is heterogeneous and incongruent because of the summation of findings from different species, different cell types, different experimental conditions, studies using different levels of technological sophistication, and different historical periods of investigation. Moreover, much of it was not developed with human cells and tissues. A first response to this situation is obvious. Ideally, effective biomedical engineering cannot be accomplished with information of such low intrinsic continuity and questionable relevance to human cellular processes. It will be difficult to effectively design when the data are characterized by inconsistencies of scale, quantification, and biology. In addition, despite the overall tremendous volume of data, for any specific segment (e.g., cell type-specific), it is likely to be incomplete as well.

One essential goal exists that must be achieved if the great potential of biochemical pathway research in biomedical engineering will be realized. For each of the main areas of medical therapy challenges (e.g., cancer, cardiovascular disease, neurological disease), a technically sound, self-consistent body of quantitative data that captures the dynamics of all key biochemical determinants of cellular behavior, their modes of regulation, and their connectivity must be developed. Such a store of information could be used by cell and tissue engineering efforts worldwide. It is unlikely that any one entity or institution will have the expertise and resources to accomplish such a lofty goal alone. So, the field of biomedical engineering must adopt a general consensus that collective development of a comprehensive base of knowledge that covers all areas of cell and tissue biochemistry relevant to human health and disease is a critical undertaking for continued advances in biomedicine. Individual efforts should strive to build self-consistent datasets that can be integrated easily with others. Successful integration will require the establishment of standardized methods of biochemical quantification and analysis, and a mandate for using human cells and tissues for interrogation whenever possible. Studies in animal models will still have an important role to play, but it will be important to precisely define their relationship to processes in humans.

4. CHANGING TIMES FOR BIOCHEMICAL PATHWAY RESEARCH

Despite this rich history of life science discoveries and biomedical applications, including groundbreaking advances in chemistry, enzymology, protein crystallography, macromolecular regulation, spectroscopy, and medicine to give but a small sampling, many would say that biochemistry is no longer an active field of investigation. Instead, it has gone the way of the Latin language: once spoken in most of the Eurocentric world, now only seen in the foundations of languages that are derivative of it. So may appear the current fate of biochemistry.

Until the 1990s, biochemistry was a major field of investigation in every major research institute and medical school. However, because of its molecular genetic methods for isolation and production of enzymes that were much more efficient than the biochemical method of assay-based protein fractionation, the rapidly maturing field of molecular biology eclipsed biochemistry. One unforeseen effect of the advances in molecular technology was a change in the life sciences focus from enzymology and biochemical regulation to molecular modification and gene expression regulation. Research on the actions of biochemical pathway metabolites and small biomolecules in the regulation of cell behavior took a backseat to the rapid emergence of the ability to evaluate regulation and gene product function by molecular genetic approaches (e.g., site-directed mutagenesis; polymerase chain reaction).

The effects of advances in molecular biology on biochemical research were compounded by the lack of comparable advances in biochemical methods. The core principles and conceptual approaches to biochemical research have not changed fundamentally in the last century. Biochemical research is grounded in the fundamental concepts of separation chemistry, analytical chemistry, and protein fractionation. Although incremental improvements in these technologies have occurred, there have been no recent fundamental conceptual or technological advances, as in the field of molecular biology (e.g., DNA sequencing, gene cloning, polymerase chain reaction, gene microarrays).

Ancillary fields, like protein crystallography, that have a connection to the field of biochemistry have gone their own way, integrating new technologies and conceptual approaches (e.g., nuclear magnetic resonance and computational approaches) to make advances in their own right. In contrast, biochemical pathway research has all but stopped. Even laboratories that previously focused on enzymology and biochemical pathway regulation have begun studying gene regulation of involved enzymes. Of course, such studies are quite relevant to understanding biochemical pathway function, integration, and regulation. Unfortunately, the greater challenges of quantifying and evaluating the biochemical features of these systems have given way to more yielding molecular analyses.

Here, it is important to acknowledge that biochemical research is inherently difficult because of the nature of the molecules and processes under study. Protein enzymes are inherently more complex in structure and information coding than DNA and RNA; and, unlike nucleic acids, they have an unpredictable molecular structure, have functions that cannot be predicted from knowing their molecular structure, cannot be synthesized directly with their own structure as a template, can be intractable to purification, and lack a method for routine comprehensive sequencing. Many of the biomolecules acted on by enzymes are difficult to detect, because of low concentration or challenging separation properties, and interconvert or turnover rapidly. These properties combined to make biochemical pathway questions impenetrable at times given current methodologies.

These challenges that currently confront biochemical research may yield to new interdisciplinary approaches that are brought to bear in a focused way on general problems in biochemical pathway research. Many examples of cellular biochemical pathways exist in which many, if indeed not all, of the substrates, metabolites, products, cofactors, and enzymes are known in detail. In addition, the kinetics properties of the involved enzymes have been determined *in vitro*; and their molecular expression properties in cells are known. However, the metabolite concentrations and reaction rates *in vivo* are unknown, and it is not possible to measure changes in them as a cell changes with time or as it is induced to transit between different physiological states. Recent developments in identification of cellular biomolecules by mass spectrometry (MS) after their separation by different methods of liquid chromatography (LC) have potential for application to problems of this type. However, currently, the difficult to control, unpredictable nature of molecular ionization poses a major limitation to development of the quantitative MS approaches needed. This limitation might be overcome by chemically preparing isotopic standards for all evaluated pathway biomolecules that can be added to cell extracts prior to LC separation and MS analysis, which is no small undertaking, of course, but is the scope of the problems that must be solved for continued advances in understanding biochemical pathways.

5. NEW TOOLS FOR BIOCHEMICAL PATHWAY RESEARCH

The previous two sections make the case that, for biochemical pathway research to have a greater impact in biomedical engineering, new approaches to the acquisition and management of biochemical data must be adopted that are compatible with engineering methodology. However, it will be difficult to make these conceptual shifts without parallel advances in new tools and technologies for biochemical pathway research that can provide the necessary advances in biochemical data.

In particular, three general categories of technology development are envisioned that will be critical for realization of the proposed revisions in biochemical pathway research practice. The first category is technologies for high-volume cellular biochemical steady-state and kinetics measurements. The second category is improved technologies for making such measurements for individual cells. The third category is technologies for performing these evaluations with temporal resolution (i.e., "biochemical time lapse"). The focus on these three categories of technology development is in keeping with the goal of quantifying changes in diverse pathway components (e.g., enzymes, substrates, metabolites, products) simultaneously. Such measurements under steady-state conditions will be useful. However, it will be the ability to monitor the kinetics of pathway component changes after small or large perturbations from steady state or as a function of cell lifetimes that will yield the greatest advances in knowledge of cell and tissue function.

Currently, the gene expression microarray tool equivalent does not exist for pathway enzymes and metabolites. Although new microarray tools are in development for proteins, complexity comparable with that of gene microarrays is not in the foreseeable future. In addition, just as in many cases mRNA expression is not a valid metric for a given protein's production level, so too, the cellular concentration of a given enzyme will often misrepresent its cellular activity. Unlike mRNA-protein nonequivalencies, which have no solution other than direct protein quantification, protein-activity nonequivalencies can be resolved by the development of specific activity probes. Activity probes are enzyme substrate analogues that give a conveniently detected and quantified signal (e.g., fluorescence) when enzymatically transformed into a product analogue. When provided to cells in tracer amounts, activity probes can provide direct quantitative information about the cellular activity level of their targeted enzyme. The applicability of activity-probe analyses falls off rapidly as the number of enzymes in the biochemical pathway/network of interest increases, because a chemically specific and spectrally distinct probe is required for each enzyme. Thus, increased chemistry ingenuity, focused on the development of diverse libraries of well-defined, specific activity probes for cellular pathways of interest, will also be an important factor.

If the signal basis is fluorescence, then the activity-probe approach is quite compatible with a longstanding mainstay technology for single-cell fluorimetry, flow cytometry. Several brands of highly sophisticated flow cytometers are commercially available. These instruments are capable of simultaneous multiprobe measurements. They are also effective for isolation of individual cells based on their quantitative fluorescence profile. This feature allows for subsequent orthogonal analyses (e.g., proliferation capability) of biochemically profiled cells. However, although flow cytometry provides many of the answers for high-throughput, quantitative, single-cell analyses when combined with the activity-probe approach, it has two significant shortcomings. It lacks single-cell temporal resolution, and it loses cell lineage relationship information.

Determination of how biochemical pathway activities are integrated with cell lifetime and cell developmental programs is the present frontier of cell biochemistry research. In the last hundred years, cell function research has largely been based on the average properties of studied cell populations. Evaluating the distributions of single-cell function will be significantly more informative regarding the structure and kinetics of cellular biochemical systems, which is particularly true in the case of tissue cell systems that are composed of collections of patterned, lineage-related cells with distinctive, but interdependent, processes and functions. Although these challenges are formidable, questions in this area have begun to yield to recent ingenuity in integrating existing technologies to produce novel ones specifically designed for these analyses. For example, capillary electrophoresis has been coupled to temporally defined "chemical cytometry" in a manner that provides quantitative data on the specific protein content of single sister cell products of individual cell divisions (9,10). As technologies of this power are integrated with improvements in metabolite separation resolution and analytical chemistry sensitivity, the envisioned new era of biochemical pathway research will take off, with major advances in biomedical engineering not far behind.

6. AN EXAMPLE OF CELL ENGINEERING ENABLED BY BIOCHEMICAL PATHWAY RESEARCH

It may be instructive to consider a case study of how biochemical pathway research can beneficially inform biomedical engineering in the particular case of cell engineering. A recent example can be found in the field of adult stem cell biological engineering (11). Biomedical engineering falls under the heading of the newly emerging field of biological engineering. Biological engineering is here defined as the engineering discipline concerned with application of engineering methods and principles to biological discovery for the benefit of humanity. Beyond engineering for problems in medicine, like the applied field of biomedical engineering, biological engineering is a novel engineering discipline that, in addition to invention and manipulation, sets discovery across all scales of biology as a problem to solve.

A major goal of adult stem cell biological engineering is to gain sufficient knowledge of adult stem cell function, so that these unique tissue cells can be engineered in a manner that supports continued investigation of their special properties while providing useful biotechnology (e.g., drug screening) and biomedical applications (e.g., cell replacement therapy). Adult stem cells are found in all major human tissues and organs. These cells are a small fraction of total tissue cells that are responsible for the renewal, regeneration, and repair of tissues. They exist in a relatively undifferentiated state, but divide to yield daughter cells with "asymmetric" differentiation and cell kinetics properties. Specifically, one daughter retains the properties of its stem cell predecessor. It remains relatively undifferentiated and retains division capacity for the life span of the body.

In contrast, the other daughter cell becomes a nonstem cell. It serves as the precursor for a differentiating cell lineage that terminates as a population of permanently nondividing cells. This nondividing lineage may be composed of the initial nonstem cell daughter that undergoes differentiation and terminal arrest without further division. However, in most tissues, the terminal cell lineage is characterized by multiple transient cell divisions associated with differentiation of cell phenotype. For some tissues, a single type of mature, terminal, arrested cell type is produced. Adult stem cells that produce this type of single-phenotype differentiating lineage are called unipotent (e.g., lens stem cells, sperm stem cells). In other tissues, adult stem cells produce progeny lineages that undergo branching differentiation programs to produce several distinct functional cell types that construct the tissue. These adult stem cells are described as multipotent.

The property of adult stem cells to divide continuously while maintaining their cell kinetics capacity and potency, and simultaneously producing differentiating progeny cells, is called asymmetric self-renewal. Asymmetric cell kinetics underlie this defining adult stem cell property. Adult stem cell kinetics are defined as asymmetric, because, whereas adult stem cells have capacity to divide for the life span of the body, their differentiating progeny undergo a terminal cell cycle arrest. *In vivo*, the cell cycle arrest is followed by expiration of mature cells and their subsequent removal from tissues by apoptosis programs or physical wear. By fueling the continuous production of new differentiating cells that progress to arrest and removal, adult stem cells effect tissue renewal. A similar program is thought to be responsible for replacement and repair of damaged tissue during wound healing. However, in the injury setting, it is postulated that adult stem cells must also undergo symmetric divisions to create additional tissue units to replace lost tissue. Symmetric adult stem cell divisions produce two adult stem cells that are identical to each other and their predecessor. These expanded adult stem cells, under controls present in the wound environment, are thought to restart their asymmetric self-renewal program for restoration of normal tissue architecture and function. Adult stem cells must also undergo regulated symmetric divisions during growth of the body from neonatal through adult development. Tight regulation of such divisions is recognized as imperative to avoid development of benign masses or malignant tumors (8).

As a result of their remarkable regenerative properties, adult stem cells are an exciting prospect for cell engineering. These cells are ideal for *in vitro* tissue engineering research, organ culture efforts, and drug metabolism and toxicology research. They are ideal targets for gene therapy, and they are the key to development of biomedical applications toward effective cell replacement therapies. Yet, progress in all of these areas has been incredibly slow because adult stem cells are rare tissue cells, impossible to prospectively identify, and difficult to isolate. In particular, expansion and propagation of any adult stem cell in culture has been a well-known intractable problem in stem cell biology (8). Prior to the research that will now be described, it had been especially difficult to establish clonally pure cultures of adult stem cells. The ability to prepare such populations opens up the possibility to attain the level of completeness needed in our understanding of adult stem cell function to effectively engineer them for biomedical applications.

The first successful engineering of adult stem cell expansion was the result of integrating molecular biological and biochemical pathway knowledge in the context of a carefully designed model cell culture system that was found fortuitously to exhibit asymmetric self-renewal [reviewed in (8)]. Conceiving an approach to the expansion problem required several steps of logic. First, it had been shown that restoration of normal expression of the well-studied p53 tumor suppressor protein in established cultured cell lines induced asymmetric self-renewal. This finding led to an examination of the cell division kinetics of explanted tissue cells that had been cultured only for short periods. If these cells were derived from mice that produced normal p53, they exhibited asymmetric self-renewal. However, if they were derived from transgenic mice engineered to have a p53 gene knockout, they exhibited symmetric self-renewal, which promoted exponential proliferation of cells from these mice.

Given a sufficient period of culture, all explanted adult mammalian tissue cells are known invariably to undergo a complete cell division arrest called senescence. Asymmetric self-renewal by adult stem cells in culture was postulated to be an important factor that led to senescence. As the stem cell number does not increase during asymmetric self-renewal, whereas the number of differentiating progeny cells *does* increase, adult stem cells were predicted to be "lost" with continued culture as a simple consequence of dilution among accumulating differentiating progeny cells.

Based on these concepts, the idea was advanced that expanding functional adult stem cells in culture would require converting them from asymmetric self-renewal to symmetric self-renewal. For engineering the expansion of adult stem cells based on this principle, it would be ideal to effect a reversible change in self-renewal. A reversible method would allow expanded adult stem cells to be subsequently returned to asymmetric self-renewal for study and evaluation for potential medical and biotechnology applications. A method for reversible "suppression of asymmetric cell kinetics" (SACK) was available because of prior research on biochemical pathways that control p53-dependent asymmetric cell kinetics. The ability of p53 to induce asymmetric self-renewal had been shown to depend on its ability to reduce the synthesis of IMPDH, the rate-limiting enzyme for guanine ribonucleotide biosynthesis (see Fig. 2). Knowledge of the involvement of this biochemical pathway led to the discovery that the purine nucleoside xanthosine (Xs) could reversibly convert p53-dependent cultured model cells from asymmetric self-renewal to symmetric self-renewal. The basis for this regulation is the ability of Xs supplementation to maintain guanine nucleotide pools even though IMPDH is under regulation by p53.

SACK engineering is based on knowledge of asymmetric self-renewal gained from studies with cultured mouse cell models that exhibit p53-dependent asymmetric self-renewal. This manipulation was applied to develop several independent murine cell strains by using parent cells of diverse tissue origin (e.g., adult breast cells and fetal fibroblast cells) and two different gene promoter systems for controlling p53 gene expression. Each of these different cell strains exhibits the same p53-dependent asymmetric self-renewal because of the same biochemical pathway. The findings with these cell strains led to the general proposal that an integrated molecular-biochemical system composed of p53, IMPDH, and guanine ribonucleotide

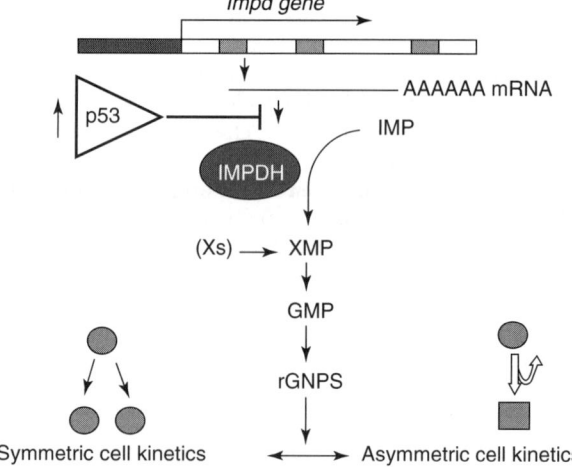

Figure 2. A molecular-biochemical pathway for engineering adult stem cell expansion. The p53 gene product controls the decision between asymmetric and symmetric cell kinetics by molecular regulation of the synthesis of the IMPDH enzyme. Expression of p53 is associated with a concomitant reduction in expression of IMPDH messenger RNA and protein. IMPDH is rate-determining for conversion of inosine monophosphate (IMP) into guanine ribonucleotides (rGNPS). Higher levels of cellular rGNPs promote symmetric cell kinetics, which promote adult stem cell expansion; and reduced levels induce asymmetric cell kinetics characterized by production of a nondividing cell or progenitor for a terminal cell lineage (represented by square) at each adult stem cell division. By promoting the IMPDH-independent formation of xanthosine monophosphate (XMP), the SACK agent xanthosine (Xs) expands rGNP pools sufficiently to convert adult stem cells from asymmetric cell kinetics to symmetric cell kinetics. (This figure is available in full color at http://www.mrw.interscience.wiley.com/ebe.)

metabolites regulated the symmetry of adult stem cell kinetics. This knowledge was tested for its completeness for designing an approach to expansion of adult stem cells.

As a first test of the SACK engineering, an attempt was made to expand adult stem cells from adult rat liver (11). Much interest exists in liver stem cells that can produce mature hepatocytes for applications in cell therapy development, drug metabolism analyses, and preclinical toxicity testing. In addition to hepatocytes, the liver contains several other resident cell types that are also of interest (e.g., stellate cells, Kupffer cells, sinusoidal endothelial cells, and cholangiocytes). If other hepatic cell types are renewed by asymmetrically cycling liver stem cells, the SACK method should lead to their expansion as well.

In these studies, single isolated liver cells were cultured in the presence of Xs and expanded to become stable clonal cell strains. Cell strains established in the presence of Xs show evidence of asymmetric self-renewal when Xs was withdrawn. The two best-characterized cell strains, which had the highest degree of asymmetric self-renewal, had *in vitro* properties indicative of the two liver epithelial cell types, hepatocytes and cholangiocytes, respectively. Previously, there had been only limited success in culturing primary cholangiocytes, and stable cholangiocyte stem cell lines had not been reported. The SACK-engineered cell strains exhibit Xs-dependent asymmetric self-renewal and Xs-controlled expression of respective mature hepatic differentiation markers.

The success with the SACK engineering approach demonstrated its potential to become a general method for the expansion of diverse types of adult stem cells. The cellular biochemistry of nucleotide metabolism is one of the most conserved cellular processes across diverse species. It is noteworthy that biochemical pathway knowledge established in murine cells led to success in rat tissues. This initial success in translation across species is a harbinger for success in future attempts to engineer the expansion of human adult stem cells based on the same biochemical pathway principles.

7. THE FUTURE OF BIOCHEMICAL PATHWAY RESEARCH IN BIOMEDICAL ENGINEERING

The success in engineering the expansion of adult rat liver stem cells serves to highlight the advantage gained from developing a biomedical engineering strategy that integrates advances in molecular biology and biotechnology with those from biochemical pathway research. This intellectual advance has not occurred in several other active areas of biomedical engineering research, including cell engineering targeted to gene regulation, cell cycle regulation, and signal transduction. These processes are central to normal tissue function, and defects in them are important elements in the etiology of many illnesses. A major shortcoming in current biomedical engineering research is that elucidation of cellular regulation mechanisms is characteristically viewed as a problem in macromolecular interactions. Little attention is paid to the small cellular molecules that fuel and modify the macromolecular components of these systems.

The major impetus for recent interests in engineering cell regulation processes is the perception that genomics has given a complete "parts list" of all the important components in cell regulation networks. Of course, even the least informed students of biology now appreciate that a significant proportion of cell regulation information is coded in postgenomic sites such as RNAs and proteins. Accordingly, to fill in these missing parts, new technologies for comprehensive cataloguing of RNA expression are maturing; analogous technologies for protein and protein modification systematics are in development (i.e., proteomics); and other methods for identifying and quantifying all members of a class of information bearing macromolecules are said to be on the horizon (e.g., glycomics). However, little attention has been given to comprehensive quantitative assessment of small cellular molecules that act as cofactors for enzymes, metabolites in biochemical pathways, allosteric regulators, substrates for macromolecular synthesis, storehouses of chemical energy, and substrates for macromolecular modification reactions. Unfortunately, often in science and engineering, expediency is the mother of failed inventions.

Currently, most efforts to engineer cell regulation based on genomic and gene expression profiling data is in the mathematical modeling stage. In general, modeling features common to nonbiological disciplines have been applied to modeling gene transcription networks, protein interaction networks, and growth factor signaling networks. Measurements of enzymatic activities and biomolecular interactions are included in these models, but concentrations and conversion rates of involved small molecules are not. For example, modeling of signaling networks based on hydrolysis of guanosine-5′-triphosphate (GTP) or adenosine-5′-triphosphate (ATP), with either phosphate release or phosphate transfer, does not include the concentrations and dynamics of GTP and ATP pools. An untested assumption is made that these pools are saturating and stable, which is certainly unlikely for GTP, a small nucleotide pool whose levels change significantly with changes in cellular state (12).

Mathematical models based on data that omits incorporation of information on critical small cellular molecules are unlikely to be faithful predictors of cellular behavior for which small biomolecules are most certainly important determinants. Thus far, these types of mathematical models have not been put to this practical test. Therefore, it is a powerful statement that one mathematical model that has been put through such testing, and does well, is based on both pathway enzymes and metabolites. Curto et al. (13) developed a comprehensive mathematical model for purine nucleotide metabolism that incorporates the dynamics of pathway metabolites as well as the interregulation and intrinsic kinetics properties of involved enzymes. This model showed excellent performance in predicting known clinical outcomes for specific physiological perturbations. Thus, it may have application in diagnosis of disturbances in purine metabolism from several measurements of metabolites in patients; and it may be effective in predicting clinical responses to agents that have a known effect on particular components of the purine nucleotide biosynthesis system. This model is one of the best examples of biomedical engineering enabled by biochemical pathway research. If others emulated the principles provided by this excellent example, then progress in cell and tissue engineering would be rapidly accelerated.

The biochemical knowledge and principles of biochemistry are fundamental to our understanding of life processes across diverse species. In the fast-paced modern world of genomic science, much is made of the recognition that many examples of gene and protein sequences exist that are conserved even among organisms that are quite different in form. Yet, myriad biochemical processes are invariant throughout nature, indicating that the common denominator of all life forms is chemistry not gene sequence. Thus, the future of biochemical pathway research seems as vibrant as ever.

At the moment, many emerging new fields based on the high-throughput cataloguing principles of genomics (e.g., proteomics and glycomics) go forth without consideration of cellular chemistry. However, eventually, all of these constructions will be found to be lacking for successful engineering, because they omit key determinants required for effective design and manipulation of complex cellular systems. In the end, the vast majority of the information encoded in the genome is chemical. The macromolecules that currently engender the most excitement in cell engineering efforts are only ends to the means for chemical reactions that build, energize, and renew life forms. Life is, in its essence, a collection of self-sustaining chemical reactions.

One of the currently more esoteric forms of cell engineering, the design and synthesis of minimal genomes provides one final illustration of the impact that biochemical pathway research can have in biomedical engineering. In these efforts, gene assembly, gene ablation, and *in silico* gene analysis approaches are applied to design and produce minimal genomes based on the calculated minimal requirements for a living cell (14,15). In some design, no *a priori* consideration is given to what biochemical processes might be required for life. For instance, a genome chosen for minimization may be dissected by random insertion mutagenesis. The pattern of insertions that result in a viable organism is used to decipher the minimal set of essential gene elements (14). Whereas the final outcome of such an undertaking holds intrigue for biologists, its utility for biomedical engineering is dubious. In other studies (15), significant effort is placed in developing experimentally testable computer models that incorporate detailed biochemical data along with molecular data. These studies demonstrate that comprehensive incorporation of molecular and biochemical information yields increased power in developing robust models that can be used to predict cellular properties in related organisms for which experimental data is not available. The models developed provide an effective means to explore cellular dimensions that are otherwise inaccessible with current technology. So far, this comprehensive approach that embraces data from biochemical research has been limited to microbial organisms. However, its future acceptance as the preferred approach to engineering of human cells and tissues will yield tremendous benefits in medicine as well as significant growth in our grasp of the fundamental nature of life.

Acknowledgments

Many thanks to R. Taghizadeh, A. M. Nichols, and Dr. J.-F. Paré for carefully reviewing the manuscript and providing helpful suggestions for its completion.

BIBLIOGRAPHY

1. A. L. Lehninger, *Bioenergetics, The Molecular Basis of Biological Energy Transformations*, 2nd ed. Menlo Park, CA: W. A. Benjamin, Inc., 1973.
2. A. L. Lehninger, *Principles of Biochemistry*. New York: Worth Publishers, 1982.
3. L. Stryer, *Biochemistry*. New York: W. H. Freeman, 1995.
4. J. L. Sherley, An emerging cell kinetics regulation network: integrated control of nucleotide metabolism and cancer gene function. In: K. W. Pankiewicz and B. M. Goldstein, eds., *Inosine Monophosphate Dehydrogenase: A Major Therapeutic Target*. Washington, DC: American Chemical Society, 2003.
5. V. A. McKusick, Medical genetics. In: A. M. Harvey, R. J. Johns, V. A. McKusick, A. H. Owens, and R. S. Ross, eds., *The Principles and Practice of Medicine*, 20th ed. New York: Appleton-Century-Crofts, 1980.
6. G. N. Stephanopoulos, A. A. Aristidou, and J. Nielsen, *Metabolic Engineering: Principles and Methodologies*. San Diego, CA: Academic Press, 1998.
7. J. W. Hyle, R. J. Shaw, and D. Reines, Functional distinctions between IMP dehydrogenase genes providing mycophenolate resistance and guanine prototrophy to yeast. *J. Biol. Chem.* 2003; **278**:28470–28478.
8. J. L. Sherley, Asymmetric cell kinetics genes: the keys to expansion of adult stem cells in culture. *Stem Cells* 2002; **20**:561–572.
9. H. Ahmadzadeh and S. Krylov, On-column labeling reaction for analysis of protein contents of a single cell using capillary electrophoresis with laser-induced fluorescence detection. *Methods Mol. Biol.* 2004; **276**:29–38.
10. K. Hu, H. Ahmadzadeh, and S. N. Krylov, Asymmetry between sister cells in a cancer cell line revealed by chemical cytometry. *Anal. Chem.* 2004; **76**:3864–3866.
11. H.-S. Lee, G. G. Crane, J. R. Merok, J. R. Tunstead, N. L. Hatch, K. Panchalingam, M. J. Powers, L. G. Griffith, and J. L. Sherley, Clonal expansion of adult rat hepatic stem cell lines by suppression of asymmetric cell kinetics (SACK). *Biotech. Bioeng.* 2003; **83**:760–771.
12. M. L. Pall, GTP: a central regulator of cellular anabolism. *Curr. Top. Cell. Regul.* 1985; **25**:1–20.
13. R. Curto, E. O. Voit, A. Sorribas, and M. Cascante, Validation and steady-state analysis of a power-law model of purine metabolism in man. *Biochem. J.* 1997; **324**:761–775.
14. C. A. Hutchison, S. N. Peterson, S. R. Gill, R. T. Cline, O. White, C. M. Fraser, H. O. Smith, and J. C. Venter, Global transposon mutagenesis and a minimal Mycoplasma genome. *Science* 1999; **286**:2165–2169.
15. C. H. Schilling, M. W. Covert, I. Famili, G. M. Church, J. S. Edwards, and B. O. Palsson, Genome-scale metabolic model of Helicobacter pylori 26695. *J. Bacteriol.* 2002; **184**:4582–4593.

BIOCHEMICAL PROCESSES/KINETICS

EBERHARD O. VOIT
Georgia Institute of Technology
and Emory University
Atlanta, Georgia

1. INTRODUCTION

Biochemical processes are responsible for the proper metabolic functioning of cells and organisms. They include transport steps, chemical conversions, and mechanisms of signal transduction. Individual biochemical process steps are typically arranged in biochemical pathways, which convert an initial substrate into a desired metabolite, such as an amino acid, carbohydrate, or lipid. Although textbooks may give the impression that pathways neatly organize the biochemical machinery within a cell, the truth is that pathways are massively interconnected, and their widespread use lies in the fact that they are simply convenient representations. Thus, although glycolysis is depicted as a more or less linear chain of events, in reality, many branches that siphon material off the pathway or supply the pathway with material from other pathways exist.

The analysis of biochemical processes and pathway systems employs approaches from three complementary fields: thermodynamics, kinetics, and stoichiometry. Although the prime focus here is on kinetics, it is useful to sketch out how these three contributors elucidate different aspects of the biochemical conversions that govern the functioning of every cell. Briefly, thermodynamics addresses questions of energy associated with the substrates and products of a biochemical reaction. These energetic considerations constrain the range of reactions that are possible and permit computations of efficiency of a biochemical system. Kinetics is primarily interested in temporal issues, asking how fast a reaction might proceed and which factors may affect the turnover rate. Stoichiometry accounts for the mass distribution within a biochemical network and ensures that all material entering a metabolite pool is appropriately balanced by material leaving this pool.

2. THERMODYNAMIC CONSTRAINTS

As biological systems are part of the physical world, they must obey the laws of physics and, of particular interest here, the laws of thermodynamics (for a good introduction, see Ref. 1). These laws constrain the range of possible chemical reactions to those that lead to a lower level of energy. Thus, for a chemical reaction A → B to take place, B has to have a lower energetic state than A. Classical thermodynamics characterizes such states, usually in closed systems that neither receive nor release energy or matter. For biological systems, the mandatory transition to lower-energy states appears to be quite counterintuitive. For example, plants use low-energy compounds like CO_2 and water to generate high-energy compounds like sugars, proteins, and fats. Even so, organisms are bound to operate within the constraints of thermodynamics, which they accomplish, generally, by taking up energy from the outside, either as sunlight or as chemical energy in the form of food, and by coupling thermodynamically infeasible, energy-consuming processes with thermodynamically favorable, energy-releasing mechanisms. Although the total energy of the closed system, which in this context must include all organisms as well as earth, sun, and the rest of the universe, decreases, the energy within an individual organism can easily rise to a higher level, for instance, as the consequence of taking in food. Expressed differently, biological systems are open systems that dissipate energy. They can grow, thereby using a lot of energy, or they can maintain their size in a more or less stationary state, which still requires some energy, although less, from outside the system. In contrast to the equalized energy state that any closed system ultimately approaches, each biological system operates in a nonequilibrium stationary state, where influxes and effluxes exchange matter and energy with the environment and are more or less in balance over time. Thus, as an extension of the classical thermodynamics of closed systems, nonequilibrium thermodynamics deals with questions of energy that are linked to transport and metabolism in open, dissipative systems, which require a net influx of energy. Nonequilibrium thermodynamics allows the estimation of energy requirements for specific biochemical systems, the extent and stoichiometry of reactions or pathways, and the degree of coupling among them. It also characterizes the efficiency of coupled reactions, which may be computed as the ratio of free energy consumed in one direction over the energy liberated in the opposite direction. As a specific example, it permits the estimation of the number of moles of oxygen needed for the phosphorylation of one mole of ADP to ATP.

3. THE BASICS OF KINETICS

The defining feature of the thermodynamic approach is its focus on energy. Thermodynamic computations characterize the energetic possibility or likelihood of a reaction, but they typically give no indication of whether this reaction will occur on a time scale of seconds or years. For cells and organisms, these considerations of timing are of great importance. They are at the heart of kinetics, which addresses the temporal aspects of a reaction while implicitly assuming that the reaction is thermodynamically possible. Typical questions of kinetics are: How fast does the reaction proceed? What is the half-life of a metabolite? What affects the speed or rate of the reaction? Kinetics approaches these questions by formulating rate laws that describe how the velocity of a reaction is affected by substrates and modulators.

In the simplest case, the velocity v of the reaction is directly proportional to the concentration X of the substrate of this reaction. This situation translates immediately into the mathematical formulation

$$v(X) = -kX. \qquad (1)$$

The negative sign indicates that substrate X is lost in this reaction, and the positive rate constant k quantifies the turnover per time unit. It is important to note that the rate $v(X)$ represents the change in substrate concentration over time, even though time is not explicit in Equation 1. The connection to the temporal domain becomes clear when the rate of the reaction, is formulated that is the "change over time," as the derivative of the function describing the substrate concentration, X, with respect to time, namely

$$v = \frac{dX(t)}{dt} = \dot{X} = -kX(t), \quad X(0) = X_0. \tag{2}$$

Thus, without much effort, the formulation of a simple reaction as a differential equation is arrived at. The right-hand side is typically expressed as $-kX$, as in Equation 1, which does not show t explicitly, but time is present implicitly in X. The expression $X(0) = X_0$ is called the initial condition of the differential equations. The differential equation itself describes how the substrate concentration changes, whereas the initial condition specifies the substrate concentration at a chosen time $t = 0$.

The representation of substrate loss is easily augmented with a process description for the generation of product Y. As no material is lost in this ideal situation, the production of Y equals the degradation of X, but with opposite sign, because one increases and the other decreases. Thus, the process can be written as a small system of two differential equations, namely:

$$\begin{aligned} \dot{X} &= -kX \quad X(0) = X_0, \\ \dot{Y} &= +kX \quad Y(0) = 0. \end{aligned} \tag{3}$$

The initial condition for Y is set to zero, indicating that no product is present at the beginning of the experiment at $t = 0$.

In the particular case of Equation 3, one can "solve" the differential equation. That is, one can deduce from the differential equation what specific function of time the substrate concentration X is. The solution is

$$X = X_0 \cdot \exp(-k \cdot t), \tag{4a}$$

$$Y = X_0 \cdot (1 - \exp(-k \cdot t)), \tag{4b}$$

where X_0 is again the initial substrate concentration at $t = 0$ and the exponential function describes how the substrate disappears during the reaction. Equation 4a really corresponds to Equation 2 as can be seen if Equation 4a is differentiated with respect to time. The left-hand side becomes dX/dt, whereas the right-hand side becomes $-k \cdot X_0 \cdot \exp(-k \cdot t)$, which is the same as $-kX$ according to Equation 4a. All the material disappearing from the substrate pool X is "collected" in Equation 4b, so that the total mass in the reaction is constant. Initially, at $t = 0$, $Y = 0$. Over time, Y increases, first very fast, then slower, until it approaches the final state $Y = X_0$ where all substrate has been converted into product.

If a reaction is more complicated, for instance, because it involves the effects of inhibitors and other modulators, one cannot write down an explicit function describing the dynamics of X as a function of time. However, it is usually possible to formulate the process as a differential equation as in Equation 2, where one describes which components of the biochemical system affect the rate of change in a metabolite of interest. Furthermore, many numerical algorithms exist that compute very precise approximations of the solution, which cannot be obtained otherwise. For these reasons, the formulation of biochemical reactions as differential equations is the preferred representation, and it is crucial for the analysis of biochemical pathways and networks.

As a simple example, consider the formation of product X_3 from two substrate molecules of type X_1 and one molecule of type X_2. According to the law of mass action, the differential equation for X_3 is set up as

$$\dot{X}_3 = k_3 X_1^2 X_2 \tag{5a}$$

with an appropriate turnover rate k_3. Note that X_1 appears with a power of 2, which reflects that two molecules of type X_1 are used in the reaction. Thus, although the reaction is often written as the "sum" $2X_1 + X_2 \rightarrow X_3$, the substrates enter the rate law as products.

As the speed of formation of X_3 is the same as the speed of disappearance of X_2, one obtains

$$\dot{X}_2 = -\dot{X}_3 = -k_3 X_1^2 X_2. \tag{5b}$$

Substrate X_1 disappears twice as fast, because each reaction yielding one molecule of X_3 consumes two molecules of X_1. The equation for X_1 is therefore

$$\dot{X}_1 = -2k_3 X_1^2 X_2. \tag{5c}$$

In this fashion, differential equations can be constructed for metabolic networks of considerable complexity, as will be discussed later.

Although kinetics has a different focus than thermodynamics, it is worth noting that the two are not independent. A bridge is statistical mechanics, which goes back more than a century to Ludwig Boltzmann (1844–1906) and Svante Arrhenius (1859–1927) and describes the probability that one of very many substrate molecules enters a new (product) state within a given period of time. This stochastic process for each molecule can be combined and averaged for thousands of molecules and leads to the so-called master equation, which characterizes the average number $\bar{x}$ of remaining substrate molecules at any given time. The result is

$$\bar{x} = x_0 \cdot \exp(-k \cdot t), \tag{6}$$

which is exactly of the form in Equation 4a.

One can also derive the rate law in Equations 2–4 and 6 from the transition state theory, which supposes that both substrate and product have energy states that are locally minimal, so that not only the product but also the substrate has a certain degree of stability (2). To be converted into product, the substrate has to overcome an intermediate state of even higher energy so that activation energy must be infused to allow the reaction to proceed. The situation is shown in Fig. 1. It is now assumed that the energy states of all molecules in the system fluctuate somewhat in a random fashion so that some substrate molecules are activated enough to jump over the threshold and become product. To a lesser degree, which clearly depends on the energy differential between substrate and product and on the activation energy, some product molecules may have enough energy to revert to substrate. Based on thermodynamic principles, it is possible to compute the average number of molecules jumping in either direction, and these probabilistic considerations again lead to the rate law discussed before.

3.1. Enzymes

Up to this point, only reactions that occur spontaneously have been considered. Fortunately, these are in the minority in living cells, because otherwise all reactions that could happen would happen on a continual basis, maybe slowly, but surely and without a means of controlling them, which, in reality, does not happen. Instead, many reactions can only exceed the energy level E_A necessary to overcome the intermediate state (Fig. 1) if they are aided by enzymes, which are proteins that catalyze the process by temporarily forming a physical complex with the substrate that energetically exceeds the required activation energy. Once the reaction has occurred, the enzyme is released in its original form. Enzymes thus make reactions possible that would otherwise not occur and are therefore crucially important controllers for any living organism. As a result of its 3D structure, each enzyme is more or less specific for one or a few substrates that fit into the enzyme's binding site, which renders targeted control of metabolism possible. This control is exerted through the modulation of enzyme activity by inhibitors, cofactors, or even the products for whose generation the enzyme is ultimately responsible. A typical example of the latter is endproduct inhibition of the enzyme catalyzing the first step of a linear pathway (Fig. 2). This type of feedback offers a very efficient, ubiquitous mechanism of preventing the generation of unneeded product. Enzymes and their modulators provide the cell with a multitude of effective,

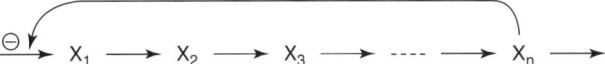

Figure 2. Generic linear pathway with feedback inhibition by the endproduct of the pathway.

fine-tuned means of control. A prominent mechanism of endproduct inhibition is allosteric modulation, where the endproduct, or some other modifier, binds to the enzyme outside its active site. The binding causes a conformational change in the active site of the enzyme, thereby reducing its activity.

About a century ago, Henri, Michaelis, and Menten proposed a detailed mechanism for the action of an enzyme-catalyzed process. As illustrated in Fig. 3, they postulated that the substrate, S, forms an intermediate complex (ES) with the enzyme, E. This complex subsequently breaks apart irreversibly to yield product, P, while simultaneously releasing the unchanged enzyme molecule. The proposed mechanism also allows for the possibility that the complex (ES) could revert back to substrate and enzyme.

Employing rate equations of the type of Equations 2, 3, and 5, one can directly write down differential equations for the process by studying what produces and what degrades a given pool. For instance, S disappears with a rate of k_1, but some S may be recouped from the complex (ES) with rate k_{-1}. Thus, the overall dynamic of S is

$$\dot{S} = -k_1 S \cdot E - k_{-1} \cdot (ES). \quad (7a)$$

The equations of (ES) and P are constructed in the same fashion, yielding

$$(\dot{ES}) = k_1 S \cdot E - (k_{-1} + k_2) \cdot (ES) \quad (7b)$$

$$\dot{P} = k_2 (ES). \quad (7c)$$

Once the reaction has started, it is customary to assume that the intermediate complex (ES) remains constant in concentration. This so-called quasi-steady-state assumption has been discussed frequently in the literature. If it is valid, the system of differential equations 3 can be simplified, and it becomes possible to express the rate of product formation, $\dot{P}$, as a function of the substrate concentration. Thus, this "solution" is not a pair of functions $S(t)$, $P(t)$ in the traditional sense of solving a differential equation, but a relationship between a rate $v_P = \dot{P}$ and the substrate concentration. This relationship is known as the (Henri–)Michaelis–Menten rate law (HMMRL)

$$v_p = \frac{V_{\max} S}{K_M + S}. \quad (8)$$

It has a hyperbolic shape (Fig. 4) that indicates that the reaction speed increases with the substrate concentration,

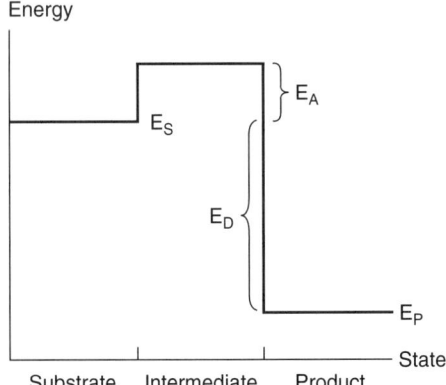

Figure 1. The substrate of the reaction $S \rightarrow P$ has a higher energetic state (E_S) than the product (E_P). However, for the reaction to occur, activation energy E_A is needed to overcome the intermediate transition state. If the energy differential E_D between substrate and product is small, the reaction also occurs, to some degree, in the reverse direction.

$$S + E \underset{k_{-1}}{\overset{k_1}{\rightleftharpoons}} (ES) \overset{k_2}{\rightarrow} P + E$$

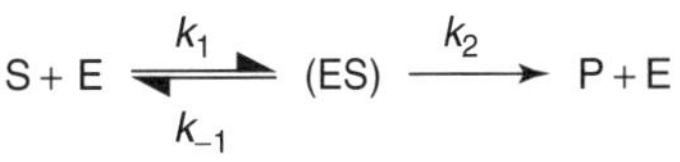

Figure 3. Diagram of an enzyme-catalyzed reaction mechanism according to Henri, Michaelis, and Menten.

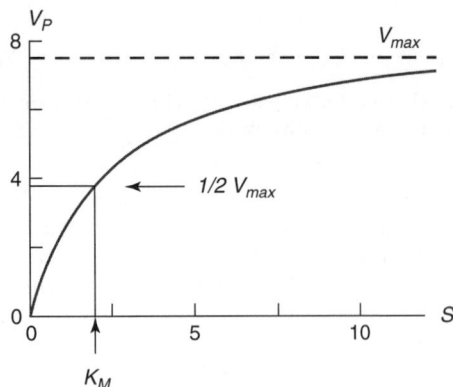

Figure 4. Henri–Michaelis–Menten rate law with $V_{max} = 7.5$ and $K_M = 2$. If the substrate concentration is further increased, the reaction rate approaches V_{max}. (This figure is available in full color at http://www.mrw.interscience.wiley.com/ebe.)

but that it eventually reaches saturation. That is, if much substrate is available, the addition of more substrate makes little difference.

The rate law has two parameters, namely the maximum velocity V_{max} and the Michaelis constant K_M. The maximum velocity designates the fastest possible turnover rate of the reaction, which occurs if a large amount of substrate exists. The Michaelis constant designates the substrate concentration where the speed of the reaction is half the maximum velocity (see Fig. 4). The K_M is considered a more or less constant property of the enzyme under a variety of conditions, whereas V_{max} can vary considerably from one situation to another. The K_Ms of hundreds of enzymes have been measured and are catalogued in databases such as BRENDA (http://www.brenda.uni-koeln.de).

The HMMRL has been discussed many times in the literature, and although considerable doubt exists whether its underlying assumptions are satisfied within a living cell, this rate law is one of the most recognized functions in biochemistry, primarily because it is simple and accurately captures many kinetic data that were obtained in purified assays. One unexpected advantage of this form is that the rate law becomes linear if it is plotted as $1/v$ against $1/S$. The linearity of this so-called Lineweaver–Burk plot allows the estimation of V_{max} and K_M from measured data via simple linear regression.

The HMMRL has been extended to more complicated mechanisms involving more than one substrate or a substrate and an inhibitor. For example, using the same principles as we used for the development of the HHMRL, one can formulate a rate law that accounts for an inhibitor I that competes with the substrate for binding to the enzyme. Setting up the differential equations, employing the quasi-steady-state assumption and expressing the speed of product formation as a function of substrate and inhibitor concentrations yields the rate law

$$v_P(S, I) = \frac{V_{max} S}{K_M \left(1 + \frac{I}{K_I}\right) + S}. \tag{9}$$

If the inhibitor concentration is 0, the term I/K_I vanishes, and the rate law reduces to HMMRL. With increasing inhibitor concentration, the rate of product formation decreases.

In contrast to competitive inhibition, which results in "slowed-down" Michaelis–Menten-type kinetics, allosterically inhibited enzymes can lead to sigmoidal rate functions. The allosteric inhibition mechanism is important, for instance, for the previously mentioned feedback by endproduct in a long chain of reactions. As every reaction step changes the initial substrate somewhat, the endproduct may no longer have much similarity to the substrate with respect to the binding site for which the enzyme is specific. Thus, competitive inhibition is no longer effective. Instead, the endproduct binds to a different binding site on the enzyme, which thereby changes its affinity in the primary site. The most famous description of allosteric regulation is the Hill function, which has the form

$$v_P = \frac{V_{max} S^n}{K_M^n + S^n}. \tag{10}$$

The Hill coefficient n determines the steepness of the rate law and has been associated with the number of subunits in an allosteric enzyme. Typical values of n are 2 and 4. For $n = 1$, the Hill rate law is equivalent with HMMRL. For higher n, the enzyme is said to display cooperativity, caused by interactions between subunits. Specifically, cooperativity occurs when the binding of a ligand to one of several identical binding sites on an enzyme increases (positive cooperativity) or decreases (negative cooperativity) the affinity for ligands at the unbound sites. Probably the best-studied example is hemoglobin, which consists of two alpha- and two beta-chain subunits and exhibits cooperativity that leads to sigmoid oxygen-binding curves.

The extension of HMMRL to rate functions that involve several substrates and modulators is possible with the same principles, but rapidly leads to unwieldy expressions, because the number of possible interactions between substrates, enzymes, and modulators grows combinatorially. Thus, depending on the mechanism, a rate law for a reaction with two substrates and two products may involve a dozen or more parameters in various combinations. The interested reader is referred to Schultz (4), who went through the trouble of dissecting some of these reactions. The increase in complexity of rate laws is a problematic trend, because many of the governing parameters are very difficult to measure, if at all possible. Furthermore, the problems grow very rapidly if not a single reaction is of interest, but an entire network.

4. SYSTEMS OF REACTIONS

If individual rate functions become intractable, it is clear that the mathematical treatment of entire systems of biochemical reactions is no longer feasible in the detailed manner provided by the law of mass action and the Michaelis–Menten mechanism. Of course, modern computers can easily handle hundreds of equations, but the problems are not merely numerical in nature; rather, the complexity of such simulation studies can cloud real insights when dozens of variables and parameters exist.

For instance, it becomes difficult to distinguish what is important and what is not, which factors contributing to the system response are most influential, or which parameters may vary several-fold without causing significant changes in output. These difficulties necessitate alternative representations that combine biological validity with a simpler mathematical implementation that permits analyses of larger systems and efficient interpretations of the computational results. Two complementary approaches have been developed and dominate the field. The first is a reduction of complexity to the stoichiometry of biochemical networks. This strategy leads to linear representations, which are easily scaled up and analyzable for very large systems. The second strategy is the replacement of the rational functions in the tradition of the HMMRL with simpler expressions, such as products of power-law functions. This strategy requires more complex mathematics than the stoichiometric approach but allows the analysis of fully regulated systems; it comes in several variations.

4.1. Stoichiometric Systems

The stoichiometric approach to analyzing and manipulating biochemical systems focuses on the conservation of mass and explicitly establishes constraints that reactions and pathways have to satisfy (e.g., Ref. 5). In addition to the constraints dictated by thermodynamics, this approach enforces conservation laws among all molecular species involved in the system. Thus, the incoming and outgoing carbon and nitrogen atoms have to balance for every chain of reactions, and all reactions have to balance at each metabolite pool. The approach has two important mathematical consequences. First, stoichiometric constraints reduce the degrees of freedom in the system, which is a welcome benefit for larger systems. Maybe more importantly, stoichiometric models are linear and therefore permit an incomparably richer repertoire of analytical methods than any nonlinear approaches.

The stoichiometry of a biochemical system is like the "wiring diagram" of some electronic gadget. It describes which metabolite is converted into which other metabolite(s), which graphically may be represented as a directed graph, which in turn corresponds mathematically to a set of balance equations that are typically written as linear ordinary differential equations. At the core is the stoichiometric matrix S, which consists of one row for each metabolite and one column for each reaction. If a reaction generates a metabolite, the corresponding element is $+1$, and if a reaction uses a metabolite as substrate, the matrix element is -1. If a metabolite and a reaction are unrelated, the corresponding matrix element is zero. If two substrate molecules are used to form one product molecule, the loss in substrate is coded as -2. Generally, the element S_{ij} represents the stoichiometric coefficient of the ith chemical species in the jth reaction. The dynamics of the system variables, which are collected in vector $\mathbf{X}$, is represented by the stoichiometric matrix, multiplied to the vector $\mathbf{v}$ of reaction rates. In typical notation, a stoichiometric model thus has the form

$$\frac{d\mathbf{X}}{dt} = \mathbf{S} \cdot \mathbf{v}. \qquad (11)$$

As an example, consider the dynamics of a small generic system consisting of four metabolites and seven reactions. The pathway and the stoichiometric matrix are shown in Fig. 5. Note that regulatory signals are ignored in this approach.

As a result of their linearity, stoichiometric models permit a variety of straightforward analyses, which show whether a given substrate can possibly be converted into some product of interest, how the flux distribution changes on a mutation that disables some enzymatic step, or how to optimize the flux distribution with respect to a desired output flux or metabolite. These types of analyses are collectively referred to as metabolic flux analysis (6). Combined with the consideration of thermodynamic and other physicochemical constraints, stoichiometric modeling is the foundation of flux balance analysis (7).

4.2. Regulated Systems

The greatest advantage of stoichiometric approaches, namely linearity, is at once their most significant limitation. It is known that linearity precludes system responses such as saturation, stable oscillations, and chaos, which are relevant in metabolic networks. More problematic is that regulation cannot be built into these types of stoichiometric models without destroying their linear nature. It is therefore important to look for approaches that combine some of the advantages of linear analysis with the capability of representing regulatory features.

The transition from stoichiometric to regulated systems is formally not difficult, but has far-reaching consequences. As before, one can set up a model as $d\mathbf{X}/dt = \mathbf{S} \cdot \mathbf{v}$. However, the novelty is that the fluxes in $\mathbf{v}$ are now considered functions of metabolites, enzymes, and modulators. In most cases, these functions are nonlinear, which permits the modeling of dynamic phenomena that cannot be captured with linear systems. As an example, consider the simple pathway in Fig. 6, which consists of a linear chain of reactions, one feedback inhibition, and one feedback activation; describing equations are shown in the

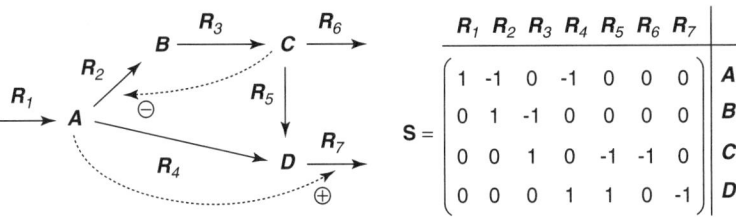

Figure 5. Small generic network consisting of four metabolites and seven reactions. Incoming and outgoing fluxes are represented respectively as $+1$ or -1 entries in the stoichiometric matrix S. Signals, such as the feedback inhibition by C or the feedforward activation by A, are not modeled in this stoichiometric representation.

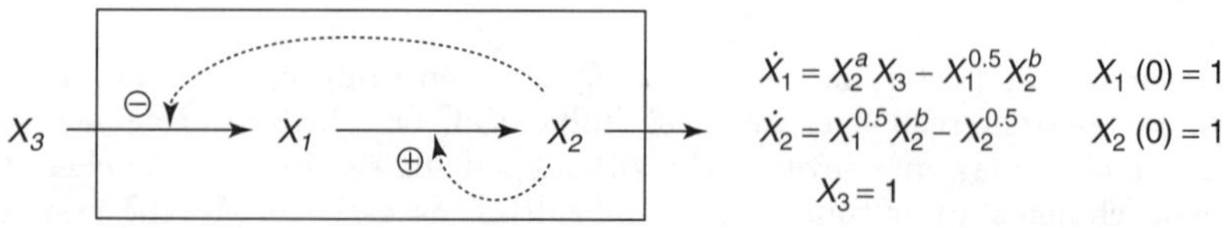

Figure 6. Linear pathway with two regulatory signals and constant input, X_3. The describing equations allow for elimination of the feedback inhibition signal ($a = 0$) or the feedback activation signal ($b = 0$). Responses of the system are shown in Fig. 7.

$$\dot{X}_1 = X_2^a X_3 - X_1^{0.5} X_2^b \quad X_1(0) = 1$$
$$\dot{X}_2 = X_1^{0.5} X_2^b - X_2^{0.5} \quad X_2(0) = 1$$
$$X_3 = 1$$

right panel of the figure. Without regulation (i.e., without inhibition and activation; $a = b = 0$), material flows through the system in quite a predictable fashion. For instance, doubling input ($X_3 = 2$) at time $t = 5$ in a persistent manner leads to a simple monotonic transient response toward a new steady-state (Fig. 7a). The system response becomes more difficult to predict, if regulation is included (e.g., $a = -2$, $b = 0.5$). Now, if the input ($X_3 = 2$) is persistently doubled at $t = 5$, the response consists of a damped oscillation toward the stable steady-state, where X_1 is actually *lower* than before, even though more material enters the system (Fig. 7b). Intriguingly, if the input is reduced to $X_3 = 0.75$, the system enters a stable (limit cycle) oscillation (Fig. 7c), which would have been very difficult to anticipate without a quantitative analysis. In other words, it is not easily possible to infer "global" behavior of a complex system from knowledge of its "local" properties, which is a very important observation that holds for many nonlinear systems.

The permission to use nonlinear functions vastly expands the repertoire of possible behaviors, but it also makes the mathematical analysis more complicated. In particular, an explicit closed-form solution for the system of dynamic equations, no longer exists like the exponential function in the simplest univariate linear case. Not even an explicit solution exists for the system of algebraic steady-state equations, $d\mathbf{X}/dt = \mathbf{S} \cdot \mathbf{v} = 0$, except under fortuitous conditions. The task then becomes to develop a mathematical representation for rate laws that retain some of the advantages of stoichiometric systems, but are capable of capturing the pertinent aspects of regulation. This choice should balance validity of representation, biological support, theoretical support, as well as mathematical and computational tractability.

Several observations and theoretical considerations have led to the insight that power-law functions of the form $v = k \cdot X^g$ might offer a very effective compromise. Mathematically, they are backed by the famous theorem of Brook Taylor (1685–1731), which is applied in a logarithmic coordinate system (2). This theorem guarantees that the power-law function is the correct representation at some point of choice, which is called the operating point, and that it is a very good representation within a reasonable interval around this point. How far one might justifiably stray is a matter of the desired accuracy and depends highly on the phenomenon in question. Another mathematical advantage of the power-law representation is that it is easily extended to any number of variables, with an increase in complexity that is much more favorable than for rate laws such as HMMRL. Specifically, the univariate power-law function $v = k \cdot X^g$ simply becomes

$$v = k \cdot X_1^{g_1} \cdot X_2^{g_2} \ldots X_n^{g_n}. \quad (12)$$

Thus, in the power-law approach to kinetics, each flux in the stoichiometric formulation is written as $v = k \cdot X_1^{g_1} \cdot X_2^{g_2} \ldots X_n^{g_n}$, where k is a positive rate constant and the exponents g_i are kinetic orders that quantify the effect of each variable X_j on flux v. In the special case of mass action, the flux in the reaction $2A + B \rightarrow C$ is given as $v = k \cdot A^2 \cdot B$, where A appears in the second power because two molecules of A are required for the reaction to proceed, and B appears in the first power because one molecule is needed. In the general power-law representation, the kinetic orders may or may not be nonnegative integers. They may assume any real value, with the magnitude characterizing the strength of the effect and the sign indicating whether the effect is inhibiting (negative sign) or

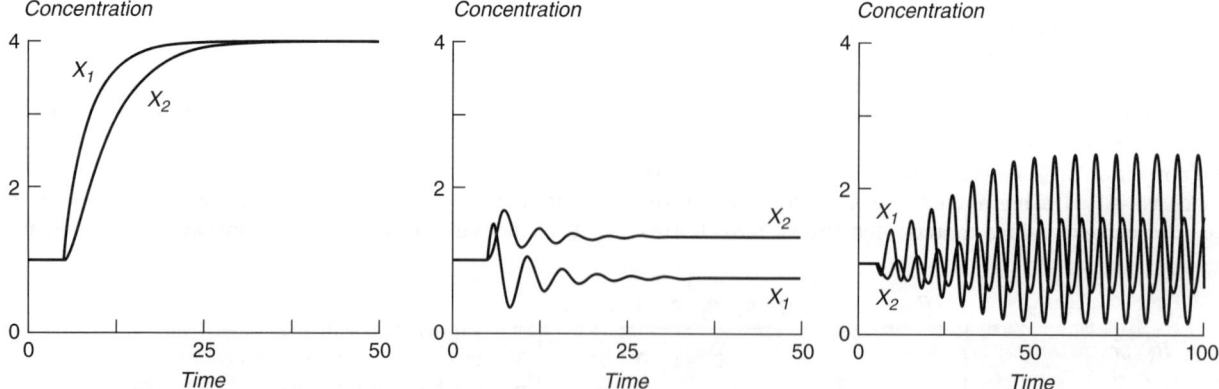

Figure 7. Responses of the pathway in Fig. 6. Left Panel: The system is not regulated ($a = b = 0$); input, X_3, is doubled at time $t = 5$. Center Panel: The system is regulated and input is changed as in left panel. Right Panel: Same system as in center panel, but input is decreased to 0.75. (This figure is available in full color at http://www.mrw.interscience.wiley.com/ebe.)

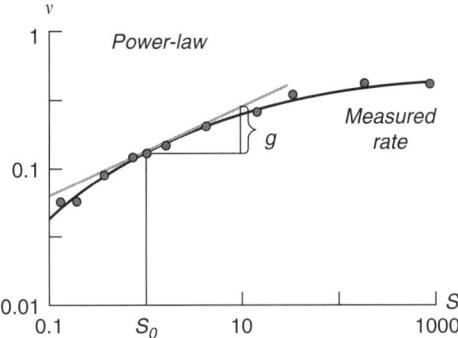

Figure 8. Visualization of the value of the kinetic order g in the rate function $v = k \cdot X_1^{g_1} \cdot X_2^{g_2} \ldots S^g \ldots X_n^{g_n}$. In the underlying biochemical experiment, all variables X_j are held constant. Metabolite S is varied and some flux rate v in the system is measured (dots). The kinetic order g may be computed as the slope of the smoothed data plot at a chosen operating point S_0. Note that the coordinates are logarithmic. As a variant of this method, the kinetic order may also be taken as the slope of the regression line through all data points, or the most relevant set of data points in log-log plot. (This figure is available in full color at http://www.mrw.interscience.wiley.com/ebe.)

activating (positive sign). For instance, $g_i = -0.1$ signifies a weak inhibitory effect. The value of a kinetic order may be visualized in a different fashion. Suppose one executes an experiment where metabolite X is varied and all other metabolites and effectors are kept constant. If one measures a flux v somewhere in the system and plots $\log(v)$ versus $\log(X)$, then the kinetic order g associated with X in flux v is the slope of the plot at a metabolite concentration of choice (Fig. 8). As an alternative, one may use the slope of the regression line through all data points in the same log-log representation. Or, as a variation on the latter, one may limit this regression to the most relevant subset of data points.

The use of power-law functions is the hallmark of biochemical systems theory (BST), which was designed as a framework for analyzing fully regulated biochemical and genetic networks (2,3,8). Combining different concepts from before, a biochemical system may therefore be written as $d\mathbf{X}/dt = \mathbf{S} \cdot \mathbf{v}$, where all fluxes have the format $v = k \cdot X_1^{g_1} \cdot X_2^{g_2} \ldots X_n^{g_n}$. In streamlined notation, the generalized mass action (GMA) representation within BST is

$$\dot{X}_i = \sum_{k=1}^{p} \pm v_{ik}(X_1, X_2, \ldots, X_n)$$
$$= \sum_{k=1}^{p} \pm \gamma_{i,k} \prod_{j=1}^{n} X_j^{f_{i,k,j}}, \qquad (13)$$

where the parameters $\gamma_{i,k}$ are positive rate constants and the exponents f_{ikj} are real-valued kinetic orders. It has been shown with mathematical rigor that this form is structurally very rich and can capture any smooth nonlinearity (9).

As an alternative to this representation, one may split the stoichiometric matrix naturally into a difference of two terms, one associated with all production processes of a metabolic species and one associated with its consumption, $\mathbf{S} = \mathbf{S}^+ - \mathbf{S}^-$. For any metabolic species that is produced through only one reaction and consumed through only one reaction, the GMA form has only one positive and one negative term, and thus the differential equation immediately reduces to the difference of two products of power functions. By contrast, if a metabolite X_i is produced or consumed through several independent reactions, one can collect all incoming fluxes into one aggregate influx V_i^+ and all outgoing fluxes into one aggregate efflux V_i^- and approximate these aggregates with a single product of power functions each. As a consequence, the rate of change of this metabolite can be written as a single difference between two power-laws, one representing the aggregated production, or total influx, and one representing the aggregated consumption, or total efflux, of the metabolite. All metabolites affecting either one of the influxes or one of the effluxes are still present in this formulation, but instead of several product of power-law function only one difference is left. The result is the S-system format

$$\dot{X}_i = V_i^+(X_1, X_2, \ldots, X_n) - V_i^-(X_1, X_2, \ldots, X_n)$$
$$= \alpha_i \prod_{j=1}^{n} X_j^{g_{i,j}} - \beta_i \prod_{j=1}^{n} X_j^{h_{i,j}} \qquad (14)$$

with rate constants and kinetic orders defined in analogy to those of the GMA form.

The S-system form has a somewhat simpler form and additional mathematical advantages. In particular, the characterization and analysis of the steady state of the system is greatly simplified, because setting the equations in Equation 14 equal to zero, moving the beta-terms to the left-hand side, and taking logarithms results in a system of linear algebraic equations, for which many analytical and computational tools exist (3). Thus, the differential equations themselves are highly nonlinear, as in the GMA case, but many features associated with the important steady-state point can be analyzed with methods of linear mathematics. In other words, the S-system form combines the richness of nonlinearity with some advantages of linear mathematics. It might seem that the S-system form is less accurate, because it does not track every flux separately. However, careful analysis has shown that this is not the case. Furthermore, as the GMA form, the S-system form is rich enough to capture any smooth nonlinearity (9).

A slightly different approach is to set up the biochemical model by formulating each reaction rate as a linear equation in the logarithm of the metabolites. More specifically, all components of the system, namely intracellular (x) and extracellular (c) metabolite concentrations, enzyme activities (e), and flux rates (v) are formulated in relation to a reference state. Defining variables with tilde as raw variables divided by the corresponding reference quantities, each rate in this so-called lin-log model is represented as

$$v = \tilde{e} \cdot (\mathbf{I} + \mathbf{E}_x \cdot \ln \tilde{x} + \mathbf{E}_c \cdot \ln \tilde{c}), \qquad (15)$$

where $\mathbf{I}$ is the unit vector of appropriate length and $\mathbf{E}_x$ and $\mathbf{E}_c$ are reference matrices of elasticities, which characterize the effect of a variable on a flux (10). Similar to the

GMA representation, the lin-log model combines linear with nonlinear features, but in a different manner. The lin-log model is much newer than GMA and S-systems, and careful comparisons between these forms with respect to accuracy of representation and mathematical tractability are not yet available. It is clear from its mathematical structure that the lin-log model is more accurate than power-law models for very high substrate concentrations, but that it fails for low concentrations, where the rate becomes negative.

As another combination of the linear and nonlinear worlds, one may define a system by representing the derivative of the logarithm of a metabolite as the sum of all effectors. Expressed differently, this strategy represents the change in each variable as the sum of products between the variable itself and all other variables, one at a time. This Lotka–Volterra form (11)

$$\dot{X}_i = X_i \cdot \sum_{j=1}^{n} (b_0 + b_{ij} X_j) \quad (16)$$

has been used very widely in ecological modeling, but not much for kinetic analyses.

4.3. Systems Analysis

Now that we have developed streamlined and scalable representations of biochemical systems, one must ask what to do with them. The multitude of possibilities may be divided into analyses close to the steady state as opposed to dynamic analyses.

The steady state is often the normal operating state of a natural system. In this state, material is entering and leaving the system, and internally numerous conversions occur, but all fluxes are in balance and all metabolites are found at a constant concentration. Typical questions associated with this state are: Can we compute this state? Is there more than one state of this type? If the system is perturbed, will it return to this state? For S-systems, lin-log, and Lotka–Volterra systems, the steady state is easy to compute, whereas a numerical search algorithm or integrator is needed to compute the state(s) in a GMA system or a system based on Michaelis–Menten rate laws.

Whether the system recovers from a (small) perturbation is a matter of stability analysis. A system does return from a small perturbation, although not necessarily from a large one, if the steady state is stable, whereas a system moves away from an unstable state. A second class of question concerns mutations or other persistent changes: If a parameter is changed (mutated), does the system assume new features? For instance, does it start to oscillate? These types of questions require a bifurcation analysis, which is typically quite complex.

A third and very important complex of questions addresses sensitivities, gains, and control. Examples are the following: How is the steady state affected if a given parameter is increased by a small percentage? What is the effect of increasing or decreasing one of the inputs? Which parameter has the most influence on the flux through a system. All these questions can be addressed with any of the above methodological frameworks.

For the latter type of question, a specific set of methods, called metabolic control analysis (MCA) (12), has been developed. As it is only concerned with pathway features close to a steady state, it does not evoke differential equations at all. Instead, this approach is purely algebraic and therefore simpler to apply if the situation of interest is suitable for this type of analysis. MCA has led to a number of rules, called summation theorems and connectivity theorems, that allow the researcher to determine which parameter is most significantly controlling a process, where the major bottlenecks of a pathway are located, and whether all controlling influences have been accounted for in an analysis. As only elementary computations are involved in this approach, it is relatively easy to execute specific metabolic control analyses. The limitations are that a few exceptions exist where this method is not valid or at least not straightforward and that the method does not address dynamic features. Recent extensions have addressed some of these limitations.

A specific task associated with the steady state is the optimization of metabolites or fluxes in a biotechnological setting. The generic question here is: How is it possible to improve yield within a steady-state culture of microorganisms by strategically redistributing metabolic fluxes while ensuring that viability of the culture is not jeopardized? On a large scale, such optimization tasks have been executed with stoichiometric models (6). Mathematically, this type of optimization is also easy with S-systems, but the limitation in this case is that much more kinetic information is needed than in the stoichiometric case (13). Optimization in the case of GMA systems is the subject of current research.

In addition to steady-state analyses, the dynamic models above allow simulation studies that elucidate the temporal behavior of biochemical systems. Typical in this category are what-if scenarios. How does the pathway respond over time if substrate is reduced to 20%? What happens if a gene is upregulated, which leads to increased activity in some enzyme? How does the organism respond to a mutation, in which one branch of the pathway is deleted? Where are the best drug targets located within a diseased metabolic system? What is the likelihood of side effects if a drug affects a particular reaction? It is easy to imagine that the scope of such simulations is endless (see Ref. 8).

4.4. Research Frontiers

Much effort in the immediate future will be devoted to specific pathways, their functioning, and their manipulation and optimization. Insights gained from these studies will be applied in metabolic engineering, drug targeting, disease analysis and treatment, food production, and improved environmental stewardship.

An important part of the analysis of specific pathways will be the identification of models from experimental data. In the past, this identification has almost always started at the bottom and worked its way up. Specifically, many pieces of detailed information on enzymes and metabolites in the pathway of interest were used to construct rate functions for all steps, and the integration of this local information was used to generate global responses, which were compared with observations. This bottom-up approach will continue to be at the forefront of modeling for some time. Complementing this approach will be a top-down approach, where observed *in vivo* time course data on gene expression, protein prevalence, and metabolite profiles are the starting point. These dynamic data will be mined in a sense that model parameters are obtained that let the model reproduce the observed dynamics. As dynamic data contain rich information on intact cells or organisms, less mixing of data will occur from different organisms and conditions, from experiments *in vitro* with data obtained *in vivo*. This analysis of time data is presently a daunting challenge, even for relatively small systems.

Beyond specific applications and techniques, an urgent need exists to discover design and operating principles in biological systems. These principles are expected to exist, because nature is usually efficient and frugal, using well-working designs and procedures as often as effectively possible. If no such general designs existed, every system would be genuinely different, and it would hardly be possible to make extrapolations between organisms or different physiological situations, which is considered unlikely, and one of the research frontiers in the area is therefore the discovery and characterization of such principles. The strategy is to determine with mathematical and computational means what the role of specific signals or mechanisms is and why nature employed the observed design and not a different, hypothetically possible design, which is accomplished with comparative systems analyses, where two systems that only differ in one aspect (for instance, a feedback signal) are analyzed side by side. Any differences in their dynamic or steady-state behavior can therefore be attributed to the differing signal. The discovery of design and operating principles will yield fundamental insights into the functioning of cells and organisms and form the foundation for rational manipulations of biological systems.

BIBLIOGRAPHY

1. D. Jou and J. E. Llebot, *Introduction to the Thermodynamics of Biological Processes*. Englewood Cliffs, NJ: Prentice-Hall, 1990.
2. M. A. Savageau, *Biochemical Systems Analysis. A Study of Function and Design in Molecular Biology*. Reading, MA: Addison-Wesley, 1976.
3. M. A. Savageau, Biochemical systems analysis. *J. Theor. Biol.* 1969; **25**:365–379.
4. A. R. Schulz, *Enzyme Kinetics. From Diastase to Multi-Enzyme Systems*. Cambridge, UK: Cambridge University Press, 1994.
5. R. Heinrich and S. Schuster, *The Regulation of Cellular Systems*. New York: Chapman and Hall, 1996.
6. G. N. Stephanopoulos, A. A. Aristidou, and J. Nielsen, *Metabolic Engineering. Principles and Methodologies*. San Diego, CA: Academic Press, 1998.
7. J. S. Edwards and B. Ø. Palsson, How will bioinformatics influence metabolic engineering? *Biotechn. Bioeng.* 1998; **58**(2-3):162–169.
8. E. O. Voit, *Computational Analysis of Biochemical Systems. A Practical Guide for Biochemists and Molecular Biologists*. Cambridge, UK: Cambridge University Press, 2000.
9. M. A. Savageau and E. O. Voit, Recasting nonlinear differential equations as S-systems: a canonical nonlinear form. *Mathem. Biosci.* 1987; **87**:83–115.
10. J. J. Heijnen, Approximative kinetic formats used in metabolic network modeling. *Biotechn. Bioeng.* 2005; **91**(5):534–545.
11. M. Peschel and W. Mende, *The Predator-Prey Model: Do we Live in a Volterra World?* Berlin: Akademie-Verlag, 1986.
12. D. A. Fell, *Understanding the Control of Metabolism*. London: Portland Press, 1997.
13. N. V. Torres and E. O. Voit, *Pathway Analysis and Optimization in Metabolic Engineering*. Cambridge, UK: Cambridge University Press, 2002.

BIOCOMPATIBILITY OF ENGINEERING MATERIALS

ROBERT BAIER
State University of New York
Buffalo, New York

1. TECHNICAL BACKGROUND ABOUT THE BIOLOGICAL ENVIRONMENTS HOSTING ENGINEERING MATERIALS

1.1. Adhesion in the Environment of Blood: The Need is "Prevention"

In the development of true artificial hearts, major continuing problems are the unwanted bioadhesion of thrombotic and coagulum masses, as well as scale-forming minerals from the bloodstream itself (1). Materials selection is severely limited to those elastomers that can withstand the 100 million or more flexing cycles necessary during the forecast period of service of such devices (2). Challenges to surface science are great in that the desirable surface qualities must be imparted to each elastomer material without degrading its mechanical properties in any way.

During even short-term implantation of artificial heart components, and especially of flexing bladders, it became immediately obvious that deposition of unwanted blood-borne materials is unavoidable (3). Adjustments in flow dynamics, velocity profiles, and other mechanical design features can minimize, but not eliminate, problems with these deposits. Implantation times of days to weeks, and now to months and years, became routine in many clinical centers as the devices matured in engineering design and continue, still, to improve. However, a serious further consequence is surface deposition of both organic matter and inorganic matter, predominantly on the flexing surfaces of the elastomeric heart bladders. Surface deposition remains a major problem and might be considered the limiting parameter for further progress in development of this most important artificial internal organ.

Application of the increasingly powerful surface diagnostic methods now available demonstrates that blood (either in its unmodified state or when deliberately anticoagulated with various natural or synthetic reagents) first deposits on any and every foreign material surface a proteinaceous "conditioning" film dominated by the glycoproteinaceous macromolecule, fibrinogen (3–5). Fibrinogen is not the most abundant protein component dissolved in the contacting fluid phase nor does it have the fastest diffusion to the walls. It is, in fact, one of the most highly hydrated macromolecules known in the protein category, and it can undergo significantly different degrees of configurational change as it dehydrates and accommodates itself to the varying potential binding sites on an enormous variety of solid and semisolid (hydrogel) materials. Transmission electron microscopic views of such test materials exposed to fresh-flowing blood for only a minute show that the fibrinogen-dominated layer accumulates in varying densities on different materials to a general thickness of about 200 Angstroms (6).

The next event in blood contact with nonphysiologic surfaces, following the acquisition of the mandatory protein-dominated conditioning film, is preferential—in fact, nearly exclusive—primary attachment of the typically three-micrometer-diameter disk-shaped blood cells called platelets. The platelets remain disk-like or partially rounded, occasionally with a few protruding pseudopods, when the original surface free energy of the substratum material is between 20 and 30 mN/m. They neither spread to flat forms that are attractive for further attachment of other platelets nor exude granules and soluble products that trigger unwanted, at least for implanted biomaterials, coagulation events in their near vicinity. For all materials with higher or lower critical surface tensions (outside the range of minimal bioadhesive potential either on the lower-energy side, as typified by Teflon™ and other fluorocarbon materials, or on the higher-energy side, as typified by polyacrylates and nylon and other common engineering polymers), the initially adherent platelets respond to the more tightly bound and "denatured" fibrinogen-rich conditioning film more aggressively. The platelets flatten, become "sticky" to their arriving siblings, disgorge biochemicals to the surrounding fluid, and, through these activities, trigger the building of pyramid-shaped thrombotic deposits. They also stimulate chemotaxis, that is, directed movement toward the surface, of one of the five major types of white blood cells, the segmented polymorphonucleocytes.

At a slight distance from the engineering materials' surfaces, the conversion of fibrinogen to fibrin is apparent in the vicinity of activated, flattening platelets. The emerging fibrin net entangles and entraps red blood cells to form the typical red clots. On the majority of currently available materials, not a surface chemically controlled to be in the nonbioadhesive range, the platelet layer and associated debris is so tenaciously bound that the arriving white cells are not able to successfully carry out their functions of digestion, engulfment, or displacement of the surface-bound mass. Instead, they become part of the "incompatibility" problem rather than the resolution of the problem. The white thrombi that trigger coronary occlusion are composed mainly of platelets and white cells that responded to the adverse surface conditions of atherosclerotically modified blood vessel walls, ultimately occluding those vessels by this platelet-buildup process.

Deliberate modification of the surface properties of biomaterials, exposed to flowing blood, secures a more benign outcome. Although the mandatory initial protein-conditioning film still deposits, it remains more loosely attached and subject to displacement by other arriving species. The platelets also do continue to arrive at and attach to this fibrinogen-dominated interface-conversion layer, but three-dimensional aggregation of these early cell-like colonizers (which retain their discoid or well-rounded forms) is rare. Although no obvious platelet distress is noted in electron micrographs of specimens examined at this stage in their blood exposure histories, such as degranulation or disgorgement of their internal biochemicals, there certainly is some influence of the binding process on the platelets. They do broadcast whatever signal is necessary to cause rapid and abundant recruitment of the segmented polymorphonuclear leukocytes to the debris-strewn surface zone. From this point, significant contrast exists to the case where bioadhesive surface properties of the original materials induce substantial protein denaturation and platelet spreading, however. The white cells arriving at low-surface-energy materials (within the "biocompatible" zone of 20 to 30 mN/m critical surface tension) effectively weaken the tentative bonds between the protein "carpet" and substratum through their exogenous production of digestive enzymes. The observed end result, usually within a few hours of the initiation of this process, is spalling or shedding of the accumulated biomass from the engineering material's surface into the flowing stream. Left behind is a protein-rich layer, "processed" by the events just described, that enters a state of long-term dynamic equilibrium with the flowing solution phase. It is rare, with the many now successfully implanted biomedical devices, to observe sustained colonization of low-surface-energy materials by cellular layers.

Considerable similarities exist among bioadhesive phenomena in blood and in other biological phases. Knowledge of successful coating materials that perform satisfactorily in other environments allows a direct extrapolation from a secure database to provide medical-grade silicone surface layers for the blood contacting phases of artificial hearts. The expected outcome was attained, with minimization of thrombogenic deposits on these materials over the early and late courses of their trials in animals, and now excellent clinical results in human patients. Complications of unwanted mineral scale formation, dominated by calcium phosphate deposits, first noted in the growing male calf model used for most artificial heart tests, were controllable by modest use of a common anticoagulant that interferes with the vitamin K/calcium metabolism.

1.2. Adhesion in the Environment of Tissue: The Need is "Promotion"

The human implantation of "integrated" biomedical or dental prosthetic devices provides challenges to engineering materials science and technology among the most severe encountered in the bonding art. Custom-fitted, strong, usually metallic devices such as dental and orthopedic implants provide good examples of the challenges to be met and the extreme complexity of the problems posed.

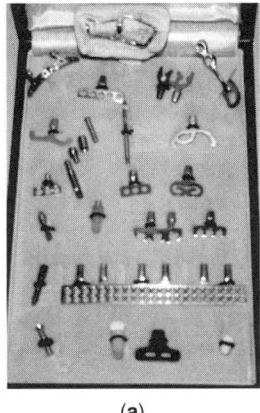
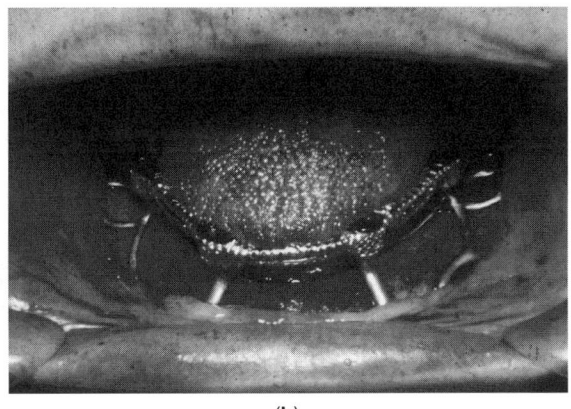

Figure 1. (a) The broad variety of dental implants clinically used over the past 30 years, most with poor outcomes because of poor bone and tissue biocompatibility. (b) A full subperiosteal dental implant, in place on a patient's exposed mandibular bone. (This figure is available in full color at http://www.mrw.interscience.wiley.com/ebe.)

These appliances can present contrasting adhesive requirements over very short distances along the same structure. As illustrated in Fig. 1, subperiosteal implants are designed to rest over the remaining mandibular bone in a patient whose teeth have been lost in that area. The remaining tissue "flap" must cover and adhere tightly to the supporting saddle-like base. Extending from the base are posts that protrude, hopefully with a bacterial-tight, infection-free seal through the tissue, into the nonsterile, intraoral region. While remaining free of adherent deposits, the posts are expected to provide mounting sites for "artificial teeth" or other load-bearing, functional, prosthetic dental restorations.

The portion of the implant buried in the tissue must be very closely approached by the host tissue in order to provide the immobile, firm support required. The bonding integrity of the tissue phase immediately adjacent to the permucosal post must be nearly perfect to prevent bacterial seepage, infection, and resulting inflammatory responses. The structural parts residing in the saliva-bathed oral cavity should resist colonization by biological materials, especially plaque-forming bacteria to improve the prospects for continued service while contributing to good oral hygiene. The distance over which the adhesive properties must change so drastically might be as little as a few micrometers along the surface of such implants. These conditions must be achieved and maintained in environments that are bloody, nonsterile, and contaminant-rich.

Although osseointegrated dental implants of the permucosal type have been used clinically for over 2 decades, the results obtained with subperiosteal implants have been ambiguous at best. Many failures have been noted as a result of loss of tissue bonding for the implanted portions, serious plaque and debris accumulations on the exposed segments, and very poor healing around the protruding posts at the epithelial junction. It is not uncommon for the implants to extrude completely from their placement sites within only a few weeks. Such trials in animals, and premature applications of otherwise promising devices in humans, have often proceeded too rapidly.

Knowledge of the zones of surface energy that can favor or minimize adhesion in biological environments has made it increasingly possible to design and develop improved prosthetic materials for a number of important applications. For example, basic knowledge developed by standard surface analytical methods revealed the intrinsic surface properties of natural dental enamel (7). Data for the same surfaces, after varying treatments (for example, with citric acid or phosphoric acid etching), opened the way to obtaining long-term retention of a whole generation of pit and fissure sealant materials and dental cements (8), which have entered general dental and orthodontic practice (9,10).

Glow discharge (plasma) treatments of biomedical devices of even such physically large dimensions as artificial hips can be quite beneficial in improving their receptivity to "bone cements" and other adhesive phases used to bind them securely within the prepared femoral shafts. Figure 2 provides a photograph of an artificial hip completely enveloped in the cleansing, sterilizing, plasma sheath within a glass chamber. Such glow discharge treatment devices were first designed and fabricated for implant treatment purposes in our laboratory (11) and are now widely commercially available as low-temperature sterilization systems.

During evaluation of the tissue response to variously surface-treated biomedical implants in rabbits, cross-sectional views obtained by standard histological methods (of the subdermal fascial zone, after implant removal at 10 or 20 days) showed dramatically increased cellularity when glow-discharge-treated or relatively high-surface-energy test specimens were removed. There was obvious "ripping" of the tissue caused during the mechanical removal of the implants. Tenacious tissue bonding to the higher surface energy, but not low-surface-energy, implants was the norm.

The "healing" response involves rapid recruitment of fibroblast/fibrocyte cells to the vicinity of the implant surface and their organization into adhesive capsular tissue adjacent to high-surface-energy engineering materials. The cellular response observed within the fascial tissue reorganized against low-surface-energy engineering material implants comprises only very thin, but sometimes dense, scar-like zones. The deeper tissue includes only a few elongated nuclei from responding fibroblasts, enmeshed within a loose connective tissue matrix. Scanning electron micrographs of the interfacial layers of tissue illustrate the weakness of the cell-poor subadjacent

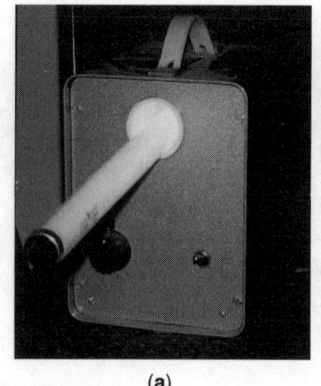 

Figure 2. (a) A portable radio-frequency-glow-discharge-treatment (RFGDT) device, with a disconnected fluorescent bulb being remotely activated/illuminated by the radiofrequency energy induced in the device's cylindrical chamber. (b) A capacitance-coupled RFGDT device cleaning and activating the surface of an orthopedic hip femoral stem implant prior to its cementation into bone. (This figure is available in full color at http://www.mrw.interscience.wiley.com/ebe.)

tissue and its ease of separation and fragmentation even during the specimen handling procedure.

Shifting the point of observation from the tissue side of the interface to the low-surface-energy engineering material implants, themselves, scanning electron micrographs show that modest mechanical forces disrupt the tissue/implant bond quite readily. Bond "failure" is predominantly an "adhesive type," through or at the glycoproteinaceous "conditioning film" layer, rather than by cohesive failure in either of the bulk phases. Scanning electron microscopic views in regions of textured, low-energy substrata make it further apparent that "conditioning film"- level clean separations of the host tissue from low-surface-energy foreign bodies can occur even when surface textural variations are present. Increased or "special" types of surface texture are not the answer to implant-to-tissue bonding failures *in vivo*.

"Adhesive failure" of the type discussed here does not mean that the base, initial implant is actually re-exposed by mechanical failure of the tissue bond to its face. Rather, the failure is in the integrity of the glycoproteinaceous "glue line," preventing cellular layers organizing beyond it (even if they were present) from obtaining permanent liaison with the original substratum.

Further examination of the tissue faces adjacent to low-surface-energy implants, in the views provided by the scanning electron microscope, shows that rounded fibroblast cells dominate the surface zone of the host tissue. The spheroidal nature of the cellular population corresponds with the low cell density and poor intercellular bonding obtained in this zone. Indeed, the tissue layers can peel like the layers of an onion, one or two from the other, in thin sheets, which further illustrates the poor mechanical integrity of the capsule with which low-surface-energy implants become invested during their implantation periods.

In important and clear contrast to the circumstances just described are those noted when the identical bulk materials have their surface energies elevated by glow-discharge-treatment. Figure 3 shows a histologic cross-sectional view of this more desirable tissue response. At the implant boundary, almost no scar-like amorphous separating layer is present. Immediately beneath this interfacial zone and for a large distance from it, the tissue response is of the preferred flat fibroblastic/fibrocytic form. This tight interfacial capsular response to high-surface-energy engineering material implants shows that aggressive cellular activity is underway, often allowing healing within periods less than 2 weeks. Evidence is usually abundant for cells in the doubling process of mitosis. Some inclusions in this zone are adipose cells with lipid storage deposits.

Higher magnification views of the tissue interface region, mechanically separated from a high-surface-energy engineering material implant, reveal no scar-like, fibrous separating layers. Rather, the cells fragment in a manner consistent with cohesive failure of the implant/tissue bond by disruption within the cells themselves. Confirming this interpretation, scanning electron micrographs illustrate the surface condition of the high-surface-energy materials at 2-days "healing" time to be dominated by membrane fragments and internal cellular debris left behind by cohesive failure in the tissue mass itself. Thus, simple modification of the surface energy state of equally smooth implants leads to circumstances as divergent as no practical adhesion at all versus the strongest practical adhesion of which the system is capable (that is, the cohesive strength of the weaker of the two bound phases, in this case, the biological tissue itself). Scanning electron micrographs obtained at low and high magnifications for the tissue faces separated from the high-surface-energy engineering material implants contrast with the views of rounded, poorly interconnected cells surrounding the low-energy test plates. It is immediately apparent that the higher energy surfaces induce not only abundant cellularity, but also morphologic features of cell flattening, spreading, and tight integration within a fibrous matrix. The same cells from the same hosts over the same time periods differentiate to strikingly different populations depending on the surface energies of the substrata against which they reside. The question remains for future work to determine whether a more graded response can be selected, as might be desired for new biomedical device requirements. Can one obtain partial encapsulation and modest adhesive strength or must it be an all-or-none phenomenon?

2. TECHNICAL BACKGROUND ABOUT THE VARIETY OF ENGINEERING ENVIRONMENTS HOSTING BIOLOGICAL MATERIALS

2.1. Adhesion in Food Handling Equipment: The Need is "Prevention"

In traditional engineering studies, such as those dealing with the rate and consequences of biological fouling of

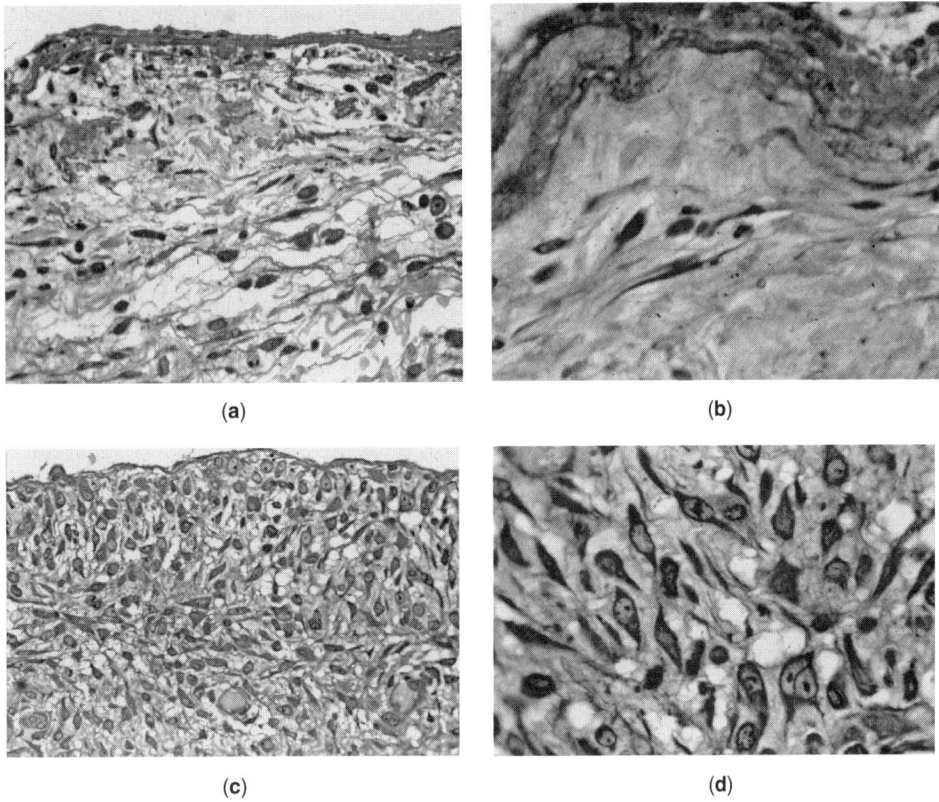

Figure 3. (a) (b) Hemotoxylin and Eosin (H&E)-stained cross sections, at 40× and 100× original magnifications, illustrating the usual type of poorly cellular and functionally nonadhesive tissue capsule ("foreign body reaction") formed around most implanted materials. (c) (d) H&E-stained sections, at original magnifications of 40× and 100×, showing the surface and deeper zone cellularity characterizing strongly adhesive tissue layers attached to materials first exposed to RFGDT (This figure is available in full color at http://www.mrw.interscience.wiley.com/ebe.).

heat exchange materials (as in power plant condenser tubes), ignorance of the early microfouling or "conditioning" events is usually dismissed with the label "induction period," which is that time after first installation of engineering materials or after their cleaning and re-exposure to fouling-prone cooling waters from rivers, lakes, or seas (usually amounting to a few weeks) during which the traditional engineering-control parameters of pressure drop and temperature differential do not change significantly. In fact, these parameters often do change in the manner indicating improved, although certainly temporary in most cases, heat exchange performance. In these engineering systems, gains in basic understanding of the performance and control of important energy conversion devices is highly dependent on improved knowledge of the actual, initial microfouling events. The first adhesive layers potentiate the systems for grosser biofouling, with its consequent deterioration of friction factors, head losses, and heat exchange coefficients.

A typical example is surface fouling by dairy products. There continue to be pressing needs for energy conservation and increased food availability in many parts of the world. Nowhere are such problems so closely coupled as in the pasteurization of dairy products, especially milk, where deposition and bacterial growth on the heat exchanger plates of common pasteurizing devices severely limit operational time. The surface fouling requires excessive consumption of energy and plate-cleaning reagents. Copious amounts of fresh rinse waters must be passed through the devices before the pasteurizing process can be reinitiated for only a few more hours. The cleaning cycle must begin again, promptly, and requirements for its efficacy are ill-defined.

Scanning electron microscope views of the adsorbed films formed on heat exchange plates after only several minutes of contact with milk reveal the complex nature of the deposits. Questions about the actual composition of such critical early deposits, so thin as to be immeasurable by most other analytical tools, are immediately answerable by internal reflection infrared spectroscopy (12). Data can be obtained for the intact film, unmodified and never contacted with or extracted with any solvents or mechanical scraping devices. Milk deposits on test plates, taken from simple laboratory models of pasteurizing units, are shown by the technique to be composed of each of the major categories of material present in the milk itself: protein, lipids (fats and oils), and carbohydrates (13). These components might be bound into chemical complexes or present as single components in a simple mixture. Concern remains about which of these components, if any, constitute the primary interface conversion layer—permanently bound—and which might be transient components subject to displacement as the fouling film thickens and matures.

After simple extraction of the distilled-water-soluble components of this film, scanning electron microscopy shows that some of the original film deposits can be easily removed. Ellipsometric measurements show that the film's thickness does diminish. Contact potential measurements reveal changes in surface electrical properties toward the starting state of the "clean" materials. These observations support the direct spectroscopic evidence provided elsewhere for the nearly quantitative removal by distilled water rinsing alone of all the carbohydrate portion of the original film deposit (13). The ready degree of water solubility suggests that the composition of the extracted matter is mainly of the lactose sugar abundantly present in milk. The film, further extracted by brief exposure to the lipid solvent, acetone, and re-examined by the same technique, is seen to lose most of its granular and globular deposits to exhibit a much more homogeneous appearance. Sequential application of ellipsometric and contact potential methods proves that the film components lost are, indeed, those illustrated directly by internal reflection spectroscopy to have been the lipid, or fatty acid and fatty-ester-rich "cream components" of the complex milk fluid. The remaining deposit is irreversibly bound and resistant to all forms of removal short of mechanical abrasion or chemical hydrolysis with strongly basic or acidic cleaning agents.

Analytically, this film is a thin proteinaceous remnant that other experiments show is, in fact, the first strongly bound material of all that randomly arrive at the test plate surface. Close inspection of the characteristic infrared spectra for similar films usually reveals their retention of some hydrocarbon moieties and some sugar moieties, suggesting that the first deposited protein does indeed contain side chains having both hydrocarbon character and carbohydrate character. Such features are typically observed with proteins of the class called glycoproteins. Separate experiments provide the strong suggestion that the main protein involved in the initial fouling events of equipment processing dairy products is the protein called beta lactoglobulin, neither the most abundant nor the most rapidly diffusing species in the complex pool of potentially adsorbable candidates available. Note the similarity of this result to the events during early blood contact with implant materials!

2.2. Adhesion in Oceanic Environments: "Prevention" Needs Dominate

In marine commerce, continuing problems exist regarding the biofouling of ship bottoms, propellers, and cooling water piping. There, microfouling films are generally precursors to macrofouling aggregations of barnacles, tube worms, algae, bryozoa, and hydroides that can diminish ship efficiency by 40% in as little as 6 months in tropical waters. Bacterial slimes, alone, accumulating on even highly toxic marine paints, also can exhibit these and other consequences of "biodeterioration."

When the goal has been to characterize, and eventually minimize, biofouling deposits in power plant heat exchangers/condensers, or on ship bottoms, special flow cells built according to a very simple design have served this purpose (14). The flow rates are controlled as required while allowing access to and use of flat diagnostic test plates (15,16). Using such flow cells in large numbers, with the test plates exposed for typically 1, 3, 6, 9, and 12 days in tropical or subtropical oceanic waters, it has been possible to document and decipher the initial sequence of fouling events taking place in natural waters (17). These events are quite comparable with the processes demonstrated to be occurring in the oral cavity, in tissue culture, in blood, and other biofouling circumstances.

The early colonizing film of micro-organisms is common on substrata of every quality so far considered and in every water condition so far tried. The materials evaluated to date include a variety of inert and toxic metals and metal alloys and a large range of organic coatings useful for adjusting the critical surface tensions of metallic support plates. Water qualities tested have ranged from natural ocean water available at given test sites to that same water modified by intermittent chlorination or by electrolytic enrichment in copper and aluminum ions. Both such treatments have been correlated with partial suppression of the biofouling film growth in past trials.

Beneath the first layer of pioneer colonizing bacteria is, without exception, a spontaneously adsorbed protein-dominated "conditioning" film. This interface conversion layer is first recruited from the macromolecular constituents of the flowing water itself. Many such components are available within the ubiquitous 5 to 10 parts per million background concentration of "humic" or "gelbstoffe" substances in all natural coastal waters where fouling is most intense.

After the pioneer (mostly short rod-like) bacteria attach and secrete their slimy exudates into their zone of influence, a second population of attaching organisms enters the fouling layer on top of the pre-adsorbed conditioning film. Continuing deposition of macromolecules is often noted. This second wave of attaching species is dominated by prosthecate micro-organisms that can use the very scarce nutrients in natural ocean waters by enhancing their surface-to-volume ratios. These prosthecate micro-organisms are oligotrophic, living in very nutrient-poor conditions. Their preferential attachment to and growth on the boundary layers represents an escape from oligotrophy; not completely consistent with their presumed resistance to growth in media of higher organic content. When the substratum colonized by these organisms is free of leachable toxic atoms, and when the water is similarly free of poisons that can interfere with bacterial metabolism and growth, these pioneer attached species secrete mucilagenous polymers around themselves. Filamentous growths then proliferate over the entire surface, making it an almost perfect filter medium for collection of other oceanic organisms and debris, as well as an excellent culture medium for the growth and multiplication of the attached and entrapped species. As noted with dental plaques of long standing, subsequent mineralization then occurs.

Collections of fouling debris on inert surfaces over which ocean water passed at a shear rate of 1000 inverse seconds for a period of only a few days illustrate the

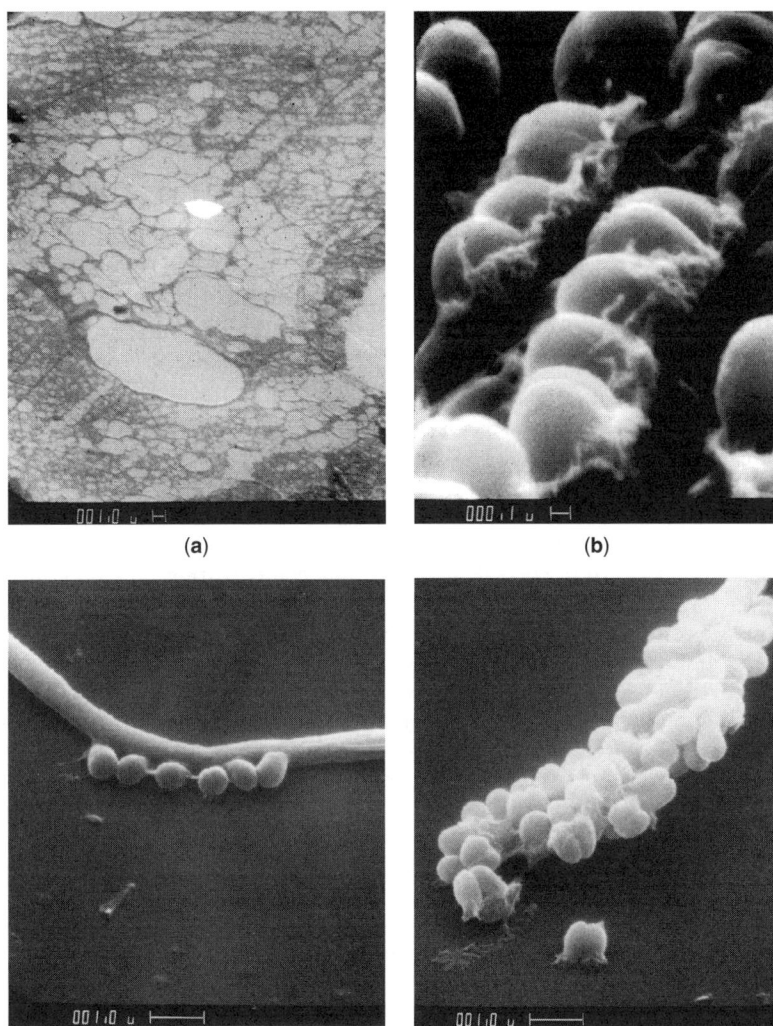

Figure 4. (a) Scanning electron micrograph illustrating spontaneous dehiscence of glycoproteinaceous "conditioning film" from methyl-siliconized substratum, remaining "plaque"-free while placed in thriving culture of bioadhesive coccoidal bacteria (CC5A). (b) Scanning electron micrograph illustrating the preformed bioadhesive tufts of glycoprotein at the tips of CC5A coccoidal bacteria harvested from human dental plaque. (c) (d) Scanning electron micrographs of CC5A coccoidal bacteria colonizing the filaments of Bacterionema matruchotti, in a cocultured system, giving rise to a "corn cob" structure also sometimes seen in the complex bacteria biofouling films on the sides of oceanic vessels. (This figure is available in full color at http://www.mrw.interscience.wiley.com/ebe.)

diversity of organisms that can be present in these films. Of special interest is the "corn cob" structure made up of small coccoidal organisms colonizing a filamentous form, prompting the further characterization of these films as the "dental plaque of the ocean" (see Fig. 4). Indeed, the events of biological adhesion are the same wherever one finds them, even though they occur at different rates in accord with the well-understood concentration dependence of film deposition and particle attachment processes.

It is both interesting and important to note that the strength of biological adhesion in natural circumstances can also be controlled by adjustments of the critical surface tensions of substrata. This control technique has allowed impressive successes in mitigating biofouling of ship bottoms (18,19) and has been considered for application to power plant heat exchanger tubes to minimize use of toxic chlorine (20).

Of greater concern to the owners and operators of commercial ships is the macrofouling community composed of barnacles, tube worms, encrusting bryozoa, and bushy hydroid growths that so seriously impede the efficient movement of originally smooth ship hulls through the waters between merchant ports. Even when killed by various toxic agents or parasites, their remaining, strongly cemented, empty shells still cause enormous increases in drag coefficients and excessive consumption of increasingly precious petroleum fuels. Barnacle attachment begins with a barely visible, soft-bodied larval form called the "cyprid." Cyprids randomly arrive at potential settlement and growth sites and responsively activate contact-pulloff probing of the contacting surfaces until they attach to a site from which they have insufficient contractile strength to release their own cement bond. The critical dependence of the success of the cyprid colonization process on the previous presence of bound protein films was highlighted in the work of Crisp and Meadows (21). Investigations of the potential further influence of the critical surface tensions of the subjacent solid supports are still required.

The tentative correlation developed from early studies was given as a qualitative plot of the degree of biological fouling versus the critical surface tension of the material (22). The same sharp minimum in strength of biological adhesion was noted as with similar data from studies of both prokaryotes and eukaryotes (23).

When mechanical challenge is added to the test environment, applied most directly by a simple wipe of the retrieved test plates, it is obvious that the strength of attachment of the biological deposits is inversely correlated with critical surface tension down to a minimum value at about 22 mN/m. In fact, the most successful formulation tested in seawater has been identical to that used in fabricating artificial hearts.

2.3. Adhesion and Biotechnology: "Promotion" Needs Dominate

The many burgeoning fields of biotechnology require significant advances in techniques for retaining, selectively, specific cells in specific geometries on the surfaces of fixed-film processing units and bioreactors. Here, "technology transfer" from the medical and dental areas may be the most rapid route to success.

For example, cellular responses to contact with implants suggest that normal behavior and configuration differ for the dominant cells involved, the fibroblasts, and provide comparison data for other cell lines that may behave differently. It is generally known that fibroblastic/fibrocytic cells can and do show two different morphological habits depending on the materials they are cultured against in laboratory experiments. Cells cultured on intrinsically low-energy surfaces like those of paraffin, waxes, common plastic culture dishes (not specially treated to enhance cell adhesion or growth), or silicone rubber remain rounded. They do not develop strong anchorage to the obligatory predeposited proteinaceous films (contributed from the serum supplements to most tissue culture media, usually). These rounded cells are poorly adhesive and slow to grow on such low-energy substrata. At appropriately high magnifications, the cells present the same view to the observer as provided by bulbous, barely clinging, rain droplets on a freshly waxed automobile hood. A point of continuing confusion is that events of cellular response to low-energy materials take place within an aqueous phase, while the observable features parallel those predicted directly from tests of pure liquids on the same surfaces surrounded by air. The lack of "inversion" of the wetting/spreading relationships as circumstances change from the gaseous to the aqueous surrounding media seems at odds with general thermodynamics (24). This apparent contradiction in findings is easily resolved when it is noted that the predeposited "conditioning" films of complex glycoproteins, prior to cell attachment to the surfaces, provide an "interface conversion" function that can restore thermodynamic consistency to the observations actually made.

The same cells, when cultured against intrinsically high-energy surfaces (like those of clean glass, laboratory vessels, glass microscope slides, many carefully prepared organic-free metals, or plastic culture dishes deliberately treated to increase their surface polarity) do spread into flat polygons and grow to cover the entire exposed substrata area. Fibroblastic cells, in particular, can also grow in multilayers in flattened configurations one over the other, synthesizing extracellular cementing materials (such as the ubiquitously distributed fibrous protein, collagen) between themselves. It is this latter pattern of cell growth, and fabrication of a fiber-reinforced composite matrix, that is observed during the normal healing of wound sites in tissues such as skin and its subjacent layers (25). Cellular growth on and strong attachment to bioprocessor surfaces is most desirable for devices whose performance depend on permanent fixation and immobilization of functional cells at the placement sites. The pattern of cellular response involving only rounded cells poorly retained on low-energy materials is most desired in cases where the devices function best without permanently attached biomass covering their surfaces.

Observations of the bioadhesive events in the human oral cavity, and in laboratory simulations thereof, are also instructive. They illustrate how to follow the initial events from the solid surface out through the initial adhesive layers to the stage where microbial mats are present and thriving.

Figure 4 shows a scanning electron micrograph of a test plate brought to a "nonadhesive" initial surface condition and then exposed to the rich biological environment of dental plaque bacteria in laboratory culture (26). The spontaneously deposited layer of glycoproteinaceous material is made most apparent in this view by its peeling and imperfect coating qualities on this substratum. A close-up view of the specific coccoidal organisms, taken from human teeth by micromaniputation and grown up in the laboratory, also is shown in Fig. 4 where it can be noted that the individual organisms are ellipsoidal in shape with preformed adhesive tufts at one end. These tufts have been demonstrated to be "organelles" specialized for adhesion to foreign surfaces. "Corn cob" structures formed by coccoidal cellular colonization of filamentous bacteria (26) also are illustrated in Fig. 4. As shown in Fig. 5, uniform, microbial biofilms are obtained on clean test plates incubated in cultures of the coccoidal organisms for 24 hours. Scanning electron micrographs of the surfaces of these biofilms, attached to receptive substrata on top of thinner predeposited glycoprotein-dominated "conditioning" films, show the organisms to pack in a close array, "shoulder-to-shoulder" or side-by-side like eggs in a large crate. Internal reflection infrared spectroscopy through the plates actually serving as the substrata for closely-packed, richly colonized bacterial "monolayers" provide unambiguous spectral evidence that the preformed "sticky" tufts comprise predominantly glycoproteinaceous materials and not the glycerol phosphate types of polymers (lipoteichoic acids) hypothesized by other workers (27). Surface-modified test plates, dominated by closely packed methyl groups, exposed to identical growing cultures of organisms specialized by evolution, apparently, for adhesion to solid substrata such as teeth, emerge from the culture as clean and shining as they go in. Inspection of the plates makes it immediately clear that no active antiadhesion principle is at work, Rather, a "passive" or benign initial substratum surface condition prevents the prerequisite "conditioning" film from ever disposing itself in the uniform carpet-like array required for successful bacterial attachments. Applied thin coatings (demonstrated by ellipsometry to be about 175 Angstroms deep) of covalently bound polydimethylsiloxane (equivalent to

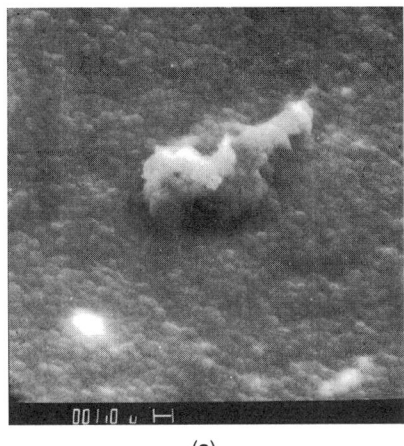 

Figure 5. (a) Scanning electron micrograph illustrating strongly adhesive full-thickness, dense CC5A bacterial "plaque" on test material with critical surface tension of 30–40 mN/m and adsorptive properties similar to human enamel. (b) Scanning electron micrograph illustrating weakly retained tissue layers formed around the same test material modified to display critical surface tension of 20–30 mN/m. (This figure is available in full color at http://www.mrw.interscience.wiley.com/ebe.)

"medical-grade silicone rubber" in many of its properties) are sufficient to impart long-lasting "abhesive" (the opposite of adhesive) surface qualities. These films, dominated by closely packed methyl groups, exhibit empirically measured critical surface tensions (using contact angle techniques) of about 22 mN/m (28). Thus, it is seen that simple modification of the surface properties of immersed solid materials, specifically to low-surface-energy ranges between 20 and 30 mN/m, prevents successful interface conversion by the spontaneously depositing glycoproteinaceous macromolecules. This induction of a poor "primer coat" inhibits the practical, longer-term adhesion of biological cells, even those cells specialized for attachment, as also illustrated in Fig. 5.

This example is just one of the circumstances in which adjustment, or accidental exposure, of surface chemical arrays with surface energies in the narrowly confined zone between 20 and 30 mN/m have correlated with minimal biological adhesive responses (29).

3. ACTIVE FIELDS OF SURFACE MODIFICATION FOR IMPROVED BIOCOMPATIBILITY

Although few new engineering materials have been successfully introduced as biocompatible implant components since von Recum's 1986 publication of scientific, technical, and clinical testing data in Handbook of Biomaterials Evaluation (30), the fields of surface–protein interaction and surface modification to enhance biocompatibility have remained significantly active. Four main themes exist, around which research has been focused for more than 30 years, and for which agreement is still lacking: (1) principles governing biomolecule interactions at foreign interfaces (31); (2) the intrinsic roles of surface hydrophilicity and hydrophobicity in determining biocompatibility (32); (3) the influence of surface micromechanics as reflected in crosslinking (33) or crystallinity (34) and (4) the phenomena of cellular and substrata macromolecules and their excluded volumes determining contact or adhesion (35).

When model substrata of modulated hydrophilic/hydrophobic ratios were employed to study macrophages (35) and 3T3 cells (36), and both "soft" and "rigid" surfaces of each type—as employed in intraocular lens polymers—were examined for endothelial cell adhesion-induced damage (37), direct relationships were not found. More detailed consideration must be given to the structure and reactivity of water at biomaterials surfaces if the remaining biocompatibility issues are to be resolved (38), because strong association of biology with surfaces inevitably requires dehydration of the mutual interfaces between substrata and proteinaceous macromolecules. Although the events of surface dehydration may take longer on such water-loving substrata as those carrying grafted chains of hydrophilic polyethylene oxide, no known successful long-term biomedical device exists, contact lens or implant, nor any ship bottom coating, that retains its initial hydrophilic properties for a sufficient time to be usefully described as resistant to biofouling. Some theoretical efforts have attempted correlations of the complex hydrophilic/hydrophobic effects on contact angles and cell adhesion, attributing stronger cell adhesion to surface hydrogen-bonding (39), while experimental observations continue to repeat findings that simple hydrophobic surfaces dominated by closely packed methyl groups are "almost non-reactive with the blood" (40). These surface-specific influences are confirmed for even very thin plasma-polymerized interface layers (41), and chemically modified surfaces that express closely packed methyl groups inhibit both cell attachment and cell proliferation without actually reducing amounts of adsorbed proteins (42). It is only the strength of interactions remaining, once water has been displaced from the interface, that dictates subsequent events of cellular wetting, spreading, and adhesion on any given substratum. Long-term and continuing "easy-release" of depositing biomass, in response to interfacial shear forces or other mechanical perturbations, is the general outcome for materials of simple, low-surface-energy quality exposed in flowing biological media.

Particularly important studies of human monocyte/macrophage adhesion, motility, and induced foreign body giant cell formations on silane-modified surfaces have revealed a methylated-surface-inhibited cellular spreading pattern, reduction of adhesion, and suppression of foreign body giant cell formation (43–45) that should contribute to biocompatibility of implants for which benign, nonadhesive responses are sought. On the other hand, some recent success is claimed in reducing bacteria and leukocyte adhesion to phospholipid-coated polymer surfaces under dynamic flow conditions (46), not in concordance with the "easy-release" criterion just noted. These competing hypotheses require more study and cross-comparison if we are to fulfill bioengineering goals of reducing capsular thickness and enhancing angiogenesis around implant drug-release systems (47), or mediating the extracellular matrix production associated with foreign body reactions to implanted biomaterials (48). Regarding infection control, preventing bacterial adhesion onto surfaces by using the low-surface-energy approach continues to show good promise for nontoxic systems (49).

Remaining unresolved, also, are issues of surface texture control of cellular attachment and function for engineering biomaterials, as distinguished from the effects of surface energy (50). Quantitative analyses of cell proliferation and orientation on uniformly grooved substrata, as well as those with different roughness and porosity, show that these influences change with increasing time of adhesion (51–53). Considerable opportunity exists to make further progress in this regard.

4. SUMMARY AND CONCLUSIONS

Having accepted and met the challenge of establishing "biocompatibility" with prosthetic devices, in the sense of preventing biological adhesion in some cases and promoting that adhesion in others, a good bit of the mystique associated with the placement of new synthetic materials into biological and engineering environments can be overcome. The stage is set, therefore, for meeting the challenges of developing new processes for surface conversion of a wider diversity of synthetic engineering materials and the architecturally superior materials of natural origin that might be taken from animal or human donors and used in life- and limb-saving applications in unrelated human recipients.

Work with substitute internal organs such as the artificial heart and blood vessel grafts, and with food handling equipment, highlights the importance of fluid flow parameters as well as surface parameters in the attainment of biofouling-free conditions for long periods. A major observation that applies in all facets of bioadhesion research is that initial biological colonization of all boundary phases must be accepted. Control of the steady-state quantity of bound material must be predicated on the application of certain mechanical removal forces, usually produced by shear-stress fields, set up adjacent to these boundaries by the flowing streams that deliver the deposited matter in the first place. The relative strength of adhesion can be selected by adjustment of the material's critical surface tension and its influence on the integrity of the first-bound "conditioning" films. Thus, the preferred plane of parting of the accumulating surface layers can be identified. Successful control of biofouling actually depends, then, on balancing the rate of deposition with a similar rate of re-entrainment. Of great importance is the ability to monitor and adjust the flow rates of potentially fouling streams within test apparatus used to evaluate the fouling tendencies of various material/fluid combinations.

Wettability profiles of conditioning-film-coated implants of the intermediate, high-energy, and low-energy types reveal that the nonadhesive surface state is best described by a critical surface tension remaining within the zone between 20 and 30 mN/m, despite the presence of abundant (but loosely organized) overcoating material. Conversely, tenacious bioadhesion is correlated with critical surface tensions converging in the zone between 30 and 40 mN/m, as a result of tighter, more-coherent overlying films bound to such substrata. Continued expression of the initial surface properties of practical biomaterials through such complex films of spontaneously deposited and then selectively retained biological matter is an intriguing example of nature's "biomolecular engineering" that might be emulated by the next generation of university-trained bioengineers.

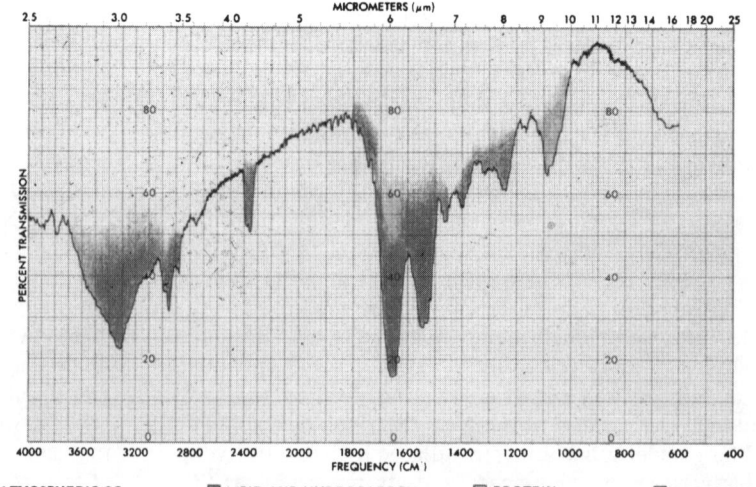

Figure 6. Infrared spectrum of the typical thin capsule immediately surrounding implanted biomaterials, revealing the substance to be predominantly glycoproteinaceous. (This figure is available in full color at http://www.mrw.interscience.wiley.com/ebe.)

Histologic cross sections of tissue capsules adjacent to intermediate-surface-energy test plates show intermediate cellular responses. Rich fibroblast-dominated cellular populations immediately adjacent to the implants trail off to more amorphous and loose connective tissue zones at greater distances. As a final reminder of the enormous dependence of all bioadhesive phenomena on the presence and properties of proteinaceous interface conversion layers, Fig. 6 presents an infrared spectrum of the typical thin film on the surfaces of any implanted material. The protein-dominated composition of this layer is obvious from such spectral records. Increased attention must be given to studies of the particular molecules—their sizes, shapes, elemental, and monomeric constituents; their origins; their lifetimes; and their two-dimensional and three-dimensional configuration—if further advances are to be made in the art of interfacing manmade synthetic or engineering materials with biological systems.

As in common observations of acquired pellicle-to-biofilm transitions during plaque formation in the human mouth, an acquired preparticulate "conditioning" film is seen in all other biological settings, also, to be preferentially composed of a minority constituent of the total adsorbabie material pool available. Common features among such conditioning films, recruited from a variety of biological phases are (1) that the molecules are usually truly macromolecules, that is, molecules having very high molecular weights, (2) are glycoproteinaceous in character, as contrasted with pure proteins, carbohydrates, lipoproteins, lipids, or proteoglycans (mucoid substances), and (3) are very highly hydrated in their solution states. Many experiments suggest that the driving force for such highly specific deposition of the same types of films on a large variety of solid surfaces is the event of dehydration, of both the original surface and the macromolecules, during their spontaneous surface localization (54).

Extrapolation of this now secure base of understanding to emerging areas of bioengineering and biotechnology should pay handsome dividends.

BIBLIOGRAPHY

1. W. S. Pierce, J. H. Donachy, G. Rosenberg, and R. E. Baier, Calcification inside artificial hearts: inhibition by warfarin-sodium. *Science* 1980; **208**:601–603.
2. J. W. Boretos, W. S. Pierce, R. E. Baier, A. F. Leroy, and H. J. Donachy, Surface and bulk characteristics of a polyether urethane for artificial hearts. *J. Biomed. Mater. Res.* 1975; **9**:327–340.
3. R. E. Baier, The organization of blood components near interfaces. *Ann. N. Y. Acad. Sci.* 1977; **283**:17–36.
4. R. E. Baier and R. C. Dutton, Initial events in interactions of blood with a foreign surface. *J. Biomed. Mater. Res.* 1969; **3**:191–206.
5. R. E. Baier, Key events in blood interactions at nonphysiologic interfaces—a personal primer. *Artif. Organs* 1978; **2**:422–426.
6. R. C. Dutton, A. J. Webber, S. A. Johnson, and R. E. Baier, Microstructure of initial thrombus formation on foreign materials. *J. Biomed. Mater. Res.* 1969; **3**:13–23.
7. R. E. Baier, Occurrence, nature, and extent of cohesive and adhesive forces in dental integuments. A. Lasslo and R. P. Quintana, eds., *Surface Chemistry and Dental Integuments*. Springfield, IL: Charles C. Thomas, 1973.
8. M. G. Buonocore, Bonding to hard dental tissues. R. S. Manly, ed., *Adhesion in Biological Systems*. New York: Academic Press, 1970.
9. A. J. Gwinnett and A. Matsui, A study of enamel adhesives: the physical relationship between enamel and adhesive. *Arch. Oral Biol.* 1967; **12**:1615–1620.
10. A. J. Gwinnett, The scientific basis of the sealant procedure. *J. Prevent. Dent.* 1976; **3**:15–28.
11. R. E. Baier and V. A. DePalma, Electrodeless glow discharge cleaning and activation of high-energy substrates to insure their freedom from organic contamination and their receptivity for adhesives and coatings, Cornell Aeronautical Laboratory Report No. 176, 1970: 17.
12. N. J. Harrick, *Internal Reflection Spectroscopy*. New York: Interscience Publishers, 1967.
13. R. E. Baier, Modification of surfaces to reduce fouling and/or improve cleaning. B. Hallstrom, D. B. Lund, and C. Tragardh, eds., *Fundamentals and Applications of Surface Phenomena Associated with Fouling and Cleaning in Food Processing*. Sweden: Division of Food Engineering, Lund University, 1981.
14. V. A. DePalma and R. E. Baier, Flow cell and method for continuously monitoring deposits on flow surfaces, U.S. Patent 4, 175,233, 1979.
15. V. A. DePalma and R. E. Baier, Microfouling of metallic and coated metallic flow surfaces in model heat exchange cells. Proc. Ocean Thermal Energy Conversion (OTEC) Biofouling and Corrosion Symp., PNL-SA-7115, U.S. Department of Energy, 1978: 89–106.
16. R. W. King, A. E. Meyer, R. C. Ziegler, and R. E. Baier, New flow cell technology for assessing primary biofouling in oceanic heat exchangers, Proc. Eighth Ocean Energy Conference, U.S. Department of Energy, 1981.
17. R. E. Baier, V. A. DePalma, A. E. Meyer, R. W. King, and M. S. Fornalik, Control of heat exchange surface microfouling by material and process variations. J. M. Chenoweth and H. Impagliazzo, eds., *Fouling in Heat Exchange Equipment*, HTD-vol. 17. New York: The American Society of Mechanical Engineers, 1981.
18. R. E. Baier, Influence of the initial surface condition of materials on bioadhesion, Proc. Third Int. Cong, on Marine Corrosion and Fouling, Evanston, IL: Northwestern University Press, 1973: 633–639.
19. S. C. Dexter, Influence of substrate wettability on the formation of bacterial slime films on solid surfaces immersed in natural sea water, Proc. Fourth Int. Cong. on Marine Corrosion and Fouling, Boulogne, France: Cent de Rech et D'Etud Oceanogr., 1977: 137–144.
20. S. C. Dexter, J. D. Sullivan, J. Williams III, and S. W. Watson, Influence of substrate wettability on the attachment of marine bacteria to various surfaces. *Appl. Microbiol.* 1975; **30**:298–308.
21. D. J. Crisp and P. S. Meadows, Adsorbed layers: the stimulus to settlement in barnacles. *Proc. Roy. Soc. B* 1963; **158**:364–387.
22. D. W. Goupil, V. A. DePalma, and R. E. Baier, Physical/chemical characteristics of the macromolecular conditioning film in biological fouling. Proc. Fifth Int. Cong. on Marine Corrosion and Fouling, Madrid, Spain, 1980: 401–410.
23. R. E. Baier, Substrata influences on the adhesion of microorganisms and their resultant new surface properties. G. Bitton and K. C. Marshall, eds., *Adsorption of Microorganisms to Surface*. New York: Wiley-Interscience, 1980.

24. R. E. Baier, Comments on cell adhesion to biomaterial surfaces: conflicts and concerns. *J. Biomed. Mater. Res.* 1982; **16**: 173–175.
25. R. E. Baier, Surface chemistry in epidermal repair. H. I. Maibach and D. T. Rovee, eds., *Epidermal Wound Healing.* Chicago, IL: Year Book Medical Publishers, 1972.
26. C. Mouton, H. S. Reynolds, E. A. Gasiecki, and R. J. Genco, *In vitro* adhesion of tufted oral streptococci to Bacterionema matruchotti. *Curr. Microbiol.* 1979; **3**:181–186.
27. R. E. Baier, Adhesion to different types of biosurfaces. S. A. Leach, ed., *Dental Plaque and Surface Interactions in the Oral Cavity.* Arlington, VA: Information Retrieval, Inc., 1980.
28. R. E. Baier, E. G. Shafrin, and W. A. Zisman, Adhesion: mechanisms that assist or impede it. *Science* 1968; **162**:1360–1368.
29. R. E. Baier, Surface properties influencing biological adhesion, R. S. Manly, ed., *Adhesion in Biological Systems.* New York: Academic Press, 1970.
30. A. F. von Recum, *Handbook of Biomaterials Evaluation.* New York: Macmillan Publishing Company, 1986.
31. A. S. Hoffman, Principles governing biomolecule interactions at foreign interfaces. *J. Biomed. Mater. Res. Symp.* 1974; **5**(1): 77–83.
32. A. S. Hoffman, Letter to the Editor: a general classification scheme for "hydrophilic" and "hydrophobic: biomaterials surfaces. *J. Biomed. Mater. Res.* 1986; **20**:ix–xi.
33. A. Chilkoti, G. P, Lopez, B. D. Ratner, M. J. Hearn, and D. Briggs, Analysis of polymer surfaces by SIMS. 16. Investigation of surface cross-linking in polymer gels of 2-Hydroxyethyl Methacrylate. *Macromolecules* 1993; **26**:4825–4832.
34. A. Park and L. G. Cima, In vitro cell response to differences in poIy-L-lactide crystallinity. *J. Biomed. Mater. Res.* 1996; **31**:117–130.
35. A. J. Lentz, T. A. Horbett, L. Hsu, and B. D. Ratner, Rat peritoneal macrophage adhesion to hydroxyethyl methacrylate-ethyl methacrylate copolymers and hydroxystyrene-styrene copolymers. *J. Biomed. Mater. Res.* 1958; **19**:1101–1115.
36. T. A. Horbett, J. J. Waldburger, B. D. Ratner, and A. S. Hoffman, Cell adhesion to a series of hydrophilic-hydrophobic copolymers studied with a spinning disc apparatus. *J. Biomed. Mater. Res.* 1988; **22**:383–404.
37. N. B. Mateo and B. D. Ratner, Relating the surface properties of intraocular lens materials to endothelial cell adhesion damage. *Invest Ophthalmol. Vis. Sci.* 1989; **30**(5):853–860.
38. E. A. Vogler, Structure and reactivity of water at biomaterial surfaces. *Adv. Colloid Interface Sci.* 1998; **74**(1–3):69–117.
39. R. J. Good, M. Islam, R. E. Baier, and A. E. Meyer, The effect of surface hydrogen bonding (acid–base interaction) on the hydrophobicity and Hydrophilicity of copolymers: variation of contact angles and cell adhesion and growth with composition. *J. Dispersion Sci. Technol.* 1998; **19**(6–7):1163–1173.
40. B. D. Ratner, Blood compatibility—a perspective. *J. Biomater. Sci. Polymer. Edn.* 2000; **11**(11):1107–1119.
41. S. D. Johnson, J. M. Anderson, and R. E. Marchant, Biocompatibility studies on plasma polymerized interace materials encompassing both hydrophobic and hydrophilic surfaces. *J. Biomed. Mater. Res.* 1992; **26**:915–935.
42. T. O. Collier, C. R. Jenney, K. M. DeFife, and J. M. Anderson, Protein adsorption on chemically modified surfaces. *Biomed. Sci. Instrumentation* 1997; **33**:178–183.
43. C. R. Jenney, K. M. DeFife, E. Colton, and J. M. Anderson, Human monocyte/macrophage adhesion, macrophage motility, and IL-4-induced foreign body giant cell formation on silane-modified surfaces in vitro. *J. Biomed. Mater. Res.* 1998; **41**:171–184.
44. C. R. Jenney and J. M. Anderson, Alkylsilane-modified surfaces: inhibition of human macrophage adhesion and foreign body giant cell formation. *J. Biomed. Mater. Res.* 1999; **46**:11–21.
45. T. O. Collier, C. H. Thomas, J. M. Anderson, and K. E. Healy, Surface chemistry control of monocyte and macrophage adhesion, morphology, and fusion. *J. Biomed. Mater. Res.* 2000; **49**:141–145.
46. J. D. Patel, Y. Iwasaki, K. Ishihara, and J. M. Anderson, Phospholipid polymer surfaces reduce bacteria and leukocyte adhesion under dynamic flow conditions. *J. Biomed. Mater. Res.* 2005; **73A**:359–366.
47. B. D. Ratner, Reducing capsular thickness and enhancing angiogenesis around implant drug release systems. *J. Controlled Release* 2002; **78**:211–218.
48. P. Puolakkainen, A. D. Bradshaw, T. R. Kyriakides, M. Reed, R. Brekken, T. Wight, P. Bornstein, B. Ratner, and E. H. Sage, Compromised production of extracellular matrix in mice lacking secreted protein, acidic and rich in cystein (SPARC) leads to a reduced foreign body reaction to implanted biomaterials. *Amer. J. Pathol.* 2003; **162**:627–635.
49. J. Tsibouklis, M. Stone, A. A. Thorpe, P. Graham, V. Peters, R. Heerlien, J. R. Smith, K. L. Green, and T. G. Nevell, Preventing bacterial adhesion onto surfaces: the low-surface-energy approach. *Biomaterials* 1999; **20**:1229–1235.
50. E. T. den Braber, J. E. de Ruijter, H. T. J. Smits, L. A. Ginsel, A. F. von Recum, and J. A. Jansen, Effect of parallel surface microgrooves and surface energy on cell growth. *J. Biomed. Mater. Res.* 1995; **29**:511–518.
51. E. T. den Braber, J. E. de Ruijter, H. T. J. Smits, L A. Ginsel, A. F. von Recum, and J. A. Jansen, Quantitative analysis of cell proliferation and orientation on substrata with uniform parallel surface-micro-grooves. *Biomaterials* 1996; **17**:1093–1099.
52. A. F. von Recum, C. E. Shannon, C. E. Cannon, K. J. Long, T. G. van Kooten, and J. Meyle, Surface roughness, porosity and texture as modifiers of cellular adhesion. *J. Tissue Eng.* 1996; **2**(4):241–253.
53. T. G. van Kooten and A. F. von Recum, Cell adhesion to textured silicone surfaces—the influence of time of adhesion and texture on focal contact and fibronectin fibril formation. *J. Tissue Eng.* 1999; **5**:223–240.
54. R. E. Baier, Book review of "Colloids and Interfaces in Life Sciences" by Willem Norde. *Biofouling* 2004; **20**:130–131.

BIOCOMPUTATION

DAN NICOLAU Jr.
University of Melbourne
Melbourne, Australia
and
Swinburne University of Technology
Melbourne, Australia

DAN NICOLAU
Swinburne University of Technology
Melbourne, Australia

Biocomputation is a term that has been understood to refer to several areas of research at the boundary between computation and biology. Among these are computational biology and bioinformatics (the use of computation and mathematics to solve problems in biology), natural computing (the use of naturally inspired computing models to solve mathematical problems, e.g., evolutionary algorithms, neural networks, etc.), biological computation (the investigation of how nature processes information at the cellular and subcellular level), and what may be termed biomolecular computation (the attempt to exploit biomolecules and cells to perform computations). The latter two of these are related and complementary, two parts of a greater effort to understand and use the computational properties of biological entities. We deal here strictly with biocomputation in the last sense, of harnessing the power of biosystems to compute (i.e., to solve mathematical problems). The most important example of such efforts is DNA computing, in which one uses DNA molecules and their interactions to perform computations. Recently, however, some other promising models of biocomputation have been proposed (e.g., membrane computing).

The use of biological entities and processes for computation is motivated by a number of factors. The first factor is that the enormous pace of progress in current computing architectures and technologies cannot continue indefinitely, and, indeed, cannot continue for much longer. The principal reason is that this pace of development has been mainly due to improvements in miniaturization of electronic components and circuits, and current technologies are approaching some "natural limits" in this respect. The current generation of microelectronic components have dimensions in the hundreds of nanometers, whereas the next generation will see this reduced to tens of nanometers or less. Functional electronic components cannot be made smaller than atomic dimensions, of course; thus, although many improvements can be made to current architectures and materials technology, it is clear that, in the absence of radical new ideas, the progress of microelectronics will grind to a halt. In light of this, biocomputation has been proposed as an alternative or complement to "silicon," along with quantum computation and other ideas.

Secondly, many problems of theoretical and practical interest exist whose solution using classic computational devices (digital computers) appears to be, in many cases, impractical. Roughly speaking, this is because all algorithms for these problems may reduce to an exhaustive search through a set of solutions whose size grows exponentially with the size of the problem (these are called NP-complete or NP-hard problems), which implies that instances of such problems of modest size exist whose solution on a classic computer will require an impractically large amount of time. Although classic computers process information very rapidly, the fact that they do so in a sequential (one operation at a time) manner makes them unsuitable for problems where such a brute-force approach is required. On the other hand, computations in nature do not proceed sequentially, but exhibit massive parallelism. For example, the neurons in the brain process information concurrently rather than "waiting" on one another, as would be the case with a sequential algorithm. In addition, it seems that biological systems are not only capable of computing, but of doing so with incredible competence in terms of energy efficiency, data storage capacity, and speed.

1. INTRODUCTION

It is difficult to identify the beginnings of computation, but it is clear that thousands of years ago, humans were able to count and perform simple arithmetic; driven by inventiveness and necessity, we have become increasingly adept at (and reliant on) this crucial activity. It would not be too much to say that our dominant position as a species is due, in great part, to our ability to compute in one form or another. "Computers" (i.e., artificial devices used to perform calculations) have a long history, beginning with the abacus and continuing with functional and wished-for devices because of Pascal, Babbage, and others. It is the modern-day electronic computer, however, that has truly revolutionized computing and society. Although the concept of computation is a very broad one, the current computing technologies (digital computers) have been so enormously successful that we have come to identify them with the concept itself. This success is attributable, in great part, to three achievements made during the twentieth century.

The first achievement was the development, by Alan Turing, of the theory of computation as it is understood today. Turing defined computation by considering a so-called "Turing machine," a theoretical computing device used as a model for mathematical calculation. A Turing machine consists of a line of cells known as a "tape" that can be moved back and forth, an active element known as the "head" that can change the machine's two properties (known as "state" and "color") of the "active" cell underneath it, and a set of instructions for how the head should modify the "active cell" and move the tape. At each step, the machine can change the color, state, or both for the active cell. This abstract description resembles (and indeed, is modeled on) the computing approach of a person (who writes symbols on a piece of paper, etc.). Intuitively, a Turing machine is a complete model for computation as we understand it. A Universal Turing machine is one that can be programmed, via a finite length of tape, to perform the actions of any Turing machine whatsoever (universality refers informally to the ability to compute any function, as long as one can provide an algorithm). Turing machines can further be divided into deterministic (if only one instruction is associated with a given tape state) and nondeterministic (if more than one instruction is permitted for any such state). Informally, nondeterministic machines can "go more than one way" at each step whereas deterministic ones are only allowed one legal transition for each internal state. Electronic computers are deterministic from a theoretical point of view (at each step, the next state of the system can be predicted given the current state and current input). Turing left the physical implementation of a Turing machine unspecified, but soon after his seminal work, methods and ideas for doing this were discovered.

The second crucial development was the proposal by John von Neumann of what has become the dominant computer design in use today. The heart of this architecture is a "processing unit," which has associated with it a stored program (a set of instructions) and computes using repeated cycles of fetching and executing instructions from the program; which represents an implementation of a deterministic Turing machine. Finally, it is necessary to implement the operations needed by such a machine physically—for which silicon provided the medium. It is the fortuitous combination of these three factors, along with some advances in physics, chemistry, and mathematics, that have given this computing paradigm power and the ability to progress at great pace.

The technologies used to build semiconductor components have certainly greatly improved over the last 50 years; many variations on the core technology have come and gone. The von Neumann architecture also has undergone some changes—a degree of parallelism has been added to the processing unit (modern CPUs can execute one instruction while fetching another and access data at the same time)—and certainly the data and instruction lengths have increased. However, both of these are still recognizable—we still use close relatives of the von Neumann architecture and we still use semiconductor-based gates and memory elements.

On the other hand, it is clear that many problems exist that do not lend themselves to current electronic computers (in particular, because they do not lend themselves to efficient computation by deterministic Turing machines); at the same time, it is also becoming clear that the current technologies cannot continue to improve at the same pace indefinitely. As early as 1965, Gordon Moore proposed what has become knows as Moore's Law: the number of transistors manufactured on a silicon chip appears to double every year. This law has held up admirably since 1965 (although, at the turn of the millennium, the doubling period had increased to around 1.5 years), but it cannot do so forever. There are approximately 4 silicon atoms per nanometer, and so this is an absolute physical limit. Before that point is reached, other physical and process problems are likely to appear. These considerations, along with some others, motivate the search for new ideas. To this end, one need only look to nature to discover powerful computing paradigms (and implementations).

In investigating biocomputation in the natural world, we encounter what appears to be a paradox of sorts: On the one hand, the "operations" performed on information in biological systems are carried out rather slowly compared with their equivalents in electronic computers. For instance, the legendary "speed of thought" is not actually large—in computer engineering terms, operations in the brain have frequencies of the order of Hertz or less. By contrast, today's microprocessors can perform billions of instructions per second, and so have frequencies of GHz. On the other hand, it is clear that even very simple biological entities are extremely proficient at many types of computation. Biological systems are adept at, for example, pattern recognition, a task that has proven intractable for human algorithms and computers in any but the simplest of cases. Organisms are capable of an incredible degree of self-regulation; moreover, they achieve this with a decentralized approach, in contrast to our own highly centralized computing architectures.

This apparent incompatibility can be resolved by considering the types of "problems" to which biological systems compute solutions. In general, these are problems whose solution requires only simple operations, but for which very large numbers of candidate solutions exist. An example of this is pattern identification—where there are plenty of contending patterns, and the identification of a specific one (e.g., the shape of a predator) is required. The key for many such tasks is to perform many operations at the same time. Although the speed or accuracy of each operation may not be impressive, the enormous parallelism of molecularity (there are billions of molecules in even a microscopic volume) ensures the correct solution is found among many candidates. Accuracy is not as critical—often, an approximate solution is sufficient. By contrast, the problems typically solved with electronic computers usually require the rapid and accurate use of a series of sequentially applied sophisticated and abstract operations. Calculus problems fall in this category—the sum of many elements is what is required to approximate a definite integral, and each of these must be known very accurately. Of course, parallelism still helps, but it is the accuracy and speed of the operations that makes the problem tractable.

It should be no surprise then that although our computational devices are extremely efficient at solving problems requiring an accurate but sequential approach, they are not so adept at solving problems requiring a massively parallel approach. Conversely, biological entities are endowed with tools useful for solving problems such as pattern identification, control of a complex system with many variables (e.g., metabolism), and so on. In other words, although electronic computers' architectures are "tuned" into the structure of calculus problems, for example, nature's computational machines appear to be tuned into the structure of combinatorial problems.

This realization has led to the attempt to gather inspiration from natural computing mechanisms for implementation of algorithms to solve human problems on electronic computers. Actually, biologically motivated computing paradigms existed early in the history of modern computing but were sidelined by artificial approaches that delivered better results in the short and middle term. Recently, for the reasons mentioned above as well as because of a better understanding of biological systems, these paradigms have seen a revival. The best-known examples are neural networks and evolutionary algorithms, which use abstracted data structures and machinery from nature (e.g., models of neurons and the process of natural selection) to solve human problems such as face identification in a photograph, optimization, and many others. Of course, the simulation of, for example, a very simple neuronal network on a sequential computer loses all the computational power of parallel processing, but may keep some of the advantages of the networks' attunement to combinatorial problems.

Although these fields are intimately related with biocomputation as we have defined it above, they do not deal with radically new methods of computation but rather with new algorithms. The same computers are used to execute these algorithms, based on the same architectures and the same materials. In order to capture the computational power, efficiency, and data storage capacity of biological systems, as well as to harness the suitability of nature's algorithms, we must turn to actually implementing physically different computers, with biomolecular components and biomolecular operations, all done on the mesoscale. Although we do not currently have the technological capacity to build artificial biocomputer components, billions of years of evolution have provided us with readymade components: biomolecules and cells for information storage and biochemical reactions for operations. Several approaches to biomolecular computation have been proposed—among these, DNA computing is the most well-developed theoretically, and so far, the only one that has seen successful and nontrivial physical implementations.

2. DNA COMPUTING

Although molecular electronics has been investigated for the last two decades, the field of DNA computing was effectively launched in 1994 by an experiment performed by Leonard Adleman (1). Adleman used DNA to efficiently compute the solution to the well-known Traveling Salesman Problem, thus demonstrating for the first time the principle of biomolecular computation. Since then, DNA computing has received a great deal of attention because of its promised problem-solving efficiency, data storage capacity, energy efficiency, and new mathematical outlook on computation.

2.1. Using DNA for Computation

In what follows, we provide an overview of the structure and properties of DNA that make DNA computing possible in principle, as well as some recent mathematical results and a brief comparison of the DNA computing with the accepted computing paradigm. For the mathematical theory of DNA computing, the reader is referred to Păun et al. (2). As this is a very young and active field, the best source of information on current developments is the literature itself, which abounds with reviews suited to most levels of expertise. The bibliography (see below) lists some articles of interest, and again the text by Păun et al. (2) provides a very readable and informal introduction to the theoretical aspects of DNA computing.

2.1.1. Structure of DNA. DNA is the molecule whose properties make DNA computing possible, both theoretically and practically. In this section, we concisely review the structure of DNA, as well as what these properties are and what "operations" can be carried out with DNA. Inevitably, this description will be both simplified and brief. For the purposes of biocomputation, many chemical and biochemical details can be omitted. For a thorough treatment of DNA, its structure and properties, we refer the reader to Griffiths et al. (3).

Deoxyribonucleic acid (DNA) is a polymer molecule made of monomers called deoxyribonucleotides. In cells, it performs two critical functions: First, it encodes the information needed for the production of proteins and second, it carries out self-replication, such that an exact copy of the genetic information it contains is passed down to "offspring" cells. It is the chemical structure of DNA that allows these duties to be performed successfully.

The monomers that make up a DNA molecule are called *deoxyribonucleotides*—usually referred to simply as "nucleotides." Chemically, each of these consists of three "components": a sugar, a phosphate group, and a nitrogenous base. Both the sugar and the base have carbon atoms; the phosphate group is attached to an atom of the sugar, whereas another carbon atom of the latter is attached to the base. Nucleotides are divided into *purines* and *pyrimidines*—the only difference between these is in the structure of their respective bases. There are two purines, *adenine* and *guanine* (abbreviated A and G, respectively), and two pyrimidines, *cytosine* and *thymine* (abbreviated C and T, respectively).

The manner in which nucleotides form bonds is critical not only for the functions of DNA and, hence, cells, but also of crucial interest for biocomputation, as will be seen shortly. Essentially, nucleotides can form links of two types. The first of these is the phosphodiester bond, which is a strong covalent bond formed between the phosphate group of one nucleotide and the hydroxyl group of the sugar of another nucleotide. The second and more important type for DNA computing consists of hydrogen (weak) bonds between the bases of two nucleotides. Importantly, A and T link together and C and G link together, but the other combinations do not occur, which is known as Watson–Crick complementarity.

These two kinds of bonds combine to produce the famous double-helix structure of DNA discovered by Watson and Crick (4). The phosphodiester bonds are needed to form what is called *single-stranded* DNA, a polymer made of many nucleotides covalently linked together. Watson–Crick complementarity is responsible for weakly bonding two such single strands to produce the double helix. The contribution of many hydrogen bonds over the length of the two double strands ensures that this structure is stable (although single hydrogen bonds are themselves quite weak). *In vivo*, the structure of DNA is more complicated because of packing (needed to make a long DNA molecule fit in a small cell) and other effects—but for the purposes of harnessing the properties of DNA for computing, this is not important.

2.1.2. Operations on DNA Molecules. We have briefly described the structure of DNA, which tells us what the "units" in a DNA computer would be, but in order to carry out a computation, even theoretically, we need to be able to perform some operations on these units. In the case of DNA computing, these operations are completely different, both in practice and in spirit, to the traditional ones. Nevertheless, the point is that the combination of these operations (made possible by the structure and properties of DNA) and the large number of DNA molecules available may make DNA computing practical and useful. Describing the operations one can carry out on DNA molecules *in vitro* in any detail is a task far beyond the scope of this article. In what follows, we summarize the most important of these in very approximate terms and in light of their usefulness for computations. For a detailed biochemical description, we refer the reader to Griffiths et al. (3), although a thorough treatment of the mathematical nature of biochemical operations for DNA computing is given in Păun et al. (2).

First, it is possible to separate double-stranded DNA into its single strands, which can be done by heating the solution above a certain denaturation temperature (generally between 85 °C and 95 °C). Alternatively, the use of certain reagents (e.g., formamide) can lower this temperature. Cooling the solution slowly leads to the complementary base pairs "locating" each other and thus to the reformation of the double-stranded structure (*annealing*, or if the DNA was previously denatured, *reannealing*). Sometimes annealing is referred to as *hybridization*, although this term is also used to describe a different "fusing" operation involving strands of different origin.

Another operation that can be performed is changing the length of a DNA strand—either shortening or lengthening it. In the latter case, *polymerases* can be used to add a sequence of nucleotides to an existing DNA strand in certain conditions. Roughly speaking, it is possible to produce both single-stranded and double-stranded DNA chains of arbitrary nucleotide sequence via a series of such "concatenations." Artificially synthesized single-stranded DNA molecules are called *oligonucleotides* (or simply *oligos*). On the other hand, *nucleases* can be used to degrade the structure of DNA. These can either extract nucleotides sequentially from the ends of a DNA strand (*exonucleases*) or break the phosphodiester bonds linking nucleotides together (*endonucleases*). Both operations are quite flexible. In the case of exonucleases, it is possible to select the end of the DNA strand from which the nucleotides in question are to be removed, and there are exonucleases available for single-strand DNA cleaving, double-strand DNA cleaving, or both. In the case of endonucleases, one may also "cut" either single strands or double strands; additionally, roughly speaking, it is possible to make the cut at any position in the strand.

Ligation is a process by which strands of DNA may be linked together. This reaction is made possible by enzymes called *ligases*. Ligation is normally done in a series of steps involving annealing as well as ligation and depends, among other things, on the way in which DNA strands were cut or cleaved. *Modifying enzymes* can be used to modify the chemical composition of nucleotides in a sequence. Both ligation and the modification of nucleotides are technically involved procedures, but the key point is that they are available and can be quite flexible.

Finally, the operation that, more than any other, brings DNA computing from the realm of mathematics to that of the laboratory is the *polymerase chain reaction* (PCR), discovered by Mullis in 1985. PCR is essentially capable of amplifying a very small amount of a particular DNA strand (as low as one single molecule) to produce, in principle, exponentially larger quantities of this strand. One needs only to supply the so-called *primers*—the end-sequences or flanking sequences of the DNA strand we wish to amplify—which is not a significant restriction, as, in general, these are known in advance. PCR can start, as mentioned, with very few or even a single strand of interest, mixed with millions of unwanted DNA strands, and in a short time produce many copies of the sequence of interest only. This sensitivity and efficiency have led to PCR revolutionizing genetic engineering. It is used in forensic analysis, genomic sequence determination, genetic engineering, and other areas. We mention that it is possible biochemically to *sequence*, with some degree of effort, a given DNA strand (i.e., determine its composition) with PCR (for example, the Human Genome Project was made possible because of this), which is of some importance for biocomputation.

In the context of DNA computing, PCR means that we can use relatively small amounts of DNA (the order of milligrams to grams) to produce and filter through all possible candidate solutions to a problem. Although there may be many of these candidates, and thus few DNA strands encoding each, we can use PCR at the end of the experiment to amplify only the sequences of interest—those encoding the "answer," which is really a simplified view of DNA computing that demands some qualification. We hope this point will be made more clear through the description of Adleman's seminal experiment in DNA computing as well as some recent efforts (see below).

From a strictly mathematical point of view, the basic operations one needs to perform DNA computing can be summarized by six abstract "commands." Each command makes use of the concept of "test tube"—a virtual multiset N of DNA molecules.

- Amplify: given a test tube, N, produce two copies of it.
- Merge: given two test tubes, N_1 and N_2, containing DNA strands, form the union of their contents, U.
- Detect: given a test tube, N, determine whether it contains any DNA strands.
- Sequence-separate: given a test tube, N, and a sequence of elements, s, from the set $\{A, C, G, T\}$, produce two test tubes—$V(N, s)$ containing the DNA strands that do contain the given sequence and $W(N, s)$ containing the ones that do not. This operation should preserve the multiplicity of the respective DNA strands.
- Length-separate: given a test tube, N, and an integer, n, produce a test tube, $T(N, n)$, containing only those DNA strands in the given test tube whose length is less than or equal to n, preserving multiplicity.
- Position-separate: given a test tube and a sequence s of elements from the set $\{A, C, G, T\}$, produce two test tubes—$B(N, s)$ containing all the DNA strands in the original tube whose sequences begin with the given sequence and $E(N, s)$ containing all the DNA strands in the original tube whose sequences end with the given sequence.

We begin to see now how DNA computing can be carried out. The DNA strands are the "variables" and the operations defined above form a basic "programming language." The main difference from the traditional computing paradigm is that the variables evolve in parallel according to the given "rules" instead of sequentially. Rather than going too far into mathematical details, we hope to illustrate the *modus operandi* of a calculation by describing Adleman's original experiment (see below).

2.1.3. Theoretical Results in DNA Computing. Too many results have already been obtained in this field to list thoroughly. We simply mention a few important ones. From a mathematical point of view, a very general result is due to Păun et al. (2; p. 67) — that Watson–Crick complementarity (the property in the double helix DNA structure in which A and T link together and C and G link together, but the other combinations do not occur – see above) assures universality of DNA computing having sufficient input/output capacities. According to the currently accepted Church–Turing Thesis, every computation can be performed by a Turing machine; this result implies, in broad terms, that DNA computing is capable of performing any computation also. The power of a number of variations of DNA computing, such as DNA computing on surfaces, has been explored and shown to be logically equivalent to that of solution-phase DNA computing.

Many algorithms have been proposed to solve specific problems using DNA computing (in particular, NP-complete problems). The first and most general of these was because of Lipton (5), who showed that using the DNA computing paradigm as discussed above, a most important NP-complete problem, 3-SAT, can be solved in linear time. This result also implies that any NP-complete problem can be solved in linear time with DNA computing (as any such problem can be reduced to 3-SAT). Thus, it is possible to break the Data Encryption Scheme (DES); solve the Hamiltonian Path Problem and Traveling Salesman Problem; and the "Knapsack" Problem; expand symbolic determinants; decide graph connectivity; efficiently carry out matrix multiplication; and solve many other problems efficiently. Where these problems are NP-complete, it is possible to solve them in polynomial time; if they are solvable in polynomial time by a deterministic Turing machine, the reduction in time complexity implies that problems in *P* may be solvable in subpolynomial time by a DNA computer.

2.1.4. Comparing DNA Computing with the Current Computing Paradigm. Computing as we understand it today was defined by Turing (6). For more than half a century, this paradigm has been incarnated with incredible success in the electronic computer. Biocomputation, on the other hand, challenges this paradigm and has begun to force the rethinking of our concepts of computing. Some intuitive comparisons immediately present themselves. Turing thought of a computer as a "clerk" that writes symbols down on a tape. The symbols on the tape, together with the "state of mind of the computer," define the state of the computation at any time, which is, essentially, in the case of a deterministic Turing machine, a sequential set of operations. This is precisely how electronic computers function. Although each operation may be performed very quickly by electronic circuits, the system must wait for the result of one step to begin the next, roughly speaking.

In DNA computing, by contrast, instead of a diligent clerk performing the calculations sequentially, we have a large number of DNA molecules evolving in parallel. Individually, the DNA molecules have no more intelligence than this clerk (in fact, they have far less). Their power lies simply in their large number and the massively parallel fashion in which they evolve. Whether this model leads to more or less efficient computations than the traditional one depends on the problem as well as the speed of various steps and the error rates involved. Whatever these realities may be, in principle, the DNA computing model is a serious rival of the traditional model, at least from a computer theoretical point of view.

It is intuitively clear, for example, that some problems are more suited theoretically for DNA computing. If NP-complete problems cannot be solved in polynomial time on a sequential computer and exhaustive search is the only strategy that one can use, then the massive parallelism of DNA computing will at some point be more powerful than the sequential calculation procedure of electronic computers, no matter how quickly each of the steps performed in these can be carried out. Examples of such problems abound. At the same time, Watson–Crick complementarity guarantees the universality of DNA computing, in the sense that a "DNA computer" can be programmed to calculate any "computable" number, subject to the computer having sufficient capabilities for handling inputs and outputs (2).

Of course, unacceptably high error rates, slow reaction times, or expense may mean that DNA computing is impractical *in vitro* for "real" problems. Even if this is the case (and it is too early at present to tell), DNA computing has already been of value in forcing computer scientists and mathematicians to rethink the paradigm of computation.

2.2. DNA Computing Implementations and Algorithms

In what follows, we briefly describe the current state of DNA computing in terms of *in vitro* implementations and proposed algorithms, beginning with Adleman's experiment. We go into some detail here, not only because this experiment "launched" the field but also because it illustrates very well some of the concepts discussed above. An overview of the current position of DNA computing is also given, listing the main results to date.

2.2.1. Adleman's Experiment and DNA Computing. Generally, in the physical sciences, important new results are established through one or a series of experiments with spectacular results. An example was the experimental verification of the remarkable prediction by Einstein's theory of relativity that light is "bent" by gravitational fields. In mathematics, progress is made generally only via rigorous proofs of theorems, conjectures, and so on (i.e., there is no experimental aspect and thus no uncertainty in results). Computer science research often lies at the boundary between mathematics and experimental science. On the one hand, the mathematical theory of computation is well established. On the other, many new results are initially obtained or suggested by "software experimentation" and proved rigorously later. Often, new ideas in computer science are brought to the fore through a *demonstration*, which was the case with DNA computing, a field in which undoubtedly the seminal experiment (demonstration) was that performed by Leonard Adleman in 1994 (1).

Adleman's experiment solves the so-called Hamiltonian Path Problem (HPP) for a given directed graph. This problem can be formulated as follows: Given a directed graph, G, and two vertices, does a path exist connecting the two vertices that visits every other vertex, but only once? The significance of solving this problem is that HPP is a so-called NP-complete problem (see above). It is clear that HPP can be solved by an exhaustive search—simply try all the possible paths in the graph, checking, in each case, whether the path satisfies the conditions above. If such a path is found, then the graph has a Hamiltonian path. If, on the other hand, no such path exists, then we can answer HPP in the negative. The difficulty associated with this approach is that the number of candidate paths grows exponentially with the number of nodes in the graph, which makes the problem rapidly become *intractable*, in the sense that an inordinately large amount of time may be required for even a graph of modest size. Although various algorithms have been devised to "prune" the tree of candidate solutions, none of these escape the exponential complexity problem. Indeed, no NP-complete problems are *tractable* at present, in the sense of being solvable in a time polynomial with the dimensions of the problem—all known algorithms amount to exhaustive search. The question of whether this is in fact possible is considered by many to be one of the single most important problems in mathematics today (and a 1 million dollar prize has been offered by the Clay Institute for its solution).

Adleman's experiment solved this problem for a very small graph; in fact, it is possible to find the Hamiltonian path simply by inspection (the graph used by Adleman is shown in Fig. 1). However, with the *massive parallelism* and *complementarity* of DNA computing, HPP can be solved efficiently in this way, even in cases where there are many nodes. Although the graph used by Adleman was very modest in size and it took days to solve HPP on this graph in the laboratory, in principle, the method could be applied to larger graphs, for which HPP cannot be solved practically on electronic computers. Moreover, the success of Adleman's experiment suggests that other NP-complete problems, some of great practical importance, could be solved using DNA computing, which is essentially why this single experiment was so significant. We now briefly describe the experiment itself.

Adleman's solution of the HPP was based on the following algorithm:

Input: A directed graph G consisting of n vertices, together with two vertices, v_{in} and v_{out}.

Step 1: Randomly generate the multiset S of all paths in

G in large quantities.

Step 2: Eliminate all paths from this multiset, that do not begin with v_{in} and end with v_{out}.

Step 3: From the remaining multiset, eliminate all paths not of length n.

Step 4: From the remaining multiset, for each vertex v (not including v_{in} and v_{out}), eliminate all paths that do not contain vertex v.

Output: If any elements remain in S, then a Hamiltonian path exists in G; otherwise, no such path exists.

This algorithm effectively carries out an exhaustive search. What makes it efficient in the context of DNA computing is that there are potentially an extremely large number of DNA molecules in the solution. Each of the molecules, consisting essentially of a random sequence of nucleotide "strings," can be thought of as coding for one possible path through the graph. By eliminating the strands that do not encode a Hamiltonian path (using methods from biochemistry), one can eventually reach a state where either all strands remaining are encoding Hamiltonian paths or where no strands remain. Of course, the elimination procedures eliminate entire classes of illegal paths at once (e.g., all paths not beginning with v_{in} can be removed in one step). Thus, although the algorithm is nondeterministic (many possibilities exist for each step and there is no way to predict precisely how the algorithm will evolve for each input), this massive parallelism compensates for the unpredictability of the algorithm's evolution.

In this case, although it is not known how many copies of each path will be generated at Step 1, with very large numbers of DNA molecules to represent these topics will guarantee that *all* paths will be generated. Moreover, Watson–Crick complementarity can be used to ensure that the DNA molecules in the solution encode only paths in G rather than any sequence of nodes.

In Adleman's experiment, each vertex of the graph was associated with a 20-mer (20-nucleotide) strand of DNA. As there are 7 nodes in the graph he used (see Fig. 1), there were seven of these strands. For example, the stand encoding node 2 was s_2 = TATCGGATCGGTATATCCGA. Let the Watson–Crick complement of s_i be $h(s_i)$, then $h(s_2)$ = ATAGCCTTAGCCATATATGGCT. Then, dividing each of the s_i in two halves, denoted $s_i's_i''$, an edge from node i to j will be encoded by the strand $h(s_i's_j'')$ (if such an edge exists). Thus edges also are represented by 20-mer strands of nucleotides. Note that because each half of $h(s_i's_i'')$ represents one node, the edges $e_{i \to j}$ and $e_{j \to i}$ where $i \neq j$ will be encoded differently (this is crucial because G is a directed graph, so that some edges do not run in both directions).

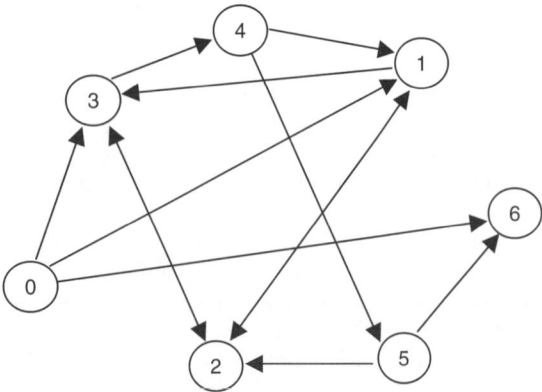

Figure 1. The graph used in Adleman's experiment.

In the experiment, large quantities of nucleotides representing the s_i (nodes) and the $e_{i \to j}$ (edges) were mixed together in a ligation reaction (see above). These combined to produce random long strains of DNA molecules encoding all paths in the graph G (because of Watson–Crick complementarity, the s_i acted as "joints" to link together only nucleotides representing edges that were both connected to the same node). Some additional reactants were used (ligase buffers) to aid the reaction, and the mixture was incubated for around 4 hours at room temperature.

Following this step (Step 1 in the algorithm above), the resulting "soup" was then progressively "trimmed" until only nucleotides representing the Hamiltonian 0123456 in G (see Fig. 1) remained. In terms of the "programming language" discussed above, the algorithm used by Adleman could be written:

Input: A test tube, N, containing the DNA strands produced in the ligation reaction (representing all legal paths through the graph G).

Step 1: Position-separate N using the sequence s_0, keeping $B(N, s_0)$.

Step 2: Position-separate $B(N, s_0)$ using the sequence s_6, keeping $E(B(N, s_0), s_6)$.

Step 3: Length-separate $E(B(N, s_0), s_6)$ using $n = 140$, giving test tube $T(E(B(N, s_0), s_6), 140)$.

Step 4: for $i = 1$–5, let $T = V(T, s_i)$.

Step 5: Detect using test tube T.

In Step 3, recall that the oligos s_i have length 20 and there are 7 nodes in G. After Step 4, all DNA strands originally present in N whose sequences did not begin with v_{in} and end with v_{out}, were longer than the length of a potential Hamiltonian, and did not contain each of the other five vertices are eliminated by progressive "carving" of the solution set. Should any DNA strands remain, Step 5 will detect this. If none are left, we conclude that the graph does not have a Hamiltonian path. Note that the success of this algorithm rests on (1) the original test tube N containing all the legal paths through G and (2) each of the steps being error-free. Of course, these are not trivial assumptions *in vivo* and need to be demonstrated. For example, Step 2 above was implemented by a PCR using the primers representing s_0 and s_6, and this reaction can, in some instances, have a significant error rate. Subsequent steps used similar biochemical techniques (for details, we refer the reader to Refs. 1 and 2 and our discussion above). Theoretically, however, this algorithm solves the HPP for any given directed graph.

We mention here that the number of DNA molecules used in this experiment far exceeded the needed quantity. For each edge, around 10^{13} encoding molecules were present in the solution; hence, given the small number of paths in the graph Adleman used, many molecules encoding the Hamiltonian would have been produced by the ligation reaction. In principle, only one such molecule would have been enough to solve the problem. For pragmatic reasons, it is preferable to have as many copies of the solution to begin with as possible, but certainly in this experiment, the quantity of nucleotides used was much larger than needed. The implications of this are that (1) problems of a much larger size could be practically solved using similar quantities and (2) conversely, much smaller quantities could be used for small problems.

In addition, although Adleman's experiment required 7 days of laboratory work, this is not a true reflection of the effectiveness of this method. There are two points to be considered here. The first is that the purpose of the experiment was to demonstrate a principle; the procedure used was rather a crude one. Certainly, a great deal of optimization could be carried out. In particular, the PCR procedures used can be automated to some extent and made far more efficient. The second and more fundamental point is that this procedure requires no more than linear time in the number of nodes of the graph G. The most time-consuming part of the experiment was Step 4 listed above (simply because it had to be carried out for each node). Were we to double the number of nodes, this step would simply take twice as long to complete, even in the absence of any optimization, whereas Steps 1, 2, and 3 would not be considerably more complex from a mathematical point of view. On the other hand, using a sequential computer, doubling the number of nodes may mean that exponentially longer time will be required to solve the problem. Thus, DNA computing's massive parallelism ensures that this NP-complete problem remains tractable in time. The "price" for this is that, in the worst case, exponentially greater amounts of DNA may be needed to guarantee the generation of all solutions (paths, in this case). However, as discussed already, this may not amount to very much for many practical problems, because in theory, a single molecule is needed in the final "soup" to indicate the existence of a solution. The main point is that the combination of massive parallelism and complementarity make the efficient solution of NP-hard problems possible, at least mathematically.

Since Adleman's experiment, biomolecular computation has received a great deal of attention in the literature. The avalanche of theoretical results has not been matched by *in vivo* implementations, however. Adleman's original experiment has been replicated by a few groups. Some progress has been made toward solving small instances of 3-SAT, notably on a surface (7). DNA-based addition has been implemented. More recently, Adleman solved a larger instance of the HPP, involving around a million possible solutions (8).

2.2.2. DNA Computing on Surfaces. As mentioned above, the practical success or failure of DNA computing implementations as defined by Adleman's experiment ("solution-phase" DNA computing) depends on the ability to maintain the error rates of the various biochemical operations below relatively small limits. As the idea behind DNA computing is to organize massively parallel and, hence, efficient searches through the complete solution space, DNA computing requires the representation of this space using DNA strands. In Adleman's experiment, this representation was guaranteed by the ratio of DNA strands available to the number of candidate solutions, which was very large. As the complexity of the problem (expressed in terms of the dimensions of the input) increases, this ratio decreases for a given initial amount of DNA (i.e., the average number of strands encoding one candidate solution becomes smaller), which places demands on the maximum acceptable error rate or, equivalently, on the minimum amount of DNA needed. Thus, DNA computing as defined suffers from a "scalability" problem, which has prompted the search for means to better control the error rates in DNA computing operations (e.g., PCR, hybridization). One avenue for improving experimental control during DNA computing experiments is to immobilize the DNA strands on a surface before manipulation, which is referred to as "DNA computing on surfaces" and was first proposed by Liu et al. (7).

First, the nucleotides encoding the solutions are generated in the same way as for solution-phase DNA computing—possibly with some redundancy because the biochemical operations performed on them will not be 100% accurate (such that each base represents one or more bits). These nucleotides are then immobilized on a surface (such as glass, silica, or thermally grown oxides on silicon wafers) in a nonaddressed fashion (i.e., without regard to spatial organization). Following this step, operations equivalent in computational power to solution-phase DNA computing operations can be performed. Specifically, it has been proposed (7) that a "mark-destroy-unmark" scheme is sufficient for many computations. Roughly speaking, one first "marks" the DNA strands of interest (say, those whose jth base represents a 0) by hybridizing them with complementary strands (followed by polymerization). Thus each strand binds to its complement if the latter is present on the surface. At the next step, those strands on the surface that are not bound to their respective complements can be destroyed (i.e., degraded) with exonucleases (see above). One can then "unmark" the remaining DNA strands simply by washing with distilled water (because, in the absence of double-strand stabilizing salts, the complements will denature from the surface-bound oligos, leaving the single-stranded DNA attached). These steps can be repeated a number of times until all that remains on the surface are the oligos representing the solution to the problem instance. One can test whether this set is empty (and extract the solution if not) using PCR for solution-phase DNA computing by either cleaving the remaining DNA from the surface (if single-stranded) or by "unmarking" the strands as described and amplifying the complements (Fig. 2).

There are advantages and disadvantages to this method. The operations on the oligos are far more efficient in surface-based DNA computing because less strands are lost at each step (and so, less solution set representation redundancy is needed initially) because of the immobilization of the oligos at the surface and thus to reduced interference between strands. Other advantages include ease of purification and the ability to use more established biochemical techniques and to automate the process. A high information density is achievable (one base per bit or better) because of the single-base mismatch approach. However, relatively speaking, these gains come at the price of a massively reduced information density (from 3D storage to 2D storage). Additionally, the number of operations per second is limited by the slower enzyme kinetics at the surface while the hybridization efficiency cannot be expected to be as high as in solution. Finally, the surface-based method does not eliminate scaling problems because discrimination of single-base mismatches becomes more difficult as the strand length increases and the operations are not error-free in any case. The most serious of these limitations is the loss of information density—one must either increase the surface area (e.g., by using microbeads instead of a planar surface) or attempt to employ a local three-dimensional surface chemistry.

From a mathematical point of view, surface-based DNA computing is a competitor to solution-phase DNA computing. It is known (9) that surface-based DNA chemistry supports general circuit computation on many inputs in parallel efficiently and that the number of parallel operations needed to decide the satisfiability of a Boolean circuit is proportional to the size of the circuit.

3. OTHER BIOCOMPUTATIONAL MODELS

The realization that many biological processes can be regarded as computations has led to a number of proposals

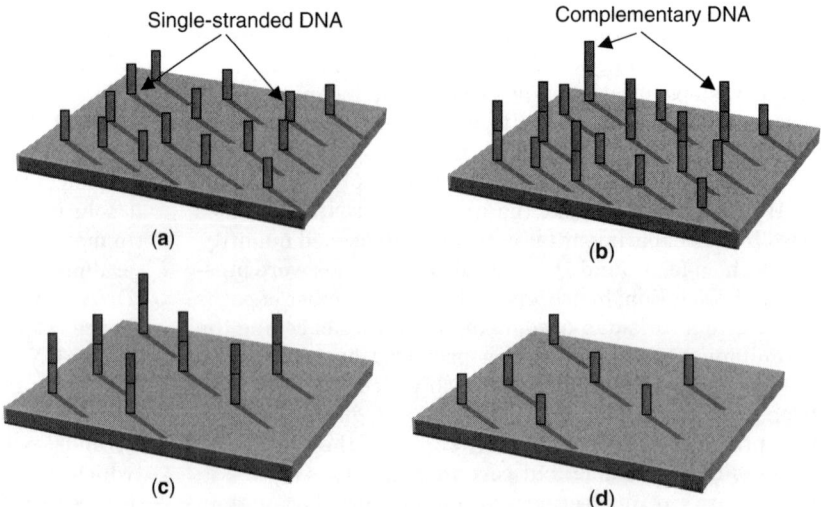

Figure 2. DNA computing on surfaces. Mark-Destroy-Unmark cycle: (a) immobilization, (b) hybridization, (c) exonuclease treatment, (d) unmarking. (This figure is available in full color at http://www.mrw.interscience.wiley.com/ebe.)

regarding the harnessing of biology for performing efficient calculations. In a sense, biocomputation is a subset of natural computing, a field dealing with computing occurring in nature and computing using ideas inspired by nature. Interpreting complex natural phenomena as computational processes enhances both the understanding of these phenomena and of the essence of computation. Examples of more classic computational strategies inspired by nature include artificial neural networks and evolutionary computing—these aim simply to "simulate" *in silico* processes imitating those occurring in nature in the hope of exploiting their suitability to certain problems. At the other end of the scale, we have what has been called here biocomputation: biomolecular computing (DNA computing), biological circuits, membrane computing, computing with mobile bioagents, and other schemes that aim to actually implement using biomolecular components and strategies *in vitro*, which is illustrated schematically in Fig. 3, where the vertical scale indicates the strength of the biocomputational (i.e., *in vivo*) flavor. It is interesting to note that all of the computational schemes inspired by nature that have been proposed so far have been supplied by biological systems—suggesting that perhaps computation is something particular to life.

One important and recent computing paradigm involves membrane computing or "P-systems", proposed by Păun (10). It is based on the notion of membrane (cell wall-like) structure that is used to enclose computing cells in order to make them independent computing units. Also, a membrane serves as a communication channel between a given cell and other cells "adjacent" to it. This model comes from the observation that the processes that take place in the complex structure of a living cell can be considered as computations. Since these computing devices were introduced, several variants have been considered. A good introduction to P-systems can be found in Păun and Calude (11). Briefly, a membrane contains a number of "objects," possibly with some multiplicity, together with a set of rules (with priority relations) controlling the

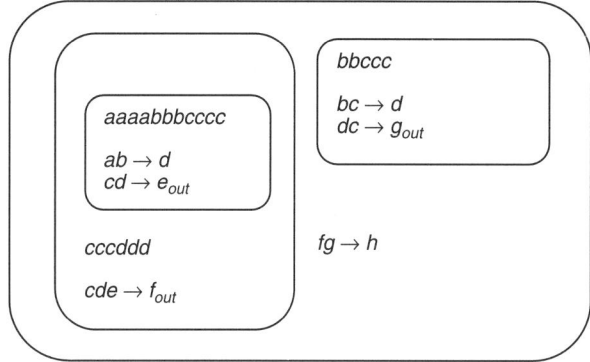

Figure 4. A P-system with multiple nested membranes.

combination of objects to produce other objects (e.g., objects a and b can combine to produce c via the rule $ab \to c$). Membranes can contain other membranes (leading to a tree of membranes) and, in the general case, rules can dissolve membranes and send objects outside of the membrane and membranes can also replicate. It is not difficult to see how this scheme imitates the functioning of a living cell, with the membrane representing the cell wall, the objects representing chemical species in the cell, and the rules representing the biochemical reactions between these. A P-system is illustrated in Fig. 4.

This computing paradigm is extremely strong from a mathematical point of view, both in terms of computational power and in terms of efficiency, which is principally because P-systems evolve in parallel on two levels: in the membrane, a computation is being carried out in parallel, whereas different membranes can evolve in parallel also (including replicating themselves and being destroyed). It is known (10) that NP-complete problems can be solved in linear time by P-systems with active membranes, and various other theoretical results exist for variants of P-systems. Unfortunately, implementing even an extremely simple version of such a system *in vivo* has not even been

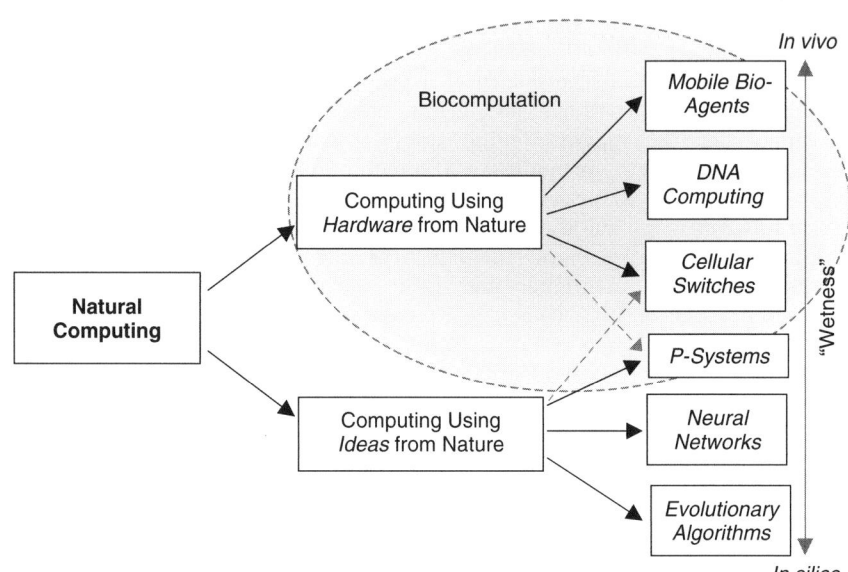

Figure 3. Biocomputation as a field of natural computing. (This figure is available in full color at http://www.mrw.interscience.wiley.com/ebe.)

attempted and is unlikely to be achieved in the near future because of the incredible complexity of living cells.

A biologically related and more realistic biocomputing scheme involves the construction of biological circuits using "gene switches." Gardner et al. demonstrated the construction of a genetic toggle switch in *Escherichia coli* (12). Elowitz and Leibler (13) built a simple synthetic "genetic clock" with genetic expression in cells, which could be considered the first successful attempts at showing artificial computation to be possible in a cell. Such biological circuits are unlikely to replace the functions of electronic circuits or solve complex problems, but they do hold a great deal of promise with respect to controlling artificially living cell systems, analogously to the way in which electronic systems control electrical and mechanical systems.

A different avenue for computation using biological systems involves the use of the natural directional mobility of some biological entities, which may be coupled with the ability of cells to communicate with one another and interact with the environment. Nakagaki et al. (14) demonstrated the solving of a maze by an autonomous amoeboid organism (a nontrivial if not NP-hard mathematical problem). This demonstration shows that cellular materials can display a type of primitive intelligence and are capable of relatively sophisticated computations. Nicolau and Nicolau (15) suggested that the directional motion of actin on myosin (protein molecular motors responsible for many processes including muscular contraction and cell division) in the *in vitro* motility assay can be regarded as a computation and, because of the large numbers of molecular motors that can be made to move simultaneously in microstructures, may be able to be used to efficiently compute the solutions to difficult combinatorial problems (e.g., NP-complete problems).

4. PERSPECTIVES

Aside from the services DNA computing may render to the theory of computation and the way we view this activity, the central question facing the field of DNA computing is whether it will practically change the way that we compute or merely remain a "curiosity," which is by no means a vague question—in fact, it appears that the answer depends quite precisely, as suggested above, mainly on (1) the error rates with which the operations can be performed and (2) the limits with which problems can be represented by DNA strands.

The advantages of DNA computing over the traditional paradigm are clear. The first of these is the processing speed of the "soup": In Adleman's model, the number of operations per second was estimated at around 10^{18}— around a million times more rapid than the fastest supercomputer available today. Secondly, DNA has an enormous capacity for data storage, as attested to by the incredible complexity of even the simplest life form. An information density of one bit per cubic nanometer is probably achievable, which is some billions of times more dense than current computer memories. It is estimated that a single DNA strand could hold more memory than all the hard drives ever manufactured to date. Finally, DNA computing would be far more energy efficient than electronic circuits are. Computers generally perform at the level of a billion operations per Joule, whereas a DNA computer could achieve an energy efficiency some billions of times better than this.

For the first of the above conditions, a rough mathematical analysis is possible, as undertaken for example by Adleman et al. (16), in relation to the amounts of DNA needed to have a given chance of breaking the DES, which is one of the most commonly used algorithms for secure communications. If an error rate as low as 10^{-4} (i.e., one error in 10,000 operations, on average) can be attained, then the amounts required are rather low, of the order of 1 g (which is actually not a small quantity by the standards of molecular biology but is still very reasonable in general). On the other hand, if the error rate were one in a hundred operations, then the required amount would be larger than the mass of the Earth. Even with a one in one thousand error rate, one would still require kilograms to break a 56-bit DES—an amount far too large to make routine DNA computing feasible on real-life problems. Such quantities would not only be prohibitively expensive but pose serious problems for the machinery performing the mixing, and so on. Thus, the feasibility of DNA computing will depend in the strongest terms on the error rate attainable *in vitro*. It is difficult to say even with current techniques what these error rates are expected to be in practice; and one can only speculate on what improvements future developments can bring.

We can also say something about the second condition above. As DNA computing is known to solve NP-complete problems in linear time, we would expect, in general, to only require strand lengths that grow linearly with the input size—suggesting that length should not be an insurmountable problem, which in fact, is not strictly true. In keeping with the DES example, DNA strands of the order of 10,000 nucleotides are expected to be required for breaking a 56-bit code. Although a realistic problem size, oligonucleotides longer than around 15,000–20,000 bases are subject to breakage because of even modest fluid shear forces. Although it may be possible to improve on both the representation efficiency and the maximum stable strand length, these figures indicate an order-of-magnitude limit on the length of a problem that can safely be encoded. Without significant new developments, this order-of-magnitude limit would pose a clear problem.

In addition, arguments such as these apply to "combinatorial problems," which appear to be naturally suitable to DNA computing. Other problems are likely to be no more tractable experimentally. Finally, it should be mentioned that, even in principle, the time efficiency of DNA computing comes at a price—exponentially larger amounts of DNA are needed as the input size of an NP-complete problem grows linearly, in the worst case. No matter how small the error rate can be made, eventually these amounts will become prohibitively large, which, of course, is the case with any computational device—the question is whether DNA computing will be superior to electronic computing for these or other problems.

This discussion would suggest a bleak outlook for the possibility of implementing DNA computing. On the contrary, theoretical and experimental developments so far give us plenty of reasons for optimism. It is clear that DNA computing is a very promising field. It is also becoming clear that new techniques will be needed to improve our control over the DNA molecules *in vitro* in order for this promise to be fulfilled. A consensus is slowly emerging that, in the future, DNA computers may well complement their electronic counterparts in some way, rather than replacing them. Additionally, DNA computing may prove instrumental in nanotechnology and related fields.

Other biocomputational techniques are, at the time of writing, too immature to allow any but the crudest form of speculation about the future. Ten years ago, it is improbable than anyone would have been able to foresee Adleman's experiment and the astonishing rate of progress in the field since. Thus, it is surely foolish to predict the next 10 years. Nevertheless, one can say with some confidence that it is likely that biocomputation will have some role in the future of computer science. It is clear that biological systems are extremely proficient at performing certain kinds of computations—fortuitously, those appear to be problems that digital computers are not proficient at. Attempts at implementing primitive "biohardware" and solving simple problems using biological systems have already met with some success, and the pace of development can be expected to continue in the near term.

Acknowledgments

The authors would like to thank Kristi Hanson, Luisa Filipponi, Gerardin Solana, and Audrey Riddell for helpful comments and suggestions during the preparation of the manuscript.

BIBLIOGRAPHY

1. L. M. Adleman, Molecular computation of solutions to combinatorial problems. *Science* 1994; **266**:1021–1024.
2. G. Păun, G. Rozenberg, and A. Salomaa, *DNA Computing: New Computing Paradigms*. New York: Springer-Verlag, 1998.
3. A. J. F. Griffiths, J. H. Miller, D. T. Suzuki, R. C. Lewontin, and W. M. Gelbart, *An Introduction to Genetic Analysis*, 7th ed. New York: Freeman, 2000.
4. J. D. Watson and F. H. Crick, A structure for deoxyribose nucleic acid, *Nature* 1953; **171**:737–738.
5. R. J. Lipton, Using DNA to solve NP-complete problems. *Science* 1995; **268**:542–545.
6. A. M. Turing, Computing machinery and intelligence. *Mind* 1950; **59**:433–460.
7. Q. Liu, L. Wang, A. G. Frutos, A. E. Condon, R. M. Corn, and L. M. Smith, DNA computing on surfaces. *Nature* 2000; **403**:175–179.
8. R. S. Braich, N. Chelyapov, C. Johnson, P. W. K. Rothemund, and L. Adleman, Solution of a 20-variable 3-SAT problem on a DNA computer. *Science* 2002; **296**:499–502.
9. W. Cai, A. E. Condon, R. M. Corn, E. Glaser, Z. Fei, T. Frutos, Z. Guo, M. G. Lagally, Q. Liu, L. M. Smith, and A. Thiel, The power of surface-based DNA computation. Proc. 1st Annual International Conference on Computational Molecular Biology, 1997.
10. G. Păun, Computing with membranes. *J. Comp. Syst. Sci.*, 2000; **61**(1):108–143.
11. G. Păun and C. Calude, *Computing with Cells and Atoms*. London, UK: Taylor and Francis, 2001.
12. T. S. Gardner et al., Construction of a genetic toggle switch in Escherichia coli. *Nature* 2000; **403**:339–403.
13. M. B. Elowitz and S. Leibler, A synthetic oscillatory network of transcriptional regulators. *Nature* 2000; **403**:335–338.
14. T. Nakagaki, H. Yamada, and A. Toth, Maze-solving by an amoeboid organism. *Nature* 2000; **407**:470–471.
15. D. V. Nicolau, Jr. and D. V. Nicolau, Computing with the Actin-Myosin molecular motor system. *SPIE Proc.* 2002; **4937**:219–225.
16. L. M. Adleman, P. W. K. Rothemund, S. Roweiss, and E. Winfree, On applying molecular computation to the Data Encryption Standard. Proc. 2nd Annual Meeting on DNA Based Computers, 1996: 28–48.

READING LIST

R. J. Lipton, DNA solution of hard computational problems. *Science* 1995; **268**:542–545.

BIOELECTRICITY AND BIOMAGNETISM

SHOOGO UENO
MASAKI SEKINO
MARI OGIUE-IKEDA
University of Tokyo
Tokyo, Japan

1. INTRODUCTION

Interactions between living organisms and electromagnetic fields have long attracted attention from numerous scientists. However, it is only recently that scientists have begun to study the interactions systematically, and to apply electromagnetic fields to the study of biological phenomena. Applying an interdisciplinary approach, these studies cover a wide range of fields from medicine and biology to physics and engineering (1). A technique of localized and vectorial transcranial magnetic stimulation (TMS) has enabled us to obtain noninvasive functional mapping of the human brain (2–6). Recent studies have shown that TMS potentially has therapeutic effects for several diseases such as mental illnesses, ischemia, and cancer (7–20). The development of bioimaging technologies such as electroencephalography (EEG), magnetoencephalography (MEG), and magnetic resonance imaging (MRI) enabled the identification of the locations of human brain functions (21–26). Despite these technologies, however, it is still difficult to understand the dynamics of brain functions, which include millisecond-level changes in functional regions and dynamic relations between brain neuronal networks. Development of new bioimaging methods for visualizing neuronal electrical activities and electrical conductivities in the brain is an attractive research field (27–33). Technological advances of superconducting magnets have enabled us to study diamagnetic forces on biological macromolecules and cells. The diamagnetic force causes magnetic orientation of macromolecules and cells, which has potential applications in regenerative medicine (34–43). This chapter reviews recent advances in biomagnetics and bioimaging techniques such as TMS, EEG, MEG, MRI, cancer therapy, magnetic control of cell orientation and cell growth, and magnetoreception in animals.

2. TRANSCRANIAL MAGNETIC STIMULATION

TMS of the human brain was developed by Barker et al. in 1985 (2). When a strong electric current is applied to a coil over the head for 0.1 ms, a pulsed magnetic field of 1 T is produced. This magnetic field induces eddy currents in the brain, which stimulate the neurons. Geometry of the coil is an important factor affecting spatial distribution of the eddy currents. Most of the presently used stimulators have a circular coil or a figure-eight coil. The figure-eight coil consists of a pair of circular coil elements and was proposed by Ueno et al. (3). Pulsed electric currents were applied to the two coil elements in the opposite directions. This geometry enables focal and vectorial stimulation because the eddy current converges below the intersection of the coil elements and flows in a specific direction. Calculations of the eddy current distributions in the human brain generally require numerical methods because the human head consists of multiple tissues with different electric conductivities. Eddy current distributions were obtained for circular coils and figure-eight coils using the finite element method (7,8). Figures 1a and 1b show the magnetic field distributions on the surface of the head for a 100-mm circular coil and a 75-mm figure-eight coil. Figures 1c and d1 show the eddy current distributions for these coils. The brain surface under the intersection of the figure-eight coil exhibited high current density. The circular coil induced more widespread eddy currents compared with the figure-eight coil.

TMS is a useful method to examine dynamic brain function without causing any pain, producing so-called "virtual lesions" for a short period of time. The authors were able to noninvasively evaluate the cortical reactivity and functional connections between different brain areas. An associative memory task involving pairs of Kanji (Chinese) pictographs and unfamiliar abstract patterns were studied (4). The subjects were 10 Japanese adults fluent in Kanji, so only the abstract patterns represented novel material. During memory encoding, TMS was applied over the left and right dorsolateral prefrontal cortex (DLPFC). A significant reduction in subsequent recall of new associations was seen only with TMS over the right DLPFC. This result suggests that the right DLPFC contributes to encoding of visual-object associations.

TMS has a wide range of clinical and preclinical applications such as identification of the location of the motor cortex and treatments of neurological and psychological

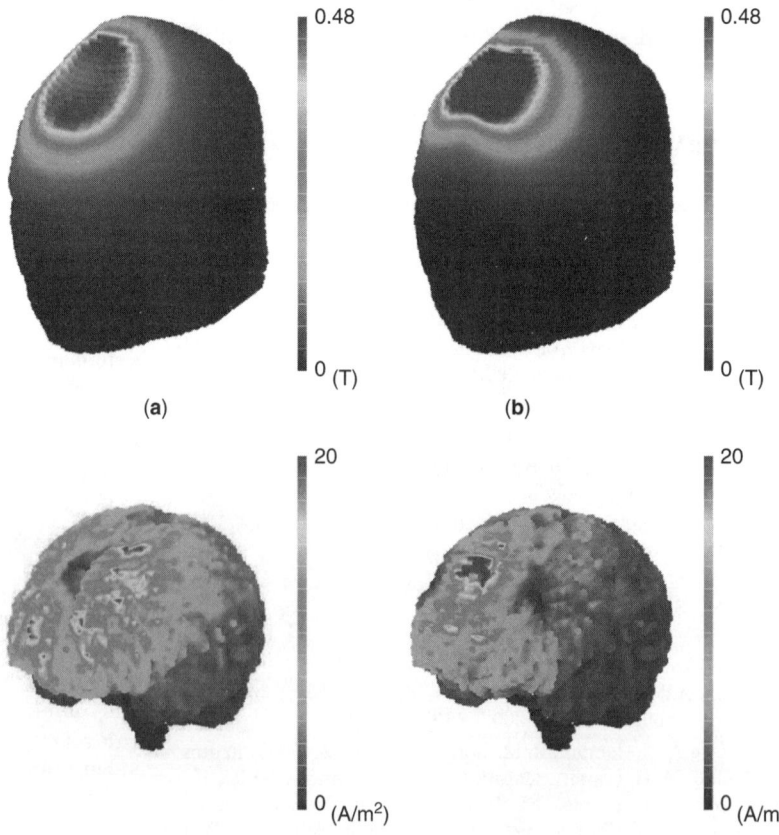

Figure 1. (a) Magnetic field distribution on the surface of the head model for transcranial magnetic stimulation (TMS'1) using a 100-mm circular coil. (b) Magnetic field distribution for a 75-mm figure-eight coil. (c) Eddy current distribution on the surface of the brain for TMS using the 100-mm circular coil. (d) Eddy current distribution for the 75-mm figure-eight coil. (This figure is available in full color at http://www.mrw.interscience.wiley.com/ebe.)

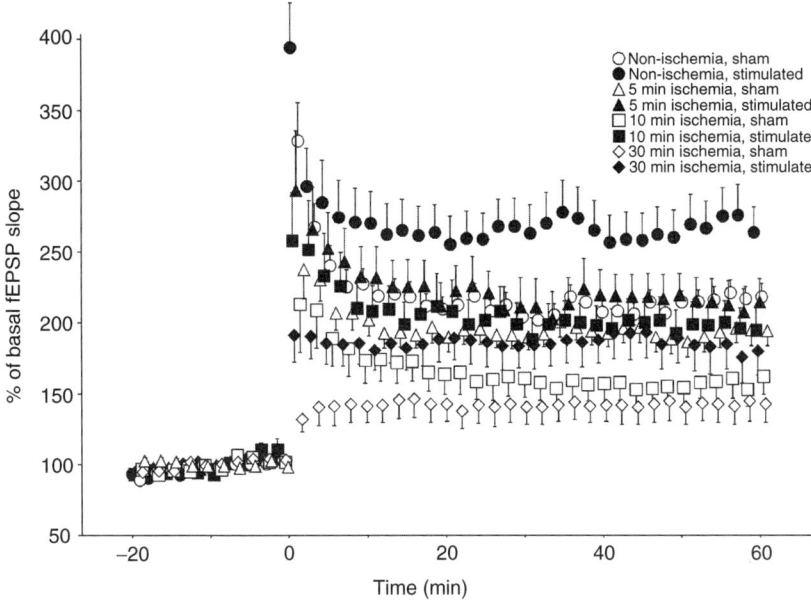

Figure 2. LTP in the 0.75-T rTMS group and sham control groups in each ischemic condition. The LTP in the 0.75-T rTMS group was enhanced compared with the LTP of the sham control group in each ischemic condition. Tetanus stimulation was applied at time = 0 min. Error bar = ±1 SEM.

diseases. The localized TMS is particularly useful for noninvasive mapping of the brain function for presurgical planning (5,6). Recording of motor evoked potentials caused by TMS shows us the location of the motor cortex. This technique enables us to avoid motor deterioration after a brain surgery. For practical purposes, the functional mapping using TMS requires equipment for measuring the location of the coil and software for visualizing the stimulation targets on individual MRI. Stereotactic TMS devices allow precise mapping of the spatial and temporal representation of brain activities that react to magnetic stimuli such as sensory, motor, language, and cognitive functions.

Recent studies showed that repetitive TMS (rTMS) has therapeutic effects on neurological and psychological diseases. Electroconvulsive therapy (ECT), in which electric currents are applied to the brain, improves severe mental illnesses such as depression. Effects of TMS on depression have been investigated because TMS has a potential to give a comparable therapeutic effect to ECT with less invasiveness. In several studies, TMS improved schizophrenia (9) and Parkinson's disease (10) as well as depression (11).

We investigated the effect of rTMS on long-term potentiation (LTP) in the rat hippocampus (12,13). Rats were magnetically stimulated at a rate of 1000 pulses/day for 7 days by a circular coil, in which the peak magnetic fields at the center of the coil were 0.50, 0.75, 1.00, and 1.25 T. LTP enhancement was observed in the 0.75-T rTMS group, and LTP suppression was observed in the 1.25-T rTMS group, whereas no change was observed in the 0.50-T and 1.00-T rTMS group. These results suggest that the effect of rTMS on LTP depends on the stimulus intensity. In another study, LTP was suppressed after the hippocampal slices were exposed to ischemic conditions (14). The LTP in the 0.75-T rTMS group was enhanced compared with the LTP of the sham control group in each ischemic condition, suggesting that rTMS resulted in acquisition of ischemic tolerance in the hippocampus (Fig. 2).

The effect of rTMS on injured neurons was investigated in the rat brain after administration of the neurotoxin MPTP (l-methyl-4-phenyl-1,2,3,6-tetrahydropyridine) (15). The rats received rTMS (10 trains of 25 pulses/s for 8 s) 48 h after MPTP injection. Tyrosine hydroxylase (TH) and NeuN expressions were investigated in the substantia nigra. The functional observational battery-hunched posture score for the MPTP-rTMS group was significantly lower and the number of rearing events was higher compared with the MPTP-sham group. These results suggest that rTMS reactivates the dopaminergic system in the brain.

3. ELECTROENCEPHALOGRAPHY AND MAGNETOENCEPHALOGRAPHY

Activities of neurons and muscles originate from exchange of ions between intracellular space and extracellular space. Electric current associated with the transfer of ions generates electric field and magnetic field around the activating cell. Electroencephalography (EEG) and electromyography (EMG) are techniques to record electric potentials at electrodes attached to the surface of the head and muscles, respectively. The electric potentials obtained from these techniques reflect activities of a group of neurons and muscle cells unlike in the case of single cell recording. Berger reported the first EEG measurement of the human brain in 1929. Researchers recognized his result after Adrian performed a precise EEG measurement to validate his result in 1935. Electrodes for EEG recordings are usually arrayed at 21 points on the scalp according to the International 10-20 Electrode System. For some research purposes, more than 60 electrodes are used for detailed mapping of potentials, whereas EEG recordings with a few electrodes meet some monitoring purposes. Neuronal activities can be observed using magnetic sensors as well as surface electrodes.

Magnetoencephalography (MEG) is a method for measuring magnetic fields as weak as 5×10^{-15} T (i.e., 5 fT by superconducting quantum interference devices (SQUIDs) arrayed on the scalp). Cohen succeeded in recording MEG corresponding to alpha-waves from the human brain in 1972 (21). The main advantages of EEG and MEG over other imaging modalities are their complete noninvasiveness and their high temporal resolution of milliseconds.

The relation between neuronal electric currents and the resulting electric potentials or magnetic fields on the surface of the head is one of the most important topics in EEG and MEG. Estimation of the surface potentials or magnetic fields from a given neuronal current source is called the forward problem, and estimation of the neuronal current source from measured potential or magnetic fields is called the inverse problem. The inverse problem is an ill-posed problem whose solution is underspecified. In order to specify a current source from the measured potentials or magnetic fields, a model should be employed to restrict the solution according to *a priori* knowledge, or a mathematical constraint should be imposed in the inverse problem. Methods for solving the inverse problems are classified by their models or mathematical constraints. Signals of EEG or MEG occur as a result of a spatiotemporal coherence of electrical activities of a group of neurons in a part of the brain. From an electromagnetic point of view, the source of electrical activities of neurons can be approximated by a current dipole. In the case of MEG, magnetic field $B(i)$ generated from a current dipole and recorded by a sensor i is given by the following equation based on the Biot–Savart's law:

$$\begin{aligned} B(i) &= \frac{\mu_0}{4\pi} \frac{\mathbf{Q} \times (\mathbf{R}(i) - \mathbf{L})}{|\mathbf{R}(i) - \mathbf{L}|^3} \cdot \mathbf{n}(i) \\ &= \frac{\mu_0}{4\pi} \frac{(\mathbf{R}(i) - \mathbf{L}) \times \mathbf{n}(i)}{|\mathbf{R}(i) - \mathbf{L}|^3} \cdot \mathbf{Q} \\ &= \frac{\mu_0}{4\pi} \mathbf{g}(i) \cdot \mathbf{Q}, \end{aligned} \quad (1)$$

where $\mathbf{L}$ is the position vector of the current dipole, $\mathbf{Q}$ is the current dipole moment, $\mathbf{R}(i)$ is the position vector of the sensor i, $\mathbf{n}(i)$ is the direction of the sensor i, and μ_0 is the magnetic permeability of free space. For multiple sensors, Equation 1 is extended to

$$\mathbf{B} = \begin{bmatrix} B(1) \\ \vdots \\ B(m) \end{bmatrix} = \frac{\mu_0}{4\pi} \begin{bmatrix} \mathbf{g}(1) \\ \vdots \\ \mathbf{g}(m) \end{bmatrix} \cdot \mathbf{Q} = \mathbf{G}(\mathbf{L}) \cdot \mathbf{Q}, \quad (2)$$

where m is the number of sensors and $\mathbf{g}$ is the gain matrix giving the relation between the current dipole and the measured magnetic fields. An extension of Equation 2 for the case of multiple current dipoles leads to

$$\mathbf{B} = \begin{bmatrix} \mathbf{G}(\mathbf{L}_1) & \cdots & \mathbf{G}(\mathbf{L}_p) \end{bmatrix} \begin{bmatrix} \mathbf{Q}_1 \\ \vdots \\ \mathbf{Q}_p \end{bmatrix}, \quad (3)$$

where p is the number of current dipoles. Equations 2 and 3 have nonlinearity in terms of the relation between the magnetic field and the position of current dipole. The position and moment of current dipole can be estimated from the measured magnetic field using nonlinear fitting algorithms such as the Marquardt algorithm (22). In this approach, the algorithm minimizes the error of mean square developing between the measured magnetic field and the magnetic field given by the above forward model. The fitting approach is mathematically defined as

$$\sum_{i=1}^{m} (B(i)_{\text{cal}} - B(i)_{\text{meas}})^2 \to \min, \quad (4)$$

where $B(i)_{\text{cal}}$ is the magnetic field calculated from the forward model and $B(i)_{\text{meas}}$ is the measured magnetic field. In another approach for solving the inverse problem, multiple current dipoles are arrayed on the surface of the brain (or inside the brain), and their dipole moments are estimated under appropriate constraints. In this framework, an infinite pattern of dipole moments exists that explains the measured magnetic field because the number of equations in this problem is much smaller than the number of unknown parameters. This kind of problem requires some performance function in order to choose one pattern of dipole moments (or one solution to this problem), which is the most reasonable in electromagnetic and physiological point of view. Divergence of solution, in which extraordinarily large dipole moments with opposite directions juxtapose to each other, gives rise to a physiologically unreasonable solution. The minimum norm method is widely used for avoiding the divergence (23). In this method, minimization of the norm of the dipole moments results in one appropriate solution to the inverse problem.

We measured the event-related magnetic field P300 m with MEG and the event-related potential P300 with electroencephalography (EEG) (24). Multiple equivalent current dipoles were estimated from the P300 and P300 m waveforms obtained from visual and somatosensory oddball paradigm tasks. Estimated sources from P300 m were located on both sides of the occipito-temporal gyrus for visual stimuli and on the post central gyrus for somatosensory stimuli. The sources of P300 m were modality specific. The equivalent current dipoles of P300 were located on the cingulate gyrus and the thalamus in addition to the locations estimated from P300 m. However, the dipoles of P300 m in MEG were not located on the cingulate gyrus and the thalams. The discrepancy between EEG and MEG was because of the difficulty to measure radially oriented dipoles in MEG.

4. MAGNETIC RESONANCE IMAGING

Magnetic resonance imaging (MRI) is a method to obtain spatial distribution of nuclear magnetic resonance (NMR) signals using gradient magnetic fields and Fourier transform. The basic principle of MRI was proposed by Lauterbur in 1973 (25). A MRI system consists of a magnet, gradient coils, and radio-frequency (RF) coils. The magnet applies a strong static magnetic field B_0 to nuclear spins, which causes a gap in energy levels of the spin states. The frequency of NMR signal, ω_0, is proportional to the intensity of magnetic field. This relation is given by

$$\omega_0 = \gamma B_0, \quad (5)$$

where γ is the gyromagnetic ratio (2.7×10^8 rad/s/T for ^{1}H nucleus). The energy of a photon at the NMR frequency, $\varepsilon = \hbar\omega_0$, equals to the gap in the energy levels of the spin states. Presently used magnets for clinical MRI systems are either superco6nducting magnets or permanent magnets, producing magnetic fields between 0.2 T and 3.0 T. The gradient coils produce linear gradient magnetic fields in x, y, and z directions. The gradient magnetic fields enable image reconstruction based on slice selection, frequency encode, and phase encode. The RF coils produce electromagnetic fields B_1 at the NMR frequency and receive NMR signals from the nuclear spins. The principles of MRI are described in detail in Ref. 26.MRI visualizes a variety of tissues properties such as brain activities, blood flow, metabolism, diffusion of water molecules, neuronal fiber tracts, temperature, and elasticity in addition to anatomy. Figure 3 shows magnetic resonance images of the human brain and the rat brain, magnetic resonance angiogram, and magnetic resonance spectra of a skeletal muscle. Magnetic resonance spectroscopy is useful for investigating metabolism in tissues. Visualization of electric and magnetic phenomena in living bodies has recently been realized as discussed below.

Magnetic resonance imaging of electrical phenomena in living bodies is useful for quantitative evaluations of biological effects of electromagnetic fields. Magnetic field in an object causes a shift in the resonant frequency (44,45) and a change in the phase of magnetic resonance signals (46). Stationary electric current causes an increase in the apparent diffusion coefficient (47). Spatial distributions of externally applied magnetic field and electrical current can be estimated from these changes in magnetic resonance signals. These methods have potential medical applications such as imaging of current distributions in electrical defibrillation (48). Detection of weak magnetic fields induced by neuronal or muscular electrical activities using MRI is a potentially effective method for functional imaging of the brain. However, the detection of these weak magnetic fields requires extremely high sensitivity. A numerical analysis of the theoretical limit of sensitivity for detecting weak magnetic fields generated in the human brain was performed (27). The theoretical limit of sensitivity was approximately 10^{-8} T. The effect of neuronal electrical activities on magnetic resonance signals was investigated in several experimental studies (28,29). These studies will lead to a new method for visualizing brain function with a spatial resolution of millimeters and a temporal resolution of milliseconds.

Estimation of impedance distributions of biological tissues is essential for various analyses in biomedical engineering, such as obtaining current distributions in electric stimulation and magnetic stimulation, calculating the absorption of electromagnetic waves from mobile phones, and current source estimations in EEG and MEG.

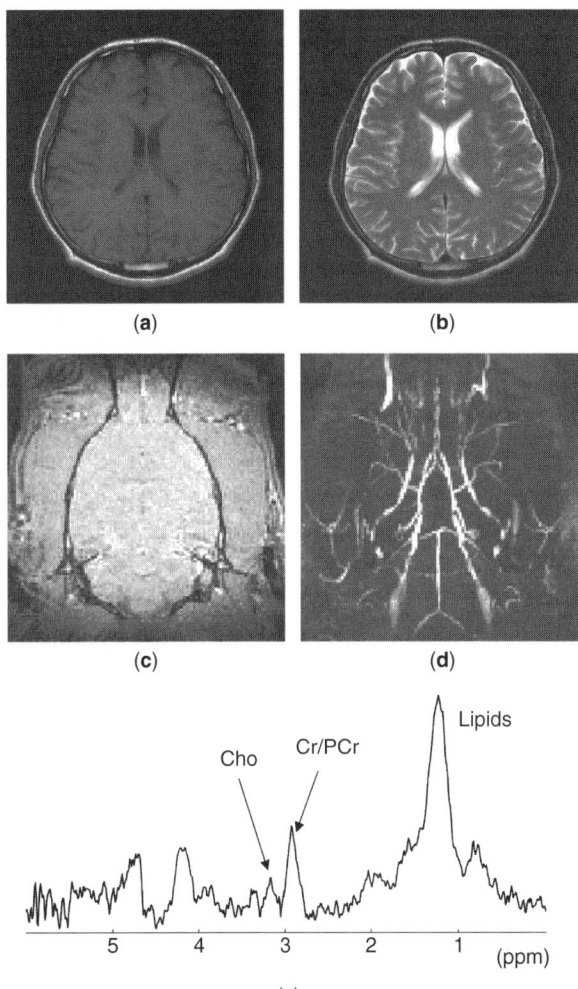

Figure 3. (a) T_1-weighted image of the human brain. (b) T_2-weighted image of the human brain. (c) T_1-weighted image of the rat brain. (d) Magnetic resonance angiogram of the rat cerebral arteries. (e) ^{1}H magnetic resonance spectrum of the rat gastrocnemius.

Impedance-weighted magnetic resonance images were obtained during applications of external oscillating magnetic fields, which induce impedance-dependent eddy currents in a sample (30). In another study, spatial distribution of electrical impedance was obtained from the electrical current distributions using an iterative algorithm (49). The ADC reflects electrical conductivity of a tissue, which enables an estimation of anisotropic conductivity of the tissue (31–33). Conductivity distributions were obtained in the rat brain and the human brain using MRI. The estimation of conductivity was based on the proportionality between the self-diffusion coefficient of water and conductivity. Figure 4 shows conductivity maps of the human brain. The signals in the corpus callocum exhibited high anisotropy because of alignment of neuronal fibers. Regions with high conductivity anisotropy were found in the white matter.

A distinctive signal inhomogeneity occurs in images of objects with high dielectric constant. This phenomenon, dielectric resonance, particularly appears in scanners

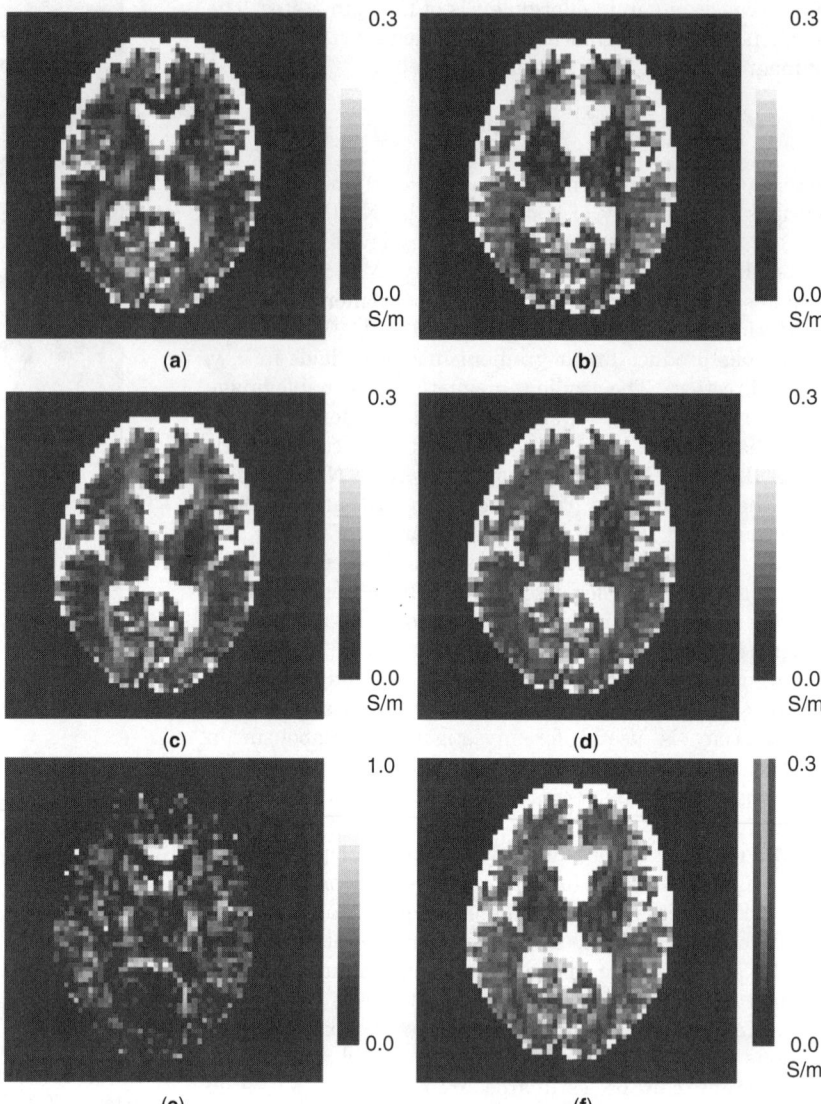

Figure 4. (a)–(c) Images of electrical conductivities in the human brain in the superior-inferior, right-left, and anterior-posterior directions. (d) and (e) Images of the mean conductivity (MC) and the anisotropy index (AI). (f) Color map of conductivity. The intensities of red, green, and blue are proportional to the conductivities in the superior-inferior, right-left, and anterior-posterior directions, respectively. (This figure is available in full color at http://www.mrw.interscience.wiley.com/ebe.)

with high static fields. Spatial distributions of magnitude and phase of magnetic resonance signals in cylindrical objects were obtained by theoretical calculations and experiments (50). As diameter of the object approaches to the wavelength of electromagnetic fields in the object, the center of the object exhibited high magnitude of signals.

5. CANCER THERAPY

Magnetic force acting on magnetic materials moves the materials along magnetic field gradients. A new method to destruct targeted cells was developed using magnetizable beads and pulsed magnetic force (16,17). TCC-S leukemic cells (51) were combined with magnetizable beads. After combination, the cell/bead/antibody complexes were placed on a magnet for enough aggregation. The aggregated beads were then stimulated 10 times at 5 sec intervals by a circular-shaped coil, which produced a magnetic field of 2.4 T at the center of the coil. After the stimulation, the viability of the aggregated and stimulated cell/bead/antibody complexes was significantly decreased, and the cells were destructed by the penetration of the beads into the cells or rupturing of the cells by the beads, as shown in Fig. 5. In principle, the magnetic force acting on any particular material is proportional to the magnetic field, magnetic field gradient, and the magnetic susceptibility of the material; however, the magnetic force acting on the individual nanoscale magnetic particles inside the beads are too weak to affect them, because the magnetic susceptibility of each individual particle is very low. In contrast, when the nanoscale particles inside the beads are closely assembled, the magnetic mass susceptibility is sufficiently high to force the attachment of the beads to the cells by the magnetic force. Thus, the magnetic force acting on the aggregated beads was strong enough to shift the beads and damage the cells.

Repetitive magnetic stimulation has antitumor effects by activating immune functions (19,20). Magnetic stimulations

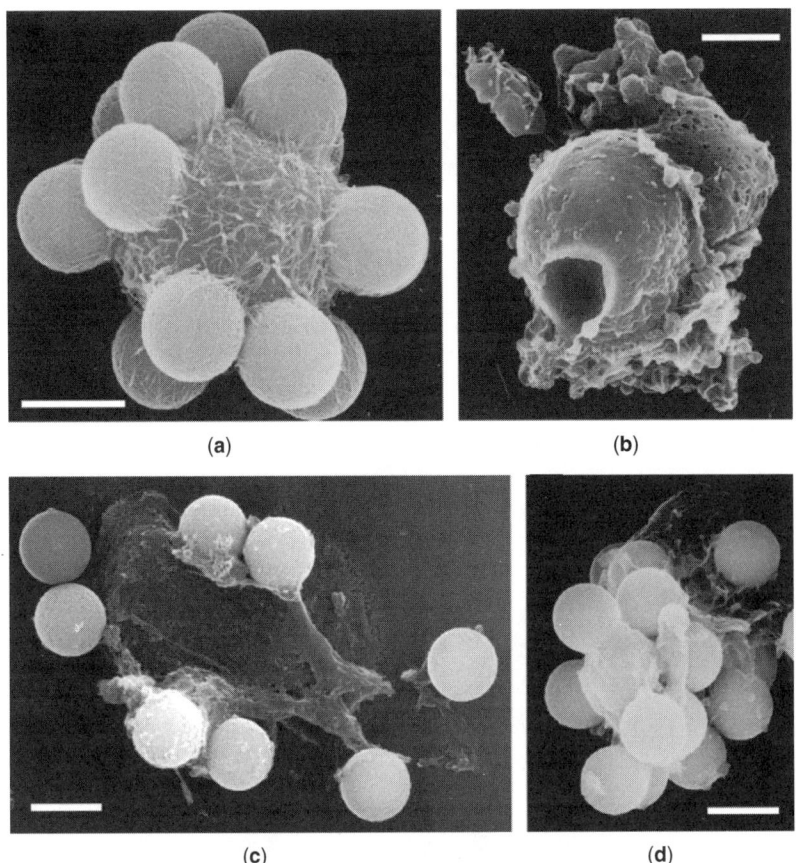

Figure 5. Scanning electron micrographs of the cell/bead/antibody complexes with and without pulsed magnetic stimulation. (a) The nonstimulated cell/bead/antibody complex was not damaged. (b) The stimulated cell/bead/antibody complexes were damaged by penetration of the beads. (c) and (d) The stimulated cell/bead/antibody complexes were damaged by rupturing by the beads. Scale bars = 4.5 µm.

were applied from a circular coil with a peak intensity of 0.25 T and a frequency of 25 pulses/sec. Tumor growth study showed a significant tumor weight decrease and survival probability increase because of the application of magnetic stimulation (Fig. 6a and 6b). An immunological assay was also performed to examine the effects of the magnetic stimulation on immune functions. *In vivo* study, the productions of TNF-α (tumor necrosis factor-α), which plays a tumor-suppression role in tumor immunity mainly by TNFR1 – TRADD - FADD – Caspase-8 - Caspase-3 apoptosis pathways (52,53), and IL-2 (interleukin-2), which is produced by T cells, activates the proliferation and functional development of T cells (helper T cells and cytolytic T cells) and B cells (54,55), were measured in the spleen after exposure of the magnetic stimulation 3 or 7 times (Fig. 6c and 6d). TNF-α production significantly increased in the stimulated group. *In vitro* study, isolated spleen cells (lymphocytes) were exposed to the magnetic stimulation and a proliferation assay was performed. The proliferation activity of the lymphocytes was upregulated in the exposed samples. These results indicate that the immune functions might be activated by repetitive magnetic stimulation exposure, resulting in a tumor weight decrease.

6. MAGNETIC ORIENTATION

When diamagnetic materials such as fibrin, collagen are exposed to static magnetic fields of T (tesla) order, these materials align either parallel or perpendicular to the direction of the magnetic field depending on the magnetic anisotropy of the materials (34–37). Such materials have variations in magnetic energy with their angles from the direction of externally applied magnetic field. The magnetic orientation appears when the variation of the magnetic energy exceeds the energy of thermal motion. The condition for the magnetic orientation of fibrous polymers such as fibrin is given by

$$NB^2 \Delta\chi \sin 2\theta > kT, \qquad (6)$$

where N is the number of polymer molecules, B is the intensity of magnetic field, $\Delta\chi$ is the anisotropy of magnetic susceptibility, θ is the angle between the molecular axis and the magnetic field, k is the Boltzmann constant, and T is temperature. Biological cells such as osteoblasts and smooth muscle cells are oriented parallel to the direction of the magnetic field because of the magnetic torque acting on the diamagnetic components in cells (37–40). Figure 7 shows magnetically oriented biological macromolecules and cells. Fibrin fibers are oriented parallel to the magnetic field, and collagen are oriented perpendicular to the magnetic field (34–37). Osteoblasts, smooth muscle cells, Schwann cells, red blood cells, and platelets are also oriented in magnetic fields (37–42). In a recent experiment, Schwann cells with magnetically oriented collagen were oriented along the magnetically oriented collagen (41). The magnetic control of biological cells may translate into

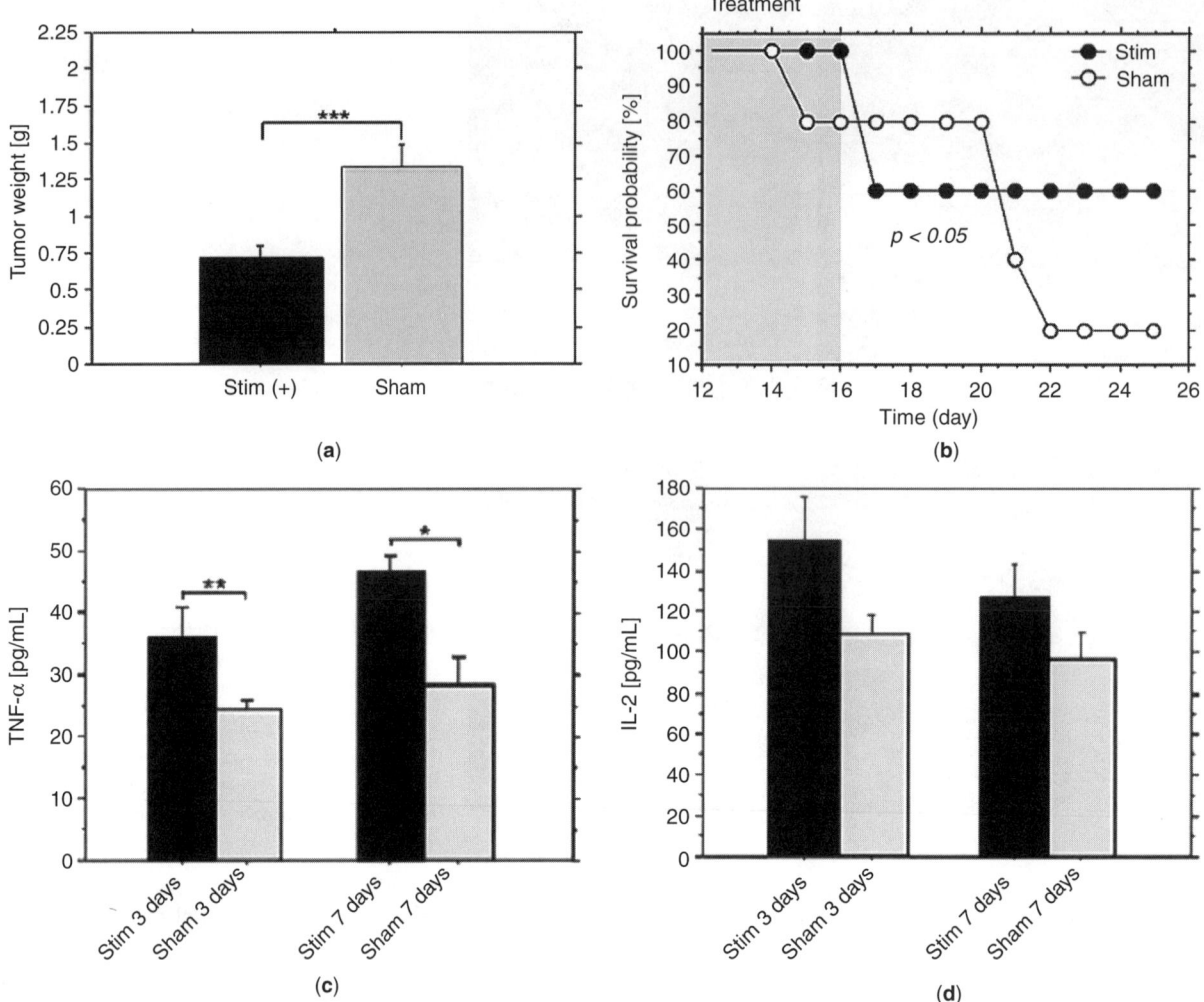

Figure 6. The effects of repetitive magnetic stimulation on tumors and immune functions. (a) The tumor weights were decreased because of the application of magnetic stimulation. (b) The survival probability of mice was increased because of the application of magnetic stimulation. (c) TNF-α production significantly increased in the stimulated group. (d) IL-2 production increased in the stimulated group; however, no significant differences were observed.

potentially viable tissue and medical engineering applications including nerve regeneration.

The introduction of bone formation to an intentional orientation is a potentially viable clinical treatment for bone disorders. The effects of static magnetic fields of 8 T on bone formation were investigated in both *in vivo* and *in vitro* systems (36,37). After 60 h of exposure to the magnetic field, cultured mouse osteoblastic MC3T3-E1 cells were transformed to rod-like shapes and were oriented in the direction parallel to the magnetic field. Although the magnetic field exposure did not affect cell proliferation, it upregulated cell differentiation and matrix synthesis as determined by alkaline phosphatase and Alizarin red stainings, respectively. The magnetic fields also stimulated ectopic bone formation in and around subcutaneously implanted bone morphogenetic protein-2 (BMP-2) containing pellets in mice, in which the orientation of bone formation was parallel to the magnetic field. Strong magnetic fields have the potency to stimulate bone formation as well as to regulate its orientation in both *in vitro* and *in vivo* models. The authors propose that the combination of strong magnetic fields and a potent osteogenic agent such as BMP may possibly lead to an effective treatment of bone fractures and defects.

Schwann cells aid in neuronal regeneration in the peripheral nervous system by guiding the regrowth of axons. Schwann cells provide a supportive role in the peripheral nervous system, forming a layer or myelin sheath along single segments of an axon. Schwann cells were exposed to 8T magnetic fields, or the cells were cultivated in magnetically oriented collagen gels (41). After 60 h of exposure, Schwann cells oriented parallel to the magnetic fields. In contrast, the Schwann cells and collagen mixture aligned along the magnetically oriented collagen fibers perpendicular to the magnetic field after 2 h of exposure. Although further experiments are necessary to determine whether magnetic field exposure promotes Schwann cell proliferation and to clarify the mechanisms

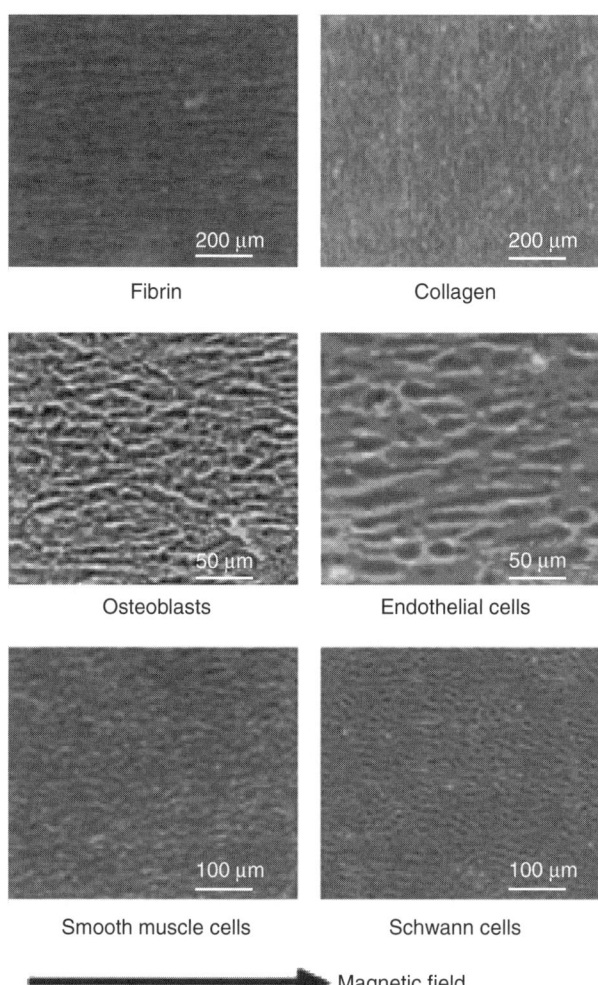

Figure 7. Magnetic orientation of biological macromolecules and cells. Fibrin, osteoblasts, endothelial cells, smooth muscle cells, and Schwann cells were oriented in the direction parallel to the magnetic field. Collagen was oriented in the direction perpendicular to the magnetic field.

for the apparent orientation, the magnetic control of Schwann cell alignment is potentially useful for nerve regeneration techniques in tissue engineering and regenerative medicine.

7. MAGNETORECEPTION IN ANIMALS

Orientation, navigation, and homing by animals ranging from bacteria through higher vertebrates are likely to depend on events occurring in the individual cells that detect magnetic fields (56–59). Magnetoreception was observed in sea turtles (60), newts (61), fish (57), birds (62,63), insects (64,65) and so on. All magnetic field sensitivity in living organisms is thought to be a result of a highly evolved, finely-tuned sensory system based on single-domain, ferromagnetic crystals. Edmonds showed that a very sensitive magnetic compass is formed by the incorporation of a small quantity of ferromagnetic single-domain crystals, such as magnetite within a nematic liquid crystal (66). Winklhofer et al. localized high concentration of Fe^{3+} in the upper-beak skin of homing pigeons, and identified the materials of magnetite nanocrystals as a core of a magnetic-field receptor (63). In rainbow trout, the magnetic crystal in the area of olfactory lamellae was found and the arrangement of several magnetic crystals in a chain of about 1 µm has been confirmed (57). Although no experimental evidence exists whether animals have separate magnetite-based magnetoreceptor cells that are specialized for magnetic field direction and intensity, it can be predicted that animals should be able to reconstruct the total field vector accurately and with high sensitivity (59).

BIBLIOGRAPHY

1. S. Ueno, Biomagnetic approaches to studying the brain. *IEEE Eng. Med. Biol.* 1999; **18**:108–120.

2. A. T. Barker, R. Jalinous, and I. L. Freeston, Noninvasive magnetic stimulation of human motor cortex. *Lancet* 1985; **2**:1106–1107.

3. S. Ueno, T. Matsuda, and M. Fujiki, Functional mapping of the human motor cortex obtained by focal and vectorial magnetic stimulation of the brain. *IEEE Trans. Magn.* 1990; **26**:1539–1544.

4. C. M. Epstein, M. Sekino, K. Yamaguchi, S. Kamiya, and S. Ueno, Asymmetries of prefrontal cortex in human episodic memory: effects of transcranial magnetic stimulation on learning abstract patterns. *Neurosci. Lett.* 2002; **320**:5.

5. T. Krings, B. R. Buchbinder, W. E. Butler, K. H. Chiappa, H. J. Jiang, B. R. Rosen, and G. R. Cosgrove, Stereotactic transcranial magnetic stimulation: correlation with direct electrical cortical stimulation. *Neurosurgery* 1997; **41**:1319–1325.

6. T. Krings, H. Foltys, M. H. Reinges, S. Kemeny, V. Rohde, U. Spetzger, J. M. Gilsbach, and A. Thron, Navigated transcranial magnetic stimulation for presurgical planning—correlation with functional MRI. *Minim. Invasive Neurosurg.* 2001; **44**: 234–239.

7. M. Sekino and S. Ueno, Comparison of current distributions in electroconvulsive therapy and transcranial magnetic stimulation. *J. Appl. Phys.* 2002; **91**:8730–8732.

8. M. Sekino and S. Ueno, FEM-based determination of optimum current distribution in transcranial magnetic stimulation as an alternative to electroconvulsive therapy. *IEEE Trans. Magn.* 2004; **40**:2167–2169.

9. H. M. Haraldsson, F. Ferrarelli, N. H. Kalin, and G. Tononi, Transcranial magnetic stimulation in the investigation and treatment of schizophrenia: a review. *Schizophr. Res.* 2004; **71**:1–16.

10. A. Pascual-Leone, J. Valls-Sole, J. P. Brasil-Neto, A. Cammarota, J. Grafman, and M. Hallett, Akinesia in Parkinson's disease. II. Effects of subthreshold repetitive transcranial motor cortex stimulation. *Neurology* 1994; **44**:735–741.

11. T. A. Kimbrell, J. T. Little, R. T. Dunn, M. A. Frye, B. D. Greenberg, E. M. Wassermann, J. D. Repella, A. L. Danielson, M. W. Willis, B. E. Benson, A. M. Speer, E. Osuch, M. S. George, and R. M. Post, Frequency dependence of antidepressant response to left prefrontal repetitive transcranial magnetic stimulation (rTMS) as a function of baseline cerebral glucose metabolism. *Biol. Psychiat.* 1999; **46**: 1603–1613.

12. M. Ogiue-Ikeda, S. Kawato, and S. Ueno, The Effect of transcranial magnetic stimulation on long-term potentiation in rat hippocampus. *IEEE Trans. Magn.* 2003; **39**:3390–3392.
13. M. Ogiue-Ikeda, S. Kawato, and S. Ueno, The effect of repetitive transcranial magnetic stimulation on long-term potentiation in rat hippocampus depends on stimulus intensity. *Brain Res.* 2003; **993**:222–226.
14. M. Ogiue-Ikeda, S. Kawato, and S. Ueno, Acquisition of ischemic tolerance by repetitive transcranial magnetic stimulation in the rat hippocampus. *Brain Res.* 2005; **1037**: 7–11.
15. H. Funamizu, M. Ogiue-Ikeda, H. Mukai, S. Kawato, and S. Ueno, Acute repetitive transcranial magnetic stimulation reactivates dopaminergic system in lesion rats. *Neurosci. Lett.* 2005; **383**:77–81.
16. M. Ogiue-Ikeda, Y. Sato, and S. Ueno, A new method to destruct targeted cells using magnetizable beads and pulsed magnetic force. *IEEE Trans. Nanobiosci.* 2003; **2**:262–265.
17. M. Ogiue-Ikeda, Y. Sato, and S. Ueno, Destruction of targeted cancer cells using magnetizable beads and pulsed magnetic force. *IEEE Trans. Magn.* 2004; **40**:3018–3020.
18. S. Yamaguchi, M. Ogiue-Ikeda, M. Sekino, and S. Ueno, The effect of repetitive magnetic stimulation on the tumor generation and growth. *IEEE Trans. Magn.* 2004; **40**:3021–3023.
19. S. Yamaguchi, M. Ogiue-Ikeda, M. Sekino, and S. Ueno, Effect of magnetic stimulation on tumor and immune functions. *IEEE Trans. Magn.*, in press.
20. S. Yamaguchi, M. Ogiue-Ikeda, M. Sekino, and S. Ueno, Effects of pulsed magnetic stimulation on tumor development and immune functions in mice. *Bioelectromagnetics*, in press.
21. D. Cohen, Magnetoencephalography: detection of the brain's electrical activity with a superconducting magnetometer. *Science* 1972; **175**:664–666.
22. D. W. Marquardt, An algorithm for least-squares estimation of non-linear parameters. *J. Soc. Indust. Appl. Math.* 1963; **11**:431–441.
23. M. S. Hamalainen and R. J. Ilmoniemi, Interpreting magnetic fields of the brain: minimum-norm estimates. *Med. Biol. Eng. Comput.* 1994; **32**:35–42.
24. T. Maeno, A. Kaneko, K. Iramina, F. Eto, and S. Ueno, Source modeling of the P300 event-related response using magnetoencephalography and electroencephalography measurements. *IEEE Trans. Magn.* 2003; **39**:3396–3398.
25. P. C. Lauterbur, Image formation by induced local interactions: examples employing nuclear magnetic resonance. *Nature* 1973; **242**:190–191.
26. P. T. Callaghan, *Principles of Nuclear Magnetic Resonance Microscopy.* Oxford, UK: Oxford University Press, 1993.
27. T. Hatada, M. Sekino, and S. Ueno, FEM-based calculation of the theoretical limit of sensitivity for detecting weak magnetic fields in the human brain using magnetic resonance imaging. *J. Appl. Phys.* 2005; **97**:10E109.
28. H. Kamei, K. Iramina, K. Yoshikawa, and S. Ueno, Neuronal current distribution imaging using magnetic resonance. *IEEE Trans. Magn.* 1999; **35**:4109–4111.
29. J. Xiong, P. T. Fox, and J. H. Gao, Directly mapping magnetic field effects of neuronal activity by magnetic resonance imaging. *Hum. Brain Map.* 2003; **20**:41–49, 2003.
30. S. Ueno and N. Iriguchi, Impedance magnetic resonance imaging: a method for imaging of impedance distributions based on magnetic resonance imaging. *J. Appl. Phys.* 1998; **83**:6450–6452.
31. M. Sekino, K. Yamaguchi, N. Iriguchi, and S. Ueno, Conductivity tensor imaging of the brain using diffusion-weighted magnetic resonance imaging. *J. Appl. Phys.* 2003; **93**:6730–6732.
32. M. Sekino, Y. Inoue, and S. Ueno, Magnetic resonance imaging of mean values and anisotropy of electrical conductivity in the human brain. *Neurol. Clin. Neurophysiol.* 2004; **55**:1–5.
33. D. S. Tuch, V. J. Wedeen, A. M. Dale, J. S. George, and J. W. Belliveau, Conductivity tensor mapping of the human brain using diffusion tensor MRI. *Proc. Natl. Acad. Sci. USA* 2001; **98**:11697–11701.
34. J. Torbet, J. M. Freyssinet, and G. Hudry-Clergeon, Oriented fibrin gels formed by polymerization in strong magnetic fields. *Nature* 1981; **289**:91–93.
35. M. Iwasaka, M. Takeuchi, S. Ueno, and H. Tsuda, Polymerization and dissolution of fibrin under homogeneous magnetic fields. *J. Appl. Phys.* 1998; **83**:6453–6455.
36. H. Kotani, M. Iwasaka, S. Ueno, and A. Curtis, Magnetic orientation of collagen and bone mixture. *J. Appl. Phys.* 2000; **87**:6191–6193.
37. H. Kotani, H. Kawaguchi, T. Shimoaka, M. Iwasaka, S. Ueno, H. Ozawa, K. Nakamura, and K. Hoshi, Strong static magnetic field stimulates bone formation to a definite orientation in vivo and in vitro. *J. Bone Miner. Res.* 2002; **17**:1814–1821.
38. M. Iwasaka and S. Ueno, Optical absorbance of hemoglobin and red blood cell suspensions under magnetic fields. *IEEE Trans. Magn.* 2001; **37**:2906–2908.
39. A. Umeno and S. Ueno, Quantitative analysis of adherent cell orientation influenced by strong magnetic fields. *IEEE Trans. Nanobiosci.* 2003; **2**:26–28.
40. M. Ogiue-Ikeda and S. Ueno, Magnetic cell orientation depending on cell type and cell density. *IEEE Trans. Magn.* 2004; **40**:3024–3026.
41. Y. Eguchi, M. Ogiue-Ikeda, and S. Ueno, Control of orientation of rat Schwann cells using an 8-T static magnetic field. *Neurosci. Lett.* 2003; **351**:130–132.
42. T. Higashi, A. Yamagishi, T. Takeuchi, N. Kawaguchi, S. Sagawa, S. Onishi, and M. Date, Orientation of erythrocytes in a strong static magnetic field. *Blood* 1993; **82**:1328–1334.
43. M. Ogiue-Ikeda, H. Kotani, M. Iwasaka, Y. Sato, and S. Ueno, Inhibition of leukemia cell growth under magnetic fields of up to 8T. *IEEE Trans. Magn.* 2001; **37**:2912–2914.
44. M. Sekino, T. Matsumoto, K. Yamaguchi, N. Iriguchi, and S. Ueno, A method for NMR imaging of a magnetic field generated by electric current. *IEEE Trans. Magn.* 2004; **40**:2188–2190.
45. Y. Manassen, E. Shalev, and G. Navon, Mapping of electrical circuits using chemical-shift imaging. *J. Magn. Reson.* 1988; **76**:371–374.
46. M. Joy, G. Scott, and M. Henkelman, In vivo detection of applied electric currents by magnetic resonance imaging. *Magn. Reson. Imaging* 1989; **7**:89–94.
47. K. Yamaguchi, M. Sekino, N. Iriguchi, and S. Ueno, Current distribution image of the rat brain using diffusion weighted magnetic resonance imaging. *J. Appl. Phys.* 2003; **93**:6739–6741.
48. R. S. Yoon, T. P. DeMonte, K. F. Hasanov, D. B. Jorgenson, and M. L. G. Joy, Measurement of thoracic current flow in pigs for

the study of defibrillation and cardioversion. *IEEE Trans. Biomed. Eng.* 2003; **50**:1167–1173.

49. H. S. Khang, B. I. Lee, S. H. Oh, E. J. Woo, S. Y. Lee, M. H. Cho, O. Kwon, J. R. Yoon, and J. K. Seo, J-substitution algorithm in magnetic resonance electrical impedance tomography (MREIT): phantom experiments for static resistivity images. *IEEE Trans. Med. Imaging* 2002; **21**:695–702.

50. M. Sekino, H. Mihara, N. Iriguchi, and S. Ueno, Dielectric resonance in magnetic resonance imaging: signal inhomogeneities in samples of high permittivity. *J. Appl. Phys.* 2005; **97**:10R303.

51. Y. Kano, M. Akutsu, S. Tsunoda, H. Mano, Y. Sato, Y. Honma, and Y. Furukawa, In vitro cytotoxic effects of a tyrosine kinase inhibitor STI571 in combination with commonly used antileukemic agents. *Blood* 2001; **97**:1999–2007.

52. B. B. Aggarwal, Signalling pathways of the TNF superfamily: a double-edged sword. *Nat. Rev. Immunol.* 2003; **3**:745–756.

53. A. Ashkenazi, Targeting death and decoy receptors of the tumour-necrosis factor superfamily. *Nat. Rev. Cancer* 2002; **2**:420–430.

54. B. H. Nelson, IL-2, regulatory T cells, and tolerance. *J. Immunol.* 2004; **172**:3983–3988.

55. K. A. Smith, Interleukin-2: inception, impact, and implications. *Science* 1988; **240**:1169–1176.

56. M. M. Walker, C. E. Diebel, C. V. Haugh, P. M. Pankhurst, J. C. Montgomery, and C. R. Green, Structure and function of the vertebrate magnetic sense. *Nature* 1997; **390**: 319–426.

57. C. E. Diebel, R. Proksch, C. R. Green, P. Neilson, and M. M. Walker, Magnetite defines a vertebrate magnetoreceptor. *Nature* 2000; **406**:299–302.

58. J. L. Kirschvink, M. M. Walker, and C. E. Diebel, Magnetite-based magnetoreception. *Curr. Opin. Neurobiol.* 2001; **11**:462–467.

59. M. M. Walker, T. E. Dennis, and J. L. Kirschvink, The magnetic sense and its use in long-distance navigation by animals. *Curr. Opin. Neurobiol.* 2002; **12**:735–744.

60. K. J. Lohmann, S. D. Cain, S. A. Dodge, and C. M. Lohmann, Regional magnetic fields as navigational markers for sea turtles. *Science* 2001; **294**:364–366.

61. J. B. Phillips, M. E. Deutschlander, M. J. Freake, and S. C. Borland, The role of extraocular photoreceptors in newt magnetic compass orientation: parallels between light-dependent magnetoreception and polarized light detection in vertebrates. *J. Exp. Biol.* 2001; **204**:2543–2552.

62. W. Wiltschko, U. Munro, R. Wiltschko, and J. L. Kirschvink, Magnetite-based magnetoreception in birds: the effect of a biasing field and a pulse on migratory behavior. *J. Exp. Biol.* 2002; **205**:3031–3037.

63. G. Fleissner, E. Holtkamp-Rotzler, M. Hanzlik, M. Winklhofer, N. Petersen, and W. Wiltschko, Ultrastructural analysis of a putative magnetoreceptor in the beak of homing pigeons. *J. Comp. Neurol.* 2003; **458**:350–360.

64. D. Strickman, B. Timberlake, J. Estrada-Franco, M. Weissman, P. W. Fenimore, and R. J. Novak, Effects of magnetic fields on mosquitoes. *J. Am. Mosq. Control Assoc.* 2000; **16**:131–137.

65. T. J. Slowik, B. L. Green, and H. G. Thorvilson, Detection of magnetism in the red imported fire ant (Solenopsis invicta) using magnetic resonance imaging. *Bioelectromagnetics* 1997; **18**:396–399.

66. D. T. Edmonds, A sensitive optically detected magnetic compass for animals. *Proc. Biol. Sci.* 1996; **263**:295–298.

BIOENERGETICS AND SYSTEMIC RESPONSES TO EXERCISE

BART VERHEYDEN
FRANK BECKERS
ANDRÉ E. AUBERT
University Hospital
Gasthuisberg
Leuven, Belgium

1. BIOENERGETICS AND SYSTEMIC RESPONSES TO EXERCISE

1.1. Introduction

Physiology is a branch of biological science concerned with the function of organisms and their systems (1). Exercise physiology is a discipline that traditionally has focused on the study of how exercise alters the structure and function of the human body. This research field encloses a multidisciplinary area including the study of aging, cell biology, developmental biology, epidemiology, genomics, immunology, molecular biology (biochemistry, genetics, and biophysics), clinical medicine, neurosciences, pediatric pharmacology, physiology, preventive medicine, and public health (2). In the last 50 years, tremendous achievements have been made in the field of exercise physiology because of the application of new or improved research techniques (implying bioengineering aspects) or the application of techniques on new conditions (3). Because exercise is a whole body commitment, the entire body must respond and adapt. However, it is impossible to discuss the responses and adaptations related to all areas that are involved in the study of exercise physiology in one chapter. It is not the intention of this section to give a complete and detailed description of every mechanism underlying physiological changes caused by exercise. The goal is to cover some metabolic considerations of exercise and to elaborate the most important and largest cardiovascular, respiratory, and muscular responses and adaptations to sports exercise.

1.2. Exercise Metabolism

Metabolism can be defined as the total sum of all chemical reactions (catabolic and anabolic) occurring in a living organism (1). During exercise, the metabolism gradually increases and is associated with a global increment in energy expenditure. For example, during heavy exercise, the body's total energy expenditure may increase fifteen to twenty-five times above expenditure at rest. The body uses carbohydrate and fat nutrients consumed daily as the fuel to provide the necessary energy to maintain cellular activities both at rest and during exercise. During exercise, the primary nutrients used for energy are fats and carbohydrates, with protein contributing for a small amount of the total energy (1,4–6). Regulation of fuel selection during exercise is under complex control and is dependent on several factors, including diet and the intensity and duration of exercise. In general, carbohydrates are used as the major fuel source during high-intensity exercise. During prolonged exercise, a general shift occurs from carbohydrate metabolism toward fat metabolism. Proteins contribute less than 2% of the fuel used during exercise of less than one hour in duration. During prolonged exercise (three to five hours in duration), the total contribution of protein to fuel supply may reach 5–15% during the final minutes of work via liver gluconeogenesis. During exercise, increased energy consumption is necessary to provide adinosine triphosphate (ATP) to the working skeletal muscles (7). ATP is often called the universal energy donor and serves to couple the energy released from the breakdown of foodstuffs (carbohydrates, fats, and proteins) into a usable form of energy required by all cells. Although ATP is not the only energy-carrying molecule in the cell, it is the most important one, and without sufficient amounts of ATP, most cells die quickly. The structure of ATP consists of three main parts: (1) an adenine portion, (2) a ribose portion, and (3) three linked phosphates. The formation of ATP occurs by combining adenosine diphosphate (ADP) and inorganic phosphate (Pi) and requires a rather large amount of energy. Some of this energy is stored in the chemical bond joining ADP and Pi. Therefore, this bond is called a high-energy phosphate bond. When the enzyme ATPase breaks this bond, energy is released and serves as the immediate source of energy for muscular contraction:

$$\text{ATP} \xrightarrow{\text{ATPase}} \text{ADP} + \text{Pi} + \text{energy}$$

A mole of ATP, when broken down, has an energy field of 7.3 kcal. In fact, energy-liberating reactions are linked to energy-requiring reactions.

1.3. Bioenergetics

Because muscle cells store limited amounts of ATP, metabolic pathways must exist in the cell with the capability to produce ATP rapidly during exercise. Indeed, muscle cells can produce ATP by any one or a combination of three metabolic pathways, depending on the type of exercise (intensity and duration): (1) immediate formation of ATP by phosphocreatine (PC) breakdown, (2) formation of ATP via the degradation of glucose or glycogen (called glycolysis), and (3) oxidative formation of ATP (8). Formation of ATP via the PC pathway and glycolysis does not involve the use of O_2. These systems are referred to as anaerobic, meaning they are not dependent on O_2. The third, is referred to as the aerobic energy system. Energy transduction in this system is dependent on the presence of O_2. The combination of stored ATP and PC is called the ATP-PC system or the "phosphagen system." As rapidly as ATP is broken down to ADP + Pi at the onset of exercise, ATP is reformed via the PC reaction in which Pi of CP is given to ADP:

$$\text{PC} + \text{ADP} \xrightarrow{\text{Creatine kinase}} \text{ATP} + \text{Creatine}$$

The reaction is catalyzed by the enzyme creatine kinase. PC reformation requires ATP and occurs only during the recovery from exercise (9,10). The ATP-PC system will be

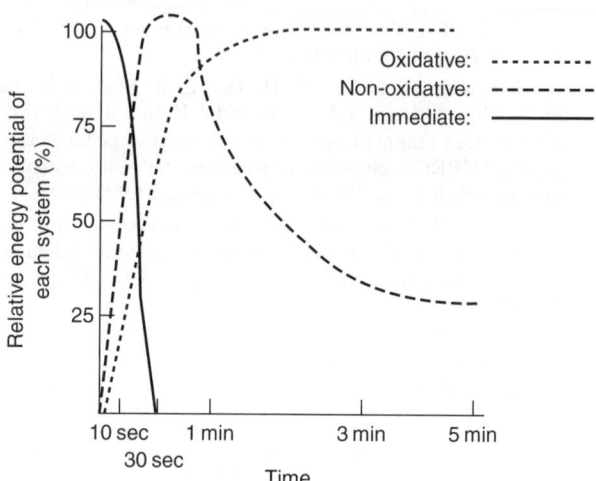

Figure 1. Energy potential for a muscle as a function of activity duration. Aerobic (oxidative) energy systems are slow to respond compared with the anaerobic (nonoxidative and immediate) energy delivery systems concerning the rate of ATP production.

very important in activities that require only a few seconds to complete and thus need a rapid supply of ATP. In power events, where the activity lasts a few seconds or less, the working muscle will be mostly dependent of this phosphagen system (1,11) (Fig. 1). Glycolysis is an anaerobic pathway that occurs in the cytoplasm of the cell and is used to transfer bond energy from glucose and glycogen to rejoin Pi to ADP. This process involves a series of enzymatically catalyzed, coupled reactions for the breakdown of glucose or glycogen to form two molecules of pyruvic acid or lactic acid (12,13). In this reaction, two distinct phases can be subtracted: (1) energy investment phase, and (2) energy generation phase (Fig. 2). The net gain of glycolysis is two ATP if glucose is the substrate, and three if glycogen is the substrate because glycogen-glycolysis requires just one ATP in the energy investment phase instead of the two ATP for glucose. The hydrogens that are removed during glycolysis can bond on pyruvic acid to form lactic acid. The enzyme that catalyzes this reaction is lactate dehydrogenase (LDH):

$$\text{Pyruvic acid} \xrightarrow[\text{LDH}]{\text{NADH} + \text{H}^+ \quad \text{NAD}} \text{Lactic acid}$$

Hydrogens are frequently removed from nutrient substrates in bioenergetic pathways and are transported by "carrier molecules." Two biologically important carrier molecules are nicotinamide adenine dinucleotide (NAD) and flavin adenine dinucleotide (FAD). Both NAD and FAD transport hydrogens (NADH and FADH) and their appropriate energy is used for later generation of ATP in the mitochondria via aerobic processes, at least if enough O_2 is available. In the presence of O_2 in the mitochondria, pyruvate can participate in the aerobic production of ATP. Thus, glycolysis can also be considered the first step in the aerobic degradation of carbohydrates. For rapid, forceful, exercises lasting from a few seconds to approximately one

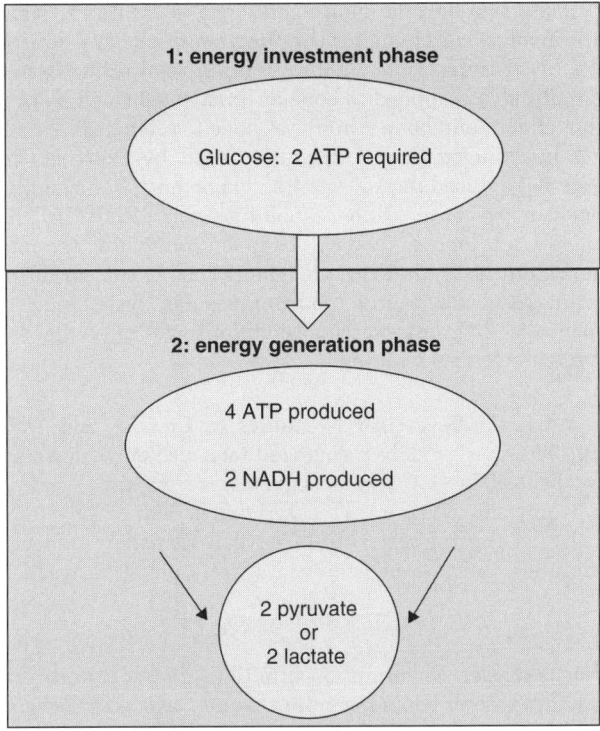
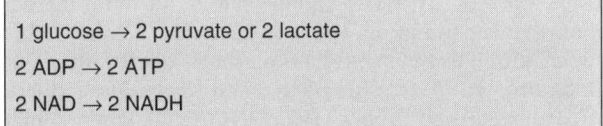

Figure 2. Illustration of the two existing phases in the glycolysis and its products. In the energy investment phase, 2 ATP are required to generate an additional 4 ATP in the energy generation phase.

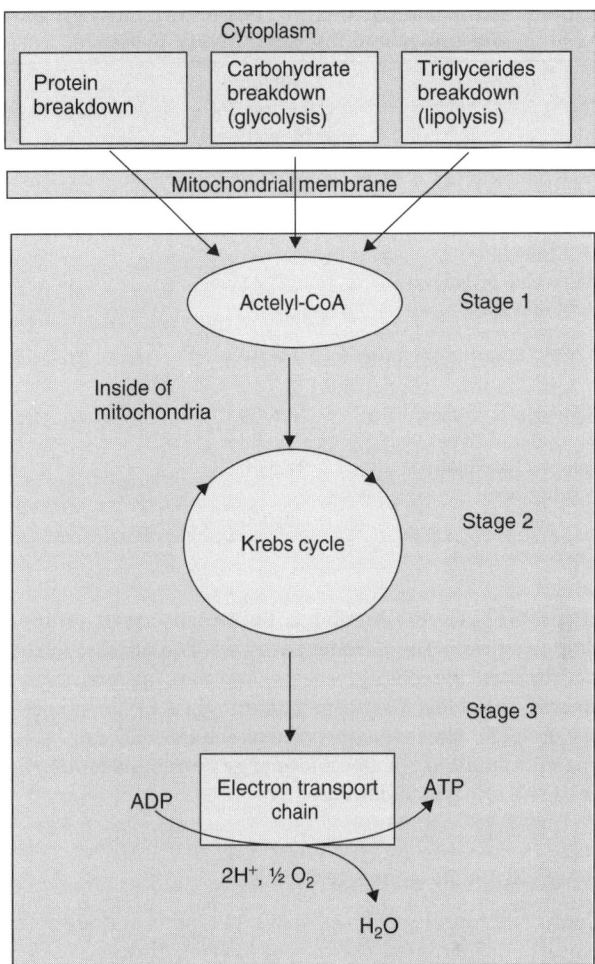

Figure 3. The three stages of oxidative phosphorylation in the mitochondria. Actelyl-CoA is the breakdown product of three foodstuffs (carbohydrates, fats, and proteins). Actelyl-CoA is oxygenated in the Krebs cycle. Finally, in the electron transport chain, ATP is generated.

minute, muscles depend mainly on nonoxidative or glycolytic energy sources as well as on immediate energy sources (Fig. 1). The aerobic ATP production occurs in the mitochondria and can be considered as a three-stage process (Fig. 3) (14). Stage one is the generation of a key two-carbon molecule, acetyl-Coenzyme A (Acetyl-CoA). Acetyl-CoA can be formed from the breakdown of either carbohydrates, fats, or proteins. Stage two is the oxidation of Acetyl-CoA in the Krebs cycle. The primary function of the Krebs cycle is to remove hydrogens (containing the potential energy in the food molecules) from various substrates involved in the cycle. The energy from removed hydrogens can be used to combine NAD and hydrogen for NADH formation, which later provides energy in the electron transport chain for ADP and Pi for the reformation of ATP in the third stage. In this stage, oxygen serves as the final hydrogen acceptor at the end of the electron transport chain (i.e., water is formed, $H_2 + O \rightarrow H_2O$). The process of aerobic production of ATP is termed oxidative phosphorylation. The aerobic metabolism of one molecule of glucose and glycogen results in the production of 38 and 39 ATP molecules, respectively. For activities lasting two minutes or more, oxidative mechanisms become increasingly important (Fig. 1). The maximal energy capacity of immediate and nonoxidative energy systems is small compared with that of the oxidative energy system (Table 1). Immediate and nonoxidative energy systems, however, are important because they are activated very rapidly when muscles start to contract. The maximal rates (power) at which the various systems provide energy for muscle contraction are shown in Table 1, together with the maximal capacities for energy release. The oxidative energy system is activated more slowly and produces energy at a lower rate even when fully activated (15). In this comparison, immediate and nonoxidative energy systems are revealed to have superior, although short-lived, power capacities. Thus, the three energy systems in the muscle provide a means to sustain short, intense bursts of activity as well as more sustained activities of lesser intensity (16). In conclusion, the energy required for muscle contraction or other forms of biologic work is produced by aerobic and anaerobic pathways. ATP breaks down inside muscle cells and becomes a source of energy. Although not all ATP is formed aerobically, the amount of energy yielded from

Table 1. Estimation of Maximal Power and Capacity of the Aerobic and Anaerobic Energy Delivery Systems

	Maximal Capacity (Total kcal Available)	Maximal Power (kcal.min^{-1})
Anaerobic energy system		
ATP-CP system (immediate)	11.4	36
Anaerobic glycolysis (nonoxidative)	15	16
Aerobic energy system		10
Glycogen in muscle	2000	
Glycogen in liver	280	
Fat (triglyceride in adipose)	141000	
Body proteins	24000	

anaerobic ATP production is extremely small (Table 1) compared with the aerobic pathway. The aerobic system, which uses glucose, glycogen, and fats as energy substrates, provides very large amounts of ATP for muscular energy (17). However, the rate at which ATP can be produced is small in the aerobic energy system in comparison with the anaerobic delivery.

1.4. Systemic Responses to Exercise

Continued muscle contractions for extended periods of time require an increased delivery of oxygen. However, oxygen delivery to the cell is critical; hence, the capacity to provide oxygen to the tissues usually determines the intensity of physical activity that an individual can perform, which largely depends of the cardiovascular system that provides the means to transport the oxygen to the contracting skeletal muscle. To improve the capability of the cardiovascular function to deliver oxygen to the working musculature, some chronic and acute exercise adaptation will occur. The lung function is to provide a way for oxygen to be transferred between atmospheric air and the blood and for the majority of metabolically produced carbon dioxide to be removed from the body. As a result of the increased oxygen consumption to fuel mitochondrial respiration, together with an increased carbon dioxide production by the working musculature during exercise, lung function increases directly with the rate of metabolism. As the hydrogen ion content in blood influences blood acid-base balance, the lungs are also important for regulating blood pH during exercise. During muscle contractions, a sudden increase occurs in ATP use and the stimulation of metabolic adaptations that enable the muscle fibers to increase the regeneration of ATP. Different types of muscle fibers in several motor units pose different metabolic capacities. Therefore, the contribution of different muscle fibers to the metabolic demand of exercise is very specific and is associated with acute adaptations of muscle motor unit and fiber-type function during different exercise conditions and after different types of exercise training. Exercise will also induce a series of acute and chronic neuro-endocrine adaptations, which are involved in changing the function of skeletal muscle, the heart, lungs, liver, kidneys, brain, and other tissues. Finally, stress applied to bone at junctions formed by muscle, tendon, and bone stimulates bone to alter itself, resulting in changed bone structure caused by exercise. The exercise-induced improvements in bone have had ramifications in preventive and rehabilitative medicine. In what follows, a brief overview will be given concerning the most important acute and chronic adaptations of the cardiovascular, respiratory, and (neuro)-muscular systems in response to different exercise conditions and after different types of exercise training.

1.4.1. Cardiovascular Responses to Exercise and Training.
The most widely recognized measure of cardiovascular fitness is the aerobic capacity or maximal oxygen consumption (VO_{2max}). This variable is defined physiologically as the highest rate of oxygen transport and consumption that can be achieved at peak physical exertion.

$$VO_2 = HR \times SV \times (CaO_2 - CvO_2)$$

VO_2 is oxygen consumption in milliliters per minute; HR is heart rate in beats per minute; and $CaO_2 - CvO_2$ is the arteriovenous oxygen difference in milliliters of oxygen per deciliter of blood (18). At the workload beyond which an increase in further work does not result in an increase of oxygen consumption, the individual has attained his or her maximal oxygen uptake (i.e., there is a leveling off or plateauing in VO_2). This point is an important indicator of cardiovascular fitness (19). A metabolic equivalent or MET approximates 3.5 milliliters of oxygen per kilogram of bodyweight per minute (3.5 mL/kg/min). This expression of resting oxygen consumption is relatively constant for all persons, regardless of age, bodyweight, or fitness level. Multiples of this value are often used to quantify relative levels of energy expenditure and aerobic fitness (20). Much of the circulatory data increases from rest during exercise, with physically active individuals having higher maximal values than sedentary persons, as shown in Table 2. As there is little variation in maximal heart rate and maximal systemic arteriovenous oxygen difference ($CaO_2 - CvO_2$) with training, VO_{2max} virtually defines the pumping capacity of the heart.

1.4.1.1. Acute Cardiovascular Responses to Exercise.
The heart and circulation respond to the requirements of metabolism during exercise by increasing blood flow to active areas and decreasing it to less critical areas. Heart rate (HR) generally increases progressively as a function of exercise intensity. A roughly linear relationship exists between HR and workload or power output (21). The equation (220 − age) provides an approximation of the maximal HR, but the variance for any given age is considerable (standard deviation ± 10 bpm). During exercise, an increase in stroke volume (SV) resulting from both the Frank–Starling mechanism (enhanced venous return results in larger cardiac preload increasing the force of myocard contraction) and a decreased end systolic volume also occurs. The latter is because of increased

Table 2. Adaptation to Exercise in a Sedentary Man Versus a World-Class Endurance Athlete.

Parameter	Sedentary Man		Physical Active Man
	Rest ↓ Max. ex.		Rest ↓ Max. ex.
Oxygen consumption (L/min)	X 10	↔	X 16
Cardiac output (L/min)	X 3	↔	X 6
Heart rate (beats/min)	X 3	↔	X 4
Stroke volume (mL/beat)	X 1	↔	X 1
Arteriovenous oxygen difference (mL/dL blood)	X 4	↔	X 4

Values present the gain in circulatory data that are found from rest to maximal exercise (max. ex.).

ventricular contractility, secondary to catecholamine-mediated sympathetic stimulation (Fig. 4). Up to exercise levels of 50% VO_{2max}, the increase in cardiac output (CO = HR × SV) is accomplished through increases in both HR and SV. At higher exercise intensities, a further increase of cardiac output results almost solely from the continued increase in HR (Fig. 5) (22). The arterial and mixed venous oxygen content at rest are approximately 20 mL and 15 mL of oxygen per 100 dL of blood, respectively. At the muscle, an increased extraction of oxygen from arterial blood supply occurs when exercise intensity increases from moderate to heavy loads. As the workloads approach the point of exhaustion, the venous oxygen content is decreased to 5 mL/dL blood or lower, which indicates a threefold increase of the arteriovenous oxygen difference from

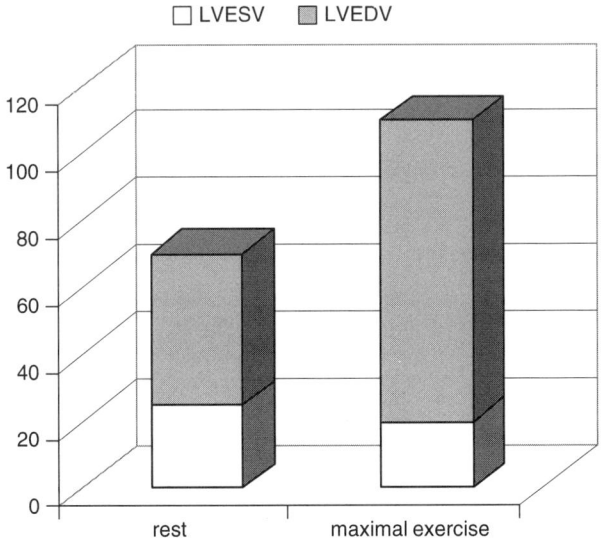

Figure 4. Changes in SV from rest to maximal exercise. LVESV, left ventricular end-systolic volume; LVEDV, left ventricular end-diastolic volume.

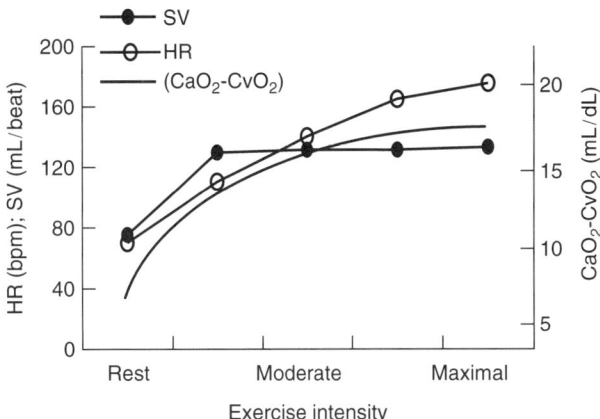

Figure 5. Arbitrary determined SV, HR, and arteriovenous oxygen difference during different levels of intensity. At higher exercise intensities, cardiac output mainly increases because of increases in HR instead of increases in SV.

5 mL/dL to 15 mL/dL blood (Fig. 5). A linear increase occurs in the systolic blood pressure with increasing exercise, with maximal values reaching 170 mmHg to 220 mmHg. Diastolic blood pressure generally remains unchanged or decreases slightly from rest to maximum exercise. During exercise, a shunting of blood occurs so that the working muscles receive as much as 85–90% of the cardiac output through vasodilatation, while blood flow to the visceral organs is simultaneously decreased by vasoconstriction. Blood flow to the myocardium (coronary arteries) is increased in proportion to the increased metabolic activity of the heart, whereas blood flow to the brain is maintained via autoregulation (23). Increased skin blood flow occurs during exercise to facilitate heat dissipation. When maximal exertion is approached, blood does not supply the skin any longer in order to meet the metabolic requirements of working muscles, and consequently, core temperature may quickly rise (24).

1.4.1.2. Cardiovascular Adaptations to Prolonged Exercise. Cardiovascular adaptations to physical activity generally depend on the type and intensity of exercise. Two types of exercise can be distinguished: endurance and resistance. Endurance training is best for improving the capacity of cardiovascular function and results in adaptive changes in many aspects of cardiovascular function. These changes are listed in Table 3. Improvements in maximal oxygen consumption with training depend on current fitness, type of training, and age and result in a maximum increase in VO_{2max} of approximately 20% (1). At the level of the heart, long term endurance training will increase the left ventricular diastolic cavity dimensions, wall thickness, and mass (25–27). These changes are described as the "athlete's heart." Endurance training results in an increased volume load (increased left ventricular volume), whereas resistance or strength training induces a pressure load (increased left ventricular wall thickness) of the heart (28). The volume load during endurance training reduces resting and submaximal exercise HR and increases the SV at rest, submaximal, and maximal

Table 3. Cardiovascular Adaptations Resulting from Endurance Training During Submaximal Exercise (submax. ex.) and Maximal Exercise (max. ex.)

Factor	Rest	Submax Ex.	Max. Ex.
Heart rate	↓	↓	=
Stroke volume	↑	↑	↑
Arteriovenous O_2 difference	↑ =	↑	↑
Cardiac output	= ↓	↓ =	↑
VO_2	=	=	↑
Work capacity	-----	-----	↑
Syst. blood pressure	= ↓	↓ =	=
Diast. blood pressure	= ↓	↓ =	=
Mean arterial blood pressure	= ↓	↓ =	=
Total peripheral resistance	=	↓ =	=
Coronary blood flow	↓	↓	↑
Brain blood flow	=	=	=
Visceral blood flow	=	↑	=
Inactive muscle blood flow	=	=	=
Active muscle blood flow	= ↓	↓ =	↑
Skin blood flow	=	↑ =	=
Blood volume	↑	-----	-----
Plasma volume	↑	-----	-----
Red cells mass	↑ =	-----	-----
Heart volume	↑	-----	-----

Symbols: ↑ increase; ↓ decrease; =, no change; ----- not applicable. Adapted from reference (1).

exercise (1). Aerobic exercise training is suggested to increase the parasympathetic modulation of the heart via the vagal nerve (29–31). Consequently, a shift toward vagal predominance in autonomic HR control has been demonstrated with endurance training. Probably, this phenomenon partly explains the observed lower resting HRs after prolonged endurance exercise training. The latter might have beneficial effects regarding the risk for (lethal) cardiac events like tachycardia and other arrhythmias (32). The decline of HR at rest and during submaximal exercise does not reduce cardiac output because SV increases significantly (up to 20%) through an increase of end-diastolic volume and cardiac contractility (force at a given sarcomere length) after endurance exercise training. The result is a more efficient pressure-time relationship.

All these mechanisms are significantly accompanied by large increases in cardiac output. Exercise physiologists have traditionally regarded oxygen transport, particularly increased maximal cardiac output, as the primary mechanism of improvement in VO_{2max} after sustained endurance training. Also, some cardiovascular morphologic characteristics (central blood volume and total hemoglobin per total blood volume) appear to increase with regular endurance exercise. Increased cardiac output, SV, central blood volume, and total hemoglobin are generally called the central adaptations to exercise that improve VO_{2max} or cardiovascular fitness (Fig. 6). However, because the arteriovenous oxygen difference increases slightly with training, improvements in VO_{2max} may also be achieved by some peripheral adaptations to exercise training (Fig. 6). Indeed, studies involving the biopsy removal of muscle tissue have shown an improvement in the metabolic preparedness of the muscle cells as a result of structural and enzymatic adaptations. The oxidative capacity of skeletal muscles is increased through increases in the volume density of skeletal muscle mitochondria (mitochondria form reticulum), as well as increases in muscle oxidative enzymes (33). Additionally, elevated myoglobin concentrations are found together with increased muscle capillary density after endurance training interventions.

Actually, circulatory and biochemical changes with training are linked. Skeletal muscle vascularity (blood vessels in skeletal muscles) increases with endurance training, which facilitates diffusion of oxygen, substrates, and metabolites in exercising muscles. Increased vascularity also decreases total peripheral resistance, which is one of the mechanisms that increases cardiac output with endurance training. Training also decreases muscle blood flow during submaximal exercise. The trained muscle has an increased oxygen extraction capacity because of improved diffusion capability and muscle respiratory capacity (1). Decreased muscle blood flow at submaximal work allows more blood to be directed to the viscera and

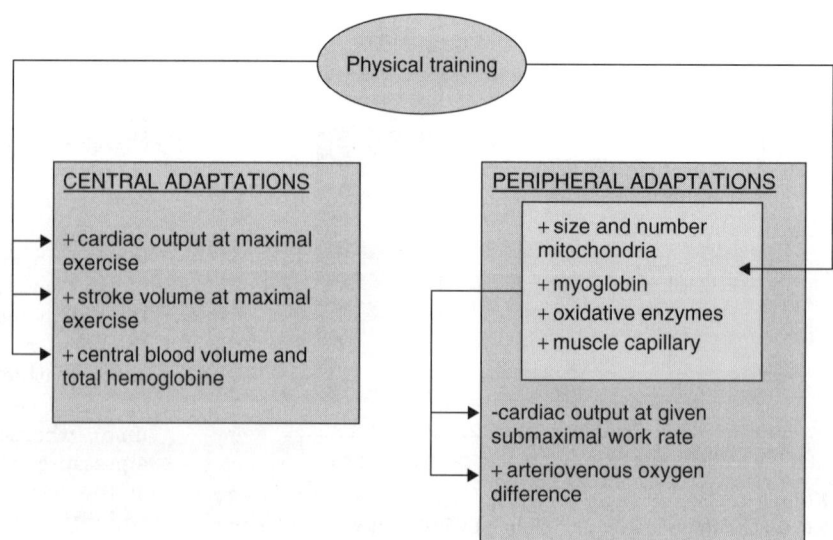

Figure 6. Mechanisms responsible for the increase in VO_{2max} with physical training (+, increase; −, decrease).

skin. The result appears to be a decreased cardiac output and muscle blood flow at any given submaximal workload. The muscle needs and receives a smaller blood flow than before training, which means that a better use of available blood flow occurs, resulting in economy and reserve capacity with respect to cardiac work (18). The hemodynamic advantages of a more efficient distribution of cardiac output and increased oxidative capacity of skeletal muscle is reflected by decreased lactate release and a lowered respiratory exchange ratio (RER) (lower RER also because of greater percentage of fat oxidized in trained at given submaximal workload) at any given level of exercise, with increases in arteriovenous oxygen difference at submaximal and maximal exercise.

1.4.2. Pulmonary Responses to Exercise and Training. Ventilation is the process of providing sufficient airflow through the respiratory passageways filling the gas-exchange areas in an attempt to accommodate to the cellular needs of oxygen delivery and carbon dioxide removal (34). As a result of homeostasis that characterizes the human body, the respiratory system will respond very specifically to different types of acute and chronic exercise.

1.4.2.1. Acute Pulmonary Responses to Exercise. Respiratory rate can increase five- to six-fold during maximal exercise with respiratory rates of 50 to 60 breaths per minute. Accordingly, tidal volume is substantially altered from rest, with five-fold to seven-fold increases during maximal exercise. The net effect on minute ventilation is a 30- to 40-fold increase over resting airflow values. During rest, central and peripheral chemoreceptors influence the intrinsic breathing pattern established by the medulla (34). For a few seconds before exercise up to the first 10 to 20 seconds of exercise, an initial moderate increase in respiration rate occurs as the immediate response to the cellular demands (35–37). This initial phase is relatively short in duration and results mainly from neurogenic stimulation from the cerebral cortex (central command) as an individual prepares for exercise (34). The next stage yields a rapid progressive increase in ventilation until a steady state is reached (during submaximal intensity exercise). This more rapid increment in respiration rate is the result of an increased central command together with an increase in neural stimuli to the medulla by activation of muscle-joint receptors. Additionally, central and peripheral chemoreceptors, reacting to an increasing PCO_2 and decreasing pH in the blood or cerebrospinal fluid, as well as mechanoreceptors, become proportionally more important. As exercise progresses to maximal effort, ventilation progressively and continuously increases as the need for oxygen delivery increases. Whereas the control mechanisms from the initial phase continue to be an active force for ventilation, inputs from central and peripheral chemoreceptors lead them to play a greater role in the control of the ventilatory response. Central and peripheral chemoreceptors provide feedback about blood homeostasis of oxygen, carbondioxide, hydrogen ion concentration, and temperature (36,38,39). The steep decrease on the cessation of exercise reflects the reduction in regulatory inputs from "central command" and mechanoreceptors, whereas the slower decrease appears to be related to the matching of pulmonary ventilation with blood chemical stimuli as the resting state is restored (36,40). Changes in tidal volume are the major contributor to increases in pulmonary minute ventilation in light- to moderate-intensity exercise. At higher exercise intensities, tidal volume tends to plateau once approximately 60% of vital capacity is reached (38,41). At the point of tidal volume plateau, further increases in pulmonary minute ventilation can be attributed to continued increases in breathing rate (36,40,42). The increase in airflow is linearly related to the metabolic demand of increased oxygen consumption and carbon dioxide elimination accompanying light- to moderate-intensity exercise. If exercise intensity reaches or exceeds 55–65% of maximal aerobic capacity, the increase in pulmonary ventilation is mainly related to the removal of CO_2 (35,38,41,42). At this breakpoint ("anaerobic threshold," "ventilatory threshold," "lactate inflection point") the metabolic energy-producing pathways shift toward anaerobic glycolysis. The byproducts (specifically lactate acid) from the energy metabolism accumulate at a rate greater than that at which those substrates can be eliminated (43). Buffering of the hydrogen ion from lactic acid through the bicarbonate system yields nonmetabolic-produced carbon dioxide and becomes a major driver of ventilation to maintain blood homeostasis (42):

$$La^- + H^+ + NaHCO_3 \rightarrow NaLa + H_2CO_3$$

$$H_2CO_3 \rightarrow H_2O + CO_2$$

The resulting effect is a rapid production of CO_2 in the cardiovascular system, as CO_2 is produced both metabolically (Krebs cycle and catabolism of acetyl-CoA) and nonmetabolically through lactic acid buffering (1,42,44). Therefore, a disproportionate increase in ventilation occurs to remove the produced CO_2. Increased H^+ also directly drives chemoreceptors to increase ventilation during exercise.

During exercise, an increase in pulmonary capillary blood flow, together with a linear increase of pulmonary capillary blood volume, occurs (45). Exercise pulmonary capillary transit time decreases with approximately 50%; however, this does not significantly affect the exchange of oxygen and carbon dioxide in the pulmonary capillary bed (34,45).

1.4.2.2. Pulmonary Adaptations to Prolonged Exercise. Responses and adaptations of the respiratory system are considerably less remarkable than in other body systems. This lack of adaptation to prolonged activity is not very surprising because tremendous reserves are available accompanying the lungs, even without physical training. The respiratory system is not considered a limiting factor to exercise performance in the majority of persons (46). This notification is supported by research that has shown no relationship among level of physical training, training methods, and adaptations in lung capacity and volumes, although some adaptations can be notified in the change in pulmonary ventilatory dynamics. As a result of prolonged exercise training, minute ventilation is increased during maximal-exercise intensity and decreased during submaximal-exercise intensity. At rest, during submaximal and maximal exercise, tidal volume is increased in trained individuals. In contrast, respiration rate is decreased during rest and submaximal exercise and increased during maximal exercise. Additional increases in respiratory muscle strength and endurance are also reported (47,48). These adaptations, in which a smaller amount of air is ventilated to provide the same amount of oxygen, are specific to the method of training, but consistent across the intensity of exercise (49).

1.4.3. Muscular Responses to Exercise and Training.
The skeletal muscle is a highly adaptable tissue. Repetitive use of the skeletal muscles (physical training) results in some adaptive changes in its structural functional and properties. In this section, the principles by which muscle structure and function adapt to increases and decreases in habitual levels and types of physical activity will be summarized. Therefore, the focus will be on the chronic adaptations to exercise.

1.4.3.1. The Principle of Myoplasticity.
The capacity of skeletal muscles for adaptive changes to nutritional and endocrine factors caused by exercise training is governed by the principle of myoplasticity. Skeletal muscle adaptations are characterized by modifications of morphological, biochemical, and molecular variables that alter the functional attributes of fibers in specific motor units (1). Acute changes in the microenvironment lead to altered rates of protein synthesis and degradation (muscle synthesis and degradation). The molecular basis for the adaptations in skeletal muscle proteins to exercise training is an altered gene expression. When a protein structure is altered in this way, we say that the muscle's phenotype has been changed. Phenotype is the outward or observable characteristics of muscle; it reflects the underlying genes (genotype) and their regulation by several factors, including exercise training.

1.4.3.2. Adaptations in Muscle Structure to Endurance Training.
During muscle contraction, oxygen must be supplied to and used by the contracting muscle fibers. When oxygen is released from hemoglobin at the tissue level, it must be transferred from the capillary to within the muscle fiber and then to within the mitochondria. Despite convincing evidence from animal research, endurance-training research with human subjects has not shown an increase in myoglobin concentrations (50). However, the major chronic skeletal muscle metabolic adaptations to training for long-term muscular endurance are related to mitochondrial respiration, muscle glycogen concentration, and the concentration and activity of glycolitic enzymes (Table 4). The increase of mitochondrial proteins and many of the enzymes of beta oxidation for fatty acids result in an approximately two-fold increase of the oxidative metabolism in the skeletal muscle (3). More specifically, an increase in the mitochondrial membrane surface area is found, which improves the capacity for the exchange of metabolites between the cytosol and mitochondria (3). The increased mitochondrial volume provides a greater concentration of mitochondrial oxidative enzymes that, in turn, increases VO_{2max}. Also, some evidence is found of increased muscle glycogen concentrations in long-term endurance-trained athletes (51).

Table 4. Chronic Skeletal Muscle Metabolic Adaptations Resulting from Exercise Training for Long-Term Muscular Endurance

Metabolic Pathway	Adaptation
Mitochondrial respiration	↑ number and size of mitochondria
	↑ concentration of glycolitic enzymes
Glycogen	↑ concentration
Glycolysis	↑ activity of glycolitic enzymes

Next to these metabolic adaptations, some structural adaptations to chronic endurance exercise also occur in the skeletal muscle. As already mentioned, skeletal muscle vascularity (blood vessels in skeletal muscles) increases with endurance training. Because the transit time for red blood cells is decreased because of the improved capillary, the diffusion of oxygen, substrates, and metabolites in the muscles is facilitated. The trained muscle has an increased oxygen extraction capacity because of improved diffusion capability and muscle respiratory capacity (1). The result appears to be a decreased cardiac output and muscle blood flow at any given submaximal workload. Indeed, during submaximal exercise, muscle blood flow is decreased following endurance training. Decreased muscle blood flow allows more blood to be directed to the viscera and skin. The muscle needs and receives a smaller blood flow than before training. Thus, a better use of available blood flow occurs, resulting in economy and reserve capacity with respect to cardiac work (18). During maximal aerobic exercise training, however, blood flow increases in the active motor units. Finally, increased vascularity also decreases total peripheral resistance, which is one of the mechanisms that increases cardiac output with endurance training. The functional significance of all these changes is seen primarily during sustained exercise in which there will be a delay in the onset of metabolic acidosis, an increase in the capacity to oxidize free fatty acids and other fuels, and a conservation of carbohydrates (1).

1.4.3.3. Adaptations in Muscle Structure to Resistance Training.
The major adaptation that occurs with resistance training is an increase in the cross-sectional area of the muscle, called muscle hypertrophy (3). The principle mechanism for muscle hypertrophy is cellular hypertrophy (increased cross-sectional area) and not hyperplasia (increased number of cells). The functional significance of morphological changes is primarily a greater capacity for strength and power development. Both type I (slow twitch, oxidative) and type II (fast twitch, explosive, anaerobe) fibers increase in cross-sectional area resulting from high-intensity resistance training (Fig. 7) (52). During more intense exercise, muscle hypertrophy is greater than that for endurance activities (49). In addition to fiber areas, the angle in fiber pinnation can also adapt to training.

Individuals that only perform resistance training without endurance training risk reduction of endurance

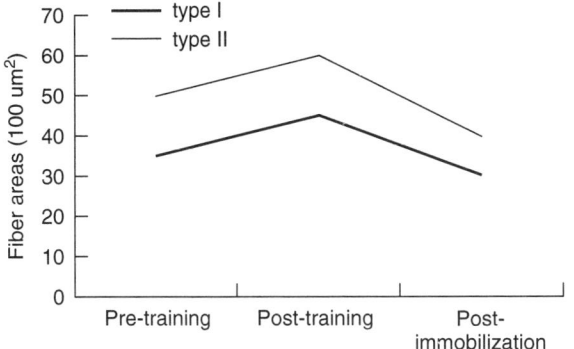

Figure 7. Cross-sectional area for type I (slow) and type II (fast) fibers following training and immobilization.

capacity. Performing some endurance training in the training program can counteract this problem. Thus, from a practical point of view, when training stimuli are combined, a decrease in metabolic capacity is not inevitable. It is common to find increases in strength of 30–40%, which exceed the amount of muscle hypertrophy. Especially at the beginning of resistance training, additional gain in strength is found without any hypertrophy occurring, which can be ascribed to the optimization of recruitment patterns (neurological training) (53,54). Independent of any improvements that occur, one becomes more adept in effectively using muscle mass (53,54). The gain in strength without hypertrophy of the muscle is given by the concept of specific force, which involves the maximal force being normalized to the physiological cross-sectional area of the muscle. This measurement allows comparing intrinsic capacity with develop force, regardless of the size of the muscle (1). Increasing evidence from animal studies supports morphologic adaptations of the neuromuscular junction (NMJ) in response to increased activity (55). In conclusion, the neural component of adaptation to exercise may be important to better understand increased exercise tolerance and performance after training.

1.5. Benefits from Regular (Recreational) Physical Activity: Public Health Perspectives

Numerous health benefits are associated with regular participation in intermittent, moderate-intensity physical activity. The Surgeon General's report, Physical Activity and Health (56) stated some conclusions that merit more attention here. Significant health benefits can be obtained by including a moderate amount of physical activity on most, if not all, days of the week. "Through a modest increase in daily activity, most people can improve their health and quality of life." Additional health benefits can be gained through greater amounts of physical activity. These statements imply that health and fitness benefits associated with physical activity most likely follow a dose-response relationship, which is suggested to be bell-shaped (57). Although the optimal dose of exercise has yet to be defined, the dose-response relationship between physical activity and various health benefits supports the need to encourage the public to engage in at least moderate amounts and moderate intensities of daily physical activity [i.e., activities that are approximately 3–6 METs or the equivalent of walking at 3 to 4 mph (15–20 minutes to walk 1 mile)] for most healthy adults.

Numerous laboratory-based studies have quantified the many health and fitness benefits (e.g., physiologic, metabolic, and psychological) associated with regular physical activity (46). In addition, an increasing number of prospective epidemiologic studies support the notion that a physically active lifestyle and a moderate to high level of cardiorespiratory fitness independently lower the risk of various chronic diseases. However, until now, the optimal weekly training load in terms of intensity, frequency, and duration of exercise that is required to improve health-related fitness parameters remains forthcoming (58–63). Nevertheless, the bulk of the epidemiologic evidence support the hypothesis that regular activity increases longevity (64). Benefits of regular physical activity or exercise might be outlined as represented in Table 5 (58).

1.6. Conclusion

In this section, a general overview is provided of both exercise metabolism and chronic and acute systemic responses and adaptations to sports or recreational exercise. During exercise, the body uses three types of substrates (fats, carbohydrates, and proteins) as fuel for the construction of ATP in the working musculature. When the ATP bond breaks down, the energy that is released serves as the immediate source of energy for muscular contraction. Three metabolic pathways exist in the muscle cells to produce ATP rapidly during exercise: (1) immediate formation of ATP by PC breakdown, (2) formation of ATP via the degradation of glucose or glycogen (glycolysis), and (3) oxidative formation of ATP. Formation of ATP via the PC pathway and glycolysis are referred to as anaerobic, meaning not dependent on O_2. The third is referred to as the aerobic energy system. The energy transduction from this system depends on the presence of O_2. The proportional importance of these three pathways in the total energy delivery is dependent on the type of exercise (intensity and duration). The aerobic system provides large amounts of ATP for muscular energy at a slow rate (long duration and low-intensity exercise). The anaerobic delivery, however, provides smaller amounts of ATP but at a higher rate (short duration, high-intensity exercise). As a response to physical exercise, a series of acute and chronically systemic alterations occur to provide a more efficient energy delivery in the working musculature. The capacity to provide and to use oxygen in the tissues (aerobic capacity) largely depends on the cardiovascular system. The aerobic capacity can be quantified by the maximal oxygen consumption (VO_{2max}). This index depends on HR, SV, and the arteriovenous oxygen difference. At these levels, major acute and chronic alterations will occur because of physical endurance exercising. Another parameter that partly determines endurance capacity is the pulmonary functioning. However, alterations at this level are much less pronounced than those in the cardiovascular system. Finally, the skeletal muscle is also a highly adaptable tissue because of the principle of myoplasticity that allows

Table 5. Benefits from Regular Physical Activity or Exercise

1) Improvements of cardiovascular and respiratory function:
   a. Increased maximal oxygen uptake caused by both central and peripheral adaptations
   b. Lower minute ventilation at a given submaximal intensity
   c. Lower myocardial oxygen cost for a given submaximal intensity
   d. Lower HR and blood pressure at a given submaximal intensity
   e. Increased capillary density in skeletal muscle
   f. Increased exercise threshold for the accumulation of lactate in the blood
   g. Increased exercise threshold for the onset of disease signs or symptoms
2) Reduction in coronary artery disease risk factors
   a. Reduced resting systolic/diastolic pressures
   b. Increased serum high-density lipoprotein cholesterol and decreased serum triglycerides
   c. Reduced total body fat, reduced intra-abdominal fat
   d. Reduced insulin needs, improved glucose tolerance
3) Decreased mortality and morbidity
   a. Primary prevention (i.e., interventions will prevent an acute cardiac event)
      i. Higher activity or fitness levels are associated with lower death rates from coronary artery disease
      ii. Higher activity or fitness levels are associated with lower incidence rates for combined cardiovascular disease, cancer of the colon, and type 2 diabetes
   b. Secondary prevention (i.e., interventions after a cardiac event to prevent another)
      i. All cause of mortality is lower in patients involved in a cardiac rehabilitation exercise training program as shown by meta analysis
4) Other postulated benefits
   a. Decreased anxiety and depression
   b. Enhanced feelings of well being
   c. Enhanced performance of work

the muscle to alter its phenotype in function of very specific stimuli. Therefore, both endurance and resistance training will lead to relatively large muscular alterations and adaptations. In fact, because exercise is a whole body commitment, the entire body will respond and adapt to exercise. Cardiovascular, pulmonary, endocrine, muscular, immune, gastrointestinal, renal skeletal, and skin responses to exercise are all responsible for an efficient energy delivery during exercise.

2. BIOMEDICAL ENGINEERING CONTRIBUTIONS TO THE FIELD OF EXERCISE PHYSIOLOGY

2.1. Introduction

The term "sports medicine" is defined as multidisciplinary, including the physiological, biomechanical, psychological, and pathological phenomena associated with exercise and sport (58), and also includes associated medical specialities, allied health professionals, and applied sciences. Figure 8 shows a schematic representation of the relationship between exercise physiology and bioengineering. Somewhat arbitrarily, two large subdivisions are made: biomechanics and systems physiology. The first will be discussed in separate chapters of this encyclopedia. The latter has been discussed previously from the physiological point of view, and bioengineering implications will now be presented. Biomedical implications have become indispensable in the evaluation and measurement of physical fitness parameters and systemic/metabolic adaptations to exercise.

At any time (whether at rest or during exercise), systems generate signals, which can be recorded with appropriate sensors. Sensors convert signals of one type, such as hydrostatic fluid pressure, into an equivalent signal of another type, for example, an electrical signal. As such, biomedical sensors serve as the interface between a biologic and an electronic system. They must function in such a way as to not adversely affect either of the systems and not interfere with exercise. Therefore, most of these techniques will be noninvasive (i.e., no indwelling catheters or ionizing radiation).

The term "physical fitness" has been defined in many ways, always referring to the capacity for movement. Although many different literal definitions of physical fitness exist, there is relative uniformity in the operational definition of physical fitness being a multifactor construct that includes several components. Health-related physical fitness typically includes cardiorespiratory endurance, body composition, muscular strength and endurance, and flexibility. Measurements of physical fitness are a common and appropriate practice in preventive and rehabilitative exercise programs to promote health. Appropriate measurement techniques enable measurement and evaluation of health-related physical fitness in presumably healthy adults before, during, and after exercise training. Most of the equipment used is in common with general physiologic measurements, with some restrictions and specific requirements. Possibilities are rather limited for measurements on the track or in the field because of portability, weight restrictions, and ruggedness. During exercise, for example, ECG can be measured with a telemetry system or a Holter recorder (tape recorder for ECG for a total duration of 24 hours), or only HR with pulse systems.

Studies in the laboratory for condition assessment or pre- and post-exercise measurements give rise to less restrictions. An overview of the most important parameters and the corresponding measuring devices is given in Table 6. Some of the less well-known methods will be briefly described.

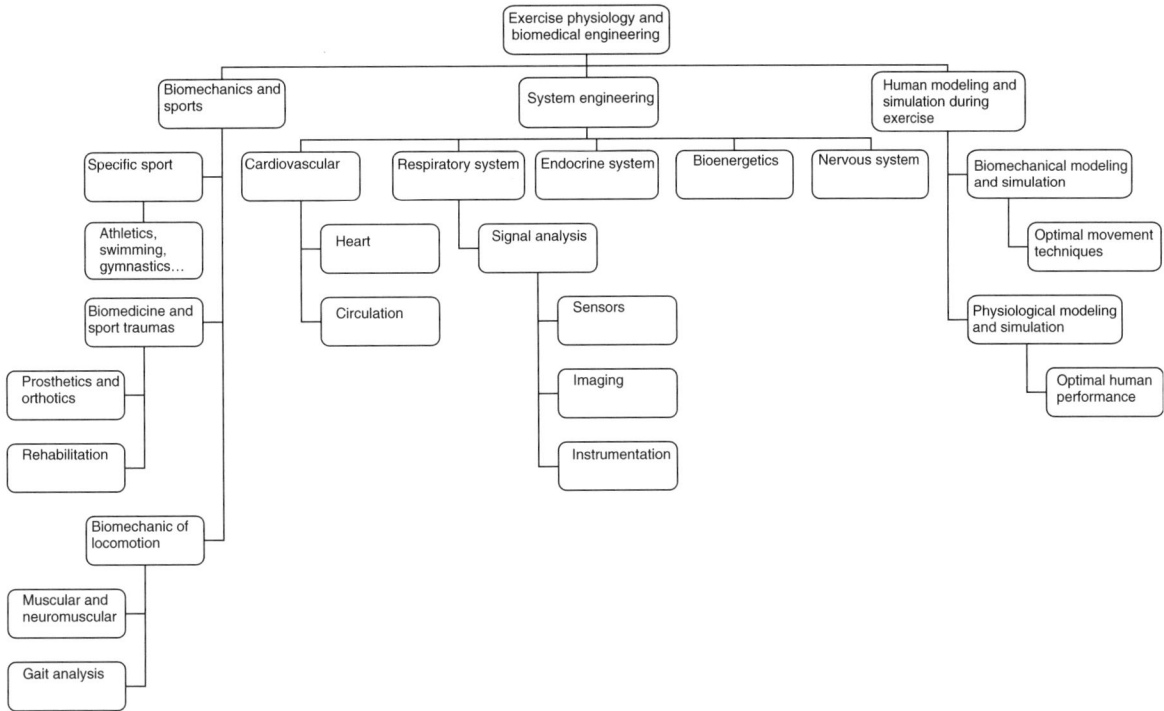

Figure 8. Overview of exercise physiology and its relationship to biomedical engineering. Data analysis is in common for all systems generating any kind of signal.

2.2. Biomedical Engineering Contributions

2.2.1. Bioelectrical Impedance (BIA). BIA involves passing a small electric current through the body and measuring the impedance or opposition to current flow. Fat-free tissue is a good conductor of electrical current, whereas fat is not. The resistance to current flow is thus inversely related to the fat-free mass and total body water, both of which can be predicted by this technique (65).

2.2.2. Pulse Oxymetry. This method relies on the detection of the time variance photoplethysmographic signal, caused by changes in arteriolar blood volume

Table 6. Schematic of Systemic Measurement Methods in Sport Physiology

Cardiovascular		
	Electrical	ECG, heart rate
	Mechanical	Blood pressure: intermittent (arm cuff), continuous (finger plethysmograpth)
	Blood flow	Cardiac output (rebreathing)
		Doppler ultrasound
		Magnetic resonance (phase velocity mapping)
	Dimensions	Echocardiography
		Magnetic resonance imaging
		Exercise nuclear imaging
Pulmonary		
	Pulmonary minute volume	Pneumotachometer, spirometer
	Breathing rate	Impedance
	O_2 saturation	Pulse oximetry
Body composition	Densitometry	Plethysmography: dual chamber plethysmograph (air displacement, body volume)
		Bioelectrical impedance (BIA)
		Dual energy x-ray absorptiometry (DXA)
		Magnetic resonance spectroscopy
Activity		Accelerometer
Temperature		Thermistor
Metabolism	Chemical components	Near-infrared interactance (NIR)

associated with cardiac contraction. Saturated O_2 is derived by analyzing only the time variant changes in absorbance caused by the pulsating arterial flow at the same red and infrared wavelengths used in a conventional invasive (in the blood flow)-type oximeter (66). A pulse oximeter consists of a pair of small and inexpensive red and infrared LEDs and a single highly sensitive silicon photodetector. The majority of the sensors are of the transmittance type. The sensor is usually applied at the earlobe or fingertip.

2.2.3. Magnetic Resonance Velocity Mapping (67).

Magnetic resonance imaging is firmly established in imaging static organs and even moving targets such as the cardiovascular system with an appropriate triggering system (cardiac and breathing cycle). Conventional images represent a map of the amplitude of the radio-frequency signal emitted by sensitive nuclei in the imaging plane under the influence of a static field and applied magnetic field gradients and radio-frequency pulses. It is possible to encode velocity in the phase of the signal so that the phase map becomes a velocity map. The encoding of velocity is achieved by a combination of magnetic field gradients that leave the phase in each pixel of the image proportional to velocity in a chosen direction, either through the image plane or within it. Clinical applications have centred on the measurement of flow pulmonary circulation, in shunts, in valvular regurgitation and stenosis. Especially important for exercise physiology is the measurement of flow in native coronary arteries in a complete noninvasive way (68). This method could therefore be implemented in exercise physiology. However, a limitation of the method is the maximum velocity that can be measured (about 6 m/s).

2.2.4. Magnetic Resonance Spectroscopy (69).

Conventional magnetic resonance imaging measures signals emitted by hydrogen (proton) nuclei from small pixels that have been selected by spatial variations in frequency and phase. However, when using frequency changes for spatial encoding and imaging, the ability to discriminate much of the important information about the chemical environment among the nuclei is lost. Water, fat, and other chemicals (aminoacids) combine to produce a single net signal from each pixel.

Magnetic resonance spectroscopy can extract information about the chemicals that reside on the frequency scale between water and fat in both a quantitative manner and a qualitative manner, and a plot representing chemical composition within a pixel is generated.

An RF pulse is applied to the sample (which is in a large, continuous, main magnetic field). The reflected signal from the sample is measured and Fourier transformed. The peaks (or resonances) (Fig. 9) on the x-axis correspond to specific chemical components. The signal intensity (amplitude on the y-axis) and linewidth provide the area that can be used to quantitate the amount of the observed chemical. As these effects are very small, and in order to resolve the different chemical species, it is necessary to achieve very high levels of homogeneity of the main mag-

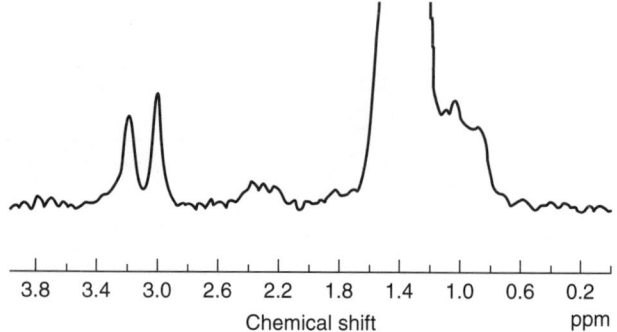

Figure 9. *In vivo* proton MR spectrum from the calf muscle of a healthy volunteer. Normal choline and creatine/phosphocreatine (at 3.2 ppm and 3.0 ppm, respectively). A high concentration of lipids (at 1.4 ppm) is present in the healthy muscle (amplitude of the y-axis has been expanded in order to visualize the smaller peaks at 3.2 and 3.0 ppm).

netic field, approximately one part in 10^8–10^{10}. In human studies, three nuclei are of particular interest:

1. 1H or proton spectroscopy, mainly employed in studies of the brain with prominent peaks developing at choline-containing compounds, myo-inositol, and lactate. Other applications are in the study of creatine/phospocreatine and intramyocellur lipids. Recently, it has been observed that acetyl groups, an intermediary in metabolism, can also be studied with exercise (70) and to evaluate intramuscular lipid metabolism (71). This technique enables the distinction between separate lipid compartments within muscle. Intramuscular cytoplasm serves as a readily available fuel source during exercise. Studies have shown a decrease in response to intermittent and long-duration exercise (71), but not with sprinting (72). These results suggest that both exercise duration and workload are important factors in determining the reduction in intramuscular cytoplasm (73).

    Recently, a post-exercise peak has been shown (74) that presumably reflects acetyl carnitine. As acetyl carnitine has a role in the interplay between carbohydrate and fat metabolism, future studies of this peak will likely contribute to an enhanced understanding of muscle substrate selection by exercise.

2. 13C spectroscopy (75) is used for studying changes in muscle glycogen and metabolic fluxes in the brain and the liver. It was shown that exhaustion during exercise (soccer-specific performance) is related to the capacity to use muscle glycogen (76).

3. 31P spectroscopy (77) detects compounds involved in energy metabolism: creatine phospate, ATP, and inorganic phospate. It is also possible to determine intracellular pH (modified Henderson–Hasselbach equation) because the inorganic phosphate peak position is pH-sensitive (78).

Phosphorous-containing molecules (phosphagens) are of interest because of their role in the energy flux to the cell.

Thus, 31P magnetic spectroscopy has been used to investigate the effects of exercise and recovery on phosphagen metabolism.

At rest, muscle energy metabolism is supplied by oxidative phosporylation, which is maintained at a relatively low rate. The energy state of the cell *in vivo* is correlated with the ratio of PC to inorganic phosphate and can thus be used in metabolic studies. In healthy muscle, at rest, this ratio is 6–12 (79). This ratio is used to study recovery rates and decreased acidification in long-distance runners during and after exercise, attributed to their greater capacity for aerobic metabolism (80). Therefore, this method is very suited to show differentiation between endurance-trained, sprint-trained, and untrained subjects (81).

In summary, it can be stated that 31P spectroscopy has generated a tremendous improvement in the study of muscle bioenergetics.

Great expectations for magnetic resonance spectroscopy exist, a powerful technique that noninvasively monitors biochemistry. These techniques will undoubtedly contribute to an enhanced understanding of normal metabolism and changes associated with exercise.

BIBLIOGRAPHY

1. G. A. Brooks, T. D. Fahey, and T. P. White, *Exercise Physiology. Human Bioenergetics and its Applications*. Mountain View: Mayfield Publishing, 1995.
2. J. A. Carson and F. W. Booth, Molecular biology of exercise. In: W. E. Garett and D. T. Kirkendall, eds., *Exercise and Sport Science*. Philadelphia, PA: Lippincot Williams & Wilkins, 2000, pp. 251–264.
3. R. A. Roberts and S. O. Roberts, *Exercise Physiology: Exercise, Performance and Clinical Applications*. St. Louis, MO: Mosby-Year Book, 1997.
4. P. W. Lemon, K. E. Yarasheski, and D. G. Dolny, The importance of protein for athletes. *Sports Med.* 1984; **1**:474–484.
5. P. W. Lemon, D. G. Dolny, and K. E. Yarasheski, Moderate physical activity can increase dietary protein needs. *Can. J. Appl. Physiol.* 1997; **22**:494–503.
6. P. D. Gollnick, Metabolism of substrates: energy substrate metabolism during exercise and as modified by training. *Fed. Proc.* 1985; **44**:353–357.
7. P. C. Tullson and R. L. Terjung, Adenine nucleotide metabolism in contracting skeletal muscle. *Exerc. Sport Sci. Rev.* 1991; **19**:507–537.
8. S. P. Bessman and A. Fonyo, The possible role of the mitochondrial bound creatine kinase in regulation of mitochondrial respiration. *Biochem. Biophys. Res. Commun.* 1966; **22**:597–602.
9. C. Marconi, D. Pendergast, J. A. Krasney, D. W. Rennie, and P. Cerretelli, Dynamic and steady-state metabolic changes in running dogs. *Respir. Physiol.* 1982; **50**:93–110.
10. P. E. Di Prampero, U. Boutellier, and P. Pietsch, Oxygen deficit and stores at onset of muscular exercise in humans. *J. Appl. Physiol.* 1983; **55**:146–153.
11. G. A. Gaesser and G. A. Brooks, Muscular efficiency during steady-rate exercise: effects of speed and work rate. *J. Appl. Physiol.* 1975; **38**:1132–1139.
12. G. F. Graminski, Y. Kubo, and R. N. Armstrong, Spectroscopic and kinetic evidence for the thiolate anion of glutathione at the active site of glutathione S-transferase. *Biochemistry* 1989; **28**:3562–3568.
13. P. Cerretelli, D. Pendergast, W. C. Paganelli, and D. W. Rennie, Effects of specific muscle training on VO_2 on-response and early blood lactate. *J. Appl. Physiol.* 1979; **47**:761–769.
14. W. C. Stanley and R. J. Connett, Regulation of muscle carbohydrate metabolism during exercise. *FASEB J.* 1991; **5**:2155–2159.
15. S. K. Powers and E. T. Howley, *Exercise Physiology: Theory and Applications to Fitness and Performance*. Madison, WI: Brown and Benchmark, 1997.
16. F. Lipmann, The roots of bioenergetics. *Ciba Found. Symp.* 1975; **31**:3–22.
17. D. K. Mathews and E. L. Fox. In: *The Physiological Basis of Physical Education and Athletics*. Philadelphia, PA: WB Saunders, 1997.
18. B. A. Franklin, Cardiovascular responses to exercise and training. In: W. E. Garrett and D. T. Kirkendall, eds., *Exercise and Sport Science*. Philadelphia, PA: Lippincot Williams and Wilkins, 2000, pp. 107–116.
19. W. T. Phillips, B. J. Kiratli, M. Sarkarati, G. Weraarchakul, J. Myers, B. A. Franklin, I. Parkash, and V. Froelicher, Effect of spinal cord injury on the heart and cardiovascular fitness. *Curr. Probl. Cardiol.* 1998; **23**:641–716.
20. P. K. Wilson, Etiologic and initial cardiovascular characteristics of participants in a cardiac rehabilitation program. *Am. Correct Ther. J.* 1973; **27**:122–125.
21. J. H. Wilmore and A. C. Norton, *The Heart and Lungs at Work: A Primer of Exercise Physiology*. Champaign, IL: Beckman Instruments, 1975.
22. J. H. Mitchell and G. Blomqvist, Maximal oxygen uptake. *N. Engl. J. Med.* 1971; **284**:1018–1022.
23. E. G. Zobl, F. N. Talmers, and R. C. Christensen, Effect of exercise on the cerebral circulation and metabolism. *J. Appl. Physiol.* 1965; **20**:1289–1293.
24. L. B. Rowell, Cardiovascular aspects of human thermoregulation. *Circ. Res.* 1983; **52**:367–379.
25. R. Fagard, A. Aubert, J. Staessen, E. V. Eynde, L. Vanhees, and A. Amery, Cardiac structure and function in cyclists and runners. Comparative echocardiographic study. *Br. Heart J.* 1984; **52**:124–129.
26. R. Fagard, A. E. Aubert, R. Lysens, J. Staessen, L. Vanhees, and A. Amery, Noninvasive assessment of seasonal variations in cardiac structure and function in cyclists. *Circulation* 1983; **67**:896–901.
27. R. H. Fagard, Impact of different sports and training on cardiac structure and function. *Cardiol. Clin.* 1992; **10**:241–56.
28. A. Roy, M. Doyon, J. G. Dumesnil, J. Jobin, and F. Landry, Endurance vs. strength training: comparison of cardiac structures using normal predicted values. *J. Appl. Physiol.* 1988; **64**:2552–2557.
29. A. E. Aubert, D. Ramaekers, Y. Cuche, R. Leysens, H. Ector, and F. Van de Werf, Effect of long-term physical training on heart rate variability. *IEEE Comp. Cardiol.* 1996; **16**:17–20.
30. A. E. Aubert, F. Beckers, and D. Ramaekers, Short-term heart rate variability in young athletes. *J. Cardiol.* 2001; **37**(Suppl 1):85–88.
31. A. Aubert, B. Seps, and F. Beckers, Heart rate variability in athletes. *Sports Med.* 2003; **33**:889–919.
32. D. Ramaekers, H. Ector, A. E. Aubert, A. Rubens, and W. F. Van de, Heart rate variability and heart rate in healthy volunteers. Is the female autonomic nervous system cardioprotective? *Eur. Heart J.* 1998; **19**:1334–1341.

33. J. Henriksson and J. S. Reitman, Quantitative measures of enzyme activities in type I and type II muscle fibres of man after training. *Acta. Physiol. Scand.* 1976; **97**:392–397.
34. D. D. Brown, Pulmonary responses to exercise and training. In: A. Garozzo and D. T. Kirkendall, eds., *Exercise and Sport Science*. Philadelphia, PA: Lippincott Williams and Wilkins, 2000, pp. 117–134.
35. M. L. Fox and S. Keteyian, *Fox's Physiological Basis for Exercise and Sport*. Boston, MA: WCB McGraw-Hill, 1998.
36. B. J. Whipp, Peripheral chemoreceptor control of exercise hyperpnea in humans. *Med. Sci. Sports Exerc.* 1994; **26**:337–347.
37. R. L. Pardy, S. N. Hussain, and P. T. Macklem, The ventilatory pump in exercise. *Clin. Chest Med.* 1984; **5**:35–49.
38. N. L. Jones and E. J. M. Campbell, *Clinical Exercise Testing*. Philadephia, PA: WB Saunders, 1997.
39. F. L. Eldridge, Central integration of mechanisms in exercise hyperpnea. *Med. Sci. Sports Exerc.* 1994; **26**:319–327.
40. N. L. Jones, Dyspnea in exercise. *Med. Sci. Sports Exerc.* 1984; **16**:14–19.
41. W. D. McArdle, F. I. Katch, and V. L. Katch, *Exercise Physiology: Energy, Nutrition and Human Performance*. Baltimore, MD: Williams and Wilkins, 1996.
42. K. Wassermann, J. E. Hansen, D. Y. Sue, B. J. Whipp, and R. Casaburi, *Principles of Exercise Testing and Interpretation*. Philadelphia, PA: Lea and Febiger, 1994.
43. G. A. Brooks, Current concepts in lactate exchange. *Med. Sci. Sports Exerc.* 1991; **23**:895–906.
44. J. A. Davis, Anaerobic threshold: review of the concept and directions for future research. *Med. Sci. Sports Exerc.* 1985; **17**:6–21.
45. J. A. Dempsey, J. B. Wolffe memorial lecture. Is the lung built for exercise? *Med. Sci. Sports Exerc.* 1986; **18**:143–155.
46. L. A. Talbot, E. J. Metter, and J. L. Fleg, Leisure-time physical activities and their relationship to cardiorespiratory fitness in healthy men and women 18–95 years old. Med. Sci. Sports Exerc. 2000; **32**:417–425.
47. E. Byrne-Quinn, J. V. Weil, I. E. Sodal, G. F. Filley, and R. F. Grover, Ventilatory control in the athlete. *J. Appl. Physiol.* 1971; **30**:91–98.
48. M. Miyamura, T. Tsunoda, N. Fujitsuka, and Y. Honda, Ventilatory response to CO_2 rebreathing at rest in the ama. *Jpn. J. Physiol.* 1981; **31**:423–426.
49. D. H. Clarke, Adaptations in strength and muscular endurance resulting from exercise. *Exerc. Sport Sci. Rev.* 1973; **1**:73–102.
50. I. Jacobs, P. A. Tesch, O. Bar-Or, J. Karlsson, and R. Dotan, Lactate in human skeletal muscle after 10 and 30 s of supramaximal exercise. *J. Appl. Physiol.* 1983; **55**:365–367.
51. D. L. Costill, W. J. Fink, M. Hargreaves, D. S. King, R. Thomas, and R. Fielding, Metabolic characteristics of skeletal muscle during detraining from competitive swimming. *Med. Sci. Sports Exerc.* 1985; **17**:339–343.
52. P. D. Gollnick, R. B. Armstrong, C. W. Saubert, K. Piehl, and B. Saltin, Enzyme activity and fiber composition in skeletal muscle of untrained and trained men. *J. Appl. Physiol.* 1972; **33**:312–319.
53. V. J. Caiozzo, J. J. Perrine, and V. R. Edgerton, Training-induced alterations of the *in vivo* force-velocity relationship of human muscle. *J. Appl. Physiol.* 1981; **51**:750–754.
54. E. F. Coyle, D. C. Feiring, T. C. Rotkis, R. W. Cote, III, F. B. Roby, W. Lee, and J. H. Wilmore, Specificity of power improvements through slow and fast isokinetic training. *J. Appl. Physiol.* 1981; **51**:1437–1442.
55. M. R. Deschenes, J. Covault, W. J. Kraemer, and C. M. Maresh, The neuromuscular junction. Muscle fibre type differences, plasticity and adaptability to increased and decreased activity. *Sports Med.* 1994; **17**:358–372.
56. United States Department of Health and Human Services. *Physical Activity and Health: A Report of Surgeon General*. Washington, DC: U.S. Department of Health and Human Services, Centers for Disease Control and Prevention, National Center for Chronic Disease Prevention and Health Promotion, 1996.
57. K. Iwasaki, R. Zhang, J. H. Zuckerman, and B. D. Levine, Dose-response relationship of the cardiovascular adaptation to endurance training in healthy adults: how much training for what benefit? *J. Appl. Physiol.* 2003; **95**:1575–1583.
58. American College of Sports Medicine Position Stand. The recommended quantity and quality of exercise for developing and maintaining cardiorespiratory and muscular fitness, and flexibility in healthy adults. *Med. Sci. Sports Exerc.* 1998; **30**:975–991.
59. B. F. Hurley and J. M. Hagberg, Optimizing health in older persons: aerobic or strength training? *Exerc. Sport Sci. Rev.* 1998; **26**:61–89.
60. A. J. Schuit, L. G. van Amelsvoort, T. C. Verheij, R. D. Rijneke, A. C. Maan, C. A. Swenne, and E. G. Schouten, Exercise training and heart rate variability in older people. *Med. Sci. Sports Exerc.* 1999; **31**:816–821.
61. P. K. Stein, A. A. Ehsani, P. P. Domitrovich, R. E. Kleiger, and J. N. Rottman, Effect of exercise training on heart rate variability in healthy older adults. *Am. Heart J.* 1999; **138**:567–576.
62. A. Loimaala, H. Huikuri, P. Oja, M. Pasanen, and I. Vuori, Controlled 5-mo aerobic training improves heart rate but not heart rate variability or baroreflex sensitivity. *J. Appl. Physiol.* 2000; **89**:1825–1829.
63. R. Perini, N. Fisher, A. Veicsteinas, and D. R. Pendergast, Aerobic training and cardiovascular responses at rest and during exercise in older men and women. *Med. Sci. Sports Exerc.* 2002; **34**:700–708.
64. I. M. Lee, C. C. Hsieh, and R. S. Paffenbarger, Jr., Exercise intensity and longevity in men. The Harvard Alumni Health Study. *JAMA* 1995; **273**:1179–1184.
65. M. B. Snijder, B. E. Kuyf, and P. Deurenberg, Effect of body build on the validity of predicted body fat from body mass index and bioelectrical impedance. *Ann. Nutr. Metab.* 1999; **43**:277–285.
66. M. W. Wukitsch, Pulse oximetry: historical review and Ohmeda functional analysis. *Int. J. Clin. Monit. Comput.* 1987; **4**:161–166.
67. S. R. Underwood, D. N. Firmin, R. S. Rees, and D. B. Longmore, Magnetic resonance velocity mapping. *Clin. Phys. Physiol. Meas.* 1990; **11**(Suppl A):37–43.
68. F. Stahlberg, L. Sondergaard, and C. Thomsen, MR flow quantification with cardiovascular applications: a short overview. *Acta Paediatr. Suppl.* 1995; **410**:49–56.
69. G. B. Matson and M. W. Weiner, Spectroscopy. In: D. D. Stark and J. W. Bradley, eds., *Magnetic Resonance Imaging*. St. Louis, MO: Mosby-Year Book, 1992.
70. R. Kreis, B. Jung, S. Rotman, J. Slotboom, and C. Boesch, Non-invasive observation of acetyl-group buffering by 1H-MR spectroscopy in exercising human muscle. *NMR Biomed.* 1999; **12**:471–476.

71. J. Rico-Sanz, M. Moosavi, E. L. Thomas, J. McCarthy, G. A. Coutts, N. Saeed, and J. D. Bell, In vivo evaluation of the effects of continuous exercise on skeletal muscle triglycerides in trained humans. *Lipids* 2000; **35**:1313–1318.
72. J. Rico-Sanz, J. V. Hajnal, E. L. Thomas, S. Mierisova, M. Ala-Korpela, and J. D. Bell, Intracellular and extracellular skeletal muscle triglyceride metabolism during alternating intensity exercise in humans. *J. Physiol.* 1998; **510**(Pt 2): 615–622.
73. K. Brechtel, A. M. Niess, J. Machann, K. Rett, F. Schick, C. D. Claussen, H. H. Dickhuth, H. U. Haering, and S. Jacob, Utilisation of intramyocellular lipids (IMCLs) during exercise as assessed by proton magnetic resonance spectroscopy (1H-MRS). *Horm. Metab. Res.* 2001; **33**:63–66.
74. R. Kreis, B. Jung, S. Rotman, J. Slotboom, and C. Boesch. Non-invasive observation of acetyl-group buffering by 1H-MR spectroscopy in exercising human muscle. *NMR Biomed.* 1999; **12**:471–476.
75. U. K. Decking, 13C-MR spectroscopy–how and why? *MAGMA* 1998; **6**:103–104.
76. J. Rico-Sanz, M. Zehnder, R. Buchli, M. Dambach, and U. Boutellier, Muscle glycogen degradation during simulation of a fatiguing soccer match in elite soccer players examined noninvasively by 13C-MRS. *Med. Sci. Sports Exerc.* 1999; **31**:1587–1593.
77. M. Horn, Experimental spectroscopy. General methodology and applications for 31P. *MAGMA* 1998; **6**:100–102.
78. J. T. Chen, T. Taivassalo, Z. Argov, and D. L. Arnold, Modeling *in vivo* recovery of intracellular pH in muscle to provide a novel index of proton handling: application to the diagnosis of mitochondrial myopathy. *Magn. Reson. Med.* 2001; **46**:870–878.
79. Z. Argov, M. Lofberg, and D. L. Arnold, Insights into muscle diseases gained by phosphorus magnetic resonance spectroscopy. *Muscle Nerve* 2000; **23**:1316–1334.
80. T. Yoshida, The rate of phosphocreatine hydrolysis and resynthesis in exercising muscle in humans using 31P-MRS. *J. Physiol. Anthropol. Appl. Human Sci.* 2002; **21**:247–255.
81. L. Johansen and B. Quistorff, 31P-MRS characterization of sprint and endurance trained athletes. *Int. J. Sports Med.* 2003; **24**:183–189.

BIOHEAT TRANSFER MODEL

JING LIU
Chinese Academy of Sciences &
Tsinghua University
Beijing, China

1. INTRODUCTION

Bioheat transfer models have significant applications in a wide variety of clinical, basic, and environmental sciences (1,2). In particular, understanding the heat transfer in biological tissues involving either raising or lowering of temperature is a necessity for many therapeutic practices such as cancer hyperthermia (3), burn injury (2,4), brain hypothermia resuscitation (5), disease diagnostics (6), thermal comfort analysis (7), and cryosurgery (8–10) and cryopreservation (11).

In a hyperthermia process, whose primary objective is to raise the temperature of the diseased tissue to a therapeutic value, typically, 42–43°C, and then thermally destroy it, various apparatuses such as the microwave (12), the ultrasound (13), and the laser (14) have been used to deposit heating for treating the tumor in the deep biological body. However, in some other cases, the high temperature should be avoided. For example, skin burns caused by exposing to heat in a flash fire, laser irradiation, or contact with hot substances (2,4) are the most commonly encountered hazards in daily life or in industry. In these situations, the prediction of variation of temperature in both space and time is requested to evaluate burn injury of tissues, which would serve to find ways either to induce or prevent such thermal hazards to the target tissues.

In contrast to the principle of hyperthermia therapy, cryosurgery realizes its clinical object by a controlled destruction of tissues through deep freezing and thawing (15). Applications of this treatment are wide in dermatology and gynecology, glaucoma, lung tumor, urology, orthopedics, otology, neurosurgery, ophthalmology, management of cancer, and other specialties because of its outstanding virtues such as quick, clean, relatively painless, good homeostasis (arresting of bleeding), and satisfaction of the little scar. An accurate understanding of the extent of the irregular shape of the frozen region, the direction of ice growth, and the temperature distribution within the ice balls during the freezing process is a basic requirement for the successful operation of a cryosurgery.

Except for the above extreme cases, quantification on the thermal processes of biological body in a mild temperature range also found significant applications in many medical issues, such as disease diagnostics (6), thermal comfort analysis (16), and thermal parameter estimation (17,18). For example, temperature mapped at the skin surface of biological body is a unique index to reflect the disease statue, which has led to the rapid progress of disease diagnostics by infrared thermometer, because the body surface temperature is often determined by the blood circulation underneath the skin, the local metabolism, the radiation emissivity and humidity of skin, the convective heat transfer coefficient, and the relative humidity and temperature of the surrounding air. Changes in any of these parameters will induce alteration from the normal temperature or heat flux range at the skin surface, reflecting the physiological or pathological status of human body (6). To better understand the disease processes that go with thermal abnormalities, and consequently to use thermal imaging information not only to detect pathology but also to manage disease better, accurate correlations between skin thermal information and human pathophysiology need to be clearly understood.

During some adverse pathological or emergent clinical situations such as circulation arresting, or cardiac operations, especially for procedures requiring reduced perfusion or circulatory arrest (5), it has been commonly accepted that a mild or moderate hypothermia (>30°C) is by far the most principal means of neurologic protection. In fact, brain cooling has been proven to be an efficient way to restrain the cerebral oxygen and metabolic demands and thus prevent cerebral tissues from being damaged. Comprehension on the hypothermia behavior involved would further advance the brain resuscitation. To numerically simulate the thermal development of head subject to hypoxia and then to reveal its effect to the oxygen transport in the cerebrovascular network, establishing a mathematical model is very necessary.

All of these medical issues caused a high demand for modeling the bioheat transfer process. In this section, aiming to provide a fundamental knowledge for the readers to grasp the basic rule of theoretically tackling the above bioengineering issues, an overview on the bioheat transfer model and its typical applications from low to high temperature will be presented. Considering that the thermoregulation mechanisms of the biological bodies were often neglected for simplicity in most of the former studies, this discussion also falls into such a category. Besides, covering various model forms is not the current objective. Therefore, only the most commonly encountered basic bioheat transfer models in tissue level will be discussed. For those in micro-scale or requiring specific derivation, readers are referred to Refs. 1, 9, 11 and 19 for more detail.

2. BIOHEAT TRANSFER MODEL WITHOUT PHASE CHANGE

2.1. Classic Pennes Bioheat Equation

Generally, due to highly nonhomogeneous thermal properties, it is extremely difficult to quantify the thermal behavior of a biological body by means of distinguishing its temperature in local tissue from the vascular network. A simple yet intuitive way is to introduce the temperature T to characterize the overall thermal state in a specific position of the tissue. Until now, the classic Pennes equation has been commonly accepted as the best practical approach for modeling bioheat transfer, due to its simplicity and validity. This is because most of the other models still lack sound experimental grounding and are generally very complex. Therefore, in view of its simplicity and wide applicability, the Pennes equation kept being proved as a generalized model for characterizing the heat transport process in the skin tissues.

As is well known, the arterial temperature decreases due to conduction of heat from arterial blood to the surrounding tissue, and in the capillary bed, the condition of very slow flow with a superposed oscillating component favors almost complete thermal equilibrium between the bloodstream and the surrounding tissue. As the precapillary and capillary beds are the major sites for exchange of heat in tissue, it is reasonable to assume the equality in postcapillary blood and tissue temperatures. It was based on this justification that the Pennes equation (20) was established and its generalized form can be written as

$$\rho C \frac{\partial T(\mathbf{X},t)}{\partial t} = \nabla \cdot k(\mathbf{X}) \nabla [T(\mathbf{X},t)]$$
$$+ w_b(\mathbf{X}) C_b [T_a - T(\mathbf{X},t)]$$
$$+ Q_m(\mathbf{X},t) + Q_r(\mathbf{X},t), \quad \mathbf{X} \in \Omega, \quad (1)$$

where ρ and C are, respectively, the density and the specific heat of tissue; c_b denotes the specific heat of blood; $\mathbf{X}$ contains the Cartesian coordinates x, y, and z; Ω denotes the analyzed spatial domain; $k(\mathbf{X})$ is the space-dependent thermal conductivity; and $w_b(\mathbf{X})$ is the space-dependent blood perfusion and can generally be measured through a thermal clearance method. This value represents the blood flow rate per unit tissue volume and is contributed mainly from microcirculation including the capillary network plus small arterioles and venules; T_a is the blood temperature in the arteries supplying the tissue and is often treated as a constant at 37°C; $T(\mathbf{X},t)$ is the tissue temperature; $Q_m(\mathbf{X},t)$ is the metabolic heat generation; and $Q_r(\mathbf{X},t)$ is the distributed volumetric heat source due to externally applied spatial heating.

Clearly, after introducing the intuitive concept of Pennes' blood perfusion term, analysis on bioheat transfer was significantly simplified. It should be noted that for those tissues embedded with large blood vessels, energy equations for the blood flow in a single vessel will be needed and then combined with the Pennes equation to predict the whole temperature fields.

2.2. Particularities of Geometry and Parameters in the Bioheat Transfer Model

The geometric shape, dimensions, thermal properties, and physiological characteristics for tissues, as well as the arterial blood temperature, were usually used as the input to the Pennes equation for a parametric study. For example, the biological tissue can be stratified as three layers (including skin, fat, and flesh layer, respectively). In each layer, the thermal parameters were treated as constant yet different from each other. A typical geometry for such simplified structure can be shown as Fig. 1, although real anatomical geometry can also be dealt

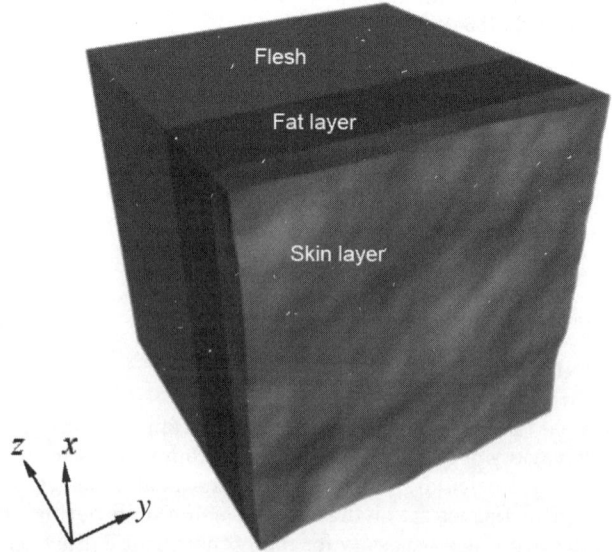

Figure 1. Simplified three-layer model of tissue (z denotes the tissue depth from the skin surface, and three layers are the skin, fat, and flesh layer, respectively). (This figure is available in full color at http://www.mrw.interscience.wiley.com/ebe.)

with. Physiologically, in the skin and fat layers, the blood perfusion and metabolic rate are both zero. The bioheat transfer equation can be directly used to characterize the thermal process of the biological bodies subject to various external or internal factors such as the convective interaction with a heated or cooled fluid, radiation by fire or laser, contact to a heating or freezing apparatus, electromagnetic trauma, or a combination among them. Such issues can be treated using different boundary conditions as well as spatial heating or freezing patterns. Meanwhile, the bioheat transfer equation should be modified by incorporating more complex yet real situations including coordinate systems and the calculation domain.

The flexibility of the Pennes equation in a variety of bioheat transfer analysis can be illustrated by the following examples. When cerebral circulation arrests a transient state, the blood perfusion may gradually reduce in a short period of time, but the brain metabolism continues until it runs out of the metabolic substrate. In such a case, Equation 1 can be modified by incorporating these properties, although completely characterizing them is still a difficulty. Meanwhile, the ischemic region of the brain can be reflected by reducing the blood perfusion rate to a lowered value or a certain percentage of its normal value. In some brain cooling via ventilating cold blood through the vessels, the arterial temperature in the Pennes equation was intentionally lowered to accommodate to the practical situation.

The metabolic heat generation generally depends on the local tissue temperature. But in most cases, it was treated as a constant. The baseline metabolic rate is chosen as that which enables the patient to maintain his/her own body temperature in a normal state. The most widely adopted approach in estimating the metabolic term has been to set it as the product of the oxygen consumption and its caloric value. It is assumed that the metabolic rate changes according to the temperature coefficient ϕ, i.e. (5),

$$Q_m = Q_{m0} \cdot \phi^{[T-37]/10}, \quad (2)$$

where Q_{m0} is the reference metabolic rate at 37°C and ϕ is usually set as 3.0.

It should be emphasized that the thermal parameters involved in the bioheat transfer model for the living tissues would vary under different statuses like tumor formation, vascular disorders, hemodynamic dysfunction, and during or after external heating or freezing. Therefore, when performing a thermal analysis, caution choosing the appropriate parameters should always be taken.

2.3. Boundary and Initial Conditions

The boundary condition for the heat transfer occurring at the skin surface generally is composed of three parts: convection, radiation, and evaporation; i.e. (6),

$$-k\frac{\partial T}{\partial n}\bigg|_{skin} = h_f(T_s - T_f) + \sigma\varepsilon(T_s^4 - T_f^4) + Q_e, \quad (3)$$

where h_f is the convection heat transfer coefficient; T_s and T_f are the skin and surrounding air temperature, respectively; ε is the skin emissivity; σ is the Stefan–Boltzmann constant; Q_e is the evaporative heat losses due to sweat secretion; and (refer to Refs. 7 and 21)

$$Q_e = Q_{dif} + Q_{rsw}, \quad (4)$$

$$Q_{dif} = 3.054(0.256T_s - 3.37 - P_a), \quad (5)$$

$$Q_{rsw} = 16.7h_f W_{rsw}(0.256T_s - 3.37 - P_a), \quad (6)$$

where Q_{dif} is the heat loss by evaporation of implicit sweat secretion when the skin is dry; Q_{rsw} is the heat loss by evaporation of explicit sweat secretion; W_{rsw} is the skin humidity; $0 \leq W_{rsw} \leq 1$ and $W_{rsw} = 0, 1$, respectively, mean that the skin is dry and entirely wet; P_a is the vapor pressure in ambient air; $P_a = \phi_a P_a^*$, where ϕ_a is the relative humidity of surrounding air; and P_a^* is the saturated vapor pressure at surrounding air temperature.

The nonlinear boundary condition in Equation 3 due to the occurrence of the $\sigma\varepsilon T_s^4$ term can be solved through iteration until an acceptable error was obtained. The whole initial temperature field for the biological bodies was often treated as uniform at 37°C just like that of the body core for simplicity or obtained through solving the bioheat equation at steady state by incorporating the corresponding boundary conditions.

Solving the generalized bioheat transfer model and the boundary and initial conditions, the relative contribution of each thermal factor to skin temperature distribution can be evaluated. For example, Fig. 2 depicts a comparison of the heat fluxes due to radiation, convection, and evaporation and the sum of these heat fluxes at the skin surface (6). It indicates that the thermal radiation and evaporation of skin is evident. In addition, heat loss due to evaporation when the skin is dry is much less than that while the skin is partially wet. Consequently, the nonhomogeneous skin humidity can also result in abnormal temperature distribution at the skin surface. All of these complexities should be considered in some specific clinical practices such as accurate thermal diagnosis. But for a general analysis, the evaporative and radiative heat transfer at the skin surface were often omitted or just attributed to the apparent convective heat transfer term for simplicity.

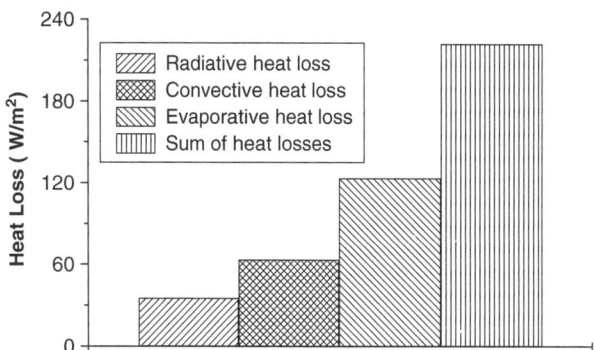

Figure 2. The heat fluxes at skin surface due to radiation, convection, and evaporation ($\varepsilon = 0.9$, $W_{rsw} = 0.2$, $\phi = 40\%$, $T_f = 25°C$, and $h_f = 10\,\text{W/m}^2 \cdot °C$).

2.4. Extension of the Classic Bioheat Transfer Model

Until now, most thermal medical analyses are based on the well-known Pennes equation. The history of bioheat transfer art and science can be merged to modify and improve this model. Among all of the efforts ever made, the blood perfusion term in the Pennes equation has been substantially studied, which led to several conceptually innovated bioheat models such as the Wulff continuum model (22), the Chen–Holmes model addressing both the flow and the perfusion properties of blood (23), and the Weinbaum–Jiji three-layer model to characterize the heat transfer in the peripheral tissues (24).

Except for the extensive works performed on modeling the blood flow heat transfer, some other efforts were made to better understand the mechanisms of the metabolic term in the Pennes equation (25). Meanwhile, attentions were also paid to the heat conduction term. To date, nearly all of the studies on the bioheat conduction were based on the Fourier law, which implies an instantaneous thermal energy deposition in medium; i.e., any local temperature disturbance causes an instantaneous perturbation in temperature at each point in the medium. Modifications on this term have led to several studies on wave-like heat transfer in living tissues (26).

2.5. Solutions to the Classic Bioheat Transfer Model

It is often desirable to obtain as flexible as possible a way to analytically solve the most widely used Pennes equation. Such solutions are mainly based on the Green's function method (27), which is beneficial for dealing with the nonhomogeneous problem with the spatial or transient heating source and initial temperature distribution, as well as complex boundary conditions. This is because the Green's function obtained for the differential equation is independent of the source term and the transient or space-dependent boundary conditions.

When the analytical solutions were not available, numerical approaches such as the finite difference method (FDM), finite element method (FEM), and boundary element method (BEM) all have been used. Among these, the FDM is convenient for compiling the computer code but is not convenient to osculate the coordinate of the complex biological shape. As an alternative, the FEM has good adaptability to the complex shape (28). Compared with these, the BEM has a unique virtue to provide a complete problem solution in terms of boundary values only, with substantial savings in computer time and data preparation (29). But severe restrictions exist in the traditional BEM for solving the bioheat transfer equation (BHTE). One is that the fundamental solution to the BHTE with the nonhomogeneous blood perfusion term is hard to obtain. Also, in most cases, it is inconvenient to alter the program by incorporating a new fundamental solution when the user wishes to study a slightly different bioheat equation. Furthermore, the nonhomogeneous term accounting for the spatial heating needs to be included in the ordinary BEM formulation by means of domain integrals, which makes the technique time consuming and loses the attraction of its "boundary-only" character. In this side, the dual reciprocity boundary element method (DRBEM) can avoid the above restrictions (30). Overall, the geometry and properties of biological bodies vary drastically which make the analysis of bioheat transfer complex. And the numerical calculations are therefore often requested. Except for the solution to the direct problems with input model parameters, solution to the inverse problems based on partially measurable variables of the model to infer the unknown system parameters also consists of one extremely important issue for hyperthermia treatment (31). In fact, such problems seem more complicated and yet common in reality. As requested by many clinical practices, optimization on the bioheat transfer problem to find out the best treatment protocol is strongly needed.

3. BIOHEAT TRANSFER MODEL IN CRYOMEDICAL ENGINEERING

3.1. Particularities of Modeling With Phase Change

For a cryosurgery or cryopreservation, it is in the cooling and rewarming processes that injury was induced. To obtain an optimal output of the biological materials, quantitative evaluation on the temperature history during the phase change is highly desirable. Knowing the phase change front and the transient temperatures is critical for adjusting the freezing to realize the specific object either leading to damage or preserving the cells (8–10). A major difference between simulation of cryosurgery and that in hyperthermia lies in that a phase change process occurred in the former case. As a result, the thermal conductivities for the frozen tissue and blood-perfused region were chosen differently; for example, a smaller value for tissue was often taken as $k_l = 0.5$ W/m $\cdot$ °C, whereas in the frozen region, a larger value as $k_s = 2$ W/m $\cdot$ °C was used (32). Clearly, blood perfusion and metabolic heat generation in the unfrozen region exist, whereas they disappear after being frozen. Considering that biological tissues in cryosurgery experience a wide range of temperature change, their thermal properties (including specific heat and thermal conductivity) are expected to change significantly over the freezing/thawing process. That is to say, the thermal parameters of the biological tissues are generally temperature-dependent. Several previous studies have shown that inclusion of temperature dependence has a significant effect on phase change predictions (33,34). But for simplicity, constant assumptions were often adopted or only space-dependent parameters were taken into concern. For example, to avoid the expensive and intensive numerical iteration, a multi-segmental constant thermal conductivity has been used to approximate the temperature-dependent case (34). Furthermore, due to the nonideal solution property, phase change temperature for the biological tissues usually occurs in a wide range (35), say between −1°C (upper limit) and −8°C (lower limit), not just fixed at 0°C, as assumed in most calculations.

At this time, only a few efforts have been made analyzing the rewarming behavior and the role of the blood flow to the phase change heat transfer process. In most modeling on the rewarming process during cryosurgery, it has often been assumed that when tissue is thawed, blood will reflow to the originally frozen and then completely thawed region. This, however, only partially holds true. In fact, if the blood vessels were destroyed due to irreversible freezing injury, blood will no longer be able to, or need certain time to, reflow to this area. For such cases, the heat transfer models with blood reflow behavior (36) should be used to accurately predict the thawing behavior.

3.2. Models for Characterizing Different Phases Separately

For the heat transfer in unfrozen tissue, the classic model was often used; i.e.,

$$\rho_u C_u \frac{\partial T_u(\mathbf{X},t)}{\partial t} = \nabla \cdot k_u \nabla [T_u(\mathbf{X},t)] + w_b C_b [T_a - T_u(\mathbf{X},t)] + Q_m, \quad \mathbf{X} \in \Omega_u(t), \quad (7)$$

where subscript u indicates the unfrozen phase.

In the frozen region, due to the absence of blood perfusion and metabolic activities, the heat balance is given by

$$\rho_f C_f \frac{\partial T_f(\mathbf{X},t)}{\partial t} = \nabla \cdot k_f \nabla [T_f(\mathbf{X},t)], \quad \mathbf{X} \in \Omega_f(t), \quad (8)$$

where subscript f indicates frozen tissue.

For ideal biological tissues, the temperature continuum and energy balance conditions at the moving solid–liquid interface are given as follows (assuming that the density of tissue ρ is same constant for both liquid and solid phases):

$$T_f(\mathbf{X},t) = T_u(\mathbf{X},t) = T_m, \quad \mathbf{X} \in \Gamma_{m.i.}, \quad (9)$$

$$k_f \frac{\partial T_f(\mathbf{X},t)}{\partial n} - k_u \frac{\partial T_u(\mathbf{X},t)}{\partial n} = Q_l V_n, \quad \mathbf{X} \in \Gamma_{m.i.}, \quad (10)$$

where n denotes the unit outward normal; Q_l, T_m are, respectively, the latent heat and freezing point of tissue; $\Gamma_{m.i.}$ is the moving boundary, i.e., moving interface resulted by phase change; and V_n is the normal velocity of the moving interface.

Due to the high nonlinearity of Equations 7–10, a complex iteration at the moving boundary is inevitable by directly discretizing these governing equations.

3.3. Effective Heat Capacity Model

To avoid the iteration at the moving boundary, the effective heat capacity method is also often adopted. Since first proposed by Bonacina et al. (37), the effective heat capacity method has been used by many investigators to solve phase change problems. The advantage of this method lies in that a fixed grid can be used for the numerical computation, and that the nonlinearity at the moving boundary can thus be avoided. The essence of the effective heat capacity method is to approximate the latent heat by a generalized effective heat capacity over a small temperature range near the freezing point (38). Following this strategy, the numerical solution can be carried out on the fixed grid throughout the calculation process, which is much easier to implement.

To apply the effective heat capacity method, Equations 7 and 8 must be substituted by a uniform energy equation that can be constructed as follows (38):

$$\Lambda \frac{\partial T(\mathbf{X},t)}{\partial t} = \nabla \cdot k(T) \nabla [T(\mathbf{X},t)] + w_b(T) C_b [T_a - T(\mathbf{X},t)] + Q_m(T), \quad \mathbf{X} \in \Omega, \quad (11)$$

where $\Lambda = \rho C(T) + Q_l \delta(T - T_m)$ and

$$\rho C(T) = \begin{cases} \rho C_f, & T < T_m \\ \rho C_u, & T > T_m \end{cases}, \quad \delta(T - T_m)$$

is the Dirac function;

$$k(T) = \begin{cases} k_f, & T < T_m \\ k_u, & T > T_m; \end{cases} \quad Q_m(T) = \begin{cases} 0, & T < T_m \\ Q_m, & T > T_m; \end{cases} \text{ and}$$

$$w_b(T) = \begin{cases} 0, & T < T_m \\ w_b, & T > T_m. \end{cases}$$

As the phase change of real biological tissue does not take place at a specific temperature but within a temperature range, it is reasonable to substitute a large effective heat capacity over a temperature range (T_{ml}, T_{mu}) for the latent heat, where T_{ml} and T_{mu} are, respectively, the lower and upper phase transition temperatures of tissue. Introducing the effective heat capacity $\tilde{C}$, effective thermal conductivity $\tilde{k}(T)$, effective metabolic heat generation $\tilde{Q}_m$, and effective blood perfusion $\tilde{w}_b(T)$, respectively, as (assuming that k_u, k_f, C_u, and C_f are all constant)

$$\tilde{C}(T) = \begin{cases} \rho C_f, & T < T_{ml} \\ \frac{\rho Q_l}{(T_{mu} - T_{ml})} + \frac{\rho C_f + \rho C_u}{2}, & T_{ml} \leq T \leq T_{mu} \\ \rho C_u, & T > T_{mu}, \end{cases} \quad (12)$$

$$\tilde{k}(T) = \begin{cases} k_f, & T < T_{ml} \\ (k_f + k_u)/2, & T_{ml} \leq T \leq T_{mu} \\ k_u, & T > T_{mu}, \end{cases} \quad (13)$$

$$\tilde{Q}_m(T) = \begin{cases} 0, & T < T_{ml} \\ 0, & T_{ml} \leq T \leq T_{mu} \\ Q_m, & T > T_{mu}, \end{cases} \quad (14)$$

$$\tilde{w}_b(T) = \begin{cases} 0, & T < T_{ml} \\ 0, & T_{ml} \leq T \leq T_{mu} \\ w_b, & T > T_{mu}. \end{cases} \quad (15)$$

Then Equation 11 can be rewritten as

$$\tilde{C} \frac{\partial T}{\partial t} = \nabla \cdot \tilde{k} \nabla T + \tilde{w}_b C_b (T_a - T) + \tilde{Q}_m, \quad \mathbf{X} \in \Omega. \quad (16)$$

Consequently, through introducing the effective heat capacity, the complex nonlinear phase change problems are simplified as nonhomogeneous ones that can be easily dealt with by a general numerical method.

3.4. Moving Heat Source Model

Except for the above modeling approach, the three equations used to characterize the solid phase, the liquid phase, and the solid–liquid interface separately can in fact be equivalently transformed to a single heat conduction model with a moving heat source term (39), such that for a cryopreservation process, the one-dimensional model reads as

$$\rho C \frac{\partial T(\mathbf{X},t)}{\partial t} = \nabla \cdot [k\nabla T(\mathbf{X},t)]$$
$$+ \rho Q_l \frac{ds(\mathbf{X},t)}{dt}\delta[\mathbf{X}-s(\mathbf{X},t)], \quad t>0, \qquad (17)$$

with a condition at the phase change front:

$$T(\mathbf{X},t) = T_m, \quad \mathbf{X} = s(\mathbf{X},t), \quad t>0, \qquad (18)$$

where $\delta[\mathbf{X}-s(\mathbf{X},t)]$ is the delta function and $s(\mathbf{X},t)$ is the moving interface.

The second term in the Equation 17 appears as a moving heat source q_r. With such a relatively simple expression, the equations can be analytically solved using the Green's function method (39).

Except for characterizing the freezing process, the moving heat transfer modeling approach can also be extended to deal with more complex phase change problems. In skin rewarming after cryosurgery, if certain spatial heating apparatus such as microwave or ultrasound are applied for a better thawing effect, the moving heat source should be expressed as

$$q_r(\mathbf{X},t) = \rho Q_l \frac{ds(\mathbf{X},t)}{dt}\delta[\mathbf{X}-s(\mathbf{X},t)] + Q_r(\mathbf{X},t), \qquad (19)$$

where Q_r is the external volumetric heating.

In fact, the physical meaning for the spatial heat source can be much wider. It can be sensed as an apparent heat product contributed from both the metabolism as well as the blood heat transfer effect. During cryosurgery, effects of the blood perfusion heat transfer and the metabolic rate on the phase change can be approximately studied through defining the appropriate heat source in the energy equation. Assuming that the perfusion and metabolic activity will cease once freezing at that location is initiated, then the corresponding moving heat source in the energy equation can be expressed as follows:

$$q_r(\mathbf{X},t) = \rho Q_l \frac{ds(\mathbf{X},t)}{dt}\delta[\mathbf{X}-s(t)] + q_m H[\mathbf{X}-s(t)], \qquad (20)$$

where $q_m = w_b C_b (T_a - T) + Q_m$ represents heat generation due to blood perfusion and metabolic rate and H is the heavy side function. Analysis from this approach can reflect a certain *in vivo* freezing situation.

3.5. Solution to the Phase Change Heat Transfer Model

The main difficulties encountered in cryosurgical simulation are the unknowns on the extent of the irregular shape of the frozen region, the direction of ice growth, and the temperature distribution within the ice balls during the freezing process. Such moving boundary problems are highly nonlinear. Exact solutions are only possible for some simple one-dimensional problems (32,39). Clearly, the complexity of multidimensional phase change problems during cryosurgery suggests numerical approaches. In the long-term development of cryosurgery technology, several numerical models to solve the phase change problems of biological tissues have been proposed. Generally, the existing numerical schemes can be divided into two basic approaches (38): One is based on the front-tracking technique, whereas another is on the non-front-tracking technique including enthalpy formulation and the effective heat capacity method. Most of the previous numerical efforts were mainly focused on the one- or two-dimensional heat transfer models.

Numerical calculations on the phase change problems of biological tissues can be done by the FDM, FEM, the finite volume method, or BEM. As explained, the BEM has an advantage over others due to its requiring only discretization on the boundaries of the domain. However, the traditional BEM for problems involving nonlinearities may still yield difficulty in deriving the so-called fundamental solution and thus requires additional domain discretization. This difficulty can be resolved by the DRBEM, applying the reciprocity relationship twice. Since its introduction by Nardini and Brebbia (40), the DRBEM has been successfully applied to various heat transfer problems including phase change problems for both single moving boundary and multi-moving-boundary cases. Recently, this method was extended to solve the multidimensional phase change problem of biological tissues during cryosurgery (30).

Depicted in Fig. 3 is a typical temperature distribution for the freezing by three probes with an identical insertion depth into the tissue (38). It was shown that the temperature responses at tissues surrounding the three probe tips are much different from the rest of the tissues, and that three identical valleys appear in the temperature distributions during freezing, whereas three identical peaks were produced at the same positions during heating. Figure 4 gives the location and size of the ice ball produced by the freezing of probes. Compared with the case of a single freezing, the ice ball formed by the three-probes freezing is much larger. It must be noticed that the domain of the ice ball should exceed the area of the tumor, which is for completely destroying the target. The use of multiple probes permits overlapping the requested frozen/heated areas in the treatment of large tumors and provides a method of destroying the tissue to the desired size and shape in complex tumor ablation. Meanwhile, overfreezing might also cause an irreversible injury to the neighboring healthy tissues. Therefore, numerical calculations can provide a very informative prediction on the tissue temperature responses and thus help to optimize the treatment parameters before the tumor operation. For the method to optimize the cryosurgical protocols, readers are referred to Refs. 41–43 for more detail.

4. ROLE OF SINGLE LARGE BLOOD VESSEL TO HEAT TRANSFER

The role of the large blood vessel to the bioheat transfer during tumor hyperthermia has been well documented in

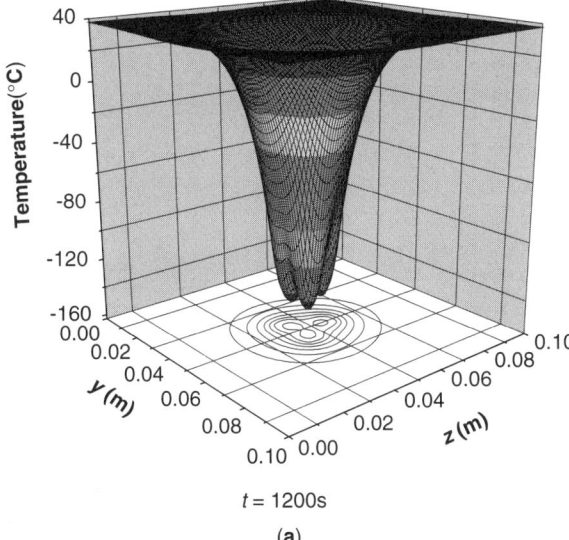

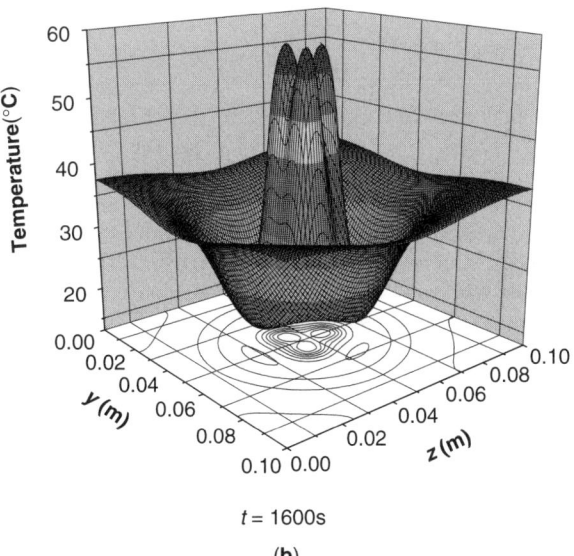

Figure 3. Tissue temperature distributions at cross section of $x = 0.027$ m for the case of three-probes with identical insertion depth. (This figure is available in full color at http://www.mrw.interscience.wiley.com/ebe.)

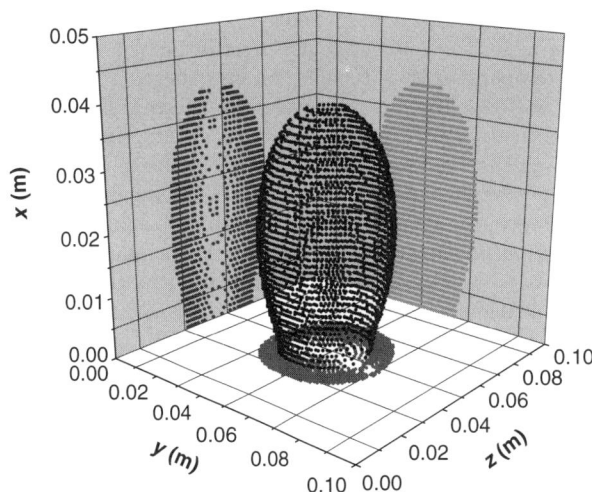

Figure 4. The location and size of an ice ball after freezing for $t = 1200$ s. (This figure is available in full color at http://www.mrw.interscience.wiley.com/ebe.)

the literature (44). However, few efforts were made on understanding their effects in the cryosurgical treatment up to now. Without any doubt, large blood vessels play extremely important roles over the heat transfer process, leading to tissue freezing, just like that in the hyperthermia clinics.

As is well known, tumor growth and survival critically depend on its blood vessel network. The process of neovascularization is a universal characteristic of solid cancers larger than a few millimeters, which results in the fact that tumors are often situated near some large blood vessels. In addition, some malignancies, such as pancreatic tumors, encase the aorta and other major vessels. The presence of tumor involvement of vessels will, in most cases, make the patient ineligible for curative resection. Therefore, the use of cryosurgery or hyperthermia in these cases often appears as an attractive choice. However, it would be a plague to implement cryosurgery or hyperthermia when a tumor is too close to a critical blood vessel or such a vessel transits the tumor. There are mainly two reasons can resolve this mechanism: On the one hand, the heating/cooling nature of the flowing blood in the large vessels can produce steep temperature gradients in frozen/heated tissues, resulting in inadequate cooling/heating temperatures and then contributing to the non-killing of the tumor during cryosurgery/hyperthermia; on the other hand, cutting off the circulation of blood vessels and/or bleeding due to ruptures of large blood vessels by the ice ball or burning injury during the ablation procedure may cause undesired damage to healthy tissues or organs (45). Besides, there is still much concern about the effect of cryotherapy/hyperthermia on major vessels in a young patient with anticipated subsequent growth, although blood vessels do seem to tolerate some freezing/heating. To implement an effective cryosurgery/hyperthermia for the case of large blood vessels embedded in or close to the tumor, the effects of large blood vessels to the transient temperature distributions of tissues subject to controlled freezing or heating must be well understood.

The Pennes bioheat equation has collectively considered the thermal effect of blood flow. However, it has the inherent limitation that it cannot simulate the effects of widely spaced thermally significant blood vessels. Such vessels are distributed throughout the body and can perturb the temperature field of tissue. In particular, when large blood vessel are present, the convective effect of the blood flow may significantly affect the surrounding tissues frozen or heated by the medical applicator, thus possibly forming steep temperature gradients between large blood vessels and the tissues near the vessels. If such temperature gradients are present, the surgical protocol must be revised to provide adequate freezing or heating and to avoid damaging normal tissue.

Many vascular heat transfer models currently exist that account for the convective effects of large blood vessels on the temperatures of tissues. Among these, Chato's work (46) was the first to investigate the thermal behaviors of blood vessels. By introducing several simplifications, he analytical solved the temperature fields for a single vessel and a counter-current vessel pair embedded in nonperfused tissues. Weinbaum–Jiji's modified bioheat transfer equation (24), which did not explicitly include large blood vessels, is also used as an alternative equation to study bioheat transfer problems. Kotte et al. established the discrete vasculature thermal model, in which the vessel network is described as a structured tree of vessel segments (47). Chen and Roemer (44) developed several vascular models and then studied the effects of large blood vessels on the tissue temperature distributions during simulated hyperthermia. Due to its important applications in tumor hyperthermia, the study on thermal behavior of large blood vessels had always been a focus in the bioheat transfer field. Unfortunately, few attentions had been paid to the effects of large blood vessels on the phase change heat transfer in living tissues subject to freezing. Until recently, Zhang et al. (45) made the first attempt to develop a theoretical model as well as simulating experiments for cryogenic heat transfer in biological tissues embedded with large blood vessels. They proposed a conceptual model for characterizing the heat transfer in two-dimensional cylindrical tissues with a single blood vessel, and they analytically solved it. Using this model, they investigated the influences of the blood vessel entrance temperature, the vessel diameter, the blood flow velocity, and the vessel length to the tissue temperature distributions. However, due to the complexity of such problems, only a two-dimensional steady-state case with a single blood vessel transmitting the tissues was considered.

Focusing on several most typical vascular models, Deng and Liu theoretically investigated the effects of counter-current large blood vessels to the transient tissue temperature distributions during cryosurgery treatment (48). In such models, the whole tissue domain consists of unfrozen tissue, frozen tissue, and large blood vessel domains. The thermal model combines the Pennes bioheat transfer equation describing for perfused tissues and the energy equation for single or counter-current large blood vessels with a constant Nusselt number. Here, the temperature of blood in a large vessel, which varies along the flow direction, is governed by the convective heat transfer equation (44,48)

$$C_b \frac{\partial T_b}{\partial t} = \frac{hP}{S}(T_{wb} - T_b) - C_b v \frac{\partial T_b}{\partial z}, \quad (21)$$

where $h = N_u \cdot k_b/D$ is the convective heat transfer coefficient between the blood and tissue, D is the diameter of the vessel, P is the perimeter of the vessel, S is the cross-sectional area of the vessel, v is the mean blood velocity along the vessel, Nu is the Nusselt number, and T_{wb} is the wall temperature of the vessel. Here, conduction inside the vessel in the z direction (flow direction) is neglected for large flow rate, and a constant Nusselt number is often assumed.

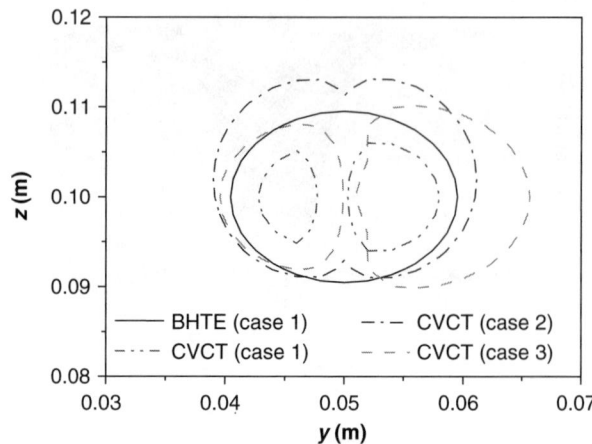

Figure 5. Ice fronts at certain cross sections of tissues subject to four typical freezing configurations (BHTE: Pennes bioheat transfer equation; CVCT: model with counter-current flow vessels close to the tumor; case 1, 2, 3: three typical cryoprobe configurations). (This figure is available in full color at http://www.mrw.interscience.wiley.com/ebe.)

Applying the finite difference formulation, the effects of the blood flow in large vessels to the three-dimensional phase change processes in biological tissues embedded with large blood vessels can be clearly revealed. Figure 5 illustrates the effects of large blood vessels and cryoprobe configurations on the ice ball parameters (48). It demonstrated a difference of the freezing fronts in tissues for different vascular models and cryoprobe configurations. Clearly, the blood vessel has significant effects on the temperature profiles during a phase change process.

5. PROSPECTIVE OF BIOHEAT TRANSFER MODELING

The authors have presented here an overview on the modeling of bioheat transfer either in high or low temperature. Overall, geometrical irregularity of the structure of blood vessels is the major obstacle in developing a general bioheat transfer model. For a specific clinical process, quantitative information of the vascular structure of the target is normally unavailable except for a simple structure such as skin. There is still limited knowledge on the heat transfer in many organs such as liver, renal, brain, kidney, and prostate. Meanwhile, the externally applied three-dimensional heating or freezing patterns by the medical apparatus should also be carefully quantified, which is a basic requirement for an accurate temperature prediction.

Clearly, establishing the bioheat transfer model with a generalized purpose still has plenty of space to explore. One possible way may be to characterize the anatomical structure. Morphometry (vessel number density, length and diameter, etc.) of arterial/venous vasculature has been known as a function of the vessel generation order, i.e., the vessel bifurcation branch. If such vascular structure can be characterized through a certain mathematical way like the scaling law, then a general modeling on the biological heat transfer can be possible. Previously, fractals have been developed especially to describe the self-similarity of the vasculature using a noninteger dimension. In particular, it was even used to study the bioheat transfer (49) as well as the scaling and allometry of metabolic rates in the organ (50). All previous efforts may warrant further investigations in the near future.

BIBLIOGRAPHY

1. C. K. Charny, Mathematical models of bioheat transfer. *Adv. Heat Transfer* 1992; **22**:19–155.
2. K. R. Diller, Modeling of bioheat transfer processes at high and low temperatures. *Adv. Heat Transfer* 1992; **22**:157–357.
3. R. B. Roemer, Engineering aspects of hyperthermia therapy. *Annu. Rev. Biomed. Eng.* 1999; **1**:347–376.
4. S. Karaa, J. Zhang, and F. Q. Yang, A numerical study of a 3D bioheat transfer problem with different spatial heating. *Math. Comput. Simul.* 2005; **68**:375–388.
5. Y. Ji and J. Liu, Numerical study on the effect of lowering temperature on the oxygen transport during brain hypothermia resuscitation. *Comput. Biol. Med.* 2002; **32**:495–514.
6. Z. S. Deng and J. Liu, Computational study on temperature mapping over skin surface and its implementation in disease diagnostics. *Comput. Biol. Med.* 2004; **34**:495–521.
7. P. O. Fanger, *Thermal Comfort*. New York: McGraw-Hill, 1970, pp. 19–67.
8. B. Rubinsky, Cryosurgery. *Annu. Rev. Biomed. Eng.* 2000; **2**:157–187.
9. J. C. Bischof, Quantitative measurement and prediction of biophysical response during freezing in tissues. *Annu. Rev. Biomed. Eng.* 2002; **2**:257–288.
10. A. A. Gage, Selective cryotherapy. *Cell Preservation Technol.* 2004; **2**:3–14.
11. J. O. M. Karlsson and M. Toner, Long-term storage of tissues by cryopreservation: Critical issues. *Biomaterials* 1996; **17**:243–256.
12. L. Zhu, L. X. Xu, and N. Chencinski, Quantification of the 3-D electromagnetic power absorption rate in tissue during transurethral prostatic microwave thermotherapy using heat transfer model. *IEEE Trans. Biomed. Eng.* 1998; **45**:1163–1172.
13. P. M. Meaney, R. L. Clarke, G. R. Ter Haar, and I. H. Rivens, 3-D finite-element model for computation of temperature profiles and regions of thermal damage during focused ultrasound surgery exposures. *Ultrasound Med. Biol.* 1998; **24**:1489–1499.
14. B. M. Kim, S. L. Jacques, S. Rastegar, S. Thomsen, and M. Motamedi, Nonlinear finite-element analysis of the role of dynamic changes in blood perfusion and optical properties in laser coagulation of tissue. *IEEE J. Select. Top. Quantum Electron* 1996; **2**:922–932.
15. A. A. Gage and J. Baust, Mechanism of tissue injury in cryosurgery. *Cryobiology* 1998; **37**:171–186.
16. S. D. Burch, S. Ramadhyani, and J. T. Pearson, Analysis of passenger thermal comfort in an automobile under severe winter conditions. *ASHRAE Trans.* 1991; pt **1**:247–257.
17. M. M. Chen, K. R. Holmes, and V. Rupinskas, Pulse-decay method for measuring the thermal conductivity of living tissues. *ASME J. Biomech. Eng.* 1981; **103**:253–260.
18. G. T. Anderson, J. W. Valvano, and R. R. Santos, Self-heated thermistor measurements of perfusion. *IEEE Trans. Biomed. Eng.* 1992; **39**:877–885.
19. B. Rubinsky, Microscale heat transfer in biological systems at low temperatures. *Exper. Heat Transfer* 1997; **10**:1–29.
20. H. H. Pennes, Analysis of tissue and arterial blood temperatures in the resting human forearm. *J. Appl. Physiol.* 1948; **1**:93–122.
21. J. Liu and C. Wang, *Bioheat Transfer (in Chinese)*. Beijing: Science Press, 1997, pp. 285–290.
22. W. Wulff, The energy conservation equation for living tissues. *IEEE Trans. Biomed. Eng.* 1974; **BME-21**:494–497.
23. M. M. Chen and K. R. Holmes, Microvascular contributions in tissue heat transfer. *Ann. NY. Acad. Sci.* 1980; **335**:137–150.
24. S. Weinbaum and L. M. Jiji, A new simplified bioheat equation for the effect of blood flow on local average tissue temperature. *ASME J. Biomech. Eng.* 1985; **107**:131–139.
25. C. C. Wang, Relationship between the basal metabolic rate and temperature regulation of human being and animals. Proc. Int. Symp. Macro-& Microscopic Heat and Mass Transfer in Biomedical Eng., Sept. 2–6, 1991.
26. J. Liu, X. Chen, and L. X. Xu, New thermal wave aspects on burn evaluation of skin subjected to instantaneous heating. *IEEE Trans. Biomed. Eng.* 1999; **46**:420–428.
27. Z. S. Deng and J. Liu, Analytical study on bioheat transfer problems with spatial or transient heating on skin surface or inside biological bodies. *ASME J. Biomech. Eng.* 2002; **124**:638–649.
28. I. Chatterjee and R. E. Adams, Finite element thermal modeling of the human body under hyperthermia treatment for cancer. *Int. J. Comput. Appl. Tech.* 1994; **7**:151–159.
29. C. L. Chan, Boundary element method analysis for the bioheat transfer equation. *ASME J. Biomech. Eng.* 1992; **114**:358–365.
30. Z. S. Deng and J. Liu, Modeling of multidimensional freezing problem during cryosurgery by the dual reciprocity boundary element method. *Eng. Anal. Boundary Elements* 2004; **28**:97–108.
31. C. T. Liauh, S. T. Clegg, and R. B. Romer, Estimating three-dimensional temperature fields during hyperthermia: Studies of the optimal regularization parameter and time sampling period. *ASME J. Biomech. Eng.* 1991; **113**:230–238.
32. Y. Rabin and A. Shitzer, Exact solution to one-dimensional inverse-Stefan problem in nonideal biological tissues. *ASME J. Heat Transfer* 1995; **117**:425–431.
33. B. Han, A. Iftekhar, and J. C. Bischof, Improved cryosurgery by use of thermophysical and inflammatory adjuvants. *Technol. Cancer Res. Treatment* 2004; **3**:103–111.
34. Z. S. Deng and J. Liu, Numerical simulation of selective freezing of target biological tissues following injection of solutions with specific thermal properties. *Cryobiology* 2005; **50**:183–192.
35. H. Budman, J. Dayan, and A. Shitzer, Controlled freezing of nonideal solutions with applications to cryosurgical processes. *ASME J. Biomech. Eng.* 1991; **113**:430–437.
36. Y. T. Zhang and J. Liu, Numerical study on three-region thawing problem during cryosurgical re-warming. *Med. Eng. Phys.* 2002; **24**:265–277.
37. C. Bonacina, G. Comini, A. Fasano, and M. Primicero, Numerical solution of phase-change problems. *Int. J. Heat Mass. Transfer* 1973; **16**:1825–1832.
38. Z. S. Deng and J. Liu, Numerical simulation on 3-D freezing and heating problems for the combined cryosurgery and hyperthermia therapy. *Numer. Heat Transfer, Part A: Applicat.* 2004; **46**:587–611.
39. J. Liu and Y. X. Zhou, Analytical study on the freezing and thawing process of biological skin with finite thickness. *Heat Mass Transfer* 2002; **38**:319–326.

40. D. Nardini and C. A. Brebbia, A new approach to free vibration analysis using boundary elements. In: C. A. Brebbia, ed., *Boundary Elements in Engineering*. New York: Springer-Verlag, 1982.
41. D. C. Lung, T. F. Stahovich, and Y. Rabin, Computerized planning for multiprobe cryosurgery using a force-field analogy. *Comput. Methods Biomech. Biomed. Eng.* 2004; **7**:101–110.
42. R. G. Keanini and B. Rubinsky, Optimization of multiprobe cryosurgery. *ASME J. Heat Transfer* 1992; **114**:796–801.
43. J. G. Baust and A. A. Gage, Progress toward optimization of cryosurgery. *Technol. Cancer Res. Treatment* 2004; **3**:95–101.
44. Z. P. Chen and R. B. Roemer, The effects of large blood vessels on temperature distributions during simulated hyperthermia. *ASME J. Biomech. Eng.* 1992; **114**:473–481.
45. Y. T. Zhang, J. Liu, and Y. X. Zhou, Pilot study on cryogenic heat transfer in biological tissues embedded with large blood vessels. *Forschung im Ingenieurwesen-Eng. Res.* 2002; **67**:188–197.
46. J. C. Chato, Heat transfer to blood vessels. *ASME J. Biomech. Eng.* 1980; **102**:110–118.
47. A. N. T. J. Kotte, G. M. J. Van Leeuwen, and J. J. W. Lagendijk, Modelling the thermal impact of a discrete vessel tree. *Phys. Med. Biol.* 1999; **44**:57–74.
48. Z. S. Deng and J. Liu, Numerical study on the effects of large blood vessels on 3-D tissue temperature profiles during cryosurgery. *Numer. Heat Transfer, Part A: Applicat.* 2006; **49**:47–67.
49. J. W. Baish, A fractal-based theory of bioheat transfer. AICHE Symp. Series (Heat Transfer—Philadelphia), 1989:406–410.
50. J. R. Lees, T. C. Skalak, E. M. Sevick, and R. K. Jain, Microvascular architecture in a mammary carcinoma: Branching patterns and vessel dimensions. *Cancer Res.* 1991; **51**:265–273.

BIOIMPEDANCE

SVERRE GRIMNES
ØRJAN G. MARTINSEN
University of Oslo
Oslo, Norway

Bioimpedance describes the passive electrical properties of biological materials and serves as an indirect transducing mechanism for physiological events, often in cases where no specific transducer for that event exists. It is an elegantly simple technique that requires only the application of two or more electrodes. According to Geddes and Baker (1), the impedance between the electrodes may reflect "seasonal variations, blood flow, cardiac activity, respired volume, bladder, blood and kidney volumes, uterine contractions, nervous activity, the galvanic skin reflex, the volume of blood cells, clotting, blood pressure and salivation."

Impedance Z [ohm, Ω] is a general term related to the ability to oppose ac current flow, expressed as the ratio between an ac sinusoidal voltage and an ac sinusoidal current in an electric circuit. Impedance is a complex quantity because a biomaterial, in addition to opposing current flow, phase-shifts the voltage with respect to the current in the time-domain. *Admittance* Y [siemens, S] is the inverse of impedance ($Y = 1/Z$). The common term for impedance and admittance is *immittance* (2).

The conductivity of the body is ionic (electrolytic), because of for instance Na^+ and Cl^- in the body liquids. The ionic current flow is quite different from the electronic conduction found in metals: Ionic current is accompanied by substance flow. This transport of substance leads to concentrational changes in the liquid: locally near the electrodes (electrode polarization), and in a closed-tissue volume during prolonged dc current flow.

The studied *biomaterial* may be living tissue, dead tissue, or organic material related to any living organism such as a human, animal, cell, microbe, or plant. In this chapter, we will limit our description to human body tissue.

Tissue is composed of cells with poorly conducting, thin-cell membranes; therefore, tissue has capacitive properties: the higher the frequency, the lower the impedance. Bioimpedance is frequency-dependent, and impedance *spectroscopy*, hence, gives important information about tissue and membrane structures as well as intra- and extracellular liquid distributions. As a result of these capacitive properties, tissue may also be regarded as a *dielectric* (3). Emphasis is then shifted to ac *permittivity* (ε) and ac losses. In linear systems, the description by permittivity or immittivity contains the same information. It must also be realized that permittivity and immittivity are *material constants*, whereas immittance is the directly measured quantity dependent on tissue and electrode geometries. In a heterogeneous biomaterial, it is impossible to go directly from a measured immittance spectrum to the immittivity distribution in the material. An important challenge in the bioimpedance area is to base data interpretation on a better knowledge of the immittivity of the smaller tissue components (2–11).

1. TYPICAL BIOIMPEDANCE DATA

Figure 1 shows the three most common electrode systems. With two electrodes, the current carrying electrodes and signal pick-up electrodes are the same (Fig. 1, left). If the electrodes are equal, it is called a bipolar lead, in contrast to a monopolar lead. With 3-(tetrapolar) or 4-(quadropolar) electrode systems, separate current carrying and signal pick-up electrodes exist. The impedance is then

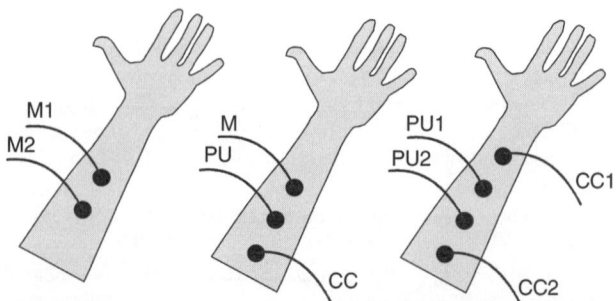

Figure 1. Three skin surface electrode systems on an underarm. Functions: M = measuring and current carrying, CC = current carrying, PU = signal pick-up.

transfer impedance (12): The signal is not picked up from the sites of current application.

The 4-electrode system (Fig. 1, right) has separate pick-up (PU) and current carrying (CC) electrodes. With ideal voltage amplifiers, the PU electrodes are not current carrying, and therefore, their polarization impedances do not introduce any voltage drop disturbing measured tissue impedance. In the 3-electrode system (Fig. 1, middle), the measuring electrode M is both a CC and signal PU electrode.

1.1. A 4-Electrode Impedance Spectrum

Figure 2 shows a typical transfer impedance spectrum (Bode plot) obtained with the 4-electrode system of Fig. 1 (right). It shows two *dispersions* (to be explained later). The transfer impedance is related to, but not solely determined by, the arm segment between the PU electrodes. As we shall see, the spectrum is determined by the sensitivity field of the 4-electrode system as a whole. The larger the spacing between the electrodes, the more the results are determined by deeper tissue volumes. Even if all the electrodes are skin surface electrodes, the spectrum is, in principle, not influenced by skin impedance or electrode polarization impedance.

For many, it is a surprise that the immittance measured will be the same if the CC and PU electrodes are interchanged (the reciprocity theorem).

1.2. A 3-Electrode Impedance Spectrum

Figure 3 shows a typical impedance spectrum obtained with three skin surface electrodes on the underarm (Fig. 1, middle). Notice the much higher impedance levels than found with the 4-electrode system. The measured zone is under M and comprises electrode polarization, skin

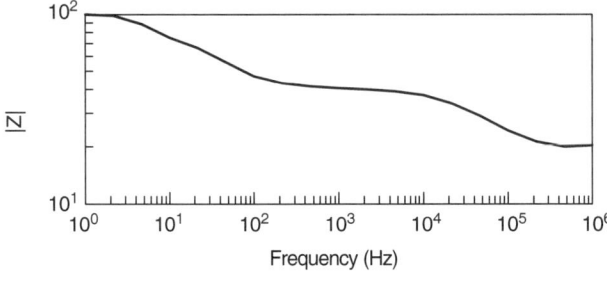

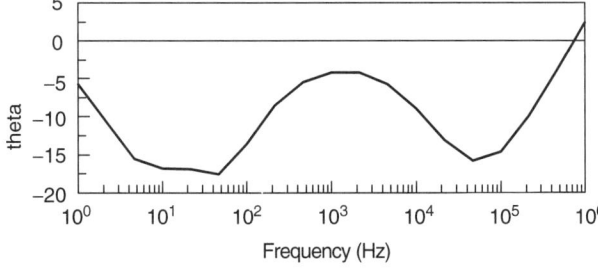

Figure 2. Typical impedance spectrum obtained with four equal electrodes attached to the skin of the underarm as shown on Fig. 1 (right). All electrodes are pregelled ECG-electrodes with skin gel-wetted area $3\,cm^2$. Distance between electrode centers: 4 cm.

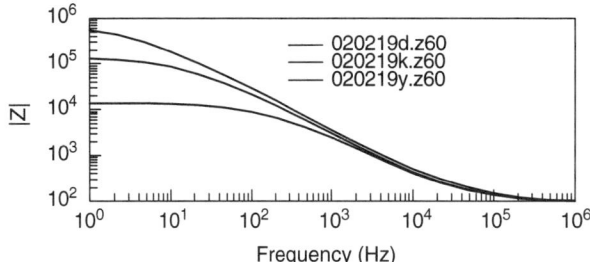

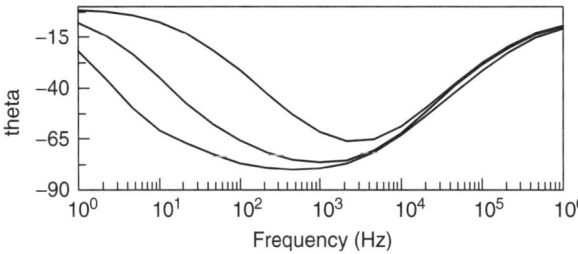

Figure 3. Typical impedance spectrum obtained with the 3-electrode system on the underarm as shown on Fig. 1 (middle). The parameters are: at the time of electrode onset on dry skin, after 1 h, and after 4 h of contact.

impedance, and deeper layer impedance, all physically in series. The electrode polarization impedance is a source of error; electrode impedance is not tissue impedance. At low frequencies (< 1000 Hz), the result is dominated by the high impedance of the human skin with negligible influence from the polarization impedance of the electrode. At high frequencies (> 100 kHz), the results are dominated by deeper layer tissues.

Figure 3 also shows the effect of contact electrolyte penetration into the initially dry skin (three curves: at the moment of electrode onset on dry skin, after 1 h, and after 4 h). The electrode polarization contribution can be judged by studying Fig. 11. Initially, the electrode polarization impedance has negligible influence on the LF results; however, at HF, around 20% of the measured impedance is from the M electrode itself.

Also, the immittance measured will be the same if the CC and PU electrodes are interchanged (the reciprocity theorem).

2. FROM MAXWELL TO BIOIMPEDANCE EQUATIONS

The Maxwell equation most relevant to bioimpedance is:

$$\nabla \times \boldsymbol{H} - \partial \boldsymbol{D}/\partial t = \boldsymbol{J} \qquad (1)$$

$$\boldsymbol{D} = \varepsilon_o \boldsymbol{E} + \boldsymbol{P} \qquad (2)$$

where $\boldsymbol{H}$ = magnetic field strength [A/m], $\boldsymbol{D}$ = electric flux density [coulomb/m^2], $\boldsymbol{J}$ = current density [A/m^2], $\boldsymbol{E}$ = electric field strength [V/m], ε_o = permittivity of vacuum [farad (F) /m], and $\boldsymbol{P}$ = electric polarization, dipole moment pr. volume [coulomb/m^2].

If the magnetic component is ignored, Equation 1 is reduced to:

$$\partial \boldsymbol{D}/\partial t = -\boldsymbol{J} \qquad (3)$$

Equations 1–3 are extremely robust and also valid under *nonhomogeneous, nonlinear, and anisotropic* conditions. They relate the time and space derivatives at a point to the current density at that point.

Impedance and permittivity in their simplest forms are based on a basic capacitor model (Fig. 4) and the introduction of some restrictions:

a) Use of sufficiently small voltage amplitude v across the material so the system is *linear*. b) Use of *sinusoidal* functions so that with complex notation a derivative (e.g., $\partial E/\partial t$) is simply the product $j\omega E$ (j is the imaginary unit and ω the angular frequency). c) Use of $D = \varepsilon E$ (space vectors), where the permittivity $\varepsilon = \varepsilon_r \varepsilon_o$, which implies that D, P, and E all have the same direction, and therefore, that the dielectric is considered isotropic. d) No fringe effects in the capacitor model of Fig. 4.

Under these conditions, a lossy dielectric can be characterized by a complex dielectric constant $\varepsilon = \varepsilon' - j\varepsilon''$ or a complex conductivity $\sigma = \sigma' + j\sigma''$ [S/m], then $\sigma' = \omega \varepsilon''$ (2). Let us apply Equation 3 on the capacitor model where the metal area is A and the dielectric thickness is L. However, Equation 3 is in differential form, and the interface between the metal and the dielectric represents a discontinuity. Gauss law as an integral form must therefore be used, and we imagine a thin volume straddling an area of the interface. According to Gauss law, the outward flux of D from this volume is equal to the enclosed free charge density on the surface of the metal. With an applied voltage v, it can be shown that $D = v\varepsilon/L$. By using Equation 3, we then have $\partial D/\partial t = j\omega v \varepsilon/L = J$, and $i = j\omega v \varepsilon A/L = v j \omega C$.

We now leave the dielectric and take a look at the external circuit where the current i and the voltage v (time vectors) are measured and the immittance determined. The admittance is $Y = i/v$ [siemens, S]. With no losses in the capacitor, i and v will be phase-shifted by 90° (the quadrature part). The conductance is $G = \sigma' A/L$ [S], and the basic equation of bioimpedance is then (time vectors):

$$Y = G + j\omega C \tag{4}$$

Three important points must be made here:

First, Equation 4 shows that the basic impedance model actually is an *admittance* model. The conductive and capacitive (quadrature) parts are physically in parallel in the model of Fig. 4.

Second, the model of Fig. 4 is predominantly a dielectric model with *dry* samples. In bioimpedance theory, the materials are considered to be *wet*, with double layer and polarization effects at the metal surfaces. Errors are introduced, which, however, can be reduced by introducing 3- or 4-electrode systems (Fig. 1). Accordingly, in dielectric theory, the dielectric is considered as an *insulator* with dielectric losses; in bioimpedance theory, the material is considered as a *conductor* with capacitive properties. Dry samples can easily be measured with a 2-electrode system. Wet, ionic samples are prone to errors and special precautions must be taken.

Third, Equations 1–3 are valid at a point. With a homogeneous and isotropic material in Fig. 4, they have the same values all over the sample. With inhomogeneous and anisotropic materials, the capacitor model implies values averaged over the volume. Then, under *linear* (small signal) conditions, Equation 4 is still correct, but the measured values are difficult to interpret. The capacitor is basically an *in vitro* model with a biomaterial placed in the measuring chamber. The average anisotropy can be measured by repositioning the sample in the capacitor. *In vivo* measurements, as shown in Fig. 1, must be analyzed from sensitivity fields, as shown in the next chapter.

From Equation 3, the following relationship is easily deduced (space vectors):

$$J = \sigma E. \tag{5}$$

Equation 5 is not valid in anisotropic materials if σ is a scalar. Tissue, as a rule, is anisotropic. Plonsey and Barr (5) discussed some important complications posed by tissue anisotropy and also emphasized the necessity of introducing the concept of the *bidomain*. A bidomain model is useful for cardiac tissue, where the cells are connected by two different types of junctions: tight junctions and gap junctions where the interiors of the cells are directly connected. The intracellular space is one domain and the interstitial space the other domain.

3. GEOMETRY, SENSITIVITY AND RECIPROCITY

Resistivity ρ [$\Omega \cdot$m] and *conductivity* σ [S/m] are *material constants* and can be extended to their complex analogues: impedivity [$\Omega \cdot$m] and admittivity [S/m]. The resistance of a cylinder volume with length L, cross-sectional area A, and uniform resistivity ρ is:

$$R = \rho L/A. \tag{6}$$

Equation 6 shows how bioimpedance can be used for volume measurements (plethysmography). Notice, however, that, for example, a resistance increase can be caused either by an increased tissue length, a reduced cross-sectional area, or an increased resistivity. Tissue dielectric and immittivity data are listed by Duck (13).

Figure 5 illustrates typical resistance values for body segments (2), valid without skin contribution and without current constrictional effects caused by small electrodes. By using the term "resistance," we indicate that they are not very frequency-dependent. Notice the low resistance of the thorax (13 Ω) and the high resistance of one finger (500 Ω).

Figure 6 shows the effect of a constrictional zone caused by one small electrode. Such an electrode system is called *monopolar* because most of the measuring results are a result of the impedance of the small electrode proximity zone.

Figure 4. The basic capacitor model.

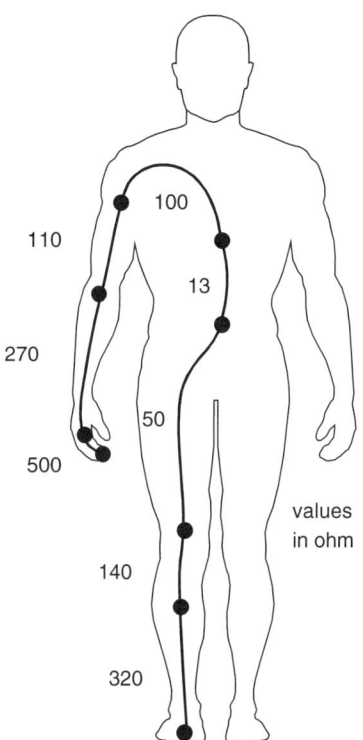

Figure 5. Typical body segment resistance values. From (2), by permission.

3.1. Sensitivity Field of an Electrode System

Intuitively, it is easy to believe that if a small tissue volume changes immittivity, the influence on the measurement result is larger the nearer that tissue volume is to the electrodes, which is indeed the case and is illustrated by an equation based on the work of Geselowitz (14):

$$R = \int_V \rho \mathbf{J}_{CC} \cdot \mathbf{J}_{reci} dv, \quad (7)$$

where R is the transfer resistance [Ω] measured by a 4-electrode system; ρ is the local resistivity in the small volume dv (if ρ is complex, the integral is the transfer impedance $\mathbf{Z}$); $\mathbf{J}_{CC}$ [1/m²] is the local current density space vector in dv caused by the CC electrodes carrying a *unity* current; and $\mathbf{J}_{reci}$ [1/m²] is the local current density space vector in dv caused by the PU electrodes *if they also carried a unity current* (reciprocal excitation).

The product $\mathbf{J}_{CC} \cdot \mathbf{J}_{reci}$ is the local dot vector product, which may be called the local *sensitivity* S [1/m⁴] of the electrode system:

$$S = \mathbf{J}_{CC} \cdot \mathbf{J}_{reci}. \quad (8)$$

Unity current is used so that sensitivity is a purely geometrical parameter not dependent on any actual current level. S is a scalar with positive or negative values in different parts of the tissue; the spatial distribution of S is the *sensitivity field*.

The implications of Equation 7 are important, and at first sight counter intuitive: In a 4-electrode system, *both* electrode pairs determine the sensitivity, not just the PU electrodes as one may intuitively believe. There will be zones of negative sensitivity in the tissue volume between the PU and CC electrodes. The zones will be dependent on, for example, the distance between the PU and CC electrodes. The PU and CC current density fields enter Equation 7 in the same way, and the interchange of the PU and CC electrodes do not change the value of R. Equation 7 is therefore based on the *reciprocity* theorem (14).

In a monopolar or bipolar electrode system, Equation 7 simplifies to

$$R = \int_V \rho \mathbf{J}^2 dv \quad S = \mathbf{J}^2, \quad (9)$$

where $\mathbf{J}$ [1/m²] is the local current density caused by a unity current passed through the electrode pair.

The analysis of the sensitivity field of an electrode system is of vital importance for the interpretation of measured immittance. Figure 7 (top) illustrates, for instance, the effect of electrode dimensions in a bipolar electrode system. As the gap between the electrodes narrows, the local sensitivity in the gap increases according to $S = \mathbf{J}^2$. However, the volume of the gap zone also becomes smaller, and the overall contribution to the integral of Equation 7 is not necessarily dominating. If local changes (e.g., from a pulsating blood artery) is to be picked up, the artery should be placed in such a high-sensitivity zone.

Figure 7 (bottom) illustrates the effect of electrode-electrode center distance in a bipolar system. As distance is increased, the sensitivity in deeper layers will increase but still be small. However, large volumes in the deeper layers will then have a noticeable effect, as small volumes proximal to the electrodes also have.

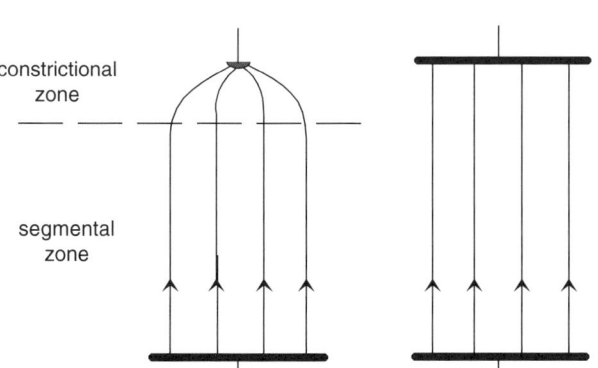

Figure 6. Left: Monopolar system with one electrode much smaller than the other. Increased resistance from the constrictional current zone with increased current density. Right: Segment resistance with uniform current density, large bipolar electrodes. From (2), by permission.

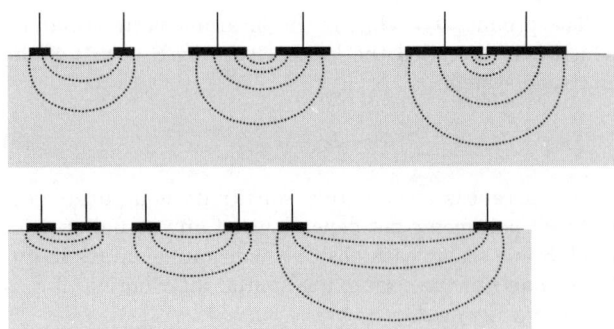

Figure 7. The measuring depth as a function of electrode dimensions (top) and electrode spacing (bottom). From (2), by permission.

4. ELECTRICAL MODELS

Bioimpedance is a measure of the passive properties of tissue, and, as a starting point, we state that tissue has resistive and capacitive properties showing relaxation, but not resonance, phenomena. As shown by Equation 4, *admittance* is

$$Y = G + j\omega C_p$$
$$\varphi = \arctan(\omega C_p / G)$$
$$|Y|^2 = G^2 + (\omega C_p)^2, \quad (10)$$

where φ is the phase angle indicating to what extent the voltage is time-delayed, G is the parallel conductance [S], and C_p the parallel capacitance [F].

The term ωC_p is the capacitive *susceptance*.

Impedance is the inverse of admittance ($Z = 1/Y$); the equations are:

$$Z = R - j/\omega C_s$$
$$\varphi = \arctan(-1/\omega R C_s)$$
$$|Z|^2 = R^2 + (1/\omega C_s)^2. \quad (11)$$

The term $-1/\omega C_s$ is the capacitive *reactance*.

The values of the series (R_s, C_s) components values are not equal to the parallel ($1/G$, C_p) values:

$$Z = R_s - j/\omega C_s = G/|Y|^2 - j\omega C_p/|Y|^2. \quad (12)$$

Equation 12 illustrates the serious problem of choosing, for example, an impedance model if the components are physically in parallel: R_s and C_s are both frequency-dependent when G and C_p are not. Implicit in these equations is the notion that impedance is a series circuit of a resistor and a capacitor, and admittance is a parallel circuit of a resistor and a capacitor. Measurement results must be given according to one of these models. A model must be chosen, no computer system should make that choice. An important basis for a good choice of model is deep knowledge about the system to be modeled. An electrical model is an electric circuit constituting a substitute for the real system under investigation, as an equivalent circuit.

One ideal resistor and one ideal capacitor can represent the measuring results on one frequency, but can hardly be expected to mimic the whole immittance spectrum actually found with tissue. Usually, a second resistor is added to the equivalent circuit, and one simple and often surprisingly effective addition is also to replace the capacitor C by a more general CPE (Constant Phase Element). A CPE is not a physical device but a mathematical model, you cannot buy a CPE as you buy a resistor ($\varphi = 0°$) or a capacitor ($\varphi = 90°$). A CPE can have any constant phase angle value between $0°$ and $90°$, and mathematically, it is a very simple device (2). Figure 8 shows a popular equivalent circuit in two variants.

One such circuit defines *one dispersion* (15), characterized by two levels at HF and LF, with a transition zone where the impedance is complex. Both at HF ($Z = R_\infty$) and at LF ($Z = R + 1/G_{var}$), the impedance Z is purely resistive, determined by the two ideal resistors. The circuit of Fig. 8 (left) is with three ideal, frequency-independent components, often referred to as the Debye case, and the impedance is:

$$Z = R_\infty + \frac{1}{G_{var} + G_{var} j\omega\tau} \quad \tau = C/G_{var}.$$

The diagram to the right in Fig. 8 is with the same two ideal resistors, but the capacitor has been replaced by a CPE (2). The equivalent circuit of a CPE consists of a resistor and a capacitor, both frequency-dependent so that the phase becomes frequency-independent.

$$Z = R_\infty + \frac{1}{G_{var} + G_1(j\omega\tau)^\alpha}$$
$$j^\alpha = \cos(\alpha\pi/2) + j\sin(\alpha\pi/2) \quad (13)$$

In Equation 13, the CPE admittance is $G_1(j\omega\tau)^\alpha$; and τ may be regarded as a mean time constant of a tissue volume with a distribution of different local time constants. τ may also be regarded just as a frequency scaling factor; $\omega\tau$ is dimensionless and G_1 is the admittance value at the characteristic angular frequency when $\omega\tau = 1$. α is related both to the constant phase φ of the CPE according to j^α

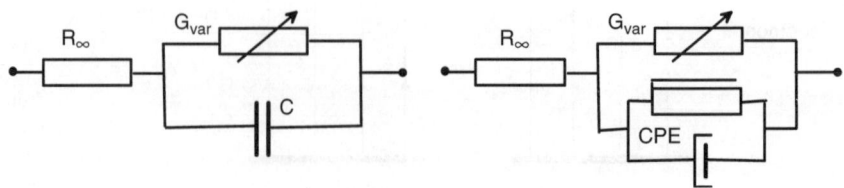

Figure 8. Two versions of a popular equivalent circuit.

and $\varphi = \alpha \cdot 90°$ and to the frequency exponent in the term ω^α. This double influence of α presupposes that the system is Fricke-compatible. According to Fricke's law, the phase angle φ and the frequency exponent m are related in many electrolytic systems so that $\varphi = \mathrm{m} \cdot 90°$. In such cases, m is replaced by α (2,16).

A less general version of Equation 13 was given by Cole (17):

$$Z = R_\infty + \frac{\Delta R}{1 + (j\omega\tau)^\alpha} = R_\infty + \frac{1}{\Delta G + \Delta G(j\omega\tau)^\alpha}. \quad (14)$$

The Cole model does not have an independent conductance G_var in parallel with the CPE; ΔG controls both the parallel ideal conductance and the CPE, which implies that the characteristic frequency is independent of ΔG, in the same way that $\tau = C/G_\mathrm{var}$ in Equation 11 would be constant if both C and G_var varied with the same factor. The lack of an independent conductance variable limits the application of the Cole equation (18). All of Equations 12–14 are represented by circular arc loci if the impedance is plotted in the complex plane, but one of them must be chosen.

Instead of Bode plots, a plot in the complex Argand or Wessel (2) plane may be of value for the interpretation of the results. Figure 9 shows the data of Fig. 2 as a Z and a Y plot. The two dispersions are clearly seen as more or less perfect circular arcs. The LF dispersion is called the α-dispersion, and the HF dispersion is called the β-dispersion (15). β-dispersion is caused by cell membranes and can be modeled with the Maxwell–Wagner structural polarization theory (15,19). The origin of the α-dispersion is more unclear.

Figure 8 is a model well-suited for skin impedance: The R_∞ is the deeper tissue series resistance and the parallel combination represents the stratum corneum with independent sweat duct conductance in parallel. Figure 10

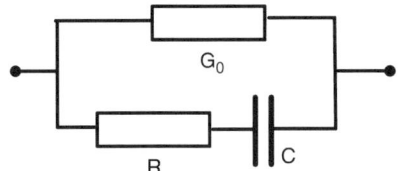

Figure 10. Tissue or suspension equivalent circuit.

shows another popular model better suited for living tissue and cell suspensions. The parallel conductance G_0 is the extracellular liquid, the capacitance is the cell membranes, and the R is the intracellular contributions.

5. ELECTRODES AND INSTRUMENTATION

The electrode is the site of charge carrier transfer, from electrons to ions or vice versa (20–22). The electrode proper is the contact zone between the electrode metal (electronic conduction) and the electrolyte (ionic conduction). As ionic current implies transport of substance, the electrolytic zone near the metal surface may be depleted or filled with electrolyte species. A double layer will be formed in the electrolyte at the electrode surface. This double layer represents an energy barrier with capacitive properties. Both processes will contribute to electrode polarization immittance. Figure 11 shows the impedance spectrum of a commercial, pregelled ECG electrode of the type used for obtaining the results in Figs. 2 and 3.

5.1. Electrode Designs

Skin surface electrodes are usually made with a certain distance between the metal part and the skin [Fig. 12 (top)]. The enclosed volume is filled with contact electrolyte, often in the form of a gel contained in a sponge. The

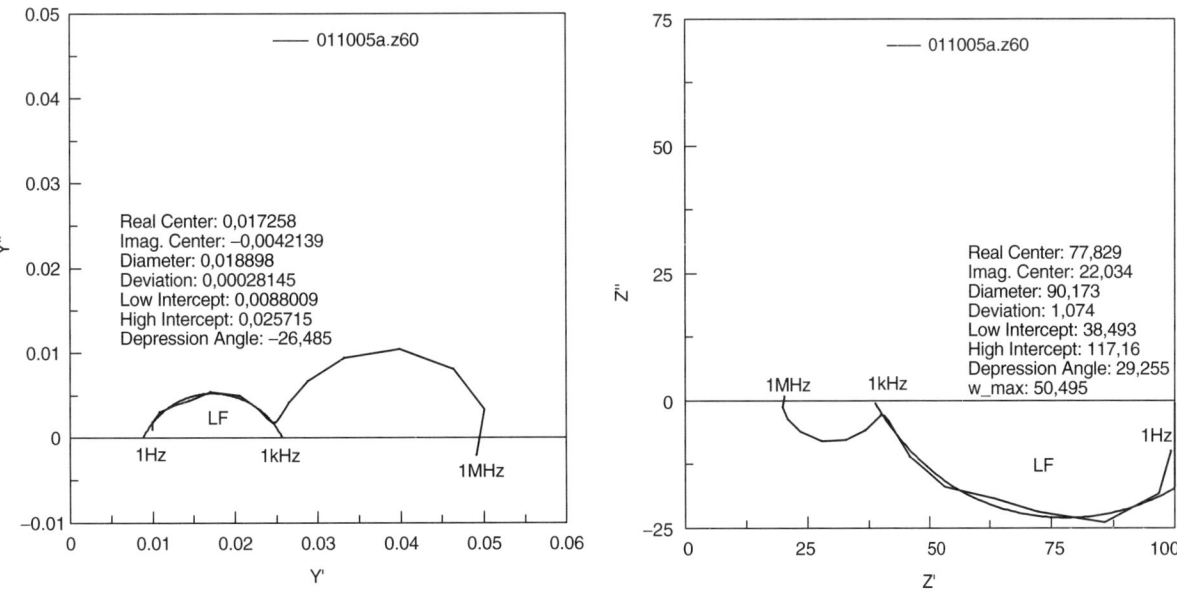

Figure 9. The data in Fig. 2 shown in the complex Wessel diagram as Y and Z plots. Circular arc fits for the same LF dispersion are shown in both plots.

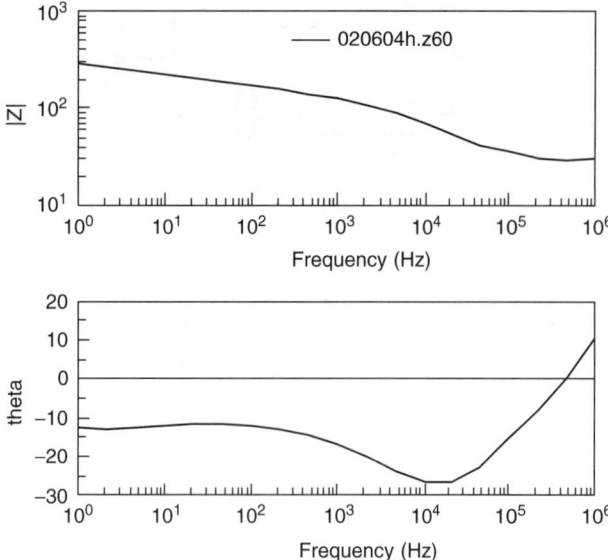

Figure 11. Electrode polarization impedance, one pregelled ECG commercial electrode of the type used in Figs. 2 and 3. The positive phase at HF is caused by the self-inductance of the electrode wire.

shaft. Some are of a coaxial type with a thin center lead isolated from the metal shaft.

With modern technology, it is easy to fabricate microelectrodes with small dots or strips with dimensions in the micrometer range that are well-suited for single-cell discrimination.

5.2. Instrumentation and Software

Bridges achieve high precision. The frequency range is limited (23), although modern self-balanced bridge designs have extended it below 10 Hz. Now, lock-in amplifiers are the preferred instrumentation for bioimpedance measurements (2). The lock-in amplifier needs a synchronizing signal from the signal oscillator for its internal synchronous rectifier. The output of the lock-in amplifier is a signal not only dependent on input signal *amplitude*, but also *phase*. A two-channel lock-in amplifier has two rectifiers synchronous with the in-phase and quadrature oscillator signals. With such an instrument, it is possible to measure complex immittance directly. Examples of commercially available instruments are the SR model series from Stanford Research Systems, the HP4194A, and the Solartron 1260/1294 system with an extended LF coverage (10 µHz–32 MHz). Bioimpedance is measured with small currents so that the system is linear: An applied sine waveform current results in a sine pick-up signal. A stimulating electrode, on the other hand, is used with large currents in the nonlinear region (7), and the impedance concept for such systems must be used with care.

Software for Bode plots and complex plane analysis are commercially available; one example is the Scribner ZView package. This package is well-suited for circular arc fits and equivalent circuit analysis.

5.3. Safety

In a 2- and 3-electrode system, it is usually possible to operate with current levels in the microampere range, corresponding to applied voltages around 10 mV rms. For most applications and direct cardiac, these levels may be safe (24,25).

With 4-electrode systems, the measured voltage for a given current is smaller, and for a given signal-to-noise ratio, the current must be higher, often in the lower mA range. For measuring frequencies below 10 kHz, this is unacceptable for direct cardiac applications. LF mA currents may also result in current perception by neuromuscular excitation in the skin or deeper tissue. Dependent on the current path, these LF current levels are not necessarily dangerous, but are unacceptable for routine applications all the same.

6. SELECTED APPLICATIONS

6.1. Laboratory-on-a-Chip

With microelectrodes in a small sample volume of a cell suspension, it is possible to manipulate, select, and characterize cells by rotation, translocation, and pearl chain formation (26). Some of these processes are monitored by bioimpedance measurements.

stronger the contact electrolyte, the more rapid the penetration into the skin. There are two surface areas of concern in a skin surface electrode: The area of contact between the metal and the electrolyte determines the polarization impedance; the electrolyte wetted area of the skin (the effective electrode area, EEA) determines the skin impedance.

Figure 12 (bottom) shows needle electrodes for invasive measurements. Some types are insulated out to the tip; others have a shining metal contact along the needle

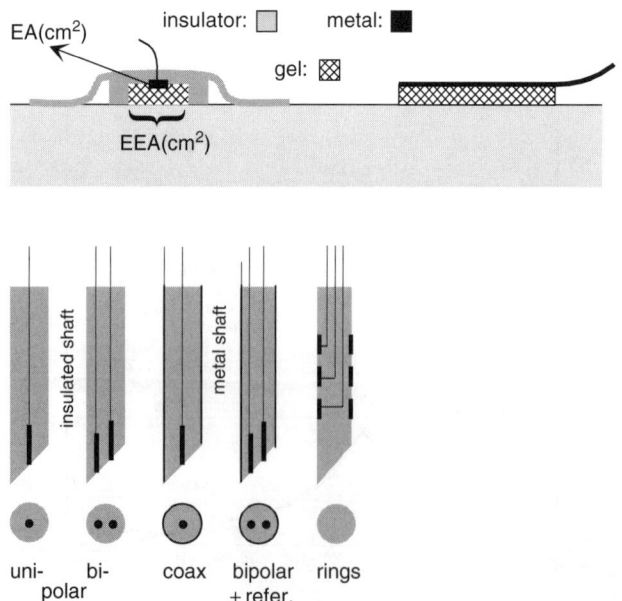

Figure 12. Top: skin surface types. Bottom: invasive needles. From (2), by permission.

6.2. Cell Micromotion Detection

A monopolar microelectrode is convenient to study cell attachment to a surface. Many cell types need an attachment to flourish, and it can be shown that measured impedance is more dominated by the electrode surface the smaller it is. Cell micromotion can be followed with nm resolution on the electrode surface (27).

6.3. Cell Suspensions

The Coulter counter counts single cells and is used in hospitals all over the world. The principle is based on a cell suspension where the cells and the liquid have different impedivities. The suspension is made to flow through a capillary, and the capillary impedance is measured. In addition to rapid cell counting, it is possible to characterize each cell on passing (28).

6.4. Body Composition

Bioimpedance is dependent on the morphology and impedivity of the organs, and with large electrode distances, it is possible to measure segment or total body water, extra- and intracellular fluid balance, muscle mass, and fat mass. Application areas are as diversified as sports medicine, nutritional assessment, and fluid balance in renal dialysis and transplantation. Body composition instruments represent a growing market. Many of them are single-frequency instruments, and with the electrode systems used, it is necessary to analyze what they actually measure (29).

6.5. Impedance Plethysmography

See that entry in this encyclopedia.

6.6. Impedance Cardiography (ICG) and Cardiac Output

A tetrapolar system is used with two band electrodes around the neck, one band electrode corresponding to the apex of the heart, and the fourth further in caudal direction A more practical system uses eight spot electrodes arranged as four double electrodes, each with one PU and one CC electrode. The amplitude of the impedance change $\Delta \mathbf{Z}$ as a function of the heartbeat is about 0.5% of the baseline value. The $\Delta \mathbf{Z}$ waveform is similar to the aorta blood pressure curve. The *first time derivative* $d\mathbf{Z}/dt$ is called the impedance cardiographic curve (ICG). By adding information about patient age, sex, and weight, it is possible to estimate the heart stroke volume and cardiac output. The resistivity of blood is flow-dependent (11), and as long as the origin (heart-, aorta-, lung-filling/emptying) of the signal is unclear, the cardiac output transducing mechanism will also be obscure. Sensitivity field analysis may improve this status (30).

6.7. Skin Moisture

The impedance of the stratum corneum is dependent on its water content. By measuring the skin susceptance, it is possible to avoid the disturbance of the sweat duct parallel conductance (31), and hence assess the hydration state of the stratum corneum. Low-excitation frequency is needed to avoid contribution from deeper, viable skin layers.

6.8. Skin Fingerprint

Electronic fingerprint systems will, in the near future, eliminate the need for keys, pincodes, and access cards in a number of daily-life products. With a microelectrode matrix or array, it is possible to map the fingerprint electrically with high precision using bioimpedance measurements. Live finger detection is also feasible to ensure that the system is not fooled by a fake finger model or a dead finger (32).

6.9. Impedance Tomography

By applying many electrodes on the surface of a body, it is possible to map the distribution of immittivity in the volume under the electrodes (33–35). One approach is to use, for example, 16 electrodes, excite one pair and arrange a multichannel measurement of the transfer impedance in all the other unexcited pairs (Fig. 13). By letting all pairs be excited in succession, one complete measurement is performed. By choosing a high measuring frequency of, for example, 50 kHz, it is possible to sample data for one complete image in less than one tenth of a second, and live images are possible. The sensitivity in Equation 9 clearly shows that it is more difficult to obtain sharp spatial resolution the larger the depth from the skin surface. In practice, the resolution is on the order of centimeters; therefore, other advantages are pursued (e.g., the instrumentation robustness and the simplicity of the sensors).

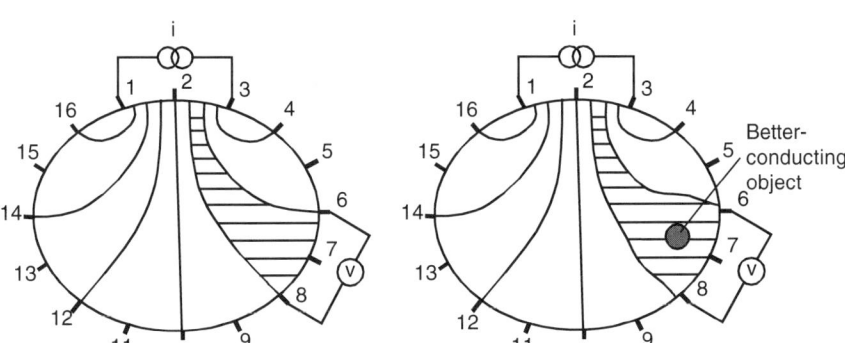

Figure 13. Principle of a tomography setup. From (2), by permission.

6.10. Monitoring Tissue Ischemia and Death

Large changes in tissue impedance occur during ischemia (36,37), tissue death, and the first hours afterward (2). The changes are related to changed distribution of intracellular and extracellular liquids, variations in the gap junctions between the cells, and, in the end, the breakdown of membrane structures.

7. FUTURE TRENDS

Basic scientific topics: immittivity of the smaller tissue components, sensitivity field theory, impedance spectroscopy and tissue characterization, nonlinear tissue properties, and single-cell manipulation.

Instrumentation: ASIC (Application Specific Integrated Circuit) design as a new basis for small and low-cost instrumentation, also for single-use applications. Telemetry technology to improve signal pick-up and reduce noise and influence from common-mode signals.

Applications: microelectrode technology; single-cell and microbe monitoring; electroporation; electrokinetics (e.g., electrorotation); tissue characterization; monitoring of tissue ablation, tissue/organ state, and death/rejection processes.

BIBLIOGRAPHY

1. L. A. Geddes and L. E. Baker, *Principles of Applied Biomedical Instrumentation*. New York: John Wiley, 1989.
2. S. Grimnes and Ø. G. Martinsen, *Bioimpedance and Bioelectricity Basics*. San Diego, CA: Academic Press, 2000.
3. R. Pethig, *Dielectric and Electronic Properties of Biological Materials*. New York: John Wiley, 1979.
4. S. Ollmar, Methods for information extraction from impedance spectra of biological tissue, in particular skin and oral mucosa—a critical review and suggestions for the future. *Bioelectrochem. Bioenerg.* 1998; **45**:157–160.
5. R. Plonsey and R. C. Barr, A critique of impedance measurements in cardiac tissue. *Ann. Biomed. Eng.* 1986; **14**:307–322.
6. J. Malmivuo and R. Plonsey, *Bioelectromagnetism*. Oxford, UK: Oxford University Press, 1995.
7. R. Plonsey and R. C. Barr, *Bioelectricity. A Quantitative Approach*. New York: Plenum Press, 2000.
8. K. R. Foster and H. P. Schwan, Dielectric properties of tissues and biological materials: a critical review. *CRC Crit. Rev. Biomed. Eng.* 1989; **17**(1):25–104.
9. J.-P. Morucci, M. E. Valentinuzzi, B. Rigaud, C. J. Felice, N. Chauveau, and P.-M. Marsili, Bioelectrical impedance techniques in medicine. *Crit. Rev. Biomed. Eng.* 1996; **24**:223–681.
10. S. Takashima, *Electrical Properties of Biopolymers and Membranes*. New York: Adam Hilger, 1989.
11. K. Sakamoto and H. Kanai, Electrical characteristics of flowing blood. *IEEE Trans. BME* 1979; **26**:686–695.
12. F. F. Kuo, *Network Analysis and Synthesis*, international edition, New York: John Wiley, 1962.
13. F. A. Duck, *Physical Properties of Tissue. A Comprehensive Reference Book*. San Diego, CA: Academic Press, 1990.
14. D. B. Geselowitz, An application of electrocardiographic lead theory to impedance plethysmography. *IEEE Trans. Biomed. Eng.* 1971; **18**:38–41.
15. H. P. Schwan, Electrical properties of tissue and cell suspensions. In: J. H. Lawrence and C. A. Tobias, eds., *Advances in Biological and Medical Physics*, vol. V. San Diego, CA: Academic Press, 1957, pp. 147–209.
16. H. Fricke, Theory of electrolytic polarisation. *Phil. Mag.* 1932; **14**:310–318.
17. K. S. Cole, Permeability and impermeability of cell membranes for ions. *Cold Spring Harbor Symp. Quant. Biol.* 1940; **8**:110–122.
18. S. Grimnes and Ø. G. Martinsen, Cole electrical impedance model—a critique and an alternative. *IEEE Trans. Biomed. Eng.* 2005; **52**(1):132–135.
19. H. Fricke, The Maxwell–Wagner dispersion in a suspension of ellipsoids. *J. Phys. Chem.* 1953; **57**:934–937.
20. J. R. Macdonald, *Impedance Spectroscopy, Emphasizing Solid Materials and Systems*. New York: John Wiley, 1987.
21. J. Koryta, *Ions, Electrodes and Membranes*. New York: John Wiley, 1991.
22. L. A. Geddes, *Electrodes and the Measurement of Bioelectric Events*. New York: Wiley-Interscience, 1972.
23. H. P. Schwan, Determination of biological impedances. In: W. L. Nastuk, ed., *Physical Techniques in Biological Research*, Vol. 6. San Diego, CA: Academic Press, 1963, pp. 323–406.
24. IEC-60601, Medical electrical equipment. General requirements for safety. International Standard, 1988.
25. C. Polk and E. Postow, eds., *Handbook of Biological Effects of Electromagnetic Fields*. Boca Raton, FL: CRC Press, 1995.
26. R. Pethig, J. P. H. Burt, A. Parton, N. Rizvi, M. S. Talary, and J. A. Tame, Development of biofactory-on-a-chip technology using excimer laser micromachining. *J. Micromech. Microeng.* 1998; **8**:57–63.
27. I. Giaever and C. R. Keese, A morphological biosensor for mammalian cells. *Nature* 1993; **366**:591–592.
28. V. Kachel, Electrical resistance pulse sizing: Coulter sizing. In: M. R. Melamed, T. Lindmo, and M. L. Mendelsohn, eds., *Flow Cytometry and Sorting*. New York: Wiley-Liss Inc., 1990, pp. 45–80.
29. K. R. Foster and H. C. Lukaski, Whole-body impedance—what does it measure? *Am. J. Clin. Nutr.* 1996; **64**(suppl): 388S–396S.
30. P. K. Kaupinen, J. A. Hyttinen, and J. A. Malmivuo, Sensitivity distributions of impedance cardiography using band and spot electrodes analysed by a 3-D computer model. *Ann. Biomed. Eng.* 1998; **26**:694–702.
31. Ø. G. Martinsen and S. Grimnes, On using single frequency electrical measurments for skin hydration assessment. *Innov. Techn. Biol. Med.* 1998; **19**:395–399.
32. Ø. G. Martinsen, S. Grimnes, K. H. Riisnæs, and J. B. Nysæther, Life detection in an electronic fingerprint system. *IFMBE Proc.* 2002; **3**:100–101.
33. K. Boone, Imaging with electricity: report of the European concerted action on impedance tomography. *J. Med. Eng. Tech.* 1997; **21**:6,201–232.
34. P. Riu, J. Rosell, A. Lozano, and R. Pallas-Areny, Multifrequency static imaging in electrical impedance tomography. Part 1: instrumentation requirements. *Med. Biol. Eng. Comput.* 1995; **33**:784–792.
35. J. G. Webster, *Electrical Impedance Tomography*. New York: Adam Hilger, 1990.
36. M. Schaefer, W. Gross, J. Ackemann, M. Mory, and M. M. Gebhard, Monitoring of physiological processes in ischemic heart muscle by using a new tissue model for description of

the passive electrical impedance spectra. Proc. 11th Int. Conf. on Electrc. Bioimpedance, Oslo, Norway, 2001: 55–58.

37. M. Osypka and E. Gersing, Tissue impedance and the appropriate frequencies for EIT. *Physiol. Meas.* 1995; **16**:A49–A55.

BIOINFORMATICS

LEI LIU
ALI E. ABBAS
University of Illinois at Urbana-Champaign
Urbana, Illinois

1. INTRODUCTION

In the past two decades we have witnessed revolutionary changes in biomedical research and biotechnology and an explosive growth of biomedical data. High throughput technologies developed in automated DNA sequencing, functional genomics, proteomics, and metabolomics enable us to produce such high volume and complex data that the data analysis becomes a big challenge. Consequently, a promising new field, Bioinformatics has emerged and is growing rapidly. Combining biological studies with Computer Science, Mathematics and Statistics, Bioinformatics develops methods, solutions, and software to discover patterns, generate models, and gain insight knowledge of complex biological systems.

Before we discuss further of the field, let us briefly review the basic concepts in molecular biology, which are the foundations for bioinformatics studies. The genetic information is coded in DNA sequences. The physical form of a gene is a fragment of DNA. A genome is the complete set of DNA sequences that encode all the genetic information for an organism, which is often organized into one or more chromosomes. The genetic information is decoded through

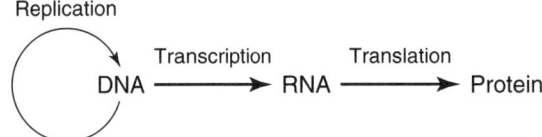

Figure 1. Central dogma of molecular biology. (This figure is available in full color at http://www.mrw.interscience.wiley.com/ebe.)

complex molecular machinery inside a cell composed of two major parts: transcription and translation, to produce functional protein and RNA products. These molecular genetic processes can be summarized precisely by the Central Dogma shown in Fig. 1. The proteins and active RNA molecules combined with other large and small biochemical molecules, organic and inorganic compounds form the complex dynamic network systems that maintain the living status of a cell. Proteins form complex 3-D structures that carry out functions. The 3-D structure of a protein is determined by the primary protein sequence and the local environment. The protein sequence is decoded from the DNA sequence of a gene through the genetic codes as shown in Table 1. These codes have been shown to be universal among all living forms on earth.

The high throughput data can be generated at many different levels in the biological system. The genomics data are generated from the genome sequencing that deciphers the complete DNA sequences of all the genetic information in an organism. We can measure the mRNA levels using microarray technology to monitor the gene expression of all the genes in a genome known as transcriptome. Proteome is the complete set of proteins in a cell at a certain stage, which can be measured by high throughput 2-D gel electrophoresis and Mass Spectrometry. We also can monitor all the metabolic compounds in a cell known as metabolome in a high throughput fashion. Many new terms ending with "ome" can be viewed as the complete set of entities in a cell. For example, the

Table 1. The Genetic Code

| First Position | | Second | Position | | Third Position |
	T	C	A	G	
T	TTT Phe [F]	TCT Ser [S]	TAT Tyr [Y]	TGT Cys [C]	T
	TTC Phe [F]	TCA Ser [S]	TAC Tyr [Y]	TGC Cys [C]	C
	TTA Leu [L]	TCG Ser [S]	TAA *Stop* [end]	TGA *Stop* [end]	A
	TTG Leu [L]	TCC Ser [S]	TAG *Stop* [end]	TGG Trp [W]	G
C	CTT Leu [L]	CCT Pro [P]	CAT His [H]	CGT Arg [R]	T
	CTC Leu [L]	CCC Pro [P]	CAC His [H]	CGC Arg [R]	C
	CTA Leu [L]	CCA Pro [P]	CAA Gln [Q]	CGA Arg [R]	A
	CTG Leu [L]	CCG Pro [P]	CAG Gln [Q]	CGG Arg [R]	G
A	ATT Ile [I]	ACT Thr [T]	AAT Asn [N]	AGT Ser [S]	T
	ATC Ile [I]	ACC Thr [T]	AAC Asn [N]	AGC Ser [S]	C
	ATA Ile [I]	ACA Thr [T]	AAA Lys [K]	AGA Arg [R]	A
	ATG Met [M]	ACG Thr [T]	AAG Lys [K]	AGG Arg [R]	G
G	GTT Val [V]	GCT Ala [A]	GAT Asp [D]	GGT Gly [G]	T
	GTC Val [V]	GCC Ala [A]	GAC Asp [D]	GGC Gly [G]	C
	GTA Val [V]	GCA Ala [A]	GAA Glu [E]	GGA Gly [G]	A
	GTG Val [V]	GCG Ala [A]	GAG Glu [E]	GGG Gly [G]	G

"interactome" refers to the complete set of protein-protein interactions in a cell.

The theory of evolution is also a fundamental base for many aspects of Bioinformatics, especially on sequence and phylogenetic analyses. According to the Darwin's theory of evolution, mutation and natural selection is the driving force during the evolution. In 1980s, the neutral theory of molecular evolution was proposed by Kimura (1) based on the observation that the mutation rates were not even on different parts of genomes. The places that change rapidly might not be under the natural selection. The assumptions on how genes are changing have profound influence on the analysis of biological sequences.

Bioinformatics is needed at all levels of high throughput systematic studies to facilitate the data analysis, mining, management, and visualization. But more importantly, the major task is to integrate data from different levels and prior biological knowledge to achieve system level understanding of biological phenomena. Since Bioinformatics touches on many areas of biological studies, it is impossible to cover every aspect in a short chapter. In this article, the authors will provide a general overview of the field and focus on several key areas including: sequence analysis, phylogenetic analysis, protein structure, genome analysis, microarray analysis, and network analysis.

Sequence analysis often refers to sequence alignment and pattern searching in both DNA and protein sequences. This area can be considered as the "classical" Bioinformatics, which can be dated back to 1960s, long before the word "Bioinformatics" appeared. It deals with the problems such as how to make an optimal alignment between two sequences; how to search sequence databases quickly with an unknown sequence. Phylogenetic analysis is closely related to sequence alignment. The idea is to use DNA or protein sequences comparison to infer evolution history. The first step in this analysis is to perform multiple sequence alignment. Then a phylogenetic tree is built based on the multiple alignments. The protein structure analysis involves the prediction of protein secondary and tertiary structures from the primary sequences. So far the analyses focus on individual sequences or handful of sequences. The next three areas are involved in system wide analysis. Genome analysis mainly deals with the sequencing of a complete or partial genome. The problems include genome assembly, gene structure prediction, gene function annotation, and so on. Many techniques of sequence analysis are used in genome analysis, but many new methods were developed for the unique problems. Microarray technologies provide an opportunity for biologist to study the gene expression at a system level. The problems faced in the analysis are completely different from sequence analysis. Many statistical and data mining techniques are applied in the field. Network analysis is another system level study of biological system. Biological networks can be divided into three categories: metabolic network, protein-protein interaction network, and genetic network. The questions in this area include network modeling, network inference from high throughput data, such as microarray, and network properties study. In the following several sections, we will make a more in-depth discussion of each area.

2. SEQUENCE ALIGNMENT

2.1. Pair-wise Sequence Alignment

Sequence alignment can be described by the following problem. Given two strings of text, X and Y, (that may be DNA or amino acid sequences) find the optimal way of inserting dashes into the two sequences so as to maximize a given scoring function between them. The scoring function depends on both the length of the regions of consecutive dashes and on the pairs of characters that are in the same position when gaps have been inserted. The following example from Abbas and Holmes (2) illustrates the idea of sequence alignment for two strings of text. Consider the two sequences, COUNTING, and NTIG shown in Fig. 2a. Figures 2b–d, show possible alignments obtained by inserting gaps (dashes) at different positions in one of the sequences. Figure 2d shows the alignment with the highest number of matching elements. The "optimal alignment" between two sequences depends on the scoring function that is used. As we shall see, an optimal sequence alignment for a given scoring function may not be unique.

Now we have discussed what is meant by an optimal sequence alignment, we need to explain the motivation for doing it. Sequence alignment algorithms can detect mutations in the genome that lead to genetic disease, and also provide a similarity score, which can be used to determine the probability that the sequences are evolutionarily related. Knowledge of evolutionary relation between a newly identified protein sequence and a family of protein sequences in a database may provide the first clues about its three-dimensional structure and chemical function.

```
(a)   Sequence 1   C O U N T I N G
      Sequence 2   N T I G

(b)   Sequence 1   C O U N T I N G     Possible Alignment
      Sequence 2   - - - N T I G -     (Shifting Sequence 2)

(c)   Sequence 1   C O U N T I N G     Possible Alignment
      Sequence 2   - - - N T - I G     (Shifting Sequence 2 and inserting a gap)

(d)   Sequence 1   C O U N T I N G     Possible Alignment
      Sequence 2   - - - N T I - G     (Shifting Sequence 2 and inserting a gap)
```

Figure 2. Possible alignments of two sequences.

Furthermore, by aligning families of proteins that have the same function (and may have very different sequences) we can observe a common subsequence of amino acids that is key to its particular function. These subsequences are termed protein motifs. Sequence alignment is also a first step in constructing phylogenetic trees that relate biological families of species.

A dynamic programming approach to sequence alignment was proposed by Needleman and Wunsch (3). The idea behind the dynamic programming approach can be explained using the two sequences, CCGAT and CA-AT, of Fig. 3a. Suppose we have an optimal alignment for the two sequences and an additive scoring system for their alignment. If we break this alignment into two parts (Fig. 3b), we have two alignments: the left is the alignment of the two sequences CCGA and CA-A, and the right is the alignment of the last elements T-T. If the scoring system is additive, then the score of the alignment of Fig. 3b is the sum of the scores of the four base-alignment on the left (CCGA and CA-A) plus the score of the alignment of the pair T-T on the right. If the alignment in Fig. 3a is optimal then the four-base alignment in the left hand side of Fig. 3b must also be optimal. If this were not the case (for example if a better alignment would be obtained by aligning A with G) then the optimal alignment of Fig. 3c would lead to a higher score than the alignment shown in Fig. 3a. The optimal alignment ending at any stage is therefore equal to the total (cumulative) score of the optimal alignment at the previous stage plus the score assigned to the aligned elements at that current stage.

The optimal alignment of two sequences ends with the last two symbols aligned, the last symbol of one sequence aligned to a gap, or the last symbol of the other sequence aligned to a gap. In our analysis x_i refers to the ith symbol in sequence 1 and y_j refers to the jth symbol in sequence 2 before any alignment has been made. We will use the symbol $S(i,j)$ to refer to the cumulative score of the alignment up until symbols x_i and y_j, and the symbol $s(x_i, y_j)$ to refer to the score assigned to matching elements x_i and y_j. We will use d to refer to the cost associated with introducing a gap.

1. If the current stage of the alignment matches two symbols, x_i and y_j, then the score, $S(i,j)$, is equal to the previous score, $S(i-1, j-1)$, plus the score assigned to aligning the two symbols, $s(x_i, y_j)$.
2. If the current match is between symbol x_i in sequence 1 and a gap in sequence 2 then the new score is equal to the score up until symbol x_{i-1} and the same symbol y_j, $S(i-1, j)$, plus the penalty associated with introducing a gap, $-d$
3. If the current match is between symbol y_j in sequence 2 and a gap in sequence 1 then the new score is equal to the previous score up until symbol y_{j-1} and the same symbol x_i, $S(i, j-1)$, plus the gap penalty $-d$

The optimal cumulative score at symbols x_i and y_j is:

$$S(i,j) = \max \begin{cases} S(i-1, j-1) + s(x_i, y_j) \\ S(i-1, j) - d \\ S(i, j-1) - d. \end{cases}$$

The previous equation determines the new elements at each stage in the alignment by successive iterations from the previous stages. The maximum at any stage may not be unique. The optimal sequence alignment (s) is the one that provides the highest score. This is usually performed using a matrix representation, where the cells in the matrix are assigned an optimal score, and the optimal alignment is determined by a process called trace back (4,5).

The optimal alignment between two sequences depends on the scoring function that is used. This brings the need for a score that is biologically significant and relevant to the phenomenon being analyzed. Substitution matrices, present one method of achieving this using a "log-odds" scoring system. It lists the likelihood of change from one amino acid or nucleotide to another in homologous sequences during evolution. One of the first substitution matrices used to score amino acid sequences was developed by Dayhoff et al (6) and called Percent Accepted Mutation (PAM) Matrix, which was derived from a relatively small set of closely related proteins. Other matrices such as the BLOSUM50 matrix (7) were also developed and use databases of more distantly related proteins.

The Needleman- Wunsch (N-W) algorithm and its variation (4) provide the best *global* alignment for two given sequences. Smith and Waterman (8) presented another dynamic programming algorithm that deals with finding the best *local* alignment for smaller subsequences of two given sequences rather than the best global alignment of the two sequences. The local alignment algorithm identifies a pair of subsegments, one from each of the given sequences, such that there is no other pair of subsegments with greater similarity.

2.2. Heuristic Alignment Methods

Heuristic search methods for sequence alignment have gained popularity and extensive use in practice because of the complexity and large number of calculations in the dynamic programming approach. Heuristic approaches search for local alignments of subsegments and use these alignments as "seeds" in which to extend out to longer sequences. The most widely used heuristic search method available today is BLAST (Basic Local Alignment Search Tool) by Altschul et al (9). BLAST alignments define a measure of similarity called MSP (Maximal Segment Pair) as the highest scoring pair of identical length subsegments from two sequences. The lengths of the subsegments are chosen to maximize the MSP score.

```
C C G A T          C C G A    T          C C G A T
| |   | |          | |   |    |          |   | | |
C A - A T          C A - A    T          C - A A T
   (a)                (b)       +           (c)
```

Figure 3. Overview of the dynamic programming approach.

2.3. Multiple Sequence Alignments

Multiple sequence alignments are alignments of more than two sequences. The inclusion of additional sequences can improve the accuracy of the alignment, find protein motifs, identify related protein sequences in a database, and predict protein secondary structure. Multiple sequence alignments are also the first step in constructing phylogenetic trees.

The most common approach for multiple alignments is progressive alignment, which involves choosing two sequences and performing a pairwise alignment of the first to the second. The third sequence is then aligned to the first and the process is repeated until all the sequences are aligned. The score of the multiple alignment is the sum of scores of the pairwise alignments. Pairwise dynamic programming can be generalized to perform multiple alignments using the progressive alignment approach; however, it is computationally impractical even when only a few sequences are involved (10). The sensitivity of progressive alignment was improved for divergent protein sequences using CLUSTAL-W (11), available at (*http://clustalw.genome.ad.jp/*).

Many other approaches to sequence alignment have been proposed in the literature. For example, a Bayesian approach was suggested for adaptive sequence alignments (12,13), (Zhu et al 1998). Another approach to sequence alignment that has found great success is hidden Markov models. We will refer to this approach in more detail in the section on genome annotation. The data that is now available from the human genome project has suggested the need for aligning whole genome sequences where large-scale changes can be studied as opposed to single-gene insertions, deletions, and nucleotide substitutions. MuMMer (14) follows this direction and performs alignments and comparisons of very large sequences.

3. PHYLOGENETIC TREES

Biologists have long built trees to classify species based on morphological data. The main objectives of phylogenetic tree studies are (1) to reconstruct the genealogical ties between organisms and (2) to estimate the time of divergence between organisms since they last shared a common ancestor. With the explosion of genetic data in the last few years, molecular based phylogenetic studies have been used in many applications such as the study of gene evolution, population subdivisions, analysis of mating systems, paternity testing, environmental surveillance, and the origins of diseases that have transferred species.

From a mathematical point of view, a phylogenetic tree is a rooted binary tree with labeled leaves. A tree is binary if each vertex has either one or three neighbors. A tree is rooted if a node, R, has been selected and termed the root. A root represents an ancestral sequence from which all other nodes descend. Two important aspects of a phylogenetic tree are its topology and branch length. The topology refers to the branching pattern of the tree and the branch length is the "evolutionary" time between the splitting events. Figure 4a shows a rooted binary tree with six leaves. Figure 4b shows all possible distinct rooted topologies for a tree with 3 leaves.

The data that is used to construct trees is usually in the form of contemporary sequences and is located at the leaves. For this reason trees are represented with all their leaves "on the ground level" rather than at different levels.

The tree-building analysis consists of two main steps. The first step, estimation, uses the data matrix to produce a tree, $\tilde{T}$, that estimates the unknown tree, T. The second step provides a confidence statement about the estimator $\tilde{T}$. This is often performed by bootstrapping methods.

Tree-building techniques can generally be classified into one of four types: distance-based methods, parsimony methods, maximum likelihood methods, and Bayesian methods. For a detailed discussion of each of these methods see Li (15). Now we will give a brief overview of each of the tree-building methods, more details of which can be found in Abbas and Holmes (2).

Distance-Based methods first calculate an "evolutionary distance", d_{xy}, between each two sequences X and Y in a multiple sequence alignment. The pairwise distance for N sequences results in a distance matrix of dimension $N \times N$, which is symmetric about its diagonal if the distance d_{xy} is symmetric. One of the widely used distance-based methods of phylogenetic tree construction the Jukes-Cantor model (16) that provides an estimate of the evolutionary distance between X and Y as

$$d_{xy} = -\frac{3}{4}\log\left(1 - \frac{4}{3}\left(1 - \left(\frac{\#AA}{K} + \frac{\#CC}{K} + \frac{\#GG}{K} + \frac{\#TT}{K}\right)\right)\right),$$

where K denotes the number of characters (columns) in the dataset, and $\#AA$ denotes the number of times a letter A in sequence X is matched with a letter A in sequence Y. Once the distance matrix is calculated, the phylogenetic tree is estimated using a clustering technique. Clustering is the task of segmenting the sequences into a number of homogeneous subgroups or clusters. The most commonly applied clustering methods are the unweighted pair group

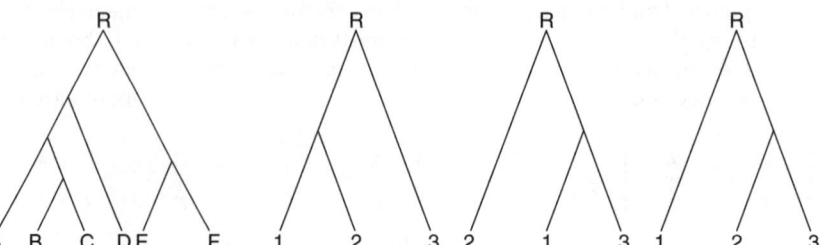

Figure 4. (a) Rooted tree with six leaves. (b) All possible topologies for three leaves.

method with arithmetic mean (UPGMA), and the neighbor-joining method (17).

The Parsimony Method for constructing phylogenetic trees is based on the assumption that "evolution is parsimonious" which means that there should be no more evolutionary steps than necessary. As a result, the phylogenetic tree that is selected is the one with the minimum number of substitutions between ancestor and descendants. Maximum likelihood methods select the tree that has the highest probability of producing the observed data. Under this model the likelihood for each possible tree is separately computed for each sequence (row) in the data set. This requires computing the likelihood of all the possible trees and so the method is computationally expensive and requires efficient search procedures. For more details see Felsenstein (18). The Bayesian estimation methods start the tree construction with a very wide prior distribution on the space of all trees. The approach then uses Gibbs sampling and Monte Carlo Markov Chains to compute a posterior probability distribution on the tree conditioned on the dataset. To facilitate this task, Huelsenbeck and Ronquist (19) developed a software package called Mr. Bayes to perform Bayesian inference of phylogenetic trees using use MCMC simulation.

Tree-building methods can be compared using several criteria such as accuracy, consistency, efficiency, and robustness. To clarify some of these issues, we refer the reader to Holmes (20) where a geometric analysis of the problem is provided and these issues are further discussed. The second part of the tree-building analysis is concerned with how "close" we believe the estimated tree is to the true tree. Felsenstein (21) suggested the use of the bootstrap to answer this question of how much confidence should we have in the estimated trees. Another method builds on a probability distribution on the space of all trees. The difficult part of this problem is that there are exponentially many possible trees. A nonparametric approach using a multinomial probability model on the whole set of trees would not be feasible as the number of trees is (2N-3)!!. The Bayesian approach defines parametric priors on the space of trees, and then computes the posterior distribution on the same subset of the set of all trees. This analysis enables confidence statements in a Bayesian sense (22).

4. PROTEIN FOLDING, SIMULATION, AND STRUCTURE PREDICTION

The structure of a protein greatly influences its function. Knowledge of protein structure and function can help determine the chemical structure of drugs needed to reverse the symptoms that arise due to its malfunction. The bonds in a molecular structure contribute to its overall potential energy. We shall neglect all quantum mechanical effects in the following discussion and consider only the elements that contribute largely to the potential energy of a structure (as suggested by (23)).

1. **Pair Bonds:** This is a bond that exists between atoms physically connected by a bond and separated by a distance b. It is like a spring action where energy is stored above and below an equilibrium distance, b_0. The energy associated with this bond is $U(b) = \frac{1}{2}K_b(b-b_0)^2$, where b_0 can be determined from the X-ray of the crystal structure showing the electron density maps, and K_b can be determined from spectroscopy.

2. **Bond Angles:** This bond exists when an angular deviation from an equilibrium angle, θ_o, occurs between three atoms. The bond angle energy associated with the triplet is $U(\theta) = \frac{1}{2}K_\theta(\theta-\theta_0)^2$.

3. **Torsion Angles:** This bond exists when a torsion angle, ϕ, exists between the first and fourth atoms on the axis of the second and third atoms. The energy associated with this bond is $U(\phi) = K_\phi(1-\cos(n\psi+\delta))$, where δ is an initial torsion angle.

4. **Non-bonded pairs:** Bonds also exist between atoms that are not physically connected in the structure. These bonds include

   a. Van der Waal forces, which exist between non-bonded pairs and contribute to energy, $U(r) = \varepsilon[(\frac{r_0}{r})^{12} - 2(\frac{r_0}{r})^6]$, r_0 is an equilibrium distance and ε a constant.

   b. Electrostatic Interactions, which contribute to an energy of $U(r) = \alpha\frac{q_iq_j}{r}$, and

   c. Hydrogen bonds, which result from Van Der Waals forces and the geometry of the system, and contribute to the potential energy of the structure.

The total potential energy function of a given structure can thus be determined by the knowledge of the precise position of each atom. The three main techniques that are used for protein structure prediction: homology (comparative modeling), fold recognition and threading, and *Ab initio* folding.

4.1. Homology or Comparative Modeling

Comparative modeling techniques predict the structure of a given protein sequence based on its alignment to one or more protein sequences of known structure in a protein database. The approach uses sequence alignment techniques to establish a correspondence between the known structure "template" and the unknown structure. Protein structures are archived for public use in an Internet-accessible database known as the Protein Data Bank. *http://www.rcsb.org/pdb/* (24).

4.2. Fold Recognition and Threading

When the two sequences exhibit less similarity, the process of recognizing which folding template to use is more difficult. The first step in this case is to choose a structure from a library of templates in the protein databank. This is called fold recognition. The second step "threads" the given protein sequence into the chosen template. Several computer software programs are available for protein structure prediction using the fold recognition and threading technique such as PROSPECT (25).

4.3. Ab Initio (New Fold) Prediction

If no similarities exist with any of the sequences in the database, the ab initio prediction method is used. This method is one of the earliest structure prediction methods, and uses energy interaction principles to predict the protein structure (23,26,27). Some of these methods include optimization where the objective is to find a minimum-energy structure (a local minimum in the energy landscape has zero forces acting on the atoms and is therefore an equilibrium state).

Another type of analysis uses Molecular dynamics uses equations of motion to trace the position of each atom during folding of the protein (28). A single structure is used as a starting point for these calculations. The force acting on each atom is the negative of the gradient of the potential energy at that position. Accelerations, a_i, are related through masses, m_i, to forces, F_i, via Netwon's second law ($F_i = m_i a_i$). At each time step, new positions and velocities of each of the atoms are determined by solving equations of motion using the old positions, old velocities, and old accelerations. Beeman (29) showed that new atomic positions and velocities could be determined by the following equations of motion

$$x(t + \Delta t) = x(t) + v(t)\Delta t + [4a(t) - a(t + \Delta t)]\frac{(\Delta t)^2}{6},$$

$$v(t + \Delta t) = v(t) + [2a(t + \Delta t) + 5a(t) - a(t - \Delta t)]\frac{\Delta t}{6},$$

where $x(t)$ = position of the atom at time, t, $v(t)$ = velocity of the atom at time, t, $a(t)$ = acceleration at time, t, and Δt = time step in the order of 10^{-15} seconds for the simulation to be accurate.

In 1994, the first large-scale experiment to assess protein structure prediction methods was conducted. This experiment is known as CASP (Critical Assessment of techniques for protein Structure Prediction). The results of this experiment were published in a special issue of *Proteins* (1995). Further experiments were developed to evaluate the fully automatic web servers for fold recognition. These experiments are known as CAFASP (Critical Assessment of Fully Automated Structure Prediction). For a discussion on the limitations, challenges, and likely future developments on the evaluation of the field of protein folding and structure prediction, we refer the reader to (30).

5. GENOME ANALYSIS

Analysis of completely sequenced genomes has been one of the major driving forces for the development of bioinformatics field. The major challenges in this area include genome assembly, gene prediction, function annotation, promoter region prediction, identification of single nucleotide polymorphism (SNP), and comparative genomics of conserved regions. For a genome project, one must ask several fundamental questions: how can we put the whole genome together from many small pieces of sequences? where are the genes located on a chromosome? and what are other features we can extract from the completed genomes?

5.1. Genome Assembly

The first problem is pertaining to the genome mapping and sequence assembly. During the sequencing process, large DNA molecules with millions of base pairs, such as a human chromosome, are broken into smaller fragments (~100 kb) and cloned into vector such as bacterial artificial chromosome (BAC). These BAC clones can be tiled together by physical mapping techniques. Individual BACs can be further broken down into smaller random fragments of 1-2 kb. These fragments are sequenced and assembled based on overlapping fragments. With more fragments sequenced, there will be enough overlaps to cover most of the sequence. This method is often referred as "shotgun sequencing". Computer tools were developed to assemble the small random fragments into large contigs based on the overlapping ends among the fragments using similar algorithms as the ones used in the basic sequence alignment. The widely used ones include PHRAP/Consed (31,32) and CAP3 (33). Most of prokaryotic genomes can be sequenced directly by the "shotgun sequencing" strategy with special techniques for gap closure. For large genomes, such as human genome, there are two strategies. One is to assemble large contigs first and then tile together the contigs based on the physical map to form the complete chromosome (34) Another strategy is called Whole Genome Shotgun Sequencing (WGS) strategy, which assemble the genome directly from the "shotgun sequencing" data in combination with mapping information (35). WGS is a faster strategy to finish a large genome, but the challenge of WGS is how to deal with the large number of repetitive sequences in a genome. Nevertheless, WGS has been successfully used in completing the *Drosophila* and human genomes (36,37).

5.2. Genome Annotation

The second problem is related to deciphering the information coded in a genome, which is often called genome annotation. The process includes the prediction of gene structures and other features on a chromosome and the function annotation of the genes. There are two basic types of genes in a genome: RNA genes and protein encoding genes. RNA genes produce active RNA molecules such as ribosomal RNA, tRNA, small RNA. Majority of genes in a genome are protein encoding genes. Therefore, the big challenge is how to find the protein encoding region in a genome. The simplest way to search for a protein encoding region is to search for open reading frames (ORF), which is a contiguous set of codons between two stop codons. There are six possible reading frames for a given DNA sequence. Three of them start at the first, second, and third base. The other three reading frames are at the complementary strand. The longest ORFs between the start codon and the stop codon in the same reading frame provide good, but not sufficient evidence of a protein encoding region. Gene prediction is generally easier and more accurate in prokaryotic than eukaryotic organisms due to the intron/exon structure in eukaryote genes. In prokaryotic organisms (bacteria and archaea), the translation and transcription are highly coupled. The protein is synthesized from the RNA even before the transcription of the gene is finished. The coding region spans the whole RNA without interruption. On the other hand, in eukaryotic organisms, transcription is more complicated. After the initial RNA product is generated from the gene, it goes through a process called splicing to get rid of non-coding regions (intron) and concatenate the coding regions (exon) together. Computational methods of gene prediction based on Hidden Markov Model (HMM) have been quite successful, especially in prokaryote genome. These methods involve training a gene model to recognize genes in a particular organism. Because of the variations in codon usage, a model must be trained for each new genome. In prokaryote genome, genes are packed densely with relatively short intergenic sequences. The model reads through a sequence with unknown gene composition and find the regions flanked by start and stop codons. The codon composition of a gene is different from that of an intergenic region and can be used as a discriminator for gene prediction. Several software tools, such as GeneMark (38) and Glimmer (14) are widely used HMM methods in prokaryotic genome annotation. Similar ideas are also applied to eukaryote gene prediction. Because of the intron/exon structure, the model is much more complex with more attention on the boundary of intron and exon. Programs such as GeneScan (39) and GenomeScan (40) are HMM methods for eukaryote gene prediction. Neural network based methods have also been applied in eukaryote gene prediction, such as Grial (41). Additional information for gene prediction can be found using expressed sequence tags (ESTs), which are the sequences from cDNA libraries. Because cDNA is derived from mRNA, a match to an EST is a good indication that the genomic region encodes a gene. Functional annotation of the predicted genes is another major task in genome annotation. This process can be also viewed as gene classification with different functional classification systems such as Enzyme Commission Numbers (EC number) system, protein families, metabolic pathways, and Gene Ontology. The simplest way is to infer annotation from the sequence similarity to a known gene, e.g. BLAST search against a well-annotated protein database such as SWISS-PROT. A better way can be a search against protein family databases (e.g. Pfam (42)), which are built based on profile HMMs. The widely used HMM alignment tools include HMMER (43) and SAM (44). All automated annotation methods can produce mistakes. More accurate and precise annotation requires experimental verification and combination of information from different sources.

Besides the gene structures, other features such as promoters can be better analyzed with a finished genome. In prokaryotic organisms, genes involved in the same pathway are often organized in an operon structure, in which the genes contiguous on the chromosome and transcribed together. Finding operons in a finished genome provides information on the gene regulation. For eukaryotic organisms, the completed genomes provide upstream sequences for promoter region search and prediction. Promoter region prediction and detection has been a very challenging bioinformatics problem. The promoter regions are the binding sites for transcription factors (TF). Promoter prediction is to discover the sequence patterns which are specific for TF binding. Different motif finding algorithms have been applied including scoring matrix method (45), Gibbs sampling (46), and Multiple EM for Motif Elicitation (MEME) (47). The results are not quite satisfactory. Recent studies using comparative genomics methods on the problem have produced some promising results and demonstrated that the promoters are conserved among closely related species (48). In addition, microarray studies can provide additional information for promoter discoveries (see, the section on microarray analysis).

5.3. Comparative Genomics

With more and more genomes being completely sequenced, comparative analysis becomes increasingly valuable and provides more insights of genome organization and evolution. One comparative analysis is based on the orthologous genes, called clusters of orthologous groups (COG) (49). Two genes from two different organisms are considered orthologous genes if they are believed to come from a common ancestor gene. Another term, paralogous genes, refers to genes in one organism and related to each other by gene duplication events. In COG, proteins from all completed genomes are compared. All matching proteins in all the organisms are identified and grouped into orthologous groups by speciation and gene duplication events. Related orthologous groups are then clustered to form a COG that includes both orthologs and paralogs. These clusters correspond to classes of functions. Another type of comparative analysis is based on the alignment of the genomes and studies the gene orders and chromosomal rearrangements. A set of orthologous genes that show the same gene order along the chromosomes in two closely related species is called a synteny group. The corresponding region of the chromosomes is called synteny blocks (50). In closely related species, such as mammalian species, the gene orders in synteny regions are generally conserved with many rearrangements. The chromosomal rearrangements include inversion, translocation, fusion and fission. By comparing completely sequenced genomes, for example, human and mouse genomes, we can reveal the rearrangement events. One challenging problem is to reconstruct the ancestral genome from the multiple genome comparisons and estimate the number and types of the rearrangements (51). Detailed comparisons of genomes often reveal not only the differences between species but much more insight of evolution and function and regulation of genes. Recent publication of chimpanzee genome and its comparison to human genome shows approximately thirty-five million single-nucleotide changes, five million insertion/deletion events, and various chromosomal rearrangements (52).

6. MICROARRAY ANALYSIS

Microarray technologies allow biologists to monitor genome-wide patterns of gene expression in a high throughput fashion. Gene expression refers to the process of transcription. Gene expression for a particular gene can be measured as the fluctuation of the amount of messenger

RNA produced from the transcription process of that gene in different conditions or samples.

DNA microarrays are typically composed of thousands of DNA sequences, called probes, fixed to a glass or silicon substrate. The DNA sequences can be long (500–1500 bp) cDNA sequences or shorter (25–70 mer) oligonucleotide sequences. The probes can be deposited with a pin or piezoelectric spray on a glass slide, known as spotted array technology. Oligonucleotide sequences can also be synthesized in situ on a silicon chip by photolithographic technology (i.e., Affymetrix GeneChip). Relative quantitative detection of gene expression can be carried out between two samples on one array (spotted array) or by single samples comparing multiple arrays (Affymetrix GeneChip). In spotted array experiments, samples from two sources are labeled with different fluorescent molecules (Cy3 and Cy5) and hybridized together on the same array. The relative fluorescence between each dye on each spot is then recorded and a composite image may be produced. The relative intensities of each channel represent the relative abundance of the RNA or DNA product in each of the two samples. In Affymetrix GeneChip experiments, each sample is labeled with the same dye and hybridized to different arrays. The absolute fluorescent values of each spot may then be scaled and compared with the same spot across arrays. Figure 5 gives an example of a composite image from one spotted array.

Microarray analyses usually include several steps: image analysis and data extraction, data quantification and normalization, identification of differentially expressed genes, and knowledge discovery by data mining techniques such as clustering and classification. Image analysis and data extraction is fully automated and mainly carried out using a commercial software package or a freeware depending on the technology platforms. For example, Affymetrix developed a standard data processing procedures and software for its GeneChips (for detailed information *http://www.affymetrix.com*); GenePix is widely used image analysis software for spotted arrays. For the rest of steps, the detailed procedures may vary depending on the experiment design and goals. We will discuss some of the procedures below.

6.1. Statistical Analysis

The purpose of normalization is to adjust for systematic variations, primarily for labeling and hybridization efficiency, so that we can discover true biological variations as defined by the microarray experiment (53,54). For example, as shown in the self-hybridization scatter plot (Fig. 6) for a two-dye spotted array, variations (dye bias) between dyes is obvious and related to spot intensities. To correct, the dye bias, one can apply the following model:

$$\log_2(R/G) -> \log_2(R/G) - c(A),$$

where R and G are the intensities of the dyes; A is the signal strength ($\log_2(R*G)/2$); M is the logarithm ratio ($\log_2(R/G)$); c(A) is the locally weighted polynomial regression (LOWESS) fit to the MA plot (55,56).

After correction of systematic variations, we want to determine which genes are significantly changed during the experiment and to assign appropriately adjusted p-values to the genes. For each gene, we wish to test the null hypothesis that the gene is not differentially expressed. P-value is the probability of finding a result by chance. If P-value is less than a cut-off (e.g., 0.05), one would reject the null hypothesis and state that the gene is differentially expressed (57). Analysis of variance (ANOVA) is usually used to model the factors for a particular experiment. For example,

$$\log(m_{ijk}) = \mu + A_i + D_j + V_k + \varepsilon_{ijk},$$

where m_{ijk} is the ratio of intensities from the two dye-labelled samples for a gene; μ is the mean of ratios from all

Figure 5. An image from a spotted array after laser scanning. Each spot on the image represents a gene and the intensity of a spot reflects the gene expression. (This figure is available in full color at http://www.mrw.interscience.wiley.com/ebe.)

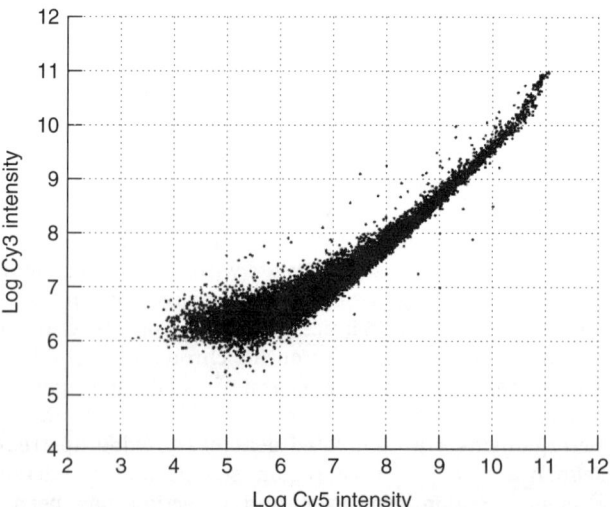

Figure 6. Self-hybridization scatter plot. Y-axis is the intensity from one dye; X-axis is the intensity from the other dye. Each spot is a gene.

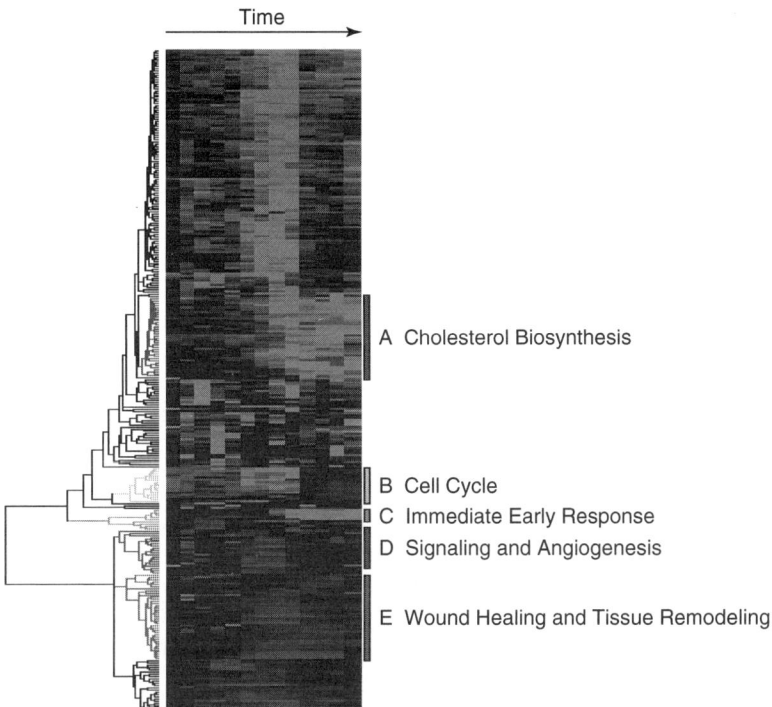

Figure 7. Hierarchical clustering of microarry data. Rows are genes. Columns are RNA samples at different time points. Values are the signals (expression levels), which are represented by the color spectrum. Green represents down-regulation while red represents up-regulation. The color bars beside the dendrogram show the clusters of genes which exhibit similar expression profiles (patterns). The bars are labeled with letters and description of possible biological processes involving the genes in the clusters. (Reprinted from Eisen et al, (64)). (This figure is available in full color at http://www.mrw.interscience.wiley.com/ebe.)

replicates; A is the effect of different arrays; D is the dye effects; V is the treatment effects (58). Through F-test, we will determine if the gene exhibits differential expression between any V_k. For a typical microarray, there are thousands of genes. We need to perform thousands of tests in an experiment at the same time, which introduce the statistical problem of multiple testing and adjustment of p-value. Many methods exist for such as purpose including Bonferroni adjustment and False discovery rate (FDR) (59).

For Affymetrix GeneChips analysis, even though the basic steps are the same as spotted microarrays, because of the difference in technology, different statistical methods were developed. Besides the statistical methods provided by Affymetrix, several popular methods are packaged into software such as dChip (60) and RMA (53) in Bioconductor (*http://www.bioconductor.org*). With rapid accumulation of microarray data, one challenging problem is how to compare microarray data across different technology platforms. Some recent studies on data agreements have provided some guidance (61–63).

6.2. Clustering and Classification

Once obtained from the statistical test a list of significant genes, we would apply different data mining techniques to find interesting patterns. At this step the microarray data set is organized as a matrix. Each column represents a condition; each row represents a gene. An entry is the expression level of the gene under the corresponding condition. If a set of genes exhibit the similar fluctuation under all of the conditions, it may indicate that these genes are co-regulated. One way to discover the co-regulated genes is to cluster genes with similar fluctuation patterns using various clustering algorithm. Hierarchical clustering was the first clustering method applied to the problem (64). The result of hierarchical clustering forms a two-dimensional dendrogram as shown in Fig. 7. The measurement used in the clustering process can be either a similarity such as Pearson's correlation coefficient or a distance such as Euclidian distance.

Many different clustering methods have been applied later on, such as, k-means (65), self-organizing map (66), and support vector machine (67). Another type of microarray study involves classification techniques. For example, we can use the gene expression profile to classify cancer types. Golub et al (68) first reported using classification techniques to classify two different types of leukemia as shown in Fig. 8. Many commercial software packages, e.g., GeneSpring and Spotfire, offer the use of these algorithms for microarray analyses.

7. COMPUTATIONAL MODELING AND ANALYSIS OF BIOLOGICAL NETWORKS

Biological system is a complex system involving hundreds of thousands of elements. The interaction among the elements forms an extremely complex networks. With the development of high throughput technologies in functional genomics, proteomics, and metabolomics, one can start looking into the system level mechanisms governing the interactions and properties of biological networks. Network modeling has been used extensively in social and economical fields for many years (69). Many methods can be applied to biological network studies.

The cellular system involves complex interactions between proteins, DNA, RNA, and smaller molecules and can be categorized in three broad subsystem, metabolic

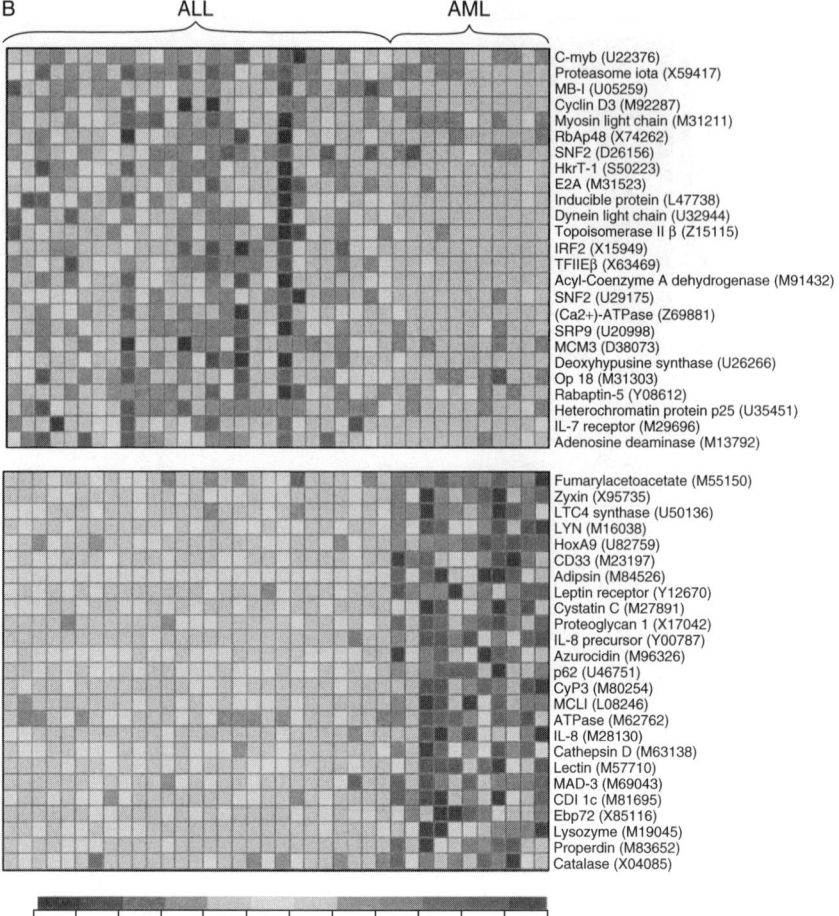

Figure 8. An example of microarray classification. Genes distinguishing acute myeloid leukemia (AML) and acute lymphoblastic leukemia (ALL). The 50 genes most highly correlated with the ALL-AML class distinction are shown. Each row corresponds to a gene, with the columns corresponding to expression levels in different samples. Expression levels for each gene are normalized across the samples such that the mean is 0 and the SD is 1. Expression levels greater than the mean are shaded in red, and those below the mean are shaded in blue. The scale indicates SDs above or below the mean. The top panel shows genes highly expressed in ALL, the bottom panel shows genes more highly expressed in AML. (Reprinted from Golub et al, (68)). (This figure is available in full color at http://www.mrw.interscience.wiley.com/ebe.)

networks or pathway, protein networks, and genetic or gene regulatory networks. *Metabolic networks* represent the enzymatic processes within the cell, which provide energy and building blocks for cells. It is formed by the combination of a substrate with an enzyme in a biosynthesis or degradation reaction. Considerable information about metabolic reactions has been accumulated through many years and are organized into large databases, such as KEGG (70), EcoCyc (71) and WIT (72). *Protein networks* refer to the signaling networks where the basic reaction is between two proteins. Protein-protein interactions can be determined systematically using techniques such as yeast two-hybrid system (73) or derived from the text mining of literatures (74). *Genetic networks or regulatory networks* refer to the functional inference of direct causal gene interactions (75). One can conceptualize gene expression as a genetic feedback networks. The networks can be inferred from the gene expression data generated from microarray or proteomics studies in combination with computation modeling.

Metabolic networks are typically represented as a graph with vertex being all the compounds (substrates) and the edges being reactions linking the substrates. With such representation, one can study the general properties of the metabolic networks. It has been shown that metabolic networks exhibit typical property of small world or scale-free network (76,77). The distribution of compound connectivity follows a power law as shown in Fig. 9. There are nodes serving as hubs in the networks. Such property makes the networks quite robust to random deletion of nodes, but vulnerable to selected deletion of nodes. For example, deletion of hub nodes will cause the network collapse very quickly. A recent study also shows that the metabolic networks are organized in modules based on the connectivity, which are correlated with functional classification (78).

Flux analysis is another important aspect in metabolic network study. Building on the stoichiometric network analysis, which only uses the well-characterized network topology, the concept of elementary flux modes was introduced (79,80). An elementary mode is a minimal set of enzymes that could operate at steady state, with the enzymes weighted by the relative flux they need to carry out the mode to function. The total number of elementary modes for given conditions has been used as a quantitative measure of network flexibility and as an estimate of fault-tolerance (81,82).

A system approach to model regulatory networks is essential to understand their dynamics. Recently several high-level models have been proposed for the regulatory

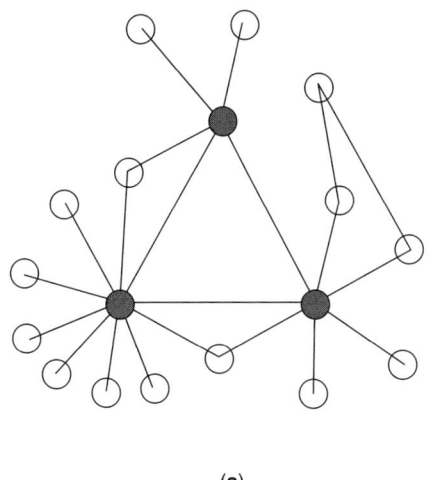 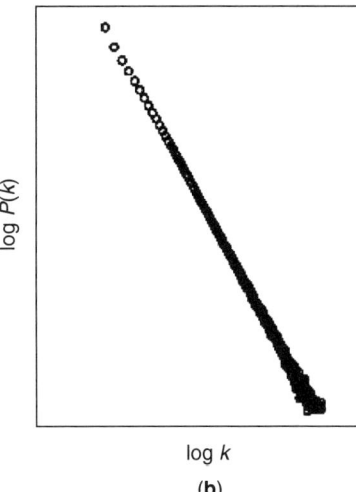

Figure 9. A. In the scale-free network most nodes have only a few links, but a few nodes, called hubs (filled circle), have a very large number of links. B. The network connectivity can be characterized by the probability, $P(k)$, that a node has k links. $P(k)$ for a scale-free network has no well-defined peak, and for large k it decays as a power-law, $P(k) \approx k^{-\gamma}$, appearing as a straight line with slope $-\gamma$ on a log–log plot (Reprinted from Jeong et al, (76)).

network including Boolean models, continuous systems of coupled differential equations, and probabilistic model. *Boolean networks* assume that a protein or a gene can be in one of two states, active or inactive, represented by 1 or 0. This binary state varies in time and depends on the state of the other genes and proteins in the network through a discrete equation:

$$X_i(t+1) = F_i[X_i(t), \ldots, X_N(t)].$$

Thus the function F_i is a Boolean function for the update of the ith element as a function of the state of the network at time t (75). Figure 10 gives a simple example.

Gene expression patterns contain much of the state information of the genetic network and can be measured experimentally. We are facing the challenge of inferring or reverse engineering the internal structure of this genetic network from measurements of its output. Genes with similar temporal expression patterns may share common genetic control processes and may therefore be related functionally. Clustering gene expression patterns according to a similarity or distance measure is the first step toward constructing a wiring diagram for a genetic network (84).

Differential equations can be an alternative model to the Boolean network and applied when the state variables X are continuous and satisfy a system of differential equations of the form

$$\frac{dX_i}{dt} = F_i[X_1(t), \ldots, X_N(t), I(t)],$$

where the vector $I(t)$ represents some external input into the system. The variables X_i can be interpreted as representing concentrations of proteins or mRNAs. Such model has been used to model biochemical reactions in the metabolic pathways and gene regulation (75).

Bayesian networks are provided by the theory of graphical models in statistics. The basic idea is to approximate a complex multi-dimensional probability distribution using a product of simpler local probability distributions. Generally, a Bayesian network model is based on a directed acyclic graph (DAG) with N nodes. In genetic network, the nodes may represent genes or proteins and the random variables X_i levels of activity. The parameters of the model are the local conditional distributions of each random variable given the random variables associated with the parent nodes, whose product yields the joint distribution of all of the random variables:

$$P(X_1, \ldots, X_N) = \prod_i P(X_i | X_j : j \in N^-(i)),$$

where $N^-(i)$ denotes all the parents of vertex i. Given a data set D representing expression levels derived using DNA microarray experiments; it is possible to use learning techniques with heuristic approximation methods to

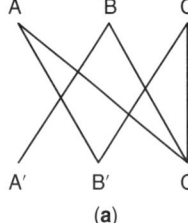

Input			Output		
A	B	C	A'	B'	C'
0	0	0	0	0	0
0	0	1	0	1	0
0	1	0	1	0	0
0	1	1	1	1	1
1	0	0	0	1	0
1	0	1	0	1	1

(c)

Figure 10. Target Boolean network for reverse engineering. (a) The network wiring and (b) logical rules determine (c) the dynamic output. The challenge lies in inferring (a) and (b) from (c) (Reprinted from Liang et al, (83)).

infer the network architecture and parameters. Because data from microarray experiments are still limited and insufficient to completely determine a single model, people have developed heuristics for learning classes of models rather than single models, for instance, for a set of co-regulated genes (75) Bayesian networks have been used to combine heterogeneous data sets and applied recently to genetic networks using microarray data (85,86).

In this chapter we reviewed some major development in the field of bioinformatics and introduced some basic concepts in the field covering six areas: sequence analysis, phylogenetic analysis, protein structure analysis, genome analysis, microarray analysis, and network analysis. Due to the limited space, some topics have been left out. One of such topics is text mining, which using Natural Language Processing (NLP) techniques to extract information from the vast amount of literatures in biological research. Text mining has become an integral part in bioinformatics. With the continuing development and maturing of new technologies in many system level studies, the way that we conduct biological research is undergoing revolutionary change. Systems biology is becoming a major theme and driving force. The challenges for bioinformatics in the post-genomics era lie on the integration of data and knowledge from heterogeneous sources and system level modeling and simulation providing molecular mechanism for physiological phenomena.

BIBLIOGRAPHY

1. M. Kimura, *The neutral theory of molecular evolution*. Cambridge University Press, 1983.
2. A. Abbas, S. Holmes, Bioinformatics and Management Science. Some Common Tools and Techniques. *Operations Research* 2004; **52**(2):165–190.
3. S. B. Needleman, C. D. Wunsch, A General Method Applicable to the Search for Similarities in Amino Acid Sequence of Two Proteins. *J. Mol. Biol.* 1970; **48**:443–453.
4. O. Gotoh, An Improved Algorithm for Matching Biological Sequences. *J. Mol. Biol.* 1982; **162**:705–708.
5. S. Durbin, S. Eddy, A. Krogh, and G. Mitchison, *Biological Sequence Analysis: Probabilistic models of proteins and nucleic acids*. Cambridge, U.K.: Cambridge University Press, 1998.
6. M. O. Dayhoff, R.M. Schwartz, B. C. Orcut, *A model of evolutionary change in proteins*. Atlas of Protein Sequence and Structure. Vol 5, supplement 3. National Biomedical Research Foundation. Washington, D.C., 1978, pp. 345–352.
7. S. Henikoff, J. G. Henikoff, Amino Acid Substitution Matrices from Protein Blocks. *Proc. Natl. Acad. Sci. USA* 1992; **89**:10915–10919.
8. T. F. Smith and M.S. Waterman, Identification of Common Molecular Subsequences. *J. Mol. Biol.* 1981; **147**:195–197.
9. S. F. Altschul, W. Gish, W. Miller, E. Myers, and J. Lipman, Basic Local Alignment Search Tool. *J. Mol. Biol.* 1990; **215**:403–410.
10. J. D. Lipman, S. F. Altschul, and J. D. Kececioglu, A Tool for Multiple Sequence Alignment. *Proc. Natl. Acad. Sci.* 1989; **86**:4412–4415.
11. J. D. Thompson, D. G. Higgins, and T. J. Gibson, CLUSTAL W: improving the sensitivity of progressive multiple sequence alignment through sequence weighting, position specific gap penalties and weight matrix choice. *Nucleic Acids Research* 1994; **22**:4673–4680.
12. C. E. Lawrence, S. F. Altschul, M. S. Boguski, J. S. Liu, A. N. Neuwald, J. Wootton, Detecting subtle sequence signals: A Gibbs sampling strategy for multiple alignment. *Science* 1993; **262**:208–214.
13. J. Zhu, J. S. Liu, and C. E. Lawrence, Bayesian Adaptive Sequence Alignment Algorithms. *Bioinformatics* 1998; 14, 25–39.
14. A. L. Delcher, D. Harmon, S. Kasif, O. White, and S. L. Salzberg, Improved microbial gene identification with GLIMMER. *Nucleic Acids Res.* 1999; **27**(23):4636–4641.
15. W. H. Li, 1997. *Molecular Evolution*. Sinauer Assoc, Boston.
16. T. Jukes, C. Cantor, Evolution of Protein Molecules, in: H. N. Munro, ed., *Mammalian Protein Metabolism*. New York: Academic Press, 1969, pp. 21–132.
17. N. Saitou, M. Nei, The neighbor-joining method: a new method for reconstructing phylogenetic trees. *Mol. Biol. Evol.* 1987; **4**(4):406–425.
18. J. Felsenstein, Evolutionary trees from DNA sequences: a maximum likelihood approach. *J. Mol. Evol.* 1981; **17**(6):368–376
19. J. Huelsenbeck, F. Ronquist, 2002. Mr. Bayes. Bayesian Inference of Phylogeny. http://morphbank.ebc.uu.se/mrbayes/links.php
20. S. Holmes, 2002. Bootstrapping Phylogenetic Trees. To appear in *Statistical Science*. Submitted in (2002).
21. J. Felsenstein, Confidence limits on phylogenies: An approach using the bootstrap. *Evolution* 1985; **39**:783–791.
22. S. Li, D. K. Pearl, H. Doss. Phylogenetic tree construction using MCMC. *J. Am. Statistical Association* 2000; **95**:493–503.
23. M. Levitt, S. Lifson, Refinement of Protein Confirmations using a Macromolecular Energy Minimization Procedure. *J. Mol. Biol.* 1969; **46**:269–279.
24. H. M. Berman, J. Westbrook, Z. Feng, G. Gillil, T. N. Bhat, H. Weissig, I. N. Shindyalov, and P. E. Bourne, The Protein Data Bank, *Nucleic Acids Research* 2000; **28**:235–242.
25. Y. Xu, D. Xu, Protein threading using PROSPECT: Design and evaluation. *Proteins: Structure, Function, and Genetics* 2000; **40**:343–354.
26. M. Levitt and A. Warshel, Computer Simulation of Protein Folding. *Nature* 1975; **253**:694–698.
27. G. Nemethy and H. A. Scheraga, Theoretical Determination of Sterically Allowed Conformations of a Polypeptide Chain by a Computer Method. *Biopolymers* 1965; **3**:155–184.
28. M. Levitt, Molecular Dynamics of Native Protein: Computer Simulation of the Trajectories. *J. Mol. Biol.* 1983; **168**:595–620.
29. D. Beeman, Some Multi-step Methods for Use in Molecular Dynamics Calculations. *J. Comput. Phys.* 1976; **20**:130–139.
30. P. E. Bourne, CASP and CAFASP experiments and their findings. *Methods Biochem. Anal.* 2003; **44**:501–507.
31. D. Gordon, C. Abajian, and P. Green, Consed: a graphical tool for sequence finishing. *Genome Res.* 1998; **8**(3):195–202.
32. D. Gordon, C. Desmarais, and P. Green, Automated finishing with autofinish. *Genome Res.* 2001; **11**(4):614–625.
33. X. Huang and A. Madan, CAP3: A DNA sequence assembly program. *Genome Res.* 1999; **9**(9):868–877.
34. R. H. Waterston, E. S. Lander, and J. E. Sulston, On the sequencing of the human genome. *Proc. Natl. Acad. Sci. USA* 2002; **99**(6):3712–3716.

35. E. W. Myers, et al., A whole-genome assembly of *Drosophila* 2000; **287**(5461):2196–2204.
36. M. D. Adams, et al., The genome sequence of *Drosophila melanogaster*. *Science* 2000; **287**(5461):2185–2195.
37. J. C. Venter, et al., 2001. The sequence of the human genome. *Science* 2001; **29**:1304–1351.
38. A. V. Lukashin, M. Borodovsky, GeneMark.hmm: new solutions for gene finding. *Nucleic Acids Res.* 1998; **26**(4):1107–1115.
39. C. Burge, and S. Karlin, Prediction of complete gene structures in human genomic DNA. *J. Mol. Biol.* 1997; **268**:78–94.
40. R.-F. Yeh, L. P. Lim, and C. B. Burge, Computational inference of homologous gene structures in the human genome. *Genome Res.* 2001; **11**:803–816.
41. Y. Xu and C. E. Uberbacher, 1997. Automated Gene Identification in Large-Scale Genomic Sequences. *J. Comp. Biol.* 1997; 4:325–338.
42. A. Bateman, L. Coin, R. Durbin, R. D. Finn, V. Hollich, S. Griffiths-Jones, A. Khanna, M. Marshall, S. Moxon, E. L. L. Sonnhammer, D. J. Studholme, C. Yeats, and S. R. Eddy, The Pfam Protein Families Database. *Nucleic Acids Research* 2004; **32**:D138–D141.
43. S. Eddy, Profile hidden Markov models. *Bioinformatics* 1998; **14**:755–763.
44. A. Krogh, M. Brown, I. S. Mian, K. Juolander, and D. Haussler, Hidden Markov models in computational biology applications to protein modeling. *J. Mol. Biol.* 1994; **235**:1501–1531.
45. G. D. Stomo and G. W. Hartzell, Identifying protein-binding sites from unaligned DNA fragments. *Proc. Natl. Acad. Sci.* 1989; **86**:1183–1187.
46. C. E. Lawrence and A. A. Reilly, An expectation maximization (EM) algorithm for the identification and characterization of common sites in unaligned biopolymer sequences. *Proteins Struct. Funct. Genet.* 1990; **7**:41–51.
47. L. T. Bailey and C. Elkan, 1994. Fitting a mixture model by expectation maximization to discover motifs in biopolymers. *Proceedings of the Second International Conference on Intelligent Systems for Molecular Biology* pp. 28-36.
48. M. Kellis, B. W. Birren, and E. S. Lander, Proof and evolutionary analysis of ancient genome duplication in the yeast *Saccharomyces cerevisiae*. *Nature* 2004; **428**:617–624.
49. R. L. Tatusov, E. V. Koonin, and D. J. Lipman, A genomic perspective on protein families. *Science* 1997: 631–637.
50. S. J. O'Brien, M. Menotti-Raymond, W. J. Murphy, W. G. Nash, J. Wienberg, R. Stanyon, N. G. Copeland, N. A. Jenkins, J. E. Womack, and J. A. M. Graves, The promise of comparative genomics in mammals. *Science* 1999; **286**:458–481.
51. G. Bourque and A. P. Pevzner, Genome-scale evolution: reconstructing gene orders in the ancestral species. *Genome Res.* 2002; **12**:26–36.
52. Chimpanzee Sequencing and Analysis Consortium. Initial sequence of the chimpanzee genome and comparison with the human genome. *Nature* 2005; **437**:69–87.
53. B. M. Bolstad, R. A. Irizarry, M. Astrand, and T. P. Speed, A Comparison of Normalization Methods for High Density oligonucleotide Array Data Based on Bias and Variance. *Bioinformatics* 2003; **19**(2):185–193.
54. Peter Bajcsy, Lei Lice, and Mark Band, *DNA microarray image processing in "DNA Array Image Analysis": Nuts & Bolts*. Gerda Kanberova ed., DNA Press, 2005.
55. Y. H. Yang and N. Thorne, 2003. Normalization for Two-color cDNA Microarray Data. Science and Statistics: A Festschrift for Terry Speed. In: D. Goldstein eds., *IMS Lecture Notes, Monograph Series*. Vol 40, pp. 403–418.
56. Y. H. Yang, S. Dudoit, P. Luu, D. M. Lin, V. Peng, J. Ngai, and T. P. Speed. Normalization for cDNA microarray data: a robust composite method addressing single and multiple slide systematic variation. *Nucleic Acids Research* 2002; **30**(4):e15.
57. G. K. Smyth, Y. H. Yang, and T. P. Speed. Statistical issues in microarray data analysis. In: Functional Genomics: Methods and Protocols. In: M. J. Brownstein and A. B. Khodursky eds., *Methods in Molecular Biology*, Volume 224, Totowa, NJ: Humana Press, 2003, pp. 111–136.
58. M. Kerr and G. Churchil, Analysis of variance for gene expression microarray data. *J Comp Biol.* 2000; **7**:819–837.
59. Y. Benjamini and Y. Hochberg, Controlling the false discovery rate: a practical and powerful approach to multiple testing. *J. Royal Statistical Society, Series B.* 1995; **57**(1):289–300.
60. C. Li and W. J. Wong, Model-based analysis of oligonucleotide arrays: expression index computation and outlier detection. *Proc. Natl. Acad. Sci.* 2001; **98**:31–36.
61. H. Wang, X. He, M. Band, C. Wilson, and L. Liu, A study of inter-lab and inter-platform agreement of DNA microarray data. *BMC Genomics* 2005; **6**(1):71.
62. A. C. Culhane, G. Perriere, and D. G. Higgins, Cross-platform comparison and visualisation of gene expression data using co-inertia analysis. *BMC Bioinformatics* 2003; **4**:59.
63. A. Jarvinen, S. Hautaniemi, H. Edgren, P. Auvinen, J. Saarela, O. Kallioniemi, and O. Monni, Are data from different gene expression microarray platforms comparable? *Genomics* 2004; **83**:1164–1168.
64. M. B. Eisen, P. T. Spellman, P. O. Brown, and D. Botstein, Cluster analysis and display of genome-wide expression patterns. *Proc. Natl. Acad. Sci. USA* 1998 95(25):14863–14868.
65. A. Ben-Dor, R. Shamir, and Z. Yakhini, Clustering gene expression patterns. *J. Comp. Biol.* 1999; **6**(3/4):281–297.
66. P. Tamayo, D. Solni, J. Mesirov, Q. Zhu, S. Kitareewan, E. Dmitrovsky, E. S. Lander, and T. R. Golub, Interpreting patterns of gene expression with self-organizing maps: Methods and application to hematopoietic differentiation. *Proc. Natl. Acad. Sci. USA.* 1999; **96**(6):2907–2912.
67. O. Alter, P. O. Brown, and D. Bostein, Singular value decomposition for genome-wide expression data processing and modeling. *Proc. Natl. Acad. Sci. USA* 2000; **97**(18):10101–10106.
68. T. R. Golub, D. K. Slonim, P. Tamayo, C. Huard, M. Gaasenbeek, J. P. Mesirov, H. Coller, M. L. Loh, J. R. Downing, M. A. Caligiuri, C. D. Bloomfield, and E. S. Lander, Molecular Classification of Cancer: Class Discovery and Cass Prediction by Gene Expression Monitoring. *Science* 1999; **286**:531–537.
69. R. V. Sole, R. Ferrer-Cancho, J. M. Montoya, and S. Valverde, Selection, tinkering, and emergence in complex networks. *Complexity* 2003; **8**:20–33.
70. M. Kanehisa, A database for post-genome analysis. *Trends Genet.* 1997; **13**:375–376.
71. I. M. Keseler, J. Collado-Vides, S. Gama-Castro, J. Ingraham, S. Paley, I. T. Paulsen, M. Peralta-Gil, and P. D. Karp, EcoCyc: A comprehensive database resource for Escherichia coli. *Nucleic Acids Research* 2005; **33**:D334–D337.
72. R. Overbeek, N. Larsen, G. D. Pusch, M. D'Souza, E. Selkov, N. Kyrpides, M. Fonstein, N. Maltsev, and E. Selkov,WIT: integrated system for high-throughput genome sequence analysis and metabolic reconstruction. *Nucleic Acids Res.* 2000; **28**(1):123–125.
73. S. Fields and O. K. Song, A novel genetic system to detect protein-protein interactions. *Nature* 1989; **340**:245–246.
74. N. Daraselia, A. Yuryev, S. Egorov, S. Novichkova, A. Nikitin, and I. Mazo, Extraction of human protein interactions from

75. P. Baldi and G. W. Hatfield, *Microarrays and Gene Expression*. Cambridge, UK: Cambridge University Press, 2001.

76. H. Jeong, B. Tombor, R. Albert1, Z. N. Oltvai, and A. L. Barabási, The large-scale organization of metabolic networks. *Nature* 2000; 407:651–654.

77. A. Wagner and D. A. Fell, The small world inside large metabolic networks. *Proc. R. Soc. Lond. B.* 2001; **268**:1803–1810.

78. R. Guimera and A. L. Nunes Ameral, Functional cartography of complex metabolic networks. *Nature* 2005; **433**: 895–900.

79. S. Schuster, C. Hilgetag, J. H. Woods, and D. A. Fell, Reaction routes in biochemical reaction systems: algebraic properties, validated calculation procedure and example from nucleotide metabolism. *J. Math. Biol.* 2002; **45**(2):153–181.

80. S. Schuster, D. A. Fell, and T. Dandekar, A general definition of metabolic pathways useful for systematic organization and analysis of complex metabolic networks. *Nature* 2000; **18**: 326–332.

81. J. Stelling, S. Klamt, K. Bettenbrock, S. Schuster, and E. D. Gilles, Metabolic network structure determines key aspects of functionality and regulation. *Nature* 2002; **420**:190–193.

82. T. Cakir, B. Kirdar, and K. O. Ulgen. Metabolic pathway analysis of yeast strengthens the bridge between transcriptomics and metabolic networks. *Biotechnol. Bioeng.* 2004; **86**:251–260.

83. S. Liang, S. Fuhrman, and R. Somogyi, REVEAL, a general reverse engineering algorithm for inference of genetic network architectures. *Pacific Symposium on Biocomputing* 1998; **3**:18–29.

84. R. Somogyi, S. Fuhrman, and X. Wen, *Genetic network inference in computational models and applications to large-scale gene expression data*. Cambridge, MA MIT Press, 2001.

85. Segal et al., 2003.

86. O. G. Troyanskaya, K. Dolinski, A. B. Owen, R. B. Altman, and D. Botstein, A Bayesian framework for combining heterogeneous data sources for gene function prediction (in *Saccharomyces cerevisiae*). *Proc. Natl. Acad. Sci.* 2003; **100**:8348–8353.

FURTHER READING

S. F. Altschul, T. L. Madden, A. A. Schaffer, J. Zhang, Z. Zhang, W. Miller, D. J. Lipman, Gapped BLAST and PSI-BLAST: a new generation of protein database search programs. *Nucleic Acids Res.* 1997; 25:3389–3402.

P. Baldi, Y. Chauvin, T. Hunkapillar, M. McClure, Hidden Markov Models of Biological Primary Sequence Information. *Proceedings of the National Academy of Sciences of the USA* 1994; **91**:1059–1063.

P. Baldi and S. Brunak, *Bioinformatics: The Machine Learning Approach*. 2nd ed. Cambridge, MA: MIT Press, 2001

P. Bork, T. Dandekar, Y. Diaz-Lazcoz, F. Eisenhaber, M. Huynen, and Y. Yuan, Predicting Function: From Genes to Genomes and Back. *J. Mol. Biol.* 1998; **283**:707–725.

J. Bower and H. Bolouri, *Computational Modeling of Genetic and Biochemical Networks*. Cambridge, MA: MIT Press, 2001.

N. Bray, I. Dubchak, and L. Pachter, AVID: A global alignment program. *Genome Res.* 2003; **13**(1): 97–102.

P. O. Brown and D. Botstein, Exploring the new world of the genome with DNA microarrays. *Nature Genetics* 1999; **21**: 33–7.

M. Brudno, C. B. Do, G. M. Cooper, M. F. Kim, E. Davydov, E. D. Green, A. Sidow, and A. Batzoglou, LAGAN and Multi-LAGAN: efficient tools for large-scale multiple alignment of genomic DNA. *Genome Res.* 2003(a); **13**(4):721–31.

M. Brudno, S. Malde, A. Poiakov, C. Do, O. Couronne, I. Dubchak, and A. Batzoglou, Glocal Alignment: Finding Rearrangements during alignment. *Bioinformatics*. Special Issue on the Proceedings of the ISMB 2003, 2003(b); **19**:54i–62i,

S. H. Bryant and S. F. Altschul, Statistics of sequence-structure Threading. *Current Opinion in Structural Biology* 1995; **5**:236–244.

F. E. Cohen, Protein misfolding and prion diseases. *J. Mol. Biol.* 1999; **293**:313–320.

P. Diaconis and S. Holmes, Random walks on trees and matchings. *Electronic J. Probability* 2002; **7**:1–17.

J. C. Doyle, Robustness and dynamics in biological networks. In: *The First International Conference on Systems Biology*. New York/NY: Japan Science and Technology Corporation, MIT Press, 2000

S. Dudoit, J. Fridlyand, and T. P. Speed, Comparison of discrimination methods for the classification of tumors using gene expression data. *J. Am. Statist. Assoc.* 2002; **97**:77–87.

S. Eddy, G. Mitchison, and R. Durbin, Maximum Discrimination Hidden Markov Models of Sequence Consensus. *J. Comput. Biol.* 1995; **2**:9–23.

S. R. Eddy, Non-coding RNA genes and the modern RNA world. *Nature Reviews Genetics*. 2001; **2**:919–929.

B. Efron, E. Halloran, and S. Holmes, Bootstrap confidence levels for phylogenetic trees. *Proc. National Academy Sciences* 1996; **93**:13429–34.

J. S. Farris, The logical basis of phylogenetic analysis, In: N. Platnick and V. Funk, eds., *Advances in Cladistics*. vol. 2, 1983, pp. 7–36.

A. N. Fedorov and T. O. Baldwin, Contranslational Protein Folding. *J. Biol. Chem.* 1997; **272**(52):32715–32718.

J. Felsenstein, PHYLIP, 1993. (Phylogeny Inference Package) version 3.5c., Distributed by the author. Department of Genetics, University of Washington, Seattle. Available: *http://evolution.genetics.washington.edu/phylip.html*

D. Fischer, C. Barret, K. Bryson, A. Elofsson, A. Godzik, D. Jones, K. J. Karplus, L. A. Kelley, R. M. MacCallum, K. Pawowski, B. Rost, L. Rychlewski, and M. Sternberg, CAFASP-1: critical assessment of fully automated structure prediction methods. *Proteins* 1999; 3:209–17.

W. M. Fitch and E. Margoliash, Construction of phylogenetic trees. *Science* 1967; **155**:279–284.

L. R. Foulds and R. L. Graham, The Steiner problem in Phylogeny is NP-complete. *Advanced Applied Mathematics* 1982; **3**:43–49.

N. Friedman, M, Linial, I. Nachman, and D. Peter, Using Bayesian Networks to analyze expression data. *J. Comp. Biol.* 2000; 7:601–620.

M. Gardner, *The Last Recreations*. NY: Copernicus-Springer Verlag, 1997.

S. Geman and D. Geman, Stochastic relaxation, Gibbs distribution and the Bayesian restoration of images. *IEEE Transactions on Pattern Analysis and Machine Intelligence* 1984; **6**:721–741.

K. D. Gibson and H. A. Scheraga, Revised algorithms for the build-up procedure for predicting protein conformations by energy minimization. *J. of Comp. Chem.* 1987; **9**:327–355.

P. A. Goloboff, 1995 SPA. (S)ankoff (P)arsimony (A)nalysis, version 1.1. Computer program distributed by J. M. Carpenter, Dept. of Entomology, American Museum of Natural History, New York.

S. Gribaldo and P. Cammarano, The root of the universal tree of life inferred from anciently duplicated genes encoding components of the protein-targeting machinery. *J. Mol. Evol.* 1998; 47(5):508–516.

E. Haeckel, Generelle, Morphologie der Organismen: *Allgemeine Grundzuge der organischen FormenWissenschaft, mechanisch begrundet durch die von Charles Darwin reformirte Descendenz-Theorie.* Berlin: Georg Riemer, 1866.

S. Hannenhalli and P. A. Pevzner, Transforming cabbage into turnip: polynomial algorithm for sorting signed permutations by reversals. *STOC* 1995: 178–189.

J. V. Helden, B. Andre, and J. Collado-Vides, Extracting regulatory sites from the upstream region of yeast genes by computational analysis of oligonucleotide frequencies. *J. Mol. Biol.* 1998; 281:827–842.

E. Hooper, *The River.* Boston: Little, Brown, 1999.

S. Karlin and S. F. Altschul, Methods for assessing the statistical significance of molecular sequences features by using general scoring schemes. *Proc. Nat. Ac. Sci. USA* 1990; **87**(6) 2264–2268.

J. M. Keith, P. Adams, D. Bryant, D. P. Kroese, K. R. Mitchelson, D. A. E. Cochran, G. H. Lala, A simulated annealing algorithm for finding consensus sequences. *Bioinformatics* 2002; 18: 1494–1499.

BLAST-like alignment tool. *Genome Res.* 2002; **12**(4):656–64.

S. Kirkpatrick, C. D. Gelatt, Jr., and M. P. Vecchi, Optimization by simulated annealing. *Science* 1983; **220**:671–680.

I. Korf, P. Flicek, D. Duan, and M. R. Brent, Integrating genomic homology into gene structure prediction. *Bioinformatics* 2001; **17**:S140–S148.

M. Levitt, Protein Folding by Restrained Energy Minimization and Molecular Dynamics. *J. Mol. Biol.* 1983; **170**:723–764.

D. H. Ly and D. J. Lockhart, R. A. Lerner, and P. G. Schultz, Mitotic misregulation and human aging. *Science* 2000; **287**:1241–1248.

B. Ma, J. Tromp, and M. Li, PatternHunter: Faster And More Sensitive Homology Search. *Bioinformatics* 2002; **18**:440–445.

B. Ma, Z. Wang, and K. Zhang, Alignment between Two Multiple Alignments. In *Combinatorial Pattern Matching*: 14th Annual Symposium, CPM 2003, Morelia, Michoacán, Mexico, June 25-27. *Lecture Notes in Computer Science.* Volume 2676 / 2003, Springer-Verlag Heidelberg, 2003

D. Maddison and W. Maddison. MacClade. 2002. Sinauer. Available : *http://phylogeny.arizona.edu/macclade/*

H. McAdams and L. Shapiro, Circuit Simulation of Genetic Networks. *Science* 1995; **269**:650–656.

N. Metropolis, A. Rosenbluth, M. Rosenbluth, A. Teller, and E. Teller. Simulated Annealing. *J. Chem. Phys.* 1953; 21:1087–1092.

E. Mjolsness, D. H. Sharp, and J. Rinetz. A connectionsit model of development. *J. Theor. Biol.* 1991; 152:429–453.

L. B. Morales, R. Garduno-Juarez, and D. Romero, Applications of simulated annealing to the multiple-minima problem in small peptides. *J. Biomol. Struc. Dyn.* 1991; **8**:721–735.

B. Morgenstern, Dialign2: improvement of the segment-to-segment approach to multiple sequence alignment. *Bioinformatics* 1999; **15**:211–218.

J. L. Mountain and L. L. Cavalli-Sforza, Inference of human evolution through cladistic analysis of nuclear DNA restriction polymorphisms. *Proc. Natl. Acad. Sci. USA* 1994; **91**:6515–6519.

U. Muckstein, I. L. Hofacker, and P. F. Stadler, Stochastic pairwise alignments. *Bioinformatics* 2002; **18**(sup. 2):S153–S160.

C. Notredame, D. Higgins, and J. Heringa, T-Coffee: A novel method for multiple sequence alignments. *Journal of Molecular Biology* 2000; **302**:205–217.

M. C. Peitsch, ProMod and Swiss-Model: Internet-based tools for automated comparative protein modeling. *Biochem. Soc. Trans.* 1996; **24**:274–279.

P. A. Pevzner, *Computational molecular biology, an algorithmic approach.* Cambridge, MA: MIT Press, 2000.

U. Pieper, N. Eswar, V. A. Ilyin, A. Stuart, and A. Sali, ModBase, a database of annotated comparative protein structure models. *Nucleic Acids Res.* 2002; **30**:255–259.

G. N. Ramachandran and V. Sasisekharan, Conformation of polypeptides and proteins. *Adv Protein Chem.* 1968; **23**:283–438.

B. Rannala, Z. Yang, Probability distribution of molecular evolutionary trees: a new method of phylogenetic inference. *J. Mol. Evol.* 1996; **43**:304–311.

F. M. Richards, 1991. The Protein Folding Problem. *Scientific American,* pp. 54-63, January.

T. Schlick,. Optimization methods in computational chemistry. In *Reviews in Computational Chemistry,* III, VCH Publishers, 1992, 1-71.

I. Schmulevich, E. Dougherty, S. Kim, and W. Zhang, Probabilistic Boolean Networks: A rule-based uncertainty model for gene regulatory networks. *Bioinformatics* 2002; **18**:261–274.

E. Schröder, Vier combinatorische Probleme. *Z. Math. Phys.* 1870; **15**:361–376.

C. E. Shannon, A mathematical theory of communication. *Bell sys. Tech. Journal* 1948; **27**(379–423):623–656.

M. E. Snow, Powerful simulated annealing algorithm locates global minima of protein folding potentials from multiple starting conformations. *J. Comput. Chem.* 1992; **13**:579–584.

R. Stanley, *Enumerative Combinatorics.* Vol. I, 2nd ed. Cambridge University Press, 1996

J. Stuart, E. Segal, D. Koller, and S. Kim, A Gene Co-Expression Network for Global Discovery of Conserved Genetic Modules. *Science* 2003; **302**(5643):249–55.

D. L. Swofford. PAUP. Phylogenetic analysis using parsimony. V4.0. 2001. Available from Sinauer Associates. Boston, Massachusetts

A. Tozeren and S.W. Byers, *New Biology for Engineers and Computer Scientists.* Englewood Cliffs, NJ: Prentice Hall, 2003.

L. S. Wang, R. Jansen, B. Moret, L. Raubeson, and T. Warnow, Fast phylogenetic methods for the analysis of genome rearrangement data: An empirical study. *Proc. of 7th Pacific Syrnposium on Biocomputing* 2002.

J. D. Watson and F. H. Crick, A Structure for Deoxyribose Nucleic Acid. *Nature* 1953; (April).

K. P. White, S.A. Rifkin, P. Hurban, and D. D. Hogness, Microanalysis of *drosphila* development during metamorphosis. *Science* 1999; **286**:2179–2184.

H. Winkler, *Verbeitung und Ursache der Parthenogenesis im Pflanzen und Tierreiche.* Jena: Verlag Fischer, 1920.

J. Xu and A. Hagler, Review: Chemoinformatics and drug discovery. *Molecules* 2002; 7:566–600.

Z. Yang and B. Rannala, Bayesian phylogenetic inference using DNA sequences: a Markov chain Monte Carlo method. *Mol. Biol. Evol.* 1997; **14**:717–724.

BIOLOGICAL DATABASE INTEGRATION

ZINA BEN MILED
YANG LIU
NIANHUA LI
OMRAN BUKHRES
Indiana University Purdue
University Indianapolis
Indianapolis, Indiana

1. INTRODUCTION

Recent technological advances and the availability of new experimental data have led to the emergence of many biological databases. These databases are distributed, heterogeneous, autonomous, and often only accessible via the Internet. Currently, about 400 biological databases are available online (1), and the number is growing rapidly. Only 2 years ago, the same survey (2) listed about half that many databases. These databases often focus on a specific subject area, and individually, they only represent a fraction of all available biological data.

The biological databases cover a wide range of subjects and data types including gene sequences, gene expression data, protein structures, protein sequences, and metabolic pathways. Although these databases are distributed, the biological data they hold are often semantically interrelated. To allow scientists to issue multi-database queries—queries that simultaneously invoke multiple databases—support for the integration of these databases is needed and an integrated view of the data must be available. Substantial research efforts (3–7) have been devoted to addressing this research issue. Because of the unique characteristics of biological data, traditional database integration methods, which have been developed for business data stored in relational databases, may not be applicable to biological databases. In the business domain, the databases are often well structured and adhere to a well-defined schema, whereas in the biological domain, most databases are unstructured. Furthermore, several databases store data that are highly heterogeneous. These characteristics make the traditional query formulation algorithms, schema mapping algorithms, and query processing optimization algorithms nonapplicable.

Although this article focuses mostly on biological databases, other biomedical databases, such as drug compound databases in the pharmaceutical industries and patient record databases in the health-care industries face similar issues.

In the following sections, some of the most popular online biological databases are first described, followed by a discussion of the challenges associated with the integration of biological databases. The discussion is intended to provide a deeper understanding of the limitation of some current databases and database integration systems. The section on "Integration" presents an overview of the different integration models and their advantages and disadvantages. The following section describes some currently available commercial and academic integration systems for biological databases. The purpose of this section is to offer an overview of the capabilities, limitations, and potential usage of these systems. The last section summarizes state-of-the-art research and the available technology in the field. It also discusses future trends in the area of distributed databases.

2. OVERVIEW OF BIOLOGICAL DATABASES

Biological databases have grown in size rapidly during the past two decades. Scientists use these databases to exchange information and to disseminate knowledge (Table 1). They contain valuable information and are essential in conducting research studies. Most public domain biological databases were originally stored in plain text files. As the size and complexity of the data increased, these databases migrated to a database management system. Users can access them by submitting queries to their Web interfaces or by downloading the entire database, in text format, to a local machine. Direct access to the database through the database management system is often not allowed. Some of the most popular databases are described below. These databases are classified into two groups: general-purpose databases and subject-specific databases. The latter group tends to include information related to a specific subject such as the study of a given organism or the study of a given biological process (e.g., metabolic pathways). The former group of databases tends to include a broad range of information that is specific neither to a given organism nor to a given subject matter. However, general-purpose databases may sometimes focus on a specific biological data type such as protein sequences or nucleotide sequences.

2.1. General-Purpose Databases

Sequence data are an important source of biological knowledge. This importance was one of the drivers behind the creation of the International Nucleotide Sequence Database Collaboration effort that encompasses three major nucleotide sequence databases: GenBank (8) at the National Center for Biotechnology Information, the European Molecular Biology Laboratory (EMBL) (9), and the DNA Databank of Japan (DDBJ) (10). These three databases synchronize their data on a regular basis.

The Protein Information Resource (PIR) (11) database is an example of a general-purpose database that is organized around protein information. PIR includes not only protein sequence data, but also protein family classification and protein annotations. The protein entries in PIR are classified into a hierarchy of protein families that consist of families and superfamilies. In addition to detailed information on protein sequences, PIR also contains protein family information such as signature features. As of April 2003, PIR counts more than one million nonredundant sequences.

Another example of a general-purpose database that focuses on storing protein sequences and related information is SWISS-PROT (12). The sequence data in SWISS-PROT include translations of DNA sequences from the EMBL Nucleotide Sequence database, results of protein sequencing experiments submitted by researchers, and

Table 1. Examples of Biological Databases

Database	Subject	Content						
		DNA	Protein	Protein Family	Gene/Protein Annotation	Genome	Pathway	Bibliography
GenBank EMBL DDBJ	General	X			X	X		X
PIR	General		X	X	X			X
SWISS-PROT	General		X		X			X
Pfam	General		X	X				
PubMed/Medline	General							X
GDB	Specific organism (human)	X	X	X		X		X
MGD	Specific organism (mouse)	X	X	X	X	X		X
TAIR	Specific organism (Arabidopsis)	X			X	X		X
FlyBase	Specific organism (Drosohpila)	X	X		X	X		X
SGD	Specific organism (Saccharomyces)	X	X		X	X	X	X
OMIM	Specific organism (human)				X			X
BRENDA	Specific protein (enzyme)		X	X				X
MetaCyc	Specific pathway (metabolic pathway)		X				X	

sequences extracted from PIR. To reduce the amount of redundancy, slightly variant sequences, such as polymorphisms, are identified among the entries in the database (12). SWISS-PROT also contains human expert annotations of the protein sequence, as well as cross-references to other databases such as GenBank and the Online Mendelian Inheritance in Man (OMIM) (13). As previously mentioned, GenBank is a major repository for nucleotide sequences, and the OMIM database, which will be discussed later in this section, stores information related to human genes.

The Protein Data Bank (PDB) (14) is a database that includes three-dimensional structures of biological macromolecules and related information such as synonyms, enzyme classification, and source organism. As of April 2003, there were about 20,747 structures in PDB. New structures are deposited by scientists in PDB on a continuous basis.

Pfam (15) is a database that focuses on protein families. It includes the multiple sequence alignment of proteins in each family. In addition, Pfam includes hidden Markov model-based profiles that can be used to identify whether a new sequence belongs to a given protein family. The profile for each family is constructed by using a representative set of protein sequences. This profile is then used to search for member sequences in other databases. Once the member sequences for a given family are retrieved, the hidden Markov model-based profile is used to align all sequences in the family. For each family, the database also includes functional annotation and literature references.

PubMed (16) (previously called Medline) is a bibliographic database covering the fields of biomedicine and health. The database contains more than 12 million references and abstracts (17). This database is widely used, and it is cross-referenced by many biological databases.

2.2. Subject-Specific Databases

Many subject-specific databases are organism-specific databases. For example, the Human Genome Data Base (GDB) (18), Mouse Genome Databases (MGD) (19), The Arabidopsis Information Resource (TAIR) (20), FlyBase (21), and Saccharomyces Genome Database (SGD) (22) include genome information related to human, mouse, *Arabidopsis thaliana*, fruit fly, and budding yeast, respectively. The data in these databases consist of, among others, genome maps (e.g., cytogenetic and physical maps), phenotype information, and protein structure and function. For example, TAIR contains expression data obtained from microarray experiment, as well as information related to metabolic pathways and enzymes. Most data are deposited by scientists or extracted from other databases. Each gene in SGD is associated with related data such as links to protein and structure information from other databases and calculated data such as protein length. Another example of an organism-specific database is OMIM (13). This database includes expert annotations of human genes and genetic disorders.

BRENDA (23) is a database that includes information related to a specific type of protein, namely enzyme. The information collected in BRENDA is extracted from more than 40,000 references that cover more than 6000 organisms. The data in BRENDA are organized under seven main categories: enzyme nomenclature (e.g., EC number, synonyms), enzyme structure (e.g., molecular weight, three-dimensional (3-D) structure), enzyme–ligand interactions (e.g., inhibitors, cofactor), functional parameters (e.g., specific activity, pH range), molecular properties (e.g., temperature stability, storage stability), organism-related information (e.g., source tissue, organ), and bibliographic data.

The last example of subject-specific databases that is discussed in this section is MetaCyc (24). This metabolic pathway database integrates pathway information from various literature databases. Specifically, MetaCyc includes descriptions of pathways, enzymes, and reactions. This database includes more than 400 pathways and 1000 enzymes for a wide range of organisms.

2.3. Motivation for Database Integration

The example databases discussed here indicate that the scientific community can benefit from the integration of various databases. As mentioned, some databases link their records to records in other databases. For example, SWISS-PROT includes cross-references to both GenBank and OMIM. Also, the subject-specific databases often extract data from various general-purpose databases to which they sometimes add annotations. Support for the integration of these databases can allow scientists to issue multi-database queries that are beyond the scope of individual databases. Furthermore, this integration combined with more flexible access methods can reduce the need for subject-specific databases. One example query discussed in Ref. 25 that was used to motivate the need for integration of biological databases consists of retrieving "all mammalian gene sequences for proteins identified as being involved in signal transduction" as well as related "annotations and literature citations." This query may be answered by using the following steps. First, all proteins involved in signal transduction have to be retrieved from a pathway database such as KEGG (26). The resulting proteins, along with mammalian organism names, can then be used to query protein databases (e.g., SWISS-PROT). The gene sequences and annotations of the proteins generated in the second step can be obtained from nucleotide sequence databases (e.g., GenBank). The abstracts of related citations can be retrieved from PubMed. Additional annotations can also be found in organism-specific databases such as OMIM for human and MGD for mouse. This example illustrates how tedious the process of information retrieval can be when the databases are not integrated. Support for an integrated access to the distributed, heterogeneous, biological databases can greatly facilitate knowledge extraction. The next section highlights some challenges facing the integration of these biological databases.

3. INTEGRATION CHALLENGES

Several challenges face the efficient and meaningful integration of distributed biological databases. These challenges can be caused by the inherent characteristics of the biological data or can be an artifact of the biological databases in which the data are stored.

3.1. Data Complexity

Some challenging characteristics of the biological data include the fact that not only is it doubling in size every 2 years (27), but it is complex as well. There are several types of biological data ranging from sequences to more complex data such as tertiary structures and network pathways. Therefore, a unique data representation is not applicable to all biological data within a given database. The biological data are often represented by using several complex data types such as free text, strings, lists, sets, deeply nested records (28), and graphs (29). For example, abstracts and annotations may be represented by using free text. Strings are used to represent sequences. Literature references may consist of records that include three fields: title (string), author (list of author names), and journal (a set of fields including journal title, page numbers, volume, number, and year). Metabolic pathways may be represented by using complex networks, where the chemical compounds are the nodes of the network and the edges represent chemical reactions (29). In addition, the data can be temporal (as in the case of gene expression time series) (30). The data can also be spatial. Examples of this case are secondary and tertiary structures of proteins.

Biological data are also semistructured. Semistructured data, also referred to as unstructured data, are data whose structure or schema is not explicitly defined (31). Biological data represented in HTML format are an example of semistructured data. These data are readable by humans, but they can be easily processed by a computer. Several online biological databases return results in HTML format.

The relationships between biological objects can also be complex. These relationships can be hierarchical, unidirectional, bidirectional, or many-to-many. Examples of the first type of relationship include the hierarchical classification of proteins in families and superfamilies and the hierarchical taxonomic tree of species (32). A gene that *encodes* a protein or a protein that *regulates* a gene expression are examples of unidirectional relationships. Bidirectional relationships are commonly used to depict protein–protein interactions. Metabolic pathways can give rise to many-to-many relationships. A given protein may appear in multiple pathways, and one pathway may contain multiple proteins. The importance of the explicit recording of these relationships is highly dependent on the specific research question that is being addressed. However, these relationships are often not explicitly included in the biological databases (27,28).

3.2. Data and Databases Heterogeneity

Heterogeneity is one major challenge for the integration of multiple biological databases. This heterogeneity is present in several contexts (3,5): syntactic, semantic, data model related, and schematic.

Syntactic heterogeneity originates because different databases may use synonyms to refer to the same concept (27). For example, the scientific species name *Escherichia coli* can be used by one database, whereas one of its abbreviations (*E. Coli* or *E. coli*) can be used by another database.

Resolving cases in which two synonyms are used to represent the same concept is less difficult than resolving cases in which one name may have different meanings in different contexts (semantic heterogeneity (33)). An example of this type of heterogeneity as it relates to the concept of gene was presented in (4). This example compares the use of *gene* in GDB that refers to a "DNA fragment that can be transcribed and translated into a protein" with its use in GenBank that refers to a "DNA region of biological interest with a name and that carries a genetic trait or phenotype."

The last two types of heterogeneities—data model-related heterogeneity and schematic heterogeneity—are a result of the choices of database management systems that are used to manage the data.

A data model is an abstraction of the entities (objects) and their relations in the database. It describes how the attributes of an entity are structured, the constraints imposed on the data, and how the entities (objects) are related to each other. There are two widely used data models: the relational model and the object-oriented model. Under the relational data model, the data are organized into a table format, where each column of the table represents an attribute of an object, and each row (record) of the table represents an instance of the object. In each table, one or more attributes are designated as the primary key that is used to uniquely identify each record.

The object-oriented model is based on concepts that are similar to ones used in object-oriented programming languages (34). For example, encapsulation, inheritance, and polymorphism are also applicable to the object-oriented data model. Each data object is defined as a class that has multiple attributes. These attributes can describe features of the object or can be a reference to other classes. The object protocol model (OPM) (35) is an example of an object-oriented model that is used by GDB (36).

Relational data models have often been deemed to be inappropriate for the biological domain. For example, in Ref. 35, it is argued that the relational data model can force objects to be scattered among multiple tables and thus lead to large databases. Similar observations were made in Ref. 37 about pharmacogenomic databases. An independent study (38) concluded that relational data models do not easily support hierarchical data types (e.g., trees, graphs), and they can make structural updates to the database difficult. As discussed in the previous subsection, hierarchical data types are important in the biomedical research field. For example, graphs represent metabolic pathways and proteins may be classified in a tree structure of families and superfamilies, which seems to make the case for an object-oriented data model. However, object-oriented database technology is not as mature as the relational database technology (34,38).

When databases that use different data models are integrated, a mechanism that translates the databases to a common data model is needed. Even if the databases use the same data model, heterogeneities can still occur at the schema level depending on how the data are organized (3).

A database schema is a description of the database that is specified by using a given data model (i.e., relational or object oriented). In a relational model, the schema consists of the structure of the tables in the database, the attribute in these tables, and the relationships among the attributes within a table and across the tables. Differences in schema definition lead to schematic heterogeneities. For example, GDB and TAIR are organism-specific databases that use the object-oriented data model. However, they represent their data by using different schema. In the TAIR schema, "gene," "polymorphic marker," and "map" are siblings. However, in the GDB schema, "gene" and "polymorphic marker" are children of "genomic segment," which is a sibling of "map." This example shows how similar concepts can be organized differently in different schema.

3.3. Database Accessibility

The accessibility approaches vary among the biological databases. Almost all community databases provide access through a Web interface. Results returned by these interfaces may be in HTML (e.g., PDB, SWISS-PROT) or XML (e.g., PIR). Alternatively, some databases may return the result of a query in plain text files (e.g., one output format option in EMBL is text files). Internally, these databases may be implemented by using an off-the-shelf database management system such as Oracle or a custom-built data search and retrieval software such as ACeDB. Read and write access to the database for update and maintenance is performed directly through the database management system or the custom-built software. Read-only user access to the database is supported by an application interface. In addition to being read only, these interfaces limit access to a set of prespecified queries. This aspect makes integration difficult because of the limitations imposed on the queries and the fact that the query result is formatted for human inspection rather than to support any further automated processing.

3.4. Data Redundancy

The data in various biological databases can be redundant and conflicting. Redundancy in biological databases originates as a side effect of the multiple data acquisition sites as well as the multiple ways in which data are extracted from one database and stored in another database (38). There are several major general-purpose databases, and although these databases may exchange information, at any given point in time, data in these databases may not be fully synchronized. Subject-specific databases that extract information from other databases and add their annotations can also be a source of conflict. For example, a protein family database may contain a description of proteins and their families, which may vary from one database to another (39).

4. INTEGRATION

In a database integration system, the databases participating in the integration are referred to as *component databases*. In general, the integration of distributed databases involves the following steps: transforming the schema of the component databases to a common data model, establishing the relationship between the objects in different databases, decomposing multi-database queries into database-specific subqueries, accessing the component databases to retrieve the data, and combining the result of the subqueries (40). Depending on the integration method used, one or more of these steps may not be necessary. The complexity, dynamicity, syntactic and semantic heterogeneities of the biological data, and the limited accessibility, redundancy, and variation in data models and schema of the biological databases, have made the integration of biological databases both challenging and interesting.

Different approaches to the database integration of biological databases have been proposed. These approaches include federated databases, data warehousing, and link driven. When the component databases are Web databases, the latter approach is also called hypertext linking or hypertext navigation. One main advantage of federated databases is that they do not require that component databases be physically combined into one database (3). Database federation is used to integrate autonomous, heterogeneous, and distributed databases (41). This class of integration systems can be further divided into two subclasses: tightly coupled and loosely coupled.

In a tightly coupled federated databases system, a global data schema is constructed and the queries are expressed on this unified schema (40). The global schema integrates all component schema. Therefore, the component databases are transparent to the user. TAMBIS (42) is an example of a tightly coupled federated database.

In a loosely coupled federated database system, there is no global schema, and the user needs to specify the component databases in the query by using a multi-database query language (41). One disadvantage of loosely coupled federated databases is that the component databases and any heterogeneity that may be present among their schema are not transparent to the user. This approach is used by BioKleisli (43).

The second type of integration approaches is data warehousing. The component databases are mapped to a single database called the data warehouse. Creating a data warehouse is a process that consists of two steps (3). First the schemas of the component databases are mapped onto the global schema (i.e., the schema of the data warehouse). Second, data are extracted from the component databases and used to populate the data warehouse. An example system that uses this approach is genomics unified schema (GUS) (44). Data warehouses have not seen a wide spread use in the biological domain because they do not scale well to many databases. Adding a new database to the warehouse may lead to a redesign and repopulation of the data warehouse.

The third type of integration approach, link driven, is currently the most popular in the biological domain. This approach does not rely on a common data model or a global schema and is therefore easy to implement. In this approach, a physical link is created between related records from different databases. These links can be stored in the form of index files as in sequence retrieval system (SRS) (45) or simply in the form of hypertext links as in Entrez (17). Access to this type of integration system allows the user to navigate from one database to another using these predefined static links. The disadvantage of this approach is that it limits the scope of the queries that can be issued. The previous two types of integration approaches (i.e., federated databases and data warehouses) do not suffer from such a limitation, but their implementations are more complex.

In the following subsections, the steps involved in processing multi-database queries, namely data model translation, schema integration, query translation, data retrieval, and result assembly, are discussed. The role of each step in the three types of integration approaches (federated databases, data warehouses, and link driven) is explained.

4.1. Data Model Translation

Each component database may adopt a different data model that is used to define its local schema. In most integration approaches, the local database schema needs to be translated to the component schema that is based on a common data model (40). Data model translation is only needed for federated databases and data warehouses.

As mentioned, the relational data model has serious limitations when it comes to representing biological data. The object-oriented model, although less mature, is better suited for expressing the complex relationships between biological objects. OPM is an example object-oriented data model that has been used by GDB and other databases (46). Once this technology becomes more mature, more community biological databases are expected to migrate to the object-oriented model.

4.2. Schema Integration

After a common data model has been selected, the local schema of the component databases is translated to a component schema that uses this common data model. For example, schema transformation from the relational model to the OPM model and vice versa is described in Ref. 35. The next step consists of integrating the component schema by creating a global schema and mapping the component schema to the global schema. As stated in Ref. 40, schema integration faces several challenges. First, syntactic and semantic variations among the related entities in the component databases need to be disambiguated. Second, the relations among the entities need to be explicitly specified in the global schema. Schema integration for biological databases requires extensive domain expertise.

Consider, for example, integrating the TRANSFAC (*http://transfac.gbf.de/TRANSFAC/*) and the SWISS-PROT databases. For the purpose of this example, only a subset of the content of these two databases will be considered. Namely, in TRANSFAC, it is possible to retrieve a list of genes and their transcription factors. Also in SWISS-PROT, it is possible to retrieve a list of proteins and the genes that encode these proteins. One query of interest may be to retrieve all transcription factors, their coding genes, and the genes regulated by these transcription factors. Another query of interest may be to retrieve the genes that encode proteins and the transcription factors of these genes. For the first query, the global schema of choice may consist of *transfactor*, *gene-T*, and *gene-S*, where *transfactor* is the attribute obtained by joining the *transfactor* attribute from TRANSFAC and the *protein* attribute from SWISS-PROT. The attributes *gene-T* and *gene-S* are obtained directly from TRANSFAC and SWISS-PROT, respectively. *Gene-T* corresponds to the gene whose transcription factor is *transfactor*, and *gene-S* corresponds to the gene that encodes this transcription factor. For the second query, a more appropriate global schema is *gene*, *transfactor-T*, and *protein-S*, where the attribute *gene* in this case is result of the joining of the attribute *gene* from TRANSFAC and the attribute *gene* from SWISS-PROT. The attributes *transfactor-T* and *protein-S* are from TRANSFAC and SWISS-PROT, respectively. They correspond to the transcription factor of *gene* and the protein that is encoded by *gene*, respectively.

Schema integration is important in tightly coupled federated databases and data warehousing, but it is usually not used in the loosely coupled federated databases or in the link-driven approach.

4.3. Query Translation

Query translation is a sequence of transformation steps that are applied to a multi-database query when certain types of integration approaches are used. In a data warehouse, a multi-database query is issued directly against the data warehouse. Therefore, query translation is not necessary. In a loosely coupled federated databases system, the multi-database query is expressed by using a query language that embodies the schema definitions of the component databases. In this case, the multi-database query translation is also not needed. The same is true for the link-driven approach. Query translation is, however, a major part of tightly coupled federated databases. Tightly coupled federated databases are often implemented by using a mediator-wrapper architecture (47). The mediator accepts multi-database queries that are expressed by using the global schema. Each multi-database query is translated into a set of component database-specific subqueries (40). These subqueries are expressed by using the component schema. The wrappers in the mediator-wrapper architecture are responsible for retrieving the data from the component databases. In the process of performing this task, the wrappers will translate the subquery expressed on the component schema to a subquery expressed on the local schema, which can be directly executed by the corresponding component database.

4.4. Data Retrieval

Data retrieval is relatively simple in data warehouses and link-driven integration systems. In the first case, the data are accessed by using the database management system (DBMS) that is managing the data warehouse. In the second case, the records that need to be retrieved are indicated by a direct link. For federated databases, data are retrieved from the component databases by using a wrapper (also called the data driver). The wrapper construction for component databases that are directly accessed through their DBMSs is straightforward. DiscoveryLink (48) simplifies issues related to data retrieval by requiring that component databases be accessible through a well-structured query language and that the returned data be at a minimum formatted by using a relational table. However, most biological databases are only accessible through a Web interface. Wrapper construction for these databases is a complex task for two reasons. First, the Web databases use the query by example approach. That is, a query is usually constructed by entering values in predefined data fields rather than by using a query language such as structured query language (SQL). Second, the returned Web pages are often human readable and not meant to be machine readable (i.e., not appropriately tagged for automated data extraction). Thus, locating and identifying fields in the result pages is not trivial. The wrapper system proposed in W4F (49) can be used to retrieve information from both traditional databases as well as Web databases. W4F consists of three layers: the retrieval layer, the extraction layer, and the mapping layer. The retrieval layer is responsible for retrieving HTML pages from Web data sources. The extraction layer uses predefined parsing rules to extract relevant data from the returned pages. The mapping layer transforms the extracted information into the target structure to be used by the integration system.

4.5. Result Assembly

The final step in the integration process is the assembly of the result returned by the component databases (40). This step is only necessary for federated databases. In a data warehouse, the query result is generated in an integrated form directly from the data warehouse. In the link-driven approach, the records retrieved from the different databases are already coupled using links, and in most cases, data conflicts and data redundancies are addressed in advance when these links are established. When a federated databases system is used, the query result returned from individual wrappers needs to be integrated. This step offers an opportunity for resolving conflicting and redundant data because it is often not possible to resolve these data inconsistencies until the actual data have been retrieved from the component databases (27). Consider the databases BRENDA (*http://www.brenda.uni-koeln.de*) and EMP (*http://emp.mcs.anl.gov*). BRENDA, as mentioned, includes citations that are related to enzymes. EMP also includes citations that are related to enzymes. For example, a query on the Enzyme with EC number 1.1.1.1 generated 101 citations from BRENDA and 288 citations from EMP. These citations include duplicates. If the two databases are integrated, duplicate citations, which are query specific, have to be removed, and this can only be performed during query execution.

Developing efficient techniques to resolve data conflicts is the subject of active research. One possible approach to addressing this issue is to associate a confidence level with each dataset so that when two conflicting records from two different databases are retrieved, the confidence level can be used to select one record.

5. DATABASE INTEGRATION SYSTEMS

As mentioned, there are three different types of integration approaches: link driven, federated databases, and data warehouses (Table 2). With the exception of GUS, the data warehouse approach has not been widely adopted in the biological research field. This section presents a review of some integration systems that are based on the first two types of integration approaches. The intent is not to be comprehensive but to offer example systems for each approach.

Entrez from NCBI uses the link-driven approach. It integrates, among others, nucleotide sequence data, protein sequence data, gene expression data, protein structures, protein domain information, genome maps, and literature (17). These component databases are stored by using the ASN.1 format flat files. These component databases are derived from other databases. For example, as described in Ref. 50, structure data are imported from PDB. Once the data are retrieved from the source database, it is validated and converted into the ASN.1 format. Links between different types of data (e.g., from sequences to literatures, or from sequences to structures) are established during this process. The *neighbors* is a unique feature of Entrez. They are precomputed, and they establish links among similar data of the same type, such as similar sequences, or similar documents (17). For example, sequence neighbors are established by using the BLAST sequence alignment algorithm. The Entrez system allows only one database to be searched at a time. However, when a query is issued against a database specified by the user, the returned result includes the target records from the specified database as well as links to their related and neighbor records from the same or other databases (51).

SRS (45) also uses the link-driven approach. It integrates component databases in flat file format. The integrated data includes, among others, nucleotide sequence, protein sequence, protein 3-D structure, protein domain information, and metabolic pathways. More than 100 SRS servers are installed all over the world. The main server at the European Bioinformatics Institute integrates more than 130 databases and 10 applications (e.g., homology search tools). SRS uses many indices to support the link-driven approach. In this system, each database record in any component database is processed separately as a collection of data fields (52). An index is created for each data field (data field index). This index classifies each record in the database according to a set of keywords from a controlled vocabulary. Each database will have one index per data field. A different type of indices (link index) is used to link individual databases. A link index is created for each pair of databases that includes cross-references to each other. These links are bidirectional. Databases that do not directly reference each other can still be connected by traversing links through intermediate databases. SRS has a unique approach to addressing syntactic and semantic heterogeneities. As mentioned, for a given database and a given data field, each database record is classified by using keywords from a controlled vocabulary. Therefore, the set of keywords assigned to similar data fields should overlap. SRS relies on this overlap to retrieve related records.

The remainder of this section presents example integration systems that are based on the federated databases approach, which can be either loosely coupled or tightly coupled. BioKleisli uses the loosely coupled federated databases approach. The CPL-Kleisli (43), the language used in BioKleisli, can be used to query various databases including relational databases, object-oriented databases, and ASN.1 flat files. The system provides different wrappers for different types of databases. Each wrapper enables data retrieval from one type of database (44). The results returned by the wrappers are expected to be in a common data exchange format, which addresses data model related heterogeneities. The component databases are not transparent to the user. A multi-database query includes explicit specifications of the target component databases. Furthermore, the multi-database query also includes explicit specifications of subqueries results assembly. The task of resolving syntactic and semantic heterogeneities is left to the user, and it is handled on a query-by-query basis. It is possible to add another layer to BioKleisli that combines the component schema into a global schema that will make the component databases transparent to the user and yield a tightly coupled federated databases system. This addition was implemented in K2/Kleisli (44) and TAMBIS. This latter tightly coupled federated databases system is discussed below.

DiscoveryLink (48) is an example of a tightly coupled federated databases system. The global schema in DiscoveryLink uses the relational data model. DiscoveryLink was designed to integrate heterogeneous databases, particularly databases from the biological and pharmaceutical domains. As in the case of BioKleisli, the system includes various wrappers in which each wrapper enables

Table 2. Multiple Approaches to the Integration of Biological Databases

Data Warehousing	Link-driven Approach	Federated Database
Remote data copied to local server	Links information between databases	Builds a homogenizing layer on top of different sources
• Sufficient for integration of low number of databases	• Links are static and often unidirectional	• Uses semantic relationships of the data
• Requires a deep understanding of data schema	• Links may not exist between related data	• Supports complex queries
• Lacks scalability	• Links limit the scope of queries	• Low maintenance
	• Poor scalability	• Good scalability
		• Complex design
GUS	Entrez, SRS	BioKleisli, DiscoveryLink, TAMBIS, BACIIS

data retrieval from one type of database (e.g., relational databases and flat text files). These wrappers do not include database-specific information, such as data field names, and therefore each wrapper can be used in conjunction with several databases of the same type. The local schema of each component database is mapped onto the component schema that is expressed by using the relational data model. This mapping addresses the data model-related heterogeneity. Syntactic heterogeneities can be addressed in DiscoveryLink by defining a translation table that includes the relationships among the attributes in different databases. Semantic heterogeneities are not addressed in DiscoveryLink.

To resolve semantic heterogeneity, in particular for tightly coupled federated databases, ontologies have recently emerged as an efficient framework for capturing biological knowledge (53,42). Comprehensive reviews on ontology-based knowledge representation in biological applications are provided in Refs. 54 and 55. An ontology is a formal description of the concepts and entities in a given domain. A core part of the ontology is a vocabulary of terms and their meaning (54). An ontology is ideally suited for resolving both syntactic and semantic variations in the biological domain. Semantic variations can be resolved by expressing the concepts in an ontology at different degrees of granularity (27). Furthermore, by expressing the relationships among the various concepts, the ontology organizes the concepts in a structured way, thus reducing the number of possible interpretations of these concepts (54). The ontology can also be complemented with a dictionary of synonyms to resolve syntactic variations.

An example ontology that may see widespread use in databases is the gene ontology (GO) (56). GO is a controlled vocabulary that consists of three hierarchies describing molecular function, cellular component, and biological process. The terms in GO can be used as attributes of gene products by collaborating databases, thus facilitating uniform query formulation across multiple databases. Currently, databases such as SWISS-PROT, the MGD, the Rat Genome Database, and FlyBase (database for the *Drosophila melanogaster*), have already adopted GO or have created terminology indices that are based on GO terms.

TAMBIS is an integration system that is built on top of BioKleisli. It extends BioKleisli by also resolving semantic heterogeneities. In TAMBIS, component database-specific collection programming language (CPL) queries are mapped onto a global schema. An ontology is used as the global schema. This ontology is also used for query construction and query validation. The ontology used in TAMBIS includes domain knowledge and allows the query to be expressed by the user without any knowledge of the underlying component databases.

BACIIS (53) is another example of a tightly coupled federated databases system, which is based on a mediator-wrapper architecture. Like TAMBIS, BACIIS uses an ontology for a global schema. In BACIIS, the component schema, which describes each component database, uses the concepts defined in the ontology. This approach isolates the ontology from the impact of changes to the component databases.

6. SUMMARY

The number of biomedical databases has been growing at a fast pace. These databases represent a valuable resource that is used by scientists on a daily basis for the discovery of new knowledge and for the validation of information. Scientists may have to access several of these databases in the course of a single study. This task is made difficult because the databases are heterogeneous, autonomous, and distributed. Additionally, automated integrated access to the databases can be a challenge because accessibility is often only offered through a human-readable rather than a machine-readable Web interface.

To address these issues, several approaches and systems have been proposed. These systems either use a federated, data warehouse, or link-driven approach. They differ in the way and the level at which they address syntactic, semantic, schematic, and data model-related heterogeneities.

Continued research effort in the area is needed to provide scientists with flexible integration systems that can answer multi-database queries and make the heterogeneities of the component databases transparent to the user. Specifically, XML standards for the delivery of query results have to be developed and adopted by the community databases. These standards can facilitate wrapper induction (i.e., the automated wrapper creation process). In addition, the confidence in the quality of the data has to be explicitly defined in the databases by using a unified metric. Finally, the need for adopting a common ontology or a controlled vocabulary is critical to the efficient integration of the distributed biological databases.

Acknowledgment

The authors are supported by the National Science Foundation under grants CAREER DBI-0133946 and DBI-0110854. We would like to thank Dr. Hiroki Yokota for his feedback on this manuscript.

BIBLIOGRAPHY

1. A. D. Baxevanis, The Molecular Biology Database Collection: 2003 update. *Nucleic Acids Res.* 2003; **31**(1):1–12.
2. A. D. Baxevanis, The Molecular Biology Database Collection: An updated compilation of biological database resources. *Nucleic Acids Res.* 2001; **29**(1):1–10.
3. P. Karp, A strategy for database interoperation. *J. Computat. Biol.* 1995; **2**(4):573–586.
4. S. Schulze-Kremer, Integrating and exploiting large-scale, heterogeneous and autonomous databases with an ontology for molecular biology. In: R. Hofestädt and H. Lim, eds. *Molecular Bioinformatics, Sequence Analysis—The Human Genome Project*. Aachen: Shaker Verlag, 1997, pp. 43–56.
5. W. Sujansky, Heterogeneous database integration in biomedicine. *J. Biomed. Inform.* 2001; **34**(4):285–298.
6. V. M. Markowitz and O. Ritter, Characterizing heterogeneous molecular biology database systems. *J. Comput. Biol.* 1995; **2**(4):547–556.
7. A. C. Siepel, A. N. Tolopko, A. D. Farmer, P. A. Steadman, F. D. Schilkey, B. D. Perry, and W. D. Beavis, An integration

platform for heterogeneous bioinformatics software components. *IBM Syst. J.* 2001; **40**(2):570–591.

8. D. A. Benson, I. Karsch-Mizrachi, D. J. Lipman, J. Ostell, B. A. Rapp, and D. L. Wheeler, GenBank. *Nucleic Acids Res.* 2002; **30**(1):17–20.

9. G. Stoesser, W. Baker, A. Broek, M. Garcia-Pastor, C. Kanz, T. Kulikova, R. Leinonen, Q. Lin, V. Lombard, R. Lopez, R. Mancuso, F. Nardone, P. Stoehr, M. A. Tuli, K. Tzouvara, and R. Vaughan, The EMBL Nucleotide Sequence Database: Major new developments. *Nucleic Acids Res.* 2003; **31**(1): 17–22.

10. Y. Tateno, K. Fukami-Kobayashi, S. Miyazaki, H. Sugawara, and T. Gojobori, DNA Data Bank of Japan at work on genome sequence data. *Nucleic Acids Res.* 1998; **26**(1):16–20.

11. C. H. Wu, H. Huang, L. Arminski, J. Castro-Alvear, Y. Chen, Z. Z. Hu, R. S. Ledley, K. C. Lewis, H. W. Mewes, B. C. Orcutt, B. E. Suzek, A. Tsugita, C. R. Vinayaka, L. S. Yeh, J. Zhang, and W. C. Barker, The Protein Information Resource: An integrated public resource of functional annotation of proteins. *Nucleic Acids Res.* 2002; **30**(1): 35–37.

12. B. Boeckmann, A. Bairoch, R. Apweiler, M. C. Blatter, A. Estreicher, E. Gasteiger, M. J. Martin, K. Michoud, C. O'Donovan, I. Phan, S. Pilbout, and M. Schneider, The SWISS-PROT protein knowledgebase and its supplement TrEMBL in 2003. *Nucleic Acids Res.* 2003; **31**(1):365–370.

13. V. A. McKusick, *Mendelian Inheritance in Man. Catalogs of Human Genes and Genetic Disorders*, 12th ed. Baltimore, MD: Johns Hopkins University Press, 1998.

14. J. Westbrook, Z. Feng, S. Jain, T. N. Bhat, N. Thanki, V. Ravichandran, G. L. Gilliland, W. Bluhm, H. Weissig, D. S. Greer, P. E. Bourne, and H. M. Berman, The Protein Data Bank: Unifying the archive. *Nucleic Acids Res.* 2002; **30**(1):245–248.

15. A. Bateman, E. Birney, R. Durbin, S. R. Eddy, K. L. Howe, and E. L. Sonnhammer, The Pfam protein families database. *Nucleic Acids Res.* 2000; **28**(1):263–266.

16. J. McEntyre and D. Lipman, PubMed: Bridging the information gap. *CMAJ.* 2001; **164**(9):1317–1319.

17. D. L. Wheeler, D. M. Church, S. Federhen, A. E. Lash, T. L. Madden, J. U. Pontius, G. D. Schuler, L. M. Schriml, E. Sequeira, T. A. Tatusova, and L. Wagner, Database resources of the National Center for Biotechnology. *Nucleic Acids Res.* 2003; **31**(1):28–33.

18. S. I. Letovsky, R. W. Cottingham, C. J. Porter, and P. W. Li, GDB: The Human Genome Database. *Nucleic Acids Res.* 1998; **26**(1):94–99.

19. J. A. Blake, J. T. Eppig, J. E. Richardson, and M. T. Davisson, The Mouse Genome Database (MGD): Expanding genetic and genomic resources for the laboratory mouse. *Nucleic Acids Res.* 2000; **28**(1):108–111.

20. S. Y. Rhee, W. Beavis, T. Z. Berardini, G. Chen, D. Dixon, A. Doyle, M. Garcia-Hernandez, E. Huala, G. Lander, M. Montoya, N. Miller, L. A. Mueller, S. Mundodi, L. Reiser, J. Tacklind, D. C. Weems, Y. Wu, I. Xu, D. Yoo, J. Yoon, and P. Zhang, The Arabidopsis Information Resource (TAIR): A model organism database providing a centralized, curated gateway to Arabidopsis biology, research materials and community. *Nucleic Acids Res.* 2003; **31**(1):224–228.

21. The FlyBase Consortium, The FlyBase Database of the Drosophila Genome Projects and community literature. *Nucleic Acids Res.* 2003; **31**(1):172–175.

22. S. Weng, Q. Dong, R. Balakrishnan, K. Christie, M. Costanzo, K. Dolinski, S. S. Dwight, S. Engel, D. G. Fisk, E. Hong, L. Issel-Tarver, A. Sethuraman, C. Theesfeld, R. Andrada, G. Binkley, C. Lane, M. Schroeder, D. Botstein, and J. Michael Cherry, Saccharomyces Genome Database (SGD) provides biochemical and structural information for budding yeast proteins. *Nucleic Acids Res.* 2003; **31**(1):216–218.

23. I. Schomburg, A. Chang, and D. Schomburg, BRENDA, enzyme data and metabolic information. *Nucleic Acids Res.* 2002; **30**(1):47–49.

24. P. D. Karp, M. Riley, S. M. Paley, and A. Pellegrini-Toole, The MetaCyc Database. *Nucleic Acids Res.* 2002; **30**(1):59–61.

25. R. Robbins, Report of the invitational DOE workshop on genome informatics, 26–27 April 1993; Genome informatics I: Community databases. *J. Computat. Biol.* 1994; **1**(3):173–190.

26. H. Bono, H. Ogata, S. Goto, and M. Kanehisa, Reconstruction of amino acid biosynthesis pathways from the complete genome sequence. *Genome Res.* 1998; **8**:203–210.

27. Practical data integration in biopharmaceutical R&D: Strategies and technologies. White paper. 3rd Millennium, Inc., Cambridge, MA. (2002). Available: *http://www.3rdmill.com/pdf/3rd_Millennium_Biopharma-data-integration.pdf*.

28. S. B. Davidson, C. Overton, and P. Buneman, Challenges in integrating biological data sources. *J. Computat. Biol.* 1995; **2**(4):557–572.

29. M. Y. Becker and I. Rojas, A graph layout algorithm for drawing metabolic pathways. *Bioinformatics* 2001; **17**(5): 461–467.

30. V. Honavar, A. Silvescu, J. Reinoso-Castillo, C. Andoff, and D. Dobbs, Ontology driven information extraction and knowledge acquisition from heterogeneous, distributed biological data sources. Proc. IJCAI-2001 Workshop on Knowledge Discovery from Heterogeneous, Distributed, Autonomous, Dynamic Data and Knowledge Sources, 2001.

31. P. Buneman, Semi-structured data. Proc. 16th ACM SIGACT—SIGMOD—SIGART Symp. on Principles of Database Systems, 1997: 117–121.

32. A. Mackey, Relational modeling of biological data: Trees and graphs. *The O'Reilly Network.* (2002). Available: *http://www.oreillynet.com/pub/a/network/2002/11/27/bioconf.html*.

33. R. Hull, Managing semantic heterogeneity in databases: A theoretical perspective. Proc. 16th ACM Symp. on Principles of Database Systems (PODS), 1997: 51–61.

34. K. Aberer, The use of object-oriented data models for biomolecular databases. Proc. Conf. on Object-Oriented Computing in the Natural Sciences (OOCNS) '94, Heidelberg, Germany, 1994.

35. I. A. Chen and V. M. Markowitz, An overview of the object-protocol model (OPM) and OPM data management tools. *Inform. Syst.* 1995; **20**(5):393–418.

36. K. H. Fasman, S. I. Letovsky, R. W. Cottingham, and D. T. Kingsbury, Improvements to the GDB Human Genome Data Base. *Nucleic Acids Res.* 1996; **24**(1):57–63.

37. D. L. Rubin, F. Shafa, D. E. Oliver, M. Hewett, and R. B. Altman, Representing genetic sequence data for pharmacogenomics: an evolutionary approach using ontological and relational models. *Bioinformatics* 2002; **18**(suppl. 1): 207–215.

38. D. Frishman, K. Heumann, A. Lesk, and H. W. Mewes, Comprehensive, comprehensible, distributed and intelligent databases: Current status. *Bioinformatics* 1998; **14**(7): 551–561.

39. E. Shoop, K. A. Silverstein, J. E. Johnson, and E. F. Retzel, MetaFam: A unified classification of protein families. II. Schema and query capabilities. *Bioinformatics* 2001; **17**(3): 262–271.
40. C. T. Yu and W. Meng, *Principles of Database Query Processing for Advanced Applications*. New York: Morgan Kaufmann, 1998.
41. S. Busse, R. D. Kutsche, U. Leser, and H. Weber, Federated information systems: Concepts, terminology and architectures. Technical Report 99-9, Technische Universitat Berlin, 1999.
42. C. A. Goble, R. Stevens, G. Ng, S. Bechhofer, N. W. Paton, P. G. Baker, M. Peim, and A. Brass, Transparent access to multiple bioinformatics information sources. *IBM Syst. J.* 2001; **40**(2):532–552.
43. S. B. Davidson, C. Overton, V. Tannen, and L. Wong, BioKleisli: A digital library for biomedical researchers. *J. Digital Libraries* 1997; **1**(1):36–53.
44. S. B. Davidson, J. Crabtree, B. P. Brunk, J. Schug, V. Tannen, G. C. Overton, and C. J. Stoeckert, Jr., K2/Kleisli and GUS: Experiments in integrated access to genomic data sources. *IBM Syst. J.* 2001; **40**(2):512–531.
45. E. M. Zdobnov, R. Lopez, R. Apweiler, and T. Etzold, The EBI SRS server-recent developments. *Bioinformatics* 2002; **18**(2): 368–373.
46. I-M. A. Chen, A. S. Kosky, V. M. Markowitz, and E. Szeto, Constructing and maintaining scientific database views in the framework of the object-protocol model. Proc. 9th Int. Conf. on Scientific and Statistical Database Management, IEEE, New York, 1997: 237–248.
47. G. Wiederhold, Mediators in the architecture of future information systems. *IEEE Computer*. 1992; **25**(3):38–42.
48. L. M. Haas, P. M. Schwarz, P. Kodali, E. Kotlar, J. E. Rice, and W. C. Swope, DiscoveryLink: A system for integrated access to life sciences data sources. *IBM Syst. J.* 2001; **40**(2): 489–511.
49. A. Sahuguet and F. Azavant, Building light-weight wrappers for legacy Web datasources using W4F. Proc. 25th Int. Conf. on Very Large Data Bases, Edinburgh, Scotland, U.K., September 7–10, 1999.
50. J. Chen, J. B. Anderson, C. DeWeese-Scott, N. D. Fedorova, L. Y. Geer, S. He, D. I. Hurwitz, J. D. Jackson, A. R. Jacobs, G. J. Lanczycki, C. A. Liebert, C. Liu, T. Madej, A. Marchler-Bauer, G. H. Marchler, R. Mazumder, A. N. Nikolskaya, B. S. Rao, A. R. Panchenko, B. A. Shoemaker, V. Simonyan, J. S. Song, P. A. Thiessen, S. Vasudevan, Y. Wang, R. A. Yamashita, J. J. Yin, and S. H. Bryant, MMDB: Entrez's 3D-structure database. *Nucleic Acids Res*. 2003; **31**(1):474–477.
51. J. A. Epstein, J. A. Kans, and G. D. Schuler, WWW Entrez: A hypertext retrieval tool for molecular biology. Proc. the 2nd Int. World Wide Web Conf., Chicago, IL, October 1994.
52. T. Etzold, A. Ulyanov, and P. Argos, SRS: Information retrieval system for molecular biology data banks. *Methods Enzymol*. 1996; **266**:114–128.
53. Z. Ben Miled, N. Li, G. M. Kellet, B. Sipes, and O. Bukhres, Complex life science multidatabase queries. *Proc. IEEE* 2002; **90**(11):1754–1763.
54. S. Schulze-Kremer, Ontologies for molecular biology and bioinformatics. *Silico Biol*. 2002; **2**(3):179–193.
55. R. Stevens, C.A. Goble, and S. Bechhofer, Ontology-based knowledge representation for bioinformatics. *Brief. Bioinformatics* 2000; **1**(4):398–414.
56. The Gene Ontology Consortium, Gene Ontology: Tool for the unification of biology. *Nature Genet*. 2000; **25**:25–29.

BIOLOGICAL NEURAL CONTROL

NORBERTO M. GRZYWACZ
MÓNICA PADILLA
University of Southern
California
Los Angeles

Neural control refers to the manipulation of inputs to particular structures of the nervous system to cause desirable output behavior. There are two kinds of neural control mechanisms, namely, closed loop and open loop. As an example of the former, the hypothalamus receives sensory information from the body to control its temperature. If the environmental temperature rises, then the hypothalamus makes the skin sweat to cool the body. This temperature control is called closed loop, because after the output behavior is modified, new sensory data are taken to see whether the output behavior should be increased or reduced. Hence, closed-loop neural control depends on feedback information about the effects of output behavior. In contrast, the nervous system can sometimes perform open-loop control. This form of control does not rely on feedback to correct erroneous outputs, and thus, the nervous system must rely on knowledge of system behavior to compute its output. Therefore, open-loop neural control requires accurate models of the system. Although such control can be less precise, it has the advantage of speed, because it does not require feedback for computations. One example of open-loop neural control occurs in the first 100 ms of the eye trying to do smooth pursuit of a target in the visual field. This period of pursuit is open loop, because no visual feedback is available because of the delays in the visual system. Thereafter, visual feedback (and other sources of information) is available to close the loop, which improves performance.

The first section shows that open-loop and, especially, closed-loop mechanisms of control are pervasive in the nervous system. Then, the section on general principles of neural control distills these examples into a small set of theoretical principles. These principles are applicable to both biological and artificial forms of neural control, which is important because in many modern circles, neural control refers only to applications with artificial neural networks. Although this article focuses on biological aspects of neural control, the principles learned in the general-principles section straddle the boundary to artificial applications. Finally, the last four sections illustrate these principles in computational detail for four different neural systems.

1. CONTROL MECHANISMS IN THE NERVOUS SYSTEM

In this section, we illustrate closed-loop and open-loop control mechanisms in the nervous system. We do it through a few examples, but many others exist, such as mechanisms for development of sensory and motor functions (1), and for attention and context in sensory systems (2). However, for the sake of brevity, we will focus here on seven examples:

1. *Action potentials.* Action potentials are all-or-none voltage events that transmit information between distant neurons. The firing of an action potential is controlled in closed loops by negative and positive feedback. All cells have a resting potential (polarization in their membrane). However, only neurons and muscle cells can fire or signal through action potentials. The resting potential in the membrane of the cell is maintained by Na^+, K^+, and Cl^- channels (membrane proteins), and by a Na^+ and K^+ pump (an energy-consuming protein). When depolarization occurs (the intracellular potential becomes more positive), Na^+ channels open, which allows an inward current into the cell. This current produces more depolarization, which causes even more channels to open and, thus, more inward current to flow. This positive feedback cycle is what makes the membrane potential reach the action-potential peak. In turn, the delayed opening of K^+ channels mediates a negative feedback. Their effective closed-loop control is shown in Fig. 1. The outward current that takes place when K^+ channels open makes the cell repolarize and the action potential end (2).

2. *Sensory Adaptation.* This is an example of a closed-loop control. In an important case of this example, the gain of sensory systems adapts according to the mean intensity of the input signals (3). Other known cases include adaptation of spatial and temporal properties of sensory processing (4). Sensory neurons have different adaptation rates and respond differently depending on their function. Slowly adapting neurons respond to prolonged and constant stimuli. These types of neuron respond strongly at the initiation of the stimulus, but if the stimulus is continued, then they reduce the amount of response and maintain a steady state until the stimulus is stopped. This decrease in response is called adaptation. In contrast, rapidly adapting neurons sense things like velocity and acceleration of the stimulus. These neurons respond at the beginning and at the end of the stimulus, but they cease to respond during a constant stimulus.

   The receptors in the visual system provide an illustration of how gain control occurs, by adapting slowly to conditions of dark or light. The response of the photoreceptors is determined by the intensity of the input light. Light triggers an enzymatic cascade, which causes the level of cyclic guanosine monophosphate (cGMP) to fall. When this process happens, cGMP-gated channels in the photoreceptor's membrane close, which stops the influx of both Na^+ and Ca^{2+}. The photoreceptor then hyperpolarizes (becomes more negative). During light adaptation, the photoreceptors slowly depolarize to a steady-state potential by reopening of cGMP-gated channels. These changes in the photoreceptor take place because of a decrease in internal Ca^{2+} concentration, which affects the function of different enzymes in the phototransduction cascade. In contrast, during dark adaptation, the opposite process occurs (2).

   Adaptation can occur at different levels of processing. For instance, in the case of the visual system, adaptation also happens at the level of the cortex (see References 5–8, among other places). One example of adaptation in the visual cortex occurs when a subject sees a grating with high contrast. The initial contrast threshold is measured before the person is exposed to this grating. When one measures the threshold of contrast again after approximately 1 min of adaptation, the threshold value is higher. It takes tens of seconds for the subject to readapt again to low contrast images and go back to its previous threshold levels of detection (9). A more detailed example of sensory adaptation appears in the section on detailed example 3.

3. *Motor Systems.* These systems mediate rlocomotion, and eye movements among other things. Motor systems are controlled in open and closed loops. We already discussed how smooth-pursuit eye movements occur in both open- and closed-loop modes. Another example of a motor closed-loop control is the pupillary reflex. The size of the pupil is regulated by the hypothalamus according to the amount of light available. In the dark, the size of the pupil increases reflexively to allow more light in, whereas at high illumination, the pupil's size decreases to reduce the amount of light that enters the eye.

   Figure 2a provides a general schematic view of the locomotion system. Central pattern generators maintain the rhythmic movements that produce the locomotion (e.g., as in walking or swimming). (A simple generator that has been widely studied is the one used in the lamprey's swimming. This example will be discussed in detail in the section on detailed example 2.) In humans, the locomotion pattern generator is in the spinal cord, which controls the extension and flexion of the limbs. To initiate the movement and to regulate the speed of the

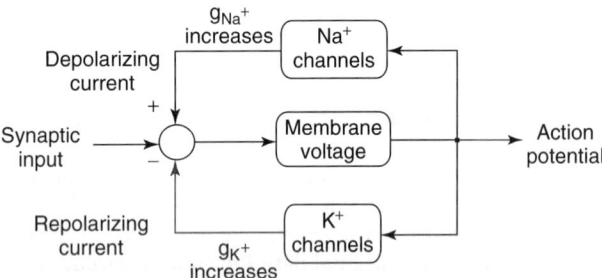

Figure 1. Positive and negative feedbacks mediate action potentials in neurons. Red and blue lines represent positive and negative effects, respectively. An excitatory (synaptic) input causes the membrane voltage to become more positive, which causes Na^+ channels to open, increasing the Na^+ conductance (g_{Na}), which in turn, because there is an excess of Na^+ in the extracellular space, depolarizes the neuron. The depolarization of the membrane potential has the opposite effect through K^+ channels. They open (with a delay), increasing K^+ conductance (g_K), but because there is an excess of K^+ in the intracellular space, the K^+ current is outward. (This figure is available in full color at http://www.mrw.interscience.wiley.com/ebe.)

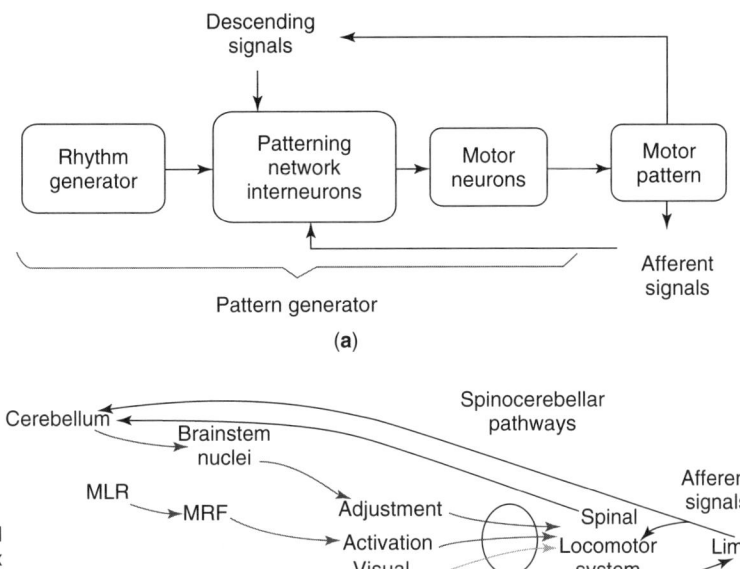

Figure 2. Locomotion system. (a) The general strategy of locomotion. (b) Neurobiological details of the mammalian locomotion system. MLR is the mesencephalic locomotor region, and MRF is the medial reticular formation. (This figure is available in full color at http://www.mrw.interscience.wiley.com/ebe.)

locomotion, descending (feedforward) signals are necessary to gate the generator. Initially, then, locomotion is regulated in an open-loop control (the early descending signals). Later, sensory feedback is necessary to update the locomotion according to changes in the terrain and obstacles that can be encountered, which gives rise to the closed-loop control. Three types of sensory information are used to regulate stepping. First, inputs from the vestibular system in the ear control balance. Second, visual inputs provide information about the path and obstacles. Third, somatosensory inputs from receptors of the muscle and skin inform about the motion and about deformations of the terrain. These somatosensory receptors are classified as proprioceptors and exteroceptors. Exteroreceptors, as the name indicates, adjust stepping to external stimuli and are located in the skin. Input from proprioceptors participates in the regulation of the automatic stepping and is located in muscles and joints (2).

Figure 2b shows with more detail where descending and afferent (sensory) signals come from, go to, and interact in mammals. Because accurate visual information is necessary for locomotion, the vestibulo-ocular reflex allows eye movements to update for the position of the body. The vestibular system has another important role in humans. Locomotion in humans has similarities with locomotion in other mammals, but because humans are bipedal, their locomotion requires special maturation of the balance control by the vestibular system. In turn, the cerebellum has various control schemes to maintain a steady locomotion rhythm. The cerebellum compares internal feedback signals of the intended movement with feedback signals of the actual movement (sensory information). When the movement is different from the intended movement, the cerebellum generates corrective signals. These signals are feedforward or are anticipatory and sent to the brainstem and cerebral cortex. The role of the latter is to plan the movement.

4. *Regulation of Body Temperature*. We already mentioned a role for the hypothalamus in the control of pupillary reflex. This organ also contributes to the regulation of body temperature, in a clear example of a closed-loop control. There is a set point for the internal body temperature (37°C). Information about the external temperature is given by thermoreceptors in the skin. Different sensors respond to either cold or hot external temperatures. When the external temperature rises about the set point, sweat starts (dissipation) and heat production reduces to maintain the internal body temperature. When the external temperature decreases, the body starts shivering (muscle contractions) and heat-production increases. The hypothalamus directs these temperature-controlling actions. Fever occurs when the temperature set point has been changed to a higher value and the body tries to maintain the temperature at this new value (10).

5. *Regulation of Body Weight*. Early theories for body-weight control defined a control loop similar to what was described for the control of body temperature. These theories had just one set point and one closed-loop control from the hypothalamus. However, the hypothalamus receives input from different sources to regulate body weight and food intake. Moreover, the forebrain and the hindbrain are involved in controlling these two processes. Hence, body weight is regulated by several closed-loop control mechanisms. Different factors provide positive- and negative-feedback controls to maintain energy balance

and metabolic fluxes. Therefore, different steady states can be set depending on life cycle, lactation, and body rhythms, which means that there might be different body-weight and body-fat set points. Between steady states, periods of non-steady states can occur in which the body weight fluctuates. Two main hormones are involved in the regulation of body weight, namely, leptin and ghrelin. Other metabolic signals contribute to body-weight regulation, but these two are the main ones. These hormones interact with the hypothalamus and brainstem to maintain a particular steady state of energy balance and food intake. It is when there is overstimulation or understimulation of these pathways that eating disorders take place (obesity or anorexia, for example). The presence of leptin signals indicates that there is abundant energy, whereas the presence of ghrelin indicates an insufficiency, consequently, requiring eating. The pathways of both hormones arrive to several receptors in the hypothalamus, mainly in the dorsomedial hypothalamic nucleus (DMN) (10,11).

6. *Learning*. A closed-loop control takes place for learning, as will be shown in more detail in the section on detailed example 4. The process of learning is part of the fine-tuning of synaptic connections between neurons that takes place after development. A connection that performs such fine-tuning is called a Hebbian synapse (12). It has an adaptive process, in which synaptic gain varies according to the presynaptic input and to the results sensed from actions taken (e.g., the postsynaptic potential), that is, from feedback. These synaptic modulations can last for a short period, short-term memory, or be long lasting, long-term memory. Then, different types of learning give rise to different types of memory. One simple case in which feedback helps is in associative learning. An example of associative learning is when an animal is in a cage with a lever at one wall. After activating the lever, the result of this action is receiving food. Receiving the food serves as reinforcement, so the animal learns to associate the lever with food. This example is also called trial-and-error learning (2).

7. *Control of Neuron Number in Development*. Neural development has many forms of control processes, some using the Hebbian mechanisms described for learning (13). Here, we illustrate a different kind of developmental control. The amount of neurons that form during development is a controlled process. This process is determined by a previous program stored in the genetic information of each species and by feedback from target cells connected with the neurons. At different stages of the development, different types of neurons develop. Neurons form from ectodermic cells, which after being recruited, start acquiring differentiated properties aided by signaling from neighboring cells (inducing factors), with some cells becoming neurons and others becoming glia. Neurons then start migrating to different zones in the developing embryo, depending on their future function or fate, and they start forming connections with target cells. Several neurons are genetically preprogrammed to die at a certain moment of development, a process known as programmed cell death. Almost half of the neurons initially generated are lost through this process. This way, the number of final neurons is controlled in an open-loop control. The survival of neurons is also regulated by a closed-loop control given by feedback from target cells and from the environment. Target cells secrete a variety of neurotrophic factors, and elimination or augmentation of these factors activates neuronal death. For example, immediately after innervation, retinal ganglion cells become dependent on neurotrophins for their survival (2,14).

2. GENERAL PRINCIPLES OF NEURAL CONTROL THEORY

Although the systems described in the preceding section are from different areas of the nervous system, their functions share a few basic control principles. In control theory, the process to be controlled is called the plant. For instance, in the hypothalamic control of temperature through sweat, the plant is the set of sweat glands and their sympathetic innervation. The plant typically includes actuators, which are devices whose manipulation achieves control objectives. In the case of the nervous system, these devices are neural hormones or synapses from the innervation. Moreover, a neural-control system includes the controller, which is a device that drives the actuators. In the temperature-control system, for instance, the controller is the hypothalamus, which sets the control parameters with its neural networks. Neural networks are collections of interacting neurons that perform computations collectively (see the section on detailed example 2). Many computational models of neural networks exist, some being biologically realistic and others more artificial. Besides neural networks, the nervous system can perform control by chemical means. Manipulations of enzymatic cascades or of ionic channels in membranes can control the responses of individual cells (see the section on detailed example 1). Finally, neural-control systems require sensors if implementing closed-loop computations. These sensors may be organs providing information about external (e.g., visual, see the section on detailed example 3) or internal (e.g., blood pressure) variables of the body. The sensors may also be neural networks performing error-correction calculations.

A major mechanism in many neural networks is that their neurons have synapses whose weights can change to perform learning tasks (as illustrated in the preceding section). Therefore, many models of learning are of the neural-network kind (see the section on detailed example 4). However, because learning involves obtaining skills from different, unexpected examples, it is a statistical process. Many models of learning are thus, statistical, involving concepts of Bayesian probability (see the section on detailed example 3). As we discuss in the section on detailed example 4, there is a deep relationship between statistical and neural-network models of learning.

An important concept in learning and other neural closed-loop control models is stability. If the feedback is too sluggish and the actuator is too strong, then by the time the sensors report an error, it may be large. The actuator then compensates the plant vigorously, but before the sensors report that the system arrived to the desired point, the plant may overshoot this point by much. If these delayed overcompensations continue, then the system oscillates wildly. However, if the actuator is weakened or the delay is reduced, then the oscillations are terminated and the desired point is reached. Hence, the behavior of the system, stable or oscillatory, depends on its parameters. It turns out that one can classify the possible behaviors into four groups (see the section on detailed example 2). In the behavior called the stable fixed point, the neural control brings the system to a single value and small perturbations of the system by noise do not cause the system to change much. For example, this is the situation in thalamic control of temperature, which remains relatively fixed over time. Another neural form of stable situation occurs in certain kinds of oscillation. These oscillations converge to a stable limit cycle; that is, the system's frequency is robust against noise. An example of stable limit cycles occurs in locomotion, as when one walks at a steady pace. In contrast, to those stable states of neural control, certain parameters of the nervous system can cause unstable fixed points and limit cycles. For instance, an overly strong synaptic connection may cause instability. However, although unstable fixed points and limit cycles have theoretical interest, they are unobservable, as noise destroys them.

Until now in this section, we have emphasized theoretical concepts that are common to all neural-control systems. However, some differences also exist, and it is thus interesting to classify these control systems into a few major groups. Here, we follow the Suykens and Bersini classification and identify four main neural-control strategies (15,16).

1. *Neural Adaptive Control.* In adaptive control, the neural controller is trained for instance to track specific reference inputs (see the section on detailed example 3). The system does this in indirect adaptive control by deriving a neural network or statistical model online from input–output measurements on the plant (Figure 3a). To model this derivation, one may use a cost function based on the model and the controller. (A cost function measures how much the system pays for making particular errors in the task.) Alternatively, in direct adaptive control, one may define the cost function with respect to the

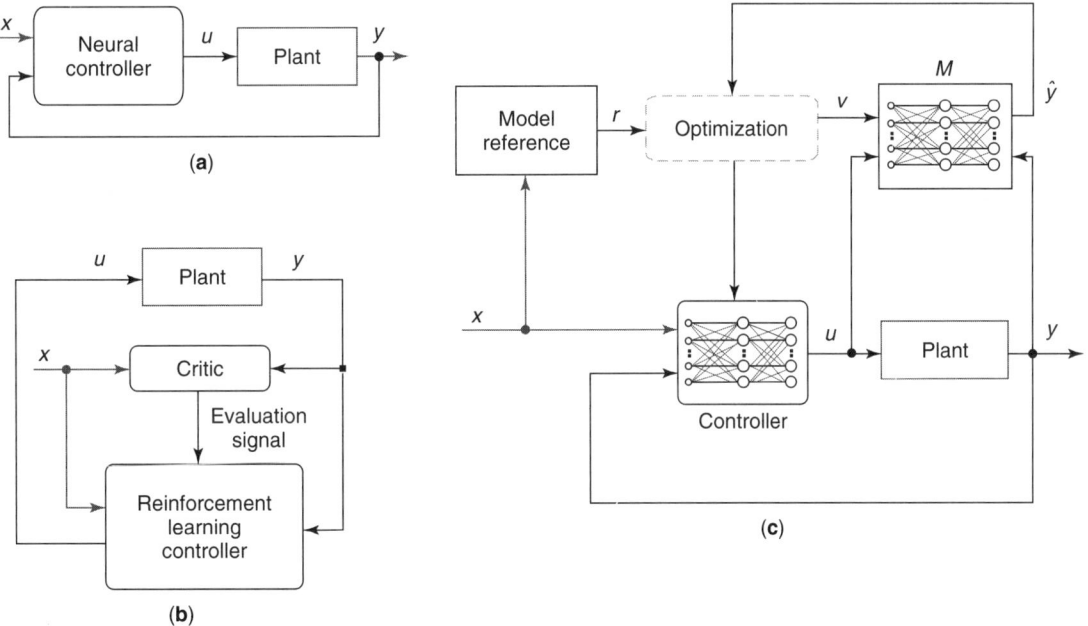

Figure 3. Three neural control strategies. (a) Neural adaptive control. This is a form of neural closed-loop control. The controller compares the external input x with the plant's output y to generate the actuator signal u. (b) Reinforcement learning control. Reinforcement learning begins with a scheme identical to adaptive control but adds a critic. The critic compares the input with the plant's output to send an evaluation signal to the controller. (c) Model-predictive control. This form of control also begins with the scheme of adaptive control. However, corrections to the controller, which is a neural network (or a statistical system), do not come from a supervisor but from a model. This model receives the external input and feeds an output r onto an optimization processor. In turn, this processor sends its output v to a neural network (or a statistical system) M. M adjusts its computations to send a prediction $\hat{y}$ of the plant's output back to the optimization processor. With this feedback loop, the system computes the optimal value of this prediction and then sends it to the controller. The prediction is valid only for a short time into the future (the time horizon). (This figure is available in full color at http://www.mrw.interscience.wiley.com/ebe.)

real plant, instead of based on a model. The system adapts the controller over time.

2. *Learning Control.* This control process maps situations to actions to maximize a scalar reward called a reinforcement signal (see the section on detailed example 4). In supervised learning, the reinforcement controller is rewarded or punished by a supervising element (a strong version of the critic in Fig. 3b) through several trials on the systems. The supervisor may provide different types of information, from the exact state of the input to just whether the system's decision is correct. In all these cases, the system may control the plant directly without modeling it (17). Inspection of Fig. 3a and b shows that there is a tight relationship between learning and adaptive control. They are related closed-loop neural-control strategies.

3. *Neural Optimal Control.* Classic optimal-control theory deals with nonlinear dynamical systems (that is, nonlinear systems of differential equations) (18–20). In optimal control, one time-dependent variable is at the control of the system. The choice of this variable minimizes a cost function. For example, early, fast eye movements follow nonlinear dynamical equations of motion (21). Because the goal of the movements is to track as closely as possible moving targets in scenes, one can formulate a cost function in terms of tracking error. Consequently, early, fast eye movements can fall in the category of optimal control, because they are nonlinear and must minimize a cost function. Neural optimal control is a variation of classic optimal-control strategies (22) in which neural networks or related statistical-learning techniques are black-box neural-controller models. The system performs the optimization by setting the free parameters of the controller. Because the system does not apply explicit feedback, neural optimal control is useful in open-loop situations.

4. *Model-Predictive Control.* This is a control strategy based on solving an online optimal-control problem. A receding-horizon approach is used for this form of control. In this approach, one solves an open-loop, optimal-control problem for the current state of the system over some future interval (the time horizon). One then uses this solution to try to control the system toward the desired solution at a short interval later. The procedure is repeated over time, with the new states of the system as initial conditions. Consequently, different from simple optimal control, not only are the plant and the controller in a closed loop but also is the input to the controller. In neural model-predictive control, one makes explicit use of neural-network or equivalent statistical models in the control scheme (Fig. 3c). The control signal of the controller is determined from predictions of the neural-network model for the plant over a given time horizon. The Kalman filter approach in the section on detailed example 4 is an illustration of model-predictive control.

3. DETAILED EXAMPLE 1: SINGLE-NEURON FIRING

As seen in example 1 of the section on control mechanisms in the nervous system, a closed loop governed by a chain reaction of different ionic channels can control a neuron's membrane voltage. They are activated or inactivated at different levels of voltage, and the resulting currents depend on ion concentrations inside and outside the cell. Furthermore, a neuron's membrane voltage is controlled by chemical means (ligands and neurotransmitters). Besides the basic Na^+ and K^+ channels that participate in the maintenance of the membrane resting potential and in the generation of all-or-none single action potentials, neurons contain a large variety of ion channels (23). They have different properties and appear in different concentrations depending on the site on the cell. One can model a patch of cell membrane as an equivalent electrical circuit. In this model, conductances with associated batteries represent the ionic channels. This model also includes a membrane capacitance and the Na^+-K^+ pump modeled as a current flow. The full equivalent circuit of a cell's membrane considers different ion channels as shown in Fig. 4. Currents and conductances of K^+, Na^+, and Ca^{2+} channels change according to ion concentrations, voltage levels, and neurotransmitters. Normally, the Cl^- conductance is not variable as can be seen in the figure. However, some neurons also have voltage-gated Cl^- channels.

The equations that describe the currents that flow through each of these channels are

$$I_k = g_k(V_m - E_k), \quad (1)$$

$$I_{Na} = g_{Na}(V_m - E_{Na}), \quad (2)$$

$$I_{Ca} = g_{Ca}(V_m - E_{Ca}), \quad (3)$$

and

$$I_{Cl} = g_{Cl}(V_m - E_{Cl}). \quad (4)$$

The different ion channels and different concentrations allow different types of firing to take place depending on the function of the neuron. Each basic type of ion channel has variants. They differ in the speed of activation, voltage-activation range, and sensitivity to various ligands (23). In addition, many neurons have channels that are significantly permeable to either both Na^+ and K^+ or both Na^+ and Ca^{2+}, and they activate slowly by hyperpolarization (negative voltage). There are at least four types of voltage-activated K^+ channels. And two or more types of Na^+ voltage-activated channels exist in the

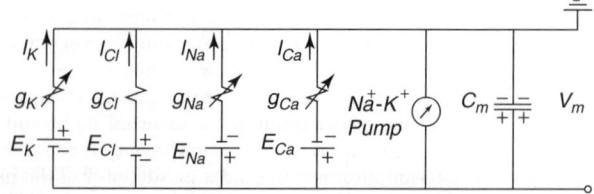

Figure 4. Example of an equivalent electrical circuit model of a patch of neural membrane.

nervous system. Most neurons have voltage-gated Ca^{2+} channels, and there are at least five subtypes of them. To expand, we now consider different types of K^+ channels. The delayed rectifier, the anomalous rectifier, and the A channel are potassium channels that are voltage activated. The delayed rectifier is activated by the depolarization caused by an action potential and is responsible for the repolarization in the action potential as explained in example 1 of the section on control systems in the nervous system. The anomalous rectifier is activated by hyperpolarization, and as a result, a small outward potassium current is produced. Activation of the A channel requires prior hyperpolarization and then depolarization. There are also messenger-modulated K^+ channels (M and S channels), which are activated by depolarization and are closed by different substances (peptides or serotonin, for example).

There are also Ca^{2+}- and Na^+-activated channels, which open when the intracellular concentrations of these ions rise. In the case of Ca^{2+}-activated channels, different subtypes exhibit different values of conductances. Calcium concentration of a resting cell is extremely low. Hence, the flux of Ca^{2+} through voltage-gated channels can increase this concentration significantly and modulate the opening and closing of other channels. It can enhance the probability of opening of Ca^{2+}-activated K^+ channels. It can also inactivate Ca^{2+} channels that are sensitive to intracellular concentration of Ca^{2+}. Calcium influx in a cell can have then two opposing effects, repolarization or generative depolarization. Depolarization can be accomplished by the positive charge that Ca^{2+} carries into the cell, which can mediate in some cases the firing of slow action potentials. Repolarization occurs because of the opening of K^+ channels and the closing of Ca^{2+} channels, both of which create a net outward current.

The existence of many types of channels is responsible for a wide variety of types of firing. One of the most prevalent is repetitive firing (Fig. 5a). It can take place in neurons of different parts of the nervous system by constant depolarization. However, by holding the membrane potential at different values, different firing patterns can take place in the same cell. Moreover, neurotransmitters may mediate different firing patterns. An example of how the response of a neuron can be controlled by an external input that changes the membrane voltage is a neuron from the nucleus tractus solitarius. An injection of a depolarizing (excitatory) current triggers an immediate train of action potentials, when one initially holds the membrane potential at the resting voltage. However, a delay in the spike train can occur when one first holds the cell at a hyperpolarized membrane potential as shown in Fig. 5a, which allows A-type K^+ channels that are normally inactive at the resting potential to be activated by the applied depolarization current. These channels generate a transient outward current that drives the voltage away from threshold. Because of the transient current, the neuron reaches the threshold needed for firing later.

Another example of firing modulated by hyperpolarization occurs in thalamic neurons. Depolarizing a cell from the resting potential results in tonic, steady firing. However, if before the depolarization, one hyperpolarizes the cell, then one triggers a transient burst of action potentials. The hyperpolarization causes voltage-gated Ca^{2+} channels to recover from inactivation. These channels produce a long Ca^{2+} action potential in which the normal Na^+ action potentials ride (Fig. 5b). (The system naturally produces the hyperpolarization in both this and the preceding example by an inhibitory synaptic input, that is, by negative feedback.)

Neurotransmitters can sometimes alter the firing of a neuron in response to input stimuli. Therefore, in effect, these transmitters can mediate positive or negative feedback on how action potentials take place (2). Examples of this type of feedback control are the firing patterns of sympathetic ganglion cells and of hippocampus pyramidal neurons. Both types of cells are regulated by neurotransmitters. In the case of sympathetic ganglion cells, the neurotransmitter acetylcholine (ACh) closes M-type K^+ channels that normally activate with depolarization. This closure allows the neuron to fire several action potentials in response to the same stimulus as shown in Fig. 5c. Without the action of ACh, the neuron fires a single action potential in response to a stimulus, because the outward current provided by the K^+ channels prevents additional firing. In the case of hippocampus pyramidal neurons, which are important in short-term memory, the neurotransmitter norepinephrine modulates Ca^{2+}-activated K^+ channels. This modulation allows the neuron to fire more action potentials, without changing the resting potential of the cell (2,24).

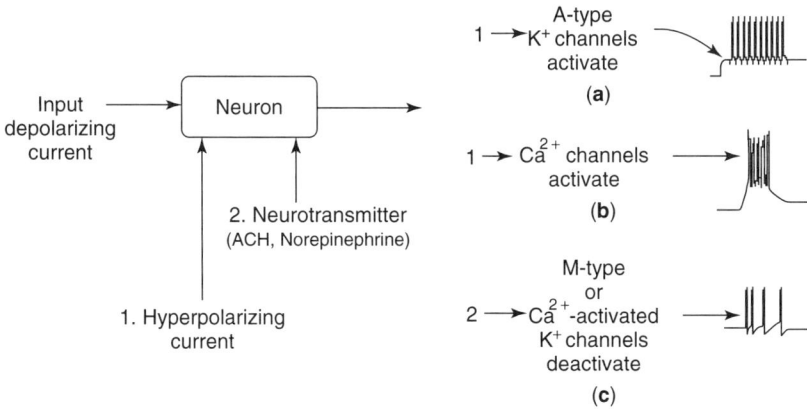

Figure 5. Examples of responses of different neurons to the same depolarizing current when other signals are present (hyperpolarization or neurotransmitters). (a) Example based on a neuron from the nucleus tractus solitarius, where hyperpolarization delays firing. (b) Example based on a thalamic neuron, where hyperpolarization causes bursts of action potentials. (c) Example based on sympathetic ganglion cells or hippocampal pyramidal neurons, where neurotransmitters increase firing. (This figure is available in full color at http://www.mrw.interscience.wiley.com/ebe.)

Neurons in these examples receive excitatory input to fire action potentials. Other neurons can fire even without a synaptic input. These neurons fire spontaneously different patterns of brief bursts of action potentials. The activity of these types of neurons is controlled by chain activation of ionic channels. For instance, a thalamocortical relay neuron produces multiple bursts of action potentials through three types of voltage-gated ion channels. These channels activate and deactivate by receiving feedback from the voltage in the cell, which maintains the cycle of bursts. Channels called H-type activate during hyperpolarization, which enable an inward current and lead to a gradual depolarization. Then, voltage-gated Ca^{2+} channels that open at low levels of depolarization activate, which allows a faster inward current. This current generates the necessary depolarization for voltage-gated Na^+ channels to open and cause the burst of action potentials. The H-type channels close during the bursts because of the high depolarization, and the calcium channels inactivate, which allows K^+-dependent hyperpolarization to take place. After the hyperpolarization takes place, H-type channels can be activated again to restart the process as shown in Fig. 6a.

Other cells that do not need excitatory inputs to fire trains of action potentials are pacemaker cells. For instance, pacemaker cells in the heart rhythmically fire action potentials without a synaptic input. However, although this is an autonomous system, synapses control the frequency of pacemaker firing. This way, the frequency of the beating of the heart can be controlled according to the needs of the body (25). Figure 6b illustrates how a closed-loop system controls the firing frequency of the pacemaker cells. According to the state of the body (rest, exercising, or running for example), the heart should be beating at a certain frequency. The autonomous nervous system sends neurotransmitters to increase or decrease the frequency of the firing of action potentials of the pacemaker cells. The pacemaker cells or the whole system act as the plant. The heart beating frequency is adjusted then, and the new value is sent to the autonomous nervous system to see if more adjustment is needed.

4. DETAILED EXAMPLE 2: LOCOMOTION

As shown in the section on control mechanisms in the nervous system, locomotion is a rhythmic motor activity. It is generated by a central pattern generator (26). Modeling of mammalian locomotion generally involves many neurons. In contrast, invertebrates or primitive vertebrates, where structures are well known, allow simpler modeling of central-pattern generators. Such models consider a representative of each type of neuron involved in the control of the motion (27), as in the example that we will discuss in this section.

One example of locomotion that has been widely studied in vertebrates is the swimming pattern of the lamprey. Its motor generator is a bilateral ipsilateral (same side) system. The lamprey swims by alternating activation of motor neurons on the two sides of the body. Figure 7a shows the neural-network elements that control the rhythmic locomotion present in a segment of each side of the lamprey's body. The complete body has several segments, each with its given phase lag necessary for undulatory swimming (28). On each side of a segment's network, there are excitatory interneurons (E) that provide ipsilateral excitatory input. Furthermore, there are two classes of inhibitory interneurons (L and C). The C interneurons inhibit the contralateral (other side) E, C, and L neurons, to ensure that when one side of the network is active, the muscles on the other side are inactive. These types of interneurons crossing the midline of the spinal cord are necessary to maintain the left–right coordination of the locomotion. The E and C interneurons provide input to the motor neurons (M). The L neurons inhibit the C neurons in the same side, removing the inhibition to the contralateral side of the network, which thus allows that side of the segment to become active. Connections from the reticulospinal (R) neurons in the brainstem provide feedforward excitatory input to neurons on the same side of the central pattern generator unit. The reticulospinal system is the main pathway between the brain and the spinal cord. This pathway

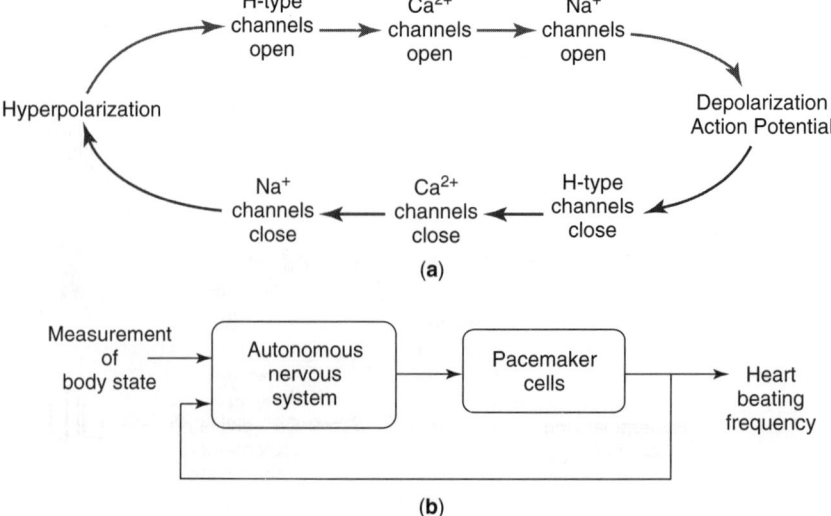

Figure 6. Examples of neural cascades generating synaptic-independent firing. (a) A series of channel openings and closings mediates cyclic-burst firing in thalamocortical relay neurons. (b) Neural signals from the body, setting the desired rhythm of pacemaker firing are compared with the actual rhythm in a feedback loop of the autonomic nervous system to set the correct firing. (This figure is available in full color at http://www.mrw.interscience.wiley.com/ebe.)

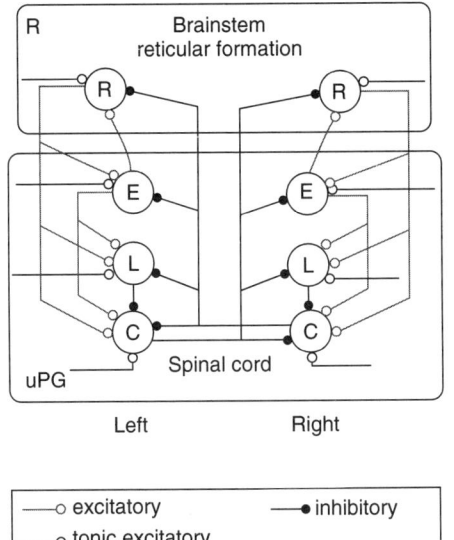
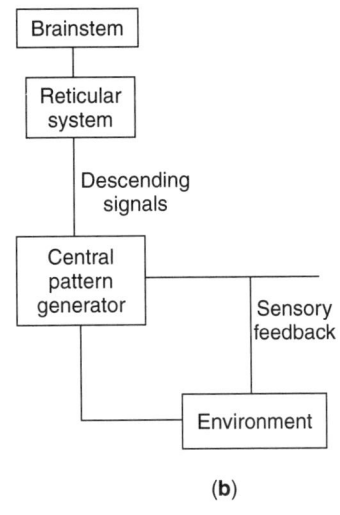

Figure 7. (a) Neural network model of the Lamprey's locomotion system. The symbol uPG represents "unit Pattern Generator" (the blue rectangle), R the reticular neurons, E the excitatory neurons, L the lateral inhibitory interneurons, and C the crossed inhibitory interneurons. (b) Schematic of the Lamprey's locomotion system including sensory feedback. (This figure is available in full color at http://www.mrw.interscience.wiley.com/ebe.)

participates in the control of the speed of the lamprey's locomotion.

Sensory feedback is not as relevant in the lamprey as in other animals, except in the case of swimming in cross current. Then sensory feedback is necessary to counteract the effects of the current. A schematic representation of this locomotion system including the sensory feedback is shown in Fig. 7b.

Jung et al. discussed a network model that simulates the dynamic behavior of the swimming pattern of the lamprey (29). This model defined a unit pattern generator (uPG) composed by interneurons E, L, and C (Fig. 7a). Moreover, the model posited interactions with the reticular system defined by neurons R as shown in Fig. 7a. Each neuron was defined with basic properties as a living neuron and was assumed to receive tonic synaptic drive and weighted input from other connecting neurons. The output function of each neuron as a function of voltage was a static sigmoidal function, with points of threshold and saturation. A set of initial parameter values according to findings of previous studies was used for the analysis of the behavior of this neural network. The Jung et al. study investigated the effects of changing the tonic drives to different neurons of the uPG (29). They also probed variations of the inhibitory and excitatory interconnection strengths among these neurons. Finally, they investigated the effects of varying feedforward and feedback interconnection strengths between uPG and R neurons.

The mathematical formulation of the Jung et al. model begins with each neuron having a potential that is a function of the resting conductance and of synapses received from other neurons (29). The time derivative of this membrane potential is given by

$$\frac{dv_i}{dt} = G_R^i(V_R^i - v_i) + G_T^i(V_T^i - v_i) + \sum_j G_{ji} h(v_j)(V_{syn}^j - v_i), \quad (5)$$

where the index i labels the ith neuron, v_i is the potential, G_R^i is the resting conductance, V_R^i is the resting potential, G_T^i is the tonic excitatory synaptic conductance, V_T^i is its reversal potential, G_{ji} is the maximal synaptic conductance (connection strength) from neuron j to neuron i, V_{syn}^j is the corresponding synaptic reversal potential, and h is a sigmoidal function of potential. According to h, when a neuron's potential is higher than the threshold, the neuron affects neurons connected to it. The neuron's output is given by

$$h(v) = \begin{cases} -20v^7 + 70v^6 - 84v^5 + 35v^4 & for\ 0 \leq v \leq 1 \\ 0 & for\ v < 0 \\ 1 & for\ v > 1, \end{cases} \quad (6)$$

such that $h(v)$ is uniquely determined by the following conditions:

$$h(0) = h'(0) = h''(0) = h'''(0)$$
$$h(1) = 1 \quad (7)$$
$$h'(1) = h''(1) = h'''(1) = 0.$$

A value of 0 in the output function means that the neuron's voltage is below threshold, whereas a value of 1 means that the neuron has reached saturation.

Jung et al. performed a study of the behavior of the network model with bifurcation analysis. It revealed a complex dependence on parameters. Several types of bifurcation took place, as different parameters of the system were varied. A bifurcation is a point where a change in the number and stability of fixed points and limit cycles occurs. Initially, Jung et al. studied only the responses of the uPG neurons without considering the feedforward and feedback connections among uPG and R neurons (29). This process was similar to looking at the responses of the isolated spinal cord *in vitro* under different conditions. Later, the effects of varying the connections between R neurons and the spinal uPG were studied with the network model. Interestingly, this model by itself with a set of initial parameters for different conductances and resting membrane voltages causes the membrane voltage of each neuron to oscillate periodically. The phase relationships between the different neurons are similar to those encountered experimentally. Following Jung et al. we now describe the results of the analysis of the neural network for the following two cases:

1. *The Isolated uPG Network.* The uPG network by itself without considering the R neurons can have different stable states, with fixed points or limit cycles. The switch between these regimes is a function of the tonic drive to each neuron (G_T^i). Considering tonic drive applied only to neuron C, different dynamic behaviors were found, with a one-parameter bifurcation analysis. Four different states occurred: states of stable and unstable fixed points, and stable and unstable limit cycles. Sometimes two stable states were possible in the same range as shown in Fig. 8a. These states occurred while tonic drives to neurons E and L were at their default values. When the tonic drives to these neurons also changed, different stable states and limit cycles occurred.

   In another range of tonic drive to the C neuron, a stable behavior of oscillatory membrane voltage of the left and right C neurons is possible (Fig. 8b). Here, the voltages in neurons in the left and right have the same waveform but opposite phase. Such opposite-phase oscillation is what the animal needs for locomotion. Changing the tonic drive to two neurons at a time while leaving the other parameters at default values was also considered. While maintaining E at its default value, larger tonic drive to C than to L interneurons or only tonic drive to C interneurons is sufficient for opposite-phase stable oscillations. But a drive too high to the C or L neuron can prevent them. To obtain oscillations with L at its default value, one must apply a minimum tonic drive to E and C neurons. However, a drive too high to either E or C causes loss of oscillations. By changing the tonic drive for L and E, one shows that tonic drive to neuron E is not necessary for stable oscillations. Nevertheless, if its drive is not balanced with the tonic drive to neuron L, no oscillations occur.

   Jung et al. also studied the interconnections between neurons in the uPG network. The roles of ipsi- and contralateral inhibitory connections were studied by changing the value of the synaptic conductance G_{ji}. The inhibitory effect of the L neurons in the activity of C neurons is important to obtain cyclic activity. Changing different parameter values shows that ipsi- and contralateral inhibitions are essential for obtaining stable oscillatory output. In addition, Jung et al. found different possible states from excitatory interconnections by changing synaptic conductances. One can also reach different states with different inhibition from C neurons.

2. *The RN-uPG Network.* Bifurcation analysis was also performed to determine how feedforward and feedback interactions of uPG and R neurons alter the dynamic behavior of the system. Feedforward gain refers to the maximal synaptic conductance of the R-to-uPG connections. Feedback gain refers to the maximal conductance from uPG to R. Neurons R receive both excitatory positive feedback from neurons E and inhibitory negative feedback from neurons C. Strong feedforward annihilates the oscillation, but this can be prevented up to a point (critical value of the feedforward) by increasing the negative feedback. The frequency of oscillation changes with increments of the feedforward value. For a fixed value of feedback, increases in feedforward strength can increase or decrease oscillatory frequency. At different values of feedback, the pattern can be different, which accounts for experimental observations, where frequency of swimming movements varies with signals from the brainstem.

5. DETAILED EXAMPLE 3: SENSORY ADAPTATION

As described in the section on control mechanisms in the nervous system, one of the most important properties of biological sensory systems is adaptation. This property is ubiquitous to them (30,31). It allows them to adjust to variations in the environment and thus to deal better with it. In that section, we pointed out that sensory adaptation might happen synaptically (as in the visual cortex). We also pointed out that it might happen biophysically at the level of single cells (as in photoreceptors). Hence, one may use biophysical or neural-network modeling techniques as in the last two sections to model sensory adaptation. However, it can also be modeled in a third way, which is introduced in this section to expand the examples presented in this article.

Because sensory adaptation is a form of neural adaptive control, one can conceptualize it as in Fig. 3a. Here,

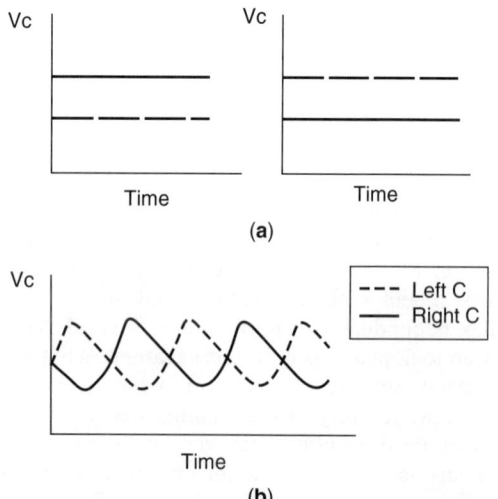

Figure 8. Membrane voltages for interneurons C with different tonic drives. (a) Example of bistable behavior. (b) Example of bistable limit cycles necessary for locomotion.

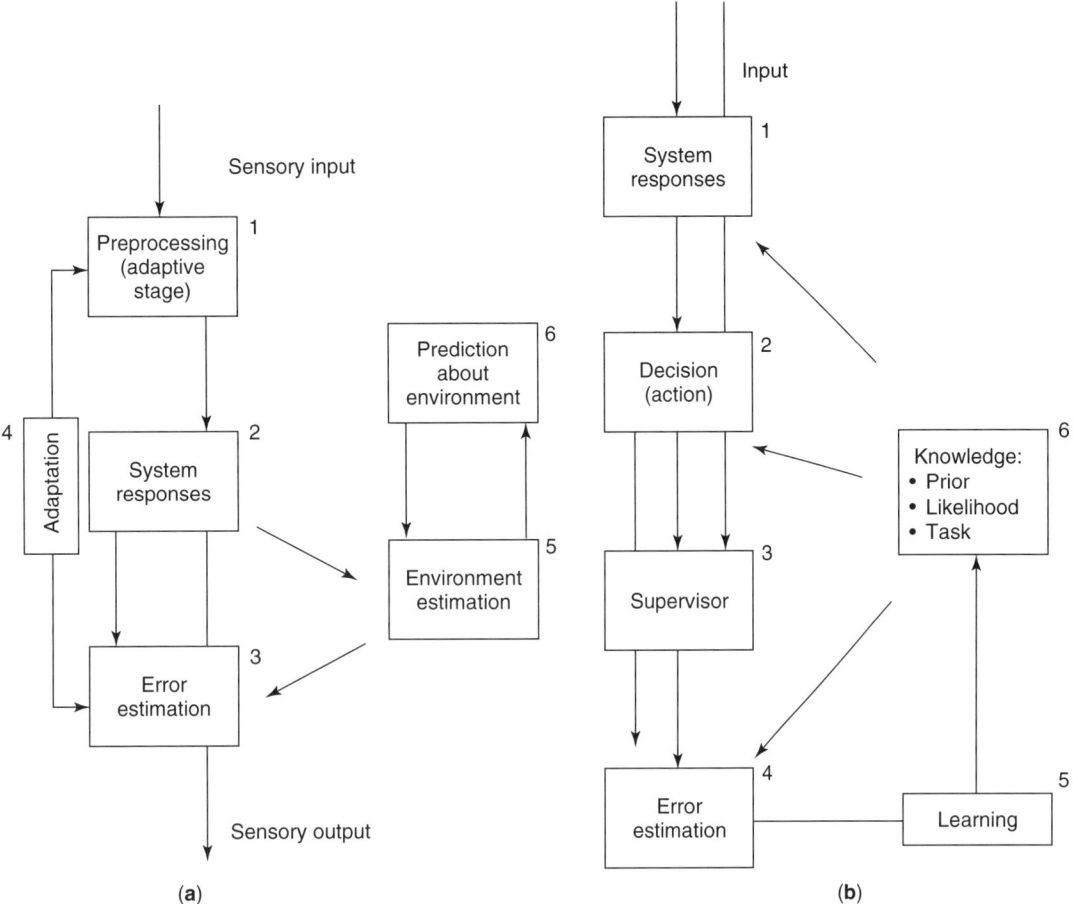

Figure 9. Frameworks for (a) sensory adaptation and (b) supervised learning.

we modify and generalize that figure to sketch modern terminology of statistical models of adaptation. The result is Fig. 9a, which is a version of the model-predictive control shown in Fig. 3c (32,33). In Fig. 9a, the plant is as simple as in Fig. 3a (Box 2), but the neural controller contains several explicit subcomponents (Boxes 1, 3, 4, 5, and 6). The sensory input is processed by an adaptive stage (Box 1), whose output is processed even more to yield the system responses (Box 2). The system then estimates with how much error these responses code important attributes from the environment (Box 3). Next, the system adapts the preprocessing stage to minimize this error (Box 4). To estimate the error, the system needs to have knowledge about the environment. The trouble is that, as the environment changes, its parameters must be estimated for the system to know what knowledge to apply. In this model, the system estimates them from its responses (Box 5) and from predictions that it makes about the environment based on past responses (Box 6). If current responses are statistically consistent with these predictions, then the estimates do not change. Otherwise, they change, but they do so slowly to take into account the tendency of the environment to remain stable.

To formalize the model in Fig. 9a, we use modifications of the notations introduced by Grzywacz and Balboa (32) and by Burgi et al. (34). Vectors are denoted by bold symbols, time t_k is denoted by subscript k, and time series from t_0 on are denoted by an inverted hat accent. In other words, we define $\check{\mathbf{z}}_k = \{\mathbf{z}_k, \mathbf{z}_{k-1}, \ldots, \mathbf{z}_1\}$. Let $\mathbf{I}$ be the input to the sensory system and $\mathbf{H}^1(\mathbf{I}), \mathbf{H}^2(\mathbf{I}), \ldots, \mathbf{H}^N(\mathbf{I})$ be the N relevant task-input attributes to extract from $\mathbf{I}$. For instance, these attributes could be things like contrasts and positions of occluding borders in a visual image. The input is first preprocessed by an adaptive stage (Box 1 in Fig. 9a), whose output $\mathbf{O}$ is transformed by the task-coding stage (Box 2) to $\mathbf{R}^{H1}(\mathbf{O}), \mathbf{R}^{H2}(\mathbf{O}), \ldots, \mathbf{R}^{HN}(\mathbf{O})$. These functions are estimates of the values of the task-input attributes, and one can interpret them physiologically as the responses of the system. The system tries to make the error in each of these estimates small. This error is the discrepancy between $\mathbf{H}^i(\mathbf{I})$ and $\mathbf{R}^{Hi}(\mathbf{O})$. The system cannot know exactly what this discrepancy is, because it does not have access to the input, only to its estimates. However, the system can estimate the expected amount of error (Box 3). Grzywacz and Balboa (32) showed that the mean Bayesian expected loss over all possible responses is

$$E_i(\mathbf{A}, \Lambda) = \int_I \int_{\mathbf{R}^{Hi}} P(\mathbf{R}^{Hi}|\mathbf{I}, \mathbf{A}) P(\mathbf{I}|\Lambda) L(\mathbf{I}, \mathbf{R}^{Hi} : \Lambda, \mathbf{A}), \quad (8)$$

where $\mathbf{A}$ are the adaptation parameters of the adaptive stage, Λ are the hyperparameters of the prior distribution that the system selected, and L is the loss function, which measures the cost of deciding that the ith attribute is $\mathbf{R}^{Hi}$

given that the input is I. One can give intuitive and practical interpretations of the probability terms in this equation. The first term, the likelihood function $P(\boldsymbol{R}^{Hi}|\boldsymbol{I},\boldsymbol{A})$, embodies the knowledge about how sensory mechanisms encode the input. In other words, $P(\boldsymbol{R}^{Hi}|\boldsymbol{I},\boldsymbol{A})$ tells how the system responds when the stimulus is I and the adaptation parameters are $\boldsymbol{A}$. This is not an arbitrarily free probabilistic function, but it has strong constraints from the underlying neural system. Such probabilistic dependence on I and $\boldsymbol{A}$ is already present in the $\boldsymbol{O}$ variables, which then pass it to $\boldsymbol{R}^{Hi}$. The second term, $P(\boldsymbol{I}|\Lambda)$, is the prior distribution of the input when the environment is Λ.

Because the goal in the model is to estimate N attributes, the total error is

$$E(\boldsymbol{A}, \Lambda) = \sum_{i=1}^{N} E_i(\boldsymbol{A}, \Lambda). \quad (9)$$

A way to find $\boldsymbol{A}$ and Λ would be to find the values that minimize $E(\boldsymbol{A},\Lambda)$. However, this way would fail to recognize the knowledge that the environment does not change or it tends to change slowly. In other words, if we know what Λ was a short while ago, then it is highly likely that Λ did not change. Hence, we propose that the system estimates Λ from past responses (Box 5 in Fig. 9a). With Λ given, then we can estimate $\boldsymbol{A}$ (Box 4) as

$$\boldsymbol{A}(\Lambda) = \arg\min_{\boldsymbol{A}^*} E(\boldsymbol{A}^*, \Lambda). \quad (10)$$

In other words, $\boldsymbol{A}$ is the adaptation argument that minimizes the error for a fixed value of Λ.

The goal is thus to estimate Λ from past responses. Moreover, responses are related to the environment through the adaptation state of the system. Therefore, one must in general take into account the past adaptation states to estimate Λ. The most general way to relate Λ to past responses and adaptation states is to calculate $P(\Lambda_k|\breve{\boldsymbol{R}}_k^{Hi}, \breve{\boldsymbol{A}}_k)$. Using the Bayes rule and a few algebraic manipulations (34,35), we get

$$P(\Lambda_k|\breve{\boldsymbol{R}}_k^{Hi}, \breve{\boldsymbol{A}}_k) = \frac{P(\boldsymbol{R}_k^{Hi}, \boldsymbol{A}_k|\Lambda_k, \breve{\boldsymbol{R}}_{k-1}^{Hi}, \breve{\boldsymbol{A}}_{k-1})}{P(\boldsymbol{R}_k^{Hi}, \boldsymbol{A}_k|\breve{\boldsymbol{R}}_{k-1}^{Hi}, \breve{\boldsymbol{A}}_{k-1})} P(\Lambda_k|\breve{\boldsymbol{R}}_{k-1}^{Hi}, \breve{\boldsymbol{A}}_{k-1}). \quad (11)$$

One way to use this equation to get Λ_k is to set it from maximum *a posteriori* estimation:

$$\Lambda_k(\breve{\boldsymbol{R}}_k^{Hi}, \breve{\boldsymbol{A}}_k) = Arg\max_{\Lambda_k^*} P(\Lambda_k^*|\breve{\boldsymbol{R}}_k^{Hi}, \breve{\boldsymbol{A}}_k). \quad (12)$$

However, other estimations, with different loss functions are possible (36).

Equation 11 can be understood as a form of Kalman filtering. The rightmost term represents the *prediction stage*, in which we find the likely values of the current environment as predicted by the past measurements (Box 6 in Fig. 9a). In turn, the numerator in the division represents the *measurement stage*. The new measurements are combined using the Bayes theorem (arrow from Box 2 to Box 5) and, if consistent, reinforce the prediction (arrow from Box 6 to Box 5) and decrease the uncertainty about the current environment; inconsistent measurements may increase the uncertainty. That $\boldsymbol{A}$ appears as an argument of the probability function of the measurement stage is a mathematical consequence of the Bayes theorem. However, Grzywacz and de Juan (33) argue that $\boldsymbol{A}$ is not a random variable, which simplifies Equation 11. (Finally, the denominator is just a normalizing term.)

The measurement stage of Equation 11 relates the environment to the responses. Consequently, for a sensory system, this stage depends on the underlying biological mechanisms. In turn, the prediction stage of Equation 11 must somehow take into account that environments are often stable or do not change rapidly. The predictive distribution function can be expressed as (34,35)

$$P(\Lambda_k|\breve{\boldsymbol{R}}_{k-1}^{Hi}, \breve{\boldsymbol{A}}_{k-1}) = \int_{\Lambda_{k-1}} P(\Lambda_k|\Lambda_{k-1}) P(\Lambda_{k-1}|\breve{\boldsymbol{R}}_{k-1}^{Hi}, \breve{\boldsymbol{A}}_{k-1}). \quad (13)$$

The importance of this equation is that its rightmost term is identical to the leftmost term in Equation 11, except for being one step back in time. Hence, if one knows how to model the measurement stage of Equation 11 and one knows the first term of the integral of Equation 13, one can solve both equations recursively. The first term of the integral of Equation 13, is the knowledge of how slowly the environment changes. It is, therefore, a second prior term necessary in adapting systems. The first prior term, $P(\boldsymbol{I}|\Lambda)$, appeared in Equation 8. Incidentally, as we now treat the environment as a function of time, it is necessary to rewrite Equations 8, 9, and 10, as

$$E_i(\boldsymbol{A}_k, \Lambda_k) = \int_I \int_{\boldsymbol{R}^{Hi}} P(\boldsymbol{R}^{Hi}|\boldsymbol{I}, \boldsymbol{A}_k) P(\boldsymbol{I}|\Lambda_k) L(\boldsymbol{I}, \boldsymbol{R}^{Hi} : \Lambda_k, \boldsymbol{A}_k), \quad (14)$$

$$E(\boldsymbol{A}_k, \Lambda_k) = \sum_{i=1}^{N} E_i(\boldsymbol{A}_k, \Lambda_k), \quad (15)$$

and

$$\boldsymbol{A}_k(\Lambda_k) = \arg\min_{\boldsymbol{A}_k^*} E(\boldsymbol{A}_k^*, \Lambda_k). \quad (16)$$

Recent papers provided specific examples of how to apply these equations to model different kinds of retinal adaptation (32,33,37,38). The power of these examples was that they accounted for surprising features of the data. One example was an explanation for the behavior of the extent of the lateral inhibition mediated by retinal horizontal cells as a function of background light intensity. The example explained both the fall of the extent from intermediate to high intensities and its rise from dim to intermediate intensities. Another example was an application of the Kalman-filtering framework to retinal contrast adaptation. It was shown that this application could account for surprising features of the data. For example, it accounted for the differences in responses to increases and decreases of mean contrasts in the environment. In addition, it accounted for the two-phase decay of contrast gain when the mean contrast in the environment rose suddenly.

6. DETAILED EXAMPLE 4: LEARNING

Learning and sensory adaptation have many common functional properties. In both cases, the nervous system changes when exposed to novel ensembles of inputs. Furthermore, as a comparison of Fig. 3a and b shows, changes in both processes occur as an error-correction process in the neural controller. Consequently, it is not surprising that models like those described in the preceding section do a good job in accounting for properties of learning. Extending the models of the last section to learning is the main subject of this section. However, it is important to point out that these models became dominant in learning theory since the 1990s, and another important class of models was dominant beforehand and is still influential. The origin of that class was the perceptron, an artificial network of neurons capable of simple pattern-recognition and classification tasks (39). It was composed of three neural layers, where signals only passed forward from nodes in the input layer to nodes in the hidden layer and finally out to the output layer. There were no connections within any layer of the perceptron. Its main limitation was to form only linear discriminate functions, that is, classes that could be divided by a line or hyperplane (40). This limitation disappeared with the advent of the backpropagation method (41,42) and the multilayer perceptron (43). With the backpropagation method, for instance, the perceptron neural network could be trained to produce good responses to a set of input patterns, which rekindled the interest in neural networks. Other work on them included, for instance, Boltzmann machines (44,45), Hopfield networks (46), competitive-learning models (45,47), multilayer networks (43), and adaptive resonance theory models (48). Neural networks were also intriguing to the neurophysiologist, because their main cellular mechanism, namely, synaptic plasticity, occurs during learning in the brain (Example 6 of the section on control mechanisms in the nervous system).

However, in the last 15 years, an approach to learning similar to the models described for sensory adaptation in the last section began gaining rapidly in popularity. The new approach is the so-called statistical learning theory in its simplest form. However, it accepts a Bayesian generalization, which, for simplicity, will be what we will refer to as statistical learning theory here. One reason that this theory is gaining in popularity is that despite important achievements in some specific applications with neural networks (49,50), they have at least five problems: First, the theoretical results obtained did not contribute much to general learning theory. Second, no new learning phenomena were found in experiments with neural networks (51). Third, procedures like backpropagation are sensitive to local minima of the learning cost function and there is no fundamental way to avoid them. Fourth, the convergence of the gradient-based method is slow. Fifth, the sigmoid function used in backpropagation and other neural networks has a scaling factor that affects the quality of the learning, but there is no fundamental way to choose this factor. In other words, neural networks are not efficient and not well-controlled learning machines. In contrast, statistical learning theory does not suffer from these problems, because it has a solid foundation (51). Furthermore, one can find instantiations of statistical learning theory that are equivalent to neural networks (52,53). In other words, statistical learning theory is better and more general than neural-network models.

The rest of this section is devoted to explaining the key ideas in statistical learning theory. Its cornerstone is that learning is a problem of function approximation, a notion explained in Fig. 10. Here, we use this figure in the context of supervised learning (learning with a supervisor or a strong critic, Fig. 3b), but we extend the arguments to unsupervised learning later. Suppose that a neural structure receives an input x and must use it to perform a task $f(x)$ (the red line on the figure). Suppose also that this structure does not know in the beginning how to perform the task, but it can learn it. For example, to learn, the neural structure may get random examples from the possible inputs. Then a critic or a teacher (e.g., another neural structure) would tell our neural structure what are the appropriate task outputs (green open circles). The problem is to find an approximation to the task function (blue line) that can generalize. In other words, this approximation must predict good task outputs when the neural structure receives novel inputs (points different from the green open circles). As Fig. 10 shows, if the number of examples is small and the task is complex, the approximation is

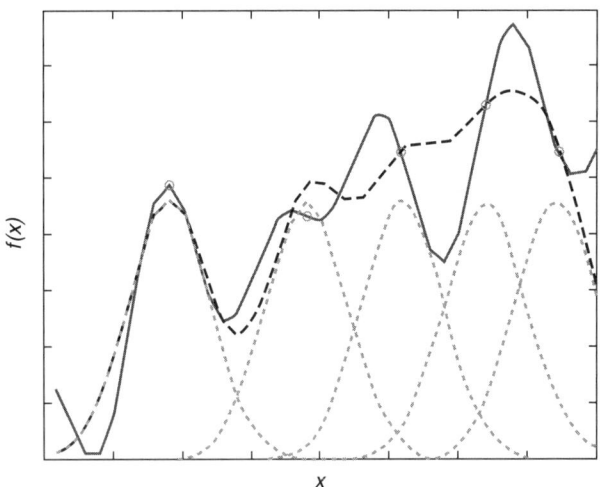

Figure 10. Supervised learning as a problem of function approximation. The variable x represents the input, whereas $f(x)$ is a task that the system must perform (red line). Green circles are examples of correct tasks provided to the system by a supervisor. Learning would be to find a function that approximates the correct performance of the task, e.g., the blue line. The approximated function may match the supervisor's examples exactly (three right circles) or may match them approximately if there is reason to suspect the supervisor's reliability (two left circles). If the number of examples is small, as in this figure, the approximation is crude (compare blue and red curves), but it gets better with more examples. There are several methods to achieve the approximation. One method is to sum estimator functions (called Parzen estimators) centered on the examples and whose width fall as the number of examples increases. In this figure, we used weighted Gaussian estimators (dashed green lines). Weights were found to minimize the difference between the blue and the red curves. (This figure is available in full color at http://www.mrw.interscience.wiley.com/ebe.)

crude. However, the approximation improves with the number of examples. Much work in statistical learning theory addresses the rate of improvement and addresses what approximations converge rapidly to the desired function.

From Fig. 10, one can understand why a mathematical description like that in Equation 8 is a good starting point for statistical learning theory. Its approximations involve choosing from a given set of functions $f(x,\alpha)$, the one that best approximates the supervisor's response. In other words, approximations involve the choice of parameters α, as we did in Equations 8–10. Moreover, to quantify which parameters are best, one must define and minimize the expected loss as in Equation 8. The necessity of working with expected loss to model learning originates because the input comes from a random distribution, which is expressed in Equation 8 as $P(I|A)$. However, learning is more general than sensory adaptation for two reasons: First, learning may involve a supervisor to provide information to the system (Fig. 9b). Second, in learning, one may have to find the optimal arbitrary action (or decision rule). This is different from what Equation 8 expresses, because in that case, the system is only estimating input attributes not arbitrary actions.

We use Fig. 9b to show how the supervisor and the learning of the decision rule generalize the model of adaptation. As in Fig. 9a, Fig. 9b has a plant as simple as in Fig. 3b (Box 2), but a neural controller with several explicit subcomponents (Boxes 1, 4, 5, and 6). In addition, Fig. 9b has a supervisor (Box 3), as indicated by Fig. 3b. Learning begins with the input being processed, which yields neural responses (Box 1). With these responses and with the knowledge of how to perform the task (Box 6), the system can make a decision (Box 2). This decision is then sent to the supervisor along with a copy of the input, and the supervisor provides a critique (Box 3). The system uses this critique to estimate the expected error in the decision (Box 4). To do so, the error-estimating stage inspects both the critique and the decision. Moreover, the estimation of expected error applies knowledge about how to perform the task, about the prior (the distribution of inputs), and about the likelihood function (the encoding process and its noise—Box 6). Finally, the system modifies (learns) this knowledge to minimize the expected error (bottom-leftward arrow to Box 5). Learning may also involve modifying the decision rule and the encoding process (middle and top leftward arrows, respectively).

The generalization of Equations 8–10, for the problem of supervised learning based on Fig. 9b is

$$E(\boldsymbol{G},\boldsymbol{C},\boldsymbol{D}) = \int_I \int_R \int_S P(\boldsymbol{S}|\boldsymbol{I},d(\boldsymbol{R}:\boldsymbol{D}))P(\boldsymbol{R}|\boldsymbol{I},\boldsymbol{G})P(\boldsymbol{I}|\boldsymbol{C}) \quad (17)$$
$$\times L(\boldsymbol{S},\boldsymbol{I},d(\boldsymbol{R}:\boldsymbol{D}))$$

and

$$(\boldsymbol{G},\boldsymbol{C},\boldsymbol{D}) = \mathrm{argmin}_{G^*,C^*,D^*} E(\boldsymbol{G}^*,\boldsymbol{C}^*,\boldsymbol{D}^*), \quad (18)$$

where $\boldsymbol{G}$, $\boldsymbol{C}$, and $\boldsymbol{D}$ are sets of parameters to be learned, $\boldsymbol{S}$ is the input of the supervisor, and $d(\boldsymbol{R}:\boldsymbol{D})$ is the decision rule.

We now list the seven assumptions underlying these equations and highlight their differences and similarities with other models of supervised learning.

1. The prior distribution may be learned, as indicated by parameter $\boldsymbol{C}$ in the distribution of $\boldsymbol{I}$; see $P(\boldsymbol{I}|\boldsymbol{C})$ in Equation 17. Although other models learn the prior as in this equation, not all do. A system may choose not to learn the prior to be consistent with Vapnik's restricted-information principle (51). This principle states that "when solving a given problem, try to avoid solving a more general problem as an intermediate step" (51). Below, we consider three different strategies of learning, one that obeys this principle (strategy A), and two that learn the prior (strategies B and C). Only strategy A is standard statistical learning theory (51), whereas the other two are part of its Bayesian generalization. The reason why we consider alternative strategies is that many interesting forms of learning in the brain do not have access to $\boldsymbol{I}$. For instance, suppose that one wants to learn to score a point in a basketball game. The visual input is processed by the retina and lateral geniculate nucleus before reaching the learning centers in the cortex (2,54). This situation is similar to that in Fig. 9b, where a preprocessing stage intervenes between the input and the learning center. Hence, in this model, the signal available to the learning center is $\boldsymbol{R}$ not $\boldsymbol{I}$. Nevertheless, mechanisms that have direct access to $\boldsymbol{I}$ are possible in the brain. They constitute particular cases of the model in Fig. 9b, because its Box 1 can be the identity operator.

2. The system may try to learn the likelihood function, as embodied by parameter $\boldsymbol{G}$ in term $P(\boldsymbol{R}|\boldsymbol{I},\boldsymbol{G})$ of Equation 17. This form of learning is different from learning the prior, because the latter represents variables external to the system. Therefore, when learning the prior, the system tries to obtain information about something not known *a priori*. In contrast, the system knows the family of functions $P(\boldsymbol{R}|\boldsymbol{I},\boldsymbol{G})$, because it is part of the system. Consequently, learning these functions is fine-tuning the encoding, so that $P(\boldsymbol{R}|\boldsymbol{I},\boldsymbol{G})$ becomes optimal for the task and inputs. Such fine-tuning is automatic if the prior is known, because the system can find the optimal likelihood function (and decision rule) by directly minimizing Equation 17. In general, learning the likelihood function is like the sensory adaptation described in the preceding section (parameter $\boldsymbol{A}$ in Equation 8). The only difference is that learning the likelihood function has an effect that is more permanent. This is why we use a different parameter set here ($\boldsymbol{G}$). Importantly, this set does not control a free family of functions, but they do have strong constraints from the neural system. Such constraints often ensure that the learning process converges.

As pointed out in assumption 1, we consider below three different strategies of learning. Strategy A obeys Vapnik's restricted-information principle, whereas strategies B and C learn the likelihood function and assume it known, respectively. The former strategy does not try to learn the likelihood function. In turn, assuming a known likelihood function is akin to saying that learning is performed with a fixed, known preprocessing of the input (Box 1 in Fig. 9b). To know this processing, the system must learn it beforehand, probably through development or evolution.

3. The optimal decision rule (action) is possibly learned, as indicated by its parameter D; see $d(R:D)$ in Equation 17. The mechanism for learning the decision rule is fundamentally different from those for learning the prior and the likelihood function. The system may learn the former without knowing the latter (see assumptions 1 and 2), but the system must always specify a decision rule. However, learning it has an important similarity with learning the likelihood function. If the prior is known, then the setting of the optimal decision rule is automatic (assumption 2). Consequently, it only makes sense to learn the decision rule (or the likelihood function), when the prior is not known. Strategy A allows for the learning of decision rules without knowing the priors. In turn, strategy B learns priors and decision rules (and possibly likelihood functions) simultaneously. Another situation in which the system does not need to learn the decision rule occurs when it is obvious (strategy C). For instance, if a task is to determine which of two inputs is stronger, often the best decision rule could simply be to pick the input generating the larger response.

4. The supervisor is possibly probabilistic ($P(S|I,d(R:D))$), which is done to indicate that the supervisor may possibly be somewhat unreliable. Figure 10 illustrates a consequence of suspecting the supervisor to be unreliable. The learned task (blue line) may not pass exactly through the data points provided by the supervisor (green circles). However, to make sense, a supervisor cannot be too unreliable and must be less unreliable than the system-responses stage in Fig. 9b. Assuming an unreliable supervisor is not necessary, because its function can be written deterministically, as for instance, $P(S|I,d(R:D)) = \delta(S = S(I,d(R:D)))$. Strategy A for solving Equations 17 and 18 does not require knowing the supervisor function. In turn, strategies B and C assume it known. We justify these latter strategies by the typical approach of psychophysical experiments, which informs the subject what the supervision input is. For instance, the supervisor will provide a tone only if the subject's decision is the correct interpretation of the input.

5. The supervisor bases its critique on both the inputs (I) and the decisions ($d(R:D)$); see $P(S|I,d(R:D))$ in Equation 17. This is different from many models of learning, in which the supervisor computes its critiques from the inputs alone (51). Those models are reasonable, because the supervisor may provide direct information about what the input is. Furthermore, the analysis of the learning problem is simpler if the supervisor only uses the inputs. However, many types of supervised learning are ignored if one does not consider the decision. For instance, take a subject making visual-size judgments and the supervisor only providing "too-large" or "too-small" answers. It must wait for the judgments before critiquing. Finally, another reason to use a supervisor that also considers decisions is that, as in assumption 1, only input-dependent supervisors are particular cases of the model expressed in Equation 17, because $P(S|I,d(R:D))$ includes the possibility $P(S|I)$).

6. The decision rule depends on responses (R); see $d(R:D)$ in Equation 17. Other models of learning typically assume decision rules that depend on the inputs (I – (51)). Although these types of rule are easier to analyze, we choose the form $d(R:D)$ for the reasons stated in the discussion of assumption 1.

7. The general form of the loss function in supervised learning depends on both inputs and supervisor's critiques; that is, $L = L(S,I,d(R:D))$. This process allows the learning to depend on just the supervisor's output ($L(S,I,d(R:D)) = L(S,d(R:D))$), or both on it and the inputs. An example of a system that learns just based on the supervisor's output is one that tries to minimize the amount of negative critique received. In turn, a sports player being coached has to learn both from the supervisor and from the result of actions. For instance, a coach may teach the player the right technique to shoot a basketball. However, the latter will only learn completely after shooting several balls and seeing if they score.

Having completed the setting of supervised learning, we briefly turn our attention to unsupervised learning. From Equation 17, one can see that unsupervised learning is a particular case of the general supervised-learning formulation. If one does not have a supervisor, then $P(S|I,d(R:D))$ is uninformative; that is, it is a constant. Furthermore, the loss function cannot depend on S; that is, $L(S,I,d(R:D)) = L(I,d(R:D))$. (Therefore, learning cannot be done by supervision but through consequences of errors of action. An example of such a consequence is a baby tumbling when trying to learn how to walk.) Without the supervisor and with such a loss function, Equation 17 turns into

$$E(G,C,D) = \int_I \int_R P(R|I,G)P(I|C)L(I,d(R:D)). \quad (19)$$

An extreme form of unsupervised learning occurs when the system receives inputs passively, i.e., without reacting. Because it does not react, there are no actions with each to err. Nevertheless, learning can occur in such situations, such as when the brain fine-tunes during development under abnormal-rearing conditions (2,45,47). That there is no action does not mean that we cannot model such learning with an action-error loss function as in Equation 19. Such learning must be thought of as preparing the system for future action, and thus, one must consider it with an appropriate loss function. For instance, if passively viewing a display with mostly horizontal line segments for several experimental sessions, the brain may ready itself for orientation-discrimination tasks performed with nearly horizontal segments. Positing an orientation-discrimination loss function would force the brain to learn the orientation prior and the orientation likelihood function.

How does one minimize Equations 17 and 19 to implement learning? As explained in the discussions of assumptions 2 and 3 after Equation 18, only certain kinds of learning make sense. For instance, learning the likelihood function only makes sense when the system does not know $P(I|C)$. Similarly, one can compute the decision rule automatically if one knows $P(I|C)$. Only the prior cannot be computed automatically if one knows the other two functions. Of these latter functions, we will show below that one can try to learn the decision rule but not the likelihood function without paying attention to the prior (invoking Vapnik's restricted-information principle—assumption 1). Alternatively, one can try to learn the various unknown functions simultaneously. The following three strategies (algorithms) for minimization of Equations 17 and 19 are the simplest ones to address the sensible kinds of learning.

A. *Learning the Decision Rule Without Attempting to Learn the Prior.* To see how it is possible to learn the decision rule without knowledge of the prior, it is useful to consider the general setting of the learning problem (51). Define the probability density function $P(z)$ on space Z. Consider the set of functions $Q(z, \alpha)$, where α is a set of parameters. The goal of learning is to minimize the risk functional

$$E(\alpha) = \int_z P(z)Q(z:\alpha). \qquad (20)$$

What makes the minimization of this equation a learning problem is that we do not know $P(z)$. If we knew it, then the minimization would not require examples as in Fig. 10. Instead of $P(z)$, the system is given a sample of independent examples $\{z_1, \ldots, z_k\}$. The simplest way to minimize Equation 20 with an unknown $P(z)$ is to use the so-called empirical-risk minimization (ERM) principle. In it, this equation is replaced by the empirical risk functional

$$E_{emp}(\alpha) = \frac{1}{k}\sum_{i=1}^{k} Q(z_i:\alpha). \qquad (21)$$

One then approximates the function $Q(z, \alpha_0)$ that minimizes Equation 20 by the function $Q(z, \alpha_k)$ that minimizes Equation 21. (Other principles that we can use instead of ERM are, for instance, structural-risk minimization and minimal-description length (51).)

How do we apply ERM (or a version of it) specifically to supervised learning as expressed in Equation 17? Equations 20 and 21 teach us that we must identify the pertinent sample of independent examples (z). Moreover, we must ascertain that they have an unknown distribution $P(z)$. This sample is as in Fig. 10, except that in our model, the learning structure does not have direct access to the input, namely, I (see assumption 1 after Equation 18). Instead, the system must work through a sample of its responses and of supervisor inputs, namely, $\{(R_1, S_1), \ldots, (R_k, S_k)\}$. This sample corresponds to z in Equations 20 and 21. In what sense are $P(R)$ and $P(S)$ unknown [as is $P(z)$]? Because in this strategy A, the prior, the likelihood function, or both are unknown, so are the distributions of I and R. Because the distribution of S depends on these variables, its distribution is also unavailable to the system. Consequently, we can adapt Equation 21 to the learning of the decision rule as follows:

$$E_{emp}(D) = \frac{1}{k}\sum_{i=1}^{k} L(S_i, d(R_i:D)) \qquad (22)$$

and

$$D = \mathrm{argmin}_{D^*} E_{emp}(D^*). \qquad (23)$$

Equation 22 assumes that the loss function depends on only the supervisor critiques not on the inputs (see the discussion of assumption 7), because the system has no access to I. Hence, strategy A attempts to minimize the magnitude of errors reported by the supervisor, without regard to chosen actions. A consequence of the inability of this strategy to take into account errors of action is that it is not applicable to unsupervised learning (the minimization of Equation 19).

Equations 22 and 23 cannot be adapted to the case of learning the likelihood function without attempting to learn the prior, because the likelihood function depends on I. Therefore, one cannot write an empirical-risk functional as in Equation 22, i.e., a functional for which one can measure all independent variables.

B. *Simultaneous Learning of the Decision Rule (or the Likelihood Function or Both) and the Prior.* This learning strategy is more complex than the preceding one, but it is more thorough, because it can take into account errors of action. Therefore, this strategy is also applicable to unsupervised learning. The complexity of the process follows from the dependence of the supervisor input on the system's decisions (assumption 5) and of the responses on the system's choices of parameter G. The trouble is that the optimal decision rule and the likelihood function at every instant depend on the sample up to that point. Consequently, one cannot find a single decision rule for the entire sample as in strategy A. Similarly, one cannot find a single likelihood function for the entire sample. In other words, the sample must be treated as a time series (as in Equation 11).

To represent the measurement sample in simultaneous learning, we use the notation for time series introduced in the section on detailed example 3. The pertinent sample is thus $\{(\check{R}_k, \check{S}_k, \check{G}_k, \check{C}_k, \check{D}_k)\} = \{(R_k, S_k, G_k, C_k, D_k), \ldots, (R_1, S_1, G_1, C_1, D_1)\}$. This sample resembles z in Equations 20 and 21, but it is different in including parameters G_i, C_i, and D_i. Parameter D_i is included, because it is necessary to interpret the supervisor's input. Similarly, we include parameter G_i to interpret the responses at different times. In addition, the sample includes the string of C_i to provide information about the system's estimates of the prior at every instant. (Strictly speaking, we may not need all these strings in the sample, because the decision rule or the likelihood function may be known.) In what sense is the distribution of this sample unknown [as is $P(z)$]? Strategy B assumes that the system does not know $P(I|C)$ and one or both of $d(R:D)$ and $P(R|I,G)$. Hence, the distribution of the variables (including S) and parameters in our sample are also unknown (see discussion of strategy A).

With these considerations, we can apply a version of ERM to learn the prior simultaneously with the decision rule, the likelihood function, or both. Here, we consider the general case of learning the three of them together, but learning pairs is similar and easier. Assume that the system knows the pertinent sample up to time $k-1$; i.e., the learning process is complete until then. Next, the system makes a new measurement R_k, which has probability $P(R_k|I_k, G_{k-1})$. This measurement must use parameter G at time $k-1$. With this measurement, the best decision by the system uses parameter D_{k-1}, that is, $d(R_k : D_{k-1})$. This decision is optimal, because without a new input from the supervisor, it is impossible to update the decision rule with strategy B. (At time 1, one may pick, for instance, parameters G_0 and D_0 at random, or based on some previous knowledge we may have about them.) If the system makes this decision, then the supervisor responds with input S_k, which has probability $P(S_k|I_k, d(R_k : D_{k-1}))$. Now, we can use the system's response and the supervisor's input to build a version of the empirical risk functional from Equation 17 as

$$E_{emp}(C_k) = \frac{1}{k} \sum_{i=1}^{k} \int_I P(S_i|I, d(R_i : D_{i-1})) P(R_i|I, G_{i-1})$$
$$\times P(I|C_k) L(S_i, I, d(R_i : D_{i-1})).$$
(24)

By minimizing this functional, we obtain the optimal C_k. Because we now have an optimal estimate of the prior at time k, we can estimate the optimal G_k and D_k automatically (assumptions 2 and 3). Consequently, we learn the prior [$P(I|C_k)$], likelihood function [$P(R|I, G_k)$], and optimal decision rule [$d(R : D_k)$] up to time k. With these functions in hand, one can proceed iteratively to later times.

To apply this strategy to unsupervised learning, one can adapt Equation 24 for Equation 19; i.e.,

$$E_{emp}(C_k) = \frac{1}{k} \sum_{i=1}^{k} \int_I P(R_i|I, G_{i-1}) P(I|C_k) L(I, d(R_i : D_{i-1})).$$
(25)

C. *Learning the Prior Without Attempting to Learn Both the Decision Rule and the Likelihood Function.* This strategy is only possible if the system knows the likelihood function. Because this function is not being learned, it cannot be disambiguated from the prior, because responses depend on both the prior and the likelihood function. With knowledge of this function, the system can learn the prior by first estimating $P(R|C)$ from $\{R_1, \ldots, R_k\}$ with the empirical distribution function (51). If we know $P(R|I)$, then we can estimate $P(I|C)$ from the integral equation

$$P(R|C) = \int_I P(R|I) P(I|C).$$
(26)

Although this equation yields a computationally inexpensive method, it has a limitation. The resolution with which one knows $P(R|I)$ limits the resolution of the learned function. If we know $P(R|I)$ reliably for N values of R, then we can determine at most N values of $P(I|C)$.

If in addition, the decision rule is known to the system, then the estimate of the prior can be refined. With a supervisor, the system would first estimate $P(S,R|C)$ from $\{S_1, \ldots, S_k\}$ and from $\{R_1, \ldots, R_k\}$. Again, the system would include the empirical distribution function for this estimation (51). Then the system would solve

$$P(S,R|C) = \int_I P(S|I, d(R)) P(R|I) P(I|C)$$
(27)

for $P(I|C)$. However, if there was no supervisor, then knowing the decision rule would not add information to what is available by $P(R|I)$.

Finally, strategy C can be adapted for unsupervised learning, because Equation 26 does not depend on the supervisor's critique.

The three strategies outlined here make different testable predictions for learning in the nervous system. For instance, if a neural structure applies strategy A but not the other strategies, then this structure does not learn the prior. Therefore, a change of task without a change of prior would cause the learning process to start from scratch. In contrast, in strategies B and C, learning would not start from scratch, because the system would possess the prior after learning. The same experiment could distinguish between strategies B and C. In the latter, the decision rule is always optimal for the prior assumed by the system (assumption 3 after Equation 18). Hence, if the system learned the prior, then changing the task would not make the system less optimal. However, in strategy B, although the system would not start from scratch, it would have to learn the optimal decision rule for the new task.

The literature is inconclusive on which of these strategies our brain uses. For instance, perceptual-learning data suggest different strategies depending on conditions. Dosher and Lu (55) interpreted their data saying that perceptual learning was a retuning of connections from neural responses to the decision process to emphasize relevant aspects of signals. In Fig. 9b, such retuning would affect the arrow from Box 1 to Box 2, modifying $d(\boldsymbol{R}{:}\boldsymbol{D})$ in Equation 17. In other words, the Dosher and Lu interpretation supports learning of the decision rule, without knowledge of the prior, i.e., strategy A. Other studies support this interpretation by showing task-specific forms of perceptual learning [e.g., see Saffell and Matthews (56)]. If learning does not transfer across tasks with identical priors, then it does not acquire priors, thus using that strategy. However, if one repeats this experiment with different tasks requiring much shared information about the priors, then task transfer occurs (57). In this case, because decision rules must be relearned, the transfer suggests learning of priors, which implies strategy B or C. Support for these strategies comes from experiments showing no learning in a difficult task, but improved performance in an easy task after completion of the hard-task training (57). That no learning occurs with the difficult task shows that the system is not improving the decision rule. Other experiments show increased transfer as learned tasks become easier, which suggests that learning strategies shift with conditions (57–59). Finally, a relevant experimental condition shows learning of a task involving a specific prior (e.g., an orientation) that does not transfer to the same task with a different prior [another orientation—e.g., Ramachandran and Braddick (60)]. Such experiments have suggested that prior learning occurs. However, one can explain them with the Lu and Dosher model, which retunes the input to the decision process to emphasize specific neural responses (55).

The procedures for using Equations 26 and 27 (strategy C) or minimizing Equations 22 (strategy A), 24, and 25 (strategy B) are complex and subject to intense research. We recommend the book by Vapnik (51) as a good starting point to learn about some of these estimation procedures. For example, estimating $P(\boldsymbol{R}|\boldsymbol{C})$ for Equation 26 is called the density-estimation problem. Mathematically this problem is ill-posed in the sense that it does not have a unique solution. To make the problem well-posed, one uses regularization constraints, which essentially make $P(\boldsymbol{R}|\boldsymbol{C})$ as smooth as possible (61). A judicious choice of constraints leads to a solution with the so-called Parzen estimators (62). This solution boils down to centering appropriately weighted kernel functions (e.g., Gaussians) on the data and shrinking the widths of these kernels as the number of examples increase. (For instance, the learned task in Fig. 10, blue line, is a sum of weighed Gaussians; dashed green lines, centered on the horizontal positions of the examples, green open circles.)

We can use similar kernel procedures for the problem of finding decision rules (63). The most common decision rules involve classification (e.g., pattern recognition) or real-valued functions (e.g., regression). Another popular modern technique to learn these kinds of decision rules is the support vector machine (SVM) (64). The SVM purposely maps the input vector into a high-dimensional space to simplify analysis. For instance, to segment (i.e., to classify parts of) a map of a circular island with surrounding waters, one needs to draw a circle. However, if one projects this map into three-dimensional space by elevating each point in proportion to its distance from the center of the island, then one can "cut" the island away from the water with a plane. This mathematical procedure, is easier because the plane is a linear function.

Acknowledgments

We would like to thank Dr. Zhong-Lin Lu, Dr. David Merwine, Ms. Susmita Chatterjee, Mr. Joaquín Rapela, and Mr. Jeff Wurfel for discussions during the writing of this article. This work was supported by National Eye Institute Grants EY08921 and EY11170.

BIBLIOGRAPHY

1. J. R. Lackner and P. A. DiZio, Aspects of body self-calibration. *Trends Cogn. Sci.* 2004; **4**(7):279–288.
2. E. R. Kandel, J. H. Schwartz, and T. M. Jessell, *Principles of Neural Science*, 4th ed. New York: McGraw-Hill, 2000.
3. V. Torre, J. F. Ashmore, T. D. Lamb, and A. Menini, Transduction and adaptation in sensory receptor cells. *J. Neurosci.* 1995; **15**(120):7757–7768.
4. U. Hillenbrand and J. L. Van Hemmen, Adaptation in the corticothalamic loop: computational prospects of tuning the senses. *Phil. Trans. R. Soc. Lond.* 2002; **357**:1859–1867.
5. C. W. G. Clifford and M. R. Ibbotson, Fundamental mechanisms of visual motion detection: modles, cells and functions. *Prog. Neurobiol.* 2003; **68**:409–437.
6. H. Tabata, K. Yamamoto, and M. Kawato, Computational study on monkey VOR adaptation and smooth pursuit based on the parallel control-pathway theory. *J. Neurophysiol.* 2002; **87**:2176–2189.
7. A. Reeves, Vision adaptation. In: L. M. Chalupa and J. S. Werner, eds., *The Visual Neurosciences*, Cambridge, MA: The MIT Press, 2004, pp. 851–863.
8. M. A. Webster, Pattern-selective adaptation in color and form perception. In: L. M. Chalupa and J. S. Werner, eds., *The Visual Neurosciences*, Cambridge, MA: The MIT Press, 2004, pp. 936–947.
9. S. Yantis, *Visual Perception: Essential Readings*. London: Psychology Press, 2001, pp. 172–189.
10. R. M. Berne and M. N. Levy, *Physiology*, 3rd ed. St. Louis, Mosby Yearbook, 1993.
11. L. L. Bellinger and L. L. Bernardis, The dorsomedial hypothalamic nucleus and its role in ingestive behavior and body weight regulation: lessons learned from lesioning studies. *Physiol. Behav.* 2002; **76**:431–442.
12. D. O. Hebb, *The Organization of Behavior: A Neuropsychological Theory*. New York: Wiley, 1949.
13. L. M. Chalupa and E. Gunhan, Development of on and off retinal pathways and retinogeniculate projections. *Prog. Retinal Eye Res.* 2004; **23**:31–51.
14. U. Guha, W. A. Gomes, J. Samanta, M. Gupta, F. L. Rice, and J. A. Kessler, Target-derived BMP signaling limits sensory

neuron number and the extent of peripheral innervation in vivo. *Development* 2004; **131**(5):1175–1186.
15. J. A. K. Suykens and H. Bersini, Neural control theory: an overview. *J. Control* (special issue). 1996; **96**(3):4–10.
16. J. A. K. Suykens, J. Vandewalle, and B. De Moor, *Artificial Neural Networks for Modelling and Control of Non-Linear Systems*, Dordrecht, The Netherlands: Kluwer Academic Publishers, 1996.
17. A. G. Barto, R. S. Sutton, and C. W. Anderson, Neuronlike adaptive elements that can solve difficult learning control problems. *IEEE Trans. Syst. Man Cybernet.* 1983; SMC-**13**(5):834–846.
18. K. S. Narendra and K. Parthasarathy, Identification and control of dynamical systems using neural networks. *IEEE Trans. Neural Networks* 1990; **1**(1):4–27.
19. K. S. Narendra and K. Parthasarathy, Gradient methods for the optimization of dynamical systems containing neural networks. *IEEE Trans. Neural Networks*. 1991; **2**(2):252–262.
20. P. Werbos, Minimisation methods for training feedforward neural networks. *Neural Networks*. 1990; **7**(1):1–11.
21. D. A. Robinson, Integrating with neurons. *Annu. Rev. Neurosci.* 1989; **12**:33–45.
22. A. E. Bryson and Y-C. Ho, *Applied Optimal Control; Optimization, Estimation and Control*. Waltham, MA: Blaisdell Pub. Co., 1969.
23. B. Hille, *Ionic Channels of Excitable Membranes*. 2nd ed. Sunderland, MA: Sinauer Associates Inc. Publishers, 1992.
24. J. G. Nicholls, A. R. Martin, and B. G. Wallace, *From Neuron to Brain*, 3rd ed. Sunderland, MA: Sinauer Associates Inc. Publishers, 1992, pp. 93–99.
25. R. M. Berne and M. N. Levy, *Cardiovascular Physiology*. 6th ed. St Louis, MO: Mosby Yearbook, 1992.
26. F. Brocard and R. Dubuc, Differential contribution of reticulospinal cells to the control of locomotion induced by the mesencephalic locomotor region. *J. Neurophysiol.* 2003; **90**:1714–1727.
27. M. Rabinovich, A. Selverston, L. Rubchinsky, and R. Huerta, Dynamics and kinematics of simple neural systems. *Chaos*. 1996; **6**(3):288–296.
28. A. A. V. Hill, M. A. Masino, and R. L. Calabrese, Intersegmental coordination of rhythmic motor patterns. *J. Neurophysiol.* 2003; **90**:531–538.
29. R. Jung, T. Kiemel, and A. H. Cohen, Dynamic behavior of a neural network model of locomotor control in the lamprey. *J. Neurophysiol.* 1996; **75**(3):1074–1086.
30. J. Thorson and M. Biederman-Thorson, Distributed relaxation processes in sensory adaptation. *Science* 1974; **183**:161–172.
31. S. B. Laughlin, The role of sensory adaptation in the retina. *J. Exp. Biol.* 1989; **146**:39–62.
32. N. M. Grzywacz and R. M. Balboa, A Bayesian framework for sensory adaptation. *Neural Computat.* 2002; **14**:543–559.
33. N. M. Grzywacz and J. de Juan, Sensory adaptation as Kalman filtering: theory and illustration with contrast adaptation. *Network: Comput. Neural Syst.* 2003; **14**:465–482.
34. P. Y. Burgi, A. L. Yuille, and N. M. Grzywacz, Probabilistic motion estimation based on temporal coherence. *Neural Comput.* 2000; **12**:1839–1867.
35. Y-C. Ho and R. C. K. Lee, A Bayesian approach to problems in stochastic estimation and control. *IEEE Trans. Automat. Control.* 1964; **9**:333–339.
36. J. O. Berger, *Statistical Decision Theory and Bayesian Analysis*. New York: Springer-Verlag, 1985.

37. R. M. Balboa and N. M. Grzywacz, The role of early retinal lateral inhibition: more than maximizing luminance information. *Visual Neurosci.* 2000; **17**:77–89.
38. R. M. Balboa and N. M. Grzywacz, The minimal local-asperity hypothesis of early retinal lateral inhibition. *Neural Computat.* 2000; **12**:1485–1517.
39. F. Rosenblatt, *Principles of Neurodynamics: Perceptron and Theory of Brain Mechanisms*. Washington, DC: Spartan Books, 1962.
40. M. Minsky and S. Papert, *Perceptrons*. Cambridge, MA, MIT Press, 1969.
41. Y. LeCun, *Learning Processes in an Asymmetric Threshold Network. Disordered Systems and Biological Organizations*. Les Houches, France: Springer, 1986, pp. 233–240.
42. D. E. Rumelhart, G. E. Hinton, and R. J. Williams, *Learning Internal Representation by Error Propagation. Parallel Distributed Processing: Explorations in Macrostructure of Cognition*, vol. I. Cambridge, MA: Badford Books, 1986, pp. 318–362.
43. C. M. Bishop, *Neural Networks for Pattern Recognition*. Oxford, UK: Oxford University Press, 1995.
44. D. H. Ackley, G. E. Hinton, and T. J. Sejnowski, A learning algorithm for Boltzmann machines. *Cognit. Sci.* 1985; **9**:147–169.
45. P. Dayan and L. F. Abbott, *Theoretical Neuroscience: Computational and Mathematical Modeling of Neural Systems*. Cambridge, MA: MIT Press, 2001.
46. J. Hopfield, Neural networks and physical systems with emergent collective computational abilities. *Proc. Natl. Acad. Sci. USA* 1982; **79**(8):2554–2558.
47. V. D. Malsburg, Self-organization of orientation sensitive cells in the striate cortex. *C. Kybernetik*. 1973; **14**:85–100.
48. C. Dang, Y. Leung, X. B. Gao, and K. Z. Chen, Neural networks for nonlinear and mixed complementary problems and their applications. *Neural Networks* 2004; **17**(2):271–283.
49. S. Grossberg, Adaptive pattern classification and universal recoding: I. parallel development and coding of neural feature detectors. *Biologic. Cybernet.* 1976; **23**:121–134.
50. M. Stodolski, P. Bojarczak, and S. Oswski, Fast second order learning algorithm for feedforward multilayer neural networks and its applications. *Neural Networks* 1996; **9**(9):1583–1596.
51. V. N. Vapnik, *The Nature of Statistical Learning Theory*, 2nd ed. New York: Springer-Verlag, 2000.
52. T. Poggio and F. Girosi, Regularization algorithms for learning that are equivalent to multilayer networks. *Science*. 1990. **247**:978–982.
53. H. Abdi, Linear algebra for neural networks. In: N. J. Smelter, and P. B. Baltes, eds., *Encyclopedia of the Social and Behavioral Sciences*. London, UK: Elsevier Science, 2001.
54. K. P. Kording and D. M. Wolpert, Bayesian integration in sensorimotor learning. *Nature*. 2004; **427**(6971):244–247.
55. B.A. Dosher, Z. L. Lu, Mechanisms of perceptual learning. *Vision Res.* 1999; **39**:3197–3221.
56. T. Saffell and N. Matthews, Task-specific perceptual learning on speed and direction discrimination. *Vision Res.* 2003; **43**:1365–1374.
57. M. Padilla and N. M. Grzywacz, Perceptual learning in a motion-segmentation task. in preparation.
58. M. Ahissar and S. Hochstein, Task difficulty and the specificity of perceptual learning. *Nature.* 1997; **387**:401–406.

59. Z. Liu and D. Weinshall, Mechanisms of generalization in perceptual learning. *Vision Res.* 2000. **40**:97–109.
60. V. S. Ramachandran and O. Braddick, Orientation-specific learning in stereopsis. *Perception*. 1976; **2**:371–376.
61. A. N. Tikhonov, On solving ill-posed problem and method of regularization. *Doklady Akademii Nauk USSR*. 1963; **153**: 501–504.
62. E. Parzen, An approach to time series analysis. *Ann. Math. Statist*. 1961; **32**:951–989.
63. T. Poggio and S. Smale, The mathematics of learning: dealing with data. *Notices the AMS*, 2003; **50**(5):537–544.
64. C. Cortes and V. Vapnik, Support vector networks. *Machine Learn*. 1995; **20**:1–25.

BIOLOGICAL NEURONAL NETWORKS, MODELING OF

M. GIUGLIANO
Brain Mind Institute—EPFL
Lausanne, Switzerland
and
University of Bern
Bern, Switzerland

M. ARSIERO
University of Bern
Bern, Switzerland

1. INTRODUCTION

In recent decades, since the seminal work of A. L. Hodgkin and A. F. Huxley (1), the study of the dynamical phenomena emerging in a network of biological neurons has been approached by means of mathematical descriptions, computer simulations (2,3), and neuromorphic electronic hardware implementations (4). Several models[1] have been proposed in the literature, and a large class of them share similar qualitative features. Specifically, a reduction of the biological complexity is usually operated, and the study of electrophysiological phenomena, emerging uniquely from the interactions of several neurons, is undertaken by describing the activity of each cell of a population through differential equations. Alternative approaches propose simplified average descriptions, introducing state variables and descriptors that do not directly relate to biophysical observables. A well-known example of such an approach is represented by the Wilson–Cowan population model (5). In this model the average firing rate of an entire assembly of neurons is described by a single heuristic mathematical equation. Similarly, the synaptic connectivity and signals transduction dynamics are usually introduced only in abstract terms.

[1]With the aim of providing a deeper introduction to the interested readers, and to let them run their own computer simulations, by e-mail request, we make some programming code (i.e., ANSI-C) employed in the reported simulations, freely available which will ultimately show how to translate the mathematical models discussed here into simple numerical implementations, to be simulated on a personal computer.

Today, thanks to the availability of fast computers and cheaper computational resources, it is possible to focus on large networks and study them in detail. To reach such goals, extended computer simulations are implemented in terms of the numerical integration of the mathematical equations for each model neuron, and, similarly, to the computational techniques of *molecular dynamics*, they are employed to make quantitative predictions and to explore the model's parameters space.

For these reasons, the recent literature increasingly reported realistic mathematical descriptions and simulations, whose results can be directly related to the cellular and subcellular biophysical details. In the following, we focus on the class of spiking neuronal models (2,6), which mimic the temporal evolution of the neuron membrane voltage as well as the generation of action potentials (i.e., the *spikes*), characterizing real neurons in an experimental context. These models are considered an intermediate description level between highly simplified abstract models, whose direct experimental verification is possible only under approximate terms, and multicompartmental models of the single-neuron electrophysiology, whose simulation often require prohibitive CPU loads, even for a network of small size. Thus, most of the conclusions and predictions obtained by the chosen modeling approach may be directly tested, comparing the result of computer simulations with single- and multi-electrode electrophysiological recordings, *in vivo* and *in vitro*.

Among the models of spiking neurons, two important classes can be outlined: models that accurately describe the subcellular biophysical phenomena characterizing the nonlinear voltage-dependent ionic permeability properties of the neuronal membranes (7) and models that provide a simplified phenomenological description. In the first case, the voltage-dependent ionic permeability properties of the neuronal membranes are characterized. In the second case, simplifying hypotheses on the biophysics of the neurons and synapses are considered. Such a reduction is performed with the aim of obtaining analytically and computationally tractable descriptions while retaining enough biological details to predict emerging phenomena that depend on the concerted interactions between cells.

The first class of models is constituted by detailed descriptions of the integrative and excitable properties of single neurons, including the ionic flows through the plasmatic membrane, following the original approach introduced by Hodgkin and Huxley (1). Such a theoretical framework has been greatly expanded by the discovery of new voltage-dependent ion currents and the availability of highly sophisticated experimental data, such as the *whole-cell* patch-clamp, the *single-channel*, and simultaneous *somato-dendritic* recordings (8), as well as the three-dimensional reconstruction and simulation of the neuronal morphologies. These models are characterized by a common repertoire of the excitable biological processes and by the description of the voltage-dependent membrane ionic conductances. Neurons are usually assembled in model networks, and the neurochemical transduction steps, which lead to the synaptic transmission between cells, are mimicked as additional postsynaptic brief transient inputs, triggered by the emission of presynaptic action potentials (i.e., excitatory and inhibitory postsynaptic currents—*EPSC/IPSC*). These contributions affect the subsequent temporal evolution of the simulated postsynaptic membrane voltage in terms of emission of action potentials and integration of additional incoming currents.

As mentioned, quite a different strategy to model the emergent properties of a large network of neurons consists of neglecting some biological details, by simplifying the dynamical properties of single neurons and synapses, which leads to phenomenological descriptions useful to define the effective *characteristics* of the network as a whole, still preserving some realism and allowing a direct comparison with the experimental data. Even if the further simplification of these reduced descriptions led to considerable advances and it had a strong impact on the development of the theory of formal neural computation and statistical learning theories (9,10), the search for realistic intermediate mathematical descriptions is still an open issue in neuroengineering.

2. MODELING THE SINGLE-NEURON ELECTROPHYSIOLOGY

In the following, we review two single-compartment models of the electrophysiology of a neuron: the conductance-based model, proposed by C. Morris and H. Lecar, and the leaky Integrate-and-Fire phenomenological model. Under the appropriate conditions, both models reproduce realistic discharge patterns as observed *in vitro* and *in vivo*. However, the second model is considerably simpler and can be effectively incorporated in large-scale computer simulations. Thus, these networks can be studied in detail, and it is possible to relate single-cell properties to the collective network activity. Moreover, a closed form for the input-output single-neuron transfer properties can be derived, in the case of the Integrate-and-Fire neurons, which is relevant for the statistical analysis of the network collective activity, under *extended mean-field* approaches, as discussed extensively in the literature (11,12). Such approaches are related to the hypothesis that individual cells in a large homogenous network of neurons cannot be distinguished, in a statistical sense. In fact, because of the very large number of synaptic connections and the presence of inhomogeneities and noise sources, neurons roughly tend to experience the same input current. Of course, each neuron will instantaneously receive a different realization of the same process, but its descriptors (i.e., current mean, variance, and correlation time length) are assumed to be the same. In other words, each neuron experiences the same mean field, extended to the regime of input fluctuations. Under such hypotheses, the characterization of the discharge properties of a single neuron becomes statistically representative of the others as a whole.

2.1. A Conductance-Based Model Neuron

In the mammalian central nervous system, neurons are characterized by an impressive morphological complexity of the dendritic branchings, where most of the incoming synaptic inputs from other cells are established. However, no conclusive evidence yet exists that such a complex anatomical morphology is playing a substantial role in determining the neuronal discharge response properties *in vivo*, other than maximizing the membrane surface available for receiving synaptic contacts.

As a consequence, many authors choose to focus on the single-compartment modeling of neuronal excitability (13,14), which implies that the spatial character of the dynamic distribution of the voltage across the neuronal membrane is neglected, assuming it to be characterized by the same value $V(t)$ (i.e., the space-clamp hypothesis) (3). The mathematical models of this kind are usually described by a set of coupled ordinary differential equations, as opposed to the multicompartmental descriptions, where systems of *partial* differential equations are used to predict the spatial and temporal evolution of the membrane voltage $V(x,y,z,t)$.

By assuming almost perfect dielectric properties for the phospholipidic bilayer that constitutes the neuronal membrane, charge-conservation considerations lead to a law of temporal evolution for the transmembrane voltage $V(t)$.

$$C_m \frac{dV}{dt} = I_{ext} - f(V), \quad I_{ext} = I_{syn} + I_{stim}$$

The previous equation states that a change in V is the result of external and intrinsic ionic current contributions. We denoted, with C_m, the equivalent electric capacitance per membrane area unit ($\mu F/cm^2$); with $f(V)$, the current densities (mA/cm^2) related to the ion-selective and voltage-dependent ionic permeability of the membrane; and with I_{ext}, the total external contribution to the membrane current density (mA/cm^2). The last includes synaptic inputs from other neurons, injected stimulus currents I_{stim}, and the metabolic contributions from electrogenic transport mechanisms, as the ion-pumps. Because, on the time scale characterizing the single action potential, metabolic contributions are usually small and time-invariant, they are neglected in the foregoing description. The kinetic description of the voltage-dependent intrinsic ionic currents $f(V)$ takes advantage of the model originally proposed by Morris and Lecar (15). This description leads to a system of two ordinary differential equations, and it is considerably simpler when compared with more detailed models (13). Such a modeling approach is referred to as *conductance-based*, because the evolution of the membrane voltage in time is determined by the interplay of several ion conductances (i.e., accounting for leakage currents, K^+ and Na^{2+} voltage-dependent permeabilities) uniformly distributed throughout the membrane. For our particular choice, the expression of the total ionic current density $f(V)$ is

$$f(V) = I_{leak} + I_K + I_{Na}.$$

We assume an Ohmic dependence of individual currents on the membrane voltage, so that the dependence of each term on V can be made explicit as

$$I_{leak} = \bar{g}_{leak}(V - E_{leak})$$

$$I_K = \bar{g}_K n (V - E_K)$$

$$I_{Na} = \bar{g}_{Na} m_\infty(V)(V - E_{Na}),$$

$$m_\infty(V) = \frac{1}{2}\left[1 + tgh\left(\frac{V - V_1}{V_2}\right)\right].$$

In the previous equation, the *Nernst* equilibrium voltages for the mixed passive leakage currents, the potassium ions, and the sodium ions have been indicated with E_{leak}, E_K, and E_{Na}, respectively. Moreover, $\bar{g}_{leak}$, $\bar{g}_{Na}$, and $\bar{g}_K$ are the maximal conductances related to the specific membrane permeabilities.

Thus, the final form of the differential equation can be rewritten as

$$C_m \frac{dV}{dt} = I_{ext} - \bar{g}_{leak}(V - E_{leak}) - \bar{g}_K n(V - E_K) \quad (1)$$
$$- \bar{g}_{Na} m_\infty(V)(V - E_{Na})$$

and completed by the equation satisfied by the state variable $n(t)$, which represents the fraction of K^+-sensitive ion channels that are in an *open state* (i.e., their ion permeability is nonzero).

$$\frac{dn}{dt} = \frac{(n_\infty(V) - n)}{\tau(V)} \quad (2)$$

where $n_\infty(V) = \frac{1}{2}[1 + tgh(\frac{V-V_3}{V_4})]$ and $\tau(V)^{-1} = \tau_n^{-1} cosh(\frac{V-V_3}{2V_4})$.

The last equation describes the voltage dependence of the molecular conformational states of the *delayed-rectifier K^+* channels, participating in the generation of action potentials and accounting for the after-discharge absolute refractoriness. As opposed to the channels selective to Na^+ ions, whose conductance is assumed to be as instantaneously dependent on V (i.e., $m = m_\infty(V)$), the evolution of n is much slower.

At the steady state, as $I_{ext} = 0$, the membrane voltage asymptotically tends to a resting value $V = V_0$, which is mainly determined by the leak current I_{leak}.

$$C_m \frac{dV}{dt} \sim I_{ext} - \bar{g}_{leak}(V - E_{leak})$$

This is a result of the voltage-dependence of the sodium and potassium conductances, whose resting contribution to the net membrane current is weak (i.e., $m_\infty(V_0) \simeq 0$ and $n \simeq n_\infty(V_0) \simeq 0$). Therefore, under such conditions, the temporal evolution of V, for small external currents, is approximately a passive RC-like response. However, for stronger depolarizing currents, the sigmoidal voltage-dependence of $m_\infty(V)$ instantaneously activate the sodium conductances in such a way that I_{Na} affects V as well. For voltages that are depolarized enough with respect to V_1 (see Equation 1), V quickly starts approaching the reversal potential E_{Na} (i.e., $E_{Na} > V_0$):

$$C_m \frac{dV}{dt} \sim I_{ext} - \bar{g}_{leak}(V - E_{leak}) - \bar{g}_{Na}(V - E_{Na}).$$

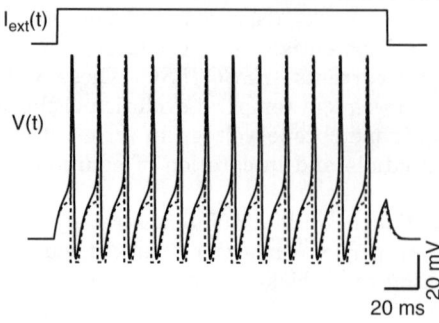

Figure 1. Temporal evolution of the simulated membrane voltage response $V(t)$ (lower traces) evoked by an intracellular somatic DC current injection $I_{ext}(t)$ (upper trace), for the single-compartment conductance-based (Equations 1 and 2) (continuous line) and the *Integrate-and-Fire* model neurons (Equation 4) (dashed line).

With much slower reaction times, also, the potassium conductance turns on and becomes nonzero, resulting in a stronger opposite current contribution I_K, which makes V bouncing back to more negative voltages, toward E_K (i.e., $E_K < E_{Na}$):

$$C_m \frac{dV}{dt} \sim I_{ext} - \bar{g}_{leak}(V - E_{leak}) - \bar{g}_K(V - E_K).$$

The interplay of these currents underlies the generation of an action potential, characterized by a very fast increase of V and by a delayed hyperpolarizing drive that reset the voltage to negative values (Fig. 1—thick line), which constitutes a biophysical description of the electrical excitable properties of a neuronal membrane, and accounts for the existence of a minimal current stimulation to induce repetitive action potentials generation and for the refractoriness that follows each spike. Therefore, small amplitude current stimuli produce passive membrane responses, and if a threshold voltage is overcome, a train of action potentials is generated, resulting from the oscillation of $V(t)$. Finally, the relationship between the spike frequency and the current stimulus amplitude is nonlinear, and it can be determined only in a numerical way.

2.2. The Leaky Integrate-and-Fire Model

The choice of the accuracy level of a description, aimed at gaining insights on the essence of a physical phenomena, is a well-known problem in the natural sciences. In our case, the choice of conductance-based single-compartment descriptions, or of other simpler caricatures of the excitable behavior of a neuron, depends on the degree of abstraction and on the phenomena of interest. In the context of the network activity, emerging in a population of synaptically interacting neurons, a reduction of the single-cell mathematical description is possible and required to reduce the CPU loads needed by a large network computer simulation. Similar to the complex systems studied in statistical physics, when the phenomena of interest are the result of the interaction of a large number of elements, most of the details of the single element can be incorporated into an effective description. Although a similar description may fail in faithfully reproducing the entire range of experimental observations at the level of the single unit, it provides a very good abstraction to account for the collective behavior of the whole population while preserving experimentally measurable correlates.

In the context of neuronal modeling, reduced models can, in some cases, preserve a link to the biophysical properties that can be experimentally measured, and they retain enough details to quantitatively account for the collective activity emerging in a network. In the following, and for the network simulations reported here, we consider a reduced description of neuronal excitability, originally introduced by Lapicque (16) and referred to as a leaky *Integrate-and-Fire* model neuron. We show how to derive such a model from the Morris–Lecar neuron (Equations 1 and 2), ignoring one of the two equations, under appropriate hypotheses. The key point in operating a similar reduction is the assumption that the action potential generation and the refractoriness mechanisms are highly stereotyped phenomena.

Although we are interested in preserving the integrative-capacitive electrical properties of nervous cells, we can neglect Equation 2, reducing the accuracy in the activation variable n. In other words, because of the invariance of the shape of action potentials, which are identified by the time of occurrence and described by repolarization-hyperpolarization voltage amplitudes and by the refractory period, we claim that the precise description of an action potential is not relevant in predicting the collective properties of a network. Such an observation can be efficiently described with a lower degree of fidelity, considerably reducing the CPU times of a computer-simulation.

Let us consider the dynamics of n to be instantaneous, compared with the passive time constant associated with Equation 1. Then $n \simeq n_\infty(V)$, and we can rewrite such an equation as

$$C_m \frac{dV}{dt} \cong I_{ext} - \bar{g}_{leak}(V - E_{leak}) - \bar{g}_K n_\infty(V)(V - E_K)$$
$$- \bar{g}_{Na} m_\infty(V)(V - E_{Na}). \quad (3)$$

This equation is a good approximation for the description of the membrane voltage, below the excitability threshold, when V is approximately constant. This condition may correspond to the statistical average value of V, under a given activity regime, or, for instance, to its resting value, when $I_{ext} = 0 \, mA/cm^2$. The study of the second case implies that the right-hand term of the last equation is zero when V is at the resting membrane potential V_0. We can thus perform a Taylor-series expansion of Equation 3, whose leading term is a reduced linear description of the subthreshold dynamics, proportional to a constant $\bar{g}$, with the meaning of an effective passive membrane ion conductance.

$$C_m \frac{dV}{dt} \cong I_{ext} - \bar{g}(V - V_0)$$

Then, because we are regarding the emission of an action potential as a highly stereotyped phenomenon, a fixed-voltage threshold V_{th} must be artificially reintroduced. Therefore, as soon as V crosses V_{th}, a spike is emitted

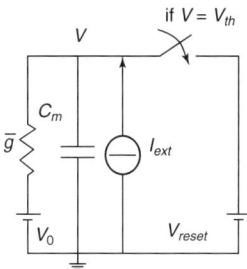

Figure 2. Sketch of the electrical equivalent circuit of the single-compartment leaky *Integrate-and-Fire* model employed in the simulations of large biological neuronal networks: For the sake of simplicity, the absolute refractoriness was not represented. The condition $V(t) = V_{th}$ corresponds to the emission of an action potential at time t and to the reset of V.

and the voltage is reset to an hyperpolarizing voltage V_{reset} (i.e., $V_{reset} < V_{th}$). Furthermore, on the emission of a spike, V is refractory to any external perturbation during a subsequent time interval of duration τ_{ref}, mimicking the latency of $n(t)$ in Equation 2. These steps identify an *Integrate-and-Fire* model (Fig. 2), whose description can be summarized as follows:

$$C_m \frac{dV(t)}{dt}$$
$$= \begin{cases} \bar{g}(V_0 - V) + I_{ext}(t) & \text{if } V(t) \leq V_{th} \\ 0, \, V(t) = V_{reset} & \text{if } V(t_0^-) = V_{th} \\ & \text{and } t \in (t_0^+; \, t_0^+ + \tau_{ref}). \end{cases} \quad (4)$$

The qualitative comparison between the temporal evolution of membrane voltage of the Morris–Lecar model neuron and the Integrate-and-Fire model neuron is reported in Fig. 1, where a DC pulse current injection results in the same output mean firing rate for both models.

As anticipated, the operated reduction led to a mathematical model that can be treated analytically and analyzed in detail, compared with the conductance-based models (17). In particular, the relationship between the input current I_{ext} and statistics of the model discharge response can be studied and general expressions can be derived under a realistic input current drive (17), which holds for a deterministic, DC current waveform, as well as for a stochastic input drive (i.e., a delta-correlated gauss-distributed noise), mimicking the superposition of a large number of independent fast synaptic currents experienced by a neuron embedded in a large network. Under such conditions and in a regime of asynchronous firing, every postsynaptic neuron experiences a synaptic current that is approximately gauss-distributed, as predicted by the central limit theorem (18). In this case, by indicating with μ and σ^2 the *infinitesimal* expected value and variance of the gauss-distributed current I_{ext}, therefore measured as currents per unit of time, the expression of the output mean firing rate ν for the leaky Integrate-and-Fire model can be expressed as follows (19):

$$v = \Phi(\mu, \sigma)$$

$$= \begin{cases} 0 & \text{if } \sigma = 0, \mu < \bar{g}(V_{th} - V_0) \\ [\tau_{ref} + \tau \ln(\frac{a}{b})]^{-1} & \text{if } \sigma = 0, \mu \geq \bar{g}(V_{th} - V_0) \\ [\tau_{ref} + \tau\sqrt{\pi}\int_a^b e^{x^2}(1 + erf(x))dx]^{-1} & \text{if } \sigma > 0, \end{cases}$$
(5)

where $erf(x)$ is the error function, $\tau = C_m/\bar{g}$ is the membrane time constant, and the expressions of a and b are indicated below:

$$a = (C_m(V_{reset} - V_0) - \mu\tau)/(\sigma\sqrt{\tau}),$$
$$b = (C_m(V_{th} - V_0) - \mu\tau)/(\sigma\sqrt{\tau}).$$

Thanks to Equation 5, most of the basic features characterizing neuronal excitability can be immediately understood in quantitative terms. For instance, the role of the refractoriness that follows each spike (i.e., τ_{ref}), with respect to the limiting of the maximal mean output frequency, either in the deterministic ($\sigma^2 = 0$) or in the noisy regimes, is apparent from the linear dependence of $v \sim \tau_{ref}^{-1}$, as μ is sufficiently large. Under these conditions, the spike response evoked by I_{stim} becomes insensitive on both the mean injected current μ and the fluctuations amplitude σ. Furthermore, the minimal current amplitude $\bar{g}(V_{th} - V_0)$ (i.e., the *rheobase*), which is required to evoke a sustained spiking response, under DC current stimulation, is playing a different role as a *noise-dominated regime* is considered, where neurons may fire as a consequence of the input variance, even for negative mean stimuli ($\mu < 0, \sigma^2 > 0$).

Finally, as opposed to the conductance-based models, one of the main advantages of the Integrate-and-Fire neurons (Equation 4) is that their response properties to a synaptic/external input drive can be related in a simple way to a small set of effective parameters (V_{th}, V_{reset}, $\bar{g}$, C_m—see Equation 4), which can be experimentally identified (20) and related to the biophysical properties of the cells.

We also note that a direct experimental identification of the parameters of the Integrate-and-Fire model has been recently attempted *in vitro* for cortical rat neurons. Surprisingly, the Integrate-and-Fire dynamics can indeed quantitatively predict the discharge response properties of real cortical neurons with high accuracy (20,21).

3. MODELING THE CHEMICAL SYNAPTIC TRANSMISSION

As our goal is modeling the electrical activity emerging from neuronal interactions after the introduction of the model of electrophysiological activity in single cells, one of the most fundamental steps is the description of the synaptic communication. The most common form of communication between neurons involves the chemical synaptic release of neurotransmitter molecules (e.g., glutamate, GABA, glycine, acetylcholine). Such a release lasts for a fraction of a millisecond, and it is normally triggered by a presynaptic action potential, which propagates through the axon down to the synaptic boutons. The close spatial proximity of the postsynaptic membrane to the presynaptic boutons lets the neurotransmitter molecules diffuse in the synaptic cleft and activate the molecular recognition devices, corresponding to postsynaptic membrane receptors. These receptors are highly selective and constitute, in most of the cases, a class of ligand-gated ion channels. Reminiscent of the description of the voltage-dependent membrane ion conductances, it is possible to introduce a similar mathematical description that accounts for the transient changes of the postsynaptic receptor conductance as a function of the instantaneous concentration of the ligand molecules in the cleft.

3.1. A Kinetic Markov Model of Neurotransmitter-Gated Postsynaptic Receptors

The total postsynaptic current caused by N independent ionotropic synaptic contacts can be expressed by the ohmic formalism, already employed for the intrinsic ion currents (3,22):

$$I_{syn} = \sum_{i=1}^{N} g_i(t)(E_{syn\,i} - V),$$
(6)

where V is the postsynaptic potential and $E_{syn\,i}$ is the synaptic apparent reversal potential related to the specific ion specie the receptors are selective to (e.g., $0\,mV$ for excitatory glutamatergic synapses and $-80\,mV$ for inhibitory GABAergic synapses). The time course of each synaptic conductance $g_i(t)$ can be defined according to the state diagram of a Markov model (see Equation 7), operationally grouping the functional configurations of an ion channel population in several states, each characterized by distinct conductances. Transitions from one state to the other are usually spontaneous or they require the neurotransmitter molecules to interact with the receptors. As mentioned, $g_i(t)$ is related to the *open-state conductance(s)* of the receptors. Typically, postsynaptic receptors that can be characterized by two functional states (e.g., the open state and the closed state); but the same approach can be used in more complicated situations (22). Quantitatively, we may write:

$$g_i(t) \propto [TR_i^*] \quad \forall i = 1, \ldots, N.$$

$$R_i + T_i \underset{\beta}{\overset{\alpha}{\rightleftharpoons}} TR_i^* \quad [R_i] + [TR_i^*] = 1,$$
(7)

where T_i represents the actual concentration of neurotransmitter molecules in the cleft, and $\alpha, \beta, [R_i]$, and $[TR_i^*]$ are the forward and backward rate constants for transmitter binding and the unbound and the bound fraction of postsynaptic membrane receptors, respectively. If we define the maximal synaptic conductance $\bar{g}_{syn}$ (i.e., the absolute synaptic strength) and the fraction of ligand-gated channels in the functional open state r_i, for the generic ith afferent synapse, then, by definition

$$g_i(t) = \bar{g}_{syn} r_i(t) \quad \forall i = 1, \ldots, N.$$

The previous kinetic scheme is equivalent to a differential equation, so that, given a large number of ion channels and neglecting statistical fluctuations (23), $r_i(t)$ satisfies the following equation

$$\frac{dr_i(t)}{dt} = -\beta r_i(t) + \alpha T_i(t)(1 - r_i(t)). \quad (8)$$

Assuming that the transmitter concentration $T_i(t)$ in the synaptic cleft occurs as a pulse of amplitude T_{max} and duration C_{dur}, triggered by a presynaptic action potential, a closed solution of Equation 8 exists, and its efficient iterative calculation can be expressed as follows:

$$r_i(t + \Delta t) = \begin{cases} r_i(t)exp(-\Delta t/\tau_r) & \text{if } \bar{t}_i > (t + \Delta t) > \bar{t}_i + C_{dur} \\ + (1 - exp(-\Delta t/\tau_r))r_\infty & \\ r_i(t)exp(-\beta \Delta t) & \text{if } (t + \Delta t) > \bar{t}_i + C_{dur} \end{cases} \quad (9)$$

where r_∞ and τ_r are constants defined in Table 1, Δt is the simulation discrete-time step size, and $\bar{t}_i$ is the last occurrence time of a presynaptic action potential at the ith synapse. Such a model describes how, upon a presynaptic spike emission, the concentration T_i changes as a brief, piecewise-constant pulse that leads to the *activation* of the fraction of postsynaptic receptors in the open state $r_i(t)$ (Fig. 3a–c), which induces a transient change in the membrane synaptic conductances at the postsynaptic neurons and results in excitatory (i.e., depolarizing) or inhibitory (i.e., hyperpolarizing) postsynaptic potentials (EPSPs/IPSPs) (Fig. 3a–c).

Finally, it is interesting to note that such a model implicitly accounts for saturation and summation of multiple presynaptic events (Fig. 3c), and it is efficiently computed at each simulation time step, with very small CPU loads and no approximation, by Equation 9. Although further algorithmic techniques for the fast calculation of the overall synaptic current (Equation 6) can be devised (24,25), Equations 4 and 6 for each neuron, and Equation 9 for each synapse, constitute an appropriate description level to perform extended computer simulations of the activity of a network (26) (Figs. 4 and 5).

3.2. Introducing Short-Term Homosynaptic Responses Depression

Recently, the dependence of synaptic responses to the presynaptic activity history was experimentally investigated at the central synapse and found to considerably modulate signal transmission between pairs of *in vitro* neocortical neurons (27) in a frequency-dependent manner. A phenomenological model has been proposed to quantify such dynamic behavior by means of the definition of a limited amount of *resources* available for signal transduction at each synapse (2,28). This model accounts for the so-called *homosynaptic* short-term depression, as it involves subcellular mechanisms that depend exclusively on the presynaptic activity history (i.e., not on the correlated pre- and postsynaptic activation), and it occurs over a time scale of a few hundreds of milliseconds.

Possible biophysical correlates of such short-term depression include postsynaptic receptors desensitization (Fig. 4a) as well as presynaptic ready-releasable neurotransmitter vesicle pool depletion (Fig. 4b). Both models can be represented by a three-state kinetic scheme, as sketched in Fig. 4. Compared with the model of postsynaptic receptor discussed in the previous section, an additional transition from the bound state to an inactive state R_{inact} has been added (Fig. 4a). In such a state, transmitter-gated channels are inactivated and functionally closed, and the slow recovery to the unbound state occurs with a rate γ. β represents the rate of inactivation, and α represents the probability of the ligand-receptor binding per unit of time.

In the kinetic description depicted in Fig. 4b, short-term plasticities are instead assumed to result from the presynaptic dynamics of neurotransmitter vesicles, including their exocytosis, depletion, and refilling. This description assumes that the amount of neurotransmitter released in the cleft depends on the previous synaptic activity, determining the availability of a ready-releasable vesicle pool. In particular, η represents the decay rate of the actual concentration of neurotransmitter in the cleft, because of enzymes and reuptake mechanisms, whereas μ is the rate of recovery phenomena such as the endocytosis or the docking of new vesicles to the presynaptic membrane.

The two models lead to an equivalent phenomenological description that can be considerably simplified as a single-state process and combined to the kinetic model

Table 1. Numerical parameters employed in the computer-simulations of Figs. 1–6

Symbol	Value
τ_n	$0.1\,ms$
V_1	$-1\,mV$
V_2	$15\,mV$
V_3	$10\,mV$
V_4	$14.5\,mV$
E_{Na}	$100\,mV$
E_K	$-70\,mV$
E_{leak}	$-50\,mV$
E_{syn}	$0\,mV$
V_0	$-65\,mV$
C_m	$1\,\mu F/cm^2$
$\bar{g}_{Na}$	$0.75\,mS/cm^2$
$\bar{g}_K$	$1.49\,mS/cm^2$
$\bar{g}_{leak}$	$0.5\,mS/cm^2$
$\bar{g}_{Na}$	10–100000
$\bar{g}$	$0.05\,mS$
$\bar{g}_i$	$0.05/N\,mS$
N	10–100000
T_{max}	$1\,mM$
C_{dur}	$1\,ms$
τ_{rec}	$400\,ms$
f_i	0.75
α	$2\,ms^{-1}mM^{-1}$
β	$1\,ms^{-1}$
r_∞	$(\alpha T_{max})/(\alpha T_{max} + \beta)$
τ_r	$(\alpha T_{max} + \beta)^{-1}$

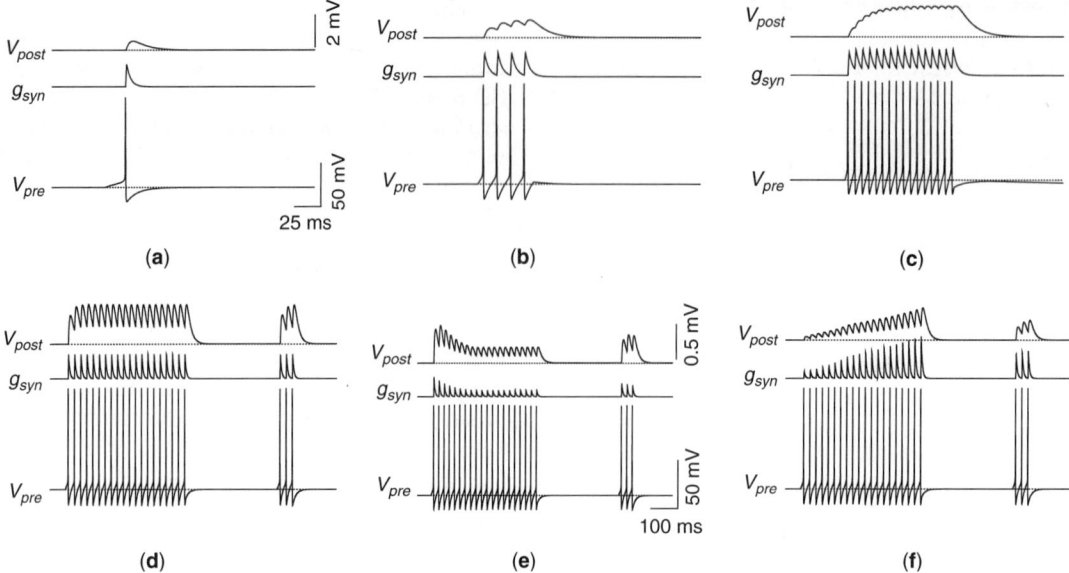

Figure 3. Model of chemical synaptic transmission and short-term plasticities: the kinetic model (Equations 7–9) was computer-simulated in the case of an excitatory connection between two model neurons (i.e., $E_{syn} = 0\,mV$). (a–d) The emission of a single presynaptic action potential, or of a train of action potentials (lower traces), evokes a transient postsynaptic receptor activation and, thus, a change in the total synaptic excitatory conductance (middle traces), which leads to a single or a series of excitatory postsynaptic potentials (EPSPs) (upper traces), which show temporal summation because of the capacitive properties of the postsynaptic membrane.

synapse introduced in the previous section. These models predict that, as result of the activity at a particular synapse, the amplitude of subsequent EPSPs/IPSPs in a train is not constant (Fig. 3d–f). Instead, the synaptic transduction of signals from the presynaptic neuron is characterized by a kind of effective *fatigue* (or *facilitation*—Fig. 3f), which result in a frequency-dependent modulation of the postsynaptic responses.

For the sake of simplicity, it can be proved the models of Fig. 4 can be equivalently rephrased by rewriting $g_i(t)$ as follows:

$$g_i(t) = \bar{g}_i z_i(t) r_i(t). \qquad (10)$$

$z_i(t)$ is a positive-state variable (29) that exponentially approaches 1 between any two subsequent presynaptic spikes. Such a process is characterized by an equivalent recovery time constant, τ_{rec}, associated to the biophysical short-term mechanisms (i.e., $\tau_{rec} \sim \varepsilon^{-1}$ or $\tau_{rec} \sim \gamma^{-1}$—see Fig. 4):

$$\frac{dz_i(t)}{dt} = \frac{(1 - z_i(t))}{\tau_{rec}}.$$

Figure 4. Markov kinetic schemes for (a) a simple form of postsynaptic receptors inactivation and (b) the presynaptic dynamics of the neurotransmitter vesicle pool, including exocytosis, depletion, refilling, and its interaction with postsynaptic receptors.

However, each time an incoming presynaptic spike induces the activation of the ith synapse, $z_i(t)$ must be reduced by a constant fraction $f < 1$ ($f > 1$, for facilitating synapses—Fig. 3f).

$$z_i \rightarrow z_i f$$

In Fig. 3d–f, the results from the computer simulations of Equations 4, 6, and 10 are reported for a two-neuron circuit, showing that, for a given presynaptic spike train, the postsynaptic response can be affected substantially by the underlying short-term synaptic dynamics. The impact on the network collective activity of the short-term synaptic responses depression may have quite dramatic consequences, which will be discussed in the following sections.

4. COMPUTER SIMULATIONS

4.1. Bistability in a Network of Excitatory Neurons

As a striking example of a network property emerging from the synaptic interactions between neurons and not characterizing the discharge response of isolated cells, we review the existence of regimes of reverberating self-sustained asynchronous activity (i.e., a network bistability) in networks of excitatory neurons. This class of phenomena has been extensively investigated and related to the evidences of *in vivo* electrophysiological experiments aimed at dissecting the cellular bases of mnemonic associations, working-memory states, and the delay-activity observed in match-to-sample behavioral task experiments (30). During these kinds of experiments, trained primates are

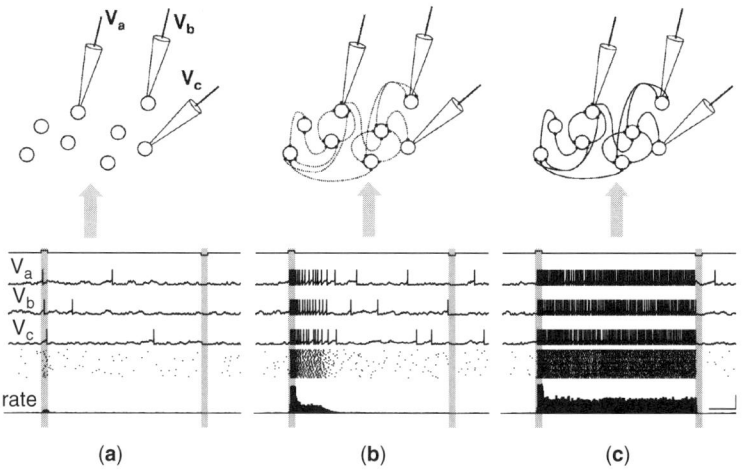

Figure 5. Bistability as a network emerging phenomenon in a population of 300 model neurons randomly and recurrently connected by excitatory synapses: Each panel contains the temporal evolution of an external background current stimulation (top trace), the membrane voltages $V_a(t), V_b(t), V_c(t)$ of three neurons taken by chance, the raster plot of the spikes emitted by a subset of the network (middle panels), and the overall population mean firing rate (bottom trace), estimated over a 10 ms sliding time window, as a peristimulus time histogram.

instructed to memorize a transient visual stimulus, and later compare it with a second stimulus occurring tens of seconds later in pair-wise associations. During simultaneous chronic recording of singe-unit cortical activity in the infero-temporal cortex, as well as in many other associative areas, the spiking activity of most of the neurons that strongly respond selectively to the first stimulus show a *delayed* enhanced activation, compared with the level of spontaneous activity, which persists in the absence of any visual stimulation (7).

The theoretical works who first explored the analogy between such *in vivo* experimental results and multistability in neuronal distributed systems (11,31,32) proposed network bistability as a paradigmatic example of collective emerging phenomena, which might act as the computational substrate of a *working memory*. Following such a hypothesis, in this section, we introduce and focus on a single homogenous population representing only a small portion of a larger neuronal population. These neurons are supposed to selectively respond to a particular visual stimulus, and, under some conditions, show enhanced electrical activity after stimulus removal, replicating the results of delayed match-to-sample experiments.

Under these perspectives, network bistability might underlie several kinds of associative computations, efficiently maintaining an internal (*working*) neuronal representation whose persistence is ensured by the strong cortical recurrent excitation (33).

We consider a homogeneous population of neurons characterized by unstructured topological arrangement of the connections (Fig. 5). Although the results we are going to discuss do not depend on the specific mathematical model neurons or synapses, we simulated a network of integrate-and-fire neurons connected by the kinetic synapses, as introduced in the previous section. Therefore, a population of N cells randomly connected with a fixed probability C_{ee} was simulated by a system of N simultaneous differential equations (see Equation 4). Individual nondepressing synapses, between any two connected neurons (i.e., up to N^2), were described by Equations 6 and 9, and it was further assumed that an additional background synaptic activity influenced each neuron of the network, which is reminiscent of an unmodeled set of afferents that are independent of the activity of the simulated network and spontaneously active. Under these conditions, the membrane voltage of each neuron randomly fluctuates, as in a random walk, and a low-rate asynchronous emission of action potentials characterizes the global activity of the network (see Fig. 5a), which appears to be a realistic model of electrophysiological recordings performed in the intact cortex of behaving animals as well as in *in vitro* cultured networks of dissociated neurons (21): The membrane voltage of cells is fluctuating (34), and a spontaneous irregular spike emission occurs (11).

The result of the computer simulations points out clearly that, if a large fraction of the neurons receives a transient depolarizing input, and therefore increases its level of activity, an assembly of coactive neurons may develop, spread to the entire population, and last indefinitely (see Fig. 5). This result is a consequence of the recruitment of the appropriate number of recurrent excitatory synaptic inputs, which can self-sustain a net recurrent input, even without external stimulation, as in a positive feedback system, which produces a bistable collective network response, resembling an elementary reverberating mechanism to actively preserve memory traces as neuronal representations. Once started, the spiking activity may last, even when the external input is removed, in a self-sustained manner. Such a phenomenon is a consequence of the synaptic interactions, as a population of uncoupled neurons (Fig. 5a), receiving the same transient external input stimulation, increasing its firing rate, and immediately relaxing back to the level of spontaneous activity. It is therefore not requested that the spike response of individual neurons is bistable, and it is usually the case *in vivo* and as it can be tested by DC current injection in the model (see Fig. 1). Under the appropriate conditions, when embedded in a population of synaptically interacting units, neurons may receive an additional (recurrent) synaptic drive that depends on their own firing, sufficient to keep such a regime indefinitely, in a self-consistent input-output relationship.

We further mention that a similar collective phenomenon may also auto-organize and emerge from a correlated activity pattern, imposed by an external stimulation, and consolidated by the activity-dependent long-term strengthening of synaptic efficacies. Within a particular range of synaptic potentiation, the activity of the network can therefore sustain two stable collective dynamical equilibria, one at the level of the global spontaneous activity (e.g., 1–5 Hz) and the other at a higher firing rate (e.g., 30 Hz), similar to what was observed in *in vivo*.

We conclude the present section underlying that, in the general case of cortical networks, the activation of inhibitory populations has been demonstrated to play a fundamental role. In particular, because of the intrinsic impact on the postsynaptic neurons, synaptic inhibition is expected to lower the output mean spiking rate, reproducing more closely the experimental single-electrode recordings (35). Most importantly, inhibitory populations have been hypothesized to participate in the generation of competition across different stimulus-selective populations.

4.2. Computations in a Network with Short-Term Depressing Synapses

As anticipated in the previous section, short-term synaptic depression may contribute to increase the dynamical range of collective phenomena developing in a network of neurons. Here, we present the results of computer simulations in feed-forward network architectures, as sketched in Fig. 6, where homosynaptic short-term depression was included. In a first simulation (Fig. 6a), the corresponding presynaptic afferent activation was modeled as independent random events by identical Poisson process whose mean activation rate v_{pre} step-changed from 20 Hz to 90 Hz. As expected, when the simulated afferent synaptic activation changes, a new steady state is quickly reached and the total net current experienced by the postsynaptic neuron increases. If the new presynaptic activity level is sufficient to depolarize enough of the postsynaptic membrane voltage, a postsynaptic integrate-and-fire neuron, receiving nondepressing synapses, discharges irregularly and indefinitely, which is a consequence of a proportional relationship between presynaptic population mean firing rate and the average resulting postsynaptic input current under an asynchronous presynaptic regime, because nondepressing synapses are performing a temporal integration of incoming spikes.

However, as short-term depression was introduced in the synapses, its firing response did not encode the presynaptic mean activity level anymore but, instead, the occurrence time of the transition in the background activity. Although the details of the computer simulation are indicated in the figure caption (see Fig. 6), an explanation of such phenomenon can be given through an approximate description: For a normal synapse, the average conductance change is given by $\Delta g_i \propto \Delta v_{pre}$, whereas for a depressing synapse, it is $\Delta g_i \propto \Delta v_{pre}/v_{pre}$ (29). As a consequence, whereas in the first case the synaptic conductance acts as a low-pass filter, in the second case, the depressing synapses act as high-pass filters, unable to respond to constant presynaptic activity level but signaling its temporal transitions.

Such kind of derivative response might be exploited by the nervous system to perform other kinds of computations. The very same model synapse actually can indeed predict the direction selectivity emerging in the neurons of

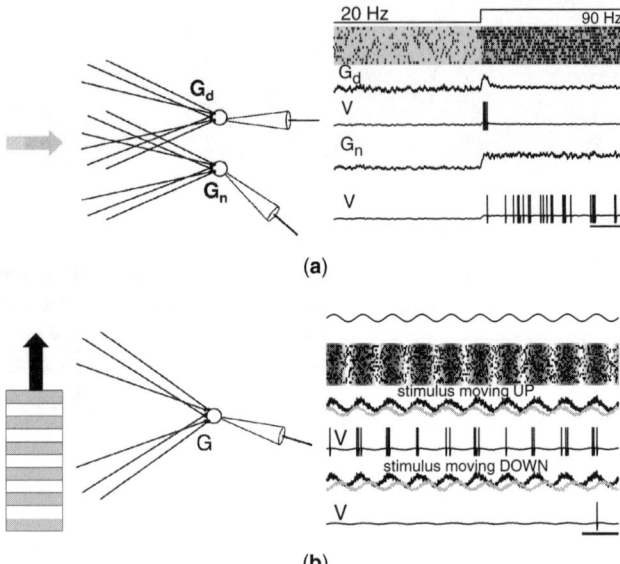

Figure 6. Short-term synaptic depression may contribute to process incoming information in feed-forward model networks. (a) Two identical integrate-and-fire models, one receiving short-term depressing synaptic afferents (i.e., $f_i < 1$) and the other receiving nondepressing synaptic inputs (i.e., $f_i = 1$), detect different features in the time course of the mean level of asynchronous background activity. Actually, if the mean level of background activity step changes from 20 Hz to 90 Hz, the previously silent neuron responds by a burst of action potentials, signaling the change, whereas the other starts to fire tonically, encoding the new input mean level in its output firing rate. The plot reports the time evolution of the mean firing rate of the presynaptic population, the raster plot of the spikes emitted by a subset of the presynaptic afferents (top panels), and the total postsynaptic conductances $G_d(t)$ and $G_n(t)$, together with the corresponding neuron membrane voltage $V(t)$.

the mammalian primary visual cortex under visual stimulation (36). By employing a retinotopic excitatory synaptic projection, arranged as an alternation of depressing and normal synaptic afferents, the postsynaptic neuron will be driven by two overlapping synaptic contributions: one in phase with the moving stimulus and the other characterized by a fixed phase-lag. Such a phase-lag is consequence of the derivative response properties of each depressing synapse and resembles what would be produced by a high-pass filter. Because of the linear superposition of such contributions at the postsynaptic neuron, there will be only one particular activation order (i.e., first depressing synapse, then nondepressing) that will result in a constructive interference between the two subpopulations.

As a consequence, the output neuron will receive a net input drive sufficiently large enough to fire, only when the two synaptic waveforms are somehow in phase, producing a response that is selective to one sliding direction only.

5. CONCLUSION: AIMS OF MATHEMATICAL MODELING AND COMPUTER SIMULATIONS

Modeling constitutes an important complementary tool to the experimental techniques *in vivo* and *in vitro* (37), similar to other scientific fields such as physics and engineering. In the context of biological sciences, such an approach is in full agreement with the typical perspectives of neuroengineering (38): Modeling is essential in addressing conceptual issues that develop from the study of the nervous system at different description levels. Many advantages exist: (i) A quantitative model can make the dynamics of a complex neurobiological system, constituted by many interacting components, more accessible; (ii) brand new phenomena may be discovered by comparing the predictions of analytic solutions and numerical computer simulations with the experimental results and, more importantly, new experiments can be designed on the bases of those predictions; (iii) experiments that are considerably difficult or impossible to perform in real biological preparations (i.e., such as the selective lesion of a particular molecular or cellular mechanism, or the exploration of unphysiological synaptic arrangements) can be effectively simulated by the use of a model. As mentioned in the introduction, one strategy consists of simulations that try to incorporate as many of the cellular details as possible. Although such an approach can be very useful, the realism of the model is a weakness and a strength at the same time: As the model is made increasingly realistic by adding more details and parameters (e.g., the distribution of a variety of active ion-channels and the multicompartmental reconstructed morphology of the dendritic branching), the simulation might turn to be as poorly understood as the studied neurobiological system itself. Equally worrisome, as all the cellular details are not yet know, is that important features may be left out, thus affecting the results. Finally, realistic simulations of a network of neurons are highly computation-intensive. Present constraints limit simulations to small-scale systems or to the subcomponents of a more complex system. Only recently has sufficient computational power been available to go beyond the simplest models.

On the other hand, when the interest is mainly focused on the network level, a simplified description of individual neurons and synapses, as those provided in this contribution, may successfully lead to a deeper understanding of the collective electrophysiological activity. More importantly, in this case, extended computer simulations can be approached, so that realistic network sizes and time scales can be studied and compared with real data (21,39).

Acknowledgments

Michele Giugliano and Maura Arsiero acknowledge support by the Human Frontier Science Program (LT00561/2001-B) and by the *EC Thematic Network in Neuroinformatics and Computational Neuroscience*, respectively.

BIBLIOGRAPHY

1. A. L. Hodgkin and A. F. Huxley, A quantitative description of membrane current and its application to conduction and excitation of nerve. *J. Physiol.* 1952; **117**:500–544.
2. L. F. Abbott and P. Dayan, *Theoretical Neuroscience*. Cambridge, MA: The MIT Press, 2001.
3. C. Koch and I. Segev, *Methods in Neuronal Modeling: From Ions to Networks*, 2nd ed. Cambridge, MA: The MIT Press, 1998.
4. R. Douglas, M. Mahowald, and C. Mead, Neuromorphic analogue VLSI. *Annu. Rev. Neurosci.* 1995; **18**:255–281.
5. H. R. Wilson and J. D. Cowan, Excitatory and inhibitory interactions in localized populations of model neurons. *Biophys. J.* 1972; **12**(1):1–24.
6. W. Gerstner and W. Kistler, *Spiking Neuron Models: Single Neurons, Populations, Plasticity*. Cambridge, UK: Cambridge University Press, 2002.
7. X. J. Wang, Synaptic Reverberation Underlying Mnemonic Persistent Activity. *Trends Neurosci.* 2001; **24**(8):455–463.
8. B. Hille, *Ion Channels of Excitable Membranes*. Sunderland, MA: Sinauer Associates, 2001.
9. V. N. Vapnik, *Statistical Learning Theory*. New York: John Wiley and Sons, 1998.
10. J. Hertz, A. Krogh, and R. G. Palmer, *Introduction to the Theory of Neural Computation*. Reading, MA: Addison-Wesley, 1991.
11. D. J. Amit and N. Brunel, Model of global spontaneous activity and local structured (learned) delay activity during delay. *Cerebral Cortex* 1997; **7**:237–252.
12. D. J. Amit and M. V. Tsodyks, Quantitative study of attractor neural network retrieving at low spike rates i, substrate-spikes, rates and neuronal gain. *Network* 1991; **2**:259.
13. J. Rinzel and G. B. Ermentrout, *Analysis of Neural Excitability and Oscillations*, 1st ed. Cambridge, MA: The MIT Press, 1989, pp. 135–169.
14. I. Segev, J. W. Fleshman, and R. E. Burke, *Compartmental Models of Complex Neurons*, 1st ed. Cambridge, MA: The MIT Press, 1989, pp. 63–96.
15. C. Morris and H. Lecar, Voltage oscillations in the barnacle giant muscle fiber. *Biophys. J.* 1981; **35**:193–213.
16. H. C. Tuckwell, *Introduction to Theoretical Neurobiology*. Cambridge, UK: Cambridge University Press, 1988.
17. N. Fourcaud and N. Brunel, Dynamics of the firing probability of noisy integrate-and-fire neurons. *Neural Comp.* 2002; **14**:2057–2110.
18. P. L. Meyer, *Introductory Probability and Statistical Applications*. Reading, MA: Addison Welsley, 1965, p. 287.
19. L. M. Ricciardi, *Diffusion Processes and Related Topics in Biology*. Berlin: Springer, 1977.
20. A. Rauch, G. La Camera, H.-R. Lüscher, W. Senn, and S. Fusi, Neocortical pyramidal cells respond as integrate-and-fire neurons to *in vivo*-like input currents. *J. Neurophysiol.* 2003; **90**:1598–1612.
21. M. Giugliano, P. Darbon, M. Arsiero, H.-R. Lüscher, and J. Streit, Single-neuron discharge properties and network activity in dissociated cultures of neocortex. *J. Neurophysiol.* 2004; June:1–20.
22. A. Destexhe, Z. Mainen, and T. J. Sejnowski, Synthesis of models for excitable membranes, synaptic transmission and neuromodulation using a common kinetic formalism. *J. Comp. Neurosci.* 1994; **1**:195–230.
23. A. F. Strassberg and L. J. DeFelice, Limitations of the hodgkin-huxley formalism: effects of single channel kinetics on transmembrane voltage dynamics. *Neural Comp.* 1993; **5**:843–855.
24. M. Giugliano, Synthesis of generalized algorithms for the fast computation of synaptic conductances with markov kinetic models in large network simulations. *Neural Comp.* 2000; **12**(4):771–799.

25. M. Giugliano, M. Bove, and M. Grattarola, Fast calculation of short-term depressing synaptic conductances. *Neural Comp.* 1999; **11**(6):1413–1426.
26. J. Reutimann, M. Giugliano, and S. Fusi, Event-driven simulation of spiking neurons with stochastic dynamics. *Neural Comp.* 2003; **15**:811–830.
27. M. V. Tsodyks and H. Markram, The neural code between neocortical pyramidal neurons depends on neurotransmitter release probability. *Proc. Natl. Acad. Sci. USA* 1997; **94**:719–723.
28. M. V. Tsodyks, K. Pawelzik, and H. Markram, Neural networks with dynamic synapses. *Neural Comp.* 1998; **10**:821–835.
29. L. F. Abbott, J. A. Varela, K. Sen, and S. B. Nelson, Synaptic depression and cortical gain control. *Science* 1997; **275**:220–223.
30. V. Yakovlev, S. Fusi, E. Berman, and E. Zohary, Inter-trial neuronal activity in infero-temporal cortex: a putative vehicle to generate long term associations. *Nat. Neurosci.* 1998; **1**:310–317.
31. D. J. Amit and M. V. Tsodyks, Effective neurons and attractor neural networks in cortical environtment. *Network* 1992; **3**:121–137.
32. D. J. Amit, *Modeling Brain Function*. Cambridge, UK: Cambridge University Press, 1989.
33. R. J. Douglas and K. A. Martin, Neocortex. In: *The Synaptic Organization of the Brain, 3rd ed*. New York: Oxford University Press, 1990, chapt. 12, pp. 389–438.
34. A. Destexhe and D. Paré, Impact of network activity on the integrative properties of neocortical pyramidal neurons *in vivo*. *J. Neurophysiol.* 1999; **81**:1531–1547.
35. N. Brunel, Persistent activity and the single cell *f-i* curve in a cortical network model. *Network* 2000; **11**:261–280.
36. F. S. Chance, S. B. Nelson, and L. F. Abbott, Synaptic depression and the temporal response characteristics of V1 cells. *J. Neurosci.* 1998; **18**(12):4785–4799.
37. P. S. Churchland and T. J. Sejnowski, *The Computational Brain*. Cambridge, MA: Bradford Books, The MIT Press, 1996.
38. M. Grattarola and G. Massobrio, *Bioelectronics Handbook*. New York: McGraw-Hill, 1998.
39. M. Giugliano, M. Arsiero, P. Darbon, J. Streit, and H.-R. Lüscher, Emerging network activity in dissociated cultures of neocortex: novel electrophysiological protocols and mathematical modelling. In: *Advances in Network Electrophysiology Using Multi-Electrode Arrays*. New York: Kluwer Academic/Plenum Publishers, 2005.

BIOMEDICAL ELECTRONICS

MART MIN
TOOMAS PARVE
Tallinn University of Technology
Tallinn, Estonia

RODNEY SALO
Guidant Corporation
St. Paul, Minnesota

1. INTRODUCTION

Electronics specifically applied to biomedical applications may be divided into two main categories: data acquisition and therapy. Data acquisition involves acquiring, processing, and recording signals from the body whereas therapy involves the application of energy or therapeutic agents (e.g., pharmaceuticals) to the body. The fact that biomedical electronics must interface with a living body generates a unique set of safety requirements. Other unique requirements develop from a need for small size and low power consumption because of limited space or a need for battery power. Implantable devices with extraordinary computational ability fitting within a volume of less than 30 ml with lifetimes of up to ten years are perhaps the most extreme example of the genre and must embody all of these characteristics.

A generic biomedical electronic system embodying all the usual components involved in data acquisition and therapy is shown in Fig. 1. Biological information is input into the system from sensors with analog or digital outputs. After signal conditioning, which may include amplification, filtering, demodulation, and multiplexing, the analog signals are digitized by an analog-to-digital (A/D) converter and output to a digital microcomputer (i.e., controller or signal processor). Digital signals are clocked into the microcomputer directly through a digital input–output (I/O) interface. The real-time microcomputer then records or processes the acquired digital data and controls therapy based on the resulting information. Changes in therapy (e.g., change in rate, energy, or site of electrical pacing; change in direction or amplitude of mechanical movement; change in the rate of drug infusion) are realized through digital and analog (after D/A conversion) control of actuators.

As, in most cases, electronics is directly connected to the living tissue/organ or patient through electrodes, safety considerations must be included in the design. Galvanic isolation between the medical-grade power supply and the instrument from one side, and between the instrument and patient from the other side, is required.

Finally, biomedical devices are commonly connected to data storage devices and to one or more information networks. Networks specific to medical applications include body area (BAN) and personal area (PAN) networks as well as wired or wireless public networks (i.e., the Internet). The system may use USB, Bluetooth, ZigBee, or some near-field communication means for both data transmission and energy transmission.

This chapter will concentrate on technologies, circuits, and systems particularly applicable to portable/handheld and wearable/implantable devices where low-power and low-voltage operation are either desirable or necessary.

2. DEVELOPMENT COURSE

The invention of the transistor in 1947 forever changed the electronics world. The manufacture of the silicon transistor in 1954 similarly affected biomedical electronics. In 1958, Swedish engineer Rune Elmqvist designed and built the first implantable pacemaker, which consisted of two bipolar silicon transistors and a few other components. The implant of this pacemaker by Swedish physician Åke Senning ushered in the era of implantable electronics. The

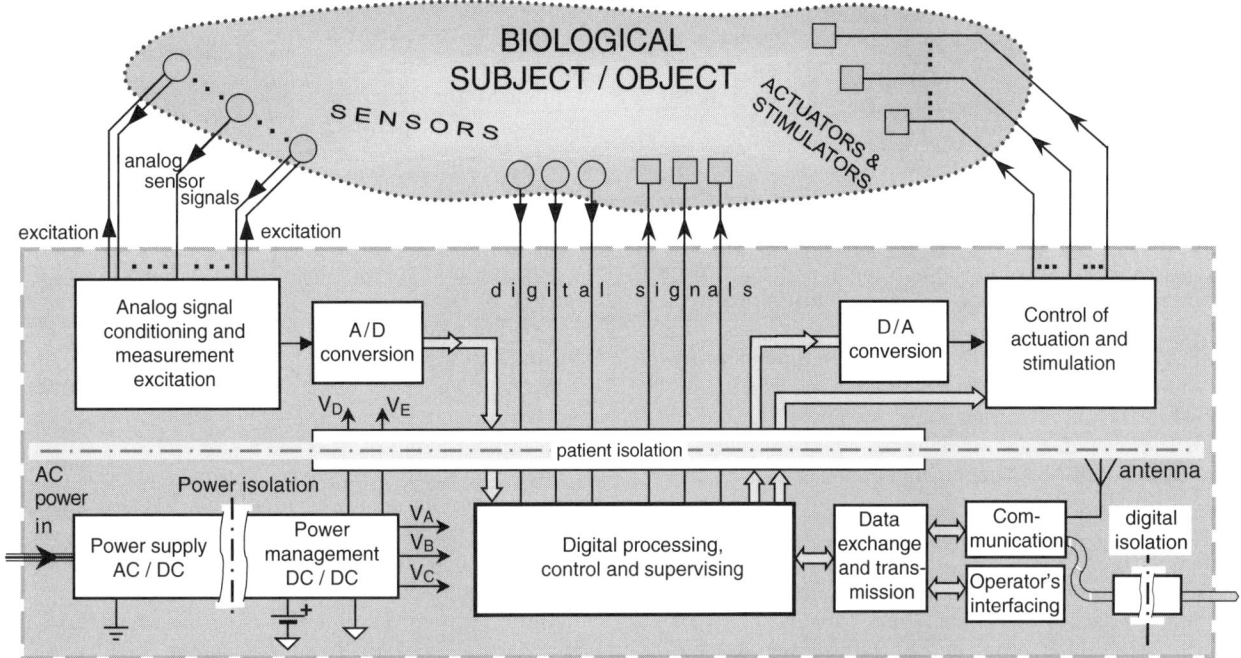

Figure 1. Generic electronic system for biomedical devices. (This figure is available in full color at http://www.mrw.interscience.wiley.com/ebe.)

later invention of the integrated circuit in 1958 by Jack Kilby (Nobel price winner in 2000) and Robert Noyce firmly established bipolar technology using silicon as the semiconductor material as an industry standard.

Bipolar transistors (BT) have low voltage noise characteristics, about $1\,nV/Hz^{1/2}$ RMS spectral density. Therefore, biomedical amplifiers using this technology exhibit low noise in applications that involve signals from relatively low-impedance (<1 kOhm) sensors at frequencies ranging from direct current (DC) to hundreds of MHz. As it is easy to manufacture bipolar transistors with similar characteristics in integrated circuits, matched circuits with a small difference in base-emitter voltage drop, low-voltage offset <1 mV, and small temperature drift ($\approx 5\,\mu V/^\circ K$) can be designed for DC applications (1). Less desirable characteristics include input currents of approximately 100 nA with accompanying current noise about $1\,pA/Hz^{1/2}$ RMS spectral density, relatively low base-to-emitter input resistance, from 1 to 100 kOhm, and current consumption per transistor more than one microampere.

Commercial field effect transistors with reversely biased p-n junction gate (JFET) with input currents three orders of magnitude smaller than bipolar transistors (i.e., picoamp range) and 100 MΩ range input resistance became available in the late 1960s. Unfortunately, all the other operating characteristics of JFETs are worse than those of bipolar transistors.

Concurrently, the insulated metal gate field effect transistor, MOSFET, came into use, primarily in digital electronics, but also in radio-frequency (RF) analog circuitry. These MOSFETs were high-speed devices with super-high-input resistance and practically zero level input current, but they were too noisy for applications in wide-band analog circuits, especially for amplification of low-frequency or DC signals. Shortly afterwards, complementary n- and p-channel MOSFETs (CMOS) revolutionized digital circuits (1) permitting the inclusion of tens and hundreds of millions of transistors in one digital chip.

At the beginning of twenty-first century, CMOS technology also dominates in analog circuitry (2). Modern CMOS analog circuit characteristics exceed even bipolar circuits with voltage offset of 100 nV, and temperature drift $25\,nV/^\circ K$. At about $3\,nV/Hz^{1/2}$ spectral density, voltage noise is somewhat higher but current noise is practically absent. As the same CMOS technology can be used for both analog and digital circuits, it is possible to design mixed signal analog/digital integrated microchips. This approach is already popular and this trend is sure to accelerate, particularly in low-voltage and super-low-power biomedical electronics. Bipolar transistors are now used only in addition to CMOS technology (Bi-CMOS) to meet particular requirements, for example, at the input when low noise level is crucial or at the output when high-level (more than 1 A) currents are to be handled at high speed.

3. BIOSIGNAL AMPLIFIER

The primary function of biomedical electronics is the recovery of small analog signals corrupted by noise and disturbances from sensors and electrodes with minimal distortions. The next task is conditioning of analog signals (amplification, filtering, demodulation, sampling, and holding) before digitizing to get the highest resolution and maximal effective number of bits (ENOB) (see also ANALOG

TO DIGITAL CONVERSION). The accomplishment of both of these tasks demands well-designed or correctly chosen biosignal amplifiers. Figure 2 illustrates a generalized differential biosignal amplifier with a signal source at the input and a load at the output.

Typically, the biosignal amplifier has a differential stage at the input to suppress common mode signal $V_{CM} = (V_{IN1} + V_{IN2})/2$, which appears at both inputs simultaneously, and to amplify the differential signal $V_d = V_{IN1} - V_{IN2}$. A typical common mode signal is an induced or radiated power line 50/60 Hz disturbance. Also, communication and other interferences can generate unwanted common mode signal, which can be much stronger than sensor signals. The extracted differential signal is amplified in the amplification stage and output to the load through the output stage, which can behave either as a voltage source with low-output impedance or as a current source with high-output impedance.

The amplifier in Fig. 2 can be either an operational amplifier with high ($G_D = 10^4$ to 10^6), but indeterminate, differential gain or an instrumentation amplifier with lower, but precisely determined, gain, typically $G_D = 1$ to 1000. Classically, the input and output signals are both voltages, but it is not uncommon for biomedical circuits, especially integrated circuits, to use current output amplifiers or operational transconductance amplifiers (OTA). In these circuits, the gain is characterized as transconductance, G_D, in units of mA/V or mS (milli-Siemens). Penetration of the common mode voltage V_{CM} to the output is characterized by the common mode gain $G_{CM} \ll 1$. The common mode rejection ratio for the complete amplifier is defined as $\text{CMRR} = 20 \cdot \log_{10}(G_D/G_{CM})$ dB. The CMRR is an important parameter. For example, in a practical case where $V_{CM} = 1$ V and $V_d = 1$ mV, the CMRR = 100 dB or 10^5 is required to limit the relative error caused by V_{CM} to 1%.

The CMRR value is determined by the exactness with which signals are subtracted in the differential stage. Therefore, this stage must be very symmetrical with respect to its two inputs, which is accomplished by making the transistors in the two branches of the circuit as identical as possible.

Modern analog biomedical electronics use low-voltage power supplies operating at 1.8 to 2.5 V. To operate with up to 1 V RMS signals is reasonable to obtain acceptable signal-to-noise ratio. Therefore, it is necessary to use the voltage range between power supply rails cleverly to avoid distorting the waveform against the common mode voltage V_{CM} at the input, and the output voltage V_{OUT} at output (see Fig. 2). Special rail-to-rail amplifier circuits have been designed to achieve a wide dynamic range at both the output and the input.

The value of input offset voltage, V_{off}, and its drift over temperature and time are important in direct current (DC) applications (1). The RMS value of spectral density of noise voltage, in nV/(Hz)$^{1/2}$, becomes critical (1) when low-level signals are to be amplified. Finally, special attention must be paid to the 1/f type increase of the spectral density of noise at frequencies lower than some kHz (See also NOISE IN INSTRUMENTATION)

Amplifiers exhibit inertia in their response to input signals causing their gains to change with signal frequency. Classic operational and transconductance amplifiers have nearly the same frequency response as integrators, that is, their gains are inversely proportional to frequency dropping off at a rate of -20 dB/decade or -6 dB/octave (1,2). An important parameter is the transition frequency f_T at which the overall gain is reduced to a unity, $G(f_T) = 1$. In voltage feedback amplifiers, gain (G) and bandwidth (BW) product is constant and $G \times \text{BW} = f_T$ or $\text{BW} = f_T/G$ (1–5). However, when the current feedback is used, the gain bandwidth product $G \times \text{BW}$ is not constant in current feedback amplifiers. These amplifiers have the advantage that higher closed-loop gains do not necessitate proportional decreases in bandwidth (6). It is especially important in programmable gain amplifiers where the BW must remain almost constant when changing gain.

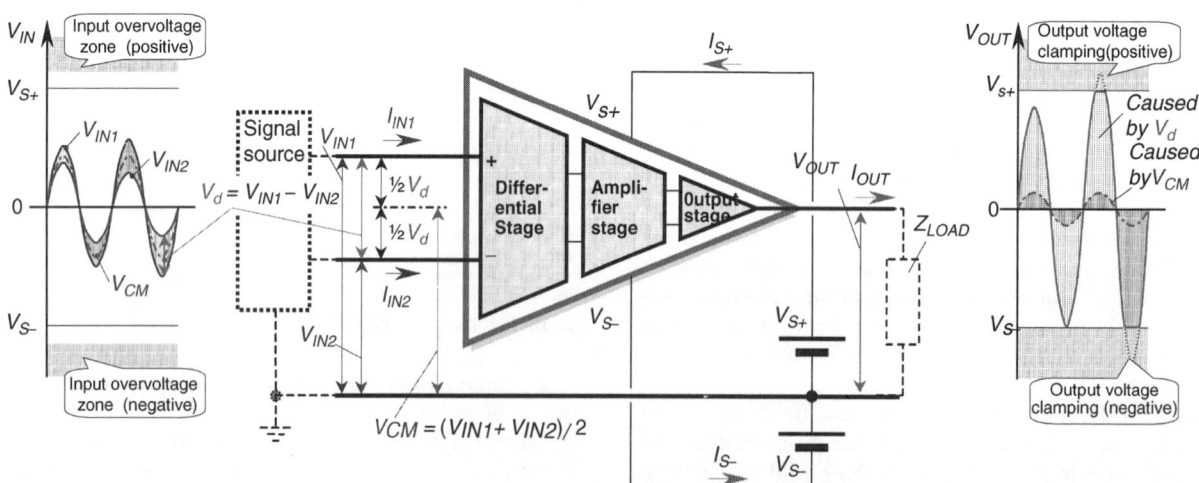

Figure 2. Differential biosignal amplifier. (This figure is available in full color at http://www.mrw.interscience.wiley.com/ebe.)

4. CMOS TECHNOLOGY

Biomedical electronics has used several integrated circuit (IC) technologies, but the basic semiconductor material has been silicon (Si) for decades. In the next 10 to 15 years, silicon will most likely remain the primary substrate for integrated circuits, and the prevailing silicon processing technology will continuously be CMOS (Complementary Metal-Oxyde-Semiconductor) for both analog circuits and digital circuits. Currently, all digital circuits and more than 90% of analog circuits in biomedical electronics are implemented in CMOS. Low-voltage, low-current digital CMOS designs using 90 nm technology and operating at microampere current consumption with 0.8 to 3.0 volt power supplies are commonplace throughout the industry. Some development is still necessary for analog circuitry, but serious efforts are underway to implement 1.0–2.5 V submicron CMOS analog circuits into pacemaker electronics (3).

CMOS technology uses n-type and p-type conductivity channels (complementary n-MOSFET and p-MOSFET), which are controllable through an electric field developed by a voltage V_{GS} between a metal gate, G, insulated from the channel by a super thin 1 to 5 nm silicon dioxide SiO_2 layer, and the source, S, at one end of the channel (Fig. 3). At the other end of the channel is drain D, which is connected to the voltage supply $+V$ when it is an n-channel and to $-V$ when it is a p-channel. So, the drain current I_D becomes controllable by the gate to source voltage V_{GS} (4).

Two types of channels in MOS technology exist (5). The enhancement-type channels (Fig. 3) do not conduct current when the control voltage is absent ($V_{GS}=0$). Charge carriers will be induced into the channel only when the forward-biased voltage $V_{GS}>V_{TH}$ is applied (V_{TH}-threshold voltage, typically 0.5–1.5 V or even lower).

The other one is a depletion-type channel, which is conductive in the absence of control voltage ($V_{GS}=0$) and a quiescent current I_{DSS} flows through the channel. The channel closes when the inverse V_{GS} is higher than a pinch-off voltage V_p (1,2).

Digital CMOS circuits use only the enhancement-type MOSFETs. Analog CMOS circuits use both types of MOSFETs, but the enhancement type is generally preferred.

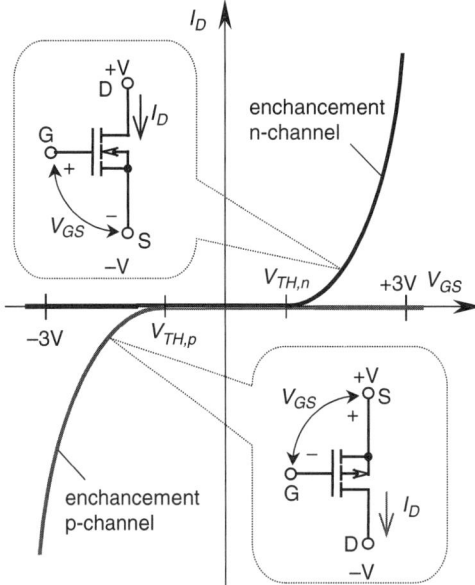

Figure 3. Enhancement-type CMOS transistors and their transconductance characteristics. (This figure is available in full color at http://www.mrw.interscience.wiley.com/ebe.)

5. CMOS CELLS FOR DIGITAL AND SWITCHED MODE ELECTRONICS

The basic digital CMOS cells are presented in Fig. 4, where complementary pairs, the enhancement-type n-channel and p-channel MOSFET transistors Q1 and Q2, are connected serially. In CMOS switches (Fig. 5), the complementary pairs are connected in parallel.

The circuit in Fig. 4a is known as a logic inverter (NOT gate). When the input V_i has a low near-to-ground potential (logic 0), then the n-channel of Q2 is closed (does not conduct) and the p-channel of Q1 is open (conducts). The output voltage V_o has a high level, very near to $+V_{CC}$ (logic 1). However, when the input V_i becomes high (logic 1), which means higher than threshold $V_{TH,n}$ of Q2 (see Fig. 3), then the channel of Q2 opens, Q1 closes, and the output V_o falls to ground potential (logic 0). Thus, the circuit performs a logic inversion $V_o = \overline{V}_i$.

In principle, current only flows through the circuit during transitions (or switching) when a brief current pulse

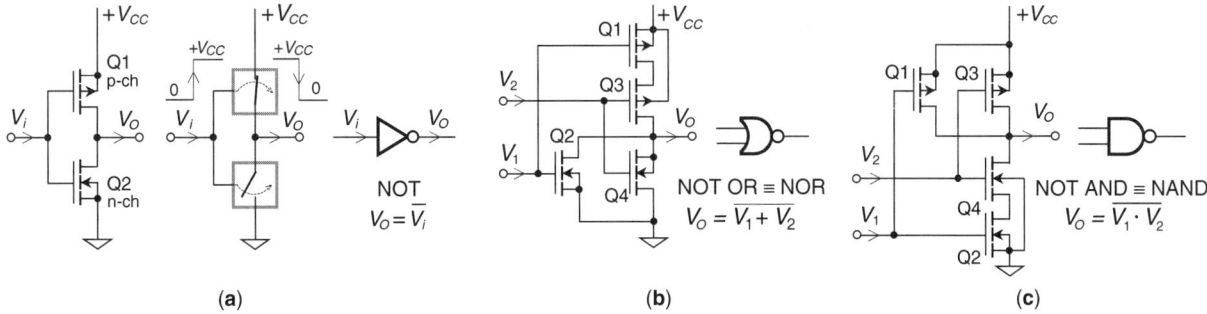

Figure 4. The basic CMOS logic circuits: inverter NOT with its equivalent circuit (a), NOR gate (b), and NAND gate (c). (This figure is available in full color at http://www.mrw.interscience.wiley.com/ebe.)

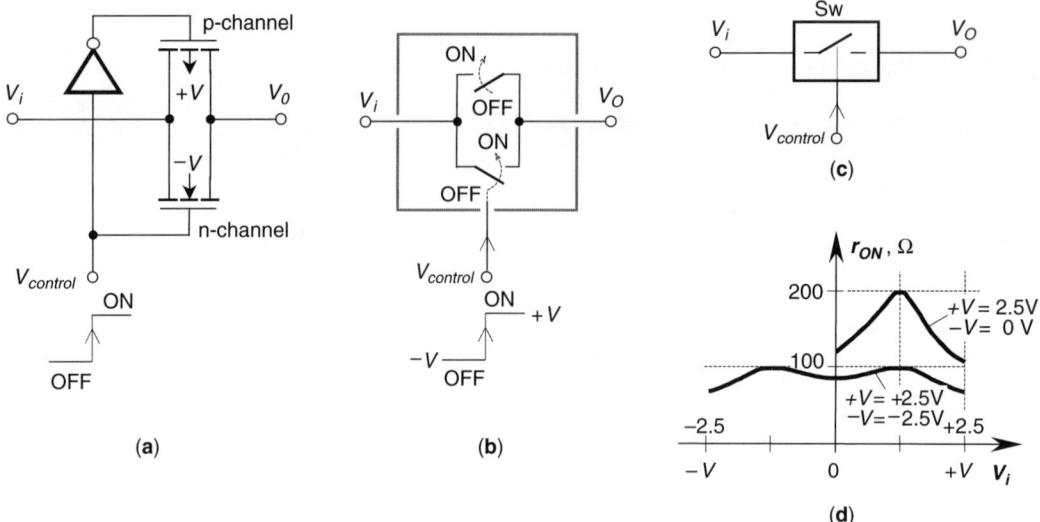

Figure 5. Bidirectional switch (a), its relay equivalent (b), circuit symbol (c), and dependence of ON resistance on input voltage (d). (This figure is available in full color at http://www.mrw.interscience.wiley.com/ebe.)

passes from $+V_{CC}$ to ground through both transistors. Therefore, current consumption depends on the pulse rate appearing at the input and dynamic properties of the circuit (faster switching circuits consume less power). In addition, some leakage current flows from the channel to a silicon substrate material through reversely biased p-n junctions denoted as arrows in Fig. 4 and Fig. 5. These p-n junctions form the interface between the channels and the semiconductor substrate when bulk silicon processing is used. When a silicon-on-insulator (SOI) process is used, these p-n junctions are absent, and current leakage is much smaller (3). The channels are not perfectly insulated, even when V_i is lower than the threshold, which results in some leakage current in the nanoampere range in any case.

The basic NOR and NAND logic gates are shown in Fig. 4b and Fig. 4c. The NOR gate has two complementary pairs of transistors connected serially (Q1 and Q2, and Q3 and Q4, respectively), whereas the n-channel transistors Q2 and Q4 are connected in parallel and p-channel transistors Q1 and Q3 in serial mode. When both V_1 and V_2 are low, then V_o has the high $+V_{CC}$ (logic 1). But when either V_1 OR V_2 will become high, then V_o falls low (logic 0).

The NAND gate in Fig. 4c also has two complementary pairs of transistors. The n-channel transistors Q2 and Q4 are connected serially and the p-channel transistors Q1 and Q2 are connected in parallel. When V_1 or V_2 is low, then V_o is high. Only when V_1 AND V_2 are both high will V_o be low (logic 0).

All other gates, including OR, AND, XOR (exclusive OR), and XNOR (exclusive OR with inverted output) and more complex components, such as triggers, registers, counters, coders, as well as computer components, such as arithmetical units, memories, peripheral interfaces, and other programmable units, can be synthesized by combining NOR, NAND, and NOT gates.

Several companies produce CMOS logic. For biomedical applications, the advanced ultra-low-power AUP family from Texas Instruments, for example, the NAND gate SN74AUP1G00, NOR gate G02, and NOT gate G04, are most suitable (7). These gates can work with power supplies from 0.8 to 3.3 V and consume 0.5 µA current at low switching frequencies. In complex digital circuits, the average standby current consumption per gate remains in the pA range because of intelligent power management, wherein the power supply of subcircuits is switched off automatically when they are not in use.

The CMOS pair transistors in a switch circuit in Fig. 5 is connected in parallel. The n-channel transistor Q1 is controlled directly, but the p-channel transistor Q2 is controlled through an inverter. As a result, both transistors open (ON state) and close (OFF state) simultaneously. The cell behaves like a bidirectional semiconductor switch (Figs. 5b and 5c) and can be used as in analog as well as in digital circuits. As the transistors are connected in parallel, the ON resistance r_{ON} depends weakly on polarity and level of the voltage V_i to be switched. It is important that V_i falls within the power supply limits ($-V$ to $+V$). Depending on size of the channel (low or high current switch) and the supply voltage levels, the ON resistance r_{ON} can be as low as few Ω or as high as a few thousand Ω. Typically, the ON resistance changes about 10% with the value of V_i (see Fig. 5d), but in ultra-low-power pacemaker circuits (3), it can depend up to 50% on the level of V_i and can reach tens of thousands of Ω.

CMOS switches play an important role in biomedical electronics. In digital circuits, it sometimes makes sense to use switch-based relay logic instead of classical logic gates shown in Fig. 4, especially when the logic function is simple. Switches are foundational components in switched capacitor and switched current filters and amplifiers (1,8,9) and synchronous demodulators (4,10). Switches are essential building blocks of analog multiplexers (Mux) and demultiplexers (Dmux), sample-and-hold (S/H) circuits, and analog-to-digital (A/D) and digital-to-analog (D/A) converters (1,9). CMOS switches with higher power and voltage capabilities function as voltage multipliers and pulse formers in pacemakers and defibrillators (8,10), as pulse width modulators (PWM) for controlling actuating power, and in isolating AC/DC and DC/DC converters as power modulators and demodulators (5).

6. CHARACTERIZATION OF MOS TRANSISTORS IN CMOS ANALOG CIRCUITS

The transconductance characteristics of CMOS transistors, $I_D = F(V_{GS})$ at $V_D = $ const, displayed in Fig. 3, are nonlinear functions and have a quadratic form when V_{GS} is greater than threshold voltage ($V_{GS} > V_{TH}$):

$$I_D = K(V_{GS} - V_{TH})^2. \tag{1}$$

Here, K is a constant that depends on the geometry of the transistor, primarily on W/L (the width/length ratio for the channel), and on the thickness of the silicon dioxide layer between the gate and channel.

The most important parameter for MOS transistors is the dynamic transconductance $g_m = \Delta I_D/\Delta V_{GS}$ at a constant value of V_D. The derivative of Equation 1 gives

$$g_m = 2K(V_{GS} - V_{TH}). \tag{2}$$

As from Equation 1 we can get $K = I_D/(V_{GS} - V_{TH})^2$, the transconductance can also be expressed as

$$g_m = 2I_D/(V_{GS} - V_{TH}). \tag{3}$$

Modern low-voltage ($V_D = 1$ to 3 V) and low-current ($I_D = 0.1$ to 1 µA) transistors have 80- to 160-nm long channels and 2- to 5-nm thick dioxide layer (the breakthrough voltage of 1-nm thick SiO_2 layer is about 1 V). The transconductance for nanoampere transistors can be derived from an experimental relationship, where g_m/I_D is about 15–25 V^{-1} (subthreshold regime, $V_{GS} < V_{TH}$). That is, when $I_D = 100$ nA, the transconductance g_m is at least 1.5 µA/V. Using a load resistance $R_L = 50$ MΩ, the voltage gain, $G = g_m \times R_L = 75$, is reachable.

7. BASIC CELLS FOR ANALOG CMOS CIRCUITS

As shown in Fig. 6a, in addition to a matched pair of p-channel (or n-channel) transistors, Q1 and Q2, a constant current source and current mirror, are important components of the differential stage (see Fig. 2) of amplifiers commonly included in CMOS analog circuits. The output current I_{out} proportional to the difference ΔV_i between the input voltages (Fig. 6a)

$$I_{out} = g_D(\Delta V_i)G_I, \tag{4}$$

where g_D, A/V is a differential transconductance.

Although more diverse types of analog circuits than digital circuits exist, there are some traditional building blocks or circuit cells that can be used for creating different analog circuits. Perhaps the most frequently used cell is a current mirror (CM), shown in Figs. 6b and 6c. The implementation in Fig. 6b is based on p-channel transistors, and the cell in Fig. 6c uses n-channel transistors. In both cases, the transistor Q4 (or Q6) operates as a source of current I controllable by bias current I_{bias}. Transistor Q3 (or Q5) ensures a linear relationship (constant current gain G_I) between the currents I and I_{bias}. The current gain G_I depends on a difference of transconductances of the transistors Q3 and Q4 (or Q5 and Q6). The current mirrors can operate as an amplifier stage in biosignal amplifiers (Fig. 2). When the current I_{bias} has a fixed value, the CM functions as a constant current source (Fig. 6d).

8. DIFFERENTIAL CMOS BUILDING BLOCKS

The differential amplifier shown in Fig. 7a corresponds to the amplifier in Fig. 2 with a current output. Both n-channel transistors Q1 and Q2 in a matched differential pair work under equal conditions with no direct connection between the differential pair and output. Therefore, this symmetrical differential amplifier is inherently more precise than the simpler differential circuit in Fig. 6a. This building block is called as an operational transconductance amplifier (OTA) and is characterized by a differential voltage input and single-ended current output. The

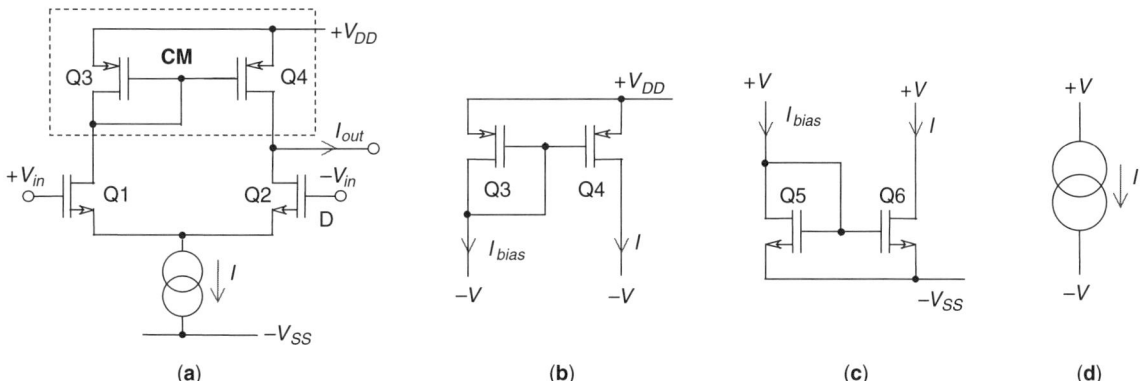

Figure 6. Differential amplifier stage (a) and current mirrors based on p-MOS (b) and n-MOS (c) transistors, and symbol of a constant current source (d). (This figure is available in full color at http://www.mrw.interscience.wiley.com/ebe.)

506 BIOMEDICAL ELECTRONICS

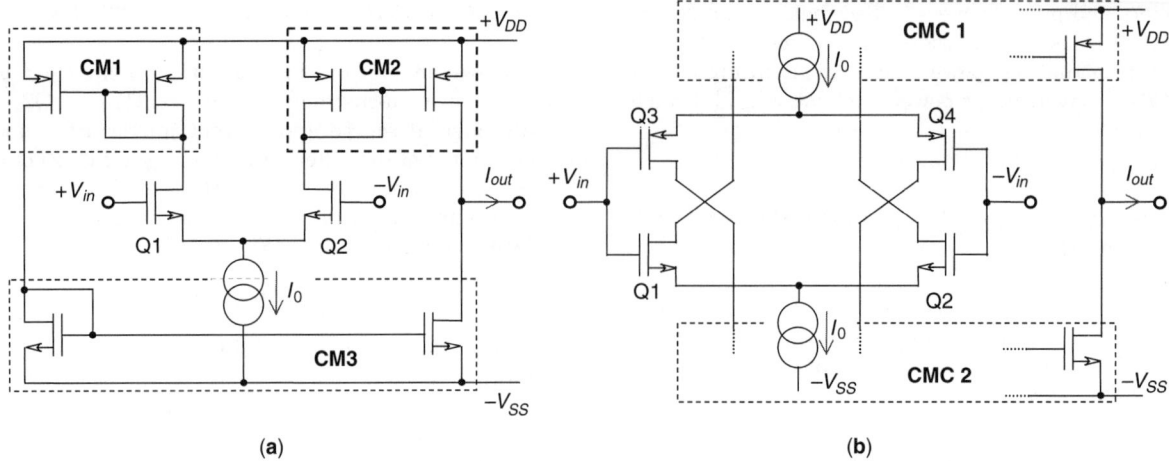

Figure 7. Operational transconductance amplifier (a) and its version with rail-to-rail differential input (b). (This figure is available in full color at http://www.mrw.interscience.wiley.com/ebe.)

main parameter of OTA is its differential transconductance g_D with respect to the input voltage difference ΔV_i.

The OTA with a rail-to-rail input is illustrated in Fig. 7b. It has two differential pairs of matched transistors (n-channel pair Q1 and Q2, and p-channel pair Q3 and Q4) at the input. This arrangement permits a symmetrical and almost rail-to-rail voltage swing between power supply voltages $-V_{SS}$ and $+V_{DD}$ at the input. This feature is especially important in low-voltage circuits, where the available range for voltage swing is narrow, especially for common-mode voltages V_{CM} (see Fig. 2) operating at the both inputs simultaneously (1,3,4,10).

The use of current-mode circuits (CMC1 and CMC2 in Fig. 7b) is characteristic to modern low-voltage circuits in which designers must deal with a very limited voltage range. By using current, the voltage responses can be kept low. Current-mode techniques provide the only solution for ultra-low-voltage (less than 1 V) analog electronics in the near future (6).

The novel circuit in Fig. 8a is called as differential difference amplifier (DDA). This amplifier has two precision differential pairs of n-channel transistors (Q1;Q2, and Q3;Q4) in parallel at the input, forming together with current mirrors (CM1, CM2, and CM3) two identical operational transconductance amplifiers OTA1 and OTA2 with equal transconductances g_D and a common current output (see Fig. 8b). The output current I_{out} is proportional to the difference $\Delta V_1 - \Delta V_2$ between two voltage

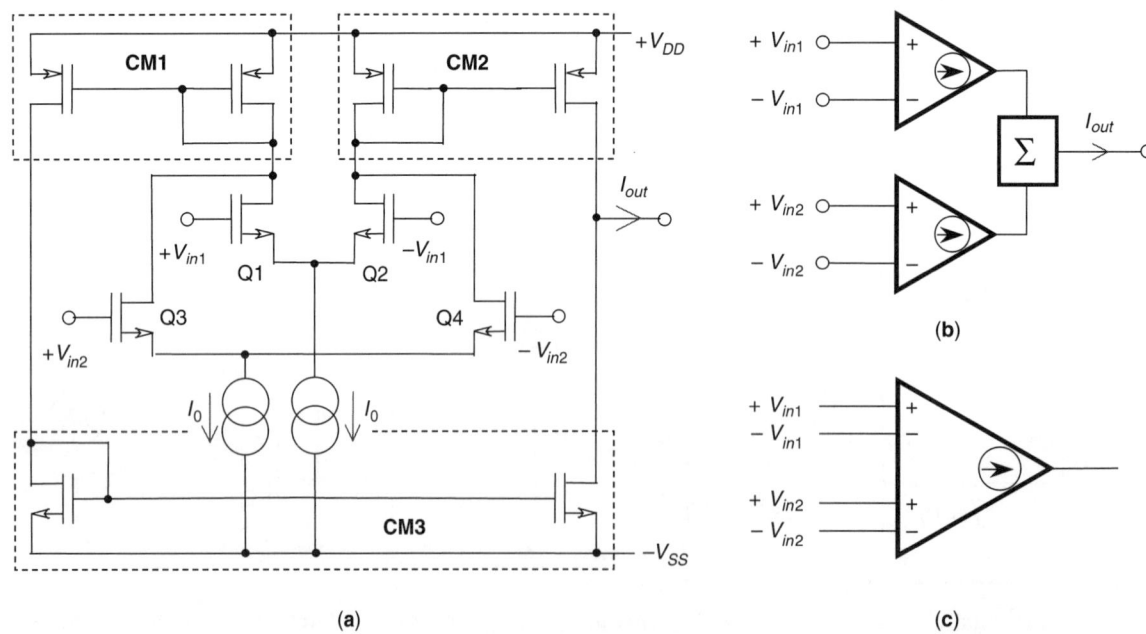

Figure 8. Differential difference amplifier (DDA), where (a) depicts the circuit diagram, (b) explains principles of operation, and (c) gives a circuit symbol. (This figure is available in full color at http://www.mrw.interscience.wiley.com/ebe.)

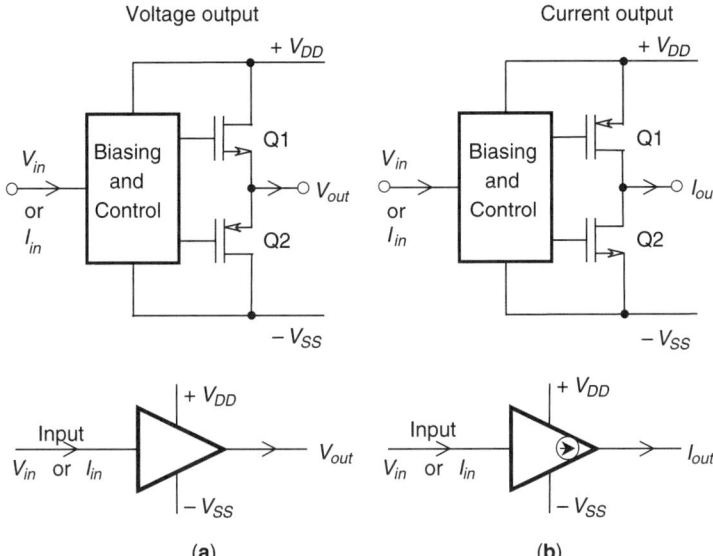

Figure 9. Output stages with low-(a) and high-(b) output impedance. (This figure is available in full color at http://www.mrw.interscience.wiley.com/ebe.)

differences ΔV_1 and ΔV_2 at the inputs of these OTAs

$$I_{out} = g_D \cdot G_I \cdot (\Delta V_1 - \Delta V_2). \qquad (5)$$

This new building block is suitable for a number of applications. An acceleration sensor signal amplifier (3) and a grounded load-current source and integrator (11) are the example medical applications.

9. OUTPUT STAGES OF CMOS ANALOG CIRCUITS

Two types of AB class CMOS output stages are illustrated in Fig. 9; in each, a small quiescent current flows through both transistors, n-channel Q1 and p-channel Q2. In a low-impedance stage of Fig. 9a, where the output voltage is taken from united sources of Q1 and Q2, the output impedance can be computed through $1/g_1$ and $1/g_2$, where g_1 and g_2 are transconductances of the transistors. Under real conditions, the output impedance may range from 100Ω to $100 \mathrm{k}\Omega$. Bipolar transistors may be added to obtain the lower output impedance (Bi-CMOS technology). This low-impedance output stage can be added to the high-impedance current output of an OTA to transform operational transconductance amplifiers to classical voltage output operational amplifiers (OA).

Figure 9b illustrates a high-impedance output stage, where the output current I_{out} is taken from the united drains of transistors Q1 and Q2. The output impedance depends strongly on the load voltage and can be as high as hundreds of $M\Omega$. This type of output stage is used in OTAs and other low-power analog circuits, for example, in rail-to-rail output voltage amplifiers (3).

10. A TRANSOR CIRCUIT

An interesting analog building block is described in Fig. 10. Two pairs of complementary transistors Ql; Q3 and Q2; Q4 together with constant current sources and current mirrors are connected to obtain a three-node (G – gate, S – source, and D – drain) circuit cell with power supply voltages $+V_{DD}$ and $-V_{SS}$. The circuit behaves as a symmetrical MOS transistor (Fig. 10b), which has properties of n-channel and p-channel MOSFETs

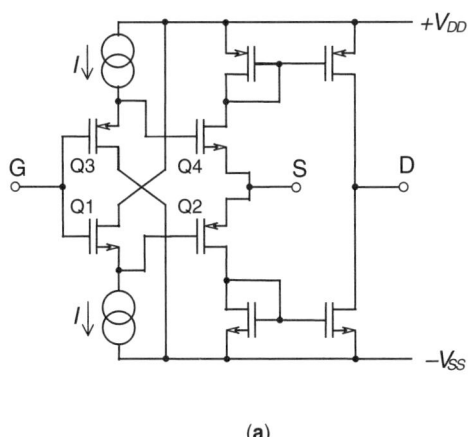
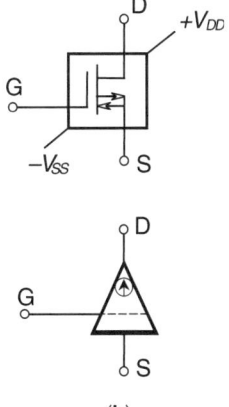

Figure 10. A transor circuit as an ideal transistor: (a) circuit diagram, and (b) circuit symbols. (This figure is available in full color at http://www.mrw.interscience.wiley.com/ebe.)

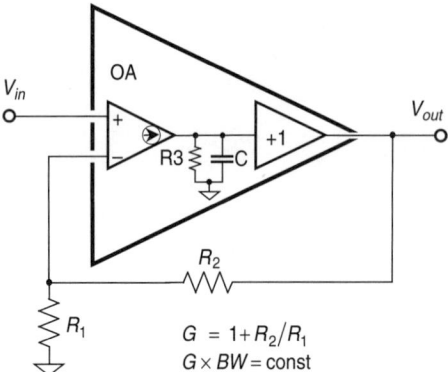

Figure 11. Voltage feedback amplifier.

at the same time (Fig. 3). In circuit theory, a similar component is known as transconductor (6) or transor – a linear voltage-controlled current source. The first implementation of a nearly ideal transor is called as "diamond transistor" (Burr-Brown OPA660 from Texas Instruments) (12). A similar component AD844 from Analog Devices is known as current conveyor. The transors can play an important role in the future integrated circuits as current sources and current feedback amplifiers (see Fig. 12).

11. VOLTAGE AND CURRENT FEEDBACK AMPLIFIERS

Integrated circuit amplifiers have poorly defined and unstable gains. This problem can be addressed by the addition of negative feedback (1–4). Figure 11 depicts a noninverting voltage feedback operational amplifier OA consisting of a differential input OTA and a low-impedance output stage (Fig. 9a) operating as a voltage follower (unity gain). The feedback factor is $\beta = R1/(R1 + R2)$. Assuming that the open-loop gain, A, is much larger than closed-loop gain with feedback, the closed-loop gain can be expressed as $G = 1/\beta = 1 + R2/R1$. The open-loop gain, A, is frequency-dependent because of a stray capacitance C (Fig. 11). Thus, the complex transfer function is $A(j\omega) = A_0/(1 + j\omega T)$, where $T = R3 \cdot C$ is a time constant. The Bode plot of the open-loop gain $A(j\omega)$, shown in Fig. 13, has a maximum $A_0 = 2 \cdot 10^4$, and a corner frequency f_C at $1/(2\pi T) = 50$ Hz. The transition frequency, where gain falls to unity ($G = 1$), is $f_T = 10^7$ Hz. One can see that bandwidth (BW) of the feedback amplifier is reciprocal to gain, which means that gain and bandwidth product $G \times BW$ is almost constant in voltage feedback systems (1–4).

The current feedback amplifier (4,6) in Fig. 12 is based on two transors (see Fig. 10). The first transor, T1, operates as a voltage amplifier, and the other, T2, as a voltage follower with near to unity gain. The open-loop frequency response has a low-frequency gain $A_0 = 2 \cdot 10^3$ and a corner frequency f_C at 10^5 Hz (Fig. 14). The feedback chain is the same as in the voltage feedback amplifier in Fig. 11, with $R2 \gg R1$. In this circuit, R1 plays a double role: it determines the feedback factor $\beta = R1/(R1 + R2)$, and also the gain of stage T1, which is the open-loop gain of the amplifier as well, $A_0 = (R3)/R1$. Changing the closed-loop gain $G = 1/\beta = 1 + R2/R1$ by changing R1 also changes the open-loop gain A_0. As a matter of fact, the product $A \times \beta$ stays almost constant. The closed-loop bandwidth also remains constant (BW = 100 kHz) when changing the gain $G = 1 + R2/R1$ (see Fig. 14), which is a major advantage of current feedback amplifiers in addition to their somewhat greater bandwidth compared with voltage feedback amplifiers under the same conditions.

12. INSTRUMENTATION AMPLIFIERS

An instrumentation amplifier (IA) is a differential amplifier with a precisely determined gain G and a high common-mode rejection ratio CMRR (a measure of the relative gain for differential voltage, ΔV_i, versus common-mode voltage, V_{CM}).

Figure 15 describes a standard voltage feedback, three operational amplifiers (OA)-based IA (1,4,5,10) with output $V_{out} = \Delta V_i(1 + 2(R2/R1))$. Gain for the common-mode voltage V_{CM} is always unity for the first differential stage on OA1 and OA2 independent of the values of R1 and R2, which determine the gain for the differential

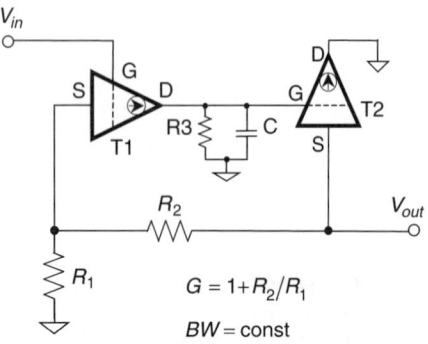

Figure 12. Current feedback amplifier.

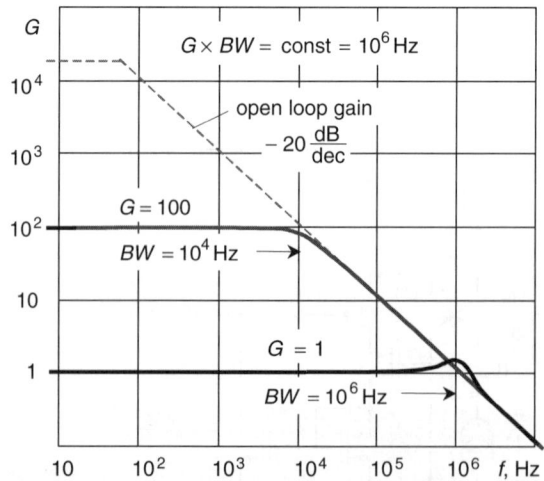

Figure 13. Frequency response of the voltage feedback amplifier. (This figure is available in full color at http://www.mrw.interscience.wiley.com/ebe.)

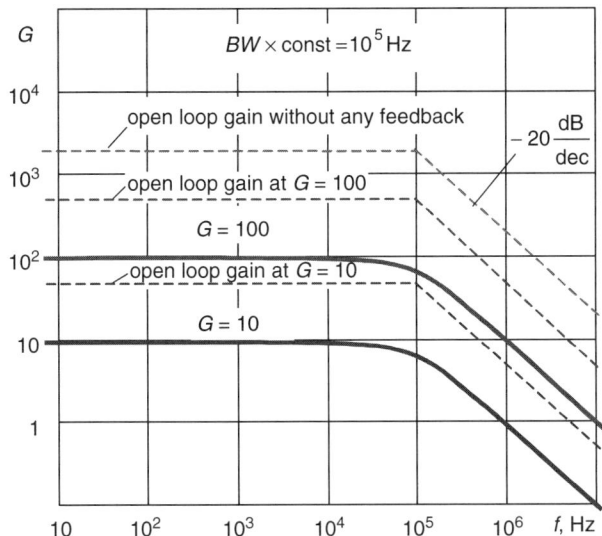

Figure 14. Frequency response of the current feedback amplifier. (This figure is available in full color at http://www.mrw.interscience.wiley.com/ebe.)

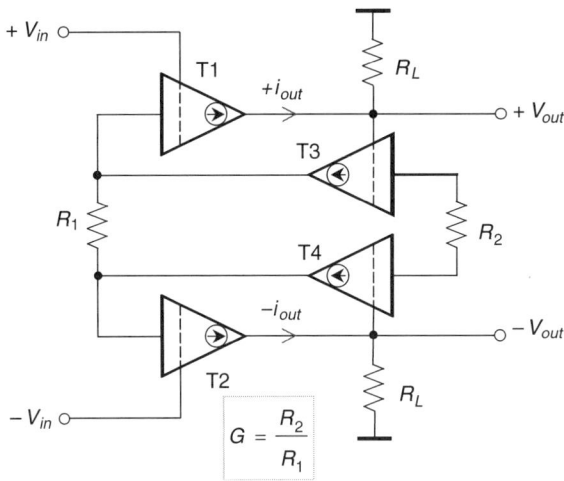

Figure 16. Current feedback instrumentation amplifier. (This figure is available in full color at http://www.mrw.interscience.wiley.com/ebe.)

signal ΔV_i. Therefore, the CMRR of the IA is G times that for differential amplifier, based on OA3. For example, when $G = 100$, and CMMR of OA3 stage is 80 dB, the CMRR for the IA is $40 + 80 = 120$ dB. As such, the instrumentation amplifiers are commercially available from several analog electronics companies (e.g., Burr Brown INA118 from Texas Instruments).

A current feedback IA in Fig. 16 is a fully differential amplifier, which is built up using two pairs of transors. T1 and T2 operate as a feed-forward differential amplifier with a gain determined by R1. Transors T3 and T4 operate as a feedback amplifier with a gain determined by R2. As a result, changing of the gain $G = R2/R1$ does not affect the bandwidth of IA (BW = const). As such, the instrumentation amplifiers AD620 and AMP01 are commercially available from Analog Devices (4).

Both types of instrumentation amplifiers IA are commonly used in biomedical instrumentation (4,10).

13. FULLY DIFFERENTIAL (SYMMETRICAL) AMPLIFIER

Fully differential amplifiers with differential input and output enable the design of symmetrical circuits that are much less sensitive to additive common-mode disturbance than the single-ended circuits. Ideally, the entirely processing chain from sensors to an A/D converter should consist of symmetrical circuits. An example of a fully differential amplifier is the AD8139 from Analog Devices, depicted in Fig. 17. An interesting feature of this amplifier is common-mode feedback (here, V_C is a common-mode voltage at the output). Independent of common-mode features, an input for biasing (shifting) both outputs to a desired level V_{bias} also exists.

14. LOW-VOLTAGE POWER SUPPLIES

Low-voltage (1 to 3 V) circuits require, low power consuming power supplies (13,14). Current solutions use CMOS low drop-out (LDO) voltage supplies illustrated in Fig. 18.

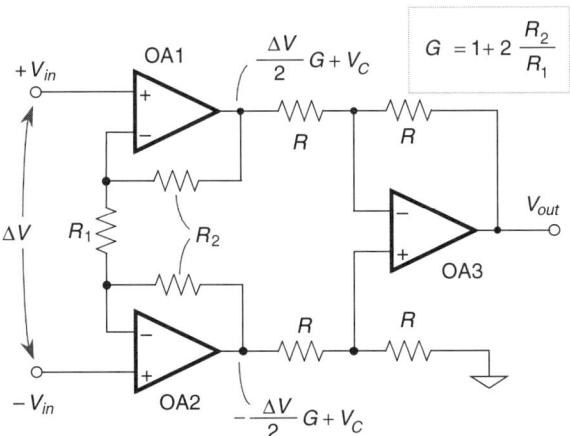

Figure 15. Voltage feedback instrumentation amplifier. (This figure is available in full color at http://www.mrw.interscience.wiley.com/ebe.)

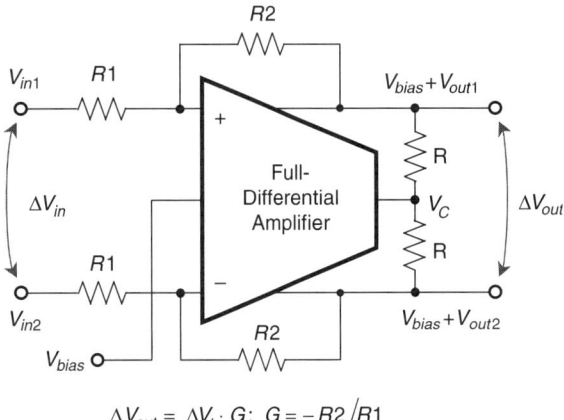

Figure 17. Differential input/output amplifier. (This figure is available in full color at http://www.mrw.interscience.wiley.com/ebe.)

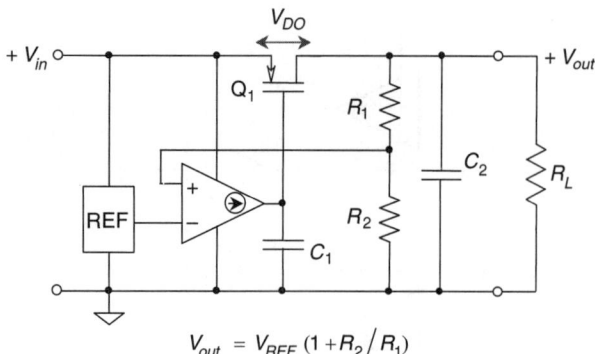

Figure 18. Low drop-out (LDO) voltage supply. (This figure is available in full color at http://www.mrw.interscience.wiley.com/ebe.)

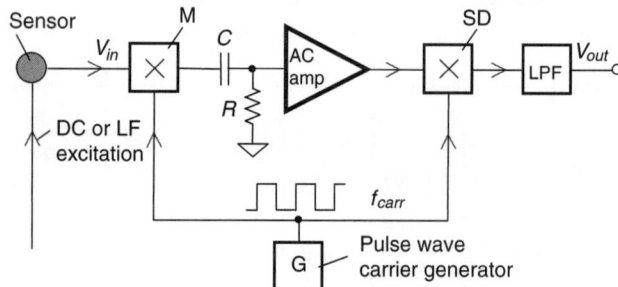

Figure 20. AC carrier-based DC amplifier. (This figure is available in full color at http://www.mrw.interscience.wiley.com/ebe.)

In an LDO supply, a regulating p-MOS transistor Q1 operates in a weak inversion regime, where only tens of millivolts (max $V_{DO} < 100\,\text{mV}$) are lost or dropped out (13). A very special regulating n-MOS transistor with negative threshold ($V_{TH} \leqslant 0$) can give even better results (14).

The low-voltage LDO needs a low reference V_{REF}. Traditional Zener diode based reference sources provide too high voltage, more than 5 V (Fig. 19a). One alternative is to use the p-n junction of a bipolar transistor (Fig. 19b). The drawbacks are that the voltage cannot be changed (about 0.6 V), and it is sensitive to temperature.

A better solution (13), a bandgap reference source (14), is shown in Fig. 19c. The voltage is given by $V_{REF} = V_{BE} + \Delta V_{BE}(R2/R3)$, where $\Delta V_{BE} = V_{BE1} - \Delta V_{BE2}$ is a voltage drop on R3 and $\Delta V_{BE}(R2/R3)$ appears on R2. The difference ΔV_{BE} achieved because of different currents flowing through transistors Q1 and Q2. The bipolar transistors are available as side products in bulk CMOS manufacturing technology.

15. CARRIER-BASED CONDITIONING OF SENSOR SIGNALS

Many biomedical signals, for example, respiratory and accelerometric responses, ECG, and EEG, are primarily low-frequency signals ($f < 100\,\text{Hz}$) (4,10,15). Many biomedical sensors, including temperature, pressure, displacement, and concentration sensors, use DC. Some sensors use DC or low-frequency (LF) excitation [e.g., bridge-type resistive sensors (Fig. 20)]. Unfortunately, the DC amplifiers have voltage offset that is temperature-dependent and drifts over time. Furthermore, because of a flicker or $1/f$ type pink noise, the noise level is much higher at low frequencies as is evident in the spectral density curve of noise power in Fig. 21. One solution is to use a higher frequency AC carrier for the sensor signal (14), which avoids errors from the DC offset and drift and reduces the noise floor to that of white noise. The noise level is proportional to the area S under the noise power spectral density curve $S_n(f)$ in V^2/Hz. This area and noise level can be reduced using narrow band filters tuned to the carrier frequency f_{carr}.

When the noise bandwidth Δf is constant, then the noise dispersion is directly proportional to the value of $S_n(f)$ at the carrier frequency. As is evident from Fig. 21, moving from f_{exc} to f_{carr} reduces the noise level significantly.

In Fig. 20, the DC or LF input voltage V_{in} from a sensor is multiplied with a rectangular wave carrier in a modulator M (chopper), and the AC product is amplified by an AC amplifier and then demodulated in a synchronous detector SD. The amplified and demodulated DC component is extracted at the output using a low-pass filter (LPF), which also determines the noise bandwidth Δf, see Fig. 21. Both modulation and demodulation are performed using analog CMOS switches (see Fig. 5) controlled digitally from a pulse wave generator.

Another modification, where the AC excitation itself is the carrier, is depicted in Fig. 22. The AC excitation is

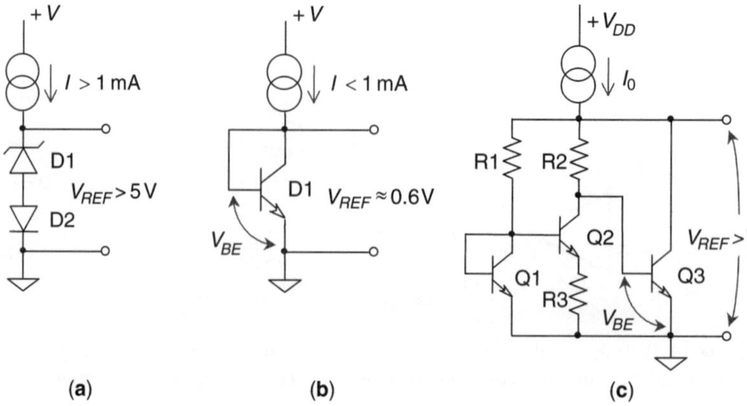

Figure 19. Reference voltage sources based on (a) diode-compensated Zener, (b) base-emitter junction; and (c) basic bandgap circuit. (This figure is available in full color at http://www.mrw.interscience.wiley.com/ebe.)

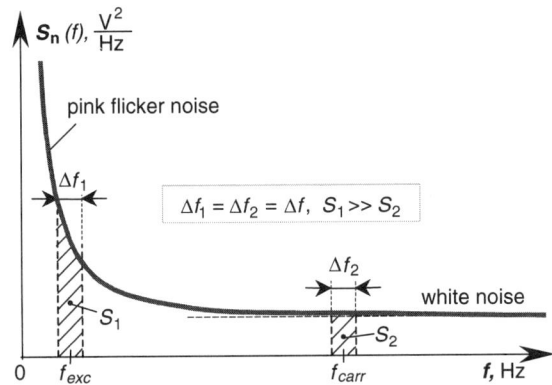

Figure 21. Noise shaping using AC carrier. (This figure is available in full color at http://www.mrw.interscience.wiley.com/ebe.)

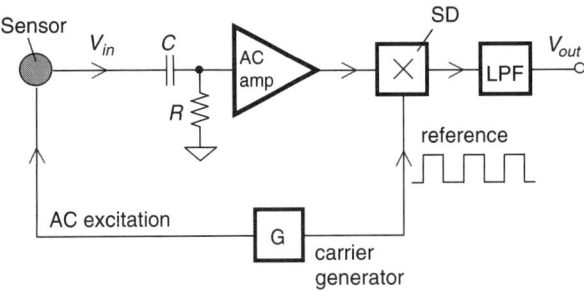

Figure 22. Using sensor excitation as AC carrier. (This figure is available in full color at http://www.mrw.interscience.wiley.com/ebe.)

modulated in the sensor by a physiological parameter. The modulated V_{in} is then amplified in the AC amplifier, demodulated synchronously in the SD, and filtered out by the LPF. Technically, the AC excitation could be a rectangular pulse waveform, but some sensors [e.g., complex bioimpedance sensors (4,10,16)] require sine wave excitation, in principal. An alternative approach is to use specific pulse-level or pulse width (PWM) modulations instead of sine waves (17,18). In some studies, simultaneous excitation at different frequencies and at different locations of the biological subject is desired. In this case, a digital arbitrary waveform generator (10) or a system of direct digital synthesizers (DDS) (10,19) may be appropriate for the generating of excitation. A digital synchronous demodulation, using some modification of the discrete Fourier transform (DFT), may be applied for demodulation of the sensor signal (19).

16. GALVANIC ISOLATION

In medical devices, a 4 kV galvanic isolation of the patient from a medical instrument is required (IEC 60601–1) (see also ELECTRICAL SAFETY). Galvanic isolation means that the parts of medical instrument are separated (insulated from each other) having in mind a direct electrical conductivity. This isolation usually resides in the digital section of instrument (see Fig. 1), but sometimes it is not possible or recommended. In this case, the AC carrier technique with modulation and demodulation may be applied, as shown in Fig. 23. In a modulator, the amplified input voltage V_{in} modulates the carrier, which passes the band-pass galvanic isolator via magnetic (Fig. 24a), optical (Fig. 24b), or capacitive connection (1,4,10). After that, the carrier is demodulated, and the modulation is filtered out and further amplified. The carrier frequency is typically higher than 100 kHz, and the signal bandwidth reaches to tens of kHz (4,10,15).

Optical isolation is possible without a carrier. In this case, both the feed-forward and feedback throughputs of optrons are required (Fig. 24c) for linearization of the nonlinear transfer characteristics (4) of optrons.

Digital transfer of pulses do not require linearity, and only a small number of couplers is needed for serial communications. Optical couplers, as in Fig. 24b, consume more energy than transformer-based couplers. New four-channel transformer couplers as ADuM24XX from Analog Devices meet the requirements with 5 kV$_{rms}$ isolation for one minute, and support a data rate from DC to 90 Mbps.

Power supplies also require isolation. Figure 25 shows a circuit diagram of a carrier-based direct current power coupler (DC/DC converter), which uses a switched-mode power modulator and demodulator, and a transformer-based magnetic power transmission. The THP 3 Series 3 W DC/DC converters from TRACO Electronic AG are examples. These devices feature an isolation of 4 kV$_{rms}$ for one minute and a carrier frequency of 150 kHz.

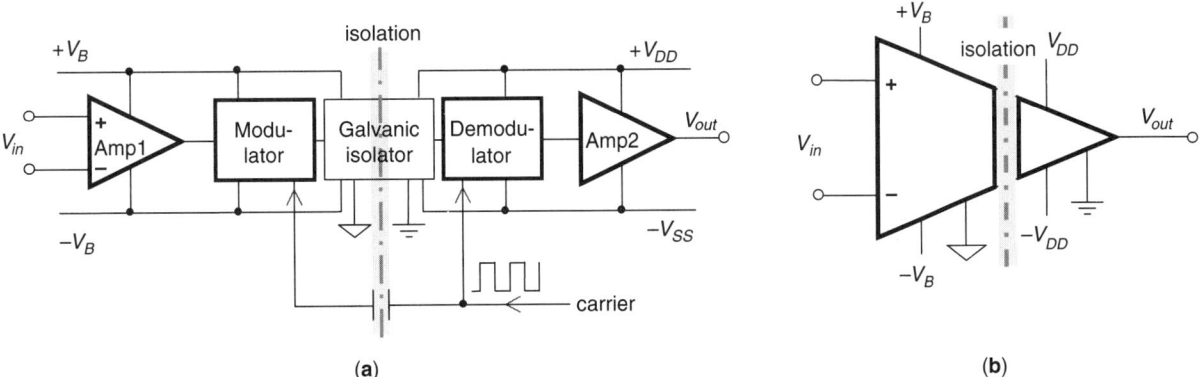

Figure 23. Galvanic isolation in an analog signal amplifier: (a) circuit diagram, (b) circuit symbol. (This figure is available in full color at http://www.mrw.interscience.wiley.com/ebe.)

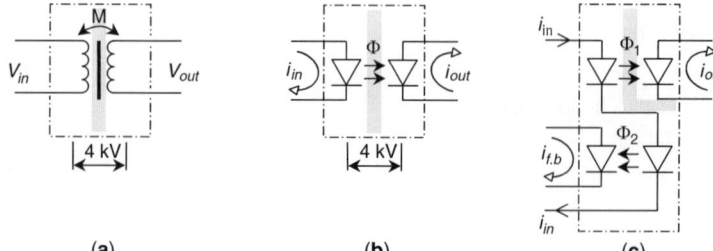

Figure 24. Magnetic (a) and optical (b, c) isolation. (This figure is available in full color at http://www.mrw.interscience.wiley.com/ebe.)

17. INPUT PROTECTION

Low-voltage electronic circuits are sensitive to overvoltage and static electricity, which particularly applies to MOS electronics with super thin gate isolation, with breakdown voltages of about a pair of volts. At the same time, electrostatic voltage can reach 10 kV and cause a short-term (less than 1 µs) current about 10 A.

Therefore, all CMOS circuits include diode-based internal protection circuits. In addition, exterior protection is required in medical applications, where pacing pulses, defibrillation shock pulses, and electro-cautery present potential hazards to circuitry (15).

Gas discharge surge arresters are most effective, operating at the largest current densities and presenting the smallest capacitance. They, however, can limit overvoltage at 50 to 70 V level. Semiconductor limiters have smaller voltages, but also much bigger capacitance (Fig. 26c and 26d). Anyway, the diode bridge circuit in Fig. 26d can successfully compete with all others, except the gas discharge surge arresters, which are expected to bear powerful electrical charges from defibrillators and other electric shock machines. Several kilovolts can be generated by an external defibrillator during some milliseconds causing about 50 A current pulses delivering up to 300 J of energy (15). Implantable defibrillators generate hundreds of volts and deliver about 30 J energy.

18. MIXED SIGNAL ANALOG/DIGITAL SYSTEMS ON CHIP

Advances in CMOS technology permit complex mixed signal analog and digital systems on the chip (SoC). Texas Instruments has developed the mixed signal microcontroller family MSP430FG43XX, diagrammed in Fig. 27, especially optimized for portable/handheld medical devices (20). The chip contains a multiplexed 12-input, 12-bit analog-to-digital (A/D) converter with a 100 kHz sampling rate. The analog section also includes two 12-bit digital-to-analog (D/A) converters and three configurable operational amplifiers (OA) with rail-to-rail inputs and outputs and possibility to configure the OAs as programmable gain inverting and noninverting units and instrumentation amplifiers. In addition, the chip has a 48-pin digital input/output port, and a serial asynchronous (UART) and synchronous (SPI) communication interfaces. Using the MSP430FG43XX family chip, the designer can make a programmable signal-chain-on-chip (SCoC) class system (signals from sensors to actuators) that can fulfil most of the tasks shown in Fig. 1.

The digital section contains a 16-bit RISC CPU (central processing unit) with a Flash memory and RAM, running

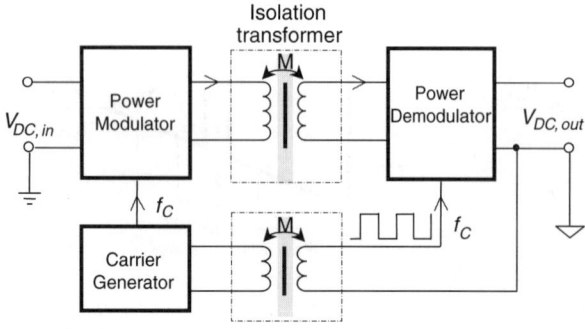

Figure 25. Isolated power supply – a DC/DC converter. (This figure is available in full color at http://www.mrw.interscience.wiley.com/ebe.)

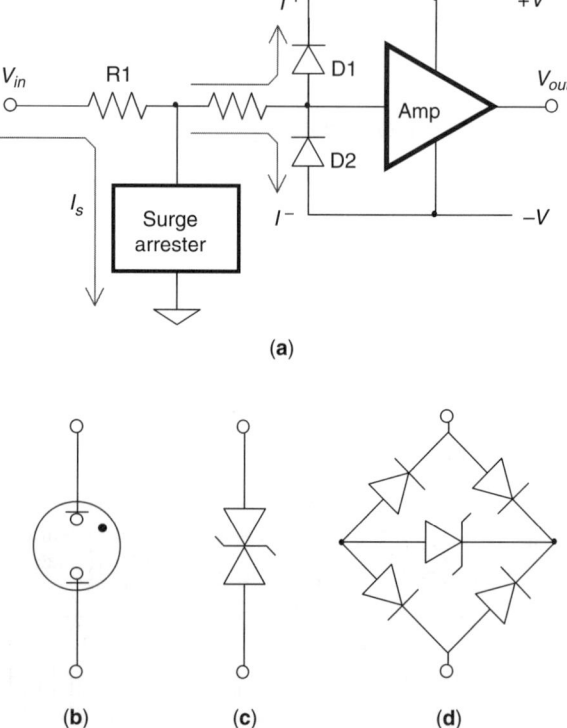

Figure 26. Input protection of medical devices: a circuit diagram (a), and gas discharge tube (b), bidirectional Zener diode (c), and a Zener in diode bridge (d) as surge arresters. (This figure is available in full color at http://www.mrw.interscience.wiley.com/ebe.)

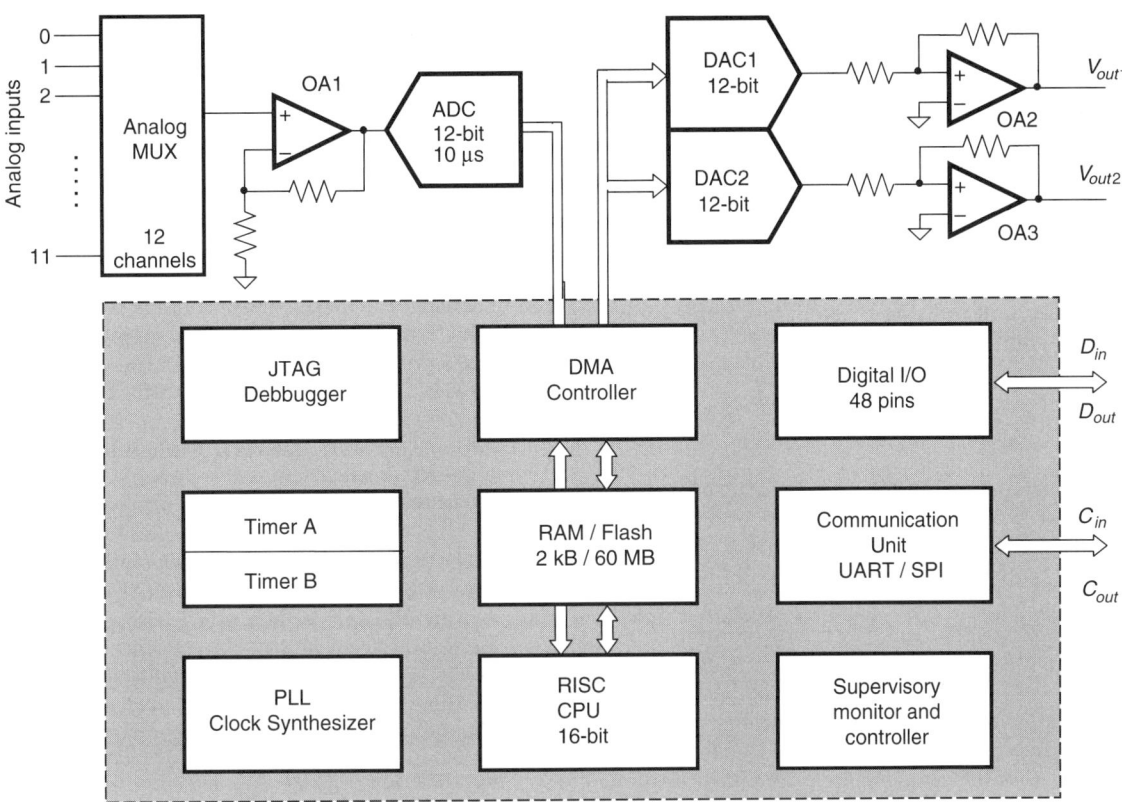

Figure 27. A signal-chain-on-chip (SCoC) based on the mixed signal microcontroller.

under a programmable clock with five power-saving modes. A special feature is the flexible direct memory access (DMA controller) providing data handling and transfer without CPU interaction. Special care has been taken to provide high reliability. The chip contains several subunits cooperating together as a supervisory controller, which fulfils the tasks of a watchdog, interrupt/reset handler, power brownout controller, supply voltage supervisor, and fail-safe clock provider. The supply voltage range is 1.8 to 3.6 V, current consumption is 0.1 µA in a RAM retention off mode, 1.1 µA in standby mode (real-time clock runs), and 300 µA in active mode (1 MHz clock, 2.2 V power supply).

Many systems-on-chip (SoC) use heterogeneous multiprocessors. The Texas Instruments OMAP platform is oriented to handheld devices with multiple communication channels and protocols. The system has a RISC controller and a digital signal processor (DSP) on the same chip, which are bridged and share common memory resources. Both the CPUs have their own real-time operation system (RTOS).

Besides public communications, the ultra-low-power RF communication is specific to medical electronics (14,21,22). For short-range communications (10–50 m) at 868 MHz (Europe) or 915 MHz (United States) and 2.4 GHz (global) bands, the 20 to 250 kbps battery-powered (3 µA standby) Zigbee IEEE 802.15.4 wireless standard is appropriate (21). Recent ITU-T Recommendation SA 1346 outlines the shared use of the 402 to 405 MHz frequency band for a medical implant communications service (MICS). It is expected that MICS will become a true global standard within several years. A 2-meter wireless link can be established between an implanted device and a base station with 10 channels, 300 kHz each, allowed at the limited output power 25 µW (22). Binary amplitude shift-keying (ASK) and phase-shift-keying (PSK) are the preferred digital data modulation methods (14).

19. TRENDS FOR FUTURE DEVELOPMENT

Electronic devices for numerous biomedical applications, including implantable and other sophisticated devices, have reached significant complicity already today (23,24). As the number of possible applications will grow, the number of dedicated solutions will grow with it.

Future trends in biomedical electronics will be determined by advances in technology and by regulatory requirements. Devices will continue to shrink in size and use less energy (25,26), which, coupled with improvements in battery technology, will increase the useful life of implanted devices and permit distributed architectures with multiple independent devices communicating over a "body bus." These devices will include diverse sensors as well as integrated and independent units for computation, data storage, and communication, which will allow a level of redundancy that can greatly increase system reliability so that malfunctions do not cripple the system or result in potentially dangerous behavior. In addition, wireless extra-corporal communication to wearable equipment or even distant data processing centers will add information processing, integration, and monitoring capability to further improve the safety and reliability of therapy devices.

Despite continued improvements in electronic design tools, biomedical electronics development costs continue to escalate and schedules continue to lengthen. A trend that will increasingly influence electronics is the coupling of hardware and software. Real-time systems, often based on embedded systems-on-chip (27), place a huge demand on software, particularly in patient-critical biomedical applications. Currently, the major limitation is software testing. Without significant improvements in testing tools and strategies, it will be difficult to meet the requirements of regulatory bodies and to bring these more complex next-generation systems to market.

20. SUMMARY

As biomedical electronics interfaces with living organisms, the highest priority must be given to safety and reliability during the design and manufacturing process. Thus, electrical isolation, fault mitigation, and output limits must be included as a matter of course. Other unique requirements develop from size, weight, and functional life limitations, which are normally addressed by using the latest low-power IC technology with high-power density batteries and packaging these components in biocompatible materials. Finally, the circuitry must function in an electrically noisy environment necessitating symmetric amplifier designs, shielding, and signal filtering as well as careful control of any electrode interface. Strict attention to these factors will greatly improve the chances of a successful design.

BIBLIOGRAPHY

1. P. Horowitz and W. Hill, *The Art of Electronics*, 2nd ed. New York: Cambridge University Press, 2001.
2. U. Tietze and C. Schenk, *Electronic Circuits: Handbook for Design and Application*. Berlin: Springer-Verlag, 2003.
3. F. Silveira and D. Flandre, *Low Power Analog CMOS for Cardiac Pacemakers: Design and Optimization in Bulk and SOI Technologies*. Boston, MA: Kluwer Academic Publishers, 2004.
4. R. B. Northrop, *Analysis and Application of Analog Electronic Circuits to Biomedical Instrumentation. -The Biomedical Engineering Series*. Boca Raton, FL: CRC Press, 2004.
5. R. C. Dorf, ed., *The Electrical Engineering Handbook*. Boca Raton, FL: CRC Press, 2000.
6. C. Toumazou, F. J. Lidgey, and D. G. Haigh, eds., *Analog IC Design: The Current-Mode Approach*. London: Peter Peregrinus, 1998.
7. C. Cockrill, S. Cohee, and T. Nanan. (2005). Texas *Instruments Little Logic Application Report SCEA029*. Texas Instruments Inc. (online) Available: http://www.ti.com/sc/psheets/scea029/scea029.pdf.
8. J. G. Webster, ed., *Design of Cardiac Pacemakers*. Piscataway, NJ: IEEE Press, 1995.
9. R. van de Plassche, *CMOS integrated analog-to-digital and digital-to-analog converters*. Boston, MA: Kluwer Academic Publishers, 2003.
10. D. Prutchi and M. Norris, *Design and Development of Medical Electronic Instrumentation: A Practical Perspective of the Design, Construction, and Test of Medical Devices*. Hoboken, NJ: J. Wiley & Sons, 2005.
11. W. G. Jung, ed., *Op Amp Applications*. Norwood, MA: Analog Devices Inc., 2002, pp. 6.179–6.180.
12. C. Henn. (2005). *Macromodels for RF op amps are a powerful design tool. Burr-Brown Application Bulletin AN-189 SBOA074 QPA660*. (online) Available: http://focus.ti.com/lit/an/sboa074/sboa074.pdf.
13. W. Jung. (2005). *Voltage references and low dropout linear regulators*. (on line). Available: http://www.analog.com/UploadedFiles/Technical_Articles/3752210Publication_V-Ref.pdf.
14. M. Sawan, Y. Hu, and J. Coulomb, Wireless smart implants dedicated to multichannel monitoring and microstimulation. *IEEE Circuit Syst. Mag.* 2005; **5**(1):21–39.
15. D. Bronzino, ed., The Biomedical Engineering Handbook, 2nd ed. vols. 1 and 2. Boca Raton, FL: CRC Press, 2000.
16. R. W. Salo, B. D. Pederson, and J. A. Hauck, The measurement of ventricular volume by intracardiac impedance. D. L. Wise, ed., *Bioinstrumentation: Research, Developments and Application*. Boston, MA: Butterworths, 1990, pp. 853–891.
17. M. Min, T. Parve, V. Kukk, and A. Kuhlberg, An implantable analyzer of bio-impedance dynamics: mixed signal approach. *IEEE Trans. Instrum. Measurement* 2002; **51**(4):674–678.
18. A. Kink, R. Land, M. Min, and T. Parve, Method and device for measurement of electrical bioimpedance. International Patent PCT WO 2004/052198 A1, June 24, 2004.
19. M. Min, R. Land, O. Märtens, T. Parve, and A. Ronk, A sampling multichannel bioimpedance analyzer for tissue monitoring. Proc. of the 26. Annual International Conference of the IEEE EMBS, 2004: 902–905.
20. Texas Instruments. *MSP430x4xx Family. User's Guide – Mixed Signal Products*, SLAU056E. Dallas, TX: Texas Instruments Inc., 2005. Available: http://focus.ti.com/lit/ug/slau056e/slau056e.pdf.
21. M. Galeev, Home networking with ZigBee. *Embedded Syst. Eur.* 2004; 20–23. Available: http://www.embedded.com/europe.
22. A. Sivard, P. Bradley, P. Chadwick, and H. Higgins, Challenges of in-body communications. *Embedded Syst. Eur.* 2005: 34–37. Available: www.embedded.com/europe.
23. Wear it well: overcoming the limitations of ambulatory monitoring with wearable technology. *Special Issue IEEE Eng. Med. Biol. Mag.* 2003; **22**(3).
24. Biomimetic systems: implantable, sophisticated, and effective. *Special Issue IEEE Eng. Med. Biol. Mag.* 2005; 24(5).
25. B. P. Wong, A. Mittal, Y. Cao, and G. Starr, *Nano-CMOS Circuit and Physical Design*. Hoboken, NJ: J. Wiley & Sons, 2005.
26. S. Luryi, J. Xu, and A. Zaslavsky, *Future Trends in Microelectronics: The Nano, the Giga, and the Ultra*. Hoboken, NJ: J. Wiley & Sons, 2004.
27. R. C. Dorf, ed., *The Engineering Handbook*, 2nd ed. Boca Raton, FL: CRC Press, 2005.

READING LIST

Biomedical circuits and systems: new wave of technology. *Special Issue IEEE Trans. Circuits and Systems* 2005; **52**(12).

P. H. Sydenham and R. Thorn, eds., *Handbook of Measuring System Design*, vols. 1–3. Hoboken, NJ: J. Wiley and Sons, 2005.

R. Pallàs-Areny and J. G. Webster, *Sensors and Signal Conditioning*, 2nd ed. New York: J. Wiley & Sons, 2001.

J.G. Webster, ed., *The Measurement, Instrumentation and Sensors Handbook*. Boca Raton, FL: CRC Press and Heidelberg, Germany: Springer-Verlag GmbH & Co, 1999.

BIOMEDICAL PRODUCTS, INTERNATIONAL STANDARDS FOR

DONALD E. MARLOWE
U.S. Food and Drug
Administration
Rockville, Maryland

People have developed standards since the beginning of recorded history, and probably earlier. Some were created by decree. Some grew from a need for people to harmonize activities or ideas with the changes in their environments. All standards were developed to solve one problem or another in an increasingly complex society. Today we develop standards to allow a supplier and a user of goods or services to describe to each other what is being supplied. We develop standards to ensure or improve public safety and health. We describe test procedures by which a product might be compared with its expected performance. To quote the Honorable William M. Daley, U.S. Secretary of Commerce, "The needs of business drive standards."

In recent times, particularly since the end of World War II and the development of the global economy, the concept of international standards has developed and evolved. In its simplest terms, an *international standard* is a standard developed by an international group of persons (individual interests, individual companies, or individual countries) who are participants on the technical committee that develops the standard.

1. "STANDARD"

What is a standard? A standard is nothing more than an agreed to way of doing something. Standards are skillfully constructed tools of industry, regulators, and commerce. They are living documents. They might describe products, specify their characteristics, or establish expected levels of performance. They might also describe the production process or the methods by which the product is tested. They might also describe management processes or desirable behavior. Standards may be the specifications in a contract. They may be a sign, a known symbol that signifies integrity. Standards are a language of trade used by buyers and sellers. They are passports to world markets. They are regulations. Standards, when they are very good, are synonymous with quality and market relevance. Standards can impart safety to products and interchangeability to parts. Standards are references, weights, and measures. They make the pieces fit.

First, a word about language. As in any other field, people in standardization communicate with each other in their own language. It is not difficult, it is not complicated and outside of a technical committee, it is not technical. It is merely colloquial. For instance, standards are not "set." They are "developed." People who develop standards are "participants." Sometimes the word *standard* is an adjective, i.e., a standard specification or a standard test method.

The taxonomy of standards can be viewed from at least three distinct perspectives: by type, by developer, or by scope.

1.1. Types of Standards

Terminology standards define the lexicon within which the industry sector develops product, tests the product, and describes its performance. *Parametric standards* include the terminology and define the important performance variables for the product and describe the test methods by which the variables will be characterized. *Performance standards* include the information described in parametric standards and establish the acceptable levels of performance for the product.

1.2. Developers of Standards

Company Standards are intellectual property, developed within the company for the company. A company standard may be a specification for the product it manufactures or describe the internal operating procedures of a particular department. This group of documents also includes standards developed by an industry consortium, a group of companies with similar interests. *Voluntary standards*, also called *consensus standards*, are developed through a consensus process. The operative words here are *consensus process*, which will be discussed next. The final developers of standards, particularly for the biomedical engineering sector, are governments. They develop *regulatory standards*. Typically, regulatory standards implement statues or establish minimum requirements when public safety is threatened by the improper design, manufacture, or use of a product.

No discussion of standards would be complete without a brief discussion of the concept of *consensus*. The development of standards is a process. Like any other process, standards development is guided by a set of rules. The adherence to the rules guarantees the integrity of the final product, i.e., the consensus standard.

Rule Number One: The process must be open, which means that a standards activity cannot be unduly exclusive. For example, an industry standards activity cannot be consensus if it purposefully excludes its smaller members, its largest member, or any member. The industry process usually admits only industry members, although they may invite others. A full consensus process will include participants from three general categories: producers, users, and those with a general interest in the outcome of the standard (this means they have no proprietary interest in the standard; they may be academics or government representatives).

Some processes will admit only those from one country or one region. The processes used by some international standards development organizations will admit only one official delegation per country. The rationale is as follows: The number or kind of participants in the process is directly related to the use or acceptance of the standard. That is, people who develop a standard, and/or people who are invested in a particular process, are more likely to use it. There are exceptions of course. One need only think of the Microsoft Windows standard (a company standard) or the Internet Standards (developed by small, highly specialized teams). These standards have virtually universal acceptance. But wide acceptance of standards developed by narrow or selective groups are the exceptions, not the norm.

1.3. Scope of Standards

The scope of standards describes either basic safety issues (*horizontal standards*) or product-specific issues (*vertical standards*). Horizontal standards indicate fundamental concepts, principles, and requirements with regard to general safety or effectiveness issues applicable to all types of, or families of, products and processes. Examples of horizontal standards are standards for risk assessment, biocompatibility, sterilization of medical products, or electrical safety. Vertical standards indicate necessary safety or effectiveness aspects of specific products or processes, which make reference, wherever possible, to the horizontal standards. Examples of vertical standards are standards for heart valve performance, anesthesia machine performance, or total joint wear.

2. HISTORY OF STANDARDIZATION

In the early 1850s, the first consensus standards were developed almost simultaneously in the United States and in the United Kingdom. The first products standardized were machine screws and nuts. In the 1850s, the machine tool industry was the biggest driver in the industrial economies of the two countries, the "dot-coms" of their era. Unfortunately, the standardized designs were different and did not permit interoperability of parts. The consequences were only of nuisance value until the armies of Britain and the United States were allied against the Germans in World War II in the deserts of North Africa. Then, bolts and screws that were nominally identical did not fit the war equipment operated by the two armies. The logistical and financial consequences to the war effort were enormous. After the war, the world was a different place than it had been for the previous century. The business community wanted to sell their products globally. An international screw thread standard was established.

3. INTERNATIONAL STANDARDS

The definition of "international standard" provided here can now be seen as a vast oversimplification. After the above discussion, we can provide a more technically correct definition and a more complete understanding. An international standard is, ultimately, one that is useful in the global marketplace to enable the stakeholders in that standard to do their business, whether it is sales, purchasing of good or services, and regulation.

3.1. Development of International Standards

Like anything else with enormous potential for good, standards can create mischief. They move trade and make the world healthier and safer. They transform chaos into order and progress. They foster prosperity and make technology available to those who do not have the means to create it. But flip the coin, and these same wonderful inventions can become instruments of unfair competition.

Before the General Agreement on Tariffs and Trade (GATT),[1] tariffs protected domestic markets from "foreign" producers. The tariff, for all practical purposes, is now passé. With tariffs no longer creating wholesale unfair advantages and imbalances in the market, the WTO turned its attention to the rising incidences of non-tariff obstacles, or barriers to trade, and negotiated the Technical Barriers to Trade Agreement (TBT).[2] The TBT is an agreement among governments. Its basic premise is that standards used in regulation should not create unnecessary barriers to trade. By the completion of the Second Triennial Review of the TBT Agreement[3] in 2000, 77 countries had certified that their trade practices conformed. But in the final analysis, a standard is currency. The definition of *international*, which is intended to confer credibility on the standard, is related to the process by which the standard is developed. Within the TBT, Annex 3 describes a "Code of Good Practice for the preparation, adoption and application of standards." Code of Practice was amplified even more to describe the process for development of international standards in Annex 4 of the Second Triennial Review. Annex 4 describes the principles by which the process for development of international standards shall operate. They are as follows:

Transparency—All essential information regarding current work programs as well as proposals for standards, guides, and recommendations under consideration, and the final results should be made easily accessible to at least all interested parties. Procedures should be established so that adequate time and opportunities are provided for written comments. The information on these procedures should be effectively disseminated.

Openness—Membership of an international standardizing body should be open on a nondiscriminatory basis to all interested parties. Any interested party with an interest in a specific standardization activity should be provided with meaningful opportunities to participate at all stages of standard development.

Impartiality and Consensus—All relevant parties should be provided with meaningful opportunities to contribute to the elaboration of an international standard so that the standard development process will not give privilege to, or favor, the interests of a particular supplier/s, country/ies, or region/s. Consensus procedures should be established that seek to take into account the views of all parties concerned and to reconcile any conflicting arguments.

Effectiveness and Relevance—International standards should be relevant and effectively respond to

---

[1] The General Agreement on Tariffs and Trade (GATT) of 30 October 1947. The GATT 1994 established the World Trade Organization (WTO).
[2] The WTO Agreement on Technical Barriers to Trade; WTO, 1994.
[3] Second Triennial Review of the Operation and Implementation of the Agreement on Technical Barriers to Trade; WTO, 13 November 2000.

regulatory and market needs, as well as scientific and technological developments.

- Coherence—To avoid the development of conflicting international standards, it is important that international standardizing bodies avoid duplication of, or overlaps with, the work of other international standardizing bodies.
- Development Dimension—Constraints on interested parties from developing countries, in particular, to effectively participate in standards development, should be taken into consideration in the standards development process.

4. INTERNATIONAL STANDARDS FOR BIOMEDICAL ENGINEERING

For biomedical engineering, the development of standards, the roles they play in the management of product risk, and the regulation of those products by governments worldwide are inexorably linked.

Regulatory authorities develop and use standards of many types at every level in the regulatory process. As the complexity of biomedical products increases, the interaction among developers, producers, and regulators to manage the risks posed by these products must increase. Information on "standards of performance" that the regulators expect these products to achieve is of great importance to the regulated industry. Standards constitute the medium of information exchange, encouraging coordination of technical discussions and information dissemination, and enabling more effective engagement among stakeholders.

4.1. Risk Management

The fundamental principle of standards development for biomedical engineering is risk management. The ideal standard contains only the clauses that address specific product risks and then provides information on risk reduction. For medical products, the risk management process has been described in ISO standard 14971.[4] An evaluation of risk following the approach described by this standard will lead through a hierarchal assessment of risk types and remediation.

4.2. Standards and Regulation

Regulation of biomedical products throughout the international economy is greatly reliant on the use of the standards by the manufacturers of the products to describe product performance, test methods, labeling, and handling procedures using standards. Standards are only a tool by which a product manufacturer can demonstrate regulatory compliance. They are not regulation. Because the use of the standards is voluntary, they give flexibility to both the product manufacturer and the regulatory agency. The product manufacturer can choose whether

Table 1. Major International Standards Development Organizations for Biomedical Products

Name of Organization	URL
Association for the Advancement of Medical Instrumentation	http://www.aami.org/
ASTM International	http://www.astm.org/
International Electrotechnical Commission	http://www.iec.ch/
International Standards Organization	http://www.iso.ch/
NCCLS	http://www.nccls.org/

to cite the standard in its application. The regulatory body can take deviation from the standard, if necessary.

The use of consensus standards by governments in the management of risk derived from biomedical products raises the issue of "harmonization" of requirements. "Why should the regulatory requirements of one government differ from those of another?" To address this concern, the governments of the three global sectors, i.e., North America, Europe, and the Asian Rim, established the Global Harmonization Task Force (GHTF). Recognizing that the statues regulating biomedical products differ from country to country, GHTF has recognized the role of standards in their regulation.[5]

Within the United States regulation of biomedical products is within the jurisdiction of the U.S. Food and Drug Administration (FDA). The U.S. FDA has long recognized the importance and use of standards to manage the risks from medical products. Currently, the FDA "Recognizes" more than 600 voluntary standards as useful in meeting some or all of the regulatory requirements for these products. A complete and current listing of these 600 standards and the regulatory limits of their utility are available on the FDA Internet website.[6]

4.3. Developers of International Standards for Biomedical Engineering

There are comparatively few developers of international standards for biomedical products. They are as shown in Table 1.

4.4. Major Areas of Standardization

The individual international standards for biomedical products are too numerous to list here. Such a list would be out-of-date before it could be published. The major categories of standards are shown in Table 2.

The current system of voluntary standards development and application may seem awkward and inefficient. However, it generally identifies the projects that need work and arrives at the right result for ensuring product safety.

---

[4]ISO 14971 Medical Devices—Application of risk management to medical devices; International Organization for Standardization, Geneva, Switzerland.

[5]Role of Standards in the Assessment of Medical Devices (including in Vitro Diagnostic Devices)(DRAFT); GHTF Study Group 1; Doc No SG1/N044R4.

[6]www.cdrh.fda.gov/science/standards/constand.htm.

Table 2. Major Areas of International Standards for Biomedical Products and the Relevant Standards Development Organizations

Anesthetic and respiratory equipment	ISO
Biological evaluation of medical devices	ASTM, ISO
Clinical laboratory testing and *in vitro* diagnostic test systems	ISO, NCCLS
Devices for administration of medicinal products and intravascular catheters	ISO
Dentistry	ISO
Electrical safety	IEC
Electromagnetic compatibility	IEC
Healthcare informatics	ISO
Implant materials	ASTM, ISO
Implants for surgery	
a. Cardiovascular devices	ISO
b. Neurological devices	ISO
c. Orthopedic devices	ASTM, ISO
d. Plastic surgery devices	ASTM
e. Spinal devices	ASTM, ISO
f. Urological devices	ASTM
Implant retrieval	ASTM, ISO
Lifts, escalators, passenger conveyors	ISO
Mechanical contraceptives	ASTM, ISO
Optics and photonics	ISO
Prosthetics and orthotics	ISO
Protective clothing	ASTM
Quality management and quality assurance	ISO
Rubber medical products	ASTM
Safety of electomedical equipment	IEC
Statistics	ASTM, ISO, NCCLS
Sterilization of health-care products	ISO
Surgical instruments	ASTM, ISO
Technical systems and aids for disabled or handicapped persons	ISO
Tissue-engineered medical products	ASTM

5. CONCLUDING REMARKS

There is always a tension among the government, user community, and regulated industry over the amount of control needed to ensure that the best technology improves public health. Such a tension is healthy. It keeps us all intellectually honest. Development of standards provides a forum in which that tension can play out in a fruitful way.

BIOMEDICAL SENSORS

JOHN T. W. YEOW
University of Waterloo
Waterloo, Ontario, Canada

1. INTRODUCTION

One of the most primitive inborn functions, which every organism on this planet possesses, is the ability to sense their surrounding environment, which should be taken as the first clue that the sensory functions we possess are not only necessary to our well-being, but are also vital to our survival. With the advent of modern technology, the art of biological and physiological sensing has brought about a revolution in diagnostic and therapeutic medical instruments. Novel clinical treatment techniques involving sensors and innovative biomedical sensor networks are appearing because of immense technological breakthroughs in biomedical sensing and its peripheral technologies. As a result, biomedical sensors are indispensable instruments in modern day medicine.

The definition of a sensor from dictionaries suggests that it is a device that gives a signal for the detection or measurement of a physical property to which it responds. The ability to detect the surrounding environment is provided by many sensing organs in our body, and, consequently, they allow us to respond to external factors. Our skin allows us to sense temperature, our ears detect sound waves, and our eyes acquire visual images. These acquired stimuli are converted into electrical signals, which are then interpreted by our brain. The external factors are evaluated and corrective responses are formulated through a feedback system. Our sensory organs provide us with useful information that allows us to interact with our surroundings more effectively. Our ability to respond to potentially hazardous external environmental factors is the key to our survival. Any highly sophisticated and integrated intelligent feedback system would consist of sensors that acquire environmental parameters, a microprocessor that interprets the information, and actuators acting on the commands from the processing unit. Therefore, sensing elements are crucial to any feedback systems if an effective control scheme is to be implemented. A feedback system is shown in Fig. 1.

Since the first commercial use of the sensor, it has found a broad range of applications (1–6). In addition to the traditional physical properties, modern sensors have the capability to measure chemical and electrical properties too. In general, a sensor can detect and generate a response proportional to the property or condition that it is measuring. The property or condition measured is called the measurand. The response output of the sensor may vary; it could be in the form of analog electrical signals or in digital formats. The relationship of measurand vs. output of a sensor is defined by the sensor manufacturer during device calibration. Therefore, the sensor output provides us with reliable information about the

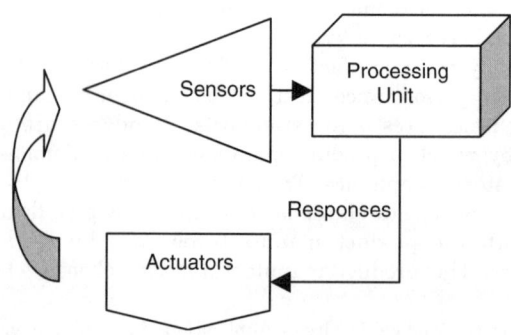

Figure 1. A feedback system incorporating sensors. Sensors are necessary components in any feedback system.

magnitude of the measurand. Biomedical sensors are a subset of specialized sensors responsible for sensing physiological or biological measurands. The biomedical measurands are used as indicators for clinical diagnostic or therapeutic purposes. Since ancient times, physicians have always relied on physiological signals for clinical diagnosis and therapeutic treatments. In traditional Chinese medicine, the arterial pulse waveform measured from the human waist by the experienced fingers of a physician will reveal the physical condition of the patient. It is understood that our biological signals or measurands carry a lot of encoded information on our physiological and psychological states. The key is to access the biomedical measurands effectively and decode them accurately, reliably, and swiftly. Modern technology has enabled us to detect subtle physiological signals or biochemistry constituents, which are then analyzed by physicians to determine our state of health.

The measurands used most in medical diagnostics can be grouped into the following categories: (1) biopotential (ECG signals); (2) flow (blood flow); (3) dimensions (imaging); (4) displacement (velocity, acceleration, and force); (5) impedance (plethysmography); (6) temperature (skin); (7) chemical concentrations (enzymes activities); (8) sound (phonocardiography); and (9) pressure (blood and eyes). The measurands are usually acquired by physical sensors, which are placed in close proximity or in physical contact with the organ or anatomical structure of interest. The measurands may be acquired *in vivo* or *ex vivo*. In either case, the sensor should be minimally invasive and minimize the effect of its presence on normal bodily functions. Stringent requirements are imposed on biomedical sensors because of the unique nature of their applications and the environment that the devices have to function in.

2. SENSOR CLASSIFICATIONS

Biomedical measurands are quantified using many different sensing principles. The interaction between the environment and the sensor alters the state of the sensing mechanism. The majority of the sensing effects rely on changes in the electrical properties of the sensing materials. The transducer mechanism in the sensor will convert the change in material properties to a signal that is proportional to the environmental factor. The signal can be presented as a visual reading, as in the case of a thermometer, or may be used to provide feedback parameters to control the quantities of medications administered by an implantable insulin pump or ingredients in a food processing plant. Sensors are often categorized by the measurands that they are quantifying. Traditionally, the medical measurands are classified into three main categories: (1) physical; (2) electrical; and (3) chemical. Physical measurands include mechanical quantities such as position, displacement, flow, and pressure and thermal quantities such as temperature and heat flow. Electrical measurands are in the form of electrostatic, magnetic fields and fluxes, and radiation intensity (electromagnetic, nuclear, etc.). Chemical sensors represent the most widely used and largest group of biomedical sensors. They measure chemical quantities such as the humidity, gas components, and ion concentration. More recently, the interest in chemical biomedical sensors is in the development of biosensors. Biological quantities, such as the concentration of enzyme substrate, antigens, and antibodies, are measured by exploiting biological agents that are naturally sensitive to the measured analytes. A biosensor contains biological agents that react with the analyte of interest. The reactions are detected by optical or electrochemical means, and the concentration or presence of the analyte of interest is deduced. Details of the sensing techniques are described in the section entitled "Biomedical Sensing Technologies."

With the tremendous improvement of technology, particularly in microelectronics and miniaturization techniques, biomedical sensors with sophisticated functions can be integrated together with microprocessor units to improve their sensing performance. The miniaturization of already commercially successful macroscopic sensors and the incorporation of integrated signal conditioning circuitry have brought a paradigm shift in the way sensors are employed. The collective cross-disciplinary efforts of engineers, physicists, chemists, biologists, and physicians have allowed rapid developments of new materials, real-time signal processing algorithms, surgical implantation techniques, and novel sensing modalities. These efforts have contributed to the vast improvement and successful implementations of biomedical sensors in a wide range of applications. The improvement in sensing technology has enabled us to measure quantities and physiological phenomenon that were difficult or impossible to quantify before. As a result, the measurands of modern-day sensors demand a broader and more comprehensive classification, more specifically, the inclusion of biophotonic and chemical biosensing principles.

3. BIOMEDICAL SENSING REQUIREMENTS

The specification of sensors depends on where and how the sensors are deployed. As mentioned earlier, biomedical sensors have special requirements because of their uniqueness in applications and their consequences during operational failures. Most often, a biomedical sensor is used as a diagnostic tool to measure the instantaneous physiological status such as heart rate, blood pressure, and temperature. These diagnostic sensors must be compact, durable, accurate, consume little power, have a fast response time, and provide disturbance rejection capability. Examples of diagnostic biomedical sensors include digital thermometers, blood pressure sensors, blood oxygen sensors, and endoscopic biomedical imaging probes. In the case of therapeutic sensors, where the devices are used to monitor bodily states continuously and have the added function of controlling actuators to normalize biological states, the design of the sensors is even more challenging (7). Many chronic diseases such as diabetes and heart diseases require state-of-the-art implantable sensors and controlling systems to help in determining and dispensing the amount of medications accurately. The danger of overcompensation or undercompensation of medications, which could be potentially fatal, always exists.

Sensors implanted inside the human body require them to have long-term reliability, stability, robustness; to have a long lifetime and to be biocompatible. As in the case of the implantable devices, invasive sensors for *in vivo* biomonitoring or bioimaging require them to be disposable, small in size, low in mass, and consume little power. Miniaturization is the enabling technology that ensures that the sensors are noninvasive and do not affect the normal functioning of the body. Sensors that are implanted or inserted into human cavities are designed to withstand sterilization procedures that include the use of ethanol, ethylene oxide (ETO) gas, gluteraldehyde, electron beam radiation, and autoclave. Issues such as sterilization and material corrosion have to meet stringent requirements set by regulatory agencies such as Association for the Advancement of Medical Instrumentation (AAMI). Biocompatibility issues, such as decomposition, oxidation, and toxic corrosion in the case of metallic implants, could cause serious health problems. Metalosis, which results in the spread of potentially toxic metallic particles in the surrounding tissues or blood stream, could cause irreparable damage. Therefore, implantable sensors should be made of biocompatible materials that are structurally stable to prevent biological rejection, prevent formation of protein precipitation, and minimize blood clotting. Materials made from a combination of organic and inorganic compounds are needed in order to prolong the life of the implanted device and keep the recipient safe from infection. In addition, they must be economical and disposable to be commercially viable.

4. BIOMEDICAL SENSING TECHNOLOGIES

This section provides an introduction to biomedical sensors based on optical, chemical, and biosensing techniques. An overview of their underlying sensing technologies and their applications is presented. The readers are referred to several optical and chemical sensing books for more detailed theoretical sensing principles (8–12).

4.1. Optical Sensors

The field of optics has always been an active area of research and development. The quest for data bandwidth in an optical communication network has fueled much of the developments in optical technologies. Optical units such as optical fibres, amplifiers, lasers, filters, and high-speed detectors that were originally developed for telecommunication networks are finding novel applications in biomedical systems. An optical sensor consists of a laser that provides the light source, a transmission medium that delivers the light beam to the sample, and a detector that collects the modulated light beam after interaction with the sample. An optical biosensor is an optical sensor with the purpose of sensing biological analytes. Examples of optical biophysical sensing devices include body temperature and blood pressure sensors. Optical biochemical sensors are used to measure blood chemical content and detect cancerous tissue.

4.1.1. Optical Biosensing Techniques. Optical biosensing relies on different spectral features or fluorescence intensity of the analytes. The outgoing beam spectrum or intensity will be different from that of the returning beam. The difference is measured by comparing absolute beam intensity, or by the interference pattern with a reference beam. The most common spectroscopic techniques are measurement of optical absorption, fluorescence, reflectance, and interference.

4.1.1.1. Optical Absorption. As an optical beam is transmitted through a sample, the intensity of the beam decreases because a portion of it is absorbed by the analytes. This technique is effective for determining the concentration of analytes that have characterized absorbance (A) well. The absorption is governed by the Beer–Lambert law:

$$I_T = I_O e^{-\alpha L}, \tag{1}$$

where, I_T is the intensity of the transmitted beam. I_O is the intensity of the incident beam. α is the absorption coefficient (uin cm^{-1}). L is the absorption path length (cm).

The absorbance of the sample is given by

$$A = \log\left(\frac{I_O}{I_T}\right) = \varepsilon L C, \tag{2}$$

where ε is the molar absorptivity (Lmol^{-1} cm^{-1}). C is the concentration of the absorbing species.

As shown in Equation 2, the absorbance A of the sample is directly proportional to the concentration C of the absorbing species. It should be noted that the bandwidth ($\Delta\lambda$) of the incident beam and absorbing and scattering characteristics of the samples may cause deviation from the perfect Beer–Lambert relationship.

4.1.1.2. Fluorescence. Fluorescence is a phenomenon that is used commonly in detection of analytes in the biological science community. This phenomenon occurs when an analyte absorbs a high-energy photon and releases a lower energy photon at a longer wavelength. The released photons cause luminescence in the visible range. Equation 2 can be represented by

$$\log_e\left(\frac{I_T}{I_O}\right) = -2.303\varepsilon L C. \tag{3}$$

When $\varepsilon L C \ll 1$, Equation 3 can be expressed as

$$I_T = I_O[1 - 2.303\varepsilon L C]$$
$$I_O - I_T = 2.303 I_O(\varepsilon L C). \tag{4}$$

Without significant scattering and low concentration of the absorbing species, Equation 4 represents the intensity of light absorbed in the medium. Intuitively, the fluorescence intensity (I_F) is expected to be proportional to the intensity of light absorbed. Therefore,

$$I_F \propto \varepsilon L C. \tag{5}$$

The fluorescence sensing technique is based on the linear relationship of the luminescence intensity to the

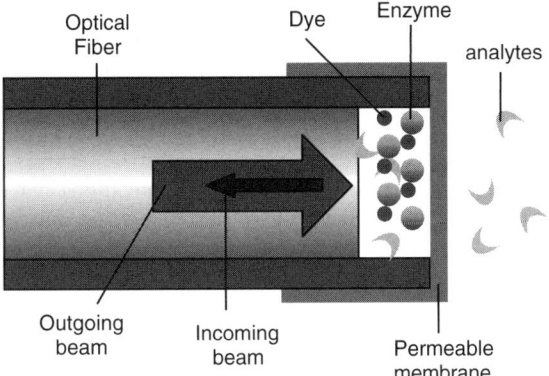

Figure 2. A typical enzymatic fiber optic biosensor. (This figure is available in full color at http://www.mrw.interscience.wiley.com/ebe.)

concentration of the absorbing species. The recent report on fiber optic biosensing for *Escherichia coli* (*E. coli*) identification uses fluorescence intensity to identify different strains of *E. coli*. The concentration of beta-glucuronidase (GUD), which is a marker enzyme of *E. coli*, is proportional to the fluorescence intensity (13). Figure 2 shows a fiber optic biosensor complete with an enzymatic reaction chamber. The analytes diffuse through the permeable membrane and react with the enzyme in the chamber. A dye that is sensitive to the change in environmental factors, such as pH, gas concentration, or any chemical byproducts, is used to provide spectral information. The spectral change of the dye is determined from the incoming beam, and the concentration of analytes is deduced.

4.1.1.3. Reflectance. Converse to the absorption technique, the intensity of light reflected from absorbing species may also be used to deduce the concentration of the species. However, the relationship between the concentration of absorbing species C and the reflectance R is described by the Kubelka–Munk (F_{KM}) function, and is nonlinear.

$$F_{KM} = \frac{(1-R)^2}{2R} = \frac{\varepsilon C}{S} \qquad (6)$$

$$R = \ln\left(\frac{I_o}{I_r}\right), \qquad (7)$$

where S is the scattering coefficient and ε is the molar absorptivity. Reflectance is less used in fiber optical biosensing than the other two optical detection techniques.

4.1.1.4. Interferometry. Interferometric biomedical sensing relies on the physical interaction between two or among more individual light waves usually of the same origin. The light interaction involves interference of multiple light waves. The Michelson interferometer is a commonly used optical instrument for producing interfering light patterns for biomedical sensing applications. One of the latest developments of interferometry techniques for biosensing applications is the application of optical coherence tomography (OCT) (14) for *in vivo* biomedical imaging. OCT is a novel sensing technique that allows high-resolution imaging of subcutaneous tissue samples. Sample tissue morphology can be created by measuring the backreflected infrared light. This interferometric-based approach shows promise for imaging minute cavities or blood vessels where conventional biopsies cannot be performed effectively. The recent development of the OCT sensing modality is the incorporation of micromachining technologies (15,16). Endoscopes incorporating micromachined-based OCT sensing modality will find applications in cardiology for guiding coronary procedures and detection of early cancer lesions in the colon and esophagus.

4.1.2. Optical Biosensing Applications. Optical biosensor represents the second largest family of biosensors after chemical biosensors. The field of medicine has benefited greatly from the enormous efforts in the research and development from the optics community. Many innovative optical noninvasive diagnostic systems used for biomedical imaging and analyte detections find their origin in the telecommunication industry.

4.1.2.1. Optical Oximetry. The knowledge of oxygen saturation (SO_2) of oxygen-carrying compounds, such as hemoglobin, myoglobin, and cytochromes aa3, is important in many clinical diagnostic and therapeutic settings (17). Optical oximeter is an optical sensor that measures SO_2 with the wavelength-dependent absorption properties of the aforementioned hemochromes. There are two modes of operating the optical oximeters: (1) transmission mode and (2) reflectance mode. Transmission-mode oximetry involves measuring the absorption of the incident light. On the other hand, reflectance-mode oximetry measures the reflectance (R) light at two different wavelengths (λ_1, λ_2), one in the red and another in the infrared range. The oxygen saturation is deduced from the following relationship:

$$SO_2 = A + B\left[\left(\frac{R(\lambda_1)}{R(\lambda_2)}\right)\right], \qquad (8)$$

where A and B are empirical constants that are dependent on the fiber geometry and physiological parameters of blood. More advanced versions of oximeter use more wavelengths to account for hematocrit variations.

4.2. Electrochemical Sensors

Electrochemical sensor is one of the broadest and most popular types of sensors used to measure chemical measurands. Most of the electrochemical sensors are used for measuring biomedical analytes such as ionic concentration in blood or bodily fluids, pH monitoring in gastric acid, and partial pressure of dissolved oxygen (pO_2) and carbon dioxide (pCO_2) (18–22). The principles of sensing for this class of sensors are based on potentiometric, amperometric, and conductivity measurements. A typical electrochemical sensor consists of two electrodes submerged in an ionic conductive material called the electrolyte. The electrodes detect the electrolyte reaction that is proportional to the concentration of the analytes.

4.2.1. Potentiometric Technique. When a piece of metal is placed in an ionic solution, opposing charges exist at the

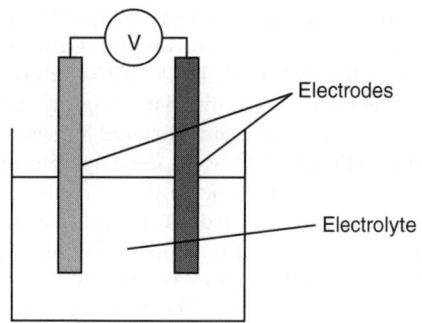

Figure 3. A typical cell in an electrochemical sensor.

boundary between the metal and the solution. The nature and concentrations of analytes in the solution and the nature of the electrodes can affect the amount of charge accumulated. Therefore, the concentrations of the analytes can be deduced by measuring the charge accumulation at the material boundary, which can be achieved by placing a high-impedance voltmeter between the two immerged electrodes in the ionic solution to measure the difference in potential. Figure 3 shows the basic components in an electrochemical cell. The potential measured between the two electrodes is caused by the redox reactions occurring at the electrode material and electrolyte interface:

$$Ox + ne^- = Re, \qquad (9)$$

where Ox denotes the oxidized species, Re the reduced species, n the number of electrons in the redox reaction, and e^- an electron.

The difference in potential between the two electrodes is given by the Nernst equation:

$$E = E^0 + \frac{RT}{nF} \ln\left(\frac{a_{Ox}}{a_{Re}}\right), \qquad (10)$$

where E^0 is the electrode potential at standard state, a_{Ox} and a_{Re} are activities of the oxidized and reduced species, n is the number of electrons in the reaction, F is the Faraday constant, R is the gas constant, and T is the operating temperature in Kelvin.

4.2.2. Amperometric Technique. Amperometric technique involves the deduction of the concentrations of analytes by their linear relationship with the current generated. As chemical analytes approach the sensing electrodes, electrons are transferred from the analytes to the electrode. The transfer of electrons will contribute to measurable current on the electrode. The current generated is dependent on the nature of the electrode, concentrations of the analytes, and the electric potential on the electrode. Amperometric sensing devices are by far the most prevalent and commercially successful electrochemical sensors to date.

4.2.3. Conductivity Technique. The conductivity sensing method measures the ability for a solution containing the analytes of interest to conduct electric current. The conductivity of the solution is directly proportional to the concentration of ions in the solution; therefore, sensors based on conductivity can be used to measure ionic concentration in bodily fluids. The disadvantage of this technique is that conductivity is nonspecific for a given ion type. Therefore, this technique is generally used with membranes that provides analyte selectivity.

4.2.4. Electrochemical Sensor Applications. Electrochemical sensors may have advantages over optical biosensors when the medium that contains the analytes is opaque or the analytes are not optically active. The incorporation of miniaturization and microelectronics technology has enhanced the performance of electrochemical sensors. This section describes the commercial applications of electrochemical sensors.

4.2.4.1. Measurement of pH Level and Gas Concentration in Bodily Fluids. In most commercially available electrochemical sensors, a membrane separates the electrode from the medium that contains the analyte of interest. Analyte diffuses through the membrane and reacts at the electrode-electrolyte interface. The reactions either develop a potential between the two electrodes or generate a current that is proportional to the concentration of the analytes. For example, the blood pH measurement is primarily based on the glass electrode covered with a selective membrane (23). The membrane in the sensor is permeable to CO_2 molecules but impermeable to water, hydrogen ions (H^+), and bicarbonate ions (HCO_3^-). A layer of sodium-carbonate ($NaHCO_3$) is used between the pH electrode and the membrane. The following equation shows the reaction of the diffused CO_2 molecules within the electrolyte (24):

$$CO_2 + H_2O \Leftrightarrow H_2CO_3 \Leftrightarrow H^+ + HCO_3^-. \qquad (11)$$

The glass electrode measure pCO_2 indirectly through:

$$\mathrm{pH} = constant - \log pCO_2. \qquad (12)$$

4.3. Biosensors

In recent years, there has been tremendous interest in the research and development of biosensors. As in the case of any sensors, biosensors contain two main elements: (1) the receptor and the (2) transducer. The uniqueness of a biosensor is that the receptor incorporates biological elements as a part of its sensing system. The biological receptor (bioreceptor) of the sensor interacts with specific analytes that are present in a sample, and the transducer converts the measurands into measurable signals. The interaction is usually chemical in nature, and the reaction mimics the kind of interactions between living biological entities. Figure 4 illustrates that the receptor of the biosensor only reacts with a selected analyte.

4.3.1. Immobilization. A transducer produces measurable signals as a result of the chemical reaction occurring at the receptor. The fact that the receptor is made of biomaterials presents a significant challenge to the fabrication process of the biosensors. High-fabrication-yield techniques have evolved throughout the years to effectively

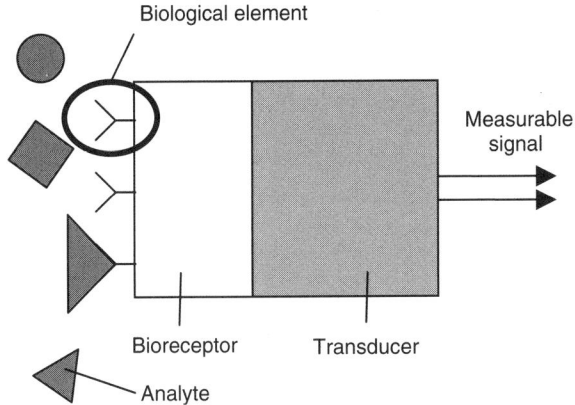

Figure 4. Sensing elements of a biosensor.

connect the bioreceptor to the transducer. The process of the connection is called immobilization. Immobilizing bioreceptors on the transducer surface remains one of the main technical obstacles to overcome (25,26). The choice of immobilization techniques depends on the applications, transducer types, and the bioreceptors involved. In general, six methods of immobilization exist:

4.3.1.1. Absorption. This technique is the most straightforward and involves minimal materials. Bioreceptors are either physically bonded by hydrogen bonds, van der Waals, or electrostatic forces. Physical absorption is usually weak and temporary. Obvious disadvantages of absorption immobilization are difficulty in reproducibility of identical biosensors and unreliability in performance. A stronger absorption is possible by forming chemical covalent bonds between the two entities. Although much stronger bonds are achieved, absorption immobilization is a short-term approach, and is suitable for proof-of-concept type biosensors.

4.3.1.2. Microencapsulation. The bioreceptors are prepared in a permeable membrane or dissoluble capsule and are physically attached to the transducer surface. The main advantages of this technique include protection of the receptors against biocontamination, thereby increasing reliability and sensor life; close and strong contact with the transducer; and robustness of sensor performance in an environment with varying temperature, pH, and chemical concentration.

4.3.1.3. Entrapment. Bioreceptors are physically entrapped within an active membrane or gel. Interaction with analytes is achieved by diffusion through the gel barrier. Potential problems occur when the gel barrier is too thick for effective diffusion of the analytes. As a result, the limited chemical reactions slow down sensor response time.

4.3.1.4. Cross-linking. This approach makes use of a bifunctional crosslinking reagent, which reacts with supporting material, such as proteins, to chemically bond bioreceptors. For example, the detection of glucose is performed by biosensors that use cross-linking as an immobilization technique. Glutaraldehyde is used as a bifunctional agent to bind the bioreceptors with lysine residues on the exterior of the proteins.

4.3.1.5. Covalent Bonding. The functional groups of the bioreceptors are covalently bonded to the support material. It is important that the catalytic functional groups of the enzyme are not involved in the binding process. Covalent binding is very strong and is therefore suitable for commercial sensors because of its long lifetime. The binding process is usually performed under mild conditions such as low temperature and low ionic strength to prevent losing the activity of the enzyme.

4.3.1.6. Natural Interaction. The exploitation of the natural biological interaction and immobilization techniques shows the most promise. Inherent problems with other immobilization techniques are avoided by using naturally occurring biological binding processes. This approach is gentle; bioreceptors are positioned in a favorable orientation, and their activities are not altered in the process. The immobilization of enzymes, antibodies, and bioreceptors has been demonstrated and discussed in the literature (25–29).

4.3.2. Biosensor Applications. This section describes the commercial applications of biosensors in various diagnostic and sensing domains. The applications of the biosensors are minimally invasive physically because they are small in size. They also present minimum disruption to biological processes because they are biocompatible. As a result, biosensors are capable of providing rapid and accurate sensing for medical diagnosis. The development of biosensors has evolved beyond the initial impetus from the health-care sector. At present, biosensing technology finds applications in industrial process control, environmental monitoring, and military applications.

4.3.2.1. Health Care (Glucose Biosensor). The glucose biosensor is, without doubt, the most commercially successful biosensor to date. The purpose of the creation of the glucose biosensor was to allow accurate measurement of blood glucose level for people with diabetes (30,31). Diabetic patients are characterized by their lack of ability or limited ability to produce adequate quantity of insulin, a polypeptide hormone produced by the beta-cells of the pancreas, which is responsible for the metabolism of blood sugar contents. The deficiency in insulin results in patients having higher than normal blood glucose levels. Therefore, a controlled amount of insulin has to be administered, usually by subcutaneous injection, to keep the blood glucose level close to normal. The desired amount of insulin injected is determined from the patient's most recent blood glucose level test. One of the obvious disadvantages is that precise control over the amount insulin administered is not possible, and hyperglycemia, or hypoglycemia is induced. Hence, a closed system incorporating an implantable glucose biosensor that detects glucose level in real-time and provides accurate glycemic control of insulin is introduced. This system independently and

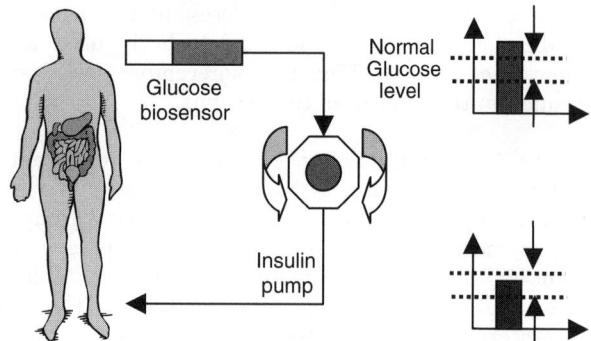

Figure 5. A closed-loop insulin control system with implantable glucose biosensor that monitors blood glucose level continuously. (This figure is available in full color at http://www.mrw.interscience.wiley.com/ebe.)

continuously monitors the blood glucose levels of the patient and responds accordingly. Patients are relieved of the stress from daily manual insulin injections. It is for the medical diagnostic application that so much current research has been directed at the development of a cheap, accurate, and implantable glucose biosensor. Figure 5 shows the operating principles of an implantable insulin pump.

Patients with renal deficiency during kidney malfunction suffer from excess waste build-up in their blood and can be potentially fatal if adequate dialysis is not performed. It is estimated that 661,330 people will require dialysis by the end of the decade (32). Therefore, an accurate and easily administered mean of urea monitoring is required. The use of a biosensor for dialysis monitoring was recently reported (33). The urea biosensor consists of two subsystems: (1) a biosensing component and (2) a photodetection system. The sensor works in conjunction with a dialysis machine to detect urea concentration in the dialysate outflow. The sensor uses a unique layer-by-layer technique (1) to chemically attach an absorbing dye, Congo Red, and (2) to immobilize urease onto the thin films. The urease reacts with urea in the dialysate to produce ammonia, resulting in a pH change of the dialysate. The absorbance spectra of Congo Red molecular changes as pH changes. Hence, the concentration of urea can be deduced by simple device calibration.

The Centers for Disease Control and Prevention estimates that there are 20,000 illnesses that can be attributed to *E. coli* infections (34). The detection of *E. coli* O157:H7 with high affinity and sensitivity has increased the use of testing devices such as biosensors. The recent development of a biosensor for bacterial detection focused on immunoassay using conductive polymers, such as polyaniline, polypyrrole, polyacetylene, and polythiophene (35,36). The conductive polymers detect biological events and translate them into sensible electrical signals. An example of an electrochemical sandwich immunoassay biosensor using polyaniline for detecting *E. coli* O157:H7 was reported (37). The biosensor consists of two types of proteins: capture protein and reporter protein. The capture protein is immobilized on a pad between two electrodes, and the reporter protein is binded to polyaniline. After sample application, the target analyte binds to the reporter protein and forms a sandwich complex with the capture protein. The polyaniline in the sandwich complex forms a molecular wire that conducts an electrical signal that is proportional to the amount of target analyte. It is reported that the biosensor could detect approximately 7.8×10^1 colony-forming unit per milliliter of *E. coli* O157:H7 in 10 min.

4.3.2.2. Food Industry (Lactose Biosensor). Demand and an increasing need exist to monitor and make online measurement of food contents for quality assurance and safety assessment. The food industry performs lactose analysis during the fermentation process for quality control. The constant and direct monitoring of specific compounds in the food has the potential to increase production yield by increasing productivity and reducing unnecessary material costs. The use of biosensors enables real-time feedback control of the production lines. Commercially available biosensors for the food industry are used to monitor glucose, galactose, sucrose, uric acid, amylase, L-lysine, L-lactate, D-lactate, and many forms of organic acids. The dairy and fermentation industry is one of the biggest markets for biosensors. The detection of lactose, for example, uses an amperometric multienzyme system (38). The lactose concentration is determined by the following cascaded biochemical and electrochemical reactions:

$$\text{Lactose} + H_2O \xrightarrow{\beta\text{-galactosidase}} \text{D-galactose} + \beta\text{-D-glucose}$$

$$\beta\text{-D-glucose} + O_2 \xrightarrow{\text{glucose oxidase}} \text{D-glucano-}\delta\text{-lactone} + H_2O_2$$

The depletion of O_2 during chemical reaction (2) or oxidation of H_2O_2 is measured by an electrode to determine the concentration of lactose. Alternatively, horseradish peroxidase can be used to oxidize H_2O_2, with 5-aminosalycilic acid as a mediator (39). The oxidized version of the mediator is reduced at the electrode to produce an amperometric signal that is proportional to the lactose concentration. Although biosensors show enormous potential in the food industry, many challenges must be overcome before gaining industry-wide acceptance. Issues such as more effective immobilization techniques, protection from strong denaturing agents, and reliable mass-production technology must be addressed.

4.3.2.3. Military (Toxicity Biosensor). Our body requires naturally occurring proteins called enzymes to act as a catalyst in many biological processes. It is, therefore, not surprising that enzymes are the most commonly used bioreceptors in biosensors to facilitate enzymatic reactions. Enzymes are obtained by purification from their sources, and precaution is taken to avoid denaturing of the enzymes, which results in the loss of enzymatic activities. This purification process can be eliminated if whole cells are used as bioreceptors. The advantages of cell-based biosensors include:

- Less susceptible to changes in environmental conditions;
- Cost effective because no purification steps are required;

- Longer lifetime because the cell is an enclosed environment;
- Enzymes already in the optimal environment to perform their duties.

Cell-based biosensors have been demonstrated previously with much success (40,41). Perhaps the most common cell-based biosensor is one that uses yeasts as the bioreceptors. Yeasts are robust, with wide physiochemical tolerance and tough cell walls. They are also very well understood; therefore, genetic alteration to increase permeability of cell membranes, selectivity in analyte recognition, and enzymatic activities is possible. The constant threat of terrorism has renewed interests in the military applications of biosensors. The ability to detect and measure environmental toxicity is crucial to safeguard property and lives. The recently reported toxicity biosensor with genetically modified *Saccharomyces cerevisiae* (type of yeasts) shows enormous potential (27). The genetically modified *Saccharomyces cerevisiae* is made bioluminescent. The intensity of the light generated reduces as the concentration of toxin increases. The *Saccharomyces cerevisiae*-based biosensor appears to perform very well in harsh toxic environments.

5. FUTURE TRENDS

The development of futuristic sophisticated biomedical sensors will continue to exploit novel technologies, as they are available to improve performance and usability. Implantable biosensors with wireless and telemetry capability are one of the latest research topics. Smart sensors with on-chip processing modules can be realized with integrated circuit fabrication techniques. Sensors with multiple functions or multiple sensors may also be combined with integrated circuits to provide a robust sensing capability. Large-scale smart-sensor networks could be deployed in various clinical theaters to enhance performance by providing more informative and accurate descriptions of the measurands.

5.1. Wireless Sensor Networks

The transfer of information to and from sensors implanted within a patient is the subject of recent research. One of the most obvious requirements of an implantable sensor is that the sensor must be noninvasive and must minimize any effects it has on normal biological functioning of the body. This important requirement motivated the use of wireless communication links with external control, or computer systems. With wireless techniques, interconnections to the sensor chip can be eliminated. Challenges of wireless sensor networks have been discussed extensively (42–46). The implementation of wireless sensors for biomedical applications has been reported (47,48). The use of a camera, which transmits images to an array of embedded smart sensors within the eye, shows tremendous potential as a visual aid to the visually impaired. The artificial retina sensor array consists of two components: (1) integrated circuits for signal processing and (2) electrical probes that transmit electrical pulses to the retina

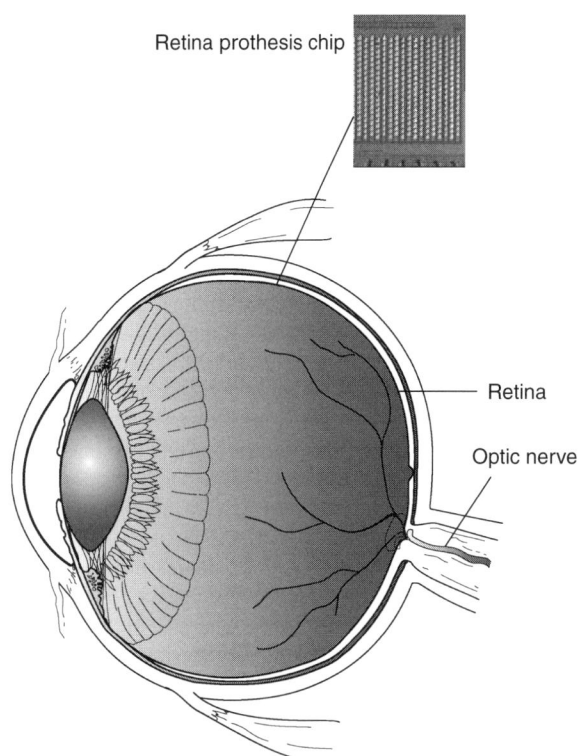

Figure 6. Wireless artificial retina embedded within the eye (© with permission from IEEE) (29).

nerves. The artificial retina has the capability to transmit and receive data. As the array of sensing devices is embedded within the eye, the challenge is to successfully network the sensors and the camera to an external processing unit via wireless communication protocols. Figure 6 shows the implantable microsensor array and its location within the eyeball.

5.2. Integrated Multisensors

Next generation sensors will benefit from the advancement in fabrication techniques, telecommunication protocols, and low-power integrated circuitry. Biomedical instruments based on micro-electromechanical systems (MEMS), to realize lab-on-a-chip devices, offer the potential for large-scale biological extraction and clinical analysis on the same silicon chip (49,50). Systems-on-chip technology provides the ability to build functioning systems on a single silicon chip to reduce development time and cost. The integration of lab-on-a-chip and systems-on-chip technologies to create a multiple-function sensor microsystem has been proposed (51). An integrated multifunction sensor incorporating multiple sensors, an application-specific integrated circuit (ASIC), a wireless transmitter, and a battery is described. The sensing device may be used to detect abnormalities by real-time monitoring of the gastrointestinal (GI) tract. The device contains sensors for measuring conductivity, pH, temperature, and dissolved oxygen. The circuitry for analog signal condition, multiplexing, analog-to-digital (ADC), digital-to-analog (ADC), oscillator, and transmitter are implemented using ASIC technology. A growing trend exists for more integrated and performance-enhanced multifunction biomedical sensors. Given the ability to detect and process more physiological and biological parameters, future sensors will be able to provide a more advanced and complete picture of our biological states.

5.3. Wearable Sensors

Often, it is desirable to have continuous monitoring of physiological signals to detect any abnormalities. In situations where sensors are used as part of a feedback mechanism to counteract a potentially debilitating disease such as diabetes, implantable sensors are the obvious choice. However, implantable sensors may not be the most economical, and certainly not the only, option. In cases where sensors serve as diagnostic tools to advance disease detection or prevention, noninvasive wearable sensors are a cost-effective and easy-to-maintain alternative. Wearable biosensors (WBS) may be used to detect elevated blood pressure or heart rate variability after taking dose-critical medication. WBS will be an indispensable instrument for patients who suffer from chronic cardiovascular diseases. WBS will continuously monitor for physiological indices (abnormal heart rate or blood pressure) that precede a heart attack. Life threatening situations or costly hospitalization can be effectively avoided. The development of WBS with advanced photoplethysmographic (PPG) techniques to deduce the patient's cardiovascular state has been reported (52). The WBS is implemented as a ring around a patient's finger. Figure 7 shows a WBS prototype using a coin-sized battery as a power source. Physiological parameters such as heart rate, oxygen saturation, and heart rate variability may be detected with the wearable PPG sensor. Challenges in WBS include minimizing effects of motion artifacts, interference with normal biological functioning (blood circulation in this case), and the power requirement. In addition, the WBS should not cause any discomfort to the wearer. Results have indicated that WBS has the potential to provide reliable and effective biosensing over a broad range of health applications.

Another notable wearable sensor technology that led to a successful spin off, Sensatex Inc., is the Georgia Tech Wearable Motherboard developed at Georgia Tech (53). The technology evolved from an initial military focus to everyday civilian medical applications. The Smart Shirt is a wearable health-monitoring system that can record heart rate, body temperature, pulse oximetry, and many other physiological parameters. The vital signals can be transmitted wirelessly to a processing unit that can perform real-time tracking of defining physiological events to trigger preventive actions before the onset of fatal symptoms. The development of the Smart Shirt technology takes into account the following design criteria: (1) wearability; (2) durability; (3) usability; (4) maintainability; (5) performance; and (6) functionality. By integrating the latest technologies in telemedicine and information processing, the Smart Shirt has the potential to make significant impact by advancing clinical studies and diagnosis of diseases that display subtle symptoms before an episode.

Another example of a successfully commercialized wearable sensor technology is a GlucoWatch® Biographer, manufactured by Cygnus Inc., that can monitor blood sugar levels for up to 13 hours. The biographer is worn on a person's forearm like a wristwatch to provide frequent, automatic, and noninvasive measurement of glucose level. The glucose is extracted through the skin by iontophoresis, and the glucose oxidase reactions are measured by an electrochemical sensor (54). A current generated by the electrochemical reaction is converted into a glucose measurement by signal processing algorithms. Although biographer represents a significant advancement in wearable technology for glucose measurement, even Cygnus recommends that it should not be used as a replacement to conventional glucose meters or implantable glucose sensors.

A huge concerted effort exists in the Europe Union, which includes over 30 european research and academic institutes of excellence with strong industrial support, in developing Intelligent Biomedical Clothes (IBC). Notably, Italy is one of the major players in this field. The Centro "E. Piaggio" at the University of Pisa has demonstrated futuristic e-Texiles, WEALTHY, which contains strain fabric sensors based on piezo-resistive yarns, and fabric electrodes realized with metal-based yarns (55). WEALTHY is a wearable and wireless instrumented garment capable of recording physiological signals such as respiration rate, electrocardiogram, electromiogram, and temperature. A miniaturized short-range wireless system can be integrated in the sensitive garment and used to transfer the signals to processing PCs, PDAs and mobile phones. The performance of the high-tech garment remains unaffected after repeated washing and is fully reusable.

6. CONCLUSIONS

Biomedical sensing technologies and their applications are presented in this chapter. Future trends of biosensing technologies seem to be moving toward a more integrated,

Figure 7. Battery-powered wearable photoplethysmographic sensor (© with permission from IEEE) (33). (This figure is available in full color at http://www.mrw.interscience.wiley.com/ebe.)

multifunctional, and real-time monitoring approach. The research and development of biomedical sensors will, without doubt, continue to flourish because of the uniqueness and importance of their applications. Technologies originally developed for optical and wireless communication networks or other exciting application areas will eventually be adapted for medical biosensing.

BIBLIOGRAPHY

1. J. Janata and A. Bezegn, Chemical sensors. *Anal. Chem.* 1988; **60**:62R–74R.
2. S. M. Sze, *Semiconductor Sensors*. New York: Wiley, 1994.
3. J. I. Peterson and G. G. Vurek, Fiber optic sensor for biomedical applications. *Science* 1984; **224**:123–127.
4. J. P. Payne and J. W. Severinghaus, *Pulse Oximetry*. Berlin, Germany: Springer-Verlag, 1986.
5. I. Fatt, Polarographic oxygen sensors. Cleveland, OH: CRC Press, 1976.
6. C. D. Fung, P. W. Cheung, W. H. Ko, and D. G. Fleming, *Micromachining and Micropackaging of Transducers*. Amsterdam: Elsevier, 1985.
7. M. R. Neuman and C.-C. Liu, Biomedical sensors in interventional systems: present problems and future strategies. *Proc. IEEE* 1988; **76**(9):1218–1225.
8. V. G. Welch, *Optical-Thermal Response of Laser Irradiated Tissue*. New York: Plenum Press, 1995.
9. B. Culshaw and J. Dakin, *Optical Fiber Sensors*, vol. I & II. New York: Artech House, 1989.
10. M. Lambrechts and W. Sansen, *Biosensors: Microelectrochemical Devices*. University of Leuven, Belgium: Institute of Physics and IOP Publishing Ltd, 1992.
11. F. S. Ligler and C. A. R. Taitt, *Optical Biosensors: Present & Future*. New York: Elsevier Scienct, 2002.
12. L. J. Blum and P. R. Coulet, *Biosensor Principles and Applications*. New York: Marcel Dekker, 1991.
13. C. Chou, S. L. Pan, L. C. Peng, C. Y. Han, and K. W. Yu, Fiber-optic biosensor for E. coli identification. *CLEO 2000*, May 7–12, 2000, pp. 501–502.
14. J. G. Fujimoto, C. Pitris, S. A. Boppart, and M. E. Brezinski, Optical coherence tomography: an emerging technology for biomedical imaging and optical biopsy. *Nature* 2000; **2**:9–25.
15. Y. T. Pan, H. K. Xie, and G. K. Fedder, Endoscopic optical coherence tomography based on a microelectromechanical mirror. *Opt. Lett.* 2001; **26**:1966–1968.
16. T. W. Yeow, V. Yang, A. Chahwan, M. Gordon, B. Wilson, and A. A. Goldenberg, Micromachined optical coherence tomography. *Sens. Actuators A* 2005; **117**:331–340.
17. S. Takatani and J. Lian. Optical oximeter sensors for whole blood tissue. *IEEE Eng. Med. Biol. Mag.* 1994; **13**(3):347–357.
18. G. Harsanyi, I. Peteri, and I. Deak, Low cost ceramic sensors for biomedical use: a revolution intranscutaneous blood oxygen monitoring? *Sens. Actuators B* 1994; **18–19**:171–174.
19. H. Suzuki, A. Sugama, and N. Kokima, Micromachined clark oxygen electrode. *Sens. Actuators B* 1993; **10**:91–98.
20. S. J. Alcock and A. P. F. Turner, Continuous analyte monitoring to aid clinical practice. *IEEE Eng. Med. Biol. Mag.* 1994; **13**(3):319–325.
21. L. C. Clark, Monitoring and control blood and tissue oxygen. *Trans. Am. Soc. Artif. Intern. Organs* 1958; **2**:41–48.
22. H. Jinghong, C. Dafu, L. Yating, C. Jine, D. Zheng, Z. Hong, and S. Chenglin, A new type of transcutaneous pCO_2 sensor. *Sens. Actuators B* 1995; **24–25**:156–158.
23. I. W. Severinghaus and A. F. Bradely, Electrodes for blood PO^2 and PCO^2 determination. *J. Appl. Physiol.* 1958; **13**:515–520.
24. G. Harsanyi, *Sensors in Biomedical Applications: Fundamentals, Technology & Applications*. Lancaster, PA: Technomic, 2003.
25. G. G. Guilbault, J.-M. Kauffmann, and G. J. Patriarche, In: R. F. Taylor, ed., *Protein Immobilisation, Fundamentals and Applications*. New York: Marcel Dekker, 1991, pp. 209–262.
26. R. F. Taylor, In: R. F. Taylor, ed., *Protein Immobilisation, Fundamentals and Applications*. New York: Marcel Dekker, 1991, pp. 286–289.
27. R. P. Hollis, K. Killham, and L. A. Glover, Design and application of a biosensor for monitoring toxicity of compounds to eukaryotes. *App. Environ. Microb.* 2000; **66**:1676–1679.
28. R. Wilson and A. P. F. Turner, Glucose oxidase: an ideal enzyme. *Biosens. Bioelectron.* 1992; **65**:238–240.
29. P. I. Hilditch and M. J. Green, Disposable electrochemical biosensors. *Analyst* 1991; **116**:1217–1220.
30. C. Podaru, C. Bostam, C. Malide, O. Neagoe, M. Simion, G. Popescu, D. Gargancinc, and N. Tzetci, An amperometric glucose biosensor. *Inter. Semi. Conf.* Oct. 9–12, 1996; **1**:101–104.
31. P. S. Grant, Y. Lvov, and M. J. McShane, Nanostructured fluorescent biosensor for glucose detection. *IEEE EMBS/BMES Conf.* Oct. 23–26, 2002; **2**:1710–1711.
32. G. Obrador, B. Pereira, and A. Kausz, Chronic kidney disease in the United States: an underrecognized problem. *Sem. Nephrol.* 2002; **22**(6):441–448.
33. C. E. Stanecki, M. J. McShane, A. M. Hannibal, A. Watts, and K. Driggers, A novel biosensor for on-line dialysis monitoring. 25th Annual International Conference, IEEE-EMB, Engineering in Medicine and Biology Society, Cancun, 2003: 3009–3011.
34. J. M. Jay, *Modern Food Microbiology*. Gaithersburg, MD: Aspen, 2000.
35. T. A. Sergeyeva, N. V. Lavrik, and A. E. Rachkov, Polyaniline label-based conductimetric sensor for IgG detection. *Sens. Actuator B* 1996; **34**:283–288.
36. D. T. Hoa, S. Kumar, T. N. Punekar, N. S. Punekar, and L. R. Srinivasa, Biosensor based on conducting polymers. *Anal. Chem.* 1992; **64**:2645–2646.
37. Z. Muhammad-Tahir and E. C. Alocilja, Fabrication of a disposable biosensor for Escherichia coli O157:H7 detection. *IEEE Sensors J.* 2003; **3**(4):345–351.
38. W. F. Scheller, E. V. Ralis, A. Makower, and D. Pfeiffer, Amperometric bi-enzyme based biosensors for the detection of lactose-characterization and application. *J. Chem. Tech. Biotechnol.* 1990; **49**:255–256.
39. I. Eshkenazi, E. Maltz, B. Zion, and J. Rishpon, A three-cascaded-enzymes biosensor to determine lactose concentration in raw milk. *J. Dairy Sci.* 2000; **83**:1939–1945.
40. G. T. A. Kovacs, Electronic sensors with living cellular components. *Proc. IEEE* 2003; **91**(6):915–929.
41. M. D. Antonik, N. P. D'Costa, and J. H. Hoh, A biosensor based an micromechanical interrogation of living cells. *IEEE Eng. Med. Biol. Mag.* 1997; **16**(2):66–72.
42. I. F. Akyildiz, W. Su, Y. Sankarasubramaniam, and E. Cayirci, A survey on sensor networks. *IEEE Commun. Mag.* 2002; **40**(8):102–114.
43. D. Estrin, R. Govindan, J. Heidemann, and S. Kumar, Next century challenges: scalable coordination in sensor networks.

ACM/IEEE Inter. Conf. on Mobile Computing Networking, 1999: 263–270.
44. S. K. S. Gupta, S. Lalwani, Y. Prakash, E. Elsharawy, and L. Schwiebert, Towards a propagation model for wireless biomedical applications. *IEEE ICC03*, 2003; **3**:1993–1997.
45. S. Shakkottai and T. S. Rappaport, Research challenges in wireless networks: a technical overview. *Sympos. Wireless Personal Multimedia Communicat.* 2002; **1**:12–18.
46. L. Schwiebert, S. K. S. Gupta, and J. Weinmann, Research challenges in wireless networks of biomedical sensors. Proc. ACM/IEEE Conf. on Mobile Computing and Networking, 2001: 151–165.
47. T. B. Tang, E. A. Johannessen, L. Wang, A. Astaras, M. Ahmadian, A. F. Murray, J. M. Cooper, S. P. Beaumont, B. W. Flynn, and D. R. S. Cumming, Toward a miniature wireless integrated multisensor microsystem for industrial and biomedical applications. *IEEE Sensors J.* 2002; **2**(6):628–635.
48. L. Schwiebert, S. K. S. Gupta, P. S. G. Auner, G. Abrams, R. Iezzi, and P. McAllister, A biomedical smart sensor for the visually impaired. *IEEE Sensors J.* 2002; **1**:693–698.
49. G. Medoro, N. Manaresi, A. Leonardi, L. Altomare, M. Tartagni, and R. Guerrieri, A lab-on-a-chip for cell detection and manipulation. *IEEE Sensors J.* 2003; **3**(3):317–325.
50. E. Verpoorte and N. F. De Rooij, Microfluidics meets MEMS. *Proc. IEEE* 2003; **91**(6):930–953.
51. L. Wang, T. B. Tang, E. A. Johannessen, A. Astaras, M. Ahmadian, A. F. Murray, J. M. Cooper, S. P. Beaumont, and D. R. S. Cumming, Integrated micro-instrumentation for dynamic gastrointestinal tract monitoring. Proc. 19th IEEE IMTC, 2002: 1717–1720.
52. H. H. Asada, P. Shaltis, A. Reisner, R. Sokwoo, and R. C. Hutchinson, Mobile monitoring with wearable photoplethysmographic biosensors. *IEEE Eng. Med. Biol. Mag.* 2003; **22**(3):28–40.
53. S. Park and S. Jayaraman, Enhancing the quality of life through wearable technology. *IEEE Eng. Med. Biol. Mag.* 2003; **22**(3):41–48.
54. M. J. Tierney, J. A. Tamada, R. O. Potts et al., The GlucoWatch Biographer: a frequent automatic and noninvasive glucose monitor. *Ann. Med.* 2000; **32**(9):632–641.
55. R. Paradiso, A. Gemignani, E. P. Scilingo, and D. De Rossi, Knitted Bioclothes for Cardiopulmonary monitoring. 25th Annual International Conference, IEEE-EMB, Engineering in Medicine and Biology Society, Cancun, 2003: 3720–3723.

BIOMEDICAL TRANSDUCERS

TATSUO TOGAWA
Waseda University
Tokorozawa, Saitama, Japan

1. INTRODUCTION

A transducer is a device that converts a measured object quantity into an electrical signal. Biomedical transducers are transducers with specific uses in biomedical applications, such as physiological measurement and patient monitoring, and in health care. The object quantities in biomedical measurements are physical and chemical quantities that reflect the physiological functions in a living body. Typical object quantities and transducers or systems for biomedical measurements are listed in Table 1. Although some quantities, such as blood composition, can be determined from a sample extracted from the body, real-time and continuous measurements can be achieved if a transducer is attached to the body, and use of a transducer attached to the body is essential when continuous monitoring of an object quantity is required.

Usually, a measurement system consists of a transducer and an electronic instrument as shown in Fig. 1 (1). The physical or chemical quantity that characterizes the physiological function in the body is detected by the transducer and converted into an electrical quantity. Once the object quantity is converted into an electrical signal, then ordinary electronic instruments, such as computers, can be used to analyze, display, or store the data.

Sometimes, measurements require active procedures to be applied to the object, such as excitation, transmission, illumination, irradiation, stimulation, application, or injection. Such procedures are considered part of the measurement process. A transducer that involves active procedures is called an active transducer, and a transducer without such functions is called a passive transducer.

Table 1. Typical Object Quantities and Transducers or Systems for Biomedical Measurements

Object quantity	Transducer or measurement system
Blood pressure	Diaphragm-type transducer
	Cuff technique
	Tonometric transducer
Blood flow in a vessel	Electromagnetic flowmeter
	Ultrasonic flowmeter
	Dilution method
	Clearance method
	Laser Doppler flowmeter
	Plethysmography
	Magnetic resonance imaging
Respiratory air flow	Rotameter
	Pneumotachograph
	Hot-wire anemometer
	Ultrasound flowmeter
	Volume-type spirometer
	Body plethysmography
	Impedance plethysmography
Body motion	Displacement and rotation transducers
	Optical motion analyzer
	Magnetic induction transducer
	Accelerometer
	Gyroscope
	Force plate
Body temperature	Contact thermometer
	Infrared radiation thermometer
Bioelectric potential	Body surface electrode
	Extracellular electrode
	Intracellular electrode
Biomagnetism	SQUID magnetometer
	Fluxgate magnetometer
Body fluid component	Electrochemical transducer
	Optical transducer
	Acoustic transducer
	Functional magnetic resonance imaging

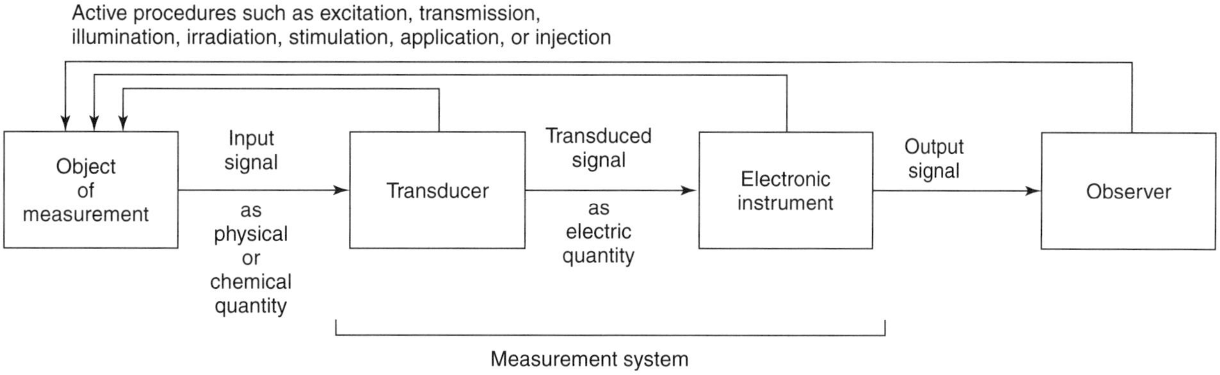

Figure 1. The general structure of a measurement system.

In principle, many physical or chemical quantities in the body can be measured by passive transducers inserted into the body. For example, if a small pressure transducer can be introduced into an artery, then blood pressure can be measured accurately. However, to realize this, a surgical procedure is required, and such an invasive procedure is not acceptable in the ordinary patient monitoring situation, unless continuous blood pressure monitoring is essential for care of the patient. When invasive blood pressure measurement is unacceptable, a noninvasive measurement is employed, in which an external pressure is applied, and the pulse rate is detected to determine the systolic and diastolic pressures.

Although use of an active procedure is unavoidable in actual measurement situations, its influence on the body should be minimized for two reasons: (1) to minimize hazards, and (2) to minimize the measurement errors because of the change in object quantity caused by the active procedure. Therefore, the level of the active procedure should be the minimum required for achievement of the measurement.

Many situations exist where different measurement methods are required, and many types of transducers have been developed. Although noninvasive transducers are desirable, their use is not possible in some cases, and use of an invasive procedure, such as one introducing a fine needle probe into tissue, is sometimes unavoidable. Even in such cases, effort is taken to minimize the effect of the invasive procedure. A transducer in which the invasive effects are reduced is called a lessinvasive transducer. In some situations, a transducer has to be placed within the body. Implantable or indwelling transducers are used in such situations. When a continuous physiological measurement is required in an unconstrained subject, measurement has to be performed by a transducer that can be worn. For patient care or health management in the home, transducers for home use are required, and in the following sections, transducers for this purpose are described.

2. NONINVASIVE AND LESS-INVASIVE TRANSDUCERS

A noninvasive transducer is a transducer that does not involve an invasive technique, such as puncturing of the skin or blood vessels for the insertion of different types of mechanical objects (2). Although a transducer may not need to puncture the skin for use, it will be regarded as invasive if it requires the application of energy to a patient above the level at which damage to the tissue can occur. In contrast, when the applied energy of a required procedure is at a low level, such that any adverse effect is negligible, it can be regarded as noninvasive. Many active procedures, such as ultrasound, visible or near-infrared light, electric current, magnetic flux, x-ray and gamma-ray, mechanical force, heat load, and inhalation of a gas, have been successfully used in noninvasive transducers.

Ultrasound is a convenient tool in noninvasive measurements, because a sharp beam can be easily available, it can transmit in the tissue fairly well, and it is absolutely safe at a level that is needed to detect back-scattered signal. When a short burst of ultrasonic wave is transmitted from a crystal into the tissue, back-scattered signals from tissue boundaries can be detected by the same crystal. The distance between the transmitter/receiver crystal and the scattering object can be determined from the traveling time of the wave. Repeating this procedure, the motion of the scattering object can be observed in real-time. With this technique, the motion of a heart valve can be monitored noninvasively from the body surface, as shown in Fig. 2.

The ultrasound scattered at a moving object causes Doppler shift, and the velocity of the object can be determined by frequency analysis of the Doppler component. As ultrasound is scattered by blood cells, blood flow velocities can be measured noninvasively. As shown in Fig. 3a, back-scattered signal from a definite sample volume can be extracted by gating the received signal with a time window. In the ordinary blood pressure measurement method, a cuff is attached to the upper arm and the sound of the vascular activity, called the Korotkoff sound, is audibly detected with a stethoscope to determine the systolic and diastolic pressures. Although this method is noninvasive, it requires a medical professional's presence, and thus involves a subjective factor. To make the measurement objective in nature, and to obtain measurement results as an electrical signal, electronic blood pressure meters are available, which also use a cuff that is attached to the patient's upper arm; but the systolic and diastolic pressures are determined by detecting the change in pulsatile pressure in the cuff with a pressure sensor, or by detecting the

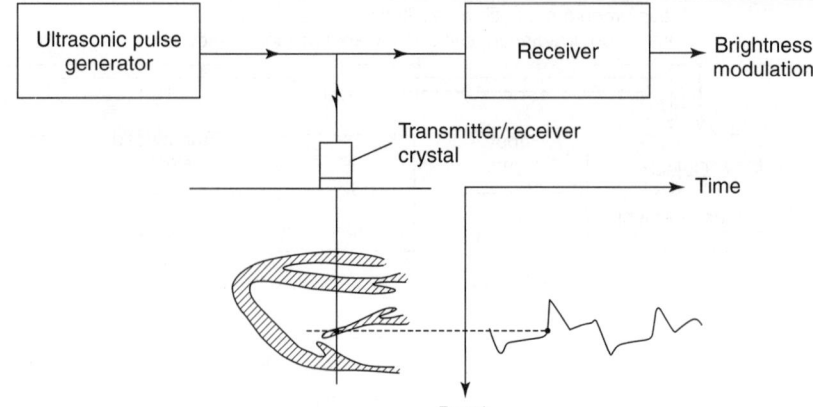

Figure 2. Ultrasonic transducer for noninvasive detection of valvular motion of the heart with M-mode display.

vascular sound with a microphone. Such a system can be regarded as a noninvasive transducer that converts intravascular pressure into an electronic signal.

When a patient's blood circulation is unstable, and a fatal event, such as hemodynamic shock, may occur, the blood pressure is monitored with an arterial catheter connected to a pressure transducer. Although this method is invasive, it is commonly used in high-risk patients. Noninvasive continuous monitoring of blood pressure has been realized with different techniques. The vascular unloading technique is based on the principle that the internal pressure in a cavity can be monitored with a controlled external pressure set to equate both pressures at each moment. This principle was realized with a pneumatic servo-controlled system, schematically shown in Fig. 4 (3). A probe with an infrared light source, a photo detector, and a cuff is attached to the finger, and the cuff pressure is controlled such that the light transmittance is constant. The change in transmittance is mainly caused by pulsation of the arteries, so if the pulsation can be compensated for by applying an external pressure, the change in the external pressure has to be the same as the intra-arterial pressure. This condition can be achieved with a fast-response pneumatic servo-control system. As a result, the arterial pressure waveforms, as well as its absolute pressure level, can be monitored continuously, giving similar results to an intra-arterial catheter connected to a manometer. Therefore, this technique is sometimes called the noninvasive catheter-manometer system (4).

Another noninvasive method of measuring intra-arterial pressure is the use of tonometry (5). The principle of this technique is based on the use of reaction force measurements. When a flat plate is pressed onto a flexible deformable boundary membrane, on which internal pressure is exerted, then the internal pressure can be measured from the outside as the reaction force, regardless of the transverse tension developed in the membrane. This principle has been applied to intraocular pressure measurements. The same principle has also been applied to arterial pressure monitoring, in which a probe is attached

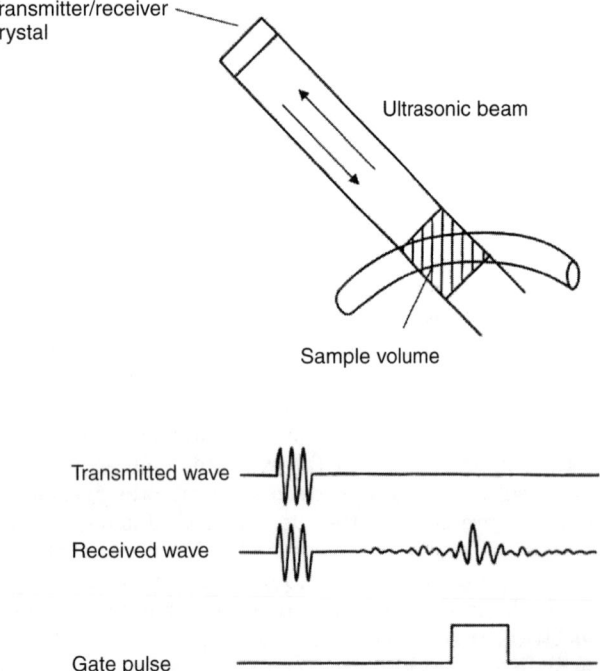

Figure 3. Ultrasound Doppler blood flow measurement with range discrimination by gating the received signal with a time window.

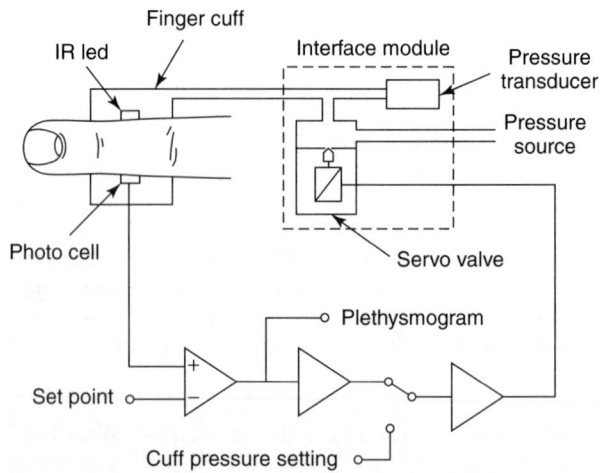

Figure 4. The vascular unloading technique for instantaneous blood pressure monitoring using a servo-control of the finger cuff pressure so as to equate to the intra-arterial pressure.

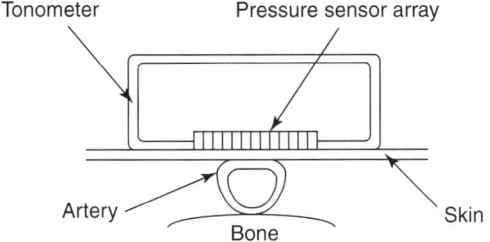

Figure 5. A tonometry transducer for blood pressure monitoring.

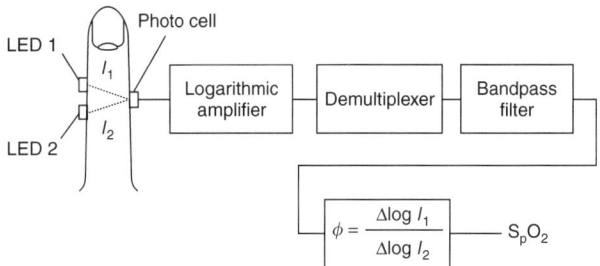

Figure 6. Principle of the pulse oximeter.

to the wrist so that the radial artery is partially occluded, as shown in Fig. 5 (3). In this example, the transducer has a pressure sensor array, and the highest recorded value among the sensor outputs corresponds to the arterial pressure.

Arterial blood oxygen saturation can be monitored noninvasively with a pulse oximeter (6). As a result of the difference in the spectral absorption of oxyhemoglobin and reduced hemoglobin, the oxygen saturation level of a blood sample can be determined by measuring the absorption at two different wavelengths, typically in the red and infrared part of the spectrum. However, human tissue contains arterial and venous blood, and hence, light absorption by arterial blood cannot be determined simply from the tissue absorption. However, the arterial component of light absorption can be estimated selectively by extracting the pulsatile component of the tissue absorption (7). Monitoring of the arterial oxygen saturation level is usually performed with a patient's finger, as shown in Fig. 6. The arterial oxygen saturation level is determined from the ratio of amplitude of the pulsatile components at two different wavelengths.

3. NONCONTACT TRANSDUCERS

Some physiological quantities can be measured without placing a transducer in contact with the body. For example, the body surface temperature can be measured with a radiation thermometer, and the temperature distribution on the body's surface can be observed with a thermal camera or a thermography system. Optical, electromagnetic, acoustic, and pneumatic techniques can be employed to realize noncontact measurement of physiological quantities.

Noncontact transducers have the advantage of being able to avoid contamination, and thus prevent infection. They are also inherently safe against electrical hazards, such as micro-shocks caused by a leakage current passing through the heart. They also avoid causing discomfort by mechanical contact with the body surface. Such features are important where the tissue is sensitive, such as on the cornea or on burned skin.

Pneumatic methods are sometimes employed to realize noncontact measurements, and Fig. 7 shows an example of a noncontact intraocular pressure measurement system (8). This technique produces a linearly increasing air pulse that impinges on the cornea. It detects the point when the corneal surface becomes flattened. At that point, the external and internal forces at the cornea are balanced, and the internal pressure can be estimated from the force developed by the air pulse.

A noncontact measurement is similar to a remote-sensing function, such as the visual, auditory, and olfactory systems by which animals can obtain extensive information from a distance. Although the performance of many available artificial-sensing techniques is still inferior to the natural senses of animals, especially in smell sensing, performance similar to those of animal senses may be artificially realized in principle, and thus further developments are expected, especially in detecting chemical quantities with sensitivity similar to that of the animal olfactory function.

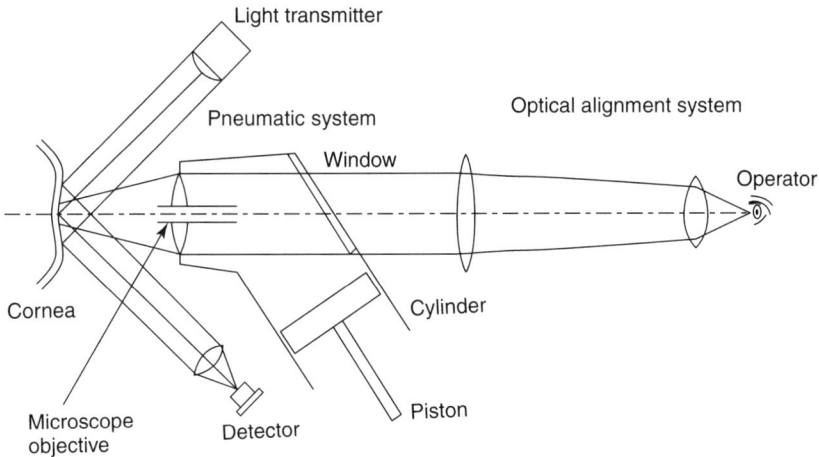

Figure 7. A tonometer for noncontact intraocular pressure measurement.

4. INDWELLING TRANSDUCERS

In the intensive care of high-risk patients, some transducers are placed in the body, because of the need for accurate and continuous monitoring of the physiological condition of the patient. Even in a situation where continuous monitoring is not necessary, the constant presence of a transducer is sometimes more convenient than having to place and remove the transducer for each measurement.

A simple way to place a transducer in the body is to use a natural orifice. For example, rectal thermometer probes are used for long-term body temperature monitoring. During anesthesia, an esophageal thermometer probe is sometimes used instead, because the esophageal temperature undergoes rapid changes in temperature, but rectal temperature does not (9). When a balloon catheter is placed in a patient's bladder, a temperature sensor can be placed at the tip of the catheter to monitor body temperature. Figure 8 shows a thermistor-tipped bladder catheter for that purpose. The temperature obtained with such a probe is close to the rectal temperature and can be a reliable monitor of body temperature (10), which is advantageous for a patient for whom urine drainage is necessary.

When monitoring is essential for patient intensive care, a transducer can be inserted in a blood vessel for periods of many days. For example, the thermo-dilution method is used to monitor cardiac output (11). Figure 9 shows a method in which a thermo-dilution catheter is inserted intravenously into the pulmonary artery. To measure the cardiac output, a small bolus of cold saline is injected into the right atrium, and the time course of the changes in blood temperature is measured, which exhibits a dilution curve for an impulsive heat input. The cardiac output is then calculated from the heat input and the dilution curve. Although the thermo-dilution method requires an invasive procedure, the cardiac output can be measured repeatedly once the thermo-dilution catheter has been inserted.

In principle, transducers can be implanted in tissue. If the data readout can be obtained without connecting a cable through the skin, then it will be safer and more convenient than an indwelling transducer using a cable connection. Even if it is unacceptable to place a transducer with a surgical procedure in ordinary patients, it will be possible to place a small transducer in artificial organ components that can be implanted in the body. Data measurements can then be taken with either electromagnetic or optical coupling. In practice, cardiac pacemakers have sensing mechanisms in which cardiac potentials can be detected from the electrode so as to control the pacing rate, and some have a capability for data output, such as the

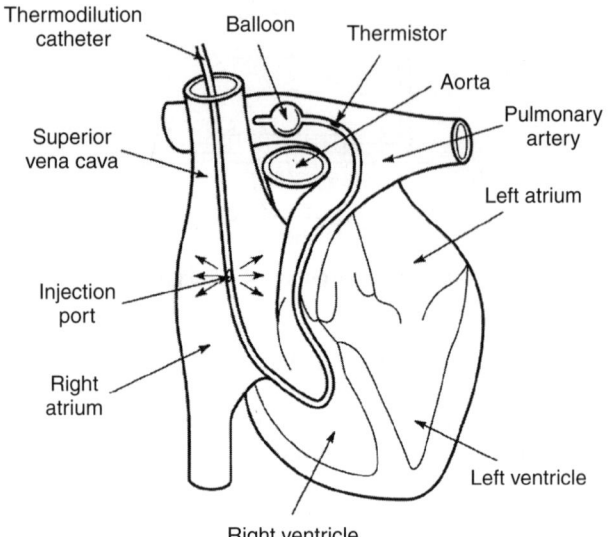

Figure 9. Thermodilution method for cardiac output measurement.

trend in heart rate (12). Rate-responsive pacemakers have a sensor that detects physiological quantity such as temperature or pH.

5. WEARABLE TRANSDUCERS

Continuous monitoring is employed for the long-term observation of physiological changes. From the long-term data obtained, rare episodes, such as an arrhythmia occurring over a short period, can be identified, and effective action, such as the use of anti-arrhythmic agents, can be taken. Such a technique is called ambulatory monitoring. To realize long-term continuous monitoring in unconstrained patients, a monitoring system has to be wearable, and thus wearable transducers are required for such purposes.

The most common monitoring item in unconstrained subjects is the electrocardiogram (ECG), such as the Holter ECG (13). In ordinary Holter ECG monitoring, three disposable electrodes, shown in Fig. 10, are attached to the chest wall, and the ECG is driven and recorded continuously for 24 or 48 hours. The data are stored in digital memory.

The ambulatory blood pressure monitoring system is also used. Although its principle is the same as the ordinary electronic blood pressure meter, intermittent measurements can be performed fully automatically in it (14). Strictly, the monitoring here is not continuous, but

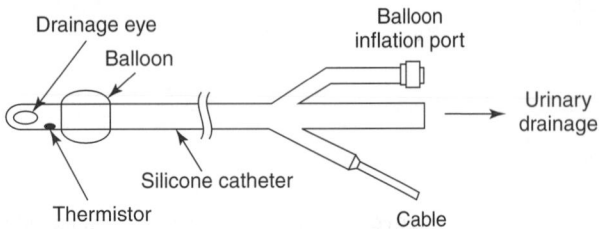

Figure 8. A thermistor-tipped bladder catheter.

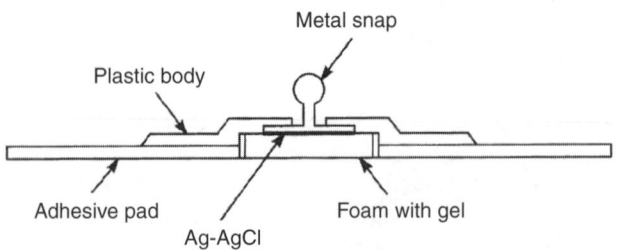

Figure 10. A cross-section of a disposable foam electrode.

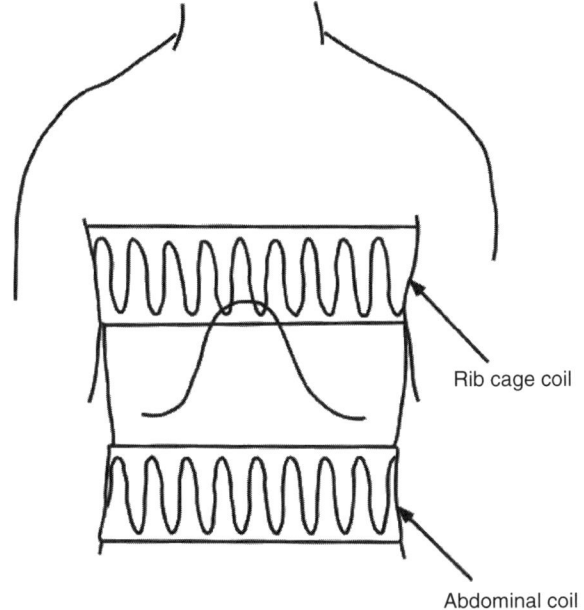

Figure 11. An inductance plethysmography for monitoring respiration.

rather intermittent. However, the measurement interval can range from every few minutes to once an hour, according to the monitoring purpose.

To monitor respiration in an unconstrained subject, the inductance plethysmography technique, as shown in Fig. 11, can be used. This technique consists of two coils arranged in a zigzag fashion, secured by two elastic bands on the rib cage and the abdomen. The inductance of each coil changes according to the cross-sectional area, and thus, the two coils record the volume changes in the rib cage and abdomen, allowing for an estimation of the tidal volume (15).

Body motion can be detected with an accelerometer attached to body segments. Accelerometers produced by silicon fabricated technology are commonly used for this purpose. For a rough estimate of the level of physical activity, the acceleration measured at one or more selected sites can be employed. For example, a rough estimate of the average energy expenditure can be made by monitoring the vertical acceleration at the waist (16), and the acceleration at the wrist can be an indicator for monitoring sleep (17).

Ring-type monitors are also increasing in popularity. A study has shown that a small electronic instrument attached to a finger ring can be used to monitor heart rate and arterial blood oxygen saturation (18).

Although chemical quantities are always difficult to monitor with wearable noninvasive transducers, it has been shown that glucose levels in interstitial fluid can be estimated with an enzymatic biosensor combined with iontophoresis, and a wristwatch-type wearable transducer system has been developed for this purpose (19).

6. TRANSDUCERS FOR HOME USE

Some medical devices, such as clinical thermometers and blood pressure meters, contain transducers and have been used in the home. Although the number of transducers for home use is limited, their potential importance has been stressed recently, for many reasons, including their ability to cope with the increase in the population of elderly citizens, their ability to obtain information for early diagnosis and disease prevention, their ability to allow for patient care at a distance, their ability to reduce periods of admission in hospital, their ability to reduce costs, and their ability to provide home care for the many patients who prefer it (20). Transducers for home use must be simple, safe, reliable, and inexpensive, and they must not disturb normal home life. Most transducers used in hospitals and medical facilities do not satisfy such requirements, and thus special considerations must be taken into account in the design of home transducers.

Blood pressure measurement is not restricted to the upper arm. As shown in Fig. 12, blood pressure monitors at the wrist position and on a finger are also used. In applications where the patients use the instrument themselves, the cuff is more convenient to attach than the upper-arm type. However, errors caused by gravitational

Figure 12. Wrist-type (left) and finger-type (right) blood pressure monitors.

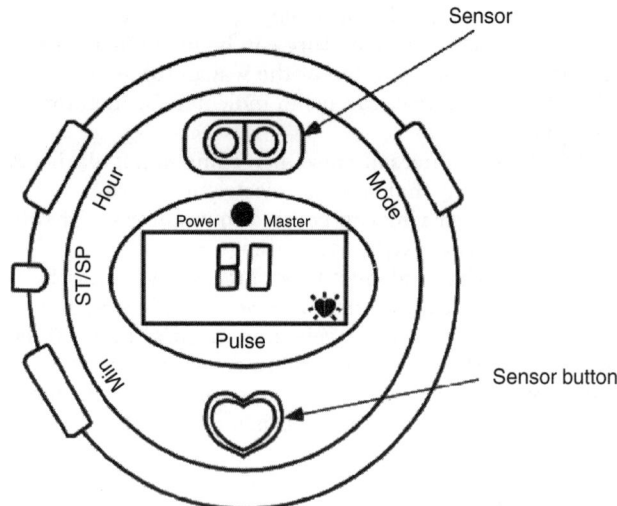

Figure 13. A wristwatch-type pulse monitor.

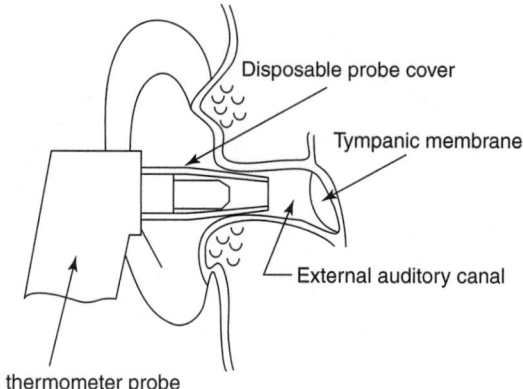

Figure 14. The tympanic thermometer probe inserted into the external auditory canal.

force, which is proportional to the relative height from the heart, is larger than that for measurements taken at the upper-arm location.

The heart rate or pulse rate is often monitored at home. Although it is possible to manually detect arterial pulses by placing a finger on the radial artery and counting the number of pulses per minute, a wristwatch-type pulse rate meter, as shown in Fig. 13, can also be used to display the pulse rate by placing a fingertip on the sensor, which consists of a small light source and a photo sensor.

Body temperature has commonly been measured at home, and frequent measurements are required in the home care of chronic diseases. Basal body temperature measurements are also required to monitor the menstrual cycle. A high degree of temperature accuracy is required for clinical thermometers, typically $\pm 0.1 \, \text{K}$ for ordinary use and $\pm 0.05 \, \text{K}$ for basal body temperature measurements. Although mercury-in-glass clinical thermometers have been used until recently, these have been replaced by electronic thermometers to avoid the risk of mercury contamination from accidental breakage.

The ordinary clinical electronic thermometer has a thermistor as the sensor, and the body temperature is displayed as a digital value. Completion of a measurement is notified by an audible beep. Although the response time of the sensor is faster than that of a mercury-in-glass thermometer, the time required for the sensor to reach a stable oral temperature is about the same, because of the fixed heat capacity of the tissue near the thermometer tip. Some models include an algorithm that can predict the final temperature, which has reduced the measurement time from about 3 min to 1 min or less.

The tympanic thermometer, which uses an infrared radiation measurement, is becoming popular. It has an infrared sensor, usually a thermopile that is installed near the probe tip, which is inserted into the auditory canal and oriented toward the tympanic membrane, as shown in Fig. 14. As the response time of the detector is very fast, and the tympanic temperature is close to the deep body temperature, a measurement can be carried out within a period of a few seconds. It was shown that the accuracy of tympanic thermometers remains in the acceptable range for ordinary clinical use (21,22).

For continuous monitoring of body temperature, devices that are used in hospitals, such as rectal or bladder thermometers, are unacceptable for use at home. However, the deep body thermometer, which uses the zero-heat-flow method, can be used without difficulty even for patient care at home. As shown in Fig. 15, a deep body thermometer probe has two temperature sensors to detect the outward heat flow from the body, and it is compensated for by a servo-controlled electric heating function. When the servo-control operates precisely, no heat can penetrate into the probe, and thus the probe can be regarded as being an ideal heat insulator. If a region of the surface of the skin is perfectly insulated, then the temperature of the surface of the skin beneath the probe equilibrates to the deep tissue temperature. It has been shown that the temperature obtained by attaching the probe at the chest, abdomen, or forehead is close to the deep body temperature value (23).

Body weight is an essential parameter for health management. The body-mass index, defined as the weight (in kilograms) divided by the square of the height (in meters), is sometimes used for this purpose. A detailed study has shown that a greater body-mass index is associated with a higher mortality (24). It is known that people with body-mass indices between 19 and 22 live the longest, and that

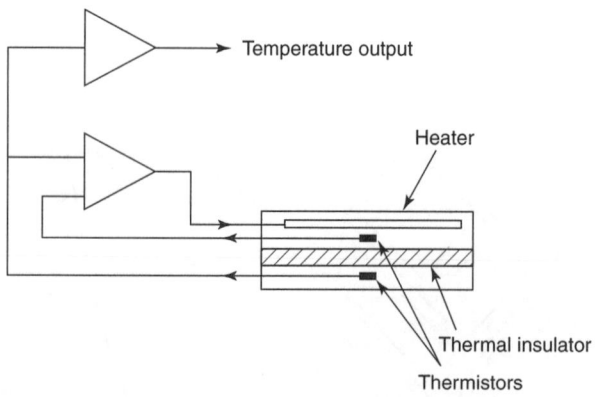

Figure 15. Schematic diagram of the deep body thermometer.

death rates are notably higher for people with indices above 25. A scale is available, in which family members' heights can be memorized, and the body-mass indices can be displayed by selecting a button.

Body composition, especially the amount of body fat, is a better determination of health than the body-mass index. Although body fat can be derived from body density, it is not easy to determine body density at home. Sometimes, body fat can be estimated from the thickness of the subcutaneous fat layer, which can be measured directly with a skin-fold caliper. A convenient method of estimating body fat is to use whole-body impedance measurements. The operating principle of this method is based on the fact that different tissues have different electrical impedances. In the low-frequency region, <1 MHz, the resistivity of skeletal muscle in the transverse direction is about 16 Ωm, and the resistivity of fat is 25 Ωm. Body impedance is usually measured with a four-electrode technique, in which a constant alternating current with a frequency of $f = 50$–100 kHz and a magnitude of $I = 0.1$–1 mA is applied between the outer electrode pair, and the alternating voltage developed between the inner electrode pair is then detected. In a commercial instrument, the four electrodes are located on a weighing scale, as shown in Fig. 16. The alternating excitation current is applied between the toes of both feet, and the voltage developed between electrodes at both heels is then detected. As the impedance between the feet and the body weight can be measured simultaneously, the approximate mass of fat can be estimated from a comparison between the measured body weight and the weight estimated from the observed impedance, assuming a lean body type. It has been shown that the impedance technique is highly reproducible for estimating lean body mass (25).

The use of near-infrared interactance to determine body composition has been studied (26). This method is based on the difference in absorption spectra between fat and lean tissue. In a commercial instrument, the specific wavelengths of $\lambda = 938$ and 948 nm are used, and the backscattered light is measured. Although the measurement can be performed at any site on the body, measurements at the biceps are mostly used to determine the whole body-fat value.

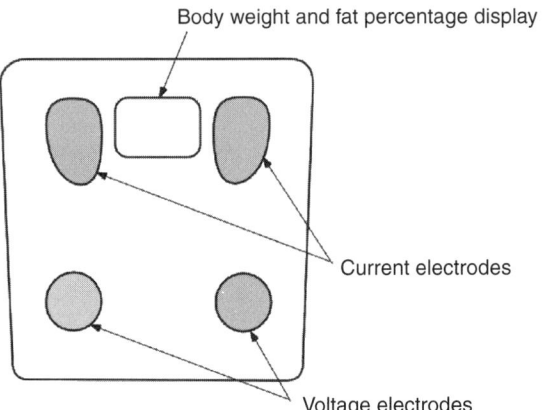

Figure 16. A body scale with impedance electrodes for body fat measurement.

7. BIOMIMICRY TECHNOLOGIES

Although many studies have been carried out on biomedical transducers and instrumentation, it may be said that existing devices have not reached their attainable limits. If their performances are compared with the sensing mechanisms of biological receptors, then many sensing organs exist superior to current artificial devices, which implies that many possibilities for improving the performance of artificial devices still exist. To achieve improvement, mimicking biological systems may be a possible approach, but it is also believed that high-tech solutions will eventually exceed the performance of biological systems, because many technologies are available that are not found in biological systems.

An example of a biological sensing function that does not have an artificial counterpart is tactile sensation. The softness of an object is sensed manually by touch with a fingertip. However, no comparable artificial device exists that can measure the elasticity or deformability of soft tissue. Although there have been attempts to estimate the visco-elastic properties of soft tissues by measuring the stress-strain relationship under high-frequency-applied shear forces (27), such systems are bulky and not as smart as a fingertip.

A remarkable feature of some receptors is the fact that standards of the quantity to be measured are contained in them. For example, a body temperature receptor can sense the absolute level of temperature so that it can provide signals to control the dissipation of heat from the body from the difference between the set point and the body temperature. Although existing temperature sensors need calibration before use, receptor cells do not need any calibration. In principle, a temperature standard can be obtained from a phase transition temperature, such as the ice point of water, but no convenient molecular standard temperature exists that undergoes a phase transition at a temperature close to body temperature. If a transducer with its own internal standard becomes available, then no temperature calibration would be necessary, and it would be highly reliable. Such a chemical indicator for pH would be a molecule that changed its light absorption at a specific pH, which would then be a molecular standard. A fiber optic pH transducer using such a chemical indicator has been attempted (28), but it still needed a calibration step.

A remarkable feature of biological receptor organs is their capability to discriminate molecular species with extremely high sensitivity. In particular, the olfactory function of some animals, such as dogs, is far above that of any chemical analyzer. Although many types of biosensors have been developed, their sensitivities are still far below that of the olfactory receptors of animals, and only precision chemical analyzers, such as mass spectrometers, can detect molecules at very low concentrations, which can be used to detect many chemical species in exhaled air (29), but a mass spectrometer is a bulky instrument employing high vacuum technology. If a simple smell sensor with sensitivity similar to that of a dog could be developed, it would be able to gather much information from a distance, as a dog does, and would be applicable to patient monitoring at home.

In ordinary clinical examinations, much important information is obtained from blood analysis. Some items, such as the blood glucose level, can be analyzed with a small volume of blood, even in the order of a few microliters. However, to obtain a blood sample, an invasive technique, such as puncturing the finger with a lancing device or a needle, has to be used. The mosquito is far smarter at taking a drop of blood than any available artificial puncturing system. A mosquito penetrates the skin by putting a probe that is about 30 μm in diameter and 2 mm long into a small blood vessel, and then removes a few microliters of blood in a period of a few minutes (30). No pain is involved for the victim, who only experiences a slight tingling sensation, if an itching sensation does not remain. If an artificial device mimicking a mosquito could be developed with micromachining technology, then blood samples could be obtained automatically.

As shown in the examples above, biomimicking technologies are one approach to attain future biomedical transducers with performances that would be similar to animal sensors, and would also be a step nearer to achieving the much higher level of sophistication that is realized in biological systems.

BIBLIOGRAPHY

1. T. Togawa, P. Å. Öberg, and T. Tamura, *Biomedical Transducers and Instruments*. Boca Raton, FL: CRC Press, 1997.
2. P. Rolfe, Preface. In: P. Rolfe, ed., *Non-invasive Physiological Measurement*, vol. 2. London: Academic Press, 1983, pp. i–ix.
3. T. Togawa, Patient Monitoring. In: J. G. Webster, ed., *Wiley Encyclopedia of Electrical and Electronics Engineering*, vol. 16, New York: John Wiley and Sons, 1999, pp. 1–10.
4. N. T. Smith, K. H. Weseling, and B. de Wit, Evaluation of two prototype devices producing noninvasive, pulsatile, calibrated blood pressure measurement from a finger. *J. Clin. Monit.* 1985; **1**:17–29.
5. J. S. Eckerle, Arterial tonometry. In: J. G. Webster, ed., *Encyclopedia of Medical Devices and Instrumentation*. New York: John Wiley and Sons, 1988, pp. 2770–2776.
6. J. W. Severinghause and P. Astrup, History of blood gas analysis. VI. Oximetry. *J. Clin. Monit.* 1986; **2**:270–288.
7. I. Yoshiya, Y. Shimada, and K. Tanaka, Spectrophotometric monitoring of arterial oxygen saturation in the fingertip. *Med. Biol. Eng. Comput.* 1980; **18**:27–32.
8. M. Forbes, G. Pico, Jr., and B. Grolman, A non-contact applanation tonometer: description and clinical evaluation. *Arch. Ophthalmol.* 1975; **91**:134–140.
9. R. J. Vale, Monitoring of temperature during anesthesia. *Int. Anesthesiol. Clin.* 1981; **19**:61–83.
10. J. K. Lilly, J. P. Boland, and S. Zekan, Urinary bladder temperature monitoring: a new index of body core temperature. *Crit. Care Med.* 1980; **8**:742–744.
11. W. Ganz, R. Donoso, H. S. Marcus, J. S. Forester, and H. J. C. Swan, A new technique for measurement of cardiac output by thermodilution in man. *Am. J. Cardiol.* 1971; **27**:392–396.
12. J. E. Waktare and M. Malik, Holter, loop recorder, and event counter capabilities of implanted devices. *Pacing Clin. Electrophysiol.* 1997; **20**(Pt 2):2658–2669.
13. W. J. Tompkins, Ambulatory monitoring. In: J. G. Webster, ed., *Encyclopedia of Medical Devices and Instrumentation*. New York: John Wiley and Sons, 1988, pp. 20–28.
14. G. Meaning, S. G. Vijan, and M. W. Millar-Craig, Technical and clinical evaluation of the Medilog ABP non-invasive blood pressure monitor. *J. Ambulat. Monitor.* 1994; **7**:255–264.
15. J. D. Sackner, A. J. Nixon, B. Davis, N. Atkins, and M. A. Sackner, Non-invasive measurement of ventilation during exercise using a respiratory inductance plethysmography. *Am. Rev. Respirat. Dis.* 1980; **122**:867–871.
16. S. B. Servais and J. G. Webster, Estimating human energy expenditure using an accelerometer device. *J. Clin. Eng.* 1984; **9**(2):159–171.
17. R. J. Cole, D. F. Kripke, W. Gruen, D. J. Mullaney, and J. C. Gillin, Automatic sleep/wake identification from wrist activity. *Sleep* 1992; **15**:461–469.
18. S. Rhee, B. H. Yang, K. Chang, and H. H. Asada, The ring type sensor: a new ambulatory wearable sensor for twenty-four-hour patient monitoring system. *Proc. 20th Ann. Int. Conf. IEEE Eng. Med. Biol.*, 1998: 1906–1909.
19. J. D. Neuman and A. P. F. Turner, Biosensors for monitoring glucose. In: P. Å. Öberg, T. Togawa, and F. A. Spelman, eds., *Sensors in Medicine and Health Care*. Weinheim, Germany: Wiley-VCH, 2004, pp. 45–78.
20. T. Togawa, Home health care and telecare. In: P. Å. Öberg, T. Togawa, and F. A. Spelman, eds., *Sensors in Medicine and Health Care*. Weinheim, Germany: Wiley-VCH, 2004, pp. 381–405.
21. T. Shinozaki, R. Dean, and F. M. Perkins, Infrared tympanic thermometer: evaluation of a new clinical thermometer. *Crit. Care Med.* 1988; **16**:148–150.
22. M. E. Weiss, A. F. Pue, and J. Smith, III, Laboratory and hospital testing of new infrared tympanic thermometer. *J. Clin. Eng.* 1991; **16**:137–144.
23. T. Togawa, Non-invasive deep body temperature measurement. In: P. Rolfe, ed., *Non-invasive Physiological Measurement*, vol. 1, London: Academic Press, 1979, pp. 261–277.
24. J. Stevens, J. Cai, E. R. Pamuk, and D. F. Williamson, The effect of age on the association between body-mass index and mortality. *New Eng. J. Med.* 1998; **338**:1–7.
25. K. R. Segal, B. Gutin, E. Presta, J. Wang, and T. B. Van Itallie, Estimation of human body composition by electrical impedance methods: a comparative study. *J. Appl. Physiol.* 1985; **58**:1565–1571.
26. J. M. Conway, K. H. Norris, and C. E. Bodwell, A new approach for the estimation of body composition: infrared interactance. *Am. J. Clin. Nutr.* 1984; **40**:1123–1130.
27. K. B. Arbogast, K. L. Thibault, B. S. Pinheiro, K. I. Winey, and S. S. Margulies, A high-frequency shear device for testing soft biological tissues. *J. Biomech.* 1997; **30**:757–759.
28. J. L. Peterson, S. R. Goldstein, and R. V. Fitzgerald, Fiber optic pH probe for physiological use. *Ann. Chem.* 1980; **52**:864–869.
29. P. Spanel, P. Rolfe, B. Rajan, and D. Smith, The selected ion flow tube (SIFT): a novel technique for biological monitoring. *Ann. Occup. Hyg.* 1996; **40**:615–626.
30. J. C. Jones, The feeding behavior of mosquito. *Sci. Amer.* 1978; **238**:112–120.

READING LIST

T. J. H. Essex and P. O. Byrne, A laser Doppler scanner for imaging blood flow in skin. *J. Biomed. Eng.* 1991; **13**:189–194.

BIOMETRICS

TING MA
YAN ZHANG
YUANG-TING ZHANG
The Chinese University of Hong Kong
Shatin, NT, Hong Kong

Biometrics deals with the automatic recognition of individuals based on statistical analysis of physiological and/or behavioral characteristics. Any human physiological or behavioral characteristic that is unique, universal, stable, and collectable could be used as a biometric characteristic.

In the modern automated world, access to a reliable authentication system becomes increasingly essential. However, traditional authentication methods based on the user's exclusive knowledge, such as password, personal identification number (PIN), or something belonging to one, such as a cardkey, smart card, or token [like identity cards (ID)], can hardly meet the requirements of the reliability of an authentication system because passwords or PIN may be forgotten and ID cards can be lost, forged, or misplaced. Compared with traditional methods, biometrics can provide enhanced security and convenience. As biometric recognition systems are increasingly deployed for many security applications, biometrics and its applications have attracted considerable interests. Recently, biometrics has emerged as one of the most reliable technologies for future human identification and verification.

1. HISTORICAL ASPECTS OF BIOMETRICS

As a term, "biometrics" is derived from the Greek words *bio* (life) and *metric* (to measure). The study of biometrics can be traced back several centuries. In the fourteenth century, fingerprint was a characteristic that distinguished young children from each other in China, as reported by explorer and writer Joao de Barros. In the 1890s, Bertillonage, a method of bodily measurement, was used by police authorities throughout the world (1). From then on, fingerprint-based biometric technology has become essentially important in forensic applications for criminal investigation, which also promote the development of biometrics.

Automated biometrics has a 40-year history; however, it has received intensive attention only in the last 25 years because of the challenging task of designation of an automated biometrics system for large population identification. In late 1960s, the Federal Bureau of Investigation (FBI) began to automatically check finger images, and by the mid-1970s, several automatic fingerprint-based identification systems had been installed (2). Based on the success of biometric technologies in law enforcement, the applications of biometrics have been extended to a wide range of civilian markets, including access control, e-business, network security, and other various security applications.

With the spread of biometrics, more and more research fundings are being devoted to this important security technology. Besides fingerprint, many other biometric technologies are also developed, including face, iris, speaker, and signature. Electrocardiogram (ECG) signals, photoplethysmography (PPG) signals, and evoked action potentials (EAPs) have been investigated recently for human recognition. Biometrics has separated from the original subset stage of signal processing, pattern recognition, image processing, computer vision, computer security, and other subjects, and it has developed into a relatively new and independent research area.

2. WHAT IS BIOMETRICS?

Biometrics deals with the automatic recognition of individuals based on statistical analysis of physiological and/or behavioral characteristics (2,3). Any human physiological or behavioral characteristic that is unique, universal, stable enough, and collectable could be used as a biometric characteristic (4).

Common physiological characteristics include fingerprint, iris, face, palm, and retina. These characteristics are based on direct measurements of part of the human body. Although the behavioral characteristics, such as voice, signature, and keystroke dynamics, are traits that are learned or acquired, they are based on indirect measurements and data are collected from an action of a human. No matter which biometric technology, physiological or behavioral, any one of them cannot be separated from the other completely. Some typical physiological and behavioral characteristics are shown in Fig. 1 (4).

3. BIOMETRIC SYSTEM STRUCTURE

There are two kinds of biometric systems: verification system and identification system. Figure 2 (5) shows the structure of a typical verification system and a typical identification system. Verification systems are based on biometric characteristics to verify whether someone really is whom he or she claims to be. Such systems usually receive two inputs: One is the claimed identity of the user requesting verification, which might be an ID number, a username, or a PIN, and the other is the biometric characteristic(s) of that user. The claimed identity is used to retrieve a particular biometric template stored in advance in a database, and this template is used as a reference to be compared against the currently offered biometric. The similarity between the template and the test determines the system output, match or no match. These systems are usually referred to as one-to-one matching.

On the other hand, identification systems answer the question, "Who am I?" Therefore, these systems only receive one input, which is the biometric characteristic(s) of the user requesting identification. A database established in advance is scanned for a matching biometric. If a matched biometric is found in the database, this person is identified. The output returned by these systems is an identity such as a username or ID number. These systems are usually referred to as one-to-many matching (5).

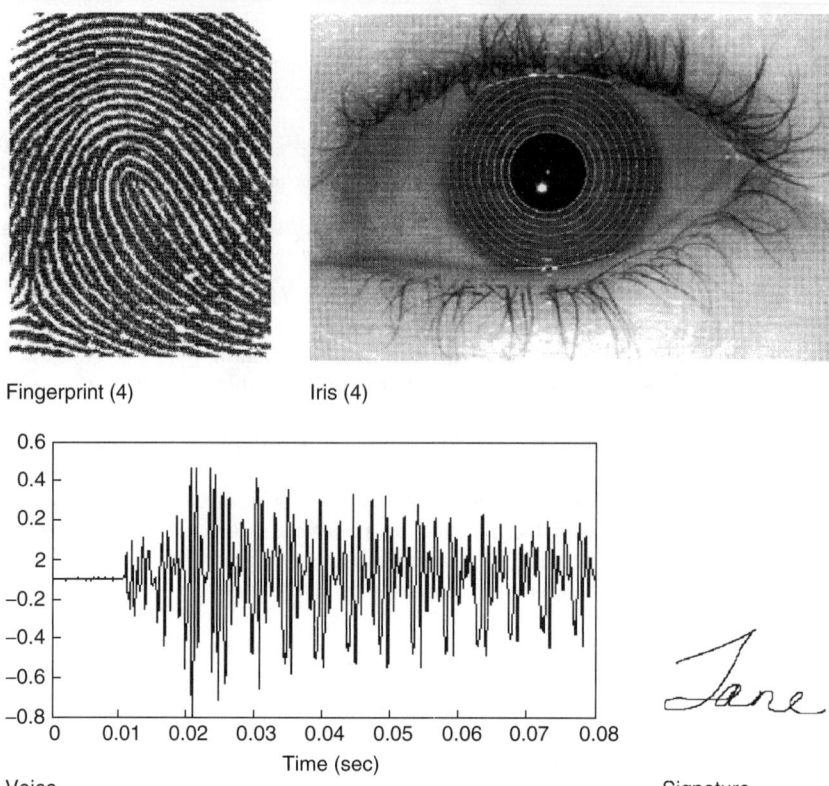

Figure 1. Some typical physiological and behavioral characteristics which are used in biometrics.

4. HOW DOES BIOMETRIC SYSTEM WORK?

No matter which kind of biometric system, it can be divided into two parts: the enrollment module and the recognition module (2,3,5,6). The enrollment module registers a user's biometric and trains the system to authenticate a given person, whereas the recognition module identifies or verifies the user's biometric. Although probably based on different biometric characteristics, all biometric systems follow the same functioning procedure. Both enrollment and identification modules consist of data collection and feature extraction. The difference is that one storage stage is involved in enrollment module, whereas comparison and matching stages are included in the identification module.

4.1. Data Collection Stage

The data collection stage is the first stage in human recognition systems. Therefore, the quality of the captured data is crucial to the performance of the whole system. The less noise, the better the performance. This stage completes data capture, together with data preprocessing sometimes, of a physiological or behavioral sample input

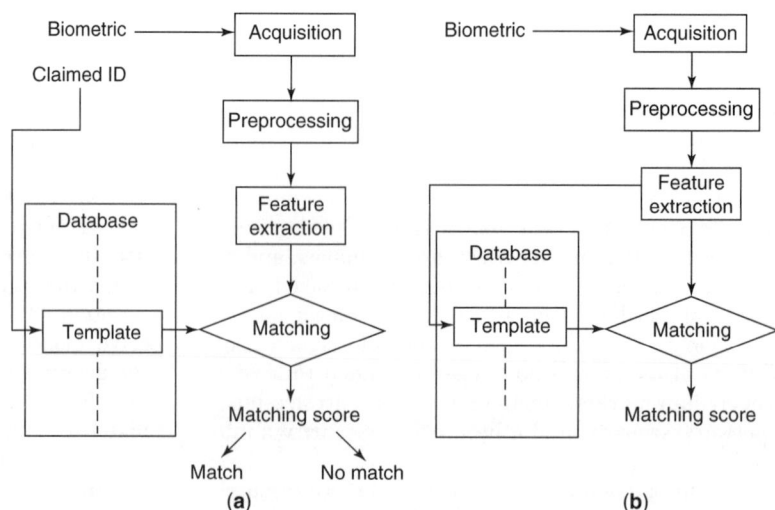

Figure 2. Typical structures of (a) a verification system and (b) an identification system, respectively (6).

into the system by users. The original signals are then translated into digital formats for computer processing. Different systems use different devices for data collection. In general, cameras or sensors are employed for physiological biometric acquisition. Video cameras are used to capture iris and face images. Some specially designed fingerprint scanners are available in the market. Newly investigated PPG-based verification technology uses an infrared optical sensor as its data acquisition device. Behavioral characteristic capturing devices are distinct from each other. A voice-based system uses only a microphone, and a signature-based system uses a writing board.

4.2. Feature Extraction Stage

The feature extraction stage typically comprises three modules: segmentation, feature extraction, and quality control as shown in Fig. 3.

The segmentation module decides whether there is a biometric signal in the received raw data, which is detected in the data collection stage. If so, it segments the biometric signal from the received signal.

Feature extraction module extracts unique biometric features from the segmented signal and then generates a feature vector that should be distinct for any two persons while similar enough for different samples of the same person. Therefore, feature extraction module must process the signal in some way to preserve or enhance the between-individual variation while minimizing the within-individual variation. The output of this module is a set of numerical features or called feature vector.

The quality control module then performs a statistical check on the extracted features. If this check is not successfully passed, the system may alert the user to resubmit the biometric pattern.

4.3. Matching Stage

In the comparison stage, the newly extracted template from a sample is compared with a registered one in the database. The comparison algorithm should tolerate insignificant changes from the same person and yet distinguish persons by the difference of the samples. For identification systems, comparison means that the new template should be compared with all registered templates in the database, whereas for verification systems, it means that the new template is only compared with a particular registered template.

4.4. Decision Stage

In the decision stage, the system determines whether the template extracted from the new sample matches the registered one. Based on the comparison results, a matching score is gained to show the consistency of the observed template. Generally, a threshold is set to give a definite answer of "yes" or "no,". When the matching score is greater than the threshold, an answer of "yes" is given; otherwise "no" is the output.

5. PERFORMANCE EVALUATION

Performance evaluation is a difficult and important aspect in biometric technologies. As described, a score indicating the similarity between the registered template and the test sample is obtained from the comparison. If the score exceeds the stated threshold, a match is returned.

Two performance measurements to rank the level of matching accuracy have been used for many years: false rejection rate (FRR) and false acceptance rate (FAR). The FRR is the probability that the real system users are rejected by the system; the FAR is the probability that the impostors are accepted. Both rates can be described as a percentage using the following simple calculations (2):

$$\text{FRR} = \frac{\text{NFR}}{\text{NAA}} \times 100\% \qquad (1)$$

and

$$\text{FAR} = \frac{\text{NFA}}{\text{NIA}} \times 100\%, \qquad (2)$$

where NFR and NFA are the numbers of false rejections and false acceptances, respectively. NAA is the number of authorized identification or verification attempts, and NIA is the number of impostor identification attempts. Although FRR and FAR are the most commonly used evaluation principle for identification systems, it is difficult to justify the performance of a biometric system simply by FRR and FAR.

Another evaluation method is the equal error rate (EER) or the crossover rate. EER, representing the accuracy level at which FRR equals FAR, is a general indicator of a system's resistance to impostors and ability to match templates with authorized users. Therefore, EER is commonly used as a representation of overall system accuracy.

Receiver operation curve (ROC) and the difference between the means of the genuine distribution and impostor distribution (d') are two other commonly used evaluation methods. ROC provides a measure of the system performance at different operating points. It is a good representation of the tradeoff between FAR and FRR and can be used to select an appropriate operating point for a particular application. d' is defined as the difference between

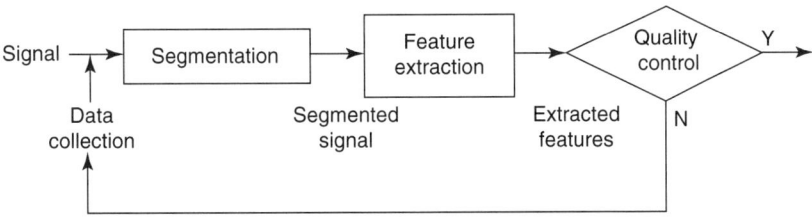

Figure 3. Typical signal processing procedure of a biometric system.

the means of the genuine distribution and impostor distribution divided by a conjoint measure of their derivation

$$d' = \frac{||M_{\text{impostor}} - M_{\text{genuine}}||}{\sqrt{(SD_{\text{impostor}}^2 + SD_{\text{genuine}}^2)/2}}, \quad (3)$$

where M_{genuine}, M_{impostor}, SD_{genuine}, and SD_{impostor} are the means and standard derivations of the genuine distribution and impostor distribution, respectively (2).

Recently, a variant of ROC plot, the detection error tradeoff (DET) curve (7), has been used as a measure for evaluation that plots FRR and FAR on both axes, giving uniform treatment to both types of error. A complete DET curve can fully describe system error tradeoffs. Compared with the ROC plot, the DET curve may move the curves away from the lower left corner when performance is high and produce linear curves, which makes system comparisons easier (8). The evaluation methods mentioned serve as a guide to understand a system's general ability, but when a particular biometric system is installed, some other factors should still be considered, including vulnerability, convenience, cost, intrusion to person, applicability, speed, storage size, and stability.

6. TYPES OF BIOMETRICS

Various biometric technologies have been proposed, investigated, and evaluated for recognition applications. Each of them has its intrinsic advantages and disadvantages and appeals to a particular application. The existing and widely used biometric technologies are summarized as follows.

6.1. Fingerprint

Fingerprint recognition is one of the earliest and most widespread techniques of biometric technologies. It uses the pattern of friction ridges and valleys on an individual's fingertips to recognize persons. Their formations depend on the initial conditions of the embryonic development and are considered unique to all persons; even the same fingers of identical twins are different from each other.

Commonly used features include: (1) the directional field (the local orientation of the ridge-valley structures), (2) the singular points (the discontinuities in the directional field), and (3) the minutiae (the details of the ridge-valley structures) (3). Figure 4 (4) illustrates these features for a sample fingerprint. The ridges and valleys form special structure characteristics of fingerprints and can be classified into six categories: right loop, left loop, twin loop, whorl, arch, and tented arch. Minutiae offer a compact representation of a fingerprint. Usually two types of minutiae characteristics are used for their robustness and stability: ridge ending (the point at which a ridge ends) and bifurcations (the point at which one ridge divides into two).

A greater variety of fingerprint devices are available than for any other biometrics at a low cost, and the relatively small size allows the sensor to be integrated into other devices. Although fingerprint recognition is one of the most mature biometric technologies and has been

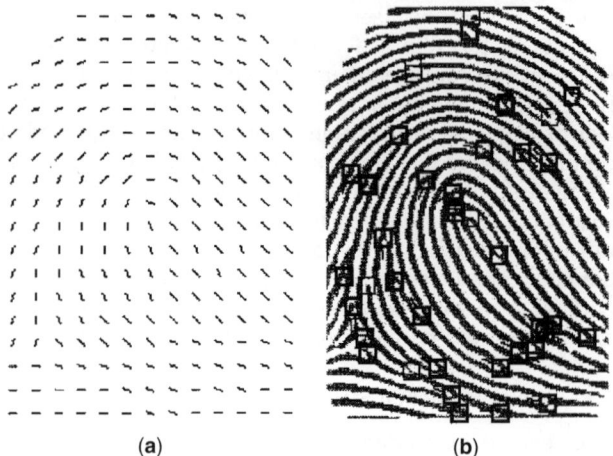

Figure 4. (a) The orientation and (b) the minutiae of a fingerprint (4).

widely used in forensic applications for criminal investigation, experiments have shown that the sensor of a fingerprint recognition system cannot provide sufficient resolution images for up to 5% of the people because of special problems, like intensive use of manual tools leading to fingerprint damaging by friction and erosion (8–10).

6.2. Palmprint

Like the fingerprint, the palmprint also has a long history as a reliable human identifier. The palm is the inner surface of the hand between the wrist and the fingers and has been found to have distinctive and stable features, which could be categorized as geometry features (e.g., width, length, and area of a palm), line features (e.g., principal lines, coarse wrinkles, and fine wrinkles), and point features (e.g., minutiae and delta points). Geometry features are easily captured; however, it is relative easy to create a fake palm, whereas point features can only be obtained from the fine resolution image, which costs the storage much. Therefore, line features, such as principal lines and coarse wrinkles, are usually adopted as the biometric features. Figure 5 illustrates three principal lines of a palm that are named as heart line, head line, and life line, respectively (11). Both location and form of principal lines keep stable from time to time and have become important biometric features. Some advantages have been achieved by using line features (12): (1) It can be used on the image with low resolution; (2) significant features can be determined even in the presence of noise; (3) a line contains more information than a point compared with fingerprint biometric features. There is an increasing application of palmprint biometrics for its stability and uniqueness. Automatic palmprint recognition systems already have been developed for law enforcement applications (13,14).

6.3. Face

Face-based identification systems were developed from the late 1980s, and the systems can be commercially available in the 1990s. Face-based identification systems analyze the characteristics of a person's face image, including

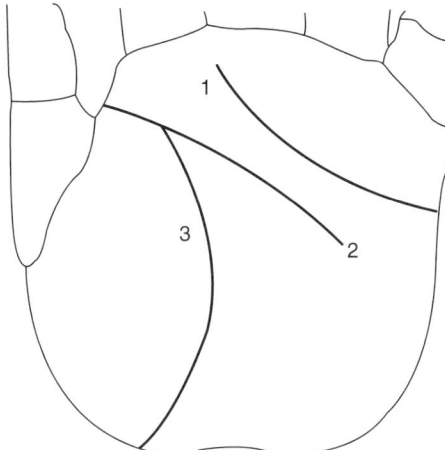

Figure 5. A palm with three principal lines, which are named as heart line, head line, and life line, respectively (11).

distances between eyes, nose, mouth, and jaw edges. Similar to fingerprint recognition technology, there are a variety of approaches to face recognition. Two primary face recognition approaches are (1) the transform approach and its variants (15): using a set of orthonormal basis vectors to represent the universe of face image domain and (2) the attribute-based approach (16): Facial attributes are first extracted from the face image and those invariant geometric properties are used as recognizing features. An advantage of face-based biometric technology is that it provides continuous personal identity recognition. Another benefit is that it uses smaller templates. Face-based biometric technology has become one of the most acceptable biometrics. However, change of face expressions and slight variants on the face can greatly affect system performance.

6.4. Iris

The use of iris for personal identification was originally proposed by ophthalmologist Frank Burch in 1936. Iris-based biometric systems (17) analyze the distinctive features of the human iris that exist in the colored tissue surrounding the pupil for human recognition. Currently, three major types of iris-based recognition systems are kiosk-based systems, physical access devices using motorized cameras, and inexpensive desktop cameras (5). Iris recognition technology is extremely powerful for high security applications because of the uniqueness of the eyes, even between the left and right eye of the same person. And its relatively high speed and easy of use make it a great potential biometric. However, it encounters some difficulties in capturing iris images if the users are not willing to hold their heads in the right spot for the scan. Besides, it is difficult to minimize the size of image capture devices and it requires expensive high-resolution image for vertex detection in feature extraction.

6.5. Infrared Facial and Hand Vein Thermograms

A thermogram-based system uses the pattern of heat radiation from the human body, which is considered a

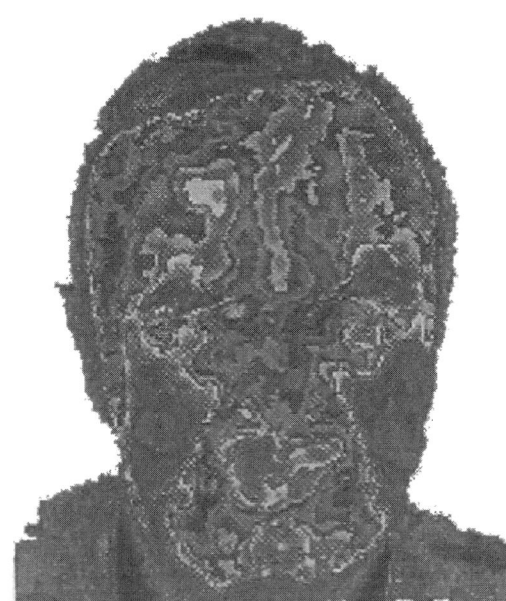

Figure 6. An image of facial thermograms for biometrics (4).

characteristic of each body, for authentication. Thermograms can be acquired by an infrared sensor, and these thermal images indicate the heat emanating from different parts of human body. Nowadays, facial thermogram (18) based recognition systems are commercially available. A hand vein thermogram-based recognition system requires near-infrared imaging technology to obtain the hand vein structure. A thermal image obtained by sensing the infrared radiation from the face of a person is shown in Fig. 6 (4), and an infrared image of the back of a clenched human fist is shown in Fig. 7 (4). A thermogram-based system seems to be acceptable by the public for it is noncontact. However, the applications of a thermogram-based system may be dramatically limited by the surrounding conditions, such as air conditioner and vehicle exhaust pipes. Furthermore, the high price of infrared sensors is

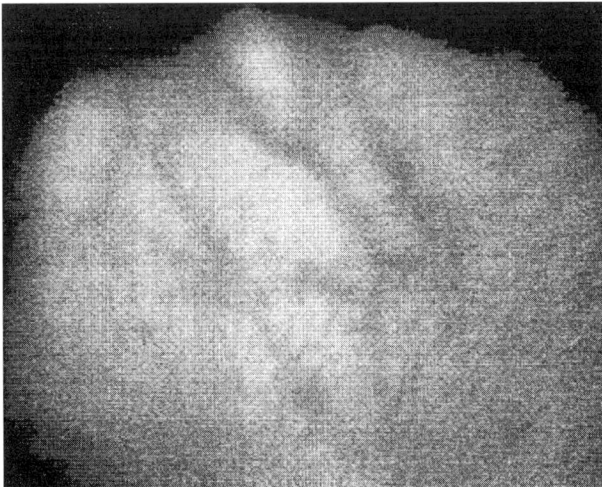

Figure 7. An image of hand veins for identification (4). (This figure is available in full color at http://www.mrw.interscience.wiley.com/ebe.)

also a factor prohibiting the widespread use of thermograms for identification.

6.6. DNA

DNA is often called the code of life and is known to be totally unique for each person except identical twins, who have an identical DNA pattern. DNA identification systems (4,19) are based on the distinctive signature at one or more polymorphic loci. Currently, most DNA-based recognition systems are used in forensic applications and their testing procedures involve the use of restriction endonucleases, which are used to cleave a person's DNA into a reproducible pattern of fragments at recognition sites, DNA probes, and Southern blot hybridization, to isolate, detect, and visualize specific target DNA fragments. DNA recognition is the most precise biometric technology and is widely accepted as absolutely accurate. But it may encounter some problems in applications, such as contamination, sensitivity, privacy issues, and even discrimination.

6.7. Retina

A retina-based biometric system analyzes the pattern of blood vessels situated on the back of the eye to recognize persons (4). Figure 8 (4) illustrates the vasculature structure of a retinal image, which is considered to be unique to each person and each eye. The first retina scan device for commercial use was made in 1984. Typically, a retina scan device is composed of a low-intensity light source, an optical coupler, and a pattern reader that can work at a high level of accuracy. The retina is claimed to be the most accurate biometric available today. And it is also a long-term and high-security biometric trait for the continuity of the retina pattern throughout life and the difficulty in mimicking the retina vasculature. Unfortunately, it does require the user to remove glasses, place their eye close to the device, and focus on a certain point, which may mislead the user into thinking that it is harmful to the eyes. And the cost of the hardware for scanning limits its applications to some particular situations.

6.8. Hand and Finger Geometry

Hand geometry (4,20) has become a popular biometric topic recently. The geometry of the human hand is not unique, but it does provide sufficient information for recognition. This technology measures and analyzes the shape of the hand such as length of fingers, width, and thickness of the hand for recognition. It is easy for users to work the system and has no negative public attitude to overcome. The size of storage is very small, which is an attractive feature for bandwidth- and memory-limited systems. However, hand image capturing devices require the cooperation of users, and they are inconvenient as they limit flexibility of the hand.

Finger geometry is a variant of hand geometry, which relies only on geometrical invariants of fingers (index and middle), and it is relatively new in the biometrics area, and thus not as mature as hand geometry.

6.9. Ear

The shape and structure of an ear may not unique, but they are distinctive enough for identity recognition, and they do not change radically over time. Ear recognition technology (4,21) is based on the comparison of the distances of salient points on the ear from a landmark location on the ear. The features used for ear recognition are shown in Fig. 9 (4). Ear-based recognition technology is promising for it is easy to extract features from the ear. However, because it is still a new research topic, no commercial system is available in the market yet.

6.10. Odor

It is well known that odor exuded from an object is an indicator of its chemical components, which makes it possible for the object to be distinguished by its odor. Odor-based recognition systems (4,22) first detect the odor of one person by an array of chemical sensors, each of which is sensitive to a certain group of odor, and then the feature vector is constructed by the characteristics of the

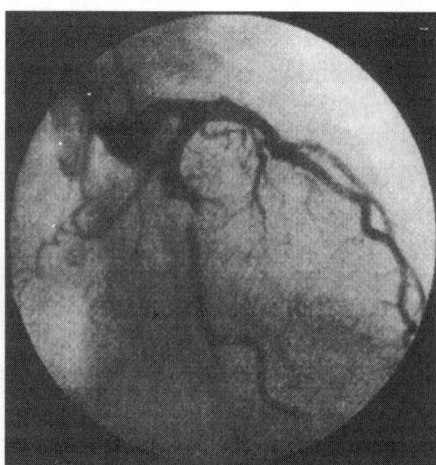

Figure 8. An image of retina for recognition (4). (This figure is available in full color at http://www.mrw.interscience.wiley.com/ebe.)

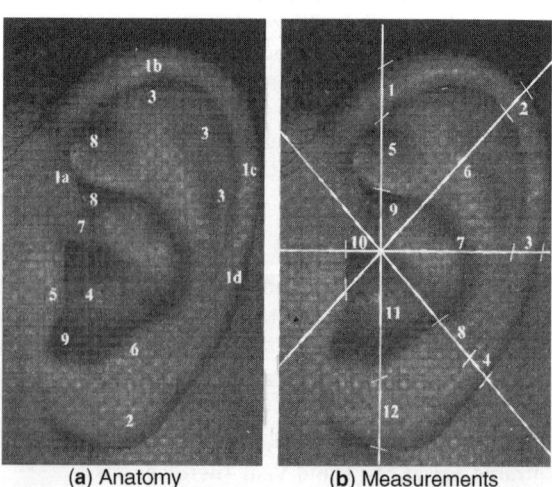

Figure 9. An ear image and the features used for ear-based verification (4).

normalized measurement from each sensor. After the detection, sensors have to be initialized by a flux of clean air. Odor-based biometric systems require odor sensors with extremely high odor sensitivity and selectivity. Furthermore, it is still not clear if the invariance in an odor could be detected by the chemical sensors in the environment in which chemical components vary under the present time.

6.11. Speaker

Speaker recognition uses the speaker-specific information included in speech waves to recognize who is speaking. This technology involves elements of both behavioral and physiological characteristics: the shape of the vocal tract that determines how a voice sounds and the user's behavior that determines what is spoken and in what tone.

Speaker recognition methods can be divided into text-dependent (23) and text-independent (24) methods. The former one requires the same passwords to be used for both enrollment and recognition. The latter one does not place any constraint on the phrase used for enrollment or recognition. Speaker recognition has the most potential for growth, not only because the voice is easy for the user to provide, but also because the required speech collection hardware is readily available on most personal computers today. However, as in many behavioral biometric systems, the performance of the matching system is contingent on users' motivation to verify. This technology is potentially more susceptible to attacks than other biometrics. Furthermore, low-quality capturing devices and ambient noise may challenge its accuracy, and the large size of the template limits the number of potential applications.

6.12. Signature

Signature verification analyzes the way that a person signs his or her name, by measuring dynamic signature features such as speed, pressure, and angle. There are two types of signature recognition systems: offline (or called static) (25) and online (or called dynamic) signature systems (26). In an offline system, the signature is written on paper and an optical scanner or a camera digitizes the signature data. Afterward, the signature is verified through examining the overall or detailed shapes of the signature. In an online system, the signature is acquired in real time with a digitized pen tablet (or an instrumented pen or other touch panel specialized hardware), which captures both the static and dynamic information of the signature during the signing process. Because the acquisition of the offline signature is somewhat complex with weak performance, these systems are limited and usually used as an aid in legal cases to identify criminals. However, an online signature verification system can use not only the shape information of the signature, but also the dynamic time-dependent information. Therefore, its performance (accuracy) is normally considered to be better than that of an offline system.

Although the study of human signatures has a long history, automatic signature verifications is still a new topic in the biometrics domain. Automatic signature technology is potentially one of the most powerful and publicly acceptable means of personal authentication currently available and this technology is not privacy invasive. More importantly, people easily accept it as an identity recognition method because people are accustomed to the signature as a means of transaction-related identity recognition and most would see nothing unusual in extending it to biometrics applications. However, signature acquisition hardware may limit the applications of signature recognition systems and network response is an important consideration. Nowadays, this technology is mainly employed in e-business and in other applications in which a signature is an already accepted method for personal authentication.

6.13. Keystroke Dynamics

Keystroke dynamics biometric technology (2,4,6,27) is a purely behavioral biometrics compared with speaker and signature technologies. This technology uses a person's distinctive typing pattern on a keyboard, such as the length of time a user held down each key and the time elapsed between keystrokes, for recognition. Although this behavioral biometrics may be not unique to each person, it offers sufficient discriminatory information for identity recognition. Generally, typical matching approaches use neural network architecture to associate identity with the keystroke dynamics features and this technology is normally deployed and integrated with passwords. Although there are limitations in the potential accuracy of keystroke dynamics verification technology, it can be operated in environments and applications where other biometrics cannot work efficiently. Some commercial keystroke dynamics verification systems are already in the market.

6.14. Gait

The use of gait as human identity is very recent, and no commercial system based on gait (4,28) exists. Gait is the distinctive and particular way that each person walks, which can be understood from a biomechanics standpoint. Typically, gait-based recognition systems extract gait features from video, which include characteristics of several different movements of articulate joints and muscles. One major merit of gait-based biometrics is its acceptability by the public, for people are adept at recognizing a person at a distance from his or her gait. Another one is its nonconcealable nature. However, the main challenge to this technology is the gait variance over a large period of time, caused by large fluctuation of body weight, major shift in body weight, or inebriety. And automatic extraction of gait features from video is also an ambitious problem to be solved in gait biometric technology.

6.15. Electrocardiogram (ECG)

Recently, a new approach based on ECG (29) for human identification is proposed. An ECG is a technique that records a well-coordinated series of electrical events that take place within the heart, caused by the smooth, rhythmic contraction of the atria and ventricles for blood circulation. This set of electrical events, which is intrinsic to the heart, is not only valuable in clinical applications but

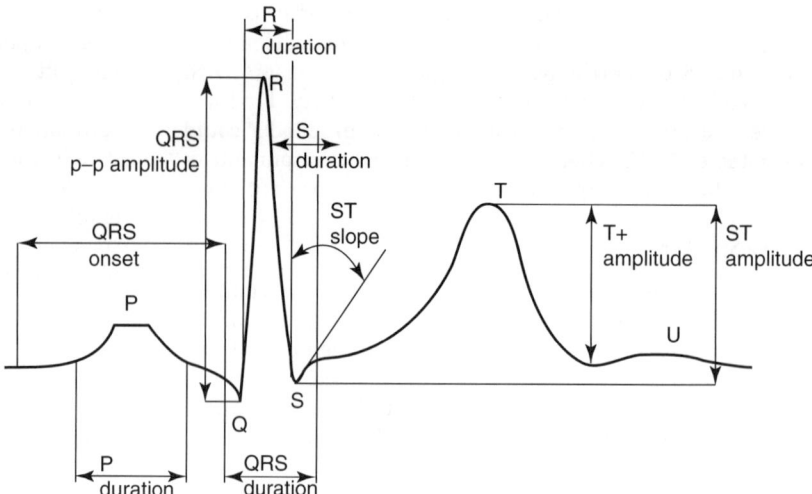

Figure 10. An ECG with the features used for biometrics (21).

also a good biometric trait for human recognition. Preliminary exploration was carried out based on the distinctive features of the ECG waveform in the time domain. A standard 12-lead rest ECG was collected, and the representative features were extracted from the waveform of ECG as shown in Fig. 10 (29). With further investigation, one recent research result has shown that the signal from one lead is enough for human identification (30). Actually, in all of these identification methods, it is difficult to implement identification on moving humans. Although biometric technology based on ECG analysis can solve this problem and can achieve a high recognition rate, which indicates that ECG is a potential biometric signal. In addition, ECG is a one-dimension signal that is easy for signal processing. However, there are still problems with this new biometric method. It is inconvenient for users to place several electrodes on the body, and several other issues, like data collection, motion effects, and the stability of ECG, are still under study.

6.16. Photoplethysmographic (PPG) Signal

The application of the PPG signal for human verification purposes is a new exploration (31). Typically, photoplethysmography uses an infrared optical sensor to detect the volume change of red blood cells in the peripheral microvascular bed associated with each pressure pulse initiated by the heart. PPG signals may not be unique to each individual, but they do provide sufficient characteristics for identity verification. The first trial of this newly proposed biometric method is based on four features, which are extracted from the PPG signals in the time domain as shown in Fig. 11 (31). The Euclidean distance was adopted as the criteria of decision-making, and the pilot investigation shows a promising verification result; i.e. 16 subjects were successfully recognized out of 17. The PPG-based biometric system uses an easily collected signal, which can be simply obtained at peripherals, such as the fingertip, by optical sensor. Besides, PPG-based biometric systems can be easily combined with other biometric techniques for personal identity recognition. However, the PPG signals vary substantially under different conditions, especially under different pressures and temperatures. Further study is needed to solve these problems.

6.17. Muscle Evoked Action Potential (EAP)

The use of muscle EAP for human identification was proposed in 2003 (32). EAP is an electrical waveform elicited in the nervous system or muscle, which is excited by an external stimulus. In the pilot study, muscle EAPs were collected from individuals at hypothenar eminence, which is activated by external stimulating the ulnar nerve. The template for each person was constructed by the wavelet coefficients of the muscle EAP on a specified scale. The decision making on human identity is based on the calculated Euclidean distance between the registered and the sample template. The preliminary result indicates great potential of this biometric technology. The major merits of muscle EAP-based biometrics are that the signal is very stable within the same person and remarkably distinctive between individuals, as shown in Fig. 12 (32). In addition, the muscle EAP is a one-dimension signal, which is easy to process. However, as a newly proposed technology, muscle EAP-based biometrics is under further investigation. One limit of using muscle EAP is that its data collection requires electrical stimulation, which affects its

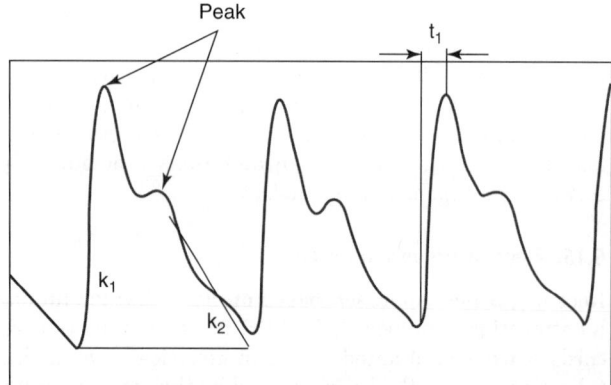

Figure 11. A typical PPG signal with features used in biometrics (23). (This figure is available in full color at http://www.mrw.interscience.wiley.com/ebe.)

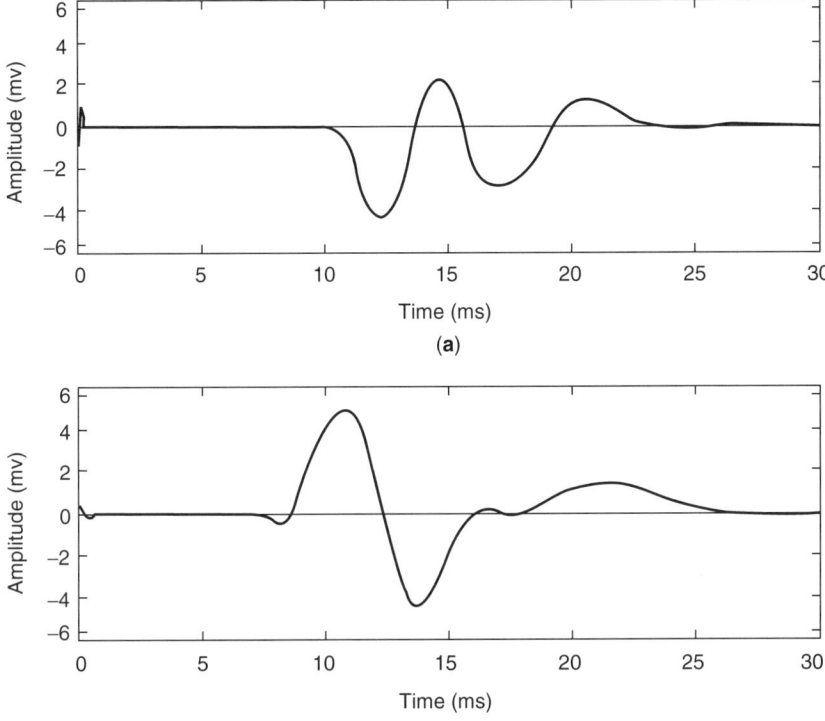

Figure 12. Muscle EAPs from two individuals which are illustrated in (a) and (b), respectively (24).

acceptability in public. Besides, the reliability of muscle EAP in the long term still needs further examination.

6.18. Multimodal Biometrics

No matter which biometric technology is employed, each technique has its advantages and limitations, and no single biometric technology can efficiently meet all desired performance requirements of an application. Currently, the performance improvement of biometric technologies can be realized mainly through two ways: exploring more accurate and reliable biometric characteristics, as well as efficient integration of existing biometric technologies, called multimodal biometrics, which combines multiple biometrics in making an identification system. Integration of two or more biometric technologies may overcome the demerits of one technology by the merits of another technology. Multimodal biometrics can achieve at least one or both advantages: the identification accuracy improvement and the identification speed improvement at matching step. Researchers have investigated some fusion technologies to improve the performance of this type of recognition system (33).

7. BIOMETRICS APPLICATIONS

Biometrics-based human recognition systems have been employed in many practical applications, which can be simply classified as follows:

- *Criminal Identification*: Criminal recognition is perhaps the earliest and largest biometric application, in which biometrics-based identity authentication systems are used to recognize the identity of a suspect or detainee. Fingerprint and palm-based recognition technologies have been widely used in this application. In turn, this application also promotes the development and widespread of biometric technologies.

- *Access Control*: Biometric technologies have been used as a means of access control in many high secure environments for decades. Nowadays, the primary applications of biometrics in access control lie in physical access control, which uses biometrics to recognize the identity of the individual entering or leaving one place, typically a room or a building.

- *Computer and Network Security*: With the rapid development of the computer network, the security of e-world has suffered growing problems, such as hacker and electronic eavesdropping, which have already threatened the prosperity and productivity of corporations and individuals using the Internet. Therefore, computer and network security based on biometric technologies has evoked great public interest. Biometric technology-based security computer systems are used to recognize the identity of individuals accessing the network, computer system, software applications, or other computer resources, and these systems remove the need for various traditional security methods, which are mainly passwords and tokens. Different from other applications, biometrics technology-based security computer systems require incorporating portable or handheld devices with the ability to be secured. Nowadays, fingerprint and voice recognition are the most promising techniques in this application.

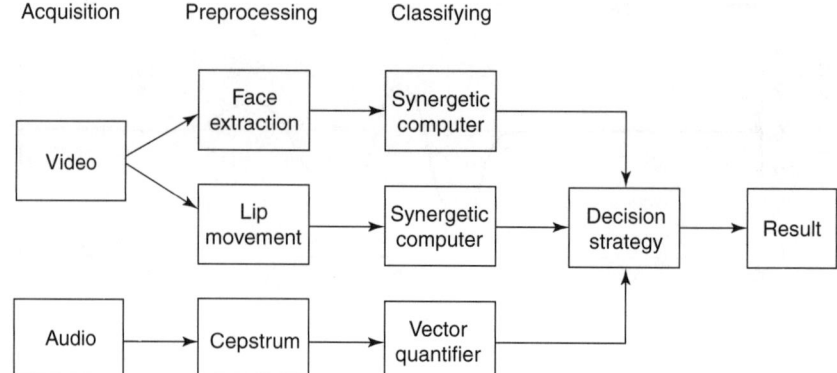

Figure 13. Schematic diagram of a BioID system (34).

- **E-Commerce Applications**: E-commerce developers explore biometric technologies to recognize the identity of individuals conducting remote transactions for goods or services. It is obvious that biometric technologies will provide higher security than traditional authentication mechanisms such as passwords and PINs. Generally, e-commerce applications involve remote user authentication, unsupervised enrollment, and transactional verification processes.
- **Citizen Identification**: Biometrics-based national identity management systems have already been in use in some countries to complement or replace traditional authentication methods like document provision and signature. These systems use biometric technologies for identity recognition for the purpose of card issuance, voting, immigration, social services, or employment background checks. Generally, these systems involve storing a biometric template on a card as a national identity document.
- **Telephone Systems**: Being frequently attacked from increasing fraud, telephone companies now use biometric technologies to defend against these onslaughts. In general, these systems involve remote user authentication, unsupervised enrollment, and verification processes. It is obvious that speaker recognition is well suited the telephone environment and will quickly become popular in this new market.
- **Time, Attendance, and Monitoring**: Time, attendance, and monitoring systems are usually deployed to restricting, registering, or controlling the presence of an individual in a given place. Although there are many advantages of biometric systems over traditional authentication methods that are usually performed by punch cards, biometric-based monitoring systems might sometimes elicit privacy problems.
- **An Application Example**: Recently, a multimodal identification system, BioID, was developed by the Dialog Communication System Company (DCS AG) for access control utility (34). Figure 13 (34) gives the schematic diagram of a BioID system, which employs three different biofeatures, face, voice, and lip movement, as well as biometric traits. At the data collection stage, the system records a sample of a person speaking, in which a 1-second sample consists of a 25-frame video sequence and an audio signal. Different signal processing methods are applied to extract features representing unique characteristics of users' speech, face, and lip movement. In detail, the speech signal is divided into several segments, each of which is represented by an audio feature vector composed by cepstral coefficients. A class of the audio feature vectors is collected as the biometric characteristics of the speech. Face recognition is completed by two steps: face localization and facial feature extraction. To detect the location of a face in an arbitrary image, BioID adopts the first image in the video sequence and uses a model-based algorithm, which is based on Hausdorff distance, to match a binary model of a typical human face to a binarized, edge-extracted version of the video image. Figure 14 (34) demonstrates the face localization process. All faces captured from different users are scaled to a uniform size, and appropriate facial features are extracted, such as the head size, to form a feature vector. The lip movement is characterized by vector fields, which represent the local movement of each image part to the next image

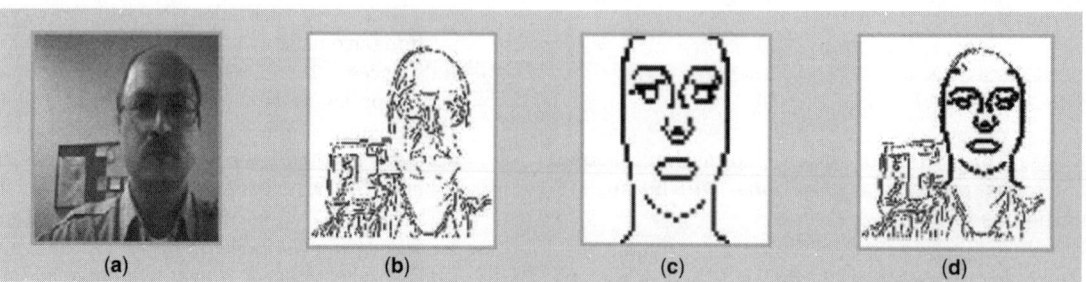

Figure 14. The face localization process of a BioID system (34). (This figure is available in full color at http://www.mrw.interscience.wiley.com/ebe.)

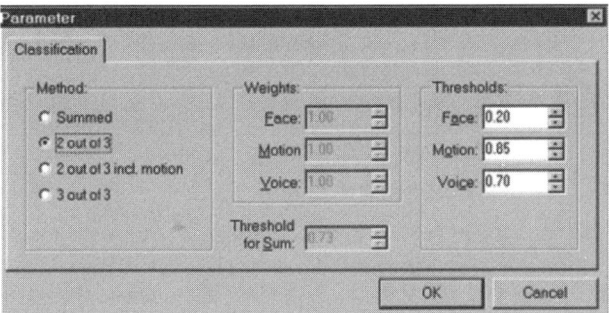

Figure 15. The available options for decision strategy of a BioID system (34). (This figure is available in full color at http://www.mrw.interscience.wiley.com/ebe.)

in the video sequence. Specifically, the first 17 images for the mouth area are cut from the video sequence to construct 16 vector fields, which are then used to create a one-dimensional feature vector as the pattern for lip movement. Finally, feature vectors representing uniqueness of speech, face, and lip movement, respectively, are collected as the biometric pattern for a user.

As shown in Fig. 13, image and audio features are extracted by the synergetic computer and vector quantifier, respectively. The template patterns for all users are constructed and stored in the system for human recognition. For a specific identification task, a sample, containing the audio and video signals, is input to the BioID system. After signal processing and feature extraction, the sample pattern is compared with all template patterns and three scores are obtained by comparing the features of voice, face, and lip movement, respectively. The decision strategy can be designed by the system operator to achieve a different security level or recognition accuracy by choosing sensor fusion methods, weights for different biometric traits, and the thresholds for comparison results. Figure 15 (34) shows the available options for decision

Figure 16. A user who is trying to access to a computer network through the BioID system (34). (This figure is available in full color at http://www.mrw.interscience.wiley.com/ebe.)

strategy. For example, in normal operation, a two-out-of-three method is chosen for sensor fusion; i.e., a person will be identified with an ID when two out of his three biometric traits are best matched to those of the corresponding template pattern without falling below threshold values set in advance. The three-out-of-three strategy can be selected for a higher security level. In addition, the identification accuracy can be improved by adjusting the weights assigned to feature vectors of face, voice, and lip movement. For example, if the system always correctly identifies a person by voice, the weight assigned to its feature vector will be enhanced. BioID can be employed in any technical system in which access control is needed. Figure 16 (34) shows one application example when a user poses in front of the PC camera and speaks his name for seeking access to a computer network.

8. CONCLUSION

Compared with traditional authentication technologies, biometrics is distinctive because it uses natural signals of human beings for identity recognition. Signal processing, such as neural network, and other relative techniques have advanced the development of biometrics considerably. Various biometric technologies have been explored by many research groups, and biometrics is emerging as one of the most reliable technologies for future human recognition. However, biometrics is still in its developing stage. The exploration of new signatures from the human body and efficient integration of existing biometric technologies are two hot topics in this area. And interwoven development of data acquisition, signal processing, classification technique, and performance evaluation criteria, would offer an opportunity to stimulate further development of biometric technology.

BIBLIOGRAPHY

1. Court Technology Laboratory (CTL) (2003). Available: *http://clt.ncsc.dni.us*.
2. D. D. Zhang, *Automated Biometrics: Technologies and Systems*. Amsterdam: Kluwer Academic Publishers, 2000.
3. D. D. Zhang, *Biometric Solutions For Authentication In An E-World*. Amsterdam: Kluwer Academic Publishers, 2002.
4. A. Jain, R. Bolle, and S. Pankanti, *Biometrics: Personal Identification in Networked Society*. Amsterdam: Kluwer Academic Publishers, 1999.
5. S. Nanavati, M. Thieme, and R. Nanavati, *Biometrics: Identity Verification in a Networked World*. New York: Wiley Computer Publishing, 2002.
6. S. Pankanti, R. M. Bolle, and A. Jain, Biometrics: The future of identification. *Computer* 2003; **33**:46–49.
7. A. Martin, G. Doddington, T. Kamm, M. Ordowski, and M. Przybocki, "The DET curve in assessment of decision task performance," in Proc. ESCA 5th Eur. Conf. Speech Comm. and Tech., EuroSpeech '97, Rhodes, Greece, 1997: 1895–1898.
8. J. Ortega-Garcia, J. Bigun, D. Reynolds, and J. Gonzalez-Rodriguez, Authentication gets personal with biometrics. *IEEE Signal Processing Mag.* 2004; **21**(2):50–62.

9. P. Peter, The smart cards are coming ... really. Available: http://www.forbes.com/technology/feeds/general/2005/02/11/generalbhsgml_2005_02_11_19213_791446171-0019-KEYWORD.Missing.html.
10. Summary of NIST standards of biometric accuracy, tamper resistance, and interoperability. Available: http://www.itl.nist.gov/iad/894.03/NISTAPP_Nov02.pdf.
11. W. Shu and D. Zhang, Palmprint verification: an implementation of biometric technology, Proc. 14th Int. Conf. on Pattern Recognition, 1998: 219–221.
12. D.D. Zhang, *Palmprint Verification in Automated Biometrics*. Boston: Kluwer Academic Publishers, 2000.
13. NEC Automatic Palmprint Identification System (2003). Available: http://www.nectech.com/afis/download/PalmprintDtsht.q.pdf.
14. Automatic Palmprint Identification System (2003). Available: http://www.printrakinternational.com/omnitrak.htm – Printrak.
15. J. Zhang, Y. Yan, and M. Lades, Face recognition: Eigenface, elastic matching, and neural nets. *Proc. IEEE*. 1997; **85**:1423–1435.
16. M. J. Lyons, J. Budynek, A. Plante, and S. Akamatsu, Classifying facial attributes using a 2-D Gabor wavelet representation and discriminant analysis. Proc. Automatic Face and Gesture Recognition, Proceedings. Fourth IEEE International Conference, 2000.
17. G. O. Williams, Iris recognition technology. *IEEE Aerospace Electronic Syst. Mag.* 1997; **12**:23–29.
18. F. J. Prokoski, R. B. Riedel, and J. S. Coffin, Identification of individuals by means of facial thermography. Proc. IEEE 1992 International Carnahan Conference on Security Technology: Crime Countermeasure, Atlanta, GA, Oct. 14–16, pp. 120–125.
19. R. Sanchez-Reillo and A. Gonzalez-Marcos, Access control system with hand geometry verification and smart cards. *IEEE Aerospace Electronic Syst. Mag.* 2000; **15**:45–48.
20. B. Lehr, DNA's lasting imprint. *IEEE Potentials*. 1989; **8**:6–8.
21. M. Burge and W. Burger, Ear biometrics in computer vision. Proc. 15th Int. Conf. on Pattern Recognition, 2000: 822–826.
22. K. Cai, T. Maekawa, and T. Takada, Identification of odors using a sensor array with kinetic working temperature and Fourier spectrum analysis. *IEEE Sensors J.* 2002; **2**:230–234.
23. K. Chen, D. Xie, and H. Chi, A modified HME architecture for text-dependent speaker identification. *IEEE Trans. Neural Networks*. 1996; **7**:1309–1313.
24. R. A. Finan and A. T. Sapeluk, Text-independent speaker verification using predictive neural networks. Proc. Fifth Int. Conf. on Artificial Neural Networks, 1997: 274–279.
25. E. J. R. Justino, A. El Yacoubi, F. Bortolozzi, and R. Sabourin, An off-line signature verification system using hidden Markov model and cross-validation. Proc. XIII Brazilian Symposium on Computer Graphics and Image Processing, 2000: 105–112.
26. L. Nakanishi, N. Nishiguchi, Y. Itoh, and Y. Fukui, On-line signature verification method utilizing feature extraction based on DWT. Proc. 2003 Int. Symp. on Circuits and Systems, 2003: 25–30.
27. J. A. Robinson, V. W. Liang, J. A. M. Chambers, and C. L. MacKenzie, Computer user verification using login string keystroke dynamics. *IEEE Trans. Systems, Man Cybern.* 1998; **28**:236–241.
28. C. BenAbdelkader, R. Cutler, and L. Davis, Stride and cadence as a biometric in automatic person identification and verification. Proc. Fifth IEEE Int. Conf. on Automatic Face and Gesture Recognition, 2002: 357–362.
29. L. Biel, O. Pettersson, L. Philipson, and P. Wide, ECG analysis: A new approach in human identification. *IEEE Trans. Instrumentation Measure.* 2001; **50**:808–812.
30. T. W. Shen, W. J. Tompkins, and Y. H. Hu, One-lead ECG for identity verification. Proc. Second Joint IEEE EMBS/BMES Conf., Huston, TX, 2002.
31. Y. Y. Gu, Y. Zhang, and Y. T. Zhang, A novel biometric approach in human verification by photoplethysmography signal. Proc. ITAB2003 4th Annu. IEEE EMBS Special Topic Conf. on Information Technology Applications in Biomedicine.
32. T. Ma, Y. Y. Gu, Y. Zhang, and Y. T. Zhang, A novel biometric approach by evoked action potentials. Proc. Int. IEEE EMBS Conf. on Neural Engineering, Italy, 2003.
33. V. Chatzis, A. G. Bors, and I. Pitas, Multimodal decision-level fusion for person authentication. *IEEE Trans. Systems, Man Cybern.* 1999; **29**:674–680.
34. R. W. Frischholz and U. Dieckmann, BiolD: A multimodal biometric identification system. *Computer*. 2000; **33**(2):64–68.

READING LIST

D. Maltoni, D. Maio, A. K. Jain, and S. Prabhakar, *Handbook of Fingerprint Recognition*. New York: Springer, 2003.

S. Liu and M. Silverman, A practical guide to biometric security technology. *IT Professional*. 2001; **3**(1):27–32.

M. Turk and A. Pentland, Eigenfaces for recognition. *J. Cognitive Neurosci.* 1991; **3**(1):71–86.

J. Daugman, High confidence visual recognition of persons by a test of statistical independence. *IEEE Trans. Pattern Analysis Machine Intell.* 1993; **15**(11):1148–1161.

D. Zhang, W. K. Kong, J. You, and M. Wong, Online palmprint identification. *IEEE Trans. Pattern Analysis Machine Intell.* 2003; **25**(9):1041–1050.

R. Brunelli and D. Falavigna, Person identification using multiple cues. *IEEE Trans. Pattern Analysis Machine Intell.* 1995; **17**(10):955–966.

A. Ross and A. K. Jain, Information fusion in biometrics. *Pattern Recognition Lett.* 2003; **24**(13):2115–2125.

BIOMOLECULAR LAYERS: QUANTIFICATION OF MASS AND THICKNESS

FLORIN FULGA
DAN V. NICOLAU
Swinburne University of Technology
Hawthorn, Australia

1. INTRODUCTION

Biomolecular films immobilized on surfaces have broad industrial and academic applications. One challenge in developing hybrid devices, like biosensors, that use biomolecular films is to design and fabricate surfaces that allow high concentrations of biomolecules and preserve their bioactivity.

The approach of building biomolecular devices starts with the preparation of thin biomolecular films with a structural control, preferably in the direction perpendicular to the thin-film plane of different substrate materials, such as silicon, oxides, gold, etc. Different techniques for the preparation of biomolecular films exist, starting with those that produce various degrees of molecular organization [e.g., Langmuir–Blodgett deposition, self-assembly from solution, electrochemical deposition, covalent binding at molecularly engineered sites, binding by biospecific interaction (e.g., biotin-avidin)], to those that produce molecularly random films [e.g., surface adsorption, bulk entrapment (absorption), nondirected covalent binding to the surface].

The biomolecular films have many characteristics that may require quantification, and a variety of tools for surface and interface analysis have been used to characterize the films microscopically, spectroscopically, and phenomenologically, such as the mass of the film, reaction heats, absorption coefficients, and so on (1). From the point of view of building biomolecular devices and irrespective of the deposition method, one of the parameters that must be quantified is the amount of biomolecules immobilized in the film. Some methods applied to determine this amount, methods that are important for applications, will be reviewed in the paper. Methods exist, based on phenomena like birefringence, that give information about the organization of the biomolecular film, but they are not the objective of this chapter. Also, it is not the purpose of this chapter to describe methods that quantify other parameters, like the range of activity measurements (e.g., enzyme activity, antibody-binding activity, orientation and steric hindrance, electron transfer efficiencies, etc.).

2. THE PHYSICAL MODEL

Whatever the technology used for the preparation of the biomolecular film and the technique used for immobilization, in order to preserve their biological function, the biomolecular films should be interfaced with a specific solution, and the techniques applied to measure different properties should be able to operate and provide reliable results in this environment.

The system under study consists of a substrate, on top of which can exist several other thin layers of organic or inorganic material, and a thin layer of biomolecules. This system is usually placed in a stationary liquid cell or a flow cell. The measurements are performed either dynamically, when studying the kinetics of biomolecular film deposition, or stationary, when measuring the final parameters of the film.

In order to interpret the data obtained from the characterization instruments, it is commonly assumed that the biomolecular film is isotropic and homogeneous, the substrate and all other layers are flat, and the thickness of the biomolecular film is constant. The film is characterized in terms of physical parameters like refraction index; density; shear modulus; or shear viscosity, thickness, and mass.

3. AVAILABLE METHODS

The first question regarding a biomolecular film is how many molecules are deposited on the substrate, or equivalently, what is the mass of the film? The monitoring of the mass changes of the biomolecular film can reveal many important physical and chemical processes. However, the small amounts of mass deposited pose important problems—the mass of a very thick biomolecular film will be barely measurable by the most advanced available balance, and only if large amounts of expensive biomolecules are used. Therefore, measuring the mass of a biomolecular film is not straightforward and is performed indirectly by several methods that measure different parameters related to the mass or thickness and with physical models based on known properties of the film.

Several reviews (2,3) point to the ideal technique for measuring biomolecular film, which:

- should not affect the biomolecular film properties;
- should quantify the amount of adsorbed biomolecules *in situ*;
- should be able to be equally used on different substrate materials;
- should give information on conformation or orientation in the adsorbed film (if possible).

A brief comparison of the available techniques used for the study of biomolecular films is given in Table 1. From the methods listed in the table, the optical methods and the acoustical methods will be reviewed because they have the required precision, they are simple and inexpensive, and they are close to satisfying the criteria for the ideal measurement techniques.

4. OPTICAL METHODS

Optical methods use an incident beam of light specifically tuned to interact with the sample, which comprises the biomolecular film. The measurement of different parameters of the outgoing beam allows the calculation of the thickness or the refractive index of the biomolecular film.

Usually, the measurement is performed in flow cells filled with aqueous solution, first in a control-experiment mode (i.e., without the biomolecules in solution). Then the actual measurements are performed with a solution of a known concentration of biomolecules and compared with the previous ones. In order to quantify the influence of the solution on the film, the measurements can also be performed after drying the film.

4.1. Ellipsometry

Ellipsometry is a convenient and accurate technique for the measurement of thicknesses and refractive indexes of thin films on solid surfaces and for the measurement of optical constants of reflecting surfaces. The technique is easily adaptable to the study of biomolecular films under liquids. The standard reference books (4,5) provide a comprehensive overview of the technique with an emphasis on applications.

4.1.1. Principle of the Method. A typical ellipsometry system comprises a sample with a biomolecular film of

Table 1. Techniques Used for the Study of Biomolecular Films

Technique	Principle of operation	Characteristics
Optical techniques • Ellipsometry • Surface plasmon resonance (SPR) • Optical waveguide lightmode spectroscopy (OWLS) • Raman spectroscopy • Infrared reflectance spectroscopy	The light beam interacts with the biomolecular film and changes in amplitude, phase, polarization, etc. These changes allow the quantification of the optical parameters and even bonding states of molecules.	• Optical methods do not perturb the biomolecular film. • Measurements *in situ* are possible. • Depending on the optical method, different surface materials are available. • Optical methods can give information on conformation or orientation in the adsorbed film can be obtained.
Labeling techniques • Fluorescence labeling • Radiolabeling	The biomolecules are tagged with a radioactive or fluorescent label, which is then detected. It is assumed that the signal is in a linear or known relationship with the number of biomolecules on the surface.	• Labeling changes the properties of biomolecules and impacts on their adsorption properties. • The surface material does not influence the ability to measure the biomolecular film. • No information on conformation or orientation in the adsorbed film can be obtained. • High concentrations of fluorophore can induce auto-fluorescence with impact on the linearity of the signal vs. concentration.
Acoustic vibration techniques • Quartz crystal microbalance (QCM) • Surface acoustic wave devices (SAW)	The resonance frequency changes with adsorbed mass on the surface of the piezoelectric material.	• Possible impact on the adsorption of biomolecules. • The surface material is limited to piezoelectric materials. • Measurements *in situ* are possible. • No information on conformation or orientation in the adsorbed film can be obtained. • For operation in air, the quantification is straightforward; but for films interfaced with a solution, there is a coupling between the oscillations in the crystal and the liquid.
Electrical methods • Impedance • Electro-kinetic methods	As biomolecules in solution will become charged as the result of the ionization, the change in electrostatic charge at the surface is detected. If the charge of an individual biomolecule is known, then the total number of biomolecules can be calculated.	• Electrical methods do not influence the biomolecular film. • Different surface materials are available. • Measurements *in situ* are possible. • No information on conformation or orientation in the adsorbed film can be obtained.

index of refraction n_f deposited on a substrate with an index of refraction n_s, immersed in a medium of index of refraction n_1, as shown in Fig. 1. The incident beam, usually from a laser, is passed through an optical system to be converted into an elliptically polarized beam. The incident angle is usually expressed with regard to the plane of incidence (i.e., the plane through the incident, reflected, and transmitted beam). The elliptically polarized light beam can be decomposed into two orthogonal components, one linearly polarized parallel to the plane of incidence, denoted by a subscript p, and one linearly polarized perpendicular to the plane of incidence, denoted with a subscript s. The incoming light on the sample has a state of polarization, with both components having different interactions with a surface, either the bare substrate surface or the surfaces the light encounters in the sample. As a result, the light beam is modulated by the other components on the same orthogonal axes, and the state of polarization has changed during interaction with the sample. This change in the polarization state of the outgoing light beam compared with the incoming light beam gives the information about thicknesses of the layers the light encounters in the sample.

The reflection of both p and s components can be described by complex reflection coefficients R:

$$R_p = |R_p|e^{i\delta_p}$$
$$R_s = |R_s|e^{i\delta_s} \quad (1)$$

It would be possible to determine the reflection coefficients R_p and R_s separately, but in order to eliminate the effect of intensity fluctuations in the incident light beam, an ellipsometer determines the ratios of the amplitudes of the p and s components, and the phase difference between p and s components rather than the absolute values. Usually, these two parameters are expressed as two angles, Ψ and Δ.

$$R_p/R_s = \tan\Psi * \exp(i\Delta), \quad (2)$$

where Ψ and Δ are measured by the ellipsometer. The reflection coefficients are dependent on the angle of incidence; the wavelength λ of the incident light; the refractive indexes of the substrate, film, and immersion medium, n_s, n_f, n_m, respectively; and the thickness d of the biomolecular film.

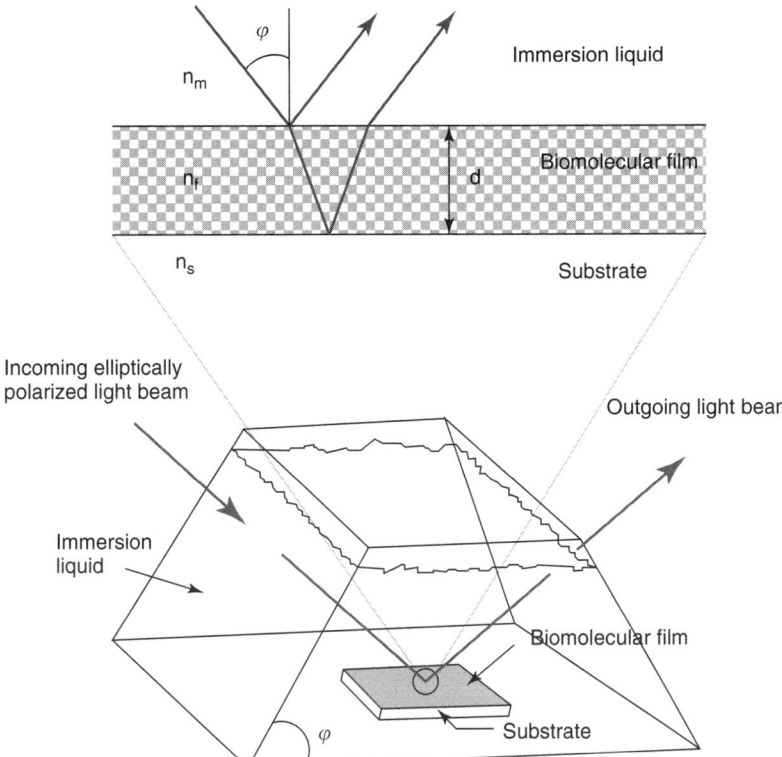

Figure 1. Typical experimental setup for ellipsometric measurements. The state of polarization of the outgoing light beam is different than the state of polarization of the incoming light beam because of the interaction with the optical system made of substrate, biomolecular film, and immersion liquid. (This figure is available in full color at http://www.mrw.interscience.wiley.com/ebe.)

In order to study a biomolecular film by ellipsometry, the reflection coefficients of the bare substrate are first measured and the complex refractive index of the substrate estimated. The substrates easiest to work with are those with high reflectance. Both smoothness and flatness are factors that must be considered in the selection of the substrate. Irregularities that are small compared with the dimensions of the light beam, which is commonly of the order of 1 mm^2, are averaged and do not affect the result.

A biomolecular film is then deposited on the substrate, and the new reflection coefficients of the combination are measured.

The value of n_m, the index of refraction of the medium, either is known, in the case of measurements performed in air, or it is measured by refractometry, in the case of measurements performed in aqueous solution containing biomolecules.

For the case shown in Fig. 1, Equation 2 becomes

$$C_1(\exp D)^2 + C_2(\exp D) + C_3 = 0, \quad (3)$$

where C_1, C_2, and C_3 are complex functions of refractive indexes, Ψ, Δ, and

$$D = -4\pi i n_f \cos(\varphi d/\lambda),$$

where φ is the angle of incidence.

For a given value of the coefficients, Equation 2 gives two solutions for $\exp(D)$, and a value for the thickness d of the film may be calculated from each solution, assuming a value for n_f, the film refractive index. For instance, for proteins, this value is usually considered to be 1.45–1.5. As the coefficients are complex, the film thicknesses calculated from this equation would also be expected to be complex. However, the correct value for the thickness should be a real quantity. Therefore, the solution of the quadratic that yields a real film thickness is the correct solution. In practice, various experimental errors will result in both solutions yielding complex values, so the thickness with the smallest imaginary component is selected.

The method of ellipsometry allows more biomolecular films adsorbed on a substrate to be measured, but, in this case, the equations become more complex and an iterative procedure for calculating the thicknesses should be used.

The measurements in liquids require a closed environment (a cell). The cell can be either a flow cell or a stationary fluid cell, as in Fig. 1. The sample is placed on the base of the cell that is filled with a liquid of known refractive index (or measured by refractometry). The light enters and leaves the cells through optically flat windows, inclined at the angle of incidence φ, with respect to the base of the cell, so that the light passes through at normal incidence. Therefore, the reflection of light at the surface of the cell window will be independent of its direction of polarization, and the polarization of the incident light will not be changed. In order to avoid any optical disturbances, the glass should be stress-free and the inner and the outer sides of each window should be parallel.

4.1.2. Calculation of the Adsorbed Mass from the Refractive Index and Thickness of an Adsorbed Biomolecular Film. Cuypers et al. (6) developed a model, starting from the Lorentz–Lorenz relation for the refractive index n_f of a mixture of substances:

$$\frac{n_f^2 - 1}{n_f^2 + 1} = A_1 N_1 + A_2 N_2 + \ldots, \qquad (4)$$

where A_i, and N_i are, respectively, the molar refractivity of the biomolecule and the number of moles of substance i per unit volume.

For a pure substance, as is the case of a homogeneous biomolecular film, we may write

$$\rho_f = M \cdot N = \frac{M}{A} \frac{n_f^2 - 1}{n_f^2 + 1}, \qquad (5)$$

where ρ_f is the density of the biomolecular film in mass per unit volume and M is the molar weight of the biomolecule. If we consider a biomolecular film of thickness d expressed in nm, we find for the mass m in micrograms per square centimeter

$$m = d \cdot \rho_f = \frac{0.1 M \cdot d}{A} \left(\frac{n_f^2 - 1}{n_f^2 + 1} \right). \qquad (6)$$

If the biomolecular film is a mixture of pure biomolecular substance and buffer solution, then the above formula should be corrected as follows:

$$m = d \cdot \rho_f = \frac{0.3 d \frac{n_f + n_l}{(n_f^2 + 2)(n_l^2 + 2)}}{\frac{A_{bio}}{M_{bio}} - V_{20} \frac{n_l^2 - 1}{n_l^2 + 2}} (n_f - n_l), \qquad (7)$$

where V_{20} is the partial specific volume of the biomolecule at 20°C and n_l is the refractive index of the pure aqueous solution. From this relationship, it follows that the molecular weight, the molar refractivity, and the partial specific volume of the adsorbed biomolecule must be known in order to obtain m from d and n_f.

De Feijter et al. (7) developed an alternative and simpler relationship for the adsorbed mass, based on the assumption that the refractive index of an aqueous biomolecular solution is a linear function of its concentration. For proteins, the linearity of a graph of the refractive index versus bulk concentration has been verified by a number of authors (8). The slope of this graph is called the refractive index increment (dn/dc), and the accepted value from a large amount of experiments is 0.188 ml/g. The model gives for the mass of the film per unit area,

$$m = \frac{d \cdot (n_f - n_l)}{dn/dc}. \qquad (8)$$

The uncertainty in determining the Ψ and Δ angles is about 0.02°. The dependence of Ψ and Δ on n_f and d is nonlinear. As an example, if the refractive index n_f is assumed to be 1.45, than the thickness d may be between several Å to 25 Å (9) or, for a refractive index of 1.455, the thickness of the film is between 63 nm and 68 nm (10).

4.2. Surface Plasmon Resonance

Surface plasmon resonance (SPR) is an optical technique that relies on the interaction under specific conditions between the incident light on a metal surface and the surface plasmons, which are quasi-particles of the electron charge density in the metal. The resonance is a result of energy and momentum transferred from incident photons to surface plasmons. The resonance is sensitive to the refractive index of the medium on the opposite side of the metal film.

4.2.1. Principle of the Method.
When a beam of light passes from material with a high refractive index n_1 (e.g., glass) into material with a low refractive index n_2 (e.g., water or biomolecular buffer solution), some light is reflected from the interface.

When the angle of incidence is greater than the critical angle θ_c, defined as

$$\theta_c = \sin^{-1}\left(\frac{n_2}{n_1}\right), \qquad (9)$$

the light is completely reflected (total internal reflection, TIR) at the interface and propagates back into the high refractive index medium.

Although the fully reflected beam does not lose any net energy across the TIR interface, the light beam leaks an electrical field intensity called an evanescent field wave into the low refractive index medium. The amplitude of this evanescent field wave decreases exponentially with distance from the interface, effectively extinguishing over a distance of about one light wavelength from the surface (11).

The penetration depth d_p is defined as the distance from the interface where the intensity of light is reduced to $1/e$ of its originally value.

$$d_p = \frac{\lambda}{2\pi} (n_1^2 \sin^2(\theta) - n_2^2)^{-1/2} \qquad (10)$$

The value of d_p is 200–300 nm.

If the surface of the glass is coated with a thin film of a noble metal (e.g., 50 nm of gold), the evanescent field interacts with the free electrons of the metal, exciting electromagnetic surface plasmon waves propagating within the conductor surface that is in contact with the low refractive index medium. The surface plasmons are quantum oscillations (or quasi-particles) of the density of electrons of the metal, and like any particle, they are defined by their momentum and energy. When the wavevector of the surface plasmon and the incident photon are equal both in magnitude and in direction for the same frequency of the waves, a resonance condition occurs and the photon is absorbed with the creation of a plasmon. The surface plasmon enhances the evanescent electric field, as described in Fig. 2.

The wavevector of the photon depends on the wavelength of the incident light and on the angle of incidence. The resonance condition can be met by varying these parameters. The resonance conditions, which occur at a specific angle of incidence (critical angle), will translate

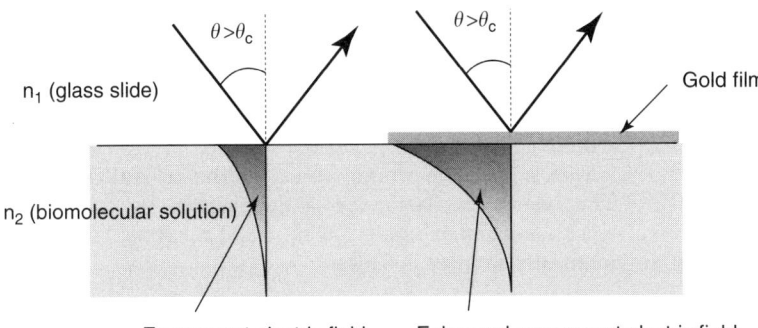

Figure 2. In the left side of the figure, TIR and the evanescent electric field. In the right side, the enhanced evanescent electric field because of the surface plasmons excitation. (This figure is available in full color at http://www.mrw.interscience.wiley.com/ebe.)

into a drastic decrease of the reflected light intensity (12,13).

Because the evanescent electric field enhanced by the plasmon penetrates a short distance into the lower refractive index medium, the conditions for SPR are sensitive to the refractive index of the medium at the gold surface. If the gold surface is immersed in an aqueous buffer and biomolecules bind to the surface, the refractive index at the surface increases, which, in turn, translates to a shift in the critical angle detected by the instrument. The shift is quantified in response units (RUs) with the equivalence $1\,\text{RU} = 10^{-4}$ degrees. The change of the refractive index is measured in real time and plotted against time. This variation of the refractive index measured in RU versus time is called a sensorgram (in Fig. 3).

The thickness of the film is calculated with the shift ΔRU and complex Fresnel calculations. Similarly to ellipsometry, the value of the refractive index is assumed known. In principle, the refractive index of the film can be extracted from the shape of the entire SPR curve. The uncertainty regarding the refractive index is a source of errors (e.g., a variation of 0.05 in the film refractive index leads to an error of approximately $\pm 7\%$ in the estimation of film thickness (14).

In general, different proteins have very similar refractive index contributions (i.e., the refractive index change is the same for a given change in protein concentration). Values for glycoproteins, lipoproteins, and nucleic acids are of the same order of magnitude. SPR thus provides a mass detector that is essentially independent of the nature of the chemical species immobilized on the surface. Moreover, because it is the evanescent field wave and not the incident light that penetrates the sample, the measurements can also be performed on opaque samples.

An empirical formula is used for the evaluation of the mass of the film,

$$\Delta m_{\text{SPR}} = C_{\text{SPR}} \cdot \Delta \text{RU}, \quad (11)$$

where C_{SPR}, having been determined experimentally for many different proteins, is

$$C_{\text{SPR}} = 6.5 \cdot 10^{-2}\,\text{ng/cm}^2, \quad (12)$$

and ΔRU is the measured change in response units (15,16).

The precision of the SPR is similar to ellipsometry, perhaps a little bit better (14).

4.3. Other Optical Methods

In this section, two other optical methods are briefly described. The first method, Optical Waveguide Lightmode Spectroscopy (OWLS), allows the estimation of the thickness of biomolecular films, similar to ellipsometry and SPR, whereas the second, Total Internal Reflection Fluorescence (TIRF), allows the measurements on mixtures of biomolecular films, at least in relative terms. When TIRF is coupled with either ellipsometry, SPR, or OWLS, it allows quantitative measurements.

4.3.1. Optical Waveguide Lightmode Spectroscopy (OWLS).
An OWLS device is based on a thin-film optical waveguide, which confines a discrete number of guided electromagnetic waves. The incident light is guided, at various angles of incidence, to the waveguide through a grating coupler. An evanescent wave that extends about 200 nm beyond the surface results when a wave is excited in the waveguide. If the waveguide is the base of a flow cell designed for biomolecular adsorption, then the angle of incidence of the incoming light is sensitive to the refractive index and the thickness of the biomolecular film adsorbed. Provided that the index of refraction is known, like in the optical methods described above, the thickness of the biomolecular film can be determined. The method is quite

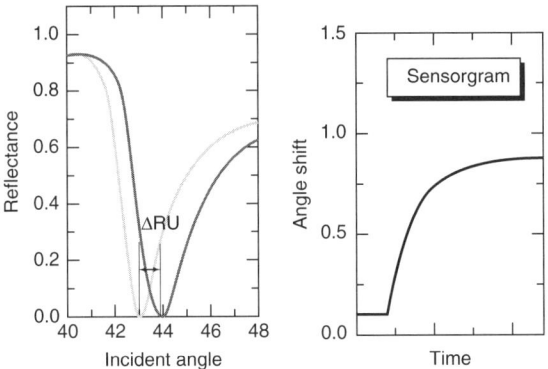

Figure 3. Illustration of the dip in reflectance curve because of surface plasmon excitation and the shift in the dip position as a biomolecular film is adsorbed (left side). Right side: Increase of the angle shift with time (the thickness of the film)—the sensorgram. (This figure is available in full color at http://www.mrw.interscience.wiley.com/ebe.)

sensitive, with precision of approximately 0.5 ng/cm², but it has the disadvantage of working only on highly transparent surfaces.

4.3.2. Total Internal Reflection Fluorescence (TIRF).

TIRF is a surface-sensitive method, which uses the evanescent wave that is formed on total reflection between two optically different media, as it was described in Section 4.2. This method allows the observation of biomolecules mixtures, for instance, when two or more biomolecules adsorb simultaneously on a surface. The Lorentz–Lorenz formula (Eq. 4, in section 4.1.2) can be used to calculate the refracting angle of a mixture of biomolecules. However, neither ellipsometry nor SPR allow the estimation of the relative concentration of each biomolecular species in the film. In the TIRF method, the biomolecules are labeled with specific fluorescent dyes, or alternatively, they are labeled one at a time followed by respective measurements.

The fluorescence intensity is given by

$$I_F = k \int_0^{d_{cell}} f(z)\phi(z)\varepsilon(z)c(z)(E^{ev}(z))^2 dz, \quad (13)$$

where k is a constant dependent on experimental factors, f is the fraction of emitted light, ϕ is the quantum yield, ε is the extinction coefficient, c is the fluorophore concentration, and E^{ev} is the electric field of the evanescent wave. The instrumental factor k is obtained by measuring the fluorescence intensity for a special solution used for calibration. Relative intensities are quantified for each component. If an independent method, for instance, ellipsometry, is used to measure the total mass of the film, the relative composition of the film is quantified pro rata.

5. NONOPTICAL METHODS

5.1. Quartz Crystal Microbalance

A quartz crystal microbalance (QCM) is a device built around a quartz crystal, which is made of a piezoelectric material. Piezoelectricity describes the generation of electric charges on opposing surfaces of a sheet induced by its mechanical deformation. The quartz crystal microbalance is making use of the converse piezoelectric effect (i.e., the generation of oscillating mechanical deformations induced by the oscillating electric potential difference applied on the electrodes deposited on the surface of the crystal). The most versatile microbalance, although not the most sensitive, is the Thickness Shear Mode (TSM) QCM. The quartz crystal is disk-shaped and cut along a specific crystallographic plane. The AT-cut (the quartz plate is cut at an angle of 35° 10′ with respect to the optical z-axis), which is predominantly used for TSM, allows good stability of the frequency of oscillation and a low temperature coefficient. An alternating potential difference is applied on the electrodes, which have been evaporated on both surfaces of the AT-cut disk-shaped quartz crystals. The transverse mechanical waves with shear displacement in the plane of the surface propagate in the crystal, interfere and generate standing waves with the frequency

$$f_n = \frac{v_p n}{2 d_q}, \quad (14)$$

where v_p is the propagation velocity of the acoustical waves, n is the overtone, and d_q is the crystal thickness. The resonance frequency is determined by the crystal thickness. For example, for $d_q = 330$ microns, the fundamental frequency is 5 MHz.

5.1.1. The Physical Model.

Essential to the functioning of a QCM is that the crystal is very sensitive to the boundary conditions at the electrodes and that the changes in these boundary conditions can be quantified and interpreted in terms of electrical parameters. Mason (17) suggested a one-dimensional wideband acoustical model that provides a basis for the theoretical modeling of complex resonators near the resonance frequency.

The system under study is composed of the quartz crystal mounted with one facing the fluid environment in a flow cell with biomolecules in an aqueous solution. These biomolecules will adhere to the gold electrode exposed to the solution following particular adsorption kinetics. The models for the optical methods described before (i.e., ellipsometry and SPR) require the variation of the refractive index and the thickness for the biomolecular film, as well as the refractive index of the medium (i.e., the aqueous solution). The phenomenological model for QCM requires the description of the mechanical properties of the film and of the medium, as in Fig. 4.

The bare quartz crystal with the gold electrodes is easily described electrically, as in Fig. 5. The equivalent of series of resistances, impedances, and capacitors represents well the motional arm near resonance. The resistance R_q accounts for the energy loss because of the viscous effects in the quartz crystal, internal friction, and damping caused by the holder; the capacitance C_q accounts for the mechanical elasticity of the quartz; the inductance L_q accounts for the mass; and the static capacitance C_0 develops from the plane-parallel capacitor formed by the electrodes and the quartz as dielectric. The biomolecular film and the medium surrounding the system are accounted for by electrical impedance Z_L in series with the motional arm. Different models for the film relate

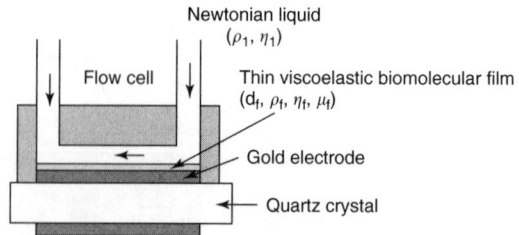

Figure 4. Schematic illustration of the geometry and parameters used to simulate the quartz crystal covered with a viscoelastic biomolecular film with a thickness d_f in a semi-infinite Newtonian fluid. The film has the elastic module μ_f, viscosity η_f, and density ρ_f. The liquid has the density ρ_l and the viscosity η_l. (This figure is available in full color at http://www.mrw.interscience.wiley.com/ebe.)

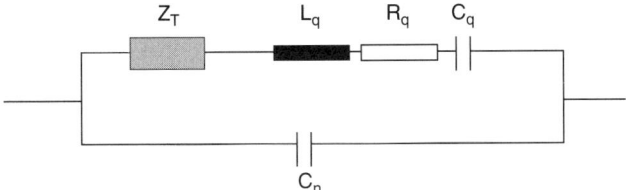

Figure 5. Equivalent electric circuit of the system. The quartz crystal near the resonance frequency is equivalent with a series combination of L_q, R_q, and C_q elements (the motional arm) in parallel with the dielectric capacitance C_0, Z_L represents the electrical impedance of the film and the fluid. Different models for the film and the fluid assume different forms of Z. (This figure is available in full color at http://www.mrw.interscience.wiley.com/ebe.)

the electrical impedance to the physical parameters of the film and the medium.

In order to measure the electrical parameters, the quartz crystal is connected to a monitoring electronic system. If we assume a model that interprets the properties of the biomolecular film through the analogy with an electrical circuit, then the thickness and mass of the film can be calculated. Two modes for measuring the electrical parameters of the system basically exist:

1. Oscillator mode. The resonator is included in an oscillator loop that compensates for the losses in the system under study. This simple and inexpensive method measures the shift in resonance frequency. In a variant of this method, QCM-D, developed by Rodahl et al. (18), it is possible to measure the energy dissipation factor, which allows the oscillator mode to be used for viscoelastic models describing the biomolecular film.
2. Impedance analysis. The system under study is coupled at a frequency generator that induces variable frequency sinusoidal waves close to the resonance frequency. The electrical impedance is measured with a spectrum analyzer (19). Assuming an equivalent electrical model, the mass of the film and the energy dissipation factor are estimated from the electrical impedance that fits the measured curve impedance versus frequency. This mode is recommended for viscoelastic films.

5.1.2. Models Describing the Biomolecular Film. Different models describing the biomolecular film translate to several relationships relating the deposited mass of the film (per unit area) to the measured frequency shift to the following:

- For homogeneous and uniform film, rigid, loss-free material, the characteristic impedance matching that of the quartz (20), the medium surrounding the film is air (or it is neglected the influence of the liquid)

$$\frac{\Delta m}{A} = -\frac{\sqrt{\mu_q \rho_q}}{2f_0^2}\Delta f = -\frac{Z_{cq}}{2f_0^2}\Delta f = -C\Delta f, \quad (15)$$

where Δm is the mass in ng; A is the area of the electrode in cm^2; f_0 is the bare crystal resonance frequency in MHz; Δf is the shift in resonance frequency in Hz measured with a frequency counter; μ_q and ρ_q are shear modulus of quartz and the density of quartz, respectively, and Z_{cq} is the quartz characteristic impedance. For $f_0 = 5$ MHz, C, the mass-sensitivity constant, is 17.7 ng cm^2 Hz^{-1}.

- The same as above, but the film is immersed in a semi-infinite liquid, an aqueous solution of the biomolecules under study, surrounding one side of the quartz; the liquid also causes a frequency shift (21). Martin et al. (22) has shown that, in certain approximations, the two frequency shifts are additive. Therefore,

$$\frac{\Delta m}{A} = -\frac{Z_{cq}}{2f_0^2}\Delta f - \sqrt{\frac{\rho_l \eta_l}{4\pi f_0}}, \quad (16)$$

where ρ_l is the density of the liquid, η_l is its viscosity. The liquid is assumed to be Newtonian, and the first layer of fluid is assumed to have the same transversal velocity as the quartz (the nonslip condition).

If the film is viscoelastic (similar to a rubber), then it is described by a complex shear modulus

$$G = \mu_f + i2\pi f \eta_f, \quad (17)$$

where μ_f is the shear modulus of the film, η_f is the shear viscosity (or loss modulus) of the film, and f is the frequency of oscillation. The assumed model for the viscoelastic film is usually a Voight or Kelvin model (23,24). The film immersed in fluid is then supposed to have a uniform thickness d_f and a uniform density ρ_f. The changes in the resonant frequency Δf and the energy dissipation factor ΔD, which are both measured, become

$$\Delta f = \frac{\text{Im}(\beta)}{2\pi t_q \rho_q}$$

$$\Delta D = -\frac{\text{Re}(\beta)}{\pi f t_q \rho_q}, \quad (18)$$

where

$$\beta = \xi_1 \frac{2\pi f \eta_f - i\mu_f}{2\pi f} \frac{1 - \alpha\exp(2\xi_1 d_f)}{1 + \alpha\exp(2\xi_1 d_f)}$$

$$\alpha = \frac{\dfrac{\xi_1}{\xi_2}\dfrac{2\pi f \eta_f - i\mu_f}{2\pi f \eta_l} + 1}{\dfrac{\xi_1}{\xi_2}\dfrac{2\pi f \eta_f - i\mu_f}{2\pi f \eta_l} - 1} \quad (19)$$

$$\xi_1 = \sqrt{-\frac{(2\pi f)^2 \rho_f}{\mu_f + i2\pi f \eta_f}}$$

$$\xi_2 = \sqrt{i\frac{2\pi f \rho_l}{\eta_l}},$$

Table 2. Comparison of the Methods Used for Biomolecular Films Characterization

Method	Benefits	Drawbacks
Ellipsometry/SPR	• The biomolecular films require no treatment with markers • The procedure is fast • The results agree well with quantitative radioimmunoassay methods (RIA) • The resolution is very good • Provides kinetic data in aqueous solution	• Requirement of reflecting, or gold surface (for SPR). These materials, in particular gold, are not always optimum surfaces for biomolecules, leading to the use of other layers on top—increased complexity of the system • Rather complex theory if the optical parameters are unknown or the systems under study are complex. Iterative procedures and approximations are needed
QCM	• The only requirement for the surface material is to be deposited as a thin film on the crystal • It is a simple, fast, and inexpensive method • Requires no markers • Provides kinetic data in aqueous environment	• The method measures the water coupled with the biomolecular film, including hydro dynamically coupled water, water associated with the hydration layer, or water trapped in cavities in the film • The results do not agree very well with other methods

where t_q is the quartz thickness. Four unknown parameters exist: μ_f, η_f, d_f, and ρ_f. Only two measurable quantities exist: Δf and ΔD. In this case, Δf and ΔD are measured at different overtones (multiples of the fundamental frequency), and the four unknown parameters are determined by a nonlinear extraction procedure.

The detection limit of the QCM is poorer than that of optical methods, such as SPR and ellipsometry, because of the larger sensor surface area and the inherently relatively reduced sensitivity. However, in recent years, the QCM has been developed from a pure mass sensor in the gas phase to a versatile tool in bioanalysis, providing information not only about binding events at surfaces, but also revealing material-specific quantities, such as elastic moduli, surface charge densities, and viscosity. Indirectly, one can also deduce conformational changes of proteins and the water content and net charge of biomolecules.

5.2. Atomic Force Microscopy (AFM)

The AFM uses a sharp tip mounted at the end of a flexible cantilever to probe topological features and mechanical characteristics of a sample. The displacement of the tip is controlled vertically and horizontally with a precision of the order of 1 Angstrom. A small area of the surface of the sample is imaged by the AFM, typically $1\,\mu m \times 1\,\mu m$ or $2\,\mu m \times 2\,\mu m$. The information this measurement provides is related to the degree of uniformity of the surface. The surface can be the gold substrate, for instance, or the layer of biomolecules.

This information is mentioned here, and it is not the purpose of this chapter to go into a detailed discussion of this method, because we want to point out that the AFM is the tool that can tell, by probing locally, if the assumption that the layers, the substrate, or the biomolecular film, assumed for all the models, for ellipsometry, SPR, and QCM, are uniform is valid for the system under study. AFM methods are treated extensively in a different chapter.

6. BENEFITS AND DRAWBACKS—COMPARISON OF THE METHODS

All the methods described have been largely used for studying the thickness and the mass of biomolecular films adsorbed on surfaces. More and more comparative studies are ongoing that are trying to see how the different methods can be used synergistically in order to get complementary results and to use the benefits and avoid or mitigate the drawbacks of each method. A brief description of the benefits and drawbacks of the methods more thoroughly examined is given in Table 2.

An important issue regarding the methods presented is to assess the extent of their complementarity with each other. For instance, is it possible to determine the thickness of biomolecular films with the QCM method and to use this value in an ellipsometer experiment to determine the index of refraction? To answer this question, a series of experiments have been performed in identical conditions (25,26). The results for the change in mass in ng/cm² measured with different methods are described in Table 3.

The results from the optical methods agree fairly well, but large differences exist with the QCM measurements, explained by the absorbed water in the biomolecular film that is accounted by QCM but not by the optical methods. The optical methods are based on the difference in the refractive index between the biomolecular film and the water displaced by the biomolecules. Therefore, the water included in the film is not included in the mass

Table 3. Comparison of the Measurements of Different Protein Layers

Method	Mussel adhesive protein (Mefp-1)	Human Serum Albumin	Fibrinogen	Hemoglobin
QCM	740 ± 20	333 ± 27	1177 ± 18	626 ± 25
Ellipsometer	130 ± 15	162 ± 35	397 ± 10	350 ± 50
OWLS		214 ± 9	451 ± 27	368 ± 32
SPR	165 ± 10			

determination as estimated by optical methods (25). On the other hand, the coupled water is sensed by the QCM and the amount can be estimated, taking as thickness of the layer the thickness given by the optical methods.

7. FURTHER DEVELOPMENTS

We expect that a lot of effort will occur in the future to solve problems common to QCM and SPR (e.g., a reproducible immobilization of the biomolecular films on the crystal or gold surface and addressing the nonspecific binding of proteins. QCM, which has found a wide range of applications in the food industry, environmental and clinical analysis because of its low cost and easy-to-use method, appears to be very suitable for biosensors. Other trends relate to the increase of the resolution, similar to that obtained with optical methods; the reusability of the crystals; and the development of multichannel devices.

The research in SPR and ellipsometry is constantly concerned with their miniaturization and related high-density arrays, focusing on the increase in spatial resolution and sensitivity in order to be able to tackle new applications, such as the study of conformational changes of biomolecules or multianalysis performed in real-time.

BIBLIOGRAPHY

1. W. Gopel, *Biosens. Bioelectron.* 1995; **10**:35.
2. B. Ivarsson and I. Lundstrom, *CRC Crit. Rev. Biocompat.*, 1986; **2**:1.
3. J. J. Ramsden, *Quarterly Rev. Biophys.* 1994; **27**:41.
4. R. M. Azzam and N. M. Bashara, *Ellipsometry and Polarized Light*. New York: North-Holland Publishing Company, 1977.
5. H. G. Tompkins, *A User's Guide to Ellipsometry*. Boston, MA: Academic Press, 1993.
6. P. A. Cuypers, J. W. Corsel, M. P. Janssen, J. M. M. Kop, W. T. Hermens, and H. C. Hemker, *J. Biol. Chem.* 1983; **258**:2426.
7. J. A. De Feijter, J. Benjamins, and F. A. Veer, *Biopolymers* 1978; **17**:1759.
8. S. Welin-Klintstrom, *Ellipsometry and Wettability Gradient Surfaces*. Linkoping University, Sweden, 1992.
9. M. Sastry, *Bull. Mater. Sci.* 2000; **23**:159.
10. F.L. McCrackin, E. Passaglia, R. R. Stromberg, and H. L. Steinberg, *J. Res. Natl. Inst. Stand. Technol.* 2001; **106**:589.
11. A. J. de Mello, In: J. Davies, ed., *Surface Analytical Techniques for Probing Biomaterial Processes*. Boca Raton, FL: CRC Press, 1996.
12. E. N. Economou, *Phys. Rev.* 1969; **182**:539.
13. E. Burstein, W. P. Chen, Y. J. Chen, and A. J. Hartstein, *J. Vac. Sci. Technol.* 1974; **11**:1004.
14. C. E. Jordan, B. L. Frey, S. Kornguth, and R. M. Corn, *Langmuir* 1994; **10**:3642.
15. E. Stenberg, B. Persson, H. Roos, and C. Urbaniczky, *J. Colloid. Interface Sci.* 1991; **143**:513.
16. B. Liedberg, I. Lundstrom, and E. Stenberg, *Sens. Actuators, B* 1993; **11**:63.
17. Physical Acoustics: Principles and Methods. In: W. P. Mason, ed., *Properties of Gases, Liquids, and Solutions*, vol. IIA. New York and London: Academic Press, 1965.
18. M. Rodahl, F. Hook, and B. Kasemo, *Anal. Chem.* 1996; **68**:2219.
19. K. Bizet, C. Gabrielli, H. Perrot, and J. Therasse, *Biosens. Bioelectron.* 1998; **13**:259.
20. G. Sauerbrey, *Z. Phys.* 1959; **155**:206.
21. K. K. Kanazawa and J. G. Gordon, *Anal. Chem.* 1985; **57**:1770.
22. S. J. Martin, V. E. Granstaff, and G. C. Frye, *Anal. Chem.* 1991; **63**:2272.
23. C. E. Reed, K. K. Kanazawa, and J. H. Kaufman, *J. Appl. Phys.* 1990; **68**:1993.
24. M. V. Voinova, M. Rodahl, M. Jonson, and B. Kasemo, *Physica Scripta* 1999; **59**:391.
25. F. Hook, B. Kasemo, T. Nylamder, C. Fant, C. Sott, and H. Elwing, *Anal. Chem.* 2001; **73**:5796.
26. F. Hook, J. Voros, M. Rodahl, R. Kurrat, P. Boni, J. J. Ramsden, M. Textor, N. D. Spencer, P. Tengvall, J. God, and B. Kasemo, *Colloids and Surfaces, B* 2002; **24**:155.

BIO-OPTICAL SIGNALS

ATA AKIN
Boğaziçi University
Bebek, İstanbul, Turkey

1. INTRODUCTION

Living organisms undergo dynamical changes when they are stimulated or diseased. In the case of cells, stimulation as well as diseases cause their intracellular chemical dynamics to constantly change. The chemical components responsible for such dynamics typically possess unique optical signatures that can be measured and quantified by light sources working in a broad range around visible spectrum from ultraviolet (UV) (< 300 nm) to infrared (IR) (1 μm to 10 μm). The concept of near-infrared spectroscopy developed when the optical properties of biochemical components associated with physiological activity were discovered (1–5). Through collaborations of biochemists and biophysicists, it soon became apparent that it is possible to exploit optics in monitoring and measuring these dynamics even when the organism is still alive. As early as the 1940s it was discovered that a range exists within the physiologically relevant optical spectrum that covers the 650–950 nm range called the "optical window" or "therapeutic window," in which from a relative lessened coefficient of absorption of water, it is possible to probe tissues to a depth close to 5 cm (5–8). The possibility of seeing inside the body without penetration led to the development of functional near-infrared spectroscopy (fNIRS) systems that particularly focused on neuroimaging of the brain and detection of cancerous tissue.

Biooptical signals are acquired by using a light source aimed at a body part with a detector or camera collecting the reflected or transmitted light through the tissues under investigation. Tissues possess mainly three optical characteristics: absorption, scattering, and fluorescence. Based on the method of illumination and measurement, relevant information can be obtained invasively or

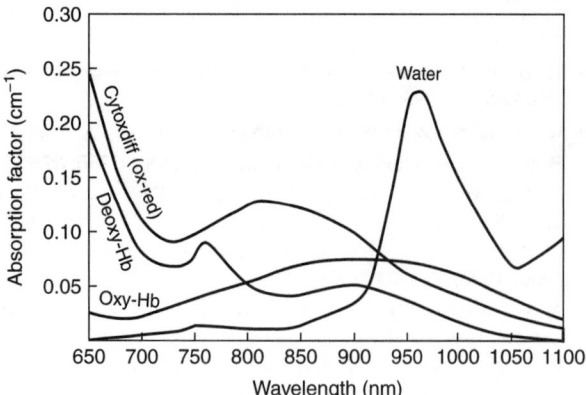

Figure 1. Absorption spectra of various tissue types in the optical window range.

noninvasively. Hence, biooptical signals are the signals received after light has interacted with tissue in one (or more) of the three ways mentioned. Most noninvasive biooptical signal applications have focused around (1) quantification of tissue oxygenation with a special emphasis on functional imaging of the brain, in which the aim is to monitor the changes in the hemodynamic activity during stimulation, and (2) optical biopsy using spectroscopical methods, in which the aim is to detect the presence of cancer tissue at an early stage.[1]

2. LIGHT–TISSUE INTERACTION

Light sources operating in the optical window range can penetrate into the tissue up to 5 cm because of the decreased absorption of water. Light within this wavelength range interacts mainly in two ways with the tissues: (1) absorption and (2) scattering (8–11). Scattering effects dominate the light–tissue interaction within this wavelength, which causes a loss in the resolution (detectability) and sensitivity analysis. Scattering happens because of a change of refractive index around the denser cell organelles like the mitochondria, nucleus, cell membrane, and endoplasmic reticulum. The main absorbers in the near-infrared region are the oxygenated and deoxygenated hemoglobin molecules, HbO_2 and Hb, respectively. Their absorption spectra can be seen in Fig. 1. The combined effects of absorption and scattering alter the path of photons traveling inside the tissue as sketched in Fig. 2a. Depending on the placement of the detector with respect to the light source, the method of measurement receives the name "reflectance" (source and detector are on the same surface: 1 and 2 in Fig. 2a) or "transmission" (source and detector are on opposite sides of the object: 3 in Fig. 2a).

2.1. Photon Migration in Tissue

There are basically three ways that photons interact with media based on the scattering coefficient of the media: (1) absorbing but no scattering, (2) both absorption and scattering, and (3) diffusion process. Even though the light–tissue interaction is listed as a diffusion process, the complexity of the physics underlying this interaction and the computational power required to calculate necessary parameters forced researchers to approximate the physics to the more simple approach listed as the first two ways below.[2]

2.1.1. Absorbing-Not Scattering Medium.
The light intensity transmitted through a medium can be expressed analytically by the Beer–Lambert law:

$$I_L = I_0 e^{-\varepsilon(\lambda)CL}, \quad (1)$$

where I_0 is the incident light intensity, I_L is the transmitted light intensity through the medium, $\varepsilon(\lambda)$ is the absorption coefficient as a function of wavelength, C is the concentration of the absorber, and L is the optical pathlength (distance from source to detector). This equation can be changed to yield direct information on C, the concentration of the absorber, as

$$OD(\lambda) = \log(I_0/I_L) = \varepsilon(\lambda)CL, \quad (2)$$

where OD is the optical density. This equation assumes a homogeneous solution (medium) with no scatterers that is not a valid approach in biological applications (12).

2.1.2. Absorbing and Scattering Medium.
This case is typical for all biological tissues. The medium is heterogeneous in terms of optical properties. Optical propagation through a medium can be described by (1) single scatterer, (2) multiple scatterers or (3) diffusion process, in which the medium is so dense that light is diffusely scattered as in Fig. 2b.

The main difference between this approach and the previous one is the presence of scatterers that increase the L value from a mere source-detector separation to the total length that a photon has to travel in a turbid media before it loses its initial direction. A slight modification to Equation 2 is applied to compensate for the prolonged L and attenuation related to the geometry of the subject matter:

$$OD(\lambda) = \varepsilon(\lambda)CLB + G, \quad (3)$$

where B is the differential pathlength factor to account for the prolongation of L because of scatterers and G is the constant attenuation factor related to the optical properties and geometry of the tissue. Although B should be determined experimentally, currently the existing technology does not allow that. Assuming that the measurements are made to monitor the *changes* of the chromophore, then L, B, and G can be taken as constants (13,14):

$$\Delta C = \frac{\Delta OD(\lambda)}{\varepsilon(\lambda)LB}. \quad (4)$$

2.1.3. Diffusion Analysis.
Whenever the particle concentration in the medium exceeds 5% or the medium thickness becomes so large that light will have to go through

---

[1]For a detailed explanation on optical spectroscopy, see NEAR-INFRARED SPECTROSCOPIC IMAGING.

[2]This part is for readers advanced in differential equations.

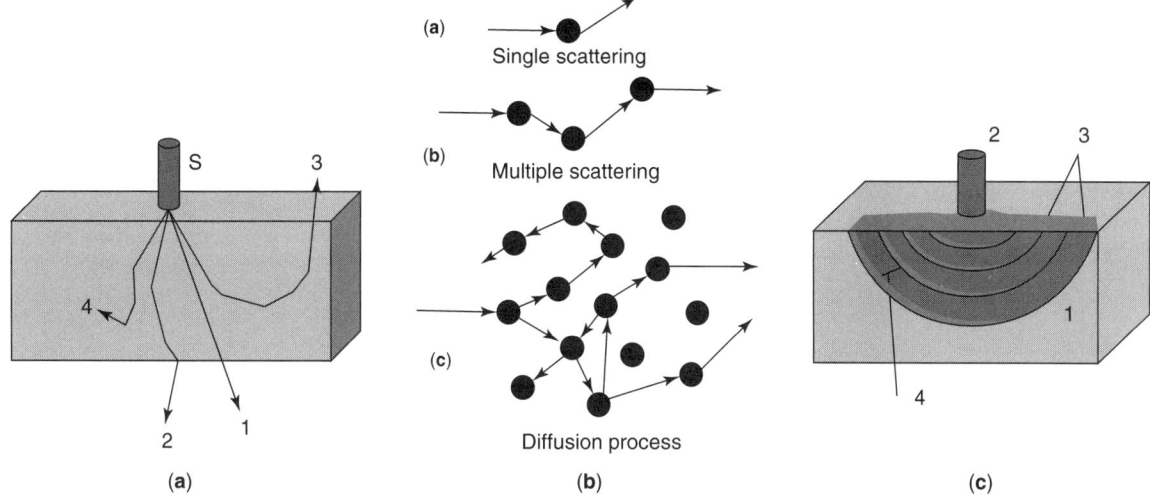

Figure 2. (a) Optical paths for various events. S: Source, 1. Ballistic photons, 2. Transmitted scattered photons, 3. Reflected scattered photons, 4. Scattered and absorbed photons. (b) Schematic representation of light propagation in (a) scattering medium, (b) multiple scattering medium, and (c) diffusion process (Adopted from Ref. 10).

repeated multiple scattering, diffusion analysis is performed. At this point, the coherence of incident light is lost and light becomes *diffused*, which gives rise to diffuse photon density waves (DPDWs). Diffusion analysis is an approximation to the transport equation (15–25).

The time-domain photon diffusion equation approximating the propagation pattern of photons is given as (18,24):

$$-\nabla \cdot D(\mathbf{r}\nabla\Phi(\mathbf{r},t) + \frac{1}{v}\frac{\partial \Phi(\mathbf{r},t)}{\partial t} + \mu_a(\mathbf{r})\Phi(\mathbf{r},t) = S(\mathbf{r},t), \quad (5)$$

where $S(\mathbf{r},t)$ ([Watts/m^3]) is the power density source, $D(\mathbf{r})$, the diffusion constant, defined as

$$D(\mathbf{r}) = \frac{1}{3\{\mu_a(\mathbf{r}) + \mu_s(\mathbf{r})[1 - g(\mathbf{r})]\}}, \quad (6)$$

with g being the average scattering angle cosine, μ_a and μ_s being the absorption and scattering inverse mean-free paths [m^{-1}], respectively, and v being the speed of light in tissue taken to be c_0/n, with $n = 1.38 \sim 1.5$ (approximating to speed of light in water) (24). Because in tissue scattering events dominate absorption events in the near-infrared regions ($10\,\text{cm}^{-1} < \mu_s < 50\,\text{cm}^{-1}$ and $0.03\,\text{cm}^{-1} < \mu_a < 0.15\,\text{cm}^{-1}$ @ $\lambda = 650$ nm), Li (24) assumes $\mu_a \ll \mu_s$ and rearranges Equation 5 as follows:

$$\nabla \Phi(\mathbf{r},t) - \frac{v\mu_a}{D}\Phi(\mathbf{r},t) - \frac{1}{D}\frac{\partial \Phi(\mathbf{r},t)}{\partial t} = -\frac{v}{D}S(\mathbf{r},t), \quad (7)$$

where $D = 1/3\mu_s'$ and $\mu_s' = \mu_s(1-g)$ is the reduced scattering coefficient (reciprocal of the random walk step, i.e., the average length it takes for a photon's direction to become random), whereas g is the single-scattering anisotropy factor, a measure of how much of the incident light is scattered in the forward direction. In typical biological tissues, the scattering is predominantly in the forward direction ($g = 0.9$). For example, in breast tissue, the scattering length is about 0.1 mm, but the random walk step is 1 mm; that is, it takes about ten scattering events for the photon direction to become random with respect to its incident direction. For the steady state, the equation is written independent of time as (26)

$$(\nabla - \mu_a/D)\Phi(\mathbf{r}) = S(\mathbf{r}). \quad (8)$$

Analytical solutions of Equations 5 and 7 exist and are used in image reconstruction algorithms in which the goal is to estimate the optical properties, μ_a and μ_s, of the medium from multiple observations[3] (21,24). Once optical properties are estimated, then it may be possible to classify the functional state of the underlying tissues as in cancer cases.

3. fNIRS

The fNIRS instrument [a.k.a functional optical imaging (fOI)] is an extension of a standard oximeter that uses two wavelengths (λ_1, λ_2) to measure the [Hb] and [HbO$_2$] concentration changes by a probe placed over the tissues. At each wavelength, a combination of Δ[Hb] and Δ[HbO$_2$] can be derived from Equation 4 based on ΔOD measurements:

$$\Delta OD(\lambda_1) = (\varepsilon_{\text{HbO}_2}(\lambda_1)\Delta[\text{HbO}_2] + \varepsilon_{\text{Hb}}(\lambda_1)\Delta[\text{Hb}])LB(\lambda_1), \quad (9)$$

$$\Delta OD(\lambda_2) = (\varepsilon_{\text{HbO}_2}(\lambda_2)\Delta[\text{HbO}_2] + \varepsilon_{\text{Hb}}(\lambda_2)\Delta[\text{Hb}])LB(\lambda_2). \quad (10)$$

These equations represent the changes in the total light attenuation signal with respect to extinction coefficients (absorption coefficients) of the corresponding tissues. L is the mean free pathlength, and $B(\lambda)$ is the correction of this pathlength with respect to wavelength (27,28). Given Equations 9 and 10, and assuming $B(\lambda_1) \cong B(\lambda_2) = B$, Δ[HbO$_2$] and Δ[Hb], the changes in [HBO$_2$] and [Hb]

---
[3]Derivation of such solutions is beyond the scope of this article. Refer to the references for reviews.

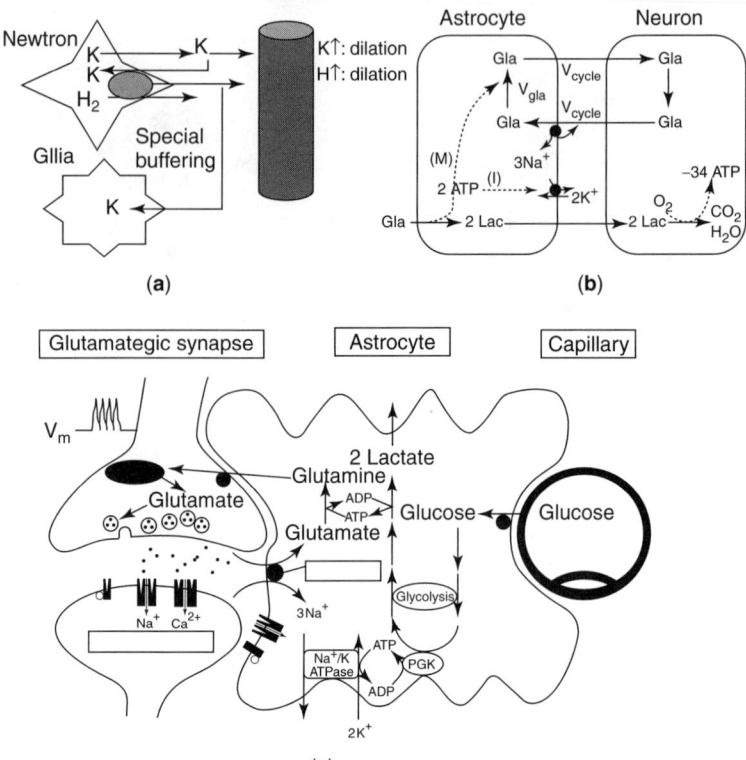

Figure 3. (a) A graphical representation of the neurovascular coupling hypothesis that includes the role of metabolites released to ECS causing vasodilation action (reproduced from Ref. 38 by permission). (b) Lactate-derived glucose utilization of astrocytes (picture is a courtesy of Ref. 36). (c) Glutamate release in synaptic cleft drives glucose utilization in astrocytes (picture is a courtesy of Ref. 39).

concentrations, respectively, can be calculated from Equations 9 and 10 as

$$\Delta[Hb] = \frac{\Delta OD(\lambda_1) - \frac{\varepsilon_{HbO_2}(\lambda_1)}{\varepsilon_{HbO_2}(\lambda_2)}\Delta OD(\lambda_2)}{LB\left[\varepsilon_{Hb}(\lambda_1) - \varepsilon_{Hb}(\lambda_2)\frac{\varepsilon_{HbO_2}(\lambda_1)}{\varepsilon_{HbO_2}(\lambda_2)}\right]}, \quad (11)$$

$$\Delta[HbO_2] = \frac{\Delta OD(\lambda_1) - \frac{\varepsilon_{Hb}(\lambda_1)}{\varepsilon_{Hb}(\lambda_2)}\Delta OD(\lambda_2)}{LB\left[\varepsilon_{HbO_2}(\lambda_1) - \varepsilon_{HbO_2}(\lambda_2)\frac{\varepsilon_{Hb}(\lambda_1)}{\varepsilon_{Hb}(\lambda_2)}\right]}, \quad (12)$$

from which total blood volume change and oxygenation can be estimated by

$$\Delta[BV] = \Delta[HbO_2] + \Delta[Hb], \quad (13)$$

$$\Delta[Oxy] = \Delta[HbO_2] - \Delta[Hb], \quad (14)$$

which presents insight on the functional activity level of the underlying tissues. In case of brain activity, an increase in metabolism in a region of the brain leads to an increase in regional cerebral blood flow as dictated by the neurovascular coupling hypothesis. The raise in the inflow of blood to a region to compensate for the demand for glucose and especially O_2 can be indirectly monitored by measuring $\Delta[HbO_2]$ and $\Delta[Hb]$.

3.1. Physiological Events During Brain Activation

Neuronal circuitry of the brain constantly requires the surrounding blood vessels to provide necessary nutrients to sustain its aggressive operational level. The rapid bursts of action potentials are products of Na^+/K^+ pumps that require adenosine triphosphate (ATP) to maintain their continuous activity. ATP is generated biochemically by (1) oxidative phosphorylation according to the chemical reaction (29):

$$3ADP + 3P_i + 1/2O_2 + NADH + H^+ \xrightarrow{Exo} 3ATP + NAD^+ + H_2O; \quad (15)$$

(2) glycolysis, in which glucose is converted into pyruvate and lactate; and (3) mitochondrial respiration consuming O_2 and pyruvate in the tricarboxylic acid cycle (30,31). The demand for O_2 and glucose is responded by increasing the regional cerebral blood flow (CBF) and volume (CBV) in a volume of brain giving rise to neurovascular coupling. Although intracellular biochemical reactions giving rise to neuronal activation are investigated in detail, there are controversies in the explanation of how increased O_2 and glucose demand actually evoke a change in the cerebrovascular dynamics (increase in blood flow). Several hypotheses have been proposed in explaining the coupling of neuronal activity to cerebrovascular dynamics and energy metabolism (32–35). These hypotheses have been tested on animal and human studies using several imaging modalities like positron emission tomography (PET), functional magnetic resonance imaging (fMRI), indicator dilution methods, and electro-physiologic mapping via needle electrodes as well as with intrinsic optical signals (IOSs) and fNIRs (36,37).

There are basically three working hypotheses in explaining the hemodynamical changes observed by various methods: (1) Local brain activity in neurons is shown to induce a local arteriolar vasodilation and hence an increase in local (regional) CBV and CBF via release of metabolites such as K^+, H^+ or adenosine as well as several neurotransmitters and messengers such as nitric oxide (NO) into the extracellular space (as shown in Fig. 3a) (40); (2) release of lactate by astrocytes during glucose utilization that is picked up by neurons and further oxidized (as shown in Fig. 3b); and (3) release of glutamate during synaptic activity stimulates the use of glucose by astrocytes that provides an indirect mechanism of coupling of neurons to hemodynamic activity (as shown in Fig. 3c) (34,41). Monitoring of the CBF and CBV can be accomplished indirectly through continuous monitoring of oxygenated and deoxygenated hemoglobin activity by near-infrared spectroscopy instruments. The dynamic coupling of blood flow and oxygenation to neural function and metabolism is achieved by vasomotor action of cerebral arteries and arterioles. Within a few seconds of the formation of the action potential, a quick adjustment of blood flow to the increased functional activity area results. The original extra/intracellular distribution of ions around neurons then has to be reestablished on a long-term basis. Pumps are activated to carry K^+ back into and Na^+ out of the cells. Increased pump activity is accompanied by the increased metabolic activity (32,34). The demand for ATP is correlated with a demand for O_2 according to Equation 15, which is the basis for neuroimaging methods that make use of the metabolic activity as an indicator of function.

3.2. Biooptical Signals of the Brain

It is possible to monitor the oxygenation levels of the brain while the subjects are performing cognitive tasks (protocols). One such instrument is the fMRI that measures the blood oxygenation level-dependent signal (BOLD signal) as a function of the concentration of deoxyhemoglobin ([Hb]) present in a tissue voxel (42). The fMRI instrument has enabled the cognitive neuroscientist to investigate the distribution of activities on the brain during stimulation, producing a map of functional activation superimposed on the anatomical regions (42). The time course of the BOLD signal is shown to depend on the concentration change of [Hb] (Δ[Hb]), hence, indirectly related to neuronal metabolism and neuronal activation.[4] An alternative to fMRI is the fNIRS with a higher temporal resolution (100 s of milliseconds compared with seconds in fMRI) to perform neuroimaging of cognitive activity. The ultimate goal in neuroimaging studies has been to understand the sequence of metabolic events taking place during stimulation and provide a model for it for health and disease. Unfortunately neither fMRI nor fNIRS have resolutions in the micrometer range, which limits the modeling of metabolic activity of a single neuron. The gross imaging techniques of fMRI and fNIRS have recently demonstrated that during activation, an increase in BOLD signal in fMRI and Δ[HbO$_2$] in fNIRS are observed (8,11,42). The time course of these signals are investigated in detail to relate them back to actual metabolic events. The existing neurovascular coupling approach explained above depends on experiments performed on cell groups.

3.2.1. Components of Biooptical Signals of the Brain. Ongoing research studies on the signal components of the brain hemodynamic activity agree on the existence of a model that includes (1) ultra low frequency that may be related to basal activity centered around 0.004 Hz, (2) very slow components related to dominant frequency of brain hemodynamic activity (centered around 0.04 Hz), (3) task-related high-frequency activity centered around 0.1 Hz, (4) respiratory activity centered around 0.2–0.25 Hz, (5) heart rate activity peaking around 1–1.5 Hz, and (6) ultra high-frequency activity sometimes denoted as fast signal centered at frequencies higher than 2 Hz (1,10). A typical fNIRS system used in monitoring brain optical signals during cognitive tasks includes a probe connected to a computer-based controller unit (see Fig. 4a). The probe used in adult brain activity monitoring studies shown in Fig. 4b is usually 7 cm by 16 cm containing four light emission diode near-infrared light sources each operating at least two different wavelengths (typically one around 735 nm for monitoring Δ[Hb] and the other around 850 nm for Δ[HbO$_2$]).

3.2.2. Protocols of Obtaining Data. The ultimate goals of neuroimaging techniques are (1) to localize the activated regions, (2) to map the distributed patterns of activation for diseased and healthy brains, and (3) to evaluate

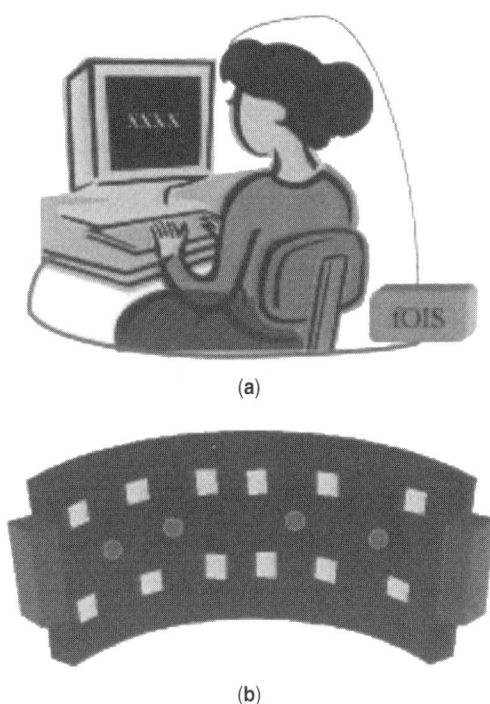

Figure 4. (a) Typical fNIRS measurement setup where subject is performing an *oddball* task. (b) The probe housing sources shown as circular objects surrounded by detectors shown as rectangular objects placed 2.5 cm away from the center of each source.

---
[4] See the article BRAIN FUNCTION, MAGNETIC RESONANCE IMAGING OF for a review.

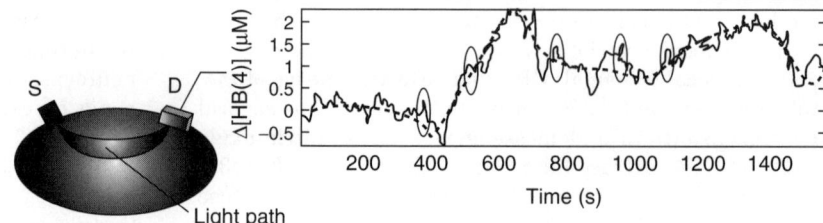

Figure 5. Typical reflectance method measurement shows the graphical representation of the light path through tissues.

the amount of physiological changes during activation. Several tasks are identified and tested in accomplishing especially the first two goals that are basically the expertise of medical imaging groups (43). So far fNIRS researchers have focused on establishing a standard in acquiring the hemodynamic signals before even quantifying the changes observed in normal and diseased persons. Hence, only the protocols aiming to answer the first goal were run by researchers. Experiments using animals, in which sensory-motor stimulations were carried out, have successfully demonstrated the use of fNIRS as an alternative to PET and fMRI techniques (1,9–12,39,44–49). So far only a few researchers have attempted to investigate the changes during cognitive activity from the prefrontal lobe. Studies have yielded results with high statistical variations leading to difficulties in the interpretations. Nevertheless, the relative ease and noninvasiveness of fNIRS data acquisition still promotes the promising nature of this technique in future clinical neuroscience environments (Fig. 5).

The challenges in quantifying the physiological changes observed in cognitive activity monitoring can be listed as follows:

1. Cognitive activity is not as localized as motor activity. Hence, probe geometry must be carefully adjusted to monitor the distributed nature of activation.
2. The signal of the cognitive activity tends to be modulated by several factors such as emotions, tiredness, boredom, alertness expectations, and habitation (see Fig. 6).
3. The signals recorded are contaminated by certain physiological noise such as dynamic fluctuations in vessels from heart rate, breathing changes, and baroreceptor reflexes.
4. Scalp, skull, and cerebrospinal fluid anatomy have varying contributions to the measured signals that must be deconvolved for optimal performance quantification (14).
5. Comparison of single-event trial data to fMRI signals have yielded controversial results suggesting mere analyses of both types of data.

4. FINAL WORD

Functional neuroimaging techniques aim to identify and localize the physiological changes taking place during brain activation. The goal is to decipher which physiological changes have dominance during sustained performance and disease. Biooptical signals of the brain that can be recorded by near-infrared spectroscopy systems offer the advantage of being noninvasive, comfortable, and portable while causing minimal disturbance during the course of activity. Nevertheless, the complexity of the incoming data challenges the researchers to identify the so longed physiological changes rapidly and correlate them to behavioral performance metrics as well as diseases.

The affects of several physiological activities along with the constant background activity of brain complicate the extraction of useful and relevant signals from these signals. Signal processing methods such as Fourier transform, wavelet transform, principal component analysis, and independent component analysis have been tried on several occasions; however, the success of such algorithms depends heavily on the initial assumptions made regarding the signal model. In certain cases, such as visual or motor stimulation tasks, biooptical signals have been recorded with great success, and in fact, the mentioned models and signal processing approaches have proven to be very useful in analyzing the signals (8,10,14,27,36,38,39,44–46,48,50,51). Thus, the complexity of decision making is forced into such models, and this requires the integration of cognitive scientists with cognitive neuroscientists, signal processing experts, mathematical modeling experts, and systems design engineers in creating the future versions of fNIRS. Such an integration of different professionals will eventually lead to a design more suitable for analyzing brain activity online and on-site.

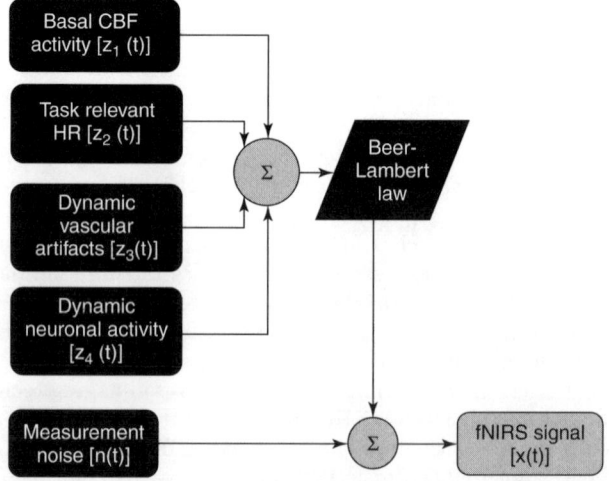

Figure 6. A hypothetical model of signal components gives rise to changes in the brain biooptical signal during cognitive activity.

BIBLIOGRAPHY

1. B. Chance, Z. Zhuang, C. Unah, C. Alter, and L. Lipton, Cognition-activated low frequency modulation of light absorption in human brain. *Proc. Natl. Acad. Sci. USA*. 1993: 3770–3774.
2. Y. Hoshi and M. Tamura, Detection of dynamic changes in cerebral oxygenation coupled to neuronal function during mental work in man. *Neurosci. Lett.* 1993; **150**:5–8.
3. Y. Hoshi and M. Tamura, Oxygen dependence of redox state of copper in cytochrome oxidase in vitro. *J. Appl. Physiol.* 1993; **75**:1622–1627.
4. F. F. Jobsis, Non-invasive infrared monitoring of cerebral and myocardial oxygen sufficiency and circulatory parameters. *Science* 1977; **198**:1264–1267.
5. G. A. Millikan, The oximeter, an instrument for measuring continuously the oxygen saturation of arterial blood in man. *Rev. Sci. Instrum.* 1942; **13**:434–444.
6. D. K. Hill and R. D. Keynes, *J. Physiol.* 1949; **108**:278–281.
7. R. Richards-Kortum and E. Sevick-Muraca, Quantitative optical spectroscopy for tissue diagnosis. *Annu. Rev. Phys. Chem.* 1996; **47**:555–606.
8. A. Villringer and B. Chance, Non-invasive optical spectroscopy and imaging of human brain function. *Trends Neurosci.* 1997; **20**(10):4435–4442.
9. B. Chance, E. Anday, S. Nioka, S. Zhou, H. Long, K. Worden, C. Li, T. Turray, Y. Ovetsky, D. Pidikiti, and R. Thomas, A novel method for fast imaging of brain function, noninvasively, with light. *Optics Express*. 1998; **2**:411–423.
10. H. Obrig, M. Neufang, R. Wenzel, M. Kohl, J. Steinbrink, K. Einhaupl, and A. Villringer, Spontaneous low frequency oscillations of cerebral hemodynamics and metabolism in human adults. *Neuroimage*. 2000; **12**:623–639.
11. H. Obrig, R. Wenzel, M. Kohl, S. Horst, P. Wobst, J. Steinbrink, F. Thomas, and A. Villringer, Near-infrared spectroscopy: Does it function in functional activation studies of the adult brain? *Intrl. J. Psychophysiol.* 2000; **35**: 125–142.
12. A. Villringer, Functional neuroimaging, optical approaches. In: A. Villringer and U. Dirnagl, eds., *Optical Imaging of Brain Function and Metabolism*, vol. 2. New York: Plenum Press, 1997, pp. 1–18.
13. D. T. Delpy, M. Cope, P. van der Zee, S. R. Arridge, S. Wray, and J. S. Wyatt, Estimation of optical pathlength through tissue from direct time of flight measurements. *Phys. Med. Biol.* 1988; **33**:1433–1442.
14. M. Firbank, E. Okada, and D. T. Delpy, A theoretical study of the signal contribution of regions of the adult head to near-infrared spectroscopy studies of visual evoked responses. *NeuroImage*. 1998; **8**:6978.
15. J. C. Hebden, S. R. Arridge, and D. T. Delpy, Optical imaging in medicine: I. Experimental techniques. *Phys. Med. Biol.* 1997; **42**:825–840.
16. S. R. Arridge and J. C. Hebden, Optical imaging in medicine: Ii. Modelling and reconstruction. *Phys. Med. Biol.* 1997; **42**:841–853.
17. A. Ishimaru, *Wave Propagation and Scattering in Random Media*. New York: Academic Press, 1978.
18. A. D. Kim and A. Ishimaru, Optical diffusion of continuous-wave, pulsed, and density waves in scattering media and comparisons with radiative transfer. *Appl. Opt.* 1998; **37**(22): 5313–5319.
19. X. Li, Fluorescence and diffusive wave diffraction tomographic probes in turbid media, Ph.D. dissertation, University of Pennsylvania, Philadelphia, PA, 1998.
20. X. D. Li, M. A. O'Leary, D. A. Boas, and B. Chance, Fluorescent diffuse photon density waves in homogeneous and heterogeneous turbid media: Analytic solutions and applications. *Appl. Opt.* 1996; **35**(19):3746–3758.
21. M. O'Leary, Imaging with diffuse photon density waves, Ph.D. dissertation, University of Pennsylvania, Philadelphia, PA, 1996.
22. M. O'Leary, D. Boas, B. Chance, and A. G. Yodh, Refraction of diffuse photon density waves. *Phys. Rev. Lett.* 1992; **69**: 2658–2666.
23. A. Ya. Polishchuk, S. Gutman, M. Lax, and R. R. Alfano, Photon-density modes beyond the diffusion approximation: Scalar wave-diffusion equation. *J. Opt. Soc. Am. A*. 1997; **14**(1): 230–234.
24. J. S. Reynolds, A. Przadka, S. P. Yeung, and K. J. Webb. Optical diffusion imaging: A comparative numerical and experimental study. *Appl. Opt.* 1996; **35**(19):3671–3679.
25. V. Twersky, Absorption and multiple scattering by biological suspension. *J. Opt. Soc. Amer.* 1970; **60**:1084–1093.
26. S. Takatani and J. Ling, Optical oximetry sensors for whole blood and tissue. *IEEE Eng. Med. Biol.* June/July 1994: 347–357.
27. G. Strangman, J. P. Culver, J. H. Thompson, and D. A. Boas, A quantitative comparison of simultaneous bold fMRI and NIRS recordings during functional brain activation. *NeuroImage*. 2002; **17**:719–731.
28. K. Uludag, M. Kohl, J. Steinbrink, H. Obrig, and A. Villringer, Cross talk in the lambert-beer calculation for near-infrared wavelengths estimated by monte carlo simulations. *J. Biomed. Opt.* 2002; **7**(1):51–59.
29. B. Chance, NMR and time-resolved optical studies of brain imaging. In: U. Dirnagl, A. Villringer, and K. M. Einhäupl, eds., *Optical Imaging of Brain Function and Metabolism*. New York: Plenum Press, 1993, pp. 1–7.
30. A. Aubert and R. Costalat, A model of the coupling between brain electrical activity, metabolism, and hemodynamics: Application to the interpretation of functional neuroimaging. *NeuroImage*. 2002; **17**:1162–1181.
31. Larry R. Squire, ed, *Fundamental Neuroscience*, 2nd ed. San Diego, CA: Academic Press, 2003.
32. P. J. Magistretti, Cellular bases of functional brain imaging: Insights from neuron-glia metabolic coupling. *Brain Res.* 2000; **886**:108–112.
33. P. J. Magistretti and L. Pellerin, Metabolic coupling during activation, a cellular view. In: A. Villringer and U. Dirnagl, eds., *Optical Imaging of Brain Function and Metabolism*, vol. 2. New York: Plenum Press, 1997, pp. 161–166.
34. P. J. Magistretti and L. Pellerin, Cellular mechanisms of brain energy metabolism and their relevance to functional brain imaging. *Phil. Trans. R. Soc. Lond. B*. 1999; **354**:1155–1163.
35. M. E. Raichle, Circulatory and metabolic correlates of functional activation in the brain. In: R. S. Frackowiak, P. Magistretti, R. G. Shulman, J. S. Altman, and M. Adams, eds., *Neuroenergetics: Relevance for Functional Imaging*, Starsbourg, France: HFSP, 2001, pp. 65–71.
36. D. Malonek and A. Grinvald, Interactions between electrical activity and cortical microcirculation revealed by imaging spectroscopy: Implications for functional brain mapping. *Science*. 1996; **272**(5261):551–555.

37. R. A. Stepnoski, A. LaPorta, F. Raccuai-Behling, G. E. Blonder, R. E. Slusher, and D. Kleinfeld, Noninvasive detection of changes in membrane potential in cultured neurons by light scattering. *Proc. Natl. Acad. Sci.* 1991; **88**:9382–9386.

38. Y. Hoshi, S. J. Chen, and M. Tamura, Spatiotemporal imaging of human brain activity by functional near-infrared spectroscopy. *Am. Lab.* Oct 2001: 35–39.

39. R. P. Kennan, S. G. Horovitz, A. Maki, Y. Yamashita, H. Koizumi, and J. C. Gore, Simultaneous recording of event-related auditory oddball response using transcranial near infrared optical topography and surface eeg. *Neuroimage.* 2002; **16**:587–592.

40. W. Kulschinky, Neuronal-vascular coupling, a unifying hypothesis. In: A. Villringer and U. Dirnagl, eds., *Optical Imaging of Brain Function and Metabolism*, vol. 2. New York: Plenum Press, 1997, pp. 167–176.

41. N. R. Sibson, A. Dhankhar, G. F. Mason, D. L. Rothman, K. L. Behar, and R. G. Shulman, Stoichiometric coupling of brain glucose metabolism and glutamatergic neuronal activity. *Proc. Natl. Acad. Sci. USA.* 1998; **95**:316–321.

42. D. Kim and K. Ugurbil, Bridging the gap between neuroimaging and neuronal physiology. *Image Anal. Stereol.* 2002; **21**:97–105.

43. R. Cabeza and L. Nyberg, Imaging cognition II: An empirical review of 275 PET and fMRI studies. *J. Cogn. Neurosci.* 2000; **12**(1):1–47.

44. W. N. J. M. Colier, V. Quaresima, G. Baratelli, P. Cavallari, M. Sluijs, and M. Ferrari, Detailed evidence of cerebral hemoglobin oxygenation changes in response to motor cortical activation revealed by a continuous wave spectrophotometer with 10 hz temporal resolution. In: B. Chance, R. R. Alfano, and A. Katzir, eds., *SPIE Proc. Optical Tomography and Spectroscopy of Tissue: Theory, Instrumentation, Model, and Human Studies II*, vol. 2979, 1997: 390–396.

45. G. Gratton, M. Fabiani, and P. M. Corballis, Can we measure correlates of neuronal activity with non-invasive optical methods. In: A. Villringer and U. Dirnagl, eds., *Optical Imaging of Brain Function and Metabolism*, vol. 2. New York: Plenum Press, 1997, pp. 53–62.

46. A. Maki, Y. Yamashata, Y. Ito, E. Watanabe, Y. Mayanagi, and H. Koizumi, Spatial and temporal analysis of human motor activity using non-invasive NIR topography. *Med. Phys.* 1995; **22**(12).

47. H. Obrig, H. Israel, M. Kohl-Bareis, K. Uludag, R. Wenzel, B. Muller, G. Arnold, and A. Villringer, Habituation of the visually evoked potential and its vascular response: Implications for neurovascular coupling in the healthy adult. *NeuroImage.* 2002; **17**:1–18.

48. P. Wobst, R. Wenzel, H. Obrig, M. Kohl, and A. Villringer, Linear aspects of changes in deoxygenated hemoglobin concentration and cytochrome oxidase oxidation during brain activation. *NeuroImage.* 2001; **13**:520–530.

49. C. B. Akgul, B. Sankur, and A. Akin, Selection of frequency bands in functional near infrared spectroscopy. *J. Comp. Neurosci.*, in press.

50. J. Mayhew, D. Johnston, J. Berwick, M. Jones, P. Coffey, and Y. Zheng, Spectroscopic analysis of neural activity in brain: Increased oxygen consumption following activation of barrel cortex. *NeuroImage.* 2000; **12**:664–675.

51. D. A. Boas, T. Gaudette, G. Strangman, X. Cheng, J. J. A. Marota, and J. B. Mandeville, The accuracy of near infrared spectroscopy and imaging during focal changes in cerebral hemodynamics. *NeuroImage.* 2001; **13**:76–90.

BIO-OPTICS: OPTICAL MEASUREMENT OF PULSE WAVE TRANSIT TIME

KALJU MEIGAS
JAANUS LASS
RAIN KATTAI
Tallinn Technical University
Tallinn, Estonia

1. INTRODUCTION

The arterial blood pressure has long been known as one of the most important medical factors reflecting the state of the cardiovascular system. A useful and convenient parameter for continuous monitoring of blood pressure is pulse wave velocity or pulse wave transit time (PWTT) between different regions of human body. It is suggested that, although it may be difficult to estimate systolic blood pressure from PWTT with acceptable accuracy, possibilities still remain to use changes of pulse arrival time as an indicator of changes in systolic blood pressure. During the exercise, arterial blood pressure increases, which is mainly achieved by increasing the contractility of the heart and vasculature resistance. Both of these values result in the increase of pressure wave speed in the vascular system, which means that the time delay between the start of electromechanical contraction and PWTT to any particular point in the body decreases. In our previous studies, we have shown that it is possible to calculate the mean arterial pressure from pulse wave transit time.

Time delay between the signal of heart electrical activity and pulse wave is involved mainly with the pulse wave propagation velocity, but another important parameter also exists, which is the isovolumic contraction time. It is the interval required for the heart to convert the electrical stimulus into a productive mechanical contraction capable of ejecting blood from the ventricles. This time interval contributes to the delay between electrocardiogram (ECG) R-peak and aortic valve opening. This period has been shown to be uncorrelated with arterial pressure and may make up a substantial part of the ECG-peripheral pulse delay.

Optical sensors are popular in many fields. They have several advantages compared with other types of sensors. One of the advantages of optical measurement methods is that many vital functions can be measured noninvasively, (i.e., no need exists to injure the skin during measurements). Moreover, the optical methods are safe to use, as direct electrical contact between the measurement device and the measured person does not occur, as the information is transmitted in an electromagnetic field. The diagnostic devices based on optics use low emitted power, which is typically less than several mW.

One common optical technique for biomedical diagnostics is the laser Doppler method. This simple optical method is also applicable for pulse wave profile, pulse wave delay time, and blood flow measurement. The method is based on recording of the Doppler frequency shift related to a moving target—blood vessel walls or small particles. The Doppler signal is detected with the use of the self-mixing that occurs in the diode laser cavity when radiation scatters back from the moving target into the laser and interferes with the field inside it. Two different ways may simultaneously be used for the self-mixing signal extraction: with the help of a photodiode accommodated in the rear facet of the diode laser package and with the help of resistor from the laser pump current. Pulse wave delay time in different regions of the human body can be calculated relative to the ECG signal.

Self-mixing method as a part of laser Doppler technique allows us to realize coherent photodetection. For example, the coherent photodetection of the radiation reflected back from the object surface is used in vibration measurements (1), surface motion measurement (2), deformation measurement (3), and distance measurements (4). The method for coherent photodetection is used in several biomedical applications of coherent imaging detection (5), Doppler interferometry (6), Doppler anemometry (7), and in principal schemes of a number of different sensors. However, several realizations are technically complicated and carefully aligned high-quality optics are required.

Self-mixing method in a laser has been known to have several advantages over conventional systems (8,9). This method enables us to simplify the optical scheme of such devices and achieve the mixing effect, where a small portion of the light reflected from the mirror is returned into the laser cavity and is mixed with the original oscillating wave inside the laser. The method of self-mixing inside the laser cavity facilitates the design of the same principle without beam splitters and external photodetector. The optical setup contains only cheap optical components: a laser diode with fiber or lenses. Self-mixing systems can be easily aligned because they have only one optical axis, they have been called self-aligning systems. Self-mixing also gives a larger modulation depth, hence a comparatively high signal-to-noise ratio.

The theories of heterodyne or homodyne reception and photodetection with self-mixing in semiconductor laser (10,11), gas laser (12), as well as theory of self-mixing in a single-mode or multilongitudinal-mode diode laser (13) are based on a presumption of coherence of light in all equipment and on the way to the target and back. The laser is used to send light, either in free space or through an optical fiber, to a movable target from which the backscattering is detected. The self-mixing effect has been explained as a spectral mode modulation inside laser caused by backscattered radiation. The theories operate with interference between irradiated and scattered back to the laser cavity radiation.

In medical applications, laser Doppler methods have been applied for different purposes, such as tissue vibration measurement (14), eye movements' registration (15), and others. Blood flow measurements have been of special interest for many years because quantitative determination of blood flow provides essential information to diagnose serious diseases and circulatory disorders in a certain part of the body, including vessels. Laser Doppler flowmetry has been developed as an effective and reliable method for noninvasive blood flow measurement (16–19). The self-mixing effect in diode lasers provides remarkable advantages in blood circulation measurement. Intra-arterial measurements of the velocity and the average flow of red blood cells were investigated by means of a fiber-coupled laser Doppler velocimeter based on the self-mixing effect (20). This technology is close to becoming commercial product for clinical application. The efficiency of the self-mixing velocimeter has been tested *in vivo* with the optical fiber inserted in the artery in upstream and in downstream blood flow conditions (16).

Blood pressure and cardiovascular pulsation are fundamental indicators of cardiovascular disease. The pulse is considered as one of the four most fundamental medical parameters. Abnormal shape and rhythm of arterial pulsation are directly connected to diverse cardiovascular disorders. Small and weak pulses can be related to heart failure, shock, or aortic stenosis. Large and bounding pulses can represent hyperkinetic states, aortic regurgitation, or abnormal rigidity of arteries. Arrhythmia can lead to a changing amplitude or irregularity of pulsation. Abnormal amplitude of pulsation can also be related to the blood pressure. Pulse wave velocity is known to be correlated with blood pressure. The pulsation of the carotid artery can be an indicator of cerebrovascular disease. In our preliminary publication, we noted that this method allows an optical no-touch measurement of skin surface vibrations, which can reveal the pulsatile propagation of blood pressure waves along the vasculature (21). The same method and equipment has been applied to detect small moving micron-size particles (22) and as an optical noninvasive method for blood flow velocity measurement (23).

2. THEORETICAL BACKGROUND

The pulse wave as a pressure wave causes changes in blood vessels radius and movement of its wall. Typically, vessel wall expansion and shrinkage results in movement on the skin surface, which can reveal the pulsatile propagation of pulse waves along the vasculature and contain such useful parameters as period of pulsation, moment of pulse wave arrival, and pulsation profile.

The laser light is coupled into a fiber and guided toward a moving object—skin near the arteries. A small part of the Doppler-shifted light scattered by the moving object is collected by the same fiber and guided back into the laser. The Doppler-shifted light interferes with the laser light present in the laser cavity and causes an intensity modulation of the light inside the laser cavity. The frequency of this intensity modulation is related to the Doppler shift.

Self-mixing effect in a semiconductor laser diode is observed as modulation of the amplitude and the spectra of the emitted light because of the optical feedback of backscattered light into the laser cavity. This external optical feedback affects some internal parameters of the laser diode such as threshold gain and lasing frequency. The light backscattered from the target comes to a laser-active cavity and causes changes in a number of carriers in the active area and in the threshold of excitation current (10–12,24). In this way, the backscattered light is active in the process of laser generation and affects inversion.

A theory that explains the phenomena observed in the self-mixing process has been developed by several authors.

An analytical model for the three-mirror cavity has been proposed and widely used to explain optical feedback-induced changes in output parameters of a diode laser. The amplitude and spectral behavior of the laser with optical feedback have been studied (25). A spectral line narrowing or broadening, depending on feedback conditions, has been shown to take place in the output spectra of the semiconductor lasers (26). A theoretical model for the shape and the amplitude of self-mixing signals has been described (17). A broadening of the apparent linewidth of the semiconductor laser modes with external cavity has been shown to be caused by the coherent nature of the feedback and multiple reflections in the external cavity (27). A detailed theoretical analysis of stability for a semiconductor laser with an external cavity has shown that instability is related to jumps of the laser frequency between external cavity modes or to feedback-induced intensity pulsations caused by the carrier-density dependence on the refractive index (28). The conditions that decide whether an external cavity laser oscillates in the mode with the lowest threshold gain have been discussed (29), where the mode coupling between the side modes has been shown to allow the external cavity laser to oscillate stable in a mode, even though it has side modes with higher gain. The results of numerical analyses of the external feedback on a single-mode semiconductor laser have demonstrated that the lasing mode with the minimum linewidth is most stable, rather than the mode with minimum threshold gain (30,31).

A three-mirror cavity model can be used for description of the self-mixing method applied for measurements in this study. The principal scheme of the method is presented in Fig. 1. Two mirrors, R1 and R2, constitute the laser cavity. The moving object (skin) can be presented as a third mirror, R3. The light reflected from the target interfered with the light at the laser front facet with different phase swift determined by the distance to the target. The mirror, R3, and one of the laser facets, R2, constitute an effective laser mirror RE, the reflectivity of which depends on distance. The dependence of effective reflection from the second laser mirror on the length of external cavity causes changes in threshold of generation and the output light power of the laser. It is clear that the laser optical output includes modulation term dependent on the feedback strength and the distance of the external reflector. It corresponds to a variation of the $\lambda/2$ displacement at the external reflector and is a repetitive function with a period of 2π rad. This model is based on coherence of light inside the external cavity.

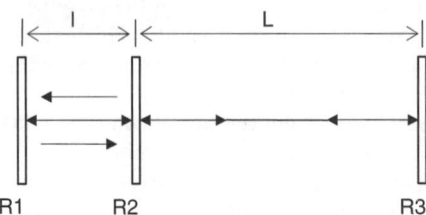

Figure 1. Schematic of a simple laser with external optical feedback: l, laser-cavity length; L, distance from laser-cavity front facet to target; R1, R2, R3, laser mirrors, and target.

According to the three-mirrors model (18,25), the field in the laser cavity can be calculated by applying the amplitude and phase criteria for the stationary state of the light propagating in the laser cavity. In such an approach, it can be seen that the diode laser undergoes threshold gain and lasing frequency variations. The changes in threshold gain, and therefore in the optical output variation ΔP caused by feedback, has been shown to have a nonmonotonic dependence on the length of the external cavity

$$\Delta P \propto \frac{1-R^2}{Rl} R_{ext} \cos \omega \tau_{ext}, \qquad (1)$$

where R is the reflectivity of the laser facets, l is the length of the laser cavity, R_{ext} is the reflectivity of the moving object (skin), ω is the lasing frequency with optical feedback, $\tau_{ext} = 2L/c$ is the round-trip time of the laser light in the external cavity, and L is a distance from the front facet of the laser cavity to the moving object (skin).

The maximum value of optical output takes place when the waves from the laser cavity and the external cavity meet each other in the same phase at the laser front facet. This condition can be satisfied if the effective optical length of the external cavity is the same or multiplied to the effective optical length of the laser cavity. The maximum optical output is realized when the following condition is satisfied:

$$\frac{L}{\eta l} = m, \qquad (2)$$

where η is a refractive index of the cavity, $m = M' - M$ difference of mode index, which means that the maximum output is attained at the external cavity length equal to integer multiples of the effective laser cavity length ηl.

The dependence of the self-mixing signal on distance between the laser and moving object and threshold gain (pump current) was investigated in experiments.

3. EQUIPMENT AND EXPERIMENTS

3.1. Self-Mixing Method

Experimental set-up for measurement of dependence of the self-mixing signal on distance is presented in Fig. 2. The Philips 1550 nm GaInAsP laser diode CQF 58 with monomode fiber with 1-m length was used in our experiments. A special potentiometer P exists for precise adjustments of the laser pump current. A special low-noise two-channel amplifier is used for the signals from photodiode PD and from the resistor in the laser current chain correspondingly. Light consists of nine longitudinal modes, with 1.11 nm between them (Fig. 3). The frequency difference between longitudinal modes of the laser-active cavity was $\Delta f = 1.25 \times 10^{11}$ Hz, the equivalent optical (taking into account refractive index) longitudinal size of the cavity was about 1.2 mm.

Measured dependence of a self-mixing interference on the distance between laser and target is presented in Fig. 4 (first 5 maximums and others). The distance

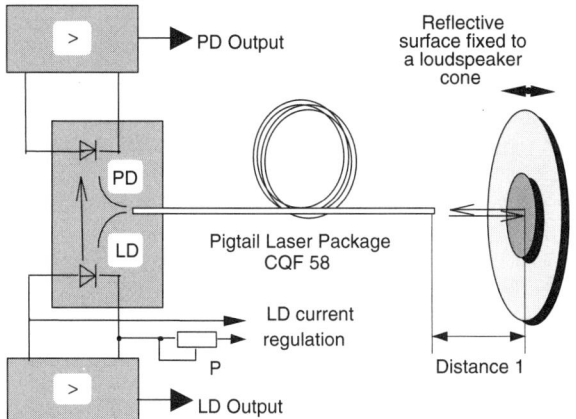
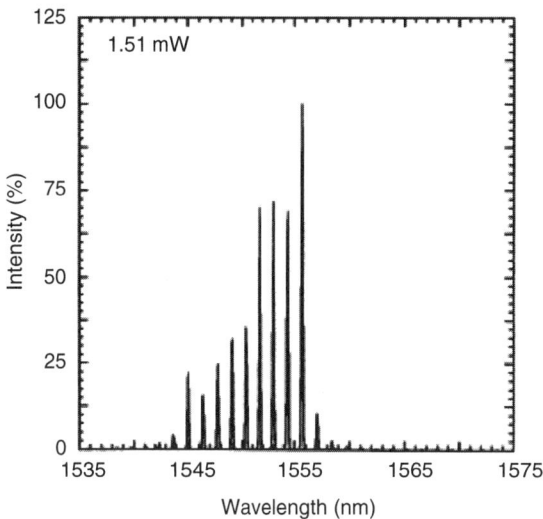

Figure 2. Schematic experimental arrangement with amplifiers for PD and LD output signals and with LD current regulation. Used laser package from Philips CQF 58, which contains an optical fiber 1 m in length, PD - photodiode (integral), and LD - laser diode.

Figure 3. The spectrum of laser diode, CQF 58 passport data.

between minimum and maximum on the curve of self-mixing modulation function was 0.89 mm.

We used a reflective surface attached to a loudspeaker cone driven by a signal generator as a target to provide phase variations of the external optical feedback. The diode laser package incorporates a photodiode accommodated in the rear facet for monitoring the laser power. This characteristic of the device is particularly well-suited to observe the self-mixing interference, and it provides a convenient internal detector. A typical output signal obtained in this case is shown in Fig. 5. The feedback was less than 10% in all experiments. The upper trace in Fig. 5 is the signal applied to achieve the periodic target movement, and the resultant intensity modulation (middle trace) is the self-mixing signal observed. Another structure of the same device has also been observed. For the signal separation and capturing, an additional resistor (R = 50 Ω) exists with intermittent potential in laser-pump current chain. On the same figure (lower trace), the self-mixing signal is presented.

Theoretical value for distances between positions of two maximums on the curve of self-mixing modulation function is about 1.2 mm. The difference between theoretical estimation of distance between maximums and measured real value 0.89 mm (Fig. 4) would be explained by a slightly different length of laser cavity used in theoretical estimation. This length was derived from passport data of light spectrum of the diode laser (Fig. 3).

The dependence of amplitude of maximums of the self-mixing signal on distance can be explained by the character of spatial broadening of a laser light beam, as well as by limited longitudinal coherence. The spatial broadening of the beam from an optical fiber was measured by the method applied in (21). The radiation angle of the fiber was 40 degrees at 90% level, and the intensity dependence on the distance from the fiber r was typical for spherical

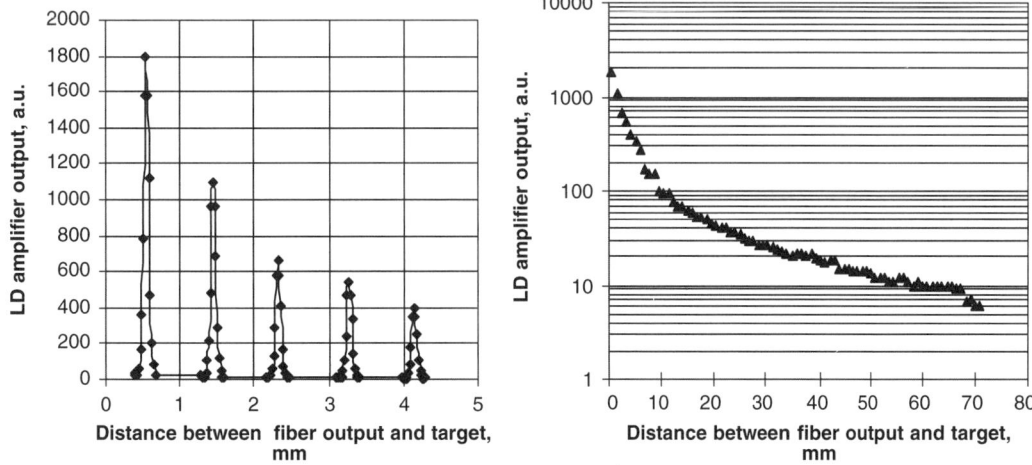

Figure 4. Dependence of a self-mixing interference on the distance: left—first 5 maximums; right—all maximums; horizontal axis—distance between optical fiber output and target; vertical axis—signal amplitude in arbitrary units.

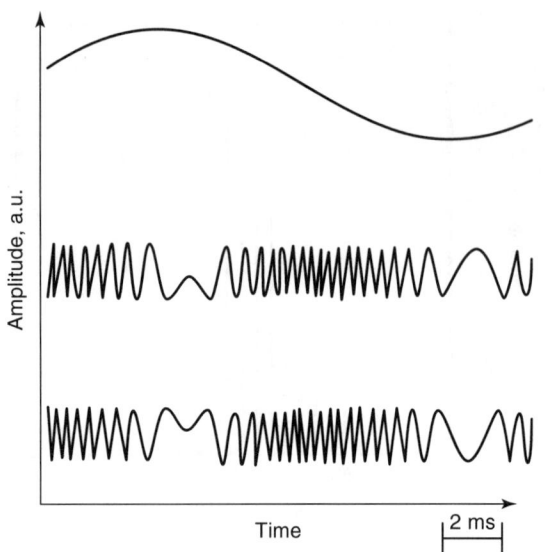

Figure 5. Typical signals observed from self-mixing. Upper trace, signal applied to achieve the periodic target movement; middle and lower traces, self-mixing signals from photodiode and resistor in laser-current chain. Horizontal axis—time; vertical axis—signal amplitude in arbitrary units.

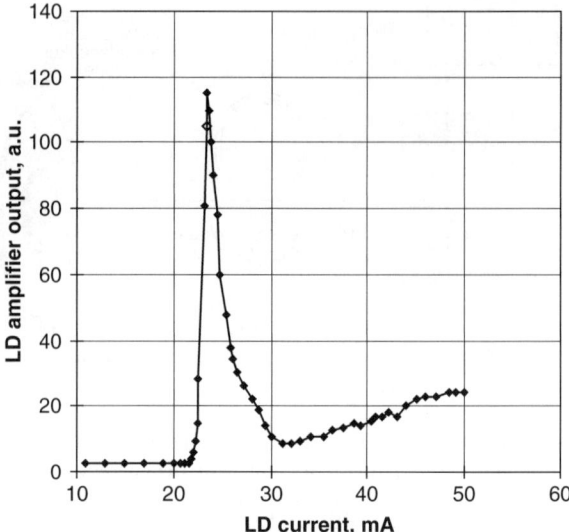

Figure 6. Mixed-signal amplitude dependent from laser current. The laser used was QF 4142 with threshold current $I_{th} = 24$ mA. Horizontal axis—laser current in mA; vertical axis—signal amplitude in arbitrary units.

wave $1/r^2$. The decreasing of the measured maximums (Fig. 4) with increasing of the distance fit well to this dependence.

The longitudinal coherence is determined by the width of the spectrum of radiated light. If the round-trip time of the light in the external cavity $\tau_{ext} = 2L/c$ is within the coherence time, the interference between the backscattered field and the field inside the laser-active cavity is strongest. If the round-trip time is larger than the laser light coherence time, the phase changes between the intracavity and backscattered light causes strong reduction of the interference and, consequently, the self-mixing signal. Introducing the longitudinal coherence length through the width of the spectrum of the radiated light $l_{coh} = c/\Delta f$, the relative reducing of the self-mixing signal is proportional to the ratio (16)

$$\frac{\langle I^2 \rangle}{I_{coh}^2} \propto \frac{1}{[1+(2L/l_{coh})^2]^{1/2}}, \quad (3)$$

where I_{coh} is the level of the signal for $\tau_{ext} \ll l_{coh}$ or $2L/c \ll c/\Delta f$.

In our experiments, the multimode laser was used, and its spectrum consists of the series of narrow lines (Fig. 3). In this case, two different coherence scales exist. One of them is determined by the width of the single longitudinal mode. On the other hand, the whole spectrum includes N modes separated by $\Delta F = c/2l$, where l is length of the laser cavity. The total intensity is obtained as a sum of N uncorrelated spectrum lines, and the longitudinal coherence length $l_{coh} = 2l$ has its minimal value 2.4 mm. The real length of the external cavity L consists of two parts: length of the optical fiber $l_f = 1$ m and distance from the fiber r and $L = l_f + r$, which is much longer than the coherence length. The relatively small changes in distance L caused by variances of r (Fig. 4) cannot affect the condition $L \gg l_{coh}$. Calculated by (3), dependence of the self-mixing signal on distance r from the fiber is not significant. In this case, the measured dependence of maximums amplitude on distance (Fig. 4) was caused by decreasing of the intensity of light with distance caused by the beam broadening only.

The self-mixed signal amplitude is a function of the laser-pump current measured. Pump current was adjusted precisely with a potentiometer, and output signals of both channels were recorded simultaneously. These characteristics are similar, and one of them is in Fig. 6. As we can see, in both cases, the self-mixed signal depends on the laser-pump current as inversely proportional, and the maximum value of the signal corresponds to the pump current near threshold.

3.2. Application of Method for Pulse Wave Delay Time Measurement

Pulse wave delay time was measured as a time interval between the ECG signal and skin vibration signal near different arteries of the human body. We used a four-lead system with standard ECG clinical device Mingograf-4 produced by Siemens-Elema for ECG recordings. Six different measurement points Mp1–Mp6 to measure pulse wave in different arteries were used. Both of the signals were recorded with a sampling rate of 10 kHz; the duration of the recording was 10 seconds. Recorded signals are given in Fig. 7. The first signal (upper curve) is ECG as a reference signal for time measurement, the second curve is a processed laser-Doppler signal (pulse profile), and the third one is a signal from a laser diode as an input signal (without processing).

A recorded laser-Doppler signal was preprocessed using a sliding window, the length of the Hanning-windowed

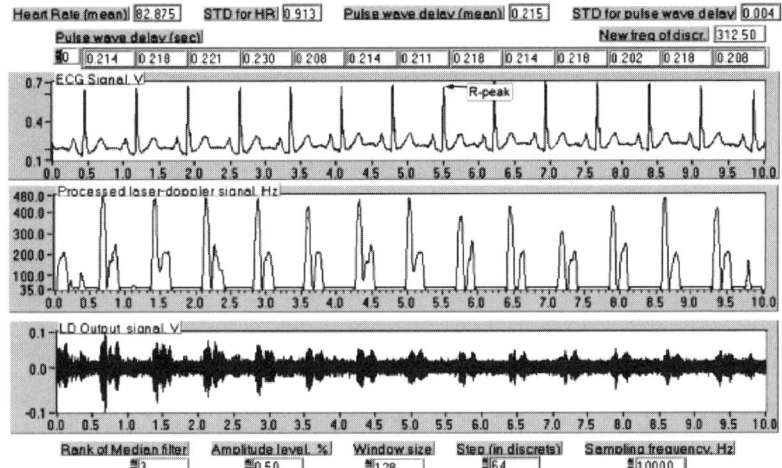

Figure 7. Recorded pulse profile, processed pulse profile, and ECG signals.

signal was 128 points, and the power spectrum was calculated after every 8 points (resolution 0.8 ms) using an FFT algorithm. From every calculated power spectrum, the frequency with maximum amplitude was detected. The maximum of every pulse profile was detected using the adaptive-level peak detector, after which the pulse vibration starting point was determined. Calculated time-dependent frequency (pulse profile) never goes exactly to zero because of the noise, which makes it difficult to determine the exact starting point of the pulse vibration. In this work, the starting point of the pulse wave was fixed to the time point where the tangent of the maximum value of the first derivative of the current pulse profile and zero frequency converge. The frequency of discretization of the ECG was reduced to 335 Hz, because no relevant information exists above 100 Hz. Thereafter, the ventricular electrical activation starting point R-peak in the ECG was located by the adaptive-level peak detector. Pulse delay was calculated as a time difference between the R-peak of the ECG and the following pulse starting point of the laser-Doppler signal.

3.3. Calculation of Pulse Wave Transit Time and its Correlation with Blood Pressure Using Different Physiological Signals

A noninvasive multiparameter registration system was developed for this study. The system enables simultaneous registration of electrocardiogram (ECG), photoplethysmogram (PPG), pulse pressure wave (PW), and bioimpedance signal (BI). Two metal electrodes placed on the left and right hand measure the ECG signal. The same electrodes were used for bioimpedance measurement. The PPG signal was measured with a standard SpO2 optical sensor from the middle finger of the left hand (infrared light channel was used in our case). PW was measured with a piezo-electric transducer located on the index finger of the left hand. Experimental arrangement for the registration system is represented in Fig. 8. This device consists of four different signal-conditioning modules (ECG, PPG, PW, and BI), which form four simultaneous measurement channels to get time synchronization between ECG and the other three physiological signals. Analogue output signals from this device are connected to the laptop computer with data acquisition card and data processing software LabVIEW. The sampling frequency for all data channels is 1 kHz.

In our study, we used a group of volunteers (34 persons) who took the bicycle exercise test. The volunteers represented different patients with different diseases with the mean age of 46 ± 14 years (14 female, 20 male). The test consisted of cycling sessions of increasing workloads during which the HR changed from 60 to 180 beats per minute. In addition, a blood pressure (NIBP) was registered with standard sphygmomanometer once per minute during the test. The NIBP measurement values were synchronized to other signals to exact time moments where the systolic blood pressure was detected (Korotkoff sounds starting point). The computer later interpolated the blood pressure signal in order to get an individual value for every heart cycle. The other signals were measured continuously during all tests. At the end of every session, a recovery period was included until the person's NIBP and HR normalized. All measurements were also done during this period.

Laboratory signal-processing software was used to detect the pulse parameters and to calculate different time delays between ECG and the other three signals, which reflect the pulse wave propagation in a vascular system. From ECG signal, the HR was calculated by measuring R-peaks and calculating RR intervals. The delay was measured as a time interval between R-peak to the corresponding pulse onset. The pulse onset was detected using different moments in pulse wave in order to find the maximum correlation point between the time delay and BP for a person and for a group. The first detection point was located at where the 10% amplitude occurred on the rising edge of the pulse signal (calculated from baseline value). The next point was located in the middle (50%) point of the pulse wave rising edge, and the last point was located at the maximum (100%) of the pulse wave amplitude. The same calculation procedure was applied for three different pulse signals (PPG, PW, BI).

An example of recorded signals is shown in Fig. 9. The first strip represents ECG signal, the second strip

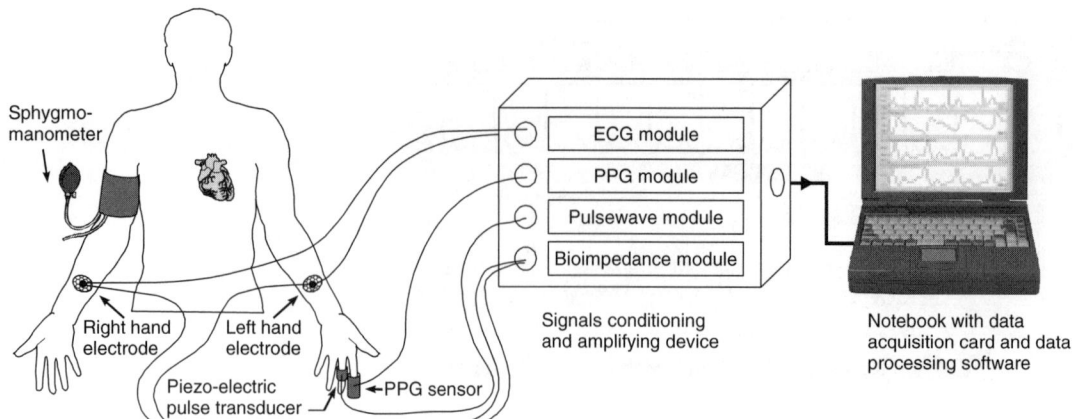

Figure 8. Experimental arrangement.

represents PPG signal, and the third and the fourth strips represent PW and BI signals, respectively. As it can be seen from the Fig. 9, the earliest detection point can be located in the BI signal (BI10) compared with PW and PPG, which means that the bioimpedance signal represents changes in blood volume near to the heart, and the time point BI10 closely matches with the beginning of the ejection phase in the heart cycle; it is also in good agreement with earlier studies (32). The next signal that can be triggered is mechanical pulse pressure wave (PW10), which represents the point when the arterial pulse pressure starts to increase in the major arteries of the index finger. The last wave front that can be detected is PPG (PPG10), which represents blood volume changes in the finger. This kind of measurement setup allow us to estimate the isovolumic contraction time (BI10) and the pulse pressure propagation time (PW10-BI10) and compare the changes in these parameters with other intervals and NIBP. It should be mentioned that the average BI10 values turned out to be higher than expected, which could mean that the BI10 does not represent the true isovolumetric phase but extends also to the beginning ejection phase. Nevertheless, the BI10 is the only parameter in this set where the isovolumic phase has an essential part.

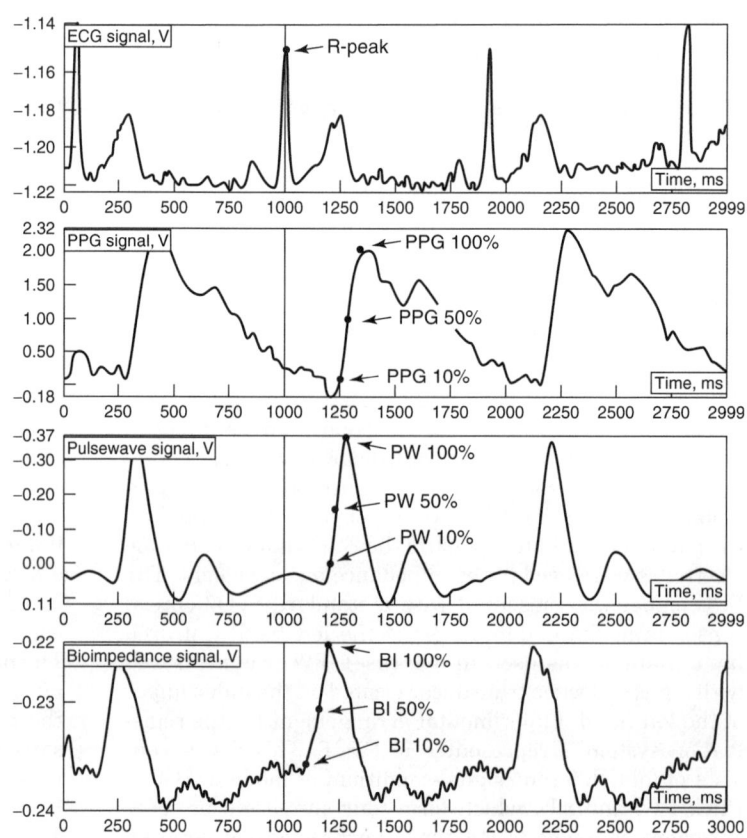

Figure 9. An example of a recording fragment.

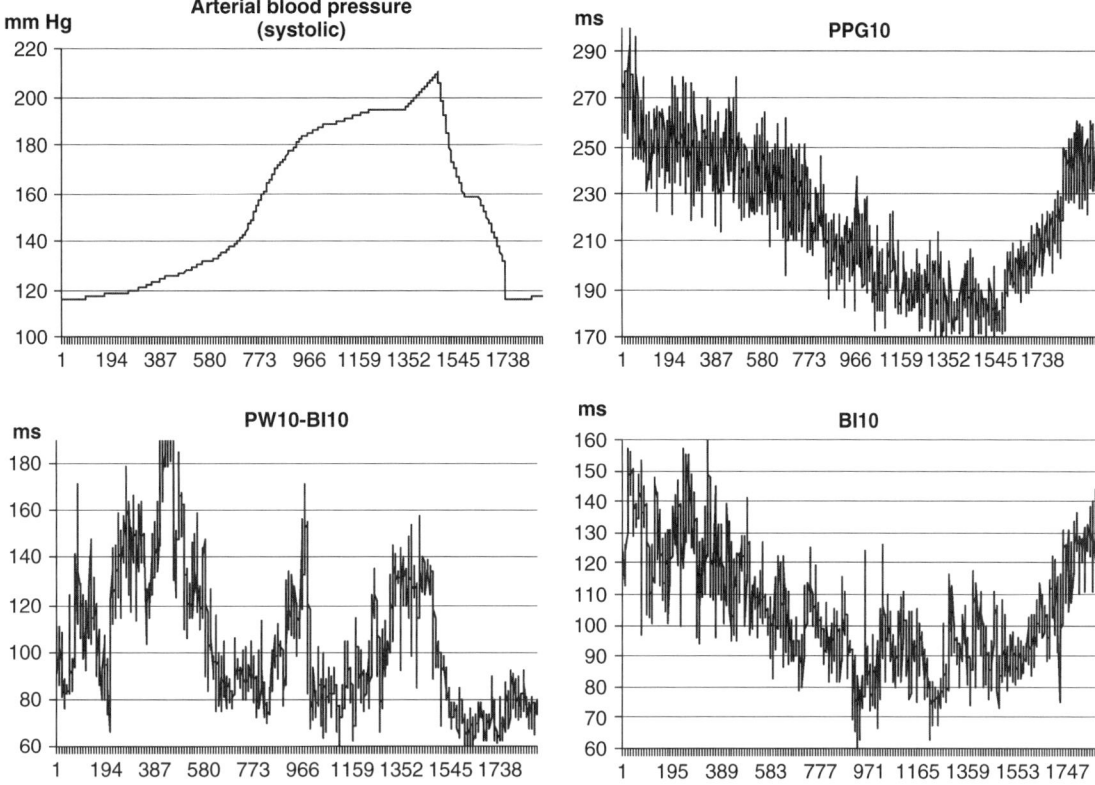

Figure 10. An example of calculated parameters during exercise. (This figure is available in full color at http://www.mrw.interscience.wiley.com/ebe.)

In Fig. 10, we can see the changes of some of the calculated parameters during the exercise for one person from the group. It can be noticed that, whereas PW10 and BI10 are inversely proportional to the systolic arterial blood pressure, the PW10-BI10 curve does not correlate very well with other lines. The same result is valid for the group (Table 1). Table 1 is the average correlation matrix for the study group for all calculated parameters. The Pearson's correlation coefficient is used for the matrix calculation. From Table 1, we can see that the best correlation with systolic blood pressure has the PPG10 parameter. It is the only parameter, the correlation of which is more than − 0.7. The dynamic range of the PPG10 parameter and systolic pressure is approximately the same (1 mmHg = 1 ms). The PPG50 and PPG100 have slightly lower correlation coefficients. The BI and PW parameters have much lower correlation with systolic blood pressure compared with PPG. From the table, it can be seen that diastolic pressure does not correlate with any of the parameters, which leads us to a conclusion that diastolic pressure cannot be estimated by the PWTT measurement. In Table 1, a parameter PPG100-PPG10 also exists, which represents the rising time of the PPG pulse wave; it can be seen that it does correlate with blood pressure changes (r = − 0.25), which means that the change in the rising time of pulse signal cannot be a very reliable parameter for arterial pressure estimation.

Table 1. Average Correlation Coefficients of Calculated Parameters for the Group

	PPG100	PPG10	PPG50	PW100	PW10	PW50	BI100	BI10	BI50	PW10-BI10	PPG100-PPG10	Systoli
PPG100												
PPG10	0.681											
PPG50	0.755	0.849										
PW100	0.480	0.520	0.532									
PW10	0.420	0.481	0.448	0.655								
PW50	0.490	0.514	0.515	0.784	0.761							
BI100	0.320	0.461	0.427	0.446	0.445	0.446						
BI10	0.375	0.448	0.416	0.351	0.309	0.339	0.328					
BI50	0.481	0.587	0.585	0.452	0.356	0.458	0.516	0.491				
PW10-BI10	− 0.077	− 0.198	− 0.168	− 0.037	0.144	0.049	0.009	− 0.030	− 0.224			
PPG100-PPG10	0.484	0.251	0.324	0.198	0.179	0.213	0.122	0.199	0.223	0.105		
Systoli	− 0.620	− 0.700	− 0.674	− 0.500	− 0.381	− 0.474	− 0.374	− 0.392	− 0.578	0.358	− 0.246	
Diastoli	0.212	0.163	0.198	0.120	0.102	0.141	0.140	0.175	0.116	0.203	0.168	− 0.046

4. EVALUATION

The laser diode used is based on double-heterostructure technology and has quite a high threshold current, approximately 80 mA. When a handheld monitoring device is considered, this high of a current consumes too much power for present-day battery technology, which means that a type of laser diode with a much lower threshold current should be used.

Semiconductor laser types, which have very low-operating currents, are based on quantum well (QW) technology. In this technique, the thickness of the active region is reduced. Because of the reduced area of the active region, a lower current is needed for stimulated emission and, thus, laser operation. The threshold current for QW lasers is typically less than 10 mA.

The problem in cardiovascular pulse measurements are the external artifacts that cause disturbances to the Doppler signal. These artifacts are easily generated because of the high sensitivity of the measuring device. Unfortunately, the errors caused by the artifacts are typically on the same frequency band as the information signal from the displacement of the arterial wall.

One possible technique to reduce the effect of the artifacts on the Doppler signal is to use two self-mixing devices to measure the pulsation. The idea is to insert another self-mixing device in the measurement probe. The first one is positioned above the radial artery, and it measures both the displacement of the arterial wall and possible errors caused by external artifacts. The second one is positioned in such a way that it measures only the error signal. If both channels are identical, it could be possible to manipulate the signals so that effect of artifacts could be minimized.

BIBLIOGRAPHY

1. P. A. Roos, M. Stephens, and C. E. Wieman, Laser vibrometer based on optical-feedback-induced frequency modulation of a single-mode laser diode. *Appl. Opt.* 1996; **34**:6754–6761.
2. V. P. Ryabukho and S. S. Ul'yanov, Spectral characteristics of dynamic speckle-fields interference signal for surfaces motion measurements. *Measurement* 1992; **10**(1):39–42.
3. X. C. Lega and P. Jacquot, Deformation measurement with object-induced dynamic phase shifting. *Appl. Opt.* 1996; **25**:5115–5121.
4. C. B. Carlisle, R. E. Warren, and H. Riris, Single-beam diode-laser technique for optical path-length measurements. *Appl. Opt.* 1996; **22**:4349–4354.
5. B. Deveraj, M. Takeda, M. Kobayashi, M. Usa, and K. P. Chan, In vivo laser computed tomographic imaging of human fingers by coherent detection imaging method using different wavelengths in near infrared region. *Appl. Phys. Lett.* 1996; **69**:3671–3673.
6. G. F. Schmid, B. L. Petrig, C. E. Riva, K. H. Shin, R. A. Stone, M. J. Mendel, and A. M. Laties, Measurement by laser Doppler interferometry of intraocular distances in humans and chicks with a precision of better than ±20 mm. *Appl. Opt.* 1996; **19**:3358–3361.
7. H. Mignon, G. Grehen, G. Gouesbet, T. H. Xu, and C. Tropea, Measurement of cylindrical particles with phase Doppler anemometry. *Appl. Opt.* 1996; **25**:5180–5190.
8. N. Schunk and K. Petermann, Numerical analysis of the feedback regimes for a single-mode semiconductor laser with external feedback. *IEEE J. Quant. Electron.* 1988; **24**(7): 1242–1247.
9. S. Schuster, T. Wicht, and H. Haug, Theory of dynamical relaxation oscillations and frequency locking in a synchronously-pumped laser diode. *IEEE J. Quant. Electron.* 1991; **27**(2):205–211.
10. R. F. Kazarinov and R. A. Suris, Heterodyne reception with injection laser. *JETF* 1974; **66**(3):1067–1078 (in Russian).
11. R. Lang and K. Kobayashi, External optical feedback effects on semiconductor injection laser properties. *IEEE J. Quant. Electron.* 1980; **16**:347–355.
12. B. Zakharov, K. Meigas, and H. Hinrikus, Coherent photodetection with the aid of a gas laser. *Sov. J. Quant. Electron.* 1990; **20**:189–193.
13. W. M. Wang, W. J. O. Boyle, K. T. V. Grattan, and A. W. Palmer, Self-mixing interference in a diode laser: experimental observations and theoretical analysis. *Appl. Opt.* 1993; **9**:1551–1558.
14. S. S. Ul'yanov, V. P. Rayabukho, and V. V. Tuchin, Speckle interferometry for biotissue vibration measurement. *Optic. Engineer.* 1994; **33**(3):908–914.
15. A. V. Skripal and D. A. Usanov, Semiconductor laser interferometry of eye movements. *Proc. SPIE* 2000; **3908**:7–12.
16. F. F. M. de Mul, L. Scalise, A. L. Petoukhova, M. Herwijnen, P. Moes, and W. Steenbergen, Glass-fiber self-mixing intra-arterial laser Doppler velocimetry: signal stability and feedback analysis. *Appl. Opt.* 2002; **41**(4):658–667.
17. M. H. Koelink, M. Slot, F. F. M. de Mul, J. Greve, R. Graaf, A. C. M. Dassel, and J. G. Aarnoudse, Laser Doppler velocimeter based on the self-mixing effect in a fiber coupled semiconductor laser: theory. *Appl. Opt.* 1992; **18**:3401–3408.
18. S. K. Özdemir, S. Shinohara, S. Takamiya, and H. Yoshida, Noninvasive blood flow measurement using speckle signals form a self-mixing laser diode: in vitro and in vivo experiments. *Opt. Eng.* 2000; **39**(9):2574–2580.
19. H. W. Jentink, F. F. M. de Mul, H. E. Suichies, J. G. Aarnoudse, and J. Greve, Small laser Doppler velocimeter based on the self-mixing effect in a diode laser. *Appl. Opt.* 1988; **2**:379–385.
20. L. Scalise, W. Steenbergen, and F. de Mul, Self-mixing feedback in a laser diode for intra-arterial optical blood velocimetry. *Appl. Opt.* 2001; **40**(25):4608–4615.
21. K. Meigas, H. Hinrikus, R. Kattai, and J. Lass, Coherent photodetection for pulse profile registration. *Proc. SPIE* 1999; **3598**:195–202.
22. K. Meigas, Method for small particle detection by laser. *Optic. Engineer.* 1998; **9**:2587–2591.
23. K. Meigas, H. Hinrikus, R. Kattai, and J. Lass, Simple coherence method for blood flow detection. *Proc. SPIE* 2000; **3915**:112–120.
24. H. Kakiuchida and J. Ohtsubo, Characteristics of a semiconductor laser with external feedback. *IEEE J. Quant. Electron.* 1994; **9**:2087–2097.
25. K. Petermann, *Laser Diode Modulation and Noise*. Dordrecht, The Netherlands: Kluwer Academic Press, 1988.
26. L. Goldberg, H. F. Taylor, A. Dandridge, J. F. Weller, and R. O. Miles, Spectral characteristics of semiconductor lasers with optical feedback. *IEEE J. Quant. Electron.* 1982; **18**(4):555–564.
27. A. Olsson and C. L. Tang, Coherent optical interference effects in external-cavity semiconductor lasers. *IEEE J. Quant. Electron.* 1981; **17**(8):1320–1323.

28. B. Tromborg, J. H. Osmundsen, and H. Olesen, Stability analysis for a semiconductor lasers in an external cavity. *IEEE J. Quant. Electron.* 1984; **20**(9):1023–1032.
29. B. Tromborg, J. Mork, and V. Velichansky, On mode coupling and low-frequency fluctuations in external-cavity laser diodes. *Quant. Semiclass. Opt.* 1997; **9**(5):831–851.
30. N. Shunc and K. Petermann, Numerical analysis of the feedback regimes for a single-mode semiconductor lasers with external feedback. *IEEE J. Quant. Electron.* 1988; **24**:1242–1247.
31. N. Yanong, K. T. V. Grattan, B. T. Meggitt, and A. W. Palmer, Characteristics of laser diodes for interferometric use. *Appl. Opt.* 1989; **28**(17):3657–3662.
32. J. Malmivuo and R. Plonsey, *Bioelectromagnetism*. New York: Oxford University Press, 1995, pp. 405–420.

READING LIST

I. Tepner, J. Lass, D. Karai, H. Hinrikus, and K. Meigas, Monitoring blood pressure from pulse wave transit time. Proc. of the 4th BSI International Workshop, June 24–26, Como, 2002; 239–242.

K. Meigas, H. Hinrikus, R. Kattai, and J. Lass, Self-mixing in a diode laser as a method for cardiovascular diagnostics. *J. Biomed. Optics* 2003; **8**(1):152–160.

BLIND SOURCE SEPARATION

ALEXANDER M. BRONSTEIN
MICHAEL M. BRONSTEIN
MICHAEL ZIBULEVSKY
Technion–Israel Institute of Technology
Haifa, Israel

1. INTRODUCTION

The term *blind source separation* (BSS) refers to a wide class of problems in signal and image processing in which one needs to extract the underlying sources from a set of mixtures. Almost no prior knowledge about the sources or about the mixing is known, hence the name *blind*. In practice, the sources can be 1D (e.g., acoustic signals), 2D (images), or 3D (volumetric data). The mixing can be *linear*, or *nonlinear* on one hand, and *instantaneous* or *convolutive*; in the latter case, the problem is referred to as *multichannel blind deconvolution* or *convolutive BSS*. In many medical applications, the instantaneous linear mixing model holds, hence, the most common situation is when the mixtures are formed by superposition of sources with different scaling coefficients. These coefficients are usually referred to as *mixing* or *cross-talk coefficients* and can be arranged into a *mixing (cross-talk) matrix*. The number of mixtures can be smaller, larger, or equal to the number of sources.

Independent component analysis (ICA), a method for finding independent components in multidimensional statistical data, is usually exploited to solve the BSS problem.

One of the fundamental assumption of this approach is that the sources are *statistically independent* (i.e., the value of any one of the sources gives no information on the values of the other sources). Hence, finding the minimally dependent components by minimizing a certain measure of dependence (usually employing a *numerical optimization* procedure) produces an estimate of the original sources. An emerging class of alternative methods is the *sparse component analysis* (SCA or SPICA), which allows one to relax the assumption of statistical independence of the sources and relies on the assumption that sources are *sparse* (i.e., have a small number of nonzero values) or *sparsely representable* in an appropriate domain (e.g., in the domain of the Fourier transform, wavelet transform, Gabor transform). Another important class of approaches to the BSS problem is based on *independent factor analysis* (IFA).

In the context of medical signal and image processing, the BSS problem develops, for example, in analysis of electroencephalogram (EEG), magenetoencephalogram (MEG), and electrocardiogram (ECG) signals and functional magnetic resonance images (fMRI). Recent research has also shown the feasibility of BSS techniques in hyperspectral analysis of tissues.

2. THE LINEAR BSS PROBLEM

In a typical scenario of an instantaneous linear BSS problem, real signals $s_i(t)$ from N sources are recorded by M sensors; t is a multi-index and refers to data of any dimension. The sensor signals $x_i(t)$ are instantaneous linear combinations of the source signals, and are possibly contaminated by additive *sensor noise* $\xi_i(t)$:

$$\begin{aligned} x_1(t) &= a_{11}s_1(t) + \cdots + a_{1N}s_N(t) + \xi_1(t) \\ &\vdots \\ x_M(t) &= a_{M1}s_1(t) + \cdots + a_{MN}s_N(t) + \xi_M(t), \end{aligned}$$

where a_{ij} are the cross-talk coefficients. In EEG, for example, source signals are resulting from electromagnetic activity of the brain cortex, and sensor signals are the electric potentials measured by the scalp electrodes. The cross-talk coefficients represent the attenuation the source signals undergo during their propagation and are related to the geometry of the head. The noise can result, for example, from electromagnetic disturbances.

In matrix notation, the linear BSS problem has the form

$$\boldsymbol{x}(t) = A\boldsymbol{s}(t) + \xi(t),$$

where $\boldsymbol{s}(t)$, $\boldsymbol{x}(t)$, and $\xi(t)$ are column vectors with values $s_i(t), x_i(t)$, and $\xi_i(t)$, respectively. When the data is discrete, the problem can be written as

$$X = AS + \Xi, \qquad (1)$$

where X and S are matrices containing the mixtures and the sources, respectively, as their rows, and Ξ is the corresponding matrix of noise. If the mixtures are

2D (images), they are parsed into vectors and treated in an essentially similar way.

The BSS problem consists of finding an estimate $Y(X)$ of S, given only the observed (possibly noisy) data X. If no *a priori* information is available, the latter is possible up to an arbitrary permutation and scaling only. In a particular case when the matrix A is square (i.e., $N=M$) and invertible, and under the assumption of zero noise, the problem is equivalent to estimating the *unmixing matrix* $W = (\mathbf{w}_1, \ldots, \mathbf{w}_N)^T = EA^{-1}$ such that

$$EY = WX,$$

where E is a permutation and scaling matrix.

3. INDEPENDENT COMPONENT ANALYSIS (ICA)

If the sources s_i are statistically independent, the ICA approach can be used to solve the BSS problem. Statistical independence implies that the joint probability density of the sources $f_s(s_1, \ldots, s_N)$ can be factorized into a product of marginal densities,

$$f_s(s_1, \ldots, s_N) = \prod_{i=1}^{N} f_{s_i}(s_i).$$

The assumption of statistical independence of sources is the guiding principle of ICA: The sources can be estimated as a linear transformation $Y = WX$ of the mixtures, yielding the "most independent" components. The sources estimated in this way are termed *independent components* (ICs).

The property of independence is stronger than *uncorrelatedness* (i.e., $\mathbb{E}s_i s_j = \mathbb{E}s_i \mathbb{E}s_j$ for $i \neq j$, where $\mathbb{E}$ denotes expectation). Uncorrelated components can be extracted by means of linear *decorrelation* (sometimes termed *sphering* or *whitening*) of the data $\mathbf{x}(t)$ (i.e., by linearly transforming $\mathbf{x}(t)$ into $\mathbf{y}(t) = P\mathbf{x}(t)$ by a whitening matrix P, such that the *covariance matrix* of $\mathbf{y}(t)$ equals units):

$$C_{yy} = \mathbb{E}(\mathbf{y}(t)\mathbf{y}^T(t)) = PC_{xx}P^T = I.$$

The covariance C_{xx} is estimated from a finite realization of $\mathbf{x}(t)$. Substituting the eigendecomposition $C_{xx} = VDV^T$ (where V is a unitary diagonalizing matrix and D is diagonal), the covariance C_{yy} can be expressed as

$$C_{yy} = PVC_{xx}V^T P^T.$$

The whitening matrix therefore equals $P = D^{-\frac{1}{2}}V^T$. This procedure is a version of *principal component analysis* (PCA) (also called *Hotteling-* or *Karhunen–Loeve transform*), in which the variances are normalized.

Geometrically, decorrelation of the data undoes the mixing up to a rotation matrix (i.e., the unmixing matrix can be found as a product $W = UP$ of the whitening matrix P and some rotation matrix U). When the sources are jointly Gaussian, independence and uncorrelatedness are equivalent. As such distribution is invariant under rotation, in case of Gaussian sources unmixing is only possibly up to arbitrary rotation (see Fig. 1). For this reason, the Gaussian case is usually excluded in the classic ICA model.

More precisely, the classic setting of the (zero-noise) ICA problem is identical to Equation 1 with the addition of the following *identifiability conditions:* (1) at most one of the sources s_i is Gaussian; (2) the number of mixtures is at least as large as the number of sources, ($M \geq N$); (3) the mixing matrix A is of full column rank; and (4) $s_i(t)$ are stationary and ergodic. Typically, a stronger version of these conditions, assuming non-Gaussianity of all the sources and independent identical distribution (i.i.d.) of the values of $s_i(t)$ over t is used.

ICA is usually performed by formulating a criterion of statistical dependence $\varphi(\mathbf{y})$, referred to as *contrast function* or *constrast* (the terms *objective*, *cost*, and *loss function* are synonymous) and minimizing it—sometimes referred to as *minimum contrast estimation*. The selection of specific contrast and numerical algorithm for its optimization gives rise to different ICA methods.

Typically, decorrelation of the data using PCA is performed as a preprocessing stage. Some methods using *orthogonal contrasts* demand explicitly uncorrelatedness of the data. If the number of sources N equals the number of mixtures M, it is usually convenient to pose the problem as estimating the unmixing matrix W:

$$W = \underset{W}{\operatorname{argmin}}\ \varphi(W\mathbf{x}).$$

If $M > N$, the first N principal components of X are used as the data.

Depending on whether all the ICs are estimated simultaneously or one-by-one, *multi-unit* or *one-unit* contrast functions are used, respectively. Another distinction is made between *online* (or *adaptive*) methods that use instantaneous data at each iteration and *batch* methods that use all the data simultaneously; this terminology is common in the neural networks community, where some fundamental works on ICA came from.

4. CONTRAST FUNCTIONS

Minimum contrast estimation is a general statistical estimation approach commonly used in ICA. Different multi-unit contrast functions can be derived from different principles such as maximum likelihood and information maximization. It can be shown that in many cases these contrasts are equivalent.

4.1. Mutual Information

Independence can be measured as the *Kullback–Leibler divergence* between the true density $f_\mathbf{y}(\mathbf{y})$ of $\mathbf{y}$ and the product of marginal densities $f_0(\mathbf{y}) = \prod_{i=1}^{N} f_i(y_i)$

$$\varphi_{MI}(\mathbf{y}) = \int f_y(\mathbf{y}) \log\left(\frac{f_y(\mathbf{y})}{f_0(\mathbf{y})}\right) d\mathbf{y},$$

also known as the *mutual information* of $\mathbf{y}$. Mutual information reflects the quantity of information shared between the elements of $\mathbf{y}$; it is always non-negative and vanishes only if y_i are statistically independent. Mutual

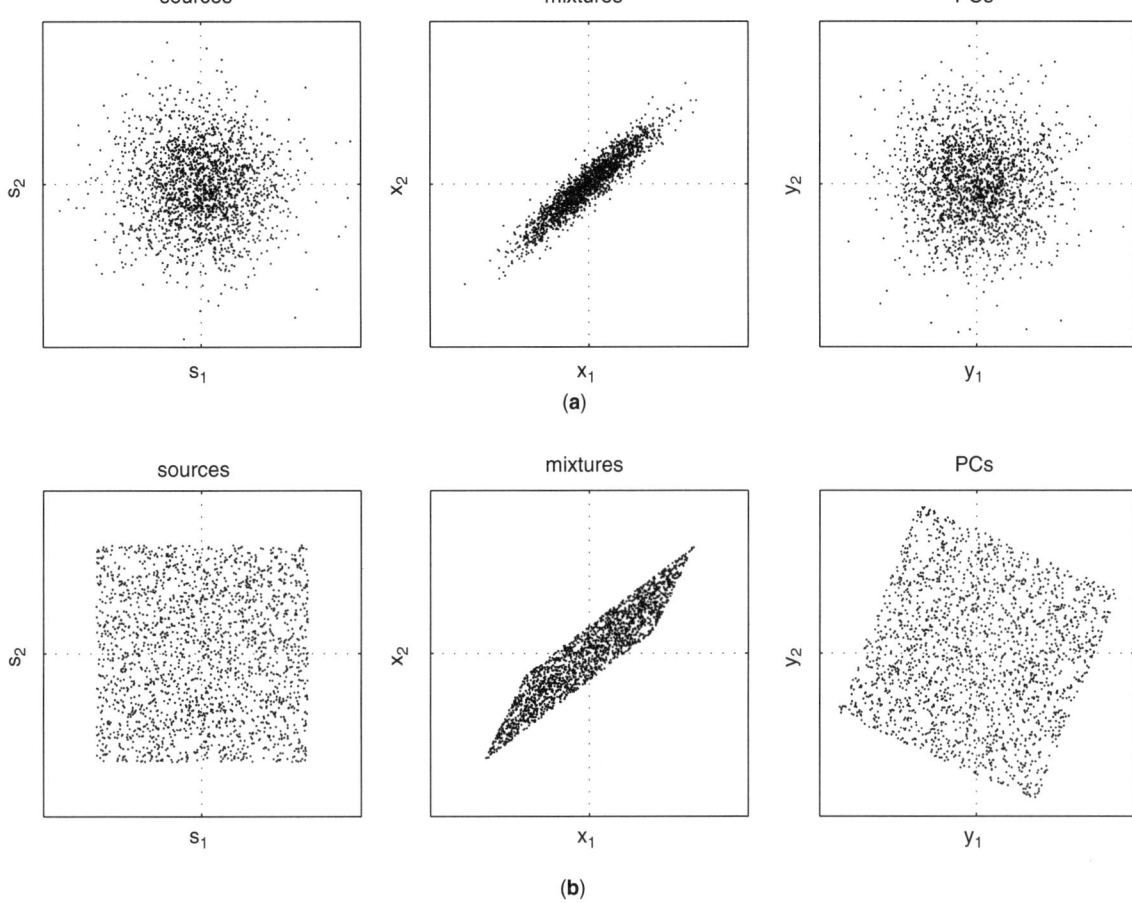

Figure 1. PCA applied to mixtures of two different kinds of sources: Gaussian (A) and uniform (B).

information can be also expressed via the *differential entropy* as

$$\varphi_{MI}(\mathbf{y}) = \sum_{i=1}^{N} H(y_i) - H(\mathbf{y})$$

$$= \sum_{i=1}^{N} H(w_i^T \mathbf{x}) - H(\mathbf{x}) - \log|\det W|,$$

with w_i denoting the ith row of W and

$$H(\mathbf{x}) = -\int f_\mathbf{x}(\mathbf{x}) \log f_\mathbf{x}(\mathbf{x}) d\mathbf{x}. \quad (2)$$

If the data is decorrelated, φ_{MI} becomes the orthogonal mutual information contrast:

$$\varphi_{MI}^{\perp}(\mathbf{y}) = \sum_{i=1}^{N} H(y_i). \quad (3)$$

4.2. Likelihood

Assuming that $s_i(t)$, $t = 1, \ldots, T$ are i.i.d., the samples of the mixtures $x_i(t)$ can be considered as independent measurements of the data. The distribution of $\mathbf{s}(t)$ in this case is given by

$$f(\mathbf{x}|W) = |\det W| \cdot f_\mathrm{s}(W\mathbf{x}) = |\det W| \cdot \prod_{i=1}^{N} f_{s_i}(\mathbf{w}_i^T x_i), \quad (4)$$

where W are the parameters of the model that need to be estimated. The *likelihood* is the probability of the observed data as a function of the parameters of the model:

$$L(X;W) = |\det W| \cdot \prod_{t=1}^{T} \prod_{i=1}^{N} f_{s_i}(\mathbf{w}_i^T x_i(t)).$$

Maximizing the likelihood function $L(X,W)$ with respect to W produces an estimate of the unmixing matrix. Usually, it is more convenient to minimize the normalized *minus log-likelihood function*

$$\ell(X;W) = -\frac{1}{T} \sum_{t=1}^{T} \sum_{i=1}^{N} \log f_{s_i}(\mathbf{w}_i^T x_i(t)) - \log|\det W|.$$

The normalized minus-log-likelihood $\ell(X;W)$ can be seen as an approximation (up to a constant) of the Kullback–Leibler divergence

$$d_{KL}(f_{W\mathrm{x}}, f_\mathrm{s}) = \int f_{W\mathrm{x}}(\mathbf{y}) \log\left(\frac{f_{W\mathrm{x}}(\mathbf{y})}{f_\mathrm{s}(\mathbf{y})}\right) d\mathbf{y},$$

on a finite sample of size T, for example,

$$\ell(X;W) \overset{T\to\infty}{\to} d_{KL}(f_{Wx}, f_s) + \text{const.}$$

Hence, the *maximum likelihood* (ML) estimation is associated with the contrast function

$$\varphi_{ML}(\mathbf{y}) = d_{KL}(f_{Wx}, f_s). \tag{5}$$

The probability densities f_{s_i} do not need to be necessarily known exactly; when approximate f_{s_i} are used, the estimation is termed as *quasi-maximum likelihood* (QML) (1). Often, if the sources are known to be sub-Gaussian or super-Gaussian, approximate f_{s_i} can be selected accordingly.

4.3. Infomax

Estimated ICs can be considered as outputs of a single-layer multiinput multioutput neural network, with $x(t)$ serving as inputs and outputs given by $y_i(t) = g_i(\mathbf{w}_i^T \mathbf{x}(t))$, where g_i are nonlinear scalar functions. When g_i are selected as the cumulative distribution functions F_{s_i} of the sources (or in other words, $g_i' = f_{s_i}$), maximizing the entropy of the outputs

$$\varphi_{IM}(\mathbf{y}) = H(g_i(\mathbf{w}_1^T \mathbf{x}(t)), \ldots, g_N(\mathbf{w}_N^T \mathbf{x}(t))), \tag{6}$$

is equivalent to maximum likelihood estimation. This approach is known as *network entropy* or *information maximization* (*infomax*) and was introduced by T. Bell and L. Sejnowski (2). The intuition of infomax is the following: $\mathbf{g}(\mathbf{s}) = (g_1(\mathbf{s}), \ldots, g_N(\mathbf{s}))$ is uniformly distributed on $[0,1]^N$ and possesses the largest entropy among all the distributions on $[0,1]^N$. Therefore, $\mathbf{g}(\mathbf{y}) = \mathbf{g}(W A \mathbf{s})$ has the largest entropy when $WA = I$.

4.4. Negentropy

Another non-Gaussianity maximization approach stems from a fundamental result of information theory, according to which a Gaussian variable possesses the largest entropy among all random variables with equal variance. A normalized version of entropy $J(\mathbf{y}) = H_0 - H(\mathbf{y})$, where H_0 is the entropy of a Gaussian vector possessing the same covariance as $\mathbf{y}$, is called *negentropy*. Negentropy is nonnegative and vanishes only if $\mathbf{y}$ has a Gaussian distribution. Minimization of the contrast function

$$\varphi_{NE}(\mathbf{y}) = -\sum_i J(y_i) \tag{7}$$

leads to ICs with maximum non-Gaussianity. Minimization of $\varphi_{NE}(\mathbf{y})$ can be approximated by maximization the sum of squared kurtoses $\sum_i k_i^2$. A. Hyvärinen et al. proposed to use the more general approximation of the form

$$J(\mathbf{y}) \approx (\mathbb{E}h(\mathbf{y}) - \mathbb{E}h(\mathbf{v}))^2, \tag{8}$$

where $\mathbf{v}$ is a normal random variable and h can be virtually any nonquadratic function.

4.5. High-Order Cumulants

Another way to derive contrast functions is by using high-order statistics. Such contrasts can be considered approximations of the information theoretic contrasts. High-order statistical information is expressed via *cumulant tensors*; for zero-mean variables x_i, x_j, x_k, x_l, the second- and the fourth-order cumulants are:

$$C_{ij}(\mathbf{x}) = \mathbb{E}(x_i x_j),$$

$$C_{ijkl}(\mathbf{x}) = \mathbb{E}(x_i x_j x_k x_l) - \mathbb{E}(x_i x_l)\mathbb{E}(x_k x_l)$$
$$- \mathbb{E}(x_i x_k)\mathbb{E}(x_j x_l) - \mathbb{E}(x_i x_l)\mathbb{E}(x_j x_k).$$

The auto-cumulants $C_{ii}(\mathbf{x}) = \mathbb{E}x_i^2 = \sigma_i^2(\mathbf{x})$ and $C_{iiii}(\mathbf{x}) = \mathbb{E}x_i^4 - 3\mathbb{E}^2 x_i^2 = k_i(\mathbf{x})$ are the *variance* and the *kurtosis* of the ith source, respectively. For statistically independent sources, the cross-cumulants vanish (i.e., $C_{ij}(\mathbf{s}) = \sigma_i^2 \delta_{ij}$, $C_{ijkl}(\mathbf{s}) = k_i \delta_{ijkl}$, where δ denotes the Krönecker symbol).

If the data is decorrelated, the mutual information contrast φ_{MI} can be approximated as:

$$\varphi_{KUR}^{\perp}(\mathbf{y}) = \sum_{ijkl \neq iiii} C_{ijkl}^2(\mathbf{y})$$
$$= -\sum_i C_{iiii}^2(\mathbf{y}) + \text{const.} \tag{9}$$

This orthogonal contrast was used already in the early works on ICA (3). The intuition behind this contrast is based on the *non-Gaussianity maximization* principle. According to the central limit theorem, the distribution of a sum of non-Gaussian sources is closer to Gaussian distribution than the distributions of the sources. Gaussian distribution has zero kurtosis, distributions with positive kurtosis are termed as *super-Gaussian* and distributions with negative kurtosis are termed as *sub-Gaussian*. The first class includes, for examples, sparse distributions. Non-Gaussianity of y_i can be measured as the absolute or the squared value of the kurtosis $K_i(\mathbf{y})$. Hence, the minimizer of $\varphi_{KUR}^{\perp}(\mathbf{y})$ corresponds to the most non-Gaussian components.

4.6. Tensorial Methods

Tensorial methods rely on eigendecomposition of the fourth-order cumulant. A fundamental result from statistics states that every matrix $\mathbf{w}_i \mathbf{w}_i^T$, where $\mathbf{w}_i^T$ is the ith row of the mixing matrix W, is an eigenmatrix of the cumulant tensor of the whitened data $\mathbf{y}$, with the corresponding eigenvalue k_i, for example,

$$C(\mathbf{w}\mathbf{w}^T) = k_i \cdot \mathbf{w}\mathbf{w}^T,$$

where

$$C(\mathbf{w}\mathbf{w}^T)_{ij} = \sum_{k,l} w_k w_l C_{ijkl}$$

is the product of the tensor C with the matrix $\mathbf{w}\mathbf{w}^T$. This method gives a direct method for estimating the mixing matrix: the observation $\mathbf{y}$ is whitened, then the fourth-order cumulant tensor C is estimated and its eigendecomposition is computed. Estimate of the mixing matrix is constructed from the eigenmatrices of C.

4.7. Joint Approximate Diagonalization of Eigenmatrices (JADE)

JADE is a particular case of tensorial methods introduced by J-F. Cardoso and A. Souloumiac (4). The tensor C has N^2 orthonormal eigenmatrices D_m and the corresponding eigenvalues λ_m, of which only N are nonzero. If the data is decorrelated, the mixing matrix is unitary (a rotation matrix), and diagonalizes the eigenmatrices of C, for example, $B_m = WD_mW^T$, with D_m a diagonal matrix. Thus, the unmixing matrix can be found by joint diagonalization of the N eigenmatrices B_m, which can be carried but efficiently using Jacobi iterations. The orthogonal contrast associated with the JADE algorithm is the joint diagonalization criterion:

$$\varphi_{JADE}^{\perp}(\mathbf{y}) = \sum_{ijkl \neq ijkk} \mathsf{C}_{ijkl}^2(\mathbf{y}). \qquad (10)$$

5. NUMERICAL ICA ALGORITHMS

Independent component analysis is often carried out by means of minimization of the contrast function φ with respect to the argument W. This task is performed by an iterative numerical procedure, called *optimization algorithm*. A generic optimization algorithm produces a sequence $W^{(0)}, W^{(1)}, \ldots, W^{(K)}$ of estimates of the function minimizer, on which the function approaches the optimal value. The iterations are stopped when, for example, the norm of the gradient $\|\nabla_\varphi(W^{(k)}\mathbf{y})\|$ is sufficiently small. Classic optimization algorithms produce $W^{(k+1)}$ by additive update, for example, by making a step in direction $V^{(k)}$ on every iteration: $W^{(k+1)} = W^{(k)} + \alpha^{(k)}V^{(k)}$. Typically, the choice of the direction depends on the gradient $\nabla\varphi$ or the Hessian $\nabla^2\varphi$ of φ at each iteration. For example, choosing $V^{(k)} = -\nabla_\varphi(W^{(k)}\mathbf{y})$ is known as *steepest* or *gradient descent*, and $V^{(k)} = -(\nabla^2\varphi(W^{(k)}\mathbf{y}))^{-1}\nabla_\varphi(W^{(k)}\mathbf{y})$ is known as *Newton* step.

5.1. Relative Optimization

A relative optimization algorithm is a class of optimization algorithm that rely on the fact that the mixing matrices form a multiplicative group. The latter implies that, starting from the observed mixtures, one can iteratively improve the source estimate (in terms of some contrast function) by finding an estimate of the unmixing matrix leading to a decrease of the contrast function, and using the obtained source estimate as the observation at the next iteration. A general relative optimization algorithm has the following structure (5):

1. Start with an initial source estimate $Y^{(0)}$.
2. For $k = 0, 1, 2, \ldots$, until convergence
   (a) Starting from $V^{(k)} = I$ and using one or more steps of some unconstrained minimization algorithm, find such a matrix $V^{(k)}$ that decreases the contrast function $\varphi(V^{(k)}Y^{(k)})$.
   (b) Update the source estimate: $Y^{(k+1)} = V^{(k)}Y^{(k)}$.
3. End of loop.

The output of the algorithm is the source estimate $Y = Y^{(K)}$ minimizing the contrast φ. The unmixing matrix is obtained as the product $W = V^{(0)} \cdot V^{(1)} \cdot \ldots \cdot V^{(K)}$.

Different algorithms can be obtained by different choices of the unconstrained minimization algorithm used in Step (a). The *relative gradient* (also referred to as the *natural gradient*) algorithm, introduced by A. Cichocki et al. (6), J-F. Cardoso and B. Laheld (7), and S-I. Amari et al. (8), although derived from different considerations, can be obtained by using a steepest descent iteration at Step (a). The *relative Newton* algorithm (1,5) is obtained by using a single Newton iteration at Step (a). An efficient block-coordinate version of this algorithm was proposed by Bronstein et al. (9). Relative optimization techniques can be used in both batch and online modes. Extensions for blind deconvolution are also available (10,11).

5.2. Fixed Point Algorithms

Fixed point algorithms were introduced by A. Hyvärinen and E. Oja (12) and are commonly known under the name of *Fast ICA*. This class of algorithms is based on fixed point iterations, a method known to have very fast convergence. The one-unit Fast ICA algorithm has the following form:

- Center and whiten the data $\mathbf{y}$.
- Start with some estimate $\mathbf{w}^{(0)}$ of a row vector of W.
- For $k = 1, 2, \ldots$, until convergence
  ○ Update $\mathbf{w}^{(k)} \leftarrow \mathbb{E}\{\mathbf{y}\varphi'((\mathbf{w}^{(k-1)})^T\mathbf{y})\} - \mathbb{E}\{\varphi''((\mathbf{w}^{(k-1)})^T\mathbf{y})\} \cdot \mathbf{w}^{(k-1)}$.
  ○ Normalize: $\mathbf{w}^{(k)} \leftarrow \mathbf{w}^{(k)}/\|\mathbf{w}^{(k)}\|$.
- End of loop.

In practice, the expectations in Step 4 are replaced by empirical averages and the separation vectors $\mathbf{w}$ are orthogonalized. φ can be virtually any twice-differentiable function and does not have to correspond to any statistically or information theoretically based contrast. Different choices of φ result in different algorithms suitable for different classes of sources. Fast ICA is a batch algorithm and generally used for estimating the ICs one by one. Extensions for the multiunit case are available. (Fig. 2)

6. SPARSE COMPONENT ANALYSIS (SCA)

The fundamental assumption of statistical independence of source used in ICA can be relaxed and replaced by the assumption of their *sparsity*, which gives rise to the *sparse component analysis* (SCA), an alternative class of methods for BSS (13,14). Sparsity implies that most of the values of the source are zero or near zero. For simplicity, consider the problem of two sources and two mixtures ($M = N = 2$). The observed mixture $\mathbf{x}(t)$ is a set of points in the plane (*scatter plot*), given by

$$\mathbf{x}(t) = \mathbf{a}_1 \cdot \mathbf{s}_1(t) + \mathbf{a}_2 \cdot \mathbf{s}_2(t),$$

where $\mathbf{a}_1, \mathbf{a}_2$ are the column vectors of the mixing matrix A. If the sources are sparse, most of the instantaneous mixture values are contributed by a single "active" source only (either $\mathbf{s}_1(t)$ or $\mathbf{s}_2(t)$). Statistical independence is

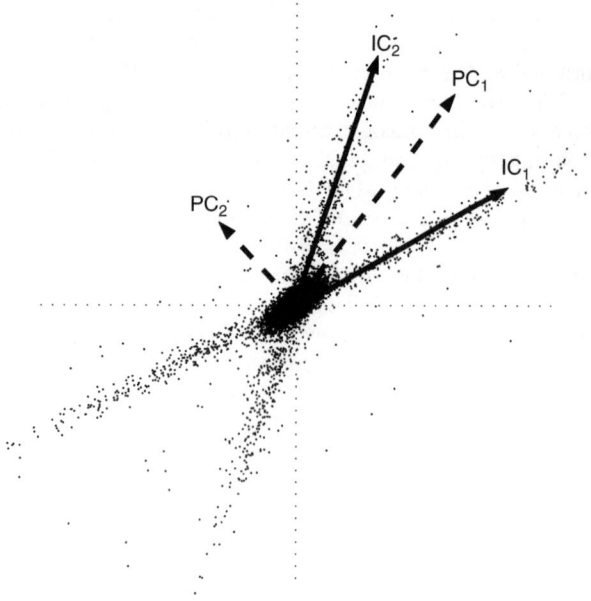

Figure 2. The difference between PCA and ICA on non-Gaussian sources.

If the sources are not originally sparse, in many cases they can be *sparsified*, for example, brought to a sparse representation by application of a linear *sparsifying transformation* given by a $T_1 \times T$ matrix Φ:

$$S' = S\Phi^T,$$

where T_1 is the dimension of the resulting representation and superscript $()^T$ denotes matrix transposition. The transformation Φ is not necessarily invertible. As the mixing is linear, applying Φ on the mixtures is equivalent to mixing the sparsified sources:

$$X' = X\Phi^T = (AS)\Phi^T = A(S\Phi^T) = AS'.$$

Unmixing is performed by first estimating the unmixing matrix W given the sparsified mixtures X' and then applying it to the original mixtures X. The sparsifying transformation is generally dependent on the source types. A possible selection of Φ is, for example, the *short-time Fourier transform* (STFT) for acoustic signals (15) and a discrete derivative for images (16). A richer family of transformations is obtained by using multiresolution representations such as the *wavelet* or the *wavelet-packet* (WP) transforms.

not necessary for this value to hold; it is enough that in the realization of the sources the nonzero values do not overlap in most of their occurrences. Consequently, $\mathbf{x}(t)$ consists mainly of points lying along directions $\mathbf{a}_1$ and $\mathbf{a}_2$ (Fig. 3). Recovering these directions allows one to estimate the mixing matrix. This BSS approach is termed *geometric separation*. The applicability of geometric separation is practically limited to problems with number of mixtures $M = 2$ or 3. However, the advantage of this method compared with ICA is that it allows one to handle BSS problems with number of sources $N > M$.

Sparsification can be employed together with QML estimation in cases when $M = N$. QML methods require that the distribution of sources is modeled at least approximately. In some cases, it is very difficult or practically impossible to write the source distribution in an analytical form. In addition, many distributions result in nonconvex objective functions that are problematic for optimization. Instead, the sources are assumed sparse (with the absolute value or its smoothed version used as a model of $-\log f_{s_i}(s)$) and an appropriate sparsification transformation is applied on the mixtures, making the problem equivalent to separation of sparse sources. This method

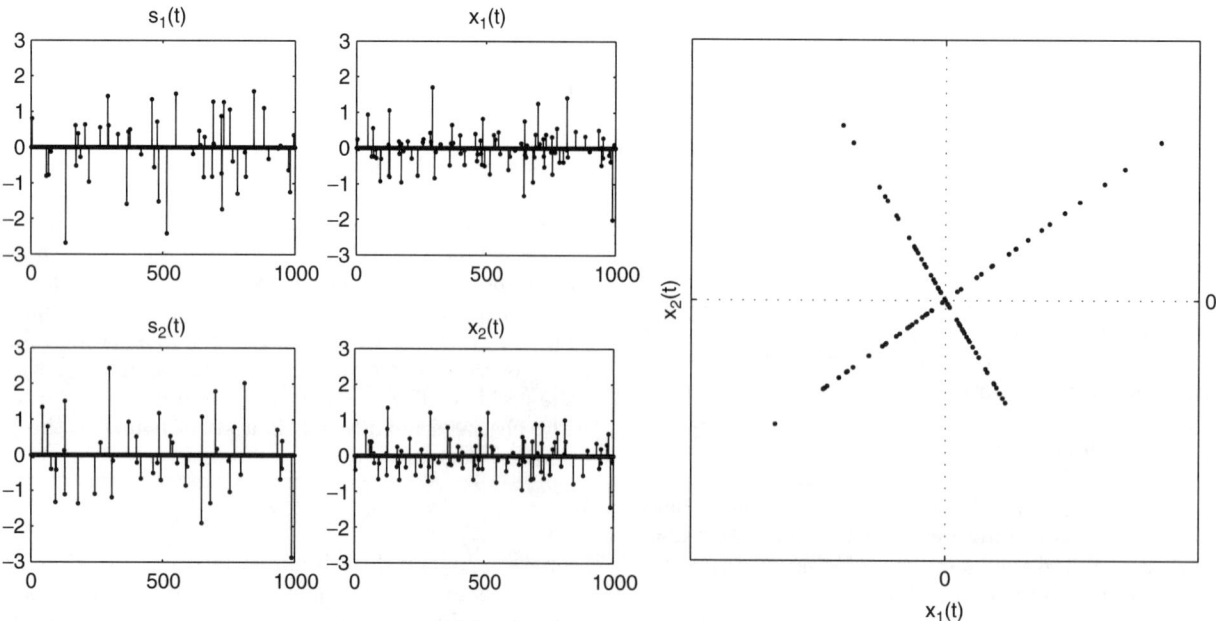

Figure 3. Sparse sources (left), their mixtures (center), and the mixtures scatter plot (right).

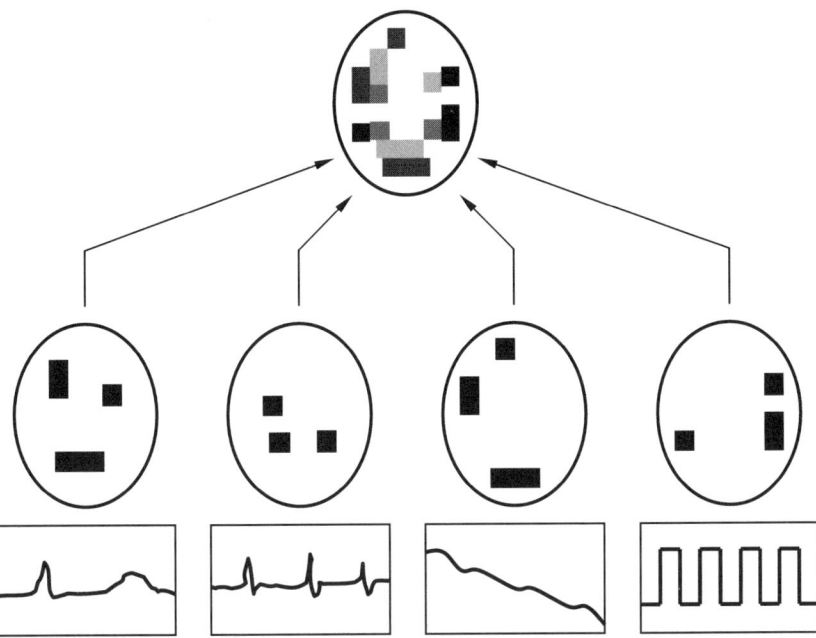

Figure 4. Linear mixing model in fMRI.

usually outperforms QML estimation without sparsification (Fig. 4).

7. INDEPENDENT FACTOR ANALYSIS (IFA)

One of the weaknesses of ICA algorithms is the fact they do not incorporate noise into the model. For this reason, the applicability of ICA is limited to low- to medium-noise cases. SCA is more robust to noise in some cases; however, it also ignores the explicit noise model. As a possible remedy, H. Attias (17) proposed the *independent factor analysis* (IFA), which can be viewed as a generalization of ICA, PCA, and ordinary factor analysis.

Similarly to ICA, the IFA framework assumes instantaneous linear mixing model. Observations $x(t)$ are assumed to be contaminated by zero-mean Gaussian noise $\xi(t)$ with some covariance matrix Λ. Both Λ and the mixing matrix A are unknown. The sources $s(t)$ are considered mutually statistically independent and their probability density functions are usually modeled as a *mixture of Gaussians* (MOG)

$$f_{s_i}(s_i|\theta_i) = \sum_{k=1}^{n_i} z_{ik} f_G(s_i - \mu_{ik}, \sigma_{ik}^2),$$

where $f_G(s - \mu, \sigma^2)$ is the Gaussian probability density function with mean μ and variance σ^2, and $\theta_i = (z_{i1}, \ldots z_{1n_i}, \mu_{i1}, \ldots, \mu_{in_i}, \sigma_{i1}^2, \ldots, \sigma_{in_i}^2)$ is the vector of parameters describing the mixture, where $\sum_k z_{ik} = 1$. The resulting model observation density is given by

$$f(x|\Lambda, A, \theta_i, \ldots, \theta_M)$$
$$= \int f(x|s)f(s)ds$$
$$= \int f_G(x - As, \Lambda) \prod_{i=1}^{M} f_{s_i}(s_i|\theta_i) ds_1 \ldots ds_N.$$

IFA consists of two steps. First, the parameters Λ, A, $\theta_i, \ldots, \theta_M$ of the model are adapted to minimize a distance function between the model and the observed probability densities of x. A typical choice for such a distance function is the Kullback–Leibler divergence. Various numerical algorithms can be used to perform the minimization; H. Attias showed the use of an *expectation-minimization* (EM) algorithm. Once the model parameters are estimated, source recovery is performed. A good restoration can be obtained by the *least mean squares* (LMS) estimator,

$$y_{LMS} = \mathbb{E}\{s|x\},$$

which minimizes $\mathbb{E}(y - s)^2$. This above conditional expected value is readily available as a byproduct of the EM algorithm. Another way to estimate the sources is the *maximum a posteriori estimator* (MAP), which maximizes the joint probability density $f(s, x)$.

8. GENERALIZATIONS OF THE BSS PROBLEM

A more general setting of the BSS problem is required in situations where the linear instantaneous mixing model cannot be assumed.

8.1. Convolutive Mixture

Convolutive mixture is a generalization of the instantaneous mixing model. The sensor signals in this case are given by

$$\begin{aligned}
x_1(t) &= a_{11}(t) * s_1(t) + \cdots + a_{1N}(t) * s_N(t) + \xi_1(t) \\
&\vdots \\
x_M(t) &= a_{M1}(t) * s_1(t) + \cdots + a_{MN}(t) * s_N(t) + \xi_M(t),
\end{aligned}$$

where $a_{ij}(t)$ are some unknown filters. This problem is called *multichannel blind deconvolution* (BD). As in the BSS case, the sources and the filters can be 1D, 2D, or 3D. It is common to distinguish between the following cases: $M, N > 1$ - *multiple-input multiple-output* (MIMO), which is the most general setting; $M > 1$, $N = 1$ - *single-input multiple-output* (SIMO); and $M = N = 1$ - *single-input single-output* (SISO). In image processing, the latter case is sometimes simply referred to as *blind deconvolution*.

A particular setting of the MIMO BD problem is a generalization of the instantaneous (delay-less) linear BSS problem to the cases when the signal propagation velocity is low. The filters model the delays in signal propagation. Such a situation is common (e.g., when the sources are acoustic signals). Often, in order to account for acoustic effects like reverberation, the filters must be more complicated than just delays.

The SISO BD problem develops in optical imaging applications, where the source image is acquired through scattering medium. The action of the medium can be modeled as a *linear shift-invariant* (LSI) system and described by convolution with some 2D filter (usually referred to as the *point spread function* or PSF). The source reconstruction requires one to undo the effect of the convolution with the PSF. If more than one observation of the same source is available degraded by different PSFs, the problem is formulated as SIMO BD.

Some theoretical methods and numerical algorithms used for BSS can be generalized to BD problems. For example, ML and QML estimators can be formulated in a way similar to the BSS case. An emerging field is the extension of SCA approaches to BD (10,11). In some cases, the MIMO BD problem can be posed as an instantaneous BSS problem. For example, in case of acoustic signals, P. Smaragdis proposed the *time-frequency domain ICA* approach (18). The sensor signals are transformed into the STFT domain

$$x_i(t) \stackrel{STFT}{\to} X_i(t, \omega),$$

where the convolutive mixtures are translated into instantaneous ones, with a frequency-dependent mixing matrix:

$$X_i(t, \omega) = A_{i1}(\omega)S_1(t, \omega) + \cdots + A_{iN}(\omega)S_N(t, \omega).$$

Assuming that in narrow frequency subbands the filters are approximately constant (i.e., that A is constant), the BD problem can be formulated as a set of instantaneous BSS problems in each subband. The difficulty in this method is the permutation and scale ambiguity, which must be resolved before the separated signals in each subband are merged. Usually, some additional information, such as the directivity pattern, is necessary for this purpose (19).

8.2. Nonlinear Mixture

Nonlinear mixture is a generalization of the linear mixing model. The mixing matrix A is replaced by an invertible nonlinear mixing function $\mathscr{A}$ from $\mathbb{R}^N$ to $\mathbb{R}^M$. The sensor signals in this case are given by

$$\mathbf{x}(t) = \mathscr{A}(\mathbf{s}(t)).$$

A particular case when the nonlinear operator is applied in a component-wise manner after linear mixing

$$\mathbf{x}(t) = \mathscr{A}(A \cdot \mathbf{s}(t))$$

is referred to as *post-nonlinear* BSS. Typically, the nonlinear BSS problem is significantly harder than the linear BSS. Several approaches, for example, based on correlation maximization and kernel learning (20), were proposed, yet the nonlinear BSS problem is still an open research field.

9. APPLICATIONS

BSS techniques have been successfully employed in biomedicine (e.g., for the analysis of EEG, MEG, ECG, and fMRI data). In these applications, the linear mixture assumption is usually justified by the physical principles of signal formation, and high signal propagation velocity allows one to use the instantaneous mixture model (21). Otherwise, nonlinear BSS or BD methods are used.

9.1. Electroencephalography (EEG)

The brain cortex can be thought of as a field of K tiny sources, which, in turn, are modeled as current dipoles. The jth dipole is characterized by the location vector $\mathbf{r}_j$ and the dipole moment vector $\mathbf{q}_j$. The electromagnetic field produced by the neural activity determines the potential on the scalp surface, sampled at a set of M sensors. Denote the ith sensor location by $\mathbf{r}'_i$, and the potential it measures by v_i. The mapping from the neural current sources to the measured scalp potentials $\{\mathbf{q}_j, r_j\} \to \{v_i, \mathbf{r}'_i\}$ is usually referred to as *the forward model* and the measured potentials as the *forward field*. The forward model is linear in the dipole moment $\mathbf{q}$. As the propagation velocity of electromagnetic waves is very large compared with head dimensions, zero propagation time can be assumed. Hence, one can express the forward field at sensor j because of dipole i as the inner product $v_j^i = \mathbf{g}_{ij}^T \mathbf{q}_j$, where $\mathbf{g}_{ij} = \mathbf{g}(r_j, r'_i)$ is the *vector kernel* or the *lead field* depending on the geometry and the electromagnetic properties of the head. By electromagnetic superposition, $v_j = \sum_i v_j^i$. Assuming T time samples of $\mathbf{v}(t) = (v_1(t), \ldots, v_M(t))^T$ are observed, the entire spatio-temporal forward model can be expressed in matrix form as

$$\begin{pmatrix} v_1(t_1) & \cdots & v_1(t_T) \\ \vdots & \ddots & \vdots \\ v_M(t_1) & \cdots & v_M(t_T) \end{pmatrix} = \begin{pmatrix} \mathbf{g}_{11}^T & \cdots & \mathbf{g}_{1K}^T \\ \vdots & \ddots & \vdots \\ \mathbf{g}_{M1}^T & \cdots & \mathbf{g}_{MK}^T \end{pmatrix} \cdot \begin{pmatrix} \mathbf{q}_1(t_1) & \cdots & \mathbf{q}_1(t_T) \\ \vdots & \ddots & \vdots \\ \mathbf{q}_K(t_1) & \cdots & \mathbf{q}_K(t_T) \end{pmatrix},$$

or $V = GQ$, where V is an $M \times T$ matrix of the measured forward field, G is the $M \times 3K$ *gain matrix*, which is assumed to be fixed in time, and Q is a $3K \times T$ matrix of current sources.

Typically, the number of sensors ranges from tens to hundreds, and the number of dipoles K in a faithful forward model is larger by several orders of magnitude. As a consequence, the *inverse problem* (i.e., the problem of determining the entire set of K current sources from the measured forward field) is underdetermined; however, many types of brain activity can be modeled as the forward field from sets of spatially fixed (either localized or widely spread around the cortical surface) concurrently acting dipoles activated by temporally independent sources. Makeig et al. were the pioneers in using ICA for separation of such sources (22). ICA may determine what temporally independent activations compose the collected scalp recordings without specifying directly where in the brain these activations occur. ICA has been shown particularly efficient for analysis *of event-related potential* (ERP) data (23) and for removing encephalographic artifacts (24).

9.2. Magnetoencephalography (MEG)

Similar to EEG, the forward model in MEG is also essentially linear. The sensors measure the vector of the magnetic field $\mathbf{b}_j$ around the scalp. The forward field at sensor j because of dipole i can be expressed as $\mathbf{b}_j^i = G_{ij} \mathbf{q}_j$, where $G_{ij} = G(r_j, r_i')$ is the *matrix kernel* depending on the geometry and the electromagnetical properties of the head. BSS can be used for separation of independent temporal components in the same way it is used in EEG (21).

9.3. Electrocardiography (ECG)

The mechanical action of the heart is initiated by a quasi-periodic electrical stimulus, which causes an electrical current to propagate through the body tissues and results in potential differences. The potential differences measured by electrodes on the skin *(cutaneous* recording) as a function of time is termed *electrocardiogram* (ECG). The measured ECG signal can be considered as a superposition of several independent processes, resulting, for example, from *electromyographic* activity (electrical potentials generated by muscles), 50 Hz or 60 Hz net interferences, or the electrical activity of the fetal heart (FECG). The latter contains important indications about the fetus health.

The transfer from bioelectrical sources to the electrodes can be considered linear and delay-less because of the high propagation velocity of electromagnetic waves. Cutaneous ECG voltage recordings, $v_1(t), \ldots, v_M(t)$, measured at M electrodes on the skin, are given as a superposition of N electrical sources $s_1(t), \ldots, s_N(t)$:

$$\begin{pmatrix} v_1(t) \\ \vdots \\ v_M(t) \end{pmatrix} = \begin{pmatrix} g_{11} & \cdots & g_{1K} \\ \vdots & \ddots & \vdots \\ g_{M1} & \cdots & g_{MK} \end{pmatrix} \begin{pmatrix} s_1(t) \\ \vdots \\ s_N(t) \end{pmatrix} + \begin{pmatrix} \xi_1(t) \\ \vdots \\ \xi_N(t) \end{pmatrix},$$

where $\xi(t)$ stands for additive noise and g_{ij} are the transfer coefficients (conductivity) from source j to electrode i. BSS methods have been successfully used for separation of interferences in ECG data. A particular success was demonstrated in separation of a fetal ECG from a mother's ECG (25).

9.4. Functional Magnetic Resonance Imaging (fMRI)

The principle of fMRI is based on different magnetic properties of oxygenated and deoxygenated hemoglobin, which allows one to obtain a *blood oxygenation level-dependent* (BOLD) signal. The observed spatio-temporal signal $q(\mathbf{r}, t)$ of magnetic induction can be considered as a superposition of N *spatially independent* components, each associated with a unique time course $\beta_k(t)$ and a spatial map $s_k(\mathbf{r})$. Each source represent the *loci* of concurrent neural activity and can be either *task-related* or *non-task-related* (e.g., physiological pulsations, head movements, background brain activity). The spatial map corresponding to each source determines its influence in each volume element (*voxel*), and is assumed to be fixed in time. Spatial maps can be overlapping.

ICA has been successfully used to separate either the independent spatial sources (26) or the independent time courses. These techniques are usually known as *spatial* and *temporal* ICA, respectively. In the spatial approach, fMRI data $q(\mathbf{r}, t_1), \ldots, q(\mathbf{r}, t_T)$ acquired at T different times can be considered as a superposition of N independent source images $s_1(\mathbf{r}), \ldots, s_N(\mathbf{r})$ mixed with different contributions:

$$\begin{pmatrix} q(\mathbf{r}, t_1) \\ \vdots \\ q(\mathbf{r}, t_T) \end{pmatrix} = \begin{pmatrix} \beta_1(t_1) & \cdots & \beta_N(t_1) \\ \vdots & \ddots & \vdots \\ \beta_1(t_T) & \cdots & \beta_N(t_T) \end{pmatrix} \cdot \begin{pmatrix} s_1(\mathbf{r}) \\ \vdots \\ s_N(\mathbf{r}) \end{pmatrix}$$

$$+ \begin{pmatrix} \xi(\mathbf{r}, t_1) \\ \vdots \\ \xi(\mathbf{r}, t_T) \end{pmatrix},$$

where $\xi(\mathbf{r}, t)$ stands for additive noise. In the temporal approach, the observed data is considered as K time signals $q(\mathbf{r}_1, t), \ldots, q(\mathbf{r}_K, t)$, consisting of linear mixtures of N independent time courses $\beta_1(t), \ldots, \beta_N(t)$:

$$\begin{pmatrix} q(\mathbf{r}_1, t) \\ \vdots \\ q(\mathbf{r}_K, t) \end{pmatrix} = \begin{pmatrix} s_1(\mathbf{r}_1) & \cdots & s_N(\mathbf{r}_1) \\ \vdots & \ddots & \vdots \\ s_1(\mathbf{r}_K) & \cdots & s_N(\mathbf{r}_K) \end{pmatrix} \cdot \begin{pmatrix} \beta_1(t) \\ \vdots \\ \beta_N(t) \end{pmatrix}$$

$$+ \begin{pmatrix} \xi(\mathbf{r}_1, t) \\ \vdots \\ \xi(\mathbf{r}_K, t) \end{pmatrix},$$

where K stands for the number of voxels. Combined *spatio-temporal* techniques maximize some measure of independence over space and time simultaneously, without necessarily achieving independence over either of them separately (27). The main advantage of BSS techniques over other fMRI analysis tools is that no need exists to assume any *a priori* information about the time course of processes contributing to the measured signals.

BIBLIOGRAPHY

1. D. Pham and P. Garrat, Blind separation of a mixture of independent sources through a quasi-maximum likelihood approach. *IEEE Trans. Sig. Proc.* 1997; **45**:1712–1725.
2. A. J. Bell and T. J. Sejnowski, An information maximization approach to blind separation and blind deconvolution. *Neural Computation* 1995; **7**(6):1129–1159.
3. P. Comon, Independent component analysis – a new concept. *Signal Proc.* 1994; **36**(3):287–314.
4. J-F. Cardoso and A. Souloumiac, An efficient technique for blind separation of complex sources. *Proc. EEE SP Workshop on Higher-Order Stat.* 1993:275–279.
5. M. Zibulevsky, Blind source separation with relative Newton method. *Proc. Int. Sympos. Independent Component Anal. Blind Signal Separation* 2003:897–902.
6. A. Cichocki, R. Unbehauen, and E. Rummert, Robust learning algorithm for blind separation of signals. *Electron. Lett.* 1994; **30**(17):1386–1387.
7. J-F. Cardoso and B. Laheld, Equivariant adaptive source separation. *IEEE Trans. Sig. Proc.* 1996; **44**(12):3017–3030.
8. S-I. Amari, S. C. Douglas, A. Cichocki, and H. H. Yang, A new learning algorithm for blind signal separation. *Advances Neural Inform. Proc. Syst.* 1996; **8**:757–763.
9. A. M. Bronstein, M. M. Bronstein, and M. Zibulevsky, Blind source separation using block-coordinate relative Newton method. *Signal Proc.* 2004; **84**(8):1447–1459.
10. A. M. Bronstein, M. M. Bronstein, M. Zibulevsky, and Y. Y. Zeevi, Blind deconvolution of images using optimal sparse representations. *IEEE Trans. Image Proc.* 2005; **14**(6):726–736.
11. A. M. Bronstein, M. M. Bronstein, and M. Zibulevsky, Relative optimization for blind deconvolution. *IEEE Trans. Signal Proc.* 2005; **53**(6):2018–2026.
12. A. Hyvarinen and E. Oja, A fast fixed-point algorithm for independent component analysis. *Neural Computation* 1997; **9**(7):1483–1492.
13. M. S. Lewicki and T. J. Sejnowski, Coding time-varying signals using sparse, shift-invariant representations. *Advances Neural Inform. Proc. Syst.* 1999; **11**:730–736.
14. M. Zibulevsky and B. A. Pearlmutter, Blind source separation by sparse decomposition. *Neural Computation* 2001; **13**(4):863–882.
15. M. Zibulevsky, P. Kisilev, Y. Y, Zeevi, and B. A. Pearlmutter, Blind source separation via multinode sparse representation. *Advances Neural Inform. Proc. Syst.* 2002; **12**.
16. A. M. Bronstein, M. M. Bronstein, M. Zibulevsky, and Y. Y. Zeevi, Sparse ICA for blind separation of transmitted and reflected images. *Int. J. Imaging Sci. Technol.* 2005; **15**(1):84–91.
17. H. Attias, Independent factor analysis. *Neural Computation* 2001; **11**(4):803–851.
18. P. Smaragdis, Blind separation of convolved mixtures in the frequency domain. *Neurocomputing* 1998; **22**:21–34.
19. N. Mitianoudis and M. Davies, Permutation alignment for frequency domain ICA using sub-space beamforming methods. *Proc. International Symposium on Independent Component Analysis and Blind Signal Separation*, 2004.
20. S. Harmeling, A. Ziehe, M. Kawanabe, B. Blankertz, and K-R. Muller, Nonlinear blind source separation using kernel feature spaces. *Proc. International Symposium on Independent Component Analysis and Blind Signal Separation*, 2001.
21. T-P. Jung, S. Makeig, M. J. McKeown, A. J. Bell, T-W. Lee, and T. J. Sejnowski, Independent component analysis of biomedical signals. *Imaging Brain Dynamics Using Independent Component Anal.* 2001; **89**(7):1107–1122.
22. S. Makeig, A. J. Bell, T-P. Jung, and T. J. Sejnowski, Independent component analysis of electroencephalographic data. *Advances Neural Inform. Proc. Syst.* 1996; **8**:145–151.
23. S. Makeig, T. P. Jung, A. J. Bell, and T. J. Sejnowski, Blind separation of auditory event-related brain responses into independent components. *Proc. Nat. Acad. Sci.* 1997; **94**:10979–10984.
24. T-P. Jung, C. Humphries, T-W. Lee, M. J. McKeown, V. Iragui, S. Makeig, and T. J. Sejnowski, Removing electroencephalographic artifacts from by blind source separation. *Psychophysiology* 2000; **37**:163–178.
25. V. Zarzoso, A. K. Nandi, and E. Bacharakis, Maternal and foetal ecg separation using blind source separation methods. *IMA J. Math. Appl. Med. Biol.* 1997; **14**(3):207–225.
26. M. J. McKeown, S. Makeig, C. G. Brown, T-P. Jung, S. S. Kindermann, and T. J. Sejnowski, Analysis of fMRI by blind separation into independent spatial components. *Human Brain Mapping* 1998; **6**(3):160–188.
27. J. V. Stone, J. Porrill, C. Buchel, and K. Friston, Spatial, temporal, and spatiotemporal independent component analysis of fMRI data. *Proc. 18th Leeds Statistical Research Workshop on Spatial-temporal modelling and its applications*, 1999. (online) Available: http://www.shef.ac.uk/pc1jvs/paperspublished.html.

READING LIST

J. Karhunen, A. Hyvärinen, and E. Oja, *Independent Component Analysis*. New York: John Wiley and Sons, 2001. – A comprehensive introduction to ICA. The book includes the fundamental mathematical background needed to understand and utilize it. It offers a general overview of the basics of ICA, important solutions and algorithms, and in-depth coverage of different applications.

S. J. Roberts and R. M. Everson, eds., *Independent Components Analysis: Principles and Practice*. Cambridge; Cambridge University Press, 2001. – A self-contained book built as a structured series of edited papers by leading researchers in the field, including an extensive introduction to ICA. The major theoretical bases are reviewed from a modern perspective, current developments such as the SCA paradigm are surveyed and many case studies of applications, including biomedical ones, are described in detail.

J-F. Cardoso, Blind signal separation: statistical principles. *Proc. IEEE* 1998; **9**(10):2009–2025. (online). Available: http://www.tsi.enst.fr/~cardoso/jfbib.html. – One of the best review articles on ICA. The author, one of the leading researchers in the field, gives a comprehensive survey on different approaches to ICA and algorithms used therein.

A. Hyvärinen, Survey on independent component analysis. *Neural Computing Surveys* 1999; **2**:94–128. (online). Available: http://www.cis.hut.fi/aapo/papers/NCS99web. – A comprehensive survey paper on ICA by a leading researcher in the field.

T-P. Jung et al., Imaging brain dynamics using independent component analysis. *Proc. IEEE* 2001; **89**(7):1107–1122. (online). Available: http://www.sccn.ucsd.edu/~scott. – A review article on biomedical applications of ICA written by the pioneers of the field. The paper is based mainly on research of S. Makeig *et al.* and highlights several important applications of ICA in EEG, MEG, ECG and fMRI.

A. Hyvärinen, Fixed-point ICA (FastICA) algorithm (language: MATLAB). (online). Available: http://www.cis.hut.fi/projects/ica/fastica.

M. Zibulevsky, Relative Newton algorithm (language: MATLAB). (online). Available: http://iew3.technion.ac.il/~mcib.

EEGLab, Toolbox for processing and visualization of electrophysiological data. Includes an implementation of the InfoMax BSS algorithm (language: MATLAB, C). (online). Available: http://www.sccn.ucsd.edu/eeglab.

BLOOD FLOW MEASUREMENT

SHARMEEN MASOOD
GUANG-ZHONG YANG
Imperial College
London, United Kingdom

1. INTRODUCTION

The main function of blood flow is to provide cells with a constant supply of oxygen and nutrients whilst carrying away waste substances such as carbon dioxide. Patterns of flow in the major blood vessels and cardiac chambers can provide insights into normal function and subsequent changes in disease. In medicine, both invasive and noninvasive techniques have been used to measure blood flow for decades. The purpose of this chapter is to outline the main techniques used in clinical blood flow measurement, and the basic principles behind these techniques.

2. INVASIVE TECHNIQUES FOR FLOW MEASUREMENT IN BLOOD VESSELS

Indicator dilution technique is one of the earliest methods used for measuring regional blood flow or cardiac output. The method uses indicators, such as radioisotopes, dyes, and cold saline, injected into the blood vessel or the right atrium and subsequently measures the concentration in the region of interest over time. Stewart was the first to use this principle to measure blood flow in 1897. He used sodium chloride (NaCl) as the indicator, continuously recording the increased conductivity caused by the NaCl, allowing a measure of the delay time and dispersion of the indicator. Subsequently, Hamilton expanded the ideas of using appearance times to estimate blood flow. These theories were later modified and developed to include detailed indicator input functions for establishing the relationship between the indicator dilution curve and blood flow (1,2). The dyes used in the past include indocyanine and iodinated contrast mediums for angiographic applications. When radioisotopes are used, an external detector can be used to measure the radioisotope washout. Thermodilution techniques comprise an injection of cold saline in the proximal part of the catheter with a thermistor mounted at its tip. The thermistor can then be used to record changes in temperature over time and thus the indicator clearance curve (2–4).

Quantitatively, the principle behind indicator dilution techniques is Fick's law of diffusion at steady state, which relates the flow of a fluid to the indicator added per unit time and the initial and measured concentrations, which correspond to the arterial and venous flows, for example,

$$Q = \frac{A}{(C_i - C_o)}, \quad (1)$$

where Q is the flow in litres/min, A is the indicator added per unit time in mg/min, and C_i and C_o are the input and output concentrations in mg/liter. This method can also be used to measure volumes and cardiac output (2).

The electromagnetic flowmeter was introduced independently by Wetterer in Germany and Kolin in the United States circa 1936. Peppe and Wetterer successfully measured the flow in the ascending aorta in 1940, and Gregg et al. subsequently used electromagnetic flowmeters to measure flow in the coronary and other arteries. A history of the development of the technique for blood flow measurement is presented in Cappelen (5) and Wyatt (6). Since then, electromagnetic flowmeters have been miniaturized and mounted in diagnostic catheters. In general, electromagnetic flowmeters can be used to measure flow in vessels from 50 mm to 1 mm in external diameter (2,4,6), and they are especially suited for reconstructive arterial surgery and other surgical procedures. The basic physical principle behind electromagnetic flowmeters is shown in Fig. 1, where the magnetic force applied to the blood vessel causes the moving charged ions in the blood to be deflected, producing an electric field E. Therefore, a voltage difference exists between the two electrodes, and $V = Ed$. The velocity of the blood can be obtained by $v = E/B$, where E is the electric field produced by the ions moving in the magnetic field B. The same relationship can be obtained by looking at the balancing forces caused by the electric and magnetic fields, for example,

$$\begin{aligned} F_B + F_E &= 0 \\ \Rightarrow qE - qvB &= 0 \\ \Rightarrow v &= E/B. \end{aligned} \quad (2)$$

The main disadvantage of electromagnetic flowmeters is that they are invasive. Nevertheless, it is possible to leave flowmeters attached to blood vessels after surgery for up to a few days to monitor recovery. They cannot, however, be left indefinitely as this can lead to sepsis and further complications.

For measuring blood flow in the microcirculation, laser Doppler velocimetry/flowmetry is a minimally invasive method that has been in use for nearly 30 years. The early work was mainly concentrated on retinal blood flow because of the ease of access to the retinal circulation (7,8). The method has since been extended to measure blood

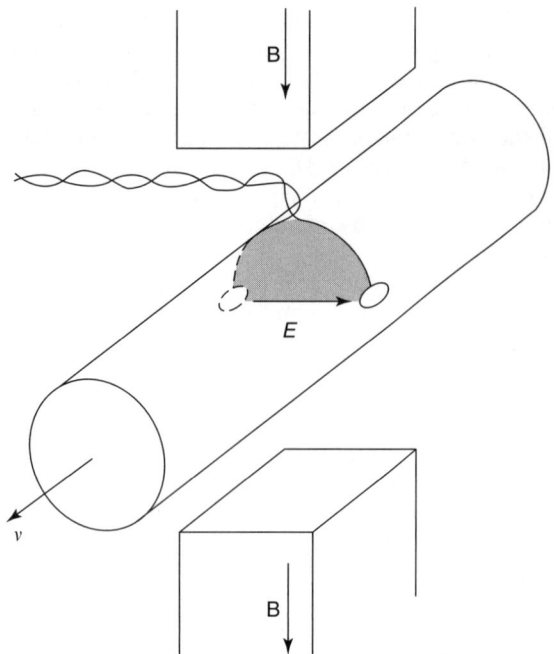

Figure 1. Principle behind electromagnetic flowmeter. The magnetic field, B, causes moving charged ions in the blood to be deflected. This produces an electric field, E, which can be measured directly with a voltmeter. The velocity of the charged ions can then be calculated as the balance between the electric and magnetic fields, $v = E/B$.

flow in other tissues including the measurement of cerebral blood flow, ocular circulation (9), cutaneous blood flow (10), and dental circulation (11). Another mode of operation of laser Doppler velocimetry allows a two-dimensional (2-D) relative measurement of perfusion in tissues. Although laser Doppler is the only modality that allows measurement of the microcirculation at high temporal resolution, only local measurements can be made and they are prone to motion artefacts. In addition, it only allows relative measurements of blood flow and perfusion, thus arbitrary units of measurement must be used.

Through imaging, it is also possible to quantify *in vivo* blood flow. Angiography allows the clinician to examine blood flow by injecting a dye into the circulation and observing its progress through the cerebral or coronary circulation, with x-ray imaging. Adding the ability to measure the blood flow by using the traditional indicator dilution method can further enhance the clinical value of the technique. For angiographic applications, the indicator dilution method is typically carried out by means of iodinated contrast mediums. Angiography can also be combined with intravascular flow velocity measurements obtained with a Doppler ultrasound guidewire.

3. NONINVASIVE TECHNIQUES FOR FLOW MEASUREMENT IN BLOOD VESSELS

Strain gauge plethysmography is a noninvasive method for measuring peripheral blood flow. The method was first described in 1909 by Hewlett and van Zwaluwenburg, but was not applied routinely because of the equipment required. The first plethysmographs were water or air-filled plethysmographs constructed of glass, brass, or other metals making them cumbersome to use (12,13). The development of mercury in plastic strain gauges led to its use in research and, subsequently, in routine patient use for assessing blood flow in the peripheral circulation. Measurement of flow is achieved by tying a strain gauge around the limb under investigation. The strain gauge consists of a stretchable tube of liquid metal such as mercury or indium-gallium alloy. Changes in length as a result of blood flow cause changes in the cross-sectional area of the liquid metal leading to variations in the electrical impedance, which can be easily monitored and recorded (3,14). An inflatable cuff can be used to interrupt venous return. As the cuff is deflated, the changes in volume induce changes in impedance in the strain gauge allowing measurement of venous blood flow. This technique is termed venous occlusion plethysmography (13).

An alternative to mercury strain gauge plethysmography is to use an optical sensor, also known as photoplethysmography (15), which overcomes the disadvantages of the mercury strain gauge: temperature sensitivity, and corrosion of mercury sensor electrodes leading to short storage times and the escape of free mercury from broken sensors contributing to environmental pollution. Optical sensors employ the principle of mode coupling in optical fibers to measure changes in dimension of the gauge. As for strain gauge plethysmography, this allows direct estimation of blood flow.

Plethysmography has been used to measure blood flow in the limbs for several decades. The clinical applications of plethysmography are concentrated on the measurement of peripheral circulation. Blood flow to the digits, measured using photoplethysmography, has been used extensively for investigating endothelial function (13,16–18). The technique is routinely applied clinically to monitor blood pressure in patients and can be used for ambulatory monitoring of hypertensive patients. Strain gauge plethysmography is especially useful for examining blood flow in the arms and legs and has found applications in measuring muscular blood flow and venous blood flow (19–21). A review of the research and clinical applications of venous occlusion plethysmography is given by Wilkinson and Webb in (13).

4. NONINVASIVE IMAGING TECHNIQUES FOR MEASURING FLOW IN BLOOD VESSELS AND CARDIAC CHAMBERS

With recent advances in imaging technologies, blood flow in the major blood vessels and chambers of the heart is now measured noninvasively on a routine basis with Doppler ultrasound and CMR imaging. Laser Doppler perfusion imaging, which permits an assessment of relative perfusion of the skin, has also been extensively used for direct measurement of blood flow.

4.1. Doppler Ultrasound Flow Imaging

Doppler ultrasound can be used to measure blood flow in real time, allowing the assessment of stenoses, cerebral blood flow, and even muscle velocities. The acoustic windows available into the body, however, restrict Doppler ultrasound, and the measurement of velocity is affected by the angle of inclination of the transducer to the blood vessel or cardiac chamber. Another limitation is that only the velocity parallel to the beam can be measured. Nevertheless, ultrasound is relatively inexpensive and portable, which can be a significant advantage in a clinical environment.

Doppler ultrasound employs the basic principle that as an object emitting waves at a certain frequency moves in relation to the observer, its frequency appears to change which is easily observed by listening to the siren of a passing fire truck; as it comes closer, the siren is higher pitched and as it moves away, the pitch becomes lower. This principle can be applied to ultrasonic scans of moving tissue. The transducer (i.e., emitter and receiver) is stationary, but the reflector is moving, which gives rise to a shift in frequency that is directly proportional to the velocity of the moving reflector.

$$f_d = \frac{-2vf_0 \cos\theta \cos(\delta/2)}{c}, \quad (3)$$

where v is the reflector's velocity, f_0 is the transmitted frequency, c is the speed of ultrasound, θ is the angle between the bisector of the transmitter and receiver beams and the direction of the movement, and δ is the angle between the transmitter and receiver beams as illustrated in Fig. 2. If the movement of the reflector is away from the transducer (i.e., positive), the shift in frequency is negative (i.e., it becomes lower). The angle between transducer and receiver is usually small enough to reduce the expression to

$$f_d = \frac{-2vf_0 \cos\theta}{c}. \quad (4)$$

It should be noted that if $\theta = 90°$, no Doppler shift occurs. Consequently, to measure the velocity of the reflector accurately, which in the case of blood is the red blood cell, the transducer should be in line with the blood vessel of interest or the angle between the vessel, and the transducer should be known (2,22).

The two main types of systems that can be used to measure blood flow velocities with Doppler ultrasound are pulsed wave and continuous wave systems. Continuous wave systems use separate emitter and receiver transducers, thus allowing continuous imaging of blood velocities. Pulsed wave systems, on the other hand, use the same transducer to send and receive the ultrasonic signal. Pulsed wave Doppler allows depth resolution and can therefore be used to measure flow in single vessels and to measure velocity variation within blood vessels with small sample volumes. Continuous wave Doppler, however, is simpler to use, less expensive, and does not have the aliasing problems associated with pulsed wave Doppler systems (22). Doppler color flow imaging can be used to map the mean velocity in the heart and blood vessels as a color map onto 2-D anatomical images. Doppler velocity information can be used to calculate two different hemodynamic indices, the first being blood flow and the second pressure gradients (23). Blood flow is calculated using the measured velocity and the cross-sectional area of the blood vessel, which can be obtained by using M-mode ultrasound. Pulsed and continuous wave Doppler also allow direct pressure measurement by using the modified Bernoulli equation as follows:

$$\Delta P = 4v^2, \quad (5)$$

where ΔP is the pressure difference measured in mmHg and v is the blood velocity measured in m/s.

Clinically, it is possible to estimate intravascular and intracardiac pressures as an indicator of disease and disease severity (22–24). As an example, Fig. 3 shows aortic pressure gradients measured in a patient with aortic stenosis. The top figure shows the aortic (Ao) and left ventricular (LV) pressures in systole during continuous catheterization. The most common pressure gradient measured during catheterization is peak-to-peak pressure difference between Ao and LV. These, however, occur at different parts of systole, with LV peak pressure occurring earlier than Ao. The peak instantaneous pressure difference is an important physiologic quantity and is defined as the maximum pressure difference between Ao and LV at a given point in time, which is equivalent to the Doppler-derived pressure gradient, as can be seen in Fig. 3, $v = 5$ m/s, corresponding to 100 mmHg using Equation 5.

In practical clinical use, the Doppler system is usually combined with a standard B-mode ultrasound scanner to incorporate anatomical information with the flow information to display a color map superimposed on an anatomical image of the region. In the example shown in Fig. 4, velocities in red indicate flow toward the transducer and

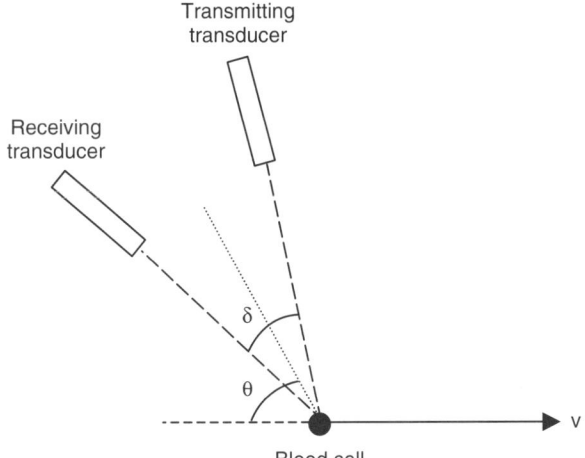

Figure 2. Principle of Doppler ultrasound imaging. Sound waves emitted from the transmitting transducer hit the moving reflector, the blood cell, at an angle. The frequency of the reflected beam is recorded by the receiving transducer, and the Doppler equations can be used to calculate the velocity of the blood cell. v is the reflector's velocity, θ is the angle between the bisector of the transmitter and receiver beams and the direction of the movement, and δ is the angle between the transmitted and received beams.

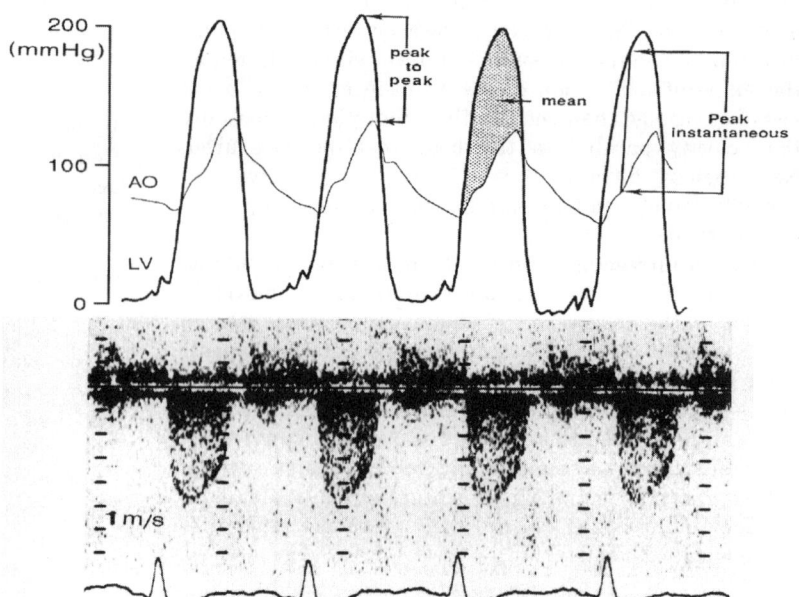

Figure 3. Doppler ultrasound measurements of aortic gradients in a patient with aortic stenosis. Top: Pressure gradients derived during continuous catheterization where Ao is the aortic pressure in systole and LV represents the left ventricular pressure in systole. Bottom: Doppler ultrasound measurement of velocities in the aorta. The peak Doppler velocity corresponds to the peak instantaneous pressure difference when converted using the modified Bernoulli equation. [Image courtesy Dr. P. Nihoyannopolous, Department of Clinical Cardiology, Hammersmith Hospital, Imperial College, London, UK.]

blue indicate flow away from the transducer. The B-mode ultrasound image also helps the operator locate the vessel of interest and direct the beam accordingly (22,24). Doppler color flow imaging has found application in a range of clinical settings, from measuring blood flow in the cardiac chambers and major vessels (25,26), to opthalmological (27) and oncological applications. In oncological imaging, color Doppler ultrasound has found relevance in investigating breast lesions, prostate cancer, ovarian cancer, and detecting metastases (28,29). Doppler ultrasound has also shown potential in gynecological imaging, allowing the uterine and ovarian blood flow to be examined (30). Color Doppler ultrasound allows the clinician to examine patterns of flow in the cardiac chambers and the major arteries, allowing diagnosis of disease based on changes in the normal flow pattern. Figure 4 shows a color Doppler image of a patient with mitral regurgitation, the regurgitant flow can be seen as the orange flow back into the left atrium. Limitations of color Doppler ultrasound are as those for conventional ultrasound, which include limited depth resolution, noise caused by speckle, and limited acoustic windows into the body. In terms of flow measurement, the major disadvantage is that only flow in a direction parallel to the transducer can be measured. Its strength lies in the high temporal resolution obtainable as compared with CMR, allowing near real-time flow imaging with time between frames as low as 5 ms. It is also portable and relatively inexpensive, giving it a considerable advantage over the expense of and dedicated space needed to house CMR systems.

4.2. Blood Flow Measurement using CMR

Compared with Doppler ultrasound flow imaging, CMR is more expensive but more versatile. The technique allows any oblique plane in the body to be imaged, thus permitting the alignment of the imaging planes with the blood vessel of interest. Velocity can be measured in all directions, and a wide variety of complex flows, such as those found in the cardiac chambers and curved vessels, can be accurately quantified. CMR allows a great deal of flexibility as images are anatomically correct and different functional and anatomical indices can be calculated. For example, in cardiac CMR, a range of indices such as chamber volumes, filling rates, and perfusion can be measured in a single examination.

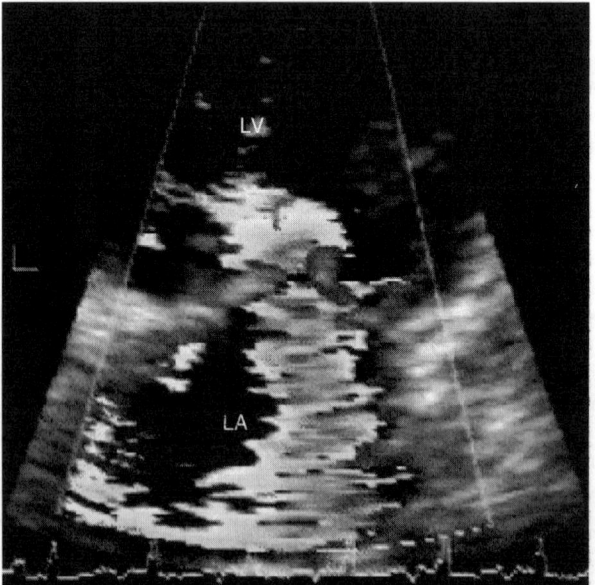

Figure 4. Color Doppler image of a patient with mitral regurgitation. LV is the left ventricle and LA is the left atrium. The blue is blood flowing into the LV, whereas the orange is the regurgitant flow back into the LA. [Image courtesy Dr. P. Nihoyannopolous, Department of Clinical Cardiology, Hammersmith Hospital, Imperial College, London, UK.] (This figure is available in full color at http://www.mrw.interscience.wiley.com/ebe.)

CMR techniques to measure blood flow were first developed in the early 1980s (31,32). The two main methods for measuring blood flow with CMR are time-of-flight and phase contrast. Time-of-flight methods are based on inflow effects. Spins perpendicular to the vessel to be imaged can be magnetically saturated, "tagged", and the time taken for the tagged blood to travel a certain distance can be used to calculate its speed. This method is routinely used in CMR angiography. The phase contrast method employs the inherent property of CMR to alter the phase shifts of the spins*, and has become established for clinical application in the major blood vessels and cardiac chambers.

4.2.1. Phase Contrast Velocity Mapping. The principle behind phase contrast velocity mapping can be derived from basic MR principles. It uses the fact that a spin moving in a gradient magnetic field will experience a motion-related phase shift. The phase acquired by a particle can be calculated from the zero and first moments of the gradients applied. For stationary spins, the phase, $\phi(t)$, is given by the zero moment, Equation 6, and for spins moving at a constant velocity by the first moment, Equation 7.

Zero gradient moment:

$$\phi(t) = \gamma r_0 \int G(t) dt \qquad (6)$$

First gradient moment:

$$\phi(t) = \gamma v_0 \int t G(t) dt, \qquad (7)$$

where γ is the gyromagnetic constant, $G(r) = (G_x, G_y, G_z)$ is the gradient magnetic field applied, r_0 is the spatial position of the spin, and v_0 is the spin's constant velocity (33). The phase shifts associated with higher orders of motion, such as acceleration, can also be calculated.

CMR sequences can be specially designed to exploit the above phase shift effect by using a "bipolar" gradient (i.e., a gradient pulse that has two symmetrical lobes). Although it may seem that this pulse achieves nothing, this is only true for stationary spins. Moving spins acquire a phase shift directly related to their velocity. This relationship between the gradient applied and the phase can therefore be used to measure the velocity of the moving spins. If a balanced bipolar gradient is applied before image acquisition, stationary tissues will not experience a net phase change. However, spins moving at a constant velocity will experience a phase change that is directly proportional to their velocity, as can be seen in Fig. 5 (34). By subtracting the phase images from a scan with and without the bipolar pulse, a direct measurement of the velocity can be obtained, as shown in Fig. 6(b) (34).

The velocity sensitivity of a bipolar gradient pulse can be easily derived using Equation 7 and Fig. 6(a) as follows:

$$\phi(t) = 2\gamma v_0 G \delta \Delta. \qquad (8)$$

*Spin is a fundamental quantum mechanical property associated with any atom. In MR imaging, the received MR signal consists of the contribution of spins of the hydrogen atoms from the water in the tissue.

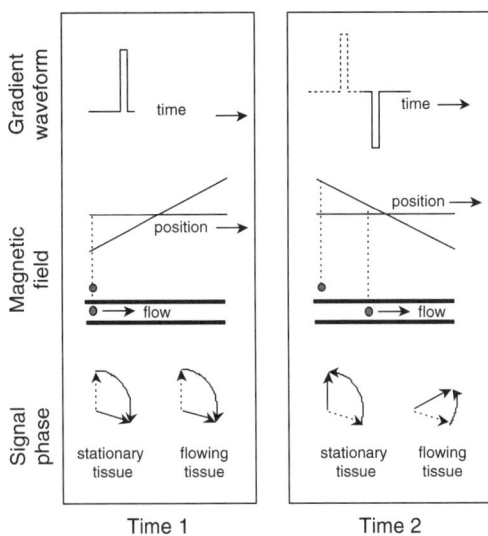

Figure 5. Principles of CMR phase contrast velocity mapping. The left-hand side shows that the initial gradient applied at Time 1 causes the same amount of phase shift in moving and stationary tissues. However, when the second half of the bipolar gradient is applied at Time 2, the phase of the stationary tissue returns to its original position, but the phase of the flowing tissue does not, which is because this tissue is now in a different spatial position experiencing a reduced phase shift related to that position. This phase shift is directly proportional to the velocity of the flowing tissue (34).

γ, G, δ and Δ are constant for any particular CMR sequence (34). Thus, the phase is directly proportional to the velocity. The range of phase that is physically measured is set to 2π, which allows the calculation of the velocity sensitivity (venc) of the gradient applied, or alternatively, the gradient needed to achieve a particular venc. The velocities in the region of interest to be imaged should cause large phase shifts to be measured accurately. However, if they are too large (i.e., greater than $\pm 180°$), velocity aliasing will occur. To avoid this, the venc value should be estimated before designing the CMR sequence. For pulsatile flow, the sequence can be designed to have higher venc during the period with higher velocities and a lower venc during the period with lower velocities to record them more accurately. For example, this would correspond to a higher venc during systole and a lower venc during diastole in a cardiovascular application. Another method that can be used to extend the dynamic range of the velocity-encoding sequence is to use phase unwrapping after image acquisition, which can be performed in only the spatial domain or both the spatial domain and temporal domain (35). The method used by Yang et al. (35) allows the effective velocity sensitivity of the sequence to be increased almost twofold, as illustrated in Fig. 7. Figure 7 (b, c, d, e) show the flow images with and without the phase unwrapping in the descending aorta and the outflow tract of the right ventricle. Figure 7(a) shows the corresponding anatomical MR magnitude images. Figure 8 shows the blood velocities recorded in the RV outflow tract, the descending aorta, and the aortic root at the regions indicated by the arrows A, B, and C. It can be

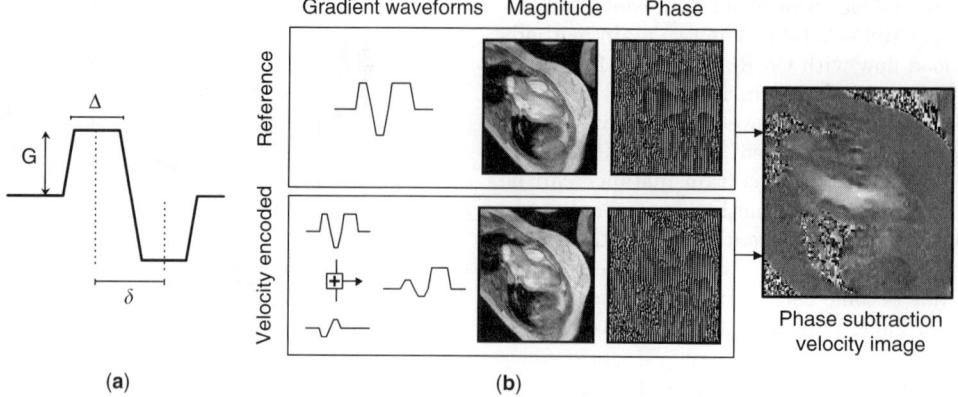

Figure 6. (a) The amplitude and timing of the bipolar gradient determine the velocity sensitivity of the sequence. G is the gradient amplitude, δ is the time between the middle of the two lobes of the pulse, and Δ is the duration of the gradient. (b) The gradient waveforms used to produce a velocity-encoded image and their corresponding magnitude and phase images. The velocity image is obtained by subtraction of the reference (top) and velocity-encoded (bottom) phase images (34).

seen that the sensitivity of the velocity mapping sequence has been increased twofold by using the phase unwrapping post-processing technique, allowing much improved velocity accuracy.

4.2.2. CMR Sequence Design. Phase contrast velocity mapping can be used with any type of CMR sequence as it only requires the addition of a bipolar gradient pulse. However, standard spin echo sequences cannot be used for CMR velocity mapping because of the outflow of blood by the time the second 180° pulse is applied. Thus, the most common sequences are gradient-recalled echo (GRE) sequences with additional flow compensation [also known as even echo rephrasing, motion artifact suppression (MAST), gradient moment nulling, or gradient moment rephrasing or refocusing (GMR)] to reduce signal loss caused by intravoxel phase dispersion (33,34). In cardiac imaging, velocity mapping sequences are usually ECG-triggered, segmented GRE sequences. A segmented sequence encodes more than one line of data per cardiac cycle to reduce the acquisition time. Cardiac imaging is limited by superposition of respiratory motion on the normal motion of the heart, which can be circumvented by using either breath-hold sequences or respiratory

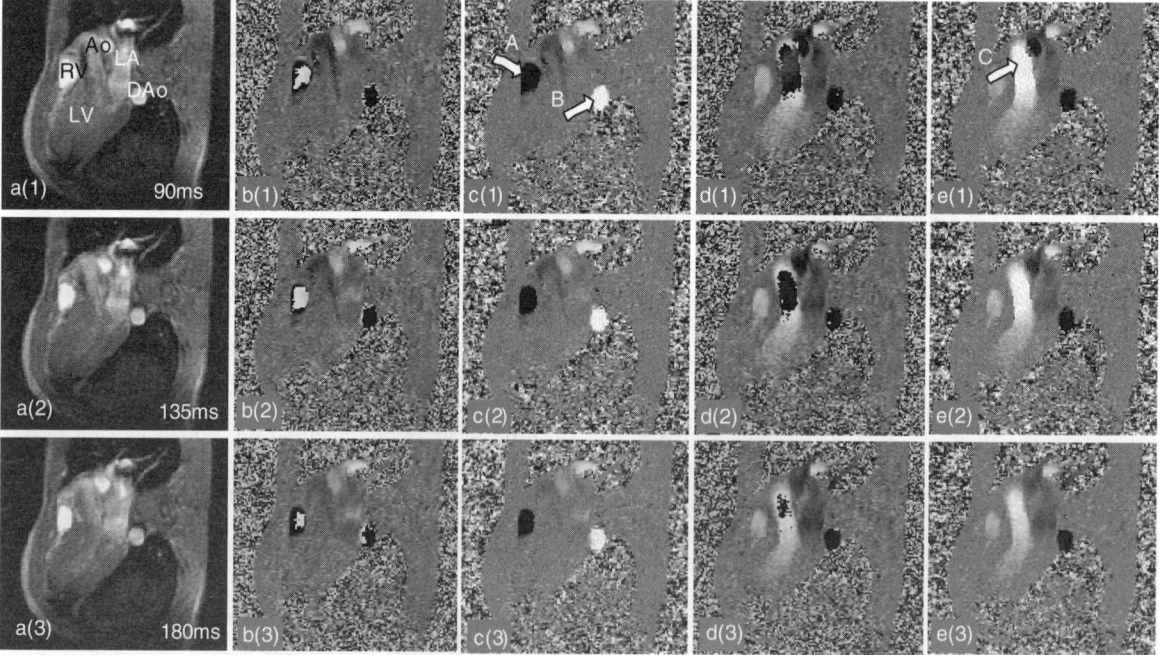

Figure 7. Phase unwrapping for dynamic range extension for phase contrast velocity mapping. Images (b, c, d, e) show the flow images with and without the phase unwrapping in the descending aorta and the outflow tract of the right ventricle. Figure 7(a) shows the corresponding anatomical MR magnitude images (35).

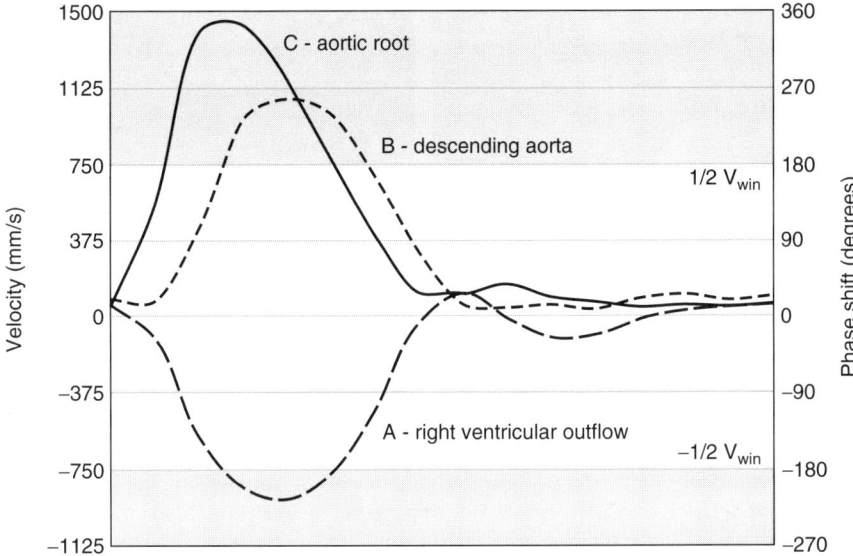

Figure 8. The blood velocities recorded in the RV outflow tract, the descending aorta and the aortic root at the regions indicated by the arrows A, B, and C in Fig. 7. It can be seen that the sensitivity of the velocity mapping sequence has been increased two-fold by using the phase unwrapping post-processing technique, allowing much improved velocity accuracy (35).

triggering. In practice, it is difficult to design breath-hold sequences for CMR velocity mapping as velocity mapping requires one reference scan plus additional scans for each velocity encoding direction, which means that up to four scans may be required for each phase of the cardiac cycle. Designing CMR velocity mapping sequences achievable in a breath-hold either entails decreased spatial or temporal resolution or using schemes such as "view-sharing" with which data from adjacent time frames can be used to synthesize in between time frames.

Typically, a nonbreath-hold flow study with three velocity directions can require up to a few minutes to acquire per image. This limitation can be overcome by using rapid CMR imaging techniques such as echo planar imaging (EPI), spiral imaging, parallel imaging, and techniques such as UNFOLD and kt-BLAST that exploit spatio-temporal correlations of the cardiac structures. Techniques such as UNFOLD allow a reduction in acquisition time by unaliasing through Fourier encoding in the temporal domain (36–38). These techniques are only useful in imaging applications where a series of images is acquired over time, and thus are ideal for cardiac imaging. In terms of sequence design, EPI and spiral imaging allow an image to be acquired in a short period by covering a number of k-space lines in a single cardiac cycle. Potentially, these techniques could be combined with velocity mapping to allow multidirectional flow data to be acquired in a single breath-hold. Further enhancement in imaging speed can be realized with the use of parallel imaging techniques, which employ the sensitivities of the receiver coils to unwrap aliased images to speed up the acquisition (39–41).

To enable complete coverage of the cardiac cycle, continuous imaging can be carried out in conjunction with restrospective gating, which involves recording the electrocardiogram (ECG) trace and then extrapolating the data back onto the cardiac cycle. It should be noted that as the design of bipolar velocity encoding gradients does not take into account higher orders of motion such as acceleration, imaging of turbulent flow can lead to signal loss. Interestingly, signal loss has proven to be useful for visual assessment of cases such as mitral regurgitation where the extent of the signal loss can help in visualizing the regurgitant jet.

4.2.3. Clinical Applications of CMR Blood Flow Imaging. The unique capability of CMR for examining detailed blood flow in any region of the body makes it an ideal tool for assessing a range of clinical conditions. It is useful for examining velocities in the thoracic aorta as the vessel is relatively large and CMR offers a multidirectional velocity profile of the whole volume of the vessel (42,43). Thus far, CMR has been used to study blood flow in all the major blood vessels, such as the pulmonary arteries and veins, and the caval veins (44).

Assessment of valvular disease can also be enhanced with CMR velocity mapping. Regurgitant flow, such as that found in mitral regurgitation, can be analyzed with CMR velocity mapping (45–47). As mentioned earlier, the turbulent regurgitant flow causes a signal void, thus providing a semiquantitative way of measuring the severity of regurgitation (48,49). CMR is especially useful for examining congenital heart disease, as no physical assumptions are made when imaging the anatomy or function. CMR velocity mapping has also become an essential tool for the study of spatial and temporal patterns of blood flow in the cardiac chambers. It is especially useful for assessing flow pattern changes in the left ventricle in diseases with ventricular remodeling such as dilated or hypertrophic cardiomyopathy.

CMR phase contrast velocity mapping can also be used to compute relative pressures in the major vessels and cardiac chambers. The Navier–Stokes equations for incompressible fluids can be used to compute the pressure differences associated with blood flow (50,51). The method has been validated against the gold standard of catheter measurements and Doppler ultrasound (50,52). Figure 9 shows an example of the pressure field found in an aortic

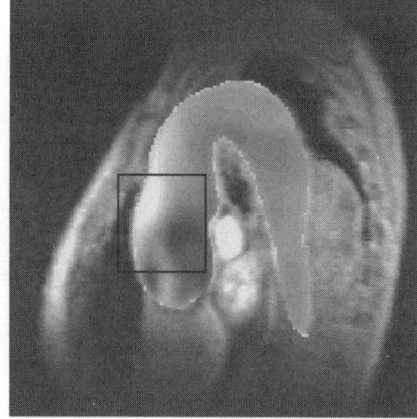 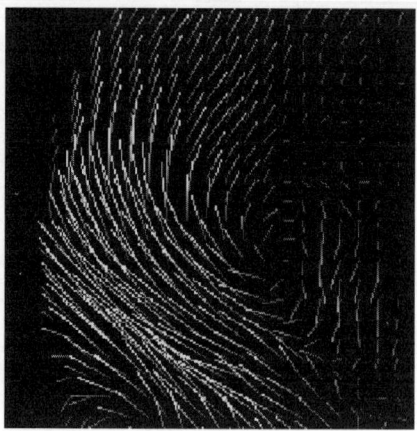

Figure 9. Pressure field in an aortic aneurism calculated from the MR velocity measurements by using the Navier–Stokes equations for incompressible fluids. (This figure is available in full color at http://www.mrw.interscience.wiley.com/ebe.)

aneurism calculated with CMR phase contrast velocity mapping (50). Although this is not a routinely used technique, an online system could be developed to allow simultaneous pressure and velocity measurement within the body using CMR velocity mapping. Finally, flow in the coronary arteries has also been studied with CMR, but it is still a challenge because of the small size of the vessels and their movement caused by cardiac and respiratory motion (53–56).

4.2.4. Flow Field Visualization and Restoration. Although CMR velocity mapping permits flexible multidirectional flow information to be measured, direct visualization and quantification of the flow data is of limited clinical use, which is because velocity is displayed as a grayscale value for each pixel and the separate velocity components can only be displayed one at a time. Thus, effective visualization and quantification of flow data has been the subject of research for several years. The two threads of the work center around the display of flow data for easy clinical diagnosis and the quantification of flow features that can aid in diagnosis. For visualization, two popular methods are streamlines and "seeds" that drift with the flow over time (57–59). Figure 10 shows an example of visualization of blood flow in the dilated left ventricle of a patient using the "seeds" method employed by Yang et al. (57).

As mentioned earlier, one important application of CMR velocity mapping is to examine blood flow in the left ventricle. This application becomes increasingly important for heart disease involving myocardial remodeling. It has been shown that significant differences exist in flow pattern topology between normal subjects and patients with remodeled left ventricles. The difference is manifested not only in general flow patterns, such as the formation and propagation of vortices, but also in its dynamics over the cardiac cycle. Analysis of ventricular flow data requires the extraction of critical points associated with salient flow features, which can be tracked throughout the cardiac cycle. Work is ongoing in developing algorithms that allow accurate identification and tracking of vortical features in 3-D flow data (60–62). Recent work centered on tracking the vortices in ventricular blood flow throughout the cardiac cycle has developed an effective framework for the restoration, abstraction, and extraction of flow features such that the associated dynamic indices can be accurately tracked and quantified (61).

4.2.5. Future of CMR Blood Flow Imaging. Current state-of-art CMR imaging is increasingly moving towards real-time imaging to enable the analysis of physiological changes in response to exercise or pharmacological stress. Stress testing with exercise or pharmacological agents has been proven to be a valuable tool for evaluating the

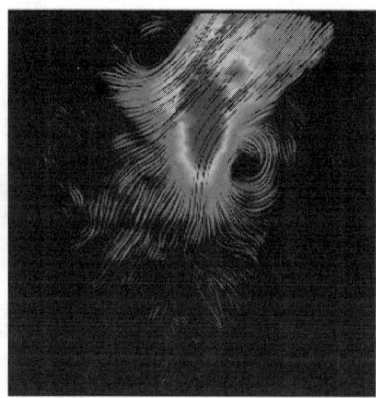 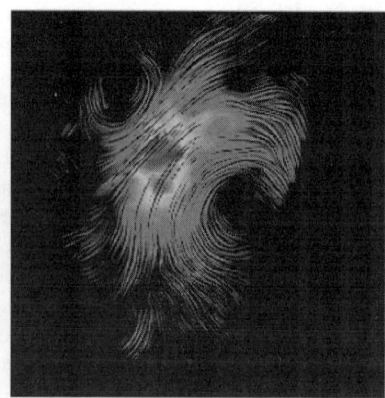 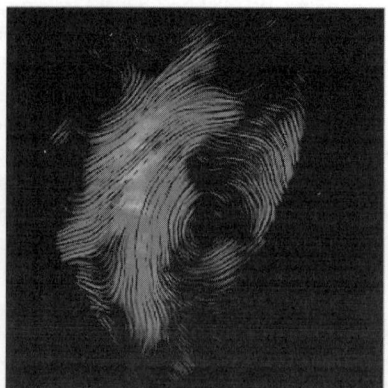

Figure 10. Visualization of blood flow using the "seeds" method. These images show blood flowing into the left ventricle of a patient with a dilated left ventricle. (This figure is available in full color at http://www.mrw.interscience.wiley.com/ebe.)

prognostic importance of functional abnormalities introduced by coronary artery disease. Previous studies involving CMR for stress testing were limited by the temporal averaging nature of conventional velocity encoding sequences, typically requiring several minutes for the acquisition of each of the velocity components. Quantitative systolic ejection indices of maximal acceleration, peak velocity, and volume of blood ejected from the left ventricle can be used to assess the functional significance of coronary artery disease and inducible myocardial ischemia. In order to have a more quantitative evaluation of dynamic changes of the ventricular function related to stress, rapid high-resolution imaging of flow and structural changes is required, which can be achieved through real-time CMR flow imaging techniques.

Real-time imaging sequences usually consist of EPI or spiral imaging. Real-time EPI and spiral sequences can be adapted to incorporate velocity mapping by adding the necessary bipolar gradients (59,63), which allows the assessment of real-time blood flow in the cardiac chambers and major blood vessels. Despite the continuing improvement of CMR imaging, velocity mapping techniques are faced with the ultimate spatio-temporal resolution limited by CMR hardware and physiological motion. One important trend in the field is to integrate computational fluid dynamics with the acquired flow data to enable visualization of a complete 4-D (3-D plus time) flow field without the need to acquire velocity scans of the whole heart (64,65). CFD blood flow simulation has important applications in cardiac imaging as it allows the calculation of flow indices that cannot be directly measured such as mass transport, wall shear, and boundary flow layer. It also allows the creation of flow fields at a far higher temporal and spatial resolution than can be provided by flow imaging. Furthermore, it permits the prediction of flow parameters and evaluation of efficacy of surgical and therapeutic measures. It is worth noting that CMR imaging is not the only modality that can be used for this work. IntraVascular UltraSound (IVUS), for example, can provide an effective and practical solution for estimating endothelial shear stress despite its invasiveness (66). As a general trend for the assessment of cardiac and vascular flow, it will become increasingly common to fuse data from different modalities as each technology has its own advantages in terms of resolution, accessibility, and the range of functional properties that can be measured. The combination of these techniques promises to provide an effective tool for investigating the relationships between morphological structure and the hemodynamic properties of flow.

5. DISCUSSION AND CONCLUSION

Blood flow measurement with CMR is an important component of the ongoing interdisciplinary research towards understanding the localized development and progression of disease. Initiatives such as the Physiome Project are working towards integrating all levels of information available to us through technological advances, from genetics to CMR imaging (67). The fusion of probabilistic atlases of cardiac anatomy and functional models of the heart with blood flow data is being pursued at several research centers (65,67), which will allow a more complete understanding of how all the different levels of physiological function come together to produce an effective working human body. It also provides further insight into the development of disease and helps in devising effective means of treatment and prevention. The measurement of blood flow in different regions of the body has become a fundamental tool for diagnosis of disease. Improvements of traditional techniques and the development of new imaging techniques are now allowing the clinician access to information hitherto unquantifiable. The cardiovascular system have been the focus of such research as both morphological and functional information is readily available with techniques such as CT, echocardiography, PET, and CMR imaging (67).

Acknowledgments

The authors would like to thank Dr. Peter Gatehouse, Prof. David Firmin, Dr. Raad Mohiaddin, Dr. Petros Nihoyannopoulos, Dr. Robert Merrifield, Dr. Pauline Ng, and Dr. Paramate Horkaeow for their input.

BIBLIOGRAPHY

1. K. Zierler, Indicator dilution methods for measuring blood flow, volume, and other properties of biological systems: a brief history and memoir. *Ann. Biomed. Eng.* 2000; **28**: 836–848.
2. W. Nichols and M. O'rourke, *MacDonald's Blood Flow in Arteries: Theoretical, Experimental and Clinical Principles*, 3rd ed. London, UK: Edward Arnold, 1990.
3. R. T. Mathie, *Blood Flow Measurement in Man*. Kent, UK: Tunbridge Wells, Castle House, 1982.
4. R. Tabrizchi and M. Pugsley, Methods of blood flow measurement in the arterial circulatory system. *J. Pharm. Tox. Methods* 2000; **44**:375–384.
5. C. Cappelen, New findings in blood flowmetry. Presented at Congress on Blood Flow Measurement, Oslo, Norway, 1968.
6. D. Wyatt, The electromagnetic blood flowmeter. *J. Scientific Instruments (J. Phys. E.)* 1968; **2**(1):1146–1152.
7. C. Riva, J. Grunwald, and S. Sinclair, Laser Doppler measurement of relative blood velocity in the human optic nerve head. *Invest. Opthalmol. Vis. Sci.* 1982; **22**:241–248.
8. C. Riva, B. Ross, and G. Benedek, Laser Doppler measurements of blood flow in capillary tubes and retinal arteries. *Invest. Opthalmol.* 1972; **11**:936–944.
9. M. Geiser, U. Diermann, and C. Riva, Compact laser Doppler choroidal flowmeter. *J. Biomed. Optics* 1999; **4**:459–464.
10. T. Winsor, D. Haumschild, D. Winsor, Y. Wang, and T. Luong, Clinical application of laser Doppler flowmetry for measurement of cutaneous circulation in health and disease. *Angiology* 1987; **38**:727–736.
11. R. Emshoff, I. Emshoff, I. Moschen, and H. Strobl, Laser Doppler flow measurements of pulpal blood flow and severity of dental injury. *Int. Endod. J.* 2004; **37**:463–467.
12. N. Raine and J. Sneddon, A simple water-filled plethysmograph for measurement of limb blood flow in humans. *Advan. Physiol. Edu.* 2002; **26**:120–128.

13. I. Wilkinson and D. Webb, Venous occlusion plethysmography in cardiovascular research: methodology and clinical applications. *Br. J. Clin. Pharmacol.* 2001; **52**:631–646.
14. C. Roberts, *Blood Flow Measurement*. London, UK: Sector, 1972.
15. E. Stenow and P. Oberg, Venous occlusion plethysmography using a fiber-optic sensor. *IEEE Trans. Biomed. Eng.* 1993; **40**:284–289.
16. S. Hansel, G. Lassig, F. Pistrosch, and J. Passauer, Endothelial dysfunction in young patients with long-term rheumatoid arthritis and low disease activity. *Atherosclerosis* 2003; **170**:177–180.
17. Y. Higashi and M. Yoshizumi, New methods to evaluate endothelial function: method for assessing endothelial function in humans using a strain-gauge plethysmography: nitric oxide-dependent and -independent vasodilation. *J. Pharmacol. Sci.* 2003; **93**:399–404.
18. C. Tentolouris, D. Tousoulis, C. Antoniades, E. Bosinakou, M. Kotsopoulou, A. Trikas, P. Toutouzas, and C. Stefanadis, Endothelial function and proinflammatory cytokines in patients in ischemic heart disease and dilated cardiomyopathy. *Int. J. Cardiol.* 2004; **94**:301–305.
19. Q. Zhang, G. Andersson, L. Lindberg, and J. Styf, Muscle blood flow in response to concentric muscular activity vs passive venous compression. *Acta Physiol. Scand.* 2004; **180**: 57–62.
20. S. Romano and M. Pistolesi, Assessment of cardiac output from systemic arterial pressure in humans. *Crit. Care Med.* 2002; **30**:1834–1841.
21. J. Stewart and L. Montgomery, Regional blood volume and peripheral blood flow in the postural tachycardia syndrome. *Am. J. Physiol. Heart Circ. Physiol.* 2004; **287**(3):1319–1327.
22. P. Fish, *Physics and Instrumentation of Diagnostic Medical Ultrasound*. Chichester, UK: Wiley, 1990.
23. H. Feigenbaum, *Echocardiography*, 5th ed. Philadelphia, PA: Lea & Febiger, 1994.
24. J. Kisslo, D. B. Adams, and R. N. Belkin, *Doppler Color Flow Imaging*. New York: Churchill Livingstone, 1988.
25. C. Gorg, J. Riera-Knorrenschild, and J. Dietrich, Pictorial review: colour Doppler ultrasound flow patterns in the portal venous system. *Br. J. Radiol.* 2002; **75**:919–929.
26. J. Sands, L. Ferrell, and M. Perry, The role of color flow Doppler ultrasound in dialysis access. *Semin. Nephrol.* 2002; **22**:195–201.
27. T. Williamson and A. Harris, Color Doppler ultrasound imaging of the eye and orbit. *Surv. Ophthalmol.* 1996; **40**: 255–267.
28. M. Blomley and R. Eckersley, Functional ultrasound methods in oncological imaging. *Eur. J. Cancer* 2002; **38**:2108–2115.
29. T. Mehta, S. Raza, and J. Baum, Use of Doppler ultrasound in the evaluation of breast carcinoma. *Semin. Ultrasound CT MR* 2000; **21**:297–307.
30. L. Valentin, Use of colour and spectral Doppler ultrasound examination in gynaecology. *Eur. J. Ultrasound* 1997; **6**: 143–163.
31. P. van Dijk, Direct cardiac NCMR imaging of heart wall and blood flow velocity. *J. Comput. Assisted Tomogr.* 1984; **8**:429–436.
32. D. Bryant, J. Payne, D. Firmin, and D. Longmore, Measurement of flow with NMR imaging using a gradient pulse and phase difference technique. *J. Comput. Assisted Tomogr.* 1984; **8**:588–593.
33. M. Haacke, R. Brown, M. Thompson, and R. Venkatesan, *Magnetic Resonance Imaging: Physical Principles and Sequence Design*. New York: John Wiley & Sons, 1999.
34. P. Gatehouse and D. Firmin, Methoden der flussmessung. In: E. Fleck, ed., *Kardiovaskuläre Magnetresonanztomographie*. Darmstadt, Germany: Steinkopff Verlag, 2002, pp. 25–29.
35. G. Yang, P. Burger, P. Kilner, S. Karwatowski, and D. Firmin, Dynamic range extension of cine velocity measurements using motion-registered spatiotemporal phase unwrapping. *J. Mag. Res. Imag.* 1996; **6**:495–502.
36. B. Madore, G. Glover, and N. Pelc, Unaliasing by fourier-encoding the overlaps using the temporal dimension (UNFOLD), applied to cardiac imaging and fMRI. *Mag. Res. Med.* 1999; **42**:813–828.
37. J. Tsao, P. Boesiger, and K. Pruessmann, k-t BLAST and k-t SENSE: dynamic MRI with high frame rate exploiting spatiotemporal correlations. *Mag. Res. Med.* 2003; **50**:1031–1042.
38. B. Madore, Using UNFOLD to remove artifacts in parallel imaging in partial-Fourier imaging. *Mag. Res. Med.* 2002; **48**:493–501.
39. K. Pruessmann, M. Weiger, M. Scheidegger, and P. Boesiger, SENSE: sensitivity encoding for fast MRI. *Mag. Res. Med.* 1999; **42**:952–962.
40. M. Bydder, D. Larkman, and J. Hajnal, Generalized SMASH imaging. *Mag. Res. Med.* 2002; **47**:160–170.
41. M. Griswold, P. Jakob, R. Heidemann, M. Nittka, V. Jellus, J. Wang, B. Kiefer, and A. Haase, Generalized autocalibrating partially parallel acquisitions (GRAPPA). *Mag. Res. Med.* 2002; **47**:1202–1210.
42. R. Mohiaddin, P. Kilner, S. Rees, and D. Longmore, Magnetic resonance volume flow and jet velocity mapping in aortic coactation. *J. Am. Coll. Cardiol.* 1993; **22**:1515–1521.
43. R. Mohiaddin, D. Firmin, and D. Longmore, Age-related changes of human aortic flow wave velocity measured noninvasively by magnetic resonance imaging. *J. Appl. Physiol.* 1993; **74**:492–497.
44. R. Paz, R. Mohiaddin, and D. Longmore, Magnetic resonance assessment of the pulmonary arterial trunk anatomy, flow, pulsatility and distendibility. *Eur. Heart J.* 1993; **14**:1524–1530.
45. P. Kilner, C. Manzara, R. Mohiaddin, D. Pennell, M. Sutton, D. Firmin, S. Underwood, and D. Longmore, Magnetic resonance heat velocity maping in mitral and aortic valve stenosis. *Circulation* 1993; **87**:1239–1248.
46. G. Aurigemma, N. Reichek, M. Schiebler, and L. Axel, Evaluation of mitral regurgitation by cine magnetic resonance imaging. *Am. J. Cardiol.* 1990; **66**:621–625.
47. S. Kozerke, J. Schwitter, E. Pedersen, and P. Boesiger, Aortic and mitral regurgitation: quantification using moving slice velocity mapping. *J. Magn. Reson. Imag.* 2001; **14**:106–112.
48. K. Nayak, B. Hu, and D. Nishimura, Rapid quantitation of high-speed flow jets. *Magn. Reson. Med.* 2003; **50**:366–372.
49. K. Nayak, J. Pauly, A. Kerr, B. Hu, and D. Nishimura, Real-time color flow MRI. *Magn. Reson. Med.* 2000; **43**:251–258.
50. G. Yang, P. Kilner, N. Wood, S. Underwood, and D. Firmin, Computation of flow pressure fields from magnetic resonance velocity mapping. *Magn. Reson. Med.* 1996; **36**:520–526.
51. R. Thompson and E. McVeigh, Fast measurement of intracardiac pressure differences with 2D breath-hold phase-contrast MRI. *Magn. Reson. Med.* 2003; **49**:1056–1066.

52. S. Caruthers, S. Lin, P. Brown, M. Watkins, T. Williams, K. Lehr, and S. Wickline, Practical value of cardiac magnetic resonance imaging for clinical quantification of aortic valve stenosis: comparison with echocardiography. *Circulation* 2003; **108**:2236–2243.
53. R. Edelman, W. Manning, D. Burstein, and S. Paulin, Corornary arteries: breath-hold MR angiography. *Radiology* 1991; **181**:641–643.
54. R. Edelman, W. Manning, E. Gervino, and W. Li, Flow velocity quantification in human coronary arteries with fast, breath-hold MR angiography. *J. Mag. Res. Imag.* 1993; **3**: 699–703.
55. J. Keegan, D. Firmin, P. Gatehouse, and D. Longmore, The application of breath hold phase velocity mapping techniques to the measurement of coronary artery blood flow velocity: phantom data and initial *in vivo* results. *Mag. Res. Med.* 1994; **31**:526–536.
56. J. Keegan, P. Gatehouse, R. Mohiaddin, G. Yang, and D. Firmin, Comparison of spiral and FLASH phase velocity mapping, with and without breath-holding, for the assessment of left and right coronary artery blood flow velocity. *J. Mag. Res. Imag.* 2004; **19**:40–49.
57. G. Yang, P. Kilner, R. Mohiaddin, and D. Firmin, Transient streamlines: texture synthesis for *in vivo* flow visualisation. *Int. J. Card. Imag.* 2000; **16**:175–184.
58. G.-Z. Yang, The role of quantitative MR velocity imaging in exploring the dynamics of *in vivo* blood flow. *IEEE Engineer. Med. Biol.* 1998; **17**:64–72.
59. S. Napel, D. Lee, R. Frayne, and B. Rutt, Visualizing three-dimensional flow with simulated streamlines and three-dimensional phase-contrast MR imaging. *J. Mag. Res. Imag.* 1992; **2**:143.
60. P. Kilner, G. Yang, A. Wilkes, R. Mohiaddin, D. Firmin, and M. Yacoub, Asymmetric redirection of flow through the heart. *Nature* 2000; **404**:759–761.
61. P. Ng and G. Yang, Vector-valued image restoration with application to magnetic resonance velocity imaging. *J. Win. Sch. Comp. Graph.* 2003; **11**(2):2338–2345.
62. G. Yang, R. Mohiaddin, P. Kilner, and D. Firmin, Vortical flow feature recognition: a topological study of *in vivo* flow patterns using MR velocity mapping. *J. Comput. Assist. Tomogr.* 1998; **22**:577–586.
63. P. Gatehouse, D. Firmin, S. Collins, and D. Longmore, Real time blood flow imaging by spiral scan phase velocity mapping. *Magn. Reson. Med.* 1994; **31**:504–512.
64. N. Wood, S. Weston, P. Kilner, A. Gosman, and D. Firmin, Combined MR imaging and CFD simulation of flow in the human descending aorta. *J. Magn. Reson. Imag.* 2001; **13**: 699–713.
65. R. Merrifield, Patient specific modelling of left ventricular morphology and flow using magnetic resonance imaging and computational fluid dynamics. In: *Computing*. London, UK: Imperial College, University of London, 2003.
66. R. Kram, J. J. Wentzel, J. A. Oomen, R. Vinke, J. C. Schuurbiers, P. J. de Feyter, P. W. Serruys, and C. J. Slager. Evaluation of endothelial shear stress and 3D geometry as factors determining the development of atherosclerosis and remodeling in human coronary arteries *in vivo*. Combining 3D reconstruction from angiography and IVUS (ANGUS) with computational fluid dynamics. *Arterioscler. Thromb. Vasc. Biol.* 1997; **17**(10):2061–2065.
67. E. Crampin, M. Halstead, P. Hunter, P. Nielsen, D. Noble, N. Smith, and M. Tawhai, Computational physiology and the physiome project. *Experiment. Physiol.* 2003; **89**:1–27.

FURTHER READING

M. Stern, In vivo evaluation of microcirculation by coherent light scattering. *Nature* 1975; **254**:56–58.

A. Shepherd and P. Oberg, Laser-Doppler blood flowmetry. In: *Developments in Cardiovascular Medicine*. Boston, MA: Kluwer Academic, 1990.

J. Briers, Laser Doppler, speckle and related techniques for blood perfusion mapping and imaging. *Physiol. Meas.* 2001; **22**:R35–R66.

C. Choi and R. Bennet, Laser Doppler to determine cutaneous blood flow. *Dermatologic Surg.* 2003; **29**:272–280.

C. Riva, Basic principles of laser Doppler flowmetry and application to the ocular circulation. *Int. Opthalmol.* 2001; **23**: 183–189.

BLOOD FLOW SIMULATION, PATIENT-SPECIFIC IN-VIVO

ROBERT MERRIFIELD
QUAN LONG
YUN XU
GUANG-ZHONG YANG
Imperial College
London, UK

1. INTRODUCTION

As blood is transported through the body, it undergoes complex interactions with the cardiovascular system. Although the blood flow patterns in healthy subjects provide an efficient transport mechanism for oxygen and nutrients, they can be severely compromised for those suffering from disease. In some cases, small changes to the cardiovascular anatomy can have significant impact on the flow patterns throughout the body. Because of the close coupling between cardiovascular anatomy and blood flow, the study of hemodynamic properties can provide great insight into the localization and development of cardiovascular disease. Flow pressure distribution, velocity, wall shear stress, and flow topology are some important indices that are of clinical interest. The development of modeling and *in vivo* quantification techniques for the analysis of these properties has been the subject of significant research over the years.

Most early techniques for investigating cardiovascular flow were based on *in vitro* experiments and theoretical models. Approaches such as the Windkessel model and the Womersley equations allowed the prediction of flow parameters with simplified vessel geometries (1). With the introduction of *in vivo* flow imaging techniques (2–11), the nature of blood flow can now be studied in greater detail and in a subject-specific manner. Imaging modalities such as Doppler ultrasound and magnetic resonance (MR) imaging have permitted the acquisition of flow velocity with extensive spatiotemporal coverage. Despite clear clinical benefits, existing flow imaging techniques do have certain limitations. The limited resolution and the presence of imaging artifacts often involve extensive postprocessing steps for restoring the intrinsic flow distributions before reliable hemodynamic indices can be derived. Over recent years, there has been a rapid surge of interest in combining computational fluid dynamics (CFD) with *in vivo* imaging techniques for studying interactions between blood flow and vessel morphology (12–20). CFD gives the ability to compute features/properties that cannot be measured, e.g., wall shear stress and mass transfer rate, but are important to studies of atherosclerosis (21) or the design of vessel prostheses (22–24). Moreover, it can also provide details of the flow that are often beyond the discrimination of existing imaging techniques. This trend is driven by our increased understanding of biomechanics, maturity of computational modeling techniques, and advancement in imaging. CFD has traditionally been used in aerodynamics and thermodynamics to perform simulations controlled by the physical laws that govern the motion of fluids. Its application to blood flow modeling has increased the scope of studying vessel-flow dynamics that is far beyond what is provided by imaging alone.

Early techniques of CFD for blood flow simulation were mainly based on simplified/idealized vessel geometries. With the improvement of imaging techniques, the use of patient-specific information for CFD has become increasingly popular. At first, models of the arterial anatomy were constructed from simple measurements taken from x-ray angiograms. More recently, sufficient development has been made to allow the simulation of arterial flows in physiologically realistic models reconstructed from MR (25–29) or ultrasound images (30). The approach of combining CFD with patient-specific morphology and flow captured with imaging has important clinical applications particularly for the design of vessel prostheses (22–24). As CFD is based on simulation rather than exclusively on measurement, significant errors can be introduced from numerical errors or inappropriate choice of boundary conditions. It is therefore unsuitable to use it as a direct replacement for flow imaging. Instead, it is desirable to exploit the principal benefits of each of these techniques. In this way, the combination of imaging and simulation will provide a unique and versatile tool for the study of vascular morphology and flow. It is envisaged that subject-specific CFD combined with imaging will have an important role in improved design of vessel prostheses and in assessing the efficacy of therapeutic measures.

2. STRUCTURAL MODELING

The principles of CFD are based on the Navier–Stokes equations derived in the nineteenth century by the French and English scientists M. Navier and Sir G. G. Stokes. The equations provide the exact solution for the laminar flow of a simple fluid. The first of these (Equation 1) represents the conservation of mass under the assumption that the fluid is incompressible. The second (Equation 2) expresses the conservation of momentum for a Newtonian fluid. Although these equations can be solved analytically in some cases (31), most flow simulations require numerical approximation methods. Here, the equations are provided in a form that is amenable to a numerical matrix solution.

$$\rho \frac{\partial}{\partial t} \int_V dV + \rho \int_S (U - U_S) \cdot n dS = 0, \quad (1)$$

$$\rho \frac{\partial}{\partial t} \int_V U dV + \rho \int_S U(U - U_S) \cdot n dS$$
$$= \int_S [-p + \mu \nabla U + \mu (\nabla U)^T] \cdot n dS, \quad (2)$$

where U and p are the instantaneous fluid velocity and static pressure at time t; ρ and μ are the density and viscosity of the fluid, respectively; and n is a unit normal vector of the surface S enclosing the volume V. U_S denotes the velocity of the surface S.

The first stage of blood flow simulation involves the acquisition of data that describe the anatomical structure. For *in vivo* modeling, this process typically involves the acquisition of a three-dimensional (3-D) image series covering the anatomical regions of interest. This imaging process is then followed by a reconstruction stage that creates a model representing the physical surfaces of the flow domain. The accuracy with which the simulation can be performed is largely dependent on the amount of detail that can be incorporated into the model. The next stage is the creation of a volume mesh that divides the flow domain into a large set of cells of similar topology and size. The dynamic nature of the cardiovascular system makes this one of the most challenging aspects of blood flow simulation. Following this stage, the boundary conditions that specify the properties of the flow at the inlets and outlets of the flow domain must be defined. Once the volume mesh and the boundary conditions have been successfully integrated, the simulation process can begin, which involves the creation of a set of equations that link the properties of the flow in each cell to those of its immediate neighbors. These equations are then solved in parallel with an iterative numerical algorithm. The solution typically consists of a list of flow parameters for each cell. This information can finally be reformatted to produce velocity maps, pressure measurements, and parameters such as wall shear stress.

2.1. Imaging

The motion of a fluid is highly sensitive to the shape and dynamics of the surrounding anatomical structure. It is therefore important to use subject-specific anatomical data for detailed blood flow simulations. Such models may be reconstructed from high-definition images acquired with modalities that permit 3-D acquisition such as computed tomography (CT) or magnetic resonance imaging (MRI). The acquired data can be computationally reconstructed into multiple two-dimensional (2-D) slices that provide complete spatial coverage of the anatomy. With CT, multiple transmitters and receivers may be used to increase the imaging speed, which allows a set of slices to be acquired within a single breath-hold, thereby reducing the overall duration of the examinations. The introduction of 16- and 64-channel CT detectors has provided a substantial improvement to both the imaging speed and the resolution. Modern scanners are now capable of acquiring complete temporal coverage of the heart with isotropic voxels in under 10 cardiac cycles. Despite the high accuracy and reproducibility that can now be achieved (32) and further speed increases promised by the forthcoming 256 channel detectors, the major limitation of CT is due to its use of ionizing radiation and its reliance on contrast agents for providing adequate signal variation between cardiovascular structures. Furthermore, CT is limited in its ability to quantify flow properties that are important in patient-specific blood flow simulation.

MR imaging, on the other hand, is a noninvasive modality with no known side effects. It is more acceptable to patients, especially when repeated examinations are required. In addition, the contrast between tissues is good for most cardiac assessment techniques without the use of contrast agents. The use of MR, however, can be unsuitable for patients with metallic implants, as the magnetic fields and radio-frequency pulses that are required can cause considerable heating in metals. Also, care must be taken not to bring ferromagnetic objects near to the scanner as they can be forcibly attracted by the strong magnetic fields. For morphological examinations, MR imaging is often considered to be the gold standard against which other modalities are validated, because of the high resolution and definition of images that may be acquired and its ability to provide complete spatial and temporal coverage of cardiac vessels and chambers as demonstrated in Fig. 1.

Cardiovascular MR (CMR) imaging is significantly complicated by the continuous and complex motion of the heart. This motion may be considered to be the combination of three principal factors. First, the heart typically changes shape in a periodic fashion over the cardiac cycle to fulfill its role as a cardiovascular pump. Second, its proximity to the diaphragm causes its displacement and deformation over the longer and less uniform respiratory cycle. Finally, its location is affected by any overall movement of the patient within the scanner. In combination, these three factors yield a rapid, nonperiodic motion

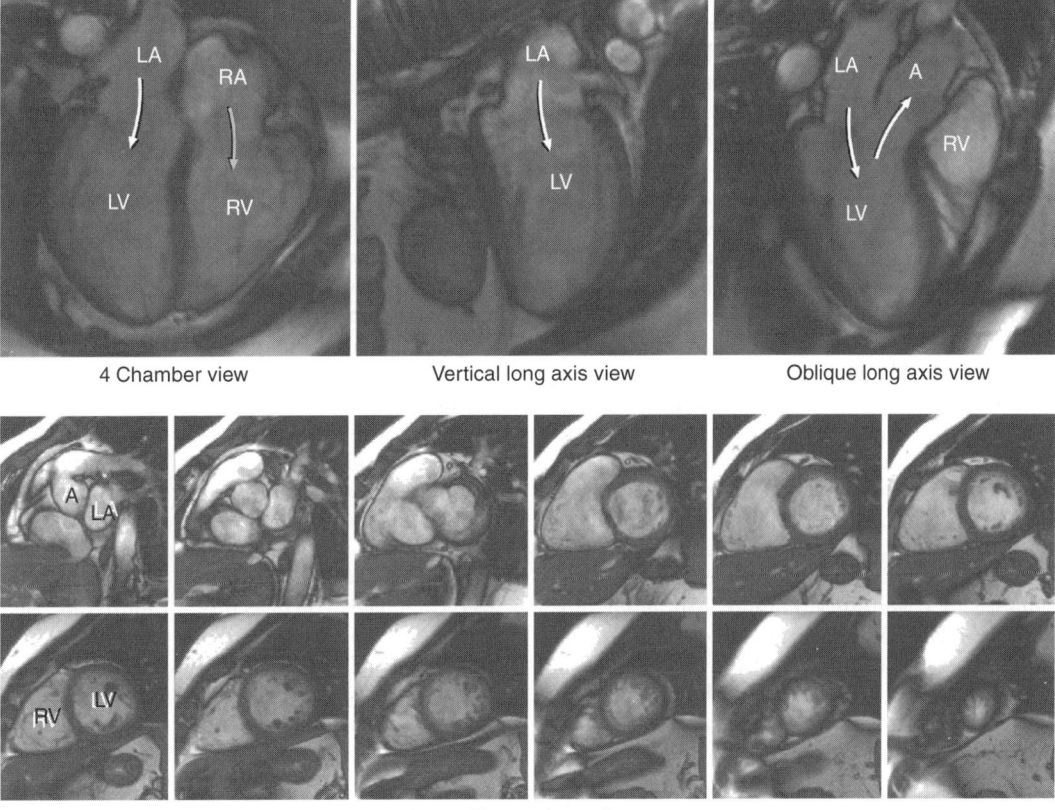

Figure 1. A demonstration of MR imaging planes that can be used for the creation of patient-specific models of the heart. MR imaging can rapidly acquire high-definition cine images of an arbitrary orientation. Shown here are images acquired using a TrueFISP sequence (echo time = 1.1 ms, repetition time 2.2 ms, slice thickness = 8 mm, pixel size = 1.4 mm) on a 1.5-T scanner with a peak gradient of amplitude of 30 mT/m and slew rate of 200 mT/m/ms. The first row shows images acquired in the long-axis plane. The four-chamber view allows both ventricles and the atria to be visualized. The oblique long-axis view portrays the inflow and outflow tracts of the LV. The short-axis stack shown in the second and third rows is typically used for the assessment of cardiac morphology as it allows complete spatial coverage of the heart. (LV = left ventricle, LA = left atrium, RV = right ventricle, RA = right atrium, A = aorta).

that presents significant challenges for cardiac imaging and analysis.

Because of the large amount of data that is required to generate each CMR image, it is common to build up images over multiple cardiac cycles. This approach uses cardiac gating to synchronize the data acquisition to the heartbeat of the patient. The heartbeat is typically monitored by a set of three or four electrodes attached to the chest. For each phase of a cine acquisition, a subset of the data domain called k-space is acquired. Complete images of k-space may then be constructed over a series of cycles. This scheme is therefore also known as segmented k-space acquisition. It relies on the assumption that the dynamic morphology of the heart remains uniform across multiple cardiac cycles. Although this assumption is not true for patients with arrhythmias, it is sufficiently reliable to produce detailed images for most subjects. One major issue related to electrocardiogram (ECG) gating is the difficulty in acquiring late diastolic images, which has motivated extensive research in the development of retrospective gating techniques. Research has also been conducted in the practicality of surface electrode ECG, particularly in high-field and fetal cardiac imaging applications. It has been demonstrated that it is possible to use data directly acquired with MR to perform cardiac gating. For example, an undersampled radial k-space acquisition has been used to measure a real-time image-derived flow-gating waveform, from which the gating times are derived (33).

Unlike cardiac motion, respiratory-induced motion has poor intercycle consistency, which makes gating at different phases of the respiratory cycle particularly difficult. For the measurement of respiratory motion, a reliable marker strongly correlated to the motion of the heart is required. Traditionally, methods of directly monitoring respiration with a belt or bellows have been used (34). An alternative method involves the monitoring of ECG demodulation, in which an optically coupled ECG sensor is used to reduce interference from MR (35). Existing research has shown that the diaphragm can have a range of motion four to five times greater than that of the chest wall (36). Because of the proximity of the heart to the diaphragm, the motion of the diaphragm can be used as a sensitive measure of respiratory motion. MR navigator echoes involve the acquisition of a signal from a column of material running perpendicular to the direction of motion based on a readout gradient along its length (37–40). The application of a Fourier transform on the measured data provides a well-defined edge that gives a scalar value of respiratory position. The acquisition takes minimal time and can be interleaved with the imaging sequence, which allows a real-time measurement of respiratory position throughout the acquisition, that is essential for prospective respiratory motion compensation.

Recently, many techniques have been developed to increase the effective imaging speed and resolution of cardiovascular MR. These techniques include parallel imaging and the use of prior information. Parallel imaging uses the spatial characteristics of the phase array coils for minimizing the acquisition time. Thus far, several techniques have been proposed. Most of them however, are originated from the SMASH (41–43) and SENSE (44–46) methods. SMASH is a partially parallel imaging technique that makes use of the multiple coils generally used in MRI to encode Fourier lines in the spatial harmonics of the receiver coil's signals, which allows the recovery of additional k-space lines in the postprocessing stage, reducing the number of k-lines that are required and thus increasing the efficiency of the imaging process. MR receiver coils have spatial variation in their sensitivity, which for a standard circular surface coil is monotonic in all directions with distance from the coil. The combination of multiple coils may be used to create a sinusoidal spatial sensitivity profile, and suitable modulations in the amplitude of this can replace the phase or frequency encoding normally produced by magnetic field gradients. The use of prior information, on the other hand, exploits the redundancy of k-space values in representing the image or series of images, which is of particular importance when the development of fast image acquisition is limited by hardware constraints. For dynamic imaging, a variety of methods have been developed, which either exploit correlations in k-space (partial Fourier, reduced field of view, parallel and prior information driven, e.g., BLAST (47)), in time (keyhole (48,49), view sharing and UNFOLD (50)), or in both k-space and time (cardiac UNFOLD, k-t BLAST (51)) also known as k-t space.

Another imaging modality that has an important role in blood flow simulation is ultrasound (52), which is an imaging modality that is showing considerable promise in the field of subject-specific flow simulation. The technique is based on the differential reflection of ultrasonic sound waves at the interfaces of anatomical structures. It has been developed from a one-dimensional scanning technique, in which an ultrasonic probe is used to measure the time between the emission of a sound wave and its reception. With this technique, it is possible to calculate the distance to the point at which a wave was reflected. As this calculation occurs at the interfaces between tissues of different density, information regarding the underlying structure of the anatomy may be retrieved. To construct a 2-D image, piezoelectric crystals may be used to generate scanning patterns that allow the 2-D position of the reflections to be deduced. The electronics required for signal emission and reception can typically be incorporated within a handheld device. Because of its low cost, speed, and accessibility, 2-D ultrasound is by far the most widely used modality for general cardiac imaging. It has recently been extended to permit the acquisition of dynamic 3-D datasets. With this technique, the scanning probe is automatically rotated in relation to the patient, which allows multiple imaging planes to be acquired. This process represents a huge development in terms of the accuracy with which subject-specific models can be created with ultrasound. In the near future, it is likely that ultrasound will play an increasingly important role in blood flow simulation as its high temporal resolution makes it particularly well suited to the modeling of the dynamic properties of the mitral valve leaflets. For the detection and evaluation of coronary artery disease, intravascular ultrasound (IVUS) (53,54) has attracted significant interests. In clinical practice, IVUS is most often used as an adjunct to balloon angioplasty to detect dissection, stent underdeployment, stent thrombosis, and predict restenosis risk. It is also used as an accessory to diagnostic angiography to evaluate lesions of uncertain severity and to detect disease that is not visible on a conventional angiogram. Another invasive technique that is showing promise for deriving vessel geometry in patient-specific modeling is rotational x-ray angiography (55). This technique acquires x-ray images from multiple angles for computing the underlying volumetric data.

2.2. Morphological Segmentation

The implicit representations of flow domains provided by medical images are not sufficient for CFD simulation. To extract the explicit anatomical structure, it is necessary to delineate the boundaries of the vessels. Over the last 25 years, the segmentation of medical images has been a widely researched area. The different techniques range from manual methods to fully automatic approaches. For many years, only those techniques that involved a large degree of user interaction proved reliable enough to be used in practice. Recently, with the continued improvement of the quality of imaging hardware and segmentation techniques, it has been possible to rely on more automated approaches.

One of the most widely used semiautomatic alternatives to manual delineation is known as the active contour technique (56,57). To create a dynamic 3-D model that is suitable for CFD simulation, it is possible to apply this technique to multiple timeframes and imaging slices. The method involves the rapid manual delineation of an approximate border, followed by the automatic deformation of the contour to its true position. The deformation process involves taking into account the surrounding image features and the curvature of the contour. The technique may be formulated as an energy minimization problem in which high curvature and poorly defined image gradients correspond to high energy.

The active contour method can be problematic when applied to cardiac delineation. Significant errors can be introduced when delineating complex edges such as the papillary muscles and trabeculations of the endocardial border. In addition, the technique cannot accurately identify poorly defined boundaries, which can be problematic in areas with strong partial volume effects such as those near the apex where acute tapering of the ventricular wall makes the identification of the endocardial and epicardial borders difficult. As a result, most implementations require a human operator to verify the quality of each contour and to correct those that are inadequate.

Balloon expansion (58) is an extension of the active contour model based on the geometric deformation of a sphere until it reaches the desired cardiac boundaries. Similarly, the method is formulated as the minimization of a cost function that constitutes both internal and external constraints. The first of these provides the expansion of the model similar to the filling of a balloon with air; the second limits expansion in regions of well-defined image features; and the final component constrains the morphology to limit its local deformation.

Active appearance models (59) may be used to provide knowledge-driven automatic segmentation of medical images by taking into account statistical variations of the anatomical structure across time and subjects (60). The technique first requires an expert observer to manually delineate tissue boundaries in a training set of images. The resulting contours may be used to construct a model of the average shape and the statistical appearance of the anatomical structure. The basic active appearance technique only models the shape and appearance of objects in 2-D images, which does not use the full spatial and temporal information that is available with CMR imaging. To improve the reliability of segmentation, 3-D or 4-D models may be constructed that capture the complete dynamic morphology of the ventricle. Active appearance models present a promising method for cardiac segmentation because of their ability to cope with the statistical variability in the shape and appearance of the anatomy.

The effectiveness of cardiac segmentation can be improved by the incorporation of surface embedding techniques (61,62). These create a one-to-one mapping between the surface of a structure and a normalized 2-D space. Shape-embedding techniques are designed so that specific regions of the anatomy are consistently mapped to particular locations within the normalized space. As such, detailed statistical models of the shape and motion of structures may be derived. The techniques have been used to segment the four-chamber model of the heart shown in Fig. 2a without user interaction.

3. MESH GENERATION

Although the laws that govern the motion of fluids within simple domains have been extensively studied and are well understood, the simulation of flow within a complex geometry such as the cardiovascular system is a challenging task. If the domain was treated as a single entity, it is infeasible that a set of equations could be constructed to simulate the flow. For this reason, the domain is typically divided into a large set of tetrahedral or hexahedral cells through a preprocessing step called meshing. A list of equations that govern the flow within each of these cells can then be constructed. As the flow within each cell is dependent on its surroundings, it is necessary to solve all equations in parallel, which requires the solution of a set of nonlinear partial differential equations using an approximate numerical approach such as the finite difference, finite volume, or finite element method.

The division of an arbitrary flow domain into a large set of volumetric elements is possibly the most challenging aspect of CFD simulation, because an interwoven set of conditions must be stringently maintained. The first condition is that every point of space within the 3-D domain must be covered by only one cell. The second is that all cells within the domain should have approximately the same size. Finally, no cells should be extensively deformed from the shape of a regular tetrahedron or hexahedron. The most basic method used for this process is called structured meshing, which involves dividing the flow domain into regular grids of cells. Alternatively, if the flow

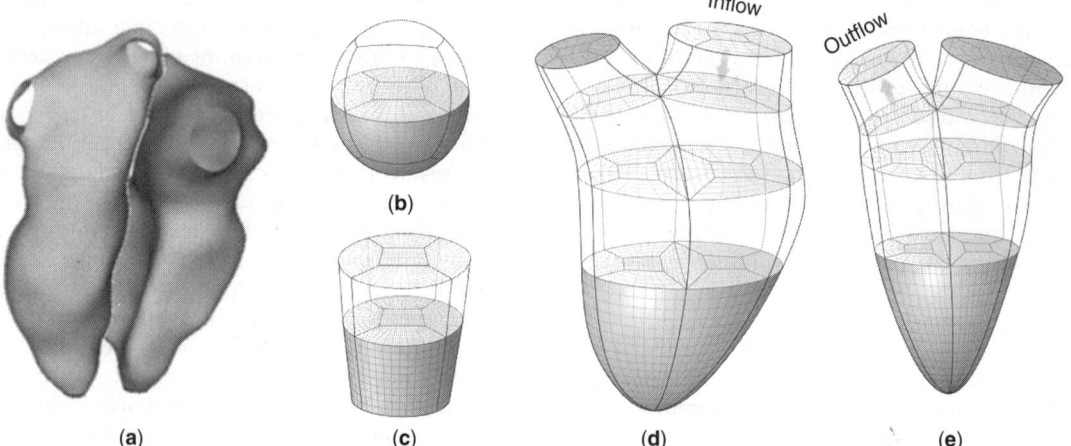

Figure 2. (a) A four-chamber cardiac model derived with the shape-embedding segmentation technique. Models such as these can now be segmented automatically by using a statistical model that captures the principal modes of variation of cardiac morphology. (b) and (c) An illustration of how structured meshes may be used to represent the complete flow domain for simple object topologies. (d) and (e) An illustration of how the dynamic morphology of the left ventricle can be represented with structured meshing.

domain has a complex topology, a more sophisticated technique called unstructured meshing may be employed.

3.1. Structured Meshes

A structured volume mesh typically comprises a large set of hexahedral cells arranged in a regular lattice. This use of a regular layout ensures that neighboring cells have a similar size and orientation. Additionally, it promotes a simple indexing scheme in which neighbors of each cell can be readily identified. As demonstrated in Fig. 2, the technique may be used to describe a wide range of morphologies. For simple objects that have the same topology as a hexahedron, a single lattice structure can be used. For more complex topologies such as cylinders (b), spheres (c), or bifurcations (d) and (e), the flow domain can be broken into a set of individual lattice structures. Using this so-called block-structured approach, it is important to take care of the interfaces between lattices. There must be an equal number of cells on both sides of each interface, and each pair of adjacent cells must share four vertices. For many topologies, it is possible to use a standard layout that ensures that these conditions are maintained.

3.2. Unstructured Meshes

Unstructured meshes represent a flexible alternative to structured meshes for modeling objects with a complex topology. A mesh is considered to be unstructured if it contains an irregular arrangement of cells. In this situation, the cells cannot be referenced by a simple indexing scheme. Instead, it is necessary to complement the list of cells with a description of the connections between them, which increases the overall complexity of the simulation and requires additional computational storage and processing. However, the lack of a regular structure does allow a wider range of topologies to be represented. Furthermore, additional cells may be introduced in regions where turbulent flow is expected such as near boundaries and bifurcations, which can help to improve the accuracy of the simulation.

Unstructured meshes are typically constructed out of tetrahedral cells as these can be easily tessellated to cover 3-D space. Although it is theoretically possible to construct any shape in terms of tetrahedra, in practice it can often yield a computationally unfeasible number of cells or highly distorted cells that are inappropriate for CFD calculation. Unstructured mesh generation can typically be broken into two main processes. The first of these processes is the placement of vertices within the flow domain so that they are relatively evenly spaced. The second is the specification of the connectivity between the vertices enabling cells to be formed. There are a range of established methods of unstructured mesh generation (63) including Delaunay triangulation (64) and advancing front methods (65).

3.3. Mesh Generation for Objects With Dynamic Morphologies

The challenging process of mesh generation is complicated if the morphology of the flow domain changes over time. It is theoretically possible to generate both structured and unstructured meshes for many such objects, but in practice the complications can be overwhelming. For dynamic objects, it is necessary to divide the temporal dimension into many discrete phases. For each of these phases, a valid mesh must be generated. The difficulty is that a further set of constraints must typically be placed on the mesh generation process to resolve its dynamic properties. The first constraint is that the number of cells within the mesh must remain constant over all phases of the simulation. The second is that the cells must be well connected in both their temporal and their spatial dimensions, which means that each cell must retain the same neighboring cells over time and that each vertex must not undergo excessive displacement between phases. These constraints have severely inhibited the development of a general-purpose meshing strategy for dynamic objects. Several successful techniques have been proposed for specific object geometries such as dynamic deformed cylinders, spheres, and bifurcations. These techniques have all been based on structured meshes and have typically required a large degree of manual supervision.

Out of the wide variety of meshing schemes that are available (66–69), one of the most commonly used techniques is based on the structured boundary conforming mesh, which is generated by first defining a curvilinear surface that encloses the flow domain. A smooth continuous volume mesh is then constructed within its interior, which may be performed algebraically with interpolation of the boundaries. Alternatively, a more flexible but computationally expensive technique involving the solution of a set of partial differential equations may be applied.

4. BOUNDARY CONDITIONS AND NUMERICAL SIMULATION

To obtain a unique solution to a CFD simulation, the Navier–Stokes equations require boundary conditions to be specified at all points on the outer surface of the fluid domain (70). Three principal types of boundary conditions are widely used. The first of these uses Dirichlet conditions, in which the velocity at each boundary point is specified. These conditions are typically used at the inlet of a fluid domain, so that the velocity of the inflow can be prescribed. The second type relies on Neumann conditions, in which the derivative of the velocity is specified instead of its actual value. These conditions are typically used at the outlet of a fluid domain so that the outflow velocity is dependent on the flow simulation. For simplicity, a fully developed outflow is normally assumed, which has the property that the acceleration of the fluid perpendicular to boundary is set to zero. Finally, impermeable walls may be represented by setting the velocity of the fluid relative to the boundary to zero.

In terms of CFD calculation, blood flow may be considered to be 3-D, time-dependant, incompressible, viscous, and laminar. Extensive work has already been performed for such flows within the fields of aerodynamics and fluid dynamics. The three most popular methods for this type of flow simulation shall be discussed. These methods are known as the finite difference, finite element, and finite volume methods. All three techniques share some common features and requirements. The first requirement is that a set of equations is used to describe variables at a finite number of points in the flow domain. The second is that a set of starting conditions is required to initialize the time-dependent simulation. Third, the values of variables must be set at the boundaries of the flow domain.

4.1. The Finite Difference Method

The finite difference method (71) was introduced by the eighteenth-century Swiss mathematician, Leonhard Euler. Although this technique was originally developed for hand evaluation, it has become a widely used method for numerical flow simulation. The technique uses a Taylor series to build a set of equations that describe the derivatives of the variables used. The derivatives are represented as the differences between the variables at discrete points in space and time. The volume mesh is used by assigning an appropriate difference formula to each vertex. These formulas relate the variables at each vertex to those of its immediate neighbors, which generates a large set of simultaneous equations that may be solved to deduce the values of the variables. Using this method of flow simulation, no guarantee exists that momentum and mass will be locally conserved.

4.2. The Finite Volume Method

The finite volume method (72) is an alternative technique that can be used to construct and evaluate the partial differential equations involved in blood flow simulation. It was specifically developed to simulate heat transfer and fluid flow while ensuring local volumetric conservation. It is now probably the most widely used CFD technique. It works by modeling the fluxes of small volumetric cells within the flow domain and ensuring that they remain constant for adjacent surfaces. One of the key benefits of the technique is that it is easily applicable to both structured and unstructured meshes.

4.3. The Finite Element Method

The finite element method (73) involves the division of the flow domain into several subdomains known as elements. The variation of variables within each of these elements is constrained. This piece-wise description is used to build a picture of how the variables vary over the whole domain. Depending on the severity of the constraints, the local conservation of mass and momentum may be enforced to a specified level. There is a fundamental difference between the finite element method and the other techniques described. Rather than producing equations that relate the variables of neighboring elements, it generates equations for each element independently, which means that the interaction between elements is only taken into account in the final stages of the simulation.

The finite element method is also different in the way that it handles boundary conditions. In this framework, the behavior of the boundary conditions must be built into the element equations, which contrasts with the finite difference and finite volume methods where the boundary conditions may simply be inserted into the solution matrix. Finite element methods tend to be flexible as they only produce numerical equations from data at known points. Consequently, there is no restriction on how the elements are connected so long as the faces of neighboring elements are aligned. This technique is therefore ideally suited for use with an unstructured mesh.

4.4. Simulation Output

The finite difference, finite volume, and finite element methods for flow simulation are capable of generating 3-D vector fields that characterize the velocity distribution

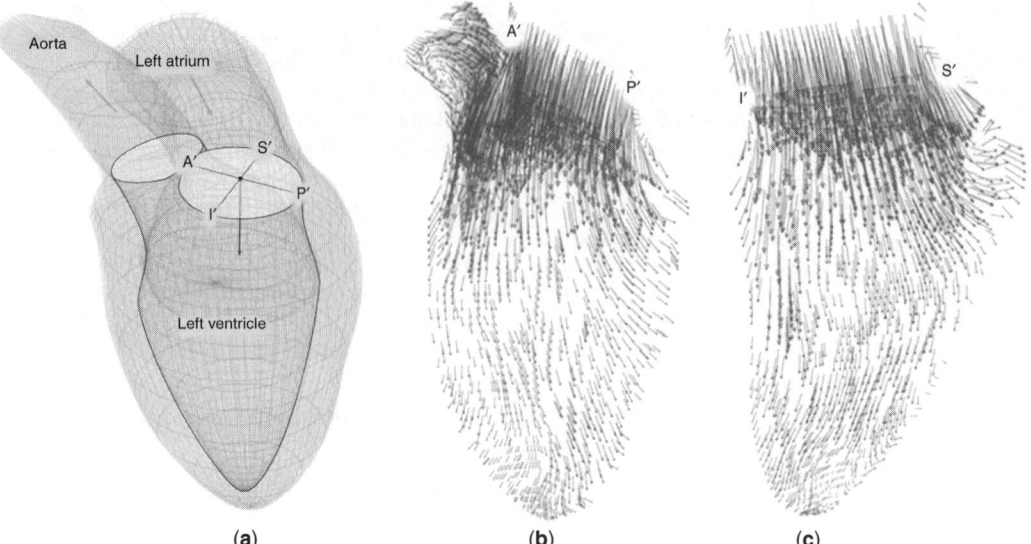

Figure 3. (a) A left ventricular model prepared for CFD blood flow simulation. The dynamics of the epicardial border and the motion of the valve planes may be incorporated into the simulation. The anterior-posterior A'-P' plane of the resulting simulated flow is represented in (b), and the inferior-superior I'-S' plane is shown in (c). The flow field is shown for a diastolic phase of the cardiac cycle with blood flowing through the mitral valve into the left ventricle.

within the flow domain. These methods can be generated with spatial and temporal resolutions that exceed those possible with flow imaging techniques. An example simulated flow field for the left ventricle is shown in Fig. 3. For this simulation, a computer model of the flow domain (a) has been constructed from a series of MR images acquired with the same imaging planes shown in Fig. 1. The model of the endocardial border of the left ventricle (LV) is constructed using the software LVtools (Cardiovascular Imaging Solutions, London, U.K.) as the union of two generalized cylinders that bifurcate to form the inflow and outflow tracts. A block structured meshing scheme has been used to provide a volumetric description of the flow domain suitable for the finite-volume based CFD solver, CFX4 (CFX international, AEA technology, Harwell). Two orthogonal planes of a resultant diastolic flow field are shown in (b) and (c). Blood can be seen passing from the left atrium, through the mitral valve and into the left ventricle. Although the simulations provide a useful tool for investigating the relationships between cardiac morphology and blood flow, the use of a simplified geometry and the introduction of numerical inaccuracies during simulation mean that the resultant flow fields cannot be assumed to be an accurate depiction of the *in vivo* flow. To remove this limitation, it will be necessary to develop modeling techniques that can represent the complex morphology of the endocardial surface and advanced meshing schemes that can handle the resulting flow-domain topologies.

Flow simulations can be highly sensitive to changes in the prescribed boundary conditions. For this reason, it is important that the boundary conditions are specified with as high an accuracy as possible. Figure 4 shows example simulations performed using the same flow-domain morphology, but with four different inflow areas at the mitral valve. It is evident that small changes to the inflow conditions can have significant effects on the resultant flow fields, which signifies the importance of defining the boundary conditions with a high level of accuracy, particularly for time-varying dynamic structures. It is therefore necessary to optimize morphological and flow imaging techniques so that detailed information about the dynamics of the inflow region can be acquired.

In the case of the left ventricle, it is not likely that fully realistic flow simulations will be possible until all key features of the ventricular anatomy are incorporated into the simulated flow domain, which is a challenging modeling task that involves detailed segmentation of the papillary muscles, trabeculations, and valve leaflets. In addition to this, the subsequent meshing process will require the development of sophisticated algorithms that can produce valid volume meshes for time-points throughout the cardiac cycle. The combination of these challenges means that CFD blood flow simulations are unlikely to be applied in a clinical setting in the near future. Despite this, the measurements that can be derived provide a unique and powerful research tool for the investigation of *in vivo* hemodynamic properties.

4.5. Coupled Modeling

Although it is possible to simulate many aspects of cardiac mechanics and flow, it remains difficult to simulate the interactions between the two. Such simulations require detailed knowledge of the tissue properties of the wall as well as sophisticated numerical algorithms to solve the coupled fluid and solid equations (74). In the wider field of computational mechanics, there has been a considerable effort to solve this type of problem. There are three main approaches to solving a coupled simulation. The first

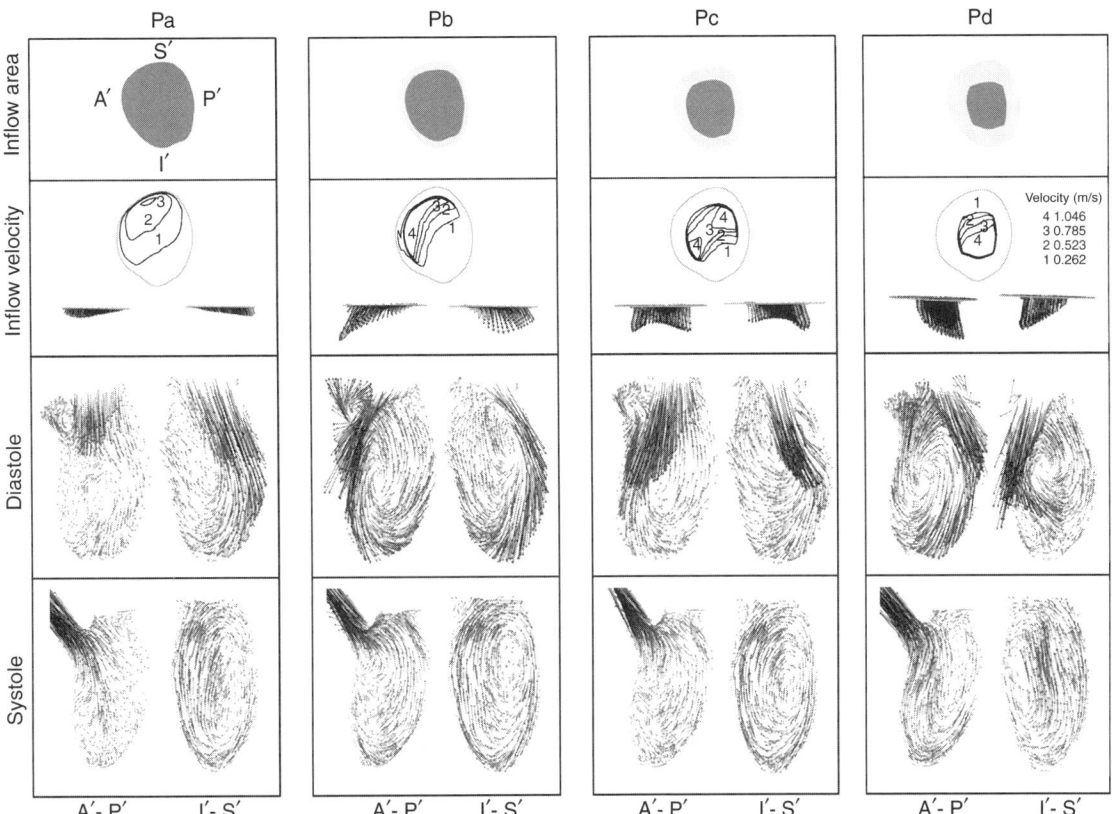

Figure 4. A demonstration of the sensitivity of CFD blood flow simulations. The diagram shows the results of four ventricular flow simulations (Pa - Pd), each with different inflow boundary conditions. It can be shown that by making small adjustments to the area of the inflow pressure boundary, the resulting flow fields can vary dramatically. As a consequence of this high level of sensitivity, it is essential that physiologically accurate cardiac models are used as the starting point for flow assessment.

is based on a simultaneous solution, the second on an iterative one, and the third is based on a hybrid technique. The simultaneous and hybrid approaches require the equations for both the wall and the flow to be solved simultaneously. As this is not catered for by conventional CFD packages, use of these approaches has been limited. The iterative approach is different because the fluid and wall equations are designed to be independent of each other. The iterative process is then performed to provide a loose coupling between the equations. This technique fits into the framework of a variety of existing CFD packages.

Figure 5 is an example of blood flow simulation using coupled modeling. In this study, a detailed model of the carotid bifurcation was derived using *in vivo* MR imaging. An iterative algorithm was employed to simulate the flow and wall parameters independently, couple the results together, and modify the morphology of the flow domain ready for the next iteration. The CFD code CFX4 (CFX international, AEA technology, Harwell) using the finite volume method was used in combination with a the finite element simulator ABAQUS (Hibbit, Karlsson and Sorensen Inc., Pawtucket, RI) for the wall. The combined technique allows the investigation of the interactions between flow and wall mechanics with the derivation of instantaneous flow fields (a) and parameters such as wall shear stress (b) and wall tensile stress (c).

5. FUTURE DIRECTIONS

Blood flow simulation is a growing research area that is likely to take many years to mature. At present, simulations tend to be based on simplified models of the anatomy. As such, it cannot be expected that the resultant flow fields are fully realistic. Although it is possible that the overall flow topology can be derived, measurements that are dependent on detailed flow patterns cannot yet be assumed to be accurate. To secure the use of CFD as a practical clinical tool, it is necessary to make improvements to all processes involved including imaging, modeling, meshing, and simulation. In addition, it is important to form a stronger relationship between blood flow imaging and CFD to enable detailed validation.

The key areas in imaging that require improvement are the imaging speed, the temporal and spatial resolution, and the contrast between different tissues. As most scanning techniques involve a tradeoff between these different properties, it is difficult to envisage how they can be simultaneously achieved. That said, the increasing use

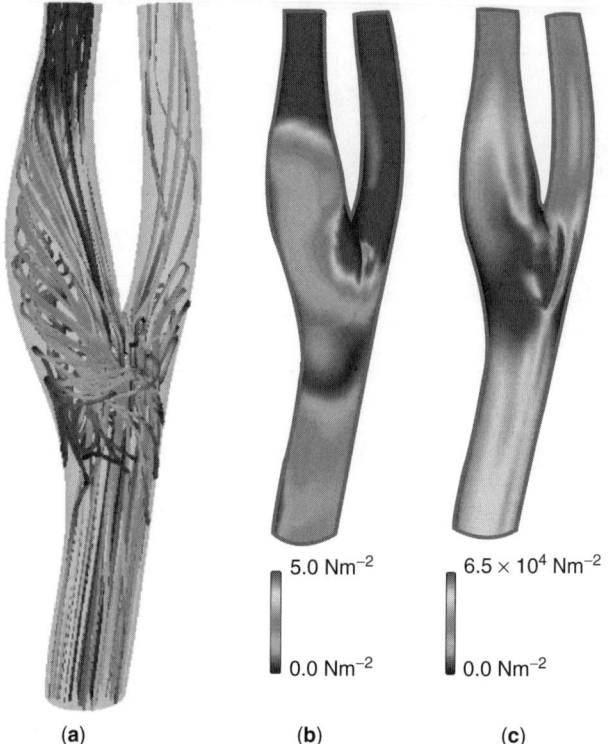

Figure 5. (a) A blood flow simulation of the carotid artery using coupled modeling, which involves the simulation of the two-way interactions between the wall of the vessel and the blood flow. The basic form of the vessel is derived using *in vivo* MR imaging and is used as a starting point for the morphology of the flow domain. The flow simulations have the capacity to modify the wall structure as well as the hemodynamic properties. (b) and (c) Wall shear stress and the wall tensile stress may be calculated from the simulated parameters. (This figure is available in full color at http://www.mrw.interscience.wiley.com/ebe.)

of parallel imaging coupled with k-space sharing techniques being developed for MR imaging will greatly improve the scan efficiency. Because of the general demand of more accurate and reproducible cardiac and vascular assessment, it is likely that we will see an increase in the number of temporal 3-D acquisition sequences that are specifically designed for anatomical model reconstruction.

In terms of modeling, there are many ways in which existing techniques could be improved. Primarily, the level of detail that is extracted could be increased substantially. As an example, it would be possible to extend existing cardiac modeling schemes to incorporate finer anatomical details such as the valve leaflets, papillary muscles, and trabeculations. In this respect, modeling schemes are not making full use of the resolution that is provided by modern imaging strategies. One key difficulty in this area is the development of fully automatic delineation schemes. Although this area has received considerable research interest over the last 25 years, only recently has image quality been sufficiently high to yield reliable and reproducible results.

Probably the greatest challenge in developing a practical tool for blood flow simulation is that of mesh generation. Existing schemes for dividing a flow domain into uniformly sized and shaped cells is well developed for static morphologies but limited for dynamic ones. The ultimate goal of being able to provide fully automatic mesh generation for arbitrary dynamic objects is still a distant prospect. As it is difficult to see how the division of an arbitrary flow domain could be performed using structured meshing, it is likely that solutions will be based on an unstructured meshing scheme that uses temporal constraints.

Validation is critical to the wider usage of CFD for blood flow simulation. Before the technique can be used for clinical studies, it must be validated quantitatively with *in vivo* flow imaging. One interesting possibility is that data from flow imaging could be used to constrain the CFD calculations, which would ensure that the overall flow patterns were accurate, whereas CFD would permit the investigation of detailed patterns and parameters that cannot be measured directly. Another important issue is that of statistical modeling. It is desirable to know the principal ways in which flow varies for normal subjects and across different patient groups. By using such a statistical model, it will become possible to derive the severity of flow patterns by comparing them with an extensive flow database.

Over the next decade, it is likely that we will see a change in focus throughout biomedicine. Although most previous work has been directed toward the diagnosis and treatment of disease, it is envisaged that the emphasis will gradually be shifted toward the early identification of disease. CFD promises to be valuable in reaching this goal. It provides us with a powerful tool for predicting changes to the cardiovascular and vascular systems before they occur. With the capability to foresee hemodynamic changes, our ability to combat disease will be dramatically enhanced.

BIBLIOGRAPHY

1. W. W. Nichols and M. F. O'Rourke, *McDonald's Blood Flow in Arteries: Theoretical, Experimental and Clinical Principles*, 4th ed. Edward Arnold, 1998.
2. P. van Dijk, Direct cardiac NMR imaging of heart wall and blood flow velocity. *J. Comput. Assist. Tomogr.* 1984; **8**:429–436.
3. D. J. Bryant, J. A. Payne, D. N. Firmin, and D. B. Longmore, Measurement of flow with NMR imaging using a gradient pulse and phase difference technique. *J. Comput. Assist. Tomogr.* 1984; **8**:588–593.
4. G. Z. Yang, P. J. Kilner, N. B. Wood, S. R. Underwood, and D. N. Firmin, Computation of flow pressure fields from magnetic resonance velocity mapping. *Magn. Reson. Med.* 1996; **36**: 520–526.
5. P. J. Kilner, G. Z. Yang, R. H. Mohiaddin, D. N. Firmin, and D. B. Longmore, Helical and retrograde secondary flow patterns in the aortic arch studied by three-dimensional magnetic resonance velocity mapping. *Circulation* 1993; **88**:2235–2247.
6. P. G. Walker, G. B. Cranney, R. Y. Grimes, J. Delatore, J. Rectenwald, G. M. Pohost, and A. P. Yoganathan, Three-dimensional reconstruction of the flow in a human left heart by using magnetic resonance phase velocity encoding. *Ann. Biomed. Eng.* 1996; **24**:139–147.

7. R. Frayne and B. K. Rutt, Measurement of fluid-shear rate by Fourier-encoded velocity imaging. *Magn. Reson. Med.* 1995; **34**:378–387.

8. J. N. Oshinski, D. N. Ku, S. Mukundan, Jr., F. Loth, and R. I. Pettigrew, Determination of wall shear stress in the aorta with the use of MR phase velocity mapping. *J. Magn. Reson. Imaging.* 1995; **5**:640–647.

9. S. Oyre, S. Ringgaard, S. Kozerke, W. P. Paaske, M. B. Scheidegger, P. Boesiger, and E. M. Pedersen, Quantitation of circumferential subpixel vessel wall position and wall shear stress by multiple sectored three-dimensional paraboloid modelling of velocity encoded cine MR. *Magn. Reson. Med.* 1998; **40**:645–655.

10. D. N. Firmin, P. D. Gatehouse, J. P. Konrad, G. Z. Yang, P. J. Kilner, and D. B. Longmore, Rapid 7-dimensional imaging of pulsatile flow. *Proc. IEEE Computers in Cardiology*, London, UK, 1993: 353–356.

11. P. J. Kilner, G. Z. Yang, R. H. Mohiaddin, D. N. Firmin, and M. Yacoub, Asymmetric redirection of flow through the heart. *Nature* 2000; **404**(6779):759–761.

12. K. Perktold, R. M. Nerem, and R. O. Peter, A numerical calculation of flow in a curved tube model of the left main coronary artery. *J. Biomech.* 1991; **24**:175–189.

13. X. Ma, G. C. Lee, and S. G. Wu, Numerical simulation for the propagation of nonlinear pulsatile waves in arteries. *J. Biomech. Eng.* 1992; **114**:490–496.

14. A. Santamarina, E. Weydahl, J. M. Siegel, Jr., J. E. Moore, Jr., Computational analysis of flow in a curved tube model of the coronary arteries: Effects of time-varying curvature. *Ann. Biomed. Eng.* 1998; **26**:944–954.

15. Q. Long, X. Y. Xu, M. W. Collins, T. M. Griffith, and M. Bourne, The combination of magnetic resonance angiography and computational fluid dynamics: A critical review. *Crit. Rev. Biomed. Eng.* 1998; **26**:227–276.

16. D. J. Doorly, Modelling of flow transport in arteries. In: S. G. Sajjadi, G. B. Nash, and M. W., Rampling, eds., *Cardiovascular Flow Modelling and Measurement with Application to Clinical Medicine.* Oxford, UK: Oxford University Press, 1999, pp. 67–68.

17. T. Yamaguchi, Computational mechanical model studies in the cardiovascular system. In: T. Yamaguchi, ed., *Clinical Application of Computational Mechanics in the Cardiovascular System.* Tokyo, Japan: Springer-Verlag, 2000.

18. N. B. Wood, S. J. Weston, P. J. Kilner, A. D. Gosman, and D. N. Firmin, Combined MR imaging and CFD simulation of flow in the human descending aorta. *J. Magn. Reson. Imaging.* 2001; **13**:699–713.

19. P. Papathanasopoulou, S. Zhao, U. Kohler, M. B. Robertson, Q. Long, P. Hoskins, X. Y. Xu, and I. Marshall, MRI measurement of time-resolved wall shear stress vectors in a carotid bifurcation model, and comparison with CFD predictions. *J. Magn. Reson. Imaging.* 2003; **17**(2):153–162.

20. N. R. Saber, N. B. Wood, A. D. Gosman, G. Z. Yang, R. D. Merrifield, P. D. Gatehouse, C. L. Charrier, and D. N. Firmin, Progress towards patient-specific computational flow modeling of the left heart via combination of MRI with CFD. *Ann. Biomed. Eng.* 2003; **31**:42–52.

21. D. P. Giddens, C. K. Zarins, and S. Glagov, The role of fluid mechanics in the localization and detection of atherosclerosis. *J. Biomech. Eng.* 1993; **115**:588–594.

22. J. Golledge, Vein grafts: Haemodynamic forces on the endothelium—a review. *Eur. J. Vasc. Endovasc. Surg.* 1997; **14**:333–343.

23. J. F. LaDisa, Jr., I. Guler, L. E. Olson, D. A. Hettrick, J. R. Kersten, D. C. Warltier, and P. S. Pagel, Three-dimensional computational fluid dynamics modeling of alterations in coronary wall shear stress produced by stent implantation. *Ann. Biomed. Eng.* 2003; **31**(8):972–980.

24. Y. Papaharilaou, D. J. Doorly, S. J. Sherwin, J. Peiro, J. Anderson, B. Sanghera, N. Watkins, and C. G. Caro, Combined MRI and computational fluid dynamics detailed investigation of flow in a realistic coronary artery bypass graft model. *Proc. Int. Soc. Mag. Reson. Med.* 2001: 379.

25. J. R. Cebral, P. J. Yim, R. Lohner, O. Soto, and P. L. Choyke, Blood flow modelling in carotid arteries with computational fluid dynamics and MR imaging. *Acad. Radiol.* 2002; **9**(11):1286–1299.

26. L. D. Jou, C. M. Quick, W. L. Young, M. T. Lawton, R. Higashida, A. Martin, and D. Saloner, Computational approach to quantifying hemodynamic forces in giant cerebral aneurysms. *Am. J. Neuroradiol.* 2003; **24**(9):1804–1810.

27. J. S. Milner, J. A. Moore, B. K. Rutt, and D. A. Steinman, Hemodynamics of human carotid artery bifurcations: Computational studies in models reconstructed from magnetic resonance imaging of normal subjects. *J. Vasc. Surg.* 1998; **28**:143–156.

28. C. A. Taylor, M. T. Draney, J. P. Ku, D. Parker, B. N. Steele, K. Wang, and C. K. Zarins, Predictive medicine: Computational techniques in therapeutic decision-making. *Comput. Aided. Surg.* 1999; **4**(5):231–247.

29. Q. Long, X. Y. Xu, B. Ariff, S. A. Thom, A. D. Hughes, and A. V. Stanton, Reconstruction of blood flow patterns in a human carotid bifurcation: A combined CFD and MRI study. *J. Magn. Reson. Imaging.* 2000; **11**:299–311.

30. D. Augst, D. C. Barratt, A. D. Hughes, F. P. Glor, Thom SAMcG, and X. Y. Xu. Accuracy and reproducibility of CFD predicted wall shear stress using 3D ultrasound images. *ASME J. Biomed. Eng.* 2003; **125**:218–222.

31. N. B. Wood, Aspects of fluid dynamics applied to the larger arteries. *J. Theor. Biol.* 1999; **199**:137–161.

32. A. Schmermund, B. Rensing. P. Sheedy, and J. Rumberger, Reproducibility of right and left ventricular volume measurements by electron-beam CT in patients with congestive heart failure. *Int. J. Cardiol. Imaging* 1998; **14**:201–209.

33. R. B. Thompson and E. R. McVeigh, Flow-gated phase-contrast MRI using radial acquisitions. *Magn. Reson. Med.* 2004; **52**(3):598–604.

34. J. N. Oshinski, L. Hofland, S. Mukundan, Jr., W. T. Dixon, W. J. Parks, and R. I. Pettigrew, Two-dimensional coronary MR angiography without breath holding. *Radiology* 1996; **201**(3):737–743.

35. J. Felblinger and C. Boesch, Amplitude demodulation of the electrocardiogram signal (ECG) for respiration monitoring and compensation during MR examinations. *Magn. Reson. Med.* 1997; **38**(1):129–136.

36. Y. Wang, S. J. Riederer, and R. L. Ehman, Respiratory motion of the heart: Kinematics and the implications for the spatial resolution in coronary imaging. *Magn. Reson. Med.* 1995; **33**(5):713–719.

37. R. L. Ehman and J. P. Felmlee, Adaptive technique for high-definition MR imaging of moving structures. *Radiology* 1989; **173**(1):255–263.

38. P. G. Danias, M. V. McConnell, V. C. Khasgiwala, M. L. Chuang, R. R. Edelman, and W. J. Manning, Prospective navigator correction of image position for coronary MR angiography. *Radiology* 1997; **203**:733–736.

39. P. Jhooti, J. Keegan, P. D. Gatehouse, A. M. Taylor, G. Z. Yang, and D. N. Firmin, Coronary artery imaging with real-time phase encode reordering for optimal scan efficiency. *Magn. Reson. Med.* 1999; **41**:555–562.

40. D. Manke, K. Nehrke, and P. Bornert, Novel prospective respiratory motion correction approach for free-breathing coronary MR angiography using a patient-adapted affine motion model. *Magn. Reson. Med.* 2003; **50**(1):122–131.

41. D. K. Sodickson and W. J. Manning, Simultaneous acquisition of spatial harmonics (SMASH): fast imaging with radiofrequency coil arrays. *Magn. Reson. Med.* 1997; **38**(4):591–603.

42. P. M. Jakob, M. A. Griswold, R. R. Edelman, W. J. Manning, and D. K. Sodickson, Accelerated cardiac imaging using the SMASH technique. *J. Cardiovasc. Magn. Reson.* 1999; **1**(2):153–157.

43. M. A. Griswold, P. M. Jakob, Q. Chen, J. W. Goldfarb, W. J. Manning, R. R. Edelman, and D. K. Sodickson, Resolution enhancement in single-shot imaging using simultaneous acquisition of spatial harmonics (SMASH). *Magn. Reson. Med.* 1999; **41**(6):1236–1245.

44. K. P. Pruessmann, M. Weiger, M. B. Scheidegger, and P. Boesiger, SENSE: Sensitivity encoding for fast MRI. *Magn. Reson. Med.* 1999; **42**(5):952–962.

45. B. Madore and N. J. Pelc, SMASH and SENSE: Experimental and numerical comparisons. *Magn. Reson. Med.* 2001; **45**(6):1103–1111.

46. M. Weiger, K. P. Pruessmann, and P. Boesiger, Cardiac real-time imaging using SENSE. SENSitivity Encoding scheme. *Magn. Reson. Med.* 2000; **43**(2):177–184.

47. S. Kozerke, J. Tsao, R. Razavi, and P. Boesiger, Accelerating cardiac cine 3D imaging using k-t BLAST. *Magn. Reson. Med.* 2004; **52**(1):19–26.

48. J. J. van Vaals, M. E. Brummer, W. T. Dixon, H. H. Tuithof, H. Engels, R. C. Nelson, B. M. Gerety, J. L. Chezmar, and J. A. den Boer. "Keyhole" method for accelerating imaging of contrast agent uptake. *J. Magn. Reson. Imaging* 1993; **3**(4):671–675.

49. R. A. Jones, O. Haraldseth, T. B. Muller, P. A. Rinck, and A. N. Oksendal, k-space substitution: A novel dynamic imaging technique. *Magn. Reson. Med.* 1993; **29**(6):830–834.

50. B. Madore, G. H. Glover, and N. J. Pelc, Unaliasing by fourier-encoding the overlaps using the temporal dimension (UNFOLD), applied to cardiac imaging and fMRI. *Magn. Reson. Med.* 1999; **42**(5):813–828.

51. J. Tsao, P. Boesiger, and K. P. Pruessmann, k-t BLAST and k-t SENSE: Dynamic MRI with high frame rate exploiting spatiotemporal correlations. *Magn. Reson. Med.* 2003; **50**(5):1031–1042.

52. M. J. Starmans-Kool, A. V. Stanton, S. Z. Zhao, X. Y. Xu, S. A. M. Thom, and A. D. Hughes, Measurement of haemodynamics in human carotid artery using ultrasound and computational fluid dynamics. *J. Appl. Physiol.* 2002; **92**:957–961.

53. R. Krams, J. J. Wentzel, J. A. Oomen, R. Vinke, J. C. Schuurbiers, P. J. de Feyter, P. W. Serruys, and C. J. Slager, Evaluation of endothelial shear stress and 3D geometry as factors determining the development of atherosclerosis and remodeling in human coronary arteries in vivo. Combining 3D reconstruction from angiography and IVUS (ANGUS) with computational fluid dynamics. *Arterioscler. Thromb. Vasc. Biol.* 1997; **17**(10):2061–2065.

54. C. J. Slager, J. J. Wentzel, J. C. Schuurbiers, J. A. Oomen, J. Kloet, R. Krams, C. von Birgelen, W. J. von der Giessen, P. W. Serruys, and P. J. de Feyter, True 3-dimensional reconstruction of coronary arteries in patients by fusion of angiography and IVUS (ANGUS) and its quantitative validation. *Circulation* 2000; **102**(5):511–516.

55. D. A. Steinman, J. S. Milner, C. J. Norley, S. P. Lownie, and D. W. Holdsworth, Image-based computational simulation of flow dynamics in a giant intracranial aneurysm. *Am. J. Neuroradiol.* 2003; **24**(4):553–554.

56. M. Kass, A. Witkin, and D. Terzopoulos, Snakes: Active contour models. *Int. J. Comput. Vision* 1988; **1**(4):321–331.

57. S. Ranganath, Contour extraction from cardiac MRI studies using snakes. *IEEE. Trans. Med. Imaging* 1995; **14**(2):56–64.

58. L. D. Cohen, On active contour models and balloons. In: *Comp. Vis. GraphIm. Proc.: Imag. Underst.*, 1991.

59. T. F. Cootes, G. J. Edwards, and C. J. Taylor, Active appearance models. In: *Eur. Conf. Comput. Vision*, 1998.

60. S. C. Mitchell, B. P. F. Lelieveldt, R. J. van der Geest, H. G. Bosch, J. H. C. Reiber, and M. Sonka, Multistage hybrid active appearance model matching: Segmentation of left and right ventricles in cardiac MR images. *IEEE Trans. Med. Imaging* 2001; **20**(5).

61. P. Horkaew and G. Z. Yang, Optimal deformable surface models for 3D medical image analysis. *IPMI* 2003: 13–24.

62. P. Horkaew and G. Z. Yang, Construction of 3D dynamic statistical deformable models for complex topological shapes. *MICCAI* 2004: 217–224.

63. W. C. Thacker, A brief review of techniques for generating irregular computational grids. *Int. J. Numer. Methods Eng.* 1980; **15**:1335–1341.

64. J. F. Thompson and N. P. Wearherill, Structured and unstructured grid generation. *Crit. Rev. Biomed. Eng.* 1992; **20**:73–120.

65. R. Lohner and P. Parikh, Three dimensional grid generation by the advancing-front method. *Int. J. Numer. Methods Fluids* 1988; **8**:1135–1149.

66. J. F. Thompson, Z. U. A. Warsi, and C. W. Mastin, Boundary-fitted coordinate systems for numerical solution of partial differential equations—a review. *J. Computat. Phys.* 1982; **47**(1).

67. J. F. Thomson, A survey of composite grid generation for general three-dimensional regions. In: *Numerical Methods for Engine-Airframe Interpolation*. New York: AIAA, 1986.

68. J. F. Thompson, Z. U. A. Warsi, and C. W. Masin, *Numerical Grid Generation: Foundations and Applications*. Amsterdam: North-Holland, 1985.

69. J. F. Thompson, W. J. Minkowycz, E. M. Sparrow, and G. E. Schneider, *Grid Generation. Handbook of Numerical Heat Transfer*. New York: AIAA, 1988.

70. Q. Long, R. Merrifield, G. Z. Yang, X. Y. Xu, P. J. Kilner, and D. N. Firmin, The influence of inflow boundary conditions on intra left ventricle flow predictions. *J. Biomech. Eng.* 2003; **125**:922–927.

71. C. D. Smith, *Numerical Solution of Partial Differential Equations: Finite Difference Methods*, 3rd ed. Oxford: Oxford University Press, 1985.

72. H. K. Versteeg and W. Malalasekera, *An Introduction to Computational Fluid Dynamics: The Finite Volume Method*. Reading, MA: Addison-Wesley, 1995.

73. M. J. Turner, R. W. Clough, H. C. Martin, and L. J. Topp, Stiffness and deflection analysis of complex structures. *J. Aeronaut. Sci.* 1956; **23**:805–824.

74. S. Z. Zhao, X. Y. Xu, and M. W. Collins, The numerical analysis of fluid-solid interactions for blood flow in arterial structures. Part 2: Development of coupled fluid-solid algorithms. *Proc. Inst. Mech. Eng. Part H* 1998; **212**:241–251.

BLOOD OXYGEN SATURATION MEASUREMENTS

XIAO-FEI TENG
YUAN-TING ZHANG
The Chinese University of Hong Kong
Shatin, Hong Kong

The percentage of hemoglobin bound with oxygen is termed the blood oxygen saturation (SO_2), which is called SvO_2 in the venous circulation and SaO_2 in the arterial circulation (1). SO_2 is a vital indicator of a patient's health, considering that a human being cannot survive for a prolonged time without a constant oxygen supply to the brain. It has to be measured in the general ward, operating room, intensive care unit, emergency department, patient transport, birth and delivery, and neonatal care. In some clinical applications, like anesthesia in a surgical procedure, it is a mandatory vital sign measurement.

Lower blood oxygenation (*hypoxemia*) would result in the onset of lower tissue oxygenation (*hypoxia*) and can lead to tissue damage. However, the recognition of the presence of hypoxemia by the human eye is rather poor. Normally, when red blood cells pass through the lungs, more than 95% of their hemoglobin is saturated with oxygen. In cases of lung disease or other types of medical conditions, fewer of the red blood cells carry their usual load of oxygen and oxygen saturation will be lower than 95%.

Blood hemoglobin oxygen saturation can be measured in two ways: chemical and optical. Before the establishment of optical methods, the common practice for reliable measurement was to draw blood from patients and to analyze the samples at regular intervals—several times a day, or even several times an hour—using large hospital laboratory equipment. The blood gas analyzer is one of the commonly used *in vitro* analysis instruments. The breakthrough of blood oxygen saturation measurement came in the mid-1970s when pulse oximetry was designed for *in vivo* measurement. *Oximetry* is a general term that refers to the measurement of blood oxygen saturation. Optical techniques are usually applied. Such measurements depend on the light-absorbing properties of the different forms of hemoglobin.

1. BIOCHEMICAL BASIS OF BLOOD OXYGEN SATURATION MEASUREMENTS

Oxygen molecules that pass through the thin alveolar-capillary membrane dissolve in blood plasma, and most of these molecules quickly bind with hemoglobin. Hemoglobin (Hb) is a protein that is contained in red blood cells. It combines with oxygen in the lungs and delivers it to all tissues requiring oxygen to maintain the viability of cells. Every hemoglobin molecule contains four Fe^{+2} heme and four globin units with each heme and globin unit having the capacity to carry one oxygen molecule. When it is saturated with oxygen, it appears red. When it gives up oxygen, its color changes to dark red. This color change is used in the application of oximetry (2).

Hemoglobin exists in several forms, most of which can carry oxygen and therefore are referred to as *functional hemoglobins*. When a hemoglobin is fully saturated with oxygen (carrying four oxygen molecules), it is called *oxyhemoglobin* (HbO_2); otherwise, it is called *deoxyhemoglobin* (HHb). The *functional oxygen saturation* (*functional* SO_2) is determined by the percentage of concentration of oxyhemoglobin to the total functional hemoglobin.

$$Functional\ SO_2 = \frac{c_{HbO_2}}{c_{HbO_2} + c_{HHb}} \times 100\% \quad (1)$$

Dysfunctional hemoglobin is defined as hemoglobin that is incapable of carrying oxygen. Methemoglobin (MetHb), carboxyhemoglobin (COHb), sulfhemoglobin, and carboxysulfhemoglobin are the four most common dyshemoglobins. MetHb occurs when the iron is oxidized to the ferric state (Fe^{+3}). Certain compounds, such as chlorates, nitrates, and phenacetin, introduced into the blood can cause this oxidation. Under certain physiological circumstances, the amount of MetHb is below 0.6% of the total hemoglobin (3). *Carboxyhemoglobin* is hemoglobin bound to carbon monoxide instead of oxygen. Carbon monoxide binds to hemoglobin with an affinity of 210 times greater than that of oxygen. Therefore, high levels of carbon monoxide will reduce the amount of oxygenated hemoglobin, leading to hypoxic injury to the brain, liver, and renal tubules. In nonsmokers, the level of COHb is usually below 2% (4). The reaction of oxyhemoglobin with hydrogen sulfide produces *sulfhemoglobin*, and reaction of sulfhemoglobin with carbon monoxide produces *carboxysulfhemoglobin*. These two components are usually not significant in human blood.

The *fractional oxygen saturation* is determined by the percentage of the concentration of oxyhemoglobin to the total hemoglobin.

$$Fractional\ SO_2 = \frac{c_{HbO_2}}{c_{total\ hemoglobin}} \times 100\% \quad (2)$$

The amount of oxygen attached to the hemoglobin is related to the amount of oxygen dissolved in the blood, which is represented by *partial pressure of oxygen* (PO_2). At the alveolar-capillary interface in the lungs, the PO_2 is typically high, and therefore the oxygen binds readily to hemoglobin. As the blood circulates to other body tissue in which the PO_2 is low, the hemoglobin releases the oxygen into the tissue because the hemoglobin cannot maintain its full bounding capacity of oxygen in the presence of lower PO_2. The relationship between partial pressure of oxygen and blood oxygen saturation is called the *oxyhemoglobin dissociation curve*.

Hemoglobin's affinity for oxygen increases as successive molecules of oxygen bind to it. Hence the curve is sigmoidal or S-shaped. Many variables can affect hemoglobin's affinity for oxygen, and thus the position of the curve. Decreasing concentrations of carbon dioxide, 2, 3-diphosphoglycerate, and hydrogen ions, decreasing temperature, and increasing pH, increase hemoglobin's affinity for oxygen, shifting the curve to the left. A left-shifted curve implies that the hemoglobin molecules would be more saturated at a lower PO_2. Fetal hemoglobin, which binds more readily with oxygen

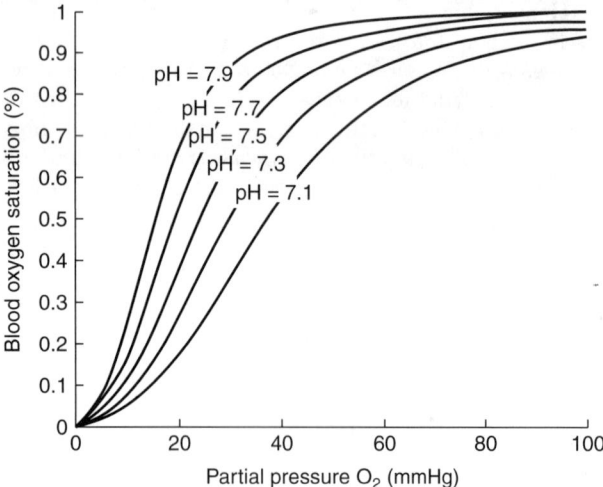

Figure 1. Left shift of oxyhemoglobin dissociation curve because of the increasing pH at a constant temperature of 37°C (2).

than adult hemoglobin, also affects the curve. The effect of increasing pH on the oxyhemoglobin dissociation curve is illustrated in Fig. 1. In Fig. 2 (5), a typical oxyhemoglobin dissociation curve is given. P50 is the partial pressure of oxygen at which 50% of hemoglobin is saturated with oxygen. Point A represents normal 75% *venous blood oxygen saturation* (SvO_2) and point B a 90% *arterial blood oxygen saturation* (SaO_2), which corresponds to a *partial pressure of arterial oxygen* (PaO_2) of about 60 mmHg. Hypoxemia occurs when PaO_2 or SaO_2 is less than the value at point B.

2. REVIEW OF CHEMICAL METHOD

The *Van Slyke* apparatus was used clinically for measuring the oxygen content of a blood sample. Oxygen, carbon dioxide, and other gases are liberated from the blood sample by a releasing agent. Carbon dioxide and oxygen are

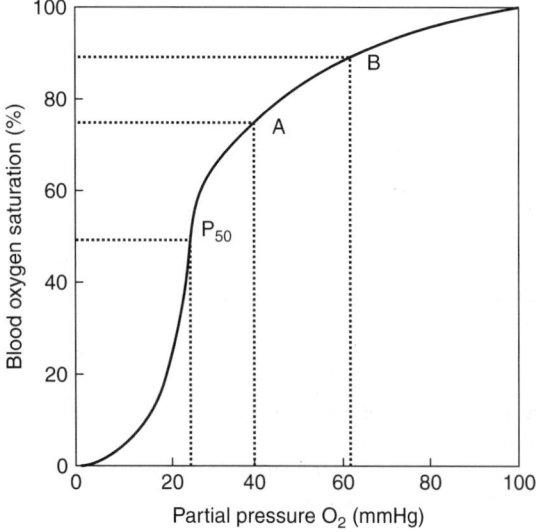

Figure 2. Oxyhemoglobin dissociation curve (5).

removed subsequently, and the pressures of the remaining gases are measured by compressing them into the same fixed volume, respectively. The difference between the two pressures is PO_2 in the blood sample, which is related to the oxygen content (6). Oxygen content (CaO_2) can be calculated as follows,

$$CaO_2 = 1.36 \times Hb \times SaO_2\% + 0.0031 \times PaO_2. \quad (3)$$

Hb is the hemoglobin with unit of $gm\,dL^{-1}$ and PaO_2 is the partial pressure of arterial oxygen with unit of torr. The constant, 1.36, is the amount of oxygen (ml at 1 atmosphere) bound per gram of hemoglobin. The exact value of this constant varies from 1.34 to 1.39, depending on the reference and the way it is derived. The constant 0.0031 represents the amount of oxygen dissolved in plasma at 1 atmosphere. The dissolved oxygen term generally can be ignored, but becomes significant at high pressures—as in a hyperbaric chamber.

Alternatively, oxygen can be extracted from the blood sample using the Van Slyke apparatus and analyzed by a gas chromatograph. Although the Van Slyke technique can provide results with accuracies of ±0.03% (6) and has been a standard for measuring blood oxygen in the past, it required a highly skilled operator and was slow and laborious by today's standards.

Another commonly used and compact type of oxygen analyzer is the *Clark electrode*. In 1954, Clark designed his PO_2 electrode, which uses the basic polarographic principles of oxidation and reduction to measure the PO_2 in a solution. The Clark electrode consists of a silver (Ag) anode and a platinum (Pt) cathode in contact with an electrolyte solution. In this system, shown in Fig. 3 (7), O_2 diffuses from blood to the electrode and is reduced (electrons lost according to equations):

$$Pt\ cathode : O_2 + 2H_2O + 4e^- \rightarrow 4OH^-$$

$$Ag\ anode : 4Ag + 4Cl^- \rightarrow 4AgCl + 4e^-.$$

Usually, a membrane is added to the polarographic electrode, otherwise there would be many problems for blood measurement. The potential applied to the polarizable electrode is increased and the current increases until a plateau develops (8). The maximum number of electrons used in the platinum cathode reaction is proportional to the PO_2 present in the bath. Therefore, by measuring the current between the two electrodes, the PO_2 in the solution is determined.

The *galvanic electrode* is another polarographic method introduced in 1975. It is similar in operation to the Clark electrode, but the cathode is made of Ag, the anode of lead (Pb), and the electrolyte solution is potassium hydroxide. In this system, O_2 is reduced at the Ag electrode while oxidizing at the Pb electrode. The galvanic electrode has no means to replenish the electrolyte solution in the electrode because the anode and the cathode do not participate in the chemical reaction. Therefore, it has only a limited lifetime, which depends on the PO_2 and exposure time (9). However, if the galvanic electrode is made with a removable membrane, the electrolyte can be replenished.

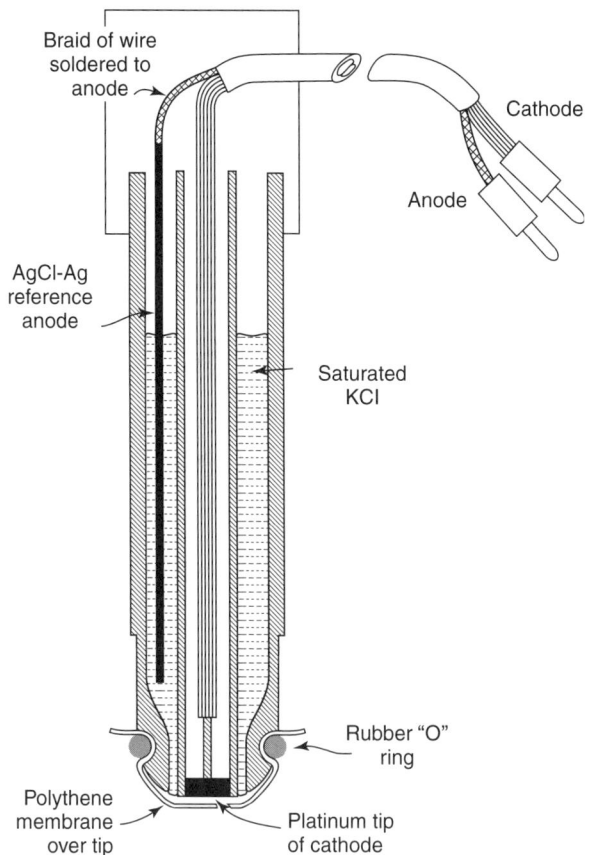

Figure 3. Chemical reactions in a Clark Electrode (5).

3. REVIEW OF OXIMETRY

Oximetry measures the percentage of hemoglobin saturated with oxygen by passing specific wavelengths of light through the blood. The history of oximetry dates back to the early 1860s. Using a spectroscope, a German Professor, Felix Hoppe-Seyler found that when blood was mixed with oxygen, there was a change in light absorption. In 1929, Glen Allan Millikan, an American physiologist, measured the color changes in the mixture of desaturated hemoglobin solutions and oxygen with a photoelectric blood oxygen saturation meter. Photoelectric cells enabled a more precise and immediate measurement of the light intensity than previous methods. They were proved to be crucial in the development of oximeters (10).

Significant advances were made in 1935, when Kurt Kramer used only red light to show that its transmission and absorption were dependent on oxyhemoglobin. In the same year, Karl Matthes introduced the two-wavelength ear saturation meter (5), which was the first *in vivo* instrument for continuous blood oxygen saturation monitoring in humans. However, the sensor was so cumbersome that its use, even as a trend monitor, was impractical (10). In the early 1940s, Millikan developed a lightweight device for monitoring oxygen sufficiency in pilots at high altitude in unpressurized cockpits, and referred to this device as an "oximeter" (8). Millikan noted that changes in oxygen saturation had no effect on light transmitted through a green filter, so this light could be used to stabilize the oximeter. In 1948, Earl Wood at the Mayo Clinic improved Millikan's oximeter by including a pressure capsule. Later, in the 1950s, Brinkman and Zijlstra developed a reflectance oximeter. These instruments led to the use of oximeters in clinical settings, but they required unique settings and individual calibrations (10).

4. PRINCIPLES OF OXIMETRY

Spectrophotometry is the basis of all oximetry. A spectrophotometer operates by passing a beam of monochromatic light of intensity I_0 at a particular wavelength through a medium and measuring the intensity I of the transmitted light. Measurement of the blood oxygen saturation using spectrophotometry is based on Beer's Law (1852, also referred to as Beer–Lambert's Law):

$$I = I_0 e^{-\varepsilon(\lambda)cd}, \qquad (4)$$

where $\varepsilon(\lambda)$ is the extinction coefficient or molar absorptivity of the absorbing substance at a specific wavelength with unit of $L\,mol^{-1}\,cm^{-1}$, c is the concentration of the absorbing substance with unit of $mol\,L^{-1}$, which is constant in the medium, and d is the light path length through the medium. The *unscattered absorbance* (A) of this process can be calculated from

$$A = -\ln(I/I_0) = \varepsilon(\lambda)cd. \qquad (5)$$

Beer's Law is valid even when more than one light-absorbing substance is in the medium. The total light absorbance of n absorbing substances is the sum of their n independent absorbances. Therefore, the unknown concentrations of n different absorbing substances in a homogeneous medium can be determined if n different wavelengths are used to measure the light absorbance and the extinction coefficients of these substances are known.

Different hemoglobin species have different optical absorbances (Fig. 4). In the red region, HHb absorbs more light than HbO_2, and vice versa in the infrared region. The extinction coefficients of both hemoglobin species are

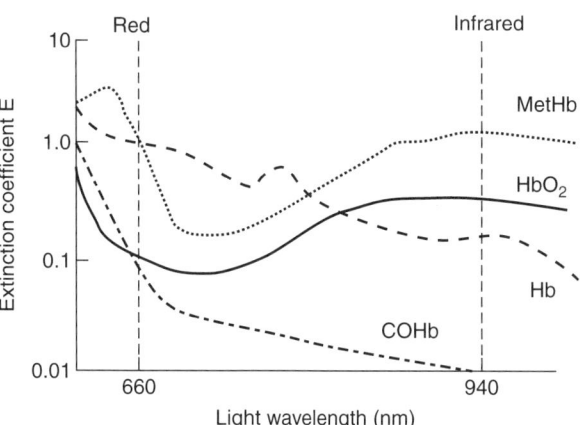

Figure 4. Optical absorption spectrum of different hemoglobin derivatives (7).

equal at the isosbestic point (805 nm), where the absorption is independent of oxygen saturation.

5. THE CO-OXIMETER

The CO-oximeter is a specifically designed spectrophotometer for analyzing *in vitro* the concentrations of HHb, HbO_2, COHb, and MetHb. In 1966, Instrumentation Laboratories released the first commercial CO-oximeter. By subtracting the absorbance readings of the blank solution from the absorbance readings of the hemolyzed plasma sample at each wavelength, the CO-oximeter can give the absorbance of the blood at each wavelength. From these absorbances, the concentration of each type of hemoglobin can be calculated. As the CO-oximeter is designed for four types of hemoglobin, as few as four wavelengths are necessary to make the calculation.

Any substances in the sample that scatter light affect the measurements because the amount of transmitted light does not solely depend on the light absorbed by hemoglobin species. CO-oximeters use hemolyzed samples to reduce the light scattering. Hemolyzed blood is a blood sample with the red blood cell membranes removed. To compensate for the errors caused by the presence of fetal hemoglobin, bilirubin, and other absorbents in the sample, some CO-oximeters use more wavelengths of light (11).

The CO-oximeter is one of the most accurate methods available for measuring the four clinically relevant hemoglobin species, although it can provide accurate readings only for the instant at which the blood is drawn. In addition, CO-oximeters are used as a standard for calibration of *in vivo* oximeters.

6. *IN VIVO* EIGHT-WAVELENGTH OXIMETER

Despite these developments, the oximeter remained an impractical instrument to use until Hewlett-Packard (HP) developed the HP 47201A eight-wavelength ear oximeter in 1976. It was designed to be self-calibrating and claimed to be unaffected by motion (12). This was the first *in vivo* device for measuring blood oxygen saturation that did not require the drawing of patient blood samples. An ear probe was coupled through a fiber-optic cable to the oximeter mainframe, which contained the light source and receivers. The device heated the ear to "arterialize" the capillary blood. The device had accuracy of $\pm 4\%$ for the range of 65–100% SaO_2, but was rendered inaccurate in the presence of jaundice, COHb, or skin pigments. Despite these limitations, this device quickly became a standard clinical and laboratory tool in pulmonary medicine. Although Hewlett-Packard's eight-wavelength oximeter was considered an accurate device, its need for bulky fiber optic-cables to carry the light source to the patient and the transmitted light back to a light sensor made it impractical (13) (Fig. 5).

7. PULSE OXIMETRY

Pulse oximetry was developed in the early 1970s by Takuo Aoyagi, while he was working in Nihon Kohden Corporation in Tokyo developing a noninvasive cardiac

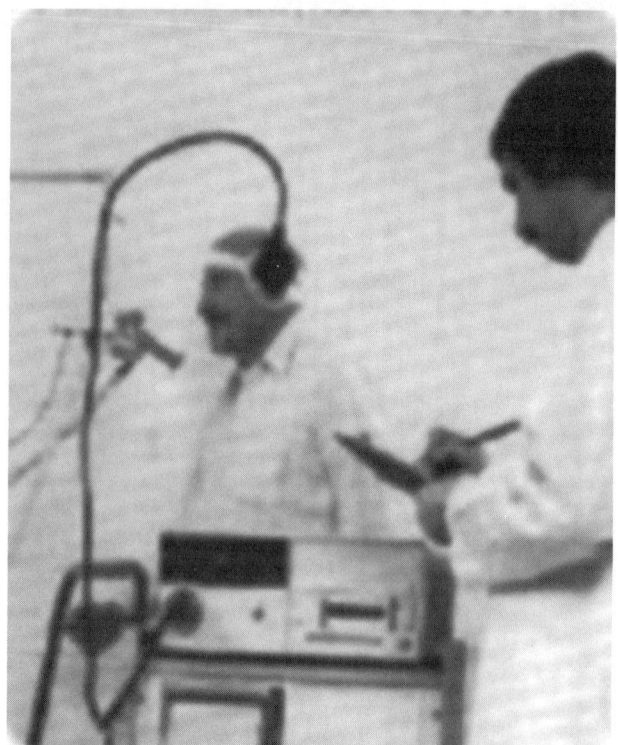

Figure 5. HP ear oximetry. *http://www.medical.philips.com* (This figure is available in full color at http://www.mrw.interscience.wiley.com/ebe.)

output measurement, using dye dilution and ear densitometer. He noticed a correlation in the difference between unabsorbed infrared and red lights and the oxygen saturation and realized that an ear oximeter based on pulsatile light absorbance could measure arterial hemoglobin saturation without heating the ear (14). After several prototypes were tested, Aoyagi and others delivered the first commercial pulse oximeter, OLV-5100, in March 1974. In 1977, the Minolta Camera Company introduced the Oximet MET-1471 pulse oximeter with a fingertip probe and fiber optic cables (15). Later, Nellcor, founded by William New from Stanford University and Jack Lloyd, produced a microprocessor-based pulse oximeter, which did not need any user calibration. It was smaller, less expensive, and accurate enough for clinical applications (16). Some compact and portable pulse oximeters available today are shown in Fig. 6. Pulse oximetry differs from previously described oximetry in that it does not rely on absolute measurements, but rather on the pulsations of arterial blood. By taking advantage of the pulsatile blood volume change, the pulse oximeter is able to eliminate the discomfort caused by the measurement of oximetry, which often required heating and squeezing of the tissue. The SaO_2 estimated by pulse oximetry is abbreviated as SpO_2.

Two types of pulse oximeter probes exist. A transmission pulse oximeter measures the amount of light that passes through the tissue on extremities such as a fingertip, toe, or earlobe. A reflectance pulse oximeter measures the amount of light reflected back to the probe on the chest or cheek. Both types use the same technology, differing only in the position (2).

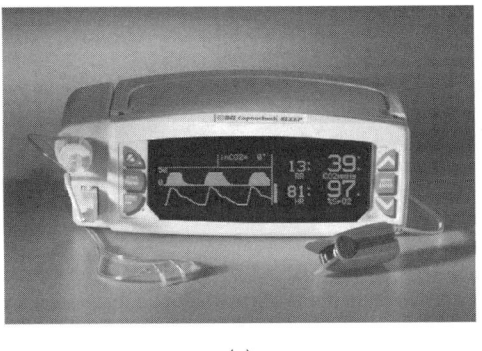 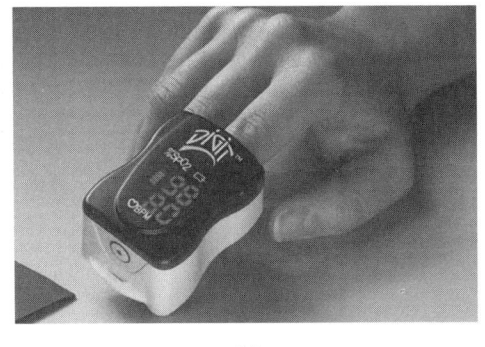 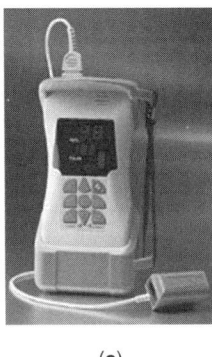

(a)   (b)   (c)

Figure 6. Compact portable and handheld pulse oximeters. http://www.sims-bci.com (This figure is available in full color at http://www.mrw.interscience.wiley.com/ebe.)

Pulse oximeters usually assume that no dysfunctional hemoglobins exist in the arterial blood. As only two types of hemoglobin, oxygenated and deoxygenated, are being measured, two wavelengths of light, typically 660 nm (red) and 940 nm (infrared), are necessary. Figure 7 schematically illustrates the series of absorbers in a living tissue sample. At the top of the figure is the pulsatile or AC component, which is attributed to the pulsating arterial blood. Typical red and infrared AC components are shown in Fig. 8. The pulsatile expansion of the arterial wall produces an increase in path length, thereby increasing the absorbance. The baseline or DC component represents the absorbances of the venous blood, nonpulsatile arterial blood, skin, tissue, and bone. The pulse oximetry traces the peak-to-peak change at each wavelength and divides this change by the corresponding DC component to obtain a normalized absorbance that is independent of the incident light intensity. It then calculates the ratio (R) of these normalized AC amplitudes:

$$R = \frac{AC_R/DC_R}{AC_{IR}/DC_{IR}}, \quad (6)$$

where the subscript R represents red light and the subscript IR represents infrared light. This ratio has been shown to be a function of SaO_2:

$$SaO_2 = \frac{\varepsilon_{HHb}(\lambda_R) - \varepsilon_{HHb}(\lambda_{IR})R}{\varepsilon_{HHb}(\lambda_R) - \varepsilon_{HbO_2}(\lambda_R) + [\varepsilon_{HbO_2}(\lambda_{IR}) - \varepsilon_{HHb}(\lambda_{IR})]R} \times 100\%, \quad (7)$$

where $\varepsilon_{HHb}(\lambda_R)$ and $\varepsilon_{HbO_2}(\lambda_R)$ are extinction coefficients of deoxyhemoglobin and oxyhemoglobin at the red wavelength and $\varepsilon_{HHb}(\lambda_{IR})$ and $\varepsilon_{HbO_2}(\lambda_{IR})$ are extinction coefficients of deoxyhemoglobin and oxyhemoglobin at the infrared wavelength.

In actual use, commonly available light emitting diodes (LEDs) are not monochromatic light sources; therefore, the extinction coefficient for hemoglobin cannot be used directly in a theoretical calculation. Furthermore, the theoretical calculation based on Beer's Law is only approximately true. For example, the two wavelengths do not necessarily have the exact same path length changes, and scattering effects have been ignored. Therefore, this ratio is converted to SpO_2 via an empirical calibration curve.

Figure 9 is an example of a pulse oximeter calibration curve. When the ratio R is one, the saturation is approximately 85%. By analyzing the relationship between oximeter R value and oxygen saturation obtained from blood samples of human subjects with a CO-oximeter, the calibration curve is obtained. During this process, subjects breathe gas mixtures containing oxygen and nitrogen to change the arterial oxygen saturation. As volunteers should not be taken below 70% SaO_2, calibration at levels below this is by extrapolation. The Finger Phantom is an

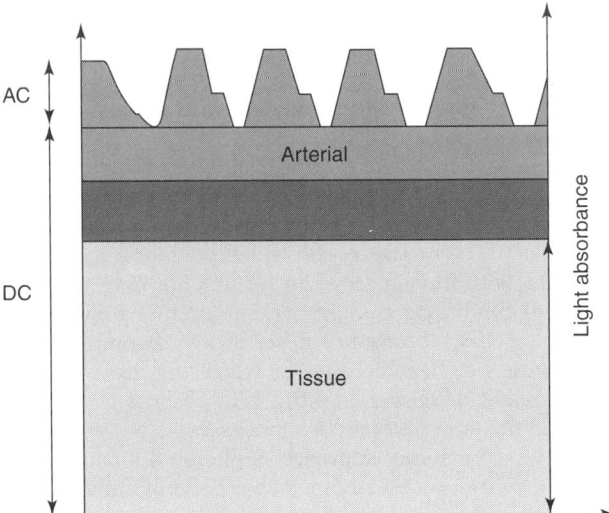

Figure 7. Light absorption by arterial blood, venous blood, bone, and tissue (7).

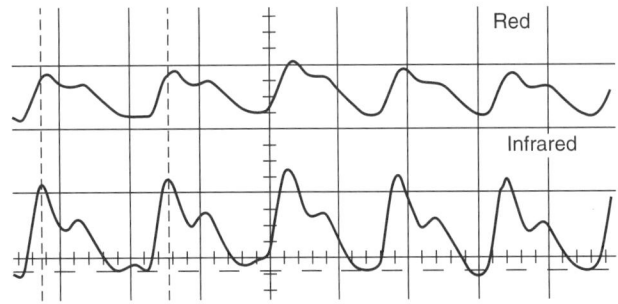

Figure 8. Transmitted red and infrared AC.

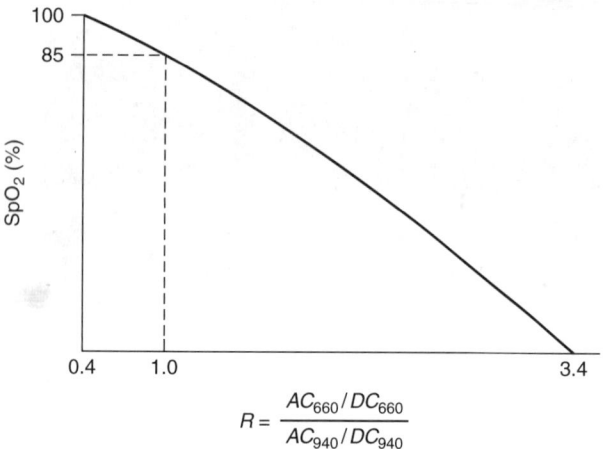

Figure 9. Typical pulse oximeter calibration curve (7).

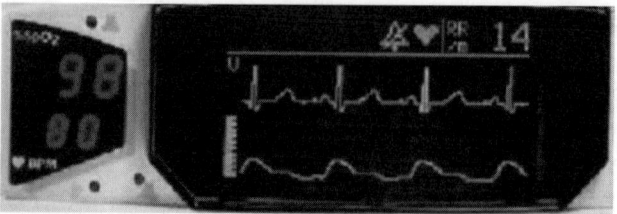

Figure 10. A common interface of pulse oximeter. The upper waveform is electrocardiogram and the lower one is plethysmogram. The numbers on the left side: the upper one is SpO_2 and the lower one is heart rate. http://www.sims-bci.com (This figure is available in full color at http://www.mrw.interscience.wiley.com/ebe.)

alternative for *in vivo* calibration (17,18). Such a device simulates the light absorption and arterial blood flow of the human finger at different SaO_2 levels.

The accuracy of pulse oximeters is usually assessed by a CO-oximeter. A meta-analysis concluded that pulse oximeters were accurate to within 2% in the range of 70–100% SaO_2, but failed to accurately record the true SaO_2 during severe or rapid desaturation, hypotension, dyshaemoglobinaemia, and low perfusion states (19). Finger probes are accurate to within 0.2–1.7%, which is superior to the values obtained using ear, nose, and forehead probes (20).

8. LIMITATIONS OF PULSE OXIMETRY

Once the pulse oximeter became widely used, it became clear that pulse oximeters did not work in all situations. The most common causes of inaccuracies are patient motion and low patient perfusion at the sensor site. As the pulse oximetry method relies on the pulsatile part of the absorption, and the pulsatile part accounts for only 2–5% of the whole transmitted signal and only 1–2% of the reflected signal (11), any movement of the patient (body movement, such as seizures or shivering), sensor, stretcher, or the transporting vehicle will give a false reading. By their very nature, pulse oximeters require a pulse of regular rhythm and a site with adequate perfusion. The presence of vasoconstrictors, such as cold, fear, hypothermia, and medications, can cause inadequate perfusion, therefore making it difficult to distinguish the true signal from background noise. Other causes of an inaccuracies include the presence of an increasing amount of dysfunctional hemoglobin, dark pigmentation, intravenous dyes, interference from radiated light, and pulsatile venous system (21,22).

All the above artifacts could cause false alarms. Several approaches have been introduced to address this problem, using alarm delay, averaging, median filtering, changing the alarm limits, and synchronization with ECG. Figure 10 shows a pulse oximeter simultaneous display of the electrocardiogram and the plethysmogram. When the heart rate obtained from plethysmogram differs significantly from that obtained from electrocardiogram, the SpO_2 may be erroneous and must be interpreted with caution. Recently, Masimo introduced an innovative approach using signal extraction technology (SET) (23,24) to extract the true signal from the artifact caused by motion and low perfusion. Masimo SET® technology enables the power of adaptive filters to be applied to real-time physiologic monitoring with a proprietary technique to accurately establish a "noise reference" in the detected physiologic signal, thus enabling the direct calculation of arterial oxygen saturation and pulse rate. As it is not bound by a conventional "red over infrared" ratio approach, the Masimo SET® system substantially eliminates the problems of motion artifact, lower peripheral perfusion, and most low signal-to-noise situations, which greatly extends the utility of SpO_2 in high-motion, low-signal, and noise-intensive environments (25,26).

9. MODELING OF PULSE OXIMETRY

The incident light passing through human tissue is not split only into absorbed light and transmitted light because of the interaction between incident light and the skin, tissues, and blood. The light is also reflected, refracted, and scattered. With photoplethysmography, light reflection at the skin surface and light absorbance because of tissue other than the pulsating blood are overcome. However, scattering caused by the skin surface, tissue, muscle, and especially blood still affects the absorbance of light. Several approaches have been taken to create models to find a relationship between SaO_2 and the ratio R of normalized absorbances for whole blood instead of hemoglobin solutions only.

An analytical theory, based on electromagnetic field theory, has been developed by Twersky to describe the scattering of light by large, low-refracting, and absorbing particles (27). The theory can be adapted for a special setting and will provide accurate results, but once the physiological conditions change, recalibration is required (2). In this theory, absorbance described by Beer's Law and attenuation of light because of scattering are treated as independent processes, and the total absorbance of whole blood is the sum of these two processes.

Another modeling approach is photon diffusion theory (28,29). Schmitt (28) examined the effects of multiple scattering by comparing the results of the photon diffusion analysis with those obtained using an analysis based on Beer's Law and found that the shape of the calibration curve is affected by tissue blood, volume, source-detector placement, and other variables that change the wavelength dependence of the attenuation coefficient of the tissue. Marble et al. (30) found that the three-dimensional photon diffusion theory can be used to model tissue optics, although pulse oximetry violates many of the requirements of this model. However, they concluded that this theory could not replace clinical calibrations.

10. FETAL PULSE OXIMETRY

Pulse oximetry is used to monitor SaO_2 of both the mother and the fetus during childbirth. The fetal pulse oximeter consists of a catheter that has an oxygen sensor at one end and is connected to a display monitor at the other end. The catheter is inserted into the vagina and moved past the cervix so that the sensor rests against the scalp or cheek of the fetus. Although the technology of conventional pulse oximetry for adults is well established, fetal pulse oximetry is challenged by at least three substantial differences: (1) sensor positioning is difficult because the fetus is not readily visible or accessible to the clinicians; (2) the oximeter must be capable of working with fetal pulses that are generally much weaker than those of adults or children; and (3) as the fetal blood oxygen saturation normally varies between 25% and 70%, conventional emitting wavelengths and empiric calibration curves for adults cannot be used in fetal pulse oximetry (31).

The sensor placement and attachment issue must be dealt with before other issues can be addressed. All sensors currently under investigation are placed on the surface of the fetus's head and are used to measure light that is re-emitted from the illuminated tissues by optical fiber. The fetal signals obtained with reflectance pulse oximetry are typically one-tenth of the adult transmission pulse oximetry signals, which is at the lower limit of signal acceptance for some commercial transmission pulse oximeters. The fetal pulse oximeters use several ways to enhance the signal processing. Fetal electrocardiogram is used to synchronize the red and infrared pulses, a weighted moving average of red-to-infrared ratio is calculated over several heartbeats and pulse oximeter verifies if the red and infrared peak and trough are in phase with each other.

Pulse oximeters for adults and neonates develop worsening accuracy as the SaO_2 drops below 70%. Within the normal breathing population, saturations below this level are generally indicative of an unhealthy condition and limited accuracy does not affect the clinical utility of the system. In the fetal environment, a threshold value of saturation separating healthy versus unhealthy is around 30%. Conventionally used wavelengths of 660 nm and 940 nm are not suitable for fetal pulse oximetry at low saturation. Instead, the wavelength pairing of one from the far-red region (735 nm) and one from the near-infrared region close to 900 nm (890 nm) is employed (32).

The calibration of pulse oximetry for fetal use is more complicated than the calibration of pulse oximetry for adult or neonatal use, as the arterial blood samples cannot be obtained from a human fetus. Animal experiments have been suggested. However, unknown or uncontrollable physiological parameters cause large variations in the calibration curve. Therefore, *in vitro* models have emerged as a practical method for exploring accurate calibration at low saturations (33).

11. RETINAL PULSE OXIMETRY

The unique transparency of the ocular media means that the eye is the only area of the body where systemic blood vessels may readily be observed. Oxygen saturation measurement from retinal vessels offers some distinct advantages over conventional pulse oximeter readings at peripheral areas of body, such as the finger or earlobe (34). First, the blood supply to the retinal arteries originates via the ophthalmic artery from the internal carotid artery, which supplies the cerebral tissues. Thus a measurement of retinal SaO_2 will indicate the cerebral oxygen saturation. Second, the retinal circulation, unlike peripheral blood flow, is not susceptible to arterial shutdown in case of shock, hypothermia, and hypovolemia, where conventional pulse oximeters fail to display any reading as the peripheral pulse disappears. Finally, the measurement of retinal arterial oxygen saturation may be of use in the understanding and potential treatment of retinal diseases.

Retinal pulse oximeters perform reflective measurements from the retinal fundus (34). An incoherent, trifurcated fiber-optic bundle, which is coupled to a haptic contact lens, guides the light to and from the eye, which enables the incident and reflected light paths to be coaxial and allows all electrical connections to be remote from the subject and the weight on the eye to be minimal. In this application, aside from the absorption spectra, the scattering at the erythrocytes in the blood must be taken into account. Furthermore, the scattering and absorption properties of the embedding retina and the fundus layers behind the vessel must be considered (35). Other noninvasive measurements of retinal oxygenation include calculation of the oxygen saturation from fundus photographs and imaging fundus spectroscope (36).

12. SvO_2 MEASUREMENTS

Venous blood oxygen saturation (SvO_2) measurements can be obtained oximetrically with appropriate catheters and monitors. For example, SvO_2 can be measured with a fiber optic artery catheter, which is currently a useful monitoring technique for the critically ill patient. These permit minute-to-minute assessment of total tissue oxygen balance (i.e., the relationship between oxygen delivery and oxygen consumption). Noninvasive measurement of SvO_2 by a vibratory oximetry sensor has been proposed recently. With an external perturbation cuff, this sensor can optically differentiate and localize the veins and arteries (37).

Several studies have suggested that SvO_2 monitoring may constitute a useful early warning of an impending low-cardiac-output state, but it is worth remembering that other causes of falling SvO_2 exist apart from a reduction in cardiac output. The determinants of SvO_2 can be approximately defined by the Fick equation:

$$SvO_2 = SaO_2 - VO_2/(CO \times 13.9 \times Hb), \qquad (8)$$

where CO is the cardiac output, VO_2 is the oxygen consumption, and Hb is the hemoglobin with unit of gm L^{-1}. (This modified equation does not allow for the presence of dissolved oxygen). The normal SvO_2 at rest is 75%, which indicates that, under normal conditions, tissues extract 25% of the oxygen delivered. When the SvO_2 is less than 30%, tissue oxygen balance is compromised, and anaerobic metabolism ensues. This condition should be viewed as a medical emergency.

Blood oxygen saturation measurement by transmission pulse oximeter has been widely used. Such measurement requires a site, such as a fingertip or an earlobe, through which light can be transmitted. When the circulation to the fingertip or the earlobe is diminished during shock or hypothermia, measurements fail. On the other hand, the reflectance pulse oximeter can be applied to any portion of the body, and thus it has potential for wider clinical application. However, reflectance pulse oximeter for fetal and retinal use is still under active development. Cerebral oxygen sufficiency monitoring is another important area of interest, which applies near-infrared spectrophotometry (NIRS) (38).

BIBLIOGRAPHY

1. Dr. Martin, *All You Really Need to Know to Interpret Arterial Blood Gases*, 2nd ed. Philadelphia, PA: Lippincott Williams & Wilkins, 1999.
2. J. G. Webster, *Design of Pulse Oximeters*. Philadelphia, PA: Institute of Physics Pub., 1997.
3. H. F. Bunn, *Hemoglobin: Molecular, Genetic, and Clinical Aspects*. Philadelphia, PA: Saunders, 1986.
4. M. W. Wukitsch, M. T. Petterson, D. R. Tobler, and J. A. Pologe, Pulse oximetry: analysis of theory, technology, and practice. *J. Clin. Monit.* 1988; **4**:290–391.
5. N. F. Jensen (2001). *Clinical arterial blood gas analysis* (online). Available: http://www.boardprep.com/pdfs/abgs.pdf.
6. A. P. Adams and C. E. W. Hahn, *Principles and Practice of Blood-Gas Analysis*, 2nd ed. New York: Churchill Livingstone, 1982.
7. I. Fatt, *Polarographic Oxygen Sensors: Its Theory of Operation and Its Application in Biology, Medicine, and Technology*. Cleveland, OH: CRC Press, 1976.
8. M. Bourke (2001). *Anaesthetic gas, blood gas analysis and oximetry* (online). Available: http://www.anesthesia.org/resident/program/physics_fluids/physics_fluids_gasanalysis_oximetry_2001.pdf.
9. B. A. Shapiro, R. A. Harrison, R. D. Cane, and R. Templin, *Clinical Application of Blood Gases*, 4th ed. Chicago, IL: Year Book Medical, 1989.
10. J. W. Severinghaus and P. B. Astrup, History of blood gas analysis, VI. Oximetry. *J. Clin. Monit.* 1986; **2**(4):270–288.
11. J. T. B. Moyle, *Pulse Oximeters*. London: BMJ, 1994.
12. E. B. Merrick and T. J. Hayes, Continuous non-invasive measurements of arterial blood oxygen levels. *Hewlett-Packard J.* 1976; **28**(2):2–9.
13. A. S. Rebuck, K. R. Chapman, and A. D'Urzo, The accuracy and response characteristics of a simplified ear oximeter. *Chest* 1983; **83**:860–864.
14. T. Aoyagi, M. Kishi, K. Yamaguchi, and S. Watanabe, Improvement of the ear piece oximeter. 13th Meeting of the Japanese Society for Medical Electronics and Biological Engineering, abstract, 1974: 90–91.
15. J. W. Severinghaus, History and recent developments in pulse oximetry. *Scand J. Clin. Lab. Invest. Suppl.* 1993; **214**:105–111.
16. T. Santamaria and J. S. Williams, Pulse oximetry. *Med. Dev. Res. Rep.* 1994; **1**(2):8–10.
17. E. C. Morillo and Y. Mendelson, Multiwavelength transmission spectrophotometry in the pulsatile measurement of hemoglobin derivatives in whole blood. *Proc. IEEE 23rd Northeast Bioeng. Conf.* May 1997: 5–6.
18. http://www.bcgroupintl.com/BiomedMain.htm.
19. L. A. Jensen, J. E. Onyskiw, and N. G. Prasad, Meta-analysis of arterial oxygen saturation monitoring by pulse oximetry in adults. *Heart Lung* 1998; **27**:387–408.
20. D. G. Clayton, R. K. Webb, A. C. Ralston, D. Duthie, and W. B. Runciman, Pulse oximeter probes: a comparison between finger, nose, ear, and forehead probes under conditions of poor perfusion. *Anaesthesia* 1991; **46**:260–265.
21. A. C. Ralston, R. K. Webb, and W. B. Runciman, Potential errors in pulse oximetry. *Anaesthesia* 1991; **46**:291–295.
22. Y. Mendelson, Pulse oximetry: theory and application for noninvasive monitoring. *Clin. Chem.* 1992; **38**(9):1601–1607.
23. J. M. Goldman, M. T. Petterson, R. J. Kopotic, and S. J. Barker, Masimo signal extraction pulse oximetry. *J. Clin. Monitor. Comput.* 2000; **16**:475–483.
24. M. K. Diab, M. E. Kiani, and W. M. Weber, Signal processing apparatus. U.S. Patent 6,263,222, July 17, 2001.
25. P. J. Witucki and S. J. Bell, Comparison of three new technology pulse oximeters during recovery from extreme exercise in adult male. *Crit. Care Med.* 1999; **27**(12):A87(224).
26. S. J. Barker, The effect of motion on the accuracy of six "motion-resistant" pulse oximeters. *Anesthesiology* 2001; **95**: A587.
27. V. Twersky, Multiple scattering of waves and optical phenomena. *J. Opt. Soc. Am.* 1962; **52**:145–171.
28. J. M. Schmitt, Simple photon diffusion analysis of the effects of multiple scattering on pulse oximetry. *IEEE Trans. Biomed. Eng.* 1991; **38**(12):1194–1203.
29. V. Ntziachristos, Oximetry based on diffuse photon density wave differentials. *Med. Phys.* 2000; **27**(2):410–421.
30. D. R. Marble, D. H. Burns, and P. W. Cheung, Diffusion-based model of pulse oximetry: in vitro and in vivo comparisons. *Appl. Opt.* 1994; **33**:1279–1285.
31. P. D. Mannheimer, M. E. Fein, and J. R. Casciani, Physio-optical considerations in the design of fetal pulse oximetry sensors. *Eur. J. Obstetr. Gynecol. Reproduct. Biol.* 1997; **72**(1): S9–S19.
32. P. D. Mannheimer, J. R. Casciani, M. E. Fein, and S. L. Nierlich, Wavelength selection for low-saturation pulse oximetry. *IEEE Trans. Biomed. Eng.* 1997; **44**(3):148–158.
33. T. Edrich, M. Flaig, R. Knitza, and G. Rall, Pulse oximetry: an improved in vitro model that reduces blood flow-related artifacts. *IEEE Trans. Biomed. Eng.* 2000; **47**(3):338–343.
34. J. P. de Kock, L. Tarassenko, C. J. Glynn, and A. R. Hill, Reflectance pulse oximetry measurements from the retinal fundus. *IEEE Trans. Biomed. Eng.* 1993; **40**(8):817–823.

35. M. Hammer, S. Leistritz, L. Leistrits, and D. Schweitzer, Light paths in retinal vessels oxymetry. *IEEE Trans. Biomed. Eng.* 2001; **48**(5):592–598.
36. D. Schweitzer, M. Hammer, J. Kraft, E. Thamm, E. Konigsdorffer, and J. Strobel, In vivo measurement of the oxygen saturation of retinal vessels in healthy volunteers. *IEEE Trans. Biomed. Eng.* 1999; **46**(12):1454–1465.
37. P. A. Shaltis and H. Asada, Monitoring of venous oxygen saturation using a novel vibratory oximetry sensor. *Proc. 24th Annual Conf. IEEE Eng. Med. Biol. Soc.* 2002; **2**:1722–1723.
38. T. Kusaka, K. Isobe, K. Nagano, K. Okubo, S. Yasuda, M. Kondo, S. Itoh, K. Hirao, and S. Onishi, Quantification of cerebral oxygenation by full-spectrum near-infrared spectroscopy using a two-point method. *Compar. Biochem. Physiol. Part A* 2002; **132**:121–132.

BLOOD SUBSTITUTES

Hiromi Sakai
Eishun Tsuchida
Waseda University
Tokyo, Japan

1. INTRODUCTION: PROBLEMS OF BLOOD TRANSFUSION SYSTEM AND EXPECTATIONS FOR THEIR SUBSTITUTES

Since the discovery of blood type antigen by Landsteiner in 1900, allogeneic blood transfusion has been developed as a routine clinical practice; it has contributed to human health and welfare. Infectious diseases such as hepatitis and HIV are now social problems, but a strict virus test by nucleic acid amplification test (NAT) is extremely effective to detect trace presences of a virus to minimize infection (though it is available in few developed countries). Even so, NAT poses problems such as detection limits during the window period and limited species of viruses for testing. Emergence of new viruses (such as West Nile virus, avian influenza, Ebola, dengue) and a new type of pathogen, prions, also threaten us. The preservation period of donated red blood cells (RBCs) is limited to 3–6 weeks. Platelets can be preserved for only a few days. Immunological responses (such as anaphylaxis and graft versus host disease) and contingencies of blood type incompatibility further limit the utility of blood products. To obviate or minimize homologous transfusion, the transfusion trigger has been reconsidered, and roughly reduced from 10 to 7–8 g/dl. Bloodless surgery and preoperational enhancement of erythropoiesis for storing autologous blood have become common. However, these epoch-making treatments are not always practical for all patients. Some developed countries with aging population are facing a decreasing number of young donors and an increasing number of aged recipients. On the other hand, in some developing countries, establishment of a safe blood donation system is difficult. Under such circumstances, research toward blood substitutes has gathered great attention and has been developed worldwide (1,2). In Japan, for example, the government has given strong support to a spectrum of projects for development of blood substitutes in the wake of two tragedies: the infection of hemophiliac patients, who had received non-pasteurized plasma products, by AIDS; and the Great Hanshin Earthquake disaster. In China, because of the lack of safe transfusion, blood substitute R&D is a national project.

Blood is separable into two fractions after centrifugation: plasma and cells. The roles of all plasma components are well characterized and their substitutes are already established (Table 1). Especially, recombinant human serum albumin (rHSA) will be commercialized soon in Japan. On the other hand, substitutes for cellular components—platelets and RBCs—are challenging (3). In this chapter, we specifically examine artificial oxygen carriers, which are substitutes for RBCs. The requisites for artificial oxygen carriers should be not only effectiveness for tissue oxygenation, but also the following:

1. No blood type antigen and no infection (no pathogens);
2. Stability for long-term storage (e.g., over 2 years) at room temperature for stockpiling for any emergency;

Table 1. Roles of Blood Components and Their Substitutes

Fraction		Components	Substitutes[*]
Plasma (55 vol%)	Plasma proteins	Albumin (maintenance of blood volume)	Plasma expanders (dextran, hydroxyethyl starch, modified gelatin, recombinant human serum albumin)
		Globulin (antibody)	Antibiotics artificial immunoglobulin
		Fibrinogen coagulation factors	Fibrin adhesive recombinant coagulation factors
	Electrolytes and other solutes	$Na^+, K^+, Ca^{2+}, Mg^{2+}, Cl^-, HCO_3^-, HPO_4^{2-}$, etc.	Electrolyte infusion
		Vitamins, amino acids, glucose, lipids, etc.	Nutrient infusion (triglyceride, amino acids, saccharides)
Cells (45 vol%)		Platelets	Artificial platelets
		White blood cells	None (antibiotics)
		Red blood cells	Artificial red cells (artificial O_2 carriers, O_2-infusions)

[*]including the materials under development.

3. Low toxicity and prompt metabolism even after massive infusion;
4. Rheological properties can be adjusted to resemble human blood; and
5. Reasonable production expense and cost performance.

Realization of an artificial oxygen carrier will bring innovative change in transfusion medicine.

2. CHEMICALLY MODIFIED HEMOGLOBIN AS AN OXYGEN CARRIER

Historically, the first attempt in the 1930s of Hb-based O_2 carrier was to simply use stroma-free Hb because Hb in RBCs binds and releases O_2. However, several problems became apparent: impurity of stroma-free Hb' dissociation into dimers that have a short circulation time; renal toxicity; high oncotic pressure; and high O_2 affinity. Since the 1970s, various approaches have been developed to overcome these problems, especially in the United States, because of military use for infusion to combat casualties (Fig. 1).

Materials included intra-molecular crosslinking using dibromosalicyl fumarate (4) or pyridoxal 5'-phosphate, polymerization using glutaraldehyde (2) or oxidized o-raffinose (5), and conjugation with water-soluble polymers such as polyethylene glycol (PEG), hydroxyethyl starch (HES), and dextran (2). The source of Hb is mostly human Hb purified from outdated donated human blood. An industrial–scale production of human Hb-based O_2 carriers requires a cooperation with blood banks, the Red Cross, and hospitals to establish a collection system of outdated donated RBCs. However, the amount is limited due to the limited number of blood donors and the fact that the hospitals are trying to use packed RBCs in a well-planned manner to reduce the discarded packed RBCs. Bovine or swine Hb can be a huge source obtainable from the cattle and hogs industries. The absence of heterologous immune reaction and prion protein has to be guaranteed. Recent biotechnology enables production of human Hb from transgenic swine blood. Moreover, a large-scale production of reombinant human Hb mutants as well as recombinant human serum albumin is possible from E. coli or yeast that should not include any pathogens from humans and mammals. For all the cases, Hbs should be strictly purified and free of pathogen via rigorous purification procedure such as ultrafiltration, pasteurization, irradiation, and solvent-detergent method (6), because the dose rate is considerably large.

In some cases of chemically modified Hbs, their structure (acellular structure) is so different from that of RBCs and caused side effects such as vasoconstriction (4). They are presumably attributable to the specific affinity of Hb to endogenous gas molecules, NO and CO, which are important messenger molecules for vasorelaxation. Although many companies have developed chemically modified Hb solutions as a transfusion alternative for elective surgery and trauma, some of them suspended clinical trials because of vasoactive properties. (See CLINICAL TRIALS in another chapter).The fact that myocardial lesion is caused by intramolecular crosslinked Hb (both chemically modified and recombinant Hb mutants) deters further development of these Hb-based O_2 carriers (7). Presently, glutaraldehyde-polymerized bovine or human Hbs and PEG-conjugated Hbs have progressed to the final stages of clinical trials (Table 2).

Oxyglobin^{TM}, a polymerized bovine Hb produced by Biopure Co. (Cambridge, MA), is now approved for veterinary use in the United States. This material can be stored in a liquid state at room temperature for years because Hb is stabilized by deoxygenation with addition of N-acetylcysteine. A more purified product, Hemopure^{TM}, with a narrower molecular weight distribution, is approved in South Africa for treating adult surgical patients who are acutely anemic and for eliminating, reducing, or delaying the need for allogeneic red blood cell transfusion in such patients (8). Because prion proteins are known to cause mad cow

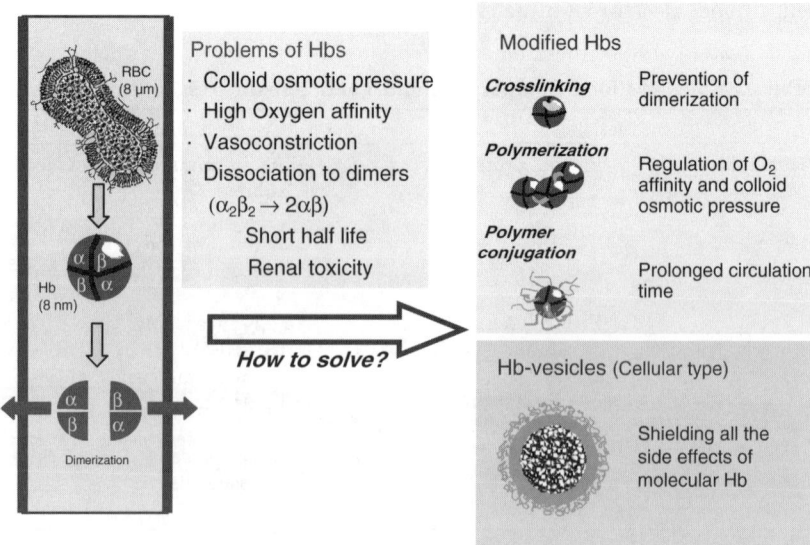

Figure 1. Chemically modified Hbs and encapsulated Hb to solve the side effects of molecular Hbs. (This figure is available in full color at http://www.mrw.interscience.wiley.com/ebe.)

Table 2. Artificial Oxygen Carriers Currently Developed for Clinical Application

Products (Group)	Composition	Indication	Present R&D Situation
PolyHeme (Northfield Labs. Inc.)	Glutaraldehyde-polymerized human Hb	Trauma	Phase III (US)
Hemopure (Biopure Corp.)	Glutaraldehyde-polymerized bovine Hb	Elective surgery	Phase III (US) approved in South Africa
PHP (Curacyte AG)	Pyridoxalated-human Hb, PEG-conjugated	Septic shock	Phase II (US)
Hemospan (Sangart Inc.)	PEG-modified human Hb	Elective surgery	Phase II (Sweden)
Hemolink (Hemosol Corp.)	o-raffinose polymerized human Hb	Elective surgery	Phase III, suspended
Oxygent (Alliance Pharm. Corp.)	Perfluorooctylbromide emulsion	Elective surgery	Phase II (US) suspended, R&D in China
Perftoran (Perftoran)	Perfluorodecalin, perfluoromethylcyclohexilpiperidine, proxanol	Hypovolemia, elective surgery	Approved in Russia
Hb-vesicles (Waseda-Keio-Oxygenix-Nipro)	Phospholipid vesicles encapsulating Hb,		Preclinical
Hemozyme (SynZyme Technol.)	Polynitroxyl human Hb		Preclinical
HemoTech (Hemobiotech Inc.)	Bovine Hb conjugated with o-ATP, o-adenosine and reduced glutathione.		Preclinical
PLA-PEG Hb nanocapsules (McGill Univ.)	Polylactide-PEG copolymer nanocapsules with Hb and enzymes		Preclinical
PolyHb-SOD-CAT (McGill Univ.)	Copolymerized Hb with SOD and catalase		Preclinical
Dex-BTC-Hb (Univ. Henri Poincare-Nancy)	Dextran conjugated Hb		Preclinical
HRC 101 (Hemosol Corp.)	Human Hb and hydroxyethyl starch conjugate		Preclinical
PEG-bHb (Beijing Kaizheng Biotech Corp)	PEG-modified bovine Hb		Preclinical
TRM-645 (Terumo Co.)	Liposome-encapsulated Hb		Preclinical
OxyVita (Oxyvita Inc.)	Zero-link polymer of bovine Hb		Preclinical
Albumin-hemes (Waseda-Nipro Corp.)	Synthetic heme-albumin composite		Preclinical
PHER O$_2$ (Sanaguine Corp)	Second generation Fluosol		Preclinical

disease (bovine spongiform encephalopathy: BSE), the key is to collect safer bovine blood exclusively from closed herds with well-documented health histories and controlled access.

PolyHeme^{TM} is a glutaraldehyde polymerized human Hb developed by Northfield Laboratories Inc. (Evanston, IL). Even though most chemically modified Hbs show vasoconstriction, it is reported that PolyHeme^{TM} does not induce vasoconstriction (9). Information on this material is more limited in the academic literature than for other products. PolyHeme^{TM} is now undergoing phase III clinical trials designed to evaluate the safety and efficacy of Polyheme when used to treat patients in hemorrhagic shock following traumatic injuries. According to the company, this is the first trial of an Hb-based O$_2$ carrier in which treatment begins in the pre-hospital setting, such as in an ambulance during transport.

Ajinomoto Co. Inc. (Tokyo) first tested PEG-conjugation to pyridoxalated Hb (PHP^{TM}) (10); Curacyte AG (Chapel Hill, NC) is continuously developing that material as an NO scavenger. PHP^{TM} has been demonstrated to reverse the vasodilatation caused by excess NO produced by inducible NO synthase. It resolves the hypotension associated with septic shock. It has completed Phase II clinical studies in distributive shock. Sangart Inc. (San Diego, CA) has developed PEG-modified human Hb (Hemospan^{TM}) with unique physicochemical properties: markedly higher O$_2$ affinity [P$_{50}$ (partial pressure of O$_2$ at which Hb is half-saturated with O$_2$) = 6 Torr]; viscosity (2.5 cP); and colloid osmotic pressure (55 Torr). It is effective for microcirculation and targeted O$_2$ transport to tissues (11). This material is now in clinical phase II trials in Sweden. Even though criticism exists that the O$_2$ affinity is too high to release O$_2$ in peripheral tissues, a comparative study of PEG-modified albumin indicated that Hemospan reliably delivers O$_2$ to tissues with no vasoconstriction or hypertension (12). This reliability suggests that the appropriate physicochemical properties for artificial O$_2$ carriers should not necessarily be merely equal to those of blood or RBCs (13).

3. IMPORTANCE OF Hb-ENCAPSULATION IN RBC FOR ARTIFICIAL RBC DESIGN

Physicochemical analyses have revealed that the cellular structure of RBCs retards O_2 release and binding of the inside Hb in comparison with a homogeneous Hb solution (14,15). However, nature has selected this cellular structure during evolution. Historically, Barcroft et al. insisted that the reasons for Hb encapsulation in RBCs were (1) a decreased high viscosity of Hb and a high colloidal osmotic pressure, (2) prevention of the removal of Hb from the blood circulation, and (3) preservation of the chemical environment in the cells such as concentration of phosphates (2,3-DPG, ATP, etc.) and other electrolytes (1). Moreover, during the long development of Hb-based O_2 carriers, numerous side effects of molecular Hb have become apparent, such as the dissociation of tetrameric Hb subunits into two dimers ($\alpha_2\beta_2 \rightarrow \alpha\beta$) that might induce renal toxicity, and entrapment of gaseous messenger molecules (NO and CO) inducing vasoconstriction, hypertension, reduced blood flow, and tissue oxygenation at microcirculatory levels (16,17), neurological disturbances, and the malfunctioning of esophageal motor function (18), and heme-mediated oxidative reactions with various active oxygen species (19). These side effects of molecular Hbs imply the importance of the cellular structure or the larger particle dimension of Hb-based O_2 carriers.

Pioneering work of Hb encapsulation to mimic the cellular structure of RBCs was performed by Chang in 1957 (2) who prepared microcapsules (5 μm) made of nylon, collodion, etc. Toyoda in 1965 (20) and the Kambara-Kimoto group (21) also covered Hb solutions with gelatin, gum Arabic, or silicone, etc. Nevertheless, it was extremely difficult to regulate the particle size that was appropriate for blood flow in the capillaries and to obtain sufficient biocompatibility. After Bangham and Horne reported in 1964 that phospholipids assemble to form vesicles in aqueous media, and that they encapsulate water-soluble materials in their inner aqueous interior (22), it was reasonable to use such vesicles for Hb encapsulation. Djordjevich and Miller in 1977 prepared liposome-encapsulated Hb (LEH) composed of phospholipids, cholesterol, fatty acids, etc. (23). In the US, Naval Research Laboratories showed remarkable progress of LEH (24). What we call Hb-vesicles (HbV) with a high-efficiency production process and their improved properties have been established by Tsuchida's group based on technologies of molecular assembly and precise analysis of pharmacological and physiological aspects (25,26) (Fig. 2).

Liposomes, as molecular assemblies, had been generally accepted as structurally unstable. Many researchers have sought to develop stabilization methods that use polymer chains (27). Polymerization of phospholipids that contain dienoyl groups was studied extensively. For example, gamma-ray irradiation induces radiolysis of water molecules and generates OH radicals that initiate intermolecular polymerization of dienoyl groups in phospholipids. This method produces enormously stable liposomes, like rubber balls, which are resistant to freeze-thawing, freeze-drying, and rehydration (1,28). However, the polymerized liposomes were so stable that they were not degraded easily in the macrophages even 30 days after injection. It was concluded that polymerized lipids would not be appropriate for intravenous injection. Selection of appropriate lipids (phospholipid/cholesterol/negatively charged lipid/PEG-lipid) and their composition

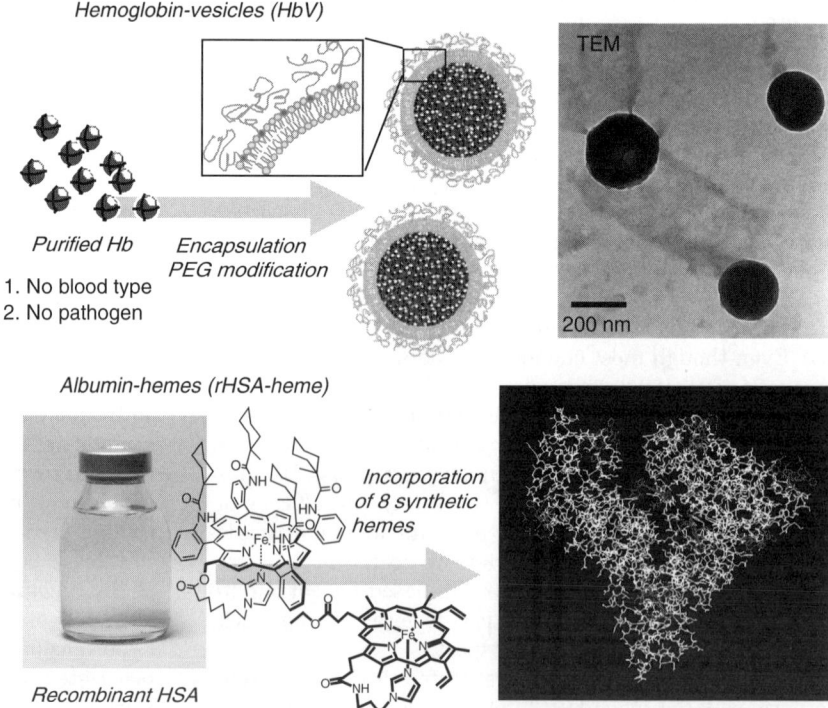

Figure 2. Hemoglobin-vesicles and albumin-hemes as new types of artificial oxygen carriers. (This figure is available in full color at http://www.mrw.interscience.wiley.com/ebe.)

are important to enhance the stability of liposomes without polymerization. Surface modification of liposomes with PEG chains is effective for dispersion stability. Using deoxygenation and PEG-modification, HbV can be stored at room temperature under deoxygenated conditions for two years (29). Moreover, storage does not induce aggregation and metHb formation. Even after injection into blood stream, HbV is homogeneously dispersed in the plasma phase and contributes to tissue oxygenation, as clarified by the microcirculatory observations (30).

One particle of HbV (ca. 250-nm diameter) contains about 30,000 Hb molecules. The HbV acts as a particle in the blood and not as a solute. Therefore, the colloid osmotic pressure of the HbV suspension is nearly zero. It requires addition of a plasma expander for a large substitution of blood while maintaining the blood volume. Candidates of plasma expanders are plasma-derived HSA, hydroxyethyl starch (HES), dextran, or gelatin, depending on the clinical setting, cost, country, and clinician. Recombinant human serum albumin (rHSA) is an alternative that will be approved for clinical use in Japan. The HbV suspended in HSA or rHSA was tested for resuscitation from hemorrhagic shock (31) and extreme hemodilution (30). Moreover, HbV with a high O_2 affinity (low P_{50}) suspended in HES was tested for oxygenation of an ischemic skin flap (32). The results imply the further application of HbV for other ischemic diseases such as myocardial and brain infarction and stroke.

Safety of HbV has been confirmed in terms of blood compatibility (33), no vasoactivity (17), biodistribution of ^{99m}Tc-labeled HbV to reticuloendothelial system (RES) (34) and prompt degradation in RES, even after a massive infusion (35,36). Based on the safety and efficacy of HbV, a joint collaboration partnership of academia, a biotech venture company and a corporation in Japan are seeking clinical trials of HbV within a few years.

4. TOTALLY SYNTHETIC OXYGEN CARRIERS

4.1. Metal Complexes and Heme Derivatives

Minoshima et al. tested the crystalline state of cobalt histidine chelate complex as an O_2 carrier that reversibly binds an O_2 molecule (37). The Kambara and Kimoto group studied heme-derivatives of imidazole complexes. However, the irreversible O_2 binding and the short lifetime of the O_2 complex could not be overcome. Because a heme is inserted into a hydrophobic pocket of a globin macromolecule (such as Hb, myoglobin, neuroglobin), stable O_2 binding requires a hydrophobic environment. Collman et al. in 1973 (38) synthesized a derivative of iron tetraphenyl porphyrin-imidazole complex that makes its O_2 binding site hydrophobic and binds O_2 reversibly in an organic solvent, but not in an aqueous solution because of the spontaneous and irreversible oxidation of heme. Tsuchida et al. in 1983 synthesized an amphiphilic derivative of iron porphyrin that can be inserted into the hydrophobic bilayer membrane of phospholipid vesicles (liposomes) (39). This system represents the first example of an entirely synthetic O_2 carrier that reversibly binds O_2 under physiological conditions.

One role of serum albumin is to provide a hydrophobic binding site to carry nutrients, metabolic wastes, or functional molecules. It was clarified that a synthetic heme derivative can be incorporated efficiently into human serum albumin (HSA) solution, thereby providing a red albumin-heme hybrid (40). In Japan, recombinant human serum albumin (rHSA) is manufactured through expression in *Pichia pastoris* yeast; the Japanese FDA will soon approve it. Combination of the heme derivative, rHSA-heme is a new class of synthetic hemoprotein that requires no blood as a raw material source (41) (Fig. 2). The in vivo tests clarified the efficacy of rHSA-heme for hemodilution and shock resuscitation (42). A physiological colloid osmotic pressure was regulated by 5-wt% HSA concentration in the blood. To increase the O_2 transporting capacity of rHSA-heme, albumin-dimer is effective to reduce the colloid osmotic pressure and to increase the heme content. The dimer can be prepared using intermolecular crosslinking at Cys-34 (43). Surprisingly, this rHSA-heme shows no vasoconstriction or hypertension even though its NO binding properties are similar to those of other modified Hbs of similar molecular size (44,45). This phenomenon is explained by characteristics of negatively charged rHSA molecules, which reduce the permeability across the negatively charged endothelial cell layers, where NO is produced for relaxation of the smooth muscle layer.

The small molecular dimension of rHSA-heme, which causes no vasoconstriction, will be appropriate to carry O_2 effectively to tissues where RBCs are difficult to reach, such as tumor tissues. The tumor vasculature is highly heterogeneous and is therefore susceptible to hypoxia. In such conditions, tumor cells become resistant to chemotherapy and irradiation. It has been confirmed that injection of rHSA-heme considerably increased the O_2 tension in an implanted tumor in a rat model (46). The succeeding irradiation therapy shows reduced tumor size and improved survival. This therapeutic possibility for cancer therapy is also supported by the trials of chemically modified Hb solutions and perfluorochemicals (47).

rHSA incorporates a protoheme IX into the hydrophobic cavity of the subdomain IB. Introduction of proximal histidine into the heme binding site by site-directed mutagenesis allows O_2 binding to the prosthetic heme group. This albumin-protoheme is a new type of synthetic O_2-carrier (48).

4.2. Perfluorochemicals

Two major discoveries exist in the study of perfluorochemicals (PFC): (1) Clark and Gollan found that mice can survive by breathing an oxygenated PFC liquid (49); (2) Geyer et al. showed that an emulsified PFC can be used to replace the blood of rats completely (50). The former Green Cross Co. (Osaka) produced a PFC solution composed of perfluorodecalin ($C_{10}F_{18}$) and perfluorotripropylamine with a mixture of Pluronic and egg-yolk lecithin as surfactants. The resulting white colored emulsion, Fluosol-DA, was approved in 1978 to undergo clinical trials (51). Because the PFC concentration in the emulsion is only 20–35 vol%, its O_2 carrying capacity is less than one-tenth that of blood at ambient O_2 pressure. Therefore, patients require inhalation of 100% O_2 gas during an operation. The US FDA approved Fluosol-DA for intracoronary administration only during percutaneous transluminal coronary angioplasty (PTCA). Because of its insufficient O_2 transporting capacity and side effects such as accumulation, pneumonia, and anaphylactic reactions, the company stopped production of Fluosol-DA in 1993.

Riess et al. showed that PFC emulsions from perfluorooctylbromide ($C_8F_{17}Br$) had four-times' higher O_2 solubility than that of Fluosol-DA (52,53). Alliance Pharmaceutical Corp. (San Diego, CA) has extensively developed a so-called second-generation PFC emulsion (Oxygent™) that is in multi-center international phase II/III trials aimed at its use as a pre-operational or peri-operational infusion for elective surgery to obviate or minimize allogeneic transfusion. In Russia, Perftoran (Moscow) developed PFC emulsion of perfluorodecalin and perfluoromethyl cyclohexylpiperidine. This material is approved in Russia for medical application (54).

5. METABOLISM OF BLOOD SUBSTITUTES AND SIDE EFFECTS

As a dose rate of blood, substitutes would be considerably larger than those of other drugs and the circulation time would be significantly shorter than RBC; their biodistribution, metabolism, excretion, and the side effects have to be characterized. Normally, free Hb released from RBC is rapidly bound to haptoglobin and removed from the circulation by hepatocytes. However, when the Hb concentration exceeds the haptoglobin binding capacity, unbound Hb is filtered through the kidney, where it is actively absorbed. When the reabsorption capacity of the kidney is exceeded, hemoglobinuria and eventually renal failure occur. The encapsulation of Hb in both RBC and liposomes completely suppresses renal excretion. However, both senescent RBCs and Hb-vesicles in the blood stream are finally captured by phagocytes in the RES (or MPS), that was confirmed by radioisotope-labeling techniques (23,34,35). Particles of Perfluorocarbon emulsions and chemically modified Hbs (such as pyridoxalated polymerized Hb) are also captured by RES (55,56). It has to be clarified whether the accumulation of these materials in phagocytic cells may lead to transient impairment of the function of RES such as elimination of other foreign elements (35). There needs to be a balance between the circulation time of the O_2 carriers and the rates of metabolism and excretion. When their circulation time is too short, they burden on the functions of RES, kidney, and other related organs.

The released heme from Hb-based O_2 carriers should be mainly metabolized by the inducible form of heme oxygenase-1 in the Kupffer cells in the liver and macrophages in the spleen. The resulting bilirubin is excreted in the bile duct. Iron deposition is confirmed as hemosiderin for the chemically modified and encapsulated Hbs. Normally, iron from a heme is stored in the ferritin molecule. This protein has 24 subunits and encloses as many as 4,500 iron atoms in the form of an aggregate of ferric hydroxide (57). Ferritin in the lysosomal membrane may form paracrystalline structures and eventually aggregate in mass with an iron content as high as 50%. These are hemosiderins composed of degraded protein and coalesced iron. Not only infusion of polymerized Hb and Hb-vesicles, but also transfusion of stored RBCs induces hemosiderin deposition in RES. As iron acts as a catalyst for Fenton reaction to produce toxic cytotoxic OH radicals from hydrogen peroxide, the level of hemosiderosis should be carefully monitored.

As for the membrane components of Hb-vesicles and perfluorocarbon emulsions, it was reported that the infused lipid components of liposomes are entrapped in the Kupffer cells, and diacylphosphatidylcholine is metabolized and reused as a component of the cell membrane, or excreted in bile, especially as fatty acids and in exhaled air (35). There is no metabolic pathway for inert parfluorocarbon, and this gradually diffuses from the RES to the blood stream and is excreted in exhaled air through the lungs. The PEG chain is widely used for surface modification of both Hb and Hb-vesicles. The chemical crosslinker of PEG-lipid or PEG-Hb is susceptible to hydrolysis to release PEG chains during metabolism. The released PEG chains, which is known as an inert macromolecule, should be excreted in the urine through the kidneys (58).

6. NEW CONCEPTS

Development of artificial O_2 carriers was originally initiated with a simple idea and an expectation that the materials that bind or dissolve O_2 can behave like RBCs in the blood stream. However, it was not easy to complete that project. During its long history of development, unexpected side effects were clarified such as capillary plugging, renal toxicity, vasoconstriction, vascular injury, and accumulation. Even after R&D of artificial O_2 carriers for decades, no material is commercially available for clinical use in Europe, Japan, or the US. Recent advanced biotechnology enables *ex vivo* RBC production from hematopoietic stem cells (59). However, problems remain of large-scale production and long-term storage for stockpiling. On the other hand, no doubts exist about a strong demand and expectation of blood-substitute development.

The importance of the sophisticated function of RBCs in concert with vascular physiology has been clarified, and new concepts are proposed in terms of the physicochemical properties of Hb-based artificial O_2 carriers. Historically, it has been regarded that the O_2 affinity should be regulated similarly to RBCs (25–30 Torr). Theoretically, this allows sufficient O_2 unloading during blood microcirculation as can be evaluated by the arterio-venous difference in O_2 saturation in accordance with an O_2 equilibrium curve. It has been expected that decreasing O_2 affinity (increasing P_{50}) increases O_2 unloading. However, small artificial O_2 carriers should release O_2 faster in arterial blood flow (14,15). It has been suggested that faster O_2 unloading from the HBOCs is advantageous for tissue oxygenation. However, this concept is controversial in light of recent findings because an excess O_2 supply would cause autoregulatory vasoconstriction and microcirculatory disorders. The new concept is that an Hb-based O_2 carriers with a high O_2 affinity (low P_{50}) should retain O_2 in the upstream artery or arteriole and release O_2 in the capillaries of the targeted tissue. This concept is recently supported by the results of PEG-modified Hbs and Hb-vesicles by the microcirculatory observations (60–62).

Because an infusion of an artificial O_2 carrier results in substitution of a large volume of blood, impact on hemorheology is great. It has been regarded that lower blood viscosity after hemodilution is effective for tissue perfusion. However, microcirculatory observation shows that, in some cases lower viscosity engenders decreased shear stress on the vascular wall, engendering vasoconstriction and reduced functional capillary density (63). Therefore, an appropriate viscosity might exist, which maintains the normal tissue perfusion level. In relation to this, solutions of Hb-based O_2 carriers with a higher molecular weight are more viscous and would be appropriate. Moreover, as mentioned above, a larger molecular dimension can reduce the vascular permeability and minimize trapping of NO and CO as vasorelaxation factors.

These new concepts suggest reconsideration of the design of artificial O_2 carriers (13,14). Actually as shown in Table 2, new products are appearing, though they are in the preclinical stage, such as zero-link polymerized Hb (64), Hb-vesicles (65), and HRC 101 with larger molecular dimensions and higher O_2 affinities. The biodegradable polylactide (PLA)-PEG copolymer -nanocapsules (80–100 nm in size) contain Hb and hemolysate-derived enzymes (66). RBCs contain radical scavenging functions by SOD and catalase, and the sophisticated metHb reducing system. Hemozyme, Hemotech, and PolyHb-SOD-CAT have antioxidative properties that would be appropriate for eliminating active oxygen species in ischemia-reperfusion injury (66).

7. ADVANTAGES OF ARTIFICIAL OXYGEN CARRIERS AND CLINICAL INDICATIONS

Advantages of artificial O_2 carriers are the absence of blood-type antigens and infectious viruses, and stability for a long-term storage that overwhelm RBC transfusion. Easy manipulation of physicochemical properties enables tailor-made O_2 carriers that suit the clinical indications. The shorter half-lives of the HBOCs in the blood stream (2–3 days) limit their use, but they are applicable for a shorter periods of use, as: (1) resuscitative fluids for hemorrhagic shock during a pre-hospital emergency situation for temporary use or bridging until packed RBCs are available; (2) fluids for pre-operative hemodilution or peri-operative O_2 supply fluids for a hemorrhage in an elective surgery to obviate or delay allogeneic transfusion; (3) a priming solution for the circuit of an extracorporeal membrane oxygenator (ECMO); (4) O_2 therapeutics to oxygenate ischemic tissues; or (5) *ex vivo* oxygenation of harvested cell cultures, reconstructed tissues, and organs for transplantation (Table 3).

Clinicians and patients await the realization of safe and functional artificial O_2 carriers and their new clinical applications in the near future. This development might require continuous interdisciplinary cooperation to overcome not only emerging problems in preclinical and clinical tests, but also the dogmas of classical blood substitutes and modern transfusion medicine.

Table 3. Expected Clinical Indication of Artificial O_2 Carriers

Transfusion Alternative	Other Expected Applications
Resuscitative fluid for shock in an emergency.	Oxygenation of local ischemic disease (brain or myocardial infarction)
Hemodilution for autologous blood preservation for elective surgery.	Tumor oxygenation for photosensitization
Prime for circuit of extracorporeal membrane oxygenator (ECMO)	Oxygenation fluid of cultured cells for tissue reconstruction
Chronic anemia	Perfusion of organs for transplantation
Infusion to rate blood type patients	
Infusion to patients who do not accept transfusion (e.g., fear of infection, religious reason)	

BIBLIOGRAPHY

1. E. Tsuchida, *Artificial Red Cells: Materials, Performances, and Clinical Study as Blood Substitutes*. Chichester: John Wiley & Sons, 1995.

2. T. M. S. Chang, *Blood Substitutes: Principles, Methods, Products, and Clinical Trials*. Basel: Karger, 1997.

3. E. Tsuchida, *Blood Substitutes: Present and Future Perspectives*. Lausanne: Elsevier, 1998.

4. E. P. Sloan, M. Koenigsberg, D. Gens, M. Cipolle, J. Runge, M. N. Mallory, and G. Rodman Jr., Diaspirin cross-linked hemoglobin (DCLHb) in the treatment of severe traumatic hemorrhagic shock: a randomized controlled efficacy trial. *JAMA*. 1999; **282**:1857–1864.

5. F. J. Carmichael, A. C. Ali, J. A. Campbell, S. F. Langlois, G. P. Biro, A. R. Willan, C. H. Pierce, and A. G. Greenburg, A phase I study of oxidized raffinose cross-linked human hemoglobin. *Crit. Care Med*. 2000; **28**:2283–2292.

6. H. Abe, K. Ikebuchi, J. Hirayama, M. Fujihara, S. Takeoka, H. Sakai, E. Tsuchida, and H. Ikeda, Virus inactivation in hemoglobin solution by heat treatment. *Artif. Cells Blood Substit. Immobil. Biotechnol*. 2001; **29**:381–388.

7. K. Burhop, D. Gordon, and T. Estep, Review of hemoglobin-induced myocardial lesions. *Artif. Cells Blood Substit. Immobil. Biotechnol*. 2004; **32**:353–374.

8. J. H. Levy, The use of haemoglobin glutamer-250 (HBOC-201) as an oxygen bridge in patients with acute anaemia associated with surgical blood loss. *Expert Opin. Biol. Ther*. 2003; **3**:509–517.

9. S. A. Gould, E. E. Moore, D. B. Hoyt, P. M. Ness, E. J. Norris, J. L. Carson, G. A. Hides, I. H. Freeman, R. DeWoskin, and G. S. Moss, The life-sustaining capacity of human polymerized hemoglobin when red cells might be unavailable. *J. Am. Coll. Surg*. 2002; **195**:445–452.

10. K. Ajisaka and Y. Iwashita, Modification of human hemoglobin with polyethylene glycol: a new candidate for blood

substitute. *Biochem. Biophys. Res. Commun.* 1980; **97**:1076–1081.

11. K. D. Vandegriff, A. Malavalli, J. Wooldridge, J. Lohman, and R. M. Winslow, MP4, a new nonvasoactive PEG-Hb conjugate. *Transfusion* 2003; **43**:509–516.

12. R. M. Winslow, J. Lohman, A. Malavalli, and K. D. Vandegriff, Comparison of PEG-modified albumin and hemoglobin in extreme hemodilution in the rat. *J. Appl. Physiol.* 2004; **97**:1527–1534.

13. A. G. Tsai, P. Cabrales, and M. Intaglietta, Oxygen-carrying blood substitutes: a microvascular perspective. *Expert Opin. Biol. Ther.* 2004; **4**:1147–1157.

14. T. C. Page, W. R. Light, C. B. McKay, and J. D. Hellums, Oxygen transport by erythrocyte/hemoglobin solution mixtures in an in vitro capillary as a model of hemoglobin-based oxygen carrier performance. *Microvasc. Res.* 1998; **55**:54–64.

15. H. Sakai, Y. Suzuki, M. Kinoshita, S. Takeoka, N. Maeda, and E. Tsuchida, O_2 release from Hb vesicles evaluated using an artificial, narrow O_2-permeable tube: comparison with RBCs and acellular Hbs. *Am. J. Physiol. Heart Circ. Physiol.* 2003; **285**:H2543–H2551.

16. N. Goda, K. Suzuki, M. Naito, S. Takeoka, E. Tsuchida, Y. Ishimura, T. Tamatani, and M. Suematsu, Distribution of heme oxygenase isoforms in rat liver. Topographic basis for carbon monoxide-mediated microvascular relaxation. *J. Clin. Invest.* 1998; **101**:604–612.

17. H. Sakai, H. Hara, M. Yuasa, A.G. Tsai, S. Takeoka, E. Tsuchida, and M. Intaglietta, Molecular dimensions of Hb-based O_2 carriers determine constriction of resistance arteries and hypertension. *Am. J. Physiol. Heart Circ. Physiol.* 2000; **279**:H908–H915.

18. J. A. Murray, A. Ledlow, J. Launspach, D. Evans, M. Loveday, and J. L. Conklin, The effects of recombinant human hemoglobin on esophageal motor functions in humans. *Gastroenterology* 1995; **109**:1241–1248.

19. A. I. Alayash, Oxygen therapeutics: can we tame haemoglobin? *Nat. Rev. Drug Discov.* 2004; **3**:152–159.

20. T. Toyoda, Artificial blood. *Kagaku* 1965; **35**:7–13 (in Japanese).

21. M. Kitajima, W. Sekiguchi, and A. Kondo, A modification of red blood cells by isocyanates. *Bull. Chem. Soc. Jpn.* 1971; **44**:139–143.

22. A. D. Bangham and R. W. Horne, Negative staining of phospholipids and their structure modification by surface-active agents as observed in the electron microscope. *J. Mol. Biol.* 1964; **8**:660–668.

23. L. Djordjevich and I. F. Miller, Lipid encapsulated hemoglobin as a synthetic erythrocyte. *Fed. Proc.* 1977; **36**:567.

24. A. S. Rudolph, R. W. Klipper, B. Goins, and W. T. Phillips, In vivo biodistribution of a radiolabeled blood substitute: ^{99m}Tc-labeled liposome-encapsulated hemoglobin in an anesthetized rabbit. *Proc. Natl. Acad. Sci. U S A.* 1991; **88**:10976–10980.

25. H. Sakai, K. Hamada, S. Takeoka, H. Nishide, and E. Tsuchida, Physical properties of hemoglobin vesicles as red cell substitutes. *Biotechnol. Prog.* 1996; **12**:119–125.

26. S. Takeoka, T. Ohgushi, K. Terase, T. Ohmori, and E. Tsuchida, Layer-controlled hemoglobin vesicles by interaction of hemoglobin with a phospholipid assembly. *Laugmuir* 1996; **12**:1755–1759.

27. H. Ringsdorf, B. Schlarb, and J. Venzmer, Molecular architecture and function of polymeric oriented systems – models for the study of organization, surface recognition, and dynamicc of biomembranes. *Angew. Chem. Int. Ed.* 1988; **27**:113–158.

28. E. Tsuchida, E. Hasegawa, N. Kimura, M. Hatashita, and C. Makino, Polymerization of unsaturated phospholipids as large unilamellar liposomes at low-temperature. *Macromolecules* 1992; **25**:2007–2212.

29. H. Sakai, K. Tomiyama, K. Sou, S. Takeoka, and E. Tsuchida, Poly(ethylene glycol)-conjugation and deoxygenation enable long-term preservation of hemoglobin-vesicles as oxygen carriers in a liquid state. *Bioconjug. Chem.* 2000; **11**:425–432.

30. H. Sakai, A. G. Tsai, H. Kerger, S. I. Park, S. Takeoka, H. Nishide, E. Tsuchida, and M. Intaglietta, Subcutaneous microvascular responses to hemodilution with a red cell substitute consisting of polyethyleneglycol-modified vesicles encapsulating hemoglobin. *J. Biomed. Mater. Res.* 1998; **40**:66–78.

31. H. Sakai, Y. Masada, H. Horinouchi, M. Yamamoto, E. Ikeda, S. Takeoka, K. Kobayashi, and E. Tsuchida, Hemoglobin-vesicles suspended in recombinant human serum albumin for resuscitation from hemorrhagic shock in anesthetized rats. *Crit. Care Med.* 2004; **32**:539–545.

32. C. Contaldo, J. Plock, H. Sakai, S. Takeoka, E. Tsuchida, M. Leunig, A. Banic, and D. Erni, New generation of hemoglobin-based oxygen carriers evaluated for oxygenation of critically ischemic hamster flap tissue. *Crit. Care Med.* 2005; **33**:806–812.

33. S. Wakamoto, M. Fujihara, H. Abe, H. Sakai, S. Takeoka, E. Tsuchida, H. Ikeda, and K. Ikebuchi, Effects of poly(ethyleneglycol)-modified hemoglobin vesicles on agonist-induced platelet aggregation and RANTES release in vitro. *Artif. Cells Blood Substit. Immobil. Biotechnol.* 2001; **29**:191–201.

34. K. Sou, R. Klipper, B. Goins, E. Tsuchida, and W. T. Phillips, Circulation kinetics and organ distribution of Hb-vesicles developed as a red blood cell substitute. *J. Pharmacol. Exp. Ther.* 2005; **312**:702–709.

35. H. Sakai, H. Horinouchi, K. Tomiyama, E. Ikeda, S. Takeoka, K. Kobayashi, and E. Tsuchida, Hemoglobin-vesicles as oxygen carriers: Influence on phagocytic activity and histopathological changes in reticuloendothelial system. *Am. J. Pathol.* 2001; **159**:1079–1088.

36. H. Sakai, Y. Masada, H. Horinouchi, E. Ikeda, K. Sou, S. Takeoka, M. Suematsu, M. Takaori, K. Kobayashi, and E. Tsuchida, Physiological capacity of the reticuloendothelial system for the degradation of hemoglobin vesicles (artificial oxygen carriers) after massive intravenous doses by daily repeated infusions for 14 days. *J. Pharmacol. Exp. Ther.* 2004; **311**:874–884.

37. T. Minoshima, T. Artificial blood—past, present and future. *Kokyu To Junkan*. 1982; **30**:758–767 (in Japanese).

38. J. P. Collman, R. R. Gagne, T. R. Halbert, J. C. Marchon, and C. A. Reed, Reversible oxygen adduct formation in ferrous complexes derived from a "picket fence" prophyrin. A model for oxymyoglobin. *J. Am. Chem. Soc.* 1973; **95**:7868–7870.

39. E. Tsuchida, H. Nishide, M. Sekine, and A. Yamagishi, Liposomal heme as oxygen carrier under semi-physiological conditions. Orientation study of heme embedded in a phospholipid bilayer by an electrooptical method. *Biochim. Biophys. Acta.* 1983; **734**:274–278.

40. E. Tsuchida, K. Ando, H. Maejima, N. Kawai, T. Komatsu, S. Takeoka, and H. Nishide, Properties of and oxygen binding by albumin-tetraphenylporphyrinatoiron(II) derivative complexes. *Bioconjug Chem.* 1997; **8**:534–538.

41. E. Tsuchida, T. Komatsu, Y. Matsukawa, K. Hamamatsu, and J. Wu, Human serum albumin incorporating Tetrakis(o-pivalamido) phenylporphinatoiron(II) derivative as a totally synthetic O_2-carrying hemoprotein. *Bioconjug. Chem.* 1999; **10**:797–802.

42. T. Komatsu, H. Yamamoto, Y. Huang, H. Horinouchi, K. Kobayashi, and E. Tsuchida, Exchange transfusion with synthetic oxygen-carrying plasma protein "albumin-heme" into an acute anemia rat model after seventy-percent hemodilution. *J. Biomed. Mater. Res. A.* 2004; **71**:644–651.
43. T. Komatsu, Y. Oguro, Y. Teramura, S. Takeoka, J. Okai, M. Anraku, M. Otagiri, and E. Tsuchida, Physicochemical characterization of cross-linked human serum albumin dimer and its synthetic heme hybrid as an oxygen carrier. *Biochim. Biophys. Acta.* 2004; **1675**:21–31.
44. T. Komatsu, Y. Matsukawa, and E. Tsuchida, Reaction of nitric oxide with synthetic hemoprotein, human serum albumin incorporating tetraphenylporphinatoiron(II) derivatives. *Bioconjug. Chem.* 2001; **12**:71–75.
45. E. Tsuchida, T. Komatsu, Y. Matsukawa, A. Nakagawa, H. Sakai, K. Kobayashi, and M. Suematsu, Human serum albumin incorporating synthetic heme: Red blood cell substitute without hypertension by nitric oxide scavenging. *J. Biomed. Mater. Res. A.* 2003; **64**:257–261.
46. K. Kobayashi, T. Komatsu, A. Iwamaru, Y. Matsukawa, H. Horinouchi, M. Watanabe, and E. Tsuchida, Oxygenation of hypoxic region in solid tumor by administration of human serum albumin incorporating synthetic hemes. *J. Biomed. Mater. Res. A.* 2003; **64**:48–51.
47. M. Nozue, I. Lee, J. M. Manning, L. R. Manning, and R. K. Jain, Oxygenation in tumors by modified hemoglobins. *J. Surg. Oncol.* 1996; **62**:109–114.
48. T. Komatsu, N. Ohmichi, A. Nakagawa, P.A. Zunszain, S. Curry, and E. Tsuchida, O_2 and CO binding properties of artificial hemoproteins formed by complexing iron protoporphyrin IX with human serum albumin mutants. *J. Am. Chem. Soc.* 2005; **127**:15933–15942.
49. L. C. Clark Jr. and F. Gollan, Survival of mammals breathing organic liquids equilibrated with oxygen at atmospheric pressure. *Science* 1966; **152**:1755–1756.
50. R. P. Geyer, Whole animals perfusion with fluorocarbon dispersions. *Fed. Proc.* 1970; **29**:1758–1763.
51. T. Mitsuno, H. Ohyanagi, and R. Naito, Clinical studies of a perfluorochemical whole blood substitute (Fluosol-DA) Summary of 186 cases. *Ann. Surg.* 1982; **195**:60–69.
52. J. G. Riess, Overview of progress in the fluorocarbon approach to in vivo oxygen delivery. *Biomater. Artif. Cells Immobil. Biotechnol.* 1992; **20**:183–202.
53. J. G. Riess, Oxygen carriers ("blood substitutes")—raison d'etre, chemistry, and some physiology. *Chem. Rev.* 2001; **101**:2797–2920.
54. E. Maevsky, G. Ivanitsky, L. Bogdanova, O. Axenova, N. Karmen, E. Zhiburt, R. Senina, S. Pushkin, I. Maslennikov, A. Orlov, and I. Marinicheva, Clinical results of Perftoran application: present and future. *Artif. Cells Blood Substit. Immobil. Biotechnol.* 2005; **33**:37–46.
55. A. G. Greenburg, The effects of hemoglobin on reticuloendothelial function. *Prog. Clin. Biol. Res.* 1983; **122**:127–137.
56. G. Lenz, H. Junger, M. Schneider, N. Kothe, R. Lissner, and A.M. Prince, Elimination of pyridoxylated polyhemoglobin after partial exchange transfusion in chimpanzees. *Biomater. Artif. Cells Immobilization Biotechnol.* 1991; **19**:699–708.
57. C. A. Finch and H. Huebers, Perspectives in iron metabolism. *N. Engl. J. Med.* 1982; **306**:1520–1528.
58. T. Yamaoka, Y. Tabata, and Y. Ikada, Distribution and tissue uptake of poly(ethylene glycol) with different molecular weights after intravenous administration to mice. *J. Pharm. Sci.* 1994; **83**:601–606.
59. M. C. Giarratana, L. Kobari, H. Lapillonne, D. Chalmers, L. Kiger, T. Cynober, M. C. Marden, H. Wajcman, and L. Douay, Ex vivo generation of fully mature human red blood cells from hematopoietic stem cells. *Nat. Biotechnol.* 2005; **23**:69–74.
60. P. Cabrales, H. Sakai, A. G. Tsai, S. Takeoka, E. Tsuchida, and M. Intaglietta, Oxygen transport by low and normal oxygen affinity hemoglobin vesicles in extreme hemodilution. *Am. J. Physiol. Heart Circ. Physiol.* 2005; **288**:H1885–H1892.
61. J. A. Plock, C. Contaldo, H. Sakai, E. Tsuchida, M. Leunig, A. Banic, M. D. Menger, and D. Erni, Is the Hb in Hb vesicles infused for isovolemic hemodilution necessary to improve oxygenation in critically ischemic hamster skin?. *Am. J. Physiol. Heart Circ. Physiol.* 2005; **289**:H2624–H2631.
62. A. G. Tsai, K. D. Vandegriff, M. Intaglietta, and R. M. Winslow, Targeted O_2 delivery by low-P_{50} hemoglobin: a new basis for O_2 therapeutics. *Am. J. Physiol. Heart Circ. Physiol.* 2003; **285**:H1411–H1419.
63. A. G. Tsai, B. Friesenecker, M. McCarthy, H. Sakai, and M. Intaglietta, Plasma viscosity regulates capillary perfusion during extreme hemodilution in hamster skinfold model. *Am. J. Physiol.* 1998; **275**:H2170–H2180.
64. H. Sakai, M. Yuasa, H. Onuma, S. Takeoka, and E. Tsuchida, Synthesis and physicochemical characterization of a series of hemoglobin-based oxygen carriers: objective comparison between cellular and acellular types. *Bioconjug. Chem.* 2002; **11**:56–64.
65. B. Matheson, H. E. Kwansa, E. Bucci, A. Rebel, and R. C. Koehler, Vascular response to infusions of a nonextravasating hemoglobin polymer. *J. Appl. Physiol.* 2002; **93**:1479–1486.
66. T. M. S. Chang, Hemoglobin-based red blood cell substitutes. *Artif. Organs* 2004; **28**:789–794.

BONE, MECHANICAL TESTING OF

M.P. HORAN
Medical University of South Carolina
Charleston, South Carolina

Y.H. AN
Medical University of South Carolina
Charleston, South Carolina
and
Clemson University
Clemson, South Carolina

1. INTRODUCTION

Bone is a complex heterogeneous material that in the human body serves the function of support, movement and protection, body mineral homeostasis, and hematopoesis. The study of bone brings together the fields of medicine and engineering in addition to the basic sciences of chemistry, biology, and physics to find ways of preventing and treating disease. As a material whose normal function and operation is integral to the daily life of the human being, bone has been the subject of countless research studies covering topics as diverse as the treatment of fractures to replacing pieces with artificial materials to building bone de novo on the lab bench. It has a hierarchical structural that presents certain behaviors under daily loading conditions in its physiologically normal state, only to offer somewhat unpredictable behaviors in the setting of a variety of bone pathologies and abnormal loading conditions. The combination of its physiologic roles, complicated geometry, and dynamic composition contribute to the challenging nature of characterizing its mechanical behavior. As such the topic of mechanical testing of bone is quite extensive and cannot be fully discussed in a single book chapter. It is therefore the authors' intent to provide a brief overview of the various mechanical tests that engineers, physicians, and researchers use to characterize this dynamic material.

2. OVERALL OBJECTIVES OF MECHANICAL TESTING OF BONE

According to Hoffler et al. (1), there are four main motives to use mechanical testing in studying bone. First, there is the "form-function" (structure-mechanics) evaluation as described by Katz (2) that can detail bone structural characteristics in relationship to the functional consequences of their variation. Next is the elucidation of the etiology and pathogenesis of disease to help understand both prevention and treatment. As Hoffler et al. point out, the functional deficit that affects mechanical integrity is most manifest and it is usually the effect that is easier to understand than the cause. Nonetheless, these two motives complement each other in either starting with the disease and searching for the effect, or starting with the deficit and seeking the factors that can contribute to it. The third motive is to investigate the behavior of bone under iatrogenic manipulation such as fracture fixation, grafting, or total joint arthroplasty. Finally, understanding the behavior of bone allows for advanced modern methods of computer simulation and research such as finite element analysis in which the computer needs to be programmed with the normal physical behavior of bone in order to perform computations with it. Hoffler et al. continue in realizing that these motives are "neither mutually exclusive or exhaustive" and simply serve the purpose of providing a basis on which to understand the broad spectrum of research on bone.

3. STRUCTURAL BASICS

Bone consists of an organic matrix of mostly collagen and inorganic hydroxyapatite (HA) crystals. Other components include noncollagenous proteins, proteoglycans, phospholipids, glycoproteins, and phosphoproteins. The inorganic and organic materials form fibrils, which are more ductile than the brittle HA crystals alone and can bear more load and add more stability over the flexible collagen. The fibrils then form isotropic lamellae arranged into one of two structures, either stacked sheets or tubes. The tubular forms are circular concentric columns of fibril sheets called Haversian osteons that provide strength along their long axis. The stacking of lamellae into sheets provides increased strength in the plane of the sheet as found in plexiform bone. Ultimately bone is formed into cortical (compact) or trabecular (cancellous or spongy) bone by the combination of mineral matrix, Haversian osteons and/or trabeculae. These structures contribute directly to the mechanical properties and make orientation, sample selection, and specimen preparation consideration vital to the successful mechanical test. These concepts are illustrated in Fig. 1.

It is also important to understand that the structure described forms a complex hierarchy that directly models its mechanical properties. Each structural level from mineral to the macroscopic bone structure affects the properties of the whole. Please refer to Table 1 for an expanded listing of factors associated with the hierarchical levels of bone. Figure 1 illustrates this structural concept. Each of these structural levels can be evaluated individually, but given the scope of this chapter, only the ultra structural methods will be considered. For more in depth review of bone structure, consider reviewing the following books, chapters, or journal articles: *Strength of Biological Materials* by Yamada (3), *Mechanical Properties of Bone* by Evans (4), *The Mechanical Adaptations of Bones* by Curry (5), *Bone Mechanics Handbook* edited by Cowin (6), *Skelatal Tissue*

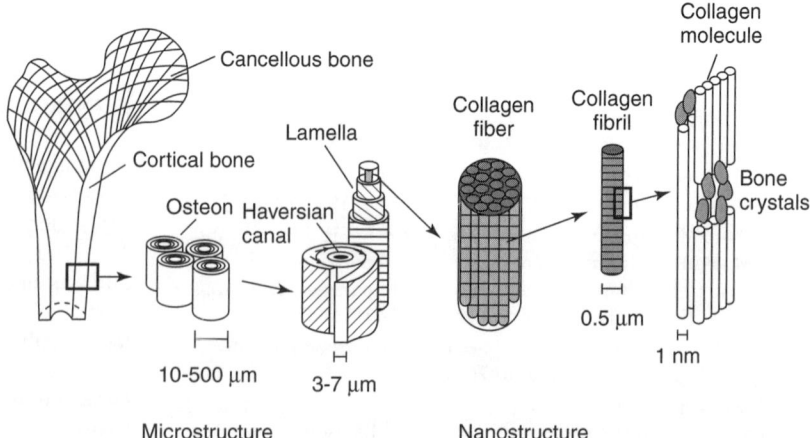

Figure 1. Hierarchical Structure of Bone: The various structural layers and components of bone are illustrated by the hierachial model. Reprinted from Rho et al. (1998) with permission.

Table 1. The Hierarchical Levels of Bone (13,14)

Level	Elements (specimens)	Main Factors Determining Bone Strength
Macrostructure	Femur, humerus, vertebrae, frontal bone, phalangeal bones, calcaneous, etc.	Macrostructure such as tubular shape, cross-sectional area, and porosity of long bone, cortical bone-covered vertebrae, or the irregular pelvic bone
Architecture	Compact bone or cancellous bone blocks, cylinders, cubes, or beams	Densities, porosity, the orientation of the osteons, collagen fibers, or trabeculae
Microstructure	Osteons, trabeculae	Loading direction, with maximum strength along their long axis
Submicrostructure	Lamella, large collagen fibrils	Collagen-HA fibrils are formed into large collagen fibers or lamellar sheets with preferred directions of maximum and minimum strengths for a primary loading direction
Ultrastructure	Collagen fibril and molecule, mineral components	HA crystals are embedded between the ends of adjoining molecules; this composite of rigid HA and flexible collagen provides a material that is superior in mechanical properties to either of them alone, more ductile than hydroxyapatite, allowing the absorption of more energy, and more rigid than collagen, permitting greater load bearing.

Mechanics by Martin and Burr (7), Nordin and Frankel (8), Albright (9), Einhorn (10), Hayes and Bouxsein (11), Whiting and Zernicke (12), and Rho et al. (13).

4. SAMPLE SELECTION AND PREPARATION

The source and selection of bone is important in designing a mechanical test. Bone can be obtained from either animals or humans, either premortem or postmortem, and from either healthy subjects or pathological subjects. Pathological subjects can either be from a natural etiology or from investigator induced pathology. Factors that affect the mechanical properties and performance of bone are numerous. During life, these include: age, subject sex, specimen porosity and structure, differences in anatomical location, weightlessness, activity levels, hormones (sex hormones, parathyroid hormone, growth hormone, and steroids) and pathology. Therefore it is necessary to obtain samples from similar sources. One of the best ways to do this is to use samples from matched locations such as a right and left tibia from the same individual. When using smaller machined samples, ensuring that they are uniformly machined will reduce the effects of size variation between samples.

Bone can be obtained, processed, stored, and prepared for testing in numerous ways. Bone can be obtained with consent from patients or from experimental animals during surgery or at necropsy or autopsy. Specimens should be taken as close to death as possible to avoid natural degradation of tissue. If samples cannot be obtained at the time of death, then freezing the source can be performed as long as samples are used within a few days of freezing. It is important to note that if histology is planned for the experiment that the sample not be frozen because the thawing of intracellular ice crystals will make microscopy impossible. Some sources may be embalmed, which is not appropriate for mechanical testing of bone properties, but can be used in testing surgical techniques or implants. Formalin fixation alters the collagenous matrix of bone and will significantly alter the mechanical properties of bones.

When obtaining specimens, it is important to note the sample location, subject sex, and pathology, if any, of the sample for future reference. Soft tissue can either be left intact or removed depending on the nature of the test being performed. It is also important to use caution when harvesting samples due to the fact that stray cuts with instruments into bone can create stress concentrations and change the outcome of tests, especially when using small samples or samples from small laboratory animals. It is also a good idea to obtain both radiographs and photographs of samples when harvested to document pathology or absence of it. Surrounding tissue changes, as seen in arthritis, fractures, or implants are also important to note as these all affect the structure and remodeling of bone.

Once a sample has been harvested, it must be processed. The least amount of processing required for a sample is to keep the tissue bathed in saline for prompt use right away. Hydration is important because simple drying in room air can decrease the elastic modulus by up to 3% (15). If a sample will not be used immediately, then it should be frozen according to accepted standards. Many studies have evaluated storage methods ranging from sterile storage at room temperature in saline or ethanol to storage in ethylene oxide to freezing at a variety of temperatures. For long term storage (greater than 3 months), the most common practice is to freeze specimens at $-20°C$ in saline to prevent dehydration. Soft tissues should be left intact as this will guard against freezer burn. If soft tissues are removed, saline soaked gauze can be substituted. Specimens should be wrapped in plastic wrap and placed in sealed plastic bags for additional protection. Although bone degradation can still occur at this low temperature and certainly enzymatic degradation continues when the bone is thawed to room temperature, the effects on mechanical properties are minor (16). Various laboratories have researched the length of time that a specimen can be frozen without significant reduction in mechanical properties and the majority have found that there is no significant change (17–19). As no data is available for time periods longer than 8 months, storage longer than this time period is not recommended.

When it is time to test the bone samples, they should be thawed in saline at room temperature for at least 3 hours. If the situation arises that the testing cannot be completed *en toto*, bone can be refrigerated for several days between tests without significant change in properties. Bone can also be refrozen and tested multiple times as complex experiments may warrant. This has been verified by both our laboratory (20) and by the work of Linde and Sørenson (21).

Once the bone is ready for testing, it must be machined both into the appropriate shape and to the desired structural level as outlined by the test being performed. Bone structure can be tested from whole bone down to single osteons or trabeculae and each requires unique preparation methods. Whole bone requires minimal processing prior to testing but microsamples require unique methods of dissection and machining. Equipment used to prepare large specimens includes saws, drills, polishers, grinders, and measuring tools (calipers, measuring tapes) as illustrated in Fig. 2. Rough cuts are made with the larger saws (handsaw, bandsaw) and the sample is machined to the desired size and shape with finer cuts from more precise saws (diamond wafering saw). Bone can also be machined into cylindrical shapes. Drill presses fitted with appropriate hollow bits are used to cut cores of the required size. The specific shapes that are commonly used in different tests will be discussed in the respective testing sections below.

In addition to testing whole bone and macroscopic sections, investigators sometimes use microscopic samples for testing. Microsamples of osteons and smaller bone structures can be obtained with specialized equipment such as that described by Ascenzi, et al. (28). These essentially consist of either using microneedles to "drill" out cores of bone with a dental drill (down to 200 micrometer diameter × 500 micrometer length), or using "splitting and scraping" techniques of creating microfractures under microscopic guidance along natural planes in the bone. Single trabeculae can also be dissected out and tested as described by Runkle et al. and Townsend et al. for tensile, compressive, buckling, torsional, and bending tests.

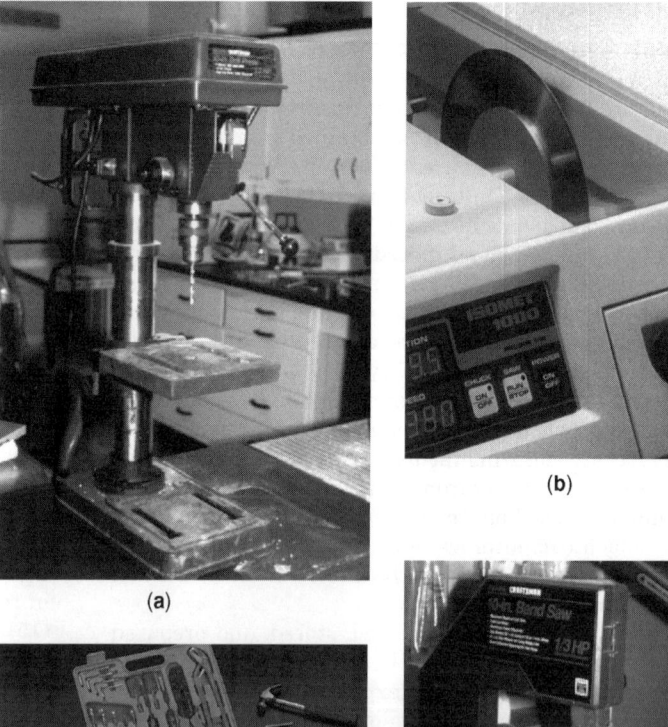

Figure 2. Lab Equipment. Some of the basic laboratory equipment necessary for specimen preparation: (a) Drill Press (b) Diamond Wafering Saw (c) Various Hand Tools (d) Band Saw.

The quality and method of machining performed on a sample can affect the outcome of a test. Uneven surfaces can induce stress concentrations and moments that will lead to inaccurate results. The contact between the machine and the specimen can induce frictional forces that will likewise alter the data. When a sample is tested in such a way that these other forces are eliminated or reduced to a negligible level, one refers to the test as a "pure" test, as in pure axial loading or pure bending. An investigator should also be sure that machining is performed in a standardized fashion because of the potential to induce sample to sample variability in shape and size. The effects of machining itself also pay a factor. High heat processes such as bandsawing and coring can burn the edges of bone to depths of 1–2 mm (22–24). The cut edges need to be ground with a polishing wheel or other similar post-machining finishing steps to remove the burns.

5. MATERIAL PROPERTIES

A basic overview of some of the important mechanical properties encountered in this type of testing is necessary before discussing the test equipment and testing methods. These include stress, strain, stiffness, the modulus of elasticity, and stress-strain curves. First, the modulus of elasticity, E, is a measure of the ductility or elasticity of a material. It will tell an engineer if a material is brittle like a ceramic or ductile like a piece of rubber. This value is a ratio of the stress (force applied to an area) and strain (ratio of change in length to original length) in a material before the yield point. The yield point is a point in which the material behavior changes from elastic deformation to plastic deformation. Elastic deformation means that if a material is loaded and then unloaded, it will return to its previous size and shape similar to a rubber band stretched and released. When a material is loaded beyond the elastic point, it is said to undergo plastic deformation, or a change in size or shape that cannot be recovered when unloaded. This is similar to molding a piece of clay.

Figure 3 shows a typical stress-strain curve for a tensile test. This type of curve is usually generated from the raw data recorded during the test, namely the load applied and the deformation recorded. Using formulas for stress and strain discussed below, these values are plotted automatically by a computer or manually by the investigator. In addition to point values for stress and strain, the following information can be obtained from plotting a stress strain curve and are illustrated in Fig. 3 :

1. The beginning of the elastic portion where the specimen is engaged by the machine.
2. The proportional limit or the limit to which stress and strain are proportional.
3. The elastic limit or the limit at which the greatest stress can be applied without leaving permanent deformation upon removal of the load.
4. The elastic range, that part of the curve where strain is directly proportional to stress.

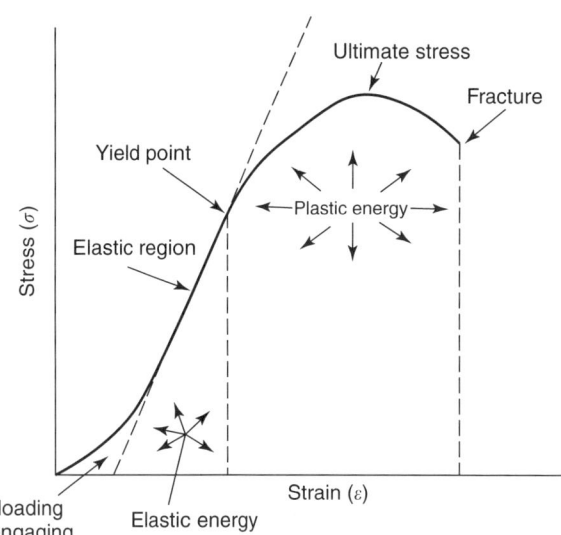

Figure 3. Stress-Strain Curve. A typical stress strain curve illustrating the important testing parameters recorded from the mechanical test.

5. The yield point, or the point at which permanent deformation occurs.
6. The range of plastic deformation, the part of the curve from the yield point to the failure point.
7. The breaking point or ultimate stress or ultimate strength.
8. The amount of energy absorbed by the specimen before failure, or the area under the curve.
9. The modulus of elasticity or the slope of the linear portion of the curve.

The stiffness of a material related to the modulus of elasticity is a measure of the amount of deformation under a given load. This is usually calculated as the slope of the linear portion of a load-displacement curve but can also be calculated as the product of the modulus of elasticity and the moment of inertia (discussed below) of the specimen. The moment of inertia is related to the shape of the specimen and computed using formulas specific to shape. As such, this can be a difficult number to calculate and the slope of the load-displacement curve works well instead.

The inverse of the stiffness is called the compliance of a material and this is sometimes reported when for specimens with very low stiffness values. Ligaments and soft tissues have a high compliance and low stiffness whereas hard bone has a high stiffness and low compliance.

The last concept to understand is that bone is a viscoelastic material. This refers to the time dependent change in mechanical properties that is encountered when bone is loaded at different rates. When bone is loaded at a high strain rate it appears stiffer and stronger. When loaded at a slower rate, it is the opposite. Most biological materials have some amount of viscoelasticity. For this reason, mechanical tests are conducted at accepted rates to standardize testing.

A more detailed discussion of mechanical properties, specific concepts related to mechanical testing, material dimensions, and loading behavior is beyond the scope of this chapter but can be found in any mechanics of materials book or reviewed in a variety of specific bone mechanics books (6,25).

6. TESTING EQUIPMENT

A biomechanics laboratory that evaluates bone properties or bone mechanics will require basic equipment to perform tests. This includes a mechanical testing machine, stress and strain recorders, data logging devices, and a variety of jigs and fixtures. Testing machines include single axis, multiaxis, micromechanical, and indentation/hardness testers. Modern single axis or multiaxial biomechanical testing machines are typically computer controlled table top units that are powered by servohydraulics (Fig. 4). These are available from numerous private commercial vendors who specialize in this field. Machines can also be screw driven, pneumatic, or electromagnetic. Single axis machines can perform compressive, tensile, or bending tests although it is possible to convert single axis machines into torsional testers with geared jigs and force translators as demonstrated in our lab and by other authors (Fig. 5) (26,27). Multiaxis machines can perform a wider range of mechanical tests and apply two or more different loads in different directions (such as an axial test and torsional test simultaneously) but they are more expensive than single axis machines. Each of these machines

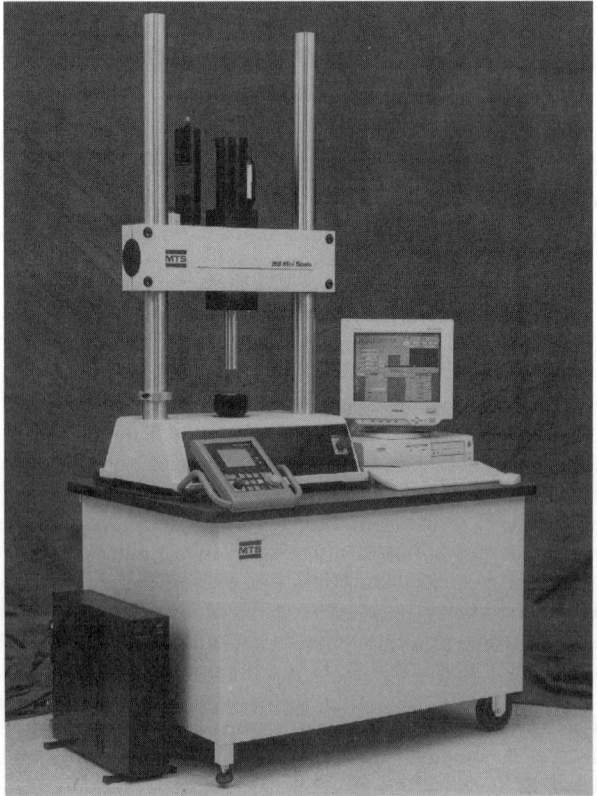

Figure 4. MTS Machine. A modern table top single axis servohydraulic mechanical testing machine (MTS Mini Bionix Model 855). (Courtesy of MTS Systems, Inc., Eden Prairie, MN).

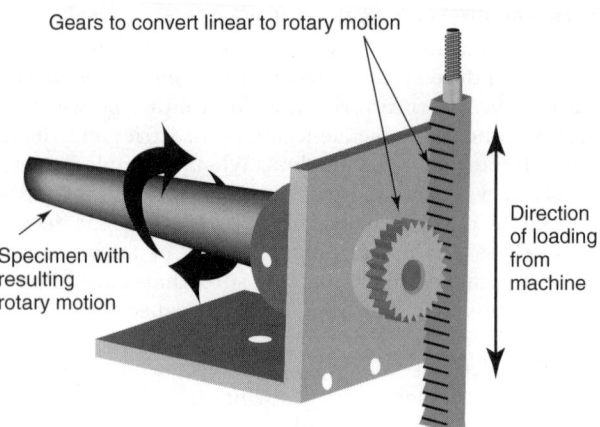

Figure 5. Rotational Tester. An example of a method to convert the common uniaxial mechanical tester into an additional torsional testing machine. This jig can be constructed at a local machine shop. (This figure is available in full color at http://www.mrw.interscience.wiley.com/ebe.)

operates by the same method in which an electrically generated function is sent to the machine, which in turn moves the actuator(s) and applies force to the test specimen. Force is recorded by a load cell and sent to the data logging equipment for software based data analysis. Most biomechanical machines do not need to exceed loads of 1000–2000N, but in our experience having a load cell that can achieve 10000N is useful in some compressive tests of vertebral bodies. Although some labs will require large load capacities, others will require micromechanical testing capabilities. Machines for this type of testing are used to test single osteons or trabeculae. Bending, compression, tension, and torsion can all be tested with these specialized testers (28,29). One final type of testing machine is the hardness tester (Fig. 6). This is used to determine resistance to indentation or abrasion. Macroindentation according to the Brinell method is used for loads of 1 kg or greater applied to bone macrosections. Microsections can be tested with the Knoop and Vickers indenters to determine mineral content. Nanoindenters are also available for testing of osteons and trabeculae on an even smaller scale. It is obviously important to decide what type of test will be performed to optimize equipment choice.

Most mechanical testing machines are used to measure the load applied to the load cell. Most machines also report the distance moved by the actuator that is applying the force. Many investigators use this to record the displacement of their specimen; however, there may be a difference in the amount of displacement in the sample versus the distance the actuator travels due to non-homogenous properties of the sample or slippage of the sample in the machine. For example, if a vertebral body is being compressed and portions of intervertebral discs are still present, the disk will act as a shock absorber and compress before the bone. If the measurement is taken simply from the actuator travel, the compression of the bone will appear greater than it actually is, as shown in Fig. 7. The use of strain gauges or extensometers applied directly onto the sample itself in the area of interest will prevent this common error (30–36). These devices yield much more

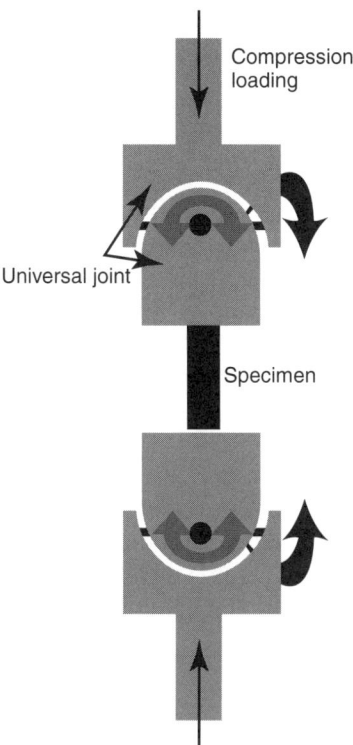

Figure 6. Hardness Tester. A Brinell-Rockwell-Vickers hardness tester (Courtesy of Nanotek, Inc., Opelika, AL).

accurate measures of the displacement and calculated strain over a defined area.

Other items that are needed to perform tests include machining equipment mentioned previously, such as saws, drills, grinders, and polishers. Basic operating tools such as scalpels, towels, gloves, saline, and storage bags and boxes should also be available (37).

Mechanical testing requires additional equipment and materials to fix specimens in the testing machines. These fixtures and materials will vary depending on the type of test and the property being investigated. Bone specimens will either have a uniform shape after machining or will have an irregular shape, as seen with whole bone specimens. Uniform shapes can be attached with standard

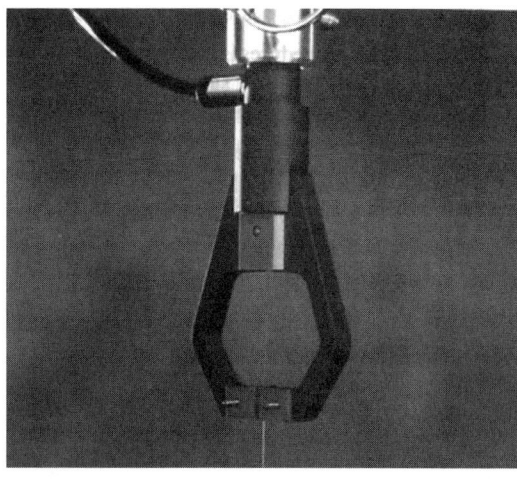

Figure 8. Tensile Test with grips. An example of a pneumatic grip attachment. Reprinted with permission from reference (25).

grips and fixtures such as pneumatic grips in a tensile test (Fig. 8), or simply placed on the platens in a compressive test. Standard grips and fixtures are commercially available. However, it is often necessary to have custom jigs and fixtures made to accommodate the specific test the investigator is trying to perform. Local machine shops can make these fixtures out of stock steel or aluminum to produce a cost effective custom appliance.

The more irregular shapes, such as the ends of long bones or vertebrae, will require potting in a hard setting material that can be mounted in the mechanical tester. Materials used for this purpose include: polymethylmethacrylate (PMMA), epoxy resins, plaster, or calcium sulfate-based materials such as dental stone (37). Once these specimens are potted, they may or may not be compatible with standard fixtures.

7. FACTORS AFFECTING THE MECHANICAL TEST

It is necessary to consider all the sources of error that might be encountered in mechanical testing of bone. The most common sources include machine compliance, the specimen fixture interface, measurement errors, and of

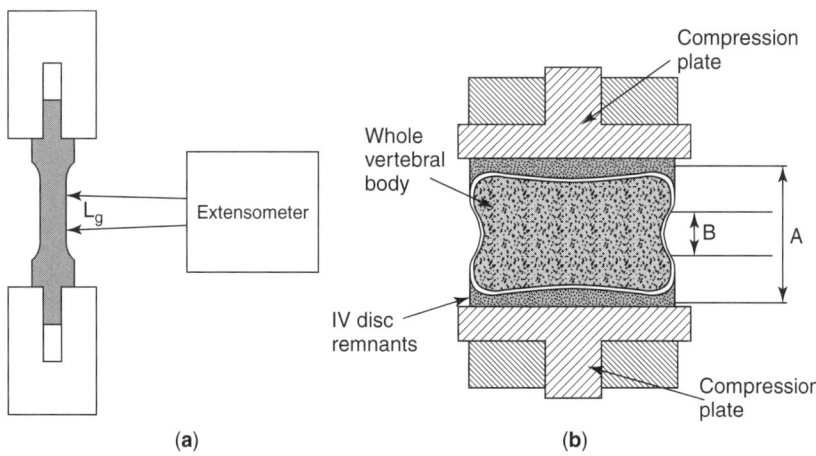

Figure 7. Compression of Vertebral Body, Extensometer. (a) An extensometer attached to a sample illustrating the defined displacement measurement region, or gauge length (L_g). (b) Without an extensometer, a compression test will only yield information about the overall displacement of a specimen, illustrated as dimension A, with no way to determine if the displacement is due to compression of the IV disc on the ends of the vertebral body or if it is from compression of the bone itself. In a test of bone displacement, an extensometer should be placed to measure the displacement listed as dimension B for accuracy. Reprinted with permission from reference (25).

course simple human error in calculation and procedure. Machine compliance comes into play because the two values often being measured, the displacement of the actuator and the force generated on the load cell, may not be the actual values being exhibited by the sample. If the machine's frame, actuator, and loading platform have a stiffness that is less than that of the sample, the data will be erroneous. To account for this effect, the compliance of the machine should be determined through the standard method described by Bensen and An (37) and Turner and Burr (22). A small load (P) is applied to the load cell without any sample in place. The stiffness of the machine (S) is calculated by dividing the load (P) by the displacement (δ) (Equation 1). The process is repeated with the sample in place and the sample stiffness is calculated. These two values are related through equation 2 to yield an actual specimen stiffness (S_{actual}). If the $S_{machine}$ is much greater than the S_{actual}, as is the case with most large biomechanical testing machines, then the error caused by the machine compliance will be minimal (22) and no correction will be necessary. For smaller microtesting machines, this becomes a more important calculation.

$$S = \frac{P}{\delta}, \qquad (1)$$

$$S_{actual} = \frac{S_{sample} S_{machine}}{(S_{sample} + S_{machine})}. \qquad (2)$$

The specimen-machine interface is also a common source of error. As with any mechanical testing, each and every attachment point of the sample to the machine constitutes the boundary conditions of the test. These boundary conditions must be part of the mechanical analysis to account for all loads applied and developed in the sample. They must also be fixed without slippage or shearing between the sample and attachment or loading point. In compression testing, there is friction that develops between the flat platens and the specimen. The moisture content, surface finish of the platen and sample, and the shape of the sample will all affect the frictional force. Small bone sections expand in a transverse direction to the axis of compressive loading and likewise contract under tensile loading. This is known as Poisson's Effect. Too high frictional forces from rough or dry surfaces will constrain this slight expansion in compression and create a shear stress within the sample. This will not only alter the stiffness of the specimen but also cause fractures to occur at 45° to the direction of loading. On the other hand, too little friction from greasy specimens will be caused by transverse strains from excessive spreading of the material (22). It is recommended that a surface finish of $2\,\mu m\,cm^{-1}$ (mirror polish) be used for the platens and that petroleum jelly be added for samples that create high frictional forces. The effect of friction in relation to Poisson's effect of expansion is greatest with smaller bone sections, but small sections that are too large will be affected by a phenomenon known as the vertical end effect described first by Linde and Hvid (38). This is seen in compression testing of trabecular specimens where the axial and lateral deformation of a specimen are largest at the ends compared to the midpoint between the plates.

Keaveny et al. (39, 40, 42) studied the common use of cubes versus cylinders in this type of testing and found that cylinders with an L/D of 2 provided better estimation of the modulus of elasticity in respect to the end effect and frictional forces. End caps or potting in PMMA has been shown to reduce this effect as well (38,41). A through discussion of this topic can be found in Keller and Liebschner (23) and Turner and Burr (22).

The fixation of the specimen in a jig is also a source of error. Slippage of a specimen is common when using biological materials that may be wet or greasy or highly ductile. This is most commonly seen in tensile and torsional tests. Figure 9 shows a load displacement curve of a tensile test in which the sample is slipping in the grips. This will result in the recording of displacement errors by the machine and is another reason for the use of extensometers in the measurement of displacement. The use of potting material or grips with serrated teeth will help reduce slippage.

The shape and alignment of the specimen are crucial to accurate results for all tests due to the fact that off center loading creates a lever arm through which the force is conducted. This in turn adds an unintended bending moment to the specimen in compression, tension, or bending tests. The moment will then add stress concentrations to the sample that effect the outcome data (Fig. 10). Zhu et al. (43) discuss a parallelism index that can be used to asses

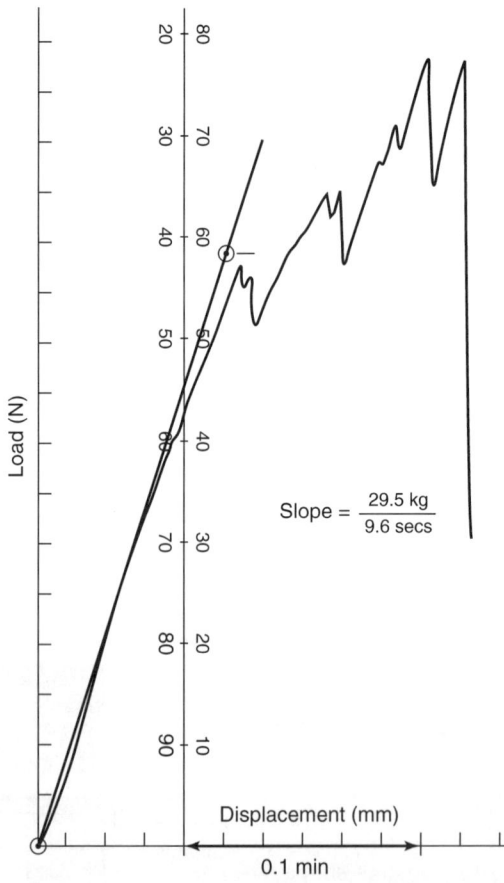

Figure 9. Slippage. A load displacement curve illustrating the jagged "catch and release" behavior of a specimen that is slipping with respect to its fixation.

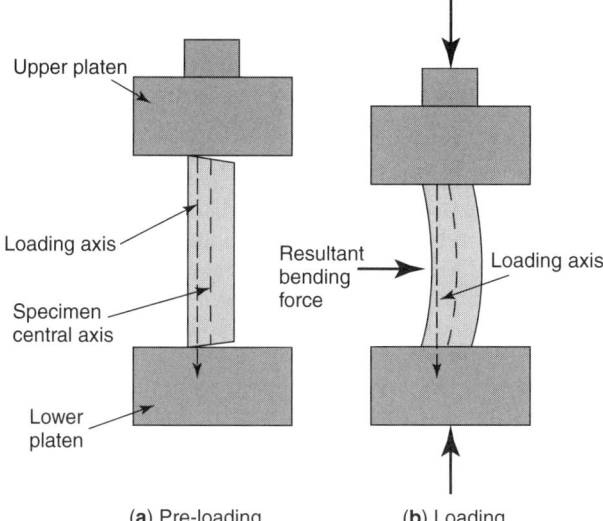

Figure 10. Uneven Loading. An Illustration of off-center loading and the resulting unintended bending moment applied to the specimen.

the alignment of the ends of the specimen for use in compressive testing. In other forms of testing, one should be sure that the load is being applied along the intended axis and any moment arms are eliminated or accounted for in the mechanical analysis.

In test specimens that may not be exactly parallel, special features can be added to jigs for compression and tensile tests. The jig fixture or platens can be fitted with a universal joint that allows bending motion in two directions or a spherical joint that allows motion in three directions (Fig. 11). This will eliminate bending or torsional moments that are developed within the specimen. With such a jig in place, the specimen is said to be loaded in pure tension or compression.

8. TYPES OF TESTS

There are numerous mechanical tests that can be performed on bone, but the most common are the compression, tensile, bending, indentation, and torsional tests. Each test requires a different specimen shape, testing jig, and testing parameters and of course, each tests a different mechanical property.

The tensile and compression tests are similar tests in that they both test mechanical properties of axial loading although they are in different directions. In pure compression and tension tests, the ends of the sample in contact with the machine are required to be parallel to prevent triaxial loading and unintended moments as previously discussed. The tensile test is the standard mechanical property measuring test used for many materials in engineering. The test specimen is typically machined into a shape with the ends wider than the central portion. This is called the dogbone shape and is illustrated in Fig. 12. It can be used to test cancellous or cortical bone with equal success. The test is run with the specimen fixed into grips or it is embedded into a potting material that is fixed into a

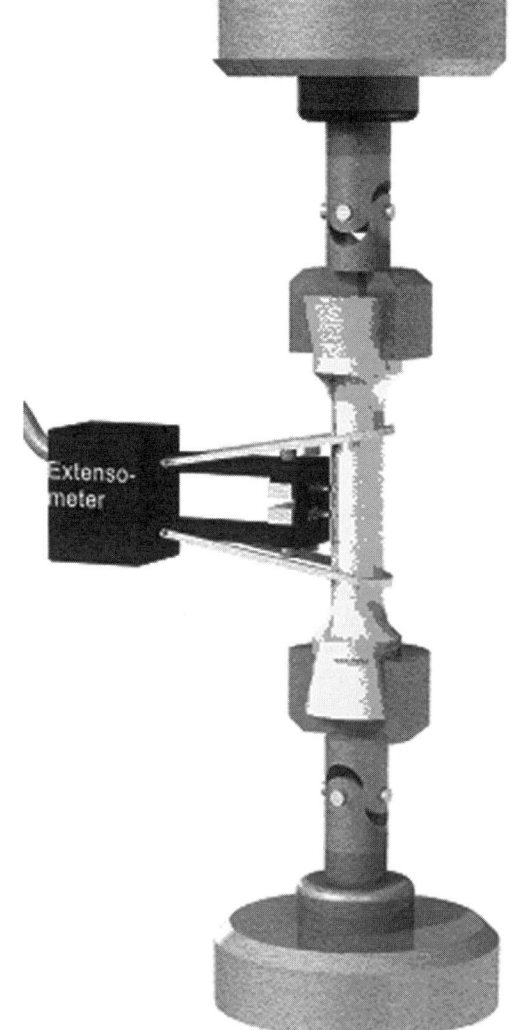

Figure 11. Universal Joint. With the use of a universal joint in compression testing, the specimen can be loaded in pure compression. Any unintended bending moments created by compression will be reduced through the ability of the joints to rotate while maintaining axial alignment.

specialized holder, usually a cylindrical pipe. Tests are run in displacement control with the actuator moving at a slow rate of 1 mm to 5 mm per minute and the load is measured from the load cell. Extensometers are most useful in this type of testing because they can be attached to the narrow central region and easily measure displacement of this region. Slippage of the test specimen is the main source of error and the machining of specimens to the dogbone shape is not an easy task. Therefore, although this test is commonly used for performing material tests in other fields, it is not as widely employed in bone testing.

Compression tests are similar to tensile tests but they are run in the opposite direction. In machined samples, the dogbone shape is not used, rather, a cylinder with an L/D ratio of at least 2 is typically used. This shape will minimize the chance of buckling and reduce the errors of the end effect described previously in addition to making test set up easier and quicker. In compression tests, the

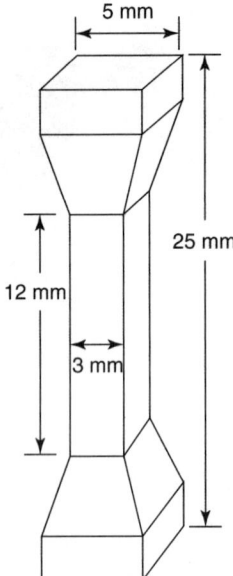

Figure 12. Dogbone Shape. The standard dogbone shape used in many types of material testing shown with standard dimensions. Reprinted with permission from Y. H. An and R. J. Friedman (eds.), Animal models in orthopaedics. Boca Raton, FL: CRC Press, 1999.

ends of the sample may or may not be fixed depending on the type of sample being used. Some studies have shown that fixing the sample in PMMA may reduce the end effect, but this error may be minimal in properly sized specimens. Unfixed specimens tested on flat platens need to have frictional forces accounted for secondary to the previously discussed changes in apparent stiffness from high frictional forces or exaggerated spreading from too little.

Calculation of material properties is straightforward in either compressive or tensile tests. In a solid cylindrical core sample being compressed, the ultimate compressive or tensile stress (σ) is calculated by:

$$\sigma = \frac{P}{A}, \quad (3)$$

where P equals the maximum load attained and A is the area of the sample. Strain (ε) can be determined for either test with the formula:

$$\varepsilon = \frac{\delta}{L}, \quad (4)$$

where δ is the change in length and L is the original length. It is important to make sure that the length used in the calculations is the gauge length of the portion of the specimen where the change in displacement was measured. In a dogbone sample, this is the narrow center area or the area between the arms of the extensometer. The modulus of elasticity (E) can be calculated by:

$$E = \frac{SL}{A}, \quad (5)$$

where S equals the specimen stiffness (see Equation 1), L is the gauge or specimen length, and A is the uniform cross section area of the specimen. Again, the length is specific to the area being investigated and measured.

It is sometimes difficult to test material properties with compression or tensile tests. Such is the case when trying to machine small bones into the required shapes for these tests. An alternative test that can be used to determine mechanical properties is the bending test. Bending tests are used on these as well as longer whole bone sections to evaluate stress, strain, and modulus of elasticity. Bending tests are also useful because many bone fractures occur as a result of bending stresses, such as a ski boot top fracture. As shown in Fig. 13a and 13b, a material loaded in bending experiences compressive stresses along one axis and tensile stresses along the opposite. In bone testing, failure usually occurs along the tensile side.

There are two types of bending tests that are used in bone testing: the three point and the four point bending tests (Fig. 13a and 13b). The three point test is simple to construct and perform, but large point shear loads are developed at the center load point. This can adversely affect the outcome of the test and therefore, the four point bending test is considered the most sound. The four point bending test will place the center section of the specimen in pure bending with a constant moment and zero shear. This is the ideal scenario in a uniform cross section specimen, such as a steel pipe, however, most long bones do not have a uniform cross section. They have a small length to diameter ratio, inconsistent cross sectional area, and inconsistent intramedullary components (marrow in some areas, trabecular bone in others). Ideally the L:D ratio should be at least 20 to reduce the bone displacement secondary to direct shear at the load points. The variance in geometry and composition causes unequal loading along the shaft of the long bone, more so than with the three point bending test. For this reason, the simpler three point bending test is often used for testing whole bone sections.

It is possible to machine sections of long bones for use in bending tests. This requires a bone of sufficient size that a strip of the appropriate size (length to thickness ratio of at least 20) can be machined from it. For most whole bones, the L:D ratio is insufficient. The deformation, strain calculation and the modulus of elasticity will be overestimated. Therefore, if the L:D ratio is less than 20, the best way to calculate these values is with a strain gauge bonded directly to the bone surface.

For either the three or four point bending test, fixtures generally consist of load supports made of rounded cylinders with their long axis mounted perpendicular to the long axis of the bone. Next the bone is loaded onto the fixture and a load is applied at the center point(s) until failure is reached. Typically, bones have an elliptical cross section such that the placement in the fixture can be important depending on the type of test. One may wish to test both the long and short diameter orientations for strength, or one may only be interested in a particular loading scenario such as a fracture from a force in a certain direction. If the orientation is not important, loading the bone with the long diameter horizontal makes for a more stable construct.

For three point bending, the calculation of the modulus of elasticity (E), stress (σ), and strain (ε) is again straightforward, although now the cross sectional moment of inertia (I) about the bending axis must be calculated. The

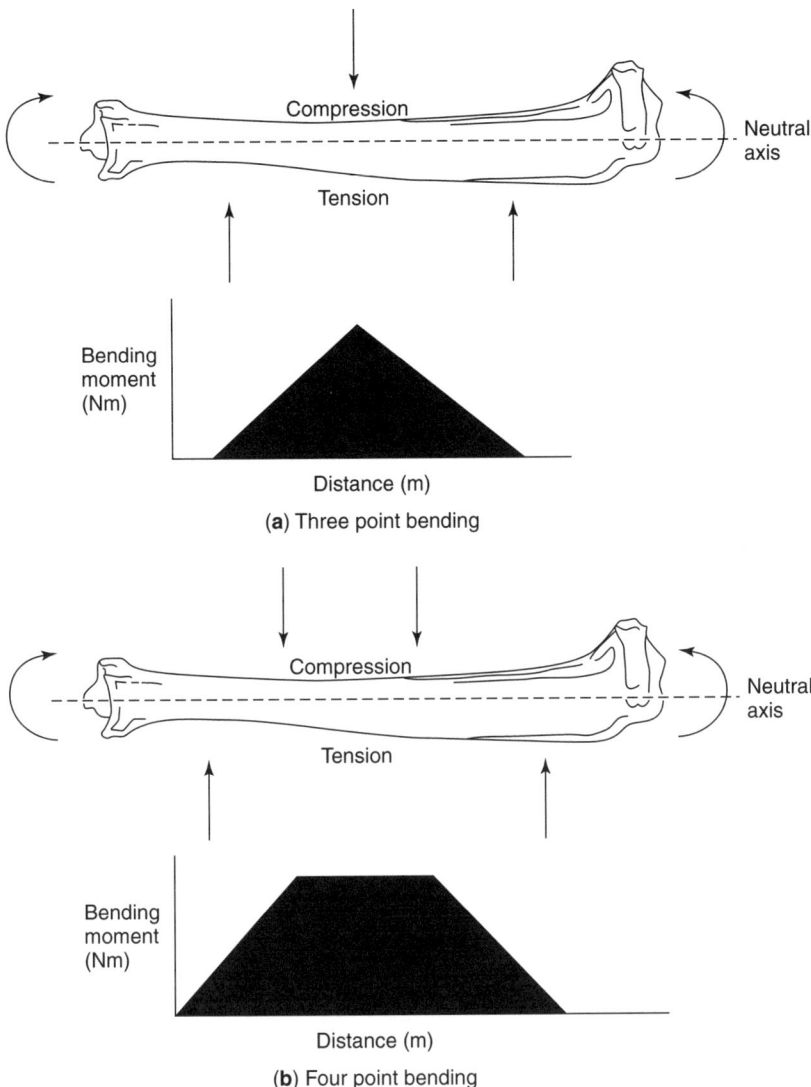

Figure 13. Bending Tests: (a) Three Point Bending (b) Four Point Bending. The two types of bending tests illustrating the compression-tension relationship of forces along the surfaces of the loaded specimens. Moment diagrams are also shown for each type of bending test illustrating the maximum load at the application point of the center load in the three point bending test and the constant moment across the loading region between the two center loads in the four point bending test. Reprinted with permission from reference (25).

equations are listed below:

$$I = \pi \frac{(a^3 b - a'^3 b')}{64}, \tag{6}$$

$$E = \frac{SL^3}{48I}, \tag{7}$$

$$\sigma = \frac{PLa}{8I}, \text{ and } \varepsilon = d\left(\frac{12a}{L^2}\right). \tag{8}$$

For four point bending, the moment of inertia is the same but equations for modulus of elasticity (E), stress (σ), and strain (ε) are as follows:

$$E = S\left(\frac{c^2}{12I}\right)(3L - 4c), \tag{9}$$

$$\sigma = \frac{Pac}{4I}, \tag{10}$$

$$\varepsilon = d\left(\frac{6a}{2c(3L - 4c)}\right), \tag{11}$$

where a, b, a' and b' are the average measurements of the external and internal anteroposterial and mediolateral diameters of the specimen as shown in Fig. 14, S is the stiffness, L is the distance between the supports, P is the applied load, c is the distance between the support and the load point in the four point model and d is the displacement of the actuator.

The cantilever test is essentially a one sided bending test. It is conducted with one end of a specimen rigidly fixed and a load applied to the unfixed end in a direction perpendicular to the axis of orientation, as shown in Fig. 15. This applies a maximum bending moment at the fixation interface. Shear is a factor in specimens that are short and the applied load is near the point of fixation since the point of maximum shear is located at the specimen fixation interface. The investigator should decide which side of the bone needs to be in compression and which needs to be in tension and orient the sample accordingly. For this reason, it is also important that the samples are all in the same rotational alignment when potted and placed in the testing jig. For this test, the same cross sectional moment of inertia (I) is used as with the

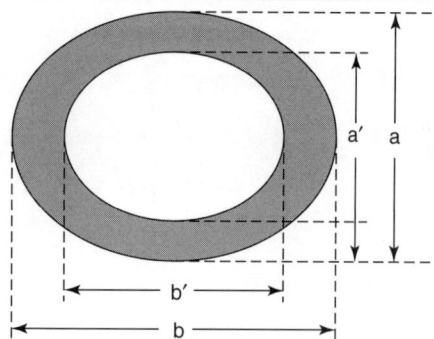

Figure 14. Measurements of Diameters for MOI. Dimensions for determining the moment of inertia for a typical long bone.

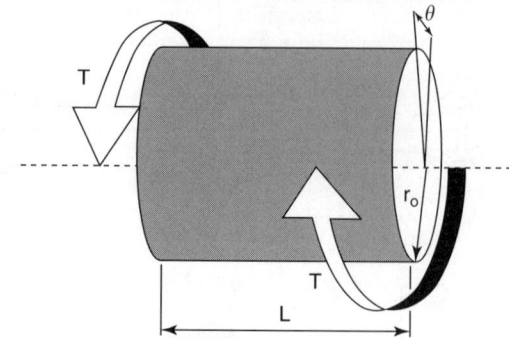

Figure 16. Torsional Test. Torsional Test of a cylindrical specimen showing the applied torque, T, and resulting rotational displacement, θ.

three and four point bending tests (Equation 6). The bending moment (M) and bending stress (σ) are calculated as listed below:

$$M = PL, \qquad (12)$$

$$\sigma = \frac{PLa}{4I} = P\left(\sin\frac{\alpha}{A}\right) + \frac{PLa(\cos\alpha)}{2I}, \qquad (13)$$

where L is the distance between the potted support and the applied load, P is the applied load, a is the average measurement of the external diameter of the specimen, α is the angle between the fracture surface and the vertical, and A is the cross sectional area of the specimen.

Torsional tests are usually used to test whole bone specimens. A torsional test puts a rotational shear stress on a bone fixed at both ends. The ends are rotated in opposite directions from each other until failure occurs (Fig. 16). Torsional tests are useful because they load the ends of the bone and place a maximum stress at the weakest point in between. This can be useful in determining weak points, the effects of bony defects, or bone repair. Generally bones are potted in standard potting material into a cup or cylinder at both ends and then placed in a torsional testing machine or a jig designed to convert a uniaxial tester into a torsional test machine. In torsional calculations the polar moment of inertia (J) about the axis of rotation is used. For most bones that resemble a tubular shaft with radii of $r_{Outside}$ and r_{Inside}, this is

$$J = \frac{\pi}{2}\left(r_{Outside}^4 - r_{Inside}^4\right). \qquad (14)$$

The maximum torque applied and angle of deformation are recorded. The shear stress (τ) and shear modulus (G) are usually calculated and their formulas are shown below:

$$\tau_{Max} = \frac{T r_{Outside}}{J}, \qquad (15)$$

$$G = \frac{TL}{\theta J}, \qquad (16)$$

where T is the applied torque, θ is the rotational displacement, L is the length of the test region, and $r_{Outside}$ is the external radius of the specimen.

The indentation test is a test used to measure the hardness of a material. The test is performed by driving an indenter of specified geometry into a material and measuring the penetration. As previously discussed, bone exists as a hierarchical structure with four main levels: macrostructure (cancellous or cortical bone), microstructure (Haversian systems, osteons), nanostructure (collagen and lamella), and subnanostructure (molecular structure). The indentation test is useful to measure the respective macroindention, microindention, and nanoindention properties of each hierarchical level (44). Hardness properties are specific to each bone level and so each level should be tested to fully characterize the hardness of a bone sample. The most common use of these tests is in examining the subchondral and trabecular bone near large joints in the body in studies of osteoarthritis and in implant studies (22). The basic tests are similar for each substructural level with some specifics regarding the indenter and specimen preparation for nanoindentation or osteopenetrometer testing. As such, only the

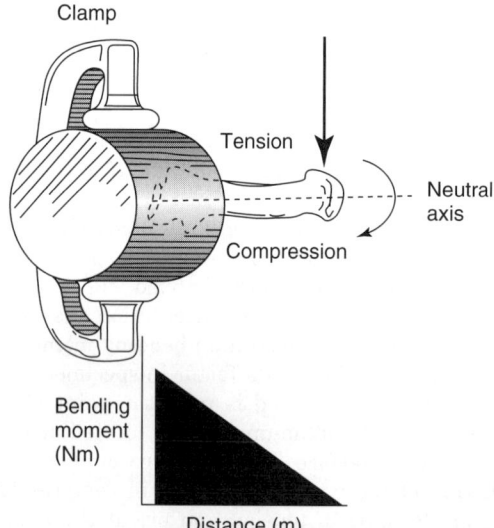

Figure 15. Cantilever Test. Illustration of a cantilever test showing a specimen with one fixed end being loaded at the opposite end. A moment diagram shows that the peak moment is at the point of fixation. Reprinted with permission from reference (25).

general principle for macroindentation is discussed here. The test consists of driving an indenter with a blunt end, either a sphere or a flat ended rod, and into a flat specimen at a rate of 1–2 mm/min to a depth of 0.2 to 0.5 mm or until a maximum load is reached (22,44). From this test a load vs. displacement curve is generated similar to the compression and tensile tests previously discussed. The modulus of elasticity (E) is determined by the equation:

$$E = S\left(\frac{1-v^2}{d}\right), \quad (17)$$

where S is the stiffness taken from the slope of the load-displacement curve, v is Poisson's ratio, and d is the diameter of the indenter. The ultimate indentation strength (σ) is determined dividing the maximum load by the cross-sectional area of the indenter represented by the equation:

$$\sigma = \frac{4P}{\pi d^2}, \quad (18)$$

where P is the maximum load obtained and d is the cross-sectional diameter of the indenter.

Some other methods of testing bone include using ultrasound, specialized tests of pure shear, fracture mechanics studying crack propagation, and fatigue testing. These tests are more specialized than the basic tests outlined here and as such are not elaborated upon in this text.

9. CONCLUSION

The intent of this chapter is to provide an overview of the reasons for performing mechanical testing of bone, the basic preparation of a testing lab and testing samples, and finally to offer a description of the core group of tests that an experimenter can use to characterize the mechanical properties of bone. It is important to remember that testing any material will always yield some form of result but it is up to the investigator to ensure that the result is a true measure of what was sought at the onset of the test. Understanding the hierarchical structure of bone, the effects of preparation and influence of external factors such as loading and the machine interface is the first step to performing a sound mechanical test. Using the basic mechanical principles, adequate, thoughtful preparation and careful planning will yield accurate results that will be reproducible and acceptable to the highest standards.

BIBLIOGRAPHY

1. C. E. Hoffler, B. R. Mc Creadie, E. A. Smith, and S. A. Goldstein, A Hiearchical Approach to Exploring Bone Mechanical Properties. In: R. A. Draughn, ed., *Mechanical testing of bone and the bone-implant interface*. Boca Raton, FL: CRC Press, 2000, pp. 133–149.
2. J. Katz, The Structure and Biomechanics of Bone. *Symp. Soc. Exp. Biol.* 1980; **34**: 137–168.
3. H. Yamada and F. G. Evans, *Strength of biological materials*. Baltimore: Williams & Wilkins, 1970.
4. F. G. Evans, *Mechanical properties of bone*. American lecture series; publication No. 881. Springfield, Ill.: Thomas, 1973.
5. J. D. Currey, *The mechanical adaptations of bones*. Princeton, N.J.: Princeton University Press, 1984.
6. S. C. Cowin, *Bone mechanics handbook*. 2nd ed. Boca Raton, FL: CRC Press, 2001.
7. R. B. Martin and D. E. Burr, Skeletal Tissue Mechanics, New York, NY: Springer, 1998.
8. M. Nordin and V. H. Frankel, Biomechanics of Whole Bones and Bone Tissue. In: M. Nordin and V. H. Frankel, eds., *Basic biomechanics of the musculoskeletal system*. Philadelphia: Lea & Febiger, 1989.
9. J. Albright, Bone: Physical Properties, In: J. A. Allbright and R. A. Brand, eds., *The Scientific Basis of Orthopaedics*. Norwalk, CT: Appleton & Lange, 1987.
10. T. Einhorn, Biomechanics of Bone, In: J. P. Bilezikian, L. G. Raisz, and G. A. Rodan, eds., *Principles of bone biology*. San Diego: Academic Press, 2002.
11. W. Hayes and M. Bouxsein, Biomechanics of Cortical and Trabecular Bone: Implications for Assessment of Fracture Risk. In: V. C. Mow and W. C. Hayes, eds., *Basic orthopaedic biomechanics*. Philadelphia: Lippincott-Raven, 1997.
12. W. C. Whiting and R. F. Zernicke, Biomechanical Concepts. In: W. C. Whiting and R. F. Zernicke, eds., *Biomechanics of musculoskeletal injury*. Champaign, IL: Human Kinetics, 1998.
13. J. Y. Rho, L. Kuhn-Spearing, and P. Zioupos, Mechanical properties and the hierarchical structure of bone. *Med. Eng. Phys.* 1998; **20**(2):92–102.
14. C. Hoffler et al., A Hierarchical Approach to Exploring Bone Mechanical Properties. In: Y. H. An and R. A. Draughn, eds., *Mechanical testing of bone and the bone-implant interface*. Boca Raton, FL: CRC Press, 2000, pp. 133–149.
15. E. D. Sedlin and C. Hirsch, Factors affecting the determination of the physical properties of femoral cortical bone. *Acta Orthop. Scand.* 1966; **37**(1):29–48.
16. L. Stromberg and N. Dalen, The influence of freezing on the maximum torque capacity of long bones. An experimental study on dogs. *Acta Orthop. Scand.* 1976; **47**(3): 254–256.
17. T. I. Malinin, O. V. Martinez, and M. D. Brown, Banking of massive osteoarticular and intercalary bone allografts–12 years' experience. *Clin. Orthop.* 1985; **197**:44–57.
18. M. M. Panjabi et al., Biomechanical time-tolerance of fresh cadaveric human spine specimens. *J. Orthop. Res.* 1985; **3**(3):292–300.
19. S. C. Roe, G. J. Pijanowski, and A. L. Johnson, Biomechanical properties of canine cortical bone allografts: effects of preparation and storage. *Am. J. Vet. Res.* 1988; **49**(6): 873–877.
20. Q. Kang, Y. H. An, and R. J. Friedman, Effects of multiple freezing-thawing cycles on ultimate indentation load and stiffness of bovine cancellous bone. *Am. J. Vet. Res.* 1997; **58**(10):1171–1173.
21. F. Linde and H. C. Sorensen, The effect of different storage methods on the mechanical properties of trabecular bone. *J. Biomech.* 1993; **26**(10):1249–1252.
22. C. H. Turner and D. B. Burr, Experimental Techniques for Bone Mechanics. In: S. C. Cowin, ed., *Bone mechanics handbook*. Boca Raton, FL: CRC Press, 2001.
23. T. Keller and M. Liebshner, Tensile and Compression Testing of Bone. In: Y. H. An and R. A. Draughn, eds., *Mechanical testing of bone and the bone-implant interface*. Boca Raton, FL: CRC Press, 2000, pp. 175–206.

24. Y. H. An and R. A. Draughn, Mechanical Properties and Testing Methods of Bone. In: Y. H. An and R. J. Friedman, ed., *Animal models in orthopaedic research*. Boca Raton, FL: CRC Press, 1999, pp. 139–150.
25. Y. H. An and R. A. Draughn, *Mechanical testing of bone and the bone-implant interface*. Boca Raton, FL: CRC Press, 2000.
26. J. A. Szivek and R. A. Yapp, A testing technique allowing cyclic application of axial, bending, and torque loads to fracture plates to examine screw loosening. *J. Biomed. Mater. Res.* 1989; **23**(A1 Suppl):105–116.
27. Z. Damián, P. A. Lomelí, and L. Núñez, A Device for Biomechanical Torsion Tests of Long Bones in an Instron Test Machine. *Journal of the Mexican Society of Instrumentation* 1997; **3**(Nr. 7):3.
28. A. Ascenzi, P. Baschieri, and A. Benvenuti, The torsional properties of single selected osteons. *J. Biomech.* 1994; **27**(7):875–884.
29. P. L. Mente and J. L. Lewis, Experimental method for the measurement of the elastic modulus of trabecular bone tissue. *J. Orthop. Res.* 1989; **7**(3):456–461.
30. W. E. Caler, D. R. Carter, and W. H. Harris, Techniques for implementing an in vivo bone strain gage system. *J. Biomech.* 1981; **14**(7):503–507.
31. D. R. Carter, Anisotropic analysis of strain rosette information from cortical bone. *J. Biomech.* 1978; **11**(4):199–202.
32. V. L. Roberts, Strain gage techniques in biomechanics. *Experimental Mechanics* 1966; 19A-22A.
33. Y. H. An and C. V. Bensen General Considerations of Mechanical Testing. In: Y. H. An and R. A. Draughn, eds., *Mechanical testing of bone and the bone-implant interface*. Boca Raton, FL: CRC Press, 2000, pp. 119–132.
34. T. M. Wright and W. C. Hayes, Strain gage application on compact bone. *J. Biomech.* 1979; **12**(6):471–475.
35. T. S. Keller and D. M. Spengler, In vivo strain gage implantation in rats. *J. Biomech.* 1982; **15**(12):911–917.
36. S. Boyd et al., Measurement of cancellous bone strain during mechanical tests using a new extensometer device. *Med. Eng. Phys.* 2001; **23**(6):411–416.
37. C. V. Bensen and Y. H. An, Basic Facilities and Instruments for Mechanical Testing of Bone. In: Y. H. An and R. A. Draughn, eds., *Mechanical testing of bone and the bone-implant interface*. Boca Raton, FL: CRC Press, 2000, pp. 87–102.
38. F. Linde and I. Hvid, The effect of constraint on the mechanical behaviour of trabecular bone specimens. *J. Biomech.* 1989; **22**(5):485–490.
39. T. M. Keaveny et al., Theoretical analysis of the experimental artifact in trabecular bone compressive modulus. *J. Biomech.* 1993; **26**(4–5):599–607.
40. T. M. Keaveny et al., Trabecular bone modulus and strength can depend on specimen geometry. *J. Biomech.* 1993; **26**(8):991–1000.
41. K. Choi et al., The elastic moduli of human subchondral, trabecular, and cortical bone tissue and the size-dependency of cortical bone modulus. *J. Biomech.* 1990; **23**(11):1103–1113.
42. T. M. Keaveny et al., Trabecular bone exhibits fully linear elastic behavior and yields at low strains. *J. Biomech.* 1994; **27**(9):1127–1136.
43. M. Zhu, T. S. Keller, and D. M. Spengler, Effects of specimen load-bearing and free surface layers on the compressive mechanical properties of cellular materials. *J. Biomech.* 1994; **27**(1):57–66.
44. B. E. McKoy, Q. Kang, and Y. H. An, Indentation Testing of Bone. In: Y. H. An and R. A. Draughn, eds., *Mechanical testing of bone and the bone-implant interface*. Boca Raton, FL: CRC Press, 2000, pp. 233–240.

BONE RESORPTION

DAVID R. HAYNES
DAVID M. FINDLAY
University of Adelaide
Adelaide, Australia

1. BONE REMODELING

Healthy bone is continually being removed and replaced in the mature skeleton by the process of remodeling. Remodeling of bone relies on the integrated activity of osteoclast (bone resorbing) and osteoblast (bone forming) cells and is required to maintain the integrity of the skeleton by removing areas of damaged matrix, as well as for calcium homeostasis (1). However, either excessive resorption or decreased bone formation, or combinations of these, result in bone loss in disease conditions. In addition, disuse or chronic unloading of bones results in bone loss. Significant advances have occurred recently in our understanding of the factors that regulate the formation and activation of the cell type uniquely responsible for bone resorption, the osteoclast. This review of bone resorption discusses the current understanding of osteoclast formation and activation and the factors that are responsible for focal or systemic bone loss in various pathologies.

2. THE OSTEOCLAST

Osteoclasts can be defined as multinucleated giant cells that have the unique ability to resorb mineralized tissue. Osteoclasts are formed by fusion of circulating precursor cells of the monocyte-macrophage lineage (2), which are recruited to sites of bone turnover and repair or, in pathology, to sites of osteolysis (3). Immature and mature cells of the monocyte-macrophage lineage, in a wide variety of tissues, apparently have the ability to develop into osteoclasts under suitable conditions (3–6). In the process of osteoclast differentiation, a loss of monocyte/macrophage markers (such as CD11a, CD11b, CD14, HLA-DR, and CD68) and an acquisition of markers specific to osteoclasts occurs. Osteoclasts characteristically express a variety of proteins, which include tartrate-resistant acid phosphatase, cathepsin K, carbonic anhydrase II, matrix metalloproteinases MMPs, osteopontin, calcitonin receptors, and vitronectin receptors (7). Bone resorption by osteoclasts requires the secretion of acid that enables the dissolution of hydroxyapatite, the mineral component of bone. In addition, matrix proteases are released that degrade the organic phase of bone matrix (principally type I collagen) at acid pH (7). The process of differentiation of precursor monocytes to mature functioning osteoclasts may take several weeks and requires an appropriate environment and the presence of several key cytokines. A large number of factors have now been identified that can influence osteoclast differentiation positively or negatively. As a result of the complex environment in which osteoclasts form, it has taken many years of research to reach our current understanding of the biological process of bone resorption (described in detail below).

When an osteoclast attaches to the surface of bone, it undergoes polarization because of cytoskeletal reorganization (8), which allows osteoclasts to adhere firmly to the bone via integrins, molecules that attach to extracellular matrix proteins of the bone matrix, to create what is known as the "sealing zone" (9). This tight junction between the bone surface and the osteoclast prevents escape of the protons or enzymes used in the degradation process (9). Vesicles containing protons, which demineralize bone, and proteases, which degrade the organic matrix, concentrate to areas of membrane within the resorption zone, where the cell surface area is greatly increased by formation of the "ruffled border." The contents of the vesicles are released into the compartment formed between the osteoclast and the bone and a "resorption lacuna" is created. Once the mineral phase is dissolved, the degraded matrix is endocytosed by the cell and secreted at the opposite membrane (10). Osteoclastic resorption is illustrated in Fig. 1.

3. FACTORS REGULATING OSTEOCLAST FORMATION AND ACTIVITY

3.1. Receptor Activator of NF-κB Ligand (RANKL) and Osteoprotegerin (OPG)

Major advances in our understanding of the process of osteoclast formation have been obtained from research over the past two decades. Rodan and Martin (11) first hypothesized that osteoblasts play a significant role in regulating bone resorption. This concept was based on numerous studies showing that pro-resorptive influences act first on osteoblastic cells (12). The importance of the osteoblast is demonstrated by the fact that coculture of osteoclast precursors with osteoblastic cells is normally required for osteoclast differentiation (13). Studies have shown that osteoclast formation requires cell-cell contact between the osteoclast precursors and osteoblastic cells, as separation of monocytes from stromal cells by a membrane filter prevents osteoclasts from forming. Furthermore, media obtained from osteoblastic cells in culture cannot support osteoclast formation. These experiments indicate that pro-osteoclastic signal(s) expressed by osteoblasts require intimate cell contact between osteoclast precursors and osteoblastic cells.

The reason for this necessary close contact between osteoblasts and osteoclast precursors has now been discovered. In 1998, the cell surface molecule, RANKL (RANK ligand, osteoclast differentiation factor, osteoprotegerin ligand), was identified as a key factor in stimulating osteoclast formation. RANKL is expressed by osteoblasts in response to pro-resorptive molecules, such as parathyroid hormone, Vitamin D, or IL-11. RANKL binds to its receptor, RANK, on the surface of osteoclast precursors, stimulating these cells to differentiate into mature, active osteoclasts capable of resorbing bone (14,15). Although the RANKL/RANK interaction is essential for physiological osteoclast formation, other factors are also necessary, and it is now clear that RANKL, together with macrophage-colony stimulating factor (M-CSF), is required for osteoclast formation.

The initial discovery of osteoprotegerin (OPG), a soluble TNF "receptor-like" molecule that is a natural inhibitor of RANKL, played an important role in subsequently identifying RANKL (14–16). OPG binds to RANKL and prevents its ligation to RANK and, thus also, to the intracellular signaling pathways that are needed for osteoclast differentiation. The importance of OPG and RANKL in regulating bone metabolism is clearly demonstrated by transgenic and gene knockout studies in mice. Severe osteoporosis is observed in animals failing to express

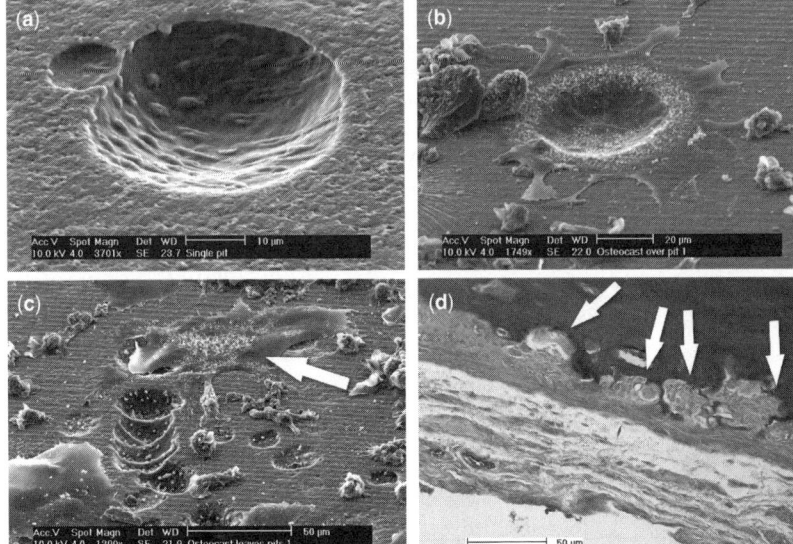

Figure 1. Panel A shows a typical lacunae created by osteoclasts *in vitro* on a dentine or bone surfaces. These lacunae range from smaller than 10 μm to over 100 μm in diameter and are 1–5 μm deep. Panel B shows an osteoclast overlying a resorption lacuna it has formed in dentine. Panel C shows a series of lacunae created sequentially by a single osteoclast. An osteoclast, indicated by an arrow, is in the process of migrating to a new section of dentine to form another lacuna. Panel D shows a section of tissue from a rheumatoid joint with synovial tissue ingrowth into bone. Lacunae of 20–50 μm in diameter, indicated by arrows, can be seen on the surface of the bone. These lacunae contain large multinucleated cells, likely to be osteoclasts.

BONE RESORPTION

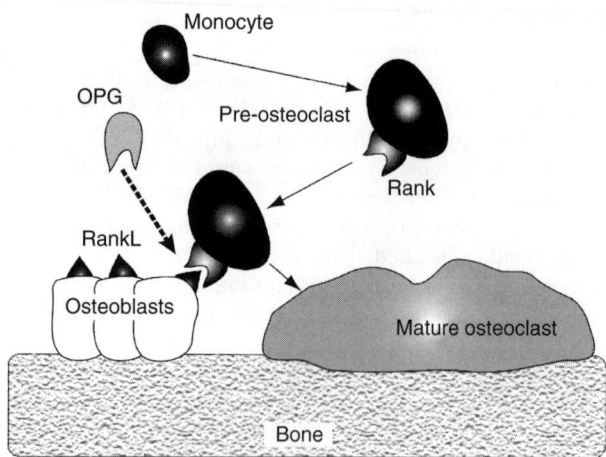

Figure 2. The diagram represents how receptor activator for NFκB (RANK), its ligand, RANKL, and the RANKL inhibitor, OPG, regulate physiological osteoclast formation. During inflammation, a reduction in OPG and an increase in production of RANKL, derived from inflammatory cells, occurs, which can result in elevated osteoclast formation.

OPG, whereas, conversely, osteopetrosis is seen in RANKL gene knockout mice. Hence, the relative levels of RANKL and OPG are likely to be important in determining whether osteoclasts will form. Figure 2 summarizes the physiological role of RANKL, RANK, and OPG in osteoclast differentiation.

3.2. Macrophage-Colony Stimulating Factor (M-CSF)

Macrophage-colony stimulating factor (M-CSF), like RANKL, also appears to be an essential cytokine regulating physiological bone metabolism. M-CSF, also known as Colony Stimulating Factor 1 (CSF-1), normally plays an essential role in osteoclast formation as is demonstrated by the fact that CSF-1 knockout mice (*op/op*) are osteopetrotic. Daily injection of CSF-1 reverses the osteopetrotic phenotype showing that osteoclast formation depends on this cytokine (17). M-CSF appears to be an essential factor for the proliferation and survival of osteoclast progenitors that occurs during differentiation of mature osteoclasts. M-CSF may also prolong the survival of mature osteoclasts, as well as modulate their bone resorbing activity.

3.3. Parathyroid Hormone (PTH)

Parathyroid hormone is a systemic hormone secreted in response to falls in blood calcium levels and is important in bone remodeling. A chronic continuous excess of PTH increases bone resorption (18) and, at high levels, PTH has been shown to inhibit osteoblast synthesis of collagen and other matrix proteins (19). Conversely, when delivered in low doses intermittently, PTH causes an increase in bone mass, an effect that may be mediated by reduced osteoblast cell death (18). Although PTH is not essential for osteoclast bone resorption, its ability to act independently of other factors, and its role in regulating blood calcium levels, make it an important factor regulating bone metabolism systemically in health and disease. In addition to parathyroid gland-derived PTH, a PTH-like molecule termed parathyroid hormone-related protein (PTHrP), first identified as an osteolytic and hypercalcaemic agent in cancer, can mimic the bone resorptive actions of PTH. PTHrP shares N-terminal homology with PTH and acts through the same receptors. It is produced and acts locally in various tissues, and its local production by metastatic tumor cells in bone has been implicated in tumor-induced osteolysis (20).

3.4. Prostaglandins (PGs)

Prostaglandins are a group of nonpeptide metabolites of arachidonic acid that act locally to affect osteoblastic cells as well as preosteoclasts and mature osteoclasts. These chemical mediators have a wide range of effects on osteoclastic cells that may depend on the stage of osteoclast differentiation. The action of prostanoids in the regulation of bone turnover is complex, and they have been shown to have opposing effects depending on the timing and location of their release. Their effects are further complicated by the fact that they typically interact with several different receptors, which may result in different biological outcomes. PGE_2, a prostanoid found at high levels in inflammation, stimulates osteoclastic bone resorption in bone organ culture systems as well as osteoclast formation in murine marrow cultures (21). Overall, prostanoids are thought to stimulate, more than inhibit, osteoclast formation and activity. PGE_2 may indirectly stimulate osteoclastogenesis *in vitro*, where it has been shown to induce expression of RANKL by osteoblasts (22).

3.5. Interleukin (IL)-1

Interleukin (IL)-1 is a proinflammatory multifunctional cytokine that stimulates bone resorption, helps recruit inflammatory cells, stimulates eicosanoid (especially PGE_2) release, and stimulates release of matrix metalloproteinases. Importantly, IL-1 induces the expression of RANKL by osteoblasts. The ability of IL-1β to stimulate bone loss in disease has been widely studied. IL-1β acts directly on preosteoclasts and osteoclasts via functional IL-1R and appears to be important in the coordinated signaling between preosteoclasts and osteoblastic cells during osteoclast formation (23). IL-1β prolongs the survival of osteoclastic cells by stimulating intracellular signaling of the PI3-kinase/Akt and extracellular signal-regulated kinase (ERK) pathways (24). In addition, it stimulates multinucleation and actin ring formation, which are important stages in osteoclast resorption (25).

3.6. Tumor Necrosis Factor (TNF)

TNFα has been implicated in bone resorption because of its association with the joint destruction seen in rheumatoid arthritis. TNFα can stimulate osteoclast formation by inducing RANKL expression. It also sensitizes osteoclast precursors to low levels of RANKL (26). Recent studies have also suggested that TNF-α may stimulate osteoclastogenesis directly, in the absence of RANKL (27). As osteoblasts do not secrete TNFα, it may not have a role in normal bone metabolism. However, during inflammatory bone loss, TNFα is likely to have an important role. Recent studies have indicated that in some human diseases osteoclasts may form in the absence of RANKL (28). However, because high levels of RANKL are usually present at sites of pathological bone loss, it is more likely that TNFα acts in synergy with RANKL by sensitizing preosteoclasts to low levels of RANKL. TNF-antagonists and antibodies are now an accepted treatment regime for active rheumatoid arthritis. Inhibition of TNF activity not only reduces the inflammatory response but may also inhibit osteoclastic bone resorption by reducing the synergistic activity of RANKL and TNFα in the joint tissues.

3.7. Interleukin (IL)-6

Osteoblastic cells supporting osteoclast formation *in vitro* express mRNA encoding IL-6 and IL-6 receptors (29). IL-6 appears to be involved in osteoclast differentiation and may also have a role in regulating the activity of mature osteoclasts (30). IL-6 production is stimulated by many inflammatory mediators, and some of the effects of IL-1 and TNF are likely to be mediated via IL-6 (30). IL-6 has become of particular interest for its role in estrogen deficiency-related bone loss, in which raised levels of proinflammatory cytokines, particularly IL-6, have emerged as factors that modify normal bone remodeling. Increased expression and secretion of IL-6, as well as IL-1 and TNF-α, exists with estrogen deficiency. Patient studies further demonstrate that IL-6 may mediate the osteoporotic effects of estrogen deficiency. Serum levels of IL-6 in postmenopausal women on hormonal replacement therapy are lower than in those women who are not on therapy (31,32). In addition, IL-6 deficient mice do not suffer from bone loss caused by estrogen depletion. IL-6 is reported to stimulate the expression of RANKL by osteoblastic cells (33). Thus, although IL-6 may not be an essential factor for osteoclast formation, it is likely to have an important role in stimulating osteoclast formation in disease in a similar way to other proinflammatory cytokines.

3.8. Interleukin (IL)-11

IL-11 is a multifunctional cytokine, closely related to IL-6, with many properties in common with IL-6, and is implicated in regulating bone metabolism in health and disease. IL-11 stimulates osteoclast formation (34) and osteoblast-mediated osteoid degradation (35). In normal bone, IL-11 appears to be produced by cells of the mesenchymal lineage. ST-2 murine osteoblastic cells express mRNA encoding IL-11 *in vitro* (36). IL-1, TNFα, PGE$_2$, parathyroid hormone (PTH), and 1 alpha,25-dihydroxyvitamin D3 (1 alpha,25(OH)2D3) induce the production of IL-11 by osteoblasts, providing further evidence that gp130-coupled cytokines, such as IL-6 and IL-11, have a central role in osteoclast development. Interestingly, like TNFα these cytokines have been reported to stimulate osteoclast formation in the absence of RANKL (37). As is the case with TNFα it is difficult to determine how important this is for human osteoclast formation *in vivo*, as IL-11 is a potent inducer of RANKL expression by osteoblasts (38).

3.9. Transforming Growth Factor (TGF)-β

Molecules belonging to the Transforming Growth Factor (TGF)-β superfamily are very important in regulating the repair of soft and mineralized tissues. They stimulate cell proliferation, differentiation, and regulate apoptosis of a variety of cells during tissue repair. TGF-β has been shown to stimulate the differentiation of osteoblast cells (39); however, the effect of TGF-β on osteoclasts is still controversial. TGF-β is clearly important in stimulating the laying down of new bone during healing but may also stimulate osteoclast activity. Connective tissues, including bone, contain large amounts of inactive TGFβ bound to the extracellular matrix that is released and becomes activated when the matrix is damaged or degraded. Although reports exist that TGF-β inhibits the proliferation and fusion of human osteoclast precursors (40), more recent reports show that TGF-β increases the formation of osteoclasts and can stimulate osteoclast formation in the absence of RANKL (41). Other members of the TGFβ superfamily, particularly some of the bone morphogenetic proteins (BMPs), have also been shown to enhance osteoclastogenesis under certain circumstances.

4. BONE RESORPTION IN BONE LOSS PATHOLOGIES

Bone loss pathologies can be broadly divided into two categories: systemic diseases, such as osteoporosis, in which bone is lost throughout the skeleton, and conditions characterized by more focal bone loss, such as the joint erosions produced by active rheumatoid arthritis (RA); tumor-induced osteolysis, including primary bone tumors such as giant cell tumors of bone and osteosarcomas and multiple myeloma or metastatic bone disease produced by tumors of breast or prostate origin; periodontal bone loss; and bone loss around orthopedic devices, primarily joint replacement components. In seeking to understand the mechanisms of these bone loss pathologies, it is becoming clear that many of the same molecular players seen in physiological osteoclast formation participate in osteoclast-mediated bone resorption in these conditions. In pathological bone loss, the process of bone resorption lacks regulation, or is not matched by bone formation, leading to a net loss of bone. Although the same molecules may be involved in health and disease, they appear to be expressed by different cell types depending on the particular pathology concerned.

The following section describes the current understanding of the mechanisms of bone loss in a variety of pathological states, focusing largely on inflammatory arthritis and aseptic loosening of orthopedic implants as examples. The factors that recruit osteoclast progenitors to sites of bone resorption, and that promote their differentiation and activity at those sites, are discussed briefly. Figure 3 summarizes, in a simple form, the major events of localized pathological bone loss, reviewed below.

4.1. Recruitment of Osteoclast Precursors—The Role of Chemokines

The accumulation of macrophages in bone or the tissues adjacent to bone is likely to be a significant initial event in

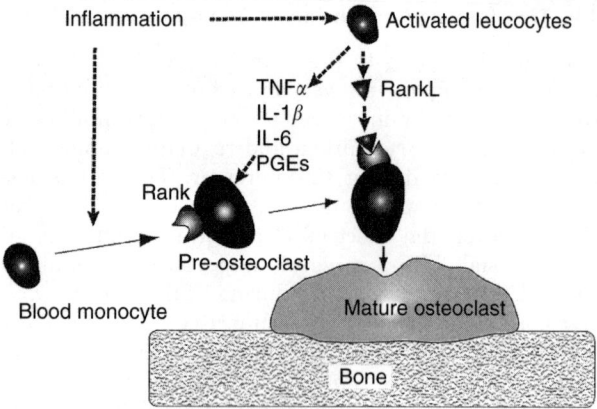

Figure 3. Diagram summarizing the major cytokines and mediators involved in the process of osteoclast formation in the tissues associated with pathological bone loss.

inflammatory-mediated bone loss, which is likely to be because of expression of locally produced chemokines that recruit peripheral blood monocytes into the tissues. The granulomatous inflammatory response seen in human pathologies is often initiated by pathogens or foreign materials, such as particles of worn implants, that cannot be easily removed or degraded. Recruitment of osteoclast precursors is required in normal bone turnover, but excessive recruitment often occurs at sites of bone loss associated with inflammation. Chemokines can be involved in the recruitment of many types of cells, but it is those that are involved in the recruitment of monocytes/macrophages and lymphocytes that are likely to be most important in peri-prosthetic, rheumatoid arthritis, and periodontal osteolysis. Chemokines have been classified into four distinct groups according to the arrangement of cystine amino acids in their structure, and chemokines and their receptors are the subject of several recent reviews (42–44). Chemokines belong to a family of molecules that are different from the more classic chemoattractants, such as the complement fragments C3a and C5a, platelet activating factor (PAF), other metabolites of arachidonic acid, and bacterial products. It is those that belong to the group known as the CC chemokines that are most likely to be involved in granuloma formation. Three important CC chemokines are chemoattractant protein (MCP)-1, macrophage inflammatory protein (MIP)-1α, and RANTES, and these chemokines are likely to be most important in focal bone loss pathologies, as they are chemotactic for both monocytes and activated T cells (43). Strong evidence exists that MCP-1 is important in granulomatous formation, as abnormalities in monocyte recruitment and granuloma formation are seen in MCP-1-deficient mice (45) as well as in mice with knockout of the chemokine receptor (CCR)2 gene, the cell surface receptor for MCP-1 (46). Several other chemotactic molecules have been implicated in studies on experimental and human granulomatous diseases, which show strong expression of MIP-1α (47) and RANTES (48) as well as MCP-1 (49). Importantly, MCP-1 and MIP-1α are implicated in the recruitment of osteoclast progenitors (50), and MIP-1α may also mediate osteoclast differentiation (51).

Chemokines are present in the interface tissue around loosening implants (52), and prosthetic wear particles have been shown to stimulate the production of molecules chemotactic for monocytes (52). Three members of the CC group of chemokines, MCP-1, MIP-1α, and RANTES, were stimulated by prosthetic particles *in vitro* (53). Similarly, other diseases of localized bone loss are associated with elevated expression of these chemokines near sites of osteolysis.

4.2. Interleukin (IL)-8

Interleukin (IL)-8 is a chemokine present in diverse inflammatory disorders, which stimulates the recruitment, proliferation, and activation of vascular and immune cells. Tissue samples obtained from patients undergoing revision hip replacement because of peri-prosthetic osteolysis have significantly higher levels of IL-8 compared with normal joint capsule tissue (54). IL-8 is also elevated in rheumatoid arthritis (55), and fibroblastic cells have been identified as the cell type producing IL-8 in peri-prosthetic tissue adjacent to areas of bone loss and in rheumatoid tissues (56). IL-8 has been reported to promote osteoclastogenesis, independently of RANKL, in metastatic cancer (57). Osteoclasts are also a source of IL-8; human osteoclasts isolated from osteoporotic femoral heads have also been shown to synthesize mRNA (messenger ribo nucleic acid) for IL-8 and release IL-8 protein in culture (55). IL-8 is present at high levels in peri-prosthetic osteolytic lesions, where its expression has been related to aseptic loosening of total hip replacement (56). Inflammatory stimuli, such as LPS, IL-1α, and TNFα significantly increase IL-8 mRNA expression and IL-8 protein release from osteoclasts. In contrast, "noninflammatory" cytokines and systemic hormones, such as IL-6, TGF-β1, and TGF-β3, do not stimulate IL-8 release (55). It has also been suggested (55) that human osteoclast-derived IL-8 may be an important autocrine/paracrine mediator of bone cell physiology and immunoregulation involved in normal as well as pathological bone remodeling.

4.3. RANKL and OPG in Pathological Bone Resorption

As mentioned previously, RANKL and OPG may be key factors in the process of pathological bone loss as well as normal bone turnover. Immunostaining for human RANKL and OPG suggested that elevated levels of RANKL and OPG protein are commonly present in diseased tissue adjacent to osteolysis. Studies on healthy and pathologic human tissues (58–61) show that a number of cell types may be important in the ectopic production of RANKL in the tissues adjacent to localized bone loss. Very high levels of RANKL protein are expressed within inflammatory cell infiltrates in the tissues adjacent to pathological bone loss in RA, peri-prosthetic loosening, and periodontal disease. Importantly, the expression of RANKL in the tissues adjacent to bone loss was significantly greater than in the relevant controls (58,60,61).

Dual staining has shown that different types of cells colocalize with RANKL protein in different bone loss pathologies. In rheumatoid arthritis and periodontal bone loss, CD3-positive lymphocytes were the predominant cell type that colocalized with RANKL (58,60,61), which is consistent with reports that lymphocytes present in the rheumatoid tissues may be the major cell type that produces RANKL in inflammatory arthritis (62,63). Inhibition of RANKL by OPG treatment *in vivo* can reduce both bone destruction and cartilage destruction in a model of adjuvant arthritis (62). These reports suggest that RANKL is also associated with CD68-positive monocyte/macrophages in rheumatoid arthritis. As monocyte/macrophages can express RANK and may also become osteoclasts, it is possible that autocrine stimulation of RANK may occur in disease.

OPG has also been detected in peri-prosthetic osteolysis and other osteolytic states (64). OPG was detected in blood vessels (on Factor VIII-positive endothelial cells) and synovium (on CD68-positive type A synoviocytes) in RA and peri-prosthetic tissues associated with osteolysis, as well as the epithelium in periodontal tissues. Whether the OPG produced by endothelial cells or the type A synoviocytes is able to block RANKL and inhibit the formation of osteoclasts in these tissues is yet to be determined. However, it is significant that the expression of OPG was greatly reduced in the blood vessels and synovium in inflammatory disease. Although the role of OPG production in blood vessels is not yet understood, the difference in OPG expression between diseased and control tissues suggests that the reduced production of OPG may contribute to the disease. A significant correlation exists between osteoclast formation *ex vivo* and the ratio of RANKL to OPG mRNA levels in cells isolated from RA tissues (65). Consistent with this fact, the ratio of RANKL to OPG mRNA was also higher in cells isolated from peri-prosthetic tissues, from which osteoclasts readily formed (66). It is interesting that cells from several of these samples formed active osteoclasts without additional osteoblast-like/stromal cells. More recently, the levels of RANKL and OPG mRNA have been compared in healthy gums with tissue from moderate or advanced periodontitis (67). Like the RA and peri-implant tissues, lower levels of OPG mRNA were found in moderate and advanced periodontitis compared with healthy tissue, while the expression of RANKL mRNA was highest in the advanced periodontitis samples. These results suggest that a high ratio of RANKL to OPG in tissues adjacent to bone may stimulate macrophages to become mature osteoclasts without the need for them to be in contact with osteoblast cells.

The effect of TNFα on OPG production by endothelial cells has also been investigated *in vitro* (27). The finding that TNFα reduces OPG detected on endothelial cells but increases OPG released into the supernatant may be relevant to inflammatory joint diseases. It is possible that, although TNFα might initially upregulate OPG production, chronic exposure to inflammatory cytokines *in situ* might exhaust the production of OPG by endothelial cells. TNF antagonists and antibodies are now an accepted treatment regime for active RA. Not only does inhibition of TNFα activity reduce the inflammatory response (68), but blocking elevated TNFα in disease may also enhance the differentiation and proliferation of osteoblasts (69). These recent findings demonstrate that the RANKL/RANK pathway of osteoclast differentiation is an important pharmaceutical target for regulating the bone destruction in inflammatory bone loss conditions such as peri-prosthetic loosening, RA, and periodontitis.

4.4. Interleukins 1, 6, 11, 17

Levels of IL-1 are raised in peri-prosthetic tissue at sites of osteolysis (70) as well as in the synovial fluid of RA patients (71). In general, IL-1 is also elevated in periodontal tissues associated with bone loss (72,73), as well as the gingival crevicular fluid (74) from inflamed periodontal tissues, compared with normal or healthier sites. The inflammatory response and bone loss in experimental periodontitis can be inhibited by antagonists to IL-1 and TNFα, demonstrating the role IL-1 may have in bone loss in this disease (68). In peri-prosthetic tissues associated with osteolysis, macrophages containing wear particles express both IL-1β and TNFα protein (75). Macrophages, foreign body giant cells and osteoclasts isolated from peri-prosthetic tissues, also express IL-1β receptor (76). The detection of IL-1β in fibroblasts, without concurrent mRNA expression, suggests that macrophages release IL-1β that then binds to fibroblasts, and possibly other macrophages (75). The production of IL-1 in these pathological tissues is likely to be important in the light of evidence that IL-1 acts as a cofactor for, and may be synergistic with, RANKL (as discussed above). IL-1 may also induce osteoclast formation independently of RANKL (25).

IL-6 and soluble IL-6 receptors are elevated in the synovial fluids of rheumatoid arthritic patients (77) and are produced by macrophages and other cells present in the peri-prosthetic tissues during osteoclast formation (78). IL-6 is also elevated in inflamed periodontal tissues, with notably higher levels in periodontitis than in gingivitis tissue (79,80). In peri-prosthetic bone loss, IL-6 is thought to have a role in regulating the inflammatory response at the bone-implant interface, as well as having an adverse effect on the process of bone remodeling. It is important to note that IL-6 has been positively related to the severity of osteolysis around prostheses, although not to the same extent as TNFα (81). Furthermore, the IL-6 receptor is present on macrophages and giant cells, possible precursors of osteoclasts, which are present in these peri-prosthetic tissues (76).

It has long been established that infection in bone or around teeth can cause extensive and rapid osteolysis. Inflammatory cells that are recruited into the area release chemokines and cytokines, such as IL-1-β and TNF-α. These cytokines can then stimulate osteolysis, as discussed in previous sections of this review. It has been shown that endotoxin (lipopolysaccharide; LPS) released from bacteria activates cells, including tissue macrophages, which causes the release of TNFα, IL-1-β, and other osteolytic cytokines. In addition, LPS has been found to directly contribute to osteoclast differentiation by interacting with Toll-like receptors on osteoclast progenitors (82).

Immunohistochemical staining has identified IL-11 in the interface and pseudocapsular tissues obtained from sites of peri-prosthetic osteolysis (83). Cells expressing IL-11 are more numerous at the bone-implant interface and pseudocapsular tissues from patients with aseptic loosening of their implant because of peri-prosthetic osteolysis than in control synovial tissues from patients undergoing primary hip replacement. The presence of IL-11 in peri-prosthetic tissue suggests that it may be another factor involved in stimulating peri-prosthetic bone resorption. In rheumatoid arthritis, IL-11 is produced predominantly by the synovial cells, and has been identified in these cells at both the RNA and protein level (84). It is also detected in the serum and synovial fluid of patients with rheumatoid arthritis (85). It is therefore likely that IL-11 production in the rheumatoid joint is involved in the bone destruction of this disease (84).

IL-17 is a T cell-derived cytokine involved in inflammation. IL-17 is present in the synovium of rheumatoid patients (86) and levels are significantly higher in the synovial fluid of patients with rheumatoid arthritis compared with osteoarthritic patients (86). IL-17 is reported to stimulate osteoclast activity and bone destruction in rheumatoid arthritis (86). Recent studies using a murine model also suggest a role for IL-17 in cartilage and bone destruction. The intra-articular administration of IL-17 into a normal mouse induced cartilage destruction, whereas blocking IL-17 with specific inhibitors protected the bone by inhibiting bone destruction (87). The effect of IL-17 on bone and cartilage may be explained by its ability to stimulate the production of the proinflammatory cytokines, IL-1, and TNF-α by macrophages, as well as IL-6, IL-8, PGE_2, and RANKL expression by stromal/osteoblast lineage cells (88).

The extent to which IL-17 is involved in pathological osteolysis is yet to be determined. As IL-17 is largely a T cell-derived cytokine, its role in peri-prosthetic osteolysis may not be as important as the factors mentioned above because T cells are typically in low numbers in these tissues (89). However, IL-17 could have an important role in other pathologies, such as rheumatoid arthritis and periodontal disease, where large numbers of T cells are present. A recent study found mRNA encoding IL-17 to be expressed by osteoblast lineages and its production by osteoblastic cells may therefore be involved in physiological and pathological bone loss in various diseases (29).

4.5. TNFα

Tumor necrosis factor (TNF)α is primarily produced by activated monocytes/macrophages in inflamed joints in rheumatoid arthritis (90,91), whereas TNF-β (lymphotoxin) is produced by activated T lymphocytes. TNFα mediates leukocyte recruitment and activation, synovial macrophage and fibroblast cell proliferation, increased prostaglandin and matrix degrading MMP activity, as well as bone and cartilage destruction. The production of TNFα in inflammation has long been recognized as an inducer of bone resorption (92). In support of this fact, a correlation has been found between the presence of TNFα and the severity of osteolysis (81). The administration of TNFα antagonists to patients with rheumatoid arthritis has been shown to reduce inflammatory symptoms and to prevent bone loss in this disease (93) and is now an important treatment modality in this disease. Indeed, experimentally, antiTNF therapy, in concert with antiIL-1 or antiRANKL therapy, can give almost complete disease remission and protection from bone loss (94). In peri-prosthetic osteolysis, abundant TNFα has been observed associated with macrophages in areas containing wear debris, as well as with fibroblasts and some endothelial cells (95). Numerous reports exists of the ability of prosthetic wear particles to stimulate macrophages to release TNFα *in vitro* (96).

4.6. Prostanoids

PGE_2 is an inflammatory mediator that is likely to be important in several bone loss pathologies. It has been identified in the peri-prosthetic pseudosynovial membrane (97) and macrophage-like cells derived from this tissue, cultured together with human osteoblasts, form bone resorbing osteoclasts after 14 days incubation (98). The addition of exogenous PGE_2 in this cell culture system caused a dose-dependent two-to three-fold increase in lacunar bone resorption compared with untreated controls (98). Conversely, although it appears to be a requirement for normal osteoclast formation, excessive PGE_2 may inhibit osteoclast formation (99). Other arachidonate metabolites may also be involved in osteoclast formation in the peri-prosthetic tissues. Anderson et al. have developed a culture system that uses prosthetic wear-like particles to stimulate osteoclast formation (100). With this model, they showed inhibition of osteoclast formation by specific inhibitors of leukotriene synthesis, indicating that other arachidonic acid metabolites, besides prostaglandins, may regulate osteoclast formation in disease conditions. In periodontal disease, a significant amount of PGE_2 can be produced in the inflamed tissues (101). Levels of PGE_2 in the gingival crevicular fluid are reduced in patients with improved clinical parameters following treatment (101), suggesting that PGE_2 may be involved in the pathogenesis of oral bone loss disease. It is likely that arachidonic acid metabolites, prostaglandins, and leukotrienes are important mediators of osteoclast formation; however, it is not known if they have a direct effect on the process or if their effects are mediated via regulation of other important factors involved in osteoclastogenesis.

4.7. M-CSF

Considering its pivotal role in osteoclast differentiation, M-CSF production by activated cells in the soft tissues adjacent to localized bone loss is likely to be an important factor in osteoclast formation. It is, therefore, important that, in addition to bone marrow stromal cells and osteoblasts, M-CSF has been associated with a variety of cell types, such as monocyte/macrophages, fibroblasts, and vascular endothelial cells. M-CSF is present in the synovial fluid (102) and in the synovial-like membrane (103) of peri-prosthetic tissues taken from patients with implant loosening because of peri-prosthetic osteolysis. High levels of M-CSF have been identified in the peri-prosthetic tissues adjacent to failed prostheses than in the synovial membrane of patients undergoing primary hip replacement (102). It is important to note that the M-CSF receptor is present on the surface of macrophages and foreign body giant cells in peri-prosthetic tissues (78). These findings support the concept that M-CSF may play an important role in the regulation of osteoclastogenesis in disease. Indeed, studies of osteoclast formation from cells isolated from peri-prosthetic tissues illustrate the importance of M-CSF and its receptor in peri-prosthetic osteolysis. High levels of M-CSF are produced during osteoclast formation *in vitro* in cocultures of arthroplasty-derived mononuclear cells and human osteoblastic cells (78). The addition of antibodies to block endogenous human M-CSF binding to its receptor markedly reduced the numbers of osteoclasts that formed in these cocultures. However, the addition of exogenous M-CSF or IL-6 to these cultures only slightly increased the numbers of osteoclasts that formed, suggesting that cells exists in peri-prosthetic tissues that are releasing sufficient M-CSF to mediate osteoclast formation (78).

Granulocyte-macrophage colony stimulating factor (GM-CSF) is another colony stimulating factor that regulates monocyte maturation. It is closely related to M-CSF and is also present at sites of osteolysis (104). GM-CSF is reported to stimulate osteoclast formation in much the same way as M-CSF. However, at similar concentrations, GM-CSF is a weaker stimulator of osteoclast formation than M-CSF and may not be as important as M-CSF in mediating osteolysis associated with diseases such as peri-prosthetic loosening, rheumatoid arthritis, and periodontitis.

4.8. TGFβ

TGF-β is thought to be associated with repair of damaged tissues. It is produced not only by osteoblastic cells but has been shown in peri-prosthetic tissues near loose hip implants, where it may modulate bone metabolism (105). TGF-β is also associated with macrophages (90). *In vitro*, both osteoblastic cells and blood monocytes express mRNA encoding TGF-β (29). The actions of TGF-β in pathological bone loss are uncertain, but its activities may depend on the types of cells and cytokines present in the soft tissues adjacent to the bone.

4.9. Martix Metalloproteinases (MMPs)

Matrix metalloproteinases (MMPs) are essential to break down the nonmineral component of bone and, therefore, the resorption of bone by osteoclasts. MMPs are also believed to play an important role in joint destruction in a range of arthritides and osteolytic diseases by contributing to degradation of cartilage extracellular matrix. Fibroblasts from the interface membrane can be stimulated by wear particles to produce metalloproteinases in culture (106). Elevated levels of matrix MMP have been reported near loose artificial hip joints (107) and in the synovial fluids from RA patients (108). The MMP-specific degrading activity of rheumatoid synovial fluid may contribute to both cartilage destruction and bone loss. Tissue inhibitors of metalloproteinases (TIMPs) specifically regulate the enzymatic activity of MMPs. The balance between MMPs and TIMPs is thought to be particularly important in determining resultant cartilage damage. A recent study found that the molar ratio of MMPs to TIMPs was 5.2-fold higher in RA patients compared with OA patients (108).

4.10. Other Factors and Cells

Recently, it has been proposed that cells other than osteoclasts can directly cause osteolysis. Fibroblasts that grow into the bone interface during peri-implant osteolysis may also cause bone dissolution by a similar mechanism to that of osteoclasts (109). Although not forming a resorptive compartment like osteoclasts, fibroblasts can secrete protons into the local environment and cause degradation of the bone surface (110). This lower pH could result in the dissolution of the mineral component of bone and allow degradation of the connective tissue component by enzymes that may also be released by fibroblasts. It is not certain if these effects are directly contributing to bone loss or if it is because of hypoxia and an acidic environment that can, in turn, promote the activation of osteoclasts (111,112). As hypoxic or acidic conditions are likely to be common in swollen and inflamed joints, they may contribute to osteoclast-mediated bone loss, although this concept remains to be tested *in vivo*.

An increase in mechanical stress or fluid flow may upregulate osteoclastogenic factors, leading to bone loss in pathologies. These mechanical influences can cause bone loss adjacent to prosthetic joints and in alveolar bone. The release of the osteolytic cytokines, IL-6, TNF-α, and IL-β from monocyte-derived macrophages is significantly increased when exposed to cyclic pressure regimes of different frequencies in culture (113). Mechanical strain or fluid sheer stress has also been reported to regulate PGE_2, TGF-β, OPG, and IL-11 expression. In addition, fluid sheer stress is reported to stimulate bone resorbing activity of osteoclasts in culture (114). Mechanical factors have been suggested as contributing to inducing peri-prosthetic osteolysis prior to the production of wear particles (115). Pressure changes or fluid flow have been suggested as possible causes of peri-prosthetic bone resorption and implant loosening (115). Cocultures of human PBMC and periodontal ligament cells, as the stromal support, have demonstrated an increase in osteoclastogenesis when cells were placed under pressure (116). In support of this fact, when periodontal ligament cells were placed under a compressive force, an increase in expression of RANKL mRNA protein resulted (116). Mechanical strain or fluid sheer stress is reported to regulate PGE2, TGF-β, OPG, and IL-11 expression and stimulate bone resorbing activity of osteoclasts in culture (114). It is likely that peri-implant fluid pressure acts in synergy with particulate debris to greatly accelerate peri-implant osteolysis (117).

A simplified description of the regulation of osteoclasts in inflammatory diseases of localized bone resorption is shown in Fig. 3.

5. OTHER BONE LOSS PATHOLOGIES

5.1. Osteoporosis

Osteoporosis (OP) is a systemic skeletal disease characterized by low bone mineral density (BMD) and microarchitectural deterioration of bone tissue, with a consequent increase in bone fragility and susceptibility to fracture. OP is common in western countries and particularly among

Caucasian women (118). Although the underlying causes of OP are not well understood, the accelerated rate of bone loss after the menopause in women clearly reduces the strength of bones and their ability to resist fracture. Factors that may cause this bone loss have been identified, and include reduced estrogen levels in women, low Vitamin D levels, poor nutrition, such as low calcium intake, insufficient exercise, and propensity to fall (119). Recently, it has been argued (120) that the rate of bone remodeling is an important factor that determines bone fragility because increased osteoclastic resorption leads to loss of trabecular elements in bone and greater cortical porosity. Abnormal expression of molecules regulating bone metabolism have also been reported, and higher levels of RANKL, relative to OPG, may exist in OP bone (121), which may be initiated by reduced estrogen levels and increased proinflammatory cytokines, such as IL-6 and TNFα (122). Understanding how the interactions of hormonal and other factors influence the structural integrity of bone is likely to greatly advance our knowledge of this disease and lead to improvements in treatments in our aging populations.

5.2. Paget's Disease

Paget's disease of bone, a disorder that affects up to 3% of the population over 60, disturbs bone remodeling and results in bone lesions. Paget's disease involves the formation of abundant new woven bone, with single or multiple skeletal lesions (123). The pathology is most commonly found in the pelvis, lumbar spine, femur, tibia, and skull. In the early stages, Paget's disease is associated with marked osteolysis, which is visible radiographically. The lesions may be caused by local stimulation of osteoclast formation as well as activity (19). Generally, the histopathological changes indicate abnormal bone remodeling (124). The early stages involve disorganized bone remodeling and abundant woven bone formation, which are later replaced with lamellar bone. Pagetic osteoclasts are larger than normal, with a greater number of nuclei, and contain viral-like nuclear inclusions (125). The preosteoclasts and osteoclasts from Paget's patients appear hyperresponsive to 1,25 dihydroxy Vitamin D3 and RANKL (126).

Increases in IL-6 levels have been reported in Paget's disease (123) and osteoclasts themselves express IL-6 mRNA as well as IL-6 receptors (125). At the same time, a marked increase in serum alkaline phosphatase occurs, as well as an increase in serum osteocalcin, indicating increased bone metabolism. Similarities exist between Paget's disease and bone metastases, as there is increased osteoclastogenesis mediated by IL-6 and RANKL in both conditions (127). In juvenile Paget's disease, the gene for OPG (TNFRSF11B) is subject to an inactivating mutation, leading to increased resorption and accelerated remodeling (128).

5.3. Osteolytic Tumors, Giant Cell Tumors (GCT)

Many tumors of bone can cause pathological bone resorption, including primary malignancies such as GCT and osteosarcoma, as well as metastatic disease from cancers such as breast, lung, and prostate. Osteolytic tumors may induce bone resorption by direct or indirect induction of osteoclastic resorption. GCT are the primary example of tumors that cause the recruitment of osteoclast progenitors and promote their differentiation and activity. GCT stromal cells have been shown to express high levels of proresorptive factors and, in particular, RANKL (129). In contrast, other primary tumors of bone or tumors metastatic to bone induce the production of RANKL by nearby stromal cells to indirectly stimulate osteoclastic resorption (130).

5.4. Multiple Myeloma

An important tumor that can cause extensive resorption and degradation of bone is multiple myeloma. Myeloma is a plasma cell dyscrasia involving expansion of a single clone of immunoglobulin-secreting cells and secondary myeloma lesions are commonly detected in the spine, pelvis, ribs, and proximal long bones. As in Paget's disease, proliferation and differentiation of myeloma cells is associated with high serum levels of IL-6. Osteoclast formation and activity is stimulated by production of cytokines and growth factors such as IL-1, IL-6, TNF-β, and M-CSF. The actual tumor cells express CD38, a plasma cell-associated antigen. They may express epithelial membrane antigen and are negative for leukocyte common antigen and CD22, a mature B cell marker (19). Recent evidence strongly suggests that RANKL expression by myeloma cells confers on them the ability to participate directly in the formation of osteoclasts in vitro and in vivo (131).

6. TREATING BONE RESORPTION IN DISEASE

A number of treatments are commonly applied to treat the skeleton-wide bone loss of osteoporosis, such as estrogen therapy, bisphosphonates, calcitonin, calcium, and Vitamin D. There have been a large number of reviews that discuss the clinical approaches to this common form of bone loss [e.g., Riggs and Parfitt (120)]; however, it is not the intention of this review to elaborate further on this topic. Although rheumatoid arthritis, peri-prosthetic loosening, periodontitis, and other conditions characterized by a local or widespread loss of bone appear to be quite different diseases, the primary mediators involved in pathological bone loss are common. Therefore, similar approaches might be useful to inhibit osteolysis in a variety of bone loss diseases. Although we now have an extensive range of anti-inflammatory drugs to control inflammation, very few treatments are available for inhibiting the debilitating bone loss seen in a variety of inflammatory diseases. The recent discoveries of the key factors involved in regulating osteoclast resorption will enable the development of new therapies to treat this problem. The RANK-RANK interaction in the formation of osteoclasts is an ideal target of therapy because it is a point at which numerous pathways for osteolysis converge. Experimental treatments for peri-implant osteolysis that are based on the inhibition of RANKL by its natural inhibitor OPG (132) have been successful, as have similar approaches in rheumatoid arthritis (62,133). Soluble RANK also appears to be useful as an antiresorptive agent (134). Suppression of TNFα has also been used to successfully treat peri-implant osteolysis (135). Suppression of osteoclast formation by various combinations of TNFα, IL-1, and RANKL inhibitors have also been used in an animal model of rheumatoid arthritis (93) and may be more successful than individual therapies. In addition, selective blockade of potassium channels may also reduce osteoclast formation in T cell-driven osteolysis in inflammatory diseases. These animal studies are promising, and preliminary studies based on blocking RANK-RANKL interaction in humans are currently underway (136). Similar therapies based on bisphosphonates, a group of drugs currently used to inhibit osteoclast activity in osteolytic bone tumors and osteoporosis, have also been used in models of peri-implant osteolysis animal studies (137), although they may not be effective in all cases.

7. CONCLUSION

An increased understanding of the factors regulating osteoclast-mediated bone resorption over the past decade mean that we can apply molecular therapies to treat excessive bone loss in a wide range of systemic and localized bone loss disease. Although further studies are required to refine these approaches, they are likely to yield exciting results that enhance the duration and quality of life of many individuals.

BIBLIOGRAPHY

1. A. M. Parfitt, Modeling and remodeling: how bone cells work together. In: D. Feldman, W. Pike, and F. H. Glorieux, eds., *Vitamin D*. 2nd ed. New York: Elsevier, 2005, pp. 711–720.
2. U. Sarma and A. M. Flanagan, Macrophage colony-stimulating factor induces substantial osteoclast generation and bone resorption in human bone marrow cultures. *Blood* 1996; **88**:2531–2540.
3. Y. Fujikawa, J. M. W. Quinn, A. Sabokbar, et al., The human osteoclast precursor circulates in the monocyte fraction. *Endocrinology* 1996; **137**:4058–4060.
4. T. Tsurakai, N. Takahashi, E. Jimi, et al., Isolation and characterization of osteoclast precursors that differentiate into osteoclasts on calvarial cells within a short period of time. *J. Cell Physiol*. 1998; **177**:26–35.
5. N. Udagawa, N. Takahashi, T. Akatsu, et al., Origin of osteoclasts: mature monocytes and macrophages are capable of differentiating into osteoclasts under a suitable microenvironment prepared by bone marrow-derived stromal cells. *Proc. Natl. Acad. Sci. USA* 1990; **87**:7260–7264.
6. J. M. W. Quinn, A. Sabokbar, and N. A. Athanasou, Cells of the mononuclear phagocyte series differentiate into osteoclastic lacunar bone resorbing cells. *J. Pathol*. 1996; **179**:106–111.
7. G. R. Mundy, Bone resorption and turnover in health and disease. *Bone* 1987; **8**:S9–S16.
8. H. Vaananen, Y. Liu, P. Lehenkari, et al., How do osteoclasts resorb bone? *Mater. Sci. Eng. C* 1998; **6**:205–209.
9. J. Bilezikian, L. Raisz, and G. Rodan, *Principles of Bone Biology*. San Diego, CA: Academic Press, 1996, Chapters 1,4–8,10,11.
10. L. T. Duong and G. A. Rodan, Regulation of osteoclast formation and function. *Rev. Endocr. Metab. Disord*. 2001; **2**:95–104.
11. G. A. Rodan and T. J. Martin, Role of osteoblasts in hormonal control of bone resorption - a hypothesis. *Calcif. Tissue Int*. 1982; **34**:311.
12. T. J. Martin and K. W. Ng, Mechanisms by which cells of the osteoblast lineage control osteoclast formation and activity. *J. Cell Biochem*. 1994; **56**:357–366.
13. N. Udagawa, N. Takahashi, T. Katagiri, et al., Interleukin (IL)-6 induction of osteoclast differentiation depends on IL-6 receptors expressed on osteoblastic cells but not on osteoclast progenitors. *J. Exp. Med*. 1995; **182**:1461–1468.
14. H. Yasuda, N. Shima, N. Nakagawa, et al., Osteoclast differentiation factor is a ligand for osteoprotegerin/osteoclast inhibitory factor and is identical to TRANCE/RANKL. *Proc. Natl. Acad. Sci. USA* 1998; **95**:3597–3602.
15. D. L. Lacey, E. Timms, H-L. Tan, et al., Osteoprotegerin ligand is a cytokine that regulates osteoclast differentiation and activation. *Cell* 1998; **93**:165–176.
16. H. Yasuda, N. Shima, N. Nakagawa, et al., Identity of osteoclastogenesis inhibitory factor (OCIF) and osteoprotegerin (OPG): a mechanism by which OPG/OCIF inhibits osteoclastogenesis in vitro. *Endocrinology* 1998; **139**:1329–1337.
17. H. Kodama, A. Yamasaki, M. Nose, et al., Congenital osteoclast deficiency in osteopetrotic (op/op) mice is cured by injections of macrophage colony-stimulating factor. *J. Exp. Med*. 1991; **173**:269–272.
18. T. Bellido, A. A. Ali, L. I. Plotkin, Q. Fu, I. Gubrij, P. K. Roberson, R. S. Weinstein, C. A. O'Brien, S. C. Manolagas, and R. L. Jilka, Proteasomal degradation of Runx2 shortens parathyroid hormone-induced anti-apoptotic signaling in osteoblasts. A putative explanation for why intermittent administration is needed for bone anabolism. *J. Biol. Chem*. 2003; **278**:50259–50272.
19. N. A. Athanasou, *Clinical Radiological and Pathological Correlation of Diseases of Bone, Joint and Soft Tissue*. London: Arnold Publishers, 2001.
20. T. J. Martin and J. M. Moseley, Mechanisms in the skeletal complications of breast cancer. *Endocr. Relat. Cancer* 2000; **7**:271–284.
21. K. Takahashi, H. Yamana, S. Yoshiki, et al., Osteoclast-like cell formation and its regulation by osteotropic hormones in mouse bone marrow cultures. *Endocrinology* 1988; **122**:1373–1382.
22. K. Tsukii, N. Shima, S-I. Mochizuki, et al., Osteoclast differentiation factor mediates an essential signal for bone resorption induced by $1\alpha,25$ dihydroxyvitamin D_3, prostaglandin E_2 or parathyroid hormone in the microenvironment. *Biochem. Biophys. Res. Commun*. 1998; **246**:337–341.
23. D. R. Haynes, G. J. Atkins, M. Loric, et al., Bidirectional signaling between stromal and hemopoietic cells regulates interleukin-1 expression during human osteoclast formation. *Bone* 1999; **25**:269–278.
24. Z. H. Lee, S. E. Lee, C. W. Kim, et al., IL-1alpha stimulation of osteoclast survival through the PI 3-kinase/Akt and ERK pathways. *J. Biochem. (Tokyo)* 2002; **131**:161–166.
25. E. J. Jimi, I. Nakamura, L. T. Duong, et al., Interleukin 1 induces multinucleation and bone-resorbing activity of osteoclasts in the absence of osteoblast/stromal cells. *Expt. Cell. Res*. 1999; **247**:84–93.
26. J. Lam, S. Takeshita, J. E. Barker, et al., TNF-α induces osteoclastogenesis by direct stimulation of macrophages

exposed to permissive levels of RANK ligand. *J. Clin. Invest.* 2000; **106**:1481–1488.

27. K. Kobayashi, N. Takahashi, E. Jimi, et al., Tumor necrosis factor α stimulates osteoclast differentiation by a mechanism independent of the ODF/RANKL-RANK interaction. *J. Exp. Med.* 2000; **191**:275–285.

28. A. Sabokbar, O. Kudo, and N. A. Athanasou, Two distinct cellular mechanisms of osteoclast formation and bone resorption in periprosthetic osteolysis. *J. Orthop. Res.* 2003; **21**:73–80.

29. G. J. Atkins, D. R. Haynes, S. M. Geary, et al., Coordinated cytokine expression by stromal and hematopoietic cells during human osteoclast formation. *Bone* 2000; **26**:653–661.

30. O. A. Adebanjo, B. S. Moonga, T. Yamate, L. Sun, C. Minkin, E. Abe, and M. Zaidi, Mode of action of interleukin-6 on mature osteoclasts. Novel interactions with extracellular Ca2+ sensing in the regulation of osteoclastic bone resorption. *J. Cell Biol.* 1998; **142**:1347–1356.

31. F. P. Cantatore, G. Loverro, A. M. Ingrosso, et al., Effect of oestrogen replacement on bone metabolism and cytokines in surgical menopause. *Clin. Rheumatol.* 1995; **14**:157–160.

32. R. H. Straub, H. W. Hense, T. Andus, et al., Hormone replacement therapy and interrelation between serum interleukin-6 and body mass index in postmenopausal women: a population-based study. *J. Clin. Endocrinol. Metab.* 2000; **85**:1340–1344.

33. T. Nakashima, Y. Kobayashi, S. Yamasaki, et al., Protein expression and functional difference of membrane-bound and soluble receptor activator of NF-κB ligand: modulation of the expression by osteotropic factors and cytokines. *Biochem. Biophys. Res. Commun.* 2000; **275**:768–775.

34. G. Girasole, G. Passeri, R. L. Jilka, et al., Interleukin-11: a new cytokine critical for osteoclast development. *J. Clin. Invest.* 1994; **93**:1516–1524.

35. P. A. Hill, A. Tumber, S. Papaioannou, et al., The cellular actions of interleukin-11 on bone resorption in vitro. *Endocrinology* 1998; **139**:1564.

36. G. J. Atkins, M. Loric, T. N. Crotti, et al., presented at the SIROT, Sydney, Australia, 1999.

37. O. Kudo, A. Sabokbar, A. Pocock, et al., Interleukin-6 and interleukin-11 support human osteoclast formation by a RANKL-independent mechanism. *Bone* 2003; **32**:1–7.

38. T. Suda, N. Takahashi, N. Udagawa, et al., Modulation of osteoclast differentiation and function by the new members of the tumor necrosis factor receptor and ligand families. *Endocrine Rev.* 1999; **20**:345–357.

39. T. A. Linkhart, S. Mohan, D. J. Baylink, et al., Growth factors for bone growth and repair: IGF, TGF-beta and BMP. *Bone* 1996; **19**(1 Suppl):1S–12S.

40. C. Chenu, J. Pfeilschifter, G. R. Mundy, et al., Transforming growth factor β inhibits formation of osteoclast-like cells in long-term human marrow cultures. *Proc. Natl. Acad. Sci. USA* 1988; **85**:5683.

41. I. Itonaga, A. Sabokbar, S. G. Sun, et al., Transforming growth factor-beta induces osteoclast formation in the absence of RANKL. *Bone* 2004; **34**:57–64.

42. P. M. Murphy, M. Baggiolini, I. F. Charo, et al., International union of pharmacology. XXII. Nomenclature for chemokine receptors. *Pharm. Rev.* 2000; **52**:145–176.

43. C. Lloyd, Chemokines in allergic lung inflammation. *Immunology* 2002; **105**:144–154.

44. M. N. Ajuebor and M. G. Swain, Role of chemokines and chemokine receptors in the gastrointestinal tract. *Immunology* 2002; **105**:137–143.

45. B. Lu, B. J. Rutledge, L. Gu, et al., Abnormalities in monocyte recruitment and cytokine expression in monocyte chemoattractant protein 1-deficient mice. *J. Exp. Med.* 1997; **185**:1959–1968.

46. J. L. Gao, T. A. Wynn, Y. Chang, et al., Impaired host defense, hematopoiesis, granulomatous inflammation and type 1-type 2 cytokine balance in mice lacking CC chemokine receptor 1. *J. Clin. Invest.* 1997; **100**:2552–2561.

47. S. Hashimoto, T. Nakayama, Y. Gon, et al., Correlation of plasma monocyte chemoattractant protein-1 (MCP-1) and monocyte inflammatory protein-1alpha (MIP-1alpha) levels with disease activity and clinical course of sarcoidosis. *Clin. Exp. Immunol.* 1998; **111**:604–610.

48. M. Petrek, P. Pantelidis, A. Southcott, et al., The source and role of RANTES in interstitial lung disease. *Eur. Respir. J.* 1997; **10**:1207–1216.

49. J. S. Friedland, R. J. Shattock, and G. E. Griffin, Phagocytosis of mycobacterium tuberculosis or particulate stimuli by monocytic cells induces equivalent chemotactic protein-1 gene expression. *Cytokine* 1993; **5**:150–160.

50. M. H. Zheng, Y. Fan, A. Smith, et al., Gene expression of monocyte chemoattractant protein-1 in giant cell tumors of bone osteoclastoma: possible involvement in CD68+ macrophage-like cell migration. *J. Cell Biochem.* 1998; **70**:121–129.

51. B. A. Scheven, J. S. Milne, I. Hunter, et al., Macrophage-inflammatory protein-1 alpha regulates preosteoclast differentiation in vitro. *Biochem. Biohys. Res. Commun.* 1999; **254**:773–778.

52. N. Ishiguro, T. Kojima, T. Ito, et al., Macrophage activation and migration in interface tissue around loosening total hip arthroplasty components. *J. Biomed. Mater. Res.* 1997; **35**:399–406.

53. Y. Nakashima, D. H. Sun, M. C. D. Trindade, et al., Induction of macrophage C-C chemokine expression by titanium alloy and bone cement particles. *J. Bone Joint Surg. Br.* 1999; **81-B**:155–162.

54. A. S. Shanbhag, J. J. Jacobs, J. Black, et al., Cellular mediators secreted by interfacial membranes obtained at revision total hip arthroplasty. *J. Arthroplasty* 1995; **10**:498–506.

55. L. Rothe, P. Collin-Osdoby, Y. Chen, et al., Human osteoclasts and osteoclast-like cells synthesize and release high basal and inflammatory stimulated levels of the potent chemokine interleukin-8. *Endocrinology* 1998; **139**:4353–4363.

56. M. S. Bendre, D. C. Montague, T. Peery, N. S. Akel, D. Gaddy, and L. J. Suva, Interleukin-8 stimulation of osteoclastogenesis and bone resorption is a mechanism for the increased osteolysis of metastatic bone disease. *Bone* 2003; **33**:28–37.

57. J. Lassus, V. Waris, J-W. Xu, et al., Increased interleulin-8 (IL-8) expression is related to aseptic loosening of total hip replacement. *Arch. Orthop. Trauma Surg.* 2000; **120**:328–332.

58. T. Crotti, M. D. Smith, R. Hirsch, et al., Receptor activator NF kappaB ligand (RANKL) and osteoprotegerin (OPG) protein expression in periodontitis. *J. Periodontal Res.* 2003; **38**:380–387.

59. D. R. Haynes, E. Barg, T. N. Crotti, et al., Osteoprotegerin expression in synovial tissue from patients with rheumatoid arthritis, spondyloarthropathies and osteoarthritis and normal controls. *Rheumatology (Oxford)* 2003; **42**:123–134.

60. T. N. Crotti, M. D. Smith, H. Weedon, et al., Receptor activator NF-kappaB ligand (RANKL) expression in synovial tissue from patients with rheumatoid arthritis, spondyloarthropathy, osteoarthritis, and from normal patients: semiquantitative and quantitative analysis. *Ann. Rheum. Dis.* 2002; **61**:1047–1054.

61. T. N. Crotti, M. D. Smith, D. M. Findlay, et al., Factors regulating osteoclast formation in human tissues adjacent to peri-implant bone loss: expression of receptor activator NFkappaB, RANK ligand and osteoprotegerin. *Biomaterials* 2004; **25**:565–573.
62. Y-Y. Kong, U. Feige, I. Sarosi, et al., Activated T cells regulate bone loss and joint destruction in adjuvant arthritis through osteoprotegerin ligand. *Nature* 1999; **402**:304–309.
63. E. Romas, O. Bakharevski, D. K. Hards, et al., Expression of osteoclast differentiation factor at sites of bone resorption in collagen-induced arthritis. *Arthritis Rheum.* 2000; **43**:821–826.
64. D. R. Haynes, E. Barg, T. N. Crotti, et al., Osteoprotegerin (OPG) expression in synovial tissue from patients with rheumatoid arthritis, spondyloarthropathies, osteoarthritis and normal controls. *Rheumatology* 2003; **43**:1–12.
65. D. R. Haynes, T. N. Crotti, M. Loric, et al., Osteoprotegerin and receptor activator of nuclear factor kappa B ligand (RANKL) regulate osteoclast formation by cells in the human rheumatoid arthritic joint. *J. Rheumatol.* 2001; **40**:623–630.
66. D. R. Haynes, T. N. Crotti, A. E. Potter, et al., The osteoclastogenic molecules RANKL and RANK are associated with periprosthetic osteolysis. *J. Bone Joint Surg. Br.* 2001; **83-B**:902–911.
67. D. Liu, J. K. Xu, L. Figliomeni, et al., Expression of RANKL and OPG mRNA in periodontal disease: possible involvement in bone destruction. *Int. J. Mol. Med.* 2003; **11**:17–21.
68. R. Assuma, T. Oates, D. Cochran, et al., IL-1 and TNF antagonists inhibit the inflammation response and bone loss in experimental periodontitis. *J. Immununol.* 1998; **160**:403–409.
69. D. R. Haynes, S. J. Hay, S. D. Rogers, et al., Regulation of bone cells by particle-activated mononuclear phagocytes. *J. Bone Joint Surg. Br.* 1997; **79-B**:988–994.
70. T. Y. Konttinen, H. Kurvinen, M. Takagi, et al., Interleukin-1 and collagenase around loose total hip prostheses. *Clin. Exp. Rheumatol.* 1996; **14**:255–262.
71. P. Kahle, J. G. Saal, K. Schaudt, et al., Determination of cytokines in synovial fluids: correlation with diagnosis and histomorphological characteristics of synovial tissue. *Ann. Rheum. Dis.* 1992; **51**:731–734.
72. J. J. Jandinski, P. Stashenko, and L. S. Feder, Localization of interleukin-1 beta in human periodontal tissues. *J. Periodontol.* 1991; **62**:36–43.
73. P. Stashenko, P. Fujiyoshi, M. S. Obernesser, et al., Levels of interleukin 1β in tissue from sites of active periodontal disease. *J. Clin. Periodontol.* 1991; **18**:548–554.
74. D. S. Preiss and J. Meyle, Interleukin-1 beta concentration of gingival crevicular fluid. *J. Periodontol.* 1994; **65**:423–428.
75. W. A. Jiranek, M. Machado, M. Jasty, et al., Production of cytokines around loosened cemented acetabular components. *J. Bone Joint Surg.* 1993; **75-A**:863–879.
76. S. D. Neale and N. A. Athanasou, Cytokine receptor profile of arthroplasty macrophages, foreign body giant cells and mature osteoclasts. *Acta Orthop. Scand.* 1999; **70**:452–458.
77. S. Kotake, K. Sato, K. J. Kim, et al., Interleukin-6 and soluble interleukin-6 receptors in the synovial fluids from rheumatoid arthritis patients are responsible for osteoclast-like cell formation. *J. Bone Miner. Res.* 1996; **11**:88–95.
78. S. Neale, A. Sabokbar, D. W. Howie, et al., Macrophage colony-stimulating factor and interleukin-6 release by periprosthetic cells stimulates osteoclast formation and bone resorption. *J. Orthop. Res.* 1999; **17**:686–694.
79. K. Yamazaki, T. Nakajima, E. Gemmell, et al., IL-4 and IL-6-producing cells in human periodontal disease tissue. *J. Oral Pathol. Med.* 1994; **23**:347–353.
80. M. Geivelis, D. W. Turner, E. D. Pederson, et al., Measurements of interleukin-6 in gingival crevicular fluid from adults with destructive periodontal disease. *J. Periodontol.* 1993; **64**:980–983.
81. S. Stea, Cytokines and osteolysis around total hip prostheses. *Cytokine* 2000; **12**:1575–1579.
82. K. Itoh, N. Udagawa, K. Kobayashi, et al., Lipopolysaccharide promotes the survival of osteoclasts via Toll-like receptor 4, but cytokine production of osteoclasts in response to lipopolysaccharide is different from that of macrophages. *J. Immunol.* 2003; **170**:3688–3695.
83. J. W. Xu, T.-F. Li, G. Partsch, et al., Interleukin-11 (IL-11) in aseptic loosening of total hip replacement (THR). *Scand. J. Rheumatol.* 1998; **27**:363–367.
84. H. Taki, E. Sugiyama, T. Mino, et al., Differential effects of indomethacin, dexamethasone, and interferon gamma (INF-γ) on IL-11 production by rheumatoid synovial cells. *Clin. Exp. Immunol.* 1998; **112**:133–138.
85. P. Trontzas, E. Kamper, F. A. Potamianou, et al., Comparative study of serum and synovial fluid interleukin-11 levels in patients with various arthridities. 1998; **31**(8):673–679.
86. M. Chabaud, J. M. Durand, N. Buchs, et al., Human interleukin-17: a T cell-derived proinflammatory cytokine produced by the rheumatoid synovium. *Arthritis Rheum.* 1999; **42**:963–970.
87. S. Kotake, N. Udagawa, N. Takahashi, et al., IL-17 in synovial fluids from patients with rheumatoid arthritis is a potent stimulator of osteoclastogenesis. *J. Clin. Invest.* 1999; **103**:1345–1352.
88. M. Chabaud, E. Lubberts, L. Joosten, et al., IL-17 derived juxta-articular bone and synovium contributes to joint degradation in rheumatoid arthritis. *Arthritis Res.* 2001; **3**:168–177.
89. S. Kotake, N. Udagawa, M. Hakoda, et al., Activated human T cells directly induce osteoclastogenesis from human monocytes. *Arthritis Rheum.* 2001; **44**:1003–1012.
90. B. Vernon-Roberts and M. A. R. Freeman, In: S. A. V. Swanson and M. A. F. Freeman eds., *The Scientific Basis of Joint Replacement*. Tunbridge Wells, Kent: Pitman Medical Publishing, 1977, pp. 112–121.
91. S. B. Goodman, P. Huie, Y. Song, et al., Cellular profile and cytokine production at prosthetic interfaces. Study of tissues retrieved from revised hip and knee replacements. *J. Bone Joint Surg.* 1998; **80-B**:531–539.
92. D. R. Bertolini, G. E. Nedwin, T. S. Bringman, et al., Stimulation of bone resorption and inhibition of bone formation *in vitro* by human tumour necrosis factors. *Nature* 1986; **319**:516–518.
93. G. Elliott, R. N. Maini, and M. Feldmann, Treatment of rheumatoid arthritis with chimeric monoclonal antibodies to tumor necrosis factor alpha. *Arthritis Rheum.* 1993; **36**:1681–1690.
94. J. Zwerina, S. Hayer, M. Tohidast-Akrad, et al., Single and combined inhibition of tumor necrosis factor, interleukin-1, and RANKL pathways in tumor necrosis factor-induced arthritis: effects on synovial inflammation, bone erosion, and cartilage destruction. *Arthritis Rheum.* 2004; **50**:277–290.
95. J. W. Xu, Y. T. Konttinen, J. Lassus, et al., Tumour necrosis factor-alpha (TNF-alpha) in loosening of total hip replacement (THR). *Clin. Exp. Rheumatol.* 1996; **14**:643–648.

96. D. R. Haynes, S. D. Rogers, S. Hay, et al., The differences in toxicity and release of bone-resorbing mediators induced by titanium and cobalt-chromium-alloy wear particles. *J. Bone Joint Surg. Am.* 1993; **75-A**:825–834.

97. M. J. Perry, F. Y. Mortuza, F. M. Ponsford, et al., Properties of tissue from around cemented joint implants with erosive and/or linear osteolysis. *J. Arthroplasty* 1997; **12**:670–676.

98. S. D. Neale, A. Sabokbar, Y. Fujikawa, et al., presented at the SIROT, Sydney, Australia, 1999.

99. J. M. W. Quinn, A. Sabokbar, M. Denne, et al., Inhibitory and stimulatory effects of prostaglandins on osteoclast differentiation. *Calcif. Tiss. Int.* 1997; **60**:63–70.

100. D. M. Anderson, R. MacQuarrie, C. Osinga, et al., Inhibition of leukotriene function can modulate particulate-induced changes in bone cell differentiation and activity. *J. Biomed. Mater. Res.* 2001; **58**:406–414.

101. E. Leibur, A. Tuhkanen, U. Pintson, et al., Prostaglandin E2 levels in blood plasma and in crevicular fluid of advanced periodontitis patients before and after surgical therapy. *Oral Dis.* 1999; **5**:223–228.

102. I. Takei, M. Takagi, H. Ida, et al., High macrophage-colony stimulating factor levels in synovial fluid of loose artificial hip joints. *J. Rheumatol.* 2000; **27**:894–899.

103. J. W. Xu, Y. T. Konttinen, V. Waris, et al., Macrophage-colony stimulating factor (M-CSF) is increased in the synovial-like membrane of the prosthetic tissues in the aseptic loosening of total hip replacement (THR). *Clin. Rheumatol.* 1997; **16**:244–248.

104. N. al-Saffar, H. A. Khwaja, Y. Kadoya, et al., Assessment of the role of GM-CSF in the cellular transformation and the development of erosive lesions around orthopedic implants. *Am. J. Clin. Pathol.* 1996; **105**:628–639.

105. T. Y. Konttinen, V. Waris, J. W. Xu, et al., Transforming growth factor-beta 1 and 2 in the synovial-like interface membrane between implant and bone in loosening of total hip arthroplasty. *J. Rheumatol.* 1997; **24**:694–701.

106. J. Yao, T. T. Glant, M. W. Lark, et al., The potential role of fibroblasts in periprosthetic osteolysis: fibroblast response to titanium particles. *J. Bone Miner. Res.* 1995; **10**:1417–1427.

107. M. Takagi, S. Santavirta, H. Ida, et al., Matrix metalloproteiases and tissue inhibitors of metalloproteinases in loose artificial hip joints. *Clin. Orthop.* 1998; **352**:35–45.

108. Y. Yoshihara, H. Nakamura, K. Obata, et al., Matrix metalloproteinases and tissue inhibitors of metalloproteinases in synovial fluids from patients with rheumatoid arthritis or osteoarthritis. *Ann. Rheum. Dis.* 2000; **59**:455–461.

109. H. Sakai, S. Jingushi, T. Shuto, et al., Fibroblasts from the inner granulation tissue from hips at revision arthroplasty induce osteoclast differentiation, as do stromal cells. *Ann. Rheum. Dis.* 2002; **61**:103–109.

110. T. Pap, A. Claus, S. Ohtsu, et al., Osteoclast-independent bone resorption by fibroblast-like cells. *Arthritis Res. Ther.* 2003; **5**:R163–R173.

111. T. R. Arnett, D. C. Gibbons, J. C. Utting, et al., Hypoxia is a major stimulator of osteoclast formation and bone resorption. *J. Cell Physiol.* 2003; **196**:2–8.

112. T. Arnett, Regulation of bone cell function by acid-base balance. *Proc. Nutr. Soc.* 2003; **62**:511–520.

113. G. M. Ferrier, A. McEvoy, C. E. Evans, et al., The effect of cyclic pressure on human monocyte-derived macrophages in vitro. *J. Bone Joint Surg. Br.* 2000; **82-B**:755–759.

114. K. Sakai, M. Mohtai, J. Shida, et al., Fluid shear stress increases interleukin-11 expression in human osteoblast-like cells; its role in osteoclast induction. *J. Bone Miner. Res.* 1999; **14**:2089–2098.

115. H. M. Van Der Vis, P. Aspenberg, D. K. Kleine, et al., Short periods of oscillating fluid pressure directed at a titanium-bone interface in rabbits lead to bone lysis. *Acta Orthop. Scand.* 1998; **69**:5–10.

116. H. Kanzaki, M. Chiba, S. Yoshinobu, et al., Periodontal ligament cells under mechanical stress induce osteoclastogenesis by receptor activator of nuclear factor κB ligand upregulation via prostaglandin E_2 synthesis. *J. Bone Miner. Res.* 2002; **17**:210–220.

117. B. Skoglund and P. Aspenberg, PMMA particles and pressure-a study of the osteolytic properties of two agents proposed to cause prosthetic loosening. *J. Orthop. Res.* 2003; **21**:196–201.

118. L. J. Melton, How many women have osteoporosis now? *J. Bone Miner. Res.* 1995; **10**:175–177.

119. R. P. Heaney, Is the paradigm shifting? *Bone* 2003; **33**:457–465.

120. B. L. Riggs and A. M. Parfitt, Drugs used to treat osteoporosis: the critical need for a uniform nomenclature based on their action on bone remodeling. *J. Bone Miner. Res.* 2005; **20**:177–184.

121. H. Tsangari, D. M. Findlay, J. S. Kuliwaba, G. J. Atkins, and N. L. Fazzalari, Increased expression of IL-6 and RANK mRNA in human trabecular bone from fragility fracture of the femoral neck. *Bone* 2004; **35**:334–342.

122. J. Pfeilschifter, R. Koditz, M. Pfohl, and H. Schatz, Changes in proinflammatory cytokine activity after menopause. *Endocr. Rev.* 2002; **23**:90–119.

123. F. R. Singer and G. D. Roodman, In: J. P. Bilezikian, L. G. Raisz, and G. A. Rodan eds., *Principles of Bone Biology*. San Diego, CA: Academic Press, 1996, pp. 969–978.

124. P. A. Revell, *Pathology of Bone*. New York: Springer-Verlag, 1986.

125. G. D. Roodman, Osteoclast function in Paget's disease and multiple myeloma. *Bone* 1995; **17**:57S–61S.

126. S. V. Reddy, N. Kurihara, C. Menaa, et al., Paget's disease of bone: a disease of the osteoclast. *Rev. Endocr. Metab. Dis.* 2001; **2**:195–201.

127. G. D. Roodman, Studies in Paget's disease and their relevance to oncology. *Semin. Oncol.* 2001; **28**:15–21.

128. P. Salmon, Loss of chaotic trabecular structure in OPG-deficient juvenile Paget's disease patients indicates a chaogenic role for OPG in nonlinear pattern formation of trabecular bone. *J. Bone Miner. Res.* 2004; **19**:695–702.

129. G. J. Atkins, D. R. Haynes, S. E. Graves, et al., Expression of osteoclast differentiation signals by stromal elements of giant cell tumors. *J. Bone Miner. Res.* 2000; **15**:640–649.

130. H. Morgan, A. Tumber, and P. A. Hill, Breast cancer cells induce osteoclast formation by stimulating host IL-11 production and downregulating granulocyte/macrophage colony-stimulating factor. *Int. J. Cancer* 2004; **109**:653–660.

131. A. N. Farrugia, G. J. Atkins, L. B. To, et al., Receptor activator of nuclear factor-kappaB ligand expression by human myeloma cells mediates osteoclast formation in vitro and correlates with bone destruction in vivo. *Cancer Res.* 2003; **63**:5438–5445.

132. J. J. Goater, R. J. O'Keefe, R. N. Rosier, et al., Efficacy of *ex vivo* OPG gene therapy in preventing wear debris induced osteolysis. *J. Orthop. Res.* 2002; **20**:169–173.

133. E. Romas, N. A. Sims, D. K. Hards, et al., Osteoprotegerin reduces osteoclast numbers and prevents bone erosion in

collagen-induced arthritis. *Am. J. Pathol.* 2002; **161**:1419–1427.

134. L. M. Childs, E. P. Paschalis, L. Xing, et al., In vivo RANK signaling blockade using the receptor activator of NF-kappaB:Fc effectively prevents and ameliorates wear debris-induced osteolysis via osteoclast depletion without inhibiting osteogenesis. *J. Bone Miner. Res.* 2002; **17**:192–199.

135. L. M. Childs, J. J. Goater, R. J. O'Keefe, et al., Efficacy of etanercept for wear debris-induced osteolysis. *J. Bone Miner. Res.* 2001; **16**:338–347.

136. P. J. Bekker, D. L. Holloway, A. S. Rasmussen, R. Murphy, S. W. Martin, P. T. Leese, G. B. Holmes, C. R. Dunstan, and A. M. DePaoli, A single-dose placebo-controlled study of AMG 162, a fully human monoclonal antibody to RANKL, in postmenopausal women. *J. Bone Miner. Res.* 2004; **19**:1059–1066.

137. M. Iwase, K. J. Kim, Y. Kobayashi, et al., A novel bisphosphonate inhibits inflammatory bone resorption in a rat osteolysis model with continuous infusion of polyethylene particles. *J. Orthop. Res.* 2002; **20**:499–505.

BRAIN FUNCTION, MAGNETIC RESONANCE IMAGING OF

KÂMIL UĞURBIL
WEI CHEN
NOAM HAREL
PIERRE-FRANCOIS VAN DE MOORTELE
ESSA YACOUB
XIAOHAN ZHU
University of Minnesota Medical School
Minneapolis, Minnesota

KAMIL ULUDAG
Max Plank Institute for Biological Cybernetics
Tübingen, Germany

1. INTRODUCTION

Most of our understanding of brain function derives from electrophysiology studies that use single and multiple unit recordings. This methodology is precise, but tedious, and the data produced are inherently undersampled because coverage over large areas of the cortex is impractical. Optical imaging techniques provide an alternative approach. However, these methods cannot penetrate more than ∼1 mm in depth on the exposed cortical surface when used for high resolution functional mapping and are restricted to a few millimeters with relatively course spatial resolution when performed noninvasively through the intact scalp and skull (1). Radioactively labeled glucose analog 2-[1-^{14}C]-deoxyglucose has been used in early seminal experiments on functional parcellation in the brain [e.g., (2)]; this method, however, is limited by the availability of differently labeled glucose analogs, and hence, the number of conditions that can be studied in a single animal. Of course, a major drawback with all of these methodologies is that they are invasive, or even terminal, therefore, unsuitable for human brain applications.

These limitations are the reasons why functional magnetic resonance imaging (fMRI), since its introduction (3–5), has come to play a dominant role in both human and animal model studies. Today, functional images in the brain can be obtained using the BOLD mechanism (3–5), measurement of cerebral blood flow (CBF) changes with arterial spin labeling (ASL) [e.g., (6–11) and references therein], intravoxel incoherent motion (12,13), and cerebral blood volume (CBV) alterations [e.g., (14–17)]. These methods have revolutionized our ability to study brain function, especially human brain function, which is endowed with unique capabilities that often cannot be studied in animal models. In this review, the authors have tried to provide a critical evaluation of most of these methods—their shortcomings, strengths, and promises. The critical evaluation is not meant, however, to detract from the accomplishments and current utility of these methods. Rather, it is provided with the view to future developments that can rectify some of the current limitations, recognizing that even the simplest and the most commonly used functional imaging study (i.e., gradient echo (GE) BOLD fMRI) performed at 1.5 Tesla magnetic field strength has been and will continue to be of immense utility.

2. BOLD-BASED FUNCTIONAL IMAGING OF THE BRAIN

The most commonly used fMRI approach, introduced in 1992 (3–5), is based on imaging regional deoxyhemoglobin perturbations that accompany modulations in neuronal activity. This contrast mechanism is referred to as blood oxygen level dependent (BOLD) contrast (18–21). It originates from the intravoxel magnetic field inhomogeneity induced by paramagnetic deoxyhemoglobin sequestered in red blood cells, which in turn are compartmentalized within the blood vessels. Magnetic susceptibility differences between the deoxyhemoglobin-containing compartments versus the surrounding space devoid of this strongly paramagnetic molecule generate magnetic field gradients across and near the boundaries of these compartments. Therefore, in images sensitized to BOLD contrast, signal intensities are altered if the regional deoxyhemoglobin content is changed. This occurs in the brain because of spatially specific metabolic and hemodynamic responses to enhanced neuronal activity. Regional cerebral blood flow (CBF) increases while oxygen consumption rate ($CMRO_2$) in the same area is elevated to a lesser degree, resulting in decreased extraction fraction and lower deoxyhemoglobin content per unit volume of brain tissue. Consequently, signal intensity in a BOLD sensitive image increases in regions of the brain engaged by a "task" relative to a resting, basal state.

2.1. Data Acquisition Considerations

In BOLD contrast, signal is acquired simply after a delay, t_e, following excitation. A refocusing pulse may or may not be applied during this delay. When refocusing pulses are not applied, one uses field gradient pulses to form an echo in imaging; consequently, such echoes are called "gradient-recalled" or "gradient" echoes, abbreviated simply as GE. This is the first technique introduced for functional mapping using the BOLD approach (3–5). In the presence of a refocusing pulse, the spin-echo (SE) signal (echo-amplitude) detected after the echo time TE decreases according to exp(-TE/T_2) where T_2 is the spin-lattice relaxation time associated with the decrease of magnetization in the transverse plane. Spin echoes can also yield functional information in the brain leading to SE based fMRI. SE fMRI is sensitive to only a subset of the processes that lead to BOLD contrast in the GE experiment, while everything that contributes to functional signals in SE fMRI also contribute to GE BOLD fMRI.

In a gradient echo, signal loss occurs through relaxation processes that contribute to T_2 as well as by signal cancellation that arises due to "dephasing" of the magnetization in the presence of magnetic field inhomogeneities. The latter is recoverable with a refocusing pulse and does not contribute to spin-echoes. In a gradient echo image, the appropriate relaxation constant of magnetization in the transverse plane is T_2^*. BOLD contrast refers to T_2^* or T_2 changes due to magnetic field inhomogeneities generated by magnetic susceptibility difference across red blood cells membranes and across the luminal boundaries of blood vessels. Oxygenated blood is diamagnetic and has similar properties as tissue. However, deoxyhemoglobin molecule is paramagnetic, and its presence leads to a large susceptibility difference between compartments that contain this molecule and others that are devoid of it.

During increased neuronal activity, deoxyhemoglobin content is altered in the brain; this change is accompanied with small but measurable decreases in T_2 and T_2^* and was experimentally documented in early days of fMRI (Fig. 1) (22). Provided that "noise" in fMRI data is dominated by the intrinsic thermal noise of MR images, it is easy to show that the optimum TE for detecting the signal intensity difference induced by this T_2 or T_2^* change in a spin- or gradient-echo signal is equal to T_2 or T_2^*, respectively (23). In the brain, T_2 and T_2^* are field dependent (Table 1) (24–27).

Image acquisition can be accomplished using "slices" where a frequency selective RF pulse is employed to restrict the signal origin to a single slice, typically 3 to 10 mm in thickness. Signal from the slice is spatially encoded in the two remaining orthogonal dimensions and acquired either by means of techniques that require multiple applications of the RF pulse [such as FLASH (28), and see reviews (29–31)], or by methods that complete the entire spatial encoding and acquisition for the slice in a

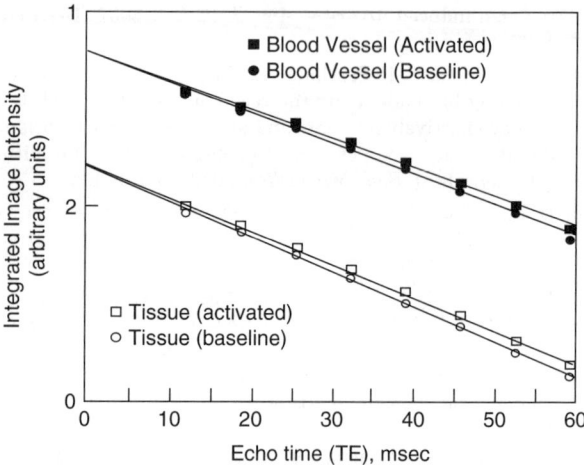

Figure 1. Change in T_2^* in the visual cortex for the human brain induced by visual activation. Log of signal intensity is plotted as a function of echo delay TE for a gradient recalled echo study of activation in the visual cortex. A linear relationship is seen as expected from the exponential signal decay with time constant T_2^*. The data are shown for the same voxels that were found to be "activated" during the visual stimulations study, both before and during the visual stimulation period. Some voxels were identified as originating from large venous vessels. Tissue areas had no visible blood vessels at resolution of vessel weighted MR images. A multi-echo sequence was used to collect the data for all echo times simultaneously. From Menon et al. (22).

single shot (such as Echo Planar imaging (EPI) or SPIRAL imaging [e.g., (32)]). Segmented versions of EPI and SPIRAL are also employed based on hardware restriction and/or resolution requirements; these segmented approaches rely on not a single but a few RF excitations, but significantly less than those employed in methods like FLASH to collect the data for a single slice [e.g., (33–35)].

In MR imaging terminology, k-space refers to the two or three-dimensional matrix of data points collected during image acquisition. When the image acquired is from a slice only, the k-space data reside in two dimensions, representing the encoding along two orthogonal directions that define a plane perpendicular to the slice selection direction. A 2-D Fourier transform then converts this into an image of the slice. In techniques like FLASH, one line along a single dimension of k-space is collected after each RF pulse. In single shot approaches, as the name implies, the entire k-space points are covered after a single RF pulse using multiple gradient echoes. Acquiring the image of a slice with single shot methods takes approximately ~20–100 msec depending on the available hardware. In contrast, techniques like FLASH typically take seconds to

Table 1. T_2 and T_2^* Values (ms) as a Function of Field Strength (24–27)

	White Matter		Gray Matter		Source
	T_2	T_2^*	T_2	T_2^*	
1.5 Tesla	74±5		87±2		25
4 Tesla	49.8±2.2		63±6.2		26
4 Tesla	57.9±3.8		67.1±6.0	41.4±5.5	25 (for T_2); 24 (for T_2^*)
7 Tesla	45.9±1.9		55.0±4.1	25.1±3.5	25 (for T_2); 24 (for T_2^*)

achieve the same task. Segmented EPI schemes are in between the single shot and FLASH approaches: They cover more than 1 line but less than the entire k-space in a single segment.

At higher magnetic fields, it is necessary to acquire single shot images faster, because the signal disappears faster subsequent to excitation (i.e., the T_2^* is shorter). The main reason for this is the increased magnetic field inhomogeneities generated in the sample as a result of compartments with different magnetic susceptibilities. These compartments can be blood vessels ranging in size from a few microns to several millimeters or significantly larger entities such as the air filled sinuses and ear cavities. These inhomogeneities are detrimental in single shot techniques but their deleterious influence can be minimized by rapid image acquisition. Thus, at 4 Tesla and 7 T, single shot images have been acquired so far in ~30 msec, limited in speed by gradient hardware. At 1.5 T, acquisition speeds in the range 50 to 100 msec are tolerable and are utilized.

Single shot techniques provide distinct advantages to the others. These advantages are the ability to acquire rapidly multiple slices to cover the whole head, and suppress image-to-image signal fluctuations. The latter advantage touches on a very interesting subject: namely, in consecutively acquired images from the brain, there exists image-to-image signal fluctuations that arise from physiologically induced processes such as blood vessel pulsation, vasomotion, and respiration (reviewed further on). Of course, the intrinsic signal-to-noise ratio (SNR) of a single image has to be sufficiently good (e.g., better than 50/1) so that these physiologically induced fluctuations can be detected. These image-to-image signal fluctuations come into single shot versus multiple shot acquisitions in a different way and are much more deleterious for the multiple shot techniques (36,37). This effect can be minimized by data processing strategies in multiple shot as well as in single shot techniques [e.g., (36–44)]. However, such strategies require intrinsically very good SNR and, for some, rapid image acquisition so as to capture the physiologically induced fluctuations accurately.

The main disadvantage of the single shot techniques is the difficulty of achieving high-resolution images, especially at high magnetic fields where the signal loss on the transverse plane is faster. Thus, columnar level functional imaging in humans (45) that requires at least 0.5 mm spatial resolution would be practically impossible to perform with single shot techniques. In this respect, multiple shot approaches or segmentation schemes are advantageous.

2.2. Mechanism of BOLD-Based Functional Imaging Signals

BOLD contrast reports on the deoxyhemoglobin content in the brain, which is determined by two parameters: (1) the deoxyhemoglobin concentration in blood and (2) the total amount of deoxyhemoglobin containing blood in a given volume of brain tissue. Deoxyhemoglobin concentration in blood is determined by $CMRO_2/CBF$ ratio (i.e., the rate at which oxygen is used compared to its delivery rate). The total amount of deoxyhemoglobin is determined by the product of deoxyhemoglobin concentration and blood volume in tissue. This aspect of the BOLD effect is therefore determined only by physiology.

Susceptibility gradients generated around blood vessels because of the presence of deoxyhemoglobin depends on vessel diameter. Consequently, the blood vessels play a critical role in generation of the MR detectable signals induced by alterations in neuronal activity. The nature of the MR detectable signals also depends on parameters other than physiology, namely magnetic field magnitude and the type of pulse sequence used.

The BOLD effect has two components: extravascular and intravascular (i.e., blood related). First, let us ignore the intravascular space and focus on the *extravascular space* only. When deoxyhemoglobin is present in a blood vessel, magnetic susceptibility of the space within the blood vessel is different than outside the blood vessel. This results in a homogeneous field within the blood vessel[1] but an inhomogeneous field outside the blood vessel. If one considers an infinite cylinder as an approximation for a blood vessel with magnetic susceptibility difference $\Delta \chi$ then the magnetic field expressed in angular frequency,[2] at any point in space, will be perturbed from the applied magnetic field ω_0 (46). Inside the cylinder, the perturbation, $\Delta \omega_B$ will be given by the equation

$$\Delta \omega_B^{in} = 2\pi \Delta \chi_O (1-Y) \omega_0 [\cos^2(\theta) - 1/3] \quad (1)$$

At any point outside the cylinder, the magnetic field will vary depending on the distance and orientation relative to the blood vessel and the external magnetic field direction, according to the equation:

$$\Delta \omega_B^{out} = 2\pi \Delta \chi_O (1-Y) \omega_0 [r_b/r]^2 \sin^2(\theta) \cos(2\phi). \quad (2)$$

In these equations, $\Delta \chi_O$ is the maximum susceptibility difference expected in the presence of fully deoxygenated blood, Y is the fraction of oxygenated blood present, r_b designates the cylinder radius, r is the distance from the point of interest to the center of the cylinder in the plane normal to the cylinder. The angles and the relevant distances are depicted in Fig. 2. Note that outside the cylinder, the magnetic field changes rapidly over a distance comparable to two or three times the cylinder radius; at a distance equal to the diameter of the cylinder from the cylinder center, $\Delta \omega_B^{out}$ is already down to 25% of its value at the cylinder boundary.

In a GE BOLD-based fMRI experiment, images are acquired after a delay TE in order to sensitize the image to

---

[1] In reality, this statement is correct strictly if we treat blood as a homogeneous medium with a magnetic susceptibility that differs from the surrounding tissue. However, within blood, deoxyhemoglobin is compartmentalized within red blood cells. This leads to magnetic field inhomogeneities around the red blood cells, thus to a non-uniform field distribution within the blood. However, the spatial scale of these inhomogeneities is sufficiently small that water diffusion in blood averages them out, effectively resulting in a uniform field.

[2] In the presence of a magnetic field B_o the angular "resonance" or Larmor frequency ω_o is given by $\omega_o = \gamma B_o$ where γ is the gyromagnetic ratio which is 2.6751965×10^4 rad s^{-1} G^{-1} for protons in a spherical water sample.

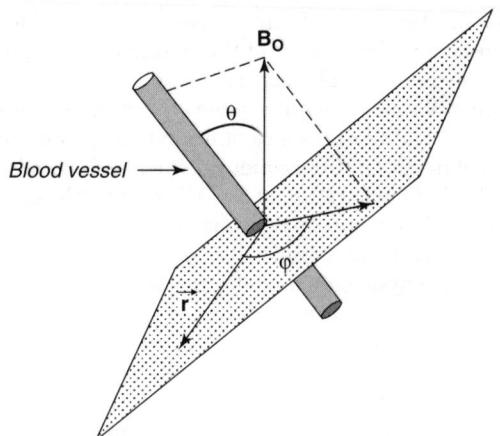

Figure 2. Diagram of a cylinder representing a blood vessel and the parameters that determine magnetic field at point outside of the cylinder when the susceptibility inside and outside the cylinder is not the same.

magnetic field inhomogeneities. In the SE approach, there is a single 180 degree "refocusing" pulse in the delay period TE. As previously discussed, intra-voxel magnetic field inhomogeneities will lead to signal loss during the evolution period TE. If the typical diffusion distances of water in tissue during the delay TE are comparable to the distances spanned by the magnetic field gradients within that tissue, then during this delay, the tissue spins will experience a time-averaged magnetic field in the extravascular space where the magnetic field gradients exist. This type of averaging due to motion of the spins is referred to as "dynamic" averaging. Dynamic averaging is what is detected in SE fMRI experiments. GE, on the other hand, contains additional contributions.

Typical TE values used in fMRI experiments depend on the field strength and the specifics of the pulse sequence, but in general range from ~ 15 to ~ 100 ms. Thus, blood vessel size compared to the diffusion distances in this ~ 15–100 ms time domain becomes a critical parameter in the BOLD effect. In this time scale, diffusion will dynamically average the gradients around small blood vessels (e.g., capillaries) that contain deoxyhemoglobin and, thus, result in a signal decay that will be characterized with a change in apparent T_2 (47–49). In a spin-echo experiment with a single refocusing pulse in the middle of the delay period (i.e., a Hahn Spin Echo), the phase accumulation that has taken place during the first half of the echo will not be reversed fully because the spins will not be able to trace back their trajectories exactly due to the presence of the diffusion. Of course, applying many refocusing pulses as in a Carr–Purcell pulse train or applying a large B_1 field (relative to the magnitude of the magnetic field inhomogeneity) for spin-locking during this delay will reduce or even eliminate this signal loss due to dynamic averaging. In a gradient echo measurement, dynamic averaging will also occur during the entire delay TE. If the imaging voxel contains only such small blood vessels at a density such that one-half the average distance between them is comparable to or less than diffusion distances (as is the case in the brain where capillaries are separated on the average by ~ 25 to $40\,\mu$m (50,51)[3], then the entire signal from the voxel will be affected by dynamic averaging.

In considering the movement of water molecules around blood vessels, we need not be concerned with the exchange that ultimately takes place between intra- and extravascular water across capillary walls. Typical lifetime of the water in capillaries exceeds 500 ms (52–54), significantly longer than the T_2 and T_2^* values in the brain tissue and longer than the period TE typically employed in fMRI studies.

For larger blood vessels, complete dynamic averaging for the entire voxel will not be possible. Instead, there will be "local" or "partial" dynamic averaging over a subsection of the volume spanned by the magnetic field gradients generated by the blood vessel. However, there will be signal loss from the voxel due to *static averaging* if refocusing pulses are not used or asymmetric spin echoes are employed. Following the excitation and rotation onto the plane transverse to the external magnetic field, the bulk magnetization vector of the nuclear spins will precess about the external magnetic field with the angular frequency $\omega_{B^{\text{out}}}$. A water molecule at a given point in space relative to the blood vessel will see a "locally" time-averaged $\omega_{B^{\text{out}}}$, $\overline{\omega}$, B^{out}, which will vary with proximity to the large blood vessel. Thus, signal in the voxel will then be described according to equation

$$S(t) = \sum_k s_{ok} e^{-TE/T_{2k}} (e^{-i\overline{\omega}_k TE}), \quad (3)$$

where the summation is performed over the parameter k, which designates small volume elements within the voxel; the time-averaged magnetic field experienced within these small volume elements is $\overline{\omega}_k$ in angular frequency units. Because $\overline{\omega}_k TE$ varies across the voxel, signal will be "dephased" and lost with increasing echo time TE. This signal loss occurs from "static averaging." In this domain, if the variation $\overline{\omega}_k$ over the voxel is relatively large and evenly distributed (e.g., gaussian distribution), signal decay can be approximated with a single exponential time constant T_2^*. In a spin-echo, the static dephasing will be refocused and thus eliminated.

Consistent with the discussion above, modeling studies (48) demonstrate that the relaxation of spins due to the presence of cylinders with a different susceptibility are given by:

$$R_2^* = 1/T_2^* = \alpha[\Delta\chi_O B_O(1-Y)]b_{vl}. \quad (4)$$

For large vessels (static dephasing regime). For small vessels only,

$$R_2^* = R_2 = 1/T_2 = \eta[\Delta\chi_O B_O(1-Y)]^2 b_{vs} p. \quad (5)$$

---

[3]Capillary density in the brain is not a constant and varies from region to region. In the cat visual cortex, the average distance between capillaries has been reported to be $\sim 25\mu$m. Pawlik G, Rackl A, Bing RJ. Quantitative capillary topography and blood flow in the cerbral cortex of cats: an *in vivo* microscopic study. Brain Res. 1981; 208(1)35–58.; in contrast in the human primary motor cortex it was reported to be $\sim 40\mu$m. Duvernoy HM, Delon S, Vannson JL, Cortical blood vessels of the human brain. Brain Res Bull 1981; 7(5):519–579.

All other relaxation mechanisms (e.g., dipole-dipole coupling) are ignored. In these equations, α and η are constants, is the external magnetic field, $\{\Delta\chi_O B_O(1-Y)\}$ is the frequency shift due to the susceptibility difference between the cylinder simulating the deoxyhemoglobin containing blood vessel and the space outside the cylinder, b_{vl} is the blood volume for *large* blood vessels (veins and venules with a radius greater than $\sim 5\,\mu m$ for 4 T) and b_{vs} is the *small* vessel blood volume (capillaries and small venules, less than $\sim 5\,\mu m$ in *radius* that permit dynamic averaging), and p is the fraction of active small vessels (i.e., filled with deoxyhemoglobin containing red blood cells).

Figure 3 illustrates R_2^* modeled for infinite cylinders in a voxel as a function of the voxel diameter and the difference in the susceptibility between the intra- and extra-cylinder volume expressed as frequency difference in Hz. When the cylinder radius is between 5 to 10 microns, a transition occurs in the dependence of R_2^* on the radius. At radii larger than this transition region, R_2^* is independent of the radius—this is the pure static averaging regime associated with large cylinders. For radii in the 10 micron or smaller range, the R_2^* decreases because the diffusion actually averages these field inhomogeneities, increasingly effectively at smaller radii.

An important prediction of modeling studies (47–49,55) is that the large and small vessel *extravascular* BOLD effects differ and the microvascular contribution varies supralinearly with the external magnetic field. In contrast, the dependence on the external magnetic field is linear for large blood vessels in the static averaging domain.

However, in considering the effect of magnetic fields on the BOLD mechanism, we must also consider purely blood effects (intravascular contribution) as opposed to extravascular contributions specified in Equation 4 and Equation 5. In the blood, hemoglobin is also compartmentalized within red blood cells. Thus, when the deoxy form is present, there are field gradients around the red cells. However, because the dimensions are very small compared to diffusion distances, the effect is dynamically averaged and becomes an "apparent" T_2 effect only. The dynamic averaging in this case also involves exchange across the red blood cell membrane that is highly permeable to water. The exchange is between two compartments, plasma and the interior of the red blood cell where the magnetic field is significantly different because of the presence of paramagnetic deoxyhemoglobin. Thus, in the presence of deoxyhemoglobin containing red blood cells, apparent T_2 of blood decreases and can be expressed as $1/T_2 = A_o + kB_o^2(1-Y)^2$ where A_o is a field independent term and k is a constant (see Ref. 56 and references therein). Therefore, the T_2 of *blood* will change when the content of deoxyhemoglobin is altered by elevated neuronal activity and this will lead to a signal change in the apparent T_2 or T_2^* weighted image. This effect will be present wherever the content of deoxyhemoglobin has changed, thus potentially both in large and small blood vessels.

2.3. Inflow Effects in BOLD-Based fMRI

Flow increases within large vessels supplying and draining the activated brain tissue is inevitable during the hemodynamic response that supplies more blood to the area of increased activity. This macrovascular flow change can lead to signal alterations in images intended to report on BOLD contrast [e.g., (22,57–60)]. This is not because BOLD contrast itself contains a direct flow effect; rather, image contrast may not be purely of BOLD origin because any repeated, slice-selective image is inherently flow sensitive if the signal within the slice does not attain full relaxation between consecutive signal excitations. In single slice studies, *allowing full relaxation in between RF pulses eliminates this problem completely*; however, this condition often is not satisfied in many studies since it leads to a loss in SNR per unit time. Consequently, such studies essentially obtain images of macrovascular flow rather than the much smaller BOLD changes. This problem was demonstrated with clarity by comparing presumably BOLD-based "functional" images with vessel images in two and three dimensions (58,61).

When full relaxation is allowed for each slice, a small macrovascular inflow problem may still be present in multislice studies due to inter slice effects. For example, blood experiencing an RF pulse in one particular slice will subsequently travel to other parts of the brain and, in turn, affect the intensity of signals from a slice sampled at a later time. The sensitivity to this problem will depend on the imaging sequence used, and how the different slices are sampled and the orientation of slices since most macrovascular flow occurs along a superior-anterior direction in the human brain. Experimental evidence, however, indicates that this type of inter-slice flow contribution to BOLD-based multislice imaging is negligible (62).

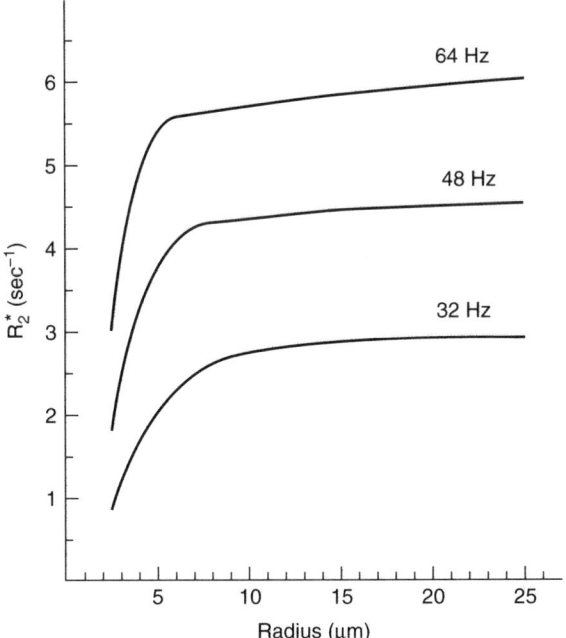

Figure 3. The susceptibility induced R_2^* in the presence of water diffusion plotted as a function of cylinder radius. Plots are shown at three different values of frequency shifts (32, 48, 64 Hz). Echo time = 40 msec, fractional "blood volume" (i.e., volume within cylinder relative to voxel volume) = 0.02. From Ogawa et al. (48).

3. PERFUSION IMAGING

BOLD-based images reflect a complex interplay between function-induced changes in oxygen consumption, blood flow and blood volume, with blood flow changes dominating the effect for the positive BOLD response. Blood-flow changes alone can be utilized to generate functional maps as well using magnetic resonance techniques, thereby simplifying the underlying mechanism involved in the functional image.

Images of perfusion or perfusion changes associated with increased neuronal activity can be obtained using ASL techniques that utilize the water protons in the blood as an endogenous "transient" tag. These methods rely on either continuous [e.g., (63–65)] [or dynamic (i.e., modulated) versions of continuous (66,67)] or pulsed [e.g., (7,68,69)] tagging approaches. All of these techniques benefit from increased T_1 encountered in higher magnetic fields. This is expected to ameliorate errors introduced by transit delays, extend coverage over the brain and yield higher CNR, and specificity to tissue. Excellent perfusion images based on continuous arterial spin tagging have already been accomplished at 3 Tesla (70) and this approach has also been used for functional mapping at 3 Tesla (71).

The tissue specificity of ASL can improve at high magnetic fields because of the fact that tagged spins require a finite amount of time to reach the capillaries and exchange with tissue water. At shorter periods, larger blood vessels in the arterial side can dominate the measurement, confounding quantitation of blood flow or blood flow changes [e.g., (72)], and appearing as "activated" in perfusion based functional images. In ASL measurements, generally the tag that can be detected in the veins subsequent to the tag's passage through the capillaries is thought to be negligible and is ignored, even though there has not been an experimental confirmation of this. At high fields, this potential contribution should vanish due to the short T_2 of venous blood.

4. SPATIAL SPECIFICITY OF BOLD AND PERFUSION-BASED FMRI

Although the neuro- and cognitive science communities have embraced fMRI with exuberance, caution must be exercised in quantitatively using fMRI data because fMRI maps are based on secondary metabolic and hemodynamic events that follow neuronal activity, and not the electrical activity itself. One of these issues is the spatial specificity of the fMRI maps (i.e., how accurate are the maps generated by fMRI compared to actual sites of neuronal activity?).

The ability to obtain accurate functional maps at the level of neuronal clusters that respond to a highly reduced attribute of the input is imperative if fMRI is to make contributions to brain science beyond its current capabilities. This is not just a question of mapping ocular orientation or dominance columns but has important implications for cognitive neurosciences. It is not known to what degree functional parcellation occurs in higher order visual areas or in other higher order cognitive functions. For example, clustering, analogous to the well-recognized columnar organization in the early visual areas has been proposed for some higher order visual areas [e.g., inferiotemporal (IT) cortex of monkey with respect to face and object recognition (73,74)] and could be a fundamental principle in the entire brain. Such fine functional parcellation of cortical territory and our inability to detect it easily in the human brain may in fact be the reasons for existing controversies.

In discussing the spatial specificity of this approach, we must distinguish single condition versus differential mapping. Differential mapping refers to functional images obtained by using two analogous but orthogonal activation states designed specifically to eliminate or suppress common signals; the functional image generated corresponds to a subtraction of the two conditions. In contrast, single condition mapping does not rely on a subtraction from a second orthogonal activated state. Mapping ocular dominance columns by stimulating one eye vs the other in an alternating fashion is differential mapping. Stimulating one eye and using a dark state as the control would correspond to single condition mapping. Even if the imaging signals are not specific enough to be confined to the territory of one column, a functional map of the columns can be obtained in the differential image if a given group of columns respond slightly differently to the stimulation of each eye. In addition, false (i.e., nonspecific) activation, often associated with draining veins, induced equally by stimulating the two different eyes would be eliminated in the difference. Phase encoded mapping of multiple areas using an activation paradigm that cycles through all possible stimulations (as employed in the visual system [e.g., (75–77)] is also a kind of differential mapping because it suppresses common, non modulating components.

Single condition mapping is significantly more demanding on the spatial accuracy of the imaging signals than differential imaging. However, ability to perform single condition mapping is important because it is not always feasible to have *a priori* knowledge of analogous but orthogonal activation conditions.

The degradation of spatial specificity might arise from imprecise spatial-coupling between neuronal activity and the physiological and metabolic events that ultimately yield the functional images. Thus, blood flow increases that accompany enhanced neuronal activity might exceed the boundaries of active neurons. Alternatively, or in addition, the lack of spatial specificity could originate from the fact that vasculature plays a crucial role in the generation of MR detectable signals from deoxyhemoglobin changes that accompany neuronal activity. In fact, it is well documented that large "draining" vessels do contribute to the T_2^* based BOLD signals (22).

4.1. Specificity of Perfusion Changes Coupled to Neuronal Activity

The two potential limitations on spatial specificity described above, in principle, equally apply to perfusion-based fMRI. Clearly, when the control of perfusion by the vasculature occurs at a coarser spatial scale than functional clustering, it will lead to spatial blurring in perfusion fMRI. In addition, when blood flow to a volume of tissue increases, flow in the large arterial and venous blood vessel that supply or drain blood from that particular volume of tissue must also increase due to conservation of mass. It is, therefore, not surprising that blood vessel dilation produced by neuronal activation propagates to more distant blood vessels (78). However, the perfusion measurement techniques by MR can be "tuned" to be selectively sensitive to capillary/tissue level flow and the large vessel effects can be minimized so as to yield accurate perfusion maps, i.e. maps that report on water delivery to the capillary bed and, by exchange across the capillary wall, to surrounding tissue.

In these MR techniques, a transient perturbation (the "label") is induced in the population of hydrogen nuclei (spins) of water in blood, outside the tissue of interest; this label is monitored as it shows up in the tissue of interest after a delay (the tagging time) that is long enough to permit arrival into the capillary and tissue but not long enough to reach a new equilibrium state. In this case, the amount of "label" detected in the tissue of interest is proportional to blood flow, and increases with elevated neuronal activity. Since arterial side is permeated with fresh blood significantly faster relative to the tissue, tagging times sufficiently long to allow the equilibrium state to be reached in the large arteries versus the tissue is different. Thus, long tagging times (~2 s) (72) eliminate arterial component from perfusion images. The long tagging times, however, may lead to the tagged spins appearing in the venous side, thus leading to false "activation" in the veins. At high magnetic fields, the spin-spin relaxation rate (T_2) of venous blood is very short (79,80) so that this effect can be selectively eliminated by a brief delay after excitations of spins but before image acquisition. Accordingly, perfusion-based fMRI maps have been shown to yield accurate images and co-localize with Mn^{+2} uptake (81), a marker of calcium dependent synaptic activity (82).

In the absence of large vessel contributions, perfusion-based fMRI can be used to examine the critical physiological question related to specificity of blood flow increases. Namely, are perfusion changes confined accurately to the region of increased neuronal activity in the spatial scale of columnar organizations? Such a study was performed recently using the iso-orientation columns in the cat visual system (8). It demonstrated that while perfusion increases that follow neuronal activation were not "perfectly" localized at the iso-orientation column level, the difference between active and neighboring inactive columns was large and permitted single condition mapping (Fig. 4) (8). This is a fundamentally important result for brain physiology and functional mapping because it demonstrates for the first time blood flow changes are regulated even at the level of iso-orientation columns and that these tissue level flow changes can be used to obtain maps with columnar specificity. This conclusion was initially considered to be in conflict with results from optical imaging techniques because CBV maps from such techniques were shown not to have columnar specificity. However, there is a very important difference between the optical imaging and MR perfusion imaging data. Namely, as discussed above, the MR methods can be set to eliminate the confounding problem coming from large vessel effects, which degrade spatial specificity. When the signal origin is restricted to the accurate capillary/tissue signals, columnar activation is detected by the MR method. However, the optical methods intrinsically do not have this selective "tuning" capability and report on all volume changes; thus they are in fact dominated by large vessel effects. Recent optical imaging studies conducted subsequent to the perfusion mapping of iso-orientation columns by MR, have confirmed this fundamental difference and demonstrated that when large vessel effects are taken out CBV based optical imaging techniques also yield columnar level mapping signals [e.g., (83–85)].

This perfusion based fMRI work by Duong et al. indicated for the first time that there must be regulation of blood flow control in the submillimeter (~300 to 400 μm) scale. Consistent with this results, the point spread

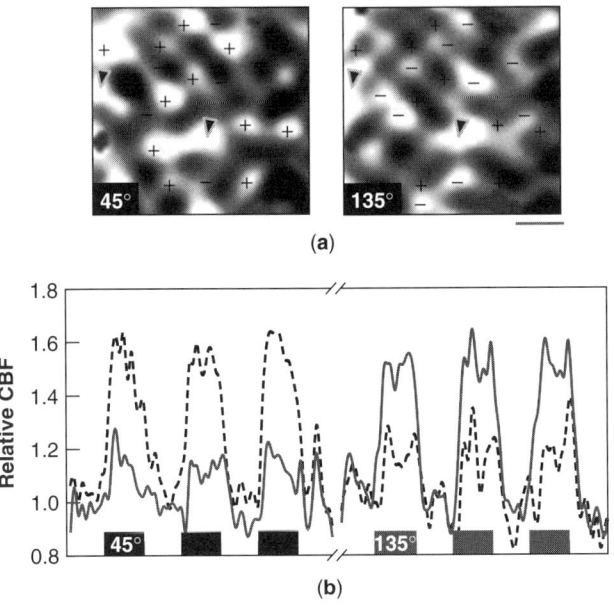

Figure 4. Activation maps of two orthogonal iso-orientation domains in the cat visual cortex obtained separately with perfusion based mapping. Panel a, shows perfusion based images obtained with two orthogonal orientations (45 and 135 degree gratings), demonstrating their complimentarity. Each map was acquired as a *single condition* map where the grating of one orientation was either moving back and forth (activation condition) or static (control condition). Bottom trace (Panel b) shows the blood flow (perfusion) response in all the voxels identified as "activated" either by the 45 degree or 135 degree gratings for the two orthogonal stimulations. All voxels identified as "activated" for 45 degree orientation display a large perfusion increase during stimulation by this orientation gratings and show a smaller but detectable perfusion increase in response to the 135 degree gratings. The opposite of this is observed for the voxels identified as activated for 135 degree gratings. A marked perfusion increase (~55%) following 45° or 135° *stimulus* was observed in the regions tuned to these orientations, while the stimulation with the orthogonal orientation lead to a 3.3 ± 0.6 fold smaller perfusion increase in the same region. Adapted from Ref. 8. (This figure is available in full color at http://www.mrw.interscience.wiley.com/ebe.)

function (PSF) of the CBF response to visual stimulation has been measured to be ~0.6 mm [full width half maximum (FWHM)] in the cat primary visual cortex (86). This number includes the neuronal contribution to the PSF as well. This PSF imposes a fundamental limitation on spatial specificity in all functional imaging techniques that directly [e.g., ASL or positron emission tomography (PET)] or indirectly (e.g., BOLD) rely on blood flow as the imaging signal. However, it also demonstrates that blood flow regulation does not exist only at the arteriolar level but capillary level regulation must occur. This follows simply from the fact that spatial separation of arterioles within the cortex is such that it cannot yield a PSF of 0.65 mm. Subsequently, CBV based fMRI studies, conducted using the extravascular contrast agent MION (see discussion further on), have also demonstrated single condition columnar resolution fMRI changes (87). Furthermore, with its superior sensitivity, CBV activation studies in animals have demonstrated that fMRI can detect and resolve laminar activity and that the larger intralaminar fMRI signal changes correspond to cortical layer 4 [(88,89) and references therein].

4.2. Hahn Spin Echo (HSE)-Based Bold Images

Virtually all current fMRI studies are carried out using GE (T_2^* weighted) BOLD technique. However, with the availability of high magnetic fields, there is increasing justification for considering HSE-based functional mapping for improved accuracy. BOLD-based functional maps with suppressed large vessel contribution can be obtained with Hahn Spin Echo approach at high but not low magnetic fields. HSE fMRI responds to apparent T_2 (as opposed to T_2^*) changes both in the extravascular space around microvasculature (48,90), and in blood itself (91–93). The former provides spatial specificity in the hundred micron spatial scale because capillaries are separated on the average by ~25 to ~40 μm (50,51) (see footnote 3). The blood effect, however, can be associated with large and small blood vessels and hence degrades spatial specificity of fMRI. However, the apparent T_2 of venous blood decreases quadratically with magnetic field magnitude (79) and is diminished from ~180 ms at 1.5 Tesla (94) to ~6 ms at 9.4 Tesla (80), significantly smaller than brain tissue T_2 and the TE values that would be used at such field strengths.

The *extravascular* HSE BOLD effect, which is microvascular in origin, is a small effect. As a result, it is reasonable to question whether it is at all detectable at any field strength or whether HSE BOLD signals originate predominantly from blood contribution. Both of these questions were specifically examined using separate experiments

Whether extravascular BOLD effect exists was examined using weak but detectable ^1H resonances of metabolites, such as N-acetyl aspartate (Fig. 5), that are

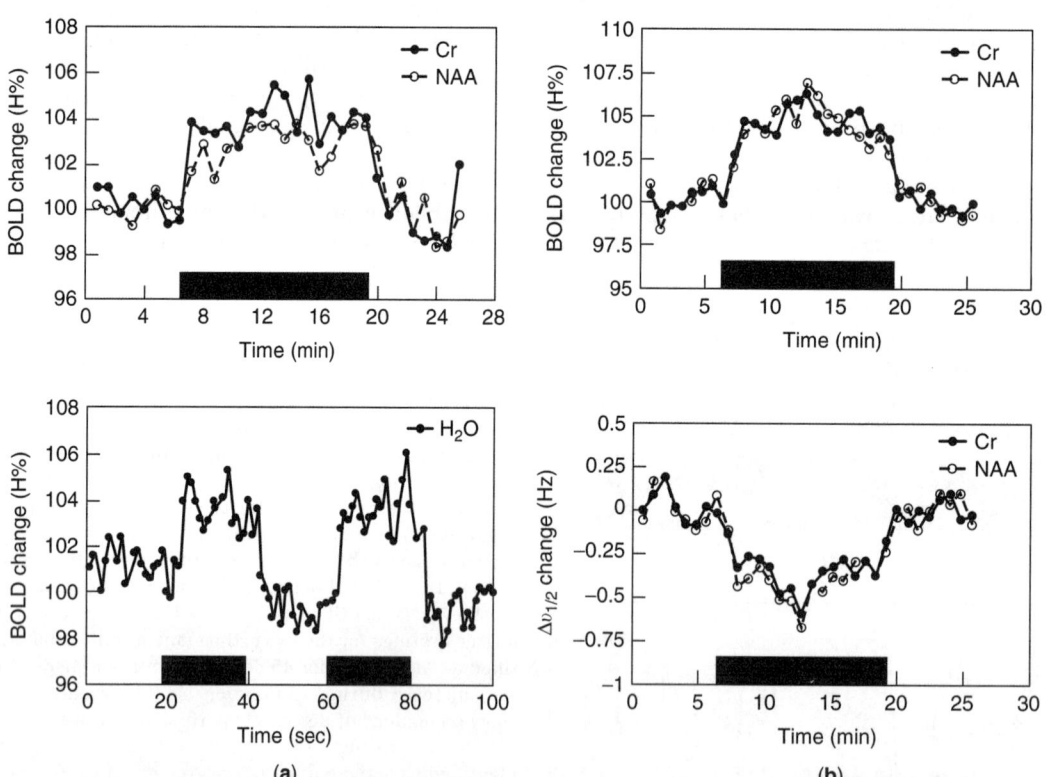

Figure 5. Demonstration of extracellular BOLD effect using intracellular metabolites. (a) Time courses of fMRS BOLD changes (%Peak Height) in the NAA and Cr metabolites (top) and water (bottom) from the same subject. (b) Time courses of BOLD and linewidth changes (top: %Height; bottom: % linewidth) in the NAA and Cr metabolites, respectively, from another subject. The dark bars indicate the task periods of visual stimulation (95).

sequestered intracellularly [95]. Any BOLD effect observed on such signals can only be *entirely* extravascular in origin. A BOLD effect on intracellular metabolites was clearly detectable at 4 and 7 Tesla, leading to the clear demonstration that *extravascular spin-echo* BOLD effects exist in the human brain.

The blood contribution to HSE BOLD fMRI was evaluated using Stejskal and Tanner gradient pair first introduced for diffusion measurements (96). This gradient pair, often referred to as "diffusion weighting" gradients, also suppresses blood significantly more than tissue spins in the brain. This is because the diffusion of blood water is much faster than in tissue, and the gradient pair dephases blood spins. The latter effect arises because flowing spins acquire a velocity dependent phase in the presence of these gradients. Flow rates are non-uniform within a blood vessel. Furthermore, the blood vessels change directions within a voxel, or there may be several different blood vessels with different flow rates and/or different orientations relative to the gradient directions. Since the blood signal detected from the voxel will be a sum of all of these, the net result can be signal cancellation due to dephasing of flowing spins (IVIM effect) (97).

The Stejskal-Tanner pulsed gradients can be used to distinguish between intra-and extra-vascular BOLD effects in functional images. Such experiments conducted at 1.5 Tesla have concluded that most of the BOLD-based signal increase during elevated neuronal activity is eliminated by bipolar gradients, leading to the conclusion that most of the fMRI signal at 1.5 Tesla arises from *intravascular* or blood related effects (98,99). At 3 Tesla, ~50% of the HSE functional signal changes has been attributed to blood using the same type of experiments (100).

The effect of the Stejskal–Tanner gradients on brain tissue signal intensity at 4 and 7 Tesla was examined and found to agree well with the modeling predictions (56). The dependence on the b value for diffusion-weighted spin-echo BOLD data, averaged for all subjects, are illustrated in Fig. 6 for an echo time of 32 ms for both at 4 and 7 T, and for echo times of 65 and 55 ms for 4 Tesla and 7 T, respectively (56). The 32 ms echo time is shorter than tissue T_2 at both fields. Blood contributions are expected to be echo time dependent and diminish with increasing echo time for TE values exceeding the blood T_2. At about 32 ms TE, blood contribution to 7 Tesla is expected to be minimal since T_2 of blood is short (24). However, for 4 T, where the T_2 of blood is ~20 ms, the blood is still expected to contribute. The 32 ms TE data in Fig. 6 emphasizes that at this echo time, signal changes associated with activation that are attributable to blood (and thus can be suppressed with diffusion gradients) are a small fraction of the total signal change at 7 T; in contrast, signal changes associated with activation at this echo time arise predominantly from blood at 4 Tesla. This would still be the case in gradient echo fMRI experiments where typical echo times employed would be approximately 30 ms. At the longer echo times, however, as expected, percent changes at both fields were only slightly reduced by the diffusion-weighting gradients. There is a relatively small but persistent reduction in stimulation-induced signal intensity change that is field independent; this is ascribed to the

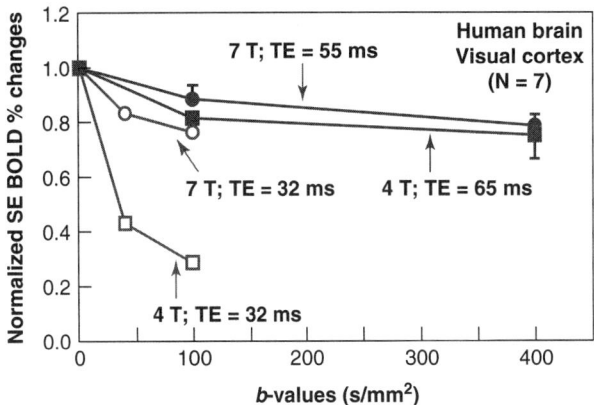

Figure 6. Experimental evaluation of blood contribution to 7 Tesla Hahn Spin Echo Functional Images: Figure shows normalized HSE BOLD percent changes as a function of b value for short and long echo times at 4 and 7 Tesla in a region of interest (ROI) defined in the b = 1 map. Percent changes were normalized to the BOLD change at $b = 1 \, s/mm^2$ for each subject and averaged for each field. Closed rectangles and circles indicate HSE data with TE of 65 ms at 4 Tesla (n = 7) and with TE of 55 ms at 7 Tesla (n = 4), respectively. Open rectangles and circles indicate HSE data with TE of 32 ms at 4 and 7 T, respectively, for 2 subjects at each field (six *repeated* measurements were made for each subject). Error bars are standard errors of the means. For long TE, the attenuation was not statistically different between 4 and 7 Tesla (P > 0.05) (56). (This figure is available in full color at http://www.mrw.interscience.wiley.com/ebe.)

elimination of so called "in-flow" effects that arise because full relaxation is not allowed between consecutive images.

At 9.4 Tesla, the effect of the Stejskal-Tanner gradients are similar to 7 Tesla results reported for the human brain above (80). In a Hahn spin-echo weighted fMRI study conducted in the rat brain (forepaw stimulation, symmetric spin-echo with one 180° pulse), we observed that the activation is not altered at all going from very small to very high b values (Fig. 7) (80).

Therefore, one can conclude that at these very high magnetic fields, unwanted blood contributions to the BOLD effect are virtually eliminated at typical TE values that correspond to gray matter T_2. This TE value would represent the optimum TE for spin-echo fMRI assuming that the "noise" in the spin-echo fMRI data (i.e., image-to-image fluctuations in signal intensity) is TE-independent; this certainly would be the case at very high resolution where intrinsic image SNR dominates the "noise" in the fMRI series and is also the case for HSE functional mapping even at low resolution (see discussion later on and Ref. 101). Thus, at these "optimal" echo times, both 7 Tesla and 4 Tesla human fMRI data as well as 9.4 Tesla rat data are equally devoid of blood-related degradations in specificity.

Even though both at 4 Tesla and 7 T undesirable blood contributions are suppressed at the optimum TE, there is still an advantage to the 7 Tesla over the 4 Tesla. This advantage is the substantially improved CNR of the HSE functional images, which increases approximately quadratically with field magnitude (Fig. 8) (25). This is expected based on modeling studies and was also verified

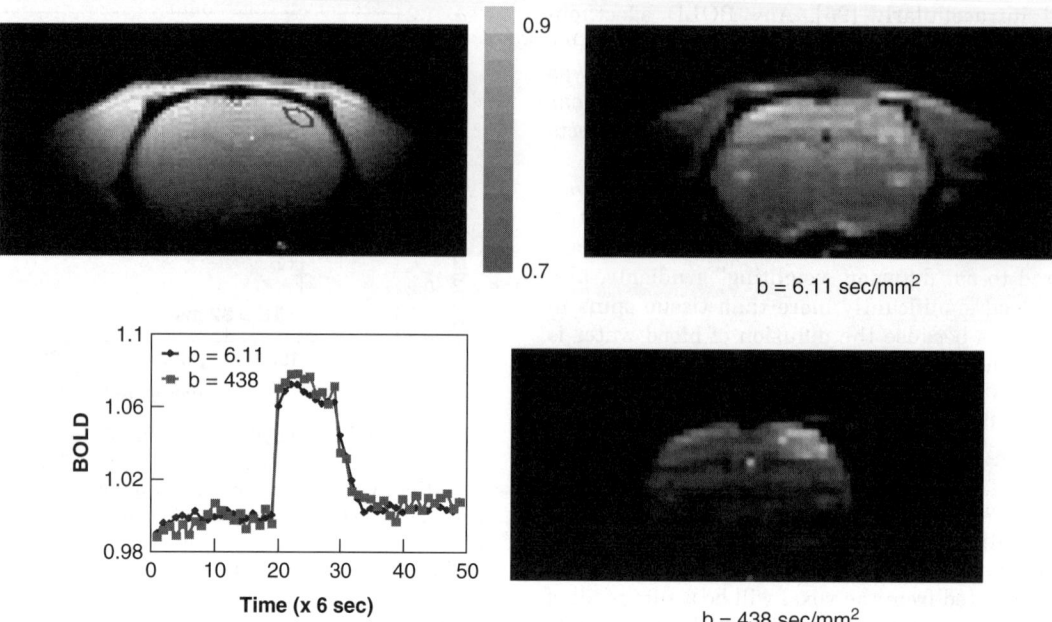

Figure 7. 9.4 Tesla diffusion-weighted spin-echo fMRI maps with b values of 6.1 (top right) and 438 s/mm^2 (bottom right) overlaid on one of the original, consecutively acquired EPI images (BOLD and diffusion weighted) collected during the functional imaging study. Coronal single-slice single-shot spin-echo EPI images of rat brain were acquired with a matrix size of 64×32, a FOV of 3.0×1.5 cm^2, a slice thickness of 2 mm, and TE = 30 msec. Somatosensory stimulation was used. Color bar indicates a maximum cross-correlation value from 0.7 to 0.9. Signal intensity (shown in background) was significantly reduced by bipolar gradients, as expected due to diffusion. Localized activation is observed at the somatosensory cortex in the contralateral side of a stimulated forepaw. Foci of activation site (color) agree very well in both fMRI maps. A Turbo FLASH image with a region of interest is shown in the upper left corner and time courses of diffusion-weighted images within the ROI are shown in the bottom left corner. If the macro-vascular contribution were significant, relative BOLD signal changes would decrease when a higher b value is used. However, relative signal changes remained the same in both images, suggesting that extravascular and micro-vascular components predominantly contribute to spin echo BOLD at 9.4 T (80). (This figure is available in full color at http://www.mrw.interscience.wiley.com/ebe.)

experimentally (25). This CNR enhancement is extremely important in view of the fact that CNR for HSE fMRI is low even at 4 Tesla. Note, however, that similarly long echo times are not useful for 1.5 Tesla for suppressing the blood effect because blood T_2 is equal to or exceeds tissue T_2. At that field magnitude, the blood contribution is substantial at TE values in the vicinity of tissue T_2 ($\sim$90 ms), and are increasing with TE (Fig. 9) until about 150 ms [(56) and references therein].

The experiments with the "diffusion" weighted gradients confirm the expectation that blood effects are negligible at high fields. However, experimentally, we are unable to judge at what level of the vascular tree we were able to eliminate the blood signals from the image using the diffusion gradients. Since the apparent T_2 of venous blood is so short, we expect that at all levels of the venous tree (venules to large veins), blood signals must be suppressed provided we use echo times that are about three fold larger than the apparent venous blood T_2. The intra-capillary blood, however, is unlikely to be fully suppressed. In the capillary, red blood cell density is non-uniform and oxygenation level varies from the arterial value to the venous value from one end of the capillary to the other end. However, this is *immaterial* as to whether capillary blood still contributes or not at high fields to functional maps; in either case, spatial resolution is dictated by the capillary distribution.

Thus, as in perfusion imaging, HSE BOLD techniques are expected to yield columnar level mapping. That they are capable of distinguishing among different layers has already been shown (25,88,102). Recent work measuring the point spread function (PSF) in the cat cortex have confirmed this by demonstrating that the full width at half maximum of the PSF is approximately the same for both CBF and HSE methods and less than a millimeter ($\sim$0.7 mm) [(86) and see discussion further on].

4.3. Gradient Echo (T2*) Bold Functional Imaging

Virtually all current fMRI studies are carried out using GE (T_2^* weighted) BOLD technique. All other fMRI methods are primarily employed by investigators interested in mechanisms and methodology development. GE BOLD fMRI, however, suffers from inaccuracies in functional mapping because of large vessel contributions. This is true at all magnetic fields because the extravascular

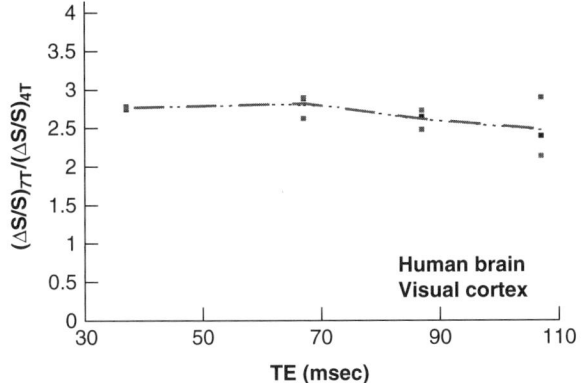

Figure 8. Magnitude of HSE functional imaging signals at 7 Tesla versus 4 Tesla, as a function of echo time. Data were obtained in the same subjects in the same slice at the two different fields. The data demonstrate approximately quadratic gains at 7 Tesla in the magnitude of the HSE fMRI signals (25). (This figure is available in full color at http://www.mrw.interscience.wiley.com/ebe.)

BOLD effect associated with large vessels is always present and increase linearly with magnetic field, as previously mentioned. However, the situation gets better even for this BOLD approach with higher magnetic fields.

The GE BOLD effect also comes from intravascular and extravascular sources. The intravascular effect originates from same mechanism as in the HSE BOLD, namely the T_2 changes induced by deoxyhemoglobin concentration perturbation when neuronal activity is altered. However, there is a second mechanism by which blood comes in to the GE BOLD effect (103,104). This mechanism is operative when blood occupies a large fraction of the volume of the voxel. When deoxyhemoglobin is present in the blood, the blood water will dynamically average the gradients surrounding the red blood cells and will behave as if it encounters a uniform magnetic field. This will differ from the magnetic field experienced by the rest of the voxel. In the immediate vicinity of the blood vessel, the magnetic field will vary and approach a constant value in tissue distant from the blood vessel. For simplicity, we can neglect the gradients near the blood vessel and consider the voxel to be composed of two large bulk magnetic moments, one associated with blood and the rest with the extravascular volume. These magnetic moments will precess at slightly different frequencies; therefore, the signal from the voxel will decrease with time as the two moments lose phase coherence. In this scenario, the signal can even oscillate as the phase between the two magnetic moments increase and then decrease. This mechanism cannot be operative when a voxel only contains capillaries since the blood volume is $\sim 2\%$ (50). This blood-related effect appears to be the main source of fMRI signals at 1.5 Tesla and explains the reason why there exists very large stimulation induced fractional changes at low magnetic fields (103,104). This effect is diminished and even eliminated at high fields as the blood signal contributions becomes smaller due to the short T_2. Note that, unlike the extravascular BOLD effect, these blood-related mechanisms do not require that ΔS is dependent on S (or equivalently, $\Delta S/S$ is independent of S), where S represents the voxel signal intensity and ΔS is the activity-induced signal change in the fMRI data.

As in GE fMRI, all blood related effects diminish dramatically with increasing magnetic field as blood T_2 decreases and blood signal vanish in a typical fMRI acquisition. However, unlike HSE, the extravascular BOLD effect for large vessels persists and continues to be a source of inaccurate functional mapping signals in GE. Nevertheless, at higher fields such as 7 Tesla, the microvascular contributions also become large enough so that its contributions is comparable to the large vessel effects (101). Figure 10a displays a histogram of the number of activated voxels, defined by a statistical threshold of $p < 0.05$, versus percent signal change for one subject, for three different runs performed on different days at 7 Tesla (101); the figure contains both HSE and GE fMRI data. An example of a GE image that was used to extract these data is illustrated in Fig. 10b. In each case, anatomical landmarks were used to aim for the selection of the same slice. The activation in the HSE data has a narrow distribution in percent signal change induced by visual stimulation, with only few voxels showing increases larger than 10%. This is consistent with the concept that a single blood vessel type contributes to these signal changes. In contrast,

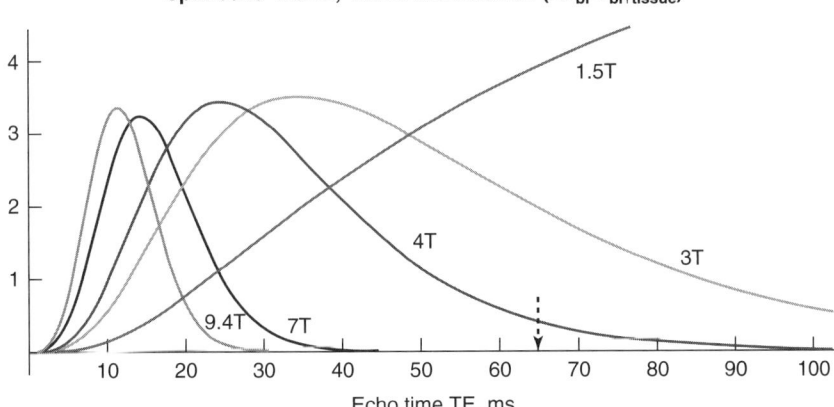

Figure 9. Simulation of the intravascular BOLD signal change, $\Delta S_{blood}/(S_{blood} + S_{tissue})$, as a function of echo time for a Hahn spin echo at 1.5, 3, 4, 7 and 9.4 T. We assumed a venous blood volume of 0.05, no stimulus-evoked blood volume change, and an increase in venous oxygenation level from 0.6 to 0.65 during stimulation. The T2 values of blood water at different field strengths and as a function of echo time were calculated from experimental data (see Ref. 56 for details) (56). (This figure is available in full color at http://www.mrw.interscience.wiley.com/ebe.)

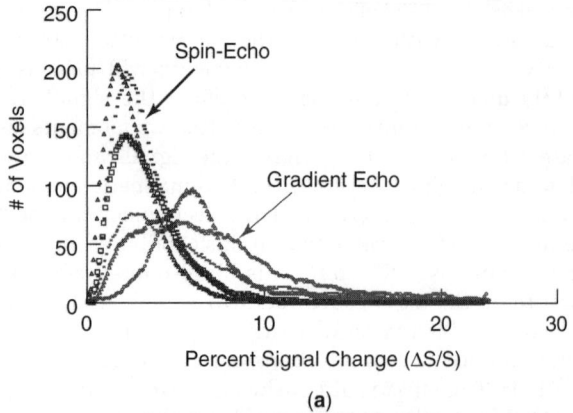

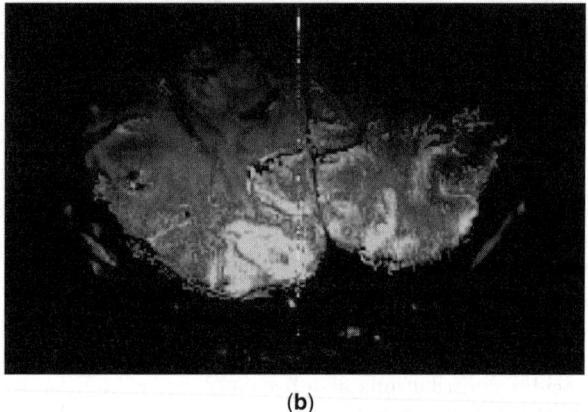

Figure 10. Stimulus-invoked percent signal change at 7 Tesla in the human visual cortex. (a) Histograms of percent changes of repeated (different days) HSE (in blue) and GE (in red) BOLD studies from the same subject, in the same anatomical location. (b) Percent signal change versus basal image signal intensity (normalized). The data were extracted from functional images obtained with $0.5 \times 0.5 \times 3$ mm^3 spatial resolution at 7 Tesla (b) (101). (This figure is available in full color at http://www.mrw.interscience.wiley.com/ebe.)

the GE BOLD measurements are characterized by a broad distribution of stimulus-induced percent changes. In the GE data, there exists a large concentration of voxels at small percent changes just as in HSE series; however, there are also a significant number of voxels displaying large signal changes ranging up to ~25%. This would be expected if different blood vessels, from capillaries to draining veins, are all contributing.

The presence of large vessel contribution even at 7 Tesla in T_2^* BOLD fMRI images in the human brain was also demonstrated by comparing perfusion based images in the human brain with T_2^* BOLD fMRI. Both because of the enhanced SNR and the long T_1, high fields provide advantages for perfusion based functional imaging. In high resolution perfusion images that were obtained at 7 Tesla (11), the activity was found to be confined to the gray matter ribbon; in contrast, the T_2^* BOLD fMRI images showed highest intensity changes not in the cortex but in the cerebral spinal fluid space (CSF) space in the sulci. Within the cortical gray matter, there are 50 to 100 micron veins separated by 1 to 1.5 millimeters that drain the capillaries. These are the blood vessels that can be seen in high resolution T_2^* weighted images as dark lines traversing the gray matter perpendicular to the cortical surface [e.g., (20)]. These small veins drain immediately into similarly sized or somewhat larger veins on the cortical surface. The draining "pial" veins located on the cortical surface are the likely source of intense but false activation seen within sulci in human GRE BOLD images at 7 Tesla (11).

Despite the presence of large vessel contributions, high resolution GE functional images at 7 Tesla look very different than high resolution GE functional images at lower fields because of the substantial microvascular contributions at the higher field strength. At lower fields, high resolution functional images look like venograms (58); only if the spatial resolution is reduced, the functional image becomes diffuse and activated areas *appear* to cover gray matter areas rather than depicting vessels (103). At higher fields, microvascular contribution increasingly becomes apparent. At 7 Tesla, the activity is diffuse and covers gray matter even at very high resolution (Fig. 10) (101) albeit together with contaminating signal from large vessels. At fields of 3 Tesla or above, if contrast-to-noise permits it would therefore useful to utilize methods that selectively suppresses large vessel contributions [e.g., (105,106)].

We should add that even with these inaccuracies, still a large number of questions about human cognition can be answered using GE fMRI even at 1.5 T. However, high-resolution images at the columnar level become impossible to attain by GE (T_2^*) BOLD fMRI at any field strength unless differential methods can be successfully applied. It was rigorously demonstrated that T_2^* BOLD fMRI method fails to generate *single-condition* functional images of iso-orientation domains in the cat cortex (107,108) at 4.7 Tesla (107). Although the image resolution was sufficiently good to detect columnar organizations associated with orientation preference in these studies, they were *not* observed. Furthermore, the images were not complementary when orthogonal orientations were used, and the highest "activity" was associated with a large draining vein, the sagittal sinus. Thus, spatial specificity in the submillimeter domain of iso-orientation columns does not exist in T_2^* BOLD fMRI. Additional limitations on the accuracy of T_2^* BOLD fMRI were provided by multiple site single unit recordings and fMRI studies on the same animal. These studies suggested that the limit of spatial specificity of T_2^* BOLD may be in the 2 to 3 mm range for *single-condition maps* (109,110). In this study, high-resolution T_2^* BOLD fMRI was performed together with single unit recording at multiple sites. The BOLD response was found to be proportional to the spiking rate when data from all recording sites was averaged and compared with the BOLD response from the entire area where it was deemed statistically significant. However, this correlation started breaking down when smaller and smaller areas were considered for the averaging of the two distinct measurements, in particular when the area started decreasing below approximately 4×4 mm^2. At 1.5 Tesla human brain studies using a phase-encoding method (77), the full width at half maximum of the pint spread function was estimated to be 3.5 mm.

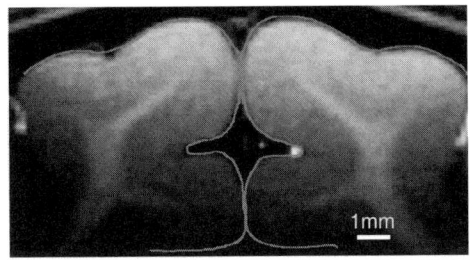
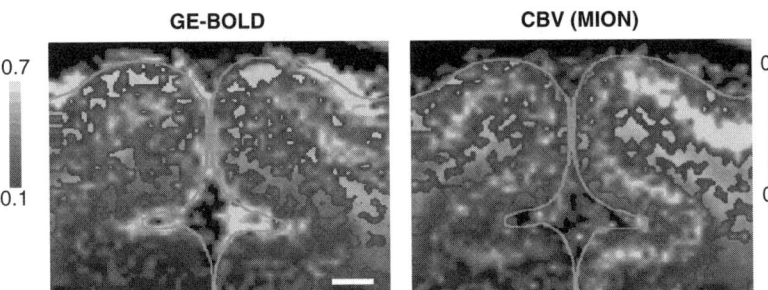

Figure 11. Improvement in the spatial specificity of fMRI signals can be obtain by using an intravascular iron oxide contrast agent sensitive to CBV changes. High-resolution fMRI images (0.15 × 0.15 × 2 mm) obtained in cat visual cortex before (i.e., BOLD contrast) and following the administration of MION (i.e., CBV-weighted, images are shown). Note that the largest BOLD signals changes are observed on the surface of the cortex where the cortical blood vessels are, while the largest CBV changes are located over the middle cortical layers, which have the highest neuronal activity increase (modified from Ref. 88). (This figure is available in full color at http://www.mrw.interscience.wiley.com/ebe.)

4.4. Functional Mapping With Initial "Dip"

In principle, it is possible to obtain functional images with better spatial specificity than perfusion if one can map the changes in cerebral oxygen consumption ($CMRO_2$). $CMRO_2$ increases must occur through increased activity of the mitochondria in the "activated" neurons, and the mitochondrial density is high in the synapse rich regions, predominantly in the dendrites (111). It might be possible to image $CMRO_2$ increases directly using the magnetic isotope of oxygen (^{17}O) [e.g., [112–116] and references therein].

$CMRO_2$ increases can also be mapped indirectly through its perturbations on the deoxyhemoglobin content. It has been shown that optical imaging of intrinsic signals in the cat and monkey visual cortex display a transient and small signal change ascribed to the deoxyhemoglobin increase before the onset a larger increase in blood flow (117–121). In this explanation, it is the lag in the blood flow response which provides the *temporal opportunity* to detect the deoxyhemoglobin increase that results from the faster response in of $CMRO_2$ elevation. The optical imaging data also showed that this deoxyhemoglobin increase yielded better maps of iso-orientation or ocular dominance columns than the signals associated with the hyperoxygenated state. Deoxyhemoglobin increase, however, can only be co-localized with elevated $CMRO_2$ at early time points after the onset of $CMRO_2$ elevation. Even if subsequent blood flow enhancement did not occur and the only response detected was a deoxyhemoglobin increase, this deoxyhemoglobin increase will co-localize with enhanced neuronal activity initially, but later will propagate down the vasculature and appear in the venous system as discussed previously.

The transient deoxyhemoglobin increase would yield a negative signal change in a BOLD weighted MR image. Such a "dip" has been observed in human fMRI experiments (122–128) and in animal models (107,108,129).

However, both in optical imaging and fMRI literature, the observation of this "dip" has been controversial. The source of this controversy in the early human fMRI studies has been the failure of several laboratories to detect this effect at 1.5 Tesla studies. However, this is a small effect at low magnetic fields but was shown to increase substantially with magnetic field magnitude (127) to yield easily detectable signals at 7 Tesla. Recently, in intra-operative studies in the human brain, the MR functional maps obtained from this early "dip" were also detected and shown to co-localize with electrical recordings, whereas T_2^* BOLD maps did not (128).

In anesthetized animal models, this small dip has not been highly reproducible presumably because of the perturbations of anesthesia on the animal physiology. When observed, it was found to yield accurate maps at the iso-orientation columnar level (107,108). However, this early negative response corresponding to deoxyhemoglobin increase is a very small effect. Furthermore, in order to obtain accurate maps from it, one must restrict the observation window to early time points of the early "dip" itself before deoxyhemoglobin disperses to areas distant from actual volume of activation due to blood flow. At these early time points of this transient "dip" pattern, the signals are even smaller. Thus, as a general high resolution MR imaging approach, it suffers from low contrast-to-noise.

5. "NOISE" IN FMRI DATA AND FIELD DEPENDENCE OF FMRI CNR

It is well recognized that the standard deviation of image-to-image fluctuations in an fMRI series (σ_{fMRI}) is usually not dominated by the thermal noise (σ_{Therm}) (37,57,130–133) that determines the intrinsic SNR of individual MR images. Instead, physiological processes contribute to σ_{fMRI} so that σ_{fMRI} can be expressed as $(\sigma_{Phys}^2 + \sigma_{Therm}^2)^{1/2}$, ignoring the potential contributions from instrument instabilities. These physiological processes include brain motion due to cardiac pulsation, magnetic field perturbations induced by alterations in lung volume, and temporal instabilities in blood flow, blood volume, erythrocyte density, and oxygen consumption caused by vascular events and/or spontaneous activities of neurons. These may be mediated through BOLD mechanisms but other mechanisms unrelated to BOLD are possible. Usually, respiration dependent contributions, and sometimes cardiac pulsation-induced instabilities, are suppressed by post-processing [e.g., (42,43) and references therein].

Recently, it has been postulated that the physiological component of the noise, σ_{Phys}, is linearly proportional to signal magnitude S (131–133) and this postulate was supported by experimental data obtained at 1.5 and 3 Tesla for GE-based BOLD fMRI. A linear dependence of σ_{Phys} on S is, of course, expected for physiological noise that is mediated through the BOLD mechanism; for example, as blood volume, blood flow and/or deoxyhemoglobin concentration in blood fluctuates, so will the basal BOLD effect. In fact, shortly after the introduction of fMRI, it was demonstrated that largest signal fluctuations in a GE fMRI time series occurs in and around large blood vessels, and that this can even be used as a mask to selectively eliminate large vessel regions in functional images (57). However, recent studies demonstrate that the physiological noise characteristics are somewhat different at 7 Tesla compared to 1.5 or 3 Tesla for GE fMRI (101). More interestingly, the noise in an HSE and GE fMRI time series at 7 Tesla also differ dramatically from each other (101); this is highly significant since one of the major benefits of the high fields is high specificity functional mapping using HSE BOLD effect. The results, however, demonstrate that HSE fMRI not only provides gains in accuracy of the functional imaging signals but also substantial gains with increasing magnetic field in contrast-to-noise ratio as well.

A linear dependence of σ_{Phys} on signal intensity S has important implications on the field dependence of CNR in functional imaging. CNR is equal to the activation-induced signal change in the fMRI time series (ΔS) divided by σ_{fMRI}. Given that the activation-induced changes in $1/T_2$ and $1/T_2^*$ are very small,[4] ΔS can be approximated as $(S_0 e^{-TE \cdot R}(\Delta R \cdot TE))$ and CNR can thus be expressed as

$$CNR = \Delta S / \sigma_{fMRI} = (S_0 e^{-TE \cdot R}(\Delta R \cdot TE)) / \sigma_{fMRI}, \quad (6)$$

where R is either R_2 or R_2^* (i.e., $1/T_2$ or $1/T_2^*$) for HSE or GE FMRI, respectively. We drop the subscript 2 on the R but it is implicit that we are talking about either R_2 or R_2^* depending on whether we are considering a HSE or a GE experiment. TE is the echo time, S_o is the signal intensity at TE = 0, and Δ represents the stimulus induced changes. Note that $S_0 e^{-TE \cdot R}$ is simply signal intensity S after the echo delay TE. For the *extravascular* BOLD effect, ΔR is calculated from Equation 4 and Equation 5, based on changes in fractional deoxyhemoglobin content in blood (1-Y), blood volume, and so on.

At the limit of very high-resolution functional imaging, for example, for mapping columnar organizations, image SNR decreases and we operate in a domain where σ_{Therm} dominates over σ_{Phys}; in this limit

$$CNR = \Delta S / \sigma_{Therm} \propto SNR_{io} \cdot e^{-TE \cdot R}(\Delta R \cdot TE). \quad (7)$$

In the above equation, SNR_{io} is the signal-to-noise ratio in the image corresponding to TE = 0. SNR_{io} increases approximately linearly field magnitude for protons at fields of 1.5 T or above (134). Based on this information and using Equation 4 and Equation 5 for ΔR, implications of Equation 7 for large vessels and microvasculature can be easily deduced as a function of TE for the *extravascular* BOLD effect. When Equation 7 is applicable, the *optimum* CNR is obtained for TE equal to T_2^* or T_2 for GE or HSE experiments, respectively[5], in which case, $e^{-TE \cdot R} = e^{-1}$. Brain tissue T_2 changes slowly with magnetic field (Table 1). Thus, provided we are not comparing extremely different field strengths such as 1.5 Tesla vs. 9.4 Tesla one can consider keeping the TE the same. Certainly the T_2 difference between 4 Tesla and 7 Tesla is small enough so the same TE can be used. For HSE fMRI at high fields, the large blood vessels do not contribute significantly and the functional signals originate predominately from the extravascular BOLD effect associated with microvasculature. Then, CNR for HSE fMRI will increase as the *cubed power* of the magnetic field magnitude, providing an enormous advantage in going to higher fields.

The situation is different for GE fMRI since T_2^* decreases approximately linearly with magnetic field and both large and small vessels contribute. Operating under conditions of TE = T_2^* means, then, that $\Delta R \cdot TE$ will be field independent for large blood vessels, and will increase as B_0 for microvasculature. Thus, under conditions where thermal noise dominates σ_{fMRI}, CNR will increase with magnetic field as SNR_{io} for large blood vessels and as $(B_0 \cdot SNR_{io})$ for microvasculature in the GE fMRI approach executed with optimal TE. Therefore, for high resolution imaging, higher magnetic fields provide major advantages for CNR even when GE fMRI is the method of choice. Not only CNR increases in a GE experiment, the CNR for microvasculature increases much more rapidly compared to the undesirable large vessels.

If we are in the limit where σ_{Phys} is much greater than σ_{Therm} (i.e., $\sigma_{fMRI} \sim \sigma_{Phys}$) then predictions of field dependence require that we define σ_{Phys} better. For BOLD-based noise one can write

$$\sigma_{Phys(BOLD)} = c \cdot S \cdot \sigma_{\Delta R} \cdot TE, \quad (8)$$

where c is a constant and S is the signal intensity $\sigma_{\Delta R}$ is the standart deviation accounting for the temporal variations in ΔR (133). For other non-BOLD sources, such as

---

[4]Signal difference between the activated and the basal state is $S_{activated} - S_{basal} = S_0(e^{-TE \cdot R_{activated}} - e^{-TE \cdot R_{basal}}) = S_0 e^{-TE \cdot R_{activated}}(1 - e^{-TE \cdot \Delta R})$ and for $TE \cdot \Delta R \ll 1$, one has the approximation, $e^{-TE \cdot \Delta R} \approx 1 - TE \cdot \Delta R$, where R is either R_2 or R_2^*.

[5]Optimum TE is calculated simply by solving the equation $\frac{\partial(CNR)}{\partial(TE)} = 0$ where CNR is given by Equation 4.

respiration and cardiac pulsation,

$$\sigma_{Phys(non-BOLD)} = c' \cdot \bar{S}, \quad (9)$$

where $\bar{S}$ is the time averaged signal intensity in the time series. The resultant physiological noise is given by $\sigma_{Phys} = (\sigma^2_{Phys(BOLD)} + \sigma^2_{Phys(non-BOLD)})^{1/2}$ (131). The CNR is calculated from Equation 6 with the substitution of σ_{fMRI} with σ_{Phys} and remembering that $S = S_0 e^{-TE \cdot \Delta R}$. For simplicity, the predictions for cases where physiological noise is only BOLD mediated or only non-BOLD mediated can be considered.

If the fMRI noise is exclusively BOLD mediated, in the limit $\sigma_{fMRI} \sim \sigma_{Phys}$, CNR is just a constant, *independent* of field magnitude, ΔR, and TE for either the HSE or GE experiment.[5] No need to buy high field magnets for functional images just to gain CNR in this case, although one can still argue a utility for high fields based on the accuracy of functional imaging signals. On the other hand, if the noise is purely of non-BOLD variety given by Equation 9, then CNR will be proportional to $\Delta R \cdot TE$, predicting an increasing CNR at longer TE values.[6] For the same TE, CNR will increase linearly for large blood vessels and quadratically for the microvasculature in a GE study. For a HSE experiment, quadratic increase in CNR with magnetic field magnitude will be applicable, since the contribution is predominantly microvascular. If we again use TE = T_2^* in the GE measurement so that we change the TE linearly with $1/B_o$, then CNR in a GE experiment will be field *independent* for *large* vessels, and increase linearly with field for voxels dominated by the microvasculature only.

Thus, when σ_{Phys} dominates, it is possible to be in a scenario where higher magnetic fields provide no gains in CNR in the context of the noise models considered so far. However, studies conducted at 7 Tesla demonstrated that fMRI "noise" in GE and HSE data are fundamentally different at this field strength, and that and unlike GE data, σ_{Phys} in 7 Tesla HSE fMRI was virtually independent of voxel volume and hence $\bar{S}$ (102) while functionally induced signal changes, ΔS were proportional to $\bar{S}$ under basal conditions. Note that independence of σ_{Phys} from $\bar{S}$ rules out the explicit formulation of BOLD and non-BOLD mechanisms given by Equation 8 and Equation 9. Although mechanisms considered by Equation 8 and Equation 9 remain valid for GE BOLD fMRI, they are not applicable to SE fMRI.

6. FMRI USING CEREBRAL BLOOD VOLUME

6.1. Exogenous Contrast Agents

Historically, the first human fMRI experiment was based on cerebral blood volume (CBV) measurement using an exogenous contrast agent (GdDTPA) (14). However, data acquisition had to be performed transiently during bolus passage of the contrast agent GdDTPA; hence the temporal resolution and repeatability of such an approach was limited. Since then, more advanced contrast agents that alter the T_2^* relaxation times, with better sensitivity to CBV, have been developed. These agents, are mainly composed of superparamagnetic particles of iron oxide, which have a blood half life time lasting up to several hours (135,136). Dextran coated Monocrystaline Iron Oxide Nanoparticles (MION) is an example of such a compound that is commercially available.

Following the intravenous injection of the contrast agent, the MR signal decreases due to the change in the magnetic susceptibility of vascular space, consequently magnetic field inhomogeneities are generated around blood vessels. During elevated neuronal activity, a local increase in CBV leads to an increase in the amount of contrast agent within a voxel, resulting in a further decrease in MRI signals. While changes in deoxyhemoglobin concentrations (i.e., BOLD) are concurrently present in a T_2^*-weighted sequence, the effects from the exogenous contrast agent, applied with a sufficient dose, will dominate, and the fMRI signal changes will be mainly weighted by CBV changes.

There are two main advantages to the use of an exogenous contrast agent for measuring CBV changes. First, there is an enhanced CNR when compared to conventional BOLD, provided sufficiently high doses can be used, and second, there is an increase in the spatial specificity of the mapping signals to the site of neuronal activity compared to GE BOLD.

The contrast agents at sufficiently high dose yield a higher CNR when compared to BOLD contrast due to the stronger intravascular versus tissue susceptibility difference induced by the particles relative to deoxyhemoglobin. The enhanced CNR, however, is heavily dependent on factors such as dose and echo time used. At high doses appropriate for animal model studies the net gain in CNR is about 5 times at low fields (1.5–2 T) (135) while at higher fields (3–4.7 T) the gain is approximately a factor of 3 relative to BOLD contrast (137). Thus, the gain in CNR when using an exogenous contrast agent is magnetic field dependent. The reason for this is that the BOLD contribution increases approximately supralinearly with the magnetic field (25,48) whereas the CBV change remains constant because the iron oxides are saturated at high fields (135). However, theoretical calculations and empirical data have shown that the benefits provided by exogenous agents will persist at fields as high as 9.4 Tesla and beyond (138).

The second advantage of the contrast agent is the increase in the spatial specificity of the MR signal compared with GE BOLD (Fig. 11). In a high-resolution fMRI study it has been shown that while the GE BOLD fMRI signal is mainly being detected at the cortical surface, a region containing large vessels, the CBV-weighted signals are centered over the supragranular and granular cortical layers (88,139). Due to the shortening of the T_2^* of blood following the injection of the contrast agent, voxels containing large vessels (i.e., larger amount of contrast agent) exhibit a larger signal reduction, when compared with

---

[6]One cannot conclude that TE can be increased indefinitely since at some TE, the condition $\sigma_{fMRI} \sim \sigma_{Phys}$ that leads to this prediction will fail, and thermal noise will dominate. But as long as $\sigma_{fMRI} \sim \sigma_{Phys}$ remains valid, longer TE's will yield higher CNR if the source of σ_{Phys} is non-BOLD type given by Equation 4.

tissue signals. The signal-to-noise ratio from these large vessel regions becomes so low to be undetectable after a relatively short echo time TE. As a result, the functional maps are "clean" from large vessel contamination even though blood volume changes occur in such large vessels (89,138). In contrast, large veins dominate the GE BOLD maps and are nonspecific to the site of neuronal activity. CBV-weighted signals have been used to image at the colunar resolution in the cat cortex (87).

With respect to the specificity of the CBV-weighted signal, another mechanism is working to reduce the vessels contribution in the functional maps as well. During activation, CBV increases leading to an increase in the content of MION in a voxel, and hence to a *negative* signal change. Countering this signal decrease is the BOLD contrast, which induces a positive signal change during activation. Therefore, the two mechanisms will compete. The effect is expected to be heavily spatially dependent. The larger BOLD effect is observed around large vessels, mainly on the surface region while functional signal changes from MION will be larger where the fractional blood volume changes are the largest. This may lead to a better cancellation of MION signals at the cortical surface, while the tissue region is less affected by this phenomenon.

MION has been used successfully in animal models such as the mouse, rat, cat, and monkey with very promising results. The advantages of the exogenous contrast agents should certainly be taken into account when high spatial and/or temporal fMRI studies are planned where CNR is a limiting factor. While exogenous CBV-weighted contrast agents exhibit obvious benefits over the currently available hemodynamic-based fMRI contrasts (e.g., BOLD, ASL), its use is currently limited to animal models due to the fact that the desired dose is not approved for human use.

6.2. Endogenous Contrast Agents

In recent years, a number of non-invasive techniques for measuring CBV have been developed that use blood as an endogenous contrast agent. These methods are based on nulling the blood signal using inversion recovery techniques.

The vascular space occupancy (VASO) method (17,140,141) is based on eliminating the blood signal in a manner that is independent of blood oxygenation and flow. Since the T_1 of blood is independent of oxygenation, VASO aims to null out blood in all compartments of the microvasculature. The remaining VASO signal is composed of brain parenchyma and CSF. As CBV increases, the VASO signal should decrease as parenchyma and CSF is displaced by blood in a given voxel. As a consequence, changes in CBV can be assessed through changes in the remaining extravascular water signal.

Similar to the MION signal, during an increase in CBV the VASO signal shows an inverse correlation with the stimulus paradigm (e.g., decreased MR signal, consistent with local vasodilatation). Although the VASO signal provides an indication of the dynamics of CBV, it does not provide any quantitative measurement or absolute values of CBV changes.

Another approach was introduced by Pike and colleague (142), who developed a new, noninvasive fMRI technique for direct quantification of venous CBV changes. Venous refocusing for volume estimation (VERVE) isolates the deoxygenated blood signal by exploiting the dependence of the transverse relaxation rate in deoxygenated blood on the refocusing interval. In an fMRI study, an increase of $16\pm2\%$ of venous CBV was measured in visual cortex on healthy young adults (142).

Thus, there are several approaches that use exogenous and endogenous contrasts for measuring blood volume changes using MRI techniques. Potentially most promising are the development of the exogenous contrast agents that can yield, as in animal studies, significant improvements in both the CNR and the spatial specificity over the conventional BOLD contrast, and which can be used in humans. Development and refinement of these agents for human use and ultimately for clinical applications will prove highly advantageous.

6.3. Mapping of Neuronal Currents Directly

The temporal and spatial resolution of BOLD and ASL fMRI is not only limited by the measurement technique but also by the coarseness of the hemodynamic response to the neuronal event. As discussed above, ongoing research is dedicated to bypass the limitations of the hemodynamics using clever experimental designs and analysis methods. In contrast, EEG and MEG are more directly related to neuronal current but lack spatial accuracy and have their own methodological limitations. At present, no single method can be used to probe brain function with mm spatial and ms temporal accuracy and therefore, it is highly desirable to measure neuronal currents directly with MRI (nc-MRI).

Recently, a few attempts have been made to measure nc-MRI on phantoms, cellcultures, healthy humans, and epileptic patients (143–150). Time varying currents inside the neurons and within the extracellular fluid induce small changes in magnetic field strength. If many neurons have the same orientation (like parallel-oriented pyramidal cells) and fire coherently then the concomitant magnetic and electric field changes can be measured noninvasively by MEG and EEG, respectively. The measurement squids are at a distance of 2 to 4 cm from the dipole source, and at this distance the magnetic field changes caused by evoked stimulus responses and spontaneous rhythmic activity (e.g., alpha oscillations) cover a range from 10^{-12} Tesla to 10^{-13} Tesla corresponding to an estimated dipole current of 10–100 nAm. Proximal to the dipole source, the corresponding magnetic field strength is approximately 10^{-9} T.

As in other magnetic field disturbances in fMRI (e.g., deoxyhemoglobin in BOLD fMRI) this leads to changes in intensity due to altered relaxation time and to a phase shift between the undisturbed and disturbed regions of the probe. Simulations have shown that phase is more sensitive to neuronal currents than intensity with a contrast-to-noise advantage of two orders (147). The only question is, Is this effect big enough to be measurable with current technology?

To simulate neuronal firing, Bodurka et al. induced an electric current of different magnitudes in a wire in a water-filled container and measured phase changes using a gradient-recalled echo (GE) sequence at 3 Tesla. They found that magnetic field strengths as small as $\Delta B_0 \sim 2 \cdot 10^{-9}$ Tesla could be detected (145). This value is close to the aforementioned estimated magnetic field strength change occurring in the brain during evoked or spontaneous activity. In a follow-up study, Bodurka and Bandettini compared the sensitivity of GRE and spin-echo (SE) sequences with detect the phase changes (144). In order to mimic physiological perturbations (e.g., respiration), a sinusodial varying current at 0.28 Hz was superimposed on to the constant current. Because SE refocuses large-scale and slow varying magnetic field gradients, SE should be more sensitive to detecting the constant current than GE. Indeed, SE turned out to be more sensitive to the constant current and less sensitive to the sinusoidal current and capable of detecting transient magnetic flux changes of 40 ms in duration and $2 \cdot 10^{-10}$ Tesla in strength.

Using simulation on dipoles of finite spatial extent, Konn et al. modeled the influence of size of the dipole, orientation and location relative to the measurement volume on detectability of neuronal currents (147). As expected, detectability is maximal for a perpendicular oriented dipole and additionally for the voxel adjacent to the dipole. This leads to the speculation that focal neuronal changes might be easier to detect than widespread activations. In summary, if proven, nc-MRI will have different sensitivities for detecting various types of neuronal currents.

To date, measurability of neuronal currents with fMRI in humans remains controversial. Xiong et al. tried to separate the slow hemodynamic signal from the fast signal caused by neuronal currents (146) using a rapid event-related paradigm with a stimulation rate of 2 s and acquiring images with different delays (-200, -100, 0, 100, 200 ms) to the stimulus onset in different runs. The rationale is to bring the BOLD signal into steady-state, and therefore any differences between the images at different delays can be attributed to neuronal currents. With this method, they found at 1.9 Tesla an nc-MRI signal change of 1% compared with baseline. However, this finding could not be reproduced by Chu et al. (149), who criticized that the experimental design in the former study was not optimal to detect neuronal currents. Fast BOLD signal changes evoked by the stimulus and signal fluctuations in different runs due to fatigue can lead to artifactual nc-MRI signals. To separate hemodynamics from neuronal currents with high sensitivity, Chu et al. (149) used two pseudorandom m-sequences at two different time scales (fast and slow). M-sequences are known to be efficient in determining the linear and nonlinear aspects of the fMRI impulse response (151). First, the BOLD impulse response containing both the hemodynamic and neuronal signal was determined using the slow m-sequence. Second, while only selecting the first 2000 ms the nc-MRI impulse response was determined with the fast m-sequence. With this method no nc-MRI was found although on the same subjects an MEG signal change during the first 2000 ms was present. The authors concluded that nc-MRI is at least one order of magnitude below the sensitivity of BOLD fMRI.

However, because spontaneous activity (e.g., alpha oscillations) and epileptic discharges can evoke neuronal currents one to two orders of magnitude larger than stimulus-evoked responses, the feasibility of nc-MRI in humans for these applications remains open. To test this, Konn et al. used eyes closed versus eyes open and visual stimulation as experimental conditions to cause large changes in alpha oscillations (152). Utilizing a short TR of 40 ms at 3 Tesla and a subsequent frequency analysis of the fMRI data, significant nc-MRI activation in 3 out of 4 subjects was observed. However, because similar activations were also found in CSF and blood vessels the authors suggest that further improvements in the methodology are needed before claiming detectability of neuronal currents by MRI. Using simultaneously acquired EEG-fMRI data in one epileptic patient, Liston et al. separated MRI changes related to neuronal current from the BOLD response utilizing generalized spike-wave discharges measured with EEG as a predictor instead of the standard hemodynamic response function (150). They could successfully show nc-MRI changes in the patient which only partially overlapped with active regions of the BOLD response.

In summary, nc-MRI as a new method might be very useful for both the study of healthy human brain function and in clinical epilepsy applications. For instance, presurgical mapping of epileptogenic foci by invasive intracortical EEG recordings could be replaced by the noninvasive nc-MRI method without loss in spatial and temporal resolution. Although in phantoms mimicked neuronal currents were measured reproducibly, an unassailable proof of detectability of neuronal currents in vivo still lacks despite the very promising latest results. Ongoing research has to optimize the stimulation paradigm, acquisition sequence, MRI parameters and analysis method in order to make nc-MRI more reliable (see Ref. 148). Also crucial will be the correction method for physiological noise, which was not employed in the aforementioned studies.

7. CONCLUSIONS

Our understanding of MR detectable functional signals in the brain has improved significantly over the years since the development of the methodology. At the same time, we have seen major advances and refinements in instrumentation, such as the introduction of ultrahigh fields, and methodologies that also substantially improve data quality. Some of these developments have not yet been incorporated into routine research or clinical use. However, we can expect this to happen in the near future resulting in further substantial advances in this field. In addition, intense research efforts continue to focus on new and novel methodologies. If history is a guide, the fantastically dynamic nature of the MR methodology, and the very large amount of effort committed to this field of research would predict that in a decade or two, MR-based functional imaging may be performed using totally different approaches and provide substantially better information than the already available rich techniques at the cutting edge today.

Acknowledgment

The part of the work reviewed here performed at the University of Minnesota was supported by NIH National Research Resource grant RR07089, the Keck Foundation and the MIND Institute.

BIBLIOGRAPHY

1. A. P. Gibson, J. C. Hebden, and S. R. Arridge, Recent advances in diffuse optical imaging. *Phys. Med. Biol.* 2005; **50**(4):R1–43.
2. C. Kennedy, M. Des Rosiers, L. Sokoloff, M. Reivich, and J. Jehle, The ocular dominance columns of the striate cortex as studied by the deoxyglucose method for measurement of local cerebral glucose utilization. *Trans. Am. Neurol. Assoc.* 1975; **100**:74–77.
3. S. Ogawa, D. W. Tank, R. Menon, J. M. Ellermann, S. G. Kim, H. Merkle, and K. Ugurbil, Intrinsic signal changes accompanying sensory stimulation: functional brain mapping with magnetic resonance imaging. *Proc. Natl. Acad. Sci. U. S. A.* 1992; **89**(13):5951–5955.
4. K. K. Kwong, J. W. Belliveau, D. A. Chesler, I. E. Goldberg, R. M. Weisskoff, B. P. Poncelet, D. N. Kennedy, B. E. Hoppel, M. S. Cohen, R. Turner, et al. Dynamic magnetic resonance imaging of human brain activity during primary sensory stimulation. *Proc. Natl. Acad. Sci. USA* 1992; **89**(12):5675–5679.
5. P. A. Bandettini, E. C. Wong, R. S. Hinks, R. S. Tikofsky, J. S. Hyde, Time course EPI of human brain function during task activation. *Magn. Reson. Med.* 1992; **25**(2):390–397.
6. J. A. Detre and J. Wang, Technical aspects and utility of fMRI using BOLD and ASL. *Clin. Neurophysiol.* 2002; **113**(5):621–634.
7. S.-G. Kim, Quantification of relative cerebral blood flow change by flow-sensitive alternating inversion recovery (FAIR) technique: application to functional mapping. *Magn. Reson. Med.* 1995; **34**:293–301.
8. T. Q. Duong, D. S. Kim, K. Ugurbil, and S. G. Kim, Localized cerebral blood flow response at submillimeter columnar resolution. *Proc. Natl. Acad. Sci. U S A.* 2001; **98**(19):10904–10909.
9. R. D. Hoge, J. Atkinson, B. Gill, G. R. Crelier, S. Marrett, and G. B. Pike, Linear coupling between cerebral blood flow and oxygen consumption in activated human cortex. *Proc. Natl. Acad. Sci. U S A* 1999; **96**(16):9403–9408.
10. A. Shmuel, E. Yacoub, J. Pfeuffer, P. F. Van de Moortele, G. Adriany, X. Hu, and K. Ugurbil, Sustained negative BOLD, blood flow and oxygen consumption response and its coupling to the positive response in the human brain. *Neuron* 2002; **36**(6):1195–1210.
11. J. Pfeuffer, G. Adriany, A. Shmuel, E. Yacoub, P. F. Van De Moortele, X. Hu, and K. Ugurbil, Perfusion-based high-resolution functional imaging in the human brain at 7 Tesla. *Magn. Reson. Med.* 2002; **47**(5):903–911.
12. A. W. Song and T. Li, Improved spatial localization based on flow-moment-nulled and intra-voxel incoherent motion-weighted fMRI. *NMR Biomed* 2003; **16**(3):137–143.
13. A. W. Song, T. Harshbarger, T. Li, K. H. Kim, K. Ugurbil, S. Mori, and D. S. Kim, Functional activation using apparent diffusion coefficient-dependent contrast allows better spatial localization to the neuronal activity: evidence using diffusion tensor imaging and fiber tracking. *Neuroimage* 2003; **20**(2):955–961.
14. J. W. Belliveau, D. N. Kennedy, R. C. McKinstry, B. R. Buchbinder, R. M. Weisskoff, M. S. Cohen, J. M. Vevea, T. J. Brady, and B. R. Rosen, Functional Mapping of the Human Visual Cortex by Magnetic Resonance Imaging. *Science* 1991; **254**:716–719.
15. J. Mandeville, J. Marota, J. Keltner, B. Kosovsky, and J. Burke, al e. CBV functional imaging in rat brain using iron oxide agent at steady state concentration. 1996; New York, New York. p 292.
16. S. G. Kim, K. Ugurbil, High-resolution functional magnetic resonance imaging of the animal brain. *Methods* 2003; **30**(1):28–41.
17. H. Lu, X. Golay, J. J. Pekar, and P. C. Van Zijl, Functional magnetic resonance imaging based on changes in vascular space occupancy. *Magn. Reson. Med.* 2003; **50**(2):263–274.
18. S. Ogawa, T-M. Lee, A. S. Nayak, and P. Glynn, Oxygenation-sensitive contrast in magnetic resonance image of rodent brain at high magnetic fields. *Magn. Reson. Med.* 1990; **14**:68–78.
19. S. Ogawa, T-M. Lee, A. R. Kay, and D. W. Tank, Brain Magnetic Resonance Imaging with Contrast Dependent on Blood Oxygenation. *Proc. Natl. Acad. Sci. USA.* 1990; **87**:9868–9872.
20. S. Ogawa and T. M. Lee, Magnetic Resonance Imaging of Blood Vessels at High Fields: in Vivo and in Vitro Measurments and Image Simulation. *Magn. Reson. Med.* 1990; **16**:9–18.
21. S. Ogawa, T. M. Lee, and B. Barrere, Sensitivity of magnetic resonance image signals of a rat brain to changes in the cerebral venous blood oxygenation. *Magn. Reson. Med.* 1993; **29**:205–210.
22. R. S. Menon, S. Ogawa, D. W. Tank, and K. Ugurbil, 4 Tesla gradient recalled echo characteristics of photic stimulation-induced signal changes in the human primary visual cortex. *Magn. Reson. Med.* 1993; **30**(3):380–386.
23. K. Ugurbil, M. Garwood, J. Ellermann, K. Hendrich, R. Hinke, X. Hu, S. G. Kim, R. Menon, H. Merkle, S. Ogawa, et al. Imaging at high magnetic fields: initial experiences at 4 T. *Magn. Reson. Q.* 1993; **9**(4):259–277.
24. E. Yacoub, A. Shmuel, J. Pfeuffer, P. F. Van De Moortele, G. Adriany, P. Andersen, J. T. Vaughan, H. Merkle, K. Ugurbil, and X. Hu, Imaging brain function in humans at 7 Tesla. *Magn. Reson. Med.* 2001; **45**(4):588–594.
25. (a) E. Yacoub, T. Q. Duong, P. F. Van De Moortele, M. Lindquist, G. Adriany, S. G. Kim, F. W. Ugurbil. *J. Comput. Assist. Tomogr.* 1984; **8**(3):369–380. (b) F. Wehril et al., *J. Comput. Assist. Tomogr.* 1984; **8**(3):369–380.
26. P. Jezzard, et al. *Radiology* 1996; **199**(3):773–779.
27. A. Hasse, J. Frahm, D. Matthaei, W. Hainicke, and K. D. Merboldt, FLASH imaging. Rapid NMR imaging using low flip angle pulses. *J. Mang. Reson.* 1986; **67**:258–266.
28. D. Chien and R. R. Edelman, Ultrafast imaging using gradient echoes. *Magn. Reson. Quarterly.* 1991; **5**(1):31–56.
29. K. Hu, X. Spin-echo fMRI in humans using high spatial resolutions and high magnetic fields. *Magn. Reson. Med.* 2003; **49**(4):655–664.
30. E. M. Haacke, J. A. Tkach, and M. R. Fast, imaging: techniques and clinical applications. *AJR Am. J. Roentgenol.* 1990; **155**(5):951–964.

31. E. M. Haacke, P. A. Wieloposki, J. A. Tkash, A comprehensive technical review of short TR, fast, magnetic resonance imaging. *Magn. Reson. Med.* 1991; **3**(2):53–170.

32. Y. Yang, G. H. Glover, P. van Gelderen, A. C. Patel, V. S. Mattay, J. A. Frank, and J. H. Duyn, A comparison of fast MR scan techniques for cerebral activation studies at 1.5 tesla [published erratum appears in Magn Reson Med 1998 Mar;39(3):following 505]. *Magn. Reson. Med.* 1998; **39**(1):61–67.

33. S. G. Kim, X. Hu, G. Adriany, and K. Ugurbil, Fast interleaved echo-planar imaging with navigator: high resolution anatomic and functional images at 4 Tesla. *Magn. Reson. Med.* 1996; **35**(6):895–902.

34. R. S. Menon, C. G. Thomas, J. S. Gati, Investigation of BOLD contrast in fMRI using multi-shot EPI. *NMR Biomed.* 1997; **10**(4-5):179–182.

35. G. H. Glover and S. Lai, Self-navigated spiral fMRI: interleaved versus single-shot. *Magn. Reson. Med.* 1998; **39**(3):361–368.

36. X. Hu, T. H. Le, T. Parrish, and P. Erhard, Retrospective estimation and correction of physiological fluctuation in functional MRI. *Magn. Reson. Med.* 1995; **34**(2):201–212.

37. X. Hu and S.-G. Kim, Reduction of Physiological Noise in Functional MRI using Navigator Echo. *Mag. Reson. Med.* 1994; **31**:495–503.

38. P. P. Mitra, D. J. Thompson, S. Ogawa, X. Hu, and K. Ugurbil, Spatio-temporal Patterns in fMRI Data Revealed by Principal Component Analysis and Subsequent Low Pass Filtering. 1995; Nice, France. p 817.

39. P. P. Mitra, S. Ogawa, X. Hu, and K. Ugurbil, The nature of spatio-temporal changes in cerebral hemodynamics as manifested in functional magnetic resonance imaging. *Magn. Reson. Med.* 1997; **37**(4):511–518.

40. S. Ogawa, P. P. Mitra, X. Hu, and K. Ugurbil, Spatio-temporal patterns revealed in denoised fMRI Data. Visualization of information processing in the human brain:. In: Hashimoto I, Okada YC, Ogawa S, editors. Recent advances in MEG and functional MRI: EEG Suppl 47; 1996. p 5–14.

41. B. Biswal, E. A. DeYoe, and J. S. Hyde, Reduction of physiological flucuations in fMRI using digital filters. *Magn. Reson. Med.* 1996; **35**:117–123.

42. J. Pfeuffer, P. F. Van De Moortele, K. Ugurbil, X. Hu, and G. H. Glover, Correction of physiologically induced global off-resonance effects in dynamic echo-planar and spiral functional imaging. *Magn. Reson. Med.* 2002; **47**(2):344–353.

43. P. F. Van De Moortele, J. Pfeuffer, G. H. Glover, K. Ugurbil, and X. Hu, Respiration-induced B0 fluctuations and their spatial distribution in the human brain at 7 Tesla. *Magn. Reson. Med.* 2002; **47**(5):888–895.

44. G. H. Glover, T. Q. Li, and D. Ress, Image-based method for retrospective correction of physiological motion effects in fMRI: RETROICOR. *Magn. Reson. Med.* 2000; **44**(1):162–167.

45. R. S. Menon, S. Ogawa, J. P. Strupp, and K. Ugurbil, Ocular dominance in human V1 demonstrated by functional magnetic resonance imaging. *J. Neurophysiol.* 1997; **77**(5):2780–2787.

46. C. S. Springer and Y. Xu, Aspects of Bulk Magnetic Susceptibility in in vivo MRI and MRS. In: Rink PA, Muller RN, eds., *New Developments in Contrast Agent Research*. Blonay, Switzerland: European Magnetic Resonance Forum; 1991. pp. 13–25.

47. J. L. Boxerman, L. M. Hamberg, B. R. Rosen, and R. M. Weisskoff, MR contrast due to intravscular magnetic susceptibility perturbations. *Magn. Reson. Med.* 1995; **34**:555–556.

48. S. Ogawa, R. S. Menon, D. W. Tank, S.-G. Kim, H. Merkle, J. M. Ellermann, and K. Ugurbil, Functional Brain Mapping by Blood Oxygenation Level-Dependent Contrast Magnetic Resonance Imaging. *Biophys J.* 1993; **64**:800–812.

49. R. P. Kennan, J. Zhong, and J. C. Gore, Intravascular susceptibility contrast mechanisms in tissue. *Magn. Reson. Med.* 1994; **31**:9–31.

50. G. Pawlik, A. Rackl, and R. J. Bing, Quantitative capillary topography and blood flow in the cerebral cortex of cats: an in vivo microscopic study. *Brain. Res.* 1981; **208**(1):35–58.

51. H. M. Duvernoy, S. Delon, and J. L. Vannson, Cortical blood vessels of the human brain. *Brain. Res. Bull.* 1981; **7**(5):519–579.

52. J. O. Eichling, M. E. Raichle, R. L. Grubb, and M. M. Ter-Pogossian, Evidence of the limitations of water as a freely diffusable tracer in brain of the Rheusus monkey. *Circ. Res.* 1974; **35**(3):358–364.

53. O. B. Paulson, M. M. Hertz, T. G. Bolwig, and N. A. Lassen, Filtration and diffusion of water across the blood-brain barrier in man. *Microvasc. Res.* 1977; **13**(1):113–124.

54. O. B. Paulson, M. M. Hertz, T. G. Bolwig, N. A. Lassen, Water filtration and diffusion across the blood brain barrier in man. *Acta. Neurol. Scand. Suppl.* 1977; **64**:492–493.

55. N. Fujita, Extravascular contribution of blood oxygenation level-dependent signal changes: a numerical analysis based on a vascular network model. *Magn. Reson. Med.* 2001; **46**(4):723–734.

56. T. Q. Duong, E. Yacoub, G. Adriany, X. Hu, K. Ugurbil, and S. G. Kim, Microvascular BOLD contribution at 4 and 7 T in the human brain: Gradient-echo and spin-echo fMRI with suppression of blood effects. *Magn. Reson. Med.* 2003; **49**(6):1019–1027.

57. S.-G. Kim, K. Hendrich, X. Hu, H. Merkle, and K. Ugurbil, Potential pitfalls of functional MRI using conventional gradient-recalled echo techniques. *NMR in Biomed.* 1994; **7**(1/2):69–74.

58. C. Segebarth, V. Belle, C. Delon, R. Massarelli, J. Decety, J.-F. Le Bas, M. Decorpts, and A. L. Benabid, Functional MRI of the human brain: Predominance of signals from extracerebral veins. *NeuroReport* 1994; **5**:813–816.

59. J. Frahm, K.-D. Merboldt, W. Hanicke, A. Kleinschmidt, and H. Boecker, Brain or vein-oxygenation or flow? On signal physiology in functional MRI of human brain activation. *NMR in Biomed.* 1994; **7**(1/2):45–53.

60. J. H. Duyn, C. T. W. Moonen, G. H. Van Yperen, R. W. De Boer, and P. R. Luyten, Inflow versus deoxyhemoglobin effects in BOLD functional MRI using gradient echoes at 1.5 T. *NMR in Biomed.* 1994; **7**(1/2):83–88.

61. V. Belle, C. Delon-Martin, I. R. Massarell, J. Decety, J. Le Bas, A. Benabid, and C. Segebarth, Intracranial gradient-echo and spin-echo functional MR angiography in humans. *Radiology* 1995; **195**(3):739–746.

62. A. M. Howseman, S. G. Grootoonk, D. Porter, J. Ramdeen, A. P. Holmes, and R. Turner, The effect of slice order and thickness on fMRI activation data using multislice EPI. *NeuroImage* 1999; **9**(4):363–376.

63. J. A. Detre, J. S. Leigh, D. S. Williams, and A. P. Koretsky, Perfusion imaging. *Magn. Reson. Med.* 1992; **23**(1):37–45.

64. W. Zhang, D. S. Williams, and A. P. Koretsky, Measurement of rat brain perfusion by NMR using spin labeling of arterial water: in vivo determination of the degree of spin labeling. *Magn. Reson. Med.* 1993; **29**(3):416–421.

65. J. A. Detre, W. Zhang, D. A. Roberts, A. C. Silva, D. S. Williams, D. J. Grandis, A. P. Koretsky, and J. S. Leigh, Tissue

specific perfusion imaging using arterial spin labeling. *NMR Biomed.* 1994; **7**(1-2):75–82.
66. E. L. Barbier, A. C. Silva, H. J. Kim, D. S. Williams, and A. P. Koretsky, Perfusion analysis using dynamic arterial spin labeling (DASL). *Magn. Reson. Med.* 1999; **41**(2):299–308.
67. E. L. Barbier, A. C. Silva, S. G. Kim, and A. P. Koretsky, Perfusion imaging using dynamic arterial spin labeling (DASL). *Magn. Reson. Med.* 2001; **45**(6):1021–1029.
68. R. E. Edelman, B. Siewer, D. G. Darby, V. Thangaraj, A. C. Nobre, M. M. Mesulam, and S. Warach, Quantitative mapping of cerebral blood flow and functional localization with echo-planar MR imaging and signal targeting with alternating radio frequency. *Radiology* 1994; **192**:513–520.
69. E. C. Wong, R. B. Buxton, and L. R. Frank, Quantitative imaging of perfusion using a single subtraction (QUIPSS and QUIPSS II). *Magn. Reson. Med.* 1998; **39**(5):702–708.
70. S. L. Talagala, F. Q. Ye, P. J. Ledden, and S. Chesnick, Whole-brain 3D perfusion MRI at 3.0 T using CASL with a separate labeling coil. *Magn. Reson. Med.* 2004; **52**(1):131–140.
71. T. Mildner, R. Trampel, H. E. Moller, A. Schafer, C. J. Wiggins, D. G. Norris, Functional perfusion imaging using continuous arterial spin labeling with separate labeling and imaging coils at 3 T. *Magn. Reson. Med.* 2003; **49**(5):791–795.
72. N. V. Tsekos, F. Zhang, H. Merkle, M. Nagayama, C. Iadecola, and S. G. Kim, Quantitative measurements of cerebral blood flow in rats using the FAIR technique: correlation with previous iodoantipyrine autoradiographic studies. *Magn. Reson. Med.* 1998; **39**(4):564–573.
73. G. Wang, K. Tanaka, and M. Tanifuji, Optical imaging of functional organization in the monkey inferotemporal cortex. *Science* 1996; **272**:1665–1668.
74. I. Fujita, K. Tanaka, M. Ito, and K. Cheng, Columns for visual features of objects in monkey inferotemporal cortex. *Nature* 1992; **360**:343–346.
75. M. I. Sereno, A. M. Dale, J. B. Reppas, K. K. Kwong, J. W. Belliveau, T. J. Brady, B. R. Rosen, and R. B. H. Tootell, Borders of Multiple Visual Areas in Humans Revealed by Functional Magnetic Resonance Imaging. *Science* 1995; **268**:889–893.
76. E. A. DeYoe, G. J. Carman, P. Bandettini, S. Glickman, J. Wieser, R. Cox, D. Miller, and J. Neitz, Mapping striate and extrastriate visual areas in human cerebral cortex. *Proc. Natl. Acad. Sci. U S A* 1996; **93**(6):2382–2386.
77. S. A. Engel, G. H. Glover, and B. A. Wandell, Retinotopic organization in human visual cortex and the spatial precision of functional MRI. *Cereb. Cortex.* 1997; **7**(2):181–192.
78. C. Iadecola, G. Yang, T. J. Ebner, and G. Chen, Local and propagated vascular responses evoked by focal synaptic activity in cerebellar cortex. *J. Neurophysiol.* 1997; **78**(2):651–659.
79. K. R. Thulborn, J. C. Waterton, P. M. Matthews, and G. K. Radda, Oxygenation dependence of the transverse relaxation time of water protons in whole blood at high field. *Biochim. Biophys. Acta.* 1982; **714**(2):265–270.
80. S.-P. Lee, A. C. Silva, and K. Ugurbil, S.-G. K. Diffusion weighted spin echo fMRI at 9.4 T: Microvascular/Tissue contribution to BOLD signal changes. *Mag. Reson. Med.* 1999; **42**(5):919–928.
81. T. Q. Duong, A. C. Silva, S. P. Lee, and S. G. Kim, Functional MRI of calcium-dependent synaptic activity: cross correlation with CBF and BOLD measurements [In Process Citation]. *Magn. Reson. Med.* 2000; **43**(3):383–392.
82. Y. J. Lin and A. P. Koretsky, Manganese ion enhances T1-weighted MRI during brain activation: an approach to direct imaging of brain function. *Magn. Reson. Med.* 1997; **38**(3):378–388.
83. S. Sheth, M. Nemoto, M. Guiou, M. Walker, N. Pouratian, and A. W. Toga, Evaluation of coupling between optical intrinsic signals and neuronal activity in rat somatosensory cortex. *Neuroimage* 2003; **19**(3):884–894.
84. R. V. Harrison, N. Harel, J. Panesar, and R. J. Mount, Blood capillary distribution correlates with hemodynamic-based functional imaging in cerebral cortex. *Cereb. Cortex* 2002; **12**(3):225–233.
85. I. Vanzetta, H. Slovin, D. B. Omer, and A. Grinvald, Columnar resolution of blood volume and oximetry functional maps in the behaving monkey; implications for FMRI. *Neuron* 2004; **42**(5):843–854.
86. J. C. Park, I. Ronen, D.-S. Kim, and K. Ugurbil, Spatial Specificity of High Resolution GE BOLD and CBF fMRI in the Cat Visual Cortex. *Proc. Intl. Soc. Mag. Reson. Med.* 2004; **12**: 10–14.
87. F. Zhao, P. Wang, K. Hendrich, and S. G. Kim, Spatial specificity of cerebral blood volume-weighted fMRI responses at columnar resolution. *Neuroimage* 2005; **27**(2):416–424.
88. N. Harel, J. Lin, S. Moeller, K. Ugurbil, and E. Yacoub, Combined imaging-histological study of cortical laminar specificity of fMRI signals. *Neuroimage* 2006 Feb 1;**29**(3): 879–887.
89. H. Lu, S. Patel, F. Luo, S. J. Li, C. J. Hillard, B. D. Ward, and J. S. Hyde, Spatial correlations of laminar BOLD and CBV responses to rat whisker stimulation with neuronal activity localized by Fos expression. *Magn. Reson. Med.* 2004; **52**(5):1060–1068.
90. J. L. Boxerman, R. M. Weisskoff, B. E. Hoppel, and B. R. Rosen, MR contrast due to microscopically heterogeneous magnetic susceptibilty: cylindrical geometry. 1993; New York, NY. *Int. Soc. of Magn. Reson. Med.* p 389.
91. K. Ugurbil, X. Hu, W. Chen, X.-H. Zhu, S.-G. Kim, and A. Georgopoulos, Functional Mapping in the human brain using high magnetic fields. *Philos. Trans. R. Soc. Lond. B. Biol. Sci.* 1999; **354**(1387):1195–1213.
92. P. C. van Zijl, S. M. Eleff, J. A. Ulatowski, J. M. Oja, A. M. Ulug, R. J. Traystman, and R. A. Kauppinen, Quantitative assessment of blood flow, blood volume and blood oxygenation effects in functional magnetic resonance imaging [see comments]. *Nat. Med.* 1998; **4**(2):159–167.
93. K. Ugurbil, G. Adriany, P. Andersen, W. Chen, R. Gruetter, X. Hu, H. Merkle, D. S. Kim, S. G. Kim, J. Strupp, X. H. Zhu, and S. Ogawa, Magnetic resonance studies of brain function and neurochemistry. *Annu. Rev. Biomed. Eng.* 2000; **2**:633–660.
94. M. Barth, E. Moser, Proton NMR relaxation times of human blood samples at 1.5 T and implications for functional MRI. *Cell. Mol. Biol. (Noisy-le-grand).* 1997; **43**(5):783–791.
95. X. H. Zhu and W. Chen, Observed BOLD effects on cerebral metabolite resonances in human visual cortex during visual stimulation: a functional (1)H MRS study at 4 T. *Magn. Reson. Med.* 2001; **46**(5):841–847.
96. E. O. Stejskal and J. E. Tanner, Spin diffusion measuremnts: spin-echoes in the presence of a time dependent field gradient. *J. Chem. Phys.* 1965; **42**:288–292.
97. D. Le Bihan, E. Breton, D. Lallemand, P. Grenier, E. Cabanis, and M. Laval-Jeantet, MR imaging of intravoxel incoherent motions: application to diffusion and perfusion in neurologic disorders. *Radiology* 1986; **161**(2):401–407.
98. A. W. Song, E. C. Wong, S. G. Tan, and J. S. Hyde, Diffusion weighted fMRI at 1.5 T. *Magn. Reson. Med.* 1996; **35**(2):155–158.

99. J. L. Boxerman, P. A. Bandettini, K. K. Kwong, J. R. Baker, T. L. Davis, B. R. Rosen, R. and M. Weisskoff, The intravascular contribution to fMRI signal change: Monte Carlo modeling and diffusion-weighted studies in vivo. *Magn. Reson. Med.* 1995; **34**(1):4–10.

100. T. Mildner, D. G. Norris, C. Schwarzbauer, and C. J. Wiggins, A qualitative test of the balloon model for BOLD-based MR signal changes at 3 T. *Magn. Reson. Med.* 2001; **46**(5):891–899.

101. E. Yacoub, P. F. Van De Moortele, A. Shmuel, and K. Ugurbil, Signal and noise characteristics of Hahn SE and GE BOLD fMRI at 7 T in humans. *Neuroimage* 2005; **24**(3):738–750.

102. F. Zhao, P. Wang, and S. G. Kim, Cortical depth-dependent gradient-echo and spin-echo BOLD fMRI at 9.4 T. *Magn. Reson. Med.* 2004; **51**(3):518–524.

103. S. Lai, A. L. Hopkins, E. M. Haacke, D. Li, B. A. Wasserman, P. Buckley, L. Friedman, H. Meltzer, P. Hedera, and R. Friedland, Identification of vascular structures as a major source of signal contrast in high resolution 2D and 3D functional activation imaging of the motor cortex at 1.5T: preliminary results. *Magn. Reson. Med.* 1993; **30**(3):387–392.

104. E. M. Haacke, A. Hopkins, S. Lai, P. Buckley, L. Friedman, H, Meltzer, P. Hedera, R. Friedland, S. Klein, L. Thompson, et al. 2D and 3D high resolution gradient echo functional imaging of the brain: venous contributions to signal in motor cortex studies [published erratum appears in NMR Biomed 1994 Dec;7(8):374]. *NMR Biomed.* 1994; **7**(1-2):54–62.

105. D. B. Rowe, Modeling both the magnitude and phase of complex-valued fMRI data. *Neuroimage* 2005; **25**(4):1310–1324.

106. R.S. Menon, Postacquisition suppression of large-vessel BOLD signals in high-resolution fMRI. *Magn. Reson. Med.* 2002; **47**(1):1–9.

107. Q. D. Duong, D.-S. Kim, and K. Ugurbil, S-G. K. Spatio-Temporal Dynamics of BOLD fMRI Signals:Towards mapping Submillimeter Cortical columns Using the early Negative Response. *Magn. Reson. Med.* 2000; **44**(2):231–242.

108. D. S. Kim, T. Q. Duong, and S. G. Kim, High-resolution mapping of iso-orientation columns by fMRI. *Nat. Neurosci.* 2000; **3**(2):164–169.

109. L. J. Toth, I. Ronen, C. Olman, K. Ugurbil, and D.-S. Kim, Spatial correlation of BOLD activity with neuronal responses. 2001. p 6.

110. K. Ugurbil, L. Toth, and D. S. Kim, How accurate is magnetic resonance imaging of brain function? *Trends. Neurosci.* 2003; **26**(2):108–114.

111. M. Wong-Riley, B. Anderson, W. Liebl, and Z. Huang, Neurochemical organization of the macaque striate cortex: correlation of cytochrome oxidase with $Na+K+ATPase$, NADPH-diaphorase, nitric oxide synthase, and N-methyl-D-aspartate receptor subunit 1. *Neuroscience* 1998; **83**(4):1025–1045.

112. I. Ronen, J. H. Lee, H. Merkle, K. Ugurbil, and G. Navon, Imaging $H2(17)O$ distribution in a phantom and measurement of metabolically produced $H2(17)O$ in live mice by proton NMR. *NMR Biomed.* 1997; **10**(7):333–340.

113. X. H. Zhu, H. Merkle, J. H. Kwag, K. Ugurbil, and W. Chen, 17O relaxation time and NMR sensitivity of cerebral water and their field dependence. *Magn. Reson. Med.* 2001; **45**(4):543–549.

114. X. H. Zhu, Y. Zhang, R. X. Tian, H. Lei, N. Zhang, X. Zhang, H. Merkle, K. Ugurbil, and W. Chen, Development of (17)O NMR approach for fast imaging of cerebral metabolic rate of oxygen in rat brain at high field. *Proc. Natl. Acad. Sci. U S A* 2002; **99**(20):13194–13199.

115. X. H. Zhu, N. Zhang, Y. Zhang, X. Zhang, K. Ugurbil, and W. Chen, In vivo (17)O NMR approaches for brain study at high field. *NMR Biomed* 2005; **18**(2):83–103.

116. N. Zhang, X. H. Zhu, H. Lei, K. Ugurbil, and W. Chen, Simplified methods for calculating cerebral metabolic rate of oxygen based on 17O magnetic resonance spectroscopic imaging measurement during a short 17O2 inhalation. *J. Cereb. Blood. Flow. Metab.* 2004; **24**(8):840–848.

117. A. Grinvald, R. D. Frostig, R. M. Siegel, and E. Bartfeld, High-resolution optical imaging of functional brain architecture in the awake monkey. *Proc. Natl. Acad. Sci. U S A* 1991; **88**(24):11559–11563.

118. I. Venzetta and A. Grinvald, Phosphorescence decay measurements in cat visual cortex show early blood oxygenation level decrease in response to visual stimulation. *Neurosci-Lett* 1998; Supp 51:S42.

119. I. Vanzetta and A. Grinvald, Increased cortical oxidative metabolism due to sensory stimulation: implications for functional brain imaging. *Science* 1999; **286**(5444):1555–1558.

120. D. Malonek, U. Dirnagl, U. Lindauer, K. Yamada, I. Kanno, and A. Grinvald, Vascular imprints of neuronal activity: relationships between the dynamics of cortical blood flow, oxygenation, and volume changes following sensory stimulation. *Proc. Natl. Acad. Sci. U S A* 1997; **94**(26):14826–14831.

121. A. Grinvald, H. Slovin, and I. Vanzetta, Non-invasive visualization of cortical columns by fMRI. *Nat. Neuroscience* 2000; **3**(2):105–107.

122. X. Hu, T. H. Le, and K. Ugurbil, Evaluation of the early response in fMRI using short stimulus duration. *Magn. Reson. Med.* 1997; **37**: 877–884.

123. R. S. Menon, S. Ogawa, X. Hu, J. P. Strupp, P. Anderson, and K. Ugurbil, BOLD based functional MRI at 4 Tesla includes a capillary bed contribution: echo-planar imaging correlates with previous optical imaging using intrinsic signals. *Magn. Reson. Med.* 1995; **33**(3):453–459.

124. D. B. Twieg, G. G. Moore, and Y. T. Zhang, Estimating Fast Response Onset Time. 1997; Vancouver, British Columbia, Canada. p. 1645.

125. E. Yacoub, T. H. Le, K. Ugurbil, and X. Hu, Further evaluation of the initial negative response in functional magnetic resonance imaging. *Magn. Reson. Med.* 1999; **41**(3):436–441.

126. E. Yacoub, T. Vaughn, G. Adriany, P. Andersen, H. Merkle, K. Ugurbil, and X. Hu, Observation of the Initial "Dip" in fMR1 Signal in Human Visual Cortex at 7 Tesla. Proc Intl Sot *Mag. Reson. Med.* 2000; **8**:991.

127. E. Yacoub, A. Shmuel, J. Pfeuffer, P. F. Van De Moortele, G. Adriany, K. Ugurbil, and X. Hu, Investigation of the initial dip in fMRI at 7 Tesla. *NMR Biomed* 2001; **14**(7-8):408–412.

128. A. F. Cannestra, N. Pouratian, S. Y. Bookheimer, N. A. Martin, D. P. Beckerand, A. W. Toga, Temporal spatial differences observed by functional MRI and human intraoperative optical imaging. *Cereb. Cortex.* 2001; **11**(8):773–782.

129. N. K. Logothetis, H. Gugenberger, S. Peled, and J. Pauls, Functional imaging of the monkey brain. *Nature Neuroscience* 1999; **2**(6):555–560.

130. J. S. Hyde, B. Biswal, A. W. Song, and S. G. Tan, Physiological and Instrumental Fluctuations in fMRI data. 1994; Madison, WI. pp. 73–76.

131. G. Kruger and G. H. Glover, Physiological noise in oxygenation-sensitive magnetic resonance imaging. *Magn. Reson. Med.* 2001; **46**(4):631–637.

132. G. Kruger, A. Kastrup, and G. H. Glover, Neuroimaging at 1.5 T and 3.0 T: comparison of oxygenation-sensitive magnetic resonance imaging. Magn Reson Med 2001; **45**(4):595–604

133. J. S. Hyde, B. B. Biswal, and A. Jesmanowicz, High-resolution fMRI using multislice partial k-space GR-EPI with cubic voxels. *Magn. Reson. Med.* 2001; **46**(1):114–125.

134. J. T. Vaughan, M. Garwood, C. M. Collins, W. Liu, L. DelaBarre, G. Adriany, P. Andersen, H. Merkle, R. Goebel, M. B. Smith, and K. Ugurbil, 7T vs. 4T: RF power, homogeneity, and signal-to-noise comparison in head images. *Magn. Reson. Med.* 2001; **46**(1):24–30.

135. J. B. Mandeville, J. J. Marota, B. E. Kosofsky, J. R. Keltner, R. Weissleder, B. R. Rosen, and R. M. Weisskoff, Dynamic functional imaging of relative cerebral blood volume during rat forepaw stimulation. *Magn. Reson. Med.* 1998; **39**(4): 615–624.

136. R. P. Kennan, B. E. Scanley, R. B. Innis, J. C. Gore, Physiological basis for BOLD MR signal changes due to neuronal stimulation: separation of blood volume and magnetic susceptibility effects. *Magn. Reson. Med.* 1998; **40**(6):840–846.

137. F. P. Leite, D. Tsao, W. Vanduffel, D. Fize, Y. Sasaki, L. L. Wald, A. M. Dale, K. K. Kwong, G. A. Orban, B. R. Rosen, R. B. Tootell, and J. B. Mandeville, Repeated fMRI using iron oxide contrast agent in awake, behaving macaques at 3 Tesla. *Neuroimage* 2002; **16**(2):283–294.

138. J. B. Mandeville, B. G. Jenkins, Y. C. Chen, J. K. Choi, Y. R. Kim, D. Belen, C. Liu, B. E. Kosofsky, and J. J. Marota, Exogenous contrast agent improves sensitivity of gradient-echo functional magnetic resonance imaging at 9.4 T. *Magn. Reson. Med.* 2004; **52**(6):1272–1281.

139. N. Harel, J. Lin, Y. Zhang, and K. Ugurbil, E. Yacoub. Tissue Specificity of fMRI Signals at Ultra-high Resolution - But Where in the Tissue ? *Proc. Of. Int. Soc. Magn. Res. Med.* 2004; **12**:200.

140. H. Lu, P. C. van Zijl, J. Hendrikse, and X. Golay, Multiple acquisitions with global inversion cycling (MAGIC): a multislice technique for vascular-space-occupancy dependent fMRI. *Magn. Reson. Med.* 2004; **51**(1):9–15.

141. H. Lu, and P. C. van Zijl, Experimental measurement of extravascular parenchymal BOLD effects and tissue oxygen extraction fractions using multi-echo VASO fMRI at 1.5 and 3.0 T. *Magn. Reson. Med.* 2005; **53**(4):808–816.

142. B. Stefanovic and G. B. Pike, Venous refocusing for volume estimation: VERVE functional magnetic resonance imaging. *Magn. Reson. Med.* 2005; **53**(2):339–347.

143. H. Kamei, K. Iramina, K. Yoshikawa, and S. Ueno, Neuronal current distribution imaging using magnetic resonance. *IEEE Trans. Magn.* 1999; **35**:4109–4111.

144. J. Bodurka and P. A. Bandettini, Toward direct mapping of neuronal activity: MRI detection of ultraweak, transient magnetic field changes. *Magn. Reson. Med.* 2002; **47**(6): 1052–1058.

145. J. Bodurka, A. Jesmanowicz, J. S. Hyde, H. Xu, L. Estkowski, and S. J. Li, Current-induced magnetic resonance phase imaging. *J. Magn. Reson.* 1999; **137**(1):265–271.

146. J. Xiong, P. T. Fox, and J. H. Gao, Directly mapping magnetic field effects of neuronal activity by magnetic resonance imaging. *Hum. Brain. Mapp.* 2003; **20**(1):41–49.

147. D. Konn, P. Gowland, and R. Bowtell, MRI detection of weak magnetic fields due to an extended current dipole in a conducting sphere: a model for direct detection of neuronal currents in the brain. *Magn. Reson. Med.* 2003; **50**(1):40–49.

148. M. Bianciardi, F. Di Russo, T. Aprile, B. Maraviglia, and G. E. Hagberg, Combination of BOLD-fMRI and VEP recordings for spin-echo MRI detection of primary magnetic effects caused by neuronal currents. *Magn. Reson. Imaging* 2004; **22**(10):1429–1440.

149. R. Chu, J. A. de Zwart, P. van Gelderen, M. Fukunaga, P. Kellman, T. Holroyd, and J. H. Duyn, Hunting for neuronal currents: absence of rapid MRI signal changes during visual-evoked response. *Neuroimage* 2004; **23**(3):1059–1067.

150. A. D. Liston, A. Salek-Haddadi, S. J. Kiebel, K. Hamandi, R. Turner, and L. Lemieux, The MR detection of neuronal depolarization during 3-Hz spike-and-wave complexes in generalized epilepsy. *Magn. Reson. Imaging* 2004; **22**(10):1441–1444.

151. G. T. Buracas, and G. M. Boynton, Efficient design of event-related fMRI experiments using M-sequences. *Neuroimage* 2002; **16**(3 Pt 1):801–813.

152. D. Konn, S. Leach, P. Gowland, and R. Bowtell, Initial attempts at directly detecting alpha wave activity in the brain using MRI. *Magn. Reson. Imaging* 2004; **22**(10):1413–1427.

BULK MODULUS

SERDAR ARITAN
Hacettepe University
Hacettepe, Turkey

Soft tissues generally exhibit nonlinearity, nonhomogeneity, anisotropy and viscoelasticity. Viscoelastic materials present some difficulties for characterising their mechanical response. Energy storage and dissipation within the complex molecular structure produces hysteresis and allows creep and relaxation to occur. Soft tissues are relatively compliant at low strains and become dramatically stiffer at high strains (2). There is a continuing need to develop methods that reveal the complete set of anisotropic material properties. These properties are important in case and control studies and in numerical implementations of constitutive equations in biological tissue. This context was developed to demonstrate these properties and to provide experience in analysis of this type of data.

1. WHAT ARE MECHANICAL PROPERTIES?

The mechanical design and function of an organism occurs at two independent levels: materials and structures. In fact, there is a considerable relationship between structure and mechanical properties. All soft tissues are composite materials. Among the common components of soft tissues, collagen and elastin fibers mostly affect the overall mechanical behavior of the tissue (3).

The result of the force acting on biological materials is deformation, which can be measured. The adequacy of the structure to resist these forces depends on its material properties, as well as on the shape of the structure. In solid mechanics, force and displacement are normalised as stress and strain. Stress (σ) is defined as,

$$\sigma = \frac{F}{A}, \tag{1}$$

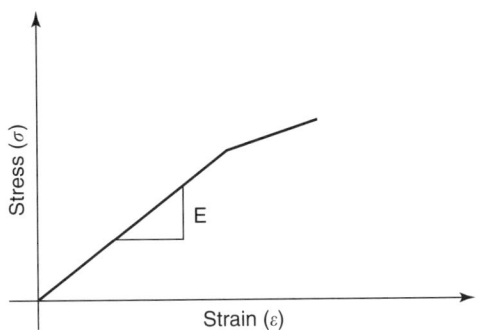

Figure 1. Graphical representation of Elastic modulus (E).

where, F is the force and A is the cross-section area to which the force is applied.

Similarly, deformation or displacement of the material (ΔL) can be expresses as a strain by dividing by the resting length of the material ($\varepsilon = \Delta L/L$). As shown in Fig. 1, the slope of the linear part of the stress-strain curve defines the elastic modulus ($E = \sigma/\varepsilon$). The elastic modulus is a indicator of the material's stiffness in terms of stress and strain (1,3–5).

The relative magnitude of strain in two planes (lateral strain to the longitudinal strain) perpendicular to the plane of applied stress is defined by the Poisson's ratio (v) of the material in those planes.

$$v_{xy} = -\varepsilon_y/\varepsilon_x \; ; \; v_{xz} = -\varepsilon_z/\varepsilon_x. \quad (2)$$

In isotropic materials, v_{xy} and v_{xz} are equal. Anisotropic materials such as bone, muscle and other biological materials show differing amounts of strain and stiffness along the materials (1,3,4).

Often, the stresses and strains of the material are developed in three-dimension. Therefore, the components of stress in a body can be defined by considering the forces acting on an infinitesimal cubical volume element (in Fig. 2) whose edges are parallel to the coordinate axes x, y and z. In equilibrium, the forces per unit area acting on the cube faces are

F1 on the yz plane,
F2 on the zx plane,
F3 on the xy plane.

These three forces are then resolved into their nine components in the x, y and z directions as follows:

F1: σ_{xx}, σ_{xy}, σ_{xz},
F2: σ_{yx}, σ_{yy}, σ_{yz},
F3: σ_{zx}, σ_{zy}, σ_{zz}.

The first subscript refers to the direction of the normal to the plane on which the stress acts, and the second subscript to the direction of the stress. In the absence of body torques the total torque acting on the cube must also be zero and this implies three equalities:

$$\sigma_{xy} = \sigma_{yx}, \; \sigma_{xz} = \sigma_{zx}, \; \sigma_{yz} = \sigma_{zy}. \quad (3)$$

These components of stress are defined by six independent quatities, σ_{xx}, σ_{yy} and σ_{zz}, the normal stress which is normal to the surface and represents direct stress, and σ_{xy}, σ_{yz} and σ_{zx}, which are tangential to the surface and represent shear stresses. These form six independent components of the stress tensor σ_{ij}:

$$\sigma = \begin{bmatrix} \sigma_{xx} & \sigma_{xy} & \sigma_{xz} \\ \sigma_{yx} & \sigma_{yy} & \sigma_{yz} \\ \sigma_{zx} & \sigma_{zy} & \sigma_{zz} \end{bmatrix}. \quad (4)$$

The state of stress at a point in body is determined when the normal components and the shear components of acting on a plane can be specified. Cauchy generalised Hooke's law to three-dimensional elastic bodies and stated that the 6 components of stress are linearly related to the 6 components of strain. The stress state is a second order tensor since it is a quantity associated with two directions. As a result, stress components have 2 subscripts. A surface traction is a first order tensor (i.e. vector) since it a quantity associated with only one direction. Vector components therefore require only 1 subscript. Mass would be an example of a zero-order tensor (i.e. scalars), which have no relationships with directions (and no subscripts) (6).

The stress-strain relationship written in matrix form, where the 6 components of stress and strain are organised into column vectors is,

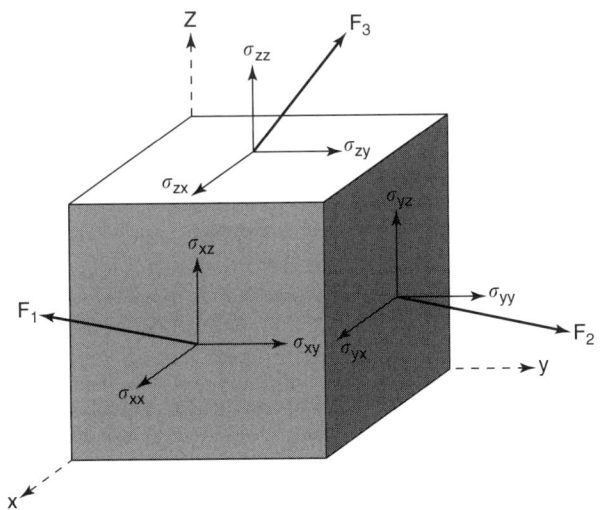

Figure 2. The stress state is represented by an *infinitesimal* cube with three stress components on each of its six sides.

$$\begin{bmatrix} \varepsilon_{xx} \\ \varepsilon_{yy} \\ \varepsilon_{zz} \\ \varepsilon_{yz} \\ \varepsilon_{zx} \\ \varepsilon_{xy} \end{bmatrix} = \begin{bmatrix} S_{11} & S_{12} & S_{13} & S_{14} & S_{15} & S_{16} \\ S_{21} & S_{22} & S_{23} & S_{24} & S_{25} & S_{26} \\ S_{31} & S_{32} & S_{33} & S_{34} & S_{35} & S_{36} \\ S_{41} & S_{42} & S_{43} & S_{44} & aS_{45} & S_{46} \\ S_{51} & S_{52} & S_{53} & S_{54} & S_{55} & S_{56} \\ S_{61} & S_{62} & S_{63} & S_{64} & S_{65} & S_{66} \end{bmatrix} \begin{bmatrix} \sigma_{xx} \\ \sigma_{yy} \\ \sigma_{zz} \\ \sigma_{yz} \\ \sigma_{zx} \\ \sigma_{xy} \end{bmatrix}, \quad (5)$$

$$\begin{bmatrix} \sigma_{xx} \\ \sigma_{yy} \\ \sigma_{zz} \\ \sigma_{yz} \\ \sigma_{zx} \\ \sigma_{xy} \end{bmatrix} = \begin{bmatrix} C_{11} & C_{12} & C_{13} & C_{14} & C_{15} & C_{16} \\ C_{21} & C_{22} & C_{23} & C_{24} & C_{25} & C_{26} \\ C_{31} & C_{32} & C_{33} & C_{34} & C_{35} & C_{36} \\ C_{41} & C_{42} & C_{43} & C_{44} & C_{45} & C_{46} \\ C_{51} & C_{52} & C_{53} & C_{54} & C_{55} & C_{56} \\ C_{61} & C_{62} & C_{63} & C_{64} & C_{65} & C_{66} \end{bmatrix} \begin{bmatrix} \varepsilon_{xx} \\ \varepsilon_{yy} \\ \varepsilon_{zz} \\ \varepsilon_{yz} \\ \varepsilon_{zx} \\ \varepsilon_{xy} \end{bmatrix}, \quad (6)$$

where, C is the stiffness matrix, S is the compliance matrix. In general, stress-strain relationships are known as constitutive relations. Normally, there are 36 stiffness matrix components. However, it can be shown that conservative materials possess a strain energy density function and as a result, the stiffness and compliance matrices are symmetric.

Therefore, only 21 stiffness components are actually independent in Hooke's law. The two elastic constants are usually expressed as the Young's modulus E and the Poisson's ratio v. However, the alternative elastic constants K (bulk modulus) and/or G (shear modulus) can also be used. Direct stresses tend to change the volume of the material (e.g. hydrostatic pressure) and are resisted by the body's bulk modulus (which depends on the Young's modulus and Poisson ratio). Shear stresses tend to deform the material without changing its volume, and are resisted by the body's shear modulus (6).

2. BULK MODULUS

When an isotropic material is exposed to hydrostatic pressures, shear stress will be zero and the normal stress will be the same. In response to the hydrostatic load, the material will change its volume. Its resistance to do so is quantified as the bulk modulus K, also known as the modulus of compression. Technically, K is defined as the ratio of hydrostatic pressure to the relative volume change (which is related to the direct strains).

$$K = -V \frac{\Delta p}{\Delta V}, \quad (7)$$

where V is the initial volume, ΔV is change in the volume and Δp is change in the pressure (4,6).

3. CHARACTERISATION AND MODELLING OF SOFT BIOLOGICAL TISSUE

A muscle is ordinarily thought of as an active system; its prime function is to generate force. Muscles are composite materials, composed of stiff strong fibres, plates or particles in a relatively complaint matrix. The properties of composite materials depend very much upon their structure. Many soft composites change their mechanical properties as part of their normal functioning. Muscle is stiffened by contraction and softened by relaxation. At present, there is not a representation of nonlinear viscoelasticity which gives an adequate description of the behavior and provides some physical insight into the origins of this behavior (7,8).

4. ULTRASOUND EXPERIMENTS AND RESULTS

Ultrasound testing is used to detect flaws where an ultrasonic source is pressed against the part to be tested, using a gel to act as a sonic coupling to the surface, and sound is passed into the material. Reflections or echoes occur from the back face of the material and any internal discontinuity will reflect the sound wave and generate a signal in the receiver. The time lags of the echoes are measured to determine the thickness of the material and the distance to the discontinuity. The propagation of sound waves through a material is governed by the mechanical properties of that material. Therefore ultrasound could be a useful aide in studies of the elastic properties of biological tissues. If distortion is produced at a point in an elastic material, it is transmitted as a sound wave or side wave. The sound wave velocity V can be shown as a function of bulk modulus K, and as the density ρ of isotropic material. The equation is written as;

$$V = \sqrt{\frac{K}{\rho}}. \quad (8)$$

Therefore, the bulk modulus can be calculated if V and ρ are known. However, the elastic constant (such as, Elastic modulus, Poisson's ratio) of human tissue cannot be theoretically estimated because of its complex structure and composition, but the constant may be function of elasticity explained above (1).

Levinson (9) used ultrasound and assumed linear elastic theory to obtain mechanical properties of seven frog sartorius specimens. Linear elastic theory is often applied as a means of defining the mechanical properties of biological materials. Levinson obtained four observable elastic constants associated with a transversely isotropic model. The average values of these constants were;
for resting,

$$C_{11} = 2.64 \, \text{Pa},$$
$$C_{13} = 3.39 \, \text{Pa},$$
$$C_{33} = 4.40 \, \text{Pa},$$

undergoing tetanic contraction,

$$C_{11} = 2.65 \, \text{Pa},$$
$$C_{13} = 3.43 \, \text{Pa},$$
$$C_{33} = 4.57 \, \text{Pa},$$

where index $_1$ and $_3$ represent the transverse and longitudinal axes. In all cases C_{44} was 0 indicating a minimal contribution from longitudinal shear. For all seven specimens, the model of transverse isotropy provided a better fit of the data than that of isotropy (9). In terms of the characterisation of ultrasonic wave propagation, this approach has been most extensively applied to bone.

Levinson et al. (10) also used sonoelastography (which uses ultrasound) to measure the propagation of shear waves induced by externally applied vibrations. Because

Table 1. Calculated Values of Elastic Modulus

Propagation (Hz)	Applied Loads (kgf)	Elastic Modulus (E, kPa)
30	0	7 ± 3
30	7.5	29 ± 12
30	15	57 ± 37
60	0	25 ± 6
60	7.5	75 ± 11
60	15	127 ± 65

Table 2. Material Properties Given by Schock et al. (1982)

Tissue	Shear Modulus (kPa)	Bulk Modulus (kPa)
Skin	28.2	53.1
Fat	18.3	12.2
Muscle	48.3	39
Skin	175	178
Fat	61.5	99.2
Muscle	147	169
Skin	1260	1280
Fat	65.2	105
Muscle	43.8	39

shear waves predominate in incompressible viscoelastic media at low frequencies, sonoelastic data should be comparable to those obtained using conventional means. They demonstrated the practicality of using sonoelastography to measure the viscoelastic properties of human muscles *in situ*. They recorded vibration propagation speeds as a function of applied load in the quadriceps muscle of ten volunteers, as shown in Table 1.

4.1. Indentor Test Results

In an indentor test, a steady load is applied to soft tissue under constant environment conditions, the amount of indentation being measured after a fixed time. Indentor tests have been used widely to determine mechanical properties of soft biological tissues, such as cartilage, skin and subcutaneous tissues. In general, biological tissues are multi-phasic and show time-dependent behavior (11–16).

Indentor tests are widely used in characterising the mechanical response of biological tissue. Advantages of using an indentor test are to have the ability to analyse samples *in situ* and perform multiple tests at several different sites. When tissue is modeled using the linear biphasic theory, the equilibrium deformation is that of a compressible elastic solid and Poisson ratio v ($0 \leq v \leq 0.5$). In indentation creep or stress-relaxation tests, compressible elastic models can then predict the three-dimensional deformation of a tissue layer at equilibrium. However, determining the material properties of tissue from such a test is a difficult computational task.

Historically, viscoelastic behavior has also been reported in soft tissues (2,11,12,14,17) most of these analyses have used a static linear elastic material model. Hayes et al. (12) obtained a theoretical solution to the problem associated with axisymmetric indentation, where the articular cartilage bounded to subchondral bone was modeled as an elastic layer fixed to a rigid boundary. The elastic modulus E can be calculated using the formula;

$$E = \frac{P(1-v^2)}{2awk(v,a/h)}, \quad (9)$$

where, P is the indentation force, v the Poisson's ratio, w the indentation depth, a the radius of the indentor, h the thickness of the test tissue, k a geometry and material-dependent factor.

Sakamoto et al. reported similar results using a new mathematical model (18). The solution of Hayes (12) has been widely used to calculate the elastic modulus of cartilage (13,19,20) and skin together with subcutaneous tissues.

Sacks et al. (21) used weighted indentors to measure, *in vivo*, static stiffness of the soft tissue of the proximal thigh of four male subjects; two paraplegic and two without disabilities. Schock et al. (22) reported *in vitro* indentor tests of static stiffness of porcine skin, fat and muscle and compared these with a finite difference analytical model. Assuming linear elasticity, they gave three sets of material properties as in Table 2.

Their conclusions stated that no single set of data led to adequate agreement between experimental and analytical results; they suggested that more accurate constitutive equations were needed. They also noted that their model was more accurate at low strains.

Krouskop et al. (23) used Doppler ultrasound motion sensing system to make non-invasive measurement of the elastic modulus of soft tissue *in vivo*. The traditional drawback of ultrasound in measuring tissue properties is that the strain rate is very much higher than the physiological one, making the results useful for detecting anisotropy but not for measurement of stiffness. They used a 10 Hz mechanical oscillator to drive the tissue; using ultrasound only to measure the resulting tissue motion. This method is used to measure dynamic viscoelastic stiffness; results can be seen in Table 3.

Chow and Odell (24) analysed wheelchair seating stress using a static, geometrically nonlinear model. They studied the deformations and stresses in the buttocks of a seated person using pressure measurement on human subjects, bench tests of the deformations of various seating materials and a nonlinear finite element model. Nakamura et al. (25) used a static, materially and geometrically nonlinear model for cadaver heel tissue, calculating moduli from 500 to 7000 kPa. They created a two-dimensional plane-stress finite element model of the foot to study the effect of various shoe sole materials.

Table 3. Tissue Properties Repeated by Krouskop et al. (1987) Given as a Elastic Modulus (E)

Contraction State	Elastic Modulus (kPa) at 10% Strain
Relaxed	6.204
Mild	35.848
Maximum	108.925

Steege et al. (26) used indentor tests of below-knee residua and a static, linearly elastic model to derive an elastic modulus of 60 kPa. Steege and Childress (27) later used a geometrically and materially nonlinear model with some improvement noted in the correlation between observed and calculated interface pressures. Reynolds (28) used intendor tests of non-amputated legs and a static, linear elastic model to derive elastic moduli in the range 50 to 145 kPa. Mak et al. (17) used intendor tests of residual and nonamputated limbs and a static, linearly elastic model and calculating moduli from 21 to 195 kPa. It is clear that the results presented, even using similar test procedures gave a very wide range of elastic moduli.

Recently, Vannah et al. (29) used indentor tests of non-amputated lower limbs and modelled geometrically and materially non-linear using large-strain finite element formulation. The resulting composite material stiffness was non-linear and could be approximated using the Jamus-Green-Simpson strain energy function. Typical values for the coefficients were $c_{10} = 2.6$ kPa, $c_{01} = 0.64$ kPa and $c_{11} = 5.57$ kPa.

Oomens et al. (15,16) investigated the time-dependent compression characteristics of *in vitro* skin and subcutaneous fat by using a confined compression test. They modeled the skin as a mixture of a solid and a fluid. A numerical procedure was presented to solve nonlinear field equations describing such a mixture. The confined compression test, however, gives quite repeatable results, but it does not represent living tissue behavior.

Zhang et al. (30) investigated the reaction of skin and soft tissue to shear force applied externally to the skin surface. They also measured skin blood flow by using laser Doppler flowmetry while variable shear forces and normal force were applied to the skin surface. They analysed the internal stress using simplified models incorporating elastic theory.

The most majority the researchers used indentor test to obtain mechanical properties of soft tissue *in vivo*. The drawback of the indentor test to the load applied on the soft tissue was limited which depending on the cross-section of indentor's tip and the tissue under the indentor has three dimensional movement of behavior. Therefore, the some of the response of the soft tissue is not observable.

5. COMPRESSION TEST RESULTS

Ursino and Cristalli (31) presented a mathematical model of the mechanical behavior of arm tissue under the effect of external pressure loads. The model was used to study stress and strain distribution across the tissue and pressure transmission to the brachial artery. They compressed the arm by inflating two independent adjacent cuffs. They calculated elastic modulus and Poisson ratio from 10 subjects. Their results were Elastic modulus E in the range 90 to 115 kPa and Poisson ratio v in the range 0.44 to 0.38. Their arm model included soft tissue, bone and vessels based on cylindrical geometrical shapes.

For a linear viscoelastic solid the creep behavior is completely specified at a given temperature (human body temperature *in vivo* studies) by a measurement of the response to a constant stress over the requires period of time. But for a nonlinear viscoelastic solid the behavior over the range stress required must be mapped out in detailed over the required period of time (32). This type of data can only be obtained specially designed testing machine. There are several of testing procedures and testing instruments in mechanical engineering for testing materials. By contrast, there is not a certain type of testing machine and procedure in the area of soft tissue biomechanics. Usually adaptation of mechanical engineering tests has been used for soft tissue research (2). Lack of testing machine and protocol to run a series of experiments, caused negligence in the *in vivo* soft tissue research area.

Aritan et al. (4,33) were designed a new testing machine and control software to investigate the response

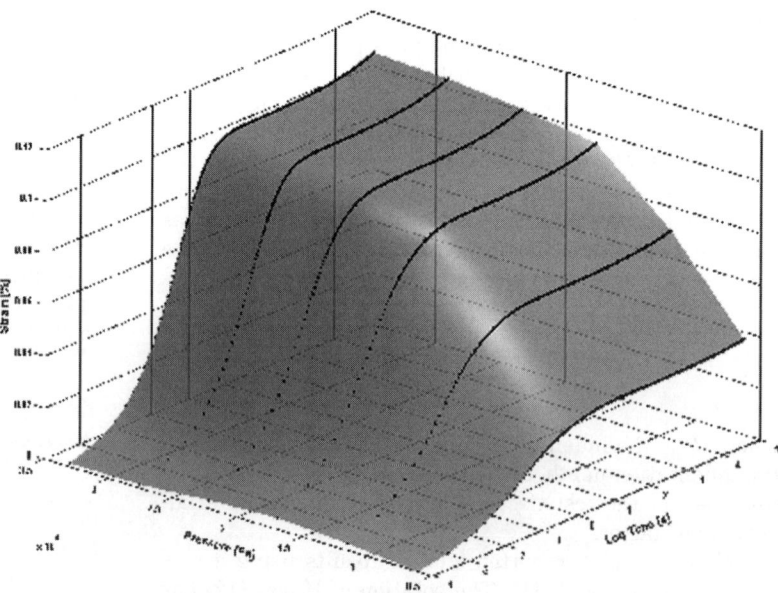

Figure 3. Elastic modulus of muscular bulk tissue in relax condition.

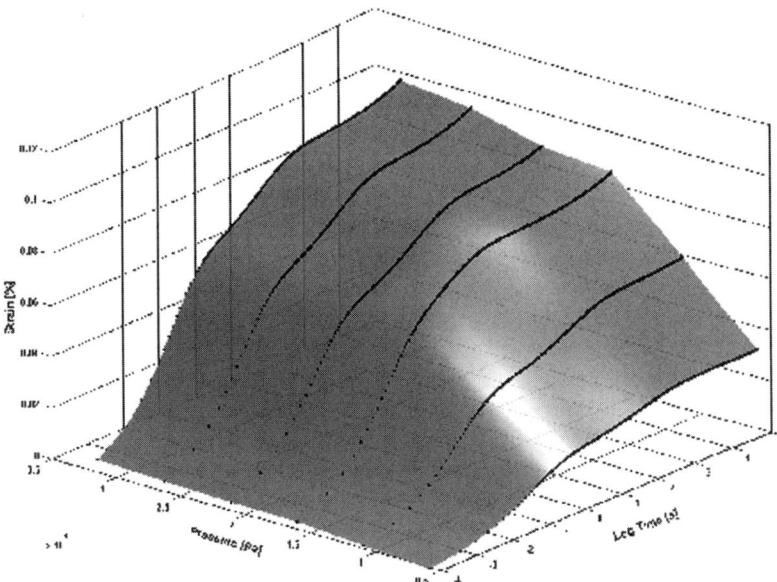

Figure 4. Elastic modulus of muscular bulk tissue in contracted condition.

of the bulk tissue under compression. Using the test machine, it was possible to collect creep, stress-relaxation and stress-strain time-dependent (viscoelastic) data from the bulk tissue of the human upper arm. Creep and stress-relaxation tests measure the dimensional stability of a material, which is very important in the theory of viscoelasticity. Creep and stress relaxation tests applied to the subject with two different protocols. In the first protocol, the subject did not contract his muscle during the test. In the second test, the subject sustained the muscle contraction at 25% of maximum voluntary contraction (MVC) during the test.

The relationship between stress and strain for a fixed time of measurement can be seen in Figs 3 and 4. These relationships produce three-dimensional surfaces, as shown in figures. In the creep test, relax test procedure showed higher elongation results than the contracted test procedure. It is obvious that, this response is caused by the increased stiffness of the contracted muscle tissue. The comparisons can be seen in Figs. 3 and 4. For each load, the creep behavior follows an exponential law typical of viscoelastic materials.

The linear increase of the applied load showed non-linear increase in the elongation. This situation can be explained in terms of engineering that the applied load is reaching the limiting load and can also be explained, in terms of soft tissue, the human upper arm reached the state of the incompressibility. There is also quite an interesting result from the experiment can be observed, the creeping of the contracted muscle is higher than the relax muscle. This response can be interpreted that since the contraction of the muscle has developed fluid in the muscle structure which causes higher creeping response than the relax muscle.

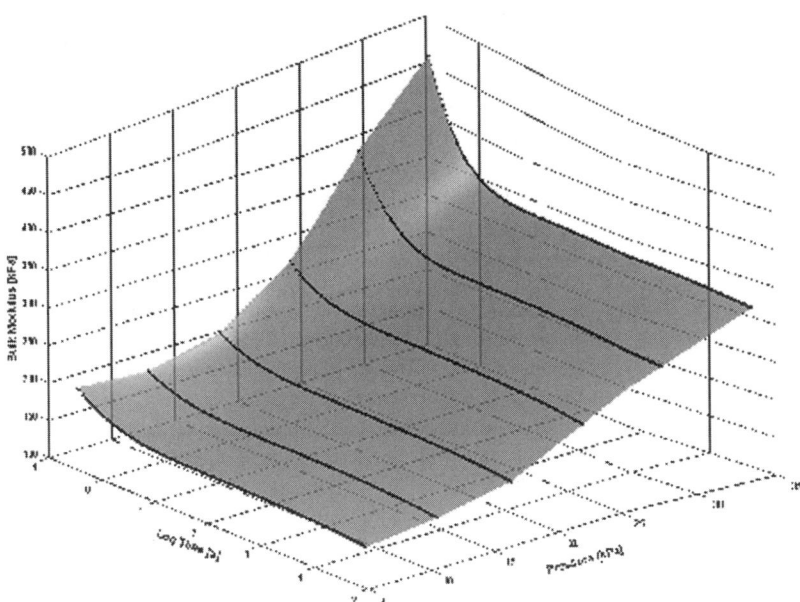

Figure 5. Bulk modulus of muscular bulk tissue in relax condition.

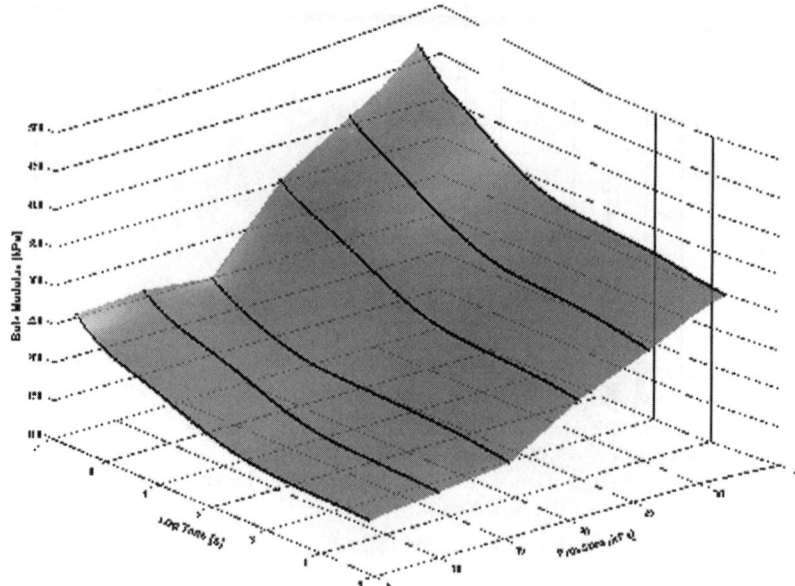

Figure 6. Bulk modulus of muscular bulk tissue in contracted condition.

In their investigation, Aritan et al. (4,33) were simulated the creep behavior of the upper arm by using four Voigt elements in series. Basically, in their study, four mechanical models were used to represent the upper arm. Each one represented one main part of the upper arm. The three obvious materials of the arm - skin, fat and muscle - were modeled in series. The effects of blood vessels and connective tissue were also modeled in series with the previous ones. It was also assumed that the load applied to the upper arm had no deformation effect on the bone. The bulk properties of the upper arm were calculated from the values of the mechanical model parameters. The bulk properties of the muscular tissue can be seen in Figs. 5 and 6.

Finally, all investigations show that there is not a single value, which defines the mechanical properties of soft tissue. Results usually depend on the research approach (*in vivo* or *in vitro*), tissue type (which part of the human body) and the numerical method applied to extract mechanical properties. Prior inverse approaches have been mostly done for planar soft tissue *in vitro*. These studies can give satisfactory results in control experiments in two dimensions but using these data for an *in vivo* three-dimensional study, tissue model demands three-dimensional data collection and analysis.

BIBLIOGRAPHY

1. E. Ackerman, L. B. Ellis, and L. E. Williams, Biophysical Science, 2nd ed. 1979. Prentice-Hall Inc, New Jersey.
2. Y. C. Fung, Biomechanics: mechanical properties of living tissues. 1981. Springer-Verlag, New York.
3. N. Ozkaya, and M. Nordin, Fundamentals of biomechanics: Equilibrium, motion and deformation. 1991. Van Nostrand Reinhold, New York.
4. S. Aritan, S. O. Oyadiji, and R. M. Bartlett, Proceedings of ESDA04 7th Biennial Conference on Engineering Systems Design and Analysis July 19-22, 2004, Manchester, United Kingdom.
5. M. B. Nigg and W. Herzog, Biomechanics of the Muscle-Skeletal System. 1994. John Wiley & Sons Ltd., West Sussex.
6. S. Timeshenko and J. N. Goodier, Theory of elasticity. Mc Graw Hill. New York. 1951.
7. S. A. Glantz, A three-element description for muscle with viscoelastic passive elements. *Journal of Biomechanics* 1977; **10**:5–20.
8. C. E. Jamison, R. D. Marangoni, and A. A. Glaser, Viscoelastic properties of soft tissue by discrete model characterization, *Journal of Biomechanics* 1968; **1**:33–56.
9. S. F. Levinson, Ultrasound propagation in anisotropic soft tissues: the application of linear elastic theory. *Journal of Biomechanics* 1987; **20**:251–260.
10. S. F. Levinson, S. Masahiko, and S. Takuso, Sonoelastic determination of human skeletal muscle elasticity. *Journal of Biomechanics* 1995; **28**:1145–1154.
11. W. C. Hayes and L. F. Mackros, Viscoelastic properties of human articular cartilage. *Journal of Applied Physiology* 1971; **3**:562–568.
12. W. C. Hayes, L. M. Keer, G. Herrmann, and L. F. Mackros, A mathematical analysis for indentation tests of articular cartilage. *Journal of Biomechanics* 1972; **5**:541–551.
13. R. Y. Hori and L. F. Mockros, Indentation test of human articular cartilage. *Journal of Biomechanics* 1976; **9**:679–685.
14. C. W. J. Oomens, D. H. Van Campen, and H. J. Grootenboer, *In vitro* compression of a soft tissue layer on a rigid foundation. *Journal of Biomechanics* 1987; **20**:923–935.
15. C. W. J. Oomens, D. H. Van Campen, H. J. Grootenboer, and L. J. DeBoer, Experimental and theoretical compression studies on porcine skin. 1985 (Perren S.M. and Schneider, E. eds. Biomechanics: current interdisciplinary research, 221–226, Dordrecht, Netherlands).
16. C. W. J. Oomens, D. H. Van Campen, and H. J. Grootenboer, A mixture approach to the mechanics of skin. *Journal of Biomechanics* 1987; **20**:877–886.
17. A. F. T. Mak, G. H. W. Liu, and S. Y. Lee, Biomechanical assessment of below-knee residual limb tissue. *Journal of Rehabilitation Research and Development* 1994; **31**: 188–198.

18. M. Sakamoto, G. Li, T. Hara, and E. Y. S. Chao, A new method for theoretical analysis of static indentation test. *Journal of Biomechanics* 1996; **29**(5):679–685.
19. C. G. Armstrong, A. S. Bahrani, and D. L. Gardner, Changes in the deformational behavior of human hip cartilage with age. *Journal of Biomechanical Engineering* 1980; **102**:214–220.
20. V. C. Mow, M. C. Gibbs, W. M. Lai, W. B. Zhu, and K. A. Athanasiou, Biphasic indentation of articular cartilage. Part II: a numerical algorithm and an experimental study. *Journal of Biomechanics* 1989; **22**:853–861.
21. A. H. Sacks, H. O'Neill, and I. Perkash, Skin blood flow changes and tissue deformations caused by cylindrical indentors. *Journal of Rehabilitation Research and Development* 1987; **22**:1–6.
22. R. B. Schock, J. B. Brunski, and G. V. B. Cochran, *In vivo* experiments on pressure sore on biomechanics: stresses and strains in indented tissues. 1982 (In: Advances in Bioengineering, ASME, 88–91, New York).
23. T. A. Krouskop, D. R. Dougherty, and F. S. Winson, A pulsed doppler ultrasonic system for making non-invasive measurement of the mechanical properties of soft tissue. *Journal of Rehabilitation Research and Development* 1987; **24**:1–8.
24. W. W. Chow and E. I. Odell, Deformations and stresses in soft body tissues of a sitting person. *Journal of Biomechanical Engineering* 1978; **100**:79–87.
25. S. Nakamura, R. D. Crowninshield, and R. R. Cooper, An analysis of soft tissue loading in the foot: A preliminary report. *Bull prosthetic and Research* 1981; **18**:27–34.
26. J. W. Steege, D. S. Schnur, and D. S. Childress, Finite element prediction of pressure at the below-knee socket interface. 1987 (In: Biomechanics of normal and prosthetic gait, ASME, BED-4: 39–44, New York).
27. J. W. Steege and D. S. Childress, Finite element modelling of the below-knee socket and limb: Phase II. 1988 (In: Modelling and control issues in biomechanical systems, ASME, BED-4: 121–9, New York).
28. D. Reynolds, Shape design and interface load analysis for below-knee prosthetic sockets. 1988 (Ph.D. Thesis, University of London, London).
29. W. M. Vannah and D. S. Childress, Indentor tests and finite element modelling of bulk muscular tissue *in vivo*. *Journal of Rehabilitation Research and Development* 1996; **33**:239–252.
30. M. Zhang, A. R. Turner-Smith, and V. C. Roberts, The reaction of skin and soft tissue to shear forces applied externally to the skin surface. *Proceedings Institute Mechanical Engineering* 1994; **208**:217–222.
31. M. Ursino and C. Cristalli, Mathematical modeling of noninvasive blood pressure estmation techniques - Part I: Pressure transmission across the arm tissue. *Journal of Biomechanical Engineering* 1995; **117**:107–116.
32. W. Maurel, Y. Wu, N. M. Thalmann, and D. Thalmann, Biomechanical models for soft tissue simulation. 1998. Springer, Berlin.
33. S. Aritan, Development and validation of a biomechanical model of the human upper arm. 1998 (Ph.D. Thesis, the Manchester Metropolitan University, Manchester, UK).